GEOLOGIC TIME SCALE

ERA	DURATION (millions of years)	PERIOD		EPOCH	DURATION (millions of years)	MILLIONS OF YEARS AGO	
CENOZOIC	65	Quaternary		Holocene (Recent)	.01	.01	First humans
		Tertiary	Neogene	Pleistocene	2.5	2.5	Dominance of mammals and flowering plants
				Pliocene	4.5	7	
				Miocene	19	26	
			Paleogene	Oligocene	12	38	
				Eocene	16	54	
				Paleocene	11	65	
MESOZOIC	160	Cretaceous			70		Last dinosaurs First Flowering plants
		Jurassic			55	135	First birds Dominance of reptiles and conifers
		Triassic			35	190	First dinosaurs First mammals
						225	Wide extinctions
PALEOZOIC	345	Permian			55		Mammal-like reptiles
		Carboniferous	Pennsylvanian		45	280	First reptiles
			Mississippian		20	325	Dominance of amphibians and ferns First seed plants
		Devonian			55	345	First amphibians Air-breathing fishes
		Silurian			30	400	First jawed fishes First vascular plants First land-dwelling invertebrates
		Ordovician			70	430	Jawless fishes
		Cambrian			70	500	First vertebrates Invertebrates widely established Appearance of numerous invertebrate fossils
PRECAMBRIAN					4 billion years	570	Fossils rare Algae

HAMMOND BARNHART

DICTIONARY OF SCIENCE

by ROBERT K. BARNHART

with SOL STEINMETZ
Managing Editor

 INCORPORATED   MAPLEWOOD, NEW JERSEY   07040-1396

FIRST EDITION

ISBN: 0-8437-1689-4

LIBRARY OF CONGRESS CATALOG CARD NUMBER: 86-045735

TABLE OF CONTENTS

PREFACE

The *Hammond Barnhart Dictionary of Science* is a record of the basic terminology of science in English. We have emphasized scientific terms and omitted most of the technical terms that are used in the application of science to our everyday lives. This book is designed to support the student in his or her introduction to a first systematic study of the physical and biological sciences.

While we have not included the entire vocabulary of any science, we have followed curriculum guides and syllabuses to explain the language which is markedly different in its use of words and in its syntax from what we are accustomed to using in day-to-day communications. By explaining the vocabulary of science, as in the Associated Terms, and by showing examples of actual usage, we have attempted to build this dictionary into a tool the student and teacher can use to augment the textbook and the lecture.

The expert help of the Editorial Committee has enabled us to formulate definitions and explanations that correspond to recent findings of scientists working in their various disciplines. The advice of teachers has directed our efforts toward the needs of the student. Precise editorial review in the offices of the publisher has contributed to a standard of excellence, even in the pictures. And surely more than any other single contribution is that which was made by the dedicated editorial staff at Barnhart Books working with Mr. Sol Steinmetz to create a new approach to the language of science.

Many others have contributed to the making of this dictionary, including Richard Whittemore, who understood the objectives of the project in presenting it to Stuart Hammond and Hugh Johnson who, in turn, had the vision to carry the project through with the guidance of Kathleen Hammond and the experienced editorial hand of Dr. George Sullivan.

Many of the ideas encompassed in the editorial policies of the *Hammond Barnhart Dictionary of Science* were developed from working closely with my father, Clarence Barnhart, and the threads of his influence will be found to run through the fabric of the dictionary. To him for his advice and counsel and to Cynthia Barnhart for her unflagging efforts and support to see the project through, I dedicate this book.

Bronxville, N.Y.   Robert K. Barnhart
May, 1986

EDITORIAL COMMITTEE OF CONTRIBUTING
SCIENTISTS

ROBERT H. MARCH

Professor of Physics and Integrated Liberal Studies, University of Wisconsin, at Madison.
Author of *Physics for Poets*.

YOSHI KAZU SASAKI

George Lynn Cross Research Professor, School of Meteorology, University of Oklahoma.
Director, Cooperative Institute for Mesoscale Meteorological Studies.
Author of *Final Report*, Meteorological Research and Development, Naval Environmental Prediction Grant; co-author of "Mesoscale Cloud Features Observed from Skylab," in *NASA Skylab Explores the Earth;* "Methods in Numerical Weather Prediction," in *Finite Elements in Fluids*.

HYRON SPINRAD

Professor of Astronomy, University of California, at Berkeley.
Author of "Faint Galaxies in Cosmology," in *Publications of the Astronomical Society of the Pacific;* co-author of "Spectroscopy of Extremely Distant Radio Galaxies," in *The Astrophysical Journal*"; Kitt Peak Spectroscopy of P/Halley," in *Asteroids, Comets, and Meteors II;* "A Third Update of the Status of the Three CR Sources: Further New Red Shifts and New Identifications of Distant Galaxies," in *Publications of the Astronomical Society of the Pacific*.

EDITORIAL STAFF

For Barnhart Books

Managing Editor: Sol Steinmetz

General Editors:
 Cynthia A. Barnhart
 Benjamin B. Normark

Senior Associate Editor:
 Anne L.Bartling

Associate Editors:
 Shirley Abramson
 Maria Bastone
 Gerald Dalgish

Editorial Assistants:
 David F. Barnhart
 Virginia M. Barnhart
 Maria R. Bastone
 Clarie Day
 George S. Waldo

Office Assistants:
 Albert S. Crocco
 Katherine E. Barnhart
 Rebecca L. Barnhart

Illustrations: John R. Barnhart

George Hendrix, Art Production
Henri A. Fluchère, Consultant

For Hammond Incorporated

Editorial Liaison:
 George E. Sullivan

CONTENTS OF THE DICTIONARY

This is a dictionary of the basic terms of the physical sciences and the biological sciences. It includes vocabulary used to introduce students to the processes and facts that describe natural phenomena. When a scientist communicates and compares his or her findings with others, it is through a specialized vocabulary used in a generally agreed-upon way. Understanding and careful use of this vocabulary avoids ambiguity and misinterpretation. The purpose of this dictionary is to explain a part of that vocabulary by definition and to familiarize the student with the relationships among various words by contrasting and comparing related terms. If this dictionary can supplement the textbook, the demonstration, and the lecture or classroom presentation, we will have accomplished our purpose.

By using all of the dictionary techniques at hand, from notes on usage and the exposition of associated terms to selection of citations, pronunciation of difficult words and etymological explanation of sources, this dictionary is designed to help students understand the mosaic of the language of science as it has developed in modern times.

Terms are defined in a variety of styles to fit the experience and knowledge of the user at that stage of learning where a term is likely to be first encountered. Information within an entry is distributed in such a way that it may be easily assimilated, and in terms we hope will not require reference to other entries. Although citations of usage are included to help distribute the burden of information, they are primarily examples of how the word is used in context and show the environment of usage.

Subject labels provide a guide to the context of terms used in defining. A subject label also serves as a quick reference to finding a meaning in a multi-definition word, and further shows that a particular definition has been reviewed by a specialist in that field.

The artwork is devoted largely to concepts rather than to illustrations of specific animals, instruments, and plants. The legends in many cases are a link between the definition and the illustration and serve to expand the definition. We have also included numerous cross-references to the pictures so that related concepts can be more easily explained.

The compiling of a specialized dictionary is, perhaps more than in most other reference books, the result of many scholars and editors working together. We are fortunate to have an outstanding group of active scientists who read, commented upon and corrected the entire manuscript and we are indebted to them for their thorough work. In addition we are indebted to a large number of science educators—teachers and curriculum coordinators and specialists—from many parts of the United States who have given us invaluable advice and made available syllabuses, study guides, etc., to help us select the terminology of introductory course materials.

The staff of editors and compilers has worked with great care to make a suitable dictionary for the student in introductory science courses. Many hours of diligent work and consultation with critics was necessary to make a new book which rests on the foundations of the Thorndike-Barnhart series of school dictionaries, most notably the *Thorndike-Barnhart Advanced Dictionary* and *The New Century Dictionary*, on the Barnhart dictionaries of New English and the new Supplements of the *Oxford English Dictionary*. No dictionary has been compiled without building upon previous reference books, and we are grateful for the right to form a part of our first-draft manuscript from these books. It is, however, the many outside critics and specialists who add authority to this dictionary, and our editors, working with them and with the resources of our quotation files, who have produced a book that reflects the modern vocabulary of science.

ARRANGEMENT OF ENTRIES

The words entered are in one alphabetical list in bold heavy type, whether they are single words, compounds, derivatives, or affixes (word elements). In a few instances words are listed within an entry as part of a group of related terms in a classification (as under *soil*), but these are not considered main entries in this book.

Derivatives

Derivative words are listed at the end of main entries. The ones listed for spelling purposes are those having easily derived meanings and are undefined.

> **albinism** (al′bə niz′əm), *n. Biology.* the absence of natural coloration of pigmentation; condition of being an albino: *Albinism ... is a recessive characteristic in man as it is in other animals* (R. Beals and H. Hoijer) —**albinistic,** *adj.*

> **aseptic** (ā sep′tik), *adj. Medicine.* **1** free from the living microorganisms causing infection: *Surgical instruments are made aseptic by boiling them.* **2** using all possible measures to exclude infectious microorganisms: *aseptic surgery.* —**aseptically,** *adv.*

Those for which the meaning is not easily derived from the main entry are defined, often with the addition of an illustrative phrase.

> **analgesia** (an′l jē′zhə), *n. Physiology.* a deadening or absence of the sense of pain without loss of consciousness. Compare **anesthesia.** [from Greek *analgēsia,* from *an-* not + *algein* feel pain]
> —**analgesic,** *adj.* of or causing analgesia: *There are several other substances closely related to aspirin that have analgesic properties ... and antipyretic properties* (Pauling, *Vitamin C*). —*n.* an analgesic drug.

> **algebra,** *n.* the branch of mathematics dealing with the relations and properties of quantities by the use of symbols and letters, negative numbers as well as ordinary numbers, and equations to solve problems involving a finite number of operations. $x + y = x^2$ is a way of stating, by algebra, that the sum of two numbers equals the square of one of them. [from Medieval Latin *algebra,* from Arabic *al-jabr* the bone setting; hence reduction (i.e., of parts of a fracture set into a whole)]
> —**algebraic** (al′jə brā′ik), *adj.* of or having to do with algebra; used in algebra.

Homographs

Homographs (different words having the same spelling—for example, *limb*[1] arm, *limb*[2] crescent) are individually entered, and the entry words are distinguished by a small superscript number after each entry word. At the end of these homographs is an etymology that explains why, though the words are coincidentally written with the same letters and usually pronounced in the same way, they are in fact two different words in English.

> **limb**[1], *n. Anatomy.* a leg, arm, wing, or other member of an animal body distinct from the head or trunk: *The limbs of seals, moles, bats and antelope look very different from each other and are adapted to the functions of swimming, digging, flying and running* (Mackean, *Introduction to Biology*).
> **2** *Botany.* a large branch.
> [Old English *lim*]

> **limb**[2], *n.* **1** *Astronomy.* the edge of the disk of a celestial body: *The problem is one of identifying the satellite's edge, or limb, whether from an occultation (in which the blocking of the light from a star can be precisely timed) or from a photograph* (Science News).
> **2** *Botany.* the expanded flat part of a structure, such as the upper part of a corolla or the blade of a leaf.
> **3** *Geology.* one of the two sides of a fold. Also called **flank.**
> [from Latin *limbus* border, edge]

Variants

Variant spellings are usually given next to the main-entry word.

> **acarpelous** or **acarpellous** (ā kär′pə ləs), *adj. Botany.* (of a flower) having no carpels.

A variant form (a different word for the main entry) is listed in its alphabetical order, though it may be listed at the end of the preferred form after the phrase. "Also called _____." Entries that are variant forms are cross-referred to the preferred form with an equal sign (=).

> **airstream,** *n.* = airflow.

> **airflow,** *n.* **1** *Meteorology.* a natural movement of air: *Most of the airflow is from a westerly direction with speeds increasing from the pole southward* (Scientific American).
> **2** *Physics.* the flow of air around and relative to an object moving in air, usually in a direction opposite to that of the object's motion or flight. Also called **airstream.**

Affixes and Combining Forms

Affixes (prefixes or suffixes) and combining forms are listed in alphabetical order with examples of how they function in English word formation. Prefixes (*di-, dia-*), suffixes (*-ase, -ide*), and combining forms (*bio-, cyto-, -logy*) help users recognize and understand the word-formation process of scientific vocabulary. This is particularly useful for the SI units of measure.

> **giga-** (jig′ə-), *combining form.* one billion (10⁹) of any SI unit: *Gigavolt = one billion volts. The distance from the sun to the earth is 150 gigametres or 94 megamiles* (London *Times*). *Symbol:* G [from Greek *gigas* giant]

> **milli-**, *combining form.* one thousandth of, as in *millimeter = one thousandth of a meter. Symbol:* m [from Latin *mille* thousand]

Inflected forms

Inflected forms are given for nouns only, as verbs and adjectives follow the rules of English and if they are not within the experience of the user, they can be found in any general dictionary. However, the nouns used in science are sometimes borrowings from Latin or Greek or are modeled on the pattern of nouns from Latin or Greek. This process produces plural forms that are irregular in English, and to help the user we have entered these irregular forms where they exist.

> **epithelium** (ep′ə thē′lē əm), *n., pl.* **-liums, -lia** (-lē ə). *Anatomy.* a thin layer of cells forming a tissue which covers the internal and external surfaces of the body, and which performs protective, secretive, or other functions. [from New Latin, from Greek *epi-* on + *thēlē* nipple]

Subentries

Some entries are explained within the main entry, because these subentries are closely related to the main entry but have a specific application. Such entries are also listed in boldface.

> **moment** *n.* **1** *Physics.* **a** a tendency to cause rotation around a point or axis: *When torques about a center of moment are equal, no rotation takes place* (E.S. Obourn, *World of Science*). **b** the product of a (specified) physical quantity and the length of the perpendicular from a point or axis. The **moment of force** about a point is the product of the magnitude of the force and the length of the perpendicular disfance from the point to the line of action of the force; also called *torque.* **2** *Statistics.* any of a series of quantities that express values derived from sums of powers of the variables in a set of data. [from Latin *momentum* moving power, movement, from *movere* to move]

> **mother**, *adj.* producing others; being the source or origin; parent, as in: **mother cell**, *Biology.* the cell from which daughter cells are formed by cell division. **mother cloud**, *Meteorology.* the cloud from which the funnel of a tornado descends: *The cloud above it, or the mother cloud, was so low, and the funnel was so*

wide, that one witness described the tornado as a "turbulent, boiling mass of blackness" (James E. Miller).
mother element, *Physics.* the radioactive element whose decay gives rise to a daughter element: *they could record the alpha particles emitted by the "mother" element 105 atoms ... and thus aid in the detection of the "daughter" element 103 (lawrencium-256) atoms* (Albert Ghiorso).
—*n. Chemistry.* a slimy, filmy substance formed on the surface of an alcoholic liquid undergoing fermentation, as in making vinegar or milk products. The substance consists of yeast and bacterial cells whose enzymes promote fermentation. Also called **mother of vinegar.**

DEFINITIONS, LABELS, AND CITATIONS

The editors have tried to keep the definitions at a level that is understandable to the user in terms of the experience he or she brings to the dictionary in a subject field. For example, definitions for functions in trigonometry are written at a more advanced technical level than those for arithmetic.

In entries with multiple definitions, each definition is brought to the margin for easy access and, with the subject label at the beginning of the specialized definitions, the user should find a desired meaning without difficulty.

intercalate (in tėr′kə lāt), *v.* **1** *Astronomy.* to put (an additional day or month) into the calendar.
2 *Anatomy.* to place or insert between: *a neuron that is intercalated in a chain of neurons.*
3 *Chemistry.* to become inserted between two or more other components: *Large numbers of ethidium bromide molecules can intercalate in a nicked duplex loop or a linear duplex* (Scientific American).
4 *Geology.* to be inserted or interleaved between the original layers or series: *Marine mud and sand ... intercalated here and there with strata of limestone* (A.C. Ramsay).
[from Latin *intercalatum* interposed, proclaimed between, from *inter-* between + *calare* proclaim]
—**intercalation,** *n.* **1** the act or process of intercalating: *Intercalation is a process whereby donor ions position themselves inside a host structure. These ions occupy otherwise vacant regions ... without changing the structure of their host* (Science News). **2** something that has been intercalated, as a layer of rock between layers of different origin.

Labels

Many technical reference books enter numerous words and definitions with a general sense that is not confined to science. We have tried to eliminate this practice in favor of including only those terms which have specific application in science. Therefore, unless a term is truly a general term of science, a subject label, or in some cases two subject labels, will appear at the head of a definition. The user will then know in what subject field a specific meaning is defined and how to interpret the words used in the definition.

node, *n.* **1** *Botany.* any joint in a stem where leaves grow out: *Usually the nodes are not visible in the older stems of mature trees, shrubs, or lianas but are always evident on the younger twigs* (Emerson, *Basic Botany*).
2 *Physics.* a point, line, or plane in a vibrating body at which there is comparatively no vibration: *Certain points known as the nodes remain always at rest* (Sears and Zemansky, *University Physics*).
3 *Astronomy.* either of the two points at which the orbit of a celestial body intersects the path of the sun or the orbit of another celestial body: *The points of intersection are known as the Moon's nodes, the one where the Moon crosses from the south side of the ecliptic to the north being called the ascending node and the other the descending node* (Duncan, *Astronomy*).
4 *Geometry.* a point at which a curve crosses itself, or a similar point on a surface.
5 *Anatomy.* **a** = node of Ranvier. **b** = lymph gland.
[from Latin *nodus* knot]

Citations

Citations are generously distributed throughout the book. The purpose of a citation is to show users the application of a given definition and to help understand the use of the term in its appropriate environment. We have drawn this material from our files collected over the past forty years. These files provide the raw material that definers use to frame definitions. Many of the sources will be familiar to users as premier publications of a specific field of study; others are more general publications which include relatively new terms or include particularly fine writing with an apt turn of phrase that illustrates use of a given term. The dictionary also includes citations from scientific articles and syllabuses of course instruction.

radioactivity, *n. Physics.* **1** the property exhibited by certain elements, of emitting radiation in the form of alpha particles, beta particles, or gamma rays as the result of spontaneous nuclear decay: *The intense radioactivity of radium and its compounds made it possible to investigate the nature of the radiation and the effects that accompany its emission ... for, as the investigators later recognized, radioactivity is a property of the atomic nucleus* (Furry, *Physics*).
2 the radiation emitted by a radioactive substance: *Radioactivity is measured in curies: one curie is equal to the radioactivity from one gram of radium (37 billion atoms disintegrating per second). Thus a millionth of a curie, acting on the body over a period, is a dangerous dose* (Scientific American).

salt, *n.* **1** the common name of sodium chloride (NaCl), a white substance found in the earth and in sea water, used as a seasoning, a preservative for food and hides, and in many industrial processes: *Salt is one of the common rock minerals of the earth ... Inexhaustible supplies are available for human use* (Finch and Trewartha, *Elements of Geography*). *Formula:* NaCl
2 *Chemistry.* a compound derived from an acid by replacing the hydrogen wholly or partly with a metal or an electropositive radical. A salt is formed when an acid and a base neutralize each other. Sodium bicarbonate is a salt. *Salts* [are] *usually defined as ionic*

compounds which in water solution yield a positive ion other than hydrogen and a negative ion other than hydroxyl. *The process by which an acid and a base unite to form water and a salt is termed neutralization* (Jones, *Inorganic Chemistry*). *Salts which contain a polyvalent ion combined with two different ions, are called double salts. Acid salts and basic salts are, therefore, double salts* (Offner, *Fundamentals of Chemistry*).

USAGE NOTES, ASSOCIATED TERMS, AND CROSS-REFERENCES

From time to time the user will find a special mark (►) at the beginning of a paragraph within an entry (*acid, binary star*). After this mark we enter editorial statements about usage and usually some restriction as to how certain terms are applied, or a statement further differentiating meaning. These editorial remarks reflect usage of the term among scientists today.

acid, *n. Chemistry.* any of a class of substances whose water solutions are generally characterized by sour taste, ability to turn litmus dye red, and ability to react with bases and certain metals to form salts. Specifically: **a** a substance that yields hydrogen ions when dissolved in water. **b** a substance that can act as a proton donor. **c** a substance that can act as an electron acceptor. Compare **base.**

—*adj.* Also, **acidic. 1** *Chemistry.* **a** of or containing acids; having the properties of an acid: *an acid solution.* **b** having a pH factor of less than 7; having a relatively high concentration of hydrogen ions (contrasted with *alkaline* especially as a characteristic of soil).

2 *Geology.* containing a large proportion of silica: *Granite is an acidic rock.* Compare **basic.**

► Traditional definitions of *acid* and *base* refer to easily recognizable physical and chemical properties. The first structural definition was that of Arrhenius in 1887: acids yield hydrogen ions in a water solution, bases yield hydroxide ions. This definition did not include reactions occurring without the intervention of water, and a more generalized definition was provided by Brönsted and Lowry in 1923: acids are proton (hydrogen ion) donors, while bases are proton acceptors, to which was added a still more generalized definition by Lewis: acids accept a pair of electrons to form a covalent bond, while bases donate such electron pairs. A substance (for example ammonia) may serve as an acid in one reaction and as a base in another. Thus, strictly, a substance is defined as an acid or base according to its behavior in a particular reaction.

ASSOCIATED TERMS: *Acid, base, salt* refer to definite physical properties of a substance; *weak, strong,* and *transition* are classifications of those properties; *pH, alkaline* are measures of the properties.

binary star, *Astronomy.* a pair of stars that revolve around a common center of gravity. Binary stars are relatively common; perhaps a third to a half of all stars are located in systems of binary stars. Also called **binary, double star.**

ASSOCIATED TERMS: Binary stars are commonly classified as visual, spectroscopic, or eclipsing. *Visual binary stars* are near enough to earth to be separately visible by a telescope; *spectroscopic binaries* are recognized only by observation

through a spectroscope; *eclipsing binaries* have orbits so nearly edgewise to the line of sight from the earth that they alternately eclipse each other.

▶ Visual binary stars should not be confused with
optical double stars, two stars in almost the same line
of sight but actually situated at a great distance from
each other and having no physical relation. Some astronomers call both of these *double stars*, distinguishing them as visual or optical.

Associated Terms

The ASSOCIATED TERMS introduce the user to broad groupings of related terms and help
explain the structure of the language of science. The vocabulary of the Associated Terms is
chiefly one of relationships and not of meaning alone, so that here the dictionary assumes
the function of a textbook and a thesaurus as well. Whereas traditional dictionaries present
words of science as isolated hard words, in the Associated Terms it is possible to present the
selected vocabulary of a structured concept.

electron (i lek′tron), *n. Physics.* an elementary particle
having a very small mass at rest (9.095×10^{-28} gram)
and a unit charge of negative electricity equal to
1.60219×10^{-19} coulombs. All atoms have electrons
surrounding a nucleus at various distances in *orbitals*
or *shells*. The hydrogen atom has one electron; the
uranium atom has 92 electrons. The electron is a member of the class of elementary particles known as
*leptons. When electrons move through a conductor
such as a wire, it is called current electricity. Whenever any electrical device such as an iron, a radio, a motor, or a light bulb is operated, there is a stream of
billions of electrons moving through the wires*
(Obourn, *Investigating the World of Science*). *Symbol:*
e [from *electr* (*ic*) + *-on*, as in *ion*]
ASSOCIATED TERMS: *Electrons, protons,* and *neutrons* are
three of the units of which *atoms* are composed. An electron
is negatively charged and has a *mass* of approximately 1/1836
of a proton. A proton is positively charged and has a mass of
approximately one *dalton.* A neutron has a zero charge and
a mass of approximately one dalton. Each *element* has a particular number of protons and, in a neutral atom, an equal
number of electrons to balance the *charge.* The number of protons in the nucleus is the *atomic number* of an element. The
mass number indicates the total amount of protons and neutrons in the nucleus. The electrons are outside the nucleus at
various *energy levels.* When electrons have absorbed energy
and shifted to higher energy levels, the atom is in an *excited
state.*

muscle, *n. Anatomy.* **1** the tissue in the body of people
and animals that can be tightened or loosened to make
the body move. Muscles contract in response to nerve
stimuli. Muscle tissue is composed of bundles of fibers
and is of two general types, striated muscle and
smooth muscle. *Muscles are sensitive to stretch and
automatically (reflexly) adjust their activity to changes
in tension or stretch* (Science News).
2 an organ consisting of a special bundle of such tissue
which moves some particular bone, part, or substance.
The biceps muscle bends the arm. The heart muscle
pumps blood.
[from French *muscle,* from Latin *musculus,* diminutive of *mus* mouse; so called from the appearance of
certain muscles]
ASSOCIATED TERMS: The human body has over 600 major

muscles. The two main types of muscles are the *skeletal muscles* and the *smooth muscles*. A third type, the *cardiac muscle* or *myocardium* has characteristics of both the skeletal and smooth muscles. Skeletal muscles are also called *striated muscles* or *voluntary muscles*. These muscles are joined to bones by *tendons*. The fixed attachment of a muscle during *contraction* is called the *origin*, while the more movable attachment is called the *insertion*. Skeletal muscles operate in pairs, one bending a limb or part (*flexor*), the other stretching it (*extensor*). Smooth muscles are also called *involuntary muscles* because they operate automatically. They are not *striated* and have only one *nucleus*. Smooth muscles, as those that move food in the stomach and intestines, are stimulated by nerves of the *autonomic nervous system*. Muscles are contracted or stimulated by the sliding action of filaments in muscle fiber made up of the proteins *actin* and *myosin*. The energy for this action is provided by the chemical *ATP* (adenosine triphosphate).

Cross-references

Throughout this dictionary there are cross-references to the Usage Notes and the Associated Terms.

> **local group** or **Local Group,** *Astronomy.* a group of about 20 relatively nearby galaxies to which the Milky Way belongs, including also the Magellanic Clouds and the Andromeda spiral: *When we remember that a single light year is equal to almost six million million miles, we can see that the Spiral is almost inconceivably remote—and yet it is one of the closest of the external systems, and is a member of what we call the Local Group of galaxies* (Listener).
> ASSOCIATED TERMS: see **galaxy.**

> **nuclear physics,** the branch of physics dealing with the structure of atomic nuclei, and the behavior of nuclear particles: *One of the best established facts in nuclear physics is that the density of nuclear matter is constant* (D. Allan Bromley). ► See the note under **particle physics.**

There are also cross-references to individual terms found after the direction to *Compare* or *Contrast* various terms. The significance of contrast is an important pedagogical device designed to help the user to isolate and fix a concept or meaning by emphasizing differences where concentrating on similarities may be confusing.

> **abyssal,** *adj.* **1** *Oceanography.* of or having to do with the depths of the ocean to which light does not penetrate approximately from 3000 to 6000 meters in depth: *In the deeper parts of the ocean, the abyssal region, there is no ... photosynthetic plant life to serve as a source of food for animals* (Hegner and Stiles, *College Zoology*). Compare **bathyal, hadal, littoral.**
> **2** = plutonic: *abyssal rock.*

> **heterolysis** (het′ə rol′ə sis), *n.* **1** *Biology.* the destruction of cells by enzymes or lysins from another organism. Contrasted with **autolysis.**
> **2** *Chemistry.* the breaking of a bond in an unsymmetrical manner, so as to yield two differently charged fragments. Contrasted with **homolysis.**

PRONUNCIATION

Since this book will be used by students who have a certain familiarity with English, we have pronounced only hard or unusual words or words that are pronounced in ways that are exceptions to the sound-letter correlation of traditional English orthography. The pronunciation system in this dictionary was developed in the Thorndike-Barnhart school dictionaries and is, in one form or another, found in most American, and some British, dictionaries today. The full key with traditional representation of the consonants and foreign sounds is given below; an abbreviated key with the vowel symbols is given at the bottom of the right-hand column of every two-page spread.

COMPLETE PRONUNCIATION KEY

a	cap	j	jam	u	cup
ā	face	k	kin	u̇	put
ã	air	l	land	ü	rule
ä	father, far	m	me	yü	use
		n	no	yu	uric
		ng	long		
b	bad			v	very
ch	child	o	rock	w	will
d	did	ō	go	y	yet
		ô	order, all	z	zero
e	best	oi	oil	zh	measure
ē	be	ou	out		
ėr	term				
f	fat	p	paper	ə	represents:
g	go	r	run		a in about
h	he	s	say		e in taken
		sh	she		i in pencil
		t	tell		o in lemon
i	pin	th	thin		u in circus
ī	five	TH	then		

foreign sounds

Y as in French *lune*, German *süss*. Pronounce ē as in *equal* with the lips rounded as for ü, as in rule.

œ as in French *peu*, German *König*. Pronounce ā as in *age* with the lips rounded for ō as in *open*.

N as in French *bon*. The N is not pronounced, but shows that the vowel before it is nasal.

H as in German *ach*, Scottish *loch*. Pronounce k without closing the breath passage.

(The mark ′ is placed after a syllable with a primary or strong accent; the mark ′ after a syllable indicates a secondary or lighter accent in a multisyllable word).

Of the special symbols in the key, the schwa (ə) represents the colorless or muted vowel occurring in unaccented syllables. It is now a standard symbol in most American dictionary keys and represents a sound that occurs with great frequency.

Another special symbol is barred *th* (TH) which is used to distinguish the voiced sound represented by *th* in *then* from the voiceless or aspirated sound represented by *th* in *thin*.

A special use of two regular symbols is that of *l* and *n*. These letters may stand alone in unaccented syllables and represent a sound associated with either *l* or *n* where a surrounding vowel sound is almost completely lost. They are called syllabic consonants.

ETYMOLOGY

Etymologies are given for basic words in a family or group of terms (*accelerate, acoustic*) and for hard words (*ascogonium, anaerobe*). They are added in brackets at the end of an entry, and explained without the usual abbreviations. Thus we use "from" rather than the symbol (<), and we write out the language names (Latin, Greek) and language processes (diminutive, frequentative) instead of using the abbreviated forms in traditional transcriptions of etymologies. We have devoted the space to make the etymologies readable so that emphasis is placed squarely on form and meaning.

accelerate, *v. Physics.* to change the velocity of (a moving object): *When an electron is accelerated, its kinetic energy, which is one half the product of its mass times the square of its velocity, increases* (J.R. Pierce). [from Latin *acceleratum* hastened, from *ad-* to + *celer* swift]

ascogonium (as′kə gō′nē əm), *n., pl.* -nia (-nē ə). *Biology.* the female sex organ in the gametophyte of ascomycetous fungi. [from New Latin *ascogonium*, from Greek *askos* skin bag + a root *gen-* to bear] —**ascogonial,** *adj.* of or having to do with an ascogonium.

ABBREVIATIONS

Most of the standard abbreviations and symbols that the user will find in introductory texts of science are entered in this dictionary. For quick reference and easy access many are also given at the end of the entries or definitions of the words they stand for. Some abbreviations, such as *DNA*, that are more commonly known than the word or words they stand for, are given standard extended treatment under the abbreviated form.

The only abbreviations not defined are those for parts of speech which we assume the user has already had enough experience to recognize (i.e. *n.* noun, *adj.* adjective, *v.* verb, *pl.* plural, etc.), and those for synonyms (SYN) and antonyms (ANT).

MEASUREMENT AND UNITS

The measurements used in this dictionary, except for main entries of customary units, are given in metric units. This is the primary measuring system of science; it is to the user's benefit to adjust to metric measurement so that no ambiguity arises in reference to scientific texts. Where the old customary units are defined, equivalent customary units are used, but descriptions of various phenomena and general concepts of science are referred to in metric measurement. The SI units (International System of Units) and the MKS (meter-kilogram-second) system provide the user with a standard system found in most scientific writing today.

Tables

Within the dictionary there are a number of small tables and scales of measurements at various entries: *metric system, circular measure, linear measure, Beaufort scale, Richter scale,* etc. There are also tables of soil classification (at *soil*) and amino acid components of proteins (at *amino acid*). These are tables of quick reference; they are also given to help the user establish relationships among quantities of a system of measure or nomenclature.

Other tables are provided after this section to show some of the common signs and symbols used in scientific notation. While almost every science has special signs and symbols, we have included only those that the user may encounter in introductory work. We have, however, separated the Periodic Table and a table of geological eras from the body of the dictionary so that these tables, which are so often referred to, may be more easily accessible.

SIGNS AND SYMBOLS

Chemistry

−	single bond *or* negative charge *or* (as a prefix) levorotatory
=	double bond *or* (in equations) forms, yields
≡	triple bond
+	positive charge *or* (as a prefix) dextrorotatory
⟶, ⟵	used to indicate a reaction in the direction specified
⇌	used to indicate a reversible reaction
↓	appears as a precipitate
↑	appears as a gas
⬡	benzene ring

Biology

♀	female
♂ or ♀	male
○	female (in genealogical charts)
□	male (in genealogical charts)
x	crossed with
F	filial generation
F_1	first filial generation
F_2, F_3, etc.	second, third, etc. filial generation
P	parental generation
+	wild type

Mathematics

+	plus *or* positive
−	minus *or* negative
×	multiplied by
·	multiplied by
÷	divided by
/	divided by
=	is equal to
≈	is approximately equal to
≡	is equivalent to
≅	is congruent to *or* is approximately equal to
∼	is similar to
≠	is not equal to
<	is less than
>	is greater than
≤	is less than or equal to
≥	is greater than or equal to
∝	is proportional to
⊥	is perpendicular to
∥	is parallel to
⟶	approaches
:	is to, divided by
±	plus or minus
!	factorial
°	degree *or* degrees
′	minute *or* minutes
″	second *or* seconds
∴	therefore
π	pi

Astronomy

☉	the sun
☾ or	the moon
●	new moon
○	full moon
)	first quarter
(	last quarter
☄	comet
✶ or ✳	fixed star
○	conjunction
□	quadrature
☍	opposition
☊	ascending node
☋	descending node

Physical constants

	Symbol	Value
speed of light in a vacuum	c	2.99792458×10^8 m/s
elementary charge	e	1.602189×10^{-19} C
rest mass of electron	Me	9.109534×10^{-31} kg
Planck's constant	h	6.626176×10^{-34} J/s
Boltzman constant	k	1.38066×10^{-23} J/K
Bohr magneton	μ_B	9.274078×10^{-24} J/T
Rydberg constant	R	$1.09737318 \times 10^{-7}$ /m
Avogadro number	N_A	6.022045×10^{23} /mol

SI units and combining forms

unit	symbol	unit	symbol	combining form	symbol
ampere	A	mole	mol	atto-	a
candela	cd	newton	N	centi-	c
coulomb	C	ohm	μ	deci-	d
farad	F	pascal	Pa	deka-	da
henry	H	radian	rad	femto-	f
hertz	Hz	second	s	giga-	G
joule	J	siemens	S	hecto-	h
kelvin	K	steradian	sr	kilo-	k
kilogram	kg	tesla	T	mega-	M
lumen	lm	volt	V	micro-	μ
lux	lx	watt	W	milli-	m
meter	m	weber	Wb	nano-	n
				pico-	p
				tera-	T

A Dictionary
of Science

A

a, *abbrev.* or *symbol.* **1** ampere *or* amperes. **2** are (100 square meters). **3** atto-.

A, *abbrev.* or *symbol.* **1** absolute temperature. **2** adenine. **3** ampere *or* amperes. **4** angstrom unit *or* angstrom units. **5** one of four major blood groups (used to identify blood for compatibility in transfusion). ► See the note under **ABO. 6** mass number.

Å, *symbol.* angstrom unit *or* angstrom units.

aa (ä′ä), *n. Geology.* a lava flow with a rough, slaggy, fragmentary surface: *It is not uncommon for a pahoehoe flow to turn into an aa flow as it moves downslope away from the vent. However, the opposite transformation never occurs* (Birkeland and Larson, *Putnam's Geology*). [from Hawaiian *'a'a*]

ab-¹, *prefix.* away from, as in *abactinal, abaxial, aboral.* [from Latin *ab* off, away]

ab-², *prefix.* belonging to or denoting an electromagnetic unit in the centimeter-gram-second system, as in *abampere, abcoulomb, abfarad,* and *abhenry.* [from *ab*(*solute*)]

AB, *symbol.* one of four major blood groups (used to identify blood for compatibility in transfusion). ► See the note under **ABO.**

ABA, *abbrev.* abscisic acid.

abactinal (ab ak′tə nəl), *adj. Zoology.* having to do with or situated on the side of a radially symmetrical animal opposite to the mouth or oral area: *the abactinal end of a sea urchin.* Compare **aboral. —abactinally,** *adv.*

abampere, *n.* the CGS unit of current, equal to 10 amperes.

abaxial (ab ak′sē əl), *adj. Biology.* situated away from, or on the opposite side of, the axis of an organ or organism. Compare **adaxial.**

abcoulomb, *n.* the CGS unit of electric charge, equal to 10 coulombs.

ABC soil, *Botany, Geology.* soil having well-defined A, B, and C horizons.

abdomen, *n. Anatomy.* **1 a** (in mammals) the part of the body between the chest and the pelvis. The abdomen contains the stomach, intestines, and other digestive organs, most of the urinary organs, and some of the genital organs. It is lined with a serous membrane called the peritoneum and separated from the chest by a diaphragm. *The trunk is naturally divided into the chest or thorax and the belly or abdomen* (T. H. Huxley). **b** the corresponding region in vertebrates other than mammals, though there is no diaphragm to separate the abdomen from the chest or thorax: *The abdomen [of the striped garter snake] is olive, with a central row of yellow blotches becoming narrower and disappearing toward the tail* (R.L. Ditmars). **2** (in insects and many other arthropods) the last of the three parts of the body containing the digestive apparatus: *Except for a few ... specialized forms, the adults of all insects are alike in having ... a fused head, a thorax ... with six legs ... and a distinct abdomen* (Storer, *General Zoology*). See the pictures at **crustacean** and **thorax.**

—abdominal, *adj.* of, in, or for the abdomen.

abduct, *v. Physiology.* to draw (an arm, finger, toe, etc.) away from the axis of the body or of an extremity. Contrasted with **adduct.** [from Latin *abductum* led away]

abduction, *n. Physiology.* the drawing away of a part from the axis of the body or of an extremity. Contrasted with **adduction.**

abductor, *n. Anatomy.* a muscle that draws an arm, finger, toe, etc., away from the axis of the body or of an extremity: *Muscles which pull a limb away from the median body line are abductors, while those that pull it back toward the median line are adductors* (Winchester, *Zoology*). Contrasted with **adductor.**

Abelian group or **abelian group** (ə bēl′yən), *Mathematics.* a group whose rule of operation is commutative: *If I is the set of integers and o is ordinary multiplication, then [I, o] is an Abelian group; a o b = b o a where a and b are arbitrary members of I.* [named after Niels H. *Abel,* 1802–1829, Norwegian mathematician]

aberration, *n.* **1** *Biology.* an abnormal structure or development; deviation from a normal type: *Changes in chromosome structure are called chromosomal mutations or aberrations* (The Effects of Nuclear Weapons). SYN: mutation.
2 *Physics.* the failure of a lens, mirror, or other optical device to bring the rays of light coming from one point to a single focus. *Spherical aberration* causes a distorted image. *Chromatic aberration* causes an image fringed with prismatic colors. Compare **achromatic.**
3 *Astronomy.* the apparent displacement of a heavenly body from its actual position, as seen by an observer on the earth. The effect is caused by the motion of the earth so that the time it takes for light from the heavenly body to travel through a telescope makes the body's image seem to be slightly displaced from its real position. *Raindrops descending vertically on a calm day strike the face of a pedestrian ... If he runs instead, the apparent slanting direction of the rain becomes more noticeable ... This is a familiar example of aberration* (Baker, *Astronomy*).
[from Latin *aberrationem,* from *aberrare* wander away]

abfarad, *n.* the CGS unit of capacitance, equal to 10⁹ farads.

cap, fāce, fäther; best, bē, tėrm; pin, five;
rock, gō, ôrder; oil, out; cup, pùt, rüle;
yü in use, *yù* in uric;
ng in bring; *sh* in rush; *th* in thin, ᴛʜ in then;
zh in seizure.
ə = *a* in about, *e* in taken, *i* in pencil, *o* in lemon, *u* in circus

abhenry, *n., pl.* **-ries** or **-rys.** the CGS unit of inductance, equal to 10^{-9} henry.

abiogenesis (ā′bī′ō jen′ə sis), *n.* = spontaneous generation: *The most difficult aspect in developing a theory of abiogenesis has been to account for the origin of the complex unit of life, the cell* (Sidney W. Fox and Angus Wood). [from *a-* not + *biogenesis*]

abiogenic (ā′bī′ō jen′ik), *adj. Biology.* not produced by living organisms; not biogenic: *abiogenic proteins. Some natural petroleumlike hydrocarbons may be abiogenic* (Lawrence Ogden). —**abiogenically,** *adv.*

abiological (ā′bī′ə loj′ə kəl), *adj. Biology.* **1** not connected with living organisms; not biological: *rocks formed by abiological processes.* **2** = abiogenic. —**abiologically,** *adv.*

abiotic (ā′bī ot′ik), *adj. Ecology.* independent of life or living things: *An ecosystem involves interactions between abiotic and biotic factors ... The abiotic environment includes physical and chemical factors which affect the ability of organisms to live and reproduce* (Biology Regents Syllabus). [from *a-* without + *biotic*]

ablastin (ab las′tin), *n. Biochemistry.* a substance in the blood that impairs the ability of certain parasitic germs to reproduce: *The results indicate that the parasites inhibited by ablastin are forming no nucleic acid and very little protein* (William H. Taliaferro). [from *a-* not + Greek *blastos* sprout]

ablate, *v. Geology.* to remove by ablation: *The low sun ... evaporates or "ablates" the ice continually* (Times Literary Supplement). [from Latin *ablatum* carried away]

ablation, *n. Geology.* the wearing down or away of a glacier, snow, or other formation by melting, evaporation, or wind erosion: *The vast amount of ablation ... which a glacier undergoes through the melting of the surface* (Guide to the Western Alps).
ASSOCIATED TERMS: see **erosion.**

ABO (ā′bē′ō′), *Immunology, Genetics.* —*n.* a system of classifying blood groups, used to determine blood compatibility in transfusions. The main groups are A, B, AB, and O.
—*adj.* of or having to do with this system of classification.
Compare **Rh factor.**
▶ The ABO classification is based on the presence or absence of two antigens, A and B. In red blood cells with only antigen A, the blood is group A and the plasma has anti-B antibodies. Red blood cells with only antigen B in the blood are of group B and the plasma has anti-A antibodies. When both antigens are present, the blood is group AB, and has neither antibody. In the blood group O both antibodies are present.

abohm (ab ōm′), *n.* the electromagnetic unit of resistance in the CGS system, equal to 10^{-9} ohm.

abomasum (ab′ə mā′səm), *n., pl.* **-sa** (-sə). *Zoology.* the fourth and true stomach of cows, sheep, deer, and other ruminants, in which the food is digested: *Food substances pass from the rumen (the first stomach in cud chewing animals) into the abomasum and intestine, where they are digested and the products absorbed, as is the case in nonruminant mammals* (John R. Porter). Compare **omasum, reticulum.** [from New Latin *abomasum,* from Latin *ab-* away + *omasum* bullock's tripe]

aboral, *adj. Zoology.* situated away from the mouth; having to do with the end away from the mouth: *The mouth [of a comb jellyfish] is situated at one end (oral), and a sense organ ... at the opposite or aboral end* (Hegner and Stiles, *College Zoology*).

abort, *v.* **1** *Medicine.* to give birth to before the embryo or fetus can live on its own; give birth prematurely: *Of eight female rhesus monkeys ... four of the females conceived but aborted their fetuses, and ... only two ... were able to carry their infants to term* (New Yorker). SYN: miscarry.
2 *Biology.* to be arrested in growth, as an organ; fail to develop: *Thorns, such as those of the rose, are aborted branches* (J. Hogg).
[from Latin *abortum* miscarried]

abortion, *n.* **1** *Medicine.* **a** birth that occurs before the embryo or fetus develops enough to live on its own, usually during the first twelve weeks of pregnancy (*spontaneous* abortion): *Many abortions ... are due to deficiency of vitamins or of other necessary food elements* (Bulletin of Atomic Scientists). SYN: miscarriage. **b** the deliberate expulsion of an embryo or fetus with the intent of destroying it (*induced* abortion): *Abortion involves a complicated moral and ethical choice that women have ... to make* (N.Y. Times).
2 *Biology.* **a** arrested development of an organ at an early stage: *... the partial or complete abortion of the reproductive organs* (Charles Darwin). **b** an imperfectly developed organ.

abortive, *adj. Biology.* imperfectly formed or developed; rudimentary: *an abortive seed or branch, an abortive organ.*

abrade, *v. Geology.* to wear down or away by abrasion; scrape off: *Glaciers abrade rocks.* [from Latin *abradere,* from *ab-* off + *radere* to scrape] —**abradable,** *adj.*

abranchiate (ā brang′kē it *or* ā brang′kē āt), *Zoology.*
—*adj.* having no gills.
—*n.* an animal having no gills.
[from *a-* without + Greek *branchia* gills]

abrasion, *n. Geology.* the action of wearing down or away by friction, especially in reducing the size of rocks by scraping of particles in a stream bed or under a glacier: *Using ... the rocks ... the ice is able to accomplish ... part of its erosion by the process of abrasion.* (Finch and Trewartha, *Elements of Geography*). [from Latin *abrasionem,* from *abradere* scrape away] ASSOCIATED TERMS: see **erosion.**

abrasive, *adj. Geology.* wearing away by rubbing; causing abrasion: *the abrasive action of younger on older rock.*

abrupt, *adj.* **1** *Botany.* suddenly tapering off; truncate: *abrupt leaves.*
2 *Geology.* (of adjacent rock formations) having a sharp boundary at the point of contact. [from Latin *abruptum* broken off]

abscise (ab sīz′), *v. Botany.* to undergo or separate by abscission: *Not all leaves abscise: Pteridophytes and many herbaceous angiosperms retain their dead leaves* (Barbara D. Webster). [from Latin *abscisum* torn off]

abscisic acid (ab sis'ik), *Biochemistry.* a plant hormone that promotes abscission of leaves and dormancy and inhibits growth. *Formula:* $C_{15}H_{20}O_4$ *Abbreviation:* ABA

abscisin (ab sis'in), *n.* = abscisic acid.

abscissa (ab sis'ə), *n., pl.* **-sas** or **-sae** (-sē). *Mathematics.* (in a system of rectangular coordinates) the horizontal coordinate of a point on a graph, usually symbolized *x;* a quantity representing perpendicular distance of a point from the vertical axis. It is positive if the point lies to the right of the vertical (*y*) axis, and negative if the point lies to the left of the vertical axis. Compare **ordinate.** [from Latin (*linea*) *abscissa* (line) cut off, from *abscindere* tear off]

abscission, *n. Botany.* the normal separation of a mature fruit, leaf, or stem from a twig by the formation of a corky layer of young cells at the base. The **abscission layer** of cells protects the twig from decay. [from Latin *abscissionem* a tearing off, from *abscindere* tear off]

absolute, *adj.* **1** not compared with anything else; based on a system of measurement independent of variable standards: *absolute pressure, absolute humidity, absolute magnitude.*
2 of or expressed in absolute temperature.
3 based on some primary units (especially those of length, mass, and time) of invariable value which are taken as fundamental: *the centimeter-gram-second system of absolute units.*
[from Latin *absolutum* loosened or freed from]

absolute alcohol, *Chemistry.* pure ethyl alcohol; ethyl alcohol containing a negligible amount of water.

absolute humidity, *Meteorology.* the mass of water vapor present in a given volume of air, usually expressed in grams of water vapor per cubic meter.

absolute magnitude, *Astronomy.* the magnitude a star would have if it were 10 parsecs (32.6 light years) from the earth, used as a standard for expressing the intrinsic brightness of stars.

absolute pressure, *Physics.* pressure measured with respect to a vacuum.

absolute scale, *Physics.* a scale in which temperatures are measured from absolute zero. The Kelvin and Rankine scales are absolute scales. Also called **absolute temperature scale.**

absolute temperature, *Physics.* temperature measured with respect to absolute zero, usually expressed in kelvins: *Absolute temperature is directly proportional to the average kinetic energy of random motion of the molecules of an ideal gas* (Physics Regents Syllabus).

absolute unit, *Physics.* a unit defined directly in terms of the fundamental units of mass, length, and time, as opposed to relative measurements: *The newton is called an absolute unit of force because its value is defined in terms of the effect of forces on the motion of an object quite independently of the position of the object in the universe or of gravitational pull on it* (Shortley and Williams, *Elements of Physics*).

absolute value, *Mathematics.* the value of a real number regardless of any accompanying sign: *The absolute value of +5, or −5, is 5.*

absolute zero, *Physics.* the lowest possible temperature; zero kelvins. It is about −273.15 degrees Celsius. *Any object with a temperature above absolute zero ... will emit radiation* (Wall Street Journal). *An object is at absolute zero when its internal energy is a minimum* (Physics Regents Syllabus). See also **zero-point energy.**

absorb, *v.* **1** *Chemistry, Physics.* **a** to take in (a liquid, gas, radiant energy, etc.) by chemical or molecular action: *Animal and vegetable matters deposited in soils are absorbed by plants* (Sir Humphry Davy). *Whenever a solid body loses some of that force of attraction by means of which it remains solid, heat is absorbed* (Michael Faraday). *The α radiation from uranium and its compounds is rapidly absorbed in its passage through gases* (Ernest Rutherford). See the picture at **adsorb.** **b** to take in without reflecting: *The rate at which sound energy is absorbed is a function of the type of wall surface and the amount of soft furnishings installed* (Science News).
2 *Physiology.* to take (digested food, oxygen, etc.) into the body's fluids and cells; take in by absorption: *Through the surface of the villi [in the small intestine], amino acids, simple sugars, and other nutrients are absorbed by diffusion or active transport into the blood.* (H. Kolb).
[from Latin *absorbere,* from *ab-* from + *sorbere* suck in]
ASSOCIATED TERMS: In the sense of definition la and b, *absorb* is often contrasted with *reflect.* The waves of radiant energy which enter a substance are said to be *absorbed* when they are not transmitted through the substance, but are changed into some other form of energy, such as heat; the waves which are transmitted or turned back without entering the substance are said to be *reflected.*

absorbance or **absorbancy,** *n. Physics.* the common logarithm of the absorptivity of a substance.

absorptance, *n.* = absorptivity.

absorption, *n.* **1** *Chemistry, Physics.* the act or process of absorbing: *When liquids are brought into contact with the leaves of plants, absorption takes place* (J.H. Balfour). *If we pass an electric current ... across the junction, there will be, by elementary principles, a continuous absorption of energy* (O. Heaviside).
2 *Physiology.* the process by which the end products of digestion, as well as other dissolved liquids and gases, enter the fluids and cells of an organism.
ASSOCIATED TERMS: *Absorption* and *adsorption* are physical and chemical processes by which substances are taken up and retained. In *absorption,* a substance penetrates and enters the inner structure of another substance, usually becoming generally distributed within the absorbing object or material. In *adsorption,* a substance is attracted to and collected on the surface of another, where it may be held weakly or strongly, but without being absorbed.

absorption band, *Physics.* any of the dark bands in an absorption spectrum: *This radiation is most intense at wavelengths very close to the principal absorption band (13 to 17 microns) of the carbon dioxide spectrum* (Scientific American).

---

cap, fāce, fäther; best, bē, tėrm; pin, five;
rock, gō, ôrder; oil, out; cup, pùt, rüle,
yü in use, *yủ* in uric;
ng in bring; *sh* in rush; *th* in thin; ᴛʜ in then;
zh in seizure.
ə = *a* in about, *e* in taken, *i* in pencil, *o* in lemon, *u* in circus

absorption coefficient, 1 *Physics.* a constant for a material indicating the degree to which it absorbs radiation, atomic particles, rays of light, etc., that pass through. **2** *Biology.* the rate at which the human body absorbs a particular substance, usually expressed in milligrams per kilogram of body weight per hour.

absorption line, *Physics.* any of the lines in an absorption spectrum: *Some 66 [chemical elements] have been found in the atmosphere of the sun by their telltale absorption lines in the solar spectrum* (Armin J. Deutsch). Compare **emission line.** See also **spectral line.**

absorption spectrum, *Physics.* a continuous spectrum broken by dark lines or bands, into which white light or other electromagnetic radiation can be separated after passing through a gas or liquid. The lines or bands indicate wavelengths absorbed by the medium through which the radiation has travelled, and they can be analyzed to determine its chemical composition. *The whole science of spectroscopy is based upon this fundamental selectivity of light absorption by atoms. Experiments have shown that hydrogen atoms always absorb the same wavelengths, and every other atom likewise has its own pattern. Like the fingerprints of men, the absorption spectra of atoms identify them completely* (Armin J. Deutsch). Compare **emission spectrum.** See also **absorption line, absorption band.**

absorptivity, *n. Physics.* the ratio of the radiant energy absorbed by a surface to the total of the energy striking the surface: *Let it be supposed that a fraction R of the incident radiation is reflected and a fraction A is absorbed. The quantities R and A are known as the reflectivity and the absorptivity of the surface respectively* (Hardy and Perrin, *Principles of Optics*).

abstrict, *v. Botany.* to separate by abstriction.

abstriction, *n. Botany.* branched spore formation in certain fungi in which portions of the spore-bearing filament (hypha) are separated by the formation of walls (septa). [from Latin *ab-* away + Late Latin *strictionem* a drawing together]

abundant number, *Mathematics.* a whole number whose divisors (other than itself) have a sum greater than the number. The number 18 is an abundant number because the sum of its divisors (9, 6, 3, 2, 1) is greater than 18. Compare **deficient number, perfect number.**

abvolt, *n.* the CGS unit of potential difference, equal to 10^{-8} volt.

abyssal, *adj.* **1** *Oceanography.* of or having to do with the depths of the ocean to which light does not penetrate approximately from 3000 to 6000 meters in depth: *In the deeper parts of the ocean, the abyssal region, there is no ... photosynthetic plant life to serve as a source of food for animals.* (Hegner and Stiles, *College Zoology*). Compare **bathyal, hadal, littoral.** **2** = plutonic: *abyssal rock.*

abyssal plain, *Geology.* a level or nearly level area in the deepest part of the ocean, such as the one lying about 1200 to 1600 kilometers off the eastern coast of North America.

Ac, *abbrev. or symbol.* **1** actinium. **2** altocumulus.

a.c. or **A.C.,** *abbrev.* alternating current.

Acadian (ə kā′dē ən), *adj. Geology.* of or having to do with the formation of mountains in or near the Devonian, especially in the northern Appalachians. [from *Acadia,* the former name of Nova Scotia, a region in southeastern Canada]

acantha (ə kan′thə), *n., pl.* **-thae** (-thē). **1** *Botany.* a prickle. **2** *Zoology.* a spine or prickly fin. **3** *Anatomy.* a spinous process of a vertebra. [from Greek *akantha* thorn, spine]

acanthoid (ə kan′thoid), *adj. Biology.* **1** shaped like a spine. **2** having spines.

acanthopterygian (ak′ən thop′tə rij′ē ən), *Zoology.* —*n.* any of a large group (Acanthopterygii) of fishes with bony skeletons and hard, spiny rays in the dorsal and anal fins, including the sunfish, perch, bass, porgy, mackerel, and swordfish.
—*adj.* of or belonging to this group.
[from Greek *akantha* thorn + *pterygia* fins]

acarid (ak′ər id), *n.* any of an order (Acarina) of small arachnids that includes the mites and ticks. [from New Latin *Acaridae,* from Greek *akari* mite]

acarpelous or **acarpellous** (ā kär′pə ləs), *adj. Botany.* (of a flower) having no carpels.

acarpous (ākär′pəs), *adj. Botany.* (of a plant) not producing fruit; sterile. [from Greek *akarpos,* from *a-* without + *karpos* fruit]

acaudal (ā kô′dəl), *adj. Zoology.* without a tail. [from *a-* not + Latin *cauda* tail]

acaulescence (ak′ô les′ns), *n. Botany.* the condition of being or appearing to be without a stem.

acaulescent (ak′ô les′nt), *adj. Botany.* (of plants) stemless or apparently stemless. [from *a-* not + *caulescent*]

accelerate, *v. Physics.* to change the velocity of (a moving object): *When an electron is accelerated, its kinetic energy, which is one half the product of its mass times the square of its velocity, increases* (J.R. Pierce). [from Latin *acceleratum* hastened, from *ad-* to + *celer* swift]

acceleration, *n. Physics.* **1** a change in velocity, either an increase (*positive acceleration*) or a decrease (*negative acceleration*): *Except in certain special cases the velocity of a moving body changes continuously as the motion proceeds. When this is the case the body is said to move with accelerated motion, or to have an acceleration* (Sears and Zemansky, *University Physics*). **2** the rate of change in the velocity of a moving body. Acceleration is expressed in meters per second per second. *Acceleration is defined as rate of change of velocity. Since velocity is not merely speed, but directed speed, acceleration may appear as changing speed or changing direction, or both* (R.H. Baker).

acceleration of gravity, *Physics.* the acceleration of a freely falling body due to gravitational force. Near the earth's surface, it is equal to about 9.8 meters per second per second. See also **standard gravity.**

accelerator, *n.* **1** *Physics.* any of various devices for greatly increasing the speed and energy of charged particles, such as protons and electrons. Also called **particle accelerator.** Accelerators are either circular, as the cyclotron, betatron, and synchrotron, or linear, as the SLAC (Stanford Linear Accelerator). Accelerators are often used to bombard atomic nuclei, causing them to release new particles and energy. *High-energy particle accelerators (once known as atom-smashers) are tools with which physicists attempt to understand* *the fundamental nature of matter* (D. Spurgeon). *The imposing building erected by the Medical Research Council* [*will*] *house the 45in. cyclotron and the eight million-volt linear accelerator—used for the treatment of malignant disease* (London *Times*).

2 *Chemistry.* any catalyst that increases the rate of a reaction, especially one that speeds up the vulcanization of rubber.

3 *Physiology.* any muscle, nerve, or secretion that increases the speed of a bodily function.

accelerometer (ak sel′ə rom′ə tər), *n.* **1** an instrument for measuring the acceleration of a moving body, such as an aircraft or rocket: *Used with great success in the "Redstone" missile, this directional system ... employs gyroscopes and accelerometers to keep track of the direction and velocity of the missile* (Newsweek).
2 an instrument for measuring vibrations: *The instrument is essentially a piezo-electric accelerometer with an output that is electrically differentiated. ... Thus, it is inherently isolated from extraneous motions such as building or operating-table vibrations* (Science).

accent, *n.* **1** a mark at the right of a number indicating minutes of a degree, two such marks indicating seconds, as in 20°10′30″ (read *20 degrees, 10 minutes, 30 seconds*).
2 one, two, or three marks at the right of a number used to denote feet, inches, and lines (twelfths of an inch), as in 3′6″7‴ (read *3 feet, 6 inches, 7 lines*).
3 a mark placed at the right of a letter so that it may be used to represent different mathematical quantities, as in *b′* (read *b prime*), *b″* (read *b double prime* or *b second*), *b‴* (read *b triple prime* or *b third*).

accept, *v.* **1** *Chemistry.* to take up and combine with: *Not all bases accept protons with equal ease. We use the term ... basic strength to designate ease of proton acceptance* (Baxter and Steiner, *Modern Chemistry*).
2 *Immunology.* to take into the body without rejecting: *A mouse whose thymus has been removed ... will accept a skin graft from an unrelated animal, whereas normal mice invariably reject such foreign grafts* (Scientific American).

acceptor, *n.* *Chemistry.* an atom, molecule, or ion that combines with another atom, molecule, or ion, especially an atom that shares two electrons in a bond with another atom but contributes neither electron: *In the modern concept of Brönsted, an acid is simply a proton donor. Accordingly, a base is a proton acceptor* (Dull, *Modern Chemistry*). Compare **donor.**

accessory bud, *Botany.* an additional bud forming in or near the leaf axil, such as a bud on a young stem of the red maple.

accessory fruit, *Botany.* any fruit having conspicuous fleshy parts surrounding an ovary or ovaries, as an apple or strawberry.

acclimate (ak′lə māt *or* ə klī′mit), *v.* = acclimatize.

acclimation (ak′lə mā′shən), *n.* = acclimatization.

acclimatization, *n.* *Biology.* the adaptation of an animal or plant to changes in climate or other environmental conditions: *There is an increase in the total volume of blood in the process of acclimatization to a warmer climate and a similar decrease in adjusting to a colder climate* (T.A. Blair).

acclimatize (ə klī′mə tīz), *v.* *Biology.* to adapt to changes in climate or other environmental conditions.

acclivity, *n.* *Geology.* an upward slope of ground; ascent. Contrasted with **declivity.** [from Latin *acclivitatem,* from *acclivis* ascending]

accommodation, *n.* *Optics.* the automatic adjustment of the focal length of the lens of the eye to see objects at varying distances. The eye muscles adjust the thickness and curvature of the lens and the diameter of the pupil to produce an accurate focus.

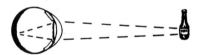

Accommodation:
Distant objects reflect light rays that pass through the lens with little bending needed to converge on the retina, so that the lens is nearly flat, providing little power of refraction. Light rays from near objects require more refraction to focus on the retina, so that the lens becomes thicker to increase its power of refraction.

accordant, *adj.* *Geology.* **1** having nearly the same elevation: *accordant summits.* **2** having similar orientation: *accordant folds.* **3** meeting or joining one another without break or noticeable irregularity: *accordant valleys.*

accrescent (ə kres′ənt), *adj.* *Botany.* growing larger after flowering: *an accrescent calyx.* [from Latin *accrescentem,* from *accrescere* grow, be added to]

accretion, *n.* **1** an increase in size by gradual external addition: *Accretion ... is illustrated by the growth or increase in size of an icicle, or of a snowball rolling down a hill* (Harbaugh and Goodrich, *Fundamentals of Biology*).
2 something formed or resulting from such growth or additions: *The deep floor of the Pacific Ocean is spotted with metallic accretions containing copper, manganese, nickel, molybdenum, and other metals* (J.E. Bardach).
[from Latin *accretionem,* from *ad-* to + *crescere* grow]

accretionary, *adj.* characterized or formed by accretion: *... the debate* [*as to*] *whether the moon's rocks are volcanic or accretionary in origin* (New Scientist).

accretion disk, *Astronomy.* a disk-shaped formation of gases or other interstellar matter around a black hole, neutron star, or other heavenly body: *This formation, called an accretion disk, may be very important in*

cap, fāce, fäther; best, bē, tėrm; pin, five;
rock, gō, ôrder; oil, out; cup, pùt, rüle,
yü in use, *yu̇* in uric;
ng in bring; *sh* in rush; *th* in thin; ᴛʜ in then;
zh in seizure.
ə = *a* in about, *e* in taken, *i* in pencil, *o* in lemon, *u* in circus

some stellar birth sequences. In our own solar system ... the planets and asteroids formed from the accretion disk as the sun grew in the embryo core (S.P. Maran).

accumbent (ə kum′bənt), *adj. Botany.* lying or leaning against something: *accumbent cotyledons.* [from Latin *accumbentum,* from *accumbere* lie down]

acellular (ā sel′yə lər), *adj. Biology.* lacking cells; not cellular: *The concept of protozoans being identified as acellular rather than unicellular. ... This problem is one of definition* (M.T. Jollie).

-aceous (-ā′shəs), *adjective suffix.* having the appearance of; of or like; containing, as in *arenaceous, cretaceous, herbaceous, sebaceous.* The suffix *-aceous* is used in botany to form adjectives of New Latin names ending in *-aceae* for various families of plants, as in *liliaceous* (New Latin *Liliaceae,* the lily family). [from Latin *-aceus*].

acephalous (ā sef′ə ləs), *adj. Zoology.* lacking a distinct head: *acephalous worms.* [from Late Latin *acephalus*]

acerate (as′ər āt *or* as′ər it), *adj. Botany.* needle-shaped and stiff, as the leaves or needles of the pine. [from Latin *acer* sharp]

acerose (as′ə rōs), *adj.* = acerate.

acerous (as′ər əs), *adj. Zoology.* without horns, antennae, or tentacles. [from Greek *akeros,* from *a-* without + *keras* horn]

acervate (ə sėr′vit *or* ə sėr′vāt), *adj. Botany.* growing in closely compacted clusters; heaped. [from Latin *acervatum* heaped up] —**acervately,** *adv.*

acetabular (as′ə tab′yə lər), *adj. Anatomy, Zoology.* of, having to do with, or like an acetabulum.

acetabulum (as′ə tab′yə ləm), *n., pl.* **-lums, -la** (-lə). *Anatomy, Zoology.* **1** the cup-shaped socket at the base of the hipbone into which the ball-shaped top part of the thighbone (femur) fits.
2 a similar structure in the body of an insect.
3 any cup-shaped structure, such as a sucker on the ventral side of a parasitic flatworm.
[from Latin *acetabulum* (literally) small cup]

acetal (as′ə tal), *n. Chemistry.* any of a class of organic compounds obtained by the reaction of aldehydes or ketones with alcohol.

acetaldehyde (as′ə tal′də hīd), *n. Chemistry.* a transparent, colorless liquid aldehyde with a characteristic smell, produced by the partial oxidation of ordinary alcohol. *Formula:* CH_3CHO Also called **aldehyde.** [from *acet(ic) aldehyde*]

acetate (as′ə tāt), *n. Chemistry.* a salt or ester of acetic acid.

acetic (ə sē′tik), *adj.* of, producing, or derived from vinegar or acetic acid. [from Latin *acetum* vinegar]

acetic acid, *Chemistry.* an acrid, pungent, colorless liquid responsible for the characteristic taste and odor of vinegar. It is used in organic synthesis, in the manufacture of cellulose acetate and pharmaceuticals, and as a solvent in various gums and resins. *Formula:* CH_3COOH

acetification (ə sē′tə fə kā′shən), *n. Chemistry.* the process of converting into acetic acid.

acetify (ə sē′tə fī), *v. Chemistry.* to turn into acetic acid.

acetous (ə sē′təs), *adj. Chemistry.* producing or containing acetic acid.

acetyl (ə sē′tl), *n. Chemistry.* the univalent group of acetic acid and its derivatives. *Formula:* CH_3CO-

acetylate (ə set′l āt), *v. Chemistry.* to add one or more acetyl groups to (an organic compound).
—**acetylation,** *n.* the act or process of acetylating.

acetylcholine (ə sē′tl kō′lēn′), *n. Biochemistry.* a substance secreted at the ends of nerve fibers of the somatic and parasympathetic nervous systems, closely associated with the transmission of nerve impulses in the body. It is a derivative of choline. *Formula:* $C_7H_{17}O_3N$

acetyl coenzyme A or **acetyl-coA** (ə sē′tl kō′ā), *n. Biochemistry.* a compound important as an intermediate in the Krebs cycle and in biochemical functions. *Formula:* $C_{25}H_{38}N_7O_{17}P_3S$

acetylene (ə set′l ēn), *n. Chemistry.* a colorless gas formed by the action of water on calcium carbide, by cracking petroleum hydrocarbons with steam or natural gas, and by other methods, and used especially as the starting point for the manufacture of many chemicals. *Formula:* C_2H_2 $(HC \equiv CH)$

acetylene series, *Chemistry.* a series of unsaturated, open-chain hydrocarbons containing a triple bond and having the general formula C_nH_{2n-2}. A hydrocarbon of this series is called an alkyne. Also called **alkyne series.**

achene (ā kēn′), *n. Botany.* any small, dry, hard fruit consisting of one seed with a thin outer covering that does not burst open when ripe, such as the sunflower seed. [from New Latin *achaenium*]
—**achenial** (ā kē′nē əl), *adj.* having to do with or resembling an achene.

Achilles tendon, *Anatomy.* the strong tendon at the back of the leg that connects the muscles in the calf to the bone of the heel. See the picture at **tendon.**

achlamydeous (ak′lə mid′ē əs), *adj. Botany.* lacking a calyx and a corolla. [from *a-* not + Greek *chlamys, chlamydos* mantle]

achondrite (ā kon′drīt), *n. Geology.* a stony meteorite that lacks chondrules: *Anchondrites comprise only six to seven percent of all meteorites ... Because they closely resemble igneous rocks, achondrites are thought to represent early stages of planetary evolution* (Science News).
—**achondritic** (ā kon′drit′ik), *adj.* of or resembling achondrites.

achordate (ā kôr′dāt), *Zoology.* —*adj.* lacking a notochord.
—*n.* an achordate animal.

achromatic (ak′rə mat′ik), *adj. Optics.* **1** refracting white light without breaking it up into the colors of the spectrum; giving an image free from chromatic aberration: *an achromatic telescope or microscope.*
2 *Biology.* consisting of material difficult to stain with the usual stains or dyes: *achromatic cells.* [from Greek *achromatos* colorless]

achromatic lens, *Optics.* a compound lens corrected for chromatic aberration, made by combining two or more lenses of different kinds of glass so that the various colors in light can be made to meet at practically a single focal point.

achromatin (ā krō′mə tin), *n. Biology.* that portion of the mitotic cell nucleus or of individual chromosomes which, under the action of staining agents, remains less highly colored than the rest.

achromatism (ā krō′mə tiz əm), *n. Optics.* freedom from chromatic aberration.

acicula (ə sik′yə lə), *n., pl.* **-lae** (-lē). a needlelike part: *The spines or bristles of some plants and the crystals of certain minerals are aciculae.* [from Late Latin *acicula* pin, diminutive of Latin *acus* needle]
—**acicular** (ə sik′yə lər), *adj.* needle-shaped. —**acicularly**, *adv.*
—**aciculate** (ə sik′yə lit), *adj.* **1** having aciculae. **2** acicular; needle-shaped.

aciculum (ə sik′yə ləm), *n., pl.* **-lums, -la** (-lə). **1** *Botany.* = acicula. **2** *Zoology.* = seta. [New Latin *aciculum,* from Late Latin *acicula* pin]

acid, *n. Chemistry.* any of a class of substances whose water solutions are generally characterized by sour taste, ability to turn litmus dye red, and ability to react with bases and certain metals to form salts. Specifically: **a** a substance that yields hydrogen ions when dissolved in water. **b** a substance that can act as a proton donor. **c** a substance that can act as an electron acceptor. Compare **base.**
—*adj.* Also, **acidic. 1** *Chemistry.* **a** of or containing acids; having the properties of an acid: *an acid solution.* **b** having a pH factor of less than 7; having a relatively high concentration of hydrogen ions (contrasted with *alkaline* especially as a characteristic of soil). **2** *Geology.* containing a large proportion of silica: *Granite is an acidic rock.* Compare **basic.**
▶ Traditional definitions of *acid* and *base* refer to easily recognizable physical and chemical properties. The first structural definition was that of Arrhenius in 1887: acids yield hydrogen ions in a water solution, bases yield hydroxide ions. This definition did not include reactions occurring without the intervention of water, and a more generalized definition was provided by Brönsted and Lowry in 1923: acids are proton (hydrogen ion) donors, while bases are proton acceptors, to which was added a still more generalized definition by Lewis: acids accept a pair of electrons to form a covalent bond, while bases donate such electron pairs. A substance (for example ammonia) may serve as an acid in one reaction and as a base in another. Thus, strictly, a substance is defined as an acid or base according to its behavior in a particular reaction.
ASSOCIATED TERMS: *Acid, base, salt* refer to definite physical properties of a substance; *weak, strong,* and *transition* are classifications of those properties; *pH, alkaline* are measures of the properties.

acid-fast, *adj. Bacteriology.* **1** retaining dye when treated with acid to remove the dye: *Tubercle bacilli are acid-fast.* **2** differentiated by such a stain from closely related forms.

acidification, *n. Chemistry.* the process of acidifying.

acidify, *v. Chemistry.* to change into an acid; make or become acid: *milk allowed to acidify.* —**acidifiable,** *adj.* —**acidifier,** *n.*

acidity, *n.* **1** *Chemistry.* **a** the degree of acid quality: *the acidity of a soil.* **b** the number of replaceable -OH groups in the molecule of a base. **2** *Physiology.* an excess of acid, especially hyperacidity. [from Latin *acidus* sour]

acid number, *Chemistry.* a number expressing the degree of acidity of a substance. It is equal to the number of milligrams of potassium hydroxide required to neutralize the free fatty acids in one gram of the substance. Also called **acid value.**

acidophil (a sid′ə fil), *n. Biology.* a cell that stains readily with acid dyes. Compare **basophil.**
—**acidophilic** (as′ə dof′ə lik), *adj.* that stains readily with acid dyes.

acid radical, *Chemistry.* a radical formed from an organic acid by the removal from the acid of the hydroxyl group (-OH): *A salt always contains a metal or a metallic radical, as well as an acid radical* (R.1. Brownlee).

acid rain, *Ecology.* rain containing a high concentration of acidity (rated on the pH scale as 5.6 or less) resulting from the emission into the atmosphere of pollutants, primarily sulfur and nitrogen oxides, which form sulfuric and nitric acids in raindrops: *Botanists have also become concerned that acid rain is causing ecological damage by killing plant tissues, altering their growth, and making nutrients unavailable* (D.W. Newsom).

acid salt, *Chemistry.* a salt formed from an acid of which only part of the hydrogen has been replaced by a metal or radical.

acidulate (ə sij′ə lāt), *v. Chemistry.* to make acid or slightly acid. —**acidulation,** *n.*

acid value, = acid number.

acinaciform (as′ə nas′ə fôrm), *adj. Botany.* scimitar-shaped: *acinaciform leaves.* [from Latin *acinaces* scimitar + English *-form*]

acinar (as′ə nər), *adj. Anatomy.* having to do with an acinus or acini.

aciniform (ə sin′ə fôrm), *adj Botany, Zoology.* clustered like berries or grapes; acinous.

acinous (as′ə nəs), *adj. Botany, Zoology.* composed of or resembling a cluster of small berries; consisting of acini. [from Latin *acinosus* grapelike, from *acinus* grape, berry]

acinus (as′n əs), *n., pl.* **acini** (as′n ī). **1** *Botany.* **a** one of the small, fleshy berries that make up such compound fruits as the blackberry. **b** the compound fruit that they compose. **c** the stone or seed of a grape or berry. **d** a berry which grows in clusters, such as grapes or currants.
2 *Anatomy.* a minute lobule; one of the small terminal sacs with constricted lumen in a lung or exocrine gland. [from Latin *acinus* berry, grape]

aclinic (ā klin′ik), *adj. Geology.* having no magnetic dip or inclination. The **aclinic line** is another name for the magnetic equator, at which a magnetic needle balances horizontally without dipping. [from Greek *aklines* not bending]

---

cap, fāce, fäther; best, bē, tėrm; pin, fīve;
rock, gō, ôrder; oil, out; cup, pùt, rüle;
yü in use, *yù* in uric;
ng in bring; *sh* in rush; *th* in thin, ŦH in then;
zh in seizure.
ə = *a* in about, *e* in taken, *i* in pencil, *o* in lemon, *u* in circus

acoelomate (ā sē′lə māt), *Zoology.* —*adj.* being without either a true or false body cavity (coelom), as the flatworms. —*n.* an acoelomate worm; flatworm.

acoelous (ā sē′ləs), *adj. Zoology.* having no digestive tract or body cavity. [from Greek *akoilos* not hollow, from *a-* not + *koilos* hollow]

acontium (ə kon′shē əm), *n., pl.* **-tia** (-shē ə). *Zoology.* any of several long, delicate threads equipped with stinging cells, arising from the septa of some sea anemones and protruding when the animal contracts. [from New Latin *acontium,* from Greek *akontion,* ultimately from *akē* point]

acotyledon (ā′kot l ēd′n), *n. Botany.* a plant without cotyledons, such as a fern or moss. SYN: cryptogam.
—**acotyledonous** (ā′kot l ēd′ə nəs), *adj.* having no cotyledons.

acoustic or **acoustical**, *adj.* **1** *Physics.* having to do with sound or the science of sound: *acoustic feedback, the acoustical performance of a room.*
2 *Anatomy, Physiology.* having to do with the sense or organs of hearing.
[from Greek *akousikos,* from *akouein* hear]

acoustic nerve or **acoustical nerve**, *Anatomy.* = auditory nerve.

acoustics, *n. Physics.* **1** *pl. in form and use.* the structural features of a room, hall, auditorium, etc., that determine how well sounds can be heard in it; acoustic qualities.
2 *pl. in form, sing. in use.* the science of sound: *Another device for focusing radio waves is borrowed from acoustics. It uses a horn to funnel the radio waves to a point* (Scientific American).

acoustic velocity, *Physics.* the rate at which a sound wave travels through a specified medium; the velocity of sound.

acquired character or **acquired characteristic**, *Biology.* a change of structure or function in a plant or animal as a result of use or disuse or in response to the environment: *So-called 'acquired characters' ... are not inherited at all, or else to such a slight degree as not to be of any great importance in heredity and evolution* (J.S. Huxley).
▶ The theory that characteristics acquired by parents during their own lifetime can be transmitted to their offspring was postulated by Lamarck in 1801–1809. The theory lost support after Darwin published his theory of evolution through natural selection in 1859.

acrasin (ə krā′sin), *n. Biochemistry.* a substance produced by certain amoebas as an attractant, causing a group of amoebas to aggregate. [from Greek *akrasia* lack of strength or control]

acro-, *combining form.* tip; end; extremity, as in *acrocentric, acrosome.* [from Greek *akros* tip]

acroblast (ak′rə blast), *n. Biology.* the structure that forms the acrosome in spermatogenesis. [from *acro-* + Greek *blastos* sprout]

acrocarpous (ak′rə kär′pəs), *adj. Botany.* producing fruit at the end or top of the main stem, as certain mosses do. [from Greek *akrokarpos* bearing fruit at the top, from *akros* tip + *karpos* fruit]

acrocentric (ak′rō sen′trik), *Genetics.* —*adj.* having the centromere near one end, so that the chromosome has one long arm and one very short arm: *an acrocentric autosome.*
—*n.* an acrocentric chromosome: *The chromosome involved is one of the four small acrocentrics—those having one arm much shorter than the other* (New Scientist). Compare **metacentric, telocentric.**

acrodont (ak′rə dont), *Zoology.* —*adj.* **1** attached by the base to the edge of the jawbone, without sockets: *acrodont teeth.* **2** having teeth so attached.
—*n.* an acrodont animal: *Some lizards are acrodonts.* [from *acro-* + Greek *odontos* tooth]

acrodrome (ak′rə drōm), *adj. Botany.* having the main veins coming together and uniting at the tip of the leaf: *an acrodrome plant.* [from *acro-* + Greek *dromos* a running]

acrogen (ak′rə jən), *n. Botany.* a plant growing only at the top or apex, such as ferns and mosses. [from *acro-* + *-gen*] —**acrog′enous**, *adj.* of or having to do with acrogens.

acromial (ə krō′mē əl), *adj.* of or having to do with the acromion.

acromion (ə krō′mē ən), *n. Anatomy.* the outer, triangular end of the scapula, to which the collarbone is connected and which forms the point of the shoulder. [from Greek *akrōmion,* from *akros* end, tip + *ōmos* shoulder]

acropetal (ə krop′ə tl), *adj. Botany.* developing from below toward the top or apex (used especially of the order in which the parts of a plant develop). [from *acro-* + Latin *petere* to seek]

acrosin (ak′rə sən), *n. Biochemistry.* an enzyme in the acrosome which digests the protein protective layer around the egg cell.

acrosome (ak′rə sōm), *n. Biology.* a minute structure at the front end of a sperm cell. It produces enzymes involved in penetrating the egg cell. *The contents of the acrosome ... are released in the form of a filament when the sperm touches an egg* (New Scientist). [from *acro-* + *-some*]
—**acrosomal** (ak′rə sō′məl), *adj.* of an acrosome: *the acrosomal filament.*

acrospire, *n. Botany.* the first sprout appearing in the germination of grain. [from *acro-* + Greek *speira* a coil]

acrylate (ak′rə lāt), *n. Chemistry.* a salt or ester of acrylic acid.

acrylic, *adj. Chemistry.* having to do with, containing, or derived from acrylic acid: *acrylic plastics, acrylic resin.* [from Latin *acris* sharp + English *-yl* a radical (as in *acetyl*) + *-ic*]

acrylic acid, *Chemistry.* a colorless, pungent liquid that polymerizes easily, is soluble in water and alcohol, and is used in making acrylic resins. *Formula:* $CH_2{:}CHCO_2H$

ACTH (ā′sē′tē′āch′), *n. Biochemistry.* a hormone of the anterior lobe of the pituitary gland which stimulates the cortex of the adrenal gland to produce cortisone and other hormones: *ACTH is chemically quite different from cortisone, since it is ... of protein nature, as are most of the building units of the body* (A.J. Birch). [from *a(dreno)-c(ortico)-t(ropic) h(ormone)*]

actin, *n. Biochemistry.* a protein component of muscle cells that acts with another protein, myosin, in muscle contraction: *When actin and myosin are mixed they form actomyosin, which can be made into threads that rapidly decompose ATP. At the same time, the threads contract* (McElroy, *Biology and Man*). [from Latin *actus* motion + English *-in*]

actin-, *combining form.* the form of **actino-** before vowels, as in *actinal, actinic, actinoid.*

actinal (ak′tə nəl), *adj. Zoology.* having to do with or situated on the side of a radially symmetrical animal on which the mouth or oral area is found. Compare **abactinal. —actinally,** *adv.*

actinic (ak tin′ik), *adj. Chemistry.* of, having to do with, or exhibiting actinism. [from Greek *aktis, aktinos* ray]

actinide (ak′tə nīd), *n. Chemistry.* any of the series of heavy, radioactive metallic elements belonging to the series with atomic numbers 89 or 90 through 103. The properties of the actinides differ only slightly with atomic number. *Because the first member of the series is actinium, this family of elements is called actinides (as the rare earths are called lanthanides). The family resemblance between the actinides and lanthanides provides a key to chemical separation and recognition of the individual transuranium elements* (Albert Ghiorso and Glenn T. Seaborg).

actinism, *n. Chemistry.* the action or property in radiant energy that produces chemical changes, for example in photography, where certain wavelengths of electromagnetic radiation create strong effects on photographic plates or film.

actinium (ak tin′ē əm), *n. Chemistry.* a radioactive, metallic, trivalent element somewhat like radium, found in pitchblende after uranium has been extracted, or obtained from radium by bombardment with neutrons. *Symbol:* Ac; *atomic number* 89; *atomic weight* 227 (the most stable isotope, with a half-life of 21.8 years); *melting point* 1050°C; *boiling point* 3200°C; *oxidation state* +3.

actinium series. *Chemistry.* the series of isotopes produced by radioactive decay of actinium.

actino-, *combining form.* 1 ray; radial; radiant; radiating; radiation, as in *actinomorphic, actinometer.* 2 ray of light, as in *actinology.* Also spelled **actin-** before vowels. [from Greek *aktis, aktinos* ray]

actinoid (ak′tə noid), *adj. Zoology.* having the form of rays; radiated: *A starfish is actinoid.*

actinolite (ak tin′ə līt), *n. Mineralogy.* a species of green amphibole containing iron, usually occurring in needle-shaped crystals.

actinology (ak′tə nol′ə jē), *n.* the branch of physics dealing with the chemical action of light.

actinometer, *n.* an instrument for measuring the degree of actinic action in radiant energy.

actinometric (ak′tə nə met′rik), *adj.* of or having to do with actinometry.

actinometry (ak′tə nom′ə trē), *n.* the measurement of intensities of radiant energy.

actinomorphic (ak′tə nə môr′fik), *adj. Botany.* having radial symmetry: *Actinomorphic ... flowers like those of the Mustard, etc., ... are capable of division in more than one direction* (H.W. Youngken). Compare **zygomorphic.**

actinomycete (ak′tə nō mī′sēt′), *n. Bacteriology.* any of a group of bacteria found in soil that are structurally similar to certain fungi. Antibiotics such as streptomycin are derived from some actinomycetes. [from *actino-* + Greek *mykes* fungus]

actinon (ak′tə non), *n.* a gaseous, radioactive isotope of radon, formed by the decay of actinium. Symbol: An

actinophage (ak tin′ə fāj), *n. Microbiology.* a bacteriophage that attacks actinomycetes.

actinost (ak′tə nost), *n. Zoology.* one of the bones in a fish immediately supporting the rays of the pectoral and ventral fins. [from *actin-* + Greek *osteon* bone]

action current, *Physiology.* the change in action potential when nerve impulses travel along the fibers of a sensory or motor nerve.

action potential, *Physiology.* a difference of electric potential on the surface of a cell, nerve fiber, or muscle fiber resulting from stimulation and associated with the transmission of a nerve impulse: *Arrival of the impulse excites the muscle fiber so that an action potential spreads over the muscle surface, resulting in contraction of the fiber* (McElroy, *Biology and Man*).

activate, *v.* 1 *Chemistry.* to make (an atom, molecule, etc.) capable of reacting or of speeding up a reaction. See **activation energy.**
2 *Physics.* to make (a substance) radioactive by bombardment with neutrons, protons, or other nuclear particles. See **activation analysis.**
3 to make (charcoal, carbon, etc.) capable of adsorbing impurities, especially in the form of gases: *Basically, charcoal is activated by keeping it at red heat over an extended period with limited access of air. This alters the normally smooth surface, creating a great network of pores with vast adsorption area* (N.Y. Times).
4 to purify (sewage) by treating it with air and bacteria: *activated sludge.*
—activation, *n.* **—activator,** *n.*

activation analysis, *Chemistry.* a method of identifying and measuring the elements in a substance by analyzing the radiation given off when the substance is bombarded with nuclear particles, especially neutrons: *The usual method of detecting very small traces of elements is by activation analysis* (New Scientist).

activation energy, *Chemistry.* the minimum energy needed in an atomic system for a particular process to occur, as for a molecule to enter into a reaction: *A mixture of hydrogen and oxygen gases remains inert at room temperatures until the necessary activation energy is introduced by means of an electric spark* (Dull, *Modern Chemistry*).

active immunity, *Immunology.* immunity from a disease due to the production of antibodies by the organism: *When the body becomes immune by working against poisons and producing its own antibodies, as with smallpox, we say the body is protected by active im-*

---

cap, fāce, fäther; best, bē, tèrm; pin, five;
rock, gō, ôrder; oil, out; cup, pùt, rüle;
yü in use, *yu* in uric;
ng in bring; sh in rush; th in thin, ᴛʜ in then;
zh in seizure.
ə = *a* in about, *e* in taken, *i* in pencil, *o* in lemon, *u* in circus

munity (Beauchamp, *Everyday Problems in Science*). Contrasted with **passive immunity.**

active mass, *Chemistry.* the molecular concentration of the substances involved in a reaction, expressed in terms of moles per liter.

active site, *Biochemistry.* the part of an enzyme or antibody at which a catalytic reaction occurs: *Some mechanism capable of discriminating between the side chains of the amino acids of the substrate must be incorporated in the enzyme molecule at or near the active site* (Scientific American).

active transport, *Biology.* the movement of substances across a cell membrane by means of chemical energy in a direction opposite to that of diffusion: *The transmission of nerve pulses, heart function, and kidney processes are coupled to a process known as active transport—the movement of sodium and potassium ions across cell membranes* (J.G. Lepp). *Active transport is a process in which cellular energy is used to move particles through a membrane. This movement is from a region of low concentration toward a region of high concentration. Carrier proteins embedded in the cell membrane aid the transport of materials* (Biology Regents Syllabus).

ASSOCIATED TERMS: *Diffusion, osmosis,* and *active transport* are physiological processes involving the movement or exchange of dissolved substances. *Diffusion* is the movement of molecules from a region of high concentration to one of low concentration. *Osmosis* is the passage of a solvent, such as water, through a semipermeable membrane from a region of low solute concentration to a region of higher solute concentration. *Active transport* occurs when neither diffusion nor osmosis are able to meet the demands of living tissues, and the cell membrane itself becomes actively involved in the transport of fluids.

activity series, = electromotive series.

actomyosin (ak′tō mī′ə sən), *n. Biochemistry.* a combination of the proteins actin and myosin, constituting the structural unit in muscle cells by means of which muscles are enabled to contract.

acuate (ak′yü it), *adj. Botany, Zoology.* needle-shaped; sharp-pointed. [from New Latin *acuatus,* from Latin *acus* needle]

aculeate (ə kyü′lē it), *adj.* **1** *Botany.* having sharp prickles growing from the bark; prickly. **2** *Zoology.* **a** having minute spines on the wing membranes, as some moths do. **b** equipped with a sting, as a wasp or bee. [from Latin *aculeatus* having stings, from *aculeus* thorn, sting]

aculeus (ə kyü′lē əs), *n., pl.* **-lei** (-lē ī). **1** *Botany.* a prickle growing from the bark, as in the rose or blackberry. **2** *Zoology.* the sting of wasps, bees, or other insects. [from Latin *aculeus* thorn]

acuminate (ə kyü′mə nit), *adj. Botany.* (of leaves) tapering to a point. [from Latin *acuminatus* sharpened]

acute angle, *Mathematics.* an angle less than a right angle; any angle less than 90 degrees. See the picture at **angle.**

acute triangle, *Mathematics.* a triangle having three acute angles.

acyclic (ā sī′klik *or* ā sik′lik), *adj.* **1** *Botany.* not cyclic; arranged spirally rather than in circles or whorls: *acyclic flower parts.*

2 *Chemistry.* having an open-chain structure: *Acyclic compounds ... contain no ring structures of the atoms, and are frequently called chain compounds* (Parks and Steinbach, *Systematic College Chemistry*).

acyl (as′əl), *n.* an acid radical formed from a carboxylic acid: *Coenzyme A is involved in the transfer of acetyl and other acyl groups* (Robert S. Harris). *Formula:* -CO

—**acylate** (as′ə lāt), *v.* to introduce an acid radical into (a substance). —**acylation,** *n.*

Adam's apple, *Anatomy.* the slight lump at the front of a person's throat, formed by the thyroid cartilage of the larynx. It is normally more evident in men than in women. Its name derives from the notion that a piece of the forbidden fruit stuck in Adam's throat. Technically it is sometimes called *laryngeal prominence* or *laryngeal protuberance.*

adapt, *v.* **1** *Biology.* to change in structure, form, or habits so as to fit different conditions; undergo adaptation: *The Australopithecines could not balance themselves as well as man and were still in the process of adapting to erect progression* (Science).

2 (of a sense organ) to become adjusted to a varying condition: *There is a detectable difference in the amount of hormone from the sinus-gland in the eye stalks of the light and dark adapted prawn* (Science News).

[from Latin *adaptare* to join or fit to]

adaptation, *n.* **1** *Biology.* a change in structure, form, or habits to fit different conditions: *Animals and plants survive by adaptation. ... This adaptation is achieved through natural selection of the best fitted genotypes* (Bulletin of Atomic Scientists). *So far only Mongoloids, in the Andes and Tibet, are believed to have shown altitude adaptation, in that they have relatively large hearts and large quantities of blood rich in oxygen-carrying red corpuscles* (Harper's).

ASSOCIATED TERMS: see **evolution.**

2 *Physiology.* adjustment of a sense organ to a varying condition: *The function of the pupil is to regulate the quantity of light entering the eye ... This process is known as adaptation* (Sears and Zemansky, *University Physics*).

—**adaptational,** *adj.* of or involving adaptation: *The modifications which insect larvae undergo may be divided into two kinds—developmental, and adaptational or adaptive* (John Lubbock). —**adaptationally,** *adv.*

adaptive, *adj. Biology.* of or characterized by adaptation; showing adaptation; enabling the organism to fit into a certain environment or situation. —**adaptively,** *adv.* —**adaptiveness,** *n.*

adaptive convergence, *Biology.* the tendency in distantly related animals or plants to assume similar characteristics, as of form, structure, or habits, under similar conditions. The similar shape of sharks (fish) and whales (mammals) is the result of adaptive convergence. *Adaptive convergence often occurs when animals of different groups come to live in a common habitat* (Storer, *General Zoology*). Compare **convergent evolution.**

adaptive radiation, *Biology.* the tendency in closely related animals or plants to evolve into strikingly different forms because of different environmental requirements. The diverse forms of bats and mice result from adaptive radiation. *The adaptive radiation of the reptiles ... developing terrestrial, marine, and fly-*

ing forms—every variation we are familiar with among modern mammals (R.M. Garrels).

adaxial (ad ak′sē əl), *adj. Biology.* situated on the side nearest to the axis of an organ or organism. Compare **abaxial.**

add, *v. Mathematics.* to unite or combine (two or more quantities) into one quantity; to find the sum: *When several numbers are added together, it is indifferent in what order the numbers are taken* (Smith, *Algebra*).

addend (ad′end *or* ə dend′), *n. Mathematics.* a number or quantity to be added to another number or quantity: *In the problem "421 + 365 = ?" 365 is the addend.* Contrasted with **augend.**

addition, *n.* **1** *Mathematics.* the operation or process usually indicated by the sign +, of combining separate numbers or quantities into one number known as the sum.
2 *Chemistry.* the process of combining several substances or of incorporating one substance into another: *Because addition reactions take place more easily than substitution reactions, unsaturated compounds tend to be more reactive than saturated compounds* (Chemistry Regents Syllabus).

additive (ad′ə tiv), *n. Chemistry.* a substance combined with or incorporated into another substance to preserve it, increase its effectiveness, etc.: *food additives. The effect of certain copper-based fungicides is greatly enhanced when they contain traces of additives—compounds of nickel, chromium or manganese* (New Scientist).
—*adj. Mathematics.* involving addition: *We construct the additive function by introducing an algebraic commutative group and a logarithmic expression* (Jean Piaget). See also **additive inverse.**

additive compound, *Chemistry.* a compound formed by addition, involving conversion of a double bond into a single bond.

additive inverse, *Mathematics.* either of two numbers whose sum is zero. The additive inverse of +5 is −5; the additive inverse of −5 is +5.

address, *n.* a number, label, or other symbol identifying a particular location where information is stored in the memory of a computer.
—**addressable,** *adj.* capable of being addressed; accessible by means of an address or addresses.

adduct, *v. Physiology.* to draw (an arm, finger, toe, etc.) inward toward the main axis of the body or of an extremity. Contrasted with **abduct.** [from Latin *adductus* brought toward]

adduction, *n. Physiology.* the drawing of a part inward toward the main axis of the body or of an extremity. Contrasted with **abduction.**

adductor, *n. Anatomy.* a muscle that adducts, such as the muscle that moves the thumb inward against the fingers. Contrasted with **abductor.**

adenine (ad′n ēn′), *n. Biochemistry.* a substance present in nucleic acid in cells. It is one of the purine bases of DNA and RNA. *The DNA molecule resembles a ladder that has been twisted into a helix. ... The rungs, which join two sugar units, are composed of pairs of nitrogenous bases: either adenine paired with thymine or guanine paired with cytosine* (Seymour Benzer). *Formula:* $C_5H_5N_5$ *Abbreviation:* A See the picture at **DNA.** [from Greek *aden* gland + English *-ine* (so

called because found originally in the pancreatic gland of an ox)]

adenohypophysis (ad′n ō hī pof′ə sis), *n. Anatomy.* the anterior, glandular part of the pituitary gland. It produces and secretes many important hormones, such as ACTH, growth hormone, prolactin, thyroid-stimulating hormone (TSH), luteinizing hormone (LH), and follicle-stimulating hormone (FSH). The posterior, neural part of the pituitary is called the **neurohypophysis.** [from Greek *aden* gland + *hypophysis* outgrowth]

adenoid (ad′n oid), *Anatomy.* —*adj.* **1** of the lymphatic glands or lymphoid tissue.
2 like a gland; glandular.
—*n.* Usually **adenoids,** *pl.* enlarged lymphoid tissue in the upper part of the throat, at the back of the nose, often hindering normal breathing and speaking.
—**adenoidal** (ad′ə noi′dəl), *adj.* **1** of lymphoid tissue; adenoid. **2** having adenoids.

adenosine (ə den′ə sēn′), *n. Biochemistry.* a substance composed of adenine and ribose (a pentose sugar), found in tissue, especially muscle tissue, and in ribonucleic acid. It is important in muscle contraction and the metabolism of sugars. *Formula:* $C_{10}H_{13}N_5O_4$

adenosine diphosphate (dī fos′fāt), *Biochemistry.* a compound of adenosine and two phosphate groups, formed from adenosine triphosphate in the muscles. *Formula:* $C_{10}H_{15}N_5O_{10}P_2$ *Abbreviation:* ADP

adenosine monophosphate (mon′ə fos′fāt), *Biochemistry.* **1** = adenylic acid. **2** a cyclic isomer of adenylic acid, functioning as a regulatory agent in many cellular and enzymatic processes. Also called **cyclic AMP.** *Abbreviation:* AMP

adenosine triphosphatase (trī fos′fə tās), = ATPase.

adenosine triphosphate (trī fos′fāt), *Biochemistry.* a compound of adenosine and three phosphate groups. The removal of one or two phosphate groups releases large amounts of energy which are used for many biochemical processes, such as muscle contraction, the metabolism of sugars, and photosynthesis. *Formula:* $C_{10}H_{16}N_5O_{13}P_3$ *Abbreviation:* ATP

adenovirus (ad′n ō vī′rəs), *n. Biology.* any of a large group of viruses that contain DNA and attack mucous tissues, especially of the respiratory tract, originally found in human adenoid tissue.
—**adenoviral,** *adj.*

adenylic acid (ad′n il′ik), *Biochemistry.* an acid composed of adenine, ribose, and phosphoric acid, formed in the body from red blood cells and muscle tissue. It is the phosphoric acid ester of adenosine, and is an intermediate in the release of energy for muscular and other cellular activity. *Formula:* $C_{10}H_{14}N_5O_7P$ Also called **adenosine monophosphate.**

---

cap, fāce, fäther; best, bē, tėrm; pin, fīve;
rock, gō, ôrder; oil, out; cup, pùt, rüle,
yü in use, *yu̇* in uric;
ng in bring; *sh* in rush; *th* in thin, ᴛʜ in then;
zh in seizure.
ə = *a* in about, *e* in taken, *i* in pencil, *o* in lemon, *u* in circus

adhesion, *n.* **1** *Physics.* the molecular attraction exerted between the surfaces of unlike bodies in contact, as a solid and a liquid.
2 *Medicine.* a growing together of body tissues that are normally separate, especially after surgery. [from Latin *adhaesionis,* from *adhaerere* to stick to, adhere]

adiabatic (ad′ē ə bat′ik), *adj. Physics.* occurring without loss or gain of heat: *No entirely adiabatic process occurs in practice, but the expansion of steam in the cylinder of a steam engine ... and the compression of the air ... in an air compressor are all approximately adiabatic* (Shortley and Williams, *Elements of Physics*). [from Greek *adiabatos* impassable] —**adiabatically,** *adv.*

adiabatic lapse rate, *Meteorology.* the rate at which air cools as it rises. When no condensation occurs, it is called the **dry adiabatic lapse rate,** and is equal to about 1 degree Celsius per 100 meters. The **wet adiabatic lapse rate,** or **saturation adiabatic lapse rate,** where condensation does occur, equals about 3 to 6 degrees Celsius per 1000 meters.

adipate (ad′ə pāt), *n. Chemistry.* a salt or ester of adipic acid.

adipic (ə dip′ik), *adj. Chemistry.* of or derived from fatty or oily substances. **Adipic acid** is a fatty acid found in beet juice or prepared synthetically, used in making polyurethane foams, food additives, etc. *Formula:* $C_6H_{11}O_4$ [from Latin *adeps, adipis* fat]

adipose (ad′ə pōs), *Biology.* —*adj.* consisting of or resembling fat; fatty. **Adipose tissue** is a type of connective tissue made up of rounded or polygonal cells containing droplets or globules of fat; the cells are found under the skin and around organs such as the kidneys. —*n.* the animal fat stored in fatty tissues. [from Latin *adeps, adipis* fat]

adipose fin, *Zoology.* a small, fatty, fin-shaped projection behind the dorsal fin of certain fishes, such as trout and salmon, that lacks supporting rays.

aditus (ad′ə təs), *n., pl.* **-tus, tuses.** *Zoology.* a canal leading inwards, as in sponges. [from Latin *aditus* entrance]

adjacent angle, *Mathematics.* one of a pair of angles having one side in common and interiors that do not intersect.

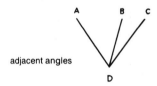
adjacent angles

Angle ADB and angle BDC are adjacent angles.

adnate (ad′nāt), *adj. Biology.* (of unlike parts of plants or animals) growing together or adhering throughout their length.
[from Latin *adnatum* grown onto]
—**adnation,** *n.* the condition of being adnate.

adnexa (ad nek′sə), *n.pl. Anatomy.* parts appended or adjunct to another or others, such as the eyelids and tear glands in relation to the eyeball. [from Latin *adnexa,* from *adnectere* connect, from *ad-* to + *nectere* bind]
—**adnexal,** *adj.* appended or adjunct to another or to other parts.

adobe (ə dō′bē), *n. Geology.* a mixture of clay and silt, usually calcareous, found in desert basins, especially in southwestern North America. It is used to make sun-dried bricks. **Adobe flats** are smooth plains covered with sandy clay deposited by streams which flow during heavy rains.

adoral (ad ôr′əl), *adj. Zoology.* of or near the mouth: *adoral cilia, the adoral zone of a protozoan.* [from *ad-* to + *oral*]

ADP, *abbrev.* adenosine diphosphate.

adradial (ad rā′dē əl), *adj. Zoology.* situated near or beside a radial part, such as the arm of a starfish. [from *ad-* to + *radial*]

adrenal (ə drē′nl), *adj. Anatomy.* **1** near or on the kidney; suprarenal. The **adrenal gland** is either one of two ductless glands, on or near the upper part of the kidneys of vertebrates, whose outer wall or cortex secretes cortisone and other important hormones and whose inner part or medulla secretes adrenaline and noradrenaline. Also called **suprarenal gland.**
2 of, belonging to, or derived from the adrenal glands: *Stimulation of the adrenal cortex releases adrenocortical hormones accompanied by a depletion of adrenal ascorbic acid* (Time).
—*n.* Usually **adrenals,** *pl.* an adrenal gland: *The adrenals [are] famous as producers of adrenalin and anti-arthritis cortisone, among other hormones* (Science News).
[from Latin *ad-* near + *renes* kidneys]

adrenaline or **adrenalin** (ə dren′l ən), *n.* **1** *Biochemistry.* a hormone secreted by the inner part or medulla of the adrenal glands. Adrenaline speeds up the heartbeat and thereby increases bodily energy and resistance to fatigue. *Formula:* $C_9H_{13}NO_3$ *When the sense organs of the animal transmit to the brain impulses which are associated with danger or other situations needing vigorous action, motor impulses are relayed to the adrenal medulla, which releases adrenaline into the blood.* (Mackean, *Introduction to Biology*). Also called **epinephrine.**
2 Adrenalin. a trademark for a levorotatory form of adrenaline extracted from animals or prepared synthetically, used in medicine to constrict blood vessels, stimulate the heart, relax muscles, and to counter allergic reactions, anesthesia, and cardiac arrest.

adrenergic (ad′re nėr′jik), *adj. Physiology.* producing or activated by adrenaline or an adrenaline-like substance: *an adrenergic nerve fiber. The intense adrenergic stimulation preceding stress-induced shock is well recognized* (Science). [from *adren*(aline) + Greek *ergon* work + English *-ic*]
ASSOCIATED TERMS: *Adrenergic* and *cholinergic* are terms used to describe opposing actions of sympathetic and parasympathetic nerve endings. Adrenergic fibers release adrenaline while cholinergic fibers release acetylcholine; adrenergic chemicals stimulate transmission of nerve impulses across synapses, causing the heartbeat to speed up, while cholinergic chemicals inhibit the transmission of such impulses, causing the heartbeat to slow down.

adrenocortical (ə drē′nō kôr′tə kəl), *adj. Biochemistry.* having to do with or derived from the outer wall or cortex of an adrenal gland: *adrenocortical steroids.*

Some of the adrenocortical hormones have anti-inflammatory effects (New Scientist).

adrenocorticotropic hormone, = ACTH.

adrenolytic (ə drē′nə lit′ik), *Physiology.* —*adj.* inhibiting the action of the adrenergic nerves or the effect of adrenaline: *an adrenolytic drug or agent.*
—*n.* an adrenolytic substance: *The so-called adrenolytics and "sympatholytics" have little more than diagnostic applications in hypertensive disease* (Morris Fishbein).

adsorb, *v. Chemistry.* to take up and hold (a gas, liquid, or dissolved substance) in a thin layer of molecules on the surface of a solid substance: *Solids ... with extremely fine pores tend to condense on their surface any ... substances with which they are in contact ... The thin films of these adsorbed substances are held so tenaciously that great pressures are required for their removal* (Shull, *Principles of Animal Biology*). Compare **absorb.** [from Latin *ad-* to + *sorbere* suck in]
—**adsorbability,** *n.* the condition or quality of being adsorbable.
—**adsorbable,** *adj.* that can be adsorbed: *adsorbable molecules.*

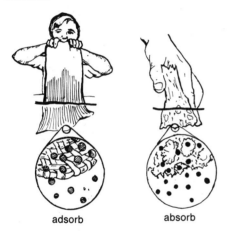

adsorb   absorb

The enlarged segment of cloth shows the molecules of dye being adsorbed and forming a bond on the surface of the fibers of the cloth. The enlarged segment of the sponge shows the molecules of water being absorbed by intermingling with the molecules of the sponge.

adsorbate, *n. Chemistry.* anything that is adsorbed.

adsorbent, *Chemistry.* —*adj.* adsorbing readily: *All colloidal matter is ... highly adsorbent* (Turk, *Introduction to Chemistry*).
—*n.* a substance that adsorbs readily.

adsorption, *n. Chemistry.* the act or process of adsorbing; condensation of gases, liquids, or dissolved substances on the surface of solids: *Adsorption plays an important part in various industrial processes ... vapor is passed through a column or tower stacked with activated carbon, and the vapor particles are adsorbed on the surface* (A.B. Garrett).
ASSOCIATED TERMS: see **absorption.**

adsorptive, *adj.* **1** of or having to do with adsorption. **2** able to adsorb.

adularia (aj′ə lār′ē ə), *n. Mineralogy.* a low-temperature, translucent or transparent variety of orthoclase or microcline feldspar. [from New Latin *adularia,* from *Adula,* a mountain group in the Swiss Alps]

adult, *Biology.* —*adj.* full grown; fully developed: *an adult animal or plant.* SYN: mature.
—*n.* a plant or animal grown to full size and strength.

advection (ad vek′shən), *n. Meteorology.* the transfer of heat, cold, or other properties of air by the horizontal motion of a mass of air: *It is this advection of air masses that causes most of the day-to-day weather changes and the storminess of winter climates in the middle latitudes* (Finch and Trewartha, *Elements of Geography*). [from Latin *advectionem* a conveying]
—**advective,** *adj.* of or having to do with advection.
ASSOCIATED TERMS: In meteorology, *advection* and *convection* are used especially to describe the transfer or movement of heat by masses of air. Although in physics *convection* refers to the displacement of heat by either vertical or horizontal movement of air masses, as applied to the atmosphere in meteorology, it is used in the more restricted sense of vertical movement or transfer. *Advection* denotes the much larger horizontal transfer of heat by winds and air masses.

advection fog, *Meteorology.* a fog that occurs when a body of relatively warm, moist air moves over a cold surface: *Advection fogs often form in the Newfoundland Banks area, where warm air blows over the cold water of the Labrador Current* (G.F. Taylor).

advective (ad vek′tiv), *adj. Meteorology.* of or having to do with the horizontal moving of masses of air, involving the transfer of heat or other properties of air.

adventitia (ad′ven tish′ē ə), *n. Biology.* a membranous structure covering but not properly belonging to an organ, especially a blood vessel. [from Latin *adventicia,* neuter plural of *adventicius* external]

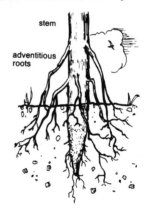

stem

adventitious roots

adventitious (ad′ven tish′əs), *adj. Biology.* appearing out of the normal or usual place, as roots on stems, buds on leaves, or hair where it does not usually grow:

cap, fāce, fäther; best, bē, tèrm; pin, fīve;
rock, gō, ôrder; oil, out; cup, pùt, rüle,
yü in use, yủ in uric;
ng in bring; sh in rush; th in thin, ŦH in then;
zh in seizure.
ə = a in about, e in taken, i in pencil, o in lemon, u in circus

Adventitious roots commonly form on stem cuttings of large numbers of house and greenhouse plants (Emerson, *Basic Botany*). [from Latin *adventicius* external, extraneous]

adventive (ad ven'tiv), *Biology.* —*adj.* introduced into a new environment; not native or established, though growing with cultivation.

—*n.* a plant or animal not native to the environment: *Many foreign weeds are adventives, until they become naturalized* (Norman Taylor).

adverse, *adj. Botany.* turned toward the stem.

aecial stage (ē'shē əl), *Biology.* the phase in the life cycle of certain rust fungi in which aecia are produced.

aeciospore (ē'shē ə spôr), *n. Biology.* a spore produced in an aecium, having two nuclei.

aecium (ē'shē əm), *n., pl.* **-cia** (-shē ə). *Biology.* a cup-shaped structure which produces chains of spores (aeciospores), formed by the fusion of spores of opposite mating types of certain rust fungi. They usually appear on the lower surface of the leaf of the host plant. [from New Latin *aecium,* from Greek *aikia* injury, from *aikēs* unseemly]

aeolian (ēō'lē ən), *adj.* = eolian: *Aeolian sandstone is the consolidated sand of dunes* (Fenton, *The Rock Book*).

aeolotropic (ē'ə lə trop'ik), *adj. Crystallography.* = anisotropic. [from Greek *aiolos* changeful + *tropikos* turning]

aeolotropy (ē'ə lot'rə pē), *n. Crystallography.* = anisotropy.

aeon (ē'ən *or* ē'on), *n.* = eon: *"Aeon" is being increasingly used by earth and planetary scientists as a convenient short synonym for "billion years"* (Science News).

aerate, *v. Chemistry.* **1** to expose to and mix with air: *Water in some reservoirs is aerated and purified by being sprayed into the air.*
2 to charge or mix with a gas, such as carbon dioxide, often under pressure.
3 to expose to chemical action with oxygen; oxygenate by respiration: *The blood ... has been thus aerated in the lungs* (Erasmus Darwin). *So long as the animal is alive the muscles are well aerated by the blood supply* (New Scientist).
[from Latin *aer* air + English *-ate*]
—**aerator,** *n.* any apparatus used for aerating.

aeration, *n. Chemistry.* the act or process of aerating: *One use of aeration occurs when air is artificially introduced into sewage wastewater or effluent in order to create aerobic conditions and foster biological and chemical purification.* (Paul Sarnoff).

aerenchyma (ār eng'kə mə), *n. Botany.* a type of tissue containing large air spaces, found especially in the stems of many aquatic plants: *Hydrophytes contain more air-space tissue (aerenchyma) in their leaves, stems, and roots than mesophytes* (Albert R. Grable). [from Greek *aēr* air + *enchyma* infusion]

aerial, *adj.* **1** *Botany.* growing in the air instead of in soil: *In many species upright branches arise from the rhizomes and grow above the ground, forming aerial stems* (Emerson, *Basic Botany*).

2 *Meteorology.* of the air; atmospheric: *aerial currents.* [from Latin *aer* air + English *-ial*]

aero-, *combining form.* **1** air; of the air, as in *aerometer* = air meter (for measuring the density of air).
2 atmosphere; atmospheric, as in *aerology* = study or science of the atmosphere.
3 gas; of gas or gases, as in *aerodynamics* = the dynamics of gases.
[from Greek *aēr, aeros* air]

aeroallergen (ār'ō al'ər jən), *n. Immunology.* any allergen carried by the air, such as pollen or mold spores.

aerobe (ār'ōb), *n. Bacteriology.* an aerobic bacterium or other microorganism; an organism that can live only where there is atmospheric oxygen. Contrasted with **anaerobe.** [from Greek *aēr* air + *bios* life]

aerobic, *adj.* **1** *Bacteriology.* living and growing only where there is atmospheric oxygen: *A city plant uses aerobic bacteria, those that work in the presence of oxygen, to transform waste matter into harmless liquid or gas* (Atlantic).
2 *Biology.* having to do with or occurring in the presence of atmospheric oxygen: *The second stage [of cellular respiration] is called the aerobic stage since it requires molecular oxygen* (Otto and Towle, *Modern Biology*). Contrasted with **anaerobic.**
—**aerobically,** *adv.*

aerobic respiration, *Biology.* a process in which uncombined atmospheric oxygen (O_2) serves as the final hydrogen acceptor during the breakdown of organic compounds, such as glucose, with the result that water is formed. Carbon dioxide is also produced, and there is a high output of energy. Contrasted with **anaerobic respiration.**

aerobiology, *n.* the study of the biological components of the atmosphere, such as bacteria, viruses, spores, and pollen, and their effect on plants and animals.
—**aerobiological,** *adj.* of or having to do with aerobiology.

aerobiosis (ār'ō bī ō'sis), *n. Biology.* life sustained by an organism in the presence of oxygen. Contrasted with **anaerobiosis.**
—**aerobiotic** (ār'ō bī ot'ik), *adj.* of or characterized by aerobiosis; aerobic.

aerodynamic, *adj. Physics.* of or having to do with the forces of air or other gases in motion: *When the elastic properties are taken into account the wing [of an airplane] will distort under the aerodynamic lift which must be produced to keep the plane in the air* (G.N. Lance).
—**aerodynamically,** *adv.*

aerodynamics, *n.* the branch of dynamics that deals with forces such as pressure or resistance exerted by air or other gases in motion on both flying and wind-blown bodies. Compare **aerostatics.**
▶ Aerodynamics is often regarded as a subdivision of *hydrodynamics,* the study of fluids in motion, whose fundamental equation is *Bernoulli's principle,* which relates pressure, velocity, and elevation at points along a line of flow or streamline.

aerogel (ār'ə jel), *n. Chemistry.* a colloidal solution of a solid in which a liquid is replaced by a gas or by air, thus producing a high degree of porosity.

14

aerogenic, *adj.* **1** gas-producing: *aerogenic bacteria.* **2** having the form of air or gas: *The inhalable, or aerogenic, form of the vaccine gives greater immunity than does the injected form* (Science News Letter).

aerography (ãr og′rə fē), *n. Meteorology.* the study or description of the air or atmospheric conditions.

aerolite (ãr′ə līt), *n. Geology.* a meteorite made up entirely or almost entirely of silicates; a stony meteorite. [from *aero-* + *-lite* stone, rock]
—**aerolitic** (ãr′ə lit′ik), *adj.* made up entirely or almost entirely of stone, as a meteorite.

aerology, *n.* the branch of meteorology that deals with the properties and phenomena of the upper atmosphere.
—**aerological,** *adj.* of or having to do with aerology: *Through aerological investigations scientists have learned much about the atmosphere and the weather* (J.V. Finch). —**aerologically,** *adv.*

aeronomy, *n.* the science that deals with the physical and chemical conditions of the upper atmosphere: *Aeronomy is concerned with that part of the atmosphere where dissociation and ionization processes are important and includes the ionosphere* (William Swider). [from *aero-* + Greek *nomos* distribution]
—**aeronomical,** *adj.*

aeropause, *n. Meteorology.* the region of the upper atmosphere where aerodynamic support of flight begins to cease and the conditions of outer space are approached. It extends from about 20 to 200 kilometers above the earth.

aerophysics, *n.* the branch of physics that deals with air or the atmosphere.

aerophyte (ãr′ə fīt), *n. Botany.* a plant nourished by the air instead of the soil; epiphyte.

aeroscepsis (ãr′ō skep′sis), *n. Zoology.* the ability to perceive the state or quality of the atmosphere by means of certain sensory organs. [from *aero-* + Greek *skepsis* perception]

aerosol, *n. Chemistry.* a dispersion of very fine colloidal particles suspended in the air or in some other gas: *The stratosphere contains small dustlike particles, or aerosols, which tend to reduce the solar energy input and thermal energy emitted from the earth. Changes in the aerosol concentration can affect the earth's climate* (M.A. Calabrese). [from *aero-* + *sol* colloidal solution, from *sol*(ution)]

aerostatics, *n.* the branch of statics dealing with the static equilibrium of the air and other gases and of solid objects suspended or moving in them. Compare **aerodynamics.**

aerotactic, *adj. Biology.* moving toward or away from oxygen or air: *The bacteria under the coverslip ... are "positively aerotactic," aggregating around the edges where the oxygen concentration is higher than in the centre* (New Scientist).

aerotaxis, *n. Biology.* movement of an organism or part of an organism toward or away from oxygen or air: *Motile bacteria respond to gases* (*aerotaxis*), *light* (*phototaxis*), *and various chemical compounds* (*chemotaxis*) (New Scientist). [from *aero-* + Greek *taxis* arrangement]

aestivate, *v.* = estivate.

A.F. or **a.f.,** *abbrev.* audio frequency (the frequency corresponding to audible sound vibrations, ranging in human beings from about 15 to about 20,000 hertz).

afference (af′ər əns), *n. Physiology.* the action or excitation of afferent nerves.

afferent (af′ər ənt), *adj. Physiology.* carrying inward to a central organ or point: *Sensory, or afferent, neurons are those which conduct impulses from receptors to or toward the central nervous system* (Storer, *General Zoology*). Contrasted with **efferent.** [from Latin *afferentem* carrying to, from *ad-* to + *ferre* bring]

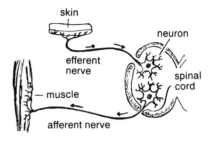

affine (ə fīn′), *adj. Mathematics.* having to do with or designating a transformation that carries parallel lines to parallel lines and finite points to finite points: *Affine geometry ... plays an important part in the mathematics of relativity* (Martin Gardner). [from Latin *affinis* related, bordering on]

affinity, *n.* **1** *Biology.* structural resemblance between species, genera, or other natural groupings that makes a common ancestry probable: *The true Reptiles and the Birds ... are nevertheless related to one another by various points of affinity* (Nicholson, *Manual of Palaeontology*).
2 *Chemistry.* the force that attracts certain elements to others and keeps them combined: *This new attraction we call chemical affinity, or the force of chemical action between different bodies* (Michael Faraday).
[from Latin *affinitatem* relation, from *affinis* related, bordering on, from *ad-* on + *finis* border]

afforestation (ə fôr′ə stā′shən), *n. Ecology.* the changing of open land into forest; establishment of a forest by seeding, planting trees, etc.

aflatoxin (af′lə tok′sən), *n. Biochemistry.* any of a group of poisonous substances produced by certain molds, that contaminate stored food crops, such as peanuts, and cause liver cancer in animals: *Aflatoxins are among the most carcinogenic and hepatotoxic natural products known, and they are highly toxic to most animals* (L.L. Wallen). [from *A*(*spergillus*) *fla*(*vus*), the first species of mold identified as producing this poison + *toxin*]

afterbirth, *n. Embryology.* the placenta and fetal membranes expelled from the uterus shortly after birth.

cap, fāce, fäther; best, bē, tèrm; pin, five;
rock, gō, ôrder; oil, out; cup, pùt, rüle,
yü in use, *yu̇* in uric;
ng in bring; sh in rush; th in thin, ŦH in then;
zh in seizure.
ə = *a* in about, *e* in taken, *i* in pencil, *o* in lemon, *u* in circus

15

afterbrain, *n. Embryology.* the portion of the hindbrain containing the medulla oblongata. Also called **myelencephalon.**

afterglow, *n. Astronomy.* **1** a high arch of radiance seen occasionally in the western sky after sunset. It is caused by the scattering effect on white light from the sun by very fine particles of dust suspended in the upper atmosphere. Compare **airglow.**
2 a luminous trail, as of a meteor.

afterimage, *n. Physiology.* a visual sensation that persists or recurs after the stimulus causing it has ceased to act: *the afterimage — the persistent colored image seen after gazing at a bright light and then averting the eye* (New Scientist).

aftershock, *n. Geology.* a smaller shock coming after the main shock of an earthquake: *Aftershocks are attributed to delayed adjustment to redistributed stress in the focal region following a major shock* (Stacey, *Physics of the Earth*).

afwillite (af wil′it), *n. Mineralogy.* a rare, colorless, monoclinic crystalline form of hydrous calcium silicate, found in Ireland, Africa, and the United States. *Formula:* $Ca_3Si_4O_4(OH)_6$ [from *A*(lpheus) *F. Will*(iams), American mining engineer who discovered it + *-ite*]

agalmatolite (ag′əl mat′ə lit), *n. Mineralogy.* any of various grayish or greenish minerals, usually micaceous, soft enough to be carved into ornaments and figurines, and also used for glazing pottery. Also called **pagodite.** [from Greek *agalma, agalmatos* statue + English *-lite*]

agamete (ā gam′ēt *or* ā gə mēt′), *n. Biology.* an asexual reproductive cell. [from Greek *agametos* not married]

agamic (ə gam′ik), *adj. Biology.* **1** = asexual: *agamic plants, agamic reproduction.*
2 (of ova) not requiring fertilization by the male: *Drones are always developed from agamic or unfertilized eggs* (Carpenter, *Animal Physiology*).
[from Greek *agamos* not married]
—**agamically,** *adv.*

agamogenesis (ā gam′ə jen′ə sis *or* ag′ə mō jen′ə sis), *n. Biology.* asexual reproduction. [from Greek *agamos* unmarried + English *genesis*]

agamogenetic (ā gam′ə jə net′ik *or* ag′ə mō jə net′ik), *adj. Biology.* propagating asexually; reproducing or being reproduced without sexual union.

agamous (ag′ə məs), *adj.* = asexual.

aganglionic (ā gang′glē on′ik), *adj. Biology.* having no ganglia.

agar (ä′gər), *n.* Also, **agar-agar. 1** a gelatinlike extract obtained from certain seaweeds, used in making culture media for bacteria and fungi, as a stabilizer in foods, and in medicine as a mild laxative.
2 a culture medium containing agar.
[from Malay *agar-agar*]

age, *n.* **1** *Geology.* a period or major stage in the history of the earth, especially a span of time shorter than an epoch: *the ice age.*
2 *Biology.* a period in the evolution of living things, especially of the animal kingdom: *the age of mammals.*

agenesis (ā jen′ə sis), *n. Biology, Medicine.* absence or incomplete development of a part or organ. [from *a*-without + Greek *genesis* origin]

agent, *n. Chemistry.* any substance or force capable of causing a change or reaction: *a catalytic agent. The element phosphorus [is] a key agent in fomenting heavy growth of smelly green algae that blankets many bodies of water* (Science News Letter).

agglomerate (ə glom′ər it), *n. Geology.* a rock composed of angular volcanic fragments fused by heat. [from Latin *agglomeratum* wound onto a ball, from *ad-* to + *glomus* ball]
—*adj. Botany.* gathered into a cluster or mass, but not cohering, as some types of flowers.
—**agglomeratic** (ə glom′ə rat′ik), *adj. Geology.* having the nature of an agglomerate.

agglutinate (ə glüt′n āt), *v. Immunology.* **1** to cause (bacteria or blood cells) to clump together.
2 (of bacteria or blood cells) to clump together; undergo agglutination: *Incompatible blood transfusions result from antigen-antibody reactions ... and cause them to agglutinate* (McElroy, *Biology and Man*).
[from Latin *agglutinatum* glued together, from *ad-* to + *gluten* glue]

agglutination, *n. Immunology.* the clumping together of bacteria or blood cells, usually by the introduction of antibodies: *the antibody is known as an agglutinin, and the clumping reaction between it and the antigen is termed agglutination* (Harbaugh and Goodrich, *Fundamentals of Biology*).

agglutinin (ə glüt′n ən), *n. Immunology.* an antibody that causes bacteria or blood cells to agglutinate. Compare **agglutinogen.**

agglutinogen (ag′lü tin′ə jən), *n. Immunology.* an antigen that stimulates the production of agglutinins: *Extensive tests show that two types of antigens (agglutinogens) called A and B occur in the red cells of different persons* (Storer, *General Zoology*). [from *agglutinin* + *-gen* something that produces]

aggrade, *v. Geology.* to raise the level of a bay, valley, or other enclosed area by filling it up with silt or other debris.
—**aggradation,** *n.* the process of aggrading.
—**aggradational,** *adj.* of or having to do with the process of aggrading: *The work of waves and currents, like that of rivers and glaciers, has two phases: degradational and aggradational, or erosional and depositional* (Finch and Trewartha, *Elements of Geography*).

aggregate (ag′rə git *or* ag′rə gāt), *adj.* **1** *Botany.* (of a flower) consisting of many florets arranged in a dense mass. **2** *Geology.* composed of different mineral fragments united into one rock by heat, as granite, or by cementation, as sandstone.
—*n. Geology.* a rock composed of several different mineral constituents that can be separated by mechanical means.
[from Latin *aggregatum* added to, from *ad-* to + *grex* flock]

aggregate fruit, *Botany.* a fruit composed of a cluster of several ripened ovaries that were separate in the flower: *Aggregate fruits are the product of all the carpel ripenings in one flower, the cluster of carpels being crowded on the ripened receptacle forming one mass,* as in the *Raspberry, Blackberry, Dewberry,* and *Strawberry* (Youngken, *Pharmaceutical Botany*).

ASSOCIATED TERMS: An *aggregate fruit* is often distinguished from a *multiple fruit* (also called a *compound fruit* or *collective fruit*), such as the mulberry or the pineapple, which is composed of a cluster of ripened ovaries produced by several flowers rather than by a single flower. Both of these types of fruit are in turn distinguished from a *simple fruit,* which has developed from a single ripened ovary, such as a tomato, apple, or acorn.

aglycone (ā glī′kōn) or **aglycon** (ā glī′kon), *n. Biochemistry.* the nonsugar part of a glycoside molecule. [from *a-* not + Greek *glykys* sweet + English *-one*]

agnathan (ag′nə thən), *adj. Zoology.* having no lower jaw; jawless: *an agnathan fish.* [from *a-* not + Greek *gnathos* jaw]

agonic line (ə gon′ik *or* ā gon′ik), *Geography.* the irregular imaginary line passing through the two magnetic poles of the earth along which the magnetic needle points directly north and south; the line of no magnetic declination. [from Greek *agōnos* without angle, from *a-* not, without + *gōnia* angle]

agonist (ag′ə nist), *n.* **1** *Physiology.* a contracting muscle which is resisted or counteracted by another muscle (the *antagonist*).

2 *Biochemistry.* a substance that can combine with a nerve receptor and initiate or produce a reaction typical for that substance: *Adenyl cyclase is an enzyme which is activated when dopamine agonists interact with the dopamine receptor on the nerve cell receiving the neurotransmitter signal* (London *Times*). [from Greek *agōnistēs* competitor, contestant]

agouti (ə gü′tē), *n. Zoology.* an alternation of light and dark bands in the hair or fur of various animals, producing a grizzled appearance: *The agouti series ... accounts for the typical mousy appearance of the wild mouse's fur.* (Scientific American). [named after the *agouti,* a rodent of tropical America usually grizzled in color, ultimately from the Tupi Indian name]

agrestal (ə gres′təl) or **agrestial** (ə gres′chəl), *adj. Botany.* growing wild in cultivated land, as weeds. [from Latin *agrestis* of the field, from *ager* field]

agro-, *combining form.* **1** field; soil, as in *agrobiology.* **2** agriculture or agricultural, as in *agroclimatology.* [from Greek *agros* open country, land]

agrobiology, *n.* the study of plant nutrition and growth in relation to the condition and constituents of the soil, especially to increase crops.
—**agrobiological,** *adj.* of or having to do with agrobiology. —**agrobiologically,** *adv.*

agroclimatology, *n.* the branch of climatology that deals with the effects of weather upon crops: *Agroclimatology can advise as to whether such crops are likely to succeed in a given locality and which varieties are the best to plant* (Scientific American).

agronomy (ə gron′ə mē), *n.* the study of soil and the improvement of crop production; science of managing farmland. [from Greek *agronomos* an overseer of land, from *agros* land + *-nomos,* related to *nemein* manage]

agrostology (ag′rə stol′ə jē), *n.* the branch of botany that deals with grasses. [from Greek *agrōstis* a kind of grass + English *-logy*]

a.h., *abbrev.* ampere-hour.

AHF, *abbrev.* antihemophilic factor.

AHG, *abbrev.* antihemophilic globulin.

A horizon, *Botany, Geology.* the uppermost layer of soil; topsoil: *The A horizon is rich in organic matter only ... where freshly deposited organic litter is decomposing. The lower portion ... is sandy and light-colored* (Weier, *Botany*). Contrasted with **B horizon** and **C horizon.**

AI, *abbrev.* **1** artificial insemination. **2** artificial intelligence.

aiguille (ā gwēl′ *or* ā′gwēl), *n. Geology.* a slender, sharply pointed peak of rock, especially in the Alps. [from French *aiguille* (literally) needle]

air, *n.* the odorless, tasteless, and invisible mixture of gases that surrounds the earth and directly or indirectly supports every form of life on earth; atmosphere. Air consists chiefly of nitrogen and oxygen, along with argon, carbon dioxide, hydrogen, and small quantities of neon, helium, and other inert gases. *The active energizing element of the air is its oxygen, which combines readily with other chemical elements and is necessary to all life* (Blair, *Weather Elements*).

ASSOCIATED TERMS: The air involved in normal breathing is called *tidal air.* It is the air ordinarily inhaled and exhaled at each breath, and amounts to about 500 millimeters in adults. Air forced out of the lungs after normal exhalation is called *supplemental air* and amounts to about 1,100 millimeters, leaving in the lungs about 1,200 millimeters of *residual air,* which cannot be forced out. Air forced into the lungs after normal inhalation is called *complemental* (or *complementary*) *air,* and can amount to about 3,000 millimeters, bringing the *vital capacity* (the maximum amount of air that can move through the lungs) up to from 4,500 to 6,500 millimeters.

air bladder, *Biology.* a sac in most fishes and various animals and plants that is filled with air. The air bladder of a fish adjusts the specific gravity of the fish to the water pressure at varying depths. In certain plant species, large air bladders, located at various places on the plant, help to hold it up when it is covered with water. The air bladder of a fish is also called **swim bladder.**

air cell, *Biology.* **1** a tiny cavity for air in an organism, as the air sac of a bird.

2 an alveolus: *the air cells of the lungs.*

3 a space between the membranes in the white of an egg, usually located at the broad end of the egg: *The air cell has a purpose. The embryo's beak lies directly beneath it, and during the later stages of incubation ... the embryo gulps air from the cell* (New Yorker).

air current, *Meteorology.* **1** a stream of air; wind. **2** = air flow.

airflow, *n.* **1** *Meteorology.* a natural movement of air: *Most of the airflow is from a westerly direction with speeds increasing from the pole southward* (Scientific American).

2 *Physics.* the flow of air around and relative to an object moving in air, usually in a direction opposite to that of the object's motion or flight. Also called **airstream.**

---

cap, fāce, fäther; best, bē, tėrm; pin, fīve;
rock, gō, ôrder; oil, out; cup, pùt, rüle;
yü in use, *yù* in uric;
ng in bring; *sh* in rush; *th* in thin; ₮H in then;
zh in seizure.
ə = *a* in about, *e* in taken, *i* in pencil, *o* in lemon, *u* in circus

---

17

airglow, *n. Astronomy.* a faint glow in the sky, barely visible to the naked eye, believed due to chemical reactions caused by the sun's rays in the upper atmosphere. The airglow consists of many faint emission lines of the common upper atmospheric gases, like oxygen, nitrogen, and the hydroxyl molecule (OH): *Even in the absence of the moon the sky is not completely black during a clear night, it is instead faintly luminescent. This is mainly ... an emission from the atmosphere known as the airglow [that] ... occurs throughout the twenty-four hours* (Bates, *The Earth and Its Atmosphere*). Compare **afterglow.**

▶ Airglow occurring in the daytime is usually called *dayglow,* while that occurring at night is called *nightglow.* The *twilight glow,* which is caused by the direct action of sunlight on atoms, is similar to the nightglow but much more intense. See also **aurora.**

air layering, *Botany.* a method of causing a branch or stem to form roots for planting, by making a cut halfway through the branch and wrapping it in moist earth or moss. New roots form in the area of the cut.

air mass, *Meteorology.* a large body of air within the atmosphere that has nearly uniform temperature and humidity at any given level and moves horizontally over great distances without changing: *Air masses are to the meteorologist what ... population masses to the statistician. Each mass ... represents a typical unit possessing certain distinctive characteristics or traits* (Neuberger, *Weather and Man*).

air plant, = epiphyte.

air pollution, *Ecology.* the contamination of the atmosphere by industrial waste gases, fuel exhaust, particulate matter such as smoke, and the like: *It ... defined community air pollution ... as the presence in the ambient atmosphere of substances put there by the activities of man in concentrations sufficient to interfere ... with his comfort, safety or health, or with full use and enjoyment of his property"* (N.Y. Times). —**air pollutant.**

air pressure, = atmospheric pressure.

air sac, *Biology.* **1** any of various air-filled spaces in different parts of the body of a bird, connected with the lungs. It aids in breathing and in regulating body temperature. **2** one of the alveoli of the lungs: *The lungs are divided into many tiny cavities, or air sacs.... One careful scientist has estimated that the lungs of a grown person contain 600,000,000 air sacs* (Beauchamp, et al. *Everyday Problems in Science*). **3** a thin-walled enlargement in the trachea of an insect, resembling a sac.

air space, *Botany.* any gas-filled space found in the tissues of plants.

airstream, *n.* = airflow.

Al, *symbol.* aluminum.

ala (ā′lə), *n., pl.* **alae** (ā′lē). **1** *Zoology.* a wing or winglike part. **2** *Anatomy.* one of the lateral cartilages of the nose. **3** *Botany.* one of the two side petals of a papilionaceous flower. [from Latin *ala* wing]

Ala, *abbrev.* alanine.

alabaster, *n. Mineralogy.* a smooth, white or delicately shaded, translucent variety of gypsum. Alabaster is a soft mineral, commonly carved into ornaments and vases. [from Latin *alabaster* vase (of alabaster), from Greek *alabastros*]

alanine (al′ə nēn), *n. Biochemistry.* a crystalline amino acid occurring in several proteins. *Formula:* $C_3H_7NO_2$ *Abbreviation:* Ala

alate (ā′lāt), *adj. Botany, Zoology.* having wings or winglike parts; winged: *alate leaves, alate quadrupeds.* [from Latin *alatus,* from *ala* wing]

albedo (al bē′dō), *n. Astronomy, Meteorology.* reflecting power, as of a planet or other heavenly body or of ice or snow: *It [Jupiter] has a high albedo, i.e., Jupiter reflects more than half of the light that falls upon it* (Science News Letter). [from Latin *albedo* whiteness, from *albus* white]

albertite (al′bər tīt), *n. Mineralogy.* a natural, jet-black, lustrous pyrobitumen, resembling asphalt. [from *Albert* County, New Brunswick + *-ite*]

albinism (al′bə niz′əm), *n. Biology.* the absence of natural coloration or pigmentation; condition of being an albino: *Albinism ... is a recessive characteristic in man as it is in other animals* (R. Beals and H. Hoijer). —**albinistic,** *adj.*

albino (al bī′nō), *n. Biology.* **1** a person or animal characterized from birth by the absence of coloring pigment in the skin, hair, and eyes so that the skin and hair are abnormally white or milky and the eyes have a pink color with a deep-red pupil and are unable to bear ordinary light: *the complete albino with solidly white coat and pink eyes* (Storer, *General Zoology*). **2** a plant having a pale, defective coloring because of its inability to synthesize chlorophyll: *Since they are unable to carry on photosynthesis, the albino seedlings starve to death when the food in the grain is exhausted* (Greulach and Adams, *Plants*). [from Portuguese *albino,* from *albo* white, from Latin *albus*]

albite (al′bīt), *n. Mineralogy.* a plagioclase feldspar, usually white, occurring in granite and other igneous rocks. *Formula:* $NaAlSi_3O_8$ [from Latin *albus* white + English *-ite*] —**albitic** (al bit′ik), *adj.* of the nature of or containing albite.

albumen (al byü′mən), *n.* **1** *Biology.* the white of an egg consisting mostly of albumin dissolved in water. **2** = albumin. [from Latin *albumen,* from *albus* white]

albumin (al byü′mən), *n. Biochemistry.* any of a class of proteins that are soluble in water and can be coagulated by heat, found in the white of an egg, milk, blood serum, and in many other animal and plant tissues and juices. [from French *albumine,* from Latin *albumen* white of an egg] —**albuminous,** *adj.*

albuminoid (al byü′mə noid), *Biochemistry.* —*n.* a substance like albumin; a protein. Albuminoids are obtained chiefly from animal connective tissues and bones, and are characterized by insolubility. *The albuminoids include the collagen of bone and cartilage from which gelatine and glue are made* (Offner, *Fundamentals of Chemistry*). Also called **scleroprotein.** —*adj.* of or like albumin.

albuminous (al byü′mə nəs), *adj. Biochemistry.* of, like, or containing albumin: *Albuminous seeds are those in which the nourishment is not stored in the embryo until germination takes place. Such seeds show a larger nourishing tissue region and a smaller embryo region* (Youngken, *Pharmaceutical Botany*).

alburnum (al bėr′nəm), *n. Botany.* the lighter, softer, and more recently formed wood between the inner bark and the harder center, or heartwood, of a tree; sapwood. [from Latin *alburnum,* from *albus* white]

alcohol, *n. Chemistry.* **1** the colorless, flammable, volatile liquid in wine, beer, whiskey, gin, and other fermented and distilled liquids that makes them intoxicating. Alcohol is commercially prepared from grain, potatoes, or molasses, and is used in medicine, in manufacturing, and as a fuel. *Alcohol for industrial purposes ... must first be "denatured," rendered unfit for drinking by the addition of wood alcohol, benzine, or other substances* (Offner, *Fundamentals of Chemistry*). It is also called **ethyl alcohol, ethanol,** and **grain alcohol.** *Formula:* C_2H_5OH
2 any of a group of similar organic compounds. Alcohols contain a hydroxyl group (-OH) and react with organic acids to form esters. Wood alcohol or methyl alcohol, CH_3OH, is very poisonous.
[from Medieval Latin *alcohol,* originally, "fine powder," then "essence," from Arabic *al-kuhl* the powdered antimony]
—alcoholic, *adj.*

alcoholic fermentation, *Biology.* the conversion of sugars to carbon dioxide and alcohol by certain yeasts, molds, and bacteria in the absence of oxygen. It is a form of anaerobic respiration.

alcosol (al′kə sol), *n. Chemistry.* a colloidal solution in alcohol. [blend of *alco(hol*) and *sol(ution*)]

aldehyde (al′də hīd), *n. Chemistry.* **1** any of a group of organic compounds containing the group -CHO, derived from various alcohols by oxidation. Formaldehyde is an aldehyde produced by the oxidation of methyl alcohol.
2 = acetaldehyde.
[from New Latin *al*(cohol) *dehyd*(rogenatum) alcohol dehydrogenated]
—aldehydic (al′də hī′dik), *adj.* **1** of the nature of aldehydes. **2** resembling the aldehydes.

aldose (al′dōs), *n. Chemistry.* a sugar containing any one of a group of organic chemical compounds having the radical CHO and the OH radical on the C atom adjacent to the CHO group. [from *ald(ehyde*) + *-ose*]

aldosterone (al dos′tə rōn), *n. Biochemistry.* a steroid hormone of the adrenal cortex which controls the salt and water balance in the body. *Formula:* $C_{13}H_{16}N_2O$
[from *ald*(ehyde) + *ster*(ol) + (horm)*one*]

alecithal (ə les′ə thəl), *adj. Embryology.* having little or no yolk. Compare **centrolecithal, homolecithal, telolecithal.** [from *a-* without + Greek *lekithos* yolk]

aleurone (ə lür′ōn *or* al′yə rōn′), *n. Botany.* a mixture of minute protein granules found in seeds and grains. The **aleurone layer** is a row of cells in the endosperm or nourishing tissue of a plant embryo, containing aleurone. [from Greek *aleuron* flour]

alexin (ə lek′sin), *n. Immunology.* any substance, such as one present in normal blood serum, that is capable of destroying bacteria or other foreign substances.

Compare **phytoalexin.** [from German *Alexin,* from Greek *alexein* to ward off]

Alfvén wave (äl vān′), *Physics.* a magnetohydrodynamic wave generated by motion within an electrically conducting plasma in a magnetic field: *The magnetic wake is caused by Alfvén waves, which are rather like the sound waves that travel ahead of objects moving at subsonic speeds* (New Scientist). [named after Hannes Alfvén, born 1908, Swedish physicist who discovered the wave]

alga (al′gə), *n., pl.* **algae** (al′jē). *Biology.* any of a large group of mostly aquatic organisms that contain chlorophyll but lack special tissues for carrying water. Some algae are single-celled; others, such as certain seaweeds, are multicellular and may be very large; and some live in symbiotic association with fungi in the form of lichens. Algae were traditionally classified as plants, but today some are often treated as monerans and others as protists. *Through photosynthesis, algae are the main producers of food and oxygen in water environments* (Otto and Towle, *Modern Biology*).
▶ The two main groups of algae are the **blue-green algae,** which consists of prokaryotic cells (without a visible nucleus), and eight divisions with eukaryotic cells (with a visible nucleus), comprising the **red algae,** the **green algae,** the **brown algae,** the **yellow-green algae,** the **euglenoids,** the **pyrrophytes** the **diatoms,** and a small group with yellow pigments . (division *Xanthophyta*). Until recently the prokaryotic blue-green algae have been thought to form a natural group with the prokaryotic bacteria. It has been found, however, that the metabolic processes of the blue-green algae resemble more closely those of the eukaryotic algae and higher plants. In addition, the internal membranes differ in the two groups. The blue-green algae are therefore an intermediate group between the prokaryotic bacteria and the eukaryotic algae.
[from Latin *alga* seaweed]
—algal, *adj.* of or resembling algae.

algebra, *n.* the branch of mathematics dealing with the relations and properties of quantities by the use of symbols and letters, negative numbers as well as ordinary numbers, and equations to solve problems involving a finite number of operations. $x + y = x^2$ is a way of stating, by algebra, that the sum of two numbers equals the square of one of them. [from Medieval Latin *algebra,* from Arabic *al-jabr* the bone setting; hence reduction (i.e., of parts of a fracture set into a whole)]
—algebraic (al′jə brā′ik), *adj.* of or having to do with algebra; used in algebra.

algin (al′jən), *n., or* **alginic acid,** *Chemistry.* a gelatinous compound found in the cell walls of brown algae, used in plastics, cosmetics, paints and varnishes, and as a food emulsifier and thickener. *Formula:* $(C_6H_8O_6)_n$

---

cap, fāce, fäther; best, bē, tėrm; pin, fīve;
rock, gō, ôrder; oil, out; cup, pùt, rüle,
yü in use, *yu* in uric;
ng in bring; *sh* in rush; *th* in thin; *ᴛʜ* in then;
zh in seizure.
ə = *a* in about, *e* in taken, *i* in pencil, *o* in lemon, *u* in circus

19

ALGOL or **Algol** (al′gol), *n.* an algebraic system for programming a computer to solve scientific problems.

▶ *ALGOL* is still used to some extent in Europe but rarely in the United States, where *FORTRAN* is the most common programming language for performing mathematical and scientific computations. [from *Algo(rithmic) L(anguage)*]

algology (al gol′ə jē), *n.* the branch of botany that deals with algae. Also called **phycology.**

algorithm (al′gə riᴛʜ′əm), *n. Mathematics.* a formal procedure for any mathematical operation, especially a set of well-defined rules for solving a problem in a finite number of steps: *If a problem is to be solved by a computer, an algorithm is indispensable, because the computer can follow only instructions that are stated in the unambiguous form of an algorithm* (Scientific American). [ultimately from al-*Khuwārizmi,* 9th-century Arab mathematician]

—**algorithmic** (al′gə riᴛʜ′mik), *adj.* of, having to do with, or according to algorithms: *an algorithmic method of solving problems.*

alicyclic (al′ə sī′klik *or* al′ə sik′lik), *Chemistry. —adj.* having the characteristics of both aliphatic and cyclic organic compounds. An alicyclic compound combines open-chained and closed-ring carbon atoms. [from *ali(phatic)* + *cyclic*]

—*n.* an alicyclic compound.

alimentary (al′ə men′tər ē), *adj. Biology.* having to do with food, nutrition, or digestion: *alimentary habits, the alimentary tract.*

alimentary canal, *Anatomy.* the tube or passage of the body of an animal through which food passes and in or by means of which food is digested and wastes are eliminated. In mammals, the mouth, esophagus, stomach, intestines, and anus are parts of the alimentary canal.

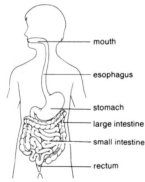

— mouth

— esophagus

— stomach

— large intestine

— small intestine

— rectum

aliphatic (al′ə fat′ik), *Chemistry. —adj.* of or belonging to a class of organic compounds in which the carbon atoms form chains with open ends rather than rings (closed chains). Aliphatic hydrocarbons include the paraffins, the olefins, and the acetylene series.

—*n.* an aliphatic compound.

[from Greek *aleiphatos* fat, oil (so called because fats are derivatives of this class of organic compounds)]

aliquant (al′ə kwənt), *Mathematics. —adj.* dividing a number or quantity with a remainder: *5 is an aliquant part of 14.* —*n.* an aliquant part. [from Latin *aliquantus* a certain (amount), from *alius* some + *quantus* how much]

aliquot (al′ə kwət), *adj. Mathematics.* dividing a number or quantity without a remainder: *3 is an aliquot part of 12.*

—*n.* **1** *Mathematics.* an aliquot part.

2 *Chemistry.* the amount of titrant needed in titration to complete the chemical change.

[from Latin *aliquot* several, from *alius* some + *quot* how many]

alisphenoid (al′ə sfē′noid), *Anatomy. —adj.* having to do with or being either of two bones at the base of the skull that form the greater wings of the sphenoid. —*n.* an alisphenoid bone. [from Latin *ala* wing + English *sphenoid*]

alizarin (ə liz′ər ən), *n. Chemistry.* an orange-red crystalline dye prepared from coal tar, formerly obtained from madder, used to form other dyes. *Formula:* $C_{14}H_8O_4$ [from French *alizari* madder]

alkalescent (al′kə les′nt), *adj. Chemistry.* tending to become alkaline; slightly alkaline.

—**alkalescence,** *n.* alkalescent quality or condition.

alkali (al′kə lī), *n., pl.* **-lis** *or* **-lies.** *Chemistry.* **1** any strong base or hydroxide that is soluble in water, neutralizes acids and forms salts with them, and turns red litmus blue. Alkalis are generally bitter-tasting in water solution and have a pH value of more than 7. Lye and ammonia are two common alkalis.

▶ In its most restricted, but usual, sense *alkali* is applied only to the hydrates of the alkali metals (lithium, sodium, potassium, rubidium, cesium, and francium). In a more general sense it is applied to the hydrates of the alkaline-earth metals and also to the alkaloids.

2 any salt or mixture of salts that neutralizes acids, found in some desert soils.

—*adj. Geology.* = alkalic: *alkali basalt, alkali granite.* [from Arabic *al-qali* the soda ash]

alkalic (al′kə lik *or* al kal′ik), *adj. Geology.* (of igneous rocks) containing much sodium and potassium, usually in the form of alkali feldspar or feldspathoid.

alkali feldspar, *Mineralogy.* one of the two major types of feldspar, containing the alkali metals sodium and potassium in varying amounts. Microcline and orthoclase are alkali feldspars. Compare **plagioclase.**

alkali flat, *Geology.* an arid plain at the bottom of an undrained basin, containing an excess of alkali in its soil.

alkali metal, *Chemistry.* any of the group of univalent metals that includes lithium, sodium, potassium, rubidium, cesium, and francium: *The alkali metals are of very low density; lithium, sodium, and potassium float on water, and the lightest of all, lithium (specific gravity 0.57) is far less dense than most woods* (J. Crowther).

alkalimetry (al′kə lim′ə trē), *n. Chemistry.* the measurement of the amount of alkali in a solution or compound.

alkaline (al′kə līn *or* al′kə lən), *adj. Chemistry.* **1a** of or having to do with alkali: *an alkaline reaction.* **b** containing much alkali: *alkaline waters.*

2 having a pH factor of more than 7; having a relatively low concentration of hydrogen ions (contrasted with *acid,* especially as a characteristic of soil).

—**alkalinity** (al′kə lin′ə tē), *n.* alkaline quality or condition.

alkaline earth, 1 any oxide of the alkaline-earth metals. **2** = alkaline-earth metal.

alkaline-earth metal, *Chemistry.* any of a group of strongly basic, bivalent metallic chemical elements, including calcium, strontium, barium, beryllium, magnesium, and radium. Some authorities omit the last three.

alkali soil, *Geology, Botany.* soil containing soluble mineral salts, in quantities sufficient to restrict or prevent the growth of plants, found usually in arid regions.

alkaloid (al′kə loid), *n. Chemistry.* any of a group of complex organic compounds of nitrogen with alkaline properties, found in or obtained from many plants; most are colorless, crystalline, have a bitter taste, and some are very poisonous. Alkaloids include such drugs as cocaine, strychnine, morphine, nicotine, caffeine, and quinine.
—**alkaloidal** (al′kə loi′dl), *adj.* of or having to do with an alkaloid.

alkane (al′kān), *n.* = paraffin. [from *alk(yl)* + *-ane*]

alkane series, = methane series.

alkene (al′kēn), *n.* = olefin.

alkene series, = ethylene series.

alkyd (al′kid), *n.* or **alkyd resin,** *Chemistry.* any of a group of sticky resins that become plastic when heated, used in paints, lacquers, etc. Alkyds are derived from phthalic acid and glycerol or from similar substances. [from *alk(yl)* + *(ac)id*]

alkyl (al′kəl), *n. Chemistry.* a univalent group obtained from aliphatic hydrocarbon derivatives by removal of a hydrogen atom. Examples of alkyl groups are methyl, ethyl, and propyl. The corresponding aromatic radical is called **aryl.** [from German *Alk(ohol)* + English *-yl*]
—**alkylate** (al′kə lāt), *v.* to introduce an alkyl group into (an organic compound).
—**alkylation,** *n.* the act or process of alkylating.

alkyne (al′kīn), *n. Chemistry.* any unsaturated hydrocarbon of the acetylene series containing a triple bond between two carbon atoms. The general formula is C_nH_{2n-2}. [alteration of *alkine,* from German *Alkin,* from *Alk(ohol)* alcohol + *-in* -ine]

alkyne series, = acetylene series.

allanite (al′ə nīt), *n. Mineralogy.* a brownish-black, monoclinic mineral of the epidote family, sometimes occurring in granite and other igneous rocks. [from Thomas *Allan,* 1777–1833, English mineralogist + *-ite*]

allantoin (ə lan′tō in), *n. Biochemistry.* an oxidation product of uric acid found in allantoic fluid, used to stimulate growth of healthy tissue in wounds, ulcers, or the like: *Man ... suffers from the handicap that he cannot convert uric acid to the more easily excreted substance allantoin, as other mammals can. Consequently if he produces too much uric acid, it accumulates in his body* (DeWitt Stetten, Jr.). *Formula:* $C_4H_6N_4O_3$

allantois (ə lan′tō is), *n. Embryology.* an appendage on the embryos of reptiles, birds, and mammals, developing as a membranous sac from the posterior part of the intestinal cavity. It is important in the formation of the umbilical cord and the placenta in mammals. [from

New Latin *allantois,* from Greek *allantoeidēs* sausage-shaped]
—**allantoic** (al′ən tō′ik), *adj.* of or relating to the allantois.

allele (ə lēl′), *n. Genetics.* one member of a pair or series of genes occupying a specific position (locus) in a specific chromosome; one of the alternative forms of a gene: *Today ... we know that there may be several alleles available within a population of organisms to occupy a particular locus in a chromosome. Normally an individual has only two alleles for any trait—one gene derived from its male parent, the other from its female parent* (Miller and Leth, *High School Biology*). Also called **allelomorph.**
▶ By convention, a capital letter usually symbolizes a dominant allele, while the lower-case form of the same letter symbolizes the recessive allele. For example: In certain pea plants, the allele for tallness (T) is dominant and the allele for shortness (t) is recessive. If two genes of an allelic pair are the same, the genotype is said to be homozygous (TT or tt). If they differ, the genotype is said to be heterozygous (Tt). [from Greek *allēlōn* of each other]
—**allelic** (ə lē′lik *or* ə lel′ik), *adj.* of or having to do with an allele or alleles: *Two genes are said to be allelic if they normally occupy the same place on a chromosome. For example, one might determine curly hair while the alternative allele might determine straight hair* (Bulletin of Atomic Scientists).

allelomimetic (ə lē′lə mi met′ik *or* ə lel′ə mi met′ik), *adj. Zoology.* of or characterized by imitativeness within a group; imitative of one another: *All the sheep in a flock, or all the fish in a school, or all the dogs in a pack, tend to do the same thing at the same time. Scientifically, this is termed allelomimetic behavior* (Science News Letter). [from Greek *allēlōn* of each other + English *mimetic*]

allelomorph (ə lē′lə môrf *or* ə lel′ə môrf), *n.* = allele. [from Greek *allēlōn* of each other + *morphē* form]
—**allelomorphic,** *adj.* = allelic.

allelopathy (al′ə lop′ə thē), *n. Botany.* the harmful effect on a plant by another plant that secretes a toxic chemical. [from Greek *allēlōn* of each other + English *-pathy* harmful effect, disease]

Allen's rule, *Zoology.* the rule that in warm-blooded animals the tail, ear, beak, and other protruding part is relatively shorter in the colder areas of the range of a species than in the warmer ones. [named after J.A. *Allen,* 1838–1921, U.S. zoologist]

allergen (al′ər jən), *n. Immunology.* any substance that causes or reveals an allergy in a particular individual or individuals.
—**allergenic** (al′ər jen′ik), *adj.* of or having to do with allergens.

---

cap, fāce, fäther; best, bē, tèrm; pin, five;
rock, gō, ôrder; oil, out; cup, pùt, rüle,
yü in use, *yu* in uric;
ng in bring; *sh* in rush; *th* in thin; ŦH in then;
zh in seizure.
ə = *a* in about, *e* in taken, *i* in pencil, *o* in lemon, *u* in circus

21

allergic (ə lėr'jik), *adj.* **1** having an allergy: *allergic to pollen or dust.* **2** of or caused by allergy: *Hay fever is an allergic reaction.*

allergology (al'ər jol'ə jē), *n.* the study of allergies.

allergy (al'ər jē), *n., pl.* **-gies.** *Medicine.* **1** an unusual sensitivity to a certain substance, such as a particular kind of pollen, food, hair, or cloth. Hay fever, asthma, and hives are common manifestations of allergy. **2** = anaphylaxis. [from New Latin *allergia,* formed from Greek *allos* different, other + *ergon* action, work]

allo-, *combining form.* other; different, as in *allograft = a graft from another or different individual.* [from Greek *allos* other]

alloantigen, *n.* = isoantigen: *In every species studied thus far, there exists a group of "strong" transplantation antigens, called alloantigens, which account for most of the rejection reaction to tissues transplanted within the same species* (C. B. Carpenter).

allogamy (ə log'ə mē), *n.* = cross-fertilization. Compare **heterogamy, syngamy.** [from *allo-* + *-gamy*]
—**allogamous** (ə log'ə məs), *adj.* of or caused by allogamy.

allogeneic (al'ə jə nē'ik) or **allogenic** (al'ə jen'ik), *adj. Immunology.* genetically different although belonging to the same species: *allogeneic tissues.* [from *allo-* + *(syn)geneic* or *-genic*]

ASSOCIATED TERMS: *Allogeneic* and *syngeneic* are terms used in tissue grafting or transplantation in connection with antigenic or immunological reactions. An *allogeneic graft* is a graft taken from a donor of the same species who differs genetically from the recipient; a *syngeneic* (or *isogeneic*) *graft* is a graft from a donor whose tissue is genetically identical with that of the recipient. See also **xenogeneic.**

allograft, *n. Immunology.* a graft of tissue from a genetically different donor of the same species as the recipient: *Grafts of normal tissue exchanged between members of the same inbred line (isografts) are accepted. Grafts exchanged between members of two different lines (allografts) are uniformly rejected within 14 days* (Scientific American).

ASSOCIATED TERMS: An *allograft* is a transfer of tissue between nonidentical members of the same species, while a *homograft* or *isograft* is a graft between identical members of the same species. If the donor and the recipient belong to different species, the graft is called a *heterograft* or *xenograft.* A graft from one part of the same individual's body to another is an *autograft.*

allomerism (ə lom'ə riz əm), *n. Chemistry.* variability in chemical makeup with similarity in crystalline form. [from Greek *allos* other + *meros* part]
—**allomerous** (ə lom'ər əs), *adj.* alike in crystalline form, but different in chemical makeup.

allometry (ə lom'ə trē), *n. Biology.* the relative growth of one part of an organism with reference to the rest, as a result of which its relative size changes with age.
—**allometric** (al'ə met'rik), *adj.* of or characterized by allometry.

allopathy (ə lop'ə thē), *n. Medicine.* a method of treating a disease by using remedies to produce effects different from those caused by the disease treated. Contrasted with **homeopathy.**
—**allopathic** (al'ə path'ik), *adj.* of allopathy; using allopathy.

allopatric (al'ə pat'rik), *adj. Biology.* of or designating two species or populations whose natural ranges do not overlap. Contrasted with **sympatric.**
—**allopatry,** *n.* the occurrence of allopatric organisms or species.
[from Greek *allos* other + *patria* clan]

allophane (al'ə fān), *n. Mineralogy.* a translucent, amorphous clay mineral, hydrous aluminum silicate, with a sky-blue, green, brown, or yellow color. [from Greek *allophanēs* appearing different, from *allos* different + *phainesthai* appear]

allopolyploid (al'ə pol'i ploid), *n. Genetics.* a polyploid organism in which the different sets of chromosomes originate from different species.

all-or-none, *adj.* characterized by either a complete action or effect or by no action or effect at all: *Nerve impulses follow the All or None Law—that is, a nerve impulse generated by a weak stimulus is just as strong as one generated by a strong stimulus* (Harbaugh and Goodrich, *Fundamentals of Biology*). *This transfer is an "all-or-none" process, the electron getting all of the photon's energy or none at all* (Sears and Zemansky, *University Physics*).

all-or-nothing, *adj.* = all-or-none: *G-6-PD is an all-or-nothing substance: the enzyme is either produced in the normal amount, or is not produced at all* (Time).

allosome (al'ə sōm), *n. Genetics.* **1** a chromosome that is not typical; an irregular or abnormal chromosome. **2** a sex chromosome. [from *allo-* + *-some*]

allosteric (al'ə ster'ik), *adj. Biochemistry.* of or involving changes in an enzyme at a site other than the active site. [from *allo-* + Greek *stereos* solid + English *-ic*]

allotetraploid (al'ō tet'rə ploid), *n.* = amphidiploid.

allotrope (al'ə trōp), *n. Chemistry.* an allotropic form of an element.

allotropic (al'ə trop'ik), *adj. Chemistry.* differing in physical and chemical properties but consisting of the same element. Ordinary oxygen gas (O_2) and ozone (O_3) are allotropic forms of oxygen. Charcoal, graphite, and diamond are allotropic forms of carbon. [from *allo-* + Greek *tropos* way]
—**allotropy** (ə lot'rə pē), *n.* the property or fact of being allotropic.

allotype, *n.* **1** *Biology.* a single specimen of the opposite sex of the specimen (holotype) on which a description of a species, variety, etc., is based.
2 *Genetics.* a variant form or type demonstrable through immunization with an isoantigen.
—**allotypic** (al'ə tip'ik), *adj.* of or having to do with an allotype: *an allotypic character.*

alloy, *n. Chemistry.* a metal made by mixing and fusing two or more metals, or a metal and a nonmetal, to obtain some desirable quality or qualities, such as hardness, lightness, and strength. Brass is an alloy of copper and zinc. —*v.* to make into an alloy.
[from French *aloi,* from *aleier* unite, combine, from Latin *alligare* bind to]

alluvial (ə lü'vē əl), *adj. Geology.* having to do with, consisting of, or formed by sand, silt, mud, etc., left by flowing water: *an alluvial plain. A delta is an alluvial deposit at the mouth of a river.*

alluvial fan or **alluvial cone,** *Geology, Geography.* a fan-shaped deposit of sand, silt, or mud formed at the point where a stream emerges from a ravine into a plain or other relatively flat area.

alluvium (ə lü′vē əm), *n., pl.* **-viums, -via** (-vē ə). *Geology.* a deposit of sand, silt, mud, or other detritus left by flowing water in river beds, flood plains, lakes, and the like. [from Late Latin *alluvium,* from Latin *alluere* wash against] Compare **eluvium.**

ally, *n., pl.* **-lies.** a related animal, plant, form, or thing: *ferns and their allies, the alkaline metals and their allies.*

—*v.,* **allied, allying. 1** to relate, as by similarity of structure or common descent: *Dogs are allied to wolves.*

2 to unite or combine: *Allied reflexes ... produce a harmonious effect, such as the muscular movements of a person in walking or in an earthworm or caterpillar when crawling* (Storer, *General Zoology*).

allyl (al′əl), *n. Chemistry.* a univalent, unsaturated, open-chain radical found especially in the liquids yielded by pressing garlic, onion, and mustard seeds. *Formula:* $CH_2 = CHCH_2$- [from Latin *allium* garlic + English *-yl*]

—**allylic** (ə lil′ik), *adj.* of an allyl.

allyl alcohol, *Chemistry.* a colorless, poisonous liquid used in the synthesis of organic chemicals. *Formula:* $CH_2:CHCH_2OH$

alpha (al′fə), *n. Astronomy.* Usually, **Alpha.** the chief or brightest star of a constellation.

—*adj. Chemistry.* **1** designating the first carbon atom adjacent to a functional group in an organic compound. EXAMPLE: $CH_3CH_2CHClCO_2H$ is alpha-chloro butyric acid, since $CH_3CH_2CH_2CO_2H$ is butyric acid and $-CO_2H$ is the functional group of acids. **2** designating the principal allotropic form of metal: *alpha iron. Symbol:* α

alpha-adrenergic receptor, = alpha receptor.

alpha decay, *Nuclear Physics.* the disintegration of a radioactive substance through the emission of alpha particles.

alpha emitter, *Nuclear Physics.* an atom which emits an alpha particle.

alpha globulin, *Biochemistry.* a type of globulin in blood plasma which has great colloidal mobility in electrically charged neutral or alkaline solutions. Compare **beta globulin** and **gamma globulin.**

alpha-helix, *n. Molecular Biology.* the single spiral structure of many protein molecules, especially as distinguished from the double helix characterizing the molecular structure of DNA. [so called because of a structural model of polypeptide chains proposed by Linus Pauling, in which the twisting of the helix is due to an angular configuration at the *alpha* carbon atom of each peptide] —**alpha-helical,** *adj.*

alphanumeric (al′fə nü mer′ik), *adj.* composed of or using both letters and numbers: *an alphanumeric code or system.* [from *alpha(bet)* + *numeric(al)*]

alpha particle, *Physics.* a positively charged particle consisting of two protons and two neutrons, released at a very high speed in the disintegration of radium and other radioactive elements: *The alpha particle was so named when its identity was not known; today, it is known to be identical with the helium nucleus or he-*

lium ion (Jones, *Inorganic Chemistry*). Compare **beta particle.**

alpha ray, *Physics.* **1** Usually, **alpha rays.** a stream of alpha particles. Also called **alpha radiation. 2** = alpha particle.

alpha receptor, *Biochemistry.* a site in a cell at which the stimulus of an adrenergic agent, especially adrenaline, produces a usually excitatory response, stepping up the activity of muscles. Also called **alpha-adrenergic receptor.** Compare **beta receptor.**

alpha rhythm or **alpha waves,** *Physiology.* the smooth, regular pattern of electrical oscillations in the brain when a person is awake and relaxed. As recorded by an electroencephalograph, alpha rhythm occurs in the frequency of 8 to 13 hertz.

alpine or **Alpine,** *adj.* **1** *Geography, Geology.* of, having to do with, or resembling the Alps or similar high mountains: *alpine terrain.* **2** *Biology.* living on mountains above the timberline: *alpine plants.*

alpine glacier, = valley glacier.

alt., *abbrev.* altitude.

altazimuth (al taz′ə məth), *n. Astronomy.* a telescope that can be moved up and down and in a circle, mounted so that its angle from north and from the horizontal can be read on scales. [from *alt(itude)* + *azimuth*]

alternate, *v. Electricity.* **1** to reverse direction at regular intervals. **2** to produce or be operated by a current that does this. See **alternating current.**

—*adj. Botany.* placed singly at different heights along the sides of a stem; not opposite: *alternate leaves.*

alternate angles, *Mathematics.* two angles, both interior or both exterior but not adjacent, formed on opposite sides of a line that crosses two other lines. If the two lines are parallel, the alternate angles are equal.

alternating current, *Electricity.* a current in which the electricity flows regularly in one direction and then the other. *Abbreviation:* a.c. or A.C. Compare **direct current.** See also **alternator.**

alternation of generations, *Biology.* the regular alternation of forms or of mode of reproduction in successive generations of an animal or plant, especially the alternation between sexual and asexual propagation in certain lower invertebrates and some plants. Also called **heterogenesis** and **metagenesis.**

alternator, *n. Electricity.* a generator for producing an alternating current.

altimeter, *n.* any instrument for measuring altitude, especially an aneroid barometer that shows the atmospheric pressure at any given altitude. Altimeters are used in aircraft.

altiplano (äl′ti plä′nō), *n. Geography.* a high, upland plain, especially in the region of the Andes, reaching altitudes of more than 4,500 meters above sea level: *Aymara ... Indians live on the high, cold, semidesert*

---

cap, fāce, fäther; best, bē, tèrm; pin, five;
rock, gō, ôrder; oil, out; cup, pùt, rüle,
yü in use, *yù* in uric;
ng in bring; *sh* in rush; *th* in thin; ᴛʜ in then;
zh in seizure.
ə = *a* in about, *e* in taken, *i* in pencil, *o* in lemon, *u* in circus

altitude

altiplanos around Lake Titicaca in Bolivia and Peru (Clifford Evans). [from American Spanish *altiplano*, from Latin *altus* high + *planus* level]

altitude, *n.* **1a** height above the earth's surface. **b** height above sea level: *The altitude of Denver is 1.6 kilometers.*
2 *Geometry.* the perpendicular distance from one vertex of a figure to the opposite side or face (the *base*).
3 *Astronomy.* the angular distance of a star, planet, or other heavenly body above the horizon. [from Latin *altitudo*, from *altus* high]
—**altitudinal** (al′tə tüd′n əl), *adj.* of or having to do with altitude.

altocumulus (al′tō kyü′myə ləs), *n., pl.* **-li** (-lī). *Meteorology.* a fleecy cloud formation consisting of rounded heaps of white or grayish clouds, often partly shaded, occurring at heights of between 2900 and 6000 meters. [from Latin *altus* high + English *cumulus*]
ASSOCIATED TERMS: see **cloud.**

altostratus (al′tō strā′təs or al′tō strat′əs), *n., pl.* **-ti** (-tī). *Meteorology.* a bluish-gray sheetlike cloud formation, ill-defined at the base, occurring at heights between 1900 and 6000 meters. [from Latin *altus* high + English *stratus*]
ASSOCIATED TERMS: see **cloud.**

altricial (al trish′əl), *adj. Zoology.* (of certain birds) born blind, usually without feathers, and thus helpless for some time after hatching: *The common song birds are all altricial, while domestic fowls, partridges, most wading birds, and the various ducks are precocial* (Shull, *Principles of Animal Biology*). [from Latin *altricem* a nurse]

alula (al′yə lə), *n., pl.* **-lae** (-lē). *Zoology.* a set of three or four quill-like feathers growing at a small joint in the middle of a bird's wing. Also called **bastard wing.** [from New Latin *alula*, diminutive of Latin *ala* wing]

alum (al′əm), *n. Chemistry, Mineralogy.* **1a** a white crystalline salt used in medicine and in dyeing. Alum is sometimes used to stop the bleeding of a small cut. *Formula:* $KAl(SO_4)_2 \cdot 12H_2O$ Also called **potash alum** and **aluminum potassium sulfate. b** a colorless crystalline salt containing ammonia, used in foam fire extinguishers, water purification, medicine, etc. *Formula:* $NH_4Al(SO_4)_2 \cdot 12H_2O$ Also called **ammonia alum** and **aluminum ammonium sulfate.**
2 any of a group of double salts analogous to and including potash alum and ammonia alum, formed by the union of a trivalent and a univalent metal.
3 a white salt made by treating bauxite with sulfuric acid, used in dyeing, medicine, water purification, and in paper and leather manufacture. *Formula:* $Al_2(SO_4)_3 \cdot 18H_2O$ Also called **aluminum sulfate.** [from Old French *alum*, from Latin *alumen, aluminis*]

alumina (ə lü′mə nə), *n. Chemistry.* the oxide of aluminum, occurring naturally as the mineral corundum, of which ruby and sapphire are varieties. Rocks rich in alumina include bauxite, clays, and emery. *Formula:* Al_2O_3 Also called **aluminum oxide.** [from New Latin *alumina*, from Latin *aluminis* alum]

aluminate (ə lü′mə nit), *n. Chemistry.* a salt in which alumina acts toward the stronger bases as an acid.

aluminium (al′yü min′ē əm), *n. British.* = aluminum.

aluminosilicate (ə lü′mə nō sil′ə kit), *n. Chemistry, Mineralogy.* a compound of aluminum and silicon with any of various metals. Feldspars and zeolites are aluminosilicates.

aluminum (ə lü′mə nəm), *n. Chemistry.* a common, very lightweight, silver-white, metallic element which is ductile and malleable, an excellent conductor of electricity, and highly resistant to tarnish and oxidation. Aluminum is the most abundant metallic element in the earth's crust, of which it makes up about seven per cent. It occurs only in combination with other elements. Aluminum is widely used in alloys and to make utensils, instruments, aircraft parts, etc. *Symbol:* Al; *atomic number* 13; *atomic weight* 26.9815; *melting point* 660°C; *boiling point* 2450°C; *oxidation state* +3. [from *alumina*]

aluminum oxide, = alumina.

alunite (al′yə nīt), *n. Mineralogy.* a hydrous potassium aluminum sulfate, occurring chiefly as a hydrothermal alteration product of various feldspathic igneous rocks. *Formula:* $KAl_3(SO_4)_2(OH)_6$ [from French *alun* alum + English *-ite*]

alveolar (al vē′ə lər), *adj. Anatomy, Zoology.* **1** of the part of the jaws where the sockets of the teeth are. **2** of, containing, or like an alveolus or alveoli.

alveolate (al vē′ə lit), *adj. Biology.* having many small cavities; pitted like a honeycomb.
—**alveolation** (al vē′ə lā′shən), *n.* alveolate condition, formation, or structure.

alveolus (al vē′ə ləs), *n., pl.* **-li** (-lī). *Anatomy, Zoology.* **1** a small cavity, pit, or cell. The air cells in the lung are alveoli. **2** the bony socket of a tooth. **3** one of the cells of a honeycomb. [from Latin *alveolus,* diminutive of *alveus* cavity]

Am, *symbol.* americium.

AM, *abbrev.* amplitude modulation.

amacrine (ā mak′rīn or am′ə krīn), *Anatomy.* —*adj.* not having long fibers; lacking axons: *The retina contains a number of kinds of nerve cell in addition to the light-sensitive rods and cones. These bipolar cells, amacrine cells and ganglia seem to be capable of more sophisticated functions than merely passing light signals on to the brain* (Derek H. Fender).
—*n.* an amacrine cell, especially one which conducts signals from the retina to the ganglia. [from Greek *a-* not + *makros* long + *inos* muscle, fiber]

amalgam (ə mal′gəm), *n. Chemistry.* an alloy of mercury with some other metal or metals. Tin amalgam is used in silvering mirrors. Silver amalgam is used as fillings for teeth. [from Medieval Latin *amalgama*]
—**amalgamate** (ə mal′gə māt), *v.* **1** to combine mercury with (another metal). **2** to enter into combination with mercury.

amber, *n.* a hard, translucent, yellow or yellowish-brown substance, the resin of fossil pine trees. Amber accumulates a negative charge of static electricity when rubbed and is considered a good electric insulator. [from Old French *ambre*, from Arabic *'anbar* ambergris]

ambergris (am′bər grēs′ or am′bər gris), *n.* a waxlike, grayish substance found floating in tropical seas, originating as a secretion in the intestines of the sperm whale. It is added to perfumes to keep them from evaporating too quickly. [from Middle French *ambre gris* gray amber]

ambulacral (am′byə lak′rəl *or* am′byə lā′krəl), *adj. Zoology.* of or having to do with an ambulacrum or the ambulacra: *a median ambulacral groove.*

ambulacrum (am′byə lak′rəm *or* am′byə lā′krəm), *n., pl.* **·cra** (-krə). *Zoology.* any of the radial areas of an echinoderm, containing a series of perforations through which the tube feet are protruded and withdrawn. [from Latin *ambulacrum* a promenade, walk, from *ambulare* to walk]

ameba (ə mē′bə), *n., pl.* **-bas, -bae** (-bē). *Biology.* any of various microscopic one-celled organisms of the genus *Amoeba* or related genera, found in fresh and salt water, in the soil, and as parasites in other animals. An ameba is essentially a shapeless mass of protoplasm enclosed by a flexible membrane and containing one or more nuclei: *Amebas move by means of their pseudopodia. The motion is a sort of flowing, with new pseudopodia reaching out and old ones disappearing back into the cytoplasm* (Otto and Towle, *Modern Biology*). Also spelled **amoeba.** [from Greek *amoibē* change]
—**amebic,** *adj.* **1** of or like an ameba or amebas. **2** caused by amebas: *amebic dysentery.* Also spelled **amoebic.**
—**ameboid,** *adj.* **1** of or like an ameba; like that of an ameba.
2 related to amebas. Also spelled **amoeboid.**

amebocyte, *n. Zoology.* a cell having an ameboid shape or motion, as in the bloodstream or in a sponge. Also spelled **amoebocyte.** [from *ameba* + *-cyte*]

ameloblast (am′ə lə blast *or* ə mel′ə blast), *n. Biology.* any of a group of columnar cells that form dental enamel on the developing teeth of vertebrates: *To make perfect enamel each enamel-forming cell must stay healthy until the work is done, for the ameloblasts, unlike most other cells of the body, cannot reproduce* (Scientific American). [from obsolete English *amel* enamel + Greek *blastos* germ]

amensalism (ā men′sə liz əm), *n. Biology.* a symbiotic relationship in which an organism affects or inhibits another without being affected or inhibited in return: *Amensalism by chemical means may be very common in nature, for plants are leaky systems, passively contributing all sorts of substances to their environment* (Weier, *Botany*). [from *a-* without + (com)*mensalism*]
ASSOCIATED TERMS: see **symbiosis.**

ament (am′ənt *or* ā′mənt), *n. Botany.* a long, slender, scaly flower spike that grows on willows, birches, etc. Its common name is **catkin.** [from Latin *amentum* thong]
—**amentaceous** (am′ən tā′shəs), *adj.* **1** consisting of or like an ament. **2** = amentiferous.
—**amentiferous** (am′ən tif′ər əs). bearing catkins: *The pussy willow is amentiferous.*
—**amentiform** (ə men′tə fôrm), *adj.* having the shape of a catkin.

americium (am′ə rish′ē əm), *n. Chemistry.* an artificial, radioactive metallic element produced by bombarding plutonium with high-energy neutrons. Americium is a transuranic element and one of the actinides. *Symbol:* Am; *atomic number* 95; *atomic weight* 243 (the most stable isotope); *melting point* 995°C;

oxidation state 6, 5, 4, 3 [from New Latin, named after *America*]

amianthus (am′ē an′thəs), *n. Mineralogy.* a variety of asbestos with long, flexible, pearly white fibers. [from Latin *amiantus,* from Greek *amiantos* (*lithos*) undefiled (stone)]

amide (am′īd *or* am′id), *n. Chemistry.* **1** a compound in which a metal is substituted for one of the hydrogen atoms of ammonia.
2a any of a group of organic compounds having the univalent group -$CONH_2$, which can be obtained by substituting the -NH_2 group for the -OH group of a corresponding acid. **b** a similar compound in which alkyl groups are substituted for the hydrogen atoms of the -NH_2 group.
[from *am*(*monia*) + *-ide*]

amido (ə mē′dō *or* am′ə dō), *adj. Chemistry.* of or containing the -NH_2 group combined with an acid radical. Compare **amino.**

amination (am′ə nā′shən), *n. Chemistry.* conversion into an amino compound: *New reaction mechanisms are postulated in which amination of phosphorylated acids gives rise directly to the corresponding amino acid* (New Scientist).

amine (ə mēn′ *or* am′ēn), *n. Chemistry.* any of a group of organic compounds formed from ammonia by replacement of one or more of its three hydrogen atoms by hydrocarbon groups. [from *am*(*monia*) + *-ine*]

amino (ə mē′nō *or* am′ə nō), *adj. Chemistry.* of or containing the -NH_2 group combined with a nonacid radical. Compare **amido.**

Amino Acids that Make up Proteins

Amino Acid	Abbreviation
alanine	Ala
argenine	Arg
asparagine	Asn
aspartic acid	Asp
cystine	Cys-Cys
cysteine	Cys
glutamic acid	Glu
glutamine	Gln
histidine	His
isoleucine	Ile
leucine	Leu
lysine	Lys
methionine	Met
phenylalanine	Phe
serine	Ser
threonine	Thr
tryptophan	Trp
tyrosine	Tyr
valine	Val

amino acid, *Biochemistry.* any of a large number of organic compounds containing a basic amino group

cap, fāce, fäther; best, bē, tèrm; pin, five;
rock, gō, ôrder; oil, out; cup, pùt, rüle,
yü in use, *yù* in uric;
ng in bring; *sh* in rush; *th* in thin, ᴛʜ in then;
zh in seizure.
ə = *a* in about, *e* in taken, *i* in pencil, *o* in lemon, *u* in circus

(-NH$_2$) and an acidic carboxyl group (-CO$_2$H). Protein is synthesized from 20 amino acids, 9 of which are essential to human life but cannot be produced by the body and must be obtained from food; these are called **essential amino acids** and include histidine, isoleucine, leucine, lysine, methionine, phenylalanine, threonine, tryptophan, and valine. The **nonessential amino acids** are those that can be synthesized by the body, such as alanine, aspartic acid, cystine, glycine, and glutamic acid. *Certain amino acids may occur in two forms: e.g., glutamic acid and glutamine. Glutamic acid has two carboxyl (COO–) groups, whereas glutamine has an amide (CONH$_2$) group in the place of one of the carboxyls* (Scientific American).

aminobenzoic acid (ə mē′nō ben zō′ik), *Biochemistry.* a substance found in yeast, especially brewers' yeast, which is part of the vitamin B complex, and apparently essential to the growth of some animal species and bacteria. It is a constituent of folic acid. *Formula:* C$_7$H$_7$NO$_2$

amitosis (ā′mi tō′sis *or* ā′mī tō′sis), *n. Biology.* a method of cell division in which the cell separates into two new cells without the appearance of the usual mitotic chromosomal configurations: *Many supposed examples of amitosis are merely distorted forms of mitosis, ... due either to faulty preparation or to natural degenerative changes in the cells* (Shull, *Principles of Animal Biology*). [from *a-* without + *mitosis*]
—**amitotic** (ā′mi tot′ik *or* ā′ mī tot′ik), *adj.* reproducing without mitosis: *In both plants and animals, amitotic division of cells usually indicates a diseased or degenerative condition* (Harbaugh and Goodrich, *Fundamentals of Biology*). —**amitotically,** *adv.*

ammeter (am′ē′tər), *n.* an instrument for measuring in amperes the strength of an electric current: *Actually, the current of electrons is measured in amperes. A current of one ampere means 6.3 billion billion (6.3 × 10^{18}) electrons per second* (Pierce, *Electrons, Waves and Messages*). [from *am(pere)* + *-meter*]

ammine (am′ēn *or* a mēn′), *n. Chemistry.* **1** any of a group of inorganic compounds containing ammonia molecules and a metallic salt. **2** (infrequently) an ammonia molecule as found in such a compound.

ammocoete or **ammocete** (am′ə sēt), *n. Zoology.* the larva of a lamprey. The anatomy of ammocoetes is typical of chordates, and they are therefore often used in zoology courses as representative. They are blind and toothless, and look like worms. They live in the sand and mud on the bottom of streams for several years. [from Greek *ammos* sand + *koitē* bed]

ammonia, *n. Chemistry.* a colorless soluble gas, consisting of nitrogen and hydrogen, that has a pungent smell and a strong alkaline reaction. Ammonia can be condensed to a colorless liquid under pressure and cold. Ammonia is an intermediate in the metabolism of nitrogen, and is used in making fertilizers, explosives, and plastics. *Astronomers ... detected molecules of ammonia gas in interstellar space near the centre of our galaxy. It marked the first time that such a complex molecule had been found in the regions between stars* (Donald Lynden-Bell). *Formula:* NH$_3$ [from Lat-

in (*sal*) *ammoniacus* (salt) of Ammon, an Egyptian god]

ammonification (ə mon′ə fə kā′shən), *n. Chemistry.* **1** the process of ammonifying or being ammonified. **2** the production of ammonia in the decomposition of organic matter, especially through the action of bacteria. In the nitrogen cycle, by which nitrogen is made available to green plants, ammonification plays an essential role.

ammonify (ə mon′ə fī), *v. Chemistry.* to combine or be combined with ammonia or ammonium compounds.

ammonite (am′ə nīt), *n. Paleontology.* the fossil shell of an extinct cephalopod mollusk of the Cretaceous period, coiled in a flat spiral and up to 180 centimeters in diameter. Also called **ammonoid.** [from New Latin *ammonites,* from Latin *cornu Ammonis* (literally) horn of Ammon]

ammonium, *n. Chemistry.* a univalent basic radical whose compounds or salts (such as ammonium hydroxide, NH$_4$OH, and ammonium chloride, NH$_4$Cl) are similar to those of the alkali metals. Ammonium never appears in a free state. *Formula:* -NH$_4$

ammonium chloride, *Chemistry.* a white, crystalline compound used in dry cells, in medicine, in fertilizers, etc. *Formula:* NH$_4$Cl Also called **sal ammoniac.**

ammonoid (am′ə noid), *n.* = ammonite.

ammonolysis (am′ə nol′ə sis), *n. Chemistry.* decomposition by the action of ammonia.

ammophilous (ə mof′ə ləs), *adj. Biology.* living or growing in sand. [from Greek *ammos* sand + *philos* loving]

amniocentesis (am′nē ō sen tē′sis), *n. Medicine.* the insertion of a long hollow needle through the wall of the uterus into the amnion to withdraw amniotic fluid in order to detect possible disorders or defects in the fetus. [from *amnion* + New Latin *centesis* surgical puncture, from Greek *kentēsis* a pricking]

amnion (am′nē ən), *n., pl.* **-nions, -nia** (-nē ə). *Embryology.* a membrane forming the inner sac which encloses the embryo or fetus of reptiles, birds, and mammals. The amnion is filled with a serous fluid that protects the embryo and keeps it moist. [from Greek *amnion,* diminutive of *amnos* lamb]
—**amnionic** (am′nē on′ik), *adj.* amniotic: *The embryo is surrounded by another membrane, the amnion, which contains the amnionic fluid, in which the embryo floats* (Winchester, *Zoology*).
—**amniote** (am′nē ōt), *n.* any of the group of vertebrates that develop amnions in their embryonic stages.
—**amniotic** (am′nē ot′ik), *adj.* **1** of or inside the amnion: *amniotic fluid.* The amnion is often called the *amniotic sac* or *amniotic membrane.* **2** having an amnion: *an amniotic cavity.*

amoeba (ə mē′bə), *n., pl.* **-bas, -bae** (-bē). = ameba.

amoebic (ə mē′bik), *adj.* = amebic.

amoebocyte (ə mē′bə sīt), *n.* = amebocyte.

amoeboid (ə mē′boid), *adj.* = ameboid.

amorphous, *adj.* **1** *Biology.* without the definite shape or organization found in most higher animals and plants: *An ameba is amorphous.* **2** *Chemistry, Mineralogy.* not consisting of crystals; uncrystallized; noncrystalline. [from Greek *amorphos,* from *a-* without + *morphē* shape]

AMP, *abbrev.* adenosine monophosphate.

26

ampere (am′pir), *n.* the SI or MKS unit of electric current. It is the constant current that if maintained in two straight parallel conductors of infinite length and negligible cross section placed one meter apart in a vacuum will produce between these conductors a force equal to 2×10^{-7} newton per meter of length. The ampere is a fundamental unit. *Symbol:* A

An **ampere-hour** is the quantity of electricity (3600 coulombs) delivered by a current of one ampere in one hour.

An **ampere turn** is a unit of magnetomotive force equal to the force produced by a current of one ampere passing through one complete loop of an electric coil. [named after André-Marie *Ampère,* 1775–1836, French physicist]

—**amperage** (am′pər ij), *n.* strength of an electric current measured in amperes.

amphi-, *combining form.* both; on both sides; of two kinds, as in *amphiaster, amphiblastula, amphicarpic.* [from Greek *amphi-* both, around]

amphiarthrosis (am′fē är thrō′sis), *n., pl.* **-ses** (-sēz). *Anatomy.* a form of articulation which permits slight motion, as that between the vertebrae. [from *amphi-* + Greek *arthrōsis* a joining]

amphiaster (am′fē as′ tər), *n. Biology.* the spindle, together with the two asters, that forms during the prophase of mitosis and can be separated from the rest of the cell intact.

amphibian, *Biology.* —*n.* **1** any of a class (Amphibia) of cold-blooded vertebrates with moist, scaleless skin that, typically, lay eggs in water where the young hatch and go through a larval or tadpole stage, breathing by means of gills; some larval forms lose their gills and develop lungs for breathing. Frogs, toads, newts, and salamanders belong to this class.

2 an animal living both on land and in water but unable to breathe under water. Crocodiles, seals, and beavers are amphibians.

3 a plant that grows on land or in water.

—*adj.* able to live both on land and in water; amphibious.

[from Greek *amphibios* living in two ways, from *amphi-* both + *bios* life]

amphibious (am fib′ē əs), *adj. Biology.* able to live both on land and in water; amphibian: *Frogs are amphibious.*

amphiblastula (am′fə blas′chù lə), *n. Embryology.* a blastula in which the cells of one hemisphere differ from those of the other, as in certain sponges.

amphibole (am′fə bōl), *n. Mineralogy.* any of a group of hydrous silicate minerals, including hornblende and actinolite. Amphiboles usually consist of a silicate of calcium, magnesium, and one or more other metals, such as iron or sodium. [from Greek *amphibolos* ambiguous, from *amphi-* both + *-bolos* struck]

▶ Amphibole is one of the five essential minerals that make up most igneous rocks; the others are quartz, feldspar, mica, and pyroxene.

amphibolite (am fib′ə lit), *n. Geology.* a rock of metamorphic origin consisting chiefly of amphibole and often containing quartz, garnet, or epidote.

amphicarpic (am′fə kär′pik), *adj. Botany.* bearing fruit at two different times, or of two different kinds.

amphicoelous (am′fə sē′ləs), *adj. Zoology.* hollowed at both ends; concave on both sides, as the vertebrae of fishes. [from *amphi-* + Greek *koilos* hollow]

amphidiploid (am′fə dip′loid), *n. Genetics.* a hybrid organism having a diploid set of chromosomes derived from each parent; a double diploid. Also called **allotetraploid.**

amphidromic (am′fə drom′ik), *adj. Oceanography.* of or having to do with tidal action occurring around a point of no tide. An **amphidromic point** is a point of no tide from which cotidal lines radiate. [from Greek *amphidromos* running around]

amphigastrium (am′fə gas′trē əm), *n., pl.* **-tria** (-trē ə). *Botany.* a small, rudimentary leaf on the underside of the stem of certain liverworts. [from New Latin *amphigastrium,* from *amphi-* + Greek *gastros* belly]

amphigean (am′fə jē′ən), *adj. Botany.* extending around the earth, as species of plants found throughout the same latitude. [from *amphi-* + Greek *ge* earth]

amphimixis (am′fə mik′sis), *n. Biology.* the union of the gametes of two organisms in sexual reproduction. SYN: fertilization. [from *amphi-* + Greek *mixis* a mingling]

amphiphloic (am′fə flō′ik), *adj. Botany.* having to do with plant stems that have the phloem on the inner and outer surfaces of the xylem. Compare **ectophloic.** [from *amphi-* + Greek *phloios* bark]

amphiprotic (am′fə prō tik), *adj.* = amphoteric.

amphithecium (am′fə thē′shē əm *or* am′fə thē′sē əm), *n., pl.* **-cia** (-shē ə *or* -sē ə). *Botany.* the outer layer of cells in the spore case of mosses, surrounding the endothecium. [from New Latin *amphithecium,* from Greek *amphi-* both + *thēkion* small box]

amphitrichous (am fit′rə kəs), *adj. Microbiology.* having a single flagellum at each end of the cell: *amphitrichous bacteria.* [from *amphi-* + Greek *trichos* hair]

amphitropous (am fit′rə pəs), *adj. Botany.* having a partly inverted ovule attached near the middle of one side and parallel with the placenta. [from *amphi-* + Greek *tropē* a turning]

ampholyte (am′fe līt), *n. Chemistry.* an electrolyte having amphoteric properties: *Amphoteric substances are often referred to as ampholytes* (Jones, *Inorganic Chemistry*). [from *ampho(teric)* + *(electro)lyte*]

—**ampholytic,** *adj.* of or having to do with an ampholyte.

amphoteric (am′fə ′ter′ik), *adj. Chemistry.* reacting either as an acid or as a base; showing characteristics of both acids and bases. Amino acids are amphoteric compounds since they contain both a basic and an acidic group. *An amphoteric hydroxide is one which reacts as a base when treated with a strong acid, and reacts as an acid when treated with a strong base*

---

cap, fāce, fäther; best, bē, tèrm; pin, fīve;
rock, gō, ôrder; oil, out; cup, pùt, rüle;
yü in use, *yu* in uric;
ng in bring; *sh* in rush; *th* in thin, ᴛн in then;
zh in seizure.

ə = *a* in about, *e* in taken, *i* in pencil, *o* in lemon, *u* in circus

(Parks and Steinbach, *Systematic College Chemistry*). [from Greek *amphoteros* both]

amplectant (am plek′tənt), *adj. Botany.* twining: *an amplectant tendril.* [from Latin *amplecti* to embrace]

amplexicaul (am plek′sə kôl), *adj. Botany.* nearly surrounding or clasping the stem, as some leaves do at their base. [from New Latin *amplexicaulis,* ultimately from Latin *amplecti* to embrace + *caulis* stem]

amplexus (am plek′səs), *n.* **1** *Zoology.* the copulatory embrace of a frog or toad. **2** *Botany.* the overlap, in vernation, of two sides of one leaf with the two sides of the leaf above it. [from Latin *amplexus* an embrace, from *amplecti* to embrace]

amplification, *n. Physics.* an increase in the magnitude or strength of an electric current or other physical quantity or force.

AMPLIFICATION

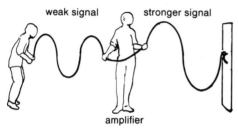

weak signal   stronger signal

amplifier

The weak signal is changed into a stronger signal with larger waves when it passes through the amplifier.

amplify, *v. Physics.* to increase the magnitude or strength of (an electric current, mechanical force, or other physical quantity).

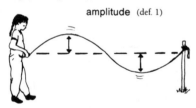

amplitude (def. 1)

The arrows indicate the amplitude of the wave (represented by the rope); the dotted line indicates the position of rest.

amplitude, *n.* **1** *Physics.* one half the extent of a vibration, oscillation, or wave. The distance between the position of rest and the highest position in the arc of a pendulum is its amplitude; thus a pendulum swinging through an angle of 90 degrees has an amplitude of 45 degrees. *The amplitude A of a simple harmonic motion is the magnitude of the maximum displacement of the particle from its equilibrium position* (Shortley and Williams, *Elements of Physics*).
2 *Electricity.* the peak strength of an alternating current in a given cycle.
3 *Mathematics* (of a complex number) the directed angle formed at the origin of a coordinate system by the rotation of the positive horizontal axis into a vector representing a complex number: *The amplitude of 3 + 3i is 45 degrees.*

4 *Astronomy.* (of a heavenly body) the arc along the horizon measured between either true east or west and a point where a line drawn from the zenith through the body would intersect the horizon.

amplitude modulation, *Electronics.* deliberate change of the amplitude of radio waves in order to transmit sound or visual images. Amplitude modulation is used for ordinary radio broadcasting and for the transmission of the picture portion of television. *Abbreviation:* AM Compare **frequency modulation.**

ampulla (am pul′ə), *n., pl.* **-pullae** (-pul′ē). **1** *Anatomy, Zoology.* the dilated portion of a canal or duct, as of the semicircular canals of the ear. **2** *Botany.* a flask-shaped organ or bladder on an aquatic plant. [from Latin *ampulla* pear-shaped bottle]

amu or **AMU,** *abbrev.* atomic mass unit.

amygdala (ə mig′də lə), *n., pl.* **-lae** (-lē *or* -lī). *Anatomy.* an almond-shaped mass of gray matter in the lateral ventricle of the brain. Also called **amygdaloid body** or **nucleus.** [from New Latin, from Latin *amygdala* almond, from Greek *amygdalē*]

amygdaloid (ə mig′də loid), *n. Geology.* a volcanic rock having small gas cavities filled with such minerals as agate, calcite, and opal. [from Greek *amygdalē* almond + English *-oid*]

amygdaloid body or **amygdaloid nucleus,** = amygdala.

amyl (am′əl), *n. Chemistry.* any of various forms of a univalent hydrocarbon radical derived from pentanes. *Formula:* $-C_5H_{11}$ Also called **pentyl.** [from Greek *amylon* starch; *amyl* was so named because its alcohol was originally obtained from spirits distilled from potato or grain starch]

amyl alcohol, *Chemistry.* **1** an acrid, oily liquid, the chief constituent of fusel oil, used in organic synthesis and as a solvent. *Formula:* $C_5H_{11}OH$ **2** any of various isomeric forms of amyl alcohol or mixtures of these isomers.

amylase (am′ə lās), *n. Biochemistry.* any of various enzymes present in saliva, pancreatic juice, etc., or in parts of plants, that helps to change starch into sugar. [from *amyl* + *-ase*]

amyloid (am′ə loid), *n. Biochemistry.* an insoluble, waxy protein developed in diseased degeneration of various organs.

amylolysis (am′ə lol′ə sis), *n. Biochemistry.* the conversion of starch into substances that can be absorbed and utilized by the body, such as certain sugars, by the action enzymes or acids.

—amylolytic (am′ə lə lit′ik), *adj.* of or having to do with amylolysis: *Germination is allowed to proceed from four to six days, during which time amylolytic, i.e. starch-digesting enzymes are developed in the barley embryos, and the conversion of the starch reserves of the grain to sugar begins* (H. J. Bunker).

amylopectin (am′ə lə pek′tin), *n. Biochemistry.* the outer part of a starch granule, consisting of almost insoluble polysaccharides of a branched molecular structure and high molecular weight. Amylopectin stains violet with iodine and does not form a gel in water. Compare **amylose.**

amyloplast (am′ə lə plast), *n. Botany.* a colorless body in plant cells which forms starch granules.

amylopsin (am′ə lop′sən), *n. Biochemistry.* the amylase in the pancreatic juice. [from *amyl* + (*pep*)*sin*]

amylose (am′ə lōs), *n. Biochemistry.* **1** the inner part of a starch granule, consisting of relatively soluble polysaccharides of unbranched, linear or spiral structure. Amylose stains blue with iodine and forms a gel in water. Compare **amylopectin. 2** any of various other polysaccharides.

An, *symbol.* actinon.

anabiosis (an′ə bī ō′sis), *n. Biology.* a state of apparently suspended animation, especially one in which certain aquatic invertebrates survive long periods of drought: *Cellular protein converts to an inactive, crystalline form during anabiosis* (James A. Pearre). [from Greek *anabiōsis,* from *ana-* again + *bioun* live]

anabolic (an′ə bol′ik), *adj. Biology.* of, involving, or exhibiting anabolism: *Proteins are synthesized in the anabolic processes of both plants and animals through the chemical combination of large numbers of units called amino acids* (Harbaugh and Goodrich, *Fundamentals of Biology*).

anabolic steroid, *Biochemistry.* any of a group of steroid hormones that promote the storage of protein and the growth of tissue, used in medicine and by athletes during training.

anabolism (ə nab′ə liz′əm), *n. Biology.* the phase of metabolism that uses energy to build up complex compounds of living matter from the simple nutritive compounds obtained from food: *Assimilation, an important part of anabolism, is the process of converting absorbed material into protoplasm* (Hegner and Stiles, *College Zoology*). Also called **constructive metabolism.** Contrasted with **catabolism.** [from Greek *ana-* upward + English *(meta)bolism*]

anabranch (an′ə branch), *n. Geology.* a stream that branches off from a river and reenters the river lower down, forming an island. [from *ana(stomosing) branch*]

anaclinal (an′ə klī′nəl), *adj. Geology.* descending in a direction opposite to that of the dip of the rock formation below, as a valley or stream. [from *ana-* again + Greek *klinein* to lean]

anadromous (ə nad′rə məs), *adj. Zoology.* going up rivers from the sea to spawn. Salmon and shad are anadromous. Contrasted with **catadromous.** Compare **diadromous.** [from Late Greek *anadromos,* from Greek *ana-* up + *dromos* a running]

anaerobe (an ār′ōb), *n. Biology.* an anaerobic bacterium or other microorganism; an organism that can live where there is no atmospheric oxygen. Contrasted with **aerobe.** [from Greek *an-* without + *aēr* air + *bios* life]

anaerobic (an′ār ō′bik), *adj. Biology.* **1** living or growing where there is no atmospheric oxygen. Anaerobic bacteria get their oxygen by decomposing compounds containing oxygen. *In swamps, peat bogs and other places where oxygen is largely excluded, certain anaerobic bacteria are able to reduce the nitrates ($-NO_3$) to nitrites ($-NO_2$) by removing one oxygen atom* (Emerson, *Basic Botany*).
2 having to do with or occurring in the absence of atmospheric oxygen: *Anaerobic metabolism is the process whereby tissue can exist with a shortage of oxygen* (Science News).
—**anaerobically,** *adv.*

anaerobic respiration, *Biology.* a process in which organic compounds are broken down and energy is released without the intake of atmospheric oxygen: *If molecular oxygen is not present, anaerobic respiration ensues in most plant cells and CO_2 and alcohol are formed* (Weier, *Botany*). *The most common type of anaerobic respiration is alcoholic fermentation* (Greulach and Adams, *Plants*).

anaerobiosis (an′ār ō′bī ō′sis), *n. Biology.* life sustained by an organism in the absence of oxygen. Contrasted with **aerobiosis.**
—**anaerobiotic** (an′ār ō′bī ot′ik), *adj.* of or characterized by anaerobiosis; anaerobic.

anagenesis (an′ə jen′ə sis), *n. Biology.* the type of evolutionary change within a species that results in linear or phyletic descent. Contrasted with **cladogenesis.** [from Greek *ana-* upward + *genesis* origin]

anal (ā′nl), *adj. Anatomy.* of, at, or near the anus or a part resembling the anus: *the anal pore of the paramecium.* See the picture at **fin.**

analcime (ə nal′sēm *or* ə nal′sim), *n. Mineralogy.* a whitish or slightly tinted isometric species of zeolite (hydrous aluminum sodium silicate) occurring in basaltic rock rich in alkali. *Formula:* $NaAlSi_2O_6 \cdot H_2O$ [from French *analcime,* from Greek *analkimos* weak, from *an-* not + *alkimos* strong]

analcite (ə nal′sīt), *n.* = analcime.

analemma (an′ə lem′ə), *n. Astronomy.* a graduated scale representing the sun's declination at any date of the year, drawn on the Torrid Zone of a terrestrial globe, often in the shape of a long figure 8 centered on the equator: *The difference in hour angle between the sun and a clock, which runs at a uniform rate, is the "equation of time." A plot of these differences through the year gives ... the so-called "analemma" which is familiar on globes of the earth* (Scientific American). [from Greek *analēmma* sundial, (literally) support, ultimately from *ana-* up + *lambanein* take]

anal fin, *Zoology.* a fin located near the anus on the underside of fishes.

analgesia (an′l jē′zhə), *n. Physiology.* a deadening or absence of the sense of pain without loss of consciousness. Compare **anesthesia.** [from Greek *analgēsia,* from *an-* not + *algein* feel pain]
—**analgesic,** *adj.* of or causing analgesia: *There are several other substances closely related to aspirin that have analgesic properties ... and antipyretic properties* (Pauling, *Vitamin C*). —*n.* an analgesic drug.

analogous, *adj. Biology.* corresponding in function, but not in structure and origin: *Analogous organs are similar in function but are not genetically related. For example, the wings of butterflies and birds* (Hegner and Stiles, *College Zoology*). Contrasted with **homologous.** [from Greek *analogos* proportionate, from *ana logon* according to due ratio]

---

cap, fāce, fäther; best, bē, tėrm; pin, five;
rock, gō, ôrder; oil, out; cup, pùt, rüle,
yü in use, *yu* in uric;
ng in bring; *sh* in rush; *th* in thin, ŦH in then;
zh in seizure.
ə = *a* in about, *e* in taken, *i* in pencil, *o* in lemon, *u* in circus

analogue, *n.* **1** *Biology.* an organ analogous to another organ. Contrasted with **homologue.**
2 *Chemistry.* a substance which resembles another structurally but differs from it in function, especially a compound of this kind that interferes in a reaction because of its structural similarity to one of the normal reactants.

analogy, *n.* **1** *Biology.* correspondence of organs or other parts in function but not in structure and origin: *Another biological principle, analogy ... Body structures are said to be analogous which have the same function, but a different embryonic background* (Winchester, *Zoology*). Contrasted with **homology.**
2 *Mathematics.* proportion; agreement of ratios.

analysis (ə nal′ə sis), *n., pl.* **-ses** (-sēz′). **1** *Chemistry.* **a** the determination of the nature or amount of one or more components of a substance. **b** the intentional separation of a compound into its parts or elements.
2 *Physics.* the resolution of light into its prismatic constituents.
3 *Mathematics.* **a** algebraic reasoning, especially as applied to geometry. **b** treatment by calculus.
[from Greek *analysis,* from *analyein* loosen up, from *ana-* up + *lyein* loosen]
—**analytic** or **analytical,** *adj.* **1** of analysis; using analysis as a method or process. **2** concerned with or based on analysis. —**analytically,** *adv.*

analytical balance, a precision balance, used especially in analytical chemistry for weighing quantities as small as 1/10,000 of a gram.

analytical chemistry, the branch of chemistry that deals with the analysis of the components making up samples of matter. See also **qualitative analysis, quantitative analysis.**

analytic geometry, the use of algebra and coordinates to solve problems in geometry. Also called **coordinate geometry.**

analyze, *v. Chemistry.* **1** to determine the nature or amount of the components of (a substance).
2 to separate intentionally (a compound) into its elements.

anamorphic (an′ə môr′fik), *adj. Optics.* having to do with or producing anamorphosis: *Anamorphic lenses ... can expand and subsequently compress the image on the negative* (Saturday Review).

anamorphose (an′ə môr′fōz), *v. Optics.* to represent by anamorphosis.

anamorphosis (an′ə môr′fə sis *or* an′ə môr fō′sis), *n., pl.* **-ses** (-sēz). **1** *Optics.* **a** a distorted image or projection that appears in natural form when reflected from a curved mirror or the like. **b** a method of producing such images or projections.
2 *Botany.* an anomalous change of form in a plant.
3 *Biology.* a gradual change of form within a group of plants or animals over a long period of time.
[from Greek *anamorphōsis* a forming anew, ultimately from *ana-* again + *morphē* form]

anandrous (an an′drəs), *adj. Botany.* lacking stamens: *Anandrous flowers are female.* [from Greek *anandros* husbandless, from *an-* without + *andros* man]

ananthous (an an′thəs), *adj. Botany.* lacking flowers: *an ananthous plant.* [from Greek *ananthēs,* from *an-* without + *anthos* flower]

anaphase (an′ə fāz), *n. Biology.* the third stage in mitosis, characterized by the movement at the two sets of daughter chromosomes to opposite ends of the spindle. It occurs after the metaphase and before the telophase. See the picture at **mitosis.** [from Greek *ana-* up + English *phase*]

anaphylaxis (an′ə fə lak′sis), *n. Medicine.* increased sensitivity to the action of a normally nontoxic drug upon injection with it for the second time, sometimes causing severe or even fatal shock. Compare **allergy.** [from Greek *ana-* back + *phylaxis* protection]

anarthrous (an är′thrəs), *adj. Zoology.* lacking joints; without joints. [from Greek *anarthros,* from *an-* without + *arthron* joint]

anastigmatic (an′ə stig mat′ik), *adj. Optics.* free from astigmatism. An **anastigmatic lens** is a compound lens in which each part is designed to compensate for the astigmatic effects of the other. [from *an-* not + *astigmatic*]

anastomose (ə nas′tə mōz), *v.,* **-mosed, -mosing.** to communicate, unite, or connect by anastomosis.

anastomosis (ə nas′tə mō′sis), *n., pl.* **-ses** (-sēz′). a cross connection between separate parts of any branching system, as the veins of leaves, the veins in the wings of insects, or rivers and their branches. [from Greek *anastomōsis,* from *ana-* back + *stoma* mouth]

anat., *abbrev.* **1** anatomical. **2** anatomy.

anatase (an′ə tās), *n. Mineralogy.* a tetragonal form of titanium oxide, occurring in brown, dark blue, and black crystals. It is polymorphous with rutile and brookite. *Formula:* TiO_2 Also called **octahedrite.** [from Greek *anatasis* extension]

anatexis (an′ə tek′sis), *n. Geology.* the process by which plutonic rock is melted and regenerated as magma. [from Greek *anatēxis* a melting, ultimately from *ana-* up + *tēkein* to thaw]

anatomy, *n. Biology.* **1** the science of the structure of animals and plants based upon dissection, microscopic observation, etc.
2 the dissecting of animals or plants to study the position and structure of their parts.
3 the bodily structure of an animal or plant: *the anatomy of an earthworm.*
[from Greek *anatomē,* from *ana-* up + *tomos* a cutting]
—**anatomic** or **anatomical,** *adj.* **1** connected with the study or practice of anatomy or dissection. **2** of anatomy; structural.
—**anatomize,** *v.* to dissect (an animal, plant, etc.) to study the structure and relation of the parts.

ancipital (an sip′ə təl), *adj. Botany.* having two sharp edges, as the stems of certain plants. [from Latin *ancipitem* two-edged, two-faced]

andalusite (an′də lü′sīt), *n. Mineralogy.* a silicate of aluminum occurring in orthorhombic crystals of various colors, trimorphous with kyanite and sillimanite. *Formula:* Al_2SiO_5 [from *Andalusia,* Spain, where first found + *-ite*]

andesite (an′də zīt), *n. Geology.* a volcanic igneous rock occurring in many varieties, in all of which a plagioclase of high sodium content is the chief constituent. [from *Andes* (mountains) + *-ite*]

androecium (an drē′shē əm *or* an drē′sē əm), *n., pl.* **-cia** (-shē ə *or* -sē ə). *Botany.* the male organ of a flower; the stamens. Compare **gynoecium**. [from Greek *andros* man + *oikion* house] —**androecious,** *adj.*

androgen (an′drə gən), *n. Biochemistry.* any hormone, such as testosterone, that induces or enhances masculine characteristics.

—**androgenic** (an′drə jen′ik), *adj.* of or having to do with an androgen.

—**androgenicity** (an′drə jə nis′ə tē), *n.* the capability of a hormone or other substance to reinforce masculine characteristics.

androgyne (an′drə jin), *n.* **1** *Botany.* an androgynous plant. **2** = hermaphrodite.

androgynous (an droj′ə nəs), *adj.* **1** *Botany.* having flowers with stamens and flowers with pistils in the same cluster.

2 *Zoology.* being both male and female; hermaphroditic.

[from Greek *andros* man + *gynē* woman]

androgyny (an droj′ə nē), *n.* = hermaphroditism.

androsterone (an dros′tə rōn), *n. Biochemistry.* a steroid hormone that reinforces male characteristics, isolated from male urine. *Formula:* $C_{19}H_{30}O_2$ [from Greek *andros* man + English *ster(ol)* + *-one*]

-ane, *noun suffix. Chemistry.* a saturated hydrocarbon of the methane or alkane series, as in *butane, propane.* [imitative formation patterned on *-ene, -ine,* etc.]

anelectric (an′i lek′trik), *Physics.* —*adj.* that cannot be electrified by friction.

—*n.* an anelectric substance.

[from *an-* not + *electric*]

anemometer (an′ə mom′ə tər), *n.* an instrument for measuring the velocity of wind. [from Greek *anemos* wind]

anemophilous (an′ə mof′ə ləs), *adj. Botany.* fertilized by pollen carried by the wind, as the flowers of grasses, sedges, and pines; wind-pollinated. [from Greek *anemos* wind + *philos* loving]

aneroid barometer (an′ə roid′), a barometer worked by the pressure of air on the elastic lid of an airtight metal box from which the air has been pumped out. A change of pressure causes a pointer attached to the lid to move along a scale. [from French *anéroïde* using no liquid, from Greek *a-* without + Late Greek *nēron* water]

anesthesia (an′əs thē′zhə), *n. Physiology.* loss of the feeling of pain, touch, cold, or other sensation. Anesthesia can be produced by ether, chloroform, nitrous oxide, halothane, procaine, or other chemical agent, or by hypnotism, acupuncture, paralysis, or disease. The entire body (**general anesthesia**) or only a certain area or areas (**local anesthesia**) may be affected. [from Greek *anaisthēsia* insensibility, from *an-* without + *aisthēsis* sensation]

—**anesthetic,** *n.* a substance that causes anesthesia.

—**anesthetize** (ə nes′thə tīz), *v.* to make (a person, animal, area of the body, etc.) unable to feel pain, touch, cold, or other sensation; produce anesthesia in.

anestrum (an es′trəm) *or* **anestrus** (an es′trəs), *n.* = diestrum. [from New Latin, from *an-* not + *estrum* or *estrus* estrous cycle]

aneuploid (an yü′ploid), *adj. Genetics.* having a number of chromosomes not a multiple of the haploid number for the species. Compare **euploid**. [from *an-* not + *euploid*]

aneurysm *or* **aneurism** (an′yə riz′əm), *n. Medicine.* a permanent swelling of an artery or vein, caused by pressure of the blood on a weakened part. Aneurysms may be congenital or caused by disease or injury. [from Greek *aneurysma* dilation, from *an-* up + *eurys* wide]

angiology (an′jē ol′ə jē), *n.* the branch of anatomy that deals with the blood and lymph vessels. [from Greek *angeion* receptacle + English *-logy*]

angiosperm (an′jē ə spėrm′), *n. Botany.* any of a group of seed plants having the seeds enclosed in an ovary or fruit; a flowering plant. Grasses, beans, strawberries, and oaks are angiosperms. The angiosperms constitute one of two groups (the other being the gymnosperms) into which the seed-bearing plants (spermatophytes) can be divided. [from Greek *angeion* vessel + *sperma* seed]

—**angiospermous** (an′jē ə spėr′məs), *adj.* having the seeds enclosed in an ovary.

angiotensin (an′jē ō ten′sin), *n. Biochemistry.* a peptide occurring in the blood that affects the caliber of blood vessels and otherwise alters blood pressure: *Scientists agree that renin, a hormone produced by the kidney, helps maintain blood pressure by converting angiotensin 1, a protein in the blood, to its active form, angiotensin 2* (Michael H. Alderman). [from Greek *angeion* receptacle + English *tens(ion)* + *-in*]

angle, *n. Mathematics.* **1** the region between two rays having a common end point. The point at which they meet is the *vertex* and the rays that form the angle are its *sides.*

2 the figure formed by two such rays. Compare **acute angle, obtuse angle, right angle.**

3 the difference in direction between two such rays, measured especially in degrees or parts of degrees. [from Latin *angulus*]

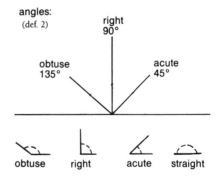

angles:
(def. 2)

angle of aberration, *Astronomy.* the angle between the actual and the apparent position of a star or other heavenly body to an observer.

---

cap, fāce, fäther; best, bē, tėrm; pin, five;
rock, gō, ôrder; oil, out; cup, pùt, rüle,
yü in use, yù in uric;
ng in bring; sh in rush; th in thin, ᴛн in then;
zh in seizure.
ə = a in about, e in taken, i in pencil, o in
lemon, u in circus

angle of dip, *Geology.* **1** the angle of downward inclination from the horizontal of the needle of a magnetic compass, ranging from 0 degrees at the magnetic equator to 90 degrees at either of the magnetic poles: *On a straight reading of the rock magnetism results in Europe the angle of dip of the magnetization suggests that this continent once lay well to the south of the Equator, and the horizontal deviation indicates a clockwise rotation of 30–40 degrees* (New Scientist). **2** the angle of downward inclination from the horizontal in a plane of stratification.

angle of incidence, *Physics.* the angle that a ray or wave of light, sound, etc., falling upon a surface, makes with a line perpendicular to that surface. The angle of incidence is always equal to and adjacent to the angle of reflection. See picture at **incidence.**

angle of reflection, *Physics.* the angle that a ray or wave of light, sound, etc., makes with a line perpendicular to a reflecting surface.

angle of refraction, *Physics.* the angle that a ray of light, etc., makes with a line perpendicular to a surface separating two media.

anglesite (ang′glə sīt), *n. Mineralogy.* native lead sulfate, formed by oxidation of galena. *Formula:* PbSO₄ [from Angles(ey), Wales, where first found + -ite]

angstrom (ang′strəm), *n.,* or **angstrom unit,** a unit of length, used for measuring wavelengths of light, equal to one ten-millionth of a millimeter (10^{-10} meter). *Symbol:* Å [named after Anders J. *Ångström,* 1814–1874, Swedish physicist]

angular, *adj. Mathematics.* **1** consisting of an angle: *an angular point.* **2** measured by an angle: *angular distance.*
—**angularity,** *n.* condition of being angular.

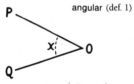

angular (def. 1)

Angle X = the angular distance of point P from point Q when measured from O.

angular acceleration, *Physics.* the rate of change of angular velocity: *While a rotating body is speeding up or slowing down ... it is said to have an angular acceleration* (Sears and Zemansky, *University Physics*).

angular aperture, *Optics.* the angular breadth of the light which an optical instrument transmits from the point viewed.

angular displacement, *Physics.* the amount of rotation of a body about an axis: *Angular displacement of a rigid body about a fixed axis is defined as the angle through which any radius of the body turns* (Shortley and Williams, *Elements of Physics*).

angular momentum, *Physics.* the product of the moment of inertia of a body and its angular velocity: *Angular momentum is ... extremely important in connection with nuclear physics. Each elementary particle of physics, such as the proton and the neutron, has an angular momentum called* spin. (Pierce, *Electrons, Waves and Messages*).

angular unconformity, *Geology.* an unconformity in which younger strata overlie an erosion surface on folded or tilted layers of rock.

angular velocity, *Physics.* the rate of angular motion about an axis, usually expressed in radians per second: *For a particle revolving in a circular orbit, the angular momentum is the mass times the angular velocity* (Scientific American).

anharmonic (an′här mon′ik), *adj. Physics.* not harmonic: *an anharmonic oscillation or vibration.*

anhedral (an hē drəl), *adj. Mineralogy.* having the molecular form of a crystal but not its external form. [from *an-*¹ lacking, not + Greek *hédrā* seat, base, surface + English -*al*¹]

anhydride (an hī′drīd *or* an hī′drid), *n. Chemistry.* **1** any oxide that unites with water to form an acid or base. Sulfur dioxide, SO₃, is the anhydride of sulfuric acid. An anhydride of a nonmetal or of an organic radical is an *acid anhydride;* an anhydride of a metal is a *basic anhydride.*
2 any compound formed by the removal of water. [from *anhydr(ous)* + -*ide*]

anhydrite (an hī′drīt), *n. Mineralogy.* a white or grayish mineral consisting of anhydrous sulfate of calcium, which commonly hydrates to form gypsum. *Formula:* CaSO₄

anhydrous (an hī′drəs), *adj. Chemistry.* containing no water of crystallization; not hydrated: *anhydrous ammonia. Many crystals, like diamond, sulfur, quartz, sodium chloride, and others, are anhydrous* (Offner, *Fundamentals of Chemistry*). [from Greek *anydros,* from *an-* without + *hydōr* water]

aniline (an′l ən), *n. Chemistry.* a colorless, poisonous, oily liquid, obtained from coal tar and especially from nitrobenzene, used in making dyes, medicines, plastics, etc. *Formula:* C₆H₅NH₂ **Aniline dyes** are derivatives of aniline used to color textiles, in making inks, paints, and varnishes, and in biology to stain cells for microscopic study. [from German *Anilin,* from *Anil* indigo, from Arabic *al-nil*]

animal, *n. Biology.* any member of a kingdom (Animalia) of organisms that ingest their food, lack chlorophyll, and generally have a capacity for voluntary motion. Many members also have more advanced types of sensation and response to stimuli. In some common classification systems, only multicellular organisms are considered animals. *Animals require complex organic materials as food, obtained only by eating plants or other animals* (Storer, *General Zoology*). [from Latin *animal* living being, from *anima* life, breath]

animal heat, *Physiology.* the temperature maintained during life in the body of a warm-blooded vertebrate animal, and necessary for its physiological functions.

animal kingdom, *Biology.* the primary division of living things that includes all animals.

animal pole, *Embryology.* that part of the surface of an egg or ovum having the least yolk, in which is found the most metabolically active protoplasm of the embryo. Contrasted with **vegetal pole.**

animal starch, = glycogen.

anion (an′ī′ən), *n. Chemistry.* **1** a negatively charged ion that moves toward the positive pole in electrolysis: *When a single molecule of an electrolyte dissociates, it yields one or more ions charged positively, called*

cations, and one or more charged negatively, known as anions (Parks and Steinbach, *Systematic College Chemistry*). **2** an atom or group of atoms having a negative charge.
[from Greek *anion* (thing) going up, from *ana-* up + *ienai* go]
—**anionic** (an'ī on'ik), *adj.* of or having to do with an anion or anions.

anion exchange, *Chemistry.* ion exchange in which the negative ions are exchanged.

aniso-, *combining form.* unequal; unlike; dissimilar, as in *anisogamete.* [from Greek *anisos,* from *an-* not + *isos* like, equal]

anisocarpic (an ī'sə kär'pik), *adj. Botany.* having fewer carpels than other floral parts. [from *aniso-* + Greek *karpos* fruit]

anisogamete (an ī'sə gam'ēt), *n. Biology.* either of a pair of conjugating gametes that differ from each other in form or size. A large, nonmotile egg and a small, motile sperm are anisogametes. Also called **heterogamete.** [from *aniso-* + *gamete*]

anisogamy (an ī sog'ə mē), *n. Biology.* the conjugation of dissimilar gametes, as among certain protozoans and algae. Also called **heterogamy.** [from *aniso-* + *-gamy*]
—**anisogamous** (an ī sog'ə məs), *adj.* characterized by anisogamy.

anisomerous (an'ī som'ər əs), *adj. Botany.* **1** having irregular or asymmetrical floral parts. **2** having unequal numbers of parts in each whorl. [from *aniso-* + Greek *meros* part]

anisometric (an ī'sə met'rik), *adj.* of unequal measurement; having nonsymmetrical parts; not isometric: *anisometric crystals.*

anisotropic (an ī'sə trop'ik), *adj.* **1** *Physics, Crystallography.* having or showing different properties when measured along different axes or directions: *There exist, however, many transparent crystalline substances which, while homogeneous, are anisotropic. That is, the velocity of a light wave in them is not the same in all directions* (Sears and Zemansky, *University Physics*). Contrasted with **isotropic.**
2 *Botany.* responding differently or unequally to external stimuli.
[from *aniso-* + Greek *tropikos* turning]
—**anisotropy** (an'ī sot'rə pē), *n.* the state or quality of being anisotropic.

ankerite (ang'kə rīt), *n.* a mineral closely related to dolomite, with the magnesium largely replaced by iron. *Formula:* Ca(Fe,Mg,Mn)(CO$_3$)$_2$ [from German *Ankerit,* from M. J. *Anker,* 19th-century Austrian mineralogist + *-it* -ite]

ankle, *n. Anatomy.* the joint that connects the foot and the leg, enabling the foot to move up and down. The ankle is formed by the articulation of the tibia and fibula (the lower leg bones) with the talus (the highest tarsal bone).

anklebone, *n. Anatomy.* the principal bone of the ankle; the astragalus or talus.

anlage (än'lä gə), *n., pl.* **-gen** (-gen) or **-ges.** **1** *Embryology.* the first clustering of embryonic cells constituting the beginning of an organ or part. SYN: primordium, blastema.

2 *Genetics.* the hereditary predisposition of an individual for a particular trait or traits.
[from German *Anlage* foundation, basis]

annabergite (an'ə ber'gīt), *n.* a mineral, hydrous arsenate of nickel, of an apple-green color, occurring in capillary crystals or incrustations as an alteration of nickel arsenides. *Formula:* (Ni,Co)$_3$(AsO$_4$)$_2$·8H$_2$O [from *Annaberg,* a region in Germany where the mineral is found + English *-ite*]

anneal (ə nēl'), *v.* **1** to make (glass or metals) less brittle by heating and then cooling: *An alloy when cast is usually inhomogeneous and must be annealed at a high temperature, so that the atoms can diffuse to give it a uniform composition throughout* (Science News).
2 *Molecular Biology.* to separate strands of DNA by heating and recombine them with complementary strands after cooling: *Revealed under high magnification, a long single strand of adenovirus DNA annealed to a shorter strand of fiber messenger RNA (mRNA) produces a duplex structure ... that contains four single-stranded loops of DNA* (R. Haselkorn).
[Old English *anǣlan,* from *an-* on + *ǣlan* to burn]

annelid (an'l id), *Zoology.* —*n.* any of a phylum (Annelida) of worms or wormlike animals characterized by a soft, elongated body composed of a series of similar ringlike segments. Earthworms, leeches, and various sea worms are annelids.
—*adj.* of or belonging to this phylum.
[from Latin *annellus,* diminutive of *annulus* ring]

annihilation, *n. Particle Physics.* the destruction of a particle and its antiparticle as a result of their collision, their energy being shared by lighter particles that emerge from the collision.
—**annihilate,** *v. Particle Physics.* **1** to cause the annihilation of: *Electrically neutral particles also have corresponding antiparticles which are annihilated with the same total energy as the original photons* (Scientific American). **2** to undergo annihilation: *When an electron and a positron meet they annihilate to form a photon* (Science News).

annual, *Botany.* —*adj.* living only one year or season: *Corn and beans are annual plants.*
—*n.* a plant that lives only one year or growing season: *There are many annuals or biennials among the dicotyledons and monocotyledons* (Weier, *Botany*).
ASSOCIATED TERMS: *Annual* plants germinate, flower, produce seed, and die within the same year or growing season; *biennial* plants, such as the carrot, germinate in one year or growing season, and flower, produce seed, and die in the next year or growing season; *perennial* plants, such as the holly, the iris, and the crocus, are those which persist and grow from year to year, usually in some reduced form during the winter months.
[from Late Latin *annualis,* from Latin *annus* year]

---

cap, fāce, fäther; best, bē, tèrm; pin, five;
rock, gō, ôrder; oil, out; cup, pùt, rüle,
yü in use, yu̇ in uric;
ng in bring; sh in rush; th in thin, ᴛʜ in then;
zh in seizure.
ə = a in about, e in taken, i in pencil, o in
lemon, u in circus

annual parallax, *Astronomy.* the apparent change or amount of change in the position of a celestial body as a result of the earth's motion around the sun. Also called **heliocentric parallax.**

annual ring, *Botany.* any of the concentric rings of wood (xylem) seen when the stem of a tree or shrub is cut across. Each ring shows one year's growth. Also called **growth ring.** See also **dendrochronology.**

annular (an′yə lər), *adj.* of or like a ring; ring-shaped. An **annular eclipse** is an eclipse of the sun in which the moon covers the sun incompletely, leaving a narrow uneclipsed ring which surrounds the dark moon. [from Latin *annulus* ring]

annulation (an′yə lā′shən), *n. Anatomy.* 1 formation of rings. 2 a ringlike structure.

annulus (an′yə ləs), *n., pl.* **-li** (-lī), **-luses. 1** *Astronomy.* the uneclipsed ring of the sun in an annular eclipse: *A ring of the sun's surface, called the annulus, the Latin word for "ring," will be visible around the black disc of the moon* (Science News Letter).
2a *Botany.* a ring of specialized cells or tissue on the spore case of a fern or moss, or on the stem of certain mushrooms. **b** *Zoology.* any ringlike part or band: ... *annuli, i.e. growth-rings, corresponding in many respects to the growth-rings in the trunks of trees, representing the years of life of the fish* (New Biology).
3 *Geometry.* the region between two concentric circles.
[from Latin *annulus* ring]

anode (an′ōd), *n. Electronics.* **1** the positive electrode of an electrolytic cell or electron tube. **2** the negative terminal of a battery that is producing current. [from Greek *anodos* a way up]

anolyte (an′ə līt), *n. Chemistry.* the part of an electrolyte close to the anode during electrolysis. [from *ano(de)* + *(electro)lyte*]

anomalistic (ə nom′ə lis′tik), *adj. Astronomy.* having to do with the anomaly of a planet: *the anomalistic period of Saturn.* —**anomalistically,** *adv.*

anomalistic month, *Astronomy.* the time required for the moon to pass from perigee to perigee. An anomalistic month averages 27 days, 13 hours, 18 minutes, and 33.1 seconds.

anomalistic year, *Astronomy.* the time between two successive passages of the perihelion by the earth in its orbit around the sun; 365 days, 6 hours, 13 minutes, and 53.1 seconds: *The anomalistic year ... is not identical with the sidereal year because the position of the earth's perihelion advances a slight amount annually due to the perturbative effects of the other planets* (Krogdahl, *The Astronomical Universe*).

anomaly, *n.* **1** *Biology.* deviation from a norm or type: *There is no greater anomaly in nature than a bird that cannot fly* (Darwin, *Origin of Species*).
2 *Astronomy.* the angle measured in the orbit of one heavenly body about another. The **true anomaly** is the angle measured between a line from the sun to a planet and a line from the sun to the planet's perihelion. The **mean anomaly** is the angle measured between a planet's perihelion and an imaginary planet having a period equal to that of the real planet but moving with a constant velocity about the sun.

3 *Geology.* an unusually high concentration of an element in soil, suggesting the presence of a nearby mineral deposit.
[from Latin *anomalia,* ultimately from Greek *an-* not + *homalos* even]
—**anomalous,** *adj.* having to do with or characterized by an anomaly; irregular; eccentric; aberrant.

anorthite (an ôr′thīt), *n. Mineralogy.* a calcium-rich plagioclase feldspar occurring as a constituent of mafic igneous rocks. *Formula:* $CaAl_2Si_2O_8$ [from *an-* not + Greek *orthos* straight + English *-ite*]
—**anorthitic** (an′ôr thit′ik), *adj.* of or containing anorthite.

anorthoclase (an ôr′thə klās *or* an ôr′thə klāz), *n. Mineralogy.* a triclinic feldspar rich in sodium as well as in potassium. Anorthoclase is related to microcline, and is found in some felsic igneous rocks. [from *an-* not + *orthoclase*]

anorthosite (an ôr′thə sīt), *n. Geology.* a coarse-grained igneous rock composed chiefly of a sodium calcium plagioclase feldspar, especially labradorite: *These unusual earth rocks, and their lunar counterparts, are called anorthosites, because they consist almost entirely of a calcium-rich feldspar called anorthite* (Paul D. Lowman, Jr.). [from French *anorthose* feldspar + English *-ite*]
—**anorthositic** (an ôr′thə sit′ik), *adj.* of or containing anorthosite: *A rock rich in magnesium called anorthositic gabbro appeared to be the commonest, though it was just one of a large number of rock types that made up the complex [lunar] highland lithology* (New Scientist).

ANS, *abbrev.* autonomic nervous system.

ant-, *prefix.* the form of **anti-** before vowels and *h,* as in *antacid, anthelix.*

antacid (ant′as′id), *n. Chemistry.* a substance that neutralizes acids or counteracts acidity, such as baking soda and magnesia.

antagonist, *n.* **1** *Physiology.* a muscle which resists or counteracts another muscle (the *agonist*), as by relaxing while the opposite one contracts.
2 *Biochemistry.* a substance which counteracts the effects of another in the body by combining with its nerve receptor or by some other chemical action.

Antarctic Circle or **antarctic circle,** *Geography.* **1** a parallel of latitude on the earth's surface, 66 degrees 32 minutes (66°32′) south of the equator, constituting the boundary of the Antarctic Zone. **2** = Antarctic Zone. See the picture at **zone.**

Antarctic Convergence, *Oceanography.* the point in the Antarctic Ocean where warmer tropical waters meet the cold polar waters. Air temperatures, cloud formations, and ocean life change abruptly at the convergence.

Antarctic Zone, *Geography.* the region between the Antarctic Circle and the South Pole, including the Antarctic Ocean and Antarctica.

antecedent, *n. Mathematics.* **1** the first term of a ratio. In the ratio 1:4, 1 is the antecedent, and 4 is the consequent. **2** the part of a conditional which states the condition upon which the consequent depends. In $p \rightarrow q$, p is the antecedent. Also called **hypothesis.** [from Latin *antecedentis* going before]

antenna (an ten'ə), *n., pl.* **-tennae** (-ten'ē) *for 1,* **-tennas** *for 2.* **1** *Zoology.* one of the long, slender, segmented feelers on the heads of insects and certain other arthropods, such as centipedes and lobsters. Insects have one pair of antennae; crustaceans have two pairs. Most antennae function primarily as organs of touch, but some are sensitive to odors and other stimuli.
2 *Electronics.* an electrical conductor designed to receive electromagnetic waves. See the picture at **satellite.**
[from Latin *antenna* sail yard]

antennule (an ten'yül), *n. Zoology.* a small antenna or similar organ, especially one of a pair of small antennae in front of a pair of larger antennae on the head of a crustacean: *the antennules of a hermit crab.*

anterior, *adj.* **1** *Anatomy.* situated toward or near the front or head; designating a forward part: *the anterior part of a fish.* ANT: posterior.
ASSOCIATED TERMS: see **dorsal.**
2 *Botany.* situated or turned away from the axis.
[from Latin *anterior,* comparative of *ante* before]
—anteriorly, *adv.*

anthelion (ant hē'lē ən *or* an thē'lē ən), *n., pl.* **-lia** (lē ə) *or* **-lions.** *Astronomy.* a bright white spot sometimes seen directly opposite and at the same altitude as the sun. [from Greek *anthēlion,* from *anti-* opposite + *hēlios* sun]

anthelix (ant hē'liks), *n.* = antihelix.

anther (an'thər), *n. Botany.* the part of the stamen of a flower that bears the pollen. It is usually a double-celled sac situated at the end of a slender, thread-like stem (the filament). [from Greek *anthēros* flowery, from *anthos* flower]

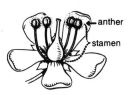

anther
stamen

antheridiophore (an'thə rid'ē ə fôr), *n. Botany.* a gametophore that bears only antheridia.

antheridium (an'thə rid'ē əm), *n., pl.* **-ridia** (-rid'ē ə). *Botany.* the part of a fern, moss, etc., that produces male reproductive cells.
—antheridial, *adj.* of an antheridium.

antherozoid (an'thər ə zō'id), *n.* = spermatozoid.

anthesis (an thē'sis), *n. Botany.* the time at which an anther or flower opens: *The life-span of pollen from typical wheat varieties is less than five minutes after anthesis* (Scientific American). [from Greek *anthēsis,* from *anthein* to blossom, from *anthos* flower]

anthocarp (an'thə kärp), *n.* = accessory fruit. [from Greek *anthos* flower + *karpos* fruit]
—anthocarpous (an'thə kär'pəs), *adj.* of or having to do with an anthocarp or anthocarps.

anthocyanin (an'thə si'ə nən), *n. Biochemistry.* any of a group of pigments in the cell sap of plants, which produce the deep red, blue, lavender, and purple colors of petals, fruits, etc.: *The flowers of the common garden four-o'clock are white, pink, or red, depending on the amount of anthocyanin present* (Emerson, *Basic Botany*). [from Greek *anthos* flower + *kyanos* blue]

anthophagous (an thof'ə gəs), *adj. Zoology.* feeding on flowers: *anthophagous beetles.* [from Greek *anthos* flower + *phagein* eat]

anthophilous (an thof'ə ləs), *adj. Zoology.* attracted to or living on flowers: *Bees are anthophilous.* [from Greek *anthos* flower + *philos* loving]

anthophore (an'thə fôr), *n. Botany.* an elongation of the receptacle of a flower, between the calyx and the corolla, forming a stalk bearing the corolla, stamens, and pistil. [from Greek *anthos* flower + English *-phore*]

anthozoan (an'thə zō'ən), *Zoology.* **—n.** any of a class (Anthozoa) of marine cnidarians with radial segments, including the sea anemone, coral, and other polyps.
—adj. of or belonging to the class of anthozoans. Compare **hydrozoan, scyphozoan.**
[from Greek *anthos* flower + *zōia* animals]

anthracene (an'thrə sēn'), *n. Chemistry.* a colorless, crystalline compound used in making alizarin dyes. It is a complex hydrocarbon obtained in distilling coal tar. *Formula:* $C_{14}H_{10}$

anthracite (an'thrə sīt), *n. Mineralogy.* coal that burns with very little smoke and flame. It is of the highest metamorphic rank and consists almost entirely of fixed carbon; commonly called **hard coal** and contrasted with **bituminous coal.** *Anthracite is regarded by many geologists as the ultimate product in the conversion of vegetable matter into coal. It is really bituminous coal transformed by tremendous pressure* (White and Renner, *Human Geography*). [from Greek *anthrax* live coal, charcoal]

anthropoid (an'thrə poid), *Zoology.* **—adj. 1** of or belonging to a suborder (Anthropoidea) of primates that includes the Old World and New World monkeys.
2 of or belonging to the group of apes that includes the chimpanzees, gorillas, orangutans, and gibbons.
—n. an anthropoid primate.
[from Greek *anthrōpos* man]

anti-, *prefix.* **1** against; counteracting; opposing, as in *antigravity, antitoxin, antitrades.*
2 preventing or inhibiting, as in *anticholinergic, antienzyme.*
3 corresponding or analogous to, as in *anticodon, antimere, antilogarithm, antineutron.* Also spelled **ant-** before vowels and *h.*
[from Greek *anti-* opposite]

anti-antibody, *n. Immunology.* an antibody that attacks other antibodies, especially antibodies produced by injected lymphocytes: *Some of the patient's antibodies could be used as a source to raise anti-antibodies—that is, antibodies that would immobilize the graft-rejecting antibodies* (Jacques M. Chiller).

antibaryon, *n. Nuclear Physics.* an antiparticle of a baryon, such as antiproton or antineutron.

cap, fāce, fäther; best, bē, tèrm; pin, fīve;
rock, gō, ôrder; oil, out; cup, pùt, rüle,
yü in use, *yu̇* in uric;
ng in bring; sh in rush; th in thin, ᴛH in then;
zh in seizure.
ə = a in about, e in taken, i in pencil, o in lemon, u in circus

antibiosis

antibiosis (an'ti bī ō'sis), *n. Biology.* **1** an association between organisms which is detrimental to at least one of them: *In studying soil microbes he was fascinated alike by symbiosis, the ability of diverse organisms to live together, and antibiosis, a state of antagonism between two forms of life* (N.Y. Times).
ASSOCIATED TERMS: see **symbiosis.**
2 the antagonistic association between bacteria or other organisms and antibiotics: *Antibiosis may play [an] ... important role in the composition and structure of plant communities* (Greulach and Adams, *Plants*).
[from *anti-* + Greek *biōsis* way of life, from *bios* life]

antibiotic (an'ti bī ot'ik), *Biology.* —*n.* a substance produced by or derived from a living organism, especially a bacterium or a fungus, that destroys or weakens harmful microorganisms. Penicillin, tetracycline, and streptomycin are antibiotics.
► *Antibiotic* as a noun was introduced about 1944 to cover all biologically-produced substances that are harmful to other organisms. Since then its meaning has gradually narrowed so that it is most commonly used in the sense of a drug used against bacterial infections.
—*adj.* of or having to do with antibiosis or antibiotics; destroying or inhibiting the growth or action of bacteria or other organisms through antibiosis.

antibody, *n., pl.* **-bodies.** *Immunology.* a protein substance produced in the blood or tissues that destroys or weakens bacteria or neutralizes poisons of organic origin. Antibodies are formed in response to specific antigens and are named according to the type of their activity such as antitoxins, agglutinins, and precipitins. *When a human host is invaded by a pathogen, the host reacts by producing substances called antibodies ... if a new infection by the same kind of pathogen occurs, the host can act immediately against it. Such resistance is called immunity* (Miller and Leth, *High School Biology*). See **B cell** and **T cell.** [from *anti-* against (disease) + *body,* in the sense of "a compact mass," a translation of German *Antikörper;* a more appropriate but rarely used name is *immune body*]

anticholinergic (an'ti kō'lə nėr'jik), *adj. Biochemistry.* inhibiting cholinergic action; preventing the liberation of acetylcholine at the nerve endings.

anticholinesterase (an'ti kō'lə nes'tə rās), *n. Biochemistry.* any substance that prevents the action of cholinesterase, such as physostigmine.

anticlinal valley, *Geology, Geography.* a valley formed by erosion of weak materials along the axis or ridge of an anticline.

anticline (an'ti klīn), *n. Geology.* an upward fold of rock strata that bends downward on both sides from its axis. Contrasted with **syncline.** See the picture at **syncline.** [from *anti-* + Greek *klinein* to lean]
—**anticlinal** (an'ti klī'nl), *adj.* of or like an anticline.

anticlinorium (an'tē klī nôr'ē əm), *n., pl.* **-noria** (-nôr'ē ə). *Geology.* a mountain range or system in which the folds of strata constitute a large composite of anticlines and synclines. [from New Latin *anticlinorium,* from *anti-* against + Greek *klinein* to slope + *oros* mountain]

anticodon (an'ti kō'don), *n. Molecular Biology.* a unit of three nucleotides in transfer RNA that binds to a corresponding codon in messenger RNA during the RNA-directed synthesis of protein: *Each kind of tRNA can bond with only one kind of amino acid. The loop opposite to the amino acid binding site has a triplet anticodon ... that can attach only to the mRNA codon for the specific amino acid* (Greulach and Adams, *Plants*).

anticryptic, *adj. Zoology.* that serves to make an animal less conspicuous to its prey, such as natural colorings or markings.

anticyclone (an'ti sī'klōn), *n. Meteorology.* **1** a storm or winds moving spirally outward from a center of high pressure. The motion of winds in an anticyclone is clockwise in the Northern Hemisphere and counterclockwise in the Southern. *Many errors in forecasting are caused by over attention to comparatively local conditions, without proper regard to the grand movement upon which the cyclones and anticyclones—the parents of rain and sunshine—are borne* (London Times).
2 an area of relatively high barometric pressure. Also called **high.**
—**anticyclonic,** *adj.* of or like an anticyclone.

antiderivative, *n. Mathematics.* any function in calculus the derivative of which is a given function: *The derivative of x^2 is $2x$; an antiderivative of $2x$ is x^2.* Also called **indefinite integral.**

antideuteron, *n. Nuclear Physics.* the antiparticle of the deuteron, consisting of an antiproton and an antineutron.

antidiuretic hormone (an'ti dī yù ret'ik), = vasopressin.

antidromic (an'ti drom'ik), *adj. Physiology.* able to conduct impulses in a direction opposite to the usual, as a nerve fiber. [from *anti-* + Greek *dromos* a course]
—**antidromically,** *adv.*

antienzyme, *n. Biochemistry.* any substance that inhibits or reduces the effects of an enzyme.

antiferromagnetic, *Physics.* —*adj.* that resists being magnetized or is nonmagnetic.
—*n.* an antiferromagnetic substance.
—**antiferromagnetism,** *n.* the property of resisting magnetization or of being nonmagnetic: *By applying external magnetic fields to samples of ... crystals at temperatures near absolute zero, the scientists discovered a new type of antiferromagnetism in which the magnetic structure as a whole grows along various axes of the crystal* (Science News Letter).

antigen (an'tə jən), *n. Immunology.* any substance that causes the body to produce antibodies to counteract this substance. Antigens include toxins, bacterial cells, foreign blood cells, etc. *When antigens gain entry into the body, they stimulate the production of antibodies, which react with the antigens and destroy or inactivate them* (McElroy, *Biology and Man*). [from *anti(body)* + *-gen* something that produces]
—**antigenic** (an'tə jen'ik), *adj.* of or like an antigen.
—**antigenically,** *adv.*
—**antigenicity** (an'tə jə nis'ə tē), *n.* the condition of being an antigen; the capacity of causing the body to produce antibodies.

antigorite (an tig′ə rīt), *n. Mineralogy.* a species of serpentine, occurring as micaceous flakes principally in serpentinites. *Formula:* $Mg_3Si_2O_5(OH)_4$ [from *Antigorio*, a valley in Italy where it is found + *-ite*]

antihelix, *n. Anatomy.* the curved ridge of cartilage within the outer rim of the human ear. Also spelled **anthelix.**

antihemophilic factor, *Biochemistry.* a protein in blood plasma that causes clotting: *Hemophilia is due to a deficiency of one of the blood-clotting proteins; the most common form, hemophilia A, involves a deficiency in factor VIII, the antihemophilic factor* (Mary Ellen Switzer). *Abbreviation:* AHF

antihemophilic globulin, = antihemophilic factor. *Abbreviation:* AHG

antilogarithm, *n. Mathematics.* the number corresponding to a given logarithm. The antilogarithms of the common logarithms 1, 2, and 3 are 10, 100, and 1000. *If a is the logarithm of x, then x is the antilogarithm of a.*

antimatter, *n. Physics.* a hypothetical form of matter identical in appearance, structure, and other properties with physical matter but with the electric charges or magnetic properties of its elementary particles reversed; matter composed of antiparticles: *When normal matter collides with antimatter, both are annihilated and tremendous amounts of energy are released* (Science News).

antimere (an′tə mir′), *n. Zoology.* one of the symmetrically corresponding parts of the body of a bilaterally or radially symmetrical animal.
—**antimeric** (an′tə mer′ik), *adj.* of or characterized by antimeres.
[from *anti-* + Greek *meros* part]

antimetabolite, *n. Biochemistry.* a substance similar in structure to a metabolite but having the effect of inhibiting or opposing the metabolite's action: *"Antimetabolites" ... have a well-established capacity to suppress the production of immune lymphocytes* (Newsweek).

antimony (an′tə mō′nē), *n. Chemistry.* a brittle, silver-white, metallic element with a crystalline texture, that occurs chiefly in combination with other elements, especially in stibnite (antimony sulfide). Antimony is used to make alloys harder, and its compounds are used to make medicines, pigments, and glass. *Symbol:* Sb; *atomic number* 51; *atomic weight* 121.75; *melting point* 630.5°C; *boiling point* 1635°C; *oxidation state* $+3$, $+5$ [from Medieval Latin *antimonium*]
—**antimonial** (an′tə mō′nē əl), *adj.* containing antimony in combination.
—**antimonic** (an′tə mon′ik), *adj.* of or containing antimony, especially with a valence of 5.
—**antimonious** (an′tə mō′nē əs), *adj.* of or containing antimony, especially with a valence of 3.

antineutrino, *n. Nuclear Physics.* the antiparticle of the neutrino: *In the normal process of emission of a negative electron from a nucleus during beta decay, a neutron changes into a proton plus the electron plus an antineutrino* (New Scientist).

antineutron, *n. Nuclear Physics.* the antiparticle of the neutron. It corresponds in mass and spin to the neutron but its magnetic moment is opposite to that of the neutron.

anting, *n. Zoology.* the practice among certain species of birds of picking up ants and dropping them among their feathers or of rubbing their feathers with them.

antinodal, *adj.* of or having to do with an antinode or antinodes.

antinode, *n. Physics.* the point or line of maximum displacement of a vibrating body. Also called **loop.**

antinucleon, *n. Nuclear Physics.* an antineutron or antiproton.

antiparallel, *adj. Physics.* parallel but oppositely directed: *The internal forces are such as to make interacting spins antiparallel to each other* (F. Bitter).

antiparticle, *n. Nuclear Physics.* a unit of matter such as a positron, antiproton, or antineutron, corresponding in mass and properties to a given elementary particle but with an opposite electrical charge, opposite magnetic properties, or opposite coupling to other fundamental forces. When an antiparticle collides with a corresponding particle, they destroy each other, thereby releasing great energy. Compare **antimatter, annihilation.**
► The existence of the first known antiparticle, the positron or antielectron, was predicted in 1928 by the British physicist Paul Dirac and confirmed in 1932 by the American physicist Carl Anderson.

antipodal (an tip′ə dəl), *adj. Geography.* of or on the antipodes; on the opposite side of the earth.
—*n.* Also called **antipodal cell.** *Botany.* any of three cells or nuclei at the end of the embryo sac opposite that of the egg apparatus in most angiosperms.
—**antipodally**, *adv.*

antipodes (an tip′ə dēz), *n. pl. Geography.* two places on directly opposite sides of the earth: *The North Pole and the South Pole are antipodes.* [from Greek *antipodes*, plural of *antipous* with feet opposite, from *anti-* opposite + *pous* foot]

antiproton, *n. Nuclear Physics.* the antiparticle of the proton. Antiprotons are created when a proton strikes a neutron. *The antiproton ... can be confused with the negative hydrogen ion (a proton with two electrons). To distinguish between the two, one has to note that the antiproton keeps its negative charge, while the hydrogen ion loses relatively easily one or even two electrons* (Bulletin of Atomic Scientists).

antiquark, *n. Nuclear Physics.* an antiparticle of a quark.

antiscorbutic acid (an′ti skôr byü′tik), = *vitamin C.*

antisepsis (an′tə sep′sis), *n. Medicine.* the process by which microorganisms are prevented from gaining access to or multiplying in body tissues injured by accident, surgery, etc.; prevention of infection.
ASSOCIATED TERMS: *Sepsis* is the condition of being infected by disease-producing microorganisms and their toxins; *antisepsis* is the prevention of sepsis by means of antiseptics;

---

cap, fāce, fäther; best, bē, tèrm; pin, fīve;
rock, gō, ôrder; oil, out; cup, pùt, rüle,
yü in use, *yu* in uric;
ng in bring; *sh* in rush; *th* in thin; ŦH in then;
zh in seizure.
ə = *a* in about, *e* in taken, *i* in pencil, *o* in lemon, *u* in circus

antiseptic

asepsis is the condition of being free of disease-producing microorganisms before infection.

antiseptic (an'tə sep'tik), *Medicine.* —*n.* a substance that prevents the growth of microorganisms that cause infection or putrefaction. Mercurochrome, alcohol, and boric acid are widely used antiseptics.
ASSOCIATED TERMS: see **disinfectant.**
—*adj.* **1** preventing infection or putrefaction by stopping the growth and activity of microorganisms. **2** of, having to do with, or using antiseptics.

antiserum (an'ti sir'əm), *n., pl.* **-serums, -sera** (-sir'ə). *Immunology.* a serum containing antibodies that are specific for certain antigens. Antitoxin and antivenin are antiserums.

anti-Stokes line (an'tē stōks'), *Physics.* a line in a spectrum of the same, or higher, frequency as that of the exciting light. It is the result of absorption of a light quantum by a vibrating molecule. [see under **Stokes' law** and **stoke**]

antisymmetric, *adj.* **1** *Mathematics.* denoting a function which is transformed into its negative when its variables are interchanged in pairs.
2 *Physics.* denoting a system in which each point has properties opposite to those of a point symmetrically related to it. —**antisymmetrically,** *adv.*

antitoxin (an'ti tok'sən), *n. Immunology.* **1** an antibody formed in response to the presence of a particular toxin and able to prevent or reduce the effects of that toxin: *Antitoxins combine with and so neutralize the poisonous toxins produced by bacteria* (Mackean, *Introduction to Biology*).
2 a serum containing antitoxins, obtained from the blood of an animal or person immunized against the corresponding toxin. It is used to inject into persons or animals to make them immune to a specific disease or to treat them if already infected.

antitrades (an'ti trādz'), *n.pl. Meteorology.* **1** (at tropical latitudes) winds that blow in a direction opposite to the trade winds on a level above them. **2** (at temperate latitudes) the prevailing westerlies.

antitragus (an tit'rə gəs), *n., pl.* **-gi** (-jī). *Anatomy.* a protuberance toward the rear of the external ear, inside the helix, opposite to the tragus. [from Greek *antitragos,* from *anti-* opposite + *tragos* tragus]

antivenin (an'ti ven'ən), *n. Immunology.* **1** an antitoxin to the venom of a snake, scorpion, or other venomous animal. **2** a serum containing antivenins, prepared from the blood of animals immunized against the venom of snakes and other animals.

antivitamin, *n. Biochemistry.* **1** any substance that inhibits or prevents the absorption of a vitamin. **2** an enzyme that destroys vitamins.

antler, *n. Zoology.* **1** a bony, hornlike growth on the head of a deer, moose, elk or other member of the deer family, and usually having one or more branches. Antlers are shed once each year and new ones grow in their place the next year. **2** branch of such a growth.

antrorse (an trôrs'), *adj. Botany, Zoology.* bent forward or upward: *antrorse hairs.* Contrasted with **retrorse.** [from New Latin *antrorsus,* from Latin *anter(ior)* in front + *versus* turned] —**antrorsely,** *adv.*

antrum (an'trəm), *n., pl.* **-tra** (-trə). *Anatomy.* a cavity, especially one in a bone, often applied to the sinus in the maxilla. [from Latin *antrum,* from Greek *antron* cave]

anus (ā'nəs), *n. Anatomy.* the opening at the lower end of the alimentary canal, through which solid waste material is eliminated from the body. [from Latin *anus*]

aorta (ā ôr'tə), *n., pl.* **-tas, -tae** (-tē'). *Anatomy.* the main artery that carries the blood from the left side of the heart and, with its branches, distributes it to all parts of the body except the lungs. See the picture at **heart.** [from New Latin *aorta,* from Greek *aortē*]
—**aortic** (ā ôr'tik), *adj.* of or having to do with the aorta. The **aortic arch** is an artery in a vertebrate embryo connecting the large ventral and dorsal arteries.

ap-, *prefix.* the form of **apo-** before vowels and *h,* as in *apastron, aphelion.*

apamin (a'pə min), *n. Biochemistry.* a substance, derived from bee venom, that is poisonous to nervous tissue, used experimentally in neurology and medicine. [from Latin *ap(is)* bee + *amin(e)*]

apastron (ap as'trən), *n. Astronomy.* the point at which the components of a binary star are farthest from each other in their orbits. Contrasted with **periastron.** [from *ap-* away from + Greek *astron* star]

apatetic (ap'ə tet'ik), *adj. Zoology.* deceptively resembling the coloration of an animal's environment or the markings of another species. [from Greek *apatētikos* fallacious, from *apatan* to deceive]

apatite (ap'ə tīt), *n. Mineralogy.* a native phosphate of lime, varying in color from white to green, blue, violet, and brown, occurring in crystals or masses, and commonly mined for use as a fertilizer: *The mineral apatite is the chief primary source of phosphorus. It occurs as a minor constituent in most types of igneous rocks* (Jones, *Minerals in Industry*). Formula: $Ca_5(PO_4,CO_3)_3(F,OH,Cl)$ [from Greek *apatē* deceit + English *-ite* (so called because its diverse forms were often wrongly identified)]

aperiodic (ā'pir ē od'ik), *adj. Physics.* having irregular vibrations; lacking oscillation.
—**aperiodicity** (ā pir'ē ə dis'ə tē), *n.* aperiodic condition.

aperture, *n. Optics.* the diameter of the opening through which light passes in a camera, telescope, or other optical instrument. The aperture of a lens is expressed in **f-numbers,** such as f/4, in which the aperture is one fourth of the **focal length.**

apetalous (ā pet'l əs), *adj. Botany.* having no petals.

apex (ā'peks), *n., pl.* **apexes, apices** (ā'pe sēz' or ap'e sēz'). **1** the highest point; tip: *The apex of a triangle or of a leaf is the point opposite the base.* SYN: vertex. **2** *Botany.* that portion of a root or shoot containing the apical and primary meristems. [from Latin *apex*]

aphanite (af'ə nīt), *n. Geology.* any compact rock so fine-grained in texture that no distinct crystals are visible to the unaided eye; cryptocrystalline rock. [from French *aphanite,* from Greek *aphanēs* not manifest, from *a-* not + *phainein* show forth]
—**aphanitic,** *adj.* of or resembling aphanite: *rocks with aphanitic texture.*

aphelion (ə fē′lē ən), *n., pl.* **-lia** (-lē ə). *Astronomy.* the point farthest from the sun in the orbit of a planet or comet: *Perihelion and aphelion are the two points on the earth's orbit respectively nearest and farthest from the sun; they are at the extremities of the major axis.* (Baker, *Astronomy*). [from *ap-* away from + Greek *hēlios* sun]

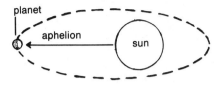

aphyllous (ə fil′əs), *adj. Botany.* lacking leaves; naturally leafless, as a cactus. [from Greek *aphyllos,* from *a-* without + *phyllon* leaf]

apical (ap′ə kəl *or* ā′pə kəl), *adj.* of, at, or forming an apex: *an apical cell.* —**apically,** *adv.*

apical dominance, *Botany.* the suppression of growth of lateral buds by the apical meristem.

apical meristem, *Botany.* the formative tissue at the tip of a root or shoot. Compare **primary meristem.**

apiculate (ə pik′yə lit *or* ā pik′yə lit), *adj. Botany.* ending with a short and abrupt point: *apiculate leaves.* [from New Latin *apiculus* tip, diminutive of Latin *apex* apex]

aplacental (ā′plə sen′təl), *adj. Zoology.* having no placenta, as monotremes and marsupials.

aplanatic (ap′lə nat′ik), *adj. Optics.* free from aberration, especially from spherical aberration. [from *a-* not + Greek *planasthai* to wander]

aplite (ap′līt), *n. Geology.* a granite of fine grain, consisting of quartz and feldspar, found in dikes. [from Greek *haplos, haplous* simple + English *-ite*]

apo-, *prefix.* **1** away from, off, as in *apogee, apogeotropic.*
2 detached; separate, as in *apogamy, apomixis.* Also spelled **ap-** before vowels and *h.*
[from Greek *apo-,* from *apo* off]

apoapsis (ap′ō ap′sis), *n. Astronomy.* the point in the orbit of a satellite where it is farthest from the celestial body around which it revolves. Compare **periapsis.** [from *apo* + *apsis*]

apocarp (ap′ə kärp), *n. Botany.* an ovary or fruit having separate or distinct carpels.
—**apocarpous** (ap′ə kär′pəs), *adj.* having the carpels distinct or separate.

apochromatic, *adj. Optics.* having neither chromatic nor spherical abberration: *an apochromatic lens.*

apocrine (ap′ə krin *or* ap′ə krīn), *adj. Physiology.* of or having to do with a type of gland whose secreting cells lose some of their cytoplasm, which becomes a part of the secretion. The **apocrine glands** are large, deep-set glands found mainly in the area of the armpits, breasts, and genitals. [from *apo-* away from + Greek *krinein* to separate]

apodeme (ap′ə dēm), *n. Zoology.* one of the plates of chitin which extend inward from the exoskeletons of arthropods, supporting their internal organs. [from New Latin *apodema,* from Greek *apo-* from, off + *demas* body, frame]

apoenzyme, *n. Biochemistry.* the protein portion of an enzyme, to which the coenzyme attaches to form an active enzyme.

apogamy (ə pog′ə mē), *n. Botany.* the production of a sporophyte directly from cells of a gametophyte instead of by the usual process of fertilization. [from *apo-* + *-gamy*] —**apogamous,** *adj.*

apogee (ap′ə jē), *n. Astronomy.* the point farthest from the earth in the orbit of the moon or any other satellite: *The satellite went into an orbit with a high point or apogee of 727 miles and a low point or perigee of 670 miles* (N.Y. Times). [from Greek *apogaion,* from *apo-* away from + *gē* earth]

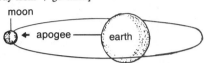

apogeotropic, *adj. Botany.* bending or turning away from the earth: *apogeotropic leaves.* —**apogeotropically,** *adv.*
—**apogeotropism,** *n.* a tendency of leaves and other plant parts to bend or turn upward and away from the earth.

apolune (ap′ə lün′), *n. Astronomy.* the point in a lunar orbit farthest from the center of the moon. [from *apo-* away from + French *lune* moon]

apomict (ap′ə mikt′), *n. Biology.* an organism that reproduces by apomixis: *Some recent work on apomicts has shown that environmental factors may affect both vegetative and seed apomixis* (D.H. Valentine).
—**apomictic** (ap′ə mik′tik), *adj.* of or characterized by apomixis. —**apomictically,** *adv.*

apomixis (ap′ə mik′sis), *n. Biology.* reproduction in which union of sexual cells or organs does not occur. Apogamy is a form of apomixis. *Since species that can reproduce only by apomixis lack sexual reproduction, they lack the variability associated with sexual reproduction and are very uniform* (Greulach and Adams, *Plants*).
[from *apo-* away from + Greek *mixis* a mingling]
—**apomictic,** *adj.* of or exhibiting apomixis: *Dandelions are apomictic.*

aponeurosis, *n., pl.* **-ses** (-sēz′). *Anatomy.* a whitish, shining, fibrous membrane, an expanded tendon, serving as a fascia for sheathing, or as the end or attachment of, certain muscles. [from Greek *aponeurōsis,* from *aponeurousthai* change into a tendon, from *apo-* + *neuron* sinew]
—**aponeurotic,** *adj.* of or having to do with aponeuroses: *aponeurotic laminae.*

apophyllite (ə pof′ə lit *or* ap′ə fil′it), *n. Mineralogy.* a mineral consisting of a hydrous silicate of potassium and calcium, sometimes with a trace of fluorine. It has

cap, fāce, fäther; best, bē, tèrm; pin, fīve;
rock, gō, ôrder; oil, out; cup, pùt, rüle,
yü in use, yù in uric;
ng in bring; sh in rush; th in thin, ᴛʜ in then;
zh in seizure.
ə = *a* in about, *e* in taken, *i* in pencil, *o* in lemon, *u* in circus

a pearly luster and is found in crystals or laminated formations. *Formula:* $KCa_4Si_8O_{20}(F,OH) \cdot 8H_2O$ [from French, from Greek *apo-* away + *phyllon* leaf]

apophyseal (ə pof′ə sē′əl), *adj.* of or having to do with an apophysis: *apophyseal joints.*

apophysis (ə pof′ə sis), *n., pl.* **-ses** (-sēz′). *Anatomy.* a natural outgrowth, projection, or swelling, especially any process of a vertebra. [from Greek *apo-* off + *phyein* grow]

apopyle (ap′ə pīl), *n. Zoology.* an opening in the radial canal of a sponge for water to pass through. [from *apo-* + Greek *pylē* gate]

aposematic (ap′ə si mat′ik), *adj. Zoology.* (of an animal's markings or coloration) serving to warn or alarm enemies. [from *apo-* + *sematic*] —**aposematically,** *adv.*

apothecium (ap′ə thē shē əm *or* ap′ə thē sē əm), *n., pl.* **-cia** (-shē ə *or* -sē ə). *Botany.* the fruiting body of various lichens and fungi, consisting of a cup-shaped receptacle to which the spore sacs (asci) are attached. [from New Latin, from Latin *apotheca* storehouse]

apothem (ap′ə them), *n. Mathematics.* the radius of the circumscribed circle of a regular polygon; the perpendicular from the center to any one of its sides. [from *apo-* + Greek *thema* thing placed]

apozymase (ap′ə zi′mās), *n. Biochemistry.* the part of zymase that is composed of protein. Compare **apoenzyme.**

apparatus, *n., pl.* **-tuses** or **-tus.** *Anatomy.* the group or system of organs of the body by which a specific natural function or process is carried on: *the digestive apparatus.*

apparent magnitude, *Astronomy.* the brightness of a star as it appears to the observer relative to that of other stars, as distinguished from the absolute magnitude: *The brightest star in the sky, Sirius, has an apparent magnitude of −1.4, while the faintest stars visible to the unaided eye have apparent magnitudes of about +6.*

apparent noon, *Astronomy.* the moment when the apparent sun's center crosses the meridian.

apparent position, 1 *Astronomy.* the position where a star or other heavenly body appears to be on the celestial sphere, measured by right ascension and declination.
2 *Optics.* the position in which an object appears to be when seen through a glass, water, or other diffracting medium, as distinguished from its true position.

apparent solar time, *Astronomy.* the measure of the day by the apparent positions of the sun, usually determined by the apparent noon: *Apparent solar time is the hour angle of the sun plus 12 hours* (Bernhard, *Handbook of Heavens*). Also called **solar time.**

apparent sun, *Astronomy.* the sun as it appears to an observer on earth as it moves eastward along the ecliptic: *The three time reckoners in use are the vernal equinox, the apparent sun, and the mean sun. The corresponding kinds of time are respectively sidereal time, apparent solar time, and mean solar (or civil) time* (Baker, *Astronomy*).

apparent time, = apparent solar time.

appendage, *n. Anatomy. Biology.* any of various external or subordinate parts of a body. Arms, tails, fins, and legs are appendages.

appendicular (ap′ən dik′yə lər), *adj. Anatomy.* of or having to do with an appendage or appendages. The **appendicular skeleton** is the skeleton of the limbs of the body.

appendix, *n., pl.* **-dixes, -dices** (-də sēz′). = vermiform appendix: *The appendix is a small, apparently useless organ jutting out from the beginning of the large intestine* (Science News Letter). [from Latin]

appestat (ap′ə stat′), *n. Physiology.* an area in the hypothalamus regarded as the center which regulates the appetite. [from *appe(tite)* + *-stat*]

apposition, *n. Botany.* the deposit of successive layers of cell-wall material in plants, thus increasing the thickness of the wall.

appressorium (ap′re sō′rē əm), *n., pl.* **-ria** (-rē ə). *Biology.* the organ of adhesion of parasitic fungi. [from New Latin, from Latin *appresus* pressed to]

apsis (ap′sis), *n., pl.* **apsides** (ap′sə dēz′). *Astronomy.* either of two points in the elliptical orbit of a heavenly body at one of which (the lower apsis) it is nearest to, and at the other (the higher apsis) it is farthest from, the body or point about which it is revolving. The **line of apsides** is the straight line that joins these two points. [from Latin *apsis* arch, orbit]

apterium (ap tir′ē əm), *n., pl.* **-ria** (-rē ə). *Zoology.* one of the featherless spaces on the skin of a bird. [from New Latin, from Greek *apteros* featherless, wingless; see under *apterous*]

apterous (ap′tə rəs), *adj.* **1** *Zoology.* wingless: *Lice are apterous insects.*
2 *Botany.* having no winglike expansions, as stems, seeds, or other plant parts. [from Greek *apteros,* from *a-* without + *pteron* wing]

aquatic, *Biology.* —*adj.* growing or living in water: *Water lilies are aquatic plants.* —*n.* a plant or animal that lives in water. [from Latin *aquaticus* watery, from *aqua* water] —**aquatically,** *adv.*

aqueous, *adj.* **1** of, containing, or like water; watery: *an aqueous solution.* The **aqueous humor** is the watery liquid which fills the space in the eye between the cornea and the lens.
2 *Geology.* produced by the action of water. Aqueous rocks are formed of sediment deposited by water.

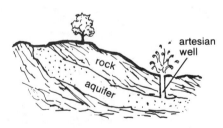

aquifer (ak′wə fər), *n. Geology.* a stratum of earth or permeable rock that contains water: *Another kind of natural reservoir is the aquifer—an underground deposit of rain water which has ... seeped through the soil and porous stone until it is trapped by an impervious layer of rock* (Harper's). Compare **artesian well.** [from Latin *aqua* water + *ferre* to carry]

Ar, *symbol.* argon.

Arabic numerals, *Mathematics.* the figures 1, 2, 3, 4, 5, 6, 7, 8, 9, 0. They are called Arabic because they were introduced into western Europe from sources of Arabic scholarship.

arabinose (ə rab′ə nōs), *n. Biochemistry.* a pentose sugar obtained from plant polysaccharides such as arabic and certain other plant gums. *Formula:* $C_5H_{10}O_5$

arachnid (ə rak′nid), *Zoology.* —*n.* any of a class (Arachnida) of arthropods closely allied to the insects and crustaceans, but distinguished by possession of eight legs, absence of wings and antennae, a body divided into two regions, and breathing mechanism of tracheal tubes or pulmonary sacs. Spiders, scorpions, mites, ticks, and daddy-longlegs belong to this class. —*adj.* of, belonging to, or having to do with the arachnids.
[from Greek *arachnē* spider, web]

arachnoid (ə rak′noid), *adj.* **1** *Zoology.* of or resembling an arachnid.
2 *Anatomy.* of or designating the delicate serous membrane enveloping the brain and spinal cord, lying between the dura mater and pia mater.
3 *Botany.* covered with or formed of long delicate cobweblike hairs or fibers.

arachnoid granulations, = Pacchionian bodies.

arachnology (ar′ak nol′ə jē), *n.* the branch of zoology that deals with the arachnids.

aragonite (ə rag′ə nit *or* ar′ə gə nīt), *n. Mineralogy.* one of the three crystalline forms of calcium carbonate, the others being calcite and vaterite. Aragonite crystallizes in orthorhombic prisms. *Formula:* $CaCO_3$ [from *Aragon,* Spain, where it was found]

arboreal (är bôr′ē əl), *adj.* **1** *Zoology.* **a** living in trees: *an arboreal animal.* **b** of or belonging to those animals living in trees: *The arboreal habits are also correlated with the development of the eyesight to a high degree* (Winchester, *Zoology*).
2 *Botany.* of or resembling a tree or trees. [from Latin *arbor* tree] —**arboreally,** *adv.*

arborescent (är′bə res′nt), *adj.* **1** *Botany.* like a tree in structure, growth, or appearance: *arborescent monocotyledons. An arborescent grass, very like a bamboo* (Charles Darwin).
2 *Anatomy.* treelike in arrangement; branching: *an arborescent network of veins.*

arboretum (är′bə rē′təm), *n., pl.* **-tums, -ta** (-tə). *Botany.* a place where trees and shrubs are grown and exhibited for scientific and educational purposes. [from Latin *arboretum,* from *arbor* tree]

arborize (är′bə rīz′), *v. Biology.* to have or produce branching formations: *The airways of the lung are called the bronchial tree because they branch or arborize like a tree* (New Scientist).
—**arborization,** *n.* a branching formation; treelike arrangement or form, as that of a dendrite.

arbor vitae (är′bər vī′tē), *Anatomy.* a treelike arrangement of white and gray nerve tissue in a section of the cerebellum. [from Latin *arbor vitae* tree of life]

arbovirus (är′bə vī′rəs), *n. Biology.* any of several viruses containing RNA, transmitted by mosquitoes, ticks, and other arthropods. Yellow fever, dengue, and equine encephalitis are caused by arboviruses. [from *ar(thropod)-bo(rne) virus*]

arc, *n.* **1** *Mathematics.* a continuous part of a circle or other curve. **2** *Electricity.* a stream of brilliant light or sparks formed as a strong electric current jumps from one conductor to another: *an electric arc.*
—*v.* to form an electric arc.
[from Latin *arcus* bow]

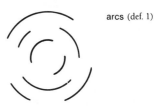
arcs (def. 1)

arch-, *combining form.* a form of **archi-,** mainly in the sense of early, primal, primitive, as in *archencephalon = primitive encephalon.*

archaebacteria (är′kē bak tir′ē ə), *n.pl. Microbiology.* a group of microorganisms that are chemically and genetically distinct from bacteria and higher living organisms. Archaebacteria exist in a warm, oxygen-free environment by ingesting carbon dioxide and hydrogen to produce methane. *The identification of archaebacteria's unique genetic structure suggests that there may be a third line of evolution. It also provides an important clue to earth's early environment* (Time). Also called **methanogen.** [from *archae-* ancient, primitive + *bacteria*]

archaeo-, *combining form.* ancient; early; primitive, as in *archaeomagnetism, archaeocyte.* [from Greek *archaios* ancient, from *archē* a beginning]

archaeocyte (är′kē ə sīt), *n. Zoology.* an undifferentiated, itinerant cell, as in the sponge embryo: *Archaeocytes are embryonic amoebocytes with blunt pseudopodia, which can produce other types of cells, particularly reproductive cells* (Hegner and Stiles, *College Zoology*). Also spelled **archeocyte.**

archaeomagnetism (är′kē ə mag′nə tiz əm), *n. Geophysics.* the natural magnetism remaining in geological or archaeological specimens, used especially as a means of determining the age of a specimen and the location of the earth's magnetic poles in antiquity.

archaeopteryx (är′kē op′tər iks), *n. Zoology.* the oldest known fossil bird, having such reptilian characteristics as claws on its wings, teeth, and a long, bony tail. The archaeopteryx lived in the more recent part of the European Jurassic. [from *archaeo-* primitive + Greek *pteryx* bird, wing]

Archean (är kē′ən), *Geology.* —*n.* **1** the earlier of the two eons of Precambrian time, the other being the Proterozoic. **2** the rocks formed during this eon.
—*adj.* of or having to do with this eon or its rocks.

cap, fāce, fäther; best, bē, tėrm; pin, five;
rock, gō, ôrder; oil, out; cup, pùt, rüle,
yü in use, *yù* in uric;
ng in bring; *sh* in rush; *th* in thin, ŦH in then;
zh in seizure.
ə = *a* in about, *e* in taken, *i* in pencil, *o* in lemon, *u* in circus

[from Greek *archaios* ancient]

► There is no recognized worldwide boundary between the Archean and Proterozoic eons. Many geologists avoid the terms, rejecting the idea of a simple twofold division of Precambrian time.

archegoniophore (är′kə gō′nē ə for), *n. Botany.* an outgrowth of the prothallium of mosses, ferns, etc., which bears archegonia.

archegonium (är′kə gō′nē əm), *n., pl.* **-nia** (-nē ə). *Botany.* the female reproductive organ in plants such as ferns and mosses. The archegonium is a multicellular, flask-shaped organ containing a single egg and corresponding to the pistil of flowering plants. [from Greek *archē* beginning + *gonos* race]
—**archegonial** (är′kə gō′nē əl), *adj.* of or having to do with an archegonium.

archencephalon (ärk′en sef′ə lon), *n. Embryology.* the part of the primitive brain from which the forebrain and midbrain develops: *The activities of the archencephalon dominate the actions of animals, but the cerebrum in man [now] controls the archencephalon* (Foster Kennedy). [from *arch-* early, primitive + *encephalon*]

archenteron (är ken′tə ron′), *n. Embryology.* the primitive intestinal or alimentary cavity of a gastrula: *Gastrulation by invagination thus results in the formation of a primitive intestine, termed the archenteron, the wall of which is the endoderm* (Harbaugh and Goodrich, *Fundamentals of Biology*). [from *arch-* early, primitive + Greek *enteron* intestine]

archeocyte (är′kē ə sīt), *n.* = archaeocyte.

Archeozoic (är′kē ə zō′ik), *n. Geology.* **1** the earlier of the two eras of Precambrian time, the other being the Proterozoic. **2** the rocks formed during this era.
—*adj.* of or having to do with this era or its rocks. [from Greek *archaios* ancient + *zōē* life]

► The term *Archean Eon* is now more common than *Archeozoic Era.* See also the note under **Archean.**

archespore (är′kə spôr), *n. Biology.* the cell or group of cells from which spores are ultimately developed. [from New Latin *archesporium,* from Greek *arche-* primal + *spora* seed]
—**archesporial** (är′kə spôr′ē əl), *adj.* having to do with or of the nature of an archespore.

archi-, *combining form.* **1** early; primitive; primal, as in *archibenthos, archipallium.* **2** chief; principal, as in *archipelago.* [from Latin *archi-,* from Greek *arche-, archi-,* related to *archē* a beginning]

archibenthal or **archibenthic,** *adj. Oceanography.* belonging to or living in the archibenthos.

archibenthos (är′kə ben′thos), *n. Oceanography.* the deep part of the ocean extending from about 200 to 1000 meters: *For the bottom dwellers, the term proposed by Alexander Agassiz, archibenthos (or archibenthic zone) is sometimes used* (Rhodes W. Fairbridge). [from Greek *archi-* primal + *benthos* depth (of the sea)]

archiblast (är′kə blast), *n. Embryology.* the yolk in an ovum, constituting the germ. [from *archi-* + Greek *blastos* sprout]

archicarp (är′kə kärp), *n. Botany.* the first stage of the fruiting body in some fungi.

Archimedes' principle, *Physics.* the principle that a body immersed in a fluid is buoyed up by a force equal to the weight of the fluid it displaces: *By applying Archimedes' principle, it is possible to find the specific gravity of an object.* (Obourn, *Investigating the World of Science*). [named after *Archimedes,* 287?–212 B.C., Greek mathematician, physicist, and inventor]

archipallium (är′kə pal′ē əm), *n. Anatomy.* the part of the pallium associated with the sense of smell.
—**archipallial,** *adj.* of or having to do with the archipallium.

archipelago (är′kə pel′ə gō), *n., pl.* **-goes** or **-gos.** *Geography, Geology.* **1** a group of many islands: *Some areas of the earth crust are neither strictly continental nor strictly oceanic ... These areas are the island archipelagoes: ... such chains have a dominantly continental character.* (Scientific American).
2 a sea having many islands in it. [ultimately from *archi-* chief + Greek *pelagos* sea]
—**archipelagic,** *adj.* of or having to do with an archipelago.

archiplasm (är′kə plaz′əm), *n. Biology.* the protoplasmic substance of the asters and spindles of a cell during mitosis. Also spelled **archoplasm.** [from *archi-* + Greek *plasma* something formed]
—**archiplasmic,** *adj.* of or of the nature of archiplasm.

archoplasm, *n.* = archiplasm.

Arctic Circle or **arctic circle,** *Geography.* **1** a parallel of latitude on the earth's surface, 66 degrees 32 minutes (66°32′) north of the equator, constituting the boundary of the Arctic Zone. **2** = Arctic Zone. See the picture at **zone.**

Arctic Zone, *Geography.* the region between the Arctic Circle and the North Pole; Frigid Zone.

arcuate (är′kyù it), *adj. Anatomy.* curved like a bow; arched: *arcuate artery, arcuate fiber, arcuate ligament.* [from Latin *arcuare* to bend, from *arcus* bow]
—**arcuately,** *adv.*

are (är or är), *n.* a unit of surface measure in the metric system, equal to 100 square meters. *Abbreviation:* a [from French *are,* from Latin *area* area]

area, 1 *Mathematics.* the extent of a surface or plane figure as measured in square units. The SI unit of area is the square meter. The area of a rectangle is the product of the lengths of two adjacent sides.
2 *Anatomy.* **a** a portion of the cerebral cortex whose function has been identified: *the sensory area, the speech area.* **b** any other section of the body or a section of one of its organs: *the pulmonary area of the heart.*
[from Latin *area* piece of level ground]

arenicolous (ar′ə nik′ə ləs), *adj. Zoology.* living in sand, as certain worms do. [from Latin *harena* sand + *colere* inhabit]

areola (ə rē′ə lə), *n., pl.* **-lae** (-lē′), **-las.** *Zoology, Botany.* **1** a ring of color about a nipple, vesicle, etc.: *the areola about a pustule.*
2 a space between the veins of a leaf or the nervures of an insect's wing.
3 a small space or interstice in a tissue.
[from Latin *areola,* diminutive of *area*]
—**areolar,** *adj.* **1** of or resembling an areola. **2** containing areolae: *areolar tissue.*
—**areolate** (ə rē′ə lit), *adj.* characterized by or having areolae.

arête (ə rāt′), *n. Geology.* a sharp ridge or spur on a mountain, especially a rocky edge sculpted by glaciers. [from French]

Arg, *abbrev.* arginine.

argentic (är jen′tik), *adj. Chemistry.* of or containing silver, especially with a valence of 2. [from Latin *argentum* silver]

argentiferous (är′jən tif′ər əs), *adj. Mineralogy.* producing or containing silver: *an argentiferous ore.*

argentite (är′jən tīt), *n. Mineralogy.* silver sulfide, an ore of silver found in veins. *Formula:* Ag_2S [from Latin *argentum* silver]

argentous (är jen′təs), *adj. Chemistry.* containing silver, especially with a valence of 1.

argillaceous (är′jə lā′shəs), *adj. Geology.* of, having to do with, or resembling clay: *argillaceous sediments.* [from Latin *argillaceus,* from *argilla* clay]

argillite (är′jə līt), *n. Geology.* a compact rock derived from clay by low-grade metamorphism. [from Latin *argilla* clay]

arginase (är′jə nās), *n. Biochemistry.* an enzyme that breaks down arginine in mammals and converts it to urea: *The immediate antecedent of urea is the amino acid arginine, and the enzyme that catalyzes its hydrolysis to form urea is arginase* (Earl Frieden).

arginine (är′jə nīn), *n. Biochemistry.* an amino acid present in plant and animal proteins. *Formula:* $C_6H_{14}O_2N_4$ *Abbreviation:* Arg [from Latin *arg(entum)* silver + *-ine* (so called from the silver color of the first salts discovered)]

argon (är′gon), *n. Chemistry.* a colorless, odorless, inert gaseous element that forms a very small part of the air. Argon is used in electric light bulbs and radio tubes. *In spite of its inertness, observations indicate that argon can be made to combine with boron fluoride to form compounds* (Offner, *Fundamentals of Chemistry*). *Symbol:* Ar; *atomic number* 18; *atomic weight* 39.948; *melting point* −189.3°C; *boiling point* −185.9°C. [from Greek *argon,* neuter of *argos* idle, from *a-* without + *ergon* work]

argument, *n. Mathematics.* an independent variable on whose value the value of a function depends.

arid, *adj. Ecology.* having insufficient rainfall to support trees or woody plants: *an arid region or climate.* [from Latin *aridus* dry]

—**aridity** (ə rid′ə tē), *n.* arid condition.

aril (ar′il), *n. Botany.* an outside covering of certain seeds, arising from the stalk of the ovule. The pulpy inner pod of the bittersweet is an aril. [from Medieval Latin *arilli* raisins]

—**arillate** (ar′ə lāt), *adj.* having an aril or arils.

arilode (ar′ə lōd), *n. Botany.* an aril not arising from the stalk; a false aril. [from New Latin *arillus* aril + Greek *eidos* form]

arista (ə ris′tə), *n., pl.* **-tae** (-tē). 1 *Botany.* the beard of grains and grasses; awn. 2 *Zoology.* a bristlelike structure, such as one at or near the end of the antenna of certain flies. [from Latin *arista* the beard (also, whole ear) of grain]

—**aristate** (ə ris′tāt), *adj.* 1 having aristae. 2 ending in a thin spine.

arith., *abbrev.* 1 arithmetic. 2 arithmetical.

arithmetic, *Mathematics.* —*n.* (ə rith′mə tik) 1 the study of numbers and their relationship; theory of numbers. 2 calculation with numbers, especially positive, real numbers, by addition, subtraction, multiplication, division, involution, and extraction of roots.
—*adj.* (ar′ith met′ik) = arithmetical.
[from Greek *arithmētikē,* from *arithmos* number]
—**arithmetical,** *adj.* of arithmetic; according to the rules of arithmetic.

arithmetic mean, = average.

arithmetic progression or **arithmetical progression,** *Mathematics.* a sequence of numbers, each of which is obtained from the preceding number of the sequence by adding or subtracting the same number. 2, 4, 6, 8, 10 form an arithmetic progression in which 2 is added to each number; 14, 9, 4, −1 is an arithmetic progression in which 5 is subtracted from each number. Compare **geometric progression.**

arkose (är′kōs), *n. Geology.* a type of sandstone derived from granite, containing feldspar, quartz, and mica: *Arkose forms both on land and sea, especially in or near deserts* (Fenton, *The Rock Book*). [from French]

arm, *n.* 1 *Anatomy.* **a** the part of a human body between the shoulder and the hand, and sometimes including the hand. **b** a forelimb or similar organ of any animal. The front legs of a bear or the tentacles of an octopus are sometimes called arms.
2 *Botany.* a main branch or limb of a tree.
3 *Geography.* an inlet from the sea or other body of water.
[from Old French *armer,* from Latin *armāre,* from *arma* weapons]

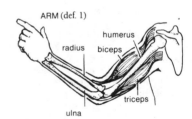
ARM (def. 1)
radius, humerus, biceps, ulna, triceps

armalcolite (är mal′kə līt′), *n. Mineralogy.* a lunar mineral composed of iron, magnesium, and titanium. *Formula:* $(Mg,Fe)Ti_2O_5$ [formed from the names of Neil *Arm(strong),* Edwin *Al(drin),* and Michael *Col(lins),* American astronauts whose spacecraft first landed on the moon (in 1969) + *-ite*]

armature, *n.* 1 *Electricity.* **a** a revolving part of an electric motor or generator, consisting of wire wound around an iron core placed between opposite poles of a magnet. See the picture at **magnet. b** a piece of soft iron placed in contact with the poles of a magnet. **c** the

---

cap, fāce, fäther; best, bē, tèrm; pin, five;
rock, gō, ôrder; oil, out; cup, pùt, rüle;
yü in use, *yu̇* in uric;
ng in bring; *sh* in rush; *th* in thin, *ᴛʜ* in then;
zh in seizure.
ə = *a* in about, *e* in taken, *i* in pencil, *o* in lemon, *u* in circus

43

movable part of an electric relay or vibrator, placed in a magnetic field which causes it to move.
2 *Biology.* any part or organ of an animal or plant serving for offense (teeth, claws) or defense (thorns). A turtle's shell is an armature.
[from Latin *armatura* armor]

aromatic (är′ə mat′ik), *Chemistry.* —*adj.* of or belonging to a class of organic compounds containing a closed chain of usually six carbon atoms. Aromatic ring systems are unsaturated yet exhibit a remarkable lack of reactivity. Benzene and its derivatives are aromatic. *Due to the pleasant odors of many of these (wintergreen oil, bitter almond oil, turpentine, etc.) they were known as aromatic compounds ... we now know of many aromatic compounds which are practically odorless* (Wertheim, *Organic Chemistry*).
ASSOCIATED TERMS: see **cyclic.**
—*n.* an aromatic compound.
[from Greek *arōmatikos* fragrant, from *arōma* fragrant herb, spice]
—**aromaticity,** *n.* the condition or degree of being aromatic.
—**aromatization,** *n.* Conversion into aromatic compounds.

array, *n.* **1** *Mathematics, Statistics.* an orderly arrangement of numbers, terms, or symbols in rows and columns.
2 *Electronics.* a group of antennas, transmitters, or other devices joined together to increase their effectiveness: *The array consists of two lines of seismic detectors, each nearly nine km long, and each having 11 equally spaced instruments, with a common time base* (New Scientist).

arrow of time, *Physics.* the direction in which time flows, either forward or, on the atomic level, forward and backward: *Accumulated starlight from the cycle subsequent to the one in which we live ... can arise, it seems, if the "arrow of time" is reversed in each cycle of the universe—that is, time flows backwards in the cycle which "follows" our own* (Nature). Compare **time reversal, time-symmetric.**

arsenate (är′sə nāt *or* ar′sə nit), *n. Chemistry.* a salt or ester of arsenic acid. **Arsenate of lead,** Pb (AsO₂)₂, is a highly toxic, white, crystalline compound used as an insecticide and herbicide.

arsenic, *n.* (är′sə nik) **1** *Chemistry.* a brittle, silver-gray, nonmetallic element which occurs chiefly in combination with other elements. Arsenic volatizes when heated and forms poisonous compounds with oxygen. It is used in making alloys, semiconductors, solders, and certain medicines. *Symbol:* As; *atomic number* 33; *atomic weight* 74.9216; *melting point* 814°C; *oxidation state* $+3$, $+5$.
2 Also called **arsenic trioxide.** a highly toxic compound of arsenic in the form of a white, odorless, tasteless powder, used especially as an insecticide and herbicide, as a pigment, and in preparing other arsenic compounds. *Formula:* As_2O_3 or As_4O_6
—*adj.* (är sen′ik) of or containing arsenic, especially with a valence of 5.
[from Latin *arsenicum,* from Greek *arsenikon*]

arsenic acid, *Chemistry.* a crystalline compound used in preparing arsenates. *Formula:* H_3AsO_4

arsenicate (är sen′ə kāt), *v. Chemistry.* to combine or treat with arsenic.

arsenide (är′sə nīd), *n. Chemistry.* a compound of arsenic and a metal or other positive element or radical.

arsenious (är sē′nē əs), *adj. Chemistry.* of or containing arsenic, especially with a valence of 3.

arsenite (är′sə nīt), *n. Chemistry.* **1** a salt or ester of arsenious acid: *Arsenites and arsenates are used as insecticides.* **2** = arsenic (def. 2).

arsenopyrite (är′sə nō pī′rīt), *n.* a silvery gray mineral, important as an ore of arsenic. *Formula:* FeAsS

arsine (är sēn′ *or* är′sēn), *n. Chemistry.* **1** a colorless, inflammable, highly poisonous gas having a garlic-like odor. *Formula:* AsH₃ **2** one of the derivatives of this gas, in which the hydrogen atoms are replaced by organic radicals. [from *ars(enic)* + *-ine*]

arterial (är tir′ē əl), *adj.* **1** of an artery or the arteries: *hardening of the arterial walls.* **2** contained in the arteries. **Arterial blood** is bright red because it has been purified and oxygenated by passing through the lungs.
—**arterially,** *adv.*

arterialize (ar tir′ē ə līz), *v. Physiology.* to convert (venous blood) into the bright-red blood of the arteries by the action of oxygen in the lungs. —**arterialization,** *n.*

arteriole (är tir′ē ōl), *n. Anatomy.* a small artery, especially one leading into capillaries: *Blood flows out from the heart to the tissues through an ever-branching system of vessels. Arteries divide into arterioles. Arterioles branch into still smaller metarterioles; and the latter give rise to the capillaries* (Harper's).
—**arteriolar** (är tir′ē ō′lər), *adj.* of or resembling an arteriole.

arteriovenous (ar tir′ē ō vē′nəs), *adj. Anatomy.* of or having to do with both an artery and a vein: *An arteriovenous aneurysm allows a vein and artery to open into one another* (Lois G. Lobb).

artery, *n., pl.* **-teries.** *Anatomy.* any of the membranous, elastic, muscular tubes forming part of the system of vessels that carry blood from the heart to various parts of the body: *The vessels carrying the fluid away from the heart are arteries, and those returning fluid from the tissue spaces toward the heart may be termed veins* (Harbaugh and Goodrich, *Fundamentals of Biology*). *The arteries have strong walls capable of withstanding considerable pressure, and they are firm enough to stand open even when empty of blood. The veins are not called upon to endure such pressures as are the arteries; their walls are comparatively thin and collapsible* (Shull, *Principles of Animal Biology*). [from Greek *artēria*]

artesian well (är tē′zhən), *Geology.* a well which rises from and above an aquifer, especially one from which the water gushes up without pumping: *Water trapped in deep-lying rock in such manner as to develop considerable hydrostatic pressure is termed artesian. A well drilled into such water-bearing rocks is known as an artesian well provided the water flows from it; if the water rises in the well but does not flow, it is termed sub-artesian* (White and Renner, *Human Geography*). See the picture at **aquifer.** [from French *artésien* of Artois, a French province where such wells were first drilled in the 1700's]

arthromere (är′thrə mir), *n. Zoology.* one of the segments or divisions of the body of a jointed or articulate animal. [from Greek *anthron* joint + *meros* portion]

arthropod (är′thrə pod), *Zoology.* —*n.* any of a phylum (Arthropoda) of invertebrate animals having segmented bodies to which jointed antennae, wings, or legs are articulated in pairs. Insects, arachnids, and crustaceans belong to this phylum. *The arthropods ... are one of the most successful manifestations of life. They show the greatest number of species of any phylum, and probably the greatest diversity of forms* (Alfred L. Kroeber).
—*adj.* of or belonging to the arthropods.
[from Greek *arthron* joint + *pous, podos* foot]

arthrospore, *n. Biology.* any of various fungal and algal spores that look like a string of beads, produced by the breaking up, or fission, of a hypha. [from Greek *arthron* joint + *spora* seed]
—**arthrosporous** (är thros′pər əs), *adj.* producing arthrospores.

articular, *adj. Anatomy.* of, belonging to, or having to do with a joint or articulation: *an articular muscle, the articular surface of a bone.*

articulate, (*adj.* är tik′yə lit; *v.* är tik′yə lāt), *Anatomy.*
—*adj.* consisting of sections united by joints; jointed. The backbone is an articulate structure.
—*v.* **1** to unite by joints: *The two bones are articulated like a hinge.* **2** to fit together in a joint: *The anklebone articulates with the bones of the leg.*
—**articulation,** *n.* **1** a joint between parts of an animal or plant. **2** act or manner of connecting by a joint or joints.
—**articulatory,** *adj.* of or having to do with articulation.

artifact, *n.* an artificial product or effect observed in a natural system, especially one resulting from investigation: *The apparent pattern in the data was an artifact of the collection method.*

artificial, *adj.* not natural or biological: *an artificial heart. Attempts to establish artificial days of 16 hours in rats and mice have failed. Although the actual time of activity of a diurnal cycle may quite easily be shifted, the general plan of 24-hour periodicity seems to be innate* (Science News). SYN: man-made, synthetic.

artificial classification, *Biology.* a classification of plant or animal organisms based on prominent points of resemblance or difference rather than on genetic or biochemical relationships: *Grouping plants by habitats (such as desert or marsh plants), or by the number of stamens or color of their flowers has ... to be an artificial classification* (Weier, Botany).

artificial gene, *Molecular Biology.* a chemically synthesized copy of a gene made by combining specific sequences of nucleotides.

artificial insemination, *Medicine.* the introduction of semen into the vagina or cervix by other than natural means: *Artificial insemination through the use of the semen of an anonymous donor (AID) has been the solution to ... sterility problems for many couples* (Americana Annual). *Abbreviation:* AI

artificial intelligence, *Electronics.* the means by which computers, robots, and other devices perform tasks which normally require human intelligence, such as limited problem solving, discrimination among objects, and response to voice command: *a system with enough artificial intelligence to recognize ... various sizes, colors, and shapes* (Science News).

artificial radioactivity, *Physics.* radioactivity of a normally stable element, induced by nuclear bombardment.

artificial selection, *Biology.* the modification of the processes of natural selection by human agency, as in breeding horses for speed, or cattle for beef or milk.

artiodactyl (är′tē ō dak′tl), *Zoology.* —*n.* any of an order (Artiodactyla) of hoofed mammals with an even number of toes, usually two or sometimes four, on each foot. Swine, camels, deer, sheep, cattle, and many other quadrupeds are artiodactyls.
—*adj.* belonging to this order; having an even number of toes on each foot.
[from Greek *artios* even-numbered + *dactylos* finger, toe]

aryl (ar′əl), *n. Chemistry.* a univalent radical occurring in aromatic hydrocarbon derivatives. Compare **alkyl.**
[from *ar(omatic)* + *-yl*]

arytenoid (ə rit′ə noid), *Anatomy.* —*adj.* having the shape of a ladle or cup (applied to two cartilages of the larynx which regulate the action of the vocal cords, and to several small glands of the larynx).
—*n.* an arytenoid cartilage: *The arytenoids ... can pivot in all directions, and also slide toward or away from each other in horizontal grooves* (Charles K. Thomas).
[from Greek *arytainoeidēs,* from *arytaina* ladle + *eidos* shape]

asbestiform, *adj. Mineralogy.* being or resembling asbestos; occurring in flexible fibers.

asbestos (as bes′təs), *n. Mineralogy.* any of several hydrous magnesium silicate minerals, often containing calcium, that occur as flexible fibers which are not combustible or heat conductive and are therefore useful as insulating materials. Excessive inhalation of asbestos fibers can damage the lungs. Chrysotile and crocidolite are types of asbestos. [from Greek *asbestos* unquenchable (originally applied to quicklime)]

ascending, *adj.* **1** extending upward; rising: *The ascending colon extends up toward the liver. The moon's ascending node is the point where the moon crosses the ecliptic going north.* **2** *Botany.* directed or growing upwards. Contrasted with **descending.**

ascidian (ə sid′ē ən), *n. Zoology.* any of a class or order (Ascidiacea) of simple or compound tunicates, usually sessile when mature, having a tough, saclike covering; a sea squirt. [from Greek *askidion,* diminutive of *askos* (literally) skin bag]

ascidium (ə sid′ē əm), *n., pl.* **-cidia** (-sid′ē ə). *Botany.* a baglike or pitcherlike part of a plant, such as the leaf of the pitcher plant. [from Greek *askidion* small bag]

ascigerous (ə sij′ər əs), *adj. Biology.* bearing asci. [from New Latin *ascus* sac, bag + Latin *gerere* to bear]

---

cap, fāce, fäther; best, bē, tèrm; pin, five;
rock, gō, ôrder; oil, out; cup, pùt, rüle,
y*ü* in use, y*ù* in uric;
ng in bring; *sh* in rush; *th* in thin, ŦH in then;
zh in seizure.
ə = *a* in about, *e* in taken, *i* in pencil, *o* in lemon, *u* in circus

ASCII (as′kē), *n.* a standardized binary code for representing letters, numbers, and symbols, used in the United States among different data processing and communications systems to facilitate the exchange of information. For example, the capital letter J in the ASCII code is 1001010; lower-case j is 1101010. [from the initials of *American Standard Code for Information Interchange*]

asco-, *combining form.* ascus; sac, as in *ascospore = spore produced in an ascus.* [from New Latin *ascus,* from Greek *askos* skin bag, sac]

ascocarp (as′kə kärp), *n. Biology.* the mature fruiting body of ascomycetous fungi.
—ascocarpous (as′kə kär′pəs), *adj.* of or resembling ascocarps; having an ascocarp.
[from *ascus* + -*carp*]

ascogonium (as′kə gō′nē əm), *n., pl.* -nia (-nē ə). *Biology.* the female sex organ in the gametophyte of ascomycetous fungi. [from New Latin *ascogonium,* from Greek *askos* skin bag + a root *gen-* to bear]
—ascogonial, *adj.* of or having to do with an ascogonium.

ascomycete (as′kō mi′sēt), *n. Biology.* any of a large class (Ascomycetes) of fungi, including yeasts, molds, mildews, etc., characterized by the formation of spores in elongated sacs (asci): *The parasitic organisms responsible for chestnut blight and Dutch elm disease are ascomycetes* (Greulach and Adams, *Plants*). Also called **sac fungus.** [from Greek *askos* skin bag + *mykēs, mykētos* fungus]
—ascomycetous (as′kō mī sē′təs), *adj.* of or belonging to the ascomycetes.

ascorbate (ə skôr′bāt), *n. Chemistry.* a salt of ascorbic acid: *White cells ... are loaded with ascorbate; in fact they serve as the body's principal storage area for the vitamin* (Rich Wentzler).

ascorbic acid (ə skôr′bik), = vitamin C. [from *a-* not + *scorb(ut)ic*]

ascospore, *n. Biology.* one of the cluster of spores formed within an ascus: *When an ascus with its eight ascospores becomes mature it may explode with such force as to send the spores a short distance into the air* (Emerson, *Basic Botany*).

ascus (as′kəs), *n., pl.* asci (as′kī *or* as′ī). *Biology.* the elongated sac or cell in which the spores of ascomycetes are formed. [from Greek *askos* skin bag]

-ase, *suffix.* enzyme, as in *proteinase = an enzyme that catalyzes the breakdown of proteins; oxidase = an enzyme that catalyzes an oxidation.* [from (*diast*)*ase*]

aseismic (ā sīz′mik), *adj. Geology.* free of seismic disturbances: *The aseismic nature of the ridge south and east of southern Africa also suggests that it is at present a dormant ridge* (Nature).

asepsis (ā sep′sis), *n. Medicine.* 1 the condition of being aseptic. 2 aseptic methods or treatment. [from *a-* without + *sepsis*]
ASSOCIATED TERMS: see **antisepsis.**

aseptic (ā sep′tik), *adj. Medicine.* 1 free from the living microorganisms causing infection: *Surgical instruments are made aseptic by boiling them.* 2 using all possible measures to exclude infectious microorganisms: *aseptic surgery.* —aseptically, *adv.*

asexual, *adj. Biology.* 1 having no sex. 2 independent of sexual processes: *asexual reproduction.* —asexually, *adv.*
ASSOCIATED TERMS: There are several types of asexual reproduction. *Binary fission* involves equal division of the cytoplasm and nuclear materials of an organism, resulting in two new organisms. *Budding* is similar, except that the division of material is unequal. In *sporulation,* specialized cells called spores are released which can develop into new organisms. *Regeneration* is the development of a new organism from a detached part of the parent organism. By *vegetative propagation,* new plants can develop from roots, stems, or leaves of a parent plant.

asexual reproduction: (def. 2)

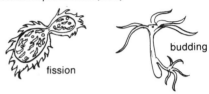

fission

budding

ash, *n.* 1 *Chemistry.* what remains of a material after it has been thoroughly burned; the incombustible residue of organic substances remaining after combustion. 2 *Geology.* fine material thrown out of a volcano in eruption.

Asn, *abbrev.* asparagine.

Asp, *abbrev.* aspartic acid.

asparaginase (as′pə raj′ə nās), *n. Biochemistry.* a bacterial enzyme that breaks down asparagine, used especially in the treatment of leukemia: *Asparaginase ... kills leukemic cells by starving them of an essential amino acid* (N. Y. Times Magazine).

asparagine (ə spar′ə jēn), *n. Biochemistry.* a crystalline amino acid, derived from aspartic acid in living systems. *Formula:* $C_4H_8N_2O_3$ *Abbreviation:* Asn

aspartic acid (ə spär′tik), *Biochemistry.* a crystalline amino acid derived from the juice of asparagus, beets, and young sugar cane. *Formula:* $C_4H_7O_4N$ *Abbreviation:* Asp

aspect, *n.* 1 *Astronomy.* a the relative position of a planet with respect to another planet, the sun, or the moon, as seen from the earth. b the relative position of any celestial body with respect to another.
2 *Geography.* the relative direction, or compass orientation, of a land slope.

asphalt, *n. Geology.* a dark, waterproof substance much like tar, a viscous liquid or solid bitumen formed by the evaporation of volatiles from petroleum. Asphalt is found in natural beds or obtained by refining petroleum. [from Greek *asphaltos*]
—asphaltic, *adj.* of the nature of or containing asphalt.

asphaltite, *n. Geology.* a dark-colored, resinous form of natural asphalt.

aspheric (ā sfer′ik), *adj. Optics.* not spherical: *Any surface having other than a plane or spherical form is said to be aspheric ... Aspheric surfaces are used to reduce the spherical aberration* (Hardy and Perrin, *Principles of Optics*).

assay, *Chemistry.* —v. (ə sā′) 1 to analyze (an ore or alloy) to find out the quantity of gold, silver, or other metal in it. 2 (of ore) to be found by analysis to contain a certain proportion of metal.

—*n.* (as′ā) **1** the determination of the proportion of gold or other metal in an ore or alloy by measuring, weighing, or calculating. **2** the determination of the strength of a drug or other substance by comparing its effects on an organism with those of a standard substance. Also called **bioassay. 3** the substance or sample analyzed or tested. **4** a list of the results of assaying an ore, drug, etc.
[from Old French *assayer,* ultimately from Latin *exigere* to weigh, prove]

assimilate, *v. Biology.* **1** to convert (nonliving substances) into living protoplasm or cells by the process of anabolism. **2** to take in (nutrients and other substances) by cells and tissues after digestion: *The human body will not assimilate sawdust.*
—**assimilation,** *n.* the process of assimilating: *Some biologists restrict the term assimilation to the making of living matter (protoplasm) from food and water ...* [*some*] *use it in its broader sense, including the formation of ... cell walls as well as protoplasm* (Greulach and Adams, *Plants*).

association area, *Physiology.* an area of the cerebral cortex where several impulses from different sense organs are correlated: *Between the receiving and effector (motor) regions of the brain are the 'association areas' of the cortex. In higher mammals, these association areas form the bulk of the cerebral cortex* (G.M. Wyburn).

associative, *adj. Mathematics.* of or having to do with a rule that the combinations by which numbers are added or multiplied will not change their sum or product. EXAMPLES: 2 + (3+4) will give the same sum as (2+3) + 4. (2×3) × 5 will give the same product as 2 × (3×5).
ASSOCIATED TERMS: According to the *associative* property of addition and multiplication, numbers to be added or multiplied may be associated or grouped in any order; according to the *commutative* property, the order of numbers to be added or multiplied may be commuted or exchanged without affecting the results; according to the *distributive* property, multiplication can be "distributed over addition" in that the sum of two numbers may be multiplied by a third number by multiplying each number separately and adding the products. See also **transitive.**

associative neuron, = interneuron.

astatic (ā stat′ik), *adj. Physics.* having no tendency to take a fixed or definite position: *A magnetic needle whose directive power has been neutralized is astatic.* [from a- not + *static*]

astatine (as′tə tēn′), *n. Chemistry.* a radioactive, nonmetallic element belonging to the halogen series. Astatine is very rare in nature and is produced artificially by bombarding bismuth with alpha particles. It is highly unstable, with many isotopes, the most stable one having a half-life of about 8 hours. *Symbol:* At; *atomic number* 85; *atomic weight* 210; *oxidation state* ±1, +3, +5, +7. [from Greek *astatos* unstable]

aster, *n. Biology.* one of two star-shaped structures formed in a cell at the beginning of mitosis. An aster consists of microtubules radiating from the centriole. [from Greek *astēr* star]

asterism (as′tə riz′əm), *n.* **1** *Astronomy.* a group of stars smaller than a constellation. The Pleiades are an asterism in the constellation Taurus.

2 *Mineralogy.* a starlike, luminous figure or effect seen in some minerals either by transmitted or reflected light due to the presence of oriented inclusions.
[from Greek *asterismos* constellation, from *astēr* star]

asteroid (as′tə roid′), *n. Astronomy.* any of thousands of small planets which revolve about the sun, chiefly between the orbits of Mars and Jupiter. The largest and first known asteroid is Ceres; it was discovered on January 1, 1801, and is about 970 kilometers in diameter. Also called **planetoid.** [from Greek *asteroeidēs* starlike, from *astēr* star]
—**asteroidal** (as′tə roi′dəl), *adj.* of or having to do with asteroids: *the asteroidal belt.* See the picture at **solar system.**

asthenosphere (as then′ə sfir′), *n. Geology.* a region of weakness in the earth's mantle, consisting of hot rock material that yields readily to prolonged strains. The movement of the rocks within the asthenosphere is believed to help equalize pressures, thereby maintaining the stability or equilibrium of the earth's crust. *Many scientists have concluded that temperatures are so high in the asthenosphere that the rock is partially ... melted, and that the molten part is the source of much of the iron- and magnesium-rich lava that erupts from volcanos at the earth's surface* (Putnam's Geology). See the picture at **converging.** [from Greek *asthenēs* weak + English *sphere*]
—**asthenospheric** (as then′ə sfer′ik), *adj.* of or having to do with the asthenosphere: *The descending lithosphere displaces asthenospheric material, forcing it upward to form new crust* (Science News).
ASSOCIATED TERMS: see **lithosphere.**

astigmatic (as′tig mat′ik), *adj. Optics.* **1** having to do with astigmatism: *A pencil that fails to unite at a single image point after refraction is said to be astigmatic, and the system is said to be afflicted with astigmatism* (Hardy and Perrin, *Principles of Optics*).
2 having astigmatism of the eye: *A third common type of defect in vision occurs when one or more of the refracting surfaces of the eye, such as the cornea or crystalline lens, are not perfectly spherical ... An eye having this characteristic is said to be astigmatic* (Shortley and Williams, *Elements of Physics*).

astigmatism (ə stig′mə tiz′əm), *n. Optics.* a structural defect of an eye or of a lens which prevents rays of light from converging at a single point, thus producing indistinct or imperfect images. Astigmatism of the eye usually refers to a defect in which the surface of the cornea is not spherical, but is more sharply curved in one plane than another, so that rays are not brought into focus at one point on the retina but seem to spread in various directions. Astigmatism of a lens, mirror, or other optical system is a form of spherical aberration resulting from the failure of rays to come to a focus at

---

cap, fāce, fäther; best, bē, tèrm; pin, five;
rock, gō, ôrder; oil, out; cup, pùt, rüle;
yü in use, *yu̇* in uric;
ng in bring; *sh* in rush; *th* in thin; ŦH in then;
zh in seizure.
ə = *a* in about, *e* in taken, *i* in pencil, *o* in lemon, *u* in circus

---

different distances along the optical axis. [from *a*-without + Greek *stigmatos* mark]

astomatous (ā stom′ə təs), *adj.* 1 *Zoology.* (of animals) having no mouth. 2 *Botany.* (of plants) having no stomata.

astragalus (ə strag′ə ləs), *n., pl.* **-li** (-lī). *Anatomy.* the uppermost bone of the tarsus; the bone of the ankle which articulates with the bones of the leg; anklebone. Also called **talus.** [from Greek *astragalos*]

astral, *adj.* 1 *Astronomy.* of the stars; stellar: *Euclidean geometry is inapplicable to astral measurements* (Saturday Review). 2 *Biology.* of or like the aster. [from Latin *astrum* star, from Greek *astron*]

astringent (ə strin′jənt), *Medicine.* —*n.* a substance, such as alum, that draws together or contracts body tissues and thus checks the flow of blood or other secretions. —*adj.* having the property of drawing together or contracting tissues. [from Latin *astringentem* drawing tight, from *ad-* to + *stringere* to bind]

astro-, *combining form.* 1 star or stars; any cosmic or heavenly body or bodies, as in *astronomy, astrophysics, astrogeology.* 2 outer space; spacecraft; space flight, as in *astronautics, astrodynamics.* [from Greek *astro-,* from *astron* star]

astrobleme (as′trə blēm′), *n. Geology.* a scar left on the earth's surface by a meteorite, usually in the form of a circle of shattered rocks or an impact crater. [from *astro-* + Greek *blema* throw of a missile, wound from such a throw]

astrochemistry, *n.* the study of the chemical composition and characteristics of celestial bodies: *The new science of astrochemistry or molecular astronomy has found dozens of chemical compounds in the interstellar clouds ... and it may indicate that interstellar clouds also house the beginnings of chemistry* (Dietrick E. Thomsen).

astrocyte (as′trə sīt), *n. Biology.* a star-shaped neuroglial cell.

astrodynamics, *n.* the branch of dynamics dealing with the motion of bodies in outer space and the forces acting upon them. —**astrodynamic,** *adj.*

astrogeology (as′trō jē ol′ə jē), *n.* the study of the rocks and other physical features of the moon or other planets or planetlike bodies; extraterrestrial geology: *When ... the geological structure of the planets has been investigated, astrogeology will not be a mere supplement to existing knowledge; terrestrial geology will, in the end, become only a division of astrogeology* (New Scientist).

astrolabe (as′trə lāb), *n. Astronomy.* 1 an ancient instrument for measuring the altitude of the sun or stars. It has been largely replaced by the sextant. 2 an optical instrument (the **prismatic astrolabe**) using a small telescope and the reflection in a prism to measure the exact position of a star. [from Greek *astrolabos* (literally) taking stars]

astronomical, *adj.* of or having to do with astronomy: *astronomical sightings.*

astronomical unit, the mean distance of the earth from the sun (about 150 million kilometers), used as a unit of measurement in expressing distances between planets and stars: *The astronomical unit ... [is] half the ma-*

jor axis of the earth's elliptical orbit (Scientific American). *Abbreviation:* A.U.

astronomical year, the period of the earth's revolution around the sun. It lasts 365 days, 5 hours, 48 minutes, and 45.51 seconds. Also called **solar year.**

astronomy, *n.* the science that deals with the constitution, motions, relative positions, sizes, etc., of the sun, moon, planets, stars, and all other heavenly bodies, as well as with the earth in its relation to them. [from Greek *astronomia,* from *astron* star + *nomos* distribution]

astrophysics, *n.* the branch of astronomy that deals with the physical characteristics of heavenly bodies, such as luminosity, temperature, size, mass, and density, and also their chemical composition.
—**astrophysical,** *adj.* of or having to do with astrophysics.

asymmetric or **asymmetrical,** *adj.* not symmetrical; lacking symmetry. An **asymmetric carbon atom** is a carbon atom combined directly with four unlike atoms or groups, resulting in an unbalanced spatial arrangement of atoms in a molecule. —**asymmetrically,** *adv.*

asymmetry (ā sim′ə trē), *n.* lack of symmetry. **Molecular asymmetry** occurs when a molecule cannot be divided into like portions. Such a molecule cannot be superimposed on its mirror image.

asymptote (as′im tōt), *n. Mathematics.* a line to which a curve comes indefinitely close but does not meet. [from Greek *asymptōtos,* from *a-* not + *symptōtos* due to coincide] —**asymptotic,** *adj.* —**asymptotically,** *adv.*

asynapsis (ā′si nap′sis), *n., pl.* **-ses** (-sēz). *Biology.* failure or absence of synapsis in meiosis. [from *a-* not + *synapsis*]

at., *abbrev.* 1 atmosphere. 2 atomic.

atavism (at′ə viz′əm), *n. Biology.* the reappearance in an animal or plant of characteristics of a remote ancestor not found in its immediate ancestors, generally as a result of a recombination of genes. [from Latin *atavus* ancestor]
—**atavistic** (at′ə vis′tik), *adj.* having to do with atavism: *... it is evident that all animals are ultimately from one primeval source. The appearance in the embryo of atavistic survivals and the fact that the process of embryological development is fundamentally the same in all animals further confirm this conclusion* (Ralph Beals and Harry Hoijer).

-ate, *noun suffix. Chemistry.* a salt or ester of an acid whose name ends in *-ic,* as in *sulfate = a salt or ester of sulfuric acid, carbonate = a salt or ester of carbonic acid.*

atlas, *n. Anatomy.* the first cervical vertebra, which supports the skull. It turns upon the second vertebra or *axis.* [from Latin and Greek *Atlas,* name of the giant in Greek mythology who supported the heavens on his shoulders]

atm. or **atm,,** *abbrev.* 1 atmosphere. 2 atmospheric.

atmolysis (at mol′ə sis), *n., pl.* **-ses** (-sez′). *n. Chemistry.* the separation of mixed gases by partial diffusion through a porous substance. [from Greek *atmos* vapor + *lysis* a loosening]

atmophile (at′mə fil), *adj. Geology.* (of an element) typically found in or forming part of the atmosphere: *Atmophile elements ... exist either in the uncombined state, such as oxygen, nitrogen, etc., or as volatiles* such as H_2O, CO_2, and the rare gases, helium, neon, argon, etc. (Rhodes W. Fairbridge). Compare **chalcophile, lithophile, siderophile.**

atmosphere, n. **1** *Meteorology, Geophysics.* the mass of gases that surrounds the earth and is held to it by the force of gravity; the air surrounding the earth: *The atmosphere has a constant proportion of oxygen, but the amount of air, and so of oxygen, diminishes on mountains* (Science News).

▶ The earth's atmosphere consists of about 78 per cent nitrogen (N_2), 21 per cent oxygen (O_2), and 1 per cent argon (Ar) by volume, with traces of carbon dioxide (CO_2), water vapor (H_2O), neon (Ne), helium (He), methane (CH_4), and other gases.

2 *Astronomy.* the mass of gases that surrounds, or may surround, any planet or other heavenly body: *the cloudy atmosphere of Venus.*

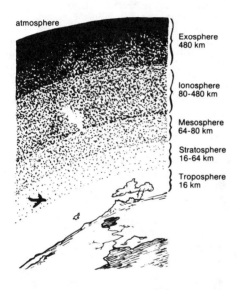

atmosphere

Exosphere 480 km

Ionosphere 80-480 km

Mesosphere 64-80 km

Stratosphere 16-64 km

Troposphere 16 km

3 *Physics.* a unit of pressure equal to 1.013×10^5 pascals. The pressure of the air on the earth's surface at sea level is about 1 atmosphere. An atmosphere is defined technically as the amount of pressure that will support a column of mercury 760 millimeters high at 0 degrees Celsius under standard gravity. *In the sea the pressure increases by about one atmosphere ... for every 10 meters of depth* (Scientific American). *Abbreviation:* at. or atm.
[from New Latin *atmosphaera,* from Greek *atmos* vapor + *sphaira* sphere]
ASSOCIATED TERMS: The *atmosphere* of the earth is the gaseous envelope surrounding the *lithosphere,* or solid part of the earth's crust, and the *hydrosphere,* or liquid portion of the earth's surface. The atmosphere itself is composed of several layers or regions. The *troposphere,* in which our turbulent weather occurs, extends to about 16 kilometers above the earth's surface, and the comparatively calm *stratosphere* extends from this upward to about 48 kilometers. The boundary between the troposphere and the stratosphere is called the *tropopause.* In the upper stratosphere is situated the *ozonosphere* that shields the earth from excessive ultraviolet radiation. Above this region are several other distinctive layers: the *mesosphere,* extending to about 80 kilometers, the *ionosphere,* extending to about 500 kilometers, and finally the *exosphere,* or outermost region of the atmosphere, where out-

er space begins. See also **chemosphere, magnetosphere, thermosphere, D region, E region,** and **F region.**

—**atmospheric,** *adj.* of, having to do with, or in the atmosphere: *atmospheric oxygen, atmospheric turbulence.* —**atmospherically,** *adv.*

atmospheric pressure, *Meteorology.* pressure caused by the weight of the air. The normal atmospheric pressure on the earth's surface at sea level is 1.013×10^5 pascals (see **atmosphere** def. 3). *Atmospheric pressure ... is probably the most important single weather element observed. It is the fundamental basis on which the weather map is constructed.* (Neuberger, *Weather and Man*). Also called **air pressure.** See the picture at **isobar.**

atmospheric tides, *Geophysics.* movement of atmospheric masses in a pattern somewhat resembling that of the tides, caused by temperature fluctuations and by gravitation.

at. no. *abbrev.* atomic number.

atoll, *n. Geography, Geology.* a ring-shaped coral island or group of islands enclosing or partly enclosing a lagoon. [from its name in the Maldive Islands]
ASSOCIATED TERMS: see **barrier reef.**

atom, *n. Chemistry, Physics.* the smallest particle of a chemical element that can take part in a chemical reaction without being permanently changed. An atom is made up of protons and neutrons in a central nucleus surrounded by electrons. The atoms of the various elements differ in mass or weight and in the number of electrons they contain. Atoms combine to form molecules; for example, a molecule of water consists of two atoms of hydrogen and one atom of oxygen. *Ordinarily, the atom as a whole appears to be electrically neutral or uncharged, so we say that the total negative charge on all the electrons is equal in magnitude to the positive charge on the nucleus. The number of electrons in a neutral atom is called the atomic number* (Shortley and Williams, *Elements of Physics*). [from Greek *atomos* indivisible, from *a-* not + *tomos* a cutting] See picture on following page.

atomic, *adj.* **1** of, belonging to, or having to do with atoms: *atomic nuclei, atomic research.*
2 using or produced by atomic energy: *atomic power, atomic radiation.*
3 *Chemistry.* separated into atoms; not combined: *atomic hydrogen.*

▶ Though **atomic** and **nuclear** are commonly regarded as synonyms, in certain established terms *atomic* is generally preferred: *atomic clock, atomic number, atomic mass, atomic theory, atomic weight.* In several terms *atomic* and *nuclear* are readily interchangeable: *atomic* (or *nuclear*) *energy, atomic* (or *nuclear*) *fission, atomic* (or *nuclear*) *bomb, atomic* (or *nuclear*) *power.* The current trend, however, is to favor the use of *nuclear* in many phrases, as in *nuclear device, nuclear*

cap, fãce, fäther; best, bē, tèrm; pin, fīve;
rock, gō, ôrder; oil, out; cup, pùt, rüle,
yü in use, *yu* in uric;
ng in bring; *sh* in rush; *th* in thin, ᴛʜ in then;
zh in seizure.
ə = *a* in about, *e* in taken, *i* in pencil, *o* in lemon, *u* in circus

atomic bomb

physics, nuclear engineering, nuclear research, nuclear radiation, nuclear reaction, nuclear reactor, and so on.

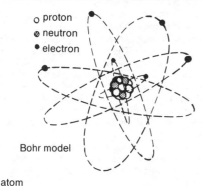

o proton
ø neutron
• electron

Bohr model

atom

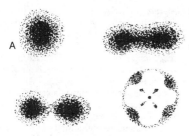

A

Schrödinger wave model

Each of these figures represents standing wave patterns which correspond to one of the Bohr orbits; A represents the ground state.

atomic bomb or **atom bomb,** a bomb in which the splitting of atomic nuclei results in an explosion of devastating force and heat, accompanied by a blinding light. The destructive force of an atomic bomb, measured in kilotons and megatons, is due to nearly instantaneous and uncontrolled successive fissions of uranium or plutonium atoms in a chain reaction, each fission releasing tremendous energy. Also called **nuclear bomb.**

atomic clock, a highly accurate instrument for measuring time, using atomic vibrations as its standard of accuracy rather than the revolution of the earth: *Atomic clocks can be made to measure time with an accuracy approaching one second per 1,000 years* (Scientific American).

atomic energy, *Chemistry, Physics.* the energy that exists in atoms. Some atoms can be made to release some of their energy, either under control (in a reactor) or uncontrolled (in a bomb). Atomic energy is generated through alteration of the nuclei of atoms chiefly by fission (splitting of heavy nuclei) or sometimes by fusion (combining of light nuclei). Also called **nuclear energy.**

atomicity (at′ə mis′ə tē), *n. Chemistry.* **1** the number of atoms contained in one molecule of an element. **2** the number of atoms or radicals which can be replaced in the molecule of a compound. **3** = valence. **4** atomic condition; state of separation into individual atoms.

atomic mass, *Chemistry, Physics.* the mass of an atom, as expressed on a scale in which the mass of the most

abundant isotope of carbon is placed at 12. Compare **atomic weight.**

atomic mass unit, = dalton. *Abbreviation:* amu or AMU

atomic number, *Chemistry, Physics.* the number of protons in the nucleus of an atom of a chemical element, used in describing the element and giving its relation to other elements in a series ranging from 1 (hydrogen) to 109 and predicted to 114. The atomic number of an element is equal to the number of electrons in a neutral atom of that element. *Abbreviation:* at. no.

atomic theory, *Chemistry, Physics.* the theory that all matter is composed of atoms, especially the theory that an atom is made up of a nucleus surrounded by electrons: *Modern chemistry begins with the Atomic Theory of John Dalton, who adapted a theory which had been current for two thousand years and more to explain the relationships between the quantities of chemical elements in chemical compounds. To him we owe the ideas of atomic weight and of the chemical formula* (F. Sherwood Taylor).

atomic weight, *Chemistry, Physics.* the relative weight of an atom of a chemical element, based on the weight of the most abundant isotope of carbon, which is taken as 12. *Abbreviation:* at.wt.

▶ Before 1962, atomic weights were based on the weight of the most abundant isotope of oxygen, which was taken as 16. On January 1, 1962, the International Union of Pure and Applied Chemistry adopted carbon-12 as the official standard for atomic weights.

ATP, *abbrev.* adenosine triphosphate.

ATPase (ā′tē′pē′ās′), *n. Biochemistry.* an enzyme that catalyzes the hydrolysis of adenosine triphosphate: *As the amino acid passes through the cell membrane, it may stimulate ATPase. ATPase in turn splits a molecule of ATP just inside the cell, in the cytoplasm* (Science News). Full name, **adenosine triphosphatase.** [from *ATP* + *-ase*]

atriopore (ā′trē ə pôr′), *n. Anatomy.* the outlet of an atrium: *Water, which is carried into the pharynx by way of the mouth, passes through the gill slits into the atrium and out of the atriopore* (Hegner and Stiles, College Zoology).

atrioventricular (ā′trē ō ven trik′yə lər), *adj. Anatomy.* of or having to do with the atria and ventricles of the heart.

atrioventricular bundle, *Anatomy.* a bundle of muscle tissue that conducts heartbeat impulses from the right atrium to the ventricles of the heart. Also called **bundle of His.**

atrioventricular node, *Anatomy.* a small mass of tissue in the right auricle of the heart, which receives the heartbeat impulses from the sinoatrial or sinus node and conducts them by way of the atrioventricular bundle. It sometimes replaces the sinoatrial node as the heart's pacemaker. Also called **A-V node.**

atrium (ā′trē əm), *n., pl.* **atria** (ā′trē ə), **atriums.** *Anatomy.* **1** either of the chambers of the heart that receives the blood from the veins and forces it into a ventricle. The heart of mammals, birds, and reptiles has two atria; that of fishes and amphibians has one atrium. Also called **auricle.** See the picture at **heart. 2** any of various cavities or sacs, especially in certain marine animals. [from Latin *atrium* main room] **—atrial,** *adj.*

atrophy (at'rə fē), *n.* **1** *Physiology, Medicine.* a wasting away of part of the body, especially through imperfect nourishment or disuse: *muscular atrophy. Progressive atrophy of the adrenal cortex ... brings on a characteristic condition of weakness which ends in death. The symptoms arising in this process are very definite and characterise the famous Addison's disease* (Science News).
2 *Biology.* arrested development of an organ of an animal or plant.
—*v.* to undergo atrophy; waste away: *The female [starling] has only one ovary, and during the nonbreeding season the sex organs of both males and females atrophy* (Scientific American).
[from Greek *atrophia,* from *a-* without + *trophē* nourishment]
—**atrophic** (ə trof'ik), *adj.* of or characterized by atrophy.

attenuate, *v.* **1** *Microbiology.* to make (microorganisms or viruses) less virulent by successive cultures, exposure to heat or chemicals, or by other means: *A live attenuated-virus vaccine against mumps was ... an advance over the killed-virus vaccines previously available, in that it confers lifelong immunity* (Frank P. Mathews).
2 *Physics.* to reduce in energy or power: *The gammas are attenuated to one-half their initial intensity only after traversing, on the average, about four inches of water or tissue* (Bulletin of Atomic Scientists).
—*adj. Botany.* gradually tapering.
[from Latin *attenuatum* made thin, from *ad-* + *tenuis* thin] —**attenuation,** *n.*

atto- (at'ō), *combining form.* one quintillionth (10^{-18}) of (any standard unit in the meter-kilogram-second system), as in *attofarad = one quintillionth of a farad, attosecond = one quintillionth of a second. Symbol:* a [from Danish *atten* eighteen]

attract, *v. Physics.* to exert a pulling force upon; to draw: *A magnet attracts iron.*
—**attractant,** *n. Physiology.* a natural or synthetic chemical substance that attracts or lures insects or other animals: *In some insects, both sexes exude attractants. The male boll weevil, for instance, can attract the female of the species from a distance of 30 feet* (John Barbour). Compare **pheromone.**
—**attraction,** *n. Physics.* the electric or magnetic force exerted by oppositely charged particles on one another that tends to draw or hold the particles together, or the gravitational force exerted by one body on another: *In the solid state, the ions are arranged in a regular array, or lattice, such that the electrical attractions between the oppositely charged ions* (K.D. Wadsworth).
—**attractive,** *adj.* attracting: *... the power of the engine is proportional to the attractive force of its magnets* (J.P. Joule).

attraction sphere, = centrosphere.

attrition, *n. Geology.* the wearing away or grinding down of rock particles: *The particles involved in attrition, notably sand grains, develop characteristic textures; for example, wind-worn sands have a distinctive "frosting"* (Rhodes W. Fairbridge).
—**attritional, attritive** (ə trī'tiv), *adj.* having to do with or characterized by attrition.
ASSOCIATED TERMS: see **erosion.**

at. wt., *abbrev.* atomic weight.

Au, *symbol.* gold. [from Latin *aurum* gold]

A.U. or **a.u.,** *abbrev.* **1** angstrom unit. **2** astronomical unit.

Å.U., A.u. or **å.u.,** *abbrev.* angstrom unit.

audibility, *n. Acoustics.* a measure of the intensity of a sound or signal, usually expressed in decibels.

audible, *adj. Acoustics.* capable of being heard; being within the range of hearing, or between 15 and 20,000 hertz for human beings.

audio (ô'dē ō), *adj. Acoustics.* using, involving, or having to do with audio frequencies: *an audio signal, an audio transformer.* [from Latin *audire* hear]

audio frequency, *Acoustics.* a frequency corresponding to audible sound vibrations, from about 15 to about 20,000 hertz or cycles per second for human beings.

audiology (ô'dē ol'ə jē), *n.* the science of hearing.
—**audiological,** *adj.* of or having to do with audiology.

audition, *n. Physiology.* the sense of hearing.

auditory, *adj. Physiology, Anatomy.* of or having to do with hearing, the sense of hearing, or the organs of hearing: *The auditory nerve transmits impulses from the ear to the brain.*

auditory canal, *Anatomy.* a duct from the eardrum to the external ear by which sound waves are transmitted from the outer to the inner ear.

auditory nerve, *Anatomy.* a sensory nerve consisting of two fibers or branches, the cochlear, that conveys sound to the brain, and the vestibular, that carries impulses involved in maintaining equilibrium. Also called **acoustic nerve** or **acoustical nerve.** See the picture at **ear.**

augend (ô'jend *or* ô jend'), *n. Mathematics.* a number or quantity to be increased by the addition of another. Compare **addend.** [from Latin *augendum,* from *augere* to increase]

Auger effect (ō zhā'), *Physics.* the movement of electrons from the outer to the inner shells of the atom, accompanied by the loss of an electron from the atom, caused by the absorption of energy by the atom. This may be followed by the further emission of electrons and X rays before the atom returns to a normal state. [named after Pierre V. *Auger,* 20th-century French physicist]

Auger electron, *Physics.* an electron emitted by an atom in the Auger effect.

augite (ô'jīt), *n. Mineralogy.* a species of monoclinic pyroxene having a greenish, brownish, or black color, occurring mostly in igneous rocks. *Formula:* $Ca(Mg,Fe)(Si,Al)_2O_6$ [from Greek *augitēs* a type of turquoise]
—**augitic** (ô jit'ik), *adj.* of, containing, or resembling augite.

---

cap, fāce, fäther; best, bē, tèrm; pin, fīve;
rock, gō, ôrder; oil, out; cup, pùt, rüle,
yü in use, *yu* in uric;
ng in bring; *sh* in rush; *th* in thin; ᴛʜ in then;
zh in seizure.
ə = *a* in about, *e* in taken, *i* in pencil, *o* in lemon, *u* in circus

aura (ôr′ə), *n., pl.* **auras, aurae** (ôr′ē). *Physics.* a field of ionized air or gas caused by the discharge of electricity from a sharp point. [from Greek *aura* breeze, breath]

aural, *adj. Anatomy, Physiology.* of, having to to with, or perceived by the ear. SYN: auditory. [from Latin *auris* ear] —**aurally,** *adv.*

aureole (ôr′ē ōl), *n.* **1** *Astronomy.* = corona.
2 *Geology.* a ring-shaped zone surrounding an igneous intrusion.
[from Late Latin *aureola* golden, from Latin *aurum* gold]

auric (ôr′ik), *adj. Chemistry.* of or containing gold, especially with a valence of three. Compare **aurous.** [from Latin *aurum* gold]

auricle (ôr′ə kəl), *n.* **1** *Anatomy.* **a** the outer projecting portion of the ear. Also called **pinna. b** = atrium (def. 1).
2 *Botany, Zoology.* an earlike part: *The auricle [is] an ear-shaped appendage occurring on the calyx of Lobelia. This term is also applied to similarly shaped lobes of foliage leaves* (Youngken, *Pharmaceutical Botany*). [from Latin *auricula,* diminutive of *auris* ear]
—**auricular** (ô rik′yə lər), *adj.* **1** having to do with an auricle of the heart: *auricular fibrillation.*
2 of or near the ear; aural.
3 shaped like an ear.

auriculate (ô rik′yə lit) or **auriculated** (ô rik′yə lā tid), *adj. Botany, Zoology.* **1** having ears, auricles, or earlike parts. **2** ear-shaped.

auriculoventricular (ô rik′yə lō ven trik′yə lər), *adj.* = atrioventricular.

auriferous (ô rif′ər əs), *adj. Mineralogy.* containing or yielding gold: *auriferous gravel.* [from Latin *aurifer,* from *aurum* gold + *ferre* carry]

aurora (ô rôr′ə or ô rōr′ə), *n., pl.* **-ras, -rae** (-rē). *Geophysics.* streamers or bands of light appearing in the sky at night, especially in polar regions. The aurora is a luminous atmospheric phenomenon due to the impact of streams of charged particles from the sun on the upper regions of the earth's atmosphere, where the particles are directed by the earth's magnetic field to the magnetic poles. The **aurora australis** (ô strā′lis) is the aurora of the southern sky, commonly called the *southern lights;* the **aurora borealis** (bôr′ē al′is) is the aurora of the northern sky, or the *northern lights.* [from Latin *aurora* dawn] —**auroral,** *adj.*

aurous (ôr′əs), *adj. Chemistry.* of or containing gold, especially with a valence of one. Compare **auric.** [from Latin *aurum* gold]

australite (ôs′trə līt), *n. Mineralogy.* a kind of tektite, a rounded lump of natural glass, found in Australia and elsewhere.

Australopithecine (ôs′trə lō pith′ə sēn), *Zoology.* —*n.* any of a group of extinct hominids found chiefly in southern Africa, having manlike teeth and a cranial capacity somewhat larger than that of the chimpanzee.
—*adj.* of or belonging to this group.
[from New Latin *Australopithecus* southern ape]

aut-, *combining form.* the form of **auto-** before vowels, as in *autecology, autoxidation.*

autecious (ô tē′shəs), *adj.* = autoecious.

autecism, *n.* = autoecism.

autecology (ô′tə kol′ə jē), *n.* the branch of ecology dealing with the biological relation between a single species, or an individual organism, and its environment.

auto-, *combining form.* self; oneself, as in *autoantibody, autointoxication.* Also spelled **aut-** before vowels. [from Greek *autos* self]

autoantibody, *n., pl.* **-bodies.** *Immunology.* an antibody that attacks the body's own cells and tissues: *an allergy to kidney tissue brought on through infection with the streptococcus ... changes kidney tissue so that it becomes antigenic and this leads to the formation of autoantibodies* (L.H. Criep).

autocatalysis, *n. Chemistry.* catalysis of a reaction by one of its own products.
—**autocatalytic,** *adj.* of or characterized by autocatalysis. —**autocatalytically,** *adv.*

autochthonous (ô tok′thə nəs), *adj.* originating or formed in the place where found: *an autochthonous blod clot, autochthonous coal.* SYN: native, indigenous. [from Greek *autochthōn*]

autoecious (ô tē′shəs), *adj. Biology.* going through all stages of the life cycle on the same host. Some parasitic fungi are autoecious. Also spelled **autecious.** Compare **heteroecious.** [from Greek *autos* same + *oikos* house]

autoecism (ô tē′siz əm), *n.* the condition of being autoecious. Also spelled **autecism.**
—**autogamic** (ô′tə gam′ik), *adj.* characterized by or adapted to autogamy.

autogamy (ô tog′ə mē), *n.* **1** *Botany.* self-fertilization; fecundation of the ovules of a flower by its own pollen.
2 *Biology.* the conjugation or fusion of closely related pairs of cells or nuclei.
—**autogamic** (ô′tə gam′ik), *adj.* characterized by or adapted to autogamy.
[from *auto-* + *-gamy*]

autogenesis (ô′tō jen′ə sis) or **autogeny** (ô toj′ə nē), *n.* = spontaneous generation. —**autogenetic,** *adj.* —**autogenetically,** *adv.*

autogenous (ô toj′ə nəs) or **autogenic** (ô tə jen′ik), *adj. Biology, Medicine.* **1** produced by or within oneself; self-generated: *an autogenous toxin, an autogenic tumor.* SYN: endogenous.
2 derived from the same individual's body: *an autogenous graft.* An **autogenous vaccine** is one prepared from bacteria present in the patient's own body.

autograft, *n. Immunology.* a tissue or organ grafted from one part of an individual's body onto another. ASSOCIATED TERMS: see **allograft.**

autoicous (ô toi′kəs), *adj. Botany.* having both male and female inflorescence on the same plant, as certain mosses. [from *aut-* + Greek *oikos* house]

autoimmune, *adj. Immunology.* caused by autoantibodies: *an autoimmune disease.*
—**autoimmunity,** *n.* the condition in which antibodies produced by an organism attack the organism's own cells and tissues.

autoinfection, *n. Biology, Medicine.* infection from within the organism; self-infection.

autointoxication, *n. Biology, Medicine.* a poisoning by or resulting from toxin formed within the body.

autologous (ô'tol'ə gəs), *adj. Immunology.* transplanted from the same individual's body: *an autologous graft.* See also **autograft.** [from *auto-* + Greek *logos* relation]

autolysate (ô tol'i sāt), *n. Biology.* a product of autolysis.

autolysin (ô tol'i sin), *n. Biochemistry.* a substance, such as an enzyme, capable of breaking down the cells or tissues of an organism within which it is produced, especially after death or in some diseased conditions.

autolysis (ô tol'ə sis), *n. Biology.* disintegration of the tissue of a plant or animal by the action of enzymes present within its own cells. Contrasted with **heterolysis.** [from *auto-* + *lysis*]

automatic, *adj. Physiology.* involuntary; reflex: *Breathing and swallowing are usually automatic.* SYN: autonomic, unconscious.

automatism (ô tom'ə tiz əm), *n. Physiology.* involuntary or reflex action.

automorphic (ô'tə môr'fik), *adj. Mathematics.* having a single value; unchanged by substitutions: *an automorphic function.*
—**automorphism,** *n. Mathematics.* a one-to-one correspondence between elements of a single set or system; isomorphism of a set with itself.

autonomic (ô'tə nom'ik), *adj.* **1** *Physiology.* **a** having to do with the autonomic nervous system: *autonomic neurons.* **b** involuntary; automatic: *the autonomic reflex system of the body.*
2 *Botany.* caused by internal stimuli; spontaneous. [from Greek *autonomos* independent] —**autonomically,** *adv.*

autonomic nervous system, *Physiology.* the part of the peripheral nervous system of vertebrates which controls involuntary actions, such as the heartbeat, gland secretions, and digestion. The system consists of ganglia and nerves that make up the pathway to the heart, stomach, intestine, and other internal organs. The autonomic nervous system is divided into the sympathetic and parasympathetic nervous systems. *Abbreviation:* ANS

auto-oxidation, *n.* = autoxidation.

autophagy (ô tof'ə jē), *n. Biology.* the process of breaking down parts of a cell by the cell's own lysosomes.
—**autophagic** (ô'tə faj'ik), *adj.* of or characterized by autophagy.

autophyte (ô'tə fīt), *n. Botany.* any plant which can manufacture its own food from inorganic substances; an autotrophic plant. [from *auto-* + Greek *phyton* plant]

autopolyploid (ô'tə pol'ə ploid), *n. Genetics.* a polyploid whose chromosomes all come from the same species and are identical: *Many garden flowers are autopolyploids.*

autoradiograph or **autoradiogram,** *n.* an image produced by a radioactive substance on a photographic film or plate in an electron-sensitive emulsion, used to detect and measure the distribution of radioactive elements in a section of tissue, mineral, etc.: *Autoradiographs of human cells made after they have been supplied with radioactive nucleic acids show that different seg-* ments of the chromosomes replicate at different times (Scientific American). —**autoradiographic,** *adj.* —**autoradiography,** *n.*

autosome (ô'tə sōm'), *n. Genetics.* any chromosome other than a sex-determining chromosome; a chromosome that is neither an X nor a Y chromosome: *All the chromosomes in the human complement other than the sex chromosomes are called the autosomes. The human therefore has 22 pairs of autosomes plus one pair of sex chromosomes* (Norman Rothwell). [from *auto-* self + *-some* body] —**autosomal,** *adj.*

autotomy (ô tot'ə mē), *n. Zoology.* a process or mechanism by which animals are able to cast off a limb or other part of the body, especially when disturbed or seeking escape. [from *auto-* + *-tomy*]

autotroph (ô'tə trōf), *n. Biology.* an organism that can manufacture its own food from inorganic substances, getting its energy either from photosynthesis or from chemosynthesis. Green plants, algae, and certain bacteria are autotrophs. Contrasted with **heterotroph.** See also **auxotroph.**
[from *auto-* + Greek *trophē* nourishment]
—**autotrophic** (ô'tə trof'ik), *adj.* **a** of or having to do with an autotroph. **b** capable of growing and multiplying in an inorganic medium. **c** independent of an outside source of vitamins, amino acids, or other complex organic molecules. Contrasted with **heterotrophic.**

autoxidation, *n. Chemistry.* **1** oxidation by direct combination with oxygen, as by exposure to air. **2** an oxidation reaction that takes place only when an additional substance induces the reaction. Also spelled **auto-oxidation.**

autumnal equinox, *Astronomy.* the equinox that occurs in the Northern Hemisphere about September 23: *The autumnal equinox marks the official beginning of autumn and the beginning of the seasons when the night is longer than the day. After still another three months* (*December 22*) *the sun will be at its farthest south; this is the winter solstice* (Krogdahl, *The Astronomical Universe*). Compare **vernal equinox.**

autunite (ô'tə nīt), *n. Mineralogy.* a yellow, radioactive mineral of a hydrated phosphate of uranium and calcium occurring in tabular crystals with a very nearly square outline. *Formula:* $Ca(UO_2)_2(PO_4)_2 \cdot 10-12H_2O$ [from *Autun,* city in France + *-ite*]

auxesis (ôk sē'sis), *n. Biology.* the growth of cells by expansion instead of division. [from Greek *auxēsis* growth, increase]

auxin (ôk'sən), *n. Botany.* any of a group of hormones synthesized in the protoplasm of the young, active parts of plants, which regulate plant growth and development: *Auxins influence division, elongation, and differentiation of plant cells. Unequal distribution of auxins causes unequal growth responses called tro-*

---

cap, fāce, fäther; best, bē, tèrm; pin, five;
rock, gō, ôrder; oil, out; cup, pùt, rüle,
yü in use, *yu̇* in uric;
ng in bring; *sh* in rush; *th* in thin; ŦH in then;
zh in seizure.
ə = *a* in about, *e* in taken, *i* in pencil, *o* in lemon, *u* in circus

pisms (Biology Regents Syllabus). [from Greek *auxein* to increase]
ASSOCIATED TERMS: see **cytokinin.**

auxochrome (ôk′sə krōm), *n. Chemistry.* any group of atoms capable of making a chromogen into a dye or pigment: *Chemical side groups that donate or accept electrons can be attached to the conjugated system to enhance the color; they are called auxochromes* (Kurt Nassau). [from Greek *auxē* an increase + *chrōma* color]

auxospore, *n. Botany.* a reproductive cell (zygote) of a diatom: *Diatoms exhibit two modes of reproduction: fission and formation of an auxospore.* [from Greek *auxē* an increase + English *spore*]

auxotroph (ôk′sə trôf), *n. Biology.* an organism, as some strains of bacteria, fungi, or algae, that has lost through mutation the capacity to synthesize one or more substances required for its nutrition. [from Greek *auxē* an increase + *trophē* nourishment]
—**auxotrophic** (ôk′sə trof′ik), *adj.* of or having to do with an auxotroph.

av., *abbrev.* **1** average. **2** avoirdupois.

avalanche (av′ə lanch), *n.* **1** *Geology.* a large mass of snow and ice, or of dirt and rocks, loosened from a mountainside and descending swiftly into the valley below.
2 *Physics.* a burst or shower of charged particles produced by the collision of an accelerated electron or other charged particle with gas molecules. The collision releases new electrons which in turn have more collisions in a self-perpetuating process. Also called **cascade.**

aventurine (ə ven′chər in), *n. Mineralogy.* a variety of quartz containing bright specks of mica or some other mineral. [from *aventurine,* type of glass which it resembles, ultimately from Italian *avventura* chance (so called for its having been discovered by chance)]

average, *Mathematics, Statistics.* —*n.* the quantity found by dividing the sum of several quantities by the number of quantities: *The average of 3, 5, and 10 is 6* $(3 + 5 + 10 = 18; 18 \div 3 = 6)$. *Abbreviation:* av. Also called **arithmetic mean.**
—*adj.* obtained by averaging; being an average: *the average wingspan, the average temperature.*
—*v.* to find the average of: *The weighted arithmetic mean ... is most commonly used in various index number formulas ... and in averaging percentages which represent unequal groups* (Parl, *Basic Statistics*).
ASSOCIATED TERMS: *Average, mean, median,* and *mode* are used in mathematics and statistics to designate an intermediate value or number. *Average* means specifically the sum of a number of quantities divided by the number, while *mean* refers more broadly to any quantity having a value halfway between the values of other quantities, as in *harmonic mean* (the middle term in a harmonic progression) and *geometric mean* (the middle term in a geometric progression). The *arithmetic mean,* or average value of a group of variable quantities, is distinguished from the *median,* which is the middle number of a series arranged according to size (as in the series 1, 2, 3, 4, 5, the median is 3). *Mode,* in turn, is the number or value that occurs most frequently in a set of numbers or a series of data (as in the series 1, 1, 2, 2, 2, 2, 3, 3, 3, 3, 4, 4, 4, the mode is 3). See also **norm, proportion, ratio.**

avidin (av′id in), *n. Biochemistry.* an albumin in raw egg white which makes biotin in humans or animals inac-

tive, causing symptoms of vitamin deficiency. [from *avid(ity)* + (*biot*)*in*]

avidity, *n. Chemistry.* **1** the relative strength of an acid or base, as determined by its degree of dissociation. **2** chemical affinity, as between an antibody and a virus.

A-V node, = atrioventricular node.

Avogadro's law or **Avogadro's hypothesis** (ä′vō-gä′droz), *Chemistry, Physics.* the statement that equal volumes of different gases, under like conditions of pressure and temperature, contain the same number of molecules. This number (called **Avogadro number** or **Avogadro's number**) is 6.023×10^{23}, indicating the number of molecules in one mole (gram molecule) of any substance. [named after Count Amedeo Avogadro, 1776-1856, Italian physicist, who stated the law]

avoirdupois (av′ər də poiz′), *n.,* or **avoirdupois weight,** a system of weights based on a pound of 16 ounces, traditionally used in English-speaking countries to weigh everything except gems, precious metals, and drugs. In this system:
27.34375 grains = 1 dram
16 drams = 1 ounce
16 ounces = 1 pound
2000 pounds = 1 ton
Abbreviation: av. [from Middle French *avoir de pois* goods of weight]

awn, *n. Botany.* one of the bristly hairs forming the beard on a head of barley, oats, or other grass.
—**awnless,** *adj.*

axial (ak′sē əl), *adj.* **1** of or forming an axis: *Wheels turn on an axial rod.*
2 on or around an axis. The **axial skeleton** is the skeleton of the head and trunk, as distinguished from the *appendicular skeleton.*
3 *Chemistry.* (of an atom in a cyclohexane ring) attached to a carbon atom in the ring with a bond that is perpendicular to the mean plane of the ring.
—**axially,** *adv.*

axial vector, = cross product.

axil (ak′səl), *n. Botany.* the angle between the upper side of a leaf or stem and the supporting stem or branch. A bud is usually found in the axil. [from Latin *axilla* armpit]

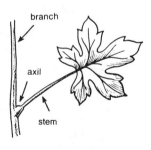

axilla (ak sil′ə), *n., pl.* **axillae** (ak sil′ē). **1** *Anatomy.* the hollow place between the arm and the wall of the thorax; armpit.
2 *Botany.* = axil.
—**axillary** (ak′sə ler′ē), *adj.* **1** of or near the axilla.
2 in or growing from an axil.

axillar (ak′sə lər), *adj.* axillary.
—*n. Zoology.* Usually, **axillars,** *pl.* a long, stiff feather growing from the axilla of a bird.

axinite (ak′sə nīt), *n. Mineralogy.* a mineral consisting essentially of an aluminum and calcium borosilicate, commonly occurring in flattened, brown crystals edged like an ax. *Formula:* $(Ca,Mn,Fe)_3Al_2$-$BSi_4O_{15}(OH)$ [from Greek *axinē* ax + English *-ite*]

axiom, *n. Mathematics.* a statement or proposition accepted as true without proof. Axioms are the basic assumptions from which all other propositions can be derived. *Examples of axioms:* A figure may be moved about in space without changing its shape or size. The positive integers 1, 2, 3, 4, etc. form an infinite sequence. The real numbers and the points of a straight line can be put into exact correspondence. —**axiomatic,** *adj.*
▶ In modern mathematics **axiom** and **postulate** are synonyms. Formerly, however, especially in Euclidean geometry, the two terms were used somewhat differently: *axioms* (which Euclid called "common notions") referred to logical properties involving equality and unequality, as in the Euclidean axioms "Things equal to the same things are equal to each other," and "If equals are added to equals, the sums are equal"; *postulates* dealt with geometric figures such as lines and circles and their relations, as in the Euclidean postulates "A straight line may be drawn between any two points," and "About any point as center a circle with any radius may be described." Unlike an axiom or postulate, a **theorem** is a proposition that has been or is to be proved, often on the basis of certain axioms or previously proved theorems.
[from Greek *axiōma,* from *axios* worthy]

axiom of choice, *Mathematics.* an axiom stating that, given a collection of sets that do not overlap, it is possible to form a new set containing one element from each of the sets: *If α consists of two sets, the set of all triangles and the set of all squares, then α clearly satisfies the axiom of choice. We merely choose some particular triangle and some particular square and then let these two elements constitute Z* (Scientific American).

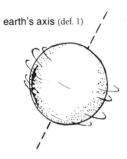

earth's axis (def. 1)

axis (ak′sis), *n., pl.* **axes** (ak′sēz′).
1 *Physics.* an imaginary or real line that passes through an object and about which an object turns or seems to turn. The earth's axis of rotation passes through the North and the South Poles.

2 *Geometry.* **a** a line about which a geometrical figure is symmetrical. Also called **axis of symmetry. b** a line passing through the vertex of a cone and the center of its base. **c** a line passing through the center of each base of a cylinder.
3 *Mathematics.* a fixed line along which distances are measured or to which positions are referred as in a graph or coordinate system. In a coordinate system, the horizontal and vertical axes meet in a common point.
4 *Anatomy.* **a** a central or principal structure extending lengthwise and having the parts of the body arranged around it: *The axis of the skeleton is the spinal column.* **b** the second cervical vertebra, on which the first vertebra (the atlas) turns.
5 *Botany.* **a** the central part or support on which parts are arranged: *The axis of a plant is the stem.* **b** the main stem and root.
6 *Crystallography.* one of the three or four imaginary lines assumed in defining the position of the planar faces of a crystal and determining the crystal system.
7 *Optics.* a straight line drawn through the optical center of a lens, and perpendicular to both its surfaces.
8 *Geology.* a line drawn along the points of maximum curvature of a fold.
[from Latin *axis* axle, axis]

axis cylinder, *Anatomy.* the central portion of a nerve fiber, the essential conducting element and the continuation of the axon of the nerve cell.

axisymmetric, *adj. Geometry.* symmetric about an axis: *an axisymmetric conical figure.*

axon (ak′son), *n. Anatomy.* the long extension of a nerve cell that carries impulses away from the body of the cell: *Each neuron is an independent living unit ... it discharges impulses to other cells via a single slender fiber, the axon, which branches profusely to make contact with numerous receiving cells* (Scientific American). See the pictures at **myelin, neuron.** [from Greek *axōn* axis]
ASSOCIATED TERMS: see **dendrite.**
—**axonal** (ak′sə nəl) or **axonic** (ak son′ik), *adj.* of or having to do with an axon or axons.

axostyle, *n. Zoology.* a slender skeletal rod in certain protozoans. [from Greek *axōn* axis + *stylos* pillar]

azeotrope (ā′zē ə trōp′ *or* ā zē′ə trōp′), *n. Chemistry.* a mixture of two or more substances which, at a certain proportion or pressure, boils at a constant temperature and, in distillation or partial evaporation, retains the same composition in the vapor state as in the liquid. [from *ā-* without + Greek *zēin* to boil + *tropos* manner]
—**azeotropic,** *adj.* of or having to do with an azeotrope. An **azeotropic mixture** exhibits either a minimum or a maximum boiling point.

cap, fāce, fäther; best, bē, tėrm; pin, fīve;
rock, gō, ôrder; oil, out; cup, pùt, rüle;
yü in use, *yù* in uric;
ng in bring; *sh* in rush; *th* in thin; ‏ŦH‎ in then;
zh in seizure.
ə = *a* in about, *e* in taken, *i* in pencil, *o* in lemon, *u* in circus

azimuth (az′ə məth), *n. Astronomy.* an arc measured clockwise from the north point of the horizon to the intersection of a vertical circle passing through a heavenly body with the horizon: *The azimuth of a star is the angular distance measured from the north point toward the east along the horizon to the vertical circle of the star, or it is the corresponding angle at the zenith.* (Baker, *Astronomy*). [from Arabic *as-sumūt* the ways] —**azimuthal** (az′ə muth′əl *or* az′ə myü′thəl), *adj.*

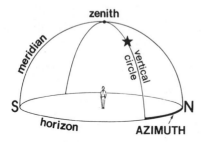

A. The azimuth of the star is 45°.

azimuth circle, = vertical circle.

azine or **azin** (az′ēn *or* az′in), *n. Chemistry.* an organic compound of a group of compounds having a ring of six atoms, at least one of which is nitrogen. The number of nitrogen atoms is indicated by a prefix, as in *diazine, triazine.* [from *az(ote)* + *-ine*]

azo (az′ō *or* ā′zō), *adj. Chemistry.* of or having to do with a compound containing the bivalent nitrogen radical -N:N-. The **azo dyes** are a large group of synthetic dyes produced from amino compounds, which can form more complex structures having a variety of dyeing properties. [from *azote*]

azole (az′ōl *or* ə zōl′), *n. Chemistry.* an organic compound of a group of compounds having a ring of five atoms, at least one of which is nitrogen. The number of nitrogen atoms is indicated by a prefix, as in *diazole, triazole.* [from *az(ote)* + *-ole*]

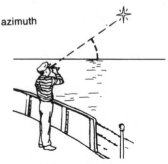

azimuth

B. Measuring the azimuth of a point on the horizon using the north star as a reference point.

azotobacter (ə zō′tə bak′tər), *n. Bacteriology.* any of a genus (*Azotobacter*) of aerobic bacteria, that fix atmospheric nitrogen for their own use.

azurite (azh′ə rīt′), *n. Mineralogy.* a blue copper ore, a hydrous carbonate of copper. *Formula:* $Cu_3(CO_3)_2(OH)_2$

azygous (az′ə gəs *or* ā′zī′gəs), *adj. Anatomy.* not one of a pair; being a single muscle, vein, etc. [from Greek *a-* without + *zygon* yoke]

B

b, *symbol.* barn.

B, *symbol.* **1** boron. **2** one of the four major blood groups (used to identify blood for compatibility in transfusion).► See the note under **ABO**. **3** flux density.

B., *abbrev. Bacillus,* a genus of bacteria.

Ba, *symbol.* barium.

babbitt (bab′it), *n.,* or **babbitt metal,** *Chemistry.* an alloy of tin, antimony, and copper, or a similar alloy, used in bearings to lessen friction. Babbitt is whitish in color and usually feels slippery when it is rubbed. [named after Isaac *Babbitt,* 1799–1862, American inventor]

baccate (bak′āt), *adj. Botany.* **1** berrylike; pulpy: *baccate fruit.* **2** bearing berries; bacciferous. [from Latin *bacca* berry]

bacciferous (bak sif′ər əs), *adj. Botany.* bearing or producing berries.

baccivorous (bak siv′ər əs), *adj. Zoology.* feeding chiefly on berries, as some birds do. [from Latin *bacca* berry + *vorare* devour]

bachelor, *n. Zoology.* any male unmated during the breeding season, such as a young male fur seal kept away from the breeding grounds by the older males.

bacillar (bə sil′ər *or* bas′ə lər), *adj.* = bacillary.

bacillary (bas′ə ler′ē *or* bə sil′ə rē), *adj. Bacteriology.* of, like, or caused by bacilli. **Bacillary dysentery** is a type of dysentery caused by certain rod-shaped aerobic bacteria.

bacillus (bə sil′əs), *n., pl.* **-cilli** (-sil′ī). *Bacteriology.* any of the rod-shaped bacteria, especially one that is a member of a genus (*Bacillus*) of aerobic bacteria that forms endospores, such as *Bacillus anthracis,* the causative agent of anthrax: *the typhoid bacillus.* See the picture at **bacteria.** [from New Latin, diminutive of Latin *bacullus* rod]

ASSOCIATED TERMS: see **coccus.**

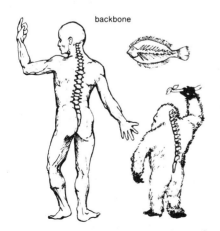

backbone

backbone, *n. Anatomy.* the main series of bones (verte-

brae) along the middle of the back in man and other vertebrates; the spinal column. SYN: spine.

backcross, *Genetics.* —*v.* to cross (a hybrid of the first generation) with either one of its parents or with an individual which is a parental type: *Using seed from pure lines of tomato plants, Drs. Burdick and Bertens irradiated it with X-rays. Plants grown from this seed were backcrossed to control plants of the same pure line. Thus, any apparent differences between seedlings from irradiated and unirradiated plants were known to be radiation-produced* (Science News Letter).

—*n.* such a cross: *He also demonstrated the efficacy of the backcross in determining the genetic constitution of an individual of doubtful ancestry* (Harbaugh and Goodrich, *Fundamentals of Biology*).

background radiation, 1 *Physics.* the low-level natural radiation from cosmic rays and trace amounts of radioactive substances present in the atmosphere: *Background radiation ... produces an annual dose of about 0.1 roentgen to a person on the earth's surface* (Collier's Year Book).

2 *Astronomy.* low-level microwave radiation coming equally from all directions in space, thought to be residual heat from the "big bang" in which the universe was born. It was discovered by Arno Penzias and Robert Wilson in 1965.

back-mutate, *v. Genetics.* to mutate back to the original form or wild type: *Mutant viruses occasionally back-mutate.* —**back-mutation,** *n.*

backscatter, *n.* **1** *Physics.* the deflection of radiation or of particles at angles greater than 90 degrees: *a neutron source ... which produces a high density neutron flux, from which the backscatter of gamma rays permits identification of many elements* (New Scientist).

2 *Geophysics.* radiation of extraneous signals by reflection from the ionosphere: *There was a certain amount of back-scatter when a low-frequency radio signal is bounced around the world. While most of the signals propagates forward, a fraction is scattered back to the transmitter as it is deflected off the ionosphere and the earth* (Newsweek).

—*v. Physics.* to deflect at an angle greater than 90 degrees.

bacteri- or **bacterio-,** *combining form.* bacteria, as in *bacterial, bacteriology.* [from New Latin *bacterium*]

bacteria (bak tir′ē ə), *n.pl.* of **bacterium.** *Bacteriology.* a large group of microscopic organisms that multiply by fission or by forming spores. Bacteria are typically

cap, fāce, fäther; best, bē, tėrm; pin, fīve;
rock, gō, ôrder; oil, out; cup, pùt, rüle,
yü in use, *yù* in uric;
ng in bring; *sh* in rush; *th* in thin, ᴛH in then;
zh in seizure.
ə = *a* in about, *e* in taken, *i* in pencil, *o* in lemon, *u* in circus

57

bactericidin

spherical, spiral, rod-shaped, or comma-shaped, and most kinds have no chlorophyll and no distinct, membrane-bound nucleus. Certain species cause diseases such as pneumonia, typhoid fever, etc.; others are concerned in such processes as fermentation and nitrogen fixation. *Formally, the distinction between viruses and bacteria is on the basis of size: they can pass through filters, through which bacteria cannot pass. But associated with this is the further distinction that, whereas bacteria can be cultured on cell-free material in the laboratory, the multiplication of virus particles is dependent on the host-organism* (A.W. Haslett). Also called **schizomycetes.**
[from New Latin, plural of *bacterium,* from Greek *baktērion,* diminutive of *baktron* rod]
ASSOCIATED TERMS: *Microorganism* is the general term for a microscopic organism. *Microbes,* or, more commonly, *germs,* usually refer to microorganisms that cause disease. *Bacteria* are one-celled microorganisms that have no distinct, membrane-bound nucleus (*prokaryotic*); some bacteria are useful to humans, and some harmful. *Bacilli* are a specific group of rod-shaped bacteria. *Protozoans* are one-celled animal-like microorganisms with a distinct, membrane-bound nucleus (*eukaryotic*). *Viruses* are infectious particles smaller than bacteria and protozoans. See also **coccus.**

bacteria:

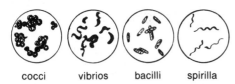

cocci   vibrios   bacilli   spirilla

—**bacterial,** *adj.* **1** of, like, or caused by bacteria: *bacterial diseases.* **2** existing in or produced by bacteria: *a bacterial enzyme.* —**bacterially,** *adv.*
bactericidin (bak tir′ə sī′dən), *n. Immunology.* an antibody that kills bacteria in the presence of a complement.
bacteriochlorophyll (bak tir′ē ō klôr′ə fil), *n. Biochemistry.* a photosynthetic pigment similar to chlorophyll, found in various bacteria: *Bacteriochlorophyll ... may have been the evolutionary precursor of chlorophyll a, the dominant photosynthetic pigment of algae and higher plants* (Greulach and Adams, *Plants*).
bacteriology, *n.* the branch of biology that deals with bacteria: *In its widest sense the physiology of bacteria embraces every aspect of bacteriology ... the growth, reproduction, metabolism and pathogenicity of bacteria are all included* (Dible, *Recent Advances in Bacteriology*). Compare **microbiology.**
► Bacteriology developed in the late 1800's with the work of Ferdinand J. Cohn, 1828–1898, who defined and classified bacteria, Louis Pasteur, 1822–1895, who discovered that diseases such as anthrax are spread by bacteria, and Robert Koch, 1843–1910, who developed the techniques of culturing bacteria.
—**bacteriological** or **bacteriologic,** *adj.* of, having to do with, or based on bacteriology: *bacteriological classification of diseases.* —**bacteriologically,** *adv.*
bacteriolysin (bak tir′ē ol′ə sin), *n. Bacteriology.* any substance, especially an antibody, that causes bacteriolysis.

bacteriolysis (bak tir′ē ol′ə sis), *n. Bacteriology.* the destruction or dissolution of bacteria.
—**bacteriolytic** (bak tir′ē ə lit′ik), *adj.* **1** destructive to bacteria. **2** of or having to do with bacteriolysis.
bacteriophage (bak tir′ē ə fāj), *n. Biology.* a virus that destroys various bacteria, normally present in the intestines, blood, etc.: *The bacteriophage particle or bacterial virus without its host cell is quite dormant and inert ... Once the particle has access to a growing bacterial host cell, however, it is transformed into one of the most active biological agents known* (J.E. Hotchin). Also called **phage.**
—**bacteriophagic** (bak tir′ē ə faj′ik), *adj.* destructive to bacteria: *bacteriophagic viruses.*
—**bacteriophagy** (bak tir′ē of′ə jē), *n.* the destruction of bacteria by a bacteriophage.
► Bacteriophages were discovered in 1917 by Félix Hubert d'Herelle, a Canadian bacteriologist, who gave them this name because of their ability to consume bacteria, from *bacterio-* + Greek *phagein* to eat.
bacteriostasis (bak tir′ē ə stā′sis), *n. Bacteriology.* arrest of growth or development of bacteria without killing them.
bacteriostat (bak tir′ē ə stat), *n. Bacteriology.* any agent that arrests growth or development of bacteria.
—**bacteriostatic** (bak tir′ē ə stat′ik), *adj.* arresting growth or development of bacteria.
bacterium (bak tir′ē əm), *n.* singular of bacteria.
bacterize, *v. Bacteriology.* to treat or change by bacterial action. —**bacterization,** *n.*
bacteroid (bak′tə roid), *n. Bacteriology.* a bacterium having an irregular or modified form, especially one found in the root nodules of nitrogen-fixing plants.
baddeleyite (bad′ə lē it), *n. Mineralogy.* natural zirconium dioxide, an uncommon ore of zirconium. *Formula:* ZrO_2 [from J. *Baddeley,* 19th-century English mineralogist + *-ite*]
bag, *n. Zoology.* **1** a sac in an animal's body: *the honey bag of a bee.* **2** an udder.
bag of waters, *Anatomy.* the membranous sac containing the amniotic fluid in which the fetus is immersed during pregnancy. Compare **amnion.**
Baily's beads, *Astronomy.* the last rays of direct sunlight observable before a total eclipse of the sun, broken up by irregularities of the moon's limb, briefly resembling a crescent-shaped string of beads. [named after Francis *Baily,* 1774–1844, English astronomer who first described the phenomenon]
bajada or **bahada** (bə hä′də), *n. Geography, Geology.* the low, graded slope between the base of mountains in arid environments and the shallow depression into which occasional runoff water settles. [from Spanish *bajada* descent, slope]
baking soda, = sodium bicarbonate.
balance of nature, *Ecology.* a condition of equilibrium existing in an ecosystem due to such factors as disease, natural enmity, food supply, etc., which keep the number of organisms relatively constant: *Civilized man, ... with his technology and agriculture, interrupts the natural cycles and disturbs the balance of nature to an extent which gives cause for concern* (Mackean, *Introduction to Biology*).
bald, *adj. Botany, Zoology.* **1** lacking the usual or natural covering, as certain animals or plants. *Bald wheat* is without a beard or awn. SYN: bare. **2** having a white

spot or blaze on the head, as some birds or mammals do.

baleen (bə lēn′), *n. Zoology.* horny plates growing downward from the upper jaws or palates of certain whales. Also called **whalebone.** [from Latin *ballena, ballaena* whale]

ball-and-socket joint, *Anatomy.* a flexible joint formed by a ball or knob fitting in a socket, such as the shoulder and hip joints, permitting rotary motion. Also called **enarthrosis.** See the picture at **socket.**

ballistic (bə lis′tik), *adj. Mechanics.* having to do with the motion or throwing of projectiles. A **ballistic trajectory** is the curved path or trajectory of a projectile, such as a ballistic missile, after the thrust or propelling force has ended. [from Latin *ballista* machine for throwing stones, from Greek *ballein* to throw]

ball lightning, *Meteorology.* a rare type of lightning, occurring as a bright reddish ball that may move rapidly along a solid object or remain floating in midair.

Balmer series (bäl′mər), *Physics, Astronomy.* a series of distinct lines in the visible spectrum of hydrogen: *These lines of the Balmer series ... are identified in stellar spectra and also in the spectrum of the sun's chromosphere* (Baker, *Astronomy*). [named after Johann *Balmer,* 1825–1898, Swiss physicist]

Banach space (ban′ək *or* ban′əн), *Mathematics.* a vector space in which the scalar multipliers are real numbers or complex numbers. [named after Stefan *Banach,* 1892–1945, Polish mathematician]

band, *n.* **1** *Electronics.* a particular range of wavelengths or frequencies, as in radio broadcasting. Bands in the radio spectrum range from extremely low frequency (below 300 hertz) to extremely high frequency (from 300 to 3000 gigahertz).
2 *Physics.* a particular range to which the energies of electrons are restricted.
3 *Geology.* a layer of rock differing in color or texture from adjacent layers.

band spectrum, *Physics.* a spectrum consisting of broad bands of molecular origin, each having a sharp edge at its low-frequency end. Compare **line spectrum, continuous spectrum.**

bank, *n. Geography, Geology.* **1** the rising ground bordering a river, lake, or other body of water. **2** a shallow place in a body of water: *the fishing banks of Newfoundland.* **3** any steep slope, especially one forming the side of a ravine or hill.

bar[1], *n. Geology.* any of various elongate coastal offshore ridges; a ridgelike accumulation of sand or gravel forming in a stream. [from Old French *barre* barrier, from Vulgar Latin *barra*]

bar[2], *n.* the CGS unit of pressure, equal to one million (10^6) dynes per square centimeter or 10^5 newtons per square meter. See also **millibar.** [from Greek *baros* weight]

bar., *abbrev.* barometer; barometric.

barb, *n.* **1** *Zoology.* **a** one of the hairlike branches on the shaft of a feather. See the picture at **feather. b** = barbel. **2** *Botany.* a hair or bristle terminating in a hook.

barbel (bär′bəl), *n. Zoology.* a long, thin, fleshy growth on the mouth or nostrils of some fish. Also called **barb.** [ultimately from Latin *barba* beard]

barbellate (bär′bə lit), *adj. Botany.* having short, stiff hairs, as some composites: *barbellate stems.* [from New Latin *barbella* stiff hair]

barbicel (bär′bə səl), *n. Zoology.* one of the very small filaments interlocking the barbules of a feather. [from New Latin *barbicella,* diminutive of Latin *barba* beard]

barbiturate (bär bich′ə rit), *Chemistry.* **1** a salt or ester of barbituric acid. **2** any of various derivatives of barbituric acid, used as hypnotics and sedatives.

barbituric acid, *Chemistry.* a colorless, crystalline acid, used as the base of various sedatives and hypnotics. *Formula:* $C_4H_4N_2O_3$ [from German *Barbitur (säure)* barbituric (acid)]

barbule (bär′byül), *n. Zoology.* one of a series of small, pointed processes fringing the barbs of a feather. See the picture at **feather.**

barchan (bär kän′), *n. Geography.* a sand dune in the shape of a crescent, with points of the crescent facing downwind. See the picture at **dune.**

bar graph or **bar chart,** a graph or chart representing different quantities by rectangles of different lengths. Compare **circle graph, line graph,** and **pictograph.**

barite (bär′it), *n. Mineralogy.* barium sulfate in its natural form, as a mineral found in many parts of the world. Also called **barytes.**

barium (bär′ē əm), *n. Chemistry.* a soft, silvery-white metallic element, which occurs only in combination with other elements, especially in barite. Barium compounds are used in making pigments, safety matches, etc. *Symbol:* Ba; *atomic number* 56; *atomic weight* 137.34; *melting point* 704°C–850°C; *boiling point* 1140°C–1637°C; *oxidation state* 2. [from New Latin, from Greek *barytēs* weight]

barium sulfate, *Chemistry.* a compound occurring as a mineral, barite, or prepared synthetically. It is used in taking X rays of the stomach and intestines, as a filler in making linoleum, etc., and as a pigment. *Formula:* $BaSO_4$

bark, *n. Botany.* the tough outside covering of the trunk, branches, and roots of trees and certain other plants: *The bark includes all the tissues outside the vascular cambium ... [It] is made up of ... cork, cork cambium, phelloderm (if present), primary cortex, and phloem* (Robbins and Weier, *Botany*).

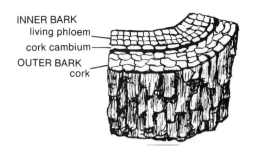

INNER BARK
living phloem
cork cambium
OUTER BARK
cork

---

cap, fāce, fäther; best, bē, tèrm; pin, fīve;
rock, gō, ôrder; oil, out; cup, pùt, rüle;
yü in use, *yu̇* in uric;
ng in bring; *sh* in rush; *th* in thin, ᴛʜ in then;
zh in seizure.
ə = *a* in about, *e* in taken, *i* in pencil, *o* in lemon, *u* in circus

Barkhausen effect (bärk′hou′zən), *Physics.* a series of abrupt changes in the magnetization of iron or other ferromagnetic substances occurring as the magnetizing force is increased or decreased. [named after Heinrich *Barkhausen*, 1881–1956, German physicist]

bar magnet, a permanent magnet in the shape of a bar or rod: *A bar magnet suspended from a string will serve as a simple compass.*

barn, *n. Physics.* a unit of area equal to 10^{-24} square centimeters, used to measure a nuclear cross section. *Symbol:* b

baro-, *combining form.* pressure, especially atmospheric pressure, as in *barometer, baroreceptor.* [from Greek *baros* weight]

baroceptor (bar′ə sep′tər), *n.* = baroreceptor.

baroclinic (bar′ə klin′ik), *adj. Meteorology.* of, having to do with, or characterized by an atmospheric condition in which surfaces of equal pressure do not coincide with surfaces of equal density. Compare **barotropic.** [from *baro-* + Greek *klinein* to slope]
—**baroclinicity** (bar′ə klə nis′ə tē), *n.* a baroclinic condition: *Extratropical cyclones are observed to form in the zone of pronounced baroclinicity* (Arnt Eliassen).

barogram (bar′ə gram), *n.* a record made by a barograph or similar instrument.

barograph (bar′ə graf), *n.* an instrument that automatically records changes in air pressure. A barograph is a recording barometer.
—**barographic,** *adj.* of, having to do with, or provided by a barograph: *barographic records.*

barometer, *n.* an instrument for measuring atmospheric pressure, used in determining height above sea level and in predicting probable changes in the weather. The two most commonly used barometers are the **aneroid barometer** and the **mercury barometer.** The barometer was invented in 1643 by Evangelista Torricelli, an Italian physicist. *Abbreviation:* bar.
—**barometric,** *adj.* of or indicated by a barometer: *barometric pressure.* —**barometrically,** *adv.*
—**barometry** (bə rom′ə trē), *n.* barometric measurement or observation.

baroreceptor (bar′ə ri sep′tər), *n. Physiology.* a nerve cell or group of cells sensitive to changes in pressure, such as blood pressure.

barotropic (bar′ə trop′ik), *adj. Meteorology.* of, having to do with, or characterized by an atmospheric condition in which surfaces of equal pressure coincide with surfaces of equal density. Compare **baroclinic.** [from *baro-* + Greek *tropos* a turning]
—**barotropy** (bə rot′rə pē), *n.* a barotropic condition; condition of zero baroclinicity.

Barr body, *Genetics.* a densely staining condensed X chromosome found in the cell nuclei of female mammals. Only one of the two X chromosomes of females is genetically active; the Barr body is the inactive member of the pair. Also called **sex chromatin.** [named after Murray L. *Barr*, 20th–century Canadian anatomist who discovered this body in 1949]

barrens, *n.pl. Geography, Geology.* an area of level land, mostly unproductive, poorly forested, and generally having sandy soil: *rock barrens, sand barrens.*

barrier island, *Geography, Geology.* a long, narrow island formed by deposition and parallel to the mainland.

barrier reef, *Geography, Geology.* a ridge of coral deposits parallel to the mainland and separated from it by a deep lagoon: *The Great Barrier Reef of Australia parallels the Queensland Coast.*
ASSOCIATED TERMS: A barrier reef is one kind of *coral reef.* Two other kinds are *fringing reefs,* which are confined to the very border of the mainland, and *atolls,* which enclose circular lagoons.

Barycentric Dynamic Time, a time scale used in astronomical studies, that differs from *Terrestrial Dynamic Time* by corrections of milliseconds, to compensate for the dependence of time measurements on the distance between the Earth and Sun. *Abbreviation:* TDB

baryon (bar′ē on), *n. Nuclear Physics.* any of a class of heavy elementary particles that includes the proton, neutron, and hyperon. Baryons are classified according to their statistical behavior as fermions, and according to their interaction as hadrons; hence they are regarded as fermionic hadrons. *The [quark] theory supposes that three quarks bind together to form a baryon* (C.B.A. McCusker). [from Greek *barys* heavy + English *-on*]
—**baryonic,** *adj.* of or having to do with baryons.

baryon number, *Nuclear Physics.* a quantum number equal to the number of baryons minus the number of antibaryons. The baryon number remains the same throughout any reaction.

barysphere (bar′ə sfir′), *n.* = centrosphere. [from Greek *barys* heavy + English *sphere*]

baryta (bə rī′tə), *n. Chemistry.* any of various compounds of barium, especially barium oxide, a whitish powder used in the manufacture of glass and barium salts, and barium hydroxide, an extremely toxic white powder or colorless crystals. [from New Latin, from Greek *barytēs* weight]
—**barytic** (bə rī′tik), *adj.* having to do with, formed of, or containing baryta.

barytes (bə rī′tēz), *n.* = barite.

basal, *adj.* **1** *Physiology.* designating the minimum level at which vital activity of an organism is maintained. **2** *Biology.* located at or growing from the base: *The basal ganglia are a group of structures at the base of the brain.*

basal body or **basal granule,** *Biology.* a minute granule at the bases of cilia and flagella and associated with their formation. Also called **kinetosome.**

basal cell, *Biology.* a variety of cell found in the deepest layer of the epithelium.

basal metabolic rate, *Physiology.* the rate at which an organism at complete rest uses up food and produces heat and other forms of energy, measured by the amount of oxygen it takes in. It is expressed in calories per hour per square meter of skin surface. *Abbreviation:* BMR

basal metabolism, *Physiology.* the amount of energy used by an organism at complete rest, measured by the basal metabolic rate and used as a standard for comparing metabolism under varying conditions.

basalt (bə sôlt′ or bā′sôlt), *n. Geology.* a hard, dark-colored rock of volcanic origin. Upon cooling it may separate into forms resembling a group of columns. *Basalt is the world's most abundant lava ... forming great lava plateaus that cover thousands of* square miles in the northwestern United States, India, and elsewhere. (Gilluly, *Principles of Geology*). [from Late Latin *basaltes*]

—**basaltic** (bə sôl'tik), *adj.* of or like basalt: *basaltic lava.*

basalt glass, = tachylyte.

basaltiform (bə sôl'tə fôrm), *adj. Geology.* having the form of basalt; columnar.

base, *n.* **1** *Chemistry.* any of a class of substances whose water solutions are generally characterized by bitter taste and a slippery feel, with an ability to turn litmus dye blue, and to react with acids to form salts. Specifically: **a** a substance that yields hydroxyl ions when dissolved in water. **b** a substance that can act as a proton acceptor. **c** a substance that can act as an electron donor. Sodium hydroxide and calcium hydroxide are common bases: *The hydroxides of the metallic elements are called bases; those of the alkali metals (potassium and sodium) are known as alkali bases, and those of the alkaline earth metals (barium, strontium, and calcium) are called alkaline earth bases* (Parks and Steinbach, *Systematic College Chemistry*).
▶ See the note under **acid.**
2a *Biology.* the part of an organ of an animal or plant nearest its point of attachment. **b** the point of attachment: *The main part of the body [of a hydra] ... is attached temporarily to some solid object at one end, called the base* (Winchester, *Zoology*).
3 *Molecular Biology.* any of the five nitrogenous (purine or pyrimidine) compounds, adenine, cytosine, guanine, thymine, and uracil, that are present in the nucleic acid of cells and that combine in certain ways to form the mechanism of the genetic code.
4 *Geometry.* a side or face of a geometrical figure. The perpendicular distance from a base to the opposite vertex is an *altitude.*
5 *Mathematics.* **a** the number on which a numeration system is based; the number of different symbols (as 0, 1, 2, 3 ...) used in a numeration system. Ten is the base of the decimal system; two is the base of the binary system. In base five, four is written 4, five is written 10, and six is written 11. **b** a number which is raised to an exponent. **c** the aggregate of the vectors or functions in terms of which other vectors or functions are to be given as sums.

base (def. 4)

base level, *Geography.* the lower limit of erosion under conditions of crustal stability: *Sea level, projected inland beneath the continent as an imaginary surface, is the base level of stream activity* (Strahler and Strahler, *Modern Physical Geography*).

base line, *Geography.* a latitudinal parallel used as a survey reference line by the U.S. Land Office Survey.

basement complex, *Geology.* a series of undifferentiated metamorphic and igneous rock underlying the oldest identifiable rocks in a specific area.

basement membrane, *Anatomy.* a fine, delicate layer underlying an epithelium.

base pair, *Molecular Biology.* the pair of bases adenine and thymine (or uracil), or guanine and cytosine, joined in the formation of a double-helical molecule of DNA or RNA: *The DNA molecule ... in many ways ... resembles a spiral staircase. The two chains correspond to the bannisters; the base pairs, to the steps* (Scientific American).

—**base-pairing,** *n.* the pairing off of adenine with thymine (or uracil), or of guanine with cytosine, in forming a nucleic acid double helix: *No matter what the class of RNA being studied, the ... base-pairing involved is identical with that which holds together the double helix of DNA* (New Scientist).

base unit, = fundamental unit.

basic, *adj.* **1** *Chemistry.* **a** being, having the properties of, or containing a base: *a basic solution.* ANT: acidic. **b** having the base in excess. **Basic slag** is slag in which the ratio of bases to acids is high. **c** = alkaline.
2 *Geology.* containing less than 52 per cent of silica: *basic rock.*

—**basicity** (bə sis'ə tē), *n. Chemistry.* **1** the quality or condition of being a base. **2** the power of an acid to combine with bases.

BASIC or **Basic** (bā'sik), *n.* a computer programming language using simple English and algebraic terms. In BASIC, numbers are written as decimals or decimal approximations; for example, the equivalent of 1/4 in BASIC is .25, and of −4 1/11, −4.09091. [from the initials of *B*eginner's *A*ll-purpose *S*ymbolic *I*nstruction *C*ode]

basic oxide, *Chemistry.* an oxide of a metal which forms a hydroxide when combined with water and neutralizes acid substances.

basidiomycete (bə sid'ē ō mī'sēt'), *n. Biology.* any of a large class of fungi that includes smuts, rusts, mushrooms, and puffballs, characterized by the formation of spores on a basidium. [from *basidium* + Greek *mykēs, mykētos* fungus]

—**basidiomycetous** (bə sid'ē ō mī sē'təs), *adj.* of or belonging to the basidiomycetes.

basidiospore (bə sid'ē ō spôr), *n. Biology.* one of the spores produced on a basidium.

basidium (bə sid'ē əm), *n., pl.* **-sidia** (-sid'ē ə). *Biology.* a small, club-shaped structure on basidiomycetes that produces spores, usually four, at the tips of minute stalks. [from New Latin, from Greek *basis* base]

—**basidial,** *adj.* of or having a basidium or basidia.

basifixed (bā'sə fikst), *adj. Botany.* attached by the base or lower end.

basilar (bas'ə lər), *adj.* of or at the base, especially of the skull: *the basilar artery.*

basilar membrane, *Anatomy.* a membranous part of the cochlea that forms the fibrous base supporting the organ of Corti.

---

cap, fāce, fäther; best, bē, tèrm; pin, fīve;
rock, gō, ôrder; oil, out; cup, pùt, rüle,
yü in use, *yù* in uric;
ng in bring; *sh* in rush; *th* in thin, ᴛʜ in then;
zh in seizure.
ə = *a* in about, *e* in taken, *i* in pencil, *o* in lemon, *u* in circus

basilic vein

basilic vein, *Anatomy.* a large vein on the inner side of the upper arm.

basin, *n. Geology.* **1** all the land drained by a river and the streams that flow into it: *the Mississippi basin.* **2** a large tract or area in which the rocks dip generally toward a central point. **3** any extensive low-lying area.

basin and range, *Geology.* a landform structure in which mountains bounded by faults are interspersed with broad, relatively flat basins.

basin range, *Geology.* a type of mountain range formed by the faulting and tilting of a block of strata.

basipetal (bā sip′ə təl), *adj. Botany.* developing from the apex downward toward the base during growth. [from Latin *basis* base + *petere* to seek]

basket cell, *Biology.* one of a layer of tiny, star-shaped cells in the cerebellum whose axons develop terminal branches in a basketlike pattern.

basophil (bā′sə fil), *n. Biology.* a cell that stains readily with a basic dye, especially a white blood cell having basophilic granules in the cytoplasm.
—**basophilic,** *adj.* staining readily with a basic dye: *basophilic cells.*

bast, *n. Botany.* **1** the inner layer of the bark of trees that contains cells carrying fluids, especially sugars in solution. Also called **phloem. 2** Also called **bast fiber.** the tough fiber in this inner layer used in making rope, matting, etc.

bastard wing, = alula.

Batesian mimicry (bāt′sē ən), *Biology.* a form of adaptation in which an edible species is protected by resembling an inedible or poisonous species shunned by predators. Also called **protective mimicry.** [named after Henry W. *Bates,* 1825–1892, English naturalist]

batholith (bath′ə lith), *n. Geology.* a great mass of granite or other plutonic, felsic igneous rock intruded below the surface, commonly along the axis of a mountain range, and exposed by erosion: *A zone of ... batholiths runs from Alaska to the southern tip of South America.* (Gilluly, *Principles of Geology*). Compare **laccolith.** [from Greek *bathos* depth + *lithos* stone] —**bath′olith′ic,** *adj.*

bathy-, *combining form.* deep; deep-sea, as in *bathyal, bathymetry.* [from Greek *bathys* deep]

bathyal (bath′ē əl), *adj. Oceanography.* of or having to do with the deeper parts of the ocean, especially the parts from 100–300 meters to 1000–3000 meters, where very little light penetrates: *The bathyal zone in the major ocean corresponds to the depths of the continental slope and continental rise.* (R.W. Fairbridge). Compare **abyssal, hadal, littoral.**
ASSOCIATED TERMS: see ocean.

bathymetry (bə thim′ə trē), *n.* **1** the science of measuring and charting the depths of seas, lakes, and other bodies of water. **2** the facts obtained by such measuring and charting. *the bathymetry of the Arabian Sea.* —**bath′y met′ric,** *adj.*

bathypelagic (bath′ə pə laj′ik), *adj. Oceanography.* of, having to do with, or inhabiting the bathyal parts of the ocean: *bathypelagic fishes.*

batrachian (bə trā′kē ən), *Zoology.* —*adj.* of or having to do with the division of vertebrates consisting of tailless amphibians, including the frogs and toads. —*n.* a tailless amphibian. [from Greek *batrachos* frog]

batture (ba tyür′), *n. Geology.* an alluvial elevation of the bed of a river, especially one that is dry or submerged according to the season: *The trees that grow out on the batture ... are half covered with water* (New Yorker). [from French *batture* shoal]

Baumé (bō mā′), *adj. Chemistry.* of, denoting, or according to an arbitrary scale used with a hydrometer to measure the specific gravity of liquids. *Symbol:* Bé. [named after Antoine *Baumé,* 1728–1804, French chemist]

bauxite (bôk′sīt), *n. Geology, Mineralogy.* a rock from which aluminum is obtained, consisting chiefly of hydrated aluminum oxide minerals (chiefly gibbsite and boehmite) and some iron oxide and silica. [from French, from Les *Baux,* in southern France, where it was discovered]

bay, *n. Geography.* a part of a sea or lake extending into the land, having a wide opening. A bay is usually smaller than a gulf and larger than a cove.
► See the note under **gulf.**

Bayesian (bā′zē ən), *adj. Statistics.* of or having to do with a method of calculating statistical probabilities from individual samples rather than from frequency-distribution data and other empirical evidence. [named after Thomas *Bayes,* 1702–1761, English mathematician]

bayou (bī′ü *or* bī′ō), *n. Geography.* a marshy inlet or outlet of a lake, river, or gulf, especially in the south central United States. [from Louisiana French, from Choctau *bayuk* creek]

BCD, *abbrev.* binary coded decimal.

B cell or **B lymphocyte,** *Biology.* a type of lymph cell that originates in bone marrow and bears immunoglobulins on the surface which function as receptors. When the receptors are stimulated by antigens, the B cell is differentiated into a plasma cell which produces the antibodies that circulate in the blood and react with the specific antigens. *One major group of lymphocytes, called B cells (because of their origin in the bone marrow), includes precursor cells that may be triggered by an invading pathogen* (Scientific American). Compare **T cell.**

B complex, = vitamin B complex.

Be, *symbol.* beryllium.

Bé., *Symbol.* Baumé.

beach, *n. Geography, Geology.* the zone of accumulated sand, stone, or gravel deposited above the water line at a shore by wave activity.

bead, *n.* **1** *Chemistry.* a globule of borax or other flux covered with a bit of the mineral to be analyzed, which is heated in a blowpipe flame as a test for the presence of metals. **2** *Astronomy.* **beads,** = Baily's beads.

beak, *n.* **1** *Zoology.* **a** a bird's bill, especially one that is strong and hooked and useful in striking or tearing. Eagles, hawks, and parrots have beaks. SYN: bill. **b** a similar, often horny, part in other animals. Turtles and octopuses have beaks.
2 *Botany.* a sharp, projecting process, or prolonged tip in some plants, such as in the seeds of the crane's-bill.

beam, *Physics.* —*n.* a concentrated stream of particles or waves, as of light or sound: *A beam of protons is normally made up of a mixture spinning in many directions, just as an ordinary unpolarized beam of light consists of vibrations in different directions, up and down, right and left and all oblique directions between these* (N.Y. Times).
—*v.* to emit or radiate (light rays, sound waves, etc.); emit in rays.

► **Beam, ray** mean a line of light or other radiation flowing from a particular point. *Beam* applies to a shaft, long and with some width, often made up of several parallel lines or rays: *the beam from a flashlight. Ray* applies to a thin line of light or the like, usually thought of as radiating, or coming out like the spokes of a wheel, from a single source: *a ray of moonlight.*

beard, *n. Biology.* a hairy or bristly growth, such as the chin tuft of a goat, the stiff hairs around the beak of a bird, or the hairs on the heads of plants, such as oats, barley, and wheat. —**bearded,** *adj.* —**beardless,** *adj.*

bearer, *n. Botany.* a tree or plant that produces fruit or flowers.

beat, *n. Physics.* the regular pulsation arising from the interference of simultaneous sound waves, radio waves, or electric currents which have slightly different frequencies: *These variations of amplitude give rise to variations of loudness which are called beats* (Sears and Zemansky, *University Physics*).

terms, *alpha rays, beta rays,* and *gamma rays*). [named after Antoine H. *Becquerel,* who first reported these rays]

bed, *n. Geology.* **1** the smallest division of a stratified series of rocks. A *stratum* may consist of beds, and a bed may consist of *layers.*
2 a seam or tabular deposit of a mineral ore: *a bed of coal.*
3 the ground under a body of water: *the bed of a river.*
—**bedding,** *n.* arrangement of rocks or other strata in beds or layers; stratification.

bedload, *n. Geology.* bits of rock, soil, or other detritus, carried along the bottom of a river or stream by the current.

bedrock, *n. Geology.* the solid rock beneath the soil and looser rocks.

bel, *n. Physics.* a unit for measuring the difference in intensity of sounds, equal to ten decibels: *The bel is used to indicate the amount of energy in the form of sound that is transmitted to one square centimeter of the ear.* (E.A. Fessenden). [named after Alexander Graham *Bell,* 1847–1922, American physicist]

bell curve or **bell-shaped curve,** *Statistics.* the symmetrical curve of a normal distribution. Also called **normal curve.**

Beaufort scale

Beaufort Number	Condition	Kilometers per Hour	Physical Effects Produced
0	calm	less than 1	smoke rises vertically
1	light air	1-5	direction of wind shown by smoke drift
2	light breeze	6-11	wind felt on face; leaves rustle
3	gentle breeze	12-19	leaves and small twigs in constant motion
4	moderate breeze	20-28	small branches sway; dust and loose paper blown about
5	fresh breeze	29-38	small trees sway; white caps on inland water
6	strong breeze	39-49	large branches sway; umbrellas used with difficulty
7	moderate gale	50-61	whole trees sway; difficulty in walking against wind
8	fresh gale	62-74	twigs break off trees
9	strong gale	75-88	slight damage to buildings
10	whole gale	89-102	trees uprooted; considerable damage to buildings
11	storm	103-117	widespread damage
12-17	hurricane	above 117	violent destruction to buildings and other fixed structures

Beaufort scale (bō′fərt), an internationally used scale of wind velocities, ranging from 0 (calm) to 12 (hurricane), used in weather maps and other meteorological work. [named after Sir Francis *Beaufort,* 1774–1857, British admiral who devised the scale]

beauty, *n. Nuclear Physics.* an informal name for the property possessed by a bottom quark: *To study the nature of the b quark ... physicists look for naked beauty, a particle in which a b quark is attached to a quark of another flavor and so can manifest its properties more or less unmasked* (Science News).

becquerel (bek′ə rel′ *or* bek rel′), *n. Physics.* the international unit of radioactivity, equal to one disintegration or other nuclear transformation per second. [named after Antoine H. *Becquerel,* 1852–1908, French physicist]

Becquerel rays, *Physics.* the invisible rays given off by radioactive substances (replaced by the more specific

belly, *n. Anatomy.* **1** = abdomen. **2** the thick, fleshy, central portion of a muscle, as distinguished from its tendinous portion.

belt, *n. Geography.* any area or region having distinctive characteristics: *The portion of the world's atmosphere near the Equator ... is known as the doldrum belt, or belt of equatorial calms* (White and Renner, *Human Geography*). SYN: zone, band, strip.

bench, *n. Geography, Geology.* a narrow stretch of high, flat land: *Many floodplains are fringed at intervals with smaller alluvial plains ... called alluvial terraces,*

cap, fāce, fàther; best, bē, tèrm; pin, five;
rock, gō, ôrder; oil, out; cup, pùt, rüle;
yü in use, *yù* in uric;
ng in bring; *sh* in rush; *th* in thin, ŦH in then;
zh in seizure.
ə = *a* in about, *e* in taken, *i* in pencil, *o* in lemon, *u* in circus

benthic

or *benches* (Finch and Trewartha, *Elements of Geography*).

benthic or **benthonic**, *adj. Oceanography.* of or having to do with the bottom of oceans or seas: *The two main divisions [of the sea] are the benthic, or sea-bottom environment, and the pelagic, or open-water environment.* (Gilluly, *Principles of Geology*). [from *benthos*]

benthos (ben'thos), *n. Biology.* the organisms living on or in the bottoms of oceans, lakes, or streams: *The marine benthos is an important ecological group because it occurs over most of the earth's surface* (R.G. Johnson). [from Greek *benthos* depth (of the sea)]

ASSOCIATED TERMS: The *benthos,* which includes such bottom dwellers as sponges, barnacles, clams, oysters, corals, and foraminifers, is distinguished from the *nekton,* or active swimming forms such as fishes, squids, seals, whales, etc., and from the *plankton,* which comprise all organisms, chiefly microscopic, that float with the ocean currents.

bentonite (ben'tə nīt), *n. Geology.* a soft, absorbent, swelling, and colloidal clay formed from the alteration of volcanic ash. It is used widely in industry as a bonding agent, sealant, thickener, filler, etc., and is one of the montmorillonite mineral clays. [from Fort *Benton,* Montana, where it was originally found + -*ite*]

benz-, *combining form.* a form of **benzo-**, as in *benzaldehyde.*

benzaldehyde (ben zal'də hīd), *n. Chemistry.* a colorless aromatic liquid obtained from the natural oil of bitter almonds or other oils, or produced artificially, used in making dyes, flavoring compounds, perfumes, and drugs. *Formula:* C_6H_5CHO

benzene (ben'zēn'), *n. Chemistry.* a colorless, volatile, flammable liquid obtained from coal tar. Benzene is a toxic, aromatic hydrocarbon, used for removing grease stains and in making dyes and synthetic rubber. *Formula:* C_6H_6 Also called **benzol.** [from *benzoin*]

benzene ring or **benzene nucleus**, *Chemistry.* a hexagonal arrangement of six carbon atoms having alternating single and double bonds between them, with each carbon atom also bonded to a hydrogen atom. This structure is found in benzene and many other compounds derived from benzene.

benzene series, *Chemistry.* a series of aromatic hydrocarbons having the general formula C_nH_{2n-6}. The simplest member of the benzene series is benzene; the second member is toluene.

benzine (ben'zēn'), *n. Chemistry.* **1** a colorless, flammable, liquid mixture of hydrocarbons obtained in distilling petroleum, used in cleaning and dyeing and as a motor fuel. **2** = benzene.

benzo-, *combining form.* of or having to do with benzene or benzoic acid, as in *benzoate, benzonitrile* (an oil obtained from benzoic acid).

benzoate (ben'zō āt), *n. Chemistry.* a salt or ester of benzoic acid. **Benzoate of soda** or **sodium bensoate** is a white, crystalline or powdery salt, C_6H_5COONa, used in medicine and as a food preservative.

benzoic acid (ben zō'ik), *Chemistry.* a white, crystalline acid occurring naturally in benzoin resin, cranberries, etc., used in medicine as an antiseptic, and as a food preservative. *Formula:* C_6H_5COOH

benzoin (ben'zō ən), *n. Chemistry.* **1** a dry, brittle, fragrant resin obtained from certain species of trees of southeastern Asia, used in perfumes, cosmetics, and incense. It is also an ingredient in cough medicine and other medicinal application.
2 a crystalline compound somewhat like camphor, derived from benzaldehyde, used medicinally and in the synthesis of organic compounds. *Formula:* $C_6H_5CH(OH)COC_6H_5$
[from French *benjoin,* ultimately from Arabic *lubān jāwī* incense of Java]

benzol (ben'zōl), *n. Chemistry.* **1** (in industrial use) = benzene. **2** a mixture containing benzene and other aromatic hydrocarbons, such as toluene, obtained from coal tar.

Bergmann's rule, *Ecology.* the principle that warm-blooded animals living in cold climates tend to be larger in size than animals living of the same species in warm climates. [named after Carl *Bergmann,* 19th-century German biologist]

bergschrund (berk'shrùnt), *n. Geology.* a deep crevasse, or a series of crevasses, at or near the head of a mountain glacier separating moving ice from immobile ice and snow. [from German *Bergschrund* (literally) mountain fissure]

berkelium (bėr'klē əm), *n. Chemistry.* a radioactive metallic element produced artificially from americium, curium, or plutonium. *Symbol:* Bk; *atomic number* 97; *atomic weight* 247; *melting point* 986°C; *oxidation state* 4,3. [from New Latin, from the University of California at *Berkeley,* where the element was first produced in 1949]

berm or **berme** (bėrm), *n. Geology.* the deposit of material, especially sand or small rocks, at the top of a beach. It is usually a nearly horizontal surface but may be of varying width according to seasonal wave action. [from French *berme*]

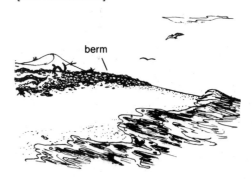

berm

Bernoulli's principle or **Bernoulli's theorem** (bėr nü'lēz), **1** *Statistics.* a theorem stating that as the number of independent trials of an event is increased, the probability of the event's occurrence increases in proportion to the trials.
2 *Physics.* the principle that the total energy per unit of mass in the streamline flow of a moving fluid is constant, being the sum of the potential energy, the kinetic energy, and the energy due to pressure. As the velocity of the fluid increases, its internal pressure decreases.
[(def. 1) named after Jacob (or James) *Bernoulli,* 1654–1705, Swiss mathematician; (def. 2) named after Daniel *Bernoulli,* 1700–1782, Swiss mathematician]

Bernoulli trials, *Statistics.* a sequence of independent trials, as in coin tossing, in which the probability of the occurrence of some event is constant from trial to trial. [named after Jacob *Bernoulli*]

berry, *n.* **1** *Botany.* **a** a simple fruit having a skin or rind surrounding the seeds in the fleshy pulp, as a grape, tomato, currant, or banana. **b** the dry seed or kernel of certain kinds of grain or other plants, as wheat, barley, or coffee. See the picture at **fruit.**
2 *Zoology.* a single egg of a lobster or fish.

beryl (ber′əl), *n. Mineralogy.* a very hard translucent or opaque mineral, usually green or blue-green, a silicate of beryllium and aluminum, used as a gem and also as the source of beryllium. Emeralds and aquamarines are transparent varieties of beryl. *Formula:* $Be_3Al_2Si_6O_{18}$ [from Latin *beryllus,* from Greek *bēryllos*]

beryllium (bə ril′ē əm), *n. Chemistry.* a grayish-white, hard, light, metallic element found in various minerals, having chemical properties similar to those of magnesium. Beryllium is used to make sturdy light alloys and to control the speed of neutrons in atomic reactors. *Symbol:* Be; *atomic number* 4; *atomic weight* 9.0122; *melting point* 1278° $\pm$ 5°C; *boiling point* 2970°C; *oxidation state* 2. [from New Latin, from Latin *beryllus* beryl]

Bessel function (bes′əl), *Mathematics.* any of a group of transcendental functions which are introduced by a differential equation, used especially in physics to represent temperature, magnetic field strength, etc., as functions of space coordinates. [named after Friedrich W. *Bessel,* 1784–1846, German astronomer]

beta (bā′tə), *n.* Usually, **Beta.** *Astronomy.* the second brightest star of a constellation.
—*adj. Chemistry.* designating the second of several possible positions of atoms or groups of atoms which are substituted in a chemical compound. *Symbol:* β

beta-adrenergic receptor, = beta receptor.

beta decay, *Physics.* the disintegration of a radioactive substance accompanied by emission of beta particles. Emission of a beta particle increases the atomic number by one, with only small loss of mass, but consequent change into a new kind of atom.

beta emitter, *Physics.* a radioactive substance which emits beta rays.

beta globulin, *Biochemistry.* a type of globulin in blood plasma intermediate between the alpha and gamma globulins in colloidal mobility in electrically charged solutions.

beta particle, *Physics.* an electron released by the nucleus of a radioactive substance in the process of disintegration: *Beta particles have the same charge and mass as electrons. They are emitted from the nuclei of radioactive atoms with tremendous speeds* (Sears and Zemansky, *University Physics*). Compare **alpha particle.**

beta ray or **beta radiation,** *Physics.* Usually, **beta rays.** a stream of beta particles.

beta receptor, *Biochemistry.* a site in a cell at which the stimulus of an adrenergic agent produces a usually inhibitory response, slowing down the activity of muscles. Compare **alpha receptor.**

beta rhythm or **beta waves,** *Physiology.* the pattern of electrical oscillations in the brain when a person is awake.

betatron (bā′tə tron), *n. Nuclear Physics.* a particle accelerator in which electrons achieve high speeds by a changing magnetic field. [from *beta* (particle) + *-tron*]

Bev or **bev,** *abbrev.* billion electron volts.
▶ This abbreviation has been generally replaced by *GeV* (gigaelectron volt).

B horizon, *Botany, Geology.* the layer of soil beneath the A horizon, in which insoluble iron oxides and clay minerals are deposited. Compare **C horizon.**

bi-, *prefix.* **1** double; doubly, as in *bicuspid, bipinnate.*
2 two; having two, as in *bilayer, bifocal.*
3 both, as in *biconcave.*
4 half, as in *bicarbonate.*
[from Latin *bi-,* from *bis* twice]

Bi, *symbol.* bismuth.

biaxial (bī ak′sē əl), *adj.* having two axes: *A biaxial crystal has two optic axes. A biaxial joint has two transverse axes at right angles to each other.*

bicapsular, *adj. Botany.* **1** having two capsules. **2** having a capsule which has two cells.

bicarbonate (bī kär′bə nit), *n. Chemistry.* a salt of carbonic acid formed by neutralizing one of two hydrogen ions.

bicarbonate of soda, = sodium bicarbonate.

biceps (bī′seps), *n., pl.* **-ceps** or **-cepses.** *Anatomy.* any muscle having two heads or origins: **a** the large muscle in the front of the upper arm, which bends the forearm. **b** the corresponding large muscle in the back of the thigh. [from Latin, two-headed, from *bi-* two + *caput* head]

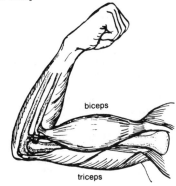

biceps

triceps

bichloride (bī klôr′īd), *n.* = dichloride.
bichloride of mercury, = mercuric chloride.
bichromate (bī krō′māt), *n.* = dichromate.
bicipital (bī sip′ə təl), *adj. Anatomy.* **1** having two heads or origins, as a muscle. **2** having to do with a biceps.
biconcave (bī kon′kāv), *adj. Optics.* concave on both sides: *biconcave lens.*

cap, fāce, fäther; best, bē, tėrm; pin, five;
rock, gō, ôrder; oil, out; cup, put, rüle,
yü in use, *yu̇* in uric;
ng in bring; *sh* in rush; *th* in thin, ᴛʜ in then;
zh in seizure.
ə = *a* in about, *e* in taken, *i* in pencil, *o* in lemon, *u* in circus

biconditional

biconditional, *n. Mathematics.* the statement or proposition "P if and only if Q," usually symbolized as *P→q: A biconditional ... is true whenever p and q have the same truth value* (Mathematics Regents Syllabus). Compare **conditional.**

bicontinuous, *adj. Mathematics.* continuous in both directions.

biconvex (bī kon′veks), *adj. Optics.* convex on both sides: *biconvex lens.*

bicornuate (bī kôr′nyu̇ it), *adj. Botany, Zoology.* having two horns or hornlike parts; crescent-shaped.

bicuspid (bī kus′pid), *Anatomy.* —*n.* a double-pointed premolar tooth that tears and grinds food. A human adult has eight bicuspids. —*adj.* having two points, fangs, or cusps.

bicuspid valve, *Anatomy.* the valve of the heart between the left atrium and left ventricle. Also called **mitral valve.**

bicyclic (bī sī′klik), *adj.* **1** *Botany.* arranged in two whorls: *bicyclic stamens.*
2 *Chemistry.* containing two rings of atoms in the molecule: *bicyclic alcohol.*

bidentate (bī den′tāt), *adj.* **1** *Botany, Zoology.* having two teeth or toothlike processes.
2 *Chemistry.* of or having to do with a molecule two of whose atoms are joined to a metal atom or ion to form a coordination complex or chelate: *a bidentate ligand.*

biennial (bī en′ē əl), *Botany.* —*adj.* (of a plant) lasting two years; germinating in one year or growing season, and flowering, producing fruit, and dying in the next year or growing season.
—*n.* a plant that lives two years or seasons. Carrots and onions are biennials.
[from Latin *biennium* a two-year period, from *bi-* two + *annus* year] —**biennially,** *adv.*
ASSOCIATED TERMS: see **annual.**

bifid (bī′fid), *adj. Botany.* divided into two parts by a cleft: *a bifid leaf.* SYN: forked. [from Latin *bifidus,* from *bi-* two + *findere* cleave]

bifocal (bī fō′kəl), *Optics.* —*adj.* having two focuses. **Bifocal lenses** have two sections of different focal lengths, the upper for distant, the lower for near vision. —*n.* a bifocal lens.

bifoliate, *adj. Botany.* having two leaves.

bifurcate, *adj. Botany, Zoology.* divided into two branches; forked.

big bang theory, *Astronomy.* the theory that the universe originated in a cosmic explosion of a hot, dense mass of matter from 10 to 18 billion years ago. Contrasted with **steady-state theory.**

bigeneric, *adj. Botany.* having the characteristics of two different genera.

bight (bīt), *n. Geography, Geology.* **1** a long curve in a coastline forming an open bay. **2** a bay formed by this curve, such as the Great Australian Bight.

bilabiate (bī lā′bē it), *adj. Botany.* having an upper and lower lip, as the corollas of flowers of the mint family.

bilateral, *adj. Biology.* **1a** having two symmetrical sides: *a bilateral organism.* **b** two-sided: *The symmetry of most animals is bilateral.*

2 having to do with or affecting two sides or parts of the body: *a bilateral reflex.* Compare **biradial.** —**bilaterally,** *adv.*

bilateral symmetry, *Biology.* a condition in which like parts are arranged in two halves, so that each half is the counterpart of the other, divisible in one plane only: *Bilateral symmetry: that is, one side of the body is in general a mirror replica of the other. This type of symmetry is characteristic of active animals* (Ralph Beals and Harry Hoijer). Contrasted with **radial symmetry.** See the picture at **symmetry.**
—**bilaterally symmetrical,** having bilateral symmetry: *Most animals (including all the higher ones) are bilaterally symmetrical.* (Shull, *Principles of Animal Biology*).

bilayer, *n. Biochemistry.* a structure consisting of two layers each of the thickness of one molecule: *The phospholipid bilayer, now confirmed as the basic molecular arrangement in membranes, may also be the structural basis of the low density lipoproteins in human blood serum* (Nature).

bile, *n. Physiology.* a bitter, greenish-yellow liquid secreted by the liver and stored in the gall bladder. It aids digestion in the duodenum by neutralizing acids and emulsifying fats. Also called **gall.**

bile acid, *Biochemistry.* any of the closely related steroid acids found in bile, commonly in combination with glycine and taurine. The most common bile acid in human bile is cholic acid.

bile duct, *Anatomy.* an excretory duct through which bile flows from the liver or gall bladder into the duodenum during digestion.

bile pigment, *Biochemistry.* one of the coloring substances in bile, such as bilirubin and biliverdin.

bile salt, *Biochemistry.* any of the sodium salts of the bile acids, important in emulsifying fats and stimulating the liver.

biliary (bil′ē er′ē), *adj.* **1** *Anatomy.* of or carrying bile: *a biliary duct, the biliary tract.* **2** *Medicine.* caused by trouble with the bile or bile duct: *Biliary colic is caused by the passage of a gallstone in the bile duct.*

bilinear, *adj. Mathematics.* **1** of or involving two lines: *bilinear coordinates.* **2** linear in two ways, especially in each of two variables or positions: *bilinear functions.*

bilious (bil′yəs), *adj.* **1** = biliary (def. 1). **2** of or having to do with diseases caused by an excess of bile: *bilious dyspepsia.*

bilirubin (bil′ə rü′bin), *n. Biochemistry.* the reddish-yellow pigment normally found in bile, formed by the breakdown of hemoglobin in the spleen, bone marrow, etc. *Formula:* $C_{33}H_{36}N_4O_6$ [from Latin *bilis* bile + *ruber* red]

biliverdin (bil′ə vėr′din), *n. Biochemistry.* a green pigment in bile produced by the oxidation of bilirubin. *Formula:* $C_{33}H_{34}N_4O_6$ [from Latin *bilis* bile + *viridis* green]

bill, *n. Zoology.* **1** the horny part of the jaws of a bird. The bills of birds differ according to how they feed and what they eat: *a duck's bill.*
2 anything shaped like a bird's bill: *the bill of a turtle.*

bilobate (bī lō′bāt) or **bilobed** (bī′lōbd′), *adj. Biology.* having or divided into two lobes: *a bilobate leaf, a bilobed ganglion.*

bimodal, *adj. Statistics.* having two modes: *a bimodal distribution.*
—bimodality, *n.* the condition or quality of being bimodal.

bimolecular, *adj. Chemistry.* **1** that is two molecules in thickness: *a bimolecular layer.* **2** having to do with or composed of two molecules: *a bimolecular reaction.* Compare **monomolecular.**

binary, *adj.* **1** consisting of two; involving two: *A binary compound is made up of two chemical components, such as an element and a radical.*
2 *Mathematics.* of, having to do with, or expressed in a system of numeration based upon two. Compare **decimal.** See also **binary notation.**
—n. = binary star.
[from Late Latin *binarius,* from Latin *bini* two at a time]

binary coded decimal, *Mathematics.* a binary number system in which each decimal digit is represented by four binary digits. In this system, $0 = 0000$, $1 = 0001$, $2 = 0010$, $3 = 0011$, $4 = 0100$, $5 = 0101$, $6 = 0110$, $7 = 0111$, $8 = 1000$, $9 = 1001$, $10 = 0001\ 0000$. *Abbreviation:* BCD Compare **binary notation, decimal system.**

binary color, = secondary color.

binary digit, *Mathematics.* either of the digits 0 or 1 used in binary notation. The binary digit is the basic unit of computer information. Also called **bit.**

binary fission, *Biology.* the division or splitting of a cell into two equal parts: *Paramecium reproduces itself by dividing in half transversely, a process known as binary fission* (New Biology).
ASSOCIATED TERMS: see **asexual.**

binary granite, *Geology.* a variety of granite composed of quartz, feldspar, and two micas, muscovite and biotite.

binary notation or **binary system,** *Mathematics.* a system of arithmetical notation which uses only two digits, 0 and 1, instead of the ordinary ten digits, 0 to 9, of decimal notation. In this system, the decimal number 1 is also noted as 1, but the decimal number 2 is noted as 10 (1 and 0), 3 as 11 (1 and 1), 4 as 100 (1 and 0 and 0), 5 as 101 (1 and 0 and 1), 6 as 110, 7 as 111, 8 as 1000, 9 as 1001, and 10 as 1010. Binary notation is the basic unit of computer information, since the digits 0 and 1 can express a choice between two possibilities, such as no or yes, off or on, etc. Compare **decimal.**

binary operation, *Mathematics.* an operation that involves two quantities, as in addition, $3 + 2 = 5$.

binary star, *Astronomy.* a pair of stars that revolve around a common center of gravity. Binary stars are relatively common; perhaps a third to a half of all stars are located in systems of binary stars. Also called **binary, double star.**
ASSOCIATED TERMS: Binary stars are commonly classified as visual, spectroscopic, or eclipsing. *Visual binary stars* are near enough to earth to be separately visible by a telescope; *spectroscopic binaries* are recognized only by observation through a spectroscope; *eclipsing binaries* have orbits so nearly edgewise to the line of sight from the earth that they alternately eclipse each other.
▶ Visual binary stars should not be confused with *optical double stars,* two stars in almost the same line of sight but actually situated at a great distance from each other and having no physical relation. Some as-

tronomers call both of these *double stars,* distinguishing them as visual or optical.

binaural (bī nôr′əl), *adj.* **1** *Physiology.* of, having to do with, or involving perception with both ears: *binaural hearing.*
2 *Electronics.* of or having to do with sound transmission or reception by two paths or channels.
[from Latin *bini* two at a time + English *aural*]
—binaurally, *adv.*

bind, *v. Chemistry.* to hold together by a strong chemical force; form a bond: *Two models ... explain how actinomycin binds to DNA ... that a hydrogen bond forms between the actinomycin molecule and the guanosine group ... [or] that a portion of the actinomycin molecule is physically locked between two layers of the DNA helix* (R.E. Marsh).

binding energy, *Physics.* the net energy necessary to break a molecule, atom, or nucleus into its smaller component parts: *The binding energy of the nucleus is the energy that must be supplied to it in order to separate it into its nucleons.* (Physics Regents Syllabus).

bine, *n. Botany.* a twining, slender stem of a climbing plant, especially of the hop. [alteration of *bind,* noun]

binocular, *adj. Zoology.* of, having to do with, or using both eyes simultaneously: *binocular vision.*
—binocularity, *n.* the simultaneous use of both eyes.

binomial (bī nō′mē əl), *n.* **1** *Mathematics.* an algebraic expression of two terms connected by a plus or minus sign. $8a + 2b$ is a binomial. Compare **monomial, trinomial.**
2 *Biology.* the scientific name of an organism consisting of two terms, the first indicating the genus and the second the species, as *Rubus occidentalis,* the blackberry, and *Equus caballus,* the horse.
—adj. consisting of two terms. **Binomial nomenclature** is the system of classifying organisms according to genus and species.
[from Late Latin *binomius* having two names, from Latin *bi-* two + *nomen* name] **—binomially,** *adv.*

binomial distribution, *Statistics.* a distribution giving the statistical probability of any particular number of recurrences of one of two possible results, such as heads or tails, in a given number of trials, each trial offering equal chances.

binomial theorem or **binomial formula,** *Mathematics.* the algebraic theorem, by Isaac Newton, for raising a binomial to any power. EXAMPLE: $(a + b)^2 = a^2 + 2ab + b^2$.

binuclear, *adj.* = binucleate.

binucleate or **binucleated,** *adj. Biology.* having two nuclei, as a cell.

bio-, *combining form.* **1** life; living things, as in *biology = the science of life.*

---

cap, fāce, fäther; best, bē, tèrm; pin, fīve;
rock, gō, ôrder; oil, out; cup, pùt, rüle,
yü in use, *yu* in uric;
ng in bring; *sh* in rush; *th* in thin, ᴛʜ in then;
zh in seizure.
ə = *a* in about, *e* in taken, *i* in pencil, *o* in lemon, *u* in circus

2 biological, as in *biochemistry* = *biological chemistry.*
[from Greek *bios* life]

bioaccumulation, *n. Ecology.* the accumulation of toxic chemicals in living things: *"The chief concern for Lake Ontario," the board reported, "is the bioaccumulation of toxic contaminants* (N.Y. Times). —**bioaccumulative,** *adj.*

bioacoustics, *n. Biology.* the study of sounds produced by or affecting living organisms.
—**bioacoustic,** *adj.* of or having to do with bioacoustics: *bioacoustic research.*

bioassay, *n.* = assay (def. 2).

bioastronautics, *n.* the science that deals with the biological aspects of travel in outer space: *Bioastronautics ... discusses some of the most important aspects of the biology of flight, including acceleration stresses, vibration and noise ... and heat protection* (Scientific American).
—**bioastronautic,** *adj.* of or having to do with bioastronautics.

biocatalyst, *n. Biochemistry.* a coenzyme, hormone, or other substance that produces a chemical reaction in a living body; biochemical catalyst.
—**biocatalytic,** *adj.* having to do with or caused by biocatalysts.

biocenose or **biocoenose** (bī′ō sē′nōs), *n. Ecology.* **1** the group of interacting organisms that inhabit a particular habitat. **2** any relationship between organisms.
[from New Latin *biocenosis, bioccenosis,* from *bio-* + Greek *koinōsis* a sharing]

biocenosis or **biocoenosis** (bī′ō sə nō′sis), *n., pl.* **-ses** (-sēz′). = biocenose.

biochemical, *adj.* of or having to do with biochemistry. —*n.* a biochemical substance. —**biochemically,** *adv.*

biochemical oxygen demand, *Bacteriology.* the amount of oxygen consumed by aerobic bacteria to decompose organic matter in a water sample, used to measure the extent of pollution by sewage, industrial waste, etc. *Abbreviation:* BOD

biochemistry, *n.* **1** the science that deals with the chemical processes of living matter; biological chemistry: *Biochemistry is concerned with ... the chemical processes by which plants germinate, photosynthesize, grow, flower, fertilize, and seed; the chemical processes by which animals reproduce and digest, absorb, utilize, and excrete food ingredients* (R.S. Harris). **2** biochemical composition or characteristics: *the biochemistry of a virus.*

biocide (bī′ə sīd), *n. Ecology.* any substance that is poisonous or destructive to living organisms: *The use of some biocides (such as pesticides and herbicides) ... has contaminated the soil, atmosphere, water supply* (Biology Regents Syllabus).
—**biocidal** (bī′ə sī′dl), *adj.* destructive of life and living things: *biocidal radiation.*

bioclimatic, *adj.* of or having to do with bioclimatology.
bioclimatology, *n. Ecology.* the study of the effects of climate on organisms, especially animals and plants.

biodegradable, *adj. Biology.* susceptible of being decomposed by biological agents, especially bacteria: *Biodegradable detergents are attacked by natural soil*

and water bacteria that destroy the foam-producing qualities of the detergent (Wall Street Journal).
—**biodegradability,** *n.* biodegradable quality or condition.
—**biodegrade,** *v.* to decompose, especially by bacterial action. —**biodegradation,** *n.*

biodynamics, *n. Biology.* the study of the effects of motion and other dynamic processes on living organisms.

bioecology, *n.* the branch of ecology dealing with relationships of organisms, especially between plants and animals.

bioelectric or **bioelectrical,** *adj. Physiology.* of or having to do with the effects of electricity on living tissues.
—**bioelectricity,** *n.* bioelectric phenomena.

bioenergetics, *n. Biology.* the study of the flow and transformation of energy in living systems.

biofacies (bī′ō fā′shē ēz), *n. Geology.* a facies distinguished by its fossil deposits of organisms. Compare **lithofacies.**

bioflavonoid (bī′ō flā′və noid), *n. Biochemistry.* any of a complex of substances, present in citrus fruits and other plant foods, that promote capillary resistance to hemorrhaging. Also called **vitamin P complex.** [from *bio-* + Latin *flāvus* yellow]

biogenesis, *n. Biology.* **1** the theory that living things can be produced only by other living things, and cannot develop spontaneously from nonliving materials. *Biogenesis* was coined in 1870 by Thomas Huxley, an English biologist. Contrasted with **abiogenesis.**
2a the production of living things from other living things. **b** = biosynthesis.
3 the supposed repetition in the development of an organism of stages in the evolution of the race or species. Also called **recapitulation.**
—**biogenetic,** *adj.* of or having to do with biogenesis.
—**biogenetically,** *adv.*

biogenic, *adj. Biology.* produced by living organisms, as certain marine sediments.

biogeochemical, *adj.* of or having to do with biogeochemistry.

biogeochemistry, *n.* the study of the effect of the earth's chemical composition upon living things.

biogeography, *n. Ecology.* the study of the geographic distribution of animals and plants. —**biogeographic** or **biogeographical,** *adj.*

biohazard, *n. Ecology.* any danger, risk, or harm resulting from exposure to infectious bacteria, viruses, or other harmful agents or their products, especially in biological research and experimentation.
—**biohazardous,** *adj.*

bioherm (bī′ō hėrm), *n. Geology.* a moundlike or dome-like mass of rock built up for the most part by marine organisms, such as coral, overlaid or enclosed by another rock. [from *bio-* + Greek *herma* sunken rock]

biological or **biologic,** *adj.* **1** of or dealing with living things and processes: *biological science.* **2** having to do with biology: *biological research.* **3** for use in or prepared by a biological laboratory: *biological serums.*
—**biologically,** *adv.*

biological clock, *Physiology.* an innate mechanism in organisms which controls the rhythm or cycle of various living functions and activities, such as photosynthesis in a plant: *The term "biological clock" is applied to the means by which living things adjust their activity patterns, without any obvious cue, to the time of day, or*

biorhythm, circadian.

biology, *n.* **1** the scientific study of life and living things, including their origin, diversity, structure, activities, and distribution: *The vastness of biology today can be seen by the various subdivisions into which the science has been broken. ... Classification of organisms ... is taxonomy. Investigation of structure is anatomy ... of inheritance from one generation to the next, genetics; of the relation of organisms to their environment, ecology ... Biology, too, can be divided according to entomology (insects) ... mycology (molds and mushrooms), bacteriology, virology, and botany (plants), and zoology (animals). The list can be greatly lengthened ...* (McElroy, *Biology and Man*).
2 the life or living things of a particular area or region: *Some of the Gulf Stream's warmer water flows into the Polar Basin ... which apparently [does] not modify appreciably either the climate or the biology* (John C. Reed).
3 the facts about an organism, biological process, etc.: *biology of aging, biology of an insect.*

bioluminescence (bī'ō lü'mə nes'ns), *n. Biology.* phosphorescence or other emission of light by living organisms. Organisms with this property include fireflies, jellyfish, and some mushrooms. **—biuminescent,** *adj.*

biolysis (bī ol'ə sis), *n. Biology.* **1** the dissolution of a living organism. **2** the decomposition of organic substances by living organisms.

biomagnification, *n. Ecology.* increase in the concentration of toxic chemicals with each new link in the food chain: *Biomagnification is ... the process by which pesticides sprayed on vegetables and grains concentrate in the fat of animals and fish that eat them and then are further concentrated in the fat of meat and fish eaters* (N.Y. Times).
—biomagnify, *v.* to undergo biomagnification.

biomass, *n. Ecology.* **1** the total amount of living material in a unit of area. A decrease of energy at each successive feeding level in a food web means that less biomass can be supported at each level. *When they sampled the reef of selected areas in attempts to estimate the biomass,* [they] *concluded that there was not enough animal matter available to support the corals* (New Scientist).
2 plant material or vegetation, especially as a source of fuel or energy: *The hydrogen production can be increased by adding biomass, essentially organic garbage, such as sewage sludge, straw and stalks* (Science News).

biomathematics, *n.* the application of mathematical principles to biological or medical studies.

biome (bī'ōm), *n. Ecology.* all plants, animals, and other organisms that make up a distinct natural community in any climatic region: *The term biome refers to the most common climax ecosystem that will form in large climatic areas* (Biology Regents Syllabus). [from Greek *bios* life + *-ōma* mass, group]
▶ Biomes may be terrestrial or aquatic. The temperate deciduous forest of the northeastern United States is a terrestrial biome. An ocean is an aquatic biome. Terrestrial or land biomes are sometimes named by the climax vegetation in the region, including tundras,

taigas, temperate deciduous forests, tropical rain forests, grasslands, and deserts.

Aquatic biomes represent the largest ecosystem on earth, and may be marine or freshwater. They are typically more stable than terrestrial biomes.

biomechanical, *adj.* of or having to do with biomechanics.

biomechanics, *n.* the science that deals with the mechanics of living organisms, especially the effects of forces applied to the body by gravity.

biomedical, *adj.* **1** of or involving both biology and medicine: *biomedical engineering.* **2** of or having to do with biomedicine: *biomedical astronautics.*

biomedicine, *n.* the medical study of the human tolerance to environmental stresses and variations, as in space travel.

biometeorology, *n.* the science that deals with the effects of atmospheric phenomena on organisms.

biometric (bī'ə met'rik) or **biometrical** (bī'ə met'rə kəl), *adj.* of or having to do with biometry. **—biometrically,** *adv.*

biometry (bī om'ə trē) or **biometrics** (bī'ə met'riks), *n.* the branch of biology that studies living things by measurements and statistics; statistical study of biological phenomena.

bionics (bī on'iks), *n.* the study of the anatomy and physiology of animals in relation to adapting or imitating biological systems in industrial design, electronic systems, devices, etc.: *In the field of bionics, or biological electronics, researchers were peering into such things as the eyes of frogs and horseshoe crabs. Their aim: to develop "eyes" and "ears" and even rudimentary "brains" for space-age robots* (A.R. Harmet).
—bionic, *adj.* of or having to do with bionics: *In building a submarine ... the bionic approach to this problem consists of studying organisms ... moving through water with the least amount of resistance* (Morley R. Kare).
[from *bio-* + *(electro)nics*]

biophile (bī'ō fīl) or **biophil** (bī'ō fil), *adj. Biochemistry.* of or designating any chemical element found in living matter, such as carbon, nitrogen, oxygen, and phosphorus.

biophysics, *n.* the branch of biology which applies the laws and methods of physics to biological problems and phenomena.
—biophysical, *adj.* of or having to do with biophysics.

biopolymer, *n. Biochemistry.* a biological polymer, such as a nucleic acid or a protein.

biopsy (bī'op sē), *n. Medicine.* **1** the surgical removal of a sample of tissue from a living body for examination and diagnosis: *The only method that assures an accurate diagnosis of cancer is a biopsy* (C.P. Rhoads).

---

cap, fāce, fäther; best, bē, tèrm; pin, fīve;
rock, gō, ôrder; oil, out; cup, pùt, rüle,
yü in use, yù in uric;
ng in bring; sh in rush; th in thin, ŦH in then;
zh in seizure.
ə = *a* in about, *e* in taken, *i* in pencil, *o* in lemon, *u* in circus

2 the scientific examination of tissue.
[from *bio-* + Greek *opsis* a viewing]

biorhythm, *n. Physiology.* innate rhythmical or cyclic changes which occur in the functions or activities of organs and organisms. Compare **biological clock.** —**biorhythmic,** *adj.*

biosphere, *n. Ecology.* **1** the portion of the earth in which life exists or which can support life. The biosphere is composed of numerous complex ecosystems. **2** the living organisms of the biosphere. —**biospheric,** *adj.* of or characteristic of the biosphere: *biospheric contamination.*

biostrome (bī′ə strōm), *n. Geology.* a blanketlike mass of sedimentary rock composed mainly of the remains of sedentary organisms. Compare **bioherm.** [from *bio-* + Greek *strōma* a spread]

biosynthesis, *n. Biochemistry.* **1** the formation of a chemical compound by a living system. Also called **biogenesis.** **2** the manufacture of a biochemical substance. —**biosynthetic,** *adj.* of or having to do with biosynthesis. —**biosynthetically,** *adv.*

biosystematics, *n. Biology.* the classification of species on the basis of genetic and ecological studies. It is an experimental form of taxonomy.

biota (bī ō′tə), *n. Biology.* the fauna and flora of a given region or period. [from Greek *biotē* life, from *bioun* to live]

biotic (bī ot′ik), *adj. Ecology.* of or having to do with life or living things; produced by or involving living organisms: *Biotic factors are all the living things that directly, or indirectly, affect the environment ... Biotic factors interact ... in nutritional relationships and symbiotic relationships* (Biology Regents Syllabus). Compare **abiotic.**

biotic potential, *Ecology.* the capacity of a group of organisms to survive and increase in numbers under optimal environmental conditions.

biotin (bī′ə tən), *n. Biochemistry.* a crystalline acid of the vitamin B complex that promotes growth, found in liver, eggs, and yeast. *Formula:* $C_{10}H_{16}N_2O_3S$ Also called **vitamin H.**

biotite (bī′ə tīt), *n. Mineralogy.* a black or dark-green mica, a hydrous silicate of aluminum and iron with magnesium and potassium, occurring in igneous and metamorphic rocks. Also called **black mica.** [from J.B. Biot, 1774–1862, French mineralogist + *-ite*]

biotope (bī′ə tōp), *n. Ecology.* a region uniform in environmental conditions and in its distribution of animal life. [from *bio-* + Greek *topos* region]

biotron (bī′ə tron), *n.* a structure or laboratory apparatus in which climatic conditions can be strictly controlled for use in studying organisms. [from *bio-* + *-tron*]

biotype, *n. Genetics.* a group of organisms having the same genotype. —**biotypic,** *adj.* of, having to do with, or resembling a biotype.

biparous (bip′ər əs), *adj.* **1** *Zoology.* (of certain animals) bringing forth two at a birth.

2 *Botany.* (of certain flower clusters) having two branches or axes.
[from *bi-* + Latin *parere* to produce]

bipartite, *adj.* **1** *Zoology.* having two parts: *A clam has a bipartite shell.* **2** *Botany.* divided into two parts nearly to the base: *a bipartite leaf.* —**bipartitely,** *adv.*

biped (bī′ped), *Zoology.* —*n.* an animal having two feet. Birds and humans are bipeds. Contrasted with **quadruped.** —*adj.* Also, **bipedal.** having two feet. [from Latin *bipedem,* from *bi-* two + *pedem* foot] —**bipedalism,** *n.* the condition of having or moving on two feet only: *Some very limited bipedalism left the hands sufficiently free from locomotor functions so that stones or sticks could be carried, played with and used* (Scientific American).

biphenyl, *n.* a colorless, crystalline hydrocarbon consisting of two phenyl groups, used in organic synthesis. *Formula:* $C_6H_5C_6H_5$ Also called **diphenyl.**

bipinnate (bī pin′āt), *adj. Botany.* doubly pinnate. A pinnate leaf with pinnate leaflets is bipinnate. —**bipinnately,** *adv.*

bipinnate leaves

bipolar, *adj.* **1** *Anatomy.* having two poles or opposite extremities: *A bipolar nerve cell has both an afferent and an efferent process.* **2** *Geography.* of or found in both polar regions of the earth. —**bipolarity,** *n.* bipolar quality or condition.

biprism, *n. Optics.* a prism with a very obtuse angle, used to split a beam of light to obtain two images from a single source.

biquadratic, *Mathematics.* —*adj.* of or containing a fourth power in mathematics: *a biquadratic equation.*

biquinary (bī kwī′nər ē), *adj. Mathematics.* of or having to do with a scale of arithmetical notation in which the base is alternatively 2 and 5.

biradial, *adj. Biology.* having both a bilateral and a radial arrangement of parts, as ctenophores: *biradial symmetry.* —**biradiality,** *n.* the condition of being biradial. —**biradially,** *adv.*

biramous (bī rā′məs), *adj. Zoology.* having or consisting of two branches.

bird, *n. Zoology.* any of a class (*Aves*) of warm-blooded vertebrates that lay eggs, having a body covered with feathers, two legs, and the forelimbs adapted to form wings by means of which most species fly.

bird of prey, *Zoology.* any of a group of birds that hunt animals to eat their flesh, including eagles, hawks, owls, vultures, etc.

birefringence (bī′ri frin′jəns), *n. Optics.* the formation of two unequally refracted rays when a ray of light passes through certain crystals. Also called **double refraction.** —**birefringent,** *adj.* characterized by birefringence.

birth canal, *Anatomy.* the canal through which the fetus passes during birth. It leads from the uterus through the cervix, vagina, and vulva.

birth rate, *Biology.* the ratio of the number of births in a given year and area to the total population of that area.

bisect (bī sekt′), *v. Geometry.* to divide into two equal parts: *a 90 degree angle is bisected into two 45 degree angles.* [from *bi-* two + Latin *sectus* cut]
—**bisection,** *n.* **1** the act of bisecting: *the bisection of an angle.* **2** the place of bisecting. **3** one of two equal parts.
—**bisector,** *n.* a straight line that bisects either an angle or a line segment.

bisect

Line DB bisects triangle ACD.

bisectrix (bī sek′triks), *n., pl.* **bisectrices** (bī sek trī′sēz). *Crystallography.* a line bisecting the angle of optic axes in a biaxial crystal.

biserrate (bī ser′it), *adj.* **1** *Botany.* doubly serrate: *a biserrate leaf.*
2 *Zoology.* serrate on both sides: *a biserrate antenna.*

bisexual, Biology. —*adj.* combining both male and female organs in one organism; hermaphroditic: *Bisexual refers to an organism that produces both eggs and sperms. It also means a flower that bears both stamens and pistils* (P. B. Sears and F. Grover).
—*n.* a plant or animal that is bisexual. Also called **hermaphrodite.** —**bisexually,** *adv.*
—**bisexuality,** *n.* bisexual condition or quality.

bismite (biz′mīt), *n. Mineralogy.* an ore of bismuth, occurring as a yellow, powdery earth. *Formula:* Bi_2O_3

biradial symmetry

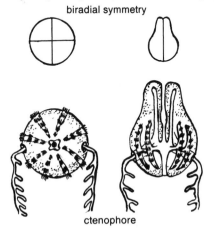

ctenophore

bismuth (biz′məth), *n. Chemistry.* a brittle, reddish-white metallic element which occurs in nature as a free metal and in various ores, used in medicine and in making low-melting alloys. *Symbol:* Bi; *atomic number* 83; *atomic weight* 208.980; *melting point* 271°C;

boiling point 1560°C; *oxidation state* 3,5. [from earlier German *Bismuth* (now Wismut)]

bismuthic (biz muth′ik), *adj. Chemistry.* of or containing bismuth, especially with a valence of five.

bismuthinite (biz muth′ə nīt), *n. Mineralogy.* a principal ore of bismuth occurring as lead-gray lustrous crystals or masses. *Formula:* Bi_2S_3

bismuthous (biz′mə thəs), *adj. Chemistry.* of or containing bismuth, especially with a valence of three.

bisulfate (bī sul′fāt), *n. Chemistry.* a salt of sulfuric acid formed by neutralizing one of its two hydrogen ions. It contains the univalent group $-HSO_4$. Compare **bisulfite.**

bisulfide (bī sul′fīd), *n.* = disulfide.

bisulfite (bī sul′fīt), *n. Chemistry.* a salt of sulfurous acid containing the univalent group $-HSO_3$.

bit, *n.* = binary digit. [from *bi(nary digi)t*]

bitangent, *Geometry.* —*adj.* touching a curved line or surface at two different points: *a bitangent line or plane.*
—*n.* a straight line which is tangent to a curve at two different points; a double tangent.

bitartrate, *n. Chemistry.* a salt of tartaric acid, which contains the univalent radical $-C_4H_5O_6$; an acid tartrate.

bitumen (bə tü′mən), *n. Mineralogy.* any of a number of natural organic materials that will burn, such as asphalt, petroleum, and naphtha. [from Latin *bitumen*]
—**bituminous** (bə tü′mə nəs), *adj.* containing or made with bitumen: *bituminous shale.*

bituminous coal, *Mineralogy.* coal in which volatile matter constitutes more than 18 per cent and which therefore burns with much smoke and a yellow flame. It is commonly called **soft coal** and contrasted with **anthracite.**

bivalent (bī vā′lənt), *adj.* **1** *Chemistry.* **a** having a valence of two. SYN: divalent. **b** having two valences, as bismuth.
2 *Biology.* double (applied to a pair of homologous chromosomes united in synapsis).
—*n. Biology.* bivalent pair of chromosomes; tetrad.
—**bivalence** or **bivalency,** *n.* bivalent quality or condition.

bivalve, *Zoology.* —*n.* **1** any mollusk whose shell consists of two parts hinged together so that it will open and shut like a book. Oysters, clams and mussels are bivalves.
2 the shell of such a mollusk.
—*adj.* **1** having a shell consisting of two parts hinged together.
2 composed of two parts hinged together: *a bivalve shell.* Contrasted with **univalve.**

bivalved, *adj.* = bivalve.

Bk, *symbol.* berkelium.

cap, fāce, fäther; best, bē, tèrm; pin, five;
rock, gō, ôrder; oil, out; cup, pùt, rüle,
yü in use, yù in uric;
ng in bring; sh in rush; th in thin, ᴛʜ in then;
zh in seizure.
ə = *a* in about, *e* in taken, *i* in pencil, *o* in lemon, *u* in circus

blackbody, *Physics.* —*n.* a theoretical surface or body capable of completely absorbing all the radiation falling on it.

—*adj.* of a blackbody: *blackbody radiation.*

black hole, *Astronomy.* a hypothetical heavenly object of extreme compactness and a strong gravitational pull that prevents the escape of light or any signal, formed by contraction and collapse of a star: *In the center of a black hole is a singularity, a place where the laws of physics may no longer apply, and where space and time seem to disappear* (Science News). *The physics involved is formidable ... the black hole slowly radiates energy and evaporates away ... until, eventually, the black hole explodes in a flash of radiation* (New Scientist).

black mica, = biotite.

bladder, *n.* **1** *Anatomy.* a soft, thin, elastic sac which stores the urine secreted by the kidneys until it is discharged.
2 *Biology.* any similar sac in animals or plants, especially: **a** a hollow sac on various plants, as bladderworts and certain seaweeds. Compare **air bladder** and **gall bladder. b** the inflated pericarp of some plants.

blade, *n.* **1** *Botany.* **a** the flat, wide part of a leaf, as distinguished from the stalk. See the picture at **leaf. b** a leaf of grass. **2** *Anatomy.* the wide, flat part of a bone: *the shoulder blade.*

blastema (blas tē′mə), *n., pl.* -mata (-mə tə), **-mas.** *Embryology.* **1** the formative substance of a germinating ovum.
2 the initial point of growth from which an organ or part is developed.
[from Greek *blastēma* sprout]
—**blastemic,** *adj.* of or resembling a blastema.

blasto-, *combining form.* germ; bud; budding, as in *blastodisk, blastogenesis.* [from Greek *blastos* sprout, shoot]

blastochyle (blas′tə kīl), *n. Embryology.* the liquid which fills the blastocoel.

blastocoel or **blastocoele** (blas′tə sēl), *n. Embryology.* the central cavity of a blastula. [from *blasto-* + Greek *koilia* belly]

blastocyst (blas′tə sist), *n. Embryology.* **1** a modified blastula, as of a placental mammal. **2** = blastula.

blastoderm (blas′tə dèrm′), *n. Embryology.* the layer of cells formed by the growth of a fertilized egg. It later divides into three layers (the germ layers), from which all parts of the animal are formed. [from *blasto-* + Greek *derma* skin]
—**blas′toder′mic,** *adj.* of or having to do with a blastoderm.

blastodisk or **blastodisc,** *n. Embryology.* a disklike aggregation of formative protoplasm at one pole of the yolk of a fertilized egg, containing the nucleus.

blastomere (blas′tə mir), *n. Embryology.* any of the cells or segments into which a fertilized egg first divides. Each blastomere has an equal number of chromosomes. [from *blasto-* + Greek *meros* part]
—**blastomeric,** *adj.* of, having to do with, or resembling a blastomere.

blastomycete (blas′tə mī′sēt), *n. Botany.* any of a group of yeastlike fungi that reproduce by multiple budding cells and cause various diseases of the skin and tissues in human beings. [from *blasto-* + Greek *mykētos* fungus]
—**blastomycetic,** *adj.* having to do with or due to blastomycetes.

blastopore, *n. Embryology.* the external opening of the cavity (archenteron) of a gastrula: *A slit-like blastopore arises at the margin of the blastoderm and invaginates into the flattened blastocoel to form the endoderm* (Winchester, *Zoology*).
—**blastoporic,** *adj.* of or having to do with a blastopore.

blastula (blas′chə lə), *n., pl.* **-las, -lae** (-lē). *Embryology.* an early stage in the development of an embryo of an animal, after the fertilization and cleavage of the ovum, consisting of a single layer of cells arranged spherically to enclose a cavity: *As cleavage continues, the cells become arranged in the form of a hollow ball, or blastula, within which a segmentation cavity, or blastocoel, appears* (Storer, *General Zoology*).
[from New Latin, from Greek *blastos* sprout, germ]
—**blastular,** *adj.* having to do with or resembling a blastula.
—**blastulation** (blas′chə lā′shən), *n.* formation of the blastula.

bleach, *Chemistry.* —*v.* to whiten by exposing to sunlight or by using chemicals. Textiles, wheat flour, sugar, and paper are examples of products which are sometimes bleached. **Bleaching powder** is a white powder used for bleaching and disinfecting, made by treating slaked lime with chlorine; it is also called **chloride of lime.**
—*n.* **1** any chemical used in bleaching, such as hydrogen peroxide, chloride of lime, and sodium hypochlorite. **2** the act or process of bleaching.

bleed, *v.* **1** *Botany.* to lose or take sap, juice, etc., from a surface that has been cut or scratched, as a tree or vine.
2 *Medicine.* to draw or let blood from.

blende (blend), *n. Mineralogy.* **1** = sphalerite. **2** any of certain other sulfides having a resinlike luster. [from German *Blende*]

blepharoplast (blef′ər ə plast), *n. Biology.* a structure at the base of the flagellum of certain cells that divides when the nucleus and cell divide. Compare **kinetoplast.** [from Greek *blepharon* eyelid + English *-plast*]

blight, *Botany.* —*n.* **1** any disease of plants that causes leaves, stems, fruits, and tissues to wither and die. Rust, mildew, and smut are blights. **2** the bacterium, fungus, or virus that causes such a disease.
—*v.* to suffer from blight; cause to wither and die: *Mildew blights roses.*

blind, *adj.* **1** not able to see. **2** *Botany, Zoology.* abortive: *a blind bud.* **3a** performed without knowledge of certain facts which are usually a part of the process: *a blind test. Analysis of the research done to establish the effectiveness of antidepressant drugs ... showed ... great deficiencies in terms of:* (*i*) *control groups;* (*ii*) *"blind" techniques* (Science). **b** used in blind testing: *By blind samples I mean that the laboratory will receive these as unknowns and will report results to the N.I.H. where the key is held* (N.Y. Times). Compare **double-blind, single-blind.**
4 *Medicine, Zoology.* closed at one end: *a blind fistula, blind pouches.*

blind spot, *Anatomy.* a point on the retina, not sensitive to light, where the optic nerve enters the eye: *Clearly there must be a 'hole' in the retina to let these optic nerves through, and where there is this hole there are no receptors and consequently a blind spot* (Science News). Also called **optic disk.**

blink microscope, *Optics.* an instrument through which two photographs may be viewed in rapid alternation to check the alteration in position or movement of a star, comet, or other heavenly body.

BL Lacertae (lə sèr′tē) or **BL Lac,** *Astronomy.* any of a group of compact celestial bodies beyond our galaxy that are the sources of intense radiation but lack clearly defined emission lines in their spectra: *The [prototype] "star" had been known as BL Lacertae, the BL being a two-letter code designating it as the ninetieth variable star discovered in the constellation Lacerta, the Lizard* (Sullivan, *Black Holes*).

block, *n.* **1** *Medicine.* an obstruction of a normal function, as in the passage of nerve or muscular impulses: *nerve block.*
2 *Geology.* a faulting in blocklike sections. **Block mountains** are formed when the land between two faults is raised and tilted.
—*v. Medicine.* to prevent normal function of (a nerve or muscular impulse), especially by the use of anesthesia.

blood, *n. Biology.* **1** the red-colored liquid in the veins, arteries, and capillaries of vertebrates, consisting of a fluid pale-yellow plasma and semisolid red and white blood cells and platelets. Blood is circulated by the heart, carrying oxygen and digested food to all parts of the body and carrying away waste materials.
2 the corresponding colored or colorless liquid in animals other than vertebrates.

blood-brain barrier, *Physiology.* a selective impermeability of the capillaries of the brain which prevents certain substances from entering the brain while allowing others to enter rapidly: *The blood-brain barrier is often a major factor determining the amount of a drug that reaches the extracellular fluid of the brain after systemic administration* (Science). Compare **blood-CSF barrier.**

blood cell, *Biology.* any of the cells or corpuscles contained in blood; an erythrocyte or leucocyte: *Blood cells are formed chiefly in the marrow contained in cavities within the bones* (Storer, *General Zoology*).

blood count, *Medicine.* a count of the number of red and white blood cells and the amount of hemoglobin in a sample (usually a cubic centimeter) of a person's blood to see if it is normal.

blood-CSF barrier, *Physiology.* a selective impermeability of the capillaries in the central nervous system which prevents most harmful substances in the blood from reaching the cerebrospinal fluid (CSF). Compare **blood-brain barrier.**

blood fluke, = schistosome.

blood group, *Immunology.* any of various groups into which human blood may be divided on the basis of the presence or absence of certain substances (antigens): *several genetically determined systems of blood groups ... found that human blood could be classed as A, B, AB, and O in accordance with the way the red blood corpuscles clump* (Ralph Beals and Harry Hoijer). Also called **blood type.** See **ABO, Rh factor.**

blood heat, *Physiology.* the normal temperature of human blood, equivalent to about 37.0°C.

blood plasma, *Biology.* the liquid part of blood, often used in transfusions in place of whole blood.

blood platelet, *Biology.* any of the colorless round or oval disks in the blood of vertebrates, smaller than red blood cells, important in coagulation: *When a blood vessel is injured, the ... tissues and ... blood platelets, floating in the plasma release substances which provide thrombin that acts upon the fibrinogen in the plasma to bring about clotting* (Storer, *General Zoology*). Also called **thrombocyte, platelet.**

blood pressure, *Physiology.* the pressure of the blood against the inner walls of the blood vessels, varying with the strength of the heartbeat, the volume of the blood, the elasticity of the arteries, and the person's health, age, and physical condition: *If the blood pressure is either too high or too low, the constancy of the internal environment and hence survival are threatened* (McElroy, *Biology and Man*).

bloodstream, *n. Physiology.* the blood as it flows through the body.

blood sugar, *Biochemistry.* glucose in the blood, the presence of which in excessive quantities is a sign of diabetes.

blood test, *Medicine.* an examination of a sample of blood to determine its contents, as for ascertaining the blood group, diagnosing illness, etc.

blood transfusion, *Medicine.* the injection of blood from the bloodstream of one person or animal into that of another.

blood type, = blood group.
—**blood-type,** *v.* to classify (the blood) according to type or (persons) according to blood group: *Blood-typing in the ABO system is based on the presence or absence of two types of antigen, A and B, on the surface of red blood cells.*

blood vessel, *Anatomy.* any of the flexible tubes in the body through which the blood circulates. Arteries, veins, and capillaries are blood vessels.

bloom, *n.* **1** *Botany.* **a** the flower of a plant or flowers collectively. SYN: efflorescence. **b** a coating like fine powder on some fruits and leaves. There is bloom on grapes and plums and on the leaves of cabbage.
2 *Zoology.* **a** a dense mass of very small, aquatic organisms, such as plankton, easily visible to the eye: *Darwin saw red water 50 miles off the southern coast of Chile. He observed that the water was colored by some organism other than the red-pigmented alga Trichodesmium, whose nontoxic "blooms" are common in the tropics and give the Red Sea its name* (Scientific American). **b** the cuticle of an eggshell: *When eggs are first laid, they have a protective film covering them which is known as the "bloom"; it seals*

cap, fāce, fäther; best, bē, tèrm; pin, five;
rock, gō, ôrder; oil, out; cup, pùt, rüle,
yü in use, yu̇ in uric;
ng in bring; sh in rush; th in thin, ᴛʜ in then;
zh in seizure.
ə = a in about, e in taken, i in pencil, o in lemon, u in circus

the pores and helps to keep odors out of the eggs (Science News Letter).

3 *Geology.* any of various supergene, efflorescent minerals occurring as a powder or crust: *cobalt bloom.*

blow, *v. Zoology.* **1** (of whales) to spout a column of hot, moist air from the blowholes, before taking in fresh air. **2** (of insects) to lay eggs in: *Some flies blow fruit.*

blowhole, *n.* **1** *Zoology.* a nostril or hole for breathing, in the top of the head of whales, porpoises, and dolphins: *Through its blowholes ... it forcefully expels air from its lungs and immediately takes a deep breath* (Scientific American).

2 *Geology.* a crack in a sea cliff through which columns of spray are ejected by incoming waves.

blowout, *n. Geography, Geology.* a shallow depression in the land surface of an arid region caused by the action of wind on loose surface material. See the picture at **dune.**

blubber, *n. Zoology.* the fat of whales and some other sea animals, lying under the skin and over the muscles, from which oil is obtained. Blubber insulates the animal from heat loss and serves as a food reserve.

blue galaxy, *Astronomy.* a galaxy beyond the Milky Way that is blue in color, generally because of the presence of young, massive blue stars.

blue-green algae, *Biology.* a division (Cyanophyta) of algae, generally bluish green because of the presence of a bluish pigment masking the chlorophyll, consisting of one or more cells that lack definite nuclei: *The blue-green algae commonly appear as a bluish green smear on moist surfaces of rocks and bark of trees, on and below the surface of damp soil* (Greulach and Adams, *Plants*).

▶ In a different classification, these organisms are called *blue-green bacteria* belonging to the division Cyanobacteria.

blue shift, *Astronomy.* a shift of the light in a celestial body towards the blue end of the spectrum, indicating movement of the light source towards the observer (Doppler shift): *During the collapse, the Doppler effect which, in our present universe, gives rise to a red shift would be reversed and any radiation produced would have been amplified by a blue shift* (Listener). Compare **red shift.**

—**blue-shift,** *v.* to shift towards the blue end of the spectrum: *During the contracting phase of each cycle, starlight and other radiation is blue-shifted ... eventually being scrambled up in the fireball between cycles* (New Scientist).

▶ Blue shifts are rare phenomena and remain largely theoretical; almost all extragalactic objects show the red shift. In 1978, however, an object (designated SS433) was discovered in the Milky Way appearing simultaneously to move toward and away from the earth at very high speeds—that is, exhibiting alternately a red shift and a blue shift (due to the ejection of beams of particles in two directions from a central star).

blue star, *Astronomy.* **1** a very bright, hot star whose light is largely of a blue color: *A great many stars, including young blue stars and white dwarfs, have surface temperatures far in excess of the sun's relatively cool 6,000 degrees C* (Scientific American). **2** a star located in a blue galaxy.

B lymphocyte, = B cell.

BMR, *abbrev.* basal metabolic rate.

BOD, *abbrev.* biochemical oxygen demand.

Bode's law (bō′dəz), *Astronomy.* an arithmetical approximation of the relative mean distances of the planets from the sun, obtained by adding 4 to the series 0, 3, 6, 12, 24, 48, 96, 192, and 384. The law applies to all the planets except Neptune and Pluto. *The spacing of the planets at their present distances from the sun ... follows a regular mathematical relationship known as Bode's Law* (Fred L. Whipple). [named after Johann E. *Bode,* 1747–1826, German astronomer who discovered the law]

body, *n.* **1** *Biology.* the whole material or physical structure of an organism, whether living or dead.

2 *Zoology.* the main part or trunk of an animal, apart from the head, limbs, or tail.

3 *Anatomy.* a small organ: *the pineal body, the pituitary body.*

4 a portion of matter; mass: *The stars and planets are celestial bodies. A lake is a body of water.*

body burden, *Biology, Medicine.* a portion of radioactive material absorbed in the body.

body cavity, = coelom.

body-centered, *adj. Crystallography.* having a cubic structure with atoms or ions at the corners and in the center of the unit cell. Contrasted with **face-centered.**

body clock, *Physiology.* an internal mechanism of the body that is supposed to regulate physical and mental functions in rhythm with normal daily activities. Swift transition through several time zones during jet flight disturbs the body clock. Compare **biological clock.**

bog, *n. Ecology.* an area of wet, spongy ground, consisting chiefly of decayed or decaying moss and other vegetable matter. A bog is created when this matter sinks to the bottom of a lake or pond, forming a slimy sediment. The water eventually evaporates or seeps out. A bog is usually a sere in a hydrarch succession that terminates in a dry-land community.

ASSOCIATED TERMS: A *bog* is a tract of low-lying ground on peaty soil which is highly acidic; a *fen* typically has peaty soil which ranges from very alkaline to slightly acidic. In contrast with a bog and fen, a *moor* is an area of high-lying land characterized by heather and other dwarf shrubs, while a *marsh* is low-lying wet land with grasses, sedges, cattails, and rushes. A *swamp* is marshy land with dense vegetation including trees. A *heath* is an uncultivated treeless habitat with low shrubby plants.

Bohr atom (bôr *or* bōr), *Physics.* the atomic structure described by the Bohr theory.

Bohr effect, *Biochemistry.* an effect by which an increase of carbon dioxide in the blood results in the dissociation of oxygen from hemoglobin and other respiratory compounds. [named after Christian *Bohr,* 19th-century Danish physiologist]

Bohr magneton, *Physics.* one half the unit of magnetic moment for electrons. [named after Niels *Bohr;* see **Bohr theory**]

Bohr model, *Physics.* a model of the hydrogen atom proposed by Niels Bohr, consisting of a positively charged nucleus and a single electron revolving in a circular orbit: *The Bohr model is not a general solution to the problem of atomic structure and has been replaced by a wave-mechanical model* (Physics Regents Syllabus). Compare **orbital.**

Bohr theory, *Physics.* a theory of the structure of atoms stating that electrons revolve around a nucleus in certain orbits of constant energy only. When an electron jumps to another orbit of less energy, energy is radiated; when it jumps to a higher energy level, energy is absorbed. [named after Niels *Bohr,* 1885–1962, Danish physicist who proposed it in 1913]

boil, *v. Chemistry.* to change from a liquid state to a gaseous state by being heated to the boiling point: *Water boils at 100 degrees Celsius.*

boiling point, *Chemistry.* the temperature at which the vapor pressure of a liquid is equal to the atmospheric pressure acting upon the surface of the liquid. The boiling point of water at sea level is 100 degrees Celsius (212 degrees Fahrenheit). *Abbreviation:* b.p. Compare **freezing point, melting point.**

bolide (bō'līd), *n. Astronomy.* a large meteor, usually one that explodes and falls in the form of meteorites: *A bolide, exploding with a loud detonation, is apt to scatter dozens of stony fragments within a small area* (Bernhard, *Handbook of Heavens*). [from Latin *bolidem,* from Greek *bolis* missile]

boll, *n. Botany.* the rounded seed pod or capsule of a plant, especially that of cotton or flax. A **boll weevil** is a beetle whose larva does great damage to young cotton bolls.

bolson (bōl'sən), *n. Geology.* a broad, shallow desert basin draining into a central, shallow lake or sink, especially such a basin surrounded by mountains in the southwestern United States and Mexico. [from American Spanish *bolsón,* from Spanish *bolsa* purse]

Boltzmann constant (bōlts'mən), a constant in physics equal to 1.38×10^{-23} joules per kelvin, a fundamental constant in the kinetic theory of gases; molecular gas constant. [named after Ludwig *Boltzmann,* 1844–1906, Austrian physicist who determined the constant]

Boltzmann distribution, *Physics.* an equation that expresses the distribution of energies for a collection of particles at a fixed temperature, based on the Boltzmann constant.

bombard, *v. Nuclear Physics.* to strike (the nucleus of an atom) with a stream of fast-moving particles, radioactive rays, etc., to change the structure of the nucleus: *Physicists at Brookhaven expect to learn more about the many kinds of particles, such as mesons and hyperons, and the various "antiparticles," that are produced in target nuclei by bombarding them with high-energy protons* (Science News Letter).

—bombardment, *n.* the act or process of bombarding: *Induced transmutations may be produced by bombardment of nuclei* (Physics Regents Syllabus).

bond, *Chemistry.* **—n.** a unit of force by means of which atoms or groups of atoms are combined or joined together in a molecule. A chemical bond usually consists of a pair of shared electrons. The number of its bonds indicates the valence of an element or radical. Also called **chemical bond.** See **covalent bond, hydrogen bond, ionic bond.**

—v. to hold together by a chemical bond: *The combined effect of pressure and firing bonds the crystals* (New Scientist). *The silicone liquid bonds into the fibers without affecting color* (Science News Letter).

bone, *n. Anatomy.* **1** one of the distinct pieces making up the skeleton of a vertebrate animal. There are more than 200 different bones in the human body.

2 the hard tissue forming the substance of the skeleton: *Bone consists of a dense organic matrix (chiefly collagen) with deposits of mineral substance ... [and] ... is, therefore, a living tissue* (Storer, *General Zoology*).

bone cell, = osteoblast.

bony fish = teleost.

book lung, *Zoology.* a breathing organ in scorpions and spiders consisting of many parallel layers arranged like leaves of a book.

Boolean or **Boolian** (bü'lē ən), *adj.* of or having to do with Boolean algebra.

Boolean algebra or **Boolian algebra,** a mathematical system dealing with the relationship between sets, used to solve problems in logic, engineering, etc. Variables consist of 0 or 1 and functions are expressed in *and, or,* and *not.* [named after George *Boole,* 1815–1864, English mathematician]

bora (bôr'ə or bōr'ə), *n. Meteorology.* a violent, dry, cold, north or northeast wind of the Adriatic and its coasts. [ultimately from Latin *boreas* north wind]

boracic (bə ras'ik), *adj.* = boric.

boracite (bôr'ə sīt), *n. Mineralogy.* a borate and chloride of magnesium, occurring in translucent to transparent, white or colorless crystals. *Formula:* $Mg_3B_7O_{13}Cl$

borane (bôr'ān), *n. Chemistry.* **1** a hydride of boron.
2 a derivative of such a compound.

borate (bôr'āt), *n. Chemistry.* a salt or ester of boric acid.

borax (bôr'aks), *n. Chemistry, Mineralogy.* a white crystalline powder and mineral having a sweetish alkaline taste, used as an antiseptic, as a cleansing agent, in fusing metals, and in making heat-resistant glass. *Formula:* $Na_2B_4O_7 \cdot 10H_2O$ Also called **sodium borate.** [from Medieval Latin, from Arabic *būraq*]

bordered pit, *Botany.* a thick hollow area in the wall of a tracheid around which a thickened cellular growth overhangs.

bore, *n. Oceanography.* a sudden, high tidal wave that rushes up a channel with great force: *The bore of the Hangchow River in China is as much as 16 feet high* (Gilluly, *Principles of Geology*). [from Scandinavian (Old Icelandic) *bāra* wave]

boreal (bôr'ē əl), *adj. Geography, Ecology.* **1** of the north; northern. The **Boreal Zone** is a region characterized by a northern type of fauna or flora. **2** of the north wind.

boric (bôr'ik), *adj. Chemistry.* of or containing boron. **Boric acid** is a white, crystalline compound occurring in nature or made from borax, used as a mild antiseptic

---

cap, fāce, fäther; best, bē, tėrm; pin, five;
rock, gō, ôrder; oil, out; cup, pùt, rüle;
yü in use, *yù* in uric;
ng in bring; *sh* in rush; *th* in thin, ⊥H in then;
zh in seizure.
ə = *a* in about, *e* in taken, *i* in pencil, *o* in lemon, *u* in circus

bornite

and in making cement, glass, soap, etc. *Formula:* H_3BO_3

bornite (bôr′nīt), *n. Mineralogy.* a brittle, reddish-brown sulfide of copper and iron, an ore of copper. *Formula:* Cu_5FeS_4 [from German *Bornit,* from Ignatius von *Born,* 1742–1791, Australian mineralogist + *-it* -ite]

boron (bôr′on), *n. Chemistry.* a nonmetallic element which occurs only in borax and other compounds, used in alloys, nuclear reactors, etc. *Symbol:* B; *atomic number* 5; *atomic weight* 10.811; *melting point* 2300°C; *oxidation state* 3. [from *bor(ax)* + *(carb)on*]

Bose-Einstein statistics (bōs′in′stīn), a statistical theory of quantum mechanics: in the distribution of a given type of nuclear particle any number of identical particles may occupy a particular quantum-mechanical state. Compare **Fermi-Dirac statistics.** [named after Satyendranath *Bose* (see the etymology under **boson**) + Albert *Einstein*]

boson (bō′son), *n. Physics.* (in quantum mechanics) any of a class of elementary particles, including photons, gluons, and weakons which conform to Bose-Einstein statistics and transmit forces between particles.
[from Satyendranath *Bose,* 1894–1974, Indian physicist + *-on*]
▶ *Bosons* have either zero or integral spin, as distinguished from *fermions,* which have fractional spin.

bot., *abbrev.* **1** botanical. **2** botany.

botanical, *adj.* **1** having to do with plants and plant life. **2** having to do with botany.

botanize, *v.* **1** to study plants in their natural environment. **2** to collect plants for study, classification, etc. **3** to explore or examine the plant life of.

botany, *n.* **1** a branch of biology that deals with plants and plant life; the study of the structure, growth, classification, diseases, etc., of plants.
2 the plant life of a particular area: *the botany of Greenland.*
3 botanical facts or characteristics concerning a particular plant or group of plants: *the botany of roses.*
[from Greek *botanē* plant]

botryoidal (bot′rē oi′dəl), *adj. Geology.* having the form of a bunch of grapes.

bottom quark, *Nuclear Physics.* a quark that has three times the mass of a charmed quark: [*The*] *six kinds of quark that form the basis of nuclear matter ... There are up quarks and down quarks* (*which compose the neutron and proton*); *charmed quarks and strange quarks; and top quarks and bottom quarks* (*sometimes called "truth" and "beauty"*) (New Scientist).

botulism (boch′ə liz′əm), *n. Medicine.* poisoning, frequently fatal, caused by a toxin (*botulin*) secreted by a certain anaerobic bacterium (*Clostridium botulinum*) sometimes present in foods not properly canned or preserved. [from Latin *botulus* sausage; originally attributed especially to sausages]

bouillon (bùl′yon), *n. Bacteriology.* a liquid, nutritive medium used for growing cultures of bacteria. Compare **broth.** [from French, from *bouillir* to boil]

boulder, *n. Geology.* a large, detached rock whose edges have become rounded or worn by the action of water or weather. [short for *boulderstone,* from Scandinavian (Swedish) *bullersten*]

boulder clay, = till.

bound, *adj.* held by a chemical bond. **Bound water** is water whose molecules are held by a hydroxyl or other group in a larger molecule. *In the freeze-drying process ... almost all the water is removed, leaving only about 2.5 per cent of bound water which prevents the protein molecules from breaking down* (New Scientist).
—*n. Mathematics.* a number which a function does not exceed (upper bound) or does not go below (lower bound).
—*v. Mathematics.* to form the boundary of.

boundary layer, *Physics.* the region next to the surface of a stationary body in which fluid flows at a much slower velocity than elsewhere.

bound charge, *Physics.* an electrostatic charge which is bound by the presence of a charge of opposite polarity. Compare **free charge.**

bounded, *adj. Mathematics.* **1** (of a function) having both an upper bound and a lower bound, always less than some fixed number and also greater than some fixed number. **2** contained within some fixed closed curve.

bound electron, *Physics.* an electron which is bound to a proton and neutron in certain atoms.

bovine (bō′vīn), *Zoology.* —*adj.* of, having to do with, or belonging to a genus (*Bos*) of ruminant mammals, that include domestic cattle, water buffaloes, and the like: *Bovine tuberculosis is a disease of cattle ...* [*having*] *living bovine bacilli* (Beaumont and Dodds, *Recent Advances in Medicine*).
—*n.* any bovine animal, such as an ox or cow.
[from Late Latin *bovinus,* from Latin *bos, bovis* ox, cow]

bowel, *n. Anatomy.* **1** the intestine. **2** a part or division of the intestine.

Bowman's capsule, *Anatomy.* a double-walled, cup-shaped structure around the glomerulus of each nephron of the kidney. It serves as a filter to remove organic wastes, excess inorganic salts, and water. [named after William *Bowman,* 1816–1892, English anatomist]

bow shock (bou), *Astronomy.* an area in outer space partly ahead of a planet, where the solar wind is disturbed by the impact of the planet's magnetic field: *The contact zone between the Earth's magnetosphere and the solar wind is referred to by space scientists as the bow shock ... it is the region where high-velocity solar wind first interacts with the magnetic field of the Earth* (Wernher von Braun).

Boyle's law, *Physics.* the statement that at a constant temperature the volume of a given quantity of gas varies inversely with the pressure to which it is subjected. Compare **Charles's law.** See also **gas laws** under **gas.** [named after Robert *Boyle,* 1627–1691, Irish scientist, who formulated the law]

b.p., *abbrev.* boiling point.

b quark, = bottom quark.

Br, *symbol.* bromine.

brace root, = prop root.

brachial (brā′kē əl), *adj.* **1** *Anatomy.* of or belonging to the arm: *the brachial artery.*
2 *Zoology.* **a** of or belonging to the forelimb of a vertebrate. **b** armlike: *the brachial appendages of a starfish.* [from Greek *brachiōn* arm]

76

brachiate (brā′kē āt), *v. Zoology.* to move by swinging from branch to branch with the arms, as gibbons and certain arboreal monkeys do.
—**brachiation** (brā′kē ā′shən), *n.* the act or practice of brachiating: *Apes employ the hook grip during the specialized mode of locomotion known as brachiation, which involves progress through the trees suspended by the hands and arms alone* (New Scientist).

brachiocephalic (brā′kē ə sə fal′ik), *adj. Anatomy.* of or involving the arm and the head. The innominate artery and vein are sometimes called the **brachiocephalic artery** and **brachiocephalic vein.**

brachiopod (brā′kē ə pod), *n. Zoology.* any of a phylum (Brachiopoda) of sea animals characterized by bivalve shell and, coiled within the shell, a pair of arms covered with cilia for sweeping tiny food organisms into the mouth. [from Greek *brachiōn* arm + *podos* foot]

brachistochrone (brə kis′tə krōn), *n. Physics.* the curve in which a body descending to a given point under an external force will reach another point in the shortest possible time; the curve of quickest descent. [from Greek *brachistos* shortest + *chronos* time]

brachium (brā′kē əm), *n., pl.* **-chia** (-kē ə). **1** *Anatomy.* the upper arm, from the shoulder to the elbow.
2 *Zoology.* the part of any limb or process corresponding to an arm.
[from Latin *bracchium* arm, from Greek *brachiōn*]

brachypterous (brə kip′tər əs), *adj. Zoology.* having very short or rudimentary wings: *brachypterous birds.* [from Greek *brachys* short + *pteron* wing]

bract, *n. Botany.* a small leaf growing at the base of a flower or on a flower stalk. See the picture at **subtend.** [from Latin *bractea* thin metal plate]
—**bracteal** (brak′tē əl), *adj.* of, like, or being a bract.
—**bracteate** (brak′tē it) or **bracted,** *adj.* having bracts.

bracteole (brak′tē ōl), *n. Botany.* a small bract situated on a secondary axis, as on a pedicel, or even on a petiole. [from Latin *bracteola,* diminutive of *bractea* thin metal plate]

bractlet, *n.* = bracteole.

Boyle's Law

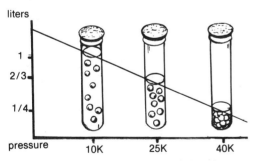

liters

1
2/3
1/4

pressure   10K   25K   40K

The volume or mass of the gas in the tubes decreases as the pressure in the tubes increases. At a constant temperature, gas compressed to one-fourth its original volume will quadruple its pressure.

bradycardia (brad′ə kär′dē ə), *n. Physiology.* slowness of the heartbeat or pulse: *Many normal persons have a slow heartbeat (bradycardia) of 40–50 beats per minute* (Harry L. Jaffe). [from Greek *bradys* slow + *kardia* heart]

bradykinin (brad′ə ki′nən), *n. Biochemistry.* a polypeptide released by blood plasma globulin when acted on by certain enzymes. It stimulates the action of smooth muscle, lowers blood pressure, and is involved in producing some of the signs of inflammation, such as swelling and pain. Compare **kallidin.** [from Greek *bradys* slow + English *kinin*]

Bragg's law, a fundamental law of X-ray crystallography, which states the condition for diffraction of an incident beam of monochromatic X rays by the successive layers of atoms or ions in a crystal. [named after W.H. *Bragg,* 1862–1942, and his son, W.L., 1890–1971, English physicists who formulated the law]

braided stream, *Geography.* a shallow stream which divides into multiple channels that reunite and subdivide again so that its paths of flow take on a braided appearance.

brain, *n.* **1** *Anatomy.* the soft, grayish and whitish mass of nerve cells and nerve fibers enclosed in the skull or head of vertebrate animals, and in man consisting of the cerebrum, cerebellum, pons, and medulla oblongata. The brain is the organ of consciousness and furnishes outgoing stimulation of muscles as a response to incoming sensory stimulation. Technically called **encephalon.**
2 *Zoology.* the part of the nervous system of invertebrates corresponding in position or function to the brain of vertebrates. See the pictures at **crustacean, thorax.**

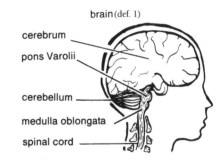

brain (def. 1)

cerebrum
pons Varolii
cerebellum
medulla oblongata
spinal cord

braincase, *n. Anatomy.* the part of the skull enclosing the brain. SYN: cranium.

brain death, *Medicine.* the prolonged and complete cessation of electrical activity in the brain as shown by flat tracings of an electroencephalograph. —**brain dead.**

brain hormone, *Biochemistry.* any of various hormones produced in the hypothalamic region of the brain, especially those acting on the pituitary gland to release other hormones: *Before the discovery of brain hor-*

cap, fāce, fäther; best, bē, tėrm; pin, five;
rock, gō, ôrder; oil, out; cup, pùt, rüle,
yü in use, *yù* in uric;
ng in bring; *sh* in rush; *th* in thin, ᴛʜ in then;
zh in seizure.
ə = *a* in about, *e* in taken, *i* in pencil, *o* in lemon, *u* in circus

mones, the pituitary ... was commonly called the body's "master gland" (Edward Edelson).

brainpan, *n.* = braincase.

brain stem, *Anatomy.* the base of the human brain lying beneath the cerebrum and the cerebellum. It consists of the midbrain, the pons, and the medulla oblongata, and connects the spinal cord with the forebrain. *Although the higher centers in the brain are capable of extensive learning, the lower centers in the brain stem and spinal cord are quite implastic* (Scientific American).

brain wave, *Physiology.* electric current produced by the rhythmic electric fluctuations between the parts of the brain: *Elaborate devices for detecting and recording electrical potentials in the brain and along nerve systems have been long used, as, for instance, in the conventional electroencephalograph, the electrical "brain wave" recorder* (N.Y. Times).

branch, *n.* **1** *Botany.* the part of a tree, shrub, or other plant growing out from the trunk; any woody part of a tree above the ground except the trunk.
2 *Geology.* a tributary stream.
—*v.* **1** to put out branches; spread in branches. **2** to divide into branches.

branched chain, *Chemistry.* a molecular pattern in which a chain of atoms branches off from another chain: *The other members of the Methane Series are similar to ethane and propane but differ in having longer and sometimes branched chains of carbon atoms* (Parks and Steinbach, *Systematic College Chemistry*).

branchia (brang′kē ə), *n., pl.* **branchiae** (brang′kē ē). *Zoology.* a gill or gill-like organ. [from Greek, plural of *branchion* gill]

branchial (brang′kē əl), *adj. Zoology.* of, having to do with, or resembling gills or gill-like organs.

branchial arch, = gill arch.

brass, *n. Chemistry.* a yellowish, malleable, ductile metal that is an alloy of copper and zinc in various proportions. Compare **bronze.**

braunite (brou′nīt), *n. Mineralogy.* a brittle, brownish-black mineral, a silicate of manganese, occurring in both crystalline and massive form. *Formula:* $Mn^{2+}Mn^{3+}_6SiO_{12}$ [from August E. *Braun,* 1809–1856, German archaeologist + *-ite*]

breadcrust bomb, *Geology.* a volcanic bomb having a compact but cracked crust and vesicular interior.

breastbone, *n. Anatomy.* the thin, flat bone in the front of the chest attached by cartilages to the ribs. Technically called **sternum.**

breathhold diving, *Zoology.* a form of diving used by seals, dolphins, etc., in which the breath is held under water for regular periods during which the heart rate and brain waves markedly slow down: *The adaptation to breathhold diving of marine animals is probably an especially well developed instance of a very general asphyxial defence mechanism common to all vertebrates from fish to man* (Science Journal).

breccia (brech′ē ə), *n. Geology.* a rock consisting of angular fragments of older rocks cemented together: *Sedimentary breccias resemble conglomerate except*

that most of their fragments are angular instead of rounded (Gilluly, *Principles of Geology*).
—**brecciated** (brech′ē ā′tid), *adj.* having the form of a breccia.
—**brecciation** (brech′ē ā′shən), *n.* brecciated condition.
[from Italian *breccia*]

breed, *v.,* **bred, breeding. 1** *Biology.* **a** to propagate sexually, especially under controlled conditions: *to breed new varieties of corn. Snapdragons, zinnias and marigolds ... have been bred "down" ... to produce miniature-sized plants* (J.L. Faust). Compare **crossbreed, interbreed. b** to produce offspring.
2 *Nuclear Physics.* to produce (fissionable material) in a breeder reactor: *With the proper array of fertile U-238 and fissionable U-235, a reactor may be able to breed more fissionable fuel than it consumes* (R.C. Cowen).
—*n. Biology.* a group of organisms having common ancestors and certain distinguishable characteristics, such as of color, size, shape, etc., especially a group within a species developed by artificial selection and maintained by controlled propagation: *Jerseys and Guernseys are breeds of cattle.*
—**breeder,** *n.* **1** an animal or plant kept for breeding.
2 = breeder reactor.

breeder reactor, a nuclear reactor that produces at least as much fissionable material as it uses. In one type of reaction it consumes uranium and produces plutonium ... *breeder reactors, in which the nuclear reaction is arranged to produce new fuel as the reactor generates power* (Barry Commoner).

breeding, *n.* **1** *Biology.* the producing of offspring.
2 *Genetics.* the producing of animals or new types of plants, especially to get improved kinds.
3 *Nuclear Physics.* production in a reactor of at least as much fissionable material as is used.

breeze, *n. Meteorology.* a wind having a velocity of 6 to 49 kilometers per hour. See **Beaufort Scale.**

bregma (breg′mə), *n., pl.* **-mata** (-mə tə). *Anatomy.* the region of the skull where the sagittal and coronal sutures join. [from Greek *bregma* front of the head]
—**bregmatic** (breg mat′ik), *adj.* having to do with the bregma.

brei (brī), *n. Biology.* tissue ground to a pulp for study or experimentation: *Extracts of yeast or spleen were as effective as breis made of these tissues* (Bulletin of Atomic Scientists). [from German *Brei* pulp]

bremsstrahlung (brems′shträ′lùng), *n. Physics.* the electromagnetic radiation produced by the deceleration (and sometimes acceleration) of charged particles, as when an electron is absorbed or emitted by an atomic nucleus. [from German *Bremsstrahlung,* literally, breaking radiation]

Brewster angle, = polarizing angle. [named after David *Brewster,* 1781–1868, Scottish physicist]

Brewster's law, *Optics.* the law relating as an equality the index of refraction of a substance and the tangent of the polarizing angle.

bridge, *n.* **1** *Anatomy.* a ridge, prominence, or similar mass of tissue connecting two parts of an organ.
2 *Chemistry.* an atom or group of atoms connecting another or others, especially within a ring.
3 *Electricity.* an apparatus for measuring the resistance of a conductor, or its inductance or capacitance.

bright-line spectrum, *Physics.* an emission spectrum composed of a pattern of bright lines on a dark background, having as its source a glowing gas that radiates in special wavelengths characteristic of the chemical elements in the composition of the gas. Compare **continuous spectrum** and **dark-line spectrum.**

brightness, *n. Optics.* **1** the intensity of light radiated or reflected by a body: *the brightness of a star.* SYN: luminance. See **albedo.**

2 the intensity of a color, as distinguished from its hue and saturation, especially as classified on a scale ranging from black (zero brightness) to white (maximum brightness).

Brillouin scattering (brē yə wåN'), *Physics.* the scattering of light by phonons: *The formula for Brillouin scattering makes it possible to measure the velocity of sound at various wavelengths in any liquid and thus to study important properties of materials* (Scientific American). [named after L. *Brillouin,* 20th-century French physicist]

Brillouin zone, *Physics.* a polyhedral zone in a crystal lattice, used to define the frequency or energy of wave motion, especially in the study of complex metals and alloys.

Brinell hardness (bri nel'), *Chemistry.* the relative hardness of a metal, determined by pressing a steel ball under a given pressure into the substance under test, expressed in kilograms per square millimeter. [named after J. A. *Brinell,* 1849–1925, Swedish engineer who devised the test]

bristle, *n. Zoology, Botany.* one of the short, stiff hairs of some animals or plants: *the bristles of a hog.*

Britannia metal or **britannia metal,** *Chemistry.* a white alloy of tin, copper, and antimony. [from Latin *Britannia* Britain]

British thermal unit, a unit for measuring heat, equal to the amount of heat necessary to raise the temperature of 0.45 kilogram of pure water 0.56 degree Celsius: *The only other heat unit in use in the United States is the British thermal unit (BTU) ... the amount of energy required to raise the temperature of one pound of water from 58.5°F to 59.5°F* (Shortley and Williams, *Elements of Physics*).

Brix (briks), *Chemistry.* —*adj.* of, relating to, or based on a hydrometric scale (**Brix scale**) for measuring the density of sugar solutions. The degrees Brix represent the percentage by weight of sugar in a solution at a particular temperature.

—*n.* the Brix scale.

[named after A.F.W. *Brix,* 1798–1890, German scientist]

broadleaf or **broad-leaved,** *adj. Botany.* having leaves that are broad and flat rather than needle-shaped: *a broadleaf tree or evergreen.*

Broca's area (brō'kəz), *Anatomy.* an area of the left hemisphere of the brain having to do with the motor aspects of speech. [named after Pierre Paul *Broca,* 1824–1880, French surgeon]

bromelin (brō'mə lin) or **bromelain** (brom'ə lān), *n. Biochemistry.* an enzyme present in the juice of pineapple which aids in the hydrolysis of proteins. [from New Latin *Bromelia* name of the pineapple genus, named after Olaf *Bromel,* 1639–1705, Swedish botanist]

bromic (brō'mik), *adj. Chemistry.* of or containing bromine, especially with a valence of 5.

bromide (brō'mīd'), *n. Chemistry.* any compound of bromine with another element or radical, especially potassium bromide: *Bromides closely resemble the chlorides in appearance and behavior. The bromide ion is colorless and highly soluble, as are most of its binary compounds* (Jones, *Inorganic Chemistry*).

brominate, *v. Chemistry.* to treat or combine with bromine. —**bromination,** *n.*

bromine (brō'mēn'), *n. Chemistry.* a dark, brownish-red element. Bromine is a nonmetallic liquid somewhat like chlorine and iodine and gives off an irritating vapor. It is used in antiknock compounds for gasoline, in drugs, and in photography. *Bromine formerly came mainly from the salt beds of Strassfurt, Germany ... We obtain most of the bromine now ... from sea water* (Offner, *Fundamentals of Chemistry*). Symbol: Br; atomic number 35; atomic weight 79.904; melting point $-7.3°C$; boiling point 58.8°C; oxidation state $\pm 1, +5$. [from Greek *brōmos* stench]

bromouracil (brō'mō yur'ə sil), *n. Biochemistry.* a compound that can take the place of thymine in DNA and cause a mutation by forming a base pair with adenine or guanine: *For purposes of tagging synthetic DNA, thymine can be replaced by 5'-bromouracil* (Arthur Kornberg). Formula: $C_4H_3N_2O_2Br$

bronchi (brong'kī), *n.pl.* of **bronchus.** *Anatomy.* the two large, main branches of the windpipe, one going to each lung: *The bronchi, which lead from the trachea into the lungs, branch and rebranch like a tree. With a microscope one can see that the smallest branches, the bronchioles, end in many small sacs, the alveoli* (Scientific American).

[from New Latin, plural of *bronchus,* from Greek *bronchos* windpipe]

—**bronchial** (brong'kē əl), *adj.* of the bronchi. The **bronchial tubes** are the bronchi and their branching tubes.

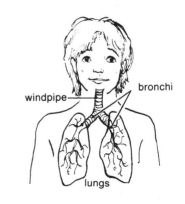

windpipe — bronchi

lungs

cap, fāce, fäther; best, bē, tèrm; pin, fīve;
rock, gō, ôrder; oil, out; cup, pùt, rüle,
yü in use, *yū* in uric;
ng in bring; *sh* in rush; *th* in thin, ŦH in then;
zh in seizure.

ə = *a* in about, *e* in taken, *i* in pencil, *o* in lemon, *u* in circus

bronchiole (brong′kē ōl), *n. Anatomy.* a very small branch of the bronchi: *The bronchioles end in microscopic compartments, or alveoli, surrounded by many blood capillaries, where the respiratory exchanges occur* (Storer, *General Zoology*).

bronchogenic (brong′kō jen′ik), *adj.* **1** *Anatomy.* having to do with or arising from the bronchi: *bronchogenic cysts*
2 *Embryology.* forming or capable of forming the bronchi.

bronchus (brong′kəs), *n.* singular of **bronchi.**

Brönsted acid or **Brønsted acid** (brœn′sted), *Chemistry.* a substance that acts as a proton donor: *The essential point here is that CH_5+ is a Brönsted acid so powerful it treats ordinary aliphatic hydrocarbons as bases and abstracts H^-ions* (Willard F. Libby). [named after Johannes N. *Brønsted,* 1879–1949, Danish chemist, who in 1923 defined acids in terms of proton transfer]
► See the note under **acid.**

Brönsted base or **Brønsted base,** *Chemistry.* a substance that acts as a proton acceptor. ► See the note under **acid.**

bronze, *n. Chemistry.* **1** a brown metal, an alloy of copper and tin.
2 a similar alloy of copper with aluminum or other metals. Compare **brass.**
► The bronzes are composed of copper with usually up to 10 per cent of tin or certain other elements. They are harder than copper, have greater tensile strength, and are highly resistant to corrosion.

brood, *Zoology.* —*n.* the young of a bird or insect hatched at one time or cared for together: *Several groups, or broods, of the periodical cicada exist in the United States. They hatch at different times* (R.E. Blackwelder).
—*v.* **1** (of birds) to sit on eggs so as to hatch them. SYN: incubate. **2** to sit on (eggs) in order to hatch: *Unmated King penguins steal eggs from unguarded nests and brood them* (Science News Letter).

brookite (brùk′it), *n. Mineralogy.* a mineral, titanium oxide, having the same composition as rutile and anatase but occurring in orthorhombic crystals of a color ranging from brown or yellow to black. *Formula:* TiO_2 [from H. J. *Brooke,* 1771–1857, English mineralogist + *-ite*]

broth, *n. Bacteriology.* a medium in which cultures of bacteria are grown. Compare **bouillon.**

brown algae, *Biology.* a division (Phaeophyta) of multicellular marine algae, generally dark brown to olive green due to the presence of a brown pigment making the chlorophyll. The kelps are brown algae.

brown coal, = lignite.

brown fat, *Histology.* fatty tissue that produces heat, found in mammals: *The human baby has brown fat tissue between the shoulder blades, around the neck, behind the breast-bone and around the kidneys—all positions duplicated in the rabbit* (Science Journal).

Brownian movement or **Brownian motion,** *Physics.* a constant, random, irregular motion often observed in very minute particles suspended in a liquid or a gas, caused by the impact of surrounding molecules: *This true movement should be distinguished from the so-called "Brownian movement," a quivering or oscil-*

lating phenomenon shown by minute particles or bodies when freely suspended in a suitable fluid. (Youngken, *Pharmaceutical Botany*).
[named after Robert *Brown,* 1773–1858, Scottish botanist, who first described it]

browridge, *n. Zoology.* the supraorbital ridge that overhangs the eye in the skulls of certain apes and hominids.

bryology (brī ol′ə jē), *n.* the branch of botany that deals with mosses and liverworts. [from Greek *bryon* moss]
—**bryological,** *adj.* of or having to do with bryology.

bryophyte (brī′ə fit), *n. Botany.* any of a division (Bryophyta) of nonflowering plants comprising the mosses and liverworts: *The bryophytes are small plants without wood, flowers, or seeds but sometimes with slightly organized stems and leaves* (Emerson, *Basic Botany*). *Simple multicellular plants, bryophytes, lack vascular tissue* (Biology Regents Syllabus). Compare **pteridophyte, tracheophyte.** [from Greek *bryon* moss + *phyton* plant]
—**bryophytic** (brī′ə fit′ik), *adj.* of or having to do with bryophytes.

bryozoan (brī′ə zō′ən), *Zoology.* —*n.* any of a phylum (Bryozoa) of minute, mosslike, aquatic invertebrates that form permanently attached colonies, reproduce by budding, and have distinct alimentary canals.
—*adj.* of or belonging to this phylum.
[from Greek *bryon* moss + *zōion* animal]

Btu, B.t.u., or **B.T.U.,** *abbrev.* British thermal unit or units.

bubble chamber, *Nuclear Physics.* a small vessel filled with a superheated, pressurized liquid, especially hydrogen or propane, through which charged subatomic particles make a bubbly track by means of which they may be examined and identified. Compare **cloud chamber, spark chamber.**

bubonic (byü bon′ik), *adj. Medicine.* having or characterized by inflammatory swelling of the lymph glands. **Bubonic plague** is caused by certain bacteria which attack the lymph glands, making them swell in the groin and some other parts of the body. [from Greek *boubōn* swelling in the groin]

buccal (buk′əl), *adj. Anatomy.* of or having to do with the cheek: *the buccal nerve.* [from Latin *bucca*]

bud, *n.* **1** *Botany.* **a** a small swelling on a plant consisting of a mass of growing tissue that develops into a flower, leaf, or branch: *Buds are short, young, undeveloped stems or specialized growing points of stems with rudimentary leaves compactly arranged upon them* (Youngken, *Pharmaceutical Botany*). **b** a partly opened flower or leaf.

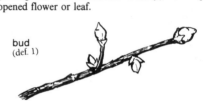

bud
(def. 1)

2 *Biology.* a small swelling (in certain organisms of simple organic structure) that develops into a new individual of the same species: *the buds of yeast plants.* Also called **gemma.** ASSOCIATED TERMS: see **asexual.**

3 *Anatomy.* a minute, bud-shaped part or organ: *a taste bud.*

—*v. Biology.* **1** to put forth buds; form buds: *Yeast cells reproduce asexually by budding.* See the picture at **asexual.**

2 *Botany.* to graft (a bud) from one kind of plant into the stem of a different kind, as a method of propagating a desired quality or variety.

bud scale, *Botany.* one of the small, often resinous or hairy, scalelike leaves forming the outside protective covering of the buds of certain woody plants in winter or in the dry season.

bud variation, *Botany.* a variation in the outgrowth of a bud from the ordinary growth of the plant, producing a sport. Many varieties in cultivated plants arise through a bud variation.

buffer, *Chemistry.* —*n.* a substance in a solution that makes the degree of acidity (hydrogen ion concentration) resistant to change when an acid or base is added.
—*v.* to treat (a solution) with a buffer: *The solution was buffered to keep the pH of the suspension constant* (New Scientist).

buffer solution, *Chemistry.* a solution which resists change in the degree of acidity on the addition of an acid or base: *Solutions which are thus prepared from a weak acid and a salt of that acid or from a weak base and a salt of that base are known as buffer solutions* (Jones, *Inorganic Chemistry*).

bug, *n. Zoology.* any of the order of hemipterous insects that are wingless or have a front pair of wings thickened at the base, and have a pointed beak for piercing and sucking. Bedbugs, lice, and chinch bugs are true bugs. *The word bug is used loosely to refer to all insects and to many things that are not insects, but only members of this order [Hemiptera] can rightfully claim this title* (Winchester, *Zoology*).

bulb, *n.* **1** *Botany.* **a** a round, underground bud from which certain plants such as onions, tulips, and lilies grow: *A true bulb ... consists of a very short, conical stem usually tipped by an apical meristem, and surrounded by a number of thick, fleshy scale leaves.* (Emerson, *Basic Botany*). Compare **corm, tuber. b** any plant that has a bulb or grows from a bulb. The narcissus is a bulb.
2 *Anatomy.* **a** = medulla oblongata. **b** a roundish dilation of any cylindrical organ or structure in an animal body.
[from Latin *bulbus,* from Greek *bolbos*]
—**bulbar** (bul'bər), *adj.* having to do with a bulb, especially the medulla oblongata: *bulbar poliomyelitis.*

bulbil, bulbel (bul'bəl), or **bulblet,** *n. Botany.* an aerial bud with fleshy scales, growing in the leaf axils or taking the place of flowers: *A bulbel or bulbil ... serves as an organ of vegetative multiplication by falling off the stem and developing into a new plant* (Youngken, *Pharmaceutical Botany*). [from New Latin *bulbillus,* from Latin *bulbus* bulb]

bulbourethral glands, = Cowper's glands.

bulbous (bul'bəs), *adj. Botany.* having bulbs; growing from bulbs: *Daffodils are bulbous plants.*

bull, *Zoology.* —*n.* **1** the uncastrated full-grown male of cattle. **2** the male of the whale, elephant, seal, walrus, and other large mammals.
—*adj.* male: *a bull moose.*

bullate (bul'āt), *adj. Botany.* having blisterlike projections, as a leaf; blistered or puckered.

bundle, *n.* **1** *Anatomy.* a group of muscle or nerve fibers bound closely together. The **atrioventricular bundle** is a muscle bundle that conducts heartbeat impulses.
2 *Botany.* an aggregation of cells for conduction and support in the stems and leaves of plants. A **vascular bundle** is a strand of vascular tissue in higher plants.

buoyancy, *n. Physics.* **1** the power of a fluid to push upward or keep a body immersed in it: *Salt water has greater buoyancy than fresh water.*
2 a body's loss in weight or tendency to float when immersed in a fluid: *Wood has more buoyancy than iron.*
—**buoyant** (boi'ənt), *adj.* **1** having buoyancy: **a** able to keep things afloat: *Air is buoyant; helium-filled balloons float in it.* **b** able to float: *Wood and cork are buoyant in water; iron and lead are not.*
2 having to do with or causing buoyancy: *If the buoyant force on an object is greater than the pull of gravity, the object will float; if less, the object will sink* (Barnard, *Science: A Key to the Future*).

bur, *n. Botany.* **1** a prickly, clinging seedcase or flower of some plants: *Animals with wool, fur, or long hair become the unintentional carriers of burs* (Emerson, *Basic Botany*). **2** a plant or weed bearing burs. Also spelled **burr.**

buran (bü rän'), *n. Meteorology.* a violent northerly wind on the Russian and Siberian steppes, carrying snow or ice particles in winter and laden with dust in summer. [from Russian]

burette or **buret** (byù ret'), *n.* a graduated glass tube, usually with a tap at the bottom. It is used for accurately measuring out small amounts of liquid or gas. *The buret and the pipet are both calibrated to deliver a certain volume rather than to contain a certain volume.* (Jones, *Inorganic Chemistry*). [from French *burette*]

burn, *v.* **1** *Chemistry.* to undergo or cause to undergo combustion; oxidize rapidly: *The body gains energy by burning food. The normal liver of the rat utilized acetic acid to synthesize fatty acids and ... burned it to carbon dioxide and water* (N.Y. Times).
2 *Nuclear Physics.* to undergo fission or fusion.

burr¹, *n.* = bur.

burr², *n. Geology.* **1** a siliceous rock with many cavities, formerly used as a millstone.
2 *British.* a tough siliceous rock occurring among calcareous or other softer formations.

bursa (bėr'sə), *n., pl.* **-sae** (-sē'), **-sas.** *Anatomy.* a sac of the body, especially one containing a lubricating fluid that reduces friction between a muscle or tendon and a bone: *Somewhat like a small collapsed balloon, the bursa has lax walls and a slick lubricated interior surface that equips it to absorb friction* (Time). [from Medieval Latin *bursa* purse, from Greek *byrsa* hide]

---

cap, fāce, fäther; best, bē, tėrm; pin, five;
rock, gō, ôrder; oil, out; cup, put, rüle,
yü in use, *yü* in uric;
ng in bring; *sh* in rush; *th* in thin; ᴛʜ in then;
zh in seizure.
ə = *a* in about, *e* in taken, *i* in pencil, *o* in lemon, *u* in circus

81

burst, *n. Physics.* a sudden increase in the radiation intensity of ionized particles, such as may be observed in a shower of cosmic rays, or in the trailing particles in an ionization chamber.

bush, *n.* **1** *Botany.* a shrub, especially one having many separate branches starting from or near the ground. Some bushes are used as hedges; others are cultivated for their fruit.
2 *Ecology.* open forest or wild, unsettled land: *the bush of Australia.*

butadiene (byü′tə dī′ēn), *n. Chemistry.* a colorless gas derived from petroleum by-products, used in making synthetic rubber. *Formula:* $CH_2=CHCH=CH_2$

butane (byü′tān), *n. Chemistry.* a colorless gas, a hydrocarbon, produced in petroleum refining, and much used as a fuel. *Formula:* C_4H_{10} [from *but(yl)* + *-ane*]

butte (byüt), *n. Geography.* a steep-sided hill with a flat surface on top, often standing alone and found primarily in arid and semiarid regions. [from French *butte* hill, mound]

butte

butyl (byü′tl), *n. Chemistry.* a univalent hydrocarbon radical obtained from butane. There are four isomeric univalent butyl radicals. *Formula:* C_4H_9 [from *but(yric acid)* + *-yl*]

butyl alcohol, *Chemistry.* any of four isomeric alcohols derived from butane and used as a solvent for resins, adhesives, varnishes, etc. *Formula:* C_4H_9OH

butylate (byü′tə lāt), *v. Chemistry.* to introduce a butyl group into (a compound). —**butylation,** *n.*

butylene (byü′tl ēn′), *n. Chemistry.* a gaseous hydrocarbon of the ethylene series, often used in making synthetic rubber. Butylene is found in three isomeric forms (*butenes*). *Formula:* C_4H_8

butyrate (byü′tə rāt), *n. Chemistry.* a salt or ester of butyric acid.

butyric (byü tir′ik), *adj. Chemistry.* of, producing, or derived from butyric acid. **Butyric fermentation** changes butter, milk, etc., into butyric acid by the action of anaerobic bacteria.

butyric acid, *Chemistry.* an oily, colorless liquid that has an unpleasant odor, formed by fermentation in rancid butter, cheese, etc. *Formula:* C_3H_7COOH [from Latin *butyrum* butter]

butyrin (byü′tər in), *n. Chemistry.* a colorless, oily liquid found in butterfat, formed from butyric acid and glycerin. *Formula:* $C_3H_5(OCOC_3H_7)_3$

butyrinase (byü′tə rə nās), *n. Biochemistry.* an enzyme in blood serum that hydrolyzes butyrin.

byr or **byr.,** *abbrev.* billion years: *Liquid water first appeared on our planet around 3.8 byr ago* (New Scientist).

bysmalith (biz′mə lith), *n. Geology.* a cylindrical body of intrusive igneous rock that cuts vertically across the adjacent sediments. [from Greek *bysma* plug + *lithos* stone]

byssaceous (bi sā′shəs), *adj. Botany.* consisting of fine threads or filaments. [from Latin *byssus* flax]

byssal (bis′əl), *adj.* of or having to do with a byssus or byssuses.

byssus (bis′əs), *n., pl.* **byssuses, bussi** (bis′ī). *Zoology.* a tuft of strong, silky filaments by which various mussels attach themselves to rocks and other objects. It is secreted by a gland in the foot. [from Latin *byssus* flax, from Greek *byssos*]

byte (bīt), *n.* a unit of eight binary digits, used as a measure of the capacity of a computer memory: *... a data memory bank of some 100 million bytes as well as a 40 million byte disk file, and 2 million bytes of system memory* (London *Times*). [perhaps from *b(inar)y (digi)t e(ight)*]

C

c.or **c,** *abbrev.* or *symbol.* **1** calorie. **2** cathode. **3** centi-. **4** centimeter. **5** *Nuclear Physics.* charm. **6** cubic. **7** curie. **8** current. **9** cycle.

C, *abbrev.* or *symbol.* **1** candela. **2** capacitance. **3** carbon. **4** Also **C.** Celsius or centigrade. **5** (electric) charge. **6** coulomb. **7** cirrus. **8** constant. **9** cytosine.

^{12}**C,** *symbol.* carbon 12.

^{13}**C,** *symbol.* carbon 13.

^{14}**C,** *symbol.* carbon 14.

ca., *abbrev.* **1** cathode. **2** centiare *or* centiares.

Ca, *symbol.* calcium.

cacodyl (kak′ə dil), *n. Chemistry.* a univalent radical consisting of arsenic and methyl, a constituent of compounds characterized by an offensive smell and highly poisonous vapors. *Formula:* $(CH_3)_2As$- [from Greek *kakōdēs* bad-smelling + English *-yl*]

cacodylate (kak′ə dil′āt), *n. Chemistry.* a salt of cacodylic acid.

cacodylic acid (kak′ə dil′ik), *Chemistry.* a crystalline, deliquescent compound, odorless and poisonous, used in the making of dyes and perfumes. *Formula:* $(CH_3)_2AsO_2H$

cadaverine (kə dav′ə rēn), *n. Biochemistry.* a poisonous, colorless, liquid ptomaine yielded by bacterial decomposition of proteins. *Formula:* $C_5H_{14}N_2$ [from Latin *cadaver* corpse]

cadmium, *n. Chemistry.* a soft, bluish-white, ductile metallic element resembling tin, which occurs only in combination with other elements and is used in plating to prevent corrosion and in making alloys. *Symbol:* Cd; *atomic number* 48; *atomic weight* 112.40; *melting point* 320.9°C; *boiling point* 767°C; *oxidation state* 2. [from Latin *cadmia* zinc ore]

cadmium sulfide, *Chemistry.* a bright yellow or orange powder used as a pigment and, because of its sensitivity to light, in such devices as transistors and solar batteries. It occurs naturally as the mineral greenockite. *Formula:* CdS

caducous (kə dü′kəs), *adj.* **1** *Botany.* falling off easily or very early, as the sepals of the poppy, which fall at once on the opening of the flower.
2 *Zoology.* falling off naturally after serving its purpose: *a caducous organ or part.*
[from Latin *caducus* falling, from *cadere* fall]

caecum (sē′kəm), *n., pl.* **-ca** (-kə). = cecum.

caesium (sē′zē əm), *n.* = cesium.

cal, *symbol.* calorie *or* calories (def. 1a); small calorie.

Cal, *symbol.* Calorie *or* Calories (def. 1b); large calorie.

calamine (kal′ə mīn), *n. Mineralogy.* = hemimorphite. [from Medieval Latin *calamina,* from Latin *cadmia* cadmium]

calamistrum (kal′ə mis′trəm), *n., pl.* **-tra** (-trə). *Zoology.* one of the curved, movable spines on the posterior legs of certain spiders, used to curl and bind the lines of silk coming from the spinnerets. [from New Latin, from Latin *calamistrum* curling iron]

calaverite (kal′ə vär′īt), *n. Mineralogy.* a telluride of gold, or of gold and silver, bronze-yellow and massive: *Next in importance to native gold as a source of the metal come the gold tellurides, chief of which is calaverite* (W. R. Jones). *Formula:* $AuTe_2$ [from *Calaveras* County, California, where it was first found + *-ite*]

calcaneum (kal kā′nē əm), *n., pl.* **-nea** (-nē ə). = calcaneus.

calcaneus (kal kā′nē əs), *n., pl.* **-nei** (-nē ī). *Anatomy.* **1** the largest bone of the row of tarsal bones; the bone of the human heel: *The tarsal bones are the talus, calcaneus, navicular, cuboid, and the three cuneiform bones. They form the heel and back part of the instep* (William V. Mayer). See the pictures at **foot, tendon.**
2 the analogous bone in the foot of other vertebrates.
[from Late Latin *calcaneus,* from Latin *calcem* heel]
—**calcaneal,** *adj.* relating to the heel or calcaneus: *a calcaneal tendon.*

calcar (kal′kär), *n., pl.* **calcaria** (kal ker′ē ə). *Botany, Zoology.* a spur or spurlike projection, as on the base of a petal or on a bird's wing or leg. [from Latin, from *calcem* heel]
—**calcarate** or **calcarated,** *adj.* furnished with a calcar.

calcareous (kal ker′ē əs), *adj.* **1** *Geology.* of or containing lime, limestone, or calcium carbonate: *calcareous sediments.*
2 *Zoology.* of or containing calcium: *The outer surface of the body [of the starfish] is covered with hard calcareous plates that provide excellent support and protection for the delicate, soft body parts beneath* (Winchester, *Zoology*).
[from Latin *calcarius,* from *calcem* lime]

calci-, *combining form.* **1** lime, calcium, or salts of lime or calcium, as in *calcification = forming or becoming hard like lime.* **2** limestone, as in *calcicole = growing on limestone.* [from Latin *calcem* lime]

calcic, *adj. Chemistry.* of, containing, or derived from lime or calcium: *calcic phosphates.*

calcicole (kal′sə kōl), *Botany.* —*adj.* growing on limestone or on chalky soil.
—*n.* a plant growing on limestone.
[from *calci-* + Latin *colere* to inhabit]

calcicolous (kal sik′ə ləs), *adj.* = calcicole.

calciferol (kal sif′ə rôl′), *n. Biochemistry.* a fat-soluble alcohol which prevents rickets: *Calciferol . . . is produced from a closely related substance, ergosterol, reg-*

cap, fāce, fäther; best, bē, tèrm; pin, fīve;
rock, gō, ôrder; oil, out; cup, pùt, rüle;
yü in use, yu̇ in uric;
ng in bring; sh in rush; th in thin, ᴛʜ in then;
zh in seizure.
ə = a in about, e in taken, i in pencil, o in lemon, u in circus

ularly present in the skin, by ultraviolet radiation (Shull, *Principles of Animal Biology*). *Formula:* $C_{28}H_{44}O$ Also called **vitamin D₂.**

calciferous (kal sif′ər əs), *adj. Biology.* yielding or containing calcite, calcium, or calcium carbonate. The **calciferous glands** of an earthworm secrete calcium carbonate. [from Latin *calcem* lime + English *-ferous*]

calcific (kal sif′ik), *adj.* of, having to do with, or involving calcification.

calcifuge (kal′sə fyüj), *n. Botany.* a plant which will not grow on limestone: *Most heaths and certain lichens are calcifuges.* [from *calci-* + Latin *fuga* flight]

calcify (kal′sə fī), *v. Physiology.* to make or become hard or bony by the deposit of calcium salts; convert or be converted into lime: *Cartilage often calcifies in older people.*

—**calcification,** *n.* **1** *Physiology.* **a** the process of calcifying: *Subsequently there occurs the impregnation of the cartilage with calcium salts. This process, termed calcification . . .* (Harbaugh and Goodrich, *Fundamentals of Biology*). Compare **ossification. b** a calcified part: *a calcification in the lung.*
2 *Geology.* the accumulation of calcium in certain soils, especially soils of cool temperate regions where leaching takes place very slowly.

calcine (kal′sīn), *Chemistry.* —*v.* **1** to heat to a high temperature without fusing in order to decompose, oxidize, etc.: *to calcine limestone, to calcine high-level radioactive wastes.* **2** to undergo calcination: *ore that calcines readily.*
—*n.* a substance obtained by calcining.
[from Medieval Latin *calcinare,* from Latin *calcem* lime] —**calcination,** *n.*

calcite (kal′sīt), *n. Mineralogy.* one of the three crystalline forms of calcium carbonate, the others being aragonite and vaterite. Calcite crystallizes in hexagonal form. It is the chief substance in limestone, chalk, and marble. *Formula:* $CaCO_3$

calcitonin (kal′sə tō′nən), *n. Biochemistry.* a thyroid hormone that lowers the level of calcium in blood plasma: *The major "reservoir" of calcium is the skeleton, calcitonin inhibiting the release of calcium ions and parathyroid hormone mobilizing the element from the bones* (New Scientist). Also called **thyrocalcitonin.** [from *calci(um)* + *ton(e)* + *-in*]

calcium (kal′sē əm), *n. Chemistry.* a soft, silvery-white metallic element. It is a part of limestone, chalk, milk, bone, etc. Calcium is used in alloys and its compounds are used in making plaster, in cooking, and as bleaching agents. *Symbol:* Ca; *atomic number* 20; *atomic weight* 40.08; *melting point* 845°C; *oxidation state* 2. [from New Latin, from Latin *calcem* lime]

calcium carbide, *Chemistry.* a heavy, gray substance that reacts with water to form acetylene gas. *Formula:* CaC_2

calcium carbonate, *Chemistry.* a compound of calcium occurring in rocks such as marble and limestone, in shells, and to some extent in plants. It is used in the manufacture of toothpastes, white paint, and cleaning powder. The common native form of calcium carbonate is the mineral calcite, though it also occurs as aragonite and vaterite. *Formula:* $CaCO_3$

calcium chloride, *Chemistry.* a compound of calcium and chlorine, used on roads to settle dust or melt ice, and in refrigeration. *Formula:* $CaCl_2$

calcium hydroxide, *Chemistry.* a white powder obtained by exposing lime to moist air or by putting water on lime. *Formula:* $Ca(OH)_2$ Commonly called **slaked lime.**

calcium oxalate, *Chemistry.* a white crystalline substance present in plant cells and urine, used in making oxalic acid and various oxalates. *Formula:* $CaC_2O_4 \cdot H_2O$

calcium oxide, *Chemistry.* a solid, white compound of calcium and oxygen, obtained by burning limestone, shells, bones, or other forms of calcium carbonate. *Formula:* CaO Commonly called **lime, quicklime;** also called **calx.**

calcium phosphate, *Chemistry.* a compound of calcium and phosphoric acid, used in medicine, in making enamels, etc. It is found in bones and as rock. *Formula:* $Ca_3(PO_4)_2$

calcrete (kal′krēt), *n. Geology.* a conglomerate of surficial sand and gravel cemented by calcium carbonate. [from *cal(careous)* + *(con)crete*]

calculate, *v. Mathematics.* to determine by mathematical means; find by adding, subtracting, multiplying, etc.: *to calculate the volume of a cylinder, calculate the velocity of light. Another example of proper use of a computer is in calculating . . .* [as] *the large quantity of mathematics involved is relatively complex, and speed and accuracy are the greatest attractions of an electronic computer* (Science News). [from Latin *calculatus* counted, from *calculus* stone used in counting, diminutive of *calcem* stone, lime]
► *Calculate* once considered a somewhat less technical term than *compute,* is now used interchangeably with *compute: Calculate* (or *compute*) *the square root of 7.*

—**calculation,** *n.* the act of calculating or the result found by calculating.

calculus (kal′kyə ləs), *n.* **1** a branch of mathematics dealing with differentiation, integration, and related topics. See also **differential calculus, integral calculus.**
2 *Medicine.* pl. **calculi.** a stone that has formed in the body because of a diseased condition. Gallstones and kidney stones are calculi.
[from Latin *calculus* stone used in counting, diminutive of *calcem* stone, lime]

calculus of variations, a branch of mathematics that studies the maximum and minimum values of functions that depend on a curve or other geometric figure rather than on a number or numbers.

caldera (kal der′ə), *n. Geology.* a large basin-shaped volcanic depression: *All the big calderas are located on a graben* (*a depressed section of crust*) (Haroun Tazieff). See the picture at **volcano.** [from Spanish *caldera* (literally, caldron]

calibrate (kal′ə brāt), *v.* to determine, check, or adjust the scale of (a thermometer, gauge, or other measuring instrument). Calibrating is usually done by comparison with a standard instrument.
—**calibration,** *n.* **1** a calibrating or being calibrated.
2 the marks made on a measuring instrument in calibrating. —**calibrator,** *n.*

caliche (kə lē′chä), *n. Geology.* **1** a crust or formation of calcium carbonate in or on soil in dry regions. **2** a rich surface deposit of sodium nitrate with a varying admixture of sand or other material, such as is found in parts of Chile. [from American Spanish, from Spanish *caliche* piece of lime, from *cal* lime, from Latin *calcem*]

California Current, *Oceanography.* a cold current originating in the northern Pacific Ocean and passing southward and then southwestward along the western coast of North America.

californium (kal′ə fôr′nē əm), *n. Chemistry.* a radioactive metallic element, produced artificially from curium, plutonium, or uranium. Californium has several isotopes, ranging in mass from 244 to 254. *Symbol:* Cf; *atomic number* 98; *atomic weight* 252 (the most stable isotope); *oxidation state* 3.
[from New Latin, from (the University of) *California,* where it was first produced in 1950]

callose (kal′ōs), *n. Biochemistry.* an amorphous substance, a carbohydrate, found in cell walls. [from Latin *callosus* callous]

callus (kal′əs), *n.* **1** *Medicine, Physiology.* a hard thickened place on the skin caused by friction or pressure. **2** *Anatomy.* a new growth to unite the ends of a broken bone. A **provisional callus** is absorbed and replaced by a **definitive callus.**
3 *Botany.* **a** a mass of cells that grows over a wound or cut end of a stem. **b** the thickening of the substance of the perforated septa (dividing walls) between sieve cells. **c** any unusually hard formation in or on a plant. [from Latin *callus*]

calm, *n. Meteorology.* a condition in which the wind has a velocity of less than one mile per hour (on the Beaufort scale, force 0). A **calm belt** is a latitudinal belt where the wind is very light or constantly shifting.

calomel (kal′ə mel), *n. Chemistry, Mineralogy.* a white, tasteless, crystalline compound of mercury and chlorine. *Formula:* Hg₂Cl₂ Also called **mercurous chloride.** [from Greek *kalos* beautiful + *melas* black]

calorescence (kal′ə res′ns), *n. Physics.* the change of nonluminous heat rays into light rays. [from Latin *calor* heat + *-escence,* as in *fluorescence*]

caloric (kə lôr′ik), *adj.* **1** having to do with heat: *the caloric effect of sunlight.* **2** of or having to do with calories: *The caloric content of grapefruit is very low.*

calorie, *n.* **1** *Chemistry, Physics.* either of two units for measuring the amount of heat: **a** the quantity of heat needed to raise by one degree Celsius the temperature of a gram of water. *Symbol:* cal Also called **small calorie, gram calorie. b** Calorie, the quantity of heat needed to raise by one degree Celsius the temperature of a kilogram of water. *Symbol:* Cal Also called **large calorie, kilocalorie, kilogram calorie.**
2 *Biology.* **a** a unit of the energy supplied by food. It corresponds to a large calorie; calorie (def. 1b). An ounce of sugar will produce about one hundred such calories. **b** the quantity of food capable of producing such an amount of energy.
[from French, from Latin *calor* heat]

calorific, *adj.* **1** *Physics.* producing heat: *calorific rays.* **2** producing calories: *calorific foods.*

calorific value, *Chemistry.* the calories or thermal units contained in one unit of a substance and released when it is burned.

calorimeter (kal′ə rim′ə tər), *n.* an apparatus for measuring the quantity of heat given off by or present in a body, such as the specific heat of different substances or the heat of chemical combination.
—**calorimetric,** *adj.* of or having to do with the calorimeter or calorimetry.
—**calorimetry,** *n.* **1** the quantitative measurement of heat. **2** studies involving the use of the calorimeter.

calvaria (kal vãr′ē ə), *n.* = calvarium.

calvarial, *adj.* of or belonging to the calvarium.

calvarium (kal vãr′ē əm), *n. Anatomy.* the upper, caplike part of the skull. [from New Latin, from Latin *calvaria* skull]

calx (kalks), *n., pl.* **calxes, calces** (kal′sēz′). = calcium oxide.

calycate (kāl′ə kit), *adj. Botany.* provided with a calyx.

calyciform (kə lis′ə fôrm), *adj. Botany.* having the form of a calyx.

calyptra (kə lip′trə), *n. Botany.* **1** a thick covering on top of the capsule in some mosses, formed by enlargement of the walls of the archegonium. **2** any hoodlike part of a flower or fruit. **3** = root cap. [from New Latin, from Greek *kalyptra* headdress, veil]

calyptrogen (kə lip′trə jən), *n. Botany.* the layer of the undifferentiated tissue meristem from which the root cap of some plants develops.

calyx (kā′liks), *n., pl.* **calyxes, calyces** (kal′ə sēz′). **1** *Botany.* the outer leaves that surround the unopened bud of a flower. The calyx is made up of sepals, either separated or joined in a cup.
2 *Anatomy, Zoology.* a cuplike structure or organ, such as one of the cuplike divisions of the pelvis of the kidney.
[from Latin, from Greek *kalyx*]

cam, *n.* a projection on a wheel or shaft that changes a regular circular motion into an irregular circular motion or into a back-and-forth motion.

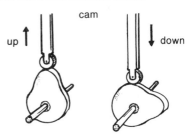

cambium (kam′bē əm), *n. Botany.* layer of soft growing tissue in a plant body which forms parallel rows of cells. The *vascular* cambium is an internal cylinder in stems and roots that develops into xylem and phloem; the *cork* cambium, or phellogen, is a cylinder in the

cap, fāce, fäther; best, bē, tèrm; pin, five;
rock, gō, ôrder; oil, out; cup, pùt, rüle;
yü in use, *yù* in uric;
ng in bring; sh in rush; th in thin, ŦH in then;
zh in seizure.
ə = *a* in about, *e* in taken, *i* in pencil, *o* in lemon, *u* in circus

85

bark of older stems and roots that develops into cork (phellem) and phelloderm. New bark and new wood develop from this actively dividing cellular tissue each year, producing the annual rings. *The cork cambium . . . forms a periderm composed mostly of cork tissue* (Raven, et al. *Biology of Plants*). See the pictures at **bark, root.** [from Medieval Latin *cambium* exchange, from Latin *cambiare* to exchange]

—**cambial,** *adj.* formed of or having to do with cambium: *In the course of the growing season each cell of the cambial sheath gives rise to a radial row of wood cells* (New Biology),

Cambrian (kam′brē ən), *Geology.* —*n.* **1** the earliest geological period of the Paleozoic era. During this period there were large numbers of primitive, invertebrate marine animals. *The Cambrian shows an assemblage quite different from that of today. Roughly fifty percent of the fossils found are remains of a form not existing today* (Garrels, *Textbook of Geology*).
2 the rocks formed during this period.
—*adj.* of or having to do with the Cambrian: *The Cambrian rocks derive their name from Cambria, the name for Wales, where, a century ago, this division of Paleozoic succession was defined* (Moore, *Introduction to Historical Geology*).

campanulate (kam pan′yə lit), *adj. Botany.* bell-shaped (applied to parts of plants, especially to the corolla). [from Late Latin *campanula,* diminutive of *campana* bell]

campylotropous (kam′pə lot′rə pəs), *adj. Botany.* having the nucellus and its integuments so curved that the micropyle is brought near the hilum of an ovule or seed. [from Greek *kampylos* curved + *tropē* a turning]

Canadian Shield, *Geography, Geology.* a vast area of Pre-Cambrian rocks, rich in minerals, occupying more than half of Canada from the Arctic Ocean to the Great Lakes, and flanking Hudson Bay in a U-shaped outline: *The Canadian Shield . . . has first importance for study of Pre-Cambrian history . . . because a more complete record has been worked out here than elsewhere on the continent* (Moore, *Introduction to Historical Geology*). Also called **Laurentian Plateau** or **Laurentian Shield.**

canal, *n. Biology.* **1** a tube, duct, or the like, in the body of an animal or plant that carries liquid, air, food, or other solid matter from one part to another. Food eaten by most animals goes through an alimentary canal.
2 a tube whose chief function is to hold some liquid or gas, especially in the body: *the semicircular canals.*
3 *Geography.* an artificial channel filled with water.

canalicular (kan′ə lik′yə lər or **canaliculate** (kan′ə lik′yə lit), *adj. Zoology, Botany.* of, having to do with, or resembling a canaliculus; minutely channeled or grooved.

canaliculation (kan′ə lik′yə lā′shən), *n. Zoology, Botany.* any minutely channeled or grooved formation, such as a canaliculus.

canaliculus (kan′ə lik′yə ləs), *n., pl.* **-li** (-lī). *Anatomy.* a small canal or duct in the body. Canaliculi connect the small cavities in the bones. [from New Latin, from Latin *canalis* groove, canal]

cancel, *v. Mathematics.* **1** to remove (a factor) common to the numerator and denominator of a fraction. EXAMPLE: The fraction $(a \cdot x)/(a \cdot y)$ can be simplified to x/y by cancelling the factor a.
2 to remove (a term or factor) from both sides of an equation. EXAMPLE: The equation $3x^2y = 6x$ can be simplified to $xy = 2$ by cancelling the factor $3x$.
—**cancellation,** *n.* the act or process of cancelling a term or factor.

cancellate (kan′sə lit), *adj.* = cancellous.

cancellous (kan′sə ləs), *adj. Anatomy.* having an open, latticed, or porous structure: *flat cancellous bones of the skull.* [from Latin *cancelli* crossbars]

cancrinite (kang′krə nīt), *n. Mineralogy.* a silicate and carbonate mineral chemically related to feldspar, often bright yellow in color, and occurring in various volcanic rocks. *Formula:* $(Na_2Ca)_4(A1SiO_4)_6CO_3 \cdot nH_2O$ [from Count *Cancrin,* 1774–1845, Russian minister + *-ite*]

candela (kan del′ə or kan dē′lə), *n.* the SI unit of luminous intensity, a fundamental unit equal to 1/60 of the radiating power of one square centimeter of a blackbody at the temperature at which platinum solidifies (1772°C). It replaced the international candle in 1948, but was not officially adopted in the United States until 1963. *Symbol:* cd Compare **candle.** [from Latin *candela* candle]

candle, *n.* a unit of luminous intensity. The **international candle** is a former unit equal to the light from five square millimeters of platinum when heated to its melting point. Since it was inconvenient to use, it was replaced by the candela. The **standard candle** is a still older unit equal to the luminous intensity of a spermaceti candle 7/8 inches long and burning 120 grains each hour. See also **candela.**

candlepower, *n.* **1** the intensity of light, measured in candelas. One candlepower is equal to 12.57 lumens.
2 the unit of luminous intensity of the former international candle or older standard candle. *Abbreviation:* c.p.

canescent (kə nes′ənt), *adj. Botany.* having grayish soft down. [from Latin *canescere* grow hoary, from *canus* hoary]

canine, *adj. Zoology.* of or belonging to the family of carnivorous mammals that includes dogs, foxes, jackals, and wolves. The coyote is a canine animal.
—*n.* **1** *Zoology.* any animal belonging to the canine family.
2 *Anatomy.* one of the four pointed teeth next to the incisors. Also called **canine tooth, cuspid.** [from Latin *caninus,* from *canis* dog]

caniniform (kā nī′nə fôrm), *adj. Anatomy.* shaped like a canine tooth: *The canine and caniniform incisor teeth of this seal function as an extremely efficient saw for cutting through thick and flinty ice* (Scientific American).

cannon bone, *Zoology.* a bone in hoofed mammals between the hock or knee and the fetlock.

canopy, *n. Ecology.* the uppermost layer in a forest or woodland, consisting of the crowns of trees or shrubs.

canyon, *n. Geography, Geology.* a narrow valley with high, steep sides, usually with a stream at the bottom: *The continental slope is often cut with deep canyons running down to the deep sea and carrying streams of sediment-laden water to be deposited at the bottom*

(New Scientist). [from Mexican Spanish *cañón* narrow passage]

capacitance, *n. Physics.* **1** the property of a capacitor that determines the amount of electrical charge it can receive and store. **2** the measure of this property. *Abbreviation:* C

capacitate, *v. Biology.* to cause (sperm) to undergo the changes in the uterus that enables them to penetrate and fertilize an egg: *The Fallopian tube fluid is still effective in capacitating sperm* (Nature).
—**capacitation,** *n.* a series of physical changes that sperms undergo before attaining the capacity to penetrate and fertilize an egg.

capacitive (kə pas′ə tiv), *adj. Physics.* of, having to do with, or due to capacitance: *Capacitive reactance is measured in ohms.*

capacitor *n. Physics.* a device for receiving and storing a charge of electricity: *A condenser, also called a capacitor, is a device . . . with the effect of stepping up, smoothing out or otherwise modifying the current* (Scientific American).

capacity, *n.* = capacitance.

capillarity (kap′ə lar′ə tē), *n. Physics.* the tendency of the surface of a liquid to rise or fall in contact with a solid, as in a capillary tube: *Capillarity depends on adhesion and surface tension.* (Obourn, *Investigating the World of Science*). Also called **capillary action.**

capillary (kap′ə ler′ē), *n. Anatomy.* a blood vessel with a very slender, hairlike opening. Capillaries join the end of an artery to the beginning of a vein. *Capillaries . . . are a network of fine blood vessels where the absorption of some products and the giving up of others through the principles of osmosis and diffusion take place* (Winchester, *Zoology*).
—*adj.* **1** of or in the capillaries: *Capillary walls are only one cell thick.*
2 like a hair; very slender: *a capillary vessel.*
3 of or having to do with capillarity: *Capillary waves are so called because they arise, not from winds as most ocean waves, but from the type of capillarity associated with surface tension phenomena* (Science News Letter).
[from Latin *capillaris* of hair, hairlike, from *capillus* hair]

capillary action, = capillarity: *Capillary action is at work when water rises in the soil, kerosene in a lamp wick and ink in blotting paper* (M.F. Vessel).

capillary attraction, *Physics.* the force that causes a liquid to rise in a narrow tube or when in contact with a porous substance. It is this force of adhesion and surface tension that allows a porous substance to soak up a liquid. A plant draws up water from the ground and a paper towel absorbs water by means of capillary attraction.

capillary repulsion, *Physics.* the force that causes a liquid to be depressed when in contact with the sides of a narrow tube.

capillary tube, *Physics.* a tube with a very slender, hairlike opening or bore: *The height of the water in the capillary tubes depends inversely upon the tube's diameter. The smaller the diameter of the tube, the higher the water will rise* (Tracy, *Modern Physical Science*).

capillitium (kap′ə lish′ē əm), *n., pl.* **-litia** (-lish′ē ə).

Botany. the tangle of hairlike filaments inside certain spore cases: *The leftover dried network of cytoplasm forms the framework of the sporangium, called the capillitium* (Emerson, *Basic Botany*). [from Latin *capillitium* head of hair, from *capillus* hair]

capitate (kap′ə tāt), *adj.* **1** *Botany.* **a** clustered into a head: *capitate flowers.* **b** having a rounded head: *a capitate stigma.*
2 *Anatomy.* having to do with or denoting a bone of the carpus or human wrist, in the center of the distal row of carpal bones.
—*n. Anatomy.* the capitate bone; magnum.
[from Latin *capitatus* headed, from *caput* head]

capitellate (kap′ə tel′it), *adj.* **1** *Botany.* growing or terminating in a small head. **2** having a capitellum or capitulum.

capitellum (kap′ə tel′əm), *n., pl.* **-tella** (-tel′ə). **1** *Anatomy.* the rounded protuberance on the outer surface of the lower end of the humerus, or upper arm.
2 *Zoology.* the portion of the body of a hydroid polyp that bears the tentacles.
[from New Latin, from Late Latin *capitellum* small head, diminutive of Latin *caput* head]

capitular (kə pich′ə lər), *adj.* **1** *Botany.* growing in a capitulum or head.
2 *Anatomy.* of or having to do with a capitulum.

capitulum (kə pich′ə ləm), *n., pl.* **-la** (-lə). **1** *Botany.* a flower head consisting of a close cluster of sessile flowers, as in red clover.
2 *Anatomy.* a protuberance of a bone, usually fitting into a hollow portion of another bone: *the capitula of the ribs.*
[from New Latin, from Latin *capitulum* small head, diminutive of *caput* head]

caprate (kap′rāt), *n. Chemistry.* a salt or ester of capric acid.

capillary attraction

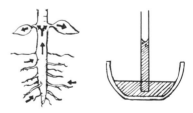

capillary attraction:
water rises in the tube or the plant draws fluid into its root system by means of capillary attraction; fluid will rise until its weight in the tube or root hair balances the surface tension of the fluid.

cap, fāce, fäther; best, bē, tėrm; pin, five; rock, gō, ôrder; oil, out; cup, pùt, rüle, *yü* in use, *yu̇* in uric;
ng in bring; *sh* in rush; *th* in thin, ᴛʜ in then; *zh* in seizure.
ə = *a* in about, *e* in taken, *i* in pencil, *o* in lemon, *u* in circus

capric acid

capric acid (kap′rik), *Chemistry.* a fatty acid found in oil and fats, especially in butter and coconut oil, used in making esters and as an intermediate. *Formula:* $C_{10}H_{20}O_2$ [from Latin *capri* goat]

caproate (kap′rō āt), *n. Chemistry.* a salt or ester of caproic acid.

cap rock, *Geology.* any of various types of overlying layers or strata, especially: **a** a layer of shale, limestone, or other material lying over and sealing in deposits, as of petroleum and natural gas. **b** (in a salt dome) an impervious body of anhydrite and gypsum, also containing calcite and sulfur.

caproic acid (kə prō′ik), *Chemistry.* a fatty acid found in oils and fats, used in analytical chemistry and in making flavors, resins, drugs, etc. *Formula:* $C_6H_{12}O_2$ [from Latin *capro-, capri* goat]

caprylic acid (kə pril′ik), *Chemistry.* a colorless liquid fatty acid with a faint but unpleasant odor, found in fats and oils, especially in butter and coconut oil, and used in making dyes, drugs, flavors, and in antiseptics and fungicides. *Formula:* $C_8H_{16}O_2$ [from Latin *capri* goat + *-yl*]

capsid (kap′sid), *n. Biology.* the protein shell surrounding the core of a virus particle: *The morphological units composing the shell have been given the name "capsomeres." The shell itself is the "capsid"* (Scientific American). [from French *capside,* from Latin *capsa* box] —**capsidal** (kap′sə dəl), *adj.*

capsomere (kap′sə mir), *n. Biology.* one of the identical units making up a capsid: *The capsomeres are packed in a regular pattern to form the shell or "capsid", as an approximate sphere* (H.T. Zwartouw). [from French *capsomère,* from *capside* capsid + *-mère* part]

capsular, *adj.* **1** of or having to do with a capsule. **2** formed or shaped like a capsule.

capsular ligament, *Anatomy.* the ligament which surrounds every moveable joint, and contains a viscid, clear lubricating liquid.

capsulate (kap′sə lit), *adj. Botany.* provided with or enclosed in a capsule: *capsulate pods.*

capsule, *n.* **1** *Botany.* **a** a dry seedcase that opens when ripe: *A capsule is a fruit developed from a compound pistil* (Emerson, *Basic Botany*). **b** the spore case of mosses and various other lower plants.
2 *Anatomy.* **a** a membrane enclosing an organ of the body: *the capsule of the kidney.* **b** either of two white layers of nerve fibers in the cerebrum.
[from Latin *capsula,* diminutive of *capsa* box]

capture, *n. Nuclear Physics.* the process by which an atomic nucleus absorbs or acquires an additional elementary particle, especially a neutron, often resulting in emission of radiation or fission of the nucleus.

carapace (kar′ə pās), *n. Zoology.* a shell or bony covering on the back of such animals as turtles, armadillos, lobsters, and crabs: *The carapace ... is really the backbone and ribs, joined into a solid mass by many bony plates* (Clifford H. Pope). [from French, from Spanish *carapacho*]

carat (kar′ət), *n.* **1** a unit of weight for precious stones, equal to 1/5 gram. **2** one 24th part of gold in an alloy: *A gold ring of 18 carats is 18 parts pure gold and 6 parts alloy.* [from Middle French, from Italian *carato,*

from Arabic *qīrāt,* from Greek *keration,* diminutive of *keras* horn]

carb-, *combining form.* the form of **carbo-** before vowels, as in *carbide, carbamic acid.*

carbamate (kär bam′āt *or* kär′bə māt), *n. Chemistry.* a salt or ester of carbamic acid.

carbamic acid (kär bam′ik), *Chemistry.* an acid occurring only in the form of salts. Its ammonium salt occurs in commercial ammonium carbonate. *Formula:* $O:C(OH)NH_2$ [from *carb(on)* + *am(ide)* + *-ic*]

carbamide (kär′bə mid *or* kär bam′id), *n.* = urea.

carbamyl (kär′bə mil), *n. Chemistry.* the univalent radical of carbamic acid. *Formula:* $-NH_2CO$

carbanion (kär ban′ī′ən), *n. Chemistry.* an organic ion carrying negatively charged carbon. Compare **carbonium.** [from *carb(on)* + *anion*]

carbene (kär′bēn), *n. Chemistry.* an organic radical containing divalent carbon, such as $-CH_2-$ (methylene), occurring as an intermediate in certain reactions.

carbide (kär′bīd), *n. Chemistry.* **1** a compound of carbon with another element, usually a metal: *iron carbide, uranium carbide.* **2** = calcium carbide.

carbinol (kär′bə nōl), *n. Chemistry.* **1** methyl alcohol or methanol. *Formula:* CH_3OH **2** an alcohol derived from it. **3** the univalent radical $-CH_2OH.$

carbo-, *combining form.* carbon, as in *carbohydrate, carbocyclic.* Also **carb-** before vowels. [from *carbon*]

carbocyclic, *adj. Chemistry.* having to do with or being a cyclic compound in which all the atoms in the ring are carbon, as in benzene.

carbohydrase (kär′bō hī′drās), *n. Biochemistry.* any enzyme that catalyzes the hydrolysis of a carbohydrate.

carbohydrate, *n. Biochemistry.* any of a large class of compounds consisting of only carbon, hydrogen, and oxygen, produced in green plants by photosynthesis and comprising a major type of food for animals. Carbohydrates include simple sugars or monosaccharides (glucose, fructose), disaccharides (lactose, sucrose), and polysaccharides (starch, cellulose). *Monosaccharides are the simplest carbohydrates and are made up of a chain of carbon atoms to which H atoms and O atoms are attached in the proportion of one carbon atom to two hydrogen atoms to one oxygen atom* (CH_2O) ... *The ultimate source of sugar in all plant cells is photosynthesis* (Raven, et al. *Biology of Plants*).

carbolfuchsin (kär′bol fŭk′sin), *n. Chemistry.* a solution of fuchsin used in staining specimens for microscopic study, in which the staining power of the dye has been enforced by the addition of carbolic acid.

carbolic acid (kär bol′ik), *Chemistry.* a very poisonous crystalline compound obtained from coal tar, used in solution as a disinfectant and antiseptic. *Formula:* C_6H_5OH Also called **phenol.**

carbon, *n. Chemistry.* a very common nonmetallic element which occurs in combination with other elements in all plants and animals. Diamonds and graphite are pure carbon in the form of crystals; coal and charcoal are mostly carbon in uncrystallized form. The atoms of carbon can link with one another in rings or chains, thus giving rise to innumerable complex compounds. *Carbon ... tends to combine with other elements by forming co-valent bonds through the sharing of electrons. It is a comparatively inactive element*

but does react with other elements at elevated temperatures (Offner, *Fundamentals of Chemistry*). *Symbol:* C; *atomic number 6; atomic weight* 12.01115; *melting point* 3600°C; *oxidation state* $+4$, $+2$. [from French *carbone,* from Latin *carbonem* coal]
—**carbonaceous,** *adj.* of, like, or containing carbon: *carbonaceous shale.*

carbon 12, *Nuclear Physics.* the most common isotope of carbon, adopted in 1962 in place of oxygen as the standard for determining atomic weights. *Symbol:* ^{12}C

carbon 13, *Nuclear Physics.* a stable, heavy isotope of carbon, having a mass number of 13, used as a tracer in physiological studies. *Symbol:* ^{13}C

carbon 14, *Nuclear Physics.* a radioactive isotope of carbon produced by the bombardment of nitrogen atoms by neutrons. Since carbon 14 decays at a uniform rate, the extent of its decay in the wood, bone, etc., of animals and plants that have died is evidence of the age of archaeological finds or geological formations in which organic matter occurs. *Symbol:* ^{14}C Also called **radiocarbon.** Compare **carbon dating.**

carbonado (kär′bə nä′dō), *n., pl.* **-does.** *Mineralogy.* an opaque, dark-colored, massive aggregate, composed of minute diamond particles, found chiefly in Brazil and used for drills. [from Portuguese *carbonado* (literally) carbonized]

carbonate, *n. Chemistry.* a salt or ester of carbonic acid, which contains the bivalent group -CO₃, such as calcium carbonate, $CaCO_3$.

carbon cycle, 1 *Biology.* the circulation of carbon in nature. Plants take in carbon dioxide from the atmosphere and convert it to carbohydrates by photosynthesis; animals eat the plants and return the carbon to the atmosphere by respiration and decay.
2 *Physics, Astronomy.* the series of thermonuclear reactions in incandescent stars that, beginning and ending with a carbon 12 atom, liberate atomic energy and transform hydrogen to helium: *The proton-proton chain . . . reactions in which protons combine to form helium with the emission of large amounts of energy, are mainly responsible for the energy production in the sun, while the carbon cycle is more important in very bright stars* (Fred Hoyle).

carbon dating, a method of determining the age of organic geological or archaeological specimens by measuring the amount of carbon 14 in them. Also called **radiocarbon dating.**

carbon dioxide, *Chemistry.* a heavy, colorless, odorless gas, present in the atmosphere or formed when any fuel containing carbon is burned, widely used for saturating water to form carbonated water (soda water), as Dry Ice, in fire extinguishers, etc. It is exhaled from an animal's lungs during respiration, and is used by plants in photosynthesis. *Microorganisms bring about the decay of organic materials everywhere, releasing most of the carbon of their organic compounds in the form of carbon dioxide* (Emerson, *Basic Botany*). *Formula:* CO_2

carbonic (kär bon′ik), *adj. Chemistry.* of or containing carbon. **Carbonic acid** is an acid made when carbon dioxide is dissolved in water. It gives the sharp taste to soda water. *Formula:* H_2CO_3

carbonic-acid gas, = carbon dioxide.

carbonic anhydrase, *Biochemistry.* an enzyme, present in the human body, which catalyzes the reaction between water and carbon dioxide.

Carboniferous (kär′bə nif′ər əs), *Geology.* —*n.* **1** the name outside of North America for a period of the Paleozoic era, after the Devonian and before the Permian. In North America this time is divided into two periods, the Mississippian and the Pennsylvanian. During the Carboniferous, the warm, moist climate produced great forests of tree ferns, rushes, and conifers, whose remains formed the great coal beds. **2** the system formed during this period, rich in coal beds.
—*adj.* **1** of or having to do with this period or system: *In the history of the earth, the Carboniferous period was one of the best for the growth of green plants. The luxuriant growth of plants . . . came near the close of a long period when shallow inland seas covered most of what is now dry land* (Emerson, *Basic Botany*).
2 carboniferous, (of rocks, etc.) containing or producing coal: *carboniferous limestone.*

carbonium (kär bō′nē əm), *n. Chemistry.* an organic ion carrying positively charged carbon. Compare **carbanion.** [from *carb(on*) + *-onium,* as in *ammonium*]

carbonize, *v. Chemistry.* **1** to change into carbon by burning: *Coal is carbonized to produce metallurgical coke.* **2** to cover or combine with carbon: *to carbonize sewing thread.*
—**carbonization,** *n.* conversion of a substance into carbon, as in the process of petrification.

carbon monoxide, *Chemistry.* a colorless, odorless, very poisonous gas, formed when carbon burns with an insufficient supply of air. It is part of the exhaust gases of automobile engines. *Carbon monoxide kills by depriving its victim of oxygen. When inhaled it combines with the hemoglobin . . . of the red blood cells* (Offner, *Fundamentals of Chemistry*). *Formula:* CO

carbon tetrachloride, *Chemistry.* a colorless, nonflammable liquid, often used in fire extinguishers and formerly widely used in cleaning fluids. Its fumes are very dangerous if inhaled. *Formula:* CCl_4

carbonyl (kär′bə nil), *n. Chemistry.* **1** a bivalent radical occurring in aldehydes, ketones, acids, etc. *Formula:* -C=O
2 any of a group of compounds of carbon monoxide united with a metal, such as nickel carbonyl.
—**carbonylation,** *n.* the introduction of a carbonyl group into a compound.
—**carbonylic,** *adj.* of or containing carbonyl.

carborane (kär′bə rān), *n. Chemistry.* any of a group of compounds of carbon, boron, and hydrogen used in chemical synthesis, especially of polymers. [blend of *carbon* and *borane*]

cap, fāce, fäther; best, bē, tėrm; pin, five;
rock, gō, ôrder; oil, out; cup, pùt, rüle,
yü in use, *yù* in uric;
ng in bring; *sh* in rush; *th* in thin, ᴛʜ in then;
zh in seizure.
ə = *a* in about, *e* in taken, *i* in pencil, *o* in lemon, *u* in circus

carboxyhemoglobin, *n. Biochemistry.* the compound formed in the blood when inhaled carbon monoxide combines with hemoglobin: *The increased amount of carboxyhemoglobin restricts the amount of oxygen that the blood can carry* (Science News).

carboxyl (kär bok′səl), *n. Chemistry.* a univalent radical existing in many organic acids, the hydrogen being replaceable by a basic element or radical, thus forming a salt. *Formula:* -CO$_2$H
—**carboxylic** (kär′bok sil′ik), *adj.* of, having to do with, or resembling carboxyl. **Carboxylic acid** is an organic acid containing the carboxyl group.

carboxylase (kär bok′sə lās), *n. Biochemistry.* an enzyme that catalyzes the addition or removal of carbon dioxide.

carboxylate, *Chemistry.* —*n.* (kär bok′sə lit) a salt or ester of carboxylic acid.
—*v.* (kär bok sə lāt) to introduce carboxyl or carbon dioxide into (a compound). —**carboxylation,** *n.*

carcinogen (kär sin′ə jən), *n. Medicine.* any substance or agent that produces or tends to produce cancer: *Malignant growths can be produced experimentally in animals by the use of certain irritating substances called carcinogens* (Newsweek).
[from Greek *karkinos* crab, cancer + English *-gen*]
—**carcinogenic** (kär sin′ə jen′ik), *adj.* producing or tending to produce cancer: *The transformation of a normal cell of the organism into a cancerous cell is the final step in a series of alterations produced inside cells which can be caused by the action of certain external agents. These agents, called "carcinogenic," can be of quite different nature: Physical, chemical, or biological* (Bulletin of Atomic Scientists).
—**carcinogenicity** (kär sin′ə jə nis′ə tē), *n.* the capacity or tendency to produce cancer.

carcinoma (kär′sə nō′mə), *n., pl.* **-mas, -mata** (-mə tə). *Medicine.* any of a class of cancers originating in epithelial tissue. Compare **lymphoma, sarcoma.** [from Latin, from Greek *karkinōma*, from *karkinos* crab, cancer]
—**carcinomatous** (kär′sə nom′ə təs), *adj.* characterized by, or of the nature of, carcinoma.

cardia (kär′dē ə), *n. Anatomy.* the orifice which connects the esophagus and the upper part of the stomach. [from Greek *kardia* heart]

cardiac, *adj. Anatomy.* 1 of or having to do with the heart: *cardiac disease, cardiac arteries.*
2 having to do with the upper part of the stomach. The **cardiac orifice** is the opening which connects the esophagus and the stomach.
[from Latin *cardiacus*, from Greek *kardiakos*, from *kardia* heart]

cardiac cycle, *Physiology.* 1 a complete pulsation of the heart, including dilation and contraction; heartbeat. 2 the time in which a complete pulsation takes place.

cardiac muscle, = myocardium.

cardinality (kär′də nal′ə tē), *n.* 1 the property of having a cardinal number. 2 the size of a set, or the number of elements it contains: *All finite sets with the same number of elements have the same cardinality* (Scientific American).

cardinal number, *Mathematics.* 1 a number that shows how many are meant. One, two, three, and four are cardinal numbers. They are the numbers used in counting. Compare **ordinal number.** 2 a number that expresses cardinality or the size of a mathematical set.

cardio-, *combining form.* of the heart; cardiac, as in *cardiology, cardiovascular.* [from Greek *kardia* heart]

cardiogenic (kär′dē ō jen′ik), *adj. Physiology.* originating in or produced by the heart. **Cardiogenic shock** is due to an impairment in the heart's output of blood.

cardioid (kär′dē oid), *n. Geometry.* a curve shaped rather like a heart, being the path of a point on the circumference of a circle when the circle rolls around a fixed circle of equal size. It is a variety of the limacon. [from Greek *kardioeidēs* heart-shaped]

cardiology, *n.* the branch of medicine dealing with the heart and the diagnosis and treatment of its diseases.
—**cardiological,** *adj.* of or having to do with cardiology.

cardiopulmonary (kär′dē ō pùl′mə ner′ē), *adj. Anatomy, Medicine.* of or having to do with the heart and lungs: *cardiopulmonary surgery.*

cardiovascular (kär′dē ō vas′kyə lər), *adj. Anatomy, Medicine.* of, having to do with, or affecting both the heart and the blood vessels: *cardiovascular diseases, the cardiovascular system.*

carina (kə rī′nə), *n., pl.* **-nae** (-nē). 1 *Zoology.* a structure or part having the form of a ridge or keel, such as the middle ridge on the breastbone of most birds.
2 *Botany.* the lower, keel-shaped pair of petals, characteristic of flowers of the pea family. [from Latin *carina* keel]
—**carinate** (kar′ə nāt), *adj.* shaped like or having a carina, or keel.

carminic acid (kär min′ik), *Chemistry.* a purplish-brown acid found in the buds of some plants, but most abundantly in the cochineal insect, used for staining microscopic specimens. *Formula:* $C_{22}H_{20}O_{13}$ [from Medieval Latin *carminium* crimson pigment made from cochineal]

carnallite (kär′nə līt), *n. Mineralogy.* a hydrous chloride of potassium and magnesium, an important source of potassium. *Formula:* $KC1MgC1_2 \cdot 6H_2O$ [from Rudolf von *Carnall*, 1804–1874, Prussian mining official + *-ite*]

carnassial (kär nas′ē əl), *Zoology.* —*adj.* of or having to do with certain teeth of carnivorous animals, adapted for tearing flesh (the last upper premolars and the first lower molars of living mammals).
—*n.* a carnassial tooth. [from Middle French *carnassier* carnivorous, ultimately from Latin *carnem* flesh]

carnitine (kär′nə tēn), *n.* = vitamin B$_t$: *Carnitine occurs in human and other animal muscles* (Newsweek). [from Latin *carnem* flesh + *-tine,* as in *creatine*]

carnivore (kär′nə vôr), *n.* 1 *Zoology.* any of an order (Carnivora) of mammals that feed chiefly on flesh, and generally have large, sharp canine teeth. Carnivores include predators and scavengers, and are typified by cats, dogs, lions, tigers, bears, and seals. Compare **herbivore, omnivore.**
2 *Botany.* a plant that eats insects. The sundew, pitcher plant, and Venus flytrap are carnivores.

[from Latin *carnivorus,* from *carnis* flesh + *vorare* devour]

► See the note under **heterotroph.**

—**carnivorous** (kär niv′ər əs), *adj.* 1 (of an animal) flesh-eating: *The carnivorous mammals that include meat as an important item of their diet . . . have strong agile bodies, with sharp claws and teeth* (Winchester, *Zoology*). Compare **herbivorous, omnivorous.**
2 (of a plant) insect-eating: *Carnivorous plants are . . . examples of the great ability of certain living things to . . . adjust . . . to unusual and hard conditions of life* (Emerson, *Basic Botany*).

carnivorean lethargy (kär′ni vôr′ē ən), *Biology.* the winter sleep of bears during which the body temperature does not drop much below normal, as it does in hibernation.

Carnot cycle (kär nō′), *Physics.* a series of thermodynamic operations consisting of isothermal expansion, adiabatic expansion, isothermal compression, and adiabatic compression, which make up the cycle of an ideal heat engine at maximum thermal efficiency. [named after Nicolas *Carnot,* 1796–1832, French physicist]

carnotite (kär′nə tīt), *n. Mineralogy.* a yellowish, radioactive vanadate mineral found in the western and southwestern United States. It is a source of uranium, radium, and vanadium. *Formula:* $K(UO_2)_2(VO_4)_2 \cdot 3H_2O$ [from Marie Adolphe *Carnot,* 1839–1920, a French inspector general of mines + *-ite*]

carotene (kar′ə tēn′), *n. Biochemistry.* a red or yellow crystalline pigment found in the carrot and other plants, and in animal tissue, and converted by the body into vitamin A: *The carotenes and xanthophylls are responsible for much of the fall leaf coloring and the color of ripe fruits* (Weier, *Botany*). *Formula:* $C_{40}H_{56}$ [from Latin *carota* carrot]

carotenoid, *n., Biochemistry.* any of a group of yellow to dark-red pigments found in various plant and animal tissues. The group includes carotene. *Amongst the exogenous pigments the carotenoids . . . have been found in the integument of many species of bugs . . . Together with the melanin pigments they are probably the most widely distributed pigments in the Insecta* (B. Nickerson).

carotid (kə rot′id), *Anatomy.* —*n.* Also called **carotid artery.** either of two large arteries, one on each side of the neck, that carry blood to the head.
—*adj.* having to do with or adjoining these arteries: *The brain's respiratory center is assumed to react to signals from tiny chemical detectors . . . in the arteries, called carotid (and aortic) bodies . . .* (Science News Letter).
[from Greek *karōtides,* from *karos* stupor (state produced by compression of carotids)]

carotinoid, *n.* = carotenoid.

-carp, *combining form.* fruit; part of a fruit, as in *ascocarp, pericarp, pseudocarp, syncarp.* [from Greek *karpos* fruit]

carpal (kär′pəl), *Anatomy.* —*adj.* of or having to do with the carpus or wrist. —*n.* one of the bones of the carpus.

carpel (kär′pəl), *n. Botany.* a modified leaf which forms a pistil or part of a pistil of a flower: *The leaf, though greatly changed in form . . . into an ovule-bearing or-*

gan is called a carpel or pistil (Youngken, *Pharmaceutical Botany*). [from Greek *karpos* fruit]

—**carpellary** (kär′pə ler′ē), *adj.* of or resembling a carpel: *carpellary leaves.*

—**carpellate** (kär′pə lāt), *adj.* having carpels: *carpellate cones.*

carpogonium (kär′pə gō′nē əm), *n., pl.* **-nia** (-nē ə). *Biology.* the single-celled female sex organ of certain algae. Its distal end is prolonged to form a tube which receives the male gamete. [from New Latin, from Greek *karpos* fruit + *gonos* offspring]

—**carpogonial,** *adj.* of or having to do with a carpogonium.

carpophagous (kär pof′ə gəs), *adj. Zoology.* fruit-eating. [from Greek *karpophagos*]

carpophore (kär′pə fôr), *n. Botany.* 1 a slender, elongation of the receptacle of a flower, supporting the carpels of some compound fruits, as in the geranium and many plants of the parsley family. 2 the stalk of a sporocarp or spore fruit.

carpospore, *n. Botany.* one of the spores produced in red algae as a result of fertilization of a carpogonium.

carpus (kär′pəs), *n., pl.* **-pi** (-pī). *Anatomy.* the wrist, or the bones of the wrist. [from New Latin, from Greek *karpos* wrist]

carrier, *n.* 1 *Chemistry.* a a catalytic agent which brings about, or helps in, the transfer of an element or group from one compound to another: *Iron may be a carrier of oxygen.* b a quantity of an element added to a radioactive isotope to facilitate handling of the isotope.
2 *Physics.* Also called **carrier wave.** a wave whose intensity and frequency are modulated in order to transmit a signal.
3 *Genetics.* an individual who carries and transmits a recessive gene.
4 *Medicine.* a person who is immune to a disease but carries and transmits the infectious agent.

carrying capacity, 1 *Ecology.* the number of individuals of a species which a given area can support indefinitely.
2 the largest amount of the biomass of an area that can exist there for an indefinite period.
3 the point at which maximum sustainable yield is achieved.

Cartesian coordinate, *Mathematics.* 1 either one of two numbers which determine the position of a point in a plane by its distance from two fixed intersecting lines.
2 any one of three numbers which determine the position of a point in space.
[named after *Cartesius,* Latinized form of the name of René *Descartes,* 1596–1650, French philosopher and mathematician]

---

cap, fāce, fäther; best, bē, tėrm; pin, fīve;
rock, gō, ôrder; oil, out; cup, pùt, rüle,
yü in use, *yu* in uric;
ng in bring; *sh* in rush; *th* in thin, ᴛʜ in then;
zh in seizure.

ə = *a* in about, *e* in taken, *i* in pencil, *o* in lemon, *u* in circus

91

Cartesian product or **Cartesian set,** *Mathematics.* the set of all ordered pairs that can be formed by matching each member of one set with each member of a second set in turn.

cartilage, *n. Anatomy.* a tough, elastic substance forming parts of the skeleton of vertebrates. Cartilage is more flexible than bone and not as hard. The external ear consists of cartilage and skin. *Certain internal parts of the connective tissue assume a consistency which is no longer fibrous, but rather that of a very hard, stiff jelly . . . called cartilage* (Norbert Wiener). SYN: gristle.
[from Latin *cartilago*]
—**cartilaginous** (kär′tə laj′ə nəs), *adj.* **1** of or like cartilage. **2** having the skeleton formed mostly of cartilage: *Sharks are cartilaginous.*

cartilage bone, *Anatomy.* bone substance or a bone that was formed in the development of the organism from cartilage.

cartilage cell, = chondrocyte.

caruncle (kar′ung kəl *or* kə rung′kəl), *n.* **1** *Botany.* a protuberance at or near the point of attachment of a seed. **2** *Zoology, Anatomy.* a fleshy process, such as the comb or wattle of a turkey or chicken. [from French, from Latin *caruncula* little piece of flesh, from *carnem* flesh]
—**caruncular** (kə rung′kyə lər), *adj.* of or resembling a caruncle.
—**carunculate** (kə rung′kyə lit), *adj.* having a caruncle or caruncles.

caryopsis (kar′ē op′sis), *n. Botany.* a small, dry seed fruit, especially of grasses. A grain of wheat is a caryopsis. [from New Latin, from Greek *karyon* nut + *opsis* appearance]

cascade, *n.* **1** *Geology, Geography.* a small waterfall, especially one of a series of waterfalls. **2** *Physics.* = avalanche.

casein (kā′sēn′), *n. Chemistry.* a protein present in milk and containing phosphorus, used in making plastics, adhesives, and certain kinds of paints. Cheese is mostly casein. *The characteristic flavour of mature Cheddar cheese must be produced by the combined effects of the breakdown products of casein, milk fat, and lactose.* (S.A. Barnett). [from Latin *caseus* cheese]

caseinogen (kā′sē in′ə jən), *n. Chemistry.* casein in a dissolved, unclotted state; the precursor of casein.

Casparian strip (kas pär′ē ən), *Botany.* a water-resistant strip or band occurring in the radial and end walls of endodermal cells. [named after R. *Caspary,* 19th-century German botanist]

casque (kask), *n. Zoology.* any helmetlike structure. The frontal boss or shield of certain birds such as the cassowary is a casque. [from French, from Spanish *casco* helmet, cask]

Cassini's division or **Cassini division,** *Astronomy.* a wide, dark gap between the bright ring and outer ring of Saturn. See the picture at **ring.** [named after *Cassini,* 1625–1712, Italian-born French astronomer, who discovered the gap in 1675]

cassiterite (kə sit′ə rīt′), *n. Mineralogy.* the dioxide of tin, found pure in nature, which is the chief source of tin. *Formula:* SnO_2 [from Greek *kassiteros* tin]

cast, *n.* **1** *Zoology.* **a** the convoluted earth and waste thrown out by an earthworm, or the sand of a lugworm: *These casts, held together by mucus, act directly in creating good soil structure* (Science News Letter). **b** the shed skin of an insect.
2 *Physiology.* a mass of fibrous, coagulated, or exuded matter that has assumed the form of a cavity in which it was molded: *a bronchial or nasal cast.*

CAT, *abbrev.* **1** clear air turbulence. **2** computerized axial tomography (X-ray photography in which images of an internal part of the body are made by a circling X-ray beam and synthesized by computer into a single cross-sectional view): *CAT scan, CAT scanner.*

cata-, *prefix.* down; downward, as in *catabolism, catadromous, catalysis.* [from Greek *kata-,* from *kata* down, against]

catabolic, *adj.* of, involving, or exhibiting catabolism: *the catabolic processes of plants and animals.* Contrasted with **anabolic.**

catabolism (kə tab′ə liz′əm), *n. Biology.* the part of metabolism that yields energy by breaking down complex molecules into simpler ones: *Catabolism serves two purposes: (1) it releases the energy for anabolism and other work of the cell, and (2) it serves as a source of raw materials for anabolic processes* (Helena Curtis). Also called **destructive metabolism, dissimilation.** [from *cata-* + *(meta)bolism*]

catabolite, *n. Biochemistry.* a product of catabolism.

catadioptric (kat′ə dī op′trik), *adj. Optics.* having to do with or involving both the reflection and refraction of light: *a catadioptric lens.* [from *cata-* + *dioptric*]

catadromous (kə tad′rə məs), *adj. Zoology.* living in fresh water but going to salt water to spawn. Eels are catadromous. Contrasted with **anadromous.** Compare **diadromous.** [from *cata-* down + Greek *dromos* a running]

catalase (kat′l ās), *n. Biochemistry.* an enzyme found in most living cells which catalyzes the separation of hydrogen peroxide into gaseous oxygen and water: *In the body, catalase occurs in all organs, with greatest concentrations in liver and red blood cells* (Science News).

catalysis (kə tal′ə sis), *n., pl.* **-ses** (-sēz′). *Chemistry.* the causing or speeding up of a chemical reaction by the presence of a catalyst. [from Greek *katalysis* dissolution, from *kata-* down + *lyein* to loose]
—**catalytic** (kat′l it′ik), *adj.* of or causing catalysis: *Metallic crystals are particularly interesting . . . because many metals show marked catalytic properties at their surfaces* (P.R. Rowland).

catalyst (kat′l ist), *n. Chemistry.* a substance that affects the rate of a chemical reaction while undergoing no permanent change in composition itself. Enzymes are important catalysts in digestion. *Catalysts change the activation energy required and thus change the rate of reaction. A catalyst does not initiate a chemical reaction* (Chemistry Regents Syllabus).

catalyze, *v. Chemistry.* to act upon by catalysis: *This enzyme* [peroxidase] *catalyzes . . . the transfer of oxygen from hydrogen peroxide or other peroxides to another substance* (Science News Letter).
—**catalyzer,** *n.* = catalyst.

cataphoresis (kat′ə fə rē′sis), *n.* = electrophoresis.
—**cataphoretic** (kat′ə fə ret′ik), *adj.* = electrophoretic.

catapleiite or **catapleite** (kat′ə plē′īt), *n. Mineralogy.* a hydrated silicate of zirconium and sodium, occurring in hexagonal opaque crystals of light yellowish-brown color. $(Na_2,Ca)ZrSi_3O_9 \cdot 2H_2O$ [from Greek *kata* together + *pleion* more (so called because it occurs along with some other minerals)]

cataract, *n. Geology, Geography.* a large, steep waterfall.

catastrophism, *n. Geology.* the theory that geological change occurs as a result of sudden, catastrophic events rather than by continuous and uniform processes. Contrasted with **uniformitarianism.**

catchment basin or **catchment area,** *Geography, Geology.* a land area, bounded by natural watersheds and usually the sea, where the precipitation is drained by one river system.

catecholamine (kat′ə kō′lə mēn′), *n. Biochemistry.* any of a group of amines that act upon nerve cells as neurotransmitters or hormones. Adrenaline, norepinephrine, and dopamine are catecholamines. *Catecholamines . . . influence the secretion of brain hormones that in turn regulate the secretion of hormones from the pituitary or master gland* (Science News). [from *catechol* a crystalline phenol + *amine*]

catecholaminergic (kat′ə kō′lə mə nėr′jik), *adj. Physiology.* activated by or producing catecholamine: *catecholaminergic neurons.* Compare **adrenergic.** [from *catecholamine* + Greek *ergon* work + English *-ic*]

catenary (kat′ə ner′ē), *n. Mathematics.* the curve formed by a heavy, perfectly flexible cord, cable, or the like, hanging freely from two fixed points not in the same vertical line. [from Latin *catēnārius* relating to a chain, from *catēna* chain]

cathepsin (kə thep′sin), *n. Biochemistry.* any of a group of proteolytic enzymes that cause disintegration of animal cells and tissues. [from Greek *kathepsein* to boil down]

cathode (kath′ōd), *n. Electronics.* 1 the negative electrode of an electrolytic cell or electron tube. 2 the positive terminal of a battery that is producing current. [from Greek *kathodos* a way down, from *kata-* down + *hodos* way] Compare **anode.**
—**cathodic** (kə thod′ik), *adj.* having to do with or resembling a cathode: *Spectacular results are being achieved with cathodic protection—that is, controlling the flow of electric currents that produce corrosion* (Wall Street Journal).

cathode rays, *Electronics.* the invisible streams of electrons from the cathode in a vacuum tube. When cathode rays strike a solid substance, they produce X rays.

cathode-ray tube, *Electronics.* a vacuum tube in which high-speed electrons are produced and passed through electromagnetic fields in the form of a beam. Cathode-ray tubes are used in reproducing images in television receivers, radar sets, and computer terminals. *A cathode-ray tube which . . . is scanned continuously by a narrow beam of electrons that covers the whole tube* (Science News). *Abbreviation:* CRT

cation (kat′ī′ən), *n. Chemistry.* 1 a positively charged ion that moves toward the negative pole in electrolysis: *A change in number of electrons changes the electrical nature of the atom . . . An atom thus changed is called an ion; with an excess of electrons it becomes an anion*

. . . with a deficit it becomes a cation (Storer, *General Zoology*).
2 an atom or group of atoms having a positive charge. [coined by Faraday from Greek *kation* going down, from *kata-* down + *ienai* go]
—**cationic** (kat′ī on′ik), *adj.* of or like a cation: *Nearly all of the long-lived fission nuclides are cationic and virtually all soils have an appreciable cation exchange capacity* (Bulletin of Atomic Scientists).

catkin, *n.* = ament.

catoptric (kə top′trik), *adj. Optics.* of or having to do with the reflection of light, especially the reflection from mirrors or polished surfaces. [from Greek *katoptrikos* pertaining to a mirror, from *katoptron* mirror] —**catoptrically,** *adv.*

Cauchy sequence (kō′shē), *Mathematics.* a sequence of elements having the property such that the numerical difference between any two of its terms is as small as desired, provided the two terms are sufficiently far out in the sequence. [named after Augustin-Louis *Cauchy*, 1789–1857, French mathematician]

caudad (kô′dad), *adv. Zoology.* toward the tail or posterior end.

caudal (kô′dl), *adj. Zoology.* 1 of, at, or near the tail or the posterior end of an animal. A **caudal fin** is a fin located at the posterior end of fishes. See the picture at **fin.**
2 taillike: *a posterior caudal tuft of longer cilia of a paramecium.*
[from Latin *cauda* tail]

caudate (kô′dāt), *adj. Anatomy.* having a tail or taillike appendage: *the caudate lobe of the liver.*

caudex (kô′deks), *n., pl.* **-dices** (-də sēz′), **-dexes.** *Botany.* 1 the woody base of a perennial plant, which sends up new herbaceous stems each year in place of the old.
2a the stem or main axis of a tree. b the stem of a palm or a tree fern.
[from Latin *caudex* trunk, stem]

caudicle (kô′də kəl), *n. Botany.* the slender, stalklike appendage of the masses of pollen in plants belonging to the orchid family. [from New Latin *caudicula,* from Latin *caudex* trunk, stem]

caulescent (kô les′nt), *adj. Botany.* having a distinct stem rising above the ground. Compare **acaulescent.** [from Latin *caulis* stalk]

cauline (kô′līn), *adj. Botany.* 1 of or belonging to a stem. 2 of or growing from the upper part of a stem, as opposed to the basal part. [from Latin *caulis* stalk]

caulocarpous (kô′lə kär′pəs), *adj. Botany.* bearing fruit repeatedly upon the same stem. [from Greek *kaulos* stalk + *karpos* fruit]

caulome (kô′lōm), *n. Botany.* the axis or stem of a plant. [from Greek *kaulos* stalk + *-ōma* a growth]

---

cap, fāce, fäther; best, bē, tèrm; pin, five;
rock, gō, ôrder; oil, out; cup, pùt, rüle,
yü in use, *yu* in uric;
ng in bring; *sh* in rush; *th* in thin, ᴛH in then;
zh in seizure.
ə = *a* in about, *e* in taken, *i* in pencil, *o* in lemon, *u* in circus

caustic

caustic, *adj.* **1** *Chemistry.* that burns or destroys tissue: *Sodium hydroxide has been called a caustic alkali, or caustic soda, because it reacts with living matter. It dissolves silk, wool, and animal tissues* (George L. Bush). SYN: corrosive.
2 *Optics.* **a** of or denoting a curved surface to which are tangent all the rays of light proceeding from a fixed point and reflected or refracted by a curved surface. **b** of or denoting the curve formed by a plane section of such a surface.
—*n.* **1** *Chemistry.* a substance that burns or destroys tissue. **2** *Optics.* a caustic curve or surface.
[from Latin *causticus,* from Greek *kaustikos,* from *kaiein* to burn]

caustic potash, = potassium hydroxide.

caustic soda, = sodium hydroxide.

cave, *n.* *Geology.* a natural underground cavity, chamber, or series of chambers, especially one with an opening in the side of a hill or mountain.

cavern, *n.* *Geology.* an underground cavity or chamber, especially one that is large or indefinite in extent, found in areas of extensive limestone.

cavernicolous (kav′ər nik′ə ləs), *adj.* *Zoology.* inhabiting caverns; dwelling in caves. [from Latin *caverna* cave + *colere* inhabit]

cavernous sinus, *Anatomy.* a venous sinus of the cranial cavity, lying on the side of the body of the sphenoid bone.

cavitation, *n.* **1** *Physics.* the formation of cavities in a fluid downstream from an object moving in it, as behind the moving blades of a propeller: *Cavitation so disrupts a liquid that even if it was formerly gas-free, it is then interspersed with gas bubbles.* (Gabriele Rabel).
2 *Medicine.* **a** the formation of cavities in any body structure, especially in tuberculous lungs. **b** any one of the cavities thus formed.

cavity, *n.* **1** *Anatomy.* an enclosed space inside the body: *the abdominal cavity.*
2 *Medicine.* a pocket of decay in a tooth.
3 *Geology.* a hole, pit, or any hollow place: *a subterranean cavity.*

Cb, *symbol.* columbium.

cc or **cc.,** *abbrev.* cubic centimeter *or* centimeters.

Cc, *abbrev.* cirrocumulus.

cd, *symbol.* candela.

Cd, *symbol.* cadmium.

Ce, *symbol.* cerium.

cecal (sē′kəl), *adj.* *Anatomy.* **1** of or having to do with the cecum; resembling the cecum: *cecal worms.* **2** ending blindly, like the cecum: *the cecal terminal of a duct.* —**cecally,** *adv.*

cecum (sē′kəm), *n., pl.* **-ca** (-kə). *Anatomy.* the first part of the large intestine, enclosed at one end: . . . *the ileum, which joins the large intestine in a kind of pouch called the cecum* (Newsweek). Also spelled **caecum.** [from Latin *caecum* blind (thing)]

ceiling, *n.* *Meteorology.* the distance between the earth and the lowest clouds: *If there is dense fog . . . or other factors . . . preventing observation of cloudiness, then the ceiling is recorded zero* (Blair, *Weather Elements*).

ceilometer (sē lom′ə tər), *n.* *Meteorology.* a device that records the height of a cloud formation: *The ceilometer beams that measure the height of clouds above many airports . . . are powerful searchlights that cast a spot of light on the base of the overcast so that an automatic instrument can calculate its height by triangulation* (Time). [from *ceil(ing*) + *-o-* + *-meter*]

celadonite (sel′ə də nīt), *n.* *Mineralogy.* a green, earthy silicate of iron, magnesium, and potassium, belonging to the mica group. [from French *céladon* a pale-green color]

celestial equator, *Astronomy.* the great circle of the celestial sphere, the plane of which is perpendicular to the axis of the earth: *The celestial equator is . . . in the same plane with the earth's equator and is the largest of the diurnal circles* (Baker, *Astronomy*). Also called **equinoctial line.** See the pictures at **equinox, north celestial pole.**

celestial horizon, = horizon (def. 1c): *The celestial horizon . . . is the horizon of astronomy as distinguished from the visible horizon* (Baker, *Astronomy*).

celestial latitude, *Astronomy.* the angular distance of a heavenly body from the nearest point on the ecliptic (either positive or negative according to whether the body is north or south of the ecliptic).

celestial longitude, *Astronomy.* the angular distance eastward from the vernal equinox to the foot of the circle of latitude drawn through a star or other celestial body.

celestial mechanics, the branch of mechanics concerned with the motions of natural or man-made celestial bodies under the influence of gravitation.

celestial meridian, *Astronomy.* the great circle on the celestial sphere which passes through the celestial poles, zenith, and nadir, intersecting the horizon at the north and south points.

celestial pole, *Astronomy.* each of the two points at which the earth's axis, if extended, would touch the celestial sphere.

celestial sphere, *Astronomy.* the imaginary sphere of infinite radius with the observer as its center, which appears to enclose the universe. To an observer on earth, the visible sky forms half of the celestial sphere. *The Celestial Sphere is the conventional representation of the sky as a spherical shell on which the celestial bodies appear projected . . . The chief convenience of the celestial sphere is in representing the positions of the stars* (Baker, *Astronomy*). See the picture at **horizon.**

celestite (sel′ə stīt), *n.* *Mineralogy.* native strontium sulfate, a mineral occurring as white, or sometimes light-blue, crystals. *Formula:* $SrSO_4$ [perhaps from Italian *celestino* sky-blue]

celiac (sē′lē ak), *adj.* *Anatomy.* of or having to do with the abdominal cavity. **Celiac disease** is a chronic digestive disorder especially of young children, characterized by abdominal swelling. Also spelled **coeliac.** [from Greek *koiliakos,* from *koilia* intestines, from *koilos* hollow]

cell, *n.* **1** *Biology.* the basic unit of living matter in all organisms (excluding viruses). Cells are of two fundamental types: *prokaryotic* cells (bacteria and blue-green algae) lack a distinct, membrane-bound nucleus, and *eukaryotic* cells (those of all other cellular organisms) possess a distinct, membrane-bound nucleus. Eukaryotic cells are generally much larger and more complex than prokaryotic cells. All cells consist of protoplasm bounded externally by a *cell membrane.* In addition, cells of bacteria, algae, fungi, and plants typically have a rigid *cell wall. The cell is a self-maintaining system with the chemical and physical mechanisms for obtaining material from its environment to satisfy nutritional and energy requirements* (D.L. Woodhouse and H.S. Sherratt). *The cells are of many different kinds—skin cells, muscle cells, bone cells, nerve cells, gland cells, and so on . . . But all the cells can be traced back to one special cell, the fertilized egg, which is the normal beginning of a new individual* (Gruenberg).

2 *Electricity, Chemistry.* a container holding materials which produce electricity by chemical action. A battery consists of one or more cells.

3 *Meteorology.* any center of high or low pressure: *In winter the cells of high pressure are centered over the cold continental areas* (Blair, *Weather Elements*).

[from Latin *cella* small room]

ASSOCIATED TERMS: In the *cytoplasm* of cells, various specialized functions occur in subcellular structures called *organelles.* Some major organelles are the *cell membrane, nucleus, nucleolus, endoplasmic reticulum, ribosome, mitochondria, Golgi apparatus, lysosome, vacuole, centriole, chloroplast,* and *cell wall.*

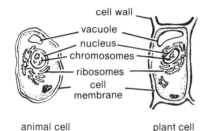

animal cell   plant cell

cell center, *Biology.* = centrosome.

cell division, *Biology.* the process by which a cell divides to form two or more cells, involving both nuclear duplication and cytoplasmic division. Compare **mitosis, amitosis.**

cell-mediated immunity, *Biology.* immunity produced by T cells attacking directly viruses, foreign tissue, etc., rather than by antibodies such as those secreted by B cells: *Cell-mediated immunity constitutes an important defense against viral infections* (Science News).

cell membrane, *Biology.* the thin membrane that forms the outer surface of the protoplasm of a cell and regulates the passage of materials in and out of the cell: *A currently accepted model of the cell membrane is the fluid-mosaic model. This suggests that the membrane is a double lipid layer in which large proteins float. Many small particles . . . diffuse through the membrane. Most larger molecules such as proteins and starches cannot diffuse into or out of cells unless they are chemically digested.* (Biology Regents Syllabus). Also called **plasma membrane.** See also **active transport.** See the picture at **cell.**

cellobiose (sel'ō bī'ōs), *n. Chemistry.* a sugar occurring as the product of the partial hydrolysis of cellulose, used as a reagent in bacteriology. *Formula:* $C_{12}H_{22}O_{11}$

celloidin (sə loi'din), *n.* a pure form of pyroxylin soluble in ether, used especially in microscopy in mounting sections on slides.

cell plate, *Botany.* a membranous plate typically arising in plant cells toward the close of cell division, eventually developing into the new cell wall that separates daughter cells.

cell sap, *Botany.* the fluid content of a plant cell.

cell theory, *Biology.* the principle that cells are the fundamental structures in the organization and life processes of all organisms, and that all cells arise from other cells. *The cell theory as set forth by Schwann and Schleiden in 1839 was primarily an anatomical theory. It proposed that cells were the building blocks of organisms composed of many cells.*

▶ There are several exceptions to the cell theory: (a) The first cell could not have arisen from a previously existing cell. (b) Viruses are not composed of cells but do contain genetic material and can reproduce in a host cell. (c) Although considered organelles, mitochondria and chloroplasts contain genetic material and can reproduce in a cell.

cellular, *adj. Biology.* **1** having to do with cells: *the cellular architecture of the brain,* **2** made up of cells: *All animal and plant tissue is cellular.*

—**cellularity,** *n.* cellular quality or condition.

cellular respiration, *Biology.* the process in living cells by which enzymes act on carbohydrates in the cells to yield energy. Atmospheric oxygen (O_2) combines with hydrogen to form water, and carbon dioxide is released as a waste product. *Cellular respiration refers to those enzyme-controlled reactions in which the potential energy of organic molecules, such as glucose, is transferred to a more available form of energy . . . stored in adenosine triphosphate molecules* (Biology Regents Syllabus). Compare **aerobic respiration.**

cellulase (sel'yə lās), *n. Biochemistry.* an enzyme that hydrolyzes cellulose, found in certain plants and insects.

cellulolytic, *adj. Biology.* capable of hydrolyzing cellulose: *cellulolytic protozoans.*

cellulose (sel'yə lōs), *n. Biochemistry.* a substance that forms a large part of the walls of plant cells; the woody part of trees and plants. Wood, cotton, flax, and hemp are largely cellulose. Cellulose is used to make paper, rayon, plastics, explosives, etc. *The chemical structure of cell walls varies considerably, but the basic substance . . . is usually cellulose, a carbohydrate rather closely related, chemically, to starch* (Emerson, *Basic Botany*). *Formula:* $(C_6H_{10}O_5)_n$ [from Latin *cellula* small room]

cellulose acetate, *Chemistry.* any of several compounds, formed from cellulose in the presence of acetic anhydride used in making textiles, camera films, lacquers, varnishes, etc.

---

cap, fāce, fäther; best, bē, tėrm; pin, five;
rock, gō, ôrder; oil, out; cup, pùt, rüle,
yü in use, *yù* in uric;
ng in bring; *sh* in rush; *th* in thin, ᴛн in then;
zh in seizure.

ə = *a* in about, *e* in taken, *i* in pencil, *o* in lemon, *u* in circus

cellulose nitrate, = nitrocellulose.

cellulosic, *Chemistry.* —*n.* any derivative of cellulose, such as cellulose acetate.

—*adj.* of or derived from cellulose: *cellulosic resin.*

cell wall, *Biology.* **1** the rigid, transparent outer covering of a plant or algal cell, made up of cellulose and other materials and surrounding the cell membrane: *Most plant cells are surrounded by a cell wall* [*that*] . . . *protects and supports the cell. It is made up of several layers and is formed by the cell itself* (Otto and Towle, *Modern Biology*). See the picture at **cell.**

2 any similar covering of a bacterial or fungal cell, differing in fundamental composition from cell walls of plant or algal cells.

celom (sē′ləm), *n.* = coelom.

Cels., *abbrev.* Celsius.

Celsius (sel′sē əs), *adj.* of or based upon the scale used in centigrade thermometers, in which the freezing point of water is 0 degrees and the boiling point 100 degrees: *The unit of temperature is the degree Celsius* (Scientific American). *Abbreviation:* C See the picture at **temperature.** [named after Anders *Celsius,* 1701–1744, Swedish astronomer, who devised the scale]

cement, *n.* **1** *Anatomy.* = cementum.

2 *Geology.* the material that occupies the spaces between individual grains of rock, binding them together. It is a chemically precipitated mineral found in clastic sedimentary rock.

cemental, *adj.* *Anatomy.* of or having to do with cement or cementum.

cement gland, *Zoology.* a gland that secretes the sticky substance by which a barnacle or other cirriped attaches itself to an object.

cementite (sə men′tīt), *n.* *Chemistry.* a hard, brittle carbide of iron, a constituent of steel. *Formula:* Fe_3C

cementum (sə men′təm), *n.* *Anatomy.* the bony tissue forming the outer crust of the root of a tooth. See the picture at **tooth.** [from New Latin, from Latin *caementum* stone chippings]

cenogonous (si nog′ə nəs), *adj.* *Zoology.* oviparous at one season of the year and ovovivaparous or viviparous at another: *cenogonous insects.* [Greek *koinos* common + *gonos* generation]

Cenozoic (sen′ə zō′ik), *Geology.* —*n.* **1** the present geological era, characterized by abundance and diversity of mammals and flowering plants. It began about 65 million years ago. *The Cenozoic saw the great spread of the mammals . . . with a peak apparently reached at some time during the Miocene and a fairly noticeable decline since that time. Among the great events of the Cenozoic . . .* [*is*] *the great Ice Age* (Garrels, *Textbook of Geology*).

2 the rocks formed in this era. —*adj.* of or having to do with the Cenozoic era or its rocks: *Cenozoic deposits.*

[from Greek *kainos* new, recent + *zōē* life]

cent., *abbrev.* centigrade.

center, *n.* **1** *Geometry.* a point within a circle or sphere equally distant from all points of the circumference or surface.

2 *Physiology.* a mass of nerve cells closely connected and acting together: *the respiratory center, the center of balance.* Also called **nerve center.**

center of buoyancy, *Physics.* a point in a floating body, corresponding to the center of gravity of the water displaced.

center of curvature, *Mathematics.* the center of the circle which serves to measure curvature at a given point: *A concave mirror is part of the inner side of an imaginary sphere . . . The center of this imaginary sphere is called the center of curvature* (E.A. Fessenden).

center of gravity, *Physics.* that point in a body around which its weight is evenly balanced: *To calculate the center of gravity of a flat body of any shape, it is necessary . . . to remember . . . that* [*the weight of*] *its line of action passes through the center of gravity in all orientations* (Sears and Zemansky, *University Physics*).

center of mass, *Physics.* that point in a body which moves as though it bore the entire mass of the body, usually identical with the center of gravity: *Many textbooks now use the term center of mass, instead of center of gravity. This is because other forces besides gravity seem to act at this point when they act on the body as a whole* (Robert F. Paton). Also called **centroid.**

center of oscillation, *Physics.* a point in a pendulum such that, if the whole mass of the pendulum were concentrated there, the time of oscillation would remain unchanged.

centi-, *combining form.* **1** 100, as in *centigrade* = *100 degrees.*

2 (in the metric system) 100th part of, as in *centimeter* = *100th part of a meter. Symbol:* c. or c [from Latin *centum* hundred]

centiare (sen′tē ār′ *or* sen′tē är′), *n.* 1/100 of an are; one square meter. *Abbreviation:* ca.

centigrade, *adj.* = Celsius. [from French, from Latin *centum* hundred + *gradus* degree]

centigram, *n.* a unit of weight in the metric system, equal to 1/100 of a gram. *Abbreviation:* cg.

centiliter *n.* a unit of volume in the metric system, equal to 1/100 of a liter. *Abbreviation:* cl.

centimeter, *n.* a unit of length in the metric system, equal to 1/100 of a meter. It is a fundamental unit of the CGS system. *Abbreviation:* cm. or c., c

centimeter-gram-second, *adj.* = CGS (having to do with a system of measurement in which the centimeter is the unit of length, the gram is the unit of mass, and the second is the unit of time).

centistoke, *n.* a unit for measuring the kinematic viscosity of a fluid, equal to 1/100 of a stoke.

centrad (sen′trad), *adv.* *Anatomy.* toward the center.

central, *adj.* **1** *Physiology.* of, relating to, or designating the central nervous system: *central anesthesia, central paralysis.* **2** *Anatomy.* of or having to do with a centrum or centrums.

central dogma, *Molecular Biology.* the theory that, with certain exceptions, genetic information is transferred by DNA serving as a template for both its own replication and the synthesis of messenger RNA, which serves in turn as a template in the synthesis of protein: *In 1970, biochemists and scientists in related fields for the first time accepted the idea that ribonucleic acid (RNA) can be a template, or pattern, for the synthesis* *of deoxyribonucleic acid (DNA). Previously, most scientists had interpreted the so-called central dogma of molecular biology to mean that the characteristic DNA-to-RNA pattern of information transfer was never reversed* (Howard M. Temin).

central force, *Physics.* a force attracting to or repelling from a center, whose strength depends solely on the distance from the center.

central nervous system, *Anatomy.* the part of the nervous system of vertebrates that consists of the brain and spinal cord: *From the central nervous system, nerves go to all parts of the body through the peripheral nervous system* (Otto and Towle, *Modern Biology*). Compare **autonomic nervous system.**

central tendency, *Statistics.* the number or point best representing a particular set of data, near the middle of a normal curve of frequency distribution.

centricipital (sen′trə sip′ə təl), *adj.* of, having to do with, or situated in the centriciput.

centriciput (sen tris′ə pət), *n. Anatomy.* the middle part of the head, between the sinciput and the occiput. [from Latin *centrum* center + *caput* head]

centrifugal (sen trif′yə gəl), *adj.* **1** *Physics.* moving or tending to move away from a center. **Centrifugal force** is the inertia, or tendency to move in one direction, which causes a body turning around a center to move away from the center. Compare **centripetal.** [from Latin *centrum* center + *fugere* flee]
2 *Physiology.* = efferent.
3 *Botany.* **a** developing from the center or apex outward or downward: *a centrifugal plant.* **b** that turns from the center toward the side of the fruit, as a radicle.

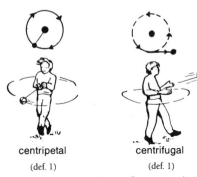

centripetal	centrifugal
(def. 1)	(def. 1)

centrifugal inflorescence, *Botany.* flowers opening in descending order from the tip.

centrifugation, *n.* the separation of materials of different densities by centrifugal force: *Centrifugation methods of separating cells and cell debris suspended in fluids are limited by the fact that the densities of these biological materials differ but little from each other and from the density of the suspension fluid* (Science News Letter).

centrifuge (sen′trə fyüj), *n.* a machine for separating two substances varying in density, as cream from milk or bacteria from a fluid, by means of centrifugal force: *The principle of a centrifuge of course is that under centrifugal force the heavier particles in a mixture tend to move farther out toward the periphery than the lighter ones* (Scientific American). Compare **ultracentrifuge.**

—*v.* to rotate in a centrifuge; subject to a centrifugal force: *Each time the solution was centrifuged, layers of fluid were separated and removed, until the different sized particles from within the bacteria were isolated in separate fractions* (Science News Letter).

centriole (sen′trē ōl), *n. Biology.* either of a pair of cylindrical bodies, composed of microtubules (or spindles), within the centrosome of a cell. Centrioles serve to organize other microtubular materials in the cell and thus help to determine the cell's overall polarity. Centrioles are common in animal cells and may also be present in flagellated protistan cells, but are lacking in the cells of higher plants. *During the reproductive process of mitosis, the centrioles, located near the nucleus of some cells, divide; each then becomes associated with the system of microtubules called the spindle ... During prophase, the two pairs of centrioles begin to migrate to opposite sides of the nucleus* (James M. Barrett). See the pictures at **centrosome, mitosis.** Compare **centrosome, aster.**

centripetal (sen trip′ə təl), *adj.* **1** *Physics.* moving or tending to move toward a center. **Centripetal force** is the force that tends to move things toward the center around which they are turning. Gravitation acts as a centripetal force. Compare **centrifugal.** See the picture at **centrifugal.**
2 *Physiology.* = afferent.
3 *Botany.* **a** developing inward or upward toward the center or apex. **b** that turn toward the axis of the fruit, as a radicle.
[from Latin *centrum* center + *petere* seek]

centro-, *combining form.* center; central, as in *centrosphere = center sphere.* [from Latin *centrum* center]

centrobaric (sen′trə bar′ik), *adj. Physics.* of or relating to the center of gravity. [from *centro-* + Greek *baros* weight]

centroid (sen′troid), *n.* **1** *Physics.* = center of mass.
2 *Geometry.* the point of intersection of the medians of a triangle.

centrolecithal (sen′trə les′ə thəl), *adj. Embryology.* (of eggs) having the yolk in the center surrounded by a layer of protoplasm. Compare **alecithal, homolecithal, telolecithal.** [from *centro-* + Greek *lekithos* yolk]

centromere (sen′trə mir), *n. Biology.* the point on a chromosome by which it is drawn to the pole during mitosis: *In most cases it would appear that the chromosomes are being pulled through a liquid by special fibers which run from the poles and are attached to a definite region of each chromosome ... the centromere, kinetochore, or simply "the spindle fiber attachment"* (Scientific American). [from *centro-* + Greek *meros* part]
—**centromeric,** *adj.* of or belonging to the centromere: *centromeric DNA.*

cap, fāce, fäther; best, bē, tėrm; pin, fīve;
rock, gō, ôrder; oil, out; cup, pùt, rüle,
yü in use, *yu* in uric;
ng in bring; *sh* in rush; *th* in thin, ᴛн in then;
zh in seizure.
ə = *a* in about, *e* in taken, *i* in pencil, *o* in
lemon, *u* in circus

centrosome (sen'trǝ sōm), n. Biology. a specialized region adjacent to the nucleus of a cell that serves as a center for the organization of microtubules (or spindles). The centrosome of an animal cell or a flagellated protistan cell typically contains two centrioles. *In the middle of the major microtubule organizing center of almost all animal cells, called the cell center or the centrosome, is a centriole pair . . . many microtubules are organized by the cell center, which therefore might be considered the "command post" of the cell* (Bruce Alberts). Also called **cell center.** Compare **centriole.**
—**centrosomic,** adj. of or having to do with a centrosome or centrosomes.

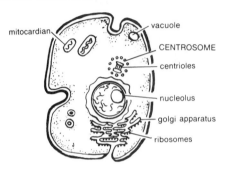

mitocardian, vacuole, CENTROSOME, centrioles, nucleolus, golgi apparatus, ribosomes

centrosphere (sen'trǝ sfir'), n. Geology. the central core of the earth: *The outer core lies beneath the mantle. At the center of the earth is the inner core. This inner core is also called the centrosphere* (S.E. Ekblaw and R.W. Burnett). Also called **barysphere.**

centrosymmetric, adj. Geometry, Physics. having symmetry with respect to a point or center: *Each of the . . . four edges, b, c, e and f, are centrosymmetric; that is, they are unaltered by a 180-degree rotation around a midpoint* (Scientific American).

centrum (sen'trǝm), n., pl. **-trums, -tra** (-trǝ). Anatomy. the body of a vertebra. The centrum is the solid part to which a bony arch and processes are attached. *Each vertebra is made up of a spool-like centrum surmounted by a neural arch to house the nerve cord* (Storer, *General Zoology*). [from Latin, center]

cephal-, combining form. the form of **cephalo-** before vowels, as in *cephalic.*

cephalad (sef'ǝ lad), adv. Zoology. toward the head or anterior end of the body.

cephalic (sǝ fal'ik), adj. Anatomy. 1 of or having to do with the head: *cephalic ganglia.*
2 near, toward, on, or in the head: *a cephalic horn.*

cephalic vein, Anatomy. a large superficial vein on the front of the arm, running from the elbow to the shoulder.

cephalin (sef'ǝ lin), n. Biochemistry. any of a group of substances containing phosphorus and resembling lecithin, found in brain tissue.

cephalization, n. Biology. the degree to which the sensory and neural organs are concentrated in the head, representing a specialization in evolutionary development.

cephalo-, combining form. head, as in *cephalopod, cephalothorax.* Also spelled **cephal-** before vowels. [from Greek *kephalē* head]

cephalocaudal, adj. Anatomy. extending from head to tail; having to do with the long axis of the body.

cephaloid (sef'ǝ loid), adj. Biology. shaped like or resembling the head.

cephalopod (sef'ǝ lǝ pod), Zoology. —n. any of the most highly developed class (Cephalopoda) of mollusks, having long, armlike tentacles around the mouth, a large head, a pair of large eyes, and a sharp, birdlike beak. Many can expel a dark, inklike fluid. Cuttlefish, squids, octopuses, and nautiluses are cephalopods.
—**adj.** of or belonging to the cephalopods.
[from *cephalo-* + Greek *podos* foot]

cephalothorax (sef'ǝ lō thôr'aks), n. Zoology. the combined head and thorax of some animals, such as crabs and spiders. See the picture at **crustacean.**
—**cephalothoracic** (sef'ǝ lō thǝ ras'ik), adj. of or having to do with the cephalothorax.

Cepheid variable, or **Cepheid** (sef'ē id), n. Astronomy. any of a class of variable stars whose changes in brightness recur in a characteristic manner with a relatively long periodicity, due to expansion and contraction of the star: *The periods of classical Cepheids range from rather over one day to about fifty days* (A. W. Haslett). [from *Delta Cephei,* such a star, from the constellation *Cepheus*]

cerargyrite (sǝ rär'jǝ rīt), n. Mineralogy. native chloride of silver. Formula: AgCl [from Greek *keras* horn + *argyros* silver]

cercaria (sǝr kār'ē ǝ), n., pl. **-iae** (-ē ē). Zoology. a second larval stage of trematode worms, in which the body is usually shaped like a tadpole. [from New Latin, from Greek *kerkos* tail]
—**cercarial,** adj. of or resembling a cercaria.

cercus (sėr'kǝs), n., pl. **-ci** (-kī). Zoology. either of a pair of small sensory appendages at the posterior end of the abdomen of certain arthropods, such as the female mosquito. [from New Latin, from Greek *kerkos* tail]

cere (sir), n. Zoology. a waxy-looking membrane through which the nostrils open near the beak of certain birds, especially parrots and birds of prey. [from Medieval Latin *cera,* from Latin, wax]

cerebellum (ser'ǝ bel'ǝm), n., pl. **-bellums, -bella** (-bel'ǝ). Anatomy. the part of the brain that controls the coordination of the muscles. It consists of a middle lobe and two lateral lobes and is located below the back part of the cerebrum. *The middle portion influences muscles of the trunk, neck, and head; each side of the cerebellum acts on muscles of the same side of the body* (Shull, *Principles of Animal Biology*). See the picture at **brain.**
[from Latin, diminutive of *cerebrum* brain]
—**cerebellar,** adj. of or having to do with the cerebellum: *the cerebellar cortex.*

cerebral (sǝ rē'brǝl), adj. Anatomy. 1 of the brain: *A cerebral hemorrhage may cause paralysis.* 2 of the cerebrum: *Cerebral dominance is the tendency for one cerebral hemisphere to be better developed.*

cerebral cortex, Anatomy. the layer of gray matter that covers the cerebrum: *The signals . . . arrive at the highest brain centre, the cerebral cortex* (New Scientist).

cerebral hemisphere, Anatomy. either of the two lobes of the cerebrum.

cerebral vesicles, *Embryology.* the three primitive, hollow dilations into which the embryonic brain of vertebrates is divided.

cerebroside (ser′ə brə sīd), *n. Biochemistry.* any of a group of lipids found in the brain and other nerve tissue: *The cerebrosides found in the nervous system are compounds of galactose and a fat-like substance* (Harbaugh and Goodrich, *Fundamentals of Biology*).

cerebrospinal (sə rē′brō spī′nl), *adj. Anatomy.* of or having to do with both the brain and the spinal cord.

cerebrospinal fluid, *Anatomy.* the clear fluid normally present within the cavities and between the membranes of the central nervous system. *Abbreviation:* CSF Also called **spinal fluid.**

cerebrovascular (ser′ə brō vas′kyə lər), *adj. Anatomy.* of or having to do with the blood vessels of the cerebrum: *cerebrovascular diseases.*

cerebrum (sə rē′brəm), *n., pl.* **-brums, -bra** (-brə). *Anatomy.* **1** the part of the human brain that controls thought and voluntary muscular movements. It consists of two lobes (the *cerebral hemispheres*) which fill most of the cranial cavity. See the picture at **brain.**
2 the corresponding part of the brain of any vertebrate: *The mammals have developed the cerebrum to the greatest size and complexity found in the vertebrates . . . The cerebrum has complex folds . . . and when the outer part folds inward the amount of gray matter is greater than it would be otherwise* (Winchester, *Zoology*).
[from Latin *cerebrum*]

Cerenkov effect or **Cerenkov radiation** (chə ren′kôf), *Physics.* light produced when electrons or other charged particles pass through a transparent solid or liquid medium faster than the speed of light in the same medium. The blue glow seen in the water of a nuclear reactor is this kind of radiation. Also spelled **Cherenkov.** [named after P.A. *Cerenkov,* 20th-century Soviet physicist, who first observed this effect]

ceric (sir′ik *or* ser′ik), *adj. Chemistry.* of or containing cerium, especially with a valence of four.

cerite (sir′īt), *n. Mineralogy.* a rare hydrous silicate of cerium and other metals. *Formula:* $(Ca,Ce)_3Si_2(O,OH,F)_9$

cerium (sir′ē əm), *n. Chemistry.* a grayish metallic element which occurs only in combination with other elements. Cerium is malleable and ductile and is used in porcelain, glass, and alloys. It is one of the rare-earth metals. *Symbol:* Ce; *atomic number* 58; *atomic weight* 140.12; *melting point* 795°C; *boiling point* 3257°C; *oxidation state* 3,4. [New Latin, named after the asteroid *Ceres*]

cerium metals, *Chemistry.* a group of closely related rare-earth elements, consisting of cerium, lanthanum, praseodymium, neodymium, promethium, samarium, and europium.

cerous (sir′əs), *adj. Chemistry.* of or containing cerium, especially with a valence of three.

ceruloplasmin (sə rü′lə plaz′min), *n. Biochemistry.* an enzyme in blood which promotes the oxidation and circulation of copper: *Ceruloplasmin, so named because of its blue color, is a protein fraction of the fluid portion of the blood* (James A. Brussel). [from Latin *caerulus* dark blue + English *plasm(a)* + *-in*]

cerumen (sə rü′mən), *n. Physiology.* a yellow, waxlike substance secreted by a modified sweat gland in the external ear. Commonly called *earwax.* [from Medieval Latin, from Latin *cera* wax]
—ceruminous (sə rü′mə nəs), *adj.* relating to or containing cerumen.

cerussite (sir′ə sīt), *n. Mineralogy.* a mineral, carbonate of lead, a common ore of lead, found in whitish crystals. *Formula:* $PbCO_3$ [from Latin *cerussa*]

cervical, *adj. Anatomy.* **1** of or having to do with the neck or a necklike part: *cervical vertebrae.*
2 of or having to do with the cervix of the uterus: *cervical cancer.*

cervix (sèr′viks), *n., pl.* **cervixes, cervices** (sər′ vi sēz′). *Anatomy.* **1** the neck, especially the back of the neck.
2 a necklike part, as of the uterus, bladder, or a tooth: *The cervix is a ring of muscle closing the lower end of the uterus where it joins the vagina* (Mackean, *Introduction to Biology*).
[from Latin]

cesium (sē′zē əm), *n. Chemistry.* a soft, silvery, metallic element ·of the alkali metal group which occurs as a minute part of various minerals. It is highly electropositive and is used in photoelectric cells. The rate of vibration of cesium atoms is used as a standard for measuring time. *Symbol:* Cs; *atomic number* 55; *atomic weight* 132.905; *melting point* 28°C; *boiling point* 705°C; *oxidation state* 1. Also spelled **caesium.** [from New Latin *caesium,* from Latin *caesius* bluish-gray; so called from the blue lines in its spectrum]

cesium 137, *Nuclear Physics.* a radioactive isotope of cesium that occurs in fission products and in fallout from nuclear explosives. It has been used in cancer research and therapy. *Cesium 137 is a relatively long-lived radioactive material similar to phosphorus* (Science News Letter).

cesium clock, an atomic clock which measures time by the vibration frequency of cesium atoms, the primary scientific standard of time since 1972. See **International Atomic time.**

cespitose (ses′pə tōs), *adj. Botany.* growing in dense tufts or clumps: *cespitose leaves or grass.* [from New Latin *caespitosus,* from Latin *caespes* turf]

cestode (ses′tōd), *Zoology.* **—n.** any of a class (Cestoda) of parasitic flatworms which may infest the intestinal tract of man and other vertebrates: *The cestodes are mostly slender and elongate, with a flat body usually of many short sections; hence the name "tapeworm. . ." There is no mouth or digestive tract; food is absorbed directly through the body wall* (Storer, *General Zoology*).
—adj. of or belonging to this class.
[from Greek *kestos* girdle]

---

cap, fāce, fäther; best, bē, tèrm; pin, fīve;
rock, gō, ôrder; oil, out; cup, pùt, rüle,
yü in use, *yu̇* in uric;
ng in bring; *sh* in rush; *th* in thin; ᴛʜ in then;
zh in seizure.
ə = *a* in about, *e* in taken, *i* in pencil, *o* in lemon, *u* in circus

cetacean (sə tā′shən), *Zoology.* —*n.* any of an order (Cetacea) of marine mammals having fishlike, almost hairless bodies, flat, notched tails, and paddle-shaped forelimbs, including whales, dolphins, and porpoises. —*adj.* of or belonging to the cetaceans. [from New Latin *Cetacea,* from Latin *cetus* large sea animal, from Greek *kētos*]

cetane (sē′tān), *n. Chemistry.* a colorless liquid hydrocarbon of the methane series, originally obtained from sperm-whale oil. *Formula:* $C_{16}H_{34}$ [from Latin *cetus* large sea animal + English *-ane*]

cetyl alcohol (sē′tl), a white, waxy crystalline alcohol extracted from sperm-whale oil, used as an emulsifier and chemical intermediate, and in pharmaceuticals, detergents, etc. *Formula:* $C_{16}H_{33}OH$ [from Latin *cetus* large sea animal, whale + *-yl*]

cevitamic acid (sē′vī tam′ik), = vitamin C. [from *ce* (for C of *vitamin C*) + *vitam(in)ic*]

Cf, *symbol.* californium.

C.F., *abbrev.* coefficient of friction.

CFC, *abbrev.* chlorofluorocarbon.

CFM, *abbrev.* chlorofluoromethane.

c.f.s. or cfs, *abbrev.* cubic feet per second.

cg. or cgm., *abbrev.* centigram *or* centigrams.

CGS or cgs, *adj.* of or designating a system of measurement based on the centimeter as the unit of length, the gram as the unit of mass, and the second as the unit of time: *The dyne is the CGS unit of force.* [from *c(entimeter)* + *g(ram)* + *s(econd)*]

► General scientific work was formerly done in the *CGS* system (replaced by SI units) in which all of the units of measurement are related except those of heat and temperature. See the note under **SI unit.**

chabazite (kab′ə zīt), *n. Mineralogy.* a colorless or flesh-colored zeolite occurring in glassy crystals, composed chiefly of silica, alumina, sodium, and calcium. *Formula:* $CaAl_2Si_4O_{12} \cdot 6H_2O$ [from French *chabazie* + English *-ite*]

chaeta (kē′tə), *n., pl.* **-tae** (-tē). *Zoology.* a bristle or seta, especially one of the bristles on the parapodia of an annelid worm. [from New Latin, from Greek *chaitē* long flowing hair]

chain, *Chemistry.* **1** a number of atoms of the same element linked together like a chain, usually within an organic molecule: *a carbon chain.*
2 a number of molecules or compounds linked together: *A protein is usually composed of one or more polypeptide chains.*

chain reaction, *Nuclear Physics.* a self-sustaining nuclear reaction occurring when a fissionable nucleus such as that of uranium absorbs a neutron and splits, releasing atomic energy and additional neutrons. These neutrons split other fissionable nuclei releasing more energy and more neutrons. *By careful design of a critical system, the neutrons can be conserved and a chain reaction produced. Uncontrolled, this chain reaction may result in an atomic explosion. Controlled, it can be used to produce useful power* (Ralph E. Lapp).

chalaza (kə lā′zə), *n., pl.* **-zas, -zae** (-zē). **1** *Zoology.* either of the two membranous twisted strings by which the yolk of a bird's egg is bound to the lining membrane at the ends of the shell and kept near the middle of the albumen.
2 *Botany.* the point on a seed where the integuments diverge from the nucellus: *The chalaza . . . is the attachment of the ovule to the scale* (Emerson, *Basic Botany*).
[from New Latin, from Greek *chalaza* hailstone] —**chalazal,** *adj.* of or containing a chalaza.

chalcanthite (kal kan′thīt), *n. Mineralogy.* native copper sulfate or blue vitriol. *Formula:* $CuSO_4 \cdot 5H_2O$ [from Greek *chalkanthon* solution of blue vitriol, from *chalkos* copper + *anthos* flower]

chalcedony (kal sed′ə nē *or* kal′sə dō′nē), *n., pl.* **-nies.** *Mineralogy.* a variety of fibrous quartz that has a waxy luster with minute crystals, and occurs in various colors and forms. A common kind is grayish or blue. Agate, onyx, carnelian, and jasper are forms of chalcedony. [from Latin *chalcedonius,* from Greek *chalkedon*] —**chalcedonic** (kal′sə dō′nik), *adj.* of or resembling chalcedony.

chalco-, *combining form.* **1** copper, bronze, as in *chalcocite.* **2** sulfur; sulfide, as in *chalcophile.* [from Greek *chalkos* copper, bronze]

chalcogen (kal′kə jən), *n. Chemistry.* any of the elements oxygen, sulfur, selenium, or tellurium (group 6A of the periodic table). [*chalco-* + *-gen*]

chalcogenide (kal′kə jə nīd), *Chemistry.* —*n.* a compound consisting of a chalcogen and another element or radical: *zinc and cadmium chalcogenides.*
—*adj.* composed of one or more of the chalcogenides: *chalcogenide glass.*

chalcophile (kal′kə fil), *adj. Geology.* having an affinity for sulfur: *Elements that . . . are associated with sulfide deposits are chalcophile* (Willard S. Moore).

chalcopyrite (kal′kə pī′rīt), *n. Mineralogy.* a yellow sulfide of copper and iron. It is an important copper ore. *Formula:* $CuFeS_2$

chalcostibite (kal′kə stib′īt), *n. Mineralogy.* a native sulfide of copper and antimony. *Formula:* $CuSbS_2$ [from *chalco-* + *stib(ium)* + *-ite*]

chalk, *n. Mineralogy.* a soft, white, gray, or yellow limestone: *Chalk consists largely of tiny shells and crystals of calcite. . . . Most deposits of chalk formed during the Cretaceous Period* (A.J. Eardley). [Old English *cealc,* from Latin *calcem* lime]

chalone (kā′lōn), *n. Biochemistry.* an internal secretion of certain tissues that is known to inhibit mitosis and is thought to reduce the activity of various tissues or organs. [from Greek *chalōn,* present participle of *chalan* to slacken]

chamber, *n. Biology.* an enclosed space in the body of an organism, as in the brain, eye, heart, or in a shell. The human heart has four chambers. SYN: cavity.

Chandler wobble or Chandler's wobble, *Astronomy.* a periodic oscillation of the earth's axis of rotation lasting about 14 months: *Because earthquakes appear to be related to Chandler's wobble, a greater understanding of the wobble could lead to scientists' predicting earthquakes* (Science News). [named after Seth Carlo Chandler, 1846–1913, American astronomer]

channel, *n. Geology, Geography.* **1** the bed or deepest portion of a stream, river, or other body of water. **2** a large strait, such as the English Channel.

character, *n. Biology.* a distinctive feature in the structure or function of an organism. The size and form of a given breed of dogs and the fragrance of a sweet pea are characters. SYN: trait, characteristic. See **acquired character.**

characteristic, *n.* **1** *Mathematics.* the whole number in a logarithm. In the logarithm 2.95424, the characteristic is 2 and the mantissa is .95424.
2 *Biology.* = character.

charcoal, *n. Chemistry.* a black, brittle form of carbon made by partly burning wood or bones in an airtight kiln. Charcoal is used as fuel and in filters. It is often specified as wood, vegetable, or animal charcoal, according to the material burned to produce it.

charge, *n. Physics.* a given quantity of electricity. A charge may be positive (if the charged body has fewer electrons than protons) or negative (if the body has more electrons than protons). The SI or MKS unit of charge is the coulomb. The charge of a single electron or proton is equal to 1.6×10^{-19} coulomb. Also called **electric charge.**
—charged, *adj.* having an electrical charge: *charged ions.*

charge conjugation, *Physics.* **1** the operation of changing the signs of all electric charges in a system, converting all particles into their antiparticles.
2 the theory that, except with respect to the weak interaction, oppositely charged particles are symmetrical or mirror images of each other. Compare **parity, time reversal.**

charge conservation, = conservation of charge: *In charge conservation, . . . if the electric charge of particles entering into a reaction is minus one, the net charge of those that emerge from the reaction must also be minus one* (Walter Sullivan).

Charles's law or **Charles' law,** *Physics.* the law that at constant pressure the volume of a given mass of gas is directly proportional to the absolute temperature. [named after Jacques A. C. *Charles,* 1742–1823, French physicist]

Charles' Law

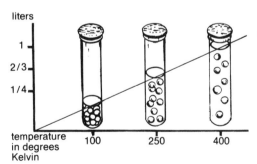

As the temperature rises, the gas expands and the volume of gas in the tubes increases. At constant pressure, a quadrupling of the temperature will increase the volume of the gas four times.

charm, *n. Nuclear Physics.* a quantum number or property of certain quarks and other particles that must be conserved in all strong interactions: *The case for charm—or the fourth quark—became much firmer when it was realized that there was a serious flaw in*

the familiar three-quark theory, which predicted that strange particles would sometimes decay in ways that they did not. In an almost magical way, the existence of the charmed quark prohibits these unwanted and unseen decays, and brings the theory into agreement with experiment* (Sheldon L. Glashow). *Abbreviation:* c

ASSOCIATED TERMS: *Charm* is a property of quarks found in certain particles such as the *J* or *psi particle.* Other properties of quarks are *strangeness, beauty,* and *truth.* Such distinctive properties are classified as *flavors.* Quarks may also be distinguished by *color.* According to the theory of *chromodynamics,* color produces the force that binds quarks together, and the carrier of this force is the *gluon.*

—charmed, *adj.* possessing or exhibiting charm: *It should be possible for charmed quarks and ordinary quarks to bind together, producing "charmed hadrons." These should be similar to ordinary hadrons, although somewhat higher in energy because the charmed quarks have higher energy than ordinary quarks* (Gerald Feinberg).

charmonium (chär mō′nē əm), *n. Nuclear Physics.* a meson composed of a charmed quark and its antiquark: *The discovery of the charmonium was an event of the utmost importance in elementary-particle physics. Nothing so exciting had happened in many years. For believers in quarks the new particle was the first experimental indication that a fourth quark existed* (Sheldon L. Glashow). [from *charm* + *-onium,* as in *muonium*]

cheek, *n.* **1** *Anatomy.* the fleshy lateral wall of the mouth in mammals.
2 *Zoology.* the lateral side of the head in vertebrates other than mammals and in invertebrates.

cheekbone, *n. Anatomy.* the zygomatic bone.

cheek pouch, *Zoology.* a pocketlike fold of skin in the cheek of various animals, especially rodents, used for holding food. Gophers and squirrels have cheek pouches.

chela (kē′lə), *n., pl.* **-lae** (-lē). *Zoology.* the prehensile claw of a lobster, crab, scorpion, etc., resembling a pincer. [from Latin *chele,* from Greek *chēlē* claw]

chelate (kē′lāt), *adj. Zoology.* having a chela (claw) or chelae.
—n. *Chemistry.* an inert complex compound or ion in which a metallic atom or ion is bound at two or more points to a molecule or ion (a ligand) so as to form a ring. The metal is called the *nuclear* atom and may be part of more than one ring. A chelate with more than one atom of metal is said to be *polynucleate. These compounds, known as chelates, are named after the Greek word for claw because of a claw-like chemical structure which literally holds iron atoms tightly in its grasp* (Science News Letter).

cap, fāce, fäther; best, bē, tèrm; pin, fīve;
rock, gō, ôrder; oil, out; cup, pùt, rüle,
yü in use, *yü* in uric;
ng in bring; *sh* in rush; *th* in thin, ŦH in then;
zh in seizure.
ə = *a* in about, *e* in taken, *i* in pencil, *o* in lemon, *u* in circus

—**v.** *Chemistry.* to join or be joined with a metallic ion so as to form a chelate: *Examples of chelated metals in vital processes are . . . magnesium in chlorophyll and iron in haemoglobin* (Science News Letter).

—**chelation** (ki lā′shən), *n. Chemistry.* the process of binding and stabilizing metallic ions by means of a chelate.

chelicera (kə lis′ər ə), *n., pl.* **-erae** (-ə rē). *Zoology.* one of the first pair of appendages near the mouth of scorpions and spiders: *Poison glands are situated in the chelicerae of the tarantulas . . . The poison they secrete passes through ducts that open on the fangs of the chelicerae* (Hegner and Stiles, *College Zoology*). [New Latin, from Greek *chēlē* claw + *keras* horn]

cheliform (kē′lə fôrm), *adj. Zoology.* having the form of a chela.

cheliped (kē′lə ped), *n. Zoology.* one of the pair of limbs, especially of a lobster or crab which bear the chelae. [from Latin *chele* chela + Greek *pedem* foot]

chem-, *combining form.* the form of **chemo-** before vowels, as in *chemosmosis.*

chem., *abbrev.* **1** chemical. **2** chemistry.

chemical, *adj.* **1** of, having to do with, or in chemistry: *a chemical formula, chemical research.*
2 made by or used in chemistry: *a chemical agent.*
3 working, operated, or done by using chemicals: *a chemical fire extinguisher.*
—*n.* any substance obtained by or used in a chemical process. Sulfuric acid, sodium bicarbonate, and borax are chemicals.
—**chemically,** *adv.* **1** according to chemistry. **2** by chemical processes.

chemical bond, *Chemistry.* the attractive force by means of which atoms are combined to function as a unit: *A chemical bond results from the simultaneous attraction of electrons to two nuclei . . . When a chemical bond is formed, energy is released. When a chemical bond is broken, energy is absorbed* (Chemistry Regents Syllabus). Also called **bond.**

chemical change, *Chemistry.* = reaction: *A chemical change results in a permanent change of properties* (Jones, *Inorganic Chemistry*).

chemical control, *Biology.* a means of regulating life processes in organisms by transmitting messages chemically from one part of the organism to another. Chemical control is achieved mainly through hormonal action. Compare **nerve control.**

chemical element, *Chemistry.* = element.

chemical kinetics, the branch of chemistry dealing with the rate of chemical reactions and the mechanisms by which reactions occur.

chemical laser, *Chemistry, Physics.* a laser that uses the energy of a chemical reaction rather than electrical energy: *Chemical lasers differ from ordinary lasers in that molecules with abnormally large amounts of energy are produced by particular chemical reactions, not by some external source of radiation* (London *Times*).

chemically pure, *Chemistry.* (of an element or compound) having a degree of purity that qualifies it for use in fine laboratory work. *Abbreviation:* c.p.

chemical reaction, *Chemistry.* = reaction: *All chemical reactions consist essentially of a rearrangement of the atoms in the molecules of the substances concerned* (K. D. Wadsworth).

chemiluminescence, *n. Chemistry.* the emission of light during a chemical reaction without an apparent rise in temperature.
—**chemiluminescent,** *adj.* producing or produced by chemiluminescence.

chemisorb (kem′ə sôrb′ *or* kem′ə zôrb′), *v.* to hold by chemisorption.

chemisorption, *n. Chemistry.* adsorption in which a single layer of molecules is held with great strength to a surface by a chemical bond.

chemistry, *n.* **1** the science that deals with the properties, composition, structure, and interactions of matter, and the energy changes that accompany these interactions: *The biggest step in the theory of chemistry was the discovery of the periodic system which affords a natural classification of all the elements* (H.R. Paneth). *Abbreviation:* chem.
2 the application of this science to a certain subject: *the chemistry of foods, the chemistry of plant cells.*
▶ Broadly, chemistry has two main branches, *organic chemistry* and *inorganic chemistry.* Other important branches of chemistry include *physical chemistry* and *biochemistry. Chemical engineering* is a branch of engineering, not of chemistry. As a field of science, chemistry may be applied or pure, and pure chemistry may be theoretical or experimental. *Chemurgy* is a branch of applied chemistry.

chemo-, *combining form.* chemical; by chemical reaction, as in *chemosynthesis* = *chemical synthesis.* Also spelled **chem-** before vowels. [from *chemical*]

chemoautotroph (kē′mō ô′tə trôf), *n. Microbiology.* a chemoautotrophic bacterium, protozoan, or other microorganism.

chemoautotrophic (kē′mō ô′tə trof′ik), *adj. Microbiology.* providing its own nourishment with energy obtained from chemical oxidations: *chemoautotrophic microorganisms.*

chemolysis (ki mol′ə sis), *n.* chemical decomposition or analysis.
—**chemolytic** (kē′mō lit′ik), *adj.* of or having to do with chemolysis.

chemoreceptor (kē′mō ri sep′tər), *n. Physiology.* a nerve ending or sense organ that reacts to chemical stimulation, as the taste buds in the tongue: *In flatworms, the simplest bilaterally symmetrical animals, chemoreceptors on the sides of the head direct the search for food* (Scientific American).
—**chemoreception,** *n.* the physiological reaction to chemical stimulation.

chemosensory (kē′mō sen′sə rē), *adj. Physiology.* of or having to do with the reception of chemical stimuli by sensory organs: *Some fishes have been found to possess almost incredible chemosensory acuity . . . [as] eels to respond to concentrations of alcohol so dilute . . . the animals' olfactory receptors could not have received more than a few molecules* (John H. Todd).

chemosmosis (kēm′oz mō′sis), *n.* a chemical reaction taking place through semipermeable membranes. [from *chem-* + *osmosis*]

—**chemosmotic** (kēm′oz mot′ik), *adj.* of or having to do with chemosmosis: *chemismotic transport of electrons.*

chemosphere (kē′mə sfir), *n. Meteorology.* a vaguely defined region of photochemical activity in the upper atmosphere, including the upper stratosphere and the mesosphere.

chemosynthesis (kē′mō sin′thə sis), *n. Biology.* the formation by cells of carbohydrates from carbon dioxide and water with energy obtained from some chemical reaction, rather than from light in photosynthesis, etc.: *Other species of bacteria . . . can synthesize simple carbohydrates from carbon dioxide and water by utilizing energy released by the oxidation of inorganic nitrogen, sulfur, or iron* (Harbaugh and Goodrich, *Fundamentals of Biology*).
—**chemosynthetic,** *adj.* of or using chemosynthesis: *chemosynthetic bacteria.*

chemotaxis (kē′mō tak′sis), *n. Biology.* movement of a cell, organism, or part of an organism toward or away from a chemical substance. Compare **chemotropism.**
—**chemotactic,** *adj.* of or produced by chemotaxis: *the chemotactic process, a chemotactic response.*

chemotaxonomy (kē′mō tak son′ə mē), *n. Biology.* the classification of organisms by their chemical constituents: *Even less is known about the chemotaxonomy of marine organisms because of the plethora of organisms to be found in the vast marine environment* (Joan Lynn Arehart).
—**chemotaxonomic** (kē′mō tak′sə nom′ik), *adj.* of or having to do with chemotaxonomy.

chemotherapy (kē′mō ther′ə pē), *n. Medicine.* the treatment of disease and infection by chemicals that have a specific toxic effect on the disease-producing organisms or malignant cells. Chemotherapy includes the treatment of bacterial infections with antibiotics and the treatment of cancer by chemicals. *Chemotherapy, or treatment with chemicals, also helps persons suffering from arthritis and certain mental illnesses, as well as infectious diseases such as pneumonia and tuberculosis* (H.C.E. Johnson).
—**chemotherapeutic** (kem′ō ther′ə pyü′tik), *adj.* having to do with treatment by chemotherapy: *chemotherapeutic drugs.*

chemotropism (ki mot′rə piz əm), *n. Biology.* the tendency of an organism or part of an organism to turn or bend in response to a chemical stimulus.

chemurgy (kem′ər jē), *n.* a branch of applied chemistry that deals with the use of farm and forest products, such as casein and cornstalks, for purposes other than food and clothing. [from *chem(istry)* + Greek *-ourgia* work]

Cherenkov effect or **Cherenkov radiation,** = Cerenkov effect.

chernozem (cher′nə zəm *or* cher′nə zyôm′), *n. Geology.* a black soil rich in humus, found in cool to temperate, semiarid regions: *Chernozem is . . . very fertile and contains abundant grass-root humus and much mineral plant food. . . . Chernozems are perhaps the most desirable soils known to man* (White and Renner, *Human Geography*). Compare **sierozem.** [from Russian *chernozyom* black earth, from *chyorni* black + *zemlya* earth, soil]

chert, *n. Geology.* a dense, sedimentary rock composed chiefly of chalcedony with an infusion of microscopic crystals: *Beds of chert are commonly associated with volcanic deposits . . . Some cherts are made up largely . . . of silica* (Gilluly, *Principles of Geology*).

chiasm (kī′az əm), *n.* = chiasma.

chiasma (kī az′mə), *n., pl.* **-mata** (-mə tə). *Biology.* **1** a crossing or intersecting: *At the optic chiasma . . . half of the fibers from each eye cross over to the other side of the brain . . . and half of the fibers stay on the same side* (F.H. George).
2 the point of interchange of two chromatids during meiosis, resulting in a cross-shaped figure: *Chromatids . . . normally break at several points and rejoin . . . crossing over . . . in the interchange . . . known as chiasma (chiasmata, plural). Chiasmata may occur at any point along the paired chromosomes* (Weier, *Botany*).
[from Greek *chiasma* a crossed arrangement, from *chiazein* to mark]

Chile saltpeter, *Mineralogy.* native sodium nitrate. *Formula:* $NaNO_3$ [so called because found abundantly in Chile]

chilopod (kī′lə pod), *n. Zoology.* any of a class (Chilopoda) of flat, elongated arthropods having one pair of long antennae, many body segments with one pair of long legs on most, and poison claws. Chilopoda vary in length from 2 or 3 centimeters to nearly 30 centimeters. Commonly called *centipede.* [from Greek *cheilos* lip + *podos* foot]

chimera or **chimaera** (kə mir′ə), *n. Biology.* an organism consisting of two or more tissues of different genetic composition, produced as a result of mutation, grafting, etc. [from Greek *chimaira* monster with a goat's body]

chinook (shə nůk′), *n. Meteorology.* **1** a warm, moist wind blowing from the sea to land in winter and spring in the northwestern United States. **2** a warm, dry wind that comes down the eastern slope of the Rocky Mountains. [named after the *Chinook* Indians of northwestern United States, from Salishan *Tsinúk*]

chiral (kī′rəl), *n. Chemistry, Physics.* not superimposable on its mirror image: *a chiral molecule.* Compare **enantiomorph.** SYN: asymmetric. [from Greek *cheir* hand + English *-al*]
—**chirality** (kī ral′ə tē). **1** the property of an object that cannot be superimposed on its mirror image. **2** = handedness.
—**chirality,** *n.* the property of being chiral.

chiropter (kī rop′tər), *n. Zoology.* any of an order (Chiroptera) of mammals having forelimbs modified as wings; a bat. [from Greek *cheir* hand + *pteron* wing]

---

cap, fāce, fäther; best, bē, tèrm; pin, five;
rock, gō, ôrder; oil, out; cup, pùt, rüle;
yü in use, *yu* in uric;
ng in bring; *sh* in rush; *th* in thin; ᵺ in then;
zh in seizure.
ə = *a* in about, *e* in taken, *i* in pencil, *o* in lemon, *u* in circus

103

chi-square or χ^2 (kī'skwār'), *adj. Statistics.* of or having to do with a test for measuring how closely the frequency distribution of observed data matches that expected in theory: *The chi-square distribution is typically used in testing independence of classification and goodness of fit* (Parl, *Basic Statistics*).

chitin (kīt'n), *n. Biology.* a horny substance forming the exoskeleton of arthropods. It is also found in the cell walls of certain fungi. *This chitin is a stiff substance rather closely related to cellulose* (Norbert Wiener). —**chitinous,** *adj.* of or like chitin: *a chitinous covering.* [from French *chitine,* from Greek *chitōn* tunic]

chiton (kīt'n *or* kī'ton), *n. Zoology.* any of an order (Polyplacophora) of marine mollusks that adhere to rocks. They have bilateral symmetry and a shell of eight overlapping plates. [from Greek *chitōn* tunic, coat of mail]

chlamydate (klam'ə dāt), *adj. Zoology.* having a pallium or mantle, as a mollusk. [from Greek *chlamydos* short mantle]

chlamydospore (klam'ə də spôr), *n. Biology.* a thick-walled, dark spore formed by smuts and certain other fungi.

chloanthite (klō an'thīt), *n. Mineralogy.* a native arsenide of nickel, of white to grayish or black color with a metallic luster. *Formula:* $NiAS_2$

chlor-, *combining form.* the form of **chloro-** before vowels, as in *chloral, chlorate.*

chloracetic acid, = chloroacetic acid.

chloramine (klôr'ə mēn), or **chloramine-T,** *n. Chemistry.* a white or yellowish crystalline powder derived from toluene, used as a strong antiseptic and oxidizing agent. *Formula:* $C_7H_7O_2NSClNa \cdot 3H_2O$

chlorargyrite (klôr är'jə rīt), *n.* = cerargyrite.

chlorate (klôr'āt), *n. Chemistry.* a salt of chloric acid containing the univalent radical -ClO_3.

chlorenchyma (klə reng'kə mə), *n. Botany.* plant tissue which contains chlorophyll: *Below the epidermis are to be found several layers of cells ... These constitute the assimilating part of the plant, give the green color to the aerial stems, and form a tissue which may be described as chlorenchyma* (Gibbs, *Botany*). [from *chlor-* + *(par)enchyma*]

chloric (klôr'ik), *adj. Chemistry.* of or containing chlorine, especially with a valence of five.

chloric acid, *Chemistry.* a colorless acid which occurs only in water solution and is a strong oxidizing agent. *Formula:* $HClO_3$

chloride (klôr'īd), *n. Chemistry.* **1** a compound of chlorine with another element or a radical. Sodium chloride is a compound of sodium and chlorine. **2** a salt or ester of hydrochloric acid.

chloride of lime, *Chemistry.* a white powder used for bleaching and disinfecting, made by treating slaked lime with chlorine. *Formula:* $CaOCl_2$ Commonly called **bleaching powder.**

chlorinated hydrocarbon, *Chemistry.* any of a class of synthetic compounds formed by a chlorine-carbon bond: *The most troublesome pollutants among pesticides are the so-called hard pesticides, principally the chlorinated hydrocarbons—DDT, dieldrin, aldrin, endrin, lindane, chlordane, heptachlor, and some of their relatives* (Science News Yearbook).

chlorine (klôr'ēn), *n. Chemistry.* a greenish-yellow, gaseous element found chiefly in combination with sodium as common salt. Chlorine is bad-smelling, poisonous, and very irritating to the nose, throat, and lungs. *Chlorine is used in water purification, sewage treatment, bleach manufacture and metal refining, among other industrial applications* (Wall Street Journal). *Symbol:* Cl; *atomic number* 17; *atomic weight* 35.453; *melting point* $-101°C$; *boiling point* $-34.11°C$; *oxidation state* ± 1, $+3$, $+5$, $+7$. [from Greek *chlōros* green]

chlorinity (klô rin'ə tē), *n. Chemistry.* a measure of the content of chlorine or other halogen in water, especially sea water.

chlorite[1] (klôr'īt), *n. Chemistry.* a salt of chlorous acid, which contains the univalent radical -ClO_2. [from *chlor(ous)* + *-ite*]

chlorite[2], *n. Mineralogy.* any of a group of green hydrous silicates of magnesium, iron, and aluminum, resembling mica. [from Greek *chlōros* pale green + English *-ite*] —**chloritic,** *adj.* having to do with or containing chlorite.

chloritoid (klôr'ə toid), *n. Mineralogy.* a native silicate of aluminum, ferrous iron, and magnesium, having a dark-green color and occurring usually in brittle laminae. *Formula:* $(Fe,Mg)_2Al_4Si_2O_{10}(OH)_4$

chloro-, *combining form.* **1** green, as in *chlorospinel = a green variety of spinel.* **2** chlorophyll, as in *chlorosis = an abnormal lack of chlorophyll.* **3** chlorine, as in *Chlorobenzene = a compound of chlorine and benzene.* [from Greek *chlōros* pale green]

chloroacetic acid, *Chemistry.* a colorless, crystalline acid, used in medicine and in organic synthesis. *Formula:* $C_2H_3ClO_2$ Also spelled **chloracetic acid.**

chlorobenzene, *Chemistry.* a colorless, flammable liquid, formed by combining chlorine with benzene, and used in organic synthesis, as a solvent, and as a chemical intermediate. *Formula:* C_6H_5Cl

chlorofluorocarbon (klôr'ə flü'ər ə kär'bən), *n. Chemistry.* any of various gaseous compounds of carbon, hydrogen, chlorine, and fluorine, used especially as refrigerants: *The theory that chlorofluorocarbons might destroy the ultraviolet light-absorbing ozone in the earth's atmosphere touched off a spate of chemical and atmospheric studies* (Science News).

chlorofluoromethane (klôr'ə flü'ər ə meth'ān), *n.* = chlorofluorocarbon.

chlorogenic acid, *Chemistry.* an acid found in most edible plants, important in plant metabolism. It is the substance that turns apples and other fruits brown after they are peeled and exposed to air. *Formula:* $C_{16}H_{18}O_9$

chlorohydrin (klôr'ə hī'drin), *n. Chemistry.* any of a group of organic compounds having both a chlorine atom and a hydroxyl (OH) radical.

chlorophyll (klôr'ə fil), *n. Biology.* the coloring matter of the leaves and other green parts of plants occurring in small bodies (chloroplasts) within the cell. Chlorophyll is essential to plants for the manufacture of carbohydrates from carbon dioxide and water in the presence of light (photosynthesis). *The spectroscopic examination of chlorophyll shows that it absorbs the light of just those wavelengths which are most effective in photosynthesis* (Emerson, *Basic Botany*).

► Chlorophyll exists in several forms, usually designated *a, b, c, d, e.* Chlorophyll *a*, which has the formula $C_{55}H_{72}MgN_4O_5$, is the principal photosynthetic pigment, except in bacteria. Chlorophyll *b*, which is also present in higher plants and green algae, differs chemically from *a* only in that it has an aldehyde group (CHO) in place of a methyl group (CH₃). Certain classes of algae contain chlorophyll *c, d,* and *e;* for example, in addition to chlorophyll *a*, the brown algae, dinoflagellates, and diatoms contain chlorophyll *c,* and the red algae chlorophyll *d.*
[from Greek *chlōros* green + *phyllon* leaf]

chlorophyllose (klôr′ə fil′ōs), *adj.* = chlorophyllous.

chlorophyllous (klôr′ə fil′əs), *adj.* having to do with or containing chlorophyll.

chloroplast (klôr′ə plast), *n. Biology.* a specialized body containing chlorophyll, found in the cells of green plants and in certain microorganisms: *The chloroplasts are bounded by an envelope of two membranes, just as are the mitochondria* (Weier, *Botany*). Compare **chromoplast.** See the picture at **leaf.** [from Greek *chlōros* green + *plastos* formed]
ASSOCIATED TERMS: see **photosynthesis.**

chlorosis (klə rō′sis), *n. Botany.* a blanching or yellowing of plants because of inadequate formation of chlorophyll, usually resulting from a lack of iron or magnesium in the soil, or from a lack of light. [from New Latin, from Greek *chlōros* green]

chlorotic (klə rot′ik), *adj.* having or affected by chlorosis: *A plant ... shut up in a dark place ... becomes chlorotic; its green color disappears* (Thomas L. Phipson). —**chlorotically,** *adv.*

chlorous (klôr′əs), *adj. Chemistry.* of or containing chlorine, especially with a valence of three.

chlorous acid, *Chemistry.* an acid occurring only in solution or in the form of its salts (chlorites). *Formula:* HClO₂

choanocyte (kō′ə nə sīt), *n. Zoology.* one of a layer of flagellated cells on the interior surface of a sponge. [from Greek *choanē* funnel + English *-cyte*]

cholate (kō′lāt), *n. Chemistry.* a salt or ester of cholic acid.

cholecystokinin (kol′ə sis′tə kī′nin), *n. Biochemistry.* the intestinal hormone that causes the gall bladder to contract and empty. [from Greek *cholē* bile + *kystis* cyst + *kinein* to move]

cholesteric (kə les′tər ik), *adj. Chemistry.* having a molecular structure characteristic of compounds containing cholesterol; consisting of a series of layers in which the molecules are arranged in close parallel, vertical lines: *The most prominent application of cholesteric liquid crystals are in thermal mapping of living systems and in color displays of electronic components* (Science News). Compare **nematic, smectic.**

cholesterol (kə les′tə rōl′), *n. Biochemistry.* a white,

crystalline substance important in metabolism, contained in all animal fats, bile, nervous tissue, blood, and also in foods such as eggs and meat: *Cholesterol, a fatty substance normally found in the body, is strongly associated with hardening of the arteries and heart disease* (Charles Marwick). *Formula:* $C_{27}H_{45}OH$ [from Greek *cholē* bile + *stereos* solid]

cholic (kō′lik), *adj. Chemistry.* having to do with bile.

cholic acid, *Chemistry.* a white, crystalline acid, related to cholesterol, produced from the nitrogenized acids of bile during its decomposition. *Formula:* $C_{24}H_{40}O_5$ Compare **deoxycholic acid.**

choline (kō′lēn′), *n. Biochemistry.* a constituent of the vitamin B complex, present in many animal and plant tissues, which prevents accumulation of fat in the liver. *Formula:* $C_5H_{15}NO_2$ [from Greek *cholē* bile]

cholinergic (kō′lə nėr′jik), *adj. Biochemistry.* producing or activated by acetylcholine: *a cholinergic agent or drug.* [from *choline* + Greek *ergon* work]

cholinesterase (kō′lə nes′tə rās′), *n. Biochemistry.* an enzyme which prevents the accumulation of acetylcholine at the nerve endings by stimulating its hydrolysis.

chondr-, *combining form.* **1** lump; grain, as in *chondrule = a small grain.* **2** cartilage, as in *chondroid = resembling cartilage.* Also spelled **chondro-.** [from Greek *chondros* lump]

chondriosome (kon′drē ə sōm), *n.* = mitochondrion.

chondrite (kon′drīt), *n. Astronomy, Geology.* a meteorite containing chondrules.
—**chondritic** (kon drit′ik), *adj.* having the peculiar granulated structure characteristic of a chondrite: *Any theory for the origin of meteorites should try to account for the chondritic stone meteorites* (Scientific American).
[from Greek *chondros* lump, grain]

chondro-, *combining form.* a form of **chondr-,** as in *chondrocyte.*

chondrocyte (kon′drə sīt), *n. Biology.* any of the cells found in the cavities in cartilage; cartilage cell.

chondrodite (kon′drə dīt), *n.* a yellow-to-red monoclinic mineral of the humite group, often occurring in granular form in crystalline marble. *Formula:* $(Mg,Fe)_3SiO_4(OH,F)_2$

chondrule (kon′drül), *n. Astronomy, Geology.* a small rounded granule or aggregate of olivine or enstatite embedded in varying numbers in some meteorites: *Most meteorites that are tuffs contain grains, or chondrules, which differ from anything formed on the earth* (Fenton, *The Rock Book*). [from Greek *chondros* lump, grain]

cap, fāce, fäther; best, bē, tėrm; pin, fīve; rock, gō, ôrder; oil, out; cup, pút, rüle, *yü* in use, *yu̇* in uric; ng in bring; sh in rush; th in thin, ŦH in then; zh in seizure. ə = *a* in about, *e* in taken, *i* in pencil, *o* in lemon, *u* in circus

105

chord, *n. Mathematics.* a line segment whose end points both lie on a given curve. [alteration of *cord*]

chords

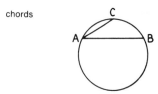

Lines AB and AC are chords.

chordamesoderm (kôr′də mes′ə dèrm), *n. Embryology.* the part of the mesoderm from which the notochord develops.

chordate (kôr′dāt), *Zoology.* —*n.* any of a phylum (Chordata) of animals that have at some stage of development a dorsal nerve cord, a notochord, and pharyngeal gill slits and that includes mammals, birds, reptiles, amphibians, fishes, and certain wormlike marine forms: *Except for a few primitive species, the chordates (102,000 species) are vertebrates; that is, their axial support is made up of small bones or vertebrae, and is known as the vertebral column, or backbone* (Hegner and Stiles, *College Zoology*).
—*adj.* of or belonging to this phylum.
[from New Latin *Chordata,* from Latin *chorda* cord]

chorioallantois (kôr′ē ō ə lan′tō is), *n. Embryology.* a baglike fetal membrane formed by the fused walls of the chorion and the allantois, especially in birds. The chorioallantois of chick embryos is used as a culture medium for viruses and cells.
—**chorioallantoic,** *adj.* of or belonging to the chorioallantois.

chorion (kôr′ē on), *n. Embryology.* the outermost membrane, enclosing the amnion, of the sac which envelops the embryo or fetus of the higher vertebrates: *In the more complex mammals ... the chorion comes into direct contact with the uterine wall, and aids in the formation of a placenta* (Harbaugh and Goodrich, *Fundamentals of Biology*).
[from Greek *chorion* membrane]
—**chorionic** (kôr′ē on′ik), *adj.* of, having to do with, or resembling the chorion: *chorionic tissue.*

chorisepalous (kôr′ə sep′ə ləs), *adj.* = polysepalous.

C horizon, *Botany, Geology.* the layer of soil beneath the B horizon, containing a mixture of decomposed rock and unchanged materials from which soil has not yet begun to form. Compare **A horizon.**

choroid (kôr′oid′), *Anatomy.* —*adj.* like the chorion; membranous. —*n.* = choroid membrane. [from Greek *choroeidēs,* from *chorion* membrane]

choroidal (kə roi′dəl), *adj.* = choroid.

choroid membrane or **choroid coat,** *Anatomy.* a delicate membrane or coat between the sclerotic coat and the retina of the eyelid.

chrom-, *combining form.* the form of **chromo-** before vowels, as in *chrominance.*

chroma (krō′mə), *n. Optics.* the degree of saturation of a color; the degree to which a color is mixed with black, white, or gray: *If you mixed steps of color between a pure red and a neutral gray of the same value,* or *lightness, you would have a series of chromas* (Faber Birren). [from Greek *chrōma* color]

chromaffin (krō′mə fin *or* krō maf′in), *adj. Biology.* having an affinity for salts of chromic acid; staining easily with chromium salts: *the chromaffin cells or tissues of the adrenal medulla.* [from *chrom-* + Latin *affinis* akin]

chromate, *n. Chemistry.* a salt or ester of chromic acid, which contains the bivalent radical $-CrO_4$.

chromatic, *adj. Optics.* of color or colors. **Chromatic aberration** is the failure of the different colors of light to meet in one focus when refracted through a convex lens: *An objective should be as free as possible from chromatic aberration, which causes colors to form around a brilliant object* (Bernhard, Bennett, and Rice). [from Greek *chrōma, chrōmatos* color]

chromatid (krō′mə tid), *n. Biology.* one half of a chromosome during the prophase or metaphase stage of cell division. The two halves, or *sister chromatids,* are joined at a specific point along their length by a *centromere. When a nucleus divides, each daughter nucleus gets one chromatid from each chromosome* (Edwin B. Matzke).
ASSOCIATED TERMS: *Chromatid, chromatin, centromere, centriole, chiasma, spindle, aster* are terms associated with the processes of *meiosis* and *mitosis.*

chromatin (krō′mə tən), *n. Biology.* a substance found throughout the nucleus of a cell which absorbs stains readily and condenses to form chromosomes during mitosis. Chromatin is composed of DNA and histone proteins. *As a nucleus starts to divide, one of the first detectable changes is seen in the chromatin, which gradually becomes more and more distinct. This is a consequence of the condensation of the chromatin into the bodies we call the chromosomes* (Norman Rothwell).
—**chromatinic** (krō′mə tin′ik), *adj.* of or having to do with chromatin.

chromatogram (krō mat′ə gram), *n. Chemistry.* the pattern of separate sections on the adsorbent in chromatography: *The mushrooms are minced, and after alcohol extraction and evaporation, a chromatogram on filter paper is prepared from the residue* (Science News Letter).

chromatography (krō′mə tog′rə fē), *n. Chemistry.* the separation and analysis of mixtures of chemical compounds by the use of an adsorbing material, so that the different compounds become adsorbed in separate sections. **Liquid chromatography** is used for mixtures in solutions. In **gas chromatography** the mixture is combined with a gas. **Paper chromatography** uses paper as the adsorbing material, while **thin-layer chromatography** uses a thin layer of filtering material.
—**chromatographic** (krō′mə tə graf′ik), *adj.* having to do with chromatography: *chromatographic analysis.*

chromatolysis (krō′mə tol′ə sis), *n. Biology.* the breakup and destruction of chromatin in a cell nucleus.
—**chromatolytic** (krō′mə tə lit′ik), *adj.* having to do with or causing chromatolysis.

chromatophore (krō′mə tə fôr), *n. Biology.* **1** one of the specialized pigment-bearing bodies in the cells of certain photosynthetic bacteria.
2 a pigment cell, especially one able to produce rapid color change in the skin of certain animals by its contraction or expansion, as in a chameleon: *Skin color depends on pigment-bearing cells called chromato-*

phores, located in the lower levels of the skin ... Its pigment is in the form of extremely tiny granules ... distributed throughout the cell or concentrated in one spot in the center (Scientific American). [from Greek *chrōma, chrōmatos* color + *-phoros* carrying]

chromatophoric (krō′mə tə fôr′ik), *adj.* having to do with or containing chromatophores.

chrome alum, *Chemistry.* an alum containing trivalent chromium, such as a violet crystalline compound, chromium potassium sulfate, used as a mordant. *Formula:* CrK(SO₄)₂·12H₂O

chromic, *adj. Chemistry.* of or containing chromium, especially with a valence of three.

chromic acid, *Chemistry.* **1** an acid occurring only in solutions or in the form of its salts (chromates). *Formula:* H_2CrO_4
2 a toxic, corrosive substance occurring in dark purplish red crystals; chromium trioxide. *Formula:* CrO_3

chromite (krō′mīt), *n.* **1** *Mineralogy.* a mineral containing iron and chromium, the commercial source of chromium. *Formula:* $FeCr_2O_4$ **2** *Chemistry.* a salt of chromium, especially one with a valence of two.

chromium (krō′mē əm), *n. Chemistry.* a grayish, hard, brittle metallic element that does not rust or become dull easily. Chromium is used to electroplate other metals, as part of stainless steel and other alloys, for making dyes and paints, in photography, etc. *Symbol:* Cr; *atomic number* 24; *atomic weight* 51.996; *melting point* 1900°C; *boiling point* 2200°C; *oxidation state* 6, 3, 2. [New Latin, from Greek *chrōma* color]

chromo-, *combining form.* **1** color; colored, as in *chromoplast = colored body.*
2 deeply staining, as in *chromomere = deeply staining part.*
[from Greek *chrōma* color]

chromodynamics, *n. Nuclear Physics.* the theory dealing with the strong force that binds quarks together: *The differentiation by "color" refers to the fact that quarks seem to behave as if they carried three different kinds of charge—not electrical charge, but something analogous to it. The study of the "color" of both matter and antimatter suggested the name "chromodynamics," which has nothing to do with real, visible color* (Malcolm W. Browne).
ASSOCIATED TERMS: see **charm.**

chromogen (krō′mə jən), *n. Chemistry.* **1** an organic substance that forms a colored compound or becomes a pigment when exposed to air. **2** a compound (not a dye) having color-forming groups and thus capable of being converted into a dye.

—chromogenic, *adj.* **1** producing color. **2** of or having to do with a chromogen. **3** developing a characteristic color: *chromogenic bacteria.*

chromomere (krō′mə mir), *n. Biology.* one of the deeply staining granules or bands of condensed chromatin lined up on a chromosome. [from *chromo-* + Greek *meros* part]

chromonema (krō′mə nē′mə), *n., pl.* **-mata** (-mə tə). *Biology.* either of a pair of spirally coiled or closely pressed threads of chromatin detectable during the early prophase of mitosis or meiosis; an elongated chromatid. [from *chromo-* + Greek *nema* thread]

chromophil (krō′mə fil), *adj. Biology.* having an affinity for color; staining readily: *chromophil granules or cells.*

chromophore (krō′mō fôr), *n. Chemistry.* a group of atoms which produce the color within the molecules of colored organic compounds. [from *chromo-* + *-phore*]

chromoplast, *n. Biology.* a yellow or red body in the cytoplasm of a plant cell containing coloring matter. The colors of flowers and fruits are largely due to the presence of chromoplasts. *Plastids are of two major types, the colored chromoplasts and the colorless leucoplasts. Certain chromoplasts are ... variously colored ... with shades of yellow and red predominating* (Harbaugh and Goodrich, *Fundamentals of Biology*). [from *chromo-* color + Greek *plastos* formed]

chromoprotein, *n. Biochemistry.* a conjugated protein containing a pigment that gives color to the compound, such as a hemoglobin or a flavoprotein.

chromosome (krō′mə sōm), *n. Biology.* any of the threadlike bodies found in the nucleus of a cell that appear when the cell divides. Chromosomes are derived from the parents and carry the genes that determine heredity. The genetic material in each chromosome is a long polynucleotide strand of DNA in association with protein. *Each species of animal or plant has a characteristic assortment of chromosomes. Most higher organisms have two identical sets of chromosomes in each nucleus; these organisms are called diploid* (Bulletin of Atomic Scientists). *The chromosomes contain all the information necessary to generate and maintain the characteristics of the entire cell* (Thomas H. Roderick). See the pictures at **cell, mitosis.**
[from Greek *chrōma* color + *sōma* body (so called because it readily absorbs dyes)]

—chromosomal (krō′mə sō′məl), *adj.* of, having to do with, or resembling a chromosome or chromosomes: *A chromosomal alteration is a change in the number of chromosomes or in the structure of the chromosomes* (Biology Regents Syllabus).

chromosome number, *Biology.* the number of chromosomes normally present in the cell nuclei of a given species of organism, usually constant for each species: *A recount indicates that the human chromosome number ordinarily is 46 (44 autosomes plus 2 sex chromosomes) and not 48 as believed for so long* (Lorus and Margery Milne).

chromosphere (krō′mə sfir′), *n. Astronomy.* a hot layer of gas around the sun or another star: *The chromosphere may be thought of as an atmosphere, roughly 10,000 kilometres deep, lying immediately above the photosphere which is the ordinary visible surface of the Sun* (New Scientist). See the picture at **sun.**

---

cap, fāce, fäther; best, bē, tèrm; pin, fīve;
rock, gō, ôrder; oil, out; cup, pùt, rüle;
yü in use, *yu* in uric;
ng in bring; *sh* in rush; *th* in thin, ᴛʜ in then;
zh in seizure.
ə = *a* in about, *e* in taken, *i* in pencil, *o* in lemon, *u* in circus

ASSOCIATED TERMS: The atmosphere of the sun is divisible into fairly distinct concentric layers or regions. The *photosphere*, which is the source of the sun's visible light, is about 540 kilometers deep. Beyond this region is the *chromosphere*, extending roughly 13,000 kilometers. The base or lowest part of the chromosphere is called the *reversing layer*; the outermost layer, the *corona*, stretches millions of kilometers in space.

—**chromospheric** (krō′mə sfer′ik), *adj.* of or having to do with the chromosphere

chromous, *adj. Chemistry.* of or containing chromium, especially with a valence of two.

chronaxie or **chronaxy** (krō′nak sē), *n. Physiology.* the minimum time required for a constant electric current to activate a nerve cell or other excitable organ, used as an index to detect changes in nervous or muscular responses. [from Greek *chronos* time + *axia* value]

chronometer (krə nom′ə tər), *n.* an instrument for measuring time, especially a timepiece adjusted to keep accurate time in all variations of temperature.

chrysalis (kris′ə lis), *n., pl.* **chrysalises, chrysalides** (krəsal′ə dēz′). **1** = pupa: *What emerges from this last larval ecdysis is called the pupa, or chrysalis. It is pale and soft at first but hardens and darkens, often approximately matching the colour of its background* (Mackean, *Introduction to Biology*).
2 the case or cocoon enveloping the pupa.
[from Latin, from Greek *chrysallis* golden sheath of a butterfly, from *chrysos* gold]

chrysalis
(def. 2)

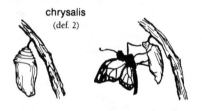

chrysoberyl (kris′ə ber′əl), *n.* a yellowish or pale-green mineral consisting of beryllium aluminate, used as a semiprecious stone; cymophane. *Formula:* $BeAl_2O_4$ [from Greek *chrysobēryllos*, from *chrysos* gold + *bēryllos* beryl]

chrysolite (kris′ə līt), *n.* = olivine. [from Greek *chrysolithos*, from *chrysos* gold + *lithos* stone]

chrysoprase (kris′ə prāz), *n. Mineralogy.* a light-green, semiprecious stone, a variety of chalcedony, with a waxy or pearly luster, used as a gem. [from Greek *chrysoprasos*, from *chrysos* gold + *prason* leek]

chrysotile (kris′ə tīl), *n. Mineralogy.* a fibrous variety of serpentine. It is the most important type of asbestos. Compare **antigorite**. [from Greek *chrysos* gold + *tilos* fiber]

chyle (kīl), *n. Physiology.* a milky liquid composed of digested fat and lymph, formed from the chyme in the small intestine and carried from there into the lacteals. [from Greek *chylos* juice, chyle, from *chein* pour]

chylomicron (kī′lō mī′kron), *n. Physiology.* one of the minute particles of emulsified fat, consisting mainly of triglycerides, found in blood and lymph during digestion of fat. [from Greek *chylos* chyle + *mikron* small]

chylous (kī′ləs), *adj.* of, having to do with, or resembling chyle.

chyme (kīm), *n. Physiology.* a pulpy semiliquid mass into which food is changed by the action and secretions of the stomach: *When digestion in the stomach is complete the pyloric sphincter relaxes ... allowing a little chyme to pass through into the first part of the small intestine called the duodenum* (Mackean, *Introduction to Biology*). [from Greek *chymos*, from *chein* pour]

chymotrypsin (kī′mə trip′sən), *n. Biochemistry.* a pancreatic proteolytic enzyme formed by the action of trypsin on chymotrypsinogen. [from *chyme* + *trypsin*]

chymotrypsinogen (kī′mə trip sin′ə jən), *n. Biochemistry.* an inactive precursor, or zymogen, of chymotrypsin. It is changed into chymotrypsin by the action of trypsin.

chymous (kī′məs), *adj.* of, having to do with, or resembling chyme.

Ci, *abbrev.* or *symbol.* **1** cirrus. **2** curie *or* curies.

cicatrix (sik′ə triks), *n., pl.* **cicatrices** (sik′ə trī′sēz). **1** *Medicine.* the new connective tissue which forms when a wound, sore, or ulcer heals; a scar.
2 *Botany.* **a** the scar left on a tree or plant especially by a fallen leaf or branch. **b** the scar on a seed where it was attached to the pod or seed container; hilum. [from Latin *cicatrix* scar]

—**cicatricial** (sik′ə trish′əl), *adj.* having to do with or resembling a cicatrix.

-cide, *combining form.* substance that kills, as in *biocide, germicide, insecticide, herbicide, fungicide.* [from Latin *-cida*, from *caedere* cut, kill]

cilia (sil′ē ə), *n.pl.* of **cilium** (sil′ə əm). *Biology.* very small, hairlike projections, composed of microtubules and capable of a whiplike motion, found on certain kinds of cells. Some microscopic organisms use cilia to move themselves or to set up currents in the surrounding water. *The linings of the air passages are covered with microscopic ... hairs, called cilia, that move back and forth. The movements of these cilia sweep dust and other useless materials up and out of the air passages* (Beauchamp, *Everyday Problems in Science*). *In the paramecium, as a result of the action of cilia, food is ingested through a fixed opening located in the oral groove* (Biology Regents Syllabus). [from Latin *cilium* eyelid, plural *cilia*]

—**ciliary** (sil′ē er′ē), *adj.* **1** *Biology.* of or resembling cilia: *Next to muscle contraction, the best understood type of cellular movement is ciliary beating ... [which helps] to sweep ova along the oviduct* (Bruce Alberts).
2 *Anatomy.* of or designating certain delicate structures of the eyeball, such as the ciliary muscle and the ciliary process.

ciliary body, *Anatomy.* a thick, wedge-shaped mass of tissue in the middle section of the eyeball, containing the ciliary muscle, the ciliary process, and related structures.

ciliary muscle, *Anatomy.* a delicate muscle around the margin of the lens of the eyeball, a principal agent in the accommodation of the lens.

ciliary process, *Anatomy.* a folded part of the choroid membrane of the eyeball, adjoining the iris.

ciliate (sil′ē it), *Biology.* —*n.* any of a class (Ciliata) of one-celled protozoans having cilia. Paramecia are ciliates.

—*adj.* Also, **ciliated.** provided with cilia: *a ciliated cell, ciliated leaves.*

—**ciliation,** *n.* **1** the condition of being ciliate. **2** = cilia.

cilium (sil′ē əm), *n.* the singular of **cilia.**

cinder cone, *Geology.* a cone-shaped mass of volcanic cinders accumulated at the vent of a volcano.

cingulate (sing′gyə lit), *adj.* of, having to do with, or surrounded by a cingulum or cingula (used especially in describing the thorax or abdomen of insects).

cingulum (sing′gyə ləm), *n., pl.* **-la** (-lə). **1** *Botany, Zoology.* a girdlelike part, such as a band or ridge on an animal. **2** *Anatomy.* a ridge at the base of the crown of a tooth.
[from Latin *cingulum* girdle, belt]

cinnabar (sin′ə bär), *n. Mineralogy.* a reddish or brownish mineral that is the chief source of mercury. *Formula:* HgS [from Latin *cinnabaris*]

cinnamate (sin′ə māt), *n. Chemistry.* a salt of cinnamic acid.

cinnamic acid (sə nam′ik), *Chemistry.* a white, crystalline acid found especially in various balsams and cinnamon. *Formula:* $C_6H_5CH=CHCO_2H$

circadian (sèr′kā′dē ən), *adj. Biology.* functioning or recurring in 24-hour cycles: *Certain marine animals ... may function in relation to circadian movements ... toward the ocean surface at night and to deeper water during daylight hours* (Robert T. Orr). Compare **infradian, ultradian.**
[from Latin *circa* about + *dies* day]

—**circadianly,** *adv.* in 24-hour cycles: *Temperature and urine flow are but two among many physiological functions which fluctuate circadianly* (New Scientist).

circadian rhythm, *Biology.* the 24-hour cycle of physiological activity in living organisms governed by the biological clock: *This circadian rhythm, as the 24- or 25-hour cycle rhythm is called, originates primarily in the organism itself* (Science News Letter).

circinate (sèr′sə nāt), *adj. Botany.* coiled from the tip toward the base. *The new leaves of a fern are circinate.* [from Latin *circinatus* made round] —**circinately,** *adv.*

circle, *n.* **1** *Geometry.* **a** a closed curve in a plane, all of whose points are equidistant from a fixed point within called the *center.* A chord that passes through the center of the circle is a *diameter.* A radius of a circle is a segment from the center to a point on the circle. **b** a plane surface bounded by such a closed curve. Compare **circumference.** See also **conic section.** See the pictures at **conic section, segment.**
2 *Geography.* **a** a parallel of latitude. **b** a meridian of longitude.

circle graph, a graph in the form of a circle divided into sectors, used to show the percentages into which any total sum is divided. Also called **pie chart.** Compare **bar graph, line graph,** and **pictograph.**

circle of curvature, *Geometry.* the circle that measures curvature at a given point. Also called **osculating circle.**

circle of latitude, 1 *Astronomy.* a great circle on the celestial sphere, perpendicular to the plane of the ecliptic, on which celestial latitudes are measured.
2 *Geography.* parallel of latitude.

circle of longitude, 1 *Astronomy.* a circle on the celestial sphere, parallel to the plane of the ecliptic, on which celestial longitudes are measured. Also called **parallel of longitude.**
2 *Geography.* = meridian (def. 1a).

circle of Willis, *Anatomy.* a circle formed by several connecting cerebral arteries at the base of the brain. [named after Thomas *Willis,* 1621–1675, English anatomist]

circuit, *n. Electricity.* the complete path, or a part of it, over which an electric current flows. A circuit usually includes the generating apparatus. When the path of the current is complete so that the electricity is free to flow, the circuit is a *closed* or *made circuit;* if interrupted at any point, it is an *open* or *broken circuit.*

circular function = trigonometric function.

circular measure, a system of measure used for angles or their corresponding arcs in a circle:
60 seconds = 1 minute
60 minutes = 1 degree
90 degrees or π 2 radians = 1 quadrant
4 quadrants or 360 degrees or 2 π radians = 1 circle

circulate, *v. Biology.* (of blood, lymph, etc.) to flow with a continuous motion through vessels or tissues in the body of an organism: *As animals get larger and more complicated they must have a blood supply circulating through the body to carry food and oxygen to remote cells and to bring back the wastes that must be eliminated and to accomplish other important functions* (Winchester, *Zoology*).

circulation, *n. Biology.* **1** the flow of the blood from the heart through the arteries and veins back to the heart: *The term circulation usually is used only in connection with animals that have a system of body spaces or vessels through which body fluids, generally blood, are forced.* (Harbaugh and Goodrich, *Fundamentals of Biology*).
2 the movement of lymph between human tissues and cells.
3 the transport of materials within cells or throughout multicellular organisms: *Intracellular circulation may be by diffusion or by transport through vascular tissue* (Biology Regents Syllabus).

—**circulatory** (sèr′kyə lə tôr′ē), *adj.* having to do with circulation. Arteries and veins are parts of the circulatory system of the human body.

circulation decimal, *Mathematics.* a decimal in which a figure or series of figures is repeated indefinitely. EXAMPLES: $1 \div 3 = .333 ..., 1 \div 7 = .142857142857$ ... Also called **repeating decimal** and **recurring decimal.**

circum-, *prefix.* around; on all sides, as in *circumterrestrial, circumvascular.* [from Latin *circum* around, originally accusative of *circus* circle, ring]

---

cap, fāce, fäther; best, bē, tèrm; pin, fīve;
rock, gō, ôrder; oil, out; cup, pùt, rüle,
yü in use, *yù* in uric;
ng in bring; *sh* in rush; *th* in thin, ᴛʜ in then;
zh in seizure.
ə = *a* in about, *e* in taken, *i* in pencil, *o* in lemon, *u* in circus

109

circumference, *n. Mathematics.* **1** the distance around a circle, equal to its diameter multiplied by pi.
2 (of a sphere) the distance around any great circle on the sphere: *The circumference of the earth is about 40,000 kilometers.*
3 any simple closed curve forming the boundary of a region. SYN: perimeter.
[from Latin *circumferentia*, from *circum* around + *ferre* to bear]

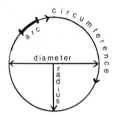

circumnutate (sėr′kəm nü′tāt), *v. Botany.* (of the growing parts of plants) to bend or move about in a spiral, circular path. —**circumnutation,** *n.*

circumscissile (sėr′kəm sis′əl), *adj. Botany.* (of fruits) bursting open by a transverse circular line, so that the upper part comes off like a lid: *circumscissile pods.*

circumscribe, *v. Geometry.* to draw (a figure) around another figure so as to touch as many points as possible: *A circle that is circumscribed around a square touches it at four points.* Compare **inscribe.** [from Latin *circumscribere,* from *circum* around + *scribere* write]

circumstellar, *adj. Astronomy.* surrounding or revolving about a star: *circumstellar dust.*

circumterrestrial, *adj. Astronomy.* surrounding or revolving about the earth: *circumterrestrial space, a circumterrestrial satellite.*

circumvascular, *adj. Anatomy.* situated or extending around a blood vessel.

cirque (sėrk), *n. Geography, Geology.* a rounded hollow or plain encircled by heights, especially one high in the mountains at the head of a stream or glacier. [from French, from Latin *circus* circle]

cirrate (sir′āt), *adj. Zoology.* furnished with cirri.

cirri (sir′ī), *n.* plural of **cirrus.**

cirriferous (sə rif′ər əs), *adj. Zoology.* bearing cirri.

cirrocumulus (sir′ō kyü′myə ləs), *n. Meteorology.* a cloud formation of very small, globular masses of white, fleecy clouds arranged in wavelike rows or groups, occurring at heights of 6000 meters and above. *Abbreviation:* Cc [from *cirrus* + *cumulus*]

cirrose (sir′ōs), *adj.* **1** *Zoology, Botany.* having or resembling a cirrus or cirri.
2 *Meteorology.* of or resembling cirrus clouds. Also spelled **cirrous.**

cirrostratus (sir′ō strā′təs), *n. Meteorology.* a thin, veil-like cloud formation of ice crystals occurring at heights of 6000 meters and above. Cirrostratus clouds consisting of a horizontal hazy layer, or in some instances inclined sheets thinning upward into light curls. *Abbreviation:* Cs

cirrous (sir′əs), *adj.* = cirrose.

cirrus (sir′əs), *n., pl.* **cirri** (sir′ī). **1** *Meteorology.* a cloud formation of thin, detached, featherlike, white clouds of ice crystals occurring at heights of 6000 meters and above: *Since cirrus are the highest of clouds, they will often be illuminated by the sun when lower clouds are submerged in the shadow of the earth* (Neuberger and Stephens, *Weather and Man*). See the pictures at **cloud, front.**
ASSOCIATED TERMS: see **cloud.**
2 *Zoology.* a slender process or appendage.
[from Latin *cirrus* curl]

cis (sis), *adj. Chemistry.* of or having to do with an isomeric compound that has certain atoms on the same side of a plane: *cis configurations.* Compare **trans.**
[from Latin *cis* on this side of]

cis-, *prefix.* **1** on this side of; near, as in *cislunar.*
2 *Chemistry.* having certain atoms on the same side of a plane: *a cis-isomeric compound.* Contrasted with **trans-.**
[from Latin *cis* on this side]

cislunar (sis lü′nər), *adj. Astronomy.* nearer than the moon; between the earth and the moon's orbit: *cislunar space.*

cissoid (sis′oid), *adj. Geometry.* contained between the concave sides of two curves which intersect each other. [from Greek *kissoeidēs* ivy-like (so called from the curve's resemblance to angles on any ivy leaf)]

cistern, *n. Anatomy.* a cavity or vessel in the body, especially any of several spaces in the brain.

cistron (sis′tron), *n. Genetics.* the smallest unit of genetic material producing a phenotypic effect: *Cistrons ... correspond roughly to the genes of classical genetics. Each cistron controls the structure of a protein* (New Scientist). [from *cis* + *tr(ans)* + *-on* (so called from the phenotypic comparisons being based on the *cis* and *trans* arrangements of genetic material in the chromosomes)]

citral (sit′rəl), *n. Chemistry.* a strong-smelling liquid aldehyde present in such oils as that of the lemon and orange and used as a flavoring and in perfumes. *Formula:* $C_{10}H_{16}O$ [from *citr(on)* + *al(dehyde)*]

citrate (sit′rāt), *Chemistry.* —*n.* a salt or ester of citric acid.
—*v.* to add a citrate to.

citric acid, *Chemistry.* a white, odorless acid with a sour taste, found in the juice of oranges, lemons, limes, and similar fruits. It is also formed in the metabolism of cells during the Krebs cycle. *Formula:* $C_6H_8O_7$

citric acid cycle, = Krebs cycle.

citrin (sit′rin), *n.* = vitamin P.

citrovorum factor (si trov′ə rəm), *Biochemistry.* a growth factor of the folic-acid group, found in liver and green vegetables. [from New Latin *citrovorum,* from Latin *citrus* citron tree + *vorare* devour]

citrulline (sit′rə lēn), *n. Biochemistry.* an amino acid, originally found in watermelon, that in living organisms is an intermediate between ornithine and arginine. *Formula:* $C_6H_{13}N_3O_3$ [from Medieval Latin *citrullus* watermelon, from Latin *citrus* citron tree]

Cl, *symbol.* chlorine.

cladistic (klə dis′tik), *adj. Biology.* having to do with or based on ancestral or evolutionary relationships: *Systematists ... developed methods of classifying animals for other purposes: ... the cladistic classifications, based on assumed lines of descent while ignoring simi-*

larity (Ronald R. Novales). [from Greek *klados* sprout, branch]

cladogenesis (klad´ō jen´ə sis), *n. Biology.* the type of evolutionary change resulting from the division or branching out of a species into isolated groups, as distinguished from phyletic or linear descent. Contrasted with **anagenesis.** [from Greek *klados* sprout, branch + English *genesis*]
—**cladogenetic** (klad´ō jə net´ik), *adj.* of or having to do with cladogenesis. —**cladogenetically,** *adv.*

cladogram (klad´ə gram), *n. Biology.* a branching diagram used to illustrate ancestral or evolutionary relationships: *Immunological distances are ... apportionable along derived cladograms and thus serve as reliable indicators of phylogenetic affinities among the taxa being compared* (Nature). Compare **dendrogram.**

cladophyll (klad´ə fil), *n. Botany.* a stem assuming the form and function of a leaf, as in asparagus.

clarain (klar´ān), *n. Geology.* a lithotype of coal occurring in bands. It often has a silky luster. [from Latin *clarus* clear + English *-ain,* as in *fusain*]

clasper, *n.* **1** *Botany.* a tendril of a climbing plant which twines round something for support.
2 *Zoology.* any of the appendages of the male of certain fishes and insects that serve to hold the female.

class, *n. Biology.* a primary group of related organisms ranking below a phylum or division, and above an order. Classes may be further divided into subclasses. *In the class Mammalia are all animals with mammary glands, including human beings* (Otto and Towle, *Modern Biology*).
ASSOCIATED TERMS: see **kingdom.**

classification, *n. Biology.* the arrangement of organisms into groups on the basis of evolutionary or structural relationship between them. In one widely used system, organisms are grouped into five kingdoms: monerans, protists, fungi, plants, and animals. Narrower classifications (in descending order) are phylum (or sometimes division in botany), class, order, family, genus, and species. *The object of all classification ... [is] to bring together those beings which most resemble each other and to separate those that differ* (Carpenter, *Zoology*). Also called **taxonomy.** See also **nomenclature.**

class interval, *Statistics.* **1** any of the arbitrary groups of equal and convenient size into which the possible values of a variable are divided: *Assuming equal distribution of the 12 items over the entire interval of 10.5 to 16.5 days, the 12 items are thereby equally divided over the three uniform class intervals: 10.5 to 12.5, 12.5 to 14.5, 14.5 to 16.5. Thus, for each uniform class interval there are four items* (Parl, *Basic Statistics*). **2** the numerical width of such a group.

clast (klast), *n. Geology.* a rock fragment broken from a preexisting rock by weathering: *The most commonly used methods in glaciated terrain include sampling of heavy clasts ... and eskers* (Stephen E. Kesler).
—**clastic** (klas´tik), *adj.* made up of clasts: *clastic rock.*

clathrate (klath´rāt), *adj.* **1** *Botany.* resembling a lattice; divided or marked by latticework: *clathrate foliage.*
2 *Chemistry.* formed by or having molecules which are interlaced in a latticelike geometrical pattern: *Enlarged clathrate crystals, containing argon trapped in a quinol lattice ...* (Science News Letter).

—*n. Chemistry.* a clathrate crystal or crystalline compound.
[from Latin *clathratus,* from *clathari* lattice]

clavate (klā´vāt), *adj. Zoology, Botany.* thickened towards the apex like a club; club-shaped: *clavate antennae, a clavate nucleus.* [from Latin *clavatus* studded with nails, from *clavus* nail]

clavicle (klav´ə kəl), *n. Anatomy.* the bone connecting the breastbone and the shoulder blade. Commonly called **collarbone.**
[from Latin *claviculla* small key, from *clavis* key; because of the shape of the bone]
—**clavicular** (klə vik´yə lər), *adj.* of the clavicle.

claviform, *adj.* club-shaped; clavate: *a claviform ironstone, a claviform foot.*

claw, *n. Zoology.* **1** a sharp, hooked nail on a vertebrate animal's foot: *a bird's claws. A tiger's claws retract, a bear's claws do not.* **2** a similar structure on the feet of insects, crustaceans, etc. **3a** a foot of an animal with sharp, hooked nails. **b** a pincer of a lobster, crab, scorpion, and various other arthropods. SYN: chela.

clay, *n. Geology.* a stiff, sticky kind of earth, composed chiefly of clay minerals, that can be easily shaped when wet and hardens after drying or baking. Bricks, dishes, and vases may be made from clay. *The feldspars and related minerals are among the most important of the rock constituents from the standpoint of soilbuilding. These unite rather readily with water and carbon dioxide, forming clays* (Emerson, *Basic Botany*).

clay mineral, any of a group of finely divided, crystalline, hydrous or hydrated aluminum silicates, found in sedimentary rocks and soils. Some clay minerals adsorb large amounts of water.

claystone, *n. Geology.* an indurated clay with the texture and composition of shale but lacking the fine lamination.

clear-air turbulence, *Meteorology.* a violent disturbance in air currents, caused by rapid changes of temperature associated with the jet stream. Clear-air turbulence is characterized by severe updrafts and downdrafts that affect aircraft flying at high altitudes. *Space scientists measured clear-air turbulence ... and concluded that jet airplane flights are smoother above 40,000 feet than between 20,000 and 40,000* (Science News Letter). *Abbreviation:* CAT

cleavage, *n.* **1** *Crystallography.* the property of a crystal or rock of splitting along planes: *Cleavage takes place parallel to crystal faces* (Garrels, *Textbook of Geology*).
2 *Biology.* any of the series of cell divisions by which a fertilized egg develops into an embryo, or the whole series: *Shortly after fertilization ... the fertilized egg begins to divide. This division, which is repeated in rapid succession until the egg is converted into many cells,*

cap, fāce, fäther; best, bē, tèrm; pin, fīve;
rock, gō, ôrder; oil, out; cup, pùt, rüle,
yü in use, *yu̇* in uric;
ng in bring; *sh* in rush; *th* in thin, ᴛʜ in then;
zh in seizure.
ə = *a* in about, *e* in taken, *i* in pencil, *o* in lemon, *u* in circus

111

is called cleavage (Shull, *Principles of Animal Biology*).

3 *Chemistry.* the splitting of a compound into simpler compounds.

cleavage nucleus, *Embryology.* the nucleus which results from the union of the male and female pronuclei, before the division of the egg into two blastomeres.

cleavage plane, *Crystallography.* a plane along which a crystal has been split, or along which it would tend to split.

cleavage spindle, *Embryology.* the karyokinetic spindle of a dividing blastomere, during the early development of the ovum.

cleistogamous (klīs tog′ə məs) or **cleistogamic** (klīs′tə gam′ik), *adj. Botany.* (of a plant, such as the violet) having small, self-pollinating flowers that do not open, in addition to its regular flowers. [from Greek *kleistos* closed + *gamos* marriage] —**cleistogamously, cleistogamically,** *adv.*

—**cleistogamy** (klīs tog′ə mē), *n.* the occurrence of cleistogamous flowers.

climate, *n. Meteorology.* the characteristic condition or course of the weather which a place has over a long period of time, or which a region averages across its areal extent, based on conditions of heat and cold, moisture and dryness, clearness and cloudiness, wind and calm: *Climate implies the summation of weather conditions over a series of years. Climate is not merely the average weather; it includes also the extremes and variability of the weather elements.* (Blair, *Weather Elements*). [from Latin *climatem,* from Greek *klima* slope (of the earth), from *klinein* to incline]

—**climatic,** *adj.* of or having to do with climate: *climatic changes, climatic regions.*

climatology (klī′mə tol′ə jē), *n.* the science that studies climate: *In climatology ... the climatologist studies the average distribution of weather elements over the earth as well as their relationship to each other and to the features of the earth's surface* (Neuberger and Stephens, *Weather and Man*). Compare **meteorology.**

—**climatological** (klī′mə tə loj′ə kəl), *adj.* of or having to do with climatology.

climax, *n. Ecology.* a self-perpetuating community in which populations remain stable and exist in balance with each other and the environment; the final stage of a succession or sere: *Illustrations of extensive and important climaxes are the tundra of the poleward margins of the Northern Hemisphere continents, the coniferous forest or taiga of the subarctic lands, and the tropical rainforest of the constantly wet lands of the low latitudes* (Finch and Trewartha, *Elements of Geography*).

climax community, = climax: *A climax community persists until a catastrophic change* (Biology Regents Syllabus).

climograph, *n. Geography.* a graph which displays the distribution of two or more climatic variables for a place during an average year.

cline (klīn), *n. Biology.* a gradual variation in a particular inherited characteristic found across a series of adjacent populations in a group of related organisms: *Gradual continuous change of characteristics ... known as clines, are exemplified by species of birds and mammals that are larger in cooler climates or darker in warm humid regions* (Storer, *General Zoology*). [from Greek *klinein* to lean, incline]

clino-, *combining form. Mineralogy.* monoclinic, as in *clinoclase, clinopyroxene.* [from *(mono)clin(ic)*]

clinoaxis, *n. Crystallography.* the diagonal or lateral axis in monoclinic crystals, forming an oblique angle with the vertical axis. Also called **clinodiagonal.**

clinochlore (klī′nə klôr), *n. Mineralogy.* a species of chlorite, occurring usually in scaly or granular aggregates, but also in monoclinic crystals. *Formula:* $(Mg,Fe^{2+},Al)_3(Si,Al)_2O_5(OH)_4$ [from *clino-* + *chlor(ite)*]

clinoclase (klī′nə klās), *n. Mineralogy.* a somewhat translucent arsenate of copper, occurring in dark-green monoclinic crystals. *Formula:* $Cu_3(AsO_4)(OH)_3$ clino- + Greek *klasis* cleavage]

clinodiagonal, *n.* = clinoaxis.

clinometer (klī nom′ə tər), *n.* an instrument for measuring deviation from the horizontal, such as the downward slope of rock strata. [from Greek *klinein* to slope + English *-meter*]

clitellum (kli tel′əm), *n., pl.* **-la** (-lə). *Zoology.* a glandular swelling around certain sections of an annelid, such as an earthworm, from which a viscous fluid is secreted which forms a cocoon for the eggs. [from New Latin, from Latin *clitellae* packsaddle]

cloaca (klō ā′kə), *n., pl.* **-cae** (-sē). *Zoology.* **1** the cavity in the body of birds, reptiles, amphibians, and most fishes, into which the intestinal, urinary, and genital canals open.

2 a similar cavity in certain invertebrates.

[from Latin *cloaca* sewer]

—**cloacal,** *adj.* having to do with or resembling a cloaca: *the cloacal duct.*

clone, *Biology.* —*n.* **1** a group of organisms produced asexually from a single ancestor. A clone may be produced by grafting, as in plants; by fission, as in single-celled organisms; and by forming buds, as hydras do. *The descendants of an individual produced asexually, even though they may number in the thousands, may be regarded as extensions of the single individual and are referred to as a clone.* (Greulach and Adams, *Plants*).

2 a cell or organism produced asexually from a single cell with which it is genetically identical: *Instead of being a mixture of genes from two parents, the clone ... is a genetic copy of its single parent* (Time).

—*v.* **1** to reproduce or propagate asexually: *Cauliflowers have been cloned ... simply by cutting slices of cauliflowers, at their market-ready stage, and putting them in nutrient solution* (New Scientist).

2 to produce a cellular constituent in quantity by means of cloning techniques: *Fragments of DNA from any source can be amplified more than a millionfold by inserting them into a plasmid or a bacterial virus (bacteriophage) and then growing these in bacterial (or yeast) cells—a process called DNA cloning* (Bruce Alberts).

[from Greek *klōn* twig]

—**clonal,** *adj.* of or like a clone: *clonal roots, clonal propagation.*

closed, *adj. Mathematics.* (of a set) producing only elements of the same set when subjected to a given operation. The set of whole numbers is closed under addition but not under division.

closed chain, *Chemistry.* an arrangement of atoms in a molecule represented in formulas and models by a ring. Compare **open chain.**

closed curve, *Mathematics.* a curve that has no end points, forming a complete loop: *A closed curve divides the plane into two parts (the inside and the outside) and you cannot get from one part to the other without crossing the curve* (Science News Letter).

closed sentence, *Mathematics.* a sentence which does not include a variable. Compare **open sentence.**

clostridium (klos trid′ē əm), *n., pl.* **-ia** (-ē ə). *Bacteriology.* any of a genus (*Clostridium*) of rod-shaped anaerobic bacteria which produce resistant spores, including the organisms causing botulism and tetanus. [from New Latin, from Greek *klōstēr* spindle] —**clostridial,** *adj.* of or having to do with a clostridium or clostridia.

closure, *n. Mathematics.* (of a set) the property of being closed.

cloud, *n. Meteorology.* a visible mass of condensed water droplets or ice particles floating in the air, usually at a height above the ground: *There is no doubt that nearly all the earth's precipitation is the result of expansion and cooling in rising air currents. The direct result of cooling due to ascent is clouds* (Finch and Trewartha, *Elements of Geography*).

ASSOCIATED TERMS: Clouds are generally classified into ten types comprising three fundamental forms: *cirrus, cumulus,* and *stratus. Cirrus* formations have a featherlike or curly appearance, *cumulus* clouds look like rounded heaps, and *stratus* clouds appear as layers or sheets. The other cloud names are derived by combinations of these three names and by the use of the prefixes *alto-,* meaning high, and *nimbus* or *nimbo-,* meaning a rain cloud. The ten forms are: *cirrus, cirrocumulus, cirrostratus; altocumulus, altostratus; stratus, stratocumulus, nimbostratus; cumulus, cumulonimbus.*

cumulus   cirrus

stratus   cumulo-nimbus

cloud chamber, *Physics.* a large vessel filled with a saturated vapor, especially a vapor of hydrogen and methyl alcohol, through which the paths of individual charged particles, such as protons and electrons, may be observed and photographed and thus be identified: *When cosmic rays traverse a cloud chamber they leave tell-tale tracks* (N.Y. Times). Compare **bubble chamber, spark chamber.**

cloud forest, *Ecology.* a dense moist forest in mountainous areas that is almost constantly covered by clouds.

cloud physics, *Meteorology.* the study of the physical processes involved in the formation, movement, action, and effects of clouds.

cluster, *n. Astronomy.* a group of stars or galaxies relatively close to each other and often found to have a common motion in space: *The globular clusters, of which about 100 are known in our galaxy, are star clusters distinguishable from galactic clusters by their apparent compactness and very great number of members* (Krogdahl, *Astronomical Universe*).

clypeiform (klip′ē ə fôrm), *adj. Botany, Zoology.* shaped like a round shield; clypeate.

clypeus (klip′ē əs), *n., pl.* **-ei** (-ē ī). *Zoology.* a shieldlike plate on the front part of the head of an insect, above the mouth. [from Latin *clypeus* round shield] —**clypeate** (klip′ē it), *adj. Botany, Zoology.* **1** shaped like a round shield or buckler: *a clypeate process.* **2** having a clypeus: *a clypeate insect.*

cm. or **cm,** *abbrev.* centimeter *or* centimeters.

Cm, *symbol.* curium.

cnidarian (nī dãr′ē ən), *Zoology.* —*n.* any of a phylum (Cnidaria) of aquatic invertebrates with radially symmetrical, saclike bodies, and a single internal cavity with one opening. Hydras, jellyfish, corals, and sea anemones belong to this phylum. —*adj.* of or belonging to this phylum. Also called **coelenterate.** [from *cnido(blast)*]

cnidoblast (nī′də blast), *n. Zoology.* a cell in the epidermis of a cnidarian, such as a hydra, in which a nematocyst is developed. Also called **cnidocyte.** [from Greek *knidē* nettle + *blastos* germ]

cnidocil (nī′də sil), *n. Zoology.* an external, irritable, ciliary part of a cnidoblast that causes the discharge of a nematocyst when it is touched. [from *cnido(blast)* + *cil(ium)*]

cnidocyte (nī′də sīt), *n.* = cnidoblast.

CNS, *abbrev.* central nervous system.

Co, *symbol.* cobalt.

^{60}Co, *symbol.* cobalt 60.

CoA, *abbrev.* coenzyme A.

coacervate (kō as′ər vāt), *n. Chemistry.* a liquid aggregate forming viscous droplets in a colloidal solution. [from Latin *coacervatum* heaped up together, from *co-* together + *acervare* to heap up, from *acervus* heap] —**coacervation,** *n.* the action of forming a coacervate by the partial mixing of two liquid substances, one or both of which are in the colloidal state: *When one takes a fairly dilute solution of a synthetic polymer and adds a trace of iron to it, the material separates into oily droplets that contain the iron. This process, called coacervation, is regarded as a "primary phenomenon" in the development of cellular structures* (James R. Newman).

cap, fāce, fäther; best, bē, tèrm; pin, fīve;
rock, gō, ôrder; oil, out; cup, pùt, rüle,
yü in use, yù in uric;
ng in bring; sh in rush; th in thin, ŦH in then;
zh in seizure.
ə = a in about, e in taken, i in pencil, o in lemon, u in circus

coaction, *n. Ecology.* the reciprocal actions or effects of animals and plants: *There are nine kinds of "coaction" possible between weak and strong organisms, ranging from symbiosis, in which there is mutual aid, to a mutual depression or dying together* (Science News Letter).

coagulant, *n. Chemistry.* a substance producing coagulation, used in precipitating solids or semisolids from a solution. The enzyme thrombin is a natural coagulant, coagulating blood and forming clots.

coagulase, *n. Biochemistry.* **1** any of several enzymes that cause coagulation, especially of the blood. Rennin is a coagulase. **2** a protein produced by a form of staphyloccus that reacts with substances in blood plasma to cause coagulation.

coagulate (kō agʹyə lāt), *v. Chemistry.* to change from a fluid to a gelatinous or semisolid state: *When a blood vessel is injured, the blood coagulates, or forms a clot, to stop up the wound and prevent loss of blood* (Storer, *General Zoology*). SYN: curdle, clot, congeal. [from Latin *coagulatus* curdled, from *coagulum* means of curdling, from *co-* together + *agere* drive]
—coagulation, *n.* **1** the act or process of coagulating: *The calcium ions which are normally found in blood are essential for the coagulating process. When they are removed from the blood or inactivated, coagulation does not take place* (K.S. Spiegler). **2** a coagulated mass.

coagulum (kō agʹyə ləm), *n., pl.* **-la** (-lə) or **-lums.** *Chemistry.* a mass of coagulated matter, as of blood. SYN: clot.

coal, *n. Geology.* a black, combustible rock composed of more than 50 per cent carbonaceous material by weight, formed in the earth from variously altered vegetable matter under special conditions of great pressure, high humidity, and lack of air. Anthracite and bituminous coal are two kinds of coal widely used as fuel.. *Coals grade from lignite, which differs little from peat, through bituminous to anthracite, which may be 90 per cent or more of carbon* (Gilluly, *Principles of Geology*).

coalification, *n. Geology.* the change by compaction and metamorphic processes in which coal beds or seams are formed from plant materials.

coal measures, *Geology.* rock strata containing beds of coal.

coal oil, = kerosene.

co-altitude, *n.* = zenith distance.

Coanda effect (kō anʹdə), *Physics.* the property or tendency of any fluid passing a curved surface to attach itself to the surface. The Coanda effect is important in fluidics and aerodynamics. *A common demonstration of the Coanda effect is seen when a falling jet of water from a tap defies gravity and runs along a spoon or jar just brought into contact with it* (Science Journal). [named after Henri Marie *Coanda,* 1885–1972, French engineer and inventor]

coarctate (kō ärkʹtit), *adj. Zoology.* (of a pupa) enclosed in an oval, horny case, and giving no external indication of the limbs or form of the insect. [from Latin *coarctatum* pressed together, from *co-* together + *arctare* to press close]

coast, *n. Geography.* the narrow zone of overlap between land and water along the margin of large landmasses within which the processes of both operate.
—coastal, *adj.* of or having to do with the coast: *a coastal plain, coastal ecology.*

coaxial (kō akʹsē əl), *adj.* **1** having a common axis: *coaxial cylinders.* A **coaxial cable** has insulated conducting material surrounding a separately insulated conducting core that carries signals resistant to disturbance by external electromagnetic fields. The arrangement permits clear transmission of multi-channel telephone signals and computer and television signals over long distances. **2** of, having to do with, or using coaxial cable.

coaxial

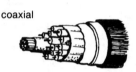

cobalt (kōʹbôlt), *n. Chemistry.* a hard, silver-white metallic element with a pinkish tint, which occurs only in combination with other elements, especially nickel and iron. Cobalt is used in steel for hardness, and in making alloys, paints, etc. *Symbol:* Co; *atomic number* 27; *atomic weight* 58.933; *melting point* 1495°C; *boiling point* 2900°C; *oxidation state* 2,3. [from German *Kobalt,* variant of *Kobold* goblin (because of miners' belief in the evil or mischievous effects of its ore)]

cobalt 60, *Physics.* a radioactive isotope of cobalt, of atomic weight 60, produced by bombarding cobalt atoms with neutrons and used as a source of gamma rays in the treatment of cancer. *Symbol:* ^{60}Co

cobalt bloom, = erythrite.

cobaltic (kō bôlʹtik), *adj. Chemistry.* of or containing cobalt, especially with a valence of three.

cobaltiferous (kōʹbôl tifʹər əs), *adj. Mineralogy.* containing or yielding cobalt.

cobaltite (kōʹbôl tīt) or **cobaltine** (kō bôl tēn), *n.* a silver-white mineral containing cobalt, arsenic, and sulfur. It is an important ore of cobalt. *Formula:* CoAsS

cobaltous (kōʹbôl təs), *adj. Chemistry.* of or containing cobalt, especially with a valence of two.

coccoid (kokʹoid), *Microbiology.* =*adj.* of or resembling a coccus. —*n.* a coccoid microorganism.

coccolith (kokʹə lith), *n. Microbiology.* any of the skeletal coverings made of calcite that protect a coccolithophore. In a fossilized state coccoliths form chalk and limestone deposits, such as the White Cliffs of Dover. [from Greek *kokkos* seed, berry + *lithos* stone]

coccolithophore (kokʹə lithʹə fôr), *n. Microbiology.* any of a group of single-celled, golden-brown microorganisms having two whiplike appendages and hard shells composed mostly of calcite. Coccolithophores, along with diatoms, are phytoplankton important to the food chain of the open ocean.

coccosphere, *n. Microbiology.* **1** a spheroidal mass of coccoliths. **2** = coccolithophore.

coccus (kokʹəs), *n., pl.* **cocci** (kokʹsī). *Bacteriology.* any bacterial cell shaped like a sphere. See the picture at **bacteria.**

ASSOCIATED TERMS: There are four basic types of bacteria, classified according to shape. *Cocci* are spherical, *bacilli* are rod-shaped, *vibrios* are comma-shaped or S-shaped, and *spirilla* have long, spirally twisted forms.
[from New Latin, from Greek *kokkos* seed, berry]
—**coccal**, *adj.* of, having to do with, or resembling a coccus; spherical or nearly spherical in form.

coccyx (kok'siks), *n.*, *pl.* **coccyges** (kok'sə jēz'), **coccyxes**. *Anatomy.* **1** a small, triangular bone forming the lower end of the spinal column in man and in tailless apes.
2 a similar part in certain animals and birds.
[from Latin *coccyx*, from Greek *kokkyx*, originally, cuckoo; because it is shaped like the cuckoo's bill]
—**coccygeal** (kok sij'ē əl), *adj.* of or having to do with a coccyx: *a coccygeal vertebra or nerve.*

cochlea (kok'lē ə), *n.*, *pl.* **-leae** (-lē ē'). *Anatomy.* a spiral-shaped cavity of the inner ear, containing the nerve endings that transmit sound impulses along the auditory nerve. See the picture at **ear.**
[from Latin *cochlea* snail]
—**cochlear** (kok'lē ər), *adj.* of the cochlea: *the cochlear duct, the cochlear nerve.*

cochleate (kok'lē it), *adj.* *Botany.* shaped like a snail shell; spiral.

cocoon, *n.* *Zoology.* **1** the silky case spun by caterpillars to live in while they are turning into adult insects. Silk is obtained from the cocoons of silkworms.
2 any similar protective covering, such as the egg case of a spider.
[from French *cocon*, from *coque* shell]

code, *n.* = genetic code.
—*v.* *Molecular Biology.* **code for,** to specify the genetic code for (the synthesis of a protein, etc.): *The structural genes that code for this enzyme were identified from research with inbred mouse strains* (Science News).

codeclination, *n.* *Astronomy.* the complement of the declination; the angular distance of a celestial body from the celestial pole. Also called **polar distance.**

codominance, *n.* *Genetics.* a condition in which both alleles of a pair are partially expressed, neither being dominant or recessive to the other.
▶ The ABO blood groups illustrate the principle of codominance. There are three genes, A, B, and O, that determine ABO blood groups. Each person has two of these genes. O is recessive to either A or B, but A and B are codominant; a person with genotype AB will have blood of group AB, neither gene being dominant. People with group O blood must have the genotype OO, while those with group A may be AA or AO, and those with group B may be BB or BO.

codominant, *n.* *Ecology.* one of the most common or important species in a plant community.
—*adj.* *Genetics.* **1** characterized by codominance; partially expressing both alleles of a pair. **2** *Ecology.* being a codominant; sharing dominance. Compare **dominant.**

codon (kō'don), *n.* *Molecular Biology.* a sequence of three nucleotides forming the genetic code for directing the placement of a particular amino acid in a polypeptide chain during protein synthesis: *A sequence of three nucleotides—known as a codon—is required to specify one amino acid. ... The codon which consists of three uracil residues, for example, specifies the amino acid phenylalanine; other codons act as punctuation marks, coding for the beginning and end of a*

protein (Science Journal). Compare **anticodon.** [from *code* + *-on*]

coefficient (kō'ə fish'ənt), *n.* **1** *Mathematics.* a number or symbol that is a factor of a term. In *3x*, *3* is the coefficient of *x*, and *x* is the coefficient of *3;* in *axy*, *a* is the coefficient of *xy.*
2 *Physics.* a ratio used as a multiplier to calculate the behavior of a substance under different conditions of heat, light, pressure, etc. The **coefficient of friction** is the ratio of the tangential force required to sustain motion to the normal force between two surfaces in contact.

coefficient of correlation, = correlation coefficient.

coefficient of friction, *Physics.* the ratio between the weight of an object being moved and the force pressing the surfaces together. The coefficient of friction varies with the different materials used. *Abbreviation:* C.F.

coelenterate (si len'tə rāt'), *n.*, *adj.* = cnidarian. [from Greek *koilos* hollow + *enteron* intestine]

coelenteron (si len'tə ron), *n.*, *pl.* **-tera** (-tər ə). *Zoology.* the internal and digestive cavity of a cnidarian (coelenterate).

coeliac (sē'lē ak), *adj.* = celiac.

coelodont (sē'lə dont), *Zoology.* =*adj.* having hollow teeth, as certain lizards.
—*n.* a coelodont animal.
[from Greek *koilos* hollow + *odontos* tooth]

coelom (sē'ləm), *n.* *Zoology.* the body cavity of most multicellular animals, having a membranous lining composed of specialized cells and containing the heart, lungs, etc. It develops within the mesoderm of the embryo. Also spelled **celom.**
[from Greek *koilōma* cavity, from *koilos* hollow]
—**coelomic** (si lom'ik), *adj.* of or having to do with the coelom: *coelomic fluid.*

coelomate (sē'lə mit), *Zoology.* —*adj.* having a coelom: *The earthworms ... is the traditional type with which students begin their study of coelomate animals* (Betty I. Roots).
—*n.* a coelomate animal.

coenenchyma (si neng'kə mə), *n.*, *pl.* **-mata** (-mə tə). *Zoology.* the common structure containing calcium and uniting the individual polyps or zooids of a compound anthozoan, such as coral. [from Greek *koinos* common + *en-* in + *chyma* what is poured]
—**coenenchymal**, *adj.* having to do with or of the nature of coenenchyma: *coenenchymal tubes.*

coenocyte (sē'nə sīt), *n.* *Botany.* a mass of cytoplasm containing many nuclei enclosed by a single cell wall, as in some green algae and many fungi.
[from Greek *koinos* common + *kytos* anything hollow]
—**coenocytic** (sē'nə sit'ik), *adj.* of or containing coenocytes: *coenocytic algae.*

---

cap, fāce, fäther; best, bē, tėrm; pin, five;
rock, gō, ôrder; oil, out; cup, pùt, rüle,
yü in use, yù in uric;
ng in bring; sh in rush; th in thin, ᴛʜ in then;
zh in seizure.
ə = a in about, e in taken, i in pencil, o in lemon, u in circus

coenosarc (sē'nə särk), *n. Zoology.* the common soft tissue which unites the individual polyps or zooids of a compound zoophyte, such as coral, and circulates food through the colony. [from Greek *koinos* common + *sarkos, sarx* flesh]
—**coenosarcal,** *adj.* having to do with or of the nature of coenosarc: *coenosarcal canals.*

coenurus (si nyùr'əs), *n., pl.* **coenuri** (si nyùr'ī). *Zoology.* a larval stage of a tapeworm, in which many heads are asexually produced within a single cyst. In the brains of sheep, coenuri cause staggers, a disease of the nerves. [from New Latin *coenurus,* from Greek *koinos* common + *oura* tail]

coenzyme (kō en'zīm), *n. Biochemistry.* an organic substance, usually containing a mineral or vitamin, capable of attaching itself to and supplementing a specific protein (the apoenzyme) to form an active enzyme system: *Coenzymes act as regulators of specific reactions in the body* (Science News Letter). Also called **cofactor.**

coenzyme A, *Biochemistry.* a coenzyme important in the metabolism of organic acids in plants and animals: *Coenzyme A is essential to the oxidation of fatty acids* (Scientific American). *Formula:* $C_{21}H_{36}N_7O_{16}P_3S$ *Abbreviation:* CoA

coenzyme Q, = ubiquinone.

coercive force, *Physics.* that intensity of a magnetic field required to reduce to zero the residual magnetism of a substance.

coercivity (kō'ėr siv'ə tē), *n. Physics.* a measure of the magnetic properties of a substance; the value of the coercive force when the substance has been initially magnetized to saturation.

coesite (kō'sīt), *n. Mineralogy.* a very dense form of silica, occurring in meteoritic craters and synthesized by subjecting quartz to very high pressure. [from Loring Coes, Jr., 20th-Century American chemist + -*ite*]

coevolution, *n. Biology.* the interactive evolution of two or more species or forms, each adapting to changes in the other. Coevolution occurs, for example, between predators and their prey, and between flowers and the insects that pollinate them.
—**coevolutionary,** *adj.* of or involving coevolution: *The coevolutionary advantages of pollen feeding not only to the Heliconius butterflies but also to the plants from which they are able to collect limited but continuous amounts of pollen* (Nature).
—**coevolve,** *v.* **1** to undergo coevolution; evolve in response to one another: *They found that the eaters and the eaten progressively evolved in close response to each other—coevolved* (Harper's).
2 to evolve at the same time: *In the birds of paradise, brilliant colors and long plumes have coevolved with extremely elaborate movements* (Natural History).

cofactor, *n.* **1** = coenzyme: *In some cases ... the enzyme protein alone is not enough to speed a chemical reaction. One or more cofactors are required* (McElroy, *Biology and Man*).
2 *Mathematics.* the minor or the additive inverse of the minor.

coffinite (kôf'ə nīt), *n. Mineralogy.* a black, hydrous silicate of uranium, like coal in appearance. [from Reuben C. Coffin, 20th-century American geologist + -*ite*]

coherent, *adj. Physics.* having waves of a similar phase, direction, and amplitude and capable of exhibiting interference: *Like water from a fire hose, coherent laser light travels in a direct stream that remains remarkably compact for long distances. Coherence is the secret of the laser's power* (National Geographic). See the picture at **laser.**

cohesion, *n.* **1** *Physics.* the attraction between molecules of the same kind: *Drops of water are a result of cohesion.*
2 *Botany.* the union of one part with another: *In the development of the flowers ... the parts of each whorl ... frequently become partly or completely united laterally. This condition is termed ... cohesion* (Youngken, *Pharmaceutical Botany*).
[from Latin *cohaesus* pressed together]
—**cohesive,** *adj. Physics.* of, causing, or characterized by cohesion: *When a substance is heated, the heat energy causes its molecules to move faster. Eventually, the molecules will move so rapidly that their cohesive forces cannot hold them together* (Louis Marick).
—**cohesively,** *adv.*

coincide, *v. Mathematics.* to occupy the same place in space: *If these two triangles △ △ were placed one on top of the other, they would coincide.*
—**coincidence,** *n.* a coinciding: *the coincidence of two triangles or circles.*

col (kol), *n. Geography, Geology.* a pass or notch which forms between circular depressions on opposite sides of a narrow alpine ridge. [from French, from Old French *col* neck, from Latin *collum*]

colatitude (kō lat'ə tüd), *n. Astronomy.* the complement of the latitude; the difference between a given latitude, expressed in degrees, and 90 degrees.

cold-blooded, *adj. Biology.* having an internal temperature that is not constant but varies with the surroundings. Such animals tend to take on the temperature of their environments. *Among the vertebrates, some are "cold-blooded" (fishes, amphibians, and reptiles) and some are "warm-blooded" (birds and mammals). A "cold-blooded" animal may very well have a higher temperature than a "warm-blooded" one ... Since the terms "warm-blooded" and "cold-blooded" are therefore flatly wrong, it is necessary to call those with internal temperature-regulating mechanisms homeothermous, and those without poikilothermous* (Simpson, *Introduction to Biology*).

colemanite (kōl'mə nīt), *n.* a mineral consisting of hydrated calcium borate occurring in colorless to white crystals. Colemanite is the principal natural source of borax. *Formula:* $Ca_2B_6O_{11} \cdot 5H_2O$ [from William T. Coleman, 1824–1893, American civic leader + -*ite*]

coleopteran (kō'lē op'tər ən), *n. Zoology.* any member of the largest order (Coleoptera) of insects, usually having two pairs of wings, the forewings being hard and sheathlike, the hind wings membranous. Beetles, weevils, and fireflies belong to this order.
[from Greek *koleos* sheath + *pteron* wing]
—**coleopterous** (kō'lē op'tər əs), *adj.* of or belonging to the coleopterans: *a coleopterous insect.*

coleoptile (kō'lē op'təl), *n. Botany.* a tubular sheath covering the terminal bud of grasses for a short time after germination of the grains: *In monocotyledonous plants, the young leaves are rolled up and completely* ensheathed in a protective tube, the coleoptile (Weier, Botany). [from Greek *koleos* sheath + *ptilon* feather]

coleorhiza (kō′lē ə rī′zə), *n.*, *pl.* **-zae** (-zē). *Botany.* the sheath enveloping the radicle or rudimentary root in the embryo of grasses and certain other plants. The root bursts through the coleorhiza in germination. [from Greek *koleos* sheath + *rhiza* root]

colicin (kō′lə sən), *n. Bacteriology.* any of various antibacterial proteins produced by certain intestinal bacteria: *Colicins are a class of proteins manufactured by many bacteria of the Escherichia coli group* (Scientific American). [from *colic*, adj., of or having to do with the colon + *-in*]

coliform (kō′lə form), *adj. Bacteriology.* of or having to do with the bacilli commonly found in the intestines of humans and other vertebrates, especially the bacillus *Escherichia coli: coliform bacteria.* [from Latin *coli* of the colon + English *-form*]

collagen (kol′ə jən), *n. Biochemistry.* the protein substance in the fibers of connective tissue, bone, and cartilage of vertebrates. Boiling with water converts collagen to gelatin. *Collagen supplies the matrix in which the calcium salts that give the bones their hardness are deposited. Collagen is the substance of cartilage and tendon* (Scientific American). [from French *collagène*, from Greek *kolla* glue]
—**collagenous** (kə laj′ə nəs), *adj.* of or having to do with collagen: *collagenous material.*

collagenase (kə laj′ə nās), *n. Biochemistry.* a pancreatic enzyme able to break down collagen.

collective fruit, = multiple fruit.

collenchyma (kə leng′kə mə), *n. Botany.* living plant tissue with cells whose walls are thickened and usually elongated: *Just beneath the epidermis, and particularly at the corners of angular stems, there occurs frequently a tough, fibrous tissue known as collenchyma* (Emerson, *Basic Botany*). Compare **sclerenchyma.** [from Greek *kolla* glue + *en-* in + *chyma* what is poured]
—**collenchymatous** (kol′ən kim′ə təs), *adj.* of, having to do with, or resembling collenchyma.

colligative (kol′ə gā tiv), *adj. Chemistry.* that depend on the relative number of particles rather than on the nature of the particles: *Colligative properties as related to solutions include changes in boiling point, freezing point, vapor pressure, and osmotic pressure* (Chemistry Regents Syllabus). [from Latin *colligatum* bound together, from *col-, com-* together + *ligare* to bind]

collimator (kol′ə mā′tər), *n. Optics.* **1** a small fixed telescope with cross hairs at its focus, used for adjusting the line of sight of other instruments.
2a the tube with a slit and lens used in the spectroscope to collect the light and project it upon the prism in parallel rays. **b** the lens or mirror itself.
3 any aperture used to limit the area or angular spread of a beam of particles or waves.

collinear (kə lin′ē ər), *adj. Geometry.* lying in the same straight line: *collinear points.*
—**collinearity,** *n.* the quality or fact of being collinear.

colloblast (kol′ə blast), *n. Zoology.* one of the cells on the tentacles of ctenophores that produce secretions helpful in the capture of prey. Also called **lasso cell.** [from Greek *kolla* glue + *blastos* sprout, germ]

color

colloid (kol′oid), *n. Chemistry.* a substance composed of particles that are extremely small but larger than most molecules (usually ranging from about 10^{-8} to 10^{-9} meter in diameter). The particles in a colloid do not actually dissolve but remain suspended in a suitable gas, liquid, or solid. *Among the colloids are polymers, such as rubber, plastics and synthetic fibers* (A.B. Garrett).
[from Greek *kolla* glue]
—**colloidal** (kə loi′dəl), *adj.* being, containing, or like a colloid: *Colloidal particles may be gaseous, liquid, or solid, and occur in various types of suspensions (imprecisely called solutions)* (Gessner G. Hawley).

colluvium (kə lü′vē əm), *n.*, *pl.* **-via** (-vē ə), **-viums.** *Geology.* loose material deposited at the base of a slope mainly by gravity, as talus. [New Latin, from Latin *colluvies* refuse, sewage, from *colluere* to wash]

colog or **colog.,** *abbrev.* cologarithm.

cologarithm, *n. Mathematics.* the logarithm of the reciprocal of a number; the negative of the logarithm. EXAMPLE: log 7/17 = log 7 + colog 17.

colon (kō′lən), *n.*, *pl.* **colons, cola** (kō′lə). *Anatomy.* the lower part of the large intestine in which solid waste is accumulated and prepared for elimination from the body. The colon extends from the cecum to the rectum. [from Latin *colon*, from Greek *kolon* part of the large intestine]
—**colonic** (kə lon′ik), *adj.* of, having to do with, or affecting the colon.

colony, *n.* **1** *Biology.* a group of organisms of the same species, living or growing together: *a colony of ants. Coral grows in colonies.*
2 *Microbiology.* a mass of microorganisms arising from a single cell, living on or in a solid or partially solid medium: *... numerous bacterial colonies can be seen without magnification. Each one of these colonies is made up of millions of cells, all descended from one individual* (Emerson, *Basic Botany*).
[from Latin *colonia*, from *colonus* cultivator, settler, from *colere* cultivate]

color, *n.* **1** *Optics.* the sensation produced by the effect of waves of light striking the retina of the eye. Different colors are produced by rays of light having different wavelengths.
2 *Nuclear Physics.* a property that renders a particle sensitive to the force that binds quarks: *The three basic quarks are subdivided still further by quantum properties called colors. By assuming three colors—red, yellow and blue—for each of the basic quarks, nine quarks become available, and it is then possible to construct all the known hadrons without violations* (Henry T. Simmons).
ASSOCIATED TERMS: see **charm.**

---

cap, fāce, fäther; best, bē, tėrm; pin, five;
rock, gō, ôrder; oil, out; cup, pùt, rüle,
yü in use, *yù* in uric;
ng in bring; sh in rush; th in thin, ᴛʜ in then;
zh in seizure.
ə = a in about, e in taken, i in pencil, o in lemon, u in circus

117

▶ Color and hue refer to the characteristic sensation produced by light waves striking the retina. *Color* can be specified in terms of brightness, wavelength, and saturation. *Hue* is the quality of a color that gives the name, such as red, blue, green, yellow, or purple, each corresponding to the dominant wavelength. White, black and gray are not regarded as hues.

colorimeter (kul'ə rim'ə tər), *n. Chemistry.* a device used in chemical analysis for comparing the color of a liquid with a standard color.
—**colorimetric** (kal'ər ə met'rik), *adj.* having to do with the comparison or measurement of colors. **Colorimetric analysis** is a method of determining the concentration of a chemical substance by measuring the intensity of its color or of a color produced by it.
—**colorimetry** (kul'ə rim'ə trē), *n.* measurement or analysis by means of a colorimeter.

color index, *Astronomy.* a measure of the color of a star, determined by the difference between its photographic magnitude and its visual magnitude.

color phase, *Zoology.* the coloration at a particular season of an animal whose fur or plumage changes in color according to season, such as the ermine or the ptarmigan.

color temperature, *Astronomy.* the temperature at which a blackbody emits light of the same color as that which is emitted by a given source, especially a celestial body: *Perhaps the most widely used one-dimensional color scale is that of color temperature for classifying light sources* (Deane B. Judd).

colostrum (kə los'trəm), *n. Physiology.* the thin, yellowish milk secreted by a mammal for the first few days after the birth of young. It is especially rich in protein and helps establish both digestion and natural immunity. *Colostrum contains whatever antibodies to disease the mother may previously have developed through having the disease herself* (Science News Letter). [from Latin *colostrum*]
—**colostral,** *adj.* of, having to do with, or caused by colostrum: *colostral immunity.*

columbite (kə lum'bīt), *n. Mineralogy.* the native ore of niobium, a black compound containing iron and manganese, and often tantalum. [from *columb(ium)* + -*ite*]

columbium (kə lum'bē əm), *n.* the former name of **niobium.** The name is still used in metallurgy. *Symbol:* Cb. [from New Latin *columbium,* from *Columbia* the United States]

columella (kol'yə mel'ə), *n., pl.* -**mellae** (-mel'ē). *Zoology, Botany.* a columnlike structure in an animal or plant, such as the small bone in the middle ear of birds, reptiles, and amphibians.
—**columellar,** *adj.* of or having to do with a columella. A columellar lip is the inner lip of a univalve shell.

columnar (kə lum'nər), *adj. Mineralogy, Biology.* having a columnlike structure; prismatic.

columnar basalt, *Geology.* basalt which has cooled in a manner that yields tightly packed columns.

columnar cell, *Biology.* an epithelial cell having a long, cylindrical or prismatic shape.

coma (kō'mə), *n., pl.* **comae** (kō'mē). **1** *Astronomy.* a cloudlike mass around the nucleus of a comet; the cometary atmosphere: *The nuclei of the smaller com-*

ets *are only a mile or so in diameter, but the visible head, or coma, may extend for thousands of miles beyond it* (New Astronomy).
2 *Optics.* the blurred appearance or hazy border surrounding an object viewed through a lens which is not free from spherical aberration: *Coma differs from spherical aberration in that a point object is imaged not as a circle, but as a comet-shaped figure, whence the term "coma"* (Sears and Zemansky, *University Physics*).
3 *Botany.* a tuft of hairs, as at the end of a seed.
[from Latin *coma,* from Greek *komē* hair]

combinatorial analysis, *Mathematics.* the branch of higher mathematics which deals with analysis by means of combinations, permutations, etc.

combinatorial topology, *Geometry.* the branch of topology which deals with forms reduced to combinations of the simplest geometric figures.

combinatorics, *n. Mathematics.* the study of the permutations and combinations of elements in finite sets: *Combinatorics involves computing the number of ways r things can be chosen from a total of n things (all different or not) with either the stipulation that: (1) order, or arrangement, counts, or (2) order does not count. Each choice in (1) is called a permutation; each choice in (2) is called a combination* (Mathematics Syllabus). *Combinatorics ... has achieved increasing importance ... in a large variety of present-day situations ... mathematical programming, statistics, production planning, and computing* (Nature).
—**combinatorial,** *adj.* of or having to do with combinatorics.

combine, *v. Chemistry.* to unite to form a compound: *Two atoms of hydrogen combine with one of oxygen to form water.*
—**combination,** *n.* **1** *Mathematics.* **a** the arrangement of individual items in groups so that each group has a certain number of items. **b** the groups thus formed. Possible two-letter combinations of *a, b,* and *c* are *ab, ac,* and *bc.*
2 *Chemistry.* the union of substances to form a compound.

combining weight, = equivalent weight.

comb jelly, = ctenophore.

combustion, *n. Chemistry.* **1** rapid oxidation accompanied by high temperature and usually by light: *The majority of the examples of combustion encountered by the average man are those of the rapid oxidation of easily ignited materials by atmospheric oxygen* (Jones, *Inorganic Chemistry*).
2 slow oxidation not accompanied by high temperature and light. The cells of the body transform food into energy by this type of combustion.
[from Latin *combustionem,* from *comburere* burn up, from *com-* up + *urere* burn]

comet, *n. Astronomy.* an object in the solar system consisting of a dense nucleus of frozen gases and dust, which develops a luminous halo and tail when near the sun. Comets follow extremely long, eccentric orbits about the sun *As the comet draws still closer to the Sun, the powerful solar wind forces dust particles to flow out of the coma and form a tail of dust* (Benedict A. Leerburger). [from Latin *cometa,* from Greek *kometes* wearing long hair, from *komē* hair]

—**cometary** (kom′ə ter′ē) or **cometic** (kə met′ik), *adj.* of, having to do with, or resembling a comet: *cometary orbits.*

comma bacillus, = vibrio.

comma tract, *Anatomy.* a tract of white nerve fibers found within the posterior external column of the spinal cord.

commensal (kə men′səl), *Biology.* —*adj.* of, having to do with, or characterized by commensalism. —*n.* a commensal organism.

commensalism (kə men′sə liz′əm), *n. Biology.* a symbiotic relationship between two organisms in which one lives in or on the other, but not as a parasite, thus leaving the other unaffected: *In commensalism, one partner benefits and the other neither benefits nor is harmed* (Otto and Towle, *Modern Biology*). [ultimately from Latin *com-* together + *mensa* table]
ASSOCIATED TERMS: see **symbiosis.**

commensurable, *adj. Mathematics.* measurable by the same standards or scale of values; expressible in terms of a common unit. A meter and a centimeter are commensurable magnitudes.

commissure (kom′ə shùr), *n.* **1** *Anatomy.* **a** any of the nerve fibers connecting corresponding parts of the brain or spinal cord: *... note the corpus callosum and anterior commissure, the great bands of nerve fibers that keep one half of the brain in communication with the other* (Robert W. Doty). **b** the line or surface along which two parts touch each other or form a connection: *the commissure of the eyelids.*
2 *Botany.* the joint or face by which one carpel coheres with another.
[from Latin *commisura* a joining]
—**commissural** (kə mish′ər əl), *adj.* having to do with or of the nature of a commissure.

common, *adj. Mathematics.* belonging equally to two or more quantities or figures. See also **common multiple.**

common denominator, *Mathematics.* a common multiple of the denominators of a group of fractions: *12 is a common denominator of 1/2, 2/3, and 3/4, because these fractions can be expressed also as 6/12, 8/12, and 9/12.*

common divisor or **common factor,** *Mathematics.* a number that will divide a group of numbers without a remainder: *2 is a common divisor of 4, 6, 8, and 10.*

common fraction, *Mathematics.* a fraction expressed as the ratio of two whole numbers. EXAMPLE: 1/2, 7/8.

common ion effect, *Chemistry.* a shift in equilibrium occurring when a substance is added to a solution of a second substance with which it has an ion in common, the volume being kept constant: *The common ion effect is observed when one of the ions of a weak electrolyte is added in excess to a solution of the electrolyte. The general effect is to render the electrolyte less ionized* (Briscoe, *College Chemistry*).

common logarithm, *Mathematics.* a logarithm in which 10 is the base.

common multiple, *Mathematics.* a number divisible by two or more numbers without a remainder: *12 is a common multiple of 2, 3, 4, and 6.*

community, *n. Ecology.* a group of organisms living together; any group of interacting organisms: *It is by no means a matter of chance that certain kinds of plant communities are found in specific locations. ... the plant communities in a region are well adapted to its physical environment* (Greulach and Adams, *Plants*).

commutative (kom′yə tā′tiv), *adj. Mathematics.* of or having to do with a property of addition and multiplication such that the order in which numbers are added or multiplied will not change the result of the operation. EXAMPLE: $2 + 3$ will give the same sum as $3 + 2$. Likewise, 2×3 gives the same product as 3×2.
ASSOCIATED TERMS: see **associative.**
—**commutativity** (kə myü′ tə tiv′ə tē), *n. Mathematics.* the property of giving the same result regardless of the order of addition or multiplication.

compact bone, *Anatomy.* dense (as opposed to spongy) bone consisting of a network of Haversian canals and the layers of tissue surrounding them.

compaction, *n. Geology.* **1** a process of rock formation by the pressing together of mineral particles, sediment, or detritus: *Eventual compaction of such ooze results in the production of limestone in the form of chalk* (Hegner and Stiles, *College Zoology*). **2** a rock or mass thus formed.

companion cell, *Botany.* any of the small cells with large nuclei adjacent to the sieve tube in the phloem of vascular plants.

comparison spectrum, *Astronomy.* a spectrum, typically an emission spectrum of helium or mercury, formed for comparison with the spectrum under observation.

compass, *n.* **1a** an instrument for indicating direction along the lines of the earth's magnetic field, usually consisting of a suspended magnetic needle that points to the North Magnetic pole. **b** an instrument showing direction, consisting of a gyroscope that points to the geographical North Pole. Also called **gyrocompass.**
2 Also, **compasses,** *pl.* an instrument consisting of two legs hinged together at one end, used for drawing circles and curved lines and for measuring distances.

compatibility, *n.* **1** *Immunology.* the ability of transferred or transplanted tissue or cells to function without being rejected: *blood compatibility.*
2 *Botany.* the ability to cross-fertilize or unite with a stock.
3 *Chemistry.* the ability of substances to mix together without impairment of function: *drug compatibility.*

compatible, *adj.* **1** *Botany.* capable of cross-fertilization.
2 *Chemistry.* able to be mixed or combined without interfering with one another: *compatible drugs.*

compensation, *n.* **1** *Biology.* the increased size or activity of one part (of an organism or organ) to make up for the loss or weakness of another part.
2 *Optics.* **a** the equalization of the retardation of two rays of light. **b** = compensator.
—**compensator,** *n. Optics.* a plate or combination of prisms for effecting compensation.

---

cap, fāce, fäther; best, bē, tèrm; pin, five;
rock, gō, ôrder; oil, out; cup, pùt, rüle,
yü in use, *yu* in uric;
ng in bring; *sh* in rush; *th* in thin, ⊤H in then;
zh in seizure.
ə = *a* in about, *e* in taken, *i* in pencil, *o* in lemon, *u* in circus

compete, *v. Biology.* to engage in competition: *If two different species compete for the same food or reproductive sites, one species may be eliminated. This usually establishes one species per niche in a community* (Biology Regents Syllabus).

competition, *n. Ecology.* the simultaneous demand by different organisms for food, living space, or other necessities: *Two or more species in the same habitat that have the same requirements for resources limited in supply are said to be in competition ... Members of the same species may also compete with one another, a situation termed intraspecific competition. This competition is important in controlling population size.* (Storer, *General Zoology*).

compital (kom′pə təl), *adj. Botany.* **1** (of veins in leaves) intersecting at various angles. **2** (of the sori of ferns) situated at the point of junction of two veins. [from Latin *compitalem* of the crossroads, from *compitum* a crossroads]

complement, *n.* **1** *Mathematics.* **a** the angle or arc by which a given angle or arc is less than 90 degrees (the additional amount of angular distance needed to make a given angle or arc equal to 90 degrees). See also **complementary angle. b** those members of a universal set that do not belong to a given subset of the universal set. EXAMPLE: In the set (1, 2, 3, 4) the complement of subset (2, 4) is (1, 3).
2 *Immunology.* a complex system of proteins and other factors found in normal blood serum, which combines with antibodies to destroy bacteria and other foreign bodies: *With the help of other body defense proteins called "complement," they then ruptured the tumor cells* (Science News).
3 *Molecular Biology.* (of a DNA or RNA molecule) having two strands that are specific for each other, so that each strand can serve as a template for synthesizing new molecules: *It is known that the genetic information within a cell ... is coded into the sequence of molecules that make up the long strands of DNA ... As necessary, the message is transcribed into complementary RNA ..., which in turn translate the information into a sequence of amino acids that carry out the instructions* (Alexander Leaf).

complemental air, *Physiology.* the air which can be drawn into the lungs by an effort after the ordinary inspiration is completed. Also called **complementary air.**
ASSOCIATED TERMS: see **air.**

complementarity, *n. Physics.* the principle that physical phenomena may be described either in terms of waves, characterized by wavelength and frequency, or in terms of the motion of particles, characterized by energy and momentum, but not fully in both ways at the same time.

complementary, *adj.* **1** of or forming a complement.
2 *Optics.* producing white or gray light in combination. **Complementary colors** are two colors which, when combined as light rays, produce white or gray. Red and green are complementary colors.

complementary air, = complemental air.

complementary angle, *Mathematics.* either of two angles which together form an angle of 90 degrees: *A 30-degree angle is the complementary angle of a 60-degree angle.* Compare **supplementary angle.**

complementary cell, *Botany.* one of the cells which make up lenticel tissue.

complementation, *n. Mathematics.* an operation to determine the complement of a subset.

complement fixation, *Immunology.* the binding of complement to an antigen-antibody mixture, making it unavailable for subsequent reaction.

complete metamorphosis, = holometabolism.

complete protein, *Biochemistry.* a protein that contains every amino acid essential for building blood and tissue. Animal proteins, unlike the proteins in most plants, are complete proteins.

complex, *v. Chemistry.* to chelate: *Extra chemicals are sometimes prescribed to complex dissolved iron* (New Scientist).
—complexation, *n.* = chelation.

complex fraction, *Mathematics.* a fraction having a fraction or mixed number in the numerator, in the denominator, or in both. EXAMPLES: 1¾/3, 1/3½, ¾/1⅞. Also called **compound fraction.**

complex number, *Mathematics.* any number of the form $a + bi$, where a and b are real numbers and i is an imaginary number such that i^2 equals -1. A complex number may be represented graphically as a point on a Cartesian coordinate system having the coordinates (a, b).

composite, *Botany.* **—** *adj.* belonging to the composite family: *a composite plant.*
—*n.* a plant of the composite family: *The daisy is a composite.*
[from Latin *compositum* put together, from *com-* together + *ponere* put]

composite family, *Botany.* a very large plant family (Compositae *or* Asteraceae) comprising the most highly developed plants, including the daisy, aster, dandelion, marigold, and lettuce. The dicotyledonous plants of this family have a close head of many small flowers or florets.

composite number, *Mathematics.* a number divisible by some whole number other than itself and one; number that has more than two factors. 4, 6, and 9 are composite numbers. Contrasted with **prime number.**

composite volcano, *Geology.* a volcano whose cone is made up of alternate layers of ash and lava.

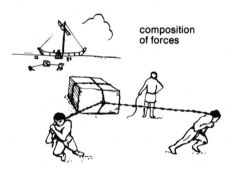

composition of forces

The combined force of the two men pulling the block at different angle results in the forward movement of the block between them. Compare the diagram at **vector.**

composition of forces, *Physics.* the joining or resolution of two or more given forces or vectors into one force or vector having an equivalent effect.

composition plane, *Crystallography.* the common plane or base between the two parts of a twin crystal.

compound, *adj.* **1** *Biology.* made up of several similar parts combined into a single structure: *a compound flower, a compound stomach.* See also **compound leaf, compound eye.** See also the picture at **fruit.**

2 *Zoology.* consisting of an intimate combination of individual animals in a colony, as coral does: *a compound tunicate.*

—*n. Chemistry.* a substance formed by chemical combination of two or more elements in definite proportions by weight: *Water is a compound of hydrogen and oxygen.*

ASSOCIATED TERMS: A compound is a substance which can be decomposed by chemical change. An *element* is a substance which cannot be decomposed by chemical change. A *mixture* consists of two or more substances differing in properties and composition. Mixtures may be *homogeneous* (as mixtures of gases) or *heterogeneous* (as a mixture of iron and sulfur). A *substance* is any form of matter all specimens of which have identical properties and composition.

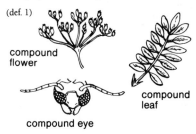

(def. 1)

compound flower

compound leaf

compound eye

compound eye, *Zoology.* the eye of certain arthropods, composed of many visual units (ommantidia), such as the large lateral eyes of insects: *The sense organs [of a crayfish] include a pair of stalked compound eyes ... composed of many individual eyes; there are about 2,500 single eyes in each ... the combined mosaic image produced by the entire group give something approximating the image produced by a lens of a higher animal's single eye* (Winchester, *Zoology*).

compound fraction, = complex fraction.

compound fruit, = multiple fruit.

compound leaf, *Botany.* a leaf composed of two or more leaflets on a common stalk. Clover has compound leaves.

compound microscope, a microscope having more than one lens, as one with an eyepiece and an objective.

compound nucleus, *Nuclear Physics.* an excited nucleus resulting from the temporary fusion of two nuclei, or of a nucleus and a nuclear particle, forming an intermediate stage in an induced nuclear reaction.

compound number, *Mathematics.* a quantity expressed in two or more kinds of units. EXAMPLES: 3 ft., 5 in.; 2 hr., 18 min., 40 sec.

compressed, *adj.* **1** *Botany.* flattened along its length; flattened laterally: *a compressed seed or leaf.*

2 *Zoology.* narrower in the lateral dimension than in the dorsoventral dimensions; of greater depth than width: *The flounder has a compressed body.*

—**compressedly,** *adv.*

compression wave or **compressional wave,** *Physics.* a longitudinal wave in which the particles move to and fro in the same direction as the motion of the wave,

causing the substance to compress: *Sound waves are compression waves.*

compressor, *n. Anatomy.* a muscle that compresses a part of the body.

Compton effect (komp'tən), *Physics.* an increase in the wavelength of X-ray or gamma-ray photons which occurs when these radiations are scattered on striking an electron. In the collision, kinetic energy is communicated to the electron and the quantum energy of the radiation is reduced. *The momentum of a photon is inversely proportional to its wavelength. The Compton effect is explained in terms of the conservation of energy and momentum in photon-particle collisions* (Physics Regents Syllabus). [named after Arthur H. Compton, 1892–1962, an American physicist, who observed it]

compute, *v. Mathematics.* to find by using mathematical processes, such as adding, subtracting, multiplying, etc. ► See note at **calculate.**

—**computation,** *n.* the act of computing or an amount computed.

—**computational,** *adj.* of or having to do with computation: *An electronic computer is used for much of the computational work* (New Scientist).

concave, *adj.* hollow and curved like the inside of a circle or sphere; curving in: *a concave mirror or lens.* Contrasted with **convex.** See the picture at **convex.** [from Latin *concavus,* from *com-* around + *cavus* hollow]

—**concavity,** *n.* **1** concave condition or quality: *The convexity of the [airplane] wing's upper surface and the concavity of its lower aid this alternate gripping and slipping of the air* (Atlantic). **2** a concave surface or thing.

concavo-convex (kon kā'vō kon veks'), *adj. Optics.* concave on one side and convex on the other. In a concavo-convex lens, the concave face has the greater curvature.

concentrate, *v. Chemistry.* to make (a substance) stronger, especially by eliminating unwanted ingredients. An acid solution is concentrated when it has very much acid in it.

—*n.* something that has been concentrated: *Lemon juice with the water removed is a concentrate.*

concentration, *n. Chemistry.* the amount of substance contained in a given quantity of a solution or mixture: *There is a certain maximum concentration of salts for any set of external conditions, beyond which there is no further increase in the rate of absorption with increase of concentration. This corresponds to a saturation of the transport system* (Science News).

concentration cell, *Chemistry.* an electrolytic cell whose difference of potential is due to the difference of concentration of the solutions in which the electrodes are immersed.

---

cap, fāce, fäther; best, bē, tėrm; pin, fīve;
rock, gō, ôrder; oil, out; cup, pùt, rüle,
yü in use, *yù* in uric;
ng in bring; *sh* in rush; *th* in thin, ᴛʜ in then;
zh in seizure.
ə = *a* in about, *e* in taken, *i* in pencil, *o* in lemon, *u* in circus

concentric

concentric, *adj. Geometry.* having the same center: *concentric circles.* —**concentrically,** *adv.*
—**concentricity** (kon′sən tris′ə tē), *n.* the quality or state of being concentric.

concentric

conceptus (kən sep′təs), *n. Biology.* an embryo or fetus: *Since the dividing cells of the conceptus "cannot be considered to be invested with individuality" while still capable of forming identical twins, arresting their development is considered contraception* (Bent G. Böving). [from Latin *conceptus* that which is conceived, from *concipere* take in, conceive]

concha (kong′kə), *n. Anatomy.* the central, hollow part of the outer ear. [from Latin *concha,* from Greek *konchē* mussel, conch]

conchiolin (kong ki′ə lin), *n. Biochemistry.* the organic part of the shells of mollusks, closely related to keratin. [from Latin *concha* conch]

conchoid (kong′koid), *n. Geometry.* a plane curve such that if a straight line is drawn from a fixed point (the pole) to the curve, the part of the line intercepted between the curve and a fixed straight line (the asymptote) is always equal to a fixed distance.
—*adj.* = conchoidal.

conchoidal (kong koi′dəl), *adj. Geology, Mineralogy.* having to do with or denoting a fracture having smooth, shell-like convexities and concavities, as in obsidian.
—**conchoidally,** *adv.*

conchology (kong kol′ə jē), *n.* the branch of zoology that deals with the shells of mollusks.
—**conchological,** *adj.* of or having to do with conchology.

conclusion, *n. Mathematics.* = consequent (def. 2).

concordant, *adj.* **1** *Geology.* (of an intrusion) having contacts parallel to the surrounding strata: *Dikes are discordant, which means that they cut across the layers, whereas sills are concordant—they more or less parallel the layering* (Birkeland and Larson, *Putnam's Geology*).
2 *Genetics.* alike with respect to a trait, especially, when comparing human twins: *If both members of a pair act in conjunction, and if both either possess the trait or are free of it, the pair is concordant, or phenotypically similar* (Monroe Strickberger).

concretion (kon krē′shən), *n. Geology.* a rounded or irregular mass formed by aggregation of solid particles or precipitation of mineral material, usually around a nucleus, common in certain types of sandstone, shale, limestone, and clay.

condensation nucleus, *Meteorology.* = nucleus.

condense, *v.* **1** *Chemistry.* **a** to change from a gas or vapor to a liquid. If steam touches cold surfaces, it condenses into water. **b** to undergo or cause to undergo

condensation (def. 1b): *Milk is condensed by removing much of the water from it.*
2 *Physics.* to increase in density and pressure; make or become more compact: *Each [theory] starts from the notion that stars condense from the matter scattered through interstellar space* (W. H. Marshall).
—**condensate** (kən den′sāt), *n. Chemistry.* a product of condensation.

—**condensation,** *n.* **1** *Chemistry, Meteorology.* **a** a changing of a gas or vapor to a liquid: *the condensation of steam into water. In the atmosphere, cooling is the only cause of any significant amount of condensation. It is important to remember this fact in considering the causes of cloudiness and precipitation* (Blair, *Weather Elements*). Compare **evaporation. b** a reaction in which two or more molecules unite to form a larger, denser, and more complex molecule, often with the separation of water or some other simple substance.
2 *Physics.* **a** an increase in density and pressure in a medium, such as air, due to the passing of a sound wave or similar wave. **b** the region in which this occurs. Compare **rarefaction.**

condenser, *n.* **1** *Physics.* = capacitor.
2 *Optics.* a strong lens or lenses for concentrating light upon a small area.
3 *Chemistry.* a device, usually of glass, for cooling a vapor so that it condenses as a liquid which can be collected.

condition, *n. Mathematics.* a requirement expressed by an open sentence. $3 + X = 5$ expresses the condition that a number you are to find, added to 3, must equal 5.

conditional, *n. Mathematics.* the statement or proposition "if *p,* then *q,*" usually symbolized as $p \rightarrow q$. Also called **implication.** Compare **biconditional.**

conditioned, *adj. Physiology.* determined or established by repeated exposure to particular stimuli, with which new responses become habitually associated: *a conditioned reflex. Habits, which are acquired by repetition, are examples of conditioned behavior. The repetition establishes pathways for nerve impulse transmission which permit rapid automatic responses to various stimuli* (Biology Regents Syllabus).

conduct, *v. Physics.* to transmit (heat, electricity, etc.); be a channel for: *Metals conduct heat and electricity.*

conductance, *n. Physics.* the power of conducting electricity as affected by the shape, length, or material of the conductor; the ease with which a substance or a solution of it permits the passage of electrical current. Its unit of measurement is the mho, or reciprocal of the ohm. Compare **resistance.**

conduction, *n.* **1** *Physics.* **a** the transmission of heat or other form of energy by the transferring of energy from one particle to another. Radiators heat the air in a room by conduction. *One primary cause of temperature changes in the lower air is conduction of heat to or from the earth's surface. Conduction is the process by which heat is transferred through matter, without transfer of the matter itself* (Blair, *Weather Elements*). **b** the transfer of an electric charge to an object by direct contact: *An object charged by conduction acquires the same kind of charge as the charging object* (Physics Regents Syllabus). Compare **induction.**

2 *Physiology.* the transmission of an impulse through muscle or nerve tissue.

ASSOCIATED TERMS: There are three basic ways of transferring heat from one point to another. *Conduction* (def. 1a) is a slow process in which the energy is transferred from molecule to molecule, without movement of the medium itself. *Convection* is the transfer of heat by the movement of heated molecules through a fluid or gaseous medium. *Radiation* is a very rapid transfer by electromagnetic waves instead of a material medium.

conductive, *adj. Physics.* of or having conductivity.

—conductivity, *n.* **1** the power of conducting heat, electricity, sound, or other form of energy: *The conductivity of a given material varies with temperature* (Sears and Zemansky, *University Physics*).
2 the characteristic conductance of a given substance; the reciprocal of the resistivity.

conductometric or **conductimetric** (kən duk′tə met′rik), *adj.* **1** *Physics.* of or having to do with the measurement of conductivity.
2 *Chemistry.* of or having to do with volumetric analysis in which the progressive change in the conductivity of a solution is used to determine the end point of a titration.

conductometry (kon′duk tom′ə trē), *n. Physics.* the measurement of the relative conductivity of different materials.

conductor, *n. Physics.* a thing that transmits heat, electricity, light, sound or other form of energy. *Copper is a good conductor of heat and electricity. Conductors are substances in which there are many free electrons. Insulators are substances in which there are few free electrons* (Physics Regents Syllabus).

conduplicate (kon dü′plə kit), *adj. Botany.* folded together lengthwise: *a conduplicate leaf in bud.*

condyle (kon′dəl *or* kon′dil), *n. Anatomy.* a rounded part that grows out at the end of a bone, articulating with another bone: *In the herbivores, the joints or condyles between the lower jaw and skull are fairly loose, allowing the sideways or back and forth movement of the lower jaw* (Mackean, *Introduction to Biology*). [from Greek *kondylos* knuckle]
—condylar (kon′də lər), *adj.* of or having to do with a condyle: *a condylar joint.*
—condyloid (kon′də loid), *adj.* having to do with or formed like a condyle.

cone, *n.* **1** *Geometry.* a surface described by a moving straight line, one point of which is fixed, the opposite end constantly touching a fixed curve. A cone in which the base is a circle and the vertex lies on the perpendicular to the base at the center is called a *right circular cone.* A cone with a circular base whose vertex does not lie on the perpendicular is called an *oblique circular cone.* See also **conic, conic section,** and **conoid.**
2 *Geology.* a cone-shaped peak formed by volcanic discharges: *Smoke and lava erupted from Parícutin as the Western Hemisphere's newest volcano rapidly built its cone* (Gordon A. Macdonald).
3 *Botany.* a scaly growth that bears the seeds on pine, cedar, fir, and other evergreen trees: *Spruces have cones that hang straight downward. Fir trees have cones that stand straight up* (K.A. Armson). Also called **strobile.** See the pictures at **gymnosperm, naked.**
4 *Biology.* one of a group of cone-shaped cells of the retina of the eye that responds to light: *The eye has two kinds of visual receptors, rods and cones. The rods are ... predominant in the eyes of nocturnal creatures. The*

cones, which are used in day vision, are found in the eyes of creatures that ... go to sleep at night (Science News Letter).

cone
(def. 1)

cone-in-cone, *n. Geology.* a structure of cones packed one inside another, found in certain sedimentary rocks. Individual cones measure from a fraction of an inch to a few feet in height.

configuration, *n.* **1** the relative position of parts; manner of arrangement: *Geographers study the configuration of the surface of the earth.*
2 *Physics.* the relative spatial position of atoms in a molecule: *In any particular molecule, atom, or nucleus, the constituent particles can usually exist in a variety of configurations* (J. Little).
—configurational, *adj.* of or having to do with configuration.

confluence (kon′flü əns), *n. Geography, Geology.* **1** the place where two or more rivers, streams, etc., come together.
2 a body of water produced in this way.
—confluent, *adj.* flowing or running together; blending into one: *confluent rivers.*

conformal, *adj.* **1** *Geography.* having the same scale in all directions at any given point: *The Mercator is a conformal projection; i.e., any very small area, such as a small bay or peninsula, is shown in practically its true shape* (Finch and Trewartha, *Elements of Geography*).
2 *Mathematics.* leaving unchanged the size of all angles.

conformation, *n. Chemistry.* the geometric or three-dimensional shape of a molecule in any of its states.
—conformational, *adj.* of or having to do with the conformation of molecules whose atoms rotate around one or more bonds: *a conformational isomer, conformational changes in DNA.* **Conformational analysis** is the determination of the conformation of molecules, especially for the purpose of correlating the preferred shapes with the physical and chemical properties of the molecules.

congener (kon′jə nər), *n.* **1** *Biology.* an organism of the same genus as another: *In Russia the small Asiatic cockroach has everywhere driven before it its great congener* (Darwin, *Origin of Species*).

---

cap, fāce, fäther; best, bē, tėrm; pin, fīve;
rock, gō, ôrder; oil, out; cup, pùt, rüle,
yü in use, *yu* in uric;
ng in bring; *sh* in rush; *th* in thin, ᴛʜ in then;
zh in seizure.
ə = *a* in about, *e* in taken, *i* in pencil, *o* in lemon, *u* in circus

2 *Physiology.* a congenerous muscle.
[from Latin *congener* of the same kind]
—**congeneric,** *adj.* of or belonging to the same genus.
Compare **conspecific.**

congenerous (kən jen′ər əs), *adj. Physiology.* having a common function; concurring in the same action, as muscles.

congenital (kən jen′ə təl), *adj. Medicine.* present at birth; acquired in the uterus, especially as a result of faulty development, infection, or injury: *a congenital deformity.* [from Latin *congenitus* born with]
ASSOCIATED TERMS: A *genetic* trait or disorder is one transmitted through the genes and is therefore *inborn, innate, hereditary* or *inherited;* a *congenital* trait or disorder is one which is acquired in the pregnancy and birth process; a *familial* trait or disorder is one occurring among several members of a family, and especially among members of the same generation. See also **acquired character.**
—**congenitally,** *adv.*

conglomerate (kən glom′ər it), *n. Geology.* a rock consisting of pebbles, gravel, or boulders, bonded by cement: *Closely allied to the sandstones are rocks called conglomerates ... held together by, a ground mass usually of sandstone, much after the manner of man-made concrete* (Finch and Trewartha, *Elements of Geography*). [from Latin *conglomeratus* rolled together]

congruent (kən grü′ənt), *adj.* **1** *Geometry.* coinciding exactly when superimposed; having the same size and shape: *congruent triangles, congruent circles.* Compare **similar.**
2 *Algebra.* producing the same remainder when divided by a given number.
[from Latin *congruentem* coming together, agreeing]
—**congruence,** *n.* the fact or condition of being congruent: *A one-to-one correspondence between the vertices of the two triangles ... is called a congruence between the two triangles* (Moise and Downs, *Geometry*).

conic (kon′ik), *adj.* Also, **conical.** of or shaped like a cone. A **conic projection** is a map projection made by projecting the earth's surface on an imaginary cone which is unrolled to a plane surface.
—*n.* **1** = conic section. **2 conics,** *pl.* the part of geometry dealing with conic sections.

conic section, *Geometry.* a curve formed by the intersection of a plane with a right circular cone. Circles, ellipses, parabolas, and hyperbolas are conic sections.

conic sections:

circle   ellipse   parabola hyperbola

conidiophore (kə nid′ē ə fôr), *n. Biology.* a specialized filament in certain fungi, bearing conidia. [from *conidium* + *-phore*]

conidiospore (kə nid′ē ə spôr), *n.* = conidium.

conidium (kə nid′ē əm), *n., pl.* **-nidia** (-nid′ē ə). *Biology.* a one-celled, asexual spore produced in certain fungi: *Each conidium contains a number of nuclei ... and when it falls on the wet surface of its proper host, it first germinates and produces as many zoöspores as it has nuclei* (Emerson, *Basic Botany*). [from New Latin *conidium,* from Greek *konis* dust]

conifer (kon′ə fər), *n. Botany.* any of a large order (Coniferales) of trees and shrubs of the gymnosperm class or division, most of which are evergreen and all of which bear cones. The pine, fir, spruce, hemlock, larch, and yew are conifers.
[from Latin *conifer* cone-bearing, from *conus* cone + *ferre* to bear]
—**coniferous** (kō nif′ər əs), *adj.* **1** bearing cones: *a coniferous tree.* **2** of conifers: *a coniferous forest.*

conjugate (kon′jə gāt), *v.* **1** *Biology.* to unite or fuse in conjugation.
2 *Chemistry.* to join (a compound) with another or others.
—*adj.* (kon′jə git) **1** *Botany.* in pairs; coupled.
2 *Mathematics.* (of two quantities, axes, points, etc. in mathematical relation) reciprocally related and interchangeable as far as certain properties are concerned.
3 *Chemistry.* expressing the relationship between an acid and a base by the difference of a proton: *In an acid-base reaction, an acid transfers a proton to become a conjugate base. This acid and its newly formed base form a conjugate acid-base pair. A base gains a proton to become a conjugate acid, forming a second acid-base pair. Each pair, made up of an acid and its base, is related by the transfer of a proton* (Chemistry Regents Syllabus).
—*n.* (kon′jə git) **1** *Biology.* an organism taking part in conjugation.
2 *Mathematics.* a conjugate axis, point, number, etc.
3 *Chemistry.* a compound joined with another or others: *a folic acid conjugate.*
[from Latin *conjugatus* yoked together, from *com-* + *jugum* yoke]
—**conjugation** (kon′jə gā′shən), *n. Biology.* **1** a process in which two one-celled organisms unite to transfer nuclear material, as in various bacteria, algae, fungi, and protozoans: *In paramecium ... there is, at intervals, a temporary union of individuals in pairs with mutual exchange of micronuclear material that is known as conjugation.* (Storer, *General Zoology*). **2** the fusion of male and female gametes to form a zygote, as in various algae.
—**conjugational,** *adj.* of or having to do with conjugation. —**conjugationally,** *adv.*

conjugated double bonds, *Chemistry.* two or more double bonds that alternate with single bonds in the molecule of a compound.

conjugated protein, *Biochemistry.* a compound consisting of a simple protein attached to a nonprotein group. Nucleoproteins are conjugated proteins.

conjugation tube, *Biology.* a tube found in certain algae that functions as a bridge between cells to facilitate conjugation.

conjunction, *n.* **1** *Astronomy.* the position of two celestial bodies when they have the same celestial longitude: *The moon is in conjunction with the sun at new moon. The planets Mercury and Venus, which lie between the earth and the sun, are said to be in inferior conjunction when one or the other of them is between the sun and the earth. Superior conjunction takes place when the sun lies between one of them and the earth* (Fletcher G. Watson). Compare **opposition, quadrature.**

2 *Mathematics.* the proposition or statement *"p* and *q"* (symbolized *p* ∧ *q*), which is true when both constituents are true, and false when either *p* or *q* or both are false. This definition is summarized in a truth table as follows:

p	q	p ∧ q
T	T	T
T	F	F
F	T	F
F	F	F

Compare **disjunction.**

conjunctiva (kon'jungk tĭ'və), *n., pl.* **-vas, -vae** (-vē). *Anatomy.* the mucous membrane that forms the inner surface of the eyelids and the front part of the eyeball. [from New Latin *(membrana) conjunctiva* conjunctive (membrane)]

connate (kon'āt), *adj.* **1** *Botany, Zoology.* congenitally united into one organ body, as leaves at the base. **2** *Geology.* entrapped in beds of sediment or rock at the time of deposition: *connate water.* [from Latin *connatus* born with, from *com-* + *nasci* be born] —**connately,** *adv.*

connective tissue, *Anatomy.* tissue that connects, supports, or encloses other tissues and organs in the body: *The tendons which unite muscles to bones consist of connective tissue; cartilage and bone are supporting connective tissues* (Hegner and Stiles, *College Zoology*).

connivent (kə nī'vənt), *adj. Botany, Zoology.* gradually converging: *The petals of certain flowers and the wings of certain insects are connivent.* [from Latin *connivere* to wink (at), shut the eyes]

conodont (kō'nə dont), *n. Paleontology.* any of certain minute fossils of a conical, toothlike form, found in Silurian and other rocks, supposed to be the jaws of annelids. [from Greek *kōnos* cone + *odontos* tooth]

conoid (kō'noid), *Geometry.* —*n.* a solid formed by the revolution of a conic section about one of its axes. —*adj.* Also, **conoidal. 1** having to do with or resembling a conoid. **2** nearly conical in shape.

consecutive points, *Mathematics.* two or more points infinitely close to one another on the same branch of a curve.

consensual (kən sen'shù əl), *adj. Physiology.* caused by reflex stimulation, as the contraction of both pupils when light strikes one eye. [from Latin *consensus* agreement]

consequent, *n. Mathematics.* **1** the second term of a ratio. In the ratio 1:4, 4 is the consequent, and 1 the antecedent. **2** the part of a conditional which states the logical conclusion of an antecedent. The proposition *q* in the conditional *p → q* is the consequent. Also called **conclusion.**
—*adj. Geography, Geology.* having a course determined by the original form or slope of the land: *consequent drainage, a consequent river.* Compare **insequent, obsequent, resequent,** and **subsequent.**

conservation *n. Physics.* the preservation of a constant quantity, as of mass, energy, or momentum: *According to the established principles of conservation in physics, energy is never created or destroyed, but is only transferred or transformed* (I. Bernard Cohen).

conservation of charge, *Physics.* the principle that the total electric charge of the universe, or of any closed system, remains constant: *Pair production is an example of conservation of charge* (Physics Regents Syllabus). Also called **charge conservation**

conservation of energy, *Physics.* the principle that the total amount of energy in the universe, or in any closed system, does not vary, although energy can be changed from one form into another: *Because energy and matter are now known to be closely related, the law of the conservation of energy has been interpreted in a new light. Energy cannot be created or destroyed. But it may be developed from matter and turned into matter* (Robert F. Paten and Ralph E. Lapp).

conservation of mass or **conservation of matter,** *Physics.* the principle that the total mass of any closed system remains unchanged by reactions within the system. Thus, in a chemical reaction, matter was thought to be neither created nor destroyed but changed from one form of the substance to another. *Careful weighing before and after chemical changes has led to the law of conservation of mass ... Scientists thought this meant that matter could not be created or destroyed. But atom-smashers can release energy from matter. Scientists now state that the total amount of energy plus the total amount of matter in the universe does not change* (E.A. Fessenden). Compare **conservation of energy.**

conservation of momentum, *Physics.* the principle that the total momentum of any system of revolving or colliding bodies remains constant in the absence of outside forces: *The third law [of motion] is implied by the law of conservation of momentum. The forces act for the same time and are equal in magnitude and opposite in direction; therefore, the total momentum remains the same* (Physics Regents Syllabus).

conservation of parity, *Physics.* the theory that in strong interactions the behavior of a group of particles and that of its mirror image cannot be differentiated in any experiment.

conserve, *v.* Usually, **be conserved.** *Physics.* to keep (some quantity, such as charge or mass) constant; to neither create nor destroy: *Recent evidence seems to confirm that leptons are conserved* (Stuart B. Palmer). See **conservation.**

consistent, *adj. Mathematics.* having at least one common solution, as of two or more equations or inequalities. The equations x + y = 4 and 2x + 2y = 8 are consistent; the equations x + y = 4 and x + y = 7 are inconsistent. The equations x + y = 5 and x − y = 3 are consistent because their graphs have exactly one point in common (i.e., they intersect in exactly one point). The equations 3x − 7y = 21 and 3x − 7y = 0 are inconsistent because their graphs are parallel.

---

cap, fāce, fäther; best, bē, tėrm; pin, fīve;
rock, gō, ôrder; oil, out; cup, pùt, rüle,
yü in use, *yu̇* in uric;
ng in bring; *sh* in rush; *th* in thin; ᴛʜ in then;
zh in seizure.
ə = *a* in about, *e* in taken, *i* in pencil, *o* in lemon, *u* in circus

conspecific

conspecific, *Biology.* —*adj.* of or belonging to the same species. Compare **congeneric.**
—*n.* an animal or plant of the same species as another: *The honeypot is thus much more vulnerable than other ant species to the crushing jaws of its conspecifics, and when death is so likely to result from combat, slavery becomes preferable* (New Scientist). Compare **congener.**

const., *abbrev.* constant.

constant, *adj. Mathematics, Physics.* retaining the same value; remaining the same in quantity, size, or other dimension: *a constant force.*
—*n.* **1** *Mathematics.* a quantity that is invariant throughout a given discussion or calculation.
2 *Physics.* a numerical quantity expressing a relation or value, as of a physical property of a substance, that remains unchanged under certain conditions: *The velocity C of electromagnetic waves in free space is probably the most important fundamental constant known to modern physics* (W.C. Vaughan). *Abbreviation:* c.

constellation, *n. Astronomy.* **1** a group of stars usually having a geometric shape. In the Northern Hemisphere, Ursa Major or the Big Dipper is the easiest constellation to locate.
2 a division of the heavens occupied by such a group: *In modern times the term constellation has come to denote not merely the stars of some conspicuous group but rather the whole area of the sky over which it is spread. Therefore constellations now signify regions upon the celestial sphere* (Krogdahl, *Astronomical Universe*).
[from Late Latin *constellationem,* from Latin *com-* together + *stella* star]

Little Dipper

north star

Big Dipper

constrictor, *n.* **1** *Zoology.* any snake that kills its prey by squeezing the prey with its coils. The boa, anaconda, and python are constrictors.
2 *Anatomy.* a nerve or muscle that compresses or narrows a part of the body: *a constrictor of the blood vessels, the constrictors of the eyelids.*
3 *Medicine, Physiology.* a chemical agent that causes a compression or narrowing.

construct, *v. Geometry.* to draw (a figure) to fulfill given conditions: *to construct a triangle having three given sides.*

constructive interference, *Physics.* the condition of the crests of two waves reaching a point at the same time so that they reinforce one another.

consumer, *n. Ecology.* an organism which ingests other organisms or food particles: *Consumers are parasitic, herbivorous, or carnivorous plants and animals which feed on other plants or animals* (Weier, *Botany*). Con-

trasted with **producer** and **decomposer.** Compare **primary consumer, secondary consumer,** and **tertiary consumer.**

contact, *n. Geology.* the surface of the boundary between adjacent rocks.

contact inhibition, *Biology.* the ceasing of cell division when the surface of one cell comes into physical contact with the surface of another cell: *It has been suggested ... that one of the factors in the control of cell division is "contact inhibition."* (Scientific American).
—**contact-inhibited,** *adj.: When normal cells are grown until they form a layer a single cell thick over the surface ... (the saturation density), cell division decreases dramatically and the cells are said to be contact-inhibited* (Nature).
—**contact-inhibitable,** *adj.: The study of surface differences between cells that are contact-inhibitable and cells that are not is being hotly pursued because one key to the governance of the cell cycle may be found here* (Daniel Mazia).

contact metamorphism, *Geology.* metamorphism caused by the heat of hot igneous rocks, often occurring at the borders of intrusions.

contagion (kən tā′jən), *n. Medicine.* the transmission of disease by direct or indirect contact between individuals. [from Latin *contagionem* a touching]
▶ Although *contagion, infection,* and *contamination* are often used in similar contexts and therefore seem to be interchangeable, these words and their corresponding derivatives have clearly distinguishable meanings. *Contagion* and *contagious* refer to the transmission of viral or bacterial agents by contact with a diseased individual, his bodily secretion, or objects touched by him, as in *Colds are contagious* (i.e., caught by contact). *Infection* and *infectious* refer to the invasion of the body by disease-causing bacteria, viruses, and the like, without reference to their transmission, as in *Hepatitis is an infectious viral disease* (i.e., caused by viral infection, whether contagious or not). *Contamination* refers in medicine to the imparting of disease through contact or association with disease-carrying agents or organisms found in dirt, sewage, etc.
—**contagious,** *adj.* transmissible or communicable by contact: *contagious dermatitis.*

contain, *v. Mathematics.* to be divisible by; be divisible by without a remainder: *12 contains 1, 2, 3, 4, 6, and 12.*

contaminant, *n.* something that contaminates: *Great advances have been made in methods of conditioning the air of museums and galleries to give constant temperature and relative humidity and to remove dust and noxious contaminants* (Harold J. Plenderleith).
▶ A *contaminant* is sometimes differentiated from a *pollutant* in that it usually refers to something impure added to the environment by man, whereas a pollutant may be natural, such as pollen or volcanic ash. Others distinguish between the two terms by considering all impurities contaminants but only harmful ones pollutants. In general, however, the two words are used synonymously though not always interchangeably.

contaminate, *v. Medicine, Biology.* to make impure or diseased by contact with dirt, germs, radioactivity, etc.: *Sicknesses caused by the consumption of contaminated water are the principal cause of infant and adult*

126

mortality in most of the Latin-American countries (Rómulo Betancourt). Compare **pollute.**

—contamination, *n.* **1** a contaminating or a being contaminated: *The contamination from oil, sewage, and the everyday products of modern life (such as pesticides) far transcend in potential danger the contamination from radioactivity* (Karl K. Turekian). **2** a thing that contaminates.

▶ See the note under **contagion.**

continent, *n. Geography.* one of the seven great masses of land on the earth. The continents are North America and South America, Europe, Africa, Asia, Australia, and Antarctica. *The deep oceans and the continents are different in their geological structure* (Gaskell and Hill).

—continental, *adj.* of, having to do with, or characteristic of a continent: *A branched crack that encircles the globe ... divides continental crust on one side from oceanic crust on the other* (Science Journal).

continental drift, *Geology.* the slow movement of the earth's landmasses (continental plates), thought to be caused by pressure that shifts them across the underlying molten material: *The main economic importance of continental drift is the way in which it controls the distribution of the Earth's surface minerals* (D. and M. Tarling, *Continental Drift*).

continental glacier, *Geology.* a sheet of ice that covers a large part of a continent, as that which covers Antarctica.

continentality, *n. Meteorology, Ecology.* the effect of a large landmass on that area's climate. The effects of continentality are found in the middle to high latitudes and produce large annual temperature ranges. *He lays stress on the distribution of land and water within the zones; climate is greatly affected by continentality and oceanity* (Nature).

continental plate, *Geology.* a crustal plate upon which a continent rests.

continental rock, *Geology.* a rock formed by processes operating on a continent. Contrasted with **marine rock.**

continental shelf, *Geology.* the shallow platform that slopes gradually out from a continent to a depth of about 100 fathoms and ends in an abrupt descent to deeper water: *If a great tidal wave would suck away the water and expose a typical continental shelf to our view, it would look like a mass of sand dunes dotted with myriads of small depressions* (Francis P. Shepard).

continental shield, *Geology.* the strongly eroded rock masses that make up the ancient cores of the present continents.

continental slope, *Geology.* the slope that extends downward from the outer edge of the continental shelf to the abyssal plain.

continental terrace, *Geology.* the continental shelf and the continental slope together: *The continental terrace ... at its seaward edge ... pitches steeply into the deep ocean basin* (Scientific American).

continued fraction, *Mathematics.* a fraction whose nu-

merator is a whole number and whose denominator is a whole number plus a fraction which has a denominator composed of a whole number plus a fraction, and so on. The value of a continued fraction may be infinite or finite.

continuity, *n. Mathematics.* **1** the property of a line, curve, or the like that extends without a break or irregularity. **2** the property of a continuous function.

continuous function, *Mathematics, Statistics.* a function which changes systematically in value as the value of the function's independent variable is charged.

continuous phase, *Chemistry.* the medium in which the particles (the disperse phase) of a colloid are distributed. Also called **dispersion medium.**

continuous spectrum, *Physics.* a spectrum whose source emits light of every wavelength in a continuous band: *Continuous spectra are produced by incandescent solids and liquids and by incandescent gases under extremely high pressure* (Physics Regents Syllabus). Compare **bright-line spectrum** and **dark-line spectrum.**

continuous variation, *Biology.* the occurrence of many small differences or intermediates between members of the same species: *Continuous variations are small additions or diminutions of certain parental characters. Discontinuous variations are sudden marked variations* (Youngken, *Pharmaceutical Botany*).

continuous wave, *Physics.* a radio or radar wave whose oscillations occur in a regular, uninterrupted pattern, at a constant amplitude and frequency.

continuum (kən tin′yü əm), *n., pl.* **-tinua** (-tin′yü ə). **1** *Physics.* a continuous whole or other quantitative concept of which the parts are indistinguishable except by reference to something outside of itself, especially the four-dimensional space-time continuum, within which it is possible to "identify" or "locate" events only by reference to three spatial coordinates and the temporal coordinate (the fourth dimension). **2** *Mathematics.* the set of all real numbers, rational and irrational. [from Latin *continuum,* neuter of *continuus* uninterrupted]

contour feathers, *Zoology.* the outer feathers which determine the contour of a bird's body.

contour interval, *Geography.* the difference in height, measured vertically, between the contour lines on a map. Intervals are usually regular and uniform, as every 100, 500, or 1000 feet.

contour line, *Geography.* a line on a map, showing height above or below sea level. All points on a contour line have the same elevation.

---

cap, fāce, fäther; best, bē, tèrm; pin, five; rock, gō, ôrder; oil, out; cup, pùt, rüle, *yü* in use, *yu* in uric; *ng* in bring; *sh* in rush; *th* in thin, ŦH in then; *zh* in seizure. ə = *a* in about, *e* in taken, *i* in pencil, *o* in lemon, *u* in circus

contour map, *Geography.* a map showing heights at regular intervals above sea level by means of contour lines.

contour map

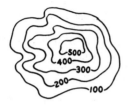

contractile (kən trak′təl), *adj.* **1** capable of contracting: *Muscle is contractile tissue.* **2** producing contraction: *Cooling is a contractile force.*
—**contractility** (kon′trak til′ə tē), *n.* the ability to contract: *If gelation is prevented ... protoplasmic contractility is lost and the cell becomes immobile* (Scientific American).

contractile vacuole, *Biology.* a vacuole in one-celled organisms that discharges its fluid by contracting: *As water diffuses through the protoplasm it takes up soluble waste products, and accumulates to form the contractile vacuole* (Harbaugh and Goodrich, *Fundamentals of Biology*). Compare **food vacuole.**

contraction, *n. Anatomy.* the shortening and thickening of tissue that permits a muscle to pull, compress, constrict, or otherwise move some part of the body.

contrapositive, *n. Mathematics.* the statement or proposition "if not *q*, then not *p*" (-*q*-*p*), in which the antecedent and the consequent of a conditional are reversed and negated. The conditional *p*→*q* and its contrapositive are logically equivalent.

control, *n.* **1** a standard of comparison for testing the results of an experiment, especially an individual or group participating in an experiment under the same conditions as a similar individual or group except for the exclusion of one factor or variable: *In the student group the age range was 18–37 years. The intention was that subjects in the latter group should act as experimental 'controls'* (P.L. Short).
2 Also called **control experiment.** an experiment made to verify the results of another experiment or experiments, using the same condition except for one factor or variable.

control group, a group which serves as a standard of comparison for testing the results of a scientific experiment on an equivalent experimental group.

control rod, *Nuclear Physics.* a mechanism containing neutron-absorbing material, used in a nuclear reactor to control the rate of a chain reaction: *Rare earths can be used in control rods for nuclear reactors* (Science News Letter).

conus (kō′nəs), *n., pl.* **coni** (-kō′nī). *Anatomy.* **1** a conical structure or organ in an animal. **2** = conus arteriosus. [from Latin *conus* cone]

conus arteriosus (är tir′ē ō′səs), *pl.* **coni arteriosi** (är-tir′ē ō′sī). *Anatomy.* **1** the anterior-most chamber of the embryonic heart in vertebrate animals. It is retained as a cone-shaped extension of the single ventricle in some adult fishes and amphibians, but in many cases it is not recognizable as a distinct chamber. **2** a conical structure in the upper left portion of the right ventricle in humans, supplying the pulmonary artery.

convection, *n.* **1** *Physics.* the transfer of heat from one place to another by the circulation of currents of heated particles of a gas or liquid: *The term convection is applied to the transfer of heat from one place to another by the actual motion of hot material. The hot-air furnace and the hot-water heating system are examples* (Sears and Zemansky, *University Physics*).
2 *Meteorology.* the rapid upward movement of air which occurs through strong heating of the earth's surface and supportive atmospheric instability.
[from Latin *convectionem*, from *convehere* carry together, from *com-* + *vehere* carry]
ASSOCIATED TERMS: see **conduction.**
—**convective,** *adj.* having to do with or resulting from convection: *convective heating.*
—**convector,** *n.* a device for heating air by convection.

converge
(def. 1)

The two lines converge at P.

converge, *v.* to cause or tend to meet in a point. A **converging lens** is one that is thicker at the middle than at the edges and converges parallel rays of light to refract and meet at a focus.
—**convergence,** *n.* **1** the act or process of converging; tendency to meet in one point: *a lens which increases the convergence of a beam of light.*
2 *Physiology.* the turning inward of the eyes in focusing on something very close to them.
3 *Meteorology.* the converging of air into a particular region.
4 *Biology.* the tendency in organisms not closely related to develop similar characteristics when living under the same conditions. See **convergent evolution** below.
5 *Geology.* movement of the earth's plates toward each other.

converging plate boundary
(def. 5)

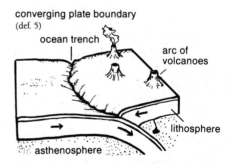

The plates move toward each other causing the surface to fold and to rise at the plate boundary.

—**convergent,** *adj.* converging; formed by convergence. **Convergent evolution** is the appearance of similar characteristics in organisms not closely related to one another: *A striking example of convergent evolution is the present similarity between cactus plants of* southwestern United States and certain Euphorbia species of Africa (Weier, *Botany*). Compare **adaptive convergence**.

convex (kon'veks), *adj.* curved out like the outside of a circle or sphere; curving out: *a convex mirror or lens.* Contrasted with **concave**. [from Latin *convexus*] —**convexity**, *n.* **1** a convex condition or quality. **2** a convex surface or thing.

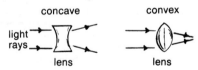

concave   convex

light rays   lens   lens

convexo-concave (kən vek'sō kon kāv'), *adj. Optics.* convex on one side and concave on the other. In a convexo-concave lens, the convex face has the greater curvature.

convexo-convex, *adj. Optics.* convex on both sides.

convexo-plane, *adj. Optics.* convex on one side and flat on the other.

convolute (kon'və lüt), *adj. Botany, Zoology.* rolled up into a spiral shape with one part over another, as the leaves in a bud or the whorls of certain sea shells. —*v.* to coil; twist. [from Latin *convolutus* rolled up, from *com-* + *volere* to roll] —**convolutely,** *adv.*
—**convolution** (kon'və lü'shən), *n. Anatomy.* an irregular fold or ridge on the surface of the brain: *the convolutions of the left cerebral hemisphere.*

coordinate, (*n., adj.:* kō ôr'də nit; *v.* kō ôr'də nāt), *n.* **1** *Mathematics.* any of two or more numbers that define the position of a point, line, or plane by reference to a fixed figure, system of lines, etc. See **Cartesian coordinate.**
2 *Geography.* the longitude or latitude of a place on the earth's surface.
—*adj.* **1** *Mathematics.* having to do with or involving the use of coordinates. A **coordinate system** is any method of representing points on a line, in space, etc., by coordinates.
2 *Chemistry.* in which one atom shares two electrons with another atom: *a coordinate bond.*
—*v. Physiology.* to act in coordination: *muscles that coordinate.*
—**coordination,** *n.* **1** *Physiology.* harmonious adjustment or working together of muscles or groups of muscles.
2 *Chemistry.* **a** the formation of coordinate bonds. **b** the group or number of atoms bound to the central atom by coordinate bonds.

Coordinated Universal Time, the international standard on which most nations base their civil time. It is derived from the scientific standard, *International Atomic Time* (TAI). *Leap seconds* are inserted in Coordinated Universal Time as needed to compensate for the gradual and irregular slowdown of the Earth's rotation; hence it is roughly coordinated with the Earth's rotation, while lagging behind TAI by a whole number of seconds that gradually increases. *Abbreviation:* UTC

coordinate geometry, = analytic geometry.

Copernican system, a system of astronomy based on the theory developed by Nicolaus Copernicus, 1473-1543, that the earth rotates on its axis and that the planets revolve around the sun: *In the Copernican system the sun was stationary at the center* (Baker, *Astronomy*). Compare **Ptolemaic system.**

coplanar (kō plā'nər), *adj. Geometry.* (of points, lines, figures) lying in the same plane. A circle is a set of coplanar points.
—**coplanarity** (kō'plā nar'ə tē), *n.* the quality or fact of being coplanar.

copolymer (kō pol'i mər), *n. Chemistry.* a compound formed by polymerizing two or more unlike substances: *Sometimes a mixture of two monomers, A and B, is polymerized [in] ... a chain such as -A-B-B-A-B-A-A-B-A- ... called a copolymer. Bakelite ... is a copolymer of phenol and formaldehyde* (Baxter and Steiner, *Modern Chemistry*).
—**copolymerize** (kō pol'i mə rī'), *v.* to polymerize two or more unlike substances: *Styrene-butadiene rubber is made by copolymerizing styrene and butadiene* (James S.Fritz).
—**copolymerization** (kō pol'i mə ri zā'shən), *n.* the act or process of copolymerizing: *The present fully synthetic fibers have low water absorption, which makes dyeing difficult and expensive and the feel of the fabric uncomfortable. This can be overcome by ... copolymerization, in which two dissimilar polymers, one supplying strength and fatigue resistance, the other, water absorption to the fiber, are combined* (Milton Harris and John F. Krasny).

copper, *n. Chemistry.* a tough, reddish-brown, ductile metallic element which occurs in various ores. Copper resists rust and is an excellent conductor of heat and electricity. *Symbol:* Cu; *atomic number* 29; *atomic weight* 63.54; *melting point* 1083°C; *boiling point* 2595°C; *oxidation state* 2,1.

copperas (kop'ər əs), *n. Chemistry.* a green, hydrated sulfate of iron, used in dyeing, in making ink, as a disinfectant, and in photography. *Formula:* $FeSO_4 \cdot 7H_2O$ [from Old French *couperose*, perhaps from Medieval Latin (*aqua*) *cuprosa* (water) of copper]

copper oxide, 1 = cupric oxide. **2** = cuprous oxide.

copper sulfate, *Chemistry.* a poisonous, blue, crystalline compound of copper and sulfuric acid, used in dyeing and printing, in electric batteries, and in sprays to destroy insects, fungi, etc. *Formula:* $CuSO_4 \cdot 5H_2O$ Also called **blue vitriol, cupric sulfate.**

copro-, *combining form.* excrement; dung, as in *coprolite, coprophagous.* [from Greek *kopros*]

coprolite (kop'rə līt), *n. Paleontology.* a stony, roundish fossil, the petrified excrement of a vertebrate.

cap, fāce, fäther; best, bē, tèrm; pin, fīve; rock, gō, ôrder; oil, out; cup, pùt, rüle, *yü* in use, *yu* in uric; *ng* in bring; *sh* in rush; *th* in thin, ᴛʜ in then; *zh* in seizure.
ə = *a* in about, *e* in taken, *i* in pencil, *o* in lemon, *u* in circus

coprophagous (kop rof'ə gəs), *adj. Zoology.* feeding on dung, as certain beetles. [from *copro-* + Greek *phagein* to eat]
—**coprophagy** (kop rof'ə gē), *n.* the condition of being coprophagous.

coprophilous (kop rof'ə ləs), *adj. Botany, Zoology.* living or growing on dung. [from *copro-* + Greek *philos* loving]

coquina (kō kē'nə), *n. Geology.* a soft, porous, whitish limestone. Coquina is a detrital limestone composed of fragments of sea shells and corals. [from Spanish *coquina* shellfish]

coracoid (kôr'ə koid *or* kor'ə koid), *Anatomy.* —*n.* **1** a beak-shaped bone between the shoulder blade and the breastbone in birds and reptiles. **2** a bony process extending from the shoulder blade to or toward the breastbone in mammals.
—*adj.* of this bone or bony process.
[from Greek *korakoeidēs* like a raven]

coral, *n. Zoology.* **1** a stony substance, mainly calcium carbonate, consisting of the skeletons of certain polyps which live in colonies in warm seas. Reefs and small islands consisting of these cnidarian coral are common in tropical seas and oceans. *Darwin saw many of the Pacific atolls ... and concluded that the coral had grown on a sinking volcano* (T.F. Gaskell and M.N. Hill).
2 any of the skeletons forming this substance.
3 a polyp that secretes a skelton of coral. Such cnidarian organisms form large branching or rounded colonies by budding.
[from Latin *corallum,* from Greek *korallion*]
—**coralline** (kôr'ə līn), *adj.* of, having to do with, or like a coral.
—*n. Zoology, Botany.* an organism resembling coral, such as some red algae whose fronds contain calcium carbonate.

corallite (kôr'ə līt), *n. Zoology.* **1** a fossil coral. **2** the coralline skeleton of a polyp.

corallum (kə ral'əm), *n., pl.* **-la** (-lə). *Zoology.* the skeleton of a compound coral, consisting of calcareous individual corallites. [from Latin *corallum* coral]

coral reef, *Geology.* a reef consisting mainly of coral: *Coral reefs are built up from the bottom in tropical areas by two different groups ... aided by a number of other lime-depositing organisms* (Shull, *Principles of Animal Biology*).
ASSOCIATED TERMS: see barrier reef.

cord, *n. Anatomy.* a nerve, tendon, or other structure in an animal body that is somewhat like a thick string: *the spinal cord, vocal cords, umbilical cord.*

cordate (kôr'dāt), *adj. Botany, Zoology.* heart-shaped; generally rounded, but pointed at one end and indented at the other: *cordate shells.* Compare **obcordate.** [from Latin *cordem* heart]

cordillera (kôr dil'ər ə *or* kôr'də lyer'ə), *n. Geology, Geography.* **1** a system of roughly parallel mountain ranges; chain of mountains.
2 Often, **Cordilleras,** *pl.* **a** the main mountain system of a continent: *the Cordilleras of the Andes.* **b** the system of mountain ranges extending from Alaska to Cape Horn. [from Spanish *cordillera*]

—**cordilleran,** *adj.* of or in a cordillera or cordilleras.

core, *n.* **1** *Botany.* the hard central part, containing the seeds, of fruits like apples and pears.
2 *Electricity.* a bar of soft iron or winding of iron wires forming the center of an electromagnet, induction coil, etc., and serving to increase and concentrate the induced magnetic field.
3 *Chemistry.* the nucleus of an atom and the inner shells of electrons: *The protons, hearts of hydrogen atoms, are used to probe the complex structure of atomic cores* (Science News Letter).
4 *Geology.* the central or innermost portion of the earth, below the mantle, probably consisting of iron and nickel. It is usually divided into a liquid *outer core* having a radius of about 2250 kilometers, and a solid *inner core* about 1300 kilometers in radius.
5 *Nuclear Physics.* the part of a nuclear reactor containing fissionable material: *The core of an atomic reactor ... is surrounded by a reflector, made of beryllium or graphite, which serves to slow down the neutrons and to prevent their loss from the reactor* (W. Kenneth Davis).
6 *Geology, Oceanography.* a cylindrical portion of rock or other material extracted from the center of a mass by cutting or drilling.

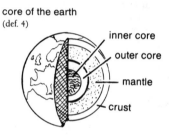

core of the earth (def. 4)

inner core
outer core
mantle
crust

Coriolis force *or* **Coriolis effect** (kôr'ē ō'lis), *Physics.* an imaginary force which affects any body moving on a rotating surface, acting at right angles to the body's direction of motion. It is simply an effect of inertia. The Coriolis forces due to the earth's rotation cause objects in motion above the earth's surface to veer to the right in the Northern Hemisphere and to the left in the Southern Hemisphere. [named after Gaspard G. *Coriolis,* 1792-1843, French mathematician, who analyzed it]

corium (kôr'ē əm), *n., pl.* **coria** (kôr'ē ə). = dermis. [from Latin *corium* leather, skin, hide]

cork, *n. Botany.* the outermost layer of the bark of woody plants: *Cork ... is a protective tissue composed of cells of tabular shape, whose walls possess a layer of waterproof substance called suberin* (Youngken, *Pharmaceutical Botany*). See the picture at **bark.**

cork cambium, *Botany.* a layer of cellular tissue or secondary meristem forming cork cells toward the outside and phelloderm toward the inside of the stem and root of many plants. Also called **phellogen.** See the picture at **bark.**

corm, *n. Botany.* a fleshy, bulblike underground stem of certain plants, such as the crocus and gladiolus, that produce leaves and buds on the upper surface and roots on the lower. A corm has smaller and thinner leaves than a bulb and it consists mostly of stem tissue.

Compare **tuber.** [from New Latin *cormus*, from Greek *kormos* stripped tree trunk]

cornea (kôr′nē ə), *n. Anatomy.* the transparent part of the outer coat of the eyeball. The cornea covers the iris and the pupil. *The human eye is ... surrounded by a tough, white, outer skin, called the sclera ... At the front of the eye, the sclera is replaced by a tough transparent membrane, called the cornea* (Hardy and Perrin, *Principles of Optics*). [from Medieval Latin *cornea* (*tela*) horny (web), from Latin *cornu* horn]
—**corneal** (kôr′nē əl), *adj.* of or having to do with the cornea: *Corneal transplants are usually successful because the cornea does not have a vascular or lymphatic system and thus there is little or no contact with the antibodies in the blood stream* (Arthur J. Snider).

cornicle (kôr′nə kəl), *n. Zoology.* **1** any small, hornlike organ or process, as one of the projections or antennae of a snail or an insect. **2** = siphuncle (def. 2). [from Latin *corniculum,* diminutive of *cornu* horn]

corniculate (kôr nik′yə lit), *adj. Botany, Zoology.* having pointed projections like horns; horned. The **corniculate cartilage** is a hornlike nodule on the tip of an arytenoid cartilage.

cornification, *n. Physiology, Zoology.* the process by which squamous epithelial cells are converted into horny material in the epidermis, nails, feathers, etc.

cornu (kôr′nyü), *n., pl.* **-nua** (-nyü ə). *Anatomy.* a horn or hornlike part or process. [from Latin *cornu* horn]
—**cornual** (kôr′nyü əl), *adj.* of a cornu or cornua; hornlike.

corolla (kə rol′ə), *n. Botany.* the petals or floral leaves of a flower, usually of some color other than green: *The corolla is the inner floral envelope, usually delicate in texture, and showing more or less brilliant colors and combinations of color* (Youngken, *Pharmaceutical Botany*). [from Latin *corolla* garland, diminutive of *corona* crown]

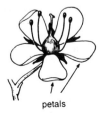

corolla

petals

corollate (kôr′ə lāt), *adj. Botany.* having or resembling a corolla.

corona (kə rō′nə), *n., pl.* **-nas, -nae** (-nē). **1** *Astronomy.* **a** a ring of light seen around the sun, moon, or other luminous body. It is usually colored and is caused by diffraction produced by thin clouds or mist in the earth's atmosphere. **b** the very hot outer atmosphere of the sun, visible as a halo of light during a solar eclipse. *Beyond the chromosphere the sun's corona, a diffuse gaseous envelope, stretches several million miles into space* (Scientific American). See the picture at **sun.**
2 *Botany.* a crownlike appendage on the inner side of the corolla in some flowers, such as the daffodil.
3 *Zoology.* the ciliated, retractile area at the anterior end of a rotifer.

4 *Electronics.* a discharge of electricity that appears as a bluish glow at the surface of a conductor at a high voltage. It is due to the ionization of the air by the voltage.
[from Latin *corona* crown]

corona (def. 1)

—**coronal** (kôr′ə nəl), *adj.* **1** of or having to do with a corona: *coronal gases.* **2** *Anatomy.* of, adjoining, or paralleling the coronal suture: *the coronal plane.*

coronagraph (kə rō′nə graf), *n. Astronomy.* a telescope or an attachment for a telescope equipped with a disk that blacks out most of the sun's corona, used for observing the sun's corona: *In the coronagraph, an artificial "moon" is arranged in such a way as to block out the solar disk, thus producing an artificial eclipse* (Steven Moll).

coronal suture, *Anatomy.* a suture extending across the skull between the frontal bone and the parietal bones.

coronary, *Aantomy.* —*adj.* of or having to do with the coronary arteries. **Coronary thrombosis** is the formation of a thrombus or blood clot in a coronary artery. A **coronary occlusion** is a blockage of a coronary artery that can result in a fatal cutting off of the blood supply to the heart muscle.
—*n.* **1** a coronary thrombosis or occlusion. **2** = coronary artery. **3** = coronary vein.

coronary artery, *Anatomy.* either of the two arteries that supply blood to the muscular tissue of the heart: *The coronary arteries arise from the base of the aorta and conduct blood back into the walls of the heart* (McElroy, *Biology and Man*).

coronary sinus, *Anatomy.* a venous sinus that opens into the right atrium of the heart and serves to drain most of the heart's veins.

coronary vein, *Anatomy.* any of the veins that drain blood from the heart and empty into the coronary sinus.

coronavirus, *n. Biology.* any of a group of spherical viruses with many minute projections which suggest the solar corona. The coronavirus group includes those for the common cold, avian infectious bronchitis, and mouse hepatitis.

coronoid (kôr′ə noid), *adj. Anatomy.* of or denoting any of various bony processes shaped like a crow's beak, as one of the lower jaw: *The coronoid process on the lower jaw, to which the main biting muscles are at-*

cap, fāce, fäther; best, bē, tèrm; pin, fīve;
rock, gō, ôrder; oil, out; cup, pùt, rüle,
yü in use, *yu* in uric;
ng in bring; *sh* in rush; *th* in thin; ᴛʜ in then;
zh in seizure.
ə = *a* in about, *e* in taken, *i* in pencil, *o* in
lemon, *u* in circus

131

tached, is remarkably small (R. F. Ewer). [from Greek *korōnē* crow + English *-oid*]

corotate, *v. Astronomy.* to rotate with, or at the same time or rate as, another body.
—**corotation,** *n.* the act or process of corotating: *How far out do you go before corotation of the sun and corona changes; that is, before the corona starts to drag behind* (Science News).

corpus (kôr'pəs), *n., pl.* **corpora** (kôr'pər ə). *Anatomy.*
1 the main body, mass, or part of a structure or organ of the body, such as the *corpus femoris,* the shaft of the femur, or the *corpus mammae,* the main mass of a mammary gland.
2 any body or organ of special character or function, such as the corpus allatum.
[from Latin *corpus* body]

corpus allatum (ə lā'təm), *pl.* **corpora allata** (ə lā'tə). *Zoology.* a gland lying behind the brain of an insect, secreting hormones that are involved in molting and metamorphosis. [from New Latin *corpus allatum* applied body]

corpus callosum (kə lō'səm), *pl.* **corpora callosa** (kə-lō'sə). *Anatomy.* the transverse band of nerve fibers connecting the cerebral hemispheres in man and other mammals: *A function was discovered for the corpus callosum, a large area of the brain; it has to do with integration of what is seen with one eye with what is seen from the other* (Science News Letter). [from New Latin *corpus callosum* callous body]

corpus cardiacum (kär dī'ə kəm), *pl.* **corpora cardiaca** (kär dī'ə kə). *Zoology.* either of a pair of organs lying at the back of the head of an insect, functioning with the corpus allatum and the glands of the prothorax. [from New Latin *corpus cardiacum* cardiac body]

corpuscle (kôr'pə səl), *n. Biology.* any of the cells that form a large part of the blood, lymph, etc. Red corpuscles carry oxygen to the tissues and remove carbon dioxide; some white corpùscles destroy disease-causing bacteria. [from Latin *corpusculum,* diminutive of *corpus* body]
—**corpuscular** (kôr pus'kyə lər), *adj.* of, like, or consisting of corpuscles. SYN: cellular.

corpuscular theory, *Physics.* the theory that light consists of a stream of particles: *By the middle of the 17th century, while most workers in the field of optics accepted the corpuscular theory, the idea had begun to develop that light might be a wave motion of some sort* (Sears and Zemansky, *University Physics*).

corpus luteum (lü'tē əm), *pl.* **corpora lutea** (lü'tē ə). *Physiology.* a yellow endocrine mass formed in the ovary from the ruptured sac (Graafian follicle) left behind after the release of a mature ovum: *The corpus luteum ... produces a hormone, progesterone. Progesterone maintains the growth of the mucous lining of the uterus* (Otto and Towle, *Modern Biology*). [from New Latin *corpus luteum* yellow body]

corpus striatum (strī ā'təm), *pl.* **corpora striata** (strī-ā'tə). *Anatomy.* either of two bodies of nerve fibers in the brain, each forming part of the undersurface of a cerebral hemisphere. [from New Latin *corpus striatum* striated body]

corrade (kə rād'), *v. Geology.* to wear down by corrasion. [from Latin *corradere* scrape together, from *com-* together + *radere* to scrape]

corrasion (kə rā'zhən), *n. Geology.* the scraping away of rock by the action of rock fragments moved by wind or water: *All scraping, grinding and scouring, sometimes called abrasion and corrasion, are weathering processes* (Finch and Trewartha, *Elements of Geography*).
ASSOCIATED TERMS: see **erosion.**
—**corrasive** (kə rā'siv), *adj.* of or characterized by corrasion.

correlation, *n.* **1** *Biology.* a mutual relation between different structures, functions, or characteristics in an animal or plant: *Although correlations are basically under hereditary control, in plants they operate through at least two main sets of internal factors: food and hormones* (Greulach and Adams, *Plants*).
2 *Statistics.* the relation or interdependence between two or more variables, as two quantities. The **correlation coefficient** is a number between −1 and 1 that indicates the degree of correlation between two quantities or sets of data. A correlation coefficient of 1 indicates a high positive correlation (for example, as one quantity increases, so does the other); a correlation coefficient of or close to zero indicates no correlation; a correlation coefficient of −1 indicates negative correlation (for example, as one quantity increases, the other decreases).
3 *Geology.* the determination of spatial and temporal relationships between geological features.
—**correlational,** *adj.* having to do with or using correlation.

correlation ratio, *Statistics.* a measure of the correlation between two sets of variables whose correlation cannot be plotted as a straight line.

correspondence, *n. Mathematics.* a one-to-one relationship between sets of objects: *A correspondence exists between the set of all seats and the set of all audience members in a theater filled to capacity.*

corrode, *v. Chemistry.* to wear away gradually by chemical action: *Moist air corrodes iron.*
[from Latin *corrodere* to gnaw or chew up, from *com-* altogether + *rodere* gnaw]
—**corrosion,** *n.* **1** the act or process of corroding; a wearing away or destruction of metals or alloys by a chemical agent or process: *Basically, corrosion occurs when positive metal ions are removed from the crystal formation in which they are loosely bound electrically and combine with negative ions in the environment, such as oxygen* (Science News Yearbook).
2 a corroded condition.
3 a product of corroding, such as rust.
—**corrosive,** *adj.* producing corrosion; tending to corrode: *acids and other corrosive materials.*
—*n.* a substance that corrodes: *Most acids are corrosives.*

corrosive sublimate, = mercuric chloride.

corrugator (kôr'ə gā'tər), *n. Anatomy.* a muscle that causes the skin to contract into wrinkles, especially one of two small muscles which contract the brow in the action of frowning. [from Latin *corrugatum* wrinkled, from *com-* together + *ruga* wrinkle]

132

cortex (kôr'teks), *n., pl.* **cortices** (kôr'tə sēz'). *Biology.*
1 the part of the tissue of roots and stems of higher plants which lies outside the vascular tissue and inside the epidermis. In woody plants it gives rise to the outer portion of the bark. *The cortex of a young root may be sloughed off as the root grows older, or it may form part of the bark around the root* (Miller and Leth, *High School Biology*). See the picture at **root**.
2 an outer layer of cells in certain algae, fungi, and lichens.
3 the outer layer or wall of an internal organ or structure, such as of the kidney or the adrenal glands. The **adrenal cortex** secretes cortisone and other important hormones.
4 the layer of gray matter which covers most of the surface of the brain: *The cortex is the seat of all conscious sensations and actions, memory, the will, and inteligence* (Storer, *General Zoology*). Also called **cerebral cortex**.
5 a layer of tightly packed cells in a shaft of hair. Compare **medulla**. See the picture at **hair**.
[from Latin *cortex* bark]
—**cortical**, *adj.* **1** of, having to do with, or involving the cortex: *cortical hormones.*
2 consisting of cortex: *cortical tissue.*

corticosteroid or **corticoid**, *n. Biochemistry.* any of a group of steroids, many of them hormones, produced by the cortex of the adrenal glands. They are important in metabolism and the ability of the body to withstand stress.

corticosterone (kôr'tə kos'tə rōn), *n. Biochemistry.* a corticosteroid that stimulates the metabolism of carbohydrates and proteins. *Formula:* $C_{21}H_{30}O_4$

cortin, *n. Biochemistry.* a secretion from the cortex of the adrenal glands containing cortisone and other hormones.

cortisone (kôr'tə sōn), *n. Biochemistry.* a steroid hormone obtained from the cortex of the adrenal glands or produced synthetically, used in the treatment of arthritis and other ailments. *Formula:* $C_{21}H_{28}O_5$

corundum (kə run'dəm), *n. Mineralogy.* an extremely hard mineral; a native form of alumina. Ruby and sapphire are varieties of corundum colored red and blue respectively by various chromophore ions. A dark-colored variety is used for polishing and scraping. *Formula:* Al_2O_3 [from Tamil *kurundam*]

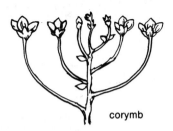
corymb

corymb (kôr'imb), *n. Botany.* a flower cluster whose outer flowers blossom first and have longer pedicels than the inner flowers, so that together they form a round, rather flat cluster on top. Cherry blossoms are corymbs. Compare **cyme** and **raceme**. [from Latin *corymbus*, from Greek *korymbos* top, cluster]
—**corymbose** (kə rim'bōs), *adj.* **1** growing in corymbs.
2 of or resembling a corymb. —**corymbosely**, *adv.*

cos, *abbrev.* cosine.

cosec, *abbrev.* cosecant.

cosecant (kō sē'kənt), *n. Trigonometry.* **1** the ratio of the length of the hypotenuse of a right triangle to the length of the side opposite an acute angle. It is the reciprocal of the sine of an angle.
2 the secant of the complement of a given angle or arc. *Abbreviation:* csc, cosec See also **secant** and **cotangent**.

cosine (kō'sīn), *n. Trigonometry.* **1** the ratio of the length of the side adjacent to an acute angle of a right triangle to the length of the hypotenuse.
2 the sine of the complement of a given angle or arc. *Abbreviation:* cos See also **sine** and **tangent**.

cosine

In the right triangle XYZ, the hypotenuse is XY and:

$$\text{the cosine of angle ZXY} = \frac{ZX}{XY}$$

$$\text{the cosine of angle ZYX} = \frac{ZY}{XY}$$

cosmic, *adj. Astronomy.* **1** of or belonging to the cosmos; having to do with the whole universe: *Cosmic forces produce stars and meteors.*
2 of or belonging to outer space, or the universe outside the earth: *cosmic radiation. The helium of mass 3 is all of cosmic origin* (E. P. George). A **cosmic year** is the period of rotation of the Milky Way, about 220 million years. —**cosmically**, *adv.*

cosmic dust, *Astronomy.* fine particles of matter in outer space, often forming clouds: *A good criterion for recognizing particles of cosmic dust on the Earth is the detection of radionuclides such as aluminium-26, formed by the action of cosmic rays on dust in free space* (Nature).

cosmic rays, *Astronomy, Physics.* radiation of great penetrating power that comes to earth from outer space, consisting of nuclei or nuclear particles traveling at very high speeds: *Cosmic rays have been bombarding the atmosphere for millions of years, and they have been constantly producing new supplies of radiocarbon* (Willard F. Libby).
► Cosmic rays that originate in outer space, such as solar and galactic rays, are usually called *primary cosmic rays* or *primaries*. When primary rays enter the earth's atmosphere, they collide with atomic nuclei and produce *secondary cosmic rays* or *secondaries*. The secondaries are chiefly mesons, such as pions and muons, produced by the collision of protons with oxy-

cap, fāce, fäther; best, bē, tėrm; pin, fīve;
rock, gō, ôrder; oil, out; cup, pùt, rüle,
yü in use, *yù* in uric;
ng in bring; *sh* in rush; *th* in thin, ᴛʜ in then;
zh in seizure.
ə = *a* in about, *e* in taken, *i* in pencil, *o* in lemon, *u* in circus

gen or nitrogen nuclei in the earth's upper atmosphere. A series of such collisions produces many new secondaries, resulting in a *cosmic-ray shower.* Most cosmic rays are deflected by the earth's magnetic field or lose energy and dissipate before reaching earth.

cosmo-, *combining form.* **1** universe, as in *cosmology, cosmochemistry.* **2** cosmic rays, as in *cosmogenic.* [from Greek *kosmos* order, world]

cosmochemistry, *n.* the branch of science dealing with the chemistry of celestial bodies and of the universe as a whole.

cosmogenic, *adj. Physics.* originating from cosmic rays: *cosmogenic isotopes of helium.*

cosmogonic (koz′mə gon′ik), or **cosmogonical,** *adj.* of or having to do with cosmogony.

cosmogony (koz mog′ə nē), *n. Astronomy.* any theory, system, or account of the origin of the earth, the solar system, or the universe: *The first crude steps could now be taken toward a cosmology describing the world at large, and toward a cosmogony accounting for the evolution of its parts* (Harold P. Robertson). [from Greek *kosmogonia,* from *kosmos* world + *gonos* birth]

cosmology (koz mol′ə jē), *n.* a branch of astronomy dealing with the history and structure of the universe as a whole: *The expansion of the entire universe is the most important single hard scientific fact of cosmology* (Allan R. Sandage).
—**cosmological** (koz′mə loj′ə kəl), *adj.* of or having to do with cosmology: *This has been extended to a "perfect cosmological principle" which says that the uniformity of the universe must always hold true, its general appearance being eternally the same* (Herbert Dingle).

cosmos (koz′məs or koz′mos), *n. Astronomy.* the universe, especially as an orderly, harmonious system: *Einstein was convinced that the cosmos is an orderly, continuous unity: gravity and electromagnetism must, therefore, have a common source. ... Einstein persisted: "I cannot believe that God plays dice with the cosmos"* (Time). [from Greek *kosmos* order, world]

costa (kos′tə), *n., pl.* **-tae** (-tē). **1** *Anatomy.* a rib.
2 *Botany.* a rib or primary vein; a midrib of a leaf or frond.
3 *Zoology.* a riblike part, as the anterior vein of an insect's wing or a ridge of a shell.
[from Latin *costa* rib]
—**costal,** *adj.* having to do with a rib or ribs: *costal cartilage.*
—**costate,** *adj.* having a rib or ribs.

costoclavicular, *adj. Anatomy.* of or affecting both the ribs and the collarbone (clavicle).

costoscapular, *adj. Anatomy.* of or affecting both the ribs and the shoulder blade (scapula).

cot, *abbrev.* cotangent.

cotangent (kō tan′jənt), *n. Trigonometry.* **1** the ratio of the length of the adjacent side (not the hypotenuse) of an acute angle in a right triangle to the length of the opposite side. It is the reciprocal of the tangent of an angle.
2 the tangent of the complement of a given angle or arc. *Abbreviation:* cot, ctn See also **secant** and **cosecant.**

cotidal (kō tī′dəl), *adj. Oceanography.* of or having to do with the coincidence in time of tidal phenomena: *A cotidal line on a map connects all those places at which high tide occurs at the same time.*

cotyledon (kot′l ēd′n), *n. Botany.* an embryonic leaf in the seed of a plant; the first leaf, or one of the first pair of leaves, growing from a seed; seed leaf. The number of cotyledons in the seed serves as an important basis of classification of angiosperms into *monocotyledons,* with one cotyledon, and *dicotyledons,* with two. See the picture at **hypocotyl.** [from Latin *cotyledon,* from Greek *kotylēdōn* cup-shaped hollow, from *kotylē* cup]
—**cotyledonary** (kot′l ēd′ner′ē), *adj.* having, or of the nature of, cotyledons.
—**cotyledonous** (kot′l ēd′n əs), *adj.* having cotyledons: *a cotyledonous leaf.*

coulee (kü′lē), *n. Geology.* **1** a deep ravine or gulch that is usually dry in summer. **2** a stream of viscous lava, especially when thick or solidified. [from French *coulée* a flow]

couloir (kü lwär′), *n. Geology.* a steep gorge or gully on the side of a mountain [from French]

coulomb (kü′lom), *n.* the SI or MKS unit of electric charge, equal to the quantity of charge which passes through a cross section of a conductor in one second when the current equals one ampere. *Symbol:* C [named after Charles A. de *Coulomb,* 1736-1806, French physicist]
—**coulombic** (kü lom′ik), *adj.* of or based on Coulomb's law: *coulombic attraction.*

Coulomb's law, *Physics.* the principle that the electrostatic attraction or repulsion between electrically charged bodies is directly proportional to the product of the electric charges on each body, and inversely proportional to the square of the distance between the bodies.

coulometry (kü lom′ə trē), *n. Chemistry.* a method used in microanalysis to determine quantities in solution by measuring the amount of electricity required to effect electrochemical deposition or other chemical change in the dissolved material. [from *coulomb* + *-metry* art of measuring]

countable, *adj.* = denumerable: *[Georg] Cantor called countable those infinite sets that can be written as a sequence X1, X2, X3 ...* (Scientific American).

countershading, *n. Zoology.* protective coloring characterized by the relatively darker coloration of an animal's exposed parts and lighter coloration of its shaded parts: *Countershading protects caterpillars against birds* (Scientific American).

country rock, *Geology.* **1** the rock adjacent to a lode, vein of ore, or dike.
2 a rock enclosing intrusive igneous bodies of any form.

couple, *n. Physics.* **1** a pair of equal, parallel forces acting in opposite directions, and tending to produce rotation. **2** = thermocouple.

covalence (kō vā′ləns) or **covalency** (kō vā′lən sē), *n. Chemistry.* **1** the total of the pairs of electrons which one atom can share with surrounding atoms: *Since carbon shares four pairs of electrons and each hydrogen shares one pair, carbon is said to have a covalence of 4, and hydrogen a covalence of 1* (Offner, *Fundamentals of Chemistry*).

2a the ability to form a bond in which two atoms share a pair of electrons. **b** Also called **covalent bond.** the bond thus formed.

—**covalent** (kō vā′lənt), *adj.* of or having to do with covalence: *When each of the atoms contribute one electron toward a shared pair which is held jointly by both atoms, the compound is known as a covalent compound* (Parks and Steinbach, *Systematic College Chemistry*).

covalent bond, *Chemistry.* a bond formed when electrons are shared by two atomic nuclei. When electrons are shared by atoms of the same element, they are shared equally and the resulting bond is a *nonpolar bond.* When electrons are shared by atoms of different elements, they are shared unequally and the result is a *polar bond.* When the two shared electrons forming a covalent bond are both donated by one of the atoms, the bond is a *coordinate covalent bond.* —**covalent bonding.**

covariance, *n. Statistics.* the average value of the product of the deviations of two variables from their respective average values.

covariant, *adj. Mathematics, Statistics.* varying with another quantity in such a way that the relationship between the two variables remain proportionally the same.

covellite (kō vel′īt), *n. Mineralogy.* a native sulfide of copper, usually occurring in masses of an indigo-blue color. It is a major copper ore. *Formula:* CuS [from Niccolò *Covelli*, 1790-1829, Italian mineralogist + *-ite*]

coverts (kuv′ərts), *n.pl. Zoology.* the smaller and weaker feathers of a bird that cover the bases of the larger feathers of the wing and tail: *All the swans' feathers I've ever seen were fine-ribbed and curly, with very long barbules, except for the wing and tail coverts* (Robertson, *Swan Song*). [from Old French *covert*, past participle of *covrir* to cover]

cow, *n. Zoology.* **1** the full-grown female of domestic cattle. **2** the female of certain other large mammals the male of which is called a bull: *an elephant cow, a buffalo cow.*

Cowper's gland (kou′pərz), *Anatomy.* either of a pair of small glands connected to the male urethra: *The Cowper's glands are each about the size of a pea. They flank the urethra and empty into it through tiny ducts* (Allgeier and Allgeier, *Sexual Interactions*). [named after William *Cowper*, 1666-1709, a British anatomist, who discovered them]

coxa (kok′sə), *n., pl.* **coxae** (kok′sē). **1** *Anatomy.* the hipbone or hip joint.
2 *Zoology.* the short segment of the leg of an arthropod by which the leg is joined to the body.
[from Latin *coxa* hip]
—**coxal** (kok′səl), *adj.* of or having to do with the coxa: *the coxal bone.*

c.p. or **C.P.,** *abbrev.* **1** candlepower. **2** chemically pure.

Cr, *symbol.* chromium.

cranial (krā′nē əl), *adj. Anatomy.* of or having to do with the cranium or skull: *cranial capacity.*

cranial nerve, *Anatomy.* any one of 12 pairs of nerves that leave the brain directly through openings in the skull. Among the cranial nerves are the olfactory, optic, and facial nerves.

craniate (krā′nē it *or* krā′nē āt), *Zoology.* —*adj.* **1** having a skull.
2 having to do with or belonging to a group of vertebrates comprising those which possess a skull and brain. The mammals, birds, reptiles, amphibians, and fishes are craniate animals.
—*n.* a craniate animal.

craniosacral, *adj.* **1** *Anatomy.* of or having to do with both the cranium and the sacrum.
2 *Physiology.* of or designating the parasympathetic nervous system.

cranium (krā′nē əm), *n., pl.* **-niums, -nia** (-nē ə). *Anatomy.* the skull of a vertebrate, especially the part of the skull enclosing the brain: *The human skull may conveniently be divided into three principal portions: the cranium or cranial vault, the face, and the lower jaw or mandible.* (Beals and Hoijer). [from Medieval Latin *cranium*, from Greek *kranion*]

crater, *n.* **1** *Geology.* **a** a bowl-shaped depression around the opening of a volcano. **b** a bowl-shaped hole or depression in the earth. An **impact crater** is a crater produced by the fall of a meteorite.
2 *Astronomy.* a round, ringlike elevation on the surface of the moon or other celestial body, resembling the crater of a volcano.
[from Latin *crater*, from Greek *kratēr* bowl]

crater lake, *Geology.* a deep lake occupying the crater or caldera of a volcano.

craton (krā′ton), *n. Geology.* a major part of the continental crust that has attained tectonic stability. [from German *Kraton*, from Greek *kratos* strength]

craw, *n. Zoology.* **1** the crop of a bird or insect. **2** the stomach of any animal.

C-reactive protein, *Biochemistry.* an abnormal protein that appears in the bloodstream in response to injury or inflammation. Tests for C-reactive protein are used in medicine to diagnose rheumatic fever. *Abbreviation:* CRP [*C*, abbreviation of *carbohydrate* (so called because a carbohydrate found in pneumococci is precipitated by this protein)]

cream of tartar, *Chemistry.* a white powder or crystalline substance obtained from the deposit in wine casks, used in medicine and cooking, especially to make baking powder, in galvanizing metals, and in foods as a buffer. *Formula:* $KHC_4H_4O_6$

creatine (krē′ə tēn), *n. Biochemistry.* a nitrogenous crystalline compound found chiefly in the muscles of vertebrates, involved with supplying energy for voluntary muscle contraction. *Formula:* $C_4H_9N_3O_2$

creatine phosphate, = phosphocreatine.

creatinine (krē at′ə nēn), *n. Biochemistry.* a constituent of urine produced by the breakdown of creatine, also found in blood, muscle, plants, soil, etc. *Formula:* $C_4H_7N_3O$

cap, fāce, fäther; best, bē, tèrm; pin, fīve;
rock, gō, ôrder; oil, out; cup, pùt, rüle;
yü in use, *yu* in uric;
ng in bring; *sh* in rush; *th* in thin, ŦH in then;
zh in seizure.
ə = *a* in about, *e* in taken, *i* in pencil, *o* in lemon, *u* in circus

135

creep, *n.* **1** *Physics.* the slow continuous deformation of a material, such as a metal, from stress or increased temperature: *Creep itself is known to be the outward effect of a multitude of atomic rearrangements which are activated by thermal motion* (M.F. Perutz).
2 *Geology.* slow movement of soil or disintegrated rock down a slope, due to gravity, frost, or ground water: *Creep is the most widespread of all downslope movements ... Distortion caused by creep is shown by many rock bodies whose previous shape or structure is known* (Gilluly, *Principles of Geology*).

cremaster (krə mas′tər), *n.* **1** *Zoology.* a pointed or rounded part with many hooklike claws at the rear of a pupa's body.
2 *Anatomy.* a muscle of the spermatic cord whose contraction elevates the testis.
[from New Latin, from Greek *kremastēr* suspender]

crenate (krē′nāt), *adj. Botany, Zoology.* having a scalloped edge: *Many leaves are crenate.* SYN: notched. [from New Latin *crenatus,* from *crena* notch]
—**crenately,** *adv.*
—**crenation** (kri nā′shən), *n.* **1** a crenate formation; a series of scallops, as on the margin of a leaf or shell, or on the edges of a red blood cell as a result of shrinkage. **2** a process of osmosis whereby red blood cells undergo shrinkage.

crenulate (kren′yə lit), *adj. Botany, Zoology.* minutely crenate, as a leaf or shell. [from Latin *crenula* notch]
—**crenulation,** *n.* **1** a crenulate condition. **2** a minute crenation.

crepuscular (kri pus′kyə lər), *adj. Zoology.* appearing or flying at twilight, as certain birds and insects do. [from Latin *crepusculum* twilight]

crescent, *Astronomy.* —*n.* the shape of the moon in its first or last quarter.
—*adj.* shaped like the moon in its first or last quarter. [from Latin *crescentem* growing, increasing]

crest, *n.* **1** *Anatomy, Zoology.* a ridge or prominence, especially one running along the surface of a bone: *the frontal crest of the skull.*
2 *Physics.* the high part or peak of a wave.

Cretaceous (kri tā′shəs), *Geology.* —*n.* **1** the last geological period of the Mesozoic era, characterized by the formation of chalk deposits. During this period flowering plants began to appear and dinosaurs became extinct.
2 the rocks formed during this period.
—*adj.* **1** of or having to do with the Cretaceous or its rocks: *The Cretaceous System is well set apart as a major division of the geologic column, because important unconformities occur at both the base and the summit in most areas* (Moore, *Introduction to Historical Geology*).
2 cretaceous, like chalk; containing chalk.
[from Latin *cretaceus,* from *creta* chalk]

crevasse (krə vas′), *n. Geology.* a deep crack or crevice in the ice of a glacier, or in the ground after an earthquake: *Irregularities of the valley floor bend and twist the brittle surface of the ice causing it to open deep transverse cracks called crevasses* (Finch and Trewartha, *Elements of Geography*).

[from French *crevasse,* from Old French *crevace* crevice]

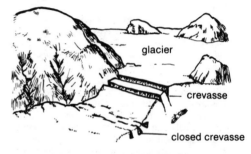

cribellum (krə bel′ləm), *n., pl.* **-bella** (-bəl′ə). *Zoology.* a platelike organ by which some spiders spin additional silk. [from Late Latin *cribellum* small sieve]

cricoid (krī′koid), *Anatomy.* —*adj.* of or denoting the ring-shaped cartilage at the lower part of the larynx.
—*n.* the cricoid cartilage.
[from Greek *krikoeidēs* ringlike]

crinoid (krī′noid), *Zoology.* —*n.* any of a class (Crinoidea) of echinoderms having a small, cup-shaped body with five branched, feathery arms, usually attached by a stalk.
—*adj.* of or belonging to this class.
[from Greek *krinoeidēs* lilylike]
—**crinoidal,** *adj.* of, having to do with, or consisting of crinoids.

crissum (kris′əm), *n. Zoology.* **1** the area surrounding the cloacal opening of a bird.
2 the feathers of this area.
[from New Latin, from Latin *crissare* move the haunches]

crista (kris′tə), *n., pl.* **cristae** (kris′tē). *Biology.* a crestlike part or process, such as one of the folds of the inner membrane of mitochondria: *Cristae in plant mitochondria frequently are sparse, irregularly arranged, and not always seemingly connected with the inner membrane of the envelope* (Weier, *Botany*).
[from Latin *crista* crest]

cristate (kris′tāt), *adj. Biology.* having a crest or crista; crested.

cristobalite (kris tō′bə līt), *n. Mineralogy.* silica occurring in small octahedral crystals as a high-temperature polymorph of quartz. [from Cerro San *Cristóbal,* near Pachuca, Mexico + *-ite*]

crith (krith), *n.* a unit of mass for gases, equal to the mass of one liter of hydrogen at standard pressure and temperature. [from Greek *krithē* barleycorn (the smallest weight)]

critical, *adj.* **1** *Physics, Chemistry.* having to do with or constituting a point at which some action, property, or condition undergoes a decisive change: *Heavy primaries of low energy can pass through a light-beam unscathed, but if their energy is increased, then beyond a certain critical value they will be split up into fragments by ordinary visible light* (E.P. George).
2 *Nuclear Physics.* having to do with or capable of sustaining a chain reaction: *To make a reactor go critical its control rods are withdrawn just enough for a chain reaction to start* (Wall Street Journal). See also **critical mass.**

—**criticality,** *n.* the point at which a nuclear reactor becomes capable of sustaining a chain reaction: *Gradually the core neared criticality and the fuel was added in smaller increments* (New Scientist).

critical angle, *Optics.* the smallest possible angle of incidence that gives total reflection: *The angle of incidence for which the refracted ray emerges tangent to the surface is called the critical angle* (Sears and Zemansky, *University Physics*).

critical constants, *Physics.* the temperature, pressure, and density of a gas above which it cannot be liquefied.

critical mass, *Nuclear Physics.* the smallest amount of fissionable material that will support a self-sustaining chain reaction.

critical point, 1 *Physics.* the point, characterized by temperature and pressure, at which the gaseous and liquid phases of a fluid substance become identical. The state of a substance at the critical point is called the **critical state.** *The end of the vaporization curve is called the critical point.* (Shortley and Williams, *Elements of Physics*). See also **critical temperature.** See the diagram at **phase.**
2 *Mathematics.* a point on a graph at which the derivative of a function is either 0 or infinite.

critical size, = critical mass.

critical temperature, *Physics.* the temperature of a substance at the critical point; the highest temperature at which a gas can be liquefied by pressure alone. The pressure exerted by a substance at the critical temperature is called the **critical pressure.**

critical velocity, *Physics.* the velocity at which the flow of a fluid changes from laminar to turbulent.

crocidolite (krō sid′ə līt), *n. Mineralogy.* an asbestiform variety of riebeckite. [from Greek *krokis, krokys* nap of woolen cloth + *-lite*]

crocoite (krō′kō īt), *n. Mineralogy.* a reddish native chromate of lead occurring in monoclinic crystals. *Formula:* $PbCrO_4$ [from Greek *krokoeis* saffron-colored + *-ite*]

crop, *n. Zoology.* a baglike swelling of a bird's or insect's food passage where food is prepared for digestion. Also called **craw.**

cross, *v., n.* = crossbreed: *Offspring resulting from the cross between members of the F_1 generation comprise the F_2 generation* (Biology Regents Syllabus).

crossbedding, *n. Geology.* a condition in which the bedding planes or original layers of sedimentary rocks are inclined, sometimes at steep angles, to the main stratification: *One kind of crossbedding forms in sand dunes ... Another kind of crossbedding forms in deltas* (Birkeland and Larson, *Putnam's Geology*).
—**crossbedded,** *adj.* characterized by crossbedding: *crossbedded river deposits.*

crossbreed, *Biology.* —*v.* to breed hybrid forms by mixing kinds, breeds, or varieties of plants or animals: *Much of the success of [Luther] Burbank with plants was through crossbreeding. The loganberry is a cross between the blackberry and the raspberry, but is different from either* (Ogburn and Nimkoff).
—*n.* **1** an individual or breed produced by crossbreeding.
2 an act or instance of crossbreeding.
—**crossbred,** *adj.* produced by crossbreeding. SYN: hybrid.

cross-fertilization, *n. Biology.* **1** the fertilization of the ovum of one individual by the sperm of another. Also called **allogamy.** Compare **self-fertilization.**
2 = cross-pollination.
3 the mating of a male of one variety with a female of another.
—**cross-fertilize,** *v.* to undergo or cause to undergo cross-fertilization.

cross hairs, *Optics.* fine strands stretched across the focal plane of an optical instrument for accurately defining the line of sight.

crossing-over, *n.,* or **crossing over,** *Biology.* a mutual exchange of genes between homologous chromosomes during meiosis, leading to a greater variability in the gene combinations: *Crossing over between chromosomes ... gives the possibility of breaking up these linkage groups as they are called, so that new combinations ... arise in the gametes* (Mackean, *Introduction to Biology*).

cross-link, *Chemistry.* —*n.* a long molecular chain joined to another chain at intervals between atomic cores for the purpose of strengthening a material, as rubber in the process of vulcanization.
—*v.* to establish cross-links: *Cross-linking increases strength and toughness in plastic film which has been irradiated* (Science News Letter).
—**cross-linkage,** *n.* the process of establishing cross-links: *Crease resistance and anti-shrink properties in textile fabrics, and increased wet-strength in paper ... were produced by a cross-linkage between adjacent cellulose or protein fibres* (R.F. Homer).

cross-match, *v. Immunology.* to determine the compatibility of a donor's and recipient's blood before transfusion. It is done by placing red cells of the donor and the recipient into the other's serum. If no agglutination occurs after cross-matching, the blood specimens are compatible.

crossover, *n. Biology.* **1** = crossing-over: *Chromosomes that have exchanged parts have undergone a crossover, and the segments involved are recognized by the genes that mark them* (Storer, *General Zoology*).
2 the characteristic inherited by crossing over.

cross-pollination, *n. Botany.* the transfer of pollen from the anther of one flower to the stigma of another. Insects and wind are agents of cross-pollination. *Once a new seed is perfected, growers must continually work to keep it free from unwanted crosspollination or from regression to its natural wild state* (Wall Street Journal). Compare **self-pollination.** See the picture at **pollination.**
—**cross-pollinate,** *v.* to undergo or cause to undergo cross-pollination.

cross product, *Mathematics.* a vector quantity whose length is the product of the lengths of two vectors and the sine of the angle between them. It is indicated by

cap, fāce, fäther; best, bē, tèrm; pin, five;
rock, gō, ôrder; oil, out; cup, pùt, rüle,
yü in use, *yù* in uric;
ng in bring; *sh* in rush; *th* in thin, ᴛʜ in then;
zh in seizure.
ə = *a* in about, *e* in taken, *i* in pencil, *o* in lemon, *u* in circus

137

a cross placed between the vectors, as in $A \times B$. Also called **vector product** or **axial vector.** Compare **dot product.**

cross reaction, *Immunology.* a reaction between an antibody and a nonspecific antigen.

—**cross-react,** *v.* to undergo a cross reaction: *Antibodies directed against antigens closely similar to normal cellular constituents of the body can cross-react with and injure the normal tissues* (Scientific American).

—**cross-reactive,** *adj.* likely to cross-react.

—**cross-reactivity,** *n.*

cross-resistance, *n. Biology.* resistance to a toxic substance which an organism transfers or develops from its resistance to a related substance: *Some of the newer antibiotics are effective against resistant staphylococci ... Some are made useless by "cross-resistance"; that is, the penicillin-resistant staphylococci quickly become resistant to the new drugs as well* (Anthony H. Rose).

—**cross-resistant,** *adj.* having or showing cross-resistance.

cross section, 1 a slice or section of an object made by cutting through it in a plane, usually at right angles to its longest axis: *a cross section of tissue.*
2 *Physics.* a quantity representing the probability of interactions between particles, expressed in units of area. Nuclear cross sections are measured in barns. *The size of the target or the "cross section" for slow neutrons is much greater than for fast ones—in fact it is about 400 times larger* (New Scientist).

crotonic acid (krō ton′ik), *Chemistry.* a colorless, crystalline substance used in organic synthesis. *Formula:* $CH_3CH:CHCO_2H$ [from Greek *krotōn* castor-oil plant]

crown, *n.* **1** *Botany.* **a** the part of a seed plant at which the root and stem join. The crown looks like the top of the root. **b** the corona of a flower. **c** the leaves and branches of a tree or shrub.
2 *Anatomy.* **a** the part of a tooth which appears beyond the gum, covered with enamel. **b** the top of the skull. See the picture at **tooth.**

crown graft, *Botany.* a graft inserted at the crown of the root of the stock.

—**crown grafting,** the act or method of making a crown graft.

CRP, *abbrev.* C-reactive protein.

cruciate (krü′shē it), *adj.* **1** *Botany.* having petals arranged in the form of a cross: *a cruciate flower.*
2 *Zoology.* crossing each other diagonally: *an insect's cruciate wings.*

crunode (krü′nōd), *n. Geometry.* a point at which a curve crosses itself. [from Latin *crux* cross + English *node*]

crus (krus), *n., pl.* **crura** (kùr′ə). *Anatomy.* **1** the leg or hind limb, especially the part between the knee and ankle. **2** any one of various parts, occurring in sets, resembling a leg. [from Latin *crus*]

crust, *n. Geology.* the solid outer layer of the earth: *the crust ... varying widely from a few miles under the ocean to perhaps 40 miles under mountains ... rests on a solid mantle extending halfway toward the center of*

the earth (Joseph Lynch). See the picture at **core.** [from Latin *crusta* rind, shell]

—**crustal,** *adj.* of or having to do with a crust, especially of the earth or the moon: *crustal rocks.*

crustacean (krus′tā′shən), *Zoology.* —*n.* any of a class (*Crustacea*) of arthropods with hard shells and jointed bodies and appendages, living mostly in the water. Crabs, lobsters, and shrimps are crustaceans.
—*adj.* of or belonging to this class.
[from New Latin *Crustacea,* from Latin *crusta* shell]

crustacean

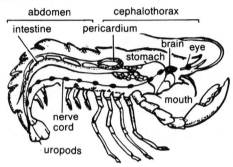

crustose (krus′tōs), *adj. Botany.* (of lichens and algae) growing in the form of crusts; sticking fast to soil, rock, or bark so as not to be detachable except in small fragments: *No organism other than a crustose lichen can maintain itself on a perfectly plain, clean rock surface* (Emerson, *Basic Botany*).
ASSOCIATED TERMS: Lichens and algae are divided into three general classes: The *crustose* types are firmly attached to soil, rock, or bark; the *foliose* types are leaflike, flattened, and easily detached; the *fruticose* types are erect or hanging, and look like tiny shrubs.

cryo-, *combining form.* low temperature; cold; freezing, as in *cryometer = a thermometer for the measurement of very low temperatures.* [from Greek *kryos* frost, cold]

cryobiology (krī′ō bī ol′ə jē), *n. Biology.* the study of the effects of low temperatures on living things: *The progress in practical cryobiology rests on an increased understanding of the characteristics of tissue as it freezes and thaws* (Science News).
—**cryobiological,** *adj.* of or having to do with cryobiology. —**cryobiologically,** *adv.*

cryochemical, *adj.* of or having to do with cryochemistry.

cryochemistry, *n.* the branch of chemistry dealing with reactions occurring at very low temperatures.

cryogen (krī′ə jən), *n. Physics.* a liquefied gas or other substance for producing very low temperatures.
—**cryogenic** (krī′ō jen′ik), *adj.* of or having to do with low temperatures or with cryogenics: *The electromagnets operated at cryogenic temperatures make use of the fact that the electrical resistivity of metals drops with decreasing temperature, thus lowering the power consumed by the magnet itself* (Z.J.J. Stekly).

cryogenics (krī′ō jen′iks), *n.* the branch of physics dealing with the behavior of matter at extremely low temperatures: *Cryogenics ... centers on the liquefying of gases. The cold temperatures provided by the liquids go to work in a myriad of tasks ranging from fuels for* *guided missiles to steel manufacture* (Wall Street Journal). Compare **low-temperature physics.**

cryolite (krī′ə līt), *n. Mineralogy.* a fluoride of sodium and aluminum found in Greenland. *Formula:* Na_3AlF_6 [from *cryo-* cold + *-lite*]

cryostat, *n. Physics.* a refrigerating unit generating very low temperatures, that is capable of liquefying a gas, such as helium, by chilling it more than − 230 Celsius.

crypt, *n. Anatomy.* a small tubular gland, cavity, or follicle. [from Latin *crypta,* from Greek *kryptē* vault]

cryptic, *adj. Zoology.* serving to hide or camouflage: *The camouflage colours which allowed the insect to merge with its background, and were hence described as cryptic coloration, provided further examples of the utility of colour to insects* (B. Nickerson) [from Greek *kryptos* hidden]

crypto-, *combining form.* hidden; covered, as in *cryptoclastic, cryptovolcanic.* [from Greek *kryptos* hidden]

cryptobiosis (krip′tō bī ō′sis), *n. Biology.* **1** the ability of an organism, such as various lower invertebrates, to survive in a state of metabolic inactivity, as at certain very low temperatures. **2** the state of suspended animation in which such organisms are able to live. [from New Latin, from *crypto-* hidden + *biosis* made of life]

cryptobiotic (krip′tō bī ot′ik), *adj. Biology.* **1** characterized by cryptobiosis; able to survive in a state of metabolic inactivity: *a cryptobiotic organism.* **2** living in concealment, as in wood, underground, etc.: *cryptobiotic insects.*

cryptoclastic (krip′tō klas′tik), *adj. Geology.* composed of particles too small to be seen by the unaided eye: *cryptoclastic rocks.* [from *crypto-* + Greek *klastos* broken off]

cryptococcal (krip′tə kok′əl), *adj. Botany.* of, having to do with, or caused by any of a group of yeastlike, often pathogenic, fungi: *cryptococcal meningitis.* [from New Latin *Cryptococcus* genus name of the fungus, from *crypto-* + *coccus,* from Greek *kokkos* seed, berry]

cryptocrystalline, *adj. Mineralogy.* having crystals too small to be resolved under an ordinary optical microscope.

cryptogam (krip′tə gam), *n. Botany.* a plant having no seeds, such as a fern or moss. [from Greek *kryptos* hidden + *gamos* marriage]
—**cryptogamic** (krip′tə gam′ik) or **cryptogamous** (kriptog′ə məs), *adj.* of or resembling cryptogams.

cryptophyte (krip′tə fīt), *n. Botany.* a plant whose buds are either beneath the soil or under water, growing on corms, bulbs, etc. [from *crypto-* + Greek *phyton* plant]

cryptovolcanic, *adj. Geology.* of or designating a circular, structurally disturbed area not directly associated with igneous rocks but having features suggesting subvolcanic activity: *Cryptovolcanic explosions expel no lava or ash* [but] *they do deform the earth surface, generally producing a circular depression from two to four miles wide with a central body of uplifted rocks* (Scientific American).

crystal, *n.* **1** *Chemistry, Physics.* one of the regularly shaped bodies with angles and flat surfaces into which many substances solidify. A crystal is a solid unit in which the atoms, ions, or molecules are arranged in a regularly repeating, characteristic pattern or network of fixed points in space, with measurable distances between them. Crystals can grow in solution by accretion. *Crystals of sugar can be distinguished from crystals of snow by their difference in form.*
ASSOCIATED TERMS: Crystallographers distinguish six basic types of crystal structure, called systems: *isometric, hexagonal, tetragonal, orthorhombic, monoclinic,* and *triclinic.*
2 *Mineralogy.* a transparent, colorless mineral that looks like ice, especially the transparent or nearly transparent form of pure quartz. Crystal is used as a gemstone.
3 *Electronics.* a crystalline material with piezoelectric or semiconducting properties.

crystals: (def. 1)

cubic
tetragonal
hexagonal
orthorhombic
monoclinic
triclinic
rhombohedral

crystal lattice, = lattice (def. 1).

crystalline, *adj.* made of crystals; solidified in the form of crystals: *Snowflakes and diamonds are crystalline.*

crystalline lens, *Anatomy.* the lens of the eye: *Immediately behind the iris is the crystalline lens ... whose surfaces can be changed by muscles around the edge in order to vary the focal length* (Shortley and Williams, *Elements of Physics*).

crystallinity, *n.* crystalline quality or character: *Varieties of color, clarity and crystallinity found in natural diamonds have been observed in the man-made diamonds* (N.Y. Times).

crystallite (kris′tə līt), *n.* **1** *Geology.* a tiny globule or prism in glassy, igneous rock that is an embryonic crystal. **2** *Crystallography.* an individual crystal in a polycrystalline substance.

cap, fāce, fäther; best, bē, tėrm; pin, fīve;
rock, gō, ôrder; oil, out; cup, pùt, rüle,
yü in use, *yu* in uric;
ng in bring; *sh* in rush; *th* in thin, ᴛʜ in then;
zh in seizure.
ə = *a* in about, *e* in taken, *i* in pencil, *o* in lemon, *u* in circus

crystallizability

3 *Chemistry.* a minute part of a polymer having a crystalline structure.

crystallizability, *n.* the quality of being crystallized; capability of forming crystals: *Crystallizability is essential for a fiber-forming polymer* (Michaela Leitner).

crystallize, *v.* to form into crystals; solidify into crystals: *Water crystallizes to form snow.*

—**crystallization,** *n.* **1** the act or process of crystallizing or quality or condition of being crystallized: *the crystallization of water by freezing.* **2** a crystallized substance or formation.

crystallographic or **crystallographical,** *adj.* of or having to do with crystallography: *crystallographic study of a mineral.* —**crystallographically,** *adv.*

crystallography, *n.* the science that deals with the form, structure, and properties of crystals. The study of crystallography is of great importance to the chemist and mineralogist, as the nature of many substances may be ascertained from an inspection of the forms of their crystals.

crystalloid, *Chemistry.* —*adj.* Also, **crystalloidal.** of crystalline form, especially as contrasted with colloid: *a crystalloid arrangement of molecules.*

—*n.* a crystalloid substance; a substance capable of crystallization that, when dissolved in a liquid, forms a true solution and will diffuse readily through vegetable or animal membranes.

crystal system, *Crystallography.* any of seven general categories, based on characteristic symmetry elements, under which the 32 classes of crystalline forms are often classified. The crystal systems include the cubic or isometric, hexagonal, tetragonal, orthorhombic, rhombohedral, monoclinic, and triclinic systems.

Cs, *symbol.* cesium.

Cs, *abbrev.* cirrostratus.

csc, *abbrev.* cosecant.

CSF, *abbrev.* cerebrospinal fluid.

ctenoid (tē′noid *or* ten′oid), *adj. Zoology.* **1** having marginal projections resembling the teeth of a comb, as the teeth or scales of certain fishes do. **2** having ctenoid scales: *a ctenoid fish.* [from Greek *ktenoeidēs* comblike]

ctenophore (ten′ə fôr), *n. Zoology.* any of a phylum (Ctenophora) of marine invertebrate animals resembling jellyfishes, having biradial symmetry, and swimming by means of eight rows of ciliated comblike plates: *Ctenophores are beautifully iridescent in sunlight, and they glow like electric light bulbs at night, due to their luminescence* (Hegner and Stiles, *College Zoology*). Also called **comb jelly.** [from Greek *ktenos* comb + *phorein* carry]

—**ctenophoran** (tə nof′ə rən), *adj.* of or having to do with the ctenophores. —*n.* = ctenophore.

ctn, *abbrev.* cotangent.

C-type virus, *Biology.* any of various strains of a cancer-producing virus (oncornavirus) consisting of a central core of RNA inside a strong protein coat or membrane. C-type viruses form by budding from the surface of infested cells and are able to alter the genetic character of the host cells. *Some viruses ... carry no DNA of their own ... Instead, these viruses, known as the C-type viruses, merely carry a simple singlestrand bit of RNA that contains all of their own virus-building instructions* (Robert Cooke).

Cu, *symbol.* copper. [for Latin *cuprum*]

cu., *abbrev.* cubic.

cube, *Mathematics.* —*n.* **1** a solid figure with 6 square faces or sides, all equal. **2** the product obtained when a number is cubed: *The cube of 4 is 64.*

—*v.* to use (a number) three times as a factor: *5 cubed (5^3) is 125, because $5 \times 5 \times 5 = 125$.* [from Latin *cubus,* from Greek *kybos* cube, die]

cube root, *Mathematics.* the number that produces a given number when cubed: *The cube root of 125 is 5.*

cubic (kyü′bik), *adj.* **1** *Mathematics.* **a** having length, breadth, and thickness; three-dimensional. A cubic centimeter is the volume of a cube whose edges are each one centimeter long. The cubic content of a room is the number of cubic meters it contains. *Abbreviation:* cu. **b** having to do with or involving the cubes of numbers. **2** *Crystallography.* = isometric. See the picture at **crystal.**

cubic equation, *Mathematics.* an equation in which the highest power of the unknown quantity is a cube.

cubic measure, a system of units used for measuring volume:

1000 cubic millimeters = 1 cubic centimeter
1000 cubic centimeters = 1 cubic decimeter
1000 cubic decimeters = 1 cubic meter
1728 cubic inches = 1 cubic foot
27 cubic feet = 1 cubic yard

cuboid (kyü′boid), *Anatomy.* —*n.* a cube-shaped bone, the outermost of the distal row of tarsal bones.

—*adj.* Also, **cuboidal.** shaped like a cube: *Cuboidal epithelium is common in salivary glands.*

cucullate (kyü′kə lāt), *adj. Botany, Zoology.* having a hood; shaped like a hood, as a leaf or nectary. [from Late Latin *cucullatus,* from Latin *cucullus* cap]

cuesta (kwes′tə), *n. Geography, Geology.* a ridge formed in resistant sedimentary strata in which one face slopes gently downward from the crest while the other, cut through the sedimentary layer by erosion, has a much steeper scarp. [from Spanish *cuesta* hill, slope]

culm (kulm), *n. Botany.* **1** the jointed stem characteristic of grasses, usually hollow. **2** the stem of any monocotyledon. [from Latin *culmus* stalk]

cultigen (kul′tə jən), *n. Biology.* **1** a cultivated plant of unknown or obscure origin, such as the cabbage. **2** any cultivated variety; cultivar: *Gardeners ... have developed many cultigens, or garden varieties, of such common flowers as irises, orchids, roses, and tulips* (Harold N. Moldenke). [from *culti(vated)* + *-gen*]

cultivar (kul′tə vär *or* kul′tə vər), *n. Biology.* a variety produced by selective breeding. Also called **cultigen.** [from *culti(vated)* + *var(iety)*]

culture, *Biology, Medicine.* —*n.* **1** the growth of microorganisms, tissue cells, or other living matter in a specially prepared nutrient medium for scientific study or medicinal use: *The method of tissue culture has also served as a direct means of studying the further development of complex structures and organ rudiments* (R.C. Parker).

2 the resulting colony or growth: *the culture is alive with the resistant strain.*
—*v.* **1** to grow (microorganisms, etc.) in a specially prepared medium: *The bacteria have been successfully cultured.*
2 to use (a substance) as a medium for culture: *to culture milk.*
[from Latin *cultura* a tending, from *colere* to till, cherish]

culture medium, *Microbiology.* a liquid, semisolid, or solid substance, consisting of nutrients and other material, in which microorganisms find nourishment and can reproduce: *The white, and even the yolk, of an egg provide a marvelous culture medium for microbes* (N. Y. Times).

cumulocirrus (kyü′myə lō sir′əs), *n. Meteorology.* a cumulus cloud resembling a cirrus, usually found at high altitudes.

cumulonimbus (kyü′myə lō nim′bəs), *n. Meteorology.* a massive cumulus cloud having peaks that sometimes resemble mountains and sometimes spread out to resemble anvils, occurring at heights of between 500 and over 6000 meters. See the pictures at **cloud, front.**

cumulostratus (kyü′myə lō strā′təs), *n. Meteorology.* a cumulus cloud with its base spread out horizontally like a stratus cloud.

cumulous (kyü′myə ləs), *adj. Meteorology.* of or like cumulus clouds.

cumulus (kyü′myə ləs), *n., pl.* **-li** (-lī). *Meteorology.* a vertical cloud formation consisting of detached clouds, rounded at the top and flat at the bottom, usually seen in fair weather and occurring at heights of between 500 and 6000 meters: *Sometimes on a hot summer day you see "thunder-heads," or cumulus clouds, towering into the sky ... formed when warm, moist air is forced upward by cool air moving in under it* (Beauchamp, et al., *Everyday Problems in Science*). See the pictures at **cloud, front.** [from Latin *cumulus* heap]
ASSOCIATED TERMS: see **cloud.**

cuneate (kyü′nē it), *adj. Botany, Zoology.* tapering to a point at the base; wedge-shaped: *a cuneate leaf, a cuneate wing.* [from Latin *cuneus* wedge]

cuneiform (kyü nē′ə fôrm *or* kyü′nē ə fôrm), *Anatomy.*
—*n.* a wedge-shaped bone, especially one of three tarsal bones.
—*adj.* of or denoting any wedge-shaped bone.
[from Latin *cuneus* wedge + English -form]

cupferron (kyüp′fer′ən), *n. Chemistry.* a crystalline compound soluble in water, alcohol, and ether, used especially as a reagent in separation of copper and iron from other metals. *Formula:* $C_6H_9N_3O_2$ [from Latin *cuprum* copper + *ferrum* iron]

cupric (kyü′prik), *adj. Chemistry.* of or containing copper, especially with a valence of two. **Cupric oxide** is an oxide of copper used as a black pigment. *Formula:* CuO [from Latin *cuprum* copper + English -ic]

cupric sulfate, = copper sulfate.

cuprite (kyü′prīt) *n.* a mineral consisting of native cuprous oxide, occurring in crystals and granular masses, and forming an important ore of copper. *Formula:* Cu_2O

cuprous (kyü′prəs), *adj. Chemistry.* of or containing copper, especially with a valence of one. **Cuprous oxide** is an oxide of copper used as a red pigment.

Formula: Cu_2O [from Latin *cuprum* copper + English -ous]

cupule (kyü′pyül), *n. Botany, Zoology.* **1** a cup-shaped membranous cover surrounding a fruit, as in the acorn.
2 any cup-shaped organ or part, such as the receptacle of certain fungi or the sucker on the feet of certain flies. [from Latin *cupula* cupola, from *cupa* tub]
—**cupulate** (kyü′pyü lit), *adj.* shaped like a cupule; having a cupule.

curie (kyùr′ē *or* kyü rē′), *n. Physics.* the unit for measuring the intensity of radioactivity. It is the quantity of a radioactive isotope which decays at the rate of 3.7×10^{10} disintegrations per second. Originally, it was the amount of radioactivity given off by one gram of radium. *The most powerful modern atomic bomb should release no more than 10 billion curies* (Scientific American). *Symbol:* Ci [named after Pierre *Curie*, 1859–1906, French chemist, co-discoverer of radium]

Curie constant, *Physics.* a constant at temperatures above the Curie point, expressed as the product of the magnetic susceptibility per unit mass of a paramagnetic substance by the absolute temperature. [see etymology under **curie**]

Curie point or **Curie temperature,** *Physics.* the temperature above which a ferromagnetic substance, such as iron, loses its magnetization or becomes paramagnetic: *As the magma — the molten rock — cools through its Curie point, the magnetic minerals it contains acquire a permanent magnetization in the direction of the Earth's magnetic field prevailing at that time* (Science Journal).

Curie's law, *Physics.* the law that the magnetic susceptibility of a paramagnetic substance is inversely proportional to the absolute temperature.

curium (kyùr′ē əm), *n. Chemistry.* a radioactive metallic element produced artificially from plutonium or americium: *Curium-242 was the source of alpha particles ... which performed the first direct chemical analysis of the lunar surface by ... alpha-proton reaction techniques* (Glenn T. Seaborg). *Symbol:* Cm; *atomic number* 96; *atomic weight* 247; *oxidation state* 3; *melting point* 1340°C. [named after Marie *Curie*, 1867–1934, and Pierre *Curie*, 1859–1906, French chemists]

current, *n. Physics.* **1** a flow of electricity. Metals are good conductors of electric current.
2 the rate or amount of such flow, usually expressed in amperes. *Abbreviation:* c. See also **alternating current, direct current.**
▶ Since the word *current* means a flow of electricity or electric charge, the use of such phrases as "current flow" and "the flow of electric current" is, strictly speaking, redundant. However, the terms which are in general use may clarify concepts.

---

cap, fāce, fäther; best, bē, tèrm; pin, five;
rock, gō, ôrder; oil, out; cup, pùt, rüle;
yü in use, *yù* in uric;
ng in bring; *sh* in rush; *th* in thin, ŦH in then;
zh in seizure.
ə = *a* in about, *e* in taken, *i* in pencil, *o* in lemon, *u* in circus

current sheet

current sheet, *Geophysics, Astronomy.* a thin broad layer of current flowing in the magnetopause, and produced by the streaming of the solar wind.

cursorial (kėr sôr′ē əl), *adj. Zoology.* 1 adapted or fitted for running: *cursorial limbs.* 2 having legs fitted for running: *The ostrich is a cursorial bird.* [from Latin *cursor* runner]

curvature, *n. Geometry.* the amount or degree of curving. For a circle, the curvature is the reciprocal of the radius. For other plane curves, the curvature at any point equals the curvature of a circle that would coincide with the curve at that point. See also **center of curvature.**

curve, *n. Mathematics.* a line, usually one that bends, especially: **a** a set of points defined by an equation. **b** the path of a moving point. **c** the line where two surfaces intersect.
  ▶ A *plane curve* is a curve that lies in a plane, a *closed curve* is one whose end points coincide, and a *simple curve* is one in which no two points coincide except possibly the end points. A circle is a simple closed curve.

curvilinear, *adj. Mathematics.* consisting of or enclosed by a curved line or lines: *The Einsteinian universe is curvilinear, with four dimensions* (N.Y. Times). **Curvilinear motion** is motion along a curve.

cusp, *n.* 1 *Astronomy.* either of the pointed ends of a crescent moon or of an inferior planet, such as Venus, in the crescent phase.
  2 *Geometry.* = spinode.
  3 *Anatomy.* a blunt or pointed protuberance on the grinding surface of a tooth or a valve of the heart. [from Latin *cuspidem* a point]

cuspid, *n.* a tooth having one cusp. Also called **canine.**

cuspidate, *adj. Botany, Zoology.* having a sharp, pointed end: *cuspidate leaves.*

cutaneous (kyü tā′nē əs), *adj. Anatomy.* of, on, or having to do with the skin: *a cutaneous gland, a cutaneous sensation.* A **cutaneous artery** branches on the inner surface of the skin. [from New Latin *cutaneus,* from Latin *cutis* skin]

cuticle (kyü′tə kəl), *n.* 1 *Zoology.* **a** the outer skin of vertebrates; epidermis. **b** any superficial skin or integument, such as the transparent membrane which envelope annelids.
  2 *Anatomy.* **a** the hard skin around the sides and base of a fingernail or toenail. **b** the outer layer of cells of a shaft of hair. See the picture at **hair.**
  3 *Botany.* a very thin film covering the outer skin of many plants: *The chief functions of ... the waxy covering of the epidermis (cuticle) are: the protection of the internal tissues of the leaf from excessive water loss, resistance to invasion by fungi, and protection from mechanical injury* (Biology Regents Syllabus). See the picture at **leaf.**
  [from Latin *cuticula,* diminutive of *cutis* skin]
  —cuticular, *adj.* of or having to do with a cuticle: *cuticular transpiration.*

cutin (kyüt′n), *n. Botany.* a waxy, waterproof substance that is the chief ingredient of the cuticle of a plant. [from Latin *cutis* skin]

cutinization (kyü′tə nə zā′shən), *n. Botany.* a modification of cell walls by which they become impermeable to water through the presence of cutin. Compare **suberization.**

cutinize (kyü′tə nīz), *v. Botany.* to modify by the formation of cutin: *Epidermal cells are usually cutinized, while cork cells tend to become suberized* (Youngken, Pharmaceutical *Botany*).

cutis (kyü′tis), *n. Anatomy.* the skin beneath the epidermis; the dermis. [from Latin *cutis* skin]

Cuvierian (kyü′vē ar′ē ən), *adj. Zoology.* of or having to do with the French naturalist Georges Cuvier, 1769–1832, or characteristic of his system of animal classification: *Cuvierian anatomical rules, Cuvierian organs, ducts, etc.*

cyan-, *combining form.* the form of **cyano-** before vowels, as in *cyanamide, cyanic.*

cyanamide (sī an′ə mid *or* sī an′ə mid), *n. Chemistry.* 1 a white, crystalline compound prepared by the action of ammonia on cyanogen chloride and in other ways. *Formula:* HN:C:NH 2 = calcium cyanamide.

cyanate (sī′ə nāt), *n. Chemistry.* a salt or ester of cyanic acid.

cyanic (sī an′ik), *adj.* 1 *Chemistry.* of or containing cyanogen. **Cyanic acid,** HOCN, is a colorless, poisonous liquid.
  2 *Biology.* blue. Flower colors in all shades of blue, and passing through violet and purple to red, are called cyanic.

cyanide (sī′ə nīd), *Chemistry.* **—n.** any of various metallic salts of hydrocyanic acid (HCN), especially potassium cyanide, a powerful poison. Cyanide is used in making plastics and extracting and treating metals.
  —v. to treat with a cyanide.

cyanite (sī′ə nīt), *n.* = kyanite.

cyano-, *combining form.* 1 blue or dark-blue, as in *cyanometry, cyanite.* 2 *Chemistry.* of or containing cyanogen or cyanide, as in *cyanoacetylene, cyanogenetic.* Also spelled **cyan-** before vowels. [from Greek *kyanos* a dark-blue color or substance]

cyanoacetylene, *n. Chemistry.* a large organic molecule discovered in cosmic gas clouds: *Cyanoacetylene* (HC_3N), *the most complex organic molecule found in space so far, has now come to light in the radio source at the galactic centre* (New Scientist). *Formula:* $HC \equiv CC \equiv N$

cyanocobalamin (sī′ə nō kō bal′ə min), *n. Biochemistry.* a dark-red, crystalline compound, one of the three forms of vitamin B$_{12}$. *Formula:* $C_{63}H_{90}N_{14}PCo$ [from *cyano-* + *cobal(t)* + *(vit)amin*]

cyanogen (sī an′ə jən), *n. Chemistry.* 1 a colorless, poisonous, inflammable gas with the odor of bitter almonds. *Formula:* C_2N_2
  2 a univalent radical (-CN) consisting of one atom of carbon and one of nitrogen, found in hydrocyanic acid and the cyanides.

cyanogenetic, *adj. Chemistry.* tending to yield or produce cyanide, as certain gases.

cyanohydrin, *n. Chemistry.* any of a group of organic compounds which contain both the CN and the OH radicals.

cyanometry, *n. Optics.* the measurement of the intensity of the blueness of light, as of the sky.

142

cyanurate (sī'ə nùr'āt), *n. Chemistry.* a salt or ester of cyanuric acid.

cyanuric acid (sī'ə nùr'ik), a white or colorless crystalline acid, formed by polymerization of cyanic acid, and used in organic synthesis. *Formula:* $C_3H_3N_3O_3$

cybernetics (sī'bər net'iks), *n.* the comparative study of the nervous system and systems of mechanical and electronic control of machinery and mechanical devices in order to better understand communication and control in both types of systems: *A fundamental goal of cybernetics is to determine why natural systems of control seem to be more versatile and adaptive in certain ways than those which man has designed and constructed* (Parkman, *The Cybernetic Society*). [coined in 1948 by Norbert Wiener, 1894–1964, an American mathematician, from Greek *kybernētikos* of a pilot, from *kybernētēs* pilot, from *kybernan* to steer] —**cybernetic**, *adj.* of or having to do with cybernetics: *cybernetic systems, cybernetic mechanisms.*

cycle, *n.* **1** any period of time or complete process of growth or action that repeats itself in the same order. The seasons of the year—spring, summer, autumn, and winter—make a cycle. *There are two very evident weather cycles of such importance that we base our reckoning of time upon them, namely, the daily and the annual cycles* (Blair, *Weather Elements*).
2 *Electricity.* a complete or double alternation or reversal of an alternating electric current. The number of cycles per second is the measure of frequency.
3 *Biology.* a recurring series of changes: *the carbon cycle, the nitrogen cycle, the estrous cycle.*
4 *Botany.* a closed circle or whorl of leaves.
5 *Physics.* a series of operations by which a substance or operation is finally brought back to the initial state: *A cycle, as applied to a wave, consists of series of changes occurring in orderly sequence by means of which the medium returns to its initial condition prior to repeating the series* (Physics Regents Syllabus).
▶ The "cycle per second" as a unit of frequency has now been replaced by *hertz,* with the symbol Hz.

cyclic (sī'klik *or* sik'lik), *adj.* **1** of a cycle; moving or coming in cycles: *In the rat, the mouse, and in man, the ovary displays cyclic activity for which the pituitary is responsible and which results in the liberation of ova from the ovaries once every five or six days in the mouse and once every month in the woman* (Science News).
2 *Botany.* arranged in whorls.
3 *Chemistry.* of, having to do with, or containing an arrangement of atoms in a ring or closed chain. —**cyclically**, *adv.*
ASSOCIATED TERMS: If all of the atoms in the ring of a cyclic compound are carbon atoms, the compound is *carbocyclic.* If the ring contains more than one type of atom, the compound is *heterocyclic.* Aromatic compounds may be either carbocyclic or heterocyclic. *Aliphatic* compounds are not cyclic; their atoms are arranged in straight or branched chains. Various cyclic compounds which share some of the characteristics of aliphatics are called *alicyclic* compounds.

cyclic AMP, = adenosine monophosphate (def. 2).

cyclic GMP, *Biochemistry.* a compound that functions as a messenger to stimulate the action of hormones and that interacts with the related chemical cyclic AMP in cellular metabolism. [*GMP,* abbreviation of *guanosine monophosphate,* a compound of guanosine (a constituent of nucleic acid) and one phosphate group]

cyclization (sī'klə zā'shən), *n. Chemistry.* the formation of one or more closed chain or ring structures by rearrangement of the atoms in a molecule.

cyclize (sī'klīz), *v. Chemistry.* **1** to establish a closed chain or ring formation in; make cyclic. **2** to undergo cyclization; become cyclic.

cyclo-, *combining form.* **1** circle; of a circle; spiral, as in *cycloid, cyclotron.*
2 *Chemistry.* cyclic, as in *cycloparaffin.*
[from Greek *kyklos* ring, wheel]

cyclogenesis, *n. Meteorology.* the formation or development of a cyclone.

cycloid (sī'kloid), *n. Geometry.* a curve traced by a point on the circumference or on a radius of a circle when the circle is rolled along a straight line and kept in the same plane.
—*adj.* **cycloid** or **cycloidal** (sī kloi'dəl), **1** like a circle. **2** *Zoology.* **a** (of the scales of certain fishes) somewhat circular, with smooth edges. **b** (of a fish) having cycloid scales. Compare **ganoid.**

cycloid (def. 1)

The cycloid is the curve described by the radius of the circle as it rolls along the straight line.

cyclone, *n. Meteorology.* **1** a storm or winds moving around and toward a calm center of low pressure, which also moves. The motion of a cyclone is counterclockwise in the Northern Hemisphere and clockwise in the Southern Hemisphere. *Cyclones usually are a few hundred miles in diameter and anticyclones generally are somewhat larger and often more eccentric* (Byers, *General Meteorology*).
2 an area of comparatively low atmospheric pressure. Also called **low.**
[from Greek *kyklōn* moving around in a circle, from *kyklos* wheel, circle]
ASSOCIATED TERMS: A *tropical cyclone* is a general term for a cyclone found in low latitudes over ocean areas, as distinguished from an *extratropical cyclone,* which is typical of middle and high latitudes. Cyclones are classified according to their intensities and regions of occurrence. A *typhoon* is a severe tropical cyclone in the western North Pacific and most of the South Pacific. A tropical cyclone originating in the North Atlantic, Caribbean Sea, Gulf of Mexico, or off the west coast of Mexico is called a *hurricane.* A *tornado,* although a very small type of cyclone, is an intense vortex of extremely powerful and destructive winds.
—**cyclonic**, *adj.* of or like a cyclone: *cyclonic rotation.*

cycloolefin (sī'klō ō'lə fin), *n. Chemistry.* any of a group of alicyclic hydrocarbons with an unsaturated ring, having the general formula C_nH_{2n-2}

cap, fāce, fàther; best, bē, tèrm; pin, five;
rock, gō, ôrder; oil, out; cup, pùt, rüle;
yü in use, *yù* in uric;
ng in bring; *sh* in rush; *th* in thin; ᵗH in then;
zh in seizure.

ə = *a* in about, *e* in taken, *i* in pencil, *o* in lemon, *u* in circus

cycloparaffin (sī'klō par'ə fin), *n. Chemistry.* any of a group of alicyclic hydrocarbons with a saturated ring, having the general formula C_nH_{2n}

cyclosis (sī klō'sis), *n. Biology.* the rotary streaming movement of protoplasm in a cell, especially of the endoplasm: *Cyclosis takes the form of migrations of various objects. Not infrequently the nucleus ... being carried along by a more or less continuous active streaming of the cytoplasm* (Emerson, *Basic Botany*). [from Greek *kyklōsis,* from *kykloun* to move around in a circle]

cyclostome (sī'klə stōm), *Zoology.* —*n.* any of a class (Cyclostomata *or* Agnatha) of slender, snakelike fishes, having a round, sucking mouth and no jaws. Cyclostomes comprise the simplest living vertebrates; lampreys and hagfishes belong to this class.
—*adj.* of or belonging to this class.
[ultimately from Greek *kyklos* ring + *stoma* mouth]

cyclothem (sī'klə them), *n. Geology.* a series of beds deposited in one sedimentary cycle. [from *cyclo-* + Greek *thema* something set down]

cyclotomy (sī klot'ə mē), *n. Mathematics.* the theory of dividing the circle into equal parts. [from *cyclo-* + Greek *-tomia* a cutting]
—**cyclotomic,** *adj.* of or having to do with cyclotomy: *cyclotomic functions.*

cyclotron (sī'klə tron), *n. Nuclear Physics.* a particle accelerator in which protons or other atomic particles are accelerated in a spiral path away from their sources by an alternating electric field in a constant magnetic field. [from *cyclo-* + *-tron*]

cylinder, *n. Geometry.* **1** a solid figure bounded by two equal, parallel circles and by a curved surface, formed by moving a straight line of fixed length so that its ends always lie on the two parallel circles. In a right circular cylinder the circles are perpendicular to the line; in an oblique circular cylinder they are not.
2 a solid bounded by two parallel planes and a curved surface formed by moving a straight line so that it constantly describes a given curve and remains parallel to its original position.
3 the surface of any such solid.
4 the volume of any such solid.
—**cylindrical,** *adj.* **1** shaped like a cylinder; having the form of a cylinder.
2 of or having to do with a cylinder. —**cylindrically,** *adv.*

cylinder (def. 1)

cylindric projection, *Geography.* a form of map projection in which the area to be displayed has been projected from the earth's spherical surface onto a cylinder.

cylindrite (sil'ən drīt), *n. Mineralogy.* a mineral composed of sulfur, lead, antimony, and tin, occurring in massive forms with concentric cylindrical structure. *Formula:* $Pb_3Sn_4Sb_2S_{14}$

cyme (sīm), *n. Botany.* a flower cluster in which there is a flower at the top of the main stem and of each branch of the cluster. The flower in the center opens first. The sweet william has cymes. Compare **corymb** and **raceme.** [from Latin *cyma,* from Greek *kyma* something swollen, sprout, from *kyein* be pregnant]
—**cymose** (sī'mōs), *adj.* having to do with or bearing a cyme or cymes: *a cymose inflorescence.*

cyme

cymophane (sī'mə fān), *n.* a variety of chrysoberyl. [ultimately from Greek *kyma* wave + *phainein* to appear]

Cys, *abbrev.* cysteine.

cyst (sist), *n. Biology.* **1** a small, abnormal, saclike growth in animals or plants, usually containing liquid and diseased matter produced by inflammation.
2 a saclike structure in animals or plants.
3 a spore in green algae during the period of dormancy, after which it germinates to produce a new plant.
4 a cell or cavity containing reproductive bodies, embryos, or bacteria in a resting stage.
5 a small, round sac, such as that enclosing an embryonic tapeworm or a pair of certain parasitic protozoans before sporulation.
[from New Latin *cystis,* from Greek *kystis* pouch, bladder]
—**cystic,** *adj.* **1** of, like, or having a cyst or cysts: *cystic disease of the breast.* **2** of the bladder or gall bladder: *the cystic duct or canal.*

cysteine (sis'tē ēn), *n. Biochemistry.* a crystalline amino acid present in proteins and derived from cystine. *Formula:* $C_3H_7O_2NS$ *Abbreviation:* Cys

cystine (sis'tēn), *n. Biochemistry.* a crystalline amino acid found in many proteins, especially keratin. *Formula:* $C_6H_{12}N_2O_4S_2$ [from *cyst* + *-ine* (so called from its discovery in gall bladder stones)]

cystocarp (sis'tə kärp), *n. Botany.* the multicellular fruiting body, consisting of a mass of asexual spores, that develops after fertilization in the red algae, sometimes contained in a special cellular envelope.

cystolith (sis'tə lith), *n. Botany.* an outgrowth of the walls of some cells, containing tiny crystals of calcium carbonate.

cyt-, *combining form.* the form of **cyto-** before vowels, as in **cytase, cyton.**

cytase (sī'tās), *n. Biochemistry.* an enzyme found especially in the cells of dates, Brazil nut, and barley, useful in the hydrolysis of cellulose.

-cyte, *combining form.* a cell, as in *leucocyte = a white (blood) cell, amebocyte = an amebalike cell.* [from Greek *kytos* receptacle, cell]

cytidine (sī'tə dēn' *or* sit'ə dēn'), *n. Biochemistry.* a white, crystalline nucleoside of ribose and cytosine, prepared from yeast ribonucleic acid. *Formula:* $C_9H_{13}N_3O_5$

cytidylic acid (sī'tə dil'ik *or* sit'ə dil'ik), *Biochemistry.* a nucleotide of ribonucleic acid which on hydrolysis yields cytosine, ribose, and phosphoric acid. *Formula:* $C_9H_{14}N_3O_8P$

cyto-, *combining form.* a cell or cells, as in *cytoplasm* = *protoplasm of a cell.* [from Greek *kytos* receptacle, cell]

cytochalasin (sī'tō kə lā'sin), *n. Biology.* any of a group of structurally related fungal by-products that inhibit the activity of contractile elements in cells, causing the arrest of cytoplasmic division, the migration of cell nuclei, and other effects on cellular processes. [from *cyto-* cell + Greek *chalasis* loosening]

cytochemical, *adj.* of, having to do with, or based on the principles of methods of cytochemistry.

cytochemistry (sī'tō kem'ə strē), *n.* **1** the branch of biochemistry dealing with the chemical composition and activity of cells.
2 the biochemical composition or characteristics of a cell or cells.

cytochrome (sī'tə krōm), *n. Biochemistry.* any of various pigments concerned with cellular respiration, important as catalysts in the oxidation process: *Nearly related chemically to haemoglobin is a pigment called cytochrome ... found in various cells of animals, particularly those of muscle* (H. Munro Fox).

cytochrome oxidase, *Biochemistry.* an oxidizing enzyme containing iron and porphyrin, found in mitochondria and responsible for the transfer of electrons from certain cytochrome molecules to oxygen molecules, resulting in the formation of water. It is also called cytochrome aa$_3$.

cytogamy (sī tog'ə mē), *n. Biology.* the union or conjugation of cells.

cytogenesis, *n. Biology.* the formation and differentiation of cells.

cytogenetics (sī'tō jə net'iks), *n.* the branch of biology dealing with the relation of cells and their constituents to heredity and variation.
—**cytogenetic,** *adj.* having to do with cytogenetics: *cytogenetic studies.*

cytokinesis (sī'tō ki nē'sis), *n.* the division of the cytoplasm of a cell following division of the nucleus.
—**cytokinetic,** *adj.* of or having to do with cytokinesis.

cytokinin (sī'tō kī'nən), *n. Biochemistry.* any of a class of plant hormones that promote cell division and growth, and delay the senescence of leaves: *The cytokinins are closely related to adenine, one of the bases found in nucleic acids, and compounds with strong cytokinin activity occur in transfer RNA* (New Scientist).

ASSOCIATED TERMS: The three most important growth-promoting hormones in plants are the *auxins,* the *cytokinins,* and the *gibberellins.*

cytology (sī tol'ə jē), *n.* **1** the branch of biology that deals with the formation, structure, and function of cells: *As a result of the refinement of the tools of cytology, the various previously separate areas of investigation of the cell and its structure have converged with a result-*

ing significant spurt in man's knowledge of cell biology (Thomas H. Roderick).
2 cellular structure and functions.
—**cytological** (sī'tə loj'ə kəl), *adj.* of or having to do with cytology: *cytological study.* —**cytologically,** *adv.*

cytolysin (sī tol'ə sin), *n.* a substance able to cause cytolysis.

cytolysis (sī tol'ə sis), *n.* the dissolution or destruction of cells: *... causing the cell to rupture, a process termed cytolysis, which permits the protoplasm of the cell to be dispersed* (Harbaugh and Goodrich, *Fundamentals of Biology*). —**cytolytic** (sī'tə lik'ik), *adj.* of or having to do with cytolysis.

cytomegalovirus (sī'tō meg'ə lō vī'rəs), *n. Biology.* a herpesvirus that attacks and enlarges epithelial cells, causing a serious congenital disease of newborn infants (cytomegalic inclusion disease) that affects the brain, liver, kidneys, and lungs. *Cytomegalovirus and herpes simplex virus have led to brain damage, deafness, blindness and other malformations of the central nervous system* (Science News). [from *cyto-* cell + *megalo-* enlarged + *virus*]

cyton (sī'ton), *n. Anatomy,* the body of a nerve cell; a neuron exclusive of its processes. [from *cyt-* + *-on*]
ASSOCIATED TERMS: see **dendrite.**

cytopathogenic, *adj. Biology.* of or having to do with the destruction of cells by a pathogenic agent, such as a virus.

cytopathy (sī top'əthē), *n. Biology.* degeneration or disease in a living cell.
—**cytopathic** (sī'tə path'ik), *adj.* of or having to do with cytopathy.

cytopharynx (sī'tə far'ingks), *n. Zoology.* a tube in certain protozoans, leading from the cytoplasm into the endoplasm: *The cytopharynx is enlarged at the base to form a vesicle called the reservoir, adjacent to which is located a contractile vacuole* (Hegner and Stiles, *College Zoology*).

cytoplasm (sī'tə plaz'əm), *n. Biology.* the living substance or protoplasm of a cell, exclusive of the nucleus or nucleoid: *The cytoplasm [has] ... a variety of complex structures ... each of which plays an important role in synthesizing chemical compounds necessary for the maintenance of cellular acticity* (Thomas H. Roderick).
—**cytoplasmic** (sī'tə plaz'mik), *adj.* having to do with cytoplasm: *The evidence indicated that another nucleic acid, ribonucleic acid (RNA), is the major cytoplasmic component which controls [protein] synthesis* (Lloyd M. Kozloff).

cytosine (sī'tə sēn'), *n. Biochemistry.* a substance present in nucleic acid in cells. It is one of the pyrimidine bases of both DNA and RNA. *The four chemical bases come in pairs—the number of units of thymine seemed to be always the same as the number of ade-*

cap, fāce, fäther; best, bē, tèrm; pin, fīve;
rock, gō, ôrder; oil, out; cup, pùt, rüle,
y<i>ü</i> in use, y<i>u</i> in uric;
ng in bring; sh in rush; th in thin, ᴛH in then;
zh in seizure.
ə = a in about, e in taken, i in pencil, o in lemon, u in circus

nine; and the number of units of cytosine the same as those of guanine (J. Bronowski). *Formula:* $C_4H_5N_3O$ · See the picture at **DNA**.

cytoskeleton, *n. Biology.* the structural framework of a cell, composed largely of actin filaments and microtubules.

cytosol (sī′tə sol), *n. Biology.* the liquid component of cytoplasm; cytoplasm without other cellular components that are associated and usually suspended in it: *Although long-chain fatty acids are at best poorly soluble in aqueous media, a mechanism to account for the apparent facility with which they traverse the cytosol (aqueous cytoplasm) has not been identified* (Science). [from *cyto(plasm)* + *sol(ution)*]
—**cytosolic,** *adj.* of or having to do with cytosol.

cytosome (sī′tə sōm), *n. Biology.* the body of a cell, exclusive of the nucleus.

cytostome (sī′tə stōm), *n. Biology.* the mouth of a protozoan, leading into the cytopharynx.

cytotaxis, *n. Physiology.* the movement of cells, or of cell masses, in reaction to external stimuli emitted by other cells.

cytotaxonomy, *n. Biology.* taxonomic classification based on the study of cellular structures, especially the chromosomes.
—**cytotaxonomic,** *adj.* of or having to do with cytotaxonomy.

cytotoxin, *n. Biochemistry.* a toxin or antibody having a specific harmful effect on certain cells.
—**cytotoxic,** *adj.* of or produced by a cytotoxin.
—**cytoxicity,** *n.*

cytotropic, *adj. Biology.* **1** characterized by cytropism. **2** that has an affinity for or is attracted by cells.

cytotropism (sī tot′rə piz əm), *n. Biology.* the moving or bending of cell masses, or of cells, toward or away from one another.

Czapek's medium (chä′peks), *Microbiology.* a culture medium consisting of agar, salt, sugar, and water, used for growing molds. [named after Friedrich *Czapek,* 20th century Czech scientist]

D

d, *abbrev.* or *symbol.* **1** deci-; one-tenth, as in *decigram.* **2** density. **3** diameter. **4** dyne.

D, *abbrev.* or *symbol.* **1** deuterium. **2** didymium. **3** displacement.

d-, *combining form. Chemistry.* **1** dextrorotatory; dextro-, as in *d-thyroxine = dextrothyroxine.*

▶ The plus sign (+) is now generally preferred to *d-.* Compare **l-.**

2 D-, (of organic compounds) having a configuration about a particular asymmetric carbon atom analogous to the configuration of dextrorotatory glyceraldehyde.

▶ To indicate the direction of optical rotation in addition to configuration, + or − (for dextrorotation and levorotation respectively) is added in parentheses, as D(-). Compare **1-.**

da, *symbol,* deka-.

dacite (dā′sīt), *n. Geology.* a volcanic igneous rock containing quartz, sodic plagioclase, and a mineral containing magnesium and iron: *Dacites occur in dikes and intrusions as well as in lava flows* (Fenton, *The Rock Book*). [from *Dacia,* a Roman province + *-ite*]
—dacitic (də sit′ik), *adj.* having to do with or having the characteristics of dacite.

dacryolin (dak′rē ə lin), *n. Biochemistry.* an albuminous protein found in tears. [from Greek *dakry* tear]

dalton (dôl′tən), *n. Chemistry.* the unit used to express atomic mass, equal to 1/12 of the mass of an atom of carbon 12 (^{12}C); about 10^{-24} grams. Also called **atomic mass unit.** [named after John *Dalton,* 1766–1844, English chemist, physicist, and meteorologist]

Dalton's law, *Chemistry.* the law that in a mixture of gases that are in equilibrium but that do not react with each other, the total pressure equals the sum of those partial pressures that each gas would exert if the others were absent. [see etymology under **dalton**]

damp, *v. Physics.* to decrease the amplitude and duration of (an oscillation or vibration): *Damping is the process of limiting the amplitude and duration of the sympathetic vibrations by friction used in some form* (Roy F. Allison).
—damper, *n.* any device for damping vibrations, as of a magnetic needle. A shock absorber is a kind of damper.

damping off, *Botany.* the decay of seedlings and cuttings caused by parasitic fungi living in water or on most soils: *Without soil sterilization or the use of fungicides, seedlings are often killed by "damping off" because of species of Pythium in the soil* (Greulach and Adams, *Plants*).

danburite (dan′bér īt), *n. Mineralogy.* a borosilicate of calcium, of a white to yellowish color, often occurring in fine crystals resembling topaz. *Formula:* $CaB_2(SiO_4)_2$ [from *Danbury,* city in Connecticut, where it occurs + *-ite*]

darcy, *n., pl.* **-cys.** *Physics.* a unit of measure of the permeability of porous substances: *Permeability is commonly measured in darcys and millidarcys* (Gilluly, *Principles of Geology*). [named after Henri *Darcy,* 1803–1858, French hydrologist]

Darcy's law, *Geology.* the law that the velocity of ground water is equal to the slope of the water table times a constant called the coefficient of permeability. This constant, P, depends on the qualities of the ground through which the water passes and is a measure of the ease with which water moves through it. [see etymology under **darcy**]

dark-adapt, *v. Biology.* to adapt to darkness: *In a night-blind person the rods in the retina take an inordinately long time to become dark-adapted* (Scientific American).
—dark-adaptation, *n.: Some night-blind people can improve their dark-adaptation by taking large doses of vitamin A* (Scientific American).

dark-field illumination, *Optics.* a method of illuminating the object under a microscope only from the sides by blocking off the rays from the center, so that the object appears bright on a dark background.
—dark-field microscope, a microscope using dark-field illumination: *In a dark-field microscope the object is illuminated only from the sides (no central light)* (Storer, *General Zoology*).

dark-line spectrum, *Physics.* a continuous spectrum composed of a pattern of dark lines on a bright background, resulting from the intervention of a cooler gas between the source of the spectrum and the observer: *Sunlight produces a dark-line spectrum, [having] passed through the atmosphere of both the sun and the earth* (Baker, *Astronomy*). Compare **bright-line spectrum.**

dark nebula, *Astronomy.* a large, dark region in space occupied by clouds of dust and gas which obscure the light of stars behind it: *One dark nebula in the milky way is so black compared to the bright areas nearby that it is called The Coalsack.*

---

cap, fāce, fàther; best, bē, tèrm; pin, fīve;
rock, gō, ôrder; oil, out; cup, pùt, rüle,
yü in use, *yu* in uric;
ng in bring; *sh* in rush; *th* in thin, ᴛH in then;
zh in seizure.
ə = *a* in about, *e* in taken, *i* in pencil, *o* in lemon, *u* in circus

147

dark reaction, *Biology.* any of the chemical reactions in photosynthesis that are not dependent on light but are controlled by enzymes sensitive to temperature changes: *The dark reactions are also referred to as carbon fixation, since the net result is the reduction of the carbon of CO_2 to a higher energy level of sugars* (Greulach and Adams, *Plants*).

dark star, *Astronomy.* a star that cannot be seen, especially a star that is one of a binary star pair, or one that is so cool that all or most of its light waves are in the infrared portion of the spectrum.

Darwinism (där′wə niz′əm), *n.,* or **Darwinian theory,** the theory of evolution through natural selection developed by Charles Darwin and Alfred Russel Wallace: in the struggle for existence, individual plant and animal forms with traits less well adapted to the environment tend to be eliminated, while the better adapted tend to survive, and the reinforced variants ultimately result in new species: *Some critics of the Darwinian theory have held that small variations commonly within the range of normal variability of the species are insignificant in bringing about major changes, and that only large variations could be significant in natural selection* (Greulach and Adams, *Plants*). Compare **Lamarckism.**
ASSOCIATED TERMS: adaptation, mutation, punctuated equilibrium, recombination, speciation, variation.

datolite (dat′ə lit), *n. Mineralogy.* a silicate containing boron and calcium, found as glassy crystals in various colors. *Formula:* $CaBSiO_4(OH)$ [from Greek *dateisthai* to divide + English *-lite*]

datum plane, *Geography.* a horizontal plane of reference used as a base level in the construction of topographic maps. The most commonly used datum plane is sea level.

daughter, *adj.* **1** *Biology.* of, belonging to, or having to do with the first generation of offspring, or resulting from a primary division or segmentation:

daughter cell: *In prophase of mitosis ... the development of a cell wall completes the formation of two identical daughter cells* (Weier, *Botany*).

daughter chromosome: *Through a series of very complex steps, each chromosome divides lengthwise, giving rise to two daughter chromosomes.* (Beals and Hoijer).

daughter nucleus: *When a nucleus divides, each daughter nucleus gets one chromatid from each chromosome* (Edwin B. Matzke). *The major result of mitosis is the formation of two daughter nuclei which are identical to each other and to the original nucleus* (Biology Regents Syllabus).

daughter virus: *Since the bacteria become infected and shortly burst, with the release of many daughter viruses, the genetic material of the virus must have entered the bacterium* (Bulletin of Atomic Scientists).
2 *Physics.* produced by or resulting from the decay of a radioactive element:

daughter atom: *The observations were made on fermium-250, the daughter atom, which has a half-life of 30 minutes* (Science).

daughter nuclide: *Each different radioactive nuclide decays into daughter nuclides at a constant rate that is not affected by physical conditions such as temperature or pressure* (Eicher and McAlester, *History of the Earth*).

daughter product: *Radioactivity is a drastic process in which elements change in kind. If radium emits helium, it can no longer be radium. The parent or original substance becomes a new element called the daughter or decay product substance* (Richards, *Modern College Physics*).
—*n. Physics.* an atom or particle produced by the decay of a radioactive element: *Because of its short life and volatility, radon—as distinguished from its daughters—does not accumulate to any great extent in plants and animals* (Andrew F. Stehney). Compare **mother, parent.**

day, *n. Astronomy.* the time occupied by one rotation of a planet or other celestial body, especially the earth: *The length of a planet's day—that is, the time required to complete one rotation about its axis—varies from 9 hours and 50 minutes for Jupiter to 243 days for Venus. The similarity in the lengths of the day for the earth and Mars is a coincidence* (Jastrow and Thompson, *Astronomy*).
▶ The earth's day can be measured in several ways: *Solar day* (or *mean solar day*) is a 24-hour day measured from midnight to midnight; astronomically, it is the interval between two successive transits of the meridian by the sun. *Lunar day* is a day of 24 hours and 50 minutes; it comprises the interval between two successive crossings of the meridian by the moon. *Sidereal day* is a day measured by the stars; it consists of one rotation of the earth in reference to any star of the vernal equinox at the meridian, and is about 4 minutes shorter than a solar day.

dayglow, *n. Astronomy.* airglow occurring in the daytime: *Successful measurements of dayglow were obtained by using ultraviolet spectrometers carried on sounding rockets* (Science News Letter).

day-neutral plant, *Botany.* a plant that grows and blooms during either short days or long days: *Day-neutral plants, such as dandelions, apparently are insensitive to the length of day as they will bloom in any season if they receive enough heat and moisture* (Harbaugh and Goodrich, *Fundamentals of Biology*). Also called **day-neutral species** or **variety.**
ASSOCIATED TERMS: Plants are often classified also into at least two other groups according to their response to day length: *short-day plants,* which flower only if exposed to a relatively short daily dose of sunlight (for example orchid, chrysanthemum), and *long-day plants,* which flower only with the long days of summer (for example, red clover, sugar beet), unless they are illuminated by artificial light.

db or **db.** or **dB,** *abbrev.* decibel *or* decibels.

d.c. or **D.C.,** *abbrev.* direct current.

deacetylate, *v. Chemistry.* to remove one or more acetyl groups from (an organic compound).
—**deacetylation,** *n.* the act or process of deacetylating.

deaerate, *v. Chemistry.* to remove air or gas from: *deaerated water.* —**deaeration,** *n.* —**deaerator,** *n.*

dealate (dē ā′lit) or dealated (dē ā′lā tid), *adj. Zoology.* (of ant queens or termites) that have dropped their wings after the nuptial flight.
—**dealation** (dē ā lā′shən), *n.* the dropping of wings after nuptial flight.

deaminate (dē am′ə nāt), *v. Chemistry.* to remove the amino group —NH₂ from (a compound): *Excess amino acids absorbed after a meal containing protein are deaminated in the liver* (Mackean, *Introduction to Biology*).
—**deamination,** *n.* the removal of the amino radical —NH₂ from a compound.

deaminize, *v.* = deaminate.

debouchure (dā bü shür′) or debouchment (di büsh′-mənt), *n. Geology.* the mouth or outlet of a river or pass: *Some rivers, like the Mississippi, combine all three stages from source to debouchure, being youthful in the upper course and old at the outlet, maturity prevailing midway* (White and Renner, *Human Geography*).

debris (də brē′), *n.* **1** *Biology.* relatively large organic particles serving as food of certain scavengers: *Sea cucumbers and some sea urchins ingest bottom mud in quantity to extract detritus for food. There is also a supply of larger organic particles, or debris, derived mainly from animals that sink after death* (Storer, *General Zoology*).
2 *Geology.* a mass of large fragments worn away from rock.
3 *Physics.* subatomic particles that arise from the bombardment of an atom.

de Broglie wave (də brô lyē′), *Physics.* a wave that is associated with an electron or other moving particle and that gives it certain properties, such as interference and diffraction: *The de Broglie wave is a way of measuring the probability of a particle's being somewhere in a given volume at a given time; its wavelength determines the area over which a particle's influence is felt, and the probability of some interaction with a target depends on this* (Science News).
[named after L.V. *de Broglie,* 20th-century French physicist who propounded the wave nature of electrons]

Debye (də bī′), *adj. Physics, Chemistry.* having to do with or designating various theories or phenomena described by the Dutch physicist and chemist P.J.W. Debye, 1884–1966, especially the Debye theory of specific heats on which a characteristic temperature of substances (the **Debye temperature,** symbol Θ) is based: *According to the modern theory of superconductivity, the pressure-induced increase of the Debye temperature, or characteristic temperature, in metals must lead to a reduction of electron-phonon interactions* (Scientific American).
The **Debye unit** is a measure of electric dipole moment in the CGS system, equal to 10^{-18} statcoulomb multiplied by one centimeter, or about 3.34×10^{-30} coulomb multiplied by one meter.

dec., *abbrev.* **1** decimeter. **2** declination.

deca- or dec-, *combining form.* = deka-.

decadic (di kad′ik), *adj. Mathematics.* of or having to do with tens; based upon ten: *the decadic system of numbers.*

decagon (dek′ə gon), *n. Geometry.* a plane figure having 10 angles and 10 sides.

decahedron (dek′ə hē′drən), *n., pl.* **-drons, -dra** (-drə). *Geometry.* a solid figure having ten faces.

decalcify (dē kal′sə fī), *v. Chemistry.* to remove lime or calcium from: *Decalcifying a tooth by the aid of a dilute mineral acid* (Tomes, *Dental Surgery*).
—**decalcification,** *n.*

decamerous (di kam′ər əs), *adj. Botany.* having ten parts in each whorl (generally written 10-merous): *decamerous flowers.* [from *deca-* ten + Greek *meros* part]

decapod (dek′ə pod), *Zoology.* —*n.* **1** any of an order (Decapoda) of crustaceans having ten legs or arms, such as a lobster, crab, or crayfish. **2** any of a suborder (Decapoda) of mollusks having ten arms or tentacles, such as a squid or cuttlefish.
—*adj.* having ten legs, arms, or tentacles.

decarbonate, *v. Chemistry.* to remove carbon dioxide or carbonic acid from (a metal, a carbonate, etc.).
—**decarbonation,** *n.* the act or process of decarbonating.

decarbonize, *v. Chemistry.* to remove carbon from: *Crude iron is often partially decarbonized in making steel.*
—**decarbonization,** *n.* the act or process of decarbonizing.

decarboxylate (dē′kär bok′sə lāt), *v. Chemistry.* to remove the carboxyl group, -COOH, from (an organic acid, etc.): *Mast cells can decarboxylate histidine at an appreciable rate* (Smith, *Allergy and Tissue Metabolism*).
—**decarboxylation,** *n.* the removal of one or more carboxyl groups: *The final key step in the formation of alcohol is the removal of CO_2 from pyruvic acid. ... Thus, when pyruvic acid loses the carboxyl group by decarboxylation it forms a new compound, acetaldehyde* (McElroy, *Biology and Man*).

decarburize, *v.* = decarbonize. —**decarburization,** *n.*

decay, *Nuclear Physics.* —*v.* (of radioactive substances) to undergo transformation through the disintegration of component nuclei, often accompanied by the emission of radiation: *The half life period T of a radioactive substance is defined as the time required for one half of the active material present at any time to decay* (Stranathan, *The 'Particles' of Modern Physics*).

cap, fāce, fäther; best, bē, tėrm; pin, five;
rock, gō, ôrder; oil, out; cup, pùt, rüle,
yü in use, *yu̇* in uric;
ng in bring; *sh* in rush; *th* in thin, ₮H in then;
zh in seizure.
ə = *a* in about, *e* in taken, *i* in pencil, *o* in lemon, *u* in circus

149

—*n.* the transformation of a radioactive substance through the disintegration of its component nuclei: *The "half-life," or rate of decay, of Carbon 14 is 5,600 years. This means it would take that many years for just half of a sample to decay* (Wall Street Journal).

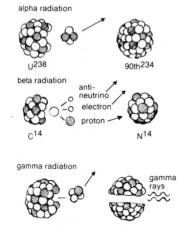

alpha radiation

U238   90th234

beta radiation

anti-
neutrino
electron
proton

C14   N14

gamma radiation

gamma
rays

decay constant, *Nuclear Physics.* the fraction of a group of unstable nuclei that will disintegrate in a given time period (second, year, etc.): *the decay constants of potassium.* Compare **half-life.**

dechlorinate, *v. Chemistry.* to free (a substance) of chlorine; remove chlorine from (water, etc.).
—**dechlorination,** *n.* the act or process of dechlorinating: *dechlorination of water by means of a carbon filter.*

deci-, *combining form.* one tenth of, as in *decigram = one tenth of a gram. Symbol:* d [from Latin *decimus* tenth, from *decem* ten]

decibel (des'ə bəl), *n. Physics.* a unit for measuring the relative intensity of sounds, equal to 1/10 of a bel: *Sound is measured in decibels. The decibel scale advances geometrically instead of arithmetically. Twenty decibels represents not twice as much noise as ten decibels, but ten times as much. The eighty-decibel level of a pneumatic drill is 100 times as noisy as the sixty-decibel level of conversation* (N.Y. Times).

decidua (di sij'ü ə), *n., pl.* **deciduae** (di sij'ü ē). *Physiology.* the part of the lining of the uterus in which a fertilized ovum is embedded and which is cast off at birth. [from New Latin (*membrana*) *decidua* deciduous (membrane)]
—**decidual,** *adj.* of or having to do with the decidua.

deciduate (di sij'ü it), *adj.* **1** *Zoology.* characterized by or having a decidua: *a deciduate mammal.*
2 *Anatomy.* composed in part by a decidua: *a deciduate placenta.*

deciduous (di sij'ü əs), *adj.* **1** *Botany.* **a** losing its leaves annually. Maples, elms, and most oaks are deciduous trees. *Evergreens are those which retain some foliage throughout the year, while deciduous trees periodically lose their leaves and are therefore bare for a portion of the year* (Finch and Trewartha, *Elements of Geography*). **b** composed of such trees or shrubs: *The Deciduous Hardwood Forest occurs chiefly in temper-*

ate regions with marked seasonal cycles (Greulach and Adams, *Plants*).
2 *Biology.* falling off at a particular season or stage of growth: *deciduous leaves, deciduous horns.*
[from Latin *deciduus* falling off, from *decidere* fall off]

decile (des'əl), *Statistics.* —*n.* one of ten equal parts of ranked data: *The range of salaries indicated by the survey extends from $24,500 or more in the highest decile to $9,500 or less in the lowest decile* (Nature). Compare **percentile.**
—*adj.* of, being, or involving deciles.

decimal (des'ə məl), *Mathematics.* —*n.* **1** a number expressed in a system of notation based upon ten, especially one containing a decimal fraction. EXAMPLES: 75.24, 3.062, .091, 5.0
2 = decimal fraction.
—*adj.* of or expressed by a decimal system: *The Hindus are distinguished in arithmetic by the acknowledged invention of the decimal notation* (Elphinstone, *History of India*).
[from Latin *decimus* tenth] —**decimally,** *adv.*

decimal fraction, *Mathematics.* a fraction whose denominator is ten or a multiple of ten, expressed by placing a decimal point to the left of the numerator. EXAMPLES: .04, .2 Also called **decimal.** Compare **common fraction, proper fraction.**

decimal place, *Mathematics.* the place to the right of the decimal point for tenths, hundredths, etc.

decimal point, *Mathematics.* the period used in expressing a decimal, as in 2.03 or .623. What is to the right of the dot is less than one. Digits to the left of the decimal point represent whole numbers, one or greater: 1.3 represents one plus three-tenths.

decimal system, *Mathematics.* any system of numeration which is based on units of ten. Compare **binary notation.**

decinormal, *adj. Chemistry.* having a concentration of one-tenth gram equivalent of solute per liter of solution: *water acidified with decinormal hydrochloric or nitric acid.*

declinate (dek'lə nit), *adj. Botany, Zoology.* bent or bending downward or aside. [from Latin *declinatum* bent away from, from *de-* away from + *clinare* to bend]

declination:

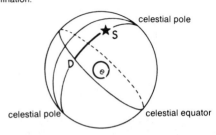

celestial pole

S
D

celestial pole

celestial equator

Line DS = declination of star S.

declination (dek'lə nā'shən), *n. Astronomy.* the angular distance of a star, planet, etc., from the celestial equator: *If the star is north of the equator, the sign of the declination is plus; if the star is south, the sign is minus* (Baker, *Astronomy*).

decompose, *v.* **1** *Chemistry.* **a** to separate (a substance) into what it is made of: *A compound is a substance which can be decomposed by a chemical change* (Chemistry Regents Syllabus). **b** (of a substance) to become separated into its parts.
2 *Biology.* to break down, especially by the action of microorganisms; decay; putrefy; rot.
—**decomposition,** *n.* **1** *Chemistry.* the act or process of decomposing. **2** *Biology.* decay; rot.

decomposer, *n. Ecology.* a fungus, saprophytic bacterium, or any of various animals that obtain nourishment by breaking down dead organic matter: *The organisms of an ecosystem can be divided into three categories: producers, consumers, and decomposers* (Weier, Botany).

decompound (dē kom'pound), *adj. Botany.* composed of things already compounded: *A decompound leaf is one whose petiole has additional petioles, each one supporting a compound leaf.*

decorticate (dē kôr'tə kit), *adj. Botany, Zoology.* without an outer covering; stripped, peeled, or husked. [from Latin *decorticatum* peeled, husked, from *de-* remove + *cortex* bark]

decrement (dek'rə mənt), *n. Mathematics.* the amount by which the value of a variable decreases. [from Latin *decrementum* decrease]

decrepitate (di krep'ə tāt), *v. Chemistry.* **1** to roast or calcine (a salt or other mineral) so as to cause crackling or until crackling stops.
2 (of salts and other minerals) to break up or make a crackling noise upon roasting. —**decrepitation,** *n.*

decumbent (di kum'bənt), *adj. Botany.* (of stems, branches, etc.) lying or trailing on the ground with the end tending to climb: *The fertile flocci were decumbent, probably from the weight of the spores* (Cooke, Fungi). [from Latin *decumbentis* lying down]

decurrent (di kėr'ənt), *adj. Botany.* extending down the stem: *decurrent leaves.*

decussate (di kus'āt), *adj. Botany.* (of leaves, etc.) arranged along the stem in pairs, each pair at right angles to the pair next above or below: *The stem has four angles, and bears decussate pairs of opposite leaves.* (Bower and Scott). [from Latin *decussatum* crossed, intersected] —**decussation,** *n.*

decussation (dē'kə sā'shən), *n. Anatomy.* a band of nerve fibers crossing the central nervous system and connecting unlike centers on the two sides.

dedifferentiate, *v. Biology.* (of a cell, tissue, organ, or organism) to undergo a process of reversion, losing its specialized form or function: *The sea squirt ... can in unfavourable surroundings dedifferentiate, or revert to an embryonic state* (J. S. Huxley, *Essays in Popular Science*).
—**dedifferentiation,** *n.* a process in which cells, tissues, etc., lose their special form or function; loss of specialization: *Dedifferentiation or return to a more embryonic condition probably underlies all types of regeneration* (American Journal of Anatomy).

deep, *n. Oceanography.* any area of the ocean more than 6,000 meters or 3000 fathoms deep: *The ocean beds have great chasms called deeps. The greatest deep is the Marianas Trench* (Gerhard Neumann).

deep scattering layer, *Oceanography.* microscopic marine organisms that form a layer deep in the ocean, causing a reflection of sonar impulses similar to those detected from reflections off the ocean floor and often confused with them.

deexcite, *v. Physics.* to reduce the high energy level of (an electron, an atom, etc.): *The density of the corona is so low that the atoms are not deexcited by collisions before they have a chance to radiate the energy* (Scientific American). —**deexcitation,** *n.*

defecate, *v. Physiology.* to discharge (food residues or waste matter) from the intestines through the anus.
—**defecation,** *n.* **1** the act or process of defecating. **2** that which has been defecated; excretion.

defibrinate, *v. Biochemistry.* remove fibrin from (blood or lymph).
—**defibrination,** *n.* the act or process of defibrinating.

deficient number or **defective number,** *Mathematics.* a whole number whose divisors (other than itself) have a sum less than this number. The number 8 is a deficient number because the sum of its divisors (1, 2, 4) is less than 8. Compare **abundant number, perfect number.**

definition, *n.* **1** *Optics.* the capacity of a lens to make the image of an object distinct to the eye: *With good definition, we can ... show features ... which had not previously been observed* (Scientific American).
2 the clarifying capacity of any other instrument or condition (such as weather): *The choice of power is determined by closeness and brightness of the two stars, together with the definition allowed by atmosphere and optics of the instrument* (Bernhard, *Handbook of Heavens*).

definitive, *adj. Biology.* fully formed: *The hair of most mammals ceases growth when it reaches a certain length, such as the body hair of humans or the fur of bears; this is called definitive hair* (Science News Letter).

definitive host, *Zoology.* the host which is inhabited by the adult stage of a parasite and in which the parasite may reproduce sexually. Contrasted with **intermediate host.**

deflation, *n. Geology.* the process by which the wind sorts and moves sand, rock particles, etc.: *Deflation is least effective in regions where soil particles are moist and adherent or are protected by a covering of vegetation. It is most effective in regions of aridity and scanty vegetation* (Finch and Trewartha, *Elements of Geography*).

deflect, *v. Physics.* to bend, turn aside, or change in direction: *In that year [1819] the Danish scientist Hans Christian Oersted observed that a pivoted magnet (a compass needle) was deflected when in the neighborhood of a wire carrying a current* (Sears and Zemansky, *University Physics*).
—**deflection,** *n.* **1** the act or process of deflecting.
2 the movement of the needle or indicator of a scientific instrument from its zero or normal position.

---

cap, fāce, fäther; best, bē, tėrm; pin, fīve;
rock, gō, ôrder; oil, out; cup, pùt, rüle;
*y*ü in use, *y*ü in uric;
ng in bring; sh in rush; th in thin, ŦH in then;
zh in seizure.
ə = *a* in about, *e* in taken, *i* in pencil, *o* in lemon, *u* in circus

deflocculate, *v. Chemistry.* to change from a flocculated state; separate into finely divided particles.
—**deflocculation,** *n.* separation into finely divided particles.

defoliate, *v. Botany.* to cause leaves to fall from (a tree or plant), especially prematurely.
—**defoliation,** *n.* a loss or shedding of leaves.

deform (di fôrm′), *v.* **1** *Physics.* to apply force to and change in shape, as by compressing, twisting, or shearing: *Moreover, since growth of the individual crystals still occurs, the internal stresses set up by this growth seriously reduce the effective creep strength of the material, and permit it to deform under very small external loads* (New Scientist).
2 *Geology.* to change the original state or size of a rock mass, especially by folding or faulting.
—**deformation,** *n.* the act or process of deforming.

deg., *abbrev.* degree or degrees.

degauss (dē gous′), *v.* = demagnetize.

degenerate (di jen′ə rāt), *v. Biology.* to evolve by losing an organ, function, or other highly developed characteristic: *The eyes of the electric eel are cloudy, degenerated, and useless in adult life* (A.W. Haslett). Also called **degrade.**
—*adj.* (di jen′ər it) **1** *Biology.* that has lost the normal or more highly developed characteristics of its kind.
2 *Genetics.* (of a codon) coding for the same amino acid as other codons; redundant: *It is implicit in a triplet code, which provides a potential surplus of triplets for 20 amino acids, that the code contains degenerate codons* (Science).
3 *Physics.* **a** being or existing in two or more quantum states with the same energy level. **b** (of a gas or other matter) having the lowest quantum states occupied; extremely compacted or dense.
—**degeneracy,** *n. Genetics.* the coding for the same amino acid by more than one codon.
—**degeneration,** *n.* **1** *Medicine.* deterioration in tissue or organs caused by disease, injury, etc.
2 *Biology.* a process of degenerating. Also called **devolution.** —**degenerative,** *adj.*

deglaciation (dē′glā shē ā′shən), *n. Geology.* the uncovering of a land mass due to the shrinkage and disappearance of a glacier.

degrade, *v.* **1** *Biochemistry.* to decompose into smaller, less complex molecules: *[A] polymerized herbicide degrades in moist soil at a controlled rate* (Science News). *Scientists mapped the molecular structure of ribonuclease, an enzyme that degrades ribonucleic acid* (Leonard Nelson). SYN: depolymerize.
2 *Geology.* to wear down by erosion. SYN: erode.
3 *Biology* = degenerate.
4 *Physics.* to change (energy) to a form less capable of doing work: *This was the line of reasoning that led to the notion of the "heat death" of the universe, the heat energy of the sun and the stars being transferred to colder bodies in the solar system and to the colder environment, until eventually the available heat becomes degraded throughout the whole of space* (I.B. Cohen).
—**degradable,** *adj.* capable of degrading or of being degraded; susceptible to decomposition: *biologically degradable detergents.* —**degradability,** *n.*

—**degradation,** *n.* the process of degrading or the condition of being degraded: *the law of degradation of energy.* SYN: decomposition.

degree, *n.* **1** *Physics.* a unit for measuring temperature: *The freezing point of water is 0 degrees (0°) Celsius or 32 degrees (32°) Fahrenheit.*
► The size of the degree Celsius (and also of the kelvin) was set in 1954 by the General Conference of Weights and Measures as 1/273.16 of the temperature of the triple point of water.
2 *Mathematics.* **a** a unit for measuring an angle or an arc of a circle. One degree is 1/90 of a right angle or 1/360 of the circumference of a circle.
► The symbol of a degree is °, that of a *minute* (1/60 of a degree) is ′, and that of a *second* (1/60 of a minute, or 1/3600 of a degree) is ″. Thus, 12°10′30″ stands for 12 degrees, 10 minutes, and 30 seconds.
b the rank of an algebraic term, determined by the sum of its exponents. EXAMPLE: a^3 and a^2b are terms of the third degree.
3 *Geography, Astronomy.* a position on the earth's surface or the celestial sphere as measured by degrees of latitude or longitude.
[from Old French *degre,* from Latin *de-* down + *gradus* grade]

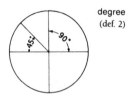

degree
(def. 2)

45° = an angle one-eighth of the circle
90° = an angle one-fourth of the circle

degree-day, *n. Meteorology.* a unit equal to one degree or deviation above or below a standard in the mean outdoor temperature for one day: *The number of heating degree-days per day is defined as the difference between 65°F and the mean daily outdoor temperature* (Science). *Plants appear to have biological clocks that ... operate on the basis of degree-days: the cumulative effect of certain temperatures multiplied by the time spent at those temperatures* (Scientific American).

degree of freedom, 1 *Physics, Chemistry.* **a** any of the ways in which a body may move; dimension. In space, there are three such dimensions. On a plane surface, there are two. **b** (of a system in equilibrium) any of the variables that may be changed, such as temperature, pressure, volume of a gas, or concentration of a solution.
2 *Statistics.* any of the independent variables that determine a value.

dehisce (dē his′), *v. Biology.* (of an organ, seed pod, etc.) to burst open, usually along a definite line, to discharge the seeds or other contents: *In ... Rhododendron and Azalea ... the anthers dehisce by small apical pores from which the pollen is shed* (Youngken, *Pharmaceutical Botany*). [from Latin *dehiscere,* from *de-* away + *hiscere* be open, gape]
—**dehiscence** (dē his′ns), *n.* the bursting open of an organ, seed pod, anther, etc., to discharge seeds, pollen, etc.

—**dehiscent** (dē his′nt), *adj.* bursting open to scatter seeds, etc.

dehiscent pods

dehydrase (dē hī′drās), *n.* **1** = dehydrogenase. **2** = dehydratase.

dehydratase (dē hī′drə tās), *n. Biochemistry.* an enzyme that causes the removal of water from a compound.

dehydrate (dē hī′drāt), *v.* **1** to remove water from.
2 *Chemistry.* to remove two atoms of hydrogen from (a compound) for each atom of oxygen removed: *A dehydrating agent is a substance which has such a pronounced affinity for water that it will tear down a molecule containing both hydrogen and oxygen atoms* (Jones, *Inorganic Chemistry*).
—**dehydration,** *n.* **1** loss or removal of water: *Terrestrial life is constantly exposed to the danger of death or injury due to dehydration, as a result many plants and animals have developed special adaptions for securing and conserving water* (C.J. Hylander).
2 *Chemistry.* the removal from a compound of two atoms of hydrogen for each atom of oxygen removed: *Reversal of hydrolysis is a reaction termed dehydration* (Storer, *General Zoology*).

dehydrochlorinase (dē hī′drə klôr′ə nās), *n. Biochemistry.* an enzyme that causes dehydrochlorination.

dehydrochlorinate, *v. Chemistry.* to remove hydrogen and chlorine from (a compound).
—**dehydrochlorination,** *n.* the removal of hydrogen and chlorine from a compound: *Mosquitoes immune to DDT by having mastered the art of inactivating it by dehydrochlorination* (New Scientist).

dehydrogenase (dē hī′drə jə nās), *n. Biochemistry.* an enzyme that causes the removal and transfer of hydrogen, as from body tissue.
ASSOCIATED TERMS: *Dehydrogenases, oxidases,* and *catalases* are specific enzymes that serve in oxidation and reduction, each kind acting on a particular substance (the *substrate*).

dehydrogenate (dē hī′drə jə nāt), *v. Chemistry.* to remove hydrogen from (a compound): *A molecule may be dehydrogenated by the action of a special dehydrogenase.* —**dehy′drogena′tion,** *n.*

deionize (di ī′ə nīz), *v. Chemistry.* to remove ions from, especially to purify (water) by changing ions into acids and using an absorbing agent to remove them.
—**dei′oniza′tion,** *n.*

deka- or **dek-,** *combining form.* ten, as in *dekagram = ten grams. Symbol:* da Also spelled **deca-** or **dec-**.
[from Greek *deka* ten]

delete (di lēt′), *v. Molecular Biology.* to remove or cause the loss of a section of DNA or other genetic material.

deletion, *n.* **1** *Molecular Biology.* the removal or loss of a section of DNA or other genetic material; also, the section deleted: *Deletions of single base pairs are the consequences of mispairing … The resulting double helix has two mispaired bases which if uncorrected by any repair process will give rise to two progeny helices, each containing a different nucleotide pair deletion* (Watson, *Molecular Biology of the Gene*). Compare **insertion.**
2 *Biology.* the loss or breaking off of part of a chromosome, resulting in a chromosomal alteration or mutation: *Chromosomes may, from time to time, undergo structural changes that result in changes in hereditary potentialities. For example, a portion of a chromosome may break off, and if it lacks a centromere (kinetochore), it will not be included in either of the two daughter nuclei at the next division. The result of such a deletion is the loss of a group of genes* (Greulach and Adams, *Plants*).
ASSOCIATED TERMS: A *deletion* (def. 2) is often contrasted with an *inversion,* in which the chromosome forms a loop, causing the original order of genes to be inverted, and *translocation,* in which parts of two nonhomologous chromosomes exchange places, often causing a great change in the expression of genes.

deliquesce (del′ə kwes′), *v.* **1** *Chemistry.* to become liquid by absorbing moisture from the air: *Calcium chloride deliquesces rapidly.*
2 *Botany.* **a** to divide into small branches or veins. **b** to become liquid; melt away: *Certain parts of some fungi deliquesce in the process of growth.*
[from Latin *deliquescere,* from *de-* + *liquescere* become fluid]
—**deliquescence** (del′ə kwes′ns), *n.* the act or process of deliquescing. Compare **efflorescence.**
—**deliquescent** (del′ə kwes′nt), *adj.* absorbing moisture from the air, sometimes becoming liquid as a result: *deliquescent calcium chloride crystals.*

delta

delta, *n. Geography, Geology.* the deposit of earth and sand, usually fanning out, that collects where a stream flows into a slower river, or into a lake or sea: *A large proportion of the sediments is cross-bedded; that is, the larger units are built of many thin, short, more or less steeply-inclined, curving beds that possibly have*

cap, fāce, fäther; best, bē, tėrm; pin, fīve;
rock, gō, ôrder; oil, out; cup, pùt, rüle,
yü in use, *yu* in uric;
ng in bring; *sh* in rush; *th* in thin, ᴛʜ in then;
zh in seizure.
ə = *a* in about, *e* in taken, *i* in pencil, *o* in lemon, *u* in circus

153

delta plain

been laid down in deltas or alluvial fans (Science News).

—*adj. Chemistry.* designating the fourth of several possible positions of atoms or groups of atoms which are substituted in a chemical compound. *Symbol:* δ Compare **alpha, beta, gamma.**

—**deltal,** *adj.* of or having to do with a delta: *the rich land of the deltal formations in the lower Mississippi River.*

delta plain, *Geography, Geology.* the nearly level lowland portion of a delta above water.

delta ray, *Physics.* an electron ejected when a fast-moving charged particle, such as an alpha particle, passes through matter.

delta wave or **delta rhythm,** *Physiology.* an electrical rhythm in the brain, especially during deep sleep, of under 6 cycles per second. In youth, these waves usually have a high amplitude; with increasing age, decreasing amplitude.

deltoid (del′toid), *n.,* or **deltoid muscle,** *Anatomy.* a large triangular muscle of the shoulder by which the arm is raised. The deltoid is an abductor muscle.

—*adj.* **1** *Biology.* shaped like the Greek letter delta; triangular in shape: *a deltoid leaf.*

2 *Anatomy.* of or having to do with the deltoid.

demagnetize, *v. Physics.* **1** to lose magnetism or be deprived of magnetic properties.

2 to neutralize the magnetism or magnetic properties of: *Because sound on tape is a kind of magnetic configuration, it can be erased (demagnetized) easily* (G. Kingsley and John Watts). Also called **degauss.**

—**demagnetization,** *n.* neutralization or loss of magnetism.

deme (dēm), *n. Biology.* a small group or population of organisms that interbreed: *The coyote-hybridization threat, together with very low red wolf numbers, requires that management be approached in terms of saving the red wolf's "gene pool." This term refers to the sum of genetic information carried by all members of a "deme," or interbreeding group* (J.H. Shaw and P.A. Jordan). [from Greek *dēmos* people, community]

demersal (di mėr′səl), *adj. Biology.* **1 a** living near the bottom of the sea. Contrasted with **pelagic. b** (of fishes' eggs) sinking to or deposited near the bottom of the sea. [from Latin *demersus* plunged down]

demonstration (dem′ən strā′shən), *n. Mathematics.* the process of proving that certain assumptions necessarily produce a certain result.

denatant (dē nā′tənt), *adj. Zoology.* swimming or migrating with the current, not against it: *denatant fishes.*

denature (dē nā′chər), *v. Biochemistry.* to change the structure of (a protein or other substance) and thus change its properties, as by adding heat, pressure, or chemicals: *In this form the protein is "denatured"; it has lost its stable, orderly architecture, is insoluble and is incapable of recombining into the original protein complex* (H. Fraenkel-Conrat).

—**denaturation,** *n.* the act or process of denaturing: *[The] distortion of enzyme molecules at high temperatures is enzyme denaturation* (Biology Regents Syllabus).

▶ The biochemical use of this term should not be confused with the commercial term *denatured alcohol,* meaning ethyl alcohol to which some ingredient has been added to render it undrinkable and therefore tax-exempt.

dendr-, *combining form.* variant form of **dendro-** before vowels, as in *dendroid.*

dendriform, *adj. Zoology.* like a tree in form; branching: *a dendriform sponge.* SYN: dendroid.

dendrite (den′drīt), *n.* **1** *Anatomy.* any of the short, branching extensions of a nerve cell that receive stimuli from other cells.

ASSOCIATED TERMS: A nerve cell, or *neuron,* differs from other cells in having *dendrites,* which conduct impulses to the cell body (*cyton*), and a single long fiber, the *axon,* which carries away the impulses from the cell body to another cell. Axons are covered with *myelin* and are either *sensory,* carrying impulses from a *sense organ* to the *central nervous system,* or are *motor,* carrying impulses from the central nervous system to a muscle.

2 *Mineralogy.* a mineral with a branching, treelike pattern, usually a manganese oxide.

3 *Crystallography.* a crystalline growth of branching form.

[from Greek *dendrites* of a tree, from *dendron* tree]

—**dendritic** (den drit′ik), *adj.* formed or marked like a dendrite; treelike. —**dentritically,** *adv.*

dendrite (def. 2)

dendro- or **dendri,** *combining form.* of a tree or trees, as in *dendrochronology, dendrology, dendriform.* Also spelled **dendr-** before vowels. [from Greek *dendron* tree]

dendrochronology (den′drō krə nol′ə jē), *n.* **1** the study of annual rings in trees to determine dates and environmental conditions in the past: *Conifer logs from houses and pavements, which in places are 28 layers thick, have been dated by dendrochronology* (New Scientist).

2 a chronology based upon such study: *By matching the patterns of rings from many trees it has been possible to set up a dendrochronology of about 2,000 years' duration that allows archeologists to date their sites within a few years* (American Scholar).

—**dendrochronological** (den′drō kron′ə loj′ə kəl), *adj.* of or having to do with dendrochronology.

dendroclimatology (den′drō klī′mə tol′ə jē), *n.* the study of rainfall and other aspects of past climatic conditions by analysis of the annual growth rings of trees: *Dendroclimatology is a subdiscipline of dendrochronology in which the annual variations in climate are reconstructed from variations in characteristics of dated rings. Tree-ring evidence for past climatic variation is unique in that the climatic information is precisely dated to the year* (Harold C. Fritts).

dendrogram, *n. Biology.* a branching diagram showing the relationship between various groups or species at different levels of similarity. Compare **cladogram.**

dendroid (den′droid), *adj. Zoology.* resembling a tree in structure or growth; arborescent: *dendroid corals.* SYN: dendriform.

dendrology, *n.* the branch of botany dealing with the study of trees.
—**dendrological,** *adj.* of or having to do with dendrology.

denitrate (dē nī′trāt), *v.* = denitrify. —**denitration,** *n.*

denitrify (dē nī′trə fī), *v. Chemistry.* **1** to remove nitrogen or its compounds (from).
2 to change (nitrates or nitrites) by reduction into a gas such as nitrogen (N_2) or ammonia: *In many soils there are denitrifying bacteria ... the denitrifying bacteria thrive in situations where oxygen is deficient, and so are likely to be abundant and active in poorly drained soils* (Greulach and Adams, *Plants*). —**deni′trifica′tion,** *n.*

denominate number (di nom′ə nit), *Mathematics.* a number used with a unit. EXAMPLES: 10kg, 20°C

denominator (di nom′ə nā′tər), *n. Mathematics.* the number below the line in a fraction, stating the size of the whole compared to the parts above the line: *In 2/3, 3 is the denominator, and 2 is the numerator.*

density, *n.* **1** *Physics.* the mass of a substance per unit volume: *The density of lead is greater than the density of wood. Although the density in stellar cores may range up to several hundred times the density of water (even the mean density of the sun somewhat exceeds that of water), quantum theory proves that at these temperatures there are no cohesive links between atoms. Accordingly, the material is neither solid nor liquid, but must be gaseous* (C.M. and H. Bondi). Compare **specific gravity.**
▶ In the metric system, density is measured in grams per cubic centimeter or in kilograms per cubic meter.
2 the quantity of anything per unit area: *If 240 people lived on one city block in one area and 80 people lived on one block in another area, the densities could be described as 240 people per block in the first case and 80 people per block in the second* (Robert W. Howe).
3 *Optics.* (of a transparent medium) the common logarithm of the opacity.

density current, 1 *Physics.* the passage by gravity of a liquid or gas through or around a fluid of slightly different density.
2 *Geology.* = turbidity current.

dent- or **denti-,** *combining form.* of or like a tooth or teeth; toothed, as in *dentoid, dentiform.* [from Latin *dentem* tooth]

dental pulp, *Anatomy.* the soft, sensitive substance containing the nerves and blood vessels that fills the cavity of a mature tooth.

dentate leaf

dentate (den′tāt), *adj. Botany, Zoology.* having toothlike projections; toothed; notched: *a dentate leaf.*
—**dentately,** *adv.*

—**dentation** (den tā′shən), *n.* **1** dentate condition or form. **2** one or more toothlike parts or projections: *the dentation on the wings of a butterfly or moth.*

denticle (den′tə kəl), *n. Zoology.* a small tooth or toothlike part. [from Latin *denticulus,* diminutive of *dentem* tooth]
—**denticular** (den tik′yə lər), *adj.* having the form of a small tooth.

denticulate (den tik′yə lit), *adj. Botany, Zoology.* finely toothed or notched: *a denticulate leaf, the denticulate scales of some fishes.* —**denticulately,** *adv.*
—**denticulation** (den tik′yə lā′shən), *n.* **1** a denticulate condition or form.
2 a denticle or denticles.

dentiform, *adj.* of the form of a tooth: tooth-shaped: *a dentiform vertebra, dentiform crystals.* SYN: dentoid.

dentin (den′tən′) or **dentine** (den′tēn′ *or* den tēn′), *n. Anatomy.* the hard, bony material beneath the enamel of a tooth, forming the main part of a tooth.
—**dentinal** (den′tə nəl *or* den tē′nəl), *adj.* of or having to do with dentin: *dentinal pulp, dentinal tissue.*

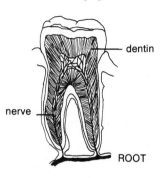
dentin
nerve
ROOT

dentition (den ti′shən), *n. Zoology.* the kind, number, and arrangement of teeth.

dentoid (den′toid), *adj.* = dentiform.

denucleate (dē nü′klē āt), *v. Biology.* to remove the nucleus or nuclei from: *to denucleate an egg cell.*
—**denucleation,** *n.* the act or process of denucleating; removal of a nucleus or nuclei.

denude, *v. Geology.* to lay (a rock, etc.) bare as by erosion: *... rapidly denuded by rain and rivers* (A. R. Wallace).
—**denudant,** *n.* any agent or agency causing denudation.
—**denudation,** *n.* the process of denuding: *Denudation of the rocks of the Alps has laid bare massive sections whose slopes can be measured as they disappear down below other rock layers* (New Scientist).

cap, fāce, fäther; best, bē, tèrm; pin, five;
rock, gō, ôrder; oil, out; cup, pùt, rüle;
yü in use, *yu̇* in uric;
ng in bring; *sh* in rush; *th* in thin; ᴛʜ in then;
zh in seizure.
ə = *a* in about, *e* in taken, *i* in pencil, *o* in lemon, *u* in circus

155

denumerable (di nü′mər ə bəl), *adj. Mathematics.* that can be put into one-to-one correspondence with the positive integers: *a denumerable set.* Also called **countable.**
—**denumerability,** *n.* denumerable quality or condition.
—**denumerably,** *adv.* by or with denumerable elements.

deoxidant, *n. Chemistry.* a deoxidizing agent.

deoxidation, *n. Chemistry.* the act or process of deoxidizing.

deoxidize (dē ok′sə dīz), *v. Chemistry.* to remove oxygen from (a compound); reduce. —**deoxidizer,** *n.*

deoxy-, *combining form. Chemistry.* containing fewer atoms of oxygen in the molecule than (the specified compound), as in *deoxycholic* (*acid*).

deoxycholic acid (dē ok′sə kō′lik), *Chemistry.* a crystalline acid occurring in the bile of certain mammals including man and containing one less hydroxyl group than cholic acid. *Formula:* $C_{24}H_{40}O_4$

deoxycorticosterone (dē ok′sə kôr′tə kos′tə rōn), *n. Biochemistry.* a steroid hormone secreted by the cortex of the adrenal gland or produced synthetically, used especially in its acetate form in the treatment of adrenal deficiency, epilepsy, and certain conditions of hypotension: *Deoxycorticosterone and the other natural adrenal hormones control electrolyte and water balance* (New Scientist). *Formula:* $C_{21}H_{30}O_3$

deoxygenate (dē ok′sə jə nāt), *v. Chemistry.* to remove free oxygen from (water, etc.).
—**deoxygenation,** *n.* the act or process of deoxygenating: *Extermination of fish through deoxygenation of water takes place sometimes ... when fish that have been landlocked in a pond or lagoon find the water getting too warm* (Science News Letter).

deoxyribonuclease (dē ok′sə rī′bō nü′klē ās), *n. Biochemistry.* an enzyme that promotes the hydrolysis of deoxyribonucleic acid. Usually called **DNAase** or **DNase.**

deoxyribonucleic acid (dē ok′sə rī′bō nü klē′ik), *Biochemistry.* a substance in all living cells, that is the genetic material passed from generation to generation. It is usually abbreviated DNA. DNA is a nucleic acid in the chromatin of all animals and plants and is also found in many viruses. It replicates and controls through messenger RNA (ribonucleic acid) the inheritable characteristics of all organisms.

A molecule of DNA consists of two parallel twisted chains of alternating units of phosphoric acid and deoxyribose, linked by crosspieces of the purine bases, adenine and guanine, and the pyrimidine bases, cytosine and thymine. See the picture at **DNA.**
▶ See the note under **ribonucleic acid.**

deoxyribonucleoprotein (dē ok′sə rī′bō nü′klē ō prō′tēn), *n. Biochemistry.* a nucleoprotein that contains DNA.

deoxyribose (dē ok′sə rī′bōs), *n. Biochemistry.* the sugar constituent of deoxyribonucleic acid: *All DNA nucleotides contain three units: phosphoric acid, a five-carbon sugar called deoxyribose, and a base, in this case a ring structure that can take up hydrogen ions* (McElroy, *Biology and Man*). *Formula:* $C_5H_{10}O_4$
See the picture at **DNA.**

dependent, *adj. Mathematics.* **1** (of a variable) determined by the values of one or more independent variables. **2** (of an equation) that can be derived from another equation. $x + y = 3$ and $2x + 2y = 6$ are dependent equations; any pair of values of x and y that satisfies one will satisfy the other.

depolarize *v. Physics.* to eliminate magnetic or other polarization.
—**depolarization** *n.* elimination of polarization, as in an electric cell. —**depolarizer,** *n.*

depolymerize (dē pol′ē mə rīz′), *v. Chemistry.* (of a polymeric molecule) to break down into smaller units:
—**depolymerization,** *n.: Glucose is a sugar that nearly all living organisms can use. The problem, then, is to convert cellulose into glucose by depolymerization* (Elwyn T. Reese).

deposit, *n. Geology.* **1** a concentrated aggregate of one or more minerals, especially one of potential economic value: *A large proportion of the deep sea deposits is organic in origin, consisting of the shells of planktonic animals.* (Science News).
2 a mass of some mineral in rock or in the ground.

deposition, *n. Geology.* **1** the laying down of sediments by floodplains, glaciers, alluvial fans, etc.: *continental and marine deposition.* SYN: sedimentation. **2** = deposit.
—**depositional,** *adj.* of or having to do with a deposition: *... depositional areas such as flood plains, old sea floor, or lake beds* (White and Renner, *Human Geography*).
ASSOCIATED TERMS: see **erosion.**

depressed, *adj. Botany, Zoology.* flattened down; broader than high.

depression, *n.* **1** *Geology.* a low place or hollow on a plain surface.
2 *Astronomy.* the angular distance of a heavenly body below the horizon.
3 *Meteorology.* an area of low atmospheric pressure: *Cyclones are low-pressure storms and commonly go by the name of lows, or depressions* (Finch and Trewartha, *Elements of Geography*).

depressor, *n. Anatomy.* a muscle that pulls down a part of the body. The **mandibular depressor** lowers the jaw to open the mouth. Contrasted with **levator.**

depressor nerve, *Physiology.* a nerve that acts to lower the blood pressure when stimulated.

deproteinize, *v. Chemistry.* to remove the protein from: *a deproteinized blood filtrate.*

depside (dep′sīd), *n. Chemistry.* any of a group of esters formed from phenolic carboxylic acids. [from Greek *depsein* to tan, from *depsa* skin, hide]

derepress (dē′ri pres′), *v. Genetics.* to induce (a gene) to operate by disengaging the repressor: *The group ... now hopes to find out how genes are repressed and derepressed—turned off and on—so that genes can be made to operate when required* (Charles S. Marwick).
—**derepressor,** *n.* the mechanism that activates a gene by disengaging the repressing mechanism, or repressor.

derivation, *n. Mathematics.* the operation of deducing a function, quantity, etc., from another, especially the deducing of a differential coefficient.

derivative, *n.* **1** *Chemistry.* a substance obtained from another by modification or by partial substitution of components: *Acetic acid is a derivative of alcohol.*
2 *Mathematics.* **a** the instantaneous rate of change of a function with respect to its variable. **b** (in differential calculus) a function deduced from another function; a differential coefficient.

derive, *v.* **1** *Mathematics.* to deduce (a function, quantity, etc.) from another or others; define in terms of more fundamental elements.
2 *Chemistry.* to obtain (a substance) from another.

derived unit, *Physics.* any unit that is derived from, or defined in terms of, one or more fundamental units. The newton is a derived unit defined in terms of the kilogram, the meter, and the second.

derma (dėr′mə), *n.* = dermis. [from Greek *derma* skin]

dermal, *adj. Anatomy.* **1** of or having to do with the dermis: *the dermal epithelium.* **2** of or having to do with the skin; cutaneous: *a dermal reflex, dermal necrosis.*

dermatogen (dər mat′ə jən), *n. Botany.* a thin layer of growing tissue from which the epidermis is developed. Also called **protoderm.**

dermatome (dėr′mə tōm), *n. Embryology.* the part of the mesoderm from which the dermis develops. [from Greek *derma* skin + *-tomia* a cutting]

dermatophyte (dėr′mə tə fīt *or* dėr mat′ə fīt), *n. Biology.* any fungus that is parasitic on the skin, hair, or nails.

dermatosome (dėr mat′ə sōm), *n. Botany.* one of the granular bodies which are thought to form the cell wall of a plant. Dermatosomes occur in rows and are united and surrounded by protoplasm.

dermis (dėr′mis), *n. Anatomy.* the sensitive layer of skin beneath the epidermis: *The thin flexible skin over the entire [frog], as in all land vertebrates, ... comprises an outer stratified epidermis and a dermis beneath, both of several cell layers* (Storer, *General Zoology*). Also called **corium, cutis.** See the picture at **skin.**

desalinate (dē sal′ə nāt′), *v. Chemistry.* to remove the salt from or reduce the amount of salt in (sea water, etc.). —**desalination,** *n.*

descending, *adj.* **1** extending or directed downward: *The descending colon extends down toward the rectum. The moon's descending node is the point extending southward.*
2 *Botany.* = determinate. Contrasted with **ascending.**

descriptive geometry, 1 the branch of geometry that solves three-dimensional problems by making projections of figures on two planes that are perpendicular to each other. **2** geometry in which projections are used in solving problems.

desegmentation, *n. Zoology.* the reduction of the number of segments by the coalescence of two or more, as in the carapace of a lobster.

desert (dez′ərt), *n. Ecology, Geography.* a dry, barren region that is usually sandy or rocky and without trees: *Most of the world's deserts are located about 30° north and south of the equator, in the so-called Horse Latitudes, [which] are continually parched by warm, dry winds that suck up any moisture* (Birkeland and Larson, *Putnam's Geology*). ▶ See the note under **biome.**
—**desertic** (di zėr′tik), *adj.* of or having to do with a desert or deserts.
—**desertification,** *n.* the process of turning into arid land or desert: *Generally thought of as the degradation of lands by natural and human means, desertification*

results in the diminution or destruction of the land's biological productivity (Science News).

desiccant (des′ə kənt), *n., adj. Chemistry.* (an agent or substance) that removes moisture: *Two of the most popular desiccants are phosphorus pentoxide and magnesium perchlorate* (New Scientist).

desmid (des′mid), *n. Biology.* any of a large family (Desmidiaceae) of single-celled green algae that live in fresh water and are typically drawn together in the middle: *Among the algae, the diatoms, desmids, and dinoflagellates, with their unusual contours and outer skeletons, take on a bizarre appearance* (McElroy, *Biology and Man*). [from Greek *desmos* chain, band]

desmosome (des′mə sōm), *n. Biology.* an adhesive part of an epithelial cell by which it adheres to adjoining cells.

desorb (dē sôrb′ *or* dē zôrb′), *v. Chemistry.* to release or be released from an absorbed or adsorbed condition.
—**desorption** (dē sôrp′shən *or* dē zôrp′shən), *n.* the process of desorbing: *Desorption from the walls of the system constitutes a major source of the residual gas that keeps dribbling into an ultra-high-vacuum system* (Scientific American).

destructive distillation, *Chemistry.* the decomposition of a complex substance, such as crude oil, coal, or wood, by subjecting it to high temperature in the absence of air or oxygen.

destructive metabolism, *Biology.* the breaking down of organic tissue into simpler substances or waste substances. Also called **catabolism.**

determinant (di tėr′mə nənt), *n.* **1** *Mathematics.* **a** a certain number of quantities arranged in a square block whose value is the sum of all the products that can be formed from them according to certain rules. **b** the square block itself.
2 *Biology.* **a** a gene: *The two types may be distinguished by a sex determinant which is possessed by the male and is lacking in the female* (E.L. Wollman and F. Jacob). **b** a hereditary factor corresponding to the function of a gene. Also called **determiner.**
—**determinantal,** *adj.* having to do with or expressed in determinants.

determinate (di tėr′mə nit), *adj. Botany.* having flowers which arise from terminal buds and thus terminate a stem or branch. The forget-me-not has determinate inflorescence. SYN: cymose. Compare **indeterminate.** See the picture at **inflorescence.**

determination, *n. Embryology.* the changes within embryonic cells which fix their future course of development.

determine, *v. Geometry.* to fix or define the position of: *Three points determine a plane.*

determiner, *n.* = determinant (def. 2).

---

cap, fāce, fäther; best, bē, tėrm; pin, fīve;
rock, gō, ôrder; oil, out; cup, pùt, rüle;
yü in use, *yù* in uric;
ng in bring; *sh* in rush; *th* in thin, ᴛʜ in then;
zh in seizure.
ə = *a* in about, *e* in taken, *i* in pencil, *o* in lemon, *u* in circus

detoxicant

detoxicant, *n. Chemistry.* a substance used to detoxify an organism: *... giving them [the birds] salt water to drink and milk as a detoxicant against the oil they have swallowed* (New Yorker).

detoxicate, *v.* = detoxify.

—detoxication, *n.* = detoxification.

detoxification, *n. Chemistry.* 1 the act or process of detoxifying. 2 the condition of being detoxified.

detoxify, *v. Chemistry.* to remove toxic or poisonous qualities from; remove a poison or its effects from: *Mosquitoes used the same enzyme as do flies to detoxify DDT* (Science News Letter).

detritus (di trī′təs), *n. Geology.* an accumulation of small fragments such as sand, silt, etc., worn away from rock: *Streams, waves, and currents are not the only agents that deposit sediments. Glaciers and winds also move and deposit detritus which may then be cemented into rock* (Gilluly, *Principles of Geology*). [from Latin, a rubbing away, from *deterere* rub away]

—detrital (di trī′tl), *adj.* composed of or involving detritus: *Late Precambrian and later iron formations were considered to be dominantly detrital* (Lawrence Ogden).

deuterate (dü′tə rāt), *v. Chemistry.* to add deuterium to (a compound).

—deuterated, *adj.* containing deuterium: *a deuterated acid.*

—deuteration, *n.* the condition of being deuterated.

deuteride, *n. Chemistry.* a compound of deuterium containing some other element or radical, as of lithium.

deuterium (dü tir′ē əm), *n. Chemistry.* an isotope of hydrogen whose atoms have about twice the mass of ordinary hydrogen. Deuterium occurs in heavy water. In nature, about one in every 6500 hydrogen atoms is deuterium. *Symbol:* D; *atomic number* 1; *mass number* 2. Also called **heavy hydrogen.** Compare **tritium.** [from New Latin *deuterium,* from Greek *deuteros* second]

deuterium oxide, = heavy water.

deuteron (dü′tə ron′), *n. Nuclear Physics.* the nucleus of an atom of deuterium, consisting of one proton and one neutron.

deutoplasm (dü′tə plaz′əm), *n. Biology.* a substance other than the nucleus and cytoplasm in a cell, especially the yolk in an egg cell. [from Greek *deut(eros)* second + *plasma* something molded]

—deutoplasmic, *adj.* of or having to do with deutoplasm.

developmental biology, the study of the development of organisms: *The question developmental biology seeks to answer concerns the way in which a single fertilized cell can grow into an organized mass of millions of cells, of many different types, all functioning together as a whole* (New Scientist and Science Journal).

deviate (dē′vē it *or* dē′vē āt), *n. Statistics.* the value of a variable measured from some standard point, usually the mean.

deviation, *n. Statistics.* the difference between the mean of a set of values and one value in the set, used to measure variation from the norm. See also **standard deviation.**

devitrify (dē vit′rə fī), *v. Chemistry, Geology.* to cause (any glassy substance) to become crystalline in structure.

—devitrification, *n.* the process of devitrifying.

devolution, *n.* = degeneration (def. 2).

Devonian (də vō′nē ən), *Geology.* —*n.* 1 the fourth period of the Paleozoic era, after the Silurian and before the Mississippian, characterized by the appearance of amphibians, wingless insects, and forests: *Another time term whose place of origin can be recognized readily is the Devonian, named for the rocks that crop out along the southwestern tip of Great Britain in Devonshire and Cornwall* (Birkeland and Larson, *Putnam's Geology*).

2 the rocks formed during this period: *Marine Devonian is found in the Appalachian trough* (Gilluly, *Principles of Geology*).

—adj. of or involving the Devonian (period or rocks).

dew, *n. Meteorology.* the fine water droplets that condense on solid surfaces when the air temperature drops below the dew point.

Dewar flask, or **Dewar** (dü′ər), *n.* a glass vessel for reducing heat loss or gain, consisting of double walls coated with silver to reflect heat and a vacuum between the walls to minimize heat conduction. It is a kind of thermos or vacuum bottle used especially to store liquefied gases. [named after Sir James *Dewar,* 1842–1923, Scottish chemist who invented the flask]

dewclaw, *n. Zoology.* 1 a small, useless inner claw or toe in some dogs and other animals, not reaching the ground in walking. 2 the false hoof of deer, hogs, and other hoofed mammals, consisting of two rudimentary toes.

dew point, *Meteorology.* the temperature to which air at a given pressure and water-vapor content must be lowered for saturation to occur; temperature at which dew forms.

dextr-, *combining form.* the form of *dextro-* before vowels, as in **dextral.**

dextral (dek′strəl), *adj.* 1 turning towards the right: *Most snail shells are dextral. There is dextral shear movement across the valley associated with the Californian fault system and ... the total displacement amounts to about 60 mm per year* (New Scientist).

2 *Physiology.* right-handed. Contrasted with **sinistral.**

—dextrality, *n. Physiology.* right-handedness. **—dextrally,** *adv.*

dextran (dek′stran), *n. Biochemistry.* any of a class of slimy white carbohydrates produced in sugar solutions by fermenting bacteria. They are used as substitutes or expanders of blood plasma because they do not transmit certain types of viruses and because they require no refrigeration. Dextrans are branched polysaccharides composed of glucose monomers.

dextranase (deks′trə nās), *n. Biochemistry.* a hydrolytic enzyme that prevents formation of dental plaque by breaking down dextran.

dextrin (dek′strin), *n. Biochemistry.* any of a class of gummy carbohydrates obtained from starch by the application of heat or by treatment with both heat and acids. They are used as adhesives, as thickening agents, and as substitutes for lactose, natural gums, etc. Dextrins are polysaccharides of intermediate chain length, composed of glucose monomers.

dextro (dek′strō), *adj.* = dextrorotatory.

158

dextro-, *combining form.* **1** toward the right, as in *dextrorotatory.* **2** dextrorotatory, as in *dextroamphetamine.* Also spelled **dextr-** before a vowel. Contrasted with **levo-**. [from Latin *dexter* right]

dextrorotation, *n. Physics, Chemistry.* rotation of the plane of polarization of light to the right.

dextrorotatory (dek'strə rō'tə tôr'ē), *adj. Physics, Chemistry.* characterized by turning the plane of polarization of light to the right, as a crystal, lens, or compound in solution. Dextrorotatory compounds are often prefixed by the symbol (+) or *d* − to distinguish them from their levorotatory forms. *By measuring the amount of rotary polarization, it is possible to determine very quickly the concentration of an optically active substance in the presence of one that is optically inactive. For example, ordinary sugar (sucrose) is dextrorotatory in an aqueous solution because of the asymmetry of its molecule* (Hardy and Perrin, *Principles of Optics*). Contrasted with **levorotatory.**

dextrorse (deks'trôrs), *adj. Botany.* rising spirally from left to right, or in a clockwise direction when viewed from directly underneath or from the base: *the dextrorse stem of a vine.* [from Latin *dextrorsum* turned to the right, from *dexter* right + *versum* turned] —**dextrorsely,** *adv.*

dextrose (dek'strōs), *n.* = glucose.

dg., *abbrev.* decigram *or* decigrams.

di-, *prefix.* **1** twofold; twice; double, as in *dicotyledon.* **2** two; having or involving two, as in *dihybrid, digraph.* **3** containing two atoms, radicals, etc., of the substance specified, as in *dioxide, disaccharide.* ▶ The corresponding prefixes for one and three are **mono-** and **tri-.**

Di, *symbol.* didymium.

dia., *abbrev.* diameter.

dia-, *prefix.* through; across; thoroughly, as in *diagenesis, diakinesis.* [from Greek *dia* through, apart]

diabase (dī'ə bās), *n. Geology.* **1** a fine-grained, dark-colored, crystalline, granular igneous rock related to basalt and gabbro, consisting essentially of augite and calcic plagioclase feldspar.
2 an altered form of basalt or dolerite.
—**diabasic** (dī'ə bā'sik), *adj.* = ophitic.

diabatic (dī'ə bat'ik), *adj. Physics.* of or involving transmission of heat. Compare **adiabatic.** [from Greek *diabatos* passable]

diacid (dī as'id) or **diacidic** (dī ə sid'ik), *adj. Chemistry.* (of a base or alcohol) able to combine with two molecules of a monobasic acid or one of a dibasic acid, forming a salt or ester: *Bases, such as barium, strontium, calcium, and magnesium hydroxides, that contain two hydroxyl ions in one molecule of the base are known as diacid bases* (Parks and Steinbach, *Systematic College Chemistry*).

diactine (dī ak'tin), *Zoology.* —*adj.* (of a sponge spicule) pointed at both ends; having two rays.
—*n.* a straight or curved spicule with two rays. [from *di-* two + Greek *aktinos* ray]

diadelphous (dī'ə del'fəs), *adj. Botany.* **1** (of stamens) united so as to form two bundles or sets. **2** (of plants) having the stamens so united. [from *di-* two + Greek *adelphos* brother]

diadromous (dī ad'rə məs), *adj. Zoology.* that migrate between salt and fresh waters: *Of freshwater and diadromous fish, carp catches declined slightly to 286,000 tons* (Donald R. Whitaker). Compare **anadromous, catadromous.**

diagenesis (dī'ə jen'ə sis), *n.* **1** *Geology.* any modification in a sediment between its deposition and its metamorphism, including its conversion into rock.
2 *Chemistry.* any breaking up into or recombination of constituent atoms or radicals.
—**di'agenet'ic,** *adj.* of, having to do with, or altered by diagenesis: *Pyrite and siderite grains are diagenetic* (Lawrence Ogden). —**diagenetically,** *adv.*

diageotropic (dī'ə jē'ə trop'ik), *adj. Botany.* tending to grow horizontally to the earth's surface, such as branches and roots. [from *dia-* through + Greek *geō-* earth + *tropikos* pertaining to a turn]
—**diageotropism** (dī'ə jē ot'rə piz'əm), *n.* the tendency to exhibit diageotropic growth.

diagnosis, *n.* **1** *Medicine.* the determination of a patient's condition or disease after thorough investigation. Compare **prognosis.**
2 *Biology.* a statement of the determining characteristics of a genus, species, etc.
[from Greek *diagnosis,* from *dia-* apart + *gignoskein* know]
—**diagnostic,** *adj.* of or having value in diagnosis.

diagonal, *Geometry.* —*n.* a line segment connecting two nonadjacent vertices in a polygon or polyhedron; line that cuts across in a slanting direction.
—*adj.* connecting two nonadjacent corners in a polygon or polyhedron: *diagonal lines.* —**diagonally,** *adv.*
▶ Straight-sided figures have *diagonals;* circles have *diameters.*

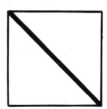

diagonal

diagram, *n. Mathematics.* a drawing or figure used in the proof of a geometrical proposition, in illustrating algebraic concepts, and in representing various data. [from Greek *diagramma,* from *dia-* apart, out + *gramma* lines (of a drawing, etc.)]

diakinesis (dī'ə ki nē'sis), *n. Biology.* the final stage in the prophase of meiosis, in which there is marked contraction in length of the chromatids.

cap, fāce, fäther; best, bē, tèrm; pin, fīve;
rock, gō, ôrder; oil, out; cup, pùt, rüle,
*y*ü in use, *y*ù in uric;
ng in bring; sh in rush; th in thin, ᴛʜ in then;
zh in seizure.
ə = *a* in about, *e* in taken, *i* in pencil, *o* in lemon, *u* in circus

dialdehyde

▶ The stages of the meiotic prophase are the *leptotene*, the *zygotene*, the *pachytene*, the *diplotene*, and the *diakinesis*.

dialdehyde (dī al'də hīd), *n. Chemistry.* a compound made up of two aldehydes.

dialysate (dī al'ə sāt *or* dī al'ə zāt), *n. Chemistry.* **1** the product of dialysis; the dialyzed part of a substance. **2** the purifying liquid through which substances from the blood pass in hemodialysis.

dialysis (di al'ə sis), *n.* **1** *Chemistry.* the separation of colloids or large molecules from dissolved substances or small molecules, (a process governed by the principle that small molecules diffuse readily through a membrane and colloids or large molecules not at all or very slightly): *If a membrane permeable to sodium chloride separates a 10 per cent solution of this salt (A) from distilled water (B), the sodium chloride diffuses from side (A) to side (B) until equilibrium is established. Should side (A) contain colloidal material, as well as salt, the colloidal matter does not diffuse through the membrane, which thus effects a separation, termed dialysis, of the colloidal matter from the dissolved material which is capable of diffusion* (Harbaugh and Goodrich, *Fundamentals of Biology*). **2** *Medicine.* the removal of poisons from the bloodstream by chemical dialysis, as of one whose kidneys do not function properly. Also called **hemodialysis**. [from Greek *dialysis* dissolution, from *dia-* apart + *lyein* to loose]
—**dialytic** (dī'ə lit'ik), *adj.* of or having to do with dialysis. —**dialytically**, *adv.*

dialyze (dī'ə līz), *v.* **1** *Chemistry, Medicine.* to separate or obtain by dialysis: *In the artificial kidney device the blood comes in contact with cellophane or another kind of semipermeable membrane through which the metabolic waste products pass into a fluid known as dialyzing solution* (N. Y. Times). **2** *Medicine.* to receive or give treatment to on a kidney dialysis machine.

diam., *abbrev.* diameter.

diamagnet (dī'ə mag'nit), *n. Physics.* a diamagnetic substance: *A superconductor ... behaves as a perfect diamagnet—that is, it excludes magnetic fields from its interior* (Science).

diamagnetic (dī'ə mag net'ik), *Physics.* —*adj.* repelled detectably by a magnet: *diamagnetic ions, a diamagnetic compound.*
—*n.* a diamagnetic substance.
—**diamagnetically**, *adv.* Compare **paramagnetic**.
—**diamagnetism** (dī ə mag'nə tiz'əm), *n.* the quality of being diamagnetic: *Some believe that the diamagnetism is a more fundamental property of a superconductor than its lack of resistance, but, in fact, a theory of superconductivity must account for both phenomena* (Science).

diameter, *n.* **1** *Mathematics.* **a** a line segment passing from one side to the other through the center of a circle, sphere, etc. **b** the length of such a line segment; measurement from one side to the other through the center; width; thickness: *The diameter of the earth is almost 13,000 kilometers.* See the picture at **sphere**.

2 *Optics.* a unit for measuring the strength of a lens of a microscope or telescope. If the lens makes the object appear twice as broad, it has magnified it two diameters, usually written 2x.
—**diametral** (dī am'ə trəl), *adj.* of, having to do with, or forming a diameter. —**diametrally**, *adv.*
—**diametric** (dī'ə met'rik), *adj.* of or along a diameter. —**diametrically**, *adv.*

diamide (dī'ə mīd' *or* dī am'īd), *n. Chemistry.* any of various compounds having two amido radicals.

diamine (dī'ə mēn' *or* dī am'ēn), *n. Chemistry.* any of various compounds having two amino radicals.

diamond, *n. Mineralogy.* an isometric crystallized form of pure carbon, used as a precious stone and as an abrasive. The varieties used as gems are colorless. Diamond is the hardest natural substance known, ranking 10 on the Mohs scale.

diandrous (dī an'drəs), *adj. Botany.* **1** having two stamens: *diandrous flowers.* **2** having flowers with two stamens: *diandrous plants.* [from *di-* two + Greek *andros* man]

diapause (dī'ə pôz), *n. Zoology.* a period of suspended development (as in some insects) during which physiological activity is very low: *Most insects survive periods when the environment is hostile by going into a resting phase called diapause* (Science News Yearbook). SYN: dormancy.
▶ A *diapause* occurs during an insect's development; *hibernation* may occur yearly after an animal is fully developed.

diapedesis (dī'ə pə dē'sis), *n., pl.* **-ses** (-sēz'). *Biology.* the movement or passage of blood cells, especially phagocytes, through the unruptured walls of the capillary blood vessels into the tissues. [from Greek *diapēdēsis* an oozing through]

diaphorase (dī af'ə rās), *n. Biochemistry.* a flavoprotein active in the transfer of electrons from the reduced form of diphosphopyridine nucleotide to molecular oxygen. [from Greek *diaphoros* different + English *-ase*]

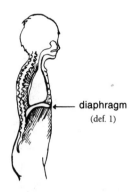

← diaphragm (def. 1)

diaphragm (dī'ə fram), *n.* **1** *Anatomy.* in mammals, a dome-shaped partition composed of muscles and tendons which separates the cavity of the chest from the cavity of the abdomen and is important in respiration: *As a result of the muscle action of the diaphragm and rib muscles, air is either inspired or expired from the lung* (Edwin Diamond).

2 *Optics.* a disk with a hole in the center for controlling the amount of light entering a camera, microscope, etc.
[from Greek *diaphragma,* from *dia-* across + *phragma* fence]

diaphysis(dī af'ə sis), *n., pl.* **-ses** (-sēz'). **1** *Anatomy.* the shaft of a long bone.
2 *Botany.* abnormal proliferation of a flower or of a flower cluster.
[from Greek *diaphysis* a growing through, point of separation]
—**diaphysial** or **diaphyseal** (dī'ə fiz'ē əl), *adj.* of or having to do with a diaphysis or diaphyses.

diapir (dī'ə pir), *n. Geology.* an anticlinal fold in which a mobile core, such as salt or gypsum, has pierced through the more brittle overlying rock: *Both young mountain ranges and mid-ocean ridges are differentiation products of the mantle and are breaking through the crust as diapirs* (Science Journal). [from Greek *diapeirein* to pierce through]
—**diapiric** (dī'ə pir'ik), *adj.* of or by diapirs: *diapiric structures, diapiric intrusion.*

diapophysis (dī'ə pof'ə sis), *n., pl.* **-ses** (-sēz'). *Anatomy.* the articular part of the transverse process of a vertebra. [from *dia-* through + *apophysis*]
—**diapophyseal** or **diapophysial** (dī'ap ə fiz'ē əl), *adj.* of or having to do with diapophysis.

diapsid (dī ap'sid), *adj. Anatomy.* having two pairs of temporal arches, as certain reptiles: *a diapsid skull.* [from *di-* two + Latin *apsis, apsidem* arch]

diarthrosis (dī'ar thrō'sis), *n., pl.* **-ses** (-sēz'). *Anatomy.* a freely movable joint or articulation, as that of the shoulder or hip joint.
—**diarthrotic** (dī'ar throt'ik), *adj.* of or having to do with diarthrosis.

diaspore (dī'ə spôr), *n. Mineralogy.* a native hydroxide of aluminum varying in color from white to violet, occurring in bauxite in scaly, crystalline masses. *Formula:* AlO(OH) [from Greek *diaspora* dispersion (so called because of its dispersion when heated)]

diastase (dī'ə stās), *n. Biochemistry.* amylase, especially certain impure extracts from molds or germinating seeds: *Diastase, an enzyme found in the kernels of barley, in the presence of water, converts starch into maltose, a disaccharide* (Youngken, *Pharmaceutical Botany*). [from Greek *diastasis* separation]

diastasis (dī as'tə sis), *n., pl.* **-ses** (-sēz'). *Physiology.* the rest period of the cardiac cycle, occurring between the diastole and the systole. [from New Latin, from Greek *diastasis* separation]

diastatic (dī'ə stat'ik), *adj. Biochemistry.* **1** of or having to do with diastase. **2** having the properties of amylase; capable of breaking down starch into dextrins and of reducing sugar.

diastem (dī'ə stem), *n. Geology.* a minor break in sedimentary rocks, accompanied by little or no erosion. [from Greek *diastēma* interval]

diastema (dī'ə stē'mə), *n., pl.* **-mata** (-mə tə). *Anatomy, Zoology.* a gap or space between teeth in a jaw of an animal. [New Latin, from Greek *diastēma* interval]

diastereoisomer (dī'ə ster'ē ō ī'sə mər), *n. Chemistry.* one of two or more stereoisomers that are not mirror images of each other. Compare **enantiomorph.**
—**diastereoisomeric** (dī'ə ster'ē ō ī'sə mer'ik), *adj.* of or having the nature of a diastereoisomer.

diastole (dī as'tl ē), *n. Physiology.* the normal, rhythmical dilation of the heart, especially of the ventricles. During diastole the chambers of the heart fill up with blood. Contrasted with **systole.** [from Greek *diastolē* expansion, from *dia-* apart + *stellein* to send]
—**diastolic** (dī'ə stol'ik), *adj.* having to do with diastole; during diastole: *diastolic pressures.*

diastrophic (dī'ə strof'ik), *adj. Geology.* of or like diastrophism.

diastrophism (dī as'trə fiz'əm), *n. Geology.* the action of the forces that have caused the deformation of the earth's crust, producing mountains, continents, etc.: *Diastrophism includes earthquakes, the rising and sinking of land, and the slow heaving up of mountains* (T. Walter Wallbank). [from Greek *diastrophē* distortion]

diathermic (dī'ə thèr' mik), *adj. Physics.* freely permeable by radiant heat; permitting the passage of heat rays. [from *dia-* through + Greek *thermē* heat]

diatom (dī'ə tom), *n. Biology.* any of a class (Bacillariophyceae) of numerous microscopic aquatic algae that have hard shells composed mostly of silica. They are either one-celled forms or an aggregation of cells forming filaments. [from Greek *diatomos* cut in half]

diatomaceous (dī'ə tə mā'shəs), *adj. Biology.* of or having to do with diatoms or their fossil remains: *skeletal remains of diatomaceous algae.*

diatomaceous earth, = diatomite: *Diatomaceous earth is quarried for commercial use in the manufacture of abrasives and polishes* (Greulach and Adams, *Plants*).

diatomic (dī'ə tom'ik), *Chemistry.* —*adj.* having two atoms in each molecule: *The primary oxidation products in such reactions are found in practice to have only one oxygen atom per molecule, whereas oxygen itself is diatomic* (K.D. Wadsworth).
—*n.* a diatomic molecule or substance.

diatomite (dī at'ə mīt), *n. Geology.* a light, porous rock consisting of the fossil remains of diatoms, used as an abrasive, insulator, filter, etc.: *Diatomite ... is a hydrous or opalescent form of silica, and the deposits are accumulations of myriads of microscopically small siliceous shells of diatoms, a group of flowerless aquatic plants of marine or fresh-water origin called algae.* (Jones, *Minerals in Industry*). ► Diatomite is also called **diatomaceous earth, kieselguhr,** and, mistakenly, **infusorial earth.**

diatreme (dī'ə trēm), *n. Geology.* a volcanic vent blasted through solid rock by exploding gases. [from *dia-* through + Greek *trēma* perforation]

diatropic (dī'ə trop'ik), *adj. Botany.* having to do with or exhibiting diatropism.

---

cap, fāce, fäther; best, bē, tèrm; pin, fīve;
rock, gō, ôrder; oil, out; cup, pùt, rüle,
yü in use, *yu* in uric;
ng in bring; sh in rush; th in thin; ᴛʜ in then;
zh in seizure.
ə = a in about, e in taken, i in pencil, o in lemon, u in circus

diatropism (dī at'rə piz əm), *n. Botany.* a tendency of certain organs in various plants to take a position transverse to the direction of the stimulus. [from *dia- + tropism*]

diazo (dī az'ō), *adj. Chemistry.* of or containing two nitrogen atoms, N_2, combined with a hydrocarbon radical: *a diazo compound.* Compare **azo.**

diazonium compound (dī'ə zō'nē əm), *Chemistry.* any of a group of compounds containing the group -N = N-.

diazonium salts, *Chemistry.* a group of compounds having the general formula (ArN:N)x, the *x* representing an acid radical.

dibasic (dī bā'sik), *adj. Chemistry.* 1 (of an acid) having two hydrogen atoms that can be replaced by two atoms or radicals of a base in forming salts: *Sulfuric acid, H_2SO_4, is a dibasic acid.* 2 having two basic atoms, each of valence 1.

dibromide (dī brō'mīd), *n. Chemistry.* a compound containing two atoms of bromine with another element or radical.

dicarboxylic acid (dī kär'bok sil'ik), *Chemistry.* an organic compound containing two carboxyl groups, -COOH.

dicentric (dī sen'trik), *adj. Biology.* having two centromeres: *dicentric chromosomes.*

dichasium (dī kā'zhē əm *or* dī kā'zē əm), *n., pl.* **-sia** (-zhē-ə *or* -zē ə). *Botany.* a cymose inflorescence in which the main axis produces a pair of lateral axes below the terminal flower, each of which similarly produces a pair, and so on. [from New Latin, from Greek *dichasis* division]
—**dichasial,** *adj. Botany.* having to do with or resembling a dichasium.

dichlamydeous (dī'klə mid'ē əs), *adj. Botany.* having both a calyx and a corolla. [from *di-* two + Greek *chlamydos* short mantle]

dichloride (dī klôr'īd'), *n. Chemistry.* a chloride containing two atoms of chlorine. Also called **bichloride.**

dichogamous (dī kog'ə məs), *adj. Botany.* having the stamens and pistils maturing at different times to ensure cross-fertilization. Contrasted with **homogamous.** [from Greek *dicho-* separately + English *-gamous*]
—**dichogamy** (dī kog'ə mē), *n.* the condition of being dichogamous.

dichotomous (dī kot'ə məs), *adj. Botany.* branching by repeated divisions into two that are nearly equal: *Dichotomous venation sets the ferns off from all modern seed plants with the one rare exception of Ginkgo* (Emerson, *Basic Botany*).

dichotomy
(def. 1)

dichotomy (dī kot'ə mē), *n.* 1 *Botany.* a branching by repeated divisions into two nearly equal parts. 2 *Zoology.* a branching in which each successive axis divides into two; repeated bifurcation, as of the veins. [from Greek *dichotomia,* from *dicha* in two + *temnein* to cut]

dichroic (dī krō'ik), *adj. Optics.* 1 (of a crystal) showing two different colors according to the direction of transmitted light, due to difference in the amount of the rays absorbed. 2 (of a solution) showing different colors at different concentrations, as a solution of chlorophyll.

dichroism (dī'krō iz'əm), *n. Optics.* the quality or condition of being dichroic: *Certain doubly refracting crystals exhibit dichroism, that is, one of the polarized components is absorbed much more strongly than the other* (Sears and Zemansky, *University Physics*). [from Greek *dichroos* two-colored + English *-ism*]

dichromate (dī krō'māt), *n. Chemistry.* a chromate whose molecules have two atoms of chromium. Also called **bichromate.**

dichromatic (dī'krō mat'ik), *adj.* 1 *Zoology.* having two color phases that are not connected with season, sex, or age. 2 *Medicine.* of, having to do with, or affected by dichromatism (def. 2).
—**dichromatism** (dī krō'mə tiz'əm), *n.* 1 *Zoology.* dichromatic quality or condition. 2 *Medicine.* color blindness in which only two of the three primary colors are seen.

dickite (dik'īt), *n. Mineralogy.* a crystalline hydrous silicate of aluminum, isomorphous with kaolinite and nacrite, found in hydrothermal veins. *Formula:* $Al_2Si_2O_5(OH)_4$ [from Allan B. *Dick,* 1833–1926, English mineralogist + *-ite*]

diclinous (dī klī'nəs), *adj. Botany.* 1 (of a plant) having the stamens and pistils in separate flowers either on the same plant (monoecious), or on separate plants of the same species (dioecious). 2 (of a flower) having only stamens or only pistils. Also called **unisexual.** Compare **monoclinous.** [from *di-* two + Greek *klinē* bed, from *klinein* to slope, bend]

dicot (dī'kot), *n.* = dicotyledon. Compare **monocot.**

dicotyledon (dī kot'l ēd'n), *n. Botany.* any of a class (Dicotyledoneae) of flowering plants of the angiosperm subdivision, having two seed leaves (cotyledons) in the embryo. The leaves typically have a network of veins and flower parts in fours or fives. Many trees and most cultivated plants are dicotyledons. *The dicotyledons, have two cotyledons in their seeds. These cotyledons are modified leaves attached to the epicotyl and hypocotyl by short stalks, and they often contain food reserves which are used in the early stages of germination* (Mackean, *Introduction to Biology*). Compare **monocotyledon.**
—**dicotyledonous** (dī kot'l ēd'n əs), *adj.* having two cotyledons.

dictyosome (dik'tē ə sōm), *n. Biology.* a subunit of the Golgi apparatus; Golgi body. [from Greek *diktyon* net + English *-some*]

dictyostele (dik'tē ə stē'lē), *n. Botany.* a hollow tube (siphonostele) whose vascular mass is broken up into a number of longitudinal strands or vascular bundles. [from Greek *diktyon* net + English *stele*]

didactyl (dī dak'təl), *adj. Zoology.* having only two fingers, claws, or toes to each limb. [from *di-* two + Greek *daktylos* finger, toe]

didymium (dī dim′ē əm), *n. Chemistry.* a mixture of rare-earth metals, usually found associated with lanthanum. It was formerly believed to be an element and consists largely of neodymium and praseodymium. Didymium, with its cerium content removed, is used to color glass and make optical filters. [from New Latin, from Greek *didymos* twin]

didymous (did′ə məs), *adj. Botany, Zoology.* growing in pairs: *didymous anthers.* [from Greek *didymos* twin]

didynamous (dī din′ə məs), *adj. Botany.* having four stamens arranged in two pairs of unequal length: *didynamous flowers.* [from *di-* two + Greek *dynamis* power]

diecious (dī ē′shəs), *adj.* = dioecious.

dielectric (dī′i lek′trik), *Physics.* —*adj.* having little or no conductivity; nonconducting. The **dielectric constant** or **coefficient** is a measure of the ability of a material to resist the formation of an electric field within it: *The dielectric constant measures the material's resistance to polarizing efforts of electric forces—the forces' tendency to separate the positive and negative charges in the atoms of the material* (Science News). —*n.* a dielectric substance, such as glass, rubber, or wood.

dielectric heating, *Physics.* heating of a nonconductor in a varying electric field, due to dielectric loss.

dielectric loss, *Physics.* loss of energy in a dielectric substance subjected to a varying electric field. This energy is converted to heat.

diencephalon (dī′en sef′ə lon), *n. Anatomy.* the posterior part of the forebrain, which connects the midbrain to the cerebrum: *It is the diencephalon that makes people sweat and turn pale in anger* (Science News). [from *di(a)-* + *encephalon*]

—**diencephalic** (dī′en sə fal′ik), *adj.* of or having to do with the diencephalon.

diene (dī′ēn *or* dī ēn′), *n.* = diolefin. [from *di-* + *-ene*]

diestrum (dī es′trəm) *or* **diestrus** (dī es′trəs), *n. Zoology.* the sexually inactive period in the estrous cycle. Also called **anestrum** or **anestrus.** [from New Latin, from *dia-* through + *estrum* or *estrus* estrous cycle]

dietetics (dī′ə tet′iks), *n.* the science that deals with the amounts and kinds of food needed by the body.

difference, *n. Mathematics.* the result obtained when one number is subtracted from another.

differential (dif′ə ren′shəl), *n. Mathematics.* the product of the derivative of a function containing one variable multiplied by the increment of the independent variable.

—*adj.* **1** *Mathematics.* having to do with or involving differentials. **2** of or exhibiting a difference in some quantity or effect, especially a difference in rate: *differential erosion, differential motion. The basis of chromatography is differential migration—that is, the flow at different rates from a starting point* (W. T. Lippincott).

differential calculus, the branch of higher mathematics that investigates differentials, derivatives, and their relations.

differential equation, *Mathematics.* an equation containing differentials or derivatives: *a differential equation on the motion of charged particles* (N.Y. Times).

differentially permeable, = semipermeable.

—**differential permeability,** the property of being semipermeable.

differentiate, *v.* **1** *Biology.* (of cells, tissues, and organs,) to undergo or cause differentiation in the process of growth or development: *There is growing evidence that an important factor in differentiation, and perhaps the most basic one, is the production of ... new types of enzymes ... and these presumably catalyze the specific biochemical processes involved in differentiation* (Greulach and Adams, *Plants*). **2** *Mathematics.* to find the differtial or derivative of, as an equation.

differentiation, *n.* **1** *Biology.* the process by which cells, tissues, and organs become modified and specialized in the course of development or growth of an organism: *The nucleic acids, deoxyribonucleic acid (DNA) and ribonucleic acid (RNA) ... regulate protein production and determine the function each cell will perform (differentiation)* (James S. Sweet).

2 *Mathematics.* the act or process of differentiating; the operation by which a function is transferred into its derivative. Compare **integration.**

diffract (di frakt′), *v. Physics.* to alter the pattern of wave motion by diffraction: *The fine-ruled grooves in an optical grating diffract light waves* (James A. Ibers). [from Latin *diffractum* broken up, from *dis-* apart + *frangere* to break]

diffraction (di frak′shən), *n. Physics.* change in the directions and intensities of a group of waves after passing by an obstacle or through an aperture. The colors of the spectrum can be produced by the diffraction of light. *If a wave train, such as that of light, strikes the edge of an object, the direction of the wave is changed. This is known as diffraction* (Robert F. Paton). Compare **refraction.**

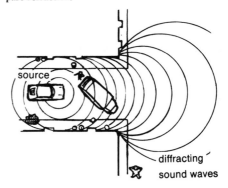

diffracting sound waves

diffraction grating, *Physics.* a set of close, fine parallel lines ruled on polished glass or metal to produce spectra by diffraction: *When a large number of identical grooves diffract a beam of light, the light is dispersed into its spectrum of wavelengths. The principle is altogether different from that of a prism, and a diffraction*

cap, fāce, fäther; best, bē, tėrm; pin, five;
rock, gō, ôrder; oil, out; cup, pùt, rüle;
yü in use, *yu̇* in uric;
ng in bring; *sh* in rush; *th* in thin; ᴛʜ in then;
zh in seizure.
ə = *a* in about, *e* in taken, *i* in pencil, *o* in lemon, *u* in circus

grating has much more resolving power than a prism of equal width (Scientific American).

diffuse, *v. Physics.* to undergo or cause diffusion; spread or dissolve into one another: *Other things being equal, oxygen diffuses through air about 10,000 times faster than through water* (A.R. Grable). SYN: scatter, disperse, disseminate.

diffuse nebula, *Astronomy.* a dark or slightly luminous galactic nebula, distinguished from planetary nebulae by the lack of a regular and distinct shape.

diffuse-porous, *adj. Botany.* (of wood) having the conducting vessels or pores distributed uniformly throughout the year's growth and the annual rings are not easily distinguishable: *In maple, birch ... and some other woods the pores are uniform in size and the wood is called "diffuse-porous." The pores of the diffuse-porous woods ... have little effect on grain or texture* (Scientific American).

diffusion (di fyü′zhən), *n. Physics.* **1** a mixing together of the atoms or molecules of gases, liquids, or solids by spreading into one another as a result of their random thermal motion: *Where the particles or molecules are free to move, as in a liquid or a gas, this migration, termed diffusion, occurs from regions of higher concentration of the molecules to regions of lower concentration* (Harbaugh and Goodrich, *Fundamentals of Biology*).
ASSOCIATED TERMS: see **active transport.**
2 the scattering of light diverted at various angles from an uneven surface.
ASSOCIATED TERMS: *Diffusion* of light is from the scattering effects of an uneven surface (or of air molecules) on rays of light. *Reflection* of light is a throwing back of rays of light from a relatively even surface. *Refraction* of light results from the bending of a beam of light in the same way.

diffusion gradient, *Physics.* the difference in concentration of a substance that results in its diffusion: *A weak solution of salt ... will contain relatively less salt and more water than a strong solution of salt. Thus the diffusion gradient for salt is from the strong to the weak solution, but for water the diffusion gradient is from the weak to the strong solution* (Mackean, *Introduction to Biology*).

digametic (dī′gə met′ik), *adj.* = heterogametic.

digastric (dī gas′trik), *Anatomy.* —*adj.* having two fleshy parts connected by a tendon: *a digastric muscle.*
—*n.* a digastric muscle, especially muscle of the lower jaw, thick at its extremities and thin and tendinous in the middle, by which the hyoid bone is raised in swallowing. [from New Latin *digastricus,* from *di-* two + Greek *gastros* belly]

digest, *v.* **1** *Biology.* to change (food) chemically into materials which the cells can assimilate, store, or oxidize and use as nourishment: *Plants as well as animals digest foods; that is, they break down complex foods into simple compounds* (Weier, *Botany*). Compare **absorb** (def. 2), **assimilate** (def. 2), **egest, ingest.**
2 *Chemistry.* to soften or decompose by combinations of heat, moisture, pressure, or chemical action: *The equipment ... extracts the sulfite waste liquids, (black liquor) from the cooked or digested chips used in pulp making by the semi-chemical process* (Wall Street Journal).

—**digestion,** *n.* **1** *Biology.* the process of digesting food: *Digestion is the process by which insoluble food, consisting of large molecules, is broken down into soluble compounds having smaller molecules* (Mackean, *Introduction to Biology*). Compare **hydrolysis. 2** *Chemistry.* decomposition of sewage, etc. by anaerobic bacteria.

—**digestive,** *adj.* of or having to do with digestion. The **digestive system** consists of the alimentary canal and its digestive glands and organs.

digit, *n.* **1** *Mathematics.* any symbol in a numeration system used to designate a quantity. In the decimal system, the digits are any of the figures 0, 1, 2, 3, 4, 5, 6, 7, 8, 9. The binary digits used in digital computers are 0 and 1.
2 *Biology.* a finger or toe. See the pictures at **homology, wing.**
[from Latin *digitus* finger]

digital, *adj.* **1** *Mathematics.* of, having to do with, or using digits, especially binary digits: *The system probably also uses a "digital" code to indicate the brightness or darkness at any point, just as the "painting by numbers" sets use a digital code to represent colour* (New Scientist).
2 *Biology.* of, having to do with, or resembling a finger or toe: *a digital organ or process.*

digitate (dij′ə tāt), *adj. Botany.* having radiating divisions like fingers: *A digitate compound leaf has finger-like leaflets coming off at the end of the petiole.*
—**digitately,** *adv.*

digitate leaf

digitigrade (dij′ə tə grād′), *adj. Zoology.* walking so that the toes are the only portion of the foot on the ground. [from New Latin *digitigrada,* from Latin *digitus* finger, toe + *gradi* to walk]
ASSOCIATED TERMS: Dogs and cats are *digitigrade* animals. Raccoons, bears, and humans are *plantigrade* (walking on the entire foot). Horses, rhinoceroses, and pigs are sometimes called *unguligrade* animals because they walk on the enlarged nails, or hoofs.

diglyceride (dī glis′ə rid), *n. Chemistry.* any of a group of fatty compounds formed when two carboxylic acid groups replace the hydrogen atoms of two hydroxyl (OH) groups in glycerol: *Because a molecule of glycerol has three reactive sites, it is possible to have monoglycerides, diglycerides, and triglycerides, depending on how many molecules of fatty acid react with each molecule of glycerol* (Scientific American).

dihedral (dī hē′drəl), *Mathematics.* —*adj.* having or formed by two intersecting plane surfaces: *Longitude is defined as a dihedral angle, or wedge angle* (Duncan, *Astronomy*).
—*n.* **1** the figure formed by two intersecting plane surfaces. **2** a dihedral angle.
[from *di-* two + Greek *hedra* seat, base]

dihybrid (dī hī'brid), *Biology.* —*adj.* having parents whose genetic make-up differs in two pairs of inheritable characters: *a dihybrid plant.*
—*n.* a dihybrid organism: *When parents differ in two pairs of characters, the F₁ offspring are termed dihybrids; in such crosses Mendel found that each pair of characters is inherited independently of the other* (Storer, *General Zoology*). Also called **dihybrid cross.** Compare **monohybrid.**

dihydrate (dī'hī'drāt), *n. Chemistry.* a compound containing two molecules of water.

dihydric, *adj. Chemistry.* containing two hydroxyl groups: *dihydric alcohols.*

dihydroxy, *adj.* = dihydric.

dikaryon (dī kar'ē ôn), *n. Biology.* 1 a cell containing two haploid nuclei which divide simultaneously. 2 a mycelium or a hypha containing such cells. [from *di-* two + Greek *karyon* kernel]

dike, *n. Geology.* a long mass of igneous rock that was intruded into a crosscutting fissure in older rock; a discordant intrusive body of tabular rock: *The evidence suggests to most investigators that the upwelling mechanism is an injection of molten deep-earth material by linear intrusions called dikes* (J.R. Heirtzler). See the picture at **volcano.**

dilatancy (dī lā'tən sē), *n. Physics, Geology.* the increase in volume of a granular mass due to the increase of space between the granules as they move: *Investigations have shown that definite changes occur in the properties of rocks that have been stressed almost to the point of rupture. During the process of deformation, microcracks and voids occur and the rock increases in volume, a phenomenon known as dilatancy* (Roper and Roper, *Earthquake*).
—**dilatant** (dī lā'tənt), *adj.* exhibiting dilatancy.

dilatation or **dilation,** *n.* 1 *Physiology.* enlargement, expansion, or stretched condition of a part or an opening of the body: *dilatation of the stomach, dilation of the pupil.* 2 *Physics, Geology.* = dilatancy.

dilate *v.* 1 *Physiology.* to make large or become wider; enlarge: *The pupil dilates as the eye accommodates to darkness.*
2 *Physics, Geology.* to increase in volume as the spaces in a granular mass extend between granules; expand: *As stressed rocks dilate and cracks open up ... shear strength ... decreases, resulting in failure or an earthquake* (Roper and Roper, *Earthquake*).

dilator, *n. Anatomy.* a muscle that dilates some part of the body.

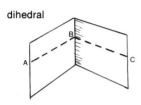

dihedral

diluent (dil'yü ənt), *n. Chemistry.* a diluting or dissolving agent: *Inert gases available as diluents of oxygen in a two-gas space cabin atmosphere include nitrogen, helium, neon, argon, krypton, and xenon* (Stanley Deutsch).

dilute (də lüt' *or* dī lüt'), *Chemistry.* —*v.* to make weaker or thinner by adding water or some other liquid: *to dilute a vaccine, to dilute alcohol in the blood.*
—*adj.* weakened or thinned by the addition of water or some other liquid: *a dilute acid.*
—**dilution,** *n.* 1 the fact or state of being diluted. 2 something diluted: *A vaccine dilution that produces a barely detectable level of antibody in the bloodstream is sufficient to prevent the virus from invading the central nervous system from the blood* (Scientific American).

diluvial (də lü'vē əl *or* dī lü'vē əl), *adj. Geology.* 1 of or having to do with a flood. 2 made up of debris left by a flood. [from Latin *diluvium* deluge]

dimension, *n.* 1 *Mathematics.* **a** Also **dimensions.** the size of a figure, especially length, width, or thickness. **b** any of the three directions of ordinary space, at right angles to each other, as at the corner of a box. **c** a factor of a term represented by a letter. α^2 and αb are both terms of two dimensions.
2 *Physics.* a fundamental physical quantity expressed as a factor in an algebraic term: *One way to spot an erroneous equation is to check the dimensions of all its terms ... In any legitimate physical equation the dimensions of all the terms must be the same* (Halliday and Resnick, *Fundamentals of Physics*).

dimer (dī'mər), *n. Chemistry.* a compound composed of two molecules of the same substance, especially one produced by polymerization: *Unfortunately, it is seldom possible to concentrate the solution much beyond 70 per cent without running the risk of forming dimers ... or causing hydrolysis to occur* (New Scientist).
—**dimeric,** *adj.* of or having to do with a dimer.
—**dimerization,** *n.* formation of a dimer.
—**dimerize,** *v.* to form a dimer; join together to form a single molecule.

dimerous (dim'ər əs), *adj.* 1 *Botany.* (of a flower) having two members in each whorl (generally written 2-merous). Compare **monomerous.**
2 *Zoology.* having two segments to each tarsus, as certain insects. Compare **trimerous.** [from *di-* two + Greek *meros* part]

dimethyl (dī meth'əl), *Chemistry.* —*adj.* (of a compound) having two methyl (CH_3) groups.
—*n.* a dimethyl compound.

dimorphic or **dimorphous,** *adj.* existing or occurring in two distinct forms; exhibiting dimorphism: *a dimorphic crystal, dimorphic organisms. Primula flowers are dimorphic. On some plants all are short-styled ... On other plants the flowers are long styled* (Gibbs, *Botany*).

dimorphism (dī môr'fiz'əm), *n.* 1 *Botany.* the occurrence of two different forms of flowers, leaves, or other parts on the same plant or in the same species, as in the disk and ray florets of the daisy, or the young and adult fo-

cap, fāce, fäther; best, bē, tèrm; pin, fīve;
rock, gō, ôrder; oil, out; cup, pùt, rüle,
yü in use, *yù* in uric;
ng in bring; *sh* in rush; *th* in thin; ŦH in then;
zh in seizure.
ə = *a* in about, *e* in taken, *i* in pencil, *o* in lemon, *u* in circus

165

liage of eucalyptus: *Dimorphism is important for cross-pollination.*

2 *Zoology.* the occurrence of two forms different in structure, coloration, etc., among animals of the same species: *Human beings exhibit distinct sexual dimorphism, with males 5–12 percent taller than females and comparably heavier as well* (Barash, *Sociobiology and Behavior*).

3 *Chemistry, Crystallography.* the property of some substances of crystallizing in two different forms. See also **polymorphism.**

dinoflagellate (din′ə flaj′ə lit), *n. Biology.* any of an order (Dinoflagellata) of tiny marine flagellates, some species of which produce a red substance poisonous to fish: *Dinoflagellates have a normal concentration of 1,000 or less per quart. But favorable ecological conditions initiate a seemingly sudden increase in their numbers. During such a "bloom," a quart of seawater may contain 60 million to 100 million dinoflagellates. This large population discolors the water; hence the name "red tide"* (M.W. Martin). [from Greek *dinos* whirling + Latin *flagellum* whip]

dinosaur, *n. Paleontology.* any of a large group of extinct terrestrial or amphibious reptiles that were chiefly herbivorous, had usually very long limbs and tails, and walked on two or four feet: *The Mesozoic Era frequently is called the Age of Dinosaurs, because these creatures were the most striking animals of the time and were world-wide in distribution* (Moore, *Introduction to Historical Geology*). [from New Latin *Dinosauria* the group name, from Greek *deinos* terrible + *sauros* lizard]

dinucleotide (dī nü′klē ə tīd), *n. Biochemistry.* a nucleotide consisting of a combination of two nucleotides. Compare **mononucleotide, trinucleotide.**

diode (dī′ōd), *n. Electronics.* an electron tube or semiconducting device having two electrodes, especially a rectifier which permits the flow of electrons in one direction only. [from *di-* two + (*electr*)*ode*]

dioecious or **diecious** (dī ē′shəs), *adj.* **1** *Botany.* having male and female flowers on separate plants of the same species. The asparagus and willow are dioecious.

2 *Zoology.* characterized by individuals that are either male or female: *Most animal species are dioecious.* Contrasted with **monoecious.** [from Greek *di-* double + *oikos* house]

—**dioecism** (dī ē′siz əm), *n.* the quality or condition of being dioecious.

diolefin (dī ō′lə fin), *n. Chemistry.* an aliphatic hydrocarbon containing two double bonds between carbon bonds: *Out of this work with butadiene and related diolefins ... have come polymers demonstrating optical as well as geometric isomerism* (Giulio Natta). Formula: C_nH_{2n-2} Also called **diene.**

Diophantine analysis (dī′ə fan′tīn), *Mathematics.* analysis of Diophantine equations to determine the presence and nature of integral solutions. [named after *Diophantus* of Alexandria, about 250 A.D., Greek mathematician]

Diophantine equation, *Mathematics.* an algebraic equation of two or more unknowns whose coefficients are integers, studied to determine the possibility or range of integral solutions: *This reduces to* $98y - 199x =$

5, a Diophantine equation with an infinite number of integral solutions (Scientific American).

diopside (dī op′sīd), *n. Mineralogy.* a species of pyroxene. Formula: $CaMgSi_2O_6$ [from French, from *di-* twice + Greek *opsis* aspect]

dioptase (dī op′tās), *n. Mineralogy.* a translucent silicate of copper, found in green, hexagonal crystals. Formula: $CuSiO_2(OH)_2$ [from French, from Greek *dia-* through + *optasia* appearance]

diopter (dī op′ter), *n. Optics.* a unit of the refractive power of a lens. It is the reciprocal of the focal length in meters. [from French, from Latin *dioptra* an optical measuring instrument, ultimately from Greek *dia-* through + *optos* visible]

—**dioptric** (dī op′trik), *adj.* of or having to do with diopters or the refraction of light: *dioptric powers.*

—**dioptrics,** *n.* the branch of optics dealing with the refraction of light, especially by lenses.

diorite (dī′ə rīt), *n. Geology.* a coarse-grained igneous rock solidified far below the earth's surface and consisting essentially of hornblende and plagioclase containing sodium: *The sedimentary rocks have been intimately intruded ... to such an extent that nearly every large exposure contains flat-lying sheets, or sills, of gabbro and diorite* (E. F. Roots). [from French *diorite,* from Greek *diorizein* distinguish]

—**dioritic** (dī′ə rit′ik), *adj.* having to do with or resembling diorite.

dioxide (dī ok′sīd), *n. Chemistry.* an oxide having two atoms of oxygen for each molecule: *carbon dioxide, sulfur dioxide.*

dip, *n.* **1** *Geology.* **a** the downward slope of a stratum or vein. **b** = angle of dip.

2 *Astronomy.* the angular distance of the visible horizon below the horizontal plane through the observer's eye: *The visible or apparent horizon ... is irregular in shape on land, and at sea it is a small circle lying below the true horizon by an angular distance called the dip* (Duncan, *Astronomy*).

dipeptidase (dī pep′tə dās), *n. Biochemistry.* an enzyme that hydrolyzes dipeptides into their constituent amino acids:

dipeptide (dī pep′tīd), *n. Biochemistry.* a peptide that, on hydrolysis, yields two amino-acid molecules: *If only two amino acids are linked, it is called a dipeptide ... If a large number are connected, it is called a polypeptide* (McElroy, *Biology and Man*).

diphosphate (dī fos′fāt), *n. Chemistry.* an ester of phosphoric acid containing two phosphate groups.

diphosphoglyceric acid (dī fos′fō gli ser′ik), *Biochemistry.* a diphosphate of glyceric acid that is an intermediate in metabolic processes in animals and plants. Formula: $C_3H_8O_9P_2$ Abbreviation: DPG

diphosphopyridine nucleotide (dī fos′fō pir′ə dēn), *Biochemistry.* the coenzyme involved in the oxidative reactions in cell metabolism. It is the primary hydrogen acceptor in a series metabolic chemical reactions in the Krebs cycle. *The special pyridine compound in the oxidation-reduction coenzyme, the diphosphopyridine nucleotide, is the vitamin nicotinamide, whose absence produces pellagra* (Scientific American). Formula: $C_{21}H_{27}N_7O_{14}P_2$ Abbreviation: DPN Also called **nicotinamide adenine dinucleotide (NAD).**

diphycercal (dif'ə sèr'kəl), *adj. Zoology.* **1** (of fishes) having a tail in which the upper and lower lobes are symmetrical and are equally divided by the end of the spine.
2 of or having to do with this kind of tail.
[from Greek *diphuēs* double + *kerkos* tail]

diphyletic (dif'ə let'ik), *adj. Biology.* **1** having two lines of descent; derived from two distinct sets of ancestors. **2** of or having to do with a classification of groups of animals on the theory that they have a diphyletic origin.

diphyllous (dī fil'əs), *adj. Botany.* having two leaves. [from *di-* two + Greek *phyllon* leaf]

diphyodont (dif'ē ə dont), *Zoology.* —*adj.* having two successive sets of teeth, as most mammals.
—*n.* a diphyodont mammal.
[from Greek *diphuēs* double + *odontos* tooth]

diploblastic (dip'lō blas'tik), *adj. Zoology.* having two germ layers, the ectoderm and endoderm, as in most cnidarians: *Because it has only two body layers, we say that Hydra is diploblastic as compared with more advanced animals, including man, which are triploblastic and have three body layers* (Winchester, *Zoology*). [from Greek *diploos* double + *blastos* germ]

diploe (dip'lō ē), *n. Anatomy.* the spongy, porous bony tissue between the hard inner and outer walls of the bones of the skull. [from Greek *diploē* a fold, from *diploos* double]
—**diploic** (də plō'ik), *adj.* of or having to do with the diploe: *diploic tissue.*

diploid (dip'loid), *Genetics.* —*adj.* having two sets of chromosomes, or double the number of chromosomes in the germ cell: *A body cell of each species of seed plant or animal has a specific number of chromosomes, the diploid or 2n number, comprised of a paternal set contributed by the sperm and a maternal set contained in the egg from which the organism developed* (Harbaugh and Goodrich, *Fundamentals of Biology*). Compare **haploid, monoploid, polyploid.**
—*n.* a diploid organism or cell.
[from Greek *diploos* double]
ASSOCIATED TERMS: *Gametes or germ cells* (the cells that unite in sexual reproduction) have the *haploid* or *n* number of chromosomes, while the *zygote* (the single cell resulting from the union of the gametes, sperm cell and egg cell) has the *diploid* or *2n* number. Through *mitosis* or cell division, the diploid number is retained in all the *somatic cells* of the adult organism. Through the process of *meiosis* or reduction division, the diploid number of chromosomes in the germ cells of the new organism is reduced to the haploid number, so that the cycle of sexual reproduction may continue in the species.
—**diploidy** (dip'loi dē), *n.* the condition of being diploid: *Organisms may differ in the degree of haploidy and diploidy of their cells, and their features of inheritance also differ. All sexually reproducing organisms, however, show the alternation of haploid and diploid phases* (McElroy, *Biology and Man*).

diplonema (dip'lə nē'mə), *n.* = diplotene.

diplont, *n. Genetics.* an organism that is diploid at all stages of its life other than the gametic stage: *Meiosis can be deferred right up to the formation of gametes. In this case ... all the cells other than the gametes are diploid and the organisms are termed pure diplonts* (Bell and Coombe, *Strasburger's Textbook of Botany*). Contrasted with **haplont.**

diplophase, *n. Genetics.* the diploid phase of an organism.

diplosis (də plō'sis), *n. Genetics.* the doubling of the number of chromosomes by fusion of two haploid sets in fertilization. Compare **haplosis.**

diplotene (dip'lə tēn), *n. Biology.* that stage in the prophase of meiosis during which the points of interchange of chromatids appear. [from Greek *diploos* double + *tainia* band]

dipole (dī'pōl), *n. Physics.* **1** two equal electric charges or magnetic poles of opposite sign which are separated by a specified distance, as in a molecule.
2 a molecule or other object having two such charges or poles: *Each water molecule is a small dipole, that is, its positive and negative charges do not coincide* (Sears and Zemansky, *University Physics*).
—**dipolar** (dī pō'lər), *adj.* **1** of a dipole. **2** of or having to do with two poles; bipolar.

dipteran (dip'tər ən), *n. Zoology.* any of a large order (Diptera) of insects that have one pair of membranous wings, the usual second pair being replaced by small club-shaped organs. Mosquitoes, gnats, and houseflies belong to this order. [from Greek *di-* two + *pteron* wing]

dipterous (dip'tər əs), *adj.* **1** *Zoology.* of or belonging to the dipterans; two-winged.
2 *Botany.* having two winglike appendages, as certain fruits and seeds do.

dipterous seed
(def. 2)
maple

direct current, *Electricity.* a current that flows in one direction only. *Abbreviation:* d.c. or D.C.
► *Direct Current* always flows in the same direction (even though it may fluctuate in magnitude), while *alternating current* flows first in one direction, then in the reverse. In the United States, *alternating current* changes at a rate of 60 complete cycles per second.

directed *adj. Mathematics.* **1** (of a line) positive in one direction and negative in the other, as a vector. **2** (of a number) marked with a plus or minus sign: *Zero is not a directed number.* **3** (of an angle) determined by rotation.
ASSOCIATED TERMS: Directed angles are important in trigonometry. A directed angle is so called because its value is determined by rotation in one direction or another. The rotation begins at an *initial side* and ends at a *terminal side*. If the rotation is counterclockwise, the angle is positive, and if clockwise, the angle is negative. Thus, *co-terminal angles*—those with the same initial and terminal sides—can have different values depending on the direction of rotation and the number of complete revolutions.

---

cap, fāce, fäther; best, bē, tèrm; pin, fīve;
rock, gō, ôrder; oil, out; cup, pùt, rüle;
yü in use, *yù* in uric;
ng in bring; *sh* in rush; *th* in thin, ᴛн in then;
zh in seizure.
ə = *a* in about, *e* in taken, *i* in pencil, *o* in lemon, *u* in circus

167

directrix (də rek′triks), *n., pl.* **directrixes, directrices** (də rek′trə sēz′). *Geometry.* a fixed straight line used in determining a conic section. For all points on the conic the distance from a fixed point (the focus) has a constant ratio to the distance from the directrix. See the picture at **eccentricity.** [New Latin, feminine of Late Latin *director* ruler]

disaccharidase (dī sak′ə rə dās), *n. Biochemistry.* any enzyme, such as invertase (sucrase), that hydrolyzes disaccharides.

disaccharide (dī sak′ə rīd), *n. Biochemistry.* any of a group of carbohydrates, such as lactose, maltose, sucrose, and various other sugars, which hydrolysis changes into two monosaccharides or simple sugars: *Ordinary table sugar and milk sugar belong to the disaccharide class of sugars* (Science News Letter).
ASSOCIATED TERMS: A *monosaccharide* cannot be divided by hydrolysis into simpler sugars, but can combine with another monosaccharide molecule to form a *disaccharide* by dehydration (loss of a water molecule). Three or four monosaccharide molecules may combine to form *trisaccharides* or *polysaccharides* such as starch and cellulose.

discharge, *n.* 1 *Electricity.* **a** the passage of an electric current through a gas. **b** the transference of electricity between two charged bodies when placed in contact or near each other. **c** the conversion of chemical energy to electric energy, as in a storage battery.
2 *Geography, Geology.* the volume of water in a stream or river passing through a cross-section of the channel in a given period of time, usually expressed in cubic meters per second.

discifloral (dis′ə flôr′əl), *adj. Botany.* having flowers in which the receptacle is enlarged into a conspicuous disk surrounding the ovary, and usually distinct from the calyx. [from Latin *discus* disk + English *floral*]

disciform, *adj. Biology.* disk-shaped; discoid: *disciform fungi.*

discoid (dis′koid) or **discoidal** (dis koi′dəl), *adj.* 1 *Biology.* having the form of a disk; disk-shaped: *Since the yolk is nonliving, cleavage is restricted to the cytoplasm and, as a result, a disc-like cap of cells is formed at the animal pole. Such a pattern of cleavage is termed meroblastic and discoidal* (Harbaugh and Goodrich, *Fundamentals of Biology*).
2 *Botany.* (of composite plants) having a flower head containing only tubular flowers in the central disk, with no ray flowers. The tansy is a discoid plant.

disconformity, *n. Geology.* a break or unconformity in a rock formation in which the strata above and below the break are parallel.

discontinuity, *n. Geology.* the boundary of a layer or formation at which marked changes in earthquake waves occur, such as the Mohorovicic discontinuity.

discontinuous phase, = disperse phase.

discontinuous variation, *Biology.* the occurence of relatively large differences between members of the same species, with no appreciable intermediate forms: *The variations in coat color are examples of discontinuous variation because there are no intermediates. ... Discontinuous variation in humans ... occur among the more serious variants, e.g. one is either an achondroplastic dwarf or one is not; intermediates do not occur*

(Mackean, *Introduction to Biology*). Contrasted with **continuous variation.**

discordant, *adj.* 1 *Geology.* (of an intrusion) cutting across the surrounding strata: *Most intrusive bodies cut across the bedding of the enclosing sedimentary rocks, hence they are called discordant, but a sill is a rock mass congealed from magma forced concordantly between sedimentary strata* (Gilluly, *Principles of Geology*).
2 *Genetics.* differing in a trait under consideration, as a pair of identical or fraternal twins. Contrasted with **concordant.**

discriminant, *n. Mathematics.* an algebraic expression used to distinguish or separate other expressions in a quantity or equation.

disepalous (dī sep′ə ləs), *adj. Botany.* having two sepals.

disequilibrium (dis ē′kwə lib′rē əm), *n. Physics, Chemistry.* lack of equilibrium: *Strong thermal disequilibrium is primarily due to the earth's being embedded in an environment partly extremely hot (the surface of the sun) and partly exceedingly cold (the depths of space)* (C.M. and H. Bondi).

disinfectant, *Medicine.* —*n.* a chemical solution, heat, or other means used to destroy or inhibit the growth of infectious microorganisms. Alcohol, tincture of iodine, and carbolic acid are disinfectants.
—*adj.* destroying or inhibiting the growth of infectious microorganisms: *a disinfectant soap.*
ASSOCIATED TERMS: *Disinfectants* may be *antiseptics* or *germicides.* Disinfectants that are too strong to use on the skin or tissues are usually called *germicides.* Disinfectants that are safe to use on the body are usually called *antiseptics.* Often antiseptics are dilute solutions of germicides.

disintegrate, *v. Nuclear Physics.* 1 to cause disintegration in.
2 to undergo disintegration. Compare **decay.**

disintegration, *n. Nuclear Physics.* a change in the structure of an atomic nucleus by its emission of particles or rays. Disintegration occurs naturally in the nuclei of radioactive substances and can be induced in any other substance by bombarding the nuclei with fast-moving particles or rays. *Helium in the rocks results from the disintegration of radioactive elements, which emit alpha-particles or helium nuclei* (E.P. George). Compare **decay.**

disintegration

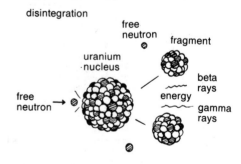

disjoint, *adj. Mathematics.* having no common members. (0, 1, 2) and (3, 4, 5) are disjoint sets. (0, 1, 2) and (2, 3, 4) are overlapping sets.

disjunct (dis jungkt′), *adj. Zoology.* having deep constrictions separating the head, thorax, and abdomen, as in ants.

168

disjunction, *n*. **1** *Mathematics*. the proposition or statement "*p* or *q*" (symbolized *p* v *q*), which is true when either one or both of the constituents are true, and false when both are false. Compare **conjunction.**

▶ There are two types of disjunction, inclusive and exclusive. The most commonly used one is the *inclusive disjunction*, which implies that one or the other or both are true; the rarer *exclusive disjunction* implies that either one or the other is true but not both.
2 *Biology*. the separation of homologous chromosomes during meiosis.

disk, *n*. **1** *Astronomy*. the circular face of a celestial body as seen by an observer: *The disk of Jupiter can be resolved with binoculars.*
2 *Botany*. a round, flat part in certain plants: *The yellow center of a daisy is a disk.* Compare **disk flower.**
3 *Anatomy, Zoology*. any round, flat part or structure, especially any of the bodies of fibrous cartilage between the vertebrae.

disk flower or **disk floret,** *Botany*. one of the florets in the central portion or disk of a daisy, aster, or other composite head flower: *The heads are usually made up of two kinds of flowers, the ray and the disk flowers* (G.H.M. Lawrence).

dislocation, *n*. **1** *Crystallography*. an imperfection in the crystal structure of a metal or other solid resulting from the absence of an atom or atoms in one or more layers of a crystal. **2** *Geology*. = displacement.

disomic (dī sō′mik), *Genetics*. —*adj*. having one extra chromosome that is homologous with a chromosome of the haploid set, especially in a haploid organism.
—*n*. a disomic cell or organism. Compare **monosomic, trisomic.**

dispermous (dī spėr′məs), *adj. Botany*. (of fruits and their cells) containing two seeds.

dispermy (dī′spėr′mē), *n. Biology*. the entrance of two spermatozoa into a single egg.

disperse, *v*. **1** *Physics*. **a** to divide (light or other electromagnetic radiation) into its different wavelengths; refract: *If a ray of white light is incident on the water surface ... it splits after passing through the water surface into a group of colored rays ... The white ray is said to be dispersed* (Shortley and Williams, *Elements of Physics*). **b** to divide or become divided by a similar separation of other waves.
2 *Chemistry*. to scatter (particles of a colloid, etc.) throughout another substance or a mixture: *Butterfat in warm, homogenized milk is an example of a liquid dispersed in another liquid* (Harbaugh and Goodrich, *Fundamentals of Biology*).

disperse phase, *Chemistry*. the particles in a colloidal system which are dispersed throughout a medium. Compare **continuous phase.**

dispersion, *n*. **1** *Physics*. **a** the separation of light or other electromagnetic radiation into its different wavelengths; refraction: *dispersion of light into its spectral colors.* **b** a similar separation of other waves: *A substance in which the velocity of a wave varies with wavelength is said to exhibit dispersion* (Sears and Zemansky, *University Physics*).

2 *Chemistry*. **a** a system composed of a dispersed substance and its medium. **b** a colloid.
3 *Statistics*. the difference in the values of one or more variables.

dispersion of light
(def. 1)

dispersion medium, = continuous phase.

dispersive medium, *Physics*. any medium that transmits waves of different frequencies at different velocities: *Glass is a dispersive medium for light. Air and water are nondispersive media for sound waves of low amplitude* (Physics Regents Syllabus).

dispersive power, *Physics*. the power of a substance to separate light rays by diffraction. It is measured by the ratio of the dispersion of the red and blue wavelengths to the deviation of the yellow produced by a prism of small angle.

dispersoid (dis pėr′soid), *n. Chemistry*. a substance consisting of finely divided particles dispersed in a medium.

displace *v*. **1** *Physics*. to change in position; cause or undergo displacement.
2 *Chemistry*. to take the place of, especially in a chemical reaction.

displacement, *n*. **1** *Physics*. **a** a change in the position of an object in space; the difference between the initial position of a body, figure, etc., and a later position: *The amplitude of the wave is the maximum displacement of a particle of the medium from the rest position* (Physics Regents Syllabus).
▶ In the literature of wave motion *displacement* is generalized to include any physical disturbance, such as changes in a transverse electric field, density of air, or density of an electron cloud.
b the weight of the volume of a liquid or gas moved from its usual position by a floating object. This weight is equal to that of the floating object.
2 *Mathematics*. a vector quantity representing the length and direction of a linear path from one point to another. Compare **distance.**

cap, fāce, fäther; best, bē, tėrm; pin, fīve;
rock, gō, ôrder; oil, out; cup, pùt, rüle;
yü in use, *yù* in uric;
ng in bring; *sh* in rush; *th* in thin, ᴛʜ in then;
zh in seizure.
ə = *a* in about, *e* in taken, *i* in pencil, *o* in lemon, *u* in circus

169

display

3 *Geology.* **a** the relative movement of the two sides of a fault. **b** the distance between the two sides.

displacement:

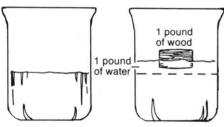

1 pound of water

1 pound of wood

(def. 1)
The one-pound block of wood displaces one pound of water in the beaker.

display, *n. Zoology.* a specialized pattern of behavior used to communicate visually, especially by the presentation of colors or plumage by male birds: *a courtship display, intimidation display.*

dissect, *v. Anatomy.* to cut apart (an animal, plant, organ, or tissue) in order to examine or study the structure. SYN: anatomize.
—**dissected,** *adj.* **1** *Botany.* cut or divided into many lobes: *dissected leaves.*
2 *Geology.* cut up by irregular valleys: *a dissected plateau.*
—**dissection,** *n.* **1** the act or process of dissecting: *At Padua, the center of medical studies, the first public dissection (as opposed to occasional private autopsies which were performed whenever a person of consequence had died under mysterious circumstances) took place in 1341* (Erwin Panofsky). **2** an animal, plant, etc., that has been dissected.

dissepiment (di sep′ə ment), *n. Botany, Zoology.* a partition in some part or organ; a septum. [from Latin *dissaepimentum,* from *dissaepire* to separate]

dissilient (di sil′ē ənt), *adj. Botany.* bursting open with force, as the dry pod or capsule of some plants. [from Latin *dissilientem* leaping apart]

dissimilation (di sim′ə lā′shən), *n. Biology.* = catabolism.

dissociate, *v. Chemistry.* to separate by dissociation: *Before a metal-ion complex can form, a water molecule must dissociate out of the metal's inner hydration sphere* (Scientific American).

dissociation, *n. Chemistry.* **1** the usually reversible changing of a substance into two or more simpler substances: *The dissociation of water into hydrogen and oxygen may ... take place in the gaseous phase at a very high temperature* (Sears and Zemansky, *University Physics*). **2** the separation of molecules of an electrolyte into constituent ions. Also called **ionization.** Hydrogen and chloride ions are formed by the dissociation of hydrogen chloride molecules in water.
—**dissociative,** *adj.* causing or connected with dissociation.

dissogeny (di soj′ə nē), *n. Biology.* a form of reproduction among the ctenophores, in which there are two periods of sexual maturity in the same individual, one in the larval stage and another in the adult form. [from Greek *dissos* double + *-geneia* origin]

dissolution, *n. Chemistry.* a making or becoming liquid; dissolving; liquefaction: *The passing of these particles into the liquid phase is known as dissolution, and the process is in many respects analogous to the evaporation of a liquid* (Jones, *Inorganic Chemistry*).

dissolve, *v. Chemistry.* to cause to pass into solution; turn into a solute.

dissolved load, *Geology.* the part of the total sediment in a stream that is in solution.

distal (dis′tl), *adj. Anatomy.* away from the center or point of origin: *Fingernails are at the distal ends of fingers.* SYN: terminal. ANT: proximal. [from *dist(ant)* + *-al*] —**distally,** *adv.*

distance, *n. Mathematics.* a scalar quantity representing the degree or amount of separation between two points. The distance D between two points (x_1, y_1) and (x_2, y_2) in a system of rectangular coordinates is given by the **distance formula:** $D = \sqrt{(x_1-x_2)^2+(y_1-y_2)^2}$ Compare **displacement** (def. 2).

distichous (dis′ti kəs), *adj. Botany.* arranged alternately in two vertical rows on opposite sides of an axis: *distichous leaves.* [from Greek *distichos* in two ranks, from *di-* two + *stichos* line]

distill or **distil,** *v. Chemistry.* **1** to heat (a liquid or solid) and condense the vapor or gas given off by cooling it, to obtain the substance or one of its constituents in a state of concentration or purity: *to distill water for drinking.*
2 to obtain by distilling: *Gasoline is distilled from crude oil. Alcoholic liquor is distilled from grain mash.* [from Latin *distillare,* from *de-* down + *stilla* drop]

distillate (dis′tl it *or* dis′tl āt), *n. Chemistry.* a distilled liquid; something obtained by distilling.

distillation (dis′tl ā′shən), *n. Chemistry.* the act or process of distilling; purification or separation of the constituents of a liquid or solid by changing it or some of its constituents to vapor and condensing the vapor to a liquid: *Petroleum products are separated from crude oil by distillation* (D.Q. Posin). *Plants for the large-scale distillation of sea water continue to be built* (New International Yearbook).

distributary (dis trib′yə ter′ē), *n. Geography.* a branch of a river that flows away from, rather than into, the main stream and never rejoins it. Compare **tributary.**

distribution, *n. Statistics.* a systematic arrangement of numerical data.

distribution curve, *Statistics.* a curve representing a frequency distribution adjusted to eliminate chance deviation due to inadequate sampling.

distributive (dis trib′yə tiv), *adj. Mathematics.* of or having to do with a rule that the same product results in multiplication when performed on a set of numbers as when performed on the members of the set individually. EXAMPLE: $3(4 + 5) = (3 \times 4) + (3 \times 5)$.
—**distributively,** *adv.*
—**distributivity,** *n.* the property of being distributive. ASSOCIATED TERMS: see **associative.**

distributive fault, *Geology.* = step fault. See the picture at **fault.**

distributive function, = probability function.

disulfide (dī sul′fīd), *n. Chemistry.* a compound consisting of two atoms of sulfur combined with another element or radical. Also called **bisulfide.**

170

diurnal (dī ėr'nl), *adj.* **1** *Biology.* occurring every day; daily: *There is a diurnal ovulation cycle in chickens.* Compare **circadian.**

2 *Astronomy.* having a daily cycle: *The rotation of the Earth on its axis, which is completed each day, results in an apparent revolution of all the heavenly bodies around the Earth in the same time. This apparent revolution is called the diurnal motion* (Duncan, *Astronomy*).

3 *Zoology.* active only in the daytime: *Diurnal animals have become adapted to meet, or to avoid, relatively high temperatures and evaporation rates, bright light and decreased conductivity of the air for odours* (Science News 28). Contrasted with **nocturnal.**

4 *Botany.* opening by day and closing by night, as certain flowers.

[from Late Latin *diurnalis,* from Latin *dies* day] —**diurnal'ity,** *n.* —**diurnally,** *adv.*

diurnal arc, *Astronomy.* the part of a circle which a celestial body appears to pass through above the horizon.

diurnal circle, *Astronomy.* the apparent circle described by a celestial body resulting from the rotation of the earth.

diurnal parallax, *Astronomy.* the parallax that a celestial body would have if viewed from two points on the earth separated by an angle of 90° with the earth's center. Also called **geocentric parallax.**

div., *abbrev.* **1** dividend. **2** division. **3** divisor.

divalent (dī vā'lənt), *adj. Chemistry.* having a valence of two; bivalent: *An element like magnesium or oxygen whose atom can hold two monovalent atoms in combination is called a divalent element* (Parks and Steinbach, *Systematic College Chemistry*).

divaricate (dī var'ə kāt), *v. Botany, Zoology.* to spread at a wide angle, as branches or insects' wings. —*adj.* (dī var'ə kit) spreading at a wide angle. [from Latin *divaricatus* spread apart] —**divarication,** *n.*

diverge, *v.* **1** to move or lie in different directions from the same point; branch off. A **diverging lens** is one that is thinner in the middle than at the edge and diverges parallel rays of light. Compare **converge.**

2 *Mathematics.* to increase indefinitely as more terms are added to a series.

—**divergence,** *n.* **1** the act or process of diverging; tendency to diverge: *a lens which increases the divergence of a beam of light.*

2 *Physiology.* a turning of both eyes outward, or of one eye when the other is fixed.

3 *Meteorology.* the spreading out of air or wind from a particular region.

4 *Biology.* the evolutionary tendency or process by which animals or plants that are descended from a common ancestor evolve into different forms when living under different conditions: *Through the process of divergence each species exploits the resources of the environment more effectively, so that the large-scale result of divergence in the inhabitants of a region is comparable to the physiological division of labour in an individual body* (Julian Huxley). See **divergent evolution** below.

5 *Geology.* the movement of the earth's plates away from each other.

—**divergent,** *adj.* diverging; formed by divergence. **Divergent evolution** is the appearance of different characteristics in animals and plants closely related by a common ancestry. Compare **convergent evolution.**

diverging plate boundaries

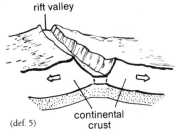

(def. 5)

The plates move away from each other, causing a widening of the ocean trench at the plate boundary.

divide, *v. Mathematics.* **1** to separate (a number or quantity) into equal parts; find out how many times one number or quantity is contained in another; perform the operation of division: *8 ÷ 4 denotes that 8 is to be divided by 4; 8 ÷ 4 = 2.*

2 (of a number or quantity) to be a divisor or factor of (another number or quantity); be contained an exact number of times in: *9 divides 36. $x + y$ divides $x^n + y^n$ when n is odd.*

divided, *adj. Botany.* cut to the midrib or base, forming distinct parts.

dividend, *n. Mathematics.* a number or quantity to be divided by another: *In 728 ÷ 16, 728 is the dividend.*

divisible, *adj. Mathematics.* leaving no remainder when divided: *15 is divisible by 3 and 5.*

division, *n.* **1** *Mathematics.* the operation of dividing one number or quantity by another; the process of finding out how many times one number or quantity is contained in the other or, more generally, of finding a quantity (the *quotient*) which multiplied by the latter (the *divisor*) will produce the former (the *dividend*); the inverse of multiplication. *48 ÷ 2 = 24 is the inverse of 2 × 24 = 48.* Compare **long division, short division.**

2 *Botany.* a major group of the plant kingdom, comparable to a phylum of the animal kingdom. The plants in a division are thought to be related by descent from a common ancestral form. *Groups of increasingly greater size are, step by step, orders, classes and, finally, divisions* (Emerson, *Basic Botany*).

division sign, *Mathematics.* any of various symbols indicating that one number is to be divided by another, especially the symbol ÷. The expression "nine divided by three" may be written 9 ÷ 3, 9/3, $\frac{9}{3}$, or $3 \overline{)9.}$

---

cap, fāce, fäther; best, bē, tėrm; pin, five;
rock, gō, ôrder; oil, out; cup, pùt, rüle;
yü in use, *yu* in uric;
ng in bring; *sh* in rush; *th* in thin; ᴛʜ in then;
zh in seizure.
ə = *a* in about, *e* in taken, *i* in pencil, *o* in lemon, *u* in circus

divisor, *n. Mathematics.* **1** a number or quantity by which another is to be divided: *In 728 ÷ 16, 16 is the divisor.*
2 a number or quantity that divides another without a remainder.

dizygotic (dī'zī got'ik) or **dizygous** (dī zī'gəs), *adj. Biology.* that have developed from two zygotes: *Identical, or monozygotic (MZ), twins have the same inherited traits, whereas fraternal, or dizygotic (DZ), twins have no more inherited traits in common than the other siblings of a family* (New Scientist).

dkg., *abbrev.* decagram *or* decagrams.

dkl., *abbrev.* decaliter *or* decaliters.

dkm., *abbrev.* decameter *or* decameters.

dl., *abbrev.* deciliter *or* deciliters.

D layer or **D region,** *Geophysics.* the lowest layer of the ionosphere, about 50 miles above the earth. Compare **E layer, F layer.** See the picture at **E layer.**

dm., *abbrev.* decimeter *or* decimeters.

DNA, *n. Biochemistry.* deoxyribonucleic acid, the genetic material of all cells and many viruses: *Genes are made of quite different long-chain molecules: the nucleic acids DNA (deoxyribonucleic acid) and, in some small viruses, the closely related RNA (ribonucleic acid)* (F.H.C. Crick). *The most important feature of DNA is that it usually consists of two very long, thin polymeric chains twisted about each other in the form of a regular double helix* (James D. Watson, *Molecular Biology of the Gene*).

DNAase (dē'en'ā'ās') or **DNase** (dē'en'ās'), *n. Biochemistry.* deoxyribonuclease, an enzyme that promotes the hydrolysis of DNA.

DNA polymerase, *Biochemistry.* an enzyme that promotes the formation of new nucleotides of DNA from DNA triphosphates by a process of replication, using DNA as a template: *Besides its copying abilities, DNA polymerase can repair strands of DNA damaged by ultraviolet light* (London *Times*).

DNA

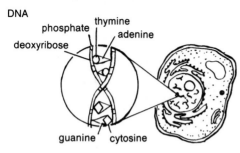

phosphate
thymine
adenine
deoxyribose

guanine   cytosine

dodecagon (dō dek'ə gon), *n. Geometry.* a plane figure having 12 angles and 12 sides. [from Greek *dōdekagōnon,* from *dōdeka* twelve + *gōnia* angle]
—**dodecagonal** (dō'de kag'ə nəl), *adj.* having 12 angles and 12 sides.

dodecahedron (dō'dek ə hē'drən), *n. Geometry.* a solid figure having 12 faces. [from Greek *dōdekaedron,* from *dōdeka* twelve + *hedra* seat, base]
—**dodecahedral** (dō'də kə hē'drəl), *adj.* having 12 faces.

dodecahedron

doldrums, *n.pl. Meteorology, Geography.* **1** certain regions of the ocean near the equator where the wind is very light or constantly shifting. **2** the calm or windless weather characteristic of these regions.

dolerite (dol'ə rīt), *n.* = diabase.

doleritic (dol'ə rit'ik), *adj.* = ophitic.

dolomite (dol'ə mit), *n. Geology.* **1** a rock consisting mainly of calcium magnesium carbonate.
2 the mineral, calcium and magnesium carbonate, that composes this rock.
[from D.G. de *Dolomieu,* 1750–1801, French geologist + *-ite*]
—**dolomitic** (dol'ə mit'ik), *adj.* containing or consisting of dolomite: *dolomitic limestone.* In **dolomitic marble** the prevalent mineral is dolomite rather than calcite.
—**dolomitize** (dol'ə mə tīz), *v.* to convert (limestone) into dolomite.
—**dolomitization,** *n.* conversion of limestone to dolomite.

domain, *n.* **1** *Physics.* any of the microscopic areas that are polarized in one direction in a ferromagnetic substance: *The magnetism of a material such as iron exists in small regions, or domains, as they are called, which are present in an unmagnetized piece of iron. Each of these domains is a small magnet having a north and south pole. . . In a magnetic field they are aligned so they give the magnetism usually observed* (Ray W. Guard).
2 *Mathematics.* the set whose members are considered as possible replacements for the variable in a given relation. Also called **replacement set.**

dome, *n.* **1** *Geology.* **a** a structure or formation in which the beds dip away approximately equally from a central point. **b** any structure or formation resembling the dome of a building.
2 *Crystallography.* a crystal form consisting of two plane surfaces forming a dihedral angle.

dominance, *n.* the quality or condition of being dominant: *Mendel expressed these findings as his Law of Dominance, which in essence states that in a cross involving a dominant and a recessive character, the dominant character only will be visible in the generation following the cross* (Harbaugh and Goodrich, *Fundamentals of Biology*). *The concept of social dominance has often been used in analyzing the structure of animal social systems, particularly nonhuman primates* (Barash, *Sociobiology and Behavior*).

172

dominant, *adj.* **1** *Genetics.* of, having to do with, or designating the allele of any pair of contrasting alleles that prevails in an organism when both alleles are present in the genotype. EXAMPLE: If a child inherits a gene for normal skin pigmentation from one parent and a gene for albinism from the other, the child will have normal pigmentation, as the gene for normal pigmentation is dominant to the gene for albinism, which is recessive. *A gene is said to be dominant when the character it represents appears in the heterozygote; its allelomorph in such a case is said to be recessive . . . The result of dominance and recessiveness in allelomorphic genes is that heterozygotic individuals do not exhibit all that they inherit from their parents, since some of the genes are recessive.* (Youngken, *Pharmaceutical Botany*).

2 *Zoology.* having high social rank, characterized by success in the competition for food or mates: *There is a clear dominance hierarchy among the adults of a baboon troop. ... If a dominant animal approaches a subordinate the latter will walk away* (Walter Fiedler).

3 *Ecology.* of or designating the most extensive and characteristic species in a plant community.

—*n.* **1** *Genetics.* **a** a dominant allele or gene. **b** an individual possessing or transmitting a dominant trait.

▶ By convention, a capital letter symbolizes a dominant allele, while a lower-case letter symbolizes the recessive allele. For example, in certain pea plants, the allele for tallness (T) is dominant and the allele for shortness (t) is recessive.

2 *Zoology.* a dominant animal: *In some cases, notably in primates, subordinates often accomplish some matings; their exclusion by the dominants is not absolute* (Barash, *Sociobiology and Behavior*).

3 *Ecology.* the most extensive and characteristic species in a plant community, determining the type and abundance of other species in the community: *The ones whose destruction brings about the greatest changes in the composition of the association are the dominants* (Emerson, *Basic Botany*).

donor, *n.* **1** *Medicine.* a person from whose body blood, tissue, or an organ is removed for transfusion, grafting, or transplant: *Normally donor grafting is possible only in identical twins* (New Scientist). Contrasted with **recipient.**

2 *Chemistry.* an atom, molecule, or ion that furnishes either a pair of electrons or a proton to form a bond with an acceptor: *A Brønsted acid acts as a proton donor.*

3 *Physics.* an impurity added to a semiconducting substance to increase the number of free electrons: *If a donor impurity is diffused into one side of a crystal and an acceptor impurity is diffused into the other, one region of the crystal will have an excess of free electrons* (Scientific American). Compare **dopant.**

dopa (dō′pə), *n. Biochemistry.* a crystalline amino acid found in various plants or made synthetically: *Administration of dopa ... was based upon the observation that dopa and/or related substances are depleted in the brain of the patient with Parkinson's disease* (Sumner M. Kalman). *Formula:* $C_9H_{11}NO_4$ [acronym formed from *d(ihydr)o(xy)p(henyl)a(lanine)*]

dopamine (dō′pə mēn), *n. Biochemistry.* a hormone found especially in the adrenal glands that is necessary for the normal nerve activity of the brain: *Dopamine, the chemical to which L-dopa is converted once it enters the body, increases blood flow in the kidneys* (Science News). [from *dopa* + *amine*]

dopant (dō′pənt), *n. Physics.* a small amount of an impurity added to a semiconductor to alter its conductive properties: *A typical integrated circuit crystal is said to be p-type, having positive charge carriers ('holes'), because it contains an acceptor dopant such as boron* (Science Journal).

Doppler broadening, *Physics.* the broadening of a line or lines of a spectrum due to the Doppler effect, as a result of a range in velocity of the radiating atoms or molecules.

Doppler effect

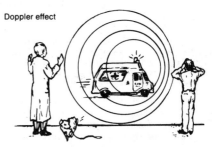

This apparent change in pitch, known as the Doppler effect, is quite evident to the bystander, although to the men...[in the ambulance] the pitch remains the same. The pitch of any sound depends upon the frequency with which the sound-waves reach the ear, and when the source of the sound approaches the observer, the waves are piled up so that more reach the ear per second than are actually given out at the source. Conversely, when the source is receding, the sound waves are spaced more widely apart so that fewer reach the ear per second.

(Bernhard, Handbook of the Heavens).

Doppler effect or **Doppler shift** (dop′lər), *Physics.* the observed change in the frequency of waves when either the source of the waves or the observer moves toward or away from the other: *The Doppler effect ... is a change in wavelength observable when the source of radiation (sound, light, etc.) is in motion. If it is moving toward the observer, the wavelength is shortened; if away, the waves are lengthened. In the case of a star moving away from us, the whole spectrum of its light is shifted toward the red, or long-wave, end* (A.R. Sandage). [named after Christian J. *Doppler*, 1803–1853, Austrian physicist]

cap, fāce, fäther; best, bē, tėrm; pin, five;
rock, gō, ôrder; oil, out; cup, pùt, rüle,
yü in use, *yù* in uric;
ng in bring; *sh* in rush; *th* in thin, TH in then;
zh in seizure.
ə = *a* in about, *e* in taken, *i* in pencil, *o* in lemon, *u* in circus

dormancy, *n. Botany.* a condition of not growing or germinating: *Some seeds will germinate immediately after harvesting, but will then slowly enter a period of non-germination, called dormancy* (A.M. Mayer).

dormant, *adj.* **1** *Botany.* (of plants, bulbs, seeds, etc.) with development suspended; not growing: *With the beginning of the cold of autumn and winter or of the drought of the dry seasons, various plants react in different ways. Some die, but the majority develop inactive or dormant buds* (Emerson, *Basic Botany*). **2** *Geology.* (of volcanoes) not active but not extinct; potentially active: *The strongest known volcanic explosions have occurred in long-dormant composite volcanoes. Some of these volcanoes seemed dead, not having erupted within historic times* (Gilluly, *Principles of Geology*).

dorsal (dôr'səl), *Anatomy.* —*adj.* of, on, or near the back or upper surface of an organ, part, or organism: *a dorsal nerve.* A **dorsal fin** is a fin or finlike part or parts on the back of most aquatic vertebrates. See the picture at **fin.**
—*n.* a dorsal part, such as a dorsal fin or vertebra. [from Latin *dorsum* the back] —**dorsally,** *adv.*
ASSOCIATED TERMS: In a typical animal with bilateral symmetry, such as a dog, the head end is *anterior,* the tail end is *posterior,* the back or upper surface is *dorsal,* the abdomen or under surface is *ventral,* and the sides are right or left *lateral.*

dorsiferous (dôr sif'ər əs), *adj.* **1** *Botany.* bearing the clusters of sporangia (sori) on the underside or back of the frond: *a dorsiferous fern.* **2** *Zoology.* = dorsiparous.

dorsiparous (dôr sip'ər əs), *adj. Zoology.* hatching the young on the back: *Certain toads are dorsiparous.* [from Latin *dorsum* back + *parere* to give birth to]

dorsiventral, *adj.* **1** *Botany.* having distinct dorsal and ventral sides: *Most foliage leaves are dorsiventral.* **2** = dorsoventral. —**dorsiventrally,** *adv.*

dorsiventral symmetry, *Botany.* symmetry in which only the sides are similar while the upper and lower surfaces are different: *Dorsiventral symmetry ... differs from true bilateral symmetry in that the front and back sides are different, so that the structure can be cut along only one plane to get similar halves. Most leaves have dorsiventral symmetry* (Greulach and Adams, *Plants*).

dorsolateral, *adj. Anatomy.* of or involving both the back and the side.

dorsoventral (dôr'sō ven'trəl), *adj. Zoology.* extending from the dorsal to the abdominal side: *dorsoventral muscles.*

dorsum (dôr'səm), *n., pl.* **-sa** (-sə). *Anatomy.* the back; outer surface of an organ or part: *the dorsum of the hand.* [from Latin *dorsum* back]

dose, *n. Medicine.* **1** the amount of medicine or other therapeutic agent to be given or taken at one time. **2** the intensity and length of exposure to heat, X rays, or other radiation: *The biological effects of radiation are measured by the dose received* (Bulletin of Atomic Scientists).

dose rate, *Medicine.* the intensity of nuclear radiation to which an organism is exposed.

dosimeter (dō sim'ə tər), *n.* an instrument for measuring the doses of X rays or radioactivity received over a given period of time.
—**dosimetric** (dō'sə met'rik), *adj.* having to do with the measurement of doses or with a dosimeter.
—**dosimetry** (dō sim'ətrē), *n.* the measurement of doses.

dot product, *Mathematics.* the product of the lengths of two vectors and the cosine of the angle between them. A dot product is indicated by a dot placed between the vectors (as in A·B). *If a and b are at right angles then $a \cdot b = 0$, and if neither a nor b is a zero vector then the vanishing of the dot product shows the vectors to be perpendicular* (I.S. Sokolnikoff). Also called **inner product** or **scalar product.** Contrasted with **cross product** or **vector product.**

double, *adj. Botany.* having more than one set of petals or sepals.

double-blind, *adj.* having to do with or designating a test or experiment in which neither the identity of what is being tested nor its possible effects are known to the subject or to the one administering the test until its completion: *The ideal way to determine that two like products are equal in therapeutic action would be to compare them in extensive, double-blind clinical studies* (Science News). Contrasted with **single-blind.**

double bond, *Chemistry.* a bond in which two electron pairs are shared between two atoms, occurring in unsaturated compounds; a bond of two valences. Compare **single bond, triple bond.**

double concave, concave on both sides: *a double concave lens.*

double convex, convex on both sides: *a double convex lens.*

double cross or **double-cross hybrid,** *Genetics.* the hybrid produced by two single crosses of inbred lines. Compare **single cross.**

double decomposition, = metathesis.

double fertilization, *Botany.* the union in seed plants of one sperm nucleus with the egg nucleus and the other sperm nucleus with the polar nuclei.

double-helical (dub'l hel'ə kəl), *adj. Molecular Biology.* consisting of two strands that coil around each other to form a double spiral or helix: *The double-helical DNA molecule is very stable at physiological temperatures* (Watson, *Molecular Biology of the Gene*).

double helix, *Molecular Biology.* **1** the structure of a molecule of DNA, made up of two spiral strands of alternating phosphate and sugar units, the strands connected by bases of adenine, guanine, thymine, and cytosine. **2** the DNA molecule itself.

double refraction, *Physics.* the splitting up of a ray of light into two rays when it passes certain crystalline substances: *The Dutch physicist, Christian Huygens (1629–1695), discovered the phenomenon of "double refraction" while studying the transmission of light* through the mineral calcite (Gilluly, *Principles of Geology*). Also called **birefringence.**

double salt, *Chemistry.* a salt that ionizes in solution as if it were two separate salts, but forms a single substance on crystallization. Alum is a double salt formed by crystallization of potassium and aluminum sulfate.

double star, *Astronomy.* **1** two stars so close together that they look like one to the naked eye. **2** = binary star.

down, *n.* **1** *Zoology.* the soft feathers of a bird: *There is no material known that is more efficient for holding heat than down* (Winchester, *Zoology*).
2 *Botany.* **a** a fine soft hair on some plants and fruits, as peach fuzz. **b** the soft, feathery pappus of some seeds, as in the dandelion and thistle.

down quark, *Nuclear Physics.* a quark having a downward spin of $+1/2$ and a charge of $-1/3$: *Ordinary neutrons and protons are made only of up and down quarks* (Science News).

downthrow, *n.* *Geology.* the downward displacement of rock on one side of a fault.

downwarp, *Geology.* —*v.* to fold or bend in a troughlike manner, as the valley of a stream: *An area of as much as 100,000 square miles rose or fell during the earthquake. Downwarping exceeded six feet along the axes of Kodiak Island and the Kenai Peninsula, and uplift amounted to 30 feet or more in broad areas to the southeast* (Richard H. Jahns).
—*n.* a troughlike condition or area: *Taylor in 1930 postulated the existence of a downwarp or trough between the Ross and Weddell seas* (Science).

DP, *abbrev.* degree of polymerization.

DPG, *abbrev.* diphosphoglyceric acid.

dpm, *abbrev. Nuclear Physics.* disintegrations per minute.

DPN, *abbrev.* diphosphopyridine nucleotide: *The primary hydrogen acceptor in the Krebs cycle is the coenzyme diphosphopyridine nucleotide (DPN). The reduced coenzyme (DPNH) is oxidized stepwise by the electron transport chain via a series of respiratory enzymes* (Efraim Racker). Compare **NAD.**

DPNH, *n.* the reduced form of DPN. [from *DPN + H(ydrogen)*] Compare **NADH.**

dps, *abbrev. Nuclear Physics.* disintegrations per second: *It [curie] refers to the quantity of a substance in which there are 3×10^{10} disintegrations (individual decay events) per second (dps)* (Eugene E. Fowler).

dr., *abbrev.* dram *or* drams.

Draconid (drak′ə nid), *n. Astronomy.* any of a shower of meteors seeming to radiate from the constellation Draco and frequently reappearing about October 9.

drag, *n. Physics.* the force acting on a body in motion through a fluid in a direction opposite to the body's motion and produced by friction: *the drag, i.e., the frictional force exerted on the plane by the air through which it moves* (Michels, *Foundations of Physics*).

drainage basin, *Geology.* the area drained by a stream and its tributaries: *A river and its tributary streams make up a drainage system; the area they drain is*

called a drainage basin (Colby and Foster, *Economic Geography*).

drainage wind, = katabatic wind.

dram, *n.* a unit of weight. In avoirdupois weight, 16 drams make one ounce. [from Old French *drame,* from Latin *drachma*]

D region, = D layer: *X-rays associated with some solar flares penetrate to the lowest stratus of the ionosphere (the D region) and increase the ionization appreciably: it appears that this process is primarily responsible for radio communications blackouts* (Hugh Odishaw).

drift, *n.* **1** = glacial drift: *It was not at first understood that the widespread glacial materials of northern Europe and North America were deposits made by great continental ice sheets. These deposits were called "drift," because it was thought that they were carried by icebergs that drifted in waters of Noah's flood* (Moore, *Introduction to Historical Geology*).
2 a current of water caused by the wind: *A general surface movement of water, which has no sharp boundaries, is known as a drift* (F.G. Walton Smith).

drift ice, *Oceanography.* masses of detached floating ice drifting with ocean currents or the wind.

dripstone, *n. Geology.* deposits of calcium carbonate built up by dripping water in caves and forming stalactites and stalagmites.

droplet infection, *Medicine.* infection spread from one individual to another by droplets of moisture expelled in coughing, sneezing, etc.

drosophila (drō sof′ə lə), *n., pl.* **-lae** (-lē). *Zoology.* any of a genus (*Drosophila*) of fruit flies, especially a species frequently used in experimental studies of heredity. [from New Latin, from Greek *drosos* dew + *philos* loving]

drug, *n. Medicine.* **1** any substance (other than food) that is used as a medicine or as an ingredient in medicine, obtained from molds, parts of plants, parts of animals, minerals, etc., or prepared synthetically. Aspirin and antibiotics are drugs. *Drugs like penicillin, streptomycin, aureomycin and the other antibiotics act specifically by killing micro-organisms* (A.J. Birch).
2 any chemical substance that affects the central nervous system, causing changes in behavior and often addiction. Drugs can be generally classified as stimulants, depressants, and hallucinogens. Amphetamine and cocaine are stimulants; alcohol, barbiturates (sleeping pills), and narcotics such as heroin and morphine are depressants; marijuana, hashish, and LSD are hallucinogens.

---

cap, fāce, fäther; best, bē, tèrm; pin, five;
rock, gō, ôrder; oil, out; cup, pùt, rüle;
yü in use, *yu* in uric;
ng in bring; *sh* in rush; *th* in thin; ŦH in then;
zh in seizure.
ə = *a* in about, *e* in taken, *i* in pencil, *o* in lemon, *u* in circus

175

drumlin, *n. Geology, Geography,* a ridge or oval hill with a smooth summit, consisting of glacial drift and shaped by glacial action: *In a few localities there are hills of considerable height which are composed of clayey till ... Some of them reach heights of 100 ft. or more ... Where conditions under the ice were favorable to the formation of one drumlin, they apparently were equally favorable to the formation of others, for commonly they are found in groups that occupy many square miles* (Finch and Trewartha, *Elements of Geography*). [apparently diminutive of Scottish and Irish *druim* back, ridge]

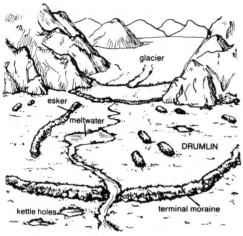

DRUMLIN

glacier

esker

meltwater

kettle holes

terminal moraine

drupaceous (drü pā′shəs), *adj. Botany.* of, having to do with, or producing drupes: *Drupaceous fruits are those in which the endocarp is always fibrous or stony in consistence, while the mesocarp is more or less succulent* (Youngken, *Pharmaceutical Botany*).

drupe (drüp), *n. Botany.* a fruit whose seed is contained in a hard pit or stone surrounded by soft, pulpy flesh: *In the peach, cherry, and plum, ... the inner part of the ovary wall hardens to form a stone-like covering around the seed, and the remainder of the ovary wall develops into the skin and the pulpy portions ... Such a fruit is a drupe* (Harbaugh and Goodrich, *Fundamentals of Biology*). See the picture at **fruit.** [from New Latin *drupa,* from Latin *druppa* very ripe olive, from Greek *dryppa* olive]

drupelet (drüp′lit), *n. Botany.* one of the small drupes composing certain fruits: *the individual parts or drupelets of blackberries and raspberries* (Emerson, *Basic Botany*).

druse (drüz), *n. Geology.* **1** a crust of small crystals lining the sides of a cavity in a rock.
2 the crystal-lined cavity. [from German *Druse*]

dry cell, an electric cell in which the chemical producing the current is made into a paste so that its contents cannot spill: *The dry cell is the most convenient small portable source of electrical energy. The positive pole is a rod of carbon and the negative pole is of metallic zinc. The ordinary fluid is replaced by a paste containing zinc chloride and ammonium chloride* (Sears and Zemansky, *University Physics*).

duct, *n.* **1** *Anatomy.* a tube in the body for carrying a bodily fluid: *tear ducts. The pancreatic ducts lead from the pancreas to the duodenum.*
2 *Botany.* = trachea.

ductile (duk′təl), *adj. Chemistry.* capable of being drawn out into a wire: *Copper and aluminium are both highly ductile metals and alloy in all proportions, but alloys of copper with from about 15 per cent to 75 per cent of aluminium are without any ductility whatsoever* (J. Crowther). Compare **malleable.**
—**ductility** (duk til′ə tē), *n.* the quality or condition of being ductile: *Most polymers lose their ductility when given even moderate exposures* (Scientific American).

ductless gland, = endocrine gland.

ductule, *n. Anatomy.* a small duct.

ductus arteriosus (duk′təs är tir′ē ō′səs), *Embryology.* a short blood vessel that connects the pulmonary artery with the aorta in the fetus: *In the heart of a human fetus the partition between the two atria is incomplete, and also there is a connection (ductus arteriosus) between the pulmonary artery and aorta* (Storer, *General Zoology*).

ductus deferens (duk′təs def′ə renz), = vas deferens.

dulosis (dü lō′sis), *n. Zoology.* the practice by some groups of ants of enslaving individuals or colonies of other ants: *From temporary social parasitism the next step is exhibited by dulosis or slavery* (A. D. Imms). [from Greek *doulōsis* slavery]
—**dulotic** (dü lot′ik), *adj.* exhibiting dulosis; practicing the enslavement of other insects.

dumortierite (dü môr′tē ə rīt), *n. Mineralogy.* an aluminum borosilicate often occurring in gneiss: *... dumortierite, a rare mineral used in spark plugs and other refractory products* (Wall Street Journal). Formula: $Al_7(BO_3)(SiO_4)_3O_3$ [from Eugène *Dumortier,* 19th-century French paleontologist + *-ite*]

dune, *n. Geology.* a mound or ridge of loose sand heaped up by the wind: *Rock surfaces are alternately heated, cooled, and fragmented, bringing about the eventual formation of sand particles. These particles are piled up, under the influence of winds, into dunes which themselves move* (R.N. Elston).

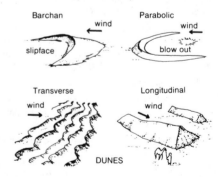

Barchan

wind

slipface

Parabolic

wind

blow out

Transverse

wind

Longitudinal

wind

DUNES

dunite (dun′īt), *n. Geology.* a granular igneous rock, composed chiefly of olivine with small amounts of chromite and other minerals. [from *Dun* Mountain, New Zealand, where the rock is found + *-ite*]

duodecimal (dü′ō des′ə məl), *Mathematics.* —*adj.* having to do with twelfths or the base of twelve; expressed by twelves. A duodecimal system of arithmetical notation uses twelve digits (such as 0, 1, 2, 3, 4, 5, 6, 7, 8, 9, x, and y) instead of the ordinary ten digits (0 to 9) of the decimal system. Compare **binary, decimal, hexadecimal.**
—*n.* one of a system of numerals, the base of which is twelve instead of ten.

duodenal (dü′ō dē′nl), *adj. Anatomy.* of or having to do with the duodenum: *a duodenal ulcer.*

duodenum (dü′ō dē′nəm), *n., pl.* **-na** (-nə). *Anatomy.* the first part of the small intestine, beginning just below the stomach and extending to the jejunum: *The duodenum ... is the shortest, the widest, and the most fixed part of the small intestine* (Goss, *Gray's Anatomy*). [from Medieval Latin *duodenum,* from Latin *duodeni* twelve each; with reference to its length, about twelve finger breadths]

duple ratio, *Mathematics.* a ratio in which the antecedent is double the consequent, as 2 to 1, or 8 to 4.

durain (də rān′), *n. Geology.* a hard, dull lithotype of bituminous coal, composed chiefly of decayed plant matter and clay. [from Latin *durus* hard + English *-ain,* as in *fusain*]

dura mater (dùr′ə mā′tər *or* dyùr′ə mā′tər), *Anatomy.* the tough, fibrous membrane forming the outermost of the three coverings (meninges) of the brain and spinal cord: *The dura mater is adherent to the inner surface of the cranium, where it serves the double function of an internal periosteum and a covering for the brain* (King and Showers, *Human Anatomy and Physiology*). [from Medieval Latin *dura mater,* literally, hard mother (that is, hard source)]

duramen (dü rā′mən), *n. Botany.* the central wood or heartwood of an exogenous tree. [from Latin *duramen* hardness; woody branch of a vine, from *durare* harden, from *durus* hard]

dust devil, *Meteorology.* a small whirlwind that picks up and carries a column of dust as it moves along, especially in the dry plains area of the western United States.

dwarf, *n., pl.* **dwarfs** or **dwarves. 1** *Biology.* an animal or plant much below the ordinary height or size of its kind or related species: *With subnormal amounts of growth hormone a child will never reach adult size; he will be a dwarf* (McElroy, *Biology and Man*).
2 *Astronomy.* any of a class of stars of small size and luminosity, including the sun; main-sequence star: *The Danish astronomer Hertzsprung had previously drawn attention to the sharp distinction between the red stars of high and low luminosity, and had named them giant stars and dwarf stars respectively* (Baker, *Astronomy*).

dwarfism, *n. Biology.* a generally underdeveloped condition of growth: *Dwarfism is sometimes dependent on light for expression, as in peas where both dwarf and tall varieties are equally tall in darkness* (Weier, *Botany*).

dwarf star, = dwarf (def. 2): *Discovery of flare-ups on two red dwarf stars raised to five the number of stars known to have flares similar to those on the sun* (Science News Letter).

Dy, *symbol.* dysprosium.

dyad (dī′ad), *n.* **1** *Biology.* one of the pairs of chromatids formed when a tetrad splits during meiosis.
2 *Mathematics.* an operator formed by the juxtaposition of two vectors, without indicating either a scalar or a vector multiplication. EXAMPLE: The dyad ab has the properties (ab)·F = a(b·F) and F·(ab) = (F·a)b. [from Latin *dyadem* two]
—**dyadic** (dī ad′ik), *adj. Biology.* having to do with a dyad or any group of two.
—*n. Mathematics.* the sum of two or more dyads.

dynamic, *adj.* **1** *Physics.* **a** having to do with energy or force in motion: *Dynamic phenomena, involving the transfer of matter by diffusion, eddying, or bulk flowing; the removal and introduction of matter by chemical reaction; and the transfer of heat* (J.F. Pearson). **b** having to do with the science of dynamics: *the dynamic laws that cause a pump to act, the dynamic theory of the tides.*
2 *Geology.* produced by earth movement or rock deformation: *dynamic metamorphism.*

dynamics, *n.* **1** *pl. in form, sing. in use.* the branch of mechanics that deals with the effects of forces on the motions of bodies. Also called **kinetics.**
▶ Dynamics is often contrasted with *statics,* the study of bodies at rest, and with *kinematics,* the study of motion without reference to forces.
2 *pl. in form, sing. in use.* any science of motion or forces: *space dynamics.*
3 *pl. in form and use.* any set of motions or forces to be studied: *the dynamics of glacier motion.*

dynamo (dī′nə mō), *n. Electricity.* a generator, usually a direct-current generator: *A dynamo converts the energy of mechanical motion into electric current. The motion may be relative motion between a coil and a ferromagnetic material or between two coils (or other conductors of electricity)* (Walter M. Elsasser). ▶ While a direct-current generator is sometimes called a **dynamo,** an alternating-current generator is more often called an **alternator.**

dynamoelectric, *adj. Physics.* having to do with the transformation of mechanical energy into electric energy, or electric energy into mechanical energy.

dynamometer (dī′nə mom′ə tər), *n.* an instrument for measuring mechanical power, as that of an automobile engine.

dyne (dīn), *n. Physics.* the unit of force in the centimeter-gram-second system. A dyne is the amount of force required to give a mass of one gram an acceleration of one centimeter per second per second. *In the clean waters of the streams where aquatic plastron-bearing insects live the surface tension of the water is high, usually between 70 and 72 dynes per centimeter* (Scientific American). [from French *dyne,* from Greek *dynamis* power]

dyne-centimeter, *n.* = erg.

---

cap, fāce, fäther; best, bē, tèrm; pin, fīve;
rock, gō, ôrder; oil, out; cup, pùt, rüle;
yü in use, *yu* in uric;
ng in bring; *sh* in rush; *th* in thin, ᴛʜ in then;
zh in seizure.
ə = *a* in about, *e* in taken, *i* in pencil, *o* in lemon, *u* in circus

dys-

dys-, *prefix.* bad, abnormal, or defective, as in *dysfunction, dysgenesis.* [from Greek *dys-*]

dyscrasite (dis′krə sīt), *n. Mineralogy.* a mineral consisting of antimony and silver. It occurs in crystals and also in massive and granular form. *Formula:* Ag₃Sb [from Greek *dyskrasia* bad mixture + English *-ite*]

dysgenesis (dis gen′ə sis), *n. Biology, Medicine.* **1** abnormal development, as of an organ. **2** lack of the ability to reproduce; sterility.

dysphotic (dis fō′tik), *adj. Biology.* growing or occurring where there is little light, as at great depths of water: *dysphotic vegetation.*

dysplasia (dis plā′zhə), *n. Biology.* abnormal development or growth of tissues, organs, or other structures. [from New Latin, from *dys-* bad + Greek *plasis* a molding]
—dysplastic, *adj.* of or having to do with dysplasia.

dysprosium (dis prō′zē əm), *n. Chemistry.* a rare-earth metallic element found in various minerals which forms highly magnetic compounds. *Symbol:* Dy; atomic number 66; atomic weight 162.51; oxidation state 3. [from New Latin *dysprosium,* from Greek *dysprositos* hard to get at]

dystrophic (dis trof′ik), *adj.* **1** *Ecology.* containing a high concentration of humus, peat, detritus, etc.: *a dystrophic lake or pond.* Compare **eutrophic.**
2 *Medicine.* characterized by dystrophy: *dystrophic muscles.*

—dystrophication, *n. Ecology.* the process of becoming dystrophic: *The sea can probably tolerate the runoff indefinitely but along the way the nitrogen creates algal "blooms" that are hastening the dystrophication of lakes and estuaries* (Ferren MacIntyre). Compare **eutrophication.**

dystrophy (dis′trə fē), *n. Medicine.* defective development or degeneration: *muscular dystrophy.* [from Greek *dys-* bad + *trophē* nourishment]

DZ, *abbrev.* dizygotic: *Twins are either fraternal or identical. If they are fraternal, they are dizygotic (DZ), arising from two separate eggs fertilized by two separate sperm* (McElroy, *Biology and Man*).

E

e, *symbol.* **1** eccentricity. **2** electron. **3** erg. **4** *Mathematics.* the quantity 2.71828+ (the base of natural logarithms).

E, *symbol.* **1** electromotive force. **2** *Physics.* energy: $E = mc^2$.

ear, *n.* **1** *Anatomy.* **a** the part of the body by which human beings and other vertebrates hear; organ of hearing. It usually consists of the external ear, the middle ear, and the inner ear. **b** the fleshy, curved outer part of the external ear. Also called **auricle**.
2 *Botany.* on a cereal plant, such as corn, wheat, oats, barley, or rye, the part that bears flowers from which grains develop.

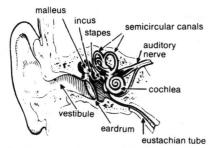

the human ear

eardrum, *n. Anatomy.* a thin membrane across the middle ear that vibrates when sound waves strike it: *Sound waves enter the outer ear, travel down the ear canal, and strike a thin membrane called the eardrum, which forms the boundary between the outer ear and the middle ear* (Shortley and Williams, *Elements of Physics*). Also called **tympanic membrane** or **tympanum**. See the picture at **ear**.

earlobe, *n. Anatomy.* **1** the soft, loosely hanging lower part of the external ear. **2** the fleshy fold beside the ear of some fowls.

earth, *n.* **1** *Astronomy.* Often capitalized **Earth**. the fifth largest planet in the solar system and the third in distance from the sun (about 150 million kilometers). The earth has a diameter of about 13,000 kilometers, but it is not perfectly round, the diameter from pole to pole being shorter by about 42 kilometers than the diameter at the equator. The earth's atmosphere extends up to about 1500 kilometers above the surface. Its surface consists of about 70 per cent water and 30 per cent land. The earth's period of rotation about its axis is 23 hours, 56 minutes, and 4.09 seconds; its period of revolution around the sun is 365 days, 6 hours, 9 minutes, and 9.54 seconds.
2 *Chemistry.* a metallic oxide from which it is hard to remove the oxygen, such as alumina: *the alkaline earths, the rare earths.*

earthquake, *n. Geology.* a shaking or sliding of a portion of the earth's crust, caused by the sudden movement of masses of rock along a fault or by changes in the size and shape of masses of rock far beneath the earth's surface. Earthquakes are often associated with volcanic activity. *Recent theories of earthquake prediction have been based on well established observations that rocks dilate, or expand slightly, when strained almost to breaking point* (Nature).
ASSOCIATED TERMS: The magnitude of an earthquake is usually measured on the *Richter scale.* Earthquakes usually occur along *faults* or along the edges of the earth's crustal *plates.* Earthquakes produce *seismic waves* which include *shear waves* and *compression waves.* The point where an earthquake originates is the *focus;* seismic waves are strongest at the *epicenter,* a point directly above the focus. *Seismology,* a branch of geophysics, is the study of seismic waves and earthquakes. See also **microearthquake**.

earth science, any of the sciences dealing with the origin, composition, and physical features of the earth, such as geology, geography, meteorology, mineralogy, and oceanography.

earthshine, *n. Astronomy.* the faint light visible on the darker part of the moon, due to the light which the earth reflects onto the moon: *The average albedo, or reflectivity, of the earth is about 35 per cent—as measured directly from satellites and indirectly from earthshine reflected off the moon* (Carl Sagan).

earwax, *n.* = cerumen.

Eastern Hemisphere, *Geography.* the half of the world that includes Europe, Asia, Africa, and Australia.

ebb tide, *Oceanography.* the flowing of the tide away from the shore. Ebb tide occurs twice every 24 hours in most parts of the world.

ebracteate (ē brak'tē āt), *adj. Botany.* without bracts: *an ebracteate flower or plant.* [from New Latin *ebracteatus,* from Latin *e-* out of + *bractea* bract]

ebullition (eb'ə lish'ən), *n. Physics.* the boiling of a liquid: *Ebullition (boiling) will begin and continue until all the water has boiled away, with no further increase in temperature* (Shortley and Williams, *Elements of Physics*). [from Latin *ebullire* boil up, from *ex-* up + *bullire* boil]

EBV, *abbrev.* Epstein-Barr virus.

EB virus, = Epstein-Barr virus.

eccentric, *adj.* **1** *Geometry.* not having the same center: *These circles ◯ are eccentric.* Contrasted with **concentric**.

---

cap, fāce, fäther; best, bē, tėrm; pin, fīve;
rock, gō, ôrder; oil, out; cup, pùt, rüle,
yü in use, *yu* in uric;
ng in bring; *sh* in rush; *th* in thin, ᴛʜ in then;
zh in seizure.
ə = *a* in about, *e* in taken, *i* in pencil, *o* in lemon, *u* in circus

eccentricity

2 *Astronomy.* **a** (of an orbit) not perfectly circular; having a positive eccentricity: *The planets are in eccentric orbits around the sun.* **b** having an eccentric orbit: *Comets are highly eccentric.*
—*n. Geometry.* a circle not having the same center as another that is within it or that intersects it.
—**eccentrically,** *adv.*

eccentric

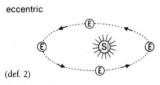

(def. 2)

The earth's orbit around the sun is eccentric.

eccentricity (ek′sen tris′ə tē), *n. Mathematics.* the ratio of the distance from the focus to any point of a conic section to the distance from the directrix. A parabola's eccentricity is 1, a hyperbola's is greater than 1, an ellipse is less than one, and a circle's is zero. Eccentricity is often used in astronomy to describe the shape of orbits. The earth's orbit has an eccentricity of .017. *The mean distance of the moon from the earth is 238,857 miles; but the eccentricity of the orbit, or ratio by which it differs from a perfect circle, causes the distance to vary from 222,000 to 253,000 miles* (Bernhard, *Handbook of the Heavens*). Symbol: e

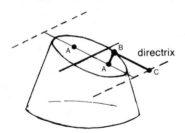

eccentricity:
points A = focus
point B = point on curve of conic section
eccentricity = lines $\dfrac{AB}{BC}$

eccrine (ek′rin *or* ek′rīn), *adj. Physiology.* of or having to do with the small sweat glands found throughout the human body surface that secrete a clear, fluid substance without a breakdown of the cells: *The glands that actually produce sweat are of two general kinds: the apocrine glands, which are usually associated with hair follicles, and the eccrine glands, which are not* (Scientific American). [from *ec-* out of (from Greek *ex-*) + Greek *krinein* to separate]

ecdysis (ek′də sis), *n., pl.* **-ses** (-sēz′). *Zoology.* the shedding of the outer covering or skin of certain animals, especially the shedding of the exoskeleton by arthropods: *Molt, or ecdysis, is therefore necessary at intervals to permit an increase in size* (Storer, *General Zoology*). [from Greek *ekdysis* a stripping]

ecdysone (ek′də sōn′), *n. Biochemistry.* a steroid hor-

mone, produced by the prothoracic glands of insects, that promotes growth and controls molting: *The insect hormone ecdysone, which controls the metamorphosis of insects, achieves its effect by acting directly on DNA* (Alexander G. Bearn). [from *ecdysis* + (*horm*)one]

ecesis (i sē′sis *or* i kē′sis), *n. Ecology.* the establishment or adaptation of a plant or animal to a new habitat. [from New Latin, from Greek *oikesis* dwelling]

echinate (ek′ə nit), *adj. Zoology.* covered with spines; spiny; bristly. [from Greek *echinos* sea urchin]

echinoderm (i kī′nə dèrm′), *n. Zoology.* any of a phylum (Echinodermata) of small sea animals usually with five arms arranged radially around a disk and a body covering strengthened and supported by calcareous plates: *The echinoderms include starfishes, crinoids, sea urchins, sea cucumbers, and a few other forms of life* (White and Renner, *Human Geography*). [from Greek *echinos* sea urchin + *derma* skin]
—**echinodermatous** (i kī′nə dèr′mə təs), *adj.* of or having to do with the echinoderms.

echo, *n. Physics.* **1** the repetition of a sound or sounds produced by the reflection of the sound waves from some obstructing surface: *Echoes may be used in determining position with relation to other objects. ... Experiments have shown that bats emit a high-pitched squeak and avoid objects by using the echoes* (Barnard, *Science*). **2** a radio wave that has been reflected. The detection of radio wave reflection is the basis of radar.

echolocation, *n. Zoology.* the process of finding the range and direction of objects by the sounds reflected from them, such as the ultrasonic sounds emitted by bats: *Man's use of instruments for echolocation, however, is very new compared with its probable use for millions of years by bats in their nocturnal search for insects, and by marine mammals in their movements through the ocean* (Robert T. Orr).

echo ranging, = echolocation.

echovirus, *n.* or **ECHO virus,** *Biology.* any of a large group of viruses containing RNA, associated with various kinds of meningitis, intestinal ailments, and respiratory illnesses in human beings; earlier called **orphan virus** because when discovered it was not known to be associated with specific diseases. [acronym formed from its full name, *e*(*nteric*) *c*(*ytopathogenic*) *h*(*uman*) *o*(*rphan*) *virus*]

eclipse

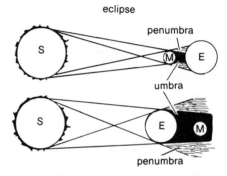

penumbra

umbra

penumbra

eclipse, *Astronomy.* —*n.* a complete or partial blocking of light passing from one heavenly body to another. A

180

solar eclipse occurs when the moon comes between the sun and the earth. A **lunar eclipse** occurs when the moon enters the earth's shadow.
—*v.* to cut off or dim the light from: *Sometimes ... there is a second, smaller dip in the light-curve. ... This behavior is explained by the periodic eclipsing of the stars by each other* (W.H. Marshall). [from Latin *eclipsis*, from Greek *ekleipsis*, from *ex-* out + *leipein* to leave]

eclipse plumage, *Zoology.* a dull or colorless plumage with which certain birds, such as male ducks, become covered at the close of the breeding season.

eclipsing binary, *Astronomy.* a variable star whose changes in brightness are due to apparent eclipse by its companion star in a binary system, which occurs when the plane of the orbit of the binary system almost coincides with the line of sight of the observer on earth. About 1100 eclipsing binaries are known to exist.

ecliptic (i klip′tik), *Astronomy.* —*n.* the apparent path of the sun and the planets among the stars in one year. It is that great circle of the celestial sphere which is cut by the plane containing the orbit of the earth.
—*adj.* **1** of the ecliptic: *The north and south ecliptic poles are the two points 90° from the ecliptic* (Baker, *Astronomy*). **2** having to do with an eclipse.

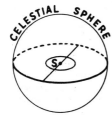

ecliptic

The inner path describes the orbit of the earth around the sun; the outer path describes the ecliptic.

eclogite (ek′lə jīt), *n. Geology.* a high-pressure metamorphic rock consisting of green pyroxene, granular red garnet, and other minerals. [from Greek *eklogē* selection + English *-ite*]

eclosion (i klō′zhən), *n. Zoology.* **1** the emergence of an insect from the pupa case. **2** the emergence of a larva from the egg. [from French *éclosion*, from *éclore* to hatch]

eco-, *combining form.* of the environment or ecology; ecological, as in *ecosphere*, *ecosystem.* [abstracted from *ecology*]

E. coli (ē′ kō′lī), *Biology.* a common rod-shaped bacterium of the intestinal tract, certain strains of which have been grown in large numbers and used extensively in experiments dealing with protein synthesis, genetic transmission, immunity, enzymology, etc. [partial abbreviation of *Escherichia coli*]

ecology (ē kol′ə jē), *n.* **1** the branch of biology that studies organisms in terms of their relationships with other organisms and with their local environment: *In its more restricted and technical sense, ecology is the study of organic diversity. It focuses on the interactions of organisms and their environments in order to address what may be the most fundamental question in evolutionary biology: "Why are there so many kinds of living things?"* (Stephen Jay Gould). **2** ecological composition or characteristics:. *There is a corresponding contrast in vegetation, with opportunities for studying the ecology of moorlands, bogs, mountains and limestone pavements* (A.W. Haslett). [from German *Ökologie*, from Greek *oikos* house + *-logia* study of]
—**ecological**, *adj.* of or having to do with ecology: *The ecological niche of an organism is the set of environmental conditions under which the particular functions of the organism could be expected to assure its survival.* (Herman H. Shugart). *The replacement of one community by another until a stable stage* (*climax*) *is reached is called ecological succession* (Biology Regents Syllabus). —**ecologically**, *adv.*

econiche (ē′kō nich′ *or* ek′ō nich′), *n. Ecology.* = niche.

ecospecies, *n. sing.* and *pl. Ecology.* a group of organisms only somewhat fertile with organisms of related groups, usually considered equivalent to a species.
—**ecospecific**, *adj.* of or having to do with an ecospecies. —**ecospecifically**, *adv.*

ecosphere, *n. Ecology.* a part of the universe suitable for life: *The term ecosystem can be applied at various levels: to the ecosystem of the Earth as a whole* (*ecosphere or biosphere*) *... or to a community in a particular environment* (Greulach and Adams, *Plants*). *To be habitable, a planet must be inside the ecosphere* (Dole and Asimov).

ecosystem, *n. Ecology.* a system of ecological relationships in a local environment, including relationships between organisms, and between the organisms and the environment itself: *An ecosystem, as we use the term, is a basic functional unit of nature comprising both organisms and their nonliving environment, intimately linked by a variety of biological, chemical and physical processes.* (Scientific American).

ecotone, *n. Ecology.* an area of transition between two communities, especially one that has characteristics of both as well as of its own. [from *eco-* + Greek *tonos* tension, tone]
—**ecotonal**, *adj.* of or having to do with an ecotone.

ecotype, *n. Ecology.* a variant type within an ecospecies: *Most wide-ranging species are composed of a continuum of ecotypes, each differing slightly in morphology and/or physiology* (Weier, *Botany*).
—**ecotypic**, *adj.* of or having to do with an ecotype.
—**ecotypically**, *adv.*

ecto-, *combining form.* to or on the outside, as in *ectoderm*, *ectoplasm.* [from Greek *ekto-*, from *ektos* outside]

ectoderm (ek′tə dėrm′), *n. Embryology.* the outer layer of cells formed during the development of the embryos of animals. Skin, hair, nails, the enamel of teeth, and the essential parts of the nervous system, grow from

cap, fāce, fäther; best, bē, tėrm; pin, five;
rock, gō, ôrder; oil, out; cup, pùt, rüle,
yü in use, *yu̇* in uric;
ng in bring; *sh* in rush; *th* in thin, ∓H in then;
zh in seizure.
ə = *a* in about, *e* in taken, *i* in pencil, *o* in lemon, *u* in circus

ectoenzyme

the ectoderm. Compare **endoderm** and **mesoderm**. [from *ecto-* + Greek *derma* skin]
—**ectodermal** (ek′tə dėr′məl), *adj.* of or having to do with the ectoderm.

ectoenzyme, *n. Biochemistry.* **1** an enzyme that exists on the outer surface of a cell membrane. **2** = exoenzyme.

ectogenesis, *n. Embryology.* the development of an embryo outside the organism in which it was formed, such as in another individual's body.

ectogenic (ek′tə jen′ik), *adj. Bacteriology.* originating or developed outside of the host, as certain disease-producing bacteria.

ectogenous (ek toj′ə nəs), *adj.* = ectogenic.

ectomere (ek′tə mir), *n. Embryology.* any of the blastomeres from which the ectoderm develops.
—**ectomeric,** *adj.* of or having to do with an ectomere.

ectoparasite (ek′tə par′ə sīt), *n. Zoology.* a parasite living on the outside of the host. Lice and fleas are ectoparasites.
—**ectoparasitic,** *adj.* of or having to do with an exoparasite or exoparasites.

ectophloic (ek′tə flō′ik), *adj. Botany.* having to do with plant stems that have the phloem only on the outside of the xylem. Compare **amphiphloic.** [from *ecto-* + Greek *phloios* bark]

ectophyte, *n. Botany.* a plant parasite living on the outside parts of the host.

ectoplasm (ek′tə plaz′əm), *n. Biology.* the semiclear, somewhat rigid outer portion of the cytoplasm of a cell. Compare **endoplasm.**
—**ectoplasmic** (ek′tə plaz′mik), *adj.* of or having to do with ectoplasm.

ectosarc (ek′tə särk), *n.* = ectoplasm. [from *ecto-* + Greek *sarkos* flesh]

ectosome, *n. Zoology.* the cortex, or outer part, of a sponge.

ectostosis (ek′tos tō′sis), *n. Physiology.* the growth of the bone around cartilage. It grows inward from without. Compare **endostosis.** [from *ecto-* + Greek *osteon* bone + New Latin *-osis* condition]

ectotherm (ek′tō thėrm), *n. Zoology.* a cold-blooded animal: *All modern reptiles, including the closest living relative of the dinosaurs, the crocodilians, are ectotherms, that is, they are dependent on external sources for control of body temperature* (J. Whitfield Gibbons). Contrasted with **endotherm.** Also called **poikilotherm.**
—**ectothermic,** *adj.* = cold-blooded.

ectotrophic (ek′tə trof′ik), *adj. Biology.* deriving nourishment from the outside, as certain fungi living on the outside surface of roots. Compare **endotrophic.**

ectozoon (ek′tə zō′ən), *n., pl.* **-zoa** (-zō′ə). *Zoology.* an animal parasite that lives on an outside part of the host. [from *ecto-* + Greek *zōion* animal]
—**ectozoic,** *adj.* having to do with ectozoons.

edaphic (i daf′ik), *adj. Ecology.* influenced by the soil, rather than by the climate: *Environmental factors ... have been catalogued as physical ... biotic ... and edaphic, which include the influence of soil [and] ... may be the result of its physical structure, its chemical composition, the conditions caused by the activities of living organisms in the soil* (Harbaugh and Goodrich,

Fundamentals of Biology). [from Greek *edaphos* bottom, ground]

edaphic climax, *Ecology.* a climax that is determined mainly by the conditions of the soil, for example whether it is moist, dry, or alkaline.

eddy, *n. Physics.* a current of water, air, smoke, etc., moving against the main current, especially when having a whirling motion, as a small whirlpool or whirlwind.

eddy current, *Physics.* an electric current induced in a conductor by a varying magnetic field, or in a moving conductor by a constant magnetic field, and causing loss of electric power or heat: *The eddy current creates an opposing force inside the metal which tries to push the metal out of the stronger magnetic field* (Wall Street Journal).

edentate (ē den′tāt), *Zoology.* —*n.* any of an order (Edentata) of mammals that are toothless or lack incisors. Armadillos, sloths, and some anteaters belong to this order.
—*adj.* **1** toothless. **2** of or having to do with the edentates.
[from Latin *edentatum* toothless, from *ex-* out + *dentem* tooth]

edentulate (ē den′chə lāt), *adj. Zoology.* without teeth or toothlike processes, as the mandibles of some insects.

edge effect, *Ecology.* the influence of two neighboring communities on the types and numbers of plants and animals in their adjoining fringes or borders. Compare **ecotone.**

effective temperature, *Astronomy.* the temperature of a star as calculated by comparison of the star's color to what the color would be under optimum radiating conditions.

effector, *n. Physiology.* **1** a muscle, gland, or organ capable of responding to a nerve impulse: *Receptors induce the transmission of nerve impulses through the nervous system, which in turn excites terminal structures, or effectors, to bring about responses* (Storer, *General Zoology).*
2 a nerve that ends in a muscle, gland, or organ, especially a nerve conducting an impulse from a nerve center: *Most of the nerves in the autonomous system are effectors, or motor nerves, but each trunk also has receptor fibers* (Foster Kennedy).

efferent (ef′ər ənt), *adj. Physiology.* carrying outward from a central organ or point. Efferent nerves carry impulses from the brain to the muscles. Contrasted with **afferent.** See the picture at **afferent.** [from Latin *efferentem* carrying out, from *ex-* out + *ferre* carry]

effloresce (ef′lə res′), *v. Chemistry.* **1** to change either throughout or on the surface to a powder by loss of water of crystallization when exposed to air. **2** to become covered with a crusty deposit when water evaporates. [from Latin *efflorescere* to bloom, blossom, from *ex-* out + *florem* flower]
—**efflorescence,** *n.* **1** a change that occurs when crystals lose their water of crystallization and become powder: *If a hydrated crystal which maintains a fairly high vapor pressure is surrounded with air in which the partial pressure due to water vapor is rather low, water will pass from the crystal into the air ... Such spontane-*

ous loss of water of hydration is known as *efflorescence* (Jones, *Inorganic Chemistry*).
2 the powder or deposit formed in this way.
3 formation of a crusty deposit when water evaporates from a solution.
—**efflorescent,** *adj.* **1** that changes from crystals into powder by losing water of crystallization when exposed to air.
2 covered with a deposit formed by efflorescence.

effluent (ef′lü ənt), *adj. Geography, Geology.* flowing out or forth, as a river from a lake.
—*n.* **1** *Geography, Geology.* a stream flowing out of a larger stream, lake, reservoir, etc. **2** *Ecology.* liquid waste matter flowing or discharged into the environment: *This effluent, or water outflow, contains considerable trash including pieces of metal, paper, organic materials and industrial chemicals* (Tracy, *Modern Physical Science*).
[from Latin *effluentem,* from *ex-* out + *fluere* to flow]

effort, *n. Physics.* a force applied on a body due to a definite cause. It is the force that directs motion in a simple machine. A heavy body on an inclined plane is said to have an effort to fall vertically.
The effort applied on a rope through a pulley (acting as a fulcrum) overcomes resistance by upward motion. Compare **resistance.**

effort

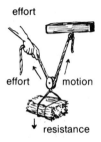

effort   motion

↓ resistance

effort arm, *Physics.* the distance between the fulcrum of a lever and the applied force or effort. Compare **load arm.**

effuse (i fyüs′), *adj.* **1** *Botany.* spread out loosely or without definite form: *effuse flowers.*
2 *Zoology.* (of certain shells) having the lips separated by a gap or groove.
[from Latin *effusum,* past participle of *effundere* pour out; see **effusion**]

effusion (i fyü′zhən), *n.* **1** *Physics.* the movement or flow of gas molecules through a small hole or holes.
2 *Medicine.* a fluid, or the escape of a fluid, such as blood or lymph, from its natural vessel into surrounding tissues or cavities.
[from Latin *effusionem* an outpouring, from *effundere* pour out, from *ex-* out + *fundere* pour]
—**effusive,** *adj. Geology.* formed by being poured out as lava on the surface of the earth: *The first is an effusive eruption, with a gentle outflow of cool dark-colored lava that pours out at a relatively slow rate of 50 to a few hundred yards an hour* (Barbara Tufty).

egest (i jest′), *v. Biology.* to discharge waste materials from the body: *In the paramecium, undigested materials are egested through a fixed opening called the anal pore* (Biology Regents Syllabus). SYN: excrete. ANT:

ingest. [from Latin *egestum,* past participle of *egerere,* from *ex-* out + *gerere* carry, lead]
—**egestion,** *n.* the act or process of egesting; the discharge of undigested or indigestible material by an organism.
—**egestive,** *adj.* of or having to do with egestion.

egesta (ē jes′tə), *n.pl. Physiology.* excrement; excreta.
—**egestatory,** *adj.* excretory.

egg, *n.* **1** *Zoology.* the round or oval body, covered with a shell or membrane, produced by the female of birds, many reptiles, amphibians, fishes, insects, and other animals: *Many land animals—including all birds—deposit the fertilized eggs, and development of the embryos occurs outside the body of the female parent.* (Miller and Leth, *High School Biology*).
2 *Biology.* = egg cell: *In the case of angiosperms ... the two gametes are very different from one another. The larger one, produced by the female parent, is the egg, and the smaller comes from the male and is called the sperm* (Emerson, *Basic Botany*).

egg apparatus, *Botany.* the group of three cells at the micropylar end of the embryo sac in seed plants. The group consists of an egg cell and two naked cells that may assist in directing the egg cell.

egg case, *Zoology.* **1** a membrane or covering around an egg. **2** = egg sac.

egg cell, *Biology.* a female reproductive cell; ovum: *There is considerable evidence that the female gametangia or gametes in plants produce chemical agents that seduce the sperms to the close proximity of egg cells* (Weier, *Botany*).

egg sac, *Zoology.* **1** the silken cocoon in which the eggs of most spiders are deposited and in which they hibernate. **2** one of the pair of egg-containing receptacles at the hinder end in certain crustaceans.

egg tooth, *Zoology.* a hard, toothlike projection in the jaw of an embryo bird, snake, or other oviparous animals, that cuts the egg membrane and shell at hatching and then drops off.

eglestonite (eg′əl stə nīt), *n. Mineralogy.* a native oxychloride of mercury, occurring in brownish isometric crystals. *Formula:* Hg_4Cl_2O [from Thomas *Egleston,* 1832–1900, American mining engineer + *-ite*]

EHF or **ehf,** *abbrev.* extremely high frequency (applied to electromagnetic waves of between 30 and 300 gigahertz).

eigenfunction (ī′gən fungk′shən), *n. Mathematics, Physics.* a solution of a differential or integral equation that satisfies certain special values of a parameter. [from German *Eigenfunktion,* from *eigen* own, characteristic + *Funktion* function]

eigenvalue (ī′gən val′jü), *n. Mathematics, Physics.* one of the restricted values of a parameter for an eigenfunction; a characteristic root: *the use of eigenvalues*

cap, fāce, fäther; best, bē, tèrm; pin, fīve;
rock, gō, ôrder; oil, out; cup, pùt, rüle,
yü in use, *yu* in uric;
ng in bring; *sh* in rush; *th* in thin, ᴛʜ in then;
zh in seizure.
ə = *a* in about, *e* in taken, *i* in pencil, *o* in
lemon, *u* in circus

Einstein equation

in the physics of continua (Science). [half translation of German *Eigenwert,* from *eigen* own + *Wert* value]

Einstein equation, *Physics.* the equation expressing the relation of mass and energy, in which the energy of a system is equated with its mass multiplied by the square of the speed of light. The equation is formulated as $E = mc^2$. Also called **mass-energy equation.** [named after Albert *Einstein,* 1879–1955, who formulated the equation in his special theory of relativity]

einsteinium (īn stī′nē əm), *n. Chemistry.* an unstable radioactive element, produced artificially: *Five new twins, or isotopes, of the artificial element, einsteinium, have been created ... by bombarding the man-made elements of berkelium and californium with helium ions and deuterons* (Science News Letter). *Symbol:* Es; *atomic number* 99; *mass number* (of most stable isotope) 254. [named after Albert *Einstein,* 1879–1955, German-born American physicist]

Einstein shift, *Physics.* a shift in the spectral lines of visible light toward the red when the light emerges from a strong gravitational field, as that of a dense star. The shift is predicted by Einstein's general theory of relativity.

ejaculate, *Physiology.* —*v.* (i jak′yə lāt) to discharge, especially semen.

—*n.* (i jak′yə lit) the semen discharged by an ejaculation: *An ejaculate of human semen contains some 100–300 million sperms* (New Scientist). [from Latin *ejaculatum* thrown out, from *ex-* out + *jacere* to throw]

—**ejaculation** (i jak′yə lā′shən), *n.* a discharge, especially of semen. —**ejac′ulato′ry,** *adj.*

ejecta (i jek′tə), *n.pl. Geology.* matter ejected, as from a volcano. [from Latin *ejecta,* neuter plural of *ejectus,* past participle of *ejicere* throw out, eject]

eka- (ek′ə- or ē′kə-), *combining form.* (of an element, especially when not discovered yet) next after (the element in the rest of the name) in that element's family in the periodic table, as in *ekalead, ekahafnium.* [from Sanskrit *eka* one]

Ekman layer, *Oceanography.* a layer of ocean water whose flow is at right angles to the wind's directions: *Thus the wind in the southern half of our square basin representing the North Atlantic transports water to the north in the thin Ekman layer* (Scientific American). [named after V. Walfrid *Ekman,* 20th-century Swedish oceanographer, who first described it]

elasmobranch (i laz′mə brangk), *n. Zoology.* any of a subclass (Elasmobranchii) of fishes whose skeletons are formed of cartilage and whose gills are thin and platelike. Sharks and rays belong to this subclass. [from Greek *elasmos* metal plate + *branchia* gills]

elastase (i las′tās), *n. Biochemistry.* an enzyme occurring in the pancreatic juice that digests elastin.

elastic, *adj. Physics.* capable of being deformed without permanently losing size, volume, or shape: *This property of returning to its original shape and size after being bent or stretched, known as the elastic behaviour of steel, is important, but is not the only factor marking the success of a building* (New Scientist). —**elastically,** *adv.*

—**elasticity** (i las tis′ə tē), *n.* elastic quality. **Elasticity of volume** is the capability of liquids and gases to regain their volume after being deformed.

elastic collision, *Physics.* a collision of two particles that leaves their total kinetic energy unchanged.

elastic fiber, *Anatomy.* a highly elastic, yellow fiber of connective tissue: *Elastic fibers bind the skin to the underlying muscles, attach many other tissues and organs to one another* (Storer, *General Zoology*). See **elastin.**

elastic limit, *Physics.* the maximum stress that an elastic material can sustain and still return to its original form.

elastic scattering, *Physics.* the scattering of particles resulting from elastic collision.

elastin, *n. Biochemistry.* a protein similar to albumin and collagen that is the main constituent of elastic fiber.

elastomer (i las′tə mər), *n. Chemistry.* any of various elastic, rubberlike substances, such as certain polymers. [from *elastic* + Greek *meros* a part]

—**elastomeric** (i las′tə mer′ik), *adj.* having the properties of an elastomer.

elater (el′ə tər), *n. Botany.* an elastic filament for discharging and dispersing spores, as in liverworts: *... filled with spores and elaters, the latter being peculiar long cells with spiral thickenings on their walls* (Emerson, *Basic Botany*). [from Greek *elatēr* driver, from *elaunein* to drive]

elaterite (i lat′ə rīt), *n. Geology.* a dark brown amorphous hydrocarbon, usually soft and elastic like rubber. [from German *Elaterit,* from Greek *elatēr* driver]

E layer or **E region,** *Geophysics,* a layer of the ionosphere that reflects low-frequency radio waves, about 80 to 145 kilometers above the earth's surface. Also called **Heaviside layer, Kennelly-Heaviside layer.** Compare **D layer, F layer.**

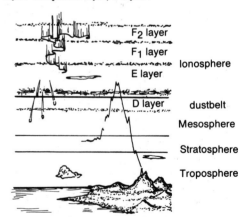

electret (i lek′trit), *n. Physics.* a nonconducting (dielectric) material with permanent polarity, the static-electricity equivalent of a magnet. [from *electr(icity)* + *(magn)et*]

electric, *adj.* of or having to do with electricity. An **electric charge** is the amount of electricity that may be discharged by an elementary particle, fuel cell, capacitor, battery, or the like. An **electric current** is the flow of electricity through a conductor or transmitter. An electric field is any space in which force due to an electric charge exists.

electrical, *adj.* 1 = electric.

2 producing or produced by electricity: *electrical energy, electrical impulses.*

3 having to do with or involving the use of electricity: *electrical devices, electrical experiments.*

electric arc, = arc (def. 2).

electricity, *n. Physics.* 1 a form of energy that can produce light, heat, magnetism, and chemical changes, and which can be generated by friction, induction, or chemical changes. Electricity is regarded as consisting of oppositely charged particles, electrons and protons, which may be at rest or moving about. *Electricity may either reside upon the surface of bodies as a charge, or flow through their substance as a current* (S.P. Thomson).

Negative electricity is the form of electricity in which the electron is the elementary unit; **positive electricity** is that in which the proton is the elementary unit. **Static electricity** is the form of electricity in which the electric charges are stationary; **dynamic electricity** is that in which the charges move in a current.

2 electric current: *An ampere-hour is the amount of electricity delivered by one ampere in one hour.*

3 the branch of physics that deals with electricity.

electro- (i lek′trō-), *combining form.* 1 electric, as in *electromagnet = an electric magnet.*

2 electrically, as in *electropositive = electrically positive.*

3 produced or operated by electricity, as in *electroplated.*

4 electron, as in *electrophile = a material attracted to electrons.*

[from Greek *ēlektron* amber, which, when rubbed, becomes electrically charged]

electroacoustics, *n.* the science that deals with the relationship of acoustic energy and electric energy.

—**electroacoustic,** *adj.* of or having to do with electroacoustics. —**electroacoustically,** *adv.*

electroanalysis, *n. Chemistry.* analysis by means of electrolysis.

—**electroanalytic,** *adj.* having to do with electroanalysis.

electrocapillarity, *n. Physics.* a change in the capillary action of a liquid, caused by the presence of an electric field on its surface. It is used in measuring very small quantities of electricity.

—**electrocapillary,** *adj.* having to do with electrocapillarity.

electrochemical, *adj.* of or having to do with electrochemistry. —**electrochemically,** *adj.*

electrochemical equivalent, *Chemistry.* the weight of an element that can be liberated in electrolysis by one coulomb of electricity.

electrochemical series = electromotive series.

electrochemistry, *n.* the branch of chemistry that deals with chemical changes produced by electricity and the production of electricity by chemical changes: *Electrochemistry ... has made possible the analysis of structure and function of living cells, all of which are in fact complicated electrochemical systems capable of transforming chemical energy and ionic transport into measurable electrical signals* (Scientific American).

electrode (i lek′trōd), *n. Electricity.* a conductor by which electricity is brought into or out of a conducting medium, as either of the two terminals of a battery or electrolytic cell. The anode and cathode of an electric cell are electrodes. *The conductors by which the current enters and leaves the solution are called electrodes* (Parks and Steinbach, *Systematic College Chemistry*). [from *electro-* + Greek *hodos* way]

electrodeposit, *v. Chemistry.* to deposit (a substance) by electrolysis. —**electrodeposition,** *n.*

electrodialysis, *n. Chemistry.* a form of dialysis in which the application of an electric current to electrodes is used to separate substances or compounds. Salt is removed from seawater on a large scale by using this method.

—**electrodialytic,** *adj.* of or having to do with electrodialysis.

electrodynamic, *adj. Physics.* 1 of or having to do with the forces produced by or affecting electric currents; electromagnetic. 2 of electrodynamics.

—**electrodynamically,** *adv.*

electrodynamics, *n.* the branch of physics that deals with the mutual influence of electric currents, the interaction of currents and magnets, and the influence of an electric current on itself. **Quantum electrodynamics** is the application of the quantum theory to electrodynamics.

electrogenesis, *n. Biology.* the production of electricity in living organisms: *The most important role of electrogenesis in animals is in ... the propagation of nerve impulses* (Barry D. Lindley).

—**electrogenic,** *adj.* producing electricity in living organisms.

electrojet, *n. Geophysics.* an electric current moving in an ionized layer in the upper atmosphere of the earth above the equator.

electrokinetics, *n.* a branch of physics that deals with electric currents or electricity in motion.

—**electrokinetic,** *adj.* of or having to do with electrokinetics.

electroluminescence, *n. Physics.* luminescence induced in crystals by the application of an electric field, as in certain phosphors: *Instead of bulbs or tubes, flat panels of glass may be used to light the walls or ceiling of a room. This method of producing light is called electroluminescence. The phosphor material is placed between two conductors, across which the voltage is applied. The lamp remains cool, and usually produces a green light* (Karl A. Staley).

—**electroluminescent,** *adj.* of or characterized by electroluminescence: *Light-emitting diodes are electroluminescent.*

electrolysis (i lek′trol′ə sis), *n. Chemistry.* the decomposition of an electrolyte or of the solvent in which it is dissolved by the action of an electric current passing

---

cap, fāce, fäther; best, bē, tėrm; pin, five;
rock, gō, ôrder; oil, out; cup, pùt, rüle;
yü in use, *yù* in uric;
ng in bring; sh in rush; th in thin, ᴛʜ in then;
zh in seizure.
ə = *a* in about, *e* in taken, *i* in pencil, *o* in lemon, *u* in circus

electrolyte

through the solution: *At the electrodes some type of chemical reaction takes place, resulting usually in deposition or solution of solid material or evolution of gas from decomposition of solvent or solute. These chemical changes are said to result from electrolysis of the solution* (Shortley and Williams, *Elements of Physics*). [from *electro-* + Greek *lysis* a dissolving]

electrolyte (i lek′trə lit), *n. Chemistry.* **1** a compound which ionizes when dissolved in a suitable liquid, or when melted, thus becoming a conductor for an electric current. Acids, bases, and salts are electrolytes. **2** a solution that will conduct an electric current: *Since its origin, the lead-acid battery has been constructed of two sets of plates, one lead, the other lead peroxide, submerged in a sulphuric acid solution called electrolyte* (Wall Street Journal). [from *electro-* + Greek *lytos* soluble, from *lyein* to loose]
  —electrolytic (i lek′trə lit′ik), *adj.* of or having to do with electrolysis or an electrolyte: *Water is the most common solvent in electrolytic solutions* (Physics Regents Syllabus).
  —electrolytically, *adv.* by means of electrolysis.

electrolytic cell, *Chemistry.* **1** the container which holds the electrolyte and the electrodes for use in electrolysis. **2** the electrolyte, its container, and the electrodes used in electrolysis.

electrolyze (i lek′trə liz), *v. Chemistry.* to decompose by electrolysis: *Humphry Davy's best research was in electrochemistry. He had a battery built with 250 plates, which was the source of the strongest electric current known at the time. He first electrolyzed water in a gold dish and proved that the process gave hydrogen and oxygen and nothing else* (Ira D. Garard, *Invitation to Chemistry*).

electromagnet, *n. Physics.* a piece of soft iron that becomes a strong magnet temporarily when an electric current is passed through wire coiled around it: *A movable electromagnet, called the armature or rotor, spins because it is alternately attracted and repelled by fixed magnets, called the field magnets* (Tracy, *Modern Physical Science*).

electromagnetic, *adj. Physics.* **1** of or caused by an electromagnet. **2** of or having to do with electromagnetism.
  —electromagnetically, *adv.* **1** by means of an electromagnet. **2** by electromagnetism.

electromagnetic field, *Physics.* the field created by the interaction of an electric field and a magnetic field when an electric current passes through a wire.

electromagnetic induction, *Physics.* the production of an electromotive force in a circuit by variation of the magnetic field with which the circuit is connected.

electromagnetic radiation, *Physics.* a radiation of electromagnetic waves: *Electromagnetic radiations are generated by accelerating charged particles.*

electromagnetic spectrum, *Physics.* the entire range of the different types of electromagnetic waves, from the very long, low-frequency radio waves, through infrared and light waves to the very short, high-frequency cosmic rays and X rays: *By using uranium, they found it possible for the first time, to generate continuous*

and coherent light waves in the infrared portion of the electromagnetic spectrum (Science News).

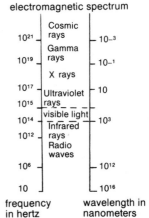

electromagnetic spectrum

frequency in hertz		wavelength in nanometers
10^{21}	Cosmic rays	10^{-3}
10^{19}	Gamma rays	10^{-1}
10^{17}	X rays	
	Ultraviolet	10
10^{15}	rays	
10^{14}	visible light	10^3
10^{12}	Infrared rays	
	Radio waves	
10^6		10^{12}
10		10^{16}

electromagnetic unit, *Physics.* any of the units in the CGS system that are based on electromagnetism. They include the *abampere* (= 10 amperes), *abcoulomb* (= 10 coulombs), *abfarad* (= 10^9 farads), and *abhenry* (= 10^9 henrys). *Abbreviation:* e.m.u. or EMU

electromagnetic wave, *Physics.* a wave of energy generated by a varying electric and magnetic field when an electric charge oscillates or is accelerated. Electromagnetic waves are light waves, radio waves, X-rays, or gamma rays, according to their frequency and wavelengths. *Whenever an electric current changes in a circuit, energy in the form of electromagnetic waves is radiated away from the circuit* (E. Mallett).

electromagnetism, *n. Physics.* **1** magnetism produced by a current of electricity. **2** the branch of physics that deals with electricity and magnetism: *Electromagnetism, together with electrostatics and such subjects as electrochemistry ... made possible all the present applications of electricity except those depending on twentieth-century developments in electronics* (Shortley and Williams, *Elements of Physics*).

electromotive, *adj. Physics.* **1** producing a flow of electricity. **2** of or having to do with electromotive force.

electromotive force, *Physics.* **1** the force resulting from differences of potential that causes an electric current. *Symbol:* E
2 the amount of energy derived from an electric source per unit of current passing through the source. Electromotive force is commonly measured in volts. *If two different metals are placed in contact, and the junction is warmed, an electromotive force is generated* (F. P. Bowden). *Abbreviation:* e.m.f. or EMF

electromotive series, *Chemistry.* a list of the metallic elements in the decreasing order of their tendencies to change to ions in solution, so that each metal displaces from solution those below it in the list and is displaced by those above it.

electron (i lek′tron), *n. Physics.* an elementary particle having a very small mass at rest (9.095×10^{-28} gram) and a unit charge of negative electricity equal to 1.60219×10^{-19} coulombs. All atoms have electrons surrounding a nucleus at various distances in *orbitals* or *shells.* The hydrogen atom has one electron; the uranium atom has 92 electrons. The electron is a member of the class of elementary particles known as *leptons. When electrons move through a conductor such as a wire, it is called current electricity. Whenever any electrical device such as an iron, a radio, a motor, or a light bulb is operated, there is a stream of billions of electrons moving through the wires* (Obourn, *Investigating the World of Science*). *Symbol:* e See the pictures at **atom, ion.** [from *electr(ic)* + *-on,* as in *ion*]

ASSOCIATED TERMS: *Electrons, protons,* and *neutrons* are three of the units of which *atoms* are composed. An electron is negatively charged and has a *mass* of approximately $1/1836$ of a proton. A proton is positively charged and has a mass of approximately one *dalton*. A neutron has a zero charge and a mass of approximately one dalton. Each *element* has a particular number of protons and, in a neutral atom, an equal number of electrons to balance the *charge*. The number of protons in the nucleus is the *atomic number* of an element. The *mass number* indicates the total amount of protons and neutrons in the nucleus. The electrons are outside the nucleus at various *energy levels*. When electrons have absorbed energy and shifted to higher energy levels, the atom is in an *excited state.*

electron beam, *Physics.* a stream of electrons moving in the same direction and with the same velocity: *The electron beam of the super-microscope has become a basic tool in research on disease* (Harper's).

electron cloud, an average region around the nucleus of an atom, where electrons are predicted to be at various states of excitement. See also **orbital, Schrödinger (wave) equation, wave-mechanical.** See picture at **atom.**

electron configuration, *Physics, Chemistry.* the number and arrangement of electrons in various orbitals or shells about a nucleus: *In an electron configuration, the number of electrons in a sublevel is indicated by a superscript following the decomposition of the sublevel. For example, calcium would be represented as $1s^2 2s^2 2p^6 3s^2 3p^6 4s^2$* (Chemistry Regents Syllabus).

electron diffraction, *Physics.* the diffraction of electrons when they pass through crystalline matter, useful in the study of the structure of materials: *Electron diffraction permits the investigation of the crystal structure of thin magnetic films prepared by evaporation in vacuum* (Scientific American).

electronegative, *adj. Physics, Chemistry.* **1** charged with negative electricity. **2** tending to pass to the anode in electrolysis. **3** tending to gain or having an attraction for electrons; nonmetallic: *Perfluoric-cyclic ether is an electronegative material. Therefore, it tries to capture any free electrons moving about in the gas* (Chester L. Dawes).

—**electronegativity,** *n.* a measure of the attraction of an atom for electrons.

electron gas, *Physics.* a concentration of electrons whose behavior is primarily determined by the interactions between the electrons rather than by other forces; any system of free electrons.

electron gun, *Electronics.* the part of a cathode-ray tube that guides the flow and greatly increases the speed of electrons. In a television set or an oscilloscope, an electron gun directs a stream of electrons to the screen. *An electron gun may supply electrons to "illuminate" the specimen in an electron microscope and so enable a chemist or a biologist to see details far smaller than a wave of light, and scarcely larger than molecules* (Pierce, *Electrons, Waves and Messages*).

electronic, *adj. Physics, Chemistry.* of or having to do with electrons or electronics: *the electronic structure of an atom or molecule.* —**electronically,** *adv.*

electronic equation, *Chemistry.* an equation showing the gain or loss of electrons in a redox reaction.

electronics, *n.* the branch of physics that deals with the production, activity, and effects of electrons in motion in a vacuum, gas, or semiconductor, especially with reference to industrial applications, as in electron tubes, transistors, and semiconductors: *Electronics ... deals with electric current mainly in the form of pulses, or signals. The current flows through electron devices, which change the current's behavior to make it work as a signal. The signals used in electronics may represent sounds, pictures, numbers, or other information* (Richard W. Henry).

electron lens, *Electronics.* the electrodes or conductors which set up electric or magnetic fields by which a beam of electrons can be focused as light rays are by an optical lens.

electron micrograph, *Physics.* a fluorescent image or photograph of an object under an electron microscope: *In an electron micrograph the clot appears as a network of protein strands of various widths* (Koloman Laki).

electron microscope, a microscope that uses beams of electrons instead of beams of light to project enlarged images upon a fluorescent surface or photographic plate. It has much higher power than any ordinary microscope. An electron microscope is focused with an electron lens.

► There are two types of electron microscopes: the *transmission electron microscope,* which passes a beam of electrons from an electron gun through the specimen and onto a fluorescent screen; and the *scanning electron microscope,* which focuses the electron beam on the specimen and scans it in a regular pattern, forming a magnified image on a television screen.

electron optics, a branch of electronics that deals with the control of beams of electrons by an electric or magnetic field in the same manner that lenses control a beam of light.

—**electron-optical,** *adj.* of or having to do with electron optics: *An electron-optical lens focuses the highly accelerated electrons into a tiny spot* (Paul H. Gleichauf).

electron tube, *Physics.* a sealed tube or container in which electrons move through a vacuum or gas, used to control the flow of electric currents. Compare **vacuum tube.**

electron volt, *Physics.* a unit of electrical energy equal to the energy gained by an electron going through a potential difference of one volt, equal to 1.6×10^{-19} joules: *The energy liberated in burning one atom of*

cap, fāce, fäther; best, bē, tèrm; pin, fīve;
rock, gō, ôrder; oil, out; cup, pùt, rüle;
yü in use, *yù* in uric;
ng in bring; *sh* in rush; *th* in thin; ͭH in then;
zh in seizure.
ə = *a* in about, *e* in taken, *i* in pencil, *o* in
lemon, *u* in circus

coal is 4 electron volts, whereas in most radioactive processes each atom contributes an amount of the order of a few million electron volts (Crammer, Atomic Energy). Symbol: eV

electroosmosis, n. Chemistry. the flow of a liquid through a porous membrane as a result of an electric field created by electrodes placed on opposite sides of the membrane.
—**electroosmotic,** adj. of or having to do with electroosmosis.

electrophile, n. Chemistry. an electrophilic substance; electron acceptor: Then by dividing heterolytic reagents into nucleophiles and electrophiles, he [Sir Christopher Ingold] saw the basis of a scheme interrelating a great range of chemical reactions (London Times).

electrophilic (i lek'trə fil'ik), adj. Chemistry. strongly attracted to electrons: an electrophilic molecule or ion. Contrasted with **nucleophilic.**

electrophoresis (i lek'trō fə rē'sis), n. Chemistry. **1** the movement of colloidal particles resulting from the influence of an electric field: When a charged particle of colloidal dimensions is suspended in a liquid between two charged electrodes, it moves towards the electrode with a charge opposite to its own ... This is known as electrophoresis (Science News). **2** a method of separating substances and analyzing molecular structure based on the rate of their movement in a colloidal suspension under the influence of an electric field: Electrophoresis, which enables us to distinguish protein fractions ... has also been applied to the differentiation of lipoproteins and mucoproteins, complexes of protein with lipids and polysaccharides respectively (F. Degering).
[from electro- + Greek phorēsis a carrying]
—**electrophoretic** (i lek'trō fə ret'ik), adj. having to do with or produced by electrophoresis: electrophoretic techniques.

electrophysiology, n. **1** the science that studies the relationship between living organisms and electricity. **2** the electrical activity associated with any part of the body and its functions: the electrophysiology of the retina.
—**electrophysiological,** adj. having to do with or produced by electrophysiology: electrophysiological measurements.

electropism (i lek'trə piz əm), n. Botany. curvature of growth in plants due to slight electric currents. [blend of electro- and tropism]

electropolar, adj. Electricity. having one end or surface positive and the other negative, as an electrical conductor.

electropositive, adj. Physics, Chemistry. **1** charged with positive electricity. **2** tending to pass to the cathode in electrolysis: Substances are frequently spoken of as being electronegative, or electropositive, according as they go under the supposed influence of a direct attraction to the positive or negative pole (Michael Faraday). **3** tending to lose electrons; metallic: Salts when ionized in solution gave an electropositive metal ion, which in many cases could be plated out on to the negative pole of an electrolytic cell (J. Crowther).

—**electropositivity,** n. the condition of being electropositive.

electroscope, n. Physics. a device that indicates the presence of minute charges of electricity and shows whether they are positive or negative. An electroscope can also detect X rays and other electromagnetic radiation.

electrostatic, adj. Physics. **1** of or having to do with static electricity: ... the removal of electrostatic charges from moving belts, paper, or cloth (Otto Frisch). **2** of or having to do with electrostatics: electrostatic experiments. —**electrostatically,** adv.

electrostatic field, Electricity. an electric field produced by charges which are stationary or at rest.

electrostatic generator, = Van de Graaff generator.

electrostatics, n. a branch of physics that deals with static electricity and charged objects.

electrostatic unit, Physics. any unit in the CGS system based primarily upon the force exerted between two electric charges. Abbreviation: e.s.u. or ESU.

electrotactic, adj. Biology. of or having to do with electrotaxis.

electrotaxis, n. Biology. the movement of organisms or cells in response to electric currents.

electrothermal or **electrothermic,** adj. Physics. of or having to do with heat and electricity, especially the conversion of electric energy into heat energy.

electrotonic, adj. Physiology. **1,** of, having to do with, or caused by electrotonus. **2** of, having to do with, or designating the electric potential found in cells and tissues as a result of metabolic activity. —**electrotonically,** adv.

electrotonus (i lek'trot'ə nəs), n. Physiology. the altered condition of a nerve or muscle during the passage of an electric current through it. [from electro- + tonus]

electrovalence or **electrovalency,** n. Chemistry. a chemical bond in which two oppositely charged ions are linked by the electrostatic attraction between them; ionic bond: This [NaCl] crystal is a collection of equal numbers of sodium ions and chloride ions, held together by the mutual attraction of positively charged particles and negatively charged particles. This sort of chemical union is called ionic valence, or electrovalence (W.H. Slabaugh). **2** the number of electrons gained or lost by an atom when it becomes an ion.
—**electrovalent,** adj. of or bonded by electrovalence: When electrovalent compounds—i.e., those made up of ions of opposite charge—are dissolved in water, the individual ions are freed from the attractions of the other ions which held them in place in the solid (Jones, Inorganic Chemistry).

electrovalency, n. = electrovalence.

electroweak theory, Physics. a theory uniting weak and electromagnetic interactions: This electroweak theory states that W and Z particles are exchanged between particles such as electrons and protons, mediating the weak interactions between them. (Gerald Feinberg). See also **Weinberg-Salam,** adj.

element, n. **1** Chemistry. a substance composed of atoms that are chemically alike and which cannot be separated into simpler parts by chemical means. Gold, iron, carbon, sulfur, oxygen, and hydrogen are among the more than 100 known elements. Every element emits or absorbs a characteristic spectrum of light (bright or dark lines at certain wavelengths) when its atoms are excited to high temperature; the elements have been *"fingerprinted" in this way in laboratories, and their prints can be matched to the spectral light from stars* (William A. Fowler). Also called **chemical element.**

► Only 103 chemical elements are presently recognized by the International Union of Pure and Applied Chemistry. However, six other elements, all radioactive, have been artificially produced since the 1960's by various groups of scientists: *element 104* (1964-1969, named either *kurchatovium* or *rutherfordium,* having up to ten isotopes, with mass numbers of 253 to 262); *element 105* (1967-1970, named *nielsbohrium* or *hahnium,* having six isotopes, with mass numbers of 255, 257, 258, 260, 261, and 262); *element 106* (1974, unnamed, having three isotopes, with mass numbers 259, 260, and 263); *element 107* (1976-1981, unnamed, having three isotopes, with mass numbers 261, 262, 267); *element 108* (1984, unnamed, having one isotope, with mass number 265); and *element 109* (1982, unnamed, having one isotope, with mass number 266). The numbers 104, 105, 106, 107, 108, and 109 are the atomic numbers of these elements.

2 *Mathematics.* **a** a member of a set. **b** a very small part of a given magnitude similar in nature to the whole magnitude. **c** one of the lines, planes, points, etc., that make up a geometrical figure.

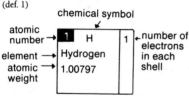

ELEMENT:
(def. 1)

elemental, *adj. Chemistry.* being an uncombined element; elementary: *elemental fluorine.*

elementary, *adj. Chemistry.* made up of only one element; not a compound: *Silver is an elementary substance.*

elementary particle, *Physics.* one of the fundamental units of which atoms and all matter are composed. Originally, elementary particles were thought to be independent, indivisible, and permanent; now many are known to convert to other particles and to be made up of smaller units. Such known elementary particles as the electron, proton, neutron, neutrino, photon, and meson are characterized by their mass, electric charge, spin, magnetic moment, and interaction. Protons, electrons, and neutrinos are stable particles; neutrons and mesons are unstable. *The historical development of elementary-particle physics might well be read as an extended lesson in skepticism. Over the course of the past century the realm of inquiry has progressed from the atom to the atomic nucleus to the hadrons that make up the nucleus to the quarks that make up the hadrons. Each of these objects has been considered for at least a time to be an elementary particle* (Scientific American). Also called **fundamental particle, subatomic particle.**

elevation, *n.* **1** *Geography.* height above sea level: *Nearby places in a highlands region may have somewhat different climates if they lie at different elevations* (Lyle H. Horn).
2 *Astronomy.* the angular distance of any heavenly body above the horizon.

ELF or **elf,** *abbrev.* extremely low frequency (applied to electromagnetic waves with frequencies below 300 hertz).

eliminate, *v. Algebra.* to get rid of (an unknown quantity) in two or more equations by combining them.

ellipse (i lips′), *n. Geometry.* a plane curve formed by the path of a point that moves so that the sum of its distances from two fixed points, called the foci, remains constant. Any conic section formed by a cutting plane inclined to the base but not passing through the base is an ellipse. The moon's path around the earth is an ellipse. See also **conic section.**

ellipse

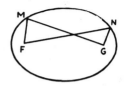

Points M and N are the foci.
Lines MG + GN = lines NF + FM

ellipsoid (i lip′soid), *n. Geometry.* **1** a solid figure of which all plane sections are ellipses or circles: *The figure obtained by rotating an ellipse is an ellipsoid.* **2** any surface of such a solid.
—adj. Also **ellipsoidal.** of an ellipsoid: *The earth is not even regular in its ellipsoid shape* (Scientific American).

elliptic or **elliptical,** *adj. Geometry.* of or shaped like an ellipse: *an elliptic cylinder.*
—ellipticity, *n.* **1** the quality of being elliptic. **2** the degree of deviation of an ellipse from a circle.

elliptical galaxy, *Astronomy.* a galaxy that appears round or elliptical and generally smooth, without the spiral arms of a spiral galaxy: *The elliptical galaxies also form a sequence, ranging from almost spherical systems to flattened ellipsoids* (Scientific American). See the picture at **galaxy.**

elliptic geometry, a system of geometry in which two or more lines passing through a point in a plane always intersect a given line in the plane. Also called **Riemannian geometry.**

El Niño (el nēn′yō), *Oceanography.* a current of warm water flowing southward along the coast of Ecuador, which, about every seven to ten years, extends to the

cap, fāce, fäther; best, bē, tėrm; pin, five;
rock, gō, ôrder; oil, out; cup, pùt, rüle;
yü in use, *yu̇* in uric;
ng in bring; *sh* in rush; *th* in thin, ᵀH in then;
zh in seizure.
ə = *a* in about, *e* in taken, *i* in pencil, *o* in lemon, *u* in circus

coast of Peru, where it causes fish and plankton to die and sea birds to migrate to find food. [from Spanish *El Niño* the Christ child (so called because the current usually appears near Christmas)]

elongation, *n. Astronomy.* **1** the angle between two celestial bodies as seen from the earth: *The elongation of the moon is the angle between the sun and the moon, which varies from 0° at new moon to 180° at full moon.* **2** the angular distance of a comet from the sun, of a satellite from its planet, etc.

eluant or **eluent** (el'yü ənt), *n. Chemistry* the solvent used in the process of elution.

eluate (el'yü it) *n. Chemistry.* the solution obtained from the process of elution.

elute (i lüt'), *v. Chemistry.* to separate or remove by elution: *The captured metal ions are then eluted and removed for final recovery by chemical precipitation* (New Scientist).

elution (i lü'shən), *n. Chemistry.* any process by which substances are separated by the action of a solvent; specifically, a chromatographic process by which adsorbed material is removed from an adsorbent by means of a solvent: *By skillful choice of solutions for elution it proved possible to resolve mixtures of the rare earths into their pure components and thus to reduce to a matter of hours an operation which had previously taken months to achieve* (K.S. Spiegler). [from Latin *elutionem* a washing out]

eluvial (i lü'vē əl), *adj. Geology.* **1** having to do with or formed by eluviation. **2** having to do with or resembling eluvium.

eluviate (i lü'vē āt), *v. Geology.* to remove by eluviation: *Some finely divided material has been carried out in suspension, or eluviated* (Finch and Trewartha, *Elements of Geography*).
—**eluviation** (i lü'vē ā'shən), *n.* the downard removal of soil particles: *These processes include ... eluviation or the removal of materials from the upper levels of the soil ...* (White and Renner, *Human Geography*).

eluvium (i lü'vē əm), *n., pl.* **-via** (-vē ə). *Geology.* an accumulation of soil or dust produced locally by the decomposition of rock or deposited by winds. Compare **alluvium.** [from New Latin *eluvium,* from Latin *ex-* out + *luere* to wash]

elytral (el'ə trəl), *adj. Zoology.* of or having to do with an elytron or elytra.

elytron (el'ə tron) or **elytrum** (el'ə trəm), *n., pl.* **-tra** (-trə). *Zoology.* **1** either of a pair of hardened front wings that form a protective covering for the hind pair; a wing case. Beetles and certain other insects have elytra. **2** a protective plate or scale on the back of certain annelid worms. [from New Latin, from Greek *elytron* sheath, from *elyein* to wrap]

EM, *abbrev.* **1** electromagnetic. **2** electron microscope.

emagram (em'ə gram), *n. Meteorology.* a chart showing temperature on a linear scale and pressure on a logarithmic scale. [from *em* the letter M + (*dia*)*gram*]

emanation, *n. Chemistry.* a heavy gaseous product resulting from the decay of a radioactive element, such as actinon from actinium, radon from radium, and thoron from thorium: *Emanations differ from each other in mass, charge, penetrating power, and ionizing power* (Chemistry Regents Syllabus).

emarginate (i mär'jə nit), *adj. Botany.* having a shallow notch at the apex, as a leaf or petal. [from Latin *emarginatum* taken off the edge]

embolic (em bol'ik), *adj.* **1** *Medicine.* having to do with an embolus. **2** *Embryology.* pushing or growing in during gastrulation.

embolus (em'bə ləs),*n., pl.* **-li** (-lī). *Medicine.* a clot, air bubble, globule of fat, etc., that is carried in the bloodstream. It sometimes blocks a blood vessel. *Sometimes clots break away from the spot where they were formed and are carried in the bloodstream to other parts of the body. This type of moving clot is called an embolus* (John W. Ferree). [from New Latin *embolus,* from Greek *embolos* peg]

emboly (em'bə lē), *n., pl.* **-lies.** *Embryology.* the process of forming a double-layered gastrula by invagination. Compare **epiboly.** [ultimately from Greek *embolē* insertion]

embryo, *n., pl.* **-bryos.** *Biology.* **1** an animal during the period of its growth from the fertilized egg until its organs have developed so that it can live independently. The embryo of a mammal is usually called a fetus in its later stages (in human beings, more than ten weeks after conception). *The higher oxygen affinity of the blood haemoglobin of the embryo in the womb, compared with that of the mother's blood, helps the transfer of oxygen from mother to foetus* (H. Munro Fox). **2** an undeveloped plant within a seed: *It is therefore the first step, the activation of the embryo, which constitutes the real germination* (A.M. Mayer). [from Medieval Latin, from Greek *embryon,* from *en-* in + *bryein* to swell]

embryos

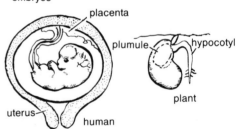

embryogenesis, *n. Embryology.* the formation and development of an embryo.
—**embryogenetic,** *adj.* of or having to do with embryogenesis.

embryogeny (em'brē oj'ə nē), *n.* = embryogenesis.
—**embryogenic** (em'brē ō jen'ik), *adj.* = embryogenetic.

embryoid (em'brē oid), *n. Biology.* a plant or animal form resembling an embryo: *The embryoids develop for a time inside the pollen grain; but eventually the pollen wall ruptures and the liberated embryoids then elongate to form root and shoot meristems* (Norman Sunderland).

embryology (em'brē ol'ə jē), *n.* the branch of biology that deals with the formation and development of embryos.
—**embryological,** *adj.* of or having to do with embryology: *embryological development.* —**embryologically,** *adv.*

embryonated, *adj. Embryology.* (of an egg) containing an embryo: *embryonated hen's eggs.*

embryonic (em′brē on′ik), *adj. Biology.* of or having to do with an embryo. The **embryonic disk** is a disklike mass of cells in the blastula from which an animal embryo develops. The **embryonic membranes** are parts of an animal embryo formed during its development, and usually including the amnion, chorion, and allantois.

embryophyte (em′brē ə fit), *n. Botany.* any of a group of plants that develop embryos: *Mosses, ferns, and flowering plants are embryophytes.*

embryo sac, *Botany.* the female reproduction structure in seed plants. When fully developed, it usually consists of the egg cell, three antipodal cells, two polar nuclei, and two naked cells or synergids: *Within the nucellus is found the embryo sac or megaspore, containing protoplasm and a nucleus* (Youngken, *Pharmaceutical Botany*).

emerald, *n. Mineralogy.* a bright-green transparent variety of beryl, whose color is due to traces of chromium. [from Old French *esmeralde,* ultimately from Greek *smaragdos*]

emergent evolution, *Biology.* evolution which according to some theorists results in characteristics that are not predictable; the theory that genuinely new properties emerge during the course of evolution.

emersed (i mèrst′), *adj. Botany.* standing out of water as part of an aquatic plant.

emersion (i mèr′zhən), *n. Astronomy.* the reappearance of the sun, moon, or other celestial body after an eclipse. [from Latin *emersionem* a dipping out, from *emergere* dip out, emerge]

emery (em′ər ē), *n. Mineralogy.* a hard, dark granular rock consisting of corundum with various iron minerals: *Emery is essentially an intimate mixture of granular corundum and magnetite, and in powdered form has long been used as emery paper, emery cloth and emery wheels* (W. R. Jones). [from Middle French *émeri,* ultimately from Greek *smyris* abrasive powder]

e.m.f. or **EMF,** *abbrev.* electromotive force.

-emia, *combining form.* condition of the blood; presence in the blood of ——, as in *toxemia* (toxic condition of the blood, presence in the blood of a toxic substance), *septicemia, uremia.* [from New Latin, from Greek *haima* blood]

emiocytosis (em′ē ō sī tō′sis), *n. Biology.* a process of cellular excretion by which the membrane of a storage granule fuses with the cell membrane and the granule is discharged through an opening in the fused membranes: *The newly formed insulin, proinsulin, and C-peptide are subsequently transferred from the Golgi apparatus to storage granules which discharge by fusing with the plasma membrane in a process known as emiocytosis* (Donald F. Steiner). Compare **endocytosis, exocytosis, phagocytosis, pinocytosis.** [from New Latin, from Greek *emein* + New Latin *cyt-* cell + *-osis* process or condition]

—emiocytotic, *adj.* of or having to do with emiocytosis: *emiocytotic release of insulin.*

emission, *n. Physics.* **1** the discharge of electrons as from an electrode, caused by heat, light, or by the impact of an electrical charge or beam of high-energy electrons. Such discharges are classified as *thermionic emission, photoemission, field emission,* and *secondary emission.* **2** any radiation of energy by electromagnetic waves.

emission line, *Physics.* any of the lines that make up an emission spectrum, each indicating a particular wavelength of radiation: *The strong emission lines at 4,577 and 3,597 angstroms were identified ... as resulting ... from radiation from quadruply ionized carbon atoms and Lyman-alpha radiation from hydrogen atoms* (Scientific American). Compare **absorption line.** See also **spectral line.**

emission spectrum, *Physics.* the spectrum into which light or other electromagnetic radiation from any source can be separated. Analysis of an emission spectrum can identify chemical elements present in the source of the radiation. *The rainbow spectrum derived from the sun's light is known as an emission spectrum. It is so named because it is produced by a glowing object and all spectra so produced are similarly classified. Emission spectra are of two types, continuous and discontinuous.* (J. Gordon Vaeth). Compare **absorption spectrum.**

emissivity, *n. Physics.* the relative ability of a surface to radiate energy. Emissivity is expressed as the ratio of the radiation emitted by a surface or body to that emitted by a perfect blackbody at the same temperature.

emittance (i mit′ns), *n. Physics.* **1** the amount of energy emitted per unit area by a radiating surface or body. **2** = emissivity.

empirical (em pir′ə kəl), *adj.* based on experiment and observation (sometimes contrasted with *theoretical*): *empirical evidence, an empirical definition.* **—empirically,** *adv.*

empirical formula, a chemical formula used to indicate the simplest ratio of the number and kind of atoms in a compound. It does not necessarily indicate the number of atoms in a molecule. For example, the empirical formula for benzene is CH, while its molecular formula is C_6H_6. Sodium chloride has only an empirical formula, NaCl, because the crystal is not made up of molecules but of Na and Cl ions. Empirical formulas are thus used to represent ionic compounds which do not exist as distinct molecular entities. Compare **molecular formula, structural formula.** See the picture at **formula.**

empiricism (em pir′ə siz′əm), *n.* the use of methods based on experiment and observation.

emplectite (em plek′tīt), *n. Mineralogy* a mineral consisting of sulfur, bismuth, and copper, occurring in thin prismatic crystals of a grayish or white color and bright metallic luster. *Formula:* $CuBiS_2$ [from Greek *emplektos* woven]

empodium (em pō′dē əm), *n., pl.* **-dia** (-dē ə). *Zoology.* a clawlike organ found between the true claws of many insects. [from New Latin, from Greek *em-* in + *podos* foot]

---

cap, fāce, fäther; best, bē, tèrm; pin, fīve; rock, gō, ôrder; oil, out; cup, pùt, rüle; *yü* in use, *yu* in uric; *ng* in bring; *sh* in rush; *th* in thin; ᴛʜ in then; *zh* in seizure.

ə = *a* in about, *e* in taken, *i* in pencil, *o* in lemon, *u* in circus

191

empty set, *Mathematics.* a set that has no members. An empty set is written as [ ] and its symbol is ∅. The set represented by [∅] is not an empty set because it has the one member, ∅. Also called **null set.**

e.m.u. or **EMU,** *abbrev.* electromagnetic unit *or* electromagnetic units.

emulsify, *v. Chemistry.* to make or convert (an oil, fat, etc.) into an emulsion: *Bile emulsifies fats in the small intestine.*
—**emulsification,** *n.* the act of emulsifying or the state of being emulsified: *Emulsification serves to increase the surface area of fats for subsequent chemical action* (Biology Regents Syllabus).
—**emulsifier,** *n.* a substance that emulsifies or that is used to produce an emulsion.

emulsion, *n. Chemistry.* a mixture of liquids that do not dissolve in each other; a colloidal mixture. In an emulsion, one of the liquids contains minute drops of the other evenly distributed throughout. *Kerosene and water form a milky emulsion which separates into its parts after a short time. While stable emulsions of two pure liquids cannot be prepared, they may be stabilized readily by the addition of a third substance, known as ... an emulsifier* (Parks and Steinbach, *Systematic College Chemistry*). Compare **suspension.** [from New Latin *emulsionem,* from Latin *emulgere* to milk out]

emulsoid (i mul'soid), *n. Chemistry.* a colloid in which a solid exhibits a strong affinity for the liquid in which it is dispersed.

enamel, *n. Anatomy.* the smooth, hard, glossy outer layer of the teeth, consisting chiefly of calcium carbonate. Enamel is the hardest substance produced by animal bodies. See the picture at **tooth.**

enantiomer (en an'tē ə mər), *n.* = enantiomorph.

enantiomorph (en an'tē əmôrf'), *n. Chemistry.* either one of a pair of molecules, crystals, or compounds that are mirror images of each other, but are not identical: *When we stand before a mirror, we see reflected a superficially bilaterally symmetrical structure, and we are misled by this apparent symmetry into treating the system as if ourselves and our reflection were identities rather than enantiomorphs (entities of opposite "handedness")* (Scientific American). [from Greek *enantios* opposite + *morphē* form]
—**enantiomorphic** or **enantiomorphous,** *adj.* having to do with or resembling an enantiomorph.

enantiotropic (en an'tē ə trop'ik), *adj. Chemistry.* (of different forms of the same substance) that can be transformed from one into the other in either direction. [from Greek *enantios* opposite + *-tropos* a turning]

enargite (en är'jit), *n. Mineralogy.* a grayish-black orthorhombic mineral, an important copper ore, composed of sulfur, copper, arsenic, and sometimes antimony. *Formula:* Cu_3AsS_4 [from Greek *enargēs* clear (so called because its cleavage is apparent)]

enarthrosis (en'är thrō'sis), *n., pl.* **-ses** (-sēz). *Anatomy.* = ball-and-socket joint. [from Greek *enarthrōsis,* from *en-* in + *arthron* joint]

encapsidate (en kap'sə dāt), *v. Biology.* to enclose (the core of a virus particle) in a protein shell to make the particle stable and transmissible: *Conventional viruses are made up of nucleic acid encapsidated in protein,* whereas viroids are characterized by the absence of encapsidated proteins (R.K. Horst). See **capsid, capsomere.**
—**encapsidation,** *n.* the process of encapsidating or the condition of being encapsidated.

encephalization, *n. Zoology.* the development and specialization of the brain as the center of the nervous system of vertebrates: *Progressive evolution of encephalization within the mammals came late in their history, in the last 50 million years of a time span of about 200 million years* (Scientific American).
The **encephalization quotient** is the ratio of brain weight to body weight, used as an index to animal development: *The herbivores display no increase in brain size through time. Their average encephalization quotient remained below 0.5 throughout the Tertiary, and they were quickly eliminated when advanced carnivores crossed the isthmus [of Panama] from North America* (Natural History).

encephalon (en sef'ə lon), *n., pl.* **-la** (-lə). *Zoology.* the brain of a vertebrate. [from New Latin, from Greek *enkephalos,* from *en-* in + *kephalē* head]

encrinite (en'krə nīt), *n. Mineralogy.* a crinoidal limestone. Much of the world's marble is composed of fragments of encrinites. [from New Latin *encrinus,* from *en-* in + *krinon* lily]

encyst (en sist'), *v. Biology.* to enclose or become enclosed in a cyst or sac.
—**encystment,** *n.* the process of encysting or the condition of being encysted.

end-, *combining form.* the form of **endo-** before vowels, as in *endamoeba, endarterial.*

endameba (en'də mē'bə), *n.* any of a genus (*Entamoeba*) of amebas or amebalike organisms that are parasitic on other forms of animal life, including the species (*E. histolytica*) that causes amebic dysentery. Also spelled **entameba.**

endarch (en'därk), *adj. Botany.* having a protoxylem developing outwards: *an endarch stele.* Compare **exarch.** [from *end-* within + Greek *archē* origin]

endarterial, *adj. Anatomy.* **1** within the artery. **2** of or having to do with the lining of an artery.

end brain, *n.* = telencephalon.

end bulb, *Anatomy.* a roundish body forming the enlarged ending of a sensory nerve fiber. End bulbs range from simple terminations to very complex sensory organs and are dispersed in the skin, mucous membranes, muscles, etc.

endemic (en dem'ik), *adj.* **1** *Medicine.* regularly found among a particular people or in a particular locality: *Relapsing fever ... is endemic ... in North Central Africa* (Science News Letter). Compare **epidemic, pandemic.**
2 *Biology.* (of plants or animals) restricted or indigenous to a certain locality. —**endemically,** *adv.*
—*n.* **1** *Medicine.* an endemic disease. **2** *Biology.* an endemic plant or animal: *Giant sequoias, Florida yews and Georgia plumes are other examples of "endemics," rare plants limited to such small ranges* (Science News).
[from Greek *endēmos* native, from *en-* in + *dēmos* people]
—**endemicity** (en'də mis'ə tē), *n.* endemic quality or character.

endite (en'dīt), *n. Zoology.* an appendage on the inner side of the limbs of an arthropod. [from *end-* within + *-ite*]

endo-, *combining form.* within; inside; inner, as in *endoderm = inner layer of cells.* Also spelled **end-** before vowels. Compare **exo-**. [from Greek *endo-,* from *endon* within]

endobiotic, *adj. Biology.* living as a parasite within the tissue of the host.

endocardium (en'dō kär'dē əm), *n. Anatomy.* the smooth membrane that lines the cavities of the heart. —**endocardial**, *adj.* **1** within the heart. **2** of or having to do with the endocardium.

endocarp (en'dō kärp), *n. Botany.* the inner layer of the pericarp of a fruit or ripened ovary of a plant. A peach stone is a hollow endocarp surrounding the seed. See the picture at **pericarp**. [from *endo-* + Greek *karpos* fruit]

endocranium (en'dō krā'nē əm), *n., pl.* **-niums, -nia** (-nē-ə). *Anatomy.* the inner surface of the skull, covering the brain. —**endocranial** (en'dō krā'nē əl), *adj.* of or belonging to the endocranium.

endocrine (en'də krən), *Physiology.* —*adj.* **1** having an internal secretion; secreting hormones directly into the blood or lymph. An **endocrine gland,** such as the thyroid and the pituitary, produces hormonal secretions that pass directly into the bloodstream or lymph without a duct. An endocrine gland is also called a **ductless gland.**
2 of or having to do with endocrine glands or the hormones they secrete: *In spring, the lengthening hours of light or rising temperature cause endocrine changes in many birds and fishes* (Science News). Contrasted with **exocrine.** Compare **apocrine, eccrine.**
—*n.* **1** an endocrine gland. **2** its secretion.
[from *endo-* + Greek *krinein* to separate]

endocrinology (en'dō krə nol'ə jē *or* en'dō krī nol'ə jē), *n.* the study of the endocrine glands and their function, especially in their relation to bodily changes and diseases.
—**endocrinologic** or **endocrinological**, *adj.* of or having to do with endocrinology.

endocytosis (en'dō sī tō'sis), *n. Biology.* a process of cellular ingestion by which the cell membrane folds inward to enclose and incorporate foreign substances: *Foreign particles such as bacteria are ingested by the process now known as endocytosis ... The cell membrane folds inward to form a pocket, the edges of which fuse to enclose the particles* (Anthony Allison). Contrasted with **exocytosis.** [from New Latin, from *endo-* within + *cyt-* cell + *-osis* process or condition]
—**endocytose**, *v.* to ingest by endocytosis: *Damaged portions of the cell surface and even potentially damaging substances that bind to it are sucked, or endocytosed, into the cell* (Garth L. Nicolson).
—**endocytotic**, *adj.* of or having to do with endocytosis: *an endocytotic process.*

endoderm (en'dō dėrm'), *n. Embryology.* the inner layer of cells formed during development of animal embryos, from which the lining of the organs of the digestive system develops: *From the endoderm comes the lining of ... the lungs, liver, and pancreas* (Shull, *Principles of Animal Biology*). Also called **entoderm, hypoblast.**

Compare **ectoderm** and **mesoderm**. [from *endo-* + Greek *derma* skin]
—**endodermal** (en'dō dėr'məl), *adj.* of or having to do with the endoderm.

endodermis (en'dō dėr'mis), *n. Botany.* a layer of cells which are united to form the inner boundary of the cortex and the sheath surrounding the vascular bundles of plants: *It is possible too that the endodermis acts as an ion-secreting membrane transporting salts into the stele, and resembles rather analogous structures such as the gastric mucosae of animals* (J.F. Sutcliffe).

endoenzyme, *n. Biochemistry.* an enzyme which acts or exists within the cell in which it originated. Contrasted with **exoenzyme.**

endogamy (en dog'ə mē), *n.* **1** *Botany.* pollination of a flower by another flower of the same plant.
2 *Biology.* the union of two closely related gametes. Compare **exogamy.**

endogenic (en'dō jen'ik), *adj.* **1** *Geology.* formed within the earth; proceeding from inside the earth: *a crater of endogenic origin.* **2** = endogenous.

endogenous (en doj'ə nəs), *adj.* **1** *Biology.* growing from within. Cells or spores developing within a cell are endogenous.
2 *Physiology.* of or having to do with the metabolism of substances within cells or tissues. Contrasted with **exogenous.**

endognath (en'dog nath), *n. Zoology.* the inner branch of the oral appendage of a crustacean. [from *endo-* + Greek *gnathos* jaw]

endolymph (en'dō limf'), *n. Anatomy.* a fluid contained in the membranous labyrinth of the ear. Compare **perilymph.**
—**endolymphatic**, *adj.* of the endolymph: *the endolymphatic duct.*

endometrium (en'dō mē'trē əm), *n., pl.* **-tria** (-trē ə). *Anatomy.* the mucous membrane which lines the uterus. [from New Latin, from *endo-* internal + Greek *mētra* womb]
—**endometrial**, *adj.* of or having to do with the endometrium.

endomitosis (en'dō mī tō'sis), *n. Biology.* a form of mitosis in which chromosomes are duplicated without an accompanying division of the nucleus.

endomixis (en'dō mik'sis), *n. Biology.* a periodic process of nuclear division and reorganization similar to conjugation that takes place in certain protozoans. [from *endo-* + Greek *mixis* a mingling]

endomorph (en'dō môrf'), *n. Mineralogy.* a mineral enclosed within another mineral. [from *endo-* + Greek *morphē* form]

---

cap, fāce, fäther; best, bē, tėrm; pin, fīve;
rock, gō, ôrder; oil, out; cup, pùt, rüle;
yü in use, *yu* in uric;
ng in bring; *sh* in rush; *th* in thin; ᴛʜ in then;
zh in seizure.
ə = *a* in about, *e* in taken, *i* in pencil, *o* in lemon, *u* in circus

endonuclease (en'dō nü'klē ās), *n. Biochemistry.* an enzyme that breaks up a sequence of DNA and RNA into discontinuous segments by attacking links within the nucleotide sequence. Compare **exonuclease.**

endoparasite, *n. Biology.* a parasite living in the tissues or cavities of organisms. The tapeworm is an endoparasite. Contrasted with **ectoparasite.**
—**endoparasitic,** *adj.* living in the tissues or cavities of organisms.

endopeptidase (en'dō pep'tə dās), *n. Biochemistry.* any of various proteolytic enzymes, such as trypsin, that break down peptide bonds inside polypeptide chains. Compare **exopeptidase.**

endoperoxide (en'dō pə rok'sīd), *n. Biochemistry.* any of a group of highly oxygenated compounds that are precursors of prostaglandins.

endophyte (en'dō fīt), *n. Botany.* a plant growing inside another plant. Endophytes are not always parasitic.
—**endophytic** (en'dō fī'tik), *adj.* of or having to do with endophytes.

endoplasm (en'dō plaz'əm), *n. Biology.* the inner portion of the cytoplasm of a cell. Compare **ectoplasm.**
—**endoplasmic** (en'dō plaz'mik), *adj.* of or having to do with endoplasm. The **endoplasmic reticulum** is a network of membranous vesicles in the cytoplasm of most cells, to which ribosomes are often attached: *The endoplasmic reticulum plays a central part in the biosynthesis of the macromolecules used to construct other cellular organelles* (Bruce Alberts).

endopodite (en dop'ə dīt), *n. Zoology.* the inner branch of a two-branched appendage in crustaceans. Compare **exopodite.** [from *endo-* + Greek *podos* foot]

end organ, *Anatomy.* a specialized structure at the distal end of a sensory or motor nerve. The retina is the end organ for vision.

endorphin (en dôr'fin), *n. Biochemistry.* any of a group of polypeptides present in the brain that control various physiological responses. Each endorphin has certain properties of morphine and acts as a natural pain suppressant or opiate of the body. Compare **enkephalin.** [from *endo-* within + (*m*)*orphin*(*e*)]

endoscope, *n. Medicine.* an optical instrument for examining the interior of a hollow organ or process, such as the urethra or rectum.

endoskeleton   exoskeleton

endoskeleton, *n. Zoology.* the internal skeleton characteristic of vertebrates and allied forms: *The larger animals, as a whole, have developed an endoskeleton in which the muscles may be outside the skeleton, as in*

man (Winchester, *Zoology*). Compare **exoskeleton.**
—**endoskeletal,** *adj.* of or having to do with an endoskeleton.

endosmosis (en'dos mō'sis), *n. Biology.* the passage of a fluid inwards through a membrane toward a solution of greater concentration: *The vacuole is surrounded by protoplasm, the adjacent surface of which functions as an osmotic membrane. When such cells are surrounded by water, water enters the cell (endosmosis) and diffuses through the protoplasm into the vacuole* (Harbaugh and Goodrich, *Fundamentals of Biology*). Contrasted with **exosmosis.**
—**endosmotic** (en'dos mot'ik), *adj.* of, or resembling, or having to do with endosmosis.

endosperm (en'dō spėrm'), *n. Botany.* food stored in the ovule of a seed plant for the early nourishment of the embryo. See the picture at **monocotyledon.**

endosperm nucleus, *Botany.* the nucleus from which the endosperm develops. The primary endosperm nucleus is formed by the fusion of two polar nuclei with the nucleus of a sperm. Successive divisions of the primary endosperm nucleus produce free endosperm nuclei which form the nutritive tissue of the endosperm.

endospore (en'dō spôr), *n.* **1** *Botany.* the inner coat or wall of a spore of certain plants. **2** *Bacteriology.* a spore formed within a cell of certain bacteria. Compare **exospore.**

endosteum (en dos'tē əm), *n., pl.* **-tea** (-tē ə). *Anatomy.* the membranous connective tissue lining the walls of the medullary cavity of a bone. [from New Latin, from *end-* within + Greek *osteon* bone]
—**endosteal** (en dos'tē əl), *adj.* **1** of or having to do with the endosteum. **2** situated in the interior of a bone or cartilage.

endostome (en'də stōm), *n. Botany.* the opening at the apex of the inner coat of an ovule.

endostosis (en'dos tō'sis), *n. Physiology.* the growth of bone within cartilage. Compare **ectostosis.** [from *endo-* + Greek *osteon* bone + New Latin *-osis* condition]

endostyle (en'dō stīl'), *n. Zoology.* a ventral, ciliated groove in the pharynx of tunicates and other chordates, used to trap food particles. [from *endo-* within + Greek *stylos* column]

endosymbiosis, *n. Biology.* a form of symbiosis in which an organism lives inside another organism.
—**endosymbiont,** *n.* an organism, usually the smaller of the two, living in endosymbiosis: *Parallel development of different species has often resulted in symbiotic relationships ... between the higher animals and the endosymbionts—examples are the bacterial species which adorn the human gut and play a useful part in digestion* (London *Times*).
—**endosymbiotic,** *adj.* having to do with endosymbiosis: *endosymbiotic algae.*

endothecium (en'dō thē'sē əm), *n., pl.* **-cia** (-sē ə). *Botany.* the inner lining of the cell of an anther. [from New Latin, from *endo-* within + Greek *thēkē* chest]

endothelium (en'dō thē'lē əm), *n., pl.* **-lia** (-lē ə). *Anatomy.* the tissue that lines blood vessels, lymphatic vessels, the heart, etc. It is a form of epithelium. [from New Latin, from *endo-* within + Greek *thēlē* nipple]
—**endothelial,** *adj.* having to do with endothelia: *endothelial cells.*

endotherm (en'dō thėrm), *n. Zoology.* a warm-blooded animal. Contrasted with **ectotherm.** Also called **homeotherm.**

endothermal, *adj.* = endothermic.

endothermic, *adj.* **1** *Chemistry.* of or having to do with a chemical change or reaction in which heat is absorbed. **2** *Zoology.* = warm-blooded.

endotoxin, *n. Bacteriology.* a poisonous substance in the cell of certain bacteria that is not released until the cell disintegrates. Contrasted with **exotoxin.**
—**endotoxic,** *adj.* of or having to do with an endotoxin.

endotrophic (en'dō trof'ik), *adj. Biology.* deriving nourishment from within, as certain fungi living within the roots of plants and deriving nourishment from their inner cells. Compare **ectotrophic.**

end plate, *Anatomy.* the expanded ending of a motor nerve in a muscular fiber. The end plate is involved in the transmission of impulses to muscles.

end point, 1 *Chemistry.* the point in a titration when some effect occurs, usually a change in color of an added indicator, showing that no more titrant should be added. It generally occurs slightly before or after the equivalence point.
2 *Mathematics.* a point in space where a line segment, etc., ends: *A point on a curve is called an end point if there are arbitrarily small neighborhoods surrounding it, each of whose boundaries has only a single point in common with the curve* (Scientific American).

end product, 1 *Chemistry.* the final product of a series of reactions or processes.
2 *Nuclear Physics.* the final member of a series of isotopes, each produced by the radioactive decay of the preceding isotope.

-ene, *noun suffix. Chemistry.* a hydrocarbon, especially of the olefin series, as in *alkene, benzene, naphthalene, toluene.* [from Latin *-enus,* from Greek *-ēnos,* adjective suffix]

energetics, *n.* **1** *pl. in form, sing. in use. Physics.* the study of energy and its transformations into various forms.
2 *pl. in form and use.* the amount of energy output or transformations in any activity or system: *the energetics of a nuclear reactor, the energetics of flapping flight.*

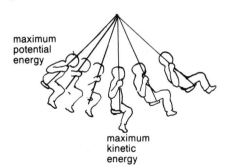

maximum potential energy

maximum kinetic energy

energy, *n. Physics.* the capacity or power for doing work possessed at any instant by a body or system. Energy may be electrical, mechanical, chemical, thermal, or nuclear, but it is always measured by the work done. Thus, one unit of energy, the newton-meter or joule, is the work done by a force of one newton in moving a body a distance of one meter. According to the law

of conservation of energy, the total energy in the universe is a constant quantity; hence energy can neither be created nor destroyed, only converted from one form to another. Matter is one form of energy, and in nuclear reactions some matter is converted to other forms, such as light and heat. The energy (E) of a particle equals its mass (m) multiplied by the speed of light (c) squared. This equation ($E = mc^2$) is called the **mass-energy equation** or the **Einstein equation.** Compare **kinetic energy, potential energy.** [from Greek *energeia,* from *en-* in + *ergon* work]

energy level or **energy state,** *Physics.* one of the usually stable states of energy of certain physical systems. In the atom, electrons cluster about the nucleus in various energy levels. *The principal energy level approximates how far the electron is from the nucleus ... When electrons have absorbed energy and shifted to higher energy levels, the atom is said to be in an "excited state"* (Chemistry Regents Syllabus).

energy of activation, *Chemistry.* the minimum amount of energy that must be added to a molecule in the ground state before it can take part in a given chemical reaction: *One significant feature of enzymes is their ability to lower the energy of activation, thus increasing the rate of chemical reaction* (McElroy, *Biology and Man*).

energy-rich bond, *Biochemistry.* a chemical bond, usually formed during metabolism, that liberates a large amount of heat on being hydrolyzed.

englacial, *adj. Geology.* of, having to do with, or being within a glacier: *an englacial boulder.* —**englacially,** *adv.*

engram (en'gram), *n. Physiology.* a physical change believed to occur in nerve tissues in response to stimuli, postulated to account for the existence and permanence of memory: *The two most popular chemical contenders for this elusive 'memory trace' or 'engram' have been ribonucleic acid (RNA) and protein* (New Scientist). [from German *Engramm,* from Greek *en-* in + *gramma* letter]

enkephalin (en kef'ə lin), *n. Biochemistry.* either of two polypeptides that are natural pain suppressants or opiates of the body. Enkephalins are produced in the pituitary gland and are related to the endorphins. *A variety of evidence suggests that the enkephalins are neurotransmitters for specific neuronal systems in the brain which mediate the integration of sensory information pertaining to pain and emotional behavior* (John W. Phillis). [from Greek *enkephalos* brain + English *-in* (chemical suffix)]

► There are two forms of enkephalin, both pentapeptides: *leu-enkephalin,* consisting of a peptide chain having the amino acid leucine at its end; and *met-enkephalin,* in which the peptide chain has the amino acid methionine at its end: *Met-enkephalin is*

cap, fāce, fäther; best, bē, tėrm; pin, five;
rock, gō, ôrder; oil, out; cup, pùt, rüle,
yü in use, *yù* in uric;
ng in bring; *sh* in rush; *th* in thin, ᴛʜ in then;
zh in seizure.
ə = *a* in about, *e* in taken, *i* in pencil, *o* in lemon, *u* in circus

195

more potent than the leu-variety, but both bind less strongly to brain receptors than does morphine (New Scientist).

enol (ē'nol), *n. Chemistry.* an organic compound, the tautomeric form of a ketone, having the formula C=COH. [from *-ene* + *-ol*]
—**enolic,** *adj.* of or having to do with an enol.

enolase (ē'nə lās), *n. Biochemistry.* a crystalline enzyme active in glycolysis, found in the muscles and in yeast.

ensiform (en'sə fôrm), *adj. Botany, Zoology.* sword-shaped: *the ensiform leaves of the iris, ensiform antennae.* SYN: xiphoid. [from Latin *ensis* sword + *forma* form]

enstatite (en'stə tīt), *n. Mineralogy.* a variety of orthorhombic pyroxene, occurring in gray, green, or brown colors. *Formula:* MgSiO$_3$ [from Greek *enstatēs* adversary (so called because of its refractory properties)]

entameba (en'tə mē'bə), *n.* = endameba.

enteric (en ter'ik) or **enteral** (en'tər əl), *adj.* = intestinal. [from Greek *enterikos,* from *enteron* intestine]

entero-, *combining form.* of the intestine or intestines; intestinal, as in *enterotoxin = intestinal toxin, enterovirus = intestinal virus.* [from Greek *enteron* intestine]

enterobacterium (en'tər ə bak tir'ē əm), *n., pl.* **-teria** (-tir'ē ə). *Bacteriology.* an intestinal bacterium, especially one belonging to a large family of rod-shaped bacteria that includes *E. coli.*

enterococcus (en'tər ə kok'əs), *n., pl.* **-cocci** (-kok'sī). *Bacteriology.* a streptococcus usually found in the human intestine.
—**enterococcal,** *adj.* having to do with or caused by an enterococcus or enterococci.

enterocoele or **enterocoel** (en'tər ō sēl'), *n. Embryology.* the coelom formed by eversion of the wall of the primitive intestinal cavity of the gastrula.

enterogastrone (en'tər ō gas'trōn), *n. Biochemistry.* a hormone of the intestinal mucosa that inhibits gastric movement and secretions. [from *entero-* + *gastr(ic)* + *(horm)one*]

enterokinase (en'tər ō kī'nās), *n. Biochemistry.* an enzyme secreted by the intestine that changes inactive trypsinogen to active trypsin.

enteron (en'tə ron), *n.* = alimentary canal. [from Greek *enteron* intestine]

enterovirus, *n. Biology.* any of a group of very small viruses containing RNA that are commonly found in the gastrointestinal tract but are associated with diseases of the nervous and respiratory systems. The poliovirus and echovirus belong to this group. —**enteroviral,** *adj.*

enthalpimetry (en thal'pim'ə trē), *n. Physics.* the measurement of total heat content generated or absorbed by a substance, used especially to follow the progress of a chemical reaction: *Enthalpimetry depends on the fact that almost all reactions are associated with the evolution or absorption of heat* (New Scientist).
—**enthalpimetric,** *adj.* of or having to do with enthalpimetry.

enthalpy (en thal'pē *or* en'thəl pē), *n. Physics.* the heat content per unit mass of a substance; the quantity of heat required to raise a unit mass of a liquid from a standard or convenient temperature, usually its freezing point, to a given temperature, and then turn it into vapor at that temperature under constant pressure. Also called **total heat.** [from Greek *enthalpein* to warm]

enthalpy of formation, *Chemistry.* the heat absorbed when one mole of a substance is formed from its elements in the phases in which they are most stable at 25°C. It is the same as the heat of formation, but opposite in sign. Also called **standard molar enthalpy of formation.**

enthalpy of reaction, *Chemistry.* the heat absorbed in a reaction when the quantities of reactants and products correspond to the number of moles indicated by the equation for that reaction. It is the same as heat of reaction, but opposite in sign.

entoblast, *n. Embryology.* any of the blastomeres from which the endoderm develops. [from Greek *entos* within + *blastos* sprout]

entoderm (en'tə derm), *n.* = endoderm. [from Greek *entos* within + *derma* skin]

entomogenous (en'tə moj'ə nəs), *adj. Biology.* (of fungi) growing upon or in insects. [from Greek *entomon* insect]

entomology (en'tə mol'ə jē), *n.* the branch of zoology that deals with insects. [from Greek *entomon* insect + English *-logy*]
—**entomological** (en'tə mə loj'ə kəl), *adj.* of or having to do with entomology.

entomophagous (en'tə mof'ə gəs), *adj. Zoology.* feeding on insects; insectivorous.

entomophilous (en'tə mof'ə ləs), *adj. Botany.* pollinated by insects, as the flowers of orchids and irises.
—**entomophily** (en'tə mof'ə lē), *n.* pollination by insects.

entozoon (en'tə zō'on), *n., pl.* **-zoa** (-zō'ə). *Zoology.* an internal parasite, especially an intestinal worm. [from Greek *entos* within + *zōion* animal]
—**entozoic,** *adj.* of or having to do with entozoa.

entrance cone, *Biology.* a conical projection formed by the protoplasm of an egg cell at the point of entrance of the sperm in fertilization. Also called **fertilization cone.** See the picture at **fertilization.**

entropy (en'trə pē), *n. Physics.* **1** (in a closed thermodynamic system) a measure of the unavailability of the thermal energy for doing mechanical work: *The entropy of a system, which fundamentally is defined in terms of its heat capacity (the amount of heat required to raise its temperature), is a measure of the lack of order, or the presence of randomness, within the system* (L.K. Runnels).
2 a measure of the degree of disorder of any system. [from German *Entropie,* from Greek *en-* in + *tropē* a turning]
—**entropic** (en trop'ik), *adj.* of or characterized by entropy: *... the entropic, increasing disorderly expansion of the physical universe* (Buckminster Fuller).
—**entropically,** *adv.*

enucleate (i nü'klē āt), *v.* **1** *Medicine.* to remove surgically (a tumor, etc.) from its capsule or cover without cutting.
2 *Biology.* to remove the nucleus of.

envelope, *n.* **1** *Botany.* a surrounding or enclosing part: *an envelope of leaves, a floral envelope.*

2 *Zoology.* any enclosing covering or integument, such as a membrane or shell.

3 *Biology.* a membrane surrounding some viruses, derived from the nuclear or cytoplasmic membrane of the host cell in which the virus was produced.

4 *Astronomy.* a nebulous mass surrounding the nucleus of a comet: *Coggia's Comet displayed magnificent envelopes* (Bernhard, *Handbook of Heavens*).

5 *Geometry.* a curve or surface touching a continuous series of curves or surfaces.

environment, *n. Biology.* all the surrounding things, conditions, and influences affecting the development of living things. Differences in environment often account for differences in the same kind of organism found in different places: *A plant's environment may be made up of soil, sunlight, and animals that will eat the plant* (J. H. Taylor).

▶ The earth's environment is made up of *biotic* and *abiotic* factors. The living or biotic environmental factors include all living organisms, the foods they consume, and their interactions among themselves and the nonliving environment. The nonliving or abiotic factors include such influences as weather and climate, living space, water, and soil. The biotic and abiotic environments interact to make up the total environment of all living things. See also **ecology.**

—**environmental,** *adj.* of or having to do with the environment.

enzymatic (en′zi mat′ik) or **enzymic** (en zī′mik), *adj. Biochemistry.* of or having to do with an enzyme or enzymes: *Enzymes are proteins, and therefore an increase in the cell's enzymic activity means that it is synthesizing these proteins* (Scientific American).

—**enzymatically** or **enzymically,** *adv.*

enzyme, *n. Biochemistry.* a complex protein substance produced in living cells that causes or accelerates chemical reactions in other substances in an organism without being permanently changed itself. Enzymes such as pepsin help break down food so that it can be digested. *The human body contains more than 1,000 types of enzymes ... Photosynthesis in plants also depends on enzymes* (Frederick B. Rudolph). [from German *Enzym,* ultimately from Greek *en-* in + *zyme* leaven]

ASSOCIATED TERMS: Enzymes act as organic *catalysts* on the *substrates* of the molecules to which their *active sites* attach. A *coenzyme* is a nonprotein substance that combines with an *apoenzyme* to form an active enzyme, or *holoenzyme.* The names of most enzymes end with *-ase,* as in *amylase, lipase, urease,* and *ribonuclease.*

enzymology (en′zi mol′ə jē), *n.* the study of enzymes and the reactions they catalyze.

—**enzymological,** *adj.* of or having to do with enzymology.

Eocene (ē′ə sēn′), *Geology.* —*n.* **1** the second epoch of the Tertiary period, after the Paleocene and before the Oligocene, during which the ancestors of many modern mammals appeared.

2 the rocks formed during this epoch.

—*adj.* of or having to do with this epoch or its rocks: *Perhaps the richest single deposit of Eocene mammals ever found, it [the Castillo Pocket] contained several primitive genera including excellent skeletons of the dawn horse Hyracotherium (eohippus)* (Henry N. Andrews).

[from Greek *ēōs* dawn + *kainos* recent]

Eogene (ē′ə jēn), *n., adj.* = Paleogene.

eolian (ē ō′lē ən), *adj. Geology.* of, produced by, or carried by the winds: *eolian dust or soil.* Also spelled **aeolian.** [from Latin *Aeolius* of or having to do with *Aeolus,* god of the winds]

eon (ē′ən or ē′on), *n. Geology.* **1** any span of time longer than an era: *the Phanerozoic eon.* **2** one billion (10^9) years. Also spelled **aeon.** [from Latin *aeon,* from Greek *aiōn* age]

eosin or **eosine** (ē′ə sən), *n. Chemistry.* a rose-red dye or stain made from coal tar, used in biology to stain cells for microscopy. *Formula:* $C_{20}H_8Br_4O_5$ [from Greek *ēōs* dawn + English *-in*]

eosinophil (ē′ə sin′ə fil) or **eosinophile** (ē′ə sin′ə fīl), *Biology.* —*n.* a cell containing granules which are easily stained by eosin or other acid dyes, especially a type of white blood cell. Compare **basophil, neutrophil.**

—*adj.* = eosinophilic.

[from *eosin* + *-phil*]

—**eosinophilic,** *adj.* **1** staining easily with eosin or other acid dyes: *eosinophilic leucocytes.*

2 caused by or having to do with eosinophils: *eosinophilic meningitis.*

epeirogeny (ep′ī roj′ə nē), *n. Geology.* deformation of the earth's crust by which continents and ocean basins are produced. [from Greek *ēpeiros* land, mainland]

—**epeirogenic** (i pī′rō jen′ik) or **epeirogenetic** (i pī′rō-je nət′ik), *adj.* of or having to do with epeirogeny.

ephapse (ef′aps), *n. Physiology.* a place where two lateral axons touch each other. [from Greek *ephapsis* a touching]

—**ephaptic** (ef ap′tik), *adj.* having to do with an ephapse: *ephaptic transmission.*

ephemeral, *adj. Geography.* transitory or of short duration: *an ephemeral stream in an arid region.*

ephemeris (i fem′ər is), *n., pl.* **ephemerides** (ef′ə mer′ə-dēz′). *Astronomy.* a table showing the computed positions of a celestial body at certain intervals of time throughout the year. [from Greek *ephēmeris* diary, calendar]

ephemeris second, *Astronomy.* the fundamental unit of ephemeris time, 1/31556925.9747 of the solar year January 0 1900 at 12 hours.

ephemeris time, *Astronomy.* a standard measure of time, used prior to 1984 in tables giving the positions of the moon and the planets in their orbits. It has been replaced by Terrestrial Dynamic Time. *Abbreviation:* E.T. or ET

epi-, *prefix.* on; up; above; among, as in *epicalyx = on the calyx, epidermis = on or above the dermis.* [from Greek, from *epi* on, at]

epibenthos (ep′ə ben′thos), *n. Biology.* the plant and animal organisms living just above the sea floor. [from *epi-* + *benthos*]

---

cap, fāce, fäther; best, bē, tėrm; pin, five;
rock, gō, ôrder; hot, out; cup, pùt, rüle;
yü in use, yü in uric;
ng in bring; sh in rush; th in thin, ᴛʜ in then;
zh in seizure.

ə = *a* in about, *e* in taken, *i* in pencil, *o* in lemon, *u* in circus

—**epibenthic,** *adj.* of or having to do with the epibenthos.

epibiont (i pib′ē ont), *n. Biology.* an organism that lives on the surface of another without feeding parasitically upon its host: *Many mosses and lichens are epibionts.*

epibiosis (i pib′ē ō′sis), *n. Biology.* a form of symbiosis in which a plant or animal derives support but not nutrition from its host. [from *epi-* + *-biosis,* as in *symbiosis*]

epiblast (ep′ə blast), *n. Embryology.* the outermost of the three layers of the wall of a blastoderm when fully formed; ectoderm. [from *epi-* + Greek *blastos* germ]
—**epiblastic,** *adj.* of or having to do with an epiblast.

epiboly (i pib′ə lē), *n. Embryology.* a process in gastrulation in which the lips of the blastopore grow over the lower hemisphere of the gastrula. Compare **emboly.** [from Greek *epibolē* a throwing or laying on]
—**epibolic** (ep′ə bol′ik), *adj.* of or having to do with epiboly.

epicalyx (ep′ə kā′liks), *n. Botany.* a ring of bracts at the base of a flower that looks like an outer calyx.

epicardium (ep′ə kär′dē əm), *n., pl.* **-dia** (-dē ə). *Anatomy.* the inner layer of the pericardium, which adheres to the heart.
—**epicardial,** *adj.* of or having to do with the epicardium.

epicarp (ep′ə kärp), *n. Botany.* the outer layer of the pericarp of a fruit or ripened ovary of a plant. The skin of a pear is its epicarp. Also called **exocarp.** See the picture at **pericarp.** Compare **endocarp.** [from *epi-* on + Greek *karpos* fruit]

epicenter, *n. Geology.* a point on the earth's surface directly above the focus or true center of an earthquake: *Grand Banks Earthquake, which occurred in 1929, caused a turbidity current that broke cables as far as 300 miles away from the epicenter of the disturbance* (Scientific American).
—**epicentral,** *adj.* of or having to do with an epicenter.

epicondyle (ep′ə kon′dəl), *n. Anatomy.* a projection or protuberance upon the condyle of a bone.

epicotyl (ep′ə kot′l), *n. Botany.* the part of the stem or axis that is above the cotyledons in the embryo of a plant: *The plant embryo consists of the hypocotyl, epicotyl, and the cotyledon(s) ... the epicotyl develops into the leaves and upper portions of the stem* (Biology Regents Syllabus). [from *epi-* + Greek *kotylē* small vessel]

epicritic, *adj. Physiology.* of or having to do with sensory nerve fibers in the skin and mouth that can make fine distinctions in touch and temperature.

epicuticle, *n. Zoology.* 1 the thin, waxy, outermost layer of an arthropod's exoskeleton.
2 the thin outer membrane surrounding a mammal's hairs.
—**epicuticular,** *adj.* having to do with or serving as an epicuticle: *Insects and arachnids possess a discrete epicuticular layer of wax and do not lose much water by evaporation* (New Scientist).

epicycle, *n.* 1 *Geometry.* a circle that rolls around the inside or outside of the circumference of another circle.

2 *Astronomy.* a small circle whose center moves around the circumference of a larger circle. In Ptolemaic astronomy, planets were believed to travel around epicycles.
[from Greek *epikyklos,* from *epi-* upon + *kyklos* circle]
—**epicyclic** (ep′ə sī′klik), *adj.* of or having to do with an epicycle.

epicycloid (ep′ə sī′kloid), *n. Geometry.* a curve traced by a point on the circumference of a circle that rolls on the outside of a fixed circle. Compare **epitrochoid.**
—**epicycloidal** (ep′ə sī kloi′dəl), *adj.* of or having the form of an epicycloid.

epidemic, *Medicine.* —*adj.* occurring extensively; affecting many people at the same time: *an epidemic outbreak of a disease.* —*n.* an epidemic disease. Compare **endemic, pandemic.** [from Greek *epidēmia* prevalence (of a disease), from *epi-* among + *dēmos* people]

epidemiology (ep′ə dē′mē ol′ə jē), *n.* 1 a branch of medicine dealing with the causes, distribution, and control of the spread of diseases in a community.
2 the factors involved in the distribution of a disease: *Recent outbreaks of enteritis, which have been associated with mussels growing near untreated sewage, have shown just how devious the epidemiology of some diseases can be* (New Scientist).
—**epidemiological,** *adj.* of or having to do with epidemiology.

epidermis (ep′ə dèr′mis), *n.* 1 *Zoology.* **a** the outer, protective layer of the skin of vertebrate animals, covering the sensitive dermis. Also called **cuticle. b** any of various outer layers of invertebrates. See the pictures at **hair, skin.**
2 *Botany.* a skinlike layer of cells in seed plants and ferns: *The chief functions of the outer cell layers of the leaf (epidermis) and the waxy covering of the epidermis (cuticle) are: the protection of the internal tissues of the leaf from excessive water loss, resistance to invasion by fungi, and protection from mechanical injury* (Biology Regents Syllabus). See the pictures at **leaf, root.**
—**epidermal** or **epidermic,** *adj.* of or having to do with the epidermis: *epidermal cells.*

epididymis (ep′ə did′ə mis), *n. Anatomy.* the part of the seminal duct lying posterior to the testicle and consisting chiefly of convoluted efferent tubules. [from Greek, from *epi-* on + *didymoi* testicles]
—**epididymal,** *adj.* having to do with the epididymis: *epididymal ducts.*

epidosite (i pid′ə sīt), *n. Geology.* a metamorphic rock composed chiefly of epidote, with quartz and some other minerals in variable amounts: *Epidosite is granular to schistose ... and commonly is very tough as well as hard* (Fenton, The Rock Book).

epidote (ep′ə dōt), *n. Mineralogy.* a hydrous silicate of calcium, iron, and aluminum, usually occurring in yellowish-green, needle-shaped crystals. *Formula:* $Ca_2(Al,Fe)_3Si_3O_{12}OH$ [from French *épidote,* from Greek *epididonai* to increase (so called because it is longer in the base of the crystal than related minerals)]

epidural, *adj. Anatomy.* situated upon or outside the dura mater.

epifauna (ep′ə fō′nə), *n., pl.* **-nas, -nae** (-nē). *Zoology.* the animals living on the surface of sediments, plants, or objects on the sea floor. Compare **infauna.**
—**epifaunal,** *adj.* of, having to do with, or belonging to the epifauna: *epifaunal organisms.*

epigeal (ep′ə jē′əl), *adj. Botany.* **1** (of cotyledons) borne above ground in germination. **2** of, producing, or characterized by epigeal cotyledons. [from Greek *epi-* upon + *gē* earth]

epigene (ep′ə jēn), *adj.* = supergene.

epigenesis (ep′ə jen′ə sis), *n.* **1** *Biology.* the development of the embryo from the substance of the egg by a series of new formations or successive differentiations: *Opposed to the preformation theory is the now generally accepted but modified theory of epigenesis, which states that the structures of the embryo, rather than existing in a preformed condition, develop from the substance of the egg* (Harbaugh and Goodrich, *Fundamentals of Biology*).
2 *Geology.* the transformation that takes place in sedimentary rocks after their lithification, exclusive of weathering and metamorphism.
—**epigenetic** (ep′ə jə net′ik), *adj.* of or produced by epigenesis: *It is an axiom of embryology that development is epigenetic: it is not the mere unfolding of a pre-existing structure, but a process of continuous interaction between the organism and its environment* (S.A. Barnett).

epiglottis (ep′ə glot′is), *n. Anatomy.* a thin, triangular plate of cartilage that covers the entrance to the windpipe during swallowing, so that food and drink do not get into the lungs. See the picture at **mouth.**
—**epiglottal** (ep′ə glot′l), *adj.* of or having to do with the epiglottis.

epigynous (i pij′ə nəs), *adj. Botany.* **1** (of the parts of a flower) appearing to arise from the top of the ovary, which is embedded in the receptacle. **2** (of a flower) having epigynous parts. Contrasted with **hypogynous, perigynous.**
[from *epi-* on + Greek *gynē* female]
—**epigyny,** *n.* the condition of being epigynous; arrangement of floral parts in which the ovary is embedded in the receptacle.

epilimnion (ep′ə lim′nē ən), *n., pl.* **-nia** (-nē ə). *Geology.* the warm, aerated upper layer of water in a lake: *The thermocline is the zone in which the warm temperatures of the epilimnion drop rapidly to the cold ones of the hypolimnion* (Scientific American). [from *epi-* upon + Greek *limnion,* diminutive of *limnē* marsh, lake]

epimer (ep′ə mər), *n. Chemistry.* an isomeric compound, common in certain sugars, that differs from its corresponding isomer in the relative positions of an attached hydrogen and hydroxyl. [from *epi-* + *(iso)mer*]
—**epimeric** (ep′ə mer′ik), *adj.* having the characteristics of an epimer; related as epimers.

epimere (ep′ə mir), *n. Embryology.* the dorsal part of a mesodermal section in chordates. [from *epi-* + *-mere*]

epimysium (ep′ə mis′ē əm), *n., pl.* **-mysia** (-mis′ē ə). *Anatomy.* the sheath of connective tissue around a muscle. [from New Latin, from Greek *epi-* on + *mys* muscle]

epinasty (ep′ə nas′tē), *n. Botany.* the bending downward of a part or organ, caused by more rapid growth or expansion along the upper surface. [from *epi-* + *nast(ic)*]
—**epinastic** (ep′ə nas′tik), *adj.* having to do with or of the nature of epinasty.

epinephrine (ep′ə nef′rən), *n.* = adrenaline. [from *epi-* on, above + Greek *nephros* kidney]

epineurium (ep′ə nůr′ē əm), *n., pl.* **-neuria** (-nůr ē ə). *Anatomy.* the thick sheath of connective tissue surrounding the trunk of a nerve. [from New Latin, from Greek *epi-* on + *neuron* sinew, nerve]
—**epineurial,** *adj.* having to do with or consisting of epineurium.

epiphragm (ep′ə fram), *n.* **1** *Zoology.* the dried mucus with which a snail closes the aperture of its shell during estivation.
2 *Biology.* a membrane closing the mouth of the spore case in certain mosses and fungi.
[from Greek *epiphragma* lid]

epiphysis (i pif′ə sis), *n., pl.* **-ses** (-sēz′). *Anatomy.* **1** the spongy end of a bone which, originally separated from the main bone by a layer of cartilage, ossifies and becomes united to the main bone. **2** = pineal gland. [from *epi-* + Greek *physis* growth]
—**epiphyseal** or **epiphysial** (ep′ə fiz′ē əl), *adj.* having to do with or resembling an epiphysis.

epiphyte (ep′ə fit), *n. Botany.* any of various plants that grow on other plants for support, but draw nourishment from the air and rain instead of from their host. Many mosses and orchids are epiphytes. Also called **aerophyte, air plant.** [from *epi-* on + Greek *phyton* plant]
—**epiphytic** (ep′ə fit′ik), *adj.* growing as an epiphyte; having the characteristics of an epiphyte: *In some instances, as in Vanilla and other epiphytic Orchids, these roots ... hang down free in the air and absorb water from rain by means of their several-layered epidermis* (Youngken, *Pharmaceutical Botany*).

epiphytotic (ep′ə fī tot′ik), *Botany.* —*adj.* (of a disease) occurring or spreading like an epidemic among plants.
—*n.* an epiphytotic disease.

epiplankton (ep′ə plangk′tən), *n. Biology.* the part of the plankton which occurs between the surface of the sea and a depth of about one hundred fathoms.
—**epiplanktonic** (ep′i plangk ton′ik), *adj.* of or having to do with epiplankton.

episome (ep′ə sōm), *n. Genetics.* a genetic particle in cells, especially bacterial cells, that can either exist autonomously in the cytoplasm or be incorporated into the chromosomes. Compare **plasmid.** [from *epi-* + *-some*]
—**episomal** (ep′ə sō′məl), *adj.* of or having to do with an episome.

---

cap, fāce, fäther; best, bē, tėrm; pin, five;
rock, gō, ôrder; oil, out; cup, pùt, rüle,
yü in use, *yu* in uric;
ng in bring; *sh* in rush; *th* in thin; ᴛʜ in then;
zh in seizure.
ə = *a* in about, *e* in taken, *i* in pencil, *o* in lemon, *u* in circus

epistasis (i pis'tə sis), *n., pl.* **-ses** (-sēz'). *Genetics.* any interaction of nonallelic genes, especially the suppression by one gene of the effect of a nonallelic gene.
—**epistatic** (ep'ə stat'ik), *adj.* of, exhibiting, or caused by epistasis.

epitaxy (ep'ə tak'sē), *n. Crystallography.* the growth of crystals of one mineral on the crystal face of another, in an orientation which retains the structural orientation of the substrate. [from *epi-* + Greek *taxis* arrangement]
—**epitaxial** (ep'ə taks'ē əl), *adj.* of or formed by epitaxy.

epithelium (ep'ə thē'lē əm), *n., pl.* **-liums, -lia** (-lē ə). *Anatomy.* a thin layer of cells forming a tissue which covers the internal and external surfaces of the body, and which performs protective, secretive, or other functions. [from New Latin, from Greek *epi-* on + *thēlē* nipple]
—**epithelial** (ep'ə thē'lē əl), *adj.* of or having to do with the epithelium: *In the normal lung, epithelial tissue defends the lung against foreign invaders such as dust and bacteria* (Edwin Diamond).

epithet (ep'ə thet), *n. Biology.* the part of a scientific name of an animal or plant which denotes a species, variety, or other division of a genus. In *Pyrus malus* (the common apple), *malus* is the specific epithet. [from Greek *epitheton* added, from *epi-* on + *tithenai* to place]

epitope (ep'ə tōp), *n. Immunology.* the part of an antigen molecule that by its structure determines the specific antibody molecule that will attach to it: *Molecules that display epitopes are called antigens.* [from *epi-* on + Greek *topos* place]

epitrichium (ep'ə trik'ē əm), *n. Embryology.* a thin membrane which overlies the epidermis and hair during fetal life, usually disappearing before birth. [from New Latin, from *epi-* + Greek *trichos* a hair]

epitrochoid (ep'i trô'koid), *n. Geometry.* a curve traced by any point fixed in relation to a circle that rolls on the outside of a fixed circle. Compare **epicycloid.** [from *epi-* + Greek *trochos* wheel]
—**epitrochoidal** (ep'i trô koi'dəl), *adj.* in the form of an epitrochoid.

epizoic (ep'ə zō'ik), *adj. Biology.* living as an external parasite or commensal.
—**epizoite** (ep'ə zō'īt) or **epizoon** (ep'i zō'ən), *n.* an epizoic organism.

epizootic (ep'ə zō ot'ik), *Zoology.* —*adj.* (of a disease) occurring or spreading like an epidemic among animals.
—*n.* an epizootic disease.

epoch (ep'ək *or* ē'pok), *n.* **1** *Geology.* one of the divisions of time into which a period is divided: *the Recent epoch of the Quaternary period.*
2 *Astronomy.* an arbitrarily chosen date or instant of time used as a reference point.
[from Greek *epochē* a stopping, fixed point in time, from *epi-* on + *echein* to hold]

epoxide (i pok'sīd), *n. Chemistry.* an epoxy compound.

epoxy (i pok'sē), *adj. Chemistry.* of or designating a large group of compounds containing a triangular structure in which oxygen acts as a bridge between two atoms already bonded together. **Epoxy resins** are syn-

thetic resins derived by polymerization of epoxy compounds and used in the manufacture of plastics, adhesives, etc.
[from *ep(i)-* on + *oxy(gen)*]

epsomite (ep'sə mīt), *n. Mineralogy.* a mineral consisting of hydrated magnesium sulfate; native Epsom salt. *Formula:* $MgSO_4 \cdot 7H_2O$

Epsom salts or **Epsom salt** (ep'səm), *Chemistry.* a bitter, colorless or white crystalline powder consisting of hydrated magnesium sulfate, used in medicine as a cathartic and to reduce inflammation. *Formula:* $MgSO_4 \cdot 7H_2O$ [named after *Epsom,* Surrey, England, where it was originally prepared from the water of mineral springs]

Epstein-Barr virus, *Biology.* a herpesvirus associated with various types of human cancers: *The Epstein-Barr (EB) virus which was first associated with Burkitt's lymphoma, then with infectious mononucleosis, then with a particular type of nasopharyngeal cancer, had recently been linked to leukemia* (Herbert Kondo). *Abbreviation:* EBV Also called **EB virus.** [named after M. A. *Epstein* and Y. M. *Barr,* 20th-century British virologists]

equal sets, *Mathematics.* sets that have the same members, regardless of their arrangements. (2, 7, 9) and (7, 9, 2) are equal sets. The set of even whole numbers and the set of whole numbers whose squares are even are equal sets.

equation, *n.* **1** *Mathematics.* a statement of the equality of two quantities, usually indicated by using the sign = between them. EXAMPLES: $(4 \times 8) + 12 = 44$; $C = 2\pi r.$
2 *Chemistry.* an expression using chemical formulas and symbols to show quantitatively the substances used and produced in a chemical change. In $HCl + NaOH = NaCl + H_2O$, HCl and NaOH are the reacting substances, and NaCl and H_2O are the resulting products, the sign = being read "produces" or "gives."

equation of state, *Physics.* an equation that relates the temperature, volume, and pressure of a gas, taking into account the forces between the molecules of the gas.

equation of time, *Astronomy.* the difference between mean solar time and apparent solar time.

equator, *n.* **1** *Geography.* an imaginary circle around the middle of the earth, halfway between the North Pole and the South Pole: *The earth's diameter is 7,927 miles at the equator and only 7,900 miles from pole to pole* (Norman D. Anderson). See the picture at **latitude.**
2 *Astronomy.* **a** a similarly situated circle on any celestial body. **b** = celestial equator. See the picture at **great circle.**
3 *Biology.* = equatorial plane.
[from Medieval Latin *aequator (diei et noctis)* equalizer (of day and night), from Latin *aequare* make equal, from *aequus* equal]
—**equatorial,** *adj.* of, at, or near the equator: *The photosphere does not rotate at a uniform rate but rotates faster at the equator—a phenomenon now known as the equatorial acceleration* (Scientific American).

equatorial plane, **1** *Astronomy.* the plane in which the celestial equator lies.
2 *Biology.* the plane lying midway between the poles of a dividing cell: *During this phase of mitosis [metaphase] the split chromosomes become located in*

the equatorial plane of the spindle (Hegner and Stiles, College Zoology).

equatorial plate, Biology. the site or position of the chromosomes in the equatorial plane during metaphase.

equi-, combining form. **1** equal, as in equidistance = equal distance. **2** equally, as in equidistant = equally distant. [from Latin aequus equal]

equiangular (ē′kwē ang′gyə lər), adj. Geometry. having all angles equal: A square is equiangular.

equidistant, adj. Mathematics. equally distant: All points of the circumference of a circle are equidistant from the center. **—equidistantly,** adv.

equilateral (ē′kwə lat′ər əl), adj. Geometry. having all sides of equal length: Every equilateral triangle is equiangular.

equilibrium (ē′kwə lib′rē əm), n., pl. **-riums, -ria** (rē ə). **1** Physics. a state of balance; condition in which opposing forces exactly balance or equal each other: The first condition of equilibrium is that if a body is in equilibrium, the resultant of all forces acting upon it is zero ... The second necessary condition for equilibrium is that the resultant of all torques acting upon the body must be zero (Glathart, College Physics). **2** Chemistry. the state of a chemical system when no further measurable change occurs in it: Ethyl alcohol and acetic acid react together to form ethyl acetate and water. But ethyl acetate and water also react together forming ethyl alcohol and acetic acid. In such reversible reactions, it is found that after a certain time, depending on the conditions, no further change appears to occur even though each pair of reactants is still present. In fact, a 'dynamic equilibrium' is set up, the concentrations of the components being such that both reactions proceed at the same rate (K. D. Wadsworth). [from Latin aequilibrium, from aequus equal + libra balance]

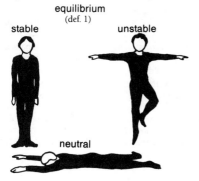

equilibrium
(def. 1)

stable   unstable

neutral

equilibrium constant, Chemistry. a constant that is characteristic of the equilibrium reached by a given reversible chemical reaction at a given temperature, equal to the product of the molar concentrations of the substances on the right side of the equation divided by the product of the molar concentrations of the substances on the left side. Symbol: K

equimolal (ē′kwə mō′ləl), adj. Chemistry. having equal molal concentrations.

equimolar (ē′kwə mō′lər), adj. Chemistry. having equal molar concentrations: The solution contained equimolar amounts of α-ketobutyric acid and sodium hydroxide (Lancet).

equinoctial (ē′kwə nok′shəl), adj. Astronomy, Geography. **1** of, having to do with, or occurring at an equinox: an equinoctial storm. **Equinoctial points** are the two imaginary points in the sky where the sun crosses the celestial equator.

2 at or near the earth's equator: Borneo is an equinoctial island.

—n. 1 = celestial equator. **2** a storm occurring at or near the equinox.

equinoctial line, Astronomy. = celestial equator.

equinox (ē′kwə noks), n. Astronomy. **1** either of the two times in the year when the center of the sun crosses the celestial equator, and day and night are of equal length in all parts of the earth, occurring in the Northern Hemisphere about March 21 (**vernal equinox**) and about September 23 (**autumnal equinox**).

2 either of the two imaginary points in the sky at which the sun's path crosses the equator.

[from Medieval Latin equinoxium, from Latin aequinoctium, from aequus equal + nox, noctem night]

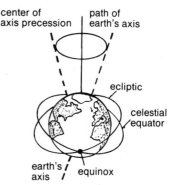

EQUINOX (def. 1)

center of axis precession   path of earth's axis

ecliptic

celestial equator

earth's axis   equinox

equipotential (ē′kwə pə ten′shəl), adj. Physics. possessing the same or equal potential: The potential distribution in an electric field may be represented graphically by equipotential surfaces. (Sears and Zemansky, University Physics).

equivalence point, Chemistry. a condition in which the substances required for a reaction are all present in precisely the amounts required for a complete reaction. Compare **end point.**

equivalent, adj. **1** Mathematics. **a** having the same set of solutions. $2x - 3 = 0$ and $3x - 2 = x$ are equivalent equations. **b** that can be put into one-to-one correspondence with each other. If the number of chairs in a room is the same as the number of people in the room, the set of chairs and the set of people are equivalent sets.

2 *Geometry.* equal in extent, but not having the same form: *A triangle and a square of equal area are equivalent.*

3 *Chemistry.* equal in combining or reacting value to a (stated) quantity of another substance. —*n.* something equivalent. —**equivalently,** *adv.*

equivalent weight, *Chemistry.* the weight of a substance which will combine with or can replace one mole of hydrogen atoms or one half mole of oxygen atoms; weight associated with the transfer of one mole of protons or electrons: *A mass equal to the atomic weight divided by the valence is called an equivalent weight* (Sears and Zemansky, *University Physics*). *Equivalent weights are simply measures of the characteristic proportions in which given elements combine, and are determined directly by ordinary chemical reaction and quantitative analysis ...* (H.R. Paneth). Also called **combining weight.**

Er, *symbol.* erbium.

era, *n. Geology.* one of the major divisions of geologic time. An era is usually divided into periods. *Three eras bounded by relatively profound, sudden, and worldwide changes in the living organisms preserved as fossils are recognized: the Paleozoic ("ancient life") Era, the Mesozoic ("middle life") Era, and the Cenozoic ("recent") Era* (Eicher and McAlester, *History of the Earth*). [from Late Latin *era,* variant of *aera* number, epoch, probably the same word as Latin *aera* counters (for reckoning), plural of *aes, aeris* copper, brass, money]

erbium (ẻr'bē əm), *n. Chemistry.* a soft, lustrous, grayish, rare-earth metallic element which occurs with yttrium or as a minute part of various minerals, used in nuclear research. *Symbol:* Er; *atomic number* 68; *atomic weight* 167.26; *oxidation state* 3. [from New Latin, from (*Ytt*)*erby,* town in Sweden where it was found]

erectile, *adj. Physiology.* that can become erect and rigid by distention, as certain animal tissues.

erection, *n. Physiology.* the state of a bodily organ or part in which the erectile tissue has become distended and rigid by the accumulation of blood.

erector, *n. Anatomy.* a muscle that erects a part.

E region, = E layer.

erepsin (i rep'sən), *n. Biochemistry.* an enzyme complex that breaks down protein molecules into amino acids, found in the intestinal and pancreatic juices. [from Latin *ereptum* set free + English (*pep*)*sin*]

erg[1], *n.* a unit for measuring work or energy in the CGS system, equivalent to the amount of work done by a force of one dyne acting through a distance of one centimeter: *An erg is a very small amount of energy: it is about the amount that a moderately slow mosquito transfers when it collides with your forehead* (Struve, *New Astronomy*). *Symbol:* e Also called **dyne-centimeter.** [from Greek *ergon* work]

erg[2], *n. Geography.* a type of desert region having extensive areas of active dunes. Compare **reg.** [from French, from Arabic (North Africa) *'irj*]

ergastoplasm (ẻr gas'tə plaz əm), *n. Biology.* cytoplasmic material that stains readily with a basic dye, especially the endoplasmic reticulum. [from Greek *ergastikos* able to work + *plasma* something formed]

ergosterol (ẻr'gos'tə rōl'), *n. Biochemistry.* a steroid alcohol or sterol derived especially from fungous growths, as those of ergot and yeast, that turns into vitamin D when exposed to ultraviolet light and can be used for the prevention or curing of rickets. *Formula:* $C_{28}H_{44}O$ [from *ergo*(*t*) + (*chole*)*sterol*]

erode, *v. Geology.* **1** to wear away gradually by erosion: *Running water erodes soil and rocks. Much of the land is eroded both because rain has washed the topsoil off the steep inclines and because the peasants and their goats have killed the forests* (Henry Giniger). **2** to form by a gradual wearing away: *The stream eroded a channel in the solid rock.* [from Latin *erodere,* from *ex-* away + *rodere* gnaw]

erose (i rōs'), *adj. Botany.* (of leaves) having the margin irregularly indented, as if gnawed away. [from Latin *erosum* gnawed away]

erosion, *n. Geology.* a gradual wearing away by the action of glaciers, running water, waves, or wind: *Forests can do much to check soil erosion and halt the spread of sand dunes that are menacing fertile farmland* (Herbert L. Edlin).
—**erosional** (i rō'zhə nəl), *adj.* produced by or causing erosion: *The energy of the sun ... drives the erosional forces that wear away the continents* (R. Dana Russell).

ASSOCIATED TERMS: *Erosion* is the general antonym of *deposition.* When rock is eroded by the scraping action of other rocks, it may be called *corrasion* if the rocks are carried by wind or water, or *abrasion* if they are moved by a glacier, although abrasion may also refer to other types of wear. Related words are *ablation,* usually referring to the melting and weathering of a glacier, and *attrition,* meaning the wearing down and decrease in size of rock particles moved by wind or water.

erosion surface, *Geology, Geography.* a surface of land shaped by the action of rain, wind, ice, and other agents of erosion. See the picture at **unconformity.**

erratic, *Geology.* —*adj.* transported from its original site, as by glacial action: *an erratic fragment.* —*n.* erratic boulder or mass of rock.

error, *n. Mathematics.* the difference between the approximate value of a quantity and its exact or true value: *In the approximate number 18.3, the maximum absolute error is 0.05, since 18.3 cannot be less than 18.25 or greater than 18.35. The relative error is the quotient of the absolute error divided by the true value* (Paul R. Rider).

erupt, *v.* **1** *Geology.* to eject material at the earth's surface, as a volcano or geyser.
2 *Physiology.* (of teeth) to break through the gums.
—**eruption,** *n.* the act or process of erupting.
—**eruptive,** *Geology.* —*adj.* of or formed by volcanic eruptions.
—*n.* a rock formed by a volcanic eruption: *Basalts, rhyolites, andesites, and tuffs are examples of eruptives* (White and Renner, *Human Geography*).

erythrite (i rith'rīt), *n. Mineralogy.* a hydrated, usually red-colored cobalt arsenate. *Formula:* $Co_3(AsO_4)_2$·$8H_2O$ Also called **cobalt bloom.** [from Greek *erythros* red]

erythroblast (i rith'rō blast), *n. Anatomy.* a nucleated cell, found in bone marrow, from which red blood cells develop. [from Greek *erythros* red + *blastos* sprout]

erythrocyte (i rith′rō sīt), *n.* = red blood cell. Contrasted with **leukocyte**. [from Greek *erythros* red + English *-cyte*]
—**erythrocytic** (i rith′rō sit′ik), *adj.* of or having to do with erythrocytes.

erythroid (i rith′roid), *adj. Physiology.* of or having to do with red blood cells.

erythropoiesis (i rith′rō poi ē′sis), *n. Physiology.* the production of red blood cells. [from Greek *erythros* red + *poiēsis* production]
—**erythropoietic** (i rith′rō poi et′ik), *adj.* having to do with the production of red blood cells.

erythropoietin (i rith′rō poi′ə tin), *n. Biochemistry.* a hormone, produced by the kidneys, that controls the production of red blood cells.

Es, *symbol.* einsteinium.

escape, *Botany.* —*v.* to grow wild; slip away from cultivation and become established in fields and by roadsides.
—*n.* a cultivated plant that grows wild: *The site was filled with "escapes" from other parts of the world-winged euonymus, garlic mustard, Japanese honeysuckle, Oriental bittersweet* (New Yorker).

escape velocity, *Physics.* the minimum speed that a moving object must reach to leave the gravitational field of the earth or other attracting body. A rocket traveling at 40,000 kilometers per hour can escape the pull of the earth without further power.

escarpment (e skärp′mənt), *n. Geography, Geology.* 1 a steep slope or descent. 2 a steep face marking the termination of high land or of stratified rocks. [from French]

Escherichia coli (esh′ə rik′ē ə kō′lī), = E. coli.

esker (es′kər), *n. Geolgraphy, Geology.* a winding ridge of sand, gravel, etc., believed to have been formed by sedimentation from streams flowing under or in glacial ice. See the picture at **drumlin.** [from Irish *eiscir*]

esophageal (i sof′ə jē′əl), *adj. Anatomy.* of or connected with the esophagus.

esophagus (i sof′ə gəs), *n., pl.* **-gi** (-jī). *Anatomy.* 1 the tubular passage for food from the pharynx to the stomach in vertebrates; the gullet. 2 an analogous tube in certain invertebrates, such as insects. See the pictures at **alimentary, mouth.** [from Greek *oisophagos,* from *oiso-* carry + *phagein* eat]

essence, *n. Chemistry.* 1 any concentrated preparation that has the characteristic flavor, fragrance, or effect of the plant, fruit, drug, or the like, from which it was obtained: *Atropine is the essence of the belladonna plant.* 2 a solution of such a substance in alcohol: *Essence of peppermint is oil of peppermint dissolved in alcohol.*

essential amino acid. See **amino acid.**

ester, *n. Chemistry.* any of a group of organic compounds produced by the reaction of an acid and an alcohol, so that the acid hydrogen of the acid is replaced by the hydrocarbon radical of the alcohol. Animal and vegetable fats and oils are esters. *Esters usually have pleasant odors. The aromas of many fruits, flowers, and perfumes are due to esters. Fats are esters derived from glycerol and long-chain organic acids* (Chemistry Regents Syllabus). [from German *Ester,* from *Essigäther* vinegar ether]

—**esterification,** *n.* formation of an ester: *Esterification is a reversible, molecular reaction which requires the presence of a dehydrating agent ... to remove the water so that the reaction can go in the desired direction* (Offner, *Fundamentals of Chemistry*).
—**esterify** (es ter′ə fī), *v.* to change or be changed into an ester.

esterase (es′tə rās), *n. Biochemistry.* any ferment or enzyme that can decompose an ester into an acid and alcohol (with the addition of a molecule of water).

estivate (es′tə vāt), *v. Zoology.* to spend the summer in a dormant or torpid condition. Some snakes and rodents estivate. Also spelled **aestivate.** Compare **hibernate.** [from Latin *aestivus* of the summer, from *aestas* summer]
—**estivation,** *n.* the act or process of estivating.

estradiol (es′trə dī′ol), *n. Biochemistry.* an estrogen that controls the growth of the sexual organs and some functions of the uterus. *Formula:* $C_{18}H_{24}O_2$

estrin, *n. Biochemistry.* 1 = estrone. 2 = estrogen.

estriol (es′trē ol), *n. Biochemistry.* an estrogen that causes or promotes estrus, found in the urine of pregnant females. It is a phenol alcohol. *Formula:* $C_{18}H_{24}O_3$

estrogen (es′trə jən), *n. Biochemistry.* any of various hormones which induce a series of physiological changes in females, especially in the reproductive or sexual organs: *The female sex hormones estrogen and progesterone ... regulate the development of secondary sex characteristics such as the development of the mammary glands and the broadening of the pelvis* (Biology Regents Syllabus). [from *estrus* + *-gen*]
—**estrogenic,** *adj.* 1 causing or promoting estrus. 2 of an estrogen or estrogens.

estrone, *n. Biochemistry.* an estrogen that causes or promotes estrus and stimulates growth of the reproductive organs. *Formula:* $C_{18}H_{22}O_2$

estrous (es′trəs), *adj. Zoology.* of, having to do with, or being in estrus: *an estrous female.*

estrous cycle, *Biology.* the recurrent bodily changes in the sexual and other organs connected with the estrus in the female of most mammals.

estrus (es′trəs), *n. Zoology.* 1 a periodically recurring state of sexual activity in most female mammals during which mating may take place. 2 = estrous cycle. [from New Latin, from Latin *oestrus* frenzy, gadfly, from Greek *oistros* gadfly]

estuary (es′chü er′ē), *n. Geography, Geology.* a broad mouth of a river into which the tide flows, especially an inlet of the sea at the lower part of a river. SYN: firth. [from Latin *aestuarium,* from *aestus* tide]
—**estuarine,** *adj.* 1 of or having to do with an estuary. 2 formed or deposited in an estuary.

---

cap, fāce, fäther; best, bē, tèrm; pin, fīve;
rock, gō, ôrder; oil, out; cup, pút, rüle;
*y*ü in use, *y*u in uric;
ng in bring; sh in rush; th in thin, ᴛн in then;
zh in seizure.
ə = a in about, e in taken, i in pencil, o in lemon, u in circus

203

e.s.u. or **ESU**, *abbrev.* electrostatic unit *or* electrostatic units.

E.T. or **ET**, *abbrev.* ephemeris time.

ethane (eth′ān), *n. Chemistry.* a colorless, odorless, flammable, gaseous hydrocarbon present in natural gas, coal gas, and crude oil. It is used as a refrigerant and as a fuel. *Formula:* C_2H_6 [from *ether*]

ethanol (eth′ə nōl), *n.* = ethyl alcohol.

ethene (eth′ēn), *n.* = ethylene.

ether (ē′thər), *n. Chemistry.* **1** a colorless, volatile, flammable, sweet-smelling liquid, produced by the action of sulfuric acid on ethyl alcohol. Because its fumes cause unconsciousness when deeply inhaled, ether was once widely used as an anesthetic. Ether is also used as a solvent for fats and resins. *Formula:* $(C_2H_5)_2O$ Also called **ethyl ether.**
2 any of a group of organic compounds consisting of two hydrocarbon groups linked by an oxygen atom. Ethers are formed by the action of acids on alcohols. [from Latin *aether,* from Greek *aithēr* upper air]

ethionine (eth ī′ə nēn), *n. Biochemistry.* an amino acid, the ethyl homologue of methionine, which inhibits the production of protein by tissue cells. *Formula:* $C_6H_{13}NO_2S$ [from *eth(yl)* + *-ionine,* as in *methionine*]

ethmoid (eth′moid), *Anatomy.* —*adj.* having to do with a bone situated in the walls and septum of the nose and containing numerous performations for the filaments of the olfactory nerve.
—*n.* the ethmoid bone.
[from Greek *ēthmoeidēs,* from *ēthmos* sieve + *eidos* form]

ethoxide, *n.* = ethylate.

ethoxyl (eth ok′səl), *n. Chemistry.* a univalent group, C_2H_5O-, consisting of an ethyl combined with an atom of oxygen.

ethyl (eth′əl), *n. Chemistry.* a univalent radical present in many organic compounds, such as ordinary alcohol. *Formula:* $-C_2H_5$ [from *eth(er)* + *-yl*]

ethyl alcohol, *Chemistry.* ordinary alcohol, a colorless, volatile liquid made by the fermentation of grain, sugar, etc.: *Formula:* C_2H_5OH Also called **ethanol.**

ethylate (eth′ə lāt), *Chemistry.* —*v.* to treat (a compound) so as to add one or more ethyl radicals.
—*n.* a compound derived from ethyl alcohol by the replacement of the hydroxyl hydrogen with a metal.

ethylene (eth′ə lēn′), *n. Chemistry.* a colorless, flammable gas with an unpleasant odor, used as an anesthetic, in making organic compounds, and for coloring and ripening citrus fruits. *Formula:* C_2H_4

ethylene glycol, = glycol.

ethylene series, *Chemistry.* a series of unsaturated, open-chain (aliphatic) hydrocarbons containing one or more double bonds and having the general formula C_nH_{2n}. Also called **alkene series, olefin series.** Compare **methane series.**

ethyl ether, = ether (def. 1).

etiolation (ē′tē ə lā′shən), *n. Botany.* a condition of plants characterized by an elongated stem, small leaves, and lack of chlorophyll, due to the absence or reduction of light. [from French *étioler* to blanch]
—**etiolate,** *v.* **1** to undergo etiolation.
2 to cause etiolation in (a plant).

etiology (ē′tē ol′ə jē), *n.* **1** the part of any science that deals with the causes of its phenomena.
2 *Medicine.* **a** the study of the causes or origins of diseases. **b** the cause or causes of a disease.
[from Greek *aitiologia,* from *aitia* cause + *-logia* -logy]
—**etiologic** (ē′tē ə loj′ik) or **etiological** (ē′tē ə loj′ə kəl), *adj.* of or having to do with etiology: *There is little or no evidence that tobacco is an etiologic (causative) factor in ... diseases of the upper gastrointestinal tract* (Science News Letter). —**etiologically,** *adv.*

eu-, *prefix.* good; well; true, as in *eukaryote, euphenics, euthenics.* [from Greek, from *eus* good]

Eu, *symbol.* europium.

eucaryote (yü kar′ē ōt), *n.* = eukaryote.

euchromatin (yü krō′mə tin), *n. Biology.* the active chromatin of a chromosome during the interphase.

euclase (yü′klās), *n. Mineralogy.* a brittle, usually light-green mineral, a silicate of aluminum and beryllium, occurring in prismatic crystals. *Formula:* $BeAlSiO_4(OH)$ [from *eu-* + Greek *klasis* breaking]

Euclidean or **Euclidian** (yü klid′ē ən), *adj.* of or according to Euclid or his principles. **Euclidean geometry** is based on the axioms and postulates described by Euclid, especially the postulate that if a point lies outside a straight line, only one line parallel to it can be drawn through the point; this postulate is not held by **non-Euclidean** geometry. *In Euclidean space the volume of a sphere increases as the cube of its radius.*

eucrite (yü′krīt), *n. Geology.* a variety of gabbro composed essentially of anorthite and augite. [from Greek *eukritos* easily distinguished]

eugeosyncline (yü jē′ō sin′klīn), *n. Geology.* a long, narrow geosyncline in which volcanic rocks abound.
—**eugeosynclinal,** *adj.* of or having to do with a eugeosyncline.

euglena (yü glē′nə), *n. Biology.* any of a genus (*Euglena*) of one-celled flagellates, often classified as algae, that are spindle-shaped and usually green, with one flagellum. They are easily grown in a culture for study. [from New Latin, from Greek *eu-* good + *glēnē* pupil of the eye]

euglenoid (yü glē′noid), *Biology.* —*adj.* of, having to do with, or like a euglena: *Euglenas possess a definite shape but are characterized by wormlike movements involving waves of contraction to which the term euglenoid movement has been applied* (Hegner and Stiles, *College Zoology*).
—*n.* a euglenoid organism.

euglobulin (yü glob′yə lin), *n. Biochemistry.* any globulin soluble in saline solutions but insoluble in water. Compare **pseudoglobulin.**

euhedral (yü hē′drəl), *adj. Mineralogy.* (of a crystal) bounded by plane faces; well-formed. Contrast with **anhedral.**

eukaryote (yü kar′ē ōt), *n. Biology.* a cell or organism having a distinct nucleus or nuclei bound by a membrane: *It has become clear that the most drastic discontinuity among living organisms is ... the separation of those organisms with an organized nucleus (eukaryotes) from those without an organized nucleus (prokaryotes)* (Peter H. Raven). Also spelled **eucaryote.** [from *eu-* good, true + Greek *karyon* nut, kernel]

—eukaryotic, *adj.* **1** containing a distinct nucleus or nuclei bound by a membrane: *eukaryotic cells.*

2 of or belonging to the group of organisms containing distinct, membrane-bound nuclei: *All multicellular plants and animals are eukaryotic.*

Euler's circles (oi′lərz), *Mathematics.* circles that show relationships between mathematical sets. [named after Leonhard *Euler,* 1707–1783, Swiss mathematician, who first devised them]

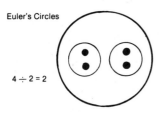

Euler's Circles

$4 \div 2 = 2$

Euler's diagram, *Mathematics.* a graphic representation of logical relations in which circles or ovals are used to represent classes of individuals to which certain predicates apply. [see etymology under **Euler's circles**]

euphotic (yü fot′ik), *adj. Ecology.* of, having to do with, or designating an upper layer of water that receives sufficient light to permit photosynthesis: *The euphotic, or illuminated, zone may be only a few centimeters deep in a very turbid river, or well over 100 meters deep in the clearest parts of the ocean* (Scientific American).

euploid (yü′ploid), *adj. Genetics.* having a number of chromosomes which is a multiple of the haploid number for the species. [from *eu-* + *-ploid,* as in *haploid, diploid*]

europium (yù rō′pē əm), *n. Chemistry.* a soft grayish, rare-earth metallic element which occurs only in combination with other elements. Europium compounds are used in the control rods of nuclear reactors. *Symbol:* Eu; *atomic number 63; atomic weight 151.96; melting point 826°C; boiling point 1489°C; oxidation state 3, 2.* [from Latin *Europa* Europe]

eury-, *combining form.* wide; broad; as in *eurythermal, euryhaline.* [from Greek *eurys*]

eurybathic (yùr′ə bath′ik), *adj. Biology.* able to live in a broad range of depths of water. [from *eury-* + Greek *bathos* depth]

euryhaline (yùr′ə hā′lin), *adj. Biology.* able to tolerate a wide range of salinity.

eurythermal (yùr′ə thèr′məl) or **eurythermic** (yùr′ə-thèr′mik), *adj. Biology.* able to tolerate a wide range of temperature.

—eurytherm (yùr′ə thèrm′), *n.* a eurythermal organism.

eurytopic (yùr′ə top′ik), *n. Biology.* able to tolerate a wide range of variations in environmental conditions. [from *eury-* + Greek *topos* place]

Eustachian tube (yü stā′kē ən *or* yü stā′shən), *Anatomy.* a slender canal between the pharynx and the middle ear, which equalizes the air pressure on the two sides of the eardrum. See the picture at **ear.** [named after Bartolommeo *Eustachio,* 1520?–1574, Italian anatomist]

eustasy or **eustacy** (yü′stə sē), *n. Geology.* world-wide changes in sea level, such as are caused by the growth or melting of glacial ice. [from *eu-* + Greek *stasis* a standing]

—eustatic (yü stat′ik), *adj.* of or characterized by eustasy: *That eustatic fall in level laid bare the bed of the north sea* (London *Times*).

eutectic (yü tek′tik), *Chemistry.* **—adj. 1** that may be easily melted, as an alloy or mixture whose melting point is lower than that of any other alloy or mixture composed of the same ingredients.

2 of or having to do with such alloys or mixtures.

—n. a eutectic alloy or mixture.

[from Greek *eutēktos* easily melting, from *eu-* well + *tēkein* melt]

eutherian (yü thir′ē ən), *Zoology.* **—adj.** of or belonging to a subclass or infraclass (*Eutheria*) comprising the placental mammals.

—n. a eutherian mammal.

[from New Latin *Eutheria,* from Greek *eu-* good, true + *thēria* beasts]

eutrophic (yü trof′ik), *adj. Ecology.* (of lakes, streams, etc.) rich in phosphates, nitrates, and other nutrients that promote the growth of algae, which deplete the water of oxygen and cause the extinction of other organisms. Compare **dystrophic.** [from Greek *eutrophos* thriving, from *eu-* well + *trephein* nourish]

—eutrophicate, *v.* to make or become eutrophic.

—eutrophication, *n.* the process of making or becoming eutrophic: *Eutrophication literally means "nourishing well," but in current usage it refers to the inadvertent nourishing of algae in lakes to the detriment of other living things* (Scientific American).

eV, *abbrev.* electron volt.

evaginate (i vaj′ə nāt), *v. Biology.* to turn inside out or cause to protrude by eversion, as a tubular organ.

—evagination, *n.* **1** the process of evaginating or state of being evaginated: *The development of the several organs from the ectoderm and endoderm is in its early stages a bending or folding of these layers, which is called invagination or evagination according as the sheets of cells are bent into, or out from, some enclosed space* (Shull, *Principles of Animal Biology*).

2 an evaginated part.

evaporate, *v. Chemistry.* **1** to change from a liquid into a vapor: *Boiling water evaporates rapidly. Moisture will always evaporate from any wet surface in contact with air that is not fully saturated with water vapor* (Emerson, *Basic Botany*). **2** to remove water or other liquid from: *Heat is used to evaporate milk.*

—evaporation, *n.* the process of evaporating or condition of being evaporated: *The effectiveness of any rainfall is controlled by the rate of evaporation* (R.N. Elston).

cap, fāce, fäther; best, bē, tèrm; pin, five;
rock, gō, ôrder; oil, out; cup, pùt, rüle,
yü in use, *yù* in uric;
ng in bring; *sh* in rush; *th* in thin, ŦH in then;
zh in seizure.

ə = *a* in about, *e* in taken, *i* in pencil, *o* in lemon, *u* in circus

evaporite (i vap′ə rit), *n. Geology.* a sedimentary deposit formed as a result of the evaporation of seawater or saline lake water: *The rich Middle East and Texan oil fields are closely associated with formations bearing thick deposits of salt and related evaporites* (New Scientist).

evapotranspiration (i vap′ō tran spə rā′shən), *n. Botany.* the loss of water to the atmosphere through its evaporation from the soil and by the transpiration of plants growing on land: *Evapotranspiration is defined as the total water lost from a soil and from the leaves of plants growing in the soil* (Grant W. Thomas).
—**evapotranspire**, *v.* to cause the loss of (water) from soil by evaporation and transpiration.

even, *adj. Mathematics.* that can be divided by 2 without a remainder: *2, 4, 6, 8, and 10 are even numbers. If n is an even integer, then n = 2k, where k is an integer.* Compare **odd.**

evening star, *Astronomy.* the planet Venus, as seen in the western sky at or soon after sunset. Compare **morning star.**

event, *n. Physics.* **1** any interaction between two or more substances, such as a collision between nuclear particles.
2 a point in space-time.

event horizon, *Astronomy.* the boundary of a black hole: *If we drop things radially into a black hole we shall never quite see them penetrate the event horizon ... Someone falling in however would pass through the event horizon in what to him would be a perfectly finite time* (New Scientist).

evergreen, *Botany.* —*adj.* having green leaves or needles all year round. In trees of this kind, the leaves or needles remain on the tree until the new ones are completely formed, as in the holly or the pine.
—*n.* an evergreen plant or tree, such as the pine, spruce, cedar, and ivy.

evoked potential, *Physiology.* the electrical response arising from the cortex of the brain upon stimulation of a sense organ: *A pattern presented to the eye gives rise to an electrical response, known as the evoked potential, the amplitude of which is an index of the effectiveness with which the input has been transmitted to the brain* (New Scientist). *Because they are objective indicators of certain brain functions evoked potentials have clinical applications. ... Evoked potentials can be used to detect deafness in infants, which is otherwise quite difficult to diagnose* (Scientific American).

evolute (ev′ə lüt), *n. Geometry.* a curve consisting of the centers of curvature of a given curve. All straight lines normal to a given curve are tangent to its evolute. Compare **involute.** See the picture at **involute.** [from Latin *evolutum* rolled out, from *ex-* out + *volvere* to roll]

evolution, *n.* **1** *Biology.* the process by which all existing organisms have developed from earlier forms through modification of characteristics in successive generations: *In 1858 Charles Darwin and Alfred Russel Wallace put forward a theoretical explanation of how evolution could have taken place and new species arisen. The theory of evolution by natural selection is the one which so far fits most, but not all, of the observed* facts and has been strengthened by discoveries since 1858 (Mackean, *Introduction to Biology*).
2 *Mathematics.* the extraction of roots from powers. Finding the square root of a number is an example of evolution. Contrasted with **involution.**
—**evolutionary**, *adj. Biology.* of or having to do with evolution: *the evolutionary history of the mammalian ear. The basic population units, the species, are successively combined in increasingly larger groups according to interpretation of their evolutionary relationships, in other words, their phylogeny* (Simpson, *Life: An Introduction to Biology*).
ASSOCIATED TERMS: Evolution (def. 1) is believed to result from *natural selection* acting upon random heritable *variation.* In the course of several generations, natural selection (sometimes called *survival of the fittest*) can modify the characteristics of a population of organisms, making the organisms more successful in their local environment. Such modifications are called *adaptations.* Hereditary characteristics are transmitted by *chromosomes* carrying *genes*, which are segments of *DNA* containing a *genetic code.* Genes can undergo *mutation* (change in structure) and *recombination* (change in arrangement), leading to evolutionary change such as the development of new species or *speciation.*

(def. 1)

Australopithecus

Homo Erectus

Homo Sapiens
Neanderthal

Homo Sapiens Sapiens
Cro-Magnon

evolve, *v. Biology.* to undergo evolution: *Species with short generation times could evolve faster than those with longer times, since over a given period of time the number of generations would be greater* (McElroy, *Biology and Man*). *According to the theory of evolution, variations among species have resulted chiefly from evolutionary adaptations to different environments. On the other hand, living things remain basically alike because they have evolved from common ancestors* (James H. Brown).

exarch (ek′särk), *adj. Botany.* having protoxylem developing toward the center: *exarch vascular tissue.* Compare **endarch.** [from *ex-* out of + Greek *archē* origin]

exasperate (eg zas′pə rit), *adj. Botany.* rough; covered with short, stiff points.

exchange reaction, *Chemistry.* a process by which atoms of the same element in two different molecules, or in two different positions in the same molecule, exchange places. It is usually studied by means of a tracer or tagged atom.

excimer (ek′sə mər), *n. Chemistry.* a substance formed by the joining together of atoms in an excited state: *The term excimer means that atoms that are energetically excited come together and form molecules ... "Two excimer" means that molecules of two different elements are involved* (Science News). [from *exci(ted)* + (*di*)*mer;* stress and articulation of the word modeled on *polymer*]

exciple (ek′sə pəl), *n. Biology.* a layer of cells partially enclosing lichens. [from Latin *excipulum* receptacle, from *excipere* take out]

—**excipular** (ek sip′yə lər), *adj.* of or having to do with an exciple.

excite, *v.* **1** *Physics.* to raise (an electron, atom, nucleus, molecule, etc.) to a higher energy level; put into an excited state: *If in a collision the energy exchange between ... an electron and atom is all energy of translation, the atom is not excited and the collision is said to be elastic* (H. E. White).
2 *Physiology.* to stimulate or irritate (an organ or tissue): *Changes which take place within the cells of the tentacles when the glands are excited* (Darwin, *Insectivorous Plants*).

—**excitable,** *adj. Physiology.* sensitive to or capable of responding to stimuli. —**excitability,** *n.* —**excitableness,** *n.*

—**excitation,** *n.* **1** *Physics.* the raising of an electron, atom, etc., to a higher energy level. The **excitation energy** of an atom, molecule, etc., is the minimum energy needed to change it from its ground state to an excited state. *The absorption of light energy by chlorophyll starts a chain of events. An important step is the excitation of an electron in the photoreactive center (trap) of an assembly of pigment molecules* (Weier, *Botany*).
2 *Physiology.* the act or process of exciting an organ or tissue: *the excitation of the rods in the retina.*

—**excitatory** or **excitative,** *adj. Physiology.* tending to excite; irritating or stimulating: *Acetylcholine is an excitatory transmitter for nerve impulses arriving at skeletal muscle junctions* (McElroy, *Biology and Man*).

excited state, *Physics.* a state of high energy in a quantized system, especially that state of energy of an atom in which the electrons have been transferred to an orbit farther from the nucleus than in the ground state: *It happens that the cyanogen molecule is excited from its ground, or lowest-energy state into its first excited state by radiation at a wavelength of 2.6 millimeters — a rather long wavelength for such a transition.* (Scientific American).

exciton (ek′sə ton or ek sī′ton), *n. Physics.* the combination of an electron and a hole in an excited semiconductor crystal: *Excitons are observed in semiconductors, where charge is carried both by electrons and by "holes," the positively charged voids formed by the absence of an electron. An exciton consists of a single electron and a single hole bound by the Coulomb force. Significantly, the exciton is an exceptionally large atom* (Scientific American). [from *excit(ation)* + -*on*]

—**excitonic,** *adj.* of or involving excitons: *There is a variety of excitonic processes, however, that give rise to electrical conductivity* (Martin Pope).

—**excitonics,** *n.* the study of excitons: *The recent developments in excitonics could have an important bearing on the study of energy-transfer mechanisms such as those involved in photosynthesis by living plants* (Scientific American).

excitor, *n. Physiology.* a nerve whose stimulation excites greater action in the part that it supplies.

exclusion principle, *Physics.* the principle that no two electrons in an atom or molecule can occupy the same energy level or have the same set of quantum numbers. It is now known to be applicable not just to electrons, but to a broad class of subatomic particles known as *fermions. One of Pauli's great discoveries in quantum mechanics is the so-called exclusion principle, which limits the states that two identical particles like the helium electrons can occupy.* (New Yorker). Also called **Pauli's principle** or **Pauli exclusion principle.**

exconjugant (eks kon′jə gənt), *n. Biology.* a microorganism that has conjugated and again become independent.

excrement (ek′skrə mənt), *n. Physiology.* waste matter, usually containing nitrogen, that is discharged from the body, especially from the intestines. Compare **excretion.** [from Latin *excrementum,* from *excretum.* See EXCRETE.]

excrescence (ek skres′ns), *n. Biology.* **1** an abnormal growth. Warts are excrescences on the skin. **2** a normal outgrowth. Fingernails are excrescences. [from Latin *excrescentem,* from *ex-* out + *crescere* grow]

—**excrescent,** *adj. Biology.* forming an abnormal growth or a disfiguring addition.

excreta (ek skrē′tə), *n.pl. Physiology.* waste matter discharged from the body, such as urine or sweat: *Dense aggregations of animals were until recently regarded as uniformly injurious and detrimental to the population, because, for instance, of the concentration of excreta in the environment of fresh-water organisms* (F.S. Bodenheimer). SYN: excretion.

—**excretal,** *adj.* having to do with or of the nature of excreta.

excrete *v. Physiology.* to separate (waste matter) from the blood or tissues and discharge it from the body: *The sweat glands excrete sweat. They generally agree that cobalt is excreted mainly in the urine after intravenous injection and that a smaller fraction may be recovered from the bile* (Science). [from Latin *excretum* sifted out, from *ex-* out + *cernere* to sift]

excretion, *n. Physiology.* **1** the act or process of discharging waste matter from the cells, tissues, and organs of the body. **2** the waste matter discharged; excreta: *Water is a necessary part of urinary excretion* (G.G. Simpson). Compare **excrement.**

---

cap, fāce, fäther; best, bē, tėrm; pin, five;
rock, gō, ôrder; oil, out; cup, pùt, rüle;
yü in use, *yù* in uric;
ng in bring; *sh* in rush; *th* in thin, ᴛʜ in then;
zh in seizure.
ə = *a* in about, *e* in taken, *i* in pencil, *o* in lemon, *u* in circus

207

—**excretory,** *adj.* excreting; having the task of discharging waste matter from the body: *an impairment of excretory function.*

▶ In many unicellular ororganisms, which lack special excretory structures, excretion is accomplished by diffusion through cell membranes. Thus, in the ameba and paramecium, carbon dioxide, ammonia, and mineral salts diffuse through the cell membrane directly into the watery environment.

excurrent (ek skėr′ənt), *adj.* **1** *Zoology.* providing an exit; giving passage outward: *Water leaves a clam shell through the dorsal opening or excurrent siphon.*
2 *Botany.* **a** having the axis prolonged to form an undivided main stem or trunk, as in the spruce tree: *When the trunk, or main stem, extends vertically upward to the tip, as it does in the Pines ... the type of branching is called excurrent; when it divides into several more or less equal divisions, as in the elm and other spreading trees, it is said to be deliquescent.* (Youngken, *Pharmaceutical Botany*). **b** projecting beyond the tip or margin, as the midrib of certain leaves.

excursion, *n.* *Physics.* **1** one of the movements of a body or particle in oscillating or alternating motion: *The particles themselves perform very small excursions, merely vibrating up and down* (Thomas H. Huxley). **2** the distance traversed by such a movement.

exergonic (ek′sər gon′ik), *adj.* *Biochemistry.* releasing energy by a chemical reaction. [from *ex-* out + Greek *ergon* work]

exfoliate (ek sfō′lē āt), *v.* *Biology, Geology.* to come off or cast off in scales or layers (of bone, skin, minerals, etc.).
—**exfo′lia′tion,** *n.* the act or process of exfoliating.
—**exfoliative** (ek′sfō′lē ā′tiv), *adj.* **1** tending to exfoliate; exfoliating: *exfoliative dermatitis.* **2** of or having to do with exfoliation: *These experiments made use of the technique of exfoliative cytology, i.e., the examination of cells which slough off or are removed from the linings of certain organs* (Harry M. Weaver).

exhalation, *n.* the act of exhaling. SYN: expiration.

exhale, *v.* **1** *Physiology.* to expel (air) from the lungs; breathe out. Also called **expire.**
2 *Biology.* to release or emit (oxygen, etc.) in the process of respiration.
Contrasted with **inhale.**
[from Latin *exhalare,* from *ex-* out + *halare* breathe]

exine (ek′sēn), *n.* = extine.

exo-, *prefix.* outside; outer, as in *exoskeleton* = *outer skeleton.* Compare **endo-.** [from Greek *exō* outside, from *ex-* out]

exobiology, *n.* the study of life on other planets or celestial bodies.
—**exobiological,** *adj.* of or having to do with exobiology.

exocarp (ek′sō kärp), *n.* = epicarp.

exocrine (ek′sə krən), *adj.* *Physiology.* **1** secreting outwardly, through a duct or into a cavity: *exocrine cells.*
2 of, produced by, or having to do with the exocrine glands. Contrasted with **endocrine.** Compare **apocrine, eccrine.**
[from *exo-* + Greek *krinein* to separate]

exocrine gland, *Physiology.* gland of external secretion that discharges its product through a duct or into a cavity. The exocrine glands include the salivary, lachrymal, sweat, mammary, and sebaceous glands, as well as the pancreas and liver which are also endocrine glands.

exocyclic, *adj.* *Chemistry.* situated outside a molecular ring: *exocyclic atoms.*

exocytosis (ek′sō sī tō′sis), *n.* *Biology.* a process of cellular excretion in which the fusion of membranes results in the discharge of a substance from the cell: *Exocytosis, in which the membrane of the storage granule fuses with the membrane of the cell, and the secretory products are lost through an opening in the fused membranes* (New Scientist). Contrasted with **endocytosis.** [from New Latin, from *exo-* outside + *cyt-* cell + *-osis* process or condition]
—**exocytose,** *v.* to discharge by exocytosis: *Upon stimulation of a mast cell ... its cytoplasmic granules are exocytosed in an energy-dependent process that involves fusion of the cell surface membrane with underlying perigranule membranes* (Science).
—**exocytotic,** *adj.* of or having to do with exocytosis.

exodermis (ek′sə dėr′mis), *n.* *Botany.* the outer layer of cortex in roots, just under the epidermis, corresponding to the hypodermis of stems.

exoenzyme (ek′sō en′zīm), *n.* *Biochemistry.* an enzyme that is active outside the cell or site of its formation. Also called **ectoenzyme.** Contrasted with **endoenzyme.**

exogamy (ek sog′ə mē), *n.* *Biology.* the union of two gametes of different descent. Compare **endogamy.**

exogenous (ek soj′ə nəs), *adj.* *Botany.* having stems that grow by the addition of layers of wood on the outside under the bark: *Exogenous stems are typical of Gymnosperms and Dicotyledons, and can increase materially in thickness due to the presence of a cambium* (Youngken, *Pharmaceutical Botany*). Contrasted with **endogenous.**

exomorphic, *adj.* *Geology.* having to do with changes in surrounding rocks by the intrusion of igneous matter.
—**exomorphism,** *n.* the state of being exomorphic.

exon (ek′son), *n.* *Molecular Biology.* a sequence of DNA that codes information for protein synthesis which is transcribed to messenger RNA: *The chicken ovalbumin gene (that is, the DNA region containing all the sequences that encode ovalbumin messenger RNA) is made up of eight exons, between each of which there is an intron having no counterpart in the messenger* (Scientific American). [from *ex(pressed unit)* + *-on* unit of genetic material]
—**exonic,** *adj.* of or having to do with exons.

exonuclease (ek′sō nü′klē ās), *n.* *Biochemistry.* an enzyme that breaks up a sequence of DNA or RNA by attacking the free ends of the nucleotide sequence: *The gap may be enlarged by an exonuclease, causing the release of mononucleotides. Subsequent polymerase activity fills in the gap* (John M. Boyle). Compare **endonuclease.**

exopeptidase (ek′sō pep′tə dās), *n.* *Biochemistry.* any of various proteolytic enzymes that break down terminal peptide bonds. Compare **endopeptidase.**

exopodite (ek sop′ə dīt), *n.* *Zoology.* the outer branch of a two-branched appendage in crustaceans. Compare **endopodite.** [from *exo-* + Greek *podos* foot]

exoskeleton, *n. Zoology.* any hard, external covering or structure which protects or supports the body, such as the shells of turtles and lobsters: *Insects, arachnids, and other arthropods have exoskeletons.* Compare **endoskeleton.** See the picture at **endoskeleton.**

—exoskeletal, *adj.* of or having to do with an exoskeleton.

exosmosis (ek′soz mō′sis), *n. Biology.* the passage of a fluid through a membrane toward a solution of lower concentration: *When the contents of a plant or animal cell include a higher concentration or percentage of water than does the material immediately surrounding it, water diffuses out of the cell, a process termed exosmosis* (Harbaugh and Goodrich, *Fundamentals of Biology*). Contrasted with **endosmosis.**

—exosmotic (ek′soz mot′ik), *adj.* of, resembling, or having to do with exosmosis.

exosphere, *n. Meteorology.* the outermost region of the atmosphere beyond the ionosphere. See the picture at **atmosphere.**

—exospheric, *adj.* of the exosphere: *exospheric temperatures.*

exospore (ek′sō spôr), *n.* **1** *Botany.* the outer coat or wall of a spore of certain plants, especially an asexual spore formed by abstriction.

2 *Bacteriology.* a spore borne outside of the parent cell of certain bacteria. Compare **endospore.**

exothermic or **exothermal,** *adj. Chemistry.* of or indicating a chemical change accompanied by a liberation of heat: *In an exothermic reaction, energy is released [and] the products have a lower potential energy than the reactants ... In an endothermic reaction, energy is absorbed [and] the products have a higher potential energy than the reactants* (Chemistry Regents Syllabus).

exotic, *adj. Geography.* (of a stream) flowing across an arid region from which it receives no additional runoff: *The Nile River is exotic along its lower course.*

exotoxin, *n. Bacteriology.* a poisonous substance secreted by a bacterium and released by it during its lifetime: *The poisonous agents produced by bacteria are often segregated into two groups—exotoxins and endotoxins. The exotoxins are composed of protein and usually do not contain nonprotein substances; most of them also stimulate the formation of antibodies* (Bruce W. Halstead).

—exotoxic, *adj.* of or having to do with an exotoxin.

expand, *v.* **1** *Physics.* (of a solid, liquid, or gas) to increase in size or volume; spread outward: *Heat can expand metal. Many stars in the sky are currently expanding, having depleted their internal hydrogen* (Scientific American). According to the theory of the **expanding universe,** the galaxies are receding from each other at speeds proportionate to their distance.

2 *Mathematics.* to express in fuller form, especially in finding a sum equivalent to a given product: *to expand the product* $(x + 4)(x - 3)$.

expansion, *n.* **1** *Physics.* the act or process of expanding: *Heat causes the expansion of gas.*

2 *Mathematics.* the fuller development of an indicated operation: *The expansion of* $(a + b)^3$ *is* $a^3 + 3a^2b + 3ab^2 + b^3$.

—expansive, *adj. Physics.* capable of expanding or of causing expansion: *the expansive force of heat.*

—expansivity, *n.* the quality, condition, or degree of being expansive: *... local variations of expansivity, viscosity, and surface tension* (Scientific American).

experiment, *n.* **1** an operation conducted to discover, test, or illustrate some fact or phenomenon, especially one in which some hypothesis is tested by the manipulation of one variable in a controlled system.

2 an apparatus or instrument used in an experiment: *On Apollo 11 was a package of three experiments—a passive seismometer, laser reflector, and solar wind sensor* (Science Journal).

—v. to conduct experiments.

—experimental, *adj.* used for, based on, or having to do with experiments. **—experimentally,** *adv.*

—experimentation, *n.* the act or process of experimenting: *The experimental method is often confused with the scientific method but actually experimentation is only a part of the latter* (Harbaugh and Goodrich, *Fundamentals of Biology*).

expiration, *n. Physiology.* a breathing out or expelling air from the lungs; exhalation.

—expiratory (ek spī′rə tôr′ē), *adj.* of or having to do with breathing out or exhalation.

expire, *v. Physiology.* to breathe out; exhale. [from Latin *expirare,* from *ex-* out + *spirare* breathe]

explant, *Biology.* **—v.** (eks plant′) to remove (living tissue) from the body to some other medium for scientific study: *In his original experiments, [he] explanted small pieces of parathyroid gland from man, mouse, and chicken* (Honor B. Fell).

—n. (eks′ plant) living tissue removed from the body for scientific study: *Dr. Kao also tried to culture brain material taken from humans in brain surgery, but the explant wouldn't grow* (Science News).

—explantation, *n.* the act or process of explanting.

explicit function, *Mathematics.* a function whose value is given in terms of the independent variable or variables.

explode, *v. Chemistry, Physics.* to burst or expand violently because of the pressure produced in a confined space by the sudden generation of one or more gases, as by gunpowder, nitroglycerin, or a nuclear reaction.

explosion, *n. Chemistry, Physics.* a violent bursting or expansion due to the pressure produced by the sudden generation of gases in confinement: *the explosion of a bomb.*

—explosive, *adj.* **1** of or involving explosion: *The idea developed that it might be possible to convert explosive energy into heat* (Bulletin of Atomic Scientists).

2 tending to explode: *Gunpowder is an explosive compound.*

—n. a chemical substance which reacts suddenly and violently on detonation by impact, friction, or heat, generating large volumes of gases, heat, etc.

---

cap, fāce, fäther; best, bē, tėrm; pin, fīve;
rock, gō, ôrder; oil, out; cup, pùt, rüle,
yü in use, *yu* in uric;
ng in bring; *sh* in rush; *th* in thin, ᵮʜ in then;
zh in seizure.
ə = *a* in about, *e* in taken, *i* in pencil, *o* in lemon, *u* in circus

exponent (ek spō'nənt), *n. Mathematics.* a number written above and to the right of a symbol or quantity to show how many times the symbol or quantity is to be used as a factor. EXAMPLES: $2^2 = 2 \times 2$; $a^3 = a \times a \times a$. Also called **index.** [from Latin *exponentem* putting forth, expounding, from *ex-* forth + *ponere* put]
—**exponential** (ek'spō nen'shəl), *Mathematics.* —*adj.* having to do with algebraic exponents; involving unknown or variable quantities as exponents. An **exponential function** is one in which an unknown or variable quantity appears as an exponent or part of an exponent, especially as an exponent of *e*, the base of natural logarithms.
—*n.* an exponential function. —**exponentially,** *adv.*
—**exponentiation** (eks'pō nen'shē ā'shən), *n.* the operation of raising one number to the power of another.

express, *v. Genetics.* Usually, (*passive*) **be expressed** or (*reflexive*) **express itself. 1** to exhibit (a genetic character or effect) in a phenotype; cause (a gene) to produce a corresponding observable character or phenotype: *The gene is present in the germplasm, as shown by its transmission to the next generation, but for some reason it has completely failed to express itself* (R.R. Gates).
2 to cause (a gene) to synthesize the specific protein it codes for by the processes of transcription and translation: [*They*] *announced that they had inserted into Escherichia coli bacteria a gene that codes for production of the hormone somatostatin and had induced the bacteria to express the gene—that is, to produce the hormone from it* (Thomas H. Maugh II).

expression, *n.* **1** *Mathematics.* any combination of constants, variables, and symbols showing or stating some operation or quantity. EXAMPLES: -1, e^x, $\sqrt{x+2}$.
2 *Genetics.* **a** the process by which a gene is expressed or expresses itself: *The expression of certain genes is controlled by the products of specific regulator genes which act as repressors* (Nature). **b** = expressivity.
—**expressivity,** *n. Genetics.* **1** the ability of a gene to manifest itself or its effects within the organism.
2 the manner or degree by which a gene is expressed or expresses itself.

exsolution, *n. Mineralogy.* the process by which a pair of minerals in solid solution separate from one another at certain temperatures.
—**exsolve,** *v.* to undergo exsolution.

exstipulate (ek stip'yə lit), *adj. Botany.* without stipules.

extension, *n.* **1** *Physics.* that property of a body by which it occupies a portion of space.
2 *Physiology.* **a** the straightening of a limb or other part by the action of an extensor. **b** the condition of being straightened in this way. —**extensional,** *adj.*

extensor (ek sten'sər), *n. Anatomy.* a muscle that when contracted extends or straightens out a limb or other part of the body: *There is a much smaller extensor muscle lying on top of the flexor which extends the abdomen* (Winchester, *Zoology*).

exterior angle, *Geometry.* **1** any of the four angles formed outside two lines intersected by a straight line.

2 an angle formed by a side of a polygon and the extension of an adjacent side: *An exterior angle of a triangle is equal to the sum of the two nonadjacent interior angles and greater than either one* (M.S. Shapiro).

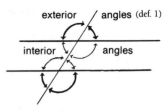

exterior / angles (def. 1)
interior angles

external, *adj.* **1** *Anatomy.* situated toward or on the outer surface; away from the median line or center of the body: *In the joints, the layer of chitin is thin and moderately flexible, but over the rest of the animal it becomes that hard external skeleton which we see on the lobster and the cockroach. An internal skeleton such as man's can grow with the animal* (Norbert Wiener). SYN: outer, exterior.
2 *Zoology.* occurring outside the body. **External fertilization** is the fertilization of an egg outside the body of the female, as in many aquatic vertebrates such as fish and amphibians, or by the introduction of a human sperm into an egg cell surgically removed from an ovary: *Some observers fear that external fertilization experiments will lead to a science fiction nightmare—a "brave new world"* (Thomas H. Maugh).

external ear, *Anatomy.* the outer part of the ear on the head, including the passage leading to the middle ear: *The ear has three parts—the external ear, the middle ear, and the inner ear* (McElroy, *Biology and Man*). Also called **outer ear.**

external respiration, *Biology.* the exchange of oxygen and carbon dioxide between the environment and the respiratory organs; breathing. Compare **internal respiration.**

exteroceptive (ek'stər ō sep'tiv), *adj. Physiology.* receiving or aroused by stimuli from outside the body: *Exteroceptive feedback yields information about events on the surface of the skin through the skin senses such as touch and heat* (New Scientist). [from Latin *exter* outside + English (*re*)*ceptive*]
—**exteroceptor** (ek'stər ō sep'tər), *n.* a sense organ that receives stimuli from outside the body, such as the ear.

extinct, *adj.* **1** *Biology.* (of a species or other taxonomic group) having no living members: *If two species compete strongly, the frequent but not inevitable outcome is that one of them becomes extinct ... On land, all the dinosaurs became extinct* (Simpson, *Life: An Introduction to Biology*).
2 *Geology.* (of a volcano) no longer active, and not expected to resume activity.
—**extinction,** *n.* the condition of being extinct: *Regression or devolution and extinction are quite as much a part of the total record* [*of evolution*] *as is progression* (Ralph Beals and Harry Hoijer).

extine (eks'tin), *n. Botany.* the outer coat of a pollen grain. Also called **exine.** [from Latin *exter* outward + English *-ine*]

extra-, *prefix.* outside; beyond; besides, as in *extraneural = outside a nerve or neuron, extrasensory = beyond the body's sense organs.* [from Latin, from *extra* outside]

extracellular, *adj. Biology.* situated or taking place outside of a cell or cells: *extracellular fluid.*

extrachromosomal, *adj. Biology.* occurring or operating outside the chromosomes: *extrachromosomal particles or elements.*

extract, *v.* **1** *Mathematics.* calculate or find (the root of a number).
2 *Chemistry.* **a** to produce (an extract). **b** to obtain (a substance) from a liquid or solid mixture.
—*n. Chemistry.* a dry substance made from a drug, plant, tissue, etc., by dissolving the active ingredients and then evaporating the solvent.
[from Latin *extractum* drawn out, from *ex-* out + *trahere* draw]

extragalactic, *adj. Astronomy.* outside our own galaxy: *The Magellanic Clouds are visible extragalactic nebulae.*

extranuclear, *adj.* **1** *Biology.* situated or existing outside the nucleus of a cell: *extranuclear determiners of hereditary.*
2 *Physics.* existing or acting outside the nucleus of an atom: *The chemical properties of an atom depend chiefly on the number of extranuclear electrons and only to a small degree on the mass of the nucleus* (Sears and Zemansky, *University Physics*).

extrapolate (ek strap′ə lāt), *v. Mathematics.* **1** to estimate the value of (a function or quantity) outside a range in which some values are known. **2** to continue (a curve) from points already plotted on a graph.
—**extrapolation,** *n.* the act or process of extrapolating.

extrapyramidal, *adj. Anatomy.* situated or occurring outside the pyramidal tracts of the brain. The **extrapyramidal system** consists of descending tracts of nerve fibers originating in the cortex and subcortical areas of the brain.

extrasolar, *adj. Astronomy.* found or existing outside the solar system: *extrasolar planets.*

extraterrestrial, *adj. Astronomy.* outside the earth or its atmosphere: *an extraterrestrial source for this high-frequency static ... was probably near the center of our Milky Way* (Ralph E. Lapp).

extratropical, *adj. Meteorology.* originating outside the tropics; occurring in the middle and higher latitudes of either hemisphere: *an extratropical cyclone.*

extravasate (ek strav′ə sāt), *v. Physiology.* to force out or escape from the proper vessels through the surrounding parts, as blood or lymph. [from *extra-* + Latin *vas* vessel]
—**extravasation,** *n.* the act or process of extravasating.

extravascular, *adj. Biology.* **1** not contained in vessels: *the extravascular circulatory system of insects.*
2 not vascular; having no blood vessels: *an extravascular structure.*

extreme, *adj. Physics.* far beyond the normal or usual range experienced: *Extreme reactions are those that occur at excessively high pressures, high temperatures, high vacuum, or under intense radiation* (R.A. Beebe and R.D. Fink).
—*n. Mathematics.* the first or last term in a proportion or series: *In the proportion, 2 is to 4 as 8 is to 16, 2 and 16 are the extremes; 4 and 8 are the means.*

extremum (ek strē′məm), *n., pl.* **extrema** (ek strē′mə). *Mathematics.* a maximum or minimum value of a function.
—**extremal,** *adj.* of or having to do with extrema.

extrinsic, *adj.* **1** *Anatomy.* originating outside the part upon which it acts: *the extrinsic muscles of the legs.*
2 *Medicine.* originating outside the body. An extrinsic allergy may be caused by a food, drug, or other substance taken into the body.
[from Late Latin *extrinsicus* from outside]

extrinsic factor, = vitamin B_{12}.

extrorse (ek strôrs′), *adj. Botany.* turned or facing outward, as anthers which face and open away from the axis of the flower. Contrasted with **introrse.** [from Late Latin *extrorsum,* from Latin *extra-* outside + *versum* turned]

extrusion, *n. Geology.* **1** the emission of lava or volcanic material at the earth's surface: *What happens when the magma is cooled by intrusion into colder rocks nearer the earth's surface or by extrusion as lava depends upon its composition, fluidity and the rate of cooling* (Herbert H. Read).
2 the lava flow or volcanic structure formed by this process: *A small flow of molten rock streaming down a mountain side, a shower of small, hot rock fragments, or a cloud of hot dust blown out of a crater is an occasional sight in many parts of the world. ... If they congeal before reaching the surface they are called intrusions. If they reach the surface they are called eruptions or extrusions* (White and Renner, *Human Geography*).
[from Medieval Latin *extrusionem,* from Latin *extrudere* thrust out]
—**extrusive,** *adj.* (of igneous rock) derived from materials ejected at the earth's surface. Also called **effusive.** Contrasted with **intrusive.**

exuviae (eg zü′vē ē *or* ek sü′vē ē), *n.pl. Zoology.* the cast-off skins, shells, feathers, etc., of animals. [from Latin, from *exuere* cast off]
—**exuvial,** *adj.* having to do with or like exuviae.

exuviate (eg zü′vē āt *or* ek sü′vē āt), *v. Zoology.* to cast off or shed skin, shell, feathers, etc. SYN: molt.
—**exuviation,** *n.* the act or process of exuviating.

eye, *n.* **1** *Anatomy.* **a** the organ of sight, occupying the anterior part of a bony socket called the orbit. The eye is made up of the sclera, cornea, choroid, ciliary body, iris, and retina. *The eye consists of a roughly spherical eyeball supported in a bony socket in the skull, a system of muscles for moving the eyeball, and tear ducts for moistening the anterior surface of the eyeball* (Shortley and Williams, *Elements of Physics*). **b** any organ that is sensitive to light. See the pictures at **compound, crustacean.**

---

cap, fāce, fäther; best, bē, tèrm; pin, fīve;
rock, gō, ôrder; oil, out; cup, pút, rüle,
yü in use, *yu* in uric;
ng in bring; *sh* in rush; *th* in thin, ₸н in then;
zh in seizure.
ə = *a* in about, *e* in taken, *i* in pencil, *o* in lemon, *u* in circus

eyeball

2 *Meteorology.* the relatively calm area at the center of a hurricane: *Warm, moist air spirals toward the storm center and flows upward in a band of clouds (the eyewall) surrounding the ... eye of the storm* (Science News).

the human eye

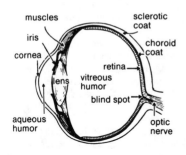

eyeball, *n. Anatomy.* the ball-shaped part of the eye without the surrounding lids and bony socket.

eye lens, *Optics.* the lens nearest the eye in an eyepiece.

eyelid, *n. Anatomy.* the movable fold of skin, upper or lower, containing muscles by means of which the eyes can be opened or shut.

eyepiece, *n. Optics.* the lens or set of lenses in a telescope, microscope, etc., that is nearest the eye of the user: *The ordinary microscope consists of two lenses, one at either end of a telescopic tube; the lens nearer the eye is called the eyepiece and the other the objective* (New Yorker).

eyepoint, *n. Optics.* the point at which the rays from the eyepiece of a telescope, microscope, or other instrument with lenses, converge, and to which the eye is applied to view an object.

eyespot, *n. Biology.* 1 a light-sensitive patch of pigment found in some unicellular and colonial green algae: *A light-sensitive eyespot lies within the chloroplast [of chlamydomonas] at the base of the flagella; with this eyespot and its flagella, the cell is able to orient itself and swim in relation to light* (J. M. Barrett).
2 a simple light receptor, consisting of a group of pigmented cells, found in certain invertebrates: *Photoreceptors sensitive to light are present in earthworms, and there are "eye spots" on various coelenterates and mollusks. From such simple structures various types of eyes have been developed* (Storer, General Zoology).

eyestalk, *n. Zoology.* the stalk or peduncle upon which the eye is borne in lobsters, shrimps, etc.

eyetooth, *n., pl.* eyeteeth. *Anatomy.* either of the two pointed, upper teeth between the incisors and the bicuspids; an upper canine tooth.

eyewall, *n. Meteorology.* a layer of turbulent funnel-shaped clouds around the eye of a storm: *Seeding the clouds outside the hurricane eyewall—the ring of strongest wind and heaviest rain may reduce the maximum wind speeds* (Jerome Spar).

F

f, *abbrev.* or *symbol.* **1** farad. **2** femto-. **3** focal length. **4** frequency. **5** fluid (ounce).

F, *abbrev.* or *symbol.* **1** farad. **2** Fahrenheit. **3** filial generation: F_1 = *first filial generation.* **4** fluorine. **5** function.

face, *n.* **1** *Anatomy.* the front part of the head, including the chin and forehead.

2 *Geometry.* any of the planes or surfaces that bound a solid figure. A cube has six faces. A polyhedron has four or more faces. *The experiments were repeated using slices of crystal, exposing only one face at a time, and it was shown that the reaction proceeded more than twice as fast on an octahedral ... face as on a cubic face* (P. R. Rowland).

face-centered, *adj. Crystallography.* having a cubic structure with an atom or ion at each corner and an atom or ion in the center of each face: *One, the so-called alpha phase, is a face-centered cubic crystal (a cube with atoms at the center of each face and at each corner); the other, the beta-prime phase, is a body-centered cubic crystal (one with an atom in the center of the cube and others at each corner)* (Scientific American).

facet (fas'it), *n.* **1** *Zoology.* one of the individual external visual units of a compound eye: *Some kinds of ants have no eyes, but most have a compound eye on each side of their heads. These large eyes may have from 6 to 1,000 or more facets, or lenses. Each facet serves as a tiny eye* (C.D. and M.H. Michener).

2 *Anatomy.* a small, smooth, flat surface, especially on a bone: *The ribs of the Sauria have only a single articular facet* (George Rolleston).

facial, *adj. Anatomy.* of the face: *a facial artery, the facial nerves, facial neuralgia, facial muscles.*

facies (fā'shē ēz), *n.* **1** *Geology.* **a** the general appearance or character of a formation in a given region with respect to fossil content, rock structure, and other characteristics: *A noteworthy feature of Cretaceous deposits in this province is the lateral change of sedimentary facies, reflecting variations in conditions of contemporary sedimentation on land, in lagoonal areas along the coast, in shallow marine waters near shore, and in somewhat deeper offshore environments* (Moore, *Introduction to Historical Geology*). **b** the part of a rock or rock face distinguished from other parts by appearance or composition. **c** a particular group of metamorphic minerals formed as a stable assemblage under a specific temperature-pressure environment.

2 *Biology.* the general appearance or makeup of the population of a species or other group: *The facies is distinguished wholly by differences in the quantity or distribution of species* (Fuller and Conrad, *Plant Sociology*).

3 *Medicine.* the appearance or expression of the face, especially as indicative of an abnormality: *During the pyrexial period the face is flushed, the eyes injected (giving a typical 'ferrety' appearance to the facies)* (Jewell and Kauntze, *Handbook of Tropical Fevers*). [from Latin *facies* form]

factor, *n.* **1** *Mathematics.* any member of a set of numbers or algebraic expressions which, when multiplied together, form a product: *5, 3, and 4 are factors of 60; 5 × 3 × 4 = 60.* Compare **prime factor.**

2 *Biology.* a gene: *The terms gene, factor, and determiner will be used as synonyms to designate the units responsible for the transmission of hereditary characters* (Harbaugh and Goodrich, *Fundamentals of Biology*).

3 *Physiology.* a substance functioning in a particular process or system, such as coagulation. Several clotting factors are present in human plasma. *The studies reveal that the brain-protecting mechanism functions through the agency of a factor secreted by the pituitary gland, the master gland situated at the base of the brain* (N.Y. Times).

—*v. Mathematics.* to separate or resolve into factors; find numbers or algebraic expressions that are factors of a given product. Factoring is similar to division. [from Latin *factor* doer, from *facere* to do]

—**factorable,** *adj. Mathematics.* that can be separated or resolved into factors.

—**factorize,** *v. Mathematics.* to resolve into factors.

—**factorization,** *n.* the process of factorizing.

factorial (fak tôr'ē əl), *Mathematics.* —*n.* the product of an integer multiplied by all the integers from one to that integer. EXAMPLE: The factorial of 4 is 4 × 3 × 2 × 1 = 24, and is symbolized 4!

—*adj.* of or having to do with a factorial: *The term factorial expression has been in some instances applied to an expression of which the factors are in arithmetical progression* (Penny Cyclopaedia).

facula (fak'yə lə), *n., pl.* **-lae** (-lē). *Astronomy.* one of the large bright patches on the surface of the sun: *Faculae ("little torches") are ... more abundant in the vicinity of sunspots and are especially conspicuous near the ... rim of the photosphere* (Baker, *Astronomy*). [from Latin *facula,* diminutive of *fax, facis* torch]

facultative (fak'əl tā'tiv), *adj. Biology.* having the power to exist under different conditions of life. A saprophytic organism that can also act as a parasite is called a

---

cap, fāce, fäther; best, bē, tèrm; pin, fīve;
rock, gō, ôrder; oil, out; cup, pùt, rüle;
yü in use, *yu* in uric;
ng in bring; sh in rush; th in thin, ᴛн in then;
zh in seizure.
ə = *a* in about, *e* in taken, *i* in pencil, *o* in lemon, *u* in circus

213

facultative parasite. Contrasted with **obligate**. [from Latin *facultas* ability, faculty]

FAD, *abbrev.* flavin adenine dinucleotide.

Fahrenheit (far'ən hīt), *adj.* of, based on, or according to a scale for measuring temperature on which 32 degrees marks the freezing point of water and 212 degrees marks the boiling point of water at standard atmospheric pressure. The **Fahrenheit thermometer** is marked off according to this scale. See the picture at **temperature**. [named after Gabriel D. *Fahrenheit*, 1686–1736, German physicist, who introduced this scale]

falcate (fal'kāt), *adj. Biology.* curved like a sickle; hooked: *a falcate cartilage.* [from Latin *falcatus*, from *falx, falcis* sickle]

falciform (fal'sə fôrm), *adj. Biology.* sickle-shaped; falcate. [from New Latin *falciformis*, from Latin *falx, falcis* sickle + *forma* form]

fall line, *Geology.* a line that marks the end of layers of hard rock, such as the edge of a plateau or piedmont, and the beginning of a softer rock layer, such as that of a plain. There are many waterfalls and rapids along the fall line.

Fallopian tube or **fallopian tube** (fə lō'pē ən), *Anatomy.* either of a pair of slender tubes through which ova from the ovaries pass to the uterus: *The female sex organs ... consist of the ovaries, fallopian tubes (oviducts), uterus (womb), and vagina (birth canal)* (McElroy, *Biology and Man*). [named after Gabriel *Fallopius*, 1523–1562, Italian anatomist, who described them]

fallout, *n.* the radioactive particles or dust that fall to the earth after a nuclear explosion: *The dangers of atomic fallout and reactor wastes have emphasized man's genetic susceptibility to excessive doses of radiation* (Arnold L. Bachman).

false, *adj. Biology.* inaccurately so called, usually because of a deceptive resemblance to another species. One name for the locust tree is "false acacia."

false ribs, *Anatomy.* ribs not attached directly to the breastbone. Human beings have five false ribs on each side. Compare **true ribs.**

false vocal cords, *Anatomy.* the upper of the two pairs of vocal cords, which do not directly aid in producing voice. Also called **superior vocal cords.** Compare **true vocal cords.**

familial, *adj. Biology.* occurring among the members of a family, especially several members of the same generation: *familial diseases.*
ASSOCIATED TERM: see **congenital.**

family, *n. Biology.* **1** a group of related organisms, ranking below an order and above a genus. In zoology, the names of families end in *-idae*, for example Felidae, the cat family; in botany, the names of families usually end in *-aceae*, for example, Rosaceae, the rose family.
ASSOCIATED TERMS: see **kingdom.**
2 *Chemistry.* a group of elements having similar properties: *In Group IA in the Periodic Table, we find the Sodium Family, a group of six similar, very active, metallic elements* (Dull, *Modern Chemistry*).

3 *Geometry.* a group of related curves or surfaces: *Family of curves is an assemblage of several curves of different kinds, all defined by the same equation of an indeterminate degree* (Charles Hutton).

fang, *n.* **1** *Zoology.* **a** a long, pointed tooth with which an animal grasps, holds, or tears its prey. **b** = chelicera: *They [spiders] have a pair of fangs at the anterior end of the ventral surface of the cephalothorax which are connected internally with poison sacs. The fangs are like hypodermic needles and when a spider bites a little poison is forced out of the poison sacs and through the fangs into the wound* (Winchester, *Zoology*). **2** *Anatomy.* the root of a tooth.

farad (far'əd *or* far'ad), *n. Physics.* the SI or MKS unit of electric capacitance. It is the capacitance of a capacitor that, when charged with one coulomb of electricity, has an electric potential of one volt. *The farad is such a large unit that it is seldom used as such; its derivatives microfarad and micromicrofarad are commonly used in practice.* (Roy F. Allison). *Symbol:* F or f [from Michael *Faraday;* see FARADAY]

faraday (far'ə dā *or* far'ə dē), *n. Physics.* a unit of quantity of electricity equivalent to about 96,500 coulombs. It is the quantity that, in electrolysis, is necessary to deposit one mole of a univalent element (usually 107.88 grams of silver). *The faraday is the Avogadro number times the electronic charge* (Jens C. Zorn). [named after Michael *Faraday,* 1791–1867, English physicist and chemist]

Faraday effect, *Physics.* the effect produced when a beam of polarized light passes through a magnetic field and is rotated in the direction of the lines of magnetic force: *A beam of linearly polarised light falls on a magneto-optical element placed near the current carrying line and, as a result of the Faraday effect, the magnetic field, which is a measure of the current flowing, produces a rotation of the polarisation which is proportional to the current* (New Scientist).

faradic (fə rad'ik), *adj. Physics.* of or having to do with induced currents of electricity.

far point, *Optics.* the point farthest from the eye at which an image is clearly formed on the retina when accommodation is relaxed: *The normal individual at 10 years of age is able in this way to focus sharply on objects lying anywhere between the far point at infinity and the near point at about 7 cm in front of the eye. The range of accommodation becomes progressively less as the individual grows older ... until in old age the power of accommodation is completely lost* (Hardy and Perrin, *Principles of Optics*).

fascia (fash'ē ə), *n., pl.* **fasciae** (fash'ē ē). *Histology.* a usually thin sheet of fibrous connective tissue covering, supporting, or binding together a muscle, part, or organ. [from Latin *fascia* band, girdle]
—**fascial** (fash'ē əl), *adj.* having to do with or consisting of fasciae: *The most basic of these techniques involves pulling the face into shape with an internal sling made of fascia, the fibrous tissue that separates and encloses the muscles of the body. Fascial slings do not restore normal muscle control, but by supporting sagging face muscles, they help to bring a certain symmetry to the face at rest* (Time).

fasciate (fash'ē āt) or **fasciated** (fash'ē ā'tid), *adj.* **1** *Botany.* compressed into a band or bundle; grown together: *fasciate stems.* **2** *Zoology.* marked with wide bands or stripes.

fasciation (fash'ē ā'shən), *n. Botany.* a malformation in plants, in which a stem or branch becomes expanded into a flat, ribbonlike shape.

fascicle (fas'ə kəl), *n.* **1** *Botany.* a close cluster, especially of flowers, leaves, or roots: [*Pine*] *leaves are needlelike, two or more growing together ... in a fascicle or group, which is sheathed at the base* (Weier, *Botany*). **2** *Anatomy.* a set of nerve or muscle fibers bound closely together.

fast, *adj. Physics.* = high-energy: *Certain important questions could only be answered by experimenting with beams of high-energy, or fast, protons* (Scientific American). See also **fast neutron.**

fastigiate (fas tij'ē it), *adj.* **1** *Botany.* having parallel, erect branches and a narrow elongated habit, as many poplars. **2** *Zoology.* fastened together in a cone-shaped bunch. [from Latin *fastigium* apex of a gable, summit]

fast neutron, *Nuclear Physics.* a neutron of relatively high energy, usually of more than 10,000 electron volts: *The neutron beam divides into three main speed ranges. The so-called fast neutrons are those with energies extending from 10 million down to about 10,000 electron volts. Below these in the hierarchy of energies are "resonance" neutrons, from 10,000 electron volts to .01 electron volt. Finally there are the thermal neutrons, with energies from .01 to .0001 electron volt* (Scientific American).

fast reactor, *Nuclear Physics.* a reactor using fast neutrons without a moderator: *The fast reactor is so called because the neutrons given off by the fissioning nuclei retain much more energy than the neutrons in a "thermal" reactor. In thermal reactors, such as Britain's advanced gas cooled reactor* (*AGR*) *or the American light water reactors, the neutrons give up most of their energy in the reactor's moderator material* (New Scientist).

fat, *n.* **1** *Chemistry.* any of a class of organic chemical compounds of which the natural fats are mixtures, comprising an important group of animal foods. Fats contain carbon, hydrogen, and oxygen, but no nitrogen, and are chiefly glycerides, compound esters of several acids. They are insoluble in water but dissolve in ether, chloroform, and benzene. *The most common lipids are fats, oils, and waxes. ... Fats occur mainly in animal tissues and as butterfat in milk and other dairy products* (Otto and Towle, *Modern Biology*). *Plants change carbohydrates into fats and store them in such seeds as peanuts, olives, cotton seed, and pecans. Plants also make proteins out of carbohydrates and minerals* (Beauchamp, Mayfield and West, *Everyday Problems in Science*).
2 *Physiology.* tissue composed mainly of fat: *It was concluded that, in the older group, there had been a replacement of muscle and other active tissues by fat, but the evidence for this view consisted in a 2 per cent reduction in specific gravity and an increase in wrinkling. Again, a comparison of fat and cell-mass would have been preferable* (A. W. Haslett). See the picture at **skin.**

▶ Fats (def. 1) and oils are chemically similar; oils contain carbon-carbon double bonds ($C = C$) that can be hydrogenated to convert the oils into fats. Fats are solid and oils are liquid at room temperature.

fat body, *Zoology.* **1** fatty tissue in insects that serves as a reserve food supply. **2** a mass of fatty tissues next to the genital glands in frogs and toads.

fat cell, *Biology.* a specialized cell for the storage of fat. Fat cells make up the adipose tissue. *The layers beneath the dermis contain numerous fat cells where fat is stored* (Mackean, *Introduction to Biology*).

fath., *abbrev.* fathom.

fathom, *n.* a unit of length in the customary system, equal to 6 feet (1.8 meters), used mostly to measure the depth of water, especially in the ocean: *The change from shallow water of less than 100 fathoms in depth to the normal deep ocean of 2,500 to 3,000 fathoms occurs sharply; this may be because the edge of the permanent continental rocks which do not exist in the deep ocean has been reached, or it may be that the steep slope represents a rubbish tip, where sediment carried down in shallow water is deposited* (T.F. Gaskell and M.N. Hill). *Abbreviation:* fath., fth. or fthm.
—*v.* to measure the depth of (water); sound.

fat-soluble, *adj. Chemistry.* that can be dissolved in fats or fat solvents: *One group, the fat-soluble vitamins, dissolves in fats. Among these are vitamins A, D, E, and K. The other group, the water-soluble vitamins, dissolves in water* (Janice M. Smith).
—**fat-solubility,** *n.* the condition of being fat-soluble.

fatty, *adj.* **1** *Biology.* of or containing fat: *fatty tissue.* SYN: adipose. **2** *Chemistry.* = aliphatic.

fatty acid, *Chemistry.* any of a group of organic acids, some of which, such as palmitic acid and stearic acid, are found in animal and vegetable fats and oils. In the body, fatty acids are absorbed through the villi into the lacteals and are transported in the lymph. *Formula:* $C_nH_{2n+1}CO_2H$

fauces (fô'sēz'), *n.pl. Anatomy.* the cavity at the back of the mouth, leading into the pharynx. [from Latin *fauces* throat]
—**faucial** (fô'shəl), *adj.* of or produced in the fauces.

fault, *Geology.* —*n.* a break in the earth's crust, with the mass of rock on one side of the break displaced: *In lowland areas large-scale structural features such as folds and faults are generally masked by soil. It is to the mountains and deserts that the geologist must go to find large rock structures clearly displayed* (Gilluly, *Principles of Geology*). *The remarkable San Andreas fault forms a continuous break in the earth's crust from beneath the Pacific Ocean at Point Arena, 100 miles north of San Francisco, southward to Cajon Pass* (Science News).

---

cap, fāce, fä̇ther; best, bē, tėrm; pin, fīve;
rock, gō, ôrder; oil, out; cup, pu̇t, rüle;
yü in use, yu̇ in uric;
ng in bring; sh in rush; th in thin, ᴛʜ in then;
zh in seizure.
ə = *a* in about, *e* in taken, *i* in pencil, *o* in lemon, *u* in circus

fault line

—*v.* of rock strata: **1** to undergo a fault or faults. **2** to cause a fault or faults in: *East and West Antarctica are separated by a great mountain range which exceeds an elevation of 15,000 feet at several points (such as Mount Markham, 15,100 feet). The range has been faulted upward, that is, pushed upward as a block, its steep slopes extending to, or almost to, sea level on either side* (Paul A. Siple).

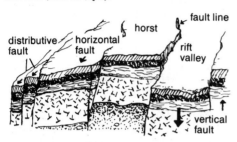

fault line, *Geology.* a line formed by the intersection of a fault plane with the surface of the earth; the trace of a fault. See the picture at **fault**.

fault plane, *Geology.* the plane along which a faulting occurs.

fauna (fô′nə), *n., pl.* **-nas, -nae** (-nē′). *Biology.* the animals or animal life of a particular place or time: *the fauna of Australia, the fauna of the Carboniferous period. Albert National Park ... is the only park in the world to shelter okapi and gorillas among its fauna* (N.Y. Times). Compare **flora.** [from New Latin, from Late Latin *Fauna*, a rural goddess, the wife of *Faunus* faun]

—**faunal** (fô′nl), *adj.* of or having to do with a fauna or faunas.

—**faunistic**, *adj.* of a fauna; faunal: *Each continental area has in fact three main floristic and faunistic elements* (New Scientist). —**faunistically**, *adv.*

f.c., *abbrev.* foot-candle.

Fe, *symbol.* iron. [for Latin *ferrum*]

feather

feather, *n. Zoology.* **1** one of the light, thin growths that cover a bird's skin, usually consisting of a partly hollow shaft bearing flat vanes formed of many parallel barbs. Compare **contour feather, pinfeather.**

2 one of the long hairs covering the neck, legs, tail, etc., of various types of dogs: *The cocker [spaniel] has long hairs, or feathers, on its ears, chest, and legs* (William F. Brown).

fecal (fē′kəl), *adj.* having to do with feces.

feces (fē′sēz), *n.pl. Physiology.* waste matter discharged from the intestines; excrement. [from Latin *faeces* dregs]

fecund (fē′kənd *or* fek′ənd), *adj. Biology.* capable of producing offspring or vegetable growth; fertile. [from Latin *fecundus*]

—**fecundate** (fē′kən dāt *or* fek′ən dāt), *v.* fertilize; impregnate. —**fe′cunda′tion**, *n.*

fecundity (fi kun′də tē), *n. Biology.* the ability to produce offspring or vegetable growth; reproductive capacity: *The slaughter, mostly by trapping, of some hundred million rabbits every year has, because of the animal's fecundity, little appreciable effect on their numbers in lush seasons* (F. Fenner and M.F. Day). SYN: fertility.

feedback, *n.* **1** *Electronics.* a process by which a system or device regulates itself by feeding back to itself part of its output; the return of a part of the output to the input: *Computers ... and related equipment use the principle of "feedback," a concept of control in which the input of machines is regulated by the machines' own output. Although the use of machines dates back to the steam engine of the 18th century and to the assembly line of the early 20th century, feedback is a new development truly unique to automation* (Walter Buckingham). *Feedback [is] circuitry by means of which a part of the electrical output from an amplifier is fed back into the input section. When the output impulses are of the same polarity as that of the input impulses, so that they are additive, the feedback is positive; when they are of opposite polarity, so that their sum at the input section is less than it would be without the feedback, the feedback is negative. Negative or inverse feedback reduces distortion and extends the frequency range of the amplifier, provided it is used properly, but reduces amplification* (Roy F. Allison).

2 *Biology.* the modification or regulation of a process by some of the results or products of the process: *The brain has to receive feedback from the muscles and joints to correct the program of impulses directed to the motor apparatus* (Scientific American). Compare **negative feedback.**

—*adj.* characterized by or using feedback: *a feedback mechanism, a feedback amplifier, a feedback oscillator.*

feedback inhibition, *Biochemistry,* a form of cellular control by which an enzyme catalyzing the production of a particular substance in the cell is inhibited at a certain level of accumulation of that substance, thereby balancing the amount produced with the amount needed: *The products of an enzyme-catalyzed reaction can also act as inhibitors if they are tightly bound to the enzyme; in this case the enzyme is inactive until the product is removed. In addition, the product may regulate a metabolic pathway by inhibiting an enzyme associated with an earlier step. This mechanism of control, called feedback inhibition, may be of considerable importance in regulating the amount of a particular product formed in a biosynthetic pathway* (McElroy, *Biology and Man*).

feedback loop, 1 the path through a feedback system or device from input to output and back to input: *The electroencephalographic signal has been found to be closely correlated with depth of anesthesia. In fact, it* has even been used in a feedback loop, controlling an anesthesia machine to maintain a constant level of anesthesia (John H. Borrowman). 2 a diagram representing such a path.

feedforward, n. Electronics, Chemistry. the control of a feedback process by anticipating any defects in the process before it is carried out: The term feedforward has been used here to cover situations in which an unknown or an intractably complex fluctuation takes place in an amplifier or possibly other transfer device. The chemical industry has also laid claim to the same term to describe a control process in which a compensating change in processing later in a cycle is anticipated by earlier measurement of system parameters. In distinction to what has been described here, their use implies all input-output-relations to be known and assumes control to be easily available (H. Seidel).

Fehling's solution or **Fehling's reagent** (fā′lingz), Chemistry. a solution of copper sulfate, potassium sodium tartrate, and sodium hydroxide, used to test a substance for the presence of sugar. [named after Hermann Fehling, 1812–1885, German chemist]

feldspar (feld′spär′), n. Mineralogy. any of a group of common rock-forming minerals composed of silicate of aluminum, combined with sodium and either potassium or calcium: Also important are the feldspars, a group of light-colored minerals including combinations of silica, aluminum, and oxygen with various other elements such as sodium, calcium, or potassium. The decomposed fragments of the feldspars, together with some fine sands and other substances, make up the bulk of ordinary clays (Finch and Trewartha, Elements of Geography). See also **alkali feldspar, plagioclase.** [alteration (influenced by spar) of German Feldspat, literally, field spar]

feldspathic (feld spath′ik), adj. of the nature of or containing feldspar: Feldspathic rocks have produced a clayey soil (Charles Darwin).

feldspathoid (feld′spə thoid), n. Mineralogy. any of a group of rock-forming minerals chemically similar to the feldspars but containing less silica: Leucite and lazurite are feldspathoids.
—**feldspathoidal,** adj. of, being, or having to do with a feldspathoid or feldspathoids.

feline (fē′lin), Zoology. —adj. of or belonging to the family (Felidae) of carnivorous mammals that includes the cats, lions, tigers, and leopards, —n. any animal belonging to the feline family. [from Latin felinus, from felis cat]

felsic (fel′sik), adj. Mineralogy. consisting of or having to do with a group of light-colored silicate minerals occurring in igneous rocks. Contrast with **mafic.** [from fe(ldspar) + l(enad) feldspathoid + s(ilica) + -ic]

felsite (fel′sīt), n. Geology. an aphanitic volcanic rock consisting chiefly of feldspar and quartz in minute crystals. [from fels(par) + -ite]

female, n. 1 Zoology. an animal belonging to the sex that produces eggs or brings forth young: In multicellular animals, sex is the total of all structural and functional characteristics that distinguish male (♂) and female (♀) (Storer, General Zoology).
2 Botany. **a** a flower having a pistil or pistils and no stamens. **b** a plant bearing only flowers with pistils.

—adj. 1 Zoology. of, having to do with, or belonging to the sex that produces eggs or brings forth young: It is also possible to make a castrated male animal, such as a steer, produce milk by injecting synthetic female hormones (New Biology).
2 Botany. **a** (of seed plants) having flowers which contain a pistil or pistils but not stamens; pistillate. **b** having to do with any reproductive structure that produces or contains elements that need fertilization from the male element.

femoral (fem′ər əl), adj. Anatomy. of, having to do with, or in the region of the femur: the femoral artery.

femto-, combining form. one quadrillionth (10^{-15}) of (any SI unit), as in femtogram = one quadrillionth of a gram. Each flash lasts 12 "femtoseconds," one such unit of time being one-thousandth of a millionth of a millionth of a second (N. Y. Times). Symbol: f Compare **atto-.** [from Danish femten fifteen]

femur (fē′mər), n., pl. **femurs, femora** (fem′ər ə). Anatomy. 1 the bone of the leg between the hip and the knee in humans. Commonly called **thighbone.** See the pictures at **knee, leg. 2** a corresponding bone in the leg or hindlimb of vertebrate animals, such as a bird. 3 the third section (from the body) of the leg of an insect. [from Latin femur, femoris thigh]

fen, n. Ecology. low, wet land; marsh. It is covered wholly or partially with shallow, often stagnant water. SYN: swamp.
ASSOCIATED TERM: see **bog.**

fenestra (fə nes′trə), n., pl. **-trae** (-trē). Anatomy. a small opening in a bone, especially an opening in the medial wall of the middle ear. The **fenestra ovalis** is an oval opening leading into the vestibule of the ear. The **fenestra rotunda** is a round opening leading into the cochlea. Compare **foramen.** [from Latin fenestra window]
—**fenestral,** adj. of or having to do with a fenestra.
—**fenestrate** (fə nes′trāt), adj. Anatomy. having fenestrae.

fenestrated membrane, Histology. the outer layer of elastic tissue in the inner coat of a large artery.

fenestration (fen′ə strā′shən), n. Anatomy. the process or condition of being perforated: Ear surgeons may treat otosclerosis by an operation called fenestration, in which they cut a tiny window in the inner ear (G.W. Beadle).

feral (fir′əl), adj. Biology. 1 having reverted from domestication back to the original wild or untamed state. 2 wild; untamed. [from Latin fera wild beast, from ferus wild]

ferment, Biochemistry. —v. (fər ment′) 1 to undergo or produce a gradual chemical change in which bacteria, yeast, etc., change sugar into alcohol and produce carbon dioxide. Enzymes help ferment animal and vegetable matter. Yeasts often ferment food, or make it

---

cap, fāce, fäther; best, bē, tėrm; pin, fīve;
rock, gō, ôrder; oil, out; cup, pùt, rüle;
yü in use, yù in uric;
ng in bring; sh in rush; th in thin, ᴛʜ in then;
zh in seizure.
ə = a in about, e in taken, i in pencil, o in lemon, u in circus

alcoholic. *Some bacteria and yeasts produce hydrogen sulfide, a gas that turns food black and gives it the odor of rotten eggs* (John T.R. Nickerson).
—*n.* (fèr′mənt) **1** a substance that causes others to ferment. **2** a chemical change caused by a ferment; fermentation.
[from Latin *fermentare*, from *fermentum* leaven]
—**fermentation,** *n.* **1** the act or process of fermenting: *Ethanol results from the fermentation of sugar.* **2** a chemical change in an organic compound caused by a ferment: *In fermentation, enzymes act as catalysts.*
fermentation tube, *Microbiology.* a tube filled with a culture medium inoculated with microorganisms, in which the gases that are formed by fermentation collect in a closed arm of the tube where they can be measured and tested for composition.
fermi (fèr′mē), *n.* a unit of length in nuclear physics, equal to 10^{-13} centimeters: *The new breed is the high-energy physicist whose domain is the tiny, teeming world within the atom's nucleus. There, the fermi is the unit of measure (one fermi equals four ten-trillionths of an inch) and the inhabitants are some 30 "strange particles" with names like pion, lambda hyperon, anti-neutrino, pi zero meson* (Newsweek). [named after Enrico *Fermi*, 1901–1954, Italian-born American physicist, who directed the first controlled chain reaction in 1942]
Fermi-Dirac statistics (fèr′mē də rak′), a statistical theory in quantum mechanics which holds that in the distribution of nuclear particles of a given type, only one of a set of identical particles may occupy a particular quantum-mechanical state. Compare **Bose-Einstein statistics.** [named after Enrico *Fermi* and Paul *Dirac,* born 1902, British physicist]
fermion (fèr′mē on), *n. Physics.* (in quantum mechanics) any of a class of elementary particles, including protons, neutrons, and electrons, which conform to Fermi-Dirac statistics. Fermions have the property that their wave function changes sign if any two particles are interchanged. All particles with spin 1/2 are fermions. *It is noteworthy that the particles whose numbers are conserved, the leptons and the baryons, are all fermions, while those whose numbers are not conserved, the pion, K-mesons, and photons, are all bosons* (Science News). [named after Enrico *Fermi*]
fermium (fèr′mē əm), *n. Chemistry.* a radioactive metallic element produced artificially from plutonium or uranium. *Symbol:* Fm; *atomic number* 100; *mass number* 254. [named after Enrico *Fermi*]
-ferous, *combining form.* producing; containing; conveying, as in *metalliferous, odoriferous.* [from Latin *-fer* (from *ferre* to bear) + English *-ous*]
ferredoxin (fer′ə dok′sin), *n. Biochemistry.* an iron-rich protein present in green plants and in anaerobic bacteria. It functions as an acceptor of electrons from chlorophyll, especially in photosynthesis. *Ferredoxins ... have no enzymic activity themselves, but nevertheless have a wide variety of applications. ... Because of the simplicity and efficiency of ferredoxin, advocates of the chemical evolution theory of the origin of life have proposed that it was among the first proteins to be synthesized on earth* (Richard Cammack). [from Latin *ferrum* iron + English *redox* + *-in*]

ferric, *adj. Chemistry.* of or containing iron, especially trivalent iron: *ferric chloride, ferric hydroxide, ferric salts.* Compare **ferrous.** [from Latin *ferrum* iron]
ferric oxide, *Chemistry.* a compound of iron and oxygen found naturally as hematite and produced chemically as a powder for use as a pigment, abrasive, etc.: *The color of ferric oxide varies from reddish-brown to red, and depends somewhat upon the heat treatment to which it has been subjected. Under the name of Venetian red, it is used as a pigment, and more highly purified forms known as rouge are used as cosmetics and for the polishing of lenses and other articles requiring a high polish* (Jones, *Inorganic Chemistry*). Formula: Fe_2O_3 Also called **iron oxide.**
ferricyanide (fer′i sī′ə nīd), *n. Chemistry.* a salt containing the trivalent group $Fe(CN)_6$.
ferrimagnet (fer′i mag′nit), *n. Physics.* a ferrimagnetic body or substance: *Ferrites and other magnetic oxides are ferrimagnets* (Scientific American).
ferrimagnetic (fer′i mag net′ik), *adj. Physics.* of, having to do with, or exhibiting ferrimagnetism: *a ferrimagnetic crystal. Because ferrimagnetic substances are electrically nonconducting ... they have proved highly useful material for the coating of magnetic tape, computer memory cores, and other important Electronic Age components* (Time).
ferrimagnetism (fer′i mag′nə tiz′əm), *n. Physics.* a form of magnetism displayed by ferrites and certain other compounds in which neighboring ions align in an antiparallel manner in a magnetic field. Ferrimagnetism is similar to but not identical with ferromagnetism.
ferrite (fer′īt), *n. Physics.* a magnetic material consisting chiefly of ferric oxide combined with a metal or metals other than iron, characterized by high electrical resistance and used especially to keep eddy current losses small: *If complete annulment is not caused, a weak ferromagnetic results, and the substance is said to be a ferrimagnetic. The most common substances of this type are the ferrites, which are mixed oxides of iron and other elements of the form Fe_2MO_4, where M represents any 2-valent metal ... The substances are poor electrical conductors and have numerous applications where magnetic materials are needed in which eddy current losses are small* (Science News). [from Latin *ferrum* iron]
—**ferritic** (fe rit′ik), *adj.* containing or resembling ferrite.
ferritin (fer′ə tin), *n. Biochemistry.* a protein containing iron, synthesized in the liver for use by the body: *Ferritin [is] a protein that contains iron particles .011 micron in diameter. These particles are readily recognized in electron micrographs* (David S. Smith).
ferro-, *combining form.* **1** alloy of iron and ___, as in *ferrochromium = alloy of iron and chromium.* **2** that contains iron; iron, as in *ferroconcrete = concrete that contains iron.* [from Latin *ferrum* iron]
ferroalloy (fer′ō al′oi), *n. Chemistry.* an alloy of iron with another element such as tungsten, manganese, chromium, or vanadium.
ferrocyanide (fer′ō sī′ə nīd), *n. Chemistry.* a salt containing the tetravalent group $Fe(CN)_6$.
ferroelectric (fer′ō i lek′trik), *Physics.* —*adj.* exhibiting electrical properties analogous to the magnetic properties of ferromagnetic materials, such as the ability to hold an electric charge: *The material used in the*

experimental ferroelectric memory systems is barium titanate (Scientific American). —*n*. a nonconducting material having ferroelectric properties.

ferromagnet (fer′ō mag′nit), *n. Physics.* a ferromagnetic body or substance: *Firstly, any ferromagnet contains strong internal magnetic fields, whether or not it is in a magnetic field; secondly, the combination of internal magnetic field and an applied field destroys superconductivity* (New Scientist).

ferromagnetic (fer′ō mag net′ik), *adj. Physics.* of, having to do with, or exhibiting ferromagnetism; able to become highly magnetic in a relatively weak magnetic field, as iron, steel, cobalt, and nickel: *If a colloidal suspension of a ferromagnetic powder is placed on the polished surface of a ferromagnetic crystal the powder will arrange itself along lines where the [magnetic] field is most intense* (W. D. Corner).

ferromagnetism (fer′ō mag′nə tiz əm), *n. Physics.* a form of magnetism in which a substance tends to take a position with the longer axis parallel to the lines of force in a magnetic field. Very high magnetic permeability and hysteresis are characteristics of ferromagnetism. *Ferromagnetism is the kind of magnetism associated with iron, cobalt and nickel in some alloys* (Richard M. Bozorth). Compare **ferrimagnetism.**

ferrous (fer′əs), *adj.* of or containing iron, especially divalent iron: *The ferrous ion, a strong absorber of light in the red region of the spectrum, tends to color the glass blue; the ferric ion, a weaker absorber in the violet, tends to color it yellow* (Robert H. Brill). Compare **ferric.**

ferruginous (fə rü′jə nəs), *adj. Geology.* **1** of or containing iron; like that of iron: *Ferruginous sandstone is characterized by having hematite or limonite as its cement* (Fenton, *The Rock Book*). **2** reddish-brown like rust. [from Latin *ferruginus,* from *ferruginem* iron rust, from *ferrum* iron]

fertile, *adj.* **1** *Biology.* **a** able to produce seeds, fruit, young, etc.: *a fertile animal or plant.* **b** able to develop into a new individual; fertilized: *Chicks hatch from fertile eggs.*
2 *Ecology.* able to produce much; producing crops easily; rich in material for plant growth: *We point out that "fertile" is not a precise designation in ecological usage, but in general it refers to sites with a high availability of plant nutrients and optimum water supply. High nutrient availability is usually a product of the parent material of the soil (the geological substrate from which soil develops), a favorable combination of temperature and moisture for decomposition and nutrient release* (Science).
3 *Botany.* **a** fruiting; capable of producing fruit; having a perfect pistil: *a fertile flower.* **b** capable of fertilizing, as an anther with well-developed pollen. **c** producing organs that bear spores, as a fern.
4 *Physics.* capable of becoming fissionable through the action of bombarding neutrons in a reactor: *Uranium-238, which makes up more than 99 per cent of natural uranium, is no use as a reactor fuel but it is "fertile", which means that neutrons can turn it into plutonium* (New Scientist).
[from Latin *fertilis,* from *ferre* to bear]
—**fertility,** *n.* the quality or condition of being fertile. SYN: fecundity.

fertilization, *n. Biology.* the act or process of fertilizing; the union of a male reproductive cell and a female reproductive cell to form a cell that is capable of developing into a new individual; union of a monoploid sperm nucleus (n) with a monoploid egg nucleus (n) to form a diploid zygote, in which the species number of homologous chromosomes (2n) is restored. Compare **external fertilization** and **internal fertilization.**

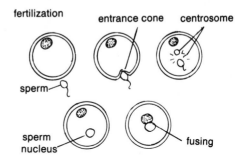

fertilization · entrance cone · centrosome · sperm · sperm nucleus · fusing

fertilization cone, = entrance cone.

fertilization membrane, *Biology.* an extension of the membrane surrounding an egg, which prevents the entry of additional sperm after fertilization has occurred: *Under normal circumstances ... the egg is doubly protected by the fertilization membrane and by the hyaline layer, either of which can stop supernumerary sperm from penetrating* (Scientific American).

fertilize, *v. Biology.* **1** to unite with (an egg cell) in fertilization: *Although a single ejaculation may contain two or three hundred million sperms, only one will actually fertilize the ovum, though the others may assist in the fertilization as a result of an enzyme they produce which helps to disperse the remaining follicle cells adhering to the surface of the ovum* (not confirmed in man so far). *In some mammals such as the rabbit, several ova are released from the ovary at the same time and will be fertilized by the corresponding number of sperms* (Mackean, *Introduction to Biology*). **2** to impregnate or inseminate; make fertile.

fetal, *adj. Biology.* **1** of or having to do with a fetus. The **fetal appendages** are the placenta, amnion, chorion, and umbilical cord. **2** like that of a fetus: *a rolled-up fetal position.*

fetus (fē′təs), *n. Biology.* an animal embryo during the later stages of its development in the womb or in the egg, especially a human embryo from about three months after fertilization until its birth. [from Latin]

Feulgen reaction (foil′gən), *Biochemistry.* a method of staining, in preparation for microscopic examination, which colors a nucleic acid purple and leaves the remainder of a cell colorless, used in the study of the nu-

---

cap, fāce, fäther; best, bē, tėrm; pin, fīve;
rock, gō, ôrder; oil, out; cup, pùt, rüle;
yü in use, *yù* in uric;
ng in bring; *sh* in rush; *th* in thin, ᴛʜ in then;
zh in seizure.
ə = *a* in about, *e* in taken, *i* in pencil, *o* in lemon, *u* in circus

219

clei of cells, especially to determine the presence and location of DNA. [named after Robert *Feulgen,* 1884–1955, German biochemist]

Feynman diagram, *Nuclear Physics.* a graphic representation of various interactions of elementary particles, such as electrons, positrons, and photons: *Electromagnetic and weak processes exhibit striking similarities when depicted in the form of Feynman diagrams. Such diagrams symbolize the interactions that underlie subnuclear phenomena, for example the collision between two particles, which physicists refer to as a scattering event* (Steven Weinberg). [named after Richard P. *Feynman,* born 1918, an American nuclear physicist who devised it]

fiber or **fibre,** *n.* **1a** any of the very fine, long threadlike pieces of which many organic and some inorganic materials consist; a threadlike part: *A muscle is made up of many fibers.* **b** a similar structure of artificial origin: *Nylon is a synthetic fiber.* SYN: filament.
2 *Botany.* **a** one of the narrow, elongated cells in the bast or sclerenchyma of plants: *When sclerenchyma fibers occur in the xylem region of fibrovascular bundles, they are termed xylem fibers, or wood fibers; when they appear in the phloem region, phloem fibers, bast fibers, or liber fibers; when in the pericycle, pericyclic fibers; when in the cortex, cortical fibers. Wood fibers are usually cut off by the cambium only and so are generally seen in the secondary xylem. They differ from the fibers of the phloem, pericycle, and cortex in that they possess bordered pits, although in many instances these are so reduced by the thickening of the wall as to appear simple. Phloem, pericyclic, and cortical fibers possess simple pits* (Youngken, *Pharmaceutical Botany*). **b** a slender, threadlike root of a plant.
3 *Physiology.* that part of an edible plant that is resistant to human digestive enzymes: *There is no concise definition of fiber, certainly none that can describe its role in human nutrition. In general, fiber can be defined as a complex of substances of plant origin that appear not to be absorbed or digested by humans. Most, but not all, of these substances are found in the plant cell wall* (David Kritchevsky).
4 *Anatomy.* a threadlike part of a nerve cell or tissue: *The axon that does so in the illustration ... originated in a secondary sensory cell group of the spinal cord and is therefore called a spinocerebellar fiber. ("Axon" and "fiber" are synonymous in neuroanatomical usage.) Many of these fibers together would compose a spinocerebellar tract or bundle* (W.J.H. Nauta and Michael Feirtag).
5 *Optics.* = optical fiber: *Fibers are an excellent means of transmitting information coded this way, because a bright pulse can represent a 1 and a dim pulse or no pulse can represent a 0* (Jay Myers).
[from Old French *fibre,* from Latin *fibra*]

fiberoptic (fī′bər op′tik), *adj. Optics.* of or having to do with fiber optics: *The snake-like tube he was operating is a new fiberoptic kind of endoscope, an instrument which allows doctors to see inside their patients, take pictures and even snip off bits of tissue for analysis* (Athens News).

fiber optics, *Optics.* **1** a bundle of fine glass or plastic fibers having refraction properties that permit them to transmit light around curves and into inaccessible locations: *The endoscopes make use of fiber optics, in which hair-thin filaments transmit light around curves with no appreciable loss of intensity* (Judith Randal). **2** the branch of optics using optical fibers to transmit light: *Technological progress in fiber optics has given the engineer an important additional tool for transferring information by means of light* (Science). See **optical fiber.**

fiberscope, *n. Medicine.* an optical instrument with a flexible bundle of fiber optics, used for inspecting internal cavities and tissues: *The fiberscope allows you to go under the skin into a body cavity such as the lung for diagnostic work without surgery* (N.Y. Sunday News). Compare **endoscope.**

Fibonacci numbers or **Fibonacci sequence** (fē′bə– nä′chē), *Mathematics.* an infinite, ordered set of numbers, 1, 1, 2, 3, 5, 8, 13, 21, 34, 55, 89, 144, etc., in which each number is the sum of the preceding two numbers: *In view of this, a very nearly correct drawing can be readily made on isometric paper, if we use four consecutive Fibonacci numbers—for instance 5, 8, 13, 21—to give lengths for segments between junction points on each of the three pairs of parallel lines* (New Scientist). *The sum of the first 10 numbers in a generalized Fibonacci series is always 11 times the seventh number* (Martin Gardner). [named after Leonardo *Fibonacci,* 13th-century Italian mathematician]

fibril (fī′brəl), *n.* **1** *Anatomy.* a small or very slender fiber, as of a muscle or nerve: *Studies using isolated fibrils from skeletal muscle are playing an important part in investigating the mechanism of contractile processes. These fibrils, which are the main cytoplasmic components of skeletal muscle cells, show a characteristic striated structure under the electron microscope, and isolated preparations suspended in a medium containing appropriate metal ions will contract on the addition of the coenzyme adenosine triphosphate* (D.L. Woodhouse and H.S.A. Sherratt).
2 *Botany.* one of the hairs on the roots of some plants; root hair.
—**fibrillar** (fī′brə lər), *adj.* of, having to do with, or like fibrils.

fibrillate (fī′brə lāt), *v. Physiology.* to undergo or cause to undergo fibrillation: *Under hypothermia the heart is especially likely to lose its regular beat and flutter uselessly (fibrillate), which may cause death* (Time).

fibrillation (fī′brə lā′shən), *n. Physiology.* a tremor in a muscle, especially a condition in the heart characterized by independent and irregular action of the muscle fibers: *Whenever disordered, ineffective cardiac rhythm (fibrillation) is detected, external defibrillation is accomplished with a 110-volt a-c shock* (Robert W. Buxton).

fibrin (fī′brən), *n. Biochemistry.* a white, tough, elastic, fibrous protein formed by the action of thrombin on fibrinogen, especially when blood clots: *The breakdown of the platelets releases into the plasma a compound which initiates a chemical reaction eventually resulting in the precipitation of a clotted protein, termed fibrin. The fibrin forms a meshwork of strands which, with the entrapped red blood cells, comprises* the clot (Harbaugh and Goodrich, *Fundamentals of Biology*).

—**fibrinous** (fī′brə nəs), *adj.* composed of or resembling fibrin.

fibrinogen (fī brin′ə jən), *n. Biochemistry.* a protein found in the blood, lymph, etc., which interacts with thrombin to form fibrin in the coagulation of blood: *Fibrinogen is the material which, when activated by the enzyme thrombin, causes blood to coagulate* (Science Journal).

fibrinoid (fī′brə noid), *n. Biology.* a homogeneous substance that stains like fibrin, found normally in the placenta and also in diseased connective tissue.

fibrinolysin (fī′brə nol′ə sən), *n. Biochemistry.* any enzyme that can cause fibrin to dissolve. Compare **plasmin.**

fibrinolysis (fī′brə nol′ə sis), *n. Biochemistry.* the dissolving of fibrin by enzymatic action. Fibrinolysis resulting from an imbalance of enzymes may cause severe bleeding, especially during surgery and childbirth. *In fibrinolysis, the dissolution pattern is very similar to that observed in clot formation. ... There are blood and tissue activators, as well as substances quite foreign to the body, all of which can speed the formation of fibrinolysin. The blood activator is not well understood, but is thought to be responsible for the increased fibrinolytic activity after exercise or in severe stress. Indeed, in cases of sudden death, the fibrinolytic mechanism is so active that the blood remains fluid* (Saturday Review). [from *fibrin* + Greek *lysis* dissolution]

—**fibrinolytic** (fī′brə nō lit′ik), *adj.* of or having to do with fibrinolysis: *a fibrinolytic enzyme.*

fibrinopeptide, *n. Biochemistry.* a protein substance formed in blood clotting: *Fibrinopeptides, short amino acid chains discarded in the process of blood clotting, are particularly rapidly evolving molecules. For them, the sequence differences have been found to be: from man to chimpanzee, 0; to gorilla, 0; to orang, 2; to gibbon, 3–5; to monkey, 5–8; to New World monkey, 9–10; and to prosimian, 18* (S.L. Washburn and E.R. McCown). *One hemoglobin chain consists of approximately 140 amino acids. Fibrinopeptides A and B, on the other hand, consist of only about 20 amino acids, which are cut out of fibrinogen and discarded during the clotting process. The hemoglobins and the fibrinopeptides also appear to be evolving individually at a uniform average rate, but their rates are quite different* (Richard E. Dickerson).

fibro-, *combining form.* fiber; fibers; fibrous tissue, as in *fibrogenic.* [from Latin *fibra* fiber]

fibroblast (fī′brə blast), *n. Histology.* a cell that gives rise to connective tissue: *One approach has been to induce the manufacture of interferon in human fibroblasts (precursor connective-tissue cells) or epithelial cells grown in large numbers in tissue culture* (Scientific American).

—**fibroblastic** (fī′brə blas′tik), *adj.* of or having to do with fibroblasts.

fibrogenic, *adj.* producing fibers: *The great fibrogenic activity of quartz compared with other minerals makes dust composition of equal importance* (New Scientist).

fibroid (fī′broid), *adj. Histology.* made up of or resembling fibers or fibrous tissue, as a tumor.

fibrous (fī′brəs), *adj.* **1** made up of or resembling fibers: *Grasses and cereals have fibrous roots.* SYN: filamentous. **2** *Mineralogy.* having a splintery or threadlike surface when fractured, or a threadlike or needlelike habit.

► See the note under **membrane.**

fibrovascular (fī′brō vas′kyə lər), *adj. Botany.* consisting of woody fibers and ducts: *the fibrovascular tissue in a plant.* The **fibrovascular bundle** is a vascular bundle surrounded by elongate fibers. Leaf veins are fibrovascular bundles.

fibula (fib′yə lə), *n., pl.* **-lae** (-lē′), **-las.** *Anatomy.* **1** the outer and thinner of the two bones in the human lower leg. It extends from knee to ankle. See the picture at **leg. 2** a similar bone in the hind leg of animals. [from Latin *fibula* clasp, brooch]

—**fibular,** *adj.* of or having to do with the fibula.

field, *n.* **1** *Physics.* **a** the space throughout which a force operates. A magnet has a magnetic field around it. *The lines of force show the direction of the field and ... the density or closeness of the lines of force indicates the strength of the field. Because a field is a vector, which has a magnitude and a direction (the direction of the lines of force), we can resolve the field into components, and we can speak of the component of the field in any direction we choose* (John R. Pierce). **b** an analog of this space in quantum mechanics: *In quantum mechanics the particles themselves can be represented as fields. An electron, for example, can be considered a packet of waves with some finite extension in space. Conversely, it is often convenient to represent a quantum-mechanical field as if it were a particle. The interaction of two particles through their interpenetrating fields can then be summed up by saying the two particles exchange a third particle, which is called the quantum of the field* (Scientific American).

2 *Optics.* the space or area in which things can be seen through the lens of a telescope, microscope, etc., usually quoted as an angle, such as (in astronomy) 1 degree on the sky. Also called **field of view.**

3 *Mathematics.* any set which has two operations called addition and multiplication, with multiplication distributive over addition. For each operation, the set is closed, associative, and commutative, has an identity element and inverses (the zero element being excluded for multiplication). *The set of all real numbers is a number field.*

field emission, *Physics.* the release of electrons from the surface of a conductor, caused by a strong electric field.

field-emission microscope, a microscope which uses the electrons emitted from the surface of a metallic conductor under influence of a strong electric field to produce a magnified image of the emitting surface on a

---

cap, fāce, fäther; best, bē, tèrm; pin, five;
rock, gō, ôrder; oil, out; cup, pùt, rüle,
yü in use, yù in uric;
ng in bring; sh in rush; th in thin, ᵀH in then;
zh in seizure.

ə = a in about, e in taken, i in pencil, o in lemon, u in circus

fluorescent screen. It has a resolution of about 20 angstroms.

field-ion microscope, a microscope similar to a field-emission microscope but using gaseous ions originating near the surface of a metal tip to produce a magnified image of the emitting surface on a fluorescent screen. It is powerful enough to permit visualization of individual atoms. Also called **ion microscope.**

field of force, *Physics.* = field (def. 1).

field of view, = field (def. 2).

field of vision, *Optics.* the space or range within which objects are visible to the immobile eye at a given time. Also called **visual field.**

field strength, *Physics.* the strength of a field at a given point, as measured by the force generated on a unit charge, unit mass, or the like, at that point.

field theory, *Physics.* **1** any theory about physical phenomena that takes into account the effects of one or more fields: *... quantum electrodynamics and other field theories* (New Scientist). **2** the branch of physics dealing with such theories: *Field theory begins from a prejudice in favor of mathematical depth. ... The emphasis is on a rigorous mathematical understanding of the theory* (Scientific American). Compare **unified field theory.**

figure, *n. Geometry.* any combination of points, lines, or planes: *Circles and triangles are plane figures; spheres and cubes are solid figures.*

filament, *n.* **1** a very fine thread; very slender, threadlike part. SYN: fiber. **2** *Botany.* the stalklike part of a stamen that supports the anther. See the picture at **stamen.** [from Late Latin *filamentum,* from Latin *filum* thread]
—**filamentous,** *adj.* having an elongated, tubelike or threadlike structure: *filamentous algae and fungi.*

filar (fī′lər), *adj.* **1** *Biology.* of or having to do with a threadlike structure. SYN: filamentous. **2** *Optics.* having threads or wires across its field of view: *a filar microscope.* [from Latin *filum* thread]

filaria (fi lãr′ē ə), *n., pl.* **-lariae** (-lãr′ē ē). *Zoology.* any of a large group of threadlike parasitic nematode worms, whose larvae develop in mosquitoes and other arthropods and are transmitted to the blood and tissues of humans and other vertebrates, causing such diseases as elephantiasis and onchocerciasis. Compare **microfilaria.** [from New Latin *Filaria* the genus name of a common type of filaria, from Latin *filum* thread]

filial generation, *Genetics.* any generation of offspring of a hybrid. *Symbol:* F; F_1, F_2, F_3, etc., mean the first, second, third, etc., filial generations. *When an individual homozygous for the dominant trait is crossed with an individual homozygous for the recessive trait, the appearance (phenotype) of the offspring, known as the F_1 generation, is like that of the dominant parent. The genotype of these offspring is heterozygous. Offspring resulting from the cross between members of the F_1 generation comprise the F_2 generation* (Biology Regents Syllabus).

fillet (fil′it), *n. Anatomy.* a band of fibers, especially a white nerve tract in the brain. [from Old French *filet,* diminutive of *fil* thread, from Latin *filum*]

film, *n.* **1** *Chemistry.* a very thin layer, sheet, surface, or coating, often of liquid. Oil poured on water will spread and make a film. *As a mulch, black polyethylene film keeps moisture in the ground and prevents the growth of weeds by shutting off light* (Hiram McCann). **2** *Biology.* a thin skin or membranous layer.

filter, *n.* **1** *Chemistry.* **a** a device for passing a liquid or gas slowly through paper, sand, charcoal, felt, or other porous media. A filter is used to remove impurities from drinking water. **b** the paper, sand, charcoal, felt, or other porous material used in such a device. **2** *Physics.* **a** a device for permitting only waves of particular frequencies to pass, eliminating or reducing the others. **b** a substance or device for absorbing or reflecting light rays of particular wavelengths.
—*v.* **1** to act as a filter for: *Charcoal filters many gases.* **2** to pass or flow very slowly: *Water filters through the sandy soil and into the well.*

filterable, *adj. Biology.* capable of passing through a filter with minute pores that arrests most microorganisms: *a filterable virus.* Also spelled **filtrable.** Compare **filter-passing.**

filter paper, *Chemistry.* porous paper, usually made of cellulose, used for filtering: *Filter paper, like blotting paper, will mop up relatively large amounts of water and can be readily divided up after the electrophoretic run to allow characterization of the separated components* (New Scientist).

filter passer, *Biology.* a bacterium or virus that is small enough to pass through a filter which arrests most microorganisms.
—**filter-passing,** *adj.* passing through a filter, as a bacterium or virus. Compare **filterable.**

filtrable, *adj.* = filterable.

filtrate (fil′trāt), *n. Chemistry.* a liquid or gas that has been passed through a filter: *An enzyme preparation from the culture filtrate of a mutant strain of Bacillus subtilis bacteria is very active in removing protein and carbohydrate stains from fabrics when it is added to detergents* (J.R. Porter).

filtration, *n. Chemistry.* the process of filtering or the condition of being filtered.

fimbria (fim′brē ə), *n., pl.* **-briae** (-brē ē). *Biology.* **1** a fringe or fringed border. **2** = pilus: *Fimbriae ... occur on the surfaces of many gram-negative bacteria. They range in diameter from 4 to 25 nm, and their length is also quite variable* (Emil Gotschlich). [from Latin *fimbria* fringe]
—**fimbriate** (fim′brē āt), *adj. Biology.* having a fimbria or fimbriae; fringed, as the petals of certain flowers.

fins of the fish:

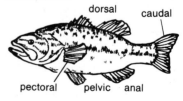

dorsal caudal

pectoral pelvic anal

fin, *n. Zoology.* one of the movable winglike or fanlike parts of a fish's or other aquatic animal's body, used for propelling, steering, and balancing the body in water: *On the back [of a bony fish] are two separate dor-*

sal fins, on the end of the tail is the caudal fin, and ventrally on the tail is the anal fin; all these are median. The lateral or paired fins are the pectoral fins behind the opercula and the ventral or pelvic fins close below (Storer, *General Zoology*).

finder, *n. Optics.* a small telescope attached to a larger one to help find objects more easily. A finder is usually low-powered and has a wide-angle lens. *For telescopes over 3 inches in aperture it is well to have a finder attached to the telescope tube* (Bernhard, *Handbook of Heavens*).

fine, *adj. Chemistry.* having a stated proportion of gold or silver in it. A gold alloy that is 925/1000 fine is 92.5 per cent gold. Compare **carat.**

finite (fī′nīt), *adj. Mathematics.* **1** (of a number) that can be reached or exceeded by counting. **2** (of a magnitude) less than infinite and greater than or equal to zero. **3** (of a set) having a limited number of elements. [from Latin *finitus,* ultimately from *finis* end]
—finitude (fin′ə tüd *or* fī′nə tüd), *n.* the quality or condition of being finite.

fiord or **fjord** (fyôrd), *n. Geography, Geology.* a long, narrow bay of the sea bordered by steep cliffs: *Other spectacular landforms of glaciated terrain are the fiords that flank many high-latitude coasts, such as in Alaska, the Canadian Arctic, Norway, Chile, and New Zealand. ... Geologists have long speculated upon the origin of fiords, and the theories range from one of tectonic origin to one of glacial scouring and overdeepening of former stream valleys. Surely, many fiords owe much of their origin to glacial erosion* (Birkeland and Larson, *Putnam's Geology*). Compare **firth.** [from Norwegian *fiord,* earlier *fjorthr*]

fireball, *n.* **1** *Astronomy.* **a** a large, brilliant meteor. **b** the bright sphere of light produced by a meteor.
2 a great, luminous cloud of hot gases, water vapor, and dust produced by a nuclear explosion.

firn, *n.* = névé. [from Swiss German *Firne* snow above glaciers]

first quarter, *Astronomy.* **1** the period of time between the new moon and the first half moon. **2** the phase of the moon represented by the first half moon after the new moon.

firth (fèrth), *n. Geography.* **1** a narrow arm of the sea. **2** the estuary of a river. [from Scandinavian (Old Icelandic) *fjörthr*. Related to FIORD.]

fish, *n., pl.* **fishes** or (*collectively*) **fish.** *Zoology.* any of of numerous vertebrates, generally cold-blooded, that live in water, have gills instead of lungs for breathing, and are usually covered with scales and equipped with fins for swimming. Some fishes lay eggs in the water; others produce living young.

fissile (fis′əl), *adj.* **1** *Geology.* (of a rock) that can be split or divided easily along close parallel planes: *Slate is fissile.*
2 *Nuclear Physics.* = fissionable: *Plutonium can be used as an alternative to the fissile isotope of uranium, U-235* (New Scientist). [from Latin *fissilis,* from *findere* to cleave]

fission (fish′ən), *n.* **1** *Biology.* a method of reproduction in which the body of the parent divides to form two or more independent individuals: *The two attached cells were originally a single cell which split to form two cells. This type of asexual reproduction is called fission, or cell division. It is a very common form of*

self-duplication in simple organisms with only one cell, such as Protococcus. The same process of cell division accounts for growth in larger and more complex plants and animals* (Robert W. Menefee). See the picture at **asexual.**
2 *Nuclear Physics.* the splitting of an atomic nucleus into two parts, especially when bombarded by a neutron. Fission releases huge amounts of energy when the nuclei of heavy elements, especially uranium and plutonium, are split. The chain reactions in atomic bombs and nuclear power plants are due to fission. *Fission means that the heavy nuclei of uranium or plutonium break into fragments about half the weight, and those fragments are unstable—that is, radioactive. While they are in the reactor, their radiations are stopped by the shield; but every few months or so the uranium must be taken out and purified* (Otto Frisch). *Unlike fission, in which the atomic nucleus is split to release energy, the basic fusion reaction requires that two nuclei of heavy hydrogen—called deuterium—fuse to make an atom of helium. Fission is the process that powered the original atom bomb. Fusion is the energy source for the much more powerful thermonuclear devices—hydrogen bombs* (Harold M. Schmeck, Jr.). Also called **nuclear fission.** See the picture at **nuclear fission.**
—v. to undergo or cause to undergo fission: *When a single uranium atom fissions, or splits in half, it immediately releases 180 million electron volts of energy. This energy release is enough to make one pound of uranium equivalent in energy content to 8,000 tons of TNT or to 1,500 tons of coal* (Harper's). *During the three-year period that the fuel normally spends in the reactor, not all the U-235 is fissioned; perhaps a quarter of it is left* (New Yorker).
[from Latin *fissionem* a splitting, cleavage, from *findere* to cleave]
—fissionable, *adj.* capable of nuclear fission; fissile: *U-235 is fissionable; that is, its atoms will split and release enormous energy* (N.Y. Times). **—n.** a fissionable substance: *Illegal manufacture of fissionables is no more difficult to detect than illegal manufacture of any other item requiring a large industrial effort* (Bulletin of Atomic Scientists).

fission-track dating, *Geology.* a method of determining the age of rocks and other geological formations by counting the characteristic tracks left by spontaneous fission of uranium 238 during the lifetime of each sample. The number of tracks is proportional to the age of the sample. *Fission-track dating is direct and visual and has proved to be applicable to many materials and over an enormous range of ages. The critical factor is the uranium content of the material to be dated. A concentration of one part per million—which is common in rocks—provides enough tracks to date an ob-*

---

cap, fāce, fäther; best, bē, tèrm; pin, five;
rock, gō, ôrder; oil, out; cup, pùt, rüle,
yü in use, *yù* in uric;
ng in bring; *sh* in rush; *th* in thin, ŦH in then;
zh in seizure.
ə = *a* in about, *e* in taken, *i* in pencil, *o* in lemon, *u* in circus

223

ject older than some 100,000 years easily (Scientific American).

fissure (fish'ər), *n.* **1** *Geology.* a long, narrow opening or crack in a rock: *limestone fissures.*
2 *Anatomy.* a natural cleft or opening in an organ or part, as those separating the convolutions of the brain, or that on the lower surface of the liver.
[from Latin *fissura,* from *findere* to cleave]
—*v. Geology.* to form a fissure: *The velocity of transmission of sound through a particular rock is directly related to the extent of fissuring* (New Scientist).

fistula (fis'chə lə), *n., pl.* **-las, -lae** (-lē'). *Medicine.* a tubelike abnormal passage connecting the surface of the body with some internal cavity or organ, caused by a wound, abscess, disease, etc.: *An injury in the fourth week, when the windpipe is budding off from the gullet, may produce the well-known defect which leaves an opening between the windpipe and gullet —so-called tracheo-esophageal fistula* (Theodore H. Ingalls). [from Latin *fistula* pipe, ulcer]

fix, *v.* **1** *Chemistry.* **a** to convert into a more reactive, usable form: *By 1946, no fewer than six reactions had been discovered by which carbon dioxide could be fixed by animal tissues, the most important being probably the condensation of carbon dioxide with pyruvic acid to give oxalo-acetic acid* (A.W. Haslett). See also **nitrogen fixation. b** to make less volatile; make more stable.
2 *Biology.* to prepare or preserve (an organism or tissue) for microscopic study: *Some method has to be found of "fixing" or killing the cells instantaneously in a way which will not change the structure that it is required to examine* (G.A. Meek).

fixation, *n. Chemistry.* **1** the process of converting into a more reactive, usable form. See also **fix** (def 1), **nitrogen fixation. 2** the process of making less volatile or more stable.

fixative, *n.* **1** *Chemistry.* a substance used to prevent something from fading or otherwise changing.
2 *Biology.* a substance used to preserve tissue for microscopic examination: *To distinguish clearly between one small part of a cell and its immediate surroundings ... with the light microscope, we use killing agents (fixatives) and strains to bring out the parts we want to examine. Literally hundreds of fixing and staining procedures are known* (McElroy, *Biology and Man*).

fixed star, *Astronomy.* a star whose position in relation to other stars appears not to change, at least without any detailed positional information. Fixed stars are so far from earth that their motion can be easily measured only over a very long period of time.

fjord (fyôrd), *n.* = fiord.

flagella (flə jel'ə), *n.* plural of **flagellum.**

flagellar (flə jel'ər), *adj.* having to do with flagella: *Flagellar movement requires energy from respiration, and the energy transfer mechanism may be quite similar to that of muscle contraction* (Greulach and Adams, *Plants*).

flagellate (flaj'ə lit), *Biology.* —*adj.* **1** having flagella: *flagellate infusoria.* **2** whiplike; flagelliform.
—*n.* any protozoan or alga having one or more flagella serving as organs of locomotion and for obtaining food. Euglenas are flagellates.

flagelliform (flə jel'ə fôrm), *adj.* like a whip; long, thin, and flexible.

flagellum (flə jel'əm), *n., pl.* **-la** (-lə), **-lums.** *Biology.* **1** a long, whiplike tail or part which is an organ of locomotion in certain cells, bacteria, protozoa, etc.: *The flagella used by some bacteria to propel themselves towards or away from a particular environment are ... yielding insights into "sensory" response at a unicellular level* (Raymond N. Doetsch). **2** a runner of a plant.
[from Latin *flagellum,* diminutive of *flagrum* whip]

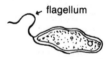

flagellum

flame cell, *Zoology.* an excretory cell in flatworms, rotifers, etc., having cilia that beat with a flamelike motion and move waste products into excretory tubes: *The excretory system comprises two longitudinal ducts connecting to a network of tubules that branch throughout the body and end in many large flame cells* (Storer, *General Zoology*).

flammable, *adj. Chemistry.* easily ignited and tending to burn rapidly.
▶ *Flammable* and *inflammable* are synonyms, not antonyms. *Inflammable* is derived from the verb *inflame. Flammable* is now the preferred term because *inflammable* has been mistakenly thought by some to mean "not flammable."

flank, *n.* **I** *Zoology.* the fleshy or muscular part of the side of an animal between the ribs and the hip.
2 *Geology.* = limb[1] (def. 3): *The sides of such folds [synclines] are called limbs, or flanks, and a line drawn along the points of maximum curvature of each bed is termed the axis* (Birkeland and Larson, Putnam's *Geology*).

flare star, *Astronomy.* any of a class of stars, usually dwarfs of low intrinsic luminosity, that exhibit at intervals brief outbursts of energy: *Certain stars have been found which, within a few minutes, increase greatly in brightness and just as quickly decline to their average brightness. This may happen only at long and irregular intervals. The name "flare stars" has been assigned to this type* (Charles P. Olivier).

flash point, *Chemistry.* the temperature at which the vapor given off from a volatile liquid will ignite spontaneously in air, in the presence of a small flame.

flavin (flā'vin), *n. Biochemistry.* **1** any of various water-soluble pigments that are coenzymes of flavoproteins: *It turned out that a second coenzyme was needed to transport electrons (plus the protons H^+) from reduced DPN to molecular oxygen flavin, a derivative of the vitamin riboflavin* (McElroy, *Biology and Man*). **2** = riboflavin.

flavin adenine dinucleotide, *Biochemistry.* a compound of adenosine diphosphate and flavin mononucleotide that constitutes the nonprotein part, or coenzyme, of various flavoproteins: *In living organisms respiratory oxidations are the work of two groups of enzymes, the dehydrogenases and oxidases ... In the respiratory dehydrogenases the prosthetic group is a derivative of nicotinamide adenine dinucleotide (NAD) or nicotinamide adenine dinucleotide phosphate (NADP); in flavoproteins, riboflavin is present as flavin adenine* dinucleotide (Storer, *General Zoology*). *Abbreviation:* FAD *Formula:* $C_{27}H_{33}N_9O_{15}P_2$

flavin mononucleotide, *Biochemistry.* a phosphoric ester of riboflavin that constitutes the nonprotein part, or coenzyme, of various flavoproteins: *Another hydrogen-transferring coenzyme is flavin adenine dinucleotide (FAD), and there is the related flavin mononucleotide (FMN). These and other flavin coenzymes, along with their characteristic enzyme proteins, are often referred to as flavoproteins* (Greulach and Adams, *Plants*). *Abbreviation:* FMN *Formula:* $C_{17}H_{21}N_4O_9P$

flavobacterium (flā'vō bak tir'ē əm), *n., pl.* **-teria** (-tir'ē ə). any of a group of rod-shaped bacteria that form yellow pigments, found in water and soil. [from New Latin *Flavobacterium* the genus name, from Latin *flavus* yellow + New Latin *bacterium*]

flavone (flā'vōn), *n. Chemistry.* 1 a colorless, crystalline compound found in various plants or produced synthetically, used as the basis of several dyes: *Nectar ... draws ethereal oils and flavones from the flower to obtain its characteristic taste, perfume and colour* (New Scientist). *Formula:* $C_{15}H_{10}O_2$ 2 any derivative of this substance. [from Latin *flavus* yellow]

flavonoid (flā'və noid), *n. Chemistry.* any of a group of natural organic compounds found chiefly as coloring matter in fruits and flowers: *Glycosides of quercetin, a mutagenic flavonoid, are present in considerable amounts in our diet from a variety of sources and, by means of hydrolysis, bacteria in the human gut readily liberate the mutagen* (Science).

flavonol (flā'və nol), *n. Chemistry.* any of various plant pigments or dyes derived from flavone.

flavoprotein, *n. Biochemistry.* any of a group of yellow respiratory enzymes containing flavin: *The flavoproteins, as their name indicates, are complex yellow nitrogenous compounds. They consist essentially of a protein component which determines the specificity of the enzyme, and a catalytic yellow nucleotide fraction whose activity depends on its iso-alloxazine content* (Dible, *Recent Advances in Bacteriology*). Also called **yellow enzyme.** [from Latin *flavus* yellow + English *protein*]

flavor, *n.* 1 *Physiology.* the sensation produced by a substance in the mouth, including its taste, smell, and feel. 2 *Nuclear Physics.* a hypothetical property, such as charm or strangeness, that distinguishes one subatomic particle from another: *The different kinds of quarks or leptons are known technically as flavors* (Robert H. March).

F layer or **F region,** *Geophysics.* a region of the ionosphere, above the E layer, which reflects high-frequency radio waves. It consists of the F_1 **layer,** about 90 to 150 miles above the earth's surface, and the F_2 **layer,** occurring from about 125 miles to over 250 miles above the earth's surface. Compare **D layer** and **E layer.** See the picture at **E layer.**

flection, *n.* = flexion.

flesh, *n.* 1 *Anatomy.* the soft tissue of the body of vertebrates, covering the bones and consisting mostly of muscles and fat. 2 *Botany.* the soft, juicy, edible part of fruits or vegetables. SYN: pulp.

—**fleshy,** *adj. Botany.* (of fruit) composed of juicy cellular tissue. SYN: pulpy. ANT: dry. See the picture at **fruit.**

flex, *v. Anatomy.* 1 to bend (a joint or limb) by the action of the flexor muscles: *As the trunk becomes more flexed, more muscle action is required to sustain the posture, and the strain on the ligaments is increased. There comes a position, however, when the ligaments will stretch no further and they take the full strain of the pull of gravity on the trunk. At this stage the muscles relax* (W.F. Floyd and F.H.S. Silver). 2 to bend or contract (a muscle or muscles). [from Latin *flexus,* past participle of *flectere* to bend]

flexion (flek'shən), *n. Anatomy.* 1 a bending of a joint in the body by the action of flexors. 2 a being bent in this way.

flexor (flek'sər), *n. Anatomy.* any muscle that when contracted bends a joint in the body: *The contraction of one group of muscles, the flexors, diminishes the angle between the skeletal segments, while the contraction of the opposite group, the extensors, increases the angle between the segments* (Harbaugh and Goodrich, *Fundamentals of Biology*).

flexure (flek'shər), *n.* 1 *Anatomy.* a bent part or thing; bend; curve. 2 *Geology.* =fold.

flight feather, *Zoology.* one of the rigid feathers of a bird's wing that are essential for flying: *The down feathers are fluffy, trapping a layer of air close to the body. The flight feathers and coverts are broad and flat and offer resistance to the passage of air* (Mackean, *Introduction to Biology*).

flint, *n. Mineralogy.* a very fine-grained, tough chalcedony which makes a spark when struck against steel. —**flinty,** *adj.* made of flint; containing flint: *... flinty concretions of Late Cretaceous age* (Moore, *Introduction to Historical Geology*).

flipper, *n. Zoology.* a broad, flat limb especially adapted for swimming. Seals have flippers.

float, *v. Physics.* to stay on top of or be held up by air, water, or other liquid: *An object floats in a fluid only if its weight is less than the weight of an equal volume of the fluid. ... To state it another way, if the density of an object is less than that of the fluid, it floats; if the density of the object is greater than that of the fluid, it sinks* (Tracy, *Modern Physical Science*). —**n.** *Geology.* a piece of ore or rock broken off a vein or strata and displaced from its original location.

floating ribs, *Anatomy.* false ribs whose anterior ends are unattached. The lowest two pairs of human ribs are floating ribs.

flocculant (flok'yə lənt), *n. Chemistry.* an agent that causes or promotes flocculation.

---

cap, fāce, fäther; best, bē, tėrm; pin, fīve;
rock, gō, ôrder; oil, out; cup, pùt, rüle,
yü in use, *yu̇* in uric;
ng in bring; sh in rush; th in thin, ᴛʜ in then;
zh in seizure.
ə = *a* in about, *e* in taken, *i* in pencil, *o* in lemon, *u* in circus

flocculate (flok′yə lāt), *v. Chemistry.* to form into flocculent masses; form compound masses of particles, as a cloud or a chemical precipitate: *The lime makes the particles clump together or flocculate, the clumps or particles behaving as the larger particles of a light soil* (Mackean, *Introduction to Biology*).
—**flocculation,** *n.* the process of flocculating; formation into flocculent masses.

floccule (flok′yül), *n. Chemistry.* something resembling a small flock or tuft of wool, as in a liquid. [from Latin *flocculus,* diminutive of *floccus* tuft of wool]

flocculent (flok′yə lənt), *adj.* **1** *Chemistry.* made up of soft, woolly masses; cloudlike; not crystalline. **2** *Zoology.* having a soft, waxy coating, as certain insects do. [from Latin *floccus* tuft of wool]
—**flocculence** (flok′yə ləns), *n.* the condition of being flocculent.

flocculus (flok′yə ləs), *n., pl.* **-li** (-lī). **1** *Astronomy.* one of the cloudy masses of very hot gases which cover the surface of the sun: *The chromosphere is mottled with bright and dark patches known as flocculi; they are masses of gas which are hotter or cooler respectively than those around them in the photographs. Calcium flocculi are generally bright; they appear especially in the vicinity of spot groups and are sometimes so closely bunched that they conceal the spots below. Hydrogen flocculi are both bright and dark, and are frequently drawn out in filaments* (Baker, *Astronomy*).
2 *Anatomy.* a small lobe on the underside of the cerebellum: *In the rat, stimulation of the flocculus evoked movement which caused the animal to take a step away from the stimulated side of the cerebellum* (Donald C. Goodman).
[from Latin *flocculus,* diminutive of *floccus* tuft of wool]

floe, *n. Oceanography.* **1** a mass or sheet of floating ice, formed by the freezing of sea water. **2** a floating piece broken off from such a mass or sheet. [from Scandinavian (Norwegian) *flo*]

floeberg (flō′bėrg′), *n. Oceanography.* a large mass of ice formed from floes heaped up by the action of the wind and waves.

flood current, *Oceanography.* the movement of water toward the coast or inland due to the movement of the tide.

floodplain, *n. Geography.* a plain bordering a river and made of sediment deposited during floods: *It is difficult to say just where delta ends and floodplain begins. However, some streams that have no deltas have well-developed floodplains* (Finch and Trewartha, *Elements of Geography*).

floor, *n. Geology.* **1** the bed or bottom of the ocean; seabed: *A U.S. oceanographic expedition had encountered petroleum hydrocarbons on the Sigsbee Knolls on the floor of the Gulf of Mexico. This discovery, made in water 11,753 ft. deep, was the first indication that oil and gas may occur in the sediments of the deep ocean floor* (Bruce C. Netschert).
2 a flat surface at the bottom: *the floor of a valley.*
3 an underlying stratum on which a seam of coal or other rock lies.

4 any large, level surface: *a forest floor of pine needles and cones.*

flora (flôr′ə), *n., pl.* **floras, florae** (flôr′ē). *Biology.* the plants or plant life of a particular region or time: *the flora of California, the flora of the Carboniferous period. On the east side of Greenland, one of the most prolific fossil floras ever discovered anywhere has revealed a wealth of plant life* (Washington University Magazine). [from New Latin, from Latin *Flora,* the Roman goddess of flowers, from *florem* flower]

floral, *adj. Botany.* of, having to do with, or resembling flowers. The **floral envelope** is the perianth of a flower.

Florence flask, a round bottle with a long neck, used in a laboratory for heating chemicals. [named after *Florence,* Italy, where such bottles were used to contain olive oil]

floret (flôr′it), *n. Botany.* one of the small flowers in a flower head of a composite plant, such as the aster: *In all instances, individual small flowers or florets are associated in groups known as spikelets* (Weier, *Botany*).

floriferous (flô rif′ər əs), *adj. Botany.* bearing flowers; freely blooming: *floriferous varieties of plants. ... the inflorescence as an aggregate of vegetative and floriferous axes* (Laurence, *Taxonomy of Vascular Plants*). [from Latin *florifer* flower-bearing]

florigen (flôr′ə jən), *n. Biochemistry.* a plant hormone that stimulates the production of flowers: *It is well established that flowering is directly controlled by a hormone, "florigen," which is produced by the leaves, and thus it is the leaves that are sensitive to day length. It has not been so well accepted that there is also an inhibitor produced by leaves that will keep the plant from flowering during an unsuitable photoperiod* (Albert J. Smith). [from Latin *floris* flower + English *-gen*]

floristic, *adj. Botany.* of or having to do with a flora or floristics. —**floristically,** *adv.*
—**floristics,** *n.* the study of the geographical distribution of plants.

Flory temperature (flôr′ē), *Chemistry.* a temperature at which a particular polymer exhibits properties that differentiate it from other polymers; a specific temperature at which any polymer exists in an ideal state for study of its properties in comparison with those of other polymers: *The Flory temperature ... became the basis for the development of hundreds of different plastics and synthetics* (Time). [named after Paul J. *Flory,* born 1910, American chemist, who discovered it]

floss, *n. Botany.* soft, silky fluff or fibers. Milkweed pods contain white floss. Ears of corn have brownish floss.

flow, *n. Physics.* the directional movement in a current or stream that is a characteristic of all fluids, as air or electricity: *the flow of lava. A flowmeter measures the rate of flow of a liquid or gas.*

flow chart, a diagram that shows the order of operations for solving a problem: *Flow charts are diagrams of algorithms* (Corcoran, *Algebra Two*).

flower, *n.* **1** *Botany.* the part of a plant that produces the seed. A flower is a shortened branch with modified leaves called petals. It consists normally of pistil, stamens, corolla, and calyx in regular series, any one or more of which may be absent. *Flowers are the reproductive structures of plants. The structures consist of pollen-bearing stamens (the male organs) and carpels*

(*the female organs*) *containing pollen-catching stigmas and ovules, the plant's "eggs." A union of the pollen with the ovules produces seeds. Most of the flowers with which we are familiar have both organs, the stamens and the carpels, in the same flower. It would be most convenient for the plant if each flower's pollen fertilized its own ovules, but many flowers cannot pollinate themselves. They are fertilized by pollen from other individuals of the same species. From an evolutionary standpoint this has advantages, for it produces a combination of different heredities and yields more variable and more flexible progeny* (Verne Grant).
2 flowers, *pl. Chemistry.* a substance in the form of a fine powder, obtained especially as the result of condensation after sublimation: *flowers of sulfur.*

flower bud, *Botany.* a bud whose stem will produce flowers only and no leaves: *A flower bud is a rudimentary shoot bearing one or more concealed and unexpanded young flowers. The drug Clove is an example of a flower bud which, if allowed to expand, would form a single flower* (Youngken, *Pharmaceutical Botany*). Compare **leaf bud, mixed bud.**

flower cup, *Botany.* **1** = calyx. **2** the cup-shaped receptacle formed by certain flowers.

flower head, *Botany.* a dense cluster of florets growing from the shortened summit of a stem, such as that on the dandelion, chrysanthemum, sunflower, or other composite plants.

flowering, *adj. Botany.* producing flowers and fruit. A **flowering plant** has its seeds enclosed in a fruit: *The angiosperms or flowering plants are the most numerous, numbering more than 200,000 species* (Greulach and Adams, *Plants*).

flow line, *Geology.* (in igneous rocks) a band of color or an arrangement of crystals that indicates the differential flow of the molten mass before solidification.

flowmeter, *n.* any of various devices for measuring the volume or rate of flow of a liquid or gas.

flow structure, *Geology.* the structure in igneous rock produced by the flow of the molten mass before solidification: *Though flow structure generally is applied only to lavas, some authors also use it for bands of mica, hornblende and other minerals in foliated granite* (Fenton, *The Rock Book*).

flueric (flü′ər ik), *adj.* = fluidic. [from Latin *fluere* to flow]

fluid, *Physics.* —*n.* any liquid or gas; any substance whose molecules flow freely past each other. Water, mercury, air, and oxygen are fluids. *The term fluid refers to a substance that does not have a fixed shape but that is able to flow and take the shape of its container; in other words, to a liquid or a gas. ... A fluid is a material substance which in static equilibrium cannot exert either tangential or tensile forces across a surface, but can exert merely pressure* (compressive force normal to a surface) (Shortley and Williams, *Elements of Physics*).
—*adj.* **1** in a state in which molecules flow freely past each other; flowing: *a fluid substance, fluid matter.* Compare **solid. 2** of or having to do with fluids. **Fluid mechanics** is the study of fluids either in motion (*fluid dynamics*) or at rest (*fluid statics*).
[from Latin *fluidus,* from *fluere* to flow]

fluid bed, a layer of solid material so finely pulverized that it has some of the characteristics of a fluid, used in ore refining and in nuclear reactors: *The fluid bed gives a high rate of heat transfer to gases passed through it and has a uniform temperature throughout the fluidized zone* (New Scientist). Also called **fluidized bed.**

fluid dynamics, *Physics.* the study of gases and liquids in motion, including both aerodynamics and hydrodynamics: *Researchers have been turning the analytical techniques of fluid dynamics to a broad variety of practical problems, ranging from traffic to air pollution and from jet noise to flash fires* (Science News Yearbook). Compare **fluid statics.**

fluidextract, *n. Pharmacology.* a tincture or solution of a vegetable drug of such strength that one cubic centimeter represents (in therapeutic effect) one gram of the standardized drug in dry form.

fluid flow, *Physics.* any characteristic movement of fluid particles, such as laminar flow and turbulent flow: *Using this method, scientists can study what they call "fluid flow," such as water running out of faucets or through pipes in your house, or the meandering of rivers and streams* (Science News Letter).

fluid flow

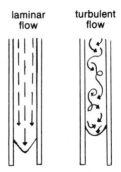

laminar flow   turbulent flow

fluidic (flü id′ik), *adj.* of or having to do with fluids or fluidics: *One thing which the study of fluidics seems to have stimulated is a fluidic approach to sensing and measuring instruments, resulting in devices which will measure such things as velocity and length without interfering with the subject of the exercise* (New Scientist).
—**fluidics,** *n.* the science or technology of using tiny jets of a gas or a liquid rather than electronic circuits for sensing, amplifying, or controlling functions: *Fluidics ... does about the same job in controlling fluids—both liquids and gases—that transistors, relays, and other electronic devices do in controlling the flow of electrons. In conditions of extreme cold or heat, or*

---

cap, fāce, fäther; best, bē, tėrm; pin, fīve;
rock, gō, ôrder; oil, out; cup, pùt, rüle,
yü in use, *yu* in uric;
ng in bring; *sh* in rush; *th* in thin, ŦH in then;
zh in seizure.
ə = *a* in about, *e* in taken, *i* in pencil, *o* in lemon, *u* in circus

227

where electric fields are undesirable, fluidics does the job better (Foster P. Stockwell).

fluidity, *n. Physics.* fluid condition or quality: *The more carbon in the iron, the lower the melting temperature and the greater its fluidity, but iron with a high carbon content is exceedingly hard and brittle* (Linton, *Tree of Culture*). *Fluidity ... is defined mathematically as the reciprocal of the viscosity* (Jones, *Inorganic Chemistry*).

fluidization, *n. Physics.* the process in which a solid is so finely ground as to take on most of the properties of a liquid.

fluidize, *v. Physics.* to give fluid properties to (a solid) by means of fluidization, so that it flows when in contact with a liquid or gaseous stream.

fluidized bed, = fluid bed.

fluid mechanics, mechanics that deals with the flow of gases and liquids.

fluid pressure, *Physics.* the pressure exerted by a confined fluid in static equilibrium, equal in all directions and perpendicular to the surfaces confining it.

fluid statics, *Physics.* the study of gases and liquids at rest. Hydrostatics is a part of fluid statics. Compare **fluid dynamics.**

flume, *n. Geography.* a deep and very narrow valley with a stream running through it. [from Latin *flumen* river, from *fluere* to flow]

fluoresce (flü′ə res′), *v. Physics.* to give off light by fluorescence; become fluorescent: *Most of the electrons strike the anode but a narrow beam passes through a small hole in the anode and continues on to the screen S which is coated on its inner surface with a substance that emits visible light (or fluoresces) when bombarded by electrons* (Sears and Zemansky, *University Physics*).

fluorescein (flü′ə res′ē ən), *n. Chemistry.* an orange-red, crystalline powder that forms a greenish-yellow, fluorescent, alkaline solution, used in making dyes, and in medicine injected intravenously especially to study the circulation in the retina and other parts of the eye: *Fluorescein ... penetrates into such areas more readily than into normal brain tissue and may be detected there during neurosurgical operations by its fluorescence under ultraviolet light irradiation* (Beaumont and Dodds, *Recent Advances in Medicine*). *Formula:* $C_{20}H_{12}O_5$

fluorescence (flü′ə res′ns), *n. Physics.* **1** a giving off of light by a substance exposed to X rays, ultraviolet rays, or certain other rays, which continues only as long as exposure to these rays continues: *Holes in prints and drawings can be invisibly repaired by the skilled workman, but the difference in fluorescence between old and new paper, or paper from different sources, as well as the effect of the adhesive used, all combine to make the site of such repairs startlingly obvious* (George Savage). **2** the property of a substance that causes this. It is an ability to transform radiation so as to emit different visible wavelengths or colors. **3** light given off in this way. [from Latin *fluor* a flowing, from *fluere* to flow]

—**fluorescent,** *adj.* that gives off light by fluorescence. Fluorescent substances glow in the dark when exposed to X rays.

fluoridate (flùr′ə dāt), *v.* to add small amounts of a fluorine compound to (drinking water), especially to decrease tooth decay. —**fluoridation,** *n.*

fluoride (flü′ə rīd′), *n. Chemistry.* a compound of fluorine and another element or radical. Fluoride occurs naturally as the mineral fluorite. *Crystals of calcium fluoride are used to bend and focus infrared light in analyzing chemical compounds* (F.C. Andrews).

fluorinate (flùr′ə nāt), *v.* **1** = fluoridate. **2** *Chemistry.* to combine or cause to react with fluorine. —**fluorination,** *n.*

fluorine (flü′ə rēn′ *or* flü′ər ən), *n. Chemistry.* a pale yellow, pungent, poisonous, gaseous element which is the most reactive of all chemical elements. It is used in small amounts in water to prevent tooth decay. *Fluorine ... burns readily but forms fireproof compounds* (Science News). *Symbol:* F; *atomic number* 9; *atomic weight* 18.9984; *melting point* $-219.62°C$; *boiling point* $-188.14°C$; *oxidation state* -1. [from *fluor(ite)*]

fluorite (flü′ə rīt′), *n. Mineralogy.* a transparent or translucent mineral, composed of calcium and fluorine, that occurs in many colors. It is used for fusing metals, making glass, etc., and is a source of fluorine. *Formula:* CaF_2 Also called **fluorspar.** [from Latin *fluor* a flowing, from *fluere* to flow]

fluorocarbon (flü′ər ə kär′bən), *n. Chemistry.* any of a group of compounds of fluorine and carbon used as solvents, lubricants, refrigerator gases, etc. Compare **halocarbon.**

fluorochrome, *n.* a fluorescent substance used to stain biological specimens: *The average pathologist has not the time to concern himself with purity of a fluorochrome and if this fluorochrome is not pure he may have to use a fractionating column or even have to reabsorb with a liver powder* (Science Journal).

fluorometer (flü′ə rom′ə tər), *n.* an instrument which measures the fluorescence given off by a substance. [from *fluor(escence)* + *-meter*]

—**fluorometric,** *adj.* of, having to do with, or using a fluorometer.

—**fluorometry** (flü′ə rom′ə trē), *n.* the measurement of fluorescent radiation with a fluorometer.

fluorspar (flü′ər spär′), *n.* = fluorite.

fluvial (flü′vē əl), *adj. Geology.* of, found in, or produced by a river: *A delta is a fluvial deposit. Another classification ... divides clastic rocks according to the agents that deposited them. Thus we have aeolian rocks, deposited by wind; fluvial rocks, laid down by running water; glacial rocks, dropped by melting glaciers, and so on* (Fenton, *The Rock Book*). [from Latin *fluvius* river, from *fluere* to flow]

fluviatile (flü′vē təl), *adj. Geology.* of or resulting from the action of a river: *fluviatile sediments.* [from Latin *fluviatilis*, from *fluvius* river]

fluvioglacial (flü′vē ō glā′shəl), *adj. Geology.* of or having to do with the combined action of rivers and glaciers: *Invasions of the Pleistocene continental glaciers are recorded over thousands of square miles by till and fluvioglacial debris* (Moore, *Introduction to Historical Geology*). Also, **glaciofluvial.**

flux, *n.* **1** *Physics.* **a** the rate of a flow of a liquid, particles, or energy across a unit surface or area per unit time: *The absolute flux of primary particles varies from 2 per square centimetre per second near the poles to 0.2 per square centimetre per second at the equator* (E.P. George). **b** = flux density: *The SI unit of flux is the weber.*
2 *Chemistry.* any substance that lowers the melting point of a substance to which it is added: *Generally, there is added to the concentrate to be smelted a suitable quantity of reducing agent to liberate the metal and a suitable quantity of a flux whose purpose is to unite with the gangue to form an easily fused glassy material, usually referred to as a slag* (Jones, *Inorganic Chemistry*).
—*v. Chemistry.* **a** to heat with a substance that helps metals or minerals melt together. **b** to melt together. [from Latin *fluxus* a flow, from *fluere* to flow]

flux density, *Physics.* the number of flux lines per unit area, proportioned to the intensity of the magnetic field. Flux density is the force exerted per unit current per unit length when the current is perpendicular to the field. *Symbol:* B Compare **magnetic induction.**

flux gate, a device used in a gyrocompass to detect the direction of the earth's magnetic field.

flux line, *Physics.* any of the imaginary lines of a magnetic field that curve from the north pole to the south pole of a magnet. Compare **line of force.**

fluxmeter, *n.* an instrument for measuring magnetic flux: *To determine the flux in an iron ring without cutting into the ring seems at first sight impossible; actually, there is an instrument called a fluxmeter which, when linked to the ring with a few turns of wire (called a secondary coil), will accurately record, in webers, all changes in flux that occur inside the ring. The operation of the fluxmeter depends on the phenomenon of electromagnetic induction* (Shortley and Williams, *Elements of Physics*).

flux unit, a unit used in radio astronomy to measure the intensity of radio waves, equal to 10^{-26} watts per square meter per hertz: *The average radio level for Cygnus X-3 is about 0.1 flux units. What his instruments detected was 22 flux units—220 times greater* (Science News). Also called **Jansky.**

flyback, *n. Electronics.* **1** the return of the tracing beam in a cathode-ray tube to its starting point after having completed its trace: *A television receiver must have a "flyback" transformer to pull its electron-scanning beam back very quickly to the starting line on the picture tube after each horizontal sweep* (Scientific American). **2** a time required for this return.

flysch (flish), *n. Geology.* a partly Tertiary, partly Cretaceous formation, consisting chiefly of sandstone, soft marls, and sandy shales, and found in the Alps and other places. [from dialectal Swiss *flysch*]

flyway, *n. Ecology.* a route usually followed by migrating birds: *That is where the wild geese are going. Up the four well-charted flyways—the Atlantic seaboard, the Mississippi valley, the wheat-belt Central flyway, the Pacific coast—geese by the tens of thousands are heading home* (Science News Letter).

fm., *abbrev.* fathom.

Fm, *symbol.* fermium.

FM or **F.M.,** *abbrev.* frequency modulation.

FMN, *abbrev.* flavin mononucleotide.

f number, *Optics.* the focal length of a lens divided by its effective diameter. An f/4 lens is one in which the diameter of the widest effective opening is 1/4 of its focal length (f). *Increased pupil size per se does not increase retinal illumination. What is important is increased pupil size relative to eye size. Image brightness can be conveniently indexed by the "f-number", familiar to photographers, which is simply the ratio of the focal length of the eye's optical system to its pupil diameter. The smaller the f-number the brighter the image* (New Scientist).

focal, *adj.* of or having to do with a focus. The **focal ratio** of a lens is the ratio of its focal length to its effective diameter: *Parallel rays entering my telescope come to a focus at a distance of 62.5 inches from the 12.5-inch objective mirror, a focal ratio of f/5* (C.L. Stong). —**focally,** *adv.*

focal length or **focal distance,** *Optics.* the distance from the optical center of a lens or mirror to the principal point of focus: *The distance f from the vertex of a mirror to its focal point is called the focal length of the mirror, and we see that the focal length of a mirror equals one-half its radius of curvature* (Sears and Zemansky, *University Physics*). *The focal length of the lens is the distance from the center, C, to the principal focus, F ... where parallel rays of light are focused by the lens. When the object is farther from the lens than the principal focus, the lens produces an inverted real image of the object, which may be shown on a screen or photographic plate* (Baker, *Astronomy*). See the picture at **focus.**

focal plane, *Optics.* a plane through the focal point of a lens.

focal point, *Optics.* the point where rays of light are focused: *When an image strikes the objective, the light waves are bent by the lens until they come to one bright point, known as the focal point* (R. William Shaw).

focus (fō′kəs), *n., pl.* **-cuses, -ci** (-sī). **1** *Optics.* **a** the point at which rays of light, heat, etc., meet, diverge, or seem to diverge after being reflected from a mirror, bent by a lens, etc.: *In the convex mirror the rays of light that are reflected diverge and thus may never cross at a real focus. If these divergent rays are projected, they will converge to a point focus behind the mirror and the focus is virtual* (Obourn, *Investigating the World of Science*). **b** = focal length.
2 *Geometry.* **a** a fixed point used in determining a conic section. A parabola has one focus while an ellipse or a hyperbola has two foci. **b** a point having a similar relation to some other curve.
3 *Geology.* the point where an earthquake originates: *The sharply pulselike records on seismograms suggests that most earthquakes begin in a very localized area, even though many miles of fault may ultimately*

---

cap, fāce, fäther; best, bē, tėrm; pin, fīve;
rock, gō, ôrder; oil, out; cup, pùt, rüle,
yü in use, *yu* in uric;
ng in bring; *sh* in rush; *th* in thin, ᴛʜ in then;
zh in seizure.
ə = *a* in about, *e* in taken, *i* in pencil, *o* in lemon, *u* in circus

be active, as in the San Francisco earthquake. The point from which the first movements appear to start is called the focus. The point on the surface vertically above the focus is called the epicenter (Greek: "above the center") (Gilluly, *Principles of Geology*).

—*v. Optics.* **1** to bring (rays of light, heat, etc.) to a focus: *Focus the light from the sun on a sheet of paper by moving the paper back and forth in back of the lens until a dazzling spot appears on the paper ... where the light of the sun has converged after passing through the lens. The light of the sun comes together at what is called the principal focus (or focal point) of the lens* (Tracy, *Modern Physical Science*).

2 to adjust (a lens, the eye, etc.) to make a clear image: *When the image can be seen most distinctly, we say that the lens is focused* (Beauchamp, *Everyday Problems in Science*).

3 to make (an image, etc.) clear by adjusting a lens, the eye, etc.: *A convex lens with greater curvature will focus an image closer to the lens than one with less curvature* (Barnard, *Science: A Key to the Future*).

[from New Latin *focus,* from Latin, hearth]

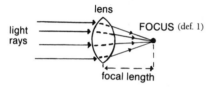

foehn (fān *or* fœn), *n. Meteorology.* a warm, dry wind that blows down the leeward slope of a mountain, especially in the Alps: *Winds of this kind are ... especially common on the northern side of the Alps in Switzerland, where they are called foehn winds, and on the eastern slope of the Rocky Mountains in Wyoming and Montana, where they are called chinooks* (Blair, *Weather Elements*). Also spelled **föhn.** [from German *Föhn*]

fog, *n.* **1** *Meteorology.* **a** a cloud of fine drops of water just above the earth's surface, in which visibility is very poor; a low cloud or thick mist: *A heavy fog is defined as one in which visibility is reduced to less than a quarter of a mile. The term "pea soup" has no standing among weathermen, but then most pea-soup fogs—the ones in London, for instance—aren't simply fogs but combinations of fog and smog, which means that they're loaded with foreign particles, mostly dust and soot* (George S. Kaufman). **b** a darkened condition of the atmosphere, or a substance in the atmosphere that causes this: *With half of Britain blacked out by a fog that cut visibility in London to 5 yards, policemen like this one were issued special smog masks* (Newsweek). **2** *Chemistry.* particles of liquid dispersed in a gas; a type of colloidal system.

fogbow (fog'bō'), *n. Optics.* an arc of faintly colored light similar to a rainbow, sometimes seen in fog.

föhn (fān *or* fœn), *n.* = foehn.

fold, *n.* **1** *Anatomy.* a doubling, as of various membranes or other parts of the body: *In other cases ... the fold may overhang and so conceal the edge of the eyelid, either completely or at some portion of its length. When the edge of the eyelid is completely covered by* the fold we have a complete Mongoloid fold (Ralph Beals and Harry Hoijer, *Anthropology).* **2** *Geology.* a bend in a layer of rock: *Folds are the typical features of most mountain chains. They range in size from microscopic crinkles to great arches and troughs fifty miles or more across* (Gilluly, *Principles of Geology*).

—*v. Geology.* to form a bend in a layer of rock: *The beds have never been folded by mountain-building forces* (E.F. Roots).

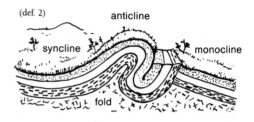

foliaceous (fō'lē ā'shəs), *adj.* **1** *Botany.* **a** of or having to do with leaves; leaflike; leafy: *a foliaceous flower or plant.* **b** = foliose: *foliaceous lichens.* **2** *Zoology.* shaped or arranged like leaves: *a foliaceous appendage.* [from Latin *foliaceus,* from *folium* leaf]

foliage (fō'lē ij), *n. Botany.* the leaves of a plant or plants: *Their empirical methods have shed little light, however, on the physiological process that transforms a plant from the vegetative (foliage-producing) to the flowering state* (Scientific American).

foliate (fō'lē it *or* fō'lē āt), *adj.* **1** *Botany.* having leaves; covered with leaves. **2** = foliated.

foliated (fō'lē ā'tid), *adj.* **1** *Botany.* having leaves. **2** *Zoology.* shaped like a leaf or leaves. **3** *Geology.* consisting of thin, leaflike layers of laminae: *foliated granite.* SYN: foliaceous.

foliation (fō'lē ā'shən), *n.* **1** *Botany.* **a** the process of putting forth of leaves. **b** being in leaf. **c** leafage; foliage. **2** *Geology.* **a** the property of splitting up into leaflike layers. **b** the leaflike plates or layers into which rocks are divided, resulting from the parallel orientation of platy or lathelike crystals: *Most metamorphic rocks have a banded or layered structure called foliation. ... The foliation appears to record widespread pervasive movement within the rock mass, during which most of the original minerals were broken, streaked out, and recrystallized into new minerals. Because movement is essential to their development, the foliated rocks are called dynamic metamorphic rocks* (Gilluly, *Principles of Geology*).

folic acid (fō'lik), *Biochemistry.* a constituent of the vitamin B complex, found in green leaves and animal tissue, and also produced synthetically, used especially in the treatment of anemia. *Formula:* $C_{19}H_{19}N_7O_6$ Also called **vitamin Bc.** [from Latin *folium* leaf]

foliolate (fō'lē ə lāt), *adj. Botany.* having to do with or consisting of leaflets (used chiefly in composition, as in *bifoliolate, trifoliolate*).

foliole (fō'lē ōl), *n.* **1** *Botany.* a division of a compound leaf; leaflet. **2** *Zoology.* a small leaflike part or organ.

foliose (fō'lē ōs), *adj. Botany.* (of lichens and algae) having a leaflike expansion of the thallus. Compare **crustose** and **fruticose.**

folivore (fō'lə vôr), *n. Biology.* an animal, especially a primate, that feeds on leaves: *The three-toed sloth is a primitive mammal that ... hangs in the forest canopy, and reaches out occasionally to pluck leaves from three favorite tree species, and is thus classified an "arboreal folivore" (a tree-dwelling leaf eater)* (Science News). Compare **frugivore.** [from Latin *folium* leaf + English *-vore,* as in *carnivore, herbivore*]
—**folivorous** (fō liv'ər əs), *adj. Biology.* leaf-eating: *Presbytis entellus and Gorilla gorilla are the only folivorous primates that do forage extensively on the ground* (Science).

follicle (fol'ə kəl), *n.* **1** *Anatomy, Zoology.* **a** a small cavity, sac, or gland in the body. Hairs grow from follicles. *It is the follicle mite, Demodex folliculorum, a tiny wormlike animal barely visible to the naked eye, which can crawl down into a hair follicle* (Bryan Silcock). See the picture at **skin. b** = Graafian follicle. The **follicle stage** involves the maturation of an egg within the follicle and the secretion of the hormone estrogen, which initiates vascularization of the uterine lining.
2 *Botany.* a dry, one-celled fruit formed of a single carpel, that splits open along one seam only, as the fruit of a milkweed.
[from Latin *folliculus* small bag, diminutive of *follis* bellows]

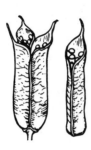

follicle
(def. 2)

follicle-stimulating hormone, *Biochemistry.* a hormone secreted by the anterior lobe of the pituitary gland that stimulates the growth of the Graafian follicles of the ovaries and promotes the production of sperm cells in the testes. *Abbreviation:* FSH

follicular (fə lik'yə lər), *adj. Biology.* of or having to do with a follicle or follicles: *the follicular cells of the Graafian follicles.*

fontanel or **fontanelle** (fon'tə nel'), *n. Anatomy.* any of the soft spots, closed by membrane and later to be filled by bone, on the head of an infant or fetus. [from French *fontanelle,* originally diminutive of *fontaine* fountain]

food, *n. Biology.* any animal or vegetable substance that can be ingested and assimilated by an organism to supply the nutrients it needs for producing energy, building tissue, and regulating physiological processes. Food supplies such nutrients as carbohydrates, proteins, fats, vitamins, and minerals. Compare **drug.**

food chain, *Ecology.* a group of organisms so interrelated that each member of the group feeds upon the one below it and is in turn eaten by the organism above it: *A food chain necessarily starts with photosynthetic green plants or green protists, since they are the organisms that acquire energy from nonorganic sources. The shortest food chains involve green plants and organisms of decay. ... More complex food chains involve animals* (Simpson, *Life: An Introduction to Biology*). *In a typical food chain, plants are eaten by animals called herbivores, which are, in turn, eaten by other animals called carnivores* (Charles F. Wurster, Jr.).

food cycle = food web: *Each stage in the flow of energy [food energy] from one population to another is known as a trophic level. The sequence of trophic levels in any ecosystem forms a food chain. All of the food chains in a particular community constitute a food cycle* (Clarence J. Hylander).

food pyramid, *Ecology.* the gradually narrowing structure representing the amount of food and nutrients passed along a food chain, as at each upward link or step the quantity of food and energy passed on becomes smaller: *In a forest, grassland, desert, lake, ocean, or any other natural community, photosynthetic plants provide the basic food supply and constitute the broad base of a "food pyramid"* (Greulach and Adams, *Plants*).

food vacuole, *Biology.* a vacuole containing food particles, found in certain protozoans, such as amoebas. It serves as a simple digestive system. *A food vacuole is a droplet of water with food particles suspended within it. As soon as one is separated from the [protozoan] pharynx, it is swept away by the rotary streaming movement of the endoplasm* (Hegner and Stiles, *College Zoology*).

food web, *Ecology.* a group of interrelated food chains in a particular community: *Food webs are very complex, even in a small community, but may be illustrated by two simplified examples. In a pond the bacteria and diatoms synthesize materials, and then in sequence small organisms are eaten by larger ones, thus: Bacteria and diatoms ⟶ small protozoans ⟶ larger protozoans ⟶ rotifers, small crustaceans ⟶ aquatic insects ⟶ fishes. The large fishes, or any intermediate organisms, by death or decay, become food for bacteria, thus completing the circuit* (Storer, *General Zoology*).

food yolk, *Biology.* the portion of the yolk of an egg that nourishes the embryo, as distinguished from the germinative portion. Compare **deutoplasm.**

foot, *n.* **1** *Anatomy.* the end part of a leg. The bones of the foot are the tarsus, metatarsus, and phalanges.
2 *Zoology.* an organ present in some invertebrates, especially the muscular, ventral protuberance of the mollusks, used for locomotion: *The snail, octopus, squid, oyster, and clam ... all have a mantle, which surrounds a part of the body called the visceral hump, that contains important internal organs. extending down ventrally from the visceral hump is a foot, in all the species that are able to move about, which is a loco-*

cap, fāce, fäther; best, bē, tèrm; pin, fīve;
rock, gō, ôrder; oil, out; cup, pùt, rüle,
yü in use, *yu̇* in uric;
ng in bring; *sh* in rush; *th* in thin, ŦH in then;
zh in seizure.
ə = *a* in about, *e* in taken, *i* in pencil, *o* in lemon, *u* in circus

foot-candle

motor organ (Winchester, *Zoology*). See the picture at **neck.**
3 a unit of length in the foot-pound-second system, equal to twelve inches (30.479 centimeters). *Abbreviation:* ft.

FOOT (def. 1)

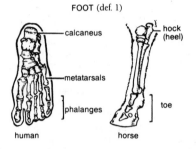

calcaneus

hock (heel)

metatarsals

phalanges

toe

human

horse

foot-candle, *n. Optics.* a unit for measuring illumination in the foot-pound-second system, equal to one lumen per square foot, or the amount of light produced by a candle on a surface at a distance of one foot. In the metric system it is equal to approximately 10.764 lux. *Abbreviation:* f.c.

foot-lambert, *n. Optics.* a unit of luminance in the foot-pound-second system, equivalent to the luminance of a perfectly diffused surface that emits or reflects one lumen per square foot. *Abbreviation:* ft.-1.

foot-pound, *n. Physics.* a unit of work or energy in the foot-pound-second system, equal to the work done by one pound of force that moves an object one foot in the direction of the force applied. In the metric system it is equal to approximately 1.356 joules. *Abbreviation:* ft.-lb.

foot-poundal, *n. Physics.* a unit of work or energy in the foot-pound-second system equivalent to the work done by a force of one poundal acting through a distance of one foot. It is equal to a foot-pound divided by the acceleration due to gravity, or about 32.2 feet per second per second. In the metric system it is equal to approximately 0.042140 joules.

foot-pound-second, *adj. Physics.* of or designating a system of units in which the foot, pound, and second are considered the basic units of length, force, and time. *Abbreviation:* fps

footwall, *n. Geology.* the side of rock under a fault, vein, lode, etc. Compare **hanging wall.**

foramen (fə rā′mən), *n., pl.* **foramina** (fə ram′ə nə) or **foramens.** *Biology.* an opening, orifice, or short passage, as in a bone or in the covering of the ovule of a plant. The **foramen magnum** (mag′nəm) is the opening in the skull for the passage of the spinal cord to the cranial cavity. The **foramen ovale** (ō vā′lē) is an opening between the right and left atria of the heart, found in the fetus but usually closed soon after birth. Compare **fenestra.** [from Latin *foramen* aperture, hole, from *forare* bore a hole]

foraminifer (fôr′ə min′ə fər), *n. Zoology.* one of the foraminifera.

foraminifera (fə ram′ə nif′ər ə), *n.pl. Zoology.* an order of usually marine protozoans that have calcareous or chitinous shells with tiny holes in them: *The foraminifera ... construct a perforated shell, usually of calcium*

carbonate, through which slender pseudopodia project (Hegner and Stiles, *College Zoology*). [from New Latin *Foraminifera,* from Latin *foramen* an opening + *ferre* to bear]

—foraminiferal, *adj.* consisting of, containing, or having to do with the foraminifera.

—foraminiferan, *n.* one of the foraminifera: *The great pyramids near Cairo, Egypt, were carved from limestone deposits made of shells of an Early Tertiary foraminiferan, Nummulites. Petroleum geologists study the foraminiferans obtained in drilling exploratory wells to identify oil-bearing strata* (Storer, *General Zoology*).

forb, *n. Botany.* any herb, excluding grasses and plants resembling grasses. [probably from Greek *phorbē* pasture, food]

forbidden lines, *Physics.* spectral lines of radiation produced by changes in energy level that occur within an atom only when the interval between atomic collisions is exceptionally long. Forbidden lines are important in the spectra of astrophysical gases at low density. *They are the so-called forbidden lines—colors which an atom is extremely reluctant to radiate and will not radiate at all unless left undisturbed by collisions for seconds or minutes at a time* (Armin J. Deutsch).

force, *n. Physics.* any influence or agency that causes a body to move or accelerate; a vector quantity that tends to produce a change in the motion of objects. Forces may be classified as gravitational, electromagnetic, (nuclear) strong, and (nuclear) weak. The unit of force in the SI or MKS system is the newton; in the CGS system it is the dyne. See **field** (def. 1), **line of force,** and **motion.**

forcipate (fôr′sə pāt), *adj. Botany, Zoology.* formed like a forceps; deeply forked; furcate: *forcipate claws.* [from Latin *forceps, forcipis* tongs]

forebrain, *n. Anatomy.* the front part of the brain, composed of the telencephalon and the diencephalon. It includes the cerebrum, thalamus, and hypothalamus. *The brain tube enlarges moderately in three regions known as the fore-, mid-, and hindbrain. This tubular structure remains in the adult as the "brain stem," but the forebrain expands enormously upward, laterally, and backward, to form the cerebrum (divided into two hemispheres), while the hindbrain develops the cerebellum* (Shull, *Principles of Animal Biology*). Also called **prosencephalon.**

foregut, *n.* **1** *Embryology.* the front section of the digestive canal of an embryo. The pharynx, esophagus, stomach, part of the duodenum, and liver develop from it. **2** *Zoology.* the front section of the digestive tract of an insect or crustacean. Compare **hindgut, midgut.**

forest, *n. Ecology.* a large group of trees growing close together with various kinds of plants: *an oak forest, a pine forest. Forests offer an interesting example of what may be accomplished by light control. A forest, depending upon its denseness, will prevent from fifty to ninety per cent of the incoming light from reaching the ground, with corresponding effects upon the temperatures of the air and soil and upon the moisture content of the soil within the forest. The specific forest climate created under the canopy of trees is favorable for the new development of some species of trees, unfa-*

vorable for others (Neuberger and Stephens, *Weather and Man*).

—forestation, *n.* the act of planting or taking care of forests.

—forestry, *n.* the science of planting and managing forests for human benefit: *Where forestry is practiced and wood becomes a cultivated crop, the parasitic fungi are often extremely difficult to control* (Emerson, *Basic Botany*).

form, *n.* **1** one of the various ways in which a thing exists, takes shape, or shows itself; manifestation; condition; type: *sulfur in the crystalline or the amorphous form, a mild form of a disease. Ice, snow, and steam are forms of water. Electricity is a form of energy.*
2 *Mathematics.* an expression of a certain type: *a bilinear or quadratic form.* The **standard form** of an equation is one that is generally accepted for the sake of uniformity.
3 *Crystallography.* the combination of all the like faces possible on a crystal of given symmetry: *A substance is said to have form when, in its solid state, it exists as crystals of one of the 32 crystal classes* (Jones, *Inorganic Chemistry*).

—formal, *adj. Mathematics.* following prescribed forms or rules: *An algorithm is a formal procedure for any operation.*

-form, *combining form.* **1** having the form of ____, as in *cruciform = having the form of a cross.* **2** having (a specified) form or forms, as in *multiform = having many forms.*

formaldehyde (fôr mal′də hīd), *n. Chemistry.* a colorless gas with a sharp, irritating odor, used in a water solution as a disinfectant and preservative: *Carbon monoxide and hydrogen have been photochemically excited with ultraviolet radiation to produce formaldehyde* (Science News). Formula: CH_2O [from *form(ic acid)* + *aldehyde*]

formalin (fôr′mə lin), *n. Chemistry.* a solution of formaldehyde in water. It may contain from 37 to 50% formaldehyde by weight.

formation, *n.* **1** the act or process of shaping, developing, or originating: *the formation of new cells.*
2 something naturally shaped or developed: *Clouds are formations of tiny drops of water in the sky. Although temperature influences the occurrence of the plant formations we have mentioned largely through its effects on water availability, low temperature is a primary factor in determining the location of tundra formations* (Greulach and Adams, *Plants*).
3 *Geology.* a series of layers or deposits of related or identical rocks; a closely related group of strata useful for mapping or description. Each formation has its own characteristic assemblage of fossils. *The basic unit of the geologic map is the formation. There are two criteria for selecting a formation: first, its contacts (i.e., the top and the bottom of a sedimentary formation) must be recognizable and capable of being traced in the field, and second, it must be large enough to be shown on the map* (Gilluly, *Principles of Geology*).

formative, *adj. Biology.* of or having to do with the formation of a new organ or organism: *formative tissue, formative yolk. In the parenchyma are scattered free formative cells, which by mitosis produce new parts in regeneration* (Storer, *General Zoology*).

form genus, *pl.* **form genera.** *Biology.* a group of animals or plants with no definite genetic relationship but classed together as a genus because they show similarities of form, structure, behavior, etc.: *The various categories of classification in this class [of fungi] are designated as form genera ... to show that the members of the genera or families do not necessarily have a natural or family relationship. For instance, some species have been named simply on the basis of the host upon which they were found, a procedure that resulted in naming hundreds of literally nonexistent "species"* (Weier, *Botany*).

formic acid, *Chemistry.* a colorless, pungent liquid that is irritating to the skin, formerly obtained from ants, spiders, nettles, etc., and now made synthetically for use in dyeing, finishing textiles, etc.: *Formic acid is a fairly strong acid ... prepared from sodium hydroxide and carbon monoxide under pressure* (Dull, *Modern Chemistry*). Formula: HCO_2H [from Latin *formica* ant]

formula, *n.,* *pl.* **-las, -lae** (-lē). **1** *Chemistry.* an expression showing by symbols the composition of a molecule or compound: *The formula for water is H_2O.* Compare **empirical formula, molecular formula, structural formula.**
2 *Mathematics.* an expression in mathematical symbols of a rule or principle: *If we let r represent the rate, d the distance, and t the time, we may express this rule by the formula $r=d/t$* (Pearson and Allen, *Modern Algebra*).
[from Latin, diminutive of *forma* form]

—formulaic (fôr′myə lā′ik), *adj.* based on or consisting of formulas: *formulaic equations.*

formulas

algebra:

$$x = \frac{-b \pm \sqrt{b^2 - 4ac}}{2a}$$

chemistry:

H_2O	H-O-H
empirical	structural

fornical (fôr′nə kəl), *adj. Anatomy.* of or having to do with a fornix: *the fornical commissure of the brain.*

fornix (fôr′niks), *n.,* *pl.* **-nices** (-nə sēz). *Anatomy.* any structure or surface resembling an arch; an arched formation or concavity, such as the one that connects the two cerebral hemispheres: *He inserted slender wire electrodes into a brain area called the fornix, between the two halves of the brain* (Scientific American). [from Latin *fornix* arch, vaulted place]

---

cap, fāce, fäther; best, bē, tėrm; pin, five;
rock, gō, ôrder; oil, out; cup, pùt, rüle,
yü in use, *yú* in uric;
ng in bring; *sh* in rush; *th* in thin; ᴛʜ in then;
zh in seizure.
ə = *a* in about, *e* in taken, *i* in pencil, *o* in lemon, *u* in circus

FORTRAN or **Fortran** (fôr′tran), *n.* an algebraic notation for programming a computer to perform mathematical and scientific computations: *The Sanskrit of computer tongues, Fortran ... was devised some years ago at Dartmouth for a time-sharing system, but its virtues have given it great currency in the world of today's personal computers* (Scientific American). Compare **ALGOL.** [from *For(mula) Tran(slation)*]

fossa (fos′ə), *n., pl.* **fossae** (fos′ē). *Anatomy.* a usually elongated shallow depression or cavity in a bone, etc.: *The probe is inserted into the skull through the nasopharynx and passed into the pituitary fossa* [*of the sphenoid bone*] *through a fine trocar and cannula* (James Fraser). [from Latin *fossa* ditch]

fossil (fos′əl), *Paleontology.* **—*n.*** anything found in the strata of the earth which is recognizable as the remains or traces of an organism of a former geological age. Fossils of ferns are sometimes found in coal. *Since land animals did not develop until late in earth history, the number and variety of their fossils cannot compare with those of sea animals* (Frederick H. Pough).

—*adj.* 1 forming a fossil; of the nature of a fossil: *the fossil remains of a dinosaur. There is persuasive geological evidence that underground deposits of petroleum, natural gas, and coal are the residues of fossil plants; hence the term "fossil fuels"* (Barry Commoner).

2 formed in the earth in past geologic ages: *Amber is a fossil resin; ozocerite is a fossil wax.*

[from French *fossile,* from Latin *fossilis* dug up, from *fodere* to dig]

ASSOCIATED TERMS: Most fossils are found in *sedimentary* rocks, different *strata* having particular kinds of fossils. Fossils may be preserved through natural prints or molds or through *petrification.* Fossils may be *petrified* in various ways, such as *replacement, permineralization,* or *carbonization.* The oldest fossils known are those of bacteria and algae.

fossil fuel, *Geology.* a combustible fossil material, such as coal, petroleum, or natural gas.

fossiliferous (fos′ə lif′ər əs), *adj. Geology.* containing fossils: *fossiliferous sandstone.*

fossilize, *v.* to change into a fossil: *Only the relatively few plants and animals that are covered by water or are in some other way protected from bacterial action can fossilize* (Emerson, *Basic Botany*). **—fossilization,** *n.*

fossorial (fo sôr′ē əl), *adj. Zoology.* **1** that digs or burrows: *a fossorial animal.* **2** adapted for digging or burrowing: *... the fossorial feet of the badger* (John G. Wood). [from Latin *fossorius,* from *fossor* digger, from *fodere* to dig]

Foucault pendulum (fü kō′), *Physics.* a spherical pendulum used to demonstrate the earth's rotation on its axis. It is suspended so that the plane of its swing slowly wheels, a phenomenon explained only by the earth's rotation. *A Foucault pendulum placed at the North Pole will appear to complete one revolution a day relative to a spot on the earth. However, it is the earth that rotates, not the pendulum* (Gerald Wick). [named after J.B.L. *Foucault,* 1819–1868, French physicist, who demonstrated this type of pendulum in 1851]

four-dimensional, *adj. Mathematics.* of or denoting a space or object each of whose points can be determined only by four coordinates: *Because it works only with numbers, the computer works as well with the four coordinates for four-dimensional space as with the three coordinates needed to locate a point in three-dimensional space. Photographing the images that result yields films of what a four-dimensional object would look like* (Kenneth Engel). See also **space-time.**

Fourier analysis (fůr′ē ā *or* fůr′ē ər), *Mathematics.* the breakdown of a periodic function into the sum of its harmonic functions by means of a Fourier series. Also called **harmonic analysis.** [named after Baron Jean Baptiste Joseph *Fourier,* 1768–1830, French mathematician and physicist]

Fourier series, *Mathematics.* an infinite series using combinations of sines and cosines of the first degree, used to approximate a function within a given domain: *The theory of Fourier series, sums of trigonometric and exponential functions named after the 19th-century French scientist Joseph Fourier ... enable mathematicians and scientists to break down complex functions into constituent components, just as musical tones can be acoustically decomposed into a sum of simple pure tones. Indeed, this analogy is so strong that the mathematical study of Fourier series is traditionally known as "harmonic" analysis* (Lynn Arthur Steen).

fovea (fō′vē ə), *n., pl.* **-veae** (-vē ē′). **1** *Biology.* a small, often round depression or pit: *the fovea of the pharynx, the fovea of the femoral head.*

2 *Anatomy.* = fovea centralis: *Vision is much more acute at the fovea than at other portions of the retina* (Sears and Zemansky, *University Physics*).

[from Latin *fovea* pit]

—foveal (fō′vē əl), *adj. Biology.* of or situated in a fovea: *Foveal vision is so much more acute than extra-foveal vision that the muscles controlling the eye always involuntarily rotate the eyeball until the image of the object toward which the attention is directed falls on the fovea* (Hardy and Perrin, *Principles of Optics*).

fovea centralis (fō′vē ə sen trā′lis), *Anatomy.* a small spot in the center of the retina where vision is sharpest: *In man, the central part of the retina, the fovea centralis or 'yellow spot' on which objects are normally focused, consists almost entirely of cones. These are responsible for colour-vision and for detailed resolution* (J.L. Cloudsley-Thompson). [from New Latin, central fovea]

f.p., *abbrev.* freezing point.

fps, *abbrev.* foot-pound-second: *the fps system of measurement.*

Fr, *symbol.* francium.

fractal (frak′təl), *n. Geometry.* any of a class of highly irregular and fragmented shapes or surfaces not represented in classical geometry: *Fractals arise in many parts of the scientific and mathematical world. Sets and curves with the discordant dimensional behavior of fractals were introduced ... by George Cantor and Karl Weierstrass* (Science News). [from Latin *fractus,* past participle of *frangere* to break]

fraction, *n.* **1** *Mathematics.* **a** the ratio of two numbers, usually shown by a horizontal or diagonal line separating the quantities and representing the division of one number by another: *½, ⅔, ¾, ⅚, and ⅞ are fractions. In the fraction ⅚, 5 is the numerator and 6 is the denominator.* **b** a similar expression in which the quantities are expressed in algebraic terms, such as a/b and xty^2/(x+2) . Compare **common fraction, complex fraction, proper fraction, improper fraction.**

2 *Chemistry.* a portion separated for collection especially by fractional distillation or precipitation: *Petroleum fractions obtained from the fractionating process may be classified in one of several rather loosely defined ways, e.g. light distillate, heavy distillate and residue* (Crabbe and McBride, *The World Energy Book*).

[from Late Latin *fractionem,* from Latin *frangere* to break]

—**fractional,** *adj.* **1** *Mathematics.* **a** having to do with fractions: *fractional powers.* **b** forming a fraction: *440 yards is a fractional part of a mile.* **2** *Chemistry.* of or designating a method for separating a mixture into its component parts based on certain differences in boiling points, solubility, etc., of these parts. **Fractional distillation** is a method of separating two or more volatile liquids having different boiling points, by heating them so as to vaporize them successively and collecting the more volatile first. **Fractional precipitation** is a method of separating elements or compounds by a series of precipitations.

fractionate, *v.* **1** *Chemistry.* to separate (a mixture) by distillation, crystallization, etc., into its ingredients or into portions having different properties: *A novel use has been shown for starch wherein mixtures of substances over a very wide molecular weight range may be fractionated by adsorption when passed over a column of starch granules, swollen to different degrees* (F. Degering).

2 *Physics.* to break up; divide: *A neutron spectrometer ... not only reduces gamma rays but also fractionates the neutrons into beams of different energies, or wavelengths* (Harold Berger).

—**fractionation,** *n.* the act or process of fractionating: *the fractionation of whole blood by a fractionator.*

—**fractionator,** *n.* a device that fractionates: *The fractionator is an elaborate centrifuge that separates whole blood into red cells, white cells platelets, plasma, and various blood protein fractions in a closed sterile system* (New Scientist).

fracture, *n. Geology.* **1** an irregular breaking or cracking of a mineral: *Fracture is developed when minerals break without cleavage or across it. There are several types of fracture: fibrous, irregular, jagged (hackly) and so on. If the broken surface suggests a clam shell, it is conchoidal* (Fenton, *The Rock Book*).

2 a break or crack in rocks due to faulting or folding: *The bottom off northern California is broken by a cliff a mile high and more than 1,000 miles long. It is one of four immense fractures apparently due to the same massive deformation of the earth's crust* (Scientific American).

—*v.* to break or crack irregularly: *Cleavage is the ability of a mineral to split, or cleave, along closely spaced parallel planes. Not all minerals have that property; many fracture along widely spaced surfaces that can be relatively smooth, but are often curved, or irregular and hackly* (Birkeland and Larson, *Putnam's Geology*).

fragmental or **fragmentary,** *adj. Geology.* formed from fragments of older rocks: *Admixture of volcanic fragmental materials with the sediments shows that there were active volcanoes, but the extent of volcanism was* much less than in the Jurassic (Moore, *Introduction to Historical Geology*). SYN: clastic.

fragmentation, *n. Biology.* a form of reproduction or propagation among certain worms, algae, etc., in which an organism splits into two or more individuals: *Although fragmentation is frequently accidental, some filamentous blue-green algae have specialized enlarged, thick-walled dead cells ... at intervals along their filaments that are associated with fragmentation* (Greulach and Adams, *Plants*).

frame of reference, *Mathematics.* a set of lines or planes used as a reference for describing position, as of a point or line: *The walls, floor, and ceiling of a room may be considered to establish, as an example, the coordinates of a frame of reference* (Kenneth F. Gantz).

frameshift, *adj. Genetics.* having to do with or causing an insertion or deletion of one or two nucleotides in the DNA chain, so that some nucleotide sequences are read in incorrect groups of three during translation: *a frameshift mutation. These mutagens are called frameshift mutagens because the reading frame of the messenger RNA (mRNA) is shifted by the addition or deletion of a base, and this effect distinguishes them from the usual mutagens that cause base pair substitutions* (Science).

francium (fran'sē əm), *n. Chemistry.* a radioactive metallic element produced artificially from actinium or thorium. Francium belongs to the alkali metals. *Symbol:* Fr; *atomic number* 87; *mass number* 223; *oxidation state* 1. [from New Latin, from *France;* named by its discoverer, the French chemist Marguerite Perey, born 1909. Compare GERMANIUM.]

fraternal twins, *Biology.* twins of the same or opposite sex coming from two separately fertilized egg cells rather than from one egg cell as identical twins do: *Fraternal twins tend to run in families and are more likely to be born to women who are taller and heavier and, more significantly, to those who are older and have borne a greater number of children. And even diet has been implicated as a possible factor; in Nigeria, for example, the high rate of dizygotic twins closely matches the pattern of consumption of a particular species of yams indigenous to that African nation* (Edwin Chen).

Fraunhofer lines (froun'hō'fər), *Physics.* dark lines in the solar spectrum caused by the absorption of some of the light by elements on the surface of the sun, such as hydrogen and calcium, and by the CH molecule: *Analysis of the Fraunhofer lines enables astrophysicists to determine the composition of the gases in the sun's outer layers* (Shortley and Williams, *Elements of Physics*). [named after Joseph von *Fraunhofer,* 1787–1826, German optician, who made the first careful study of these lines]

cap, fāce, fäther; best, bē, tėrm; pin, five;
rock, gō, ôrder; oil, out; cup, pùt, rüle;
yü in use, yu in uric;
ng in bring; sh in rush; th in thin, ᵺ in then;
zh in seizure.
ə = a in about, e in taken, i in pencil, o in lemon, u in circus

free, *adj.* **1** *Chemistry.* not combined with something else: *Oxygen exists free in air. Lightning transforms about 100 million tons of "free" nitrogen of the air into what is called "fixed" nitrogen, without which farmers would have a harder time growing their crops, or gardeners their flowers and grass* (Barbara Tufty).
2 *Biology.* not in contact with or connected to some other body or surface: *On vertebrates, tactile receptors occur over most of the exterior surface. Some are free nerve endings, and others are special corpuscles that contain the sensory nerve terminations* (Storer, *General Zoology*).
3 *Geology.* easily worked or mined; loose and soft, as rock or coal.

free charge, *Physics.* an electrostatic charge which is not bound by the presence of a charge of opposite polarity: *The conductivity of solids depends on the number of free charges per unit volume* (Physics Syllabus). Compare **bound charge.**

free electron, *Physics.* an electron that is detached from an atom and can therefore move from atom to atom, as in a conductor of electricity or heat: *In a metal ... we have a structure consisting of an array of positive ions, with a large number of electrons moving among them. These electrons are known as free electrons* (Michels, *Foundations of Physics*).

free energy, *Physics.* a thermodynamic quantity used to represent that portion of the internal energy potential of a system that is available for the performance of work.

free fall, *Physics.* the motion of a body in flight through space when it is not acted upon by any force except gravity. A body is weightless during free fall. *An interplanetary vehicle, after its free fall back to Earth, will have an approach velocity of 25,000 mph or more* (F.I. Ordway).
—free-falling, *adj.* in a state of free fall, as a ballistic missile after the termination of thrust: *The kinds of accelerations experienced in rocket flight often amount to as much as ten or twelve times that due to a free-falling object near the surface of the earth* (Manchester Guardian Weekly).

free-living, *adj. Biology.* **1** living free from and independent of a host; not parasitic: *Most of the ciliates are not parasites and are therefore said to be free-living* (Hegner and Stiles, *College Zoology*).
2 able to move about independently; not sessile: *The majority of animals are free-living ... but some water dwellers are fixed in place, or sessile* (Storer, *General Zoology*).

free radical, *Chemistry.* an organic compound in which some of the valence electrons are not paired: *The high reactivity of free radicals stems directly from the fact that they have an odd electron* (William A. Pryor).

freeze, *v. Chemistry.* to change from a liquid phase to a solid phase by being cooled to the freezing point: *Water freezes at 0 degrees Celsius.*

freeze-dry, *v.* to dehydrate by freezing the moisture content to ice and evaporating the ice by subjecting it to microwaves in a vacuum. Freeze-dried substances keep without refrigeration.

freeze-etching, *n.* a method of preparing specimens for examination under an electron microscope by freezing them and then breaking them up along natural planes of weakness to show their internal structure in three dimensions: *Since yeast cells that have been subjected to freeze-etching can recover and resume growth, if the fracturing and shadowing are omitted, it can be concluded that the structure seen in freeze-etched cells is the same as in the living state* (W.G. Rosen).
—freeze-etch or **freeze-etched,** *adj.* of or having to do with freeze-etching: *Bacteria may be quickly frozen to* −147°C *and upon thawing still survive, so bacterial cells examined by the freeze-etch technique are still alive* (Weier, *Botany*).

freeze-fracture, *v.* to prepare (a specimen) for examination under an electron microscope by freeze-etching: *When a leaf is freeze-fractured it is easy to identify the epidermal layers, the spongy mesophyll with its air spaces, and the vascular tissue* (New Scientist).

freezing point, *Chemistry.* the temperature at which the solid and liquid phases of a substance are in equilibrium at atmospheric pressure. The freezing point of water at sea level is 0 degrees Celsius or 32 degrees Fahrenheit. *The freezing point of a liquid is lowered when some other substance is dissolved in the liquid. A common example is the use of an "anti-freeze" to lower the freezing point of the water in the cooling system of an automobile engine. The freezing point of a saturated solution of common salt in water is about* −20°C (Sears and Zemansky, *University Physics*). *Abbreviation:* f.p. Compare **melting point.** See also **boiling point.**

F region, = F layer.

frenulum (fren'yə ləm), *n., pl.* **-la** (-lə). *Anatomy.* a small frenum: *Many moths have a peculiar fastening which holds the front and back wings together when they fly. This fastening serves like a hook and eye. The "hook" is on the hind wings and is called the frenulum* (Cicile H. Matschat). [from New Latin *frenulum,* diminutive of Latin *frenum* bridle]

frenum (frē'nəm), *n., pl.* **-na** (-nə). *Anatomy.* a ligament or fold of membrane which restrains the motion of the organ to which it is attached, as that which holds down the underside of the tongue: *Pedunculated cirripedes have two minute folds of skin, called by me the ovigerous frena, which serve, through the means of a sticky secretion, to retain the eggs until they are hatched* (Darwin, *Origin of Species*). [from Latin *frenum* bridle]

frequency, *n.* **1** *Physics.* **a** the number of times that any regularly repeated event, as a vibration, occurs in a given unit of time. The SI unit of frequency is the hertz. **b** the number of complete cycles per second of an alternating current or other electric wave. Different radio and television stations broadcast at different frequencies so that their signals can be received distinctly. *Radio waves reflected from a planet undergo a slight shift in frequency. If they are reflected from an object that is approaching the earth, the frequency increases; conversely, the frequency decreases if the object is receding from the earth* (Thomas Gold).
2 *Statistics.* the number of cases of the data under consideration falling within a particular class interval: *In a game of dice ... 'double-six' occurs about once in 36 times, that is, its observed frequencies cluster about 1/36* (M. Loeve, *Probability Theory*).

frequency band, *Physics.* a continuous range of frequencies of electromagnetic radiation extending between two limiting frequencies: *A frequency-band considerably wider than the range of the human ear is necessary* (Peter Bartók).

frequency distribution, *Statistics.* **1** an arrangement of data to show the number of times an event occurs in a particular way; an arrangement of data according to frequency rather than time, location, or degree: *Some frequency distributions found in practice are symmetrical about the largest or modal frequency. Others ... are skew and have a longer 'tail' on the one side than the other* (J.A. Nelder).
2 a table or graph depicting such an arrangement.

frequency modulation, *Electronics.* deliberate modulation of the frequency of the transmitting wave in broadcasting in order to agree with the changes in the sounds or images being transmitted. Frequency modulation reduces static. *In frequency modulation the amplitude of the carrier wave remains constant but the spacing between the various cycles is altered* (Scientific American). *Abbreviation:* FM Compare **amplitude modulation.** See the picture at **modulation.**

fresh, *adj.* **1** *Geology.* not salty: *The water in most rivers is fresh.* **2** *Meteorology.* fairly strong; brisk.
    Fresh breeze is a wind with a velocity of 29–38 kilometers per hour; a **fresh gale** is one of 62–74 kilometers per hour. See table under **Beaufort scale.**

freshwater, *adj.* *Ecology.* **1** of or living in water that is not salty: *He is specially interested in the group of freshwater algae known as desmids and in the feeding of small freshwater animals on algae* (New Biology).
2 of or having to do with freshwater organisms: *In freshwater biology, there are comparisons to be made between the fauna of the various streams and pools, some of which are acid, draining off peat, and others ... alkaline* (Science News).

fresnel (frez′nel *or* frā nel′), *n. Physics.* a unit of frequency equal to 10^{12} (one trillion) hertz. [named after Jean Augustin *Fresnel,* 1788–1827, French physicist]

Fresnel lens, *Optics.* a lens consisting of a central spherical section surrounded by many segmental lenses, each of different curvature, designed to direct the rays parallel and relatively free of spherical aberration: *Since Fresnel lenses are comparatively thin, their weight is less than that of an ordinary lens and they absorb a comparatively small amount of light* (Hardy and Perrin, *Principles of Optics*).

friable (frī ə bəl), *adj. Geology.* easily crumbled; crumbly: *Dry soil is friable.* [from Latin *friabilis,* from *friare* to crumble.]
—friability, *n.* the condition or quality of being friable.

friction, *n. Physics.* resistance to motion of surfaces that touch; resistance of a body in motion to the air, water, etc., through which it travels or to the surface on which it travels: *Oil reduces friction. There is no air and thus no friction in space* (Robert C. Toth). Compare **coefficient of friction.** [from Latin *frictionem,* from *fricare* to rub] **—frictionless,** *adj.*
—frictional, *adj.* **1** of or having to do with friction: *The*

stronger the shock wave, the larger the fraction of the total heat load that is transferred to the atmosphere, and the smaller the frictional component* (John V. Becker).
2 moved or caused by friction: *Much of the kinetic energy of the wind is dissipated as heat by the frictional drag of the sea-surface on the air* (R.W. James).

Frigid Zone, *Geography.* (in former climatic classification) either of the two polar regions, north of the arctic circle and south of the antarctic circle. Compare **Temperate Zone** and **Torrid Zone.**

frill, *n. Zoology.* a fringe of feathers, hair, or skin around the neck of a bird or animal.

fringing reef, *Geology.* a coral reef formed close to the shoreline: *Fringing reefs are confined to the very border of the land. They may be thousands of feet wide, but most are only a few score feet. In the aggregate, however, they are undoubtedly much bulkier than the more spectacular and individually larger barrier reefs and atolls* (Gilluly, *Principles of Geology*).
ASSOCIATED TERMS: see **barrier reef.**

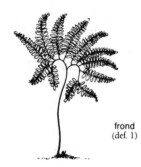

frond
(def. 1)

frond, *n.* **1** *Botany.* the leaf of a fern, palm, or cycad: *In the tropics, some ferns become treelike in form, possessing long erect trunks bearing at their summit a crown of compound leaves called fronds. In temperate regions, the land ferns have underground stems (rhizomes) from which fronds arise* (Youngken, *Pharmaceutical Botany*).
2 *Biology.* a leaflike part which includes both stem and foliage, as the thallus of a lichen.
[from Latin *frondem* leaf]

front, *n. Meteorology.* the dividing surface between two dissimilar air masses: *A front is a boundary surface, or more correctly a transition zone, separating air masses of differing character, specifically of markedly different temperatures. It is a sloping boundary and comparatively narrow, varying from five to 50 miles*

cap, fāce, fäther; best, bē, tėrm; pin, five;
rock, gō, ôrder; oil, out; cup, pùt, rüle;
yü in use, *yù* in uric;
ng in bring; *sh* in rush; *th* in thin, ͭ H in then;
zh in seizure.
ə = *a* in about, *e* in taken, *i* in pencil, *o* in lemon, *u* in circus

237

frontal

in width (Blair, *Weather Elements*). A **cold front** is the boundary line between advancing cold air and a mass of warm air under which the cold air passes. A **warm front** is the boundary line between advancing warm air and a mass of cold air over which it rises.

cold front

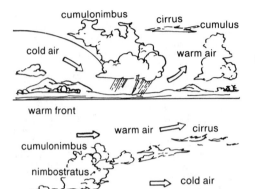

warm front

frontal, *adj.* **1** *Meteorology.* of or having to do with a front. A **frontal fog** is one that forms at a front.
2 *Anatomy.* **a** of or having to do with the front part of an organ or body. SYN: anterior. **b** of or in the region of the forehead: *the frontal bone.* See the picture at **skull.**

frontal lobe, *Anatomy.* the anterior part of either of the two lobes of the cerebrum: *The pleasure areas in the monkey brain are the septum and putamen deep in the middle of the brain. The discomfort or pain area is the orbital cortex of the frontal lobe* (Science News Letter).

frontlet, *n. Zoology.* **1** the forehead of an animal: *We can recognise the horns and frontlets of the elk* (Mayne Reid).
2 the forehead of a bird when a different color or texture of plumage: *Of two specimens, both shot at Washington, D.C., one has a whitish and the other a brown frontlet* (Elliott Coues, *Birds of the Northwest*).

frontogenesis (frun'tə jen'ə sis), *n. Meteorology.* the formation of a front, as by the convergence of dissimilar air masses: *Friction causes increased convergence, and hence frontogenesis, within the surface boundary layer, but above this layer there is a compensating divergence* (Stephen E. Mudrick). Compare **frontolysis.**

frontolysis (frun tol'ə sis), *n. Meteorology.* the dissolution of a front, as by the divergence or mingling of dissimilar air masses: *The formation of new fronts, or the regeneration and strengthening of weak and decaying fronts, is called frontogenesis. The opposite process, that of weakening or destroying existing fronts, is frontolysis* (Blair, *Weather Elements*).

frost, *n.* moisture frozen on or in a surface; feathery crystals of ice formed when water vapor in the air condenses at a temperature below freezing: *When the dew point of the air is below 32° F, water vapor in the air*

may change directly from a gas to solid ice crystals called frost. You can see this frost on the freezing unit in some refrigerators. Before frost will form on an object, temperature of the object must be below 32° F (Barnard, *Science: A Key to the Future*).

frostline, *n. Geology.* the maximum depth to which frost penetrates into the ground in a given locality, as measured during winter: *The entomologists are also relating winter soil temperatures to termite survival and have found living termites in December, most of which were in the upper six inches of soil—the same depth as the frostline* (Science News Letter).

fructification (fruk'tə fə kā'shən), *n. Botany.* **1** a forming or bearing of fruit: *At the time of fructification, watch the plants daily* (John Baxter).
2 the fruit or other reproductive material of a plant: *Nearly the whole under side of the frond is covered with the fructification* (Francis G. Heath).
3 a spore-bearing structure; sporangium or sporophore: *Prior to the reproductive stage, the plasmodium moves to a drier substratum. After a time, the plasmodium ceases moving and forms one or more spore-producing structures (sporangia or fructifications)* (Weier, *Botany*).

fructivorous (fruk tiv'ər əs), *adj.* feeding on fruits: *Some vertebrates are herbivorous ... some fructivorous, and a few even omnivorous* (White and Renner, *Human Geography*). SYN: frugivorous. [from Latin *fructus* fruit + *vorare* devour]

fructose (fruk'tōs), *n. Biochemistry.* a sugar present in many fruits, honey, etc. It is sweeter than glucose or sucrose. *Fructose, also called levulose and fruit sugar, is found in plants in relatively large amounts. It plays an important role in various metabolic processes and, like glucose, is used in the synthesis of more complex carbohydrates* (Greulach and Adams, *Plants*). Formula: $C_6H_{12}O_6$ [from Latin *fructus* fruit]

frugivore (frü'jə vôr), *n. Biology.* an animal, especially a primate, that feeds on fruit: *Those primate species that can utilize cellulose in leaves are referred to as folivores and are distinguished from frugivores, which cannot utilize cellulose. Of course, these basic categories do intergrade, since folivores may supplement their leaf diet with fruit* (Science). [from Latin *frugus* fruit + *-vore,* as in *carnivore* and *herbivore*]
—frugivorous (frü jiv'ər əs), *adj. Biology.* eating or feeding on fruit: *It was the use of fire which made it possible for our ancestors to change from the predominantly frugivorous diet of our anthropoid ancestors to the predominantly seed and root diet of our own species* (Linton, *Tree of Culture*). [from Latin *frugis* fruit + *vorare* devour]

fruit, *n. Botany.* the matured ovary of a seed plant, consisting of the seeds, the tissues connected with them, and their coverings.
ASSOCIATED TERMS: Most fruits (with the exception of *aggregate fruits* such as the raspberry, and *multiple fruits* such as the pineapple) are classified by botanists into four main types: (1) the *berries,* having a completely fleshy *pericarp,* such as the banana, grape, orange, tomato, and watermelon; (2) the *drupes,* having a hard pit and a single seed, such as the peach, cherry, apricot, and plum; (3) the *pomes,* having a central core with several seeds, such as the apple and pear; and (4) the single-seed fruits with a hard pericarp, such as pea pods, acorns, chestnuts, and the grains of wheat, corn, and rice.

fruit fly, *Zoology.* any of various small flies whose larvae feed on fruits and vegetables. Fruit flies are used in scientific studies of heredity because of the large size of the chromosomes and the rapidity of generation. *Certain animals undergo a series of dramatic changes in structure during their development—changes to which the name of metamorphosis is applied. An ideal subject for the study of metamorphosis is the common fruit fly, Drosophila* (Robert W. Menefee).

fruiting body, *Botany.* any part of a plant that produces spores: *The tiny fruiting body of the cellular slime mold is also lofted upward into the air—by as much as two whole millimeters—above the surface from which its spindly supporting stalk happens to spring* (John T. Bonner).

fruit sugar, = fructose.

frustule (frus′chül *or* frus′tyül), *n. Botany.* the siliceous bivalve shell of a diatom: *The frustule is really a box, composed of an upper and lower lid, each fitted with a girdle band around the edge* (N. Ingram Hendey). [from Late Latin *frustulum,* diminutive of Latin *frustum* piece]

frustum (frus′təm), *n., pl.* **-tums, -ta** (-tə). *Geometry.* **1** the part of a cone left after the pointed end has been cut off by a plane parallel to the base. **2** the part of any solid between two parallel planes cutting the solid. [from Latin *frustum* piece]

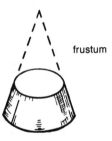

frustum

fruticose (frü′tə kōs), *adj. Botany.* having the form of a shrub; shrubby: *The fruticose, or branching, lichens may be found in the sparse soil of a rocky crevice. They look like tiny gray trees half an inch high, and often sport colorful little caps* (Ronald Rood). Compare **crustose** and **foliose.** [from Latin *fruticosus* full of bushes, shrubby, from *frutex, fruticis* bush, shrub]

Fleshy Simple Fruits

berries   drupes   pomes

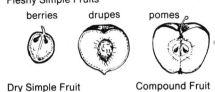

Dry Simple Fruit   Compound Fruit
maple

drupe

FSH, *abbrev.* follicle-stimulating hormone.

ft., *abbrev.* foot *or* feet.

fth. or **fthm.,** *abbrev.* fathom.

ft.-l., *abbrev.* foot-lambert.

fugacious

ft.-lb., *abbrev.* foot-pound.

fucoid (fyü′koid), *adj.* **1** *Botany.* having to do with seaweeds, especially of the fucus type.
2 *Geology.* characterized by or containing impressions of or markings like seaweeds, as sandstone: *fucoid shale.*
—n. 1 *Botany.* a seaweed of the fucus type.
2 *Geology.* a seaweed-like impression or marking: *Fucoids or fucus-like impressions occur in strata of every epoch* (David Page).

fucose (fyü′kōs), *n. Biochemistry.* a type of crystalline pentose sugar found in its levorotatory form in gum tragacanth, in various seaweeds, and in the polysaccharides of some blood groups; in its dextrorotatory form it is found in several glycosides. *Formula:* $C_6H_{12}O_5$ [from *fuc(us)* + *-ose*]

fucoxanthin (fyü′kō zan′thēn *or* fyü′kō zan′thin), *n. Biochemistry.* the brown carotenoid pigment characteristic of brown algae. *Formula:* $C_{40}H_{60}O_6$ [from *fucus* + *xanthin*]

fucus (fyü′kəs), *n., pl.* **-ci** (-sī), **-cuses.** any of an order (Fucales) of olive-brown seaweeds having flat, leathery branching fronds and often having air bladders. Commonly called **rockweed.** [from Latin *fucus* rock lichen]

fuel, *n.* **1** coal, wood, oil, or any other material that can be burned to produce useful heat or power: *The [rocket] fuel, which Rocket Research calls Monex W, has a simple recipe: the four ingredients—carbon, ammonium nitrate, aluminum and the waste material—are just blended together and they're ready to go* (Science News).
2 atomic matter producing heat by fission or fusion, as in a reactor: *We contemplated a worldwide civilian atomic power industry, which could include what we then envisioned as a complete nuclear "fuel cycle"—production of nuclear fuel, its use in reactors, and the reprocessing of spent fuel for renewed use* (Leonard Ross).

fuel cell, a device which produces electricity directly from a chemical reaction between oxygen and a gaseous fuel such as hydrogen or methane.

fuel rod, a long tube containing nuclear fuel, used in a nuclear reactor: *The uranium-oxide fuel is shaped into pellets about an inch long and half an inch in diameter, and these pellets are fitted into tubes about twelve feet long. These tubes, called fuel rods, are made of a zirconium alloy that is highly resistant to radiation damage and will not melt at the working temperature of the reactor. In a large pressurized-water reactor, there are about forty thousand fuel rods, which are packed into bundles of about two hundred rods each. Ordinary water circulates among the fuel rods* (New Yorker).

fugacious (fyü gā′shəs), *adj. Botany.* falling or fading early in the growing season: *If the calyx falls very early, it is called fugacious* (Youngken, *Pharmaceutical*

cap, fāce, fäther; best, bē, tèrm; pin, fīve;
rock, gō, ôrder; oil, out; cup, pùt, rüle,
yü in use, *yu* in uric;
ng in bring; *sh* in rush; *th* in thin, ŦH in then;
zh in seizure.
ə = *a* in about, *e* in taken, *i* in pencil, *o* in lemon, *u* in circus

Botany). Compare **deciduous, caducous.** [from Latin *fugax, fugacis* fleeting, from *fugere* flee]

fulcrum (ful'krəm), *n., pl.* **-crums, -cra** (-krə). *Physics.* the support on which a lever turns or rests in moving or lifting something: *The place where a lever is rested or supported is called the fulcrum. The distance from the weight to the fulcrum is the weight-arm, and the distance from the force (your hand) to the fulcrum is the force-arm* (Beauchamp, Mayfield and West, *Everyday Problems in Science*). See the picture at **lever.** [from Latin *fulcrum* bedpost, from *fulcire* to support]

fulgurite (ful'gyə rit), *n. Geology.* partially fused sand in the form of long, slender glass tubes formed by lightning striking unconsolidated sand or soil: *Always associated with the sky, bolts of lightning have actually been preserved in sand and rock. These "fulgurites," or "petrified lightning," are fragile, glassy tubes formed when lightning strikes in sand, melting the particles around its path and fusing them together. The hollow fulgurites, self-portraits of the lightning channels that made them, range from 1.5 to 5 centimeters in diameter. Some fossil fulgurites date back as far as 250,000,000 years* (Dava Sobel). [from Latin *fulgur* lightning]

full moon, *Astronomy.* **1** the moon seen from the earth as a whole circle: *When the earth is between the sun and the moon, you see the full moon because the light from the sun shines over the earth and on to the moon* (Beauchamp, Mayfield and West, *Everyday Problems in Science*). **2** the period when this occurs.

fulminate (ful'mə nāt), *n. Chemistry.* **1** a violent explosive. **2** a salt of fulminic acid. The fulminates, chiefly mercury and silver, are very unstable compounds, exploding with great violence by percussion or heating. *Mercuric fulminate, (Hg(ONC)_2), or fulminate of mercury, is manufactured by adding alcohol to mercury while it is being oxidized by hot nitric acid. The white precipitate obtained is unstable; when dry, it decomposes explosively on being struck. It is used in percussion caps* (Offner, *Fundamentals of Chemistry*). [from Latin *fulminatum* struck with lightning, from *fulminis* lightning, related to *fulgere* to shine]

fumarase (fyü'mə rās'), *n. Biochemistry.* an enzyme found in many plants and animals, especially in the muscles and liver of higher animals. Fumarase causes the conversion of fumaric acid to malic acid in the Krebs cycle.

fumarate (fyü'mə rāt'), *n. Chemistry.* a salt or ester of fumaric acid.

fumaric acid (fyü mar'ik), a whitish, dicarboxylic acid found in all cells and essential to their respiration. *Formula:* $C_4H_4O_4$ Compare **citric acid, succinic acid.** [from New Latin *Fumaria* the fumitory genus (a plant related to the poppy), in which this acid is found]

fumarole (fyü'mə rōl'), *n. Geology.* a hole or vent in the earth's crust in a volcanic region, from which steam and hot gases issue: *One of the geochemical techniques used for monitoring and predicting the activity of a volcano is to measure the emission rates and composition of the gas emerging from the vent in major plumes and minor fumaroles* (Peter J. Wyllie). [from Italian

fumarola, from Latin *fumus* smoke, vapor] ▶ See the note under **hot spring.**

—fumarolic, *adj.* of or formed by a fumarole: *Fleming showed the gypsum to be prominent in altered deposits at the sites of extinct or dying fumarolic activity* (G. J. Williams).

fume, *n. Chemistry.* Often, **fumes,** *pl.* vapor, gas, or smoke, especially if harmful, strong, or odorous: *zinc or lead in the form of fumes.* [from Latin *fumus* smoke]

fumigant (fyü'mə gənt), *n. Chemistry.* a poisonous compound used in its gaseous state to kill insects and other pests: *Fumigants, such as hydrogen cyanide, enter the insect's body through its respiratory tubes* (Albert A. LaPlante, Jr.).

function, *n.* **1** *Mathematics.* a quantity whose value depends on the value given to one or more related quantities: *The area of a circle is a function of its radius; as the radius increases so does the area.*

2 anything likened to a mathematical function; a factor dependent on and varying with another factor: *While the mass of a body can ordinarily be considered constant, there is ample experimental evidence that actually it is a function of the velocity of the body, increasing with increasing velocity* (Sears and Zemansky, *University Physics*).

3 *Biology.* the activity or action of any part of an organism: *Many relationships in function (physiology) parallel those of structure; in both man and anthropoids there are comparable blood groups ...* (Storer, *General Zoology*).

—functional, *adj.* of a function or functions. A **functional equation** is one in which the solution is a function rather than a number. A **functional disorder** in medicine is a disorder in the normal action of an organ for which there is no apparent organic cause or explanation. **—n.** *Mathematics.* a function whose value depends on all the values assumed by another function: *Jacques Hadamard named them functionals, and Paul Levy gave the name functional analysis to the study of analytical properties of functionals* (Edna E. Kramer).

—functionally, *adv.* in function; as regards function: *The male organs of species-hybrids are functionally weak to a higher degree than the female organs* (Sydney H. Vines).

functional analysis, *Mathematics.* the study or analysis of functionals: *The theory of the topology of metric spaces ... is not only the basis of functional analysis but also unifies many branches of classical analysis* (E.T. Copson).

functional group, *Chemistry.* a group of atoms responsible for the common properties of certain compounds, especially organic compounds, such as the carboxyl group present in organic acids: *Organic compounds can often be considered as being composed of one or more functional groups attached to a hydrocarbon group. Students should be able to recognize primary alcohols and organic acids by their functional groups* (Chemistry Regents Syllabus).

fundamental, *Physics.* **—n.** that component of a wave which has the greatest wavelength or lowest frequency: *A complex sound wave is made up of the fundamental and the overtones. The timbre of a tone is expressed in the number, intensity, and phase relations of the components, that is, the fundamental and overtones (Harry F. Olson).

—adj. of or denoting a fundamental: *There is a theory which, unfortunately, is often taken for a proven fact: it says that the fundamental frequency, or some of the lower harmonics may, by their larger volume, mask the audibility of the higher frequencies of upper partials. The idea, apparently, is that any larger volume can mask the audibility of any smaller volume. Now, this is an engineering theory, and the term masking was introduced by electrical engineers, as was the interpretation of the sound phenomenon it tries to explain* (F.A. Kuttner).

fundamental particle, = elementary particle: *Type I particles are considered fundamental particles by virtue of their relatively long lifetimes, which average a ten-billionth of a second (10^{-10} second). On the basis of their masses the particles on this list have been classified as leptons, bosons or baryons* (Scientific American).

fundamental unit, *Physics.* one of the independent units (especially those of mass, length, and time) taken as a basis for a system of units: *The fundamental units used [in the MKS system] are the meter, the kilogram, the second, the ampere, and the degree Kelvin* (Physics Regents Syllabus). *The fundamental unit of time which is used in both the English and metric systems is called the second* (J.L. Glathart). Also called **base unit.** Compare **derived unit.**

fundic, *adj. Anatomy.* of or having to do with a fundus. The **fundic glands** are the glands of the body and bottom of the stomach. *Certain cells in the mucous membrane of the stomach secrete a hormone called gastrin, which stimulates the secretion of the fundic and pyloric glands* (Hegner and Stiles, *College Zoology*).

fundus (fun′dəs), *n., pl.* **-di** (-dī). *Anatomy.* the bottom of an organ, or the part opposite or farthest from the opening: *the fundus of the stomach, eye, or uterus.* [from Latin *fundus* bottom, (piece of) ground]

fungal (fung′gəl), *adj. Biology.* of, resembling, or caused by a fungus: *The fungal component generally dominates the vegetative and reproductive characteristics of the lichens* (Greulach and Adams, *Plants*).

fungi (fun′jī), *n.* a plural of **fungus.**

fungicide (fun′jə sīd), *n.* any substance that destroys fungi: *Lethal synthesis (one approach to developing selective fungicides) involves blocking the active toxophore with a chemical grouping, such as an ester, which is removed by the fungus but not the host* (Science).
—fungicidal (fun′jə sī′dl), *adj.* that destroys fungi: *A fungicidal wash containing a silicone compound is reported to give remarkable control of mould-infested painted interiors* (New Scientist).

fungiform (fun′jə fôrm), *adj. Botany, Zoology.* having the form of a fungus: *The fungiform papillae are much smaller and more numerous than the circumvallate ones* (St. George Mivart).

fungistatic (fun′jə stat′ik), *adj. Biology.* stopping the development or growth of a fungus without killing it: *Cells of the bulbs [of certain orchids] produce fungistatic compounds with phenollike properties in response to fungal invasion* (James M. Trappe).

fungivorous (fun jiv′ər əs), *adj. Zoology.* feeding on mushrooms or fungi, as certain insects. [from *fungi* + Latin *vorare* devour]

furcate

fungoid (fung′goid), *Botany. —adj.* resembling a fungus; having spongy, unhealthful growths. *—n.* a fungoid plant.

fungous (fung′gəs), *adj.* = fungal.

fungus (fung′gəs), *n., pl.* **fungi** (fun′jī) or **funguses.** *Biology.* any of a major group (Fungi) of nonmotile, filamentous organisms that lack chlorophyll and absorb their nutrients from dead or living organisms. Fungi were traditionally classified as plants, but today they are often treated as a separate kingdom. Mushrooms, yeasts, toadstools, smuts, rusts, molds, and mildews are fungi. *The fungi, together with the heterotrophic bacteria and a few other groups of heterotrophic organisms, are the decomposers of the biosphere* (Raven, Evert, and Curtis, *Biology of Plants*). *—adj.* = fungal. [from Latin *fungus*]

funiculus (fyü nik′yə ləs), *n., pl.* **-li** (-lī). **1** *Anatomy.* a cordlike structure, such as the umbilical cord or the spermatic cord, or a bundle of nerve fibers.
2 *Botany.* the stalk by which a seed or ovule is attached to the placenta.
3 *Zoology.* a retractor muscle in bryozoans that draws the stomach aborally.
[from Latin *funiculus* little rope, diminutive of *funis* rope]

funnel, *n. Meteorology.* the narrow cloud of a tornado that extends downward from a heavy, dark mass of cumulonimbus clouds: *A tornado has a strong lifting force due to the updraft of air in the funnel* (James E. Miller).

fuoro (fyü ôr′ō), *adj. Astronomy.* of or having to do with any of a group of distant variable stars, called T Tauri stars, that exhibit eruptive activity which changes their brightness, spectrum, and apparent shape: *Although two of the fuoro stars have remained close to peak brightness for years, V 1057 appears to be sliding back to its previous condition. It has declined by a magnitude and a half in five years. So perhaps the fuoro phenomenon is not permanent, and after n number of years, the star goes back to the T Tauri class again* (Science News). [from *FU Or(i)o(nis)*, a nebula in which the first star of this type was observed]

fur, *n.* **1** *Zoology.* the hair covering the skin of certain animals. Fur grows on many mammals and usually consists of a short, soft, thick undercoat thinly covered by a longer, coarser outer coat. **2** *Botany.* any coating like fur, as on a plant.

furanose (fyür′ə nōs), *n. Biochemistry.* a sugar having a ring structure of five atoms. Hexoses may sometimes occur in the form of a furanose. [from *furan* a colorless flammable liquid, C_4H_4O, used in organic synthesis + *-ose*]

furcate (fèr′kāt or fèr′kit), *Botany, Zoology. —adj.* forked: *In the leaves of many ferns certain veins branch in a forked manner and illustrate what is*

---

cap, fāce, fäther; best, bē, tèrm; pin, five;
rock, gō, ôrder; oil, out; cup, pùt, rüle,
yü in use, *yu̇* in uric;
ng in bring; *sh* in rush; *th* in thin, ᴛʜ in then;
zh in seizure.
ə = *a* in about, *e* in taken, *i* in pencil, *o* in lemon, *u* in circus

241

furcula

known as *furcate venation* (Youngken, *Pharmaceutical Botany*).

—*v.* to form a fork; divide into branches: *Another small fissure ... furcates a short distance above* (James D. Dana).

[from Late Latin *furcatum* forked, cloven, from Latin *furca* pitchfork] —**furcation,** *n.*

furcula (fėr′kyə lə), *n., pl.* **-lae** (-lē). *Zoology.* a forked part or structure, such as the fused clavicles or wishbone of a bird. [from Latin *furcula* forked support, from *furca* pitchfork]

—**furcular,** *adj.* of or like a furcula: *a furcular bone.*

furfuraceous (fėr′fyə rā′shəs), *adj. Botany.* covered with branlike scales: *Stem somewhat flexuous, brittle, furfuraceous, then smooth* (Miles J. Berkeley). [from Late Latin *furfuraceus* branlike, from Latin *furfur* bran]

fusain (fyü′zān *or* fyü zān′), *n. Geology.* a coal lithotype characterized by black color, silky luster, fibrous structure, and friability. It occurs in patches or strands, and is soft and dusty if not mineralized. *Quite moderate variation in the amount of fusain in the coal has important effects on its utility* (Nature). *The proportion of fusain in most coals is small, generally less than 3%* (W. Francis). [from French *fusain* (originally) spindle tree, ultimately from Latin *fusus* spindle]

fused quartz or **fused silica,** = quartz glass: *Mercury has the advantage of an acoustic velocity only a quarter as large as fused quartz and a low attenuation per unit length up to a frequency of about 10 megacycles per second* (Science News).

fusel oil (fyü′zəl), *Chemistry.* an acrid, poisonous, oily liquid, consisting mainly of amyl alcohol, that occurs in alcoholic liquors when they are not distilled enough: *Leucine is the parent substance of the isoamyl alcohol of fusel oil formed during the fermentation of grain* (Muldoon, *Organic Chemistry*). [from German *Fusel* bad liquor]

fusibility (fyü′zə bil′ə tē), *n. Chemistry.* the condition or quality of being fusible: *Carbon, it is well known, gives to iron fusibility* (William C. Roberts-Austen).

fusible (fyü′zə bəl), *adj. Chemistry.* that can be fused or melted: *In the normal case, when a small piece of a fusible metal is heated to its melting point, it takes up a roughly spherical form and retains that form on cooling* (A.W. Haslett).

fusiform (fyü′zə fôrm), *adj. Biology.* rounded and tapering from the middle toward each end; shaped like a spindle: *a fusiform root, a fusiform aneurysm.* [from Latin *fusus* spindle + English *-form*]

fusing point = melting point.

fusion, *n.* **1** *Chemistry.* **a** a melting together; melting; fusing: *Bronze is made by the fusion of copper and tin.* **b** a fused mass: *Drawing out small lumps of the adhering fusion, they moulded it, before it had time to cool, into various forms* (T. Coan).

2 *Nuclear Physics.* the combining of two atomic nuclei to produce a nucleus of greater mass. Fusion releases tremendous amounts of energy and is used to produce the reaction in a hydrogen bomb. *Unlike fission, in which the atomic nucleus is split to release energy, the basic fusion reaction requires that two nuclei of heavy hydrogen—called deuterium—fuse to make an atom of helium. Fission is the process that powered the original atom bomb. Fusion is the energy source for the much more powerful thermonuclear devices—hydrogen bombs* (Harold M. Schmeck Jr.). See the picture at **nuclear fission.**

[from Latin *fusionem,* from *fundere* pour, melt]

fuzzy set, *Mathematics.* a set whose elements converge or overlap with those of other sets: *Fuzzy sets ... do not possess sharply defined boundaries* (Science).

G

g or **g.,** *abbrev.* or *symbol.* **1** gram *or* grams. **2** gravity. **3** giga.

G, *abbrev.* or *symbol.* **1** giga-. **2** specific gravity; gravitational constant. **3** guanine.

G or **g** (jē), *n., pl.* **G's** or **g's.** *Physics.* a unit of acceleration equal to about 9.8 meters per second per second. A freely falling body near the earth's surface accelerates at a rate of about 1 G. [from *G(ravity)*]

Ga, *symbol.* gallium.

GABA (gab′ə), *n. Biochemistry.* an amino acid occurring in the central nervous system, associated with the transmission of nerve impulses: *In mammals, GABA is a compound almost exclusively confined to the brain, where it is present in high concentrations* (New Scientist). [*G(amma)-A(mino)-B(utyric) A(cid)*]

gabbro (gab′rō), *n. Geology.* any of a group of fine- to coarse-grained plutonic igneous rocks, greenish-gray to black, containing pyroxene or olivine and calcic plagioclase feldspar.
  ▶ Because of its low silica content, gabbro is referred to as a basic igneous rock. Because of its high content of Mg-Fe silicates, it is also referred to as a mafic igneous rock. [from Italian *gabbro*]

gadolinium (gad′l in′ē əm), *n. Chemistry.* a highly magnetic, rare-earth metallic element which occurs in combination with other rare-earth elements in certain rare minerals. One of the isotopes of Gd has the highest neutron-absorption cross section of any element (measured in barns). *Symbol:* Gd; *atomic number* 64; *atomic weight* 157.25; *melting point* 1311°C; *boiling point* 3233°C; *oxidation state* 3. [named after Johann Gadolin, 1760–1852, Finnish chemist]

gain, *n. Electronics.* amplification of a signal, expressed as the ratio of the output of an amplifier to its input: *a voltage gain, a power gain.*

gal, *n. Physics.* a unit for measuring the acceleration of gravity, equal to one centimeter per second per second. [from *Gal(ileo)*, 1564–1642, Italian astronomer and physicist]

gal., *abbrev.* gallon. *pl.* **gal.** or **gals.**

galactan (gə lak′tan), *n. Biochemistry.* a polysaccharide found in plants during the germinative period, yielding galactose on hydrolysis.

galactic (gə lak′tik), *adj. Astronomy.* of or having to do with the Milky Way or with other galaxies: *Along the galactic plane the faint, distant stars are most dense on the sky at about galactic longitude 320° and least dense at the opposite point in longitude 140°. The former point lies amidst the magnificent star clouds in the constellation Sagittarius and marks the direction of the galactic center* (Krogdahl, *Astronomical Universe*).

galactic center, *Astronomy.* the gravitational center of our galaxy, presumed to be located in the direction of Sagittarius: *Though opaque to visible light, these clouds are quite transparent to radio radiation, and a* strong cosmic radio source (Sagittarius A) is in fact observed in the exact direction of the galactic center. It is believed to be the actual nucleus of our galaxy (Joseph Ashbrook).

galactic circle or **galactic equator,** *Astronomy.* the great circle inclined at an angle of 62 degrees to the celestial equator, whose plane nearly coincides with the central line of the Milky Way: *Galactic latitude and longitude are related to the galactic circle exactly as celestial latitude and longitude are related to the ecliptic. The origin from which galactic longitude is reckoned is the intersection of the galactic circle with the celestial equator, in α = 18ʰ 44ᵐ* (Duncan, *Astronomy*).

galactic latitude, *Astronomy.* the angular distance of a celestial body from the galactic plane.

galactic longitude, *Astronomy.* the angular distance of a celestial body measured eastward along the galactic circle from its intersection with the celestial equator.

galactic nebula, *Astronomy.* any nebula that belongs to our galaxy, classified as either bright or dark depending on its illumination: *The galactic nebulae are found in the Milky Way. The diffuse type are irregular in shape and immense in size. They may present an appearance of great shapeless clouds or may be only hazy patches, in any case showing as extended areas and thus being fainter than stars. Some of them are comparatively luminous, and others have no sources of illumination, appearing only as dark masses* (Bernhard, *Handbook of Heavens*).

galactic poles, *Astronomy.* the two opposite points on the celestial sphere, 90 degrees from the galactic circle.

galactic radio emission *Astronomy.* the radio noise emanating from interstellar charged particles in our galaxy.

galactolipid (gə lak′tō lip′id), *n. Biochemistry.* **1** a compound of galactose and a fatty substance found in tissues. **2** = cerebroside.

galactose (gə lak′tōs), *n. Biochemistry.* a white, crystalline monosaccharide, found in combined form in lactose, pectins, gums, and certain other substances: *Another gene in E. coli controls the formation of a different enzyme, beta-galactosidase. Its function is to break down a sugar, galactose, so that it can be used as a carbon source within the cell. When galactose is present, the enzyme is being formed; when galactose is absent, so, too, is the enzyme* (McElroy, *Biology and Man*). *Formula:* $C_6H_{12}O_6$

---

cap, fāce, fäther; best, bē, tèrm; pin, five;
rock, gō, ôrder; oil, out; cup, pùt, rüle;
yü in use, *yu* in uric;
ng in bring; *sh* in rush; *th* in thin; ᴛʜ in then;
zh in seizure.

ə = *a* in about, *e* in taken, *i* in pencil, *o* in lemon, *u* in circus

galactosidase (gə lak′tə sī′dās), *n. Biochemistry.* an enzyme that breaks down galactosides by hydrolysis.

galactoside (gə lak′tə sīd), *n. Biochemistry.* galactose in chemical combination with a nonsugar, a form in which it is sometimes found in nature.

galacturonic acid (gə lak′tyu̇ ron′ik), *Biochemistry.* a crystalline aldehyde acid, formed by the hydrolysis of pectins, and related to galactose, used in biochemical research. *Formula:* $C_6H_{10}O_7$

galaxy (gal′ek sē), *n. Astronomy.* **1** a system or aggregate of stars, cosmic dust, and gas held together by gravitation, often thousands of light-years in diameter. A galaxy typically contains billions of stars, and there are billions of galaxies in the universe.
2 Usually, **the galaxy** or **the Galaxy.** the galaxy containing our solar system; the Milky Way: *All the gas and dust and stars are in swirling motion, with the sun's portion of the galactic disk rotating around the center of the galaxy with a speed of about 260 kilometers per second, or almost 600,000 miles per hour. Even at this velocity it takes more than 200 million years for the sun to make one complete circuit of the galaxy* (Scientific American).
[from Greek *galaxias,* from *galaktos* milk]
ASSOCIATED TERMS: Our galaxy, the Milky Way, is a *spiral galaxy,* as distinguished from an *elliptical galaxy,* which lacks spiral arms and rotates more slowly. Our galaxy's *nebulae* (clouds of gas and dust particles) are called *galactic nebulae,* to distinguish them from the nebulae of other galaxies. The closest galaxies to the Milky Way form the *local group,* which includes the Andromeda galaxy and the Magellanic Clouds. *Radio galaxies* and *quasars* can also be studied through *radio astronomy.* The distance and motion of a galaxy is measured by its *red shift.* See also *Maffei galaxy, Seyfert galaxy.*

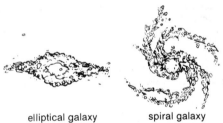

elliptical galaxy   spiral galaxy

gale, *n. Meterology.* a wind having a velocity of 50 to 102 kilometers per hour. See **Beaufort scale.**

galea (gā′lē ə), *n., pl.* **-leae** (-lē ē). *Botany, Zoology.* a structure or part that resembles a helmet, such as the upper part of a labiate flower, the membrane covering the jaws of certain insects, and a horny cap on the head of a bird. [from Latin *galea* helmet]
—**galeate** (gā′lē āt), *adj.* having or covered with a galea.

galena (gə lē′nə), *n. Mineralogy.* a metallic, gray mineral containing lead and sulfur. It is the most common lead mineral and the chief source of lead. *Formula:* PbS [from Latin]

gall, *n.* **1** *Botany.* an abnormal growth that forms on the leaves, stems, or roots of plants, caused by insects, parasitic bacteria, or fungi. The galls of oak trees are the source of tannin. *Galls are of diverse sizes and shapes and may be quite complicated in structure, their characteristics being determined by an interaction between* the host and parasite species. Gall organisms generally parasitize only one species of plant, or at most a few closely related species (Greulach and Adams, *Plants*).
2 *Physiology.* = bile.

gall bladder, *Anatomy.* a sac attached to the liver, in which excess bile is stored until needed.

gallic acid, *Chemistry.* a white, crystalline organic acid obtained especially from galls on plants, used in making ink and dyes, in tanning, etc. *Formula:* $C_7H_6O_5 \cdot H_2O$

gallium (gal′ē əm), *n. Chemistry.* a rare grayish-white metallic element similar to mercury, with a melting point slightly above room temperature and high boiling point, used in thermometers. Compounds of gallium, such as gallium arsenide, GaAs, are used as semiconductors in diodes and lasers. *Symbol:* Ga; *atomic number* 31; *atomic weight* 69.72; *melting point* 29.78°C; *boiling point* 2403°C; *oxidation state* 3. [from New Latin, from *Gallia* France]

gallon, *n.* a customary unit of volume for measuring liquids, equal to 4 quarts. The United States gallon equals 231 cubic inches or 3.7853 liters. The British gallon equals 277.420 cubic inches or 4.546 liters.

Galois field (gal′wä *or* gal wä′), *Mathematics.* a field with a finite number of elements: *Although conceived as an abstract mathematical exercise, Galois fields ... have been studied in connection with error-free codes for the transmission of information by high-speed machines* (Scientific American). [named after Evariste Galois, 1811–1832, French mathematician]

galvanic, *adj. Physics.* of, caused by or producing an electric current by chemical action; voltaic. A **galvanic current** is a direct electric current.

galvanic cell, *Electricity.* an electrolytic cell for the production of electricity by chemical action. Compare **voltaic cell.**

galvanic skin response, *Physiology.* a change or variation in the skin's electrical conductivity due to any stimulus, as detected by a galvanometer: *The galvanic skin response measures the ability of the skin to conduct electricity, which is a function of the autonomic nervous system. The latter is intimately hooked up with those mechanisms regulating the body's ability to cope with stress* (Thomas P. Hackett). *Abbreviation:* GSR

galvanism, *n. Physics.* electricity produced by chemical action. [from French *galvanisme,* from Luigi *Galvani,* 1737–1798, Italian scientist who discovered that electricity can be produced by chemical action]

galvanometer (gal′və nom′ə tər), *n.* an instrument for detecting, measuring, and determining the direction of a small electric current: *Any device used for the detection or measurement of current is called a galvanometer, and the majority of such instruments depend for their action on the torque exerted on a coil in a magnetic field* (Sears and Zemansky, *University Physics*).
—**galvanometry,** *n.* the detection, measurement, and determination of the strength of electric currents by a galvanometer.

galvanotaxis, *n. Biology.* movement in response to an electric stimulus: *The electric field is, however, effective over the whole width of the stream, and the voltage gradient across the width of a fish is sufficient to induce galvanotaxis and turn it in the direction of the positive electrode* (New Scientist).

gametangium (gam′ə tan′jē əm), *n., pl.* **-gia** (-jē ə). *Botany.* the cell or organ in which gametes are produced: *A gametangium, in the most unspecialized case, may be indistinguishable from an ordinary vegetative cell, except that several additional mitoses occur within it, giving rise to 16–32 flagellated cells* (Weier, *Botany*). [from New Latin *gametangium,* from *gaméta* gamete + Greek *angeion* vessel]

gamete (gam′ēt *or* gə mēt′), *n. Biology.* a mature reproductive cell capable of uniting with another to form a fertilized cell that can develop into a new organism; an egg or sperm cell; germ cell: *Sexual reproduction in Chlamydomonas begins with the division of the parent protoplast into eight cells which resemble the asexual spores except for their smaller size. These smaller motile cells, called gametes, are discharged, swim freely for a time and then unite in pairs. The resultant product may swim freely for a short time, during which two nuclei can be discerned in the cell. Finally it settles down, the two nuclei fuse, and a thick wall is secreted. This fusion product, called the zygote, ultimately germinates* (Greulach and Adams, *Plants*). [from New Latin *gameta,* from Greek *gametē* wife, *gametēs* husband, from *gamos* marriage]
—**gametic** (gə met′ik), *adj.* of or having to do with gametes: *the gametic stage of an organism's life.*
—**gametically** (gə met′ə klē), *adv.* as regards gametes; as a gamete.

game theory, *Mathematics.* the study of games and competitive strategies to determine or maximize the probabilities of a particular outcome: *It was not until 1926 that John von Neumann gave his proof of the minimax theorem, the fundamental theorem of game theory* (Martin Gardner). Also called **theory of games.**

gametocyte (gə mē′tō sīt), *n. Biology.* a cell that produces gametes by division.

gametogenesis (gam′ə tō jen′ə sis), *n. Biology.* the formation or development of gametes: *The process by which spermatogonia become spermatozoa and oögonia become ova is known as gametogenesis or maturation, and the resulting mature cells are called gametes* (Storer, *General Zoology*).
—**gametogenic** (gam′ə tə jen′ik), *adj.* of or having to do with gametogenesis.
ASSOCIATED TERMS: *Gametogenesis involves the process of meiosis, which occurs in the gonads. The male gonad is the testes, which produce male gametes or sperm. The female gonad is the ovary, and the gamete it produces is an ovum or egg. Fertilization occurs when a monoploid (or haploid) sperm nucleus unites with a monoploid egg nucleus, resulting in the formation of a diploid zygote.*

gametogenous (gam′ə toj′ə nəs), *adj.* = gametogenic.

gametophore (gə mē′tə fôr), *n. Botany.* a modified branch or filament bearing gametes (gametangia), as in certain liverworts.

gametophyte (gə mē′tō fīt), *n. Botany.* the individual plant or generation of a plant which produces gametes: *In plants, however, there is an alternation of generations. The gametes combine to produce a zygote, but the zygote develops into a structure which produces spores instead of gametes. If the spores are provided with the proper environment, they develop into a structure called a gametophyte. The gametophyte can then produce gametes and in this way the reproduction cycle is completed* (Robert W. Menefee).

gamma (gam′ə), *n. Physics.* **1** a unit of magnetic field strength, equal to 10^{-5} oersted: *What Mariner II found was that, at its closest approach to Venus—some 21,594 miles—there was no significant change in the interplanetary field greater than five gammas, one gamma being a hundred-thousandth of the value of the surface field at the Earth's poles* (New Scientist).
2 = gamma ray: *Prof. Kraushaar suggests, as a possible explanation for the fact that the galactic plane is so active in the production of these high energy gammas, that the rays could be produced by collisions of other kinds of cosmic rays with gas clouds known to be concentrated along the central plane of the galaxy* (Science News Yearbook).
—*adj. Chemistry.* designating the third of several possible position of atoms or groups of atoms which are substituted in a chemical compound. *Symbol:* γ Compare **alpha, beta, delta.**
[from Greek *gamma,* the third letter of the Greek alphabet]

gamma globulin, *Biochemistry.* a type of globulin in blood plasma containing many antibodies which protect against infectious diseases. Compare **alpha globulin** and **beta globulin.**

gamma radiation, *Physics.* radiation consisting of gamma rays. The emission of gamma radiation does not change the atomic or mass numbers of the elements.

gamma ray, *Physics.* an electromagnetic wave of very high frequency given off in the reactions of nuclei or nuclear particles; a high-energy photon. Gamma rays are like X rays, but have a shorter wavelength. Gamma rays are emitted when a nucleus in an excited state changes to a more stable state. See the picture at **electromagnetic spectrum.**

gamma-ray astronomy, the study of sources of celestial gamma radiation.

gamo-, *combining form. Botany.* joined; fused; united, as in *gamopetalous* = having the petals united. [from Greek *gamos* marriage]

gamont (gam′ont), *n. Biology.* a gametocyte stage in certain sporozoans; sporont. Also called **sporont.** Compare **schizont.** [from *gam(ete)* + Greek *ontos* being]

gamopetalous (gam′ə pet′ə ləs), *adj. Botany.* having the petals united: *a gamopetalous corolla.*

gamophyllous (gam′ə fil′əs), *adj. Botany.* having the leaves united: *a gamophyllous perianth.*

gamosepalous (gam′ə sep′ə ləs), *adj. Botany.* having the sepals united: *a gamosepalous calyx.*

-gamy, *combining form.* joined together; conjugation (in fertilization or reproduction), as in *allogamy* = *joined in cross-fertilization, anisogamy* = *conjugation of dissimilar gametes.* [from Greek *-gamia,* from *gamos* marriage]

---

cap, fāce, fäther; best, bē, tèrm; pin, five;
rock, gō, ôrder; oil, out; cup, pùt, rüle,
yü in use, *yu* in uric;
ng in bring; *sh* in rush; *th* in thin; ŦH in then;
zh in seizure.
ə = *a* in about, *e* in taken, *i* in pencil, *o* in lemon, *u* in circus

ganglion (gang'glē ən), *n., pl.* **-glia** (-glē ə) or **-glions.** *Anatomy.* a group of nerve cells forming a nerve center, especially outside of the brain or spinal cord: *A ganglion ... acts somewhat like a little brain for that segment and makes the proper connections for response to stimuli. An earthworm can be cut into several parts and each part will go right on crawling around for a time and reacting to stimuli much as if nothing had happened, for the nerves in the giant fibers connect the different ganglia of each part together and give it coordination* (Winchester, *Zoology*). [from Greek *ganglion* (originally) a type of swelling (ganglia were compared to tumors)]
—**ganglionic** (gang'glē on'ik), *adj.* having to do with or characterized by a ganglion or ganglia: *A group of substances known as ganglionic blocking agents lower blood pressure by interrupting the nervous pathways from the brain at the nerve-switching points called ganglia, which lie just outside the spinal column* (N.Y. Times).

ganglioside (gang'glē ə sīd'), *n. Biochemistry.* a cerebroside containing neuraminic acid, found in the surface membranes of nerve cells: *The substance apparently responsible for the chemical fixation of toxin is a ganglioside, a fatty substance found mainly in nerve tissue. There are a number of gangliosides, each somewhat different in chemical structure. They are alike in that each has two portions, one composed of water-repellent fatty acids and sphingosine and the other composed of water-soluble sugars. As a result they are readily water-soluble even though they are fatty, and this ambivalence suggests that they may have an important function in cell membranes* (W.E. van Heyningen). [from *ganglio(n)* + *(cerebro)side*]

gangue (gang), *n. Geology.* the constituents of a mineral deposit that are not of economic value, in contrast to the ore mineral to be mined: *The value of these Lake Superior deposits is further enhanced by the richness of the ore which contains very little gangue (impurities) and runs between 50% and 70% iron as mined* (Offner, *Fundamentals of Chemistry*). [from French, from German *Gang* vein of ore]

ganoid (gan'oid), *adj.* **1** (of the scales of certain fishes) hard, bony, and overlaid with enamel: *Other [scales of bony fishes] that lack such spines are termed cycloid scales, and still others have ganoid scales, which are bony and capped with ... a hard glassy, enamel-like substance* (Storer, *General Zoology*). **2** (of fishes) having ganoid scales.
—*n.* a ganoid fish. Sturgeons and garfish are ganoids. [from Greek *ganos* brightness + *eidos* form]

garigue or **garrigue** (gə rēg'), *n. Ecology.* an uncultivated land consisting of a calcareous soil overgrown with scrub oak and pine, found especially in the Mediterranean region. [from French]

garnet, *n. Mineralogy.* any of a group of hard silicate minerals occurring in many varieties. A common, deep-red, translucent variety is used as a gem and as an abrasive. *Although garnet is a frequent constituent in many gneisses and schists throughout the world, it seldom occurs in these rocks in sufficient concentration to be workable. The only important commercial deposits of abrasive garnet are those of the Adirondacks in the U.S.A., where 70% of the rock may be garnet and where crystals several feet in diameter are not uncommon* (Jones, *Minerals in Industry*). [from Old French *grenat* of pomegranate color (i.e., red), from *(pomme) grenate* pomegranate]
▶ The color of a garnet varies with its composition. Those containing manganese are dark brown; those with chromium are green, and those with calcium and aluminum are orange. The common red variety contains iron and aluminum. Its chemical formula is $Fe_3Al_2(SiO_4)_3$ and it is called almandite.

gas, *n. Physics.* a substance characterized by very low density and viscosity as compared with liquids and solids. Gases expand and contract greatly with changes in pressure and temperature. Gases are composed of molecules in constant random motion. In gases of low density, the average distance of separation of molecules is large in comparison with their diameters and the total actual volume of the gas molecules is negligible in comparison with the volume occupied by the gas. Oxygen and nitrogen are gases at ordinary temperatures. *We know that the atoms of a gas are much more widely separated than are those of a liquid or solid; hence the forces between atoms are of less consequence and the behaviour of a gas is governed by simpler laws than those applying to liquids and solids* (Sears and Zemansky, *University Physics*).
—**gaseous** (gas'ē əs), *adj.* in the form of gas; of or like a gas: *Steam is water in a gaseous condition.*
▶ The **gas laws** in physics are three laws that explain the relationship among pressure, volume, and temperature for a fixed mass of gas. These laws are *Boyle's law, Charles's law,* and *Avogadro's law.* A gas which obeys the gas laws perfectly is called an *ideal gas* or *perfect gas.* The equation PV = RT of the *gas constant* combines the gas laws into a single statement. See also **Graham's law.**

gas chromatography, *Chemistry.* a method of separating the substances in a mixture by combining it with a gas such as nitrogen and passing it through a long column of packing, such as charcoal: *Gas chromatography is a procedure whereby a volatile mixture is separated into its components by a moving inert gas passing over a sorbent ... As a method of separating the individual components of a complex mixture gas chromatography has no equal* (Scientific American).

gas constant, *Physics.* the constant of proportionality (R) in the equation PV = RT, which connects the pressure and volume of a quantity of gas with the absolute temperature. In this equation, P = the pressure of gas, V = its volume per mole, and T = its absolute temperature. The gas constant has a value of 8.314 joules per degree Kelvin or 1.985 calories per degree Celsius. Also called **universal gas constant.**

gas exchange, *Biology.* the exchange of respiratory and other gases between an organism and its environment. In monerans, protists, and fungi, gas exchange occurs by diffusion through thin, moist membranes. The stems of woody plants have lenticels which permit gas exchange. In insects such as grasshoppers gas exchange is accomplished by tracheal tubes.

gas gland, *Zoology.* a gland in the air bladder of certain fishes that secretes a gas consisting chiefly of nitrogen and oxygen: *When the blood reaches the gas gland, lactic acid is secreted into it. This raises the partial* *pressures of the dissolved gases, especially of the oxygen* (R. McNeill Alexander).

gas-liquid chromatography, *Chemistry.* a form of gas chromatography in which the mixture is passed through a liquid solvent instead of a porous solid.

gasometer (gas om′ə tər), *n. Chemistry.* a container for holding and measuring a gas, as the quantity of gas produced in a reaction.

gastric, *adj. Biology.* of, having to do with or near the stomach. The **gastric glands** are glands in the lining of the stomach that secrete gastric juice. [from Greek *gastros* stomach]

gastric juice, *Physiology.* the thin, nearly clear digestive fluid secreted by glands in the lining of the stomach. It contains pepsin and other enzymes and hydrochloric acid. *Very little absorption takes place in the stomach, but its glandular lining produces gastric juice containing the enzyme pepsin* (Mackean, *Introduction to Biology*).

gastrin, *n. Biochemistry.* a polypeptide hormone that promotes secretion of gastric juice: *Gastrin, which is normally produced in small amounts by cells of the stomach lining, stimulates the secretion of hydrochloric acid in the stomach for the digestion of food* (Neil R. Thomford).

gastrocnemius (gas′trok nē′mē əs), *n., pl.* **-mii** (-mē ī). *Anatomy.* the chief muscle of the calf of the leg, that gives it its bulging form: *The gastrocnemius flexes the femur upon the tibia and extends the foot* (Hegner and Stiles, *College Zoology*). [from Greek *gastroknemia* calf, from *gastros* belly + *knēmē* leg]

gastrocoel (gas′trə sēl), *n. Embryology.* = archenteron.

gastrodermis, *n. Zoology.* the inner cellular layer of the digestive tract of coelenterates: *The wall of the body and tentacles consists of but two cell layers, a thin external epidermis, of cuboidal cells, chiefly protective and sensory in function, and inside a thicker gastrodermis, of tall cells, serving mainly in digestion* (Storer, *General Zoology*).

gastroduodenal (gas′trō dü ə dē′nl *or* gas′trō düod′ə-nl), *adj. Anatomy.* having to do with the stomach and the duodenum: *the gastroduodenal artery, the gastroduodenal anastomosis.*

gastrointestinal (gas′trō in tes′tə nəl), *adj. Anatomy.* of or having to do with the stomach and intestines. The **gastrointestinal tract** (often *GI tract*) is the alimentary canal: *Food is moved through the GI tract by slow, rhythmic muscular contractions called peristalsis* (Biology Regents Syllabus).

gastropod (gas′trə pod), *Zoology.—n.* any of a class (Gastropoda) of mollusks having eyes and feelers on a distinct head, usually a shell that is spiral or cone-shaped, and a muscular dislike foot organ on the under surface of its body used for locomotion. Snails, slugs, and limpets are gastropods.
—adj. of or belonging to such mollusks.
[from Greek *gastros* stomach + *podos* foot]

gastrovascular (gas′trō vas′kyə lər), *adj. Anatomy.* serving for both digestive and circulatory functions: *This specimen, although incomplete, was a new species with a curious gastrovascular network of canals* (Frederick Stratten Russell).

gastrula (gas′trù lə), *n., pl.* **-lae** (-lē′). *Embryology.* a stage in the development of all many-celled animals, when the embryo is usually saclike and composed of

two layers of cells, the ectoderm and endoderm. [from New Latin, diminutive of Greek *gastros* stomach]
—gastrular, *adj.* of or having to do with a gastrula or gastrulation.

gastrulate (gas′trə lāt), *v. Embryology.* to be in or have a gastrula stage.
—gastrulation, *n.* the process of forming a gastrula: *At the end of this process, known as gastrulation, the salamander embryo looks like a rubber ball with half of it pushed in* (Scientific American).

gate, *n.* **1** *Electronics.* a circuit having one output terminal and two or more input terminals, so that the output is determined by the combination of the inputs.
2 *Physiology.* a location in the nervous system at which a sensory signal may be blocked under certain conditions.

gauge (gāj), *n. Physics.* **1** a standard measure or a scale of standard measurements of some physical property, such as size, pressure, force, temperature, radiation, etc.
2 an instrument or device for measuring some physical property: *an ionization gauge, a strain gauge, a vacuum gauge.*

gauge theory, *Physics.* one class of theory that attempts to establish relationships between fundamental forces such as gravity, electromagnetism, and the strong and weak nuclear forces: *The existence of charm has been predicted by a powerful set of theories of the fundamental forces, the gauge theories, and this discovery established those theories* (New Scientist).

gauss (gous), *n. Physics.* the unit of magnetic induction in the CGS system. One gauss is equal to 1 maxwell per square centimeter or 10^{-4} weber per square meter. The corresponding SI or MKS unit is the tesla. [named after Karl Friedrich *Gauss,* 1777–1855, German mathematician]

Gaussian curve or **Gaussian distribution,** = normal curve or distribution.

gaussmeter (gous′mē′tər), *n.* a magnetometer with the scale calibrated in gauss, used for measuring the intensity of a magnetic field: *Gaussmeter utilizes the Hall-effect principle to measure magnetic flux up to 50,000 gauss in static or a-c fields* (Science).

G.C.D., g.c.d., or **gcd,** *abbrev.* greatest common divisor.

Gd, *symbol.* gadolinium.

Ge, *symbol.* germanium.

geanticlinal (jē′an ti klī′nl), *Geology. —adj.* of or having to do with geanticline: *Many mountains owe half or more of their elevation above the sea level to geanticlinal movements* (Dana, *Manual of Geology*). *—n.* = geanticline.

geanticline (jē an′ti klīn), *n. Geology.* a general upward flexure of the earth's crust; a major upfold of subcontinental dimensions: *The crustal furrow that comprises*

---

cap, fāce, fäther; best, bē, tėrm; pin, five;
rock, gō, ôrder; oil, out; cup, put, rüle;
yü in use, *yu* in uric;
ng in bring; *sh* in rush; *th* in thin, ŦH in then;
zh in seizure.
ə = *a* in about, *e* in taken, *i* in pencil, *o* in lemon, *u* in circus

a geosyncline, commonly is bordered on one side or other by a compensating upward bend of the earth's crust—a geanticline—and it is erosion of rock materials from this uplifted belt that tends chiefly to furnish sediment deposited in the geosyncline (Moore, Introduction to Historical Geology). [from Greek ge earth + English anticline]

gear (def. 1)

gear, Physics. —*n.* **1** a wheel having teeth that fit into the teeth of another wheel. If the wheels are of different sizes, they will turn at different speeds.
2 a set of such wheels working together to transmit power or change the direction of motion in a machine.
3 any arrangement of gears or moving parts that acts as a unit in a larger mechanism.
—*v.* to connect by gears.

Gegenschein or **gegenschein** (gā'gən shīn'), *n.* Astronomy. a faint light seen near the apparent path of the sun at a point 180° from the sun: [He] offered this phenomenon in explanation of the Gegenschein—the faint diffuse glow sometimes observed at night at the antisolar point in the sky (New Scientist). [from German Gegenschein (literally) counterglow]

Geiger counter (gī'gər) or **Geiger-Müller counter** (gī'gər mYl'ər). Physics. a device which detects and counts ionizing particles, consisting of a tube with two electrodes between which a high potential difference is maintained. It is used especially to measure radioactivity, as in prospecting for radioactive mineral deposits. The radioisotopes give off radiation which can be measured by Geiger counters (Junior Scholastic). Compare **scintillation counter.** [named after Hans Geiger and W. Müller, 20th-century German physicists]

geitonogamy (gī'tə nog'ə mē), *n.* Botany. a type of fertilization in which a flower is pollinated by another flower on the same plant. [from Greek geitonos neighbor + English -gamy]

gel (jel), Chemistry. —*n.* a jellylike or solid material formed from a colloidal solution. When glue sets, it forms a gel. Compare **sol.** —*v.* to form a gel: Egg white gels when it is cooked. [from gel(atin)]

gelatin or **gelatine** (jel'ə tən), *n.* Chemistry. **1** an odorless, tasteless protein substance obtained by boiling animal tendons, bones, hoofs, etc. It forms a stiff jelly after it is dissolved in hot water and cooled. Gelatin is used for making medical capsules, as a medium for bacterial culture, in cements, jellies, etc.
2 any of various vegetable substances having similar properties.
[from French gélatine, from Italian gelatina, from gelata jelly, from gelare to jell]
—**gelatinous** (jə lat'n əs), *adj.* of, like, or containing gelatin.

gelation (jə lā'shən), *n.* Chemistry. **1** the formation of a gel from a sol: The mechanism of gelation is not well understood, but sodium silicofluoride seems to combine with the water to form hydrofluoric acid, thus

sending the latex into the highly unstable acid range (Scientific American). **2** solidification by cooling.

geminate (jem'ə nit or jem'ə nāt), *adj.* Botany. combined in a pair of pairs; coupled: geminate leaves. [from Latin geminatum doubled, from geminus twin]
—**gemination,** *n.*

gemma (jem'ə), *n.,* pl. **gemmae** (jem'ē). Botany. **1** a bud.
2 a budlike reproductive body in some plants, such as liverworts and mosses, that becomes detached from the plant and can develop into a new individual: Some of the gametophytes will show cuplike structures on their upper surface. These are called cupules. They contain special buds called gemmae which, upon being detached, may give rise directly to a new gametophyte. This is the vegetative method of reproduction (Youngken, Pharmaceutical Botany). [from Latin gemma bud]

gemmate (jem'āt), Biology. —*v.* to put forth buds; reproduce by budding.
—*adj.* having buds; reproducing by buds.
—**gemmation,** *n.* reproduction by buds; budding: In gemmation the portion abscised contains the growing tissue necessary for the bud or gemmule to develop rapidly into an adult (Harbaugh and Goodrich, Fundamentals of Biology).

gemmule (jem'yül), *n.* Biology. **1** a small bud, especially the rudimentary terminal bud of the embryo of a seed: We distinguish three parts in the embryo, corresponding to the root, stem, and leaves of the perfect plant; namely a radicle, plumule or gemmule, and one or more cotyledons (Bentley, Manual of Botany).
2 a budlike reproductive body in sponges: In freshwater sponges, masses of cells collect in the jellylike middle layer of the body wall. Hundreds of cells are in each mass, and around them is a horny layer which often contains many spicules. Such a reproductive body is called a gemmule. ... The gemmules are not shed, but when the parent's body disintegrates at the end of the season, they are left exposed on the log or stone to which the sponge was attached (Shull, Principles of Animal Biology).
[from Latin gemmula, diminutive of gemma bud]

-gen, combining form. **1** something produced or growing, as acrogen. **2** something that produces, as allergen, nitrogen. [from Greek -genēs born]

gen., abbrev. genus.

gene (jēn), *n.* Biology. the unit of transmission of hereditary characteristics in an organism. A gene is usually a segment of DNA occupying a specific place on a particular chromosome. Each gene influences the inheritance and development of some characteristic. The genes inherited from its parents determine what kind of an organism will develop from a fertilized egg cell. There are genes that determine hair color; eye color; height; number of legs, arms and feet; heart structure and so forth (Arnold L. Bachman). Also called **factor.** Compare **allele.** [from German Gen, ultimately from Greek genea breed, kind]

gene flow, Genetics. the transfer of genes from one population to another: The tolerance was hereditary, but there was little sign of "gene flow"—that is, movement of hereditary copper tolerance—into the neighbouring copper intolerant populations (New Scientist).

gene frequency, *Genetics.* the relative frequency with which a particular gene is present in a particular population of a species or other group: *The Hardy-Weinberg law, one of the most fundamental laws of genetics ... states that under certain conditions gene frequency and genotype ratios remain constant from one generation to the next in biparental (sexual) populations* (Simpson, *Life: An Introduction to Biology*).

gene mutation, *Genetics.* mutation of a single gene as a result of a change in the gene's DNA rather than a change in chromosome structure: *Usually gene mutations are recessive and the characters controlled by the gene may not be expressed* (Mackean, *Introduction to Biology*).

gene pool, *Genetics.* the total stock of genes in a breeding population, with each gene representing a number of alleles: *Natural selection is the most important process causing the population's gene pool to evolve* (Simpson, *Life: An Introduction to Biology*).

genera (jen′ər ə), *n.* a plural of **genus.**

general theory of relativity, See **relativity.**

generation, *n.* **1** *Biology.* **a** a group of organisms having the same parent or parents and comprising a single step or level in line of descent: *The growths of bacteria ... appear to have gone through as many as three bacterial generations* (Harold M. Schmeck, Jr.). **b** a form or stage of a plant or animal, with reference to its method of reproduction: *the asexual generation of a fern.* **c** the production of offspring: *Under constant conditions of temperature, nutrients, pH, etc., the time required for one cell to make two, known as the generation time, is very constant* (New Scientist). SYN: procreation.
2 *Mathematics.* the formation of a line, surface, figure, or solid especially by moving a point or line.

generative, *adj.* *Biology.* having to do with the production of offspring: *the generative organs.* SYN: reproductive.

generator, *n.* **1** *Physics.* a machine that changes mechanical energy into electrical energy and produces either direct or alternating current. SYN: dynamo. **2** *Mathematics.* = generatrix.

generatrix (jen′ə rā′triks), *n., pl.* **generatrices** (jen′ər ə trī′sēz). *Geometry.* a point, line, etc., whose motion produces a line, surface, figure, or solid: *A conical surface is generated by a moving straight line which always intersects a plane curve and always passes through a fixed point not in the plane curve. The moving line is called the generatrix* (Smith and Clark, *Modern-School Solid Geometry*). [from Latin, feminine of *generator,* from *generare* generate]

gene redundancy, *Genetics.* the presence in a cell of many copies of a single gene.

generic (jə ner′ik), *adj.* *Biology.* characteristic of a genus, kind, or class: *Cats and lions show generic differences.* —**gener′ically,** *adv.*

gene-splicing, *n.* *Molecular Biology.* the recombination of genetic material; production of recombinant DNA: *The "gene-splicing" technique consists of cutting into small pieces the long, threadlike molecules of DNA (deoxyribonucleic acid), the transmitter of genetic information, then recombining these segments with the DNA of a suitable carrier molecule and inserting the combination into an appropriate host cell, such as a bacterium. The recombinant DNA changes and con-* trols *the hereditary characteristics of the host bacteria* (Arthur J. Snider). Compare **genetic engineering.**

genetic, *adj.* *Biology.* **1** of, having to do with, or produced by genes: *The most reliable form of classification that has been devised in the field of biology is based on genetic kinship. Genetic kinship means that those organisms whose gene complexes are the most nearly alike are the most closely related to each other; and, on the contrary, those that have received very different sets of genes from their ancestors are correspondingly more remotely related to one another* (Emerson, *Basic Botany*).
ASSOCIATED TERMS: see **congenital.**
2 of or having to do with genetics: *genetic research.* —**genet′ically,** *adv.*

genetic alphabet, *Molecular Biology.* the set of symbols for the four nucleic-acid bases that combine in various ways to form the genetic code: *Most DNA consists of sequences of only four nitrogenous bases: adenine (A), thymine (T), guanine (G) and cytosine (C). Together these bases form the genetic alphabet, and long ordered sequences of them contain in coded form much, if not all, of the information present in the genes* (Scientific American).

genetic code, *Molecular Biology.* the various combinations of nucleotides which may occur in the DNA or RNA molecule of a chromosome. The genetic code determines the make-up of genes and gene products. It is the biochemical code by which the four nucleic-acid bases in the DNA molecule combine, usually in units of three (triplet codons) to specify the synthesis of particular amino acids and proteins in the cell. *The genetic code appears to be essentially the same in all organisms. This is not surprising: Variations in it from organism to organism would mean that the code had evolved by mutation, and it is almost impossible to imagine a mutation which would change the letters in a codon without being lethal* (Watson, *Molecular Biology of the Gene*).

genetic drift, *Genetics.* the fluctuation in gene frequency occurring in an isolated population, presumably due to random variations from generation to generation: *Avoiding "genetic drift" is another possible advantage of embryo freezing. Donald W. Bailey of the Jackson Laboratory in Bar Harbor, Maine, explains that a highly inbred strain of mice, or of anything else, will not remain genetically identical from one generation to the next. Random mutations gradually change the strain's genes* (Science News).

genetic engineering, *Molecular Biology.* the scientific alteration of genes or genetic material, especially through gene-splicing, to produce desirable new traits in organisms or to eliminate undesirable ones: *Research on recombinant DNA ... is commonly referred to as "genetic engineering" because it is now*

cap, fāce, fäther; best, bē, tèrm; pin, five;
rock, gō, ôrder; oil, out; cup, pùt, rüle,
yü in use, *yù* in uric;
ng in bring; *sh* in rush; *th* in thin, ᴛʜ in then;
zh in seizure.
ə = *a* in about, *e* in taken, *i* in pencil, *o* in lemon, *u* in circus

possible to cut out a segment of DNA and splice in a new segment from another organism (Harper's).

genetic load, *Genetics.* the accumulated mutations in the gene pool of a species: *We are learning more about the gene pool and genetic load, but many questions remain unanswered. We know very little about ... the relative proportions of balanced and mutational load produced by the common mutagenic agents* (Scientific American).

genetic map, *Molecular Biology.* the arrangement of genes or mutable sites on a chromosome as determined by genetic recombination experiments: *The genetic map obtained through analysis of uninterrupted matings is the same as that arrived at by analysis of frequencies of various recombinant classes. As in the chromosomal maps of higher organisms, the bacterial genes are arranged on an unbranched line. However, one important distinction exists: The genetic map of E. coli is a circle. ... The biological significance of circular genetic maps is surrounded with mystery* (Watson, *Molecular Biology of the Gene*).

genetic marker, *Genetics.* a variation in DNA that is associated with a particular gene or trait, so that it can be identified and followed from generation to generation: *In 1973 it was discovered that the great majority of patients suffering from ankylosing spondylitis have a characteristic "genetic marker," HA W-27. This marker could be used to identify persons at risk of developing ankylosing spondylitis* (Suzanne Loebl).

genetics, *n.* **1** the branch of biology dealing with the principles of heredity and variation in organisms of the same or related kinds: *Mendel's studies with peas grown in a monastery garden established the first principles of genetics, but peas were not the best organisms for genetic research* (Sheldon C. Reed).
2 genetic composition or characteristics: *The genetics of the liquid part of blood, plasma, have only been intensively studied in the last few years* (Carter, *Human Heredity*).
ASSOCIATED TERMS: Some major concepts in genetics are: dominance, segregation, recombination, intermediate inheritance, codominance, independent assortment, linkage, crossing-over, and multiple alleles.

genial (jə nī′əl), *adj. Anatomy.* of or having to do with the chin: *the lower genial process.* [from Greek *geneion* chin + English *-al*]

genic (jen′ik), *adj. Biology.* of, produced by, or like a gene; genetic: *... the total genic resources or materials of a species throughout its geographical range* (Herbert C. Hanson). *These and the host of other inherited variations, from fingerprint patterns to blood types, are manifestations of the differences that exist in the structure and arrangement of the genic material* (Evelyn M. Witkin). —**genically,** *adv.*

-genic, *combining form.* **1** producing; having to do with production, as in *carcinogenic = producing cancer.* **2** produced or formed by, as in *allogenic = produced by another.*

genicular (jə nik′yə lər), *adj. Botany.* growing on or at a node; occurring in the tissue of the node: *genicular cells.* [from Latin *geniculum* knot]

geniculate (jə nik′yə lit *or* jə nik′yə lāt), *adj. Botany, Zoology.* **1** having kneelike joints or bends: *geniculate antennae. The threads become geniculate, and unite at*

the two bends (Berkeley, *Introduction to Cryptogamic Botany*). **2** bent at a joint like the knee: *a geniculate stalk.*

geniculum (jə nik′yə lum), *n., pl.* **-la** (-lə). **1** *Botany.* a node or joint of a stem.
2 *Anatomy.* a sharp bend in any small organ, as in the facial nerve: *The point where [the facial nerve] changes its direction is named the geniculum* (Gray's Anatomy).
[from Latin *geniculum* knot, from *genu* knee]

genital, *Biology.* —*adj.* having to do with reproduction or the sex organs: *the genital region. In the female, the vulva or genital pore is located ventrally at about one third the length of the body from the anterior end* (Hegner and Stiles, *College Zoology*). —*n.* **genitals,** *pl.* = genitalia. [from Latin *genitalis,* ultimately from *gignere* beget]

genitalia (gen′ə tā′lyə), *n.pl. Biology.* the external sex organs. [from Latin]

genitourinary (jen′ə tō yùr′ə ner′ē), *adj. Anatomy, Physiology.* having to do with the genital and urinary organs or functions: *the genitourinary tract.*

genome (jē′nōm), *n. Genetics.* **1** a haploid set of chromosomes with their genes: *The most important function of the nucleus is to replicate the genome* (Daniel Mazia).
2 the sum of all the genes in such a set of chromosomes: *The human genome—the complete set of genes in an individual—consists of perhaps as many as 10 million genes* (Scientific American).
[from *gen(e)* + Greek *-oma* group, mass]

genotype (jen′ə tīp), *n. Genetics.* the genetic makeup of an organism or group as distinguished from its physical appearance or phenotype: *Moderate inbreeding is necessary to fix the successful genotype (the genetic constitution of an organism)* (Sheldon C. Reed).
—**genotypic** (jen′ə tip′ik) *or* **genotypical,** *adj.* of or having to do with a genotype: *Genotypic ratios ... refer to the transmission of the genes from one generation to the next and are somewhat different from those of the phenotypes* (Emerson, *Basic Botany*). —**genotypically,** *adv.*

gentle breeze, *Meterology.* a wind having a velocity of 12 to 19 kilometers per hour (see **Beaufort scale**).

genus (jē′nəs), *n., pl.* **genera** (jen′ər ə) *or* **genuses.** *Biology.* a group of related organisms ranking below a family and above a species. The scientific name of an organism consists of the generic name written with a capital letter and the specific name written with a small letter. EXAMPLE: *Homo sapiens. The seas around Maudheim are inhabited by three genera of seal: the crabeater, the Weddell seal, and the relatively rare Ross seal* (E.F. Roots). [from Latin *genus* race, stock, kind]
ASSOCIATED TERMS: see **kingdom.**

geo-, *combining form.* earth; of the earth, as in *geology = science of the earth, geomagnetism = magnetism of the earth.* [from Greek *geo-,* from *gē* earth]

geobotany, *n.* the study of plants in relation to their geological or geographic environment.
—**geobotanical,** *adj.* of or having to do with geobotany.

geocentric, *adj. Astronomy.* **1** viewed or measured from the earth's center. **2** having or representing the earth as a center: *In the geocentric view, upheld by Hippar-*

chus and Ptolemy, the central earth is the dominating feature (Baker, *Astronomy*). Contrasted with **heliocentric.**

geocentric parallax, = diurnal parallax.

geochemical, *adj.* of or having to do with geochemistry: *Geochemical measurements of the lunar surface ... would help to develop an understanding of the nature and origin of the moon* (Richard S. Johnston).

geochemistry, *n.* **1** the science dealing with the chemical composition and character of the earth, or of other planets or planet-like celestial bodies. **2** geochemical composition or characteristics: *the geochemistry of the moon. Recent studies have emphasized the geochemistry of clay-mineral assemblages developed in the great soil groups, in weathered tills, and in other glacial sediments* (A.V. Jopling).

geochronological (jē'ō kron'ə loj'ə kəl), *adj.* of or having to do with geochronology: *Geochronological measurements have been made on rocks of the three principal kinds which make up the Earth's crust* (New Scientist).

geochronology (jē'ō krə nol'ə jē), *n.* the science of determining the age of geological formations.

geochronometry, *n.* the measurement of geologic time by means of geochronology.

geode (jē'ōd), *n. Geology.* **1** a cavity, usually in a sedimentary rock, lined with crystals or other mineral matter. **2** a rock containing such a cavity: *Many geodes formed when seeping water deposited material from solution on the walls of cavities in bedrock; geodes of this sort often are larger than others. Some began with deposition inside shells or other fossils; when enough material was deposited it broke the shell, spread it and filled the cracks as well as the cavity* (Fenton, *The Rock Book*). [from Greek *geōdēs* earth, from *gē* earth + *eidos* form]

geodesic (jē'ə des'ik) or **geodetic** (jē'ə det'ik), *Mathematics.* —*n.* the shortest line between two points on a surface: *In the geometry of the theory of relativity the paths of light rays in space-time are the geodesics, and these play the role of the straight line. The geodesics are generally not straight* (Morris Kline). —*adj.* of or having to do with a geodesic or with geodesy: *Because the edges of the tetrahedrons join to form great circle arcs, and because "geodesic" is the term used by mathematicians to describe the shortest distances (which are great circle arcs on spheres) on curved surfaces, Fuller called his structures geodesic domes* (Robert W. Marks).

geodesy (jē od'ə sē), *n.* a branch of applied mathematics dealing with the measurement of the earth and of large areas on the surface of the earth: *More than 2,000 years ago, Eratosthenes, a Greek geometer and astronomer, first measured the curvature of the earth's surface. Then, assuming the earth to be spherical, he computed its dimensions. Though measuring techniques have been greatly refined, his reasoning is still used in modern geodesy* (Gilluly, *Principles of Geology*). [from Greek *geōdaisia*, from *gē* earth + *daiein* divide]

geographic or **geographical,** *adj.* **1** of or having to do with geography: *Thus regional geography might deal with an entire geographic realm (such as Middle America or Southeast Asia) or with a region defined by other criteria (for example the Nile Valley or the Corn Belt)* (deBlij, *Geography of Viticulture*). **2** of or having to do with a location on the earth's surface: *Chicago's geographic location.*

geographic pole, *Geophysics.* one of the hypothetical points in the Arctic or Antarctic where the earth's rotational axis meets the surface. The North Pole and South Pole are the geographic poles. Compare **geomagnetic pole.**

geography, *n.* **1** the study of the earth's surface, and the locational characteristics of climate, continents, countries, peoples, industries, and products. ▶ **Geography** studies the surface distributions of the land and water; **oceanography** continues study of the seas to their depths; **geology** studies land in depth; and **meteorology** studies the atmosphere and the "layers" above it.
2 the surface features of a place or region: *the geography of New England.*
[from Greek *geōgraphia*, from *gē* earth + *graphein* describe]

geohydrology, *n.* the branch of geology dealing with water formed or running underground.

geoid (jē'oid), *n. Geology.* a geometric figure of the earth which coincides with the mean sea level over the oceans and extends continuously through the continents. A geoid has the surface at every point perpendicular to the direction of gravity. *The geoid (or the true figure of the earth's surface, as determined by the directions of the pendulum) nearly corresponds with the spheroid on the shores of the Black Sea* (Nature). [from Greek *geoeidēs* earthlike, from *gē* earth + *eidos* form]
—**geoidal** (jē oi'dəl), *adj.* of a geoid.

geoisotherm, *n. Geology.* an imaginary line or surface passing through points in the interior of the earth that have the same mean temperature. Also called **isogeotherm.** [from *geo-* + *iso-* + Greek *thermē* heat]

geologic or **geological,** *adj.* of or having to do with geology: *A geologic formation is defined as an assemblage of rocks—igneous, sedimentary, or metamorphic—having generally like characters of lithology and forming a logically differentiated part of the rock succession for geologic mapping* (Moore, *Introduction to Historical Geology*). —**geologically,** *adv.*

geologic time, the time of existence of the earth before the ages of human history, as studied by historical geology: *The earth is very old—4,500,000,000 years or more, according to recent estimates. This vast span of time, called geologic time by earth scientists, is believed by some to reach back to the birth of the solar system* (William L. Newman).

geology, *n.* **1** the science that deals with the earth, the layers of which it is composed, and their history: *The demarcation between the study of the earth—geology—and the study of all the heavenly*

cap, fāce, fäther; best, bē, tėrm; pin, fīve;
rock, gō, ôrder; oil, out; cup, pùt, rüle,
yü in use, *yu̇* in uric;
ng in bring; *sh* in rush; *th* in thin, ŦH in then;
zh in seizure.
ə = *a* in about, *e* in taken, *i* in pencil, *o* in lemon, *u* in circus

bodies—astronomy—is obviously man-made. The answer to many questions of geology necessitates a thorough knowledge of astronomy, and vice versa (Garrels, A Textbook of Geology).
2 the study of the rocks, soils, and often physical features of the moon, planets, etc.: extraterrestrial geology, lunar geology, planetary geology.
3 the rocks, soils, mountains, oceans, or other geological features of a particular area or region: the geology of North America, the geology of Mars.
[from Medieval Latin geologia, from Greek geō- of the earth + -logia study, -logy]

geomagnetic, adj. of or having to do with the magnetism of the earth: There is some geomagnetic evidence that rock at a depth of 600 to 900 kilometers has a temperature of 1,000 to 1,500 degrees C (Scientific American).

geomagnetic pole, Geophysics. the hypothetical point in the Arctic or Antarctic where the earth's magnetic axis touches the earth's surface: The north geomagnetic pole is near Thule, Greenland, and the southern pole is in Antarctica (T. Neil Davis). Compare **geographic pole.** ► The positions of the geomagnetic poles have changed with geologic time.

geomagnetism, n. Geophysics. 1 the magnetism of the earth: The rockets, which reach a height of about 60 mi., will be instrumented to record observations in geomagnetism, cosmic rays, and aurora (Hugh Odishaw).
2 the science concerned with the magnetism of the earth: The geophysical science of geomagnetism—part of the universal science of magnetism—began in 1600 (New Scientist).

geometric or **geometrical,** adj. 1 of geometry; according to the principles of geometry: geometric proof. 2 consisting of straight lines, circles, triangles, etc.; regular and symmetrical: geometric figures. **—geometrically,** adv.

geometric mean, Mathematics. the mean of a number (n) of positive quantities produced by taking the nth root of their product: The geometric mean of the two quantities of 16 and 4 is the square root of their product, or 8.

geometric progression or **geometrical progression,** Mathematics. a sequence of numbers in which each number is multiplied by the same factor in order to obtain the following number. 2, 4, 8, 16, and 32 form a geometric progression. 3125, 625, 125, 25, 5, 1, 1/5, and 1/25 also form a geometric progression. Compare **arithmetic progression.**

geometric series, = geometric progression.

geometry, n. the branch of mathematics which studies the relationship of points, lines, angles, and surfaces of figures in space; the mathematics of space. Geometry includes the definition, comparison, and measurement of squares, triangles, circles, cubes, cones, spheres, and other plane and solid figures. [from Greek geōmetria, from gē earth + -metria measuring]

geomorphological (jē′ō môr′fə loj′ə kəl), adj. of or having to do with geomorphology.

geomorphology (jē′ō môr fol′ə jē), n. the study of the surface features of the earth or of a celestial body, with reference to their arrangement, origin, development, etc.

geophilous (jē of′ə ləs), adj. 1 Zoology. living on or in the ground, as various snails or worms. 2 Botany. growing in the ground, as various plants with deep underground buds. [from geo- + Greek phyllon leaf]

geophysical, adj. of or having to do with geophysics: The geophysical Earth is a huge, spinning, electromechanical rotor, intimately engaged in its immediate cosmic environment (Bulletin of Atomic Scientists).

geophysics, n. 1 the study of the relations between the physical processes and phenomena of the earth and the forces that change or produce them; the physics of the earth.
2 the physical processes and phenomena of the earth or of a heavenly body: planetary geophysics.

geophyte (jē′ə fīt), n. Botany. a plant which grows in earth, especially one having deep underground buds. [from geo- + -phyte]

geopressured, adj. Geology. under great pressure from geologic forces: The most speculative, but perhaps also the largest, potential source of natural gas is the geopressured zone of the Gulf Coast. This zone ... consists of large aquifers at depths of 2500 to 8000 meters. These aquifers are characterized by high temperatures (above 150°C) and pressures that are as much as twice those of conventional water at comparable depths. At such temperatures and pressures ... nearly all organic matter is eventually converted to methane (Science).

geoscience, n. any science dealing with the earth, as geology, geophysics, and geochemistry.

geosphere, n. = lithosphere.

geostationary or **geosynchronous,** Astronomy. of or having to do with a body that maintains a stationary position above a fixed point on the earth's surface, moving at the same rate as the earth rotates: A satellite in geostationary orbit by definition lies at 35,800 km above the equator, with an angular velocity which matches that of the Earth (Science Journal).

geostrophic (jē′ə strof′ik), adj. Meteorology. (of a wind or current) strongly deflected by the rotation of the earth: Most ocean currents have a motion that is geostrophic, or affected by the turning of the earth (D. James Baker, Jr.). [from geo- + Greek strophē a turning]

geosynclinal (jē′ō sin klī′nəl), Geology. —adj. of or having to do with a geosyncline: Eastern Asia was one great geosynclinal area into the Tertiary, and is characterized by simple, mild structures, high mobility, and intense magmatic activity (Lawrence Ogden). —n. = geosyncline.

geosyncline (jē′ō sin′klīn), n. Geology. a broad, elongated, downward curve or flexure of the earth's crust: It has long been recognized that geosynclines are fundamental geologic units. Furthermore, it has been a dictum of geology that they eventually evolve into mountains consisting of folded sedimentary strata. The laying down of such sediments and their subsequent folding constitute a basic geologic cycle that requires a few hundred million years (Robert S. Dietz).

geotactic, adj. of or having to do with geotaxis.
—geotactically, adv. in a geotactic manner or direction.

geotaxis (jē′ə tak′sis), n. Biology. a movement of an organism toward or away from the center of the earth; tendency to move in response to the force of gravity. [from geo- + Greek taxis arrangement]

geotectonic, *adj.* of or having to do with the structure or the arrangement of the rock strata composing the crust of the earth. —**geotectonically,** *adv.*

geothermal, *adj.* of or having to do with the internal heat of the earth: *geothermal electricity.*

geothermal gradient, *Geology.* the increase in the temperature of the earth from its surface downward to the core, estimated to be 1 degree Fahrenheit per 60 feet: *To measure the geothermal gradient on land, any facility through which the Earth's internal temperature is accessible (mine, coal field, tunnel, oil well, and so on) can be utilized* (Ki-Iti Horai).

geotropic (jē'ō trop'ik), *adj.* affected by geotropism; responding to gravity: ... *the geotropic curvature of the roots downward* (Weier, Botany). ... *shoots are negatively geotropic* (Mackean, Introduction to Biology).

geotropism (jē ot'rə pis'əm), *n. Biology.* a response by various organisms to the action of gravity. **Positive geotropism** is a tendency to move down into the earth, as plant roots do. **Negative geotropism** is a tendency to move upward: *Growing shoots tend to grow away from gravity. A response of this kind to gravity is called a geotropism and since the shoots grow away from the direction of the stimulus, the response is said to be negative* (Mackean, Introduction to Biology). [from *geo-* + Greek *tropē* a turning]

germ, *n. Biology.* **1** a microscopic organism, especially one which causes disease: *the scarlet fever germ.* SYN: microbe.

ASSOCIATED TERMS: see **bacteria.**

2 the earliest form of a living thing. SYN: seed, bud, spore. [from French *germe,* from Latin *germen* sprout]

germanium (jər mā'nē əm), *n. Chemistry.* a rare, brittle, silver-white metalloid element which occurs in zinc ores, used as a semiconductor in transistors and other electronic devices. *Symbol:* Ge; *atomic number* 32; *atomic weight* 72.59; *melting point* 937.4°C; *oxidation state* 2,4. [from New Latin, from Latin *Germania* Germany; named by its discoverer, the German chemist Clemens Winkler, 1838–1904. Compare *francium.*]

germ cell, *Biology.* **1** an egg or sperm cell; gamete: *Germ cells (the eggs produced in a female body and the sperms in a male body) contain only half the normal number of chromosomes. That is, as each germ cell is formed, only one chromosome from each pair carries over* (Sidonie M. Gruenberg). **2** a primitive cell from which an egg or sperm cell develops.

germicide (jėr'mə sīd), *n. Medicine.* any substance that kills germs, especially disease germs. Disinfectants and fungicides are germicides.

ASSOCIATED TERMS: see **disinfectant.**

—**germicidal** (jėr'mə sī'dl), *adj.* capable of killing germs: *Germicidal lamps on the walls emit ultraviolet light rays* (N.Y. Times).

germinable, *adj.* that can be germinated; ready to germinate: *a germinable plant, germinable crops.*

—**germinability,** *n.* the condition of being germinable.

germinal (jėr'mə nəl), *adj. Biology.* **1** of or like germ cells: *Germinal protoplasm, or germ plasm, is that protoplasm from which the reproductive cells, or gametes, are formed* (Harbaugh and Goodrich, *Fundamentals of Biology*). **2** in the earliest stage of development.

germinal disk, *Embryology.* the part of the blastoderm of certain vertebrate eggs where the embryo appears.

germinal vesicle, *Biology.* the enlarged nucleus of an ovum before the polar bodies are formed at the end of meiosis.

germinate, *v. Botany.* to begin to grow or develop; sprout: *Following pollination, the pollen grain germinates on the stigma and forms a pollen tube which extends into the ovule* (Biology Regents Syllabus). [from Latin *germinatum* sprouted, from *germen* sprout]

—**germination,** *n. Botany.* the act or process of germinating; a starting to grow or develop; a sprouting: *Soils which warm early in the spring aid in the germination of seeds and in the rapid growth of life in the soil* (Robert W. Howe).

—**germinative,** *adj.* capable of germinating.

germ layer, *Embryology.* any of the three primary layers of cells, the ectoderm, mesoderm, or endoderm, which become further differentiated as the embryo develops.

germ plasm, *Biology.* **1** the germ cells of an organism: *Radiation is one of the agents that cause changes called mutations in the germ plasm, the sperm and ovary borne material that transmits characteristics from generation to generation* (N.Y. Times). **2** the hereditary substance in germ cells, now known to consist of the chromosomes, that is transmitted to the offspring.

germ theory, *Biology.* the theory that infectious and contagious diseases are caused by germs.

gerontology (jer'on tol'ə jē), *n.* the branch of science dealing with the phenomena and problems of aging and old age. [from Greek *gerontos* old man]

—**gerontological** (jə ron'tl oj'ə kəl), *adj.* of or having to do with gerontology.

gestate, *v. Biology.* to carry young in the uterus during pregnancy. [from Latin *gestatum* carried about, from *gerere* to bear]

—**gestation,** *n. Biology.* **1** the carrying and developing of young in the uterus from conception to birth: *Birth usually occurs in human beings after a gestation of approximately nine months.* SYN: pregnancy. **2** the period of gestation: *The gestation ... of elephants is about two full years, so the young elephant has the protection of its huge mother's body for the first two years of its life and it is large enough when born to be able to protect itself very well* (Winchester, Zoology).

—**gestational,** *adj.* of or having to do with gestation.

getter, *n. Chemistry.* **1** a chemically active substance, such as magnesium, ignited in vacuum tubes to remove traces of gas. **2** any substance added to another to remove or neutralize traces of impurities: *Should more stringent protection against oxidation be required, a piece of titanium foil can be inserted in the package to act as a "getter" for the oxygen* (New Scientist).

---

cap, fāce, fäther; best, bē, tėrm; pin, five;
rock, gō, ôrder; oil, out; cup, pùt, rüle,
yü in use, *yù* in uric;
ng in bring; *sh* in rush; *th* in thin, ᴛʜ in then;
zh in seizure.
ə = *a* in about, *e* in taken, *i* in pencil, *o* in lemon, *u* in circus

GeV

GeV or **Gev,** *abbrev.* gigaelectron volt (one billion electron volts, a unit for measuring energy in particle physics).

geyser (gī′zər), *n. Geology.* a natural spring that spouts a column of hot water and steam into the air at frequent and irregular intervals: *Many other phenomena are associated with ground water. Springs issue from hillsides, apparently from the solid rock, and in a few places on the earth, great spouts or geysers of boiling water jet intermittently from orifices at the surface* (Garrels, *A Textbook of Geology*). [from Icelandic *Geysir,* name of such a spring in Iceland, from *geysa* to gush] ▶ See the note under **hot spring.**

geyser

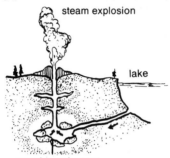

steam explosion

lake

superheated water

geyserite (gī′zə rīt), *n. Mineralogy.* a variety of opaline silica deposited about the openings of hot springs and geysers.

g-factor, *n. Nuclear Physics.* the ratio of a subatomic particle's magnetic moment to its angular momentum: *The positron's gyromagnetic ratio, called the "g-factor" ... this basic constant of the atom* (Science News). [from *g(yromagnetic) factor*]

GG or **G.G.,** *abbrev.* gamma globulin.

ghost, *n.* **1** *Biology.* a red blood cell that has undergone hemolysis: *If we suspend red blood cells in water, they swell and soon release their contents (hemoglobin) into solution. We can in this way obtain a mass of pure cell membranes and analyze them. They are called "ghosts"* (McElroy, *Biology and Man*).
2 *Optics.* a bright spot or secondary image produced by some defect in the lens or instrument.

GI or **G.I.,** *abbrev.* gastrointestinal: *the GI tract.*

giant, *n.,* or **giant star,** *Astronomy.* a very bright celestial body of large size and low density, such as Arcturus. Compare **dwarf.**

gibberellic acid (jib′ə rel′ik), *Biochemistry.* a crystalline acid first discovered in a fungus, that increases the size and the rate of growth of plants: *Treatment with a plant hormone, gibberellic acid, greatly affects phyllotaxis, or leaf arrangements, in cocklebur shoots* (Frank B. Salisbury). *Formula:* $C_{19}H_{22}O_6$ [from New Latin *Gibberella,* name of the fungus]

gibberellin (jib′ə rel′ən), *n. Biochemistry.* any of a group of hormones that are synthesized in the protoplasm of plants and that increase the rate and amount of growth. Gibberellic acid is a gibberellin. *At the present time, five principal kinds of plant hormones have been identified ... Three of the kinds of hormones—auxins, gibberellins, and cytokinins—are growth-promoting substances, although auxin inhibits its growth at high concentrations* (Greulach and Adams, *Plants*).
ASSOCIATED TERMS: see **cytokinin.**

gibbous (gib′əs), *adj.* **1** *Geometry.* curved out; convexly curved.
2 *Astronomy.* (of a celestial body) so illuminated as to be convex on both margins: *A planet is said to be gibbous when more than half of its apparent disk but not all of it is illuminated* (American Scholar).
[from Latin *gibbosus,* from *gibbus* hump]

giga- (jig′ə), *combining form.* one billion (10^9) of any SI unit: *Gigavolt = one billion volts. The distance from the sun to the earth is 150 gigametres or 94 megamiles* (London *Times*). *Symbol:* G or g [from Greek *gigas* giant]

gigacycle, *n.* = gigahertz.

gigahertz, *n.* a unit of frequency equal to one billion (10^9) hertz.

gigantism, *n.* **1** *Medicine.* an abnormal growth or size of the body or part of the body, usually due to oversecretion of growth hormone. **2** *Biology.* abnormally great vegetative growth. Compare **dwarfism.**

gigawatt, *n.* a unit of electric power equal to one billion (10^9) watts: *The development of "Q-switching" ... gives an extraordinarily high output (from megawatts to gigawatts) in a very brief period of time, one-tenth of a microsecond* (New Scientist).

gilbert, *n.* a unit of magnetomotive force in the centimeter-gram-second system, equal to $10/4\pi$ ampere turns. [named after William *Gilbert,* 1544–1603, English scientist]

gill, *n.* **1** *Zoology.* part of the body of a fish, tadpole, crab, etc., by which it breathes in water. Oxygen passes in and carbon dioxide passes out through the thin membranous walls of the gills.
2 *Biology.* the fine, thin, leaflike structure on the underside of a mushroom: *Contrast this structure to that of another great group, members of which generate spores on spoke-like surfaces that radiate on the undersides of expanded caps. These fertile structures, informally called gills, are not visible from above nor, usually, from the side, and these fungi must be upended to reveal the spore-bearing areas. Minor differences in gill construction, important in identification, may then be easily noted* (Peter Katsaros).

gill arch, *Zoology.* one of the cartilaginous arches supporting the gill of a fish or amphibian: *Each gill is supported on a cartilaginous gill arch, and its inner border has expanded gill rakers, which protect against hard particles and keep food from passing out the gill slits* (Storer, *General Zoology*). SYN: branchial arch.

gill cleft, = gill slit.

gill cover, *Zoology.* the fold of skin covering the gills of a fish. SYN: operculum.

gill filament, *Zoology.* the threadlike part of a gill filled with the blood vessels that absorb oxygen and give off carbon dioxide: *Cilia (small, threadlike outgrowths) on the gill filaments pump enormous quantities of water (several litres per animal per hour) between the gill filaments, and mucous sheets filter off plankton for use as food* (New Scientist).

gill raker, *Zoology.* one of the series of bony processes on the inner edge of a gill arch which prevents solid material from passing through the gill.

gill slit, *Zoology.* one of the openings in the pharynx of a fish, certain amphibians, etc., that serve as channels for the passage of water to the exterior. Also called **gill cleft.**

girdle, *n. Anatomy.* a bony support for the limbs: *To produce movement of the body as a whole the backward thrust of the limbs against the ground or the water must be imparted to the body. The force is usually transmitted through a skeletal structure called a girdle, which is attached to the spinal column* (Mackean, *Introduction to Biology*). The **shoulder girdle** supports the arms, and the **pelvic girdle,** the legs.

gizzard, *n. Zoology.* **1** a bird's second stomach, where the food from the first stomach is ground up. The gizzard usually contains bits of sand or gravel. **2** a muscular organ posterior to the crop in insects, earthworms, and some other animals, that serves to grind up the food.

glabella (glə bel′ə), *n., pl.* **-bellae** (-bel′ē). *Anatomy.* the small space on the forehead immediately above and between the eyebrows. [from New Latin *glabella,* from Latin *glabellus* hairless, smooth]

glabrous (glā′brəs), *adj. Biology.* without hair or down; smooth: *Nasturtiums have glabrous stems.* [from Latin *glabrum* smooth]

glacial, *adj.* **1** *Geology.* **a** of ice or glaciers; having much ice or many glaciers: *The total weight of the glacial flow is in the billions of tons, the area of glaciation including tributaries is 900 square kilometers* (V.A. Troitskaya). **b** of or having to do with a glacial period or epoch: *Gerd Luttig defined four orders of glacial time subdivisions (glacials, stadials, phases, and staffels), using curves representing ice advances and recessions, and relating maxima and minima to distances between them* (Lawrence Ogden). **c** made by the pressure and movement of ice or glaciers: *It is recognized, however, that within the regions of continental glaciation some areas have predominately the kinds of surface features that result from glacial erosion while in others the features are mainly those that result from glacial or glaciofluviatile deposition* (Finch and Trewartha, *Elements of Geography*). **2** *Chemistry.* having an icelike form in its pure state at or just below room temperature: *glacial phosphoric acid. Glacial acetic acid, about 99.5 per cent purity, is so-called because on cold days it freezes to an ice-like solid* (Parks and Steinbach, *Systematic College Chemistry*). [from Latin *glacialis,* from *glacies* ice]

glacial drift, *Geology.* clay, gravel, sand, or other rock material transported and deposited by a glacier: *The debris dumped by glaciers and by the streams flowing from them is called glacial drift. The drift assumes characteristic and easily recognized topographic forms* (Gilluly, *Principles of Geology*).

glacial epoch, *Geology.* **1** any of the times when much of the earth was covered with glaciers; an ice age. **2** the most recent time when much of the Northern Hemisphere was covered with glaciers; the Ice Age or Pleistocene.

glacial period, *Geology.* the period that includes the glacial epochs.

glaciate (glā′shē āt), *v. Geology.* **1** to cover with ice or glaciers. **2** to act on by ice or glaciers.

—glaciation, *n. Geology.* the process of covering a part of the earth's or any planet's surface with ice or glaciers: *The four recognized Ross Sea glaciations, east of the Transantarctic Mountains, occurred within the last 1.2 million years and have a correlation with the sea-level changes caused by ice sheets in the Northern Hemisphere* (Edith M. Ronne). *The possibility that the frozen polar caps of Mars were formed by glaciation of carbon dioxide has been suggested by two University of Michigan geologists* (Science News).

glacier (glā′shər), *n. Geology.* a large mass of ice moving very slowly down a mountain or along a valley, or spreading slowly over a large area of land. Glaciers are formed over many years from snow wherever winter snowfall exceeds summer melting. *Glaciers can be divided into two broad classes: ice streams and ice caps ... from the point of view of their total mass, the ice caps are the more important by far, since they occupy at least five out of the six million square miles of glaciated territory on the earth* (M.F. Perutz). [from French, from *glace* ice, from Latin *glacies*]

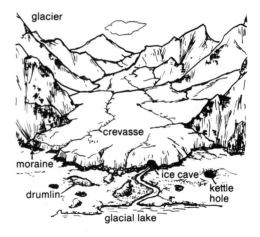

glacier
crevasse
moraine
ice cave
kettle hole
drumlin
glacial lake

glaciofluvial (glā′shē ō flü′vē əl), *adj.* = fluvioglacial.

glaciolacustrine (glā′shē ō lə kus′trin), *adj. Geology.* of or produced by glaciers and their related lakes: *glaciolacustrine deposits.* [from Latin *glacies* ice + English *lacustrine*]

glaciology, *n.* the science that deals with glaciers and glacial action: *As a relatively obscure science, normally included under hydrology, glaciology suffers from scarcity of experienced scientists and adequate techniques* (A.P. Crary).

cap, fāce, fäther; best, bē, tèrm; pin, fīve;
rock, gō, ôrder; oil, out; cup, pùt, rüle,
yü in use, yu̇ in uric;
ng in bring; sh in rush; th in thin; ᴛʜ in then;
zh in seizure.
ə = a in about, e in taken, i in pencil, o in lemon, u in circus

gladiolus (glad′ē ō′ləs), *n., pl.* **-li** (-lī). *Anatomy.* the bladelike intermediate segment of the breastbone or sternum. [from New Latin *gladiolus,* diminutive of Latin *gladius* sword]

gland, *n.* **1** *Anatomy.* **a** an organ in the body by which certain substances are separated from the blood and changed into some secretion for use in the body, such as bile, or into a product to be discharged from the body, such as sweat. The liver, the kidneys, the pancreas, and the thyroid are glands. *The superficial glands include (1) sebaceous glands, as described for the cat; (2) scent glands of various types; (3) mammary glands that produce milk; (4) sweat glands on the horse and man; and (5) lachrymal (tear) glands that moisten and cleanse the surface of the eye* (Storer, *General Zoology*). **b** any similar structure that does not secrete, such as a lymph node: *Antibodies, he found, are formed in lymph nodes, better known to the layman as glands* (Science News Letter).
2 *Botany.* any of various secreting organs or structures, generally on or near a surface: *Many plants have glands that are much more definite and clear cut in form than the rather diffuse ones thus far described. Mention has been made ... of the digestive glands on the leaves of sundew and Venus' flytrap, and of those that line the resin ducts in the wood, bark, and leaves of many of the gymnosperms. ... Other examples are those that secrete the pungent oil in orange peel* (Emerson, *Basic Botany*).
[from French *glande,* from Latin *glandem* acorn]

glandular, *adj.* of, having to do with, or consisting of a gland or glands: *glandular tissue. The liver is the largest glandular organ of the body* (Thomas H. Huxley).

glans (glanz), *n., pl.* **glandes** (glan′dēz′). *Anatomy.* the head of the penis or of the clitoris. [from Latin *glans, glandis* acorn]

glass, *n. Chemistry.* a hard, brittle, amorphous material, especially a transparent ceramic consisting mostly of silica. Glasses behave like solids, but their atomic structure is disorganized like that of a liquid. They are typically formed by the rapid cooling of a hot liquid. Obsidian and pumice are natural glasses.

glauconite (glô′kə nīt), *n.* a greenish mineral consisting essentially of a hydrous silicate of iron and potassium, occurring in greensand, clays, and seawater. [from Greek *glaukon* gray-green + English *-ite*]

glaucous (glô′kəs), *adj. Botany.* covered with a whitish, powdery or waxy substance, as plums and grapes are: *glaucous leaves.* [from Greek *glaukos* gray-green, silvery]

glaze, *n. Meteorology.* a glassy coating of ice formed by rain that freezes upon striking the ground or some other surface. Also called **sleet.**

g.l.b., *abbrev.* greatest lower bound.

gleba (glē′bə), *n., pl.* **-bae** (-bē). *Biology.* the fleshy part of certain fungi in which the spores are borne. [from Latin *gleba* lump, clod]
—**glebal,** *adj.* of or having to do with a gleba: *Inside is the spore-cup or glebal ball about one-twentieth of an inch in diameter and containing tens of thousands of spores embedded in a tough, fatty matrix* (New Scientist).

glenoid (glē′noid), *Anatomy.* —*n.* a slightly cupped cavity on a bone in which the end of another bone rests, forming a joint.
—*adj.* of or designating such a cavity: *the glenoid cavity of the scapula. A concavity in these bones at their junction furnishes the articular surface for the long wing bone (humerus) and is called the glenoid fossa* (Hegner and Stiles, *College Zoology*).

glia (glē′ə *or* glī′ə), *n.* = neuroglia.
—**glial,** *adj.* having to do with the neuroglia: *There are also other cells, the so-called glial cells, which until recently have been thought to act merely as supporting structures in the retina, and to provide pathways for the passage of energy-providing substances from the blood capillaries to the light-receptors and the neurons* (New Scientist).

gliding bacteria, = myxobacteria.

Gln, *abbrev.* glutamine.

globin (glō′bən), *n. Biochemistry.* a protein substance formed in the decomposition of hemoglobin: *Hemoglobin, on which our cells depend for oxygen, is [made up of] a conjugated protein, globin, with a prosthetic group, heme* (Simpson, *Life: An Introduction to Biology*). [from Latin *globus* globe]

globulin (glob′yə lin), *n. Biochemistry.* any of a group of proteins, found in plant and animal tissue, which are insoluble in pure water but soluble in dilute salt solutions and in weak acids and alkalis. Globulin is a protein component of blood plasma. *Some of the plasma proteins (globulin) are antibodies. They react with the invading particles (antigens) in such a way that the antigen no longer has any harmful action* (McElroy, *Biology and Man*).

glomerular (glə mer′ə lər), *adj.* **1** *Botany.* of or like a glomerule. **2** *Anatomy.* of or having to do with a glomerule or glomeruli.

glomerule (glom′ə rül′), *n. Botany.* a cyme condensed into a headlike cluster, as in the flowering dogwood. [from New Latin *glomerulus,* from Latin *glomus* ball]

glomerulus (glə mer′ə ləs), *n., pl.* **-li** (-lī). *Anatomy.* **1** a tuft of capillaries in the tubules of the kidney, contained within a Bowman's capsule and serving to filter out waste products from the blood.
2 a small rounded mass in the olfactory bulb of the brain: *The dendritic tuft of each mitral cell receives incoming axon terminals from olfactory receptor cells within a small spherical region called a glomerulus, which acts as a kind of miniature relay station within the bulb* (Scientific American).

glossopharyngeal (glos′ō fə rin′jē əl), *adj. Anatomy.* having to do with the rear part of the tongue and of the pharynx: *the glossopharyngeal nerve.*

glottis (glot′is), *n. Anatomy.* the opening at the upper part of the windpipe, between the vocal cords: *In the human, the mouth and nose communicate with the lungs through a series of special structures. The glottis is an opening in the floor of the pharynx, protected above by a lid, or epiglottis, and supported by a cartilaginous framework, the larynx* (Storer, *General Zoology*). [from Greek *glōttis,* from *glōtta* tongue]
—**glottal,** *adj.* of or having to do with the glottis.

glow discharge, *Physics.* a luminous electrical discharge in a tube containing gas at low pressure: *The type of discharge in a neon sign and a fluorescent lamp is* called a glow discharge (Sears and Zemansky, University Physics).

Glu, abbrev. glutamic acid.

glucagon (glü′kə gon), n. Biochemistry. a polypeptide hormone secreted by the islets of Langerhans that raises the blood sugar level by stimulating the breakdown of glycogen to glucose: Glucagon stimulates the release of sugar from the liver into the blood (Biology Regents Syllabus).

glucan (glü′kən), n. Biochemistry. any polysaccharide that yields glucose on hydrolysis: Polysaccharides are named for the monosaccharides of which they are composed, but the -ose suffix that designates a single sugar is replaced by the suffix -an. Thus the long chains of glucose units that make up both starch and cellulose are called glucans. Starch is an alpha-glucan; cellulose is a beta-glucan (Scientific American).

glucocorticoid (glü′kō kôr′tə koid), n. Biochemistry. any of a group of steroid hormones, such as cortisone, produced by the adrenal cortex and affecting glucose metabolism: It is known that reserpine increases blood levels of the glucocorticoids. When these hormones were directly implanted in pregnant rats, nerve cells in the fetus expressed noradrenaline production for longer than usual; however, when a glucocorticoid inhibitor was used in conjunction with reserpine, the period of noradrenaline production was normal. Thus, the glucocorticoids appear to have a direct influence on the expression of the fetal nerve cells (John A. Mutchmor). Compare **mineralocorticoid.**

gluconeogenesis (glü′kə nē′ə jen′ə sis), n. Biochemistry. the conversion of noncarbohydrate substances such as protein into glucose: The pathway of gluconeogenesis is not simply the reverse of glycolysis but involves a number of alternative steps, at which points various types of regulatory control can be exercised (Trevor W. Goodwin).

—**gluconeogenetic** (glü′kə nē′ə jə net′ik), adj. of or having to do with gluconeogenesis: In the liver and kidney a variety of induced enzymes have been demonstrated, including some key enzymes in the gluconeogenetic pathway (David Feldman).

glucosamine (glü kō′sə mēn), n. Biochemistry. an amino sugar derived from glucose by substitution of an amino group for a hydroxyl group. It is the basic structural unit of chitin, which forms the exoskeleton of arthropods. Formula: $C_6H_{13}NO_5$

glucose (glü′kōs), n. Biochemistry. the most common kind of sugar occurring in plant and animal tissues. Carbohydrates are present in the blood mainly in the form of glucose. Glucose is stored in the polysaccharide glycogen; breakdown of glycogen releases glucose for transport. The first series of reactions in anaerobic respiration involves the conversion of glucose to pyruvic acid. Of the many hexoses, two deserve particular mention: glucose and fructose. Glucose, also called dextrose and grape sugar, is the principal sugar in animal blood as well as one of the more abundant sugars in plants. Glucose assumes considerable importance in plants as the substance from which both starch and cellulose are synthesized (Greulach and Adams, Plants). Formula: $C_6H_{12}O_6$ Also called **grape sugar.** [from French, from Greek gleukos sweet wine]

glucosidase (glü′kō sə dās), n. Biochemistry. an animal or plant enzyme that hydrolyzes glucosides.

glucoside (glü′kə sīd), n. Biochemistry. a glycoside, especially one containing glucose.

glucuronic acid (glü′kyù ron′ik), Chemistry. an acid derived from glucose and present in many animals and plants: The chemical is a product of glucose when it has been burned, or oxidized, in the body. It is called glucuronic acid. It is also made synthetically and is relatively inexpensive. It is known to detoxify certain poisons in the body (Science News Letter). Formula: $C_6H_{10}O_7$ [from gluc(oside) + uronic acid]

glumaceous (glü mā′shəs), adj. Botany. like a glume; consisting of or having glumes: a glumaceous calyx, glumaceous plants.

glume, n. Botany. a chaffy bract at the base of the spikelet of grasses, sedges, etc.: Grass flowers generally grow in a head or spike. ... Each spikelet is separated from its neighbors by two small modified leaves or bracts known as glumes. They are found at the base of the spikelet and in some cases (wheat, oats, barley), they may fairly well enclose it (Weier, Botany). [from Latin gluma hull, husk]

gluon (glü′on), n. Nuclear Physics. a component of subatomic particles that has neither mass nor charge and that holds together quarks to form strongly interacting particles: In the theory of the strong force, there are particles called gluons that play a role similar to that of the intermediate vector bosons in the electro-weak theory; that is, elementary particles interact by way of the weak force by exchanging intermediate vector bosons, and particles interact by way of the strong force by exchanging gluons. Unlike the bosons, gluons have no mass, and they have already been experimentally detected (Arthur L. Robinson). According to the QCD [quantum chromodynamics] theory, the force between quarks is transmitted by particles called gluons, which are very similar to the photons, or quanta of light, that transmit ordinary electrical and magnetic forces. For example, like photons, gluons always move at the speed of light. Unlike photons, however, gluons can exert forces directly on each other (Robert H. March). [from glue + -on]

glutamate (glü′tə māt), n. Biochemistry. a salt or ester of glutamic acid.

glutamic acid (glü tam′ik), Biochemistry. a white, crystalline amino acid found in plant and animal proteins, especially in seeds and beets: A well-known product made from glutamic acid is monosodium glutamate. This saltlike substance has little flavor of its own, but it brings out the flavor of other foods (Karl S. Quisenberry). Formula: $C_5H_9NO_4$ [from glut(en) + am(ide) + -ic]

glutamine (glü′tə mēn), n. Biochemistry. a crystalline amino acid derived from the action of enzymes on gluten and found in the protein of many plants and ani-

cap, fāce, fäther; best, bē, tèrm; pin, fīve;
rock, gō, ôrder; oil, out; cup, pùt, rüle,
yü in use, yù in uric;
ng in bring; sh in rush; th in thin; ŦH in then;
zh in seizure.

ə = a in about, e in taken, i in pencil, o in lemon, u in circus

glutathione

mals. *Formula:* $C_5H_{10}N_2O_3$ [from *glutam(ic) acid* + *-ine*]

glutathione (glü′tə thī′ōn), *n. Biochemistry.* a polypeptide present in plant and animal tissues, important in physiological oxidations: *Oxidative damage to red blood cells was detected in rats kept on a selenium-deficient diet. This damage was related to reduced activity of an enzyme, glutathione peroxidase, that helps to protect hemoglobin against the injurious oxidative effects of hydrogen peroxide. The enzyme uses hydrogen peroxide to catalyze the oxidation of glutathione, thus keeping hydrogen peroxide from oxidizing the reduced state of iron in hemoglobin. Oxidized glutathione can readily be converted to reduced glutathione by a variety of intracellular mechanisms. There is some reason to believe glutathione peroxidase may even contain some form of selenium acting as an integral part of the functional enzyme molecule* (Early Frieden). *Formula:* $C_{10}H_{17}N_3O_6S$ [from *gluta(mic) acid* + Greek *theion* brimstone (sulfur)]

gluteal (glü′tē əl *or* glü tē′əl), *adj. Anatomy.* of or having to do with the gluteus: *the gluteal arteries and veins, the great gluteal muscle.*

glutelin (glü′tə lin), *n. Biochemistry.* any of a group of simple proteins found in corn, wheat, and other grains.

gluten (glüt′n), *n. Biochemistry.* the tough, sticky, elastic protein substance in the flour of wheat and other grains: *Two varieties of Panhandle wheat, Cheyenne and Nebred ... are high in protein, a quality shared by some other types. More unusual, they're high in gluten content. Gluten is the chemical part of wheat that makes dough stick together. Bakers can leave dough containing gluten in the mixing machine for several minutes beyond the usual time without its pulling apart into sticky globs* (Wall Street Journal). [from Latin *gluten* glue]

gluteus (glü tē′əs), *n., pl.* **-tei** (-tē ī). *Anatomy.* any of the three large muscles of the buttocks, especially the **gluteus maximus,** the largest and closest to the surface: *Sherwood L. Washburn of the University of California at Berkeley has expressed the view that the change from four-footed to two-footed posture was initiated by a modification in the form and function of the gluteus maximus, a thigh muscle that is powerfully developed in man but weakly developed in monkeys and apes* (John Napier). [from New Latin, from Greek *gloutos* buttock]

Gly, *abbrev.* glycine.

glyceraldehyde (glis′ə ral′də hīd), *adj. Chemistry.* a colorless, crystalline solid produced by oxidizing glycerol, important as an intermediate in carbohydrate metabolism. The D- and L- forms of glyceraldehyde are the configurational reference points for carbohydrates. *Formula:* $C_3H_6O_3$

glyceric acid (gli ser′ik *or* glis′ər ik), *Chemistry.* a colorless, syruplike acid obtained by the partial oxidation of glycerol. It is also formed during alcoholic fermentation. *Formula:* $C_3H_6O_4$

glyceride (glis′ə rīd′), *n. Chemistry.* an ester of glycerol and fatty acids. ▶ Since a molecule of glycerol has three reactive sites, it is possible to have *monoglycerides, diglycerides,* and *triglycerides,* depending on how many molecules of fatty acid react with each molecule of glycerol.

glycerin or **glycerine** (glis′ər ən), *n.* = glycerol. [from French *glycérine,* from Greek *glykeros* sweet] ▶ Although the term *glycerin* is still widely used commercially, it has generally been replaced by *glycerol* in chemistry.

glycerol (glis′ə rōl′ *or* glis′ə rol′), *n. Chemistry.* a colorless, syrupy, sweet liquid obtained from animal and vegetable fats and oils. Glycerol is an important intermediate in the metabolism of carbohydrates and fats. Commercially, it is used as a lubricant and emulsifying agent, in ointments and lotions, in antifreeze solutions and explosives, and in many other industrial applications. *Formula:* $C_3H_5(OH)_3$ ▶ See note under **glycerin.**

glycine (glī′sēn *or* glī sēn′), *n. Chemistry.* a colorless, sweet-tasting, crystalline amino acid formed when gelatin or various other animal substances are boiled in the presence of alkalis. *Formula:* $C_2H_5NO_2$ [from French *glycine,* from Greek *glykys* sweet]

glycogen (glī′kə jən), *n. Biochemistry.* a starchlike carbohydrate stored in the liver and other animal tissues. Glycogen is a polysaccharide formed from carbohydrates and changed, when the body requires it, into glucose. *Formula:* $(C_6H_{10}O_5)_n$ Also called **animal starch.** [from Greek *glykys* sweet + English *-gen*]

glycogenesis (glī′kə jen′ə sis), *n. Biochemistry.* the formation of glycogen or sugar in animals: *The metabolic effects of insulin are seen in an increase in glucose uptake, glycogenesis, glycolysis, and lipogenesis* (George Weber).

glycol (glī′kōl *or* glī′kol), *n. Chemistry.* **1** a colorless, sweet-tasting alcohol obtained from various ethylene compounds and used as an antifreeze, as a solvent, in making lacquers, etc. *Formula:* $C_2H_4(OH)_2$ Also called **ethylene glycol.**
2 any one of a class of similar alcohols containing two hydroxyl groups. [from *glyc(erin)* + *-ol*]

glycolipid (glī′kō lip′id), *n. Biochemistry.* any of a class of lipids that yield sugar and a fatty acid upon hydrolysis. Cerebrosides are glycolipids.

glycolysis (glī kol′ə sis), *n. Biochemistry.* the process by which a carbohydrate, such as glucose, is broken down to an acid: *A sugar molecule synthesized in the chloroplast by photosynthesis may move into the cytoplasm where it is changed and partially oxidized by the action of enzymes in the cytoplasm during a series of reactions called glycolysis. The oxidized product of glycolysis then moves into a mitochondrion* (Weier, Botany). *Glycolysis and respiration are the universal heritage of all eukaryotes today: one-celled protists, plants, animals and fungi* (Scientific American).
—glycolytic (glī′kə lit′ik), *adj.* of or having to do with glycolysis: *A chance observation, that cells treated with interferon showed an altered metabolism of glucose, that is increased glycolytic activity, may well be a most important clue* (New Scientist).

glycopeptide (glī kə pep′tīd), *n.* = glycoprotein.

glycoprotein (glī′kə prō′tēn), *n. Biochemistry.* any of a group of conjugated proteins which contain a carbohydrate radical and a simple protein.

glycoside (glī′kə sīd), *n. Biochemistry.* any of a large group of organic compounds which yield a sugar, often glucose, and another substance on hydrolysis in the presence of various ferments or enzymes or a dilute acid.

—**glycosidic,** *adj.* of or having to do with a glycoside: *The newborn pig lacks the ability to utilize sucrose or, in chemical terms, to hydrolyze the glycosidic bond of sucrose* (Science News Letter).

GMT, *abbrev.* Greenwich Mean Time.

gnathic (nath′ik), *adj. Anatomy.* of or having to do with the jaw. The **gnathic index** is 100 times the facial length divided by the length of the base of the cranium. [from Greek *gnathos* jaw]

gnathite (nă thīt *or* nath′īt), *n. Zoology.* one of the appendages of the mouth of an arthropod. [from Greek *gnathos* jaw]

gneiss (nīs), *n. Geology.* any of various regional metamorphic rocks consisting of coarse layers of quartz and feldspar alternating with layers of mafic minerals: *The oldest rock yet known on earth was reported found in the Godthaab district of Greenland. A metamorphic rock known as gneiss, it was dated at 3.98 billion years by the ratios of rubidium to strontium, and by isotopes of lead* (Richard M. Pearl). [from German *Gneis*]

—**gneissic** (nī′sik), *adj.* characterized by a type of foliation in which minerals within a metamorphic rock are separated into distinct layers: *On a small scale, the individual gneissic bands are fantastically contorted, with the long zig-zag folds and intricate convolutions typical of rocks that have, at least in part, recrystallized during their deformation* (E.F. Roots).

gneissoid (nī′soid), *adj. Geology.* resembling gneiss, especially when of igneous rather than metamorphic origin.

gnomon (nō′mon), *n. Geometry.* what is left of a parallelogram after a similar parallelogram has been taken away at one corner. [from Greek *gnōmon* indicator, (literally) one who knows]

gnotobiology (nō′tō bī ol′ə jē), *n.* the branch of biology dealing with gnotobiotes and gnotobiotics: *A highly specialized segment of the ultra-clean technology is gnotobiology, the raising of germfree animals, largely for research purposes* (Science Journal).

gnotobiote (nō′tō bī′ōt), *n. Microbiology.* a germ-free animal infected with one or more microorganisms in order to study the microorganism in a controlled situation. [from Greek *knōtos* known + *biotē* life]

gnotobiotic (nō′tō bī ot′ik), *adj.* **1** of or having to do with gnotobiotes or gnotobiotics.
2 free of germs or associated only with known germs: *The [tobacco] disease has been very troublesome to plant scientists because the etiology has been elusive, ephemeral, and incompletely explained. Control has been accomplished by sterile or gnotobiotic culture* (Shreve D. Woltz).

—**gnotobiotics,** *n.* the study of organisms or conditions that are either free of germs or associated only with known germs: *Gnotobiotics includes the study of both "germ-free" animals and animals whose microbial flora can be completely specified* (New Scientist).

goblet cell, *Biology.* an epithelial cell, shaped roughly like a goblet, found in the respiratory passages and in the intestines: *Through constant irritation, goblet cells are stirred into increased action; they enlarge and secrete an abundance of mucus* (Edwin Diamond).

Gödel's theorem or **Gödel's proof** (gœd′lz), *Mathematics.* the theorem that any adequate axiomatic system, such as ordinary whole number arithmetic, must contain propositions that cannot be proved or refuted using the rules of the system. Also called **incompleteness theorem.** [named after Kurt *Gödel,* born 1906, Austrian-born American mathematician]

gold, *n. Chemistry.* a shiny, bright-yellow, ductile and malleable, precious metallic element which resists alteration. *Symbol:* Au; *atomic number* 79; *atomic weight* 196.967; *melting point* 1063°C; *boiling point* 2808°C; *oxidation state* 3, 1.

golden section, *Mathematics.* the proportion resulting from the division of a straight line in such a way that the shorter part is to the longer part as the longer part is to the whole. The golden section was first formulated by Euclid and has been used widely to produce harmonious geometric figures.

Golgi apparatus or **Golgi complex** (gol′jē), *Biology.* a network or vesicles in the cytoplasm of many cells, functioning in the manufacture of proteins and carbohydrates: *In animal cells, the Golgi apparatus is generally associated with other cell constituents to form a complex region active in the elaboration of secretory products* (Weier, *Botany*). See the picture at **centrosome.** [named after Camillo *Golgi,* 1844–1926, Italian anatomist]

Golgi body, *Biology.* a unit or particle of the Golgi apparatus revealed by staining. Also called **dictyosome.**

gonad (gō′nad), *n. Biology.* an organ in which reproductive cells develop in the male or female; a sex gland. Ovaries and testes are gonads. *In animals meiosis generally takes place as the gametes are being formed in specialized organs called gonads. Sperm formation is termed spermatogenesis, and the formation of the egg, oogenesis* (Harbaugh and Goodrich, *Fundamentals of Biology*). [from New Latin *gonades,* pl., from Greek *gonē* seed, from *gignesthai* be produced]

—**gonadal** (gō nad′əl), *adj.* of or having to do with gonads: *Differences in general body conformation, which influence the structure of the pelvic region and the mammary glands; and differences in hair growth and distribution, the male generally being more hairy than the female ... are influenced by the gonadal hormones, chemical substances that are produced as development proceeds* (McElroy, *Biology and Man*).

gonadotropic (gon′ə də trop′ik) or **gonadotrophic** (gon′ə də trof′ik), *adj. Biology.* affecting or stimulating the growth or activity of the gonads: *The action of gonadotrophic hormones in promoting maturation of the female organs of reproduction, including egg development and release, is greatly enhanced by administering crude homogenates or partially purified extracts of cerebral tissue* (Science News).

---

cap, fāce, fäther; best, bē, tėrm; pin, fīve;
rock, gō, ôrder; oil, out; cup, pùt, rüle,
yü in use, *yù* in uric;
ng in bring; *sh* in rush; *th* in thin; ᴛʜ in then;
zh in seizure.
ə = *a* in about, *e* in taken, *i* in pencil, *o* in lemon, *u* in circus

gonadotropin

gonadotropin (gon′ə də trop′in) or **gonadotrophin** (gon′ə də trof′in), *n. Biochemistry.* a gonadotropic hormone: *Gonadotropins helped quail lay eggs regularly up to ten days, if periods of extended darkness were provided* (Science News Letter).

Gondwana (gond wä′nə) or **Gondwanaland,** *n. Geology.* the protocontinent of the Southern Hemisphere, a hypothetical land mass comprising India, Australia, Antarctica, Africa, and South America, which according to the theory of plate tectonics began to break up into the present continents at the end of the Mesozoic era: *Advocates of [continental] drift are challenged to say exactly how the present continents fitted together to form Pangaea, or alternatively to reconstruct the two later supercontinents Laurasia and Gondwana, which some theorists prefer to a single all-embracing land mass* (Robert S. Dietz and John C. Holden). Compare **Laurasia, Pangaea.** [named after *Gondwana,* a region of central India inhabited by the Dravidian *Gond* people. The region is noted for its unusual geological formations.]

gonidium (gō nid′ē əm), *n., pl.* **-ia** (-ē ə). *Biology.* 1 a reproductive cell produced asexually in algae, such as a zoospore.
2 one of the algal cells filled with chlorophyll that are formed in the thallus of lichens.
[from New Latin *gonidium,* from Greek *gonos* seed, what is produced, from *gignesthai* be produced]
—**gonidial,** *adj.* having to do with or containing a gonidium: *gonidial propagation, the gonidial layer, a gonidial receptacle.*

gonocyte (gon′ə sīt), *n. Biology.* a germ cell; an oocyte or spermatocyte. [from Greek *gonos* seed]

gonophore (gon′ə fôr), *n.* **1** *Botany.* a prolongation of the axis of a flower above the perianth, bearing the stamens and pistils. **2** *Zoology.* one of the generative buds in hydrozoans. [from Greek *gonos* seed + *phoros* bearing]

gonopodium (gon′ə pō′dē əm), *n., pl.* **-dia** (-dē ə). *Zoology.* an anal fin in the male of guppies and certain other fishes, modified to serve as an organ of reproduction. [from New Latin *gonopodium,* from Greek *gonos* seed + New Latin *podium* footlike structure]

gonopore (gon′ə pôr), *n. Zoology.* a reproductive opening of certain worms: *Male and female gonopores in hermaphroditic species may be separate, ... or there may be a common gonopore* (Hyman, *Invertebrates*).

gorge (gôrj), *n. Geology.* **1** a deep, narrow passage between mountains, usually steep and rocky. **2** a deep ravine or canyon, usually with a river or stream flowing through it. [from Old French *gorge* throat, ultimately from Latin *gurges* abyss]

gossan *n. Geology.* an exposed, oxidized part of a mineral vein, especially a reddish-colored outcrop of iron ore consisting largely of limonite.

Graafian follicle (grä′fē ən), *Anatomy.* one of the small, fluid-filled sacs or cavities in an ovary that contains an ovum: *The oocyte ... is surrounded by follicle cells, which provide protection and nutrients, and the whole structure is known as a Graafian follicle* (McElroy, *Biology and Man*). [named after Regnier de *Graaf,* 1641–1673, Dutch physician]

graben (grä′bən), *n., pl.* **-ben.** *Geology.* a narrow, low-lying trough of land formed between two nearly parallel faults by the depression of a block of the earth's crust: *All the big calderas are located on a graben (a depressed section of crust)* (Haroun Tazieff). Also called **rift valley.** Contrasted with **horst.** [from German *Graben* ditch]

gracilis (gras′ə lis), *n., pl.* **-les** (-lēz). *Anatomy.* a long, slender muscle on the inner surface of the thigh. The gracilis pulls the thigh in and flexes the leg. [from Latin *gracilis* slender]

gradation, *n. Geology.* the process by which the land is leveled off through the action of wind, water, etc.: *The forces of gradation ... operate largely through the work of agents such as wind, running water, moving snow and ice, and living organisms. The tendency of the forces of gradation and their processes is to bring the surface of the land to a uniform low slope or grade. This is done by tearing down all elevations, such as may be produced by the tectonic forces, and by filling up depressions* (Finch and Trewartha, *Elements of Geography*).

grade, *n.* **1** *Geology.* **a** a level or sloping surface: *Streams, or part of streams, that are thus balanced between erosion and deposition are said to be ... flowing at grade* (Garrels, *Textbook of Geology*). **b** a slope, as of a road or stream bed. SYN: gradient. **c** the amount or degree of sloping of a surface. **d** the degree or intensity of metamorphism.
2 *Mathematics.* a centesimal unit for measuring angles, equal to 1/100 of a right angle. Compare **degree.**
3 *Zoology.* **a** a group or grouping of animals at about the same point of development, though not necessarily genetically linked: *Two grades are usually recognized: the Protozoa or one-celled animals and the Metazoa or many-celled animals. Man obviously belongs to the grade Metazoa* (Beals and Hoijer, *Anthropology*). **b** a hybrid animal, especially a cross between a native and a purebred stock: *Hereford grade steer calves* (Pall Mall Gazette).
[from Latin *gradus* step, degree]

gradient (grä′dē ənt), *n.* **1** *Physics.* **a** the rate at which a variable quantity, such as temperature or pressure, changes in value: *a thermometric gradient, a barometric gradient.* **b** the curve that represents this rate: *a nearly horizontal gradient.*
2 *Mathematics.* **a** the rate at which the y-coordinate of a line or curve changes with respect to the x-coordinate: *the gradient of a function.* **b** the vector whose components along the x, y, and z axes are the partial derivatives of the function with respect to the variables: *The vector v is called the gradient of f at* P_0 (Thomas, *Calculus and Analytic Geometry*).
3 *Biology.* any system of progressively increasing or decreasing differences with respect to the rate of growth, metabolism, or other property of an organism, organ, cell, etc.: *In practice it is difficult to design an experiment with a moisture gradient and the results are often susceptible to more than one interpretation. A root system in a soil which was not uniformly moist might well show a lop-sided distribution because the dry soil has inhibited the root growth while this would have been promoted in moist soil, but such an effect could not be attributed to a directional growth re-*

sponse by the growing root tips (Mackean, *Introduction to Biology*).

4 *Geology.* a slope, especially of the surface of a stream or land: *The average gradient remains the same over considerable stretches of the river, though the bed gradient varies from deep bendways to shallow crossings* (Garrels, *A Textbook of Geology*).
[from Latin *gradientem* walking, going, from *gradus* step, degree]

graduate (graj′ù āte), *Chemistry, Physics.* —*v.* to divide into equal units by a series of lines to show degrees for measuring, as on a thermometer, barometer, etc.: *It is often necessary for the chemist to measure the volume of liquids to be used in his work. The commonest instrument so used is the graduated cylinder, on whose side are marked scratches corresponding to milliliters or multiples of milliliters. The cylinder is provided with a lip so that its contents may be poured readily into another container. Scratched in the glass of the cylinder will be the temperature (usually 20° C.) at which it contains the quantity indicated by its graduations* (Jones, *Inorganic Chemistry*).
—*n.* (graj′ù it), a container marked with degrees for measuring the volume of liquids.

graft, *v.* **1** *Botany.* **a** to insert (a shoot or bud from one kind of tree or plant) into a slit in another closely related kind of tree or plant, so that it will grow there permanently: *Grafting is essentially a method of so joining the cambiums of portions of two different but related plants that the proliferation of the cells in the two meristems will cause an interlacing of tissues and thus effect an organic union between the two parts. ... The two parts of a graft are known as the scion and the stock. The scion, which usually consists of a short portion of a branch containing several buds, is the part which will develop into the shoot. The stock may be a root, a portion of a stem which will produce roots, or the stem of a rooted plant* (Harbaugh and Goodrich, *Fundamentals of Biology*). *The cuttings, instead of being rooted, are grafted to another plant of the same species or genus, but generally of a different variety* (Greulach and Adams, *Plants*). **b** to produce or improve (a fruit, flower, etc.) by grafting. **c** to do grafting on (a plant or tree).
2 *Biology, Medicine.* to transfer (a piece of skin, bone, etc.) from one part of the body to another, or to a new body, so that it will grow there permanently.
—*n.* **1** *Botany.* **a** a grafting or a being grafted: *a successful graft.* **b** a shoot or bud used in grafting: *By cutting and grafting, the inbred characteristics of the plant are preserved, e.g. an apple tree produced by germinating a seed from a good eating apple would yield only small, sour "crabapples", but the fruit from the graft would retain their size and flavour* (Mackean, *Introduction to Biology*). **c** a place on a tree or plant where the shoot or bud is inserted: *A bad fork is apt to occur at the graft* (L.H. Bailey). **d** a tree or plant that has had a shoot or bud grafted on it.
2 *Biology, Medicine.* **a** the act or process of grafting a piece of skin, bone, etc. **b** a piece of skin, bone, etc., transferred in grafting: *For a permanent graft to replace skin burned off, victims must have some of their own skin. A badly burned person, especially a child, may not have enough intact skin to spare. So a bit of his own skin would be set to growing in a special solution. In two weeks, the piece of skin so grown may be* multiplied to ten times its original size—*a good-size piece for grafting* (Newsweek).
[from Middle French *grafe* scion, stylus (the inserted shoot was thought to resemble a writing tool), from Latin *graphium* stylus, from Greek *grapheion,* from *graphein* write]
—**graftage,** *n.* **1** the process of grafting, as on trees and plants. **2** the condition of being grafted.

graft (def. 1)

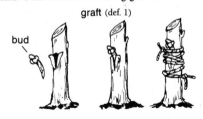

bud

graft-versus-host, *adj. Immunology.* of or denoting a condition in which transplanted cells of a donor attack the cells of the recipient's body, instead of the more common reaction in which the recipient's body rejects the transplanted cells: *The injected lymphocytes reacted immunologically against ... that part of the hybrid rats' cells that they recognized as foreign, and some of these lymphocytes produced anti-DA antibodies. This is part of what is known appropriately as a graft-versus-host reaction* (Jacques M. Chiller).

Graham's law, *Physics.* the law that under the same conditions of temperature and pressure the rate of diffusion of a gas is inversely proportional to the square root of its density. [named after Thomas *Graham,* 1805–1869, Scottish chemist who formulated this law]

grain, *n.* **1** *Botany.* **a** a single seed of wheat, corn, oats, and similar cereal grasses. In botanical usage a grain is not a seed but a fruit. *A ... grain is a dry, 1-celled, indehiscent fruit whose pericarp is always fused with the seed coat. This fruit is more likely than any other to be mistaken for a seed. (Examples: Wheat, Corn, Barley, Oats, and other members of the Gramineae or Grass Family)* (Youngken, *Pharmaceutical Botany*). **b** seeds or seedlike fruits of such plants in the mass: *to grind grain.* **c** the plants that these seeds or seedlike fruits grow on: *a field of grain.*
2 *Geology.* **a** one of the particulate units of which granular rocks are composed. **b** the plane of cleavage in coal, stone, or the like. SYN: lamination.
3a the arrangement or direction of fibers in wood, layers in stone, etc. Wood and stone split along the grain. *Distortions in the grain often cause the color to change when viewed from different angles. This lively effect is seen in mahogany, satinwood, primavera, and many other figured woods. Spirals and waviness in the grain can result in a mottle or stripe figure, as well as in fiddleback, blister, and bird's eye figures* (Morris Lieff).

cap, fāce, fäther; best, bē, tèrm; pin, five;
rock, gō, ôrder; oil, out; cup, pùt, rüle;
yü in use, yù in uric;
ng in bring; sh in rush; th in thin, ᴛʜ in then;
zh in seizure.
ə = *a* in about, *e* in taken, *i* in pencil, *o* in lemon, *u* in circus

261

b the little lines and other markings in wood, marble, etc.: *That mahogany table has a fine grain. In hardwoods an important part is played by the pores, which transport liquids in the sapwood, or living part, of the tree. The pores affect the grain and texture of the wood. Grain is defined as the pattern of light and dark resulting from differences in tissue density; texture refers to the ease with which the wood works under tools and the ultimate smoothness to which the wood can be finished* (Simon Williams). **c** the quality of a substance due to the size, character, or arrangement of its constituent particles: *a stone of coarse grain.* SYN: texture.

4 the smallest unit of weight in the avoirdupois system. One pound avoirdupois equals 7000 grains; one pound troy equals 5760 grains. [from Old French, from Latin *granum* grain, seed]
—**grainy,** *adj.* = granular.

gram, *n.* the base unit of mass in the metric system, approximately equal to the mass of one cubic centimeter of water at 4 degrees Celsius or 39.2 degrees Fahrenheit: *Since the kilogram (kg.) is somewhat large for general scientific work, it is customary to use the one-thousandth part, called the gram, as the unit of mass. The mass of 1 cubic centimeter of water at 4° C. is considered as being equal to 1 gram except in the most exact calculations* (Parks and Steinbach, *Systematic College Chemistry*). [from French *gramme,* from Late Latin *gramma* small weight, from Greek *gramma* letter]

-gram, *combining form.* **1** ___ grams, as in *kilogram =* a thousand grams. **2** ___ of a gram, as in *centigram* = one hundredth of a gram. [from Greek *gramma* small weight]

gram atom, *Chemistry.* one mole of atoms of an element.

gram-atomic mass or **gram-atomic weight,** *Chemistry.* the mass of an element in grams that equals numerically the element's atomic mass; the mass of one mole of atoms of an element.

gram calorie, *Physics, Chemistry.* = calorie: *One gram calorie is the quantity of heat which must be supplied to one gram of water to raise its temperature through one centigrade degree* (Sears and Zemansky, *University Physics*).

gram centimeter, *Physics.* a unit of energy equivalent to the work done in raising a mass of one gram vertically one centimeter. Compare **gram meter.**

gram equivalent, *Chemistry.* that quantity of an element or compound whose weight in grams is numerically equal to its equivalent weight: *The gram-equivalent weight of an element is that weight of it which combines with, liberates, or will in any wise do the same chemical work as 1.008 g. of hydrogen, or 8.0000 g. of oxygen* (Jones, *Inorganic Chemistry*).

gram formula mass, *Chemistry.* the sum of the gram-atomic masses of the atoms that make up a particular formula. The gram formula mass, calculated from the empirical formula, is used for ionic substances and network solids, since they are not molecular substances.

gramineous (grə min′ē əs), *adj. Botany.* **1** of, having to do with, or resembling grass: *a grayish-green gramineous covering.* **2** of or belonging to the grass family:

gramineous plants. [from Latin *gramineus,* from *graminem* grass]

graminivorous (gram′ə niv′ər əs), *adj. Zoology.* eating or feeding on grasses: *graminivorous birds.* [from Latin *graminem* grass + *vorare* devour]

gram meter, *Physics.* a unit of energy equivalent to the work done in raising a mass of one gram vertically one meter in height. Compare **gram centimeter.**

gram-molecular mass or **gram-molecular weight,** *Chemistry.* the mass of a substance in grams that equals numerically its molecular weight; the mass of one mole of a substance. The gram-molecular mass is the sum of the gram-atomic masses of the atoms that make up a particular molecule.

gram molecule, *Chemistry.* one mole of a substance.

gram-negative bacteria or **Gram-negative bacteria,** *Bacteriology.* bacteria that do not retain the crystal violet stain when treated with Gram's iodine solution. Gram-negative bacteria include those that cause typhoid fever and certain infections of the urinary tract. *There is almost a complete lack of gram-negative bacteria on the skin; they are most prevalent in the intestinal tract in man but do not survive on healthy skin surfaces* (John A. Ulrich). [named after Hans C. J. Gram, 1853–1938, Danish bacteriologist, who discovered a method of classifying bacteria by using stains]

gram-positive bacteria or **Gram-positive bacteria,** *Bacteriology.* bacteria that retain the crystal violet stain when treated with Gram's iodine solution and appear blue or violet. Gram-positive bacteria include those that cause scarlet fever and staphylococcal infections. [for etymology see **gram-negative bacteria**]

Gram's iodine solution or **Gram's solution,** a solution of iodine and potassium iodide in distilled water, used in Gram's method to stain bacteria. [for etymology see **gram-negative bacteria**]

Gram's method, a method of classifying bacteria by first staining them with crystal violet, treating them with Gram's iodine solution, and washing them with alcohol. Gram-positive bacteria retain the violet stain; gram-negative bacteria lose it. *Gram's method gives good results with many bacteria* (British Medical Journal), [for etymology see **gram-negative bacteria**]

Gram stain or **Gram's stain,** = Gram's method.

grana (grā′nə), *n.* plural of **granum.**

granite, *n. Geology.* a plutonic igneous rock consisting chiefly of quartz and feldspar, and usually also containing biotite. [from Italian *granito* grained, from *grano* grain, from Latin *granum*]
—**granitic** (grə nit′ik), *adj.* of or like granite: *Characteristic igneous rocks of the arc are andesitic volcanics and granitic intrusives* (Wilfred A. Elders).

granivore (gran′ə vôr), *n. Zoology.* an animal that feeds on grain or seeds. [from Latin *granum* grain + *vorare* devour]

granivorous (grə niv′ər əs), *adj. Zoology.* eating or feeding on grain or seeds: *granivorous quadrupeds.* [from Latin *granum* grain + *vorare* devour]

granodiorite (gran′ə dī′ə rīt), *n. Geology.* a plutonic igneous rock roughly intermediate between granite and diorite. [from Latin *granum* grain + English *diorite*]

granophyre (gran′ə fīr), *n. Geology.* a porphyritic igneous rock in which the groundmass is a microscopic network of quartz and potash feldspar. [from German *Granophyr,* from *Granit* granite + (*Por*)-*phyr* porphyry]

—**granophyric** (gran'ə fi'rik), *adj.* having to do with or composed of granophyre: *granophyric dikes.*

granular, *adj.* **1** consisting of or containing grains or granules: *A rock is said to have a granular texture if its mineral grains have grown to a size large enough to be seen and identified without the aid of lens or microscope. In different rocks, the average size of the grains may vary about 0.5 mm. to more than 1 cm. in diameter, but the common coarse granular rocks such as granite have grains averaging from about 3 mm. to 5 mm. in size* (Gilluly, *Principles of Geology*). SYN: grainy. **2** resembling grains or granules.

—**granularity,** *n.* granular condition or quality: *In others the chondrules are scarce and hard to distinguish from their surroundings, the entire chondrite being nearly uniform in granularity* (John A. Wood).

granulate, *v.* **1** to make into grains; form grains or granules. **2** to become granular. **3** to develop granulations. *Wounds granulate in healing.*

—**granulated,** *adj.* **1** formed into grains or granules: *granulated sugar.* **2** roughened on the surface: *granulated glass, granulated leather.*

granulation, *n.* **1** formation into grains or granules. **2** a roughening on the surface. **3** a granule on a roughened surface. **4** *Medicine.* the formation of small grainlike bodies or elevations on the surface of a wound during healing. **5** *Astronomy.* the small granular markings on the sun's photosphere.

—**granulative,** *adj.* characterized by granulation: *granulative growths.*

granule, *n.* **1** a small grain or grainlike part: *Ion-exchange resins are usually marketed in the form of coarse or spherical granules, about 1–2 millimetres in diameter, similar to greensand in softening columns* (K.S. Spiegler). **2** *Geology.* a rounded rock fragment larger than a grain of sand but smaller than a pebble. **3** *Biology.* a cellular particle, especially one that stains selectively. [from Late Latin *granulum,* diminutive of Latin *granum* grain]

granulite (gran'yə līt), *n. Geology.* a fine-grained gneiss or granite consisting chiefly of feldspar and quartz: ... *eruptive rocks, granite and granulite* (Baring-Gould, *The Deserts of Southern France*).

—**granulitic,** *adj.* of or having to do with granulite: *granulitic rock.*

granulocyte (gran'yə lō sīt), *n. Biology.* any of several types of white blood cells whose cytoplasm contains granules: *Lymphocytes in leukemia, he finds, live almost four times as long as another kind of white blood cell, the granulocytes* (Science News Letter).

—**granulocytic** (gran'yə lə sit'ik), *adj.* of or having to do with granulocytes: *granulocytic cells, granulocytic leukemia.*

granulosa cell (gran'yə lō'sə), *Anatomy.* any of the epithelial cells lining the Graafian follicle: *Among the most potent agents are progestogens such as the steroid hormone progesterone, which is secreted by the ovaries during the later part of the menstrual cycle, and it is known that progestogens can be synthesized by the granulosa cells that surround the egg* (R.G. Ed-

wards and R.E. Fowler). [from New Latin *granulosa* granular, from Late Latin *granulum* granule]

granum (grā'nəm), *n., pl.* **-na** (-nə). *Botany.* one of the disk-shaped stacks of thin platelets within the chloroplasts of plant cells. They contain chlorophyll. *The chlorophyll is arranged in orderly structures within the chloroplasts called grana and the grana in turn are separated from one another by a network of fibers or membranes. Within the grana the flat chlorophyll molecules are stacked in piles* (Scientific American). [from Latin *granum* grain]

grape sugar, = glucose.

graph, *n.* a diagram representing quantitative relationships, especially one in which a curve represents how one quantity depends on or changes with another.

—*v.* to make a graph of: *Graph the following functions* (Corcoran, *Algebra Two*). [short for *graphic formula*]

-graph, *combining form.* an instrument that traces, draws, or records, as in *seismograph* = *instrument that records earthquake data.* Compare **-graphy.** [from Greek *-graphos,* from *graphein* write]

graphic or **graphical,** *adj.* of or involving graphs and their use; showing or shown by graphs rather than calculations: *a graphic record of data.*

graphics, *n.* the science of calculating by means of graphs, diagrams, etc.

graphite (graf'īt), *n. Mineralogy.* a soft, black mineral with a metallic luster, consisting of carbon. Graphite is used for lead in pencils and for lubricating machinery. *When the atoms in a crystal are rearranged, the density of the crystal changes. The change of graphite to diamond at high pressure is an example of this process* (George R. Tilton). Also called **plumbago.** [from German *Graphit,* from Greek *graphein* write]

—**graphitic** (grə fit'ik), *adj.* having to do with or of the nature of graphite: *Extreme compression may drive off virtually all the volatile hydrocarbons, leaving a residue of nearly pure carbon* (graphite). *Pennsylvanian rocks in Rhode Island, which have been much more strongly compressed than those in Pennsylvania, contain graphitic coal that can hardly be burned* (Moore, *Introduction to Historical Geology*).

graph theory, *Mathematics.* the study of sets of points joined by lines: *While the solution to this problem is of no practical use to cartographers, the century-long endeavor to solve it generated a whole new branch of mathematics called graph theory that has been of crucial importance to the development of such fields as operations research and computer science* (Lynn A. Steen).

-graphy, *combining form.* **1** the process of tracing, drawing, or recording, as in *cardiography* = *the process of tracing or recording the electrical impulses of the heart.* **2** a descriptive science, as in *crystallography* =

cap, fāce, fäther; best, bē, tèrm; pin, five; rock, gō, ôrder; oil, out; cup, pùt, rüle, yü in use, yu̇ in uric; ng in bring; sh in rush; th in thin, ᴛʜ in then; zh in seizure. ə = *a* in about, *e* in taken, *i* in pencil, *o* in lemon, *u* in circus

263

the descriptive science of crystals. Compare -**graph.** [from Greek *-graphia*, from *graphein* write]

grass, *n. Botany.* any plant of a family (Gramineae) that includes wheat, corn, sugar cane, and bamboo. Grasses have jointed stems and long, narrow leaves, and usually a small, dry, one-seeded fruit. *In numbers of genera and species, grasses are one of the largest families of flowering plants* (Lee M. Talbot).

grassland, *n. Ecology.* a region of vegetation consisting mainly of grasses or grasslike plants.
ASSOCIATED TERMS: Grassland is one of the main kinds of natural vegetation, others being *forest, desert, tundra,* and *taiga.* Grasslands include *meadows, pampas, prairies, savannas, steppes,* and *velds.* See also **biome.**

graticule (grat′ə kyül), *n.* **1** *Geography.* a network of lines or curves representing meridians and parallels on a map or chart: *Official French maps show their graticules according to the centesimal system* (New Scientist).
2 *Optics.* a measuring scale at the focal plane of a telescope, microscope, etc.: *Sometimes, as in the thousands of hair-line graticules which were used in gun sights, the metal image is left on its glass support. By using substantially grainless photographic emulsions the lines on such graticules were quite free from graininess and could, when necessary, be made so thin that as many as 1,200 separate lines could be produced in each millimetre width* (D.A. Spencer).
[from French *graticule,* from Medieval Latin *graticula, craticula* grillwork, from Latin *cratis* grill]

grating, *n. Physics.* = diffraction grating.

graupel (grou′pəl), *n. Meteorology.* pellets of granular snow: *When riming continues to such an extent that the original crystal is unrecognizable or nearly so, the resulting snow is called graupel. Clouds must be thick or updrafts fairly strong before the crystals remain in them long enough to grow into graupel. Since graupel particles have the largest mass and the highest speed of fall of any of the particles in a snowfall, they can be a stage in the formation of rain. Much rain is melted graupel* (Charles and Nancy Knight). [from German *Graupel,* from *Graupe* small hailstone, groat]

gravel, *n. Geology.* pebbles and rock fragments coarser than sand: *Gravel is an unconsolidated deposit composed chiefly of rounded pebbles. The pebbles may be of any kind of rock or mineral and of all sizes. Most conglomerates, especially those deposited by streams, have much sand and other fine material filling the spaces between the pebbles* (Gilluly, *Principles of Geology*). [from Old French *gravele,* diminutive of *grave* sand, seashore]
—**gravelly,** *adj.* of or like gravel: *The sediments nearest the shore are gravelly; those farther out are sandy; those still farther out are muddy* (Scientific American).

gravid, *adj. Zoology.* **1** containing or carrying a fetus: *a gravid uterus, a gravid female.* **2** of or having to do with females carrying young or eggs: *the gravid period, a gravid condition.* [from Latin *gravidus* pregnant, from *gravis* heavy]

gravimeter (grə vim′ə tər), *n.* **1** a weighing device for measuring variations in the magnitude of a gravitational field: *The variations in the Earth's gravity field were measured with a ship-borne gravimeter. ... A gra-*vimeter is a very sensitive spring balance (New Scientist). **2** a device for measuring the specific gravity of a substance: *A hydrometer is a gravimeter for liquids.*

gravimetric (grav′ə met′rik), *adj.* **1** of or having to do with a gravimeter.
2 *Chemistry.* of or having to do with measurement by weight: *gravimetric analysis of a compound.*
3 *Geography.* of or having to do with the use of gravity measurements to calculate the size, shape, etc., of the earth or other planet: *The fundamental idea of the gravimetric method of mapping, and of the present world-wide gravity measuring program, is that the undulations of the geoid and its tilt at every place can be computed from the observed gravity anomalies* (Scientific American). —**grav′imet′-rically,** *adv.*

gravimetry (grə vim′ə trē), *n.* **1** the measurement of weight; determination of specific gravities by means of a gravimeter. **2** the measurement of gravitational force: *Gravimetry is concerned with the pull of gravity* (K.B. Fenton).

gravitate, *v. Physics.* to move or tend to move toward a body by gravitational force: *The sun, moon, and all the planets ... reciprocally gravitate one toward another* (Richard Bentley).

gravitation, *n. Physics.* **1** the attraction of one body for another, or the effective force of one body moving toward another; the tendency of every particle of matter toward every other particle, of which the fall of bodies to the earth or the tendency of the sun, moon, stars, and other bodies of the universe to move toward one another are instances: *Newton's law of universal gravitation states that every object in the universe attracts every other object with a force that is directly proportional to the product of their masses and inversely proportional to the square of the distance between them* (Obourn, *Investigating the World of Science*).
2 the act or process of gravitating or of causing objects to gravitate: *The simplest method of irrigation, gravitation, by which rain and river water is guided into the fields, is widespread both in the United States and in China. ... In the United States, machines prepare the land gradients: 10 years of manual labor was required to bring the Changho River to irrigate the chronically drought-ridden Linhsien region of China* (N.Y. Times).
► Though *gravity* and *gravitation* are sometimes used interchangeably, in strict usage the two words have distinct meanings. *Gravitation* means the attraction of every particle of matter for every other throughout the universe, while *gravity* means the effect of gravitation at the surface of the earth, the moon, or a planet, often thought of as combined with centrifugal force caused by the rotation of the earth, moon, or planet.
—**gravitational,** *adj.* of or having to do with gravitation: *Gravitational collapse is the drawing together of matter under the influence of the gravitational forces among atoms or larger particles. It is the way stars and planets are supposed to form from interstellar gas* (Science News).
—**gravitationally,** *adv.* by gravitation: *One of the most vexing problems for Newton was to find a rigorous proof that a sphere acts gravitationally as if all its mass were concentrated at its center* (Scientific American).

gravitational constant, the constant of gravitation, expressing the acceleration per unit of time produced by the gravitating attraction of a unit mass at the unit of distance.

gravitational field, *Physics.* any field in space in which gravitation exerts a force on an object: *It is the Sun's enormous gravitational field that holds all the planets, near and far, circling in their orbits* (Arthur C. Clarke).

gravitational mass, *Physics.* mass measured by gravitation; the mass of a body as determined by the extent to which it responds to gravitation. Gravitational mass is proportional to inertial mass (and is expressed in the same units), since the acceleration of a falling body increases in proportion to its gravitational mass and decreases in proportion to its inertial mass.

gravitational radiation, *Physics.* the propagation or radiation of gravitational waves.

gravitational wave, *Physics.* an energy-carrying wave traveling through space at the speed of light and exerting gravitational forces on any mass in its path. Gravitational waves were predicted by the general theory of relativity but have not yet been observed. *It has been postulated that ... gravitational waves may originate in the infall of matter into a singularity or singularities at the galactic centre. A black hole of as much as 10^8 solar masses could exist there* (New Scientist). Also called **gravity wave.**

gravitative, *adj. Physics.* of, having to do with, or produced by gravitation or gravity.

graviton (grav′ə ton), *n. Nuclear Physics.* a hypothetical particle constituting a unit of gravitational force: *The particle physicist now postulates that gravitation, like the other forces, is mediated by the exchange of an elementary particle—the "graviton". From observation (1) he supposes that the graviton shares with the photon the property of zero mass, but from (2) it cannot have spin 1 as the photon does* (Michael Duff). [from gravit(y) + -on]

gravity, *n. Physics.* **1a** the natural force that causes objects to move or tend to move toward the center of the earth, moon, or any planet as a result of gravitation: *Gravity is measured by the acceleration produced in a freely falling body, which is about 32 feet (981 cm.) per second per second; that is, a body falling under the influence of gravity alone increases its velocity by 32 feet per second during each second of its fall. Starting from rest, it falls 16 feet in the first second, 48 feet in the next, 80 feet in the third, and so on. This acceleration is commonly denoted by g. The weight of a body is its mass multiplied by g* (Duncan, Astronomy). **b** = acceleration of gravity: *A body starting from rest would, in the absence of air resistance, be travelling at 20 m.p.h. after one second of fall, 40 m.p.h. after two seconds, 60 m.p.h. after three seconds, and so on. This value of "g" is almost constant over the whole Earth: other planets, as we shall see later, have different gravities, most of them considerably less than that of our world. On some very tiny "planetoids", gravity is so small that it would take minutes for a falling body to descend a couple of yards* (Clarke, The Exploration of Space).

2 the force of attraction that makes objects move or tend to move toward each other; gravitation: *What is gravity? To begin with, let's simply call it a force that*

every mass in the universe exerts on every other mass in the universe (New Yorker).

3 heaviness; weight. See **center of gravity.**
[from Latin *gravitatem,* from *gravis* heavy]
▶ See the note under **gravitation.**

gravity feed, *Physics.* the process of supplying by gravity; the movement of matter by the force of gravity: *The two fuel systems in general use are the pressure feed and the gravity feed. ... In the gravity-feed system, the storage tank must be higher than the carburetor so the fuel will flow to the carburetor* (Edward F. Obert).
—gravity-feed, *v.* to supply by the action of gravity: *The ... reservoir makes use of upland topography to capture a volume of water and gravity-feed it to the towns* (New Scientist).
—gravity-fed, *adj.* supplied by the action of gravity: *a gravity-fed spring or well.*

gravity wave, *Physics.* **1** a wave between air and water or between layers of air, generated by the force of gravity: *The remarkably strong wind fields in the ionosphere ... were probably due largely to the effects of gravity waves, generated in the lower atmosphere, that propagate upwards* (William W. Kellogg). **2** = gravitational wave.

gravity wind, = katabatic wind.

gray, *n. Physics.* an SI unit for measuring absorbed doses of ionizing radiation, equal to 1 joule per kilogram. It is intended to replace the older unit *rad.* [named after Louis Harold *Gray,* 20th-century British radiobiologist]

graybody or (*especially British*) **greybody,** *n. Physics.* a body that radiates at every wavelength an amount of energy bearing a constant ratio to the amount radiated by a blackbody at the same temperature: *The observers go as far as to state that the exact blackbody intensity curve ... fits the observations better than the "greybody" approximation at long wavelengths* (Nature).

graywacke (grā′wak *or* grā′wak ə), *n. Geology.* a sedimentary rock, a coarse sandstone or fine conglomerate consisting of rounded fragments of rocks and minerals bound in a fine-grained matrix. [from German *Grauwacke,* from *grau* gray + *Wacke* wacke]

great, *adj. Biology.* distinguished by its large size from other plants or animals of the same genus: *the great lobelia, the great skua.*

great circle, 1 *Mathematics.* any circle on the surface of a sphere having its plane passing through the center of the sphere: *A great circle is a circle cut on the surface of a sphere by a plane passing through the center of the sphere, as for instance, the equator and the meridians of longitude on the earth* (Scientific American). See the picture at **sphere.**
2 *Geography.* Also spelled **Great Circle.** an arc of such a circle; the line of shortest distance between two points on the earth's surface: *A factor which favours*

---

cap, fāce, fäther; best, bē, tèrm; pin, fīve;
rock, gō, ôrder; oil, out; cup, pùt, rüle,
yü in use, *yủ* in uric;
ng in bring; *sh* in rush; *th* in thin, ᵀH in then;
zh in seizure.

ə = *a* in about, *e* in taken, *i* in pencil, *o* in lemon, *u* in circus

the planned routing of ships in the [North Atlantic] area ... is the fact that the Great Circle (shortest) route takes a ship well to the north, while the fine weather is on the whole in the south, so that considerable variations in the route taken are possible (New Scientist).

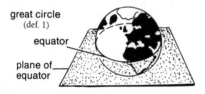

great circle
(def. 1)
equator
plane of equator

greater vestibular gland, = Bartholin's gland.

greatest common divisor, *Mathematics.* the largest number or quantity that will divide two or more others without a remainder: *The greatest common divisor of 24 and 30 is 6. Abbreviation:* G.C.D. Also called **highest common factor.**

greatest lower bound, *Mathematics.* the largest of the lower bounds of a set; the largest number less than or equal to the lowest number in a set. *Abbreviation:* g.l.b. Compare **least upper bound.**

green algae, *Biology.* a division (Chlorophyta) of mostly grass-green algae living mainly in fresh water. Green algae have definite nuclei and chloroplasts. *Although most green algae are aquatic, they are found in a wide variety of habitats, including the surface of snow, on tree trunks, in the soil, and in symbiotic relationships with lichens, protozoa, and hydra* (Raven, Evert, and Curtis, *Biology of Plants*).

green flash, *Astronomy.* a flash of vivid green light seen when the sun is just setting below the horizon. It is caused by the atmospheric dispersion of the sunlight. *The green flash can be seen much better in binoculars, and can be expected if the upper rim of the setting sun has a greenish tinge* (John B. Irwin).

green gland, *Zoology.* one of a pair of excretory organs located in the head of certain crustaceans, as the lobster: *The excretory organs [of the crayfish] are a pair of rather large bodies, the green glands ... situated in the ventral part of the head anterior to the esophagus. Each green gland consists of a glandular portion which is green in color, a thin-walled dilatation, the bladder, and a duct opening to the exterior through an excretory pore at the end of the papilla on the basal segment of the antenna* (Hegner and Stiles, *College Zoology*).

greenhouse effect, *Meteorology.* **1** the absorption and retention of the sun's radiation in the earth's atmosphere, resulting in an increase in the temperature of the earth's surface. The greenhouse effect is due to the accumulation of carbon dioxide and water vapor in the atmosphere, which allows shortwave solar radiation to reach the earth's surface but prevents reradiated longer infrared wavelengths from leaving the earth's atmosphere, thus trapping heat. *The carbon dioxide reduces the amount of heat energy lost by the earth to outer space. The phenomenon has been called the "greenhouse effect," although the analogy is inexact because a real greenhouse achieves its results less from the fact that the glass blocks reradiation in the infrared than from the fact that it cuts down the convective transfer of heat* (S. Fred Singer). **2** any similar retention of the sun's radiation by a solar collector, heavenly body, etc.: *As the rays hit the collectors they reflect off the sides and change wavelengths so the rays cannot pass back through the plastic. This "greenhouse effect" captures almost all the energy of the rays* (N.Y. Times). *How can we account for Titan's warmth? One reasonable explanation seems to be that a greenhouse effect operates there, trapping the sun's heat. This process begins when visible light passes through Titan's atmosphere and warms the surface of the satellite* (Gregory Benford).

greenockite (grē′nə kīt), *n. Mineralogy.* a mineral consisting of cadmium sulfide, found in yellow incrustations or hexagonal crystals: *Cadmium occurs in the rare mineral greenockite, CdS, and in small amounts in most zinc ores, especially zinc blend, ZnS* (Parks and Steinbach, *Systematic College Chemistry*). [from Lord *Greenock* (Charles Cathcart), 1783–1859, who discovered it + *-ite*]

greensand, *n. Geology.* a variety of sandstone containing glauconite, which gives it a greenish color: *Some minerals, such as natural greensand, are so well suited for exchange that their exchange ability can be used on an industrial scale, for instance to make hard water soft* (K.S. Spiegler).

Greenwich Time or **Greenwich Mean Time** (gren′ich), the standard time used in Great Britain and, formerly, the basis for setting standard time elsewhere. It was determined by setting noon as the time at which the sun is directly overhead at the meridian which passes through Greenwich, England. Since 1972, it has been replaced by *Universal Time,* which is based on atomic clocks. Compare **Coordinated Universal Time, UT1.**

gregarine (greg′ə rin), *n. Biology.* any of a group of protozoans that are parasitic in the intestines of insects, worms, crustaceans, and other invertebrates: *The gregarines are found mainly in invertebrata* (Claus, *Zoology*). [from New Latin *Gregarina* the genus name, from Latin *gregarius* gregarious, from *gregem* flock]

Gregorian calendar, *Astronomy.* the civil calendar now generally used, in which the ordinary year has 365 days, and leap year has 366 days. It corrected the Julian calendar by eliminating as leap years those centenary years not divisible by 400. *The average length of the year in the Gregorian calendar is* 365^{d}5^{h}49^{m}12^s *of mean solar time, which exceeds the tropical year by only 26 seconds. In its close accordance with the year of the seasons the present calendar is satisfactory* (Baker, *Astronomy*). [named after Pope *Gregory* XIII, who introduced the calendar in 1582]

greisen (grī′zən), *n. Geology.* an altered granitic rock with a coarse granular texture, consisting chiefly of quartz and mica. [from German *Greisen*]

grid, *n.* **1** *Geography.* **a** an arrangement of vertical and horizontal lines to determine the coordinates of given points or to locate points on a map, chart, etc., for which the coordinates are known. **b** the numbered squares drawn on maps and used for map references: *The first real attempt to map the distribution of biological records was based on the 10 km national grid and resulted in the "Atlas of the British Flora" in 1962* (New Scientist).

2 *Electronics.* an electrode consisting of parallel wires or a screen which controls the flow of electrons from cathode to anode in an electron tube.
[shortened from *gridiron*]

grit, *n. Geology.* **1** very fine bits of gravel or sand in sharply angular particles. **2** a coarse sandstone: *A sandstone may become a grit, or a pebbly conglomerate sandstone* (Geikie, *Textbook of Geology*).

groundmass, *n. Geology.* the crystalline, granular, or glassy base of a porphyry, in which the more prominent crystals (phenocrysts) are embedded. SYN: matrix.

ground meristem, *Botany.* a group of cells in an embryonic vascular plant that gives rise to the cortex, pericycle, and pith: *A short distance (usually a few millimeters) below apical meristem we recognize three fairly distinct primary meristematic tissues, namely protoderm, ground meristem, and procambium. These three tissues are derived directly from apical meristem* (Weier, *Botany*).

ground moraine, *Geology.* rock material deposited beneath a glacier: *Specific glacial deposits are called moraines, and the till, because much of it was held in the bottom of the ice, is called the ground moraine* (Finch and Trewartha, *Elements of Geography*).

ground state, *Physics.* that state of lowest energy of any bound physical system. For a hydrogen atom, the electron in this state is at an average distance of $.5 \times 10^{-10}$ meters from the proton. *The reactions usually encountered in organic chemistry involve the collision or interaction of molecules that lie in the ground state* (Ralph Daniels). *Because we had to pull, because we had to do work on the atom, it follows that the excited state has more energy than the ground state. But it can happen that the electron jumps back from the excited state to the ground state, and if it so jumps then the energy of the atom will diminish from the energy appropriate to the excited state to the energy appropriate to the ground state* (Hermann Bondi).

ground substance, *Biology.* **1** the material in the connective tissues, cartilage, and bone that fills the space between cells; the intercellular substance of a tissue. SYN: matrix. **2** hyaloplasm.

ground truth, geological data obtained by direct examination of features on the ground in order to verify information gathered by satellite or other airborne means: *The atmosphere may prove to be so dense that it will throw the instruments off; indeed, scientists involved in these experiments, some of whom are skeptical about the accuracy of data from an orbiting spacecraft, will be scattered around the world to get what they call "ground truth"* (Henry S.F. Cooper, Jr.).

ground water, *Geology.* water that flows or seeps downward and saturates soil or rock, supplying springs and wells. The upper level of this saturated zone is called the water table. *Georgia, Odum explained, is a good ecological microcosm for the rest of the Southeast of the United States. The finest ground water in the world lies under this state, and ground water is indispensable for cooling power plants, farm irrigation and various industrial and municipal purposes. There will probably be a scramble for ground water in the years to come* (Science News).

group, *n.* **1** *Chemistry.* **a** a combination of atoms acting as a unit in reactions, such as the hydroxyl group (-OH) and the carboxyl group ($-CO_2H$): *The -OH and -H groups of the sugar molecule may be arranged in different positions in the ring without changing the relative numbers of carbon, hydrogen, and oxygen in the formula* (Weier, *Botany*). ► See the note under **radical.** **b** in the periodic table, a vertical column that includes elements having similar properties. Group I-A contains the alkali metals; group VII-A contains the halogens. *Group O consists of six gases, helium, neon, argon, krypton, xenon, and radon. The book, like other chemistry textbooks, says that the Group O gases are inert: they form no chemical compounds with other elements* (Raymond Hull). Contrasted with **period.**
2 *Mathematics.* a set of elements that has one binary operation and includes the properties of closure, identity element, inverse, and of being associative. A *finite group* is one having a finite number of elements, as distinguished from an *infinite group* (such as the set of all integers). *In mathematics, a group is a collection of quantities called "elements" and a law of combination, such that the result of combining any two elements by that law is equal to another element of the group* (McAllister H. Hull, Jr.). *The algebraic structure called a group is a basic topic of study in modern algebra. Finite groups are of particular interest, and the major problem is to classify the ultimate building blocks that are called finite simple groups* (Irving Kaplansky).
3 *Biology.* a number of organisms belonging or classed together. The term *group* is used indefinitely for related species, genera, etc., when it is not possible, necessary, or desirable to specify the relationship or taxonomy.
4 *Geology.* a unit of stratified rocks, consisting of two or more formations usually deposited during one period.
5 = blood group: *The plan is to make use of both Group A blood and the Universal Donor Group O blood. Group O blood ... can be transfused without first testing the patient's blood* (Science News Letter).

group theory, a branch of mathematics dealing with the properties of groups. Group theory is used in physics, chemistry, and other sciences to arrange various units (such as particles or molecules) into mathematical groups. *The building blocks of group theory, analogous to the elementary particles of matter or the prime factors of integers, are called simple groups. Just as any integer can be uniquely expressed as a product of prime numbers, so any group can be uniquely represented as a "composition" of simple groups* (Scientific American). *In physics group theory underlies the classification of elementary particles. In chemistry it pro-*

cap, fāce, fäther; best, bē, tèrm; pin, fīve;
rock, gō, ôrder; oil, out; cup, pùt, rüle,
yü in use, *yù* in uric;
ng in bring; *sh* in rush; *th* in thin, ᴛʜ in then;
zh in seizure.
ə = *a* in about, *e* in taken, *i* in pencil, *o* in lemon, *u* in circus

vides a system for defining the structure of crystals. In the life sciences group theory contributed key elements to the search for the structure of the DNA molecule (Lynn A. Steen).

growth, *n. Biology.* an irreversible increase in size and weight of an organism through division, enlargement, and differentiation of cells: *Growth in higher plants is restricted largely to specific regions known as meristems. Apical meristems are found in the tips of roots and stems and are responsible for growth in length. Some plants also contain an active lateral meristem region, the cambium, located between the xylem and phloem. This is responsible for the growth in diameter of roots and stems. The growth regions contain undifferentiated cells which undergo active mitotic cell division and elongation. As a result of differentiation, the various tissues and organs are developed* (Biology Regents Syllabus).

growth curve, *Biology.* a graphic representation of the growth of a living thing or a population in relation to time: *From daily counts of the multiplication of cells in the growing colonies we can plot a typical growth curve, which turns out to be exactly like the standard growth curve of colonies of bacteria* (T.T. Puck).

growth factor, *Biology.* any substance which affects the growth of an organism, especially a vitamin.

growth hormone, *Biochemistry.* **1** a hormone secreted by the anterior lobe of the pituitary gland which regulates the growth of the body: *If excess growth hormone is produced in childhood then gigantism results, and conversely inadequate secretion of growth hormone is responsible for dwarfing* (Science News). Also called **somatotropin.**

2 Also called **growth regulator** or **growth substance.** any of various natural or synthetic substances that regulate the growth of plants: *The auxins, the first type of plant growth hormone to be discovered, were definitely established as hormones by the work of F.W. Went in 1928* (Greulach and Adams, *Plants*).
ASSOCIATED TERMS (def. 2): The three most important growth hormones are the *auxins,* the *cytokinins,* and the *gibberellins.*

growth retardant, *Biochemistry.* any synthetic chemical substance that interferes with or inhibits the normal promotion of growth by hormones in an organism: *When these growth retardants (for example, CCC, AMO-1618) are applied to plants, cell division in the subapical region practically stops, whereas the leaf-initiating activity of the apical meristem continues* (Weier, *Botany*).

growth ring, 1 *Botany.* = annual ring: *Scientists can tell much about a tree's history and age by looking at the growth rings. Each year the tree forms a new layer of wood just inside the bark* (Richard Preston, Jr.). **2** *Zoology.* one of the ringlike markings on certain fish scales, dinosaur teeth, etc., that show one year's growth: *The most accurate way to tell a rattlesnake's age is not by the number of rattles on his tail, but by the "growth rings" of his bones* (Science News Letter). SYN:annulus.

GTP, *abbrev.* guanosine triphosphate.

guanidine (gwan′ə dēn′), *n. Biochemistry.* a strongly alkaline crystalline compound formed by the oxidation of guanine. Guanidine is found in urine as a normal product of protein metabolism. It is commonly used in organic synthesis. *Formula:* $NHC(NH_2)_2$

guanine (gwä′nēn′), *n. Biochemistry.* a substance present in nucleic acid in cells. It is one of the purine bases of DNA and RNA. *The difference involves the cell's incorporation of guanine, an essential component of growth and heredity controlling chromosomes* (Science News Letter). *Guanine was first found in 1844, in the excreta of birds—forty years before it was recognized as a nucleic-acid constituent. Crystallized guanine imparts the shine to the scales of fish and reptiles* (New Yorker). *Formula:* $C_5H_5N_5O$ *Abbreviation:* G See the picture at **DNA.** [from *guano,* a source of this substance + *-ine*]

guano (gwä′nō), *n. Biology.* **1** phosphatic waste matter from sea birds or bats, often used as a fertilizer and a source of guanine. **2** any comparable fertilizer, such as one made from fish: *Fertilizers in the preparation of which fish are used, including Menhaden guano, crude and ground, guano made from fish skins, and from fish heads and bones* (Fisheries Exhibition Catalogue). [from Spanish *guano,* from Quechua *huanu* dung]

guanosine (gwä′nə sēn), *n. Biochemistry.* a white, crystalline, odorless substance, present in the pancreas and in certain plants. It is a nucleoside consisting of guanine and ribose. *Formula:* $C_{10}H_{13}N_5O_5$ [a blend of *guanine* + (*rib)ose*]

guanosine monophosphate (mon′ə fos′fāt), *Biochemistry.* a compound of guanosine and one phosphate group; guanylic acid. *Abbreviation:* GMP Compare **cyclic GMP.**

guanosine triphosphate (trī fos′fāt), *Biochemistry.* a compound of guanosine and three phosphate groups. It is a nucleotide important in protein synthesis and physiological research. *The mechanism whereby the hormone-receptor complex activates adenylate cyclase, however, has only recently been understood. It appears that another nucleotide, guanosine triphosphate (GTP), plays two critical roles in this process. GTP converts the hormone receptor to a state which, in the presence of hormone, can activate adenylate cyclase. As a separate action, GTP converts adenylate cyclase to a form where it can be activated by the hormone-receptor complex* (Ira D. Goldfine). Compare **adenosine triphosphate.**

guanylic acid (gwä nil′ik), *Biochemistry.* a nucleotide composed of guanine, ribose, and phosphoric acid, formed by the body during protein synthesis. *Formula:* $C_{10}H_{14}N_5O_8P$ Also called **guanosine monophosphate.** Compare **adenylic acid.**

guard cell, *Botany.* one of the two specialized kidney-shaped cells of the plant epidermis that control the size of the stomates by expanding and contracting: *In addition to the ordinary cells, pairs of specialized, thicker-walled, usually banana-shaped, chlorophyll-bearing cells called guard cells occur in the epidermis, and border the thousands of tiny openings, called stomata, through which diffuse gases needed and formed in various life processes* (Harbaugh and Goodrich, *Fundamentals of Biology*).

guard hair, *Zoology.* the long, coarse hair in the fur of a mammal: *The coat on many mammals is differentiated into a dense fine underfur for body insulation, and* a lesser number of heavier long guard hairs that protect against wear (Storer, *General Zoology*).

guest, *n. Zoology.* an animal that lives in the nest, shell, etc., of another animal: *In the case of the adult worm, the happiest cures are readily affected by the expulsion of the 'guest', but as regards the larvae the case is very different* (Cobbold, *Entozoa*). SYN: inquiline, commensal. Compare **host.**

gular (gyü′lər), *adj. Anatomy.* of, having to do with, or situated on the throat: *Like other members of the order, such as pelicans, boobies have a gular pouch at the base of the beak and upper throat* (C.W. Chadwick). [from Latin *gula* throat]

gulch, *n. Geology.* a deep, narrow ravine with steep sides, especially one marking the course of a stream or torrent: *The term coulee is generally applied throughout the northern tier of States to any steep-sided gulch or water channel and at times even to a stream valley of considerable length* (M.R. Campbell).

gulf, *n. Geography.* an arm of an ocean or sea extending into the land. "The distinction between *gulf* and *bay* is not always clearly marked, but in general a *bay* is wider in proportion to its amount of recession than a *gulf;* the latter term is applied to long land-locked portions of sea opening through a strait, which are never called *bays*" (Oxford English Dictionary).
▶ Gulfs often take their names from the adjoining land, as the Gulf of Mexico, the Gulf of Suez (an arm of the Red Sea).

Gulf Stream, *Oceanography.* a strong, warm ocean current that flows out of the Gulf of Mexico, north along the coast of the United States and Newfoundland, and northeast across the Atlantic toward the British Isles: *"There is a river in the sea—the Gulf Stream," said Maury, the great American oceanographer of a century ago. This statement was regarded by some as a typical American exaggeration, but modern studies essentially confirm his picture of a well-defined stream of warm water, with relatively sharp boundaries, coursing with the speed of a river across thousands of miles of ocean* (Gilluly, *Principles of Geology*).

gum, *n.* **1** *Botany.* **a** (in strict usage) any of various soft, sticky substances that are given off by certain trees and plants and that harden in the air and dissolve in water. Natural gums are polysaccharides with colloidal properties; they differ from natural resins both in their chemical composition and their solubility. Gums are chiefly used to make mucilage and as emulsifiers in foods and drugs. **b** (loosely) any similar secretion, such as resin, gum resin, etc.
2 *Anatomy.* Often, **gums,** *pl.* the firm flesh around the teeth: *Calculus is the hard substance that clings to the tooth and irritates the gums, eventually causing gum disease* (Lou Joseph).

gum ammoniac, *Botany.* a natural mixture of gum and resin used in medicine and as a cement for porcelain.

gum arabic, *Botany.* a gum obtained from acacia trees, used in making candy, medicine, and mucilage.

gumbo, *n. Geology.* a fine-grained soil that contains much clay and becomes sticky and plastic when wet. It is found in the southern, central, and western parts of the United States.

gummite, *n. Mineralogy.* a reddish-yellow mixture of hydrous oxides of uranium, thorium, and lead, usually formed by the alteration of uraninite.

gum resin, *Botany.* a natural mixture of gum, resin, and oil obtained from trees. Asafetida and myrrh are gum resins.

Gunn, *adj. Electronics.* of or based upon the Gunn effect: *a Gunn diode, a Gunn oscillator. A Gunn device generates oscillations as a result of the curious manner in which the conduction electrons behave in n-type gallium arsenide* (New Scientist).

Gunn effect, *Electronics.* the development of microwave oscillations in a small block of a semiconductor, such as gallium arsenide, when a constant direct-current voltage above a critical level is applied to opposite faces of the semiconductor material: *Gunn effect is the name given to the behavior of the electrons in certain semiconductor materials that makes them electrical oscillators rather than producers of steady current* (Science News). *If this voltage is above a certain critical value, microwave oscillations will start up as a result of the Gunn effect, and so-called "Gunn domains" will travel across the strip from one side to the other* (New Scientist). [named after John B. *Gunn,* born 1928, British physicist, who discovered the effect in 1963]

Günz (gints *or* gʏnts), *n. Geology.* the first glaciation stage of the Pleistocene in Europe, situated in the Alps: *It indicated four cold spells, which they promptly identified with the four glaciations that every geological and archaeological student has long been taught to recite—Günz, Mindel, Riss and Würm* (Sunday Times). [named after the *Günz* river, which flows through this area of glaciation]

gust, *n. Meteorology.* a sudden, violent rush of wind. A gust usually lasts less than 20 seconds and is briefer than a *squall.*

gustation (gə stā′shən), *n. Physiology.* the sense of taste; ability to taste: *Gustation ... is the perception of dissolved materials by taste buds* (Storer, *General Zoology*). Compare **olfaction.**
—gustatory (gus′tə tôr′ē), *adj.* of the sense of taste; having to do with tasting: *The taste buds comprise the gustatory region of the tongue.*
[from Latin *gustus* taste]

gutta (gut′ə), *n., pl.* **guttae** (gut′ē). **1** *Pharmacology.* a drop, as of liquid medicine. **2** *Zoology.* a drop-shaped marking, as on an insect's wing. [from Latin *gutta* drop]

gutta-percha (gut′ə pėr′chə), *n. Botany.* the latex of certain tropical trees of the Malay Peninsula related to the sapodilla. Gutta-percha is obtained by the same process as crude rubber, and is used to make surgical bandages and castings, tooth fillings, etc. [from Malay *getah percha*]

guttation (gu tā′shən), *n. Botany.* the oozing of droplets of water from plants, usually from the leaf: *Guttation occurs usually when environmental conditions are*

---

cap, fāce, fäther; best, bē, tėrm; pin, fīve;
rock, gō, ôrder; oil, out; cup, pùt, rüle;
yü in use, *yu̇* in uric;
ng in bring; *sh* in rush; *th* in thin, ᴛʜ in then;
zh in seizure.
ə = *a* in about, *e* in taken, *i* in pencil, *o* in lemon, *u* in circus

such as to check transpiration, particularly during cool nights following hot days when the air is very humid. ... The exudation of water droplets is a direct result of root pressure within a plant (Fuller and Tippo, College Botany). [from Latin gutta drop]

guttifer (gut'ə fər), n. Botany. any plant that yields gum or resin: the class of guttifers. [from New Latin guttifera, from Latin gutta a drop + ferre to bear]
—**guttiferous** (gu tif'ər əs), adj. yielding gum or resin: a guttiferous shrub.

guyot (gē'yō), n. Geology. a seamount with a platform top: These table mounts, called guyots, look like decapitated volcanoes, and that is exactly what they are. Their tops must have been cut off by some knife-edged erosive process. In the realm of the sea there is only one way that such a clean job of decapitation could have been accomplished: by the action of surf. In other words, their tops once stood above the water as islands and were sliced off by breakers cutting into them. This conclusion is confirmed by the fact that their flat tops are fringed by a narrow border of rounded cobblestones—the remains of their ancient cobble beaches (Scientific American). [named after Arnold Guyot, 1807-1884, Swiss-American geologist]

gymnocarpous (jim'nō kär'pəs), adj. Botany. having a naked fruit, as where the fruit is without hair or where the floral envelope does not adhere to the outer integument: gymnocarpous species of lichens. [from Greek gymnos naked + karpos fruit]

gymnogynous (jim noj'ə nəs), adj. Botany. having a naked ovary, as where the seeds are without a pericarp: a gymnogynous grain. [from Greek gymnos naked + English -gynous]

gymnosperm (jim'nə spėrm'), n. Botany. any of a group of seed plants having the seeds exposed, not enclosed in an ovary or fruit. The pine, fir, and spruce are gymnosperms; they bear seeds on the surface of cone scales instead of in pods. The gymnosperms constitute one of two groups (the other being the angiosperms) into which the seed-bearing plants (spermatophytes) can be divided. The most familiar gymnosperms (naked-seed plants) are the conifers, or evergreens (Scientific American). [from Greek gymnos naked + sperma seed]

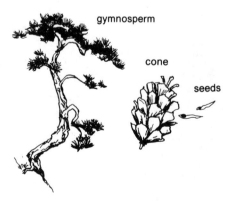

gymnosperm

cone

seeds

—**gymnospermous** (jim'nə spėr'məs), adj. belonging to the gymnosperms; having the seeds exposed:

gymnospermous conifers. **gynandromorph** (ji nan'drə môrf), n. Zoology. an organism in which one part or side of the body has male characteristics and the other female characteristics: Insect mosaics are frequently sexual mosaics. These "gynandromorphs" occur in all imaginable forms, from those which possess functional organs of both sexes to those which are a mosaic of male and female tissues and possess functional sexual organs only if the gonads and related structures are genetically all female or all male (Aloha Hannah-Alava). [from Greek gynē woman + andros man + morphē form]
—**gynandromorphic**, adj. of or like a gynandromorph: gynandromorphic moths.

gynandrous (ji nan'drəs), adj. Botany. having the stamens and pistils united in one column, as in orchids: gynandrous stamens, gynandrous species. [from Greek gynē woman + andros man]

gynobase (jī'nə bās or jin'ə bās), n. Botany. an elongation of the receptacle of a flower, bearing the gynoecium: ... an enlarged, tumid fleshy disk (the gynobase) (John Lindley). Compare **gynophore**.

gynoecium (ji nē'sē əm or ji nē'sē əm), n., pl. **-cia** (-sē ə). Botany. the female organ of a flower; the carpels or pistils: The structure of the gynoecium depends upon the number and arrangement of carpels comprising it. In the pea flower, there is a single carpel forming the gynoecium. ... In the flower of the Christmas rose there are five separate and distinct carpels comprising the gynoecium, and each carpel has its own ovary, style, and stigma (Weier, Botany). Compare **androecium**. [from Greek gynē woman + oikion house]
—**gynoecious** (jī nē'shəs or ji nē'shəs), adj. of, like, or containing gynoecia; having female organs: a gynoecious flower.

gynophore (jī'nə fôr or jin'ə fôr), n. Botany. 1 the stalk or stem of a pistil. SYN: stipe. 2 an elongation of the receptacle of a flower, bearing the gynoecium at its apex. Compare **gynobase**.

-gynous, combining form. Botany. having a certain number or type of female organs, as in monogynous = having a single pistil, gymnogynous = having a naked ovary. [from Greek -gynos, from gynē woman]

gypseous (jip'sē əs), adj. Mineralogy. like or containing gypsum: gypseous limestones.

gypsiferous (jip sif'ər əs), adj. Mineralogy. containing or yielding gypsum: ... the gypsiferous and salt-bearing formation of the Upper Silurian (Roderick I. Murchison).

gypsum (jip'səm), n. Mineralogy. a mineral used to make fertilizer and plaster of Paris; hydrous calcium sulfate. Alabaster is one form of gypsum. When heated to a temperature of from 110° C. to 120° C., gypsum ... loses more than half of its water of crystallization and is converted to the white powder, plaster of Paris, which is capable of absorbing water and setting to a comparatively hard mass. It is after conversion to plaster of Paris that gypsum has its main uses, but the uncooked or uncalcined mineral is employed in making mineral white ('terra alba'), as a filler in paper and cotton, for dusting underground passages in collieries, in paints, and is being increasingly employed in Portland cement to retard and control the time of setting (Jones, Minerals in Industry). Formula: $CaSO_4.2H_2O$ [from Latin gypsum, from Greek gypsos]

gyral (jī′rəl), *adj. Anatomy.* having to do with a gyrus: *a gyral convolution or fold.*

gyrate (jī′rāt), *adj.* **1** *Botany.* = circinate: *gyrate foliation, a gyrate cyme.* **2** *Zoology.* having convolutions: *gyrate bands.* [from Latin *gyrus* circle, from Greek *gyros*]

gyre (jir) or **gyral** (jī′rəl), *n. Oceanography.* a very large circular movement of oceanic surface water: *The circulation in this schematic ocean then divides into several gyres (rings) corresponding to the wind belts—a counterclockwise gyre in the subpolar region, a clockwise circulation in the subtropical belt above the equator, a narrow gyre on each side of the equator and a counterclockwise gyre in the subtropical region below the equator. In each gyre there is a strong, persistent current on the western side (due, as we shall see, to the rotation of the earth) and a compensating drift in the central and eastern portion* (Scientific American). *In general, the circulation of surface water in each ocean is arranged into two enormous eddies or gyrals, those of the Northern Hemisphere revolving clockwise, and those of the Southern Hemisphere counter-clockwise about comparatively motionless central areas* (White and Renner, *Human Geography*). [from Latin *gyrus* circle, from Greek *gyros*]

gyro-, *combining form.* circle; spiral, as in *gyroscope.* [from Greek *gyros*]

gyrocompass (jī′rō kum′pəs), *n.* a compass using a motor-driven gyroscope instead of a magnetic needle. It points to true north instead of the North Magnetic Pole and is not affected by nearby objects of iron or steel. *The gyrocompass, which depends for its action on the directive effect of the Earth's rotation upon the axis of a rapidly rotating, delicately balanced, massive wheel, gives directly the true azimuth of the ship's course* (Duncan, *Astronomy*).

gyrofrequency, *n. Physics.* the natural frequency of rotation of an electron or other charged particle in a constant magnetic field: *The particles in the shock wave proceed to move in tight spirals around and along the magnetic lines of force. Associated with this corkscrew movement are two important quantities: the number of revolutions per second (gyrofrequency) and the radius of the spiral* (Scientific American).

gyromagnetic, *adj. Physics.* of or having to do with the magnetic properties of a rotating charge particle. The **gyromagnetic ratio** is the ratio of an elementary particle's magnetic moment to its angular momentum. *A fundamental physical constant, the gyromagnetic ratio of the proton, has been redetermined very accurately by scientists at the National Bureau of Standards* (Science News Letter).

gyroscope, *n.* an instrument consisting of a heavy, rotating wheel or disk mounted so that its axis can turn freely in one or more directions. A spinning gyroscope tends to resist any change in the direction of its axis, no matter which way its base is turned. Gyroscopes are used to keep ships, airplanes, and guided missiles steady. *A spinning gyroscope maintains its attitude no matter how its frame is turned in space* (William J. Cromie). *The combining of a gyroscope with a TV camera in borehole photography enables geologists to measure accurately the dip and direction of a strike* (Robert M. Hamilton).

gyroscope

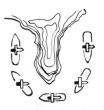

—**gyroscopic,** *adj.* of or having to do with a gyroscope.

gyrose (jī′rōs), *adj. Botany.* folded and waved; curved alternately backward and forward: *gyrose ribs of the hymenium.*

gyrus (jī′rəs), *n., pl.* **-ri** (-rī). *Anatomy.* a convolution, especially of the brain: *Localized somatic pain has been diminished by removal of the proper segment of the postcentral gyrus* (Wendell J.S. Krieg). The **gyrus of Broca** is a convolution of the left hemisphere of the brain associated with the motor aspects of speech. [from Latin *gyrus* circle]

---

cap, fāce, fäther; best, bē, tėrm; pin, five;
rock, gō, ôrder; oil, out; cup, pùt, rüle,
yü in use, *yu̇* in uric;
ng in bring; *sh* in rush; *th* in thin, ᴛʜ in then;
zh in seizure.
ə = *a* in about, *e* in taken, *i* in pencil, *o* in lemon, *u* in circus

271

H

h, *symbol.* 1 hecto-. 2 Planck's constant.

h. or H., *abbrev.* or *symbol.* 1 *Mineralogy.* hard *or* hardness. 2 hour *or* hours.

H, *symbol.* 1 hydrogen. 2 henry *or* henries. 3 magnetic field strength.

^{1}H, *symbol.* protium.

^{2}H, *symbol.* deuterium.

^{3}H, *symbol.* tritium.

H$^+$, *symbol.* hydrogen ion.

Ha, *symbol.* hahnium.

ha., *abbrev.* hectare *or* hectares.

Haber process or Haber-Bosch process (hä′bər bôsh′), *Chemistry.* a method of synthesizing ammonia by the direct combination of nitrogen and hydrogen under high pressure in the presence of a catalyst, usually iron. [named after Fritz *Haber,* 1868–1934, and Carl *Bosch,* 1874–1940, German chemists who devised the method]

habit, *n.* 1 *Botany.* the general appearance of a plant, whether climbing, erect, prostrate, etc.
2 *Mineralogy.* the characteristic form of the crystals of a particular mineral.

habitat, *n. Ecology.* the place where an organism naturally lives or grows: *A plant community can be defined as all the plants growing in a given habitat. Such a habitat may be a very small area ... or it may cover thousands of square miles of territory within a given set of climatic conditions* (Emerson, *Basic Botany*). *Man ... has been a part of many animals' habitats for millions of years, preying upon such animals as naturally as lions or wolves* (Rensberger, *The Cult of the Wild*). [from Latin *habitat* it inhabits]

hachures (hə shùrz′), *n.pl. Geography.* short lines used as shading to represent the slopes of mountains and hills on maps: *A map of land relief is sometimes ... represented by fine lines known as hachures* (White and Renner, *Human Geography*). [from French *hachure* hatch]

hadal (hā′dəl), *adj. Oceanography,* of or inhabiting the deepest parts of the ocean, at depths exceeding 6000 meters: *Colorless, totally blind, and often gigantic in size compared with their relatives from lesser depths, nearly 200 "hadal" animal species have now been dragged up from the deepest trenches in the ocean floor* (New Scientist). [from *Hades,* the lower world in Greek mythology] Compare abyssal, bathyal, littoral.

Hadley cell, *Meteorology.* an atmospheric circulation pattern involving air movement only to the north or south toward the equator with vertical uplift or subsidence: *A Hadley cell would lie at cloud heights on Venus because most of the incident solar energy is absorbed there ... On the earth a Hadley cell lies just above the surface, the place where most of the solar energy is absorbed* (Gerald Schubert and Curt Covey). [named after George *Hadley,* 1685–1768, English me-

teorologist who proposed in 1735 the existence of such cells as an explanation for the trade winds]

hadron (had′ron), *n. Nuclear Physics.* any of a class of strongly interacting elementary particles that include the baryons, antibaryons, and mesons: *Hadrons show evidence of an internal structure that is believed to be made up of the famous quarks. It is now thought that there are six varieties (the technical term is "flavors") of quark that go into the various hadronic structures* (Science News). [from Greek *hadros* thick + English *-on*]

▶ Particles that are subject to the strong interaction are called *hadrons,* while particles that do not are called *leptons.* Leptons behave like particles without structure or size, and interact with electromagnetic and gravitational fields only through the weak interaction; hadrons have a measurable size and respond to the strong interaction in addition to the weak interaction.

—hadronic, *adj.* of or having to do with a hadron or hadrons: *Hadronic atoms are hydrogen-like systems that consist of a strongly interacting particle (hadron) bound in the Coulomb field and in orbit around any ordinary nucleus.* (Clyde E. Wiegand).

haemoglobin (hē′mə glō′bən), *n.* = hemoglobin.

hafnium (haf′nē əm), *n. Chemistry.* a silvery metallic element which occurs in zirconium ores. Hafnium absorbs neutrons better than most metals and is therefore used for making the neutron-absorbing control rods of nuclear reactors. *Symbol:* Hf; *atomic number* 72; *atomic weight* 178.49; *melting point* 2150°C; *boiling point* 5400°C; *oxidation state* 4. [from *Hafnia,* Medieval Latin name for Copenhagen, where the element was discovered]

hahnium (hä′nē əm), *n. Chemistry.* an artificial radioactive element with up to six isotopes, first produced in 1970: *Hahnium was identified from the properties of its radioactive decay* (New Scientist). *Symbol:* Ha; *atomic number* 105. Also called nielsbohrium. [named after Otto *Hahn,* 1879–1968, German chemist]
▶ See the note under element.

hail, *n. Meteorology.* small, roundish pieces of ice or frozen vapor coming down from the clouds during some thunderstorms, usually in the area between latitudes 30° and 60°. *Hailstones are the balls or pellets of hail that fall in a hailstorm. Small bits of ice form in the cold air high above the earth. These particles fall into air ... are caught by the upward currents ... where they freeze again ... until the balls of ice that are formed are too heavy to be held up by air currents. These pieces of ice then fall to the earth as hail* (Beauchamp, et al. *Everyday Problems in Science*).

hair, *n.* 1 *Zoology.* a fine, threadlike, pigmented structure that grows from the skin of mammals, chiefly to provide warmth and protection: *The hairs that distinguish mammals from all other animals, unlike* feathers, are not modifications of horny scales, but are new structural elements of the skin. (Hegner and Stiles, College Zoology).

ASSOCIATED TERMS: Below the skin each hair rests in a cavity called a *follicle* which arises from a *papilla* or vascular process at the *root* of the hair. Above the skin, the *shaft* of a hair consists of an outer layer of cells called a *cuticle,* an inner layer or cortex which contains the hair pigment, and a core of loosely packed cells, the *medulla.* The *tactile hairs* on the face of many animals are surrounded by nerves that respond to touch.

b any similar outgrowth from the body of insects and other invertebrates: *The body of a bee is densely covered by hairs having short lateral barbs where pollen grains lodge easily.* (Storer, *General Zoology*). **c** a mass of such growths: *a coat of hair.*

2 *Botany.* a fine, threadlike growth from the outer layer of plants: *Epidermal appendages, such as hairs, may occur on young stems* (Weier, *Botany*).

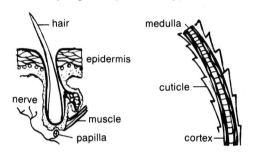

hair cell, *Anatomy.* a cell in the organ of Corti in the inner ear, having very fine hairlike processes: *They discovered the action of the sensitive hair cells in the vestibular labyrinth of the ear, and detected a relation to certain uncontrolled movements of the eye* (London Times).

half-cell, *n. Chemistry.* a single electrode immersed in an electrolytic solution, usually when an ionization process is in progress between the electrode and electrolyte.

half-life, *n., pl.* **half-lives. 1** *Physics.* the length of time it takes for half of any number of unstable nuclei or subnuclear particles to disintegrate. The half-life of a particular radionuclide is always the same, independent of temperature, chemical combination, or any other condition. It is used to measure radioactivity and as the principal characteristic in distinguishing one nuclide from another. *For every thousand atoms of carbon-14 present initially, after one half-life there will be 500 left, after a further half-life 250, and so on.* (New Scientist).

2 *Biology.* **a** the length of time it takes for half of a given substance deposited in a living organism to be metabolized or eliminated: *As one would expect, an oral dose generally provides a high level of the drug in the blood right after the dose is taken. The drug immediately begins to disappear from the blood ... The rate is usually expressed in terms of the half-life of the drug in the circulation, that is, the time it takes for the concentration of the drug to decrease to half its peak level.* (Scientific American). **b** the length of time it takes for half the species in a given group to become extinct: *Evolutionists have recently learned that groups of species also have what might be called a half-life. ... But scientists can predict only the number of species that*

will become extinct, not the identity of the individual extinct species (David M. Raup).

3 *Chemistry.* the time required for a given chemical reaction to affect half of the reactants present. It is independent of initial concentration, but dependent on temperature.

half-line, *n. Geometry.* the part of a line of infinite length that lies to one side of a given point on the line. A half-line differs from a ray by not containing its boundary point.

half-plane, *n. Geometry.* the part of a plane that lies to one side of a given line in the plane. A dihedral angle is composed of two half-planes with one edge in common.

half-reaction, *n. Chemistry.* either of the two parts of a redox reaction: *A separate equation showing gain or loss of electrons (electronic equation) can be written for each half-reaction* (Chemistry Regents Syllabus).

half-space, *n. Geometry.* the part of space that lies to one side of a given plane. The xy-coordinate plane divides space into two half-spaces.

halide (hal′īd *or* hā′līd), *n. Chemistry.* any compound of a halogen with another element or radical: *The halides include compounds of chlorine, fluorine, bromine and iodine* (Frederick H. Pough). Also called **halogenide.** [from *hal(ogen)* + *-ide*]

Hall effect, *Physics.* the development of a voltage across a current-carrying conductor placed in a magnetic field. In the Hall effect, the voltage is perpendicular to both the direction of the current and the direction of the magnetic field. The electrical properties of different materials can be measured by the Hall effect. *The Hall effect in metals and semiconductors consists basically of the appearance of a transverse voltage when the current carriers flowing along a sample are deflected by a magnetic field* (New Scientist). [named after Edwin H. Hall, 1855–1938, American physicist, who discovered this effect]

hallux (hal′əks), *n., pl.* **halluces** (hal′yə sēz). *Anatomy.* **1** the innermost digit on the hind foot of certain vertebrates. In humans it is commonly called the big toe. **2** the hind toe of birds. [from New Latin *hallux,* from Latin *allus, hallus* big toe]

halo, *n. Meteorology.* a ring of light appearing around the sun, moon, a star, or other luminous celestial body, caused by the refraction of light through ice crystals suspended in the air: *All clouds of the cirrus family are composed of ice crystals. The sun or moon shining through these ice-crystal clouds produces a halo* (Byers, *General Meteorology*). [from Latin *halos,* from Greek *halōs* disk]

halo-, *combining form.* **1** salt; salinity, as in *halophilic = salt-loving.* **2** of or containing halogen, as in *halocarbon.* [from Greek *halo-,* from *hals, halos* salt]

cap, fāce, fäther; best, bē, tèrm; pin, five;
rock, gō, ôrder; oil, out; cup, pùt, rüle,
yü in use, yù in uric;
ng in bring; sh in rush; th in thin, ᴛʜ in then;
zh in seizure.
ə = a in about, e in taken, i in pencil, o in lemon, u in circus

halobacterium (hal′ō bak tir′ē əm), *n., pl.* **-teria** (-tir′ē-ə). any of a group of rod-shaped bacteria that thrive in areas with very high salt concentrations: *Halobacteria are salt-loving cells that inhabit stagnant puddles and salt flats at the edge of tropical seas. ... They turn water orange ... and turn sunlight into chemical energy on their "purple membranes"* (Science News).

halobiotic (hal′ō bī ot′ik), *adj. Ecology.* living in a strongly salty habitat, as the sea or seashore: *halobiotic organisms.*
—**halobiont** (hal′ō bī′ont), *n.* a halobiotic organism.

halocarbon, *n. Chemistry.* any of a group of compounds of carbon and one or more halogens, used as refrigerants, propellant gases, etc.: *Concern over the tenuous, but vital, ozone layer which surrounds the Earth has centred around the possible effects ... from the halocarbons used in packaged aerosol canisters* (New Scientist). Compare **fluorocarbon.**

halocline (hal′ə klīn), *n. Oceanography.* a sharp discontinuity in the salinity of seawater, usually at a depth of about 60 meters: *There has been a marked decrease in the oxygen content of deeper waters, the report* [*on water pollution*] *states, and "if this development continues, the whole water mass below the halocline will probably turn into a lifeless 'oceanic desert' such as found in the Black Sea"* (London *Times*). [from *halo-* + Greek *klinein* incline]

halogen (hal′ə jən), *n. Chemistry.* any of the group of elements that includes fluorine, chlorine, bromine, iodine, and astatine. The halogens are electronegative, nonmetallic elements and combine directly with most metals to form salts. *The fact that each halogen has seven electrons in its valence shell accounts for the similarities in chemical properties that this family of elements displays* (Offner, *Fundamentals of Chemistry*). [from *halo-* salt + *-gen*]

halogenate (hal′ə jə nāt *or* ha loj′ə nāt), *v. Chemistry.* to combine with a halogen; add a halogen to (an organic compound).
—**halogenation,** *n.* the act or process of halogenating: *Some of the interhalogens provide convenient solvents for halogenation reactions, particularly in view of their partial ionization* (Phillips and Williams, *Inorganic Chemistry*).

halogenide (hal′ə jə nīd *or* hə loj′ə nīd), *n. Chemistry.* = halide.

halogenous (hə loj′ə nəs), *adj. Chemistry.* of, resembling, or involving a halogen: *Chlorobenzene and ethylene dibromide are halogenous compounds.*

haloid (hal′oid *or* hā′loid), *adj. Chemistry.* **1** of or like a salt: *a haloid substance.* **2** formed from a halogen: *haloid ethers.*

halophile (hal′ə fīl), *n. Biology.* an organism that grows or lives in a marsh or other region rich in salt: *Developing at the stage of evaporation when salinity has reached twenty to thirty percent, the color was caused by an explosive proliferation of pink bacteria known as halophiles* (American Scholar).
—**halophilic** (hal′ə fīl′ik) *or* **halophilous** (hə lof′ə ləs), *adj. Biology.* salt-loving; growing or living in a region rich in salt: *About a fifth of all the blue-green algae live in saline environments. A few species are truly halo-philic, or salt-tolerant; such species are found in southern France and in California in brines with a salt concentration as high as 27 percent* (Patrick Echlin).

halophyte (hal′ə fīt), *n. Botany.* a plant that grows naturally in salty soil, as on the seashore. Asparagus, saltwort, and mangrove are halophytes.
—**halophytic** (hal′ə fit′ik), *adj.* growing or adapted to grow in salty soil or in saline conditions: *Halophytic plants ... are those which thrive in salt marshes, on saline flats near the seacoast, and on the alkali flats of the interior* (Youngken, *Pharmaceutical Botany*).

haltere (hal′tər *or* hôl′tər), *n., pl.* **halteres** (hal′tirz *or* hôl′tirz). *Zoology.* either of two club-shaped organs in dipterous insects: *The hind wings of flies, for example, are reduced to knobbed stalks, the halteres, which are used as balancing organs, acting somewhat on the principle of a gyroscope* (Harbaugh and Goodrich, *Fundamentals of Biology*). [from Greek *haltēres* dumbbell-like weight for balance in leaping, from *hallesthai* to leap]

hamate (hā′māt), *Anatomy.* —*adj.* of or designating a bone of the human carpus or wrist, closest to the ulna in the distal row. —*n.* the hamate bone. Also called **unciform.** [from Latin *hamatus* hook-shaped, from *hamus* hook]

Hamiltonian, *n.,* or **Hamiltonian function,** *Physics.* a function defining the energy of a dynamic system in terms of the generalized coordinates and momenta of the system. It is equal to the total energy of the system minus the Lagrangian. *There are ... appendices—on the sign of the interaction energy in molecular orbital calculations ... and on the effective hamiltonian for the interaction of a molecule and an electromagnetic field* (David Buckingham). [named after William Rowan Hamilton, 1805–1865, Irish mathematician]

hammer, *n. Anatomy.* = malleus.

hamulus (ham′yə ləs), *n., pl.* **-li** (-lī). *Zoology.* a small hook or hooklike part, as in certain bones or in feathers. [from Latin *hamulus,* diminutive of *hamus* hook]

handedness, *n. Chemistry, Physics.* the tendency toward the right or left that distinguishes an asymmetric object from its mirror image but not from a rotated object: *The handedness of an electron depends upon the direction of its spin relative to its direction of motion: an electron with its spin oriented in the direction of motion is right-handed (just as an ordinary screw moves forward when turned to the right, or clockwise), whereas an electron with its spin against the direction of motion is left-handed* (Stuart B. Palmer). SYN: chirality.

hanging wall, *Geology.* the side of rock that lies above a fault, vein, lode, etc.: *The rock surface bounding the lower side of an inclined fault plane is known as the footwall and that above as the hanging wall* (Matthews, *Geology Made Simple*).

haploid (hap′loid), *Genetics.* —*adj.* having the gametic number of sets of chromosomes or half as many as in the somatic cells of the species; having the reduced number of chromosomes, as distinguished from the diploid number: *Sexual reproduction involves the union of haploid nuclei to give a diploid state, and meiosis is the reverse of this process. The life cycle of a sexually reproducing organism is, therefore, an* alternation of haploid and diploid states (McElroy, *Biology and Man*).

—*n.* a haploid organism or cell: *Haploids are especially useful to plant geneticists, for most mutations, or gene changes, are immediately and readily detectable in haploids* (William C. Steere).
[from Greek *haplous* single]
ASSOCIATED TERMS: see **diploid.**
—**haploidy** (hap′loi dē), *n.* the condition of being haploid.

haplont, *n. Genetics.* an organism that is haploid throughout its life except as a zygote: *Success in producing a haplont—a plant with half the usual number of chromosomes—eluded researchers until ... Indian scientists grew a haplont morning-glory* (Science News). Contrasted with **diplont.**

haplophase (hap′lō fāz′), *n. Genetics.* the haploid phase of an organism: *The point at which the haploid nuclei are reunited in conjugate association determines the duration of the haplophase* (Fischer and Holton, *The Biology and Control of Smut Fungi*). Contrasted with **diplophase.**

haplosis (hap lō′sis), *n. Genetics.* the reduction of the number of chromosomes to half during meiosis. Compare **diplosis.**

hapten, *n. Immunology.* a part of an antigen that reacts with a specific antibody but cannot stimulate the formation of that antibody without being united to a protein substance. [from Greek *haptein* to fasten]

hapteron (hap′tər on), *n., pl.* **-tera** (-tər ə). *Biology.* an organ of attachment on the stems of various aquatic plants or marine algae: *It creeps over the rocks to which it adheres by hairs or by exogenous projections known as haptera, which secrete a cement from their discoid tips* (Sculthorpe, *The Biology of Aquatic Vascular Plants*). [from New Latin *hapteron,* from Greek *haptein* to fasten]

haptic, *adj. Physiology.* of or having to do with the sense of touch. SYN: tactile.

haptoglobin, *n. Biochemistry.* an alpha globulin that is a constituent of normal blood serum and occurs in several antigenic types.

hard, *adj.* **1** *Chemistry.* **a** containing mineral salts, such as calcium carbonate, that interfere with the lathering action of soap: *hard water.* **b** that resists decomposition or biodegradation, as certain pesticides and detergents: *The foaming problem arose largely because the most commonly used "surface-active" ingredient of the hard detergents, alkylbenzene sulfonate (ABS), cannot readily be degraded by bacteria* (Scientific American).
2 *Physics.* **a** having high-energy particles that readily penetrate all types of materials; having great penetrating power: *hard X rays, the hard component of cosmic radiation.* **b** capable of being magnetized only in strong magnetic fields. Steel, iron, nickel, and cobalt are magnetically hard substances.
—**hardness,** *n.***1** the quality or condition of being hard (in any of the senses above). **2** *Physics.* resistance of any material to deformation, abrasion, etc. **3** *Mineralogy.* the comparative capacity of one mineral to scratch, or be scratched by, another mineral. The degree of hardness of a mineral is expressed with reference to the arbitrary Mohs scale, ranging from 1 (talc) to 10 (diamond).

hard palate, *Anatomy.* the bony front part of the roof of the mouth. See the picture at **uvula.**

hardpan, *n. Geography, Geology.* compact, impervious, often clayey subsoil through which roots cannot grow: *Several Brazilian attempts to establish large-scale agricultural programs in the Amazon region by clearing virgin rain forest were quickly wiped out by hardpans* (Ronald R. Dagon).

hardwood, *n. Botany.* **1** an angiospermous tree characterized by broad, flat leaves, as distinguished from a coniferous or needle-leaved tree. The oak, cherry, maple, willow, and ebony are hardwoods; the pine and fir are softwoods.
2 the wood of such a tree.
▶ The terms *hardwood* and *softwood,* when applied to wood, are somewhat misleading in that some hardwoods are comparatively soft and some softwoods are comparatively hard. For example, the wood of such hardwoods as the poplar and the tulip tree are actually soft, while that of a softwood such as the longleafed pine is very hard. However, the distinction is appropriate for hard-wooded broad-leaved trees such as the oak, hickory, osage orange, and persimmon, as contrasted with such soft-wooded conifers as the cypress, white pine, and western red cedar.

Hardy-Weinberg, *adj. Genetics.* of, having to do with, or derived from the **Hardy-Weinberg law** or **principle,** which states that the relative frequency of genes and genotypes in a large population remains constant as long as breeding is random and the environment stable from one generation to the next: *The Hardy-Weinberg principle is a theoretical one, but it shows that variation—once arisen—will persist in a population if it is found in the breeding individuals. It assumes, however, random breeding, equal survival of all genotypes to reproductive age, a large population, no additional mutations, and absence of migration of individuals in and out of the population* (McElroy, *Biology and Man*). [named after G.H. *Hardy,* English mathematician, and W. *Weinberg,* German biologist, who demonstrated this principle independently in 1908]

harmattan (här′mə tan′), *n. Meteorology.* a very dry, dust-laden wind originating in the Sahara and blowing across West Africa toward the sea, especially from November to March: *When the harmattan, a strong northeast wind, blew its dry blasts for nearly a month, tall grasses turned brown, wood shriveled, and lips and skin roughened and cracked* (National Geographic). [from a West African language, perhaps ultimately from Arabic *harām* evil thing]

harmonic, *Physics.* —*adj.* of or designating any of the frequencies of a wave or vibration that are integral multiples of the fundamental frequency: *We say that*

cap, fāce, fäther; best, bē, tèrm; pin, fīve;
rock, gō, ôrder; oil, out; cup, pùt, rüle,
yü in use, *yu* in uric;
ng in bring; *sh* in rush; *th* in thin, ŦH in then;
zh in seizure.
ə = *a* in about, *e* in taken, *i* in pencil, *o* in lemon, *u* in circus

275

harmonic analysis

[notes] *having different harmonic content differ in quality or timber* (Michels, *Foundations of Physics*).
—n. a harmonic frequency: *Most natural sounds consist of a fundamental tone, which determines the pitch, and harmonics whose number, order and relative strengths determine the quality or timbre of the sound. A harmonic one octave in frequency above the fundamental is a second harmonic; two octaves above, a fourth harmonic, etc.* (Roy F. Allison). ▶ In acoustics, harmonics are often called *overtones* or *partials*. [from Latin *harmonicus,* from Greek *harmonikos* harmonic, musical, from *harmonia* harmony]

harmonic analysis, *Mathematics.* **1** = Fourier analysis. **2** the study of Fourier series.

harmonic mean, *Mathematics.* the reciprocal of the arithmetic mean of the reciprocals of a series of values: *The harmonic mean ... is the average to be used in averaging speeds and so on* (Newman, *The World of Mathematics*).

harmonic motion, *Physics.* a vibration, as of a violin string, in which the acceleration is directly proportional to but directed away from the displacement of the body from the mean position.

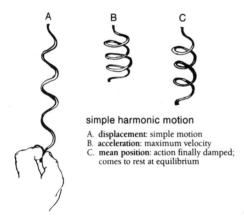

simple harmonic motion

A. **displacement**: simple motion
B. **acceleration**: maximum velocity
C. **mean position**: action finally damped; comes to rest at equilibrium

harmonic progression, *Mathematics.* the reciprocals of an arithmetic progression. Since 2, 4, 6, 8 is an arithmetic progression, 1/2, 1/4, 1/6, 1/8 is a harmonic progression.

harmonic series, *Mathematics.* an infinite series of terms in harmonic progression.

harmotome (här′mə tōm), *n. Mineralogy.* a zeolite containing barium and potassium, occurring in brittle, vitreous, usually white form. [from Greek *harmos* joint + *-tomos* a cutting]

haustellum (hô stel′əm), *n., pl.* **-la** (-lə). *Zoology.* the sucking organ or proboscis of many insects. [from New Latin *haustellum,* diminutive of *haustrum* bucket, from *haurire* to draw (water)]
—haustellate (hôs′tə lit), *adj. Zoology.* having a haustellum; fitted for sucking.

haustorial (hô stôr′ē əl), *adj. Biology.* of or having to do with a haustorium or haustoria.

haustorium (hô stôr′ē əm), *n., pl.* **haustoria** (hô stôr′ē-ə). **1** *Botany.* one of the small roots of parasitic plants that attach themselves to and penetrate the tissue of

the host and absorb nourishment: *Its parasitic tendency is so pronounced that one branch of a dodder plant may insert its haustoria into other branches of itself* (Emerson, *Basic Botany*). Commonly called **sucker.** **2** *Biology.* a specialized branch or organ of the mycelium of a fungus, by which it attaches itself to its host: *... specialized absorbing hyphae called haustoria and anchoring hyphae called rhizoids that penetrate the substrate are characteristic of many species* (Greulach and Adams, *Plants*).
[from New Latin *haustorium,* from Latin *haustor* a drainer, from *haurire* to draw (water)]

haustral (hôs′trəl), *adj. Anatomy.* of or having to do with the haustra: *haustral pouches or cells.*

haustrum (hôs′trəm), *n., pl.* **-tra** (-trə). *Anatomy.* each of the small folds, resembling sacs, in the terminal division of the colon: *Compared with the smooth small intestine, the large is sacculated along its length, the sacculations or haustra bulging between three equidistant longitudinal bands* (Lockhardt, *The Anatomy of the Human Body*). [from New Latin *haustrum,* from Latin, bucket, from *haurire* to draw (water)]

Haversian canal (hə ver′shən), *Anatomy, Physiology.* a tiny, cylindrical hollow in a bone, through which blood vessels, lymphatics, connective tissues, and nerves run: *Nearly all Haversian canals of adult dogs contain unmyelinated nerves 0.8 to 7 µm in diameter* (Science). [named after Clopton *Havers,* 1650–1702, English anatomist]

Haversian system, *Anatomy, Physiology.* the layers of tissue surrounding the Haversian canals, which together with the canals make up the structure of compact bone. Often called **osteon.**

Hayflick limit, *Microbiology.* the natural limit to the number of divisions which cells undergo in a culture: *Each species has a basic inbuilt biological clock, in the form of an appropriate Hayflick limit to the number of divisions which can take place in somatic cells* (Macfarlane Burnet). [named after Leonard *Hayflick,* born 1928, American microbiologist who discovered this limit]

haze, *n. Meteorology.* a fine mist, smoke, or dust dispersed through a part of the atmosphere. Haze makes distant objects indistinct.

H.C.F., h.c.f., or **hcf,** *abbrev.* highest common factor.

HCl, *abbrev.* hydrochloric acid.

HDL, *abbrev. Biochemistry.* high density lipoprotein (a lipoprotein that contains more protein than lipid and carries excess cholesterol out of the body tissues to the liver for excretion): *HDL is a "good" form of blood fat believed to protect, to a degree, against coronary heart disease* (Science News). Contrasted with **LDL.**

He, *symbol.* helium.

head, *n.* **1** *Anatomy.* **a** the uppermost part of the human body containing the brain, eyes, nose, ears, and mouth. **b** the corresponding part of an animal's body. See the picture at **thorax. c** the rounded end of a bone. **d** the part of a muscle closest to the origin. **2** *Astronomy.* the coma and nucleus of a comet: *The coma of a comet is the roundish nebulous region. The head region includes not only the coma, sheaths, and envelopes but also the nucleus if present. The heads have been observed to be 30,000 to 1,000,000 miles in diameter* (Bernhard, *Handbook of Heavens*).

3 *Botany.* a cluster of flowers in which the flowers or florets do not have individual stems, but grow close together from the main stem; flower head: *Botanists speak of such a flower cluster as a "head." This head is typical of the large group of flowering plants called the composites, including the daisies, sunflowers, ragweed* (Science News Letter). Compare **capitulum.**

4 *Geology, Geography.* **a** the upper limit, as of a slope or body of water; source of a river or stream: *the head of a brook.* **b** a body of water at a height above a particular level. **c** the height of such a body of water or the force of its fall: *The energy in water power comes both from the weight of the water and from the head upon it* (Hope Holway). **d** a standard pressure of water used in determining the permeability of different rocks and rock formations. **e** a projecting part of a rock, mass of ice, sandbank, etc. **f** a projecting point of land; headland: *Duncansby Head, Scotland.*

5 *Physics.* the pressure of water, gas, or other confined substance: *a full head of steam.*

headland, *n. Geography.* a point of high land jutting out into water. SYN: cape, promontory.

headwater, *n. Geography.* Often, **headwaters,** *pl.* the source or upper part of a river or stream.

heart, *n.* **1** *Anatomy.* the hollow, muscular organ that pumps the blood throughout the body of a vertebrate by contracting and relaxing: *The heart is a pair of pumps operating in parallel. The right ventricle pumps blood to the lungs; the left ventricle pumps oxygenated blood into the aorta, from which it is distributed to the rest of the body* (Victor A. McKusick).

2 *Zoology.* any organ functionally or structurally similar to the vertebrate heart: *Most of the Crustacea, including lobsters, crabs and shrimps, have a heart ... The heart propels a fluid which has cells floating in it, and these corpuscles can be seen moving round the body* (H. Munro Fox). See the picture at **thorax.**

3 *Botany.* the solid central part of a tree. See also **heartwood.**

the human heart
(def. 1)

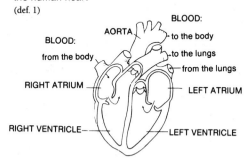

heartbeat, *n. Physiology.* a pulsation of the heart, including one complete contraction and expansion: *Most individuals with severe heart disease ... will show a relatively normal heartbeat when the EKG is made during a resting position* (Oglesby Paul).

heartwood, *n. Botany.* the central, nonliving wood in the trunk of a tree, harder and more solid than the newer sapwood that surrounds it: *In most species of trees ... the accumulation of a variety of substances, such as*

resins, oils, gums, and tannins ... commonly causes the heartwood to become deeply colored and harder than the sapwood (Greulach and Adams, *Plants*). See the picture at **wood.**

heat, *n. Physics.* **1** a form of energy that consists of the motion of the molecules of a substance, capable of being transmitted from one body to another by conduction, convection, or radiation. The effects of absorbed heat on a body are increased temperature, expansion or increased volume, and possible change of phase, as of a solid to a liquid, or of a liquid to a gas. *The energy due to molecular motion is called heat ... a form of energy, and a measurable quantity, although not a substance. It may be transformed into other expressions of energy, mechanical work, for example.* (Blair, *Weather Elements*). **2** = specific heat.

heat balance, *Meteorology.* the average equilibrium between the amount of heat received by the earth from the sun and from internal sources and that radiated from the earth into space: *... the long-term effect of carbon dioxide on the earth's heat balance* (Barry Commoner). *Heat balance in the atmosphere is a vital determinant of climate and weather* (N.Y. Times). See **greenhouse effect.**

heat capacity, *Physics.* the ratio of the quantity of heat applied to a system, body, or substance to the change in temperature the heat induces, usually expressed in calories per degree Celsius. Compare **specific heat.**

heat equator, *Meteorology.* an imaginary line drawn near the earth's equator, through the middle of the belt which represents the hottest areas of the earth, with average annual temperatures of about 26°C. Also called **thermal equator.**

heath, *n. Ecology.* a level, uncultivated area without trees, often dry, covered with low, shrubby plants. ASSOCIATED TERMS: see **bog.**

heat island, *Geography, Meteorology.* a region of consistently higher temperatures than surrounding areas, with the higher temperatures centered over an urban area.

heat lightning, *Meteorology.* flashes of lightning without thunder, often seen near the horizon. The flashes are so distant that the thunder cannot be heard. *Heat lightning is a distant cloud-to-ground or intracloud discharge that flashes on sultry summer evenings* (Natural History).

heat of combustion, *Chemistry.* the quantity of heat released per unit mass or unit volume of a substance when the substance is completely burned: *Heats of combustion of solid and liquid fuels are usually expressed in Btu/lb or in cal/gm. The heat of combustion of gases is commonly expressed in Btu/ft³* (Sears and Zemansky, *University Physics*).

---

cap, fāce, fäther; best, bē, tèrm; pin, fīve;
rock, gō, ôrder; oil, out; cup, pùt, rüle,
yü in use, *yù* in uric;
ng in bring; sh in rush; th in thin, ᴛʜ in then;
zh in seizure;
ə = *a* in about, *e* in taken, *i* in pencil, *o* in
lemon, *u* in circus

heat of formation, *Chemistry.* the quantity of heat released when one mole of a compound is formed from its elements in the phases in which they exist at ordinary temperatures: *Compounds having low or negative heats of formation are unstable; they decompose spontaneously or when heated gently* (Offner, *Fundamentals of Chemistry*). Compare **enthalpy of formation.**

heat of fusion, *Chemistry.* the quantity of heat required to convert one gram of a substance from a solid at its melting point to a liquid without an increase in temperature.

heat of reaction, *Chemistry.* the quantity of heat released by a chemical reaction. Compare **enthalpy of reaction.**

heat of vaporization, *Chemistry.* the quantity of heat required to convert one gram of a substance from a liquid at its boiling point to a gas without an increase in temperature: *The usually high heat of vaporization of water results largely because much energy is required to break its hydrogen bonds* (Greulach and Adams, *Plants*).

heat sink, *Physics.* any substance, body, or region that absorbs or dissipates heat: *Ultimately, the ice sheet grows large enough to dissipate more solar energy than it receives and it becomes a heat sink* (Richard Lewis).

heat transfer, *Physics.* the flow or transmission of heat between solids, liquids, and gases by conduction, convection, or radiation: *Nuclear boiling is marked by large rates of heat transfer and the rapid formation of bubbles from fixed points* (J.W. Westwater).

Heaviside layer (hev'ē sīd), = E layer. Also called **Kennelly-Heaviside layer.** [named after Oliver *Heaviside,* 1850–1925, British physicist who proposed the existence of this layer]

heavy, *adj.* **1** *Physics.* having a relatively large mass: *The muon is about 200 times as heavy as the electron, and is often called the heavy electron* (Science News). **2** *Chemistry.* a being or containing an isotope that possesses a greater atomic weight than another of the same element: *Heavy oxygen [oxygen 18] and dozens of compounds enriched with heavy oxygen are widely used as tracers and in measurements all over the world* (New Scientist). **b** having a relatively high atomic weight: *... announced the creation of element 107 as the heaviest transuranic species yet found* (D. Allan Bromley). **c** of or containing an element of relatively high specific gravity. A **heavy metal** is one having a specific gravity higher than 5.0; a **heavy mineral** is one having a specific gravity higher than 2.9. *Lead and cobalt are heavy metals.*

heavy chain, *Biochemistry.* either one of a pair of polypeptide chains in an antibody molecule, having an approximate molecular weight of 50,000: *Antibody molecules are ... formed from two chains of molecules called heavy chains ... [and] a smaller molecular chain called a light chain* (Ralph Snyderman).

heavy hydrogen, *Chemistry.* any isotope of hydrogen with a mass greater than that of ordinary hydrogen; deuterium or tritium: *The basic fusion reaction requires that two nuclei of heavy hydrogen—called deuterium—fuse to make an atom of helium* (Harold M. Schmeck, Jr.).

heavy water, *Chemistry.* water formed of oxygen and deuterium; deuterium oxide. Heavy water is much like ordinary water, but is about 1.1 times as heavy and has a slightly higher freezing and boiling point. It occurs in very small amounts in ordinary water. *Formula:* D_2O

hectare (hek'ter), *n.* a unit of area in the metric system, equal to 100 areas or 10,000 square meters. A hectare is equivalent to about 2.471 acres.

hecto-, *combining form.* hundred, as in *hectogram = hundred grams.* *Symbol:* h [from French, from Greek *hektaton* hundred]

hectocotylus (hek'tə kot'ə ləs), *n., pl.* **-li** (-lī). *Zoology.* an arm of the male of certain cephalopods modified to transfer sperm to the mantle cavity of the female. [from New Latin *Hectocotylus* a supposed worm genus, from *hecto-* + Greek *kotylē* small cup]

hectogram (hek'tə gram), *n.* a unit of mass equal to 100 grams. *Abbreviation:* hg.

hectoliter (hek'tə lē'tər), *n.* a unit of volume equal to 100 liters. *Abbreviation:* hl.

hectometer (hek'tə mē'tər), *n.* a unit of length equal to 100 meters. *Abbreviation:* hm.

Heinz bodies, *Biology.* refraction areas in red blood cells, found in hemolytic anemia and probably representing denatured hemoglobin. [named after Robert *Heinz,* 1865–1924, German physician who first described these bodies]

Heisenberg uncertainty principle (hī'zən bėrg), = uncertainty principle. Also called **indeterminacy principle.** [named after Werner K. *Heisenberg,* 1901–1976, German physicist]

HeLa cell (hel'ə), *Biology.* any of a strain of human epithelial cells originally obtained from cancerous cervical tissue and maintained since then in cultures for use in studying cellular processes and diseases. [named after *He(nrietta) La(cks),* a cancer patient in Baltimore from whose cells the first cultures were grown in 1951]

helical, *adj.* of, having to do with, or having the form of a helix: *Since the work of James D. Watson and F.H.C. Crick in 1953, the structure of DNA has been postulated and then known to consist of two polynucleotide strands wound around each other in a helical or spiral fashion* (Robert Haselkorn). SYN: spiral.

helices (hel'ə sēz'), *n.* a plural of **helix.**

helicity (hē lis'ə tē), *n. Nuclear Physics.* the direction of the spin of an elementary particle.

helicoid (hel'ə koid), *adj.* Also **helicoidal** (hel'ə-koi'dəl). *Zoology.* shaped like a coil; spiral. —*n. Geometry.* a surface in the form of a coil or screw.

helio-, *combining form.* **1** sun, as in *helioscope = device for looking at the sun.* **2** sunlight, as in *heliotaxis = movement in response to sunlight.* [from Greek *hēlios* sun]

heliocentric (hē'lē ō sen'trik), *adj. Astronomy.* **1** centered on the sun; moving around the sun: *"Pioneer V" was launched toward a heliocentric orbit between the Earth and Venus* (Hugh Odishaw).
2 having or representing the sun as a center: *The heliocentric view, dating formally from the time of Copernicus, establishes the solar system on an approximately correct basis* (Baker, *Astronomy*).

heliocentric parallax, = annual parallax.

heliometer (hē′lē om′ə tər), *n.* an astronomical instrument, originally for measuring the sun's diameter, but now used especially in determining the angular distance between two stars.
—**heliometric** (hē′lē ə met′rik), *adj.* having to do with a heliometer. —**heliometrically,** *adv.*

heliosphere (hē′lē ə sfir), *n. Astronomy.* the region in space over which the sun's gases and magnetic field extend: *It [a spacecraft] may even be able to detect the unknown limits of the heliosphere ... and chart the fringes of interstellar space* (Time).

heliotactic (hē′lē ə tak′tik), *adj.* having the property of heliotaxis.

heliotaxis (hē′lē ə tak′sis), *n. Biology.* movement in response to sunlight. [from *helio-* + Greek *taxis* arrangement]

heliotropic (hē′lē ə trop′ik), *adj. Botany.* turning or bending in response to sunlight: *heliotropic curvature of a plant.*
—**heliotropism** (hē′lē ot′rə piz′əm), *n.* phototropism in which a plant turns or bends in response to sunlight.

helium (hē′lē əm), *n. Chemistry.* a colorless, odorless, inert gaseous element that will not burn, occurring on earth in natural gas and in association with radioactive ores. Helium is, after hydrogen, the lightest gas and the most abundant element in the observable universe, including the sun. *Symbol:* He; *atomic number* 2; *atomic weight* 4.0026. [from New Latin, from Greek *hēlios* sun; so called because the existence of the element was inferred from observation of the solar spectrum]

helix (hē′liks), *n., pl.* **helices** or **helixes. 1** anything having a spiral, coiled form such as a screw thread, a watch spring, or a snail shell: *DNA is composed of a pair of threads that wind about each other to form a double-stranded helix* (Science News Letter).
2 *Geometry.* the curve traced by a straight line on a plane that is wrapped around a cylinder, as the thread of a screw: *In a perfect spiral the helix angle, or angle of the threads with regard to a plane perpendicular to the axis, is everywhere the same* (Scientific American).
3 *Anatomy.* the rim of the outer ear.
[from Latin, from Greek *helix* a spiral]

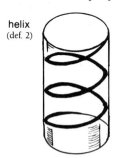

helix
(def. 2)

helminth (hel′minth), *n. Zoology.* a worm, especially an intestinal worm, such as a tapeworm, roundworm, etc. [from Greek *helminthos*]
—**helminthic** (hel min′thik), *adj.* having to do with a helminth or intestinal worm.

helminthology (hel′min thol′ə jē), *n.* the science of helminths or worms, especially of parasitic worms.

helper cell or **helper T cell,** *Immunology.* a T cell that eliminates foreign organisms: *Helper T cells bolster the activity of B cells. Others, called suppressor T cells, act to shut off that activity when it has gone far enough. The two are important in keeping the immune defense system in balance.* (Harold M. Schmeck, Jr.).

hem- or **hema-,** *combining form.* blood, as in *hemagglutination = agglutination of blood.* [from Greek *haima* blood]

hemagglutinate (hē′mə glüt′n āt), *v. Immunology.* to cause hemagglutination: *Antiserums were produced in rabbits by the intravenous injection of ... hemagglutinating units of partially purified virus* (Science).

hemagglutination (hē′mə glüt′n ā′shən), *n. Immunology.* a clumping together of the red blood cells, as by antibodies or by hemagglutinating viruses: *Use of hemagglutination, long valuable in the study of animal viruses, makes possible rapid detection ... of certain plant viruses* (L.S. McClung).

hemagglutinin (hē′mə glüt′n in), *n. Immunology.* a substance that causes red blood cells to agglutinate: *The major antigen on the surface of one type of flu virus, type A, is hemagglutinin (HA), an enormous protein with several antigenic determinants, each calling for a specific antibody* (Michael Shodell).

hemal (hē′məl), *adj. Anatomy.* **1** of or having to do with the blood or blood vessels: *the hemal system.* **2** having to do with or on the side of the body containing the heart and principal blood vessels: *the hemal cavity.*

hemapheresis (hē′mə fə rē′sis), *n. Medicine.* a process by which freshly drawn whole blood is separated into various constituents, some of which are retained while the remainder is returned to the donor's bloodstream: *A new activity for blood banks has been application of hemapheresis technology to the treatment of patients. This is termed therapeutic hemapheresis* (F. Widmann). [from New Latin *hem-* + *apheresis* separation, from Greek *aphairesis* a taking away]

hematin (hē′mə tən), *n. Biochemistry.* **1** a bluish-black pigment containing iron, formed in the decomposition of hemoglobin. *Formula:* $C_{34}H_{32}N_4O_4 \cdot FeOH$ **2** = heme.

hematite (hē′mə tīt), *n. Mineralogy.* a widely distributed mineral which is an important iron ore, occurring in crystalline, massive, or granular form, and reddish-brown when powdered: *As a rock, hematite is fine-grained and compact to loose, earthy and even fibrous* (Fenton, *The Rock Book*). *Formula:* Fe_2O_3 [from Greek *haimatitēs* bloodlike]
—**hematitic** (hē′mə tit′ik), *adj.* of, having to do with, or resembling hematite: *hematitic ores.*

hemato-, *combining form.* blood, as in *hematology = study of blood.* [from Greek *haima, haimatos*]

---

cap, fāce, fäther; best, bē, tèrm; pin, five;
rock, gō, ôrder; oil, out; cup, pùt, rüle,
yü in use, yu̇ in uric;
ng in bring; sh in rush; th in thin; ᴛʜ in then;
zh in seizure.
ə = *a* in about, *e* in taken, *i* in pencil, *o* in lemon, *u* in circus

hematoblast (hĕ′mə tō blast), *n. Anatomy.* a cell, as in bone marrow, from which a red blood cell may develop. [from *hemato-* + Greek *blastos* sprout, germ]

hematocrit (hē mat′ə krit *or* hĕ′mə tə krit), *n. Physiology, Medicine.* **1** the percentage of the volume of red blood cells in relation to the total volume in a given sample: *Each time her hemoglobin and hematocrit (red-cell concentration) readings fell alarmingly low, a blood transfusion lifted them above the danger level* (Time).
2 a centrifuge used to determine the volume of red blood cells in a given amount of blood. [from *hemato-* + Greek *kritēs* a judge]

hematogenous (hĕ′mə toj′ə nəs), *adj. Physiology.* **1** originating in or spread by the blood: *a hematogenous jaundice.*
2 derived from blood or hemoglobin: *a hematogenous pigment.*
3 having to do with the production of blood or blood cells. SYN: hematopoietic.

hematophagous (hĕ′mə tof′ə gəs), *adj. Zoology.* feeding on blood: *a hematophagous insect, such as the flea.* [from *hemato-* + Greek *phagein* eat]

hematopoiesis (hĕ′mə tō poi ē′sis), *n. Physiology.* the formation of blood or blood cells in the body. [from *hemato-* + Greek *poiēsis* a making]
—**hematopoietic** (hĕ′mə tō poi et′ik), *adj.* having to do with hematopoiesis: *the hematopoietic or blood-forming system of the body.*

heme (hēm), *n. Biochemistry.* a deep-red pigment containing iron, constituting the nonprotein part of the hemoglobin molecule: *It is the heme group that absorbs visible light and makes blood red—or blue, whenever it is not combined with oxygen* (Francis H.C. Crick). *Formula:* $C_{34}H_{32}N_4O_4Fe$ [from Greek *haima* blood]

hemi-, *prefix.* half, as in *hemisphere = half sphere.* [from Greek]

hemic (hĕ′mik), *adj. Physiology.* having to do with the blood: *hemic asthma.*

hemicellulose (hem′ē sel′yə lōs), *n. Biochemistry.* any of a group of gummy polysaccharide carbohydrates that are less complex than cellulose and are readily hydrolyzed to simple sugars: *Hemicelluloses are found in many seeds, ... in the wood and leaves of many trees, and in some fungi* (Hill, et al. Botany).

hemihedral (hem′ē hē′drəl), *adj. Crystallography.* having only half the planes required by the highest degree of symmetry: *a hemihedral crystal.* [from *hemi-* + Greek *hedra* base] —**hemihedrally**, *adv.*

hemihydrate (hem′ē hī′drāt), *n. Chemistry.* a hydrate containing half as many molecules of water as the number of molecules of the compound forming the hydrate.

hemimetabolism (hem′ē mə tab′ə liz əm), *n. Zoology.* an incomplete form of insect metamorphosis in which the pupal stage is lacking, as in cockroaches, crickets, and grasshoppers. Also called **incomplete metamorphosis.**
—**hemimetabolous** (hem′ē mə tab′ə ləs), *adj.* having to do with or characterized by hemimetabolism.

hemimorphic (hem′ē môr′fik), *adj. Crystallography.* having unlike planes or modifications at the ends of the same axis: *a hemimorphic crystal.* [from *hemi-* + Greek *morphē* form]
—**hemimorphism** (hem′ē môr′fiz əm), *n.* the state of being hemimorphic.

hemimorphite (hem′ē môr′fit), *n. Mineralogy.* a mineral, a silicate of zinc, occurring in orthorhombic crystals. *Formula:* $Zn_4(Si_2O_7)(OH)_2 \cdot 2H_2O$

hemin (hē′mən), *n. Biochemistry.* very small reddish crystals obtained when blood is heated with hydrochloric acid or glacial acetic acid and sodium chloride. The appearance of the crystals indicates the presence of blood in stains, fibers, or other material. *Formula:* $C_{34}H_{32}N_4O_4 \cdot FeCl$

hemiparasite, *n. Botany.* a partially parasitic plant: *Annual hemiparasites characteristically form haustorial connexions (root grafts) with most of the plants that surround them* (Nature). Also called **semiparasite.**
—**hemiparasitic**, *adj.* partially parasitic: *hemiparasitic flowering plants.*

hemipteran (hi mip′tər ən), *adj.* = hemipterous. —*n.* a hemipterous insect.

hemipterous (hi mip′tər əs), *adj. Zoology.* of or belonging to a large order (Hemiptera) of insects that have pointed beaks for piercing and sucking, usually two pairs of wings with forewings thickened at the base, and young that differ little from adults. Hemipterous insects include the suborder of heteropterous or true bugs such as the bedbugs, chinch bugs, and plant lice. [from *hemi-* + Greek *pteron* wing]

(def. 2) Western Hemisphere   Eastern Hemisphere

Northern Hemisphere   Southern Hemisphere

hemisphere, *n.* **1** *Geometry.* one half of a sphere formed by a plane passing through the center. See the picture at **sphere.**
2 *Geography.* half of the earth's surface: *the Western Hemisphere.*
3 *Anatomy.* = cerebral hemisphere: *Despite its linguistic superiority, the left hemisphere does not excel the right in all tasks* (Michael S. Gazzaniga).
—**hemispheric** or **hemispherical**, *adj.* **1** shaped like a hemisphere: *a hemispheric lens.* **2** of a hemisphere: *Antarctica is where cold fronts begin their hemispheric sweeps* (Newsweek). —**hemispherically**, *adv.*

hemizygote (hem'ē zī'gōt), *n. Genetics.* an animal or plant whose chromosomes have only one of a pair of genes: *The terms 'hemizygous' and 'hemizygote' ... refer to the condition of being haploid for a given gene when the genome as a whole is diploid* (H.J. Muller). Compare **heterozygote, homozygote.**
—**hemizygous** (hem'ē zī'gəs), *adj. Genetics.* of or having to do with a hemizygote.

hemocoel (hē'mə sēl), *n. Zoology.* a cavity or space between the organs of arthropods and mollusks through which the blood circulates: *Most mollusks and arthropods have an open system, blood being pumped from the heart through blood vessels to various organs but returning partly or entirely through body spaces—the hemocoel—to the heart* (Storer, *General Zoology*). [from *hemo-* + Greek *koilos* hollow]

hemocyanin (hē'mə si'ə nin), *n. Biochemistry.* a respiratory pigment found in the blood of certain crustaceans, mollusks, and arachnids. *The copper pigment, hemocyanin, serves as a vehicle for carrying oxygen in blue-blooded animals just as the iron pigment, hemoglobin, does in red-blooded creatures* (Science News Letter). [from *hemo-* + Greek *kyanos* blue substance + English *-in*]

hemocyte (hē'mə sīt), *n. Biology.* a blood cell, especially in invertebrates.

hemodialysis (hē'mə dī al'ə sis), *n.* = dialysis (def. 2): *Hemodialysis has saved the lives of many patients and has helped keep comfortable many others with chronic but not fatal renal damage* (Scientific American).

hemodynamic (hē'mə dī nam'ik), *adj. Physiology.* of or having to do with hemodynamics: *the hemodynamic effects of drugs in the bloodstream.* —**hemodynamically,** *adv.*

hemodynamics (hē'mə dī nam'iks), *n.* the branch of physiology dealing with the bloodstream and blood circulation; the study of blood pressure and flow.

hemoflagellate (hē'mə flaj'ə lāt or hē'mə flaj'ə lit), *n. Zoology.* a one-celled animal with a whiplike tail that lives as a parasite in the blood, such as a trypanosome.

hemoglobin (hē'mə glō'bən), *n. Biochemistry.* a substance in the red blood cells of vertebrates that carries oxygen from the lungs to the tissues and carries carbon dioxide from the tissues to the lungs. *Hemoglobin ... is composed of two portions—heme, an iron-containing red compound, and globin, a colorless protein* (Harold Baer).
▶ Different types of hemoglobin have been identified; these types are transmitted genetically. The most common type, found in normal adults, is *hemoglobin A.* The dominant type of hemoglobin in the fetus is called *hemoglobin F.* Other types are usually abnormal and tend to be associated with specific diseases. Thus *hemoglobin S* is the hemoglobin found in sickle-cell anemia. Also spelled **haemoglobin.** Compare **methemoglobin, oxyhemoglobin.**

hemolymph (hē'mə limf), *n. Zoology.* a nutritive fluid, analogous to blood or lymph, found circulating in the body of certain invertebrates: *The fluid hemolymph in the insect's body cavity contains no respiratory pigment* (Scientific American).

hemolysin (hē'mə li'sin), *n. Physiology.* a substance in the blood that causes hemolysis. Hemolysins may be antibodies, or they may be contained in bacterial toxins or snake venoms.

hemolysis (hi mol'ə sis), *n. Physiology.* the process of dissolution of the red blood cells, with the liberation of hemoglobin. Hemolysis in small amounts is a normal body process, usually balanced by the production of red blood cells in bone marrow. Excessive hemolysis results in anemia. [from *hemo-* + *lysis*]
—**hemolytic** (hē'mə lit'ik), *adj.* tending to dissolve the red blood cells: *hemolytic anemia.*

hemolyze (hē'mə līz), *v.* **1** to undergo hemolysis: *The cells will hemolyze when subsequently exposed to some mild form of stress* (Science Journal).
2 to cause hemolysis of (red blood cells): *Dr. Rowe recommends that frozen blood intended for transfusion be selectively hemolyzed to weed out older cells, and cells that are damaged, in order to eliminate cells that are unlikely to survive in the recipient* (Science News).

hemoprotein (hē'mə prō'tēn or hē'mə prō'tē in), *n. Biochemistry.* any of a group of protein substances, including hemoglobin, cytochromes, peroxidase, and catalase, which absorb oxygen and are pigments important in the spectral analysis of cells.

Henle's Loop (hen'lēz), *Anatomy.* a U-shaped loop formed by a tubule conveying urine when it enters the inner layer of the kidney and then turns around to pass up again into the outer layer of the kidney. Also called **loop of Henle.** [named after F.G.J. *Henle*, 1809–1885, German anatomist]

henry, *n., pl.* **-ries** or **-rys.** *Physics.* the SI or MKS unit of inductance. When a current varying at the rate of one ampere per second induces an electromotive force of one volt, the circuit has an inductance of one henry. *Symbol:* H [named after Joseph *Henry*, 1797–1878, American physicist]

heparin (hep'ər ən), *n. Biochemistry.* a mucopolysaccharide present in the liver and other animal tissues, that prevents the blood from clotting. Heparin is used in the treatment of thrombosis and other diseases. [from Greek *hēpar, hēpatos* liver]

hepatic (hi pat'ik), *adj.* **1** *Anatomy.* of or having to do with the liver. The **hepatic artery** supplies the liver with oxygenated blood. The **hepatic portal vein** carries blood from the ileum to the liver.
2 *Botany.* of or belonging to the liverworts.
—*n. Botany.* a liverwort.
[from Latin *hepaticus*, from Greek *hēpatikos*, from *hēpar, hēpatos* liver]

hepatic duct, *Anatomy.* the duct of the liver that carries the bile into the gall bladder.

hepta-, *combining form.* **1** seven, as in *heptagon.* **2** *Chemistry.* having seven atoms of one element, as in *heptane, iodine heptafluoride.* [from Greek *hepta* seven]

heptad (hep'tad), *n. Chemistry.* an element, atom, or radical with a valence of seven. [from Greek *heptas, heptados* (group of) seven, from *hepta* seven]

---

cap, fāce, fäther; best, bē, tèrm; pin, fīve;
rock, gō, ôrder; oil, out; cup, pùt, rüle,
yü in use, yù in uric;
ng in bring; sh in rush; th in thin, ᴛʜ in then;
zh in seizure.
ə = a in about, e in taken, i in pencil, o in lemon, u in circus

heptagon, *n. Geometry.* a plane figure having seven angles and seven sides. [from Greek *hepta* seven + *gōnia* angle]
—**heptagonal** (hep tag′ə nəl), *adj.* having the form of a heptagon.

heptamerous (hep tam′ər əs), *adj. Botany.* (of a flower) having seven members in each whorl. It is generally written *7-merous*. [from *hepta-* + Greek *meros* part]

heptane (hep′tān), *n. Chemistry.* one of nine isomeric hydrocarbons, including stereoisomers, of the methane series, used especially as solvents. *Formula:* C_7H_{16}

heptose (hep′tōs), *n. Biochemistry.* a monosaccharide with seven atoms of carbon in each molecule.

herb, *n. Botany.* **1** a flowering plant whose stem above ground usually does not become woody as that of a tree or shrub does. Herbs can be annual, such as corn; biennial, such as the onion; but most are perennial, such as the rose. *Some plants are trees, others are woody vines, some are herbaceous (soft-stemmed) vines, still others are erect herbs* (Fuller and Tippo, *College Botany*).
2 a plant whose leaves or stems are used for medicine, seasoning, food, or perfume. Sage, mint, and lavender are herbs. *Herbs are the aromatic leaves of plants which grow only in the temperate zone. Seeds refer to the seed, and sometimes the small fruit, of plants which grow in both the tropical and temperate zones—seeds like cardamom, celery and anise are also spices* (Science News Letter).
[from Latin *herba*]
ASSOCIATED TERMS: Plants are commonly classified into herbs, shrubs, and trees. *Herbs* (def. 1) are plants without a persistent woody stem aboveground. *Shrubs* and *trees* are perennials with woody stems, but shrubs are usually smaller than trees and have several main stems, while trees usually have a single main trunk or axis.

herbaceous (hèr′bā′shəs), *adj. Botany.* **1** of or like an herb; soft and not woody: *a herbaceous stem, herbaceous vines.*
2 having the texture, color, etc., of an ordinary leaf; green: *a herbaceous flower.*
[from Latin *herbaceus,* from *herba*]

herbal, *adj.* having to do with, consisting of, or made from herbs: *an herbal remedy.*

herbarium (hèr′ber′ē əm *or* hèr′bar′ē əm), *n., pl.* **-bariums, -baria** (-ber′ē ə *or* -bar′ē ə). *Botany.* **1** a collection of preserved plant specimens systematically arranged.
2 a place where such a collection is kept: *Specimen plants, stored in herbaria, make identification of unknown plants easier and more definite* (Weier, *Botany*).
[from Late Latin, from Latin *herba* herb]

herbicide (hèr′bə sīd), *n.* a toxic substance, especially a synthetic chemical, that destroys or inhibits plant growth.
—**herbicidal** (hèr′bə sī′dl), *adj.* that destroys or inhibits plant growth: *herbicidal compounds.*

Herbig-Haro object (hèr′big hä′rō), *Astronomy.* any of a group of small bright nebulas believed to be associated with the early stages of star formation: *Herbig-Haro objects are another of the many curious classes of objects in the sky. Small, nebular and reddish, they have a spectra consisting of emission lines of various elements with a very weak continuous spectrum in the background* (Science News). Also called **H-H object.** [named after George *Herbig,* American astronomer, and Guillermo *Haro,* Mexican astronomer, who first described such objects]

herbivore (hèr′bə vôr), *n. Zoology.* any of a large group of animals that feed chiefly on plants: *Herbivores, whether they are insects, rabbits, geese or cattle, succeed in extracting only about 50 per cent of the calories stored in the plant protoplasm* (LaMont C. Cole). [from Latin *herba* herb + *vorare* devour] ▶ See the note under **heterotroph.**

herbivorous (hèr biv′ər əs), *adj. Zoology.* feeding on grass or other plants: *The cattle, deer, rodents, and insects that eat leaves and stems of plants are said to be herbivorous; the cats, sharks, flesh flies, and many marine animals whose food is entirely or largely other animals are termed carnivorous; and those such as man, bears, rats, and others that utilize a variety of both plant and animal sources are called mixed feeders, or omnivorous* (Storer, *General Zoology*).

Hercynian (hèr sin′ē ən), *adj. Geology.* of or having to do with a series of mountain-building fractures and folds of the late Paleozoic, especially in southern Europe and northern Africa: *Hercynian orogeny.* [named after the *Hercynian* mountain range in ancient geography identified with various European ranges]

herd, *n. Zoology.* a group of animals of one kind, especially large wild animals, that keep, feed, or move together: *a herd of cattle, a herd of elephants, a herd of whales.*

hereditary, *adj. Genetics.* transmitted or capable of being transmitted by means of genes from parents to offspring: *Color blindness is hereditary. There is in each plant a pair of hereditary factors controlling flower color ... called genes* (Simpson, *Life: An Introduction to Biology*).

heredity, *n. Genetics.* the transmission of physical or mental characteristics from parent to offspring by means of genes: *It has long been known that heredity is determined by the chromosomes, the threadlike bodies in the nucleus of the cell, and by their subunits the genes* (Scientific American). *The rules by which heredity operates were first discovered and published by the Austrian monk Gregor Mendel in 1866* (Sheldon C. Reed). [from Latin *hereditatem,* from *heredem* heir]

heritability, *n. Genetics.* heritable quality or condition: *Curiously enough, the heritability of the size of the egg is the same in the fly as it is in the chicken, just as the numbers of eggs laid are strongly influenced by environment in each of them* (Grant Cannon).

heritable, *adj. Genetics.* capable of being inherited: *heritable alleles, a heritable disease. Highly heritable traits are easily subject to effective selection; slightly heritable ones require intense selection to provide significant genetic progress* (Srb and Owen, *General Genetics*).

hermaphrodite (hər maf′rə dīt), *Biology.* —*n.* an animal or plant having both male and female reproductive organs: *The term hermaphrodite is applied to ... species that contain both male and female systems* (Storer, *General Zoology*). *Among humans ... virgin birth could not happen in the case of a hermaphrodite, who would not be self-fertile. However, parthenogenesis* *might occur. This is the process by which an ovum begins to divide spontaneously, without having been fertilized by a sperm* (Time).

—*adj.* of or like a hermaphrodite: *A hermaphrodite flower is one which possesses both stamens and carpels which may or may not be functionally active* (Youngken, *Pharmaceutical Botany*).
[from Greek *Hermaphroditos*, son of Hermes and Aphrodite, who became united in body with a nymph]
—**hermaphroditic** (hər maf′rə dit′ik), *adj.* of or like a hermaphrodite: *Because both sperm and eggs may be produced in the body of one sponge, we call the sponges hermaphroditic animals as contrasted to the condition of higher forms where the two types of gametes are produced in separate individuals which are male and female animals* (Winchester, *Zoology*).
—**hermaphroditism** (hər maf′rə dī tiz′əm), *n.* the condition of a hermaphrodite: *True hermaphroditism occasionally occurs among certain primitive hagfishes and some amphibia but is even rarer among the other vertebrates. It is, however, a common condition among invertebrates of which the earthworm is a good example* (Harbaugh and Goodrich, *Fundamentals of Biology*).

hermatypic (hėr′mə tip′ik), *adj. Zoology.* that build reefs: *He* [*Thomas F. Goreau*] *experimentally demonstrated the supreme significance of the endozoic algae* (*zooxanthellae*) *present within all hermatypic corals in the necessarily high rate of calcification possessed by these reef builders* (London *Times*). [from Greek *herma* reef + *typtein* to strike, beat]

herpesvirus, *n. Biology.* any of a group of relatively large viruses that contain DNA and form inclusion bodies within the nuclei of host cells. Herpesviruses cause such diseases as chicken pox, shingles, and herpes simplex. The Epstein-Barr virus is a herpesvirus. Compare **cytomegalovirus.** [from Greek *herpēs* shingles (from *herpein* to creep) + English *virus*]

herpetology (hėr′pə tol′ə jē), *n.* the branch of zoology dealing with reptiles and amphibians. [from Greek *herpeton* reptile, from *herpein* to creep]

hertz, *n. Physics.* the SI unit of frequency, equivalent to one cycle per second. *Symbol:* Hz [named after Heinrich R. *Hertz*, 1857–1894, German physicist]

Hertzian wave, *Physics.* an electromagnetic wave, such as the wave used in communicating by radio. Hertzian waves are produced by irregular fluctuations of electricity in a conductor. *The entire electromagnetic spectrum is customarily divided, in the order of increasing frequency, into Hertzian waves, heat rays, light, ultraviolet rays, X-rays, gamma rays, and cosmic rays* (Hardy and Perrin, *Principles of Optics*).

Hertzsprung-Russell diagram (hėrt′sprung rus′əl), *Astronomy.* a graph in which the luminosity of a group of stars is plotted against their surface temperature or color: *One very valuable clue concerning how stars develop ... is known as the Hertzsprung-Russell diagram. This shows the relationship of a star's color to its brightness, and also helps to estimate the distances of stars* (Science News Letter). Also called **H-R diagram.** [named after Ejnar *Hertzsprung*, 1873–1967, Danish astronomer, and Henry N. *Russell*, 1877–1957, American astronomer, who independently devised the first of such graphs]

hesperidin (hes per′ə din), *n. Biochemistry.* a bioflavonoid that is a glucoside occurring in citrus fruit peel and in most citrus fruits when unripe: *To condition athletes before their games, Lichtman prescribes doses of ascorbic acid* (*vitamin C*) *and hespiridin, a chemical found in citrus fruits* (Newsweek). *Formula:* $C_{28}H_{34}O_{13}$ [from New Latin *Hesperides* (in early use) the orange family, from Latin, from Greek *Hesperides*, daughters of *Hesperos* having to do with the evening star, western]

hesperidium (hes′pə rid′ē əm), *n., pl.* **-ia** (-ē ə). *Botany.* a type of berry having a fleshy interior covered by a leathery, separable rind and growing on citrus trees. The fruit of the orange, lemon, grapefruit, etc., are hesperidia. [from New Latin *Hesperidium*, from Latin *Hesperides*; see HESPERIDIN]

hessite (hes′īt), *n. Mineralogy.* a telluride of silver usually occurring in gray, sectile masses. One variety also contains some gold. *Formula:* Ag_2Te [named after G.H. *Hess*, 1802–1850, Swiss chemist]

hetero-, *combining form.* of more than one kind; other; different, as in *heterogamete = different or other gamete.* [from Greek, from *heteros* different, another]

heteroatom (het′ər ō at′əm), *n. Chemistry.* an atom which substitutes for one of the carbon atoms in a hydrocarbon: *The first preparation of a sugar structure with a heteroatom other than oxygen in the ring was reported ... The substance was a sulfur-containing xylose derivative* (David M. Locke).

heterocercal (het′ər ə sėr′kəl), *adj. Zoology.* **1** having a tail in which the upper lobe is larger than the lower, and which contains the end of the spinal column: *a heterocercal fish.*
2 of or having to do with this form of tail. Contrasted with **homocercal.**
[from *hetero-* + Greek *kerkos* tail]

heterochromatic (het′ər ə krō mat′ik), *adj.* **1** *Botany.* having to do with or possessing more than one color: *heterochromatic flowers.*
2 *Optics.* having to do with light or other radiation of more than one wavelength: *If the light is heterochromatic—white, for example—the image consists of the superposed diffraction patterns produced by light of every wavelength* (Hardy and Perrin, *Principles of Optics*).
3 *Biology.* of or having to do with heterochromatin; having a deeply staining chromatin: *heterochromatic chromosomes. There is a small circular heterochromatic body in the nucleus of the female somatic cell which is absent from the male somatic cell* (G.M. Wyburn).

heterochromatin (het′ər ə krō′mə tin), *n. Biology.* a portion of chromatin that stains intensely during the interphase of mitosis. Its DNA is tightly coiled and inactive in the transcription of the genetic code.

---

cap, fāce, fäther; best, bē, tėrm; pin, fīve;
rock, gō, ôrder; oil, out; cup, pùt, rüle,
yü in use, *yù* in uric;
ng in bring; *sh* in rush; *th* in thin; ᴛʜ in then;
zh in seizure.

ə = *a* in about, *e* in taken, *i* in pencil, *o* in lemon, *u* in circus

heterochromous

Heterochromatin is known to be an effective gene repressor (G.H. Jones).

heterochromous (het′ər ə krō′məs), *adj. Botany.* having the florets of the center different in color from those of the margin, as in the daisy or aster. [from *hetero-* + Greek *chroma* color]

heterocyclic (het′ə r ə sī′klik), *Chemistry.* —*adj.* of or containing a ring made up of more than one kind of atom, especially a ring made up of several carbon atoms and one or more of another kind, such as sulfur or nitrogen: *Purines and pyrimidines are heterocyclic compounds.*
—*n.* a heterocyclic compound: *Aromatic heterocyclics are molecules in which a carbon atom of an aromatic system has been replaced by a different (hetero) atom* (New Scientist). Contrasted with **homocyclic.**

heterocyst (het′ər ə sist), *n. Botany.* a large, colorless cell found irregularly along a filament in certain species of algae: *Nitrogen-fixing blue-green algae have a problem with the oxygen produced during photosynthesis, and many strains protect their nitrogenase by compartmentation in special cells, heterocysts, which do not evolve oxygen* (John R. Postgate).

heterodont (het′ər ə dont), *adj. Anatomy.* differentiated for various uses, as into the canines, incisors, and molars of man: *heterodont teeth.* [from *hetero-* + Greek *odontos* tooth]

heterodromous (het′ə rod′rə məs), *adj. Botany.* having the spiral arrangement of the leaves on the stem in one direction and of those on the branches in the opposite direction. Contrasted with **homodromous.** [from *hetero-* + Greek *dromos* a running]

heteroecious (het′ə rē′shəs), *adj. Biology.* going through different stages of growth on different hosts, as certain parasitic fungi. [from *hetero-* + Greek *oikia* house] Compare **autoecious.**

heterogamete (het′ər ō gə mēt′), *n. Biology.* either of two gametes, differing in structure or behavior, which can unite with the other to form a zygote. A large, nonmotile egg and a small, motile sperm are heterogametes. Compare **anisogamete.**
—**heterogametic** (het′ər ō gə met′ik), *adj.* 1 of or having to do with heterogametes. 2 producing more than one type of gamete or sex chromosome. The male mammal, producing X and Y chromosomes, is heterogametic. Contrasted with **homogametic.**

heterogamous (het′ə rog′ə məs), *adj.* 1 *Biology.* a reproducing by the union of unlike gametes. b characterized by the alternation of two types of sexual reproduction in successive generations. 2 *Botany.* having flowers or florets of two sexually different kinds. Compare **anisogamous.**

heterogamy (het′ə rog′ə mē), *n. Biology.* 1 the condition of being heterogamous; reproduction by the union of unlike gametes: *The two fusing gametes are usually distinctly different in size, and this difference is regarded as evidence of a tendency toward heterogamy* (Emerson, *Basic Botany*). Compare **allogamy, anisogamy, isogamy, homogamy.** 2 = heterogenesis.

heterogeneity (het′ər ə jə nē′ə tē), *n.* the quality or condition of being heterogeneous: *To insist that species be entirely homogeneous would simply multiply the number of species and would solve no problem either*

of evolution or of classification. In practice, therefore, some heterogeneity is admitted (Shull, *Principles of Animal Biology*).

heterogeneous (het′ər ə jē′nē əs *or* het′ər ə jē′nyəs), *adj.* 1 made up of dissimilar elements or parts; unlike or varied in kind; not homogeneous: *Unlike all proteins whose structure had been determined, antibody molecules were known to be very "heterogeneous": no single sequence of amino acids—the building blocks of proteins—could represent the polypeptide chains of antibodies, as can be done, for example, in the case of a "homogeneous" protein such as insulin or hemoglobin* (Scientific American). 2 *Chemistry.* consisting of or involving more than one phase in any system or process: *A reaction may be partly homogeneous and partly heterogeneous* (C.N. Hinshelwood). 3 *Mathematics.* of different kinds and having no common integral divisor except 1. [from Medieval Latin *heterogeneus,* ultimately from Greek *heteros* other + *genos* kind]

heterogenesis (het′ər ə jen′ə sis), *n.* = alternation of generations: *Heterogenesis is the term applied to a type of life cycle that is particularly conspicuous among rotifers, water fleas, and plant lice. In this case, one or more generations that develop from unfertilized eggs (parthenogenetically) alternate with one or more generations that arise from fertilized eggs* (Hegner and Stiles, *College Zoology*).

heterogenous (het′ə roj′ə nəs), *adj. Biology.* coming from without; not arising within the body: *a heterogenous graft, heterogenous bone transplant.*

heterogonous (het′ə rog′ə nəs), *adj.* 1 *Botany.* of or denoting monoclinous flowers occurring in two or more forms, in which cross-fertilization is accomplished by the forms differing in the relative length of stamens and pistils. 2 *Biology.* characterized by heterogony.

heterogony (het′ə rog′ə nē), *n.* 1 *Biology.* alternation of generations in which a parthenogenetic generation alternates with a sexual generation. 2 *Botany.* the condition of being heterogonous. [from *hetero-* + Greek *-gonia,* from *gonos* birth]

heterograft (het′ər ə graft), *n. Immunology.* tissue from an individual grafted on a member of another species: *There has been recent interest in the use of heterograft valves from animal sources, particularly pigs* (Roy Y. Calne). Also called **xenograft.** Contrasted with **homograft.**
ASSOCIATED TERMS: see **allograft.**

heterokaryon (het′ər ō kar′ē on), *n. Biology.* a cell having several genetically different nuclei: *Protoplasts of potato and tomato are fused to give a heterokaryon, which regenerates its cell wall and divides* (New Scientist). [from *hetero-* + Greek *karyon* nut, kernel]

heterokaryosis (het′ər ō kar′ē ō′səs), *n. Biology.* the condition of having or forming heterokaryons: *The essential stages in demonstrating the parasexual cycle are (1) production of heterokaryosis, (2) production or fusion of genetically unlike nuclei, (3) recovery of the fusion or diploid nucleus, and (4) detection of segregant conidia in the diploid strain* (Edward D. Garber).

heterokaryotic (het′ər ō kar′ē ot′ik), *adj. Biology.* of or having to do with heterokaryons: *heterokaryotic character, heterokaryotic organisms.*

heterologous (het′ə rol′ə gəs), *adj. Biology.* 1 having a different relation; not corresponding: *heterologous genes, heterologous tissues.* 2 (of cells and antiserums) immunologically related but not identical. [from *hetero-* + Greek *logos* ratio, relation]

heterolysis (het′ə rol′ə sis), *n.* 1 *Biology.* the destruction of cells by enzymes or lysins from another organism. Contrasted with **autolysis.**
2 *Chemistry.* the breaking of a bond in an unsymmetrical manner, so as to yield two differently charged fragments. Contrasted with **homolysis.**
—**heterolytic** (het′ər ə lit′ik), *adj.* of or having to do with heterolysis: *In certain circumstances this polarization can go further, the pair of electrons being transferred completely to one of the atoms so that the bond dissociates giving a positive and a negative ion ... This is called heterolytic fission and usually occurs only in solution in certain types of solvents* (K.D. Wadsworth).

heteromerous (het′ə rom′ər əs), *adj.* 1 *Botany.* having the flower parts of one or more whorls differing in number from those of the others. 2 *Chemistry.* unrelated in chemical composition. [from *hetero-* + Greek *meros* part]

heteromorphic, *adj.* 1 *Biology.* of different form or size; deviating from the normal structure or type: *heteromorphic chromosome pairs, heteromorphic colonies of scyphozoans.*
2 *Zoology.* a undergoing metamorphosis, as those insects which have different forms at different stages of growth. b of or resulting from heteromorphosis: *a regenerated heteromorphic appendage.*
3 *Botany.* occurring in forms that differ in the relative length of the stamens and pistils.
[from *hetero-* + Greek *morphē* form]
—**heteromorphism,** *n.* the condition or property of being heteromorphic.

heteromorphosis (het′ər ə môr fō′sis), *n., pl.* **-ses** (-sēz). *Zoology.* the replacement or substitution of one organ or lost part by another organ or new part that is different from the part that has been removed or is in an abnormal position: *The part regenerated is not always the same as the part lost. This type of regeneration is known as heteromorphosis. For example, if an earthworm is cut in two, back of the eighteenth segment, a new tail will grow out from the posterior end instead of a new head, and the newly organized worm will starve to death* (Hegner and Stiles, *College Zoology*).

heteronomous (het′ə ron′ə məs), *adj. Biology.* having different modes of growth, as parts or organs: *heteronomous appendages, the heteronomous segments of arthropods.* Contrasted with **homonomous.** [from *hetero-* + Greek *nomos* law]

heteronuclear, *adj.* consisting of different atoms or groups of atoms: *CO is a heteronuclear molecule.* Contrasted with **homonuclear.**

heterophyllous (het′ə fil′əs), *adj. Botany.* bearing different kinds of leaves on the same plant: *a heterophyllous species.* [from *hetero-* + Greek *phyllon* leaf]
—**heterophylly** (het′ər ə fil′ē), *n.* the condition of being heterophyllous.

heteroploid (het′ər ə ploid), *Genetics.* —*adj.* having a chromosome number that is not the exact multiple of the haploid number of chromosomes of the species: *The parent mouse line ... is heteroploid and contains*

a number of biarmed chromosomes not found in normal mouse cells (M. Nobholz and W.F. Bodmer).
—*n.* a heteroploid organism.
[from *hetero-* + *-ploid*, as in *haploid, diploid*]
—**heteroploidy,** *n.* the condition of being heteroploid; a gain or loss in chromosome number.

heteropolar, *adj. Chemistry.* of or characterized by an unequal distribution of positive and negative charges, as in a semipolar bond. Contrasted with **homopolar.**

heteropteran (het′ə rop′tər ən), *adj.* = heteropterous.
—*n.* a heteropterous insect.

heteropterous (het′ə rop′tər əs), *adj.* of or belonging to the suborder (Heteroptera) of hemipterous insects that comprises the true bugs. [from *hetero-* + Greek *pteron* wing]

heteroscedastic (het′ər ə si das′tik), *adj. Statistics.* showing unequal variability; not showing the same standard deviation: *a heteroscedastic distribution.* Contrasted with **homoscedastic.** [from *hetero-* + Greek *skedastikos* that can be scattered]

heterosis (het′ə rō′sis), *n. Genetics.* an increase in vigor of growth, fertility, size, or other characteristics, resulting from crossbreeding: *The causes of heterosis may be diverse in different crosses, or different causes may be jointly effective in a single cross* (Curt Stern). Also called **hybrid vigor.** [from Greek *heterosis* alteration, from *heteros* another, different]

heterosphere (het′ər ə sfēr′), *n. Meteorology.* a part of the atmosphere, beginning at a height of about 50 miles, that is made up of distinct layers of oxygen, hydrogen, and helium.

heterosporous (het′ər ə spôr′əs *or* het′ə ros′pərəs), *adj. Botany.* having more than one kind of asexually produced spores, such as microspores and megaspores. Contrasted with **homosporous.**
—**heterospory** (het′ər ə spôr′ē *or* het′ə ros′pər ē), *n.* the production of more than one kind of asexual spores: *In heterospory the small spores give rise to male gametophytes and the large spores to female gametophytes* (Youngken, *Pharmaceutical Botany*).

heterotactic or **heterotaxic,** *adj.* characterized by or exhibiting heterotaxis.

heterotaxis (het′ər ə tak′sis), *n.* abnormal or irregular arrangement, as of parts of the body or of geological strata. Contrasted with **homotaxis.** [from *hetero-* + Greek *taxis* arrangement]

heterothallic (het′ər ə thal′ik), *adj. Biology.* having two haploid types that act as male and female: *Species of fungi that are differentiated into plus and minus strains are said to be heterothallic* (Weier, *Botany*). Contrasted with **homothallic.** [from *hetero-* + Greek *thallos* young shoot]

---

cap, fāce, fäther; best, bē, tėrm; pin, fīve;
rock, gō, ôrder; oil, out; cup, pùt, rüle,
yü in use, *yù* in uric;
ng in bring; *sh* in rush; *th* in thin, ᴛʜ in then;
zh in seizure.
ə = *a* in about, *e* in taken, *i* in pencil, *o* in lemon, *u* in circus

heterotroph (het′ər ə trof), *n. Biology.* an organism that cannot manufacture its own food, but is dependent upon complex organic substances for nutrition. Animals, fungi, protozoans, and most bacteria are heterotrophs. *The disassembly of the glucose molecule by the liver cell may be taken, however, as typical of the process by which most known aerobic heterotrophs obtain energy* (Scientific American). Compare **autotroph.**

▶ On the basis of their dependency upon other organisms for food, heterotrophs are classified into four groups: *saprophytes* (living on dead matter), *herbivores* (consuming plants), *carnivores* (consuming other animals), and *omnivores* (consuming both plants and animals).

—**heterotrophic** (het′ər ə trof′ik), *adj.* of or having to do with a heterotroph: *These, and many others, are representative of the heterotrophic bacteria—bacteria which are capable of utilising complex organic materials for their food and energy requirements* (New Scientist). *Humans are heterotrophic and therefore must ingest food* (Biology Regents Syllabus). [from *hetero-* + Greek *trophē* nourishment]

heterotypic, *adj. Biology.* of the first stages of meiosis in a cell, during which the chromosomes appear to be reduced from the diploid to the haploid number.

heterozygosis (het′ər ə zī gō′sis), *n.* = heterozygosity. [from *hetero-* + Greek *zygōsis* an equating]

heterozygosity (het′ər ə zī gos′ə tē), *n. Genetics.* the condition of being heterozygous; development from a heterozygote: *The same genes are present in the hybrids as in the inbred parents but many of them are in the heterozygous state. It is this heterozygosity which confers fitness* (Hollingsworth and Bowler, *Principles and Processes of Biology*).

heterozygote (het′ər ə zī′gōt), *n. Genetics.* an animal or plant whose chromosomes contain both genes of a contrasting pair and which, therefore, does not always breed true to type: *An organism whose two genes for any particular character are alike (WW or ww) is called a homozygote; one whose genes are different (Ww) is a heterozygote* (Shull, *Principles of Animal Biology*). *In a population exposed to malaria the heterozygote (hybrid) possessing one normal hemoglobin gene and one sickle cell hemoglobin gene has an advantage* (Scientific American). Compare **hemizygote, homozygote.**

—**heterozygous** (het′ər ə zī′gəs), *adj.* of or having to do with a heterozygote; having both genes of a contrasting pair (as for tallness and shortness) at a locus on homologous chromosomes: *Now in the great majority of cases an individual who receives a mutant gene from one of his parents receives from the other parent a corresponding gene that is "normal." He is said to be heterozygous, in contrast to the homozygous individual who receives like genes from both parents.* (Scientific American).

heulandite (hyü′lən dīt), *n. Mineralogy.* a zeolite containing calcium, found in crystals of various colors with pearly luster. *Formula:* $CaO \cdot Al_2O_3 \cdot 6SiO_2 \cdot 5H_2O$ [named after Henry *Heuland,* English mineralogist of the 1800's + *-ite*]

hexa-, *combining form.* **1** six, as in *hexagon.* **2** *Chemistry.* having six atoms of one element, as in *hexane, sulfur hexafluoride.* [from Greek *hexa-,* from *hex* six]

hexabasic, *adj. Chemistry.* **1** having six hydrogen atoms that can be replaced by basic atoms or radicals: *a hexabasic acid.* **2** having six atoms or radicals of a univalent metal.

hexachloroethane (hek′sə klor′ə eth′ān), *n.* a highly toxic, colorless, crystalline compound, used in organic synthesis and as a retarding agent and solvent. *Formula:* C_2Cl_6

hexad (hek′sad), *n. Chemistry.* an element, atom, or radical with a valence of six. [from Greek *hexas, hexados* (group of) six, from *hex* six]

hexadecimal (hek′sə des′ə məl), *adj. Mathematics.* of, having to do with, or using the decimal number 16: *The hexadecimal numeral system uses the base 16, as contrasted with the base 10 of the decimal system and base 2 of the binary system* (R. Clay Sprowls). Compare **binary, decimal, duodecimal.**

hexadic (hek sad′ik), *adj. Mathematics.* of or having to do with the number six as a base in arithmetic; increasing by sixes.

hexagon (hek′sə gon), *n. Geometry.* a plane figure having six angles and six sides. [from Greek *hex* six + *gōnia* angle]

—**hexagonal** (hek sag′ə nəl), *adj.* **1** *Geometry.* **a** having the form of a hexagon. **b** having a hexagon as base or cross section.
2 *Crystallography.* having three equal axes intersecting at 120° and lying in one plane, and a fourth unequal axis at right angles to the other three. See the picture at **crystal.**
ASSOCIATED TERMS: see **crystal.**

hexagram (hek′sə gram), *n. Geometry.* **1** a six-pointed star formed of two equilateral triangles. **2** a figure formed by six intersecting lines. [from *hexa-* + Greek *gramma* line, letter]

hexahedral, *adj. Geometry.* having six sides; like a hexahedron: *hexahedral prisms.*

hexahedron (hek′sə hē′drən), *n., pl.* **-drons, -dra** (-drə). *Geometry.* a solid figure having six faces, such as a cube. A hexahedron is one type of polyhedron. [from Greek *hexa-* + *hedra* base]

hexahydrate, *n. Chemistry.* a compound having six molecules of water in its formula.

hexamerous (hek sam′ər əs), *adj.* **1** *Zoology.* having the radiating parts or organs numbering six or a multiple of six.
2 *Botany.* (of a flower) having six members in each whorl (usually written *6-merous*): *herbs with hexamerous flowers.*
[from *hexa-* + Greek *meros* part]

hexaploid (hek′sə ploid), *adj. Genetics.* having six times the haploid set of chromosomes in the somatic cells. [from *hexa-* + *-ploid,* as in *haploid, diploid*]

hexapod, *Zoology.* —*n.* a true insect; an arthropod having six feet. Hexapods include fleas. —*adj.* having six feet: *hexapod larvae.* [from Greek *hexa-* + *pous, podos* foot]

hexokinase (hek′sō ki′nās), *n. Biochemistry.* an enzyme which catalyzes the formation of adenosine diphosphate from adenosine triphosphate.

hexosamidase (hek'sō sə min'ə dās), *n. Biochemistry.* either of a pair of enzymes, a deficiency of which causes various degenerative diseases of the central nervous system.

hexosan (hek'sə san), *n. Biochemistry.* any one of several hemicelluloses that yield hexoses on hydrolysis.

hexose (hek'sōs), *n. Chemistry.* any one of a class of simple sugars that contain six carbon atoms in the molecule, such as glucose and fructose: *Hexoses, such as pentoses, are widely distributed in plants and animals, and are the units from which the more complex carbohydrates are synthesized* (Harbaugh and Goodrich, *Fundamentals of Biology*).

HF, H.F., or **h.f.,** *abbrev.* high-frequency.

Hf, *symbol.* hafnium.

Hg, *symbol.* mercury. [for New Latin *hydrargyrum* mercury, (literally) water silver]

hg., *abbrev.* hectogram *or* hectograms.

HGH or **H.G.H.,** *abbrev.* human growth hormone.

H-H object, = Herbig-Haro object.

hiatus (hī ā'təs), *n. Anatomy.* an opening or short passage in an organ or part: *the hiatus of the facial canal.* [from Latin *hiatus* gap]
—**hiatal,** *adj.* of or involving a hiatus: *a hiatal hernia.*

hibernacle (hī'bər nak'əl), *n.* = hibernaculum (def. 2a).

hibernaculum (hī'bər nak'yə ləm), *n., pl.* **-la** (-lə). **1** *Botany.* a part of a plant, such as a bud or bulb, that protects a plant embryo during the winter.
2 *Zoology.* **a** a place where an animal hibernates during the winter: *The welfare of the hedgehog snug in its hibernaculum beneath layers of warm snow, may cause the naturalist little concern* (London *Times*). **b** an encysted winter bud of a bryozoan that germinates in the following spring.
[from Latin *hibernaculum* winter quarters, from *hibernus* wintry]

hibernate, *v. Zoology.* to spend the winter in a state like sleep or in an inactive condition, as bears, woodchucks, and some other wild animals do: *When a mammal hibernates ... breathing is almost imperceptible, and all the chemical activities of the body go on very slowly, using food stored as fat and glycogen* (Mackean, *Introduction to Biology*). [from Latin *hibernatum* wintered, from *hibernus* wintry]
—**hibernation,** *n.* the state or period of hibernating; a sleeplike, inactive, torpid condition, marked by a lowering of metabolism and body temperature, found in certain animals during winter: *True hibernation, such as is found in ground squirrels and marmots, is a state of death-like sleep* (Science News Letter). Compare **estivation, diapause.**
—**hibernator,** *n.* an animal that hibernates: *cold-blooded animals and even warm-blooded hibernators can survive a far greater reduction of their body temperature* (R. J. Hock and B. G. Covino).

hiddenite (hid'ə nīt), *n. Mineralogy.* a transparent, emerald-green variety of spodumene, sometimes used as a gem. [from William E. *Hidden,* 1832–1918, American mineralogist, who discovered it + *-ite*]

hidebound, *adj. Botany.* (of a tree) having the bark so close or unyielding as to hinder growth.

high, *adj. Biology.* of greater than usual development; more complex or advanced, especially in structure or intelligence: *the higher algae, the higher primates.*

—*n. Meteorology.* an area of relatively high barometric pressure: *If a high changes its central latitude ... in moving towards the equator the central pressure will decrease, and the system will spread out* (Science News). Also called **anticyclone.**

high-energy, *adj. Physics.* **1 a** of or having to do with the high speeds imparted by particle accelerators: *high-energy collisions.* A **high-energy particle** is an elementary particle with the energy of hundreds of millions electron volts imparted by its acceleration to very high speeds. *... the possibility that the high-energy particles that are found near the Earth have derived their energy ... in interplanetary space* (New Scientist). *Because high-energy particles have very short wavelengths, they provide the resolution required for highlighting fine details of nuclear shape and structure* (D. Allan Bromley). **b** of or having to do with high-energy physics: *high-energy research.* Compare **low-energy.**
2 *Chemistry.* yielding a very large amount of energy on hydrolysis: *The chemical bonds linking the three phosphate groups to the adenosine group are high-energy bonds* (Ellsworth S. Obourn).

high-energy physics, a branch of physics dealing with the properties and behavior of elementary particles, especially through the study of high-energy particles: *Accelerators are the chief instruments of high-energy physics* (Scientific American). Also called **particle physics.**

highest common factor, *Mathematics.* = greatest common divisor. *Abbreviation:* H.C.F.

high-field, *adj. Physics.* of or having to do with a very strong electric or magnetic field: *a high-field electromagnet.*

high-frequency, *adj. Physics.* of or having to do with a frequency ranging from 3 to 30 megahertz: *high-frequency sound waves. Abbreviation:* HF

highland, *Geography, Geology.* —*n.* Often, **highlands,** *pl.* land that is higher or hillier than the neighboring land; region high above sea level. SYN: plateau.
—*adj.* of or in a highland: *The islands [of Indonesia and the Philippines] generally have a highland center from which short rivers radiate to narrow coastal plains* (Compton Year Book).

high-level, *adj.* **1** *Meteorology.* occurring high up in the atmosphere: *high-level turbulence.*
2 *Physics.* having a high level of radioactivity: *storage high-level waste in geological formations.*

high polymer, *Chemistry.* a very large molecule composed of many repeating molecules or molecular fragments of relatively low molecular weight. A high polymer has usually a molecular weight greater than 5000. Synthetic resins such as polyester and acrylic fiber are made up of high polymers.

---

cap, fāce, fäther; best, bē, tèrm; pin, five;
rock, gō, ôrder; oil, out; cup, pùt, rüle;
yü in use, *yù* in uric;
ng in bring; *sh* in rush; *th* in thin, ŦH in then;
zh in seizure.
ə = *a* in about, *e* in taken, *i* in pencil, *o* in lemon, *u* in circus

high-pressure, *adj. Meteorology.* having a high barometric pressure: *Air generally moves from a high-pressure area into a low-pressure area* (James E. Miller).

high tide, *Oceanography.* **1** the highest level of the tide; the maximum elevation of the rising ocean on the shore. **2** the time when the tide is highest.

high water, = high tide.

hill, *n. Geography, Geology.* a raised part of the earth's surface, smaller than a mountain, usually having an elevation of less than 300 meters.

Hill reaction, *Botany.* a chemical reaction, important in photosynthesis, in which chloroplasts exposed to light yield oxygen even in the absence of carbon dioxide, provided that hydrogen acceptors such as ferric salts are present to cause the dissociation of water and the resulting liberation of oxygen. [named after Robin *Hill,* 20th-century English biochemist who reported this reaction in 1938]

hilum (hī′ləm), *n., pl.* **-la** (-lə). **1** *Botany.* **a** a mark or scar on a seed at the point of attachment to the seed vessel: *In seeds with outer coats of low permeability, much of the imbibition of water, which initiates germination, takes place through the hilum* (Greulach and Adams, *Plants*). **b** the central point of a grain of starch: *the hilum, around which a large number of curved lines arrange themselves* (Jabez Hogg). **2** *Anatomy.* a place in a gland or other organ where blood vessels, nerves, and ducts enter and leave. [from Latin *hilum* a trifle]

hindbrain, *n. Embryology.* the back part of the brain, composed of the metencephalon and myelencephalon. It includes the cerebellum, pons Varolii, and the medulla oblongata. *The brain tube enlarges moderately in three regions known as the fore-, mid-, and hindbrain. This tubular structure remains in the adult as the "brain stem" ... while the hindbrain develops the cerebellum* (Shull, *Principles of Animal Biology*). Also called **rhombencephalon.**

hindgut, *n.* **1** *Embryology.* the back section of the embryonic alimentary canal, from which the posterior part of the digestive tract develops. **2** *Zoology.* the posterior part of the digestive tract of an insect or crustacean. Compare **foregut, midgut.**

hinge joint, *Anatomy.* a joint, such as the elbow, in which a convex part of one bone fits into a concave part of another, allowing motion in only one plane.

hinge tooth, *Zoology.* any of the small, sharp projections of the shell near the hinge in bivalves.

hinterland, *n. Geology.* the land or region behind a coast. SYN: interior.

hip, *n. Anatomy.* **1** the fleshy part of the human body that covers the hip joint, extending from just below the waist to the upper thigh. **2** a similar part in animals, where the hind leg joins the body. **3** = hip joint.

hipbone, *n. Anatomy.* **1** either of the two wide, irregular bones, which, with the lower backbone, form the pelvis in man and other mammals. Each hipbone consists of three consolidated bones, the ilium, ischium, and pubis. SYN: innominate bone, coxa. **2** the upper part of the femur.

hip joint, *Anatomy.* the ball-and-socket joint formed by the upper thighbone and the hipbone.

hippocampus (hip′ə kam′pəs), *n., pl.* **-pi** (-pī). *Anatomy.* a curved ridge along the extension of each lateral ventricle of the brain. It consists mainly of gray matter. *One part of the limbic system, the hippocampus, is indisputably vital to memory. Patients whose hippocampi have been destroyed or partially removed cannot recall new information* (Time). [from New Latin *hippocampus,* from Greek *hippokampos* sea horse, from *hippos* horse + *kampos* sea monster]
—**hippocampal,** *adj.* of or having to do with the hippocampus: *In mice the hippocampal area of the brain contains the memory trace of recently learned maze behaviour* (Dwight J. Ingle).

hirsute (hėr′süt), *adj. Botany, Zoology.* covered with long, stiff hairs: *The bee assassins, the Apiomerinae, have clublike front legs that are covered with stout hairs, an arrangement that is well adapted to capturing the hirsute bee* (Scientific American). [from Latin *hirsutus*]

His, *abbrev.* histidine.

hispid, *adj. Botany, Zoology.* rough with stiff hairs, bristles, or spines: *a hispid leaf surface, the hispid hare.* [from Latin *hispidus*]
—**hispidity** (his pid′ə tē), *n.* hispid condition or appearance.

histaminase (his tam′ə nās), *n. Biochemistry.* an enzyme that can make a histamine inactive, used especially in the treatment of allergic reactions involving the release of histamine.

histamine (his′tə mēn′ *or* his′tə mən), *n. Biochemistry.* an amine derived from histidine and found in animal and plant tissues. It dilates blood vessels, stimulates gastric and other secretions, and is released by the body in allergic reactions. Histamine is used in the diagnosis and treatment of various allergies. *Histamine is a powerful stimulant of the gastric and other secretory glands, and has some stimulatory action on the intestinal musculature* (Beaumont and Dodds, *Recent Advances in Medicine*). Formula: $C_5H_9N_3$ [from *hist*(idine) + *amine*]
—**histaminic,** *adj.* of or induced by histamine: *a histaminic headache.*

histidine (his′tə dēn), *n. Biochemistry.* an essential amino acid present in many simple proteins and yielded by hydrolysis. It is sometimes used to treat peptic ulcers and arteriosclerosis. *Formula:* $C_6H_9N_3O_2$ *Abbreviation:* His [from German *Histidin,* from Greek *histion* tissue, web]

histiocyte (his′tē ə sīt), *n. Biology.* a fixed macrophage of the connective tissue, especially one that stores certain dyes: *Histiocytes are scavenger cells which remove diseased or dead cells and lymphocytes are their second cousins* (Science News Letter). [from Greek *histion* tissue + English -*cyte*]
—**histiocytic** (his′tē ə sit′ik), *adj.* composed of histiocytes: *a histiocytic tumor.*

histo-, *combining form. Biology.* tissue, as in *histochemistry* = the chemistry of tissues. [from Greek *histos* web]

histochemical, *adj.* **1** of or having to do with histochemistry. **2** using laboratory processes developed to study body tissue: *A histochemical technique with Prussian blue has shown that in Daphnia which is losing haemoglobin, iron accumulates in the wall of the excretory organs* (H. Munro Fox). —**histochemically,** *adv.*

288

histochemistry, *n.* the science dealing with the chemical composition of the tissues of the body.

histocompatibility (his'tō kəm pat'ə bil'ə tē), *n. Immunology.* the condition in which grafted or transplanted tissue is accepted by surrounding tissue without rejection: *Organ transplantation has brought urgency to the problem of the genetics of histocompatibility* (Victor A. McKusick).
—**histocompatibility antigen,** any of various proteins in the blood that stimulate the rejection of foreign tissue: *At the root of the rejection problem in transplantation surgery are the histocompatibility antigens* (New Scientist and Science Journal). Compare **HLA.**

histogen (his'tə jən), *n. Botany.* any of the areas of a plant in which tissues become differentiated.

histogenesis (his'tə jen'ə sis), *n. Embryology.* the development of tissues from embryonic germ layers.
—**histogenetic,** *adj.* of or having to do with histogenesis: *histogenetic studies of flowers.*

histogeny (his toj'ə nē), *n.* = histogenesis.

histogram (his'tə gram), *n. Statistics.* a graph of a frequency distribution in which vertical rectangles or columns are constructed with the width of each rectangle being a class interval and the height a distance corresponding to the frequency in that class interval. [from Greek *histos* web + English *-gram*]

histography, *n. Biology.* the description of bodily tissues.

histoid, *adj. Biology, Medicine.* of or like tissue, especially normal or connective tissue: *a histoid tumor.*

histological (his'tə loj'ə kəl), *adj.* of or having to do with histology: *a histological study, examination, etc.*
—**histologically,** *adv.*

histology (hi stol'ə jē), *n.* **1** the branch of biology that deals with the structure, especially the microscopic structure, of the tissues of animals and plants.
2 the tissue structure of an animal or plant.
[from Greek *histos* web, tissue]

histolysis (his tol'ə sis), *n. Biology, Medicine.* the disintegration or dissolution of bodily tissue.
—**histolytic** (his'tə lit'ik), *adj.* of or having to do with histolysis.

histone (his'tōn), *n. Biochemistry.* any of a class of proteins, such as globin, that are basic in reaction and are soluble in water. Histones occur in cell nuclei, often in combination with nucleic acid. [from German *Histon,* from Greek *histanai* to stay, stand + German *-on* -one (apparently because it halts coagulation)]

histophysiology, *n.* the branch of physiology that deals with the structure and function of tissues.

histotoxic, *adj. Medicine.* poisonous to tissues: *Histotoxic anoxia is caused by poisons that make the body tissues incapable of using the oxygen supply. The best-known chemical that causes this condition is cyanide* (W.R. Stovall).

hive, *n. Zoology.* a colony of bees.

hl., *abbrev.* hectoliter *or* hectoliters.

HLA or HL-A (āch'el'ā'), *adj. Immunology.* of or having to do with a system of tissue typing based on the presence of certain antigens in the body: *Studies of the many HLA types found in the population have also shown that some of them are associated with specific diseases: for example, a high proportion of all patients with some form of arthritis are found to have HLA B27 antigens* (London *Times*). Compare **histocom-**

patibility antigen. [from *h(uman) l(eucocyte) a(ntigen)*]

hm., *abbrev.* hectometer *or* hectometers.

Ho, *symbol.* holmium.

hoarfrost, *n. Meteorology.* the white, feathery, interlocking crystals of ice formed when dew freezes. Also called **white frost.**

hodoscope (hod'ə skōp), *n. Physics.* an instrument for tracing the paths of high-energy particles by means of an array of small radiation detectors, each with its own neon light. The flashing of light indicates the path of a particle passing through. [from Greek *hodos* way + English *-scope*]

hogback, *n. Geography, Geology.* a low, sharp ridge with steep sides of nearly equal inclination, formed from the edge of an upturned layer of rock resistant to erosion.

Holarctic (hol ärk'tik), *adj. Zoology.* of or belonging to a zoogeographical division consisting of the Nearctic division of the New World and the Palearctic division of the Old World: *The Palearctic and Nearctic regions are least separated, and their faunas have much in common, so that they are often combined as the Holarctic Region.* (Storer, *General Zoology*). [from *hol-,* variant of *holo-* whole + *Arctic*]

hole, *n. Physics.* a vacant position in a crystal left by the loss or absence of an electron, especially a position in a semiconductor which acts as a carrier of a positive electric charge: *A vacancy [in the crystal structure] may be filled when an electron from a neighboring atom moves over, transferring the deficiency or "hole" to the atom it left. The hole may move again by the transfer of an electron from a second neighboring atom, and so on. This kind of material is called o-type.* (John R. Pierce).

holmium (hōl'mē əm), *n. Chemistry.* a rare-earth metallic element which occurs in combination with other rare-earth elements in certain minerals. Its compounds are highly magnetic. *Symbol:* Ho; *atomic number* 67; *atomic weight* 164.930; *melting point* ab. 1470°C; *boiling point* ab. 2720°C; *oxidation state* 3. [from New Latin, from *(Stock)holm,* Sweden, near where it was found]

holo-, *combining form.* **1** whole; entire; all; full, as in *holobenthic, holohedral, holometabolism.* **2** wholly; entirely; fully, as in *holoblast, holoenzyme.* [from Greek *holos* whole]

holobenthic (hol'ə ben'thik), *adj. Biology.* living at or near the bottom of the sea during all stages of life.

holoblast (hol'ə blast *or* hō'lə blast), *n. Embryology.* an ovum consisting entirely or mostly of formative or germinal matter. [from *holo-* + Greek *blastos* germ]

---

cap, fāce, fäther; best, bē, tèrm; pin, fīve;
rock, gō, ôrder; oil, out; cup, pùt, rüle,
yü in use, *yù* in uric;
ng in bring; *sh* in rush; *th* in thin, ᴛʜ in then;
zh in seizure.
ə = *a* in about, *e* in taken, *i* in pencil, *o* in lemon, *u* in circus

—**holoblastic,** *adj.* (of an ovum) consisting entirely or mostly of germinal matter and therefore undergoing complete segmentation. Contrasted with **meroblastic.**

Holocene (hol′ə sēn), *Geology.* —*n.* **1** the present epoch of the Cenozoic era, after the Pleistocene epoch: *The Holocene includes only the last 10,000 years, during which the continents were largely ice free* (Eicher and McAlester, *History of the Earth*). **2** the series of rocks formed during this epoch.
—*adj.* of or having to do with this epoch or series. Also called **Recent.**
[from *holo-* + Greek *kainos* recent]

holoclastic (hol′ə klas′tik), *adj. Geology.* composed of fragments of sedimentary rock. Contrasted with **pyroclastic.**

holocrine (hol′ə krin *or* hol′ə krīn), *adj. Physiology.* **1** that produces a secretion consisting of cells which have disintegrated: *a holocrine gland.* **2** of or produced by such a gland. Compare **apocrine, endocrine.** [from *holo-* + Greek *krinein* to separate]

holoenzyme (hol′ō en′zim), *n. Biochemistry.* a fully active, complex enzyme, composed of a protein component and a coenzyme: *The molecules of many enzymes consist of two parts: the pure protein part, called the apoenzyme, and a non-protein part called the coenzyme. The active enzyme, called the holoenzyme, is the apoenzyme with the coenzyme attached to it. Often the coenzyme is a vitamin molecule or a closely related molecule* (Linus Pauling).

hologram (hol′ə gram), *n. Optics.* the photographic record of an image produced by holography: *Images are always life-size, for a hologram can't be enlarged or reduced. Its properties of minute and exact duplication make holography ideal for such scientific endeavors as data storage (holograms may replace microfiche some day), glaucoma detection, atomic research and metallurgical testing* (Maclean's).

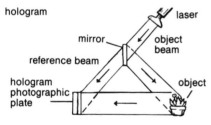

hologram

laser

mirror

object beam

reference beam

hologram photographic plate

object

The laser is split so that the mirror directs the object beam to shine on the object and the reference beam to illuminate the hologram photographic plate. Ordinary light is projected onto the plate or film and a life-sized, three-dimensional image is recorded.

holography (hō log′rə fē), *n. Optics.* a method of producing a three-dimensional image of an object by recording on a photographic plate or film the pattern of interference formed by a split laser beam and then illuminating the pattern with ordinary light: *Holography differs from photography in that it uses the coherent properties of laser light to record all the information in an arrangement of light rays on a piece of photographic film* (New Scientist).

holohedral (hol′ə hē′drəl), *adj. Crystallography.* having the full number of planes required by the maximum symmetry: *a holohedral crystal.* [from *holo-* + Greek *hedra* base]

holohedron (hol′ə hē′drən), *n., pl.* **-drons, -dra** (-drə). *Crystallography.* a holohedral crystal or form.

holometabolism, *n. Zoology.* the complete form of metamorphosis, in which an insect passes through four separate stages of growth, as embryo, larva, pupa, and imago. Also called **complete metamorphosis.** Compare **hemimetabolism.**
—**holometabolous,** *adj.* of or having to do with holometabolism: *... the larval-pupal transformation of holometabolous insects* (Science).

holophyte (hol′ə fit), *n. Biology.* an organism that produces its own food by photosynthesis. [from *holo-* + Greek *phyton* plant]
—**holophytic** (hol′ə fit′ik), *adj.* of or like a holophyte; photosynthetic.

holoplankton (hol′ə plangk′tən), *n. Biology.* organisms that live as plankton through all stages of their life cycle.

holostean (hol′ə stē′ən), *Zoology.* —*adj.* of or having to do with a group of largely extinct bony fishes that were the ancestors of the teleosts. The gar is a living example. —*n.* a holostean fish. [from *holo-* + Greek *osteon* bone]

holosymmetrical, *adj.* = holohedral.

holothurian (hol′ə thủr′ē ən), *n. Zoology.* any of a class (Holothuroidea) of echinoderms, having an elongated, leathery body covered with very small calcareous plates. The sea cucumber is a holothurian. [from New Latin *Holothuria* the sea cucumber genus, from Greek *holothourion* a kind of sea polyp]

holotrichous (hə lot′rə kəs), *adj. Biology.* uniformly covered with cilia, as certain protozoans. [from *holo-* + Greek *trichos* hair]

holotype, *n. Biology.* a single specimen of which the description of a species, variety, or lesser taxonomic group is based.
—**holotypic,** *adj.* of or based upon a holotype.

holozoic, *adj. Biology.* obtaining its food through the ingestion of complex organic substances as most animals do. [from *holo-* + Greek *zoion* animal]

homeo-, *combining form.* similar, the same; constant, as in *homeomorphism, homeotherm.* [from Greek *homoi-*, from *homoios* similar]

homeomorphic (hō′mē ō môr′fik), *adj. Mathematics.* of or characterized by homeomorphism: *homeomorphic figures or sets.*

homeomorphism (hō′mē ə môr′fiz əm), *n.* **1** *Crystallography.* similarity in crystalline form between substances of unlike chemical composition.
2 *Mathematics.* a one-to-one correspondence between the points of two geometric figures which is continuous in both directions. A topological transformation, as that produced by twisting or otherwise deforming a figure, is a homeomorphism. Compare **homomorphism.**

homeomorphous (hō′mē ə môr′fəs), *adj. Crystallography.* crystallizing in the same or a related form (used especially of substances not having an analogous chemical composition).

homeostasis (hō'mē ə stā'sis), *n. Biology.* the ability or tendency of an organism or cell to maintain internal equilibrium of temperature, fluid content, etc., by the regulation of its physiological processes and by automatic adjustments to the external environment: ... *homeostasis is the key to the regulation of metabolic processes* (E.W. Sinnott). [from *homeo-* + *stasis* position]
—**homeostatic** (hō'mē ə stat'ik), *adj.* of or having to do with homeostasis: *New blood cells have been produced to replace those lost through bleeding [and] ... help to preserve or restore the internal environment; they are homeostatic responses* (McElroy, *Biology and Man*).

homeotherm (hō'mē ə thèrm'), *n. Zoology.* a warm-blooded animal: *Man, like other mammals, is a homeotherm—his temperature varying little with alterations of the temperature of his surroundings* (Eric Neil). Also called **endotherm.**
—**homeothermal** or **homeothermic,** *adj.* warm-blooded: *The importance of environmental temperature ... is less obvious with homeothermic, or warm-blooded, animals, in which body temperature is virtually independent of environmental temperature* (Scientific American).

homeothermic, *adj.* = warm-blooded.

hominid (hom'ə nid), *Zoology.* —*adj.* belonging to the family (Hominidae) of primates of which *Homo sapiens* is the only surviving species: *Because the canines of other Miocene apes are already large and specialized, some anatomists have believed that the hominid or Man-like line of evolution must have diverged from the pongid or ape line before this specialization developed* (New Scientist).
—*n.* a hominid animal: *Africa, with its vast areas of steppe and savannah, provided the ideal environment for the evolution of ground-living primates, and it was here that the first hominids became bipedal and experimented with tool-making* (New Scientist).
[from New Latin *Hominidae* the family name, from Latin *homo, hominem* man]

hominize, *v. Biology.* to make human or manlike in the process of evolution: *hominized primates.*
—**hominization,** *n.* the process of developing human characteristics through evolution.

hominoid (hom'ə noid), *Zoology.* —*adj.* of or belonging to a superfamily (Hominoidea) of primates that includes apes and humans: *Species of Ramapithecus are among the few hominoid species ... considered as possibly close to the direct line of human ancestry* (Nature).
—*n.* a hominoid primate: *The hominoids may be divided into past and present representatives of (a) the ... gibbons, (b) the pongids or "great apes," such as the modern chimpanzees, gorillas and orangutans, and (c) the hominids or human family* (J.C. Trevor).

Homo (hō'mō), *n., pl.* **Homines** (hom'ə nēz'). *Zoology.* the genus of primate mammals comprising man, including one extant species, *Homo sapiens,* and various extinct species known from fossil remains, such as *Homo erectus* and *Homo habilis: Most of the characteristics of Homo seem to have evolved well within the Pleistocene* (J.Z. Young). [from New Latin *Homo,* from Latin *homo, hominem* man]

homo-, *combining form.* the same; similar; equal: *Homogenous = of the same origin.* [from Greek *homo-,* from *homos* same]

homocercal (hō'mə sėr'kəl), *adj. Zoology.* **1** having a tail in which the upper and lower lobes are symmetrical or almost symmetrical, the end of the spinal column coinciding with the base of the tail fin: *a homocercal fish.* **2** of or having to do with this form of tail. Contrasted with **heterocercal.**
[from *homo-* + Greek *kerkos* tail]

homochromatic, *adj. Optics.* relating to or possessing one color or hue: *The instrument is so constructed that the observer views a homochromatic field in monochromatic light of the wave length to which the instrument is adjusted* (Hardy and Perrin, *Principles of Optics*).
—**homochromatism,** *n.* the condition of being homochromatic.

homochromous (hō'mə krō'məs), *adj.* **1** *Optics.* = homochromatic. **2** *Botany.* having all the florets of the same color, as a composite flower or flower head. [from *homo-* + Greek *chroma* color]

homocyclic, *Chemistry.* —*adj.* having a ring structure in which the ring is composed of one kind of atom only: *Benzene is a homocyclic compound.* —*n.* a homocyclic compound. Contrasted with **heterocyclic.**

homocysteine (hō'mə sis'tē ēn), *n. Biochemistry.* a crystalline amino acid that is an intermediate in the metabolic conversion of methionine to cysteine: *Recent findings suggest that arteriosclerosis, a disease with high incidence in industrialized societies, is initiated by increased amounts of homocysteine formed from metabolism of diets rich in animal proteins* (Kilmer S. McCully). *Formula:* $C_4H_9NO_2S$

homodromous (hō mod'rə məs), *adj. Botany.* having the spiral arrangement of the leaves on the stem and of those on the branches in the same direction. Contrasted with **heterodromous.** [from *homo-* + Greek *dromos* a running]

homoecious (hō mē'shəs), *adj. Biology.* going through different stages of growth on the same host as a type of parasitic beetle. Compare **heteroecious.** [from *homo-* + Greek *oikia* house]

Homo erectus (hō'mō i rek'təs), *Zoology.* an extinct species of human regarded as an ancestor of *Homo sapiens: Homo erectus first appears in Africa and Java in the Early Pleistocene period, prior to 1.5 million years ago, and is known throughout the early part of the Middle Pleistocene from these areas as well as from China and Europe* (Leslie C. Aiello). See the picture at **evolution.** [from New Latin *Homo erectus* erect Man]

homogametic (hō'mō gə met'ik), *adj. Biology.* producing one type of gamete or sex chromosome. The female mammal, producing only X chromosomes, is homo-

cap, fāce, fäther; best, bē, tèrm; pin, five;
rock, gō, ôrder; oil, out; cup, pùt, rüle,
yü in use, *yù* in uric;
ng in bring; sh in rush; th in thin, ᴛʜ in then;
zh in seizure.
ə = *a* in about, *e* in taken, *i* in pencil, *o* in lemon, *u* in circus

291

gametic. *In birds, the male is homogametic, having X-chromosomes, while the female is heterogametic, with XY sex chromosomes* (Ursula Mittwoch). Contrasted with **heterogametic.**

homogamous (hō mog′ə məs), *adj. Botany.* **1** having flowers or florets that do not differ sexually.
2 having the stamens and pistils maturing at the same time, as a monoclinous flower. Contrasted with **dichogamous.**
3 *Biology.* characterized by homogamy or inbreeding.

homogamy (hō mog′ə mē), *n.* **1** *Botany.* homogamous condition, especially the condition of having the stamens and pistils of a monoclinous flower mature at the same time. Contrasted with **dichogamy.**
2 *Biology.* the mating of individuals of the same species or of similar characteristics. SYN: inbreeding.

homogenate (hə moj′ə nāt), *n. Biology.* a homogenized substance, especially tissue that has been homogenized for study or analysis: *By means of electron microscopy and centrifugation of cell homogenates at various speeds the structure and function of the various parts of all living cells can be studied* (Paul Saltman).

homogeneity (hō′mə jə nē′ə tē), *n.* the quality or condition of being homogeneous: *The steps needed to insure correct operation require that the control probe be in about as homogeneous a field as the research probe and that both probes be at the same field value. This puts severe homogeneity requirements on the magnet* (Science).

homogeneous (hō′mə jē′nē əs *or* hō′mə jē′nyəs), *adj.* **1** made up of similar elements or parts; of uniform nature or character throughout: *As the ancients knew, the melting together of two metals usually produces a homogeneous liquid and, on freezing the liquid, an apparently homogeneous solid* (J. Crowther).
2 *Chemistry.* consisting of or involving a single phase in any system or process: *The equilibrium is described as homogeneous if all of the substances involved are in a single phase and are uniformly mixed; if, on the other hand, two or more distinct phases are present, the equilibrium is described as heterogeneous* (Jones, *Inorganic Chemistry*). *One of the most rapidly expanding areas of catalysis is that of homogeneous catalysis, in which at least one of the reactants and the catalyst are present in the same phase, usually as a gas or liquid.* (Vladimir Haensel).
3 *Mathematics.* of the same degree or dimensions. A **homogeneous equation** is an equation whose terms have all the same degree. A **homogeneous function** is a polynomial in two or more variables whose terms have all the same degree.
[from Medieval Latin *homogeneus,* ultimately from Greek *homos* same + *genos* kind] —**homogeneously,** *adv.*

homogenesis (hō′mə jen′ə sis), *n. Biology.* the ordinary course of generation in which the offspring is like the parent and goes through the same cycle of development.

homogenetic (hō′mə jə net′ik), *adj. Biology.* **1** having to do with homogenesis. **2** having a common origin; derived from the same structure, however modified. SYN: homogenous.

homogenize (hə moj′ə nīz), *v.* to make homogeneous or uniform by dividing into small parts and mixing: *"Homogenized," tobacco ... generally is made of ground-up stems, tobacco dust and low-grade tobacco leaves* (Wall Street Journal). In **homogenized milk** the fat is distributed evenly throughout the milk and does not rise to the top in the form of cream. —**homogenization,** *n.* —**homogenizer,** *n.*

homogenous (hō moj′ə nəs), *adj. Biology.* corresponding in structure because of a common origin; homogenetic: *homogenous organs.* Compare **homomorphic.**

homograft, *n. Immunology.* tissue from one individual grafted on an individual of the same species: *It is the ribonucleic acid (RNA) in the cells of the liver and spleen of the donor animal that produces immunity in the recipient animal and enables it to tolerate the homograft* (New Scientist).
ASSOCIATED TERMS: see **allograft.**

Homo habilis (hō′mō hab′ə ləs), *Zoology.* an extinct species of human, believed to have been the earliest toolmaker, whose fossil fragments were discovered in the early 1960's at the Olduvai Gorge in northern Tanzania: *The smaller form of Australopithecus and a similarly gracile form of true human being called Homo habilis were thought to have been omnivorous, mixing meat with roots, nuts, eggs, shoots and fruit* (Boyce Rensberger). [from New Latin *Homo habilis* skillful man]

homoiotherm (hō moi′ə thèrm), *n. Zoology.* a warm-blooded animal.

homoiothermic (hō moi′ə thèr′mik), *adj.* = warm-blooded. Contrasted with **poikilothermic.**

homolecithal (hō′mə les′ə thəl), *adj. Embryology.* having a small amount of yolk evenly distributed throughout the egg: *Homolecithal eggs normally undergo complete or entire cleavage, by which is meant that the nucleus, the cytoplasm, and the yolk divide into two parts* (Harbaugh and Goodrich, *Fundamentals of Biology*). Compare **alecithal, centrolecithal, telolecithal.** [from *homo-* + Greek *lekithos* yolk]

homolog or **homologue** (hom′ə lôg), *n.* something that shows correspondence in structure or origin; a homologous thing or part: *Californium has the atomic number 98 and is one of the actinide elements. Chemically its properties resemble those of dysprosium, its homolog in the lanthanide series in the periodic table* (Frank P. Baranowski). *Chromosomes display a strong tendency to associate with one another in a variety of ways. Most commonly the associations are pairwise and highly specific in nature, occurring between identical chromosomes (homologs) or their multiples* (Rhoda F. Grell).

homologous (hō mol′ə gəs), *adj.* **1** *Biology.* a corresponding in type of structure and in origin: *Homologous organs may have similar functions, for example, the legs of a man and the hindlegs of a horse; or they may have different functions, for example, the arms of a man and the wings of a bird* (Hegner and Stiles, *College Zoology*). **b** derived from or involving a different organism of the same species: *a homologous antiserum, a homologous graft.*
2 *Genetics.* being normally identical in morphology and in arrangement of genetic loci: *The diploid (2n) number of chromosomes characteristic of the species*

... are present in homologous pairs (Biology Regents Syllabus). See also **homologous chromosome, homozygote.**

3 *Chemistry.* **a** belonging to or designating a series in which successive members differ regularly in formula, especially a series of organic compounds differing by multiples of $-CH_2$, as the alcohols: *A study of the behavior of one member of an homologous series indicates in general the properties of the remainder of the series* (Offner, *Fundamentals of Chemistry*). **b** of or belonging to the same group of elements in the periodic table: *Actinium and thorium are homologous.*
[from Greek *homologos* agreeing, from *homos* same + *logos* reasoning, relation]

homologous chromosome, *Genetics.* either of a pair of like chromosomes in a diploid organism, one derived from the egg and the other from the sperm, that pair during the early stage of meiosis, and are again separated into different germ cells. Homologous chromosomes contain genes for the same traits. *During the formation of gametes the process of meiosis will separate the homologous chromosomes, so that gametes will contain only one gene from each pair* (Mackean, *Introduction to Biology*).

homolographic (hom'ə lə graf'ik), *adj. Geography.* representing in equal proportions, as in a map projection in which equal areas of the earth's surface are represented by equal areas on a map or chart: *Maps which use one scale throughout are known as homolographic or equal-area projections.* (White and Renner, *Human Geography*). Compare **homolosine**. [from Greek *homalos* even, level + English *graphic*]

homologue, *n.* = homolog: *Curium has only a 3-valent form, while its predecessor, americium, shows 2-valency, as does europium, its homologue in the lower series* (H.R. Paneth).

homology (hō mol'ə jē), *n.* **1** *Biology.* correspondence in type of structure and in origin: *The appendages of the crayfish may be used to illustrate a very important biological principle, homology. Body structures which arise in similar ways in the embryo are said to be homologous. Thus, we can say that the appendages of the crayfish are homologous to one another even though they become greatly modified and assume different functions in the adult* (Winchester, *Zoology*).
2 *Chemistry.* **a** the relation of the compounds forming a homologous series. **b** the relation of the elements belonging to the same group.

homology (def. 1)

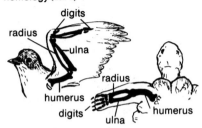

homolosine (hō mol'ə sin *or* hō mol'ə sīn), *adj. Geography.* having to do with a kind of map projection (homolosine projection) made by combining a sinusoidal projection with a homolographic projection. Compare

homolographic. [from Greek *homalos* even + English *sine*]

homolysis (hō mol'ə sis), *n. Chemistry.* the breaking of a chemical bond in a symmetrical manner, so that the two electrons in a bonding pair are separated: *The ions and electrons may recombine to produce neutral molecules in an excited state, which can undergo homolysis to produce free radicals* (Scientific American). Contrasted with **heterolysis.**
—**homolytic** (hō'mə lit'ik), *adj.* of or having to do with homolysis: *A covalent bond may in other circumstances break by homolytic fission, each of the electrons separating with one of the atoms, giving two free atoms ... or—if the atoms concerned have other atoms bound to them—free radicals* (K. D. Wadsworth).

homomorphic (hō'mə môr'fik), *adj.* **1** *Biology.* similar in external appearance, but not related in structure or origin: *homomorphic colonies of bryozoans.* Compare **homogenous.**
2 *Mathematics.* of or characterized by homomorphism: *a homomorphic transformation, a homomorphic group.*
[from *homo-* + Greek *morphē* form]

homomorphism (hō'mə môr'fiz əm), *n.* **1** *Zoology.* resemblance of form between the immature and adult stages of an animal.
2 *Biology.* the condition of being homomorphic.
3 *Mathematics.* a transformation of one set into another that preserves in the second set the operations between the members of the first set. Compare **homeomorphism.**

homonomous (hō mon'ə məs), *adj. Biology.* having the same mode of growth, as parts or organs: *The individual fingers and toes ... are homonomous structures* (Francis J. Bell). Contrasted with **heteronomous.** [from *homo-* + Greek *nomos* law]

homonuclear, *adj. Chemistry.* consisting of identical atoms or groups of atoms: O_2 *is a homonuclear molecule.* Contrasted with **heteronuclear.**

homonym (hom'ə nim), *n. Biology.* a name that has been applied previously to a different species or genus, and therefore cannot be used for a new species or genus, on the basis of the law of priority. [from Latin *homonymum*, from Greek *homōnymon*, neuter of *homōnymos* of the same name]

homoplastic, *adj. Biology.* similar but of different origin: *homoplastic organs.* [from *homo-* + Greek *plastos* something molded] —**homoplastically**, *adv.*

homopolar, *adj. Chemistry.* of or characterized by a symmetrical distribution of positive and negative electric charges between two atoms, as in the common covalent bond. Contrasted with **heteropolar.**

homopolymer, *n. Chemistry.* a polymer formed from two or more molecules of the same kind.

cap, fāce, fäther; best, bē, tėrm; pin, five;
rock, gō, ôrder; oil, out; cup, pùt, rüle,
yü in use, *yu̇* in uric;
ng in bring; sh in rush; th in thin, ŦH in then;
zh in seizure.
ə = a in about, e in taken, i in pencil, o in
lemon, u in circus

—**homopolymeric,** *adj.* of a homopolymer; formed from two or more molecules of the same kind.

—**homopolymerize,** *v.* to change or be changed by a reaction in which two or more molecules of the same kind polymerize to form a complex molecule, as in the manufacture of polystyrene. —**homopolymerization,** *n.*

homopteran (hō mop'tər ən), *adj.* = homopterous. —*n.* a homopterous insect.

homopterous (hō mop'tər əs), *adj.* of or belonging to an order (Homoptera) of insects having mouthparts adapted for piercing and sucking and wings of the same thickness and texture throughout. Aphids and cicadas belong to this order. [from *homo-* + Greek *pteron* wing]

Homo sapiens (hō'mō sā'pē enz *or* hō'mō sap'ē enz), *Zoology.* the species including all existing human beings; the only surviving species of the primate family Hominidae: *Only one species of man, Homo sapiens, has survived, although there were probably other species who lived in the past. Human beings, however, fall into several races* [*that*] *... arose through the isolation of groups by geographic or social barriers, and because of these barriers different groups of genes were isolated.* (McElroy, *Biology and Man*). See the picture at **evolution.** [from New Latin *Homo sapiens* knowing or wise man]

homoscedastic (hō'mə si das'tik), *adj. Statistics.* showing equal variability; showing the same standard deviation: *homoscedastic arrays or values.* Contrasted with **heteroscedastic.** [from *homo-* + Greek *skedastikos* that can be scattered]

homosporous (hō'mə spôr'əs *or* hō mos'pər əs), *adj. Botany.* having only one kind of asexually produced spores: *In the familiar species of fern, as in the mosses and in the algae discussed, all the meiospores are similar in appearance ... and the species are described as homosporous.* (Greulach and Adams, *Plants*).

—**homospory** (hō'mə spôr'ē *or* hō mos'pər ē), *n.* the production of only one kind of asexual spores.

homotactic or **homotaxic,** *adj.* characterized by or exhibiting homotaxis.

homotaxis (hō'mə tak'sis), *n.* similarity of arrangement, as of parts of the body or of geological strata. Contrasted with **heterotaxis.** [from *homo-* + Greek *taxis* arrangement]

homothallic (hō'mə thal'ik), *adj. Biology.* having only one haploid type, which produces self-fertile cells which can mate with one another: *a homothallic fungus.* Contrasted with **heterothallic.** [from *homo-* + Greek *thallos* young shoot]

homozygosis (hō'mə zī gō'sis), *n.* = homozygosity.

homozygosity (hō'mə zī gos'ə tē), *n. Genetics.* the condition of being homozygous; development from a homozygote: *In each generation of selection they will necessarily breed together close relatives ... The more the inbreeding is continued, the greater the extent of this homozygosity* (Hollingsworth and Bowler, *Principles and Processes of Biology*).

homozygote (hō'mə zi'gōt), *n. Genetics.* an organism whose chromosomes contain an identical pair of genes and which, therefore, always breed true to type: *the heterozygote (hybrid) possessing one normal hemoglo-*

bin gene and one sickle cell hemoglobin gene has an advantage over either homozygote (two normal genes or two sickle cell genes) (Scientific American). Compare **hemizygote, heterozygote.**

—**homozygous** (hō'mə zi'gəs), *adj.* of or having to do with a homozygote; having the same genes (as for tallness) at a locus on the homologous chromosomes: *Where the two components are the same, the cells are said to be homozygous, but where they differ, one often being dominant over the other, the term heterozygous is employed* (G. Fulton Roberts).

honey, *n. Zoology.* **1** the sweet secretion, consisting mainly of fructose and dextrose, made by bees from the nectar they collect from flowers. **2** any of various similar substances secreted by insects other than bees.

honeydew, *n.* **1** *Botany.* a sweet substance that oozes from the leaves of certain plants in hot weather. **2** *Biology.* a sweet substance on leaves, secreted by tiny insects such as aphids or leaf hoppers, or by a fungus.

honey sac or **honey stomach,** *Zoology.* the enlargement of the alimentary canal in which the bee produces and carries its honey.

hood, *n.* **1** *Zoology.* a crest or other part on a bird's or animal's head that suggests a head covering in shape or color. **2** *Botany.* any part in plants serving as a covering, especially the arched upper part of the corolla or calyx in some flowers.

hooded, *adj.* **1** *Zoology.* **a** having a crest or other structure or arrangement of colors on the head resembling or suggesting a head covering: *the hooded crow.* **b** having elastic skin at the neck that, when distended, resembles a hood, as a cobra.

2 *Botany.* **a** (of a corolla or other part) shaped like a hood; cowled. **b** (of a flower or plant) having a corolla, calyx, or other part, shaped like a hood: *the hooded violet.*

hoof, *n. Zoology.* **1** a hard, horny covering on the feet of horses, cattle, sheep, pigs, and some other animals. **2** the whole foot of such animals. See the picture at **foot.**

hook, *n.* **1** *Geography.* a curved spit or point of land: *the Hook of Holland. Sandy Hook is at the entrance of New York Bay.* **2** *Biology.* a recurved and pointed organ or appendage of an animal or plant.

horizon, *n.* **1** *Astronomy.* **a** the boundary line where the sky and the earth or sea seem to meet; the apparent or visible horizon. One cannot see beyond the horizon. *The visible or apparent horizon is irregular in shape on land, and at sea it is a small circle lying below the true horizon by an angular distance called the dip* (Duncan, *Astronomy*). **b** the plane through an observer's position at right angles to the direction of the earth's gravity. This is usually called the sensible horizon. **c** the great circle of the celestial sphere that has a plane parallel to the sensible horizon and passes through the center of the earth, or the plane of this great circle. This is called the astronomical, celestial, true, or rational horizon. **d** a horizontal reflecting surface, such as a mirror or the surface of mercury or other fluid at rest, used in measuring altitudes. This is usually called an artificial or false horizon. **e** the limit of the theoretically observable universe: *Since no signal can move faster than the speed of light, an observer can be affected only by events from which a photon would have time to reach him since the beginning of* the universe. Such events are described as lying within the observer's horizon (Scientific American).

2 Geology. **a** a stratum or series of strata differing in fossils, etc., from the deposits above or below: *Scolecodonts and conodonts are surprisingly abundant fossils in some Ordovician and Silurian strata, and the fact that the various kinds discovered at different horizons are readily distinguishable makes this group of fossils useful as stratigraphic markers* (Moore, *Introduction to Historical Geology*). **b** one of the layers in a vertical cross section of a series of soils and subsoils: *The stratum of soil gradually develops a fairly well defined system of layers, termed in soil nomenclature "horizons", and the system of horizons which makes up the stratum is called a soil "profile"* (J.H. Quastel). Compare **A horizon, B horizon, C horizon.** [from Greek *horizōn* (*kyklos*) bounding (circle), from *horos* boundary, limit]

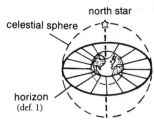

north star
celestial sphere
horizon
(def. 1)

horizontal, *adj.* **1** parallel to the plane of the horizon; at right angles to a vertical line.
2 measured in a line parallel to the horizon: *a horizontal distance.* —**horizontally**, *adv.*

horizontal fault, *Geology.* a fault that has no vertical displacement or dip. See the picture at **fault.**

hormogonium (hôr′mə gō′nē əm), *n., pl.* **-nia** (-nē ə). *Botany.* a portion of a filament in certain algae that becomes detached and develops into a new filament: *Most filamentous blue-green algae form hormogonia* (G.E. Fogg). [from New Latin, from Greek *hormos* chain, necklace + New Latin (*arche*)*gonium*]

hormonal (hôr mō′nl), *adj.* of, having to do with, or like a hormone: *Nutritional and hormonal disturbances frequently accompany and often complicate human pregnancy* (D.H.M. Woollam).
—**hormonally**, *adv*

hormone (hôr′mōn), *n. Biochemistry.* **1** a substance secreted chiefly by endocrine glands which enters the bloodstream and affects or controls the activity of some organ or tissue. Hormones control basic physiological processes such as growth, metabolism, and reproduction. Chemically, hormones may be amino acids, proteins, polypeptides, or steroids. ACTH is a polypeptide hormone that stimulates the adrenal glands to produce cortisone, a steroid hormone. Many hormones, including cortisone, are also produced synthetically. *In spring, the lengthening hours of light or rising temperature cause endocrine changes in many birds and fishes: the animals, under the influence of increased secretion of certain hormones, leave their winter quarters* (S.A. Barnett).
2 a similar substance in the protoplasm of plants. Auxins, cytokinins, and gibberellins are the chief growth-promoting plant hormones. *The young leaves and growing apex secrete at least one hormone that*

travels through the tissues, in some way suppressing the growth of all buds below. (Emerson, *Basic Botany*). Also called **internal secretion.**
[from Greek *hormōn* setting in motion, ultimately from *hormē* impulse]

hormonopoietic (hôr mō′nə poi et′ik), *adj. Physiology.* **1** producing hormones, as certain organs. **2** having to do with the production of hormones. [from *hormone* + Greek *poiētikos* productive]

horn, *n.* **1** *Zoology.* **a** a bony process, often curved and pointed and in pairs, on the heads of many ungulates: *There are four different kinds of horns among mammals: hollow horns, pronghorns, keratin-fiber horns and antlers* (Science News Letter). **b** the hard, durable, and partly transparent substance, principally keratin, of which some types of horn consist. A person's fingernails, the beak of a bird, the hoofs of horses, and tortoise shells are all made of horn.
2 *Geology.* a sharp, faceted pyramidal mountain peak bounded by cirques: *The Matterhorn is a famous example of a glacially sharpened, jagged horn.*
—**horned**, *adj. Zoology.* having a horn, horns, or a hornlike growth: *the horned coot, the horned adder.*

hornblende (hôrn′blend′), *n. Mineralogy.* a common black, dark-green, or brown silicate mineral containing aluminum and varying proportions of calcium, magnesium, and iron. Hornblende is a species of amphibole which is found in many igneous and metamorphic rocks. [from German *Hornblende*]

hornfels (hôrn′felz), *n. Geology.* a compact, fine-grained metamorphic rock composed of quartz, feldspar, mica, and other minerals, formed by the contact action of intrusive rock upon shale or other sedimentary rock. [from German *Hornfels,* from *Horn* horn + *Fels* rock]

horologic or **horological**, *adj.* of or having to do with horology.

horology (hə rol′ə jē), *n.* the science of measuring time.

horse latitudes, *Meteorology.* either of two regions characterized by high pressure and winds which are usually calm or very light. They extend around the earth at about 30 degrees north and 30 degrees south of the equator. *The heaviest traffic in the transport of angular momentum occurs at 30 degrees latitude, "the horse latitudes." At this latitude there is little or no wind movement near sea level, but strong winds blow from the west just below the stratosphere at about 40,000 feet* (Scientific American).

horsepower, *n.* a customary unit for measuring the power of engines, reactors, etc. One horsepower is equal to 550 foot-pounds per second, or to about 746 watts. *In the English system, where work is expressed in foot-pounds and time in seconds, the unit of power is one foot-pound per second. Since this unit is inconveniently small, a larger unit called the horsepower*

---

cap, fāce, fäther; best, bē, tėrm; pin, five;
rock, gō, ôrder; oil, out; cup, pùt, rüle,
yü in use, *yù* in uric;
ng in bring; *sh* in rush; *th* in thin, ᴛʜ in then;
zh in seizure.
ə = *a* in about, *e* in taken, *i* in pencil, *o* in lemon, *u* in circus

(hp) is in common use. 1 hp = 550 ft-lb/sec = 33,000 ft-lb/min. That is, a 1 hp motor, running at full load, is doing 33,000 ft-lb of work every minute it runs (Sears and Zemansky, *University Physics*).

horsepower-hour, *n.* a customary unit representing the work performed or energy consumed in working at the rate of one horsepower for one hour. It is equal to about .746 of a kilowatt-hour. *The specific fuel consumption is based on the horsepower output, and its units are pounds of fuel per horsepower-hour* (Sutton, *Rocket Propulsion Elements*).

horst (hôrst), *n. Geology.* an elongate, relatively uplifted part of the earth's crust limited by faults and standing out against its surroundings: *Geological evidence establishes that the Horlick Mountains are unmistakably a part of the antarctic horst, which appears to extend all the way from the western boundary of the Ross Sea to the eastern border of the Filchner Ice Shelf* (Science). Contrasted with **graben.** See the picture at **fault.** [from German *Horst* heap, mass, sandbank]

host, *n.* **1** *Biology.* **a** a living organism in or on which a parasite or commensal organism lives: *By absorbing blood, sap, digested food or the tissues of the living plant or animal the parasite obtains its food at the expense of its host* (Mackean, *Introduction to Biology*). **b** the dominant partner in a symbiotic or commensal relationship. Compare **guest.**
2 *Physics, Chemistry.* a crystal lattice or molecular structure that contains some foreign substance.

host-specific, *adj. Biology.* living in or on a particular species of host: *The plague bacterium is transferred from rat to rat by a particular flea which, like all other fleas, is host-specific; ordinarily it will feed on the blood of rats alone* (London *Times*).

hot laboratory or **hot lab,** a laboratory in which research with radioactive substances or with recombinant DNA is conducted, requiring special precautions: *Supersafe "hot" labs are designed for the most dangerous DNA work. Researchers use glove boxes so they never come in contact with test organisms* (World Book Science Annual).

hot spot, 1 *Geology.* any of various regions of the earth where a plume of molten material in the earth's mantle is carried upwards, heating the crust above: *Scattered around the globe are more than 100 small regions of isolated volcanic activity known to geologists as hot spots. Unlike most of the world's volcanoes, they are not always found at the boundaries of the great drifting plates that make up the earth's surface* (Scientific American).
2 *Astronomy.* an area of a planet, star, nebula, or other heavenly body that has a higher temperature than its surroundings: *They observed a region of the infrared spectrum that probes "hot spots" in Jupiter's clouds and found a series of strong absorption lines characteristic of water* (Michael J.S. Belton).
3 *Physics.* an area with a higher-than-average level of radioactivity: *In the past, under special circumstances, iodine-131 initially injected into the stratosphere or upper troposphere as a result of nuclear weapons tests has been brought down to earth unpredictably in the*

form of locally deposited fallout or "hot spots" (Bulletin of Atomic Scientists).
4 *Genetics.* a site in a portion of genetic material where mutations occur with unusually high frequency: *Nucleotide sequence data obtained from insertion and deletion experiments ... suggest that one highly mutable spot (hot spot) involves the deletion of an A from a stretch of 6 A residues* (Watson, *Molecular Biology of the Gene*).

hot spring, *Geology.* a spring producing warm water, usually at a temperature above that of the human body: *A satellite in a polar orbit can provide daily worldwide records of the changes in area or temperature of hot springs and steaming vents* (Robert W. Decker). Also called **thermal spring.**
▶ Gases and liquids escape from various openings in the earth called fumaroles, hot springs, or geysers. *Fumaroles,* usually found within craters or on the slopes of active volcanoes, emit jets of steam and other gases. *Hot springs* well forth from other openings or from volcanic fields in the wet season. *Geysers,* which spout water and steam at irregular intervals, are intermediate between fumaroles and hot springs.

hour angle, *Astronomy.* the angular distance, measured westward along the celestial equator, between the meridian of an observer and the hour circle of a heavenly body. The hour angle is used to determine solar time. *Hour angle is an angle measured from the celestial pole and is not concerned with the ecliptic, whose pole is the "ecliptic pole"* (Bernhard, *Handbook of Heavens*).

hour circle, *Astronomy.* a great circle of the celestial sphere that passes through the celestial poles. An hour circle is equivalent in the celestial system to a meridian of the earth, and, like it, is used in measuring longitude. *The hour circles are the secondary circles of the equator system; they are great circles which pass through the celestial poles and therefore at right angles to the equator.* (Baker, *Astronomy*).

hp., h.p., HP, or **H.P.,** *abbrev.* **1** high pressure. **2** horsepower.

H-R diagram, = Hertzsprung-Russell diagram.

H-substance, *n. Biochemistry.* any substance causing allergic reactions as a result of the interaction of allergens and antibodies. Histamine is the chief H-substance causing allergy in man. [from *H(istamine)*]

Hubble constant, *Astronomy.* a ratio expressing the rate at which the universe is expanding, equal to the velocity at which a typical galaxy recedes from the observer divided by the distance to that galaxy: *The new value of the Hubble constant is 53 $\pm$ 5 kilometers per second per megaparsec, or almost exactly one-tenth of Hubble's original value. The new rate of expansion pushes the age of the universe up to 18 billion years* (Scientific American). [named after Edwin P. *Hubble,* 1889–1953, American astronomer, formulator of Hubble's law]

Hubble's law or **Hubble law,** *Astronomy.* the principle that the velocity of the recession of a galaxy is proportional to its distance from the observer, as shown by the red shifts of the spectra of distant galaxies: *... in 1929, Hubble published his discovery of the apparent recession of the nebulae in accordance with what is now called Hubble's law* (Science).

hue, *n. Optics.* that property of color by which it can be distinguished as red, yellow, blue, etc., determined by one wavelength of the light: *the four parent hues—red, yellow, green, blue ... and experiments in mixing reflecting colors reveal three families—pure colors, tints and shades. These bear a relation to the three fundamental characteristics of colors—hue, saturation and brightness* (Luckiesh, *Color and Colors*). ▶ See the note under **color.**

hull, *n. Botany.* **1** the outer covering of a seed or fruit: *The hulls [of horse-chestnuts], as well as the young fruit, also contain tannin* (Morfit, *Tanning*).
2 the calyx of some fruits, such as the green leaves at the stem of a strawberry.

humate (hyü′māt), *n. Chemistry.* a salt or ester of humic acid.

Humboldt Current, *Oceanography.* = Peru Current.

humeral (hyü′mər əl), *adj. Anatomy.* of or near the humerus or shoulder: *a humeral muscle or artery.*

humerus (hyü′mər əs), *n., pl.* **-meri** (-mə rī′). *Anatomy.* the long bone in the upper part of the arm, from the shoulder to the elbow: *The upper arm bone, the humerus, acts as the fulcrum, while the forearm bone, the radius, acts as the lever arm. (There are actually two bones in the forearm, the radius and the ulna, but for simplicity we speak here of only one)* (Hardin, *Biology: Its Human Implications*). See the pictures at **arm, homology, wing.** [from Latin *humerus* shoulder]

humic acid (hyü′mik), *Chemistry.* any of various polymeric compounds or mixtures obtained from humus: *The nature of the organic portion, to which biological activity is largely confined, is so little understood that the term "humic acid" has been coined to dignify the unresolved range of polymers present* (New Scientist).

humidity, *n. Meteorology.* the amount of moisture or water vapor in the atmosphere. The **relative humidity** is the ratio between the amount of water vapor present in the air and the maximum the air could contain at the same temperature.

humite (hyü′mīt), *n. Mineralogy.* one of a group of basic magnesium silicate minerals resembling each other in physical properties and chemical composition. [from Abraham *Hume,* 19th-century English mineralogist + *-ite*]

humor, *n. Physiology.* any fluid or semifluid substance or part of the body, such as blood or lymph: *The eye ... is filled with liquid, the humors of the eye. That between the translucent cornea and the lens is the aqueous humor; the semisolid material between lens and retina is the vitreous humor* (Neal and Rand, *Chordate Anatomy*).
—humoral (hyü′mər əl), *adj.* of or having to do with any bodily humor: *Humoral immunity encompasses the processes of antibody production and numerous antibody effects* (M. Michael Sigel).

humus (hyü′məs), *n. Geology.* a dark-brown or black part of soil formed from decayed leaves and other vegetable matter, containing valuable plant foods: *The first thing to understand is that organic matter and humus are not the same thing ... Humus is, specifically, organic matter which has decomposed and is stable enough in the soil to release its nutrients to plant rootlets* (N.Y. Times). [from Latin *humus* earth]

hurricane, *n. Meteorology.* a severe tropical cyclone with a wind having a velocity of 117 or more kilometers per hour. Hurricanes typically occur in the western North Atlantic or Caribbean Sea, and in the eastern Pacific. *Accompanying a hurricane are violent winds, heavy rains, high waves and tides. In the center of a hurricane is a small core of light wind ... called the "eye" of the storm* (W.G. Osmun). Compare **typhoon.** See also table under **Beaufort scale.** [from Spanish *huracán,* from Arawak *hurakán*]
ASSOCIATED TERMS: see **cyclone.**

husk, *n. Botany.* the dry outer covering of certain seeds or fruits. An ear of corn has a husk.

Huygens' principle (hi′gənz), *Optics.* the principle that every point on a light wave front may be considered a source of secondary waves with the same speed. Huygens' principle provides a geometrical method for finding, from the known shape or position of a wave front, the shape or position of the wave front at a later time. [named after Christian *Huygens,* 1629–1695, Dutch physicist and astronomer]

hyalin (hī′ə lin), *n. Biochemistry.* a clear, homogeneous, glassy substance normally found in the matrix of cartilage, in the vitreous humor, in mucin, glycogen, etc., and pathologically in the degeneration of various tissues and cells. [from Greek *hyalos* glass]

hyaline (hī′ə lən *or* hī′ə lin), *adj.* **1** *Geology.* glassy, as certain volcanic rocks: *Rocks composed entirely of crystals of minerals are holocrystalline, partly of crystals and partly of glass, hemicrystalline, wholly of glass, hyaline or glassy* (Herbert H. Read).
2 *Anatomy, Physiology.* having to do with or consisting of hyalin: *Under normal circumstances ... the egg is doubly protected by the fertilization membrane and by the hyaline layer, either of which can stop supernumerary sperm from penetrating* (Scientific American).
—n. = hyalin.
[from Greek *hyalinos,* from *hyalos* glass]

hyaline cartilage, *Anatomy.* a translucent bluish cartilage which forms most of the fetal skeleton and is present in joints, the nose, the trachea, etc., in adults.

hyalite (hī′ə līt), *n. Mineralogy.* a colorless, transparent or translucent variety of opal, occurring in globular forms resembling drops of melted glass. [probably from German *Hyalit,* from Greek *hyalos* glass]

hyaloid (hī′ə loid), *Anatomy.* **—adj.** of or having to do with the vitreous humor of the eye. The **hyaloid membrane** is a membraneous layer of fibrils that surrounds the vitreon humor and forms the suspensory ligament that holds the lens of the eye in place.
—n. the hyaloid membrane: *the surface of the hyaloid.*
[from Latin *hyaloides,* from Greek *hyaloeidēs* glassy]

hyaloplasm (hī al′ə plaz′əm *or* hī′ə lō plaz′əm), *n. Biology.* the clear, fluid portion of the cytoplasm in which various bodies are suspended. Also called

---

cap, fāce, fäther; best, bē, tėrm; pin, five;
rock, gō, ôrder; oil, out; cup, pùt, rüle;
yü in use, *yu̇* in uric;
ng in bring; *sh* in rush; *th* in thin, ᴛʜ in then;
zh in seizure.
ə = *a* in about, *e* in taken, *i* in pencil, *o* in lemon, *u* in circus

297

ground substance. [from Greek *hyalos* glass + *plasma* something molded]

—**hyaloplasmic,** *adj.* of or having to do with the hyaloplasm.

hyaluronic acid (hī′ə lū ron′ik), *Biochemistry.* a viscous polysaccharide found in the vitreous humor, synovial fluid, connective tissues, and other parts of the body: *When an egg is discharged from the ovary, it is protected by a company of surrounding cells which are believed to be held together by a cementlike substance of which the chief ingredient is hyaluronic acid* (Scientific American). *Formula:* $C_8H_{15}NO_6$ [from Greek *hyalos* glass + English *uronic acid*]

hyaluronidase (hī′ə lū ron′ə dās), *n. Biochemistry.* an enzyme present in certain pathogenic bacteria, snake venom, sperm, etc., having the power to break down or disperse hyaluronic acid. It increases tissue permeability and is used in medicine to promote the diffusion and absorption of injections. *Hyaluronidase, known also as the "spreading factor," increases the effectiveness of antisnakebite serum when the two are given together* (Science News Letter). [from *hyaluron(ic acid)* + *-id(e)* + *-ase*]

hybrid, *n.* **1** *Biology.* the offspring of two organisms of different varieties, species, or breeds, and rarely of different genera of the same family. The loganberry is a hybrid produced by crossing the western dewberry with the red raspberry. The mule is a hybrid of a female horse and a male donkey. SYN: crossbreed.
2 *Genetics.* the offspring of two individuals that differ in at least one gene: *The offspring resulting from a cross between two individuals differing in at least one set of characters is called a hybrid* (Fuller and Tippo, *College Botany*).
—*adj.* **1** *Biology.* bred from two different species, varieties, etc.: *a hybrid rose. A mule is a hybrid animal.*
2 *Genetics.* differing from the parents in at least one gene: *A hybrid individual is one that differs in one or more heritable characters from its two parent organisms* (Sheldon C. Reed).
[from Latin *hybrida*]

—**hybridism,** *n.* **1** the production of hybrids. SYN: crossbreeding. **2** = hybridity.

—**hybridity** (hī brid′ə tē), *n.* hybrid character, nature, or condition.

—**hybridization,** *n.* the act or process of hybridizing; the production of hybrids.

—**hybridize,** *v.* to produce or cause to produce hybrids: *Perhaps the most common misconception is that grafting is a means of hybridizing plants, when as a matter of fact hybridization can occur only in the course of sexual reproduction* (Greulach and Adams, *Plants*). SYN: cross, interbreed.

hybridoma (hī brə dō′mə), *n. Biology.* a cell produced in the laboratory by the fusion of the nuclei of two different cells: *the first antibody-secreting hybridoma: a hybrid between an ordinary mouse antibody-secreting cell, which has only a limited lifespan, and an immortal tumour cell from a mouse myeloma* (New Scientist). Compare **monoclonal antibody.** [from *hybrid* + *-oma* mass, tumor, as in *myeloma* a tumor of the bone marrow]

hybrid vigor, *Genetics.* = heterosis.

hydathode (hī′də thōd), *n. Botany.* a specialized layer of cells on the tips and margins of certain leaves, from which water is secreted. [from Greek *hydōr* water + *hodos* way]

hydatid (hī′də tid), *n. Zoology.* **1** a cyst containing a clear, watery fluid, produced in animals by a tapeworm in the larval state.
2 the larva of a tapeworm in its encysted state. [from Greek *hydatidos* watery vesicle]

hydrarch (hī′drärk), *adj. Ecology.* (of a succession) that begins in water or a wet habitat, such as a pond.

hydrase (hī′drās), *n. Biochemistry.* an enzyme that catalyzes the removal or addition of water to the substrate: *The hydrases add water to organic compounds without causing a splitting* (Sumner and Somers, *The Chemistry and Methods of Enzymes*).

hydrate, *Chemistry.* —*n.* any compound produced when certain substances chemically unite with water in a definite weight ratio: *Many substances, especially salts ... combine with water in definite proportions by weight to form compounds known as hydrates* (Parks and Steinbach, *Systematic College Chemistry*).
▶ In formulas of hydrates, the addition of the water molecules is indicated by a dot, as in the formula for washing soda, $Na_2CO_3 \cdot 1OH_2O$.
—*v.* to become or cause to become a hydrate; combine with water to form a hydrate. Blue vitriol is hydrated copper sulfate.

hydration, *n. Chemistry.* the reaction of molecules of water with a substance; the incorporation of water molecules by a substance. The hydration of an ion results in the formation of hydrates.

hydraulic (hī drô′lik), *adj.* **1** having to do with water or other liquids at rest or in motion: *hydraulic flow. Because it will set under water, Portland cement is sometimes known as hydraulic cement* (Jones, *Inorganic Chemistry*).
2 operated by the pressure of water or other liquids in motion, especially when forced through an opening or openings: *a hydraulic press.*
3 having to do with hydraulics: *hydraulic engineering, hydraulic mining.*
[from Latin *hydraulicus,* ultimately from Greek *hydōr* water + *aulos* pipe] —**hydraulically,** *adv.*

hydraulics, *n.* the science dealing with water and other liquids at rest or in motion, their uses in engineering, and the laws of their actions.

hydrazine (hī′drə zēn′), *n. Chemistry.* **1** a colorless, fuming, highly poisonous liquid used in organic synthesis and as a fuel for rockets. *Formula:* N_2H_4
2 any of a group of compounds derived from this liquid by replacement of one or more hydrogen atoms with an organic radical.

hydrazoic acid, *Chemistry.* a colorless, volatile liquid obtained by treating hydrazine with nitrous acid and in various other ways. Hydrazoic acid and most of its salts are highly toxic and explosive. *Formula:* HN_3

hydrazone, *n. Chemistry.* any of various compounds containing the group $=NN=$, produced by the action of hydrazine on a ketone or an aldehyde.

hydric, *adj. Botany.* characterized by or requiring moisture: *hydric plants.* Compare **mesic, xeric.**

hydride (hī′drīd), *n. Chemistry.* a compound of hydrogen with another element or radical: *Uranium metal reacts with hydrogen at 250° to 300°C. to form a* well-defined hydride which resembles the rare-earth hydrides in many respects (Glenn T. Seaborg).

hydriodic acid (hĭ′drē od′ik), *Chemistry.* a colorless, corrosive solution of hydrogen iodide in water, used as a disinfectant. *Formula:* HI

hydro-, *combining form.* **1** of or having to do with water, as in *hydrodynamic = having to do with the force of water.* **2** having to do with any fluid, as in *hydrokinetic = having to do with the motion of fluids.* **3** containing hydrogen, as in *hydrochloric = containing hydrogen and chlorine.* [from Greek, from *hydōr* water]

hydrobromic acid (hĭ′drō brō′mik), *Chemistry.* a colorless, corrosive solution of hydrogen bromide in water, used in making organic compounds. *Formula:* HBr

hydrocarbon, *n. Chemistry.* any of a large group of organic compounds containing only hydrogen and carbon, obtained principally from petroleum and coal tar. Methane, benzene, and acetylene are hydrocarbons. Gasoline is a mixture of hydrocarbons.

hydrochloric acid (hĭ′drə klôr′ik), *Chemistry.* a clear, colorless solution of hydrogen chloride in water, that has a strong, sharp odor and is highly corrosive. It is used in food processing, in dyeing, as a reagent in organic analysis, and in many industrial processes. *Because the aqueous solution of hydrogen chloride has properties which are those of an electrolyte rather than of a covalent compound, it is known by a different name: hydrochloric acid. Crude commercial hydrochloric acid is sometimes known as muriatic acid* (Jones, *Inorganic Chemistry*). *Gastric glands ... secrete enzymes and hydrochloric acid. Hydrochloric acid provides an optimum pH for the hydrolytic activity of gastric protease* (Biology Regents Syllabus). *Formula:* HC1

hydrochloride (hĭ′drə klôr′īd), *n. Chemistry.* a compound of hydrochloric acid with an organic base.

hydrocortisone (hĭ′drō kôr′tə sōn *or* hĭ′drō kôr′tə zōn), *n. Biochemistry.* an adrenal hormone derived from cortisone, used in treating arthritis and other inflammations. *Formula:* $C_{21}H_{30}O_5$

hydrocyanic acid (hĭ′drō sī an′ik), *Chemistry.* a colorless, volatile, poisonous liquid with an odor like that of bitter almonds, used in making plastics, dyes, etc. *Formula:* HCN Also called **prussic acid.**

hydrodynamic, *adj.* having to do with hydrodynamics: *Problems encountered by swimming microorganisms are hydrodynamic, that is, related to the physical characteristics of water* (T.L. Jahn and E.C. Bovee). **—hydrodynamically,** *adv.*

hydrodynamics, *n.* the branch of physics dealing with the forces that water and other fluids in motion exert: *Chemical hydrodynamics is devoted largely to the nature of the boundary layer in streaming fluids* (Science). Compare **hydrokinetics** and **hydrostatics.**

hydrofluoric acid (hĭ′drō flü ôr′ik), *Chemistry.* a colorless, corrosive, volatile solution of hydrogen fluoride in water, used for etching glass, as a cleaning agent, etc. *Formula:* HF

hydrogen (hĭ′drə jən), *n. Chemistry.* a colorless, odorless, gaseous element that burns easily and is the lightest of all elements. It combines with oxygen to form water and is present in most organic compounds. *By far the most abundant element is hydrogen: it accounts for 93 per cent of the total number of atoms and 76 per cent of the weight of the universe's matter*

(William A. Fowler). *Hydrogen ... is able to attach itself to other atoms not only by means of its electron (a valence bond) but also by ... its unoccupied, positively charged side ... This attachment is known as the hydrogen bond* (A.M. Buswell and W.H. Rodebush). *Symbol:* H; *atomic number* 1; *atomic weight* 1.00797; *melting point* −259.14°C; *boiling point* −257.87°C; *oxidation state* ± 1. [from French *hydrogène,* from *hydro-* + *-gène* -gen]

hydrogenase (hĭ′drə jə nās), *n. Biochemistry.* an enzyme that promotes the formation of gaseous hydrogen, found in various bacteria: *The autotrophic hydrogen bacteria possess the enzyme, hydrogenase, which allows them to employ the oxidation of molecular hydrogen as the source of energy for fixation of CO_2 and all other vital processes* (Burrows, *Textbook of Microbiology*).

hydrogenate (hĭ droj′ə nāt *or* hĭ′drə jə nāt), *v.* to combine or treat with hydrogen. When vegetable oils are hydrogenated, they become solid fats. **—hydrogenation,** *n.* the act or process of hydrogenating: *Hydrogenation in the oil and fat industry is a process by which free hydrogen gas is chemically combined with oil, resulting in a rise in the melting point of the oil or fat* (Economist).

hydrogen bond, *Chemistry.* a bond in which the hydrogen atom of one molecule is attracted to the electronegative atom of another molecule, as in compounds containing nitrogen, oxygen, or fluorine. The formation of hydrogen bonds joins water molecules together. Hydrogen bonds are not as strong as covalent bonds, but they have important effects on such properties of substances as boiling point and crystal structure. *Hydrogen bonds hold life's molecules in shape, and play a vital role in enzyme catalysis, the action of drugs and anaesthetics, and taste* (New Scientist).

hydrogen bromide, *Chemistry.* a colorless, corrosive gas with a sharp odor. A solution of hydrogen bromide in water yields hydrobromic acid. *Formula:* HBr

hydrogen chloride, *Chemistry.* a fuming, poisonous, colorless gas with a strong, sharp odor. A solution of hydrogen chloride in water yields hydrochloric acid. *Formula:* HC1

hydrogen fluoride, *Chemistry.* a colorless, corrosive, poisonous gas which can cause severe burns. A solution of hydrogen fluoride in water yields hydrofluoric acid. *Formula:* HF

hydrogen iodide, *Chemistry.* a colorless, corrosive gas with a suffocating odor. A solution of hydrogen iodide in water yields hydriodic acid. *Formula:* HI

hydrogen ion, *Chemistry.* a positively charged ion of hydrogen found in all aqueous solutions of acids: *In metals the electric current is carried by electrons, whereas in the resin it is carried by the hydrogen ions or pro-*

---

cap, fāce, fäther; best, bē, tėrm; pin, five;
rock, gō, ôrder; oil, out; cup, pùt, rüle;
yü in use, *yu̇* in uric;
ng in bring; *sh* in rush; *th* in thin; ᴛʜ in then;
zh in seizure.
ə = *a* in about, *e* in taken, *i* in pencil, *o* in lemon, *u* in circus

tons, *the units of positive electricity* (K.S. Spiegler). *Symbol:* H^+ Compare **hydronium.**

hydrogenolysis (hī′drə jə nol′ə sis), *n. Chemistry.* a type of chemical decomposition in which an organic compound is broken down and changed into other compounds by the incorporation of hydrogen.

hydrogen peroxide, *Chemistry.* a colorless compound of hydrogen and oxygen. It is used in dilute solution as an antiseptic, bleaching agent, etc. *Formula:* H_2O_2 Also called **peroxide.** Compare **hydroperoxide.**

hydrogen sulfide, *Chemistry.* a flammable, poisonous gas which gives rotten eggs their characteristic odor, often found in mineral waters. *Formula:* H_2S

hydrographic (hī′drə graf′ik), *adj.* of or having to do with hydrography: *The basic research upon which hydrographic charts are prepared normally consists of hand soundings, fathometer soundings and wire dragging to locate submerged pinnacles or rocks* (Science News Letter).

hydrography (hī drog′rə fē), *n.* the science of the measurement and description of seas, lakes, rivers, and other bodies of water. Compare **hydrology.**

hydroid (hī′droid), *Zoology.* —*n.* **1** the fixed polyp form of a coelenterate as distinguished from the swimming jellyfish form. Hydroids grow in branching colonies by budding. **2** any hydrozoan.
—*adj.* of, having to do with, or like the hydrozoans: *hydroid polyps.*

hydrokinetic, *adj.* having to do with motion of fluids or with hydrokinetics.

hydrokinetics, *n.* the branch of physics dealing with the motion of fluids. Compare **hydrostatics.**

hydrolase (hī′drə lās), *n. Biochemistry.* any enzyme that catalyzes the hydrolysis of protein, starch, fat, and the like; any hydrolytic enzyme.

hydrologic or **hydrological,** *adj.* of or having to do with hydrology.

hydrologic cycle or **hydrological cycle,** = water cycle: *The hydrological cycle is continuous, though irregular, circulation of water from clouds to the earth and back to the clouds again.* (Leo A. Heindl).

hydrology (hī drol′ə jē), *n.* the science that deals with water and its properties, laws, geographical distribution, etc. Compare **hydrography.**

hydrolysis (hī drol′ə sis), *n., pl.* **-ses** (-sēz′). *Chemistry.* a process of decomposition in which a compound is broken down and changed into other compounds by taking up the elements of water: *All food digestion processes of organisms are hydrolyses. In hydrolysis the food material takes up water and then breaks down* (Youngken, *Pharmaceutical Botany*). *The reaction between a salt and water to produce an acid and a base, the reverse of neutralization ... is called hydrolysis* (Offner, *Fundamentals of Chemistry*).
—**hydrolytic** (hī′drə lit′ik), *adj.* of or producing hydrolysis: *Deoxyribonuclease and ribonuclease are hydrolytic enzymes.*

hydrolyze (hī′drə līz), *v. Chemistry.* to decompose by hydrolysis: *Nucleic acids, proteins and polysaccharides contain many bonds that hydrolyze* (G. Eglinton and M. Calvin).

hydromagnetic, *adj.* = magnetohydrodynamic.

hydromagnetics, *n.* = magnetohydrodynamics.

hydromechanical, *adj.* of or having to do with hydromechanics.

hydromechanics, *n.* the science of the mechanics of fluids or of their laws of equilibrium and motion: *In the Netherlands, hydromechanics had become urgently practical* (T. Fursdon Crang).

hydrometeorological, *adj.* of or having to do with hydrometeorology: *hydrometeorological stations.*

hydrometeorology, *n.* the branch of meteorology dealing with water in the atmosphere, particularly with hydrologic problems and phenomena resulting from such activities as irrigation, the use of hydroelectric power, flood control, and the like.

hydrometer (hī drom′ə tər), *n.* a graduated instrument for finding the specific gravity of liquids. A hydrometer is used to determine the concentration of acid in a storage battery or to test the purity of milk, the strength of alcohol or salt in a substance, etc. *A hydrometer ... consists of a glass tube with a bulb syringe for sucking up samples of the electrolyte. Within the glass tube is a calibrated float. The depth to which the float sinks is a measure of the specific gravity of the solution.* (Automotive Encyclopedia).
—**hydrometric** (hī′drə met′rik), *adj.* of or having to do with hydrometry.
—**hydrometry** (hī drom′ə trē), *n.* the determination of specific gravity, purity, etc., by means of a hydrometer.

hydronium ion (hī drō′nē əm), *n. Chemistry.* a hydrogen ion combined with a molecule of water: *The more strongly basic a solution, the weaker the concentration of the hydronium ions* (E. J. Montague). *Formula:* H_3O^+ Also called **oxonium ion.**

hydroperoxide (hī′drō pə rok′sid), *n. Chemistry.* any of a class of compounds containing the group HOO-. Hydrogen peroxide is the simplest hydroperoxide.

hydrophilic (hī′drə fil′ik), *adj. Chemistry.* of or having an affinity for water; readily absorbing water.

hydrophobic (hī′drə fō′bik), *adj. Chemistry.* of or lacking an affinity for water; tending not to absorb or repel water.

hydrophyte (hī′drə fīt), *n. Botany.* any plant that can grow only in water or very wet soil. Most algae are hydrophytes.
ASSOCIATED TERMS: *Hydrophytes* are distinguished from *xerophytes*, which are plants that can grow in deserts or very dry ground. Both in turn are distinguished from *mesophytes*, which require a moderate amount of moisture in order to grow.
—**hydrophytic** (hī′drə fit′ik), *adj.* of or like a hydrophyte.

hydroponic, *adj.* of hydroponics; produced or grown by hydroponics: *Hydroponic or "soil-less" farming has moved out of the realms of research phenomena, and into those of established farming practice* (New Scientist).
—**hydroponically,** *adv.* by means of hydroponics: *Succulent green grass ... is grown in trays hydroponically, immersed in liquid nutrients* (Science News Letter).

hydroponics (hī′drə pon′iks), *n. Botany.* the growing of plants without soil by the use of water containing the necessary nutrients: *The hydroponics farmer, having no soil to nurse, does not have to worry about drought,* floods, erosion, soil diseases or soil exhaustion (J. Durant). [from *hydro-* + Greek *ponein* to labor]

hydroquinone (hī′drō kwi nōn′ *or* hī′drō kwin′ōn), *n.* *Chemistry.* a white, sweetish, crystalline compound used as an antioxidant for fats and oils, as an inhibitor of polymerization, in photographic developers, in medicine, etc. *Formula:* $C_6H_4(OH)_2$

hydrosphere (hī′drə sfir), *n.* *Geology.* the water portion of the earth: *The solid mass of the earth (the lithosphere) is covered in part by water (the hydrosphere), and both are surrounded by a gaseous envelope (the atmosphere)* (Finch and Trewartha, *Elements of Geography*).

hydrostatic, *adj.* of or having to do with hydrostatics. The **hydrostatic equilibrium** is a state of perfect balance between the forces of gravity and pressure.

hydrostatics, *n.* the branch of physics that deals with the laws of fluids at rest and under pressure. Compare **hydrokinetics.**

hydrosulfurous acid, = hyposulfurous acid.

hydrotactic, *adj.* *Biology.* of or having to do with hydrotaxis.

hydrotaxis, *n.* *Biology.* movement toward or away from water, as of plasmodia or other organisms.

hydrothermal, *adj.* *Geology.* of or having to do with heated water, especially with the action of heated waters in dissolving and redepositing minerals: *Most ores of copper, lead, zinc, mercury, silver, and many of those of gold and tungsten are classed geologically as hydrothermal deposits; that is, they were formed by deposition from hot watery solutions* (Gilluly, *Principles of Geology*).

hydrotropic, *adj.* *Biology.* characterized by or showing hydrotropism: *Many roots have a distinct positive hydrotropic response. For this reason those of willows and cottonwoods often clog tile drains* (Emerson, *Basic Botany*). [from *hydro-* + Greek *tropos* a turning]

hydrotropism (hī drot′rə piz′əm), *n.* *Biology.* the tendency of a plant or other organism or part to turn or bend in response to moisture. Hydrotropism causes roots to grow toward water.

hydrous (hī′drəs), *adj.* *Chemistry, Mineralogy.* having water as a constituent: *a hydrous salt.*

hydroxide (hī drok′sīd), *n.* *Chemistry.* any compound consisting of an element or radical combined with one or more hydroxyl radicals. Hydroxides of metals are bases; those of nonmetals are acids.

hydroxide ion, *Chemistry.* a negatively charged ion found in all basic aqueous solutions. *Formula:* OH^-

hydroxy acid (hī drok′sē), *Chemistry.* any organic acid having a hydroxyl (-OH) group and exhibiting both alcoholic and acidic characteristics. Also called **hydroxyl acid.**

hydroxyapatite (hī drok′sə ap′ə tīt), *n.* = hydroxylapatite.

hydroxyl (hī drok′səl), *n.* *Chemistry.* a univalent group, -OH, found in all hydroxides: *If we replace one of the hydrogen atoms of a hydrocarbon with a unit of oxygen and hydrogen (OH), called a hydroxyl group, we produce a compound belonging to the class of alcohols* (McElroy, *Biology and Man*).

hydroxyl acid, = hydroxy acid.

hydroxylamine (hī drok′sə lə mēn′ *or* hī′drok sil′ə mēn′), *n.* *Chemistry.* a colorless crystalline compound, similar to ammonia but less basic, used in organic synthesis and as a reducing agent: *Viruses that have been inactivated with hydroxylamine ... not only reproduce themselves but also produce interferon* (M. Fishbein). *Formula:* NH_2OH

hydroxylapatite (hī drok′səl ap′ə tīt), *n.* *Mineralogy.* a mineral composed of calcium phosphate hydroxide that is the basic inorganic constituent of bone and tooth enamel. *Formula:* $Ca_5(PO_4)_3OH$ Also called **hydroxyapatite.**

hydroxylase (hī drok′sə lās), *n.* *Biochemistry.* an enzyme that catalyzes the hydroxylation of compounds: *Animal tissues contain hydroxylases that mediate the introduction of oxygen into specific positions of steroids* (Heftmann, *Steroid Biochemistry*).

hydroxylate, *v.* *Chemistry.* to introduce one or more hydroxyls into (a compound).

—**hydroxylation,** *n.* the introduction of one or more hydroxyls into a compound, especially in an oxidation reaction in which a hydroxyl is incorporated in an organic compound.

hydroxyproline (hī drok′sē prō′lən), *n.* *Biochemistry.* a colorless, crystalline amino acid that is a constituent of collagen, used in biochemical research. *Formula:* $C_5H_9NO_3$ *Abbreviation:* Hyp

hydrozoan (hī′drə zō′ən), *Zoology.* —*n.* any of a class (Hydrozoa) of cnidarian that includes hydras, certain colonial animals, and the Portuguese men-of-war. They are the most primitive animals that have definite tissues. Compare **anthozoan, scyphozoan.**

—*adj.* of or belonging to this class.

hygro-, *combining form.* moisture or humidity, as in *hygroscope = instrument showing variations in moisture or humidity.* [from Greek *hygro-,* from *hygros* wet]

hygrometer (hī grom′ə tər), *n.* an instrument for measuring the amount of moisture or humidity in the air. Compare **psychrometer.**

hygrometric (hī′grə met′rik), *adj.* *Meteorology.* of or having to do with the measurement of moisture in the air.

hygroscope (hī′grə skōp), *n.* an instrument that shows the variations in the amount of moisture or humidity in the air.

hygroscopic (hī′grə skop′ik), *adj.* **1** *Meteorology.* having to do with or perceptible by the hygroscope: *hygroscopic experiments.*

2 *Chemistry.* absorbing or attracting moisture from the air: *narrow hygroscopic ribbons that move under the influence of moisture changes and aid in distribution of spores* (Weier, *Botany*).

---

cap, fāce, fäther; best, bē, tèrm; pin, five;
rock, gō, ôrder; oil, out; cup, pùt, rüle,
yü in use, *yu̇* in uric;
ng in bring; *sh* in rush; *th* in thin; ᴛʜ in then;
zh in seizure.

ə = *a* in about, *e* in taken, *i* in pencil, *o* in lemon, *u* in circus

hymen (hī′mən), *n. Anatomy.* a fold of mucous membrane extending partly across the opening of the vagina. [from Late Latin, from Greek *hymēn* membrane]

hymenium (hī mē′nē əm), *n. Biology.* the spore-bearing surface of fungi, composed of numerous asci or basidia. In the common mushroom the hymenium covers the gills. [from New Latin *hymenium,* from Greek *hymēn* membrane]

hymenopteran (hī′mə nop′tər ən), *n.* a hymenopterous insect. —*adj.* of or having to do with hymenopterans.

hymenopteron (hī′mə nop′tə ron′), *n., pl.* **-tera** (-tər ə). = hymenopteran.

hymenopterous (hī′mə nop′tər əs), *adj. Zoology.* of or belonging to an order (Hymenoptera) of insects, including ants, bees, wasps, and sawflies, almost all having four transparent membranous wings, and the ovipositor of the female modified to form a stinger. The order includes all the social insects except the termites. Most kinds live in communities of complex organization and have a caste system within each group. [from Greek *hymenopteros* having membranous wings, from *hymēn* membrane + *pteron* wing]

hyoid (hī′oid), *Anatomy.* —*n.* **1** the U-shaped bone at the root of the tongue in human beings. **2** a corresponding bone or series of bones in animals.
—*adj.* of or having to do with this bone.
[from Greek *hyoeidēs* U-shaped, from *hy* upsilon (υ) + *eidos* form]

hyp., *abbrev.* **1** hypotenuse. **2** hypothesis.

Hyp, *abbrev.* hydroxyproline.

hypabyssal (hip′ə bis′əl), *adj. Geology.* of or having to do with rocks that have crystallized under conditions intermediate between the plutonic and extrusive environments: *The hypabyssal group includes the rocks of dykes, sills, and small laccoliths, etc., which occupy an intermediate position in the crust between the deep-seated plutonic bodies, and the surficial lava flows* (Tyrrell, *Principles of Petrology*). Also called **subvolcanic.**

hypanthium (hī pan′thē əm), *n., pl.* **-thia** (-thē ə). *Botany.* an enlargement of the receptacle of a flower below the calyx. [from New Latin *hypanthium,* from *hyp-* below + Greek *anthos* flower]

hyperbola (hī pėr′bə lə), *n. Geometry.* a plane curve consisting of two separate, similar branches, formed by the intersection of a plane with two similar right circular cones on opposite sides of the same vertex: *A hyperbola consists of all those points such that the difference of their distances from two fixed points, the foci, remains constant* (Corcoran, *Algebra Two*). See also **conic section.** See the picture at **conic section.** [from New Latin *hyperbola,* from Greek *hyperbolē,* from *hyper-* + *ballein* to throw]
—**hyperbolic** (hī pər bol′ik), *adj.* **1** of or having to do with hyperbolas: *a hyperbolic curve.* **2** of or having to do with a geometric system in which two or more lines can be drawn through any point in a plane and not intersect a given line in the plane: *hyperbolic geometry.*

hyperbolic function, *Geometry.* a function having a relation to a rectangular hyperbola similar to that of the ordinary trigonometric functions to a circle; a hyperbolic sine, cosine, tangent, or other function: *The tension at any point in a cable suspended by its ends and hanging under its own weight, such as an electric transmission line, may be computed in terms of hyperbolic functions* (Thomas, *Calculus and Analytic Geometry*).

hyperbolic paraboloid, *Geometry.* a curved surface of which every plane section is either a parabola or a hyperbola, the curvature being concave in one direction and convex in another.

hypercharge (hī′pər chärj′), *n. Nuclear Physics.* a quantum number equal to twice the average electric charge in an isotopic multiplet of strongly interacting particles: *In the new system for naming strongly interacting particles we shall make use of five quantities ...: atomic mass number (A), hypercharge (Y), isotopic spin (I), spin angular momentum (J) and parity (P)* (Murray Gell-Mann, et al.)

hypercomplex number (hī′pər kom′pleks), = quaternion.

hypergeometric distribution (hī′pər jē′ə met′rik), *Statistics.* a distribution showing the probable frequency of each of two kinds of elements in a sample chosen at random from a population consisting of specified quantities of the two kinds of elements.

hyperon (hī′pər on), *n. Nuclear Physics.* any of a class of highly unstable elementary particles that includes all the baryons with a mass greater than the neutron: *Hyperons ... may have positive, negative, or neutral electrical charges. The three main kinds of hyperons—xi, sigma, and lambda—are of different weights* (Ralph E. Lapp). [probably from *hyper-* above, beyond + (*prot*)*on*]

hyperparasite (hī′pər par′ə sīt), *n. Biology.* an organism parasitic upon a parasite: *Insect hyperparasites have also been hopefully cultured but not, it seems, with much success.* (New Scientist).

hyperploid (hī′pər ploid), *Genetics.* —*adj.* having or denoting a chromosome number greater than, but not a multiple of, the diploid number for the species: *hyperploid individuals.*
—*n.* an individual with a hyperploid number of chromosomes.
[from *hyper-* + (*di*)*ploid*]
—**hyperploidy** (hī′pər ploi′dē), *n.* the quality or condition of being hyperploid.

hyperpolarize (hī′pər pō′lə rīz), *v. Chemistry, Biochemistry.* to increase in polarity, as across a cell membrane: *Neurons react to an electrical current in fundamentally different ways depending on their location in the brain. Some are depolarized or excited; others are hyperpolarized or blocked* (Science News).
—**hyperpolarization,** *n.* an increase in polarity: *As a result of hyperpolarization—which is the opposite of depolarization—neurons become less responsive to stimulation. During the period of hyperpolarization, the inside of the cell has more negative charges than usual* (Science).

hypersonic, *adj. Physics.* that moves or is able to move at a speed which is five or more times faster than the speed of sound in a given medium.

hyperspace, *n. Mathematics.* space consisting of four or more dimensions: *Since most of the useful geometric notions derived from our experience in three dimensions have exact counterparts in "hyperspace" ... we can proceed by analogy. Thus we speak of distances,*

hypersthene (hī'pər sthēn), *n. Mineralogy.* a grayish, greenish-black, or brown silicate of iron and magnesium; an orthorhombic pyroxene, isomorphous with enstatite. *Formula:* $(Mg,Fe)SiO_3$ [from French *hyperstène,* from Greek *hyper-* excessive + *sthenos* strength (because it is stronger than hornblende)]

hypersurface, *n. Mathematics.* a surface in a space of more than three dimensions: *Albert Einstein ... tried to abolish the endless regress of distance by bending three-dimensional space around to form the hypersurface of a four-dimensional sphere* (Martin Gardner).

hypha (hī'fə), *n., pl.* **-phae** (-fē). *Biology.* one of the long, slender, usually branched filaments which are the structural elements of the plant body of fungi. A mass of branching hyphae constitutes the mycelium. *Fungal hyphae either ramify through decaying matter or invade the tissues of living organisms and derive nourishment therefrom* (Youngken, *Pharmaceutical Botany*). [from New Latin, from Greek *hyphē* web]
—**hyphal,** *adj.* of or having to do with a hypha or hyphae.

hypo-, *prefix. Chemistry.* indicating a compound which is less oxidized than the corresponding compound without the prefix: *Hypochlorous acid = less oxidized than chlorous acid.* [from Greek *hypo-* lower, under]

hypoblast, *n.* = endoderm.

hypochlorite (hī'pə klôr'īt), *n. Chemistry.* a salt or ester of hypochlorous acid: *The hypochlorite ion is an active oxidizing agent, and its oxidizing power is the basis of its sterilizing and bleaching action* (Pauling, *College Chemistry*).

hypochlorous acid (hī'pə klôr'əs), *Chemistry.* an unstable, weak acid existing as a yellow solution with an irritating odor, used as a bleach, disinfectant, etc. *Formula:* HC10

hypocotyl (hī'pə kot'l), *n. Botany.* the part of a plant embryo below the cotyledons, between the stem and the root: *the hypocotyl on whose lower end an embryonic root becomes organized* (Emerson, *Basic Botany*).

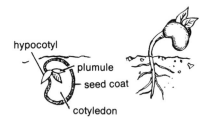

hypocotyl
plumule
seed coat
cotyledon

hypodermal, *adj. Zoology, Botany.* of or having to do with the hypodermis: *hypodermal tissue.*

hypodermis (hī'pə dėr'mis), *n.* **1** *Zoology.* a layer of tissue lying under and secreting the integument of certain insects, crustaceans, and worms.
2 *Botany.* the layer of cells just beneath the epidermis: *The ground tissue system enclosing the stele ... may have a hypodermis peripherally and an endodermis next to the stele* (Katherine Esau).

hypogastric (hī'pə gas'trik), *adj. Anatomy.* of or having to do with the hypogastrium or pubic region: *the hypogastric artery.*

hypogastrium (hī'pə gas'trē əm), *n. Anatomy.* the lower part of the human abdomen, especially the part between the right and left iliac regions. [from New Latin *hypogastrium,* from Greek *hypogastrion,* from *hypo-* under, lower + *gastros* stomach]

hypogene (hip'ə jēn), *adj. Geology.* formed beneath the surface of the earth: *Granite is a hypogene rock.* Contrasted with **supergene.** [from *hypo-* + Greek *-genēs* born]

hypogeous (hī'pə jē'əs), *adj. Biology.* living or growing below the surface of the ground, as certain cotyledons: *a hypogeous species, hypogeous fauna, a hypogeous location.* [from Latin *hypogeus,* from Greek *hypogeios* underground, from *hypo-* under + *ge* earth]

hypoglossal, *Anatomy.* —*adj.* **1** situated or occurring under the tongue: *hypoglossal paralysis.* **2** designating either of the pair of cranial nerves which control movement of the tongue.
—*n.* the hypoglossal nerve.

hypogynous (hī poj'ə nəs), *adj. Botany.* **1** situated on the receptacle below the ovary, as sepals, petals, and stamens.
2 having its parts so arranged, as the magnolia blossom: *whole hypogynous families.* Compare **epigynous, perigynous.**

hypolimnion (hī'pə lim'nēən), *n., pl.* **-nia** (-nē ə). *Geology.* the lower, cooler layer of water in a lake, below the thermocline. Compare **epilimnion.** [from *hypo-* + Greek *limnion,* diminutive of *limnē* marsh, lake]

hypophosphite (hī'pō fos'fīt), *n. Chemistry.* a salt of hypophosphorous acid.

hypophosphoric acid (hī'pō fo sfôr'ik), *Chemistry.* a crystalline acid produced by the slow oxidation of phosphorous in moist air. *Formula:* $H_4P_2O_6$

hypophosphorous acid (hī'pō fos'fər əs), *Chemistry.* a colorless acid used as a reducing agent. *Formula:* H_3PO_2

hypophyseal (hī'pə fiz'ē əl *or* hī pof'ə sē'əl), *adj. Anatomy.* of or having to do with the hypophysis: *hypophyseal growth hormone.*

hypophysis (hī pof'ə sis), *n., pl.* **-ses** (-sēz'). = pituitary gland. [from Greek *hypophysis* attachment underneath, from *hypo-* under + *physis* a growing]

hyposulfite (hī'pō sul'fīt), *n. Chemistry.* a salt of hyposulfurous acid: *sodium hyposulphite.*

hyposulfurous acid (hī'pō sul'fər əs), *Chemistry.* an acid found only in solution or as salts, used as a reducing and bleaching agent. *Formula:* $H_2S_2O_4$ Also called **hydrosulfurous acid.**

hypotenuse (hī pot'nüs), *n. Geometry.* the side of a right triangle opposite the right angle. [from Latin *hypotenusa,* from Greek *hypoteinousa,* present parti-

cap, fāce, fäther; best, bē, tėrm; pin, five;
rock, gō, ôrder; oil, out; cup, pùt, rüle;
yü in use, *yu̇* in uric;
ng in bring; *sh* in rush; *th* in thin, ᴛʜ in then;
zh in seizure.
ə = *a* in about, *e* in taken, *i* in pencil, *o* in lemon, *u* in circus

303

ciple of *hypoteinein* subtend, from *hypo-* under + *teinein* to stretch]

hypothalamic (hĭ'pō thə lam'ik), *adj. Anatomy.* of or having to do with the hypothalamus.

hypothalamus (hĭ'pō thal'ə məs), *n., pl.* **-mi** (-mī). *Anatomy.* the region of the brain under the thalamus, controlling temperature, hunger, thirst, and producing hormones which influence the pituitary gland.

hypothenuse (hĭ poth'ə nüs), *n.* = hypotenuse.

hypothesis (hi poth'ə sis), *n., pl.* **-ses** (-sēz'). **1** a proposition assumed as a basis for reasoning and often subjected to testing for its validity: *A speculation in regard to the probable cause of a phenomenon is called a hypothesis. A hypothesis gains in reliability each time a prophecy is based upon it and found to be correct* (Parks and Steinbach, *Systematic College Chemistry*). ▶ See the note under **theory.**

2 *Mathematics.* the part of a conditional which states the condition; the antecedent. A theorem in geometry consists of a hypothesis and a conclusion. [from Greek *hypothesis,* from *hypo-* under + *thesis* a placing]

hypothesize (hĭ poth'ə sīz), *v.* to make a hypothesis. SYN: theorize.

hypothetical (hĭ'pə thet'ə kəl), *adj.* of, based on, or involving a hypothesis: *a hypothetical case, a hypothetical question.*

—**hypothetically,** *adv.* in a hypothetical manner: *In my present want of information I must only speak hypothetically* (Edmund Burke).

hypoxanthine (hĭ'pə zan'thēn), *n. Biochemistry.* a crystalline, nitrogenous compound related to xanthine, found in the muscles, liver, and spleen, as well as in various plant tissues. *Formula:* $C_5H_4N_4O$

hysteresis (his'tə rē'sis), *n. Physics.* a lagging of an effect behind its cause, as when a magnetic body is subjected to a varying force, or when an electric substance is subjected to a changing intensity: *For each magnet, one can develop a recipe for varying the magnetism in such a way that the end result is an optimum shaped field. The field pattern in the gap may show long-time drifts (of the order of hours) because of magnetic hysteresis of the iron.* (Science). [from Greek *hysterein* to lag, come late, from *hysteros* later]

Hz, *symbol.* hertz.

I

i, *symbol.* **1** *Astronomy.* inclination. **2** *Mathematics.* the imaginary unit $\sqrt{-1}$, square root of minus one.

I, *symbol.* **1** iodine. **2** electric current.

I., *abbrev.* **1** inosine. **2** isotopic spin.

IAA, *abbrev.* indoleacetic acid.

-iasis, *suffix.* **1** a process, as in *odontiasis = the process of cutting teeth.* **2** a diseased condition, as in *amebiasis = a diseased condition caused by amebas.* [from New Latin, from Greek *-iasis,* a noun suffix]

-ic, *suffix. Chemistry.* indicating the presence in compounds or ions of the designated element in a higher valence or oxidation state than indicated by the suffix *-ous,* as in *ferric, chloric.*

ice, *n. Physics.* water made solid by cold; frozen water, in hexagonal crystalline form, with a melting point of 0°C or 32°F.: *When ice melts, there is no temperature rise at first because the heat has gone into breaking down the ice's molecular structure* (Science News). See **ice point.**

ice age, *Geology.* **1** any of the times when large portions of the earth were covered with glaciers, as in the late Precambrian, in the early Cambrian, and in the Permian: *In ice age times apparently there were also land bridges across the mid-Mediterranean* (Scientific American). Also called **glacial epoch.**
2 Usually, **Ice Age.** the most recent such time, during the Pleistocene, when much of the Northern Hemisphere was covered with glaciers: *The Pleistocene Ice Age, which ended only a moment in geologic time ago—12,000 years or so—produced another period of extinctions. Glaciated areas were scraped clean of plants, and they are still today being slowly revegetated* (Weier, Botany).

iceberg

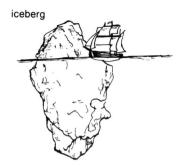

iceberg, *n. Oceanography.* a large mass of ice, detached from a glacier and floating in the sea. About 90 per cent of its mass is below the surface of the water. *Periodically, portions of the shelf* [ice] *detach themselves, creating immense, flat-topped icebergs. Icebergs of this type have been reported in length up to 100 miles* (Paul A. Siple). [from Dutch *ijsberg* (literally) ice mountain]

icecap, *n. Geology.* **1** a permanent covering of ice over large areas of land in the polar regions: *With an estimated volume of 7,500,000 cubic miles, the Antarctic icecap stores an amount of water of the order of the North Atlantic Ocean* (Richard S. Lewis). Also called **ice sheet. 2** a glacier, usually one covering a large area, that flows in all directions from its center.

ice cave, *Geology.* a cave having an accumulation of ice throughout the year. The ice forms when the temperature drops, freezing any water that seeps into the cave. See the picture at **glacier.**

ice crystal, = ice needle.

icefall, *n.,* or **ice fall,** *Geology.* a mass of glacial ice occurring on a steep grade: *Snow and ice cascade from the Khumbu icefall ... of Mount Everest* (N.Y. Times).

ice field, *Geology.* a large sheet of ice floating in the sea or resting on land, especially in a mountainous region.

ice fog, *Meteorology.* a fog of ice crystals formed during extremely low temperatures, especially in the Arctic: *Ice fog can seriously handicap airport operations in the Arctic* (Science News Letter).

ice foot, *Geology.* **1** a belt or wall of ice extending along the coast in arctic regions: *This fringe, which fills all the bays ... is really an exaggerated ice foot* (Scientific American). **2** the ice at the lower end, or foot, of a glacier.

iceland spar, *Mineralogy.* a transparent, coarsely crystalline variety of calcite, used extensively for optical purposes: *Iceland spar is the name given to the pure, transparent and colourless variety of calcite, $CaCO_3$, when it is so flawless that it can be used for polarizing light in ... optical instruments. Formerly the only source of the mineral was Iceland, hence its name,* (Jones, *Minerals in Industry*).

ice needle, *Meteorology.* a slender particle of ice floating in the air in cold weather. Also called **ice crystal.**

ice pack, *Oceanography.* a large area of masses of ice floating in the sea: *Captain James Cook sailed as far south as the Antarctic ice pack on his second voyage (1772–1775) without sighting land* (Richard S. Lewis). Compare **pack ice.**

ice point, *Physics.* the temperature at which pure water and ice are in equilibrium at atmospheric pressure, equal to 0° Celsius: *One reference temperature, the ice point, is the temperature of a mixture of air-saturated water and ice at a pressure of one atmosphere.* (Sears and Zemansky, *University Physics*). Compare **steam point.**

---

cap, fāce, fäther; best, bē, tèrm; pin, five;
rock, gō, ôrder; oil, out; cup, pùt, rüle,
yü in use, *yu* in uric;
ng in bring; sh in rush; th in thin, ᴛʜ in then;
zh in seizure.
ə = *a* in about, *e* in taken, *i* in pencil, *o* in lemon, *u* in circus

ice sheet, = icecap (def. 1).

ice shelf, = shelf ice.

ice storm, *Meteorology.* a storm in which freezing rain forms a coating of ice on the objects or surfaces that it strikes: *The so-called ice storm which produces glaze is one of the most destructive of the cool-season types of weather. It occurs when rain, near or below the freezing point, strikes surface objects the temperatures of which are below 32° and is immediately converted into ice.* (Finch and Trewartha, *Elements of Geography*).

ICF, *abbrev.* intercellular fluid.

ichthyological (ik′thē ə loj′ə kəl), *adj.* of or having to do with ichthyology.

ichthyology (ik′thē ol′ə jē), *n.* the branch of zoology dealing with fishes: *fossil ichthyology.* [from Greek *ichthys* fish + English *-logy*]

ichthyophagous (ik′thē of′ə gəs), *adj. Zoology.* feeding on fish; fish-eating: *ichthyophagous animals.* [from Greek *ichthyophagos,* from *ichthys* fish + *phagein* eat]

ichthyoplankton (ik′thē ō plangk′tən), *n. Zoology.* the small fish that float or drift in water: *Marine Resources Monitoring Assessment and Prediction ... will survey the kinds and quantities of living marine resources—such as ichthyoplankton, groundfish and pelagic fish* (Science News).

ichthyosaur (ik′thē ə sôr), *n. Zoology.* any of an order (Ichthyosauria) of extinct fishlike marine reptiles with a long beak, four paddlelike flippers, a tapering body, and a tail with a large fin: *On the land were the dinosaurs, in the sea the ichthyosaurs (looking like porpoises or sharks)* (Shull, *Principles of Animal Biology*). [from New Latin *Ichthyosauria,* from Greek *ichthys* fish + *sauros* lizard]
—**ichthyosaurian,** *adj.* of or belonging to the ichthyosaurs: *an ichthyosaurian snout.* —*n.* an ichthyosaurian reptile.

ichthyostegalian (ik′thē ə stə gal′ē ən), *adj. Zoology.* of or having to do with an order (Ichthyostegalia) of extinct vertebrates that are the most primitive amphibians known, having tails supported by fin rays and feet instead of fins. —*n.* an ichthyostegalian vertebrate. [from New Latin *Ichthyostegalia,* from Greek *ichthys* fish + *stegos* roof]

icosahedral (ī′kō sə hē′drəl), *adj. Geometry.* having twenty sides. *Almost all DNA tumor viruses have a regular, icosahedral surface consisting of viral protein subunits* (Peter K. Vogt). —**icosahedrally,** *adv.*

icosahedron (ī′kō sə hē′drən), *n.,* *pl.* **-drons, -dra** (-drə). *Geometry.* a solid having twenty faces; a twenty-sided polyhedron: *Axes of symmetry are shown for a regular icosahedron, a figure with 12 corners, 20 faces and 30 edges. Viewed along an axis at any corner, the figure can be rotated in five positions without changing its appearance* (Scientific American). [from Greek *eikosaedron,* from *eikosi* twenty + *hedra* seat, base]

-ics, *suffix.* body of facts or principles; science or field of study, as in *optics, electronics, genetics.* [originally plural of *-ic*]

ICSH, *abbrev.* interstitial cell stimulating hormone.

i.d., *abbrev.* inside diameter; internal diameter.

-ide, *suffix. Chemistry.* a binary compound of (an element), as in *oxide, chloride, sulfide.* The suffix *-ide* is added to the name of the electronegative or nonmetallic element (while that of the electropositive element remains unchanged), as in *sodium chloride, hydrogen sulfide,* etc. [from French *(ox)ide*]

ideal gas. *Physics.* a gas with molecules of negligible size that exert no force on each other; a gas that obeys the gas laws perfectly. At sufficiently high temperatures and low pressures, the behavior of real gases become increasingly close to that of an ideal gas. *This instrument makes use of the fact that the absolute temperature is proportional to the square of the speed of sound in an "ideal" gas—a gas at zero pressure* (New Scientist). Also called **perfect gas.** See also **absolute temperature.**

ideal point, *Mathematics.* a point of infinity; the point at which parallel lines intersect: *We assume ... the existence of an ideal plane of space, the locus of all ideal points* (Patterson, *Projective Geometry*).

idempotency (ī dem′pə tən sē), *n. Mathematics.* the property of remaining unchanged under multiplication by itself: *Idempotency ... corresponds to reflexiveness and occurs if an operation produces no change in the number or set on which it operates* (New Scientist). [from Latin *idem* the same + English *potency*]
—**idempotent** (ī dem′pə tənt), *adj. Mathematics.* having or characterized by idempotency: *Logical multiplication is said to be idempotent, signifying that all "powers" of a set are the same, and hence there is no need for the exponential symbolism of common algebra* (Edna E. Kramer).

identical equation, = identity (def. 2).

identical twins, *Biology.* twins of the same sex and physical appearance coming from a single fertilized egg cell rather than from two egg cells as fraternal twins do: *The occurrence of human identical twins has been known for a long time, and their potential usefulness in genetic research has been widely recognized* (Science News). See also **monozygotic.**

identity, *n. Mathematics.* **1** = identity element. **2** an equation that is satisfied by any number that replaces the letter for which the equation is defined. Also called **identical equation.** [from Late Latin *identitatem,* ultimately from Latin *idem* same]

identity element, *Mathematics.* an element that does not change any other element on which it operates; identity. The identity element for addition is 0. EXAMPLE: $28 + 0 = 28$. The identity element for multiplication is 1. EXAMPLE: $28 \times 1 = 28$.

idioblast (id′ē ə blast), *n. Botany.* a plant cell that differs greatly in form, content, size, or in other characteristic from the neighboring cells. [from Greek *idios* one's own, distinct + *blastos* germ, sprout]

idiopathic (id′ē ə path′ik), *adj. Medicine.* **1** not caused by or resulting from any disease; spontaneous; primary: *an idiopathic nervous condition.* **2** (of a disorder or disease) having no known origin: *idiopathic anemia.* [from Greek *idios* peculiar to, individual + *pathos* a suffering, disease]

idocrase (ī′də krās *or* id′ə krās), *n.* = vesuvianite. [from French *idocrase,* from Greek *eidos* form, shape + *krasis* a mixing]

Ig, *abbrev.* immunoglobulin. *Ig* is used especially in combination with a letter to designate any of various types of immunoglobulins, as in the following:

IgA, a type of antibody that comprises about 15% of Igs in normal human sera. It is the principal Ig in external secretions such as saliva, tears, and sweat. *People who are sensitive to certain plant pollens produce large amounts of IgA antibody whenever these pollens come in direct contact with the mucous membrane of the upper respiratory tract* (Buffaloe and Ferguson, *Microbiology*).

IgD, a type of antibody found in small concentrations in normal human sera whose function is poorly understood.

IgE, a type of antibody that is found only in minute concentrations in normal human sera, but which increases markedly in persons who show certain types of allergic reactivity or who contract helminthic infections: *IgE antibodies are responsible for reaginic hypersensitivity reactions* (Frank W. Putnam).

IgG, a type of antibody that comprises about 75% of Igs in normal human sera. It is the most numerous and most stable class of antibodies, and the only type that is able to cross the placenta. IgG antibodies circulate throughout the blood stream. *Some antibodies called IgG and IgM may press from the blood into the mouth in minute amounts of fluid that normally seep out of the crevices where the gingivae meet the teeth. These crevicular antibodies adhere to bacteria and enhance the action of certain white blood cells, which engulf and destroy S. mutans* (Simon W. Rosenberg).

IgM, a type of antibody that comprises about 10% of Igs in normal human sera. An IgM molecule is larger and more complex than that of any other Ig. *IgM, the phylogenetically and ontogenetically earliest immune globulin, can be produced in large amounts despite the most severe malnutrition* (Werner Dutz).

igneous (ig′nē əs), *adj. Geology.* produced by solidification from a molten state. Granite is an igneous rock. *Igneous rocks, rising molten from within the earth, are the ultimate source of all minerals* (Preston E. Cloud, Jr.). [from Latin *igneus* of fire, from *ignis* fire]

ASSOCIATED TERMS: There are three main types of rocks: *igneous, metamorphic,* and *sedimentary.* Igneous rocks, formed by the solidification (usually crystallization) of *magma,* are classified into two groups: *extrusive* and *intrusive. Basalt* is an extrusive igneous rock; *granite* is an intrusive igneous rock.

ignite, *v. Chemistry.* **1** to heat to the point of combustion or chemical change: *Generally speaking, a body can be ignited but once, whereas a body may be brought to a state of incandescence many times* (Charles Tomlinson). **2** to set fire to (a mixture of fuel and air) so as to produce power in an engine: *Spark plugs create electric sparks that ignite fuel in the combustion chambers of an engine.* [from Latin *ignitus,* past participle of *ignire* to set afire, from *ignis* fire]

—ignition, *n.* **1** the process of igniting: *ignition of a fuel mixture.* **2** a device for igniting: *a magneto ignition, an electronic ignition.* **3** *Physics.* the temperature in a plasma fusion reactor at which the energy input is equaled by the energy created by the fusion reactions. Between deuterium and tritium nuclei it is about 40,000,000 degrees Celsius.

i.hp. or **i.h.p.,** *Abbrev.* indicated horsepower.

Ile, *abbrev.* isoleucine.

ileocecal (il′ē ō sē′kəl), *adj. Anatomy.* of or having to do with both the ileum and the cecum: *... ileocecal valve, where the small bowel joins the large* (Time).

ileum (il′ē əm), *n. Anatomy.* the lowest part of the small intestine: *As the digesting food moves along by peristaltic waves it passes into the ileum, which makes up the remainder of the small intestine* (Winchester, *Zoology*). [from Latin *ileum* groin]

iliac (il′ē ak), *adj. Anatomy.* of, having to do with, or near the ilium: *iliac artery, iliac vein, the internal iliac muscle, an iliac bone.*

iliofemoral (il′ē ō fem′ər əl), *adj. Anatomy.* having to do with or connecting the ilium and the femur: the *iliofemoral ligament.*

iliosacral (il′ē ō sā′krəl), *adj. Anatomy.* having to do with the ilium and the sacrum.

ilium (il′ē əm), *n., pl.* **-ia** (-ē ə). *Anatomy.* the broad upper portion of the hipbone: *Each side of the pelvic girdle consists of an ilium, ischium, and pubis* (Shull, *Principles of Animal Biology*). See the picture at **pelvis.** [from Medieval Latin *ilium,* from Latin *ilia* loins, entrails]

Illinoian (il′ə noi′ən), *Geology.* **—adj.** of or having to do with the third glaciation of the Pleistocene in North America, beginning about 230,000 years ago, and lasting about 55,000 years: *Lakes were impounded by Illinoian ice in Pennsylvania and Illinois on the older drift and for a short time in Iowa* (J.K. Charlesworth). **—n.** the third Pleistocene glaciation in North America: *Pollen in the peat in Illinois, chiefly fir, pine, and tamarack, implies a climate cooler than today's, and suggests that possibly a transition to the succeeding Illinoian is represented* (R.F. Flint). Compare **Kansan, Nebraskan, Wisconsin.**

illite (il′īt), *n. Mineralogy.* any of a group of hydrous aluminum silicates resembling mica, found in clay: *The silica-to-alumina ratio of the illites is higher than in the micas* (G.P.C. Chambers). [from *Ill(inois)* + *-ite*]

illuminance (i lü′mə nəns), *n.* = illumination.

illumination, *n. Physics.* the amount of light per unit area of a surface, measured in lux; density of luminous flux: *experiments for determining the relative illumination of the different lights* (Andrew Ure). Also called **illuminance.**

illuvial (i lü′vē əl), *adj. Geology.* having to do with or produced by illuviation: *... a type of clay which has been enriched by illuvial accumulation* (G.W. Robinson).

illuviation (i lü′vē ā′shən), *n. Geology.* the addition to underlying levels of soil of salts, colloids, and small mineral particles moving downward by chemical or mechanical processes, most common in humid regions. [from *il-* into + *(al)luvi(al)* + *-ation*]

---

cap, fāce, fäther; best, bē, tèrm; pin, fīve;
rock, gō, ôrder; oil, out; cup, pùt, rüle,
yü in use, *yu̇* in uric;
ng in bring; *sh* in rush; *th* in thin, ᴛʜ in then;
zh in seizure.
ə = *a* in about, *e* in taken, *i* in pencil, *o* in lemon, *u* in circus

ilmenite (il′mə nīt), *n. Mineralogy.* a metallic black mineral consisting of iron, titanium, and oxygen: *The dark coloration of the lunar maria in general is undoubtedly attributable to ilmenite* (Scientific American). *Formula:* $FeTiO_3$ [from the *Ilmen* Mountains in the Urals, where it was discovered]

image, *Optics.* —*n.* the optical reproduction of an object, especially one produced by reflection in a mirror, or refraction through a lens or small hole: *Since a large part of optics is concerned with instruments that form optical images, a clear understanding of the process of image formation is essential* (Hardy and Perrin, *Principles of Optics*).
—*v.* to form an image or images of an object: *Hot stars as faint as 11th magnitude ... are routinely imaged at high resolution* (New Scientist).

imaginal (i maj′ə nəl), *adj. Zoology.* of or having to do with an imago: *... what are called 'imaginal characters'—points of structure which indicate that the larva has descended from an imago* (F. W. Myers).

imaginary, *Mathematics.*—*adj.* having to do with or being the square root of a negative number: *Such a number cannot be written in terms of ordinary real numbers, but ... the product of two imaginary numbers is a real* (Science News).
—*n.* an imaginary number or expression: *The circular functions ... are connected through the theory of imaginaries* (Arthur Cayley).
Compare **complex number.**

imago (i mā′gō), *n., pl.* **imagos, imagines** (i maj′ə nēz). *Zoology.* an insect in the final adult, especially winged, stage: *The states through which insects pass are four: the egg, the larva, the pupa, and the imago* (Kirby and Spence, *Introduction to Entomology*). [from New Latin *imago,* special use of Latin *imago* image]

imbalance, *n. Physiology.* a lack of muscular or functional balance of the eyes, the endocrine glands, etc.

imbricate (im′brə kit), *adj. Botany, Zoology.* covered with or made of scales or scalelike parts overlapping like shingles, as fish scales or leaf buds do. [from Latin *imbricatum* covered with tiles, from *imbrex* hollow tile] —**imbricately,** *adj.*
—**imbrication** (im′brə kā′shən), *n.* **1** *Botany, Zoology.* the condition of being imbricate. **2** *Geology.* the arrangement of flat pebbles or boulders by streams so that they overlap like shingles.

imidazole (im′ə daz′ōl), *n. Chemistry.* an organic crystalline base that is an antimetabolite, an inhibitor of histamine. *Formula:* $C_3H_4N_2$ [from *imid(e)* + *azole*]

imide (im′id), *n. Chemistry.* a compound containing the bivalent group -NH in combination with a bivalent acid group or two monovalent groups. [alteration of *amide*]

imido (im′i dō), *adj. Chemistry.* of or having to do with an imide or imides.

imine (im′in), *n. Chemistry.* a compound containing the -NH radical in combination with a bivalent nonacid group. [alteration of *amine;* perhaps patterned on *imide*]

imino (im′ē nō), *adj. Chemistry.* of or having to do with an imine or imines: *The proportion of an imino acid, hydroxyproline, is lower in the collagen of cold-water fish than in fish that swim in warmer waters* (Time).

imitation, *n. Biology.* = mimicry.
—**imitative,** *adj. Biology.* closely resembling: *imitative coloration in animals.*

immaculate (i mak′yə lit), *adj. Botany.* without colored marks or spots; unspotted. [from Latin *immaculatum* spotless, from *in-* not + *macula* spot, blemish]

immature, *adj. Geology.* reduced to only a slight extent by erosion, etc.

immersed, *adj.* **1** *Biology.* mostly or wholly sunk in surrounding parts, as the eyes of some beetles. **2** *Botany.* growing wholly under water.

immersion, *n. Astronomy.* the disappearance of the sun, moon, or other heavenly body in an eclipse.

immune (i myün′), *Immunology.* —*adj.* **1** protected from disease, poison, etc., by natural or acquired immunity: *In addition to immunization against smallpox, diphtheria, and scarlet fever, we can be made immune to lockjaw, typhoid fever ... bubonic plague, yellow fever, some kinds of influenza and rabies* (Beauchamp, et al. *Everyday Problems in Science*). **2** having to do with or producing immunity: *the body's immune system.*
—*n.* an immune person or animal: *After an epidemic the community remains free from that disease until the proportion of immunes declines* (Whitby and Hynes, *Bacteriology*).
[from Latin *immunis* (originally) free from obligation, from *in-* not + *munis* obliging, from *munia* duties, services]

immune body, *Immunology.* an antibody present in blood and lymph that assists in the dissolution of cells.

immune complex, *Immunology.* an aggregate formed by an antigen and its antibody. Immune complexes may cause agglutination, precipitation, and other immunological reactions. *Both autoantibodies and immune complexes are ubiquitous in mammalian serum ... they appear to be integral components of the immune system* (Scientific American).

immune response or **immune reaction,** *Immunology.* the production of lymphocytes, antibodies, and macrophages in response to disease germs, viruses, poisons, and other foreign substances: *An immune response has six major properties that distinguish it from any other type of protective mechanism: (1) it is mediated by ... antibodies, (2) it is acquired only after contact with an inciting antigen, (3) it ... is directed only at the inciting antigen, (4) it exhibits memory ... on subsequent contact with the inciting antigen, (5) it is capable of being transferred from one individual to another ... and (6) it distinguishes self [by] ... response ... only against foreign substances* (Science News). *New discoveries in the field of genetics could eventually lead to techniques for suppressing the immune reaction of patients, thus eliminating the rejection problem* (N.Y. Times).

immune serum, *Immunology.* a serum in which antibodies to a particular disease are present, especially in increased number as a result of stimulation by an antigen: *Human beings and domestic animals are now rendered immune to certain diseases by injecting the dead or attenuated organisms (vaccine) of a particular disease or the immune serum (antitoxin) from a horse or other animal that has previously been immunized* (Storer, *General Zoology*).

immunity (i myü'nə tē), *n. Immunology.* resistance of the body to disease germs, viruses, poisons, and other harmful substances, especially through the production of antibodies: *One attack of measles usually gives a person immunity to that disease. Many authorities have long held the view that "there is no immunity like convalescent immunity," meaning the immunity a person acquires after recovery from infection* (Scientific American).

▶ *Active immunity* is immunity that develops from having an immunogen in the body; thus vaccination creates active immunity to disease. *Passive immunity* is immunity resulting from the injection of serum containing antibodies. Immunity in which the body's cells attack the immunogen directly is called *cell-mediated immunity;* T cells control cell-mediated immunity. Immunity conferred by antibodies that neutralize or destroy the immunogen is called *humoral immunity;* B cells control humoral immunity.

immunize (im'yə nīz), *v. Immunology.* to give immunity to; make immune: *Vaccination immunizes people against such diseases as measles, polio, tetanus, and whooping cough.*

—**immunization,** *n.* the act or process of immunizing: *Today, through a careful program of immunization, parents are able to safeguard their children against many communicable diseases that were formerly common* (S.M. Gruenberg).

▶ *Active immunization* involves the injection of long-lasting vaccines into the body (vaccination). *Passive immunization* involves the injection of immune serum, whose effects last only a short time, or as long as the antibodies injected with the serum last.

immuno-, *combining form.* immunity; immunization, as in *immunoglobulin, immunoreaction.*

immunoadsorbent, *n. Immunology.* a chemical substance that adsorbs antigens or antibodies: *Antigens cling to the antibodies and when the immunoadsorbent is removed the antigens come with it. The antigens are then freed from the immunoadsorbent and concentrated* (Science News).

immunoassay (i myü'nō as'ā *or* i myü'nō ə sā'), *n. Immunology.* analysis of the characteristics of a bodily substance by testing its immunological or antibody-producing reactions: *Gonadotrophins from chorionic tissue are produced mostly in early pregnancy and can be assayed either biologically or by immunoassay* (New Scientist). Compare **radioimmunoassay.**

immunoblast, *n. Immunology.* an immature plasma cell.

immunocompetence, *n. Immunology.* the ability to produce or maintain immunity to disease. *the immunocompetence of lymphocytes.*

—**immunocompetent,** *adj.* capable of producing or developing immunity to disease: *The antibody-forming cells are frequently designated as immunocompetent cells* (Felix Haurowitz).

immunocyte (i myü'nə sīt), *n. Immunology.* any cell, such as a B cell or T cell, that is a part of the mechanism of immunization.

immunodeficiency, *n. Immunology.* inability of the body's immune system to produce antibodies in sufficient number or strength to fight infection: *The granulocytes and macrophages in our bloodstream are only part of our immune system ... Those children born*

immunogenic

with immunodeficiency have all the granulocytes they need, yet they die within months from infections (Ronald J. Glasser).

—**immunodeficient,** *adj.* affected by immunodeficiency: *... immunodeficient individuals, who are likely to die of overwhelming infection if their immune systems cannot be restored* (Science).

immunodiffusion, *n. Immunology.* a process of separating the components of an antigen-antibody complex by the diffusion of antigen and antibody solutions through a gel: *The diversity of the pollen wall proteins of P. deltoides is demonstrated by immunodiffusion and immunoelectrophoretic studies* (Nature).

immunoelectrophoresis (i myü'nō i lek'trō fə rē'sis), *n. Immunology.* a process of separating complex proteins according to their colloidal movement under the influence of an electric field and identifying each protein by its immune response.

—**immunoelectrophoretic** (i myü'nō i lek'trō fə ret'ik), *adj.* of or having to do with immunoelectrophoresis: *At least three and perhaps as many as six protein fractions have been identified in tears by electrophoretic and immunoelectrophoretic techniques* (Stella Y. Botelho).

immunofluorescence (i myü'nō flü'ə res'ns), *n. Immunology.* the labeling of antibodies with a fluorescent dye to reveal antigens when viewed under ultraviolet light: *Utilizing a technique of immunofluorescence, ... investigators demonstrated gamma E antibody-forming cells in lymphoid organs of the tracheal mucosa, pharynx, and intestinal tract* (Robert Keller).

—**immunofluorescent** (i myü'nō flü'ə res'nt), *adj.* of or having to do with immunofluorescence: *immunofluorescent examination.*

immunogen (i myü'nə jən), *n. Immunology.* 1 an antigenic substance; antigen: *Earlier studies had made use of haptens (small molecules that by themselves are not immunogenic) conjugated to proteins to construct immunogens* (M. M. Sigel).
2 an immunogenic substance.

immunogenetic, *adj. Immunology.* of or having to do with immunogenetics: *Several reports during the past decade have hinted that maternal blood contains fetal cells ... immunogenetic, as well as microscopic techniques, have now determined that this is indeed so* (Science News). —**immunogenetically,** *adv.*

immunogenetics, *n. Immunology.* 1 the study of the relationship between genetic factors and immunity to disease. 2 the study of biological relationships by means of serums.

immunogenic, *adj. Immunology.* making immune to a certain disease.

—**immunogenicity** (i myü'nō jə nis'ə tē), *n.* the property of causing immunity to a disease.

---

cap, fāce, fäther; best, bē, tèrm; pin, five;
rock, gō, ôrder; oil, out; cup, pùt, rüle,
yü in use, *yù* in uric;
ng in bring; sh in rush; th in thin, ᴛʜ in then;
zh in seizure.
ə = *a* in about, *e* in taken, *i* in pencil, *o* in lemon, *u* in circus

immunoglobulin (i myü′nō glob′yə lin), *n. Immunology.* any globulin that acts as an antibody; a protein in blood plasma that confers immunity: *Gamma globulin or immunoglobulin is one of the key antibody proteins which lymphocytes produce in response to microbial invasion* (Science News Yearbook). *Abbreviation:* Ig

immunohematologic (i myü′nō hem′ə tə loj′ik *or* i myü′nō hē′mə tə loj′ik), *adj.* of or having to do with immunohematology.

immunohematology (i myü′nō hem′ə tol′ə jē *or* i-myü′nō hē′mə tol′ə jē), *n.* the study of the immunological or antibody-producing properties of the blood.

immunological or **immunologic**, *adj.* of or having to do with immunology: *immunological defences.* —**immunologically,** *adv.*

immunology (im′yə nol′ə jē), *n.* the branch of biology dealing with the nature and causes of immunity from diseases: *Cooperation of lymphocytes in the evocation of the immune response is a highly important interaction, and its discovery was a milestone in modern immunology* (M. M. Sigel).

immunoreaction, *n. Immunology.* a reaction between an antibody and its antigen. —**immunoreactive,** *adj.* of or causing an immunoreaction: *immunoreactive growth hormone.*

immunosuppress, *v. Immunology.* to suppress the immune response of (an organism) against a foreign substance: *There is evidence that those individuals in whom the capacity to respond to an immunological challenge is decreased, i.e., those who are immunosuppressed, have a high incidence of cancer* (N. I. Berlin). —**immunosuppressant** (i myü′nō sə pres′ənt), *n.* an immunosuppressive drug, used especially to suppress rejection of grafted or transplanted tissue: *It is the lack of an effective immunosuppressant, free of any deleterious effect on the bone marrow, which is a major difficulty in human organ transplantation* (Science Journal). —*adj.* = immunosuppressive: *immunosuppressant chemicals or drugs.* —**immunosuppression,** *n.* suppression, especially by the use of drugs or radiation, of the immune response in order to prevent the rejection of grafts or transplants: *In order to coerce the body into accepting a donor's organ, full immunosuppression has been necessary ..., often leading unfortunately, to fatal secondary infection* (New Scientist). —**immunosuppressive,** *adj.* suppressing immunity; causing immunosuppression: *immunosuppressive drugs.* —*n.* an immunosuppressive agent or drug.

immunosurveillance, *n. Immunology.* a monitoring process by which the body's immune system detects foreign cells: *The immunosurveillance of these patients has broken down, and they are interpreting their own acetylcholine receptors as "foreign", and coating them with antibody* (New Scientist).

impact, *n. Physics.* **1** a forceful collision between inelastic bodies, resulting in a change from kinetic energy into other forms of energy, especially thermal energy. **2** the force imparted by an impact: *A four-pound bird hitting an engine on takeoff generates about 2,000 pounds of residual impact* (Science News).

—*v.* to fall or strike with an impact; collide or crash forcefully: *The Moon's craters are the scars left by impacting meteorites.*

impact crater, *Geology.* a crater produced by the impact of a meteor: *Evidently the maria are low-lying regions, in many cases enormous impact craters (judging from their roundness)* (Scientific American).

impacted, *adj.* **1** (of a tooth) wedged between the jawbone and another tooth. **2** (of a fracture) driven together so that the parts of a broken bone become locked.

impactite (im pak′tīt), *n. Geology.* a glassy kind of rock formed by partial fusion as a result of the impact of a meteorite on the surface of the earth: *Impactites ... are usually discovered near huge craters* (New Scientist).

impact parameter, *Nuclear Physics.* the distance between the path leading to the collision of two particles and the path actually taken by the particles.

imparipinnate (im par′ə pin′āt), *adj. Botany.* = odd-pinnate. [from Latin *impar* uneven + English *pinnate*]

impedance (im pēd′ns), *n. Physics.* **1** the apparent resistance in an alternating-current circuit, made up of two components, reactance and true or ohmic resistance: *impedance of a circuit.* **2** the square root of the sum of the squares of the true resistance and the difference between the inductive reactance and the capacitive reactance. **3** the ratio of the pressure in a sound wave to the product of the particle velocity and the area of a cross section of the wave at a given point: *acoustic impedance.*

impenetrability, *n. Physics.* the property of matter that prevents two bodies from occupying the same space at the same time.

impenetrable, *adj. Physics.* (of a body) excluding all other bodies from the space it occupies; having impenetrability.

imperfect fungus, *Biology.* any of an order (Fungi Imperfecti) of fungi that form only asexual spores or conidia.

impermeability (im pėr′mē ə bil′ə tē), *n.* the quality or condition of being impermeable.

impermeable, *adj.* not permitting the passage of water, gas, or other fluid: *an impermeable membrane, an impermeable shale. The transparent implant being impermeable, like all autografts, becomes invaded by the abnormal elements surrounding it* (Journal of the American Medical Association).

impervious, *adj. Geology.* = impermeable.

implant, *Biology, Medicine.* —*v.* (im plant′) to graft or set (a piece of skin, bone, etc.) into the body. —*n.* (im′plant) **1** a piece of tissue, an organ, or any living or artificial substance grafted into the body: *According to the patent, the plastics, called hydrogels, are suitable for body implants* (N.Y. Times). **2** a small radioactive tube or needle inserted in the body, especially to treat cancer: *radium implants for radiation therapy.* —**implantation** (im′plan tā′shən), *n.* **1** the introduction of tissue or an organ or any living or artificial substance into the body. **2** the passage of cells from one part of the body to another. **3** the process in which an embryo becomes attached for nourishment to the uterine wall in higher mammals, or to the yolk in other vertebrates: *Normally, implantation—the burrowing of the blastula into the lining of the womb—takes* *place on about day 20 of the woman's cycle, and some six days after fertilization has occurred* (New Scientist).

implication, *n.* = conditional.

implode, *v. Physics.* to burst or collapse violently inward. [from *im-* in + *-plode,* as in *explode*]

implosion, *n. Physics.* an imploding; a bursting or collapsing violently inward: *The whole star undergoes a very violent "gravitational collapse" in which particles are flying towards its centre at velocities near to the speed of light. In some imperfectly understood manner, part of the energy released from the gigantic implosion—the most violent process known—is converted into radio waves* (New Scientist).

impress, *v. Physics.* **1** to produce or transmit (motion, force, etc.) by pressure. **2** to produce or generate (an electromotive force or potential difference) in a conductor from some outside source, as a battery, generator, etc.

imprint (im print′), *v. Zoology.* **1** to establish (in a newborn or young animal) a behavior pattern of recognition and trust in another animal or an object identified as the parent: *In many cases, animals that become imprinted toward animals of another species never learn to recognize members of their own species* (W. J. McKeachie).
2 to be or become the object of recognition and trust of a young or newborn animal: *Goats and sheep apparently imprint on their own young in the first few moments after birth* (G. W. Barlow).
—**imprinting,** *n.* the process by which a newborn or young animal establishes a behavior pattern of recognition and trust toward another animal or an object identified as the parent: *Imprinting has been described in sheep, goats, deer and buffalo* (E. H. Hess).

improper fraction, *Mathematics.* a fraction that is equal to or greater than 1. EXAMPLES: 3/2, 4/3, 27/4, 8/5, 21/12, 4/4, 12/12. Compare **proper fraction.**

impulse, *n.* **1** *Physiology.* an electrochemical charge transmitted along a nerve cell, causing excitation of muscle, gland, and other nerve cells that it reaches: *Impulses from the eye, ear, tongue and nose travel via the cranial nerves directly to the brain* (M. E. Spencer). Also called **nerve impulse.**
2 *Physics.* **a** the product obtained by multiplying a value of force by its duration: *Impulse is a vector quantity with a magnitude equal to the product of the unbalanced force and the time the force acts. Its direction is the same as that of the force. When an unbalanced force acts on an object, there is a change of momentum which is equal to the impulse* (Physics Regents Syllabus). **b** a pulse of extremely short duration. **c** a surge of electrical power in one direction: *These impulses were recorded previously from actual signals transmitted from the control machine* (Wall Street Journal). [from Latin *impulsus* a thrust or push, from *impellere* to push, impel]

impurity, *n. Chemistry.* a foreign substance present in a very small amount or low concentration in a pure substance. An impurity may be beneficial, as in a semiconductor crystal; or it may be harmful, as an air pollutant.

In, *symbol.* indium.

-in, *suffix.* a variant of **-ine,** usually denoting neutral substances such as fats and proteins, as in *albumin, stearin.*

inactive, *adj.* **1** *Chemistry.* showing little tendency to combine with other elements or compounds, as the noble gases.
2 *Physics.* showing no optical activity in polarized light.

inbred, *adj. Biology.* developed by inbreeding; bred from closely related organisms: *Even the best inbred lines of corn yield no more than half as much as do the open-pollinated varieties from which the inbred lines are isolated* (Theodosius Dobzhansky).

inbreeding, *n. Biology.* breeding from closely related organisms: *Inbreeding tends to produce homozygous stocks, and since most genes for defects are recessive, it provides more opportunity for defective characters to appear* (Storer, *General Zoology*). Contrasted with **outbreeding.**

inbreeding depression, *Genetics.* a decline in successful breeding due to continuous inbreeding: *There will come a point when progress is hindered by the occurrence of infertility and inviability in their inbred lines. This inbreeding depression has been universally experienced by plant and animal husbandrymen.* (Hollingsworth and Bowler, *Principles and Processes of Biology*).

incandesce (in′kən des′), *v. Physics.* to glow or cause to glow with heat: *My electric lamp consists, essentially ... of an incandescing conductor of high resistance hermetically sealed in a glass vacuum chamber* (Thomas A. Edison). [from Latin *incandescere* begin to glow, from *in-* (intensive) + *candescere* become white-hot]
—**incandescence,** *n.* the condition of a body which glows and gives off light when it is heated; red-hot or white-hot condition: *It was found that electrical energy could be converted into light by sending a current through a wire of high resistance, thereby heating it to incandescence* (Hardy and Perrin, *Principles of Optics*).
—**incandescent,** *adj.* **1** heated to such a high temperature that it gives out light; glowing with heat; red-hot or white-hot: *A space charge will be developed around incandescent objects which will impede the continued emission of electrons* (Physics Regents Syllabus). **2** having to do with or containing a material that gives light by incandescence: *an incandescent lamp.*

---

cap, fāce, fäther; best, bē, tėrm; pin, five;
rock, gō, ôrder; oil, out; cup, pùt, rüle,
yü in use, *yů* in uric;
ng in bring; *sh* in rush; *th* in thin, ᴛʜ in then;
zh in seizure.
ə = *a* in about, *e* in taken, *i* in pencil, *o* in lemon, *u* in circus

Incandescent light is the light produced from an incandescent object, bulb, etc.

incenter (in′sen′tər), *n. Mathematics.* the center of the inscribed circle of a triangle: *The point of concurrency of the angle bisectors of a triangle is called the incenter of the triangle* (Moise and Downs, *Geometry*).

incenter

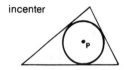

Point P is the incenter of the inscribed circle.

inch, *n.* a customary unit of length equal to 1/12 of a foot, or to about 2.54 centimeters.

inch of mercury, *Meteorology.* a customary unit used in the measurement of atmospheric pressure, equal to the pressure exerted by a one-inch column of mercury at standard gravity at a temperature of 0 degrees Celsius.

incidence, *n. Physics.* **1** the falling or striking of a ray or beam of light or other radiation upon a surface: *In equal incidences there is considerable inequality of refraction* (Sir Isaac Newton). *The incident ray, the reflected ray, and the normal to the surface at the point of incidence are in the same plane* (Physics Regents Syllabus). **2** the angle formed by this ray or beam with a line perpendicular to the surface. Also called **angle of incidence.**

—**incident,** *adj.* falling or striking upon a surface: *rays of light incident upon a mirror, the absorption of an incident photon by an atom.*
[from Latin *incidentem* happening, befalling, from *in-* on + *cadere* fall]

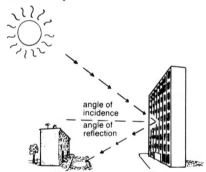

angle of incidence
angle of reflection

incised (in sizd′), *adj.* **1** *Biology.* having deep indentations or notches: *an incised leaf.* SYN: serrate. **2** *Geology.* cut or dug sharply into rock: *an incised river meander.*

incision (in sizh′ən), *n.* **1** *Botany, Zoology.* a deep indentation or notch, as on the margin of a leaf or of an insect's wing: *The incisions reach down to the rachis, or mid-rib, of the frond* (Francis G. Heath).
2 *Geology.* a sharp cut made into rock by a stream.

incisor (in si′zər), *n. Anatomy.* a tooth having a sharp edge adapted for cutting; one of the front teeth in mammals between the canine teeth in either jaw. Human beings have eight incisors.

inclination, *n.* **1** *Geology.* **a** an inclined surface; slope; slant. **b** the amount of slope or deviation from the horizontal position: *When a river approaches the sea, the inclination of its basin usually diminishes* (T.H. Huxley, *Physiography*). **c** the acute angle between the earth's magnetic field and the horizontal: *Except at the magnetic equator, the earth's field is not horizontal. The angle which the field makes with the horizontal is called the angle of dip or the inclination* (Sears and Zemansky, *University Physics*).
2 *Geometry.* **a** the angle which a line makes with a horizontal line in the same vertical plane: *If a line is parallel to the x-axis we shall regard its inclination as zero* (Rider, *First-Year Mathematics for Colleges*). **b** the smaller of the dihedral angles which a plane makes with another plane.
3 *Astronomy.* **a** the angle between the plane of the orbit of a planet and the ecliptic: *Inclination is another of the important qualities of an orbit; this is the amount by which the plane of the asteroid's orbit is tilted to the plane of the earth's orbit, the latter (plane of the ecliptic) being used as a standard* (Bernhard, *Handbook of Heavens*). **b** the angle between the plane of the orbit of a satellite and the equatorial plane of the primary. *Symbol:* i
[from Latin *inclinationem,* from *inclinare* to incline, from *in-* in + *clinare* to bend]

inclined plane, *Physics.* a plank or other plane surface set at an acute angle to a horizontal surface. It is a simple machine.

inclined plane

inclinometer (in′klə nom′ə tər), *n.* **1** an instrument for measuring the angle that the earth's magnetic field makes with the horizontal. **2** = clinometer.

included, *adj.* **1** *Botany.* **a** not protruding beyond the corolla, as a style or stamens. **b** embedded in the xylem, heartwood, etc.: *included phloem, included sapwood.*
2 *Mathematics.* taken in; contained: *An angle of a triangle is said to be included by the sides of the triangle which lie in the sides of the angle* (Moise and Downs, *Geometry*).

inclusion, *n.* **1** *Biology.* a nonliving body present in the cytoplasm of a cell, such as a starch grain or fat droplet. **2** *Crystallography, Mineralogy.* a small solid body, or a gaseous or liquid substance, contained in a crystal or a mineral mass. **3** *Geology.* a fragment of older rock embedded in an igneous mass. Also called **xenolith. 4** *Chemistry.* the material suspended in the matrix of a colloid.

inclusion body, *Biology.* a stainable body found in the cells of tissues infected with a virus, as in rabies or smallpox.

incoherent, *adj. Physics.* not coherent; having waves of irregular and random phases: *incoherent radiation. The light emitted by lasers ... is quite different from the incoherent light emitted by any other source of* *light—whether stars, candles, or electric lamps* (New Scientist).

incommensurable, *adj. Mathematics.* **1** (of lengths or numbers) having no common measure or number of which all the given lengths or numbers are integral multiples. $\sqrt{2}$ and 3 are incommensurable numbers because there is no third number which divides both of them evenly. A foot and a meter are incommensurable since the ratio between them must be expressed as an irrational number. *From the theorem of the right triangle there follows the existence of line segments that are incommensurable, that is, whose relationship with one another cannot be expressed by the natural numbers* (Scientific American). **2** = irrational.

incompatible, *adj. Biology, Medicine.* of or designating drugs, blood types, etc., that cannot be combined or used together because of undesirable chemical physiological reactions: *The red cells in less than one tablespoonful of incompatible blood may give rise to serious if not fatal reactions* (Harbaugh and Goodrich, *Fundamentals of Biology*).
—**incompatibility,** *n.* the condition of being incompatible: *Because it presents more serious difficulties for the child, the Rh factor has received greater medical attention than results of the ABO incompatibility* (Newsweek).

incomplete, *adj. Botany.* (of a flower) lacking one or more of the four kinds of floral organs, sepals, petals, stamens, and pistils: *Not all flowers have all four sets of parts. Incomplete flowers, such as some buttercups and willows, lack one or more of these sets* (Harold N. Moldenke).

incomplete metamorphosis, = hemimetabolism.

incompleteness theorem, = Gödel's theorem.

incomplete protein, *Biochemistry.* a protein that lacks one or more of the essential amino acids. A balanced diet may be attained by making foods with incomplete protein, such as wheat and beans, complement each other.

increment, *n. Mathematics.* a positive or negative change, usually small, in the value of a variable. [from Latin *incrementum* an increase, from *increscere* to increase, grow]
—**incremental,** *adj.* of, having to do with, or like an increment.

incubate, *v.* **1** *Zoology.* to sit on eggs in order to hatch them: *The female usually does the incubating, but the male often takes turns to give the female some time off the nest* (Winchester, *Zoology*). SYN: brood. **2** *Biology, Medicine.* to maintain (premature babies, eggs, cultures of microorganisms, etc.) at a temperature, humidity, and oxygen level suitable for development: *Some firms tried growing the mould in shallow trays ... each filled with the right amount of medium, aseptically inoculated with spores, and incubated for a week or so in a current of sterile air* (E. L. Smith and J.L. Crammer).
[from Latin *incubatum* incubated, from *in-* on + *cubare* lie]
—**incubation,** *n.* **1** *Zoology.* the act of incubating or condition of being incubated: *the incubation of eggs in hatcheries.* **2** *Medicine.* the stage of a disease from the time of infection until the appearance of symptoms:

The average period of incubation for mumps is 18 days.
—**incubative,** *adj.* of or having to do with incubation: *the incubative period or stage.*
—**incubator,** *n.* **1** a small box or chamber with controlled temperature and humidity, used for the artificial hatching of eggs. **2** a similar box or chamber supplied with oxygen, used for the special care of very small or premature babies. **3** a laboratory cabinet or apparatus with controlled temperature, in which cultures of microorganisms are grown: *The petri dish is left in a warm incubator for 24 hours, whereupon numerous bacterial colonies can be seen without magnification* (Emerson, *Basic Botany*).

incumbent, *adj.* **1** *Botany.* **a** (of an anther) lying flat against the inner side of the filament. **b** (of cotyledons) having the back of one against the hypocotyl: *The cotyledons are applied to each other by their faces, and ... are incumbent* (Balfour, *A Manual of Botany*). **2** *Geology.* overlying; resting upon: *an incumbent stratum.* [from Latin *incumbentem* lying upon something, from *in-* on + *-cumbere* lie down]

incus (ing′kəs), *n., pl.* **incudes** (in kyü′dēz). *Anatomy.* one of the three small bones of the middle ear of mammals, shaped somewhat like an anvil: *The vibration is transmitted mechanically by the three ear ossicles (malleus, incus, and stapes) to the membrane that covers the oval window of the inner ear* (Simpson, *Life: An Introduction to Biology*). Commonly called **anvil.** See the picture at **ear.** [from Latin *incus* anvil]

indamine (in′də mēn), *n. Chemistry.* any one of a series of basic organic compounds that form unstable bluish and greenish salts, used in making dyes. [from *ind(igo)* + *amine*]

indeciduous (in′di sij′ü əs), *adj. Botany.* **1** not deciduous, as a leaf. **2** (of a tree or plant) not losing the leaves seasonally.

indefinite integral, = antiderivative.

indehiscent (in′di his′nt), *adj. Botany.* not splitting open at maturity. Acorns are indehiscent fruits. *The typical dry, simple, indehiscent fruit is the achene, a small one-seeded fruit with a thin, tight wall* (Richard S. Cowan).

independent, *adj. Mathematics.* **1** (of a variable) causing the value of another variable to change when its own value is changed; not dependent on other variables. In statistics, the variable to be estimated or predicted is the *dependent variable,* while the variable which is used as the basis for estimating or predicting is the *independent variable.*
2 (of a system of equations) containing no equation which could be derived from another equation in the system.

cap, fāce, fäther; best, bē, tėrm; pin, five;
rock, gō, ôrder; oil, out; cup, pùt, rüle;
yü in use, *yù* in uric;
ng in bring; sh in rush; th in thin, ᴛʜ in then;
zh in seizure.
ə = a in about, e in taken, i in pencil, o in lemon, u in circus

313

independent assortment, *Genetics.* the Mendelian principle that if the genes for two different traits are located on different chromosome pairs (nonhomologous chromosomes), they segregate randomly during meiosis and, therefore, may be inherited independently of each other: *The effectiveness of independent assortment is limited by the fact that many genes are carried on a single chromosome* (Colin, *Elements of Genetics*).

indeterminacy principle, = uncertainty principle.

indeterminate (in'di tèr'mə nit), *adj. Botany.* having flowers which arise from auxiliary buds so that the tip continues to grow forming new bracts and flowers. The lily of the valley has indeterminate inflorescence. Also called **racemose.** Compare **determinate.** See the picture at **inflorescence.**

index, *n., pl.* **indexes** or **indices** (in'də sēz). **1a** a number or formula expressing some property, relationship, ratio, etc., in science. **b** the ratio of one dimension of a thing to another dimension: *The surface leads provide a better index of the activity of the whole muscle than do needle electrodes which sample the activity in only one part* (W.F. Floyd and P.H.S. Silver).
2 *Mathematics.* **a** an exponent: a^n ... *is called the nth power of a ... The integer* n *is called the index of the root* (Rider, *First-Year Mathematics for Colleges*). **b** the number indicating the root: *In* $3\sqrt{764}$ *the index is 3.* **c** the ratio of the order of a group to the order of a subgroup, expressed as an integer.
[from Latin *index* forefinger, originally, that which points out, from *indicare* point out, indicate, from *in-* in, to + *dicare* make known, proclaim]

index fossil, *Paleontology.* the fossil remains of an organism that lived in a particular geologic age, used to identify or date the rock or strata in which it is found: *Good index fossils must be abundant and widely distributed over the earth, and large enough not to be overlooked* (Shull, *Principles of Animal Biology*).

index of refraction, *Physics.* the ratio of the velocity of light in one medium, usually a vacuum or air, to its velocity in another: *Each of the different layers of silica gel in an opal has a different index of refraction, or bends light at a different angle* (Frederick H. Pough). Also called **refractive index.**

indican (in'də kən), *n. Biochemistry.* **1** a natural glucoside formed in plants yielding indigo, and which upon decomposition yields natural indigo. *Formula:* $C_{14}H_{17}NO_6$ **2** a substance found in the urine and other body fluids of animals. *Formula:* $C_8H_6NSO_4K$
[from *indico,* early variant of *indigo*]

indicator, *n.* **1** *Chemistry.* **a** a substance which, by changing color or some other property, indicates the concentration, presence or absence of a substance, or the endpoint of a chemical reaction: *The indicators in common use are litmus, methyl orange, and phenolphthalein. Litmus is changed from blue to red by acid solutions, methyl orange is changed from yellow to red, and phenolphthalein is changed from red to colorless* (Parks and Steinbach, *Systematic College Chemistry*). **b** a tracer: *a radioactive indicator.*
2 *Ecology.* **a** an organism whose existence in an area is indicative of specific environmental conditions. **b** an organism at or near the top of a food chain that reflects

or indicates the lower levels in the chain by the food residues found in its body.

indigenous (in dij'ə nəs), *adj. Ecology.* originating in a particular country; growing or living naturally in a certain region, soil, climate, etc.; native: *Cactus is indigenous only to the Americas* (Emerson, *Basic Botany*). [from Latin *indigena* a native, ultimately from *indu* in + *gignere* beget, bear]

indigo (in'də gō), *n., pl.* **-gos** or **-goes. 1** a blue dye formerly obtained from various plants, but now usually made artificially. **2** any of various plants of the pea family from which indigo was made. [from Spanish *indigo,* from Latin *indicum,* from Greek *indikon* Indian, from *Indos* India]

indium (in'dē əm), *n. Chemistry.* a rare, very soft, silvery metallic element found in nature only in combination with other elements. It is resistant to abrasion and is used as a coating on metal parts. *Symbol:* In; *atomic number* 49; *atomic weight* 114.82; *melting point* 156.61°C; *boiling point* 2000°C ($\pm$ 10°C); *oxidation state* 3. [from New Latin, from Latin *indicum.* See INDIGO.]

individual, *n. Biology.* a single organism that can exist independently: *The obvious and basic unit of life in nature is the individual. This is the unit of metabolism, of organization, of responsiveness, and of reproduction and development* (Simpson, *Life: An Introduction to Biology*).

indivisible, *adj. Mathematics.* that cannot be divided without a remainder: *Any odd number is indivisible by 2.*

indole (in'dōl), *n. Chemistry.* **1** a white, crystalline chemical compound with an unpleasant odor, produced artificially by the reduction of indigo or by other forms of syntheses, and also formed within the intestines. It is used as a reagent and in making perfumes. *Formula:* C_8H_7N
2 any of various derivatives of this compound: *A number of indole compounds have been found to give a fluorescence in the ultraviolet region of the spectra* (J.M. Sharpley).
[from *ind(igo)* + *-ole*]

indoleacetic acid (in'dōl ə sē'tik), *Biochemistry.* a plant hormone, the principal auxin, which regulates growth and development. It is synthesized in the protoplasm of the young and active parts of plants. *A chemically related hormone, indoleacetic acid, stimulates root formation in cuttings from such trees as apple and pine, which normally take hold only with great difficulty* (Scientific American). *Formula:* $C_{10}H_9O_2N$ *Abbreviation:* IAA

indolebutyric acid (in'dōl byü tir'ik), *Chemistry.* a synthetic compound used to regulate plant growth and development. *Formula:* $C_{12}H_{13}NO_2$

indophenol (in'dō fē'nōl), *n. Chemistry.* any one of various synthetic blue or green dyes derived from quinonimine. [from *ind(igo)* + *phenol*]

induce, *v.* **1** *Physics.* **a** to produce (radioactivity, fission, etc.) artificially by bombardment of a substance with neutrons, gamma rays, and other particles: *The development of particle accelerators ... has provided new sources or particles for inducing transmuting reactions* (Michels, *Foundations of Physics*). **b** to produce (an electric current, electric charge, or magnetic change) by induction: *To induce currents in a conductor, there*

must be relative motion between conductor and magnet (S.P. Thompson).
2 *Biochemistry.* to initiate or increase the production of (an enzyme or other protein): *When enough of the enzyme is formed, the cells grow as well on the disaccharide lactose as they did on glucose. In this instance, the lactose induced the formation of the enzyme needed to metabolize it* (McElroy, *Biology and Man*). [from Latin *inducere* lead in, from *in-* in + *ducere* to lead]
—**inducer,** *n. Biochemistry.* a substance whose presence initiates or increases production of an enzyme or other protein.
—**inducible,** *adj. Biochemistry.* (of enzymes) synthesized only when the material on which the enzyme acts (the substrate) is present: *Lactase, like many other enzymes, may be inducible—that is, the amount of the enzyme produced may increase as the amount of lactose ingested increases* (Paul E. Arauju).

inductance (in duk′tɔns), *n. Physics.* the property of an electric circuit by which an electromotive force is induced in it or in a nearby circuit by a change of current in either circuit. The tuner of a radio varies the inductance of its coils.

lines of induction

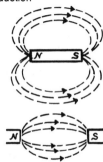

(def. 1)

induction, *n.* **1** *Physics.* the process by which an object having electrical or magnetic properties produces similar properties in a nearby object, without direct contact: *Induction can give a conductor a permanent charge, in the sense that the charge will remain on the conductor until it leaks off or is otherwise dissipated* (Scientific American). Compare **conduction.**
2 *Mathematics.* a method of proving a theorem by showing that it holds in the first instance, and that, if it holds for all the instances preceding a given one, then it must also hold for this instance: *One of the chief uses of well-ordered sets is in the method of mathematical induction which ... makes use of the well-ordering of the natural numbers* (Edna E. Kramer).
3 *Embryology.* the change in form or shape caused by the action of one tissue of an embryo on adjacent tissues or parts.
4 *Biochemistry.* the process of initiating or increasing the production of an enzyme or other protein: *In induction the protein produced by the regulator gene itself prevents the operator gene from functioning and the structural genes from producing enzymes for the metabolic sequence.* (Storer, *College Zoology*).
—**inductive,** *adj.* **1** *Physics.* having to do with or caused by electric or magnetic induction. **2**

Mathematics. of or using induction. **3** *Embryology.* capable of stimulating the formation of organs through induction: *inductive tissue.*

inductor, *n.* **1** *Embryology.* an embryonic tissue capable of influencing the differentiation of another embryonic tissue: *The tissues that stimulate the formation of organs are termed inductors* (Antone G. Jacobson). Also called **organizer.**
2 *Physics.* **a** a circuit or part of a circuit which has inductance. **b** any part of an electrical apparatus that works or is worked by induction.

induline (in′dyu̇ lēn), *n. Chemistry.* any one of a series of nitrogen compounds related to aniline dyes and producing blue or gray dyes. [from *ind(igo)* + *-ul*, a diminutive suffix + *-ine*]

indumentum (in′də men′təm), *n. Botany.* a hairy or scaly covering, often very dense: *... a thin white flaking stellate indumentum easily rubbed off and quickly falling away* (J.B. Stevenson). [from New Latin, from Latin *indumentum* clothing, from *induere* to put on]

indurate, *v. Geology.* to harden by heat, pressure, or cementation: *indurated clay.*
—**induration,** *n.* hardening by heat, pressure or cementation.

indusium (in dü′zē əm *or* in dü′zhē əm), *n., pl.* **-sia** (-zē-ə *or* -zhē ə). **1** *Botany.* **a** the membranous shield or scale covering the cluster of spore cases (sorus) of a fern. **b** a cuplike collection of hairs enclosing the stigma in certain plants: *... the fringed indusium which terminates the style of Goodeniaceae* (Lindley, *Introduction to the Natural System of Botany*).
2 *Anatomy.* a covering layer or membrane, especially the amnion.
3 *Zoology.* the covering of a larval insect.
[from Latin *indusium* tunic, from *induere* to put on]

industrial melanism, *Genetics.* the natural selection of darker variant forms of certain species of insects in industrial areas, caused by evolutionary adaptation to the sooty environment.

-ine, *suffix. Chemistry.* denoting the names of basic substances and the halogen elements, as in *aniline, fluorine.*

inelastic collision, *Physics.* a collision between two particles in which part of their kinetic energy is transformed to another form of energy: *Inelastic collisions are constantly boosting the electrons to the higher energy states appropriate to the temperature* (Scientific American).

inelastic scattering, *Physics.* the scattering of particles resulting from inelastic collision: *It is possible that the electron remains after the collision but the proton changes into several particles, one of which may still be a proton. In this case, we speak of inelastic scattering* (G. Feinberg).

cap, fāce, fàther; best, bē, tėrm; pin, fīve;
rock, gō, ôrder; oil, out; cup, pùt, rüle,
yü in use, *yu̇* in uric;
ng in bring; *sh* in rush; *th* in thin, ᴛʜ in then;
zh in seizure.

ə = *a* in about, *e* in taken, *i* in pencil, *o* in lemon, *u* in circus

inequality, *n. Mathematics.* an expression showing that two quantities are unequal. EXAMPLE: a > b means *a* is greater than *b;* a < b means *a* is less than *b;* a ≠ b means *a* and *b* are not equal. *Conditions for inequality usually include the idea of "greater than," or "less than," or "not equal to"* (Van Eugen, *Seeing Through Mathematics*).

inequilateral (in´ē kwə lat´ər əl), *adj. Biology.* not equilateral; having unequal sides: *Bivalves are all more or less inequilateral* (Samuel P. Woodward).

inert, *adj.* **1** *Chemistry.* not readily reactive with other elements; forming few or no chemical compounds. ▶ Group O elements are sometimes referred to as *inert gases.* The term *inert,* however, is no longer strictly applicable to this group, since it is possible to form compounds of krypton, xenon, and radon with fluorine and oxygen. Nevertheless, the term is still used in some contexts; for example, the electron configuration is generally referred to as the "inert gas structure" or "inert gas configuration." See also **noble.** **2** *Biology, Medicine.* having few or no active physiological, pharmacological, or other properties. [from Latin *inertem* idle, unskilled, from *in-* without + *artem* art, skill]

inertia (in ėr´shə), *n. Physics.* a tendency of all objects and matter in the universe to stay still if still, or if moving, to go on moving in the same direction, unless acted on by some outside force: *Unless a body is acted on by forces from some other body, it remains in a state of uniform motion. The natural tendency to remain in a state of uniform motion was termed inertia by Newton* (Shortley and Williams, *Elements of Physics*). [from Latin *inertia* inactivity, from *inertem* idle, inert]
—**inertial** (in ėr´shəl), *adj.* having to do with or of the nature of inertia: *water droplets tend to fly off the rim of a rapidly turning wheel ... which ... is really the inertial resistance that must be overcome to divert any mass from straight-line motion into circular motion* (Krogdahl, *Astronomical Universe*).

inertial frame, *Physics.* a frame of reference in which the law of inertia is valid. In Newtonian mechanics, all inertial frames move at constant velocity with respect to one another.

inertial mass, *Physics.* the mass of a body as determined by its momentum, as distinguished from gravitational mass: *The inertial mass of an object is proportional to the ratio of the force [acting] on an object [and] is a scalar quantity* (Physics Regents Syllabus).

infarct (in f ärkt´), *n. Medicine.* **1** a portion of dying or dead tissue, caused by the obstruction of the blood supply by an embolus, thrombus, etc. **2** the matter that fills such tissue. [from Latin *infarctum* stopped up, from *in-* in + *farcire* to stuff]
—**infarction** (in färk´shən), *n.* **1** formation of an infarct. **2** = infarct.

infauna (in´fô´nə), *n., pl.* **-nas, -nae** (-nē). *Zoology.* the animals living in the sediments on the sea floor: *While most benthic species are associated with either the epifauna or infauna, a few occur in both habitats* (Rhodes W. Fairbridge).
—**infaunal,** *adj.* of, having to do with, or belonging to the infauna: *infaunal habitats.*

infect, *v. Biology, Medicine.* to cause disease or an unhealthy condition in by introducing pathogenic organisms such as bacteria, protozoa, viruses, fungi, and worms: *Abscesses may occur on any tissue that becomes infected by bacteria* (John B. Miale). [from Latin *infectum* dyed, (originally) put in, from *in-* in + *facere* make]

infection, *n. Biology, Medicine.* **1** a causing of disease in people, animals, and plants by the introduction of pathogenic organisms such as bacteria, protozoa, viruses, fungi, and worms. Air, water, clothing, and insects are possible means of infection. Tuberculosis results from bacterial infection; measles and mumps are caused by viral infection; athlete's foot and ringworm are due to fungal infection. Elephantiasis and schistosomiasis are caused by worm infection. **2** an infectious disease: *Amebic dysentery is an intestinal infection caused by protozoans.* ▶ See the note under **contagion.**

infectious, *adj. Biology, Medicine.* **1** spread by infection; caused by bacteria, viruses, fungi, or other pathogenic organisms that invade some part of the body: *Chicken pox and German measles are infectious diseases. Most infectious diseases are communicable.* **2** causing infection: *The majority of definable diseases to which human beings are subject involve infectious microorganisms.*

infective, *adj. Biology, Medicine.* causing infection; infectious: *It has been shown that pneumonic plague patients may be directly infective* (New Scientist).
—**infectivity,** *n.* infectious quality or condition: *A group of St. Louis biochemists has taken apart and reassembled a virus without destroying its infectivity* (Scientific American).

inferior, *adj.* **1** *Botany.* **a** growing below some other part or organ: *an inferior calyx, an inferior ovary.* **b** belonging to the part of a flower that is farthest from the main stem.
2 *Anatomy.* (of organs or parts) below or posterior to others of the same kind; below or posterior to the usual or normal position: *the inferior vena cava.*
3 *Astronomy.* **a** between the earth and the sun: *Because Mercury moves in an orbit smaller than ours, it can never get into opposition, but it has two kinds of conjunction. One, which it reaches on April 5, is "inferior conjunction," when it is between us and the sun* (Science News Letter). **b** below the horizon: *the inferior passage of a star.*

inferior vena cava, *Anatomy.* the large vein carrying the blood from the lumbar, renal, hepatic, and other veins of the lower half of the body to the right auricle of the heart. Compare **superior vena cava.**

inferior vocal cords, = true vocal cords.

infinite, *adj. Mathematics.* **1** greater than any assignable quantity or magnitude of the sort in question: *There are not only, as Leibnitz had already asserted, infinite sets, but there are even what Leibnitz had denied, infinite numbers, and it can also be shown that one can operate with them, in a manner similar to that used for finite natural numbers* (Scientific American).
2 (of an assemblage) equivalent, as in value or degree, to some proper part of itself: *A collection of terms is infinite when it contains as parts other collections which have just as many terms as it has. If you can take away some of the terms of a collection, without*

diminishing the number of terms, then there are an infinite number of terms in the collection (Bertrand Russell).

3 (of a magnitude) beyond any finite magnitude. [from Latin *infinitus* without limit, from *in-* not + *finis* limit, border]

infinitesimal (in′fi nə tes′ə məl), *Mathematics.* —*adj.* approaching zero as a limit; being less than any assignable quantity or magnitude except zero. —*n.* a variable continually approaching zero as a limit. *Nonstandard analysis ... has revived the notion of the "infinitesimal"—a number that is infinitely small yet greater than zero. This concept ... has been an important tool in mechanics and geometry from at least the time of Archimedes* (Scientific American).

infinitesimal calculus, differential calculus and integral calculus, collectively: *Both analytical geometry and the infinitesimal calculus are enormously powerful instruments for solving geometrical and physical problems* (P. E.B. Jourdain).

infinity, *n., pl.* **-ties. 1** *Mathematics.* an infinite quantity or magnitude: *We have just proved that the natural numbers form an infinite set. They are said to constitute a countable or denumerable infinity* (E. E. Kramer). **2** *Geometry.* infinite distance, or the part of anything, such as a line or plane, infinitely removed from the part under consideration.

inflammable, *adj.* = flammable.

inflammation, *n. Medicine.* a reaction of tissues to injury: *The redness, heat, swelling, and pain characteristic of inflammation in man result primarily from changes in the vascular system* (Harbaugh and Goodrich, *Fundamentals of Biology*).

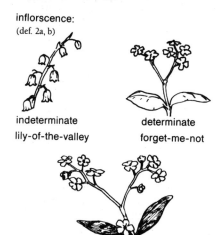

inflorscence:
(def. 2a, b)

indeterminate
lily-of-the-valley

determinate
forget-me-not

(def. 1)

inflorescence (in′flô res′ns), *n. Botany.* **1** the arrangement of flowers on the stem or axis in relation to each other. The two fundamental types of inflorescence are determinate or indeterminate.

2a a flower cluster. **b** the flower-bearing system or structure of a plant; the part of a plant bearing flowers. **c** one flower, if occurring on a plant bearing a single flower, such as the tulip.

3 a beginning to blossom; flowering stage. [from New Latin *inflorescentia* (coined by Linnaeus),

from Latin *in-* in + *florescere* begin to bloom, from *florem* flower]

—**inflorescent,** *adj.* showing inflorescence; flowering. ASSOCIATED TERMS: *Indeterminate* inflorescences (def. 1) include the *raceme, corymb, umbel, head, panicle,* and *spike. Determinate* inflorescences are sometimes called *cymose* or *cymes.* A third kind of inflorescence is the *thyrsus,* a mixed arrangement in which the main stem is indeterminate and the side branches determinate.

influent, *adj. Geography, Geology.* losing water by infiltration from the channel to the groundwater table below: *an influent stream.*

influx, *n. Geography, Geology.* the point where a river or stream flows into another river, a lake, or the sea. SYN: mouth.

infra-, *prefix.* below; beneath; beyond, as in *infrahuman, infrasonic, infrared.* [from Medieval Latin *infra-,* from Latin *infra* below, under]

infracostal, *adj. Anatomy.* located beneath the ribs: *infracostal muscles.*

infradian (in frā′dē ən), *adj. Biology.* of or having to do with biological rhythms or cycles that recur less than once per day: *Neurobiologic rhythms are organized ... in three frequency ranges: less than one per day (infradian), about one per day (circadian) and more than one per day (ultradian)* (J. Allan Hobson). [from *infra-* below + Latin *dies* day + English *-an*]

infraorbital (in′frə ôr′bə təl), *adj.* = suborbital: *an infraorbital gland.*

infraorder, *n. Biology.* a taxonomic group used in some classifications, ranking below a suborder.

infrared, *Physics.* —*adj.* of the invisible part of the spectrum whose rays have wavelengths longer than those of the red end of the visible spectrum and shorter than those of the microwaves. Most of the heat from sunlight, incandescent lamps, carbon arcs, resistance wires, etc., is from infrared rays. *Radiation occurs at invisible wavelengths longer than those of light. This infrared region has been found to extend continuously into the range of short radio waves, which it overlaps at wavelengths of the order of a millimeter* (Joseph H. Rush).

—*n.* the infrared part of the spectrum: *On the long-wavelength side of the visible lies the infrared, which may be said to merge into the radio waves* (Jenkins and White, *Fundamentals of Optics*). See the picture at **electromagnetic spectrum.**

infrared astronomy, a branch of astronomy dealing with the nature and sources of infrared radiation in space: *Infrared astronomy is beginning to study such things as intensity variations and polarizations in infrared sources* (Science News).

infrasonic, *adj.* = subsonic (def. 1).

infrasonics, *n.* the study of sound waves having a frequency below the range of human hearing (less than 10 vibrations a second or below 10 hertz).

cap, fāce, fäther; best, bē, tèrm; pin, five;
rock, gō, ôrder; oil, out; cup, pùt, rüle;
yü in use, yu̇ in uric;
ng in bring; sh in rush; th in thin, ŦH in then;
zh in seizure.
ə = a in about, e in taken, i in pencil, o in lemon, u in circus

317

infrasound, *n. Physics.* sound below the audible range: *Apparatus for detecting infrasound ... air vibrations which oscillate at less than 10 vibrations a second, or 10 hertz* (Sunday *Times*).

infraspecific, *adj. Biology.* occurring within a species.

infundibular (in′fən dib′yə lər), *adj. Biology.* funnel-shaped; infundibuliform.

infundibuliform (in′fən dib′yə lə fôrm), *adj. Biology.* funnel-shaped: *an infundibuliform flower.*

infundibulum (in′fən dib′yə ləm), *n., pl.* **-la** (-lə). *Biology.* a funnel-shaped part, organ, or structure, such as the prolongation of the third ventricle of the brain, connecting with the pituitary gland. [from Latin *infundibulum* (originally) a funnel, from *infundere* pour in, from *in-* in + *fundere* pour]

infusion, *n.* **1** *Chemistry.* **a** the act or process of extracting the active principle of a substance by steeping or soaking it in a liquid without boiling. **b** the liquid extract obtained by this means.
2 *Biology.* a liquid suspension of decaying organic matter used as a culture medium.
3 *Medicine.* the slow injection of a saline or other solution into a vein or subcutaneous tissue.

infusorial (in′fyu̇ zôr′ē əl *or* in′fyu̇ sôr′ē əl), *adj.* having to do with or containing infusorians.

infusorian (in′fyu̇ zôr′ē ən *or* in′fyu̇ sôr′ē ən), *Biology.* —*n.* any of a group of minute, usually microscopic, one-celled organisms that move by vibrating filaments, such as the paramecium. Infusorians occur in both free-living and parasitic forms. —*adj.* of or belonging to this group.
[from New Latin *Infusoria* a genus name, from Latin *infusus,* past participle of *infundere* pour in]

ingest, *v. Biology.* to take (food, etc.) into the body for digestion: *When the beetle bites a leaf or twig, it ingests the pesticide and dies almost instantly* (Wall Street Journal). *The improvement in the general state of health resulting from ingesting the optimum amount of ascorbic acid might lead to an equal additional increase in life expectancy* (Linus Pauling). ANT: egest. [from Latin *ingestum,* past participle of *ingerere* carry in, from *in-* in + *gerere* carry]

ingesta (in jes′tə), *n.pl. Biology.* substances ingested into the body, especially through the mouth. [from New Latin *ingesta,* from Latin, neuter plural of *ingestum;* see etym. under INGEST]

ingestion (in jes′chən), *n. Biology.* the act of taking food or other substance into the body for digesting: *Most species of animals secure their food by ingestion (i.e., by eating all or parts of other organisms and then digesting the food in their digestive tracts)* (Greulach and Adams, *Plants*). —**ingestive,** *adj.* of or having to do with ingestion.

ingress (in′gres), *n. Astronomy.* the entrance of a heavenly body into an eclipse, transit, or occultation.

inguinal (ing′gwə nəl), *adj. Anatomy.* of or in the region of the groin. The **inguinal canal** is a small passage through which the spermatic cord of the male enters the abdomen. [from Latin *inguinalis,* from *inguen* groin]

inhalation, *n.* the act of inhaling: *Among both men and women death rates increased with the number of cigarettes smoked per day and the degree of inhalation* (Harold S. Diehl).

inhale, *v.* **1** *Physiology.* to draw into the lungs; breathe in: *to inhale oxygen.*
2 *Biology.* to take in or absorb oxygen, etc. in the process of respiration: *Photosynthesis, in the most widely used sense, is the synthesis of carbohydrates by plants, using the energy of sunlight. Carbon dioxide is combined with hydrogen that, typically, is derived from water, and oxygen is released. Thus plants "inhale" carbon dioxide and "exhale" oxygen* (Walter Sullivan).
[from Latin *inhalare,* from *in-* in + *halare* breathe]

inherit, *v. Genetics.* to acquire (characteristics) transmitted by germ cells from one's parents or ancestors: *Sex is an inherited trait ... determined by a balance between the sex chromosomes and the autosomes* (McElroy, *Biology and Man*).

inheritance, *n. Genetics.* **1** the acquisition of characteristics transmitted by germ cells from generation to generation.
2 the sum of all the characteristics transmitted by the germ cells: *Those characteristics derived from a parent constitute one's inheritance, and the science which provides an explanation of how characters are inherited is called genetics* (Harbaugh and Goodrich, *Fundamentals of Biology*). Compare **heredity.**

inhibit, *v. Physiology, Biochemistry.* to block, check, or hinder the function or action of an enzyme, cell, organ, etc.: *Some transmitters [of nerve impulses] have an excitatory effect; others inhibit* (McElroy, *Biology and Man*). —**inhibition,** *n.* the act or process of inhibiting: *In the process of inhibition one cell or tissue mass impedes the growth of another cell or tissue* (Storer, *General Zoology*).

inhibitor (in hib′ə tər), *n.* **1** *Chemistry.* anything that checks or interferes with a chemical reaction, such as oxidation or polymerization; a negative catalyst: *Some kinds of inhibitors adsorb on the metal surface, forming a single layer of inhibitor molecules that acts as a barrier to the environment. ... Other inhibitors prevent corrosion by keeping the environment from becoming more corrosive (usually more acidic) by providing what is called a buffering action* (Jerome Kruger).
2 *Biochemistry.* a substance which inhibits the function or action of an enzyme: *In many instances the inhibition of an enzyme can be regarded as the result of chemical combination between the enzyme and the inhibitor* (W.V. Thorpe). —**inhibitory** (in hib′ə tôr′ē), *adj.* inhibiting; tending to inhibit. An **inhibiting nerve** hinders the activity of an end organ. *In contrast to excitatory chemicals that depolarize the structure which they excite, inhibitory chemicals do the reverse. They increase the polarization of the structure* (McElroy, *Biology and Man*).

inion (in′ē ən), *n. Anatomy.* a point at the external occipital protuberance of the skull. [from Greek *inion* nape of the neck; muscles at the back of the neck, from *inos* sinew]

initial side, *Trigonometry.* the ray whose amount of rotation determines the size of a directed angle: *The greater the amount of turning, the larger the angle. If*

the turning has been one complete revolution, the terminal side again coincides with the initial side, but the line is said to have been rotated through 360° (Schaaf, *Trigonometry for Home Study*).
ASSOCIATED TERMS: see **directed.**

initiator, *n. Chemistry.* a substance or molecule that initiates a reaction, such as polymerization. It differs from a catalyst in that it is usually consumed or chemically changed in the reaction.

inject, *v.* to force (a fluid, medicine, etc.) into a chamber, passage, cavity, or tissue: *inject penicillin into a muscle.* [from Latin *injectum* thrown in, from *in-* in + *jacere* to throw] —**injector,** *n.*
—**injection,** *n.* **1** the act or process of injecting: *Drugs are often given by injection.* **2** a liquid injected: *an injection of penicillin.*

ink, *n. Zoology.* a dark liquid ejected for protection by cuttlefish, squids, and other cephalopods.

inlet, *n. Geography.* a narrow strip of water running from a larger body of water into the land or between islands. SYN: arm.

inlier (in'lī'ər), *n. Geology.* an area or formation of older rocks completely surrounded by a more recent formation. Compare **outlier.**

inner core, *Geology.* the fourth layer and central portion of the earth's core: *At the center of the core lies an inner core ... about two-thirds the size of the moon and unlike the outer core is probably solid* (Charles R. Carrigan and David Gubbins). Compare **outer core.** See the picture at **core.**

inner ear, *Anatomy.* the innermost part of the ear, behind the middle ear, containing the essential organs of hearing and equilibrium: *The inner ear is filled with a fluid and ... It is here that the sound vibrations are converted to nervous impulses* (Mackean, *Introduction to Biology*). Also called **labyrinth.**

inner product, = dot product. Also called **scalar product.**

innervate (i nėr'vāt *or* in'ər vāt), *v.* **1** *Anatomy.* to supply (an organ, muscle, etc.) with nerves. **2** *Physiology.* to stimulate (an organ or muscle) by nerve impulses: *The facial nerve innervates the muscles of expression.*
—**innervation,** *n.* **1** the arrangement or distribution of nerves to an organ or other part of the body. **2** the amount or degree of stimulation of a muscle or organ by nerves.

innominate (i nom'ə nit), *n. Anatomy.* an innominate bone, artery, or vein. [from Late Latin *innominatus,* from Latin *in-* not + *nominen* name]

innominate artery, *Anatomy.* an artery arising from the arch of the aorta. It divides into the right carotid and subclavian arteries.

innominate bone, = hipbone.

innominate vein, *Anatomy.* either one of the two veins formed by the junction of the subclavian and internal jugular veins. The innominate veins unite to form the superior vena cava.

inoculant (i nok'yə lənt), *n.* a substance used in inoculating. SYN: inoculum.

inoculate (in ok'yə lāt), *v. Biology.* **1** to introduce (microorganisms) into surroundings suited to their growth: *Most contemporary biotechnologies require the addition (inoculation) of a particular microbe to a solution containing substances ... whose conversion by*

the microorganism results in a useful product (J. King and C. Orrego). **2** to treat (soil, seeds, etc.) with bacteria that will promote nitrogen fixation. [from Latin *inoculatum* engrafted, from *in-* in + *oculus* bud, eye]
—**inoculation,** *n.* **1** the act or process of inoculating: *the inoculation of cells into a culture medium.* **2** the bacteria, serums, or other agents used in inoculating.
—**inoculative,** *adj.* of or having to do with inoculation.
—**inoculator,** *n.* one that inoculates, especially an instrument used to inoculate.

inoculum (i nok'yə ləm), *n., pl.* **-la** (-lə). *Biology.* a substance containing viruses, bacteria, or other antigens used in inoculating: *Preparing an inoculum of penicillium spores ... at an antibiotic plant* (London *Times*). [from New Latin *inoculum,* from Latin *inoculare* to engraft, from *in-* in + *oculus* bud, eye]

inorganic, *adj.* **1** *Chemistry.* of or having to do with substances not containing organic groups. Chemical compounds without hydrocarbon groups are usually considered inorganic. *The inorganic compounds of plants—water, gases, salts, acids, and bases—are also constituents of nonliving things and are not characteristically biological* (Greulach and Adams, *Plants*).
—**inorganically,** *adv.*
2 *Biology.* not organic; not having the organized physical structure characteristic of animals and plants. Water and minerals are inorganic substances. *In general, the study of the inorganic is the province of the physical sciences such as chemistry, physics, astronomy, geology. The organic is the field for the biological sciences—zoology, botany, and some branches of psychology* (Ogburn and Nimkoff).
▶ Inorganic chemical compounds are frequently defined as those that do not contain carbon, as distinguished from organic compounds. However, many scientists limit organic compounds to those that contain both carbon and hydrogen (hydrocarbon groups), often along with oxygen and other elements, thus regarding as *inorganic* such compounds of carbon as carbon dioxide, calcium carbonate, and sodium cyanide.

inorganic chemistry, the branch of chemistry dealing with inorganic compounds and elements; the study of the chemical properties of all elements and compounds except organic ones, chiefly the hydrocarbons and their derivatives.

inosine (in'ə sēn), *n. Biochemistry.* a compound of hypoxanthine and ribose, obtained from the deamination of adenosine. Inosine is an important intermediate in the metabolism of animal purines. *Formula:* $C_{10}H_{12}N_4O_5$ [from Greek *is, inos* sinew, muscle + English *-ine*]

cap, fāce, fäther; best, bē, tėrm; pin, five;
rock, gō, ôrder; oil, out; cup, pùt, rüle,
yü in use, *yù* in uric;
ng in bring; sh in rush; th in thin, ŦH in then;
zh in seizure.
ə = *a* in about, *e* in taken, *i* in pencil, *o* in lemon, *u* in circus

inositol (i nō′sə tōl), *n. Biochemistry.* a white, crystalline alcohol found especially in certain muscle tissues and in plants. It occurs in several isomeric forms and is one of the B-complex vitamins. *Formula:* $C_6H_{12}O_6 \cdot 2H_2O$

inotropic (in′ə trop′ik), *adj. Physiology.* influencing or modifying the contractility of muscles: *an inotropic agent or compound.* [from Greek *is, inos* muscle + *tropos* turn, manner]

input, *n.* **1** *Physics.* **a** the work or force supplied to a machine. In mechanical work, the *input* is equal to the effort times the distance that the effort moves, while *output* is equal to the load to be moved times the distance that it is moved. To determine the efficiency of a machine, it is necessary to compute the ratio of the output to the input. **b** the power fed into an electronic device: *The power output of a transformer is necessarily less than the power input, because of unavoidable losses in the form of heat* (Sears and Zemansky, *University Physics*).
2 the information fed into a computer, data processing machine, or other automatic device: *A finite automaton is a device having: 1—A finite number of possible (internal) states; 2—A finite number of possible inputs at each moment; 3—A finite number of possible outputs at each moment* (New Scientist).
—v. to feed (information) into a computer, data processing machine, or other automatic device: *While one user is spending 5 seconds inputting some data ... the CPU [central processing unit] of a modern computer can execute as many as 1 million computations* (R. C. Sprowls).

inquiline (in′kwə lin), *n. Zoology.* an animal that lives in the nest or shell of another animal: *A true inquiline does not attack the host—that is, the primary inhabitant of the gall or nest—but ... may either monopolize the food so as to cause the death of the host or else may end by eating the host* (Klots, *Living Insects of the World*). Compare **commensal.** [from Latin *inquilinus* lodger]

ins., *abbrev.* inches.

inscribe, *v. Geometry.* to draw (one figure) within another figure so that the inner touches the outer at as many points as possible: *a triangle inscribed in a circle.* [from Latin *inscribere,* from *in-* on + *scribere* write]

insect, *n. Zoology.* any of a class (Insecta) of small air-breathing arthropods with the body divided into three sections (head, thorax, and abdomen), and having three pairs of legs and usually one or two pairs of wings sometime during the adult stage. Flies, mosquitoes, grasshoppers, gnats, bees, butterflies, and beetles are insects. *Insects are more numerous in species than all other animals taken together ... and, no doubt, hundreds of thousands remain to be discovered. ... They live in almost every conceivable type of environment ... and their structure, habits, and life cycle are correspondingly modified* (Hegner and Stiles, *College Zoology*). [from Latin *(animal) insectum,* literally, divided (animal), from *insecare* to cut into, from *in-* into + *secare* to cut (from the segmentation of its body)].
▶ In common usage, certain wingless arthropods with segmented bodies, such as spiders and centipedes, are often called insects. These, however, do not belong to the class Insecta and are therefore not scientifically classified as insects.

insect

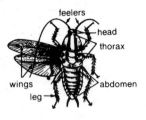

feelers
head
thorax
wings
abdomen
leg

insectarium (in′sek ter′ē əm), *n., pl.* **-ia** (-ē ə). a place for keeping, observing, and breeding insects. [from *insect* + *-arium,* perhaps patterned on *aquarium*]

insecticide, *n.* a substance for killing insects: *DDT is a stable compound and its effects last for a long time; a good property for an insecticide but potentially disastrous for the balance of nature* (Mackean, *Introduction to Biology*).

insectivore (in sek′tə vôr), *n.* **1** *Biology.* any animal or plant that feeds mainly on insects. Mantises are insectivores.
2 *Zoology.* any of an order (Insectivora) of small mammals including moles, shrews, and hedgehogs that feed chiefly on insects at night. They are the least developed of the placental animals.

insectivorous (in′sek tiv′ər əs), *adj.* **1** *Biology.* insect-eating; feeding mainly on insects: *Flying insects ... are eaten by such insectivorous birds of the treetops as warblers, vireos, orioles, and nuthatches* (Charles F. Wurster, Jr.). **2** of or belonging to the order of insectivores.
[from New Latin *insectivorus,* from Latin *insectum* insect + *vorare* devour; patterned on Latin *carnivorus* meat-eating]

inselberg (in′səl bèrg′), *n. Geology.* a prominent stump or core of a mountain left rising above a surrounding plain primarily in hot, dry regions: *The inselbergs or 'island' mountains of Tanzania and Zambia (known as kopjes in Rhodesia and South Africa) are striking features of the landscape* (Robert and Eileen Steel). [from German *Inselberg,* from *Insel* island + *Berg* mountain]

inseminate (in sem′ə nāt′), *v. Biology.* **1** to inject semen into; fertilize; impregnate. **2** to sow or plant (seeds). [ultimately from Latin *in-* in + *seminis* semen, seed]
—insemination, *n.* the act or process of inseminating. See **artificial insemination.**

insequent, *adj. Geology, Geography.* having a course that appears haphazard and exhibits no apparent relation to the form of the land: *Homogeneous surfaces tend to have insequent drainage systems.* Compare **consequent, obsequent, resequent,** and **subsequent.**

insert, *Molecular Biology.* **—v. 1** to introduce one section of DNA or other genetic material into another: *Further study revealed that the inserted sequence ... could be found at many different locations in the same gene, or in other genes. In other words, it could be transposed from one part of the bacterial chromosome to another* (Robert Haselkorn).

2 (of a DNA section) to be inserted: *The DNA might insert into an existing intervening sequence to produce two silent regions and one coding region where there was only a silent region previously* (J. A. Miller).
—*n.* an inserted sequence of DNA or other genetic material: *... when one examines the immunoglobulin gene and finds that there are three inserts separating these four functional regions* (New Scientist).

insertion, *n.* **1** *Molecular Biology.* **a** the introduction of one section of DNA or other genetic material into another. **b** the section inserted: *Partially active genes ... can be produced by crossing over between an insertion and a nearby deletion* (Watson, *Molecular Biology of the Gene*).
2 *Biology.* the place or manner of attachment of an organ, part, or tissue.
3 *Anatomy.* the part of a muscle attached to the bone or some other part that it moves: *The more proximal, or less movable, end is called the origin, and that on the more distal, or movable, part is the insertion* (Storer, *General Zoology*).

insessorial (in′se sôr′ē əl), *adj. Zoology.* **1** habitually perching. **2** adapted for perching: *insessorial feet.* [from Latin *insessor* occupant, from *insidere* sit on, from *in-* on + *sedere* sit]

insolation, *n. Meteorology.* **1** the amount of solar radiation per unit of horizontal surface in a given time: *In addition to the imposed diurnal and annual periodicities in the incoming solar radiation, one finds a slight semiannual variation of insolation in the Tropics, where the sun passes overhead twice during the year* (Abraham H. Oort).
2 the radiant energy emitted by the sun: *Solar radiation, or insolation, is the one important source of energy for the earth and its atmosphere* (C. G. Knudsen and J. K. McGuire).

insolubilize (in sol′yə bə līz′), *v. Chemistry.* to make (a substance) insoluble. —**insolubilization,** *n.*

insoluble, *adj. Chemistry.* that cannot be dissolved; not soluble: *Fats are insoluble in water.*

instar (in′stär′), *n. Zoology.* the particular stage of an insect or other arthropod between moltings: *All of the larvae died in the first, second, or third instars, or molting periods* (G. T. Hellman). [from New Latin, from Latin *instar* likeness, form]

instep, *n. Anatomy.* the arch of the human foot between the toes and the ankle, especially the upper surface of the arch.

instrument, *n.* a device for measuring, recording, or controlling. A thermometer is an instrument for measuring temperature. A compass and sextant are instruments of navigation. *The observatory's main instruments, a 6-inch telescope, a laser-camera satellite-tracking system and a 34-foot optical reflector for gamma-ray astronomy, are strung out along a knobby ridge 7,600 feet above the desert* (James C. Cornell Jr.).
—*v.* to equip with instruments, especially for recording scientific data: *Rockets can be instrumented to record observations in geomagnetism, cosmic rays, and other phenomena of the universe.* [from Latin *instrumentum*, from *instruere* to arrange]
—**instrumentation,** *n.* the mechanized use of instruments, especially for scientific or technical purposes: *... another significant forward step in the planned ex-*

pansion into new fields of instrumentation and automation (Wall Street Journal).

insula (in′sə lə), *n., pl.* **-lae** (-lē). *Anatomy.* the central lobe of the cerebrum. [from Latin *insula* island]

insular (in′sə lər), *adj.* **1** *Geography.* of or having to do with an island or islands: *Islands lying along coasts or upon shallow continental shelves, are said to possess insular situation* (White and Renner, *Human Geography*).
2 *Anatomy.* of or having to do with the islets of Langerhans.
3 *Medicine.* characterized by isolated spots, patches, or pustules, as a rash.
[from Late Latin *insularis*, from Latin *insula* island]

insulate (in′sə lāt), *v. Physics.* to keep from losing or transferring electricity, heat, sound, etc., especially by covering, packing, or surrounding with a nonconducting material: *to insulate a wire or cable. Josephson junction consists of two superconducting tin strips separated by an insulating layer of tin oxide 10 angstroms thick* (Scientific American). [from Latin *insula* island]
—**insulation** (in′sə lā′shən), *n.* **1** an insulating or being insulated. **2** material used in insulating.
—**insulator** (in′sə lā′tər), *n.* that which insulates; something that prevents the passage of electricity, heat, or sound: *Charge cannot readily travel from one part to another of a nonmetal, which hence is called an insulator or dielectric ... in nonmetals the electrons are not, in general, free to leave the atoms in which they belong* (Shortley and Williams, *Elements of Physics*). SYN: nonconductor.

insulin (in′sə lən), *n. Biochemistry.* **1** a protein hormone secreted by the islets of Langerhans that enables the body to use sugar and other carbohydrates by regulating the sugar metabolism of the body: *The metabolic effects of insulin are seen in an increase in glucose uptake, glycogenesis, glycolysis, and lipogenesis* (George Weber). *Chemically, insulin is a complex molecule composed of two polypeptide chains (A and B), joined by two disulfide linkages (A-S-S-B)* (M. Frederick Hawthorne).
2 a drug containing this hormone, obtained from the pancreas of animals for use in treating diabetes, and now also produced synthetically: *Human insulin has been produced at last by genetically engineered bacteria in a California laboratory—an achievement that catapults recombinant DNA technology into the major leagues of the drug industry* (Science News).
[from Latin *insula* island (in reference to the islets of Langerhans)]

insulinase (in′sə lə nās), *n. Biochemistry.* an enzyme secreted by the liver that is capable of inactivating insulin.

cap, fāce, fäther; best, bē, tėrm; pin, fīve;
rock, gō, ôrder; oil, out; cup, pùt, rüle,
yü in use, *yù* in uric;
ng in bring; *sh* in rush; *th* in thin, ᴛʜ in then;
zh in seizure.
ə = *a* in about, *e* in taken, *i* in pencil, *o* in lemon, *u* in circus

integer (in'tə jər), *n. Mathematics.* any positive or negative whole number, or zero: *We may classify integers as (1) positive or non-positive and (2) factorable and non-factorable* (Pearson and Allen, *Modern Algebra*). [from Latin *integer* whole, from *in-* not + *tangere* to touch]

integrability (in'tə grə bil'ə tē), *n. Mathematics.* integrable quality: *to satisfy the condition of integrability.*

integrable (in'tə grə bəl), *adj. Mathematics.* that can be integrated, as a function or differential equation: *to reduce to an integrable finite series.*

integral (in'tə grəl), *Mathematics.* —*n.* **1** the result of an integration in calculus; the quantity of which a given function is the differential or differential coefficient. **2** an expression from which a given function, equation, or system or equations can be derived by differentiating.
—*adj.* **1** having to do with an integer: *integral numbers rather than fractions or decimals.* **2** of or involving integration or integrals: *an integral equation, the integral sign.*
[from Late Latin *integralis,* from Latin *integer*]
—**integrally,** *adv.*

integral calculus, the branch of mathematics that investigates integrals, methods of finding integrals, and their applications in area, volume, etc.

integral domain, *Mathematics.* a set of elements subject to the operations of addition and multiplication, in which multiplication is commutative, which contains an identity element of multiplication, and in which at least one of a pair of elements is zero when the product of the pair equals zero.

integrality (in'tə gral'ə tē), *n. Mathematics.* integral condition or character.

integrand (in'tə grand), *n. Mathematics.* an expression to be integrated. [from Latin *integrandus,* gerundive of *integrare* make whole, integrate]

integrate (in'tə grāt), *v. Mathematics.* **1** to calculate the integral of (a given function or equation); perform the operation of integration: *By integrating the result of this summation with respect to x, we are finding the sum of all such vertical strips between the limits x = a and x = b, which clearly gives the total area desired* (William L. Schaaf). **2** to find the function, equation, or set of equations whose derivative is an integrand.
[from Latin *integratum* made whole, from *integer* whole]

integration, *n. Mathematics.* the process or operation of finding the quantity of which a given function is the differential or differential coefficient; the process of transforming a function or equation to the value of its integral.

integrodifferential (in'tə grō dif'ə ren'shəl), *adj. Mathematics.* containing or involving both integrals and differentials of a function: *integrodifferential equations.*

integument (in teg'yə mənt), *n. Biology.* **1** a natural outer covering; skin, shell, rind, etc.; tegument: *The body covering is a skin, or integument, consisting of an outer epidermis over an underlying dermis* (Storer, *General Zoology*).

2 the layer or layers of tissue surrounding an ovule: *Differentiation proceeds as outer cells of the dome develop upward to form one or two protective layers, the inner and outer integuments. The integuments do not fuse at the apex of the ovule, thus they leave a small opening known as the micropyle* (Weier, *Botany*). [from Latin *integumentum,* from *integere* to cover, enclose, from *in-* on + *tegere* to cover]
—**integumental** (in teg'yə men'tl), *adj.* of or belonging to an integument: *integumental covering.*
—**integumentary** (in teg'yə men'tər ē), *adj.* of, having to do with, or of the nature of an integument.

intensity, *n. Physics.* **1** the amount or degree of strength of electricity, heat, light, or sound per unit of area or volume: *the intensity of gravitation. Besides its wavelength composition ... light has two other aspects that are biologically important: its intensity ... and its duration* (Greulach and Adams, *Plants*).
2 the strength of a color resulting from the degree to which it is lacking its complementary color.

inter-, *prefix.* **1** between or among, as in *intercellular, interatomic, intergalactic, interglacial.* **2** one with the other, together, as in *intercrop.* [from Latin *inter-,* from *inter* among, between]

interact, *v.* to act upon each other: *There are still far too many variables and combinations of variables interacting in countless ways* (New Yorker). *Usually enzyme molecules are much larger than the molecules with which they interact* (Biology Regents Syllabus).
—**interactant,** *n.* something that interacts, such as a substance in a chemical reaction.

interaction, *n.* **1** action upon or influence on each other: *When two or more chemical reactions take place together, there is an interaction between them provided that they have reactants in common* (Science News). **2** *Nuclear Physics.* any of three types of force exerted between subatomic particles. The three types of force or interaction are often called strong, electromagnetic, and weak. *The proton and pion enter into all three types of interaction; the neutron displays only strong and weak forces. The electron and muon have identical properties. Of all the particles, the neutrino alone enters into weak interactions* (Scientific American).
—**interactive,** *adj.* acting on each other.

interatomic, *adj. Chemistry.* between atoms: *interatomic bonds.*

interbrachial, *adj. Zoology.* between the brachia or arms, as of a starfish: *the interbrachial groove.*

interbreed, *v.* **1** *Botany.* to breed by the mating of different varities or species of plants: *In the most homogeneous populations, the plants ... interbreed freely* (Weier, *Botany*).
2 *Zoology.* to breed by the mating of different varieties or species of animals: *A species is a population of individuals that are more or less alike and that are able to interbreed and produce fertile offspring under natural conditions* (Durst, *High School Biology*). SYN: crossbreed, intercross.

intercalary (in tèr'kə ler'ē), *adj.* **1** *Astronomy.* **a** inserted in the calendar to make the calendar year agree with the solar year. February 29 is an intercalary day. **b** having an added day, month, or other period of time: *an intercalary year.*

2 *Geology.* (of a stratum) lying between the normal strata.

3 *Botany.* growing or occurring at positions other than the apex, usually between the apex and the base: *Intercalary meristems may remain active for some time ... [and] contribute greatly to the rapid elongation of the stem* (Greulach and Adams, *Plants*).
[from Latin *intercalarius,* from *intercalare* to intercalate, from *inter-* between + *calare* proclaim]

intercalate (in tèr′kə lāt), *v.* **1** *Astronomy.* to put (an additional day or month) into the calendar.

2 *Anatomy.* to place or insert between: *a neuron that is intercalated in a chain of neurons.*

3 *Chemistry.* to become inserted between two or more other components: *Large numbers of ethidium bromide molecules can intercalate in a nicked duplex loop or a linear duplex* (Scientific American).

4 *Geology.* to be inserted or interleaved between the original layers or series: *Marine mud and sand ... intercalated here and there with strata of limestone* (A.C. Ramsay).
[from Latin *intercalatum* interposed, proclaimed between, from *inter-* between + *calare* proclaim]
—inter′cala′tion, *n.* **1** the act or process of intercalating: *Intercalation is a process whereby donor ions position themselves inside a host structure. These ions occupy otherwise vacant regions ... without changing the structure of their host* (Science News). **2** something that has been intercalated, as a layer of rock between layers of different origin.

intercardinal points, *Geography.* the four directions of the compass lying midway between the cardinal points; northeast, southeast, southwest, and northwest.

intercellular (in′tər sel′yə lər), *adj. Biology.* situated between or among cells: *Intercellular spaces are prominently developed ... and many cells are in contact with an intercellular space* (Weier, *Botany*). Compare **intracellular.**

intercellular fluid, *Biology.* fluid derived from blood plasma that surrounds all living cells of the body. *Abbreviation:* ICF

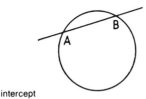

intercept

The line intercepts the cicle at points A and B.

intercept, *Mathematics.* **—v.** (in′tər sept′) to mark off or include between two points or lines: *Any two parallel chords intercept equal arcs* (C. Hutton).
—n. (in′tər sept′) **1** an intercepted part. **2** the distance from the origin to the point where a line, curve, or surface intersects a coordinate axis: *To find the equation of a straight line in terms of the intercepts which it makes on the axes* (C. Smith).
[from Latin *interceptum* caught between, interrupted, from *inter-* between + *capere* to take, catch]
—in′tercep′tion, *n. Mathematics.* the condition of be-

ing intercepted; fact of containing or being enclosed between, points, lines, boundaries or limits: *the distance between the points of interception.*

interclavicle (in′tər klav′ə kəl), *n. Zoology.* a T-shaped bone located between the clavicles in fishes, amphibia, reptiles, and birds.
—interclavicular (in′tər klə vik′yə lər), *adj.* situated between the clavicles: *the interclavicular sac, an interclavicular ligament.*

intercostal, *adj.* **1** *Anatomy.* situated between the ribs: *an intercostal muscle.* **2** *Botany.* situated between the veins of a leaf.
—n. *Anatomy.* a part situated between the ribs.
[from New Latin *intercostalis,* from Latin *inter-* between + *costa* rib] **—intercostally,** *adv.*

intercross, *Biology.* **—v.** to cross with each other; interbreed.
—n. an instance of intercrossing: *Both in the vegetable and animal kingdoms, an occasional intercross between distinct individuals is a law of nature* (Darwin, *Origin of Species*).

intercrystalline, *adj. Chemistry.* between crystals of a solid substance: *When the polymer had cooled, this hydrocarbon—which was separating the crystals—was washed away with a solvent, to expose the crystals and the intercrystalline bridges* (New Scientist).

intercrystallize, *v. Chemistry.* to crystallize together (two or more substances); to crystallize so that the atoms of one substance crystallize between the atomic layers of the other substance: *Some refractories, in particular silica, are strong to within a few degrees of their melting point because their principal constituents are intercrystallized* (J.H. Chesters). **—intercrystallization,** *n.*

intercurrent, *adj. Medicine.* (of a disease) occurring during another disease and modifying it: *an intercurrent fever, intercurrent pneumonia.*

interdigital, *adj. Anatomy.* between or joining fingers or toes: *The webbing of a duck's foot is interdigital.*

interface, *n.* **1** *Physics.* a surface lying between two bodies or spaces, and forming their common boundary: *Whenever a train of light waves, traveling in one transparent medium, strikes the surface of a second transparent medium whose index differs from that of the first ... two new wave trains are found to originate at the interface* (Sears and Zemansky, *University Physics*).

2 *Chemistry.* the area of contact between any two substances or phases that cannot be mixed: *The interface between a liquid and solid always has a certain amount of energy associated with it* (New Scientist). *Thus the molecules at the interface between the methane and*

---

cap, fāce, fäther; best, bē, tèrm; pin, fīve;
rock, gō, ôrder; oil, out; cup, pùt, rüle,
yü in use, *yù* in uric;
ng in bring; sh in rush; th in thin, тн in then;
zh in seizure.
ə = a in about, e in taken, i in pencil, o in lemon, u in circus

water molecules may crystallize into "ice" (Scientific American).

▶ Five types of chemical interfaces are possible: liquid/liquid (e.g. water and oil), solid/solid (salt and sugar), liquid/solid (mercury and glass), liquid/gas (water and carbon dioxide), and solid/gas (iron and air or oxygen).

interfacial (in'tər fā'shəl), *adj.* **1** *Geometry.* included between two faces of a solid: *the interfacial angle.*
2 *Physics, Chemistry.* **a** of or having to do with an interface: *Because of the interfacial problems associated with nearly inert biomaterials, much research during the last decade has been directed toward stabilizing the tissue-biomaterial interface by controlling either the chemical reactions or the microstructure of biomaterials* (Science). **b** formed at an interface: *Interfacial polymerization is the term applied to the formation of polymers (large molecules) through interaction of monomers (small molecules) at an interface* (New Scientist).

interfere, *v. Physics.* to cause interference: *At any point where two or more trains of waves cross one another they are said to interfere* (Sears and Zemansky, *University Physics*).

interference, *n.* **1** *Physics.* the reciprocal action of waves by which they reinforce or diminish one another: *Interference is also evidence of waves. Its characteristic pattern is formed when rays interact* (Scientific American). *Interference cancels two waves if crest meets trough, but intensifies them if the crests meet* (S.Y. Lee).
▶ *Constructive interference* occurs when two crests or two troughs come together, forming a crest or trough of increased amplitude. Maximum constructive interference occurs at points where two waves passing simultaneously through a region are in phase. *Destructive interference* occurs when a trough and a crest come together, resulting in a point of minimum disturbance. Maximum destructive interference occurs at points where the phase difference between the waves is 180° and their amplitudes are the same.
2 *Genetics.* the effect by which one crossing-over reduces or increases the probability of another crossing-over occurring along or in the vicinity of the same chromosome: *If the genes are more than 45 map units apart, interference is difficult to detect, or is nonoperative* (McElroy, *Biology and Man*).

interference fringes, *Optics.* alternate light and dark or colored bands caused by the interference of light waves: *Interference fringes represent the topographic height differences between the two states ... of the object's surface* (George W. Stroke).

interference microscope, a microscope that uses the phenomenon of interference to facilitate visualization and measurement of the surface structure of reflecting or transparent specimens, such as living cells.
—**interference microscopy,** the use of an interference microscope: *Interference microscopy, a technique in which a crystal is made visible by the fact that the light used for observation is separated into two beams ... When the two beams are recombined, their waves interfere; the interference extinguishes some of the col-*

ors in white light, giving rise to other colors that outline the crystal structures (Scientific American).

interferogram (in'tər fir'ə gram), *n.* a recording, usually photographic, of the output of an interferometer: *Holographic interferograms have been used to trap successive instants in the lifetime of such lightning-fast events as the flight of a rifle bullet* (Science News Letter).

interferometer (in'tər fə rom'ə tər), *n.* an instrument for measuring very small lengths, distances, and changes in the dimensions, density, and other properties of substances by means of the interference of two rays of light: *A kind of spectroscope that is coupled with existing telescopes, the interferometer gives a 10-times improved view of the weak absorption lines of atoms and molecules in space* (W. Tileux).
—**interferometric** (in'tər fer'ə met'rik), *adj.* of or having to do with an interferometer or interferometry: *Interferometric holograms are used to study ignition patterns inside cylinders* (E. Burgess). —**interferometrically,** *adv.*
—**interferometry** (in'tər fə rom'ə trē), *n.* the use of an interferometer: *Interferometry has commonly been used for the precise measurement and comparison of wavelengths* (Scientific American).

interferon (in'tər fir'on), *n. Biochemistry.* an antiviral protein produced by animal cells in response to infection by a virus or other chemical agent. Interferon prevents the reproduction of the virus and is able to induce resistance to other viruses in the same animal species. *There are three types of interferon, each named for the type of cell from which it originates. Leucocyte interferon ... fibroblast interferon ... and immune interferon ... made by cells produced by the immune system* (Thomas M. Roberts). *Each interferon molecule has sugar groups attached to a string of about 150 amino acids* (J. A. Miller). [from *interfere* + *-on*]

interfertile (in'tər fėr'tl), *adj. Biology.* able to interbreed: *interfertile plant species.*

interfinger, *v. Geology.* (of strata) to be inserted between and into one another laterally: *Evaporites which interfinger with the dolomite serve as a seal* (G. M. Friedman).

interfluve (in'tər flüv), *n. Geography, Geology.* a region lying between two adjacent rivers or river valleys. [from *inter-* + Latin *fluvius* river]
—**interfluvial,** *adj.* lying between two adjacent rivers or river valleys: *interfluvial land.*

intergalactic, *adj. Astronomy.* situated or taking place between galaxies: *an intergalactic gas, intergalactic space.*

intergeneric, *adj. Biology.* occurring between genera: *an intergeneric hybrid.*

interglacial, *Geology.* —*adj.* of or occurring in the period between two glacial epochs: *The Ice Age was characterized by various periods when the ice advanced (glacial) and when it retreated (interglacial)* (McElroy, *Biology and Man*).
—*n.* the period between two glacial epochs: *... the larger mammalian and reptilian fauna present in Britain since the great interglacial about 170,000 years ago* (New Scientist).

intergrade, *Biology.* —*n.* an intermediate form of an animal or plant: *the mallard-Mexican duck intergrade, the intergrades between the gray and ruddy mottled types.*
—*v.* to become alike through a continuous series of intermediate forms, as one population of animals or plants with another: *We treat as specific any form that we do not know or believe to intergrade* (Elliott Coues).
—**intergradation,** *n.* the fact or process of intergrading.

interior angle, *Geometry.* **1** any of the four angles formed inside two lines intersected by a straight line: *Given two lines cut by a transversal. If a pair of alternate interior angles are congruent, then the lines are parallel* (Moise and Downs, *Geometry*). See the picture at **exterior angle. 2** an angle formed inside a polygon by two adjacent sides: *the three interior angles of a triangle.*

interlobate, *adj. Geology.* situated between two lobes of a glacier: *an interlobate moraine.*

intermaxillary, *Anatomy.* —*adj.* situated between the maxillae: *the intermaxillary bones.*
—*n.* the premaxilla.

intermediate, *n. Chemistry.* a compound formed between the initial and final stages of a process or reaction: *Acidic resins serve as catalysts in the reaction between hydrogen peroxide and esters of fatty acids in the manufacture of chemical intermediates* (W.T. Read).

intermediate boson or **intermediate vector boson,** = weakon.

intermediate disk, = Z disk.

intermediate host, *Zoology.* the host of an adult parasite in which the parasite may reproduce only asexually. Contrasted with **definitive host.**

intermediate inheritance, *Genetics.* a form of inheritance in which heterozygous offspring are different in phenotype from their homozygous parents: *Incomplete or partial dominance as exhibited by pink snapdragons and four o'clocks is another type of intermediate inheritance* (Biology Regents Syllabus).

intermedin (in'tər mēd'n), *n. Biochemistry.* a hormone of the pituitary gland that controls pigmentation, causing changes in body color. Also called **melanocyte-stimulating hormone.** [from *intermed(iate lobe)* + *-in*]

intermetallic, *Chemistry.* —*adj.* compounded of two or more metallic elements. Intermetallic compounds are stronger and more heat resistant than ordinary metals.
—*n.* an intermetallic compound.

intermolecular (in'tər mə lek'yə lər), *adj. Chemistry.* being or occurring between molecules: *Intermolecular forces are responsible for surface tension, capillary action, adsorption and other surface phenomena* (Scientific American). Compare **intramolecular.** —**intermolecularly,** *adv.*

internal, *adj.* **1** *Anatomy.* situated inward from the surface of the body, or nearer the median line (used in the names of vessels, nerves, and the like, correlated with others called *external*): *the internal carotid artery, the internal auditory canal.*
2 *Zoology.* occurring within the body. **Internal fertilization** is fertilization in which the gamete fuse in the moist reproductive tract of the female: *Reproduction*

in most terrestrial vertebrate animals is characterized by internal fertilization (Biology Regents Syllabus).

internal ear, = inner ear.

internal energy, *Physics.* the total kinetic and potential energy associated with the motions and relative positions of the molecules of an object, as distinguished from any kinetic or potential energy of the object as a whole. An increase in internal energy either increases the kinetic energy of the random motion of an object's molecules, which results in a rise in temperature, or increases their potential energy of position, which results in a change of phase. An increase in internal energy can also raise the energy level of atoms.

internal environment, *Biology.* the fluid in which the cells of the organs and tissues of the body exist. Too much or too little water or salt in the internal environment can damage or kill cells.

internal respiration, *Biology.* the absorption of oxygen and elimination of carbon dioxide by cells: *Breathing, largely a mechanical process which serves as an adjunct of respiration, is technically termed external respiration. The oxidative process within the protoplasm constitutes internal respiration or physiologic oxidation* (Harbaugh and Goodrich, *Fundamentals of Biology*).

internal secretion, = hormone.

International Atomic Time, the primary scientific standard of time, maintained by cesium clocks at the Bureau International de la Heure near Paris, France. It is based on the definition of the second as 9,192,631,770 oscillations of this type of clock. With the aid of satellites and portable atomic clocks, the national time standards of the major nations are synchronized within a few microseconds. *Abbreviation:* TAI

international candle, a former unit for measuring the strength or intensity of light, replaced by the candela.

International Date Line, an imaginary line agreed upon as the place where each new calendar day begins. It runs north and south through the Pacific, mostly along the 180th meridian. Except at precisely noon, Greenwich Time, when it is Sunday east of the International Date Line it is Monday west of it.

international nautical mile, a unit of linear measure recommended for sea and air navigation and adopted by several countries, equal to 1,852 meters or 6,076.11549 feet.

International System of Units, the system of SI units: *The meter for length, the kilogram for mass, the second for time and the kelvin (formerly the degree Kelvin) for temperature ... together with two derived quantities (the ampere and the candela) are the building blocks of the International System of Units, or Système International* (Scientific American).

---

cap, fāce, fäther; best, bē, tėrm; pin, five;
rock, gō, ôrder; oil, out; cup, pùt, rüle,
yü in use, *yù* in uric;
ng in bring; *sh* in rush; *th* in thin; ᴛʜ in then;
zh in seizure.
ə = *a* in about, *e* in taken, *i* in pencil, *o* in lemon, *u* in circus

325

international unit, *Biology, Medicine.* a unit for measuring the activity or potency of a hormone, vitamin, enzyme, antibiotic, and the like, defined internationally for each substance on the basis of a standard amount. *Abbreviation:* IU

interneural (in'tər nùr'əl), *adj. Anatomy, Physiology.* occurring or situated between nerves: *interneural transmission of hearing.*

interneuron (in'tər nùr'on), *n. Anatomy.* a nerve cell that connects afferent and efferent nerve cells: *An interneuron is a nerve cell that is found only in the central nervous system, where it serves as a functional link between sensory neurons and motor neurons* (Scientific American). Also called **associative neuron.** —**interneuronal** (in'tər nùr'ə nəl), *adj.* of or having to do with an interneuron or interneurons: *interneuronal communication.*

internodal, *adj.* of, having to do with, or constituting an internode: *the internodal region of a plant.*

internode, *n.* **1** *Botany.* that part of a stem or branch between two of the nodes from which leaves arise: *Internodes are very short in the bud. Elongation of these internodal regions will account for the growth of the shoot in length* (Weier, *Botany*). **2** *Anatomy.* any part between two nodes or joints.

internuclear, *adj.* occurring or situated between the nuclei of atoms or cells. The internuclear distance is the distance between two nuclei in a molecule: *The covalent atomic radius is one-half the measured internuclear distance in the solid phase* (Chemistry Regents Syllabus).

internuncial (in'tər nun'shəl), *adj. Anatomy.* serving to connect nerve fibers or their paths of transmission: *Interneurons are internuncial nerve cells.* —**internuncially,** *adv.*

interoceanic (in'tər ō'shē an'ik), *adj. Geography.* situated between, or connecting, oceans: *an interoceanic canal.*

interoceptive (in'tər ə sep'tiv), *adj. Physiology.* receiving or aroused by stimuli from inside the body: *The interoceptive senses respond to changes in the organs and tissues.* Compare **exteroceptive.** [from Latin *interus* inner + English (*re*)*ceptive*]
—**interoceptor** (in'tər ə sep'tər), *n.* a sense organ stimulated inside the body, as by some chemical change in the muscles, nerves, or blood.

interphase, *n. Biology.* the period of time between the end of one cell division and the beginning of the next: *During interphase ... the nucleus shows little definable structure, except for the lightly stained chromatin and the more deeply stained nucleoli* (McElroy, *Biology and Man*).

interplanetary, *adj. Astronomy.* **1** situated or taking place between or in the region of the planets: *interplanetary gravitational attraction.*
2 within the solar system, but outside the atmosphere of any planet or the sun: *The interplanetary magnetic field comprises part of the magnetic field of the sun* (New Scientist).

interpolar, *adj. Physics.* situated between or connecting the poles, as of a voltaic battery: *interpolar electrolytic action.*

interpolate (in tèr'pə lāt), *v. Mathematics.* to find or insert (an unknown term) between two known terms in a series. [from Latin *interpolatum* freshened up, from *inter-* between + *polire* to smoothe] —**interpolation,** *n.*

interradial (in'tər rā'dē əl), *adj. Zoology.* situated between radii or rays, as are parts of an echinoderm.

interrupted, *adj.* **1** *Botany.* having the principal leaflets or flower spikes divided by a series of smaller ones. **2** *Zoology.* suddenly stopped; having a gap: *an interrupted stria.*

interscapular (in'tər skap'yə lər), *adj. Anatomy.* situated between the shoulder blades.

intersect, *v. Geometry.* to pass through or across another line or surface so as to have one or more points in common: *If a line intersects a plane not containing it, then the intersection contains only one point* (Moise and Downs, *Geometry*).
—**intersection,** *n.* **1** *Geometry.* the point where one line, surface, or solid crosses another: *The intersection of two surfaces is the locus of points common to the two surfaces. ... The intersection of two planes is a straight line determined by any two points which the planes have in common* (Smith and Clark, *Modern-School Solid Geometry*). **2** *Mathematics.* the set that contains only those elements shared by two or more sets. EXAMPLE: If set A = [1, 2, 3, 4] and set B = [3, 4, 5, 6], then the intersection of the two sets is [3, 4].
—**intersectional,** *adj.* of, having to do with, or characterized by intersection.

intersection:

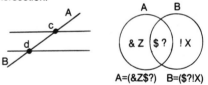

A=(&Z$?)   B=($?!X)

Points c and d are the points of intersection of line AB with the parallel lines.
The intersection of sets A and B is $?.

interseptal (in'tər sep'təl), *adj. Anatomy.* situated between septa or partitions.

intersex, *n. Biology.* an intersexual organism.

intersexual, *adj. Biology.* having sexual characteristics that are intermediate between those of the typical female and the typical male.
—**intersexuality,** *n.* the condition of being intersexual: *Spontaneous intersexuality (presence of both male and female characteristics, or of intermediate sexual features in a single individual) is found in all vertebrate classes, including man* (K. F. Liem).

interspecific or **interspecies,** *adj. Biology.* occurring or arising between species: *an interspecific hybrid.*

interstadial (in'tər stā'dē əl), *adj. Geology.* of, belonging to, or designating a period within a glacial epoch when the climate becomes temporarily warmer: *the first interstadial era of the Upper Paleolithic.* [from *inter-* between + Latin *stad*(*ium*) stage]

interstellar, *adj. Astronomy.* situated or taking place between the stars; in the region of the stars: *Radio astronomers ... found three complete molecules—*

ammonia, formaldehyde, and water—in interstellar space in 1968 and 1969 (Owen Gingerich). Previously unexplained deviations in the orbital energy of satellites with very high orbits could be due to the buffeting effect of the "interstellar wind" (New Scientist).

interstitial (in'tər stish'əl), adj. Anatomy. occupying the small, narrow spaces or interstices of tissue: interstitial cells.

interstitial cell stimulating hormone, = luteinizing hormone. Abbreviation: ICSH

interstock, n. Botany. a stock grafted between the rootstock and the scion.

interstratification, n. Geology. 1 the condition or fact of being interstratified. 2 an interposed formation or deposit.

interstratify, v. Geology. 1 to lie as strata between other strata. 2 to deposit between other strata; arrange in alternate strata.

intertidal, adj. Ecology. living or located between the high-water mark and the low-water mark: A factor of great importance in the ecology of intertidal plants and animals is the extent of wave action (F. S. Russell). —**intertidally,** adv. between the high- and low-water marks.

intertropical convergence zone, Meteorology. the zone along which tropical air masses moving in opposite directions converge.

interval, n. Mathematics, Statistics. the set of all numbers between two given numbers. A **closed interval** includes its end points. An **open interval** does not.

intervening sequence, = intron.

interventricular, adj. Anatomy. situated between the ventricles of the heart or brain: the interventricular valve or septum.

intervertebral, adj. Anatomy. situated between the vertebrae. An **intervertebral disk** is the mass of fibrous cartilage lying between adjacent vertebrae. —**intervertebrally,** adv.

intestinal, adj. Anatomy. of, in, or affecting the intestines. —**intestinally,** adv.

intestine, n. Anatomy. the part of the alimentary canal extending from the stomach to the anus; the small intestine and large intestine: An earthworm [has] ... a long intestine with pouchlike lateral extensions providing a large surface for absorption of digested portions (Storer, General Zoology). See the pictures at **alimentary, crustacean.** [from Latin intestinum, from intus inside, from in in]

intima (in'tə mə), n., pl. **-mae** (-mē). Anatomy. the innermost membrane, coat, or lining of some organ or part, especially of a blood or lymphatic vessel: In atherosclerosis there is a proliferation of smooth-muscle cells into the intima (Scientific American). [from New Latin (tunica) intima inmost (coat)]

intine (in'tin or in'tīn), n. Botany. the inner membrane of a spore or pollen grain. [from Latin intus within + English -ine]

into, prep. Mathematics. a word implying or expressing division. EXAMPLE: 5 into 30 is 6.

intorsion (in tor'shən), n. Biology. the act or fact of twisting or winding, as of the stem of a plant. [from French intorsion, from Latin intortionem, from intortus twisted]

intra-, prefix. within; inside; on the inside, as in intracellular, intracutaneous. [from Latin, from intra inside of]

intra-arterial, adj. Anatomy, Physiology. **1** within an artery or the arteries. **2** of, for, or into an artery or arteries: an intra-arterial injection.

intracellular, adj. Biology. occurring within a cell or cells: Intercellular circulation may be by diffusion or by transport through vascular tissue (Biology Regents Syllabus). —**intracellularly,** adv. within a cell or cells: Typhoid bacteria ... reproduce intracellularly (Science News Letter).

intracellular enzyme = organized ferment.

intracerebral, adj. Anatomy, Physiology. within the brain or the cerebrum: an intracerebral hemorrhage.

intracutaneous, adj. = intradermal.

intradermal, adj. Anatomy, Physiology. within the layers of the skin: Injections were given to patients by various routes from intradermal to intravenous (Science News).

intragalactic, adj. Astronomy. within a galaxy or cluster of galaxies: Whereas hot intragalactic gas in the Coma cluster is a strong source of X rays, the Hercules cluster shows little evidence of gas between galaxies and is not an X-ray emitter (Scientific American).

intramolecular, adj. Chemistry. occurring or acting within a molecule or molecules: An acceptable sunscreen must in addition be stable to light, and must not undergo any intramolecular changes that would affect its absorptive capacity (New Scientist). —**intramolecularly,** adv. within a molecule or molecules.

intramuscular, adj. Anatomy, Physiology. within or into a muscle: intramuscular injection.

intranuclear, adj. within the nucleus of an atom or cell: The larger intranuclear masses were mostly viral DNA, which differed from normal cell DNA in chemical properties and chemical structure and was produced in surplus amounts (Science News Letter).

intraocular, adj. Anatomy, Physiology. within the eyeball: intraocular pressure.

intraperitoneal, adj. Anatomy, Physiology. within the peritoneum.

intraspecific or **intraspecies,** adj. Biology. within a species; involving members of the same species: Intraspecific competition represents the keenest type because all members of a given species are competing for the same materials (Harbaugh and Goodrich, Fundamentals of Biology). Intraspecies lethal violence is not unknown in other species, but it is uncommon and occurs on a limited scale (Bulletin of Atomic Scientists).

cap, fāce, fäther; best, bē, tėrm; pin, five;
rock, gō, ôrder; oil, out; cup, pùt, rüle,
yü in use, yu in uric;
ng in bring; sh in rush; th in thin, ŦH in then;
zh in seizure.
ə = a in about, e in taken, i in pencil, o in lemon, u in circus

intratelluric (in′trə te lûr′ik), *adj. Geology.* 1 located or formed far beneath the earth's surface: *intratelluric rock.* 2 denoting the period in the formation of eruptive rock just before its appearance on the earth's surface. [from German *intratellurisch,* from *intra-* intra- + Latin *tellus, telluris* earth]

intrathoracic, *adj. Anatomy, Physiology.* situated, placed, or occurring within the thorax: *The heart and the lungs are intrathoracic organs.*

intrauterine, *adj. Anatomy, Physiology.* within the uterus.

intravascular, *adj. Anatomy, Physiology.* within a blood vessel or the blood vessels: *intravascular formation of blood cells.*

intravenous (in′trə vē′nəs), *adj. Anatomy, Physiology.* 1 within a vein or the veins. 2 of, for, or into a vein or veins.
—*n.* an intravenous injection.
[from *intra-* + Latin *vena* vein + English *-ous*] —**intravenously,** *adv.*

intravital, *adj. Biology, Medicine.* 1 occurring in or performed upon a living individual: *Intravital fluorescence microscopy ... has been of value in elucidating the functioning of various organs in animals* (New Scientist).
2 capable of staining cells without killing them: *an intravital stain.* Compare **supravital.**

intra vitam (in′trə vī′təm), *Biology.* within a living cell, tissue, or organism: *the injection of a stain intra vitam.* [from New Latin *intra vitam* (literally) within life]

intrazonal (in′trə zō′nəl), *adj. Geology, Geography.* within a zone or region: *An intrazonal type of soil is typical of a particular topographic area.*

intrinsic, *adj. Anatomy.* originating or being inside the part on which it acts: *the intrinsic muscles of the larynx.* [from Late Latin *intrinsecus* internal, from Latin, inwardly] —**intrin′sically,** *adv.*

intrinsic factor, *Biochemistry.* a substance secreted in the stomach that is needed for the utilization of vitamin B$_{12}$ by the body. Lack of it causes pernicious anemia. Compare **extrinsic factor.**

intro-, *prefix.* within; into; inward, as in *introversion, introgression.* [from Latin, from *intro* within, inwardly]

introgressant (in′trə gres′ənt), *n. Genetics.* a gene which a species acquires by introgression.

introgression (in′trə gresh′ən), *n. Genetics.* the introduction, through hybridization and chance backcrossing, of a gene of one species into another species: *Introgression can only be established by field observation, but the number of accurate and extensive field investigations has hitherto been disappointingly small* (D.H. Valentine). [from *intro-* into + (re)*gression*]

introgressive hybridization, = introgression.

intron (in′tron), *n. Molecular Biology.* a segment of DNA that does not perform a coding function: *One possibility is that some of the introns are promoters—regions of DNA where the enzyme RNA polymerase first binds to a gene before transcribing it. Introns may also play a role in control of gene expression* (T. H. Maugh II). Also called **intervening sequence.** Compare **exon.** [from *intro-* inward, inside + *-on*]

—**intronic,** *adj.* of or having to do with introns: *intronic sequences.*

introrse (in trôrs′), *adj. Botany.* turned or facing inward, toward the axis. A violet has introrse stamens. Contrasted with **extrorse.** [from Latin *introrsum,* from *intro-* inward + *versum* turned]

introversion, *n. Biology.* the act or process of turning an organ or part inward or within itself. [from *intro-* + (*re)version*]

introvert, *v. Biology.* to turn (a tubular organ or part) inward or back within itself. SYN: invaginate.

intrude, *v. Geology.* to force (molten rock) into fissures or between strata: *The sedimentary rocks have been intimately intruded by relatively basic magma, which has spread between the beds* (E.F. Roots).

intrusion, *n. Geology.* 1 the forcing of molten rock into fissures or between strata: *The Nevadan revolution was accompanied by intrusion of granitic batholiths on a huge scale. Heat and pressure derived from these intrusions ... forced their way into surrounding sedimentary formations,* (Moore, *Introduction to Historical Geology*).
2 the body of molten rock forced in and solidified in place: *Heated rocks from the depths are forced upward toward the earth's surface. If they congeal before reaching the surface they are called intrusions. If they reach the surface they are called eruptions or extrusions* (White and Renner, *Human Geography*).

intrusive, *Geology.* —*adj.* forced into fissures or between strata while molten: *intrusive rock.*
—*n.* an intrusive rock or formation: Compare **extrusive.**

intussuscept (in′təs sə sept′), *v. Physiology, Medicine.* to take within, as in telescoping one part of the intestine into another part; invaginate. [from Latin *intus* within + *susceptus,* past participle of *suscipere* take up]

intussusception (in′təs sə sep′shən), *n.* 1 *Biology.* the taking in of foreign matter, especially food, by a living organism and converting it into living tissue: *Living matter grows by intussusception, a process by which an organism takes dissimilar materials into itself, and ... makes more living matter identical with its own* (Harbaugh and Goodrich, *Fundamentals of Biology*).
2 *Physiology, Medicine.* the inversion of one part of the intestine and its slipping into an adjacent part, occurring especially in infants. SYN: invagination.
—**intussusceptive** (in′təs sə sep′tiv), *adj.* of or characterized by intussusception.

inulase (in′yə lās), *n. Biochemistry.* an enzyme that converts inulin into fructose.

inulin (in′yə lin), *n. Biochemistry.* a white, starchlike substance obtained from the roots of various composite plants, especially elecampane and dahlia, used as a substitute for starch in foods for persons who have diabetes. *Formula:* $(C_6H_{10}O_5)_n$ [from Latin *inula* elecampane]

in utero (in yü′tə rō), *Embryology.* in the uterus; before being born: *Fraternal twins interchange blood-forming tissue during the time they spend in utero* (New Scientist). [from Latin]

in vacuo (in vak′yù ō), *Physics.* in a vacuum. [from Latin]

invaginate, *v. Biology.* **1** (of a cell, organ, etc.) to fold or draw back within itself; turn inward to form a pocket: *Some cells ingest substances by invaginating.* SYN: introvert, intussuscept.

2 *Embryology.* to undergo invagination: *The ball with the invaginated region is now called the gastrula* (Winchester, *Zoology*).

[from *in-* in + Latin *vagina* sheath, vagina]
—**invagination**, *n.* **1** *Biology.* **a** the act or process of invaginating. **b** an invaginated part or condition: *Cells such as white blood cells, epithelial cells of the intestines, and others can bring substances into themselves by forming pockets, or invaginations, in the cell membrane* (McElroy, *Biology and Man*). **2** *Embryology.* the drawing inward of a portion of the wall of a blastula in the formation of a gastrula: *The invagination proceeds until the inturned cells are in contact with the opposite side of the blastula wall* (Shull, *Principles of Animal Biology*).

invariance, *n. Mathematics, Physics.* the quality or condition of being invariant: *The conservation of energy is a direct result of the principle of "time invariance." This says that the laws of nature do not change with time* (Observer).

invariant, *adj. Mathematics, Physics.* (of a quantity, property, etc.) unchanged by a transformation or other operation; unvarying; constant: *One of the simplest ways of transforming a square is to rotate it. A rotational transformation changes the coordinates of every point ... however ... its overall shape [is] ... not affected by the rotation. In the language of mathematics these properties are said to remain invariant* (Scientific American).

—*n. Mathematics.* an invariant quantity.

invasion, *n.* **1** *Ecology.* the growth of a large population of an organism in a region not before inhabited by the species: *By invasion is understood the movement of plants from an area of a certain character into one of a different character, and their colonization in the latter* (F.E. Clements).

2 *Biology, Medicine.* **a** the process by which pathogenic microorganisms enter and spread in the body. **b** the period in the development of a disease during which pathogenic microorganisms enter and spread in the body: *A fever which lasts for a long time usually starts as the period of invasion* (P. R. Cannon).

invasive, *adj. Biology, Medicine.* **1** invading healthy cells or tissues: *an invasive tumor, invasive bacteria.*

2 having to do with or involving penetration of the body by means of instruments for purposes of diagnosis or therapy: *invasive procedures.*

—**invasiveness**, *n.* the power of a pathogen or malignant cell to spread into new tissues: *We know ... that the invasiveness of some organisms is related to their ability to manufacture specific antigenic molecules* (New Scientist).

inverse (in vèrs′ *or* in′vèrs′), *Mathematics.* —*n.* **1** any operation which annuls a given operation. Subtraction is the inverse of addition. **2** one of a pair of elements in a set whose result under the operation of the set is the identity element: *For every operàtion or element there must be an inverse that, by producing the identity element, undoes whatever the operation or element did. For example, 5 times 1/5 equals 1, the identity element in multiplication* (N.Y. Times).

▶ Each real number *n* has an *additive inverse -n* such that $n + -n = 0$, and a *multiplicative inverse* $1/n$ (if *n* does not equal 0) such that $n \times 1/n = 1$.
—*adj.* of or having to do with an inverse or an inverse function: *inverse elements. For every operation there is an inverse operation such that executing an operation and then its inverse is equivalent to executing the identity operation* (Scientific American).

[from Latin *inversum* turned over, inverted]
—**inversely** *adv.*

inverse function, *Mathematics.* **1** a function which is the inverse of another function. **2** a function which replaces another function's independent variable with a value of its dependent variable.

inverse square law, *Physics.* a law stating that some quantity, such as the magnitude of a force or the intensity of radiation, varies inversely with the square of the distance from the source. Gravitation, electromagnetic radiation, sound, and many other phenomena follow inverse square laws. *The [radiation] dose may be delivered from a single point source, in which case the inverse square law applies; that is, the dose delivered is inversely proportional to the square of the distance from the source* (R. G. Wiegert).

inversion (in vèr′zhən *or* in vèr′shən), *n.* **1** *Meteorology.* an increase of air temperature with elevation instead of the usual decrease: *During a five-day stretch ... exhaust fumes from its 2,000,000 cars were trapped overhead by a temperature inversion* (Time).

2 *Mathematics.* **a** the reversal of a ratio by interchanging the positions of the numerator and denominator. Compare **reciprocal**. **b** the operation of finding the inverse of an element, function, etc.

3 *Chemistry.* the hydrolysis of certain carbohydrates, such as sucrose into glucose and fructose, in which the direction of optical rotation of the solution is reversed.

4 *Genetics.* the reversal of one segment of a chromosome. In chromosomal inversion, the linear order of the genetic material is turned around 180 degrees.
ASSOCIATED TERMS: see **deletion**.

inversion layer, *Meteorology.* a layer of air that is warmer than the air below it.

inversive, *adj.* characterized by inversion.

invert, *Chemistry.* —*v.* (in vèrt′) to decompose by inversion.

—*adj.* (in′vèrt) decomposed by inversion.

[from Latin *invertere*, from *in-* in, on + *vertere* to turn]

invertase (in vèr′tās), *n. Biochemistry.* an enzyme which acts as a catalyst to convert sucrose into glucose and fructose: *The protoplasm of the yeast plant manufactures ... invertase which passes out of the yeast cell and hydrolyzes cane sugar to dextrose* (Youngken, *Pharmaceutical Botany*). Also called **saccharase, sucrase.**

---

cap, fāce, fäther; best, bē, tèrm; pin, fīve;
rock, gō, ôrder; oil, out; cup, pùt, rüle;
yü in use, *yu̇* in uric;
ng in bring; *sh* in rush; *th* in thin, ᴛʜ in then;
zh in seizure.
ə = *a* in about, *e* in taken, *i* in pencil, *o* in lemon, *u* in circus

329

invertebrate (in vėr′tə brit *or* in vėr′tə brāt), *Biology.*
—*adj.* **1** lacking a backbone or spinal column. Sponges, worms, clams, and insects are invertebrate animals. **2** of or having to do with invertebrates: *invertebrate zoology.*
—*n.* an animal without a backbone or vertebrae. Worms and insects are invertebrates, in contrast to such vertebrates as fishes, amphibians, and reptiles.

invert sugar, *Chemistry.* a mixture of equal parts glucose and fructose obtained by the hydrolysis of sucrose. Honey contains invert sugar. Invert sugar is so called because it is produced by inversion.

inviable (in vī′ə bəl), *adj. Biology.* unable to survive, especially because of a genetic defect: *The differentiation of sperm may fail or it may be inviable* (A. D. Lees).
—**inviability,** *n.* the condition of being inviable.

in vitro (in vē′trō *or* in vit′rō), *Biology.* in an artificial environment, such as a test tube; not inside the organism: *"Nonessential" amino acids are indispensable for the growth of certain human tissue cells in vitro* (Science). [from New Latin *in vitro* in glass]

in vivo (in vē′vō), *Biology.* inside a living organism: *What actually causes human cells, in vitro or in vivo, to die out?* (A. Leaf). *The importance of protein in the prevention and treatment of pellagra had been emphasized repeatedly, but an explanation was lacking until the in vivo conversion of the amino acid tryptophan to the vitamin niacin was demonstrated* (F. J. Stare and M. F. Trulson). [from New Latin *in vivo* in the living thing]

involucral (in′və lü′krəl), *adj. Botany.* of or having to do with an involucre: *In composite flowers ... the whole head may be surrounded by green involucral bracts* (Weier, *Botany*).

involucre (in′və lü′kər), *n.* **1** *Botany.* a circle of small leaves or bracts situated close below a flower or flower cluster: *The whorl of bracts making up the involucre often has much the appearance of a complicated calyx* (Emerson, *Basic Botany*). **2** = involucrum. [from Latin *involucrum* a cover, envelope, from *involvere* roll up]

— involucre
(def. 1)

involucrum (in′və lü′krəm), *n., pl.* **-cra** (-krə). **1** *Anatomy.* a covering or sheath. **2** *Botany.* = involucre.

involuntary, *adj. Physiology.* not controlled by the will; under the control of some mechanism operating independently of the cerebral cortex: *Breathing is mainly involuntary. The autonomic nervous system is generally considered to be an involuntary system.* SYN: automatic, reflex.

involuntary muscle, = smooth muscle.

involute (in′və lüt), *adj.* **1** *Botany.* rolled inward from the edge: *an involute leaf.*
2 *Zoology.* (of shells) having the whorls closely wound.

—*n. Geometry.* a curve orthogonal to all the tangents of a given curve. The involute of a given curve is any curve whose evolute is the given curve.
[from Latin *involutum* rolled up, enveloped, from *in-* in + *volvere* to roll]

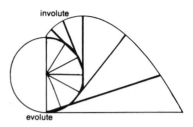

involute

evolute

involution, *n.* **1** *Mathematics.* the raising of a quantity to any power. SYN: exponentiation. Contrasted with **evolution. 2** *Biology.* degeneration; retrograde change. **3** *Botany.* **a** a rolling inward from the edge. **b** a part thus formed.

iodate (ī′ə dāt), *Chemistry.* —*n.* a salt of iodic acid. —*v.* to combine, impregnate, or treat with iodine.
—**iodation,** *n.* the act or process of iodating.

iodic (ī od′ik), *adj. Chemistry.* of or containing iodine, especially with a valence of five. Compare **iodous.**

iodic acid, *Chemistry.* a colorless or white crystalline acid, used as an astringent and disinfectant. *Formula:* HIO_3

iodide (ī′ə dīd), *n. Chemistry.* a compound of iodine with another element or radical: *sodium iodide.*

iodine, *Chemistry. n.* **1** a nonmetallic element in the form of grayish-black crystals which give off when heated a dense, violet vapor with an irritating odor resembling that of chlorine. Iodine occurs naturally only in combination with other elements and is used in medicine, in making dyes, in photography, etc. In the body, iodine occurs in trace amounts as part of the hormone thyroxine, which is secreted by the thyroid gland and controls the rate of growth. *Kelps ... are richly supplied with iodine, and consequently goiter is unknown among people who use them in large amounts in the diet* (Emerson, *Basic Botany*). *Symbol:* I; *atomic number* 53; *atomic weight* 126.9044; *melting point* 113.5; *boiling point* 184.35; *oxidation state* $\pm 1, + 5, + 7$.
2 a brown liquid containing iodine dissolved in alcohol, used as an antiseptic.
[from Greek *ioeidēs* violet in color, from *ion* violet + *eidos* form]

iodine 131, *Chemistry.* a radioactive isotope of iodine that has a half-life of eight days, emits beta and gamma rays, and is used in the diagnosis of thyroid function and brain and liver tumors, and in the treatment of hyperthyroidism and cancer of the thyroid. *Symbol:* I^{131} or ^{131}I; *atomic number* 53; *mass number* 131. Also called **radioiodine.**

iodize, *v.* to treat or impregnate with iodine or an iodide, as in photography or medicine. **Iodized salt** is table salt containing small quantities of an iodide as a dietary supplement.

iodo-, *combining form.* iodine; iodine compound, as in *iodoform = an iodine compound.* [from New Latin *iodum* iodine, from French *iode*]

iodoform (ī ō′də fôrm *or* ī od′ə fôrm), *n. Chemistry.* a yellowish, crystalline compound of iodine, used as an antiseptic, especially in surgical dressings. *Formula:* CHI_3

iodometry (ī′ə dom′ə trē), *n. Chemistry.* the determination of the amount of iodine in a mixture, usually by measuring the amount of sodium thiosulfate required to react with it.

iodopsin (ī′ə dop′sin), *n. Biochemistry.* a violet pigment that is sensitive to light, found in the cones of the retina of the eye: *The cone cells, which operate in daylight, have the violet pigment called iodopsin ... the rods ... have a red pigment, rhodopsin* (Science News Letter). [from Greek *ioeidēs* violet in color + *ōps, ōpos* eye]

iodous (ī ō′dəs *or* ī od′əs), *adj. Chemistry.* **1** of or containing iodine, especially with a valence of three. Compare **iodic. 2** resembling iodine.

ion (ī′ən *or* ī′on), *n. Chemistry, Physics.* **1** an atom or group of atoms having a negative or positive electric charge as a result of having lost or gained one or more electrons. When an acid, base, or salt dissolves in solution, some of its molecules separate into positive and negative ions. Each molecule forms two or more ions whose electric charges neutralize each other when recombined into a molecule. Cations are positive ions formed by the loss of electrons; anions are negative ions formed by the gain of electrons. *Positive sodium ions and negative chloride ions attract one another and form a crystal of table salt (NaCl: sodium chloride). This crystal is a collection of equal numbers of sodium ions and chloride ions, held together by the mutual attraction of positively charged particles and negatively charged particles. This sort of chemical union is called ionic valence, or electrovalence* (W. H. Slabaugh). **2** an electrically charged particle formed in a gas, as by the action of X rays or an electric current. [from Greek *ion,* neuter present participle of *ienai* to go]

(def. 1)

electron   nucleus   Table Salt

lost electron Na^+

gained electron

Cl^-

ion exchange, *Chemistry.* a process in which an insoluble substance, commonly a resin or zeolite, exchanges ions with the solution it is in, thereby removing a substance from the solution.

—**ion exchanger,** any substance or device by means of which ion exchange is effected: *The new process ... uses a continuous ion exchanger to decontaminate milk* (Science News Letter).

ion exchange resin, *Chemistry.* a synthetic resin which has the ability to exchange ions with a solution, used in the ion exchange process.

ionic, *adj. Chemistry.* of, having to do with, or present as ions: *Cells use a considerable amount of their energy in maintaining a correct ionic balance, by combatting the continual leakage of ions in both directions through the cell wall* (J. C. Marsden).

ionic bond, *Chemistry.* a bond between two ions with opposite charges: *When one atom gives up an electron to another, or accepts one, the two are said to be joined by an ionic bond. The atom that gains the electron thereby acquires a negative charge, and the atom that donates the electron becomes positively charged—hence the mutual attraction between the two atoms* (Scientific American). *An ionic bond is formed by the transfer of one or more electrons from metals to nonmetals* (Chemistry Regents Syllabus).

ionization (ī′ə nə zā′shən), *n. Chemistry, Physics.* separation or conversion into ions; formation of ions: *The process of separating the gas molecule into oppositely charged particles is called ionization; the free electron and the positively charged residue together constitute an ion pair* (Crammer, *Atomic Energy*).

ionization chamber, a closed, gas-filled chamber or tube containing a positive and a negative electrode. Cosmic rays or other forms of radiation passing through it are measured according to the degree of ionization they cause in the gas. The electroscope is a simple ionization chamber. *The classical detector in low-energy physics has been the ionization chamber* (Olexa-Myron Bilaniuk).

ionization potential *or* **ionization energy,** *Physics, Chemistry.* the energy required to free an electron from its atom as measured by spectroscopic or other methods: *The most important physical magnitude determining the chemical behaviour of an element is its ionization potential.* (H. R. Paneth). *Ionization energy is the amount of energy required to remove the most loosely bound electron from an atom in the gaseous phase.* (Chemistry Regents Syllabus).

ionize, *v. Chemistry, Physics.* to separate or change into ions. Acids, bases, and salts ionize when dissolved in a solution. *Atoms that lack one or more electrons are said to be ionized. If it is placed in a magnetic field, an ionized particle can move only by spiraling along one of the lines of magnetic force* (Scientific American). *Radium emanations can ionize the surrounding air, that is, electrify it* (N. Y. Times).

ionizing radiation, *Physics.* high-energy radiation, such as an X ray, that causes the formation of ions in substances through which it passes, as in the treatment of cancer and sterilization of food: *Ionizing radiation is being used to learn more about the structure of a cell, its contents, size and shape* (Science News Letter).

---

cap, fāce, fäther; best, bē, tėrm; pin, fīve;
rock, gō, ôrder; oil, out; cup, pùt, rüle;
yü in use, *yu̇* in uric;
ng in bring; *sh* in rush; *th* in thin; ᴛʜ in then;
zh in seizure.
ə = *a* in about, *e* in taken, *i* in pencil, *o* in lemon, *u* in circus

ion microscope, = field-ion microscope.

ionophore (ī on′ə fôr), n. *Biochemistry.* any of a class of organic compounds that are capable of transporting ions across lipid barriers in a cell: *The class of compounds known as ionophores has attracted increasing attention ... because of the remarkable cation selectivities shown by these substances.* (R. M. Izatt and J. J. Christensen). [from *ion* + connecting *-o-* + *-phore*]

ionophoresis (ī on′ə fə rē′sis), n. = electrophoresis.

ionosphere (ī on′ə sfir), n. *Meteorology, Geophysics.* the region of the atmosphere between the mesosphere and the exosphere. The ionosphere is composed of layers of atmosphere ionized by solar ultraviolet radiation which facilitate the transmission of certain radio waves over long distances on earth by reflections. *The ionosphere is divided into the D region (below 55 miles), the E region (from 55 to 100 miles), and the F region (above 100 miles). Whether these regions represent definite, continuous layers appears questionable* (C. G. Knudsen and J. K. McGuire). See the picture at **atmosphere.**

—**ionospheric** (ī on′ə sfir′ik), adj. in or having to do with the ionosphere: *In the polar regions geomagnetism is closely linked to auroral and ionospheric events. A strong aurora may be ... associated ionospheric disturbances* (J. C. Reed). —**ionospherically,** adv.

ion pump, a device for producing a vacuum by ionizing gas molecules and then collecting them with the aid of an electric field: *In spite of the many advantages of the diffusion pump, it is often replaced in small laboratory systems by the ion pump* (Scientific American).

iontophoresis (ī on′tō fə rē′sis), n. *Medicine.* a method of injecting ionized drugs or other substances by the use of a direct current and two electrodes attached to the body.

—**iontophoretic** (ī on′tō fə ret′ik), adj. of or having to do with iontophoresis: *iontophoretic treatment.*

IPTS, abbrev. International Practical Temperature Scale (the temperature scale in common scientific and industrial use, expressed in terms of the Celsius scale).

Ir, symbol. iridium.

IR, abbrev. **1** infrared radiation: *Snow has a high albedo (reflectivity), which delays its melting and enhances cooling. The absorption of IR causes recrystallization of the snow, reducing albedo and causing melting* (Science News). **2** Also spelled **Ir.** immune response: *The ability to respond immunologically to various antigens. ... appears to be determined by immune response (IR) genes* (J. M. Chiller).

iridium (i rid′ē əm), n. *Chemistry.* a silver-white, hard, brittle metallic element which occurs in platinum ores and is twice as heavy as lead, used as an alloy with platinum for jewelry and the points of fountain pens. *Symbol: Ir; atomic number 77; atomic weight 192.9; melting point 2410°C; boiling point 4527°C ($+$ 100°C); oxidation state 3, 4.* [from New Latin, from Latin *iris, iridis* rainbow; with reference to the element's iridescence in solution]

iridocyte (i rid′ə sīt or ī rid′ə sīt), n. = guanophore. [from Greek *iris, iridos* iris + English *-cyte*]

iris, n. *Anatomy.* the colored part around the pupil of the eye. It is a contractile disk between the cornea and the lens. The iris controls the amount of light entering the eye. *The iris ... is a ring-shaped involuntary muscle adjacent to the anterior surface of the lens* (Hardy and Perrin, *Principles of Optics*). [from Latin *iris* rainbow, from Greek]

iron, n. *Chemistry.* a very hard, strongly magnetic, silver-gray, heavy metallic element that is the commonest and most useful metal, from which steel is made. It is noted for its malleability and ductility when pure and for becoming oxidized in moist air. Iron occurs in the hemoglobin of the red blood cells where it serves to carry oxygen to all parts of the body. *The central core of the earth, whose radius is believed to be about 2200 miles, is held to consist principally of iron containing some nickel* (Jones, *Inorganic Chemistry*). *Symbol: Fe; atomic number 26; atomic weight 55.847; melting point 1535°C; boiling point 3000 °C; oxidation state 2,3.*

iron oxide, = ferric oxide.

iron pyrites, = pyrite.

ironstone, n. *Geology.* any clay, chert, or other type of rock that contains iron; iron ore.

irradiate, v. **1** to treat with some form of radiant energy, such as radiant heat from ultraviolet rays: *to irradiate fruit and vegetables to rid them of insects.* **2** to subject to the action of X rays or similar radiation: *The specimen is irradiated by 'white' x-rays, and emits characteristic secondary x-rays which fall on a rocksalt or calcite crystal mounted on an automatically scanning and counting spectrometer* (A.W. Haslett). [from Latin *irradiare,* from *in-* in, on + *radiare* to shine, from *radius* ray]

—**irradiation,** n. **1** the act of irradiating or the state of being irradiated: *Irradiation involves bombarding food molecules with relatively high amounts of energy [that] causes changes to occur in molecular structure which show up as changes in odor, flavor, color or texture* (Science News). **2** *Physics.* the intensity of the radiation falling upon a surface; the radiant energy falling per unit of time per unit area upon a surface at a specified point. **3** *Optics.* the apparent enlargement or extension of a bright object when seen against a dark background.

irrational, *Mathematics.* —adj. **1** (of an equation of expression) involving radicals or fractional exponents. **2** of or relating to an irrational number.

—n. an irrational number: *the set of real numbers (the rationals plus the irrationals).*

irrational number, *Mathematics.* any real number that cannot be expressed as an integer or as a ratio between two integers. $\sqrt{2}$ and π are irrational numbers. *An irrational number cannot be obtained by dividing an integer by a nonzero integer* (Robert W. Wirtz and Hyman Gabai).

irrefrangible (ir′i fran′jə bəl), adj. *Physics.* not refrangible; that cannot be refracted.

irregular, adj. *Botany.* (of a flower) not having all the members of the same series alike: *An irregular flower is one in which the parts of one or more whorls differ in shape and size* (Youngken, *Pharmaceutical Botany*).

irregular variable, *Astronomy.* a star that varies in brightness or magnitude but not at regularly recurring periods: *Giant and supergiant red stars whose spectra have very weak bright lines or more often none at all are likely to be irregular or semi-regular variables* (Baker, *Astronomy*).

irritability, *n. Biology.* **1** the property that living plant or animal tissue has of responding to a stimulus: *Living protoplasm is sensitive to its surroundings and is so constructed that its activities change considerably in response to differences in its environment. Such responsiveness is called irritability* (Emerson, *Basic Botany*).
2 a condition of abnormal or excessive sensitivity or excitability of an organism, organ, or part: *an allergic irritability of the skin.*

irritable, *adj. Biology.* having irritability; responding to stimuli: *Animals are irritable and respond quickly to ... changes in temperature and in light* (Hegner and Stiles, *College Zoology*).

irritant, *n.* a thing that causes irritation.
—*adj.* causing irritation: *carcinogenic agents or irritant factors.*

irritate, *v. Biology.* **1** to stimulate (an organ, muscle, tissue, etc.) to respond: *A muscle contracts when it is irritated by an electric shock.* **2** to make abnormally or excessively sensitive: *Sunburn irritates the skin.* [from Latin *iritatum* enraged, provoked]
—**irritation,** *n.* **1** an irritated condition: *irritation of the mucous membranes.* **2** something that irritates; irritant.
—**irritative,** *adj.* caused by irritation: *irritative sneezing.*

irrotational (ir′ō tā′shə nəl), *adj. Physics.* characterized by lack of rotation; not rotating: *In the physics of fluids ... so-called irrotational flow is frequently observed. In the case of a liquid drop, for example, it is possible to establish surface waves that run symmetrically around the droplet* (D. A. Bromley).

irrupt, *v. Zoology.* (of an animal species) to increase suddenly in population: *In the spring of certain years the birds have irrupted in large numbers* (A.L. Thomson).
—**irruption,** *n.* a sudden increase in the population of an animal species.

isallobar (ī sal′ə bär), *n. Meteorology.* a line on a weather map, or an imaginary line drawn in space, connecting points at which changes in atmospheric pressure are equal over a given period. [from *is-*, variant of *iso-* equal + Greek *allos* other + *baros* weight]
—**isallobaric** (ī sal′ə bär′ik), *adj.* having or indicating equal atmospheric pressure over a given period.

ischial (is′kē əl), *adj. Anatomy.* having to do with the ischium.

ischium (is′kē əm), *n., pl.* **-chia** (-kē ə). *Anatomy.* the lowest bone of the three bones forming either half of the pelvis: *Each side of the pelvic girdle consists of an ilium, ischium, and pubis. These bones in a generalized skeleton are arranged similarly to the bones of the pectoral girdle* (Shull, *Principles of Animal Biology*). See the picture at **pelvis.** [from New Latin, from Greek *ischion*]

island, *n.* **1** *Geography.* a body of land smaller than a continent and completely surrounded by water. **2** *Anatomy.* a group of cells different in structure or function from those around it.

island arc, *Geography, Geology.* an arched chain of oceanic islands lying off the coast of a continent. The Aleutian islands and the islands of the West Indies are island arcs. *Island arcs, representing the tops of mountain chains, rise steeply from the general level of the [ocean] floor* (Garrels, *A Textbook of Geology*).

island of stability, *Nuclear Physics.* a hypothetical group of superheavy elements with highly stable nuclei: *For a number of years efforts have been under way ... to create elements in the so-called "island of stability" centered on 114 by smashing heavy atoms together. Likewise, various researchers have reported evidence that such elements exist—or that they have left behind their decay products* (W. Sullivan).

islands of Langerhans (läng′ər häns), = islets of Langerhans.

islet cells, = islets of Langerhans.

islets of Langerhans (läng′ər häns), *Anatomy.* the small, scattered endocrine glands in the pancreas that secrete insulin: *The islets of Langerhans ... secrete a variety of hormones into the bloodstream. The pancreas has between one and two million islets, each islet about 200 microns in diameter and together accounting for about 2 percent of the mass of the organ. The islets are highly vascularized, and each islet cell is in close proximity to a capillary* (Scientific American). Also called **islands of Langerhans** or **islet cells.** [named after Paul Langerhans, 1847–1888, German physician and anatomist]

iso-, *combining form.* equal; alike, as in *isometric = metrically equal.* [from Greek, from *isos* equal]

isoagglutination (ī′sō ə glü′tə nā′shən), *n. Immunology.* the agglutination of the red blood corpuscles of an animal by antibodies normally present in the serum of another animal of the same species.
—**isoagglutinative** (ī′sō ə glü′tə nā′tiv), *adj.* having to do with or causing isoagglutination.

isoagglutinin (ī′sō ə glü′tə nin), *n. Immunology.* an antibody normally present in serum that will agglutinate the red blood corpuscles of another animal of the same species: *For the patient with the relatively uncommon AB blood group, it is theoretically possible to give blood of any group if it were not for the presence in the donor's blood of incompatible isoagglutinins* (Science News Letter).

isoantibody (ī sō an′tē bod′ē), *n. Immunology.* an antibody which acts against the antigen produced by another animal of the same species: *Ehrlich coined the term "isoantibodies" for antibodies synthesized by an individual that act against cells of other individuals of the same species. The antigens responsible for the development of isoantibodies are termed isoantigens* (Scientific American).

---

cap, fāce, fäther; best, bē, tèrm; pin, fīve;
rock, gō, ôrder; oil, out; cup, pùt, rüle;
yü in use, *yu* in uric;
ng in bring; sh in rush; th in thin, ŦH in then;
zh in seizure.
ə = a in about, e in taken, i in pencil, o in lemon, u in circus

isoantigen (ī′sō an′tə jən), *n. Immunology.* an antigen that stimulates the production of a specific antibody in another animal of the same species: *Organ transplants are successful between chickens of different species when the cells of the donor make a chemical called an isoantigen in the same way as do the cells of the recipient* (Science News Letter). Also called **alloantigen.**

isobar (ī′sə bär), *n.* **1** *Meteorology.* a line on a weather map connecting places having the same average barometric pressure (after allowance for height above sea level). Isobars show the distribution of atmospheric pressure at a particular time, and are used in making forecasts of the weather.
2 *Chemistry.* one of two or more kinds of atoms that have the same atomic weight, but in most cases different atomic numbers.
[from Greek *isobarēs* of equal weight, from *iso-* + *baros* weight]
—**isobaric** (ī′sə bar′ik), *adj.* **1** *Meteorology.* having or indicating equal atmospheric pressure. **2** *Chemistry.* having to do with isobars.

ISOBAR (def. 1)   ISOTHERM

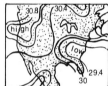

atmospheric pressure   temperature

isobath, *n. Oceanography.* a line on a map connecting points that have the same depth below the surface of a body of water: *The axis of the mid-Atlantic ridge is 2,500 meters below sea level. The 3,000-meter isobath, or contour of equal depth, lies on ocean-bottom crust that is two million years old* (J. G. Sclater and C. Tapscott). [from *iso-* + Greek *bathos* depth]

isobutane, *n. Chemistry.* a hydrocarbon, an isomer of butane, obtained from gas and gasoline, used as a refrigerant, a fuel, and in the synthesis of various chemical products. It is also known as 2-methylpropane. *Formula:* $(CH_3)_3CH$ or C_4H_{10}

isocarpic (ī′sə kär′pik), *adj. Botany.* (of a flower) having the same number of carpels as other floral parts. [from *iso-* + Greek *karpos* fruit]

isochor or **isochore** (ī′sə kôr), *n. Physics.* a curve denoting the relations between two variables, such as temperature and pressure, of a gas or quantity having a constant volume. [from *iso-* + Greek *chōra* space]

isochromatic, *adj.* **1** *Optics.* of the same color or tint, as two lines or curves in a figure formed when light waves going through a biaxial crystal interfere.
2 *Physics.* having to do with the variations involved when radiation is of constant wave length or frequency.

isochronal (ī sok′rə nəl), *adj. Physics.* **1** equal or uniform in time. **2** performed or happening in equal periods of time. **3** characterized by motions or vibrations of equal duration. [from New Latin *isochronus,* from Greek *isochronos,* from *isos* equal + *chronos* time]

isochrone (ī′sə krōn), *n. Physics.* a line connecting points on a chart where the same event occurs or would occur simultaneously. [from *iso-* + Greek *chronos* time]

isochronous (ī sok′rə nəs), *adj.* = isochronal.

isoclinal (ī′sə klī′nəl), *Geology.* —*adj.* **1** of, having to do with, or indicating equal magnetic inclination; inclining or dipping in the same direction. **2** denoting or having to do with a line on a map drawn through points on the earth's surface at which the magnetic dip is the same. **3** (of strata) of or like an isocline.
—*n.* an isoclinal line.
[from *iso-* + Greek *klinein* to dip, incline]

isocline (ī′sə klīn), *n. Geology.* a fold of strata so tightly compressed that the parts on each side of the axis dip in the same direction.

isocyanate (ī′sə sī′ə nāt), *n. Chemistry.* any of a group of compounds containing the -NCO radical, used especially in making foams and resins.

isocyclic (ī′sə sī′klik), *adj. Chemistry.* of or having to do with a ring in which all the atoms are of the same element: *Carbocyclic compounds are isocyclic.*

isodiametric (ī′sə dī′ə met′rik), *adj.* **1** *Geometry.* having equal diameters or axes.
2 *Botany.* having the diameter similar throughout: *an isodiametric cell.*
3 (of crystals) having equal lateral axes.

isodose (ī′sə dōs′), *Medicine.* —*adj.* of, having to do with, or representing surfaces or areas receiving equal doses of radiation: *The radiologist is aided in his calculations by an isodose chart superimposed over the patient's body-section diagram* (Saturday Review).
—*n.* a surface or area on a chart or on irradiated tissue, showing equal doses or intensities of radiation.

isodynamic, *adj. Physics.* **1** of or having equality of force, intensity, value, etc. **2** connecting points at which the intensity of the magnetic force is the same: *isodynamic lines on a map.*

isoelectric, *adj.* having equal electric potential; alike in potential.

isoelectric point, *Chemistry.* **1** the point at which a compound, such as a protein, has no significant excess of positive or negative charge, commonly expressed in terms of pH value.
2 the point at which a molecule or colloidal particle has no significant excess of positive or negative charges so that it does not move in either direction in an electrical field: *Every electrically responsive compound has its own isoelectric point, and this characteristic is important because it tells something about the substance's structure and chemical properties* (Scientific American).

isoelectronic, *adj. Chemistry.* having an equal number of electrons outside the nucleus: *A bond formed between boron and nitrogen is isoelectronic with the carbon-carbon bond; that is to say, both have the same number of electrons* (New Scientist).

isoenzyme (ī′sō en′zīm), *n. Biochemistry.* one of the forms of an enzyme, differing chemically but performing the same physiological function, found in a single species of an organism: *By means of starch-gel electrophoresis it is possible to separate certain enzymes into a number of components—the so-called isoenzymes* (New Scientist). Also called **isozyme.**

isogamete, *n. Biology.* either of two gametes, not differing in character or behavior, which can unite with the other to form a zygote. The gametes of certain algae are isogametes.

—**isogametic,** *adj.* of or having to do with isogametes.

isogamous (ī sog′ə məs), *adj. Biology.* having isogametes; reproducing by the fusion of two similar gametes.

isogamy (ī sog′ə mē), *n. Biology.* the fusion of two similar gametes, as in certain algae: *In plants isogamy is limited to primitive forms of the algae and fungi, and in animals to the Protozoa.* (Harbaugh and Goodrich, *Fundamentals of Biology*).

isogenous (ī soj′ə nəs), *adj. Biology.* having the same or a similar origin, as organs or parts derived from the same or corresponding tissue of the embryo.

isogeny (ī soj′ə nē), *n. Biology.* identity or similarity of origin; origination in or derivation from the same or corresponding tissues.

isogeotherm (ī′sə jē′ə thèrm), *n.* = geoisotherm.

isogon (ī′sə gon), *n. Geometry.* a figure whose angles are equal. [from Greek *isogōnios* having equal angles, from *isos* equal + *gōnia* angle]

—**isogonal** (ī sog′ə nəl), *adj.* = isogonic. —*n.* an isogonic line.

—**isogonic** (ī′sə gon′ik), *adj.* having equal angles; having to do with equal angles. SYN: equiangular. [from Greek *isogōnios,* from *iso-* + *gōnia* angle]

isogonic line, a line on a map connecting points on the earth's surface where the magnetic declination is the same.

isograft, *n.* = homograft.
ASSOCIATED TERMS: see **allograft.**

isogram, *n.* = isoline.

isohel (ī′sə hel), *n. Meteorology.* a line on a weather map or an imaginary line drawn in space, connecting points receiving equal amounts of sunshine. [from *iso-* + Greek *hēlios* the sun]

isohyet (ī′sə hī′ət), *n. Meteorology.* a line on a weather map or an imaginary line drawn in space, connecting points at which rainfall is equal: *Isohyets are drawn as nearly as may be to indicate the actual precipitation; there is no reduction to sea level as in the case of isobars and some isotherms* (Blair, *Weather Elements*). [from *iso-* + Greek *hyetos* rain]

—**isohyetal** (ī′sə hī′ə təl), *adj.* having to do with or marking equality of rainfall.

isoleucine (ī′sə lü′sēn′ *or* ī′sə lü′sən), *n. Biochemistry.* an amino acid present in casein, body tissue, etc., that is essential to the diet of animals. *Formula:* $C_6H_{13}NO_2$ *Abbreviation:* Ile

isoline (ī′sə līn′), *n. Meteorology.* a line on a chart or map indicating equality of some physical condition or quantity, as isotherms and isobars: *During the 19th century scientific atlases did much to popularize the use of ... isobars, isotherms, and isohyets ... So many types of lines were created to show the distribution of data that the general term isoline was coined* (W. M. McKinney).

isologous (ī sol′ə gəs), *adj. Chemistry.* having similar relations. In an isologous series of hydrocarbons, each member has two hydrogen atoms less than the one above it. [from *iso-* + Greek *logos* word, relation]

—**isologue** or **isolog** (ī′sə lôg *or* ī′sə log), *n. Chemistry.* an isologous compound.

isomagnetic, *adj. Physics.* **1** having or showing equality of magnetic force. **2** of or denoting a line on a map or chart connecting places that have the same magnetic properties.

isomer (ī′sə mər), *n. Chemistry.* an isomeric compound, ion, atom, etc.: *Molecules such as these sugars with the same chemical composition but with different chemical properties are called isomers* (Weier, *Botany*). *Isomers differ because their atoms are arranged differently in their molecules, just as the allotropic forms of carbon (diamond and graphite) are different because their atoms are differently arranged within their molecules* (Offner, *Fundamentals of Chemistry*). [from Greek *isomerēs* sharing equally, from *iso-* + *meros* part]

isomerase (ī som′ə rās), *n. Biochemistry.* an enzyme that catalyzes the conversion of one isomer into another.

isomeric (ī′sə mer′ik), *adj. Chemistry.* **1** composed of the same chemical elements in the same proportions by weight, and having the same molecular weight, but differing in at least one physical or chemical property because of the difference in arrangements of atoms. **2** (of the nuclei of atoms) differing in energy and behavior, but having the same atomic number and mass number.

—**isomerism** (ī som′ə riz′əm), *n.* the property or condition of being isomeric.

—**isomerize** (ī som′ə rīz′), *v.* to change into an isomer: *The only action of light upon a visual pigment is to isomerize—to straighten out—retinene to the all-trans configuration* (Scientific American). —**isomerization,** *n.*

isomerous (ī som′ər əs), *adj.* **1** having an equal number of parts, markings, etc. **2** (of a flower) having an equal number of members in each whorl.

isometric (ī′sə met′rik), *adj.* **1** *Mathematics.* having equality of measure; having to do with or related by isometry: *Two metric spaces consisting of the same elements but with different metrics may be isometric* (C.C. Krieger). **2** *Crystallography.* having three equal axes at right angles to one another. Also called **cubic.**
ASSOCIATED TERMS: see **crystal.**
3 *Geometry.* having to do with or representing an isometric projection: *an isometric drawing.* **4** *Physiology.* of or denoting lack of contraction of a muscle when under a constant tension: *The simplest form of exercise consists of tensing the muscle without shortening it, by exerting it against an immovable object—what we call isometric contraction* (E. Müller).

—**isometrically,** *adv.*

isometric projection, *Geometry.* the projection of an object upon a plane equally inclined to the object's three principal axes, so that dimensions parallel to these are represented in their true proportions.

---

cap, fāce, fäther; best, bē, tèrm; pin, five;
rock, gō, ôrder; oil, out; cup, pùt, rüle,
yü in use, *yu* in uric;
ng in bring; *sh* in rush; *th* in thin, ŦH in then;
zh in seizure.
ə = *a* in about, *e* in taken, *i* in pencil, *o* in lemon, *u* in circus

isometry (ī som'ə trē), *n. Mathematics.* a transformation of one metric space into another in which the distances between each set of points remain constant; a congruent transformation.

isomorph (ī'sə môrf'), *n.* an isomorphic organism or substance. [from *iso-* + Greek *morphē* form]

isomorphic (ī'sə môr'fik), *adj.* **1** *Biology.* having similar appearance or structure, but belonging to different species, generations, genotypes, etc.: *Their [certain seaweeds'] life-cycle is an isomorphic alternation of diploid sporophyte with haploid gametophyte* (E.J.H. Corner).
2 *Crystallography.* = isomorphous.
3 *Mathematics.* (of two sets) having a one-to-one correspondence.

—**isomorphism** (ī'sə môr'fiz'əm), *n.* the property of being isomorphic or isomorphous.

—**isomorphous** (ī'sə môr'fəs), *adj. Crystallography.* crystallizing in the same form or related forms (used especially of substances of analogous chemical composition): *Perhaps as simple an isomorphous series as any is the olivine group of minerals. The formula of this group* $(Mg,Fe)_2SiO_4$, *meaning that different specimens of olivine may have chemical compositions intermediate between the two "end members"; that is, they range from pure* Mg_2SiO_4 *to pure* Fe_2SiO_4. *It is only the iron (Fe) and the magnesium (Mg) that vary; the proportions of silicon and oxygen remain constant* (Gilluly, *Principles of Geology*).

isopach (ī'sə pak), *n. Geology.* a line drawn on a map or chart joining points of equal thickness in a particular stratum or group of strata: *The measured thickness of the pyroclastic deposits at various localities and the proposed isopachs ... generally extend south-eastwards from the eruptive centres* (Nature). [from *iso-* + Greek *pachys* thick]

isophote (ī'sə fōt), *n. Optics, Astronomy.* a line drawn on a map or chart joining points that receive equal light from a source, such as a star or the envelope of a galaxy image: *[With] isophote maps outlining the brightness of the entire sky at 15-minute intervals throughout the night ... [he] has been able to show the existence of discrete airglow cells in the upper atmosphere* (D. M. Gates). [from *iso-* + Greek *phōs, phōtos* light]

isopleth (ī'sə pleth), *n.* **1** a graph showing variations in occurrence or frequency of a phenomenon, especially in meteorology, with reference to two variables, such as time and space: *Isopleths are ... drawn for temperature, potential temperature, mixing ratio, or other elements shown by the soundings ... and are of much value to the forecaster* (Blair, *Weather Elements*).
2 *Geography.* a line on a chart or map indicating equality of some physical condition or quantity.
3 *Mathematics.* the straight line on a graph connecting the equal or corresponding values of the dependent and independent variables.
[from *iso-* equal + Greek *plēthos* number]

isoprene (ī'sə prēn'), *n. Chemistry.* a colorless, volatile, liquid hydrocarbon used in making synthetic rubber. *Formula:* C_5H_8

—**isoprenoid** (ī'sə prē'noid), —*adj.* having a chemical structure similar to that of isoprene, or resembling a connected sequence of isoprene units: *an isoprenoid hydrocarbon.* —*n.* an isoprenoid compound: *Isoprenoids ... provide the architectural base for cholesterol and many other well known biological materials* (Science News Letter).

isopropyl alcohol (ī'sə prō'pəl), or **isopropanol** (ī'sə-prō'pə nōl), *n. Chemistry.* a colorless, flammable liquid soluble in water, ether, and alcohol, used to make acetone, and as a solvent and rocket fuel. It is also known as 2-propanol. *Formula:* $CH_3CHOHCH_3$

isopyre (ī'sə pīr), *n. Mineralogy.* an impure variety of opal, sometimes used as a semiprecious stone, containing alumina, iron, and lime. [from *iso-* + Greek *pyr* fire]

isosceles (ī sos'ə lēz'), *adj. Geometry.* **1** (of a triangle) having two sides equal: *The angles at the base of an isosceles triangle are equal.* **2** (of a trapezoid) having its nonparallel sides congruent. [from Late Latin, from Greek *isoskelēs*, from *iso- skelos* leg]

isosceles triangles

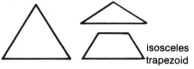

isosceles trapezoid

isoseismal (ī'sə sīz'məl or ī'sə sis'məl), *Geology.* —*adj.* **1** having to do with equal intensity of earthquake shock. **2** of or having to do with an imaginary line on the earth's surface connecting points characterized by such intensity: *Lines drawn on such a map through points of equal intensity are called isoseismal lines ... [and] generally lie in rough ovals about ... the earthquake epicenter* (Gilluly, *Principles of Geology*).
—*n.* an isoseismal line: *Isoseismals for several of the quakes ... cluster around the margins of the West China miniplate. The data help define a miniplate boundary* (Science News).
[from *iso-* + Greek *seismos* earthquake]

isosmotic (ī'sos mot'ik), *adj. Chemistry.* of or containing equal osmotic pressure: *an isosmotic medium.*

isospin (ī'sō spin'), *n. Nuclear Physics.* any of a series of quantum numbers which describe the possible charge states of an elementary particle, based on the theory that the neutron and proton are different states of the same particle: *Isospin is a number, with no simple physical meaning, that represents mathematically the difference between neutrons and protons* (Science News). Also called **isotopic spin.**

isostasy (ī sos'tə sē), *n. Geology.* **1** equilibrium of the earth's crust, believed to be due to the movement of rock material below the surface: *Where there is a high mountain chain, the sial layer extends deeper into the sima beneath, so buoying up the greater weight above; ... This condition is called isostacy* (New Biology).
2 equilibrium or stability caused by equality of pressure.
[from *iso-* + Greek *stasis* a stopping]

—**isostatic** (ī'sə stat'ik), *adj.* having to do with or characterized by isostasy: *isostatic compensation, isostatic balance.* —**isostatically,** *adv.*

isostemonous (ī'sə stē'mə nəs *or* ī'sə stem'ə nəs), *adj.* *Botany.* having the stamens equal in number to the sepals or petals. [from *iso-* + Greek *stēmōn* thread, but taken as equaling *stamen*]

isotactic (ī'sō tak'tik), *adj. Chemistry.* of or having to do with a polymer which has its attached groups of atoms in a regular order on one side of the central chain of atoms.

isoteniscope (ī'sə ten'ə skōp), *n. Chemistry.* an instrument for measuring the pressure at which a liquid and its vapor are at equilibrium at a given temperature: *Chemists routinely measure the boiling point of certain fluids with an isoteniscope* (C.L. Stong). [from *iso-* + *ten-* (stem of Latin *tenere* to hold) + *-scope*]

isotherm (ī'sə thėrm'), *n.* **1** *Meteorology.* a line on a weather map connecting places having the same average temperature. See the picture at **isobar.** **2** = isothermal curve. [from *iso-* + Greek *thermē* heat]
—**isothermal** (ī'sə thėr'məl), *adj.* **1** *Meteorology.* having or indicating equal temperatures: *The layer of air in the stratosphere is extremely thin, or rarefied. It is sometimes called the ... isothermal layer, because its temperature remains almost always the same* (E.S. Stone).
2 *Physics.* having to do with an isothermal curve or curves.
—*n.* = isothermal curve.

isothermal curve or **isothermal line, 1** *Physics.* a line or curve connecting points corresponding to pressure or volume changes, with the temperature of the substance constant.
2 *Meteorology.* = isotherm.

isotone (ī'sə tōn), *n. Nuclear Physics.* one of two or more atoms having an equal number of neutrons. [from *iso-* + Greek *tonos* tone, force, tension]

isotonic (ī'sə ton'ik), *adj.* **1** *Chemistry.* having the same osmotic pressure. A solvent isotonic with a solution will not pass through a semipermeable membrane and mix with the solution. *When two fluids contain equal concentrations of dissolved substances, they are said to be isotonic* (Storer, *General Zoology*).
2 *Physiology.* denoting or having to do with a contraction of a muscle when under a constant tension.
[from Greek *isotonos* of equal tension or tone, from *isos* equal + *tonos* tone] —**isotonically,** *adv.*
—**isotonicity** (ī'sə tō nis'ə tē), *n.* isotonic quality or condition.

isotope (ī'sə tōp), *n. Chemistry, Physics.* any of two or more forms of an element that have the same chemical properties and the same atomic number, but different atomic weights and slightly different physical properties, because of different numbers of neutrons in the atomic nucleus. Chlorine, whose atomic weight is about 35.5, is a mixture of two isotopes, one having the mass number 37, the other, 35. Hydrogen and heavy hydrogen are isotopes. *Radioactive isotopes ... are valuable as "trace elements" in research on chemical, biological, medical, industrial and other processes* (Wall Street Journal). [from *iso-* + Greek *topos* place]
► Most elements occur naturally as mixtures of isotopes. This accounts for fractional atomic masses found in reference tables. In general, the mass number of the most abundant isotope of an element can be determined by rounding off the atomic mass of the element to the nearest whole number.

The symbol for an isotope usually consists of the mass number placed over the symbol for the element. For instance, the isotope of carbon with mass number 14 is written ^{14}C.
—**isotopic** (ī'sə top'ik), *adj.* of or having to do with an isotope or isotopes; like an isotope: *Scientists frequently use deuterium to study organic and biochemical reactions. The heavy hydrogen atom serves as an isotopic tracer* (Harold C. Urey). —**isotopically,** *adv.*

isotopic spin, = isospin: *A particle such as the lambda, which is always neutral, is said to have an isotopic spin of zero* (Scientific American).

isotopy (ī sot'ə pē), *n.* the occurrence or existence of isotopes.

isotropic, *adj. Physics.* having the same properties, such as elasticity or conduction, in all directions.
—**isotropically,** *adv.* equally in all directions: *This sphere isotropically scatters the transmitted signal, so one has merely to point the transmitter and receiver antennas at it to complete the path* (J. R. Pierce).

isozyme (ī'sə zīm), *n.* = isoenzyme.

isthmian (is'mē ən), *adj. Geography.* of, having to do with, situated in, or forming an isthmus.

isthmus (is'məs), *n.* **1** *Geography.* a narrow strip of land with water on both sides, connecting two larger bodies of land: *The Isthmus of Suez separates the Red Sea, an arm of the Indian Ocean, from the Mediterranean Sea, an arm of the Atlantic.* (Colby and Foster, *Economic Geography*).
2 *Anatomy.* a narrow part of an organ connecting two larger parts or cavities.
[from Latin *isthmus*, from Greek *isthmos*]

itacolumite (it'ə kol'yə mīt), *n. Mineralogy.* a granular quartzite that is very flexible when found in thin slabs. [named after *Itacolumi*, a mountain in Brazil]

ITC, *abbrev.* intertropical convergence (zone).

-ite, *suffix.* **1** a mineral species, or a rock substance, as in *hematite.* **2** a segment of a part of a body, as in *dendrite.* **3** a fossil, as in *tribolite.* **4** a substance of organic origin, as in *dynamite.* **5** an ester or salt of an acid whose name ends in *-ous,* as in *phosphite, sulfite, nitrite.*

-itis, *suffix.* inflammation of; inflammatory disease of, as in *appendicitis, tonsillitis, bronchitis.*

-ium, *suffix.* chemical element or radical, as in *curium, sodium, ammonium.*

IUPAC, *abbrev.* International Union of Pure and Applied Chemistry.

---

cap, fāce, fäther; best, bē, tėrm; pin, fīve;
rock, gō, ôrder; oil, out; cup, pùt, rüle,
yü in use, yu in uric;
ng in bring; sh in rush; th in thin, ŦH in then;
zh in seizure.
ə = a in about, e in taken, i in pencil, o in lemon, u in circus

J

J or **j**, *symbol.* joule.

Jacob-Monod (jä′kəb mə nod′; *French* zhȧ kôb′mô-nō′), *adj. Genetics.* of or having to do with the theory that genes are controlled by the operon, first advanced by the French scientists François Jacob and Jacques Monod: *One mode of interpretation is ... the interaction of Jacob-Monod type repressors, or enzymes responsible for RNA synthesis, with control signals on the DNA* (New Scientist).

Jacobson's organ, *Zoology.* either of a pair of cavities in the roof of the mouth of many vertebrates, containing branches of the olfactory nerve. It is highly developed in most reptiles. *Snakes are equipped with a chemical sense organ, called Jacobson's organ, that tells them what is good among the available foods* (J. A. Pearre). [named after L.L. *Jacobson,* 1783–1843, Danish anatomist]

jade, *n. Mineralogy.* a hard, compact, lustrous gemstone consisting of either jadeite or nephrite. Most jade is green, but some is whitish. [from French *jade,* from Spanish (*piedra de*) *ijada* (stone of) colic, jade being anciently supposed to cure colic]

jadeite (jä′dīt), *n. Mineralogy.* a pyroxene mineral consisting of a silicate of sodium and aluminum, occurring in the form of light green monoclinic crystals. It is the tougher and more highly prized variety of jade. *Jadeite is a rare jade mineral, found chiefly in Burma, in Japan, and in carved objects in Mexico. It also occurs in California* (F. H. Pough).

Jansky (jan′skē), *n.* = flux unit. [named after Karl G. *Jansky,* 1905–1950, American electrical engineer]

Japan Current, *Oceanography.* a strong, warm current that begins near the Philippines and flows northeastward, past Japan, across the Pacific Ocean. Also called **Kuroshio Current.**

jarosite (jar′ə sīt), *n. Mineralogy.* a hydrous sulfate of iron and potassium, occurring in yellowish or brownish rhombohedral crystals. *Formula:* $KFe_3(SO_4)_2(OH)_6$ [from Barranco *Jaroso,* an area in Spain + English *-ite*]

jasper, *n. Mineralogy.* an opaque, cryptocrystalline quartz, usually red, yellow, or brown.

jaspilite (jas′pə līt), *n. Mineralogy.* a siliceous metamorphic rock consisting of interbanded layers of red jasper and hematite: *With the depletion of high-grade iron-ore deposits in the Lake Superior district ... steel companies began an intensive and costly program to mine the lower-grade taconite and jaspilite ores, and to produce from them concentrated pellets suitable for the blast furnace* (Scientific American).

jaw, *n. Anatomy.* the upper or lower bone, or set of bones, that together form the framework of the mouth of vertebrates. The upper jaw consists of the maxillae; the lower jaw, which is hinged and movable, forms the mandible.

jawbone, *n. Anatomy.* one of the bones in which the teeth are set, especially the lower jaw or mandible.

jejunum (ji jü′nəm), *n. Anatomy.* the middle portion of the small intestine, between the duodenum and the ileum. [from New Latin, from Latin *jejunum* (*intestinum*) empty (intestine)]
—**jeju′nal,** *adj.* of or having to do with the jejunum: *the jejunal tube, a jejunal ulcer.*

jellyfish, *n. Zoology.* any of a group of free-swimming marine coelenterates having an almost transparent spherical body with long, trailing tentacles that may bear stinging hairs or cells. Hydrozoans comprise most of the smaller jellyfish; scyphozoans include most of the large jellyfish.

jet[1], *n.* **1** *Physics.* a high-speed stream of fluid, such as gas, steam, or water, discharged with force especially from a small opening or nozzle. **Jet propulsion** is the propulsion of a body in one direction by a jet of fluid discharged forcefully in the opposite direction. A **jet pump** is a pump in which an accelerating fluid forces other portions of the fluid to move forward.
2 *Particle Physics.* a group of particles that emerges from a collision on nearly parallel paths.
3 *Meteorology.* = jet stream: *On Earth, equatorial jets are found in the lower stratosphere and in the oceans* (Nature).
[from Old French *giet* a throwing, from Late Latin *jectus, jactus,* from Latin *jacere* to throw]

fan jet engine:

(def. 1)

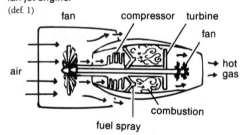

In a fanjet engine, air is forced through the compressor, where it is mixed with fuel, burned in the combustion chamber, and expelled as hot gas, providing thrust to the engine.

jet[2], *n. Mineralogy.* a hard, compact black substance, a form of lignite, that becomes glossy when polished: *Jet is a lustrous, tough, firmly compact variety of coal that breaks with a glassy, conchoidal fracture, takes a high polish, and, compared with other coals, is rather free of shrinkage cracks* (Science). [from Old French *jaiet,* from Greek *gagates* of Gagas, town in Lycia, Asia Minor]

jet stream, *Meteorology.* a current of air traveling usually from west to east, at altitudes of ten to fifteen miles. Jet streams move in wavelike fashion both vertically and horizontally, usually at a speed of 100 to 120 miles per hour but sometimes exceeding 300 miles per hour. They affect the earth's weather and are often associated with storms, tornadoes, etc. *The emphasis on obtaining meteorological data from high levels of the atmosphere will shed further light on the origin and location of the high-speed, narrow rivers of air known as the jet stream* (Science News Letter).

JH, *abbrev.* juvenile hormone.

join, *n. Mathematics.* a set which includes all the members of two or more sets. SYN: union, sum.

joint, *n.* **1** *Anatomy.* **a** the part where two or more bones of a vertebrate are joined and move on each other; the junction of two skeletal parts: *Joints may be of the hinge type, such as you have in your knees and your fingers, or they may be ball-and-socket joints, such as you have in your hips and neck.* (Beauchamp, et al., *Everyday Problems in Science*). **b** the place where parts of an invertebrate's exoskeleton are joined or fitted together. SYN: articulation. See also **pivot joint.**

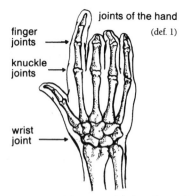

joints of the hand (def. 1)

finger joints

knuckle joints

wrist joint

2 *Botany.* a part of the stem of a plant from which a leaf or branch grows, especially when thickened, as in grasses. Also called **node.**

3 *Geology.* a fracture in a mass of rocks, usually occurring in sets of parallel planes dividing the mass into blocks: *Stresses sufficient to cause crustal fracture have developed so many times that the solid rock is nearly everywhere traversed by cracks called joints ... The joints permit the water of the ground to circulate more freely within the rocks and ... play a part in the details of shape in landform* (Finch and Trewartha, *Elements of Geography*).

Jordan curve, *Mathematics.* a curve that is closed in a plane and does not cross itself at any point. Also called **simple closed curve.** [named after M. E. Camille *Jordan,* 1838–1922, French mathematician]

Jordan curve theorem, *Mathematics.* the theorem that a Jordan curve in a plane divides the plane into two regions, an interior and an exterior. No point in the interior region can be joined to a point in the exterior region without crossing the curve.

Josephson effect, *Physics.* an effect by which, near absolute zero, an electric current can flow without resistance between two superconductors separated by a thin insulating layer. [named after Brian D. *Josephson,* born 1940, English physicist who predicted this effect in 1962]

Josephson junction, *Physics.* the junction between two superconductors separated by a thin insulating layer, exhibiting the Josephson effect: *A Josephson "junction" is an extremely fast-acting and reliable on-off switch* (G.A.W. Boehm).

joule (jül *or* joul), *n. Physics.* the SI or MKS unit of work or energy, equal to 10^7 ergs or the amount of energy corresponding to 1 watt acting for one second: *Symbol:* J or j [named after James P. *Joule,* 1818–1889, British physicist]

Joule's Law, *Physics.* **1** the law that the rate at which heat is generated by an electric current flowing through a resistance is proportional to the square of the current.

2 the law that the internal energy of a given mass of a gas depends only on its temperature.

Jovian (jō'vē ən), *adj. Astronomy.* of or having to do with the planet Jupiter: *Of the Jovian moons the outermost three revolve in an opposite or retrograde direction, from east to west. This has given rise to discussion as to whether they might not be captured asteroids and therefore not originally members of Jupiter's system* (Bernard, *Handbook of Heavens*). The **Jovian planets** Jupiter, Saturn, Uranus, and Neptune have very large masses (Jupiter has the largest mass) and are farther from the sun than the terrestrial planets (Earth, Mars, Mercury, and Venus).

J particle, = psi particle: *The most striking feature of the J particle is its very long lifetime ... [of] from a hundred to a thousand times longer than all known mesons of heavy mass* (S.C.C. Ting).

jugal bone, = zygomatic bone.

jugate (jü'gāt *or* jü'git), *adj. Biology.* occurring in pairs; connected, as a pinnate leaf in which the leaflets are in pairs. [from Latin *jugatum* joined, yoked, from *jugum* yoke]

jugular (jug'yə lər *or* jü'gyə lər), *Anatomy.* —*adj.* **1** of the neck or throat. The **jugular vein** is either of the two large veins in each side of the neck and head that return blood from the head and neck to the heart.

2 of the jugular vein: *the jugular pulse.*

—*n.* a jugular vein.

[from New Latin *jugularis,* from Latin *jugulum* collarbone, from *jugum* yoke]

jugulum (jug'yə ləm *or* jü'gyə ləm), *n. Zoology.* **1** the lower front part of the neck in birds. **2** a corresponding part in insects, fishes, etc., and certain other animals. [from New Latin *jugulum,* from Latin *jugulum* collarbone, from *jugum* yoke]

jugum (jü'gəm), *n., pl.* **juga** (jü'gə) *or* **jugums. 1** *Zoology.* a small lobe on the forewing of certain lepidopterous insects, which extends under the hind wing and holds the two wings together in flight.

2 *Botany.* a pair of opposite leaflets in a pinnate leaf. [from New Latin *jugum,* from Latin *jugum* yoke]

cap, fāce, fäther; best, bē, tėrm; pin, fīve; rock, gō, ôrder; oil, out; cup, pút, rüle, *yü* in use, *yụ* in uric; *ng* in bring; *sh* in rush; *th* in thin, ŦH in then; *zh* in seizure.

ə = *a* in about, *e* in taken, *i* in pencil, *o* in lemon, *u* in circus

juice, *n. Physiology.* a fluid secreted in the body. The gastric juice helps to digest food.

Julian calendar, *Astronomy.* a calendar in which the average length of a year was 365 1/4 days, with a leap year of 366 days every fourth year. It was introduced by Julius Caesar in 46 B.C. and used in France until 1582, and in Great Britain and its colonies until 1752, when it was replaced by the Gregorian calendar. A year in the Julian calendar was about 11 minutes and 14 seconds longer than a solar year.

jumping gene, an informal term for a transposon.

jungle, *n. Ecology.* a tropical forest with a dense undergrowth and a thick overgrowth of bushes and vines. Jungles usually develop in swamps, along riverbanks, and in tropical rain forest areas. *Jungle growth is exceedingly prolific and soon obstructs all openings* (White and Renner, *Human Geography*). [from Hindi *jangal,* from Sanskrit *jangala* desert, thicket, forest]

Jupiter, *n. Astronomy.* the largest planet in the solar system. It is the fifth in distance from the sun and has a faster rotation (once every 9 hours and 55 minutes) than any other planet. *Jupiter has 16 known satellites. The four largest moons of Jupiter are called Galilean satellites because the Italian astronomer Galileo discovered them in 1610* (Hyron Spinrad). See the picture at **solar system.**

Jura (jùr'ə), *n.* = Jurassic.

Jurassic (jù ras'ik), *Geology.* —*n.* **1** the middle period of the Mesozoic era, after the Triassic and before the Cretaceous, when dinosaurs dominated the earth and birds first appeared: *At the end of the Jurassic, after the accumulation of some thirty to forty thousand feet of sediments, smashing and folding began in this part of the geosyncline.* (Garrels, *A Textbook of Geology*). **2** the rocks formed in this period.

—*adj.* of or having to do with the Jurassic or its rocks: *Many identical fossils were found in the rocks of the Jura Mountains of Switzerland and France. The name Jurassic System, to which these strata are now referred, recalls this fact.* (Gilluly, *Principles of Geology*). [from French *jurassique,* from the *Jura* mountain range in France and Switzerland]

juvenile (jü'və nīl *or* jü'və nəl), *adj.* **1** *Geology.* having never existed on the surface of the earth or in the atmosphere; coming fresh from the earth's interior: *juvenile nitrogen, juvenile water.*
2 *Botany.* immature or undeveloped in appearance, as certain types of foliage or wood: *In many cases the juvenile leaves differ in form, attachment or arrangement from those on the adult tree* (A.B. Jackson).
—*n. Zoology.* a young bird that is able to fly but has not completely lost its down.

juvenile hormone, *Biochemistry.* a hormone of insects, secreted by the corpus allatum, that controls the metamorphosis from the larval to the adult stage: *A moth or butterfly will remain a caterpillar forever unless juvenile hormone production ceases* (N. Y. Times). *Abbreviation:* JH

K

k or **k.**, *symbol.* **1** kilo-. **2** kilogram *or* kilograms.

K, *symbol.* **1** kelvin *or* kelvins. **2** potassium [for New Latin *kalium*]. **3** a unit of storage capacity in a computer equal to 1024 (2^{10}) bytes. [from *kilo-*, prefix for 1000] **4** equilibrium constant.

kallidin (kal′ə din), *n. Biochemistry.* either of two types of kinins, one of which is bradykinin, released from blood plasma globulin by the enzyme kallikrein: *The enzyme in salivary gland that released kallidin from blood was kallikrein, a protein that had been studied in Germany even before the discovery of kallidin.* (Scientific American). [from Greek *kallos* beautiful + English *-in*]

kallikrein (kal′ə krē′ən), *n. Biochemistry.* an enzyme that breaks down proteins in blood plasma, pancreatic juice, and other body fluids, thereby releasing kallidins: *a generalized deficiency of kallikreins may explain the clinical and biochemical manifestations of cystic fibrosis* (Science). [from Greek *kallikreas* pancreas (from *kallos* beauty + *kreas* flesh) + English *-in*]

kamacite (kam′ə sīt), *n. Mineralogy.* an alloy of iron and a small amount of nickel, found in some meteorites: *Some tektites from the Philippines have been found to contain small spherules of nickel-iron (kamacite)* (E. A. King, Jr.). [from German *Kamacit,* from Greek *kamax* pole + German *-it* -ite]

kame (kām), *n. Geography, Geology.* a hill or ridge of poorly sorted gravel or sand impounded in a glacier and deposited by melting of the ice: *Upon melting of the glacier, patches of these sediments ... are left in terraces and in isolated buttes called kames and kame terraces* (Gilluly, *Principles of Geology*). [Scottish form of *comb*]

Kansan, *adj. Geology.* of or having to do with the second period of glaciation in North America, beginning about 470,000 years ago, and lasting for 156,000 years. Compare **Illinoian, Nebraskan, Wisconsin.**

kaolin (kā′ə lən), *n. Mineralogy.* a fine white clay consisting of kaolinite, used especially in making ceramics: *The purest form of clay, called kaolin, is a hydrated aluminum silicate* (Offner, *Fundamentals of Chemistry*). [from French *kaolin,* from *Kao-ling,* mountain in northern China where it was originally obtained]

kaolinite (kā′ə lə nīt), *n. Mineralogy.* a crystalline, hydrous silicate of aluminum, the chief constituent of kaolin and some other clays. Kaolinite is polymorphous with dickite and nacrite. *Formula:* $Al_2Si_2O_5(OH)_4$

kaon (kā′on), *n. Nuclear Physics.* any meson having a positively or negatively charged form with a mass of about 966 times that of an electron, and a neutral form with a mass of about 974 times that of an electron. Kaons exhibit the quantum property of strangeness and on decaying produce lighter mesons, such as pions, and sometimes neutrinos and other leptons. Also

called **K-meson.** [from *ka* the letter K (in *K-meson*) + *-on*]
—**kaonic** (kā on′ik), *adj.* containing or producing kaons: *kaonic and sigmic atoms.*

kappa particle (kap′ə), *Biology.* a protein particle containing DNA, found in certain strains of paramecia, and thought to secrete a toxic substance that kills individuals of sensitive strains: *These so-called kappa particles are probably best regarded as self-determining invaders and not as part of the normal cell* (Peacocke and Prysdale, *Molecular Basis of Heredity*).

karro or **karoo** (kə rü′), *n. Geography.* a barren tract of clayey tableland found chiefly in southern Africa. [from Afrikaans *karo*]

karst (kärst), *n. Geography, Geology.* a limestone plateau or region characterized by underground drainage, sinkholes, rolling surfaces, and caverns: *The term "karst" has long been used to define the sum of the phenomena characterizing regions where carbonate rocks, chiefly limestone, are exposed* (Science). [named after *Karst,* a desolate limestone region in northwestern Yugoslavia]

karyo-, *combining form. Biology.* cell nucleus, as in *karyology = the study of cell nuclei.* [from Greek *karyon* nut, kernel]

karyogamy (kar′ē og′ə mē), *n. Biology.* the fusion of cell nuclei, as in conjugation: *Eukaryotes ... very often have a sexual cycle of alternating karyogamy and meiosis* (A. Cronquist). [from *karyo-* + *-gamy*]

karyokinesis (kar′ē ō ki nē′sis), *n. Biology.* **1** the division of the cell nucleus, especially in mitosis. **2** the process of mitosis: *It was not until 1873 that the common method of cell division ... was discovered. ... To this method of cell division the names mitosis and karyokinesis are applied.* (Shull, *Principles of Animal Biology*). [from *karyo-* + Greek *kinēsis* motion]
—**karyokinetic** (kar′ē ō ki net′ik), *adj.* having to do with karyokinesis.

karyology (kar′ē ol′ə jē), *n.* **1** the study of cell nuclei, especially in reference to their chromosomes: *comparative karyology.* **2** the characteristic features of a cell nucleus or cell nuclei, especially in reference to the chromosomes: *the karyology of a normal cell population.*

karyolymph (kar′ē ə limf), *n. Biology.* the fluid portion within the nucleus of a cell. Also called **nuclear sap.**

karyoplasm (kar′ē ə plaz′əm), *n.* = nucleoplasm.

---

cap, fãce, fäther; best, bē, tèrm; pin, five;
rock, gō, ôrder; oil, out; cup, pùt, rüle,
yü in use, *yù* in uric;
ng in bring; sh in rush; th in thin, ᴛʜ in then;
zh in seizure.
ə = a in about, e in taken, i in pencil, o in lemon, u in circus

karyosome (kar′ē ə sōm), *n. Biology.* a body of chromatin in the nucleus of a cell, similar to but distinguished from a nucleolus.

karyotype (kar′ē ō tīp), *Genetics.* —*n.* **1** the character of a set of chromosomes as defined by their arrangement, number, size, shape, or other characteristic: *The 46 chromosomes comprise one pair of sex chromosomes ... and 22 other pairs ... of autosomes ... arranged in an approximate order of descending size to form a karyotype* (New Scientist). **2** the arrangement of chromosomes shown on photomicrographs: *A study of Mrs. J.'s chromosome pattern (called a karyotype) confirmed ... she carried an unusual chromosomal defect* (J. E. Brody).
—*v.* to classify the chromosome sets of (an organism, species, etc.) according to their arrangement, number, size, shape, or other characteristics: *Two to six weeks of growth are required before the culture is ready to be karyotyped ... and examined microscopically to determine the chromosome pattern* (Saturday Review).
—**karyotypic** (kar′ē ō tīp′ik), *adj.* of or having to do with karyotypes or karyotyping: *karyotipic patterns.*
—**karyotypically,** *adv.*

katabatic wind (kat′ə bat′ik), *Meteorology.* a wind produced by the gravitational flowing of cold, dense air down the slopes of mountains, plateaus, or other elevations: *Violent katabatic winds often descend from the glacier-covered interiors of Greenland and Antarctica* (Blair, *Weather Elements*). Also called **drainage wind, gravity wind.** [from Greek *katabatikos,* from *katabainein* to go down]

kb or **kbar,** *symbol.* kilobar.

kc, *symbol.* kilocycle *or* kilocycles.

kcal, *symbol.* kilocalorie *or* kilocalories.

keel, *n. Botany, Zoology.* a longitudinal ridge, as on a leaf or on the breastbone of a bird. Also called **carina.**

Kekulé formula or **Kekulé structure** (kā′kü lē), *Chemistry.* a ring of six carbon atoms having alternating single and double bonds between them; the benzene ring. [named after Friedrich A. *Kekulé,* 1829–1896, German chemist who discovered the ring structure of the benzene molecule]

K electron, *Physics.* one of the electrons in the K-shell.

kelvin, *n. Physics.* the SI unit of temperature, a fundamental unit equal to 1/273.16 of the temperature of the triple point of water: *The meter for length, the kilogram for mass, the second for time and the kelvin (formerly the degree Kelvin) for temperature ... together with two derived quantities (the ampere and the candela) are the building blocks of the International System of Units, or Système International* (Scientific American). *Symbol:* K [named after William Thomson, Lord *Kelvin,* 1824–1907, British physicist and mathematician who proposed the Kelvin scale for measuring temperatures]
▶ For measuring differences in temperature, a kelvin is the same as a degree Celsius. The two differ only in having different reference points—0 degrees Celsius is arbitrarily defined as the freezing point of water, but 0 kelvins is absolute zero. Water freezes at 273.15 kelvins and boils at 373.15 kelvins. As on the Celsius scale, the difference between the temperatures is 100.

One can convert any Celsius temperature to kelvins by adding 273.15.

Kelvin, *adj. Physics.* of or based on temperatures measured in kelvins: *Another scale frequently used in science is the Kelvin (Absolute) scale* (Chemistry Regents Syllabus).

Kennelly-Heaviside layer, = E layer. [named after Arthur E. *Kennelly,* 1861–1939, American electrical engineer, and Oliver *Heaviside,* 1850–1925, British physicist, who proposed the existence of this layer]

Keplerian (kep lir′ē ən), *adj. Astronomy.* of or having to do with the German astronomer Johann Kepler (1571–1630) or his laws and theories: *the Keplerian laws of planetary motion. The Keplerian universe ... was systematically ignored by Galileo* (Listener).

Kepler's law, *Astronomy.* any of three laws, formulated by Johann Kepler, describing the motions of the planets around the sun: **1** the orbits of the planets are ellipses with the sun always at one of the foci. **2** a line connecting a planet and the sun will sweep over equal areas in equal times as the planet moves about its orbit. **3** the square of the period of revolution of a planet is proportional to the cube of its distance from the sun.

Kepler's Second Law

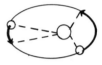

(def. 2)

kerat-, *combining form.* the form of **kerato-** before vowels, as in *keratitis* = *inflammation of the cornea.*

keratin (ker′ə tən), *n. Biochemistry.* a complex, insoluble protein, the chief structural constituent of horn, nails, hair, feathers, etc.: *Some other keratins are the horny coverings on fish eggs and the horny skeletons of bath sponges and sea fans (coelenterates)* (Storer, *General Zoology*). [from Greek *keratos* horn]
—**keratinization,** *n.* the process of keratinizing: *Turnover time of the normal epidermal cell is 28 or 29 days ... This rapid trip does not allow time for adequate maturation and keratinization of the epidermis* (K. A. Arndt and T. B. Fitzpatrick).
—**keratinize** (ker′ə tə nīz), *v.* to make or become keratinous; change into keratin: *The epidermis is made up of spongy tissue called the Malpighian layer which is covered by a thin horny or keratinized layer* (B. M. Jones).
—**keratinous** (kə rat′n əs), *adj.* of or containing keratin; horny: *keratinous tissue.*

keratinophilic (kə rat′n ō fil′ik), *adj. Biology.* growing on keratinous material; attracted to substances such as hair, feathers, etc.: *keratinophilic fungi.*

kerato-, *combining form.* **1** horny substance, as in *keratosis* = *horny growth on the skin.* **2** cornea, as in *keratochromatosis* = *discoloration of the cornea.* [from Greek *keratos* horn]

keratogenous (ker′ə toj′ə nəs), *adj.* giving rise to horny tissue: *the keratogenous zone of a hair.*

kernel, *n.* **1** *Botany.* **a** the softer part inside the hard shell of a nut or inside the stone of a fruit. **b** the body of a seed within its coating; the embryo, and, when present, the albumen or endosperm. SYN: core, nucleus. **c** a grain or seed like that of wheat or corn.

2 *Physics.* an atom exclusive of the valence electrons; the nucleus and all electrons except the valence electrons. The kernel is represented by the symbol for the element and the valence electrons are represented by dots, as in N a, . N . *Within any one period the electrons in the outer orbitals are arranged around a kernel* (Chemistry Regents Syllabus).

3 *Mathematics.* the set of all the elements that a homomorphism or other function maps into the identity element of another set.

kernite (kėr′nīt), *n. Mineralogy.* a colorless or white mineral, an important ore of boron compounds. *Formula:* $Na_2B_4O_7 \cdot 4H_2O$ [from *Kern,* county in California, where it is found]

kerogen (ker′ə jən), *n. Geology.* the solid bituminous substance found in oil shale, which yields oil when the shale is calcined. [from Greek *kēros* wax + English -*gen*]

Kerr effect (kär *or* kėr), *Physics.* **1** an electrooptical effect in which double refraction is produced in certain transparent substances to which an electric field has been applied.
2 a magnetooptical effect in which a rotation of the plane of polarization of a light beam occurs when the beam is reflected from the surface of magnetic tape or other magnetized material. [named after John *Kerr,* 1824–1907, Scottish physicist]

keto (kē′tō), *adj. Chemistry.* of or containing a ketone or ketone group. A **keto acid** is a compound containing both a ketone and a carboxyl group.

keto-, *combining form. Chemistry.* ketone or ketone bodies, as in *ketonemia = an excess of ketone bodies in the blood.* [from *keto(ne)*]

ketogenesis (kē′tə jen′ə sis), *n. Biochemistry.* the production of ketone bodies.

ketogenic (kē′tə jen′ic), *adj. Biochemistry.* producing ketones or ketone bodies.

ketone (kē′tōn), *n. Chemistry.* any of a group of organic compounds consisting of a carbonyl (-CO) radical attached to two univalent hydrocarbon radicals, or to derivatives of either of these, as acetone: *A ketone may be looked upon as a hydrocarbon derivative in which an oxygen atom has replaced two hydrogen atoms from a carbon atom not located at the end of the chain* (Offner, *Fundamentals of Chemistry*). [from German *Keton,* from French *acétone* acetone]
—**ketonic** (ki ton′ik), *adj.* of or having to do with ketones.

ketone bodies, *Biochemistry.* a group of ketones resulting from incomplete oxidation of fatty acids. Their presence in urine is a symptom of diabetes.

ketose (kē′tōs), *n. Biochemistry.* a sugar containing the bivalent –CO radical.

ketosteroid (kē′tō ster′oid), *n. Biochemistry.* any of a group of steroid hormones which originate in the adrenal glands and testes. Their presence in urine is used to determine the functioning conditions of these organs.

kettle, *n.,* or **kettle hole,** *Geography, Geology.* a depression in the surface of a ground moraine, caused by the melting of a block of subsurface ice after the moraine had formed. See the pictures at **drumlin, glacier.**

key, *n.* **1** *Biology.* a guide to the identification of a group of plants or animals, having the outstanding determining characteristics arranged in a systematic way: *An identification key is usually so constructed that contrasting characters are presented so that one of two options may be selected* (Harbaugh and Goodrich, *Fundamentals of Biology*).
2 *Botany.* = samara.
3 *Geology.* a low island or reef, especially one of the islets in the West Indies and south of Florida.

kg, *symbol.* kilogram *or* kilograms.

kgf, *symbol.* kilogram-force.

khamsin (kam sēn′), *n. Meteorology.* an oppressively hot southerly wind that blows from the Sahara toward Egypt and the Red Sea at intervals, usually in spring and early summer. [from Arabic *khamsin,* from *rih al-kahmsin* (the) wind (of) the fifty (days)]

kHz, *symbol.* kilohertz.

kidney, *n. Anatomy.* one of the pair of organs in the bodies of vertebrate animals that separate waste matter from the blood and pass it off through the bladder as urine. In human beings, the kidneys are elongated ovals about ten centimeters long, indented on one side. They are located on either side of the backbone at waist level and above. Besides excreting waste products, the kidneys control the concentration of most of the constituents of body fluids.
ASSOCIATED TERMS: Human kidneys consist of three layers: the *cortex,* the *medulla,* and the *pelvis.* In the medulla and cortex, the *renal* artery brings blood through its branches to microscopic filtration units called *nephrons.* Within each nephron water, salts, urea, amino acids, and glucose are filtered in the *renal corpuscle* from the *glomerulus* into the *Bowman's capsule.* As these materials move through the *uriniferous tubule* of each nephron, the minerals, amino acids, glucose, etc. are reabsorbed by *active transport* into the blood through capillaries associated with the tubule. The fluid remaining in the tubule is urine, which flows from the kidney through the *ureter* to the *urinary bladder.*

kieselguhr (kē′zəl gùr), *n.* = diatomite. [from German *Kieselguhr,* from *Kiesel* gravel + *Guhr* earthy sediment]

kilo-, *combining form.* a unit of 1000 or 10^3, especially in the metric system, as in *kilogram = 1000 grams, kilometer = 1000 meters. Symbol:* k [from French *kilo-,* from Greek *chilioi* a thousand]

kiloampere, *n.* a unit of electric current equal to 1000 amperes.

kilobar, *n.* a unit of pressure equal to 1000 bars or 14,500 pounds per square inch. *Symbol:* kb

kilocalorie, *n. Chemistry, Physics.* the quantity of heat needed to raise by one degree Celsius the temperature of a kilogram of water. *Symbol:* kcal Also called **Calorie.**

kilocycle, *n.* 1000 cycles per second, used formerly to express the frequency of radio waves, now expressed in kilohertz. *Symbol:* kc

---

cap, fāce, fàther; best, bē, tėrm; pin, five;
rock, gō, ôrder; oil, out; cup, pùt, rüle;
yü in use, *yu* in uric;
ng in bring; *sh* in rush; *th* in thin, ᴛʜ in then;
zh in seizure.
ə = *a* in about, *e* in taken, *i* in pencil, *o* in lemon, *u* in circus

kilodyne, *n.* a unit of force equal to 1000 dynes.

kilogauss, *n.* a unit of magnetic induction equal to 1000 gauss.

kilogram, *n.* **1** the SI or MKS unit of mass, a fundamental unit equal to 1000 grams or the mass of a liter of water at a temperature of 4°C.: *The standard for the unit of mass, the kilogram, is a cylinder of platinum-iridium alloy ... This is the only base unit still defined by an artifact* (R.D. Huntoon). *Symbol:* kg **2** = kilogram-force.

kilogram calorie, = Calorie.

kilogram-force, *n.* a unit of force equal to the weight of a one-kilogram mass measured at sea level under standard gravity. *Symbol:* kgf

kilohertz, *n.* a unit of frequency equal to 1000 hertz, used to express the frequency of radio waves. *Symbol:* kHz

kilohm (kil'ōm'), *n.* a unit of electrical resistance equal to 1000 ohms.

kilojoule, *n.* a unit of work or energy equal to 1000 joules: *The number of kilojoules which can be obtained from a sample of food is a measure of its possible value as a source of energy in the body. ... The number of kilojoules needed by ... a lumberjack doing eight hours' work per day [is] from 23,000 to 25,000 kJ ... and a child of six years about 8000 kJ* (Mackean, *Introduction to Biology*). *Symbol:* kJ or kj

kiloliter, *n.* a unit of volume equal to 1000 liters. *Symbol:* kl

kilometer, *n.* a unit of length equal to 1000 meters. *Symbol:* km

kiloparsec, *n.* a unit of distance, used to compute the distance of stars, equal to 1000 parsecs or 3262 light-years: *The gravitational force of the galaxy causes an object 10 kiloparsecs from the center of the galaxy to deviate from linear motion about 100 kilometers per year* (Science). *Symbol:* kparsec

kilorad, *n.* a unit of radiation equal to 1000 rads. *Symbol:* kr or krad

kilovolt, *n.* a unit of electromotive force equivalent to 1000 volts. *Symbol:* kv

kilowatt, *n.* a unit of electrical power equal to 1000 watts: *A common misconception is that there is something inherently electrical about a watt or a kilowatt ... but the power consumption of an incandescent lamp could equally well be expressed in horsepower, or an automobile engine rated in kilowatts* (Sears and Zemansky, *University Physics*). *Symbol:* kw

kilowatt-hour, *n.* a unit of electrical energy equal to the work done by one kilowatt acting for one hour; 3,600,000 joules. *Symbol:* kwh or kwhr

kimberlite, *n.* Geology. a mantle-derived peridotic rock found in South Africa, Zaire, Siberia, and some other areas, usually partly altered to serpentine. *Greatly weathered kimberlite, or residual clay from it, is the "blue ground" from which diamonds are mined* (Fenton, *The Rock Book*). [named after *Kimberley,* a city in South Africa]

kinematic (kin'ə mat'ik), *adj.* having to do with pure motion or with kinematics. The kinematic viscosity of water is much less than that of air. —**kinematically,** *adv.*

kinematics, *n.* **1** the branch of physics that deals with the characteristics of different kinds of pure motion, that is, without reference to mass or to the causes of the motion.
2 the kinematic features or characteristics of something: *The general kinematics of plates—their growth and consumption—requires some form of mass convection, or mass-transfer circuits, in the mantle* (John F. Dewey).
[from Greek *kinēmatos* motion, from *kinein* to move]

kinesiology (ki nē'sē ol'ə jē), *n.* the science which investigates and analyzes the anatomy, physiology, and mechanics of bodily movement, especially in humans. [from Greek *kinēsis* motion + English *-logy*]

kinesis (ki nē'sis), *n., pl.* **-neses** (-nē'sēz). *Biology.* an involuntary movement or reaction of an organism in response to a particular stimulus: *The term kinesis is used to describe undirected locomotor reactions where the speed of movement or frequency of turning depends on the intensity of stimulation.* (Storer, *General Zoology*). [from Greek *kinēsis* motion]

kinesthesia (kin'əs thē'zhə), *n. Physiology.* the sense of perception of the movement, position, balance of one's own body or body parts. Kinesthesia is mediated by the action of receptors located in the muscles, joints, and tendons. [New Latin, from Greek *kinein* to move + *aisthēsis* perception]

kinesthesis (kin'əs thē'sis), *n.* = kinesthesia.

kinesthetic (kin'əs thet'ik), *adj.* of or having to do with kinesthesia: *Humans have three orienting mechanisms: 1) vision, 2) the balancing apparatus in the inner ear, 3) the "kinesthetic" sense, which reports tension and pressure in the skin, muscles and viscera* (Time). —**kinesthetically,** *adv.*

kinetic, *adj.* of, caused by, or producing motion; having to do with kinetic energy or kinetics: *Under such circumstances, the kinetic impulse—the driving energy of the water suddenly introduced into the pipe—reaches the other end of the pipe almost instantaneously* (Saturday Review). [from Greek *kinētikos,* from *kinein* to move] —**kinetically,** *adv.*

kinetic energy, *Physics.* the energy which a body has because it is in motion. It is equal to one half the product of the mass of a body and the square of its velocity. *When a sliding or rolling body slows down, it loses kinetic energy; it does work against the force of friction. When a body falls toward the ground it gains kinetic energy. The work is done on the body by the force of gravity.* (Pierce, *Electrons, Waves and Messages*). See the picture at **energy.**

kinetic molecular theory, = kinetic theory of gases. *Abbreviation:* KMT

kinetic potential, = Lagrangian.

kinetics, *n.* **1** = dynamics (def. 1).
2 a branch of chemistry dealing with the rates of

change of concentrations in chemical reactions or other processes under conditions of nonequilibrium: *Chemical kinetics ... treats of the velocity of chemical reactions* (C.L. Speyers).

3 the kinetic aspects or characteristics of a physical or chemical change: *the kinetics of osmosis, the kinetics of evaporation.*

kinetic theory of gases, *Physics.* the theory that molecules of a gas are in a state of rapid motion, constantly colliding with the walls of any containing vessel and with one another, thereby causing changes in their velocity and direction. Also called **kinetic molecular theory.**

kinetic theory of heat, *Physics.* the theory that the temperature of a body is equivalent to the average kinetic energy of its constituent molecules, and that additional heat results in an increase in the energy.

kinetic theory of matter, *Physics.* the theory that the constituent particles of matter are in energetic motion.

kinetin (kī′nə tən), *n. Biochemistry.* a cytokinin that stimulates cell division in plants: *Cultured plant cells normally need the hormone kinetin in order to divide and grow* (W. G. Rosen). *Formula:* $C_{10}H_9N_5O$

kinetochore (ki net′ə kôr), *n.* = centromere. [from Greek *kinētos* movable + *choros* place]

kinetoplast (ki net′ə plast), *n. Biology.* a structure lying at the base of a flagellum in certain protozoans, especially trypanosomes: *The third type of cytoplasmic structure we have to consider as potentially genetically active are the kinetoplasts ...* (R.J. Cole). [from Greek *kinētos* movable + English *-plast*]

kinetosome (ki net′ə sōm), *n. Biology.* a basal granule in a cell, associated with the formation of cilia, flagella, and certain other motile structures: *The kinetosome, a barely visible granule about a thousandth of a millimeter in diameter, is the structure from which the vibrating cilia grow in cells that bear these appendages* (T.M. Sonneborn). Also called **basal body.** [from Greek *kinētos* movable + English *-some*]

kingdom, *n. Biology.* one of the primary divisions of living things. One widely used system of classification recognizes five kingdoms: monerans, protists, fungi, plants, and animals.
ASSOCIATED TERMS: The usual taxonomic classifications, from most general to most specific, are as follows: *kingdom, phylum, class, order, family, genus, species,* and *subspecies* or *variety.* The name of any group from a kingdom down to a genus is a single capitalized word: kingdom Animalia, phylum Chordata, subphylum Vertebrata, class Mammalia, order Primates, suborder Anthropoidea, family Hominidae, genus *Homo.* The name of a species is usually two words or *binomial,* consisting of the genus name (capitalized) and the species (noncapitalized): species *Homo sapiens.* For a subspecies or variety, a third italicized, noncapitalized word is added: subspecies *Homo sapiens sapiens.* See also **classification.**

kinin (kī′nən), *n. Biochemistry.* **1** any of a group of polypeptides, such as bradykinin, that form in tissues and typically cause dilation of blood vessels and contraction of smooth muscles: *Kinins ... are released as local hormones, where needed* (H.O.J. Collier). **2** = cytokinin. [from Greek *kinein* to move + English *-in*]

kininogen (kī nin′ə jən), *n. Biochemistry.* an inactive precursor of a kinin.

Kirchhoff's law (kėrk′hof), *Physics.* **1** the law that at a given temperature the ratio of the absorptive power to the emissive power for a given wavelength is the same for all bodies and is equal to the emissive power of a blackbody at that temperature: *By Kirchhoff's Law, a good absorber is a good emitter and since a black hole is a perfect absorber it ought to radiate "blackbody" radiation. This is completely in opposition to the classical theory of black holes which allows them to emit nothing* (G. Gibbons).

2 either of two fundamental laws in electricity from which the relationships among current, voltage, and resistance in direct current circuits may be derived: **a** Also called **Kirchhoff's first law.** the law that the algebraic sum of the currents entering any circuit junction is equal to zero. **b** Also called **Kirchhoff's second law.** the law that the algebraic sum of all the potential drops and applied voltages around a complete circuit is equal to zero.
[named after Gustav R. *Kirchhoff,* 1824–1887, German physicist who enunciated this law]

kJ or **kj,** *symbol.* kilojoule.

kl, *symbol.* kiloliter *or* kiloliters.

Klein bottle, *Geometry.* a continuous, one-sided surface formed by bending the small end of a tapering tube, passing it through one side, and then joining its flared opening to the opening of the wide end: *The Klein bottle has no edges, no inside or outside. What seems to be its inside is continuous with its outside, like the two apparent "sides" of a Möbius surface* (M. Gardner). Compare **Möbius strip.** [named after Felix *Klein,* 1849–1925, German mathematician]

km, *symbol.* kilometer *or* kilometers.

K-meson, *n.* = kaon.

km.p.s. or **kmps,** *abbrev.* kilometers per second.

KMT, *abbrev.* kinetic molecular theory.

knee, *n.* **1** *Anatomy.* **a** the joint between the thigh and the lower leg, formed by the joining of the thighbone and shinbone. It is covered in front by the patella, or kneecap. **b** any joint in a four-footed animal corresponding to the human knee or elbow, as in the hind leg of a horse.
2 *Botany.* a knob on the root of a tree that grows in a swamp or wet soil: *A new theory was formulated on the formation of cypress "knees"—that they result from a response of the cambium (growth layer) of a root growing in water or poorly aerated soil to chance exposure to air in spring* (H.W. Rickett).
3 a curve or angle that joins two relatively straight parts of a characteristic curve in a graph: *Some of the cosmic ray studies in the polar regions were: ... the alti-*

---

cap, fāce, fäther; best, bē, tėrm; pin, fīve;
rock, gō, ôrder; oil, out; cup, pùt, rüle,
yü in use, yù in uric;
ng in bring; sh in rush; th in thin; ŦH in then;
zh in seizure.
ə = a in about, e in taken, i in pencil, o in
lemon, u in circus

kneecap

tude of the *"knee" in the cosmic ray intensity curve and its variation during the solar cycle* (J. C. Reed).

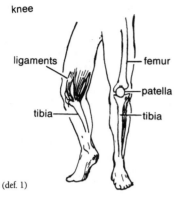

knee

ligaments

femur

patella

tibia

tibia

(def. 1)

kneecap, *n. Anatomy.* the flat, movable bone at the front of the knee; the patella. See the pictures at **knee, leg.**

knot, *n.* **1** *Botany.* **a** the hard mass of wood formed where a branch grows out from a tree, which shows as a roundish, cross-grained piece in lumber or a board: *The base of every branch becomes more deeply embedded as, year after year, new layers of wood are formed. It is these embedded branches that take the form of knots in logs and in sawn lumber* (Emerson, *Basic Botany*). **b** a joint where leaves grow out on the stem of a plant. **c** a node on a stem, such as one of the joints in grasses.
2a a unit of speed equal to one nautical mile per hour. **b** = nautical mile.

known quantity, *Mathematics.* an algebraic quantity of a given value, often designated by one of the first letters of the alphabet, as *a, b, c,* etc. Compare **unknown quantity.**

Koch's postulates (kohs), *Microbiology.* a set of conditions required to establish that a particular microorganism causes a disease: (1) the microorganism must be present in every case of the disease; (2) it must be capable of isolation and cultivation in pure culture; (3) it must produce the disease when inoculated in pure culture into susceptible animals; (4) it must be recovered from the infected animal and regrown in pure culture. [named after Robert *Koch,* 1843–1910, German physician who formulated the postulates]

kolm (kōlm), *n. Geology.* a carbonaceous material occurring in a certain shale found in Sweden and Norway. It is rich in uranium and radium. [from Swedish]

kparsec, *symbol.* kiloparsec.

kr, *symbol.* kilorad.

Kr, *symbol.* krypton.

krad, *symbol.* kilorad.

Krebs cycle, *Biochemistry.* the cycle of intracellular chemical reactions by means of which organisms convert food chemicals into physical energy. The cycle consists of a series of reactions, catalyzed by enzymes, by which acetic acid formed in the metabolism of fats, proteins, and carbohydrates is oxidized to carbon dioxide and water with release of energy. *Because this is a repeating process and because organic acids are involved, it has been called the organic acid cycle of respiration ... more frequently the Krebs cycle ... such a cycle occupies a central position in aerobic respiration* (Weier, *Botany*). Also called **citric acid cycle, tricarboxylic acid cycle.** [named after Hans A. *Krebs,* 1900–1981, German-born English biochemist who discovered the cycle]

Kronecker delta (krō'nek ər), *Mathematics.* a function of two variables that is equal to 1 when the variables have the same value, and equal to 0 when the variables have different values. [named after Leopold *Kronecker,* 1823–1891, German mathematician]

krypton (krip'ton), *n. Chemistry.* a colorless, gaseous element that is largely inert, but may combine with fluorine under certain conditions. It is chiefly used in fluorescent lamps and lasers. *Krypton and Xenon ... are present in the atmosphere in extremely small proportions* (Jones, *Inorganic Chemistry*). *Symbol:* Kr; *atomic number* 36; *atomic weight* 83.80; *melting point* $-156.6°C$; *boiling point* $-152.9°C$. [from Greek *krypton,* neuter of *kryptos* hidden]

K-shell, *n. Physics.* the lowest energy level, occupied by the orbits of electrons around the nucleus of an atom: *an electron in the filled, innermost shell (the K shell) of an atom* (G. Feinberg and M. Goldhaber). Compare **L-shell, M-shell.** See also **K electron.**

kurchatovium (kėr'chə tō'vē əm), *n.* = rutherfordium. [named after Igor *Kurchatov,* 1903–1960, Soviet nuclear physicist] ▶ See **element.**

Kuroshio Current (kü'rō shē'ō), = Japan Current. [from Japanese *kuroshio,* from *kuro* black + *shio* tide]

kurtosis (kėr tō'sis), *n. Statistics.* the relative degree of flatness or peakedness in the region about the mode of a curve describing a frequency distribution: *The kurtosis is useful in determining if a frequency distribution differs from the normal error curve* (W. L. Gore). [from Greek *kyrtōsis* a bulging, convexity, from *kyrtos* bulge, swelling]

kv, *symbol.* kilovolt *or* kilovolts.

kw, *symbol.* kilowatt *or* kilowatts.

kwh or **kwhr,** *symbol.* kilowatt-hour *or* kilowatt-hours.

kyanite (kī'ə nīt), *n.* a silicate of aluminum, usually occurring in blue, blade-shaped crystals. It is trimorphous with andalusite and sillimanite. *Formula:* Al_2SiO_5 Also spelled **cyanite.**

L

l or l., *abbrev. or symbol.* 1 length. 2 line. 3 liter *or* liters. 4 locus.

l-, *combining form. Chemistry.* 1 levorotatory; levo-: *l-glucose = levoglucose.* ▶ The minus sign (−) is now generally preferred to *l*-.
2 L-, (of organic compounds) having a configuration about a particular asymmetric carbon atom analogous to the configuration of levorotatory glyceraldehyde: *an L-amino acid.* ▶ To indicate the direction of optical rotation in addition to configuration, + or − (for dextrorotation and levorotation respectively) is added in parentheses, as L(−). Compare **d-**

L or L., *abbrev. or symbol.* 1 lake. 2 length. 3 *Botany.* Linnaeus. 4 longitude. 5 *Astronomy.* luminosity (of stellar bodies).

La, *symbol.* lanthanum.

label, *Chemistry.* —*v.* to infuse or treat (a substance) with a radioactive isotope or fluorescent dye so that its course or activity can be traced through a series of reactions, usually in a living organism: *When epinephrine labeled with carbon-14 ... was infused into ... subjects, essentially all of the radioactivity was recovered in the urine* (Science). SYN: tag.
—*n.* a radioactive isotope or fluorescent dye added to a compound or other substance to trace its course or activity: *The strategy is to label biologically important molecules with a radioactive atom that emits high-energy electromagnetic radiation. The labeled molecule is administered to a person, and external detectors record the location of the label over time* (Science News). SYN: tracer, tag.

labellate (lə bel′āt), *adj. Botany, Zoology.* having a labellum or liplike part. SYN: labiate.

labellum (lə bel′əm), *n., pl.* **-bella** (-bel′ə). 1 *Botany.* the middle petal of an orchid, usually different in shape and color from the other two and suggestive of a lip.
2 *Zoology.* a liplike part, as at the tip of the proboscis of various insects, used for lapping up liquids. [from Latin *labellum,* diminutive of *labium* lip]

labia (lā′bē ə), *n.* plural of **labium.**

labial (lā′bē əl), *adj. Biology.* of the lips or labia: *vibrations of the labial palps of hawk moths.*
—**labially,** *adv.*

labia majora (mə jôr′ə), *Anatomy.* the outer folds at the opening of the vulva. [from New Latin *labia majora* (literally) greater lips]

labia minora (mi nôr′ə), *Anatomy.* the inner folds at the opening of the vulva. [from New Latin *labia minora* (literally) lesser lips]

labiate (lā′bē it *or* lā′bē āt), *adj.* 1 *Botany.* **a** having the corolla or calix divided into two parts suggesting lips. **b** belonging to or having to do with the mint family. 2 *Zoology.* having lips or liplike parts.
—*n.* a labiate plant.

labile (lā′bəl), *adj.* having the tendency to undergo displacement in position or change in nature, form, or composition; fluctuating widely: *a labile chemical solution.* SYN: unstable. [from Latin *labilis,* from *labi* to slip, lapse]
—**lability** (lə bil′ə tē), *n.* labile quality, form, or character.
—**labilization** (lā′bə lə zā′shən), *n.* the process of making or condition of being labile.
—**labilize** (lā′bə līz), *v.* to make labile: *to labilize a chemical compound.*

labio-, *combining form.* lip or lips; labial, as in *labioglossolaryngeal = of the lips, tongue, and larynx.* [from Latin *labium* lip]

labium (lā′bē əm), *n., pl.* **-bia** (-bē ə). 1 *Botany.* a liplike portion of the corolla of certain flowers.
2 *Zoology.* **a** the organ that constitutes the lower lip of certain invertebrates, especially insects. **b** the inner margin of the opening of a gastropod's shell. Contrasted with **labrum.**
3 *Anatomy.* any of the folds at the opening of the vulva. See **labia majora, labia minora.**
[from Latin *labium* lip]

laboratory animal, any animal commonly used for experiments in a laboratory, such as guinea pigs and mice: *A laboratory animal, such as a rabbit ..., reacts to any protein from a source outside its body in the same way it does to invading disease germs* (F.W. Emerson).

Labrador Current, *Oceanography.* a current of cold water that rises in the Arctic Ocean and flows along the coast of Labrador past Newfoundland, where it meets the Gulf Stream.

labradorite (lab′rə dô rīt′), *n. Mineralogy.* a triclinic feldspar that is an essential constituent of basalts and gabbros: *Minerals in the plagioclase group contain sodium and calcium. Labradorite is a kind of plagioclase that sometimes shows a beautiful display of colors when light strikes it* (E.E. Wahlstrom). [from the Labrador peninsula, where it is found + *-ite*]

labrum (lā′brəm), *n. Zoology.* 1 the upper lip of insects and certain other arthropods. 2 the outer margin of the opening of a gastropod shell. Contrasted with **labium.** See the picture at **mouth.** [from Latin *labrum* lip, related to *labium* labium]

labyrinth (lab′ə rinth′), *n. Anatomy.* 1 any part of the body consisting of intricate passageways.

---

cap, fāce, fäther; best, bē, tèrm; pin, fīve;
rock, gō, ôrder; oil, out; cup, pùt, rüle,
yü in use, *yu* in uric;
ng in bring; *sh* in rush; *th* in thin, ᴛʜ in then;
zh in seizure.

ə = *a* in about, *e* in taken, *i* in pencil, *o* in lemon, *u* in circus

labyrinthodont

2 the inner ear, especially its interconnecting canals and cavities. The **bony labyrinth** is the system of canals and cavities within the bones of the inner ear which contains the **membranous labyrinth,** or system of canals suspended in the labyrinthine fluid. *The labyrinth contains a fluid in which changes of pressure ... cause muscular adjustments which maintain equilibrium* (Harbaugh and Goodrich, *Fundamentals of Biology*). [from Greek *labyrinthos* maze]
—**labyrinthine** (lab′ə rin′thin), *adj.* of or belonging to a labyrinth. The **labyrinthine fluid** is the fluid separating the bony and the membranous labyrinth of the inner ear.

labyrinthodont (lab′ə rin′thə dont), *Zoology.* —*adj.* **1** having teeth with a labyrinthlike internal structure.
2 of or having to do with a group of extinct amphibians resembling crocodiles and having a labyrinthlike tooth structure.
—*n.* a labyrinthodont amphibian: *Surviving modern amphibians (frogs, toads, newts, salamanders, and their relatives) [are] ... specialized descendants of early amphibians called labyrinthodonts* (Eicher and McAlester, *History of the Earth*). [from Greek *labyrinthos* labyrinth + *odontos* tooth]

laccate (lak′āt), *adj. Biology.* having the appearance of being lacquered: *laccate leaves.*

laccolith (lak′ə lith), *n. Geology.* an intrusive body of igneous rock that has spread on rising from below causing the overlying strata to bulge upward in a domelike formation: *Laccoliths are occasionally circular, or more often elliptical, in ground plan* (Tyrrell, *Principles of Petrology*). Compare **batholith, lopolith.** See the picture at **volcano.** [from Greek *lakkos* pond + English *-lith*]

lachrymal (lak′rə məl), *adj. Anatomy, Physiology.* of, having to do with, or producing tears. The **lachrymal glands** are two glands, one above each eye, that produce tears. *Mostly a salt solution, lachrymal fluid also contains substances that fight bacteria* (G.W. Beadle). Also spelled **lacrymal.** [from Medieval Latin *lachrimalis,* from Latin *lacrima* tear]

lacinia (lə sin′ē ə), *n.* **1** *Botany.* a slash, as in a leaf or petal. **2** *Zoology.* the apex of an insect's maxilla. [from New Latin, from Latin *lacinia* small piece]
—**laciniate** (lə sin′ē it *or* lə sin′ē āt), *adj.* **1** *Botany.* cut into deep and narrow irregular lobes; slashed; jagged. **2** *Anatomy.* shaped or formed like a fringe, as a ligament.
—**laciniation** (lə sin′ē ā′shən), *n.* a lobe or projecting segment; laciniate formation.

lacmoid (lak′moid), *n. Chemistry.* a dark-violet, crystalline dye used as an indicator in titration. [from *lacmus* litmus, from Dutch *lakmoes*]

lacrymal, *adj.* = lachrymal.

lact-, *combining form.* the form of **lacto-** before vowels, as in *lactate, lactide.*

lactalbumin, *n. Biochemistry.* an albumin found in milk, resembling serum albumin and having a molecular weight of about 17,400: *The typical milk proteins, casein, lactalbumin and lacto-globulin, are believed to be mainly products of the serum globulin of the blood* (F. H. Malpress).

lactase (lak′tās), *n. Biochemistry.* an enzyme present in certain yeasts and in the body, capable of decomposing lactose into glucose and galactose.

lactate, *n. Chemistry.* a salt or ester of lactic acid. —*v. Physiology.* to secrete milk: *the milk yield of lactating cows.*

lactate dehydrogenase, = lactic dehydrogenase.

lactation, *n. Physiology.* **1** the secretion or formation of milk by the mammary glands: *The pituitary gland ... secretes at least seven hormones into the bloodstream and so influences many bodily processes including growth, metabolism, water balance, reproduction, lactation* (Science News).
2 the period during which the mammary glands secrete milk.
—**lactational,** *adj.* of or having to do with lactation: *lactational fluids.*
—**lactationally,** *adv.*

lacteal (lak′tē əl), *adj. Physiology.* **1** of or resembling milk; milky: *a lacteal secretion.* **2** carrying chyle, a milky liquid formed from digested food: *the lacteal vessels.*
—*n. Anatomy.* any of the tiny lymphatic vessels that carry chyle from the small intestine to be mixed with the blood.

lactescence (lak tes′ns), *n.* **1** a milky appearance; milkiness. **2** *Botany.* an abundant flow of sap from a plant when wounded. The sap is commonly white, but sometimes red or yellow.
—**lactescent,** *adj.* **1** becoming milky; having a milky appearance. **2** producing or secreting milk. **3** (of plants and insects) producing a milky fluid.

lactic acid, *Biochemistry.* a colorless, odorless acid formed by the action of lactobacilli in sour milk, the fermentation of vegetable juices, etc., and produced by muscle tissue during exercise: *Lactic acid production is associated with fatigue* (Biology Regents Syllabus). *Formula:* $CH_3CHOHCOOH$

lactic dehydrogenase or **lactic acid dehydrogenase,** *Biochemistry.* an enzyme produced in animal tissue that causes the dehydrogenation of lactic acid to pyruvic acid in glycolysis: *Most cells have an enzyme, called lactic dehydrogenase ... which interconverts lactic and pyruvic acids, the pyruvic acid coming from the breakdown of sugars* (New Scientist). *Abbreviation:* LDH Also called **lactate dehydrogenase.**

lactide (lak′tīd), *n. Chemistry.* a compound formed by heating lactic acid, and regarded as an anhydride of that acid.

lactiferous (lak tif′ər əs), *adj.* **1** *Zoology.* secreting or conveying milk or a milky fluid. **2** *Botany.* yielding a milky juice. [from Latin *lactifer* producing milk or juice, from *lac, lactis* milk + *ferre* to bear]

lacto-, *combining form.* **1** milk, as in *lactoglobulin.* **2** lactic acid, as in *lactobacillus.* Also spelled **lact-** before a vowel. [from Latin *lac, lactis* milk]

lactobacillus (lak′tō bə sil′əs), *n., pl.* **-cilli** (-sil′ī). any member of a genus (*Lactobacillus*) of aerobic bacteria that produces lactic acid with the fermentation of sugar.

lactogen, *n. Biochemistry.* a hormone that stimulates the secretion of milk, such as prolactin: *A substance resembling a growth hormone—possibly placental lactogen—was found in maternal plasma* (H. F. Root).

lactogenic, *adj. Biochemistry.* that stimulates the secretion and flow of milk.

lactoglobulin or **lactoprotein,** *n. Biochemistry.* a protein found in milk whey, having a molecular weight of about 42,000: *About four fifths of this [milk] protein is casein, and the rest consists mostly of lactalbumin and lactoglobulin* (M. Fairman).

lactone (lak'tōn), *n. Chemistry.* any of a group of cyclic anhydrides produced by the loss of a molecule of water from the hydroxyl (-OH) and carboxyl (-COOH) radicals of hydroxy acids.

lactose (lak'tōs), *n. Biochemistry.* a crystalline sugar, a disaccharide present in milk. On hydrolysis it is converted to glucose and galactose. *Milk souring usually occurs because 'lactic acid bacteria' ... using the milk sugar as a source of energy, reproduce themselves rapidly, spreading through the bulk of the milk. They convert the lactose to lactic acid, which is responsible for the sour flavour and the clotting* (J.A. Barnett). *Formula:* $C_{12}H_{22}O_{11}$ Also called **milk sugar.**

lacuna (lə kyü'nə), *n., pl.* **-nas, -nae** (-nē). *Anatomy.* a space, cavity, depression, etc., in the anatomical structure of an animal or plant, such as a cavity in bones or tissues, or a depression in the surface of lichens: *Individual bone cells occupy small spaces, or lacunae, left between the lamellae as the mineral is deposited* (Storer, *General Zoology*). [from Latin *lacuna* hole, from *lacus* cistern, pond, lake]
—**lacunal** (lə kyü'nl), *adj.* of, having to do with, or containing a lacuna or lacunas.

lacustrine (lə kus'trən), *adj.* **1** *Geology.* **a** of, having to do with, or originating in lakes. **b** of or having to do with strata that originated by deposition at the bottom of a lake: *Another type of plain is the lake or lacustrine plain, one of the flattest land forms extant. The low relief is consequent upon ... streams ... carrying sediment which they deposited on the lake bottoms and which was redistributed smoothly by waves and currents* (White and Renner, *Human Geography*).
2 *Biology.* living or growing in lakes: *lacustrine water fleas.*
[ultimately from Latin *lacus* pond, lake]

ladder polymer, *Chemistry.* a complex polymer made up of double-stranded chains of molecules connected by bonds at regular intervals like the rungs connecting the two sides of a ladder. DNA is a ladder polymer. *The use of conventional straight-chain polymers seems to be restricted by an upper temperature limit of about 550°C, but the ladder polymers (so-called because of their integral cross-linked structure) offer more exciting possibilities* (New Scientist).

lag, *Physics.* —*v.* to fall behind (a voltage, current, etc.) in speed of response to alternations: *The current maxima, minima, etc., occur at later times than do those of the voltage. The current is said to lag the voltage* (Sears and Zemansky, *University Physics*).
—*n.* the retardation, or amount of retardation, in any current or movement.

lag fault, *Geology.* a type of overthrust fault in which the upper part of the stratigraphic section has moved less than the lower part.

lagomorph (lag'ə môrf), *n.* any member of an order (Lagomorpha) of mammals consisting of the rabbits and hares and the pikas. Lagomorphs are similar to rodents, but have short tails and two pairs of upper inci-

sor teeth, of which the second pair, just behind the first, is smaller. [from Greek *lagōs* hare + *morphē* form, shape]

lagoon, *n. Geography.* **1** a pond or small lake connected with a larger body of water.
2 shallow water separated from the sea by low sandbanks: *The lagoon behind may be brackish or fresh and may be slowly filled by silt from the land or by sand blown inland from the beach* (Gilluly, *Principles of Geology*).
3 the body of water enclosed by an atoll: *In the Pacific and Indian Oceans there are many coral atolls that are ring-like islands, enclosing a lagoon* (T.F. Gaskell and M.N. Hill). [from Italian *laguna*, from Latin *lacuna* pond]
—**lagoonal,** *adj.* of or having to do with a lagoon: *lagoonal areas, lagoonal deposits.*

Lagrangian (lə grän'jē ən), *n.,* or **Lagrangian function,** *Physics.* a function equal to the difference between the total kinetic energy and the total potential energy of a dynamic system: *Field theorists like to start work by writing down a mathematical summary statement called a Lagrangian, which expresses the energies of a general group of particles under the influence of whatever force is being considered* (D. E. Thomsen). Also called **kinetic potential.** Compare **Hamiltonian.** [named after Joseph Louis *Lagrange*, 1736–1813, French mathematician and astronomer]

Lagrangian point, *Astronomy.* a location between heavenly bodies where centrifugal and gravitational forces neutralize each other so that an object in that location remains in stable equilibrium with respect to the other bodies: *the Lagrangian points: positions ... stable for secondary bodies moving in the same plane as the primary planet if both bodies were revolving around the sun in circular orbits undisturbed by other planets* (Scientific American). Also called **libration point.**

lahar (lä'här), *n. Geology.* a flowing mass of volcanic fragments on the flanks of a volcano. [from Javanese *lahar*]

lake, *n. Geography.* a large body of water entirely or nearly surrounded by land: *Lakes are particularly numerous in plains of severe ice scour. Many of them lie in rock basins eroded by the ice with its characteristic disregard for uniform gradient.* (Finch and Trewartha, *Elements of Geography*). Compare **lacustrine.** [from Old French *lac*, from Latin *lacus*]

Lamarckian, *Biology.* —*adj.* of the French biologist Jean de Lamarck; of Lamarckism: *... the Lamarckian postulate that characters acquired by parents during their own lives can be passed on to their offspring* (Science).
—*n.* a supporter of Lamarckism: *The controversy between Darwinians and Lamarckians has raged for ... a century* (National Observer).

---

cap, fāce, fäther; best, bē, tèrm; pin, fīve;
rock, gō, ôrder; oil, out; cup, pùt, rüle,
yü in use, *yu* in uric;
ng in bring; *sh* in rush; *th* in thin; *ŦH* in then;
zh in seizure.
ə = *a* in about, *e* in taken, *i* in pencil, *o* in lemon, *u* in circus

Lamarckism or **Lamarckianism,** *n. Biology.* the theory of organic evolution proposed by Jean de Lamarck (1744–1829): *They have reopened the debate between Darwinism and Lamarckism—which holds that acquired habits can be incorporated into the genetic program and thus reappear in later generations* (N.Y. Times).

lambda (lam′də), *n.* **1** *Nuclear Physics.* a heavy, unstable elementary particle, a form of hyperon, having a neutral charge, a mass 2183 times that of the electron, a spin of 1/2, and decaying very rapidly, usually to a nucleon and a pion. Also called **lambda particle** or **lambda hyperon.**
2 *Microbiology.* a virus which infects the bacterium *Escherichia coli,* important for its ability to incorporate the bacterium's genes into its own system and transfer them to cells of other organism: *In particular, lambda has played a crucial role in our understanding of how genes are controlled at the molecular level, and of the events that lead up to copying out of these genes (in the form of messenger RNA) in the process of transcription* (New Scientist). Also called **lambda virus.** [from the Greek letter *lambda* (λ), perhaps so called because of a resemblance in shape]

lambda particle or **lambda hyperon,** = lambda (def. 1).

lambda point, *Physics.* the temperature, about 2.2 kelvins, below which liquid helium becomes a superfluid: *Upon being cooled to 2.2 degrees Kelvin, helium, although remaining liquid, undergoes as an abrupt jump in the liquid's specific heat ... Because the graph showing this abrupt change has the shape of the inverted Greek letter lambda, the point of transition (2.2 degrees) became known as the "lambda point."* (Scientific American).

lambda virus, = lambda (def. 2).

lambdoid (lam′doid), *adj. Anatomy.* of or designating the suture between the occipital bone and the parietal bones of the skull. [from Greek *lambdoeides,* from *lambda* lambda + *eidos* shape]

lambert (lam′bərt), *n. Optics.* a unit of brightness, equivalent to the brightness of a perfectly diffusing surface that emits or reflects one lumen per square centimeter. [named after Johann H. *Lambert,* 1728–1777, German physicist]

lamella (lə mel′ə), *n., pl.* **-mellae** (-mel′ē), **-mellas. 1** a thin plate or layer, especially of flesh or bone.
2 *Zoology.* one of the thin scales or plates composing some shells, as in bivalve mollusks.
3 *Botany.* **a** one of the erect scales appended to the corolla of some flowers. **b** an erect sheet of cells on the midrib of a leaf in mosses.
4 *Biology.* one of the thin radiating plates or gills forming the spore-bearing layer of a mushroom.
[from Latin *lamella,* diminutive of *lamina* thin plate]
—**lamellar** or **lamellate,** *adj.* **1** consisting of or arranged in lamellae. **2** = lamelliform.
—**lamellation,** *n.* lamellar arrangement or structure.

lamellibranch (lə mel′ə brangk), *Zoology.* —*n.* any member of a class (Lamellibranchia) of mollusks having thin, platelike gills and a headless body enclosed in a shell whose two parts are connected by a hinge. Lamellibranchs include oysters, clams, and scallops,

have a wedge-shaped foot, and are bilaterally symmetrical within a mantle secreted by the shell.
—*adj.* of or belonging to this class of mollusks.
[from Latin *lamella* lamella + Greek *branchia* gills, branchia]

lamelliform, *adj.* having the shape or structure of a lamella or thin plate.

lamina (lam′ə nə), *n., pl.* **-nae** (-nē′), **-nas. 1** *Anatomy.* a thin layer of bone, membrane, or the like.
2 *Botany.* the flat, wide part of a leaf; blade.
3 *Geology.* a thin layer of sediment: *A type of record which occurs widely throughout the world is that of varves, the laminae in certain clays and sands* (G.H. Dury).
[from Latin *lamina* thin piece of metal or wood]
—**laminal,** *adj.* = laminar.

lamina propria (lam′ə nə prō′prē ə), *pl.* **laminae propriae** (lam′ə nē prō′prē ē). *Anatomy.* the thin layer beneath the epithelium of an organ: *The minute structure of the olfactory mucosa in birds ... is basically made up of three histological portions: the epithelium, the basal lamina, and the lamina propria* (P.P.C. Graziadei). Also called **basement membrane.** [from New Latin *lamina propria* (literally) lamina proper]

laminar, *adj.* **1** consisting of or arranged in laminae.
2 *Physics.* smooth; streamlined; not turbulent. **Laminar flow** is a steady flow of a fluid near a solid body, such as the flow of air over or about an airfoil: *In general, laminar flow occurs at low velocities, between close boundaries, when the fluid is very viscous, and the fluid is of low density* (D.A. Gilbrech). See the picture at **fluid flow.**

laminated, *adj.* **1** consisting of or arranged in laminae.
2 formed in a succession of layers of material.

lamination, *n.* **1** a laminated structure; arrangement in thin layers. **2** a thin layer.

lampbrush chromosome, *Biology.* a type of large chromosome found especially in the immature eggs of amphibians, consisting of two long strands that form many brushlike loops along the main axis of the chromosome.

lanate (lā′nāt), *adj. Botany, Zoology.* having a woolly covering or surface. [from Latin *lanatus,* from *lana* wool]

lanceolate (lan′sē ə lāt), *adj. Botany, Zoology.* shaped like the head of a lance; tapering from a rounded base toward the apex: *a lanceolate leaf.* [from Latin *lanceola,* diminutive of *lancea* lance]

land, *n. Geography, Geology.* **1** the solid part of the earth's surface: *dry land.* **2** ground or soil: *Only 10 per cent of Africa's land gets enough rain to support food crops* (Boyce Rensberger).

land breeze, *Meteorology.* a breeze blowing from the land toward the sea.

land bridge, *Geography.* a neck of land connecting two land masses: *Nearly all the authorities agree that a land bridge existed at various times across Bering Strait* (L. C. Eiseley).

landform, *n. Geography.* the physical characteristics of land; irregularities of land: *The principal groups of landforms ... (a) plains, (b) plateaus, (c) hill lands, and (d) mountains* (Finch and Trewartha, *Elements of Geography*).

landlocked, *adj.* **1** *Geography.* shut in, or nearly shut in, by land: *the landlocked Great Lakes.*
2 *Ecology.* living in waters shut off from the sea: *Landlocked salmon must spend their lives in fresh water instead of making the migration to salt water.*

landmass, *n. Geography.* a large, unbroken area of land: *To explain the presence of a freshwater amphibian in Antarctica, it must be presumed that that continent was connected to some other continental landmass in Triassic times* (E. H. Colbert).

landslide, *n. Geology.* **1** a sliding down of a mass of soil or rock on a steep slope. **2** the mass that slides down.

landspout, *n. Meteorology.* a funnel-shaped cloud resembling a waterspout but occurring on land. A landspout may be produced by certain severe whirling storms of small extent.

land wind, *Meteorology.* a wind blowing from the land toward the sea.

langbeinite (lang′bī nīt), *n. Mineralogy.* an evaporate mineral, a sulfate of potassium and magnesium, that occurs in potassium salt deposits and is mined as a source of potassium sulfate fertilizer. *Formula:* $K_2Mg_2(SO_4)_3$ [from A. *Langbein,* 19th-century German chemist + *-ite*]

langley, *n., pl.* **langleys.** *Physics.* a unit of solar radiation equal to 1 calorie per square centimeter. [named after Samuel P. *Langley,* 1834–1906, American astronomer and physicist]

lanthanide (lan′thə nīd), *n. Chemistry.* any of the rare-earth elements. [from *lanthanum* (first of the rare-earth elements) + *-ide*]

lanthanide series, *Chemistry.* the rare-earth elements.

lanthanum (lan′thə nəm), *n. Chemistry.* a soft, malleable, ductile, rare-earth metallic element which occurs in various minerals. It is used in making alloys. *Symbol:* La; *atomic number* 57; *atomic weight* 138.91; *melting point* 920°C; *boiling point* 3469°C; *oxidation state* 3. [from New Latin, from Greek *lanthanein* lie hidden]

lanuginous (lə nü′jə nəs), *adj. Botany, Zoology.* covered with lanugo or soft, downy hairs.

lanugo (lə nü′gō), *n. Botany, Zoology.* a growth of fine soft hair, as on the surface of a leaf or fruit, on the body of an insect, or on the skin of a newborn child. [from Latin *lanugo* down]

lapidicolous (lap′ə dik′ə ləs), *adj. Ecology.* (of certain beetles) living under or among stones. [from Latin *lapidis* stone + *-cola* inhabitant, related to *colere* inhabit]

lapilli (lə pil′ī), *n.pl., sing.* **-pillus** (-pil′əs). *Geology.* small pyroclastic rock particles ejected from volcanoes. [from Latin, diminutive of *lapis, lapidis* stone]

lapis lazuli (lap′is laz′yə lī *or* lap′is laz′yə lē). *Geology.* a deep blue, opaque, semi-precious rock composed mainly of lazurite. [from Medieval Latin, from Latin *lapis* stone + Medieval Latin *lazulum* lapis lazuli, from Arabic *lāzuward*]

lappet (lap′it), *n. Zoology.* a loose fold of flesh or membrane, such as the lobe of the ear or a bird's wattle.

lapse, *n. Meteorology.* a decrease of temperature of the atmosphere with increase of altitude. [from Latin *lapsus* a fall, from *labi* to slip, fall]

lapse rate, *Meteorology.* the rate of decrease of an atmospheric element, such as temperature, with increase in altitude: *The average conditions in the troposphere are specified by a lapse rate of 3°F. per 1,000 feet* (D. Brunt).

large calorie, *Chemistry, Physics.* = Calorie.

large intestine, *Anatomy.* the wide, lower part of the intestines, between the small intestine and the anus, where water is absorbed and wastes are eliminated. The human large intestine is about five feet long and consists of the cecum, colon, and rectum. See the picture at **alimentary.**

larva (lär′və), *n., pl.* **-vae** (-vē *or* -vī). *Zoology.* **1** the wormlike early form of an insect that undergoes metamorphosis, from the time it hatches from the egg until it becomes a pupa. A caterpillar is the larva of a butterfly or moth. A grub is the larva of a beetle. Maggots are the larvae of flies. *Wireworms (the larvae of click beetles) and leatherjackets (the larvae of daddylonglegs) ... are two of the insect pests that starlings deal with* (Sunday *Times* Magazine).
2 an immature form of certain animals that is different in structure from the adult form and must undergo a change or metamorphosis to become like the parent. A tadpole is the larva of a frog or toad.
[from New Latin, from Latin *larva* ghost, mask]
—larval, *adj.* of, having to do with, or in the form of a larva or larvae.

larviparous (lär vip′ər əs), *adj. Zoology.* giving birth to young insects that have already passed from the egg to the larval stage; giving birth to larvae rather than eggs. [from *larva* + Latin *parere* to give birth]

laryngeal (lə rin′jē əl), *Anatomy.* **—***adj.* of, having to do with, or in the larynx. **—***n.* a part of the larynx, such as a laryngeal cartilage or nerve.

larynx (lar′ingks), *n., pl.* **larynges** (lə rin′jēz), **larynxes.** *Anatomy.* **1** the upper end of the human windpipe, containing the vocal cords and acting as an organ of voice.
2 a similar organ in other mammals, or the corresponding structure in other animals.
3 (in birds) either of two modifications of the trachea, one at the top and one at the bottom.
[from Greek *larynx, laryngos*]

lase (lāz), *v. Physics.* **1** to act as a laser; emit coherent light of a single wavelength: *The ruby "lases" and emits a pulse of red light with an energy of up to a megawatt* (New Scientist).
2 to subject to the action of a laser.
[back formation from *laser*]

laser (lā′zər), *n. Physics.* any device that emits a very narrow and intense beam of coherent light or other radiation of a single wavelength either continuously or in pulses; an optical maser. A laser uses light to stimulate the emission of more light by excited atoms, mole-

---

cap, fāce, fäther; best, bē, tėrm; pin, five;
rock, gō, ôrder; oil, out; cup, pùt, rüle,
yü in use, *yù* in uric;
ng in bring; *sh* in rush; *th* in thin, ŦH in then;
zh in seizure.
ə = *a* in about, *e* in taken, *i* in pencil, *o* in lemon, *u* in circus

L-asparaginase

cules, or other physical systems. Its light may be used to cut or melt hard materials, remove diseased body tissues, and transmit communications signals, among other functions. *A pulsed ruby laser piercing a sapphire crystal ... generates energy so intense that it can bore a sixteenth of an inch hole in the sapphire in a thousandth of a second* (Science News Letter). See the picture at **hologram.** [acronym formed from *l*(*ight*) *a*(*mplification* by) *s*(*timulated*) *e*(*mission* of) *r*(*adiation*) on the analogy of the earlier *maser*]

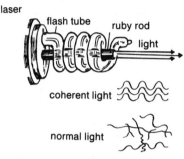

laser

flash tube   ruby rod

light

coherent light

normal light

L-asparaginase (el′as pə raj′ə nās), *n. Biochemistry.* an enzyme that causes the breakdown of L-asparagine. The enzyme is present in animal and plant tissues as well as in bacteria, and is used chiefly against leukemia cells, which require L-asparagine for their growth.

L-asparagine (el′ə spar′ə jēn), *n. Biochemistry.* the levorotatory form of asparagine: *The enzyme L-asparaginase holds hope for cancer therapy because it breaks down the amino acid L-asparagine, needed by all cells for protein synthesis* (Science News).

lasso cell, = colloblast.

lat., *abbrev.* latitude.

latent, *adj. Botany.* dormant or undeveloped: *Buds are latent when they are not externally visible until stimulated to grow.*

latent heat, *Physics, Chemistry.* the heat required to change a solid to a liquid or a vapor, or to change a liquid to a vapor, without a change of temperature. It is also the heat released in the reverse processes. Compare **heat of fusion** and **heat of vaporization.**

latent period, 1 *Physiology.* the time elapsing between a stimulus and the response to it. **2** *Medicine.* the incubation period of a disease; the time elapsing between an infection and the appearance of symptoms.

laterad (lat′ər ad), *adv. Anatomy.* toward the side. [from Latin *latus, lateris* side + *ad* to]

lateral, *adj.* of, at, or from the side; in the direction of or toward the side: *lateral roots. A river is fed by lateral streams.* [from Latin *lateralis,* from *latus, lateris* side]

ASSOCIATED TERMS: see **dorsal.**

lateral bud, *Botany.* a bud produced from the side of the stem instead of the stem apex or a leaf axil.

lateral chain, = side chain.

lateral line, *Zoology.* the row of connected sensory pores on the heads and sides of fishes, cyclostomes, and certain amphibians by which they detect changes in water pressure or current.

lateral moraine, *Geology.* a low ridge-like mass of loose rock pieces carried on, or deposited at, the side of a mountain glacier.

laterite (lat′ə rīt), *n. Geology.* a reddish soil rich in iron or aluminum or both, formed under tropical conditions by the decomposition of rock, and commonly found in parts of India, southwestern Asia, and Africa. [from Latin *later* brick, tile + English *-ite*]
—**lateritic** (lat′ə rit′ik), *adj.* of or containing laterite: *lateritic ore. Neither extensive clearing of forests nor large-scale cultivation of tropical lands offer as much promise as one might think, because much of the soil in such regions is lateritic and turns hard as the result of an oxidizing effect when it is put to the plow* (Scientific American).

laterization, *n. Geology.* the process by which laterite is formed: *While laterization has converted large areas of the humid tropics into exploitable mineral deposits, the process has reduced much larger areas almost to desert* (R. R. Dagon).

laterize, *v. Geology.* to change into laterite; make lateritic; *Under forest conditions, one hectare (2.5 acres) of soil loses only two pounds of soil a year to erosion; when the trees are cut, the same area loses thirty-four tons of soil a year. Such exposed soil may readily be laterized, or turned to stone* (A. Anderson).

latex (lā′teks), *n., pl.* **latices** (lat′ə sēz′), **latexes.** *Botany.* a usually milky liquid secreted by many plants, including milkweeds, poppies, and plants yielding rubber. It hardens in the air, and the latex from some plants is used to make chicle, rubber, and other products. *Latex ... is a complex fluid which varies in composition in different plants and can include various salts, carbohydrates, organic acids, amino acids, lipids, proteins, alkaloids, pigments, terpenes,* and *rubber* (P.G. Mahlberg). [from Latin *latex* a liquid]

laticifer (lā tis′ə fər), *n. Botany.* a plant cell which secretes latex. [from Latin *latex, laticis* a liquid + *ferre* to bear]
—**laticiferous** (lat′ə sif′ər əs), *adj.* that secretes latex.

latifoliate (lat′ə fō′lē āt *or* lat′ə fō′lē it), *adj. Botany.* having broad leaves. [from New Latin *latifoliatus,* from Latin *latus* broad + *folium* leaf]

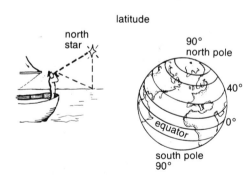

latitude

north star

90°
north pole

40°

0°

equator

south pole
90°

(def. 1)
To determine the latitude, the observer measures the angle formed at his position between the apparent horizon and the north star.

latitude (lat′ə tüd), *n.* **1** *Geography.* a distance north or south of the equator, measured in degrees. A degree of latitude is about 69 statute miles or 60 nautical

miles. **b** a place or region having a certain latitude: *Polar bears live in the cold latitudes.* **2** *Astronomy.* **a** = celestial latitude. **b** = galactic latitude. *Abbreviation:* lat. Compare **longitude.** [from Latin *latitudo* width, from *latus* wide]
—**latitudinal,** *adj.* (lat′ə tüd′n əl), of or relating to latitude. —**latitudinally,** *adv.*

lattice, *n.* **1** *Crystallography.* the three-dimensional, regularly repetitive arrangement of molecules, atoms, or ions in a crystal: *The halite lattice is built on a simple pattern in which sodium particles and chlorine particles occupy alternate corners of a continuously repeated set of cubes* (I.G. Gass).
2 *Physics.* = space lattice.
3 *Mathematics.* a partially ordered set in which any two elements have a least upper bound and a greatest lower bound: *Boolean algebras were the first lattices to be studied. They were introduced by Boole in order to formalize the calculus of propositions* (N. Jacobson).

latus rectum (lā′təs rek′təm), *pl.* **latera recta** (lā′tər ə rek′tə). *Geometry.* a chord of a conic section passing through a focus and perpendicular to the axis of the conic section. [from New Latin *latus rectum* (literally) straight side]

Laurasia (lô rā′zhə), *n. Geology.* the protocontinent of the Northern Hemisphere, a hypothetical landmass comprising North America, Europe, and Asia, which according to the theory of plate tectonics began to break up into the present continents during the Cenozoic era. Compare **Gondwana, Pangaea.** [from *Lau(rentian* Mountains of North America) + *(Eu)rasia*]
—**Laurasian,** *adj.* of or having to do with Laurasia: *Turtles are found in Triassic formation in Laurasia. None are found in Gondwanaland before Cretaceous times. This suggests a Laurasian origin* (Scientific American).

Laurentian (lô ren′shən), *adj.* **1** *Geography.* **a** of or having to do with the Canadian upland region extending from Labrador past Hudson Bay and north to the Arctic: *The Laurentian area, sometimes called the "Canadian Shield," is the most important mineral-producing section of Canada* (Colby and Foster, *Economic Geography*). **b** of or having to do with the St. Lawrence River and the regions through which it flows: *the Laurentian watershed.*
2 *Geology.* of or having to do with certain granites intrusive in the oldest Precambrian rocks of southern Canada.
[from *Laurentius,* Latin form of Lawrence]

Laurentian Plateau or **Laurentian Shield,** = Canadian Shield.

lava (lä′və *or* lav′ə), *n. Geology.* **1** the molten rock flowing from a volcano or fissure in the earth: *When lava cools, various minerals in it acquire a permanent magnetization that is parallel to the earth's magnetic field* (W.A. Elders). *Lava erupting from the mid-Atlantic ridge near the coast of Iceland created the island of Surtsey* (Walter Sullivan).
2 the rock formed by the cooling of this molten rock. Some lavas are dense and glassy or crystalline; others are light and porous.
[from Italian, from Latin *labes* a fall, falling down, from *labi* to slide, fall]

lawrencium

lava cave, *Geology.* a cave formed by the partial collapse of the roof of a lava tube.

lava cone, *Geology.* a volcanic cone built entirely or mainly of lava flows.

lava dome, *Geology.* a dome-shaped volcano built up of many lava flows issued from the central vent of a volcano: *The fourth type volcano, lava domes, are sometimes accompanied by the most destructive eruptions known to man. These domes are created from the plugs of very viscous or pasty lava that extrudes much like toothpaste from the vent of the volcano. Mt. Pelee on the island of Martinique is a lava dome volcano* (B. Tufty).

lava field, *Geology.* a large area of cooled lava.

lava flow, *Geology.* **1** the flow of lava from a volcano or fissure: *In 1960, nonexplosive lava flows from Kilauea covered the entire town of Kapoho, and more than 1000 acres were added to the coastline where lava poured into the ocean* (R. W. Decker). **2** the site of a former lava flow.

lava tube, *Geology.* a hollow subterranean channel in a lava flow formed by withdrawal of the lava after solidification of the superficial crust.

law, *n.* **1** a statement of what always occurs under certain conditions; description of a relation or sequence of phenomena invariable under the same conditions: *the laws of motion, Mendel's laws. Physicists (and other scientists) have a shorthand way of describing natural phenomena. Such shorthand descriptions are termed laws. For instance, there is a law, called the law or principle of Archimedes, describing how far into water a floating body will sink. In physics laws are usually put into mathematical form* (Harper's).
2 a mathematical rule or relationship on which the construction of a curve, a series, etc., depends.

law of large numbers, *Statistics.* a rule or theorem that a large number of items chosen at random from a population are bound, on the average, to have the characteristics of the population.

law of parsimony, = Occam's razor: *The goal of all science, as Miller sees it, is to account for the greatest number of facts with the fewest possible assumptions, in conformity with what is sometimes referred to as the Law of Parsimony* (Gerald Jonas).

law of reflection, *Physics.* **1** the law that the incident ray, the reflected ray, and the normal to the reflecting surface at the point of incidence are in the same plane. **2** the law that the angle of incidence is equal to the angle of reflection.

lawrencium (lô ren′sē əm), *n. Chemical.* a radioactive, metallic element of the actinide series. Lawrencium is short-lived and artificially produced by bombarding californium with boron ions. *Symbol:* Lr; *atomic num-*

---

cap, fāce, fäther; best, bē, tėrm; pin, five;
rock, gō, ôrder; oil, out; cup, pùt, rüle,
yü in use, *yù* in uric;
ng in bring; *sh* in rush; *th* in thin, ŦH in then;
zh in seizure.
ə = *a* in about, *e* in taken, *i* in pencil, *o* in lemon, *u* in circus

353

ber 103; *atomic weight* 256. [named after Ernest Orlando *Lawrence,* 1901–1958, American physicist]

Lawson criterion, *Physics.* **1** the requirement that for a nuclear fusion reaction to yield a gain in energy the product of the confinement time of plasma (highly ionized gas), measured in seconds, and the density of the plasma, measured in ions per cubic centimeter, must exceed a certain number, usually 10^{14}: *The fundamental criterion for a successful fusion reactor is that it should confine the hot fuel long enough so that a sufficiently large fraction will react and thereby release appreciably more energy than was invested in fuel heating. This is known as Lawson's criterion* (H. P. Furth).
2 Also called **Lawson number.** a number which indicates the gain in energy obtained in a fusion reaction: *The Lawson number, expressed in seconds per cubic centimeter, specifies the break-even condition on the assumptions that no more than a third of the energy released must be fed back to sustain the reaction* (Scientific American).
[named after D.J. *Lawson,* a British physicist, who formulated the criterion in the 1960's]

lax, *adj.* **1** (of tissue, soils, etc.) loose in texture; loosely cohering. **2** *Botany.* loose or open; not compacted.

layer, *n.* **1** *Geology.* a bed or stratum of rock.
2 *Meteorology.* a well-defined atmospheric area or region, as the *E* layer.
3 *Anatomy.* tissue of uniform thickness spread over a definite area: *the granular or pyramidal layers of the cerebral cortex.*
4 *Botany.* **a** a branch of a plant bent down and covered with earth so that it will take root and form a new plant while still attached to the parent stock. **b** a plant propagated by layering.
—**layered,** *adj.* arranged in or having layers: *In general, the layered rocks are much more open-textured than the crystallines* (Garrels, *A Textbook of Geology*).
—**layering,** *n. Botany.* a method of forming new plants by placing a shoot or twig of a plant in the ground so that it will take root while still attached to the parent stock.

lazulite (laz′yə līt), *n. Mineralogy.* a hydrous phosphate of aluminum, magnesium, and iron, often found in azure-blue monoclinic crystals. *Formula:* $(Mg,Fe)Al_2(PO_4)_2(OH)_2$ [from Medieval Latin *lazulum* lapis lazuli + English *-ite*]

lazurite (laz′yə rīt), *n. Mineralogy.* a mineral that is the chief component of lapis lazuli; one of a group of minerals chemically similar to feldspar. *Formula:* $(Na,Ca)_8(Al,Si)_{12}O_{24}$ [from Medieval Latin *lazur* azure + English *-ite*]

lb., *abbrev. pl.* **lb.** or **lbs.** pound. [for Latin *libra* pound]

l.c.d. or **L.C.D.,** *abbrev.* least common denominator; lowest common denominator.

l.c.m. or **L.C.M.,** *abbrev.* least common multiple; lowest common multiple.

LDL, *abbrev. Biochemistry.* low density lipoprotein (a lipoprotein containing more lipids than protein and carrying cholesterol from the liver to various tissues): *LDLs carry cholesterol in their core and in man represent the major mode of transporting cholesterol from*

the liver to cells of various tissues (N.Y. Times). Contrasted with **HDL.** Compare **VLDL.**

leach, *v. Geology, Chemistry.* **1** to dissolve out soluble parts from (ashes, rocks, ores, or other matter) by running water or other liquid through slowly: *to leach wood ashes for potash.*
2 to lose soluble parts as water or other liquid passes through: *Its chemically inert particles ... have strong surface adherence and do not leach off when exposed to fluids* (Science News Letter).
—**leachable,** *adj.* that can be leached.
—**leachate** (lē′chāt), *n.* the solution or soluble material that has been leached out of a substance: *Leachate from open refuse dumps and poorly engineered landfills has contaminated surface and ground waters* (R. D. Vaughan).

lead (led), *n. Chemistry.* a soft, heavy, malleable, bluish-gray metallic element which occurs naturally in galena and is used in radiation shields, as a solder, in alloys, etc.: *Lead occurs in the atmosphere virtually entirely as particulate matter originating from natural and artificial pollution sources* (J.W. Winchester). *Symbol:* Pb; *atomic number* 82; *atomic weight* 207.19; *melting point* 327.5°C; *boiling point* 1740°C; *oxidation state* 2,4.

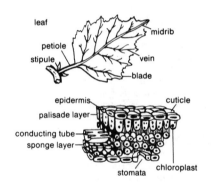

leaf, *n. Botany.* one of the thin, usually flat, green parts of a tree or other plant, that grows on the stem or up from the roots. Leaves are essential organs of most plants and use the carbon dioxide of the air or water in which they live and the light from the sun to carry on photosynthesis. *A conspicuous vascular strand commonly occupies the central part of the leaf and constitutes the main or midvein. From the midvein, smaller secondary veins depart* (Greulach and Adams, *Plants*).

ASSOCIATED TERMS: Most leaves consist of two main parts, the *blade,* or *lamina,* and the *petiole* or *leafstalk.* The leaves of many plants also have two small flaps called *stipules.* The leaves of *broadleaf* plants often have *teeth* that contain *hydathodes,* tiny structures that can release excess water. The *veins* of most broad leaves are either *pinnate* or *palmate,* and the *leaflets* of a *compound leaf* may be arranged in a pinnate or palmate pattern. Protection of a leaf is provided by the *epidermis* and its *cuticle,* which contain openings called *stomates* or *stomata.* Each *stomate* or *stoma* is surrounded by two curved guard cells that can dilate and contract, thus regulating the size of the stomate. Photosynthesis occurs inside the leaf blade in the *palisade cells* and *spongy cells,* within which are numerous *chloroplasts* containing molecules of *chlorophyll.* See also **bud, bulb, bract, tendril.**

leaf bud, *Botany.* a bud producing a stem with leaves only and no flowers; a bud which develops into a leafy branch: *There are buds that produce only stems and leaves (leaf buds)* (Weier, *Botany*).

leaflet, *n. Botany.* **1** a small or young leaf. **2** one of the separate blades or divisions of a compound leaf: *Buds occur in the axils of leaves, but not in the axils of leaflets; ... leaves stand in different planes on the stem, whereas leaflets lie in a single plane* (Weier, *Botany*).

leaf mold, *Biology.* **1** the partially decomposed leaves which form a surface layer in wooded areas. **2** a mold which attacks foliage.

leaf mosaic, *Botany.* an arrangement of the leaves on a tree, bush, or vine in which petiole length and position enable each leaf to receive the maximum amount of sunlight.

leaf scar, *Botany.* a mark left on a twig where the stem of a fallen leaf was attached.

leafstalk, *n. Botany.* a petiole.

leap second, *Astronomy.* a second of time as measured by an atomic clock, that is added or omitted each year by international agreement to compensate for changes in the earth's rotation. [patterned on *leap year*]

leap year, *Astronomy.* a year in the Gregorian calendar having 366 days, the extra day being February 29. A year is a leap year if its number can be divided exactly by four, except years at the end of a century, which must be exactly divisible by 400. *The years 1968 and 2000 are leap years; 1900 and 1969 are not.*

least common denominator, *Mathematics.* the least common multiple of the denominators of a group of fractions: *30 is the least common denominator of 2/3, 4/5, and 1/6.* Also called **lowest common denominator.**

least common multiple, *Mathematics.* the smallest quantity that is divisible by two or more given quantities without a remainder. 12 is the least common multiple of 3 and 4 and 6. Also called **lowest common multiple.**

least squares, *Statistics.* a method of determining the trend of a group of data when that trend can be represented on a graph by a straight line: *From these distances and the memory-stored geographical coordinates of the stations, the computer was further instructed to determine the center of the quake and then apply a least-squares solution to improve its accuracy* (J. J. Lynch).

least upper bound, *Mathematics.* the smallest of the upper bounds of a set; the smallest number greater than or equal to the largest number in a set. *Abbreviation:* l.u.b. Compare **greatest lower bound.**

lechatelierite (lə shä'tə lir'īt), *n. Geology.* the glassy form of silica produced by the heat of lightning or meteoric impact: *The invariable signs of flow and common presence of lechatelierite (pure silica glass) indicate that tektites were formed by rapid fusion at very high temperatures* (R. S. Dietz). [from Henry Louis *Le Chatelier*, 1850–1936, French chemist + *-ite*]

Le Chatelier's principle (lə shä'tə lirz'), *Physics, Chemistry.* the principle that if a stress, such as a change in concentration, pressure, or temperature, is applied to a system at equilibrium, the equilibrium is shifted in a way that tends to relieve the effects of the stress. For example, when a chemical system at equilibrium is disturbed, chemical reaction occurs and equilibrium is re-established at a different point, with new concentrations of reactants and products. [named after Henry Louis *Le Chatelier*; see **lechatelierite**]

lecithin (les'ə thən), *n. Biochemistry.* any of a group of nitrogenous fatty substances containing phosphorus, present in all plant and animal tissues but found especially in nerve cells and brain tissue: *Lecithin, a component of bile that helps dissolve cholesterol ...* (Science News). [from Greek *lekithos* egg yolk]

lecithinase (les'ə thə nās *or* lə sith'ə nās), *n. Biochemistry.* an enzyme capable of hydrolyzing lecithin or its components.

lee side, *Geology.* the side of a rock, dune, glacier, etc., that is away from the direction in which the wind is blowing.

lee tide, *Oceanography.* a tide running in the direction toward which the wind blows.

lee wave, *Meteorology.* any wavelike disturbance caused by some obstacle in the flow of air, especially a stationary atmospheric wave formed on the sides of mountains away from the wind and often marked by the presence of an apparently motionless cloud.

left atrioventricular valve, = mitral valve.

left-brain, *n. Anatomy.* the left hemisphere of the cerebrum: *The brain's two large cerebral hemispheres ... are now commonly known as left-brain and right-brain. The reason is that ... the two sides of the brain appear to possess such independent capacities and mental properties that each merits a separate name* (Roger W. Sperry).

left-handed, *adj. Physics, Chemistry.* **1** characterized by a direction or rotation to the left; counterclockwise.
2 = levorotatory: *One of the many unsolved mysteries of life is that living organisms possess only left-handed forms of molecules—such as the amino acids—that can also exist as right-handed isomers* (New Scientist).

leg, *n.* **1** *Anatomy.* one of the limbs on which animals support themselves and walk, especially the part of the vertebrate limb between the knee and the ankle.
2 *Geometry.* either of two sides of a right triangle that is not the hypotenuse.

legume (leg'yüm *or* li gyüm'), *n. Botany.* **1** any plant of the pea family; any of the group of dicotyledonous plants which bear pods containing a number of seeds, such as beans and peas. Legumes can absorb nitrogen from the air and convert it into nitrates by means of bacteria present in nodules on the roots of the plants.
2a the pod of such a plant. **b** the fruit or edible portion of such a pod: *Peas, beans, clover, alfalfa, and lespedeza are sometimes collectively called "legumes" for their fruits are, indeed, legumes. Other examples are wisteria, locust, and peanut* (Greulach and Adams, *Plants*).

---

cap, fāce, fäther; best, bē, tèrm; pin, fīve;
rock, gō, ôrder; oil, out; cup, pùt, rüle,
yü in use, *yu* in uric;
ng in bring; sh in rush; th in thin; ᴛʜ in then;
zh in seizure.
ə = a in about, e in taken, i in pencil, o in lemon, u in circus

3 a dry, several-seeded fruit, characteristic of plants of the pea family. It is formed by a single carpel, which is dehiscent by both sutures and so divides into two valves, the seeds being borne at the inner or ventral suture only.
[from Middle French *légume,* from Latin *legumen*]
—**leguminous** (li gyü′mə nəs), *adj.* **1** of or bearing legumes. **2** of or belonging to the pea family.

leishmania (lēsh man′ē ə), *n. Microbiology.* any member of a genus (*Leishmania*) of protozoan flagellates that are parasitic in vertebrate tissues and cause several diseases in humans: *Leishmanias ... are responsible for oriental sore and potentially fatal kala-azar* (New Scientist). [named after William B. *Leishman,* 1865–1926, British pathologist]

lek, *n. Zoology.* **1** a meeting ground used by certain species of birds during the breeding season for display and courtship: *As many as forty or fifty, or even more birds congregate at the leks* (Darwin, *Descent of Man*).
2 the display and courtship exhibited by such birds: *Baker and Parker comment that leks are normally held in wide open spaces, where predators can easily be seen before they can do any harm* (New Scientist). [from Swedish *lek* game, sport]

L electron, *Physics.* one of the electrons in the L-shell.

lemma[1] (lem′ə), *n., pl.* **lemmas, lemmata** (lem′ə tə), *Mathematics.* a subsidiary or auxiliary proposition. [from Greek *lēmma* something taken or assumed, from *lambanein* to take]

lemma[2] (lem′ə), *n. Botany.* the lower bract of the pair enclosing the flower in a spikelet of grass: *Each floret is in turn protected by two additional bracts, the lemma and the palea* (Weier, *Botany*). [from Greek *lemma,* from *lepein* to peel]

lemniscate (lem nis′kit), *n. Geometry.* a closed curve consisting of two symmetrical loops meeting at a node and generally resembling a horizontal figure 8. [from Latin *lemniscata,* feminine of *lemniscatus* adorned with ribbons, from *lemniscus* a hanging ribbon]

lemniscus (lem nis′kəs), *n., pl.* **-nisci** (-nis′ī or -nis′kī). *Anatomy.* a band of fibers, especially a band of nerve fibers in the mesencephalon. [from New Latin *lemniscus,* from Latin *lemniscus* a hanging ribbon, from Greek *lēmniskos*]

length, *n. Mathematics.* **1** (of a line segment) the distance between the end points. **2** (of a polygon) the sum of the lengths of the constituent line segments. **3** (of a curve) the limit of the sum of the lengths of an inscribed series of connected line segments as the lengths of the individual segments approach zero.

lens (lenz), *n.* **1** *Optics.* **a** a piece of glass or other transparent material which focuses or spreads the rays of light passing through it to form an image. Lenses have two opposite surfaces, either both curved or one plane and one curved, and are used alone or in combination in optical instruments. **b** a combination of two or more of these pieces, as used in a camera, microscope, telescope, etc.
2 *Anatomy.* a transparent oval body in the eye directly behind the iris, that focuses light rays upon the retina: *The lens in the human eye is a converging lens* (John

B. Walsh). Also called **crystalline lens.** See the picture at **accommodation.**
3 *Physics.* a device able to focus radiations other than those of light, such as an electron lens.
4 *Geology.* a layer of uniform sedimentary material that becomes progressively thinner along its edges. [from Latin *lens, lentis* lentil (which has a biconvex shape)]

(def. 1)

lentic, *adj. Ecology.* of or living in standing or stagnant water, such as ponds and swamps: *The stalks of inhabitants of swift currents tend to be much shorter than those of their lentic relatives* (N. Polunin). Contrasted with **lotic.** [from Latin *lentus* slow + English *-ic*]

lenticel (len′tə sel′), *n. Botany.* a usually lens-shaped body of cells formed in the corky layer of bark, which serves as a pore for the exchange of gases between the plant and the atmosphere. [from New Latin *lenticella,* from Latin *lens, lentis* lentil]
—**lenticellate** (len′tə sel′āt), *adj.* producing lenticels.

lenticular (len tik′yə lər), *adj.* **1** *Optics.* having the form of a lens, especially a biconvex lens. **2** of, having to do with, or shaped like a lens. A **lenticular cloud** is a stationary cloud shaped like a lens, formed especially above hills and mountains. The **lenticular nucleus** is a lens-shaped gray nucleus in the corpus striatum at the base of the brain. A **lenticular galaxy** has a lenslike appearance. [from Late Latin *lenticularis,* from Latin *lenticula,* diminutive of *lens, lentis* lentil]

Lenz's law (len′zəz or lent′səz), *Physics.* a law stating that the electromotive force induced by any change in physical conditions is always in such a direction that a current set up by the electromotive force will oppose the change. [named after Heinrich F.E. *Lenz,* 1804–1865, German physicist who formulated the law]

Leonid (lē′ə nid), *n., pl.* **Leonids, Leonides** (lē on′ə dēz′). *Astronomy.* one of the meteors in a meteor shower occurring on or near November 15. The Leonids seem to come from the constellation Leo.

lepidolite (lə pid′ə līt), *n. Mineralogy.* a species of mica containing lithium, commonly occurring in pegmatites as lilac, rose-colored, or grayish-white scaly masses. *Formula:* $K(Li,Al)_3(Si,Al)_4O_{10}(F,OH)_2$ [from Greek *lepis, lepidos* scale + English *-lite*]

lepidopteran (lep′ə dop′tər ən), *Zoology.* —*n.* a lepidopterous insect. —*adj.* = lepidopterous.

lepidopterology (lep′ə dop′tə rol′ə jē), *n.* the branch of entomology dealing with the lepidopterous insects.
—**lepidopterological,** *adj.* of or having to do with lepidopterology.

lepidopterous (lep′ə dop′tər əs), *adj. Zoology.* of or belonging to a large order (Lepidoptera) of insects including butterflies and moths. The larvae are wormlike and are called caterpillars. The adults have four broad, membranous wings more or less covered with small, sometimes colorful, overlapping scales, and a coiled proboscis for sucking. [from New Latin

Lepidoptera the order name, from Greek *lepis, lepidos* scale + *pteron* wing, feather]

lepidote (lep′ə dōt), *adj. Botany.* covered with scurfy scales. [from Greek *lepidōtos,* from *lepis, lepidos* scale]

leptocephalus (lep′tə sef′ə ləs), *n., pl.* -li (-lī). *Zoology.* the narrow-headed, transparent larva of various freshwater eels. [from New Latin *leptocephalus,* from Greek *leptos* thin + *kephalē* head]

lepton (lep′ton), *n. Nuclear Physics.* **1** any of a class of light elementary particles that includes the neutrinos, the electron, and the muon, and their antiparticles. Leptons partake of the weak interaction and have a spin quantum number of 1/2.
2 any weakly interacting particle, regardless of mass. [from Greek *lepton,* neuter of *leptos* thin, small]
—**leptonic,** *adj.* of or having to do with leptons: *leptonic charge, leptonic decays.*

leptonema (lep′tə nē′mə), *n.* = leptotene.

leptotene (lep′tə tēn), *n. Biology.* the stage of the prophase of meiosis just before the homologous chromosomes unite. Each chromosome, which is threadlike in appearance, actually consists of two duplicate threads (sister chromatids). [from Greek *leptos* thin + *tainia* band]

lesion (lē′zhən), *n. Biology, Medicine.* an abnormal change in the structure of an organ or body tissue, caused by disease or injury: *Two of the patients showed benign, symmetric lesions on the lower and upper lips attributable to chronic irritation from cigarette and pipe smoking* (Theodore Cornbleet). [from Latin *laesionem* injury, from *laedere* to strike]

lethal gene or **lethal factor,** *Biology.* any gene, either dominant or recessive, whose expression results in the premature death of an organism bearing it: *Various species of plants and animals carry lethal factors which, when homozygous, stop development at some stage, and the individual dies* (Storer, *General Zoology*).

Leu, *abbrev.* leucine.

leuc- or **leuk-,** *combining form.* variant form of **leuco-** before vowels, as in *leucine, leukemia.*

leucine (lü′sēn), *n. Biochemistry.* a white, crystalline amino acid essential in nutrition, produced especially by the digestion of proteins by the pancreatic enzymes: *Leucine ... is found in cheese and in muscle, and among the products of protein hydrolysis ... Leucine and isoleucine ... promote growth, and the formation of red blood corpuscles* (Muldoon, *Organic Chemistry*). *Formula:* $C_6H_{13}NO_2$ *Abbreviation:* Leu

leucite (lü′sīt), *n. Mineralogy.* a white or grayish mineral, a silicate of potassium and aluminum, found in alkalic volcanic rocks. Leucite belongs to a group of minerals that are chemically similar to feldspar. *Formula:* $KA1Si_2O_6$ [from obsolete German *Leucit,* from Greek *leukos* white + German *-it* -ite]
—**leucitic** (lü sit′ik), *adj.* of or like leucite.

leuco (lü′kō), *adj. Chemistry.* of or designating a reduced, colorless form of a dye (such as indigo) which is fixed on a fiber and then reconstituted into the dye by the action of oxidizing agents: *a leuco base.* [from *leuco-*]

leuco- or **leuko-,** *combining form.* **1** white; colorless, as in *leucocyte* = *white cell.* **2** leucocyte, as in *leukemia* = *uncontrolled increase of leucocytes.* Also spelled

leuc- or **leuk-** before a vowel. [from Greek *leukos* white]

leucocyte or **leukocyte** (lü′kə sīt), *n.* = white blood cell: *The white cells, leucocytes, engulf the bacteria and secrete chemicals that kill them* (Mackean, *Introduction to Biology*). Contrasted with **erythrocyte.** [from Greek *leukos* white + English *-cyte*]
—**leucocytic** or **leukocytic** (lü′kə sit′ik), *adj.* of or having to do with leucocytes.

leucocytosis or **leukocytosis** (lü′kō sī tō′sis), *n. Medicine.* an abnormal increase in the number of leucocytes circulating in the bloodstream: *Mercurial poisoning is sometimes accompanied by a leukocytosis* (A. Grollman).
—**leucocytotic** or **leukocytotic** (lü′kō sī tot′ik), *adj.* of or having to do with leucocytosis.

leucoplast (lü′kə plast), *n. Botany.* one of the colorless plastids in the cytoplasm of plant cells that functions in the formation and storage of starch.

leukemia (lü kē′mē ə), *n. Medicine.* a form of cancer characterized by the uncontrolled increase and spread of nonfunctional and undeveloped leucocytes. [from New Latin, from *leuk-* + *-emia*]

lev-, *combining form.* the form **levo-** before vowels, as in *levarterenol.*

levan (lev′ən), *n. Biochemistry.* a naturally occurring polysaccharide of fructose, found in the leaves of various grasses: *Levans are found widely distributed throughout the plant kingdom ... and generally serve as reserve polysaccharides in place of, or in addition to, starch* (W. Pigman). *Formula:* $(C_6H_{10}O_5)_n$ [from Latin *laevus* left (because of its levorotatory properties)]

levarterenol (lev′är tir′ə nol), *n. Biochemistry.* levorotatory noradrenaline, a crystalline compound found in the adrenal glands and synthesized for use in medicine as a vasoconstrictor. *Formula:* $C_8H_{11}NO_3$ [from *levo-* + *arterenol,* variant name of *noradrenaline*]

levator (lə vā′tər), *n., pl.* **levatores** (lev′ə tôr′ēz). *Anatomy.* a muscle that raises some part of the body, such as the one that opens the eye. Contrasted with **depressor.** [from New Latin, from Latin *levare* to raise]

levee (lev′ē), *n. Geology.* a raised bank along a river, occurring naturally as the result of deposits left during successive floods: *The Yellow River in China has high levees.* [from French *levée,* from *lever* to raise, from Latin *levare*]

level, *n.* = energy level: *After the emission of the beta ray, each residual nucleus of Si²⁸ is left in an excited level at about I · 78 Mev above its ground level* (R.D. Evans).

---

cap, fāce, fäther; best, bē, tėrm; pin, fīve;
rock, gō, ôrder; oil, out; cup, pùt, rüle;
yü in use, *yu* in uric;
ng in bring; *sh* in rush; *th* in thin; ᴛʜ in then;
zh in seizure.
ə = *a* in about, *e* in taken, *i* in pencil, *o* in lemon, *u* in circus

lever

lever (lev'ər or lē'vər), *n. Physics.* a bar which turns on a fixed support called a fulcrum and is used to transmit effort and motion. It is a simple machine. With a lever, a weight can be raised or moved at one end by pushing down at the other end. [from Old French *leveor,* from *lever* to raise, from Latin *levare,* from *levis* light]
—**leverage,** *n.* **1** the action of a lever. **2** the advantage or power gained by using a lever.

ASSOCIATED TERMS: A lever works by means of a *fulcrum,* an *effort* or *force,* and a *load* or *resistance.* The distance between the fulcrum and the effort is called the *effort arm;* the distance from the load to the fulcrum is called the *load arm.* A lever is in *equilibrium* when the effort and the load balance each other. For a lever in equilibrium, the effort multiplied by the effort arm is equal to the load mulitplied by the load arm. These products are called the *moments* or *torques* of the effort and load. A lever's *mechanical advantage* indicates how many pounds of load can be moved by each pound of effort. *First-class levers,* such as a seesaw, have the fulcrum between the effort and load. *Second-class levers,* such as a wheelbarrow, have the load between the effort and the fulcrum. *Third-class levers,* such as a forearm, have the effort between the load and the fulcrum.

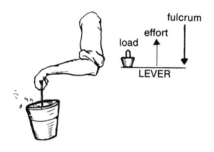

LEVER

levo (lē'vō), *adj. Chemistry.* turning or turned to the left; levorotatory: *Dextro and levo ... testosterone molecules are ... each a mirror image of the other* (Science News Letter).

levo-, *combining form.* **1** toward the left, as in *levorotatory.* **2** levorotatory, as in *levoglucose = levorotatory glucose.* Also spelled **lev-** before a vowel. [from Latin *laevus* left]

levoglucose, *n. Biochemistry.* a form of glucose, levorotatory to polarized light.

levorotation, *n. Chemistry, Physics.* rotation of the plane of polarization of light to the left when the observer is looking toward the source of light.

levorotatory (lē'vō rō'tə tôr'ē), *adj. Physics, Chemistry.* characterized by turning the plane of polarization of light to the left or counterclockwise, as a crystal, lens, or compound in solution. Levorotatory compounds are often prefixed by the symbol (−) or *l−* to distinguish them from their dextrorotatory forms. *Only levorotatory amino acids act in body biochemistry. It is probable that compounds derived from them ... should also be levorotatory* (Louise Campbell).

levulin (lev'yə lin), *n. Biochemistry.* a substance resembling dextrin, obtained from the roots of certain composite plants. It forms levulose on hydrolysis. *Formula:* $(C_6H_{10}O_5)_n$ [from *levul(ose)* + *-in*]

levulose (lev'yə lōs), *n.* = fructose.

Lewis acid, *Chemistry.* a substance that acts as an electron-pair acceptor. [named after G.N. *Lewis,* 1875–1946, American chemist] ► See the note under **acid.**

Lewis base, *Chemistry.* a substance that acts as an electron-pair donor. ► See the note under **base.**

Leyden jar (līd'n), an electrical condenser consisting of a glass jar coated inside and outside, for most of its height, with tin or aluminum foil, and sealed with a stopper containing a metal rod which is connected to the internal coating. [named after *Leiden,* Netherlands, where it was invented]

LF, L.F., or **l.f.,** *abbrev.* low-frequency.

L-form, *n. Bacteriology.* an atypical or mutant form of certain bacteria that usually lacks a cell wall, appears in various shapes, and resembles a mycoplasma: *The L-forms are completely insensitive to penicillin and are, in fact, often produced under the influence of the antibiotic* (New Scientist). [from L(*ister* Institute of Preventive Medicine), in London, where these bacterial forms were first observed]

LH, *abbrev.* luteinizing hormone.

LH-RH, *abbrev.* luteinizing hormone-releasing hormone: *LH-RH plays a key role in the onset of puberty, is the mediator responsible for the release of the ovulatory quota of LH [luteinizing hormone], and is necessary for normal implantation and maintenance of pregnancy* (Andrew V. Schally).

Li, *symbol.* lithium.

Lias (lī'əs), *n., adj.* = Liassic.

Liassic (lī as'ik), *Geology.* —*n.* **1** the earliest epoch of the European Jurassic period, characterized by clayey rocks with fossils. **2** the rocks of this epoch.
—*adj.* of or having to do with the Liassic.
[from French *liasique,* from *lias* a limestone rock]

liber (lī'bər), *n. Botany.* the bast or phloem of a plant. [from Latin *liber* inner bark]

liberate, *v. Chemistry.* **1** to set free from combination: *liberate a gas.* **2** to give off; emit; release: *liberate heat. If the potential energy of the products is lower than the potential energy of the reactants, energy has been liberated* (Chemistry Regents Syllabus).
—**liberation,** *n.* the act or process of liberating: *liberation of energy.*

librate (lī'brāt), *v. Astronomy.* to undergo libration: *The Trojan asteroids ... oscillate or librate about points on Jupiter's orbit 60 degrees ahead or behind the planet* (New Scientist).

libration (lī brā'shən), *n.* **1** *Astronomy.* a real or apparent oscillatory motion of a planet or satellite in its orbit: *The changes in the moon's orbit run in cycles; because of this, the visible surface of the moon undergoes rocking motions, or librations, which bring small areas near the edges of the observable disk into view* (C. Payne-Gaposchkin and K. Haramundanis).
2 *Physics.* any oscillatory motion associated with rotation, as that of a molecule.
[from Latin *libra* balance]
—**librational,** *adj.* of or characterized by libration; oscillatory: *Another idea pertains to attitude control in satellites, where stabilization under gravity is troubled by oscillations referred to as "librational motion"* (R.E. Rosensweig).

libration point, = Lagrangian point.

libriform (lī′brə fôrm), *adj. Botany.* having the form of or resembling liber or bast; elongated, thick-walled, and woody, as certain cells.

lichen (lī′kən), *n. Biology.* any member of a large group (Lichenes) of organisms that look somewhat like moss and grow in patches on trees, rocks, etc. Lichens consist of a fungus and an alga growing together symbiotically, the alga providing the food, and the fungus usually providing the water and protection. *Representatives of three major classes of fungi enter into the formation of lichens. The majority of forms are derived from the Ascomycetes, or sac fungi* (Scientific American). [from Latin, from Greek *leichēn*]
—**lichenized,** *adj.* forming a lichen; being one of the components of a lichen: *a lichenized fungus or alga.*

lichenology (lī′kə nol′ə jē), *n.* the branch of biology dealing with lichens.

lichenometry (lī′kə nom′ə trē), *n. Biology, Geology.* the measurement of the diameter of lichens to establish their age or the age of the area in which they grow: *Lichenometry has proved to be particularly useful in certain non-forested arctic and alpine regions* (Scientific American).
—**lichenometric** (lī′kə nə met′rik), *adj.* of, having to do with, or based on lichenometry: *lichenometric dating.*

lichenous (lī′kə nəs), *adj. Biology.* of, like, or covered with lichens.

lid, *n. Botany.* **1** the upper part of a seed vessel that bursts open transversely. **2** the operculum or coverlike part on the theca of a moss.

Lie algebra (lē), *Mathematics.* a system of vector fields on a topological space in which independent quantities are reduced to groupings whose relationships are then subject to algebraic operation: *Lie algebras and representation theory ... find applications in high-energy physics* (E.C. Zeeman). [named after Marius Sophus *Lie,* 1842–1899, Norwegian mathematician who developed this system]

Lie group, *Mathematics.* a topological group with a continuous operation in which it is possible to label the group elements by a finite number of coordinates.

lienal (lī ē′nəl), *adj. Anatomy.* of or having to do with the spleen. [from Latin *lien* the spleen]

life, *n.* **1** *Biology.* the form of existence that organisms like animals and plants have and that inorganic objects or organic dead bodies lack; animate existence, characterized by growth, reproduction, metabolism, and response to stimuli: *Ultimately ... all animal life is dependent upon that of plants, and were there no green plants in existence animal life on earth would be inconceivable* (Bell and Coombe, *Strasburger's Textbook of Botany*).
2 *Physics.* the average length of existence of a particle or state: *Each group [of radioactive substances] is arranged in the order of diminishing half-value period, and begins with the member of longest life* (R.W. Lawson). Compare **half-life.**

life cycle, *Biology.* **1** the successive stages of development that a living thing passes through from a particular stage in one generation to the same stage in a later generation: *A life cycle is the series of events from one stage during a life span through a reproductive phase*

until a state similar to the original is reached (Harbaugh and Goodrich, *Fundamentals of Biology*). **2** = life history: *A considerable number of native species that succeed in arid regions are winter annuals which complete their life cycles during the time of winter rains, thus enduring the dry summer as seeds* (Emerson, *Basic Botany*).

life expectancy, *Biology.* the length of time an organism is expected to live, as determined by statistical probability studies.

life form, *Ecology.* the lifelong or mature form of an organism, especially a plant.

life history, *Biology.* **1** the successive stages of development of an organism from its inception to death. **2** one series of such stages, often equivalent to a life cycle.

life science, any of the sciences dealing with living matter and biological processes, such as botany, zoology, biochemistry, and microbiology.
—**life scientist,** a specialist in one or more of the life sciences.

life zone, *Ecology.* a region that generally has a uniform type of plant or animal life, and a single type of climate. Because of environment, the plants in one life zone usually differ from the plants in another zone. *During the growth of the last continental ice sheet ... the climatic belts and the life zones were gradually pressed southward in front of the ice* (New Yorker).

ligament, *n. Anatomy.* a band of strong, flexible, white tissue which connects bones or holds organs of the body in place: *Ligaments are composed of tough elastic connective tissue. Ligaments connect the ends of bones at moveable joints such as the elbow, fingers, knee, and vertebral column* (Biology Regents Syllabus). See the picture at **knee.** [from Latin *ligamentum,* from *ligare* to bind]
—**ligamentous,** *adj.* having to do with, of the nature of, or forming a ligament: *ligamentous tissue.*

ligamentum (lig′ə men′təm), *n., pl.* **-ta** (-tə). = ligament.

ligand (lig′ənd *or* lī′gənd), *n. Chemistry.* an ion, molecule, or group that forms complex compounds (chelates) by establishing a coordinate bond with the ion of a metal: *It is now recognized that a bond between a ligand and a transition metal is made more stable if the ligand does not transfer too much electron density to the metal* (J. A. Kerr and L. Crombie). [from Latin *ligandus,* gerundive of *ligare* to bind]

ligase (lī′gās *or* lig′ās), *n. Biochemistry.* an enzyme that catalyzes the union of two molecules, usually with the simultaneously breaking of a pyrophosphate link in a triphosphate, such as adenosine triphosphate: *All newly made DNA is synthesized in a discontinuous manner. These discontinuous segments are then joined by an enzyme, ligase, to produce a continuous strand* (J. C. Copeland). Also called **synthetase.**

---

cap, fāce, fäther; best, bē, tėrm; pin, five;
rock, gō, ôrder; oil, out; cup, pùt, rüle;
yü in use, *yu* in uric;
ng in bring; *sh* in rush; *th* in thin, ᴛʜ in then;
zh in seizure.
ə = *a* in about, *e* in taken, *i* in pencil, *o* in lemon, *u* in circus

light

light[1], n. 1 Physics. a form of radiant energy consisting of electromagnetic waves that travel freely through space in a vacuum at a speed of 299,792 kilometers per second. The wavelength of light waves is measured in nanometers; the frequency of light waves is measured in hertz. The brightness of light is measured in candelas. We know that light consists of electromagnetic waves emitted by atoms and molecules. The waves are transverse, which means that their oscillations are at right angles to the direction in which the light travels (Scientific American).
2 the part of light that causes the sensation of vision when electromagnetic radiation acts upon the retina of the eye. Visible light ranges from about 400 nanometers (deep violet, the shortest wavelength) to about 700 nanometers (deep red, the longest wavelength).

light[2], adj. 1 Chemistry. not heavy; of low density or specific gravity. A light metal is one having a specific gravity lower than 3, such as aluminum and magnesium. A light mineral is one having a specific gravity lower than 2.85.
2 Meteorology. having a velocity of 11 kilometers per hour or less (on the Beaufort scale, force 1 and 2): a light wind.

light-adapt, v. Physiology. to adapt to increasing light or brightness: As the human becomes light-adapted ... colour discrimination becomes possible (New Scientist). Contrasted with dark-adapt.
—light-adaptation, the adjustment of the pupils of the eyes to increasing light or brightness following dark-adaptation.

light air, Meteorology. a condition in which the wind has a velocity of 1–5 kilometers per hour (force 1 on the Beaufort scale).

light breeze, Meteorology. a condition in which the wind has a velocity of 6–11 kilometers per hour (force 2 on the Beaufort scale).

light chain, Biochemistry. either one of the pair of short polypeptide chains in an antibody molecule, having an approximate molecular weight of 22,000: Each antibody molecule is made up of four polypeptide chains. Two of them are known as the light chains, and the remaining pair—which are about twice as long—as the heavy chains (New Scientist).

light curve, Astronomy. a graph that shows the variations in magnitude of a star's light at different times.

light-day, n. Astronomy. the distance that light travels in a vacuum in one day; about 26 billion (2.6×10^{10}) kilometers: If a quasar's radiation varies with a period of, say, four days, then the radius of the quasar can be no greater than four light-days (D. E. Thomsen).

light microscope, Optics. a conventional microscope, using rays of light, as distinguished especially from the electron microscope.

light-month, n. Astronomy. the distance that light travels in a vacuum in one month; nearly 800 billion (8×10^{11}) kilometers: A light-month is equal to about 10^{17} centimeters, or roughly 100 times the diameter of the solar system (G. Burbidge and Fred Hoyle).

lightning, n. Meteorology. a flash of light in the sky caused by a discharge of electricity between clouds, or between one part of a cloud and another part, or be-

tween a cloud and the earth's surface. The sound made by heat from the discharge is thunder.

light quantum, = photon.

light reaction, Biology. any of the chemical reactions in photosynthesis that are dependent on light; photochemical reaction: Light reactions ... occur within layered membranes inside the chloroplasts (Biology Regents Syllabus). Compare dark reaction.

light water, Chemistry. ordinary water, H_2O, as distinguished from heavy water (deuterium oxide, D_2O): In pressure-vessel reactors such as the United States designs, enriched uranium is used as fuel and light water is used both as moderator and coolant (London Times).

light-week, n. Astronomy. the distance that light travels in a vacuum in one week; about 180 billion (1.8×10^{11}) kilometers.

light-year, n. Astronomy. the distance that light travels in a vacuum in one year; about 9.5 trillion (9.5×10^{12}) kilometers or almost a third of a parsec. Light-years are used to measure distances to or between stars. ... Sirius, the brightest star in the sky, about nine light-years away (Clarke, The Exploration of Space).

ligneous (lig'nē əs), adj. Botany. of or like wood; woody. [from Latin ligneus, from lignum wood]

lignicolous (lig nik'ə ləs), adj. Ecology. 1 living or growing on wood, as fungi. 2 living in wood, as shipworms. [from Latin lignum wood + colere to inhabit]

lignification (lig'nə fə kā'shən), n. Botany. the act of lignifying or condition of being lignified.

ligniform (lig'nə fôrm), adj. Botany. having the form of wood; resembling wood.

lignify (lig'nə fī), v. Botany. to change into wood or woody tissue, as cells whose walls have been thickened and indurated by the deposit of lignin: The guard cells of some ... plants are lignified (woody) and cannot move at all (Emerson, Basic Botany).

lignin (lig'nən), n. Botany. an organic substance which, together with cellulose, forms the essential part of woody tissue, making up the greater part of the weight of dry wood.

lignite (lig'nīt), n. Geology. a brownish-black coal in which decomposition of vegetable matter has proceeded farther than in peat but not so far as in bituminous coal. Lignite usually has the original form of the wood clearly visible. Also called brown coal.
—lignitic (lig nit'ik), adj. of, having to do with, or containing lignite.

lignivorous (lig niv'ər əs), adj. Zoology. eating wood, as the larvae of many insects. [from Latin lignum wood + vorare devour]

lignocellulose (lig'nə sel'yə lōs), n. Botany. lignin combined with cellulose, forming an essential constituent of woody tissue.

ligroin (lig'rō in), n. Chemistry. a volatile, flammable liquid mixture of hydrocarbons obtained by the fractional distillation of petroleum, used as a solvent. Also called petroleum ether. [origin unknown]

ligula (lig'yə lə), n., pl. -lae (-lē), -las. 1 Zoology. the terminal or dorsal part of the labium of an insect. 2 = ligule. [from Latin ligula strap, tonguelike part]
—ligular, adj. of or like a ligula.

360

ligulate (lig′yə lit *or* lig′yə lāt), *adj. Botany, Zoology.* having a ligule or ligula; strap-shaped: *The ligulate ... corolla is ... usually tubular at the base, the remainder resembling a single petal* (Youngken, *Pharmaceutical Botany*).

ligule (lig′yül), *n. Botany.* any of several strap-shaped organs or parts, as the flattened corolla in the ray florets of composites or the projection from the top of the leaf sheath in many grasses. [from Latin *ligula* strap, tonguelike part]

limacon or **limaçon** (lim′ə son), *n. Geometry.* a curve generated from a circle by adding a constant length to all the radius vectors drawn from a point of its circumference as an origin. The limacon has three varieties, one of which is the cardioid. It was invented and named by Pascal. [from French *limaçon* snail, from Latin *limax, limacis*]

limb¹, *n.* **1** *Anatomy.* a leg, arm, wing, or other member of an animal body distinct from the head or trunk: *The limbs of seals, moles, bats and antelope look very different from each other and are adapted to the functions of swimming, digging, flying and running* (Mackean, *Introduction to Biology*).
2 *Botany.* a large branch.
[Old English *lim*]

limb², *n.* **1** *Astronomy.* the edge of the disk of a heavenly body: *The problem is one of identifying the satellite's edge, or limb, whether from an occultation (in which the blocking of the light from a star can be precisely timed) or from a photograph* (Science News).
2 *Botany.* the expanded flat part of a structure, such as the upper part of a corolla or the blade of a leaf.
3 *Geology.* one of the two sides of a fold. Also called **flank.**
[from Latin *limbus* border, edge]

limbate (lim′bāt), *adj. Biology.* having a border; bordered, as a flower having an edging of a different color from the rest. [from Late Latin *limbatus* bordered, edged, from Latin *limbus* edge, limbus]

limbic, *adj.* **1** *Biology.* of or having the character of a limbus or border.
2 *Anatomy.* of or having to do with the limbic lobes or the limbic system: *the limbic cortex. The limbic areas responsible for emotion (hippocampus, amygdala, and septal areas) were directly linked* (Saturday Review).

limbic lobe, *Anatomy.* either of two lobes of the brain, one in each hemisphere: *Surgeons who used electric needles to destroy small bundles of cells in the cingulum area of the limbic lobe, or feeling brain, in order to treat schizophrenics and chronic alcoholics reported significant improvement in 50% of their patients* (Peter Stoler).

limbic system, *Anatomy.* a group of interconnected neural structures in the rudimentary cortex of the brain, which surround the midline surfaces of the cerebral hemispheres and pass into the brain stem. The limbic system is believed to control emotional patterns of behavior. [*Addictive*] *drugs induce addiction by a direct action on the basic conditioning mechanisms of the limbic system* (J. R. Smythies).

limbus (lim′bəs), *n., pl.* **-bi** (-bī). *Biology.* a border or edge differentiated by color or formation, as in some flowers and shells: *In between, forming a border or "limbus" around the brain-stem and therefore called*

the *"limbic system," is a complex collection of brain structures essentially similar in man and other mammals* (Harper's). [from Latin *limbus* border, edge]

lime, *n. Chemistry.* a solid, white compound of calcium and oxygen obtained by burning limestone, shells, bones, or other forms of calcium carbonate. Lime is used in making mortar and on fields to improve the soil. *The presence or absence of ample supplies of lime in a soil make great differences in its character and productivity* (Emerson, *Basic Botany*). Formula: CaO Also called **calcium oxide.**

limen (lī′mən), *n. Physiology.* the threshold of a stimulus. [from Latin *limen* threshold]

limestone, *n. Geology.* a rock consisting mostly of calcite (calcium carbonate), much used for building and in the steel and chemical industries. Limestone yields lime when calcined. Marble is a metamorphosed limestone. *Limestone is used in blast furnaces to remove impurities from iron ore* (Science News Letter).

limicoline (lī mik′ə lin *or* lī mik′ə lin), *adj. Zoology.* of or having to do with certain shore birds or wading birds, such as the plovers, snipes, and sandpipers. [from Late Latin *limicola* dweller in mud, from *limus* mud + *colerĕ* inhabit]

limicolous (lī mik′ə ləs), *adj. Ecology.* living in mud. [from Late Latin *limicola* dweller in mud]

liminal (lim′ə nəl), *adj. Physiology.* of or having to do with a limen or threshold, especially of a sensory stimulus.

limit, *n. Mathematics.* a value toward which terms of a sequence or values of a function approach indefinitely near. [from Latin *limitem* boundary]

limiting nutrient, *Ecology.* a substance which when added to or removed from a body of water will stimulate or retard eutrophication: *Phosphates have been labeled as the culprit—the "limiting nutrient"—in eutrophication of waterways ... and phosphate levels act as the accelerator or decelerator of plant growth.* (Science News).

limit point, *Mathematics.* the limit of any sequence of points in a set. The limit point of the sequence 1, 1/2, 1/3, 1/4 ... 1/*n* ... is zero.

limnetic (lim net′ik), *adj. Ecology.* **1** inhabiting the open water of lakes, as various animals and plants. **2** having to do with life or organisms in the open water of lakes.

limnological (lim′nə loj′ə kəl), *adj.* of or having to do with limnology: *a limnological study of the Great Lakes.*

limnology (lim nol′ə jē), *n. Ecology.* the scientific study of inland bodies of fresh water, such as lakes and ponds. [from Greek *limnē* lake, marsh + English *-logy*]

---

cap, fāce, fäther; best, bē, tèrm; pin, five;
rock, gō, ôrder; oil, out; cup, pùt, rüle,
yü in use, *yu* in uric;
ng in bring; *sh* in rush; *th* in thin, ŦH in then;
zh in seizure.
ə = *a* in about, *e* in taken, *i* in pencil, *o* in lemon, *u* in circus

limonite (li'mə nīt), *n. Mineralogy.* any of various minerals consisting of hydrated ferric oxide, found in gossans, lakes, and marshes: *The various shades of yellows and browns of clays, sands, and other rocks are due mainly to the presence of limonite* (Jones, *Minerals in Industry*). [from German *Limonit*, from Greek *leimōn* meadow]

linac (lin'ak), *n.* = linear accelerator. [from *lin(ear) ac(celerator)*]

line, *n.* **1** *Mathematics.* the path traced by a moving point. It has length, but no thickness. See the picture at **segment. 2** *Astronomy.* a circle of the terrestrial or celestial sphere: *the equinoctial line.*
—**linear,** *adj.* **1** of or having to do with a straight line. **2** proportional: *There is a linear relationship between the size of a star and its luminosity.*

linear accelerator, *Nuclear Physics.* a device for accelerating charged particles in a straight line through a vacuum tube or series of tubes by means of alternating negative and positive impulses from electric fields: *In linear accelerators the particles pass just once down a straight tube, and no guiding magnets are required* (Scientific American).

linear algebra, *Mathematics.* the study of the algebraic properties of vectors, especially of a set of vectors over a field of scalars or real numbers.

linear equation, *Mathematics.* an equation whose terms involving variables are of the first degree, so called because the graph of such an equation is a straight line: $y = 4x + 3$ *is a linear equation.*

linear function, *Mathematics.* a function in which the variables are of the first degree.

linearity (lin'ē ar'ə tē), *n.* the quality or state of being linear; a linear arrangement or form.

linearize, *v.* to represent in linear form, or by means of lines. —**lin'eariza'tion,** *n.*

linear measure, 1 the measurement of length. **2** a system of units, such as a foot and mile, used for measuring length.

Metric System
10 millimeters = 1 centimeter
100 centimeters = 1 meter
1000 meters = 1 kilometer
1 kilometer = 3,280.8 feet (or about 0.6 miles)

English System	
12 inches = 1 foot	8 furlongs = 1 mile
3 feet = 1 yard	1,760 yards = 1 mile
5 1/2 yards = 1 rod	5,280 feet = 1 mile
40 rods = 1 furlong	3 miles = 1 league

linear programming, *Mathematics.* a method of solving operational problems by stating a number of variables simultaneously in the form of linear equations and calculating the optimal solution within the given limitations.

line graph, *Mathematics.* a graph in which points representing quantities are plotted and then connected by a series of short straight lines or a smooth curve.

line of apsides (ap'sə dēz), *Astronomy.* the straight line that joins the two extreme points in the elliptical orbit of a planetary body.

line of force, *Physics.* a line in a field of electric or magnetic force that indicates the direction in which the force is acting: *In a region where the intensity is large, the lines of force will be closely spaced and, conversely, in a region where the intensity is small the lines will be widely separated* (Sears and Zemansky, *University Physics*).

lineolate (lin'ē ə lāt), *adj. Botany, Zoology.* marked with minute lines: *a lineolate moth.* [from Latin *lineola,* diminutive of *linea* line]

line segment, *Mathematics.* a part of a line between two given points on the line.

line spectrum, *Physics.* a spectrum produced by a luminous gas or vapor in which distinct lines characteristic of an element are emitted by its atoms. Compare **band spectrum, continuous spectrum.**

line squall, *Meteorology.* a thunderstorm or other severe local storm appearing along a cold front.

line storm, *Meteorology.* a storm or gale occurring at or near an equinox; equinoctial storm.

linguiform (ling'gwə fôrm), *adj. Botany, Zoology.* tongue-shaped; lingulate: *a linguiform foot.* [from Latin *lingua* tongue + *forma* form]

lingulate (ling'gyə lāt), *adj. Botany, Zoology.* tongue-shaped; ligulate; linguiform: *lingulate antennae.* [from Latin *lingulatus,* from *lingula,* diminutive of *lingua* tongue]

linkage, *n. Genetics.* the association of two or more genes on the same chromosome so that they are transmitted together: *The ratios obtained when linkage occurs often do not conform to those expected* (Harbaugh and Goodrich, *Fundamentals of Biology*). Compare **independent assortment.**

linkage group, *Genetics.* a group of genes that are transmitted together: *Extensive studies of the linkage relations in various animals and plants have shown that the genes occur in linkage groups* (Storer, *General Zoology*).

linked, *adj. Genetics.* exhibiting linkage: *Because ... genes [that lie on the same chromosomes] tend to appear in the same combinations in future generations as they were in the original parents of an experiment the genes on the same chromosome are said to be linked* (Winchester, *Zoology*).

Linnean or **Linnaean** (li nē'ən), *adj.* **1** *Biology.* of or having to do with the Swedish naturalist Carolus Linnaeus (1707–1778) or his system of classification. The Linnean system of naming animals uses two words, the first for the genus and the second for the species. *Linnaean classification is a special type of binomial classification because of the unequal importance of the two words* (N.W. Pirie). **2** *Botany.* of the earlier system of plant classification introduced by Linnaeus, dividing plants into 24 classes.

linoleic acid (lin'ə lē'ik), *Biochemistry.* an unsaturated fatty acid essential to the human diet, found as a glyceride in linseed and other oils: *Coconut oil contains very little, if any, linoleic acid, an essential nutrient of particular concern from the standpoint of infant nutrition* (A. V. Gemmill). *Formula:* $C_{18}H_{32}O_2$

linolenic acid (lin'ə lē'nik), *Biochemistry.* an unsaturated fatty acid found in plant oils and essential to animal diet: *The investigation of the fatty acids present in food has shown that a particular kind called linolenic acid inhibits thrombosis and other coronary diseases* (Observer). *Formula:* $C_{18}H_{30}O_2$

lip, *n.* **1** *Botany.* **a** either of the two parts (upper and lower) into which the corolla or calyx of certain plants is divided. **b** (in an orchid) the labellum: *One petal always has a special shape, and is called the lip ... The lip has special marks to guide insects to the nectar inside the flower* (Alfred C. Hottes).
2 *Zoology.* a labium or labrum.

lip-, *combining form.* the form of **lipo-** before vowels, as in *lipase.*

lipase (lī′pās *or* lip′ās), *n. Biochemistry.* any of a class of enzymes occurring in pancreatic and gastric juices, certain seeds, etc., that can change fats into fatty acids and glycerin.

lipid (lip′id), *n. Biochemistry.* any of a group of organic compounds including the fats, oils, waxes, and sterols. Lipids are characterized by an oily feeling, solubility in fat solvents such as chloroform, benzene, or ether, and insolubility in water. *Lipids are responsible for the familiar water-repellent character of the surfaces of plants, animals, and insects. In chemical composition the surface lipids differ from the internal lipids. Collectively they are called waxes because of their peculiar physical properties, although in strict chemical terms, wax refers to esters of long-chain alcohols with long-chain acids* (Science).
▶ Lipids, like carbohydrates, contain carbon, hydrogen, and oxygen. However, the ratio of hydrogen to oxygen is much greater than 2:1, and is not constant from one lipid to another. Some lipids are the product of the dehydration synthesis of three molecules of fatty acids and one molecule of glycerol. Lipids function primarily as sources of stored energy and as components of cellular structures such as cell membranes.

lipide (lip′īd), *n.* = lipid.

lipo-, *combining form.* **1** fat; fatty, as in *lipogenesis.* **2** lipid, as in *lipoprotein.* Also spelled **lip-** before a vowel. [from Greek *lipos* fat]

lipocaic (lip′ə kā′ik), *n. Biochemistry.* a substance found in the pancreas that prevents the accumulation of fat in the liver: *A number of facts leads one to question whether or not lipocaic can be classified as a hormone* (H.J. Deuel). [from *lipo-* + Greek *kaiein* to burn]

lipochrome, *n. Biochemistry.* any of a group of fat-soluble pigments, such as the carotenoids, occurring in animals and plants. They produce the bright red or yellow plumage of birds. [from *lipo-* + Greek *chroma* color]

lipogenesis, *n. Biochemistry.* the formation of fat in the body: *Recent experiments on the control of fat biosynthesis, or lipogenesis, indicate the possibility of an unknown inhibitory factor, which will prevent synthesis taking place under certain conditions such as starvation* (New Scientist).

lipogenic, *adj. Biochemistry.* tending to produce fat: *a lipogenic substance.*

lipoic acid (li pō′ik *or* lī pō′ik), *Biochemistry.* a compound involved in the oxidation of keto acids in carbohydrate and protein metabolism. *Formula:* $C_8H_{14}O_2S_2$ Also called **thioctic acid.**

lipoid (lip′oid *or* lī′poid), *Biochemistry.* —*adj.* like fat or oil. —*n.* **1** = lipid. **2** any of a group of nitrogenous fatlike substances, such as the lecithins.

lipoidal (lip oi′dəl *or* lī′poi dəl), *adj.* = lipoid.

lipolysis (li pol′ə sis), *n. Biochemistry.* the breakdown or dissolution of a fat, as by the action of lipase: *Stimulation of the sympathetic nerves to adipose tissue causes lipolysis and mobilization of free fatty acids* (J.W. Hinman).
—**lipolytic** (lip′ə lit′ik), *adj.* of, having to do with, or of the nature of lipolysis.

lipoprotein, *n. Biochemistry.* any of a class of proteins, one of the components of which is a lipid. Alpha-lipoprotein has a high density and contains more protein than lipid; beta-lipoprotein has a low density and contains more lipid than protein.

liposome (lip′ə sōm *or* lī′pə sōm), *n. Biology.* **1** a droplet of fat or lipid in the cytoplasm of a cell.
2 an artificial membranous capsule made by enclosing an aqueous core with a phospholipid layer or layers: *It was discovered, at the clinical research centre at Northwick Park Hospital, in Harrow, that vaccines made into liposomes stimulated the immune system much more effectively* (John Newell).
—**liposomal,** *adj.* of or having to do with liposomes: *The researchers are now tagging the liposomes with antibodies that should help direct the liposomal-packaged enzymes to the appropriate target cells.* (Science News).

lipotropic, (lip′ə trop′ik *or* lī′pə trop′ik), *adj. Biochemistry.* **1** having an affinity for lipids, especially fats and oils. **2** preventing or reducing the accumulation of fat: *a lipotropic hormone.*

lipotropin (lip′ə trō′pin *or* lī′pə trō′pin), *n. Biochemistry.* a hormone of the pituitary gland that promotes the breakdown of fat in the body: *In schizophrenics an enzyme defect might lead to an imbalance in the endorphins produced from the original large pituitary chemical* (*known as beta lipotropin*) (London *Times*).

lipotropism (lip′ə trop′iz əm *or* lī′pə trop′iz əm), *n. Biochemistry.* lipotropic tendency or property.

lipped, *adj. Botany.* = labiate.

liquefaction (lik′wə fak′shən), *n. Physics.* **1** the process of changing into a liquid, especially of changing a gas by the application of pressure and cooling.
2 the condition of being changed into a liquid: *Even an initially cold earth would have been likely to pass through a stage of at least partial liquefaction, so that estimates of the age of the earth are in fact estimates of its age since 'resolidification'* (A.W. Haslett).

liquid, *Physics.* —*n.* a substance that is intermediate between a solid and a gas. A liquid is similar to a gas in that its molecules are relatively free to move with respect to each other, but it is unlike a gas and similar to a solid in that the volume of a given mass remains nearly constant in spite of changes in pressure. Most liquids assume the shape of the container in which

---

cap, fāce, fäther; bèst, bē, tèrm; pin, fīve;
rock, gō, ôrder; oil, out; cup, pùt, rüle,
yü in use, *yu* in uric;
ng in bring; *sh* in rush; *th* in thin, ᴛʜ in then;
zh in seizure.
ə = *a* in about, *e* in taken, *i* in pencil, *o* in lemon, *u* in circus

they are put and seek the lowest level. Water is a liquid at room temperature.

—*adj.* in the form of a liquid; neither solid nor gaseous: *liquid oxygen, liquid nitrogen.*
[from Latin *liquidus,* from *liquere* be fluid]

▶ **Liquid** and **fluid** are not exact synonyms. *Liquid* applies only to a substance that is normally neither a solid nor a gas: *Water and oil are liquids. Fluid* applies to anything that flows, including gases: *Air and water are fluids.*

liquid air, *Physics.* an intensely cold liquid formed by putting air under very great pressure and then cooling it. It is used as a refrigerant and as a source of nitrogen and oxygen. *Physicists first produced extremely cold temperatures in the 1870's with the development of liquid air* (M. J. Hiza).

liquid crystal, *Physics, Chemistry.* a substance that flows like a liquid but has some of the properties of a crystal within a certain temperature range. Many types of brightly colored liquid crystals have the property of changing colors because of small changes in temperature. *By sandwiching liquid crystals between panes of glass, engineers can design glare-free display devices, and windows that can be made cloudy or clear at the flick of a switch* (Michael Cusack).

liquid measure, 1 the measurement of liquids. **2** a system of units, such as pint and gallon, used for measuring the volume of liquids.

Metric system
10 milliliters = 1 centiliter
100 centiliters = 1 liter
100 liters = 1 hectoliter
English system
1 pint = 28.875 cubic inches
2 pints = 1 quart or 57.75 cubic inches
4 quarts = 1 gallon or 231 cubic inches

liquid membrane, *Chemistry.* a thin film either of oil forming a surface around a globule of water, or of water around a globule of oil, the film being stabilized by the surrounding molecules of an agent that reduces surface tension. A liquid membrane is either a barrier or a permeable film depending on its chemical composition or surface structure. *Liquid membranes, liposomes, and loaded erythrocytes have capsulelike structures with walls that differ greatly from those of classical microcapsules* (Curt Thies).

liquid oxygen, *Chemistry.* an intensely cold, transparent liquid formed by putting oxygen under very great pressure and then cooling it. It is used as a rocket fuel, in explosives, etc.

-lite, *combining form.* **1** rock, as in *rhyolite.* **2** mineral, as in *cryolite.* [from French *-lite,* from Greek *lithos* stone]

liter (lē′tər), *n.* the basic unit of volume or capacity in the metric system, equal to the volume of one kilogram of pure water at the temperature of its greatest density, 4°C, or 1000 cubic centimeters. *Abbreviation:* l. [from French *litre,* from Medieval Latin *litra,* from Greek *litra* pound]

literal equation, *Mathematics.* an equation in which one or more of the constant coefficients are letters, as in the equation ab + c = d, solve for b.

lith-, *combining form.* the form of **litho-** before vowels, as in *lithic, lithify.*

-lith, *combining form.* stone; rock; mineral, as in *batholith, laccolith.* [from Greek *lithos* stone]

lithia (lith′ē ə), *n. Chemistry.* a white oxide of lithium, soluble in water, and forming an acrid and caustic solution. *Formula:* Li_2O [from New Latin, from Greek *lithos* stone]

lithic, *adj.* **1** *Geology.* consisting of stone or rock: *At this depth the charcoal appears in small pockets and is associated with lithic tools, flint chips, and animal bones* (Science).
2 *Chemistry.* of, having to do with, or consisting of lithium.

lithification or **lithifaction,** *n. Geology.* the process by which sediment is changed into hard rock by pressure, heat, etc.: *Eventually, layers build up and a lithification, or stone-forming process, takes place* (David E. Jensen).

lithify, *v. Geology.* to change into rock: *A widespread layer in which seismic velocities ranged from 3.2 to 4.5 km. per second was thought to be a mixture of lithified volcanic ash* (Lawrence Ogden). SYN: petrify.

lithium (lith′ē əm), *n. Chemistry.* a soft, silver-white metallic element which occurs in small quantities in various minerals. Lithium is the lightest of all metals and is used in lubricants, ceramics, and chemical processes. *Symbol:* Li; *atomic number* 3; *atomic weight* 6.941; *melting point* 179° C; *boiling point* 1317° C; *oxidation state* 1. [from New Latin, from Greek *lithos* stone]

litho-, *combining form.* stone or rock, as in *lithology, lithophyte.* Also spelled **lith-** before a vowel. [from Greek *litho-,* from *lithos* stone]

lithofacies (lith′ə fā′shē ēz), *n. Geology.* a facies distinguished on the basis of lithologic characteristics. Compare **biofacies.**

lithoid (lith′oid), *adj.* of the nature or structure of stone.

lithologic or **lithological,** *adj. Geology.* **1** of or having to do with lithology: *Lithologic divisions of Devonian rocks ... are very numerous and most of them are recognized ... within 100 miles or less of the type outcrops from which these units are named* (Moore, *Introduction to Historical Geology*).
2 concerning the nature or composition of rock: *Lithologic similarity is useful in comparing two areas a few miles or perhaps a few tens of miles apart* (Garrels, *A Textbook of Geology*). —**lithologically,** *adv.*

lithology (li thol′ə jē), *n.* **1** the study of rocks and their composition. **2** the character or composition of a rock or rock formation: *a succession of beds with different lithologies.*

lithophile (lith′ə fīl), or **lithophilic** (lith′ə fil′ik), *adj. Geology.* having an affinity for silicates: *lithophilic elements.* Compare **atmophile, chalcophile, siderophile.**

lithophilous (li thof′ə ləs), *adj.* **1** *Botany.* growing on rocks: *The aquatic plants, including the algae, mosses, and flowering plants which live attached to rocks, comprise the lithophilous benthos* (Youngken, *Pharmaceutical Botany*).
2 *Zoology.* (of certain insects) living in stony places.

lithophyte (lith′ə fīt), *n. Botany.* a plant that grows among stone or rock.
—**lithophytic** (lith′ə fit′ik), *adj.* growing among stone or rock.

lithosol (lith′ə sol), *n. Geology.* any of a group of shallow soils with no distinct profiles, made up of imperfectly weathered rock fragments. [from *lith-* + Latin *solum* soil]

lithosphere (lith′ə sfir), *n. Geology.* **1** the solid portion of the earth as opposed to the atmosphere and the hydrosphere: *The earth's crust (lithosphere) is composed mainly of rocks* (Finch and Trewartha, *Elements of Geography*). Also called **geosphere.** See the picture at **converge.**
2 the rigid outer layer of the earth, comprising the crust and upper mantle.
ASSOCIATED TERMS: According to the model of plate tectonics, the crust and mantle of the earth are divided into the *lithosphere,* the *asthenosphere,* and the *mesosphere.* The *lithosphere,* which includes the crust and upper mantle, extends to a depth of about 100 kilometers of the earth; the *asthenosphere* extends from 100 to 250 kilometers into the mantle; and the *mesosphere,* or lower mantle, is the layer of hard rock below 250 kilometers to about 2900 kilometers where the core or central part of the earth begins.
—**lithospheric** (lith′ə sfir′ik), *adj.* of or having to do with the lithosphere: *Relative motions of the large lithospheric plates that make up the earth's outer shell appear to be the basic cause of earthquakes* (C. L. Drake).

lithotype, *n. Geology.* one of the four microscopically recognizable constituents of bonded coal; vitrain, clarain, durain, or fusain.

litmus (lit′məs), *n. Chemistry.* a blue coloring matter obtained from various lichens, used as an acid-base indicator. It turns red in acid solutions at pH 4.5, and blue in alkaline solutions at pH 8.3. [from Scandinavian (compare Old Icelandic *litmosi* moss used for dyeing, from *litr* color, dye + *mosi* moss)]

litmus paper, *Chemistry.* unsized paper treated with litmus, used as an indicator of pH (the ion concentration in solutions).

littoral (lit′ər əl), *Geology, Oceanography.* —*adj.* of, belonging to, or found on or near the shore: *the littoral zone, littoral drift.*
—*n.* a region along the shore or coast, especially the region or zone between the limits of high and low tides. [from Latin *litoralis,* from *litus* shore]

live-bearer (līv′ber′ər), *n. Zoology.* a fish that brings forth live young instead of eggs; a viviparous fish.

liver, *n. Anatomy.* **1** the large, reddish-brown organ in vertebrate animals that secretes bile and is active in the absorption and storage of vitamins, minerals, and sugar (which it changes into glycogen). The liver frees the blood of its waste matter and manufactures blood proteins: *The liver is a ... multi-purpose organ whose excretory functions include the breakdown of red blood cells and the production of urea following amino acid deamination* (Biology Regents Syllabus).
2 a gland in some invertebrates that secretes into the digestive tract.

living fossil, *Biology.* a plant or animal that is one of the last living species of an extinct stock known only from fossil remains: *A living fossil is defined as an organism that has survived beyond its era. A standard example is the tuatara of New Zealand, which looks like a lizard but is in fact the "sole survivor of an order of reptiles which flourished in the great Age of Reptiles and is now extinct except for this one species"* (Scientific American).

lm, *symbol.* lumen.

Ln, *symbol.* lanthanide.

load, *n.* **1** *Physics.* **a** the weight or force supported by a structure or any part of it. **b** the external resistance overcome by a machine under a given condition, measured by the power required. See the picture at **lever.** **2** *Genetics.* = genetic load.

load arm, *Physics.* the distance from the load to the fulcrum of a lever. Compare **effort arm.**

loam, *n. Geology.* soil that is between sandy soil and clayey soil in texture; rich fertile earth in which much humus is mixed with clay and sand: *There are many kinds of loam, depending on the relative amounts of the various components, as clay loam, sandy loam, etc.* (Emerson, *Basic Botany*).
—**loamy,** *adj.* of or resembling loam.

lobar (lō′bər or lō′bär′), *adj. Anatomy, Medicine.* of, having to do with, or affecting a lobe or lobes: *lobar pneumonia.*

lobate, *adj. Biology.* **1** having a lobe or lobes. **2** having the form of a lobe: *The liver is lobate.*
—**lobation,** *n.* lobe formation or condition.

lobe, *n. Biology.* a rounded projecting part. The brain, liver, etc., are divided into lobes. The lobe of the ear is the lower rounded end. *Just as the two lobes of the pituitary gland have a different origin, so have they different functions* (Bernard Donovan). [from Greek *lobos*]

lobular (lob′yə lər), *adj. Biology.* **1** having the form of a lobule or small lobe. **2** of or having to do with lobules: *a lobular vein.*

lobulate (lob′yə lit), *adj.* = lobulated.

lobulated (lob′yə lā′tid), *adj. Biology.* consisting of or separated into lobules: *lobulated kidneys.*

lobulation (lob′yə lā′shən), *n. Biology.* separation into lobules: *An automatic system for analyzing blood cells ... should give quantitative information about components of the cells' structure, such as the degree of lobulation of the nucleus* (Scientific American).

lobule (lob′yül), *n. Biology.* **1** a small lobe. **2** a part of a lobe.

local group or **Local Group,** *Astronomy.* a group of about 20 relatively nearby galaxies to which the Milky Way belongs, including also the Magellanic Clouds and the Andromeda spiral: *When we remember that a single light year is equal to almost six million million miles, we can see that the Spiral is almost inconceivably remote—and yet it is one of the closest of the external systems, and is a member of what we call the Local Group of galaxies* (Listener).
ASSOCIATED TERMS: see **galaxy.**

locomotion, *n. Biology.* the ability of an organism to move from place to place. Locomotion increases the probability of survival among animals and many protists. The chief advantage of locomotion is the oppor-

---

cap, fāce, fäther; best, bē, tėrm; pin, five;
rock, gō, ôrder; oil, out; cup, pùt, rüle;
yü in use, *yu̇* in uric;
ng in bring; *sh* in rush; *th* in thin, ᴛʜ in then;
zh in seizure.
ə = *a* in about, *e* in taken, *i* in pencil, *o* in lemon, *u* in circus

365

tunities it gives to obtain food, seek shelter, avoid predators, and mate. Locomotion in amebas is by pseudopods. In humans, locomotion is accomplished by the interaction of muscles and jointed appendages.
—**locomotive,** *adj.* **1** able to move from place to place: *Locomotive algae move by means of flagella.* **2** of or having to do with locomotion: *The locomotive structures used by a paramecium are cilia.*

locular (lok′yə lər), *adj. Biology.* having one or more locules (used chiefly in compounds, such as *bilocular, trilocular*).

loculate (lok′yə lāt *or* lok′yə lit), *adj. Biology.* having locules.

loculated (lok′yə lā′tid), *adj.* = loculate.

loculation (lok′yə lā′shən), *n. Biology.* separation into locules.

locule (lok′yül), *n. Biology.* a small cavity or cell in animal or plant tissue, separated from another locule by a septum, as in an ovary, fruit, or anther. Also called **loculus.** [from New Latin, from Latin *loculus,* diminutive of *locus* place]

loculicidal (lok′yə lə sī′dəl), *adj. Botany.* (of dehiscent seed capsules) splitting lengthwise through the back or dorsal suture of each locule or carpel, as in the seed pod of an iris. Compare **septicidal, septifragal.** [from New Latin *loculus* locule + Latin *-cidere* to cut]
—**loculicidally,** *adv.*

loculose, *adj. Botany.* divided into locules or cells.

loculus (lok′yə ləs), *n., pl.* **-li** (-lī). = locule.

locus (lō′kəs), *n., pl.* **-ci** (-sī *or* -kī). **1** *Mathematics.* a curve, surface, or other figure that contains all the points, and only those points, that satisfy a given condition. The locus of all the points 1 meter distant from a given point is the surface of a sphere having a radius of 1 meter. *The locus of points which are equidistant from the ends of a line segment is the perpendicular-bisector of the line segment* (Herberg, *A New Geometry for Secondary Schools*).
2 *Genetics.* the position of a gene on a chromosome: *Each pair of chromosomes ... constitutes a pair ... Each member of the homologous pair carries, at identical positions, or loci, a gene affecting the expression of the same trait* (Weier, *Botany*).
[from Latin *locus* place]

locusta (lō kus′tə), *n., pl.* **-tae** (-tē). *Botany.* the inflorescence or spikelet of grasses. [from New Latin *locusta,* from Latin *locusta* locust (supposedly from the shape)]

lodestone, *n. Mineralogy.* a piece of magnetite that possesses polarity and attracts iron and steel: *A lodestone or a steel magnet experiences a torque that tends to orient it in a particular direction on the earth. This led to the important invention, sometime before the 12th century A.D., of the mariner's compass* (Shortley and Williams, *Elements of Physics*).

lodicule (lod′ə kyül), *n. Botany.* one of the small scales in the flowers of most grasses, close to the base of the ovary: *Flowering begins with the swelling of two small scale-like structures at the base of the flower, the lodicules* (New Scientist). [from Latin *lodicula,* diminutive of *lodix, lodicis* coverlet]

loess (lō′is *or* les), *n. Geology.* a yellowish-brown marl or loam, usually deposited by the wind. [from German *Löss*]
—**loessal** or **loessial,** *adj.* of or having to do with loess: *Loessial soils are extremely productive of crops when well watered* (White and Renner, *Human Geography*).

log or **log.,** *abbrev.* logarithm.

logarithm (lô′gə riŦH′əm *or* log′ə riŦH′əm), *n. Mathematics.* **1** the power to which a base (usually 10) must be raised to produce a given number. If the base is 10, the logarithm of 1000 is 3; the logarithm of 10,000 is 4; the logarithm if 100,000 is 5. *The usefulness of logarithms, which youngsters learn about in high school, is that multiplication and division can be turned into addition and subtraction, while squaring and taking roots become multiplication and division* (Science News Letter). See also **common logarithm, natural logarithm.**
2 one of a system of such exponents used to shorten calculations in mathematics: *Data on the atmospheric path and orbit ... The calculations were carried out with four-place logarithms to minutes of arc* (Science). [from New Latin *logarithmus,* from Greek *logos* proportion + *arithmos* number]
—**logarithmic** (lô′gə riŦH′mik *or* log′ə riŦH′mik), *adj.* of or expressed in a logarithm or logarithms: *The logarithmic function $y = \log x$. Because of the large range of intensities over which the ear is sensitive, a logarithmic rather than an arithmetic intensity scale is more convenient* (Sears and Zemansky, *University Physics*).
—**logarithmically,** *adv.*

logic, *n.* **1** the principles of reasoning and inference; science of reasoning or the science of proof.
2a the nonarithmetical operations in a computer, such as comparing, selecting, matching, and sorting, where binary or yes-or-no decisions are involved: *Data processing, in the broadest definition, means the handling of information by arithmetic rules and logic, in. In practically all cases, the information is arithmetic and the machines add, subtract, multiply, and divide. The more complex machines can perform logic operations: "and," "not," "or"* (Evan Herbert). **b** the electronic circuitry performing such operations: *The computer logic is so fast that it has to loaf at several intervals while the input and output devices—the peripherals—are printing information* (New Yorker). [from Greek *logikē* (technē) reasoning (art), from *logos* word, from *legein* speak]
—**logical,** *adj.* having to do with logic; according to the principles of logic: *Gating circuits are the foundation for proper sequential operation of the computer, and they are peculiarly adapted for the performance of mathematical functions. The mathematical functions themselves can be performed on the basis of three logical steps, known as 'And', 'Or', and 'Not'. Circuits for carrying out these logical operations are quite simple* (Science News).

lognormal (lôg′nôr′məl *or* log′nôr′məl), *adj. Statistics.* having a normal or symmetrical logarithmic distribution. [from *log(arithm)* + *normal*]

-logy, *combining form.* science or study of, as in *endocrinology* = science dealing with endocrine glands. [from Greek *-logia,* from *logos* word, discourse, from *legein* speak]

loin, *n. Zoology.* Usually, **loins,** *pl.* the part of the body of an animal between the ribs and the hipbones. The loins are on both sides of the spinal column and nearer to it than the flanks. Compare **lumbar.**

loment (lō′ment), *n. Botany.* a leguminous fruit which is contracted in the spaces between the seeds, breaking up when mature into one-seeded segments, as in the tick trefoil. [from New Latin *lomentum*]
—**lomentaceous** (lō′men tā′shəs), *adj. Botany.* **1** of or like a loment. **2** bearing loments.

lomentum (lō men′təm), *n., pl.* **-ta** (-tə). = loment.

long. or **lon.,** *abbrev.* longitude.

long-day plant, *Botany.* a plant that flowers only when exposed to light for a relatively long period of time each day: *In long-day plants, the long photo-period activates the synthesis of the specific gibberellin (one of the five groups of compounds regulating plant growth) needed for the formation of the flower stalk* (Richard M. Klein). Contrasted with **short-day plant.**
ASSOCIATED TERMS: see **day-neutral plant.**

long division, *Mathematics.* a method of dividing numbers in which each step of the division is written out. It is used to divide large numbers.

longicorn (lon′jə kôrn), *adj. Zoology.* **1** having long antennae. **2** of or belonging to a family of beetles that often have very long antennae. [from Latin *longus* long + *cornu* horn]

longipennate (lon′jə pen′āt), *adj. Zoology.* having long wings. [from Latin *longus* long + English *pennate*]

longitude (lon′jə tüd), *n.* **1** *Geography.* distance east or west on the earth's surface, measured in degrees from a certain meridian, usually the meridian through Greenwich, England. **2** *Astronomy.* **a** = celestial longitude. **b** = galactic longitude. *Abbreviation:* long., lon. Compare **latitude.** [from Latin *longitudo* length, from *longus* long]
—**longitudinal** (lon′jə tüd′n əl), *adj.* **1** of length; in length: *longitudinal measurements.* **2** running lengthwise: *Valleys which run parallel with mountain ranges are called longitudinal valleys* (E. D. Wilson). See the pictures at **dune, wave. 3** of or having to do with longitude. —**longitudinally,** *adv.*

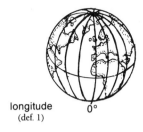

longitude
(def. 1)

longitudinal wave, *Physics.* a wave in which individual particles of a medium move back and forth in the same direction the wave moves. Compare **compression wave.** Contrasted with **transverse wave.**

long-period variable, *Astronomy.* a variable star whose period from one peak of brightness to the next is over 100 days: *The long-period variables ... are amongst the largest of known stars. They are indeed super-giants, and the biggest of them have a volume 25,000,000 times that of the sun. Yet they are extremely tenuous, and ... as various scientists have noted, little more than*

red-hot vacua (Bernhard, *Handbook of Heavens*). Also called **long-period variable star.**

loop, *n.* **1** *Physics.* **a** the portion of a vibrating string, column of air in an organ pipe, etc., between two nodes. **b** the middle point of such a part.
2 *Electricity.* a complete or closed electric circuit.
—*v. Electricity.* to join (conductors) so as to form a loop.

loop of Henlé, = Henlé's loop.

lophophore (lō′fə fôr), *n. Zoology.* an organ for gathering food, situated near the mouth of brachiopods, bryozoans, and certain other animals, and consisting of rows of ciliated tentacles that set up currents which carry tiny particles of food into the mouth. [from Greek *lophos* crest + English *-phore*]

lopolith (lop′ə lith), *n. Geology.* a centrally sunken body of intrusive igneous rock: *Most lopoliths ... are marvellously differentiated into many thin sheets and bands of contrasting mineral composition* (Gilluly, *Principles of Geology*). Compare **batholith, laccolith.** [from Greek *lopos* bent backward, convex + *lithos* stone]

lore (lôr or lōr), *n. Zoology.* **1** a space between the eye and the side of the superior mandible of a bird. **2** an area between the eye and the nostril of a snake or other reptile. **3** a corresponding space or area in fishes. Also called **lorum.** [from New Latin *lorum,* from Latin *lorum* strap, thong]
—**loreal** (lôr′ē əl or lōr′ē əl), *adj.* of or having to do with a lore.

Lorentz force (lôr′ents), *Physics.* a force acting on an electrically charged particle moving through a magnetic field: *Lorentz force ... tends to make a charged particle entering a magnetic field travel at right angles to both the direction of the original motion and the lines of force* (Scientific American). [named after Hendrik Antoon *Lorentz,* 1853–1928, Dutch physicist]

Lorentz transformation, *Physics.* any of a series of equations used in the special theory of relativity to show the relationship between quantities measured in one frame of reference and those measured in another frame moving with respect to it.

lorica (lō rī′kə), *n., pl.* **-cae** (-kē or -sē), *Biology.* **1** a hard, thickened body wall, as of a rotifer. **2** a protective case or sheath, as of a protozoan. [from Latin *lorica* leather cuirass, from *lorum* strap, thong]
—**loricate** (lôr′ə kāt), *adj.* having a lorica. —*n.* a loricate animal.

lorum (lô′rəm or lō′rəm), *n., pl.* **-ra** (-rə). = lore.

loss, *n. Electricity.* the reduction in power, measured by the difference between the power input and power output, in an electric circuit, device, or system, corre-

cap, fāce, fäther; best, bē, tèrm; pin, five;
rock, gō, ôrder; oil, out; cup, pùt, rüle;
yü in use, *yu* in uric;
ng in bring; *sh* in rush; *th* in thin, ᴛH in then;
zh in seizure.
ə = *a* in about, *e* in taken, *i* in pencil, *o* in lemon, *u* in circus

sponding to the transformation of electric energy into heat.

—lossy, *adj.* tending to lose or dissipate electrical energy: *At optical frequencies a metal transmission line structure would be very lossy and only transparent dielectric materials such as glass can be considered* (Science Journal).

lotic (lō'tik), *adj. Ecology.* of, living in, or designating rapidly flowing water, such as a stream. Contrasted with **lentic.** [from Latin *lotus* (a past participle of *lavare* to wash) + English *-ic*]

Love wave, *Geology.* a seismic wave that travels across the earth's surface at a velocity of 4 to 4 1/2 kilometers per second: *Two of them travel through the body of the earth: the primary waves (P waves) and the slower secondary waves (S waves). The other two move along the surface (that is, through the top 20 miles or so): Love waves and Rayleigh waves* (Scientific American). [named after A.E.H. *Love,* 1863–1940, English mathematician who discovered it]

low, *adj. Biology.* not advanced in organization or development: *low organisms, the lower primates.* Compare **high.**

—n. *Meteorology.* an area of relatively low barometric pressure: *Winds in a low rotate counterclockwise in the Northern Hemisphere [and] clockwise in the Southern Hemisphere* (James E. Miller). Also called **cyclone.**

low-energy, *adj. Physics.* of or having to do with energy of less than roughly 100 million electron volts; not high-energy: *low-energy electrons or neutrons, low-energy physics.*

lower, *adj.* Usually spelled **Lower.** *Geology.* being or relating to an earlier or older division of a period, system, or the like: *Lower Cretaceous.* Compare **upper.**

lower bound, *Mathematics.* a number less than or equal to every number in a given set of real numbers. Contrasted with **upper bound.** Compare **greatest lower bound.**

Lower Carboniferous, *Geology.* **1** the name outside North America for the Mississippian period.
2 the rocks formed in this period: *The term Carboniferous ... was first applied to coal-bearing rocks in England* (1822) ... *called Lower Carboniferous, corresponding closely to the Mississippian, and ... Upper Carboniferous, being approximately equivalent to the Pennsylvanian* (Moore, *Introduction to Historical Geology*).

lowest common denominator, = least common denominator.

lowest common multiple, = least common multiple.

low-frequency, *adj. Physics.* of or having to do with a frequency ranging from 30 to 300 kilohertz: *low-frequency sound waves. Abbreviation:* LF

lowland, *Geology.* **—n.** Often, **lowlands,** *pl.* a land that is lower and flatter than the neighboring country: *Upland and lowland are relative terms, and refer to the flow of water through the ground* (E. S. Deevey, Jr.). **—adj.** of or in a lowland: *lowland hills.*

low level, *adj.* **1** low in content; containing a relatively small amount of something: *Mice were given low-level doses of streptomycin* (Science News Letter).

2 *Meteorology.* occurring at a low level of altitude: *low-level turbulence.*
3 *Physics.* having a low level of radioactivity: *Much of this refuse from fission is low-level and short-lived* (Newsweek).

low-pressure, *adj. Meteorology.* having a low barometric pressure: *a low-pressure trough.*

low-temperature physics, the branch of physics dealing with matter and natural phenomena at temperatures below 90 kelvins, the boiling point of oxygen. At extremely low temperatures, thermal, electric, and magnetic properties of most materials are found to be unusual, of which perhaps the best known are the superfluidity of helium and the superconductivity of mercury, gallium arsenide, and silicon carbide. Compare **cryogenics.**

low tide, *Oceanography.* **1** the lowest level of the tide; the minimum elevation of the ocean at the shore: *The earth turns rapidly upon its axis, and soon the part of the ocean that was nearest the moon (where the water bulges out) turns away from the moon. Then the water settles back, and we have low tide* (Beauchamp et al., *Everyday Problems in Science*). **2** the time when the tide is lowest.

low water, = low tide.

loxodrome (lok'sə drōm), *n. Geography.* a loxodromic curve or line: *On the earth's surface the counterpart of the logarithmic spiral is the loxodrome (or rhumb line): a path that cuts the earth's meridians at any constant angle except a right angle. Thus if you were flying northeast and always kept the plane heading in exactly the same direction as indicated by the compass, you would follow a loxodrome that would spiral you to the North Pole* (Martin Gardner).

loxodromic curve or **loxodromic line,** *Geography.* a line on the surface of a sphere cutting all meridians at the same angle, such as that formed by the path of a ship whose course is constantly directed to the same point of the compass in a direction oblique to the equator: *countercurrents that move in a sort of spiral or loxodromic curve.* Also called **rhumb line.** [from Greek *loxos* oblique + *dromos* course]

Lr, *symbol.* lawrencium.

LRF, *abbrev.* luteinizing hormone-releasing factor.

LSD, = lysergic acid diethylamide.

L-shell, *n. Physics.* the energy level next in order after the K-shell, occupied by electrons whose principal quantum number is 2: *They have obtained pictures of neon and argon atoms taken under conditions that show the electron distribution in the L shell (the second shell out from the nucleus) enlarged 500 million times* (Science News). Compare **M-shell.** See also **L electron.**

LTH or **LtH,** *abbrev.* luteotrophic hormone.

Lu, *symbol.* lutetium.

l.u.b., *abbrev.* least upper bound.

luciferase (lü sif'ə rās), *n. Biochemistry.* an enzyme found in the cells of luminescent organisms, which acts on luciferin to produce luminosity.

luciferin (lü sif'ər in), *n. Biochemistry.* a chemical substance found in the cells of luminescent organisms, such as fireflies, which, when acted on by luciferase, undergoes oxidation and produces heatless light: *Scientists at John Hopkins University finally succeeded in reproducing luciferin (firefly light). Its com-*

plex molecules consist of 26 atoms of five different elements (A. Richard Harmet).

lucifugous (lü sif′yə gəs), *adj. Zoology.* avoiding light, as bats and cockroaches do: *such lucifugous creatures as owls.* [from Latin *lucifugus,* from *lux, lucis* light + *fugere* to flee]

lumbar (lum′bər), *Anatomy.* —*adj.* of or having to do with the loins; belonging to the area between the lowest ribs and the pelvic girdle: *Among some of the earlier men the spine is only feebly curved in the lumbar region (that is, the small of the back). As a result, the spine of such forms differs distinctively from that of modern man, characterized by a decided lumbar curve* (Beals and Hoijer, *Anthropology*).
—*n.* a lumbar vertebra, artery, nerve, etc.: *The lower oblique processes of the last lumbar are unusually far apart* (Thomas H. Huxley).
[from Latin *lumbus* loin]

lumbosacral (lum′bō sā′krəl), *adj. Anatomy.* of or having to do with both the lumbar and sacral regions or parts of the body: *Lumbosacral subluxation* [*is*] *a condition in which there is forward slipping of the fifth lumbar vertebra on the sacrum* (J. S. Batchelor).

lumbrical (lum′brə kəl), *Anatomy.* —*n.* any one of four small muscles in either the hand or the foot that help move the fingers or toes.
—*adj.* of or having to do with these muscles: *a filament to the second lumbrical muscle.*
[from New Latin *lumbricalis,* from Latin *lumbricus* worm]

lumen (lü′mən), *n., pl.* **-mina** (-mə nə), **-mens. 1** *Physics.* the SI unit of luminous flux, equal to the amount of light given out through a solid angle by a source of one candela radiating equally in all directions: *The lumen is seen to be a unit that is basically of the nature of power* (*rather than of energy*) (Shortley and Williams, *Elements of Physics*). *Symbol:* lm
2 *Anatomy.* the space within a tubular organ, such as a blood vessel.
3 *Botany.* the central cavity or space within the wall of a cell: *The lumina of both tracheids and vessels are larger, forming small conductive tubes through which water moves freely* (Emerson, *Basic Botany*).
[from Latin *lumen, luminis* light]

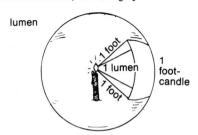

lumen

1 foot

1 lumen

1 foot-candle

1 foot

lumen-second, *n. Physics.* an SI unit of quantity of light, equal to the amount of light transmitted in one second by a luminous flux of one lumen; one lumen multiplied by one second: *The unit of photographic exposure is ... the lumen-second per square meter or, more frequently, its numerical equivalent, the meter-candle-second* (Hardy and Perrin *Principles of Optics*).

luminal (lü′mə nəl), *adj. Anatomy, Botany.* of or belonging to a lumen: *... the luminal walls of these intestinal cells* (American Naturalist).

luminance (lü′mə nəns), *n. Physics.* the intensity of light in relation to the area of its source: *The luminance of an extended source is the intensity per unit area of the source* (Sears and Zemansky, *University Physics*).

luminesce (lü′mə nes′), *v. Physics.* to exhibit luminescence: *Green plants were found to luminesce like fireflies, although on a small scale* (Science News Letter).

luminescence (lü′mə nes′ns), *n. Physics.* an emission of light occurring at a temperature below that of incandescent bodies. Luminescence includes phosphorescence and fluorescence. *Luminescence comes about through the ability of certain substances to absorb light of relatively high frequency and re-emit it in instalments of discrete lower frequencies* (New Scientist). Compare **bioluminescence.**
—**luminescent,** *adj.* **1** giving out light without much heat. **2** having to do with luminescence.

luminosity, *n.* **1** *Physics.* the relative effectiveness of light of a particular wavelength in producing brightness: *A color specification in terms of three excitation values ... can be converted into terms of dominant wavelength, purity, and luminosity* (Hardy and Perrin, *Principles of Optics*).
2 *Astronomy.* the amount of energy in the form of light emitted by the sun or a star: *The classic method for sorting out the evolutionary history of a group of stars is to plot the temperature of each star against its luminosity* (D. N. Limber).
3 *Nuclear Physics.* the number and density of accelerated particles in two intersecting beams: *The high-energy, higher-luminosity storage rings would ... be very useful for the study of the weak interaction* (Dietrick E. Thomsen).

luminous, *adj. Physics.* **1** transmitting or reflecting light: *The sun and stars are luminous bodies.* **2** of or having to do with the rate at which light is transmitted or reflected: *luminous efficiency.* [from Latin *luminosus,* from *lumen, luminis* light]

luminous energy, *Physics.* energy in the form of light; radiant energy that produces the sensation of vision.

luminous flux, *Physics.* the rate at which a quantity of light is transmitted. The SI unit of luminous flux is the lumen.

luminous intensity, *Physics.* a measure of the strength of a source of light, equivalent to the luminous flux given out per unit solid angle in a given direction.

lumisome (lü′mə sōm′), *n. Biology.* a light-emitting particle in the cells of animals that emit light; a unit particle of bioluminescence: *"Lumisomes" contain all the various molecules which have previously been identified as part of the bioluminescent system. So the list of sub-cellular organelles increases, and lumisomes*

cap, fāce, fäther; best, bē, tèrm; pin, five;
rock, gō, ôrder; oil, out; cup, pùt, rüle;
yü in use, yù in uric;
ng in bring; sh in rush; th in thin, ₮H in then;
zh in seizure.
ə = *a* in about, *e* in taken, *i* in pencil, *o* in lemon, *u* in circus

369

take their place with lysosomes, peroxisomes and the rest (New Scientist). [from Latin *lumen, luminis* light + English *-some*]

lunabase (lü′nə bās′), *n. Astronomy.* the region comprising the low, flat surfaces of the moon; the lunar maria: *The* [*lunar*] *mountains end with steep terminal slopes of 20° or so, which cannot be due to erosion as a result of melting by subsequent lavas from the mare region* (*the dark lunabase*) (Ian Ridpath). Contrasted with **lunarite.**

lunar, *adj. Astronomy.* **1** of or on the moon: *lunar soil, lunar mountains.* **2** measured by the moon's revolutions: *a lunar year.* [from Latin *lunaris,* from *luna* moon]

lunar cycle, = Metonic cycle.

lunar day, *Astronomy.* **1** the interval between two successive crossings of the same meridian by the moon: *Tidal rhythms are related to the lunar day of 24 hours, 50 minutes, and so, ... obviously the time of day of exposure at low tide will have a considerable effect upon the littoral fauna* (J.L. Cloudsley-Thompson). **2** the time it takes the moon to make one rotation on its axis; about 27 days, 7 hours, and 43 minutes.

lunar eclipse, *Astronomy.* the total or partial cutting off of the light of the full moon by the earth's shadow. It occurs when the sun, earth, and moon are in, or almost in, a straight line.

lunarite (lü′nə rīt), *n. Astronomy.* the region comprising the upland surfaces of the moon. Contrasted with **lunabase.**

lunar month, *Astronomy.* the period of one complete revolution of the moon around the earth; the interval between one new moon and the next: *The proper lunar month, which is called the synodical month, is the period between one new moon and the next, an average of 29 days, 12 hours, 44 minutes, and 2.8 seconds* (Paul Sollenberger).

lunar year, *Astronomy.* a period of 12 lunar months, about 354 1/3 days: *The Greeks had begun to compensate for the defect of the lunar year, by the occasional addition of an intercalary month* (Connop Thirlwall).

lunate (lü′nāt), *Anatomy.* —*adj.* crescent-shaped: *a lunate bone.*
—*n.* a crescent-shaped bone of the human wrist, in the proximal row of carpal bones.
[from Latin *lunatus,* past participle of *lunare* to bend into a crescent, from *luna* moon]

lunation, *n. Astronomy.* the time from one new moon to the next, about 29 1/2 days; lunar month.

lune (lün), *n. Geometry.* a crescent-shaped figure on a plane or a sphere bounded by two arcs of circles. [from French *lune,* from Latin *luna* moon]

lung, *n. Anatomy.* **1** either one of a pair of saclike, spongy organs by means of which the blood receives oxygen and is rid of carbon dioxide in vertebrates that breathe air: *The lungs form a soft, spongy mass which completely fills the chest cavity except for the small space occupied by the heart and the esophagus, or food tube. The lungs are passive; they respond to variations in air pressure, but do not themselves contain muscular fibers to initiate movement* (Charles K. Thomas). See the picture at **bronchi.**

2 a similar organ in certain invertebrates, such as snails and spiders.

lunisolar, *adj. Astronomy.* having to do with the mutual relations or joint action of the moon and sun: *lunisolar attraction.*

lunitidal, *adj. Oceanography.* having to do with the movements of the tide dependent on the moon. A **lunitidal interval** is the time between the transit of the moon and the next lunar high tide.

lunula (lü′nyə lə), *n., pl.* **-lae** (-lē). — **lunule.**

lunule (lü′nyül), *n. Botany, Zoology.* a crescent-shaped mark or part, such as the white area of a fingernail near the root: *whitish lunules on the tail-feathers* (John Stark). [from Latin *lunula,* diminutive of *luna* moon]

luster, *n. Mineralogy.* the appearance of the surface of a mineral due to the reflection of light: *Cut a piece of lead or zinc, and observe the luster of its fresh surface* (Thomas H. Huxley).
▶ The luster of a mineral is usually classified according to the following types: metallic, vitreous (like glass), resinous, pearly, adamantine (brilliant like diamonds), silky, earthy, and dull.

luteal (lü′tē əl), *adj. Biology.* of or having to do with the corpus luteum: *luteal tissue, the luteal hormone.*

lutecium (lü tē′sē əm), *n.* = lutetium.

lutein (lü′tē in), *n. Biochemistry.* **1** a yellow pigment obtained from the corpus luteum and found in egg yolks. It is a carotenoid similar to xanthophyll. **2** = xanthophyll.
[from (corpus) *lute*(*um*) + *-in*]
—**luteinization,** *n.* the act or process of luteinizing: *luteinization of the Graafian follicle cells.*
—**luteinize,** *v. Physiology.* to form lutein or cells of a corpus luteum: *This preparation ... does luteinize the follicles produced by the gonad stimulating fraction* (American Journal of Physiology).

luteinizing hormone, *Biochemistry.* a hormone produced by the anterior lobe of the pituitary gland, which in the female stimulates the development of the corpus luteum and in the male stimulates the interstitial cells of the testes to produce testosterone. *Abbreviation:* LH Also called **interstitial cell stimulating hormone.**

luteinizing hormone-releasing factor or **luteinizing hormone-releasing hormone,** *Biochemistry.* a hormone secreted by the hypothalamus that causes the release of luteinizing hormone: *Still another hormone, luteinizing hormone-releasing factor* (*LH-RH or LRF*) ... *serves as a master switch over luteinizing hormone, the sex hormones and ovulation* (Joan Arehart-Treichel). *Abbreviation:* LRF, LH-RH

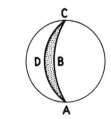

lune

The shaded area on the sphere is a lune bounded by the arcs CDA and CBA.

luteotrophic or **luteotropic,** *adj. Biochemistry.* affecting or stimulating the corpus luteum.

luteotrophic hormone or **luteotropic hormone,** = prolactin: *The feedback action of progesterone is ... to stimulate the release of luteotrophic hormone (LtH), responsible for the maturation of the corpus luteum which in turn produces both more progesterone and oestrogen* (Peter Bishop). *Abbreviation:* LTH, LtH

luteotrophin or **luteotropin,** *n.* = prolactin.

lutetium (lü tē′shē əm), *n. Chemistry.* a rare-earth metallic element which usually occurs in nature with ytterbium. *Symbol:* Lu; *atomic number* 71; *atomic weight* 174.97; *melting point* 1652°C; *boiling point* 3327°C; *oxidation state* 3. Also spelled **lutecium.** [from New Latin, named after *Lutetia,* ancient Latin name for Paris (because a French scientist discovered it)]

lux (luks), *n., pl.* **lux** or **luxes.** the SI unit of illumination, equivalent to the amount of light falling on a surface which is situated, at all points, one meter from a point source of one candela. A lux equals one lumen per square meter. *Symbol:* 1x [from Latin *lux, lucis* light]

lvs., *abbrev. Botany.* leaves.

lx, *symbol.* lux.

lycopene (lī′kə pēn), *n. Biochemistry.* a red pigment, a carotenoid, occurring in ripe fruit, especially tomatoes. [from New Latin *Lycop* the genus name of the tomato plant + English *-ene*]

Lyman-alpha line, *Physics, Astronomy.* an emission line of hydrogen in the extreme ultraviolet range of the spectrum, with a wavelength of 1216Å. It is a strong line in the spectra of stars, nebulae, quasars, and active galaxies. [named after E. M. *Lyman,* 20th-century American physicist]

lymph, *n. Biology.* a nearly colorless liquid in the tissues of the body, derived from parts of the blood which have filtered through blood capillary walls. Lymph is conveyed back to the bloodstream by the lymphatic vessels. It has a slightly alkaline quality and serves to bathe and nourish the tissues. [from Latin *lympha* clear water]

lymphangial (lim fan′jē əl), *adj. Physiology.* of or having to do with the lymphatic vessels. [from New Latin *lympha* lymph + Greek *angeion* vessel]

lymphatic, *adj. Physiology.* of, carrying, or relating to lymph: *the lymphatic system.* —**lymphatically,** *adv.*

lymphatic gland, = lymph gland.

lymphatic system, *Anatomy.* the network of small vessels, resembling blood vessels, by which lymph circulates throughout the body carrying food from the blood to the cells, picking up fats from the small intestines, and carrying body wastes to the blood.

lymphatic vessel, = lymph vessel.

lymph cell or **lymph corpuscle,** = lymphocyte.

lymph gland or **lymph node,** *Anatomy.* any of the small oval bodies occurring along the paths of the lymph vessels, active in filtering out harmful microorganisms from the lymph and as a source of lymphocytes: *Throat infections, for example, may be accompanied by swelling of lymph nodes in the neck* (Storer, *General Zoology*).

lympho-, *combining form.* lymph, as in *lymphocyte = a lymph cell.* [from New Latin, from Latin *lympha* clear water]

lymphoblast (lim′fə blast), *n. Biology.* a lymphocyte in an early stage of development. [from *lympho* + Greek *blastos* germ sprout]
—**lymphoblastic,** *adj.* of a lymphoblast: *lymphoblastic marrow.*

lymphocyte, *n. Biology.* any of the nearly colorless cells of the blood and lymphatic system, produced by lymph glands. Lymphocytes have a nucleus and are chiefly responsible for immunity. They defend the body against infection by producing antibodies and other substances that destroy viruses and other foreign bodies. B cells and T cells are lymphocytes. *There are three major types of leukocytes, namely, lymphocytes, monocytes, and the granular cells termed granulocytes* (Harbaugh and Goodrich, *Fundamentals of Biology*). Also called **lymph cell** or **lymph corpuscle.**
—**lymphocytic** (lim′fə sit′ik), *adj.* of or having to do with a lymphocyte or lymphocytes.

lymphoid, *adj. Biology.* of or having to do with lymph, lymphocytes, or lymphoid tissue: *lymphoid cells.*

lymphoid tissue, *Histology.* the tissue that forms most of the lymph glands and thymus gland, consisting of connective tissue containing lymphocytes.

lymphokine (lim′fə kin), *n. Biochemistry.* any of various substances of low molecular weight secreted by T cells that have been activated by antigens. Lymphokines are involved in cell-mediated immunity and other processes affecting cells such as stimulation of mitosis in lymphocytes and inhibition of macrophages. [from *lympho-* + Greek *kinein* to move]

lymphoma (lim fō′mə), *n. Medicine.* any of a class of cancers originating in lymphoid tissue. Compare **carcinoma, sarcoma.**

lymphopoiesis (lim′fō poi ē′sis), *n.* the formation of lymphocytes or of lymph. [from *lympho-* + Greek *poiesis* formation, composition]

lymph system, = lymphatic system.

lymph vessel, *Anatomy.* a small tube or canal through which lymph is carried from different parts of the body: *Major lymph vessels have lymph nodes which contain phagocytic cells which filter bacteria and dead cells from the lymph.* (Biology Regents Syllabus). Also called **lymphatic vessel.**

lyonization, *n. Genetics.* the inactivation of one X-chromosome in female mammals according to the Lyon principle.

Lyon principle or **Lyon hypothesis,** *Genetics.* the theory that in all normal female mammals, one X-chromosome becomes genetically inactive early in the development of the embryo and remains nonfunctional for the rest of that individual's life. [from Mary F. *Lyon,* 20th century English geneticist]

lyophile (lī′ə fīl), *adj.* = lyophilic.

---

cap, face, fäther; best, bē, tèrm; pin, five;
rock, gō, ôrder; oil, out; cup, pùt, rüle;
yü in use, *yu* in uric;
ng in bring; *sh* in rush; *th* in thin, ⟨TH⟩ in then;
zh in seizure.
ə = *a* in about, *e* in taken, *i* in pencil, *o* in lemon, *u* in circus

371

lyophilic

lyophilic (lī′ə fil′ik), *adj. Chemistry.* characterized by strong attraction between the colloid and the dispersion medium of a colloidal system: *Lyophyllic colloidal systems are affected very little by electrolytes* (W.N. Jones). Contrasted with **lyophobic.** [from Greek *lyein* to loosen + English *-phil* + *-ic*]

lyophobic (lī′ə fō′bik), *adj. Chemistry.* characterized by a lack of attraction between the colloid and the dispersion medium of a colloidal system. Contrasted with **lyophilic.** [from Greek *lyein* to loosen + English *phobic*]

lys-, *combining form.* the form of **lyso-** before vowels, as in *lysate, lysin.*

Lys, *abbrev.* lysine.

lysate (lī′sāt), *n. Biochemistry.* the product resulting from the destruction of a cell by a lysin or lysins: *A lysate ... may contain over a hundred thousand million bacteriophage particles* (Science News Letter).

lyse (līs), *v. Biochemistry.* to bring about the dissolution of cells by lysins; subject to lysis: *These viruses prevent bacterial growth and division, eventually lysing the cells* (Science).

lysergic acid (lī sėr′jik *or* lə sėr′jik), *Chemistry.* an acid derived from ergot alkaloids, used in medical research for its psychotomimetic effects. *Formula:* $C_{16}H_{16}N_2O_2$ [from *lys(is)* + *erg(ot)* + *-ic*]
—**lysergic acid diethylamide** (dī eth′ə lam′īd), a compound of lysergic acid that produces psychotic symptoms like those of schizophrenia. *Formula:* $C_{20}H_{25}N_3O$ Commonly known by the abbreviation **LSD.**

lysin (lī′sn), *n. Biochemistry.* an antibody that can dissolve bacteria, red blood cells, and other cellular elements. Lysins are developed in blood serum.

lysine (lī′sēn), *n. Biochemistry.* a basic amino acid essential for growth, formed by the hydrolysis of various proteins. *Formula:* $C_6H_{14}N_2O_2$ *Abbreviation:* Lys

lysis (lī′sis), *n., pl.* **lyses** (lī′sēz′). *Biology.* the destruction of a cell by dissolution of the cell wall or membrane, as by a lysin: *Certain antibiotics can cause physical disintegration or 'lysis' of cells when added to growing cultures of sensitive bacteria* (Nature). [from Greek *lysis* a loosening, from *lyein* to loosen]

lyso-, *combining form.* **1** dissolution; destruction, as in *lysozyme.* **2** hydrolysis, as in *lysosome.* Also spelled **lys-** before a vowel. [from Greek *lysis* a loosening, from *lyein* to loosen]

lysocline (lī′sə klīn′), *n. Oceanography.* a boundary in the sea below which certain chemical substances dissolve, as from the greater pressure: *When marine organisms die and sink to about 4,000 meters, they cross the "lysocline," below which calcium carbonate redissolves because of the high pressure* (Ferren MacIntyre). [from *lyso-* + *-cline* (as in *thermocline* and *syncline)*]

lysogenic (lī′sə jen′ik), *adj. Biology.* **1** having to do with lysis; capable of causing or undergoing lysis: *There are some bacterial strains that it is impossible to deprive of lysogenic power* (F.M. Burnet).
2 carrying a prophage within the cell: *The relationship between the lysogenic bacterium and the carried phage has been found to be subtle. In certain cases it appears that the very fact that the cell carries the phage renders it immune from attack by the same phages* (J.E. Hotchin).

lysogenization (lī soj′ə nə zā′shən), *n. Bacteriology.* the process of lysogenizing; fusion of the genetic material of a prophage with that of a host bacterium: *lysogenization of Escherichia coli by lambda and other temperate phages.*

lysogenize (lī soj′ə nīz), *v. Bacteriology.* to make lysogenic; cause (bacteria) to carry a prophage within the cell: *What would be the consequences if the reagent ... somehow infected and lysogenized E. coli in someone's gut as the result of an accident?* (Nature).

lysogeny (lī soj′ə nē), *n. Biology.* **1** the production of a lysin or lysins.
2 the initiation of the process of lysis.
3 = lysogenization: *In lysogeny the viral nucleic acid does not usurp the functions of the cell genes, but is incorporated into the bacterial chromosome to become an integral part of it* (New Scientist).

lysolecithin (lī′sə les′ə thin), *n. Biochemistry.* a substance that is highly destructive of red blood cells, obtained by the action of snake venom or lecithin: *The other detergent in bile is lysolecithin ... This substance breaks up cells by removing the lipids from their membranes* (Horace W. Davenport).

lysosomal (lī′sə sō′məl), *adj. Biology.* of or having to do with lysosomes: *Uninhibited action of lysosomal enzymes causes cell death* (London *Times*).

lysosome (lī′sə sōm), *n. Biology.* a particle in the cytoplasm of most cells that contains hydrolytic enzymes enclosed in a membrane: *The lysosome contains the digestive enzymes that break down large molecules, such as those of fats, proteins and nucleic acods, into smaller constituents that can be oxidized by the oxidative enzymes of the mitochondria* (Scientific American).

lysostaphin (lī′sə staf′in), *n. Biochemistry.* an enzyme that destroys staphylococcal bacteria by destroying their cell walls: *Enzymes may offer an alternative to antibiotic therapy in treating staph infections that cannot be blocked by known antibiotics* (Science News). [from *lyso-* + *staph(ylococci)* + *-in* (chemical suffix)]

lysozyme (lī′sə zīm), *n. Biochemistry.* an enzymelike substance that is capable of destroying many kinds of bacteria. It is found in egg white, human tears, saliva, and most body fluids. *Because of its resemblance to enzymes and its capacity to dissolve, or lyse, the cells, he called it "lysozyme"* (Scientific American). [from *lyso-* + English *(en)zyme*]

lytic, *adj. Biology.* of or having to do with lysis or a lysin: *a lytic enzyme, a lytic agent.*

lytta (lit′ə), *n., pl.* **lyttae** (lit′ē). *Zoology.* a long, worm-shaped cartilage in the tongue of dogs and other carnivorous animals. [from New Latin *lytta,* from Greek *lytta,* variant of *lyssa* rabies, rage]

M

m or **m.,** *abbrev.* or *symbol.* **1** magnetic quantum number. **2** mass. **3** meridian. **4** meter *or* meters. **5** mile. **6** milli-. **7** minim. **8** minute. **9** mist. **10** modulus. **11** moon. **12** mountain. **13** muscle. **14** molality.

m-, *prefix.* Usually italicized. *Chemistry.* meta- (def. 5a), as in *m*-dichlorobenzene.

M or **M.,** *abbrev.* or *symbol.* **1** magnetization. **2** mass. **3** mega- (one million). **4** mountain. **5** molarity.

ma., *abbrev.* milliampere.

mÅ., *abbrev.* milliangstrom.

maar (mär), *n. Geology.* the crater of a coneless volcano formed by an explosion but with no flow of lava. [from German *maar*]

machine, *n. Physics.* **1** an arrangement of fixed and moving parts capable of transmitting power or motion: *It is impossible to get more work or energy out of a machine than the work or energy that is put into it. In fact ... the machine never performs as much work as the amount of energy that is put into it* (E.A. Fessenden). See **simple machine.**
2 any mechanical, electric, or electronic device or apparatus for doing work; engine. Compare **mechanism.**

machine language, a set of instructions in binary code that a computer can use without conversion or translation: *User-oriented languages go through one or more stages of translation to convert them into a form of notation, known as machine language, which the machine can process directly* (Van Court Hare, Jr.).

machine-readable, *adj.* capable of being sensed and used directly by a computer: *Many libraries that have been converting their catalogue records to machine-readable form in order to make it possible to search them to answer user requests are finding it desirable to publish these records in books* (Scientific American).

Mach number (mäk), *Physics.* a number expressing the ratio of the speed of an object to the speed of sound in the same medium: *A Mach number less than unity indicates a subsonic speed and one greater than unity, supersonic speed. The transonic region covers, very roughly, the range of Mach numbers between about 0.75 and 1.1.*) (O.G. Sutton). [named after Ernst *Mach,* 1838–1916, Austrian physicist]

Mach's principle or **Mach's Principle,** *Physics.* the principle that the inertial and other properties of a system anywhere in the universe are determined by the interaction of that system with the rest of the universe, including its most distant parts: *It is one of the most cherished beliefs of many cosmologists and relativists that the universe is not rotating. It has been known for centuries that the "fixed stars" (or better, the most distant galaxies) define a frame of reference in which there are no detectable effects of rotation, and this has been asserted as a principle—Mach's Principle— rather than a coincidence* (London *Times*).

Maclaurin's series (mə klô′rinz), *Mathematics.* a Taylor's series in which the reference point (*a*) is replaced by zero. [named after Colin *Maclaurin,* 1698–1746, Scottish mathematician]

macle (mak′əl), *n. Mineralogy.* **1** a twinned crystal: *Crystals of ice, like macles of snow, were observed to form near the bottom* (Matthew F. Maury). **2** a dark spot in a mineral. [from French *macle*]

macro- or **macr-,** *combining form.* **1** large or long, as in *macromolecule = large molecule.* **2** abnormally or unusually large, as in *macrocephalic = having an abnormally large head.* [from Greek *makro-,* from *makros* long, large]

macroanalysis, *n. Chemistry.* the analysis of quantities of matter the size of a gram or over: *The economy of time afforded by many micro procedures favours their adoption even when the amount of sample available is sufficient for macroanalysis* (C.R.N. Strouts). Contrasted with **microanalysis.**

macroclimate (mak′rō klī′mit), *n. Meteorology, Ecology.* the climate of a large region: *As in all local climates strong departure from the macroclimate develops in the climates of cities in calm cloudless weather* (Geographical Journal). Contrasted with **microclimate.**
—macroclimatic, *adj.* of or having to do with a macroclimate: *macroclimatic forest soils.*

macrocyclic, *adj.* **1** *Chemistry.* containing a ring structure of large size, usually with over 15 atoms: *a macrocyclic compound.* **2** *Botany.* having a long or complex life cycle: *a macrocyclic rust fungus.*

macrocyte (mak′rə sīt), *n. Biology.* an abnormally large red blood cell, found especially in pernicious anemia. Contrasted with **microcyte.**
—macrocytic, *adj.* of or having to do with a macrocyte or macrocytes; characteristic of a macrocyte: *macrocytic red blood cells.*

macrodome, *n. Crystallography.* a dome whose planes are parallel to the longer lateral axis in an orthorhombic crystal.

macroenvironment, *n. Ecology.* the environment of a large area inhabited by a species of plant or animal: *The macroenvironment is similar throughout a large area occupied by a specific type of plant community such as the eastern deciduous forest as regards amount of rainfall and evaporation, temperature ranges, and other general climatic factors ... influenced by latitude*

---

cap, fāce, fäther; best, bē, tèrm; pin, fīve;
rock, gō, ôrder; oil, out; cup, pùt, rüle,
yü in use, *yu* in uric;
ng in bring; *sh* in rush; *th* in thin, ᴛH in then;
zh in seizure.
ə = *a* in about, *e* in taken, *i* in pencil, *o* in
lemon, *u* in circus

and altitude (Greulach and Adams, *Plants*). Contrasted with **microenvironment.**

—**macroenvironmental,** *adj.* of or having to do with the macroenvironment: *macroenvironmental measurements of wind speed.*

macroevolution, *n. Biology.* the evolution of organisms on a large scale, resulting in new classifications at the species level or in large groupings: *Goldschmidt considers that there are two different genetic processes: (1) microevolution, based on accumulation of small genetic changes within a species, leads to production of subspecies differing in size, proportions, tone of color, etc., but not to the formation of other species; and (2) macroevolution, by a repatterning of the entire genetic system, produces sudden abrupt changes of form ... to yield new species or higher groups* (Storer, *General Zoology*).

—**macroevolutionary,** *adj.* of or having to do with macroevolution.

macrofauna, *n., pl.* **-nas** or **-nae.** *Zoology.* **1** the macroscopic animals of a particular region or time: *... the macrofauna, represented by larger arthropods as, for example, millipedes, centipedes, woodlice, many insects, slugs and snails, earthworms, and some vertebrates* (David Madge).
2 the animals found in a macrohabitat. Contrasted with **microfauna.**

—**macrofaunal,** *adj.* of or having to do with a macrofauna.

macrofibril, *n. Biology.* a group of microfibrils: *At a magnification of × 30 000 under the electron microscope, the surface of a plant cell wall may show a remarkably regular pattern of so-called macrofibrils of cellulose.* (R.D. Preston).

macroflora, *n., pl.* **-floras** or **-florae.** *Botany.* **1** the macroscopic plants of a particular region or time. **2** the plants found in a macrohabitat. Contrasted with **microflora.**

—**macrofloral,** *adj.* of or having to do with a macroflora.

macrofossil, *n. Paleontology.* a macroscopic fossil; a plant or animal fossil that is large enough to be seen and examined without a microscope. Contrasted with **microfossil.**

macrogamete, *n. Biology.* the larger, usually the female, of two conjugating gametes of an organism which reproduces by the union of unlike gametes. Also called **megagamete.** Contrasted with **microgamete.**

macrogametocyte, *n. Biology.* a gametocyte that gives rise to macrogametes. Contrasted with **microgametocyte.**

macroglobulin, *n. Biochemistry.* an immunoglobulin of very high molecular weight, usually upwards of 900,000.

macrohabitat, *n. Ecology.* a habitat of any size containing a large number and variety of organisms. Contrasted with **microhabitat.**

macromere (mak′rə mir), *n. Embryology.* a blastomere of large size.

macromolecule, *n. Chemistry.* a large and complex molecule made up of many smaller molecules linked together, as in a resin or polymer: *Nature commonly constructs many vital compounds, including the en-*

zymes and other proteins essential to life, by linking amino acids of various types into large chainlike molecules called macromolecules (S.W. Fox and R.J. McCauley).

—**macromolecular,** *adj.* of or having to do with macromolecules: *macromolecular plastics.*

macronucleus, *n., pl.* **-clei** or **-cleuses.** *Zoology.* the larger of two types of nuclei present in various ciliate protozoans, which controls metabolic functions within the cell.

—**macronuclear,** *adj.* of or having to do with a macronucleus or macronuclei.

macronutrient, *n. Biochemistry.* a nutrient element of which relatively large amounts are necessary for plant growth, such as calcium, phosphorus, sulfur, and potassium: *Most plants obtain three of the macronutrients—carbon, hydrogen and oxygen—from the air and all the other nutrients from the soil* (C. J. Pratt).

macrophage (mak′rō fāj), *n. Biology.* any of various large phagocytes, especially those of the reticuloendothelial system: *Man ... is equipped with powerful natural defenses against the bacillus. Chief among them are the body's macrophages, or "wandering cells," which can engulf and destroy the tubercle bacillus* (Scientific American). Compare **histiocyte.** Contrasted with **microphage.**

—**macrophagic** (mak′rə faj′ik), *adj.* of or having to do with macrophages.

macrophagous (mak rof′ə gəs), *adj. Zoology.* feeding on large pieces of food.

macrophysics, *n.* the branch of physics concerned with bodies that are large enough to be observed and measured. Contrasted with **microphysics.**

—**macrophysical,** *adj.* of or having to do with macrophysics.

macrophyte (mak′rə fit), *n. Botany.* a macroscopic plant, especially one living in water. Contrasted with **microphyte.**

—**macrophytic** (mak′rə fit′ik), *adj.* having to do with or caused by macrophytes.

macropterous (ma krop′tər əs), *adj. Zoology.* having very large wings or fins. Contrasted with **micropterous.** [from Greek *makropteros,* from *makros* long + *pteron* wing]

macroscale (mak′rə skāl′), *n. Chemistry.* a scale or standard involving large units or measurements, as in macroanalysis: *Such pulses of heat easily ignite many kindling fuels—thin materials such as dried leaves or shredded newspaper, or material like rotten wood which on the macroscale appear to be solid but which may be considered as consisting of an extended network of thin, porous inflammable material* (Bulletin of Atomic Scientists).

—*adj. Meteorology.* of or designating large-scale atmospheric phenomena, such as hurricanes. Compare **microscale, mesoscale.**

macroscopic (mak′rə skop′ik), *adj.* **1** visible to the naked eye; large enough to be seen without magnification: *macroscopic fossils, macroscopic crystals.* Also called **megascopic.**
2 of or involving large units or measurements: *The early history of our knowledge of matter was naturally dominated by macroscopic concepts—witness the comparatively rapid perfecting of thermodynamics. With the growth of atomic theory the game became to*

explain these well-understood notions in terms of molecules (R.O. Davies). —**macroscopically,** adv.

macrosmatic (mak'roz mat'ik), adj. Zoology. having the organs of smell well developed: Today, mammals are divided into macrosmatic (dog, cat) and microsmatic (man) classifications, according to whether they possess a high or low odour sensitivity (Herbert E. Heist). [from macr- + osmatic]

macrosporangium (mak'rə spə ran'jə əm), n., pl. -gia (-jē ə). = megasporangium.

macrospore, n. = megaspore.

macrostructure, n. the large-scale or macroscopic structure of bodies or objects, especially the structure of a metal that is visible without or with low magnification: Structure here is to be interpreted in its widest sense, from atomic structure to macrostructure (P.E. Evans). Contrasted with **microstructure.**
—**macrostructural,** adj. of or having to do with macrostructure.

macrotaxonomy, n. Biology. the taxonomy of the larger groupings of organisms, such as family or order: Comparative anatomy has two objectives: to estimate the degree of relationship among groups of organisms (macrotaxonomy), and to reconstruct and interpret the history of animal structure (D. Dwight Davis).

macruran (mə krŭr'ən), Zoology. —n. a macrurous crustacean. —adj. = macrurous.

macrurous (mə krŭr'əs), adj. Zoology. of or belonging to a suborder of crustaceans that have ten legs and well-developed abdomens, including the lobsters, prawns, and shrimps. [from New Latin Macrura the suborder name, from Greek makros long + oura tail]

macula (mak'yə lə), n., pl. -lae (-lē). 1 Anatomy. a a spot on the skin which is unlike the surrounding tissues. b = macula lutea: The cones and rods are separable primarily through their function in color vision and twilight vision, but also through rod areas and predominantly cone areas—those areas that are nearest the macula, a rod-free area centrally placed at the back of the eye (F.H. George). c an area of sensory hair cells in the utricle of the ear: The macula is the sensor organ of the utriculus—the part of the ear that responds to gravity (Science News). 2 = macule (def. 1). [from Latin macula spot, stain]

macula lutea (lü'tē ə), pl. **maculae luteae** (lü'tē ē). Anatomy. a yellowish spot surrounding the fovea centralis in the retina of certain vertebrates. It is the point where vision is clearest. [from New Latin macula lutea (literally) yellow spot]

macular (mak'yə lər), adj. Anatomy. 1 of or having to do with a macula, especially the macula lutea: macular degeneration. 2 having maculae; spotted.

maculate (mak'yə lit), adj. Botany. marked with spots; spotted; speckled.

maculation (mak'yə lā'shən), n. Biology. the pattern of spots or other markings on an animal or plant.

macule (mak'yül), n. Anatomy, Medicine. 1 a spot, stain, or blotch on the skin which is not raised above the surface, as in certain diseases. 2 = macula (def. 1a). [from Middle French macule, from Latin macula spot, stain]

Magellanic Cloud

madrepore (mad'rə pôr), n. any of an order (Madreporaria) of stony corals that often form reefs in tropical seas. [from Italian madrepora, from madre mother + poro pore]
—**madreporic** or **madreporian,** adj. of or having to do with madrepores.

madreporite (mad'rə pə rīt or mad'rə pôr'īt), n. Zoology. the external opening of the stone canal in an echinoderm: On the aboral surface of the central disc of the starfish is a hard round structure called the madreporite which . . . admits the proper amount of sea water into the system ... The stone canal leads from the madreporite to the ring canal, from which a radial canal radiates out into each of the five arms (Winchester, Zoology).

maelstrom (māl'strəm), n. Oceanography. a powerful or turbulent whirlpool, especially a swift, dangerous current in the Arctic Ocean between two islands of the Lofoten group off the coast of Norway. [from earlier Dutch maelstrom, from malen to grind + strom stream]

Maffei galaxy (mä fā'ē), Astronomy. either of two large galaxies which are relatively close to the Milky Way and possibly part of the local group. One is an elliptical and the other a spiral galaxy; they are less than 12 million light-years away from earth. The Maffei galaxies, obscured by the Milky Way dust, appear as small, diffuse patches (Hyron Spinrad).
ASSOCIATED TERMS: see galaxy.
[named after Paolo Maffei, Italian astrophysicist who discovered them in 1971]

mafic (maf'ik), adj. Mineralogy. consisting of or designating any of a group of dark minerals occurring in igneous rocks, composed chiefly of magnesium and iron. Contrasted with **felsic.** [from ma(gnesium) + Latin f(errum) iron + English -ic]

mag., abbrev. 1 magnetism. 2 magnitude.

magcon (mag'kon), n. Astronomy, Geology. a concentration of magnetic material on the surface of a moon or planet: Magnetic concentrations ("magcons") in the lunar surface could, if of sufficient extent and field strength, interact locally with the solar wind when the magcons are on the daytime lunar side (Nature). [from mag(netic) con(centration), patterned on mascon]

Magellanic Cloud (maj'ə lan'ik), Astronomy. either of two galaxies appearing as faintly luminous areas in the heavens south of the equator. They are the two galaxies nearest to our own in the local group. The Magellanic Clouds are irregular in shape and present a straggling appearance that contrasts sharply with the great spiral in Andromeda and other island universes. (Bernhard, Handbook of Heavens). [named after Ferdinand Magellan, 1480?–1521, Portuguese navigator]

cap, fāce, fäther; best, bē, tėrm; pin, fīve;
rock, gō, ôrder; oil, out; cup, pùt, rüle,
yü in use, yù in uric;
ng in bring; sh in rush; th in thin, ᴛʜ in then;
zh in seizure.
ə = a in about, e in taken, i in pencil, o in lemon, u in circus

375

maggot, *n. Zoology.* the legless, wormlike larva of any of various kinds of two-winged flies, often living in decaying matter. [Middle English *magot*]

magic number, *Nuclear Physics.* the number of nuclear particles in a nucleus that has only closed or completed shells of protons and neutrons. The numbers 2, 8, 20, 28, 50, 82, and 126 are magic numbers. A nucleus containing such a number of protons and neutrons is highly stable. *Nuclei containing magic numbers of protons or neutrons are all more nearly spherical than their neighbours, indicating that the particles in them are arranged in a particularly symmetrical manner.* (J. Little).

magic square, *Mathematics.* a square figure formed by a series of numbers so arranged in parallel and equal ranks that the sum of each row or line taken perpendicularly, horizontally, or diagonally, is constant.

magma (mag′mə), *n. Geology.* the molten material beneath the earth's crust from which igneous rock is formed: *There are two general kinds of magma ... One is the type represented by basalt—a dark, heavy material ... The other is a lighter material, rich in silica ...; its most typical variety is rhyolite. A basaltic magma is usually hotter and more viscous than a rhyolitic one* (Scientific American). [from Greek *magma* unguent, from *massein* to knead, mold]

—**magmatic,** *adj.* of or having to do with magma: *Magmatic movement, however caused, has been a factor in the formation of the Rift Valleys* (Geographical Journal).

—**magmatism,** *n.* the movement, solidification, or activity of magma.

magnesite, *n. Mineralogy.* natural magnesium carbonate, occurring either in compact, white masses or in crystalline form, used in producing carbon dioxide and steel. *Formula:* $MgCO_3$

magnesium (mag nē′zē əm *or* mag nē′zhəm), *n. Chemistry.* a light, silver-white metallic element that is very ductile and malleable and burns with a dazzling white light. Magnesium is the lightest of all industrial metals and is found in nature only in combination with other elements. *Green plants cannot grow without magnesium, since this is essential for the formation of chlorophyll* (Clarence J. Hylander). *Symbol:* Mg; *atomic number* 12; *atomic weight* 24.312; *melting point* 650°C; *boiling point* 1107°C; *oxidation state* 2. [from New Latin, from *magnesia* an alkaline powder used in medicine, from Greek *Magnēsia* (*lithos*) (stone) from Magnesia, ancient region in Thessaly, Greece]

magnesium carbonate, *Chemistry.* a white, odorless compound, occurring in dolomite or naturally as magnesite. *Formula:* $MgCO_3$

magnesium chloride, *Chemistry.* a colorless, crystalline, salt found in salt water, used as a source of magnesium. Magnesium chloride readily absorbs moisture thus liquefying easily. *Formula:* $MgCl_2$

magnesium sulfate, *Chemistry.* a colorless crystalline salt used in matches, explosives, etc., and in the hydrated form as Epsom salts. *Formula:* $MgSO_4$

magnet, *n. Physics.* a stone, piece of metal, or other solid that has the property, either natural or induced, of attracting iron or steel. A lodestone is a natural magnet. Bar magnets and horseshoe magnets are steel magnets which retain their magnetism for a long time. *Invisible lines extend between the poles of a magnet and affect pieces of iron placed in their paths. These are magnetic lines of force. All the lines of force taken together make up what is called the magnetic field* (Beauchamp, et al. *Everyday Problems in Science*). [from Latin *magnes, magnetis,* from Greek *Magnēs* (*lithos*) (stone) from Magnesia; see MAGNESIUM]

—**magnetic,** *adj.* **1** having the properties of a magnet: *the magnetic needle of a compass. The magnetic poles are not in line with the geographic poles* (Tracy, *Modern Physical Science*).
2a of or having something to do with a magnet or magnetism: *the magnetic meridian.* **b** producing, caused by, or operating by means of magnetism: *a magnetic circuit.*
3 capable of being magnetized or of being attracted by a magnet: *magnetic nickel.* —**magnetically,** *adv.*

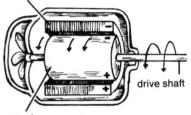

field magnet

drive shaft

armature

The armature acts as an electromagnet when current is applied, revolving between the poles of the field magnet and transmitting power to a shaft as it turns.

magnetic anomaly, *Physics, Geology.* any deviation from the general pattern of the earth's magnetic field: *The magnetic evidence for seafloor spreading is based on the pattern of the linear magnetic anomalies which lie over and to the sides of mid-oceanic ridges* (Nature).

magnetic axis, *Physics.* the straight line joining the poles of a magnet.

magnetic bottle, *Physics.* any arrangement of magnetic fields for confining plasma or highly ionized gas, especially in a controlled thermonuclear reaction.

magnetic bubble, *Physics.* a small, circular, movable magnetic domain within a piece of magnetized material in which the magnetization is opposite to that of most of the material.

magnetic circuit, *Physics.* the closed path taken by a magnetic flux: *A closed path of magnetic material ... is called a magnetic circuit, since the magnetic flux runs through the magnetic material like the electric current in an electric circuit* (Shortley and Williams, *Elements of Physics*).

magnetic declination, *Geography.* the angle between magnetic north and true north at a given point on the earth's surface.

magnetic domain, *Physics.* an area or portion of a ferromagnetic substance in which all the electrons spin in the same direction, forming a miniature magnet: *A substance will act as a large magnet when all or most of its magnetic domains (little magnets) line up in the same direction* (Michael Cusack).

magnetic equator, *Geography, Geophysics.* the line around the earth at which a magnetic needle balances horizontally without dipping: *The magnetic equator, or aclinic line, is the place on the earth where the attraction of the north and south magnetic poles is equal* (Palmer H. Craig).

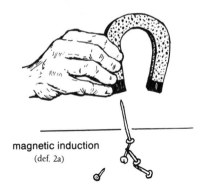

magnetic induction
(def. 2a)

Magnetic induction is produced by the force of the magnetic field radiating from the poles of the magnet through the open space to the nail which is attracting the tacks without touching the horseshoe magnet.

magnetic field, *Physics.* **1** the space around a magnet or electric current in which magnetic force is felt: *Magnetic fields can be produced by permanent magnets, but often they are produced by an electric current, that is, a flow of electrons, through a coil of copper wire.* (Pierce, *Electrons, Waves and Messages*). *The earth ..., other planets, and stars as well, have magnetic fields that extend millions of miles into space. The behavior of the magnetic fields around the planets and stars is one of the great unsolved mysteries of modern science.* (Tracy, *Modern Physical Science*). **2** the magnetic force present in such a space: *The average strength of the magnetic field inside a magnetized bar of iron is 20,000 gauss; at the center of the iron atom it is as high as 330,000 gauss* (Scientific American).

magnetic field strength, *Physics.* a vector quantity used in describing magnetic phenomena in terms of their magnetic field and magnetization. The SI unit of magnetic field strength is the ampere-turn per meter, and the CGS unit is the oersted.

magnetic flux, *Physics.* **1** the total number of magnetic lines of force passing through a specified area, generally expressed in maxwells or webers. **2** the result of dividing the magnetomotive force by the reluctance.

magnetic flux density, = magnetic induction.

magnetic force, *Physics.* **1** the force exerted between magnetic poles; the force which produces magnetization. **2** a force associated with the motion of an electric charge: *Magnetic forces exist between two charges only if both charges are in motion, the magnitude of the force being proportional to the product of the speeds of the charges* (Shortley and Williams, *Elements of Physics*).

magnetic induction, *Physics.* **1** the amount of magnetic flux in a unit area taken perpendicular to the direction of the magnetic flux. Also called **magnetic flux density.**
2 the process by which a substance (iron, steel, etc.) becomes magnetized by a magnetic field.

magnetic meridian, *Geography, Geophysics.* a line passing through both magnetic poles that represents the force exerted by the earth's magnetic field.

magnetic moment, *Physics.* **1** the product of the pole strength of a magnet and the distance between the poles: *In the case of the earth and the sun, the "strength of magnetism" (magnetic moment) of these two bodies was proportional to their angular momentum—the latter expression meaning, in effect, the time for which a specified braking force would have to be applied at the same distance from their centres, to bring them to rest.* (A.W. Haslett).
2 a magnetic force between two opposite charges in an electron or other elementary particle, associated with the spin and orbital motion of the particle: *Since the nucleus carries a charge, its rotation corresponds to an electric current, and it behaves like a tiny magnet. The result is known as the magnetic moment of the spinning body* (Scientific American).

magnetic monopole, *Physics.* a hypothetical particle having only one pole of magnetic charge instead of the two possessed by ordinary magnetic bodies: *Magnetic monopoles would be for magnetism what electrons and protons are for electricity; basic units of charge.* (Dietrick E. Thomsen). Also called **monopole.**

magnetic needle, a slender bar of magnetized steel used as a compass. When mounted so that it turns easily, it points approximately north and south toward the earth's magnetic poles.

magnetic north, *Geography, Geophysics.* the direction shown by the magnetic needle of a compass, differing in most places from the true north. See the picture at **north celestial pole.**

magnetic pole, 1 *Physics.* either one of the two poles of a magnet: *The force in magnets is concentrated near the ends; these places of magnetic concentration are called magnetic poles* (Obourn, *Investigating the World of Science*).
2 *Geography, Geophysics.* Often spelled **Magnetic Pole,** either one of the two variable points on the earth's surface toward which a compass needle points. The **North Magnetic Pole** is in the Arctic, currently at about 75.5 degrees North Latitude and 100 degrees West longitude. The **South Magnetic Pole** is in Antarctica, at about 66.5 degrees South latitude and 140.5 degrees East longitude. *At any point on earth, orienta-*

cap, fāce, fäther; best, bē, tèrm; pin, five;
rock, gō, ôrder; oil, out; cup, pùt, rüle,
yü in use, *yù* in uric;
ng in bring; sh in rush; th in thin, ŦH in then;
zh in seizure.
ə = *a* in about, *e* in taken, *i* in pencil, *o* in
lemon, *u* in circus

tion of the magnetic field is related to the position of the earth's magnetic poles (Walter Sullivan).

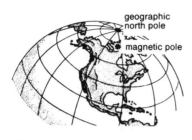

geographic north pole

magnetic pole

(def. 2)

magnetic quantum number, *Physics.* a quantum number corresponding to the component of an angular momentum in a particular direction, usually that of a magnetic field. *Abbreviation:* m

magnetic resonance, *Physics.* the absorption of microwaves, radio waves, or other forms of energy by atoms, molecules, or nuclei subjected to a magnetic field: *Laser magnetic resonance is similar to such other magnetic resonance measurement methods as electron spin resonance (electron paramagnetic resonance) and nuclear magnetic resonance. The molecule or molecular species being detected absorbs radiation corresponding to a transition between rotational energy levels* (James A. Kerr).

magnetic storm, *Geophysics, Astronomy.* a large disturbance or fluctuation of the earth's magnetic field, associated with solar flares: *The great disturbances of the earth's magnetic field known as "magnetic storms" invariably produce storms in the ionosphere — large, rapid fluctuations of electron density* (Scientific American).

magnetic variation, *Geophysics.* a variation in the earth's magnetic field. Compare **declination.**

magnetism, *n. Physics.* **1** the properties or qualities of a magnet; manifestation of magnetic properties: *Rocks tend to become magnetized by the earth's magnetic field. Their magnetism, in general, lines up so that it points like a compass needle to the magnetic poles* (R.W. Burnett).
2 the branch of physics dealing with magnets and magnetic properties.

magnetite, *n. Geology.* an important iron ore that is strongly attracted by a magnet. Magnetite that possesses polarity is called lodestone. *Changes in temperature or pressure alter the magnetic properties of magnetite and should generate small changes in the magnetic field nearby* (Robert W. Decker). *Formula:* $(Fe^{2+}, Mg)Fe^{3+}_2O_4$

magnetize, *v. Physics.* to give the properties of a magnet to. An electric coil around a bar of iron can magnetize the bar. *The fact that some substances are capable of being permanently magnetized has been known since the times of Ancient Greece, and Chinese navigators were using pieces of lodestone, or magnetite, as an aid to navigation as early as the 11th century* (Science News).

—mag′netiz′able, *adj.* that can be magnetized: *a magnetizable material or medium.*

—mag′netiza′tion, *n.* a magnetizing or a being magnetized; the extent to which a substance is magnetized: *Magnetization ... depends on the strength of the magnetic field ... in which the body is placed. The magnetization is the magnetic moment per unit volume, or the pole strength per unit cross-sectional area* (W.D. Corner).

magneto-, *combining form.* magnetism; magnetic forces; magnets, as in *magnetometer = an instrument for measuring magnetic forces.*

magnetochemical, *adj.* of or having to do with magnetochemistry.

magnetochemistry, *n.* the science dealing with the relationship of magnetic properties to chemistry.

magnetoelectric, *adj.* of electricity produced by magnets or magnetic properties.

magnetoelectricity, *n. Physics.* electricity produced by magnetic properties.

magnetograph, *n. Physics.* a device for detecting variations in the direction and intensity of magnetic fields, especially in the earth's magnetic field.

magnetohydrodynamic, *adj.* of or having to do with magnetohydrodynamics: *We find many important magnetohydrodynamic phenomena in the ionosphere and above it, such as magnetic storms and aurorae, and also the radiation belts around the Earth* (Hannes Alfvén).

magnetohydrodynamics, *n. Physics.* the study or application of the interaction of magnetic fields and electrically conducting liquids and gases: *An electric system was developed that generated 1,000 watts for five seconds using magnetohydrodynamics, in which an ionized gas moves in a magnetic field to make the current* (Science News Letter). *Abbreviation:* MHD

magnetometer (mag′nə tom′ə tər), *n. Physics.* an instrument for measuring the intensity and direction of magnetic forces.
—magnetometric (mag′nē tə met′rik), *adj.* of or having to do with a magnetometer or magnetometry.
—magnetometry (mag′nə tom′ə trē), *n.* the measurement of magnetic forces with a magnetometer.

magnetomotive force, *Physics.* the force producing a magnetic flux. *Abbreviation:* mmf

magneton (mag′nə ton), *n. Physics.* any of various units of magnetic moment: *We obtain a value of 3·14 nuclear magnetons for the magnetic moment of the Li^7 nucleus* (Physical Review).

magnetooptic or **magnetooptical,** *adj.* of or having to do with magnetooptics.

magnetooptics, *n.* the branch of optics which studies the influence of magnets on light.

magnetopause, *n. Geophysics.* the upper limits of the magnetosphere: *The magnetopause is observed to move back and forth with great rapidity* (L. J. Cahill, Jr.).

magnetoresistance, *n. Physics.* a change in the electrical resistance of a metal, semiconductor, etc., with the increase or decrease of the strength of a magnetic field applied to it.
—**magnetoresistive,** *adj.* having or characterized by magnetoresistance. *a magnetoresistive element.*

magnetosheath, *n. Geophysics, Astronomy.* a thin region that surrounds the magnetosphere of a planet and acts as an elastic medium transmitting the kinetic pressure of solar wind onto the planet's magnetic field: *The spacecraft left the tumultuous magnetosheath and entered the magnetosphere proper, the main body of the planet's magnetic field* (Science News).

magnetosphere, *n.* **1** *Geophysics.* a region in which the earth's magnetic field dominates the movement of ionized plasma: *The magnetosphere is not a static shield encircling the earth. Instead it expands and contracts with changes in the intensity of the solar wind* (T. Neil Davis).
2 *Astronomy.* any similar region around a celestial body: *If the pulsars are magnetic neutron stars spinning in a vacuum . . . such a star must have a magnetosphere, a plasma-filled region surrounding the star* (Jeremiah P. Ostriker).
—**magnetospheric,** *adj.* of, having to do with, or characteristic of the magnetosphere: *The Moon, in contrast to the Earth, has practically no magnetic field so that it has no complex magnetospheric envelope, no radiation belts and no ionosphere* (Science Journal).

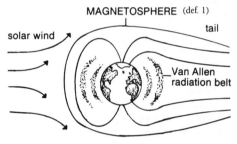

MAGNETOSPHERE (def. 1)

solar wind • tail • Van Allen radiation belt

magnetostatic, *adj.* of or having to do with magnetostatics: *magnetostatic fields.* —**magnetostatically,** *adv.*

magnetostatics, *n.* the branch of physics concerned with magnetic fields that are stationary or unchanging: *In magnetostatics it is convenient to define a magnetic pole strength, q, analogous to electric charge* (J.C. Anderson).

magnetostriction, *n. Physics.* a change in the dimensions of a ferromagnetic substance when subjected to a magnetic field: *Magnetostriction is the periodic contraction of a nickel rod in an alternating magnetic field. The end of the rod emits high frequency waves* (Gabriele Rabel).
—**magnetostrictive,** *adj.* of, having to do with, or utilizing magnetostriction: *a magnetostrictive effect, device, property, etc.*

magnetotactic, *adj. Biology.* exhibiting magnetotaxis: *magnetotactic bacteria.*

magnetotail, *n. Geophysics, Astronomy.* the elongated part of a planet's magnetosphere that extends away from the sun in a form resembling the tail of a comet.

magnetotaxis, *n. Biology.* movement of an organism in response to a magnetic field: *Several species of aquatic bacteria which orient in the Earth's magnetic field and swim along magnetic field lines in a preferred direction* (*magnetotaxis*) *have been observed in marine and fresh-water sediments of the Northern Hemisphere* (Richard B. Frankel).

magnification, *n. Optics.* **1** the ratio of the size of an image to the size of an object.
2 = magnifying power.
▶ A number and the symbol × (for times) is used to indicate magnification. For example, a 20× lens is one that magnifies an object by 20 times. The magnification of a microscope can also be expressed in diameters, so that a 20× magnification is one that magnifies the image by 20 times the diameter of the object. Compound microscopes can magnify specimens from 40× to 2000×.

magnifying glass, a lens or combination of lenses that causes close objects to appear larger. A magnifying glass usually consists of a lens whose two sides are curved to form a double convex lens.

magnifying power, *Optics, Astronomy.* the ratio of the size of an image when viewed through an optical instrument to the size of the object when seen with the naked eye: *Through the eyepiece the observer sees the imaged stars at a greater angular distance apart than they appear to the unaided eye, and this magnifying power is independent of the aperture and equals the ratio of the focal length of objective and eyepiece* (Duncan, *Astronomy*). Compare **resolving power.**

magnitude, *n.* **1** *Astronomy.* the degree of brightness of a star. The brightest stars are of the first magnitude. Those just visible to the naked eye are of the sixth magnitude. Each unit of magnitude is 2.512 times brighter than the next. *Absolute magnitude is defined as the magnitude a star would have at a distance of 10 parsecs* (Krogdahl, *The Astronomical Universe*).
2 *Mathematics.* **a** a number given to a quantity so that it may be compared with similar quantities. **b** = absolute value. **c** *Geometry.* the measure or extent of a particular line, area, volume, or angle.
3 *Geology.* the force of an earthquake: *The method most commonly employed to measure the magnitude of an earthquake is the Richter Magnitude Scale ... Magnitudes are expressed in whole numbers and decimals, the values most commonly varying between 4 and 8* (P. and J. Roper). Abbreviation: mag.
[from Latin *magnitudo,* from *magnus* large]

magnon (mag'non), *n. Nuclear Physics.* the unit particle or quantum of a spin wave: *Phonons and magnons are typical examples of quasiparticles that are bosons* (Scientific American). [from *magn*(*etic*) + *-on*]

magnum (mag'nəm), *n. Anatomy.* the capitate bone. [from Latin *magnum* great]

mail, *n. Zoology.* the protective shell or scales of certain animals, such as the tortoise or the lobster.

---

cap, fāce, fäther; best, bē, tėrm; pin, fīve;
rock, gō, ôrder; oil, out; cup, pùt, rüle;
yü in use, *yu* in uric;
ng in bring; *sh* in rush; *th* in thin, ᴛʜ in then;
zh in seizure.
ə = *a* in about, *e* in taken, *i* in pencil, *o* in lemon, *u* in circus

—**mailed**, *adj.* (of animals) having a protective covering armor or mail.

main sequence, *Astronomy.* the group of stars which, when plotted according to luminosity and spectral class on the Hertzsprung-Russell diagram, fall in a narrow diagonal band from the upper left to the lower right. Most stars fall in this sequence, showing that there is a correlation between luminosity and spectral type, and, by extension, luminosity and size.

—**main-sequence star,** any star in the main sequence, such as the sun: *Within the cores of main-sequence stars, nuclear reactions take place that convert hydrogen into helium and release huge amounts of energy (these reactions are the same as those in hydrogen bombs). The energy is emitted by the stars as light, heat, and other forms of radiation* (Natural History). Also called **dwarf.**

major arc, *Mathematics.* an arc that is greater than half a circle.

major axis, *Mathematics.* the axis of an ellipse that passes through its foci.

malachite (mal′ə kīt), *n. Mineralogy.* a green mineral that is a carbonate of copper, used as an ore and for making ornamental articles. It is found in supergene deposits. *Formula:* $Cu_2CO_3(OH)_2$ [from French, from Greek *malache* mallow (from the similar color)]

malacological (mal′ə kə loj′ə kəl), *adj.* of or having to do with malacology.

malacology (mal′ə kol′ə jē), *n.* the branch of zoology that deals with mollusks. [from French *malacologie,* short for *malacozoologie,* from *malaco-* (from Greek *malakos* soft) + *zoologie* zoology]

malacostracan (mal′ə kos′trə kən), *Zoology.* —*adj.* of or having to do with a subclass (Malacostraca) of crustaceans usually having many appendages on the thorax and abdomen. Lobsters, crabs, and shrimps belong to this subclass.

—*n.* a malacostracan crustacean.
[from New Latin *Malacostraca* the subclass name, from Greek *malakos* soft + *ostrakon* shell]

malar (mā′lər), *Anatomy.* —*adj.* of or having to do with the cheekbone or cheek.

—*n.* a malar bone. See the picture at **skull.**
[from Latin *mala* jaw, cheek]

male, *n.* 1 *Zoology.* an animal belonging to the sex that produces sperm by which it fertilizes the eggs of the female.
2 *Botany.* **a** a flower having a stamen or stamens and no pistils. **b** a plant bearing only flowers with stamens.

—*adj.* 1 *Zoology.* of, having to do with, or belonging to the sex that produces sperm: *Bucks, bulls, and roosters are male animals.*
2 *Botany.* **a** (of seed plants) having flowers which contain stamens but not pistils; staminate: *If ... the flower ... possesses one or more stamens but no pistil, it is described as staminate or male* (Youngken, *Pharmaceutical Botany*). **b** able to fertilize the female. The same plant may have both male and female flowers.

male-sterile, *adj. Physiology.* not having or producing male reproductive cells.

malic (mal′ik or mā′lik), *adj.* 1 *Botany.* of, having to do with, or obtained from apples.

2 *Chemistry.* of or having to do with malic acid. [from French *malique,* from Latin *malum* apple]

malic acid, *Chemistry.* a colorless, crystalline acid found in apples and numerous other fruits, used in making various salts and esters, for aging wine, etc. *Formula:* $C_4H_6O_5$

malic dehydrogenase, *Biochemistry.* an enzyme that catalyzes the oxidation of L-malic acid to oxaloacetic acid in the Krebs cycle: *Malic dehydrogenase, which oxidizes malic acid and is also presumed to be a metal enzyme ... play[s] a part in the burning of sugars in the body* (Science News Letter).

malleable (mal′ē ə bəl), *adj. Chemistry.* capable of being hammered or pressed in various shapes without being broken. Gold, silver, copper, and tin are malleable. Compare **ductile.**

—**malleability** (mal′ē ə bil′ə tē), *n.* the quality or condition of being malleable.

malleolar (mə lē′ə lər), *adj. Anatomy.* of or having to do with the malleolus.

malleolus (mə lē′ə ləs), *n., pl.* **-li** (-lī). *Anatomy.* the bony part that sticks out on either side of the ankle. [from Latin *malleolus,* diminutive of *malleus* hammer]

malleus (mal′ē əs), *n., pl.* **mallei** (mal′ē ī). *Anatomy.* the outermost of the three small bones in the middle ear of mammals, shaped like a hammer. Compare **incus, stapes.** See the picture at **ear.** [from Latin *malleus* hammer]

malolactic (mal′ō lak′tik), *adj. Biochemistry.* of or having to do with the bacterial conversion of malic acid to lactic acid in wine: *malolactic fermentation.*

malonate (mal′ə nāt), *n. Chemistry.* a salt or ester of malonic acid: *Straight-chain lipids are created in living organisms from simple two-carbon and three-carbon compounds: acetate and malonate, shown here as their acids* (Scientific American).

malonic acid (mə lō′nik or mə lon′ik), *Chemistry.* a white, crystalline dicarboxylic acid readily decomposed by heat, derived by oxidation from malic acid. *Formula:* $C_3H_4O_4$ [from French *malonique,* alteration of *malique* malic]

Malpighian body or **Malpighian corpuscle** (mal pig′ē ən), = renal corpuscle. [named after Marcello *Malpighi,* 1628–1694, Italian anatomist]

Malpighian tube, tubule, or **vessel,** *Zoology.* a tube-shaped gland in insects and arachnids that is connected to the alimentary canal and serves as an excretory organ: *Insects and certain other Arthropoda possess tubular excretory organs consisting of blind tubes termed Malpighian tubes, which lie in the body cavity and open only into the intestine* (Harbaugh and Goodrich, *Fundamentals of Biology*).

maltase (môl′tās), *n. Biochemistry.* an enzyme that changes maltose to glucose, present in saliva, intestinal secretion, yeast, etc.

Malthusian (mal thü′zhən or mal thü′zē ən), *adj. Biology.* of or having to do with the Malthusian theory: *In some underdeveloped areas, where the rate is highest, the Malthusian prediction that population would eventually outrun food supplies seems close to reality* (Christopher J. Pratt).

—**Malthusianism,** *n.* = Malthusian theory.

Malthusian theory or **Malthusian principle**, *Biology.* the theory that the world's population tends to increase faster than the food supply unless this tendency is checked by famine, epidemics, wars, and the like: *Darwin noted that the Malthusian principle applies even more clearly to most organisms than to man. He saw that the inevitable decimation of offspring is a possible mechanism for evolution by natural selection* (Simpson, *Life: An Introduction to Biology*). [named after Thomas R. *Malthus*, 1766–1834, English clergyman and economist who proposed this theory in *An Essay on the Principle of Population* (1798)]

maltose (môl'tōs), *n. Biochemistry.* a white crystalline sugar made by the action of various enzymes on starch. It is formed in the body during digestion. *Formula:* $C_{12}H_{22}O_{11} \cdot H_2O$ Commonly called **malt sugar.** [from French *maltose,* from English *malt*]

mamma (mam'ə), *n., pl.* **mammae** (mam'ē'). *Anatomy.* = mammary gland. [from Latin *mamma* breast]

mammal, *n. Zoology.* any of a class (Mammalia) of warm-blooded vertebrate animals usually having hair, the females of which secrete milk from mammary glands to nourish their young. Human beings, horses, cattle, dogs, lions, bats, and whales are all mammals. *Mammals have internal fertilization. The embryo develops internally within the uterus. The eggs of mammals have relatively little yolk and therefore are very small ...* (Biology Regents Syllabus). [from New Latin *Mammalia,* ultimately from Latin *mamma* breast] —**mammalian** (ma mā'lē ən *or* ma māl'yən), *adj.* of or belonging to the mammals. —*n.* one of the mammals.

mammalogy (ma mal'ə jē), *n.* the branch of zoology that deals with the study of mammals.

mammary (mam'ər ē), *adj. Anatomy.* of or having to do with the mammae or breasts.

mammary gland, *Anatomy.* a gland in the breast of mammals, enlarged in females and capable of producing milk which issues through the nipple. Also called **mamma.**

mammato-cumulus (ma mā'tō kyü'myə ləs), *n., pl.* **-li** (-lī). *Meteorology.* a cumulus cloud with rounded protuberances on the lower surface. It is usually a sign of rain. [from Latin *mammatus* (literally) having mammae + English *cumulus*]

mammilla (ma mil'ə), *n., pl.* **-millae** (-mil'ē). *Biology.* **1** the nipple of the female breast. **2** any nipple-shaped organ or protuberance. [from Latin *mamilla,* diminutive of *mamma* breast] —**mammillary** (mam'ə ler'ē), *adj.* **1** *Biology.* of, having to do with, or like a nipple or breast. **2** *Mineralogy.* having or forming smoothly rounded masses, such as certain mineral aggregates.

mandible, *n.* **1** *Anatomy.* **a** a jaw, especially the lower jaw. **b** the bone of the lower jaw; jawbone. See the picture at **skull. 2** *Zoology.* **a** an organ in the mouth of many invertebrates for seizing and biting food, especially either of a pair of such organs in an arthropod. See the picture at **mouth. b** the upper or lower part of a bird's beak. [from Late Latin *mandibula,* from Latin *mandere* to chew]

mandibular (man dib'yə lər), *adj. Zoology.* of, having to do with, or like a mandible.

mandibulate (man dib'yə lit), *Zoology.* —*adj.* **1** having a mandible or mandibles. **2** adapted for chewing. —*n.* a mandibulate animal, especially an insect.

manganese (mang'gə nēz'), *n. Chemistry.* a hard, brittle, grayish-white or reddish-gray metallic element that resembles iron but is not magnetic and is softer. It is used chiefly in making alloys of steel, fertilizers, paints, and industrial chemicals. *Symbol:* Mn; *atomic number* 25; *atomic weight* 54.9380; *melting point* 1245°C; *boiling point* 1962°C; *oxidation state* 2, 3, 4, 7. [from Italian *manganese,* alteration of Medieval Latin *magnesia*]

manganese dioxide, *Chemistry.* a black crystal, or brownish-black powder, used in making dyes, dry-cell batteries, as an oxidizing agent, etc. *Formula:* MnO_2

manganese spar, *Mineralogy.* rhodonite or rhodochrosite.

manganic (man gan'ik *or* mang gan'ik), *adj. Chemistry.* **1** of or resembling manganese. **2** containing manganese, especially with a valence of six.

manganite (mang'gə nīt), *n. Mineralogy.* a hydrated oxide of manganese, occurring in steel-gray or iron-black masses or crystals. *Formula:* γ-MnO(OH)

mangrove, *n.,* or **mangrove swamp,** *Ecology.* a type of tropical or subtropical coastal vegetation, usually dominated by mangrove trees or shrubs (genera *Rhizopora* and *Avicennia*), which have extensive root systems.

manifold, *n. Mathematics.* a topological space or surface: *One way to study a manifold is to try to break it into simple pieces resembling triangles: if the procedure is successful, it is said that the manifold has been triangulated* (Irving Kaplansky).

mannan (man'an *or* man'ən), *n. Biochemistry.* a polysaccharide found in plants which yields mannose upon hydrolysis.

mannite (man'īt), *n.* = mannitol.

mannitic (mə nit'ik), *adj. Biochemistry.* of, containing, or derived from mannitol.

mannitol (man'ə tōl *or* man'ə tol), *n. Biochemistry.* a white, odorless, crystalline alcohol obtained from glucose, seaweed, or a variety of ash tree, occurring in three optically different forms. *Formula:* $C_6H_{14}O_6$ [from French *mannite,* from *manna* a sweet secretion from a variety of ash tree containing mannitol]

mannose (man'ōs), *n. Biochemistry.* a simple sugar, a hexose, obtained from mannan or by oxidation of mannitol. *Formula:* $C_6H_{12}O_6$

---

cap, fāce, fä ther; best, bē, tèrm; pin, five; rock, gō, ôrder; oil, out; cup, pùt, rüle; *yü* in use, *yu* in uric; *ng* in bring; *sh* in rush; *th* in thin, ᴛʜ in then; *zh* in seizure. ə = *a* in about, *e* in taken, *i* in pencil, *o* in lemon, *u* in circus

manometer (mə nom′ə tər), *n. Physics.* an instrument for measuring the pressure of gases or vapors. [from Greek *manos* thin, loose + English *-meter*]

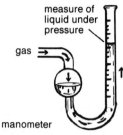

measure of liquid under pressure

gas

manometer

If the pressure of the gas is greater than the pressure of the atmosphere, the liquid in the tube will rise. The gas pressure is equal to the sum of the atmospheric pressure and the pressure of the liquid

mantissa (man tis′ə), *n. Mathematics.* the decimal part of a logarithm. In the logarithm 2.95424, the characteristic is 2 and the mantissa is .95424. [from Latin *mantissa* an addition]

mantle, *n.* **1** *Zoology.* **a** the fold of the body wall of a mollusk that lines the shell and secretes the material of which the shell is formed. **b** a pair of similar folds that secrete the shell of a brachiopod. **c** the soft tissue that lines the shell of a tunicate or barnacle. **d** the folded wings and back feathers of a bird that enclose the body like a cloak.
2 *Anatomy.* the part of the brain that includes the convolutions, corpus callosum, and formix.
3 *Geology.* the layer of the earth, or another planet, that lies between the crust and the core. See the picture at **core.**

mantle plume, *Geology.* a large upwelling of molten material from the earth's mantle: *Mantle plumes are ... hypothesized by some scientists to account for, among other things, the creation of volcanic island chains in the Pacific* (Science News). Also called **plume.**

mantle rock, = regolith.

manubrium (mə nü′brē əm), *n., pl.* **-bria** (-brē ə). *Anatomy.* **1** a process or part of a bone or other bodily structure that is shaped like a handle.
2a the broad upper division of the sternum of mammals, with which the two first ribs articulate. **b** a small, tapering, curved or twisted process of the malleus of the ear.
[from Latin *manubrium* a handle, from *manus* hand]

manus (mā′nəs), *n., pl.* **-nus.** *Anatomy.* the distal part of the forelimb of a vertebrate, including the carpus or wrist, and the forefoot or hand. [from Latin *manus* hand]

manyplies (men′ē plīz′), *n.* = omasum.

many-valued, *adj. Mathematics.* having many values; multivalued.

MAO, *abbrev.* monoamine oxidase.

map, *n.* **1** *Geography.* a graphic representation of selected features of the earth's surface or of part of it.
2 *Genetics.* = genetic map.

3 *Mathematics.* correspondence of one or more elements in one set to one or more elements in the same or another set.
—v. 1 *Geography.* to make a graphic representation of (a feature or region).
2 *Genetics.* to place (a gene or genes) in a particular arrangement on a chromosome.
3 *Mathematics.* to cause an element in (one set) to correspond to an element in the same or another set.
—mappable (map′ə bəl), *adj.* that can be mapped: *a genetically mappable gene.*

map projection, *Geography.* a systematic grid configuration that permits the earth's spherical surface, or a part of it, to be represented on a flat surface.

marble, *n. Geology.* a metamorphic rock formed by recrystallization of limestone or dolomite, usually white or variegated and capable of taking a high polish. It is widely used as a decorative stone.

marcasite (mär′kə sīt), *n. Mineralogy.* a whitish-yellow mineral with a metallic luster, similar in form to pyrite but having a different crystallization and lower specific gravity: *Marcasite is common in the chalk beds of Dover, England, and in the Joplin mining district of Missouri, Kansas and Oklahoma* (Fenton, *The Rock Book*). Formula: FeS_2 Also called **white iron pyrites.** [from Medieval Latin *marcasita,* from Arabic *marqashītā*]

marcescent (mär ses′ənt), *adj. Botany.* withering but not falling off, as a part of a plant. [from Latin *marcescentem* withering away, from *marcere* be faint, languid]
—marcescence (mär ses′əns), *n.* marcescent condition.

mare (mär′ā), *n., pl.* **maria** (mär′ē ə) or **mares.** *Astronomy.* **1** any of certain broad, dark areas on the surface of the moon: *A catastrophic event hit both the Earth and the Moon, melting the lunar surface—or at least surfaces of the mares* (Science Journal).
2 a similar dark region on Mars or any other planet: *Other conspicuous features on Mars are the so-called maria, dark areas easily distinguished from the surrounding, desert-colored, lighter expanses* (F.I. Ordway).
[from Latin *mare* sea]

marine, *adj.* of the sea or ocean; found in or produced by the sea: *Scypha is a small vase-shaped sponge about an inch in height which lives ... only in salt water and thus may be spoken of as a marine animal in contrast to a fresh-water animal* (Winchester, *Zoology*). *Except in the trade wind belts, marine climates usually have greater humidity and cloudiness than continental climates* (Blair, *Weather Elements*). [from Latin *marinus* of the sea, from *mare* sea]

marine biology, a branch of biology dealing with the living organisms of the sea.

marine cliff, *Geology, Geography.* a steep rock cliff created and maintained by the action of waves.

marine rock, *Geology.* a rock formed by processes operating in an ocean. Contrasted with **continental rock.**

marine science, the science dealing with the sea and its environment, including marine biology, oceanography, and similar specializations.

marine terrace, *Geology, Geography.* a flat area of land created by the action of waves and later uplifted to form a platform above the existing sea level.

Markarian galaxy (mär kär'ē ən), any of a class of galaxies characterized by blue color and a strong emission in the ultraviolet part of the spectrum, including some of the brightest known objects in the universe: *The Markarian galaxies, under their broad umbrella of excess ultraviolet, contain many different types: Seyferts, Zwicky objects, Haro blue galaxies, many giant galaxies, many dwarf galaxies and close galaxies; even quasars might be numbered among them.* (Science News). [named after B.E. *Markarian*, an Armenian astronomer who discovered the first such galaxies in 1968]

marker, *n.* = genetic marker.

Markov chain (mär'kôf), *Statistics.* a succession of random events each of which is determined by the event immediately preceding it: *Markov chains have found application in a wide variety of sciences ... It would have been hard for Markov to foresee that electronic computers would one day work out probabilities in ... sciences in the manner indicated by him* (Philip J. Davis). [named after Andrei *Markov*, 1856–1922, Russian mathematician]

Markov process, *Statistics.* any process based on a Markov chain.

marl, *n. Geology.* a loose, crumbly rock material or earthy deposit containing clay and calcium carbonate, used as fertilizer.

marlite (mär'līt), *n.* = marlstone.
—**marlitic** (mär lit'ik), *adj.* of or like marlstone.

marlstone, *n. Geology.* a rock consisting of a hardened mixture of clay, calcium carbonate, and other minerals: *... the marlstone formations of the Colorado region* (Gerald L. Farrar).

marly (mär'lē), *adj. Geology.* of, like, or full of marl.

marrow, *n. Anatomy.* the soft, vascular tissue that fills the cavities of most bones and is the source of red blood cells and many white blood cells: *The thymus cells enormously facilitated the ability of the bone marrow cells to produce antibody but did not themselves become antibody-producing cells* (Hugh O. McDevitt).

Mars, *n. Astronomy.* the planet next in order beyond the earth and the fourth in distance from the sun. Its orbit about the sun lies between those of the earth and Jupiter and takes 687 days to complete, at an average distance from the sun of about 141,500,000 miles. It is the seventh largest planet in the solar system, with a mean diameter of 4,220 miles. *Although the question of whether life exists on Mars was not answered conclusively, the preliminary results from the three biology experiments carried by each lander tended toward the answer no.* (Joseph Ashbrook). See the picture at **solar system.**

marsh *n. Ecology.* low land covered at times by water; soft, wet land. Such plants as reeds, rushes, and sedges grow in marshes. *A swamp — more a popular than a scientific term — is a forested wetland that can be either a bog or a fen, usually with water standing or flowing through or over it. A marsh, on the other hand, has few or no woody plants but does have grasses and reeds in its silty soil* (Jane E. Brody).
ASSOCIATED TERMS: see **bog.**
—**marshy,** *adj.* **1** of or like a marsh. **2** living in marshes.

marsh gas, = methane.

marsupial (mär sü'pē əl), *Zoology.* —*n.* any of an order (Marsupialia) of mammals having a pouch covering the mammary glands on the abdomen, in which the female nurses and carries her incompletely developed young. Kangaroos, opossums, and wombats belong to this order. *Today, pouched animals, or marsupials, are found only in North and South America, Australia and Tasmania, and a few islands to the north of them. This was not always so ... Fossil marsupials have been discovered in the gypsum paving the streets of Paris* (Science News Letter).
—*adj.* **1** of marsupials. **2** having a pouch for carrying the young. Compare **placental.**

marsupium (mär sü'pē əm), *n., pl.* **-pia** (-pē ə). *Zoology.* **1** a pouch or fold of skin on the abdomen of a female marsupial for carrying its young. **2** a similar pouch in certain fishes and crustaceans. [from Latin *marsupium*, ultimately from Greek *marsipos* pouch]

Martian (mär'shən), *adj. Astronomy.* of the planet Mars: *The Martian year lasts six hundred and eighty-seven days* (New Yorker). [from Latin *Martius* of Mars]

mascon (mas'kon), *n. Astronomy, Geology.* **1** a massive concentration of dense material lying below the lunar surface and characterized by a higher-than-average gravity: *Lunar mascons (mass concentrations), which make the Moon exert an uneven gravitational pull on craft in lunar orbit, are probably the Moon's "maria" (dry seas) themselves, rather than some unseen force or feature beneath them* (Science Journal).
2 a similar feature on the planets or their moons: *We have already noted the detection on the Martian surface of mascons, analogous to the lunar areas of especially high gravity* (New Scientist). Compare **magcon.** [from *mas*(*s*) *con*(*centration*)]

maser (mā'zər), *n. Physics.* a device which amplifies or generates electromagnetic waves, especially microwaves, with great stability and accuracy. Masers operate at temperatures near absolute zero. Their primary uses are in long-distance radar and radio astronomy. Compare **laser.** [from *m*(*icrowave*) *a*(*mplification by*) *s*(*imulated*) *e*(*mission of*) *r*(*adiation*)]

masked, *adj.* **1** *Zoology.* **a** marked on the face or head as if wearing a mask. **b** having the wings, legs, etc., of the future form indicated in outline beneath the integument, as certain insect pupae.
2 *Chemistry.* prevented by a preliminary reaction from taking part in a normal reaction: *a masked sulfate group.*

mass, *n. Physics.* **1** a measure of the quantity of matter a body contains; the property of a physical body which gives the body inertia. Mass is a property not dependent on gravity and is obtained by either dividing the weight of the body by the acceleration of gravity or comparing an unknown mass with a known mass,

---

cap, fāce, fäther; best, bē, tėrm; pin, five;
rock, gō, ôrder; oil, out; cup, pùt, rüle,
yü in use, *yù* in uric;
ng in bring; *sh* in rush; *th* in thin, ᴛʜ in then;
zh in seizure.
ə = *a* in about, *e* in taken, *i* in pencil, *o* in lemon, *u* in circus

mass defect

as on a balance. *The unit of mass in the mks system is the mass of the standard kilogram. The unit of mass in the cgs system is 1/1000 as great as the mass of the standard kilogram and is called one gram. There is no mass standard in the English gravitational system of units* (Sears and Zemansky, *University Physics*).
2 = mass number.
▶ *Mass* and *weight* are different physical properties. *Weight* is the force on a body due to the pull of gravity; the weight of an object becomes smaller if it is moved away from the earth or from another planet into outer space. The *mass* of a body, however, remains constant regardless of the forces acting on the body.

mass defect, *Nuclear Physics.* the difference between the total mass of the neutrons and protons comprising the nucleus of an atom, and the total mass of the nucleus as a whole, the difference being the equivalent in mass of the energy that holds the nucleus together: *The "mass defect," of course, refers to the fact that the mass of the nucleus is slightly smaller than the sum of the masses of the particles combined in it; the difference is the "defect"* (Scientific American).

mass-energy equation, *Physics.* an equation expressing the equivalence of mass and energy, formulated by Albert Einstein in 1905: $E = mc^2$, where E = energy; m = mass; c = the velocity of light. Also called **Einstein equation.**

massive, *adj.* **1** *Mineralogy.* composed of compact material with an irregular form; not macroscopically crystalline: *Some sandstones are well cemented ... massive sandstone, while others ... disintegrate readily upon exposure* (Finch and Trewartha, *Elements of Geography*). **2** *Geology.* without definite structural divisions.
—**massivity,** *n.* the condition of being massive.

massless, *adj. Physics.* lacking mass; having a mass of zero: *According to the uncertainty principle, virtual photons can have a nonzero mass, unlike real photons, which are massless* (Scientific American).

mass number, *Physics.* a number that indicates the atomic weight of an isotope rounded off to the nearest integer. It is equal to the sum of the protons and neutrons in the nucleus. EXAMPLES: The mass number of ordinary hydrogen is 1, of carbon 12, and of oxygen 16, indicated by a superscript before (or sometimes after) the symbol for the element: ^1H, ^{12}C, ^{16}O. Similarly, isotopes of uranium are written ^{235}U and ^{238}U. *Symbol:* A

mass spectrograph, an apparatus for determining the mass numbers of isotopes by passing streams of ions through electric and magnetic fields which separate ions of different masses. The results are recorded on a photographic plate.

mass spectrometer, an apparatus similar to the mass spectrograph except that its results are recorded electrically: *It is possible by using a mass spectrometer to determine the proportions of heavy nitrogen, N^{15}, and the normally more abundant isotope, N^{14}, in a given sample* (Nature).
—**mass spectrometry,** the use of the mass spectrometer, especially to identify the chemical components of a substance.

mass spectroscope, any of various devices utilizing magnetic fields, electric fields, or both, to determine the masses of isotopes by producing a mass spectrum. The mass spectrograph and the mass spectrometer are two types of mass spectroscopes. *In the simplest mass spectroscope, electrons bombard a gas at low pressure ... if the original beam contains ions of various masses, it spreads into a number of beams called a mass spectrum. This is similar to what happens when a beam of white light passes through a prism and forms an optical spectrum* (Alfred O. Nier).
—**mass spectroscopy,** the use of a mass spectroscope: *Mass spectroscopy is used to measure the relative abundance of naturally occurring isotopes and to make precise determinations of nuclear masses* (R. L. Thornton).

mass spectrum, *Physics.* a band of charged particles of different masses formed when a beam of ions is passed through the deflecting fields of a mass spectroscope: *Before the mass spectrum of a compound can be obtained the compound has to be vaporised. With improvements in instrumentation, it is now possible to obtain the mass spectra of ... involatile substances as ... amino acids, peptides, carbohydrates* (New Scientist).

mast cell, *Biology.* a type of cell in connective tissue that has a distinctly granular cytoplasm and is associated with the release of histamine and the anticoagulant heparin. [*mast*, from German *masten* fatten]

master gland, *Anatomy.* the pituitary gland, especially the anterior lobe: *The anterior pituitary is often referred to as the "master gland" because it secretes several hormones required by other glands* (McElroy, *Biology and Man*).

masticate, *v. Zoology.* to reduce (solid food) to small pieces or to a pulp by grinding with the teeth; to chew. [from Late Latin *masticatum* chewed, from Greek *masticham* gnash the teeth]
—**mastication,** *n.* the act or process of masticating: *Mastication is the first process in the digestion of food ... Mastication mixes the food with saliva, which reacts chemically with the food* (A. C. Gryton).

mastigoneme (mas tig′ə nēm), *n. Biology.* a stiff lateral appendage on the flagellum of certain algal cells: *It is not known just how the mastigonemes become fixed to the shaft of the flagellum, but they appear to attach to specific sites along the flagellar sheath* (Max H. Hommersand). [from Greek *mastix, mastigos* whip + *nema* thread]

mastigophoran (mas′tə gof′ər ən), *Zoology.* —*n.* any of the class (Mastigophora) of protozoans comprising the flagellates, some of whose bodies contain chlorophyll, such as the euglena, and are often classified as algae. —*adj.* of or having to do with the mastigophorans. [from New Latin *Mastigophora* the class name, from Greek *mastix, mastigos* whip + *phoros* thing that bears (from *pherein* to bear, carry)]

mastoid (mas′toid), *Anatomy.* —*n.* a projection of bone behind the ear of many mammals. —*adj.* of, having to do with, or near the mastoid. [from Greek *mastoeides* breast-shaped, from *mastos* breast + *eidos* shape, form]

math., *abbrev.* **1** mathematical. **2** mathematics.

mathematical or **mathematic,** *adj.* of or having to do with mathematics: *a mathematical theorem, mathematical models, a set of mathematical equations.*
—**mathematically,** *adv.* according to mathematics.

mathematical logic, = symbolic logic.

mathematics, *n.* **1** the science dealing with the measurement, properties, and relationships of quantities, as expressed in numbers or symbols. Mathematics includes arithmetic, algebra, geometry, calculus, etc. *The facts of mathematics can be used empirically just as the facts of science can be interpreted mathematically* (London *Times*). *Abbreviation:* math.
2 mathematical aspects, processes, or operations: *to work out the mathematics of a statistical theory.*

mathematize (math′ə mə tīz), *v.* to formulate something into mathematical terms: *to mathematize a chemical or physical process.*
—**mathematization,** *n.* formulation in mathematical terms.

mating, *n. Biology.* the act or fact of pairing for sexual reproduction: *the mating of birds.*
—*adj.* of or involving mating: *the mating season, the mating flight of winged ants.* The **mating call** of an animal is a special call or sound it makes to attract a mate: *The hi-fi of frog and toad mating calls has shed some light on the origin and evolution of mechanisms that prevent one species from breeding with another species* (Science News Letter).

matrix (mā′triks), *n., pl.* **matrices** (mā′trə sēz′), **matrixes. 1** *Anatomy.* the formative part of an organ, such as the skin beneath a fingernail or toenail.
2 *Biology.* the intercellular substance of a tissue. Also called **ground substance.**
3 *Geology.* **a** the rock in which minerals, gems, or fossils are embedded: *By etching away the limestone matrix in dilute acid, the silicified fossils, which are not affected by the acid, are freed from the rock* (Raymond C. Moore). **b** = groundmass.
4 *Mathematics.* a set of elements in a rectangular array, subject to operations such as multiplication or addition according to specified rules. Matrices are used to represent relations between quantities and are usually enclosed in parentheses or square brackets, as

$$\begin{pmatrix} 1\,2\,3 \\ 4\,5\,6 \end{pmatrix} \quad \text{or} \quad \begin{bmatrix} 1\,2\,3 \\ 4\,5\,6 \end{bmatrix}$$

5 *Statistics.* an ordered table or two-dimensional array of variables.
[from Latin *matrix* womb, from *mater* mother]

matrix algebra, *Mathematics.* an algebra whose elements are matrices in which an unoccupied space represents a zero.

matrix mechanics, *Physics.* a formulation of quantum mechanics using spectroscopic data and matrix algebra, developed by Werner Heisenberg. It is mathematically equivalent to the theory of wave mechanics.

matter, *n. Physics.* anything which has mass and can exist ordinarily as a solid, liquid, or gas. Animals and plants are organic matter; minerals and water are inorganic matter. All matter is made up of atoms. *The mass of matter does not change as the body changes state. Thus, when 1 kilogram of ice melts, 1 kilogram of water is formed; and when 1 kilogram of water is converted into steam, the latter weighs 1 kilogram* (Parks and Steinbach, *Systematic College Chemistry*). *Einstein's formula postulated the revolutionary con-*

cept that matter and energy are actually different manifestations of one underlying entity: matter is frozen energy; energy is fluid matter (William L. Laurence). [from Latin *materia* substance, matter, growing layer in trees, from *mater* mother]
► *Matter* is that which all objects are made of; *material* is matter of a certain kind, such as wood, metal, or plastic; *substance* is matter with characteristic properties and composition, such as a chemical element or compound. Glass is a material, but salt and oxygen are substances.

maturation, *n. Biology.* **1** the final stages in the preparation of germ cells for fertilization, including meiosis and various changes in the cytoplasm: *Maturation involves the reduction in the number of chromosomes to one half that of the somatic cells* (Youngken, *Pharmaceutical Botany*).
Maturation division is the name of either of the two divisions of meiosis.
2 the last stage of differentiation in cellular growth: *The period of maturation begins when the mould's active growth is finished ... a more stable period when volatile acids tend to accumulate and there is a mellowing of texture and development of flavour* (J.A. Barnett).
—**maturational,** *adj.* of or having to do with maturation.

mature, *adj.* **1** *Biology.* ripe, full-grown, or developed: *a mature plant, a mature fruit. Grain is harvested when it is mature.*
2 *Geology.* **a** so long subjected to erosion as to show mainly smooth slopes: *mature land.* **b** fully adjusted to rock formations: *a mature stream.*
[from Latin *maturus* ripe]
—**maturity,** *n.* **1** *Biology.* mature condition; full development: *Rabbits reach sexual maturity at four months, produce an average of six litters per year with an average of five or six young per litter* (F. Fenner and M.F. Day).
2 *Geology.* a stage in the evolutionary erosion of land areas where the flat uplands have been widely dissected by deep river valleys.

Maunder minimum, *Astronomy.* a period of irregular and relatively low solar activity between about 1645 and 1715: *During a 70-year period in the late 17th and 18th centuries, sunspots and other signs of activity all but vanished from the sun. Historical evidence for this period, called the "Maunder minimum," coincided with a very clear picture of the minimum shown as a radiocarbon anomaly in tree rings formed at the time* (Science News). [named after E. Walter *Maunder,* a 19th-century British solar astronomer who first described it in 1890]

---

cap, fāce, fäther; best, bē, tėrm; pin, fīve;
rock, gō, ôrder; oil, out; cup, pùt, rüle,
yü in use, *yu̇* in uric;
ng in bring; sh in rush; th in thin, ŦH in then;
zh in seizure.
ə = *a* in about, *e* in taken, *i* in pencil, *o* in lemon, *u* in circus

385

maxilla (mak sil′ə), *n., pl.* **maxillae** (mak sil′ē), **maxillas.**
1 *Anatomy.* the jaw or jawbone, especially the upper jawbone in mammals and most vertebrates. See the picture at **skull.**
2 *Zoology.* either of a pair of appendages just behind the mandibles of insects, crabs, etc. See the picture at **mouth.**
[from Latin *maxilla* jaw]
—**maxillary,** *adj., n., pl.* **-laries.** *Anatomy.* —*adj.* of or having to do with the jaw or jawbone, especially the upper jawbone. —*n.* = maxilla.

maxilliped (mak sil′ə ped), *n. Zoology.* one of the anterior limbs of crustaceans and other arthropods that are modified into accessory mouthparts. [from *maxilla* + Latin *pedem* foot]

maxillofacial (mak sil′ə fā′shəl), *adj. Anatomy, Medicine.* of or having to do with the lower half of the face: *maxillofacial prosthetics.*

maximin (mak′sə min), *n. Mathematics.* a strategy in game theory that maximizes the minimum gain that a player can guarantee himself: *Maximin is like a philosophy of complete pessimism except that the decision is based on the decision-maker's possible payoffs rather than losses* (New Scientist). Contrasted with **minimax.** [from *maxi(mum)* + *min(imum)*]

maximum (mak′sə məm), *n., pl.* **-mums, -ma** (-mə). *Mathematics.* a value of a function greater than any other value of the function in a given interval. Contrasted with **minimum.** [from Latin *maximum,* neuter of *maximus* greatest, superlative of *magnus* great]

maximum sustainable yield, *Ecology.* the largest possible yield that can be obtained from a natural resource indefinitely, especially the largest number of animals that can be taken each year from a population without impairing the ability of the population to maintain itself. Compare **carrying capacity.**

maxwell (maks′wel *or* maks′wəl), *n. Physics.* the unit of magnetic flux in the CGS system; the flux through one square centimeter normal to a magnetic field whose intensity is one gauss: *In the electromagnetic system a line of induction is called a maxwell, and magnetic induction is expressed in maxwells per square centimeter. One maxwell per square centimeter is called one gauss* (Sears and Zemansky, *University Physics*). [named after James Clark *Maxwell,* 1831–1879, Scottish physicist]

Maxwell-Boltzmann distribution, *Physics.* a mathematical formula describing the distribution of velocities among gas particles in equilibrium and the statistical probabilities of impact associated with these: *In a system containing a large number of molecules appreciable deviations from the Maxwell-Boltzmann distribution will have a very small probability of occurrence* (R.C. Tolman). [named after James Clerk *Maxwell,* 1831–1879, and Ludwig *Boltzmann,* 1844–1906, the physicists who formulated the theory]

mb., *abbrev.* millibar *or* millibars.

mc, *abbrev.* millicurie *or* millicuries.

Md, *symbol.* mendelevium.

Me, *abbrev.* methyl.

meadow, *n. Ecology.* a low, moist grassland.

mean, *Mathematics.* —*adj.* having a value intermediate between the values of other quantities: *a mean diameter.* SYN: average.
—*n.* **1** a quantity having a value intermediate between the values of other quantities, especially the average obtained by dividing the sum of all the quantities by the total number of quantities. **2** either the second or third term of a proportion of four terms. EXAMPLE: In the proportion 2/3 = 4/6, the means are 3 and 4.
ASSOCIATED TERMS: see **average.**
[from Old French *meien,* from Latin *medianus* of the middle, from *medius* middle of]

meander (mē an′dər), *n. Geography.* a looplike, winding turn occurring in a river or stream that flows across nearly level terrain.

mean deviation, *Statistics.* a measure of dispersion obtained by taking the average of the absolute values of the differences between individual numbers or scores and their mean.

mean distance, *Astronomy.* **1** the average of the aphelion and perihelion of a planet from the sun, one of the data necessary to determine the orbit of a planet: *The earth's mean distance from the sun is about 92,900,000 miles* (Baker, *Astronomy*). **2** the average of the greatest and least distance of any heavenly body, such as a star or a satellite, from the focus of its orbit.

mean free path, *Physics.* the average distance a molecule of a gas or other substance can travel before it collides with another molecule. The distance will vary according to the altitude in the case of a gas. *In a gas the mean free path may be as large as a few hundred angstroms, or much larger than the size of the atom itself* (Scientific American).

mean proportional, *Mathematics.* the means in a proportion when they are equal. EXAMPLE: In a:b = b:c, b is the mean proportional.

mean solar day, *Astronomy.* a day of twenty-four hours, measured from midnight to midnight; civil day.

mean solar time, = mean time.

mean sun, *Astronomy.* a hypothetical sun in various astronomical calculations that moves uniformly along the celestial equator at the mean speed with which the real sun apparently moves along the ecliptic. The apparent speed of the real sun varies due to the eccentricity of the earth's orbit.

mean time, *Astronomy.* time according to the hour angle of the mean sun, constituting the "ordinary time" or "clock time" of daily life. Also called **mean solar time.**

measure, *v.* to determine the dimensions, amount, capacity, duration, pressure, etc., of something, especially by the use of standard units of length, area, volume, weight, time, temperature, etc. The Celsius scale is used to measure temperature. The ampere is a unit for measuring the flow of electric current.
—*n.* **1 a** a unit or standard of measuring, such as a meter, a kilometer, a gram, or a liter. Some other common measures are a kilogram, ampere, mole, and second.
b a system of measurement: *liquid measure, dry measure, square measure.*
2 *Mathematics.* a number or quantity found a certain number of times without remainder in another number. Also called **factor.**

[from Old French *mesurer,* from Late Latin *mensurare,* from Latin *mensura* a measure, from *metiri* to measure]

—**measurement,** *n.* **1** the act or process of measuring: *weight measurements, a measurement of mass.*
2 dimension, quantity, or capacity found by measuring: *The measurements of the room are 6 by 8 meters.*
3 a system of measuring or of measures: *Metric measurement is used internationally.*

meatus (mē ā′təs), *n., pl.* **-tuses** or **-tus.** *Anatomy.* a passage, duct, or opening in the body, as in the ear. [from Latin *meatus* path, from *meare* to pass]

mechanical, *adj.* **1** of, dealing with, or worked by machinery. **2** *Physics.* having to do with or in accordance with mechanics.

mechanical advantage, *Physics.* the ratio of resistance or load to the force or effort that is applied in a machine. Mechanical advantage is shown by the number of times a machine increases the force exerted on it.

mechanical efficiency, *Physics.* the ratio of the power of an engine actually produced to the power it could theoretically produce.

mechanical energy, *Physics.* **1** the energy transmitted by a machine or machinery; energy in the form of mechanical power: *The largest single loss ... occurs when heat energy is converted into mechanical energy* (New Scientist).
2 the kinetic plus the potential energy of a body: *Provided there are no frictional or other dissipative effects, the total mechanical energy of a system of bodies remains constant* (Shortley and Williams, *Elements of Physics*).

mechanical mixture, *Chemistry.* a mixture in which the several ingredients have not entered into chemical combination, but still retain their identity and can be separated by mechanical means.

mechanics, *n.* **1** the branch of physics dealing with the action of forces on solids, liquids, and gases at rest or in motion. Mechanics includes kinetics, statics, and kinematics. **2** mechanical part or process: *the mechanics of mitosis.*

mechanism, *n.* **1a** a machine. **b** the working parts of a machine.
2 *Biology.* a system of parts working together to perform a specific function: *the mechanism of photosynthesis.*

mechanochemistry (mek′ə nō kem′ə strē), *n.* **1** the study of the means by which chemical energy can be converted directly into mechanical energy.
2 mechanochemical process or activity: *the molecular mechanochemistry of the active sites of enzymes.*
—**mechanochemical,** *adj.* of or having to do with mechanochemistry; capable of converting chemical energy directly into mechanical energy.

mechanoreception (mek′ə nō ri sep′shən), *n. Physiology.* the process by which a mechanoreceptor responds to mechanical stimuli such as pressure or touch.
—**mechanoreceptive,** *adj.* capable of responding to mechanical stimuli.
—**mechanoreceptor,** *n.* a specialized sensory receptor that responds to mechanical stimuli: *The mechanoreceptors in muscle respond to ... stimuli such as stretching and pressure* (Scientific American).

media (mē′dē ə), *n., pl.* **-diae** (-dē ē). *Anatomy.* the middle layer of the wall of a blood or lymphatic vessel. It consists mainly of cylindrical muscle fibers. [from Latin *media,* feminine of *medius* middle]

medial (mē′dē əl), *adj. Mathematics.* having to do with a mean or average. [from Latin *medialis* (of the) middle, from Latin *medius* middle]

medial moraine, *Geology.* the moraine that occurs when the lateral moraines of two glaciers meet and merge.

median (mē′dē ən), *adj.* **1** *Biology.* **a** of or in the middle; middle: *the median vein of a leaf.* **b** having to do with or designating the plane that divides something into two equal parts, especially one dividing a symmetrical animal into right and left halves.
2 *Mathematics.* of a median; having as many above as below a certain number: *The median age of the population was found to be 21 (that is, there were as many persons above 21 as below it), while the average age was found to be 25.*
—*n.* **1** *Mathematics.* the middle number of a sequence having an odd number of values. If the sequence has an even number of values, the median is the average of the two middle values. EXAMPLES: The median of 1, 3, 4, 8, 9 is 4. The median of 1, 3, 4, 8, 9, 10 is 6.
2 *Geometry.* **a** (of a triangle) a line from a vertex to the midpoint of the opposite side. **b** (of a trapezoid) the line joining the midpoints of the nonparallel sides.
ASSOCIATED TERMS: see **average.**
[from Latin *medianus,* from *medius* middle]
—**medianly,** *adv.*

medium (mē′dē əm), *adj., n., pl.* **-diums, -dia** (-dē ə). **1** *Physics.* a substance or agent through which anything acts or an effect is produced: *Copper wire is a medium of electric transmission.*
2 *Biology.* a substance in which something can live; an environment: *Water is the natural medium of fish.*
3 *Microbiology.* **a** a nutritive substance, either liquid or solid, in or upon which microorganisms are grown for study; a culture medium. **b** a substance used for displaying or preserving organic specimens.
[from Latin *medium,* neuter of *medius* middle]

medulla (mi dul′ə), *n., pl.* **-dullas, -dullae** (-dul′ē). *Biology.* **1** = medulla oblongata.
2 a the inner part or tissue of an organ or structure: *The medulla, the region of the kidney most susceptible to bacterial infection, becomes more resistant when fluids are forced through it* (Science News). **b** the marrow of bones. **c** the pith of plants.
3 = medullary sheath.
4 the central core of loosely packed cells in a shaft of hair in most mammals. Compare **cortex.** See the picture at **hair.**
[from Latin *medulla* marrow]

---

cap, fāce, fäther; best, bē, tėrm; pin, five;
rock, gō, ôrder; oil, out; cup, pùt, rüle,
yü in use, *yu̇* in uric;
ng in bring; *sh* in rush; *th* in thin, ᴛʜ in then;
zh in seizure.
ə = *a* in about, *e* in taken, *i* in pencil, *o* in lemon, *u* in circus

387

medulla oblongata (ob′long gä′tə), *Anatomy*. the part of the brain that controls breathing and other involuntary functions. It consists of nerve fibers and nerve centers and is located at the top end of the spinal cord. See the picture at **brain**. [from New Latin *medulla oblongata* prolonged medulla]

medullary (med′ə ler′ē *or* mi dul′ər ē), *adj. Biology*. of, having to do with, or like a medulla. The **medullary layer** of a typical lichen is a brood consisting of loosely interwoven fungal hyphae and some algal cells. A **medullary ray** is one of the radiating vertical bands of woody tissue which divide the vascular bundles and connect the pith with the bark in the stems of exogenous plants.

medullary sheath, 1 *Botany*. a narrow ring comprising the innermost layer of woody tissue that surrounds the pith in certain plants.
2 *Anatomy*. = myelin sheath: *The development of a medullary sheath ... brought with it a great increase in the speed and economy of nerve-signal traffic* (Scientific American).

medullated (med′ə lā′tid), *adj. Anatomy*. having a medullary sheath; myelinated: *a medullated nerve fiber.*

medullin (mə dul′ən), *n. Biochemistry*. a prostaglandin isolated from the medulla of the kidney, used in the treatment of high blood pressure.

medusa (mə dü′sə *or* mə dü′zə), *n., pl.* **-sas, -sae** (-sē *or* -zē). *Zoology*. a jellyfish, especially a free-swimming hydrozoan. [from New Latin *Medusa* the genus name, from Latin *Medusa* a mythological monster with snakes for hair (so called because one species has feelers that look like the snake hair of Medusa)]
—**medusan,** *adj.* of or having to do with a medusa. —*n.* = medusa.

medusoid (mə dü′soid), *Zoology*. —*adj.* of or like a medusa: *The bracts, swimming bells, and gonophores are constructed on a medusoid plan* (Shull, *Principles of Animal Biology*). —*n.* = jellyfish.

meg-, *combining form.* the form of **mega-** before vowels, as in *megohm.*

mega-, *combining form.* **1** large, as in *megaspore = large spore.* **2** one million, as in *megacycle = one million cycles. Symbol:* M Also spelled **meg-** before vowels. [from Greek *megas, megalou* great]

megabar, *n.* a unit of pressure equal to one million bars.

megacryst, *n. Geology*. a large mineral crystal of metamorphic origin in a finer-grained matrix.

megacurie, *n.* a unit of radioactivity equal to one million curies.

megacycle, *n.* **1** one million cycles.
2 a unit of frequency equal to one million cycles per second, used formerly to express the frequency of radio waves, now expressed in megahertz. *Abbreviation:* mc.

megadyne, *n.* a unit of force equal to one million dynes.

megagamete, *n.* = macrogamete.

megagametophyte, *n. Botany*. a female gametophyte that develops from a megaspore. Compare **microgametophyte.**

megagram, *n.* = metric ton.

megahertz *n.* a unit of frequency equal to one million hertz, used to express the frequency of radio waves: *The frequency band between about 10 megahertz and 20 gigahertz is technically suitable for space communication* (New Scientist). *Symbol:* MHz Compare **megacycle** (def. 2).

megajoule *n.* a unit of work or energy, equivalent to a million joules.

megakaryocyte (meg′ə kar′ē ə sīt), *n. Biology*. a large cell of bone marrow with a lobulated nucleus that gives rise to blood platelets: *Each platelet is only a fragment of the cytoplasm of a giant cell known as a megakaryocyte* (Marjorie B. Zucker).

megaloblast (meg′ə lō blast), *n. Biology*. a nucleated red blood cell, of abnormally large size, found in the blood of anemic persons. [from Greek *megas, megalou* great + *blastos* sprout, germ]
—**megaloblastic** (meg′ə lō blas′tik), *adj.* of or having megaloblasts; characterized by the presence of megaloblasts in the blood: *megaloblastic anemia.*

megalops (meg′ə lops), *n. Zoology*. the final larval stage of a crab. The megalops is a small organism with large, stalked eyes that swims to the shore and digs a hole in the sand where it molts into a tiny crab. [from New Latin *megalops,* from Greek *megalōpos* large-eyed]

megaparsec, *n.* a unit of distance, equivalent to one million parsecs: *A parsec is equal to 3.26 light years, and a megaparsec is a million parsecs, or 3,260,000 light years* (Science News Letter).

megarad, *n.* a unit for measuring absorbed doses of radiation, equivalent to one million rads.

megascopic (meg′ə skop′ik), *adj.* = macroscopic (def. 1).

megaseism (meg′ə sī zəm *or* meg′ə sī səm), *n. Geology*. a great or severe earthquake. [from *mega-* + Greek *seismos* earthquake]

megasporangium (meg′ə spə ran′jē əm), *n., pl.* **-gia** (-jē ə). *Botany*. a sporangium containing megaspores: *The megasporangium usually produces only four megaspores* (Emerson, *Basic Botany*). Also called **macrosporangium.**

megaspore, *n. Botany*. **1** a spore of comparatively large size from which a female gametophyte develops. Compare **microspore.**
2 the embryo sac in seed plants: *Within the nucleus is found the embryo sac or megaspore* (Youngken, *Pharmaceutical Botany*). Also called **macrospore.**

megasporophyll (meg′ə spôr′ə fil), *n. Botany*. **1** a sporophyll bearing only megasporangia. **2** = carpel.

megathere (meg′ə thir), *n. Zoology*. any of an extinct group of huge, plant-eating mammals of the Pleistocene, resembling the sloths, the fossil remains of which have been found in South America. [from New Latin (coined by Cuvier) *Megatherium,* from Greek *megas* large + *thērion* wild animal]

megatherium (meg′ə thir′ē əm), *n.* = megathere.

megavolt (meg′ə vōlt′), *n.* a unit of electromotive force equivalent to one million volts.

megohm (meg′ōm′), *n.* a unit of electrical resistance equal to one million ohms.

meiosis (mī ō′sis), *n., pl.* **-ses** (-sēz′). *Biology*. the process by which the number of chromosomes in reproductive cells of sexually reproducing organisms is reduced to half the original number, resulting in the production of gametes or spores. Meiosis consists essentially of two cell divisions. In the first, the homologous chromosomes separate equally into the two new cells so that each cell contains the haploid number of chromosomes, or half the diploid number. In the second cell division, the pairs of chromosomes split, one of each kind of chromosome going to the four new cells. Thus each new cell again contains the haploid number of chromosomes. *As a result of meiosis, diploid (2n) primary sex cells divide and form monoploid (n) cells which mature into specialized reproductive cells* (Biology Regents Syllabus). Also called **reduction division.** Compare **mitosis.** [from Greek *meiōsis* a lessening, from *meioun* lessen, from *meion* less]

meiosis:

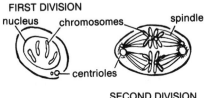

FIRST DIVISION

nucleus   chromosomes   spindle

centrioles

SECOND DIVISION

nucleus

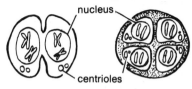

centrioles

meiospore (mī'ə spôr), *n. Biology.* a spore resulting from the meiotic division of a spore cell: *The spores of bryophytes and of the ferns and other vascular plants are all meiospores, as are some of the spores produced by algae and fungi* (Greulach and Adams, *Plants*).

meiotic (mī ot'ik), *adj. Biology.* of or having to do with meiosis: *the period between the two meiotic divisions.*

melanic (mə lan'ik), *Biology.* —*adj.* having to do with or showing melanism: *The murk of nineteenth century Manchester fostered the melanic form ... of the peppered moth* (Nature).

—*n.* an animal showing melanism: *There is a difference in the proportion of melanics found among populations of various moth species that share the same environment* (Scientific American).

melanin (mel'ə nin), *n. Biochemistry.* any of a class of dark pigments, especially the black pigment in the skin, hair, and eyes of human beings and many animals, or one developed in certain diseases. Large amounts of melanin help protect the skin from sunburn and improve vision in bright sunlight. [from Greek *melanos* black]

melanism (mel'ə niz əm), *n. Biology.* darkness of color resulting from an abnormal development of melanin in the skin, hair, and eyes of a human being, or in the skin, coat, or plumage of an animal. Compare **industrial melanism.**

—**melanistic** (mel'ə nis'tik), *adj.* characterized by melanism.

melanite (mel'ə nīt), *n. Mineralogy.* a velvet-black variety of calcium garnet containing titanium. [from Greek *melanos* black + English -*ite*]

melanization (mel'ə nə zā'shən), *n. Biology.* the process of melanizing or the condition of being melanized: *In certain moths the increased melanization of individuals found in industrial areas suggests that a possible genetical factor is concerned in the pigmentation process* (B. Nickerson).

melanize (mel'ə nīz), *v. Biology.* to produce melanism in; make dark or black: *In locusts ... the gregarious insects are highly melanized and have a greater metabolic rate than the non-pigmented solitary forms* (Science News).

melanoblast (mə lan'ə blast *or* mel'ə nō blast), *n. Biology.* an immature pigment cell, especially a precursor of melanocytes and melanophores: *... melanoblasts, the cells characteristic of both mouse and human black cancers* (Science News Letter). [from Greek *melanos* black + *blastos* germ, sprout]

melanocyte (mə lan'ə sīt *or* mel'ə nō sīt), *n. Biology.* a cell which synthesizes melanosomes, found in the skin, the choroid of the eye, and the hair; a mature melanin-forming cell. [from Greek *melanos* black + English -*cyte*]

melanocyte-stimulating hormone, *Biochemistry.* a pituitary hormone that stimulates melanocytes, causing darkening of the skin. *Abbreviation:* MSH Also called **intermedin.**

melanogenesis (mel'ə nō jen'ə sis), *n. Biology.* the formation and development of melanin.

melanophore (mə lan'ə fôr *or* mel'ə nə fôr), *n. Biology.* a cell containing melanin, especially a contractile pigment cell in lower vertebrates: *The black-pigmented cells, or melanophores, of their skin expands laterally when the animal is cold, thus darkening the body and increasing the rate at which it absorbs radiant energy* (Charles M. Bogert).

melanophore-stimulating hormone, = melanocyte-stimulating hormone.

melanosome (mə lan'ə sōm *or* mel'ə nə sōm), *n. Biology.* a specialized particle, located in the cytoplasm of a melanocyte, in which melanin is formed.

melatonin (mel'ə tō'nin), *n. Biochemistry.* a hormone of the pineal gland, thought to be involved in the regulation of various physiological activities: *Melatonin ... affects the maturing of the sex glands in certain species of young animals* (Will G. Ryan). [from *mela(nin)* + *(sero)tonin*]

M electron, *Physics.* one of the electrons in the M-shell.

melt, *v. Chemistry, Physics.* to change from a solid to a liquid by heating or being heated: *Alloys were made by melting together two or more of these metals (e.g. copper and tin to make bronze, tin and lead to make low-melting solders), or by smelting mixed ores* (J. Crowther). SYN: fuse.

---

cap, fāce, fäther; best, bē, tèrm; pin, fīve;
rock, gō, ôrder; oil, out; cup, pùt, rüle,
yü in use, *yù* in uric;
ng in bring; *sh* in rush; *th* in thin, ŦH in then;
zh in seizure.
ə = *a* in about, *e* in taken, *i* in pencil, *o* in lemon, *u* in circus

—n. **1** the act or process of melting: *Oxygen when injected into an open hearth can speed the melt of a particular batch of steel by as much as 50%* (Wall Street Journal).

2 *Geology.* molten material: *As the details of lava streams and deposits of volcanic ash became better understood, the British geologists turned their attention to the conduits through which the molten rock, or magma, as such natural silicate melts are called, had reached the surface* (Gilluly, *Principles of Geology*).

meltemi (mel tä′mē), *n. Meteorology.* the northerly summer wind in the eastern Mediterranean and the Aegean Sea. [from New Greek *meltemi,* from Turkish *meltem*]

melting point, *Chemistry.* the temperature at which a solid substance melts, especially under a pressure of one atmosphere. Different substances usually have different melting points. The melting point of ice is 0°C. Tungsten has the highest melting point of all metallic elements. *The freezing point of a substance is the same as its melting point* (Beauchamp, *Everyday Problems in Science*). *Abbreviation:* m.p. Also called **fusing point.**

meltwater, *n. Geology.* water formed from melting ice or snow, especially from a glacier: *A valley glacier ... is far wider and thicker than the corresponding stream of meltwater* (Science News). See the picture at **drumlin.**

member, *n.* **1** *Biology.* a part or organ of a plant, animal, or human body, especially a leg, arm, wing, or branch. **2** *Mathematics.* **a** a quantity that belongs to a set; element of a set. **b** the expression on either side of an equation.
[from Old French *membre,* from Latin *membrum* limb, part]
—membral, adj. Biology. of or having to do with a member, especially a member of the body.

membrane, *n. Biology.* a thin, soft, pliable sheet or layer of animal or plant tissue lining, covering, separating, or connecting some part of the organism: *Living cells are enclosed in membranes through which they obtain their food from the surrounding medium and through which they excrete unwanted materials.... Cell membranes are able to differentiate between particles in solution and transfer some kinds of them, while excluding others* (K.S. Spiegler). *Membranes consist of a double sheet of lipids, or fatty molecules, layered with one or two sheets of protein* (Edouard Kellenberger).
[from Latin *membrana,* from *membrum* member]
► Membranes may be *fibrous, mucous,* or *serous.* Fibrous membranes, such as the periosteum which covers the surface of bones, consist of connective tissue made up mostly of fibers. Mucous membranes, which form the lining of the nose, throat, mouth, and other passages open to the outside, consist of tissue moistened by secretions of mucous glands. Serous membranes, which line closed cavities of the body such as the peritoneum and the pericardium, are moistened by serous fluid.

membrane bone, *Anatomy.* a bone that originates in membranous connective tissue, instead of being developed or preformed in cartilage: *Membrane bones,* which are found only in the face and head regions, originate as membranes which have the shape of the bones to be formed. As the embryo grows these membranes develop centers of ossification which form bone from the membrane through impregnation with calcium salts (Winchester, *Zoology*).

membranous, *adj. Biology.* **1** of or like membrane: *membranous wings.*
2 characterized by the formation of a membrane or a layer like a membrane. In **membranous croup,** a deposit similar to a membrane forms in the throat and hinders breathing.

membranous labyrinth, *Anatomy.* the membranous part of the inner ear: *The wall of the membranous labyrinth contains the end organ of the auditory nerve, which carries sound messages to the brain* (Lloyd Ackerman).

memory, *n.* **1** any device or apparatus in which data may be stored and from which it may be retrieved, especially in a high-speed electronic computer: *a semiconductor memory.*
2 the amount of data that a memory can store: *a 10 million-K memory.*

menarche (mə när′kē), *n. Physiology.* the beginning of menstruation; first menstrual period: *No significant association was found between cervical cancer and ... median age at menarche* (Lancet). [from Greek *mēn, mēnos* month + *arche* beginning]

mendelevium (men′dl ē′vē əm), *n. Chemistry.* a rare, radioactive, metallic element, produced artificially from einsteinium. Mendelevium is chemically similar to thulium. *Element 101, mendelevium, with a lifetime of 1.5 hours, was discovered by bombarding an invisibly small amount of einsteinium-253 with helium ions in a cyclotron* (Isadore Perlman). *Symbol:* Md; *atomic number* 101; *mass number* 258. [named after Dmitri I. Mendeleev, 1834–1907, Russian chemist]

Mendelian (men dē′lē ən), *Genetics. —adj.* **1** of or having to do with Gregor Johann Mendel or Mendel's laws: *Any heredity is now considered Mendelian if it is dependent on chromosomes* (Shull, *Principles of Animal Biology*). *The first demonstration of a human trait that behaved according to the Mendelian rules was provided in 1903 by William C. Farabee and William E. Castle, who showed that albinism in three generations of a Negro family behaved as a Mendelian recessive* (Sheldon C. Reed).
2 inherited in accordance with Mendel's laws: *Phenylketonuria is a metabolic defect transmitted genetically as a Mendelian recessive factor* (Saturday Review).
—n. a follower of Mendel or supporter of his theories.

Mendelism (men′də liz əm), *n.* the doctrines of Gregor Johann Mendel, especially Mendel's laws.

Mendel's laws (men′dəlz), *Genetics.* the laws or principles governing the inheritance of certain characteristics by plants and animals, as formulated by the Austrian botanist Gregor Johann Mendel (1822–1884) in experiments with peas. Mendel's laws state that each characteristic is inherited independently, that characteristics show dominant and recessive forms, and that successive generations of crossbred offspring exhibit inherited characteristics in different combinations, each combination in a specific proportion of individuals. *Briefly, ... Mendel's laws mean that in-*

dependent genes come together or segregate freely in the generations, and that one is dominant over the other (Emerson, *Basic Botany*). See also **dominance, independent assortment, segregation,** and **linkage.**

meninges (mə nin′jēz), *n., pl.* of **meninx** (mē′ningks). *Anatomy.* the three protective membranes that surround the brain and spinal cord (dura mater, arachnoid membrane, pia mater). [from New Latin *meninges,* from Greek *mēningos* membrane]

meniscus (mə nis′kəs), *n., pl.* **-niscuses, -nisci** (-nis′ī). **1** *Physics.* the curved upper surface of a column of liquid. It is concave when the walls of the container are moistened, convex when they are dry. *Water has the greatest surface tension of any liquid known, so that at any given matrix potential the radius of curvature of a water meniscus will be greater than it could be for another liquid* (H.L. Penman).
2 *Optics.* a lens, convex on one side and concave on the other. A meniscus is thicker in the center so that it has a crescent-shaped section.
3 *Anatomy.* a crescent-shaped piece of cartilage with a joint, such as the knee joint.
[from New Latin, from Greek *mēniskos,* diminutive of *mēnē* moon]

meniscus (def. 1)

dry glass   wet glass

menopause (men′ə pôz), *n. Physiology.* the permanent cessation of the menses, occurring normally between the ages of 45 and 55. [from Greek *mēn, mēnos* month + *pausis* pause]

menses (men′sēz), *n.pl. Physiology.* the discharge of bloody fluid from the uterus that normally occurs approximately every four weeks between puberty and menopause. The discharge results from the shedding of the thickened lining of the uterus when fertilization does not take place. [from Latin *menses,* plural of *mensis* month]

menstrual (men′strü əl), *adj. Physiology.* of or having to do with the menses: *the menstrual cycle or period.*
▶ The menstrual cycle consists of four stages: (1) maturation of an egg (ovum) within the Graafian follicle and the secretion of the hormone estrogen, which initiates vascularization and thickening of the lining of the uterus; (2) ovulation, or the release of the egg from the follicle; (3) formation of the corpus luteum, which secretes the hormone progesterone to enhance vascularization of the uterine lining; (4) menstruation, or the shedding of the uterine lining when fertilization does not take place.

menstruate, *v. Physiology.* to have a menstrual discharge or flow; undergo menses.

—**menstruation,** *n.* the act or period of menstruating: *The interval from ovulation to menstruation is about two weeks* (Sidonie M. Gruenberg).

menstruum (men′strü əm), *n. Biochemistry.* a solvent used to extract the principles from plant and animal tissues. [from Medieval Latin *menstruum,* neuter of Latin *menstruus* menstrual (so called from the supposed belief of the alchemists in the dissolving capacity of menstrual discharge)]

mensuration (men′shə rā′shən), *n.* the branch of mathematics that deals with finding measurements, such as lengths, areas, and volumes. [from Late Latin *mensurationem,* ultimately from Latin *mensura* measure]

mer., *abbrev.* meridian.

Mercalli scale (mer kä′lē), *Geology.* a scale for measuring the magnitude of an earthquake, ranging from 1 to 12. Compare **Richter scale.** [named after Giuseppe *Mercalli,* 19th-century Italian scientist who devised the scale]

mercaptan (mər kap′tan), *n. Chemistry.* any organic compound of a series having the general formula RSH, resembling the alcohols and phenols, but containing sulfur in place of oxygen, especially ethyl mercaptan, a colorless liquid having an unpleasant odor. Also called **thiol.** [from German *Mercaptan,* from New Latin *mer(curium) captan(s)* that catches mercury]

Mercator projection (mer kā′tər), *Geography.* a map projection made with parallel straight lines instead of curved lines for latitude and longitude. In a Mercator projection the areas near the poles appear disproportionately large, because the longitudes increase in distance from each other as they approach the poles. *The Mercator projection ... is conformal, i.e., taking any small area, the shape of the regions is the same as on the globe* (Erwin Raisz). Compare **Mollweide projection.** [named after Gerhardus *Mercator,* 1512–1594, Flemish cartographer]

mercuric (mər kyùr′ik), *adj. Chemistry.* **1** of mercury. **2** containing mercury, especially with a valence of two: *a mercuric compound.*

mercuric chloride, *Chemistry.* an extremely poisonous white substance, used in taxidermy and in metallurgy, and in medicine as a germicide and disinfectant. *Formula:* $HgCl_2$ Also called **bichloride of mercury, corrosive sublimate.**

mercurous (mər kyùr′əs *or* mèr′kyər əs), *adj.* **1** of mercury. **2** containing mercury, especially with a valence of one: *a mercurous compound.*

Mercury, *n. Astronomy.* the smallest planet in the solar system and the one nearest to the sun. Its orbit about the sun takes 88 days to complete, at a mean distance of almost 36,000,000 miles. Mercury goes around the sun about four times while the earth is going around

---

cap, fāce, fäther; best, bē, tèrm; pin, fīve;
rock, gō, ôrder; oil, out; cup, pùt, rüle;
yü in use, *yù* in uric;
ng in bring; *sh* in rush; *th* in thin, ᴛʜ in then;
zh in seizure.
ə = *a* in about, *e* in taken, *i* in pencil, *o* in
lemon, *u* in circus

once. *Mercury has a rotation period equal to two-thirds its orbital period, or 59 days* (Hyron Spinrad). See the picture at **solar system.** [from Latin *Mercurius* a Roman god serving as messenger to the other gods]

mercury, *n. Chemistry.* a heavy, silver-white metallic element that is liquid at ordinary temperatures. It occurs naturally in the mineral cinnabar and combines with most other metals to form amalgams. *Symbol:* Hg; *atomic number* 80; *atomic weight* 200.59; *melting point* -38.87°C; *boiling point* 356.58°C; *oxidation state* 1, 2. [from Medieval Latin *mercurius,* from Latin *Mercurius* the Roman god Mercury]

mercury barometer, a barometer in which the height of a column of mercury enclosed in a glass tube is measured to determine atmospheric pressure.

-mere, *combining form.* part; division; segment, as in *metamere.* [from Greek *meros* portion, share]

mericarp (mer′ə kärp), *n. Botany.* any of the two or more one-seeded parts of a schizocarp. [from Greek *meros* part + *karpos* fruit]

meridian, *n.* **1** *Geography.* **a** an imaginary great circle passing through any place on the earth's surface and through the North and South Poles. **b** the half of such a circle from pole to pole. All the places on the same meridian have the same longitude. *A half-circle, extending from one pole to the other, is known as a meridian. There are 359 meridians—dividing the distance around the earth into 360 equal parts* (White and Renner, *Human Geography*).
2 *Astronomy.* the highest point that the sun or any star reaches in the sky: *There is one hour-circle that is very easy to visualize—the meridian. When you face due south you can imagine it running from the horizon through the zenith or overhead point, on to the north celestial pole, and down to the horizon* (Bernhard, *Handbook of Heavens*). Also called **celestial meridian.** See the picture at **azimuth.**
[from Latin *meridianus* of noon, from *meridies* noon, south, formed from *medius* middle + *dies* day]

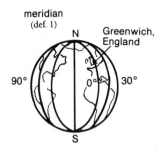

meridian
(def. 1)

N

Greenwich, England

90°   0°   30°

S

meridional (mə rid′ē ə nəl), *adj.* **1** *Geography.* of or having to do with a meridian.
2 *Meteorology, Geology.* along a meridian; in a north-south direction: *a meridional flow of air, a meridional chain of weather stations.*

meristem (mer′ə stem), *n. Botany.* the undifferentiated, growing cellular tissue of the younger parts of plants: *Meristems may be classified as primary and secondary... The principal primary meristems are the apical meristems of stems and roots and the fascicular cambium of dicotyledons and gymnosperms ... Meristems*

that arise from permanent tissues are called secondary meristems (Hill, Overholts, and Popp, *Botany*). [from Greek *meristos* divisible, divided, from *merizein* divide, from *meros* part]
—**meristematic** (mer′ə stə mat′ik), *adj.* of or having to do with the meristem: *The succulent and meristematic portions of plants, such as leaves, and the tips of stems and roots* (Harbaugh and Goodrich, *Fundamentals of Biology*).

meristic (mə ris′tik), *adj. Biology.* **1** having or composed of segments; segmental. SYN: metameric. **2** of or having to do with the number of arrangement of body parts or segments: *meristic variation.*
[from Greek *meristos* divisible] —**meris′tically,** *adv.*

meroblast (mer′ə blast), *n. Embryology.* an ovum whose contents consist of considerable nutritive as well as formative or germinal matter. [from Greek *meros* part + *blastos* germ, sprout]
—**meroblastic** (mer′ə blas′tik), *adj.* (of an ovum) containing nutritive as well as germinal matter and therefore undergoing only partial segmentation, as in birds, reptiles, and most fishes: *meroblastic cleavage.* Contrasted with **holoblastic.**

merogony (mə rog′ə nē), *n. Embryology.* the development of an embryo from a portion of an egg: *In merogony, a portion of the egg without the egg nucleus is fertilized* (N.Y. Times). [from Greek *meros* part + *-gonia* a begetting]

merosymmetrical (mer′ə si met′rə kəl), *adj. Crystallography.* being partially symmetrical.

merosymmetry (mer′ə sim′ə trē), *n. Crystallography.* partial symmetry.

Mertensian mimicry (mèr ten′zē ən), *Zoology.* close resemblance of a noxious animal to one that is less noxious: *Batesian mimicry, in which a defenseless organism bears a close resemblance to a noxious one; Mullerian mimicry, in which noxious species tend to resemble each other, and Mertensian mimicry, in which a mimic is more strongly protected than its model.* (Science News). [named after R. *Mertens,* a herpetologist who first described this type of protective mimicry]

mes-, *combining form.* the form of **meso-** before vowels, as in *mesencephalon, mesovarium.*

mesa (mā′sə), *n. Geography, Geology.* an isolated, high plateau with a flat top and steep sides, common in dry regions of the western and southwestern United States: *Usually mesas are portions of larger plateaus from which they have been detached by the formation and widening of arroyos, or canyons.... Features of the same origin of smaller size often are called buttes* (Finch and Trewartha, *Elements of Geography*). SYN: tableland. [from Spanish *mesa* table, from Latin *mensa*]

mesa

mesencephalic (mes′en sə fal′ik), *adj. Embryology.* of or having to do with the mesencephalon.

mesencephalon (mes'en sef'ə lon), *n. Embryology.* the middle section of the embryonic vertebrate brain. Also called **midbrain.**

mesenchymal (mes eng'kə məl), *adj. Embryology.* having to do with, consisting of, or derived from mesenchyme: ... *the interaction between the mesenchymal and epithelial components in the development of the pancreas, skin, liver, salivary glands, lungs, and other organs* (H. Tiedemann).

mesenchyme (mes'eng kim), *n. Embryology.* that portion of the mesoderm, consisting of cells set in a gelatinous matrix, from which the connective tissues, bone, cartilage, vascular system, and lymphatic vessels develop. [from New Latin *mesenchyma,* from Greek *mesos* middle + *enchyma* infusion]

mesenteric (mes'ən ter'ik), *adj. Anatomy.* of or having to do with a mesentery: *mesenteric arteries.*

mesenteron (mes en'tə ron), *n., pl.* **-tera** (-ter ə). *Embryology.* the interior of the primitive intestine (archenteron), bounded by endoderm. Compare **midgut.** [from New Latin *mesenteron,* from Greek *mesos* middle + *enteron* enteron]

mesentery (mes'ən ter'ē), *n. Anatomy.* a fold or membrane that envelops and supports an internal organ, attaching it to the body wall or to another organ: *The organs ... are supported from the mid-dorsal wall of the coelom by thin mesenteries, also formed of peritoneum* (Storer, *General Zoology*). [from Greek *mesenterion,* from *mesos* middle + *enteron* intestine]

mesial (mē'zē əl or mes'ē əl), *adj. Anatomy.* having to do with, situated in, or directed toward the middle line of a body. SYN: median. [from Greek *mesos* middle] —**me'sially,** *adv.*

mesic[1] (mes'ik or mez'ik; mē'sik or mē'zik), *adj. Botany.* characterized by or requiring a medium supply of moisture: *mesic plants.* Compare **hydric, xeric.** [from *mes-* medium + *-ic*]

mesic[2], *adj. Nuclear Physics.* of or having to do with a meson or mesons. A **mesic atom** is an atom in which an electron has been replaced by a negatively charged meson. [from *mes(on)* + *-ic*]

meso- (mes'ə- or mez'ə-; mēs'ə- or mēz'ə-), *combining form.* **1** middle; halfway; medium; intermediate, as in *mesoderm = middle layer of cells.* Also spelled **mes-** before vowels.
2 *Chemistry.* designating an optically inactive isomeric form of a compound: *meso-tartaric acid.*
[from Greek *mesos*]

mesoblast, *n. Embryology.* the undifferentiated mesoderm of an embryo. [from *meso-* + Greek *blastos* germ, sprout]
—**mesoblastic,** *adj.* of or having to do with the mesoblast: *the mesoblastic layer.*

mesocarp, *n. Botany.* the middle layer of the pericarp of a fruit or ripened ovary, such as the fleshy part of a peach or plum. See the picture at **pericarp.** [from *meso-* + Greek *karpos* fruit]

mesoderm, *n.* **1** *Embryology.* the middle layer of cells formed during the development of the embryos of animals: *Since the chordates are triploblastic animals there is a third body layer, the mesoderm, that must be produced in order to lay the foundation for the body parts that are to develop later* (Winchester, *Zoology*).

2 *Anatomy.* the tissues derived from this layer of cells, such as the muscles, bones, circulatory system, connective tissue, etc.: *The ectoderm is the skin and external covering of the body, the mesoderm includes connective tissues, muscles, tendons and so forth, and the endoderm represents the lining of the digestive tract and the ancillary systems connected with it* (J. L. Crammer). [from *meso-* + Greek *derma* skin]
—**mesodermal,** *adj.* of or having to do with the mesoderm.

mesogaster (mes'ə gas'tər or mē'sə gas'tər), *n. Embryology.* one of the two mesenteries of the stomach. [from New Latin *mesogastrium,* from Greek *mesos* middle + *gastēr, gastros* stomach]
—**mesogastric,** *adj.* of or having to do with the mesogaster.

mesogastrium (mes'ə gas'trē əm or mē'sə gas'trē əm), *n.* = mesogaster.

mesoglea or **mesogloea** (mes'ə glē ə or mē'sə glē' ə), *n. Zoology.* a gelatinous or fibrous layer that connects the outer and inner cell layers of a coelenterate. [from *meso-* + Greek *gloia* glue]
—**mesogleal** or **mesogloeal,** *adj.* consisting of, having to do with, or resembling mesoglea.

mesognathous (mi sog'nə thəs), *adj. Anatomy.* having jaws that are of moderate size and project only slightly. [from *meso-* + Greek *gnathos* jaw]

mesomerism, *n.* = resonance (def. 1).

mesometeorological, *adj.* of or having to do with mesometeorology: *mesometeorological systems.*

mesometeorology, *n.* the branch of meteorology that deals with atmospheric phenomena of an intermediate range, affecting several local areas: *Mesometeorology had come into prominence ... because of emphasis on analysis of the detailed structure of severe local storms* (F.W. Reichelderfer).

meson (mes'on or mez'on), *n. Nuclear Physics.* any of a group of unstable, strongly interacting subnuclear particles consisting of a quark bound to an antiquark. Kaons and pions are mesons. [alteration of the earlier name *mesotron,* from *meso-* + *-tron*]
► From 1939 to 1947 the term *meson* was used to refer to the kind of particle now known as the *muon.* An earlier (1937) name, *mesotron,* and a later name, *mu-meson,* was used the same way. Later researchers found, however, that the muon was not a strongly interacting particle and could therefore not be classified as a meson. The first known meson (as defined above) was discovered in 1947. Since then many mesons have been produced artifically. See, for example, **psi particle** and **upsilon.**
—**mesonic,** *adj.* = mesic[2].

mesonephric, *adj.* of or having to do with the mesonephros.

---

cap, fāce, fäther; best, bē, tėrm; pin, fīve;
rock, gō, ôrder; oil, out; cup, pùt, rüle,
yü in use, *yu̇* in uric;
ng in bring; sh in rush; th in thin, ᴛʜ in then;
zh in seizure.
ə = a in about, e in taken, i in pencil, o in lemon, u in circus

mesonephric duct, *Embryology.* the embryonic genital or reproductive tract of males in vertebrate animals. Also called **Wolffian duct.** Compare **Müllerian duct.**

mesonephros (mes′ə nef′ros *or* mē′sə nef′ros), *n. Embryology.* the middle division of the primitive kidney of vertebrate embryos, the pronephros and the metanephros being the anterior and posterior divisions, respectively. The mesonephros becomes part of the permanent kidney in fishes and amphibians. Also called **Wolffian body.** [from *meso-* middle, midway + Greek *nephros* kidney (so called because it develops between the pronephros and metanephros)]

mesopause, *n. Meteorology, Geophysics.* an area of the atmosphere between the mesosphere and the ionosphere.

mesophase, *n. Chemistry.* a semicrystalline phase of some substances, such as liquid crystals: *Compounds capable of forming mesophases (semicrystalline states) are widely distributed naturally throughout the plant kingdom. Examination of a variety of plant tissues revealed many substances that form mesophases in such plants as Spanish moss and dandelions* (Science News).

mesophile, *Biology.* —*n.* a mesophilic organism. —*adj.* = mesophilic.

mesophilic, *adj. Biology.* requiring moderate temperatures for development: *They studied ... flagella from mesophilic bacteria that live under more temperate conditions* (Science News Letter).

mesophyll, *n. Botany.* the inner green tissue of a leaf, lying between the upper and lower layers of epidermis. [from *meso-* + Greek *phyllon* leaf]

mesophyte, *n. Botany.* a plant that grows under conditions of average moisture and dryness: *Many mesophytes can adjust to, and grow in, water-logged soils or nonaerated nutrient solutions* (Albert R. Grable). ASSOCIATED TERMS: see **hydrophyte.**
—**mesophytic,** *adj.* of or having to do with a mesophyte.

mesoscale, *adj. Meteorology.* of or having to do with atmospheric phenomena intermediate between microscale and macroscale: *Mesoscale meteorology [is] the meteorology of areas 10 to 20 miles in diameter, or about the size of many urban areas* (Science News).

mesosome, *n. Bacteriology.* a structure on bacterial cell membranes which has to do with the formation of the cross wall when a cell divides: *The mesosome [is] a structure which can be observed in electron micrographs of bacteria.... It is the point at which DNA is attached to the cell membrane* (New Scientist).

mesosphere, *n.* **1** *Meteorology, Geophysics.* a region of the atmosphere between the stratosphere and the ionosphere. Most of the ozone in the atmosphere is created in the mesosphere and there is almost no variation in the temperature. *Winds at heights of 80 to 100 km in the Earth's atmosphere—in the so-called mesosphere—are subject to large tidal variations with periods of both 12 and 24 hours* (New Scientist). See the picture at **atmosphere.**
2 *Geology.* the lower part of the earth's mantle, extending below a depth of 250 to about 2900 kilometers: *Beneath the lithosphere is the asthenosphere, about 400 miles thick, and below the asthenosphere is the*

mesosphere, the term used by geophysicists for the rest of the mantle down to the earth's core (Earl Cook). ASSOCIATED TERMS: see **lithosphere.**
—**mesospheric,** *adj.* of or having to do with the mesosphere: *mesospheric circulation pattern.*

mesothelium (mes′ə thē′lē əm *or* mē′sə thē′lē əm), *n., pl.* **-liums, -lia** (-lē ə). **1** *Embryology.* the part of the mesoderm that lines the primitive body cavity of a vertebrate embryo.
2 *Anatomy.* the tissue derived from this part, lining the pleural, pericardial, peritoneal, and scrotal cavities of the body.
[from New Latin *mesothelium,* from *meso(derma)* mesoderm + *(epi)thelium* epithelium]

mesothermal, *adj.* of or having to do with a moderate or intermediate temperature range, as of a climate, habitat, geological process, etc.

mesothoracic, *adj. Zoology.* of or having to do with the mesothorax.

mesothorax, *n. Zoology.* the middle of the three divisions of the thorax of an insect, typically bearing the first pair of wings and the middle pair of legs.

mesothorium, *n. Chemistry.* either of two radioactive isotopes, **mesothorium 1** (*atomic number:* 88; *half-life:* 6.7 years), an isotope of radium, formed from thorium and yielding **mesothorium 2** (*atomic number:* 89; *half-life:* 6.13 hours), an isotope of actinium. Both isotopes have an atomic weight of 228, and are intermediate between thorium and radiothorium.

mesotron, *n.* the original name of the muon. ► See the note under **meson.**

mesovarium, *n. Anatomy.* the fold of peritoneum that suspends the ovary.

Mesozoic (mes′ə zō′ik), *n. Geology.* **1** the geological era before the present (Cenozoic) era. It is characterized by the development of mammals, flying reptiles, birds, and flowering plants, and the appearance and death of dinosaurs. It comprises the Triassic, Jurassic, and Cretaceous periods. The Mesozoic began about 225 million years ago and ended about 65 million years ago. *The diversity and numbers of the reptiles in the Mesozoic must be compared with that of the mammals today; one of the wonders of earth history is the remarkable parallelism of types in two groups so widely separated in time* (Garrels, *A Textbook of Geology*).
2 the rocks formed in this era.
—*adj.* of this era or these rocks.
[from *meso-* + Greek *zōē* life]

messenger RNA, *Molecular Biology.* a form of RNA (ribonucleic acid) that carries genetic information from the DNA in the nucleus of a cell to the ribosomes in the cytoplasm, specifying the particular protein to be synthesized: *Molecules of messenger RNA vary greatly in size depending largely upon the length of the code message they carry. They may code for one or several proteins* (Weier, *Botany*). *Abbreviation:* mRNA Compare **ribosomal RNA, transfer RNA.**

Met, *abbrev.* methionine.

met-, *prefix.* the form of **meta-** before vowels, as in *methemoglobin.*

meta-, *prefix.* **1** between; among, as in *metacentric.*
2 change of place or state, as in *methathesis, metabolism.*
3 behind; after, as in *metathorax.*

4 beyond, as in *metagalaxy.*
5 *Chemistry.* **a** of, bonded to, or substituting for two carbon atoms separated by one other carbon atom in a benzene ring. *Meta-* is usually italicized, as in *meta*-dichlorobenzene. Also written as **1,3-.** *Abbreviation: m-* Compare **ortho-, para-. b** similar in chemical composition to, as in *metaphosphate, metaprotein.* Also spelled **met-** before vowels.
[from Greek *meta* with, after]

metabolic (met′ə bol′ik), *adj. Biology.* having to do with or produced by metabolism: *The hormone was used to treat both hypothyroidism and metabolic insufficiency* (Science News Letter). —**metabolically,** *adv.*

metabolism (mə tab′ə liz′əm), *n. Biology.* **1** the sum of the physiological processes by which an organism maintains life. In metabolism organic compounds are broken down to yield energy, which is then used by the body to build up new cells and tissues, provide heat, and engage in physical activity. Growth and action depend on metabolism.
2 the metamorphosis of an insect.
[from Greek *metabolē* change, from *meta-* after + *bolē* a throwing]

metabolite (mə tab′ə līt), *n. Biology.* a substance produced by metabolism: *His investigations have been concerned particularly ... with the mechanism of action of carcinostatic and other metabolites that involve specific biochemical pathways concerned in nucleic acid metabolism* (Science).

metabolize, *v.* **1** to alter by or subject to metabolism: *The food eaten between the weigh-in and the match cannot be metabolized in time to restore energy lost* (Science News Letter).
2 to function in or undergo metabolism: *The [bacterial] species does not stop metabolizing as soon as it encounters anaerobic conditions; it continues to metabolize for as long as six days, even though the new metabolic product is unusual and potentially harmful* (Scientific American).

metacarpal, *Anatomy.* —*adj.* of or having to do with the metacarpus.
—*n.* a bone of the metacarpus: *Five metacarpals support the palm of the hand and 14 phalanges form the framework of the fingers* (Harbaugh and Goodrich, *Fundamentals of Biology*).

metacarpus (met′ə kär′pəs), *n., pl.* **-pi** (-pī). *Anatomy.* **1** the part of the hand, especially the bones, between the wrist and the fingers, comprising five bones in man.
2 the corresponding part of a forefoot of a quadruped, between the carpus and the phalanges.
[from New Latin, from Greek *metakarpion,* from *meta* after + *karpos* wrist]

metacentric, *Genetics.* —*adj.* having the centromere near the center, so that the chromosome's arms are of equal length.
—*n.* a metacentric chromosome. Compare **acrocentric, telocentric.**

metachromatism (met′ə krō′mə tiz əm), *n. Chemistry.*
1 change or variation of color, especially as a result of change of temperature or other physical condition.
2 the taking on of different colors by different substances when stained by the same dye.

metagalactic, *adj. Astronomy.* of or having to do with the metagalaxy.

metagalaxy, *n. Astronomy.* the universe outside the Milky Way, including the whole system of external galaxies.

metagenesis (met′ə jen′ə sis), *n. Biology.* **1** = alternation of generations.
2 regular alternation of sexual and asexual generations, such as occurs in some cnidarians (coelenterates).
—**metagenetic,** *adj.* having to do with, characterized by, or resulting from metagenesis.

metagnathous (mə tag′nə thəs), *adj. Zoology.* (of a bird) having the upper and lower tips of the mandibles crossed, as the crossbills. [from *meta-* + Greek *gnathos* jaw]

metal, *n. Chemistry.* **1** any of a class of elements, such as iron, gold, sodium, copper, lead, magnesium, tin, and aluminum, which usually have a shiny surface, are generally good conductors of heat and electricity, and can be melted or fused, hammered into thin sheets, or drawn into wires. Metals form alloys with each other and react with nonmetals to form salts by losing electrons. In the salt the metal takes on a positive charge. *A metal consists of an arrangement of positive ions which are located at the crystal lattice sites and are immersed in a "sea" of mobile electrons ... This mobility of electrons distinguishes the metallic bond from an ionic or covalent bond* (Chemistry Regents Syllabus). *Very few metals ... exist in the earth's crust in metallic form. [Most] are present as chemical compounds ... which usually bear no resemblance to metals and from which the metal is obtained by a 'reduction' process of some kind, whose difficulty increases with the chemical activity of the metal* (J. Crowther).
2 any alloy of these, such as steel, bronze, and brass.
[from Greek *metallon* (originally) mine]
—**metallic,** *adj.* of, having to do with, or consisting of a metal or metals: *a metallic chemical element, metallic reactions, metallic bonding.*

metallogenetic (mə tal′ə jə net′ik), *adj. Geology.* producing metals.

metallography (met′l og′rə fē), *n.* the study of metals and alloys, chiefly with the aid of a microscope: *Metallography primarily is used to explain why metals behave as they do under various circumstances and different combinations* (Science News Letter). [from Greek *metallon* metal + *graphein* to draw]

metalloid (met′l oid), *Chemistry.* —*n.* an element having properties of both a metal and a nonmetal. Arsenic and antimony are examples of metalloids.
—*adj.* of, having to do with, or resembling metalloids: *metalloid elements.*

metallurgic or **metallurgical,** *adj.* of or having to do with metallurgy: *metallurgic chemistry, metallurgical science.* —**metallurgically,** *adv.*

---

cap, fāce, fäther; best, bē, tèrm; pin, fīve;
rock, gō, ôrder; oil, out; cup, pùt, rüle;
yü in use, *yu̇* in uric;
ng in bring; sh in rush; th in thin, ᴛʜ in then;
zh in seizure.
ə = *a* in about, *e* in taken, *i* in pencil, *o* in
lemon, *u* in circus

metallurgy (met'l ėr'jē), *n.* the science and technology of metals. It includes the study of their properties and structure, the separation and refining of metals from their ores, the production of alloys, and the shaping and treatment of metals by stamping, heat, and rolling: *Modern metallurgy borders also on physics and chemistry, mining and mineral dressing, applied mechanics and physical chemistry* (Science News). [from New Latin *metallurgia*, ultimately from Greek *metallon* metal + *ergon* work]

metamathematical, *adj.* of or having to do with metamathematics: *Metamathematical statements are statements about the signs and expressions of a formalized mathematical system* (E. Nagel and J. R. Newman).

metamathematics, *n.* the branch of mathematics that deals with the logic and consistency of mathematical proof, formulas, and equations.

metamer (met'ə mər), *n. Chemistry.* a compound that is metameric with another.

metamere (met'ə mir), *n. Zoology.* any of a longitudinal series of similar parts or segments composing the body of various animals, such as the earthworm. Also called **somite.** [from *meta-* + *-mere*]

metameric (met'ə mer'ik), *adj.* **1** *Chemistry.* (of compounds) having the same molecular weight and the same elements combined in the same proportion but with their radicals in different positions, and hence differing in structure and chemical properties. EXAMPLE: $CH_3CH_2CH_2CHO$, an aldehyde; $CH_3CH_2COCH_3$, a ketone. SYN: isomeric.
2 *Zoology.* of or having to do with a metamere or metameres: *metameric animals, metameric segmentation.* SYN: segmental. —**metamerically,** *adv.*

metamerism (mə tam'ə riz əm), *n.* **1** *Zoology.* the condition of consisting of metameres; segmentation into metameres, typically with some repetition of organs: *Animals exhibiting metamerism are composed of a linear series of body segments* (Shull, *Principles of Animal Biology*).
2 *Chemistry.* the state of being metameric. SYN: isomerism.

metamorphic (met'ə môr'fik), *adj.* **1** *Biology.* characterized by metamorphosis; having to do with change of form: *the metamorphic stages of insects.*
2 *Geology.* changed in structure by heat, moisture, and pressure. Slate is a metamorphic rock that is formed from shale, a sedimentary rock. *Metamorphism affects both igneous and sedimentary rocks and so generally that nearly every common rock has a well-known metamorphic equivalent ... A sandstone metamorphosed becomes a quartzite, a rock of extreme hardness and resistance to erosion* (Finch and Trewartha, *Elements of Geography*).
—*n. Geology.* a metamorphic rock: *... metamorphics as the altered equivalents of sedimentary and igneous rocks* (Garrels, *A Textbook of Geology*).

metamorphism (met'ə môr'fiz əm), *n.* **1** *Biology.* change of form; metamorphosis: *an insect in its second stage of metamorphism.*
2 *Geology.* a change in the structure of a rock caused by pressure, heat, and moisture, especially when the rock becomes harder and more coarsely crystalline: *Contact metamorphism, includes changes caused by*

direct contact with hot igneous rocks as well as by heat, water, steam and other gases that come from them. Dynamic metamorphism includes the changes caused by earth movements and compression as well as those which are caused by downward pressure and the action of water (Fenton, *The Rock Book*).

metamorphose (met'ə môr'fōz), *v. Biology, Geology.* to change in form or structure by metamorphosis or metamorphism: *It is the process of losing the larval organs and of gaining the missing adult organs which is called metamorphosis* (Shull, *Principles of Animal Biology*). *The fact that mineral grains of the granite have not been distorted by compressive earth forces proves that intrusion of these rocks did not antedate the Permian folding, for otherwise they would have been much metamorphosed* (Moore, *Introduction to Historical Geology*).

metamorphosis (met'ə môr'fə sis), *n., pl.* **-ses** (-sēz'). *Biology.* a marked change in the form, and usually the habits, of an animal in its development after the embryonic stage. Tadpoles become frogs by metamorphosis; they lose their tails and grow legs. *Most insects go through several distinct stages as they develop from egg to adult ... called metamorphosis.* (Otto and Towle, *Modern Biology*). [from Greek *metamorphōsis* transformation, ultimately from *meta* (change) over + *morphē* form]

metamorphosis of the ladybug:

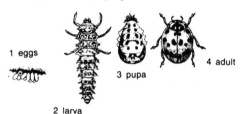

1 eggs

2 larva

3 pupa

4 adult

metanephric, *adj. Embryology.* of or having to do with the metanephros.

metanephros (met'ə nef'ros), *n. Embryology.* the posterior division of the primitive kidney of vertebrate embryos, the mesonephros and pronephros being the middle and anterior divisions, respectively. The adult kidney develops from the metanephros. *The reptiles, birds, and mammals, including man, pass through the same developmental sequence, but in their ultimate differentiation the mesonephros is replaced by a true kidney, the metanephros* (Harbaugh and Goodrich, *Fundamentals of Biology*). [from *meta-* + Greek *nephros* kidney (so called because it develops after the pronephros and mesonephros)]

metaphase, *n. Biology.* **1** the second stage in mitosis, after the prophase, characterized by the arrangement of the chromosomes along the middle of the spindle: *As they line up at the middle of the spindle at metaphase, the centromere of each chromatid faces opposite poles, and each becomes attached to certain of the fibers composing the spindle* (Norman V. Rothwell). See the picture at **mitosis.**
2 a similar phase during the first division of meiosis: *The homologous chromosomes ... arrive at metaphase as pairs of chromosomes, called bivalents, instead of single chromosomes longitudinally double* (McElroy,

—**metaphasic,** *adj.*

metaphloem (met′ə flō′əm), *n. Botany.* the primary phloem which is formed from the procambium after the protophloem; the last primary phloem to be developed: *The metaphloem consists of sieve tubes, companion cells, parenchyma, and sclerenchyma, the latter in the nature of fibers or scleroids* (Youngken, *Pharmaceutical Botany*).

metaphosphoric acid, *Chemistry.* a transparent, glacial substance which forms phosphoric acid when dissolved in water. *Formula:* (HPO_3).

metaplasia (met′ə plā′zhə), *n. Biology.* the direct formation of one type of adult tissue from another, as of bone from cartilage. [from New Latin *metaplasia,* from Greek *metaplassein* to transform]

metaplasm (met′ə plaz əm), *n. Biology.* carbohydrates, pigment, or other lifeless matter in the protoplasm of a cell.
—**metaplasmic,** *adj.* of or having to do with the metaplasm.

metaprotein, *n. Biochemistry.* any of a group of products of the hydrolytic decomposition (resulting from the action of acids or alkalis) of proteins. Metaproteins are insoluble in water but soluble in acids or alkalis.

metasomatic, *adj. Geology.* having to do with or resulting from metasomatism. —**metasomatically,** *adv.*

metasomatism (met′ə sō′mə tiz əm), *n. Geology.* the process by which the chemical constitution of a rock is changed by the action of fluids: *the metasomatism of schist into granite.* [from *meta-* + Greek *somatos* body]

metastable (met′ə stā′bəl), *adj.* **1** Chemistry. neither stable nor unstable; changing easily into a more or a less stable state: *There is no sharp distinction between a metastable and a labile solution, for if a metastable solution is cooled sufficiently, or its concentration is increased adequately, it passes over into the labile condition. The concentration at which this transition takes place is called the metastable limit* (Parks and Steinbach, *Systematic College Chemistry*). **2** Physics. having or characterized by an exceptionally long half-life: *The newly formed material may now be in a metastable state. According to thermodynamics it should revert to its original condition. But it has been frozen into permanence: at the lower temperature the change takes place with almost infinite slowness* (H. Tracy Hall).

metastasis (mə tas′tə sis), *n., pl.* **-ses** (-sēz). *Medicine.* the transfer, as through the blood vessels or the lymphatics, of a disease, cells, or microorganisms from one organ or part to another, especially such a transfer of cancerous cells. [from Greek *metastasis* removal, from *methistanai* to remove, change, from *meta-* changed, over + *histanai* to place]
—**metastasize** (mə tas′tə sīz), *v.* (of a disease, cells, or microorganisms) to spread by or undergo metastasis: *Cancer cells metastasize ... through the body ... by entering the blood vessels and traveling through the blood stream ... by being carried in the lymph ... by direct contact with another organ* (C.P. Rhoads).
—**metastatic** (met′ə stat′ik), *adj.* of, having to do with, or characterized by metastasis: *metastatic breast cancer.* —**metastatically,** *adv.*

metatarsal (met′ə tär′səl), *Anatomy.* —*adj.* of or having to do with the metatarsus: *metatarsal bones.* The **metatarsal arch** is the arch formed by bones across the ball of the foot. See the picture at **foot.**
—*n.* a bone of the metatarsus.

metatarsus (met′ə tär′səs), *n., pl.* **-si** (-sī). *Anatomy.* **1** the part of the foot, especially the bones, between the ankle and the toes. **2** the corresponding part of a hind foot of a quadruped, between the tarsus and the phalanges.

metathesis (mə tath′ə sis), *n., pl.* **-ses** (-sēz). *Chemistry.* the interchange of radicals between two compounds in a chemical reaction. Also called **double decomposition.** [from Greek *metathesis* transposition, ultimately from *meta-* changed, over + *tithenai* to place, set]
—**metathetic** (met′ə thet′ik), *adj.* of the nature of or containing metathesis.

metathoracic, *adj. Zoology.* of or having to do with the metathorax: *The flies ... each have a pair of clubbed threads ... in place of the metathoracic wings* (Hegner and Stiles, *College Zoology*).

metathorax (met′ə thôr′aks), *n., pl.* **-raxes, -races** (-rə sēz). *Zoology.* the posterior of the three segments of the thorax of an insect, bearing the third pair of legs and the second pair of wings.

metaxylem (met′ə zi′ləm), *n. Botany.* the primary xylem that is formed from the procambium after the protoxylem; last primary xylem to be developed: *When elongation has ceased, metaxylem ... is formed* (Emerson, *Basic Botany*).

metazoal (met′ə zō′əl), *adj.* = metazoan.

metazoan (met′ə zō′ən), *Zoology.* —*n.* any of a large subkingdom (Metazoa) of animals having a body made up of many cells arranged in tissues, and developing from a single cell.
—*adj.* of or belonging to the metazoans: *The first appearance of metazoan life [was] somewhat more than 600 million years ago* (Scientific American).
[from New Latin *Metazoa,* ultimately from Greek *meta-* after + *zoia,* neuter plural of *zoion* animal]
► The subkingdom Metazoa formerly comprised all multicellular animals, in contrast with the Protozoa, or one-celled animals. Today most biologists no longer classify protozoans as animals, and many now use the name Metazoa to refer to all animals except sponges, the sponges being placed in a subkingdom Parazoa.

metencephalic (met′en sə fal′ik), *adj.* of or having to do with the metencephalon.

metencephalon (met′en sef′ə lon), *n. Embryology.* the anterior part of the hindbrain, which gives rise to the cerebellum and pons.

meteor (mē′tē ər), *n. Astronomy.* a mass of rock or metal that enters the earth's atmosphere from outer space with enormous speed. Friction with air molecules in

cap, fāce, fäther; best, bē, tèrm; pin, fīve;
rock, gō, ôrder; oil, out; cup, pùt, rüle;
yü in use, *yù* in uric;
ng in bring; sh in rush; th in thin, ᴛʜ in then;
zh in seizure.
ə = a in about, e in taken, i in pencil, o in lemon, u in circus

the atmosphere causes meteors to become so hot that they glow and usually burn up before reaching the earth's surface. Most meteors are small, the size of a pea or a grain of dust. Occasionally, large meteors survive the atmospheric friction and hit the surface of the earth, at which point they are called *meteorites.* Commonly called **falling star** or **shooting star.** [from Greek *meteoron* (thing) in the air, from *meta-* after + *aeirein* to lift]

—**meteoric** (mē'tē ôr'ik), *adj.* **1** *Astronomy.* of, having to do with, or consisting of a meteor or meteors: *meteoric dust.*

2 *Meteorology.* of the atmosphere: *Wind and rain are meteoric phenomena.* —**meteorically,** *adv.*

meteor., *abbrev.* meteorology.

meteoric shower, = meteor shower.

meteoric water, *Geology.* ground water that filters through the soil after a rain or seeps into the ground from rivers and lakes: *Meteoric water makes up most of the 2 million cubic miles of ground water in the earth* (Ray K. Linsley).

meteorite (mē'tē ə rīt'), *n. Astronomy, Geology.* a meteor that has reached the earth without burning up: *The meteorites are of two types: stony and metallic. The latter consist mainly of iron, plus substantial amounts of nickel. The iron meteorites survive the trip through the atmosphere better than the stony ones; they are also more easily distinguished from terrestrial rocks—hence more of them are found* (Scientific American).

—**meteoritic** (mē'tē ə rit'ik), *adj.* of or having to do with a meteorite or meteorites: *... those who defend the meteoritic origin of all lunar craters* (New Scientist).

meteoroid (mē'tē ə roid'), *n. Astronomy.* any of the many bodies that travel through interplanetary space and become meteors when they enter the earth's atmosphere.

meteorol., *abbrev.* meteorology.

meteorological or **meteorologic,** *adj.* **1** of or having to do with the atmosphere, atmospheric phenomena, or weather: *Weathermen classifying meteorological traits of hurricanes find each one is different, making understanding difficult* (Science News Letter).

2 of or having to do with meteorology: *meteorological research.* —**meteorologically,** *adv.*

meteorology (mē'tē ə rol'ə jē), *n.* **1** the science dealing with the atmosphere and atmospheric conditions or phenomena, especially as they relate to weather: *The most challenging problem in meteorology is the nature of the long-term global wind patterns that, taken together, constitute the general circulation of the atmosphere* (Paul M. Sears).

2 meteorological condition; atmosphere, atmospheric phenomena, or weather: *Little can be said about the meteorology of deserts* (Science News).

[from Greek *meteorologia,* from *meteoron* (thing) in the air + *-logos* treating of]

meteor shower or **meteoric shower,** *Astronomy.* a large number of meteors entering and burning up in the earth's atmosphere, occurring when the earth encounters a meteor swarm or debris from a comet: *The Perseid meteor shower is intense in mid-August each year.*

meteor swarm, *Astronomy.* a large group of meteoroids that orbit together about the sun, considered to be the remains of a disintegrated comet: *Comets and meteor swarms also revolve around the sun. Their orbits are, in general, more elongated than those of the planets, and they extend to greater distances from the sun* (Baker, *Astronomy*).

meter, *n.* the SI or MKS unit of length, equal to the distance traveled by light in a vacuum in 1/299,792,458 of a second. Before 1983, it was defined as 1,650,764.73 wavelengths of an orange-red line in the spectrum of an isotope of krypton (^{86}Kr) in a vacuum. *Symbol:* m [from Latin *metrum* a measure, from Greek *metron*]

-meter, *combining form.* **1** meter (39.37 inches), as in *kilometer = one thousand meters, millimeter = one thousandth of a meter.* **2** device for measuring, as in *barometer, thermometer.*

meter candle, = lux.

meter-kilogram-second, *adj.* = MKS.

metestrus (met es'trəs), *n. Zoology.* the period of regressive changes following the estrous cycle. During the metestrus the vaginal walls are filled with white blood cells. [from *met-* + *estrus*]

methane (meth'ān), *n. Chemistry.* a colorless, odorless, flammable gas, the simplest of the hydrocarbons. Methane is formed naturally by the decomposition of plant or other organic matter, as in marshes, petroleum wells, volcanoes, and coal mines. It is obtained commercially from natural gas. *Several bacteria produced methane as a major product from the oxidation of simple organic compounds coupled with carbon dioxide reduction* (Robert G. Eagon). Formula: CH_4 Also called **marsh gas.** [from *methyl*]

methane series, *Chemistry.* a homologous series of saturated, open-chain (aliphatic) hydrocarbons, with the group formula C_nH_{2n+2} (*n* standing for the number of carbon atoms in the molecule). The methane series includes a larger number of known compounds than any other chain series, many of which, as methane, ethane, propane, and butane, are commercially important. Also called **alkane series, paraffin series.** Compare **ethylene series.**

methanogen (mə than'ə jən), *n. Bacteriology.* any of the archaebacteria: *These organisms are in fact a separate form of life, genetically and structurally distinct from either present-day bacteria or the more complex cells of plants and animals. Referred to informally as methanogens, they were known to exist in nature only in oxygen-free environs—e.g., in deep, hot (170°F) geothermal springs and in decaying material in the bottom mud of seas and bays—where they utilize carbon dioxide and evolve methane as a waste product* (Harold Sandon). [from *methane* + *-gen*]

methemoglobin (met hē'mə glō'bin), *n. Biochemistry.* a brownish substance formed in the blood when the hematin of oxyhemoglobin is oxidized, either by spontaneous decomposition of the blood or by the action of an oxidizing agent, as potassium chlorate: *Among the mountain dwellers, who live where the air is thin, the blood contains a large amount of methemoglobin, a physiologically inactive substance that may be rapidly transformed into active hemoglobin when the tissues are in need of oxygen* (Henri V. Vallois). [from *met-* + *hemoglobin*] methionine (me thī′ə nēn), *n. Biochemistry.* an essential amino acid that provides the sulfur needed for animal metabolism, occurring in various proteins, such as those in casein, yeast, and egg whites: *The amount of free methionine in the blood increases throughout the course of infection: it is produced by attack by the parasite on the host proteins and is used for the synthesis of its own proteins* (New Scientist). *Formula:* $C_5H_{11}NO_2S$ *Abbreviation:* Met [from *met-* + Greek *theion* sulfur + *-ine*]

methyl (meth′əl), *n. Chemistry.* a univalent hydrocarbon radical, occurring in methane, members of the methane series, and many other organic compounds. *Formula:* $-CH_3$ [from French *methyle,* back formation from *methylene* methylene]

methylene (meth′ə lēn), *n. Chemistry.* a bivalent hydrocarbon radical occurring almost always in combination, and regarded as derived from methane. *Formula:* $-CH_2-$ [from French *methylene,* from Greek *methy* wine + *hylē* wood, substance]

methylene blue, *Chemistry.* a dark-green, crystalline compound used as a dye, as a stain in bacteriology, in medicine as an antidote for cyanide poisoning, and as a reagent in oxidation-reduction processes: *The methylene blue test on a milk sample measures one type of bacterial activity* (Science News). *Formula:* $C_{16}H_{18}ClN_3S \cdot 3H_2O$

methylic (mə thil′ik), *adj. Chemistry.* containing, relating to, or derived from methyl.

Metonic cycle (mi ton′ik), *Astronomy.* a cycle of 19 years or 235 lunar months, after which the phases of the moon recur on the same days of the solar calendar as in the previous cycle. [named after *Meton,* Athenian astronomer of the fifth century B.C.]

metric or metrical, *adj.* 1 of or having to do with the meter or the metric system: *metric measurements, metric weights, a metric unit.*
2 *Mathematics.* of, having to do with, or defining distance or any quantity analogous to distance: *metric geometry, metric space. Space-time is a metric four-dimensional manifold* (J. Rice).
—*n. Mathematics.* a metric measure or function: *Mathematicians call a measure of length (or distance) a metric and frequently derive from it the higher dimensional measures of area and volume. The relation between distance and volume (or, in technical terms, between metrics and measures) is of crucial importance in differential geometry, that part of geometry that is concerned with the properties of curved, higher dimensional spaces* (Lynn A. Steen). —metrically, *adv.*

metric system, a decimal system of measurement based on the meter as its unit of length, the gram as its unit of mass, and the liter as its unit of volume. ▶ See the note under SI unit.

METRIC SYSTEM
linear measure
1/1000 meter = 1 millimeter
1/100 meter = 1 centimeter
1/10 meter = 1 decimeter
10 meters = 1 decameter
100 meters = 1 hectometer
1000 meters = 1 kilometer

square measure
1 square meter = 1 centiare
100 centiares = 1 are
100 ares = 1 hectare
10,000 ares = 1 square kilometer
cubic measure
1/1000 liter = 1 milliliter
1/100 liter = 1 centiliter
1/10 liter = 1 deciliter
10 liters = 1 decaliter
100 liters = 1 hectoliter
1000 liters = 1 kiloliter
mass
1/1000 gram = 1 milligram
1/100 gram = 1 centigram
1/10 gram = 1 decigram
10 grams = 1 decagram
100 grams = 1 hectogram
1000 grams = 1 kilogram
1000 kilograms = 1 metric ton

metric ton, a unit of mass in the metric system, equal to 1000 kilograms. *Abbreviation:* M.T. Also called megagram, tonne.

metrology (mi trol′ə jē), *n.* 1 the science of measurement or of measures and weights.
2 a system of measures and weights.
[from Greek *metron* measure + English *-logy*]

-metry, *combining form.* measurement; process or science of measurement, as in *spectrometry = measurement of spectra.* [from Greek *-metria* a measuring, from *metron* a measure]

Mev or MeV, *abbrev.* megaelectron volt (one million electron volts, used as a unit for measuring energy in nuclear physics).

mf or mf., *abbrev.* millifarad *or* millifarads.

mg or mg., *abbrev.* milligram *or* milligrams.

Mg, *symbol.* 1 magnesium. 2 megagram.

mh or mh., *abbrev.* millihenry.

MHD, *abbrev.* 1 magnetohydrodynamic. 2 magnetohydrodynamics.

mho (mō), *n., pl.* mhos. a unit of electrical conductance, equal to the conductance of a body through which one ampere of current flows when the potential difference is one volt; siemens. It is the reciprocal of the ohm. [reversed spelling of *ohm*]

MHz, *symbol.* megahertz.

mi., *abbrev.* mile *or* miles.

mica (mī′kə), *n. Mineralogy.* any of a group of minerals that can be divided readily into thin, partly transparent, and usually flexible layers. Mica is a hydrous potassium silicate, highly resistant to heat and is used in electric fuses and other electrical equipment. *There are several varieties of mica ... Of the three, muscovite is far and away the most important, commercially; phlogopite has some very important uses and biotite*

cap, fāce, fäther; best, bē, tėrm; pin, five;
rock, gō, ôrder; oil, out; cup, pùt, rüle,
yü in use, *yu̇* in uric;
ng in bring; *sh* in rush; *th* in thin, ŦH in then;
zh in seizure.
ə = *a* in about, *e* in taken, *i* in pencil, *o* in lemon, *u* in circus

micelle

is utilized to a considerably lesser extent (Jones, *Minerals in Industry*). [from Latin *mica* grain, crumb] —**micaceous** (mī kā′shəs), *adj.* having to do with, containing, or resembling mica: *micaceous minerals.* SYN: flaky.

micelle (mi sel′), *n. Chemistry.* **1** an electrically charged colloidal particle made up of polymeric molecules. **2** any structural unit consisting of polymeric molecules, as in cellulose and starch. [from New Latin *micella,* from Latin *mica* grain, crumb]

Michaelis constant (mi kā′lis *or* mī′kə lis), *Biochemistry.* a measure of the affinity of an enzyme for its substrate, equal to the substrate concentration at which the enzyme-catalyzed reaction proceeds at half of its maximal rate: *The extent of binding of a substrate to enzyme can be evaluated by measurement of the Michaelis constant, Km. If the natural and synthetic enzyme are identical they should have identical Km values and that, in fact, was observed* (Robert B. Merrifield). [named after Leonor *Michaelis,* 1875–1949, American chemist]

micr-, *combining form.* the form of **micro-** before vowels, as in *microhm, microsmatic.*

micra (mī′krə), *n.* microns; a plural of **micron.**

micro-, *combining form.* **1** small; very small; microscopic, as in *microorganism = a microscopic organism.*
2 abnormally small, as in *microcyte = an abnormally small red blood cell.*
3 one millionth of, as in *microfarad = one millionth of a farad. Symbol:* μ
4 done with, or involving the use of, a microscope, as in *microcrystal = a crystal visible only under a microscope.* Also spelled **micr-** before vowels.
[from Greek *mikros* small]

microampere, *n.* a unit of electric current equal to one millionth of an ampere. *Symbol:* μA

microanalysis, *n., pl.* **-ses.** *Chemistry.* the analysis of very small quantities of matter: *Microanalysis and techniques of purification ... have permitted the isolation of ... biologically active substances from large masses of inert materials* (A.J. Birch). Contrasted with **macroanalysis.**
—**microanalytic** or **microanalytical,** *adj.* of or having to do with microanalysis.
—**microanalyze,** *v.* to carry out microanalysis on (a substance).

microanatomy, *n. Biology.* **1** the anatomy of microscopic structures. Compare **histology.**
2 the microscopic structure of a part of an animal or plant: *the microanatomy of the retina.*
3 = microdissection.

microbar, *n. Physics.* a unit of pressure equal to one millionth of a bar.

microbarn, *n. Nuclear Physics.* a unit of area equal to one millionth of a barn: *One microbarn is ... about a millionth of the cross section of an atomic nucleus* (Science News).

microbarograph, *n.* an instrument for recording very small fluctuations of atmospheric pressure.

microbe, *n. Biology.* a microorganism, especially a bacterium that causes disease or fermentation: *Pasteur ... furthered (though he did not originate) the idea that microbes were associated with diseases of larger organ-*

isms, including man (Durst, *High School Biology*). [from French *microbe,* from Greek *mikros* small + *bios* life]
—**microbial** or **microbic,** *adj.* of, having to do with or caused by microbes: *Microbial control of insects would be materially advanced if convenient and inexpensive culture methods could be developed* (Edward A. Steinhaus).

microbiological, *adj.* **1** of or having to do with microbiology: *microbiological research.*
2 of or having to do with microorganisms: *The soil has an extensive microbiological population.* —**microbiologically,** *adv.*

microbiology, *n.* the branch of biology dealing with microorganisms: *Koch's technique has made possible physiological and other experiments on all kinds of living microorganisms and has been instrumental in the rise of the more strictly biological science of microbiology* (Simpson, *Life: An Introduction to Biology*). Compare **bacteriology.**

microbiota (mī′krō bī o′tə), *n. Biology.* the microscopic or submicroscopic organisms living in an area; microscopic life.

microbody, *n.* = peroxisome.

microcalorie, *n. Physics.* one millionth of a calorie.

microchemical, *adj.* of or having to do with microchemistry: *microchemical reactions.*

microchemistry, *n.* **1** a branch of chemistry dealing with very small amounts or samples. **2** chemical analysis involving very small amounts or samples. **3** *Biochemistry.* the chemistry of cells and of microorganisms.

microcirculation, *n. Physiology.* circulation of the blood or lymph through very small vessels, such as arterioles, capillaries, and venules: *It is the microcirculation which serves the primary purpose of the circulatory system: ... to maintain the environment in which the cells can exist and perform their interrelated tasks* (Benjamin W. Zweifach).
—**microcirculatory,** *adj.* of or having to do with microcirculation: *the microcirculatory system.* Compare **microvascular.**

microclimate, *n. Meteorology, Ecology.* the climate of a very small, specific area such as a glacier, valley bottom, cornfield, or animal burrow: *A special feature of the environment at high altitudes [is] the sudden and extreme temperature changes that produce a variety of microclimates, which change from hour to hour in the course of a day* (Lawrence W. Swan). Contrasted with **macroclimate.**
—**microclimatic,** *adj.* of or having to do with a microclimate or microclimatology: *... the rainless season when micro-climatic factors such as dew, shade and irrigation have a predominant influence* (New Scientist).

microclimatology, *n.* the branch of climatology dealing with the climatic conditions of small areas.

microcline, *n. Mineralogy.* a triclinic potash feldspar similar to orthoclase, white, yellow, red, or green in color. *Formula:* $KAlSi_3O_8$ [from German *Mikroklin,* from Greek *mikros* small + *klinein* to incline]

micrococcal, *adj.* relating to or caused by micrococci.

micrococcus (mī′krə kok′əs), *n., pl.* **-cocci** (-kok′sī). *Biology.* any of a genus of spherical parasitic or saprophytic bacteria, aggregating in various ways. Certain micrococci cause disease; others produce fermenta-

400

tion. [from New Latin *Micrococcus* the genus name, from Greek *mikros* small + *kokkos* berry, seed, grain]

microcolony, *n.* **1** *Microbiology.* a colony of microorganisms, as those growing in a culture: *Many soil bacteria grow as microcolonies within the larger pores of soil aggregates* (L.E. Casida).
2 *Ecology.* a group of organisms living in a microhabitat: *a microcolony of birds.*

microcontinent, *n.* *Geology.* an oceanic ridge or plateau thought to have been broken off and separated from a continent during continental drift: *The Indian Ocean, too, plainly has an involved history with presently active and inactive ridges, "microcontinents" like Madagascar and the Seychelles, and signs that its bordering continents formerly migrated apart in vigorous fashion* (P. Stubbs and G. Wick).

microcrystal, *n.* *Crystallography.* a crystal that is visible only under a microscope; minute or microscopic crystal: *Tooth enamel consists of microcrystals of apatite enclosing organic substances* (New Scientist).
—**microcrystalline,** *adj.* formed of or containing microcrystals: *microcrystalline cellulose.*

microculture, *n.* *Biology.* a culture of microscopic organisms, tissue, etc.

microcurie (mī′krō kyu̇′rē), *n.* a unit of radioactivity equal to one millionth of a curie. *Symbol:* μCi

microcyte (mī′krə sīt), *n.* *Biology.* an abnormally small red blood cell. Contrasted with **macrocyte.**
—**microcytic,** *adj.* of or having to do with microcytes: *microcytic anemia.*

microdissect, *v.* *Anatomy.* to dissect (cells, tissues, etc.) with the aid of a microscope: *a technique for microdissecting individual nerve cells and small clumps of glia* (Brian Tiplady).
—**microdissection,** *n.* dissection done under a microscope.

microearthquake, *n.* *Geology.* a small earthquake, of magnitude of 2 or less on the Richter scale.

microecology, *n.* **1** a branch of ecology dealing with environmental conditions in very small areas: *When we make a garden we make a piece of ground suitable for the growth of certain plants that would be useful to us. In doing this, we are working with microclimates or with microecology* (E. Laurence Palmer).
2 = microenvironment: *The microecology of the blowfly's gut has not been well explored* (Nature).
—**microecological,** *adj.* of or having to do with microecology.

microelectrophoresis, *n.* *Chemistry.* a technique for observing the electrophoresis of very small individual particles through a microscope or ultramicroscope.
—**microelectrophoretic,** *adj.* having to do with or produced by microelectrophoresis: *He has developed an elegant microelectrophoretic technique that makes it possible to determine the base composition of very small amounts of RNA* (Scientific American).

microelement, *n.* a chemical element present only in very small amounts. Compare **trace element.**

microenvironment, *n.* *Ecology.* the environment of a very small area, especially the immediate environment of a given species: *The Himalayan microenvironment shows great temperature contrasts between localities only a few inches apart. In the sunshine, flying insects buzz about in temperatures in the 90's. In the shade, they often drop to the ground benumbed with cold*

(Lawrence W. Swan). Contrasted with **macroenvironment.**
—**microenvironmental,** *adj.* of or having to do with a microenvironment.

microevolution, *n.* *Biology.* the evolution of organisms on the level of subspecies due to a succession of small genetic variations: *The changes within a population have been termed microevolution, and they can indeed be accepted as a consequence of shifting gene frequencies* (Science). Contrasted with **macroevolution.**
—**microevolutionary,** *adj.* of or having to do with microevolution: *Genetic drift is the principal agent responsible for the variations among villages, tribes or clans. In fact, at the ... microevolutionary levels on which we worked the differences attributable to natural selection were not large* (L. L. Cavalli-Sforza).

microfarad, *n.* a unit of electrical capacity equal to one millionth of a farad. *Symbol:* μH

microfauna, *n., pl.* **-nas** or **-nae.** *Zoology.* **1** the microscopic animals of a particular region or time: *fossil microfauna, the microfauna of the soil.*
2 the animals found in a microhabitat. Contrasted with **macrofauna.**
—**microfaunal,** *adj.* of or having to do with a microfauna: *There have often been significant changes in the microfaunal population of the sea at magnetic reversals* (Scientific American).

microfibril (mī′krō fī′brəl), *n.* *Biology.* a microscopic fibril, as in human hair or the cell wall of a plant: *They found that hair roots consist of fine, densely packed filaments (now called microfibrils)* (R.D.B. Fraser).
—**microfibrillar** (mī′krō fī′brə lər), *adj.* of, having to do with, or consisting of microfibrils.

microfilament, *n.* *Biology.* an extremely fine and very long fiber in the cytoplasm of cells, associated with protoplasmic movement and cytoplasmic division: *Cytological and biochemical studies have demonstrated that microfilaments are virtually identical to actin, one of the major contractile proteins found in all skeletal, cardiac, and smooth muscle cells.* (R. Goldman).
—**microfilamentous,** *adj.* of or having to do with microfilaments: *the actin and myosin of the microfilamentous system.*

microfilaria, *n., pl.* **-lariae.** the minute larva of a filaria: *tiny young larvae, called microfilariae.*

microflora, *n., pl.* **-floras** or **-florae.** *Botany.* **1** the microscopic plants or plantlike organisms of a particular region or time.
2 the plants found in a microhabitat: *Soil and water, by virtue of the diverse microfloras they contain, function as biological incinerators* (New Scientist). Contrasted with **macroflora.**
—**microfloral,** *adj.* of or having to do with microflora: *microfloral growths of yeast.*

cap, fāce, fäther; best, bē, tėrm; pin, fīve;
rock, gō, ôrder; oil, out; cup, pu̇t, rüle;
yü in use, *yu̇* in uric;
ng in bring; sh in rush; th in thin, ᴛʜ in then;
zh in seizure.
ə = *a* in about, *e* in taken, *i* in pencil, *o* in lemon, *u* in circus

microfossil, *n. Paleontology.* a microscopic fossil, as of foraminifera or a pollen grain: *Evidence of microfossils representing tiny primitive plants and animals is rather widespread in rocks 1,000,000,000 to 2,500,000,000 years old* (Richard H. Jahns). Contrasted with **macrofossil.** Compare **micropaleontology.**

microfungus, *n., pl.* **-gi** or **-guses.** *Biology.* a microscopic fungus.

microgamete, *n. Biology.* the smaller, typically the male, of two gametes of an organism that reproduces by the union of unlike gametes: *When the pollen tube makes contact with the archegonium, the microgametes and other nuclei of the tube are discharged into the egg cell* (Greulach and Adams, *Plants*). Contrasted with **macrogamete.**

microgametocyte, *n. Biology.* a gametocyte that gives rise to microgametes. Contrasted with **macrogametocyte.**

microgametophyte, *n. Botany.* a male gametophyte that develops from a microspore. Compare **megagametophyte.**

microgauss, *n.* a unit of magnetic induction equal to one millionth of a gauss.

microglia (mī krog′lē ə), *n. Biology.* tissue formed by small neuroglial cells of the central nervous system that function chiefly as phagocytes.

microgram, *n.* a unit of mass equal to one millionth of a gram. *Symbol:* μg

microhabitat, *n. Ecology.* a habitat, usually within a small area, containing a small number and variety of organisms: *Some species of woodlice ... are forced to leave one microhabitat which has become intolerably hot in search of another* (E.B. Edney). Contrasted with **macrohabitat.**

microhenry, *n., pl.* **-ries** or **-rys.** a unit of electrical inductance equal to one millionth of a henry. *Symbol:* μH

microhm (mī′krōm), *n.* a unit of electrical resistance equal to one millionth of an ohm.

microinject, *v. Biology.* to inject (a substance) into a microscopic body by means of a micropipette.
—**microinjection,** *n.* the act or process of microinjecting: *Biologists can manipulate cells, introduce chemicals by microinjection* (Scientific American).

microliter, *n.* a unit of volume equal to one millionth of a liter.

micromanipulation, *n. Biology.* the manipulation of very small bodies or structures under a microscope, as in microdissection and microinjection: *The micromanipulation of chromosomes of living human cells in vitro is a potential means of obtaining transplantable genetic material* (Nature).
—**micromanipulative,** *adj.* of or involving micromanipulation: *micromanipulative techniques.*
—**micromanipulator,** *n.* an instrument or device used in micromanipulation.

micromere, *n. Embryology.* a small blastomere.

micrometeorite, *n. Astronomy.* a tiny particle of meteoric dust, so small that it does not burn as it falls to the earth from outer space: *Tiny meteorites, specks of dust only a few thousandths of an inch in diameter, are constantly bombarding the earth's atmosphere. Estimates as to the amount of this material that falls on the earth's surface run as high as 1,000 tons per day. These micrometeorites, as they are called, are believed to contribute a small share of the ionization of the ionosphere* (Scientific American).
—**micrometeoritic,** *adj.* of, having to do with, or caused by micrometeorites: *micrometeoritic material, micrometeoritic craters.*

micrometeoroid, *n. Astronomy.* **1** a very small meteoroid. **2** = micrometeorite.

micrometeorology, *n.* the branch of meteorology that deals with the atmospheric phenomena of very small areas: *Micrometeorology, the study of weather in the very immediate vicinity ... for instance, just how much moisture there is within an inch or two of a growing plant* (Science News Letter).
—**micrometeorological,** *adj.* of or having to do with micrometeorology.

micrometer¹ (mī′krō mē′tər), *n.* a unit of length used in microscopic measurements, equal to one thousandth of a millimeter or one millionth (10^{-6}) of a meter. In terms of other measurements, one micrometer is equivalent to 10,000 angstrom units. *Symbol:* μm [from *micro-* + *-meter*] ► See the note under **micron.**

micrometer² (mī krom′ə tər), *n.* an instrument for measuring very small distances, angles, objects, etc. Certain kinds are used with a microscope or telescope. [from French *micromètre,* from *micro-* micro- + *-mètre* -meter]
—**micrometry** (mī krom′ə trē), *n.* the measurement of minute objects with a micrometer.

micromicrocurie (mī′krō mī′krō kyùr′ē), *n.* one millionth of one millionth of a curie. ► Since 1963, the preferred term according to the International Committee on Weights and Measures is **picocurie.**

micromicrofarad (mī′krō mī′krō far′əd), *n.* one millionth of one millionth of a farad. ► Since 1963 the preferred term according to the International Committee on Weights and Measures is **picofarad.**

micromole, *n. Chemistry.* one millionth of a mole: *Enzyme activity is expressed in micromoles of substrate metabolized per hour per gram* (Science).
—**micromolar,** *adj.* of or expressed in micromoles.

micron (mī′kron), *n., pl.* **-crons, -cra** (-krə). = micrometer¹. [from Greek *mikron,* neuter of *mikros* small]

► In 1968, the General Conference of Weights and Measures abolished further use of the term *micron,* with the symbol μ attributed to it, to designate the millionth part of the meter. The symbol μ was restricted to the decimal submultiple 10^{-6} (one millionth), and the preferred name for the millionth part of a meter became *micrometer,* with the symbol μm or μm.

micronucleus, *n., pl.* **-clei** or **-cleuses.** *Zoology.* the smaller of two kinds of nuclei of ciliate protozoans, containing chromatin materials necessary for reproduction: *The nuclei are 2 in number, a large macronucleus concerned with vegetative functions, and a smaller micronucleus that is important in reproduction* (Hegner and Stiles, *College Zoology*). Contrasted with **macronucleus.**
—**micronuclear,** *adj.* of or having to do with a micronucleus or micronuclei: *the micronuclear membrane.*

micronutrient, *n. Biochemistry.* **1** = trace element.

2 a vitamin or mineral used in small amounts by an organism but considered necessary to its proper functioning: *Micronutrients are vitamins and minerals, as opposed to macronutrients, which are fats, carbohydrates and proteins* (N.Y. Times Magazine).

microorganism, *n. Biology.* a microscopic or submicroscopic organism. Bacteria, protozoans, and algae are microorganisms. *Microorganisms ... are critical links in the earth's giant ecological system* (Brooks D. Church). Also called **microbe.**
—**microorganismal,** *adj.* of or produced by microorganisms: *Every puff of wind, every drop of water, and every handful of dust contains microorganismal life in one form or another* (Joshua Lederberg).

micropaleontology, *n.* the branch of paleontology which studies microfossils: *Micropaleontology ... deals with fossils of ... minute aquatic animals* (Scientific American).
—**micropaleontological,** *adj.* of or having to do with micropaleontology.

microparasite, *n. Biology.* a parasitic microorganism.
—**microparasitic,** *adj.* of, having to do with, or caused by microparasites: *microparasitic diseases.*

microphage (mī′krə fāj′), *n. Biology.* **1** a small phagocyte. Contrasted with **macrophage.**
2 a neutrophil, especially a neutrophilic granulocyte.
—**microphagic** (mī′krə fā′jik), *adj.* of or having to do with microphages.

microphotograph, *n.* = photomicrograph.

microphyll (mī′krə fil), *n. Botany.* **1** a small leaf. **2** a leaf having a single unbranched vein.
—**microphyllous,** *adj.* of or being a microphyll.

microphysical, *adj.* of or having to do with microphysics: *microphysical phenomena.*

microphysics, *n.* **1** the branch of physics concerned with matter or phenomena occurring on a microscopic or submicroscopic scale, especially with molecules, atoms, and nuclear particles. Contrasted with **macrophysics.**
2 the microphysical properties or composition of something: *More knowledge of the microphysics of cloud particles is likely to lead to more predictable results* (Bulletin of Atomic Scientists).

microphyte (mī′krə fīt), *n. Botany.* a microscopic plant or plantlike organism, especially a fungus. Contrasted with **macrophyte.**
—**microphytic** (mī′krə fit′ik), *adj.* having to do with or caused by microphytes: *microphytic diseases.*

micropipette, *n.* **1** a very small pipette used in microinjection.
2 a pipette used in measuring very small volumes of liquids.

microplankton, *n. Biology.* plankton consisting of microscopic organisms.

micropopulation, *n. Ecology.* the population of microorganisms living in a particular habitat: *the micropopulation of soil and water.*

micropore, *n. Physics, Geology.* a very small pore.
—**microporous,** *adj.* having very small pores: *These have high absorbency and a microporous structure that makes them extremely suitable for use as molecular sieves* (New Scientist).

micropterous (mī krop′tər əs), *adj. Zoology.* having small wings or fins. Contrasted with **macropterous.**
[from Greek *mikropteros,* from *mikros* small + *pteron* wing]

micropulsation, *n. Geology.* a short-period fluctuation in the earth's magnetic field: *... geomagnetic micropulsations, named for their very small amplitude or energy* (Charles Wright).

micropylar (mī′krə pī′lər), *adj. Botany, Zoology.* having to do with or characteristic of a micropyle: *the micropylar end of the embryo sac.*

micropyle (mī′krə pīl), *n.* **1** *Botany.* the minute opening in the outer layer or layers of an ovule, through which a pollen tube enters: *The micropyle, a circular opening in the integument, at the apex of the ovule, is directed inward toward the cone axis* (Greulach and Adams, Plants).
2 *Zoology.* any of the minute holes in the membrane covering the ovum of certain animals, through which spermatozoa enter: *The egg [of a butterfly] contains yolk, which nourishes the developing larva, and a small hole, micropyle, at the top, which admitted a sperm when the egg was fertilized on leaving the female's oviduct and which now allows air to reach the embryo* (Mackean, Introduction to Biology).
[from French *micropyle,* from Greek *mikros* small + *pylē* gate]

microquake, *n.* = microearthquake.

microradiograph, *n.* an enlarged radiographic image of a small specimen, used in examination of metal structure, body tissue, etc.
—**microradiographic,** *adj.* of or having to do with a microradiograph or microradiography: *a microradiographic image.*
—**microradiography,** *n.* the technique of producing microradiographs.

micros., *abbrev.* microscopy.

microscale, *n. Chemistry.* a scale or standard involving very small units or measurements, as in microanalysis.
—*adj. Meteorology.* of or designating small-scale atmospheric phenomena, such as water-vapor condensation and accretion of raindrops or snowflakes: *the microscale processes involved in clear-air turbulence.* Compare **macroscale, mesoscale.**

microscope, *n.* **1** an optical instrument consisting of a lens or combination of lenses for magnifying things that are invisible or indistinct to the naked eye. Microscopes are used to examine or study one-celled organisms such as bacteria and algae, specimens of tissue and blood cells, and the structures of nonliving things, such as crystals and metals. The simplest microscope is an ordinary magnifying glass, with a focal length of about 25 centimeters (10 inches). The typical microscope is a compound microscope, consisting of two lenses: the *objective,* which produces an enlarged in-

---

cap, fāce, fäther; best, bē, tėrm; pin, fīve;
rock, gō, ôrder; oil, out; cup, pùt, rüle,
yü in use, *yu* in uric;
ng in bring; *sh* in rush; *th* in thin, ᴛʜ in then;
zh in seizure.
ə = *a* in about, *e* in taken, *i* in pencil, *o* in lemon, *u* in circus

403

verted image of the specimen, and the *eyepiece* or *ocular,* which magnifies the image produced by the objective. *Optical microscopes magnify because light rays reflected from the object bend as they pass through one or more lenses. The bent rays form the image larger than the original* (Peter Gray).
2 an instrument that uses electronic or other processes to magnify things: *The acoustic microscope, which produces dramatic images with high-frequency sound waves ... can now depict structures as small as 0.2 microns in diameter within an intact cell* (Science News). Compare **electron microscope, field-emission microscope, field-ion microscope, phase-contrast-microscope, stereomicroscope, ultramicroscope.** ▶ See the note under **magnification.**
[from New Latin *microscopium,* from Greek *mikros* small + *skopein* look at]
—**microscopic** (mī'krə skop'ik), *adj.* **1** that cannot be seen without using a microscope: *microscopic fungi.*
2 of or belonging to a microscope: *a microscopic lens.*
3 occurring beyond the visible range; invisible: *The word microscopic ... is reserved for phenomena that take place on a scale much too small to be detected by any mechanical measuring instrument and too small to be seen by the best microscope, a scale in fact of the order of atomic sizes.* (Shortley and Williams, *Elements of Physics*). —**microscopically,** *adv.*
—**microscopy** (mī kros'kə pē), *n.* the use of a microscope; microscopic investigation: *(By means of phase-contrast microscopy) it has been learned that in the human more than one spermatozoon ... passes on into the ooplasm* (Scientific American).
microsecond, *n.* a unit of time equal to one millionth of a second. *Symbol:* μs
microsection, *n.* a very small section of animal tissue, mineral, or the like, prepared for microscopic examination.
microseism, *n. Geology, Meteorology.* a faint, recurrent earth tremor lasting from 1 to about 9 seconds, detectable only with a seismograph. Microseisms are not related to earthquakes, but are due to atmospheric disturbances. *Microseisms that travel about half a mile a second are the clue to changes in Great Lakes' weather detected in New York* (Science News Letter).
—**microseismic,** *adj.* having to do with or of the nature of a microseism.
microsmatic (mī'kroz mat'ik), *adj. Zoology.* having poorly developed organs of smell: *microsmatic higher primates.* Contrasted with **macrosmatic.** [from *micr- + osmatic*]
microsomal (mī'krə sō'məl), *adj.* of or having to do with a microsome or microsomes: *a microsomal enzyme, microsomal protein synthesis.*
microsome (mī'krə sōm), *n. Biology.* a particle believed to consist of fragmented endoplasmic reticulum and attached ribosomes. Microsomes are obtained by ultracentrifugation of cellular matter. [from New Latin *microsoma,* from Greek *mikros* small + *soma* body]
microspectrophotometer (mī'krō spek'trō fō tom'ə tər), *n.* a spectrophotometer used for the examination of light reflected by very small specimens: *... direct measurements of the light reflected by individual cone cells in the retina, using a special microspectrophotometer* (Lorus J. Milne).
—**microspectrophotometry** (mī'krō spek'trō fō tom'ə trē), *n.* the use of the microspectrophotometer.
microspectroscope (mī'krō spek'trə skōp), *n.* a combination of the microscope and the spectroscope, used for the examination of minute traces of substances, light, etc.
—**microspectroscopic** (mī'krō spek'trə skop'ik), *adj.* of or having to do with the microspectroscope.
microsporangium (mī'krō spə ran'jē əm), *n., pl.* **-gia** (-jē-ə). *Botany.* a sporangium containing microspores, homologous with the sac containing the pollen in flowering plants: *A microsporangium contains several hundred microspores* (Emerson, *Basic Botany*).
microspore, *n. Botany.* a spore of comparatively small size from which a male gametophyte develops. Compare **megaspore.**
microsporophyll (mī'krə spôr'ə fil), *n. Botany.* a leaf or other structure bearing microsporangia.
microstructure, *n.* the small-scale or microscopic structure of bodies or objects, such as cells or minerals: *Scientists ... have been studying the microstructure of the insect eye* (Scientific American).
—**microstructural,** *adj.* of or having to do with microstructure.
microtektite, *n. Geology.* a microscopic variety of tektite found deep in ocean sediments: *Microtektites ... are reckoned to be the fine-grained components of the so-called "strewn fields" of larger tektites* (New Scientist and Science Journal).
microthermometer, *n.* a thermometer for measuring minute variations of temperature.
microtome (mī'krə tōm), *n.* an instrument for cutting extremely thin sections of tissues for microscopic examinations. [from *micro- + Greek -tomos* that cuts]
—**microtomic** (mī'krə tom'ik), *adj.* of or having to do with microtomy.
—**microtomy** (mī krot'ə mē), *n.* the preparation of objects for microscopic examination, especially tissue examination by sectioning done with a microtome.
microtubule (mī'krō tü'byül), *n. Biology.* a minute tubular filament in cells, consisting of the globular protein tubulin and functioning especially in forming and maintaining distinctive cellular shapes: *The microtubules are very fine tubes averaging 250 angstroms in diameter. They are found in cilia, in the tail of sperm cells, in the miotic spindle of a dividing cell and in the cytoplasm of many types of cell* (Scientific American).
—**microtubular,** *adj.* of or having to do with microtubules: *Flagella or cilia are the source of motion in all single cells. They have a common structure: external spiral microtubular filaments of a peculiar cablelike cross section and a common protein chemistry* (Philip Morrison).
microvascular, *adj. Biology.* of or having to do with the very small vessels of the microcirculatory system, such as the capillaries.
microvillus (mī'krō vil'əs), *n., pl.* **-villi** (-vil'ī). *Biology.* a microscopic hairlike projection growing on the surface of certain types of epithelial cells: *Partly digested food is finally broken down on the surface of the small intestine ... by microvilli, where adsorbed enzymes complete the process of hydrolysis* (Science Journal).

microvolt, *n.* a unit of electric voltage equal to one millionth of a volt.

microwatt, *n.* a unit of power equal to one millionth of a watt.

microwave, *n. Physics.* a high-frequency electromagnetic wave, usually one millimeter to one meter in wavelength: *The biological effects of microwaves include structural or functional changes to the eyes, testicles, bone marrow, cardiovascular and central nervous system, all due to the heating caused by microwave absorption* (New Scientist).

microzoon (mī′krə zō′on), *n., pl.* **-zoa** (-zō′ə). *Zoology.* a microscopic animal, especially a protozoan.

midbrain, *n.* = mesencephalon.

Middle, *adj. Geology.* of or having to do with an intermediate principal division of a period, system, or the like, between the upper and lower divisions: *Middle Cambrian.*

middle ear, *Anatomy.* the hollow space between the eardrum and the inner ear; tympanum. In humans it contains three small bones which transmit sound waves from the eardrum to the inner ear.

middle lamella, *Botany.* the primary layer of a plant cell wall, composed chiefly of calcium pectate, on which, in older cells, secondary layers of cellulose are deposited.

midgut, *n.* **1** *Embryology.* the middle part of the digestive canal of an embryo, from which the middle section of the digestive tract develops.
2 *Zoology.* the middle section of the digestive tract of an insect or crustacean: *The digestive organs of the woodroach, like those of higher animals, are divided into segments, each of which performs a special task. Food is stored in the foregut and digested in the midgut, the insect's stomach* (Scientific American). Compare **foregut, hindgut.**

midlatitudes, *n. Geography.* the zone of latitudes lying between the subtropics and the subarctic zone; the area between about 35° and about 55°, in both the Northern and the Southern Hemispheres.

midleg, *n. Zoology.* one of the middle or second pair of legs of an insect.

midnight sun, *Astronomy.* the sun seen both day and night in the arctic and antarctic regions during summer: *The midnight sun is an example of a circumpolar star* (Baker, Astronomy).

mid-ocean ridge or **mid-oceanic ridge,** *Geology.* a series of large mountain ranges on the ocean floor, running through the middle of the Atlantic Ocean, and across the Pacific and Indian Oceans. Their existence is thought to support the theory of continental drift.

midpoint, *n. Geometry.* a point that divides a line segment into two equal parts.

midrib, *n. Botany.* the central vein of a pinnate leaf. See the picture at **leaf.**

midriff, *n. Anatomy.* **1** the diaphragm separating the chest cavity from the abdomen. **2** the middle portion of the human body. [Old English *midhrif,* from *midd* mid + *hrif* belly]

migmatite (mig′mə tīt), *n. Geology.* a common type of hybrid rock formed of a complex mixture of igneous and metamorphic rocks, characterized by gneissic bands and crosscutting veins. [from Greek *migmatos* compound + English *-ite*]

migrate, *v.* **1** *Zoology.* to go from one region to another with the change in the seasons. Most birds migrate to warm climates to spend the winter.
2 *Botany.* to spread from one localized area to another or larger area: *The wind helps trees migrate by carrying their seeds beyond the forest ... Forests can migrate over fairly level land but not across oceans or mountain ranges* (Martin H. Zimmermann).
—**migration,** *n.* **1** *Zoology.* the seasonal movement of animals from one place or region to another; the act or process of migrating: *Those almighty instincts that propel the migrations of the swallow and the lemming* (Thomas De Quincey).
2 *Physics.* **a** a movement of one or more atoms from one place to another within the molecule. **b** the movement of ions between the two electrodes during electrolysis.
—**migrational,** *adj.* of or having to do with migration: *the migrational instincts of birds.*
—**migrator** or **migrant,** *n.* **1** an animal that migrates. **2** a migratory bird.
—**migratory,** *adj.* **1** moving from one place to another; that migrates; migrating: *migratory birds.*
2 of or having to do with migration: *the migratory pattern of elephants.*

mil, *n.* a customary unit of length equal to 0.001 of an inch. [short for Latin *millesimum* thousandth, from *mille* thousand]

mile, *n.* **1** a measure of distance equal to 5,280 feet. Also called **statute mile. 2** = nautical mile. [Old English *mil,* from Latin *milia* (*passuum*) a thousand (Roman paces), plural of *mille* thousand]

miliolite (mil′ē ə līt), *n. Paleontology.* a fossil foraminifer whose minute shells occur in immense numbers in some strata and are the chief constituent of certain limestones. [from New Latin *Miliola* the genus name (diminutive of Latin *milium* millet) + English *-ite*]

milk, *n.* **1** *Physiology.* the white liquid secreted by the mammary gland for the nourishment of the young, especially cow's milk, which is the chief source of butter, cheese, and cream. Milk is composed of water, fats, milk sugar (lactose), casein, proteins, minerals, and vitamins.
2 *Botany.* any kind of liquid resembling this, such as the white juice of a plant, tree, or nut: *the milk of the coconut.*

milk sugar, = lactose.

milk teeth, = primary teeth.

Milky Way, *Astronomy.* **1** a broad band of faint light that stretches across the sky at night. It is made up of countless stars and luminous clouds of gas, too far away to be seen separately without a telescope.

---

cap, fāce, fäther; best, bē, tėrm; pin, five;
rock, gō, ôrder; oil, out; cup, pùt, rüle,
yü in use, *yù* in uric;
ng in bring; *sh* in rush; *th* in thin; ᴛʜ in then;
zh in seizure.
ə = *a* in about, *e* in taken, *i* in pencil, *o* in lemon, *u* in circus

2 Also called **the Galaxy** or **Milky Way Galaxy.** the galaxy in which these stars and gas clouds are found. The earth, sun, and all the planets around the sun are part of the Milky Way.
[translation of Latin *Via Lactea*]

millepore (mil'ə pôr), *n. Zoology.* any one of various corallike hydrozoans covered with very small openings and bearing tentacles with powerful stings. [from New Latin *millepora,* from Latin *mille* thousand + *porus* pore]
—**milleporine** (mil'ə pôr'in), *adj.* having to do with the millepores or having their characteristics.

millerite (mil'ə rit), *n. Mineralogy.* a nickel sulfide, usually occurring in brassy or bronze crystals or in incrustations. *Formula:* NiS [from German *Millerit,* from W.H. *Miller,* 1801–1880, British mineralogist + *-it* -ite]

milli-, *combining form.* one thousandth of, as in *millimeter = one thousandth of a meter. Symbol:* m [from Latin *mille* thousand]

milliampere (mil'ē am'pir), *n.* a unit for measuring electric current equal to one thousandth of an ampere. *Abbreviation:* ma.

milliangstrom, *n.* a unit of length equal to one thousandth of an angstrom. *Symbol:* mÅ

millibar, *n.* a unit of atmospheric or barometric pressure equal to 1000 dynes per square centimeter; one thousandth of a bar. Standard atmospheric pressure at sea level is about 1013 millibars or 1033.2 grams per square centimeter.

millibarn, *n. Nuclear Physics.* a unit of area equal to one thousandth of a barn.

millicurie, *n.* a unit of radioactivity equal to one thousandth of a curie: *One millicurie of radium expels thirty million alpha particles per second* (Science News). *Symbol:* mCi

millidegree, *n.* one thousandth of a degree: *... procedures capable of lowering the temperature range accessible for research to within millidegrees of absolute zero* (Scientific American).

millifarad, *n.* a unit of electric capacitance equal to one thousandth of a farad. *Abbreviation:* mf

milligal (mil'ə gəl), *n.* a unit of gravitational acceleration equal to one thousandth of a gal.

milligram, *n.* a unit of mass equal to one thousandth of a gram. *Abbreviation:* mg

millihenry, *n., pl.* **-ries** or **-rys.** a unit of inductance equal to one thousandth of a henry. *Abbreviation:* mh

milliliter, *n.* a unit of volume equal to one thousandth of a liter; about one cubic centimeter. *Abbreviation:* ml

millimeter, *n.* a unit of length equal to one thousandth of a meter. *Abbreviation:* mm or mm.
—**millimetric,** *adj.* of or having to do with the millimeter or any system of measurement based on it.

millimeter of mercury, *Meteorology.* a unit used in the measurement of atmospheric pressure, equal to the pressure exerted by a one-millimeter column of mercury at standard gravity at a temperature of 0 degrees Celsius. 760 millimeters of mercury equals one atmosphere.

millimole, *n. Chemistry.* one thousandth of a mole.

milliohm, *n.* a unit of electrical resistance equal to one thousandth of an ohm.

millirad, *n.* a unit for measuring absorbed doses of ionizing radiation, equal to one thousandth of a rad. *Abbreviation:* mrad

millirem, *n.* a unit for measuring absorbed doses of radiation, equal to one thousandth of a rem. *Abbreviation:* mrem

milliroentgen, *n.* a unit of intensity of X rays or gamma rays, equal to one thousandth of a roentgen. *Abbreviation:* mr

millisecond, *n.* a unit of time equal to one thousandth of a second: *delayed in time by several hundred milliseconds* (Science News). *Abbreviation:* ms or msec.

millivolt, *n.* a unit of potential difference or electromotive force equal to one thousandth of a volt. *Abbreviation:* mv *Symbol:* mV

milliwatt, *n.* a unit of power equal to one thousandth of a watt. *Abbreviation:* mw *Symbol:* mW

mimesis (mi mē'sis *or* mī mē'sis), *n.* = mimicry. [from Greek *mimēsis,* from *mimeisthai* imitate, from *mimos* imitator, mime]

mimetic (mi met'ik *or* mī met'ik), *adj. Biology.* having to do with or exhibiting mimicry: *In back crosses (where the hybrids are mated with the original mimetic species) the offspring exhibit poor mimicry* (New Scientist). [from Greek *mimētikos,* from *mimeisthai* imitate, from *mimos* imitator, mime] — **mimetically,** *adv.*

mimic (mim'ik), *Biology.* —*v.* to resemble (something) closely in form, color, or other aspect: *Some insects mimic leaves. The Leptalides ... fly in the same parts of the forest, and generally in company with the species they mimic* (H.W. Bates).
—*adj.* exhibiting mimicry; mimetic: *An edible mimic species gains some measure of protection from predators by virtue of its close resemblance to a model species which is unpalatable* (Nature).
[from Latin *mimicus,* from Greek *mimikos,* from *mimos* imitator, mime]

mimicry

mimicry (mim'ik rē), *n. Biology.* the close outward resemblance of an animal or plant to its surroundings or to some different animal or plant, especially for protection or concealment: *A final method of protection that might be mentioned is protective mimicry. Insects develop body shapes and colors that so closely resemble their surroundings that detection by sight is very difficult* (Winchester, *Zoology*). *Some plants are barely recognizable as plants at all, and these are the mimicry plants which are one of the most famous plants in the world* (E. Palmer).

Mindel (min′dəl), *n. Geology.* the second glaciation stage of the Pleistocene in Europe: *The first cave bears probably evolved at a time near the end of the second great ice advance of the Pleistocene epoch. That was the Mindel glaciation which began some 700,000 years ago* (Björn Kurtén). Compare **Günz, Riss,** and **Würm.** [named after the *Mindel* River, Bavaria, a locality of the glaciation]

mineral, *n.* **1** *Geology, Mineralogy.* an inorganic substance found in nature, having a uniform or restricted chemical composition and a regular crystalline form: *Perhaps the most important and significant limitation placed on the definition of a mineral is that it must be a chemical element or compound ... Minerals, with few exceptions, possess the internal ordered arrangement characteristic of the solid state ... and assume regular geometric forms as crystals* (Dana's Manual of Mineralogy). *A mineral is any naturally occurring compound of the element in question, but an ore is a mineral which contains that element in a quantity sufficient to make its extraction profitable* (Jones, Inorganic Chemistry).
2 (broadly, especially in industry) any natural substance obtained by mining or quarrying, such as coal, ore, salt, and stone. Gold, silver, and iron are metallic minerals; quartz, feldspar, and calcite are nonmetallic minerals.
—*adj.* **1** of, resembling, or containing a mineral or minerals: *a mineral substance, mineral deposits.* **2** not animal or vegetable; inorganic: *of mineral origin or composition.*
[from Medieval Latin *minerale,* from *minera* a mine]

mineralocorticoid (min′ər ə lə kôr′tə koid), *n. Biochemistry.* any of a group of steroid hormones produced by the adrenal cortex and affecting mineral metabolism: *The mineralocorticoids are concerned with the distribution of body fluid and ion balance.* (New Scientist). Compare **glucocorticoid.**

mineralogical, *adj.* of or having to do with mineralogy: *Chemical studies, solid state physics considerations, and synthetic minerals are now the frontiers of mineralogical research* (James T. Wilson). —**mineralogically,** *adv.*

mineralogy, *n.* **1** the science of minerals: *Mineralogy is an integrated field of study intimately related to geology on the one hand and to physics and chemistry on the other* (Dana's Manual of Mineralogy).
2 mineral composition or characteristics: *The mineralogy of lignite overburden is an important consideration in reclaiming mine spoils* (J.B. Dixon).

mineral oil, *Chemistry.* any oil derived from minerals, as petroleum.

mineral salt, *Chemistry.* **1** a salt which occurs as, or is derived from, a mineral: *Over-irrigation led to the depositing of mineral salts on the soil to such an extent that crop yields were reduced sharply* (Science News Letter). **2** a salt of an inorganic acid.

mineral spring, *Geology.* a spring that yields mineral water.

mineral water, *Geology.* spring water containing various mineral salts, elements, or gases. The mineral salts or elements may include Epsom salt, calcium, iron, magnesium, boron, fluorine, and others. The gases are usually carbon dioxide and hydrogen sulfide.

mineral wax, = ozocerite.

minim (min′əm), *n.* a unit of liquid measure, equal to one sixtieth of a dram, or about 0.0616 milliliter. [from Latin *minimus* smallest]

minimax (min′ē maks), *n. Mathematics.* a strategy in game theory that minimizes the maximum risk to which a player will be liable: *It has been proved that the maximin of all possible strategies, including mixed ones, is always equal to the minimax of all possible strategies of the opponent, again including mixed ones.* (S. Vajda). [from *mini*(*mum*) + *max*(*imum*)]

minimum, *n., pl.* **-mums, -ma** (-mə). *Mathematics.* a value of a function less than any other value of the function in a given interval: *Others again decrease continually; and so have no minimum ... But, on the other hand, some ... decrease to a certain finite magnitude, called their Minimum, or least state* (Hutton, Course of Mathematics). *The maximum and minimum in the retinal response may "set" brightness discriminators in the brain, and provided that there are no intervening maxima and minima (that is, visible contours) the apparent brightness of adjacent areas would not deviate from that set by the maximum or the minimum* (Floyd Ratliff). [from Latin *minimum,* neuter of *minimus* smallest, superlative of *minor* smaller, lesser]

minor, *n. Mathematics.* (of an element in a determinant) the determinant of next lower order obtained by crossing out the row and column containing the given element. See also **cofactor.** [from Latin *minor* lesser]

minor arc, *Mathematics.* an arc that is less than half a circle.

minor axis, *Mathematics.* the shorter axis of an ellipse, perpendicular to and bisecting the major axis.

minor element, = trace element.

minor planet, = asteroid: *Most such objects, technically referred to as "minor planets," circle the sun in a belt between the orbits of Mars and Jupiter* (Science News).

minuend (min′yü end), *n. Mathematics.* a number or quantity for which another number or quantity (the *subtrahend*) is to be subtracted. EXAMPLE: in 100 − 23 = 77, the minuend is 100. [from Latin *minuendus* to be made smaller, from *minus* less]

minus, *prep. Mathematics.* decreased by; less: *5 minus 2 leaves 3.*
—*adj.* **1** *Mathematics.* **a** showing subtraction. The **minus sign** (−) indicates that the quantity following it is to be subtracted or is a negative quantity. **b** having the minus sign; negative: *a minus quantity.*
2 *Botany.* of or having to do with the strain of heterothallic fungi which acts as the female in reproduction.
—*n.* **1** *Mathematics.* **a** the minus sign (−): *There is a minus before the number to be subtracted.* **b** a negative quantity. [from Latin *minus* less, neuter of *minor* lesser]

---

cap, fāce, fä́ther; best, bē, tėrm; pin, five;
rock, gō, ôrder; oil, out; cup, pùt, rüle,
yü in use, *yủ* in uric;
ng in bring; *sh* in rush; *th* in thin; ₮H in then;
zh in seizure.
ə = *a* in about, *e* in taken, *i* in pencil, *o* in lemon, *u* in circus

minute (min'it), *n.* **1** a unit of time equal to one sixtieth of an hour; sixty seconds. *Abbreviation:* min. ▶ See the note under **second.**
2 Also called **minute of arc.** one sixtieth of a degree. 10° 10' means ten degrees and ten minutes.

Miocene (mī'ə sēn), *Geology.* —*n.* **1** the fourth epoch of the Tertiary period of the Cenozoic, after the Oligocene and before the Pliocene. During the Miocene, grasses developed and grazing mammals flourished. **2** the rocks formed in this epoch.
—*adj.* of this epoch or these rocks: *the Miocene flora, Miocene deposits.*
[from Greek *meiōn* less + *kainos* new, recent]

miracidium (mī'rə sid'ē əm), *n., pl.* **-ia** (-ē ə). *Zoology.* the minute, ciliated aquatic larva which hatches from the egg of a fluke and infects the snail, the intermediate host, in the development of the fluke. [from New Latin *miracidium,* from Greek *meirakidion* little boy, diminutive of *meirakion* boy]

mirror image, something that has the exact likeness of another thing, but reversed, like an image in a mirror. If a thing is symmetrical, it can be superposed on its mirror image, but if it is asymmetrical, it cannot. *The two ridges are roughly mirror images of each other, showing that the motion was uniform on each side* (Scientific American). *Probably most interest was aroused by detection ... of a second kind of neutrino and its mirror-image counterpart, an antiparticle* (A.O. Williams, Jr.).

miscode, *v. Molecular Biology.* to provide with a wrong or faulty genetic code: *The mutant DNA miscoded a single amino acid in the sequence of structural protein in the membrane. As a result of this seemingly minor alteration, the entire membrane was defective* (Barbara J. Culliton).

misevolution, *n. Biology.* abnormal evolution of a cell, viral particle, etc.

missense, *adj. Molecular Biology.* involving or resulting from the insertion of a different amino acid from that which is usual in a polypeptide or protein molecule. Compare **nonsense.**

Mississippian, *Geology.* —*n.* **1** the fifth period of the Paleozoic era, after the Devonian and before the Pennsylvanian; Lower Carboniferous (the name used outside of North America). **2** the rocks formed during this period.
—*adj.* of or having to do with the Mississippian: *By Mississippian time, at least four separate continental blocks, which now form Asia and part of Europe, lay close together in the Northern Hemisphere* (Eicher and McAlester, *History of the Earth*).

mist, *n. Meteorology.* **1** a suspension of very fine water droplets in the air above the ground. SYN: haze. **2** a cloud of very fine drops of any liquid in the air.

mistral (mis'trəl *or* mis träl'), *n. Meteorology.* a strong, cold, dry, northerly wind common in the Mediterranean provinces of France and neighboring regions. [from French *mistral,* from Provençal (originally) dominant, from Latin *magistralis,* from *magister* master]

mitochondria (mī'tə kon'drē ə), *n., pl. of* **mitochondrion** (mī'tə kon'drē ən). *Biology.* minute specialized structures in the cytoplasm of cells, containing genetic material and many enzymes important for cell metabolism. Mitochondria produce most of the energy required by the cells and appear in the form of granules, rods, or filaments, but often change their shape although their number in each cell remains about the same. *Hibernating animals convert their mitochondria to the type possessed by cold-blooded animals when the hibernating season arrives* (John G. Lepp). See the picture at **centrosome** [New Latin, from Greek *mitos* thread + *chondros* lump]
—**mitochondrial,** *adj.* of or having to do with mitochondria: *mitochondrial DNA.*

mitochondrion (mit'ə kon'drē ən), *n.* singular of **mitochondria:** *The mitochondrion ... is frequently a spherical body about 1–2 microns in diameter* (New Scientist). Also called **chondriosome.**

mitogenetic (mī'tə jə net'ik), *adj.* = mitogenic.

mitogenic (mī'tə jen'ik), *adj. Biology.* that promotes or induces mitosis: *mitogenic agents.*

mitosis (mi tō'sis *or* mī tō'sis), *n. Biology.* **1** the process by which the nucleus of a cell divides to form two new nuclei, each containing the same number of chromosomes as the original nucleus; nuclear division. Mitosis is typically divided into four stages: *prophase,* in which the elongated chromatin threads of the nucleus condense to form microscopically visible chromosomes, each of which consists of two genetically identical sister chromatids; *metaphase,* in which the nuclear membrane disappears and the chromosomes line up near the middle of the cell; *anaphase,* in which one chromatid of each pair (now called a daughter chromosome) moves toward each end of the cell; and *telophase,* in which the daughter chromosomes lose their visible identity and again appear as chromatin threads, and two new nuclear membranes form around the two chromatinic bodies. During telophase, the cytoplasm typically divides (by a process called cytokinesis), and two new cells are formed.
2 cell division in which mitotic nuclear division occurs. Compare **meiosis.**
[from New Latin *mitosis,* from Greek *mitos* thread]
—**mitotic** (mī tot'ik), *adj.* of or having to do with mitosis: *As a result of mitotic cell division, daughter cells have one-half the number of chromosomes of the original cell* (Biology Regents Syllabus). —**mitotically,** *adv.*

mitosis:

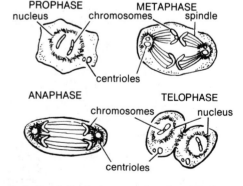

mitral valve (mī'trəl), *Anatomy.* the valve of the heart between the left atrium and left ventricle, which prevents the blood from flowing back into the atrium:

Damage to the mitral valve ... is one of the commonest pathological conditions of the heart (Arthur Selzer). Also called **left atrioventricular valve.** Compare **tricuspid valve.** See the picture at **valve.**

mixed, *adj.* **1** *Botany.* (of an inflorescence) combining both determinate and indeterminate arrangements, as a thyrsus does. **2** *Anatomy.* consisting of both sensory and motor fibers: *Nerves ... can be sensory nerves, motor nerves, and mixed nerves* (Biology Regents Syllabus). **3** *Physiology.* secreting more than one substance: *a mixed gland.*

mixed bud, *Botany.* a bud producing both foliage and flower. Compare **flower bud, leaf bud.**

mixed number, *Mathematics.* a number consisting of a positive or negative integer and a fraction. EXAMPLES: 1 1/2, 16 2/3, – 25 9/10.

mixed tide, *Oceanography.* a tide that does not flow and ebb regularly twice a day: *Some Pacific Islands have mixed tides, such as two high tides daily, with only a little ebb between, and then a very low tide* (Robert O. Reid).

mixture, *n.* *Chemistry.* the product of two or more substances mixed together, but not chemically combined. Mixtures can be separated mechanically by distillation, freezing, melting, etc. Compare **compound.** [from Latin *mixtura,* from *miscere* to mix]

MKS or **MKSA,** *adj.* of or designating a system of measurement based on the meter, the kilogram, the second, and the ampere: *The joule is the MKS unit of work.* [from *m(eter)* + *k(ilogram)* + *s(econd)* + *a(mpere)*] ▶ See the note under **SI unit.**

ml, *abbrev.* milliliter *or* milliliters.

mm or **mm.,** *abbrev.* millimeter *or* millimeters.

mm² or **mm.²,** *abbrev.* square millimeter.

mm³ or **mm.³,** *abbrev.* cubic millimeter.

mmf or **m.m.f.,** *abbrev.* magnetomotive force.

Mn, *symbol.* manganese.

MNS (em′ en′ es′), *n.* *Immunology.* a system of classifying blood groups based on genetically controlled antigens found within the membranes of red blood cells.
—*adj.* of or having to do with this system of classification. Compare **ABO.**

Mo, *symbol.* molybdenum.

mobile, *adj.* *Chemistry.* tending to be naturally fluid; moving or flowing easily: *Mercury is a mobile metal.*
—**mobility,** *n.* natural fluidity; ease of movement or flow.

Möbius strip or **Möbius band** (mœ′bē əs *or* mō′bē əs), *Geometry.* a continuous, one-sided surface formed by turning one side of a rectangle 180 degrees and then joining it to the opposite side: *The unique topological property of the Möbius strip is that it has one surface and one edge* (Carol Gibson). *If you cut the bottle in half vertically, you get two Möbius bands, one a mirror image of the other* (Martin Gardner). Also spelled **Moebius strip** or **band.** Compare **Klein bottle.** [named after August F. *Möbius,* 1790–1868, German mathematician]

modal, *adj.* *Statistics, Physics.* of or having to do with a mode or modes.

mode, *n.* **1** *Mathematics, Statistics.* the number that occurs most frequently in a set of numbers or series of data: *The presence of two or more modes usually means that the sample is not homogeneous, i.e., that*

two or more distinct distributions have been combined (O. L. Davies). ASSOCIATED TERMS: see **average.**
2 *Physics.* any one of various patterns in which vibration may occur. In a freely vibrating system, oscillation is restricted to certain characteristic patterns of motion at certain characteristic frequencies. *Waves of given frequency can go through a pipe in many patterns, called modes, each with a different wavelength and velocity. When the pipe is small, only one mode can travel* (Scientific American).
3 *Geology.* the actual mineral composition of a rock, stated quantitatively in percentages by weight or volume. Compare **norm.**
[from Latin *modus* measure, manner]

model, *n.* a simplified description or conception of a system, used to understand the system or as the basis for further study or investigation of its characteristics: *a mathematical model of the global atmosphere, the mechanistic model of the universe. A Michigan researcher has created a computer model of a bacterium, the ubiquitous Eschericia coli. Fed on machine language input, the model so far has been able to grow, function and reproduce itself just like its protoplasmic counterpart* (Science News).

moderate (mod′ər it), *adj.* *Meteorology.* denoting a breeze with a velocity of 13–18 miles per hour (on the Beaufort scale, force 4), or a gale with a velocity of 32–38 miles per hour (Beaufort force 7).
—*v.* (mod′ər āt), *Nuclear Physics.* to slow down or lower the energy of (a particle, especially a neutron): *Most of today's power reactors use graphite* (carbon) *or water* (hydrogen and oxygen) *to moderate neutron speed* (Robert C. Cowen).

moderator, *n.* *Nuclear Physics.* a material, such as graphite, used in a reactor to reduce the speed of neutrons, making them more efficient in splitting atomic nuclei.

modification, *n.* *Biology.* a change in an organism resulting from external influences, and not inheritable.

modiolus (mō dī′ə ləs *or* mə dī′ə ləs), *n., pl.* -**li** (-lī). *Anatomy.* the central conical axis around which the chochlea of the ear winds. [from New Latin *modiolus,* from Latin, nave of a wheel]

modular (moj′ú lər), *adj.* *Mathematics, Physics.* of or having to do with a modulus or moduli.

modular arithmetic, a form of arithmetic dealing with the remainders after a set of numbers are divided by a single number, the modulus: *Modular arithmetic is a concept with which we are all familiar. For example, if it is ten o'clock, then three hours later it will be one o'clock* (and not, as in simple arithmetic, thirteen o'clock). *A clock or watch is an example of what is known as a "modulo 12" system. The "12" means that there are only 12 integers* (S. J. Colley).

---

cap, fāce, fäther; best, bē, tèrm; pin, five;
rock, gō, ôrder; oil, out; cup, pùt, rüle,
yü in use, *yu* in uric;
ng in bring; *sh* in rush; *th* in thin, ᴛʜ in then;
zh in seizure.
ə = *a* in about, *e* in taken, *i* in pencil, *o* in lemon, *u* in circus

modulate (moj'ù lāt), *v. Physics.* **1** to vary the amplitude, frequency, or phase of electromagnetic waves, especially carrier waves.
2 to vary the velocity of electrons in an electron beam.
—**modulation** (moj'ù lā'shən), *n.* **1** the act or process of modulating. **2** the quality or condition of being modulated.
—**modulator** (moj'ù lā'tər), *n.* a device that modulates.

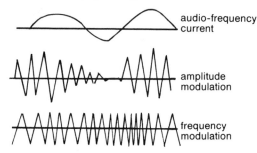

audio-frequency current

amplitude modulation

frequency modulation

modulo (moj'ə lō), *adv. Mathematics.* with respect to a (specified) modulus. EXAMPLE: 18 is congruent to 42 modulo 12, because both 18 and 42 leave 6 as a remainder when divided by 12. [from Latin *modulo,* ablative of *modulus*]
modulus (moj'ù ləs), *n., pl.* **-li** (-lī). **1** *Physics.* a quantity expressing the measure of some function, property, or the like, especially under conditions where the measure is unity: *Graphite fibers, with a tensile strength of 250,000 pounds per square inch and an elastic modulus of 50 million pounds per square inch, are the highest strength-to-weight ratio engineering materials man has yet made commercially* (Science News).
2 *Mathematics.* **a** a number by which two given numbers can be divided to leave the same remainders. EXAMPLE: the modulus of 7 and 9 is 2, written as 7 = 9 (modulus 2). **b** the factor by which a logarithm in one system is multiplied to change it to a logarithm in another system. **c** = absolute value.
[from Latin *modulus* standard of measurement, module, diminutive of *modus* measure, mode]
Moebius strip or **Moebius band,** = Möbius strip.
mofette or **moffette** (mō fet'), *n. Geology.* **1** an opening or fissure in the earth in a late-stage volcanic area, from which gases, especially carbon dioxide, emanate.
2 an emanation from such an opening. [from French *mofette,* from Italian *mofetta*]
Moho (mō'hō), *n.* = Mohorovicic discontinuity.
Mohorovičić discontinuity (mō'hə rō'və chich), *Geology.* the boundary between the earth's crust and mantle, the depth of which varies from approximately 10 to 12 kilometers under ocean basins, to 40 to 50 kilometers under the continents. [named after Andrija *Mohorovicic,* 1857–1936, Yugoslav geophysicist who discovered the discontinuity in 1909 from studies of earthquake records]
Mohs scale or **Mohs' scale,** (mōz), *Mineralogy.* an empirical scale for classifying minerals on the basis of relative hardness, determined by the ability of harder minerals to scratch softer ones. Values for the Mohs scale are as follows: talc 1; gypsum 2; calcite 3; fluorite

4; apatite 5; orthoclase 6; quartz 7; topaz 8, corundum 9; diamond 10. [named after Friedrich *Mohs,* 1773–1839, German mineralogist who invented the scale]
moisture, *n. Meteorology.* slight wetness; water or other liquid suspended in very small drops in the air or spread on a surface. Dew is moisture that collects at night on the grass.
mol, *symbol.* mole.
molal (mō'ləl), *adj. Chemistry.* (of a solution) having one mole of solute in 1000 grams of solvent. Compare **molar[2].**
—**molality** (mō lal'ə tē), *n.* the molal concentration of a solution. It is expressed as the number of moles of solute in 1000 grams of solvent. *Abbreviation:* m
molar[1], *Anatomy.* —*n.* a tooth with a broad surface for grinding, having somewhat flattened points. The twelve permanent back teeth in humans are molars.
—*adj.* of or having to do with the molars. Compare **premolar.**
[from Latin *molaris* grinding, from *mola* mill]
molar[2], *adj. Chemistry.* **1** of or having to do with a mole or moles.
2 having one mole of solute in a liter of solution. Compare **molal.** [from *mole*]
—**molarity,** *n.* the molar concentration of a solution. It is expressed as the number of moles of solute in a liter of solution. *Abbreviation:* M
molar heat capacity, *Chemistry.* the amount of heat required to raise the temperature of one mole of a substance by one degree Celsius.
mold[1], *n. Geology, Paleontology.* an impression or cavity made in earth or rock by the inner or outer surface of a fossil shell or other organic structure. [from Old French *modle,* from Latin *modulus* module]
mold[2], *n. Biology.* **1** a woolly or furry fungous growth, often greenish-blue, black, or whitish in color, that appears on the surface of food and other animal or vegetable substances when they are left too long in a warm, moist place or when they are decaying.
2 any fungus producing such a growth.
[Middle English *moulen* grow moldy]
—**moldy,** *adj.* covered with mold.
mold[3], *n. Ecology.* loose, broken, or crumbly earth, especially fine, soft soil rich in decayed leaves, manure, or other organic matter, suitable for the cultivation of plants. SYN: topsoil, humus. [Old English *molde*]
mole, *n.* the fundamental SI unit for measuring the amount of a substance, equal to the amount of a given substance that contains the same number of elementary entities (such as atoms, molecules, etc.) as there are atoms of carbon in 12 grams of carbon 12. One mole of atoms contains Avogadro's number (6.02 × 10^{23}) of atoms. *A mole of molecules of any gas occupies a volume of 22.4 liters at STP. Symbol:* mol [from German *Mol,* short for *Mol(ekül)* molecule]
molecular, *adj.* having to do with, caused by, or consisting of molecules: *Molecular substances may exist as gases, liquids, or solids, depending on the attraction that exists between the molecules. Generally, molecular solids are soft, electrical insulators, poor heat conductors, and have low melting points* (Chemistry Regents Syllabus).
molecular astronomy, the study of the chemical molecules found in interstellar space.

molecular beam, *Physics.* a stream of molecules moving in about the same direction and at approximately the same speed.

molecular biology, 1 the branch of biology dealing with the formation, structure, and activity of macromolecules essential to life, such as nucleic acid and protein molecules, and their role in cell replication and the transmission of genetic information: *Particularly relevant to cancer and aging are the revolution in molecular biology, which has brought us to a consideration of utilizing genetic engineering to cure some diseases* (Harper's). Compare **molecular genetics.**
2 biology on the molecular level: *the molecular biology of the gene.*

molecular formula, a chemical formula that shows the number and kinds of atoms in a molecule without indicating how the atoms are arranged. EXAMPLES: CH_4 (methane), H_2O_2 (hydrogen peroxide). Compare **empirical formula, structural formula.**
▶ Compounds which have the same chemical formula but different structures are called isomers. For example, the compounds CH_3CH_2CHO (propanal) and CH_3COCH_3 (acetone) are isomers, both having the molecular formula C_3H_6O.

molecular fossil, *Paleontology.* a molecule of organic material extracted from rocks used to study the early evolution of life on earth. *The exciting studies of ... "molecular fossils," have not, however, firmly established the biogenic nature of these substances* (Science Journal).

molecular gas constant, = Boltzmann constant.

molecular genetics, the branch of genetics dealing with heredity and variation in the sequence of nucleotide bases of the genetic code: *As molecular genetics and classical genetics begin to merge, long held concepts of classical genetics become less definitive. The actual interactions of gene products shed new light on our concepts and dominance and recessiveness. Dominance and recessiveness may, in fact, be relative terms* (Biology Regents Syllabus). Compare **molecular biology.**

molecular sieve, *Chemistry.* any of various substances, especially zeolites, whose crystalline structure contains regularly spaced holes of molecular size that can be used to filter out relatively large molecules: *New ... air separation methods used molecular sieves to separate nitrogen from the oxygen in the air* (Marshall Sittig).

molecular weight, *Chemistry.* the sum of the atomic weights of all the atoms in a molecule: *Recent reports indicated that the proteins of membranes may have molecular weights ranging from 15,000 to 200,000* (Robert L. Hill).

molecule, *n. Chemistry.* the smallest particle into which an element or compound can be divided without changing its chemical and physical properties. A molecule of an element consists of one or more like atoms. A molecule of a compound consists of two or more different atoms. *The tendency of atoms to bind themselves together into molecules by exchanging or sharing their valence, or outer electrons, is a well-known phenomenon* (Scientific American). [from New Latin *molecula,* diminutive of Latin *moles* mass]

mole fraction, *Chemistry.* the ratio of the number of moles of a component in a solution to the total number of moles of all components present.

mollusc, *n.* = mollusk.

molluscan (mə lusʹkən), *Zoology.* —*adj.* of or having to do with mollusks: *The chambered nautilus, famed in poetry for its beauty, does have a complete shell, one which scientists call in many ways the "most primitive" of molluscan shells* (Science News Letter). —*n.* = mollusk.

mollusk or **mollusc** (molʹəsk), *n. Zoology.* any of a large phylum (Mollusca) of invertebrate animals having a soft, unsegmented body, usually covered with a hard shell secreted by a covering mantle, and a muscular foot. Snails, clams, oysters, whelks, and mussels belong to this phylum. Slugs, octopuses, and squids are mollusks that have no shell. [from French *mollusque,* from New Latin *Mollusca* a Linnean order, (originally) neuter plural of Latin *molluscus* soft-bodied, from *mollis* soft]

molluskan, *adj., n.* = molluscan.

Mollweide projection (môlʹvīʹdə), *Geography.* a homolographic map projection in which the surface of the earth is represented as an ellipse, with the equator and parallels of latitude as straight lines. Compare **Mercator projection.** [named after Karl *Mollweide,* 1774–1825, German mathematician and astronomer who devised the projection]

molt, *Zoology.* —*v.* to shed feathers, skin, hair, shell, antlers, or other growths, before a new growth. Birds, snakes, insects, and crustaceans molt. —*n.* **1** the act or process of molting. **2** skin, hair, antlers, or other growths, shed in molting. Also spelled **moult.** [alteration of Middle English *mouten,* Old English *-mutian,* ultimately from Latin *mutare* to change]

molten, *adj. Chemistry.* made liquid by heat; melted: *molten steel. Molten rock is a conductor of electricity, whereas solidified lava is an insulator* (Robert W. Decker).

molybdenite (mə libʹdə nīt), *n. Mineralogy.* a soft, native sulfide of molybdenum that resembles graphite. It is the chief ore of molybdenum. *Molybdenite is a soft,*

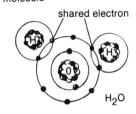

molecule

shared electron

H_2O

cap, fāce, fäther; best, bē, tèrm; pin, fīve;
rock, gō, ôrder; oil, out; cup, pùt, rüle,
yü in use, *yù* in uric;
ng in bring; sh in rush; th in thin, ᴛH in then;
zh in seizure.
ə = *a* in about, *e* in taken, *i* in pencil, *o* in lemon, *u* in circus

molybdenum

lead-grey mineral and ... will mark paper (Jones, *Minerals in Industry*). Formula: MoS_2

molybdenum (mə lib'də nəm), *n. Chemistry*. a heavy, hard, grayish or silver-white metallic element, much used to strengthen and harden steel. Molybdenum is highly resistant to heat and conducts electricity easily. It is found in the mineral molybdenite and is a necessary trace element in plant metabolism. *Symbol:* Mo; *atomic number* 42; *atomic weight* 95.94; *melting point* 2617°C; *boiling point* 4612°C; *oxidation state* 6. [from New Latin *molybdenum,* from Greek *molybdaina* ore or lead, from *molybdos* lead]

molybdic (mə lib'dik), *adj. Chemistry*. of or containing molybdenum, especially with a valence of six.

molybdous (mə lib'dəs), *adj. Chemistry*. of or containing molybdenum, especially with a valence of less than six.

moment *n.* **1** *Physics*. **a** a tendency to cause rotation around a point or axis: *When torques about a center of moment are equal, no rotation takes place* (E.S. Obourn, *World of Science*). **b** the product of a (specified) physical quantity and the length of the perpendicular from a point or axis. The **moment of force** about a point is the product of the magnitude of the force and the length of the perpendicular distance from the point to the line of action of the force; also called *torque.*
2 *Statistics*. any of a series of quantities that express values derived from sums of powers of the variables in a set of data.
[from Latin *momentum* moving power, movement, from *movere* to move]

moment of inertia, *Physics*. a measure of the resistance of a body to angular acceleration, equal to the sum of the products of each particle of a rotating body multiplied by the square of the distances of the particles from the axis of rotation: *The moon's ... moment of inertia varies according to the axis of hypothetical rotation one calculates it for* (Science News).

momentum (mō men'təm), *n., pl.* **-tums, -ta** (-tə). *Physics*. a measure of the motion of a body, equal to the mass multiplied by the velocity: *A falling object gains momentum as it falls. Momentum is a vector quantity ... Its direction is the same as that of its velocity* (Physics Regents Syllabus). Compare **kinetic energy.** See also **conservation of momentum.**
[from Latin *momentum* moving power, from *movere* to move]
ASSOCIATED TERMS: see **motion.**

mon-, *combining form.* the form of **mono-** before vowels, as in *monacid, monatomic.*

monacid, *adj., n.* = monoacid.

monadelphous (mon'ə del'fəs), *adj. Botany*. having stamens united by their filaments into one group, as in various legumes and mallows. [from Greek *monos* one, single + *adelphos* brother]

monadnock (mə nad'nok), *n. Geography, Geology*. an isolated hill or mountain of resistant rock standing in an area that is almost level from erosion. [named after Mt. *Monadnock,* in New Hampshire]

monandrous (mə nan'drəs), *adj. Botany*. **1** having only one stamen: *a monandrous flower.*
2 having such flowers: *a monandrous plant.*

[from Greek *monandros,* from *monos* single, one + *anēr, andros* husband]
—**monandry** (mə nan'drē), *n. Botany*. the condition of having but one perfect stamen.

monanthous (mə nan'thəs), *adj. Botany*. (of a plant) single-flowered; bearing one flower on each stalk. [from *mon-* single + Greek *anthos* flower]

monatomic, *adj. Chemistry*. **1** having one atom in the molecule: *Group O elements are monatomic gases. The atoms of these elements have complete outer shells, which result in an electron configuration that is stable* (Chemistry Regents Syllabus).
2 having one replaceable atom or group of atoms.
3 = univalent.

monaxial (mon ak'sē əl), *adj.* **1** *Physics*. having but one axis. SYN: uniaxial.
2 *Botany*. having flowers growing directly from the main axis.

monaxon (mon ak'son), *n. Zoology*. a tiny, rod-shaped spicule found in sponges.

monazite (mon'ə zīt), *n. Mineralogy*. a phosphate of cerium and related rare-earth metals, found in small reddish or brownish crystals: *Monazite exists in various granite formations, but mostly it has been mined in sand and gravel deposits along ocean beaches and river banks* (New Yorker). Formula: $(Ce,La,Nd,Th)PO_4$ [from German *Monazit,* from Greek *monazein* be solitary (from *monos* alone) + German *-it* -ite]

monecious (mə nē'shəs), *n.* = monoecious.

moneran (mə ner'ən), *Biology*. —*n.* any member of the kingdom (Monera) comprising the prokaryotes. Bacteria and blue-green algae are monerans.
—*adj.* of or having to do with monerans.
[from New Latin *Monera,* from Greek *monērēs* individual, solitary, from *monos* alone, single]

moniliform (mō nil'ə fôrm), *adj. Biology*. resembling a string of beads, as certain roots or pods, which have a series of swellings alternating regularly with contractions. [from Latin *monile* necklace]

mono-, *combining form.* **1** one; single, as in *monocarp, monochromatic.*
2 containing one atom, group, etc., of the substance specified, as in *monoacid, monoamine.* Also spelled **mon-** before vowels.
[from Greek *mono-,* from *monos* single]

monoacid, *Chemistry*. —*adj.* **1** (of a base or alcohol) having one hydroxyl (-OH) group that can be replaced by an atom or radical of an acid to form a salt or ester.
2 having one acid atom of hydrogen per molecule.
—*n.* an acid containing only one replaceable hydrogen atom. Also spelled **monacid.**

monoacidic, *adj.* = monoacid.

monoamine (mon'ō ə mēn' *or* mon'ō am'in), *n. Chemistry*. an amine containing one amino group, especially one that functions as a neurotransmitter: *Monoamines play a considerable part in regulating the central nervous system, and nervous disorders result when the supply of these substances is upset. Three of the key monoamines are noradrenaline, dopamine and 5-hydroxytryptamine* (New Scientist).

monoamine oxidase, *Biochemistry*. an enzyme present in the cells of most animal and plant tissue which oxidizes and destroys monoamines such as norepinephrine and serotonin: *The enzyme that breaks down phenylethylamine is known: it is a monoamine oxidase*

412

known as monoamine oxidase B. Although there is some evidence that that enzyme is abnormally weak in the bloodstream of schizophrenics there is no evidence for a similar deficiency in their brains (London *Times*). *Abbreviation:* MAO

monobasic, *adj.* **1** *Chemistry.* **a** (of an acid) having only one atom of hydrogen that can be replaced by an atom or radical of a base in forming salts. **b** having one basic hydroxyl (-OH) radical per molecule. **c** (of a salt) having one basic atom or radical which can replace a hydrogen atom of an acid.
2 *Biology.* being the sole type of its group; monotypic.

monocarp (mon′ə kärp), *n. Botany.* a plant that bears fruit only once during its lifetime. [from *mono-* + Greek *karpos* fruit]

monocarpellate or **monocarpellary,** *adj. Botany.* having or consisting of a single carpel: *The legume is ... monocarpellate and typically dehisces along two sides when ripe* (Greulach and Adams, *Plants*).

monocarpic (mon′ə kär′pik), *adj. Botany.* producing fruit but once, then dying: *Sometimes monocarpic* (*sets seed yearly and dies*) *new rosettes are formed around the old plant* (N.Y. Times).

monochasium (mon′ə kā′zhē əm *or* mon′ə kā′zē əm), *n.,* *pl.* **-sia** (-zhē ə *or* -zē ə). *Botany.* a cyme in which the main axis produces only a single branch. [from New Latin *monochasium,* from Greek *monos* one + *chasis* chasm, separation]

monochromatic, *adj.* **1** *Optics.* **a** having or showing one color only. **b** (of light or other radiation) consisting of a single wavelength: *monochromatic X rays. The use of monochromatic light, for example sodium light, does not improve the acuity of the eye to more than a very small extent* (H. Hartridge). **c** producing light or radiation of a single wave length.
2 *Physics.* (of a beam of subatomic particles) consisting of particles that all have the same energy.
—**monochromaticity,** *n.* the condition or extent of being monochromatic: *the monochromaticity of laser light.*

monochromator (mon′ə krō′mā′tər), *n.* a device for producing monochromatic beams of light, particles, etc.

monoclinal (mon′ə klī′nəl), *Geology.* —*adj.* **1** (of strata) dipping or sloping in one direction and at the same angle. **2** of or having to do with strata that dip in the same direction: *monoclinal valleys, a monoclinal ridge.*
—*n.* = monocline.
[from *mono-* + Greek *klinein* to slope, bend]
—**monoclinally,** *adv.*

monocline (mon′ə klīn), *n. Geology.* a monoclinal rock formation or fold, such as the oblique portion of a belt of strata at the place where it changes from one horizontal position to another of different level: *The beds in a monocline may dip at angles ranging from a few degrees to ninety degrees* (M.P. Billings). See the picture at **fold.**

monoclinic (mon′ə klin′ik), *adj. Crystallography.* characterized by three unequal axes with one oblique intersection. See the picture at **crystal.**
ASSOCIATED TERMS: see **crystal.**

monoclinous (mon′ə klī′nəs), *adj. Botany.* **1** (of a plant) having both stamens and pistils in the same flower. **2** (of a flower) having both stamens and pistils. Compare **diclinous.** [from French *monocline,* from Greek *monos* one + *klinē* bed, from *klinein* to slope, bend]

monocyclic

monoclonal (mon′ə klō′nəl), *Biology.* —*adj.* of or derived from a single clone; having to do with a group of cells arising by cell division from a single cell: *Milstein and Köhler developed a hybridoma that made a specific antibody. They called this type of antibody monoclonal because it was produced by a clone of one type of B-lymphocyte* (Michael Shodell).
—*n.* = monoclonal antibody.

monoclonal antibody, *Biology.* an antibody produced in the laboratory by fusing genetically distinct cells and cloning the resulting hybrids so that each hybrid cell produces the same antibody: *To identify the various types of nerve cells, the scientists made monoclonal antibodies, which react only to specific molecules on a cell. They fused white blood cells, each bearing a specific antibody, with cancerous cells that reproduce rapidly. These hybrid cells made clones, or multiple copies, of themselves* (George Adelman). Compare **hybridoma.**

monocot (mon′ə kot), *n.* = monocotyledon. Compare **dicot.**

monocotyledon (mon′ə kot′l ēd′n), *n. Botany.* any of a class (Monocotyledoneae) of flowering plants of the angiosperm subdivision, having a single cotyledon in the embryo. Monocotyledons typically have leaves with parallel veins and flower parts in units of three. They include the grasses, palms, lilies, irises, etc. *Botanists are impressed with the fact that the Polynesians divided the plant world into monocotyledons and dicotyledons* (*plants with single-leaved and double-leaved seed embryos*) *long before the 17th-century English naturalist John Ray began the classification of plants in Europe* (Donald S. Marshall).
—**monocotyledonous** (mon′ə kot′l ēd′n əs), *adj.* having only one cotyledon: *The roots of monocotyledonous plants lack cambiums, while those of conifers and dicotyledonous plants possess lateral meristems* (Harbaugh and Goodrich, *Fundamentals of Biology*).

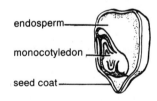

monocyclic (mon′ə sī′klik), *adj.* **1** *Biology.* having a single whorl or series of parts, as certain crinoids with a single circlet of basal plates: *When the members of a series* (*calyx, corolla, etc.*) *are in one whorl, the series is said to be monocyclic* (Sidney H. Vines).

---

cap, fāce, fäther; best, bē, tèrm; pin, five;
rock, gō, ôrder; oil, out; cup, pùt, rüle,
yü in use, *yù* in uric;
ng in bring; *sh* in rush; *th* in thin, TH in then;
zh in seizure.
ə = *a* in about, *e* in taken, *i* in pencil, *o* in lemon, *u* in circus

413

2 *Chemistry.* having a molecular structure with a single ring (of atoms): *a monocyclic terpene, monocyclic compounds.* SYN: mononuclear.

3 *Geology.* having undergone a single cycle of erosion: *a monocyclic landscape.*

monocyte (mon′ə sīt), *n. Biology.* one of the major types of leucocytes in the blood, being the largest in size and comprising about 3 to 8 per cent of the total white blood cells. It is a phagocyte with a single, well-defined nucleus. *Granulocytes and monocytes engulf and digest microbes* (Science News).
—**monocytic,** *adj.* having to do with or characteristic of a monocyte: *monocytic leukemia.*

monodactylous (mon′ə dak′tə ləs), *adj. Zoology.* having a single digit, toe, or claw. [from Greek *monodaktylos,* from *monos* single + *daktylos* finger]

monoecious (mə nē′shəs), *adj.* **1** *Botany.* having the stamens and the pistils in separate flowers on the same plant: *The gametophytes of many mosses are monoecious; that is, both antheridia and archegonia are produced by the same gametophyte* (Weier, *Botany*).
2 *Zoology.* having both male and female organs in the same individual; hermaphroditic: *In many monoecious species the spermatozoa are produced first and later the ova, but in some species this condition is reversed* (Shull, *Principles of Animal Biology*). Also spelled **monecious.**
[from *mono-* + Greek *oikos* house]

monogenesis, *n. Biology.* **1** the theory that many living things derive from a single common ancestor. **2** asexual reproduction. Contrasted with **polygenesis.**

monogenetic, *adj.* **1** *Zoology.* having a single host during the life cycle, as certain trematode worms.
2 *Geology.* formed by a single continuous process or condition: *monogenetic soil.*

monogenic, *adj.* **1** *Genetics.* **a** controlled by a single gene: *a monogenic recessive.* **b** producing offspring of only one sex.
2 *Zoology.* **a** reproducing by only one method. **b** = monogenetic (def. 1).

monogerm, *adj. Botany.* being or containing a single seed which develops into an isolated plant: *monogerm sugar beet varieties.*

monohybrid (mon′ə hī′brid), *Genetics.* —*adj.* having parents whose genetic makeup differs in only one pair of alternative inheritable characters, such as tall versus short: *a monohybrid plant.*
—*n.* a monohybrid organism. Also called **monohybrid cross.** Compare **dihybrid.**

monohydric, *adj. Chemistry.* **1** containing one replaceable atom of hydrogen. **2** = monohydroxy.

monohydroxy, *adj. Chemistry.* having one hydroxyl (-OH) group, as an alcohol.

monolayer, *n.* **1** *Chemistry.* a monomolecular layer.
2 *Biology.* a single layer, especially a layer of cells in a culture.

monomer (mon′ə mər), *n. Chemistry.* a molecule or compound that can combine with others to form a polymer. Vinyl chloride is the monomer from which polyvinyl chloride is produced. *Most monomers are more reactive, chemically, than the corresponding polymers and cannot, with impunity, be treated in the*

same way (New Scientist). [from *mono-* + Greek *meros* part]
—**monomeric** (mon′ə mer′ik), *adj.* of or like a monomer: *Each giant molecule is formed by long sequences of monomeric units* (Giulio Natta).

monomerous (mə nom′ər əs), *adj. Botany.* (of a flower) having one member in each whorl (sometimes written *l-merous*). Compare **dimerous, trimerous.** [from Greek *monomerēs,* from *monos* single + *meros* part]

monomial (mō nō′mē əl), *n.* **1** *Mathematics.* an expression in algebra consisting of a single term; a number, a variable, or a product of numbers and variables. The expressions 4, $2x^2$, 3ax, and 3a/b are monomials. Compare **binomial, trinomial.**
2 *Biology.* the scientific name of an organism consisting of a single term.
—*adj.* consisting of a single term.
[from *mono-* + *(bi)nomial*]

monomolecular, *adj. Chemistry.* **1** that is one molecule in thickness: *a monomolecular layer or film.*
2 having to do with or composed of a single molecule or molecules: *a monomolecular unit, monomolecular reactions.* —**monomolecularly,** *adv.*

monomorphic, *adj. Biology.* having only one form or genotype; having the same form throughout development: *a monomorphic species; monomorphic with regard to chromosome structure.*
—**monomorphism,** *n.* monomorphic condition or character.
—**monomorphous,** *adj.* = monomorphic.

mononuclear, *adj.* **1** *Biology.* having only one nucleus: *mononuclear cells.* SYN: mononucleate.
2 = monocyclic (def. 2).

mononucleate, *adj. Biology.* having only a single nucleus.

mononucleotide (mon′ə nü′klē ə tīd), *n. Biochemistry.* a nucleotide consisting of a molecule each of a sugar, a phosphoric acid, and a purine or pyrimidine base. Compare **dinucleotide, trinucleotide.**

monopetalous (mon′ə pet′ə ləs), *adj. Botany.* **1** having the corolla composed of united petals, as in the morning-glory; gamopetalous. **2** (of a corolla) having only a single petal.

monophagous (mə nof′ə gəs), *adj. Zoology.* eating only one kind of food.
—**monophagy** (mə nof′ə jē), *n.* the habit of feeding on only one kind of food.

monophyletic (mon′ə fī let′ik), *adj. Biology.* having to do with or descending from a single, common ancestral form, usually resembling the existing group: *monophyletic species. At present, most taxonomists agree that the largest group of plants which is definitely monophyletic is the family* (Weier, *Botany*). Contrasted with **polyphyletic.**

monophyllous (mon′ə fil′əs), *adj. Botany.* **1** consisting of one leaf: *a monophyllous calix.* **2** having only one leaf. [from Greek *monophyllos,* from *monos* single + *phyllon* leaf]

monoploid (mon′ə ploid), *Genetics.* —*adj.* **1** having a single set of chromosomes in the germ cell or gamete; haploid. **2** having the basic haploid chromosome number in a polyploid series.
—*n.* a monoploid organism or cell.
[from *mono-* + *-ploid,* as in *haploid, diploid*]

monopodial (mon′ə pō′dē əl), *adj. Botany.* of or resembling a monopodium.

monopodium (mon′ə pō′dē əm), *n., pl.* **-dia** (-dē ə). *Botany.* a single main axis which continues to extend at the apex in the original line of growth, producing lateral branches beneath, such as the trunk of a pine tree. [from New Latin *monopodium,* from Greek *monos* single + *pous, podos* foot]

monopole, *n.* = magnetic monopole: *In the theory of electromagnetism the monopole is complementary to the electron. Just as the electron is characterized by an electric charge, so does the monopole have a magnetic charge ... Because of the symmetry the monopole brings to electromagnetism, most physicists have been confident that sooner or later one would be found* (Scientific American).

monosaccharide (mon′ə sak′ə rīd), *n. Biochemistry.* any of a class of sugars, such as glucose and fructose, that cannot be decomposed by hydrolysis: *The basic unit of the carbohydrate is the monosaccharide* (Biology Regents Syllabus). *The monosaccharides and disaccharides can be further combined to form polysaccharides, among which are the starches, celluloses, and pentosans found in plants* (Harbaugh and Goodrich, *Fundamentals of Biology*).
ASSOCIATED TERMS: see **disaccharide.**

monosepalous (mon′ə sep′ə ləs), *adj. Botany.* **1** having the sepals united; gamosepalous. **2** having only one sepal: *a monosepalous calyx.*

monosome (mon′ə sōm), *n.* **1** *Genetics.* **a** an unpaired sex chromosome. **b** an individual having less than the usual number of chromosomes, especially a diploid that is missing one chromosome from the set; a monosomic individual. **2** *Biochemistry.* a single, isolated ribosome: *Many of the ribosomes of the cell are bound together in aggregates, called polyribosomes, which behave as physical units and can be separated from the free ribosomes, or monosomes* (New Scientist).
—**monosomic** (mon′ə sō′mik), *adj. Genetics.* having less than the usual number of chromosomes; having or being a diploid that is missing one chromosome.
—**monosomy** (mon′ə sō′mē), *n. Genetics.* the condition of being monosomic: *In mammals, cases of monosomy (the presence of only one chromosome of a pair) has never been reported in any but the sex-determining chromosomes* (A. Donald Merritt).

monospermous or **monospermal,** *adj. Botany.* containing or producing only one seed.

monostele (mon′ə stēl or mon′ə stē′lē), *n.* = protostele.

monostichous (mə nos′tə kəs), *adj. Botany.* arranged in a single vertical row on one side of an axis, as flowers. [from Greek *monostichos* made up of one row, from *monos* single + *stichos* row, line]

monostomous (mə nos′tə məs or mon′ə stō′məs), *adj. Zoology.* having a single mouth or mouthlike part: *Many jellyfish are monostomous.* [from Greek *monostomos,* from *monos* one + *stoma* mouth]

monostylous (mon′ə stī′ləs), *adj. Botany.* having only one style.

monosubstituted, *adj. Chemistry.* having a single substituent: *a monosubstituted amine.*

monosymmetric, *adj. Botany.* bilaterally symmetrical; zygomorphic. —**monosymmetrically,** *adv.*

monosynaptic (mon′ə si nap′tik), *adj. Biology.* involving or consisting of a single synapse: *The simplest reflex pathway in the spinal cord ... is a monosynaptic, or two-neuron, arc consisting of a fiber from a sensory neuron forming a synapse with a motoneuron* (Scientific American).

monotone, *adj.* = monotonic.

monotonic (mon′ə ton′ik), *adj. Mathematics.* either never increasing or never decreasing as a function: *a monotonic quantity. Every bounded monotonic sequence of real numbers is convergent* (E.T. Copson).
—**monotonically,** *adv.* without ever increasing or decreasing as a function.

monotypic, *adj. Biology.* being the sole representative of its group, especially a genus having only one species: *Zoologists tell us that a race is simply a recognizable division of what they call a polytypic species. A species is either a single population (monotypic) or a string of related populations (polytypic)* (Atlantic).

monovalent (mon′ə vā′lənt), *adj.* **1** *Chemistry.* having a valence of one; univalent.
2 *Biology.* **a** containing antigens from a single strain of a microorganism: *a monovalent vaccine or serum.* **b** (of an antibody or antigen) having only one site of attachment: *a monovalent agglutinin.*

monoxide (mə nok′sīd), *n. Chemistry.* an oxide containing one oxygen atom in each molecule.

monozygotic (mon′ə zī got′ik), *adj. Biology.* produced by the splitting of a single fertilized ovum or embryonic cell mass: *Identical twins are monozygotic.* Compare **dizygotic.**

monsoon (mon sün′), *n. Meteorology.* **1** a seasonal wind of the Indian Ocean and southern Asia, blowing from the southwest from April to October and from the northeast during the rest of the year.
2 the rainy season during which this wind blows from the southwest, usually accompanied by heavy rains. [from earlier Dutch *monssoen,* from Portuguese *monçao,* from Arabic *mausim* season]

montane (mon′tān), *adj. Ecology.* **1** of, having to do with, or being the zone of plant growth on mountains below the subalpine zone.
2 of or having to do with the plants and animals of this zone: *Changes in composition of montane plant communities during the Holocene were relatively minor because these communities adjusted quickly to climatic changes* (Science). [from Latin *montanus,* from *montem* mountain]

Monte Carlo (mon′ti kär′lō), *Mathematics, Physics.* of or having to do with a computational technique that uses random samples and other statistical methods for finding solutions to mathematical or physical problems: *To evaluate the relative probabilities of these two processes, we made a Monte Carlo study that included cascade protons from the struck nucleus in the spark*

---

cap, fāce, fäther; best, bē, tėrm; pin, five;
rock, gō, ôrder; oil, out; cup, pùt, rüle;
yü in use, *yù* in uric;
ng in bring; *sh* in rush; *th* in thin, *ᴛʜ* in then;
zh in seizure.
ə = *a* in about, *e* in taken, *i* in pencil, *o* in lemon, *u* in circus

chamber (taken to the aluminium), inelastic pion production, and the relativistic fragments of the incident nucleus (Nature). [named after *Monte Carlo*, Monaco, noted as a gambling resort; so called from the element of chance in Monte Carlo procedures]

month, *n. Astronomy.* **1** the time it takes the moon to make one complete revolution around the earth; the time from one new moon to the next, about 29 1/2 days. Also called **lunar month, synodical month.**
2 one twelfth of a solar year, about 30.41 days. Also called **solar month.**
3 the time it takes the moon to make one revolution around the earth in relation to the stars, about 27 1/3 days. Also called **sidereal month.**

monticule (mon′tə kyül), *n. Geology.* **1** a minor cone of a volcano. **2** a little mound or hillock. [from Middle French *monticule,* from Late Latin *monticulus,* diminutive of Latin *montem* mountain]

montmorillonite (mont′mə ril′ə nīt), *n. Mineralogy.* any of a group of mineral clays, hydrous and hydrated silicates of aluminum and certain other elements, used because of its absorbent structure for various industrial purposes: *Until recently called montmorillonite ... it is now also known as smectite, a word descriptive of a layered structure. Its microcrystals are extremely fine-grained, thin-layered and lacking in regular outlines. The layers are not tightly bound one to another and so water, numerous elements and organic matter can enter the spaces between layers* (Scientific American). [from *Montmorillon,* town in western France where it is found + *-ite*]
—montmorillonitic, *adj.* having to do with or belonging to the montmorillonites: *a montmorillonitic clay.*

monzonite (mon′zə nīt), *n. Geology.* an igneous rock composed of nearly equal amounts of plagioclase and orthoclase containing sodium and other minerals, intermediate in composition between syenite and diorite. [from German *Monzonit,* from *Monzoni,* a mountain in Tyrol + *-it* -ite]
—monzonitic (mon′zə nit′ik), *adj.* of or consisting of monzonite.

moon, *n. Astronomy.* **1** a celestial body that revolves around the earth from west to east once in approximately 27 1/3 days at a mean distance of 384,403 kilometers. It is a natural satellite of the earth that reflects the sun's light and is held in orbit by the earth's gravity. The moon's diameter is about 3475 kilometers, its volume is about 1/50 that of the earth, and its mass is about 1/80 that of the earth. The force of the moon's gravity on the earth causes tides in the ocean. *There are countless craters on the Moon, ranging in size from the great basins of 185 miles (300 kilometers) diameter to pits in rocks that can be seen only with a microscope ... On the near side of the Moon there are more than three hundred thousand craters exceeding 0.6 miles (1 kilometer) in diameter* (von Braun and Ordway).
2 the moon as it looks at a certain period of time in its cycle: *When the sun and moon are in conjunction, the illuminated hemisphere of the moon will face directly away from the earth; at this phase the moon is new. About two weeks later when the moon is at opposition, the illuminated hemisphere of the moon faces directly*

toward the earth; at this phase the moon is full (Krogdahl, *The Astronomical Universe*).
3 a natural satellite of any planet: *the moons of Jupiter.*

moonquake, *n. Astronomy, Geology.* a quake or series of vibrations on the moon similar to an earthquake: *About 3,000 moonquakes are detected each year. All of them are very weak by earth standards. The average moonquake releases about as much energy as a firecracker* (Bevan M. French).

moor, *n. Ecology.* a tract of high-lying open wasteland covered with peat, usually with heather or coarse grasses growing on it.
ASSOCIATED TERMS: see bog.

moraine (mə rān′), *n. Geology, Geography.* a mass or ridge of earth, rocks, etc., deposited at the sides or end of a glacier. The material is scraped up by the glacier as it moves along. *Ice transports rock materials in a jumbled up mixture and drops them, upon melting, in an unsorted and irregular fashion. Such deposits are known as glacial drift, till, or moraine. If this is spread over an area as a more or less regular sheet it is called ground moraine; if it is piled up in an irregular ridge along a line which was once the end position of a glacier, it is called a terminal moraine* (White and Renner, *Human Geography*). [from French *moraine*]
—morainal or **morainic,** *adj.* of or having to do with a moraine: *morainal rock.*

MORAINE

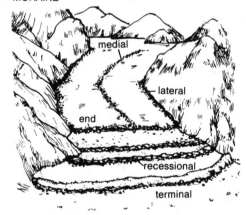

medial

lateral

end

recessional

terminal

morass, *n. Ecology.* a piece of low, soft, wet ground. SYN: marsh, swamp. [from Dutch *moeras*]

morning star, *Astronomy.* the planet Venus, as seen in the eastern sky before or at sunrise: *Venus, the familiar morning and evening star, is the brightest of the planets* (Baker, *Astronomy*).

morph (môrf), *n. Biology.* a variant form of a species of organism: *There were also differences between the various "morphs", i.e., winged or wingless, viviparous or oviparous* (New Scientist). [from Greek *morphē* form]

morphactin (môr fak′tin), *n. Biochemistry.* any of a group of compounds derived from fluorine and carboxylic acid that affect the morphogenesis of higher plants, inhibiting plant growth when applied in certain concentrations or promoting the development of ovaries into fruit without pollination. [from *morph(ogenesis)* + *act(ive)* + *-in*]

416

morphogenesis, *n. Biology.* the origin and evolution of morphological characteristics; the growth and differentiation of cells and tissues during development: *Among aspects of plant morphogenesis are ... differentiation, regeneration, and abnormal development* (Greulach and Adams, *Plants*).
—**morphogenetic,** *adj.* of or having to do with morphogenesis: *the morphogenetic effects of light on plant cells.* —**morphogenetically,** *adv.*
—**morphogenic,** *adj.* = morphogenetic.

morphological, *adj.* of or having to do with morphology; relating to form: *morphological changes. Taxonomy defines morphological categories and not identities nor biochemical similarities* (N.W. Pirie).
—**morphologically,** *adv.* from a morphological point of view; as regards morphology: *The grass family is morphologically very different from most other families of flowering plants* (Bernard N. Bowden).

morphology, *n.* **1** the branch of biology that deals with the form and structure of organisms without regard to function: *Studies in comparative morphology ... make it possible to trace the derivation of appendages of vertebrates from lateral folds on the bodies of lower chordates to the fins of sharks and bony fishes* (Storer, *General Zoology*).
2 the form and structure of an organism or one of its parts: *the morphology of a cell or a tumor, the morphology of vertebrates.*
3 the study of forms in any science, as in meteorology and geology: *The more we learn about the morphology, or form, of hurricanes, and their evolution, the better shall we be able to predict their effects on human life* (R.W. James).
[from Greek *morphē* form + English *-logy*]

morphometry (môr fom′ə trē), *n. Geography, Geology.* the measurement of the external form or shape of a natural object: *the morphometry of the Lake of Geneva.*

morphosis (môr fō′sis), *n., pl.* **-ses** (-sēz). *Biology.* the manner of formation or development of an organism or part. [from Greek *morphōsis,* from *morphoun* to form, from *morphē* form]

morula (môr′yù lə), *n., pl.* **-lae** (-lē). *Embryology.* the spherical mass of blastomeres forming the embryo of many animals, just after the segmentation of the ovum and before the formation of a blastula: *In the majority of mammals, the result of cleavage is a solid mass of cells, a morula, in which some cells are superficial and others lie inside, completely cut off from the surface by the enveloping cells* (B. I. Balinsky). [from New Latin *morula,* from Latin *morum* mulberry]
—**morulation,** *n.* the formation of a morula.

mosaic (mō zā′ik), *n. Biology.* an organism or part having adjacent cells or tissues of different genetic type; an individual exhibiting mosaicism. Also called **chimera.**
—**mosaicism** (mō zā′ə siz əm), *n.* the presence of different or antagonistic genetic characteristics in adjacent cells or tissues of the body, chiefly due to faulty cell division: *Mosaicism may be a recessive trait and appear only in organisms that inherit the trait from both parents, or it may show up with the greater frequency characteristic of a dominant trait* (Scientific American).

moss, *n. Botany.* **1** any of a class (Mucopsida) of very small, soft, green or brown bryophytic plants that grow close together in clumps or like a carpet on the ground, on rocks, or on trees. Mosses have small stems and numerous, generally narrow leaves.
2 any of various other plants which grow close together like mosses: *Many groups of plants contain members that are commonly called "mosses" (reindeer "mosses" are lichens, club "mosses" and Spanish "moss" are vascular plants, and sea "moss" and Irish "moss" are algae)* (Raven, *Biology of Plants*). [Old English *mos* bog, (later) moss]

Mössbauer effect (mes′bou ər or mœs′bou ər), *Physics.* the recoilless, elastic emission of gamma rays by radioactive nuclei of crystalline solids, and the subsequent absorption of the emitted rays by other nuclei. The Mössbauer effect is widely used in spectroscopic studies of chemical bonds, the composition of molecules, the properties of atomic motions, etc. *The application of Mössbauer effect to the study of biological molecules ... now occupies a secure place among the many physical methods used for such studies. In recent years it has played a useful role in determining the nature of several enzymes, and has provided clues about the oxygen-carrying function of hemoglobin* (George Lang). [named after Rudolf L. *Mössbauer,* born 1929, German physicist who discovered this effect in 1957]

mother, *adj.* producing others; being the source or origin; parent, as in:

mother cell, *Biology.* the cell from which daughter cells are formed by cell division.

mother cloud, *Meteorology.* the cloud from which the funnel of a tornado descends: *The cloud above it, or the mother cloud, was so low, and the funnel was so wide, that one witness described the tornado as a "turbulent, boiling mass of blackness"* (James E. Miller).

mother element, *Physics.* the radioactive element whose decay gives rise to a daughter element: *... they could record the alpha particles emitted by the "mother" element 105 atoms ... and thus aid in the detection of the "daughter" element 103 (lawrencium-256) atoms* (Albert Ghiorso).
—*n. Chemistry.* a slimy, filmy substance formed on the surface of an alcoholic liquid undergoing fermentation, as in making vinegar or milk products. The substance consists of yeast and bacterial cells whose enzymes promote fermentation. Also called **mother of vinegar.**

mother liquor, *Chemistry.* the liquid that remains after the removal of crystallizing substances from a solution.

---

cap, fāce, fäther; best, bē, tėrm; pin, fīve;
rock, gō, ôrder; oil, out; cup, pùt, rüle;
yü in use, *yu* in uric;
ng in bring; sh in rush; th in thin, ᴛʜ in then;
zh in seizure.
ə = *a* in about, *e* in taken, *i* in pencil, *o* in lemon, *u* in circus

417

mother-of-pearl, *n. Zoology.* the hard, smooth, glossy lining of certain mollusk shells, as of the pearl oyster, mussel, and abalone. It changes colors as the light changes. Also called **nacre.**

motile (mō'tl), *adj. Biology.* moving or able to move by itself; capable of spontaneous movement: *motile flagella, motile cells, motile spores. Those [algae] which are not attached are often motile ... following a jerky zigzag course across the field of view of the microscope* (Emerson, *Basic Botany*). [from Latin *motum* moved] —**motility** (mō til'ə tē), *n.* motile quality or condition.

motion, *n. Physics.* a change of position of a particle, body, or other object: *Particles of atomic size have kinetic energy arising from several different kinds of motion. All atoms are constantly in motion* (Helen M. Davis). See the picture at **pulley.**

   Newton's laws of motion state that (1) an object remains at rest or in uniform motion unless acted upon by an unbalanced force; (2) an unbalanced force acting on an object causes an acceleration which is directly proportional to the force, inversely proportional to the mass of the object, and in the same direction as the force; (3) if one object exerts a force on a second, the second exerts a force on the first that is equal in magnitude and opposite in direction. Another way of stating this third law is that for every action there is an equal and opposite reaction. See the picture at **equilibrium.** [from Latin *motionem,* from *movere* to move]
ASSOCIATED TERMS: Motion in a straight line is *rectilinear;* motion along a curved line is *curvilinear.* The rate of motion of a body in a particular direction is its *velocity. Acceleration* is a change in the velocity of a body. A decrease in velocity is called *negative acceleration* or *deceleration.* The *momentum* of a body in motion is its *mass* times its *velocity. Friction* is the resistance of a body in motion to the medium through which it travels or to the surface on which it moves. See also *potential energy, kinetic energy,* and *perpetual motion.*

motoneuron (mō'tə nur'on), *n. Physiology.* a motor neuron: *Nerve impulses generated by the motoneuron activate the muscle to which the stretch receptor is attached* (Scientific American).

motor, *adj. Physiology.* **1** (of nerves or nerve fibers) conveying or imparting an impulse from the central nervous system to a muscle or organ which results in motion or activity: *a motor neuron.* SYN: efferent. See the pictures at **afferent, reflex.**
2 (of muscles, impulses, centers, etc.) concerned with or involving motion or activity: *Furthermore, they were able to show by direct electrical stimulation that the basal lobes ... are "motor areas"* (New Scientist).
3 of or designating the effect of stimuli from the central nervous system causing motion or activity: *motor coordination. 'Motor activity' is an essential factor in perception* (George M. Wyburn).
4 of or having to do with any activity involving muscular movement: *a motor response or reflex.* [from Latin *motor* mover, from *movere* to move]

moulin (mü län'), *n. Geology.* a nearly vertical shaft or cavity worn in a glacier by surface water falling through a crack in the ice. [from French *moulin* (literally) a mill]

mound, *n. Geography, Geology.* a small or low rounded hill of earth. SYN: knoll, hillock.

mount, *n. Geography.* a mountain (used with an initial capital before a proper name): *Mount McKinley, Mount Rushmore. Abbreviation:* Mt.

mountain, *n. Geography, Geology.* a very high hill; a natural elevation of the earth's surface rising high above the surrounding level, usually higher than 600 meters. *Abbreviation:* mt.

mountain chain, *Geography, Geology.* **1** a connected series of mountains or mountain ranges. **2** = mountain system.

mountain range, *Geography, Geology.* a row of connected mountains; large group of mountains.

mountain system, *Geography, Geology.* a group of geographically related mountain ranges. Compare **cordillera.**

mouth, *n.* **1** *Anatomy.* **a** the opening through which a person or animal takes in food; cavity containing the tongue and teeth. **b** the opening to any cavity or canal in an organ or organism.
2 *Geography.* **a** an opening resembling a mouth and providing an entrance or exit: *the mouth of a cave, canyon, volcano,* etc. **b** the part of a river or the like where its waters are emptied into the sea, another river, or some other body of water: *the mouth of the Ohio River.*

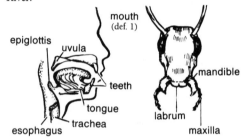

mouthbreeder, *n. Zoology.* any of various small fishes that hold their eggs and their hatched young in their mouths: *The African freshwater cichlid ... lays its eggs and then picks them up and broods them in its mouth* (Barbara N. Benson).

moutonnée (mü'tə nā'), *adj. Geology.* (of rocks) rounded and smoothed by glacial action. Compare **roche moutonnée.** [from French *moutonnée* (literally) rounded like a sheep's back]

moving cluster, *Astronomy.* a galactic cluster close enough to earth so that its relative motion can be easily measured. The Hyades form a moving cluster.

m.p., *abbrev.* melting point.

mpm or **m.p.m.,** *abbrev.* meters per minute.

mr, *abbrev.* milliroentgen.

mrad, *abbrev.* millirad.

mrem, *abbrev.* millirem.

mRNA, *abbrev.* messenger RNA: *mRNA in DNA may also be the basis for the timing cycle that regulates cell activity* (Jacob Kastner).

ms, *abbrev.* millisecond.

m/s, *abbrev.* meters per second.

msec, *abbrev.* millisecond.

M-shell, *n. Physics.* the energy level next in order after the L-shell, occupied by electrons whose principal quantum number is 3. See also **M Electron.**

m.s.l., *abbrev.* mean sea level.

mt., *pl.* **mts.** *abbrev.* mountain.

Mt., *abbrev. Geography.* Mount: *Mt. Everest.*

M.T., *abbrev.* metric ton.

muc-, *combining form.* the form of **muco-** before vowels, as in *mucic acid, mucin, mucoid.*

mucic acid (myü′sik), *Chemistry.* a white, crystalline, dibasic acid formed by oxidizing lactose or some other carbohydrate with nitric acid. *Formula:* $C_6H_{10}O_8$

mucilage (myü′sə lij), *n. Botany.* any of various sticky, gelatinous secretions present in various plants such as seaweeds. [from Late Latin *mucilago* musty juice, from Latin *mucus* mucus]
—**mucilaginous** (myü′sə laj′ə nəs), *adj.* containing or secreting mucilage: *The heterocysts ... have thick cell walls and are surrounded by a mucilaginous envelope that retards the diffusion of oxygen into the cell* (Scientific American).

mucin (myü′sin), *n. Biochemistry.* any of various mucoproteins forming the chief constituents of mucous secretions.

muco-, *combining form.* mucus; mucous, as in *mucoprotein, mucolytic.* [from Latin *mucus* slime, mucus]

mucoid (myü′koid), *Biochemistry.* —*adj.* of or resembling mucus.
—*n.* any of a group of glycoproteins resembling mucin. They occur in bone and connective tissue, and in the vitreous humor and cornea of the eye.

mucolytic (myü′kə lit′ik), *adj. Biochemistry.* that dissolves, liquefies, or disperses mucus: *mucolytic agents or enzymes.* [from *muco-* + Greek *lytikos* loosening]

mucopeptide (myü′kō pep′tīd), *n. Biochemistry.* a polymeric substance consisting of chains of mucopolysaccharides and peptides, found especially in the cell walls of many bacteria: *The cell envelope, the outer portion of the bacterial cell, is a complex structure consisting of an inner plasma membrane and a rigid mucopeptide layer, the cell wall proper, that confers strength and shape* (Scientific American). Also called **murein.**

mucopolysaccharide (myü′kō pol′ē sak′ə rīd), *n. Biochemistry.* a carbohydrate compound, such as heparin, containing amino sugar and sugar acids, found mainly in the connective tissue but also present in mucous tissue and synovial fluid: *The blood group to which each of us belongs is determined by a presence ... of a mucopolysaccharide with specific antigenic qualities* (New Scientist).

mucoprotein, *n. Biochemistry.* any one of various viscous protein compounds, such as mucin, containing a mucopolysaccharide in their molecular structure, and occurring in connective tissue and other body tissues.

mucosa (myü kō′sə), *n., pl.* **-sae** (-sē). = mucous membrane. [from New Latin (*membrana*) *mucosa* mucous (membrane)]
—**mucosal,** *adj.* of, having to do with, or characteristic of a mucous membrane: *The mucosal epithelium consists of 2 types of cylindrical cells forming a single layer: (1) absorptive cells, which are narrow, and (2) goblet cells, which contain a large, oval vacuole and which produce mucus* (Hegner and Stiles, *College Zoology*).

mucous (myü′kəs), *adj. Physiology.* 1 of or like mucus: *a mucous substance.* 2 containing or secreting mucus: *a mucous cell.*

mucous membrane, *Anatomy.* the lining of the nose, throat, digestive tract, and other passages and cavities of the body that are open to the air. It consists of tissue containing glands that secrete mucus. Also called **mucosa.** ▶ See the note under **membrane.**

mucro (myü′krō), *n., pl.* **mucrones** (myü′krō nēz). *Biology.* a sharp point; spinelike part: *the projecting mucro of a leaf.* [from Latin *mucro* point, sharp edge]
—**mucronate** (myü′krə nit *or* myü′krə nāt), *adj.* having a sharp point: *a mucronate shell, feather, or leaf.*
—**mucronation** (myü′krə nā′shən), *n.* 1 mucronate condition or form. 2 a mucronate process.

mucronulate (myü kron′yə lāt *or* myü kron′yə lit), *adj. Biology.* having a small mucro or abruptly projecting point, as a leaf.

mucus (myü′kəs), *n. Physiology.* a viscid, slimy substance, consisting chiefly of mucin, that is secreted by and moistens and protects the mucous membranes. [from Latin *mucus* slime, mucus]

mud, *n. Geology.* 1 soft, sticky, wet earth or debris produced by rain, springs, riverbeds, etc. 2 a mixture of water, clay, and silt.

mudbank, *n. Geology.* a bank or shoal of mud beside or rising from the bed of a river, lake, or sea.

mudflat, *n. Geology.* a stretch of low-lying, muddy land along a shore or island, usually submerged at high tide and left uncovered at low tide.

mudflow, *n. Geology.* 1 a flowing mass of fine-grained earth material.
2 a landslide of mud following spring thaws or heavy rain.
3 = lahar.

muff, *n. Zoology.* 1 a tuft or crest on the heads of certain birds.
2 a cluster of feathers on the side of the face of certain breeds of fowls. [from Dutch *mof,* from French *moufle* mitten]

MUFF (def. 1)

cap, fāce, fäther; best, bē, tėrm; pin, fīve;
rock, gō, ôrder; oil, out; cup, pùt, rüle,
yü in use, yủ in uric;
ng in bring; sh in rush; th in thin, ₮H in then;
zh in seizure.
ə = a in about, e in taken, i in pencil, o in
lemon, u in circus

mulch, *Ecology.* —*n.* a natural or artificial layer of loose material, such as straw, leaves, manure, and grass spread on the ground around trees or plants and serving to protect the roots from cold or heat to prevent evaporation of moisture from the soil, to check weed growth, to decay and enrich the soil itself, or to keep fruit clean.

—*v.* to cover (ground) with straw or leaves; spread mulch under or around (a tree or plant).

[perhaps Middle English *molsh* soft, Old English *melsc* mellow, sweet]

mull (mul), *n. Ecology.* a moist, granular, well-aerated forest humus that gradually blends with the mineral soil below and is conducive to plant growth: *Mull is usually rich in calcium and is less acid than other kinds of soil* (David Madge). [from German *Mull*]

Mullerian duct or Müllerian duct (myü lir′ē ən), *Embryology.* the embryonic genital or reproductive tract of the female in vertebrate animals, corresponding to the mesonephric duct of the male. [named after Johannes *Müller*, 1801-1858, German anatomist] Contrast **Wolffian duct.**

Müllerian mimicry, *Zoology.* a form of protective mimicry in which the mimicking animal is as inedible or disagreeable to predators as the animal it mimics. [named after Fritz *Müller*, 1821-1897, German naturalist]

mullite (mul′it), *n. Mineralogy.* an orthorhombic silicate of aluminum that is resistant to corrosion and heat, commonly used in artificial melting processes. Mullite is very rare in nature, but common as a synthetic constituent of ceramic products. *Formula:* $A1_6Si_2O_{13}$ [from *Mull*, island in the Hebrides, Scotland + *-ite*]

multicellular, *adj. Biology.* having or consisting of many cells: *The transport of material within cells and/or throughout multicellular organisms is circulation* (Biology Regents Syllabus).

—**multicellularity,** *n.* the condition of having many cells.

multicostate (mul′ti kos′tāt), *adj. Biology.* having many ribs or ridges.

multidentate (mul′ti den′tāt), *adj. Biology.* having many teeth or toothlike processes.

multifactorial, *adj. Genetics.* involving, dependent on, or controlled by a number of genes. Compare **multiple factors.** Contrasted with **unifactorial.** —**multifactorially,** *adv.*

multiflorous (mul′ti flôr′əs), *adj. Botany.* bearing many flowers.

multifoliate, *adj. Botany.* having many leaves or leaflets.

multilobate, *adj. Botany.* having many lobes: *a multilobate leaf.*

multilocular (mul′ti lok′yə lər), *adj. Biology.* having many chambers or cells.

multinomial, *adj., n.* = polynomial.

multinucleate or multinucleated. *adj. Biology.* having two or more nuclei: *the multinucleate cells of striated muscle.*

multiparous (mul tip′ər əs), *adj.* **1** *Biology.* producing two or more at a birth.

2 *Botany.* (of a cyme) having many axes.

[from New Latin *multiparus,* from Latin *multus* much, many + *parere* bring forth]

multiped, *Zoology.* —*adj.* having many feet. —*n.* a multiped animal. [from Latin *multipedem,* from *multus* many + *pedem* foot]

multiple, *n.* **1** *Mathematics.* a number into which another number can be divided an integral number of times without a remainder: *12 is a multiple of 3.*

2 = multiple star: *The orbital statistics of binaries and multiples have always been a problem. As our survey showed, there is a wide range in separation distances, from much less to much more than the size of the solar system* (W. K. Hartmann).

[from French *multiple,* from Late Latin *multiplus* manifold]

multiple alleles, *Genetics.* a group of three or more allelic genes, only two of which can be present at one time in the body cells of a diploid organism.

multiple factors, *Genetics.* hypothetical combinations of two or more pairs of genes which act together to produce a trait, such as size or skin pigmentation, or the variations in it.

multiple fruit, *Botany.* a fruit composed of a cluster of ripened ovaries produced by several flowers, as the mulberry. Also called **collective fruit, compound fruit.** ASSOCIATED TERMS: see **aggregate fruit.**

multiple star. *Astronomy.* a group of three or more stars comprising one gravitational system, and usually appearing to the naked eye as a single star.

multiplet (mul′tə plit), *n. Physics.* **1** two or more closely associated lines in a spectrum, exhibiting characteristic differences, as of frequency.

2 two or more elementary particles that are similar in properties such as mass and spin but have different charges: *The various possible permutations group the hadrons in what are called multiplets, thus bringing some order out of their chaos* (Science News).

multiplicand (mul′tə plə kand′), *n. Mathematics.* a number or quantity to be multiplied by another: *In 5 times 497, the multiplicand is 497.* [from Latin *multiplicandus,* gerundive of *multiplicare* to multiply]

multiplication, *n. Mathematics.* the operation that, for integers, consists of finding a number (the *product*) by adding together as many instances of one number (the *multiplicand*) as there are units in another number (the *multiplier*), or of calculating this addition briefly. The multiplication of other numbers and quantities is analogous to the multiplication of integers.

multiplication sign, *Mathematics.* either the symbol X or a centered dot (·), used to indicate the operation of multiplying. EXAMPLES: $3 \times 4 = 12$; $3 \cdot 4 = 12$.

multiplicative (mul′tə plə kā′tiv), *adj. Mathematics.* of, having to do with, or involving multiplication.

multiplicative inverse, *Mathematics.* the reciprocal of a given number. EXAMPLE: 2/3 is the multiplicative inverse of 3/2.

multiplier *n.* **1** *Mathematics.* a number by which another number is to be multiplied: *In 83 multiplied by 5, the multiplier is 5.*

2 *Physics.* an instrument or device used for intensifying by repetition intensity of a force, current, etc.

multiply, *v. Mathematics.* to find a product by multiplication. [from Old French *multiplier,* from Latin *multiplicare,* from *multus* many + *-plex* -fold]

multipolar, *adj.* having more than one pole or process; having several poles: *multipolar electromagnetic radiation, multipolar spindles in mitosis. Bipolar cells have one dendrite and one axone; multipolar cells have a multiple dendrite and single axone* (Storer, *General Zoology*).

multivalence, *n.* multivalent quality.

multivalent (mul'ti vā'lənt), *adj.* **1** *Chemistry.* **a** having a valence of three or more. **b** having more than one valence. SYN: polyvalent.
2 *Biology.* of or involving the association of three or more chromosomes during the first division of meiosis.
3 *Immunology.* (of an antigen or antibody) having several sites of attachment.

multivariate, *adj. Statistics.* having or involving two or more variables: *Through use of multivariate statistical techniques, abundance variations of foraminifera and other marine microfossils (coccoliths, diatoms, radiolaria) can provide quantitative estimates of past oceanographic parameters, such as sea-surface temperature and salinity* (Joseph J. Morley).

multivoltine (mul'ti vōl'tēn), *adj. Zoology.* that produces several broods in a single season: *a multivoltine moth or other insect.* [from *multi-* many, several + Italian *volta* a turning, turn + English *-ine*]

mu-meson, *n.* an earlier name of the muon. ▶ See the note under **meson.** [from *mu* the 12th letter of the Greek alphabet used as an arbitrary designation in a series + *meson*]

muon (myü'on), *n. Nuclear Physics.* an unstable lepton with a mass about 207 times that of the electron. Muons are the chief components of cosmic radiation on the earth's surface. They are formed by the decay of pions and in turn decay to form high-energy electrons. [from *mu*(*-meson*) + *-on*] ▶ See the note under **meson.**
—muonic (myü on'ik), *adj.* containing or producing muons: *Extensive studies of the sizes and shapes of nuclei have been made using muonic atoms* (New Scientist).

muonium (myü on'ē əm), *n. Nuclear Physics.* a short-lived particle consisting of a positively charged muon bound to a single electron: *The positive muon has been observed to collect a negative electron when coming to rest in pure gases, and the resulting "atom" muonium, has been studied with great precision* (Leon M. Lederman).

muramic acid (myù ram'ik), *Biochemistry.* a lactic acid derivative found in the mucopeptide of the cell walls of bacteria and blue-green algae: *Muramic acid occurs in all bacteria and bacteria-like forms, and is derived from a sugar. It has attached to it an acid group which serves as a useful hook on which to add other molecules* (Howard Rogers). [from Latin *murus* wall + English *am*(*ine*) + *-ic*]

murein (myùr'ē ən *or* myùr'ēn), *n.* = mucopeptide. [from *mur*(*amic acid*) + *-ein* (as in *protein*)]

muriform (myùr'ə fôrm), *adj. Botany.* resembling or suggesting a wall made of bricks arranged in courses: *muriform cellular tissue.* [from Latin *murus* wall + English *form*]

muscle, *n. Anatomy.* **1** the tissue in the body of people and animals that can be tightened or loosened to make the body move. Muscles contract in response to nerve stimuli. Muscle tissue is composed of bundles of fibers

muscular

and is of two general types, striated muscle and smooth muscle. *Muscles are sensitive to stretch and automatically (reflexly) adjust their activity to changes in tension or stretch* (Science News).
2 an organ consisting of a special bundle of such tissue which moves some particular bone, part, or substance. The biceps muscle bends the arm. The heart muscle pumps blood.
[from French *muscle,* from Latin *musculus,* diminutive of *mus* mouse; so called from the appearance of certain muscles]
ASSOCIATED TERMS: The human body has over 600 major muscles. The two main types of muscles are the *skeletal muscles* and the *smooth muscles.* A third type, the *cardiac muscle* or *myocardium* has characteristics of both the skeletal and smooth muscles. Skeletal muscles are also called *striated muscles* or *voluntary muscles.* These muscles are joined to bones by *tendons.* The fixed attachment of a muscle during *contraction* is called the *origin,* while the more movable attachment is called the *insertion.* Skeletal muscles operate in pairs, one bending a limb or part (*flexor*), the other stretching it (*extensor*). Smooth muscles are also called *involuntary muscles* because they operate automatically. They are not *striated* and have only one *nucleus.* Smooth muscles, as those that move food in the stomach and intestines, are stimulated by nerves of the *autonomic nervous system.* Muscles are contracted or stimulated by the sliding action of filaments in muscle fiber made up of the proteins *actin* and *myosin.* The energy for this action is provided by the chemical *ATP* (adenosine triphosphate).

(def. 1)

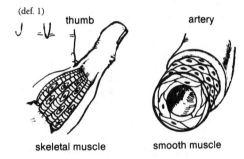

skeletal muscle   smooth muscle

muscle plasma, = plasma (def. 1b).

muscle spindle, *Anatomy.* a sensory end organ that is attached to a muscle and is sensitive to stretching: *Impulses from the muscle spindle ... excite the motoneuron, impulses from which cause contraction of the muscle* (Scientific American). Also called **stretch receptor.**

muscular, *adj. Anatomy.* **1** of the muscles; influencing the muscles: *muscular contraction.*
2 consisting of muscle: *muscular tissue.*
—muscularity, *n.* muscular development or strength.
—musculature (mus'kyə lə chər), *n.* the system or arrangement of muscles.

---

cap, fāce, fäther; best, bē, tėrm; pin, five;
rock, gō, ôrder; oil, out; cup, pùt, rüle,
yü in use, *yù* in uric;
ng in bring; *sh* in rush; *th* in thin, ŦH in then;
zh in seizure.
ə = *a* in about, *e* in taken, *i* in pencil, *o* in lemon, *u* in circus

421

mutagen (myü′tə jən), *n. Genetics.* an agent that causes mutation in an organism: *What sort of molecule could function as a specific mutagen, a reagent for a particular one of the bacterium's complement of genes* (Joshua Lederberg).
—**mutagenic** (myü tə jen′ik), *adj.* of or having to do with a mutagen; able to cause mutation: *The different types of mutations induced by one mutagenic agent ... are also produced to some extent by any other* (Bulletin of Atomic Scientists).
—**mutagenicity** (myü′tə jə nis′ə tē), *n.* mutagenic quality; ability to cause mutation: *Concern [was] felt by many scientists about the potential mutagenicity of the ever increasing number of synthetic chemicals in the environment* (Nature).

mutagenize (myü′tə jə nīz), *v. Genetics.* to cause or induce mutation in: *Their approach was systematically to make individual tests on each of several hundreds of colonies from a heavily mutagenized stock of E. coli* (Science Journal).

mutant (myü′tənt), *Genetics.* —*n.* a new genetic character or variety of plant or animal resulting from mutation: *Mutants in the T4 strain of virus (one of a group of viruses that attack bacteria) are easily isolated, and their recombinations can be detected, even in extremely low frequency, by a selective technique* (Sheldon C. Reed).
—*adj.* that is the result of mutation: *If the template theory is correct, a mutant gene may produce a mutant enzyme—an enzyme whose structure and properties are changed in some way* (N. H. Horowitz).

mutarotation (myü′tə rō tā′shən), *n. Chemistry.* a gradual change of optical rotation taking place in freshly prepared solutions of certain compounds, especially reducing sugars. [from Latin *mutare* to change + English *rotation*]
—**mutarotate,** *v.* to exhibit mutarotation: *Monosaccharides commonly mutarotate.*
—**mutarotational,** *adj.* of or characterized by mutarotation: *mutarotational forms of mannose.*

mutase (myü′tās), *n. Biochemistry.* any enzyme that promotes molecular rearrangement in a substance, especially one that causes the transfer of a phosphate group from one carbon atom to another in the same molecule. [from Latin *mutare* to change + English -ase]

mutate (myü′tāt), *n. Genetics.* to undergo or cause to undergo mutation: *Following infection of the bacterial cell, both the cell and the virus could mutate, changing their resistance and infectious characteristics* (Caroline G. Dudley).

mutation, *n. Genetics.* **1** a change within a gene or chromosome of animals or plants resulting in the appearance of a new, inheritable feature or character: *Mutation is the antithesis of heredity. Heredity insures the basic continuity and coherence of the organic form; mutation creates a diversity of genes, and thereby is ultimately responsible for the great diversity of living beings on the earth. The mutation process, in short, supplies the building blocks of evolution* (Theodosius Dobzhansky).

2 a new genetic character or new variety of plant or animal formed in this way; mutant: *It is not possible to avoid all of the environmental factors that may cause cancer and ... the same natural genetic processes producing mutations that led to the evolution of our species are also responsible for the phenomenon we call cancer ... and indeed, mutations, good and bad, have in effect dictated the course of evolution* (Harper's).
3 the process of forming a mutation on the molecular level: *Three classes of mutations result from introducing defects in the sequence of bases ... In one class, a base pair is simply changed from one into another (i.e. G-C to A-T). In the second class, a base pair is inserted (or deleted). In the third class, a group of base pairs is deleted (or inserted)* (James D. Watson). [from Latin *mutationem,* from *mutare* to change] ASSOCIATION TERMS: see **evolution.**
—**mutational,** *adj.* of or having to do with mutation: *Through a number of mutational steps, ... it has been possible to obtain a thousand-fold increase in penicillin yield* (Bulletin of Atomic Scientists). —**mutationally,** *adv.*

muticous (myü′tə kəs), *adj. Biology.* without awns or spines; awnless; spineless. [from Latin *muticus*]

muton (myü′ton), *n. Genetics.* the smallest genetic unit capable of causing mutation: *The muton ... may be as small as ... a few nucleotide pairs of the chromosomal nucleic acid* (New Scientist). [from *mut(ation)* + -*on*]

mutual inductance, *Physics.* a measure of the inductive effect of the magnetic fields produced by two circuits on each other.

mutualism, *n. Biology.* a relationship between two organisms in which they benefit each other: *Some insect species are always found in association with a certain fungus, and some fungi only with a certain insect. Such complete interdependence is called mutualism.* (S.W.T. and L.R. Batra). ASSOCIATED TERMS: see **symbiosis.**
—**mutualistic,** *adj.* involved in or characterized by mutualism: *mutualistic partners.*

muzzle, *n. Zoology.* the projecting part of the head of an animal, including the nose, mouth, and jaws. [from Old French *musel,* from *muse* muzzle]

mv or **mV,** *abbrev.* millivolt.

Mv, *symbol.* mendelevium.

mw or **mW,** *abbrev.* milliwatt.

Mw, *abbrev.* megawatt.

m.y. or **my,** *abbrev. Geology.* million years: *Professor Allen suggests that the North Atlantic began as a branching rift valley, similar to the present African one, between 120 and 130 my ago* (New Scientist).

mycelial (mī sē′lē əl), *adj. Biology.* of or having to do with the mycelium.

mycelium (mī sē′lē əm), *n., pl.* **-lia** (-lē ə). *Biology.* **1** the vegetative part of a fungus, consisting of one or more white, interwoven filaments or hyphae: *For many species of toadstools, these mycelia are perennial, remaining in the soil and producing "fruit" year after year* (Science News Letter).
2 a similar mass of fibers formed by some higher bacteria.
[from New Latin *mycelium,* from Greek *mykēs* mushroom, fungus]

mycobacterial, *adj. Bacteriology.* having to do with or caused by mycobacteria: *mycobacterial infections.*

mycobacterium (mī'kō bak tir'ē əm), *n., pl.* **-teria** (-tir'ē-ə). *Bacteriology.* any of a genus (*Mycobacterium*) of aerobic, rod-shaped bacteria. One species causes leprosy; certain other species cause tuberculosis in man, cattle, and fowl; all are acid-fast. *Many types of mycobacteria occur as free-living organisms in soil, water, and dairy products* (R.G. Krueger). [from New Latin *Mycobacterium* the genus name, from Greek *mykēs* fungus + New Latin *bacterium* bacterium]

mycology (mī kol'ə jē), *n.* **1** the branch of biology that deals with fungi.
2 the fungi of a particular region or country: *The African mycology is remarkable for the varied forms it produces among the puffballs and allied genera* (Lindley, *The Vegetable Kingdom*).
3 facts about a particular fungus: *the mycology of rusts and mildews.*
[from Greek *mykēs* fungus + English *-logy*]

mycophagous (mī kof'ə gəs), *adj. Zoology.* that feed on fungi. [from Greek *mykēs* fungus + *phagein* eat]

mycoplasma (mī'kō plaz'mə), *n., pl.* **-mas, -mata** (-mə-tə). *Bacteriology.* any of a group of extremely small and primitive bacteria that lack rigid cell walls and are gram-negative. They were formerly called **pleuropneumonialike organisms** (PPLO): *Many mycoplasmas live parasitically or saprophytically in association with animals or plants* (E. J. Braun and W. Sinclair). [from New Latin *Mycoplasma* the genus name, from Greek *mykēs* fungus + New Latin *plasma* plasma]

mycorrhiza (mī'kə rī'zə), *n. Biology.* the symbiotic association of the mycelium of certain fungi with the roots of certain higher plants, living in close relationship with the surface cells. [from Greek *mykēs* fungus + New Latin *-rrhiza,* from Greek *rhiza* root]

mycosis (mī kō'sis), *n., pl.* **-ses** (-sēz). *Biology.* **1** the presence of parasitic fungi in or on any part of the body.
2 a disease caused by such fungi.
[from Greek *mykēs* fungus + English *-osis*]
—**mycotic** (mī kot'ik), *adj.* of or having to do with mycosis.

mycotoxin (mī'kə tok'sən), *n. Biology.* a poison produced by a fungus: *The mycotoxins enter the food chain and accumulate in man through his use of milk, eggs, poultry or meat* (Daily Telegraph).

myelencephalon (mī'ə len sef'ə lon), *n.* **1** *Embryology.* the posterior part of the hindbrain, which comprises the medulla oblongata; afterbrain.
2 *Anatomy.* the brain and spinal cord as a whole.
[from Greek *myelos* marrow + New Latin *encephalon* vertebrate brain]

myelin (mī'ə lən), *n. Biochemistry.* a soft, whitish, fatty substance that forms a sheath about the core of certain nerve fibers: *Myelin ... is analogous in function to the insulation on wire; in its absence the conduction of nerve impulses along the axon is disrupted* (Scientific American). [from German *Myelin,* from Greek *myelos* marrow]
—**myelinated** (mī'ə lə nā'tid), *adj.* covered or surrounded by myelin: *myelinated nerve fibers.*

—**myelination** (mī'ə lə nā'shən), *n.* the sheathing of nerve fibers; formation of myelin: *Myelination ... leads to lower levels of excitability and more mature function in the brain* (Science News Letter).

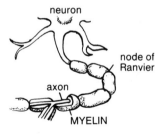

myelin sheath, *Anatomy.* the layer of myelin around the core of a nerve fiber. Also called **medullary sheath.**

myeloblast (mī'ə lə blast), *n. Biology.* a bone-marrow cell in its early stages; a rudimentary myelocyte. [from Greek *myelos* marrow + *blastos* sprout]
—**myeloblastic,** *adj.* having to do with myeloblasts: *myeloblastic leukemia.*

myelocyte (mī'ə lə sīt), *n. Biology.* an ameboid blood cell present in bone marrow and causing the development of leucocytes. [from Greek *myelos* marrow + English *-cyte*]
—**myelocytic** (mī'ə lə sit'ik), *adj.* of or having to do with myelocytes.

myelogenic (mī'ə lə jen'ik), *adj. Physiology.* originating or produced in the marrow.

myeloid (mī'ə loid), *adj. Physiology.* of, having to do with, or resembling marrow: *myeloid leukemia.*

myo-, *combining form.* muscle, as in *myocardium, myoneural.* [from Greek *myos* muscle]

myoblast (mī'ə blast), *n. Biology.* a muscle cell in its early stages; a cell which develops into a myocyte: *Embryonic muscle tissue is composed primarily of two specialized cell types: (1) fibroblasts, which lay down the connective-tissue framework of the muscle, and (2) myoblasts, which become the contractile muscle cells themselves* (Scientific American). [from *myo-* + Greek *blastos* sprout]

myocardial, *adj. Anatomy.* of or having to do with the myocardium: *a myocardial infarction or aneurysm resulting from a blood clot blocking a coronary blood vessel.*

myocardium (mī'ə kär'dē əm), *n. Anatomy.* the muscle tissue of the heart: *Heart muscle, like any other tissue in the body, dies when deprived of its blood supply, and the portion of the myocardium ... that dies and is replaced by a scar when its circulation is cut off ...* (Newsweek). [from New Latin *myocardium,* from Greek *myos* muscle + *kardia* heart]

---

cap, fāce, fäther; best, bē, tėrm; pin, fīve;
rock, gō, ôrder; oil, out; cup, pùt, rüle;
yü in use, *yù* in uric;
ng in bring; sh in rush; th in thin, ŦH in then;
zh in seizure.
ə = a in about, e in taken, i in pencil, o in lemon, u in circus

myofibril (mī′ə fī′brəl), *n. Histology.* a striated fibril of a muscle fiber: *The striations [of the fiber] arise from a repeating variation in the density, i.e., the concentration of protein along the myofibrils* (Scientific American).

myogenic (mī′ə jen′ik), *adj. Physiology.* **1** arising from the muscles: *myogenic contractions.*
2 producing or forming muscle: *myogenic cells.*

myoglobin (mī′ə glō′bin), *n. Biochemistry.* a protein similar to hemoglobin and present in muscle cells, that takes oxygen from the blood and stores it for future use. [from *myo-* + (*hemo*)*globin*]

myology (mī ol′ə jē), *n.* the scientific study of the structure, functions, and diseases of muscles.

myoneural (mī′ə nùr′əl), *adj. Anatomy.* having to do with both muscle and nerve, especially with nerve endings in muscle tissue: *The impulses activate myoneural junctions ... and cause the effector (muscle) cells to contract* (Elbert Tokay).

myopia (mī ō′pē ə), *n. Optics.* near-sightedness; an abnormal condition of the eye in which only objects close to the eye produce distinct images because parallel rays of light are brought to a focus before they reach the retina. [from New Latin *myopia,* from Greek *myōps,* from *myein* to shut + *ōps* eye]

myosin (mī′ə sin), *n. Biochemistry.* one of two protein components of muscle cells important in the elasticity and contraction of muscles. The other is actin. *Myosin is especially abundant: about half the dry weight of the contractile part of the muscle consists of myosin.* (H. E. Huxley).

myotome (mī′ə tōm), *n.* **1** *Embryology.* the part of an embryonic segment that differentiates into skeletal muscle.

2 *Anatomy.* a muscle supplied by a spinal nerve. [from *myo-* + Greek *tomos* cut, slice]

myrmecophile (mèr′mə kə fil), *n. Zoology.* a myrmecophilous insect: *Certain beetles are known as myrmecophiles.*

myrmecophilous (mèr′mə kof′ə ləs), *adj. Zoology.* **1** fond of or living with ants. Myrmecophilous insects live in anthills.
2 benefited by ants, such as plants that are cross-fertilized by them. [from Greek *myrmēx, myrmēkos* ant + *philos* loving]

myxameba or **myxamoeba** (mik′sə mē′bə), *n., pl.* **-bas, -bae** (-bē). *Biology.* a slime mold at the stage when it is an amebalike free-swimming swarm spore and before it fuses to form a plasmodium: *The myxamoebae ... now begin to congregate about a central point forming a mound of cells easily visible to the naked eye* (New Scientist). [from New Latin *myxamoeba,* from Greek *myxa* mucus + New Latin *amoeba*]

myxobacteria (mik′sə bak tir′ē ə), *n. pl.* of **myxobacterium** (mik′sə bak tir′ē əm). *Bacteriology.* a group of saprophytic bacteria that secrete a slime over which they swarm to form extensive colonies: *The myxobacteria ... are widely distributed in the sea and in fresh water, and in cow-dung and similar places on land* (Listener). Also called **gliding bacteria.** [from New Latin *Myxobacteria,* from Greek *myxa* mucus + New Latin *bacteria*]

myxomycete (mik′sō mī sēt′), *n. Biology.* any one of the slime molds that grow on damp soil and decaying vegetable matter. [from New Latin *Myxomycetes* the group name, from Greek *myxa* mucus, slime + *mykēs, mykētos* fungus, mushroom]

myxovirus (mik′sə vī′rəs), *n. Biology.* any of a group of viruses that contain RNA and agglutinate red blood cells, including the viruses which cause influenza and mumps. [from Greek *myxa* mucus + English *virus*]

MZ, *abbrev.* monozygotic.

N

n, *abbrev.* or *symbol.* **1** an indefinite number. **2** nano-. **3** negative. **4** neutron. **5** *Chemistry.* (of a solution) normal. **6** *Genetics.* the gametic (haploid) number of chromomes, as opposed to 2n (the somatic or diploid number), 3n (triploid number), etc. **7** principal quantum number.

n-, *prefix. Chemistry.* (of a hydrocarbon) normal. It is usually italicized, as in *n*-butane.

N, *abbrev.* or *symbol.* **1** newton. **2** nitrogen. **3** *Chemistry.* (of a solution) normal.

N-, *prefix. Chemistry.* (of groups in organic compounds) attached to nitrogen. In N-methyl acetamide, the methyl group is attached to the nitrogen atom in the acetamide molecule.

Na, *symbol.* sodium. [for Latin *natrium*]

NAA, *abbrev.* neutron activation analysis.

nacre (nā′kər), *n.* = mother-of-pearl. [from Middle French *nacre,* from Italian *nacchera,* ultimately from Arabic *naggārah* drum]
—**nacreous** (nā′krē əs), *adj.* of or like nacre.

nacreous cloud, *Meteorology.* an iridescent cloud resembling a cirrus, seen only in northern latitudes some 15 to 20 miles above earth: *At sunset nacreous clouds display all the colors of the spectrum increasing in brilliance as the sky darkens* (Donald T. Sanders). Compare **noctilucent cloud.**

nacrite (nā′krīt), *n. Mineralogy.* a clay mineral consisting of hydrous silicate of aluminum. It is polymorphous with kaolinite and dickite. *Formula:* $Al_2Si_2O_5(OH)_4$ [from *nacre* + *-ite*]

NAD, *abbrev.* nicotinamide adenine dinucleotide.

NADH, *symbol. Biochemistry.* the reduced form of nicotinamide adenine dinucleotide. [from *NAD* + *H*(ydrogen)]

nadir (nā′dər), *n. Astronomy.* the point in the heavens directly beneath the place where one stands; point opposite the zenith. [from old French *nadir,* from Arabic *nazir* opposite to (the zenith)]

NADP, *abbrev.* nicotinamide adenine dinucleotide phosphate.

NADPH, *abbrev. Biochemistry.* the reduced form of nicotinamide adenine dinucleotide phosphate. [from *NADP* + *H*(ydrogen)]

nail, *n. Anatomy.* the hard layer of horn on the upper side of the end of a finger or toe in man or a claw or talon in other vertebrates. Nails are a modified form of epidermis.

naked, *adj.* **1** *Botany.* **a** (of seeds) not enclosed in a case or ovary; having no pericarp, such as the seeds of a pine. **b** (of flowers) without a calyx or corolla. **c** (of stalks or branches) without leaves. **d** (of stalks or leaves) free from hairs; smooth; glabrous.
2 *Zoology.* lacking hair, feathers, shell, or other natural covering.

naked (def. 1a, b)

pine cone   lizard tail flower

stamen

pistil

wing

seed

nanism (nā′niz əm), *n. Biology.* (in animals and plants) abnormally small size or stature; dwarfishness. [from French *nanisme,* from Latin *nanus* dwarf (from Greek *nanos*) + French *-isme* -ism]

nannofossil (nan′ə fos′əl), *n. Paleontology.* a very small or microscopic fossil: *Up to 60 percent of rock consists of nannofossils, relics of the smallest kind of plankton* (Science News). Also spelled **nanofossil.** [from *nanno-,* variant of *nano-* + *fossil*]

nannoplankton (nan′ə plangk′tən), *n. Biology.* very small or the smallest plankton, including one-celled plants and animals: *Ocean sediments are commonly soft oozes made up of minute skeletons of the small organisms called foraminifera and nannoplankton* (Scientific American). Also spelled **nanoplankton.**

nano-, *combining form.* **1** one billionth (10^{-9}) of any measurement, as in *nanosecond = a billionth of a second.* Symbol: n
2 very small; dwarf, as in *nanoplankton = a very small plankton.*
[from Greek *nanos, nannos* dwarf]

---

cap, fāce, fäther; best, bē, tėrm; pin, five;
rock, gō, ôrder; oil, out; cup, pùt, rüle;
*y*ü in use, *y*ù in uric;
ng in bring; *sh* in rush; *th* in thin, ᴛʜ in then;
zh in seizure.
ə = *a* in about, *e* in taken, *i* in pencil, *o* in lemon, *u* in circus

nanofossil, *n.* = nannofossil.

nanometer (nan′ə mē′tər), *n.* a billionth (10⁻⁹) of a meter: *Usually points closer than one nanometre (10⁻⁹m) can be separated with the electron microscope, whereas with the best light microscope resolution rarely approaches 200nm* (Science Journal). *Abbreviation:* nm

nanoplankton, *n.* = nannoplankton.

nanosecond (nan′ə sek′ənd), *n.* a billionth (10⁻⁹) of a second. *Abbreviation:* ns

naphtha (naf′thə), *n. Chemistry.* any of several highly volatile and flammable liquid mixtures of hydrocarbons distilled from petroleum, coal tar, and natural gas, used as fuel, as solvents, and in making various chemicals. [from Greek *naphtha* (originally) a flammable liquid issuing from the earth]

naphthalene (naf′thə lēn), *n. Chemistry.* a white, crystalline hydrocarbon distilled from coal tar or petroleum, used as an intermediate and in making fungicides, solvents, preservatives, etc. *Formula:* $C_{10}H_8$

napiform (nā′pə fôrm), *adj. Botany.* shaped like a turnip; large and round above and slender below: *a napiform root.* [from Latin *napus* turnip + English *form*]

nappe (nap), *n.* **1** *Geometry.* either of the two equal parts of a conical surface which join at the vertex to form a cone.
2 *Geology.* a large, sheetlike body of rock that has been moved far from its original position.
[from French *nappe* sheet]

narcotic, *n. Medicine.* any drug that produces dullness, drowsiness, sleep, or an insensible condition, and lessens pain by dulling the nerves. Opium and drugs made from it are powerful narcotics. Narcotics are used in medicine in controlled doses, but taken in excess cause systemic poisoning, delirium, paralysis, or even death. SYN: opiate, anodyne. [from Greek *narkōtikos,* from *narkoun* to benumb, from *narkē* numbness]

nares (nār′ēz), *n.pl.* of **naris** (nār′is). *Anatomy.* nostrils. [from Latin]

narrows, *n. Geography.* the narrow part of a river, strait, sound, valley, or pass.

nasal, *adj. Anatomy.* of, in, or from the nose: *nasal bones, the nasal cavity.* See the picture at **skull.** [from Latin *nasus* nose] **—nasally,** *adv.*

nascent (nas′ənt *or* nā′sənt), *adj. Chemistry.* **1** having to do with the state or condition of an element at the instant it is set free from a combination.
2 (of an element) being in a free or uncombined state: *Hydrogen produced in solution by the action of a metal on an acid acts to reduce many soluble oxidizing agents. When so produced and used, hydrogen is usually described as nascent* (Jones, *Inorganic Chemistry*).
[from Latin *nascentem,* present participle of *nasci* be born]

nasopharyngeal (nā′zō fə rin′jē əl), *adj. Anatomy.* of or having to do with the nose and pharynx or with the nasopharynx.

nasopharynx (nā′zō far′ingks), *n. Anatomy.* the upper pharynx; the part of the pharynx above the soft palate that is continuous with the nasal passages.

nastic (nas′tik), *adj. Botany.* of or having to do with movement or growth of cellular tissue on one surface more than on another, as in the opening of petals or young leaves: *Nastic movements are among a plant's more beautiful motions: a typical example is the opening of a flower. They are the result of differing responses of different parts of the plant structure to the same external stimulus* (Scientific American). [from Greek *nastos* pressed together]

natatorial (nā′tə tôr′ē əl), *adj. Zoology.* **1** adapted for swimming: *Fins and flippers are natatorial organs.*
2 characterized by swimming: *Fishes are natatorial animals.*
[from Late Latin *natatorialis,* from Latin *natator* swimmer, from *natare* to swim]

native, *adj.* **1** *Biology.* originating, grown, or produced in a particular region: *... The Sino-Malayan ants introduced with the ship traffic into Oahu have radically exterminated all native ants from the lowlands of this Hawaiian island ... The more widely living, the more vital immigrants, pushed out the less vital native species* (F.S. Bodenheimer). SYN: indigenous.
2 *Chemistry, Geology.* found chemically uncombined in nature: *native copper.* **Native metals** include gold, silver, platinum, copper, iron, lead, mercury, iridium, and palladium.
—n. *Biology.* an animal or plant originating in a particular region: *The zinnia plant is a native of Mexico.*

natural, *adj.* **1** *Chemistry.* (of an element or other substance) occurring in nature; not synthetic or artificial: *Most chemical elements are natural, but many radioactive elements are produced artificially. Wood and petroleum are natural substances.*
2 *Biology.* **a** not acquired or conditioned: *natural immunity, a natural reflex.* **b** (of plants) not introduced artificially.
3 *Mathematics.* having 1 as the base of the system (applied to a function or number belonging or referred to such a system). Natural numbers are those that are positive integers, such as 1, 2, 3, and so on. Natural sines, cosines, tangents, and other functions are those taken in arcs whose radii are 1. They are not expressed as logarithms, but as the actual value of a ratio (of two sides of a right triangle) in which the hypotenuse of the triangle is taken as unity.

natural bridge, *Geology.* a rock formation shaped like a bridge and spanning a ravine or valley, especially one made by water working its way slowly through soft rock.

natural frequency, *Physics.* the frequency at which a body vibrates naturally, according to set wave patterns within itself: *If the rhythm of the externally imposed alternating voltage coincides with the very high natural frequency of the quartz plate, the maximum effect, Resonance, is obtained, and the vibrations thus produced have enormous and astounding effects* (Gabriele Rabel).

natural gas, *Chemistry.* a combustible mixture of hydrocarbons formed naturally in the earth, consisting primarily of methane, with smaller amounts of ethane, butane, propane, and nitrogen: *The gas is merely tapped or pumped from the natural gas wells which are usually found near coal and petroleum deposits. An occasional well will yield a gas that is practically pure methane* (Offner, *Fundamentals of Chemistry*).

natural history, 1 the study of animals, plants, minerals, and other things in nature, especially popular study in the field rather than scientific study in the laboratory: *Besides such common knowledge there are many items of natural history known to certain people, such as songs, habits, manner of nesting, and other details about birds or other conspicuous animals* (Storer, *General Zoology*).
2 a collection of facts about the development of some natural object or process: *the natural history of an insect, fossil, etc., the natural history of the physical sciences.*

naturalize, *v. Biology.* to introduce and establish (animals or plants) in another region; establish as native: *The English oak has been naturalized in parts of Massachusetts.*

natural logarithm, *Mathematics.* a logarithm which has as a base the irrational number *e*, whose value is approximately equal to 2.71828183+, used in analytical work.

natural number, *Mathematics.* a positive integer. The natural numbers are 1, 2, 3, 4, etc.

natural resources, *Ecology.* materials supplied by nature that are useful or necessary for life. Minerals, timber, land, and water power are natural resources.

natural science, 1 any science dealing with the objects, phenomena, or laws of nature or the physical world. Biology, geology, physics, and chemistry are natural sciences.
2 all these sciences and their branches collectively.

natural selection, *Biology.* the process in nature by which organisms best adapted to their environment tend to survive and to transmit their genetic characters; differential reproduction of different genetic types within populations, subspecies, or species: *This preservation of favourable variations and the rejection of injurious variations, I call Natural Selection* (Darwin, *Origin of Species*).
ASSOCIATED TERMS: see **evolution.**

nauplial (nô′plē əl), *adj.* having the character of a nauplius.

nauplius (nô′plē əs), *n., pl.* **-plii** (-plē ī). *Zoology.* the first stage of development of certain crustaceans, such as the shrimp, after leaving the egg; a larval form with an unsegmented body, three pairs of appendages, and a single median eye: *Larval life for the rock barnacle lasts about three months ... At first, the larva, a little swimming creature called a nauplius, is almost indistinguishable from the larva of any other crustacean* (New Yorker). [from Latin *nauplius* a kind of shellfish, from Greek *nauplios*]

navel, *n. Anatomy.* the mark or scar in the middle of the surface of the abdomen, where the umbilical cord was attached before birth. It is usually a puckered depression. Also called **umbilicus.** [Old English *nafela*]

navicular (nə vik′yə lər), *Anatomy.* —*adj.* resembling a boat in shape, as certain bones. SYN: scaphoid.
—*n.* **1a** a bone of the human foot between the talus and the metatarsals. **b** a bone of the human wrist shaped like a comma. It is one of the eight short bones that make up this joint.
2 a small, oval bone in the foot of a horse.
[from Late Latin *navicularis* of a ship, from Latin *navicula,* diminutive of *navis* ship]

Nb, *symbol.* niobium.

Nd, *symbol.* neodymium.

Ne, *symbol.* neon.

neap, *adj. Oceanography.* of, having to do with, or being a neap tide.
—*n.* = neap tide: *At London Bridge the tidal range at average spring tides is 21 feet and at average neaps it is 15 feet* (New Scientist).
[Old English *nēp*]

neap tide, *Oceanography.* a tide that occurs when the difference in height between high and low tide is least; lowest level of high tide. Neap tide comes twice a month, in the first and third quarters of the moon. Compare **spring tide.** See the picture at **tide.**

Nearctic (nē ärk′tik *or* nē är′tik), *adj. Ecology.* having to do with the region including temperate and arctic North America and Greenland, especially with reference to the distribution of animals. Compare **Palearctic.** [from Greek *neos* new + English *arctic*]

near point, *Optics.* the point nearest to the eye at which an image is clearly formed on the retina when there is maximum accommodation: *The range of accommodation gradually diminishes with age as the crystalline lens loses its flexibility. For this reason the near point gradually recedes as one grows older* (Sears and Zemansky, *University Physics*). Compare **far point.**

Nebraskan, *adj. Geology.* of or having to do with the first period of glaciation in North America, beginning about 1,200,000 years ago, and lasting about 64,000 years. Compare **Illinoian, Kansan, Wisconsin.**

nebula (neb′yə lə), *n., pl.* **-lae** (-lē′) **-las.** *Astronomy.* a cloudlike cluster of stars or a hazy mass of dust particles and gases which occurs in interstellar space and which may be either dark or illuminated by surrounding stars. **Galactic nebulae** are clouds of gas and dust particles within our galaxy. **Extragalactic nebulae** are galaxies outside the Milky Way. *Now that these* [spiral] *nebulae are well established as stellar systems outside our own, we shall henceforth call them galaxies* (D. W. Sciama). [from Latin *nebula* mist, cloud]
—**nebular,** *adj.* of or having to do with a nebula or nebulae: *The nebular system contains all those nebulae some at least of whose characteristics can be measured and studied* (G.C. McVittie).

nebular hypothesis, *Astronomy.* any of various theories which maintain that the solar system developed from a nebula, especially Laplace's theory that the planets of the solar system were formed from matter produced by the cooling and contracting of a hot, rotating nebula. Compare **planetesimal hypothesis.**

nebulosity (neb′yə los′ə tē), *n. Astronomy.* cloudlike matter; nebula: *nebulosities in the Milky Way.*

nebulous (neb′yə ləs), *adj. Astronomy.* of or like a nebula or nebulae.

---

cap, fāce, fäther; best, bē, tėrm; pin, fīve;
rock, gō, ôrder; oil, out; cup, pùt, rüle;
yü in use, yù in uric;
ng in bring; sh in rush; th in thin, ŦH in then;
zh in seizure.
ə = a in about, e in taken, i in pencil, o in
lemon, u in circus

427

neck, *n.* 1 *Anatomy.* **a** the part of the body that connects the head with the shoulders. **b** a slender or constricted part of a bone or organ. SYN: cervix. **c** the part of a tooth between the crown and the root.
2 *Zoology.* a long siphon occurring in certain mollusks: *the neck of a clam.*
3 *Geography.* **a** a narrow strip of land: *a canal across a neck of land.* SYN: isthmus, peninsula. **b** a narrow strip of water. SYN: strait.
4 *Geology.* the erosional remnant of a pipelike vent of an extinct volcano, filled with solidified lava: *Volcanic necks are found where old volcanoes have been worn away. The name also may be used for the columns of igneous rock left as magmas worked their way upward to form laccoliths.* (Fenton, *The Rock Book*).

(def. 2)

beak

neck

foot

necrophagia (nek′rə fā′jē ə), *n. Zoology.* the practice or habit of feeding on dead bodies or carrion. [from New Latin *necrophagia,* from Greek *nekros* dead body + *phagein* eat]
—**necrophagous** (ne krof′ə gəs), *adj.* feeding on dead bodies or carrion: *necrophagous organisms.*

necrosis (ne krō′sis), *n., pl.* **-ses** (-sēz). 1 *Biology, Medicine.* the death or decay of cells or body tissues. It may result from a degenerative disease, stoppage of the oxygen supply, infection, or destructive burning or freezing.
2 *Botany.* a disease of plants characterized by small black spots of decayed tissue: *Tobacco necrosis is caused by a virus.*
[from New Latin *necrosis,* from Greek *nekrōsis,* ultimately from *nekros* dead body]
—**necrotic** (ne krot′ik), *adj.* of or showing necrosis: *necrotic disease.*

nectar, *n. Botany.* a sweet liquid found in many flowers, which attracts insects and birds that carry out pollination. Bees gather nectar and make it into honey. [from Latin *nectar,* from Greek *nektar*]

nectary (nek′tər ē), *n. Botany.* the gland of a flower or plant that secretes nectar.

needle, *n.* 1 *Botany.* a thin pointed leaf, as of a pine tree: *The commoner conifers are easily recognized by their leaves which are often called needles.* (Emerson, *Basic Botany*).
2 *Mineralogy, Chemistry.* a crystal or spicule like a needle in shape: *One such isotope, in the form of germanium oxide needles, can be inserted in the body and left there* (N.Y. Times).
3 *Geology.* a pinnacle of rock tapering to a point.

negation, *n. Mathematics.* the statement or proposition "not *p*" (symbolized ∼ *p*), denying that something is true or asserting that it is false.

negative, *adj.* 1 *Mathematics.* **a** less than zero: *−5 is a negative number.* **b** lying on the side of a point, line, or plane opposite to that considered positive.
2 *Physics, Chemistry.* **a** of or having to do with the kind of electricity produced on resin when it is rubbed with silk, or that is present in a charged body which has an excess of electrons: *Thus it came to be established that there was a type of particle, called an electron, which always possessed the same negative charge e and the same mass m, which was somehow contained in atoms, and whose behaviour was responsible for the phenomenon of electricity* (H.J. Bhaba). **b** having a tendency to gain electrons: *Oxygen and other nonmetals have a negative valence.* **c** of or designating the part of an electric cell from which the current flows into the wire: *the negative pole, a negative electrode.*
3 *Biology.* moving or turning away from light, the earth, or any other stimulus.
—*n. Mathematics.* a negative or minus quantity, sign, symbol, etc.

negative acceleration, *Physics.* acceleration in a direction opposite to the velocity; a decrease in velocity: *Negative acceleration ... means that a smaller space is traveled during each successive unit of time* (Robert F. Paton).

negative angle, *Mathematics.* an angle formed by a line rotating in a clockwise direction. Compass directions in maritime navigation are given by means of negative angles.

negative catalyst, *Chemistry.* a substance that retards a reaction without itself being permanently affected.

negative feedback, *Biology.* feedback that keeps input and output in equilibrium, associated with endocrine regulation: *The negative feedback mechanism operates on the principle that the level of one hormone in the blood stimulates or inhibits the production of another hormone* (Biology Regents Syllabus).

negative staining, *Biology.* the immersing of small organisms, such as viruses and bacteria, in a stain which does not color them, so that their forms appear clearly defined against the colored background.

negatron (neg′ə tron), *n. Physics, Chemistry.* an electron with a negative charge, as contrasted with a positron or positively charged electron. The term is used only to distinguish the two members of an oppositely charged pair. *High-energy gamma rays may create matter by completely disappearing and forming a pair of electrons, one a positively charged positron and one a negatively charged ordinary electron, or negatron* (Ralph E. Lapp). [from *nega(tive)* + *-tron,* as in *positron*]

nekton (nek′ton), *n. Biology.* the relatively large organisms, such as fish, that possess the power to swim freely in oceans and lakes, independent of water movements, in contrast to plankton. [from Greek *nēkton,* neuter of *nēktos* swimming, from *nēchein* to swim]
ASSOCIATED TERMS: see **ocean** and **benthos.**
—**nektonic** (nek ton′ik), *adj.* of or having to do with nekton.

N electron, *Physics.* one of the electrons in the N-shell.

nematic (ni mat′ik), *adj. Chemistry.* (of liquid crystals) having the molecules arranged in loose parallel, vertical lines: *Nematic liquid crystals consist of rodlike organic molecules ... They maintain a parallel or nearly parallel arrangement, although each molecule can ro-*

tate around an axis pointing in its direction of alignment (George H. Heilmeier). Compare **cholesteric** and **smectic.** [from Greek *nēma, nēmatos* thread]

nematocyst (nem′ə tə sist *or* nə mat′ə sist), *n. Zoology.* a coiled, threadlike stinging process of cnidarians, discharged to capture prey and for defense. [from Greek *nēma, nēmatos* thread + English *cyst*]

nematode (nem′ə tōd), *n. Zoology.* any of a class or phylum (Nematoda) of slender, unsegmented, cylindrical worms, often tapered near the ends. Parasitic forms such as the hookworm, pinworm, and trichina belong to this group. Nematodes are commonly called *roundworms.* [ultimately from Greek *nēma, nēmatos* thread, from *nein* spin]

nematology (nem′ə tol′ə jē), *n.* the branch of parasitology that deals with nematodes.
—**nematological** (nem′ə tə loj′ə kəl), *adj.* of or having to do with nematology.

nemertean (ni mèr′tē ən), *Zoology.* —*n.* any of a phylum (Nemertea, Nemertina, or Rhynchocoela) of long, ribbonlike, unsegmented marine worms with a complete digestive system, mouth, and anus. Nemerteans are brightly colored, sometimes growing up to 90 feet long and have a long extensible proboscis. Nemerteans are commonly called *ribbon worms.*
—*adj.* of or belonging to the nemerteans: *Nemertean worms live in the sea on both coasts of the United States* (Science News Letter). Also called **nemertine, rhynchocoel.**
[from Greek *Nēmertēs,* a sea nymph]

nemertine (ni mèr′tīn), *n., adj.* = nemertean.

nemoral (nem′ər əl), *adj. Ecology.* having to do with, inhabiting, or frequenting woods or groves. [from Latin *nemoralis,* from *nemus, nemoris* grove]

neo-, *combining form.* **1** new or recent, as in *neogenesis, neoglaciation, Neocene.*
2 *Ecology.* of the New World, as in *Neotropical.*
3 *Chemistry.* **a** of or designating a new compound related to or isomeric with a specified older compound, as in *neohesperidin, neo-Vitamin A.* **b** (of a hydrocarbon) containing a carbon atom that is directly bonded to four other carbon atoms, as in *neopentane, neohexane.* [from Greek *neos* new]

neocortex (nē′ō kôr′teks), *n., pl.* **-tices** (-tə sēz). *Anatomy.* the dorsal part of the cerebral cortex, which is specially large in higher mammals and is the most recent in development: *Between the time of Australopithecus and Neanderthal man, the brain underwent rapid changes, gaining the large neocortex which was to provide the human species with its ability to manipulate symbols* (Science News). Also called **neopallium.**
—**neocortical,** *adj.* of or having to do with the neocortex.

neo-Darwinism, *n. Biology.* a theory of evolution that combines Darwinism and the findings of modern genetics. Neo-Darwinism postulates that evolution depends both on the operation of natural selection and upon the variability caused by mutations within a population.

neodymium (nē′ə dim′ē əm), *n. Chemistry.* a yellowish, rare-earth element found in monazite and various other rare-earth minerals. The rose-colored salts of neodymium are used to color glass. *Neodymium, number 60, ... was discovered by von Welsbach in 1885, when*

he separated salts of a supposed element called didymium into two fractions, one of which was praseodymium and the other neodymium (Scientific American). *Symbol:* Nd; *atomic number* 60; *atomic weight* 144.24; *melting point* 1010°C; *boiling point* 3127°C; *oxidation state* 3. [from *neo-* + (*di*)*dymium*]

Neogaea (nē′ə jē′ə), *n. Ecology.* the Neotropical region, considered with reference to the geographical distribution of plants and animals. [from New Latin *Neogaea,* from Greek *neos* new + *gaia* earth]
—**Neogaean** (nē′ə jē′ən), *adj.* of or having to do with the Neogaea.

Neogene (nē′ə jēn), *Geology.* —*n.* **1** the later of two periods of the Cenozoic era, comprising the Miocene, Pliocene, Pleistocene, and Recent epochs. The earlier period is the Paleogene.
2 the system of rock strata formed during this period.
—*adj.* of or having to do with this period or system. [from *neo-* + Greek *genēs* born]
► See the note under **Tertiary.**

neogenesis (nē′ō jen′ə sis), *n. Biology.* the formation of new tissue; regeneration: *the neogenesis of human hair follicles.*
—**neogenetic** (nē′ō jə net′ik), *adj. Biology.* of or having to do with neogenesis.

neoglacial, *adj. Geology.* of or having to do with neoglaciation: *Carbon-14 dates of vegetation, peat or soil overrun by glacier ice provide direct ages for neoglacial advances, and dating of organic matter in recessional or advance deposits associated with neoglacial moraines may provide important limiting ages* (George H. Denton and Stephen C. Porter).

neoglaciation, *n. Geology.* the formation of new glaciers, or readvance of older glaciers, particularly during the so-called Little Ice Age in the northern hemisphere between 1600 and 1850.

neon (nē′on), *n. Chemistry.* a rare element that is a colorless, odorless, inert gas, forming a very small part of the air. Tubes containing neon are used in electric signs or lamps, giving off a fiery red glow. *Symbol:* Ne; *atomic number* 10; *atomic weight* 20.183. [from New Latin *neon,* from Greek *neon* (neuter) new]

neoplasia (nē′ə plā′zhə), *n. Biology, Medicine.* **1** the development of new tissue.
2 the formation of a neoplasm or abnormal tissue: *Littermates raised in cages developed tumors in the predicted time, while other littermates raised in a terrarium simulating the natural habitat did not develop neoplasia* (Werner Dutz).
[from New Latin, from *neo-* + Greek *plasis* molding, formation]

neoplasm (nē′ō plaz əm), *n. Biology, Medicine.* a new, abnormal growth of tissue, such as a tumor: *Some of the more enthusiastic workers in this field claimed that all neoplasms and all leukemias are due to viruses*

cap, fāce, fäther; best, bē, tèrm; pin, fīve;
rock, gō, ôrder; oil, out; cup, pùt, rüle,
yü in use, *yù* in uric;
ng in bring; *sh* in rush; *th* in thin; ᴛH in then;
zh in seizure.
ə = *a* in about, *e* in taken, *i* in pencil, *o* in
lemon, *u* in circus

429

(William Dameshek). [from *neo-* + Greek *plasma* something formed]

Neotropical (nē′ō trop′ə kəl), *adj. Ecology.* of or belonging to the region that includes most of the Caribbean, tropical North America, and all of South America: *the Neotropical flora and fauna.*

neotype (nē′ō tīp′), *n. Biology.* a type specimen selected to replace a holotype: *If the type strain has been lost, an attempt is made to find a new strain that conforms to the original description. Such a strain may be designated as a neotype, which then represents the species* (W.E. Moore and L.V. Holdeman).

nephanalysis (nef′ə nal′ə sis), *n., pl.* **-ses** (-sēz′). *Meteorology.* **1** analysis of the cloud formations over a large area, using weather charts drawn especially from photographs taken by weather satellites. **2** a chart of such cloud formations. [from Greek *nephos* cloud + English *analysis*]

nepheline (nef′ə lēn), *n. Mineralogy.* a silicate of aluminum, sodium, and sometimes potassium. Nepheline occurs in various volcanic rocks and is chemically related to feldspar. *Formula:* (Na,K)Al SiO₄ Also called **nephelite.** [from French *néphéline,* from Greek *nephelē* cloud]

nephelinite (nef′ə lə nīt), *n. Geology.* a heavy, dark-colored, volcanic rock, essentially a basalt containing nepheline and pyroxene but no feldspar and little or no olivine.

nephelite (nef′ə līt), *n.* = nepheline.

nephelometer (nef′ə lom′ə tər), *n.* **1** *Chemistry.* **a** an instrument for measuring the concentration of suspended matter in a liquid dispersion by measuring the amount of light scattered by the dispersion. **b** a similar device for estimating the number of bacteria in a suspension. **2** *Meteorology.* an instrument for measuring the comparative cloudiness of the sky. [from Greek *nephelē* cloud + English *-meter*]
—**nephelometric** (nef′ə lə met′rik), *adj.* of or having to do with nephelometry: *Nephelometric standards provide four different range of turbidities found in nephelometric studies* (Science).
—**nephelometry** (nef′ə lom′ə trē), *n.* the measurement of the concentration of suspended matter in a liquid by means of a nephelometer.

nephology (ni fol′ə jē), *n.* the branch of meteorology that deals with clouds. [from Greek *nephos* cloud + English *-logy*]
—**nephological** (nef′ə loj′ə kəl), *adj.* having to do with nephology; relating to clouds or cloudiness.

nephridium (ni frid′ē əm), *n., pl.* **-ia** (-ē ə). *Zoology.* a primitive excretory organ in some invertebrates and lower vertebrates such as the mollusks and certain annelids, analogous in function to the kidneys of higher animals, and in some cases serving also in reproduction. [from New Latin *nephridium,* from Greek *nephros* kidney]

nephrite (nef′rīt), *n. Mineralogy.* a type of jade, a silicate of calcium and either magnesium or iron, varying in color from white to dark green. [from German *Nephrit,* from Greek *nephros* kidney + German *-it* -ite (so called because it was once believed to protect the wearer against kidney disease)]

nephritic (ni frit′ik), *adj.* = renal. [from Late Latin *nephriticus,* from Greek *nephritikos,* from *nephros* kidney]

nephrogenic (nef′rə jen′ik), *adj.* **1** *Physiology.* produced in the kidney. **2** *Embryology.* developing into kidney tissue. [from Greek *nephros* kidney]

nephron (nef′ron), *n. Anatomy.* any of the numerous functional units of the kidney, serving to filter waste matter from the blood. A human kidney has more than one million nephrons. Each nephron consists of a Bowman's capsule, glomerulus, and tubule. [from German *Nephron,* from Greek *nephros* kidney]

nephrostome (nef′rə stōm), *n. Zoology.* the ciliated funnel-shaped end of the nephridium in some invertebrates and lower vertebrates. [from Greek *nephros* kidney + *stoma* mouth]

Neptune, *n. Astronomy.* the fourth largest planet in the solar system, eighth in distance from the sun. It is so far from the earth that it cannot be seen without a telescope. Its orbit lies between those of Uranus and Pluto and takes 164.8 years to complete, at a mean distance from the sun of 4,486,100,000 kilometers. See the picture at **solar system.** [from Latin *Neptunus* god of the sea, related to *nebula* cloud, mist]
—**Neptunian** (nep tü′nē ən), *adj.* **1** of the planet Neptune: *Neptunian asteroids, the two Neptunian moons.* **2** Also, **neptunian.** *Geology.* resulting from or produced by the action of water, particularly oceanic water. Contrasted with **volcanic** or **plutonic.**

neptunium (nep tü′nē əm), *n. Chemistry.* a radioactive metallic element which occurs in minute amounts in uranium ore and is produced artificially by bombardment of an isotope of uranium with neutrons. A less stable isotope of neptunium (atomic weight 239) disintegrates rapidly to form an isotope of plutonium that can be used for nuclear fission. *Symbol:* Np; *atomic number* 93; *mass number* (of most stable isotope) 237; *oxidation state* 3, 4, 5, 6. [named after *Neptune,* the planet whose orbit lies beyond Uranus; similarly, neptunium comes after uranium in the periodic table]

neritic (ni rit′ik), *adj. Oceanography.* of or having to do with that part of the ocean floor between the low tide mark and a depth of about 200 meters: *Life is most abundant in the shallower parts of the ocean or neritic region of the continental shelves* (Harbaugh and Goodrich, *Fundamentals of Biology*). [from Latin *nerita* a sea mussel; so called because such life generally inhabits this depth]

nervate (nėr′vāt), *adj. Biology.* having veins or ribs: *nervate leaves.* SYN: veined.
—**nervation,** *n.* the arrangement of the veins or ribs in a leaf or an insect's wings. SYN: venation.

nerve, *n.* **1** *Anatomy.* a fiber or bundle of fibers made up of neurons through which impulses, especially of sensation and motion, pass between the brain or spinal cord and the eyes, ears, muscles, glands, and other parts of the body. The nerve fibers are enclosed within sheaths and are joined to each other with connective tissue. *Most nerves contain both motor and sensory fibers, though one or other may predominate* (Mackean, *Introduction to Biology*). See also **nervous system, ganglion.** See the pictures at **crustacean, skin.**

2 *Biology.* **a** a vein of a leaf. **b** a rib of an insects' wing. SYN: nervure. **c** the pulp of a tooth. See the picture at **tooth.**
[from Latin *nervus* sinew, tendon; sense influenced by Greek *neuron,* meaning both "sinew" and "nerve"]

nerve cell, *Biology.* **1** = neuron. **2** the cell body of a neuron, excluding its fibers.

nerve center, *Anatomy.* a group of nerve cells closely connected with one another and acting together in the performance of some function.

nerve control, *Biology.* a means of regulating life processes in multicellular animals by transmitting messages through a system of neurons. Compare **chemical control.**

nerved, *adj. Biology.* **1** having nerves. **2** = nervate. **3** having nervures.

nerve fiber, *Biology.* any of the long, threadlike processes or fibers of a neuron; an axon or a dendrite.

nerve growth factor, *Biochemistry.* a protein that stimulates the growth of sympathetic and sensory nerve cells: *Nerve growth factor (NGF) ... is a typical body protein ... [and] experiments suggest ... that NGF is at the root of some neurological and malignant diseases* (Science News).

nerve impulse, *Physiology.* an impulse transmitted along a nerve fiber, consisting of a wave of excitation in the fiber accompanied by a brief, transient change in electric potential: *The rate of discharge of nerve impulses to the motor unit is one of the factors concerned in the gradation of muscle action ... A whole muscle consists of a very large number of motor units* (W.F. Floyd and P.H. Silver).

nerve net, *Zoology.* a network of nerve cells distributed throughout the tissues of cnidarians, echinoderms, and certain other lower forms, in which it makes up a primitive nervous system.

nerve trunk, *Anatomy.* several nerve fibers bound together by a tough sheet of tissue.

nervous, *adj. Biology.* of or having to do with nerves or neurons: *nervous tissue, a nervous impulse. Automatic nervous actions of this kind do not usually involve the higher brain at all* (J.A.V. Butler).

nervous system, *Anatomy.* the system of nerve fibers, nerve cells, and other nervous tissue in a person or animal by means of which impulses are received and interpreted. The human central nervous system and that of the other vertebrates consists of the brain and spinal cord, to and from which impulses are carried by the peripheral nervous system. The peripheral nervous system consists of the cranial nerves (12 pairs in human beings), the spinal nerves (31 pairs in human beings), and the autonomic nervous system (the ganglia and nerves that regulate the involuntary muscles, viscera, and glands). *The nervous system keeps us in contact with the world outside our bodies by receiving messages from the sense organs, such as the eyes and ears* (Herbert H. Jasper).
ASSOCIATED TERMS: see **neuron.**

nervure (nèr′vyùr), *n. Biology.* **1** the principal vein of a leaf. **2** a rib of an insect's wing. [from French *nervure,* from *nerf* nerve, from Latin *nervus*]

nest, *n. Zoology.* **1** a structure or place, usually shaped somewhat like a bowl, built by birds out of twigs, straw, etc., as a place for laying eggs and rearing young: *a robin's nest.*
2 a structure or place used by insects, fishes, turtles, rabbits, or the like, for depositing eggs or young: *a wasp's nest.*
3 the birds, insects, etc., living in a nest.

nettle, *n. Botany.* **1** any of a genus (*Urtica*) of widely distributed herbs having sharp bristles on the leaves and stems that sting the skin when touched. **2** any of various related or similar plants. [Old English *netele*]

network solid, *Chemistry.* a solid consisting of covalently bonded atoms linked in a network which extends throughout the sample without any discrete particles: *Generally, network solids are hard, electrical insulators, poor conductors of heat, and have high melting points. Examples of network solids are diamond (C), silicon carbide (SiC), and silicon dioxide (SiO_2)* (Chemistry Regents Syllabus).

neur-, *combining form.* the form of **neuro-** before vowels, as in *neural, neuraminic acid.*

neural (nùr′əl), *adj. Biology.* **1** of or having to do with a nerve, neuron, or nervous system: *a neural stimulus or response. The neural processes involved in learning a given operation, such as finding the way through a maze, occur in all or most parts of the cortex* (S.A. Barnett). **2** having to do with or situated in the region or side of the body containing the brain and spinal cord: *A group of cells running the length of the dorsal surface of the embryo ... form a groove ... called the neural groove* (A.M. Winchester). SYN: dorsal. [from Greek *neuron* nerve]
—**neurally,** *adv.* by a nerve or nerves: *All states of mind are neurally conditioned* (William James).

neural arch, *Anatomy.* the arch on the dorsal side of a vertebra: *Each vertebra is made up of a spool-like centrum surmounted by a neural arch to house the nerve cord* (Storer, *General Zoology*).

neural canal, *Anatomy.* the canal formed by the vertebral foramina, enclosing and protecting the spinal cord.

neural crest, *Embryology.* the part of the ectoderm, situated on either side of the neural tube, that develops into the spinal ganglia and the ganglia of the autonomic nervous system.

neural plate, *Embryology.* the thickened dorsal plate of ectoderm that differentiates into the neural tube and the neural crest.

neural spine, *Anatomy.* a bony process on the dorsal side of a vertebra.

neural tube, *Embryology.* the dorsal tube, formed from the neural plate, that differentiates into the brain and spinal cord.

---

cap, fāce, fäther; best, bē, tèrm; pin, fīve;
rock, gō, ôrder; oil, out; cup, pùt, rüle;
yü in use, *yu* in uric;
ng in bring; *sh* in rush; *th* in thin, ᴛʜ in then;
zh in seizure.
ə = *a* in about, *e* in taken, *i* in pencil, *o* in lemon, *u* in circus

neuraminic acid

neuraminic acid (nùr'ə min'ik), *Biochemistry.* an amino acid, a constituent of carbohydrate and protein complexes, found in animal cells and tissues. *Formula:* $C_{19}H_{17}NO_8$ [from *neur-* + *amino* + *-ic*]

neuraminidase (nùr'ə min'ə dās), *n. Biochemistry.* a hydrolytic enzyme that breaks down mucoproteins, found chiefly among microorganisms inhabiting the respiratory and intestinal tracts: *The enzyme, known as neuraminidase, can attack the mucus of saliva, nasal secretion, sputum or egg white ...* (Science Journal). [from *neuramin(ic)* acid + *-ide* + *-ase*]

neurilemma (nùr'ə lem'ə), *n. Anatomy.* the delicate membranous outer sheath of peripheral nerve fibers. [from *neur-* + Greek *eilēma* covering; the *-lemma* spelling from confusion with Greek *lemma* husk]

neuro-, *combining form.* nerve; nerve tissue; nervous system, as in *neurology = study of the nervous system.* Also spelled **neur-** before vowels. [from Greek *neuron*, nerve, sinew]

neuroanatomy, *n.* the branch of anatomy that deals with the anatomical structure of the nervous system.
—**neuroanatomical**, *adj.* of or having to do with neuroanatomy: *Neuroanatomical research reveals that human brains are very much alike in structure, complexity, and functioning* (John C. Lilly).

neurobiology, *n.* the branch of biology that deals with the nervous system.
—**neurobiological**, *adj.* of or having to do with neurobiology.

neuroblast, *n. Embryology.* a cell in vertebrate embryos, derived from the ectoderm of the neural plate, that develops into a nerve cell. [from *neuro-* + Greek *blastos* germ, sprout]

neurochemical, *Biochemistry.* —*n.* a chemical substance that affects the nervous system or some part of it: *Several neurochemicals, when applied to a specific area of the brain, appear to control killing behavior in laboratory rats* (Science News).
—*adj.* of or having to do with neurochemistry: *neurochemical mechanisms.*

neurochemistry, *n.* the branch of biochemistry that deals with the chemical makeup and effects of the nervous system: *A significant development in neurochemistry has been the finding that thiol groups and disulphides are involved in the functioning and activity of neurones* (Nature).

neuroendocrine, *adj. Physiology.* of or having to do with the nervous system and the endocrine glands: *The birds' neuroendocrine apparatus can apparently respond to external stimuli to mediate an annual cycle of gonad regeneration, acceleration and culmination* (W.L.N. Tickell).

neuroendocrinology, *n. Physiology.* the study of the interactions between the central nervous system and the endocrine system.

neurofibril, *n. Biology.* any of the fibrils found in a nerve cell; a fibril of a neuron: *Neurofibrils are very long and less than a millionth of an inch thick* (Scientific American).

neurogenic, *adj. Biology.* **1** originating in the nerves or nervous tissue: *a neurogenic tumor. Broadly speaking there are two types of heart beat in the animal kingdom: one in which the beat originates in nerve cells lo-* cated in the wall of the heart (*neurogenic*), and the other in which the beat arises in the actual muscles of the heart (*myogenic*) (New Biology). **2** stimulated by nerves: *neurogenic contraction of muscles.*

neuroglia (nù rog'lē ə), *n. Biology.* the delicate connective tissue forming a supporting network for the conducting elements of nervous tissue in the brain and the spinal cord. [from *neuro-* + Late Greek *glia* glue]
—**neuroglial**, *adj.* of or having to do with neuroglia: *Neuroglial cells ... perform the function of waste-removal, decomposing the material of dead cells into their basic components—of which they absorb some and transport the remainder to the walls of capillaries to be carried away by the blood stream* (New Scientist).

neurohemal organ (nùr'ō hē'məl), *Biology.* an organ of the circulatory system composed of nerve endings, found especially in insects and thought to have to do with neurosecretion. [from *neuro-* + *hemal*]

neurohormone, *n. Biochemistry.* a hormone secreted by or stimulating nerve cells: *Spawning in starfish is controlled by a neurohormone released by the radial nerves* (James S. Kittredge).
—**neurohormonal**, *adj.* **1** of or having to do with neurohormones: *neurohormonal agents such as serotonin and norepinephrine.* **2** having to do with both nerves and hormones: *a neurohormonal reaction.*

neurohumor, *n. Biochemistry.* a neurohormone or neurotransmitter: *Under nervous stimulation the hypothalamus secretes so-called neurohumors into blood vessels along the narrow stalk that connects it to the pituitary; these secretions cause the pituitary to produce the thyroid-stimulating hormone* (Scientific American).
—*adj.* of or having to do with a neurotransmitter or its effect.

neurohypophysis (nùr'ō hī pof'ə sis), *n. Anatomy.* the posterior portion of the pituitary gland, consisting of the main body, the neural lobe, and a stalk by which the gland is attached to the hypothalamus. The anterior part is called the **adenohypophysis**.
—**neurohypophyseal** or **neurohypophysial** (nùr'ō hī pə fiz'ē əl), *adj.* of or having to do with the neurohypophysis: *The resulting sense-impulse is conveyed to the brain of the females, who secrete a neurohypophyseal hormone which triggers off the egg-laying process* (New Scientist).

neuroleptic (nùr'ə lep'tik), *n. Medicine.* a tranquilizing drug.
—*adj.* of or for the relief of nervous tension or anxiety; tranquilizing: *a neuroleptic drug.* [from *neuro-* + Greek *leptikos* seizing]

neurological, *adj.* of or having to do with neurology: *neurological abnormalities.*

neurology, *n.* the study of the nervous system and its diseases: *The most spectacular current progress in neurology is in localization and analysis of functions and connections in the brain* (W.J.S. Krieg).

neuromuscular, *adj. Biology.* of or having to do with the relationship of nerves to the muscles; involving both nerves and muscles: *Neuromuscular and psychological conditions have been linked for some time* (Science News).

432

neuron or **neurone** (nùr′on), *n. Biology*. one of the impulse-conducting cells of which the brain, spinal cord, and nerves are composed. A neuron consists of a cell body containing the nucleus, and usually several dendrites and a single axon. *The nervous system of the body contains more than 10 billion neurons, or nerve cells, most of them being in the brain. A single neuron of the cerebral cortex may be so interconnected as to influence the behavior of a hundred other cell bodies. This suggests that the interconnections ... between neurons may number around 1000 billion* (V.L. Parsegian, *This Cybernetic World*). Also called **nerve cell.** [from Greek *neuron* nerve]

ASSOCIATED TERMS: The neuron is the basic cellular unit of the *nervous system*. The nervous system is composed of three structurally different types of neurons: *sensory neurons, interneurons,* and *motor neurons.* Since nerves are bundles of neurons, they can be *sensory nerves, motor nerves,* or *mixed nerves.* An *impulse* is an electrochemical charge initiated by a *stimulus* and generated along a neuron to *receptors* or *effectors. Neurotransmitters* transmit the impulses across *synapses.* See **dendrite** for other associated terms.

—**neuronal** (nùr′ə nəl *or* nù rō′nəl), *adj.* of or having to do with a neuron or neurons: *Memory ... depends on an active neuronal mechanism which never rests, and when we remember an event (i.e. think about it) we are in some way merely picking out one of the millions of neuronal circuits available for attention* (New Scientist).

—**neuronic** (nù ron′ik), *adj.* = neuronal.

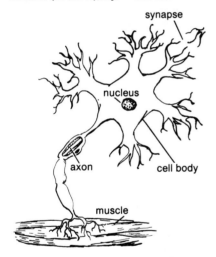

neurophysiology, *n.* the branch of physiology that deals with the nervous system: *[Galvani's] discovery that a muscle could be made to contract by an electrical current laid the foundation for the study of animal electricity, an important part of neurophysiology* (Caroline A. Chandler).

—**neurophysiological,** *adj.* of or having to do with the physiological functions of the nervous system.
—**neurophysiologically,** *adv.*

neuropile (nùr′ə pīl) or **neuropil** (nùr′ə pil), *n. Anatomy.* an area in a ganglion consisting of a network of closely interwoven nerve fibers: *The cell bodies of invertebrate neurons are gathered in a rind around the outside of the ganglion. This cellular region is devoid of axons and synapses. Each neuron sends its axon into a central region, the neuropile, where interconnections between neurons are made* (E.R. Kandel and I. Kupfermann). [from *neuro-* + Greek *pilē(ma)* felt]

neuroplasm, *n. Biology.* the protoplasm of the nerve cells and their fibrillae: *Studies of metabolic dynamics within neurons reveal that the substance—neuroplasm—within the axon (long, tiny extensions of a nerve cell) is synthesized in the nerve cell body and moves as a gel down the axon at about 1 mm per day* (E. L. Cockrum).

neuropteran (nù rop′tər ən), *Zoology. —n.* any of an order (Neuroptera) of carnivorous insects having a complete metamorphosis, four large delicate wings, and mouthparts adapted for chewing. Lacewings and ant lions belong to this order.
—*adj.* of or belonging to the neuropterans.
[from Greek *neuron* nerve, sinew + *pteron* wing]

neuroscience, *n.* any of the sciences dealing with the nervous system, such as neurology and neurochemistry, or these sciences collectively: *Perception and recognition seem to occur in steps ... and a goal of modern neuroscience is to map the brain cell circuits that make the connections between conscious and unconscious thoughts* (N.Y. Times).
—**neuroscientist,** *n.* a specialist in a neuroscience or the neurosciences.

neurosecretion, *n. Physiology.* **1** the secretion of substances, such as hormones, by nerve cells: *Neurosecretion may be of major importance in the coordination of the annelid body* (R.P. Dales).
2 the substance secreted: *It is possible that neurosecretions are the 'oldest' hormones in the animal kingdom* (New Scientist).
—**neurosecretory,** *adj.* of or having to do with neurosecretion: *The concept pictures the neurosecretory apparatus as a connecting link between the two major control systems of the body: the nervous system and the anterior pituitary, the master gland of the endocrine system* (William Etkin).

neurosensory, *adj. Physiology.* of or having to do with the sensory functions or activity of the nervous system: *neurosensory cells, neurosensory physiology.*

neurotoxin, *n. Biochemistry.* a toxin that can damage or destroy nerve tissue. The venom of the coral snake is a neurotoxin. *All of these neurotoxins in one way or another disrupt the microchemical mechanisms that transmit nerve impulses* (Scientific American).
—**neurotoxic,** *adj.* being or caused by a neurotoxin; toxic to nerve tissue: *... the neurotoxic action characteristic of the cobra poison* (New Scientist).

neurotransmitter, *n. Biochemistry.* a substance that transmits impulses between nerve cells: *Norepinephrine is a neurotransmitter, a substance responsible for carrying a signal across the gap between two neurons.* (Science News).

cap, fāce, fäther; best, bē, tèrm; pin, fīve;
rock, gō, ôrder; oil, out; cup, pùt, rüle,
yü in use, *yu* in uric;
ng in bring; *sh* in rush; *th* in thin, ᴛʜ in then;
zh in seizure.
ə = *a* in about, *e* in taken, *i* in pencil, *o* in lemon, *u* in circus

neurotropic, *adj. Biology.* drawn to or having an affinity for nervous tissue: *The virus of rabies is a neurotropic virus* (New Yorker). [from *neuro-* + Greek *tropē* a turning]

neuston (nü′stŏn), *n. Biology.* minute organisms living on the surface film of a body of water: *In the neuston ... a characteristic microflora develops* (A. Mayr-Harting). Compare **plankton.** [from German *Neuston,* from Greek *neustos* swimming, from *nein* to swim]
—neustonic (nü stŏn′ik), *adj.* of or having to do with neuston: *the neustonic zone of a pond.*

neuter, *adj.* **1** *Zoology.* without sex organs or with sex organs that are not fully developed: *Worker bees are neuter.*
2 *Botany.* having neither stamens nor pistils; functionally asexual.
—n. an animal or plant that is neuter.
[from Latin *neuter,* from *ne* not + *uter* either]

neutercane (nü′tər kān), *n. Meteorology.* a storm that draws its force from both tropical and cold-front disturbances in the atmosphere. [from *neuter* + (*hurri*)*cane*]

neutral, *adj.* **1** *Chemistry.* neither acid nor alkaline: *a neutral salt.* **2** *Physics.* neither positive nor negative in electric charge: *Neutral atoms have equal numbers of electrons and protons.* **3** *Biology.* not developed sexually; neuter.
—neutrality, *n.* the condition of being neutral.

neutral current, *Physics.* a weak interaction between nuclear particles in which no charge is exchanged: *A neutral-current process is one in which two particles interact without exchanging a unit of electric charge. If a neutrino strikes a proton and bounces off, and the neutrino remains a neutrino and the proton remains a proton, that's a neutral-current interaction* (Science News).

neutralization, *n. Chemistry.* a process of neutralizing; a chemical reaction between an acid and a base: *The process by which an acid and a base unite to form water and a salt is termed neutralization* (W. N. Jones).

neutralize, *v. Chemistry.* to make (a solution) neutral by adding a base (if the solution is acidic) or an acid (if the solution is basic).

neutrino (nü trē′nō), *n., pl.* **-nos.** *Physics.* any of three elementary particles having no electric charge and a mass too close to zero to be measured. Neutrinos are leptons with a spin of 1/2. Each of the three types of neutrinos is associated with another particle. The most familiar type, the *electron neutrino,* is emitted in the beta decay of a nucleus. The other two types are the *muon neutrino* and the *tau neutrino.* [from Italian *neutrino,* from *neutro* neuter + *-ino,* diminutive suffix]

neutron (nü′trŏn), *n. Physics.* an elementary particle with no electric charge that occurs in the nucleus of every atom except the lightest isotope of hydrogen (^{1}H) and has about the same mass as a proton. Neutrons are used to bombard the nuclei of various elements to produce fission and other nuclear reactions. *It is the neutron which figures in the transmutations which give atomic power. Neptunium and plutonium were formed by bombarding uranium 238 with neu-*

trons (Helen M. Davis). See the picture at **atom.** [from *neutr*(*al*) + *-on*]
ASSOCIATED TERMS: see **electron.**
—neutronic, *adj.* of or having to do with a neutron or neutrons: *neutronic scansion.*

neutron activation analysis, *Chemistry.* a method of analyzing the composition of a substance by bombardment with neutrons to produce radioisotopes that may be identified by their characteristic radiation. *Abbreviation:* NAA

neutron capture, *Nuclear Physics.* the absorption of a neutron by the nucleus of an atom, resulting in the emission of a gamma ray or rays or in fission of the nucleus: *Fission is brought about by neutron capture and this results in fission fragments, liberation of energy, and release of two or more neutrons* (Chemistry Regents Syllabus).

neutron flux, *Physics.* **1** a flow of neutrons, as in a nuclear reactor.
2 a measure of this flow, expressed by the number of neutrons crossing a unit area in a unit time.

neutron poison, *Physics.* an element that readily absorbs large numbers of neutrons, used in a nuclear reactor to control a chain reaction.

neutron star, *Astronomy.* a celestial body that is often the source of powerful X rays and consists of a mass of very densely packed neutrons probably formed by the collapsed atoms of a moderately large star: *Astrophysicists now believe that pulsars are neutron stars—the last stage in the life of a heavyweight star in which the matter is so tightly compressed that the protons and electrons have combined to form neutrons* (New Scientist). Compare **pulsar.**

neutrophil (nü′trə fil), *Biology.* **—n.** a very abundant type of white blood cell that is capable of absorbing and destroying harmful matter and so protects the body against infection, making up about 50 to 75 per cent of the total number of white blood cells: *Neutrophils identified by light microscopy exhibited a round to oval shape with lobed nucleus and an abundant, granule-containing cytoplasm* (Science). Compare **basophil, eosinophil.**
—adj. = neutrophilic.

neutrophilic (nü′trə fil′ik), *adj. Biology.* readily staining with either acidic or basic dyes: *neutrophilic granulocytes.*

névé (nā vā′), *n. Geology.* **1** the crystalline or granular snow that has not been compressed into ice on the upper part of a glacier. **2** a field of such snow. Also called **firn.** [from French *névé,* ultimately from *nivis* snow]

new moon, *Astronomy.* **1** the moon when seen as a thin crescent with the hollow side on the left. **2** the moon when its dark side is toward the earth, appearing almost invisible.

newton, *n. Physics.* the SI or MKS unit of force, equal to 100,000 dynes. It is the force required to give an acceleration of one meter per second per second to a mass of one kilogram. *Symbol:* N [named after Isaac Newton, 1642–1727, English physicist]

Newton's laws of motion. See **motion.**

NGF, *abbrev.* nerve growth factor.

Ni, *symbol.* nickel.

niacin (nī′ə sin), *n.* = nicotinic acid: *In the case of one vitamin, niacin, the anti-pellagra vitamin, mental illness can result if people do not get enough of the vita-*

niccolite (nik′ə līt), *n. Mineralogy.* a mineral, nickel arsenide, of a pale copper-red color and metallic luster. It usually occurs massive. *Formula:* NiAs [from New Latin *niccolum,* Latinization of *nickel*]

niche (nich), *n. Ecology.* the place and function of an organism within an ecosystem: *If two different species compete for the same food or reproductive sites, one species may be eliminated. This usually establishes one species per niche in a community. The niche is the organism's role in the community* (Biology Regents Syllabus). [from Middle French *niche,* from Old French *nichier* to nest, ultimately from Latin *nidus* nest]

nickel, *n. Chemistry.* a hard, silvery-white metallic element found in igneous rocks and in metallic meteorites, much used in alloys and in electroplating. Nickel is malleable, ductile, and magnetic, but not easily oxidized. *Symbol:* Ni; *atomic number* 28; *atomic weight* 58.71; *oxidation state* 2, 3. [from Swedish *nikel,* from German *Kupfernickel* (literally) copper devil; so-called because the ore resembles copper but yields none]

nickeliferous (nik′ə lif′ər əs), *adj. Mineralogy.* containing or yielding nickel.

nickelous (nik′ə ləs), *adj. Chemistry.* containing nickel, especially with a valence of two.

nicotinamide (nik′ə tin′ə mīd), *n. Biochemistry.* the amide of nicotinic acid; a crystalline compound of the vitamin B complex, convertible into nicotinic acid in living organisms.

nicotinamide adenine dinucleotide, = diphosphopyridine nucleotide. *Abbreviation:* NAD

nicotinamide adenine dinucleotide phosphate, = triphosphopyridine nucleotide. *Abbreviation:* NADP

nicotinic acid, *Biochemistry.* one of a group of vitamins of the vitamin B complex, essential for growth and health in many animals and important in protein and carbohydrate metabolism. It is found in meat, liver, wheat germ, milk, eggs, and yeast, and is used to treat and prevent pellagra. Synthetic nicotinic acid is made by oxidation of the toxic alkaloid nicotine ($C_{10}H_{14}N_2$) found in tobacco. *Formula:* C_5H_4NCOOH Also called **niacin.**

nictitating membrane, *Zoology.* a transparent inner eyelid that can draw over the eye to protect and moisten it, present in birds and certain other animals: *Each eye [of a frog] has a fleshy opaque upper eyelid and a lesser lower lid. Beneath these is a transparent third eyelid, or nictitating membrane* (Storer, *General Zoology*). [from *nictitate* to wink, alteration of Latin *nictatum* winked, blinked]

nidamental (nid′ə men′təl), *adj. Zoology.* secreting or forming a covering or protection for an egg or eggs: *a nidamental gland.* [from Latin *nidamentum* materials for a nest, from *nidus* nest]

nidate (nī′dāt), *v. Embryology.* to become implanted in the uterus: *A woman cannot abort until the fertilized egg cell has nidated and thus become attached to her body* (Eleanor Mears). [from Latin *nidus* nest]
—**nidation,** *n.* the implantation of a fertilized egg in the lining of the uterus.

nidicolous (nī dik′ə ləs), *adj. Zoology.* **1** that remains in the nest for some time after hatching: *Nidicolous species ... are born or hatched in a relatively helpless or*

dependent state (Gilbert Gottlieb). Compare **altricial.** Contrasted with **nidifugous.**
2 that shares the nest of another species of animal. [from Latin *nidus* nest + *colere* inhabit]

nidificate (nid′ə fə kāt), *v. Zoology.* to build a nest or nests. [from Latin *nidificare,* from *nidus* nest + *facere* make]
—**nidification,** *n.* the process or the manner of building a nest.

nidifugous (nī dif′yə gəs), *adj. Zoology.* that leaves the nest a short time after hatching; precocial: *Birds ... may be divided into two main types, those having nidifugous or 'nest-quitting' young, and those having nidicolous or 'nest-dwelling' young* (A. L. Thomson). Compare **precocial.** [from Latin *nidus* nest + *fugere* take flight, flee]

nidify (nid′ə fī), *v.* = nidificate.

nidulant (nij′ə lənt), *adj. Botany.* **1** lying free, or partially embedded, in a nestlike receptacle, as sporangia.
2 lying loose in a pulp, as seeds. [from Latin *nidulantem,* present participle of *nidulari* build a nest, from *nidus* nest]

nidus (nī′dəs), *n., pl.* **-di** (-dī) or **-duses.** *Zoology.* a nest or breeding place, especially one in which insects, snails, and certain other small animals deposit their eggs. [from Latin *nidus* nest]

nielsbohrium (nēlz bō′rē əm), *n.* = hahnium. [named after *Niels Bohr,* 1885-1962, Danish physicist] ▶ See note under **element.**

nightglow, *n. Astronomy.* airglow that occurs at night: *These studies show that the nightglow is faintest at the zenith overhead and grows in intensity down the sky until it reaches a maximum about 10 degrees above the horizon* (Scientific American).

nightside, *n. Astronomy.* the side of a planet, moon, or other celestial body, that faces away from the sun and is thus in darkness: *Temperatures on the nightside of the planet [Mars] were very low, dropping down to −85°F* (J.E. Tesar).

nilpotent (nil′pō′tənt), *adj. Mathematics.* having a power equal to zero: *a nilpotent element. When an expression raised to the square or any higher power vanishes, it may be called nilpotent* (Benjamin Peirce). [from Latin *nil* nothing + *potentem* having power, potent]

nimbostratus (nim′bō strā′təs), *n., pl.* **-ti** (-tī). *Meteorology.* a cloud formation consisting of a dark-gray layer of clouds and occurring at heights under 2700 meters. These clouds usually produce prolonged rain or snow. See the picture at **front.** [from Latin *nimbus* cloud + *stratus* a spreading out]

nimbus (nim′bəs), *n., pl.* **-buses, -bi** (-bī). *Meteorology.* **1** = cumulo-nimbus. **2** (loosely) any rain cloud, especially a low, dark layer of clouds such as a nimbostratus. [from Latin *nimbus* cloud]

---

cap, fāce, fäther; best, bē, tèrm; pin, five;
rock, gō, ôrder; oil, out; cup, pùt, rüle;
yü in use, *yù* in uric;
ng in bring; *sh* in rush; *th* in thin; ᴛʜ in then;
zh in seizure.
ə = *a* in about, *e* in taken, *i* in pencil, *o* in lemon, *u* in circus

niobic (nī ō′bik), *adj. Chemistry.* of or containing niobium, especially with a valence of five.

niobium (nī ō′bē əm), *n. Chemistry.* a soft, ductile metallic element of steel-gray color and brilliant luster, found in nature with tantalum, which it resembles in chemical properties. Niobium is used in making stainless steel and in other alloys, in the cores of nuclear reactors, and in making superconducting magnets. It was formerly called **columbium.** *Symbol:* Nb; *atomic number* 41; *atomic weight* 92.906; *melting point* 2500 $+50°C$; *boiling point* 4742°C; *oxidation state* 3,5. [from New Latin *niobium,* from *Niobe,* daughter of Tantalus (for whom the element tantalum was named)]

niobous (nī ō′bəs), *adj. Chemistry.* of or containing niobium, especially with a valence of three.

nipple, *n. Anatomy.* the small projection on the center of a breast or udder containing the outlets of the milk ducts through which the young of a mammal obtains milk.

nit[1]**,** *n. Zoology.* **1** the egg of a louse or similar parasitic insect. **2** a very young louse or similar insect. [Old English *hnitu*]

nit[2]**,** *n.* the MKS unit of luminance, equal to one candela per square meter. *Abbreviation:* nt. [from Latin *nit(ere)* to shine]

niter (nī′tər), *n. Chemistry.* **1** potassium nitrate, especially when it occurs naturally as a white salt in the soil and encrusted on rocks.
2 sodium nitrate, especially as it occurs in natural deposits.
[from Old French *nitre* sodium carbonate, from Latin *nitrum,* from Greek *nitron,* of Semitic origin]

nitrate (nī′trāt), *Chemistry.* —*n.* **1** a salt or ester of nitric acid, important in the nitrogen cycle.
2 potassium nitrate or sodium nitrate when used as fertilizers.
—*v.* **1** to combine or treat with nitric acid or a nitrate.
2 to change into a nitrate.
—**nitration,** *n.* the process of nitrating; introduction of the nitro group (-NO$_2$) into a compound.

nitrate bacteria, *Biology.* nitrobacteria that convert nitrites to nitrates by oxidation. Also called **nitric bacteria.** Compare **nitrite bacteria.**

nitric (nī′trik), *adj. Chemistry.* of or containing nitrogen, especially with a valence of five. Compare **nitrous.**

nitric acid, *Chemistry.* a clear, colorless, fuming liquid that eats into flesh, clothing, metal, and other substances, usually obtained by oxidation of ammonia or by treating sodium nitrate with sulfuric acid. It is used in nitration, organic synthesis, metallurgy, medicine, etc. *Formula:* HNO$_3$

nitric bacteria, = nitrate bacteria.

nitride (nī′trīd), *Chemistry.* —*n.* a compound of nitrogen with a more electropositive element or radical, such as phosphorus, boron, or a metal. —*v.* to transform into a nitride, as the surface of steel. [from *nitr(ogen)* + *-ide*]

nitrification (nī′trə fə kā′shən), *n. Chemistry.* **1** a process of nitrifying or being nitrified; the conversion of the nitrogen in ammonia into nitrites or nitrates by the oxidizing action of nitrobacteria.

2 the act or process of combining or treating a substance with nitrogen.

nitrify (nī′trə fī), *v. Chemistry.* **1** to oxidize (ammonia compounds, etc.) to nitrites or nitrates, especially by bacterial action.
2 to impregnate (soil, etc.) with nitrates.
3 to combine or treat with nitrogen or one of its compounds.

nitrifying bacteria, = nitrobacteria.

nitrile (nī′trəl *or* nī′trēl), *n. Chemistry.* any of a group of organic cyanides containing the univalent radical -CN. The nitriles form acids on hydrolysis, with the elimination of ammonia.

nitrite (nī′trīt), *n. Chemistry.* a salt or ester of nitrous acid, important in the nitrogen cycle.

nitrite bacteria, *Biology.* nitrobacteria that convert ammonia to nitrites. Also called **nitrous bacteria.** Compare **nitrate bacteria.**

nitro (nī′trō), *adj.* **1** designating or containing the univalent group -NO$_2$. **2** containing niter. **3** = nitric.

nitro-, *combining form.* **1** formed by the action of nitric acid, as in *nitrobenzene.*
2 indicating the presence of the -NO$_2$ radical, as in *nitrocellulose.*

nitrobacteria, *n.pl. Biology.* any of various bacteria living in soil that take part in the nitrogen cycle, oxidizing ammonium compounds into nitrites, or nitrites into nitrates.

nitrogen, *n. Chemistry.* a colorless, odorless, tasteless gaseous element that forms about four fifths of the atmosphere by volume and is a necessary part of all animal and plant tissues: *Nitrogen ... was recognized as an important plant nutrient in manures and other organic matter long before anybody understood the complex cycle by which unreactive atmospheric nitrogen is converted by legumes and soil bacteria into ammonia and soluble nitrogen salts* (Christopher J. Pratt). *Symbol:* N; *atomic number* 7; *atomic weight* 14.0067; *melting point* $-209.9°C$; *boiling point* $-195.8°C$; *oxidation state* $+1, +2, +3, +4, +5.$ [from French *nitrogène,* from *nitre* niter + *-gène* -gen]

nitrogenase (nī troj′ə nās *or* nī′trə jə nās), *n. Biochemistry.* an enzyme of nitrogen-fixing bacteria that activates the conversion of nitrogen to ammonia in the process of nitrogen fixation: *By means of a complex enzyme system, nitrogenase, [certain bacteria] channel metabolic energy to produce ammonia from the atmospheric nitrogen molecule at normal temperature and pressure* (London *Times*).

nitrogen balance, *Physiology.* the difference between the amount of nitrogen taken into the body, soil, etc., and the amount excreted or lost.

nitrogen cycle, *Biology.* the circulation of nitrogen and its compounds by living organisms in nature. Nitrogen in the air passes into the soil, where it is changed to nitrates by bacteria and used by green plants and then by animals. Decaying plants and animals, and animal waste products, are in turn acted on by bacteria and the nitrogen in them is again made available for circulation. *The nitrogen cycle is an example of a material cycle involving decomposers and other soil bacteria which, in part, break down and convert nitrogenous wastes and the remains of dead organisms into materials usable by autotrophs* (Biology Regents Syllabus).

nitrogen fixation, 1 *Biology.* the conversion of atmospheric nitrogen into nitrates by various microorganisms, especially certain bacteria and blue-green algae found in the soil and bacteria present in nodules on the roots of leguminous plants. Nitrogen fixation constantly brings atmospheric nitrogen into biological circulation for use in combined form by plants and other organisms. *At equilibrium, the rate at which nitrogen is added to the soil through biological nitrogen fixation and rain water is equal to the rate at which nitrogen is lost by leaching and denitrification* (F.J. Stevenson).
2 *Chemistry.* the combination of free atmospheric nitrogen with other substances, as in making explosives and fertilizers.

nitrogen-fixer, *n. Biology.* a nitrogen-fixing microorganism: *The free-living nitrogen-fixers are indirectly dependent on plants for their energy or ... obtain energy directly from sunlight* (Scientific American).

nitrogen-fixing, *adj. Biology.* capable of converting atmospheric nitrogen into nitrates: *Rhizobia are nitrogen-fixing bacteria that live symbiotically in nodules on the roots of leguminous plants.*

nitrogenize, *v. Chemistry.* to combine with nitrogen or one of its compounds.
—nitrogenization, *n.* the process of nitrogenizing.

nitrogenous (nī troj′ə nəs), *adj. Chemistry.* of or containing nitrogen or a compound of nitrogen: *The nucleic acids are built up from nucleotides—units which yield on analysis nitrogenous bases (pyrimidines or purines), a pentose sugar and a phosphate radical* (George M. Wyburn).

nitroso (nī trō′sō), *adj. Chemistry.* indicating the presence of the univalent radical -NO: *a nitroso compound, nitroso alcohols or acids.*

nitrous, *adj. Chemistry.* of or containing nitrogen, especially with a valence of three. Compare **nitric.**

nitrous acid, *Chemistry.* a weak acid occurring only in solution or in the form of its salts: *Nitrous acid is formed by the reaction of sodium nitrite with hydrochloric acid, which is present in the stomach* (Leonard C. Labowitz).

nitrous bacteria, = nitrite bacteria.

nival (nī′vəl), *adj. Ecology.* of, having to do with, or growing in or near snow: *nival species.* [from Latin *nivalis,* from *nivis* snow]

nivation (nī vā′shən), *n. Geology.* erosion of the ground beneath a snowbank, caused chiefly by alternate freezing and thawing. [from Latin *nivis* snow + English *-ation*]

nm, *abbrev.* nanometer.

NMR, *abbrev.* nuclear magnetic resonance.

No, *symbol.* nobelium.

no., *pl.* **nos.** number [for Latin *numero* by number].

nobelium (nō bē′lē əm), *n. Chemistry.* a very heavy radioactive element produced artificially by bombarding curium with carbon ions. *Symbol:* No; *atomic number* 102; *mass number* (of the most stable isotope) 255; *half-life* 3 minutes. [named after Alfred B. *Nobel,* 1833–1896, Swedish inventor and founder of the Nobel prizes]

noble, *adj. Chemistry.* relatively inactive; not reacting or combining readily with other elements; inert: *a noble element, a noble liquid.* The **noble gases** include helium, neon, argon, krypton, xenon, and radon. The **noble metals** include metals that are highly resistant to oxidation and corrosion, such as gold and platinum.

noctilucent cloud (nok′tə lü′sənt), *Meteorology.* a cloud that shines at night, resembling a cirrus and usually seen in the summer at great heights and in high latitudes: *There has been considerable dispute in the past about whether the noctilucent clouds are composed of ice particles or dust particles* (New Scientist). Compare **nacreous cloud.** [from Latin *nox, noctis* night + *lucentem* shining]

noctuid (nok′chù id), *Zoology.* **—adj.** of or belonging to a very large family (Noctuidae) of usually dull-colored, nocturnal moths including most of those attracted to lights at night. Many of their larvae are injurious to crops, such as the cutworms, army worms, and cotton worms. *The noctuid moth ... can detect the ultrasonic cries of the bat. By rapidly falling to the ground when the sounds of the bat are detected, the moth can escape an attack which would otherwise be fatal* (J. Richard Thomson).
—n. a noctuid moth.
[from New Latin *Noctuidae* the family name, from Latin *noctua* a night owl, from *nox, noctis* night]

nocturnal, *adj. Zoology.* active in the night: *Nocturnal species are frequently primitive. This is particularly apparent in the case of insects* (J.L. Cloudsley-Thompson). [from Latin *nocturnus* of the night, from *noctem* night] **—nocturnally,** *adv.*

nodal, *adj.* having to do with or like a node or nodes: *The cirri are borne ... upon certain joints of the stem, hence termed nodal* (George Rolleston). *We have not only nodal points in a vibrating string, but we may have nodal lines in a vibrating plate* (W. Lees).

node, *n.* **1** *Botany.* any joint in a stem where leaves grow out: *Usually the nodes are not visible in the older stems of mature trees, shrubs, or lianas but are always evident on the younger twigs* (Emerson, *Basic Botany*).

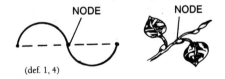

NODE   NODE

(def. 1, 4)

2 *Physics.* a point, line, or plane in a vibrating body at which there is comparatively no vibration: *Certain points known as the nodes remain always at rest* (Sears and Zemansky, *University Physics*).

cap, fāce, fäther; best, bē, tèrm; pin, five;
rock, gō, ôrder; oil, out; cup, pùt, rüle,
yü in use, *yu* in uric;
ng in bring; *sh* in rush; *th* in thin, ᴛʜ in then;
zh in seizure.
ə = *a* in about, *e* in taken, *i* in pencil, *o* in lemon, *u* in circus

3 *Astronomy.* either of the two points at which the orbit of a celestial body intersects the path of the sun or the orbit of another celestial body: *The points of intersection are known as the Moon's nodes, the one where the Moon crosses from the south side of the ecliptic to the north being called the ascending node and the other the descending node* (Duncan, *Astronomy*).

4 *Geometry.* a point at which a curve crosses itself, or a similar point on a surface.

5 *Anatomy.* **a** = node of Ranvier. **b** = lymph gland. [from Latin *nodus* knot]

node of Ranvier (räɴ vyā′), *Anatomy.* the site of a local constriction at varying intervals in the myelin and axon of a myelinated nerve cell. The node is formed at the point where two Schwann cells meet. *The ion movements can occur only in the gaps in the myelin sheath ... called the nodes of Ranvier, and the nerve impulse has to jump from one node to the next* (New Scientist). See the picture at **myelin.** [named after L.A. Ranvier, 1835–1922, French histologist]

nodical (nod′ə kəl), *adj. Astronomy.* of or having to do with nodes.

nodose (nō′dōs), *adj. Biology.* having or characterized by many nodes or protuberances; jointed or knobby at intervals.

—**nodosity** (nō dos′ə tē), *n.* the condition of being nodose; nodose state or appearance.

nodular, *adj.* having or occurring in the form of nodules: *Beneficial soil bacteria of the genus Rhizobium ... penetrate young rootlets and then multiply, forcing the root wall outward into a nodular swelling* (Noel D. Vietmeyer).

nodulate, *v. Botany.* to form nodular growth in: *Its surface is uneven and nodulated like that of a raspberry* (Robert B. Todd).

—**nodulation,** *n.* the formation of nodular growths or nodules: *the nodulation of white clover.*

nodule, *n.* **1** *Anatomy.* a small mass of tissue or aggregation of cells.

2 *Botany.* a small swelling on the root of a leguminous plant that contains nitrogen-fixing bacteria.

3 *Geology.* a small, rounded mass of ore or rock that is usually harder than the sediment or rock enclosing it: *An estimated 1.7 trillion tons of mineral deposits, spongy chunks about the size of potatoes litter the deep-sea floor. The coved lumps, called nodules, contain nickel, copper, manganese, cobalt and traces of other substances* (Paul Hofmann).

nodulose, *adj. Biology.* having little knots or knobs; consisting of small nodules.

noise, *n.* **1** *Physics.* a group of waves or vibrations which are not periodic and which cover a wide range of frequencies.

2 any fluctuation or disturbance which is not part of a desired signal or which interferes with its intelligibility or usefulness.

nomenclature (nō′mən klā′chər *or* nō men′klə chər), *n.* **1** a set or system of terms used in a particular science: *the nomenclature of mineralogy.* SYN: terminology.

2 *Biology.* the procedure of assigning names to the kinds and groups of animals or plants listed in a taxonomic classification: *Some rules of nomenclature were*

used by Linnaeus, and others were proposed later. (Storer, *General Zoology*). SYN: taxonomy.

▶ The Linnaean system of *binomial nomenclature,* using New Latin names for the genus and species of animals or plants, has been adopted by scientists throughout the world. In this system, the genus name is a single capitalized Latin word in the nominative singular, and the species name is a lower-case single or compound Latin word, usually an adjective agreeing grammatically with the genus name. Thus, the English sparrow is named *Passer domesticus.* When a species has a subspecies or variety, a *trinomial nomenclature* is employed, e.g. *Passer domesticus domesticus* for the English sparrow of continental Europe. See also **classification.**
[from Latin *nomenclatura,* from *nomen* name + *calare* to call]

—**nomenclator** (nō′mən klā′tər), *n.* a person who devises a nomenclature.

—**nomenclatorial** or **nomenclatural,** *adj.* of or having to do with nomenclature: *nomenclatorial references, nomenclatural rules.*

nomogram (nom′ə gram) or **nomograph** (nom′ə graf), *n. Mathematics.* a chart from which one can determine by alignment of scales the value of a dependent variable for any given value of the independent variable: *The interrelation of these parameters and their relationship with the magnetic field parameters are then expressed by a family of curves, or nomograms, intended for use in the solving of design problems* (Technology). [from Greek *nomos* law + English *-gram*]

nonagon (non′ə gon), *n. Geometry.* a plane figure having nine angles and nine sides. [from Latin *nonus* ninth + Greek *gōnia* angle]

nonconductor, *n. Physics.* a substance that does not readily conduct heat, electricity, sound, etc. Rubber is a nonconductor of electricity. *In a nonconductor, or insulator, all the valence electrons are tied up in the chemical bonds which hold the crystal together. Such electrons are immobilized and cannot contribute to electrical conductivity* (Scientific American).

nonconformity, *n. Geology.* an unconformity in which an erosion surface on plutonic or metamorphic rock has been covered by younger sedimentary or volcanic rock: *A nonconformity generally indicates a deep or long-continuing erosion before subsequent burial* (Plummer and McGeary, *Physical Geology*).

nondisjunction, *n. Biology.* the failure of a pair of chromosomes to separate and go to different cells when the cell divides: *The XYY deviance is caused by the nondisjunction of sex chromosomes in male meiosis* (J. Richard Thomson).

nonelectrolyte, *n. Chemistry.* a substance which in water solution does not conduct an electric current. Sugar is a nonelectrolyte.

non-Euclidean, *adj. Mathematics.* of, based upon, or in accordance with a geometric system which does not hold the Euclidean postulate that a given line has exactly one parallel through a point external to it. Non-Euclidean spaces are either hyperbolic (in which a line has an infinite number of parallels through a point) or elliptic (in which a line has no parallel through a point). *The relation between distance and volume ... is of crucial importance in differential geom-*

etry, that part of geometry that is concerned with the properties of curved, higher dimensional spaces. (Because of their unconventional, "non-Euclidean" structure, these spaces frequently serve as models for research in cosmology and relativistic physics) (Lynn A. Steen).

nonharmonic, adj. Physics. not harmonic; without harmony: When as in the case of a drum or a bell, [overtones] are not integral multiples of the fundamental they are called nonharmonic (Shortley and Williams, Elements of Physics).

nonionic (non'ī on'ik), adj. Chemistry. that does not ionize in solution: nonionic substances.

nonlinear, adj. Mathematics. **1** not linear; not containing or contained in a line: a nonlinear set of points.
2 (of an equation) containing a variable with an exponent other than one: the integration of nonlinear differential equations.
—**nonlinearity,** n. the property of being nonlinear; lack of linear relation or proportion between two related quantities.

nonmagnetic, adj. Physics. not having the properties of a magnet; that cannot be attracted by a magnet: a nonmagnetic ore.

nonmetal, n. Chemistry. a chemical element lacking the physical and chemical properties of a metal; nonmetallic element. A nonmetal forms acidic oxides and is electronegative in solution.
—**nonmetallic,** adj. not metallic; lacking the physical and chemical properties of a metal. Carbon, oxygen, sulfur, and nitrogen are nonmetallic elements.

nonpolar, adj. Chemistry. not polar; having no electric dipole moment: When electrons are shared between atoms of the same element, they are shared equally and the resulting bond is nonpolar. An example of a nonpolar covalent bond is found in the fluorine molecule (Chemistry Regents Syllabus).

nonprotein, Biochemistry. —n. a substance that is not a protein or does not contain protein. —adj. having no protein or proteins: a nonprotein molecule.

nonsense, adj. Molecular Biology. **1** that does not specify a particular amino acid in the genetic code: Nonsense codons may have the function of terminating the polypeptide chain (Watson, Molecular Biology of the Gene).
2 that results from the presence of nonsense sequences: a nonsense mutation, a nonsense protein. Compare **missense.**

nonsexual, adj. Biology. **1** having no sex; asexual: nonsexual organisms. **2** done by or characteristic of asexual animals: the nonsexual reproduction of hydras by means of budding.

nonstandard analysis, Mathematics. the study of infinitely large and infinitely small numbers: Nonstandard analysis adopts an "infinitesimal" approach. For example, it conceives of a line as consisting of an infinite number of points separated by infinitely small distances (Steven Moll).

nonstriated (non strī'ā tid), adj. not striped; smooth: nonstriated or smooth muscle.

noradrenaline or **noradrenalin,** n. Biochemistry. a hormone secreted by the adrenal medulla and functioning as a vasoconstrictor and as a neurotransmitter at the endings of sympathetic nerves. Formula: $C_8H_{11}NO_3$

Also called **norepinephrine.** Compare **levarterenol.** [from nor(mal) + adrenaline]

norepinephrine, n. = noradrenaline. [from nor(mal) + epinephrine]

norite (nôr'īt), n. Geology. a granular igneous rock solidified below the earth's surface, containing pyroxene in orthorhombic form and plagioclase containing calcium. Norite is a variety of gabbro, often associated with ore deposits. [from Nor(way), where it is found + -ite]

norm, n. **1** Mathematics. **a** an average; mean. **b** a quantity defined on a vector space with properties like those of the modulus of a complex number. **c** (of a matrix) the square root of the sum of the squares of the absolute values of the elements.
2 Geology. a hypothetical mineral composition of a rock calculated by assigning the compounds present to certain relatively simple minerals in accordance with prescribed rules. Compare **mode.**
[from Latin norma rule, pattern]

normal, adj. **1** Chemistry. **a** (of a solution) containing one gram equivalent of a dissolved substance per liter of solution: A solution of one-tenth normal strength is designated as N/10 (Strouts, Analytic Chemistry). Symbol: N **b** (of hydrocarbons) consisting of a straight unbranched chain of carbon atoms: normal paraffins. **c** not found in association: normal molecules.
2 Geometry. being at right angles: The word normal is used here with its geometrical meaning, to indicate a direction perpendicular to a plane (Shortley and Williams, Elements of Physics).
—**n.** Geometry. a line or plane that is at right angles to another: The angle between any wave front and the surface is equal to the angle between the corresponding ray and the normal to the surface, since the sides of the angles are mutually perpendicular (Sears and Zemansky, University Physics).
[from Latin normalis, from norma rule, pattern]
—**normally,** adv.

normal curve, Statistics. **1** the bell-shaped curve representing a normal distribution.
2 = normal distribution: The calculus of errors also established the "normal curve" as the distribution of the frequencies of possible measurements of a single quantity, and the mean of these measurements as the true value of the quantity (Morris Kline). Also called **Gaussian curve, probability curve.**

normal distribution, Statistics. a frequency distribution in which the quantities are so distributed that they represent the hypothetical distributions of various samplings. The normal distribution is a mathematical model represented by a symmetrical, bell-shaped, asymptotic curve whose maximum height is at the mean. Also called **Gaussian distribution.**

---

cap, fāce, fäther; best, bē, tèrm; pin, fīve;
rock, gō, ôrder; oil, out; cup, pùt, rüle;
yü in use, yu̇ in uric;
ng in bring; sh in rush; th in thin, ᵀH in then;
zh in seizure.
ə = a in about, e in taken, i in pencil, o in lemon, u in circus

normality, *n. Chemistry.* the number of gram equivalents of a dissolved substance per liter of solution.

normal pressure, 1 *Physics.* pressure equivalent to one atmosphere. **2** *Meteorology.* atmospheric pressure equivalent to 1000 millibars.

normal salt, *Chemistry.* a salt formed from an acid of which all of the hydrogen has been replaced.

normal temperature, *Physics.* zero degrees Celsius; 273 kelvins. Also called **standard temperature.**

north, *n. Geography.* the direction of the north magnetic pole, to which a compass needle points; direction to the right as one faces the setting sun.

north celestial pole, *Astronomy.* the point where the northern end of the earth's axis intersects the celestial sphere. It is located just over 1 degree from Polaris (the North Star) and is one of two points on the celestial sphere about which all stars seem to rotate: *The zenith at the north geographical pole is of special importance; this point on the celestial sphere is called the north celestial pole. Opposite it is the south celestial pole. Midway between the celestial poles is the celestial equator; it divides the celestial sphere into northern and southern hemispheres* (Krogdahl, *The Astronomical Universe*).

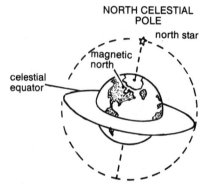

NORTH CELESTIAL POLE

northeaster, *n. Meteorology.* a wind or storm from the northeast.

northern, *adj. Astronomy.* of or in the northern half of the celestial sphere: *a northern star or constellation.*

northern circle, = tropic of Cancer.

Northern Hemisphere, *Geography.* the half of the earth that is north of the equator. See the picture at **hemisphere.**

northern lights, = aurora borealis: *The northern lights can shift about 300 to 3,000 feet a second. This speed is less than would be expected if proton bombardment were directly responsible for the formation of auroras* (Science News Letter).

north geographic pole, *Geography.* the northern end of the earth's axis; the point on the earth's surface from which every direction is south. Also called **North Pole.**

northing (nôr′thing *or* nôr′ᴛʜing), *n. Astronomy.* declination measured northward.

north magnetic pole, *Geography.* the point on the earth's surface toward which a magnetic needle points; the north pole of the earth's magnetic field. Its location varies slightly from year to year but is about 1760 kilometers from the north geographic pole, near Bathurst Island.

north pole, 1 *Physics.* the pole of a magnet that would attract that end of a compass needle that normally points north. **2** = north magnetic pole. **3** = north celestial pole. **4 North Pole.** = north geographic pole.

northwester, *n. Meteorology.* a wind or storm from the northwest.

nose, *n. Anatomy.* **1** the part of the human face or head just above the mouth which contains the nostrils. **2** any similar structure in vertebrate animals.

nosogeography (nos′ō jē og′rə fē), *n. Medicine.* the study of disease in relation to geographical factors. [from Greek *nosos* disease + English *geography*]

nosology (nō sol′ə jē), *n.* **1** the branch of medicine dealing with the classification of diseases. **2** a classification of disease. [from Greek *nosos* disease + English *-logy*] —**nosological** (nos′ə loj′ə kəl), *adj.* having to do with nosology. —**nosologically,** *adv.*

nostril, *n. Anatomy.* either of the two external openings in the nose. Air is breathed into the lungs, and odors are detected by the olfactory bulbs in the upper parts of the nostrils.

notation, *n.* **1** *Mathematics.* a set of signs or symbols used to represent numbers, quantities, or other values: *In arithmetic we use the Arabic notation (1, 2, 3, 4, etc.).* **2** the representing of numbers, quantities, or other values by symbols or signs: *Chemistry has a special system of notation.*

notochord (nō′tə kôrd), *n. Zoology.* **1** a flexible, rodlike structure of cells running lengthwise in the back of invertebrate chordates. It forms the main supporting structure of the body. *Just under the spinal chord is a flexible tubelike body which is called the notochord. ... In the higher chordates the notochord is present in the embryo only, since it is replaced by the bony vertebral column in the adult form* (Winchester, *Zoology*). **2** a similar structure in the embryos of all chordates. [from Greek *nōton* back + Latin *chorda* cord] —**notochordal** (nō′tə kôr′dəl), *adj.* of or having to do with the notochord: *the notochordal region.*

Notogaea (nō′tə jē′ə), *n. Zoology, Ecology.* a zoogeographic region that includes Australia, New Zealand, South America, and tropical North America. [from Greek *notos* south (wind) + *gaia* earth, land] —**Notogaean,** *adj.* of or having to do with the Notogaea.

notum (nō′təm), *n., pl.* **-ta** (-tə). *Zoology.* the dorsal part of an insect's thoracic segment. [from New Latin, from Greek *noton* the back]

nova (nō′və), *n., pl.* **-vae** (-vē), **-vas.** *Astronomy.* a star that suddenly becomes much brighter and then gradually fades, to its normal brightness, over a period of several weeks, months, or sometimes years: *[Stars] occasionally boil up to a state of instability that results in their exploding as a nova* (William A. Fowler). [from Latin *nova,* feminine singular of *novus* new]

novaculite (nō vak′yə līt), *n. Geology.* a very tough sedimentary rock composed mostly of microcrystalline quartz, used especially for hones and for grinding wheels. [from Latin *novacula* razor + English *-ite*]

Np, *symbol.* neptunium.

ns, *abbrev.* nanosecond.

N-shell, *n. Physics.* the energy level next in order after the M-shell, occupied by electrons whose principal quantum number is 4. See also **N electron.**

nt., *abbrev.* **1** nit (unit of luminance). **2** newton.

nth (enth), *adj. Mathematics.* last in the sequence 1, 2, 3, 4 ... n; being of the indefinitely large or small amount denoted by *n. To the nth power* means to any required power. [from *n(umber)* + the numerical suffix *-th,* as in *tenth*]

nucellar (nü sel′ər), *adj. Botany.* of or having to do with the nucellus; produced in or derived from the nucellus: *The nucellar epidermis is sometimes highly resistant and may proliferate into a nucellar cap with relatively thick walls* (Katherine Esau).

nucellus (nü sel′əs), *n., pl.* **-celli** (-sel′ī). *Botany.* the central mass of cells within the integument of the ovule containing the embryo sac: *Each ovule consists of a massive nucellus, or megasporangium with a single enveloping integument* (Greulach and Adams, *Plants*). [from New Latin *nucellus,* from Latin *nux, nucis* nut]

nuclear, *adj.* **1** of or having to do with nuclei or a nucleus, especially the nucleus of an atom: *nuclear mass, nuclear radiation. It was suspected from the beginning that radioactivity is a nuclear process and that the emission of a charged particle from the nucleus of an atom results in leaving behind a different atom, occupying a different place in the periodic table* (Sears and Zemansky, *University Physics*).
2 of or having to do with atoms or atomic energy: *nuclear power.* ► See the note under **atomic.**

nuclear bomb, = atomic bomb.

nuclear chemistry, the branch of chemistry dealing with atoms and atomic nuclei, and their relation to chemical processes and reactions, especially reactions producing new elements.

nuclear energy, = atomic energy: *When fission occurs, the nucleus of the atom splits into two (occasionally more) lighter elements, with release of nuclear energy* (Science News Letter).

nuclear envelope, = nuclear membrane: *In eukaryotic cells, the nucleus is a large, often spherical body, usually the most prominent structure within the cell. It is surrounded by two lipoprotein membranes, which together make up the nuclear envelope.* (Helena Curtis, *Biology*).

nuclear fission, = fission (def. 2): *The uranium nucleus on capturing a neutron split into nuclei of intermediate mass number; this process is called nuclear fission.* (Shortley and Williams, *Elements of Physics*).

nuclear force, *Physics.* the force which holds protons and neutrons together in the nucleus. It differs from electromagnetic forces, is of short range, and is governed by the strong interaction. *Nuclear forces operate when the distance between nucleons is less than* 10^{-15} *meters. Nuclear forces exceed all other types by many orders of magnitude* (Physics Regents Syllabus).

nuclear fuel, *Physics.* material containing a fissionable substance that will sustain a chain reaction.

nuclear fusion, = fusion (def. 2): *The principle of any hydrogen explosive is nuclear fusion. Hydrogen atoms (some of them created from lithium during the explosion) are brought together, and when they fuse far more energy is released than in fission* (Wall Street Journal). *Nuclear fusion ... is generally believed [to be] the source of the sun's heat* (Science News). See the picture at **nuclear fission.**

nuclear magnetic resonance, *Physics.* an effect in which nuclei in various electrical environments interact with an external magnetic field when various radiation frequencies are applied to the nuclei, used especially in spectroscopic studies of the structure of molecules: *Nuclear magnetic resonance is being used to measure the rate of venous blood flow in the forearm* (Science News). *Abbreviation:* NMR

nuclear membrane, *Biology.* the double-layered membrane enclosing the nucleus of a cell: *The nucleus is surrounded by a nuclear membrane and contains a clear nucleoplasm within. Arranged in the form of a network within the nucleoplasm will be the chromatin* (Winchester, *Zoology*). Also called **nuclear envelope.**

nuclear particle, *Physics.* a particle within, or emitted by, an atomic nucleus. Nuclear particles include neutrons, protons, deuterons, and alpha particles. *If the energy of the incident particle is high enough it can enter the target nucleus and cause it to break up with the emission of a number of nuclear particles* (P.E. Hodgson).

nuclear physics, the branch of physics dealing with the structure of atomic nuclei, and the behavior of nuclear particles: *One of the best established facts in nuclear physics is that the density of nuclear matter is constant.* (D. Allan Bromley). ► See the note under **particle physics.**

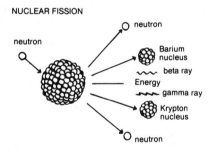

NUCLEAR FISSION

neutron · neutron · Barium nucleus · beta ray · Energy · gamma ray · Krypton nucleus · neutron

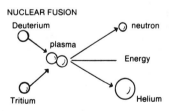

NUCLEAR FUSION

Deuterium · plasma · neutron · Energy · Tritium · Helium

nuclear reaction, *Physics.* a reaction involving a change in the nucleus of an atom. Some nuclear reactions can result in the release of a tremendous amount of energy: *Nuclear reaction results from the splitting of the atoms*

cap, fāce, fäther; best, bē, tėrm; pin, fīve;
rock, gō, ôrder; oil, out; cup, pùt, rüle,
yü in use, *yú* in uric;
ng in bring; *sh* in rush; *th* in thin, ᴛʜ in then;
zh in seizure.
ə = *a* in about, *e* in taken, *i* in pencil, *o* in
lemon, *u* in circus

... if allowed to reach maximum intensity, it results in atomic explosions (N.Y. Times). Compare **chain reaction.**

nuclear reactor, = reactor: *Nuclear reactors are the machines for converting the energy of nuclear fission into forms that can be turned to useful purposes* (Scientific American).

nuclear sap, = karyolymph: *Within the nucleus is a colorless fluid, the nuclear sap (nucleoplasm), in which there is a substance that has a strong affinity for certain dyes; this is known as chromatin* (Hegner and Stiles, *College Zoology*).

nuclease (nü′klē ās), *n. Biochemistry.* any of a group of enzymes that hydrolyze nucleic acids.

nucleate, *v.* **1** to form a nucleus or nuclei; form into a nucleus or around a nucleus: *Inclusions can nucleate, multiply and grow dendritically* (Scientific American). **2** to form a nucleus in; provide a nucleus: *Findeisen proposed as early as 1938 that for ice crystals to appear in supercooled clouds a nucleating or seeding agent might be required* (B.J. Mason).
—*adj.* having a nucleus: *nucleate organisms.*
—**nucleation,** *n.* the formation of a nucleus or nuclei.

nucleator, *n. Chemistry.* a substance or agent that produces nuclei in gases or liquids: *seeding clouds with artificial nucleators.*

nuclei, *n.* a plural of **nucleus.**

nucleic acid, *Biochemistry.* any of a group of compounds occurring chiefly in combination with proteins in living cells and viruses, consisting of linked nucleotides: *Nucleic acids are the chemicals that control cellular heredity and function* (Science News Letter).

nuclein, *n. Biochemistry.* any of a class of substances present in the nuclei of cells, consisting chiefly of proteins, phosphoric acids, and nucleic acids.
—**nucleinic,** *adj.* of, having to do with, or resembling a nuclein or nucleins.

nucleo-, *combining form.* **1** nucleus, as in *nucleophile, nucleosynthesis.* **2** nucleic acid, as in *nucleocapsid, nucleotide.* [from Latin *nucleus* kernel]

nucleocapsid, *n. Biology.* the structure of a virus or virion, consisting of the capsid or protein coat and the nucleic acid which it encloses: *The RNA of the influenza virus is found in five to seven discrete pieces, each in its own nucleocapsid, and its total mass is about 4 million daltons. Each of the pieces ... is an intact gene that controls at least one characteristic of the virus* (Science). [from *nucleo-* + *capsid*]

nucleogenesis, *n. Chemistry, Physics.* the formation of atomic nuclei: *the nucleogenesis of uranium isotopes, nucleogenesis in stars.* Compare **nucleosynthesis.**

nucleohistone, *n. Biochemistry.* a nucleoprotein whose protein component is a histone.

nucleoid (nü′klē oid), *n. Biology.* that part of a prokaryotic cell which contains the DNA; the nucleic-acid portion of a cell: *Distinct DNA-containing regions known as nucleoids occupy the rest of the cell. The DNA is so densely packed into a nucleoid that most cytoplasmic particles are excluded, but the nucleoid is not surrounded by a membrane* (Ursula Goodenough).

nucleolar, *adj. Biology.* having to do with or of the nature of a nucleolus: *The centromere is a small round body that divides the chromosome into "arms" and is believed to play the important role of "nucleolar organizer" in cell division* (Science News Letter).

nucleolus, *n., pl.* **-li** (-lī). *Biology.* a small structure, usually round, found within the nucleus of a cell, containing a high concentration of ribonucleic acid: *The most conspicuous body within the nucleus is the nucleolus. ... the site at which ribosomal subunits are constructed* (Helena Curtis, *Biology*). See the picture at **centrosome.**

nucleon (nü′klē on), *n. Physics.* a proton or neutron, especially as a component of an atomic nucleus: *It seems that each nucleon (a general name for either proton or neutron) cannot interact with all the other particles in the nucleus, but only with its neighbours* (J. Little).
—**nucleonic,** *adj.* **1** of or having to do with a nucleon or nucleons: *nucleonic wave functions.* **2** of or having to do with nucleonics: *An important tool of nucleonic research is the mass spectrograph* (Electronics).
—**nucleonics,** *n.* **1** the study of the behavior and characteristics of nucleons or atomic nuclei: *Between the expanding microcosm of nucleonics and the macrocosm of astronomy we are learning to see anew* (Newsweek). **2** the development of instruments for use in nuclear research.

nucleophile (nü′klē ə fīl), *n. Chemistry.* a nucleophilic substance with an unshared electron pair which in a reaction leads to sharing of the electrons by the formation of a covalent bond: *In the hydrolysis of polypeptides, bond breakage is usually initiated by a nucleophile, a compound with electrons to spare* (Hans Neurath).

nucleophilic (nü′klē ə fil′ik), *adj. Chemistry.* strongly attracted to the nuclei of atoms and so donating or sharing electrons: *nucleophilic ions or molecules. Putrid odors, on the other hand, are caused by molecules that have an excess of electrons and are called nucleophilic, because they are strongly attracted by the nuclei of adjacent atoms* (Scientific American). Contrasted with **electrophilic.**

nucleoplasm (nü′klē ə plaz′əm), *n. Biology.* **1** the protoplasm in the nucleus of a cell. **2** = karyolymph.
—**nucleoplasmic** (nü′klē ə plaz′mik), *adj.* having to do with or resembling nucleoplasm.

nucleoprotein, *n. Biochemistry.* any of a group of substances present in the nuclei or nucleoids of cells and in viruses, consisting of proteins in combination with nucleic acids. On hydrolysis, nucleoproteins yield amino acids, purine and pyrinidine bases, phosphoric acid, and pentose sugars.

nucleoside (nü′klē ə sīd), *n. Biochemistry.* any of a group of compounds of a nitrogen base (purine or pyrimidine) and sugar (pentose), similar to a nucleotide but lacking phosphoric acid: *Adenosine is the subunit compound, or nucleoside, of a sugar molecule and adenine, one of the four bases or side groups found in nucleic acids* (G. Allan Robinson).

nucleosome (nü′klē ə sōm), *n. Biochemistry.* the basic structural unit of chromatin, consisting of a DNA chain coiled around a core of histones: *Each nucleo-*

some is roughly spherical, about 100 Å (10 nm) in diameter, and consists of 8 histone molecules and about 200 base pairs of DNA (Bruce A.J. Ponder) [from *nucleo-* + *-some*]
—**nucleosomal** (nü′klē ə sō′məl), *adj.* of or having to do with a nucleosome or nucleosomes: *nucleosomal DNA.*

nucleosynthesis, *n. Physics, Astronomy.* the process by which chemical elements more complex than hydrogen are formed, especially in the evolution of stars: *Helium is one of the most abundant elements in the universe, and it is the first nuclear species not an isotope of hydrogen that can be made when nucleosynthesis starts with free protons and neutrons* (Science News).
—**nucleosynthetic,** *adj.* of or having to do with nucleosynthesis: *The relative abundance of ^{244}Pu was the factor determining whether any exceptionally large nucleosynthetic event occurred just before the solar system formed* (Nature).

nucleotidase (nü′klē ə tī′dās), *n. Biochemistry.* an enzyme which splits phosphoric acid from a nucleotide, producing a nucleoside.

nucleotide (nü′klē ə tīd), *n. Biochemistry.* any of a group of compounds consisting of a sugar, phosphoric acid, and a purine or pyrimidine base. Nucleotides are the constituent molecules of nucleic acids (DNA and RNA). *As a direct consequence of the base-pairing mechanism, it becomes evident that DNA carries information by means of the linear sequence of its nucleotides. Each nucleotide—A, C, T, or G—can be considered as a letter in a simple four-letter alphabet that is used to write out biological messages in a linear "ticker-tape" form* (Alberts, *Molecular Biology of the Cell*).

nucleus (nü′klē əs), *n., pl.* **-clei** (-klē ī) or **-cleuses. 1** *Physics.* the central part of an atom, consisting of protons and neutrons. The nucleus carries a positive charge and forms a core containing most of the mass of an atom around which electrons orbit. See the picture at **ion.**
2 *Biology.* a mass of specialized protoplasm found in the cells of most organisms, except blue-green algae, bacteria, and other monerans, without which the cell cannot grow and divide. A nucleus is different in structure from the rest of the cell. It consists of complex arrangements of proteins, nucleic acids, and lipids surrounded by a delicate membrane, and typically containing such structures as chromosomes and nucleoli. The nucleus controls growth, cell division, and other activities, and contains DNA, a nucleic acid which passes on the genetic characteristics of the cell. *If the nucleus is removed, the cell may live for a long time, dying later for want of the governmental functions of the nucleus* (Daniel Mazia). See the pictures at **cell, mitosis, neuron.**
3 *Chemistry.* **a** the fundamental, stable arrangement of atoms in a particular compound: *the benzene nucleus.*
b = ring.
4 *Astronomy.* the dense central part of a comet's head: *The visible nucleus is really the brightest portion of the inner coma, the true cometary nucleus being a rather small object, typically perhaps 1–5 miles across* (Hyron Spinrad).

5 *Anatomy.* a specialized mass of gray matter in the brain or spinal cord.
6 *Meteorology.* a speck of dust or other particle upon which water vapor condenses, as to form a drop. Also called **condensation nucleus.**
[from Latin *nucleus* kernel, from *nux, nucis* nut]

nuclide (nü′klīd), *n. Physics.* a particular type of atom having a characteristic nucleus and a measurable life span: *Carbon 14 is a radioactive nuclide of carbon.* Compare **isotope.**

nudibranch (nü′də brangk), *n. Zoology.* any of a suborder (Nudibranchia) of marine gastropod mollusks that usually have external adaptive gills and no shell when adult. [from Latin *nudus* naked + Greek *branchia* gills]

nullipennate (nul′ə pen′āt), *adj. Zoology.* having no flight feathers, as the penguin. [from Latin *nullus* not any + *penna* feather]

null set, = empty set: *In current terminology the "solution set" of a false statement is not the null set but an element (member) of the null set, and since the null set has no members the statement is true of nothing* (Martin Gardner).

number, *n. Mathematics.* **1** a unit or amount of units, especially the count or sum of a group, series, etc.; quantity. The symbols used to represent numbers are called numerals. The number of one's fingers is ten, represented by the numeral 10 (in the decimal system). Compare **complex number, imaginary number, irrational number, natural number, prime number, rational number, real number, whole number.** See also **integer.**
2 a figure or mark that stands for a number; a numeral. 1 and 10 are numbers.
[from Old French *nombre,* from Latin *numerus*]

number theory, *Mathematics.* the study of integers and their relationships. Number theory is especially concerned with the properties of the natural numbers 1, 2, 3, 4. ... , using theorems and prime numbers to describe these properties. Also called **theory of numbers.**

numeral (nü′mər əl), *n. Mathematics.* a figure, letter, or word standing for a number; a group of figures, letters, or words standing for a number. 1, 5, 10, 50, 100, 500, and 1000 are Arabic numerals. I, V, X, L, C, D, and M are Roman numerals. [from Late Latin *numeralis,* from Latin *numerus* number]

numeration system or **numeral system,** *Mathematics.* any system of counting or naming numbers, such as the decimal system and the binary system.

numerator (nü′mə rā′tər), *n. Mathematics.* the number above the line in a fraction which shows how many parts are taken: *In 3/8, 3 is the numerator and 8 is the denominator. In a proper fraction, the numerator is smaller than the denominator; in an improper fraction, the numerator is greater.*

---

cap, fāce, fäther; best, bē, tėrm; pin, fīve;
rock, gō, ôrder; oil, out; cup, pùt, rüle;
yü in use, *yu̇* in uric;
ng in bring; *sh* in rush; *th* in thin, *ŦH* in then;
zh in seizure.
ə = *a* in about, *e* in taken, *i* in pencil, *o* in lemon, *u* in circus

numeric (nü mer′ik), *adj.* = numerical.

numerical (nü mer′ə kəl), *adj. Mathematics.* **1** of or having to do with numbers; in or by numbers: *It [a sorting machine] can sort cards in numerical or alphabetical order at the rate of 1,000 a minute* (Wall Street Journal).
2 shown by numbers or digits, not by letters: *10 is a numerical quantity; bx is an algebraic quantity.* —**numerically,** *adv.*

numerical analysis, *Mathematics.* the study of methods of approximating the solution of various types of mathematical problems, especially through the use of computers.

numerical taxonomy, the branch of taxonomy dealing with the quantitative evaluation of similar characters and the arrangement of taxonomic units on the basis of similarities: *In the second edition of their work on numerical taxonomy, Peter Sneath and Robert Sokal stated: "Numerical taxonomy is a revolutionary approach to biological classification. ... Instead of qualitatively appraising the resemblance of organisms on the basis of certain favored characters, a taxonomist using this new methodology will attempt to amass as many distinguishing characters as possible, giving equal weight to each* (John G. Lepp).

nut, *n. Botany.* **1** (strictly) a dry, one-seeded, indehiscent fruit similar to an achene, but larger and having a harder and thicker pericarp, such as an acorn, walnut, hazelnut, or beechnut.
2 (loosely) a dry fruit or seed with a hard, woody or leathery shell containing an edible kernel, such as a peanut, almond, or coconut.

nutant (nü′tənt), *adj.* **1** *Botany.* drooping; nodding.
2 *Zoology.* **a** sloping in relation to the parts behind it or with the axis of the body: *a nutant head.* **b** bent or curved toward the anterior extremity of the body: *a nutant horn.*
[from Latin *nutantem* nodding, swaying]

nutate (nü′tāt), *v.* to undergo nutation; to twist, wobble, or oscillate: *If one boom is slow in erecting ... the satellite will be unbalanced and will nutate like a wobbling top* (New Scientist).

nutation (nü tā′shən), *n.* **1** an act or instance of wobbling or oscillating; a twisting or oscillating movement.
2 *Botany.* a twisting or rotation of the growing tip of a plant due to consecutive differences in growth rates of different sides: *Nutation is the spiral twisting of a stem as it grows* (Scientific American).
3 *Astronomy.* a slight oscillation of the earth's axis, which makes the motion of the precession of the equinoxes irregular: *When the sun or the moon is above or below the equatorial plane of the earth each exerts a gravitational pull on the equatorial bulge of the earth, which tends to displace the earth's axis. Thus the axis precesses like a top pushed a little to one side. Motions due to such forces are called nutations* (Scientific American).
[from Latin *nutationem,* from *nutare* to nod, sway]
—**nutational,** *adj.* of, having to do with, or exhibiting nutation.

nutrient (nü′trē ənt), *Biology.* —*n.* a substance that provides nourishment; nourishing substance: *The usual path of mineral nutrients from the soil to the plant is from the solid particles of soil to the water in the soil and thence into the root* (Christopher J. Pratt).
—*adj.* that provides nourishment; nourishing: *The host cells (bacterial cells) are commonly grown either in liquid cultures of nutrient medium (such as meat broth), or on the surface of a similar medium made into a solid jelly with gelatin or agar* (J.E. Hotchin).
[from Latin *nutrientem* nourishing, from *nutrire* nourish]
▶ Food includes nutrients and fiber or roughage. Nutrients include usable carbohydrates, protein, lipids, vitamins, minerals, and water. Vitamins, minerals, and water are small molecules and can be absorbed without digestion. Carbohydrates, lipids, and proteins require digestion. Nutritional requirements vary with an individual's age, sex, and activities. Malnutrition results from an improper or poor intake, absorption or use of nutrients by the body.
ASSOCIATED TERMS: see **nutrition.**

nutriment (nü′trə mənt), *n. Biology.* that which is required by an organism for life and growth; nourishment; food.

nutrition (nü trish′ən), *n. Biology.* **1** a series of processes by which food is taken in and used by living organisms for growth, energy, etc.: *Multicellular eukaryotes are divided into three kingdoms, based primarily on their mode of nutrition: fungi absorb organic molecules from the surrounding medium, plants manufacture them by photosynthesis, and animals ingest them in the form of other organisms* (Helena Curtis, *Biology*).
2 nourishment; feeding; sustenance: *In summary, there has up to the present time been an appreciable improvement in nutrition, ... a marked increase in the production of ground-nuts from the new methods, and an incidental increase in the acreage of two cereal crops, millets and sorghums* (A.W. Haslett).
[from Late Latin *nutritionem,* from *nutrire* nourish]
ASSOCIATED TERMS: Nutrition includes those activities by which organisms obtain and process the *nutrients* needed for energy, growth, repair, and regulation. Two basic types of nutrition are *autotrophic* and *heterotrophic.* The two forms of autotrophic nutrition are *photosynthesis* and *chemosynthesis* in plants and certain monerans and protists. Heterotrophic nutrition involves *ingestion, digestion,* and *egestion* of food, and is the usual form of nutrition among fungi, protozoans, and animals.
—**nutritional,** *adj.* having to do with nutrition: *nutritional growth.* —**nutritionally,** *adv.*
—**nutritious,** *adj.* valuable as food; nourishing: *nutritious fruits.*
—**nutritive,** *adj.* **1** having to do with foods and the use of foods. Digestion is part of the nutritive process. **2** = nutritious.

nymph (nimf), *n. Zoology.* an immature form of a type of insect in which the pupal stage is lacking (hemimetabolous insect). A nymph resembles the adult but lacks fully developed wings. *Each summer lacy-winged adult May flies emerged from the lake in hordes as the May fly nymphs at the lake bottom matured* (Barry Commoner). [from Old French *nymphe,* from Latin *nympha* female spirit of nature, pupa of an insect, from Greek *nymphē*]

O

o-, *prefix.* Usually italicized. *Chemistry.* ortho- (def. 2a), as in *o*-dichlorobenzene.

O, *symbol.* **1** oxygen. **2** one of the four major blood groups (used to identify blood for compatibility in transfusion). ▶ See the note under **ABO.**

oasis (ō ā′sis), *n., pl.* **-ses** (-sēz′). *Geography.* a fertile spot in the desert where there is water and some vegetation: *Oases are not strictly a part of deserts, as the climate is artificially altered* (R. N. Elston). [from Late Latin, from Greek]

ob-, *prefix.* inversely; contrary to the usual position, as in *obcompressed, obconic, obovate.* [from Latin *ob-,* from *ob* against]

obcompressed (ob′kəm prest′), *adj. Biology.* compressed or flattened in the opposite of the usual direction: *obcompressed ovary and seeds.*

obconic or **obconical,** *adj. Botany.* conical, with the base upward or outward; inversely conical: *Carpels ... immersed in a large, obconical receptacle* (Daniel Oliver). **—obconically,** *adv.*

obcordate (ob kor′dāt), *adj. Botany, Zoology.* heart-shaped, with the apex serving as base or point of attachment; inversely cordate: *an obcordate leaf, an obcordate thorax.*

obduct (ob dukt′), *v. Geology.* to push (one crustal plate of the earth) on top of another. Compare **subduct.** [from Latin *obductum,* from *obducere* cover over] **—obduction,** *n.* the process of obducting; thrusting of a crustal plate upwards over the margin of an adjacent plate: *The Alpine orogeny ... was probably preceded by subduction or obduction of the Tethyan plate along the European continental margin* (Nature).

object glass, = objective.

objective, *n. Optics.* a lens or combination of lenses that first receives light rays from the object and forms the image viewed through a telescope or microscope: *The instrument is provided with a rotating nosepiece to which are permanently attached three objectives of different magnifications* (Sears and Zemansky, *University Physics*). Also called **object glass.**

oblanceolate (ob lan′sē ə lit *or* ob lan′sē ə lāt), *adj. Botany.* shaped like a lance head but with the tapering end at the base: *an oblanceolate leaf.*

oblate (ob′lāt *or* o blāt′), *adj. Geometry.* flattened at the poles. An **oblate spheroid** or **ellipsoid** is produced by rotating an ellipse through 360° about its shorter axis. *The meridians are not circles, but ellipses; and the earth is therefore not a sphere, but an oblate spheroid—flattened at the poles and bulging at the equator* (Baker, *Astronomy*). Compare **prolate.** [from New Latin *oblatum,* from Latin *ob-* inversely + (*pro*)*latum* prolate]

obligate (ob′lə git *or* ob′lə gāt), *adj. Biology.* able to exist under or restricted to only one set of environmental conditions, as a parasite which can survive only by living in close association with its host: *Viruses ... are*

dead *in the sense that they lack any internal metabolism, but they come alive on entering the living cell. For this reason they are obligate parasites, able to reproduce only by taking over the enzymatic machinery of the host cell* (Scientific American). Contrasted with **facultative.**

oblique (ə blēk′), *adj.* **1** *Mathematics.* **a** neither perpendicular to nor parallel with a given line or surface; slanting. **b** (of a solid figure) not having the axis perpendicular to the plane of the base.
2 *Botany.* having unequal sides: *an oblique leaf.*
3 *Anatomy.* situated obliquely; not transverse or longitudinal: *oblique muscles, fibers, or ligaments.*
—n. 1 *Mathematics.* something oblique, such as a line or figure.
2 *Anatomy.* an oblique muscle: *The superior and inferior obliques roll the eyeball about the visual axis* (Scientific American).
[from Latin *obliquus*] **—obliquely,** *adv.* **—obliqueness,** *n.*

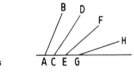

oblique lines
(def. 1a)

AB, CD, EF, and GH are oblique lines.

oblique angle, *Mathematics.* any angle that is not a right angle or a multiple of a right angle. Acute angles and obtuse angles are oblique angles.

oblong (ob′lông), *Geometry.* **—adj.** deviating from a square or circular form by being elongated in one direction; having the opposite sides parallel and the adjacent sides at right angles but not square: *An oblong spheroid is prolate, as opposed to an oblate spheroid.*
—n. an oblong figure, or something oblong in form, especially a rectangle of greater length than width.
[from Latin *oblongus,* from *ob-* toward + *longus* long]

obovate (ob ō′vāt), *adj. Botany.* inversely ovate: *an obovate leaf.*

obovoid (ob ō′void), *adj. Botany.* inversely ovoid: *obovoid fruits.*

obsequent (ob′sə kwənt), *adj. Geology, Geography.* flowing in a direction opposite to the dip or tilt of the surface strata: *obsequent streams.* Compare **consequent, insequent, resequent, subsequent.**

---

cap, fāce, fäther; best, bē, tėrm; pin, fīve;
rock, gō, ôrder; oil, out; cup, pùt, rüle;
yü in use, *yu* in uric;
ng in bring; sh in rush; th in thin, ᴛʜ in then;
zh in seizure.
ə = *a* in about, *e* in taken, *i* in pencil, *o* in lemon, *u* in circus

observatory, *n.* **1** a place or building with a telescope or other equipment for observing the stars and other celestial bodies: *Whenever possible, observatories are located on isolated mountaintops in regions of predominantly clear weather* (Helmut Abt).

2 a place or building for observing facts or happenings of nature: *a meteorological observatory.*

3 an artificial satellite designed to gather information about the earth, sun, or other heavenly body: *Observatories are larger satellites, ... either carrying many (up to 30) small experiments or a few heavy (up to 1000 lb.) complex experiments. An observatory weighs from 1000 to 4000 lb.* (John E. Naugle).

obsidian, *n.* *Geology.* a hard, dark, glassy rock that is formed when rhyolitic lava cools; volcanic glass: *The first glass was obsidian, molded by the heat of volcanic materials that spread over the sandy soil, fusing it into a smooth amorphous solid* (Science News Letter). [from Latin *obsidianus (lapis)* (stone) of *Obsius,* Roman explorer supposed to have discovered it]

obsolescent, *adj.* *Biology.* gradually disappearing; imperfectly or slightly developed: *an obsolescent organ, obsolescent tubercles.* Compare **obsolete.**

—**obsolescence,** *n.* the gradual disappearance or atrophy of an organ or part.

obsolete, *adj.* *Biology.* imperfectly developed and of little use; indistinct, especially in comparison with the corresponding character in other individuals or related species: *They [eggs of long-tailed tit] are sometimes entirely white, or with the spots almost obsolete* (F. O. Morris). SYN: vestigial. Compare **obsolescent.**

obtected (ob tek′tid), *adj.* *Zoology.* having the appendages covered or protected by a hard shell or horny case, as the pupae of many insects, especially butterflies. [from Latin *obtectum* covered up, from *ob-* against + *tegere* to cover]

obtruncate (ob trung′kāt), *v.* *Biology.* to cut or lop off the head or top from: *to obtruncate trees. The female [viper] ... obtruncates the male* (Richard Tomlinson).

obtuse angle, *Mathematics.* an angle larger than a right angle but less than 180 degrees. See the picture at **angle.**

obtuse triangle, *Mathematics.* a triangle having one obtuse angle.

obvolute (ob′və lüt), *adj.* (of two leaves in a bud) folded together so that one half of each is exterior and the other inferior, as in the poppy. [from Latin *obvolutum* wrapped around, from *ob-* against, over + *volvere* wind, roll]

Occam's razor (ok′əmz), a principle devised by the English philosopher, William of Occam (or Ockham), which states that entities must not be multiplied beyond what is necessary. In a scientific evaluation, Occam's Razor is the choice of the simplest theory from among the theories which fit the facts we know. *The decision as to whether a heat-sensitive process is biological or not must be based on additional evidence. In the end, however, the judgment is based on Occam's razor: the traditional principle that the hypothesis most likely to be correct is the one that accounts for the maximum number of observations with the minimum number of assumptions* (Norman H. Horowitz). Also spelled **Ockham's razor.** Also called **law of parsi-**

mony. [*razor* refers to the idea in this principle of shaving an argument to its simplest terms]

occipital (ok sip′ə təl), *Anatomy.* —*adj.* of, belonging to, or situated in the back part of the head or skull: *occipital artery, occipital nerve, occipital condyle.*

—*n.* **1** = occipital bone. **2** any occipital part. [from Medieval Latin *occipitalis,* from Latin *occiput* occiput] —**occipitally,** *adv.*

occipital bone, *Anatomy.* the compound bone forming the lower back part of the skull. See the picture at **skull.**

occipital lobe, *Anatomy.* the posterior lobe of each cerebral hemisphere: *Damage to the occipital lobe of the brain involves disturbance of visual perception* (Margaret Knight).

occiput (ok′sə pət), *n., pl.* **occipita** (ok sip′ə tə). *Anatomy.* **1** the back part of the head or skull. **2** = occipital bone. Compare **centriciput, sinciput.** [from Latin *occiput,* from *ob-* behind + *caput* head]

occlude, *v.* **1** *Biology, Medicine.* to obstruct or close, as an artery.

2 *Chemistry.* absorb and retain (gases and other substances): *Platinum occludes hydrogen.*

3 *Meteorology.* to undergo occlusion: *The cold front overtakes the warm front and the system is said to occlude* (S. Petterssen). [from Latin *occludere* close up, from *ob-* up + *claudere* to close]

occluded front, *Meteorology.* the front formed when a cold air mass overtakes a warm air mass and displaces it upward in a system of low barometric pressure: *Occluded fronts ... move more slowly than ordinary fronts and therefore bring persistent bad weather over the affected area* (Neuberger and Stephens, *Weather and Man*).

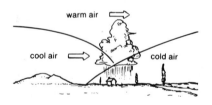

warm occluded front

warm air

cool air

cold air

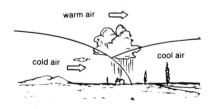

cold occluded front

warm air

cold air

cool air

occlusion, *n.* **1** *Biology, Medicine.* an obstruction or closure of a passageway or vessel.

2 *Chemistry.* the absorption and retention of a gas or other substance, as by a metal: *the occlusion of lime and magnesia by ferric oxide.*

3 *Meteorology.* **a** the process in which a cold air mass overtakes and forces upward a warm air mass in a cyclone, thereby meeting a second cold air mass originally in front of the warm air mass. Occlusion increases the intensity of a cyclone. **b** the contact between these two cold air masses.

occult (ə kult′), *v. Astronomy.* to cut off, or be cut off, from view by interposing some other body, as one celestial body hiding another by passing in front of it: *On the night of April 7–8* [*1976*] *the planet Mars passed directly in front of the 3.2-magnitude star Epsilon Geminorum ... on the average, Mars occults ... a star this bright only once in four or five centuries* (Joseph Ashbrook). SYN: eclipse.
—**occultation** (ok′ul tā′shən), *n.* the action or fact of hiding the light of one heavenly body by another passing between it and the observer: *Occultations of stars by the moon show an instantaneous disappearance or reappearance of the stellar image in a striking manner ...* (Bernhard, *Handbook of Heavens*).

ocean, *n. Geography.* **1** the great body of salt water that covers almost three fourths of the earth's surface; the sea: *The ocean is a potential source of unvarying power in some locations where there is a comparatively large difference in temperature of the water between the surface and the bottom* (J. J. William Brown).
2 any of its four main divisions; the Atlantic, Pacific, Indian, and Arctic oceans. The waters around the Antarctic continent are considered by some to form a separate ocean. *Most of the Arctic is ocean; the Polar Basin* (which includes the Arctic Ocean and several seas) *covers almost 4,000,000 square miles* (John C. Reed).
ASSOCIATED TERMS: The two main divisions of the ocean environment are the *benthic* (sea-bottom) environment and the *pelagic* (open-sea) environment. In addition, the ocean is usually subdivided into four regions or levels of depth: the *hadal* (the deepest part, below 6000 meters), the *abyssal* (too deep for light to penetrate, 3000 to 6000 meters), the *bathyal* (very little light, 100–300 to 1000–3000 meters), and the *littoral* (on or near the seashore). The life in the ocean is usually divided into various groups, including the *plankton,* minute organisms that float or drift about, consisting of *phytoplankton* (plants and plantlike organisms such as diatoms and dinoflagellates) and *zooplankton* (animals and animal-like organisms such as copepods and radiolarians); the *nekton,* consisting of fish and other organisms that swim about, the *benthos,* organisms such as sponges and starfish that live on the ocean bottom, and the *neuston,* minute organisms floating on the surface film of the water.

oceanic, *adj.* of, in, or belonging to the ocean: *oceanic islands, an oceanic climate. In the deep sea ... the rocks over vast areas are covered with oceanic sediments which may be thousands of feet in thickness* (T. F. Gaskell and M. N. Hill).

oceanity or **oceanicity,** *n. Meteorology, Ecology.* the degree to which a particular climate is influenced by proximity to the ocean or an ocean: *The term "oceanicity" traditionally has conveyed to ecologists the impression of a climate with ... a high annual precipitation* (Nature). Compare **continentality.**

oceanization, *n. Geology.* the transformation of continental crust into oceanic crust: *The Soviet tectonician Belousov has gone so far as to invoke extensive 'oceanisation' of continental crust to account for the ocean basins* (A. Hallam).

oceanographic, *adj.* of or having to do with oceanography: *oceanographic experiments.*

oceanography, *n.* the science that deals with the oceans and seas: *Basic studies in oceanography include investigations of marine life; the analyses of ocean water; examination of the ocean floor; investigation and measurement of currents, waves and tides; and analyses of the interaction of atmosphere and sea* (W. V. Burt and S. A. Kulm).

oceanological, *adj.* of or having to do with oceanology: *oceanological equipment.*

oceanology, *n.* **1** = oceanography.
2 ocean technology, including the political and economic aspects of the seas and oceans: *... the whole field of oceanology, from offshore oil exploration to physiological research in diving* (R. Barton).

ocellus (ō sel′əs), *n., pl.* **ocelli** (ō sel′ī). *Zoology.* one of the rudimentary, single-lens eyes found in certain invertebrates, especially one of the simple eyes situated between the compound eyes of insects: *Many insects have accessory simple eyes, called ocelli, which may function in increasing the irritability of the organism to light* (Harbaugh and Goodrich, *Fundamentals of Biology*). [from Latin *ocellus,* diminutive of *oculus* eye]

Ockham's razor, See **Occam's razor.**

ocrea or **ochrea** (ok′rē ə), *n., pl.* **ocreae** or **ochreae** (ok′rē ē). *Botany.* a tubular stipule or stipules sheathing the stem above the node, as in buckwheat. [from New Latin *ocrea,* from Latin *ocrea* a legging]
—**ocreate** or **ochreate** (ok′rē it), *adj.* having an ocrea or ocreae; sheathed: *ocreate stipules.*

octa-, *combining form.* **1** eight, as in *octagon.* **2** *Chemistry.* having eight atoms of a specified substance, as in *octane.* [from Greek *okta* eight]

octad (ok′tad), *n. Chemistry.* an element, atom, or radical with a valence of eight. [from Greek *oktas, oktados* (group of) eight, from *okta* eight]

octagon (ok′tə gon), *n. Geometry.* a plane figure having eight angles and eight sides. Also written *8-gon,* as in *the angles of a convex 8-gon.* [from Greek *oktagōnos,* from *okta* eight + *gōnia* angle]
—**octagonal,** *adj.* having eight angles and eight sides: *an equiangular octagonal figure.* —**octagonally,** *adv.*

octahedral (ok′tə hē′drəl), *adj. Geometry.* **1** having eight plane faces: *an octahedral solid.* **2** of or like an octahedron: *The direction ... at right angles to an octahedral plane is called an octahedral direction* (Science News).

octahedrite (ok′tə he′drīt), *n.* **1** *Mineralogy.* = anatase. **2** *Geology.* a common type of metallic meteorite.

octahedron (ok′tə hē′drən), *n., pl.* **-drons, -dra** (-drə). *Geometry.* a solid figure having eight faces. [from Greek *oktaedron,* from *okta* eight + *hedra* seat, base]

---

cap, fāce, fäther; best, bē, tèrm; pin, fīve;
rock, gō, ôrder; oil, out; cup, pùt, rüle;
yü in use, *yu̇* in uric;
ng in bring; *sh* in rush; *th* in thin, ᴛʜ in then;
zh in seizure.
ə = *a* in about, *e* in taken, *i* in pencil, *o* in lemon, *u* in circus

octal

octal (ok′təl), *adj. Mathematics.* of, having to do with, or based upon the number eight, as a numbering system based upon units of eight: *octal digits, an octal notation. Heathkit's H-8 computer's monitor lets you enter machine language based on octal (base-8) numbers* (Computer Readout).

octamerous (ok tam′ər əs), *adj. Botany.* (of a flower) having eight members in each whorl. It is generally written *8-merous.* [from *octa-* + Greek *meros* part]

octane, *n. Chemistry.* any of various isomeric hydrocarbons that occur in petroleum and belong to the methane series: *The rate of oxidation of the hydrocarbons is very sensitive to the number of carbon atoms and to their arrangement in the hydrocarbon. Thus octane (C_8H_{18}) oxidizes at about 2,000 times the rate of propane under similar conditions* (K. D. Wadsworth).

octant, *n. Mathematics.* 1 one eighth of a circle; a 45-degree angle or arc.
2 one of the eight parts into which a space is divided by three planes intersecting at one point.

octapeptide (ok′tə pep′tīd), *n. Biochemistry.* a polypeptide chain made up of eight amino acids.

ocular, *adj. Biology.* 1 of or having to do with the eye: *an ocular muscle.*
2 like an eye; eyelike: *ocular markings or spots.*
—*n. Optics.* the eyepiece of a telescope or microscope: *An ocular ... is a magnifier used for viewing an image formed by a lens or lenses preceding it in an optical system* (Sears and Zemansky, *University Physics*). [from Latin *oculus* eye]

oculomotor (ok′yə lō mō′tər), *Anatomy.* —*adj.* of or having to do with movement of the eyeball in its socket: *an oculomotor muscle, an oculomotor nerve.*
—*n.* either of a pair of cranial nerves that supply most of the muscles moving the eyeball.
[from Latin *oculus* eye + *motor* that which moves]

oculus (ok′yə ləs), *n. Anatomy.* an eye, especially a compound eye. [from Latin *oculus* eye]

OD or o.d., *abbrev.* 1 optical density. 2 outside diameter. 3 outside dimension.

odd, *adj. Mathematics.* leaving a remainder of 1 when divided by 2: *3, 5, and 7 are odd numbers.* Compare **even.**

odd-pinnate (od′pin′āt), *adj. Botany.* pinnate with an odd terminal leaflet; unevenly pinnate. SYN: imparipinnate.

odometry (ō dom′ə trē), *n.* measurement by some mechanical device of distances traveled.

odonate (o′dn āt), *Zoology.* —*adj.* of or belonging to an order (Odonata) of insects having chewing mouthparts, hind wings as large as or larger than the forewings, and large compound eyes. Dragonflies and damsel flies belong to this order. *Some strong-flying insects such as dragonflies have been captured at sea more than 450 km from the nearest land, yet there is less odonate fauna in Japan than on the adjacent Asiatic mainland* (Science).
—*n.* an odonate insect.
[from New Latin *Odonata* the order name, from Greek *odous, odontos* tooth]

odonto-, *combining form.* tooth; teeth, as in *odontology = the scientific study of the teeth.* Also spelled **odont-** before vowels. [from Greek *odous, odontos* tooth]

odontoblast (ō don′tə blast), *n. Biology.* one of a layer of cells that produce dentine as a tooth develops. [from *odonto-* + Greek *blastos* germ, sprout]

odontoid (ō don′toid), *Anatomy.* —*adj.* of or having to do with a toothlike projection of the second cervical vertebra, upon which the first cervical vertebra rotates: *the odontoid ligaments.*
—*n.* the odontoid process.
[from Greek *odontoeides,* from *odous, odontos* tooth + *eidos* form]

odontolite (ō don′tə līt), *n. Geology.* a fossil tooth or bone colored blue by iron phosphate (vivianite). [from *odonto-* + *-lite*]

odontology (ō′don tol′ə jē), *n.* the branch of anatomy dealing with the structure, development, and diseases of the teeth.

odontophore (ō don′tə fôr), *n. Zoology.* a structure in the mouth of most mollusks (other than bivalve mollusks) over which the radula is drawn backward and forward in the process of breaking up food. [from Greek *odontophoros,* from *odous, odontos* tooth + *-phoros* carrying, from *pherein* to carry]

O electron, *Physics.* one of the electrons in the O-shell.

oersted (ėr′sted), *n.* a unit of magnetic field strength in the CGS system, equivalent to the intensity of the force of one dyne acting in a vacuum on a magnetic pole of unit strength at a distance of one centimeter. [named after Hans Christian *Oersted,* 1777–1851, Danish physicist]

ogive (ō′jīv), *n.* 1 *Statistics.* a distribution curve in which the frequencies are cumulative. In an ogive of wage levels in which 10 people earn less than $50, 60 people $50 to $100, and 40 people $100 to $150, the abscissas of the curve would be 50, 100, and 150 and the ordinates 10, 70, 110.
2 *Geology.* any curved dark band extending across the surface of a glacier, convex in the direction of flow.
[from the earlier meaning "a pointed arch," from Middle French *ogive,* an architectural term for the diagonal rib of a vault]

ohm (ōm), *n.* the SI or MKS unit of electrical resistance. A wire in which one volt produces a current of one ampere has a resistance of one ohm. *Symbol:* Ω [named after Georg S. *Ohm,* 1787–1854, German physicist]

ohmage (ō′mij), *n.* the electrical resistance of a conductor, expressed in ohms.

ohmic (ō′mik), *adj. Physics.* that is consistent with Ohm's law; exhibiting behavior or properties in accordance with Ohm's law: *It is the ohmic resistance of the gas that generates the heat on passage of the current* (Lyman Spitzer, Jr.).
—*ohmically, adv.* in accordance with Ohm's law; resulting from ohmic resistance.

Ohm's law (ōmz), *Physics.* a law stating that the current in amperes in an electric circuit is directly proportional to the electromotive force in volts and inversely proportional to the resistance in ohms. When Ohm's law is used for an alternating-current circuit, resistance is replaced by impedance, also in ohms. Ohm's law is usually expressed mathematically as $E = IR$, in which the electromotive force (E), measured in volts, equals the electric current (I) in amperes multiplied by the resistance (R) in ohms. Ohm's law holds chiefly for metallic conductors at constant temperature found in electric circuits.

oil, *n. Chemistry.* **1** any of several kinds of fatty or greasy liquids that are lighter than water, that burn easily, and that will not mix or dissolve in water but will dissolve in alcohol. Mineral oils, such as petroleum and kerosene, are used for fuel; vegetable oils, such as olive oil and peanut oil, are used in cooking and medicine and in many other ways; animal oils include various fish oils used in shortenings, soaps, and detergents. Essential or volatile oils, such as oil of peppermint and turpentine, are distilled from plants, leaves, flowers, and other parts of plants, and are thin and evaporate very quickly. ► See the note under **fat.**
2 a substance that resembles oil in some respect. Sulfuric acid is called oil of vitriol.
3 = petroleum: *The oil and gas seldom are distributed uniformly throughout the total extent of the rock in which they occur but are gathered together in limited areas, called pools* (Finch and Trewartha, *Elements of Geography*).

oil of vitriol, = sulfuric acid.

Okazaki fragment (ō kə zä′kē), *Molecular Biology.* a fragment of DNA that is synthesized during replication and is afterwards linked with other fragments to form the long double-helical strand of the typical DNA molecule. [named after Reiji *Okazaki,* 1930–1975, Japanese biologist who first identified the fragments]

-ol¹, *suffix. Chemistry.* **1** a compound containing a hydroxyl group (-OH); an alcohol, as in *carbinol, phenol.* **2** a phenol or phenol derivative, as in *thymol.* [from (*alcoh*)*ol*]

-ol², *suffix. Chemistry.* a variant of **-ole,** as in *terpinol.*

old, *adj. Geology.* (of topographical features) well advanced in the process of erosion to base level.

-ole, *suffix. Chemistry.* **1** a compound containing a five-part ring, as in *pyrrole.* **2** an ether or aldehyde, as in *anethole.* Also spelled **-ol.** [short for Latin *oleum* oil]

olefin (ō′lə fin), *n. Chemistry.* any of a group of unsaturated aliphatic hydrocarbons containing one or more double bonds. Olefins having one double bond are often called *alkenes;* those with two double bonds are *diolefins.* Ethylene and propylene are important olefins. The names of olefins end in *-ene* or *-ylene. Through the use of heterogeneous catalysts it has been found possible to polymerize a number of olefins such as propylene, alpha butylene, alpha pentylene, alpha hexylene and styrene to high molecular weight products which are crystalline* (E. L. Kropa). Compare **paraffin.**
[from French (*gaz*) *oléfiant* oil-forming (gas), ethylene + English *-in*]
—**olefinic,** *adj.* of or having to do with the olefins: *the olefinic double bond.*

olefin series, = ethylene series.

oleic acid (ō lē′ik *or* ō lā′ik), *Chemistry.* an oily, unsaturated acid obtained by hydrolyzing various animal and vegetable oils and fats, much used in making soaps. *Formula:* $C_{17}H_{33}COOH$ Compare **linoleic acid.** [from Latin *oleum* (olive) oil + *-ic*]

olein (ō′lē in), *n. Chemistry.* **1** an ester of oleic acid and glycerin, a yellow, oily liquid that is one of the most abundant natural fats. Lard, olive oil, and cottonseed oil are mostly olein. *Formula:* $(C_{17}H_{33}COO)_3C_3H_5$ **2** = oleic acid.

oleoresin (ō′lē ō rez′ən), *n.* a natural or prepared mixture of essential oils and resins, such as that obtained from a plant by means of a volatile solvent. The balsams are oleoresins.
—**oleoresinous,** *adj.* **1** of, like, or containing an oleoresin: *oleoresinous deposits.* **2** consisting of a mixture of resins and drying oils heated and dissolved in turpentine or petroleum products: *Spar varnish is an oleoresinous varnish.*

oleum (ō′lē əm), *n. Chemistry.* a solution of sulfur trioxide and concentrated sulfuric acid, used as an agent in chemical processes: *It is then called oleum or fuming sulfuric acid, because this kind of sulfuric acid fumes when exposed to the air* (J. S. Fritz).

olfaction (ol fak′shən), *n. Physiology.* **1** the process or function of smelling: *Smell, or olfaction—"taste at a distance"—depends in man on slender neurons with directly exposed tips that lie in mucous membrane high in the nasal cavity* (Storer, *General Zoology*).
2 the sense of smell: *The astonishingly high acuity of the sense of smell has led many to regard olfaction as a mysterious sense* (D. G. R. Ottoson).

olfactory (ol fak′tər ē), *adj.* having to do with smelling; of or involving the sense of smell. The nose is an olfactory organ. *Olfactory stimuli are important in regulating the social behavior of many fishes and mammals* (George W. Barlow). [from Latin *olfactum* smelled, from *olere* emit a smell + *facere* make]

olfactory bulb, *Anatomy.* the enlarged distal end of the olfactory lobe. It is the point at which the olfactory nerve begins. *Each primary receptor cell is connected to a nerve fibre which runs without interruption to an extension of the brain known as the olfactory bulb* (Science Journal).

olfactory lobe, *Anatomy.* a lobe on the lower surface of the brain's frontal lobe, responsible for the sense of smell. The olfactory lobe is better developed in many animals than in man.

olfactory nerve, *Anatomy.* the first cranial nerve, a sensory nerve that carries the sensation of smell from the mucous membranes of the nose to the olfactory lobe of the brain.

oligo-, *combining form.* few; small; less, as in *oligomer, oligonucleotide.* [from Greek *oligos* few]

Oligocene (ol′i gō sēn′), *Geology.* —*n.* **1** the third epoch of the Tertiary period of the Cenozoic era, after the Eocene and before the Miocene, during which the first apes appeared and modern mammals became dominant.
2 the series of rocks formed in this epoch.
—*adj.* of or designating the Oligocene or its rocks: *By the Oligocene Epoch, dogs and cats had appeared, along with three-toed horses about as large as sheep* (Samuel P. Ellison, Jr.).
[from *oligo-* few + Greek *kainos* new, recent]

---

cap, fāce, fäther; best, bē, tėrm; pin, fīve;
rock, gō, ôrder; oil, out; cup, pút, rüle;
yü in use, *yu̇* in uric;
ng in bring; *sh* in rush; *th* in thin, ᵀH in then;
zh in seizure.
ə = *a* in about, *e* in taken, *i* in pencil, *o* in lemon, *u* in circus

oligoclase (ol'i gō klās), *n. Mineralogy.* a plagioclase containing more sodium than calcium, occurring in light gray, yellow, or greenish crystals. [from *oligo-* less + Greek *klasis* a breaking (so called because it was thought to be less perfect in cleavage than albite, another feldspar)]

oligodendrocyte (ol'i gō den'drə sīt), *n. Biology.* one of the cells comprising the oligodendroglia: *The primary function of the oligodendrocytes is the formation of the myelin sheath* (J. A. G. Rhodin).

oligodendroglia (ol'i gō den drog'lē ə), *n. Biology.* neuroglia consisting of cells similar to but smaller than astrocytes, found in the central nervous system and associated with the formation of myelin. [from *oligo-* + *dendro-* + (*neuro*)*glia*]
—**oligodendroglial,** *adj.* of the oligodendroglia: *oligodendroglial cells.*

oligomer (ə lig'ə mər), *n. Chemistry.* a compound with few molecular units, in contrast with a polymer or a monomer: *Most enzymes are oligomers, that is they consist of more than one such polypeptide chain. And most oligomeric enzymes are composed of several identical monomers or sub-units; enzymes comprising two, four or six such sub-units are the most common of all.* (New Scientist). [from *oligo-* + *-mer,* as in *polymer*]
—**oligomeric,** *adj.* of or characteristic of an oligomer: *an oligomeric structure or substrate.*

oligonucleotide (ol'i gō nü'klē ə tīd), *n. Biochemistry.* a substance composed of a relatively small number (usually two to ten) of nucleotides as compared to a polynucleotide: *Since the sequence of nucleotides in many RNA molecules is known, specific oligonucleotides with three or four monomers may be designed to bind to a selected part of the molecule by complementary base pairing* (Science Journal).

oligophagous (ol'i gof'ə gəs), *adj. Zoology.* eating few kinds of food; not polyphagous: *an oligophagous insect.* [from *oligo-* + Greek *phagein* to eat]
—**oligophagy** (ol'i gof'ə jē), *n.* the property of being oligophagous.

oligosaccharide (ol'i gō sak'ə rīd), *n. Biochemistry.* a carbohydrate that on hydrolysis yields a relatively small number (usually two to ten) of monosaccharides as compared to a polysaccharide: *Neomycin is an oligosaccharide with four sugar units linked together in a linear fashion* (W. Thomas Shier).

oligotrophic (ol'i gō trof'ik), *adj. Ecology.* not providing nutrition, as a lake with scant vegetation: *Most upwelling areas have major currents flowing through them advecting plankton and nutrients from eutrophic (rich) to relatively oligotrophic (poor) areas* (Science). [from Greek *oligotrophia,* from *oligos* few, little + *trephein* nourish]
—**oligotrophy** (ol'i got'rə fē), *n.* the condition of being oligotrophic: *the change from oligotrophy to eutrophy.*

olivenite (ō liv'ə nīt), *n. Mineralogy.* a mineral, an arsenate of copper, usually occurring in olive-green crystals or masses. *Formula:* $Cu_2(AsO_4)(OH)$ [from German *Olivenit,* from *Oliven-* of olive + *-it* -ite]

olivine (ol'ə vēn), *n. Mineralogy.* a greenish or yellow mineral, a silicate of magnesium and iron found in mafic and ultramafic igneous rocks: *The rigid mantle of the earth is thought to consist largely of olivine, an insulator under normal conditions. At the high temperatures within the earth, its properties change enough for it to become a conductor. The electrical conductivity of the earth's mantle plays an important role in many geophysical phenomena* (Science News Letter). *Formula:* $(Mg,Fe_2)SiO_4$ Also called **chrysolite.**

-oma (-ō'mə), *suffix, pl.* **-omas** (-ō'məz) *or* **-omata** (-ō'mə tə). a growth, tumor, or neoplasm, as in *adenoma, carcinoma.* [from Greek *-ōma, -ōmatos,* a noun suffix]

omasum (ō mā'səm), *n., pl.* **-sa** (-sə). *Zoology.* the third stomach of a cow or other ruminant. It receives the food when swallowed the second time, after having been chewed as a cud. Also called **manyplies.** Compare **abomasum, reticulum, rumen.** [from Latin *omasum* bullock's tripe]

omega (ō meg'ə), *n. Nuclear Physics.* **1** Also called **omega meson.** a highly unstable and short-lived neutral meson having a mass of 784 Mev (1534 times that of the electron) and usually decaying into a positive, negative, and neutral pion.
2 Also called **omega minus** or **omega hyperon.** a negatively-charged subatomic particle having a mass of 1672 Mev (3272 times that of the electron) and decaying into a xi particle and a pion or a lambda particle and a kaon.

omentum (ō men'təm), *n., pl.* **-ta** (-tə). *Anatomy.* a fold of the peritoneum connecting the stomach with certain of the other viscera. The great omentum is attached to the stomach and enfolds the transverse colon; the lesser omentum lies between the stomach and the liver. *The omentum is rich in lymphatic tissue* (Science News). [from Latin *omentum*]

ommatidial (om'ə tid'ē əl), *adj.* of or having to do with ommatidia.

ommatidium (om'ə tid'ē əm), *n., pl.* **-ia** (-ē ə). *Zoology.* one of the radial elements or segments making up the compound eye, as of insects or crustaceans: *The compound eye of insects and other arthropods, for instance, consists of hundreds or thousands of elementary structures called ommatidia. These resemble simple eyes* (Scientific American). [from New Latin *ommatidium* from Greek *omma, ommatos* eye + Latin *-idium,* a diminutive suffix]

ommatophore (ə mat'ə fôr), *n. Zoology.* a movable eyestalk, as in certain snails. [from New Latin *ommatophorus,* from Greek *omma, ommatos* eye + *-phoros* carrying, from *pherein* to carry]

omnifocal (om'nə fō'kəl), *adj. Optics.* having continuously varying focal lengths: *Omnifocal lenses eliminate a typical sharp break between lens segments of bifocals* (Chicago Tribune).

omnivore (om'nə vôr), *n. Zoology.* an organism that eats both animal and vegetable food. Compare **carnivore, folivore, frugivore, herbivore.** ▶ See the note under **heterotroph.**
—**omnivorous** (om niv'ər əs), *adj.* eating both animal and vegetable food: *Most primates are omnivorous ... they have neither the high development of the canines and incisors characteristic of carnivores nor the exces-*

sive molar specialization of herbivores (Ralph Beals and Harry Hoijer).

-on, *suffix.* **1** *Physics.* **a** a nuclear particle, as in *neutron, hadron, parton.* **b** any unit particle or quantum of energy, as in *photon, graviton, exciton.*
2 *Genetics.* a unit of genetic material, as in *operon, cistron.*
3 *Chemistry.* **a** a variant of **-one,** used for a compound that is not a ketone, as in *diuron.* **b** inert gas, as in *radon.*
[from -on, as in ion, electron, and proton]

onco-, *combining form.* tumor, as in *oncogenesis, oncology.* [from New Latin *onco-,* from Greek *onkos* mass, bulk]

oncogen (ong′kə jen), *n. Biology, Medicine.* a tumor-producing virus or other agent; oncogenic substance or organism.

oncogene (ong′kə jēn′), *n. Biology.* a tumor-producing gene: *Human cancers are initiated by oncogenes, altered versions of normal genes* (Scientific American).

oncogenesis (ong′kə jen′ə sis), *n. Medicine.* the process of forming or producing tumors.

oncogenic, *adj. Medicine, Biology.* having to do with or producing tumors; tending to cause tumors, either benign or malignant, especially in test animals: *oncogenic viruses.* SYN: carcinogenic.
—**oncogenicity,** *n.* the condition or tendency of producing tumors: *the oncogenicity of a chemical.*

oncological (ong′kə loj′ə kəl), *adj.* of or having to do with oncology.

oncology (ong kol′ə jē), *n.* the branch of medicine dealing with the study of tumors: *The mechanism by which cancer cells spread throughout the body is one of the active areas of research in the field of oncology* (N. Karle Mottet).

oncornavirus (ong′kor nə vī′rəs), *n. Biology.* any of a group of tumor-producing viruses that contain RNA: *[He] spoke of the oncornaviruses particularly in fowls, mice and cats, in which they can lead to leukemia or sarcoma* (Nature). [from onco- + RNA + virus]

-one, *suffix. Chemistry.* a ketone or any similar compound or compounds, as in *acetone, progesterone.* [from Greek -ōnē, a feminine suffix]

one-celled, *adj.* = unicellular: *The bacterium Escherichia coli, a rod-shaped, one-celled organism normally found in the human intestinal tract, is widely used in research on heredity* (Atlantic).

one-to-one, *adj. Mathematics.* matching every element in a set with one and only one element in another set. EXAMPLE: { 1, 2, 3, 4, 5 } and { 10, 20, 30, 40, 50 } are two sets showing a one-to-one correspondence.

ontogenetic (on′tō jə net′ik), *adj. Biology.* of or having to do with ontogeny: *ontogenetic development or evolution.* —**ontogenetically,** *adv.*

ontogeny (on toj′ə nē), *n. Biology.* the development of an individual organism, or the history of its development: *According to Baer's law, the common features of large groups of animals develop earliest during ontogeny. In the light of the evolutionary theory, however, these features are the ones that are inherited from the common ancestor of the animal group in question; therefore, they have an ancient origin* (Balinsky, An Introduction to Embryology). [from Greek ontos being + -geneia origin]

onyx, *n. Mineralogy.* a semiprecious variety of cryptocrystalline quartz, having straight bands of different colors and shades. It is hard, strong, and takes a high polish. **Onyx marble,** commonly found in caves, is not an onyx but a variety of calcite, that resembles onyx and is used for ornamental stonework. [from Greek *onyx* nail, claw]

oo-, *combining form.* egg or eggs; ovum, as in *oology = the science of bird's eggs.* [from Greek *ōion* egg]

ooblast (ō′ə blast), *n. Biology.* a primitive or formative ovum not yet developed into a true ovum. [from oo- + Greek *blastos* germ, sprout]

oocyst (ō′ə sist), *n. Zoology.* a cyst in sporozoans that contains developing sporozoites, present within a host organism.

oocyte (ō′ə sīt), *n. Biology.* an ovum or egg cell in the stage that precedes maturation; a female gametocyte: *The oocytes are those cells in which meiosis takes place.* (McElroy, Biology and Man).

oogamete (ō′ə gam′ēt or ō′ə gə met′), *n.* = heterogamete.

oogamous (ō og′ə məs), *adj. Biology.* heterogamous; reproducing by the union of a large nonmotile egg and a small, motile sperm.

oogamy (ō og′ə mē), *n. Biology.* the condition of being oogamous; heterogamous condition. SYN: heterogamy.

oogenesis (ō′ə jen′ə sis), *n. Biology.* the formation and development of an ovum or ova: *In animals meiosis generally takes place as the gametes are being formed in specialized organs called gonads. Sperm formation is termed spermatogenesis, and the formation of the egg, oogenesis.* (Harbaugh and Goodrich, Fundamentals of Biology).
—**oogenetic** (ō′ə jə net′ik), *adj.* of or having to with oogenesis.

oogonial (ō′ə gō′nē əl), *adj.* of or having to do with oogoniums: *... processes by which oogonial cells gradually become built up from relatively simple beginnings to the often very large and complex egg cells ripe for fertilization* (New Scientist).

oogonium (ō′ə gō′nē əm), *n., pl.* **-niums, -nia** (-nē ə). **1** *Biology.* a primitive germ cell that divides and gives rise to the oocytes: *After a number of divisions ... the spermatogonia and oogonia cease to divide and begin to increase considerably in size. At this stage they are called primary spermatocytes ... and oocytes* (L. Doncaster).
2 *Botany.* the female reproductive organ in various thallophytes, usually a rounded cell or sac containing one or more oospheres.
[from New Latin *oogonium,* from oo- egg + Greek -gonos producing]

cap, fāce, fäther; best, bē, tėrm; pin, five;
rock, gō, ôrder; oil, out; cup, pùt, rüle;
yü in use, *yù* in uric;
ng in bring; sh in rush; th in thin, ᴛʜ in then;
zh in seizure.

ə = a in about, e in taken, i in pencil, o in lemon, u in circus

oolite

oolite (ō′ə līt), *n. Geology.* **1** a rock, usually limestone, composed of rounded concretions of calcium carbonate resembling the roe of fish: *Southern Florida is like a very slightly tilted cookie tray, with low, coastal edges where dunes and outcrops of oolite, a soft, limy rock full of fossil shells, are elevated enough for building* (Philip Wylie).
2 any of the rounded concretions of which this rock is composed, each having a concentric layered structure. [probably an adaption of French *oolithe,* from Greek *oion* egg + *lithos* stone]

oological (ō′ə loj′ə kəl), *adj.* of or having to do with oology. —**oologically,** *adv.*

oology (ō ol′ə jē), *n.* the branch of zoology that deals with the study of eggs, especially the eggs of birds.

oophyte (ō′ə fīt), *n. Botany.* the generation or form of a plant that bears the sexual organs in the alternation of generations, as in ferns, mosses, and liverworts.
—**oophytic** (ō′ə fit′ik), *adj.* of or having to do with the oophyte.

ooplasm (ō′ə plaz′əm), *n. Biology.* the cytoplasm of an egg.
—**ooplasmic,** *adj.* of or having to do with ooplasm.

Oort Cloud (ürt *or* ôrt), *Astronomy.* a hypothetical swarm of comets traveling in elliptical orbits around the solar system at distances ranging from 6 trillion to 21 trillion kilometers from the sun: *The comets in the Oort Cloud spend most of their time so far from the sun that they are easily perturbed by the gravitational influences of other stars* (Stephen P. Maran). [named after Jan Hendrik *Oort,* 20th-century Dutch astronomer, who first proposed the existence of this swarm of comets]

oosphere (ō′ə sfir), *n. Botany.* a female reproductive cell contained in an oogonium which when fertilized becomes an oospore.

oospore (ō′ə spôr), *n. Botany.* the fertilized female cell or zygote within an oogonium from which a sporophyte develops; a heavy-walled spore resulting from the fusion of heterogametes. [from *oo-* + Greek *sporos* seed, spore]
—**oosporic,** *adj.* of or having to do with the oospore.

ootheca (ō′ə thē′kə), *n., pl.* **-cae** (-kē *or* -sē). *Zoology.* an egg case or capsule of certain mollusks and insects, especially cockroaches and mantises. [from *oo-* + Greek *thēkē* receptacle]
—**oothecal,** *adj.* of or having to do with an ootheca or oothecae.

ootid (ō′ə tid), *n. Biology.* a haploid cell resulting from the meiotic division of a secondary oocyte: *In the second meiotic division, the cytoplasm with one nucleus forms the ootid, and the other nucleus is passed out as the second polar body. ... With slight change in nuclear position the ootid becomes a female gamete, or ovum* (Storer, *General Zoology*). Compare **spermatid.**

ooze, *n. Geology.* a soft mud or slime, especially at the bottom of a pond or river or on the ocean bottom: *Whereas the deep-sea bed elsewhere consisted of thick deposits of fine oozes and clays, here the bed was mainly composed ... of sand and silt* (B. C. Heezen).

opacity, *n. Optics.* **1** the quality or condition of being impervious to light rays or other forms of radiant energy.

2 (of a transparent medium) the reciprocal of the transparency.

opal, *n. Mineralogy.* a mineral, an amorphous form of hydrous silica, somewhat like quartz, found in many varieties and colors, certain of which reflect light with an iridescent play of color. Black opals show brilliant colored lights against a dark background; some are so dark as to seem almost black. Milk opals are milky white with rather pale lights. Fire opals are similar with more red and sometimes yellow flashes of color. *For diffusing light, ground glass and opal glass are commonly used. ... The diffusing property of this type of glass is due to minute colloidal particles ... All opals cause the transmitted light to become noticeably yellow by virtue of the selective effect of the scattering* (Hardy and Perrin, *Principles of Optics*). [from Latin *opalus,* from Greek *opallios,* from Sanskrit *upala* gem]
—**opalescence,** *n.* the quality of reflecting an iridescent play of colors like that of an opal.
—**opalescent** or **opaline,** *adj.* having a play of colors like that of an opal.

open chain, *Chemistry.* an arrangement of atoms in a molecule represented in a structural formula by a chain with open ends rather than by a ring. Compare **closed chain.**

open sentence, *Mathematics.* a sentence (an equation or inequality) which contains one or more variables and which in its present form is neither true nor false. EXAMPLE: $x + 2 = 6$. *The members of a replacement set of an open sentence which make the sentence true constitute the solution set of the open sentence* (Mathematics Regents Syllabus).

operand (op′ə rand), *n. Mathematics.* the quantity or expression that is to be subjected to a mathematical operation: *In the mathematical operation of division, the dividend and divisor are two operands* (P. B. Jordain).

operation, *n. Mathematics.* something done to one or more numbers or quantities according to specific rules. Addition, subtraction, multiplication, and division are the four commonest operations in arithmetic.

operator, *n.* **1** *Mathematics.* a symbol indicating or expressing an operation to be carried out: *the arithmetic operators $+, -, \times,$ and $\div$.*
2 *Molecular Biology.* a chromosomal region that regulates the structural genes of an operon by interacting with a specific repressor: *According to a now classic hypothesis in genetics, there are two classes of genes. The first consists of structural genes ... The second class of genes, called operators, control the first* (Science News). Also called **operator gene.**

operator gene, = operator (def. 2).

opercular, *adj.* of or having to do with an operculum: *... the projecting movable opercular plates that close the orifice when the animal withdraws into its shell* (New Scientist). —**opercularly,** *adv.*

operculate or operculated, *adj.* of or having an operculum.

operculum (ō pėr′kyə ləm), *n., pl.* **-la** (-lə), **-lums.** *Biology.* a lidlike part or organ; any flap covering an opening, as the lid of the spore case in mosses, the plate of some gastropods that closes the opening of the shell, or the gill cover of a fish: *Snails retire into their shells, closing the doorlike operculum to shut out the air and keep some of the sea's wetness within* (New Yorker). [from Latin, from *operire* to cover]

operon (op'ər on), *n. Molecular Biology.* a unit of gene activity consisting of a sequence of genetic material that functions in a coordinated manner by means of an operator, promoter, one or more structural genes, and a terminator: *An operon is a cluster of linked, contiguous genes that functions as a unit in coding for the production of proteins* (Robert G. Eagon). [from French *opéron,* from *opérer* to operate, effect + *-on*]

ophiolite (of'ē ə lit *or* ō'fē ə lit), *n. Geology.* any of a group of mafic and ultramafic igneous rocks, ranging from basalt to peridotite and including those rich in serpentine: *A diverse category of rocks collectively termed ophiolites occurs commonly in various stages of fragmentation and deformation in mountain belts, such as the northern Appalachians. They have presented a paradox: They seem to have been formed at high temperatures but placed in their present positions at low temperatures* (Science News). [from Greek *ophis* snake + English *-lite*]

ophiology (of'ē ol'ə jē *or* ō'fē ol'ə jē), *n.* the branch of zoology dealing with snakes. [from Greek *ophis* snake + English *-logy*]

ophiophagous (of'ē of'ə gəs *or* ō'fē of'ə gəs), *adj.* feeding on snakes: *ophiophagous reptiles.* [from Greek *ophis* snake + *phagein* to feed]

ophitic (ō fit'ik), *adj. Geology.* (of igneous rocks) having a texture characterized by crystals of plagioclase embedded in a matrix of augite. Also called **diabasic.** [from Greek *ophitēs* (*lithos*) serpentine (stone), from *ophis* snake]

ophthalmic (of thal'mik), *adj. Anatomy.* of or having to do with the eye. SYN: ocular, optic. [from Greek *ophthalmos* eye]

ophthalmology (of'thal mol'ə jē), *n.* the branch of medicine dealing with the structure, function, and diseases of the eye.

opisthognathous (op'is thog'nə thəs), *adj. Zoology.* 1 having receding jaws. 2 (of a jaw) receding. [from Greek *opisthen* behind + *gnathos* jaw]

opposite, *adj. Botany.* 1 placed in pairs on different sides along a stem; not alternate: *opposite leaves.*
2 in front of an organ, coming between it and its axis, as a stamen in front of a sepal or petal.
[from Latin *oppositum* placed against, from *ob-* against + *ponere* to place] —**oppositely,** *adv.* —**oppositeness,** *n.*

opposition, *n. Astronomy.* the position of two celestial bodies when their longitude differs by 180 degrees, especially such a position of a body with respect to the sun. The moon is in opposition with the sun when it is full moon. Compare **conjunction, quadrature.** See also **aspect.**

opsin (op'sin), *n. Biochemistry.* a protein formed in the retina, one of the constituents of the visual pigments, such as rhodopsin: *It had been known that a key molecule in the retina, 11-cis-retinal, which is bound to complex proteins called opsins, carries a light-absorbing group that constitutes the basis of colour recognition, but not how that group could be "tuned" to absorb more than one colour* (Gordon Wilkinson). [from Greek *ops, opos* eye + English *-in*]

opsonic (op son'ik), *adj. Immunology.* of or having to do with opsonin or opsonization. The **opsonic index** is the ratio of the opsonizing power of blood, obtained by measuring the number of bacteria destroyed by the phagocytes in the blood serum of a given individual compared to the number destroyed in normal or standard blood serum.

opsonin (op'sə nin), *n. Immunology.* type of antibody in blood serum that weakens bacteria and other foreign cells so that the phagocytes can destroy them more easily. [from Greek *opson* a relish (as meat, fish) + English *-in*]

opsonize (op'sə nīz), *v. Immunology.* 1 to increase the opsonins in, as by immunization.
2 to make (bacteria) more susceptible to destruction by phagocytes.
—**opsonization,** *n.* the process of making bacteria more susceptible to destruction by phagocytes: *Phagocytosis is a prime mechanism of defense against bacterial infections. It has been known since the 1930's that antibody and complement render bacteria susceptible to phagocytosis, a process called opsonization* (Manfred M. Mayer).

optic, *adj. Anatomy.* of or having to do with the eye and its structures: *the optic or visual center.* [from Greek *optikos,* from *op-* see]

optical, *adj.* 1 Anatomy. of the eye or the sense of sight; visual: *Nearsightedness is an optical defect. The optical path for the right eye is longer than that for the left; hence the image presented to the right eye is abnormally small* (Scientific American).
2 Optics. a made to assist sight: *Telescopes and microscopes are optical instruments.* b of vision and light in relation to each other: *the optical cortex.* c having to do with optics: *optical experiments.*
—**optically,** *adv.* by optical means; in or by means of reflected or transmitted light: *The mass storage disc is designed to read stored data optically and convert it into electronic signals* (New Scientist).

optical activity, *Chemistry.* the property of a substance, due to asymmetrical molecular structure, by which it rotates the plane-polarized light passing through it. Compare **optical rotation.** See also **dextrorotatory, levorotatory.**

optical astronomy, the branch of astronomy that deals with observation of the heavens using visible light, as distinguished from radio astronomy, X-ray astronomy, etc.

optical center, *Optics.* a point in the axis of a lens so situated that all rays pass through the axis and the lens without being refracted.

optical density, *Optics.* the degree in which a refractive medium retards transmitted rays of light; the degree of opacity of a translucent medium. *Abbreviation:* OD

optical fiber, *Optics.* a fiber that acts as a light guide in fiber optics: *Optical fibers have been widely used ... where the light path must be flexible, for example in such medical instruments as endoscopes* (Scientific American).

---

cap, fāce, fäther; best, bē, tėrm; pin, fīve;
rock, gō, ôrder; oil, out; cup, pùt, rüle;
yü in use, yů in uric;
ng in bring; sh in rush; th in thin, ʇʜ in then;
zh in seizure.
ə = a in about, e in taken, i in pencil, o in lemon, u in circus

optical isomer, *Chemistry.* = stereoisomer.

optical maser, = laser.

optical pair, *Astronomy.* a pair of stars that appear close together because they are in nearly the same direction in space. One such star may really be much closer to earth than the other.

optical rotation, *Chemistry.* a measure of the extent to which a given quantity of substance rotates the plane of plane-polarized light. Compare **optical activity.**

optic axis, 1 *Optics.* **a** the line in a double refracting crystal in the direction of which no double refraction occurs. Crystals having a single such line are uniaxial; crystals having two such lines are biaxial. **b** the line in a lens or an optical instrument about which the behavior of light is symmetric.
2 *Anatomy.* an imaginary line passing through the midpoints of the cornea and the retina.

optic chiasma or **optic chiasm,** *Anatomy.* the structure forming the intersection or crossing of the optic nerve fibers at the base of the brain.

optic disk, *Anatomy.* the round spot on the retina where the optic nerve enters the eye: *At the exit point of the optic nerve ... no rods or cones are to be found by microscopic examination ... and, although careful experimental studies (Helson, 1929) have shown some sensitivity there, the optic disk is often called the blind spot* (Deane B. Judd).

optic nerve, *Anatomy.* the nerve of sight, which goes from the brain to the eyeball and terminates in the retina.

optics, *n.* the branch of physics dealing with the properties and phenomena of those electromagnetic waves with wavelengths greater than X rays and smaller than microwaves. Optics includes the ultraviolet, visible, and infrared parts of the spectrum.

oral, *adj. Anatomy.* of the mouth: *In the paramecium ... food is ingested through a fixed opening located in the oral groove* (Biology Regents Syllabus). The **oral cavity** in mammals contains the teeth, tongue, and the openings from the salivary glands.

orbicular (ôr bik′yə lər), *adj.* **1** like a circle or sphere; rounded. SYN: circular, spherical.
2 *Botany.* having the shape of a flat body with a nearly circular outline: *an orbicular leaf.* [from Latin *orbiculus,* from *orbis* circular]

orbicularis (ôr bik′yə lăr′is), *n. Anatomy.* a muscle surrounding an opening of the body, such as that of the mouth or the eye; sphincter. [from New Latin, from Latin *orbiculus* orbicular]

orbiculate or **orbiculated,** *adj.* = orbicular. —**orbiculately,** *adv.*

orbit, *n.* **1** *Astronomy.* **a** the path of the earth or any one of the planets about the sun: *the orbit of Mars.* **b** the path of any celestial body about another celestial body: *the moon's orbit about the earth. The most important elements of the orbit are: P the period of revolution; e the eccentricity, which measures the elongation of the ellipse; a the semi-major axis, i.e. half the length of the ellipse; and i the inclination of the plane of the orbit to a reference plane, perpendicular to the line of sight, and called the plane of the sky* (W. H. Marshall). **c** the path of an artificial satellite about any celestial body: *to put a weather satellite into orbit about the earth.*
2 *Physics.* the curved path of an electron about the nucleus of an atom in the Bohr model of atomic structure.
▶ It is no longer believed that electrons follow well-defined paths about the nucleus. Electron configurations are now described in terms of *orbitals.*
3 *Anatomy.* **a** the bony cavity or socket in which the eyeball is set. **b** the eye; eyeball.
—*v.* **1** to move or revolve in an orbit: *Having been given a big push to 18,000 miles per hour, the satellite will continue to orbit on its own* (Hanson W. Baldwin). *Earlier investigations [showed] how muonic atoms behave when the muons orbit within the nucleus* (Scientific American).
2 to travel in an orbit around (a celestial body): *Mars orbits the sun in an ellipse with an eccentricity of .093; its distance from the sun varies significantly* (Joseph Miles Chamberlain).
3 to put into an orbit: *to orbit a satellite.*
4 (of a satellite, etc.) to go into orbit; achieve orbital velocity: *The ... satellite which failed to orbit on May 27 probably travelled 7,500 miles into the south Atlantic* (London Times).
[from Latin *orbita* wheel track, from *orbis* wheel, circle]

orbital, *adj.* **1** *Astronomy.* of an orbit: *orbital speed, the moon's orbital motion.*
2 *Anatomy.* of or having to do with the orbit of the eye.
3 *Physics.* of or having to do with an orbital or orbitals: *The orbital model [of the atom] differs from the Bohr model in that it does not represent electrons as moving in planetary orbits around the nucleus* (Chemistry Regents Syllabus).
—*n. Physics.* the wave function of an electron moving in a molecule or atom, corresponding to the orbit or path of an electron in earlier theory. An orbital indicates the average region of most probable location of an electron. Electrons occupy orbitals that may differ in size, shape, or orientation in space.

orbital index, *Anatomy.* the ratio of length to height of the orbit of the eye.

orbital steering, *Biochemistry.* a process by which enzymes are thought to facilitate chemical reactions by guiding the atoms in the reacting compounds at precise angles to enable them to join together: *They call this process "orbital steering," since the electron orbitals of the reacting atoms are manoeuvred into the proper juxtaposition* (New Scientist).

orbital velocity, *Astronomy, Physics.* **1** the velocity at which a body revolves about another body: *To any pair of revolving bodies at a specified distance from each other there is associated an orbital velocity* (Krogdahl, The Astronomical Universe).
2 the velocity a body, such as an artificial satellite, must achieve and maintain to go into or remain in orbit.

orbiter, *n.* something that orbits, especially an artifical satellite.

order, *n.* **1** *Biology.* a primary group of related organisms ranking below a class and above a family. The rose family (Rosaceae), the pea family (Leguminosae), and several others, belong to one order. *In the Linnaean classification, structural characters, rather than habits of external forms, were used as a basis. Six class-*

es were employed, four of them vertebrate ... and two invertebrate. These classes were divided into orders, the orders into genera, and the genera into species (Shull, *Principles of Animal Biology*).
ASSOCIATED TERMS: see **kingdom.**
2 *Mathematics.* **a** the degree (of an algebraic term); the sum of the exponents to which the variables in a term are raised. **b** the number of elements in a group: *a group of the nth order.* **c** the number of rows and columns in a matrix. **d** the number of times a variable is successively differentiated. **e** (of a differential equation) the order of the highest derivative which appears. **f** = order of magnitude.
3 *Chemistry.* a classification of chemical reactions according to the number of molecules entering the reaction: *The thermal decomposition of acetaldehyde CH_3—CHO $\longrightarrow$ CH_4 + CO and of several other carbon compounds, also appears to have an order which is neither 1 nor 2, but somewhere in between* (K. D. Wadsworth).

ordered pair, *Mathematics.* any two numbers written in a meaningful order, so that one can be considered as the first and the other as the second of the pair.

order of magnitude, *Mathematics.* a range of magnitudes of a quantity extending from some value of the quantity to ten times that value. EXAMPLES: 250 is an order of magnitude greater than 25. 10 is two orders of magnitude greater than 1/10. Orders of magnitude are used in referring to extremely large units, as those in astronomy, or to extremely small units, as those in nuclear physics.

ordinal, *adj.* **1** *Mathematics.* showing order or position in a series (as *first, second, third, fourth*).
2 *Biology.* having to do with an order of organisms: *the specific, generic, or ordinal names of organisms.*
—*n.* = ordinal number.
[from Late Latin *ordinalis,* from Latin *ordinem* order]

ordinality, *n. Mathematics.* the condition or property of being expressible in order.

ordinal number, *Mathematics.* a number that shows order or position in a series. First, second, third, etc., are ordinal numbers; one, two, three, etc., are cardinal numbers.

ORDINATES

ORDINATE of D is DA on Y-axis

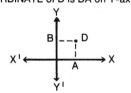

ordinate (ôrd′n it *or* ôrd′n āt), *n. Mathematics.* (in a system of Cartesian coordinates) the vertical coordinate of a point on a graph, usually symbolized *y;* a quantity representing perpendicular distance of a point from the horizontal (*x*) axis. It is positive if the point lies above the horizontal axis, and negative if the point lies below the horizontal axis. *A practical example is found in the method often used for designating an address in a city, the streets of which intersect at right angles. Thus, for a residence at 234 East 116th Street, 234 may be regarded as the abscissa and 116 as the ordinate* (Duncan, *Astronomy*). [from Latin *ordinatum* arranged, from *ordinem* order]

Ordovician (ôr′də vish′ən), *Geology.* —*n.* **1** a geological period, the second in the Paleozoic era, after the Cambrian and before the Silurian. The Ordovician is characterized by the appearance of jawless fishes and the development of many trilobites, brachiopods, and other invertebrates.
2 the rocks formed in this period.
—*adj.* of the Ordovician or its rocks: *The territory that was the earth's south polar region in the Upper Ordovician period is now the central Sahara* (Science News).
[from Latin *Ordovices* ancient Celtic tribe in Wales]

ore, *n. Geology.* **1** a mineral aggregate or rock containing a high enough concentration of one or more economically valuable metals or minerals to make mining it profitable: *Over 100 uranium-bearing minerals are known ... Of all these uranium minerals only nine are considered to be ores* (J. K. Fockler).
2 a natural substance yielding a nonmetallic material, such as sulfur.
[Old English *ar* brass]

organ, *n. Biology.* any part of an animal or plant that is composed of various tissues organized to perform some particular function. The eyes, stomach, heart, and lungs are organs of the body. Stamens and pistils are organs of flowers: *Different kinds of tissues may combine to form an organ which has some specific function to perform ... Likewise, organs are combined to form organ systems; for example, the digestive system is formed by such organs as the stomach, large and small intestine, liver, pancreas, etc.* (Winchester, *Zoology*). [from Latin *organum,* from Greek *organon* instrument, related to *ergon* work]

organelle (ôr′gə nel′), *n. Biology.* a specialized structure or part of a cell, such as a vacuole, having a special function and considered as analogous to an organ in multicellular organisms. [from New Latin *organella,* from Latin *organum* organ]
▶ Some major organelles are: *cell membrane, cell wall* (in plants and algae), *cytoplasm, nucleus,* and *nucleolus* (in eukaryotes), *nucleoid* (in prokaryotes), *endoplasmic reticulum, ribosome, mitochondrion, Golgi apparatus, lysosome, vacuole* (e.g. *food vacuole, contractile vacuole), centriole* (in animal cells), *chloroplast* and *plastids* (in plant cells).

organic, *adj.* **1** *Chemistry.* of or having to do with compounds containing carbon: *an organic substance, an organic reaction.* **Organic compounds** form the chemical basis of living things and are more numerous than inorganic compounds: *The major categories of organic compounds found in living things are carbohydrates, lipids, proteins, and nucleic acids* (Biology Regents Syllabus).

---

cap, fāce, fäther; best, bē, tėrm; pin, five;
rock, gō, ôrder; oil, out; cup, pùt, rüle,
yü in use, yù in uric;
ng in bring; sh in rush; th in thin, ᴛH in then;
zh in seizure.
ə = a in about, e in taken, i in pencil, o in lemon, u in circus

organic acid

2 *Biology.* **a** of or derived from living organisms: *organic remains.* **b** of the bodily organs; vital; affecting the structure of an organ: *When some organ of the body fails to do its work, we say the disease is organic. Heart trouble, caused by the improper working of the heart, is an organic disease* (Beauchamp, *Everyday Problems in Science*). —**organically,** *adv.*

organic acid, *Chemistry.* any carbon compound that displays typical acidic properties, especially one containing the carboxyl radical –COOH: *Carboxylic acids are the most common type of organic acids* (James S. Fritz).

▶ In the IUPAC system of nomenclature, organic acids are named from the corresponding hydrocarbons by replacing the final "-e" with the ending "-oic" and adding the name "acid". The first two members of this series, methanoic acid, HCOOH, and ethanoic acid, CH_3COOH, are known by the common names, formic acid and acetic acid.

organic chemistry, the branch of chemistry that deals with compounds of carbon, such as foods and fuels.

organism, *n. Biology.* a living body made up of separate parts, such as ribosomes, vacuoles, cells, tissues, and organs, which work together to carry on the various processes of life: *There is good reason to think that organisms were able to make most of the enzymes of which we are aware before they had evolved far enough to leave any fossil record.* (N. W. Pirie).

organizable, *adj. Biology.* that can be converted into living tissue: *the organizable part of a plant.*

organization, *n. Biology.* **1** the way in which the parts of an organism are arranged to work together: *Living things are set off from nonliving, not by being constructed from peculiar substances, but because of the unique organization which permits life processes to go on effectively* (Emerson, *Basic Botany*). SYN: constitution.
2 a living thing made up of related parts, each having a special function; organism: *A tree is an organization of roots, trunk, branches, leaves, and fruit.*

organize, *v. Biology.* **1** to furnish with organs; provide with an organic structure; make organic.
2 to assume organic structure; become living tissue.

organized, *adj. Biology.* having organs or organic structure: *A rose is a highly organized plant.*

organized ferment, 1 *Biochemistry.* an enzyme that is active only within a particular cell. Also called **intracellular enzyme.**
2 *Biology.* a living organism, such as yeast or other fungi, which is used to cause fermentation.

organizer, *n.* = inductor: *One of the most interesting developments ... is the discovery or organizers ... substances which exert marked control over regions of the embryo adjacent to them* (Hegner and Stiles, *College Zoology*).

organo-, *combining form.* **1** *Chemistry.* organic, as in *organomercurial, organometallic.* **2** *Biology.* organ or organs, as in *organology, organoleptic.*

organ of Corti (kôr′tē), *Anatomy.* a part of the structure of the inner ear that lies along the basilar membrane of the cochlea. It contains over 15,000 hair cells which transmit sound vibrations to the nerve fibers. [named after Alfonso *Corti,* 1822–1888, Italian anatomist]

organogenesis, *n. Biology.* the origin or development of the organs of an animal or plant: *Genes engaged in flower organogenesis have widely different alternative reaction modes* (J. Heslop-Harrison).
—**organogenetic,** *adj.* having to do with organogenesis: *We still do not fully comprehend the organogenetic movements involved in gonadal differentiation and the mechanisms controlling cellular specialization* (A. D. Jost). —**organogenetically,** *adv.*

organography (ôr′gə nog′rə fē), *n.* the description of the organs of living beings; descriptive organology.

organoleptic (ôr′gə nō lep′tik), *adj. Physiology.* using various sense organs to determine flavor, texture, or other quality: *Organoleptic tests of the samples after deep-fat frying did not show significant effect on flavor due to antibiotic treatment* (Jean I. Simpson). [from *organo-* + Greek *leptos* fine, delicate] —**organoleptically,** *adv.*

organology, *n.* the branch of biology that deals with the structure and function of animal and plant organs.

organomercurial, *Chemistry..* —*n.* any one of various highly toxic organic compounds that contain mercury: *Not all E. coli strains resistant to inorganic mercury carry resistance to the organomercurials, although the two resistances may well be linked in some way* (London *Times*).
—*adj.* having to do with or containing organomercurials: *organomercurial seed dressings.*

organometallic, *Chemistry.* —*adj.* consisting of an atom of a metal in combination with one or more alkyl radicals: *... the thousands of organometallic compounds using mercury, arsenic, bismuth, and other metals to form medicines and drugs indispensable to our present day life* (Parks and Steinbach, *Systematic College Chemistry*).
—*n.* an organometallic compound.

orientation, *n.* **1** *Chemistry.* **a** the relative position of atoms or radicals in complex molecules: *Even when a hydrogen molecule collides in a suitable orientation with an iodine molecule, the energy being sufficient for activation, whether reaction occurs may depend on the timing of internal vibration of each of the molecules* (K. D. Wadsworth). **b** the determination of the position of atoms and radicals to be substituted in a substance.
2 *Geography.* the alignment of a linear feature on the earth's surface with respect to a reference direction or location, often with respect to north.

origin, *n.* **1** *Anatomy.* the main or more fixed attachment of a muscle, which does not change position during the muscle's contraction: *In the case of limb muscles the origin is usually the end nearest the trunk while the distal end of the muscle is the insertion* (Winchester *Zoology*).
2 *Mathematics.* the intersection of the horizontal axis and the vertical axis in a coordinate system: *A coordinate system has a fixed reference point called the origin* (Shortley and Williams, *Elements of Physics*). [from Latin *originem,* from *oriri* to rise]

ornithic (ôr nith′ik), *adj. Zoology.* of or characteristic of birds. [from Greek *ornithikos,* from *ornis, ornithos* bird]

ornithine (ôr′nə thin), *n. Biochemistry.* an amino acid formed by hydrolyzing arginine. *Formula:* $C_5H_{12}O_2N_2$ [from Greek *ornis, ornithos* bird (so called because the amino acid is found in the excrement of birds)]

ornithoid (ôr′nə thoid), *adj. Zoology.* having a certain structural resemblance to a bird: *an ornithoid lizard.*

ornitholite (ôr nith′ə līt), *n. Paleontology.* the fossilized remains of a bird. [from Greek *ornis, ornithos* bird + English *-lite*]

ornithological, *adj.* of or having to do with birds. —**ornithologically,** *adv.*

ornithology (ôr′nə thol′ə jē), *n.* the branch of zoology dealing with the study of birds.

orogen (ôr′ə jən), *n.* a mountain formation: *Orogens can be examined for signs of the descent of the lithosphere into a trench* (Science).

orogenesis (ôr′ə jen′ə sis), *n.* = orogeny.

orogenetic (ôr′ə jə net′ik), *adj.* = orogenic: *We do not know how much carbon dioxide may have been emitted by the volcanoes of Venus in some recent orogenetic spasm* (Science News).

orogenic (ôr′ə jen′ik), *adj. Geology.* having to do with the formation of mountains: *The next phase in the development of a geosyncline is characterized by rising mountains and is known as the orogenic phase* (Science Journal).

orogenics (ôr′ə jen′iks), *n. Geology.* = orogeny.

orogeny (ô roj′ə nē), *n. Geology.* the formation of mountains, as by the folding of the earth's crust: *The Moine Thrust was formed about 400 million years ago, during the Caledonian orogeny, when a great range of mountains was born* (Peter Francis). [from Greek *oros* mountain + *-geneia* a being born, from *-genēs* born]

orographic (ôr′ə graf′ik), *adj.* **1** *Geography.* of or having to do with orography.
2 *Meteorology.* produced by the forced ascent of warm air into cooler regions because of a mountain range lying in its path: *The most ideal condition for producing orographic rainfall is when a high and relatively continuous mountain barrier lies close to a coast and the winds from off a warm ocean meet the barrier at right angles* (Finch and Trewartha, *Elements of Geography*). —**orographically,** *adv.*

orography (ô rog′rə fē), *n.* the branch of physical geography that deals with the formation and features of mountains.

orpiment (ôr′pə ment), *n. Mineralogy.* a bright-yellow mineral, arsenic trisulfide, found in soft, foliated masses or prepared synthetically as a yellow powder, used as a pigment. *Formula:* As_2S_3 [from Old French *orpiment,* also or *pigment,* learned borrowing from Latin *auripigmentum,* from *aurum* gold + *pigmentum* pigment]

ortho (ôr′thō), *adj. Chemistry.* (of a salt or acid) having the highest number of water molecules within a series: *The most highly hydrated form of an oxide which acts as an acid when so hydrated is referred to as the ortho form of the acid* (Jones, *Inorganic Chemistry*).

ortho-, *combining form.* **1** straight, upright, as in *orthoclase, orthocenter.* **2** *Chemistry.* **a** of, bonded to, or substituting for two adjacent carbon atoms in a benzene ring. *Ortho-* is usually italicized, as in *ortho*-dichlorobenzene. Also written **1,2-.** *Abbreviation: o-* Compare **meta-, para-.** **b** ortho, as in *orthophosphoric acid.* [from Greek *orthos* straight]

orthocenter (ôr′thə sen′tər), *n. Mathematics.* the point at which the altitudes of a triangle intersect.

orthoclase (ôr′thə klās *or* ôr′thə klāz), *n. Mineralogy.* a common feldspar, a silicate of aluminum and potassium, that often occurs in granite and is used in making glass, ceramics, and abrasives. *Formula:* $KAlSi_3O_8$ [from German *Orthoklas,* from Greek *ortho-* + *klasis* cleavage]

orthogenesis (ôr′thə jen′ə sis), *n. Biology.* the theory that evolution of new species follows inherent tendencies and is not influenced by environment or natural selection: *Weidenreich was an adherent of the theory of orthogenesis, according to which evolutionary changes are directed from within the organism by a sort of inner urge.* (Scientific American). —**orthogenetic** (ôr′thə jə net′ik), *adj.* of, having to do with, or exhibiting orthogenesis.

orthogonal (ôr thog′ə nəl), *adj. Mathematics.* **a** having to do with or involving right angles; rectangular. SYN: right-angled. **b** (of certain functions) having an integral that is equal to zero or to one. **c** (of vectors) having a dot product equal to zero. **d** consisting of orthogonal elements.
—*n.* an orthogonal line or plane: *... reading in both directions along all orthogonals and main diagonals* (Scientific American). [from Greek *orthogōnios,* from *orthos* right + *gōnia* angle]
—**orthoganality** (ôr′thog ə nal′ə tē), *n.* orthogonal quality or state.
—**orthogonally,** *adv.* in an orthogonal manner; at right angles: *First note that each cell of the checkerboard ... has eight neighboring cells, four adjacent orthogonally, four adjacent diagonally* (Martin Gardner).

ortho-hydrogen, *n. Physics, Chemistry.* a form of hydrogen consisting of molecules whose pairs of nuclei have spins in the same direction: *Three quarters of hydrogen gas is ortho-hydrogen* (Offner, *Fundamentals of Chemistry*).

orthonormal, *adj. Mathematics.* having to do with or consisting of normal orthogonal elements: *an orthonormal set or system.*

orthophosphoric acid, = phosphoric acid (def. 1).

orthopteran (ôr thop′tər ən), *Zoology.* —*n.* an orthopterous insect. —*adj.* = orthopterous.

orthopterous (ôr thop′tər əs), *adj. Zoology.* of or belonging to an order (Orthoptera) of insects characterized by longitudinally folded, membranous hind wings covered by hard, narrow forewings, and having mouthparts adapted for chewing. Orthopterous insects have an incomplete metamorphosis. The order includes crickets, grasshoppers, locusts, and cockroaches.

orthorhombic (ôr′thə rom′bik), *adj. Crystallography.* having three unequal axes intersecting at right angles. Also called **rhombic.** See the picture at **crystal.**

---

cap, fāce, fäther; best, bē, tèrm; pin, fīve;
rock, gō, ôrder; oil, out; cup, pùt, rüle,
yü in use, *yu* in uric;
ng in bring; *sh* in rush; *th* in thin, TH in then;
zh in seizure.
ə = *a* in about, *e* in taken, *i* in pencil, *o* in lemon, *u* in circus

orthotropic

orthotropic (ôr′thə trop′ik), *adj. Botany.* of, having to do with, or exhibiting orthotropism; growing vertically upward or downward, as a stem or root. —**orthotropically,** *adv.*

orthotropism (ôr thot′rə piz əm), *n. Botany.* a tendency to grow in a vertical direction, upward or downward.

orthotropous (ôr thot′rə pəs), *adj. Botany.* (of an ovule) having a straight nucellus and not inverted, with the chalaza at the base and the micropyle at the opposite end.

os^1 (os), *n., pl.* **-ossa** (os′ə). *Anatomy.* a bone, as in *os nasale* (nasal bone), *os parietale* (parietal bone). [from Latin *os, ossis* bone]

os^2 (os), *n., pl.* **ora** (ôr′ə or ōr′ə). *Anatomy.* a mouth or opening, as in *os uteri* (opening of the uterus). [from Latin *os, oris* mouth]

os^3 (ōs), *n., pl.* **osar** (ō′sär). *Geology.* an esker. [from Swedish *ås* ridge (of a hill or roof); *åsar,* plural]

Os, *symbol.* osmium.

oscillate, *v. Physics.* **1** to have or produce oscillations. **2** to swing from one limit to another. [from Latin *oscillatum* swung, rocked, from *oscillum* a swing]

oscillation, *n. Physics.* **1** the variation of a quantity from one limit to another, as the voltage of an alternating current: *In practice the particle will execute oscillations around the phase-stable orbit, circulating on or around the orbit, indefinitely, neither gaining nor losing energy* (L. L. Green).
2 a single swing from one limit to another: *A complete vibration or oscillation means one round trip, say from a to b and back to a, ...* (Sears and Zemansky, *University Physics*).

OSCILLATION

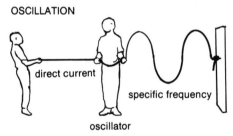

direct current

specific frequency

oscillator

Direct current passes through the oscillator which changes it to a wave of a specific frequency.

oscillogram (ə sil′ə gram), *n.* a record made by an oscillograph.

oscillograph (ə sil′ə graf), *n.* an instrument for recording electric oscillations, as of currents and voltages: *The electrical potentials can be led off from the muscles with the aid of suitable electrodes, amplified, and then recorded either photographically or by an ink-writing oscillograph* (W. F. Floyd and P. H. S. Silver).

oscilloscope (ə sil′ə skōp), *n. Electricity.* an instrument for representing the oscillations of a varying voltage or current on the fluorescent screen of a cathode-ray tube: *Oscilloscopes are invaluable tools in studying the behavior of electric circuits. They can be used to dis-*

play phenomena which occur in a few billionths of a second. (John R. Pierce).

osculant (os′kyə lənt), *adj.* **1** *Biology.* intermediate between two or more groups (applied as to genera or families, that connect or link others together).
2 *Zoology.* adhering closely; embracing.
[from Latin *osculantem* kissing, from *osculum* a kiss]

oscular (os′kyə lər), *adj. Zoology.* of or having to do with an osculum.

osculate (os′kyə lāt), *v.* **1** *Geometry.* **a** to have three or more points coincident with: *A plane or a circle is said to osculate a curve when it has three coincident points in common with the curve.* **b** (of two curves, surfaces, etc.) to osculate each other.
2 *Biology.* to share the characters of two or more groups; be intermediate.
[from Latin *osculatum* kissed, from *osculum* a kiss]

osculating circle, = circle of curvature.

osculation (os′kyə lā′shən), *n. Geometry.* a contact, as between two curves or surfaces, at three or more common points.

osculum (os′kyə ləm), *n., pl.* **-la** (-lə). *Zoology.* a mouth or mouthlike opening, as of a sponge or tapeworm. [from Latin *osculum* a kiss, (literally) little mouth, from *os, oris* mouth]

-ose, *suffix. Biochemistry.* **1** type of sugar or other carbohydrate, as in *fructose, lactose.*
2 a primary protein derivative, as in *proteose.*
[from French *-ose,* in *glucose* glucose]

O-shell, *n. Physics.* the energy level next in order after the N-shell, occupied by electrons whose principal quantum number is 5. See also **O electron.**

-osis (-ō′sis), *suffix, pl.* **-oses** (-ō′sēz). **1** act or process of, or state or condition of, as in *osmosis, ostosis.* **2** an abnormal condition, as in *thrombosis, trichinosis.* [from Latin *-osis,* from Greek *-ōsis*]

osmatic (oz mat′ik), *adj. Physiology.* of, having to do with, or possessing the sense of smell. [from Greek *osmē* odor]

osmeterium (oz′mə tir′ē əm), *n. pl.* **-teria** (-tir′ē ə). a forked process of the first thoracic segment of certain butterfly larvae, noted for the disagreeable odor it emits.

osmic (oz′mik), *adj. Chemistry.* **1** of osmium. **2** containing osmium, especially with a valence of four.

osmic acid, = osmium tetroxide.

osmium (oz′mē əm), *n. Chemistry.* a hard, heavy, bluish-white metallic element which occurs with platinum and iridium. Osmium is the heaviest or densest known element and is used for electric-light filaments and phonograph needles. *Symbol:* Os; *atomic number* 76; *atomic weight* 190.2; *melting point* 2700°C; *boiling point* 5300°C; *oxidation state* 3, 4. [from New Latin, from Greek *osmē* odor]

osmium tetroxide, *Chemistry.* an oxide of osmium used in electron microscopy to stain and fix biological material, especially animal tissues. *Formula:* OsO_4

osmol (oz′mōl′ or os′mōl′), *n. Chemistry.* a unit of osmotic pressure, equal to the pressure exerted by one mole of ions in one kilogram of solvent.

osmolal (oz mō′lal or os mō′lal), *adj. Chemistry.* of, having to do with, or expressed as an osmolality: *the osmolal concentration of a substance.*

osmolality (oz′mə lal′ə tē *or* os′mə lal′ə tē), *n. Chemistry.* osmotic pressure expressed in osmols or milliosmols per kilogram of solvent: *The problem is that both cow's milk and baby foods contain large quantities of soluble protein ... It is the business of the kidneys to regulate the solute concentrations (osmolality)* (New Scientist).

osmolar (oz mō′lər *or* os mō′lər), *adj. Chemistry.* of, having to do with, or expressed as an osmolarity.

osmolarity (oz′mə lar′ə tē *or* os′mə lar′ə tē), *n. Chemistry.* osmotic pressure expressed in osmols or milliosmols per liter of solution.

osmometer (oz mom′ə tər *or* os mom′ə tər), *n.* a device for measuring osmotic pressure: *Water molecules pass through fairly rapidly, so that the cell often behaves like an osmometer* (John E. Harris).
—**osmometric,** *adj.* of or having to do with osmometry or an osmometer. —**osmometrically,** *adv.*
—**osmometry** (oz mom′ə trē *or* os mom′ə trē), *n.* the measurement of osmotic pressure.

osmoregulation, *n. Biology.* the process by which the osmotic activity of a living cell is increased or decreased by the organism in order to maintain the most favorable conditions for the vital processes of the cell and the organism: *From the point of view of osmoregulation it is the concentration of the salts which matters, but the proportions of the various ions may be important physiologically if they differ very much from those in the body fluids* (H. B. N. Hynes).
—**osmoregularity,** *adj.* of or having to do with osmoregulation: *Chimaeras resemble sharks in ... having a peculiar osmoregulatory system in which urea is retained in the tissues in concentrations which would kill most vertebrates* (Colin Patterson).

osmosis (oz mō′sis *or* os mō′sis), *n. Chemistry, Physics.*
1 the tendency of two solutions with different concentrations of dissolved substances, separated by a semipermeable membrane, to become equally concentrated by passage of the solvent of the less concentrated solution through the membrane. Osmosis is the chief means by which nutrients dissolved in fluids pass in and out of plant and animal cells.
2 the diffusion or spreading of fluids through a membrane or partition till they are mixed.
[from New Latin, from Greek *ōsmos* a thrust]
—**osmotic** (oz mot′ik *or* os mot′ik), *adj.* of or having to do with osmosis: *osmotic absorption of water from the soil.*
—**osmotically,** *adv.* by osmotic action; by the process of osmosis: *osmotically active particles.*
ASSOCIATED TERMS: see **active transport.**

osmotic pressure, *Chemistry, Physics.* the pressure exerted by dissolved material in a solution on a semipermeable membrane separating the solution from another solution or from pure solute, such as water: *Perhaps the inflation of the cells is caused by a change in osmotic pressure, resulting in an intake of water either from the environment or from neighboring cells* (Joseph J. Maio).

osmous (oz′məs *or* os′məs), *adj. Chemistry.* containing osmium, especially with a lower valence than that contained in osmic compounds.

osseous (os′ē əs), *adj.* **1** *Anatomy.* bony: *osseous tissue.*
2 *Geology.* containing or consisting of bones: *osseous breccia.* [from Latin *osseus,* from *os* bone]

ossicle (os′ə kəl), *n. Anatomy.* **1** a small bone, especially of the ear: *As vibrations strike against this drum they set in motion the chain of bones or ossicles of the middle ear* (Simeon Potter).
2 a small bony or bonelike part: *The individual portions of the skeleton of a starfish, sea urchin, or sea cucumber are termed ossicles* (Harbaugh and Goodrich, *Fundamentals of Biology*).

ossicular (o sik′yə lər), *adj. Anatomy.* of or having to do with an ossicle or ossicles. The **ossicular chain** is a series of three small bones, the malleus, incus, and stapes, located in the middle ear of mammals and connecting the tympanic membrane with the vestibule.

ossiferous (o sif′ər əs), *adj.* containing or yielding bones, as a cave or a geological deposit.

ossific, *adj. Physiology.* making or forming bone; ossifying.

ossification, *n. Physiology.* **1** the process of changing into bone.
2 the condition of being changed into bone.
3 a part that is ossified; bony formation.

ossify, *v. Physiology.* **1** to change into bone; become bone: *The soft parts of a baby's skull ossify as he grows older.*
2 to harden like bone; make or become fixed.

osteal (os′tē əl), *adj. Anatomy.* bony; osseous.

osteo-, *combining form.* bone, as in *osteogenesis = the development or formation of bone.* [from Greek *osteon* bone]

osteoblast (os′tē ə blast), *n. Biology.* a bone-forming cell; bone cell.

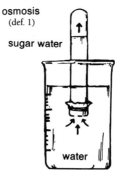

osmosis
(def. 1)

sugar water

water

The solution in the tube rises as water passes through the membrane and mixes with the sugar solution. The mixing of the two solutions is a demonstration of the process of osmosis.

—**osteoblastic** (os′tē ə blas′tik), *adj.* forming bone; osteogenetic: *It is postulated by C. A. L. Bassett that areas of electronegativity are associated with osteoblastic activity — the deposition of new bone — while*

cap, fāce, fäther; best, bē, tèrm; pin, fīve;
rock, gō, ôrder; oil, out; cup, pùt, rüle,
yü in use, *yu̇* in uric;
ng in bring; *sh* in rush; *th* in thin, ŦH in then;
zh in seizure.
ə = *a* in about, *e* in taken, *i* in pencil, *o* in lemon, *u* in circus

areas of positivity or electrical neutrality are associated with osteoclastic activity — the removal of existing bone (Robert J. Pawluk).

osteoclast (os′tē ə klast), *n. Biology.* one of the large multinuclear cells found in growing bone which absorb bony tissue, as when canals and cavities are formed. [from German *Osteoklast,* from Greek *osteon* bone + *klastos* broken]
—**osteoclastic,** *adj.* of or having to do with osteoclasts.

osteocyte (os′tē ə sīt), *n.* one of the branched cells that lie in the lacunae of bone tissue: *Dynamic interchange of calcium and phosphorus between the bone structure and the fluids that bathe it depends in part on the osteocytes that stimulate, and osteoclasts that destroy, bone formation, but the whole complex process is controlled by the parathyroid hormone* (Peter Bishop). [from *osteo-* + *-cyte*]

osteogenesis (os′tē ə jen′ə sis), *n. Biology.* the formation or growth of bone.
—**osteogenetic,** *adj.* of or having to do with osteogenesis; ossific.

osteogenetic cell, = osteoblast.

osteogenic, *adj.* = osteogenetic.

osteoid (os′tē oid), *adj. Biology.* bonelike; bony.

osteological, *adj.* of or having to do with osteology.
—**osteologically,** *adv.*

osteology (os′tē ol′ə jē), *n.* **1** the branch of anatomy that deals with bones. **2** the bony structure or system of bones of an animal or of a major part of an animal, such as the head or trunk. [from Greek *osteon* bone]

osteon, *n.* = Haversian canal.

osteosis (os′tē ō′sis), *n. Biology.* the process or condition of forming bone; bone formation. Also spelled **ostosis.**

ostiole (os′tē ōl), *n. Biology.* a very small orifice or opening, such as those in certain algae and fungi through which the spores are discharged. [from Latin *ostiolum,* diminutive of *ostium* door, opening]

ostium (os′tē əm), *n., pl.* **-tia** (-tē ə). *Biology.* an opening or mouthlike hole, as in the heart of an arthropod: *In a woman the ostium of the oviduct is closely connected to the ovary* (Winchester, *Zoology*). [from Latin *ostium* door, opening]

ostosis (os tō′sis), *n.* = osteosis.

otocyst (ō′tə sist), *n. Zoology.* an organ in many invertebrates, probably of sense of direction and balance, containing fluid and otoliths, once supposed to be an organ of hearing.
—**otocystic** (ō′tə sis′tik), *adj.* of or having to do with an otocyst or otocysts.

otolith (ō′tə lith), *n. Anatomy.* a calcareous body in the inner ear of lower vertebrates and some invertebrates, helpful in maintaining equilibrium.
—**otolithic** (ō′tə lith′ik), *adj.* of or having to do with an otolith or otoliths.

ounce, *n.* **1** a customary unit of weight, 1/16 of a pound in avoirdupois, and 1/12 of a pound in troy weight. **2** a customary unit of volume for liquids, equal to about 29.6 milliliters; fluid ounce. [from Old French *unce,* from Latin *uncia* twelfth part]

-ous, *suffix. Chemistry.* indicating the presence in a compound of the designated element in a lower valence than indicated by the suffix *-ic,* as in *stannous, ferrous, sulfurous.*

outbreeding, *n. Biology.* breeding from organisms that are not closely related. Contrasted with **inbreeding.**

outcrop, *n. Geology.* **1** a coming (of a rock, stratum, etc.) to the surface of the earth: *the outcrop of a vein of coal.* **2** a part that comes to the surface; such rock exposed at the surface or covered only by soil: *an outcrop rich in iron. Not far from our camp was an outcrop of sandstone and schist with secondary white mica. A cliff of this had broken away and buried the main irrigation flume* (F. Kingdon-Ward).

OUTCROP (def. 2)

outer core, *Geology.* the third of the four layers of the earth, lying between the mantle and the inner core. See the picture at **core.**

outer ear, = external ear.

outer space, *Astronomy.* **1** space immediately beyond the earth's atmosphere. **2** space between the planets or between the stars; space beyond the solar system; interplanetary or interstellar space.

outlier (out′lī′ər), *n. Geology.* a part of a stratified formation left detached through the removal of surrounding parts by denudation. An outlier is surrounded by rocks older than itself. Contrasted with **inlier.**

output, *n.* **1** *Physics.* **a** the work done by a machine. **b** the power delivered by an electronic device. **2** the information delivered by a computer or other automatic device.
—*v.* (of a computer) to put out or deliver (information). Compare **input.**

outwash plain, *Geography, Geology.* a broad, flat plain created by the deposition of sediment carried by the meltwater of a glacier.

ova (ō′və), *n.* plural of **ovum.**

oval (ō′vəl), *Geometry.* —*adj.* shaped like an ellipse; ellipsoidal: *the oval path of the earth around the sun.*
—*n.* a plane figure or curve shaped like an ellipse: *the arc of an oval.*

ovalbumin (ōv′al byü′mən), *n. Biochemistry.* the albumin of egg white.

oval window, *Anatomy.* a membrane in the ear of a mammal or other vertebrate connecting the middle ear with the inner ear: *A short external auditory canal at the posterior-lateral angle of the head leads to the tympanic membrane or eardrum. From the latter a bone ... transmits sound waves across the middle-ear cavity to the "oval window" of the inner ear* (Storer, *General Zoology*).

ovarian (ō vãr′ē ən), *adj.* of or having to do with an ovary.

ovary, *n., pl.* **ovaries. 1** *Zoology.* the organ of a female or hermaphroditic animal in which eggs are produced: *In the rat, the mouse, and in man, the ovary displays*

cyclic activity for which the pituitary is responsible and which results in the liberation of ova from the ovaries once every five or six days in the mouse and once every month in the woman. (Bernard Donovan).
2 Botany. the enlarged lower part of the pistil of a flowering plant, enclosing the ovules: *Each of these different fruits bear one or more seeds which contain the embryos of future plants. Now, in the flower the seeds are found in a small structure called the ovary. So, among all plants bearing flowers, the "fruit" is basically a transformed and ripened ovary containing seeds* (Science News Letter).
[from New Latin *ovarium*, from Latin *ovum* egg]

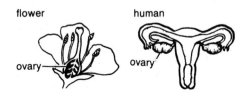

flower

human

ovary

ovary

ovate (ō′vāt), *adj. Biology.* egg-shaped: *an ovate leaf, ovate scales, an ovate berry.*

overburden, *n. Geology.* clay, rock, etc., which has to be removed to get a deposit of ore: *These ores occur beneath 250 feet of waste overburden, two tons of which must be removed for each ton of ore recovered* (Saturday Review).

overlapping, *adj. Mathematics.* having one or more members in common. The sets (0, 1, 2) and (2, 3, 4) are overlapping sets. Compare **disjoint.**

overthrust fault, *Geology.* a fault in which one section of crust has ridden up over another.

overvoltage, *n. Physics.* the difference between the theoretical voltage and the actual required voltage of an electrode in an electrolytic process.

overwinter, *v. Botany.* to stay through the winter enduring cold and snow: *Dahlia tubers can overwinter right in the ground* (Sunset).

ovi-, *combining form.* ovum; egg, as in *oviform, ovisac.* [from Latin *ovi-,* from *ovum* egg]

oviduct (ō′və dukt), *n. Anatomy.* the tube through which the ovum or egg passes from the ovary: *Each ovary leads into an oviduct, termed the fallopian tube. The ciliated, flared-out opening of each tube partially surrounds its ovary in a way that facilitates the entrance of ova into the tube at the time of ovulation, when the ovum is released from the ovary* (J. M. Barrett). [from New Latin *oviductus,* from Latin *ovum* egg + *ductus* duct]

oviferous (ō vif′ər əs), *adj. Biology.* producing or bearing eggs: *the oviferous tube of certain fishes.*

oviform (ō′və fôrm), *adj. Biology.* egg-shaped: *an oviform vesicle.*

oviparous (ō vip′ər əs), *adj. Zoology.* producing eggs that develop after leaving the body. Birds, amphibians, and most reptiles, fishes, and insects are oviparous. Compare **ovoviviparous, viviparous.** [from Latin *oviparus,* from *ovum* egg + *parere* bring forth] —**oviparously,** *adv.* —**oviparousness,** *n.*
—**oviparity** (ō′və par′ə tē), *n.* the condition of being oviparous.

oviposit (ō′və poz′it), *v. Zoology.* to deposit or lay eggs, especially by means of an ovipositor, as an insect: *The females oviposit on seaweeds, or in the cavities of empty shells* (Samuel P. Woodward). [from Latin *ovum* egg + *positum* placed, put, from *ponere* to place, put]
—**oviposition** (ō′və pə zish′ən), *n.* the laying of eggs by means of an ovipositor: *It has been found that attack by Codling moth is reduced if apple-trees are illuminated at night, because oviposition is inhibited by light in this species* (J. L. Cloudsley-Thompson).
—**ovipositor,** *n.* (in certain female insects) an organ or set of organs at the end of the abdomen, by which eggs are deposited: *When the flowers are open the female Pronuba forces her ovipositor through the wall of the ovary of the yucca flower and deposits eggs* (Emerson, *Basic Botany*).

ovisac (ō′və sak), *n. Biology.* **1** a sac, cell, or capsule containing an ovum or ova: *The eggs are carried about in the ovisacs until they are hatched* (Thomas H. Huxley). **2** = Graafian follicle.

ovogenesis (ō′və jen′ə sis), *n.* = oogenesis.

ovogenetic (ō′və jə net′ik), *adj.* = oogenetic.

ovoid (ō′void), *Biology.* —*adj.* **1** oval with one end more pointed than the other; egg-shaped; ovate. *The ovoid female cones, when broken in cross section, plainly show the seeds and the cells in which they are located* (Scientific American). **2** of this form with the broader end at the base, as a pear or avocado is.
—*n.* an egg-shaped organ or part.

ovoidal (ō voi′dəl), *adj.* = ovoid.

ovotestis (ō′və tes′tis), *n., pl.* **-tes** (-tēz). *Zoology.* a combined male and female reproductive organ, as in the snail. [from New Latin *ovotestis,* from Latin *ovum* + *testis* testis]

ovovitellin (ō′vō vī tel′in), *n. Biochemistry.* a protein contained in the yolk of eggs. [from Latin *ovum* egg + *vitellin*]

ovoviviparous (ō′vō vī vip′ər əs), *adj. Zoology.* producing eggs that are hatched within the body of the parent, so that young are born alive but without placental attachment, as certain reptiles and fishes and many invertebrate animals: *Viviparous animals in which practically the whole nourishment of the young is furnished by the egg itself are said to be ovoviviparous* (Shull, *Principles of Animal Biology*). [from Latin *ovum* egg + English *viviparous*] —**o′vovivip′arously,** *adj.*
—**ovoviviparity** (ō′vō vī′və par′ə tē), *n.* the condition of being ovoviviparous.

ovular, *adj. Biology.* of an ovule; being an ovule: *ovular membranes.*

ovulate (ov′yə lāt), *v. Zoology.* **1** to produce ova or ovules; discharge ova from the ovary: *It [the pituitary gland] ... sends out the special hormones to act upon

---

cap, fāce, fäther; best, bē, tėrm; pin, fīve;
rock, gō, ôrder; oil, out; cup, půt, rüle,
yü in use, *yủ* in uric;
ng in bring; sh in rush; th in thin, ᴛʜ in then;
zh in seizure.
ə = a in about, e in taken, i in pencil, o in lemon, u in circus

a woman's ovaries (the follicle stimulating hormone and the luteinizing hormone) so she will ovulate once a month, and will begin producing progesterone necessary to maintain pregnancy should conception occur (Alton Blakeslee).
2 (of an ovum) to be discharged: At birth a female will have all the eggs she will get for reproduction. The eggs are developed only to the prophase stage of cell division until puberty, when the cyclic rise in estrogen triggers one egg, once a month, to ovulate or resume cell division (Science News).
—**adj.** (ov'yə lit) Botany. of or designating a type of cone in coniferous plants that bears megasporangia: Ovulate cones of pine are much larger and considerably more complex in structure than the pollen-bearing cones (Raven, Biology of Plants).

ovulation (ov'yə lā'shən), n. Biology. the formation or production of ova or ovules, and especially their discharge from the ovary: There is a monthly cycle of ovulation in a woman that somewhat corresponds to the yearly cycle of ovulation in the frog, but a woman normally releases only one egg from her ovaries during ovulation while the frog may release hundreds (Winchester, Zoology). ► See the note under **menstrual.**

ovulatory (ov'yə lə tor'ē), adj. Biology. of or having to do with ovulation: the ovulatory cycle.

ovule (ō'vyül), n. **1** Botany. **a** the part of a plant that develops into a seed. In flowering plants the ovule contains one egg, which after fertilization develops into an embryo. The carpel is the unit structure which produces ovules. It is usually made up of an ovary containing at least one ovule, plus a style, and a stigma (Emerson, Basic Botany). **b** a young seed.
2 Zoology. a small ovum, especially when immature or unfertilized; the ovum before its release from the ovarian follicle. [from New Latin ovulum, diminutive of Latin ovum egg]

ovum (ō'vəm), n., pl. **ova.** Biology. a female germ cell or gamete, produced in the ovary. After the ovum is fertilized, it is called a zygote. The zygote undergoes cell division to become an embryo. The nonmotile gamete, the egg or ovum ... produced by the female is usually much larger than the sperm, due to a larger quantity of cytoplasm and in some cases to an additional nonliving material called yolk (Harbaugh and Goodrich, Fundamentals of Biology). SYN: macrogamete. [from Latin ovum egg]

oxalacetic acid (ok'sə lə sē'tik), = oxaloacetic acid.

oxalate (ok'sə lāt), n. Chemistry. a salt or ester of oxalic acid.

oxalic acid (ok sal'ik), Chemistry. a colorless, crystalline, poisonous organic acid found in many vegetables and other plants, such as wood sorrel, spinach, rhubarb, tomatoes, grapes, and sweet potatoes. It is also produced in the body, and is prepared commercially by heating sodium formate with sodium hydroxide. Formula: $C_2H_2O_4$ [from Latin oxalis wood sorrel (plant with an acid juice), from Greek oxalis, from oxyr sour]

oxaloacetate (ok'sə lō as'ə tāt), n. Chemistry. a salt or ester of oxaloacetic acid: The carbon atom [in photosynthesis] originating from the carbon dioxide molecule is thought to be passed from oxaloacetate to an acceptor molecule, thereby forming the PGA [phosphoglyceric acid] and also converting oxaloacetate into the three-carbon pyruvic acid (New Scientist).

oxaloacetic acid (ok'sə lō ə sē'tik), Chemistry. an unstable, colorless, crystalline acid that combines with acetic acid at the beginning and end of the Krebs cycle: Acetone, in low concentrations, is a normal blood component resulting from non-enzymatic breakdown of oxaloacetic acid (Stuart Eriksen). Formula: $C_4H_4O_5$ [from oxalic (acid) + acetic (acid)]

oxazine (ok'sə zēn), n. Chemistry. one of a series of isomeric compounds consisting of one oxygen, one nitrogen, and four carbon atoms arranged in a ring. Formula: C_4H_4ON

oxbow, n. Geography. a crescent shaped lake or swamp formed when a meander of a river or stream is cut off from the main channel through deposition along the banks and a realignment of the main flow.

oxidant, n. = oxidizer.

oxidase (ok'sə dās or ok'sə dāz), n. Biochemistry. any of a group of enzymes that cause or promote oxidation: Cytochrome oxidase ... specifically catalyzes the oxidation of cytochromes by molecular oxygen (P. H. Mitchell). Compare **oxygenase.** Also called **oxidation enzyme.**
—**oxidasic,** adj. of or having to do with oxidase.

oxidation, n. Chemistry. the act or process of oxidizing; combination with oxygen; removal of hydrogen from a compound or of an electron from an atom or molecule: The rusting of iron is a common example of the original meaning of oxidation. Iron combines with oxygen in the presence of moisture to form rust. ... Oxidation also occurs inside the human body when food molecules combine with inhaled oxygen to slowly produce carbon dioxide, water, and energy. The combustion of natural gas and other fossil fuels is a form of rapid oxidation (Harriet V. Taylor). Oxidation is any process that removes electrons from a substance. Reduction is the reverse process: the addition of electrons (C. C. Delwiche).

oxidation enzyme, = oxidase: Oxidation enzymes are required in all living matter to help cells burn oxygen (Wall Street Journal).

oxidation number, Chemistry. **1** a positive or negative number that indicates the ionic charge of an atom and that is equal to its valence: The oxidation number, although arbitrary, is a convenient notation for keeping track of the number of electrons involved in a chemical reaction. In assigning oxidation numbers, electrons shared between two unlike atoms are counted as belonging to the more electronegative atom; electrons shared between two like atoms are divided equally between the sharing atoms (Chemistry Regents Syllabus). **2** = oxidation state.

oxidation potential, Chemistry, Physics. a measure, in volts, of an element's tendency to oxidize or lose electrons, used to predict how the element will react with another substance. The oxidation potential of hydrogen is arbitrarily assigned as zero. The standard oxidation potential of potassium is +2.92, and gold, −1.42.

oxidation-reduction, *n. Chemistry.* a reaction involving the transfer of electrons from one atom or ion to another atom or ion. Compare **redox.**

oxidation-reduction potential, *Chemistry.* the electromotive force resulting from oxidation-reduction, usually measured relative to a standard hydrogen electrode: *Electron-donor and electron-acceptor molecules can be characterized by the quantity called oxidation-reduction potential, which can be positive or negative and is usually expressed in volts* (Scientific American).

oxidation state, *Chemistry.* the number of electrons lost or gained by an element or atom in a compound, expressed as a positive or negative number indicating the ionic charge and equal to the valence. Oxygen has an oxidation state of −2; hydrogen has an oxidation state of +1. Also called **oxidation number.**

oxidative (ok'sə dā'tiv), *adj. Chemistry.* having the property of oxidizing: *cellular oxidative processes.* —**oxidatively,** *adv.*

oxidative phosphorylation, *Biochemistry.* the process in cell metabolism by which mitochondria oxidize organic molecules and store the energy by converting adenosine diphosphate (ADP) to adenosine triphosphate (ATP): *The final stage of the breakdown of carbohydrates and fats ... is normally linked intimately with the production of ATP in the process known as oxidative phosphorylation* (New Scientist).

oxide (ok'sīd), *n. Chemistry.* a compound of oxygen with another element or radical. The oxides of sulfur and nitrogen can be used to form sulfuric and nitric acids. *Iron rust (limonite) is an iron oxide; sapphire and ruby are aluminum oxides* (Frederick H. Pough). —**oxidic** (ok sid'ik), *adj.* of or involving an oxide or oxides.

oxidizable (ok'sə dī'zə bəl), *adj.* that can be oxidized: *oxidizable metals.*

oxidize, *v. Chemistry.* **1** to combine with oxygen; undergo oxidation. When a substance burns or rusts, it is oxidized.
2 to cause to lose hydrogen by the action of oxygen. SYN: dehydrogenate.
3 to change (atoms, ions, or molecules) to a higher valence by the loss of electrons; remove an electron from an atom or molecule: *When an iron nail is dipped into a concentrated solution of copper sulfate, the iron atoms in the nail become oxidized by the loss of two electrons.*
4 to become oxidized: *Iron oxidizes in water.*

oxidizer, *n. Chemistry.* **1** a substance that oxidizes another substance; an oxidizing agent. Also called **oxidant. 2** a substance that supports the combustion of a fuel.

oxido-reducing, *adj. Chemistry.* of, having to do with, or causing oxidation-reduction: *an oxido-reducing reaction, an oxido-reducing enzyme.* Also called **redox.**

oxidoreductase (ok'sə dō ri duk'tās), *n. Biochemistry.* an enzyme that catalyzes a reaction in which one molecule of a compound is oxidized and the other reduced; an oxido-reducing enzyme.

oxime (ok'sēm), *n. Chemistry.* any of a group of compounds having the general formula $R_2C=NOH$, where R is an alkyl group or hydrogen. Oximes are

formed by treating aldehydes or ketones with hydroxylamine.

oxonium compound (ok sō'nē əm), *Chemistry.* a compound formed when an organic compound containing an atom of oxygen reacts with a stong acid. [from *ox(ygen)* + *(amm)onium*]

oxonium ion, = hydronium ion.

oxy (ok'sē), *adj. Chemistry.* of or containing oxygen: *the oxy form of hemoglobin.* [from *oxy-*]

oxy-, *combining form. Chemistry.* **1** of or containing oxygen, as in *oxyacetylene, oxyhydrogen, oxysalt.*
2 of acid, as in *oxyphilic.*
3 containing hydroxyl, as in *oxytetracycline.*
[from *oxy(gen)*]

oxyacetylene (ok'sē ə set'ə lēn), *adj. Chemistry.* of or using a mixture of oxygen and acetylene. Compare **oxyhydrogen.**

oxyacid (ok'sē as'id), *n. Chemistry.* **1** an acid that contains oxygen, such as sulfuric acid. Also called **oxygen acid.**
2 any organic acid containing both a carboxyl and a hydroxyl group.

oxychloride (ok sē klor'īd), *n. Chemistry.* **1** a combination of oxygen and chlorine with another element.
2 a compound of a metallic chloride with the oxide of the same metal.

oxygen, *n. Chemistry.* a colorless, odorless, and tasteless gaseous element that forms about one fifth of the atmosphere by volume. Oxygen is present in a combined form in water, carbon dioxide, iron ore, and many other substances. In its uncombined form, oxygen usually occurs in diatomic molecules (O_2). Animals, plants, and most other organisms cannot live without oxygen. *Oxygen is our chief source of energy, being responsible for the respiration of living organisms and the combustion of fuels* (Scientific American). *Oxygen is a by-product of photosynthesis, and most photosynthesis occurs in the sea, especially in areas of coastal upwelling* (Galen E. Jones). *Symbol:* O; *atomic number* 8; *atomic weight* 15.9994; *melting point* −218.4°C; *boiling point* −182.962°C; *oxidation state* −2. [from French *oxygène,* from Greek *oxys* sharp + *genēs* born]

oxygen acid, = oxyacid.

oxygenase (ok'sə jə nās), *n. Biochemistry.* an enzyme that promotes the use of atmospheric oxygen by an organism: *Oxygenases ... reduce O_2 but do it by incorporating oxygen atoms into organic substrates. On the other hand, oxidases reduce O_2 ... and do not incorporate it into organic compounds* (M. F. Mallette).

---

cap, fāce, fäther; best, bē, tėrm; pin, fīve;
rock, gō, ôrder; oil, out; cup, pùt, rüle,
yü in use, *yu* in uric;
ng in bring; *sh* in rush; *th* in thin, ᴛʜ in then;
zh in seizure.
ə = *a* in about, *e* in taken, *i* in pencil, *o* in
lemon, *u* in circus

463

oxygenate (ok′sə jə nāt), *v. Chemistry.* **1** to treat or combine with oxygen: *The left atrium receives oxygenated blood from the lungs through the pulmonary vein* (Biology Regents Syllabus).
2 = oxidize.

oxygenation (ok′sə jə nā′shən), *n. Chemistry.* **1** the act or process of oxygenating: *... physiochemical problems associated with oxygenation of the blood* (Science News). **2** = oxidation.

oxygen debt, *Physiology.* depletion of oxygen stored in the tissues and red blood cells due to a burst of exercise. Oxygen is stored after the exercise is completed. *Although the other tissues can accumulate an oxygen debt, it is essential for the brain to maintain its oxygen supply* (New Scientist).

oxygen demand, = biochemical oxygen demand.

oxygenic (ok′sə jen′ik), *adj. Chemistry.* having to do with or containing oxygen.

oxygenize, *v.* = oxygenate.

oxygenless (ok′sə jen lis), *adj.* **1** having no oxygen: *Sea creatures never before known may lurk in deep, nearly oxygenless basins beneath the Gulf of California* (Science News). **2** = anaerobic: *oxygenless fermentation, oxygenless digestion of sewage.*

oxyhemoglobin (ok′si he′mə glo′bən), *n. Biochemistry.* a substance in arterial blood consisting of hemoglobin combined with oxygen. Oxyhemoglobin gives arterial blood its bright red color.

oxyhydrogen (ok′si hi′drə jən), *adj. Chemistry.* of or using a mixture of oxygen and hydrogen. An oxyhydrogen torch is used for welding or cutting metals. Compare **oxyacetylene.**

oxysulfide (ok′si sul′fid), *n. Chemistry.* a sulfide compound in which one part of the sulfur is replaced by oxygen.

oxytocic (ok′si tō′sik), *Medicine.* —*adj.* hastening childbirth, especially by stimulating the contraction of the uterine muscles.
—*n.* an oxytocic drug or medicine.

oxytocin (ok′si tō′sn), *n. Biochemistry.* a hormone of the pituitary gland effecting contraction of the uterus in childbirth and stimulating lactation. *Formula:* $C_{43}H_{66}N_{12}O_{12}S_2$ [from Greek *oxys* sharp + *tokos* birth]

ozocerite (ō zō′kə rit *or* ō zō′sə rit), *n. Geology.* a waxlike fossil resin of brownish-yellow color and aromatic odor, consisting of a mixture of natural hydrocarbons, and sometimes occurring in sandstones. Also called **mineral wax.** [from German *Ozokerit,* from Greek *ozein* to smell (because it is aromatic) + *kēros* bees wax]

ozokerite (ō zō′kə rit), *n.* = ozocerite.

ozone (ō′zōn), *n. Chemistry.* a form of oxygen with a sharp, pungent odor, produced by electricity and present in the air, especially after a thunderstorm. It is a strong oxidizing agent and is produced commercially for use in water purification, air-conditioning, and as a bleaching agent. *Ozone is the principal substance that shields the surface of the earth from ultraviolet radiation from outer space* (Peter H. Raven). *Formula:* O_3 [from German *Ozon,* from Greek *ozein* to smell]

ozone layer, = ozonosphere.

ozone shield, = ozonosphere.

ozonic (ō zon′ik *or* ō zō′nik), *adj. Chemistry.* of, having to do with, or containing ozone: *an ozonic compound.*

ozonide (ō′zə nid), *n. Chemistry.* any of a group of organic compounds formed when ozone is added to an unsaturated hydrocarbon.

ozoniferous (ō′zn if′ər əs), *adj. Chemistry.* containing ozone.

ozonization (ō′zə nə zā′shən), *n. Chemistry.* **1** the act or process of ozonizing. **2** the condition of being ozonized.

ozonize (ō′zə nīz), *v. Chemistry.* **1** to charge or treat with ozone; cause to react with ozone: *to ozonize an element or compound.* **2** to convert (oxygen) into ozone.

ozonolysis (ō′zə nol′ə sis), *n. Chemistry.* **1** the process of treating a hydrocarbon with ozone followed by hydrolysis. **2** decomposition following treatment with ozone. [from *ozone* + Greek *lysis* a loosening]

ozonometry (ō′zə nom′ə trē), *n. Meteorology.* the process of measuring the amount of ozone in the atmosphere.

ozonosphere (ō zon′ə sfir), *n. Meteorology.* a region of concentrated ozone in the outer stratosphere and the mesosphere. It shields the earth from excessive ultraviolet radiation. Also called **ozone layer** and **ozone shield.**

P

p, *abbrev.* or *symbol.* **1** parental. **2** parity. **3** pico-. **4** positive. **5** pressure. **6** proton.

p-, *prefix.* Usually italicized. *Chemistry.* para- (def. 3a), as in *p*-dichlorobenzene.

P, *symbol.* phosphorus.

Pa, *symbol.* **1** pascal. **2** protactinium.

PABA, *abbrev.* para-aminobenzoic acid.

Pacchionian bodies or Pacchionian glands (pak'ē ō'ne-ən), *Anatomy.* clusters of small villi extending from the arachnoid membranes enveloping the brain. Cerebrospinal fluid passes through them. Also called **arachnoid granulations.** [named after Antonio *Pacchioni*, 1665–1726, Italian anatomist]

pacemaker, *n. Anatomy.* an area of specialized tissue in the heart, near the top of the wall of the right atrium, that sends out the rhythmic impulses which regulate the contractions of the heart muscles: *One complication of coronary heart disease long recognized by physicians had been a slow or temporary arrest of the pacemaker of the heart leading to cessation of all heart action for a few seconds or longer* (Paul Oglesby). Also called **sinoatrial node.**

pachynema (pak'i nē'mə), *n.* = pachytene.

pachytene (pak'ə tēn), *n. Biology.* the third stage of the prophase of meiosis, during which the homologous chromosomes attain a thick form and four chromatids (two from each homolog) become clearly apparent: *As soon as synapsis is complete all along the chromosomes, the cells are said to have entered the pachytene stage of prophase, where they remain for days* (Alberts, *Molecular Biology of the Cell*). [from Greek *pachys* thick + *tainia* band]

Pacinian body or Pacinian corpuscle (pə sin'ē ən), *Anatomy.* any of the numerous oval, seedlike nerve endings, especially in the skin, muscles, tendons, and joints, that function as sensory receptors. [named after Filippo *Pacini*, 1812–1883, Italian anatomist]

pack ice, *Oceanography.* a large area of sea ice consisting of a mixture of floating ice fragments packed or squeezed together: *Pack ice forms and retreats with the seasons, but it is always there* (N. Y. Times). Compare **ice pack.**

pad, *n. Zoology.* **1** a cushionlike mass of fatty tissue forming a fleshy, elastic part on the bottom side of the feet of dogs, foxes, camels, and some other animals. **2** a tarsal cushion of an insect's foot. Also called **pulvillus.** **3** any cushionlike part of an animal body.

paedogenesis (pē'dō jen'ə sis), *n.* = pedogenesis[1].

paedomorphism (pē'dō môr'fiz əm), *n.* = pedomorphism.

pagodite (pə gō'dīt), *n.* = agalmatolite.

pahoehoe (pä hō'ē hō'ē), *n. Geology.* a type of basaltic lava that has hardened into a smooth, undulating, shiny surface: *Specially fine pahoehoe may be seen among the explosive cones and ash beds of the Craters of the Moon, Idaho. Although the youngest of these flows are several hundred years old, they still glow with iridescent tones of blue, purple and bronze* (Fenton, *The Rock Book*). Compare **aa.** [from Hawaiian]

pair bond, *Zoology.* a monogamous bond or union between a male and female of a species: *Ritual "Dance" of the wandering albatross is actually the agonistic responses of an unmated male and female that are not familiar with each other ... until eventually the birds are at ease in each other's presence and a pair bond is established* (Scientific American).
—**pair bonding,** the act or condition of forming a pair bond.

pair production, *Physics.* the production of a subatomic particle and its antiparticle from some other form of energy, such as the production of an electron-positron pair from the energy contained in a gamma ray.

palaeo-, *combining form.* a variant form of **paleo-,** as in *palaeobotany, palaeogeography,* etc.

palatal, *adj.* of or having to do with the palate: *Swallowing when the palate is irritated is a palatal reflex.*

palate (pal'it), *n. Anatomy.* the roof of the mouth. The bony part in front is the **hard palate,** formed by parts of the maxillary and palatine bones, and the fleshy part in back is the **soft palate,** formed by several muscles. [from Latin *palatum*]

palatine (pal'ə tīn), *Anatomy.* —*adj.* **1** of, having to do with, or in the region of the palate: *In human beings, a palatine tonsil can be seen on each side of the back of the mouth just above the throat and below the roof of the mouth* (Paul R. Cannon). **2** designating or having to do with either of the two bones (**palatine bones**) forming the hard palate.
—*n.* a palatine bone.

pale-, *combining form.* the form of **paleo-** before vowels, as in *Palearctic.*

palea (pā'lē ə), *n., pl.* **-leae** (-lē ē). *Botany.* **1** one of the inner, scalelike, usually membranous bracts enclosing the stamens and pistil in the flower of grasses.
2 one of the bracts at the base of the individual florets in many composite plants.
3 the scales on the stems of certain ferns.

Palearctic (pā'lē ärk'tik *or* pā'lē är'tik), *adj. Ecology.* of or belonging to the northern division of the Old World (Europe, Africa north of the Tropic of Cancer, and Asia north of the Himalayas), especially in reference

---

cap, fāce, fäther; best, bē, tėrm; pin, fīve;
rock, gō, ôrder; oil, out; cup, pùt, rüle,
yü in use, *yu* in uric;
ng in bring; sh in rush; th in thin, ŦH in then;
zh in seizure.
ə = *a* in about, *e* in taken, *i* in pencil, *o* in lemon, *u* in circus

paleo-

to the distribution of animals: *the Palearctic region, Palearctic birds.* Compare **Nearctic.**

paleo- (pā′lē ō-), *combining form.* **1** old, ancient; prehistoric, as in *paleogeography, paleomagnetism.* **2** of or having to do with fossils, as in *paleochronology.* **3** early or earliest, as in *Paleocene.* Also spelled **palaeo-**, or **pale-** before vowels. [from Greek *palaio-*, from *palaios* ancient]

paleobiochemistry, *n.* a branch of paleontology that deals with the biochemical constituents of fossil organisms: *In one of the first applications of paleobiochemistry it has been found that hydrocarbon compounds in rocks 3 billion years old may be composed of fossilized chlorophyll* (Franklin J. Tobey, Jr.).

paleobotanic or **paleobotanical,** *adj.* of or having to do with paleobotany: *A surprising paleobotanical discovery was made some years ago when specimens that looked like ordinary ferns proved to have true seeds, instead of spores* (Moore, *Introduction to Historical Geology*). —**paleobotanically,** *adv.*

paleobotany, *n.* the branch of paleontology dealing with fossil plants.

Paleocene (pā′lē ə sēn′), *Geology.* **1** the earliest epoch of the Tertiary period, before the Eocene, during which shallow inland seas drained and the first primates appeared.
2 the rocks formed in this epoch.
—*adj.* of or having to do with this epoch or its rocks: *Paleocene mammals.*
[from *paleo-* + Greek *kainos* recent]

paleochronology, *n. Paleontology.* the dating of fossil animals and plants, as by counting the ridges on fossil shells and corals.

paleoclimate, *n. Geology.* a prehistoric climate: *A new independent method has been developed to study paleoclimates based on deep ice cores from the ice sheets in Antarctica and Greenland* (W. Dansgaard). —**paleoclimatology,** *n.* the study of the climate of prehistoric times.

paleocortex, *n. Anatomy.* a portion of the cortex in the brain of vertebrates, having to do with the sense of smell.

paleocrystic (pā′lē ō kris′tik), *adj. Geology.* consisting of or containing ice supposed to have remained frozen since early ages. [from *paleo-* + Greek *kryst(allos)* ice + English *-ic*]

paleoecological, *adj.* of or having to do with paleoecology. —**paleoecologically,** *adv.*

paleoecology (pā′lē ō i kol′ə jē), *n. Paleontology.* the study of the relationship of living things to environment and each other in prehistoric times; the ecology of prehistoric life: *Instances of the immediate practicality of paleoecology were beginning to be recognized — for example, in arid regions where ancient experience in the laying out of irrigation systems, drainage and salinity control pointed the way for modern planning* (William A. Ritchie).

paleofauna (pā′lē ə fô′nə), *n., pl.* **-nas, -nae** (-nē). *Paleontology.* the fossil fauna of a geological formation or period.

paleoflora (pā′lē ə flôr′ə), *n., pl.* **-floras, -florae** (-flôr′e or flôr′ē). *Paleontology.* the fossil flora of a geological formation or period.

Paleogene (pā′lē ə jēn), *Geology.* —*n.* **1** the first of two periods of the Cenozoic era, consisting of the Paleocene, Eocene, and Oligocene epochs.
2 the system of rock strata formed during this period.
—*adj.* of or having to do with this period or system. Also called **Eogene.**
► See note under **Tertiary.**

paleogeographic or **paleogeographical,** *adj.* of or having to do with paleogeography: *... paleogeographic maps showing the distribution of land and sea since the early Cambrian* (Nature). —**paleogeographically,** *adv.*

paleogeography, *n. Geology.* the geography of prehistoric or geological time.

paleomagnetic, *adj.* of or having to do with paleomagnetism: *paleomagnetic measurements.*

paleomagnetism, *n. Geology.* the study of the direction of the residual magnetism in ancient rocks to determine the movement of the magnetic poles or of the rocks: *Recent studies of paleomagnetism ... have brought renewed interest in the theory of continental drift* (George R. Tilton).

paleontologic (pā′lē on′tl oj′ik), *adj.* = paleontological: *Paleontologic and stratigraphic evidence has accumulated which indicates that many deposits of banded iron ore once thought to be Pre-Cambrian are actually Paleozoic in age* (Lawrence Ogden).

paleontological (pā′lē on′tl oj′ə kəl), *adj.* of or having to do with paleontology.

paleontology (pā′lē on tol′ə jē), *n.* the science of the forms of life existing in prehistoric time, as represented by fossil animals and plants: *Paleontology, the study of indications of prehistoric life, contributes much critical information to the biological sciences even as they, in turn, contribute to paleontology* (Harbaugh and Goodrich, *Fundamentals of Biology*). [from *paleo-* + Greek *on, ontos* a being]

paleotemperature, *n. Geology.* the temperature of oceans and seas in prehistoric times, obtained by measuring or analyzing the chemical components of fossil sediments: *Fossil bones found in certain environments, such as caves where paleotemperatures might be estimated from O^{18}/O^{16} ratios in stalagmites and stalactites, can apparently be fairly accurately dated* (Jeffrey L. Bada).

Paleotropical, *adj. Ecology.* belonging to the tropical (and subtropical) regions of the Old World or Eastern Hemisphere, especially in reference to the distribution of animals.

Paleozoic (pā′lē ə zō′ik), *Geology.* **1** the era after the Proterozoic and before the Mesozoic, characterized by the development of the first fishes, land plants, amphibians, reptiles, insects, and forests of fernlike trees: *The Paleozoic can be generalized as a time of shrinking seas, with gradually increasing crustal unrest.* (Garrels, *A Textbook of Geology*).
2 the rocks formed in this era: *In most places, the rocks beneath the Cretaceous are Paleozoics, strongly folded in Georgia, Alabama, Arkansas, and Oklahoma, but nearly horizontal elsewhere* (Moore, *Introduction to Historical Geology*).
—*adj.* of or having to do with the Paleozoic or its rocks: *Toward the end of the Paleozoic era, forests of trees and patches of undergrowth became extensive* (Emerson, *Basic Botany*).

paleozoological, *adj.* of or having to do with paleozoology.

paleozoology, *n.* the branch of paleontology dealing with fossil animals.

palindrome, *n. Molecular Biology.* a segment of double-stranded DNA having a point from which one of the strands going in one direction has the same nucleotide sequence as the complementary strand going in the opposite direction: *Mirror-image sequences, called palindromes, of a different type are known in bacteria, but such an arrangement of nucleotides has not been described in globin or other mammalian genes* (Science News)
—**palindromic**, *adj.* consisting of a palindrome or palindromes.

palingenesis (pal'in jen'ə sis), *n.* **1** *Embryology.* the development of characteristics in the early stages of an organism which recapitulate without change the characteristics of their ancestors. Compare **ontogeny**.
2 *Geology.* the formation of new magma by the melting or fusion of rocks deep within the crust.
[from Greek *palin* again + *genesis* birth, genesis]
—**palingenetic** (pal'in jə net'ik), *adj.* of, having to do with, or based on palingenesis. —**palingenetically**, *adv.*

palisade, *n.* **1** *Geography, Geology.* Usually, **palisades**, *pl.* a line of high steep cliffs. **2** *Botany.* = palisade layer. [from Middle French *palissade*, ultimately from Latin *palus* stake]

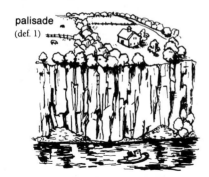

palisade
(def. 1)

palisade cell, *Botany.* any of the cells forming the palisade layer.

palisade layer, *Botany.* a layer of elongated cells between the upper and lower epidermis of leaves. The palisade layer has many chloroplasts and takes part in photosynthesis. See the picture at **leaf**.

palisade mesophyll or **palisade parenchyma**, = palisade layer.

palladic (pə lā'dik), *adj. Chemistry.* of or containing palladium, especially with a valence of four.

palladium (pə lā'dē əm), *n. Chemistry.* a light, silver-white metallic element which occurs in nature with platinum. Palladium is ductile and malleable and is used in making scientific instruments, in alloys with precious metals such as gold and silver, and as a catalyst. *Symbol:* Pd; *atomic number* 46; *atomic weight* 106.4; *melting point* 1552°C; *boiling point* 2927°C;

oxidation state 2, 4. [from New Latin, from the asteroid *Pallas* (which was discovered in 1802, a year before the discovery of the element)]

palladous (pə lā'dəs), *adj. Chemistry.* of or containing palladium, especially with a valence of two.

pallium (pal'ē əm), *n., pl.* **-liums**, **-lia** (-lē ə). **1** *Anatomy.* the cortex of the brain. **2** *Zoology.* **a** the mantle of mollusks and brachiopods. **b** the mantle or stragulum of a bird.

palmar (pal'mər), *adj. Anatomy.* of or having to do with the palm of the hand. [from Latin *palma* palm]

palmate (pal'māt), *adj.* **1** *Botany.* shaped somewhat like a hand with the fingers spread out, especially: **a** having divisions or leaflets which are all attached at one point at the end of the petiole, as in the clover: *a palmate compound leaf.* **b** having several large veins radiating from one point at the end of the petiole, as in the maple leaf: *palmate venation.*
2 *Zoology.* having the front toes joined by a web; web-footed. —**palmately**, *adv.*

palmate leaf
(def. 1)

—**palmation**, *n.* **1** palmate formation or structure. **2** one division of a palmate structure.

palmitic acid (pal mit'ik), *Chemistry.* a white crystalline acid, solid at ordinary temperatures, contained as a glyceride in palm oil and in most solid fats: *The most important of the fatty acids are palmitic, stearic, and oleic acids. Hard fats have comparatively large amounts of palmitic and stearic acids* (Leone R. Carrol). *Formula:* $C_{16}H_{32}O_2$

palmitin (pal'mə tin), *n. Chemistry.* a colorless crystalline solid, a glyceride of palmitic acid, present in palm oil and, in association with stearin and olein, in solid animal fats. *Formula:* $C_{51}H_{98}O_6$

palp, *n.* = palpus.

palpate (pal'pāt), *adj. Zoology.* having a palpus or palpi.

palpebra (pal'pə brə), *n., pl.* **-brae** (-brē). *Anatomy.* an eyelid. [from Latin]
—**palpebral**, *adj.* of or having to do with an eyelid or the eyelids.

cap, fāce, fäther; best, bē, tèrm; pin, fīve;
rock, gō, ôrder; oil, out; cup, pùt, rüle,
yü in use, yù in uric;
ng in bring; sh in rush; th in thin, ᴛʜ in then;
zh in seizure.
ə = a in about, e in taken, i in pencil, o in lemon, u in circus

palpus

palpus (pal′pəs), *n., pl.* **-pi** (-pī). *Zoology.* one of the jointed feelers attached to the mouth of insects, spiders, lobsters, and other arthropods; palp. Palpi are organs of touch or taste. [from Latin *palpus* soft touch, pat]

PALPI

paludal (pə lü′dəl *or* pal′yə dəl), *adj. Geology, Ecology.* of or having to do with a marsh; marshy.

palustrine (pə lus′trin), *adj.* = paludal.

palynologic (pal′ə nə loj′ik), *adj.* = palynological.

palynological (pal′ə nə loj′ə kəl), *adj.* of or having to do with palynology. —**palynologically,** *adv.*

palynology (pal′ə nol′ə jē), *n.* the study of plant spores and pollen, especially in fossil form. [from Greek *palynein* to strew + English *-logy*]

pampas (pam′pəz), *n.pl. Geography, Ecology.* the vast, grassy, treeless plains of South America, especially in Argentina. [from Spanish, from Quechua *pampa* a plain]
—**pampean** (pam pē′ən), *adj.* of or having to do with the pampas.
ASSOCIATED TERMS: see **grassland.**

pamprodactylous (pam′prō dak′tə ləs), *n. Zoology.* having all four toes turned forward, as certain birds. [from Greek *pam-* all, pan- + *pro* before + *daktylos* finger or toe]

pan, *n. Geography, Geology.* **1** a natural hollow or depression in the ground, especially one in which water stands temporarily and evaporates, leaving a crust of salt: *A dry pan, or waterhole, which ... was densely covered with weeds* (H. Rider Haggard). **2** = pan ice.

pan-, *combining form.* all; of all; entirely, as in *panchromatic = entirely chromatic.* [from Greek, from *pan,* neuter of *pas* all]

pancreas (pang′krē əs *or* pan′krē əs), *n. Anatomy.* **1** a large gland in vertebrates near the stomach that secretes insulin and glucagon into the blood and secretes **pancreatic juice,** a digestive juice which contains various enzymes, into the small intestine: *He ... was diagnosed as having a chronically diseased pancreas, the organ that makes insulin, the hormone chiefly responsible for enabling the body to store and use sugar, and that makes enzymes that digest food* (Science News). **2** a similar organ in certain invertebrates. [from Greek *pankreas,* from *pan-* all + *kreas* flesh]
—**pancreatic,** *adj.* of the pancreas: *pancreatic secretions, pancreatic cells.*

pancreatin (pang′krē ə tən *or* pan′krē ə tən), *n. Biochemistry.* **1** any of the enzymes of pancreatic juice, such as amylase, lipase, and trypsin. **2** a preparation extracted from the pancreas of animals, used to aid digestion.

pancreozymin (pang′krē ō zī′min *or* pan′krē ō zī′min), *n. Biochemistry.* an intestinal hormone that stimulates secretion of enzymes by the pancreas.

pandemic (pan dem′ik), *Medicine.* —*adj.* spread over an entire country or continent, or the whole world. An epidemic disease may be endemic or pandemic. *Two specifically human diseases usually associated with high mortality rates are pandemic influenza and cholera* (F. Fenner and M. F. Day).
—*n.* a pandemic disease: *The most recent great pandemic of bubonic plague in the world began somewhere in China at the end of the last century* (J. L. Cloudsley-Thompson).
[from Greek *pandēmos,* from *pan-* all + *dēmos* people]

Pangea or **Pangaea** (pan jē′ə), *n. Geology.* a hypothetical supercontinent that included all the land masses of the earth before the Triassic period, when continental drift began with the breaking away of the northern group (Laurasia) from the southern group (Gondwana): *The theory of continental drift ... in its extreme form holds that all the continents were once joined in a single great land mass. Named Pangaea, this universal continent was somehow disrupted, and its fragments — the continents of today — eventually drifted to their present locations* (Robert S. Dietz and John C. Holden). [from Greek *pan-* all + *gaia* land]
—**Pangean** or **Pangaean,** *adj.* of or having to do with Pangea: *He ... also cites evidence of small remaining portions of the Pangaean crust in the form of xenoliths in the basaltic lavas of Ascension Island* (Martin Prinz).

pan ice, *Geology.* blocks or pieces of ice formed along the shore, and afterwards loosened and driven by winds or currents.

panicle (pan′ə kəl), *n. Botany.* **1** a loose, diversely branching flower cluster, produced when a raceme becomes irregularly compound: *a panicle of oats.* **2** any loosely branching cluster in which the flowers are borne on pedicles. [from Latin *panicula,* diminutive of *panus* a swelling]
—**panicled,** *adj.* having or forming panicles.

paniculate (pə nik′yə lit), *adj. Botany.* growing in a panicle; arranged in panicles.

panmixia (pan mik′sē ə), *n. Biology.* indiscriminate crossing of breeds without selection.

pannose (pan′ōs), *adj. Botany.* having the appearance or texture of felt or woolen cloth. —**pannosely,** *adv.*

pantothenic acid (pan′tə then′ik), *Biochemistry.* a yellow oily acid, a constituent of the vitamin B complex, found in plant and animal tissues, especially liver, yeast, bran, and molasses. *Formula:* $C_9H_{17}NO_5$ [from Greek *pantothen* from every side]

pantropic (pan′trop′ik), *adj. Medicine.* drawn to or having an affinity for many kinds of tissues: *a pantropic virus.*

papaveraceous (pə pav′ə rā′shəs), *adj. Botany.* belonging to the poppy family of plants. [from New Latin *Papaveraceae* the poppy family, from Latin *papaver* poppy]

papilionaceous (pə pil′ē ə nā′shəs), *adj. Botany.* having a zygomorphic corolla somewhat like a butterfly in shape, as most leguminous plants. A papilionaceous flower consists of a large upper petal (vexillum), two lateral petals (alae), and two narrow lower petals below these, forming the carina or keel. [from Latin *papilionem* butterfly + English *-aceous*]

papilla (pə pil′ə), *n., pl.* **-pillae** (-pil′ē). *Biology.* **1** a small, nipplelike projection.

2 a small vascular process at the root of a hair or feather. See the picture at **hair.**

3 one of certain small protuberances concerned with the senses of touch, taste, or smell: *the papillae on the tongue.*

4 a pimple or pustule. [from Latin]
—**papillary** (pap′ə ler′ē), *adj.* **1** of or like papilla. **2** having papillae.
—**papillate** (pap′ə lāt), *adj.* covered with papillae.
—**papillose** (pap′ə lōs), *adj.* having many papillae.

papovavirus (pə pō′və vī′rəs), *n. Biology.* any of a group of viruses containing DNA, associated with or causing cancerous tumors and other growths, such as warts, including the polyoma virus. [from *pa*(*pilloma*) a kind of tumor + *po*(*lyoma*) + *va*(*cuolation*), the order in which the viruses became known + *virus*]

pappose (pap′ōs), *adj. Botany.* **1** having a pappus. **2** downy.

pappus (pap′əs), *n., pl.* pappi (pap′ī). *Botany.* an appendage to a seed, often made of down or bristles, which aids in the seed's dispersal by the wind: *A far more efficient means of dispersal occurs in such achenes as those of dandelion, lettuce, thistle, and salsify. The pappus on one of these fruits acts as a parachutelike float, and under favorable conditions is sufficiently buoyant to carry the achene many miles, even a slight breeze sustaining it almost indefinitely* (Emerson, *Basic Botany*). [from Latin *pappus* down on seeds]

papular (pap′yə lər), *adj. Biology.* of, having to do with, or covered with small papillae or pimples. [from Latin *papula* swelling, pimple]

par., *abbrev.* parallel.

par-, *prefix.* the form of **para-** before vowels and *h,* as in *paramylum, parenteral, parhelion.*

para-, *prefix.* **1** beside; near, as in *parathyroid, parameter.*

2 related or similar to, as in *parainfluenza, paramedical.*

3 *Chemistry.* **a** of, bonded to, or substituting for two carbon atoms that are opposite each other in a benzene ring. *Para-* is usually italicized, as in *para-*xylene. Also written **1,4-.** *Abbreviation: p-* Compare **meta-, ortho-.**
b modification or isomer of (a substance); related to, as in *paraformaldehyde.* [from Greek, from *para* beside, near]

para-aminobenzoic acid (par′ə ə mē′nō ben zō′ik *or* par′ə am′ə nō ben zō′ik), *Biochemistry.* a yellow crystalline acid, a constituent of the vitamin B complex, present in yeast and in bran. It is used in the treatment of rheumatic fever and various skin conditions. *Abbreviation:* PABA *Formula: p*-$H_2NC_6H_4CO_2H$

parabiosis (par′ə bī ō′sis), *n. Biology.* the natural or surgical union of two animals in such a way that there is an exchange of blood.
—**parabiotic** (par′ə bī ot′ik), *adj.* of or having to do with parabiosis. —**parabiotically,** *adv.*

parablast (par′ə blast), *n. Embryology.* the nutritive yolk of an ovum or egg. [from *para-* + Greek *blastos* germ, sprout]

—**parablastic,** *adj.* of, having to do with, or derived from the parablast.

parabola (pə rab′ə lə), *n. Geometry.* a plane curve formed by the intersection of a right circular cone with a plane parallel to a side of the cone: *The orbits of many comets are nearly parabolas. All parabolas, like all circles, have the same form, but not the same size* (Baker, *Astronomy*). See also **conic section.** [from Greek *parabolē* parabola, juxtaposition]
—**parabolic** (par′ə bol′ik), *adj.* having to do with or resembling a parabola. See the picture at **dune.**
—**parabolically,** *adv.*

parabola

parabolic mirror or **parabolic reflector,** *Optics.* a concave mirror the reflecting surface of which has the shape of a paraboloid, capable of focusing rays parallel to its axis to a point without spherical aberration: *Parabolic mirrors or reflectors are used in certain telescopes and solar collectors.*

paraboloid (pə rab′ə loid), *n. Geometry.* **1** a solid or surface generated by the revolution of a parabola about its axis.

2 a conoid of which sections made by planes parallel to a given line are parabolas.
—**paraboloidal** (pə rab′ə loi′dəl), *adj.* having to with or resembling a paraboloid: *The mirror of the new telescope will ... be a paraboloidal shape* (A. W. Haslett).

paraffin (par′ə fin), *n. Chemistry.* any of a group of saturated aliphatic hydrocarbons characterized by a straight or branched carbon chain and having the general formula C_nH_{2n+2}. Ethane, propane, and butane are paraffins. The names of paraffins end in *-ane.* Also called **alkane.** Compare **olefin.** [from German *Paraffin,* from Latin *parum* not very + *affinis* related (so called because of small affinity for other substances)]
—**paraffinic** (par′ə fin′ik), *adj.* derived from, related to, or like the paraffins: *paraffinic hydrocarbons found in petroleum.*

paraffin series, = methane series.

paragenesis (par′ə jen′ə sis), *n. Mineralogy.* the occurrence together of several related minerals, from whose textures a sequence of formation may be determined.
—**paragenetic** (par′ə jə net′ik), *adj.* of or having to do with paragenesis. —**paragenetically,** *adv.*

para-hydrogen, *n. Physics, Chemistry.* a hydrogen molecule in which the spins of the nuclei are opposite.

---

cap, fāce, fäther; best, bē, tėrm; pin, fīve;
rock, gō, ôrder; oil, out; cup, pùt, rüle,
yü in use, *yu* in uric;
ng in bring; *sh* in rush; *th* in thin, ᴛʜ in then;
zh in seizure.
ə = *a* in about, *e* in taken, *i* in pencil, *o* in
lemon, *u* in circus

paraldehyde (pə ral′də hīd), *n. Chemistry.* a colorless liquid, obtained by the action of sulfuric acid on acetaldehyde, used as a hypnotic and sedative, as a solvent, and in the manufacture of organic compounds. *Formula:* $C_6H_{12}O_3$

parallactic (par′ə lak′tik), *adj.* of or having to do with parallax: *a parallactic angle, parallactic displacements for nearby stars.*

parallax (par′ə laks), *n.* **1** *Optics.* the apparent change in the position of an object when it is seen or photographed from two different points which are not on a direct line with the object. Parallax is used in surveying, etc., to determine distance of objects.
2 *Astronomy.* the angle between the straight lines that join a celestial body to two different points of observation, equal to the difference in the directions in which the body is seen from the two points: *The relative distances of a series of objects could be specified by angles instead of a linear measure ... Since in most astronomical work it is the angle which is measured, it is generally more convenient to speak of the parallax of an object rather than its distance* (W. H. Marshall). See also **annual parallax, diurnal parallax.**
[from Greek *parallaxis* alteration, from *para-* beside + *allassein* to change]

parallax
(def. 1)

parallel, *adj.* **1** *Geometry.* (of straight lines or planes) lying or extending alongside one another, always equidistant and (in Euclidean geometry) never meeting however far extended, or (in projective geometry) meeting at infinity.
2 *Geography.* (of curved lines, surfaces, etc.) always equidistant at corresponding points: *A parallel is a straight line on the globe, but if the parallel becomes curved in the process of projecting it onto a map, then a relatively straight river, railway, or boundary line following that parallel must be curved on the map* (White and Renner, *Human Geography*).
—n. *Geography.* **1** any of the imaginary circles around the earth parallel to the equator, marking degrees of latitude.
2 the markings on a map or globe that represent these circles.
[from Greek *parallēlos,* from *para allēlōs* beside one another]

parallelepiped (par′ə lel′ə pī′pid), *n. Geometry.* a prism or polyhedron all of whose faces are parallelograms: *A rectangular parallelepiped is a right rectangular prism.* [from Greek *parallēlos* parallel + *epipedon* a plane surface]

parallel of declination, *Astronomy.* any imaginary circle whose plane is parallel to the celestial equator: *The hour circles run from celestial pole to celestial pole and are therefore oriented in an exactly north-south direc-*

tion. The ... parallels of declination circle parallel to the celestial equator and consequently run east-west (Krogdahl, *The Astronomical Universe*).

parallel of latitude, *Geography.* any imaginary circle on the earth's surface, parallel to the equator, by which degrees of latitude are represented: *A parallel of latitude, drawn through points equally distant from the equator on all meridians, may be constructed for any degree, minute, or second of latitude* (Finch and Trewartha, *Elements of Geography*).

parallel of longitude, *Astronomy.* = circle of longitude.

parallelogram, *n. Geometry.* a four-sided plane figure whose opposite sides are parallel and equal: *In a parallelogram, any two opposite angles are congruent. If two sides of a quadrilateral are parallel and congruent, the quadrilateral is a parallelogram.*

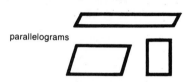

parallelograms

paramagnet, *n. Physics.* a paramagnetic substance.

paramagnetic, *Physics.* **—adj.** of or having to do with a class of substances, such as liquid oxygen, whose capability for being magnetized is slightly greater than that of a vacuum or unity, but much smaller than that of iron. Such substances, when placed in a magnetic field, are magnetized parallel to the line of force in the field and proportional to the intensity of the field.
—n. a paramagnetic substance. **—paramagnetically,** *adv.*
—paramagnetism, *n.* the quality of being paramagnetic: *If the paramagnetism exceeds the diamagnetism, the substance is paramagnetic. If the paramagnetism is small or absent altogether, the substance is diamagnetic* (Sears and Zemansky, *University Physics*).

paramecium (par′ə mē′shē əm *or* par′ə mē′sē əm), *n., pl.* **-cia** (-shē ə *or* -sē ə) *or* **-ciums.** *Zoology.* any of a genus (*Paramecium*) of protozoans shaped like a slender slipper and having a groove along one side leading into a gullet. Paramecia are free-swimming ciliates that live in almost all fresh water. *One strain of paramecium—the 'killer' strain—can produce a toxin which destroys another strain—the 'sensitive' strain* (G. M. Whyburn). [from New Latin, from Greek *paramēkēs* oblong, from *para-* on one side + *mēkos* length]

parameter (pə ram′ə tər), *n.* **1** *Mathematics.* **a** a constant in a particular calculation that varies in other cases, especially a constant occurring in the equation of a curve or surface, by the variation of which the equation is made to represent a family of such curves or surfaces. **b** an independent variable in terms of which each coordinate of a point is expressed, independently of the other coordinates.
2 *Physics.* a measurable factor which helps with other such factors to define a system: *Various individual experiments have climbed past various obstacles to reach positions close to the break-even level. In fact, in some instances two of the three essential parameters (density, temperature and confinement time) have already been achieved* (Scientific American).

—**parametric** (par′ə met′rik), *adj.* of, having to do with, or in the form of a parameter: *parametric equations of a curve.* —**parametrically,** *adv.*

paramorph (par′ə môrf), *n. Mineralogy.* a mineral that has the form of another mineral, acquired by a change in atomic or molecular structure without a change in chemical composition.

—**paramorphism** (par′ə môr′fiz əm), *n.* the change of one mineral to another having the same chemical composition but a different molecular structure.

paramylum (par am′ə ləm), *n. Biochemistry.* a starchlike food reserve found in certain one-celled organisms: *In each chloroplast of some species of Euglena there is ... a center for the formation of a starchlike substance called paramylum. Paramylum bodies may also be free in the cytoplasm in the form of disks, rods, and links* (Hegner and Stiles, *College Zoology*). [from par- + Latin *amylum* starch, from Greek *amylon*]

paramyosin (par′ə mi′ə sin), *n. Biochemistry.* a fibrous form of myosin that freezes muscle tension: *Many mollusks have special muscles, usually containing a high proportion of paramyosin, which can maintain powerful contractions over long periods with a low energy expenditure* (Graham Hoyle).

paramyxovirus (par′ə mik′sə vī′rəs), *n. Biology.* any of a group of viruses containing RNA that includes the viruses causing mumps and various respiratory diseases: *Myoviruses (influenza viruses) and paramyxoviruses (the viruses of measles, mumps and related diseases) have a set of genes that encodes the manufacture of several viral proteins* (John J. Holland).

paraphysis (pə raf′ə sis), *n., pl.* **-ses** (-sēz′). *Botany.* any of the erect, sterile filaments often occurring among the reproductive organs in certain ferns, mosses, fungi, and the like. [from New Latin, from Greek *para-* alongside + *physis* growth]

parapodium (par′ə pō′dē əm), *n., pl.* **-dia** (-dē ə). *Zoology.* one of the paired, jointless metameric processes or rudimentary limbs of certain annelids, that serve as organs of locomotion and sometimes of sensation or respiration: *The lateral appendages or parapodia are formed by outpocketings of the lateral body walls; they are usually conspicuous and variously provided with fleshy structures such as cirri, scales, and gills* (Hegner and Stiles, *College Zoology*). [from New Latin, from Greek *para-* + *pous, podos* foot]

paraselene (par′ə sə lē′nē), *n., pl.* **-nae** (-nē). *Astronomy.* a bright moonlike spot on a lunar halo.

—**paraselenic** (par′ə sə len′ik), *adj.* of or having to do with a paraselene.

parasexual, *adj. Genetics.* of, involving, or designating a process by which recombination of genes from different individuals occurs without meiosis.

parasite, *n. Biology.* an organism that lives on or in another organism from which it gets its food, always at the expense of the host, which is often injured by the relationship. Lice and tapeworms are parasites. Mistletoe is a parasite on oak trees. *There are several examples of the successful control of plant and insect pests by means of insect or microbial parasites, the most spectacular being the control of prickly pear in Australia by an imported Argentine moth Cactoblastis cactorum* (F. Fenner and M. F. Day). [from Greek

parasitos feeding beside, *para-* beside + *sitos* food]

—**parasitic,** *adj.* **1** of or like a parasite; living on others: *The white blood cells constitute a gendarmery which is always ready to repel parasitic invasion by engulfing the microscopic invaders and then digesting them* (Scientific American).
2 caused by parasites: *a parasitic disease, parasitic gastritis.* —**parasitically,** *adv.*

—**parasitical,** *adj.* = parasitic.

—**parasitism,** *n.* **1** the relationship between two organisms in which one obtains benefits at the expense of the other, often injuring it: *Parasitism is the mode of life for many species; for survival as a species the parasite should not unduly injure its host. Parasitism, along with some diseases, is one factor in the regulation of populations of host animals* (Storer, *General Zoology*).
2 parasitic infestation.
ASSOCIATED TERMS: see **symbiosis.**

parasite cone, *Geology.* a cinder cone on the side of a volcano. See the picture at **volcano.**

parasitology, *n.* the branch of biology or medicine dealing with parasites and parasitism.

parasympathetic (par′ə sim′pə thet′ik), *adj. Physiology.* of or having to do with the parasympathetic nervous system: *The other subdivision, the parasympathetic, is associated with some of the cranial and sacral nerves* (Harbaugh and Goodrich, *Fundamentals of Biology*).

parasympathetic nervous system, *Physiology.* the part of the autonomic nervous system that produces such involuntary responses as dilating blood vessels, increasing the activity of digestive and reproductive organs and glands, contracting the pupils of the eyes, slowing down the heartbeat, and other responses opposed to the action of the sympathetic nervous system.

parathormone (par′ə thôr′mōn), *n. Biochemistry.* the hormone produced by the parathyroid glands, which regulates the way the body uses calcium and phosphorus. Also called **parathyroid hormone.** [blend of *parath(yroid)* and *hormone*]

parathyroid (par′ə thi′roid), *Anatomy.* —*adj.* of, having to do with, or obtained from the parathyroid glands: *Parathyroid extract can increase survival following irradiation by more than 50%* (Science News Letter).
—*n.* **1** a parathyroid gland. **2** = parathormone.

parathyroid gland, *Anatomy.* any of several (usually four) small endocrine glands in or near the thyroid gland that secrete a vital hormone (parathormone) which enables the body to use calcium and phosphorus.

parathyroid hormone, = parathormone.

parenchyma (pə reng′kə mə), *n.* **1** *Botany.* the fundamental tissue in higher plants, composed of living unspecialized cells that may, under certain conditions, develop into other types of cells. Most of the tissue in

cap, fāce, fäther; best, bē, tèrm; pin, fīve;
rock, gō, ôrder; oil, out; cup, pùt, rüle,
yü in use, *yu̇* in uric;
ng in bring; *sh* in rush; *th* in thin, ᵺ in then;
zh in seizure.
ə = *a* in about, *e* in taken, *i* in pencil, *o* in lemon, *u* in circus

471

the softer parts of leaves, the pulp of fruits, the pith of stems, etc., is parenchyma.

2 *Zoology.* **a** the tissue of an animal organ or part that is special or essential to it, as distinguished from its connective or supporting tissue. **b** the soft, undifferentiated tissue composing the general substance of the body in some invertebrates, such as sponges and flatworms. **c** the undifferentiated cell substance of endoplasm of a protozoan.
[from Greek, anything poured in, from *para-* beside + *en-* in + *chyma* what is poured]
—**parenchymal,** *adj.* = parenchymatous.
—**parenchymatous** (par′eng kim′ə təs), *adj.* having to do with or of the nature of parenchyma: *The apex may become parenchymatous, or it may be converted into some specialized structure, such as a carpel or a hydathode* (Shirley C. Tucker).

parent, *n.* **1** *Biology.* an organism that produces offspring: *As in Mendel's pea experiments, it does not matter which parent — male or female — carries a particular character. Whether the vestigial-wing character is the male or female parent, the F_1 is always all normal-winged, and the F_2 contains three normals and one vestigial* (Simpson, *Life: An Introduction to Biology*).
2 *Nuclear Physics.* a nuclide that upon radioactive disintegration yields another nuclide: *The experiment ... was also utilized to prove that radium E was the parent of the α ray product radium F* (Ernest Rutherford). Compare **daughter.**
—*adj.* **1** being the source or origin; producing or yielding others, as in:
parent substance, *Chemistry.* a compound that yields one or more derivatives.
parent element, *Physics.* an element that yields an isotope or daughter element through radioactive decay or nuclear bombardment.
2 = parental.

parental, *adj. Genetics.* of or belonging to the generation in which hybrids are produced by cross-breeding. *Symbol:* P —**parentally,** *adv.*

parenteral (pə ren′tər əl), *adj. Physiology.* not entering by means of or passing through the alimentary canal; not intestinal. An intravenous injection provides parenteral nourishment. [from *par-* + *enteral*]

parhelic (pär he′lik), *adj.* having to do with or resembling a parhelion. A **parhelic circle** (or **ring**) is a horizontal halo or circle of light that appears to pass through the sun.

parhelion (pär hē′lē ən *or* pär hē′lyən), *n., pl.* **-helia** (-hē′lē ə *or* -hē′lyə). *Astronomy.* a bright spot of light, often showing the colors of the spectrum, that is sometimes seen on either side of the sun on a solar halo; sundog. Parhelia are caused by the refraction of sunlight through ice crystals suspended in the atmosphere. [from Greek *parēlion,* from *para-* beside + *hēlios* sun]

paridigitate (par′ə dij′ə tāt), *adj. Zoology.* having the same number of toes on each foot. [from Latin *par, paris* equal + English *digitate*]

paries (par′ē ēz), *n., pl.* **parietes** (pə rī′ə tēz). *Biology.* a wall or structure enclosing, or forming the boundary of, a cavity in an animal or plant body. [from Latin *paries* wall, partition]

parietal (pə rī′ə təl), *adj.* **1** *Anatomy.* **a** of the wall of the body or of one of its cavities: *parietal secretions of the stomach.* See **parietal cell** and **parietal lobe.** **b** of or having to do with a parietal bone. See the picture at **skull.**
2 *Botany.* belonging to, connected with, or attached to the wall of a hollow organ or structure, especially of the ovary or of a cell (used especially of ovules): *parietal placentation.*
—*n.* either of two bones that form part of the sides and top of the skull.
[from Late Latin *parietalis,* from Latin *paries* wall]

parietal cell, *Biology.* a cell of the mucous membrane of the stomach that secretes hydrochloric acid.

parietal lobe, *Anatomy.* the middle lobe of each cerebral hemisphere, which lies just behind the top of the head.

paripinnate (par′i pin′āt), *adj. Botany.* (of a leaf) pinnate with an even number of leaflets; pinnate without an odd terminal leaflet. [from Latin *par, paris* equal + English *pinnate*]

parity (par′ə tē), *n. Physics.* (in quantum mechanics) the behavior of a wave function in an atomic or other physical system when it is reflected to form its mirror image. If the sign of the function remains unchanged, parity is even; if the sign is changed, parity is odd. Parity is conserved in ordinary mechanical and electrical processes; it is not conserved in weak interactions of nuclear particles. *Parity relates a reaction to its mirror image; that is, the parity rule asserts that a particle has no intrinsic right- or left-handedness* (Gary Mitchell). [from Late Latin *paritas* equality, from *par, paris* equal]

parotid (pə rot′id), *adj. Anatomy.* near the ear. The **parotid glands,** one in front of each ear, supply saliva to the mouth through the **parotid ducts.**
—*n.* either parotid gland.
[from Greek *parōtidos,* from *para-* beside + *ōtos* ear]

parous (par′əs), *adj. Biology.* **1** having given birth one or more times: *the parous uterus.*
2 bearing offspring of a particular number or in a particular way (usually combined with a prefix, as in *biparous, oviparous, viviparous*). [from Latin *parere* give birth]

parsec (pär′sek), *n. Astronomy.* a unit used in computing the distance of stars, equal to that of a star whose annual parallax is one second of arc, or about 206,265 times the mean distance of the earth from the sun: *These hydrogen clouds have diameters of several parsecs, one parsec being the distance light travels in 3.26 years* (Science News Letter). [from *par(allax of one) sec(ond)*]

part, *n.* **1** *Mathematics.* each of several quantities into which a whole may be divided; fraction: *A tenth is one of ten equal parts. A decimeter is a tenth part of a meter. Cement may be strenghtened by adding latex in proportions of up to 0.2 part latex to 1 part cement.*
2 *Biology.* a portion of an organism; member, limb, or organ. Stamens and pistils are floral parts.

parted, *adj. Botany.* divided into distinct lobes by depressions extending from the margin nearly to the base, as a leaf; partite.

parthenocarpic (pär′thə nō kär′pik), *adj. Botany.* produced without seeds or fertilization: *Of course, such parthenocarpic fruits contain no viable seed* (P. W. Brian). —**parthenocarpically,** *adv.*

472

parthenocarpy (pär′thə nō kär′pē), *n. Botany.* the development of fruit without seeds or fertilization: *Gibberellic acid sprays ... will induce parthenocarpy in the tomato* (New Scientist). [from Greek *parthenos* virgin + *karpos* fruit]

parthenogenesis (pär′thə nō jen′ə sis), *n.* **1** *Biology.* reproduction without any male element, as the development of eggs in certain insects from virgin females without fertilization by union with one of the opposite sex: *Parthenogenesis, or virgin birth, is common among insects but rare among higher forms of life* (Scientific American).
2 *Botany.* the development of one of the sexual cells of a plant without previous fusion with a cell of the opposite sex: *A considerable number of angiosperms produce seeds regularly without pollination. The common dandelion is the most familiar example. When applied to this kind of seed production, the term parthenogenesis implies "a beginning without fertilization"* (Emerson, *Basic Botany*).
[from Greek *parthenos* virgin + English *genesis*]
—**parthenogenetic** (pär′thə nō jə net′ik), *adj.* having to do with or exhibiting parthenogenesis: *parthenogenetic development.*
—**parthenogenetically,** *adv.* in a parthenogenetic manner; by means of parthenogenesis.

parthenogenone (pär′thə noj′ə nōn), *n. Biology.* an organism born through parthenogenesis: *The human population, by virtue of its enormous size, could indeed contain a few parthenogenones* (Science Journal). [from *parthenogenesis*]

partial, *adj. Mathematics.* (of a function of two or more variables) relative to only one of the variables involved, the rest being for the time supposed constant.

partial fraction, *Mathematics.* one of the fractions into which a given fraction can be resolved, the sum of such simpler fractions being equal to the given fraction.

particle, *n. Physics.* **1a** any very small part or unit of matter, such as a molecule or atom: *Since it is inconvenient to work with individual particles (atoms, molecules, ions, electrons, etc.), chemists have chosen a unit containing many particles for comparing amounts of different materials. The mole is a unit which contains 6.02×10^{23} particles* (Chemistry Regents Syllabus). **b** = elementary or subatomic particle.
2 a minute mass of matter that, while still having inertia and attraction, is treated as a point without length, breadth, or thickness.
[from Latin *particula,* diminutive of *partem* part]

particle accelerator, any of several machines, such as the betatron, cyclotron, or synchrotron, that greatly increase the speed and energy of protons, electrons, and other atomic particles and direct them in a steady stream at a target. The accelerated particles are used to bombard the nuclei of atoms, causing the nuclei to release new particles. Also called **accelerator.**

particle beam, a concentrated flow of charged nuclear particles: *Particle beams have been essential research tools for [studying] the structure of the atom and its constituent particles* (N. Y. Times).

particle physics, a branch of physics dealing with subatomic particles, especially through the study of high-energy collisions: *One of the basic hypotheses of particle physics is that nature should be symmetrical with regard to three basic characteristics of particle in-*

teraction; electric charge, parity (right or left handedness), and direction of motion in time (Science News). Also called **high-energy physics.**
► Particle physics differs from *nuclear physics* by being concerned chiefly with the smaller particles that make up the nuclei of atoms, especially the many unstable particles discovered by the use of particle accelerators. These particles exist very briefly before decaying into lighter particles. Particle physicists divide such subatomic particles into three major groups: *leptons,* which include the electron, the muon, and the neutrinos; *quarks,* which combine to form strongly interacting *hadrons* such as baryons and mesons; and *bosons,* such as photons, gluons, and weakons, which transmit forces between particles.

particulate (pär tik′yə lit *or* pär tik′yə lāt), *Physics.*
—*adj.* of, having to do with, or consisting of small, separate particles: *Dust, smoke or other particulate material is often of greater significance when it is inhaled as an aerosol (the generic word for a suspension of solid or liquid in air) than it would be if an equivalent amount were eaten or applied to the skin* (New Scientist).
—*n.* a very small, separate particle, such as a particle of dust or fiber: *The major source of atmospheric particulates include combustion of coal, gasoline, and fuel oil; cement production; lime kiln operation; incineration; and agricultural burning* (Frances A. Wood).

parting, *n.* **1** *Geology.* **a** a layer of rock or clay lying between two beds of different formations. **b** a surface along which a rock is naturally separable into layers.
2 *Mineralogy.* the breaking of a mineral along parallel planes of weakness, resulting from strain or twinning.

partite (pär′tīt), *adj.* **1** = parted. **2** *Zoology.* divided to the base, as a wing.

partition, *n.* **1** *Anatomy, Zoology.* something that separates, such as a wall, septum, or other separating membrane in a plant or animal body.
2 *Mathematics.* a way of expressing a number as a sum of positive whole numbers.

parton (pär′ton), *n. Nuclear Physics.* a hypothetical subatomic particle forming a constituent of a composite particle: *It takes very high energy even to get evidence of the individuality of partons, let alone to pull them apart* (Science News). *Feynman assumes the proton or neutron is made up of subparticles, or partons, and interprets the electron-collision data to find the properties of the partons. If the electron hits a parton hard enough, the forces binding it to its companions become negligible* (Robert March). [from *part(icle)* + *-on*]

parturient (pär tùr′ē ənt *or* pär tyùr′ē ənt), *adj. Biology.*
1 bringing forth young; about to give birth to young.
2 having to do with childbirth.

---

cap, fãce, fãther; best, bē, tèrm; pin, fīve;
rock, gō, ôrder; oil, out; cup, pùt, rüle,
yü in use, *yu* in uric;
ng in bring; sh in rush; th in thin; ᴛʜ in then;
zh in seizure.
ə = *a* in about, *e* in taken, *i* in pencil, *o* in lemon, *u* in circus

parturition (pär'tù rish'ən, pär'tyù rish'ən, *or* pär'-chù rish'ən), *n. Biology, Medicine.* the act or process of giving birth to young. [from Latin *parturitionem* be in labor, ultimately from *parere* to bear]

parvovirus (pär'vō vī'rəs), *n. Biology.* any of a group of viruses that contain DNA and are found in various animals, especially a virus transmitted by dog feces and causing a serious and often fatal disease of dogs: *Studies ... suggest that hepatitis A virus is also parvovirus* (Science). [from Latin *parvus* little + English *virus*]

PAS, para-aminosalicylic acid.

pascal (pas kal' *or* pas'kəl), *n.* the SI or MKS unit of pressure, equal to one newton per square meter. *Symbol:* Pa [named after Blaise *Pascal,* 1623–1662, French physicist and philosopher]

Pascal's law, *Physics.* the principle in hydrostatics that in a fluid at rest the pressure is the same in all directions and that, except for the differences of pressure produced by the action of gravity, pressure applied to a confined fluid is transmitted equally in all directions: *The operation of Pascal's law and its applications are involved in the scientific study of liquids and of gases* (M.F. Vessel).

Pascal's triangle, *Mathematics.* a triangular arrangement of numbers in which each number is the sum of the two numbers to the right and the left of it in the row above: *The whole scheme is conveniently summarized in a handy table known as Pascal's triangle ... made up of the coefficients of the binomial expansion, each successive row representing the next higher power* (Scientific American).

Pascal's Triangle

```
              1
            1   1
          1   2   1
        1   3   3   1
      1   4   6   4   1
    1   5  10  10   5   1
 ...1  6  15  20  15   6   1...
```

pass, *n. Geography.* **1** a narrow road, path, way, or channel, especially through mountains: *the Khyber pass, the Donner pass in California. Passes generally occur at low points on mountain watersheds, or in valleys between mountain ridges* (R.M. Glendinning). **2** a navigable channel, such as at the mouth or in the delta of a river.

passerine (pas'ə rin), *Zoology.* —*adj.* of or belonging to the very large order (Passeriformes) of perching birds, including more than half of all birds, such as the warblers, sparrows, chickadees, wrens, thrushes, and swallows: *Among passerine birds the raven has the widest range* (Alfred R. Wallace).
—*n.* a passerine or perching bird.
[from Latin *passerinus,* from *passer* sparrow]

passive immunity, *Immunology.* immunity to a disease due to antibodies of a serum obtained from another organism: *Passive immunity ... is not so lasting as the kind acquired by getting the infection itself* (Science News Letter). Contrasted with **active immunity.**

passive transport, *Biology.* the movement of substances across a cell membrane as a result of the kinetic energy of the particles in motion: *Diffusion, a form of passive transport, is a process in which the net movement of ions or soluble molecules is from a region of higher concentration to a region of lower concentration* (Biology Regents Syllabus).

Pasteur effect (pas tèr'), *Biology.* an effect by which certain organisms and tissues shift from fermentation to respiration in the presence of certain amounts of oxygen. [named after Louis *Pasteur,* 1822–1895, French chemist]

pasteurization, *n.* the process of pasteurizing: *Pasteurization ... will kill all pathogenic bacteria which may be present, although many harmless forms survive* (Ernest Jawetz).

pasteurize (pas'chə rīz *or* pas'tə rīz), *v.* to heat (milk, wine, beer, or other liquid) hot enough and long enough to kill harmful bacteria and to prevent or stop fermentation. Commercial pasteurization of milk usually involves heating it to 72 degrees Celsius for 15 seconds. [named after Louis *Pasteur,* who developed pasteurization]
▶ *Pasteurize* is distinguished from *sterilize,* which implies the killing of all the bacteria.

patagium (pə tā'jē əm), *n., pl.* **-gia** (-jē ə). *Zoology.* **1** a wing membrane, as of a bat.
2 a fold of skin extending along the side of the body of certain gliding mammals and reptiles, as the flying squirrel.
3 the fold of skin between the upper arm and forearm of birds.
4 a small, flat sclerite above the wing base of many insects.
[from New Latin, from Latin *patagium* gold border of a tunic]

patella (pə tel'ə), *n., pl.* **-tellas, -tellae** (-tel'ē). **1** *Anatomy.* the kneecap. See the picture at **knee. 2** *Biology.* a structure in the form of a shallow pan or cup, such as the spore-bearing structure of certain lichens. [from Latin *patella,* diminutive of *patina* pan]
—**patellar,** *adj.* having to do with the kneecap.

patelliform, *adj. Biology.* having the form of a patella; shaped like a shallow pan, kneecap, or limpet shell.

path., *abbrev.* **1** pathological. **2** pathology.

pathogen (path'ə jən), *n. Biology, Medicine.* any agent capable of producing disease, especially a living microorganism: *Transmission of disease by pathogens on droplets borne by the wind has received a great deal of study for many years* (Saturday Review). [from Greek *pathos* disease + English *-gen*]

pathogenesis, *n. Medicine.* the origin or development of disease.

pathogenetic, *adj.* = pathogenic.

pathogenic, *adj.* having to do with pathogenesis; originating or producing disease.

pathogeny (pa thoj'ə nē), *n.* = pathogenesis.

pathological or **pathologic,** *adj.* **1** of pathology; dealing or concerned with diseases: *pathological studies.* **2** due to or accompanying disease: *a pathological condition of the blood cells.* —**pathologically,** *adv.*

pathology, *n. Biology, Medicine.* **1** the study of the causes and nature of diseases, especially the structural and functional changes brought about by diseases. Human pathology is a branch of medicine. *Abbreviation:* path.

2 unhealthy conditions and processes caused by a disease, especially changes in the tissues and organs of the body: *The great difficulty in effecting a solution rests in the unique pathology of cancer* (Alexander Haddow).
[from Greek *pathos* disease + English *-logy*]

-pathy, *combining form.* **1** disordered condition; disease, as in *arthropathy = disease of the joints.* **2** treatment of disease in or by, as in *hydropathy = treatment of disease by the use of water.* [from Greek *-patheia,* from *pathos* suffering]

patulous (pach′ə ləs), *adj. Botany.* **1** spreading slightly, as a calyx. **2** bearing the flowers loose or dispersed, as a peduncle. [from Latin *patulus* open, from *patere* lie open]

Pauli exclusion principle (pou′lē) or **Pauli's principle,** = exclusion principle: *In contrast to bosons, fermions are required to obey the Pauli exclusion principle, which prohibits any two identical fermions in a given population from occupying the same quantum state* (Henry T. Simmons). [named after Wolfgang *Pauli,* 1900–1958, Austrian physicist who suggested this principle]

paw, *n. Zoology.* the foot of an animal having claws or nails. Cats, dogs, monkeys, and bears have paws.

Pb, *symbol.* lead [for Latin *plumbum*].

p.c., *abbrev.* per cent.

Pd, *symbol.* palladium.

p.e. or **P.E.,** *abbrev. Statistics.* probable error.

peak, *n.* **1** *Geography.* the highest point on a mountain that has a generally conical or pyramidal shape.
2 *Mathematics.* the greatest frequency or highest value of a varying quantity, as on a graph.

pearl, *n.* **1** a hard, smooth, white or nearly white gem with a soft shine, formed inside the shell of certain oysters or in other similar mollusks by secretions of calcium carbonate with layers of animal membrane around a grain of sand, parasitic worm, or other foreign matter. **2** = mother-of-pearl. SYN: nacre.

peat, *n. Geology.* a kind of heavy turf made up of unconsolidated, partly decomposed vegetable matter, such as a sphagnum moss, found in peat bogs.

peat bog, *Geology.* a bog in which peat has formed and accumulated through partial decomposition of mosses, sedges, trees, etc.

pectase (pek′tās), *n. Biochemistry.* an enzyme found in various fruits that has the property of changing pectin into pectic acid and methyl alcohol.

pectate (pek′tāt), *n. Chemistry.* a salt or ester of pectic acid.

pecten (pek′tən), *n., pl.* **-tines** (-tə nēz′). *Zoology.* a comblike part or projection, especially a membrane in the eyes of most birds and some reptiles and fishes that projects from the choroid coat into the vitreous humor and has parallel folds that suggest the teeth of a comb. [from Latin *pecten* comb]

pectic (pek′tik), *adj. Biochemistry.* of, having to do with, or derived from pectin: *pectic enzymes from fruits.*

pectic acid, *Biochemistry.* a transparent gelatinous acid, insoluble in water, formed by the hydrolysis of certain esters of pectin. *Formula:* $C_{17}H_{24}O_{16}$ [from Greek

pēktikos curdling, congealing, from *pēgnynai* make stiff]

pectin (pek′tən), *n. Biochemistry.* any of various water-soluble colloidal substances related to carbohydrates, which, when combined with acid and sugar, yields a jelly. Pectin occurs in the rind of citrus and other fruits. *When the cells have become mature the pectin commonly unite, chemically, with calcium to form calcium pectate, a compound which is strong enough to cement the walls firmly together* (Emerson, *Basic Botany*). [from *pectic (acid) + -in*]

pectinate (pek′tə nāt), *adj. Biology.* formed like a comb; having straight, narrow, closely set projections or divisions like the teeth of a comb. [from Latin *pecten, pectinis* comb]

—**pectination,** *n.* **1** the condition of being pectinate. **2** a comblike structure; pecten.

pectines (pek′tə nēz′), *n.* plural of **pecten.**

pectoral, *adj. Anatomy.* of, in, or on the breast or chest: *pectoral muscles, the pectoral cavity.* [from Latin *pectoralis,* from *pectus* chest]

pectoral arch, = pectoral girdle.

pectoral fin, *Zoology.* either of a pair of fins in fishes, usually just behind and in line with the gills, corresponding to the forelimbs of higher vertebrates. See the picture at **fin.**

pectoral girdle, *Anatomy.* the bony or cartilaginous arch supporting the forelimbs of vertebrates, formed in man by the scapulae and clavicles: *The shoulder blades, which form part of the pectoral girdle, are not fused to the spine in mammals, but bound by muscles to the back of the thorax* (Mackean, *Introduction to Biology*). Also called **pectoral arch.**

peculiar galaxy, *Astronomy.* any galaxy having unusual properties or an unusual shape: *Many of the peculiar galaxies look as though they were normal in shape at one time, but were subsequently distorted in appearance by some unusual event ... either a collision with another galaxy, or a gigantic explosion within the galaxy that literally blew it apart* (Jastrow and Thompson, *Astronomy*).

ped, *n. Geology, Geography.* a natural aggregate of soil particles. Peds range from 1.3 to 15 centimeters. [from Greek *pedon* soil]

pedate (ped′āt), *adj.* **1** *Zoology.* **a** having a foot or feet: *pedate larvae.* **b** having tubular, somewhat footlike organs, as many echinoderms do.
2 *Botany.* (of a leaf) parted or divided in a palmate manner with the two lateral lobes divided into smaller segments.
[from Latin *pedem* foot]

pedicel (ped′ə səl), *n.* **1** *Botany.* a small stalk or stalklike part; an ultimate division of a common peduncle, supporting one flower only. The main flower stalk when

---

cap, fāce, fäther; best, bē, tėrm; pin, fīve;
rock, gō, ôrder; oil, out; cup, pút, rüle,
yü in use, *yu* in uric;
ng in bring; sh in rush; th in thin, ᴛʜ in then;
zh in seizure.
ə = *a* in about, *e* in taken, *i* in pencil, *o* in lemon, *u* in circus

small, a secondary stalk that bears flowers, or each of the secondary or subordinate stalks that immediately bear the flowers in a branched inflorescence is a pedicel. *All those parts — calyx, corolla, stamens, and carpels — are attached to the receptacle, the somewhat specialized summit of the pedicel* (Emerson, *Basic Botany*).

2 *Zoology.* any small stalklike structure in an animal. [from New Latin *pedicellus,* from Latin *pedem* foot]
—pedicellate (ped′ə sə lit *or* ped′ə sə lāt), *adj.* having a pedicel or pedicels: *pedicellate flowers.*

pedicle (ped′ə kəl), *n. Biology.* a small stalk; pedicel or peduncle: *A vertebra has a body, and above this a pedicle on either side* (A. Brazier Howell).

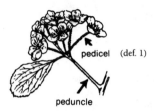

pedicel   (def. 1)

peduncle

pedipalp (ped′ə palp), *n., pl.* **pedipalpi** (ped′ə-pal′pī). *Zoology.* one of a pair of short, leglike appendages near the mouth and fangs of spiders and other arachnids, used as aids in feeding and as copulatory organs: *The pair of pedipalpi are short, six-jointed, and leglike, with enlarged bases that form "maxillae" used to squeeze and chew the food; in mature males the tip becomes a specialized container for the transfer of sperm* (Storer, *General Zoology*). [from New Latin, from Latin *pedem* foot + New Latin *palpus* palp]

pedocal (ped′ə kal), *n. Geology.* a type of soil characteristic of arid or semiarid regions, which has built up under short grass or sparse vegetation. High in calcium carbonate and low in iron content, a pedocal is an alkaline soil. [from Greek *pedon* soil + English *cal(cium*)]
—pedocalic (ped′ə kal′ik), *adj.* having to do with or characteristic of pedocals: *Pedocalic soils evolve in an arid, semiarid or sub-humid climate* (White and Renner, *Human Geography*).

pedogenesis[1] (pē′dō jen′ə sis), *n. Zoology.* reproduction by animals in the larval state: *In this type of reproduction, known as pedogenesis, the young of certain species are produced from non-fertilized eggs while in other species the young are produced only from fertilized eggs* (Harbaugh and Goodrich, *Fundamentals of Biology*). Also spelled **paedogenesis.** [from Greek *paidos* child + New Latin *genesis*]

pedogenesis[2] (ped′ō jen′ə sis), *n. Geology.* the manner in which soil originates and develops: *the role of climate in pedogenesis.* [from Greek *pedon* soil + New Latin *genesis*]
—pedogenetic (ped′ō jə net′ik), *adj.* having to do with or characterized by pedogenesis: *pedogenetic soil profiles.*
—pedogenic (ped′ō jen′ik), *adj.* = pedogenetic: *Over periods of geological time, the nitrogen level may undergo further change due to variations in climate or to alterations in soil composition through pedogenic processes* (F.J. Stevenson).

pedological (pē′də loj′ə kəl), *adj.* of or having to do with pedology: *The soil scientist would study pedological topics and the physical properties of soils in relation to civil engineering demands* (London *Times*).

pedology (pi dol′ə jē), *n. Geology.* the science dealing with the origin, classification, and utilization of soils. Also called **soil science.** [from Greek *pedon* soil + English *-logy*]

pedon (ped′on), *n. Geology, Geography.* a small volume of soil representing the surrounding soil and its horizons. [from Greek *pedon* soil]

peduncle (pi dung′kəl), *n. Biology.* stalk; stem; stalklike part (of a flower, fruit cluster, or animal body), such as a stalk that bears the fructification in some fungi, the stalk of a lobster's eye, or a white bundle of nerve fibers connecting various parts of the brain: *The main stalk of an inflorescence is known as the peduncle, and the stalk of each flower is a pedicel* (Emerson, *Basic Botany*). [from New Latin *pedunculus,* diminutive of Latin *pedem* foot]
—peduncular (pi dung′kyə lər), *adj.* of or having to do with a peduncle.
—pedunculate (pi dung′kyə lit) *or* **pedunculated,** *adj.* having a peduncle; growing on a peduncle; stalked: *pedunculate flowers, pedunculate fins.*

pegmatite (peg′mə tīt), *n. Geology.* a very coarse-grained igneous rock occurring in sills and dikes, and usually containing crystals of the common minerals found in granite, but sometimes containing rare minerals rich in such elements as lithium, uranium, and tantalum: *Feldspar is mined almost entirely from bodies of pegmatite, a special kind of rock containing large crystals of quartz and feldspar* (Ernest E. Wahlstrom). [from Greek *pēgmatos* something joined]
—pegmatitic (peg′mə tit′ik), *adj.* consisting of, characteristic of, or resembling pegmatite: *pegmatitic deposits in Norway.*

pelagic (pə laj′ik), *adj. Oceanography.* **1** of the ocean or the open sea; oceanic: *In the open parts of the ocean, the pelagic region, plants and animals differ considerably from those along the seashore* (Harbaugh and Goodrich, *Fundamentals of Biology*).
2 living on or near the surface of the open sea or ocean, at some distance from land, as certain animals and plants do: *The oceanic breeding grounds of pelagic birds ...* (A. Dillard). Compare **benthic.**
[from Greek *pelagikos,* from *pelagos* sea] **—pelagically,** *adv.*

P electron, *Physics.* one of the electrons in the P-shell.

Peléean or **Pelean** (pə lā′ən *or* pə lē′ən), *adj. Geology.* having to do with or designating a kind of volcano characterized by violent eruptions which create a circular hill of extremely viscous lava around the vent: *Peléean eruptions.* [named after Mount *Pelée,* a volcano in Martinique, noted for its catastrophic eruption in 1902]

Pele's hair (pē′lēz *or* pā′lāz), *Geology.* volcanic glass occurring in fine hairlike threads, commonly supposed to have been formed from drops of lava by the wind; naturally spun basaltic glass. [named after *Pele,* the Hawaiian goddess of Mount Kilauea, a volcano in Hawaii]

pelite (pē′līt), *n. Geology.* a rock composed especially of particles of mud or clay. [from Greek *pēlos* potter's clay + English *-ite*]

pellicle (pel′ə kəl), *n. Biology.* a very thin skin; external membrane: *The blood in each capillary of the lung is separated from the air by only a delicate pellicle* (Thomas H. Huxley). [from Latin *pellicula,* diminutive of *pellis* skin]
—**pellicular** (pə lik′yə lər), *adj.* having the character or quality of a pellicle.

peloria (pə lôr′ē ə), *n. Botany.* regularity or symmetry of structure occurring abnormally in flowers normally irregular or unsymmetrical: *In irregular flowers, those nearest to the axis are oftenest subject to peloria* (Darwin, *Origin of Species*). [from New Latin, from Greek *pelōros* monstrous]
—**peloric,** *adj.* of or characterized by peloria: *peloric flowers.*

peltate (pel′tāt), *adj. Botany.* **1** (of a leaf) having the petiole attached to the lower surface of the blade at or near the middle (instead of at the base or end). **2** (of stalked parts) having a similar attachment. [from Latin *pelta* shield]

Peltier effect (pel tyā′), *Physics.* a heating or cooling effect produced by the passage of an electric current through the junction of two dissimilar metals, the heating or cooling depending on the direction of the current: *The Peltier effect at the junction between the liquid and solid phases has been used to remove the latent heat of crystallization in crystal growth* (R.W. Ure). [named after Jean *Peltier,* 1785–1845, French physicist who discovered this effect]

pelvic, *adj. Anatomy.* of, having to do with, or in the region of the pelvis: *the pelvic cavity, pelvic fascia.*

pelvic arch, = pelvic girdle.

pelvic fin, *Zoology.* either of a pair of fins in fishes, usually behind and below the pectoral fins, corresponding to the hind limbs of higher vertebrates. Also called **ventral fin.** See the picture at **fin.**

pelvic girdle, *Anatomy.* the bony or cartilaginous arch supporting the hind limbs of vertebrates: *The fusion of the pelvic girdle to the spine is very effective in transmitting force from the legs to the body* (Mackean, *Introduction to Biology*). Also called **pelvic arch.**

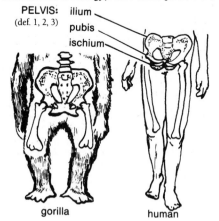

PELVIS: (def. 1, 2, 3) ilium pubis ischium

gorilla   human

pelvis (pel′vis), *n., pl.* **-vises, -ves** (-vēz′). *Anatomy.* **1** the basin-shaped cavity in human beings formed by the hipbones and the end of the backbone.
2 the corresponding cavity of any vertebrate.
3 the bones forming this cavity.

4 a basinlike cavity in the kidney which collects urine before its passage into the ureter.
[from Latin *pelvis* basin]

pen., *abbrev.* peninsula.

pendular, *adj. Physics.* **1** of or having to do with a pendulum. **2** resembling the movement of a pendulum; oscillating.

pendulum (pen′jə ləm *or* pen′dyə ləm), *n. Physics.* a weight so hung from a fixed point that it is free to swing to and fro through a regular arc under the influence of gravity. The movement of the works of a tall clock is often timed by a pendulum. *The utility of the pendulum as a timekeeper is based on the fact that the period is practically independent of the amplitude. Thus, as a clock runs down and the amplitude of the swings becomes slightly smaller, the clock will still keep very nearly correct time* (Sears and Zemansky, *University Physics*). [from New Latin, from Latin *pendulus* hanging loosely, from *pendere* hang]

pendulum

peneplain or **peneplane** (pē′nə plān′ *or* pen′ə plān′), *n. Geology, Geography.* a formerly mountainous or hilly area reduced nearly to a plain by prolonged erosion: *There is evidence of several peneplains during Cenozoic time in the Appalachians, indicating crustal uplift, erosion, renewed uplift, erosion, etc.* (Garrels, *A Textbook of Geology*). [from Latin *paene* almost + English *plain* or *plane*]
—**peneplanation,** *n.* the forming of a peneplain by erosion: *Peneplanation, especially in a region of much disturbed hard rocks, is judged to demand very prolonged work of erosive processes* (Moore, *Introduction to Historical Geology*).

penetrance (pen′ə trəns), *n. Genetics.* the measurement, expressed in percentages, of the ability of a gene to manifest itself or its effects: *Penetrance refers to the regularity with which a gene produces a detectable effect* (Hegner and Stiles, *College Zoology*).

penetrometer (pen′ə trom′ə tər), *n.* an instrument designed to measure the density, compactness, or penetrability of a substance. A marine penetrometer records the firmness of the sediment at the bottom of the ocean. *Take readings with a penetrometer — a long stick with a cone at the end of it, which is used*

---

cap, fāce, fäther; best, bē, tėrm; pin, five;
rock, gō, ôrder; oil, out; cup, pùt, rüle,
yü in use, *yú* in uric;
ng in bring; *sh* in rush; *th* in thin, ᴛʜ in then;
zh in seizure.
ə = *a* in about, *e* in taken, *i* in pencil, *o* in lemon, *u* in circus

to measure the bearing strength of the soil (Henry S.F. Cooper, Jr.).

penicillate (pen′ə sil′it or pen′ə sil′āt), adj. Biology. having or forming a small tuft or tufts of hairs, scales, or feathers. [from Latin penicillus small brush or tail, diminutive of penis tail]

penicillin (pen′ə sil′in), n. Biochemistry. any of a group of closely related antibiotic compounds produced by certain strains of molds belonging to the genus Penicillum, used against various harmful bacteria, such as some strains of staphylococci, gonococci, and pneumococci: Although penicillin is the least poisonous antibiotic available, a few persons become sensitive, or allergic to it (H. W. Florey).

penicillinase (pen′ə sil′ə nās), n. Biochemistry. an enzyme that destroys penicillin, produced by many forms of bacteria and used clinically to neutralize allergic reactions to penicillin.

penicillium (pen′ə sil′ē əm), n., pl. **-cilliums, -cillia** (-sil′-ē ə). any of a genus (Penicillium) of green and bluish-green ascomycetous fungi that grow as molds on citrus fruits, cheeses, etc., including several species used to produce penicillin and certain other antibiotic drugs. [from New Latin, from Latin penicillus small brush or tail]

peninsula (pə nin′sə lə or pə nin′syə lə), n. Geography. a piece of land almost surrounded by water, or extending far out into the water. Florida is a peninsula. [from Latin paeninsula, from paene almost + insula island] —**peninsular,** adj. in, of, or like a peninsula.

penis (pē′nis), n., pl. **-nises, -nes** (-nēz). Anatomy. the male organ of copulation. In mammals it is also the male urinary organ. In some of the more complex animals, including most mammals, the external genital organ of the male, the penis ..., conducts the sperms into the female reproductive organs, and is the functional equivalent of the pollen tube of plants (Harbaugh and Goodrich, Fundamentals of Biology). [from Latin penis penis, tail]

penna (pen′ə), n. Zoology. a contour feather of a bird, as distinguished from a down feather or plume. [from Latin penna feather, wing]

pennate (pen′āt), adj. 1 Zoology. having wings; having feathers. 2 = pinnate.

Pennsylvanian, Geology. —n. the sixth period of the Paleozoic era, after the Mississippian and before the Permian, characterized by coal-, oil-, and gas-bearing deposits and cyclic sedimentation; Upper Carboniferous (the name used outside of North America). —adj. of or having to do with the Pennsylvanian or its rocks: The Mississippian and the next following Pennsylvanian Period are the only widely recognized major geologic time divisions that are "made in America" (Moore, Introduction to Historical Geology).

pensile (pen′səl), n. Zoology. (of birds) building a hanging nest: Vireos look like warblers but have a heavier bill and are somewhat more sluggish. They build semi-pensile nests in the crotches of trees (New Yorker). [from Latin pensilis hanging down, from pendere hang]

pent-, combining form. the form of **penta-** before vowels, as in pentacid, pentane, pentyl.

penta-, combining form. **1** five, as in pentahedron, pentaploid. **2** Chemistry. having five atoms of a specified substance, as in pentabasic. [from Greek penta-, from pente five]

pentabasic, adj. Chemistry. having five atoms of hydrogen replaceable by basic atoms or radicals: a pentabasic acid.

pentacid (pen tas′id), adj. Chemistry. capable of combining with five molecules of a monobasic acid. [from pent- + acid]

pentagon, n. Geometry. a plane figure having five sides and five angles. Pentagons are sometimes written as 5-gons. [from Greek pentagōnon, from penta- + gōnia angle] —**pentagonal,** adj. **1** having five sides and five angles: a pentagonal figure. **2** contained by pentagons: the pentagonal dodecahedron. —**pentagonally,** adv.

pentahedral (pen′tə hē′drəl), adj. Geometry. having five faces.

pentahedron (pen′tə hē′drən), n., pl. **-drons, -dra** (-drə). Geometry. a solid figure having five faces. [from penta- + Greek hedra base, seat]

pentamer (pen′tə mər), n. Chemistry. a polymer consisting of five molecules: They found the pentamer, as expected, to be an irregular ten-membered ring of alternate phosphorus and nitrogen atoms, with two chlorine atoms on every phosphorus (New Scientist).

pentamerous (pen tam′ər əs), adj. Botany. (of a flower) having five members in each whorl (generally written 5-merous).

pentane (pen′tān), n. Chemistry. any of three colorless, flammable, isomeric hydrocarbons derived from petroleum and used as solvents. Formula: C_5H_{12} [from pent- + -ane]

pentapeptide, n. Biochemistry. a polypeptide composed of five amino acids.

pentaploid (pen′tə ploid), Biology. —adj. having five times the haploid number of chromosomes characteristic of the species. —n. a pentaploid organism. [from penta- + -ploid, as in diploid]

pentavalent (pen′tə vā′lənt), adj. Chemistry. having a valence of five; quinquevalent: pentavalent antimony compounds.

pentosan (pen′tə san), n. Biochemistry. any of a group of polysaccharides that yield pentoses when hydrolyzed. Pentosans occur in most plants and in humus.

pentose (pen′tōs), n. Biochemistry. any of a class of monosaccharides that contain five atoms of carbon in each molecule, are constituents of nucleic acids, and are not fermented by yeast. Pentoses are produced in animal tissues, and are obtained from pentosans by hydrolysis. Deoxyribose and ribose are pentoses. The nucleus of all cells contains a considerable amount of nucleic acid, a complex molecule containing a pentose sugar, phosphoric acid and a series of organic bases (John E. Harris).

pentoxide (pen tok′sīd), n. Chemistry. a compound containing five atoms of oxygen combined with another element or radical: phosphorus pentoxide.

pentyl (pen′təl), n. = amyl.

penumbra (pi num′brə), n., pl. **-brae** (-brē), **-bras.** Astronomy. **1** the partial shadow outside of the complete shadow formed by the sun, moon, etc., during an eclipse: The earth's shadow has two parts. The outer region, the penumbra, is that from which our globe only partly hides the sun. Inside this is a core called the umbra, where the sun is completely hidden (Science News Letter). See the picture at **eclipse.**

2 the grayish outer part of a sunspot: *A complete and fully formed spot shows a dark central portion, known as the umbra, surrounded by a not-so-dark area called the penumbra* (Krogdahl, *The Astronomical Universe*). [from Latin *paene* almost + *umbra* shadow]
—**penumbral,** *adj.* having to do with or like a penumbra.

pepsin (pep'sin), *n. Biochemistry.* an enzyme in the gastric juice of the stomach that helps to digest meat, eggs, cheese, and other protein: *Lastly, acid is necessary for the action of pepsin. This protein-splitting enzyme from the stomach is included in rennet and attacks the casein* (J.A. Barnett). [from Greek *pepsis* digestion]

pepsinogen (pep sin'ə jən), *n. Biochemistry.* the substance present in the gastric glands from which pepsin is formed during digestion: *Pepsinogen is activated (converted into pepsin) by the hydrochloric acid* (Shull, *Principles of Animal Biology*).

peptic, *adj. Physiology.* **1** having to do with or promoting digestion; digestive: *peptic glands.*
2 of, having to do with, or secreting pepsin: *peptic juice.*
3 caused by the digestive action of gastric juice: *peptic ulcer.*
[from Greek *peptikos,* from *peptos* cooked, digested]

peptidase (pep'ti dās), *n. Biochemistry.* an enzyme that breaks down peptides or peptones into amino acids: *This breakdown process is catalysed by enzymes known as proteases and peptidases, which act by breaking the peptide linkages with the simultaneous addition of water molecules* (F. Fowden).

peptide (pep'tīd), *n. Biochemistry.* any compound of two or more amino acids in which the carboxyl group of one acid is joined with the amino group of another.

peptide bond, *Chemistry.* a bond formed by the removal of water from two adjacent molecules of amino acid. The amino acids of proteins are linked by peptide bonds in protein synthesis.

peptize (pep'tīz), *v. Chemistry.* to change (as a gel) into a colloidal solution or form: *Materials called protective colloids, or peptizing agents, are added to the mixture; and they apparently coat the suspended particles and so prevent their coalescing* (Offner, *Fundamentals of Chemistry*).

peptone (pep'tōn), *n. Biochemistry.* any of a class of short-chain pepsins resulting from the hydrolysis of long-chain proteins. In the human stomach, peptones are produced from proteins by the action of the enzyme pepsin. [from German *Pepton,* from Greek *peptos* cooked, digested]
—**peptonic** (pep ton'ik), *adj.* having to do with or containing peptones.

per-, *prefix. Chemistry.* **1** the maximum or a large amount of, as in *peroxide.* **2** having the indicated element in its highest or a high valence, as in *perchloric acid.* [from Latin *per-,* from *per* through, thoroughly, to the end]

peracid, *n. Chemistry.* an acid containing a greater proportion of oxygen than others made up of the same elements. EXAMPLE: perchloric acid ($HClO_4$) is a peracid in its relation to chloric acid ($HClO_3$).

perborate (per bôr'āt), *n. Chemistry.* a salt of perboric acid, having either the univalent radical $-BO_3$ or the bivalent radical $=B_4O_3$, as perborax (sodium perborate).

perboric acid (per bôr'ik), *Chemistry.* an acid occurring only in solution or in the form of its salts. *Formula:* HBO_3

per cent or **percent,** *n. Mathematics.* parts in each hundred; hundredths. 5 per cent is 5 of each 100, or 5/100 of the whole. 5 per cent (5%) of 40 is the same as 5/100 × 40, or .05 × 40. *Per cent* is used to express many proportions: *shrinkage of less than one per cent.* Abbreviation: pct. Symbol: % [from Medieval Latin *per centum* by the hundred]
—**percentage,** *n.* **1** a rate or proportion of each hundred; part of each hundred. **2** the result obtained by multiplying a given quantity (the *base*) by the fractional or decimal equivalent of a per cent.

percentile (pər sen'tīl), *n. Mathematics, Statistics.* **1** one of a set of points on a scale arrived at by dividing a group into a hundred equal parts in order of magnitude: *The nth percentile is the measure at or below which n% of the measures fall* (Mathematics Regents Syllabus).
2 any of the parts thus formed: *A student in the ninetieth percentile of the class on a particular test is in the top ten per cent.* Compare **quartile, quintile.**
[probably from *per cent,* patterned on *quartile, sextile*]

perchlorate (pèr klôr'āt), *n. Chemistry.* a salt or ester of perchloric acid: *Perchlorate prevents the thyroid from concentrating iodide* (R.B. Greenblatt).

perchloric acid (pèr klôr'ik), *Chemistry.* a colorless, syrupy liquid used as an oxidizing agent. It is stable when diluted, but its concentrated form is highly explosive when in contact with oxidizable substances. *Formula:* $HClO_4$

percoid (pèr'koid), *Zoology.* —*adj.* **1** resembling a perch. **2** of or belonging to a large suborder (Percoidea) of spine-finned teleost fishes, including the freshwater perches, basses, and sunfishes, and certain saltwater fishes, such as the mackerels and tuna. —*n.* a percoid fish. [from Latin *perca* (from Greek *perkē* perch) + Greek *eidos* form]

percussion, *n. Medicine.* the tapping of a part of the body to determine by the quality of the sound the condition of the organs underneath. [from Latin *percussionem,* from *per-* thoroughly + *quatere* to strike, beat]

cap, fāce, fäther; best, bē, tèrm; pin, five;
rock, gō, ôrder; oil, out; cup, pùt, rüle,
yü in use, *yù* in uric;
ng in bring; *sh* in rush; *th* in thin, ŦH in then;
zh in seizure.
ə = *a* in about, *e* in taken, *i* in pencil, *o* in lemon, *u* in circus

perennate (per′ə nāt *or* pə ren′āt), *v. Botany.* to last or live through a number of years, as a perennial plant: *In most plants the seed is a perennating organ* (J. L. Harper). [from Latin *perennare,* from *perennis* perennial]

perennial (pə ren′ē əl), *Botany.* —*adj.* (of a plant) lasting through a number of years; having underground parts that live more than two years: *Perennial plants are those whose roots or underground stems and roots live indefinitely, as trees, shrubs, and perennial herbs* (Youngken, *Pharmaceutical Botany*).
—*n.* a perennial plant. Roses are perennials; so are many weeds and most wild grasses. *The annuals, of which, for instance, rice and corn are representative ... stand in contrast to perennials, the vegetative parts of which live on year after year* (Finch and Trewartha, *Elements of Geography*). [from Latin *perennis,* from *per-* through + *annus* year] —**perennially,** *adv.*

perfect, *adj.* 1 *Botany.* **a** (of a flower) having both stamens and pistils; monoclinous. Most angiosperms have perfect flowers. **b** of or designating the stage in the life cycle of a fungus at which sexual spores are produced.
2 *Mathematics.* (of a set) such that every neighborhood of each point contains at least one other point of the set: *An empty set is perfect.*

perfect gas, = ideal gas.

perfect number, *Mathematics.* a positive integer which is equal to the sum of its factors (other than itself). 6, being the sum of its factors 1, 2, and 3, is a perfect number, as is 28 (1, 2, 4, 7, and 14). Compare **abundant number** and **deficient number.**

perfusate (pər fyü′zāt), *n.* the fluid introduced in a perfusion.

perfuse, *v. Biology, Medicine.* to pass a substance through (an organ or other part of the body), especially by way of an artery or vein: *Sloviter and his colleagues perfused isolated rat brains with ethanol marked with a radioactive tracer. Since they could find no radioactivity in the brain's amino acids, they concluded that the perfused rat brain "does not metabolize ethanol at a measurable rate"* (Science News). [from Latin *perfusum* poured over, from *per-* + *fundere* pour]
—**perfusion,** *n.* the act or process of perfusing an organ, etc.: *The gland might be kept alive by perfusion, and supplied with suitable synthetic substances which it could convert to cortisone* (A.J. Birch).

peri-, *prefix.* 1 around; surrounding, as in *perimeter, periscope.* 2 near, as in *perihelion.* [from Greek]

perianth (per′ē anth), *n. Botany.* the envelope of a flower, including the calyx and the corolla: *The term perianth is ... frequently used to describe flowers, such as the tulip, in which the two outer whorls, though present, are morphologically indistinguishable* (Weier, *Botany*). [from *peri-* around + Greek *anthos* flower]

periapsis (per′ə ap′sis), *n. Astronomy.* the point in the orbit of a satellite at which it is closest to a primary. [from *peri-* + *apsis*]

periastron (per′ē as′tron), *n., pl.* **-tra** (-trə). *Astronomy.* the point at which the two components of a binary star come closest to each other in their orbits. Contrasted with **apastron.** [from *peri-* around + Greek *astron* star]

periblast (per′ə blast), *n. Embryology.* the part of the cytoplasm surrounding the nucleus of a cell or ovum. [from *peri-* around + Greek *blastos* germ, sprout]

periblem (per′ə blem), *n. Botany.* the layer of meristem in the growing ends of stems and roots of plants from which the cortex develops. [from Greek *periblema* garment, (literally) thing thrown over, ultimately from *peri-* around + *ballein* to throw]

pericardial, *adj. Anatomy.* 1 around the heart: *pericardial sinus.*
2 of or having to do with the pericardium: *the pericardial cavity, a pericardial murmur.*

pericardium (per′ə kär′dē əm), *n., pl.* **-dia** (-dē ə). *Anatomy.* the membranous sac enclosing the heart. It contains a serous fluid and part of the aorta. See the picture at **crustacean.** [from New Latin, from Greek *peri-* around + *kardia* heart]

pericarp (per′ə kärp), *n. Botany.* 1 the walls of a ripened ovary or fruit of a flowering plant, sometimes consisting of three layers, the exocarp, mesocarp, endocarp. Also called **seed vessel.**
2 a part that holds the spores in certain algae, such as one surrounding the cystocarp of red algae. [from *peri-* around + Greek *karpos* fruit]

PERICARP:
(def. 1)

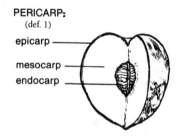

epicarp
mesocarp
endocarp

perichondrial, *adj.* of or having to do with the perichondrium.

perichondrium (per′ə kon′drē əm), *n., pl.* **-dria** (-drē ə). *Anatomy.* a membrane of fibrous connective tissue covering the surface of cartilages except at the joints. [from New Latin, from Greek *peri-* around + *chondros* cartilage]

pericranial, *adj.* of or having to do with the pericranium.

pericranium (per′ə krā′nē əm), *n., pl.* **-nia** (-nē ə). *Anatomy.* the membrane covering the bones of the skull; external periosteum of the cranium. [from New Latin, from *peri-* around + Greek *kranion* skull]

pericycle (per′ə sī′kəl), *n. Botany.* the outer portion of the stele of a plant, lying between the vascular tissues internally and the innermost layer of the cortex externally, and consisting mainly of parenchyma cells: *Often the pericycle of roots is a single layer of cells ... In stems there are usually several layers of pericycle cells* (Hill, Overholts, and Popp, *Botany*).
—**pericyclic,** *adj.* of or having to do with the pericycle: *pericyclic fibers.*

pericynthion (per′ə sin′thē ən), *n.* = perilune. [from *peri-* near + *Cynthia* goddess of the moon]

periderm (per′ə dėrm), *n. Botany.* the cork-producing tissue of stems, together with the cork layers and other tissues derived from it: *Each periderm has a few layers*

of cork cells which make it impermeable (*the phellem*), a cork cambium (*phellogen*) and sometimes a few layers of parenchyma on its inner surface (*the secondary cortex or phelloderm*) (T.C. Whitmore). [from *peri-* + Greek *derma* skin]
—**peridermal** (per′ə dėr′məl), *adj.* of or having to do with the periderm.

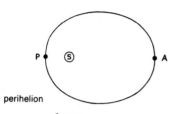

perihelion

S = sun
A = planet at aphelion
P = planet at perihelion

peridium (pə rid′ē əm), *n., pl.* **-ia** (-ē ə). *Biology.* the outer coat or envelope enclosing the sporophore of certain fungi. [from New Latin, from Greek *peridion,* diminutive of *pera* leather bag]

peridot (per′ə dot), *n. Mineralogy.* a yellowish-green, transparent variety of olivine, used as a gem. [from French *péridot*]

peridotite (pə rid′ə tīt), *n. Geology.* any of several coarse-grained ultramafic igneous rocks consisting of olivine with an admixture of various other minerals, such as pyroxene, or sometimes phlogopite, chromite, and spinel.
—**peridotitic** (per′ə də tit′ik), *adj.* having to do with, resembling, or consisting of peridotite: *This difference between ocean and continent ... is perhaps the most fundamental of all, and we believe that below the Mohorovičić discontinuity the rocks, probably peridotitic in character, are world-encircling* (T.F. Gaskell and M.N. Hill).

perigeal (per′ə jē′əl) or **perigean** (per′ə jē′ən), *adj.* of or having to do with perigee: *perigeal distances, the perigean diameter of the moon.*

perigee (per′ə jē), *n. Astronomy.* the point closest to the earth in the orbit of the moon or any other earth satellite: *... the speed is greatest at perigee, where the moon is nearest the earth, and is least at apogee, farthest from the earth* (Baker, *Astronomy*). [from *peri-* around + Greek *gē* earth]

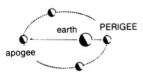

earth   PERIGEE

apogee

periglacial (per′ə glā′shəl), *adj. Geology.* **1** bordering a glacier or glaciers: *the periglacial areas of southern England.* **2** of or characteristic of a periglacial region: *periglacial deposits, a periglacial climate.*

perigynous (pə rij′ə nəs), *adj. Botany.* **1** situated on the edge of a saucer-shaped or cup-shaped receptacle surrounding the pistil: *perigynous stamens, sepals, or petals.* Compare **epigynous, hypogynous.**
2 (of a flower) having its parts so arranged. [from *peri-* around + Greek *gynē* female]

perihelion (per′ə hē′lē ən or per′ə hē′lyən), *n., pl.* **-helia** (-hē′lē ə or -hē′lyə). *Astronomy.* the point closest to the sun in the orbit of a planet or comet. Contrasted with **aphelion.** [from *peri-* around + Greek *hēlios* sun]

perikaryon (per′ə kar′ē on), *n. Biology.* the part of a nerve cell containing the nucleus; cell body of a neuron: *In the invertebrates the fibre is not so long and does not usually end on a blood vessel; the perikaryon instead provides more or less all the available storage space* (New Scientist). [from *peri-* around + Greek *karyon* kernel]

perilune (per′ə lün), *n. Astronomy.* that point in a lunar orbit closest to the center of the moon: *The orbital velocity at perilune ... was about 7200 km/hr relative to the moon* (Science Journal). Also called **pericynthion.** Compare **apolune.** [from *peri-* near + French *lune* moon]

perilymph (per′ə limf′), *n. Anatomy.* a fluid between the bony and the membranous labyrinths of the ear. Compare **endolymph.**

perimeter, *n. Geometry.* **1** the outer boundary of a figure within a plane: *the perimeter of a circle.* **2** the distance around such a boundary. The perimeter of a square equals four times the length of one side.

perimysium (per′ə mizh′ē əm or per′ə miz′ē əm), *n., pl.* **-mysia** (-mizh′ē ə or -miz′ē ə). *Anatomy.* the thin connective tissue which surrounds a muscle and also divides its fibers into bundles. [from New Latin *perimysium,* from Greek *peri-* around + *mys, myos* muscle]

perinatal, *adj. Medicine.* of or having to do with the period around childbirth including the five months preceding birth and the first month after. *perinatal mortality.* [from *peri-* around + *natal*]
—**perinatology** (per′ə nā tol′ə jē), *n.* the medical study of the period around childbirth.

perineal (per′ə nē′əl), *adj.* of or having to do with the perineum.

perineum (per′ə nē′əm), *n., pl.* **-nea** (-nē′ə). *Anatomy.* **1** the area of the body between the genitals and the anus. **2** the region included in the opening of the pelvis, containing the roots of the genitals, the anal canal, the urethra, etc. [from Late Latin]

period, *n.* **1** *Geology.* one of the subdivisions of an era. A period is divided into epochs. *These minor breaks are used to delimit secondary divisions in the geologic time scale called periods. All the rocks below the massive limestone, representing continuous deposition, are assigned to the Cambrian period. The next group above is called the Ordovician* (Garrels, *A Textbook of Geology*).
2 *Astronomy.* **a** the time it takes a celestial body to make a complete revolution around another body. **b** the interval of time between two similar phases of a regularly recurring event: *The period of a variable star*

cap, fāce, fäther; best, bē, tèrm; pin, fīve;
rock, gō, ôrder; oil, out; cup, pùt, rüle,
yü in use, *yù* in uric;
ng in bring; *sh* in rush; *th* in thin; *TH* in then;
zh in seizure.
ə = *a* in about, *e* in taken, *i* in pencil, *o* in
lemon, *u* in circus

481

periodate

is defined as the time in which the object goes from one minimum to another, or from one maximum to another. On an average, these cluster variables run twice through this range, from dim to bright, in slightly more than a day. (Bernhard, Handbook of Heavens).
3 Physics. the interval of time between the recurrence of like phases in a vibration or other periodic motion or phenomenon: The period is the time required for the completion of a cycle. It is the reciprocal of the frequency (Physics Regents Syllabus)
4 Chemistry. a sequence of elements arranged in order of increasing atomic number and forming one of the horizontal rows in the periodic table: The horizontal rows of the periodic table are called periods, or rows. The properties of elements change systematically through a period. (Chemistry Regents Syllabus). Contrasted with **group.**
5 Mathematics. **a** a digit or a group of digits set off by commas. EXAMPLE: In 6,527,308 there are three periods, 6 being the millions period and 527 the thousands period. **b** the smallest interval of the independent variable required for a function to begin to repeat itself: Mathematicians have found that the values of trigonometric ratios have periods, or repeat again and again (Elbridge P. Vance).
—**periodic,** adj. repeating or recurring at regular intervals; cyclic: the periodic motion of a pendulum, periodic variations of the element.

periodate (pèr i′ə dāt), n. Chemistry. a salt or ester of periodic acid.

periodic acid (pėr′ī od′ik), Chemistry. a colorless crystalline acid containing iodine with a valence of 7, its highest valence. Formula: H_5IO_6 [from per- + iodic]

periodic function, Mathematics. a function which repeats the same values at regular intervals. A trigonometric function is a periodic function.

periodicity (pir′ē ə dis′ə tē), n. **1** the tendency of a variable property or phenomenon to repeat itself over regular intervals.
2 Chemistry. the tendency of elements having similar positions in the periodic table to have similar properties.

periodic law, Chemistry. the law that the properties of elements change at regular intervals when the elements are arranged in the order of their atomic number: In 1869 Dmitri Ivanovitch Mendeleeff, a Russian chemist, announced his Periodic Law: The properties of the elements are periodic functions of their atomic weights (Offner, Fundamentals of Chemistry).

periodic table, Chemistry. a table in which the elements, arranged in the order of their atomic numbers, are shown in related groups: The periodic table is a useful map of the main properties of the elements. Empirically we can predict those properties which vary smoothly from neighbour to neighbour (H.R. Paneth).

periodontal (per′ē ə don′tl), adj. Anatomy. encasing or surrounding a tooth: a periodontal membrane. See the picture at **tooth.** [from peri- around + Greek odontos tooth]

periosteal (per′ē os′tē əl), adj. of, having to do with, or connected with the periosteum.

periosteum (per′ē os′tē əm), n., pl. -tea (-tē ə). Anatomy. the dense fibrous membrane covering the surface of bones except at the joints, and serving as an attachment for muscles and tendons: Within the periosteum are bone cells that function in growth and repair. (Storer, General Zoology). [from New Latin, from peri- around + Greek osteon bone]

peripheral, adj. **1** having to do with, situated in, or forming an outside boundary.
2 Anatomy. of the surface or outer part of a body or organ; external.
3 Physiology. perceived or perceiving near the outer edges of the retina: peripheral vision.
4 of the peripheral nervous system: The peripheral nerves are those near the surface of the body (Science News Letter). —**peripherally,** adv.

peripheral nervous system, Anatomy. the part of the vertebrate nervous system that is made up of all the nerves outside the central nervous system: The Peripheral Nervous System consists of 31 pairs of nerves that leave the spinal cord, and 12 pairs of nerves that connect the brain with various parts of the body. These parts include the eyes, nose, ears, lungs, heart and digestive system (William U. Gardner).

periphery, n. **1** Geometry. **a** the circumference of a circle or other closed curve. **b** the sum of the sides of a polygon. **c** the surface of a solid figure. **d** the length of, or the boundary line of, any closed plane figure.
2 Anatomy. the region in which nerves end.
[from Greek peri- around + pherein carry]

periphytic (per′ə fi′tik or per′ə fit′ik), adj. of or having to do with a periphyton: Due to the rich development there of periphytic algae growing on the bark of flooded trees as well as phytoplankton in protected areas, they have an abundance of food (New Scientist).

periphyton (per′ə fi′ton or pə rif′ə ton), n. Ecology. a complex colony of microscopic algae, insect larvae, small crustaceans, and other organisms that form a thick layer, as on the bottom of swamps and marshes, covering the stems of plants. The periphyton is a dominant biological complex in the Florida Everglades. [from peri- around + phyton]

periplasm (per′ə plaz′əm), n. Biology. the region near or immediately within a cell wall.
—**periplasmic,** adj. of or contained in the periplasm: the periplasmic space, periplasmic enzymes.

periplast (per′ə plast), n. Biology. a cell wall or cell membrane.
—**periplastic,** adj. of or occurring within a periplast.

periproct (per′ə prokt), n. Zoology. the part of the body of various invertebrates which surrounds the anus: the periproct of an echinoderm. [from peri- + Greek proktos anus]

perisarc (per′ə särk), n. Zoology. the external horny or chitinous covering of certain hydrozoans. [from Greek peri- around + sarx, sarkos flesh]

peristalsis (per′ə stal′sis), n., pl. -ses (-sēz). Physiology. the wavelike muscular contractions of the alimentary canal or other tubular organ by which its contents are moved onward. [from New Latin, from Greek peri- around + stellein to wrap]
—**peristaltic** (per′ə stal′tik), adj. of or having to do with peristalsis: peristaltic contractions.

peristome (per'ə stōm), *n.* **1** *Botany.* the one or two rings or fringes of toothlike appendages around the mouth of the capsule or theca in mosses.
2 *Zoology.* any special structure or set of parts around the mouth or oral opening in various invertebrates. [from New Latin *peristoma,* from *peri-* around + *stoma* mouth]
—**peristomial** (per'ə stō'mē əl), *adj.* of or having to do with a peristome.
perithecial (per'ə thē'shē əl), *adj.* of or having to do with the perithecium.
perithecium (per'ə thē'shē əm), *n., pl.* **-cia** (-shē ə). *Botany.* the fruiting body of certain ascomycete fungi, usually a rounded or flask-shaped receptacle with a narrow opening, enclosing the asci or spore sacs. [from New Latin, from Greek *peri-* around + *thēkē* case, receptacle]
peritoneal, *adj.* of the peritoneum: *Dialysis means the interchange of chemical substances across a membrane. In peritoneal dialysis the membrane is the peritoneum or abdominal wall* (Science News Letter).
peritoneum (per'ə tə nē'əm), *n., pl.* **-nea** (-nē'ə). *Anatomy.* the thin, transparent, serous membrane that lines the walls of the abdomen and covers the organs in it. [from Late Latin, ultimately from Greek *peri-* + *teinein* to stretch]
peritricha (pə rit'rə kə), *n.pl. Microbiology.* bacteria with the cilia (organs of locomotion) around the entire body. [from New Latin *Peritricha* the order name, from Greek *peri-* around + *thrix, trichos* the hair]
—**peritrichous** (pə rit'rə kəs), *adj.* **1** having a band of cilia around the body. **2** of or having to do with peritricha. —**peritrichously,** *adv.*
perivisceral (per'ə vis'ər əl), *adj. Anatomy.* surrounding and containing viscera. The perivisceral cavity is the general body cavity containing the alimentary canal and its appendages. *A perivisceral sinus surrounds the digestive tract in the cephalothorax* (Hegner and Stiles, *College Zoology*).
perlite (per'līt), *n. Geology.* a form of hydrated obsidian or other vitreous rock broken up by minute spherical cracks. Also spelled **pearlite.** [from French, from German *Perlit,* from *Perle* pearl + *-it* -ite]
—**perlitic** (per lit'ik), *adj.* of or having to do with perlite.
permafrost, *n. Geology.* a layer of permanently frozen subsoil, sometimes reaching a depth of 300 meters or more, found throughout most of the arctic regions. [from *perma(nent*) + *frost*]
permanent magnet, a magnet that retains its magnetism after the magnetizing current or force is removed: *A flux may exist in the iron even in the absence of any external field; when in this state the iron is called a permanent magnet* (Sears and Zemansky, *University Physics*).
permanent tooth, one of the set of teeth that replace the milk teeth and become permanent in adults. Human beings have 32 permanent teeth.
permanganate (pər mang'gə nāt), *n. Chemistry.* a salt of an acid containing manganese. A solution of potassium permanganate is used as an antiseptic.
permanganic acid (per'man gan'ik), *Chemistry.* an acid that is unstable except in dilute solutions. Its aqueous solution is used as an oxidizing agent. *Formula:* $HMnO_4$

permeability, *n.* **1** *Chemistry.* the condition of being permeable: *The damage may be in the form of changing the permeability of cell walls* (Science News Letter).
2 *Physics.* **a** the property of a material which changes the flux density in a magnetic field from its value in a vacuum. **b** the ratio of magnetic induction to the intensity of the magnetic field.
permeable, *adj. Chemistry.* that can be permeated; allowing the passage or diffusion of liquids or gases through it: *Most membranes are impermeable to colloidal particles, and for this reason the colloidal components of protoplasm remain within the cell membranes.* (Harbaugh and Goodrich, *Fundamentals of Biology*).
Permian (per'mē ən), *Geology.* —*n.* **1** the last period of the Paleozoic, after the Pennsylvanian and before the Triassic. It was characterized by the end of the trilobites, the spread of the reptiles, and widespread basaltic vulcanism. *From Devonian through Permian, the Paleozoic can be generalized as a time of shrinking seas, with gradually increasing crustal unrest.* (Garrels, *A Textbook of Geology*).
2 the rocks formed in this period.
—*adj.* of the Permian or these rocks.
per mill or **per mil** (pər mil'). by the thousand; in thousandths: *The sulfur isotopic values ... averaged around 5.3 per mil* (Science).
permineralization, *n. Geology.* a process of fossil petrifaction in which minerals replace bones, shells, or woody tissue without changing the original shape of the fossil: *As the substance dissolves, minerals replace it. In permineralization ... the actual bone or shell remains, strengthened by the minerals* (S. P. Welles).
—**permineralize,** *v.* to fossilize by permineralization: *Permineralized corals that preserve both shape and structure* (C.L. and M.A. Fenton).
permittivity (per'mə tiv'ə tē), *n.* = dielectric constant.
permutation, *n. Mathematics.* **1** a changing of the order of a set of things; arranging in different orders.
2 such an arrangement or group. The permutations of *a, b,* and *c* are *abc, acb, bac, bca, cab, cba. Mathematicians were discovering permutation groups in other fields. The set of six motions that carry an equilateral triangle into itself, for example, forms a group* (Scientific American).
[from Latin *permutationem,* from *permutare* to change through, from *per-* through + *mutare* to change]
—**permute,** *v.* to subject to permutation; alter the order of; rearrange in a different order: *When the columns are permuted in any manner, or when the lines are permuted in any manner, the determinant retains its original value* (Arthur Cayley).

---

cap, fāce, fäther; best, bē, tėrm; pin, fīve;
rock, gō, ôrder; oil, out; cup, pút, rüle,
yü in use, *yu̇* in uric;
ng in bring; *sh* in rush; *th* in thin, ᴛʜ in then;
zh in seizure.
ə = *a* in about, *e* in taken, *i* in pencil, *o* in
lemon, *u* in circus

peroxidase (pə rok′sə dās), *n. Biochemistry.* any of a group of enzymes found widely in plant, animal, and bacterial cells, which catalyzes the transfer of oxygen from organic peroxides to another substance which is able to receive it: *Peroxidases are common in plants and are responsible for discolorations seen in the tissues of many barks and fruits when they are bruised or broken and exposed to the air* (Youngken, *Pharmaceutical Botany*).

peroxide, *n. Chemistry.* 1 an oxide of an element or radical that contains the greatest possible, or an unusual, proportion of oxygen, especially one containing an oxygen atom joined to another oxygen atom: *Unsaturated organic compounds in plastics or in natural oils may be oxidized by oxygen to form peroxides that lead to discoloration or rancid products* (Frederick L. Crane). 2 = hydrogen peroxide.

peroxisomal (pə rok′sə sō′məl), *adj.* of or having to do with the peroxisome: *peroxisomal oxidases, peroxisomal respiration.*

peroxisome (pə rok′sə sōm), *n. Biochemistry.* an organelle found in the cytoplasm of most cells having a nucleus. Peroxisomes contain a variety of enzymes, especially the enzyme catalase and several oxidases that produce hydrogen peroxide: *... peroxisomes, whose major function is thought to be the protection of cells from oxygen* (Scientific American). [from *peroxi*(de) + *-some*]

perpendicular, *Geometry.* —*adj.* at right angles to a given line, plane, or surface: *A line and a plane are perpendicular if they intersect and if every line lying in the plane and passing through the point of intersection is perpendicular to the given line* (Moise and Downs, *Geometry*). Also called **normal.**
—*n.* a line or plane at right angles to another line, plane, or surface.
[from Latin *perpendicularis,* from *perpendiculum* plumb line, from *per-* thoroughly + *pendere* hang]
—**perpendicularly,** *adv.*
—**perpendicularity,** *n.* perpendicular position or direction: *The perpendicularity of these lines to each other is the difference of a right angle* (Isaac Watts).

perpetual calendar, *Astronomy.* a table or device designed to show the day of the week on which a date will fall in any given year.

perpetual motion, *Physics.* the motion of a hypothetical machine which being once set in motion should go on forever by creating its own energy, unless it were stopped by some external force or the wearing out of the machine.

per second per second, *Physics.* during one second, of a series of seconds (in connection with a constant acceleration measured in intervals of one second each). EXAMPLE: the acceleration of gravity is 9.81 meters (about 32 feet) per second per second, which means that the velocity of a freely falling body increases by 9.81 meters (32 feet) per second during each successive second of fall.

Perseid (pėr′sē id), *n. Astronomy.* any of the meteors in a meteor shower that occurs in mid-August. The Perseids appear to radiate from the constellation Perseus.

persistent, *adj.* 1 *Zoology.* permanent; not lost or altered during development: *persistent horns.* 2 (of toxic chemicals, especially insecticides) hard to decompose; chemically stable and therefore degradable only over a long period of time.

perspiration, *n. Physiology.* the salty fluid secreted by sweat glands through the pores of the skin; sweat: *Perspiration is only incidentally excretory, its primary function being temperature regulation. Evaporation of the sweat (98% water and 2% salts and urea) occurs when heat is absorbed from skin cells. This absorption of heat lowers body temperature* (Biology Regents Syllabus). [from Latin *perspirare* to sweat, from *per-* through + *spirare* breathe]

perturbation, *n.* 1 *Astronomy.* a disturbance in the motion of a planet or other celestial body in orbit caused by the gravitational attraction of a body or bodies other than its primary.
2 *Physics.* a small change in any physical system, especially a change from a relatively simple system to a more complex one.
3 *Mathematics.* a function which produces a small modification in the values of a given function.
—**perturbational** or **perturbative,** *adj.* of or having to do with perturbations.

Peru Current, *Oceanography.* a cool ocean current of the Pacific which flows northward along the west coast of South America. Also called **Humboldt Current.**

pervious, *adj.* giving passage or entrance; permeable: *Wood is fairly strong in tension and compression but all too pervious to water and vapor pressure* (Scientific American). [from Latin *pervius,* from *per-* through + *via* way]

pes (pēz), *n., pl.* **pedes** (pē′dēz or ped′ēz). 1 *Anatomy.* the terminal segment of the hindlimb of a vertebrate animal, corresponding to the human foot.
2 *Botany.* a footlike part or organ; base of support. [from Latin *pes, pedis* foot]

pesticide, *n. Ecology.* any substance that controls the spread of harmful or destructive organisms, especially an insecticide, herbicide, or fungicide.

petal, *n. Botany.* one of the parts of a flower that are usually colored; one of the leaves of a corolla. A daisy has many petals. *In a flower that is just beginning to open there are to be seen five white, pink, or blue petals attached in a circle just inside the sepals* (Emerson, *Basic Botany*). [from Greek *petalon* leaf]

petaled or petalled, *adj.* having petals: *six-petaled.*

petalite (pet′ə līt), *n. Mineralogy.* a rare mineral, a silicate of aluminum and lithium, occurring in pegmatites in white masses, often tinged with black, gray, red, or green. [from Greek *petalon* leaf + English *-ite*]

petaloid (pet′ə loid), *adj. Botany.* having the form of a petal; resembling petals in texture and color, as certain bracts: *All the stamens show a great tendency to become petaloid* (Nature).

petiolar (pet′ē ə lər), *adj.* 1 of or having to do with a petiole: *petiolar buds, a petiolar sheath, appendage, etc.* 2 proceeding from a petiole; supported by a petiole: *a petiolar tendril.*

petiolate (pet′ē ə lāt), *adj.* having a petiole: *a petiolate leaf.*

petiole (pet′ē ōl), *n.* **1** *Botany.* the slender stalk by which a leaf is attached to the stem; leafstalk.
2 *Zoology.* a stalklike part. A petiole connects the thorax and abdomen of a wasp.
[from Latin *petiolus* stalk]

petiole →

stem →

(def. 1)

petiolule (pet′ē ə lül *or* pet′ē ol′yül), *n. Botany.* a small or partial petiole, such as belongs to the leaflets of compound leaves.

Petri dish or **petri dish** (pē′trē *or* pā′trē). a shallow circular glass dish with a loose cover, used in the preparation of bacteriological cultures. [named after Julius Petri, 1852–1922, German bacteriologist who invented this dish]

petrifaction or **petrification,** *n. Geology.* the process of turning organic materials into rock by the replacement of the organic matter with minerals: *By far the most common mode of preservation is by petrification .... In this process, the original material is dissolved, and minerals from ground water precipitated in its place.* (Garrels, *A Textbook of Geology*).

petrify, *v. Geology.* to turn into rock by petrification; change (organic material) into rock: *Petrified tissues are sometimes so perfectly preserved that even cell structures are clearly visible when thin sections are examined with the microscope* (Emerson, *Basic Botany*). [from French *pétrifier,* from Latin *petra* stone]

petro-, *combining form.* **1** rock; rocks, as in *petrology = the science of rocks.* **2** petroleum, as in *petrochemical = (a chemical) derived from petroleum.* [from Greek *petra* rock]

petrochemical, *Chemistry.* —*n.* a chemical made or derived from petroleum or natural gas.
—*adj.* of or having to do with petrochemicals or petrochemistry: *petrochemical processes.*

petrochemistry, *n.* the study or science of the chemical properties and derivatives of petroleum; the chemistry of petroleum: *the petrochemistry of Alaskan rocks.*

petrogenesis, *n. Geology.* the origin of rocks, especially as a subject of scientific study: *lunar petrogenesis.*
—**petrogenetic** (pet′rō jə net′ik), *adj.* of or having to do with petrogenesis: *petrogenetic studies.* —**petrogenetically,** *adv.*

petrographic or **petrographical,** *adj.* of or having to do with petrography. —**petrographically,** *adv.*

petrography, *n.* **1** the branch of geology that deals with the description and classification of rocks. **2** the description and classification of rocks: *An enormous amount of work is in progress on the mineralogy and petrography of uranium and thorium minerals and other potentially important nuclear materials, such as lithium* (J.T. Wilson).

petroleum, *n. Chemistry.* a dark, oily, flammable liquid found in the earth's crust, consisting mainly of a mixture of various hydrocarbons. Gasoline, kerosene, and paraffin are derived from petroleum. *Coal and petroleum are derived from materials which were originally*

organic, but they have been so changed during a great lapse of time that they are now considered to belong to the mineral kingdom (Finch and Trewartha, *Elements of Geography*). [from Medieval Latin, from Greek *petra* rock + Latin *oleum* oil]

petrologic or **petrological,** *adj.* having to do with or relating to petrology: *Petrological observations ... indicate that most of the oceans are floored with igneous rocks of fairly uniform basaltic composition* (New Scientist). —**petrologically,** *adv.*

petrology, *n.* **1** the branch of geology that deals with rocks, including their origin, composition, structure, changes, etc.
2 the nature or composition of rocks: *Volcanic activity might be different in volume and petrology before, during and after the quiescent interval* (Nature).

petrous (pet′rəs *or* pē′trəs), *adj. Anatomy.* designating or having to do with the very dense, hard portion of the temporal bone (or, in certain animals, an analogous separate bone) which forms a protective case for the internal ear.

pF, *abbrev.* picofarad.

PGA, *abbrev.* phosphoglyceric acid.

ph., *abbrev.* **1** phase. **2** phosphor.

pH (pē′āch′), *symbol.* a measure of the acidity or alkalinity of a solution in terms of the relative concentration of hydrogen ions in the solution. The pH scale commonly used ranges from 0 to 14, pH7 (the hydrogen-ion concentration in pure water) being taken as neutral, 6 to 0 increasingly acid, and 8 to 14 increasingly alkaline. Most soils are in the range between pH3 and pH10. Technically, pH is defined as the negative logarithm to the base 10 of the hydrogen-ion concentration of a solution. Thus, if the hydrogen-ion concentration of a solution is 10^{-6} (one millionth) of a mole of hydrogen ions per liter, the pH is 6; if it is 10^{-8} molar, the pH is 8; and so on. *The pH is very important in cell metabolism. Big changes in the acidity of the cells of an organism can greatly affect metabolism. The pH of blood in mammals, for example, is kept very close to 7.35. A shift as small as 0.2 can result in death. Organisms have evolved chemical systems called buffers, which tend to keep the pH relatively constant* (McElroy, *Biology and Man*). [from p(*otential for*) H(*ydrogen*)]

Ph, *symbol. Chemistry.* phenyl.

phage (fāj), *n.* = bacteriophage.

-phage, *combining form.* that eats or devours, as in *bacteriophage = that devours bacteria.* [from Greek *phagein* eat]

phagocyte (fag′ə sīt), *n. Biology.* a cell, such as a white blood cell, occurring in body fluids or tissues and capable of absorbing and destroying waste or harmful material, such as disease-producing bacteria: *Part of the*

cap, fāce, fäther; best, bē, tėrm; pin, fīve;
rock, gō, ôrder; oil, out; cup, pùt, rüle,
yü in use, *yu* in uric;
ng in bring; *sh* in rush; *th* in thin, ₮H in then;
zh in seizure.
ə = *a* in about, *e* in taken, *i* in pencil, *o* in lemon, *u* in circus

phagocytize

mechanism of protection against disease is the destruction of bacteria by certain white cells in the blood, the *phagocytes* (Linus Pauling). [from Greek *phagein* eat + English *-cyte*]

—**phagocytic** (fag'ə sit'ik), *adj.* of or having to do with a phagocyte or phagocytes: *phagocytic cells.* The **phagocytic index** is the average number of bacteria destroyed by each phagocyte during an incubation of phagocytes, bacteria, and serum.

phagocytize or **phagocytose,** *v. Biology.* to engulf or destroy like a phagocyte: *Little is known of the routes by which disposition is made of the debris of degeneration during embryonic development. Necrotic cells often fuse together and are phagocytized* (John W. Saunders, Jr.).

phagocytosis (fag'ə sī tō'sis), *n. Biology.* the process by which a phagocyte engulfs or absorbs foreign matter or bacteria so as to destroy them: *As performed by such primitive cells as the protozoon amoeba, phagocytosis or engulfment represents a normal feeding mechanism* (F. Spencer). Compare **emicytosis, endocytosis, exocytosis, pinocytosis.**
[from *phagocyte* + *-osis*]

—**phagocytotic** (fag'ə sī tot'ik), *adj.* of or having to do with phagocytosis.

phagosome (fag'ə sōm), *n. Biology.* a pouchlike structure formed by a cell membrane to hold bacteria and other foreign substances which the cell has ingested: *The cell membrane folds inward to form a pocket, the edges of which fuse to enclose the particles in "phagosomes"* (Anthony Allison).

phagotrophic (fag'ə trof'ik), *adj.* = heterotrophic.

phalange (fā'lənj *or* fə lanj'), *n. Anatomy.* any bone of the fingers or toes; phalanx. See the picture at **foot.** [from French, from Latin *phalanx, phalangis*]

phalangeal (fə lan'jē əl), *adj. Anatomy.* of, having to do with, or resembling a phalanx or phalanges.

phalanx (fā'langks), *n., pl.* **phalanxes, phalanges** (fəlan'jēz). **1** *Anatomy.* any bone in the fingers or toes: *The rest of the foot consists of five metatarsal bones which articulate with the phalanges of the corresponding toes* (Harbaugh and Goodrich, *Fundamentals of Biology*).
2 *Botany.* a bundle of stamens united by their filaments in plants with two or more groups of stamens. [from Latin *phalanx,* from Greek]

phallus (fal'əs), *n., pl.* **phalli** (fal'ī), **phalluses.** **1** *Anatomy.* the penis or clitoris. **2** *Embryology.* the structure, derived from the genital tubercle, that differentiates into the penis and the clitoris. [from Latin *phallus,* from Greek *phallos*]

phanerogam (fan'ər ə gam), *n. Botany.* an old term for plants that produce seeds. They are now called spermatophytes. [from French *phanérogame,* from Greek *phaneros* visible + *gamos* marriage]

phanerophyte (fan'ər ə fīt), *n. Botany.* a perennial plant whose buds are high enough above the ground to project freely into the air.

Phanerozoic (fan'ər ə zō'ik), *Geology.* —*adj.* of or having to do with the eon comprising the Paleozoic, Mesozoic, and Cenozoic eras and the phenomena of that age: *During the 19th century a detailed stratigraphical scheme developed for those rocks with obvious fossils*

— *that is, the Phanerozoic systems beginning with the sudden record of animal life at the base of the Cambrian System* (New Scientist).
—*n.* the Phanerozoic eon: *At most times during the Phanerozoic, the sea covered much more of the continents than it does at present* (New Scientist). Contrasted with **Precambrian.** [from Greek *phaneros* visible + English *zoic*]

pharmaco-, *combining form.* drug or drugs, as in *pharmacology, pharmacogenetics.* [from Greek *pharmakon* drug, poison]

pharmacodynamic (fär'mə kō dī nam'ik), *adj.* of or having to do with pharmacodynamics; involving the powers or effects of drugs.
—**pharmacodynamics,** *n.* the branch of pharmacology dealing with the powers or effects of drugs in an organism.

pharmacogenetic, *adj.* of or having to do with pharmacogenetics; involving the interaction of genetics and drugs.
—**pharmacogenetics,** *n.* the study of the interaction of genetics and drugs: *A body of knowledge, called pharmacogenetics, was accumulating that showed that the fate of a drug in the body, or even the nature and extent of its therapeutic effect, depends in certain cases upon a discrete genetic trait* (Sumner M. Kalman).

pharmacognostic (fär'mə kog nos'tik), *adj.* having to do with pharmacognosy: *pharmacognostic studies of herbal remedies.*

pharmacognosy (fär'mə kog'nə sē), *n.* the branch of pharmacy dealing with drugs or medicines in their natural or unprepared state; the knowledge of drugs. [from *pharmaco-* + Greek *gnōsis* knowledge]

pharmacokinetics, *n.* the study of the way the body takes up, distributes, and eliminates drugs: *The magnitude of the antitumour effect depends on a complex interaction between cell kinetics and pharmacokinetics* (Kurt Heilmann).

pharmacological or **pharmacologic,** *adj.* of or having to do with pharmacology: *pharmacological tests.*
—**pharmacologically,** *adv.* in a pharmacological manner; in the way of or as a drug: *the pharmacologically powerful compound histamine.*

pharmacology (fär'mə kol'ə jē), *n.* **1** the science of drugs, their properties, preparation, uses, and effects.
2 the characteristics or properties that make a particular drug effective: *It is impossible to study the pharmacology of a drug intelligently unless one knows what that drug is and has it available in pure form* (A. Goldstein).

pharyngeal (fə rin'jē əl *or* far'in jē'əl), *adj.* having to do with or connected with the pharynx: *The external pharyngeal region of all embryonic chordates bears a series of gill clefts* (Winchester, *Zoology*).

pharynx (far'ingks), *n., pl.* **pharynxes, pharynges** (fərin'jēz). *Anatomy.* the tube or cavity that connects the mouth with the esophagus. In mammals, the pharynx contains the opening from the mouth, the opening of the esophagus, of the larynx, and of the passages from the nose. *The length of the pharynx varies slightly as the larynx is raised or lowered in speech. The diameter may be narrowed by the large constrictor muscles which bind the pharynx to the vertebral column, and* *which are used chiefly in swallowing* (Charles K. Thomas). [from Greek *pharynx, pharyngos*]

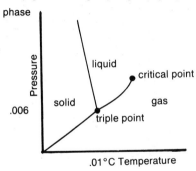

phase diagram
(def. 3a)
The phase diagram is a schematic representation of the three phases or states of water. At a pressure of .006 atmospheres and a temperature of .01°C, the triple point is reached where all three states of water (solid, gas, and liquid) coexist at equilibrium.
The critical point is reached when the difference between water in a liquid state and a vapor disappears.

phase, *n.* **1** *Astronomy.* the apparent shape of the illuminated part of the disk of the moon or of a planet at a given time. The first quarter and last quarter are two phases of the moon. *Eclipses of the sun and moon take place respectively when the moon is new and full. These phases recur every month, but eclipses are less frequent* (Baker, *Astronomy*).

2 *Physics.* a particular stage or point in a recurring sequence of movements or changes, considered in relation to a starting point of normal position (used with reference to circular motion, simple harmonic motion, or an alternating current, sound vibration, etc.): *Points on a periodic wave having the same displacement from their equilibrium position and moving in the same direction are said to be in phase* (Physics Regents Syllabus). *The current in all parts of a series circuit is in the same phase. That is, it is a maximum in the resistor, the inductor, and the capacitor at the same instant, zero in all three at a later instant, a maximum in the opposite direction at a still later instant, and so on* (Sears and Zemansky, *University Physics*).

3 *Chemistry.* **a** any of the three forms or states, solid, liquid, or gas, in which substances can exist, depending essentially on the concentration of atoms and molecules: *Change of phase of a substance is accompanied by the absorption or release of heat ... In general, phase changes* (*solid to liquid, or liquid to gas*) *are reversible and, in a closed system, equilibrium may be attained* (Chemistry Regents Syllabus). **b** a homogeneous part of a heterogeneous system, separated from other parts by definite boundaries, as ice in water. Compare **continuous phase, disperse phase.**

4 *Zoology.* one of the states, especially of coloration, of fur or plumage, characteristic of certain animals at certain seasons or ages; color phase: *Ermine is the fur of a weasel in its winter phase.*

5 *Biology.* one of the distinct stages in meiosis or mitosis. *Abbreviation:* ph.
[from Greek *phasis* phase, appearance, from *phanein* to show, appear]

phase angle, 1 *Astronomy.* the angle formed by the earth and the sun as seen from a planet: *The phase angle is greatest when the planet is near quadrature* (Baker, *Astronomy*).

2 *Physics.* an angle representing two quantities which show differences in phase: *The current is said to lag behind the voltage by ... a phase angle of 60 electrical degrees* (Shortley and Williams, *Elements of Physics*).

phase-contrast microscope, a microscope which uses the differences in phase of light passing through or reflected by the object under examination, to form distinct and contrasting images of different parts of the object: *The phase-contrast microscope ... has made it possible to observe in living cells structures which previously could be seen only if the cells were killed and stained* (Scientific American).

phase modulation, *Electronics.* a means of electronic modulation in which the phase of the carrier wave is varied in order to transmit the information contained in the signal.

phase rule, *Chemistry.* a rule stating the boundaries of thermodynamic equilibrium in a system of reactants, expressed by the formula $F = C - P + 2$, where F represents the number of degrees of freedom of a substance, C the number of its components, and P the number of its phases: *Gibbs' famous Phase Rule explained simply and logically the physical relationships among the different states of a substance, such as water, ice, and water vapor* (George Rosen).

Phe, *abbrev.* phenylalanine.

phellem (fel′əm), *n.* *Botany.* the outer layer of cork cells of the periderm, formed by the phellogen; (broadly) cork. [from Greek *phellos* cork + *-em,* as in *phloem*]

phelloderm (fel′ə dèrm), *n.* *Botany.* a layer of parenchymatous cells often containing chlorophyll, formed in the stems and roots of some plants from the inner cells of the cork cambium.

—**phellodermal** (fel′ə dèr′məl), *adj.* of or having to do with the phelloderm.

phellogen (fel′ə jən), *n.* = cork cambium.

phen-, *combining form. Chemistry.* a benzene derivative, as in *phenol, phenyl.* Also spelled **pheno-** before consonants. [from French *phén-,* from Greek *phainein* to show, appear (so called because early benzene derivatives were by-products from the making of illuminating gas)]

phenacite (fen′ə sīt) or **phenakite** (fen′ə kīt), *n.* *Mineralogy.* a mineral, a colorless, yellow, or brown silicate of beryllium, occurring in quartzlike transparent or translucent crystals. *Formula:* Be_2SiO_4 [from Greek *phenax, phenakos* a cheat + English *-ite* (so called because it was mistaken for quartz)]

cap, fāce, fäther; best, bē, tèrm; pin, five;
rock, gō, ôrder; oil, out; cup, pùt, rüle;
ÿü in use, yü in uric;
ng in bring; sh in rush; th in thin, TH in then;
zh in seizure.
ə = a in about, e in taken, i in pencil, o in lemon, u in circus

phenetic (fē net'ik), *adj. Biology.* of or having to do with a system of classification of organisms based on overall or observable resemblances among the organisms to be classified: *A good example of new phenetic data was provided by C. G. Sibley who summarized his thousands of pieces of data on proteins in bird muscle and eggs. The protein data were treated in the same way as the more classical morphological data; a classification was derived based on degree of similarity.* (Science). Compare **cladistic.** [from *phen(otype)* + *-etic*, as in *phyletic*] —**phenetically,** *adv.*
—**phenetics,** *n.* the phenetic system of classification: *Phenetics ... emphasizes the degree of similarity independent of the way in which similarity was achieved* (T. J. M. Schopf and P. L. Ames).

phenetole (fen'ə tōl), *n. Chemistry.* a colorless, volatile, aromatic liquid, the ethyl ether of phenol. *Formula:* $C_6H_5OC_2H_5$ [from *phen-* + *et(hyl)* + *-ole*]

pheno-, *combining form.* **1** the form of **phen-** before consonants, as in *phenobarbital* (a barbiturate). **2** visible; conspicuous, as in *phenocryst, phenotype.*

phenocopy (fē'nə kop'ē), *n. Genetics.* a phenotype that simulates the traits characteristic of another genotype: *In genetical language, they are inborn adaptations of a type of which very exact phenocopies can be made* (P. B. Medawar).

phenocryst (fē'nə krist), *n. Geology.* any of the large or conspicuous crystals in a porphyritic rock. [from French *phénocryste*, from *phéno-* pheno- + Greek *kryst(allos)* crystal]

phenol (fē'nol *or* fi nol'), *n. Chemistry.* **1** any of a series of aromatic hydroxyl derivatives of benzene, of which carbolic acid is the first member. **2** carbolic acid.

phenolate (fē'nə lāt), *n. Chemistry.* a salt of phenol.

phenolic (fi nō'lik *or* fi nol'ik), *adj. Chemistry.* of, having to do with, or derived from a phenol: *a phenolic hydroxyl.*

phenological (fē'nə loj'ə kəl), *adj.* of or having to do with phenology or the objects of its study: *The most important feature of the phenological year was the mild winter* (Nature). —**phenologically,** *adv.*

phenology (fi nol'ə jē), *n.* **1** a branch of science dealing with the relationship between climate and periodic occurrences in nature, such as the migration of birds and the ripening of fruit.
2 phenological relationship: *Because the phenology of the Douglas-fir tussock moth and its hosts is synchronous, bud burst has been increasingly used as a field indicator of egg hatch* (Boyd E. Wickman).
[from *pheno(mena)* + *-logy*]

phenolphthalein (fē'nōl thal'ēn), *n. Chemistry.* a white or pale-yellow crystalline compound that turns bright red in the presence of an alkaline substance, used as an indicator of alkalinity or acidity. *Formula:* $C_{20}H_{14}O_4$

phenol red, *Chemistry.* a bright- to dark-red crystalline powder, used as an acid-base indicator and as a diagnostic reagent in medicine. Phenol red turns yellow when acid is added to it. *Formula:* $C_{19}H_{14}O_5S$

phenotype (fē'nə tīp), *n. Genetics.* the external appearance or discernible characteristics of an organism resulting from the interaction of its genotype and its environment: *The phenotype changes continuously so long as the organism remains alive. A series of photographs, taken at intervals from infancy to old age, illustrates the changes in the phenotype of a person. On the other hand, the genotype of a person is relatively constant* (Theodosius Dobzhansky).
—**phenotypic,** *adj.* of or having to do with phenotypes: *It should also be emphasized that certain clones flowered freely without being subject to cold nights, indicating that flowering in this species may be considered a phenotypic response of the genotype to its environment, just as are morphological characters* (Science).
—**phenotypically,** *adv.* in a phenotypic manner; by external appearance or discernible characteristics: *One of the most important consequences of genetic dominance is that a gene for a certain character, e.g., blue eyes, may be present without manifesting iself phenotypically* (Bulletin of Atomic Scientists).

phenoxide (fi nok'sīd), *n.* = phenolate.

phenyl (fen'l *or* fē'nl), *n. Chemistry.* a univalent radical ($-C_6H_5$), formed by removing one hydrogen atom from a benzene molecule.

phenylalanine (fen'l al'ə nēn' *or* fē'nl al'ə nēn'), *n. Biochemistry.* an amino acid which results from the hydrolysis of protein and is normally converted to tyrosine in the body: *The disease, phenylketonuria, or PKU, develops in infants born without an enzyme necessary to metabolize phenylalanine, a chemical found in all proteins.* (Science News Letter). *Formula:* $C_9H_{11}NO_2$ *Abbreviation:* Phe

phenylene (fen'lēn), *n. Chemistry.* a bivalent radical ($-C_6H_4-$) formed by removing two hydrogen atoms from a benzene molecule.

phenylic (fi nil'ik), *adj. Chemistry.* of or derived from phenyl.

phenylthiocarbamide (fen'l thī'ō kär'bə mīd), *n. Chemistry.* a crystalline substance that is very bitter, slightly sweet, or tasteless, depending on the heredity of the taster (once used in various genetic tests). *Formula:* $C_7H_8N_2S$ *Abbreviation:* PTC

phenylthiourea (fen'l thī'ō yù rē'ə), *n. Chemistry.* = phenylthiocarbamide.

pheresis (fə rē'sis), *n.* = hemapheresis.

pheromonal (fer'ə mō'nl), *adj.* of or having to do with pheromones: *Of all the likely influences, a pheromonal communication, by subliminal odour, seemed biologically the most likely* (Alex Comfort).

pheromone (fer'ə mōn), *n. Physiology.* a substance secreted externally by certain animal species, especially insects, that causes a specific response when detected by other members of the species: *The queen substance of honeybees, which inhibits ovary development ... in workers, is a pheromone* (New Scientist). *Mating cecropia moths are attracted to each other by chemicals called sex pheromones* (John G. Lepp). [from *phero-* (from Greek *pherein* to carry) + *(hor)mone*]

phi (fī *or* fē), *n. or* **phi meson,** *Nuclear Physics.* a meson having zero charge and a mass 1995 times that of the electron. On decaying, it usually produces two kaons and three pions. [from *phi* (φ) the 21st letter of the Greek alphabet]

-phil or **-phile,** *combining form.* **1** having an affinity for; strongly attracted to, as in *eosinophil.*
2 a thing having an affinity for; substance strongly attracted to, as in *electrophile* = a substance strongly attracted to electrons.

3 organism that loves or is strongly attracted to, as in *sarcophile = animal organism that loves flesh.* [from French *-phile* or New Latin *-philus,* ultimately from Greek *philos* loving]

-philia, *combining form.* tendency toward or affinity for, as in *hemophilia = tendency to bleed.* [from Greek *philia* affection, from *philos* loving]

phlebology (fli bol′ə jē), *n.* a branch of medicine dealing with the study of veins. [from Greek *phleps, phlebos* vein + English *-logy*]

phlebotomy (fli bot′ə mē), *n.* **1** the opening of a vein to let blood; bleeding as a therapeutic device. Also called **venesection. 2** the transfusion or collection of blood, as at a blood bank.

phlegm (flem), *n. Physiology.* the thick mucus discharged into the mouth and throat during a cold or other respiratory disease. [from Greek *phlegma* clammy humor (resulting from heat), from *phlegein* to burn]

phloem (flō′em), *n. Botany.* the tissue in a vascular plant through which the sap containing dissolved food materials passes downward to the stems and roots; bast: *Phloem parenchyma and ray cells ... store food, and phloem fibers are strengthening and supporting cells* (Fuller and Tippo, *College Botany*). Compare **xylem.** See the pictures at **bark, root.** [from German *Phloem,* from Greek *phloos* bark]

phloem ray, *Botany.* a ray or plate of phloem between two medullary rays. Compare **xylem ray.**

phlogopite (flog′ə pīt), *n. Mineralogy.* a magnesium mica, usually of a brownish-yellow or brownish-red color, that occurs in kimberlites. [from Greek *phlogōpos* fiery + English *-ite*]

phon (fon), *n. Physics.* a unit for measuring the level of loudness of sound. The loudness level in phons of any tone is equal to the intensity level in decibels of a 1000-hertz tone that seems equally loud. Thus a tone of 60 decibels with a frequency of 1000 hertz has a loudness level of 60 phons. *According to data given at the congress, 120 phon represent pain level, 140 phon may drive people insane. Manhattan's street noise was stated to reach 80 phon* (N. Y. Times). [from Greek *phōnē* sound]

phonation (fō nā′shən), *n. Physiology.* the production or utterance of vocal sound, especially speech.
—**phonate** (fō′nāt), *v.* to utter (speech sounds); sound vocally.

phonon (fō′non), *n. Physics.* a particle or quantum of thermal energy in the form of sound or vibration in a crystal lattice, analogous to the photon: *When electrons penetrate a crystal they set up units of vibrations called phonons* (Herbert Kondo). [from Greek *phōnē* sound + English *-on*]

-phore, *combining form.* a thing that bears or carries; carrier, as in *chromophore, conidiophore, oophore, photophore.* [from Greek *-phoros,* from *pherein* to bear, carry]

phoresy (phôr′ə sē), *n. Zoology.* the nonparasitic transportation of one species by another, especially among arthropods. [from Greek *phorēsis* a carrying, from *pherein* carry]

phosgenite (fos′jə nīt), *n. Mineralogy.* a mineral, a chloride and carbonate of lead, occurring in white or yellowish tetragonal crystals. *Formula:* $Pb_2Cl_2CO_3$ [from Greek *phōs, photos* light + English *-gen* + *-ite*]

phosphatase (fos′fə tās), *n. Biochemistry.* an enzyme which splits carbohydrate and phosphate compounds.

phosphate (fos′fāt), *n. Chemistry.* a salt or ester of phosphoric acid. It is present in rocks and in the remains of organisms. Phosphates are necessary to the growth of organisms and have extensive use as fertilizers. Bread contains phosphates.

phosphate rock, = phosphorite.

phosphatic (fos fat′ik), *adj. Chemistry.* of or containing phosphoric acid or phosphates: *The presence of phosphatic nodules ... in some of the lowest azoic rocks, probably indicates life at these periods* (Darwin, *Origin of Species*).

phosphatide (fos′fə tīd), *n.* = phospholipid.
—**phosphatidic** (fos′fə tid′ik), *adj.* of, having to do with, or formed from phospholipids.

phosphatization, *n. Chemistry.* the act of phosphatizing or the condition of being phosphatized; conversion into phosphates.

phosphatize (fos′fə tīz), *v. Chemistry.* **1** to convert into a phosphate; reduce to the form of a phosphate: *These fossils are phosphatized more or less completely* (Science). **2** to treat with phosphates.

phosphene (fos′fēn), *n. Physiology.* a luminous image, as of rings of light, produced by mechanical excitation of the retina, as by pressing the eyeball when the lid is closed, or by electrical stimulation of the visual cortex: *"Seeing stars" is seeing phosphenes, an experience that can be induced by a blow on the head or by other mechanical means* (Gerald Oster). [from Greek *phōs* light + *phainein* to show, make appear]
—**phosphenic,** *adj.* of or produced by phosphenes: *phosphenic light spots.*

phosphide (fos′fīd), *n. Chemistry.* a compound of phosphorus with a basic element or radical.

phosphine (fos′fēn), *n.* **1** a colorless, extremely poisonous gas, a phosphorus hydride, with an odor like that of garlic or decaying fish. It is spontaneously flammable in air. *Formula:* PH_3 **2** any of various organic compounds derived from this gas.

phosphite (fos′fīt), *n. Chemistry.* a salt or ester of phosphorous acid.

phosphocreatine (fos′fō krē′ə tēn), *n. Biochemistry.* an energy-giving substance in muscle tissue, consisting of creatine and phosphoric acid: *The energy for the resynthesis of phosphocreatine, following contraction, comes directly from the oxidation of glucose in the muscle cells* (G. A. Baitsell). *Formula:* $C_4H_{10}N_3O_5P$ Also called **creatine phosphate.**

phosphoglyceric acid, a phosphate of glyceric acid in two isomeric forms, that is an intermediate compound in carbohydrate metabolism: *In most plants a central role is played by phosphoglyceric acid, the three-carbon compound that is the first stable product of*

---

cap, fāce, fäther; best, bē, tèrm; pin, fīve;
rock, gō, ôrder; oil, out; cup, pùt, rüle;
yü in use, *yu* in uric;
ng in bring; *sh* in rush; *th* in thin, ᴛH in then;
zh in seizure.
ə = *a* in about, *e* in taken, *i* in pencil, *o* in lemon, *u* in circus

phospholipid

carbon dioxide fixation (Scientific American). *Abbreviation:* PGA

phospholipid (fos'fō lip'id), *n. Biochemistry.* any of a class of fatty substances present in cellular tissue and consisting of esters of phosphoric acid. Also called **phosphatide.**

phosphonium (fos fō'nē əm), *n. Chemistry.* a univalent radical or ion (PH_4- or PH_4^+), analogous to ammonium.

phosphoprotein, *n. Biochemistry.* any of a group of proteins consisting of a simple protein combined with a phosphorus compound other than nucleic acid or lecithin.

phosphor (fos'fər), *n. Physics.* a phosphorescent substance, especially one that emits light when exposed to ultraviolet rays, X rays, or certain other types of energy: *Phosphor bands are practically invisible in normal light* (D. Potter). [from New Latin *phosphorus*, from Greek *phōsphoros* light-bearer]

phosphoresce (fos'fə res'), *v. Physics.* to be luminous without noticeable heat; exhibit phosphorescence: *When the Pelagia phosphoresces, it seems like a great globe of fire in the water* (E. Forbes).

phosphorescence (fos'fə res'ns), *n. Physics.* **1** the act or process of emitting light without burning or by very slow burning without noticeable heat: *the phosphorescence of fireflies.*

2 light emitted by a substance as a result of the absorption of certain rays, such as X rays or ultraviolet rays, and continuing for a period of time after the substance has ceased to be exposed to these rays: *The rate of fluorescence is generally many orders of magnitude faster than that of phosphorescence* (N. J. Turro).

—**phosphorescent** (fos'fə res'nt), *adj.* being luminous without noticeable heat; exhibiting phosphorescence.

phosphoric (fo sfôr'ik *or* fo sfor'ik), *adj. Chemistry.* having to do with or containing phosphorus, especially with a valence of five: *a phosphoric group.*

phosphoric acid, *Chemistry.* **1** a colorless, odorless acid containing phosphorus, obtained chiefly by the decomposition of phosphates, and used in making fertilizers, as a reagent, etc.: *Ammonium sulfate reacts with phosphate rock to produce a pure phosphoric acid* (Science News). *Formula:* H_3PO_4 Also called **orthophosphoric acid. 2** = metaphosphoric acid. **3** = pyrophosphoric acid.

phosphorite (fos'fə rīt), *n. Geology.* a rock consisting chiefly of calcium phosphate. It is a primary source of phosphorus for agriculture.

—**phosphoritic** (fos'fə rit'ik), *adj.* of or having to do with phosphorite.

phosphorolysis (fos'fə rol'ə sis), *n. Biochemistry.* a reaction in which the elements of phosphoric acid are absorbed by a compound.

—**phosphorolytic** (fos'fə rə lit'ik), *adj.* of or producing phosphorolysis.

phosphorous (fos'fər əs *or* fos fôr'əs), *adj. Chemistry.* **1** having to do with or containing phosphorus, especially with a valence of three. **2** = phosphorescent.

phosphorous acid, *Chemistry.* a colorless, unstable, crystalline acid. Its salts are phosphites. *Formula:* H_3PO_3

phosphorus (fos'fər əs), *n. Chemistry.* a nonmetallic element which exists in several allotropic forms. The two common forms are a white or yellow poisonous, waxy substance which burns slowly at ordinary temperatures and glows in the dark, and a reddish-brown powder, nonpoisonous, and less flammable. *Phosphorus is an essential for both animal and vegetable life. Man gets the phosphorus needed from compounds in the vegetables eaten. Plants get it from the soil, principally from phosphates.* (Science News Letter). *Symbol:* P; *atomic number* 15; *atomic weight* 30.9738; *melting point* (of white phosphorus) 44.1°C; *boiling point* (of white phosphorus) 280°C; *oxidation state* $+3$, $+5$. [from Greek *phōsphoros* bringing light, from *phōs* light + *pherein* bring]

phosphorylase (fos fôr'ə lās), *n. Biochemistry.* an enzyme which assists in the formation of glucose (in the form of phosphate) from glycogen and a phosphate: *Use of muscle to do work and its recovery depends upon the chemical action of an enzyme, phosphorylase, which is found in muscle. So long as work is paced slow enough for this chemical action to be complete, fatigue does not occur* (Science News Letter).

phosphorylate (fos'fər ə lāt), *v. Chemistry.* to convert into a phosphorus compound.

—**phosphorylation** (fos'fər ə lā'shən), *n.* conversion into a phosphorus compound: *The mitochondria are the site of oxidative phosphorylation, which is the main mechanism by which the energy of respiration is stored* (Scientific American).

phot (fōt), *n. Optics.* a unit of illumination in the CGS system, equivalent to one lumen to a square centimeter. [from Greek *phōs, photos* light]

photic (fō'tik), *adj.* **1** *Biology.* of or having to do with light, especially as it affects or stimulates organisms: *... the rather feeble photic requirements of the purple bacteria* (G. E. Hutchinson).

2 *Oceanography, Ecology.* penetrated by light; receiving sufficient light from the sun to allow photosynthesis: *The depth of the photic zone depends on the transparency of the water ... Thus, clear blue oceans far from land have a far deeper photic layer than have the green or brownish waters near the shore* (J. E. Bardach and M. D. Bradley).

photo-, *combining form.* light, as in *photochemical, photobiology.* [from Greek *phōs, photos* light]

photoautotroph, *n. Biology.* an autotroph that obtains its nourishment and energy from light: *The majority of blue-green algae are aerobic photoautotrophs: their life processes require only oxygen, light, and inorganic substances* (Scientific American).

—**photoautotrophic,** *adj.* providing its own nourishment and obtaining energy from light.

photobiological, *adj.* of or having to do with photobiology or biological processes that use radiant energy: *the photobiological domain.* —**photobiologically,** *adv.*

photobiology, *n.* the branch of biology dealing with the effects of light or radiant energy on living organisms and biological processes, such as photosynthesis: *The cure of rickets by ultraviolet light constitutes one of the most interesting chapters in photobiology* (Science).

photocathode, *n. Electronics.* a cathode which emits electrons when stimulated by radiant energy.

photocell, *n.* = photoelectric cell.

photochemical, *adj.* of or having to do with the chemical action of light: *Rays that are absorbed by the eye exert a photochemical or abiotic effect, a heat effect, or a fluorescent effect* (R. I. Pritikin and M. L. Duchon).

photochemistry, *n.* the branch of chemistry dealing with the chemical action of light: *What goes on in the light box is photochemistry, i.e., chemical reactions which are energized and promoted by light* (Harper's).

photoconduction, *n.* *Physics.* the ability of an electrical conductor to conduct electricity upon exposure to light.
—**photoconductive,** *adj.* of or having to do with photoconduction: *photoconductive properties. Photoconductive material produces a much stronger current of electricity than photoemissive material when a unit of light strikes it* (Kenneth Harwood).
—**photoconductivity,** *n.* the property of being photoconductive.

photocurrent, *n.* *Physics.* the electric current produced by the movement of a stream of electrons emitted by certain substances, usually in a photoelectric cell, when exposed to light or certain other radiations: *It is possible to make cells where ... almost all the photocurrent is generated by photons with an energy smaller than the activation energy* (New Scientist).

photodetector, *n.* *Electronics.* a semiconductor device that detects radiant energy, especially infrared radiation, by photoconductive or photovoltaic action: *Photodetectors depend on the action of light quanta on a single electron rather than on the absorption and distribution of energy over an entire macroscopic body* (S. S. Charschan).
—**photodetection,** *n.* the detection of radiant energy.

photodisintegration, *n.* *Nuclear Physics.* the breaking down of the nucleus of an atom by bombardment with high-energy gamma rays.

photodissociate, *v.* *Chemistry.* to dissociate by the action of light: *Water may also have been more readily photodissociated on Venus as a result of a higher tropopause temperature, which would allow the water to diffuse to higher altitudes more easily than on Earth* (Tobias Owen).

photodissociation, *n.* *Chemistry.* dissociation of a compound by the absorption of radiant energy, such as light and ultraviolet rays: *Nature's only oxygen manufacturing process other than the self-restricting photodissociation of water vapor is photosynthesis* (Saturday Review). Compare **photolysis.**

photodynamic, *adj.* *Biology.* having to do with or involving a toxic response to light, especially ultraviolet light: *the photodynamic inactivation of viruses and bacteria.*

photoelastic, *adj.* of, having to do with, or exhibiting the characteristics of photoelasticity: *The transparent epoxy-resin coat is photoelastic; strain alters its optical characteristics, so an observer can "see" stresses developing* (New Scientist).

photoelasticity, *n.* *Physics.* optical changes in a transparent dielectric, such as glass, due to compression or other stresses: *The double refraction produced by stress is the basis of the science of photoelasticity. The stresses in opaque objects such as girders, boiler plates, gear teeth, etc., can be analyzed by constructing a transparent model ... and examining it in polarized light* (Sears and Zemansky, *University Physics*).

photoelectric, *adj.* *Physics.* having to do with electricity or electrical effects produced by the action of light, such as the emission of electrons, generation of voltage, etc.: *An independent estimate was made of the distance of the Andromeda nebula by photoelectric measurements made in three colours—yellow, blue, and ultra-violet—on some of the brightest stars in the nebula. From this information the luminosity and distance of a star can be calculated* (A. W. Haslett). A **photoelectric cell** is an electric cell that varies the flow of current according to the amount of light falling upon its sensitive element. Variations or interruptions of the light can be used to activate mechanisms that set off alarms, measure light intensity by electric meter, etc. *This light falls on a photoelectric cell which produces an electric current corresponding to the intensity of the light* (Pierce, *Electrons, Waves and Messages*).

photoelectric effect, *Physics.* any effect involving the emission of electrons from a body exposed to light or to gamma radiation. Photoelectric effects include photoconductive, photoemissive, and photovoltaic phenomena.

photoelectron, *n.* *Physics.* an electron emitted from a surface exposed to light, as in electroemission: *The rate of emission of photoelectrons depends on the intensity of the incident radiation* (Physics Regents Syllabus).

photoemission, *n.* *Physics.* the emission of electrons from a metal subjected to the action of light or other suitable radiation: *With light of high frequency (high energy), photoemission takes place even if the light is dim, although the resulting current is very weak* (Arthur L. Joquel, II).
—**photoemissive,** *adj.* emitting electrons when subjected to the action of light: *The photocathode is photoemissive—that is, it emits electrons when struck by light* (Kenneth Harwood).

photogenic, *adj.* **1** *Biology.* phosphorescent; luminescent. *Certain bacteria are photogenic.*
2 produced or caused by light: *photogenic excitation of nerves.*

photogeological or **photogeologic,** *adj.* of or having to do with photogeology: *photogeological surveys.* —**photogeologically,** *adv.*

photogeology, *n.* a branch of geology dealing with the study and interpretation of aerial or satellite photographs to identify and map geological formations: *lunar photogeology.*

photokinesis, *n.* *Physiology.* movement caused by light: *The photokinetic action of light absorbed by photosynthetic pigments has led to the conclusion that photokinesis may be linked with photosynthesis* (W. Nultsch).

---

cap, fāce, fäther; best, bē, tèrm; pin, fĭve;
rock, gō, ôrder; oil, out; cup, pùt, rüle,
yü in use, *yu* in uric;
ng in bring; sh in rush; th in thin, ŦH in then;
zh in seizure.
ə = *a* in about, *e* in taken, *i* in pencil, *o* in lemon, *u* in circus

491

—**photokinetic,** *adj.* of or having to do with photokinesis.

photolysis (fō tol'ə sis), *n. Chemistry.* decomposition of a substance resulting from the action of light: *Another possible source of oxygen in our atmosphere is photolysis, the ultraviolet dissociation of water vapor in the outer atmosphere followed by the escape of the hydrogen from the earth's gravitational field* (Preston Cloud and Aharon Gibor). Compare **photodissociation.**

—**photolytic,** *adj.* having to do with or producing photolysis: *As far as life on the earth is concerned, the most important photolytic reaction in nature is the one that creates a canopy of ozone in the upper atmosphere* (Gerald Oster).

—**photolyze,** *v.* to break down by the action of light; cause to undergo photolysis: *Ozone ... is photolyzed ... into singlet oxygen and oxygen atoms* (Science News).

photometry, *n.* the branch of physics dealing with measurements of the intensity of light, light distribution, illimination, luminous flux, etc.: *The measurement ... is called photometry, and the instruments used in this measurement are called photometers* (Shortley and Williams, *Elements of Physics*).

photomicrograph, *n.* an enlarged photograph of a microscopic object, taken through a microscope: *A photograph of a gnat's eyebrow is a photomicrograph* (Kodak Handbook News). Also called **microphotograph.**

—**photomicrographic,** *adj.* **1** of, having to do with, or used in photomicrography: *a photomicrographic camera.* **2** obtained or made by photomicrography: *... a photomicrographic study of moth and butterfly wing scales* (Science News Letter).

—**photomicrography,** *n.* the photographing of microscopic objects on a magnified scale by attaching a camera to a microscope: *Photomicrography is used to show chromosomes splitting during cell division and abnormal chromosome arrangements* (Science News).

photomicroscope, *n.* an apparatus consisting of a microscope, a camera, and a light source, used to photograph microscopic objects.

photon (fō'ton), *n. Physics.* a quantum of radiant energy, moving as a unit with the velocity of light. It is considered to be an elementary particle. *The fundamental particle of electromagnetic energy is the photon, which, under certain circumstances, can be split into an electron and a positron* (Hugh F. Henry). *When you see different colors your eyes are being hit by streams of photons which differ in their rates of vibration* (Ralph E. Lapp). Also called **light quantum.** [from *photo-* + *-on*]

photonegative, *adj. Biology.* showing avoidance of light; photophobic: *Normally the beetles are photonegative, tending to stay in the dark half of the chamber* (New Scientist).

photooxidation, *n. Chemistry.* oxidation induced by the chemical action of light.

photoperiod, *n. Physiology.* the length of time during which a plant or animal is exposed to light each day, considered especially with reference to the effect of the light on growth and development: *Preliminary findings from manipulation of daily photoperiods suggest ... a useful tool for the study of the physiology of the orientation of migration with birds* (Science).

—**photoperiodic** or **photoperiodical,** *adj.* of or having to do with a photoperiod or photoperiodism: *This type of photoperiodic behaviour, whose flowering is accelerated by days of less than a certain critical period of illumination, is characteristic of a great many plants* (J. Heslop-Harrison). —**photoperiodically,** *adv.*

photoperiodism, *n.* the response of an organism to the length of its daily exposure to light, especially as shown by changes in vital processes: *Photoperiodism ... is largely responsible for the separation of many wild flowers into spring, summer, and fall blooming classes* (Science News Letter).

photophilous (fō tof'ə ləs), *adj. Biology.* (of an organism) flourishing in strong light, especially sunlight; light-loving: *the photophilous characteristics of lichens.*

photophobia, *n. Biology.* extreme sensitivity to light: *The mink also lacked adequate pigmentation in the iris of the eye and so exhibited photophobia* (Robert W. Leader).

—**photophobic,** *adj.* showing photophobia; shrinking from light; photonegative: *The mole is photophobic* (James Neylon).

photophore (fō'tō fôr), *n. Zoology.* a luminous, cup-shaped organ on the bellies of certain deep-sea crustaceans and fishes: *In some deep-sea fish the ventral photophores have chambers, in which light is produced, with strongly reflecting walls. These open into structures with reflecting surfaces that distribute the light after passing through a layer of coloured tissue* (Frederick S. Russell). [from *photo-* + *-phore*]

photophoresis (fō'tō fə rē'sis), *n. Physics.* the unidirectional movement of small particles, suspended in gas or falling in a vacuum, produced by a beam of light: *Photophoresis is the movement of small, suspended particles ... which is produced by shining a beam of light at them. It is a result of uneven heating of the particles* (New Scientist). [from *photo-* + Greek *phorēsis* a carrying]

photophosphorylation (fō'tō fos'fər ə lā'shən), *n. Chemistry.* phosphorylation induced by the presence of radiant energy in the form of visible or nonvisible light: *Red and far-red light can be used to control the rate of photophosphorylation, a key step in photosynthesis* (Science News).

photopic (fō top'ik or fō tō'pik), *adj. Optics.* having to do with or able to see in light of a sufficient intensity to permit color differentiation. [from Greek *phōs, photos* light + *ōps* eye]

photopigment, *n. Biochemistry.* a pigment whose characteristics are changed by the action of light: *Color vision demands not only three receptors containing different photopigments but a nervous system to process this information* (R. L. De Valois).

photopositive, *adj. Biology.* attracted to light; photophilous.

photoreaction, *n. Chemistry.* any chemical reaction induced by the presence of light: *The photoreactions involving chlorophyll, that is, the conversion of light energy to chemical energy, takes place in this membrane system* (W. W. Thomson).

photoreception, *n. Biology.* the detection and absorption of light by cells of organisms having pigments sensitive to radiant energy. Photoreception is essential to photosynthesis and phototropism in plants, and to vision in animals.

—**photoreceptive,** *adj.* of or having to do with photoreception: *The retina's photoreceptive cells undergo molecular changes when stimulated by light* (A. R. Harmet).

—**photoreceptor,** *n.* a nerve ending which is sensitive to light: *Rod photoreceptors of the eye are responsible for colourless vision of low intensities* (New Scientist).

photorespiration, *n. Botany.* respiration induced or stimulated by the presence of light: *A particular problem in photosynthesis is the role of photorespiration. Some plants lose up to half the carbon dioxide fixed by photosynthesis, but this loss seems to serve no useful function* (New Scientist).

photosensitive, *adj. Biology.* readily stimulated to action by light or other radiant energy.

—**photosensitivity,** *n.* the quality of being readily stimulated to action by light or other radiant energy: *Chlortetracycline induces photosensitivity by an unknown mechanism in anyone who takes the drug, and sunburn can result in those then exposed sufficiently to actinic energy* (Richard L. Sutton, Jr.).

—**photosensitization,** *n.* the act or process of photosensitizing: *The photosensitization can be produced by certain chemicals that come into contact with the skin* (Scientific American).

—**photosensitize,** *v.* to make sensitive to the action of light; make photosensitive.

photosphere, *n. Astronomy.* the visible, gaseous, and intensely bright layer surrounding the sun or other stars: *The photosphere [of the sun] is some 400 kilometers thick and ranges in temperature from 4,500 degrees Kelvin at the top, where it joins the overlying chromosphere, to perhaps 8,000 degrees K. at the bottom* (Scientific American). See the picture at **sun.**

—**photospheric,** *adj.* of or having to do with the photosphere or a photosphere: *photospheric light.*

photosynthesis (fō'tə sin'thə sis), *n. Biology.* the process by which plant cells make carbohydrates by combining carbon dioxide and water in the presence of chlorophyll and light, and release oxygen as a by-product. Photosynthesis is the source of most of the oxygen in the air. *Photosynthesis thus occupies a primary place in the economy of life. It is the process by which the energy of the Sun is captured and converted to the uses of the living cell. It is, in addition, the beginning process in the transfer of atoms from the inorganic world to the organic* (Harper's). *The primary process in photosynthesis ... is the absorption of light quanta by the chlorophyll molecule. Since chlorophyll absorbs in the red region of the spectrum ... we should suspect that the red quanta are the effective ones in photosynthesis* (McElroy, Biology and Man). *Amino acids are the main products of bacterial photosynthesis* (D. I. Arnon).

—**photosynthesize,** *v.* **1** to carry on photosynthesis: *A red alga is best adapted to photosynthesize in the bluish-green light of deep water* (G. E. Fogg). **2** to produce by photosynthesis: *Plants photosynthesize protein as well as carbohydrates directly under light* (Time).

—**photosynthetic,** *adj.* **1** of or having to do with photosynthesis: *When the illumination was intense the photosynthetic rate approximately doubled for every 10°C. rise in temperature* (H. Lees). **2** carrying on photosynthesis: *In any large body of water microscopic floating algae, or phytoplankton, are the primary producers of organic matter. Like flowering plants these organisms are photosynthetic and able to use inorganic substances—water, carbon dioxide, and mineral ions—and light energy for the synthesis of organic substances* (New Biology).

—**photosynthetically,** *adv.* by means of photosynthesis.

ASSOCIATED TERMS: Photosynthesis is one type of *autotrophic* nutrition. Most cells which carry on photosynthesis contain *chloroplasts.* These chloroplasts contain *pigments* which include *chlorophyll.* The two major chemical reactions in photosynthesis are the *light reaction* and the *dark reaction.* The light reaction occurs within the *grana* inside the chloroplasts, which contain the enzymes and pigments necessary for this reaction. The dark reactions occur in the *stroma* within the chloroplasts.

phototactic, *adj.* of, having to do with, or characterized by phototaxis: *A phototropic or phototactic response might become dependent on the oxygen supplied by photosynthesis* (Scientific American).

phototaxis, *n. Biology.* the tendency of an organism to move in response to light: *... negative phototaxis (movement away from light) in a number of species* (E. B. Edney).

phototroph (fō'tō trof), *n. Biology.* an organism that uses light to break down carbon dioxide for its metabolism: *Some [chlorophyll-bearing protozoa] appear to be obligate phototrophs* (R. P. Hall). [from *photo-* + Greek *trophē* nourishment]

—**phototrophic** (fō'tō trof'ik), *adj.* using light to break down carbon dioxide for its metabolism: *phototrophic bacteria.*

phototropic, *adj.* **1** *Biology.* turning or bending in response to light: *Most plants are phototropic.* **2** *Optics.* sensitive to changes in amount of radiation: *Phototropic glass [is] valuable in greenhouses because it darkens automatically when the sun shines on it* (Simeon Potter). —**phototropically,** *adv.*

phototropism (fō to'trə piz'əm), *n. Biology.* a tendency of plants to turn in response to light: *Light also brings about special responses such as phototropism (the growing toward light seen in plants) and photoperiodism (the developmental responses of living organisms to varying lengths of exposure to light)* (Clarence J. Hylander).

photovoltaic, *adj.* **1** generating an electric current when acted on by light or a similar form of radiant energy, as a photoelectric cell is: *Solar cells ... use a photovoltaic principle for converting the energy of the sun's photons into electrical energy* (George P. Sutton). **2** = photoelectric.

phragmoplast (frag'mə plast), *n. Biology.* the cytoplasmic mechanism in cell division responsible for the formation of the cell plate: *The phragmoplast, in which the cell plate develops to divide the protoplast in two*

cap, fāce, fàther; best, bē, tèrm; pin, fīve;
rock, gō, ôrder; oil, out; cup, pùt, rüle;
yü in use, *yu* in uric;
ng in bring; *sh* in rush; *th* in thin, ͫ in then;
zh in seizure.
ə = *a* in about, *e* in taken, *i* in pencil, *o* in lemon, *u* in circus

phreatophyte

as the final act of mitosis, is likewise thickly populated with cytotubules (Myron C. Ledbetter). [from Greek *phragmos* fence + English *-plast*]

phreatophyte (frē at'ə fīt), *n. Botany.* any plant that obtains water by the deep penetration of its roots into the water table: *25 million acre feet yearly are lost to useless plants, called phreatophytes, that act in effect as pumps, shooting their roots ever deeper to suck needed water out of the ground* (Wall Street Journal). [from Greek *phrear, phreatos* well + English *-phyte*]

phrenic (fren'ik), *adj. Anatomy.* of or having to do with the diaphragm: *The phrenic nerve carries to the diaphragm the stimuli which make it work* (Andrew C. Ivy). [from Greek *phrēn* diaphragm]

phthalein (thal'ēn), *n. Chemistry.* any of a series of organic dyes obtained by the action of phenols on phthalic anhydride. Compare **phenolphthalein.**

phthalic acid (thal'ik), *Chemistry.* any of three isomeric acids derived from phthalic anhydrides. *Formula:* $C_8H_6O_4$ [shortened from *naphthalic,* from *naphthal(ene)*]

phthalic anhydride, *Chemistry.* a white, crystalline acid anhydride obtained by catalytic oxidation of naphthalene, used in making various resins, pharmaceutical intermediates, and insecticides.

phthalocyanine (thal'ə sī'ə nēn), *n. Chemistry.* **1** a bright blue-green crystalline compound derived from phthalic anhydride. *Formula:* $C_{32}H_{18}N_8$
2 any of various bright blue or green pigments that are metal derivatives of this compound.

phycocyanin (fī'kō sī'ə nin), *n. Biochemistry.* a blue pigment found in association with chlorophyll, especially in the cells of the blue-green algae: *Phycocyanin was found to be the key to why plants blossom in accordance with the length of daylight and darkness* (Science News Letter). [from Greek *phykos* seaweed + *kyanos* blue]

phycoerythrin (fī'kō ə rith'rin), *n. Biochemistry.* a red pigment found in the red algae and, in association with phycocyanin, in the blue-green algae. [from Greek *phykos* seaweed + *erythros* red]

phycology (fī kol'ə jē), *n.* = algology. [from Greek *phykos* seaweed + English *-logy*]

phycomycete (fī'kō mī sēt'), *n. Biology.* a phycomycetous fungus.

phycomycetous (fī'kō mī sē'təs), *adj. Biology.* of or belonging to a division (Zygomycota) of fungi, whose members live as parasites or saprophytes and resemble algae. [from New Latin *Phycomycetes* the former class name, from Greek *phykos* seaweed + *mykēs, mykētos* fungus, mushroom]

phyla (fī'lə), *n.* plural of **phylum.**

phyletic (fī let'ik), *adj.* of or having to do with the development or evolution of a species or other group: ... *the experimental and natural observation of animals at their own phyletic level* (New Scientist). Compare **monophyletic.** [from Greek *phyletēs* tribesman, from *phylē* tribe] **—phyletically,** *adv.*

phyllite (fil'īt), *n. Geology.* a low-grade metamorphic rock, intermediate between shale and schist and consisting chiefly of mica or chlorite. [from Greek *phyllon* leaf + English *-ite*]

phylloclade (fil'ə klād), *n. Botany.* a flattened or enlarged stem or branch, resembling or performing the function of a leaf, as in the cactus. [from Greek *phyllon* leaf + *klados* branch, sprout]

phyllode (fil'ōd), *n. Botany.* an expanded and, usually, flattened petiole resembling a leaf and having the same functions, the true leaf blade being absent or much reduced in size, as in many acacias. [from Greek *phyllōdēs* leaflike, from *phyllon* leaf + *eidos* form]

phyllome (fil'ōm), *n. Botany.* **1** a leaf of a plant or any organ homologous with a leaf, or regarded as a modified leaf, such as a sepal, petal, or stamen. **2** all the leaves of a plant, taken as a whole; foliage. [from Greek *phylloma* foliage, from *phyllon* leaf]

phyllotactic (fil'ə tak'tik), *adj.* of or having to do with phyllotaxy: *the phyllotactic spiral of leaves.*

phyllotaxis (fil'ə tak'sis), *n.* = phyllotaxy.

phyllotaxy (fil'ə tak'sē), *n. Botany.* **1** the distribution or arrangement of leaves on a stem: *The phyllotaxy, or pattern of appendage arrangement, changes from 1/2 in the vegetative shoot [in plants with terminal flowers] to carpel arrangements of 2/7, 3/7, 3/8, and 4/10* (Shirley C. Tucker).
2 the laws collectively which govern such distribution. [from Greek *phyllon* leaf + *taxis* arrangement]

phylogenetic (fī'lə jə net'ik), *n.* of or having to do with phylogeny; evolutionary. **—phylogenetically,** *adv.*

phylogeny (fī loj'ə nē), *n. Biology.* **1** the evolutionary development of a species or higher grouping of organisms, or the history of its development: *The relation between metamorphosis and evolution can be considered a special case of the old idea that ontogeny (the life history of an organism) recapitulates phylogeny (its evolutionary history)* (Earl Frieden).
2 the evolutionary development of the organs or other parts of an organism: *the phylogeny of the intestinal tract in frogs.*
[from Greek *phylon* race + *-geneia* origin]

phylum (fī'ləm), *n., pl.* **-la** (-lə). *Biology.* a primary division of a kingdom, ranking above a class: *The major phylum-level patterns of animals were apparently determined in one evolutionary radiation during late Precambrian and Cambrian time. Since then extensive evolutionary change has occurred within the phyla, but few new animal phyla have appeared* (Eicher and McAlester, *History of the Earth*).
▶ Botanists now usually prefer the term *division* in place of *phylum.* Thus, Annelida and Arthropoda in the animal kindgom are generally classified as phyla, while Pterophyta and Bryophyta of the plant kingdom are now usually classified as divisions.
ASSOCIATED TERMS: see **kingdom.**
[from New Latin, from Greek *phylon* race, stock]

phys., *abbrev.* physics.

physical, *adj.* of or having to do with physics; involving application of the concepts and laws of physics: *When the information cannot be found in the published literature, the chemist will establish values for the physical properties (such as melting point, boiling point, density, and viscosity) of the appropriate chemicals* (J. F. Pearson).

physical change, *Chemistry.* a change in the size or form of a substance without any alteration in the composition of its molecules or without its producing or becoming a new substance: *A physical change may result*

494

in a more or less temporary alteration of a few of the properties of a substance, but no change in composition results from it and most of the altered properties usually regain their former value when original conditions are reestablished (Jones, *Inorganic Chemistry*).

physical chemistry, a branch of chemistry that deals with the basic laws of the properties of substances as formulated by physics and their relations to chemical composition and changes: *Traditionally, pure chemistry has become divided into three major parts: organic chemistry ... inorganic chemistry, ... and physical chemistry, which deals with the study of chemical systems by physical methods* (New Scientist).

physical geography, a branch of geography that deals with the natural features of the earth's surface, such as landforms, climate, winds, and ocean currents: *Physical geography has a humanized perspective, for it is usually an analysis of the whole natural equipment of a region, or some element of it, in terms of its resource potentialities for human use* (Finch and Trewartha, *Elements of Geography*). Also called **physiography.**

physical science, physics, chemistry, geology, astronomy, meteorology, and other sciences dealing with inanimate matter. Contrasted with **life science.**

physicochemical, *adj.* **1** of or having to do with the physical and chemical properties of substances: *Direct proof of ageing at the molecular level is derived from studies of the progressive change in the physicochemical properties of collagen with age* (Peter Alexander). **2** having to do with physical chemistry: *... a physicochemical treatment of the principles of chemical equilibrium* (Science). —**physicochemically,** *adv.*

physics, *n.* **1** the science that deals with the properties and interrelationships of matter and energy, excluding chemical and biological change. Physics includes the study of force, motion, heat, light, sound, electricity, magnetism, radiation, and atomic energy. The branches of physics include mechanics, thermodynamics, optics, electronics, and acoustics. **2** the physical characteristics, processes, etc., of any phenomenon: *The author starts with the basic physics of radiation and rightly emphasizes the importance of using frequency ... as an independent variable* (New Scientist).

physio-, *combining form.* physical, as in *physiology, physiography.* [from Greek *physis* nature]

physiographic or **physiographical,** *adj.* of or having to do with physiography: *Antarctica may be divided into two major physiographic portions, East and West Antarctica* (Paul A. Siple).

physiographic climax, *Ecology.* a climax that is determined mainly by the physical features or topography of the area, such as its having grass, woods, or hills.

physiography (fiz′ē og′rə fē), *n.* = physical geography.

physiological, *adj.* **1** having to do with physiology: *Digestion is a physiological process.* **2** having to do with the normal or healthy functioning of an organism: *Food and sleep are physiological needs.* —**physiologically,** *adv.*

physiology, *n.* **1** the branch of biology dealing with the normal functions of living things or their parts: *animal physiology, plant physiology. Research in medical physiology, including such topics as the action of endocrines and of drugs, continues to be extremely active*

and also to follow, in part, biochemical lines (Simpson, *Life: An Introduction to Biology*). **2** all the functions and activities of a living organism or of one of its parts: *Human beings also show daily rhythmic variations in their physiology, such as changes in body temperature, excretion of water and salts, hormone levels in the blood, division of cells in the skin, etc.* (Brian C. Goodwin). [from Greek *physiologia* natural science, from *physis* nature + -*logia* study of, -logy]

phyt-, *combining form.* the form of **phyto-** before vowels, as in *phytane.*

phytane (fī′tān), *n. Biochemistry.* a complex hydrocarbon, a product of the breakdown of chlorophyll, the presence of which in oilbearing rocks is thought to be evidence of the existence of living matter 3 billion years ago: *Two specific hydrocarbons, a C_{20} hydrocarbon, phytane, thought to be derived from a hydrocarbon side chain of the green pigment of plants (chlorophyll), and a C_{19} hydrocarbon, pristane, which is present in marine organisms, seem to occur ubiquitously in ancient sediments* (Eugene D. McCarthy). *Formula:* $C_{20}H_{42}$

-phyte, *combining form.* growth or plant, as in *epiphyte.* [from Greek *phyton* plant, shoot, from *phyein* grow, beget]

phyto-, *combining form.* plant or plants, as in *phytotoxic = toxic to plants.* [from Greek *phyton*]

phytoalexin (fī′tō ə lek′sin), *n. Botany.* any substance produced by a plant to counteract disease: *One particular phytoalexin can be induced simply by adding very low concentrations of a peptide* (Science Journal).

phytochemical, *adj.* of or having to do with phytochemistry: *Photosynthesis is a phytochemical process.* —**phytochemically,** *adv.*

phytochemistry, *n.* the branch of chemistry dealing with the growth, metabolism, and products of plants. Phytochemistry studies the various plant hormones, the absorption of inorganic nutrients to form sugars, proteins, etc., the chemicals, drugs, and vitamins obtained from plants, and such basic processes as photosynthesis and nitrogen fixation.

phytochrome (fī′tə krōm), *n. Botany.* a bluish, light-sensitive pigment in plants which absorbs red or infrared rays and acts as an enzyme in controlling growth and other photoperiodic responses: *Flowering, stem lengthening, and other aspects of plant growth are controlled by light acting on a blue pigment phytochrome which differs distinctly from the predominant green chlorophyll of leaves* (S. B. Hendricks).

phytogeographical, *adj.* of or having to do with phytogeography: *Botanists have always regarded the British Isles as a phytogeographical whole* (New Scientist). —**phytogeographically,** *adv.*

cap, fāce, fäther; best, bē, tèrm; pin, fīve;
rock, gō, ôrder; oil, out; cup, pùt, rüle,
yü in use, *yu* in uric;
ng in bring; sh in rush; th in thin, ŦH in then;
zh in seizure.
ə = *a* in about, *e* in taken, *i* in pencil, *o* in
lemon, *u* in circus

495

phytogeography (fī'tō jē og'rə fē), *n.* the science of plant distribution: *Phytogeography ... deals with the plant cover of the world — with its composition, its local productivity, and particularly its distribution* (N. Polunin).

phytohemagglutinin (fī'tō hē'mə glü'tə nin), *n. Biochemistry.* any one of various protein substances that are extracted from plants and cause blood cells to change in shape, divide, or clump together: *phytohemagglutinin, a mucoprotein that ... stimulates enzyme synthesis* (Science News).

phytohormone (fī'tō hôr'mōn), *n.* = plant hormone.

phytopathogenic, *adj. Biology.* causing plant disease, especially by parasitic destruction of the host: *... mixed populations of nonparasitic and phytopathogenic organisms* (Science News).

phytopathology, *n. Biology.* **1** the science that deals with the diseases of plants.
2 *Medicine.* the study of diseases caused by plant parasites and fungi.

phytophagous (fī tof'ə gəs), *adj. Biology.* feeding on plants: *Fully half the known species of insects are phytophagous, feeding on the tissues or juices of plants.* (Storer, *General Zoology*). SYN: herbivorous.

phytoplankter (fī'tō plangk'tər), *n. Biology.* any of the organisms in a phytoplankton.

phytoplankton (fī'tō plangk'tən), *n. Biology.* the plankton of any body of water which consists of plants or plantlike organisms, usually algae: *The phytoplankton serves as food for tiny sea animals ... which in turn are eaten by fish, birds or other sea-going animals* (Science News Letter). Compare **zooplankton.**

phytosociology, *n.* the branch of plant ecology dealing with the interrelations among the plants of various areas.

phytosterol (fī tos'tə rōl), *n. Biochemistry.* any of several plant alcohols that have the properties of sterols, such as ergosterol. Compare **zoosterol.**

phytotoxic (fī'tō tok'sik), *adj. Botany.* toxic or injurious to plants: *Whilst a protective fungicide must be toxic to the fungal spore it must not injure the host plant, that is, be phytotoxic* (New Scientist).
—phytotoxicity (fī'tō tok sis'ə tē), *n.* a toxic or poisonous quality injurious to plants: *Fungicides ... must be selected and timed so that disease control is achieved with minimal phytotoxicity* (Nature).
—phytotoxin, *n.* **1** a substance that is toxic or injurious to plants. **2** a toxin derived from a plant.

pi (pī), *n.* **1** *Mathematics.* the ratio of the circumference of any circle to its diameter, usually written as π and equal to 3.141592+.
2 *Physics, Chemistry.* an electron, orbital, molecule, etc., having one unit of angular momentum.
[from the name of the 16th letter of the Greek alphabet (π), used as abbreviation of Greek *periphereia* periphery]

pia mater (pī'ə mā'tər), *Anatomy.* the delicate, vascular membrane which is the innermost of three membranes enveloping the brain and spinal cord. [from Medieval Latin *pia mater* thin or tender mother]

piceous (pis'ē əs *or* pī'sē əs), *adj.* **1** *Chemistry.* of, having to do with, or resembling pitch. **2** *Zoology.* of the color of pitch; pitch-black. [from Latin *piceus,* from *picem* pitch]

pico- (pē'kō-), *combining form.* **1** one thousand billionth (10^{-12}) of an (SI unit); a trillionth part of something, as in *picofarad = one trillionth of a farad. Symbol:* p **2** very small, as in *picornavirus.* [probably from Spanish *pico* small number, peak]

picocurie, *n.* one trillionth of a curie (unit of radioactivity): *The level of 150 picocuries over a three-month period is considered by scientists to be the tolerable limit for humanity* (London *Times*). Formerly called **micromicrocurie.**

picofarad, *n.* one trillionth of a farad: *Radio engineers have been using picofarads for some years now in preference to the old-fashioned microfarads* (New Scientist).

picogram, *n.* one trillionth of a gram.

picomole, *n.* one trillionth of a mole.

picornavirus (pi kôr'nə vī'rəs), *n. Biology.* any of a group of viruses containing ribonucleic acid, including the poliovirus, rhinovirus, and similar viruses. [from *pico-* + *RNA* + *virus*]

picosecond, *n.* a trillionth of a second: *Laser pulses lasting about ... one picosecond can now be measured accurately for the first time, making it possible to measure picosecond events in atoms and molecules* (Science News).

picrate (pik'rāt), *n. Chemistry.* a salt or ester of picric acid.

picric acid (pik'rik), *Chemistry.* a very poisonous, yellow, crystalline, intensely bitter acid, used in explosives, in dyeing, and in medicine. *Formula:* $(NO_2)_3C_6H_2OH$ [from Greek *pikros* bitter, sharp]

picrite (pik'rīt), *n. Geology.* any of a group of dark igneous rocks of granular texture, composed chiefly of olivine and augite. [from Greek *pikros* bitter + English *-ite* (so called for the bitter taste of the magnesium it contains)]

picrotoxic (pik'rō tok'sik), *adj.* of or derived from picrotoxin.

picrotoxin (pik'rō tok'sin), *n. Biochemistry.* a bitter, very poisonous, crystalline chemical compound obtained from the seeds of several plants, similar to strychnine in action. Formula: $C_{30}H_{34}O_{13}$

pictograph, *n.* a diagram or chart presenting statistical data by using pictures of different colors, sizes, or numbers. Compare **bar graph.**

pie chart, = circle graph.

piedmont (pēd'mont), *Geology, Geography.* **—adj.** lying or formed at the base of a mountain range: *a piedmont alluvial plain, a piedmont glacier.*
—n. a piedmont area or region.
[named after *Piedmont,* region in Italy at the base of the Alps, from French *pied* foot + *mont* mountain]

piezo- (pē ā'zō- *or* pē āt'sō-), *combining form.* pressure, as in *piezochemistry, piezometer.* [from Greek *piezein* to press, squeeze]

piezochemistry, *n.* chemistry that deals with the effects of pressure on chemical processes and materials.

piezoelectric, *Physics.* —*adj.* of, having to do with, or affected by piezoelectricity: *A piezoelectric material produces electricity when subjected to stress or strain* (Science Journal).
—*n.* a piezoelectric material: ... *the inherent advantage of surface acoustic waves in piezoelectrics* (New Scientist). —**piezoelectrically,** *adv.*
—**piezoelectricity,** *n.* electricity or polarity induced by pressure, as in certain crystals, such as quartz, vibrating in an alternating electrical field: *Several researchers attribute piezoelectricity in bone directly to the collagen in bone, and think that all connective tissues may be piezoelectric and that piezoelectricity may play a fundamental role in physiology—perhaps as a way for organs to adjust to environmental stress* (Chica Minnerly).

piezomagnetic, *adj.* of, having to do with, or affected by changes of magnetization induced by pressure: *Laboratory experiments have shown that when magnetic rocks are subjected to pressure, their magnetic properties change. This phenomenon, called the piezomagnetic effect, has been discussed as a possible means of predicting earthquakes* (Science News).

piezometric (pē ā'zə met'rik *or* pē āt'sə met'rik), *adj.* of, having to do with, or done by piezometry.
—**piezometry** (pē ə zom'ə trē *or* pē'āt som'ə trē), *n.* the measurement of pressure or something connected with pressure.

pigeonite (pij'ə nīt), *n. Mineralogy.* a variety of clinopyroxene similar to augite but with less calcium, occurring in basalt, gabbro, and other rocks: *The large grayish crystals [of lunar rock] are plagioclase ... the brownish-orange matrix material is pigeonite* (John A. Wood). [from *Pigeon* Point, Minnesota + -ite]

pigment, *n. Biochemistry.* any natural substance occurring in and coloring the tissues of an organism. *The color of hair, skin, and eyes is due to pigment* (Barton Lyle Hodes). [from *pigmentum,* from *pingere* to paint]
—**pigmentary** (pig'mən ter'ē), *adj.* of, having to do with, containing, or consisting of pigment: *pigmentary activity, a pigmentary substance.*

pigmentation, *n. Biology.* **1** the deposit of pigment in the tissues of a living organism causing coloration or discoloration. **2** the coloration of an organism: *Amongst the endogenous pigments the melanins have received most attention, for the black coloration conferred by them is found in practically all groups of animals* (B. Nickerson).

pileate (pī'lē it *or* pil'ē it), *adj. Botany.* having a pileus or cap, as certain fungi do.

pileated (pī'lē ā'tid *or* pil'ē ā'tid), *adj. Zoology.* having the feathers on the top of the head conspicuous; having a crested pileum: *a pileated woodpecker.*

pileum (pī'lē əm), *n., pl.* **pilea** (pī'lē ə). *Zoology.* the top of a bird's head between the bill and the nape; the forehead and crown. [from Latin *pileum,* neuter of *pileus* skullcap]

pileus (pī'lē əs), *n., pl.* **pilei** (pī'lē ī). *Botany.* the broad umbrellalike fruiting structure forming the top of certain fungi, such as mushrooms; cap. It is supported by a stalk or stem (the stripe) and bears radiating plates (gills) on the under side. [from Latin *pileus* skullcap]

pillow lava, *Geology.* submarine lava, usually of a basaltic kind, found in the form of round, closely packed masses.

pilus (pī'ləs), n., pl. **-li** (-lī). **1** *Biology.* a hairlike part or structure, especially on the surface of a cell or microorganism. Also called **fimbria. 2** *Anatomy.* a hair. [from Latin *pilus* hair]

pi-meson, *n.* = pion.

pinacoid (pin'ə koid), *n. Crystallography.* a crystal form consisting of two parallel faces. [from Greek *pinax, pinakos* slab]

pinch effect, *Physics.* the constriction of plasma by the magnetic field of an electric current, used in controlling thermonuclear fusion: *The so-called "pinch effect," the contraction occurring in a gas carrying an electric charge due to its own magnetic field, was suggested as the mechanism to control thermonuclear reactions for peaceful purposes, and also to account for solar flares* (Science News Letter).

pincushion distortion, *Optics.* a distortion produced by a lens causing the sides of a square to curve inwards, suggesting the shape of a pincushion.

pineal (pin'ē əl), *adj. Anatomy.* of or having to do with the pineal body. [from French *pinéal,* from Latin *pinea* pine cone, from *pinus* pine tree]

pineal body or **pineal gland,** *Anatomy.* a small, somewhat conical structure present in the brain of all vertebrates having a cranium. It secretes various chemical substances, such as melatonin, and appears to function in various animals as a light-sensing organ, as a biological clock, or as a ductless gland whose secretions regulate the activity of sex glands.

pineal eye, *Anatomy.* an eyelike projection of the pineal gland on the head of some reptiles.

pinene (pī'nēn), *n. Chemistry.* a terpene found in oil of turpentine and other essential oils, occurring in two isomeric forms, and used in making resins or as a solvent. *Formula:* $C_{10}H_{16}$ [from *pine* (tree) + -ene]

pingo (ping'gō), *n., pl.* **-gos** or **-goes.** *Geology.* an arctic mound or conical hill, consisting of an outer layer of soil covering a core of solid ice. [from an Eskimo word]

pinion, *n. Zoology.* the last joint of a bird's wing. [from Middle French *pignon*]

pinna (pin'ə), *n., pl.* **pinnae** (pin'ē), **pinnas. 1** *Zoology.* **a** a feather, wing, or winglike part. **b** a fin, flipper, or similar part.
2 *Anatomy.* the auricle of the ear.
3 *Botany.* one of the primary divisions of a pinnate leaf, especially in ferns; leaflet.
[from Latin *pinna* feather, wing, fin]

pinnacle, *n. Geology.* a high peak or point of rock. [from Late Latin *pinnaculum,* diminutive of Latin *pinna* wing]

pinnate (pin'āt), *adj.* **1** *Biology.* resembling a feather; having lateral parts or branches on each side of a common axis, like the vanes of a feather: *The tail is pinnate at the point* (C. C. Blake). *The budding polyps are*

---

cap, fāce, fä̀ther; best, bē, tèrm; pin, five;
rock, gō, ôrder; oil, out; cup, pùt, rüle;
yü in use, *yu̇* in uric;
ng in bring; *sh* in rush; *th* in thin, ᴛʜ in then;
zh in seizure.
ə = *a* in about, *e* in taken, *i* in pencil, *o* in lemon, *u* in circus

pinnatifid

sometimes confined to two opposite sides of a branch, and pinnate forms result (James D. Dana).

2 *Botany.* (of a leaf) having a series of leaflets on each side of a stalk, the leaflets being usually opposite, sometimes alternate. Also spelled **pennate.**
[from Latin *pinnatus* feathered, from *pinna* feather]
—**pinnately,** *adv.*

pinnate leaf
(def. 2)

pinnatifid (pi nat′ə fid), *adj. Botany.* divided or cleft in a pinnate manner, with the divisions extending halfway down to the midrib, or somewhat further, and the divisions or lobes narrow or acute: *a pinnatifid leaf.*
[from Latin *pinnatus* feathered + *-fid,* a root of *findere* to cleave]

pinnatilobate (pi nat′ə lō′bāt), *adj. Botany.* lobed in a pinnate manner: *a pinnatilobate leaf.*

pinnation (pi nā′shən), *n. Botany.* a pinnate condition or formation.

pinnatipartite (pi nat′ə pär′tīt), *adj. Botany.* parted in a pinnate manner: *a pinnatipartite leaf.*

pinnatiped (pi nat′ə ped), *adj. Zoology.* having lobate toes: *a pinnatiped bird.*

pinnatisect (pi nat′ə sekt), *adj. Botany.* divided in a pinnate manner; cut down to the midrib, but with the divisions not articulated so as to form separate leaflets: *a pinnatisect leaf.*

pinnigrade (pin′ə grād), *Zoology.* —*adj.* moving on land by means of finlike parts or flippers, as seals and walruses.
—*n.* a pinnigrade animal.
[from Latin *pinna* feather, wing + *gradus* step, from *gradi* to walk]

pinniped (pin′ə ped), *Zoology.* —*adj.* **1** of or belonging to a suborder (or, in some classifications, an order) of carnivorous mammals that includes seals and walruses. **2** having feather feet.
—*n.* a pinniped animal: *The pinnipeds (the group of aquatic mammals including seals, sea lions and walruses) have milk that is often like heavy cream, containing 40 to 50 percent fat* (Scientific American).
[from Latin *pinna* wing, fin + *pedem* foot]

pinnule (pin′yül), *n.* **1** *Zoology.* a part or organ resembling a small wing or fin, or a barb of a feather: **a** a small, finlike appendage or short, detached fin ray in certain fishes, as the mackerel. **b** each of the lateral branches of the arms in crinoids.
2 *Botany.* one of the secondary or ultimate divisions of a pinnate leaf, especially in ferns; a subdivision of a pinna: *If a pinna be further divided, its divisions are called pinnules. In some ferns the pinnules are further*

subdivided into secondary pinnules (Youngken, *Pharmaceutical Botany*).
[from Latin *pinnula,* divinutive of *pinna* feather, wing, fin]

pinocytosis (pi′nō sī tō′sis), *n. Biology.* a process of taking in fluids by a cell, in which the cell membrane folds back into itself to engulf the fluid and then closes up to become a vesicle: *Glucose, which amoebae normally absorb only in trace amounts, enters freely by means of pinocytosis* (Scientific American). Compare **emiocytosis, endocytosis, exocytosis, phagocytosis.**
[from Greek *pinein* to drink + New Latin *cyt-* cell + *-osis*]
—**pinocytotic** (pi′nō sī tot′ik), *adj.* of or having to do with pinocytosis: *pinocytotic vesicles.*

pint, *n.* a customary unit of measure equal to 1/2 quart, 2 cups, or 16 ounces. [from Old French *pinte*]

pion (pī′on), *n. Nuclear Physics.* any of three lightest members of the meson family of subatomic particles, having masses about 270 times as large as that of the electron. Positive and negative pions decay into neutrinos and muons; neutral pions decay into quanta of light or radiation. *The pion is the only metastable meson that can be produced directly by protons of energy less than 1200 Mev* (D. E. Nagle). Also called **pi-meson.** [from *pi*(-*meson*) + *-on*]
—**pionic** (pī on′ik), *adj.* containing or producing pions.

pipe, *n.* **1** *Botany.* any one of various tubular or cylindrical natural formations, such as the stem of a plant.
2 *Zoology.* a tubular organ, passage, canal, or vessel in an animal body.
3 *Geology.* **a** a vein of ore of a more or less cylindrical form. **b** a mass, more or less cylindrical and often extending far into the ground, of bluish kimberlite within which diamonds may be embedded, found especially in parts of South Africa and the Soviet Union.

pipette or **pipet** (pī pet′), *n.* a slender pipe or tube used for transferring or measuring small quantities of liquids or gases. The most common type is a small glass tube that widens into a bulb at the middle, into which liquid may be sucked, and in which it may be retained by closing the top end with a stopper or finger. *The buret and pipet are both calibrated to deliver a certain volume rather than to contain a certain volume* (Jones, *Inorganic Chemistry*). [from French *pipette,* diminutive of *pipe* pipe]

piscivorous (pi siv′ər əs), *adj. Zoology.* feeding solely or chiefly on fish; fish-eating. [from Latin *piscis* fish + *vorare* devour]

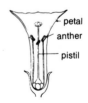

pistil (pis′tl), *n. Botany.* **1** the part of a flower that produces seeds, consisting, when complete, of an ovary, a style, and a stigma: *The central whorl of floral organs, the pistil or pistils of the flower, is formed of one or more carpels* (Harbaugh and Goodrich, *Fundamentals of Biology*).
2 such organs taken collectively, when there are more

498

than one; gynoecium. [from New Latin, from Latin *pistillum* pestle]

—pistillate (pis′tl āt), *adj.* **1** having a pistil or pistils. **2** having a pistil or pistils but no stamens.

pitch[1], *n. Chemistry.* **1** a black, sticky material formed as a residue in the distillation of coal tar or petroleum. **2** = asphalt. [Old English *pic*, from Latin *picem*]

pitch[2], *n. Acoustics.* the characteristic of sound that depends on the frequency of vibration of the sound waves. High-pitched sounds have higher frequencies than low-pitched sounds. [from Middle English *picchen* to throw, thrust]

pitchblende, *n.* = uraninite. [from German *Pechblende,* from *Pech* pitch + *Blende* blende]

pith, *n.* **1** Botany. **a** the central column of spongy tissue in the stems of most herbaceous plants and of some trees: *The roots of most dicotyledonous plants lack a pith, the xylem extending to the center* (Harbaugh and Goodrich, *Fundamentals of Biology*). **b** a similar tissue occurring in other parts of plants, as that lining the skin of an orange. **2** Zoology. the soft inner substance of the spinal column, a feather, etc. See the picture at **feather.** [Old English *pitha*]

pituitary (pə tü′ə ter′ē), *Anatomy.* —*n.* = pituitary gland: *The pituitary, a small gland buried at the base of the brain, has been called the master gland because its hormones affect so many other glands and body functions* (Science News Letter).
—*adj.* of or having to do with the pituitary gland. [from Latin *pituitarius* relating to phlegm, from *pituita* phlegm (so called because it was believed that the gland channeled mucus to the nose)]

pituitary body, = pituitary gland.

pituitary gland, *Anatomy.* a small, oval endocrine gland that is situated at the base of the brain in most vertebrates, in a cavity of the sphenoid bone, and secretes hormones that promote growth, stimulate other glands, and regulate many other bodily functions. The pituitary gland has two main sections: the anterior lobe (*adenohypophysis*), which secretes several hormones, including growth hormone, prolactin, ACTH, follicle-stimulating hormone, luteinizing hormone, and thyroid-stimulating hormone; and the posterior lobe (*neurohypophysis*), which secretes the hormones oxytocin and vasopressin. A short stalk (*infundibulum*) connects the pituitary gland to the hypothalamus, the part of the brain which produces releasing hormones that govern the secretion of hormones by the anterior lobe. Also called **hypophysis.**

pivot joint, *Anatomy.* a joint in which one bone rotates around another; a joint permitting only rotating movement: *The outer bone of the forearm rotates on its axis about the inner bone by means of a pivot joint at the elbow.* Compare **trochoid.**

pk., *abbrev.* peak.

place, *n.* **1** Mathematics. the position of a figure in a numeral or series, in decimal or any other notation. **2** Astronomy. the position of a celestial body at any instant.

placebo (plə sē′bō), *n., pl.* **-bos** or **-boes.** *Medicine, Pharmacology.* a substance containing no active ingredients of therapeutic effect but prescribed as medicine of psychological effect to satisfy a patient, or used as a control in testing the effectiveness of new drugs: *In carrying out such a test, the best experiments are those in which the subjects are divided into two groups, in a random way, with the substance being tested ... administered to the subjects in one group, and a placebo (an inactive material ...) administered to those of the other group* (Linus Pauling). [from Latin *placebo* I shall please]

placeholder, *n. Mathematics.* a symbol that holds the place for numerals that are being considered: *Astronomers are able to make such a measurement [the distance from the earth to the sun] accurate only to the nearer 100,000 miles. The zeros in the number 92,800,000 are therefore not significant. They are merely placeholders* (Obourn, *The World of Science*).

placenta (plə sen′tə), *n., pl.* **-tae** (-tē), **-tas. 1** Embryology. the spongy vascular organ in most mammals by which the fetus is attached to the wall of the uterus and nourished: *Through the [umbilical] cord ... the placenta furnishes oxygen and food material to the fetus and receives waste material from it. The placenta is expelled following the birth of the baby—hence the term "afterbirth"* (S. M. Gruenberg). See the picture at **embryo.**
2 Botany. **a** the tissue of the ovary of flowering plants to which the ovules are attached, usually the enlarged or modified margins of the carpellary leaves: *The ovules are attached to the placentae within the ovary. A placenta may have the form of a knob, ridge, or conical projection extending inward from the wall of the ovary, or sometimes it may be only a slightly specialized portion of the inner surface of the wall* (Emerson, *Basic Botany*). **b** a structure that bears the sporangia in ferns.
[from New Latin, from Latin *placenta* flat cake, from Greek *plakounta,* from *plakos* flat surface]
—**placental,** *adj.* **1** of or having to do with the placenta: *the placental membrane.* **2** having a placenta: *The Marsupials stand ... below the placental mammals* (Darwin, *Origin of Species*).

placentation (plas′ən tā′shən), *n.* **1** Embryology. **a** the formation and disposition of the placenta in the uterus. **b** the structure of the placenta. **2** Botany. the disposition or arrangement of the placenta or placentae in the ovary: *The various types of placenta arrangement (placentation) are grouped according to their relative complexity* (Youngken, *Pharmaceutical Botany*).

placer, *n. Geology.* a deposit of sand, gravel, or earth in the bed of a stream, containing particles of gold or other valuable minerals with high specific gravities. [from Spanish *placer,* ultimately from Latin *platea* broad street]

place value, *Mathematics.* the value of a digit as determined by its place in a number. In 438.7, the place values of the digits are 4×100, 3×10, 8×1, and $7 \times 1/10$: *The place values in binary numeration are*

cap, fāce, fäther; best, bē, tėrm; pin, fīve;
rock, gō, ôrder; oil, out; cup, pùt, rüle,
yü in use, *yu* in uric;
ng in bring; *sh* in rush; *th* in thin, ᵺ in then;
zh in seizure.
ə = *a* in about, *e* in taken, *i* in pencil, *o* in lemon, *u* in circus

ones, twos, fours, eights, sixteens, thirty-twos, six-ty-fours, one hundred twenty-eights, and so on (Irwin K. Feinstein).

plage (pläzh), *n. Astronomy.* a bright and intensely hot area in the sun's chromosphere, usually accompanying a sunspot. Plages shine primarily in the emission lines of calcium and hydrogen and also emit very short ultraviolet rays. *A dynamic complex of processes involving sunspots and their magnetic fields, bright local patches called plages, jets of hot gas called spicules and major eruptions in the form of prominences and flares* (Scientific American). [from French *plage* beach, bright area]

plagioclase (plā′jē ə klās), *n. Mineralogy.* one of the two major types of feldspar, containing sodium or calcium (usually both) and having its two prominent cleavage directions oblique to one another: *Labradorite is a kind of plagioclase that sometimes shows a beautiful display of colors when light strikes it* (Ernest E. Wahlstrom). *Formula:* (Na,Ca)Al(Si,Al)Si$_2$O$_8$ Compare **alkali feldspar.** [from Greek *plagios* oblique + *klasis* cleavage]

plagiotropic (plā′jē ə trop′ik), *adj. Botany.* of, having to do with, or exhibiting plagiotropism: *plagiotropic organs.* —**plagiotropically,** *adv.*

plagiotropism (plā′jē ot′rə piz əm), or **plagiotropy** (plā′-jē ot′rə pē), *n. Botany.* a turning of the organs of certain plants to an oblique position more or less divergent from the vertical, as the result of reacting differently to the influences of light, gravitation, and other external forces. [from Greek *plagios* oblique, slanting + *tropē* a turning]

plain, *n. Geology.* **1** a comparatively flat stretch of land; tract of level or nearly level land: *A plain is an area of relatively slight local relief, and therefore with gentle slopes... . It is useful to distinguish between smooth plains and rolling plains* (White and Renner, *Human Geography*).
2 any broad, level expanse, such as a part of the sea floor or a lunar sea: *Whether or not the maria were ever seas—now at least, they are ... what we call plains* (Bernhard, *Handbook of Heavens*).

planar, *adj. Geometry.* of, having to do with, or situated in a plane: *This planar amide group is a rigid part of the polypeptide chain; the amide group can be only slightly distorted from the planar configuration* (Scientific American). *If we allow five colours instead of four, any conceivable planar or spherical map can certainly be coloured as required* (New Scientist).
—**planarity** (plə nar′ə tē), *n.* the quality or condition of being planar; flatness: *the planarity of the molecular structure of butadiene.*

planarian (plə nār′ē ən), *Zoology.* —*n.* any of a family (Planariidae) of freshwater flatworms that are bilaterally symmetrical and have cilia and an intestine divided into three main branches. They are turbellarians. *A planarian can regenerate missing body parts* (J. A. McLeod).
—*adj.* of or designating a planarian.
[from New Latin *Planaria* the earlier genus name, from feminine of Late Latin *planarius* on level ground, from Latin *planus* level]

planation (plā nā′shən), *n. Geology.* the process of erosion and deposition by which a stream produces a nearly level land surface: *The floodplain is widened by further lateral planation* (A. N. Strahler). [from Latin *planum* level surface + English *-ation*]

Planck's constant (plangks), *Physics.* a universal constant relating the energy (E) of a quantum of electromagnetic radiation to the frequency of the radiation (ν), so that E = hν, where h is Planck's constant and has the dimensions of energy multiplied by time. *Symbol:* h; *approximate value:* 6.626 × 10^{-34} joule second. [named after Max *Planck,* 1858–1947, German physicist, who formulated it]

plane, *Geometry.* —*n.* a surface such that if any two points on it are joined by a straight line, the line will be contained wholly in the surface: *There is no such thing as a pair of skew planes: every two planes in space either intersect or are parallel* (Moise and Downs, *Geometry*).
—*adj.* **1** being wholly in a plane: *a plane figure.*
2 of or having to do with such figures: *Plane trigonometry treats of the relations and calculations of the sides and angles of plane triangles* (Charles Hutton).
[from Latin *planum* level space]

plane angle, *Geometry.* an angle formed by the intersection of two straight lines in the same plane.

plane geometry, the branch of geometry that deals with figures lying in one plane: *The Basic Elements of plane geometry are the point, the line, and the plane* (Holmes Boynton).

plane-polarized, *adj. Physics.* (of light) polarized so that all the vibrations of the waves take place in one plane: *Our recent work has led to the synthesis of polymers that, when placed in solution, have the property of rotating the plane of polarization—of a beam of plane-polarized light* (Giulio Natta).

planet, *n. Astronomy.* **1** one of the celestial bodies (except comets) that move around the sun in nearly circular paths. The planets of our solar system include Mercury, Venus, Earth, Mars, Jupiter, Saturn, Uranus, Neptune, Pluto, and the asteroids (minor planets) between Mars and Jupiter. *Like the moon, the planets are 'dead' bodies ... incapable of contributing any characteristic radiation of their own, apart from the appropriate thermal radiation corresponding with the temperature at or near their visible surface* (A. J. Higgs).
2 any similar body revolving around a star other than the sun: *Already, we are more than casually speculating about life among the stars, of strange yet sentient beings that may inhabit planets orbiting distant suns* (Wernher Von Braun and Frederick I. Ordway).
[from Late Latin *planetes,* from Greek *planētai* (*asteres*) wandering (stars), from *planasthai* wander]
ASSOCIATED TERMS: see **star.**

planetarium (plan′ə tār′ē əm), *n., pl.* **-ia** (-ē ə), **-iums.** *Astronomy.* **1** an apparatus that shows the movements of the sun, moon, planets, and stars by projecting lights on the inside of a dome.
2 a room or building with such an apparatus.
3 a mechanical model of the solar system, designed to illustrate the motions of the planets.
[from New Latin, from Latin *planetarius* having to do with planets, from *planetes* planet]

planetary, *adj.* **1** *Astronomy*. of or like a planet; having to do with planets: *planetary probes, planetary research. Three laws of planetary motion describing the orbits of the planets around the sun were published by Johannes Kepler in the 1600s.* See **Kepler's law.**
2 *Physics.* like that of a planet: *The orbital model differs from the Bohr model in that it does not represent electrons as moving in planetary orbits around the nucleus* (Chemistry Regents Syllabus).

planetary nebula, *Astronomy.* a nebula consisting of an envelope of gas many billions of miles in diameter surrounding a very hot star. There are over one hundred known planetary nebulae in the Milky Way galaxy. The Ring Nebula in the northern constellation Lyra is a planetary nebula.

planetary wave or **planetary wind**, *Meteorology.* a strong current of air, 10,000 to 15,000 meters above the earth's surface, that circles the earth in the Northern Hemisphere, flowing generally from west to east: *Low in the atmosphere, the momentum carried by gravity waves is far less than that carried by the general circulation and by its superimposed planetary waves. It would be rash, however, to conclude that the momentum of the gravity waves is of correspondingly little consequence* (Nature).

planetesimal (plan′ə tes′ə məl), *Astronomy.* —*adj.* of or having to do with small bodies in the developing solar system which, according to the planetesimal hypothesis, moved in planetary orbits and gradually united to form the planets.
—*n.* one of these minute bodies: *The planetesimals were about three to 150 feet in diameter, so the sun's tremendous outpouring of radiation had a great effect on the matter surrounding the planetesimals and on the tiny planets themselves* (Science News Letter).
[from *planet* + (*infinit*)*esimal*]

planetesimal hypothesis, *Astronomy.* a theory of the evolution of the solar system, formulated in 1901 by the American astronomer Forest Ray Moulton (1872–1952) and the American geologist Thomas C. Chamberlin (1843–1928). According to this theory, the planets and asteroids of the solar system were formed by the accretion of planetesimals ejected from the sun and a star passing close to it. Compare **nebular hypothesis.**

planetoid (plan′ə toid) *n.* = asteroid: *While the motion of all is direct, the planetoids differ from the major planets in having more eccentric orbits ... more highly inclined to the ecliptic* (Krogdahl, *The Astronomical Universe*).
—**planetoidal,** *adj.* of, having to do with, or like an asteroid: *planetoidal bodies.*

planetology, *n.* the scientific study of the planets: *General planetology [is] a branch of astronomy that deals with the study and interpretation of the physical and chemical properties of planets* (New Scientist).

plane wave, *Physics.* the simplest kind of wave, moving through space or a medium in a straight line: *In order to describe completely a plane wave it is necessary to specify not only the magnitude of the wave and its direction of propagation, but also the orientation of the electric (or magnetic) field in the plane perpendicular to the direction of propagation* (John B. Walsh).

plani-, *combining form.* flat; level; plane; horizontal, as in *planimetric, planisphere.* Also spelled **plano-.** [from Latin *planus* flat]

planimetric (plan′ə met′rik), *adj.* **1** *Geometry.* of or having to do with the measurement of plane surfaces.
2 (of a map) indicating only horizontal positions, without regard to elevation; not topographic: *Planimetric maps ... show the relative positions of points but do not indicate their elevations above or below sea level* (Gilluly, *Principles of Geology*).
—**planimetry** (plə nim′ə trē), *n.* the measurement of plane surfaces.

planirostral (plā′nə ros′trəl), *adj. Zoology.* having a broad, flat beak.

planisphere (plan′ə sfir), *n. Astronomy.* a projection or representation of the whole or part of a sphere on a plane, especially a map of half or more of the celestial sphere with an adjustable device to show the part of the heavens visible at a given time.

planispiral (plā′nə spī′rəl), *adj.* = planospiral.

plankter (plangk′tər), *n. Biology.* any of the organisms in a plankton: *The sediments are made up in part of the remains of the innumerable plankters that swarm in fertile seas* (Science News). [from Greek *planktēr* wanderer, from *planktos* wandering]

plankton (plangk′tən), *n. Biology.* the small organisms that float or drift in water, especially at or near the surface. Plankton includes small crustaceans, algae, and protozoans, and serves as an important source of food for larger animals, such as fish. *Plankton, the basic food for larger animals, has been estimated to be 14 times as great in the polar waters as in tropical oceans* (Paul A. Siple). See also **phytoplankton, zooplankton.** [from Greek, neuter of *planktos* wandering, from *plazesthai* to wander]
—**planktonic** (plangk ton′ik), *adj.* of, belonging to, or characteristic of plankton: *A large proportion of the deep sea deposits is organic in origin, consisting of the shells of planktonic animals* (T. F. Gaskell and M. N. Hill).
ASSOCIATED TERMS: see **ocean.**

plano-, *combining form.* flat; level; plane; horizontal, as in *plano-convex, planospiral.* Also spelled **plani-.** [from Latin *planus* flat]

planoblast (plan′ə blast), *n. Zoology.* the free-swimming form of certain hydrozoans. [from Greek *planos* wandering + *blastos* germ, sprout]

plano-concave, *adj. Optics.* flat on one side and concave on the other: *a plano-concave lens.*

plano-convex, *adj. Optics.* flat on one side and convex on the other. *a plano-convex lens.*

planospiral, *adj. Zoology.* coiled in one plane: *The shell of the nautilus is planospiral.*

---

cap, fāce, fäther; best, bē, tèrm; pin, five;
rock, gō, ôrder; oil, out; cup, pùt, rüle;
yü in use, *yu* in uric;
ng in bring; sh in rush; th in thin, ᴛʜ in then;
zh in seizure.
ə = a in about, e in taken, i in pencil, o in lemon, u in circus

plant, *n. Biology.* any member of a kingdom (Plantae) of multicellular organisms that produce their own food by photosynthesis. Plants are primarily adapted to a terrestrial existence and have thick cell walls that contain cellulose. *Almost anyone asked to tell the differences between plants and animals would mention that animals move while plants do not. Even in the restricted sense of locomotion this is not quite true. Some animals, such as corals, sponges and sea squirts, spend their entire adult lives in a fixed position, while many plants are able to move freely from place to place* (Scientific American). *It is through their role as primary producers of food for man and other animals that plants assume a special place among the elements of our environment, for from them directly or indirectly come all of our foods* (Greulach and Adams, *Plants*). [Old English *plante,* from *planta* sprout]
► Formerly, living organisms were considered to belong to either the plant or the animal kingdom. Under this system of classification, animals were distinguished from plants chiefly by their (generally) autonomous movement and their heterotrophic nutrition. According to this distinction, the plant kingdom included such diverse organisms as bacteria, algae, and fungi, as well as the complex multicellular autotrophs, but in the five-kingdom system of classification employed by most modern biologists, the term plant is restricted to the relatively complex multicellular photosynthetic organisms. See also **classification.**

plantar (plan′tər), *adj. Anatomy.* of or having to do with the sole of the foot: *a plantar wart.* [from Latin *plantaris,* from *planta* sole of the foot]

plantar arch, *Anatomy.* the main arch of the foot, extending from the heel to the ball of the foot.

plant hormone, *Biochemistry.* any of various organic compounds produced by plants which regulate their growth and other physiological activities. Plant hormones include auxins, gibberellins, cytokinins, and abscisic acid. *Plant hormone production is most abundant in actively growing areas such as the cells at the tips of roots and stems, buds, and seeds* (Biology Regents Syllabus).

plantigrade (plan′tə grād), *Zoology.* —*adj.* **1** walking on the whole sole of the foot, as bears, raccoons, and people do: *a plantigrade carnivore or mammal.* **2** characterized by placing the whole sole on the ground: *a human-like, or plantigrade foot.*
—*n.* a plantigrade animal.
[from Latin *planta* sole of the foot + *gradus* step, from *gradi* to walk]
ASSOCIATED TERMS: see **digitigrade.**

plantimal (plan′tə məl), *n. Biology.* a living cell or cells formed in the laboratory by the fusion of animal and plant cells: *Three separate research groups have now successfully fused animal cells with plant cells to form the first ... "plantimals"* (Science News). [from *plant* + (*an*)*imal*]

plant kingdom, *Biology.* the primary division of living things that includes all plants, especially as distinguished from the animal kingdom. Also called **vegetable kingdom.**

plantlet, *n. Botany.* **1** an undeveloped or rudimentary plant. **2** a small plant.

plant pathology, the study of plant diseases.

plant physiology, the study of the life processes or functions of plants.

planula (plan′yə lə), *n., pl.* **-lae** (-lē). *Zoology.* the flat, ciliated, free-swimming larva of a cnidarian. [from New Latin *planula,* diminutive from Latin *planus* flat, level]
—**planular,** *adj.* **1** of flattened form. **2** having to do with a planula or like a planula.
—**planulate,** *adj.* = planular.

plaque (plak), *n. Biology.* **1** a patch of fibrous tissue or fatty matter on the wall of an artery: *The accumulation of cholesterol in the blood vessels forms plaques which may eventually restrict the flow of blood* (N. Y. Times).
2 any formation of acid-producing bacteria and certain carbohydrates on tooth enamel: *An essential first step in* [dental] *caries production appears to be the formation of "plaque" on the hard, smooth enamel surface* (Ernest Jawetz).
3 a small area of viral infection in a cell culture, useful in assaying the total number of viral particles present in a sample.
[from Middle French *plaque,* from Middle Dutch *plak* flat board]

plasma, *n.* **1** *Biology.* **a** the clear, almost colorless liquid part of blood or lymph, in which the blood cells and platelets float. It consists of water, salts, proteins, and other substances, and it makes up the largest part of the blood. Plasma can be kept indefinitely by freezing or drying and is often used in transfusions in place of whole blood. *In the capillaries some of the liquid part of the blood, the plasma, oozes through the walls into the spaces surrounding the cells* (Beauchamp, *Everyday Problems in Science*). Also called **blood plasma.**
b the fluid contained in muscle tissue. Also called **muscle plasma.**
c protoplasm, especially as distinct from the nucleus.
2 *Physics.* a highly ionized gas, consisting of almost equal numbers of free electrons and positive ions: *Plasma, which is believed to make up 99% of the universe, is a form of matter distinct from solids, liquids, or normal gases. It is a gaseous state that is electronically neutral in the aggregate but which contains ions, free electrons, and neutral particles, and is electrically conductive* (Philip Kopper).
3 *Mineralogy.* a faintly translucent, green variety of chalcedonic quartz.
[from Greek *plasma* something formed or molded, from *plassein* to mold]

plasmablast (plaz′mə blast), *n. Biology.* the parent or stem cell of a plasma cell. [from *plasma* + Greek *blastos* sprout, germ]

plasma cell, *Biology.* an antibody-producing cell into which a B-cell develops upon reaction with a specific antigen: *A person's first exposure to an allergen triggers the formation of specialized blood cells called plasma cells. These cells produce millions of the specific IgE molecules that will recognize and react with that allergen. The IgE* [immunoglobulin E] *molecules spread through the bloodstream and attach to the special receptor sites on the membranes of mast cells and basophils throughout the body* (Robert N. Hamburger).

plasmacyte (plaz′mə sīt), *n.* = plasma cell.

plasmagene (plaz′mə jēn), *n. Biology.* a genetic element found in the cytoplasm of cells, regarded as being a hereditary factor corresponding in function to the genes found in the chromosome. Compare **episome, plasmid.**

plasmalemma (plaz′mə lem′ə), *n.* = plasma membrane. [from *plasma* + Greek *lemma* husk]

plasma membrane, *Biology.* the thin membrane that forms the outer surface or boundary of a protoplast, enclosing the cytoplasm and controlling the kinds and amounts of substances that enter or leave the cell: *The surface of a cell is formed by the plasma membrane that not only serves as a barrier between the internal and external environments of the cell but also permits the selective transfer of substances in and out of the cell* (Robert L. Hill). Also called **cell membrane, plasmalemma.**

plasmapause, *n. Astronomy.* the upper limits of the plasmasphere.

plasma physics, a branch of physics dealing with highly ionized gas, especially as it appears in a wide range of cosmic phenomena and as it is used in controlled thermonuclear reactions: *Plasma physics owes much to astronomical studies of the motions of ionized and magnetized solar gases* (Scientific American).

plasmasphere, *n. Astronomy.* an envelope of highly ionized gas about a planet: *Recent research in space science has resulted in an improved understanding of both the morphology and dynamics of the Earth's plasmasphere, and more basically the extent to which the plasmasphere influences and reflects the dynamics of the entire magnetosphere* (C. R. Chappell).

plasmatic, *adj.* of, containing, or resembling plasma: *the outer surface of a plasmatic membrane.*

plasmid (plaz′mid), *n. Biology.* a genetic element or unit of DNA which replicates in a cell independently of the chromosomal DNA, especially in the cytoplasm of bacteria. Plasmids are used in recombinant DNA research. *Researchers are fascinated by plasmids' small size ... They are on an equal footing with viruses—independent organisms that need a host cell for survival* (Science News). *Some bacteria contain, besides their chromosome, an extra piece of DNA called a plasmid. Plasmids are a serious medical difficulty because they can render bacteria resistant to antibiotics and they can be transferred from one bacterium to another* (London *Times*). Compare **episome.**

plasmin (plaz′min), *n. Biochemistry.* an enzyme in the blood which dissolves blood clots by attacking fibrin: *Normally a blood clot is dissolved by the action of the fibrinolytic enzyme plasmin which is formed from an inactive precursor known as plasminogen* (Martin C. G. Israels). Compare **fibrinolysin.**

plasminogen (plaz min′ə jən), *n. Biochemistry.* the inactive form of plasmin; a proenzyme in the blood from which plasmin is formed by certain blood activators.

plasmodesmata (plaz′mō dez′mə tə), *n.pl. Botany.* narrow strands of cytoplasm in plant cells that form connections between the plasma membranes of adjacent cells. [from New Latin, from Greek *plasma* something molded + *desma* band]

plasmodium (plaz mō′dē əm), *n., pl.* **-dia** (-dē ə). **1** a mass or sheet of protoplasm formed by the fusion, or by the aggregation, of a number of amebalike bodies, as in the slime molds: *Plasmodia are able to move by a slow irregular flowing motion... . They take up solid pieces of food of various kinds as they move about ... Undigested parts are left behind as plasmodium moves away. All this action is definitely animallike* (Emerson, Basic Botany). **2** any of a group of parasitic protozoans, including the organisms which cause malaria. [from New Latin, from Greek *plasma* something molded]

plasmogamy (plaz mog′ə mē), *n. Biology.* the fusion of the cytoplasm of two or more living cells.

plasmoid (plaz′moid), *n. Physics.* a tightly packed, luminous pellet of plasma ions, formed when plasma moves across a magnetic field.

plasmolysis (plaz mol′ə sis), *n. Botany.* the contraction of protoplasm away from the wall of a living cell, due to the withdrawal of liquid by osmosis when the cell is in a liquid of greater solute density than the cell sap. —**plasmolytic** (plaz′mə lit′ik), *adj.* having to do with, showing, or causing plasmolysis. —**plasmolytically,** *adv.*

—**plasmolyze** (plaz′mə līz), *v.* to contract by plasmolysis: *When the evaporating power of the air is very high ... the mesophyll cells plasmolyze and die* (Harbaugh and Goodrich, Fundamentals of Biology).

plasmon (plaz′mon), *n.* **1** *Physics.* a quantum or quasiparticle associated with the propagation of plasma: *Graphite is known to have a sharp plasmon of about 7eV [electron volts]* (John E. Houston). **2** *Genetics.* the aggregate of extranuclear genetic material in an organism: *Together with the nuclear genotype, the plasmon can play its part in the process of adaptation especially ... in the building up of barriers to crossing* (K. Mather). [from *plasma* + -*on*]

-plast, *combining form.* a small body or structure, especially of living matter; particle or granule, as in *bioplast, blepharoplast, mesoplast, chloroplast.* [from Greek *plastos* formed, molded]

plastic, *n.* **1** *Biology.* capable of forming, or being organized into, living tissue: *plastic lymph, a plastic exudation.* SYN: formative. **2** *Physics.* (of a substance) able to be deformed in any direction and to retain its deformed condition permanently without rupture.

plastic deformation or **plastic flow,** *Physics.* the alteration of the shape of a solid by the application of a sufficient and sustained stress: *Plastic deformation of solids is brought about when planes of atoms in the component crystals slip along adjacent planes, in a few directions favored by the crystalline atomic arrangement* (Alan Holden).

plasticity, *n.* **1** *Physics.* the capability of being deformed permanently by externally applied forces without failure.

---

cap, fāce, fäther; best, bē, tèrm; pin, fīve;
rock, gō, ôrder; oil, out; cup, pùt, rüle,
yü in use, *yu̇* in uric;
ng in bring; sh in rush; *th* in thin, ŦH in then;
zh in seizure.

ə = *a* in about, *e* in taken, *i* in pencil, *o* in lemon, *u* in circus

plastid

2 *Chemistry.* a property of particles which permits them to be displaced without removal from the sphere of attraction.

plastid, *n. Biology.* **1** any of various small, often pigmented organelles in the cytoplasm of a plant cell, such as a leucoplast, chromoplast, or chloroplast:
2 an individual mass or unit of protoplasm, such as a cell or one-celled organism.
[from German *Plastid,* from Greek *plassein* to form]

plastral (plas'trəl), *adj.* of, having to do with, or like a plastron.

plastron (plas'trən), *n. Zoology.* the ventral part of the shell of a turtle or tortoise: *The carapace and plastron are joined so that there are openings for the turtle's head, legs, and tail* (Clifford H. Pope). [from Middle French *plastron* breastplate]

-plasty, *combining form.* molding; formation; plastic surgery, as in *mammoplasty, neoplasty, osteoplasty.* [from Greek *-plastia,* from *plastos* molded]

plate, *n.* **1** *Anatomy, Zoology.* a platelike part, organ, or structure, such as a lamina. Some reptiles and fishes have a covering of horny or bony plates.
2 *Electronics.* the positive electrode in a vacuum tube. It is the electrode toward which the electrons flow, originally made in the form of a flat plate, but now usually cylindrical.
3 *Geology.* one of the series of vast segments or thin, mobile blocks that make up the crust of the earth according to the theory of plate tectonics: *Earthquakes occur because each plate is rigid and moves against another plate with great resistance* (Richard M. Pearl). *The plates, which are layers of rock several miles thick and thousands of miles long, are driven sideways by underlying forces which may be convection currents kept in motion by the earth's internal heat* (London Times).

Plate techtonics: (def. 3)
plates of the earth

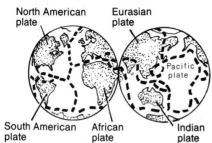

North American plate
Eurasian plate
Pacific plate
South American plate
African plate
Indian plate

plateau (pla tō'), *n., pl.* **-teaus** or **-teaux** (-tōz'). *Geology, Geography.* a large, comparatively level area, especially a plain in the mountains or at a height considerably above sea level: *Geologically the term plateau is restricted to land forms of horizontal structure. But geographically, any area characterized by a fairly level landscape and exhibiting great relief may be regarded as a plateau* (White and Renner, *Human Geography*). Also called **tableland.** [from French *plateau,* Old French *platel,* diminutive of *plat* flat]

platelet, *n. Biology.* one of numerous small disks which float in the blood plasma and are involved in clotting. It is one of the three cellular elements of blood. *Platelets are little fragments of cytoplasm that ... become detached from the cytoplasm of very large cells called megakaryocytes in the bone marrow in such a way that each platelet is completely covered with cell membrane. They have no nuclear components* (A. W. Ham and D. H. Cormack).

plate tectonics, *Geology.* a theory that the earth's crust is divided into a series of vast, platelike parts that move or drift as distinct masses: *Plate tectonics has been adequately tested in other parts of the world and has shown its ability to predict, amongst other things, the direction of the movement accompanying earthquakes* (Science Journal).

platinic (plə tin'ik), *adj. Chemistry.* of or containing platinum, especially with a valence of four.

platinoid (plat'ə noid), *Chemistry.* —*adj.* resembling platinum.
—*n.* any of a group of metals resembling platinum in several of its properties and commonly found in association with it, such as palladium, iridium, and osmium.

platinous, *adj. Chemistry.* of or containing platinum, especially with a valence of two.

platinum, *n. Chemistry.* a heavy, silver-white, precious metallic element with a very high melting point, that is ductile, malleable, resistant to acid, and does not tarnish easily. It is used as a catalyst, for chemical and industrial equipment, in dentistry, etc. *Symbol:* Pt; *atomic number* 78; *atomic weight* 195.09; *melting point* 1772°C; *oxidation state* 2, 4. [from New Latin, from Spanish *platina,* from *plata* silver]

platy (plā'tē), *adj. Mineralogy.* consisting of or readily separable into platelike layers, as mica.

platyhelminth (plat'ə hel'minth), *n.* any of a phylum (Platyhelminthes) of worms having soft, usually flat, bilaterally symmetrical bodies and no body cavity; flatworm. Tapeworms, turbellarians, and flukes belong to this phylum. [from Greek *platys* flat + *helminthos* worm]

playa (plä'yə), *n. Geography, Geology.* **1** the basin floor of an undrained desert which contains water at irregular periods.
2 (in the southwestern United States) a plain of silt or mud, covered with water during the wet season. [from Spanish *playa* beach]

pleiotaxy (plī'ə tak'sē), *n. Botany.* the condition of having more than the usual number of floral whorls. [from Greek *pleiōn* more + *taxis* arrangement]

plateaus
plain
ocean
mountains

pleiotropic (plī'ə trop'ik), *adj. Genetics.* producing change in more than one trait; having several phenotypic effects: *The mutation was pleiotropic* (Scientific American). [from Greek pleiōn more + *tropos* a turning] —**pleiotropically,** *adv.*

—**pleiotropy** (plī ot′rə pē), *n.* or **pleiotropism** (plī ot′rə piz′əm), the condition of being pleiotropic: *Pleiotropism does not interfere with a Mendelian analysis of inheritance* (B. J. Williams).

Pleistocene (plī′stə sēn′), *Geology.* —*n.* the epoch before the present period; the ice age: *During the Pleistocene, when much of the ocean's water was withdrawn from the basins to lie in great ice sheets on the land, extensive parts of the continental shelves must have been exposed to subaerial and glacial erosion* (Birkeland and Larson, *Putnam's Geology*).
—*adj.* of or having to do with the Pleistocene or its deposits: *Alaska has not yet supplied the answer, but mammoth, horse, and bison grazed there in late Pleistocene times; their bones lie in the frozen mud and gravels* (Loren C. Eiseley)
[from Greek *pleistos* most + *kainos* recent]

plenum (plē′nəm), *n., pl.* **-nums, -na** (-nə). *Physics.* **1** a condition in which the pressure of air or other gas within an enclosed space is greater than that of the outside atmosphere. **2** an enclosed space filled with matter: *In a perfect plenum, motion would be impossible* (R. Hall). Contrasted with **vacuum.** [from Latin *plenum* (*spatium*) full (space)]

pleochroic (plē′ə krō′ik), *adj. Optics.* showing different colors because of selective absorption of light when viewed in two or three different directions, as certain double-refracting crystals. A **pleochroic halo** is one of a series of circles of different color found in certain minerals as a result of emission of alpha particles by a constituent radioactive element. [from Greek *pleōn* more + *chroa* color]
—**pleochroism,** *n.* the quality of being pleochroic.

pleomorphic (plē′ə môr′fik), *adj. Biology.* able to change shape, as certain bacteria and protozoa: *Many bacteria are pleomorphic; that is, their cell shape is not always constant. An organism that is rod-shaped under optimal conditions of growth may appear branched, highly elongated, or without definite form under adverse circumstances* (Buffaloe and Ferguson, *Microbiology*). Compare **polymorphic.** [from Greek *pleōn* more + *morphē* form]
—**pleomorphism,** *n.* the condition of being pleomorphic.

pleopod (plē′ə pod), *n.* = swimmeret. [from Greek *plein* to swim + *podos* foot]

plerome (plir′ōm), *n. Botany.* the innermost region of an apical meristem, composed of actively dividing cells: *Procambium or plerome originates the primary or first vascular bundles, the cambium, and sometimes the pith* (Youngken, *Pharmaceutical Botany*). [from Greek *plērōma* a filling, from *plēroun* to fill]

pleura¹ (plùr′ə), *n., pl.* **pleurae** (plùr′ē). *Anatomy.* a thin membrane in the body of a mammal, covering each lung and folded back to make a lining for the thorax or chest cavity: *The exterior of the lungs and interior of the thorax ... are lined by smooth peritoneum, the pleura* (Storer, *General Zoology*). [from Greek *pleura* side, rib]

pleura² (plùr′ə), *n.* plural of **pleuron.**

pleural (plùr′əl), *adj. Anatomy.* of or having to do with the pleura of the lung: *The pleural membrane ... produces pleural fluid which lubricates the surfaces in the regions of contact between the lungs and thorax* (Mackean, *Introduction to Biology*).

pleuron (plùr′on), *n., pl.* **pleura** (plùr′ə). *Zoology.* a lateral part of the body of an arthropod, especially of a thoracic segment of an insect. [from New Latin, from Greek *pleuron* rib, side]

pleuropneumonialike organism, *Bacteriology.* a former name for **mycoplasma.** *Abbreviation:* PPLO

plexiform (plek′sə fôrm), *adj. Anatomy.* shaped like or resembling a plexus. The **plexiform layer** is either one of two layers of the retina made up of interconnected dendrites and functioning as a region of cellular interaction.

plexus (plek′səs), *n., pl.* **-uses** or **-us.** *Anatomy.* a closely interwoven network of nerve fibers, blood vessels, etc., usually named from its relation to or location in some part or organ, as in the *cardiac plexus, gastric plexus, hepatic plexus, plevic plexus,* etc. The solar plexus is a large network of nerves situated at the upper part of the abdomen. [from Latin *plexus,* from *plectere* to twine]

plica (plī′kə), *n., pl.* **-cae** (-sē). *Anatomy.* a fold or folding of skin or membrane: *The mucous membrane lining of the intestinal wall has many plicae* (Carl C. Francis). [Medieval Latin *plica* fold, from Latin *plicare* to fold]
—**plical** (plī′kəl), *adj.* having to do with or resembling a plica.

plicate (plī′kāt), *adj. Botany.* (of a leaf) folded along its ribs like a closed fan: *Birch leaves are plicate, folded several times lengthwise, like a fan* (N.Y. Times). [from Latin *plicatum* folded] —**plicately,** *adv.*
—**plicateness,** *n.*

Pliocene (plī′ə sēn′), *Geology.* —*n.* **1** the latest epoch of the Tertiary period, during which the first manlike apes appeared. The Pliocene was also marked by the rising of mountains in western America, and the migration of mammals between continents. **2** the rocks formed in this epoch.
—*adj.* of or having to do with the Pliocene or its rocks: *The northern part of Yucatán is basically a limestone shelf of Pliocene or post-Pliocene origin, possibly having been formed 40,000 to 50,000 years ago but probably more recently* (Donald Ediger).
[from Greek *pleiōn* more + *kainos* recent]

plot, *n.* **1** a map, chart, or diagram: *the plot of a course or route.*
2 *Mathematics.* a diagram or chart showing the relation between two variable quantities; a graph.
—*v.* **1** to mark (the position of something) on a map, diagram, or chart: *Positions are plotted on a huge plexiglass screen* (Newsweek).
2 *Mathematics.* **a** to determine the location of (a point) by means of its coordinates. **b** to form (a curve) by connecting points marked out on a graph. **c** to represent (an equation or function) by means of a curve drawn through points on a graph.

cap, fāce, fäther; best, bē, tėrm; pin, fīve; rock, gō, ôrder; oil, out; cup, pùt, rüle, *yü* in use, *yu* in uric; *ng* in bring; *sh* in rush; *th* in thin, ᵺ in then; *zh* in seizure.
ə = *a* in about, *e* in taken, *i* in pencil, *o* in lemon, *u* in circus

plug, *n. Geology.* **1** a cylindrical mass of igneous rock formed in the throat or conduit of a volcano: *If the tuff of a cone ... were swept away, we should find a central lava plug* (A. Geikie).
2 = salt dome.

plumage (plü′mij), *n. Zoology.* the covering of feathers of a bird: *The assemblage of feathers on a bird at any one time is called a plumage, and the process of feather replacement is known as molt* (Storer, *General Zoology*). [from Old French *plumage,* from *plume* feather, from Latin *pluma*]

plumate (plü′māt), *adj. Zoology,* resembling a feather, as a hair or bristle that bears smaller hairs, or an insect's antenna covered with fine hairs.

plumbago (plum bā′gō), *n. Mineralogy.* a former name for graphite. [from Latin *plumbago* lead ore, from *plumbium* lead]

plumbic (plum′bik), *adj. Chemistry.* of or containing lead, especially with a valence of four. [from Latin *plumbum* lead]

plumbous (plum′bəs), *adj. Chemistry.* of or containing lead, especially with a valence of two.

plume, n. = mantle plume.

plumose (plü′mōs), *adj. Biology.* **1** having feathers or featherlike growths; feathered: *plumose but flightless birds, plumose antennae.*
2 like a feather; feathery. **—plumosely,** *adv.*
—plumosity (plü mos′ə tē), *n.* the condition of being plumose.

plumule (plü′myül), *n. Biology.* the rudimentary terminal bud of the embryo of a seed, sometimes containing immature leaves. It is situated at the end of the hypocotyl, and is either within or enclosed by the cotyledon or cotyledons. *The terminal bud or epicotyl is ... sometimes known as plumule ... because of the featherlike appearance of the young leaves* (Emerson, *Basic Botany*). See the pictures at **embryo, hypocotyl.** [from Latin *plumula,* diminutive of *pluma* feather]

pluricellular, *adj.* = multicellular. [from Latin *plus, pluris* more + English *cellular*]

pluripotent (plü rip′ə tənt), *adj. Biology.* capable of developing, growing, or producing in a number of ways: *The bone marrow cells are called pluripotent stem cells. ... These cells are thought to give rise to committed stem cells that make red cells, white cells or platelets—or intermediary products thereof* (Joan Arehart-Treichel).
—pluripotency, *n.* the quality or condition of being pluripotent: *This condition of multiple potency on the parts of the early egg-cell has been termed pluripotency* (J. Needham).
—pluripotential, *adj.* = pluripotent.

plus, *prep. Mathematics.* increased by; added to: *3 plus 2 equals 5.*
—adj. 1 *Mathematics.* **a** showing addition. The **plus sign** (+) indicates that the quantity following it is to be added or is a positive quantity. **b** having a plus sign; positive: *a plus quantity.*
2 *Botany.* of or having to do with the strain of heterothallic algae or fungi that acts as the male in reproduction.
—n. Mathematics. 1 = plus sign. **2** a positive quantity. [from Latin *plus* more]

Pluto (plü′tō), *n. Astronomy.* the smallest of the nine major planets in the solar system and the one farthest from the sun. It was discovered in 1930 by the American astronomer Clyde W. Tombaugh. Pluto travels around the sun in an elliptical orbit that takes approximately 248 earth years to complete. It has an estimated diameter of about 3000 kilometers. In 1978 astronomers discovered a satellite of Pluto: *The total mass of both Pluto and its moon, named Charon, is only 0.0017 of the Earth's mass* (Michael J. S. Belton). See the picture at **solar system.** [named after *Pluto,* the Greek and Roman god of the lower world]

pluton (plü′ton), *n. Geology.* a body of plutonic rock: *The intrusion of large plutons may be associated with orogenic movements* (M. P. Billings). [back formation from *plutonic*]

plutonic (plü ton′ik), *adj. Geology.* **1** having to do with a class of intrusive igneous rocks that have solidified far below the earth's surface: *Coarse-grained igneous rocks are called plutonic rocks, and ... there is good reason to believe that the plutonic rocks were not spewed out to the surface like the lavas, but solidified deep underground* (Gilluly, *Principles of Geology*). Contrasted with **volcanic.**
2 Usually spelled **Plutonic.** of or having to do with the theory that the present condition of the earth's crust is mainly due to igneous action.
[from Latin *Pluto, Plutonem* god of the lower world]

plutonium (plü tō′nē əm), *n. Chemistry.* a silvery-white, radioactive, metallic element produced artificially from uranium and found in minute quantities in pitchblende and other uranium ores. It is used as a source of energy in nuclear reactors. Plutonium has six different allotropic forms and about 15 isotopes. The most important is the fissionable isotope Pu-239. *Plutonium has its drawbacks ... it is highly poisonous, thus requiring costly safety measures in plants where it is fabricated and handled ... Plutonium's high toxicity is due to its slight radioactivity. Thus, small bits that lodge in the body could result in radiation-induced cancer* (Wall Street Journal). *Symbol:* Pu; *atomic number* 94; *mass number* (of its most stable isotope) 244; *melting point* 639.5°C; *boiling point* 3235°C; *oxidation state* 3, 4, 5, 6. [from *Pluto,* the planet; patterned on *neptunium* and *uranium*]

pluvial (plü′vē əl), *adj.* **1** *Geology.* of, characterized by, or resulting from prolonged rainfall: *There is evidence of at least two great 'Pluvial' periods of heavy rainfall in the Pleistocene* (W. H. Pearsall).
2 *Meteorology.* of or having to do with rainfall or precipitation: *yearly pluvial measurements.*
—n. Geology. a period of prolonged rainfall; pluvial period: *With the onset of the last pluvial, man was able to penetrate the deserts once more* (A. J. Arkell).
[from *pluvialis,* from *pluvia* rain]

pluviometric (plü′vē ə met′rik), *adj.* of or having to do with pluviometry: *pluviometric readings or observations.* **—pluviometrically,** *adv.*

pluviometry (plü vē om′ə trē), *n. Meteorology.* the measurement of rainfall or precipitation.

Pm, *symbol.* promethium.

PNdB or **PNdb,** *abbrev.* perceived noise decibel.

pneumatic (nü mat′ik), *adj. Zoology.* containing or connected with air cavities, as the bones of birds or the swim bladder of fishes: *a pneumatic appendage,* *process, etc.; the pneumatic duct of fishes.* [from Greek *pneumatikos* of the air or breath, from *pneuma* air, breath, from *pnein* breathe]

pneumatocyst (nü mat′ə sist), *n. Zoology.* an air sac, as in a hydrozoan. [from Greek *pneuma, pneumatos* air, breath + English *cyst*]

pneumatolysis (nü′mə tol′ə sis), *n. Geology.* the alteration of rocks and formation of minerals by the chemical action of vapors given off from igneous magmas. [from Greek *pneuma, pneumatos* air, breath + *lysis* a loosening]
—**pneumatolytic** (nü′mə tō lit′ik), *adj.* of, having to do with, or formed by pneumatolysis: *pneumatolytic action.* —**pneumatolytically,** *adv.*

pneumatophore (nü mat′ə fôr), *n.* **1** *Botany.* a structure supposed to serve as a channel for air, arising from the roots of various trees that grow in swampy places.
2 *Zoology.* a hollow structure containing gas in certain hydrozoans, serving as a float. [from Greek *pneuma, pneumatos* air, breath + English *-phore*]

pneumogastric (nü mə gas′trik), *Anatomy.* —*adj.* **1** of or having to do with the lungs and the stomach or abdomen.
2 of or having to do with a vagus nerve.
—*n.* = vagus nerve.
[from Greek *pneumōn* lung + English *gastric*]

pneumogastric nerve, *Anatomy.* either of the vagus nerves.

pneumostome (nü′mə stōm), *n. Zoology.* a small opening through which air passes to and from the mantle or respiratory cavity of gastropods: *Most land snails have a mantle cavity which acts as a lung, and communicates with the outside through a small aperture called the pneumostome* (New Scientist). [from Greek *pneuma* air + *stoma* mouth]

Po, *symbol.* polonium.

pocket, *n.* **1** *Geology.* **a** a small mass of ore, or the cavity containing it: *manganese ore found in pockets in schistose rock.* **b** a natural underground cavity containing water.
2 *Zoology.* a pouch in an animal body, especially the abdominal pouch of a marsupial or the cheek pouch of a squirrel, chipmunk, etc.

pod, *n. Botany.* the bivalve shell or case in which plants such as beans and peas grow their seeds.

podetium (pō dē′shē əm), *n., pl.* **-tia** (-shē ə). *Biology.* **1** a stalklike or shrubby outgrowth of the thallus of certain lichens, bearing the apothecium or fruiting body.
2 any stalklike elevation.
[from New Latin, from Greek *pous, podos* foot]

podite (pod′īt), *n. Zoology.* a segment of the limb of a crustacean or other arthropod. [from Greek *pous, podos* foot]
—**poditic** (pə dit′ik), *adj.* of or belonging to a podite.

podium (pō′dē əm), *n., pl.* **-dia** (-dē ə) **-diums. 1** *Zoology.* an animal structure that serves as a foot.
2 *Botany.* a footstalk or other supporting part.
[from Latin *podium,* parapet, from Greek *podion* foot of a vase, diminutive of *pous, podos* foot]

podsol (pod′sol), *n.* = podzol.
—**podsolic** (pod sol′ik), *adj.* = podzolic.

podzol (pod′zol) or **podsol** (pod′sol), *n. Geology.* a white or gray acidic soil that is highly leached, found in certain cool, moist climates, especially northern Russia: *Underneath the layer of raw humus the A horizon of*

a mature podzol is leached of its iron and readily soluble minerals, and by eluviation, it has lost most of its clay and colloidal constituents also.* (Finch and Trewartha, *Elements of Geography*). [from Russian *podzol,* from *pod-* under + *zola* ashes]
—**podzolic** (pod zol′ik), *adj.* having to do with or characteristic of a podzol.
—**podzolization** (pod′zə lə zā′shən), *n.* the process of becoming podzolized; formation of a podzol.
—**podzolize,** *v.* to cause to be or resemble a podzol by its having a layer from which the minerals have been leached: *The dunes are partly podzolized, and they sustain pines, dwarfed redwoods, and shrubs* (Scientific American).

poikilotherm (poi′kə lō thèrm′), *n. Zoology.* a cold-blooded animal: *Most living reptiles are poikilotherms ... whose temperatures fluctuate with that of the environment* (John H. Ostrom). Also called **ectotherm.** [from Greek *poikilos* variegated + *thermē* heat]
—**poikilothermal,** *adj.* = poikilothermic.

poikilothermic or **pikilothermous,** *adj. Zoology.* having a body temperature that varies with the environment; cold-blooded. Contrasted with **homoiothermic.**

point, *n.* **1** *Mathematics.* something that has position but not extension. Two lines meet or cross at a point.
2 *Geography.* **a** each of the 32 positions indicating direction marked at the circumference of the card of a compass. **b** the interval between any two adjacent points of a compass; 11 degrees 15 minutes.
3 *Geology.* a piece of land with one end extending out into the water; cape.

point charge, *Physics.* a charge concentrated at one point. For simplicity of calculation, charged objects may be treated as point charges when the objects are small compared to the distance between them.

point of fusion, = melting point.

point of tangency, *Geometry.* the point where a straight line touches a circle: *The radius at the point of tangency makes a right angle with the tangent* (Rothwell Stephens).

points of the compass, *Geography.* the 32 directions marked on a compass. North, south, east, and west are the four main, or cardinal, points of the compass.

point source, *Physics.* a source of light waves, radio waves, or other electromagnetic radiation which is so highly concentrated that it can be considered to come from a single point.

poise (poiz), *n.* a unit of measurement of viscosity in the CGS system: *The unit of viscosity is that of force times distance divided by area times velocity, or, in the cgs system, 1 dyne-sec/cm². A viscosity of 1 dyne-sec/cm² is called a poise* (Sears and Zemansky, *University Physics*). [from the name of Jean Louis Marie Poiseuille, 1799–1869, French physiologist]

---

cap, fāce, fäther; best, bē, tėrm; pin, fīve;
rock, gō, ôrder; oil, out; cup, pùt, rüle,
yü in use, *yu̇* in uric;
ng in bring; *sh* in rush; *th* in thin, ᴛʜ in then;
zh in seizure.
ə = *a* in about, *e* in taken, *i* in pencil, *o* in lemon, *u* in circus

Poisson distribution (pwä sôn′), *Statistics.* a distribution differing from the normal curve that can be applied to distributions that are not continuous, as when the variable cannot have all values but is limited to particular values. The Poisson distribution is often used in bacteriological experiments when the variable is limited to the number of cells in an area. [named after Denis *Poisson,* 1781–1840, French mathematician]

Poisson ratio. *Physics.* a measure of the elasticity of a material, equal to the ratio of the lateral strain to the longitudinal strain when the material is subjected to tensile stress.

polar, *adj.* **1** *Geography.* of or near the North or South Pole: *the polar regions, a polar wind.* A **polar orbit** is an orbit of a satellite that passes over or near the poles of a planet instead of the equator.
2 *Physics.* of or having to do with the poles of a magnet, electric battery, etc.: *the polar region of a magnet.*
3 *Chemistry.* **a** having to do with or characterized by a dipole: *polar molecules. Since the bonds between the atoms of the water molecule are both polar and covalent, they are often called polar covalent* (A. B. Garrett). **b** ionizing when dissolved or fused: *polar bonds or linkages.* SYN: ionic.
4 *Geometry.* having to do with or reciprocal to a pole.
5 *Biology.* of or having to do with the poles of a nerve cell, an ovum, etc.: *polar nuclei, the polar points of the spindle.*

polar axis, *Optics, Astronomy.* the axis in the mounting of an equatorial telescope that is parallel to the earth's axis.

polar body or **polar cell,** *Biology.* one of the small cells that arise from an oocyte during meiotic division: *As a result of meiosis, four haploid cells are produced: three small polar bodies, which eventually disintegrate, and one large egg, which will be functional* (McElroy, *Biology and Man*).

polar cap, *Astronomy.* a region of ice around the pole of a planet, especially Mars: *A major Viking [spacecraft] discovery was that the residual polar caps—the small caps that remain each summer when the caps of large extent have vanished—are of water ice* (Wernher Von Braun and Frederick I. Ordway).

polar circle, *Geography.* either of two circles of the earth parallel to the equator, one of which is everywhere distant 23 degrees 30 minutes (23°30′) from the North Pole and the other equally distant from the South Pole; the Arctic or Antarctic Circle: *Geography teaches that the polar zones are separated from the temperate zones by the Polar Circles ... these being the lines beyond which during the summer the sun never disappears* (Gabriele Rabel).

polar coordinate, *Mathematics.* either one of the two coordinates for determining the position of a point in a plane. The two coordinates are the length of the line segment drawn to the point from a fixed point and the angle which this line segment makes with a fixed line or axis.

polar distance, *Astronomy.* the complement of the declination of a celestial body. Also called **codeclination.**

polar easterly, *Meteorology.* a prevailing wind which blows from east to west between each of the poles and about 60 degrees north or south latitude.

polar front, *Meteorology.* the boundary region between the cold polar winds and the warmer winds of tropical origin: *Irregularities of flow of the nature of waves along the polar front are thought to initiate depressions* (Blair, *Weather Elements*).

polarimeter (pō′lə rim′ə tər), *n.* an instrument for measuring the polarization of light.

polarimetric (pō′lər ə met′rik), *adj. Optics.* of or having to do with polarimetry.

polarimetry (pō′lə rim′ə trē), *n. Optics.* the process of measuring or analyzing the polarization of light: *With the proper filter to transmit this line alone, the mercury arc is used extensively as a source for such purposes as polarimetry* (Hardy and Perrin, *Principles of Optics*).

polarity, *n. Physics.* **1** the possession of two opposed poles. A magnet or battery has polarity: *The polarity of the earth's magnetic field reverses itself in cycles of 30,000 to more than 2,000,000 years* (Walter Sullivan). **2** a positive or negative polar condition, as in electricity: *Reversing the polarity of an applied voltage, for instance, means interchanging the positive and negative conditions of the terminals to which it is applied* (Roy F. Allison).

polarizability (pō′lə rī′zə bil′ə tē), *n.* the quality of being polarizable.

polarizable (pō′lə rī′zə bəl), *adj.* capable of being polarized.

polarization, *n.* **1** *Optics.* **a** a state, or the production of a state, in which rays of light exhibit different properties in different directions, as when they are reflected from glass in a particular way, or when they are passed through a crystal of tourmaline that confines the light vibrations to a single plane: *Sky polarization depends on the angle between the sun's rays to a particular point in the sky and an observer's line of sight to that point* (Scientific American). **b** the state in which all the vibrations of the waves take place in only one plane: *All light sources of practical importance emit radiation that exhibits little evidence of polarization. For this reason, the optical instruments based on the phenomena of polarization are provided with some element, such as a Nicol prism, for polarizing the natural light supplied by the source* (Hardy and Perrin, *Principles of Optics*).

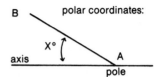

The polar coordinates of point B are the length of line AB and the size of angle X.

2 *Physics.* **a** the process by which gases produced during electrolysis rise to a reverse electromotive force. **b** the acquisition of electric charges of opposite sign: *Many crystalline substances exhibit "polarization", an effect where electric charges of opposite sign appear on opposite faces of the crystal* (New Scientist). **c** the alignment of the spin axes of elementary partiicles.

polarize, *v. Optics, Physics.* to cause polarization in: *Some natural crystals, such as tourmaline, can polarize light. Tourmaline transmits the components that lie in one vibration-direction and holds back others by absorbing them internally* (Richard T. Kriebel).

polarizing angle, *Optics.* the angle of incidence at which the maximum polarization of incident light takes place. Also called **Brewster angle.**

polaron (pō'lə ron), *n. Physics.* a conducting electron trapped by polarization charges in an ionic crystal lattice: *... the excitation known as a polaron, the presence of which can be detected by irregularities in the shape of a conduction band* (Nature). [from *polar*(*ization*) + *-on*]

polar valence, = electrovalence.

pole¹, *n.* **1** a measure of length; rod; 5 1/2 yards or 5.0292 meters. **2** a measure of area; square rod; 30 1/4 square yards or 25.289 square meters. [Old English *pal,* from Latin *palus* stake]

pole², *n.* **1** *Astronomy.* **a** either end of the earth's axis: *The spinning of the earth on an axis establishes two fixed points on the earth's surface, known as the poles* (White and Renner, *Human Geography*). **b** either celestial pole.
2 *Physics.* each of the two opposite points on a magnet at which the magnetic forces are manifested: *If a short magnet is placed in a uniform magnetic field equal forces act on its poles in opposite directions, tending to twist it into line with the field* (J. Little).
3 *Geometry.* **a** either end of the axis of any sphere. **b** the origin or fixed point in a system of polar coordinates. See the picture at **polar coordinate.**
4 *Biology.* **a** each extremity of the main axis of an organism, nucleus, or cell, especially an egg cell. **b** each extremity of the spindle formed in a cell during mitosis. **c** the point on a nerve cell where a process originates.
[from Latin *polus* end of an axis, the sky, from Greek *polos*]

pole of cold, *Geography.* the place in either polar region where the lowest winter temperature occurs.

pole of inaccessibility, *Geography.* the point in either polar region that is farthest from the coast in every direction.

poliovirus, *n. Biology.* any of a group of viruses containing ribonucleic acid that cause polio. Compare **enterovirus.**

pollen, *n. Botany.* a fine, yellowish powder consisting of grains or microspores, each of which contains a mature or immature male gametophyte. In flowering plants, pollen is released from the anthers of flowers and fertilizes the pistils. *In the great forests of pine and other evergreen trees the pollen is so abundant at shedding time that it causes the appearance of a yellow haze in the air* (Emerson, *Basic Botany*). [from Latin *pollen* fine flour]

pollen sac, *Botany.* the baglike structure in the anther of a flower in which the pollen is produced.

pollen tube, *Botany.* a tube that grows from a pollen grain. Male reproductive cells move through the pollen tube into the ovule. *Following pollination, the pollen grain germinates on the stigma and forms a pollen tube which extends into the ovule* (Biology Regents Syllabus).

pollex (pol'eks), *n., pl.* **pollices** (pol'ə sēz). *Anatomy.* the innermost digit of the forelimb; the thumb or a part corresponding to it. [from Latin *pollex* thumb, big toe]

pollinate, *v. Botany.* to transfer pollen from its site of formation to a receptive surface where it may germinate.

pollination (pol'ə nā'shən), *n. Botany.* the act or process of pollinating. For flowering plants, pollination involves the transfer of pollen from the anthers to the stigmas of flowers for fertilization, as by insects or the wind. *In the case of insect pollination, adaptation may be very complex, indicating a long association between insect vector and plant. ... Wind pollination is the common type in plants with inconspicuous flowers, as in grasses* (Weier, *Botany*). Compare **cross-pollination, self-pollination.**
—**pollinator,** *n.* an insect or other agent that pollinates plants: *In the long course of evolution the flowers of plants have become adapted through natural selection to the characteristics of their pollinators* (Verne Grant).

SELF-POLLINATION   CROSS-POLLINATION

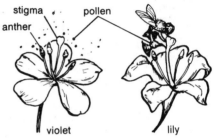

violet   lily

pollinium (pə lin'ē əm), *n., pl.* **-ia** (-ē ə). *Botany.* an agglutinated or coherent mass or body of pollen grains, characteristic of plants of the milkweed and orchid families. [from New Latin, from *pollen, pollinem* fine flour]

pollutant, *n. Ecology.* a polluting agent or medium: *a water pollutant, an air pollutant.* ► See the note under **contaminant.**

pollute, *v. Ecology.* to make impure by the introduction of harmful or hazardous matter into the environment. Compare **contaminate.**
—**pollution,** *n.* the act or process of polluting part of the environment: *the pollution of water by industrial wastes, air pollution.*

polonium (pə lō'nē əm), *n. Chemistry.* a radioactive metallic element found in pitchblende or produced artificially by bombarding bismuth with neutrons. It decays into an isotope of lead by the emission of alpha particles. *Symbol:* Po; *atomic number* 84; *mass number* (of

cap, fāce, fäther; best, bē, tèrm; pin, five;
rock, gō, ôrder; oil, out; cup, pùt, rüle;
yü in use, *yu̇* in uric;
ng in bring; *sh* in rush; *th* in thin, ŦH in then;
zh in seizure.
ə = *a* in about, *e* in taken, *i* in pencil, *o* in lemon, *u* in circus

poly-

its most stable isotope) 209; *oxidation state* 2, 4. [from *Polonia,* Latin name of Poland, homeland of Marie Curie, who discovered this element with Pierre Curie in 1898]

poly-, *combining form.* **1** more than one or several; many; multi-, as in *polynomial.* **2** *Chemistry.* polymeric, as in *polyester, polystyrene.* [from Greek *poly-,* from *polys* much, many]

polyacid, *adj. Chemistry.* equivalent in combining capacity to an acid radical of valence greater than unity as a base.

polyadelphous (pol'ē ə del'fəs), *adj. Botany.* having the stamens united in three or more bundles or groups. [from *poly-* many + Greek *adelphos* brother]

polyadenylic acid (pol'ē ad'n il'ik), *Biochemistry.* a polynucleotide formed from adenylic acid, found in various forms of RNA: *Poliovirus virion RNA contains a single covalently bound sequence of polyadenylic acid which is approximately 49 nucleotides long* (Science).

polyamide (pol'ē am'īd), *n. Chemistry.* a compound containing two or more amide (-CONH-) radicals, especially a polymeric amide: *Polyamides rank first from the industrial point of view, their most conspicuous class being the nylons* (Science News).

polyamine (pol'ē am'ēn *or* pol'ē ə mēn), *n. Chemistry.* a compound containing more than one amino group.

polyandrous (pol'ē an'drəs), *adj.* **1** *Botany.* having many stamens.
2 *Zoology.* of or characterized by polyandry. Compare **polygynous.**
[from *poly-* many + Greek *andros* man, husband]

polyandry (pol'ē an'drē), *n.* **1** *Zoology.* a mating pattern in which a single female mates with more than one male.
2 *Botany.* the condition of being polyandrous.

polyatomic, *adj. Chemistry.* containing more than two atoms, especially many replaceable hydrogen atoms: *One of the highlights of molecular astronomy ... was the discovery of the first polyatomic organic molecule in interstellar space, formaldehyde* (Science News).

polybasic, *adj. Chemistry.* (of an acid) having two or more hydrogen atoms that can be replaced by basic atoms or radicals.

polybasite (pol'ē bā'sīt), *n. Mineralogy.* a blackish mineral with a metallic luster, consisting essentially of silver, sulfur, and antimony. It is a valuable silver ore. *Formula:* (Ag, Cu)$_{16}$Sb$_2$S$_{11}$ [from German *Polybasit,* from *poly-* many + *Base* base + *-it* -ite]

polycarpic (pol'ē kär'pik), *adj.* = polycarpous.

polycarpous (pol'ē kär'pəs), *adj. Botany.* consisting of many or several carpels.

polychasial (pol'ē kā'zē əl), *adj.* of or having to do with a polychasium.

polychasium (pol'ē kā'zē əm), *n., pl.* **-sia** (-zē ə). *Botany.* a form of cymose inflorescence in which each axis produces more than two lateral axes. [from New Latin, from *poly-* many + Greek *chasis* division]

polychromatic, *adj. Optics.* having many or various colors: *Dispersion is the separation of polychromatic light into its component wavelengths* (Physics Regents Syllabus).

polyconic projection, *Geography.* a system of map projection in which each parallel of latitude is represented as if projected on a cone touching the earth's surface along that parallel.

polycotyledon (pol'ē kot'l ēd'n), *n. Botany.* a plant of which the seed contains more than two cotyledons: *Nearly all cone-bearing trees, such as pine, spruce, and hemlock, are polycotyledons, because their seeds have many seed leaves* (Vernon Quinn).
—**polycotyledonous** (pol'ē kot'l ēd'n əs), *adj.* having more than two cotyledons in the seed, as many gymnosperms.

polycrystalline, *adj. Crystallography.* **1** composed of many crystals: *Most solids are polycrystalline.*
2 composed of crystals with different space lattices: *a polycrystalline metal.*

polycyclic, *adj.* **1** *Biology.* having many rounds, turns, or whorls, as a shell.
2 *Electricity.* having many cycles or circuits.
3 *Chemistry.* having more than one ring in a molecule: *polycyclic hydrocarbons. He has found in the tar no fewer than 17 hydrogen-carbon compounds of the polycyclic group* (i.e., *with several carbon rings in the molecule*) (*Time*). Compare **polynuclear.**

polydactyl (pol'ē dak'təl), *Zoology.* —*adj.* having many or several fingers or toes, especially more than the normal number.
—*n.* a polydactyl animal.

polydisperse, *adj. Chemistry.* having to do with or made up of heterogeneous particles, as in a colloidal system.

polyembryony (pol'ē em'brē ə nē *or* pol'ē em'bri'ə nē), *n. Biology.* the formation or development of more than one embryo from a single seed or egg: *Some parasitic insects furnish examples of what is known as polyembryony; each of their eggs produces not one but many larvae, as many as 395 having been reported from a single egg* (Hegner and Stiles, *College Zoology*).

polyene (pol'ē ēn), *n. Chemistry.* any of a group of compounds characterized by a number of double bonds: *a cyclic polyene.*

polyethylene glycol, *Chemistry.* any of various colorless and odorless polymers of ethylene glycol. Polyethylene glycols are used as solvents, intermediates, waxes, etc. *Formula:* HOCH$_2$(CH$_2$OCH$_2$)$_n$CH$_2$OH

polygamous (pə lig'ə məs), *adj.* **1** *Botany.* bearing both unisexual and hermaphrodite flowers on the same plant or on different plants of the same species: *A polygamous plant, such as certain kinds of ash, buckwheat, and maple, has both perfect and imperfect flowers on the same plant* (Harold N. Moldenke).
2 *Zoology.* **a** having several mates at the same time: *The war is perhaps, severest between the males of polygamous animals* (Darwin, *Origin of Species*). **b** characterized by polygamy: *a polygamous species.*
—**polygamously,** *adv.*
—**polygamy** (pə lig'ə mē), *n.* the condition of being polygamous.

polygene (pol'ē jēn'), *n. Genetics.* any of a group of genes whose individually small but cumulatively powerful operation controls a complex hereditary trait, such as height: *Most variation in human populations clearly is the product of polygenes, that is systems of related genes no one of which individually has a discernible effect but which in their totality may produce* *a wide range of variation* (Ralph Beals and Harry Hoijer).

polygenesis, *n. Biology.* the theory that a species or other group originated from several independent ancestors or germ cells: *A second debate revolved around the supposed origin of living organism: the argument of monogenesis v. polygenesis. The polygenesist's position was that "savages" were descended from creatures separately from the progenitors of "civilized" men* (Paul Bohannon).
—polygenetic, *adj.* **1** *Biology.* **a** of or having to do with polygenesis: *a polygenetic theory of origin.* **b** = polyphyletic.
2 *Geology.* having more than one origin; formed in several different ways or at several different times: *a polygenetic topography.* **A polygenetic volcano** is one formed by several volcanic eruptions.

polygenic (pol′ē jĕ′nik), *adj. Genetics.* of or having to do with polygenes. **Polygenic inheritance** is the presence of complex hereditary characters that have developed from a relatively large number of genes operating collectively.

polygon (pol′ē gon), *n. Geometry.* a closed plane figure with three or more sides and angles: *According to a theorem first proved by the great German mathematician David Hilbert, any polygon can be transformed into any other polygon of equal area by cutting it into a finite number of pieces* (Martin Gardner). [from *poly-* many + Greek *gōnia* angle]
—polygonal (pə lig′ə nəl), *adj.* **1** having the form of a polygon; many-sided: *polygonal figures.*
2 formed of or containing polygons: *The polygonal pattern ... is typical of permanently frozen ground* (James A. West).

polygynous (pə lij′ə nəs), *adj.* **1** *Botany.* having many pistils or styles. Compare **polyandrous. 2** *Zoology.* of or characterized by polygyny. [from *poly-* many + Greek *gynē* woman, wife]

polygyny (pə lij′ē nē), *n.* **1** *Zoology.* a mating pattern in which a single male mates with more than one female: *In an evolutionary sense the single male (harem) and multimale polygynous systems are undoubtedly very close. While monogamy and polygamy, for example, are never found in the same genus, both single and multimale groups are found in several genera* (Terborgh, *Five New World Primates*).
2 *Botany.* the condition of being polygynous.

polyhalite (pol′ē hal′īt), n. *Mineralogy.* a mineral consisting essentially of a hydrated sulfate of calcium, magnesium, and potassium, occurring usually in fibrous masses, and of a brick-red color caused by the presence of hematite. It is an ore of potassium. *Formula:* $K_2MgCa_2(SO_4)_4 \cdot 2H_2O$ [from German *Polyhalit,* from *poly-* many + Greek *hals* salt + German *-it* -ite]

polyhedral, *adj. Geometry.* **1** of the form of a polyhedron; having many faces or sides: *a polyhedral figure.*
2 formed by three or more planes meeting at a point: *a polyhedral angle.*

polyhedron, *n., pl.* **-drons, -dra** (-drə). *Geometry.* a solid figure having many (four or more) plane faces; a many-sided solid. [from *poly-* many + Greek *hedra* seat, side]

polyhydroxy (pol′ē hī drok′sē), *adj. Chemistry.* having several hydroxyl (-OH) radicals.

polymer (pol′ə mər), *n. Chemistry.* a compound formed by polymerization. Starch, cellulose, and proteins are natural polymers; nylon and polyethylene are synthetic polymers. A polymer is essentially a large molecule made up by the linking of many smaller molecules or monomers. An **addition polymer** is one formed by the direct linkage of monomer units without the removal of any atoms; a **condensation polymer** is formed when small molecules are removed from the monomers during linkage, such as a water molecule for each linkage of two amino acids. Organic polymers include polysaccharides, polypeptides, and hydrocarbons. Compare **copolymer, homopolymer.** [from *poly-* many + Greek *meros* part]

polymerase (pol′ə mə rās′), *n. Biochemistry.* an enzyme that polymerizes nucleotides to form nucleic acid. Compare **DNA polymerase, RNA polymerase.**

polymeric (pol′ē mer′ik), *adj. Chemistry.* of, having to do with, or characteristic of a polymer; composed of a polymer or polymers: *polymeric hydrocarbons. Genetics tells us that the information that directs amino acid sequences in proteins is carried by long-chain polymeric molecules, the deoxyribonucleic acids (DNA)* (James D. Watson). **—polymerically,** *adv.*

polymerization (pol′ə mər ə zā′shən *or* pə lim′ər ə zā′shən), *n. Chemistry.* **1** a reaction in which many small molecules unite to form a large or more complex molecule with a higher molecular weight and different chemical properties, as in the formation of polypeptides from amino acids or of polyethylene glycol from ethylene glycol: *Another type of plastic can be made by polymerization, a reaction in which unsaturated molecules join together by a shifting of bonds to form large molecules, without splitting out any by-products* (Offner, *Fundamentals of Chemistry*).
2 the conversion of one compound into another by such a process: *the polymerization of acetylene to benzene.*

polymerize (pol′i mə rīz′ *or* pə lim′ə rīz′), *v. Chemistry.* to make or become polymeric; undergo or cause to undergo polymerization: *Formaldehyde, which can be prepared cheaply from natural gas or coal, can ... be polymerized into a tough and durable plastic* (Scientific American).

polymerous (pə lim′ər əs), *adj.* **1** *Biology.* composed of many parts, members, or segments.
2 *Botany.* having many members in each whorl: *a polymerous flower.*

polymorph (pol′ē môrf), *n.* **1** *Biology.* a polymorphic organism.
2 *Crystallography.* one of the polymorphic crystal forms of a substance.
[from *poly-* many + Greek *morphē* form]

---

cap, fāce, fäther; best, bē, tèrm; pin, five;
rock, gō, ôrder; oil, out; cup, pùt, rüle,
yü in use, *yù* in uric;
ng in bring; *sh* in rush; *th* in thin, ᴛʜ in then;
zh in seizure.
ə = *a* in about, *e* in taken, *i* in pencil, *o* in lemon, *u* in circus

—**polymorphic**, *adj.* characterized by polymorphism; having, assuming, or passing through many or various forms, stages, etc.: *Species which, like Obelia, exhibit several forms of body are said to be polymorphic* (*literally of many forms*). *In Obelia, as in many other hydroids, polymorphism is accompanied in the life cycle by an alternation of asexual and sexual reproduction* (Schull, *Principles of Animal Biology*). Compare **pleomorphic.**

—**polymorphism**, *n.* **1** *Biology.* **a** the existence of different forms or types, as in a species, genus, etc., independent of sexual variations. **b** the different forms or types encountered: *There are many differences among individuals that are totally under genetic control, that is, they are not subject to even the small physiological adaptation mentioned for skin color. These genetic differences are called genetic polymorphisms when the alternative versions of the genes determining them each occur within a population with a substantial frequency* (Scientific American). **c** the occurrence of more than one independent stage in the life cycle of a species.
2 *Chemistry.* the occurrence of a substance in two or more forms; allotropy.
3 *Crystallography.* the property of crystallizing into two or more forms; occurrence of crystals that are structurally different but chemically identical.

—**polymorphous**, *adj.* = polymorphic.

polynomial (pol'ē nō'mē əl), —*n.* **1** *Mathematics.* an algebraic expression consisting of one or more terms; a monomial or the sum of monomials. EXAMPLES: x^2y, $ab + x^2y$, and $pq - p^2 + q$ are polynomials. Compare **binomial, trinomial.**
2 *Biology.* the scientific name of a variety, species, or the like, consisting of more than two terms. Compare **binomial, monomial.**

—*adj. Biology.* consisting of more than two terms, as those indicating the genus, species, subspecies, variety, etc: *Modern humans are sometimes distinguished from their ancestors by the polynomial name Homo sapiens sapiens.*
[from *poly-* many + (*bi*)*nomial*]

polynuclear, *adj.* **1** *Biology.* having several nuclei. SYN: multinucleate.
2 *Chemistry.* **a** having two or more closed rings in a molecule: *polynuclear hydrocarbons.* Compare **polycyclic. b** containing more than one central atom: *a polynuclear metal complex.*

polynucleotide, *n. Biochemistry.* a polymeric compound consisting of a number of nucleotides; a nucleic acid composed of several mononucleotides: *The nucleic acids are made by joining up four kinds of nucleotide to form a polynucleotide chain* (Francis H. C. Crick).

polynya or **polynia** (pol'ən yä'), *n. Oceanography.* a space of open water in the midst of ice, especially in the arctic seas. [from Russian *polyn'ya*]

polyoma (pol'ē ō'mə), *n. Microbiology.* a virus containing DNA, associated with various tumors formed in rodents. It is a form of papovavirus. *A number of viral agents of cancer in lower animals, such as the polyoma virus in mice ... provide models for study of related human diseases* (James G. Shaffer). [from New Latin, from *poly-* many + *-oma*]

polyp (pol'ip), *n.* **1** *Zoology.* a small cnidarian attached at the base of its tubular body, with a mouth at the other end surrounded by fingerlike tentacles to gather in food. Polyps often grow in colonies, and include hydras, sea anemones, and corals. Some polyps produce a swimming stage called a medusa. *Living coral, the polyps that form the limestone coral reefs, actually make up only a very thin crust on the surface of the reefs* (Science News).
2 *Anatomy, Medicine.* a rounded or oblong outgrowth from the mucosal surface of a vocal cord, nasal passageway, colon, etc.
[from Greek *polypous*, from *poly-* many + *pous* foot]

polypedon (pə lip'ə don), *n. Geology, Geography.* a group of pedons.
ASSOCIATED TERMS: see **soil.**

polypeptide (pol'ē pep'tīd), *n. Biochemistry.* a peptide containing many molecules of amino acids: *In the living cell polypeptides, which are built up from amino acid molecules, form the principal chains in the structure of proteins* (Scientific American).

polyphagia (pol'ē fā'jē ə), *n.* = polyphagy.

polyphagy (pə lif'ə jē), *n. Zoology.* the habit of feeding on many different kinds of food: *Predaceous insects, spiders, birds and so on, exhibiting various degrees of polyphagy* (New Biology). [from Greek *polyphagia*, from *polyphagos* voracious, from *polys* much + *phagein* to eat]

—**polyphagous** (pə lif'ə gəs), *adj.* eating many different kinds of food: *Finally, as a corollary of the above two tendencies, they suggested that polyphagous aphids (feeding on a wide variety of plants with no particular favourites) would be more successful vectors than those feeding on a few species or only one* (New Scientist).

polyphyletic (pol'ē fī let'ik), *adj. Biology.* having to do with or descending from more than one ancestral form, as a group of animals: *A polyphyletic or multiple origin of modern man ... cannot yet be definately excluded as a possibility but is is very unlikely in view of what we know about the evolution of other mammals* (Ralph Beals and Harry Hoijer). Contrasted with **monophyletic.** Compare **polygenesis. —polyphyletically,** *adv.*

polyphyllous (pol'ē fil'əs), *adj. Botany.* **1** having distinct or separate leaves. **2** having or consisting of many leaves. [from Greek *polyphyllos* having many leaves, from *poly-* many + *phyllon* leaf]

polyploid (pol'ē ploid), *Genetics.* —*adj.* having three or more sets of chromosomes: *Polyploid plants ... tend to be larger than the original diploids from which they are derived, and this has been attributed to increased water content of their cells* (L. S. Penrose).
—*n.* a polyploid organism or cell: *Most animal polyploids reproduce either hermaphroditically ... or parthenogenetically* (J. B. Jenkins).
[from *poly-* many + *-ploid,* as in *haploid, diploid*]

—**polyploidy** (pol'ē ploi'dē), *n.* the condition of being polyploid: *Polyploidy is the condition in which organisms display a multiple of their basic chromosome number. Induction of polyploidy by use of the chemical colchicine has become important in the creation of new varieties* (Tom Stevenson and John G. S. Marshall).

polypoid (pol′ē poid), *adj. Zoology.* of or resembling a polyp.

polyribonucleotide, *n. Biochemistry.* a polynucleotide consisting of a number of ribonucleotides.

polyribosomal (pol′ē rī′bə sō′məl), *adj.* of or having to do with polyribosomes: *In the unaltered cytoplasm surrounding these areas ribosomes were gathered into polyribosomal aggregates, indicating very active synthesis of protein* (Science Journal).

polyribosome (pol′ē rī′bə sōm), *n. Biochemistry.* a cluster of ribosomes that are linked by messenger RNA and function as a unit in synthesizing proteins: *During the process of active protein synthesis, the m-RNA may accommodate as many as forty 70S ribosomes, forming a structure called a polyribosome or polysome. Present indications are that the polysome is formed by the attachment of individual ribosomes to the m-RNA as the latter is synthesized* (Paul S. Sypherd).

polysaccharide (pol′ē sak′ə rīd), *n. Biochemistry.* any of a class of carbohydrates that can be decomposed into two or more monosaccharides by hydrolysis: *When many molecules of simple sugar are combined ... a polysaccharide is produced* (Shull, *Principles of Animal Biology*).

polysepalous (pol′ē sep′ə ləs), *adj. Botany.* **1** having the sepals distinct or separate. **2** having numerous sepals.

polysome (pol′ē sōm), *n.* = polyribosome.

polysomic (pol′ē sō′mik), *Biology.* —*adj.* **1** having more than the normal number or sets of chromosomes: *Polysomic forms arise in a diploid through two daughter chromosomes passing to the same pole at mitosis, or at meiosis* (C. D. Darlington). **2** = polyploid.
—*n.* a polysomic organism or cell.

polyspermous (pol′ē spėr′məs), *adj. Botany.* containing or producing many seeds.

polysynaptic, *adj. Anatomy.* of or involving two or more synapses: *a polysynaptic response, polysynaptic reflexes.*

polytene (pol′ē tēn), *adj. Biology.* **1** having a structure made up of many strands like a rope: *Polytene chromosomes develop by means of a process in which the chromosomes repeatedly replicate but do not separate* (Scientific American). **2** of or having to do with polytene chromosomes: *polytene threads.* [from *poly-* many + Greek *tainia* band]

polytypic, *adj. Biology.* having several variant forms, especially subspecies or varieties: *Homo sapiens would be classified as a polytypic species, in which the same grade was attained independently by different hominid groups* (Margaret Mead). Compare **monotypic.**

polyunsaturated (pol′ē un sach′ə rā′tid), *adj. Chemistry.* (of a fatty acid or oil) having many double bonds; highly unsaturated, as many vegetable oils: *Polyunsaturated fats (oils) are liquid at room temperature and do not appear to be linked to cardiovascular diseases* (Biology Regents Syllabus).
—**polyunsaturate,** *n.* a polyunsaturated substance.

polyvalence, *n.* = multivalence.
—**polyvalent,** *adj.* = multivalent.

pome (pōm), *n. Botany.* a simple fleshy fruit whose outer portion is formed by the floral parts surrounding the ovary. Its core contains several seeds enclosed in a capsule. Pomes are characteristic of the rose family, to which the apple, pear, and quince belong. *In strawber-*

population

ry, the matured ovary is an achene; in apples and pears, it is a pome; in peaches, apricots, blackberries, and cherries, it is a drupe (Weier, *Botany*). See the picture at **fruit.** [from Latin *pomum* fruit, apple]

pomology (pō mol′ə jē), *n.* the branch of science that deals with fruits and fruit growing. [from New Latin *pomologia,* from Latin *pomum* fruit + *-logia* -logy]

pond, *n. Geography, Ecology.* a body of still, fresh water, usually smaller than a lake and shallow enough for sunlight to reach the bottom: *In many regions, ponds have a great variety of animal and plant life. ... All the living things in a pond depend on one another in some way* (George K. Reid). [originally variant of *pound* enclosed place, Old English *pund-*]

pongid (pon′jid), *Zoology.* —*adj.* of or having to do with the family (Pongidae) of apes that includes the chimpanzee, gorilla, and orangutan.
—*n.* a pongid ape: *Hominids have smaller teeth than pongids, especially the canines and incisors* (Ian Tattersall).
[of African origin; compare Kikongo (a Bantu language) *mpongo, mpongi* ape]

pons (ponz), *n., pl.* **pontes** (pon′tēz). *Anatomy.* **1** a part that connects two parts, as a bridge of connective tissue within an organ. **2** = pons Varolii. [from Latin *pons* bridge]

pons Varolii (ponz və rō′lē ī), *Anatomy.* a band of nerve fibers in the brain, just above the medulla oblongata, consisting of transverse fibers connecting the two lobes of the cerebellum, and longitudinal fibers connecting the medulla with the cerebrum: *The pons varolii is a passageway for nerve impulses which pass between the various parts of the brain* (Foster Kennedy). See the picture at **brain.** [from New Latin *pons Varolii* bridge of Varolii, named after Costanzo *Varolii,* about 1543–1575, Italian anatomist]

popliteal (pop lit′ē əl or pop′lə te′əl), *adj. Anatomy.* of, having to do with, or in the hollow part of the leg back of the knee: *Vascular surgeons are now frequently called on to deal with popliteal aneurysms and with ... popliteal occlusions* (Michael E. DeBakey and L. Engel). [from New Latin *popliteus (musculus*) (muscle) of the ham, from Latin *poplitem* ham, back of the knee]

population, *n.* **1** *Biology.* **a** the aggregate of organisms which inhabit a particular area or region: *An expanding population could very easily lead to overcrowding* (Otto and Towle, *Modern Biology*). *In the last two centuries, world population has increased more than four times ... By the end of this century, the figure is expected to exceed 6 billion* (Peter Matthiessen). **b** a (specified) portion of such an aggregate; a group of organisms of the same kind: *the deer population of*

---

cap, fāce, fäther; best, bē, tėrm; pin, fīve;
rock, gō, ôrder; oil, out; cup, pùt, rüle;
yü in use, yu in uric;
ng in bring; sh in rush; th in thin, ŦH in then;
zh in seizure.
ə = *a* in about, *e* in taken, *i* in pencil, *o* in lemon, *u* in circus

North America, the fish and insect populations of a lake.

2 *Statistics.* the entire group of items or individuals from which the samples under consideration are presumed to come: *The measurements involved in every scientific experiment constitute a sample of that unlimited set of measurements which would result if one performed the same experiment over and over indefinately ... Almost always one is interested in the sample only insofar as it is capable of revealing something about the population from which it came* (Scientific American). Also called **universe.**

3 Also spelled **Population.** *Astronomy.* one of two numbered groups (Population I and II) into which the stars of the various galaxies have been divided for classification, as on the basis of color, position, or chemical composition: *The type I population is represented by our region of the galaxy and was accordingly the first to be recognized* (Baker, *Astronomy*).

4 *Physics.* the atoms or particles occupying any particular energy level: *The resulting gap in the atom, due to the incomplete population of the K-shell, may now be filled by the transition of an electron from the L-level to the K-level* (R. W. Lawson).

population genetics, a branch of genetics dealing with the frequency and distribution of genes, mutants, genotypes, etc., among populations of organisms: *Population genetics is now based upon an increasing input of laboratory and field observations under an array of environments ... Much of this work involves the documentation and interpretation of genetic variability in natural populations* (Peter A. Parsons).

population inversion, *Physics.* the condition of having more atoms or particles at a higher energy level than is normally found in nature: *Laser action can take place in a medium if the number of atoms in some excited state is greater than the number in a state of lower energy. As this is the opposite of the normal, thermal, case the situation is known as a "population inversion"* (New Scientist).

population parameter, *Statistics.* a quantity that is constant for a particular distribution of a population but varies for other distributions.

pore, *n.* **1** *Biology.* a very small opening in tissue for perspiration, absorption, etc., such as any of the openings in the skin where the ducts of the sweat glands are or the openings in leaves which allow for the passage of water and carbon dioxide. See the picture at **skin.**

2 *Geology.* a space in rock or soil not occupied by mineral matter and allowing for the passage or absorption of fluids.

3 *Astronomy.* one of many dark spots on the surface of the sun: *The sun is dotted with tiny, stable sunspots ... called pores* (Science News Letter).

poriferan (pô rif′ər ən), *Zoology.* —*n.* any member of the phylum (Porifera) that comprises the sponges. —*adj.* of or having to do with the sponges. [from Latin *porus* hole, passage + Latin *ferre* to carry]

porosity, *n.* **1** a porous quality or condition.

2 a porous part or structure: *A commercial oil-bearing sandstone can have varying porosities* (G. Anderson).

3 *Geology.* the ratio of the volume of all the pores or gaps in a material to the volume of the whole.

porous, *adj.* having or full of pores; permeable by water, air, etc.: *Porous carbon is used to filter liquids. Sponges are porous.*

porphyrin (pôr′fər in), *n. Biochemistry.* any of a group of derivatives of pyrrole, especially an iron-free decomposition product of hematin or a magnesium-free decomposition product of chlorophyll. Porphyrins occur as basic substances especially in body tissue, blood, and urine. *An iron porphyrin is present in the blood pigment hemoglobin; indeed, all organisms that require free oxygen contain some kind of porphyrin* (P. H. Abelson). [from Greek *porphyra* purple]

porphyritic (pôr′fə rit′ik), *adj. Geology.* **1** of, having to do with, containing, or resembling porphyry. **2** of the nature or structure characteristic of porphyry; containing distinct crystals embedded in a compact groundmass. —**porphyritically,** *adv.*

porphyroid (pôr′fə roid), *n. Geology.* a rock resembling porphyry or of porphyritic structure, especially a sedimentary or igneous rock that has been altered by some metamorphic agency and has taken on a more or less perfectly developed porphyritic structure.

porphyropsin (pôr′fə rop′sin), *n. Biochemistry.* a purple carotenoid pigment found in the rods of the retinas of freshwater fishes and certain frogs, analogous to rhodopsin: *The supposition that the lamprey's primary visual pigment was porphyropsin had indicated a biochemical break between invertebrates and vertebrates* (Science News Letter). from Greek *porphyra* purple + *opsis* sight]

porphyry (pôr′fər ē), *n. Geology.* any igneous rock in which coarse euhedral crystals are scattered through a groundmass of finer-grained minerals or a glassy matrix. [from Greek *porphyros* purple]

porrect (pə rekt′ or pô rekt′), *adj. Zoology.* stretched out; extended, especially forward: *porrect mandibles.* [from Latin *porrectum,* from *porrigere* to stretch forth, from *por-* forth + *regere* to stretch, direct]

portal (pôr′tl), *Anatomy.* —*adj.* **1** of or having to do with the transverse fissure of the liver, through which the blood vessels enter: *the portal canals or areas of the liver.* **2** of or having to do with the portal vein.

—*n.* **1** = portal vein. **2** a place of entry to or exit from any part of the body: *Blood continues to flow without clotting or sealing off the portals to artery or vein* (Harold M. Schmeck, Jr.). [from Latin *porta* gate]

portal circulation or **portal system,** *Physiology.* any system of blood vessels that begins and ends in capillaries, as in the hepatic portal venous system, which begins with the capillaries of the gastrointestinal tract and ends in capillaries contained in the liver.

portal vein, *Anatomy.* the large vein carrying blood to the liver from the veins of the stomach, intestine, and spleen. Also called **hepatic portal vein.**

positive, *adj.* **1** *Mathematics.* **a** counting up from zero; plus: *5 is a positive number; −5 is a negative number. Five above zero is a positive quantity.* **b** lying on the side of a point, line, or plane opposite to that considered negative.

2 *Physics, Chemistry.* **a** of or having to do with the kind of electricity produced on glass by rubbing it with silk; lacking electrons: *All atoms seem to be made of* two kinds of electricity—positive electricity and negative electricity. Every atom of every substance contains both kinds. In every atom the positive and the negative charges just balance each other; so the atom is neutral, or uncharged (Beauchamp, *Everyday Problems in Science*). **b** having a tendency to lose electrons, and thus to become charged with positive electricity, as a chemical element or radical. **c** of or designating the part of an electric cell toward which the current flows: *the positive pole, a positive electrode.*

3 *Biology.* moving or turning toward light, the earth, or any other stimulus: *If a plant organ reacts by turning toward the source of a stimulus, it exhibits a positive tropism* (Emerson, *Basic Botany*).

—*n. Mathematics.* a positive or plus quantity, sign, symbol, etc.

positive acceleration, *Physics.* an increase in velocity.

positive angle, *Mathematics.* an angle formed by a line rotating in a counterclockwise direction.

positive feedback, *Biology.* a type of feedback that tends to increase the output: *Positive feedback, where the result of an action is used to effect an increase in that action, explains the explosive buildup characterizing the growth of a population* (Frank S. Beckman). Compare **negative feedback.**

positive lens, *Optics.* a convex lens that converges light.

positron (poz′ə tron), *n. Physics, Chemistry.* an elementary particle having the same magnitude of mass and charge as an electron, but exhibiting a positive charge, present in cosmic rays and also emitted in beta decay; a positive electron: *A positron is the antiparticle of an electron. It carries the same minuscule mass as an electron, but its charge is electrically opposite. If electrons and positrons get too close, they annihilate each other, converting their substance and energy into high energy gamma rays. Yet 30 years ago physicists found that for a fleeting moment, an electron and positron can exist together, bound by the attraction of their opposite charges. They called this pseudo-atom positronium* (Science). [from *posi(tive*) + *-tron,* as in *electron*]

positronium (poz′ə trō′nē əm), *n. Physics.* an electrically neutral system consisting of an electron and a positron bound together for less than a millionth of a second: *Normally, when a particle and an antiparticle meet, they annihilate each other. But it has been known for a long time that an electron and a positron on the way to annihilation can become bound together for a fleeting instant in a state that acts something like an atom and is called positronium* (Dietrick E. Thomsen).

post-, *prefix.* **1** after in time; later, as in *post-operative = occurring after an operation.* **2** after in position or space; behind, as in *postnasal = behind the nasal cavity.* ANT: pre-. [from Latin *post-,* from *post* after, behind]

postembryonic, *adj. Biology.* following the embryonic stage of life or growth.

posterior, *adj.* **1** *Anatomy.* situated behind or farther back; hinder: *the posterior process of the sphenoid bone.* SYN: caudal.

ASSOCIATED TERMS: see **dorsal.**

2 *Botany.* situated on the side nearest the axis: *the posterior or upper lip of a corolla.* SYN: adaxial, superior.

postfrontal, *adj. Anatomy, Zoology.* **1** behind the forehead; at the back of the frontal bone: *the post-frontal process, a postfrontal suture.* **2** toward the rear of the frontal lobe of the cerebrum.

postganglionic, *adj. Anatomy.* lying within or behind a ganglion. A postganglionic neuron of the autonomic nervous system has its cell body in a ganglion, with its axon extending to an organ. Compare **preganglionic.**

postglacial, *Geology.* —*adj.* coming after the glacial period or ice age: *the postglacial climate, a postglacial valley or stream.* SYN: Recent.

—*n.* a postglacial period or deposit: *Over extensive areas the ice-front discharged into the ocean ..., as indicated by the marine post-glacials* (A. L. DuToit).

postorbital, *adj. Anatomy.* situated behind the socket or orbit of the eye: *postorbital cartilage.*

postulate (pos′chə lit), *n. Mathematics.* something taken for granted or assumed as a basis for reasoning; fundamental principle; necessary condition: *One postulate of plane geometry is that a straight line may be drawn between any two points.* [from Latin *postulatum* a demand, from *postulare* to demand] ▶ See **axiom.**

postvertebral, *adj. Anatomy.* behind the vertebrae: *Postvertebral muscles hold the body erect and allow the back to be extended.*

potamoplankton (pə tam′ō plangk′ton), *n. Biology.* plankton found in rivers or streams. [from Greek *potamos* river + English *plankton*]

potash feldspar, *Mineralogy.* a feldspar having a high potassium content.

potassic (pə tas′ik), *adj.* of, having to do with, or containing potassium.

potassium (pə tas′ē əm), *n. Chemistry.* a soft, silver-white metallic element that occurs in nature only in compounds, is essential for the growth of plants, and oxidizes rapidly when exposed to the air. Potassium is one of the most abundant elements in the earth's crust, and is the lightest metal known except lithium. Potassium belongs to the alkali metals. Among its most important compounds is potassium carbonate, commonly called *potash,* and potassium nitrate, commonly called *saltpeter. Potassium is present in most types of rocks in combination with other elements, chiefly ... in the form of potassium aluminum silicates, as in such primary minerals as orthoclase feldspar, muscovite and biotite micas, and others* (Jones, *Minerals in Industry*). *Symbol:* K; *atomic number* 19; *atomic weight* 39.102; *melting point* 63.65°C; *boiling point* 774°C; *oxidation state* 1. [from New Latin, from English *potash,* earlier *pot-ashes*]

potential, *n. Physics.* **1** a function or quantity expressing the work required to move a unit from a standard reference point to some point in a field of force. The **electric potential** at any point in an electric field is the

---

cap, fāce, fäther; best, bē, tėrm; pin, fīve;
rock, gō, ôrder; oil, out; cup, pùt, rüle,
yü in use, *yu̇* in uric;
ng in bring; *sh* in rush; *th* in thin, ᴛʜ in then;
zh in seizure.
ə = *a* in about, *e* in taken, *i* in pencil, *o* in
lemon, *u* in circus

work required to move a unit positive charge from infinity to that point. *The potential at a point in an electric field is defined as the ratio of the potential energy of a test charge to the magnitude of the charge, or as the potential energy per unit charge* (Sears and Zemansky, *University Physics*). **2** = potential difference.

potential difference, *Physics.* the difference in electric potential between two points. In a circuit, the potential difference across a conductor equals the resistance of the conductor multiplied by the current. *Potential difference is measured in volts by an instrument called a voltmeter. ... An ordinary flashlight cell gives a potential difference, or voltage, of 1.5 volts* (Tracy, *Modern Physical Science*). Compare **electromotive force.**

potential energy, *Physics.* the energy which a body has because of its position or structure rather than as a result of its motion. A coiled spring or a raised weight has potential energy; when the spring is released or the weight is dropped, potential energy becomes kinetic energy. *A storage battery has potential energy which is changed into kinetic energy of electricity when the battery is in use* (Paul McCorkle). See the picture at **energy.**

potential evapotranspiration, *Meteorology.* the ideal maximum rate of water loss from a place through evaporation and transpiration, given sufficient precipitation.

pouch, *n.* **1** *Zoology.* a fold of skin that functions as a sac or pocket: *A kangaroo carries its young in a pouch. Chipmunks carry food in their cheek pouches.*
2 *Botany.* a baglike cavity or cyst in a plant.
3 *Anatomy.* any pocketlike space in the body: *the pharyngeal pouch.*

pound, *n.* a customary unit of weight equal to 16 ounces in avoirdupois (453.6 grams in the metric system) and 12 ounces in troy weight (373.24 grams in the metric system). *Abbreviation:* lb. [Old English *pund,* from Latin *pondo* by weight]

poundal (poun′dl), *n. Physics.* a unit of force equal to the force necessary to give a mass of one pound an acceleration of one foot per second per second: *A force is the product of a mass and an acceleration, so if W pounds of gas are accelerated to a velocity of v feet per second in one second, they will exert a force of Wv poundals, or $\frac{Wv}{g}$ pounds* (D. Hurden).

power, *n.* **1** *Physics.* the rate at which work is done, a scalar quantity expressed in joules per second or watts, in ergs per second, in foot-pounds per minute, etc.; work per unit of time: *The metric system uses the watt as a unit of power. Since the watt is a small unit, it is usually more practical to measure power in terms of a larger unit, the kilowatt (kw.), which is 1,000 times greater. There are 746 watts in one horsepower, or 1 hp. = 0.746 kw.* (Obourn, *Investigating the World of Science*).
2 *Mathematics.* the product of a number multiplied by itself a certain number of times: *16 is the 4th power of 2 (2 × 2 × 2 × 2 = 16).*
3 = magnifying power.

power series, *Mathematics.* a series in which some quantity is raised to successively higher powers. The simplest power series is $1 + x + x^2 + x^3 + x^4 + x^5$ and so on.

poxvirus, *n. Biology.* any of a group of large, complex viruses containing DNA, including those causing smallpox and cowpox: *Poxviruses kill the cells they infect. So a smallpox virus might very well use one of its enzymes, a DNase, to stop DNA synthesis in the host cell and thereby kill the host cell* (Science News Letter).

PPLO, *abbrev.* pleuropneumonialike organism.

p.p.m. or **ppm,** *abbrev.* parts per million: *The hair contained 10.38 parts per million of arsenic, compared with the normal content of hair of about 0.8 p.p.m.* (New Scientist).

p.p.t. or **ppt,** *abbrev.* parts per thousand.

Pr, *Symbol.* praseodymium.

prairie, *n. Geography, Ecology.* a broad, flat, grassy plain of the semiarid temperate and tropical latitudes, characterized by tall grasses.
ASSOCIATED TERMS: see **grassland.**

Prandtl number (prän′təl), *Physics.* a numerical ratio used in calculations of heat transfer in a fluid system, defined as the ratio of the viscosity of a substance to its thermal conductivity. The lower the Prandtl number of a substance, the higher is its convection capacity. Thin liquids have a very small Prandtl number, oils have a very large one. [named after Ludwig *Prandtl,* 1875–1953, German physicist]

praseodymium (prā′zē ō dim′ē əm), *n. Chemistry.* a yellowish-white rare-earth metallic element which occurs with neodymium. Its green salts are used to tint ceramics. *Symbol:* Pr; *atomic number* 59; *atomic weight* 140.907; *melting point* 935°C; *boiling point* 3127°C; *oxidation state* 3. [from New Latin, from Greek *prasios* leek green + New Latin *(di)dymium* didymium (so called because praseodymium occurred in the green fraction of didymium)]

pre-, *prefix.* **1** before in time; earlier, as in *Precambrian = before the Cambrian.* **2** before in position or space; in front of, as in *premolar = in front of the molars.* ANT: post-. [from Latin *prae-,* from *prae* before]

prebiological, *adj.* = prebiotic.

prebiotic, *adj. Biology.* **1** before the appearance of living things: *... assumes that the earth's atmosphere in prebiotic times contained methane, nitrogen, and water* (Leonard Nelson).
2 existing or occurring in prebiotic times: *prebiotic molecules. Studies on prebiotic synthesis—the way in which the chemicals of life may have originated on the primitive earth ...* (New Scientist).

Precambrian or **Pre-Cambrian** (prē′kam′brē ən), *Geology.* **—***n.* **1** the earliest era in geological time, encompassing all the time before the Cambrian period: *The Pre-Cambrian covers 85% of the total length of geological time and is the longest geological division* (Nature). *There are many suggestions of life in the Pre-Cambrian, but diligent search has produced few fossils recognizable as specific forms of life* (Garrels, *A Textbook of Geology*).
2 the rocks formed in the Precambrian: *The Pre-Cambrian are generally characterized by their complexity of structure and absence of fossils.*
—*adj.* of or having to do with the Precambrian or its rocks: *One of the oldest Precambrian sediments yet analyzed is the Soudan shale of Minnesota, which was formed about 2.7 billion years ago* (Scientific American).

precession (prē sesh′ən), *n.* **1** *Physics.* the rotation of a spinning rigid body that has been tipped from its vertical axis by an external force acting on it. The wobble of a spinning top is an example of precession.
2 *Astronomy.* **a** a slow gyration of the earth's axis around the pole of the ecliptic, caused mainly by the pull of the sun, moon, and other planets on the earth's equatorial bulge: *planetary precession. The [southern] cross has shifted southward in the sky due to the earth's precession* (I. M. Levitt). **b** = precession of the equinoxes. See the picture at **equinox.**
[from Late Latin *praecessionem* a going before, from Latin *praecedere* go before, precede]
—**precessional,** *adj.* of or caused by precession: *If the moon were always in the plane of the equator, or if the earth were absolutely spherical, there would be no precessional effect* (Armand Spitz and Frank Gaynor).
precession of the equinoxes, *Astronomy.* the slow, gradual westward motion of the equinoxes due to the precessional movement of the earth's axis. The precession of the equinoxes occurs at a rate of about 50″ (50 seconds) per year, making a complete circle around the sky in a period of about 26,000 years. Over this period, it causes the equinoxes to occur earlier each successive sidereal year.
precipitant (pri sip′ə tənt), *n. Chemistry.* a substance that causes another substance to be separated out of a solution as a solid; a reagent that makes a substance insoluble.
precipitate (pri sip′ə tāt), *v.* **1** *Chemistry.* **a** to separate out (a substance) from a solution as a solid: *The new materials that were formed were not soluble in water; so they separated out as solids. Chemists say that such insoluble materials are precipitated, because they settle to the bottom* (Beauchamp, *Everyday Problems in Science*). **b** to be deposited from solution as a solid: *Columbium carbide or vanadium nitride, which are highly soluble when the steel is hot, precipitate out of the ferrite during cooling in the form of a fine dispersion of hard particles* (Robert W. Cahn).
2 *Meteorology.* **a** to condense (water vapor) from the air in the form of rain, dew, snow, etc. **b** to be condensed as rain, dew, snow, etc.
—*n.* (pri sip′ə tit or pri sip′ə tāt) **1** *Chemistry.* a substance, usually crystalline, separated out from a solution as a solid; an insoluble compound formed in a solution by the addition of a reagent.
2 *Meteorology.* moisture condensed from vapor by cooling and deposited in drops as rain, dew, etc.
[from Latin *praecipitatum* thrown headlong, from *praecipitem* headlong, from *prae-* forth + *capitem* head] —**precipitator,** *n.*
precipitation, *n.* **1** *Chemistry.* **a** the process of separating out of a substance from a solution as a solid; a making insoluble by addition of a reagent. **b** the substance separated out from a solution as a solid; precipitate.
2 *Meteorology.* **a** the process of depositing moisture in the form of rain, dew, snow, etc.: *Hail is perhaps the most destructive form of precipitation. A heavy hail storm has been known to strip all leaves from plants and trees and to destroy a crop completely* (Neuberger and Stephens, *Weather and Man*). **b** something that is precipitated, such as rain, dew, or snow: *The distribution of precipitation is even more irregular than that of temperature. In general, the rainfall decreases from equatorial to polar regions and from coastal areas to the interiors of continents* (Blair, *Weather Elements*). **c** the amount that is precipitated: *For the continental United States, precipitation averages about 30 in. (76 cm) per year ... which, on an average basis, is more than adequate* (Paul D. Haney).

precipitin (pri sip′ə tin), *n. Immunology.* an antibody that produces a precipitate when it is combined with a specific soluble antigen. **A precipitin test** or **reaction** is a method of determining the identity of a substance by testing whether it reacts with a specific precipitin.
precipitinogen (pri sip′ə tin′ə jən), *n. Immunology.* an antigen that induces the production of a precipitin.
precocial (pri kō′shəl), *adj. Biology.* (of certain birds) born with down or feathers, and with the eyes open, thus requiring little parental care after hatching: *The common song birds are all altricial, while domestic fowls, partridges, most wading birds, and the various ducks are precocial* (Shull, *Principles of Animal Biology*). [from Latin *praecocem* maturing early, precocious, from *praecoquere* to mature or ripen early, from *prae-* before + *coquere* ripen]
precocious, *adj. Botany.* flowering or fruiting early, as before the appearance of leaves.
precursor, *n. Biochemistry.* an early stage or substance which precedes or gives rise to a more important or definitive stage or substance: *Proinsulin was identified ... as a substance within the pancreas that is a precursor of insulin* (Science News). *The structure of carbon produced in each case depends on the starting material—the precursor* (New Scientist).
predation (prē dā′shən), *n. Zoology.* the act, habit, or practice of preying on another animal or animals; predatory behavior: *At the openings of these shafts the voles are exposed to predation by owls* (Scientific American). *Any food chain, after the first plant-eating animal, is a succession of predations* (Storer, *General Zoology*). [from Latin *praedationem*, from *praedari* to prey upon, from *praeda* prey]
predation pressure, *Ecology.* the impact of predation on a given environment, constituting a continuing factor in the balance between predator and prey.
predator (pred′ə tər), *n. Zoology.* an animal that preys upon another or others: *Predators must be nocturnal if their prey comes out only at night* (J. L. Cloudsley-Thompson). *It is a sure sign that a small animal is biologically successful if it has a large number of different predators. If it was not successful it would not survive under heavy and varied predation* (J. Green). Compare **scavenger.**
—**predatory** (pred′ə tôr′ē), *adj.* **1** living by preying upon other animals. Lions and foxes are predatory mammals; hawks and owls are predatory birds.
2 feeding upon and destructive, as to crops, trees, or buildings: *predatory insects.*

---

cap, fāce, fäther; best, bē, tèrm; pin, fīve;
rock, gō, ôrder; oil, out; cup, pút, rüle,
yü in use, *yu* in uric;
ng in bring; *sh* in rush; *th* in thin, ᴛʜ in then;
zh in seizure.
ə = *a* in about, *e* in taken, *i* in pencil, *o* in lemon, *u* in circus

prefrontal, *adj. Anatomy.* of or having to do with the anterior portion of the frontal part of the brain: *prefrontal lobotomy.*

—*n. Zoology.* a bone in the anterior part of the skull of certain vertebrates, especially in amphibians and reptiles.

preganglionic, *adj. Anatomy.* lying in front of or preceding a ganglion: *In the ganglia, the neurons from the cord, called preganglionic neurons, meet at a synapse the neurons that continue to the various organs. The neurons that run from the ganglia to the organs are called postganglionic neurons* (Foster Kennedy).

prehallux (prē hal′əks), *n. Zoology.* a rudimentary structure found on the inner side of the tarsus of some mammals, reptiles, and amphibians, supposed to represent an additional digit. [from *pre-* + *hallux*]

prehensile, *adj. Zoology.* adapted for seizing, grasping, or holding on. Many monkeys have prehensile tails. [from Latin *prehensus,* from *prehendere* to grasp]

—**prehensility** (prē′hen sil′ə tē), *n.* the character of being prehensile.

prehnite (prā′nit *or* pren′it), *n. Mineralogy.* a mineral consisting of a hydrous silicate of aluminum and calcium, occurring in crystalline aggregates, and usually of a pale-green color. *Formula:* $CA_2Al_2Si_3O_{10}(OH)_2$ [from Colonel van *Prehn,* a Dutch governor of Cape Colony in the late 1700's + *-ite*]

premaxilla (prē′mak sil′ə), *n., pl.* **-maxillae** (-mak sil′ē). *Anatomy.* one of a pair of bones of the upper jaw of vertebrates, situated in front of and between the maxillary bones.

—**premaxillary** (prē mak′sə ler′ē), *adj.* **1** in front of the maxillary bones. **2** of or having to do with premaxillae. —*n.* = premaxilla.

premolar, *Anatomy.* —*n.* one of the permanent teeth between the canine teeth and the molars: *If you are blessed with a complete set of teeth, you have 8 incisors, 4 canines, 8 bicuspids or premolars, and 12 molars* (Simeon Potter).

—*adj.* of or having to do with the premolars.

premorse (pri môrs′), *n. Botany.* having the end abruptly truncate, as if bitten or broken off, as certain roots. [from Latin *praemorsum* bitten in front, from *prae-* before + *mordere* to bite]

prenuclear, *adj. Biology.* lacking a visible nucleus: *These other organisms have cells with nuclei and specialized organellas or specialized intracellular structures: they are called eukaryotic* (*truly nucleated*), *whereas bacteria and blue-green algae are prokaryotic* (*prenuclear*) (Scientific American).

preon (prē′on), *n. Nuclear Physics.* any of various hypothetical constituents of a quark or a lepton that determines its particular character: *The rationale for the preon model begins with the observation that every quark and lepton can be identified unambiguously by listing just three of its properties: electric charge, color, and generation number* (Scientific American). [probably from *pre*(*quark*) + *-on*]

preoral, *adj. Zoology.* situated in front of the mouth: *preoral somites, preoral ganglia.* —**preorally**, *adv.*

prepatellar, *adj. Anatomy.* situated in front of the patella: *the prepatellar bursa.*

prepotency, *n. Genetics.* the marked power of one parent, variety, or strain, to transmit a special trait or traits to the progeny: *Breeders of domesticated animals sometimes refer to the prepotency of one parent, implying that it has ability to transmit characteristics more fully or to the exclusion of the other parent. Any actual difference results from dominance or heterozygosity in the animals mated* (Storer, *General Zoology*).

—**prepotent**, *adj.* of, having to do with, or exhibiting prepotency. —**prepotently**, *adv.*

prepuce (prē′pyüs), *n. Anatomy.* **1** the fold of skin covering the end of the penis; foreskin. **2** a similar fold of skin covering the end of the clitoris. [from Middle French *prepuce,* from Latin *praeputium*]

prepupa (prē pyü′pə), *n., pl.* **-pae** (-pē), **-pas.** *Zoology.* **1** the inactive stage before pupation in the development of many insects.

2 the form of an insect in this stage: *We inject hemolymph from these prepupae into young larvae* (Scientific American).

—**prepupal**, *adj.* of, having to do with, or in the form of a prepupa: *Eggs were deposited on a medium containing cortisone ... and allowed to develop to the prepupal stage* (Nature).

preputial (prē pyü′shəl), *adj. Anatomy.* of or having to do with the prepuce.

pressure, *n. Physics.* **1 a** the continued or unopposed action of a weight or force: *Pressure is not a vector quantity and no direction can be assigned to it, but the force exerted by the fluid at one side of the plane, on the fluid at the other side of the plane, is at right angles to the plane whatever the orientation of the plane* (Sears and Zemansky, *University Physics*). **b** the force per unit of area: *There is a pressure of 27 pounds to the square inch in this tire.* **c** = electromotive force.
2 = atmospheric pressure: *The pattern of wind-flow depends on the variation of pressure over the earth; if one is known, the other can be inferred. Thus, if we could explain the wind-flow we should at the same time have an explanation of the pressure-differences over the surface of the globe* (Science News).

pressure gradient, 1 *Meteorology.* the rate at which atmospheric pressure decreases by units of horizontal distance: *In the Southeast, a flat pressure gradient will be noted, and in the Central Gulf States there will be a high-pressure center* (N. Y. Times).
2 *Physics.* the rate of decrease of pressure with distance or depth: *The "nitrogen" must have been deposited in the swimbladder at the bottom—that is, against a pressure gradient of 5 to 7 atm* (Science).

pressure head, *Physics.* the pressure of a fluid, expressed in terms of the height of a column of the fluid that would exert an equivalent pressure.

pressure point, *Physiology.* **1** a point on the body where pressure applied to a blood vessel can check bleeding. **2** a point in the skin where the terminal organs of nerves are located, making it extremely sensitive to pressure.

pressure ridge, *Geology.* a ridge of ice in arctic waters caused by lateral pressure.

prevailing westerlies, *Meteorology.* the variable westerly winds that blow between 30 degrees and 60 degrees latitude, both north and south of the equator, accompanying the general west-to-east movement of atmospheric cyclones and anticyclones.

primary, *adj.* **1** *Chemistry.* characterized or formed by the replacement of a single atom or group. For instance, in primary monohydric alcohol (CH_3CH_2OH), one of the hydrogen atoms of methyl alcohol (CH_3OH) has been replaced by an alkyl group ($-CH_3$). All primary alcohols have the functional group $-CH_2OH$. In primary sodium phosphate (NaH_2PO_4), one of the hydrogen atoms of phosphoric acid (H_3PO_4) has been replaced by a sodium atom. Compare **secondary, tertiary.**
2 *Geology.* **a** (of rocks) formed directly from magma. **b** (of minerals) formed at the same time as the enclosing rock. Compare **secondary.**
—*n.* **1** *Astronomy.* **a** a celestial body around which another revolves: *The moon is a secondary that revolves in an orbit around the earth, a primary* (Eric D. Carlson). **b** the brighter star of a binary star system: *Most of the mass, about two-thirds, is in the dark secondary. On the basis of this ratio, the secondary's mass is about 20 times that of the sun, the bright primary's about 10 times that of the sun* (Science News).
2 *Physics.* a cosmic-ray particle traveling at high velocity that is commonly stopped on entering the earth's atmosphere by collision with an atom of other matter, the collision resulting in the formation of a number of secondary particles: *The primaries enter the atmosphere at such tremendous speeds that when they collide with atoms and molecules of atmosphere they, in effect, smash the atmospheric particles and create new cosmic rays called "secondaries." It is the secondaries that are usually observed and measured* (Hugh Odishaw).
3 = primary feather.
4 = primary color.
Compare **secondary.**

primary color, *Optics.* any of a group of colors which, when mixed together in various combinations, can yield all other colors. For pigments, red, yellow, and blue are the primary colors, and are sometimes called *subtractive primaries.* For light, the primary colors are red, green, and blue, which are sometimes called *additive primaries.*

primary consumer, *Ecology.* an animal in a food chain, such as a rabbit or deer, that eats grass and other green plants: *The productivity of the producers, the photosynthetic plants, must be greater than that of the primary consumers* (B. K. Sladen). SYN: herbivore.

primary feather, *Zoology.* one of the large flight feathers growing on the distal section of a bird's wing. Also called **primary.** See the picture at **wing.**

primary meristem, *Botany.* the meristem that persists in an actively growing state from the time of its original development in the embryo within the seed: *The principal primary meristems are the apical meristems of stems and roots and the fascicular cambium of dicotyledons and gymnosperms* (Hill, *Botany*).

primary root, *Botany.* the single root which develops from the embryo itself: *Roots that form first, and grow directly from the stem, are called primary roots. Branches of the primary roots are called secondary roots, and branches of these are tertiary roots* (Clarence J. Hylander).

primary teeth, *Anatomy.* the first set of teeth of a young mammal, which are shed and replaced by permanent teeth. Also called **milk teeth.**

primary wave, *Geology.* an earthquake wave traveling through solids faster than the secondary wave, and causing rocks to vibrate parallel with the wave: *Primary (P) waves ... are compression-and-expansion waves like those of sound* (Scientific American).

primate (prī′mit *or* prī′māt), *n.* *Zoology.* any member of the highest order (Primates) of mammals, comprising the anthropoids and the prosimians: *Man, apes, monkeys, lemurs, and the tarsier were long since grouped together by Linnaeus, at the foundation of modern biology, as constituting the order of Primates, one of the dozen or so first subdivisions of the Mammals as a class* (Alfred L. Kroeber). [from Latin *primatem* of first rank, from *primus* first]

primatological (prī′mə tə loj′ə kəl), *adj.* of or having to do with primatology.

primatology (prī′mə tol′ə jē), *n.* the study of the origin, structure, development, and behavior of primates.

prime, *Mathematics.* —*adj.* **1** having no common integral divisor but 1 and the number itself: *a prime quotient. No even number greater than 2 is prime, because all such even numbers have 2 as a proper factor* (Pearson and Allen, *Modern Algebra*).
2 having no common integral divisor but 1: *2 is prime to 9.*
—*n.* **1** = prime number: *The only prime in the geometric sequence 2, 4, 8, 16, 32, ... is 2. But there is at least one prime between every pair of numbers in this sequence. The prime 3 is between 2 and 4, 5 is between 4 and 8, 11 is between 8 and 16, 17 is between 16 and 32, and so on* (Corcoran, *Algebra Two*).
2 a one of the equal parts into which a unit is divided, especially one of the sixty minutes in a degree. **b** the mark (′) indicating such a part, also used to distinguish one letter, quantity, etc., from another. B′ is read "B prime."
[Old English *prim* the first period (of the day), from Latin *prima* (*hora*) first (hour of the Roman day)]

prime conductor, *Electricity.* a conductor that collects and retains positive electricity.

prime factor, *Mathematics.* a factor that has no other integral factors except itself and 1; factor that is itself a prime number.

prime meridian, *Geography.* the meridian from which the longitude east and west is measured. It passes through Greenwich, England, and its longitude is 0 degrees.

prime mover, *Physics.* **1** the first agent that puts a machine in motion, such as wind or electricity: *Previous to the employment of steam as a motive force, water was the prime mover* (John Yeats).

---

cap, fāce, fäther; best, bē, tèrm; pin, fīve;
rock, gō, ôrder; oil, out; cup, pùt, rüle,
*y*ü in use, *y*ů in uric;
ng in bring; sh in rush; th in thin; ℾH in then;
zh in seizure.
ə = *a* in about, *e* in taken, *i* in pencil, *o* in lemon, *u* in circus

2 a machine, such as a water wheel or steam engine, that receives and modifies energy supplied by some natural source: [*The*] *Windmill is a machine that is a member of the class known as prime movers. It uses the energy of the wind to produce power* (A. D. Longhouse).

prime number, *Mathematics.* a whole number not divisible without a remainder by any whole number other than itself and 1. The numbers 2, 3, 4, 5, 7, 11, and 13 are prime numbers; 4, 6, and 9 are composite numbers.

primine (prī′min), *n. Botany.* the outer integument of an ovule. [from French *primine* (originally) outer coat, from Latin *primus* first]

primitive, *adj. Biology.* **1** at a very early or embryonic stage of development: *the primitive ectoderm, a primitive segment of the body.* **2** undeveloped; undifferentiated; rudimentary: *primitive vertebrae.*
—*n. Mathematics.* an algebraic or geometrical expression from which another is derived, especially the original equation from which a differential equation is obtained.

primordial, *adj.* existing at the very beginning; earliest; original: *primordial cells, primordial rock. There might also be quite a number of very much smaller black holes ... formed ... by the collapse of highly compressed regions in the hot, dense medium that is believed to have existed shortly after the "big bang" ... Such "primordial" black holes are of greatest interest for the quantum effects I shall describe here* (S. W. Hawking).

primordium (prī môr′dē əm), *n., pl.* **-dia** (-dē ə). *Embryology.* the first cells in the earliest stages of the development of an organ or structure. [from Latin, neuter of *primordius* original, from *primus* first + *ordiri* begin]

principal energy level, *Physics, Chemistry.* = shell (def. 4a).

principal quantum number, *Physics.* the quantum number which specifies the shell occupied by an electron: *The principal quantum number (n) represents the principal energy level* [*and*] *... is the same as the period number in the periodic table* (Chemistry Regents Syllabus).

principle, *n.* **1** a rule of science explaining how something works: *the principle of the lever. Traditional theories of gravitation ... are based on ... the principle of equivalence because it equates what might be called gravitational charge with mass* (Dietrick E. Thomsen). **2** one of the elements that compose a substance, especially one that gives some special quality or effect: *the bitter principle in a drug.*

prion (prē′on), *n. Biology.* a type of infectious agent about 100 times smaller than a normal virus, thought to be the cause of scrapie and other degenerative diseases of the nervous system: *The prion is known to be capable of initiating the production of new prions, at least in certain mammalian cells ... The evidence gathered so far, however, indicated the prion has no nucleic acid at all* (Stanley B. Prusiner).

[from *pr*(*oteinaceous*) *i*(*nfectious particle*) + *-on*]

prism (priz′əm), *n.* **1** *Geometry.* a solid figure whose bases or ends have the same size and shape and are parallel to one another, and each of whose sides is a parallelogram.
2 *Optics.* a transparent body of this form, often of glass and usually with triangular ends, used for separating white light passing through it into its spectrum or for reflecting beams of light: *The prism bends light of the shorter wavelengths more than light of the longer wavelengths, thus spreading a narrow beam of white light out into the visible spectrum* (Max Kozloff).
3 *Crystallography.* a crystal form consisting of three or more planes parallel to the vertical axis of the crystal.

[from Greek *prisma* something sawed off, prism, from *priein* to saw]
—**prismatic,** *adj.* of, formed by, or resembling a prism: *prismatic crystals, the prismatic spectrum.*
—**prismatically,** *adv.*

prisms

(def. 1, 2)

prismatic colors, *Optics.* the seven colors formed when white light is passed through a prism; red, orange, yellow, green, blue, indigo, and violet. These are the colors of the spectrum.

prismatic layer, *Zoology.* the middle layer of the shell of most mollusks: *The prismatic layer is made up of prisms. The color pattern of the shell usually depends upon this layer* (William J. Clench).

prismoid (priz′moid), *n. Geometry.* a solid like a prism except that one end is smaller than the other.
—**prismoidal** (priz moi′dəl), *adj.* of or having to do with the form of a prismoid.

pristane (pris′tān), *n. Biochemistry.* a complex hydrocarbon that is a product of the breakdown of chlorophyll and is found together with phytane in oil-bearing rocks. *Formula:* $C_{19}H_{40}$ [from Latin *pristis* shark, whale (so called because the substance was originally found in the fats of large sea animals)]

pro-, *prefix.* **1** before; preceding; earlier; precursor, as in *proenzyme, proinsulin, procambium.* **2** in front of; anterior, as in *procephalic, prothorax, pronephros.* [from Greek *pro-*, from *pro* before, forward]

Pro, *abbrev.* proline.

probabilistic, *adj. Mathematics, Physics.* having to do with or based on probability; subject to or involving chance or random phenomena: *Clausius, Maxwell, and Boltzmann showed how to derive the laws of gases from probabilistic assumptions about the behavior of individual molecules* (Scientific American). —**probabilistically,** *adv.*

probability, *n. Mathematics, Statistics.* **1** the likelihood that an event will occur, as measured by the relative frequency of the occurrence of events of the same kind. Probability is estimated as the ratio $\frac{p}{p+q}$, where p is the probable number of occurrences and q is the probable number of nonoccurrences. *When a statistician says the probability that a native of the U. S. has A-type blood is 4/10, he means that four out of 10 peo-*

ple have this type. This meaning of probability has become almost the standard usage in science (Scientific American).

2 a branch of mathematics dealing with the study of probability: *Probability tries to express in numbers, statements of the form: An event A is more (or less) probable than an event B* (T. H. Hildebrandt).

probability curve, = normal curve.

probability density, *Statistics.* **1** = probability distribution (def. 2). **2** a probability distribution that is a continuous function.

probability deviation, = probable error.

probability distribution, *Statistics.* **1** = probability function. **2** a function whose integral over any interval gives the probability that a random variable specified by it has values within that interval: *On the old atomic theory [electrons] were considered to move in orbits round the nucleus. In the newer wave-mechanical picture of the atom the orbits are replaced by a probability distribution of electrons. One particular electron cannot be definitely located at a particular point at a given time; all that can be said is that there is a certain probability of its being there* (Science News).

probability function, *Statistics.* a function which assigns to each possible value of a random variable the probability that this value will occur: *A probability function is mathematically symmetrical if its variables (which, in the case of the electron, take account of both orbital motion and spin direction) can be interchanged without changing its sign (plus or minus)* (Scientific American). Also called **distributive function.**

probability theory, *Mathematics.* the study of quantities having random distributions, forming the theoretical basis of statistics: *Probability theory as a mathematical discipline is concerned with the problem of calculating the probabilities of complex events consisting of collections of "elementary" events whose probabilities are known or postulated* (Scientific American).

probable error, *Statistics.* the amount by which the arithmetic mean of a sample is expected to vary because of chance alone: *The probable error means that in sampling data some error is bound to occur* (Emory S. Bogardus). Also called **probability deviation.**

proboscis (prō bos′is *or* prō bos′kis), *n., pl.* **-boscises, -boscides** (-bos′ə dēz). *Zoology.* **1** an elephant's trunk. **2** a long, flexible snout, like that of the tapir. **3** the tubelike mouth parts of some insects, such as flies or mosquitoes, developed for piercing or sucking. **4** any similar organ: **a** an organ of many worms, such as annelids and nemerteans, usually turned inside out and opening above the mouth: *Since the piece of liver is too large for ingestion as it is the Planaria throws out the proboscis from its mouth ... and attaches it to the meat* (Winchester, *Zoology*). **b** the tongue of certain gastropods. [from Greek *proboskis*, from *pro-* forth + *boskein* to feed]

procambial (prō kam′bē əl), *adj.* of or having to do with the procambium.

procambium (prō kam′bē əm), *n. Botany.* any undifferentiated tissue from which the vascular bundles are developed: *At about the same time the outer layers of the procambium become transformed into primary phloem, and between the phloem and the xylem there re-*

mains a narrow zone of meristematic cells, the *cambium* (Emerson, *Basic Botany*).

procaryote (prō kar′ē ōt), *n.* = prokaryote.

—procaryotic (prō kar′ē ot′ik), *adj.* = prokaryotic: *Fossil evidence indicates that cellular life has existed on earth for almost 4 billion years. During 2–3 billion of these years, only procaryotes were present. The first eucaryotes evidently appeared about 1.5 billion years ago. Thus eucaryotes are relatively recent arrivals* (Buffaloe and Ferguson, *Microbiology*).

procephalic (prō′sə fal′ik), *adj. Anatomy.* having to do with or forming the forepart of the head: *procephalic processes, the procephalic lobe.*

process, *n.* **1** *Biology.* a natural or involuntary sequence of changes or phenomena, leading to some particular end: *the process of decay, the photosynthetic process.* **2** *Anatomy.* a natural extension, prominence, or outgrowth; a projecting part: *the process of a bone, the spinous process of a vertebra.* **3** *Chemistry.* a test or reaction.

proctodaeum (prok′tə dē′əm), *n. Embryology.* an invagination of ectodermal tissue in the embryo which forms part of the anal passage. [from New Latin, from Greek *prōktos* anus + *odaios* (thing) that is on the way]

proctology (prok tol′ə jē), *n.* the branch of medicine dealing with the structure, function, and diseases of the rectum and anus. [from Greek *prōktos* anus + English *-logy*]

procumbent (prō kum′bənt), *adj. Botany.* (of a plant or stem) lying or trailing along the ground but not sending down roots. SYN: prostrate. [from Latin *procumbentem* leaning forward, from *pro-* forward + *-cumbere* lie down]

producer, *n. Ecology.* any of various organisms, such as green plants and certain algae which produce the organic compounds they need by means of photosynthesis or chemosynthesis and are commonly a source of food for other organisms: *The energy for a community is derived from the organic compounds synthesized by green plants. Autotrophs are therefore considered the producers* (Biology Regents Syllabus). Contrasted with **consumer** and **decomposer.**

product, *n.* **1** *Mathematics.* a number or quantity resulting from multiplying two or more numbers together: *40 is the product of 8 and 5.* **2** *Chemistry.* a substance obtained from one or more other substances as a result of chemical reaction.

proenzyme, *n. Biochemistry.* a protein formed in the cells of an organism, the inactive precursor of an enzyme, which is converted into an active enzyme by a further reaction: *Plasminogen is the plasma proenzyme which, on conversion to its active form, plasmin, is considered responsible for lysis of fibrin deposits* (Nature). Also called **zymogen.**

---

cap, fāce, fäther; best, bē, tėrm; pin, five;
rock, gō, ôrder; oil, out; cup, put, rüle,
yü in use, *yu* in uric;
ng in bring; sh in rush; th in thin, ᴛʜ in then;
zh in seizure.
ə = *a* in about, *e* in taken, *i* in pencil, *o* in
lemon, *u* in circus

proestrus (prō es′trəs), n. Zoology. the period just before estrus in the estrous cycle. It is characterized by development of follicles in the ovary and growth of the mucous membrane lining the uterus.

profile, n. 1 an outline or representation of the distinctive features or chief characteristics of something, as of a natural process or phenomenon, especially in the form of a graph, curve, or diagram: If we took gravity readings all over the earth and corrected them to sea level, we would have a gravity profile of the geoid (Scientific American).
2 Geology. a a vertical section of soil, showing the sequence of the various horizons: The stratum of soil gradually develops a fairly well defined system of layers, termed in soil nomenclature "horizons", and the system of horizons which makes up the stratum is called a soil "profile" (J. H. Quastel). b an outline of part of the earth's surface, taken in a vertical plane and showing the course of a valley, river, etc.: Profiles and contour maps of its base reveal that in all the valleys so far investigated there is a large channel beneath the flood-plain (G. H. Dury).

progestational (prō′jes tā′shə nəl), adj. Physiology. of, having to do with, or inducing the changes in the lining of the uterus during the part of the menstrual cycle preceding gestation or pregnancy. A progestational hormone is the natural hormone progesterone or any similar hormone that induces progestational changes. [from pro- before + gestation + -al]

progesterone (prō jes′tə rōn), n. Biochemistry. a steroid hormone secreted by the corpus luteum that makes the lining of the uterus more ready to receive a fertilized ovum. Formula: $C_{21}H_{30}O_2$ [from proge(stin) + ster(ol) + -one]

progestin (prō jes′tin), n. Biochemistry. any of a class of natural or synthetic hormones, such as progesterone, that make the lining of the uterus more receptive to a fertilized ovum; a progestational hormone. Also called progestogen. [from pro- before + gest(ation) + -in]

progestogen (prō jes′tə gən), n. = progestin.

proglottic (prō glot′ik), adj. of or having to do with the proglottis.

proglottid (prō glot′id), n. = proglottis.

proglottis (prō glot′is), n., pl. -glottides (-glot′ə dēz). Zoology. one of the segments or joints of a tapeworm, containing both male and female sexual organs. [from New Latin, from Greek proglōttis tip of the tongue, from pro- + glōtta tongue (so called from the shape of the segments)]

progression, n. 1 Mathematics. a sequence of quantities in which there is always the same relation between each quantity and the one succeeding it. The sequence 2, 4, 6, 8, 10 is an arithmetical progression. The sequence 2, 4, 8, 16, 32 is a geometric progression.
2 Astronomy. the apparent or actual motion of a planet from west to east.

prohormone, n. Biochemistry. the inactive precursor of a hormone; a natural substance from which a hormone develops: In the body, the molecule is first synthesized in the parathyroid gland as a prohormone, a long chain of 106 amino acids that are chemically linked together (Earl A. Evans).

proinsulin, n. Biochemistry. a single-chain protein that is a precursor of insulin in the pancreas and converted by an enzyme into insulin: Proinsulin contains a sequence of 22 amino acids joining chain A and chain B. When proinsulin is reduced, its disulfide bridges will spontaneously re-form. The conversion of proinsulin to insulin involves the removal of this link of 22 amino acids between the two chains (Claude A. Villee, Jr.).

project (prə jekt′), v. Geometry. to draw lines through (a point, line, figure, etc.) and reproduce it on a line, plane, or surface.
—projection, n. 1 the projecting of a figure, etc., upon the line, plane, or surface: The projection of a line into a plane is the set of all points of the plane which are projections of points of the line (Moise and Downs, Geometry).
2 a representation of a spherical surface on a plane surface by passing lines from various points to their intersection with a plane. The terrestrial and celestial spheres are usually represented on maps by means of projections.

projective geometry, the branch of geometry that deals with those properties of geometric figures that are unchanged after projection: Scientific geometry began with the Euclidean system (concerned with figures, angles and so on), developed in the 17th century the so-called projective geometry (dealing with problems of perspective) (Scientific American).

prokaryote (prō kar′ē ōt), n. Biology. a cell or organism without a distinct nucleus or nuclei; prokaryotic cell or organism: Bacteria and blue-green algae are prokaryotes (prenuclear cells); their genetic material, DNA, is not confined within an organized nucleus, and their respiratory and photosynthetic machinery (if it is present) is similarly dispersed (Richard E. Dickerson). Also spelled procaryote. Contrasted with eukaryote.
[from pro- before + karyon nut, kernel]
—prokaryotic (prō kar′ē ot′ik), adj. not having a distinct nucleus or nuclei; belonging to the group of organisms which lack a visible, membrane-bound nucleus: Bacteria are prokaryotic cells, that is, their genetic material is distributed throughout the cytoplasm (Scientific American).

prolactin (prō lak′tin), n. Biochemistry. a hormone from the anterior part of the pituitary gland that induces the mammary glands to give milk and affects the activity of the corpus luteum: The presence of squabs can stimulate a dove's pituitary glands to secrete more prolactin even before the stage in the cycle when the squabs normally appear (Daniel S. Lehrman). [from pro- + Latin lactem milk]

prolamin (prō′lə min) or prolamine (prō′lə mēn), n. Biochemistry. any of a group of simple proteins that are soluble in dilute alcohol but insoluble in water and absolute alcohol, such as gliadin. [from prol(ine) + am(ide) + -in]

prolan (prō′lan), n. Biochemistry. either of two hormones secreted by the pituitary gland that influence the activity of the gonads. They are present in urine during pregnancy. [from German Prolan, from Latin proles offspring]

prolate (prō'lāt), *adj. Geometry.* elongated in the direction of the polar diameter: *If [the sun] changed from a sphere to a prolate spheroid or an oblate squashed one, then it would be a body which with the same mass would attract the Earth differently, so that the Earth would leave its orbit* (Hermann Bondi). [from Latin *prolatum* extended, from *pro-* forth + *latum* brought]

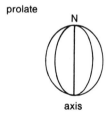

prolate

N

axis

proline (prō'lēn'), *n. Biochemistry.* a colorless, crystalline, nonessential amino acid, found in the levorotatory form in most proteins. *Formula:* $C_5H_9NO_2$ *Abbreviation:* Pro [from German *Prolin,* contraction of *Pyrrolidin* pyrrolidine]

promethium (prə mē'thē əm), *n. Chemistry.* a radioactive rare-earth metallic element which is a product of the fission of uranium, thorium, and plutonium. *Symbol:* Pm; *atomic number* 61; *mass number* (of the most stable isotope) 145, (of the most abundant isotope) 147; *oxidation state* 3. [named after *Prometheus,* a Titan in Greek mythology who stole fire from heaven and taught human beings its use]

prominence, *n. Astronomy.* a cloud of gas which erupts from the sun and is seen either as a projection from, or a dark spot on, the surface of the sun: *Spectacular upsurgings of gases in the chromosphere, known as prominences, are sometimes seen: these may shoot out to distances of the order of a few hundred thousand miles* (A. J. Higgs).

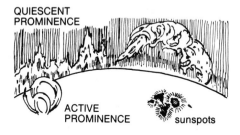

QUIESCENT PROMINENCE

ACTIVE PROMINENCE

sunspots

promontory (prom'ən tôr'ē), *n.* **1** *Geography.* a high point of land extending from the coast into the water. SYN: headland.
2 *Anatomy.* a part that bulges out; a projecting knoblike process or growth.
[from Latin *promontorium,* from *pro-* forward + *montem* mountain]

promote, *v. Chemistry.* to make (a catalyst) more active or effective; act as a promoter of: *The catalyst is preferably promoted by the addition of small amounts of aluminum oxide and potassium oxide* (Anthony Standen).

promoter, *n.* **1** *Chemistry.* a less active substance that increases the activity or effectiveness of a catalyst: *The introduction of small quantities of foreign substances*

will sometimes cause the adsorbing surface to become catalytic. Substances so used are called promoters (Jones, *Inorganic Chemistry*).
2 *Molecular Biology.* the part of an operon, situated between the operator and the structural gene, at which transcription begins: *Yet another control region for the three-gene lactose operon ... called the promoter, seems to be where mRNA begins to form to copy the code of the three-gene operon* (Victor K. McElheny).

promycelial (prō'mī sē'lē əl), *adj.* of or having to do with the promycelium.

promycelium (prō'mī sē'lē əm), *n., pl.* **-lia** (-lē ə). *Botany.* the filamentous product of the germination of a spore.

pronate (prō'nāt), *v. Anatomy.* to rotate (the hand, forelimb), etc., so that the palm is down or toward the back. [from Late Latin *pronatum* bent forward, from Latin *pronus* prone]
—**pronation** (prō nā'shən), *n.* **1** the act of rotating the hand or forelimb so that the palm turns down or toward the back.
2 a similar movement, as of the foot, hindlimb, or shoulder.
—**pronator** (prō nā'tər), *n.* a muscle that effects or assists in pronation.

pronephric (prō nef'rik), *adj.* of or having to do with the pronephros: *the pronephric duct.*

pronephros (prō nef'ros), *n. Embryology.* the most anterior part of the renal organ of vertebrate embryos. Compare **mesonephros** and **metanephros.** [from *pro-* anterior + Greek *nephros* kidney]

pronuclear, *adj.* of or having to do with the pronucleus.

pronucleus, *n. Biology.* the nucleus of a sperm or of an ovum, just before these unite in fertilization to form the nucleus (synkaryon) of the zygote: *At last, one ... sperm that penetrates the tough zona pellucida, the glass-clear membranous shell of the ovum, joins its pronucleus with that of the egg. At that moment conception takes place* (James C. G. Conniff).

proof, *n. Mathematics.* a logical argument that establishes the validity of a statement. In geometry, theorems have proofs, but postulates are given without proof.

propagate, *v. Physics.* to transmit, spread, or convey (motion, light, sound, etc.) in some direction or through some medium: *... the effects of chopping the beam into pulses and then trying to propagate toward the target several thousand pulses per second* (J. Parmentola and K. Tsipis).
—**propagation,** *n. Physics.* the act or process of propagating, especially the process by which a disturbance, as the motion of electromagnetic or sound waves, is transmitted through a medium such as air or water: *The speed of propagation of longitudinal shock waves in shelf ice—about 3,780 metres per second (12,400 ft/sec.)—was nearly the same as that in much colder*

---

cap, fāce, fäther; best, bē, tėrm; pin, five;
rock, gō, ôrder; oil, out; cup, pùt, rüle;
yü in use, *yu* in uric;
ng in bring; *sh* in rush; *th* in thin, ᴛH in then;
zh in seizure.
ə = *a* in about, *e* in taken, *i* in pencil, *o* in lemon, *u* in circus

ice at 2,700 metres (8,800 ft) above sea-level, and of the same order of magnitude as that found in temperate glaciers (E. F. Roots).

propagule (prop′ə gyül), n. Botany. a bud or other offshoot able to develop into a new plant: Few plants average as few as ten seeds or other propagules per year, and many produce thousands (Emerson, Basic Botany). [from New Latin propagulum, from Latin propago shoot, slip]

propane (prō′pān), n. Chemistry. a heavy, colorless, flammable gas, a hydrocarbon of the methane series. Propane occurs in crude petroleum and in natural gas and is used as a fuel, refrigerant, or solvent. Formula: C_3H_8 [from prop(ionic acid) + -ane]

propanol (prō′pə nōl), n. Chemistry. the normal isomer of propyl alcohol.

properdin (prop′ər din), n. Biochemistry. a substance present in blood plasma, capable of inactivating or killing various bacteria and viruses and of dissolving red blood cells: Properdin is a normal serum constituent and differs from the antibody in many respects (Science). [from pro- + Latin perdere destroy + English -in]

proper fraction, Mathematics. a fraction in which the numerator is smaller than the denominator. EXAMPLES: 1/8, 3/4, 199/200.

proper motion, Astronomy. the apparent angular motion of a star across the sky after allowing for precession, nutation, and aberration, due to real motions of the star itself: It is extremely improbable that two unconnected stars will have the same proper motion, and so the components of an optical double will be found to be moving independently. They are thus distinguished from true binaries, the proper motions of whose components differ only by a small amount due to the orbital motion (W. H. Marshall).

prophase (prō′fāz′), n. Biology. 1 the first stage of mitosis, that includes the formation of the spindle and the condensation of the chromosomes: During prophase the nucleoli are large at first, but gradually they become smaller. They also free themselves from the chromosomes to which they are attached, and finally disappear. The nuclear membrane breaks down and disappears in late prophase (McElroy, Biology and Man). See the picture at **mitosis.**
2 the first stage in meiosis, from the point at which the chromosomes appear (leptotene) to the point where the chromatics contract (diakinesis): A very small fraction (about 0.3) of the total DNA of each meiotic cell is replicated later than the bulk of the DNA in a discrete period during meiotic prophase (G. H. Jones). Compare **anaphase, interphase, metaphase, telophase.** See the picture at **meiosis.**
—**prophasic,** adj. of or having to do with prophase. [from pro- + phase]

propionate (prō′pē ə nāt), n. Chemistry. a salt or ester of propionic acid: sodium propionate, ethyl propionate.

propionic acid (prō′pē on′ik), Chemistry. a fatty acid with a pungent odor, found in perspiration and produced synthetically from ethyl alcohol and carbon monoxide. It is used in the form of its propionates as an ingredient in perfumes. Propionic acid and calcium propionate for bread are vital food preservatives (Science Journal). Formula: $CH_3CH_2CO_2H$ [from pro- before, precursor + Greek pion fat (so called because it is the first of the true carboxylic acids)]

propionyl (prō′pē ə nəl), n. Chemistry. a univalent radical (CH_3CH_2CO-) contained especially in propionic acid.

proplastid (prō plas′tid), n. Biology. a protoplasmic body that is a precursor of a plastid: The term "proplastid" is generally reserved for the plastids which are found in meristematic cells and which give rise to the other plastid forms during differentiation (W. W. Thomson).

proportion, n. Mathematics. 1 a statement of equality between two ratios. EXAMPLE: 4 is to 2 as 10 is to 5. $a : b = c : d$ and $4 : 2 = 10 : 5$ are direct proportions; $a : 1/b = c : 1/d$ and $2 : 1/3 = 1/2 : 1/12$ are indirect proportions.
2 = ratio.
[from Latin proportionem, from the phrase pro portione in relation to the part]
—**proportional,** adj. having the same or a constant ratio: The corresponding sides of mutually equiangular triangles are proportional. —**proportionally,** adv.

proprioception (prō′prē ə sep′shən), n. Physiology. the perception of internal bodily conditions, such as the state of muscular contraction: The occupational therapist is ... concerned with fine movements, especially of the hand, and she is trained in techniques for testing proprioception, thereby learning whether the child has a normal understanding of his body image and where he is in space (London Times).
—**proprioceptive** (prō′prē ə sep′tiv), adj. 1 receiving stimuli from within the body: The proprioceptive impulses keep the brain informed of the state of muscular contraction (W. Ross Ashby).
2 of or having to do with such stimuli: The fibers that activate the muscle connect to an association neuron in the spinal-reflex center; the association neuron is connected in turn to a particular type of sensory cell, the proprioceptive neuron, the terminal fibers of which are embedded in the muscle (R. W. Sperry). [from Latin proprius one's own + English (per)ceptive]
—**proprioceptor** (prō′prē ə sep′tər), n. a sense organ that receives stimuli from within the body: The movement is sensed by position or movement receptors within the body (proprioceptors), which send impulses back to the central nervous system (Donald M. Wilson).

prop root, Botany. a root that supports a plant by growing downward into the ground from above the soil, as in corn and the mangrove: Sometimes prop roots grow out of the stalk several inches above the ground to help support the plant against the wind (G. F. Sprague).

propyl (prō′pəl), n. Chemistry. the univalent radical, C_3H_7, of propane. [from prop(ionic acid) + -yl]

propyl alcohol, Chemistry. a colorless, flammable liquid used in organic synthesis and as a solvent, intermediate, and antiseptic. Formula: $CH_3CH_2CH_2OH$ Compare **propanol.**

propylene (prō′pə lēn), n. Chemistry. a colorless, gaseous hydrocarbon homologous with ethylene, used in organic synthesis. Formula: C_3H_6 **Propylene glycol**

is a colorless, viscous, liquid used as an antifreeze, as a solvent, and in organic synthesis.

propylic (prō pil′ik), *adj. Chemistry.* of, having to do with, or containing propyl.

propylite (prop′ə lit), *n. Geology.* a volcanic rock with triclinic feldspars greatly altered by hydrothermal action to carbonate, chlorite, and other minerals. Propylite is an andesitic rock, and it occurs in regions of silver deposits. [from Greek *propylon,* variant of *propylaion* gateway + English *-ite* (so called because it was formed at the start of the Tertiary period)]
—**propylitic** (prop′ə lit′ik), *adj.* of or like propylite.
—**propylitization** (prop′ə lit ə zā′shən), the process of forming propylite.

prosencephalic (pros′en sə fal′ik), *adj.* of or having to do with prosencephalon.

prosencephalon (pros′en sef′ə lon), *n., pl.* **-la** (-lə). *Anatomy.* the anterior segment of the brain, consisting of the cerebral hemispheres, or their equivalent, and certain adjacent parts; the diencephalon and telencephalon. Also called **forebrain.** [from French *prosencéphalon* forebrain, from Greek *pros* forward + *enkephalon* (literally) within the head, from *en* in + *kephalē* head]

prosenchyma (pros eng′ki mə), *n. Botany.* a type of tissue characteristic of the woody and bast portions of plants, consisting of long, narrow cells with pointed ends that sometimes form ducts or vessels. [from Greek *pros* toward + English *(par)enchyma*]
—**prosenchymatous** (pros′eng kim′ə tes), *adj.* consisting of or having to do with prosenchyma.

prosimian (prō sim′ē ən), *n. Zoology.* any member of a suborder (Prosimii) of primates, including all early fossil primates and modern lemurs, loris, tarsiers, and tree shrews: *The very important exploratory function of the hands of simians superseded the tactile sense of the prosimians (such as the lemurs) where the sensation is concentrated on the feelers or vibrissae on the face and wrists* (New Scientist). Compare **anthropoid.** [from *pro-* before, earlier + Latin *simia* ape]

prosoma (prə sō′mə), *n. Zoology.* the anterior or cephalic segment of the body in certain animals, such as cephalopods, lamellibranchs, and cirripeds. [from *pro-* anterior + Greek *sōma* body]
—**prosomal,** *adj.* of or belonging to the prosoma.

prostacyclin (pros′tə si′klin), *n. Biochemistry.* a derivative of prostaglandin that inhibits the aggregation of blood platelets and dilates blood vessels.

prostaglandin (pros′tə glan′dən), *n. Biochemistry.* any of a group of hormonelike substances produced in the tissues of mammals by the action of enzymes on certain fatty acids, found in high concentrations in seminal fluid of the prostate gland and thought to have a variety of important functions in reproduction, nerve-impulse transmission, muscle contraction, regulation of blood pressure, and metabolism: *The prostaglandins influence heart rate, blood pressure, the firing rate of nerves, and the contraction of the uterus and smooth muscle* (James S. Sweet). *Abbreviation:* PG [from *prosta(te)* gland + *-in*]

prostate (pros′tāt), *Anatomy.* —*n.* a large gland surrounding the male urethra just below the bladder. Its secretions, which transport sperm cells, make up a large part of the semen. *In later life, the prostate tends to become larger. It may press upon the urethral tube,*

interfering with the passage of urine. Surgeons may then have to remove either all, or part of, the gland (R. F. Escamilla).
—*adj.* designating or having to do with the prostate. [from Greek *prostatēs* one standing in front, ultimately from *pro-* before + *stenai* to stand]
—**prostatic** (pros tat′ik), *adj.* of or having to do with the prostate: *the prostatic duct, prostatic calculi.*

prosthetic (pros thet′ik), *adj. Biochemistry.* of, having to do with, or designating the nonprotein component of a conjugated protein or enzyme. Compare **coenzyme.** [from Greek *prosthetikos* of the nature of addition, giving additional power, from *prosthetos* added]

prosthion (pros′thē on), *n. Anatomy.* the lowest or most forward point of the upper jaw. [from Greek, neuter of *prosthios* foremost, front]

prostomial (prō stō′mē əl), *adj.* of, having to do with, or situated on the prostomium.

prostomium (prō stō′mē əm), *n. Zoology.* the part of the body in front of the mouth of mollusks, worms, and certain other invertebrates: *To tear a piece off, the worm grasps the edge between the prostomium and the lower border of the mouth, the pharynx being pushed forwards at the same time so as to provide a point of resistance for the prostomium* (Betty I. Roots). [from New Latin, from *pro-* anterior + Greek *stoma* mouth]

prostrate, *adj. Botany.* lying or trailing along the ground: *a prostrate stem or plant.* SYN: procumbent.

prot-, *combining form.* the form of **proto-** before vowels, as in *protactinium, protamine.*

protactinium (prō′tak tin′ē əm), *n. Chemistry.* a rare, heavy radioactive metallic element of the actinide series, which occurs in uranium ores and decays into actinium by emitting alpha rays. It can also be produced artificially by bombardment of thorium with alpha particles. *Symbol:* Pa; *atomic number* 91; *mass number* (of the most stable isotope) 231; *oxidation state* 4, 5. [from New Latin *prot-* + *actinium*]

protamine (prō′tə mēn′), *n. Biochemistry.* any of a group of basic proteins of low molecular weight which occur combined with nucleic acids in the sperm of many fish. Protamines are not coagulated by heat, are soluble in water, and form amino acids when hydrolyzed. *The protamines are the simplest of a series of compounds, the most complex of which are the typical proteins such as occur in blood serum, egg white, or muscle* (D. L. Woodhouse and H. S. A. Sherratt). [from *prot-* + *amine*]

protease (prō′tē ās), *n. Biochemistry.* any of various enzymes, such as pepsin and trypsin, that break down proteins into simpler compounds; a proteolytic enzyme: *Proteases of different kinds digest proteins, breaking them down into proteoses, polypeptides, and*

cap, fāce, fäther; best, bē, tėrm; pin, fīve;
rock, gō, ôrder; oil, out; cup, pùt, rüle,
yü in use, yu in uric;
ng in bring; sh in rush; th in thin, ᴛʜ in then;
zh in seizure.
ə = a in about, e in taken, i in pencil, o in lemon, u in circus

peptones, some of them carrying the hydrolytic process as far as amino acids. (Youngken, *Pharmaceutical Botany*).

protective coloring or **protective coloration**, *Biology.* a coloring of some animals and plants that blends with their environment, protecting them from being discovered by their enemies: *There are numerous methods of protection used by fish in an effort to survive ... One of the most universal is protective coloration. Fish are so colored that they are hard to see in their natural environment.* (Winchester, *Zoology*).

protective mimicry, = Batesian mimicry.

protein (prō'tēn' or prō'tē in), *n. Biochemistry.* any of a group of complex organic compounds containing carbon, hydrogen, oxygen, nitrogen, and usually sulfur. Proteins are polymers composed of chains of amino acids joined by peptide linkages and are essential to the structure and functioning of all cells and are components of viruses. Enzymes, globulins, and antibodies are all proteins. Parts of the human body such as skin, nails, and connective tissue are composed of protein. Such basic foods as meat, milk, eggs, fish, and beans are rich in protein. *A typical protein is a molecular chain containing about 200 amino acid subunits linked together in a specific sequence* (Scientific American). [from French *protéine*, from Greek *proteios* of the first quality, from *prōtos* first]

proteinaceous (prō'tə nā'shəs or prō'tē ə nā'shəs), *adj. Biochemistry.* of, containing, or resembling protein: *Beans, peas, and lentils are proteinaceous legumes.*

proteinase (prō'tə nās or prō'tē ə nās), *n. Biochemistry.* any of various proteases that hydrolyze proteins, either breaking them down to their constituent amino acids or linking them together to form polypeptides of various lengths.

proteinoid (prō'tə noid or prō'tē ə noid), *n. Biochemistry.* a proteinlike polypeptide made by polymerization of a number of animo acids: *Proteinoids might represent the first evolutionary step along the road that led to the true proteins and eventually to life itself* (New Scientist).

proteolysis (prō'tē ol'ə sis), *n. Biochemistry.* the hydrolysis or breaking down of proteins into simpler compounds, as in digestion.

proteolytic (prō'tē ə lit'ik), *adj. Biochemistry.* of, having to do with, or causing proteolysis: *Peptidase and proteinase are proteolytic enzymes.*

proteose (prō'tē ōs), *n. Biochemistry.* any of a class of water-soluble compounds derived from proteins, as by the action of the gastric and pancreatic juices.

Proterozoic (prot'ər ə zō'ik), *Geology.* —*n.* **1** the later of two divisions of Precambrian time, during which sponges, sea worms, and other forms of sea life appeared.

2 the rocks formed during this time.

—*adj.* of or having to do with the Proterozoic or its rocks.

[from Greek *proteros* prior + *zōē* life]

▶ The Proterozoic has often been considered an *era*, following the *Archeozoic Era*. It is now more commonly considered an *eon*, following the *Archean Eon*. See also the note under **Archean**.

prothallial (prō thal'ē əl), *adj.* of or having to do with a prothallium.

prothallium (prō thal'ē əm), *n., pl.* **-thallia** (-thal'ē ə). *Botany.* **1** the gametophyte of pteridophytes, such as ferns, horsetails, and club mosses.

2 the analogous rudimentary gametophyte of seed-bearing plants.

[from Greek *pro-* before + *thallos* sprout]

prothoracic, *adj.* of or having to do with the prothorax: *the prothoracic cavity, a prothoracic gland.*

prothorax, *n. Zoology.* the anterior division of an insect's thorax, bearing the first pair of legs: *With its jointed arm bent and held outward in a characteristic devotional attitude and its head raised upon a long and erect "neck" or prothorax, the mantis does appear to be praying* (Science News Letter).

prothrombin (prō throm'bin), *n. Biochemistry.* a protein in blood plasma, essential to clotting, from which thrombin is derived: *When an injury calls blood-clotting factors into action, prothrombin, through a rapid series of biochemical events, becomes the clot-producing enzyme, thrombin* (Science News).

protist (prō'tist), *n., pl.* **protists, protista** (prō tis'tə). *Biology.* any member of the kingdom (Protista) that comprises most of the one-celled organisms with a visible nucleus (eukaryotes). Protozoans, eukaryotic algae, and slime molds are protists. Certain simple multicellular organisms are also considered protists, while the only one-celled eukaryotes that are excluded are unicellular fungi such as yeasts. *In the case of the protists, the entire organism usually consists of a single autonomous cell throughout life* (J. McCabe). *Plant-like protists probably appeared several times through symbiotic unions between free-living, autotrophic prokaryote blue-green algae and various heterotrophic eukaryote protists* (Lynn Margulis). See also **classification.**

[from Greek *prōtistos* the very first, superlative of *prōtos* first]

—**protistan** (prō tis'tən), *adj.* of or having to do with protists. —*n.* = protist.

protistology, *n.* a branch of biology dealing with the protists.

protium (prō'tē əm or prō'shē əm), *n. Chemistry.* the ordinary isotope of hydrogen. *Symbol:* ^1H; *atomic number* 1; *mass number* 1. Compare **deuterium, tritium.** [from Greek *prōtos* first]

proto-, *combining form.* **1** first in time: earliest, as in *protobiont, protogalaxy.*

2 first in formation; primary, as in *protoplasm, protophloem.*

3 *Chemistry.* first or lowest in a series; having the smallest proportion of (an element or radical), as in *protosulfate, protoxide.* Also spelled **prot-** before vowels.

[from Greek *prōto-*, from *prōtos* first]

protobiont (prō'tō bī'ont), *n. Biology.* a hypothetical living cell that gave rise to all cells: *Oparin ... has suggested that the first cells, which he called protobionts, arose when a boundary or membrane formed around one or more macromolecules possessing catalytic activity, presumably proteins* (A. L. Lehninger).

protocontinent, *n.* = supercontinent.

protoderm (prō'tə dèrm), *n.* = dermatogen.

protogalaxy, *n. Astronomy.* a hypothetical first stage in the evolution of a galaxy, consisting of a mass of contracting gas (mainly hydrogen and helium) from which stars condensed and formed the galaxy: *These aggregations of matter detached themselves from the rest of the universe, began to develop as independent units ("protogalaxies") and thus set off on the road to becoming galaxies* (Jan H. Oort).

proton (prō′ton), *n. Physics.* a subatomic particle charged with one unit of positive electricity, found in the nuclei of atoms and having a mass about 1836 times that of an electron. The number of protons in each nucleus is the atomic number of the element. An atom of ordinary hydrogen contains one proton and one electron. *The number of protons in the nucleus determines the kind of matter the atom forms, whether hydrogen, uranium ... or any other* (Science News Letter). See the picture at **atom.** [from Greek *prōton,* neuter of *prōtos* first]
ASSOCIATED TERMS: see **electron.**

protonate (prō′tə nāt), *v. Chemistry.* **1** to provide with a proton; transfer a proton to (a molecule, atom, etc.). **2** to receive an additional proton: *a compound that protonates in the presence of a strong acid.*
—**protonation** (prō′tə nā′shən), *n.* an act or result of protonating; the transfer or addition of a proton: *The hydride complexes are usually obtained by reduction or protonation of suitable metal complexes. All the positively charged species listed ... were obtained by protonation, usually in very strong acid* (Science).

protonema (prō′tə nē′mə), *n., pl.* **-mata** (-mə tə). *Botany.* a threadlike structure in mosses from which the more visible, leafy portion grows: *Buds from the protonema then grow into new moss plants with leaves and rhizoids* (Rolla M. Tryon). [from New Latin, from *proto-* first + Greek *nema* thread]
—**protonemal,** *adj.* of or having to do with a protonema.

protonic (prō ton′ik), *adj.* of or having to do with a proton or protons: *protonic reagents.*

proton-proton chain, *Nuclear Physics.* a nuclear reaction believed to be the source of energy of the sun and other hydrogen-rich stars. It begins with the fusion of two protons to form a deuteron, which then changes in two stages into helium, releasing two protons to join the chain again. The reaction liberates tremendous energy.

proton synchrotron, a synchrotron designed to accelerate protons.

protophloem (prō′tō flō′em), *n. Botany.* the first phloem tissue to develop, lying closest to the outer part of the stem; the primitive phloem of a vascular bundle: *Protophloem, in its early formation, consists mainly of sieve cells and sieve tubes which are elongated and narrow.* (Youngken, *Pharmaceutical Botany*). Compare **photoxylem.**

protoplanet, *n. Astronomy.* a hypothetical first stage in the formation of a planet, consisting of gases and dust rotating as a mass around a star. The earth and other planets may have originated in this way.

protoplasm, *n. Biology.* **1** the chemically active mixture of proteins, fats, and many other complex substances suspended in water which forms the living matter of all cells, and in which metabolism, growth, and repro-

duction are manifested. **2** a former name for cytoplasm.
—**protoplasmic,** *adj.* of or having to do with protoplasm.

protoplast, *n. Biology.* the living unit of protoplasm within a cell (such as a plant or bacterial cell) that has a cell wall. The protoplast consists of the nucleus, cytoplasm, and plasma membrane.
—**protoplastic,** *adj.* of or having to do with the protoplast.

protopodite (prō top′ə dīt), *n. Zoology.* the section or joint of an appendage which attaches to the body of a crustacean. [from *proto-* + Greek *podos* foot]

protoporphyrin (prō′tə pôr′fər in), *n. Biochemistry.* a porphyrin or pigment in cells that is a basic part of hemoglobin in red blood and chlorophyll in green plants. *Formula:* $C_{34}H_{34}N_4O_4$

protostar, *n. Astronomy.* a hypothetical first stage in the formation of a star, consisting of a contracting mass of gases and dust rotating in space: *The photo also shows a typical dark globule—a dust cloud that is probably contracting to form a future protostar* (Science Journal).

protostele (prō′tə stēl or prō′tə stē′lē), *n. Botany.* the solid stele of most roots, and of the first-formed portion of some primitive stems. Also called **monostele.**
—**protostelic** (prō′tə stē′lik), *adj.* of or having to do with protostele.

prototroph (prō′tə trof), *n. Microbiology.* a microorganism possessing the nutritional requirements and capabilities characteristic of its species or strain: *Nutritional mutants are ... called auxotrophs, and the parents from which they are derived are called prototrophs* (T. D. Brock). [from *proto-* + Greek *tropē* nourishment]
—**prototrophic** (prō′tə trof′ik), *adj.* of or designating prototrophs: *prototrophic Escherichia coli.*

prototype, *n. Biology.* a primitive form; archetype: *The genes produced in each plant or animal, which transmit the heritable characters to the individuals of the succeeding generation, may on occasion be imperfect copies of the prototype; the result will be an alteration of the character or characters controlled by that gene in future generations* (Atlantic).

protovirus, *n. Biology.* a primary type of virus that serves as a model for others of the same kind; prototype of a virus: *It was hypothesized that normal cells manufacture RNA, the messenger molecule, which moves to neighbouring cells in the form of a protovirus, or template, and stimulates the production of a new form of DNA* (Peter Stoler).

protoxide (prō tok′sīd), *n. Chemistry.* that member of a series of oxides which has the smallest proportion of oxygen. [from *prot-* + *oxide*]

---

cap, fāce, fäther; best, bē, tèrm; pin, fīve;
rock, gō, ôrder; oil, out; cup, pùt, rüle,
yü in use, *yu̇* in uric;
ng in bring; *sh* in rush; *th* in thin, ᵺ in then;
zh in seizure.
ə = *a* in about, *e* in taken, *i* in pencil, *o* in lemon, *u* in circus

protoxylem

protoxylem (prō'tə zī'lem), *n. Botany.* the first xylem tissue to develop, lying closest to the pith; the primitive xylem of a vascular bundle. Compare **protophloem.**

protozoa (prō'tə zō'ə), *n. Biology.* protozoans; plural of **protozoon.**

protozoan (prō'tə zō'ən), *Biology.* —*n.* any of a phylum or subkingdom (Protozoa) of chiefly microscopic, one-celled animal-like organisms that reproduce by fission, budding, or dividing into spores. Protozoans have a well-defined nucleus (eukaryotes) and many forms move by means of cilia, flagella, or pseudopodia. Amebas and paramecia are protozoans.
—*adj.* of or belonging to the protozoans: *When a protozoan cell, such as Ameba or Paramecium, divides, the two daughter cells separate and two individuals are formed* (Winchester, *Zoology*).
[from Greek *protōs* first + *zōon* animal]

protozoological, *adj.* of or having to do with protozoology.

protozoology (prō'tə zō ol'ə jē), *n.* the branch of biology that deals with protozoans: *Antony van Leeuwenhoek ... is known as the father of protozoology. ... In 1674, at the age of 42, he discovered free-living Protozoa in a freshwater pond, including what must have been Euglena viridis and several species of ciliates. The same year he described parasitic Protozoa for the first time in material obtained from the gall bladder of a rabbit* (Hegner and Stiles, *College Zoology*).

protozoon (prō'tə zō'on), *n., pl.* **-zoa** (-zō'ə). = protozoan.

proustite (prüs'tīt), *n. Mineralogy.* a mineral consisting of a sulfide of arsenic and silver, occurring in crystals or masses of a red color: *The ruby silver minerals, proustite and pyrargyrite, are of minor importance as sources of silver* (Jones, *Minerals in Industry*). *Formula:* Ag_3AsS_3 Compare **pyrargyrite.** [from Joseph-Louis *Proust,* 1754–1826, French chemist who discovered the mineral + *-ite*]

prove, *v. Mathematics.* to demonstrate the validity of (a statement or calculation): *Prove indirectly that the null set ∅ is a subset of every set* (Pearson and Allen, *Modern Algebra*). See also **proof.**

proventriculus (prō'ven trik'yə ləs), *n., pl.* **-li** (-lī). *Zoology.* **1** the soft, anterior stomach of a bird, which secretes gastric juices. It lies between the crop and gizzard (ventriculus): *The stomach consists of 2 parts, an anterior proventriculus with thick glandular walls which secrete the gastric juice, and a thick muscular gizzard which grinds up the food with the aid of small pebbles swallowed by the bird* (Hegner and Stiles, *College Zoology*).
2 the digestive chamber between the crop and stomach in insects: *Infected fleas suffer from the plague bacilli they carry, for these eventually block the valve between the proventriculus and midgut of the insect, which slowly starves to death* (New Scientist).
3 a muscular crop in worms.
[from New Latin, from Latin *pro-* before, anterior + *ventriculus* ventricle]

provirus (prō vī'rəs), *n. Microbiology.* the precursor or latent form of a virus, capable of becoming an active, replicating virus when incorporated into the genetic material of a host cell: *The provirus may suddenly de-* velop into virus and the bacterium give rise to a group of virus particles (Scientific American).

provitamin, *n. Biochemistry.* the precursor of a vitamin, capable of assuming the activity of a vitamin through chemical change within the body. Carotene is a provitamin that is converted by the body into vitamin A.

proximal, *adj. Anatomy.* toward the point of origin or attachment, as of a limb, bone, or other structure.

proximate analysis, *Chemistry.* the separation and identification of the constituent compounds of a complex compound or mixture. An example of proximate analysis is the separation of wheat flour into starch, sugar, gluten, ligneous fiber, etc. Compare **ultimate analysis.**

pruinose (prü'ə nōs), *adj. Biology.* covered with a frostlike bloom or powdery secretion: *a pruinose stem, a pruinose shell.* [from Latin *pruinosus* frosty, from *pruina* hoarfrost]

Prussian blue, *Chemistry.* a deep-blue pigment, essentially a cyanogen compound of iron. *Formula:* $C_{18}Fe_7N_{18}$

prussiate (prus'ē āt), *n. Chemistry.* **1** a salt of prussic acid; cyanide. **2** = ferricyanide. **3** = ferrocyanide.

prussic acid (prus'ik), = hydrocyanic acid.

ps, *abbrev.* picosecond.

psephite (sē'fīt), *n. Geology.* any coarse fragmental sedimentary rock. Compare **conglomerate.** [from Greek *psephos* pebble + English *-ite*]

pseud-, *combining form.* a form of **pseudo-** sometimes used before vowels, as in **pseudaxis.**

pseudaxis (sü dak'sis), *n. Botany.* a false stem or axis; sympodium.

pseudo-, *combining form.* **1** false; spurious, as in *pseudomorph = a false form.*
2 imitative of; resembling, as in *pseudo-random.* Also spelled *pseud-* before vowels.
[from Greek *pseudēs* false]

pseudoallele (sü'dō ə lēl'), *n. Genetics.* one of a set of mutants that function as alleles of a single gene but differ structurally in that crossing-over can occur between them: *A pair of pseudoalleles in Drosophila melanogaster, which have been demonstrated to be physiologically different, have each been shown to be physiologically indistinguishable from one of two nonallelic genes in Drosophila virilis. This is further evidence that pseudoalleles may be initial steps in the evolution of genes* (Arthur G. Steinberg).

pseudoaquatic, *adj. Botany.* not really aquatic, but growing in wet places.

pseudobrookite (sü'dō brúk'īt), *n. Mineralogy.* an iron oxide of titanium that resembles brookite, found in some igneous rocks, such as andesite. *Formula:* Fe_2TiO_5

pseudocarp (sü'də kärp), *n. Botany.* a fruit that includes other parts in addition to the mature ovary and its contents, such as the apple, pineapple, or pear.
—**pseudocarpous** (sü'də kär'pəs), adj. of or having to do with a pseudocarp.

pseudocholinesterase (sü'dō kō'lə nes'tə rās), *n. Biochemistry.* an enzyme present in the liver, brain, and blood plasma that is chemically similar to and acts on the same esters as the enzyme cholinesterase.

pseudocoel (sü'də sēl), *n. Zoology.* a body cavity in some primitive animals, similar to the coelom except that it is unlined: *In roundworms . . . a cavity termed the*

528

pseudocoel, existing between the digestive tract and muscles, contains a body fluid which is set in circulation by the wriggling movements characteristic of such animals (Harbaugh and Goodrich, *Fundamentals of Biology*).

pseudoglobulin, *n. Biochemistry.* any globulin that is soluble in pure water: *Differences in solubility are also the basis for classifying proteins as albumins or globulins and globulins in turn as euglobulins and pseudoglobulins* (Scientific American).

pseudomonad (sü′də mō′nad), *n. Bacteriology.* any of a genus (*Pseudomonas*) of motile, aerobic, gram-negative bacteria, some of which are pathogenic. [from New Latin *Pseudomonas, Pseudomonadis,* from *pseudo-* + Latin *monas, monadis* monad]

pseudomonas (sü′də mō′nəs), *n., pl.* **-monades** (-mō′nə-dēz′ *or* -mon′ə dēz′). = pseudomonad: *Urinary tract infections caused by pseudomonas include everything from the serious kidney disease to cystitis, a bladder infection that is difficult to cure and recurs at periodic intervals* (Science News Letter).

pseudomorph, *n. Mineralogy.* a mineral which has the outward form of another mineral.
—**pseudomorphic** or **pseudomorphous,** *adj.* of, having to do with, or resembling a pseudomorph.

pseudopod (sü′də pod), *n.* = pseudopodium: *In the ameba, food is ingested by means of pseudopods* (Biology Regents Syllabus).

pseudopodal (sü dop′ə dəl), *adj.* = pseudopodial.

pseudopodial (sü′də pō′dē əl), *adj.* **1** having to do with a pseudopodium or with pseudopodia. **2** forming or formed by pseudopodia: *pseudopodial movement.*

pseudopodium (sü′də pō′dē əm), *n., pl.* **-dia** (-dē ə). *Biology.* **1** a temporary protrusion of the protoplasm of a one-celled organism, serving as a means of locomotion and a way of surrounding and thereby absorbing food: *Amoeboid movement has a peculiar fascination when it is observed in an active amoeba under the microscope. As the finger-like pseudopodium . . . of the amoeba advances, one sees the cytoplasm . . . flowing steadily through a central channel in the cell* (Douglas Marsland).
2 the posterior extremity of a rotifer, serving chiefly as a swimming organ.
[from New Latin, from Greek *pseudēs* false + *podos* foot]

pseudo-random, *adj. Mathematics, Statistics.* of or involving the use of numbers produced by a definite mathematical process but satisfying the tests for randomness of a statistical sampling or distribution: *There are hundreds of ways in which computers can generate . . . pseudo-random digits* (Martin Gardner).

P-shell, *n. Physics.* the energy level next in order after the O-shell, occupied by electrons whose principal quantum number is 6. See also **P electron.**

psi or **p.s.i.,** *abbrev.* pounds per square inch.

psilomelane (sī lom′ə lān), *n. Mineralogy.* any mixture of black manganese oxide minerals. [from Greek *psilos* bare + *melanos* black]

psilophyte (sī′lə fīt), *n. Biology.* any of a division (Psilophyta) of simple vascular plants with upright branches that appeared in the early Paleozoic and were the first of the land plants: *Devonian rocks contain a peculiar group of leafless land plants (psilophytes), which are quite unlike any others, but give evidence in their gen-*

eralized characters of ancestral relationship to various later forms of vegetation (Moore, *Introduction to Historical Geology*). [from Greek *psilos* bare + *phyton* plant]
—**psilophytic** (sī′lə fit′ik), *adj.* of or having to do with the psilophytes: *psilophytic flora.*

psi particle, *Nuclear Physics.* an unstable, neutral subatomic particle of large mass and long lifetime. The psi particle is a hadron thought to be composed of a charmed quark and a charmed antiquark. *The main significance of the discovery of the psi particle was that it provided compelling evidence for the existence of a fourth kind of quark, which had earlier been named the "charmed" quark* (Scientific American). Also called **J particle.**

psittaceous (si tā′shəs), *adj.* = psittacine.

psittacine (sit′ə sēn), *adj. Zoology.* of, having to do with, or resembling a parrot or parrots: *Parrots and birds of the psittacine family, such as parakeets ...* (Science News Letter). [from Greek *psittakos* parrot]

psoadic (sō ad′ik), *adj. Anatomy.* of or having to do with the psoas muscles: *the psoadic plexus.*

psoas (sō′əs), *n. Anatomy.* either of two muscles of the loin, the larger *psoas major* and the smaller *psoas minor.* [from New Latin *psoas,* plural, from Greek *psoa* loin muscle]

psychro-, *combining form.* **1** cold; coldness; low temperature, as in *psychrotolerant, psychrophilic.* **2** relative humidity, as in *psychrometry.* [from Greek *psychros* cold]

psychrometric (sī′krə met′rik), *adj. Meteorology.* of or having to do with psychrometry: *psychrometric tables.*

psychrometry (sī krom′ə trē), *n. Meteorology.* measurement of the relative humidity of the air.

psychrophile (sī′krə fil), *n. Microbiology.* a psychrophilic organism.

psychrophilic (sī′krə fil′ik), *adj. Microbiology.* able to grow at low temperatures: *psychrophilic bacteria. Psychrophilic microorganisms have been defined as organisms with the ability to grow well at 0°C* (J. L. Ingraham).

psychrotolerant, *adj. Microbiology.* able to endure cold: *Slime is caused by cold-resistant, or psychrotolerant, bacteria which are not a health hazard but which produce an objectionable smell and change the colour of the meat* (Science Journal).

Pt, *symbol.* platinum.

PTC, *abbrev.* phenylthiocarbamide.

pteridine (ter′ə dēn), *n. Biochemistry.* **1** a yellow crystalline compound obtained from pigments in the wings of butterflies, characterized by a fused pyrazine and pyrimidine ring system. *Formula:* $C_6H_4N_4$

cap, fāce, fäther; best, bē, tėrm; pin, fīve;
rock, gō, ôrder; oil, out; cup, put, rüle,
yü in use, *yu* in uric;
ng in bring; sh in rush; th in thin, ᴛʜ in then;
zh in seizure.
ə = a in about, e in taken, i in pencil, o in
lemon, u in circus

2 any of various compounds derived from it, occurring in nature especially as insect pigments and in vitamins of the B group such as riboflavin.
[from German *Pteridin,* from Greek *pteron* wing (in reference to the butterfly wings)]

pteridological (ter′ə də loj′ə kəl), *adj.* having to do with pteridology.

pteridology (ter′ə dol′ə jē), *n.* the branch of botany that deals with ferns. [from Greek *pteridos* fern + English *-logy*]

pteridophyte (ter′ə dō fit′), *n. Botany.* any member of a division (Pteridophyta) of vascular plants that includes the ferns, horsetails, and club mosses. Pteridophytes formed the chief vegetation of the earth during the Carboniferous period. They show a distinct alternation of generations in their life cycles. Compare **bryophyte, tracheophyte.** [from Greek *pteridos* fern + *phyton* plant]
—**pteridophytic** (ter′ə də fit′ik), *adj.* of or having to do with pteridophytes: *pteridophytic sporophytes.*

pteridosperm (ter′ə də spėrm), *n. Botany.* any of a group of fossil plants having the external aspect of ferns, but bearing true seeds: *Fernlike pteridosperms are the earliest to start protecting their seeds* (New Scientist). Also called **seed fern.**

pterin (ter′in), *n. Biochemistry.* a pigment or other compound containing a ring system such as that of pteridine. Folic acid and xanthopterin are pterins. [from Greek *pteron* wing]

pteropod (ter′ə pod), *Zoology.* —*adj.* of or belonging to a group of mollusks with lateral portions of the foot expanded into winglike lobes used in swimming.
—*n.* a pteropod mollusk.
[from Greek *pteron* wing, feather + *pous, podos* foot]
—**pteropodous** (tə rop′ə dəs), *adj.* **1** = pteropod. **2** characteristic of a pteropod.

pterygoid (ter′ə goid), *Anatomy.* —*adj.* **1** winglike. **2** of or having to do with either of two processes of the sphenoid bone.
—*n.* a pterygoid muscle, nerve, or other related part.
[from Greek *pterygoeridēs* winglike, from *pterygos* wing + *eidos* form]

pterygoid plate, *Anatomy.* either of the two sections, the lateral section or the medial section, that make up a pterygoid process.

pterygoid process, *Anatomy.* **1** either of two processes of bone descending from the junction of the body and the great wing of the sphenoid bone of the skull. Each process consists of a lateral and a medial section.
2 any of the pterygoid plates.

pteryla (ter′ə lə), *n., pl.* **-lae** (-lē). *Zoology.* one of the definite areas on the skin of a bird on which feathers grow: *Only certain portions of the pigeon's body bear feathers; these feather tracts are termed pterylae, and the featherless areas apteria* (Hegner and Stiles, *College Zoology*).
[from New Latin, from Greek *pteron* feather + *hylē* wood]

pterylosis (ter′ə lō′sis), *n. Zoology.* the arrangement or disposition of the feathers of a bird in definite areas. [from New Latin]

Ptolemaic (tol′ə mā′ik), *adj. Astronomy.* of or having to do with the astronomer Ptolemy (A.D. 100?–165?) or his observations and theories. The **Ptolemaic system** of astronomy taught that the earth is the fixed center of the universe and that the sun, moon, and other celestial bodies move around the earth. *In his great book, the Syntaxis, usually known by its later Arabian title, the Almagest, Ptolemy improved the theory of epicycles as applied to the sun and the moon, and extended it to the motions of the five planets. The whole theory is known, therefore, as the Ptolemaic system; its chief feature is the resolution of the complex observed motions into uniform circular ones centered at the earth* (Baker, *Astronomy*).

ptomaine or **ptomain** (tō′mān *or* tō mān′), *n. Chemistry.* any of several chemical compounds produced by bacteria in decaying matter. Ptomaines are a group of basic, nitrogenous, organic compounds, some of which are poisonous. *Defectively canned foods sometimes contain ptomaines.* [from Italian *ptomaina,* from Greek *ptoma* corpse]

ptyalin, (tī′ə lən), *n. Biochemistry.* an enzyme contained in the saliva of human beings and of certain other animals, having the property of converting starch into dextrins and maltose, thus aiding digestion. Also called **salivary amylase.** [from Greek *ptyalon* saliva, from *ptyein* to spit]

Pu, *symbol.* plutonium.

puberulent (pyü ber′yə lənt), *adj. Botany.* covered with fine, short down; minutely pubescent. [from Latin *puber* downy + *-ulent,* as in *pulverulent*]

pubes[1] (pyü′bēz), *n.* **1** *Anatomy.* **a** the hair appearing on the lower abdomen at puberty. **b** the lower abdomen. **2** *Biology.* a soft, downy growth on plants and some insects.
[from Latin *pubes* mature, adult]

pubes[2] (pyü′bēz), *n.* plural of **pubis.**

pubic (pyü′bik), *adj. Anatomy.* having to do with or in the region of the pubis: *the pubic bone.* The **pubic symphysis** is the place in front of the pelvis where the pubis of one side is joined to the other.

pubis (pyü′bis), *n., pl.* **-bes** (-bēz). *Anatomy.* the part of either hipbone that, with the corresponding part of the other, forms the front of the pelvis: *The pelvic structure is made up of three sets of paired bones, the ilium, the ischium and the pubis* (Scientific American). See the picture at **pelvis.** [from New Latin (*os*)*pubis* (bone) of the groin]

pubofemoral (pyü′bō fem′ər əl), *adj. Anatomy.* of, attached to, or common to the pubis and the femur: *pubofemoral fascia.*

pudendal (pyü den′dəl), *adj. Anatomy.* of, having to do with, or in the region of the pudendum or pudenda: *the pudendal canal, pudendal nerves.*

pudendum (pyü den′dəm), *n., pl.* **-da** (-də). *Anatomy.* the external genitals, especially of the female. [from Latin]

puerperal (pyü ėr′pər əl), *adj. Physiology.* of or having to do with childbirth: *puerperal fever, puerperal sepsis.* [from Latin *puerpera* woman in childbirth, from *puer* child + *parere* give birth]

puerperium (pyü′ər pir′ē əm), *n. Physiology.* **1** the state of having just given birth. **2** the period from childbirth to the time when the uterus regains its normal size, normally about six weeks. [from Latin, from *puerpera* woman in childbirth]

pulley, *n. Physics.* a wheel with a grooved rim in which a rope, cable, etc., can run to change the direction of the pull and so lift a load. It is a simple machine.

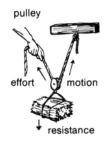

pulley

effort · motion

↓ resistance

pullulate (pul′yə lāt), *v. Biology.* **1** (of a seed) to sprout; germinate. **2** (of a plant or animal) to breed. [from Latin *pullulare,* from *pullulus* sprout, bud]
—**pullulation** (pul′yə lā′shən), *n.* **1** sprouting; germination. **2** generation or reproduction by budding.

pulmocutaneous (pul′mō kyü tā′nē əs), *adj. Zoology.* having to do with or supplying the lungs and skin. The **pulmocutaneous artery** in amphibians carries blood to the respiratory parts of the lungs, skin, and mouth. [from Latin *pulmo, pulmonen* lung + English *cutaneous*]

pulmonary, *adj.* **1** *Anatomy.* **a** of, having to do with, or situated in the lungs: *the pulmonary cells or tissues. The pulmonary vessels are those which carry the blood from the heart to the lungs and back again* (J. Harris). **b** occurring in or affecting the lungs: *pulmonary respiration. Pneumonia is a pulmonary disease.*
2 *Zoology.* having lungs or lunglike organs: *pulmonary arachnids or mollusks.* SYN: pulmonate. [from Latin *pulmonarius,* from *pulmonem* lung]

pulmonary artery, *Anatomy.* the artery which carries venous blood directly from the right ventricle of the heart to the lungs: *When the ventricle contracts, the auriculoventricular valve is closed, the pulmonary valve is opened, and blood flows through the pulmonary artery into the lungs* (Victor A. McKusick).

pulmonary circulation, *Physiology.* the part of the human circulatory system that carries the blood from the heart to the lungs and back again: *The customary use of the term "pulmonary circulation" for that part of the circulatory system which involves the right ventricle and the lungs, and of the term "systemic circulation" for the left ventricle and the balance of the circulatory system, seems to imply that the body has two distinct blood circuits. In actuality there is only one circuit; the systemic apparatus is one arc of it and the pulmonary apparatus is the other* (Julius H. Comroe, Jr.).

pulmonary valve, *Anatomy.* a valve consisting of three crescent-shaped flags at the opening of the pulmonary artery. Compare **semilunar valve.**

pulmonary vein, *Anatomy.* one of the four veins which carry oxygenated blood directly from the lungs to the left atrium of the heart.

pulmonate (pul′mə nāt *or* pul′mə nit), *Zoology.* —*adj.* **1** having lungs or lunglike organs.

2 of or belonging to a group of gastropod mollusks that have lunglike sacs and include most land snails and slugs and some aquatic snails.
—*n.* a plumonate gastropod.

pulp, *n.* **1** *Botany.* **a** the soft, fleshy part of any fruit or vegetable: *the pulp of an orange.* **b** the soft residue left when most of the liquid is pressed out of vegetables, fruit, etc. **c** the soft pith in the interior of the stem of a plant.
2 *Anatomy.* the soft inner part of a tooth, containing blood vessels and nerves: *Surrounding the pulp is a heavy layer of material, the dentine, which comprises the bulk of the tooth.* (Harbaugh and Goodrich, *Fundamentals of Biology*). See the picture at **tooth.**

pulsar (pul′sär), *n. Astronomy.* a rapidly spinning neutron star that is the source of powerful radio waves and of other radiant energy emitted in short, intense bursts or pulses at very precise intervals: *Pulsars have been found in all parts of the sky, but lie primarily in the Milky Way near the symmetry plane of the galaxy* (J. P. Ostriker). [from *pulse* + *-ar,* as in *quasar*]

pulsate, *v. Physiology.* **1** to expand and contract rhythmically, as the heart or an artery. SYN: beat, throb.
2 to exhibit a pulse or pulsation: *When alternating current is changed to direct current by means of a rectifier, the resulting direct current is often pulsating, which means that the magnitude of the direct current varies in pulses, corresponding to the frequency of the alternating current* (P. H. Craig and E. C. Easton). *It was apparently a pulsating star, which increased in brilliance as its surface area enlarged, and then sank to fainter luminosity as it collapsed to a smaller size* (G. W. Gray).
[from Latin *pulsatum* beaten, pushed, from *pulsus* a beating, pulse]
—**pulsation,** *n.* **1** a beating or throbbing, usually rhythmic, as of an artery or the heart.
2 a single beat or throb.
3 *Physics.* a periodically alternating increase and decrease in a quantity, such as a volume or voltage: *This would suggest some slow pulsation taking place in the sun which gradually alters the strength of these solar streamers* (E. P. George).

pulse, *n.* **1** *Physiology.* **a** the regular beating of the arteries caused by the rush of blood into them after each contraction of the heart. **b** a single beat in pulsation.
2 *Physics.* **a** a sudden variation or alteration in a quantity, such as a voltage, that is normally constant. **b** a short burst of radiant energy, especially a short succession of sound waves, radio waves, or the like: *A pulse of sound arriving by any particular path produces a train of waves through irregularities in the material through which it is passing* (T. F. Gaskell and M. N. Hill).

---

cap, fāce, fäther; best, bē, tėrm; pin, fīve;
rock, gō, ôrder; oil, out; cup, pút, rüle,
yü in use, *yu* in uric;
ng in bring; *sh* in rush; *th* in thin, ŦH in then;
zh in seizure.
ə = *a* in about, *e* in taken, *i* in pencil, *o* in lemon, *u* in circus

531

pulse wave

—*v. Physics.* to emit, generate, or modulate by means of pulses; produce (waves, energy, etc.) in the form of pulses: *High-powered monochromatic beams of laser light . . . may be modulated or pulsed in times as short as 10⁻¹¹s* (Nature).

—**pulsed,** *adj.* produced in the form of pulses; consisting of pulses: *The natural mode of communication with or between machines is by means of pulsed electrical signals* (Pierce, *Electrons, Waves, and Messages*).

pulse wave, *Physiology.* the wave of raised tension and arterial expansion which starts from the aorta with each ventricular systole contraction and travels to the capillaries.

pulverulent (pul ver′yə lənt *or* pul ver′ə lənt), *adj. Geology.* **1** consisting of fine powder. **2** crumbling to dust: *pulverulent rock.* [from Latin *pulverulentus* full of dust, from *pulvis* dust]

pulvillus (pul vil′əs), *n., pl.* **-villi** (-vil′ī). *Zoology.* a cushionlike pad or process on the foot of an insect, such as the fly, by which it can adhere to a vertical or other surface, such as a wall or ceiling. [from Latin *pulvillus* small pillow]

pulvinus (pul vi′nəs), *n., pl.* **-ni** (-ni). *Botany.* any cushionlike swelling at the base of a leaf or leaflet at the point of junction with the axis. The pulvinus is composed of thin-walled cells interspaced with extensive air chambers. [from Latin *pulvinus* cushion]

pumice (pum′is), *n. Geology.* a light, porous, glassy lava. Pumice is used, especially when powdered, for cleaning, smoothing, and polishing. [from Latin *pumicem*]

pumicite (pum′ə sīt), *n.* = pumice.

pump, *n.* **1** *Physics.* **a** an apparatus or machine which forces fluids into or out of things by alternate suction or compression: *Of all the pumps that have been developed, two are particularly important in physics laboratories: the rotary oil pump for pressures as low as 10⁻⁴ mm of mercury, and the mercury or oil diffusion pump for pressures as low as 10⁻⁸ mm of mercury* (Sears and Zemansky, *University Physics*). **b** any medium or device used for raising atoms or molecules to an excited state.

2 *Physiology.* **a** a molecular mechanism in living cells by which ions are transported across a cell membrane: *a diffusion pump, an enzyme pump.* **b** = sodium pump.

—*v. Physics.* **1** to excite (atoms or molecules) by a light source or by electron bombardment: *As in lasers, a variety of methods can conceivably create population inversion between nuclear states. The preferred method is to pump by neutron capture, which is immediately followed by a cascade of capture gamma-ray emissions* (George C. Baldwin).

2 to bring to or create a state of energy in (a laser) that excites the atoms in its active medium, such as helium: *Many lasers . . . are excited (or "pumped," as we say in the laboratory) by light. Others may be made to lase by radio waves, or by an electric current, or by chemical reactions* (National Geographic Magazine).

punctuated equilibrium, *Biology.* a modification of Darwinism which maintains that natural selection acts on a species to keep it stable rather than to alter it, and that the emergence of a new species is a separate event

that emphasizes the general equilibrium in nature: *Many paleontologists favoured the model of punctuated equilibrium. In this view species normally evolve slowly, and most changes occur when small populations diverge rapidly into new forms during short intervals of time* (P. J. Wyllie).

punctuational, *adj. Biology.* of or having to do with punctuated equilibrium; characterized by speciation as a separate or episodic occurrence over long periods of little or no change in nature: *Third, does evolution occur by gradual change or by episodes of rapid change interspersed with periods of stasis (punctuational evolution)*? (Robert E. Ricklefs).

—**punctuationalism,** *n.* = punctuated equilibrium: *The alternative theory is called . . . "punctuationalism." According to this, the diversity of life has come about as a result of sporadic adaptations by small, well-defined groups confronted by a new environment, interspersed with long periods of little or no change* (Manchester Guardian Weekly).

—**punctuationalist,** *n.* a supporter of punctuated equilibrium: *A convinced punctuationalist, he contrasts bivalves and mammals to support . . . the hypothesis that rate of evolution is determined by rate of speciation* (Science).

pungent, *adj. Botany.* piercing; sharp-pointed; *pungent leaves.* [from Latin *pungentem* piercing, pricking, from *punctum* point]

pupa (pyü′pə), *n., pl.* **-pae** (-pē), **-pas.** *Zoology.* **1** the intermediate stage between the larva and the adult in the metamorphosis of many insects: *Finally, the larva seems to have eaten its fill and goes into a quiescent stage called a pupa. It may spin threads of silk around itself to form a cocoon in which it spends its pupal state, or its outer body covering may simply harden to form a pupa case.* (Winchester, *Zoology*).

2 an insect in this stage. Most pupae are inactive and some, such as those of many moths, are enclosed in a tough case or cocoon.

[from Latin *pupa* girl, doll]

—**pupal** (pyü′pəl), *adj.* of, having to do with, or in the form of a pupa: *At the end of the pupal stage the full-grown wasp gnaws its way out of the nest* (Carl D. Duncan).

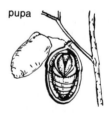

pupa

puparium (pyü pãr′ē əm), *n., pl.* **-ia** (-ē ə). *Zoology.* the hard pupal case formed, especially by certain dipterous insects, from the outermost larval skin. [from New Latin, from *pupa* pupa]

pupate (pyü′pāt), *v. Zoology.* to become a pupa; go through a pupal stage: *About half the British species [of butterflies] pupate hanging head downwards* (Mackean, *Introduction to Biology*).

—**pupation,** *n.* **1** the act of pupating: *Materials secreted by a gland-like mass associated with the brain of these insects have a controlling influence upon the process* of pupation (Harbaugh and Goodrich, *Fundamentals of Biology*).

2 the state of being a pupa; pupal stage: *Then they go through a change called pupation, and in about ten days they come out of the cocoon as full-grown wasps* (Carl D. Duncan).

pupil, *n. Anatomy.* the opening in the center of the iris of the eye, circular in man and most vertebrates and appearing as a black spot, which is the only place where light can enter the eye. The size of the pupil is regulated by expansion and contraction of the iris. [from Latin *pupilla,* originally, little doll, diminutive of *pupa* girl, doll (so called from the tiny reflection of oneself that may be seen in the eye of another person)]
—**pupillary** (pyü′pə ler′ē), *adj.* of or having to do with the pupil of the eye: *pupillary reflex.*

pupiparous (pyü pip′ər əs), *adj. Zoology.* bringing forth young that are already pupae: *Many ticks are pupiparous.* [from New Latin *pupa* pupa + Latin *parere* give birth to]

pure, *adj.* **1** concerned with theory rather than practical use; not applied; theoretical: *pure mathematics, pure research, pure science.*

2 *Genetics.* **a** homozygous, and therefore breeding true for at least one hereditary character. **b** produced by inbreeding, inbred: *Selection within pure lines . . . is ineffective because it is biologically meaningless* (Srb and Owen, *General Genetics*).

3 *Biology.* containing only one species or clone: *a pure culture of mycorrhizal fungi.*

purine (pyùr′ēn′), *n. Biochemistry.* **1** a colorless, crystalline organic base containing nitrogen and related to uric acid. Purine is characterized by a fused imidazole and pyrimidine ring system: *The purines and pterins contribute a major source of colour to the wings of butterflies; the former as a silvery white pigmentation and the latter as white, red, yellow, and orange pigments, which are also found in the integument of wasps* (B. Nickerson). Formula: $C_5H_4N_4$

2 any of a group of compounds naturally derived from it, such as the nucleic acid components adenine and guanine. Compare **pyrimidine.**
[from German *Purin,* a blend of Latin *purus* pure + New Latin *uricus* uric (acid)]

Purkinje cell (pèr kin′jē), *Biology.* any of the large branching cells, with cone-shaped bodies, that make up the intermediate layer of the cerebellar cortex: *Each Purkinje cell sends its message out of the cerebellum through a long threadlike axon* (Scientific American). [named after Johannes E. *Purkinje,* 1787–1869, Bohemian physiologist]

Purkinje fiber, *Biology.* any of the modified muscle fibers of the heart that make up a network by which muscle impulses are conducted through the heart: *From this "pacemaker" the impulse spreads over the two atria and reaches the atrioventricular node, whence it travels through a network of cells called Purkinje fibers to the rest of the heart* (Isaac Harary).

Purkinje shift, *Physiology.* a rapid decrease in the sensation of brightness produced by light of long wavelength, when illumination is reduced to a low level: *The fish eye behaves in the same way as the human eye when the illumination level is reduced: sensitivity increases, colour discrimination is reduced, and a "Purkinje shift" occurs* (Richard Fatehchand).

pus, *n. Physiology, Medicine.* a thick, yellowish-white, opaque fluid consisting of white blood cells, bacteria, serum, etc., found in sores, abscesses, and other infected tissue of the body. [from Latin]

putrefaction (pyü′trə fak′shən), *n. Biology.* the action or process of putrefying; decomposition of organic matter, especially by the action of bacteria and fungi, accompanied by the production of foul-smelling gases. SYN: decay, rot.
—**putrefactive** (pyü′trə fak′tiv), *adj.* **1** causing putrefaction: *putrefactive bacteria.*
2 characterized by or having to do with putrefaction: *putrefactive fermentation.*

putrefy (pyü′trə fī), *v. Biology.* to break down or cause to break down organic matter by the action of bacteria and fungi, producing foul-smelling gases; undergo putrefaction. SYN: decay, rot. [from Middle French *putrefier,* from Latin *putrifieri,* from *puter* rotten + *fieri* become]

putrescine (pyü tres′ēn), *n. Chemistry.* a colorless, foul-smelling ptomaine, formed in the decay of animal tissue. *Formula:* $H_2N(CH_2)_4NH_2$ [from Latin *putrescere* grow rotten, from *putere* to stink, from *puter* rotten]

pycnidial (pik nid′ē əl), *adj. Biology.* of or having to do with a pycnidium.

pycnidium (pik nid′ē əm), *n., pl.* **-ia** (-ē ə). *Biology.* a fruiting body in some basidiomycetous fungi. [from New Latin, from Greek *pyknos* thick, dense]

pycnium (pik′nē əm), *n., pl.* **-nia** (-nē ə). *Biology.* **1** a spermagonium, especially of certain rust fungi. **2** = pycnidium. [from New Latin, from Greek *pyknos* thick, dense]

pygidial (pī jid′ē əl), *adj. Zoology.* of or having to do with a pygidium: *a pygidial bladder.*

pygidium (pī jid′ē əm), *n., pl.* **-ia** (-ē ə). *Zoology.* the caudal part or terminal segment of the body in insects, crustaceans, and other invertebrates. [from New Latin, from Greek *pygidion,* diminutive of *pygē* rump]

pyknosis (pik nō′sis), *n., pl.* **-ses** (-sēz). *Biology.* a degenerative condition of cells in which the nucleus condenses and shrinks into a dense mass that stains more deeply than usual: *pyknosis of nerve cells.* [from New Latin, from Greek *pyknos* thick, dense]

pyloric (pī lôr′ik), *adj. Anatomy.* of, having to do with, or situated in the region of the pylorus: *the pyloric end of the stomach.*

pylorus (pī lor′əs), *n., pl.* **-lori** (-lôr′ī′). *Anatomy.* the opening that leads from the stomach into the intestine. [from Greek *pylōros* (originally) gatekeeper, from *pylē* gate]

pyr-, *combining form.* the form of **pyro-** before some vowels, as in *pyrone.*

---

cap, fāce, fäther; best, bē, tèrm; pin, fīve;
rock, gō, ôrder; oil, out; cup, pùt, rüle,
yü in use, yù in uric;
ng in bring; sh in rush; th in thin, ⱦⱨ in then;
zh in seizure.
ə = a in about, e in taken, i in pencil, o in lemon, u in circus

533

pyramid

pyramid, n. 1 *Geometry*. a solid figure having a polygon for a base and triangular sides which meet in a point. 2 *Crystallography*. a crystal whose faces intersect the vertical axis and one or two of the lateral axes.
3 *Anatomy*. a part of pyramidal form: **a** a mass of longitudinal nerve fibers on each side of the medulla oblongata. **b** one of the conical-shaped masses making up the medullary substance of the kidney.
[from Greek *pyramidos*]
—**pyramidal** (pə ram′ə dəl), *adj*. shaped like a pyramid; triangular: *The pyramidal cells in one hemisphere of the brain may activate the symmetrical region of the other hemisphere* (Scientific American).
—**pyramidally,** *adv*.

(def. 1)

pyran (pī′rən), n. *Chemistry*. 1 either one of two isomeric compounds, each having a ring of five carbon atoms and one oxygen atom: *Pyran resembles RNA in some features of molecular structure, which is perhaps the reason that it stimulates the output of interferon* (Scientific American).
2 any derivative of either isomer containing a ring of five carbon atoms and one oxygen atom. *Formula:* C_5H_6O
[from *pyr(one*) + -*an,* variant of -*ane*]
pyranometer, n. an instrument used to measure the intensity of direct and indirect solar radiation.
pyrargyrite (pī rär′jə rīt), n. *Mineralogy*. a mineral consisting of a sulfide of silver and antimony, and showing, when transparent, a deep ruby-red color by transmitted light. *Formula:* Ag_3SbS_3 Compare **proustite.** [from German *Pyrargyrit,* from Greek *pyr* fire + *argyros* silver]
pyrazine (pir′ə zēn), n. *Chemistry*. 1 a weakly basic white crystalline organic compound. *Formula:* $C_4H_4N_2$ 2 any of various compounds derived from it.
pyrenoid (pī′rə noid or pī rē′noid), n. *Botany*. a small protein structure found in the chloroplast of certain algae, associated with the formation and accumulation of starch. [from Greek *pyrēn* stone of a fruit]
pyridine (pir′ə dēn), n. *Chemistry*. a liquid organic base with a pungent odor, occurring especially in coal tar, and serving as the parent substance of many compounds. It is used as a solvent and waterproofing agent, and in making various drugs and vitamins. *Formula:* C_5H_5N [from *pyr(role*) + -*id(e*) + -*ine*]
pyridine nucleotide, *Biochemistry*. either one of the two oxidizing coenzymes diphosphopyridine nucleotide

and triphosphopyridine nucleotide: *It was found that in addition to coenzyme A the oxidation of fatty acids required various other factors, such as certain pyridine nucleotides, groups of enzymes and adenosine triphosphate* (Scientific American).
pyridoxal (pir′ə dok′səl), n. *Biochemistry*. a crystalline aldehyde, one of the forms of pyridoxine found in certain enzymes, important in the synthesis of amino acids and especially transamination. *Formula:* $C_8H_9NO_3$
pyridoxamine (pir′ə dok′sə mēn), n. *Biochemistry*. a crystalline amine, one of the forms of pyridoxine found in certain enzymes, important in protein metabolism. *Formula:* $C_8H_{12}N_2O_2$
pyridoxine (pir′ə dok′sēn), n. *Biochemistry*. vitamin B_6, essential to human nutrition, found especially in wheat germ, yeast, fish, and liver. *Formula:* $C_8H_{11}NO_3$ [from *pyrid(ine*) + *ox(ygen*) + -*ine*]
pyrimidine (pī rim′ə dēn *or* pi rim′ə dēn′), n. *Biochemistry*. 1 a liquid or crystalline organic base with a strong odor, whose molecular arrangement is a six-membered ring containing atoms of nitrogen. *Formula:* $C_4H_4N_2$ 2 any of various compounds naturally derived from it, such as the nucleic acid components cytosine, thymine, or uracil. Compare **purine.**
[from German *Pyrimidin,* from *Pyridin* pyridine]
pyrite (pī′rīt), n. *Mineralogy*. 1 a common yellow mineral with a metallic luster, a compound of iron and sulfur, which resembles and is often mistaken for gold; fool's gold. It is used to make sulfuric acid. *Formula:* FeS_2
2 **pyrites,** pl. any of various compounds of sulfur and a metal, such as tin pyrites, an ore of tin.
[from Greek *pyritēs* (*lithos*) (stone) of fire]
pyro-, combining form. 1 fire; heat; high temperatures; as in *pyrogenic, pyroelectricity.*
2 formed by heat, as in *pyroacid.* Also spelled **pyr-** before some vowels.
[from Greek, from *pyr* fire]
pyroacid, n. *Chemistry*. any of various acids obtained by subjecting other acids to heat.
pyrobitumen, n. *Mineralogy*. any of a class of natural hydrocarbons that yield bitumens when heated.
pyroclastic (pī′rō klas′tik), adj. *Geology*. composed chiefly of fragments of volcanic origin: *pyroclastic flows. Agglomerate and tuff are pyroclastic rocks.* Compare **holoclastic.**
pyroelectric, adj. *Physics*. of or having to do with pyroelectricity.
pyroelectricity, n. *Physics*. 1 the electrified state or electric polarity produced by a change in temperature, as in certain crystals.
2 the study of such phenomena.
pyrogenic, adj. *Geology*. produced by fire: *Volcanic rock is pyrogenic.*
pyrolusite (pī′rō lü′sīt), n. *Mineralogy*. native manganese dioxide, used as a source of manganese and in making various chemicals, such as chlorine and oxygen. *Formula:* MnO_2 [from German *Pyrolusit,* from Greek *pyr* fire + *lousis* washing]
pyrolysis (pī rol′ə sis), n. *Chemistry*. decomposition or transformation of a compound by exposure to high temperatures: *He established by pyrolysis that the*

strength of binding varies considerably in hydrocarbons (New Scientist).

—**pyrolytic** (pī′rə lit′ik), *adj.* of, involving, or produced by pyrolysis. **Pyrolytic carbon** is a strong, heat-resistant form of graphite obtained from products of hydrocarbon pyrolysis. —**pyrolytically,** *adv.*

pyromorphite (pī′rə môr′fīt), *n. Mineralogy.* a mineral consisting of chloride and phosphate of lead, occurring both in crystals and masses that are green, yellow, brown, or whitish in color. *Formula:* $Pb_5(PO_4)_3Cl$ [from German *Pyromorphit,* from Greek *pyr* fire + *morphē* form]

pyrone (pī′rōn), *n. Chemistry.* either of two isomeric compounds, one of which is the source of various natural yellow dyes. *Formula:* $C_5H_4O_2$ [from German *Pyron,* probably from *pyr-* + *-on* -one]

pyronine (pī′rə nēn), *n. Chemistry.* any of a group of red dyes used in microscopic stains, especially to indicate the presence of RNA. [from German *Pyronin,* probably irregular from *pyro-* + *-in* -ine]

pyrophoric (pī′rə fôr′ik), *adj.* 1 *Chemistry.* having the property of catching fire through exposure to air; igniting spontaneously in air, as certain compounds of phosphorus.
2 *Biology.* (chiefly of an insect or marine organism) emitting light; phosphorescent; bioluminescent.
[from Greek *pyrophoros* fire-bearing]

pyrophosphatase (pī′rō fos′fə tās), *n. Biochemistry.* an enzyme that promotes the hydrolysis of esters of phosphoric acid: *Pyrophosphatase is widely distributed in mammalian tissues* (H. D. Kay).

pyrophosphate, *n. Chemistry.* a salt or ester of pyrophosphoric acid.

pyrophosphoric acid, *Chemistry.* a crystalline acid used as a catalyst and, in the form of its salts, in medicine. *Formula:* $H_4P_2O_7$

pyrophyllite (pī′rə fil′īt), *n. Mineralogy.* a hydrous silicate of aluminum found in metamorphic rock. It resembles talc and is usually whitish or greenish in color. *Here the hydrostatic forces are transmitted to the specimen via a soft compressible stone such as pyrophyllite* (Science Journal). *Formula:* $AlSi_2O_5(OH)$ [from German *Pyrophyllit,* from Greek *pyr* fire + *phyllon* leaf]

pyrosulfuric acid, *Chemistry.* a thick, fuming liquid acid, a powerful oxidizing and dehydrating agent. *Formula:* $H_2S_2O_7$

pyroxene (pī′rok sēn′), *n. Mineralogy.* any of a group of silicate minerals, usually calcium, magnesium and iron silicate, often found in igneous rocks: *Pyroxenes occupy the space between the feldspars. Rocks that show this mineral pattern are described as having an ophitic texture; when the minerals present are small but visible without magnification, such rocks are specifically called diabases* (C. J. Schuberth). *Most abundant [on the moon] were the pyroxenes* (Laurence H. Nobles).

[from French *pyroxène,* from *pyro* + Greek *xenos* stranger]

pyroxenite (pī rok′sə nīt), *n. Geology.* a rock composed in large part, of pyroxene. It is an ultramafic igneous rock.

pyroxylin (pī rok′sə lin), *n. Chemistry.* any of several explosive or flammable compounds consisting of cellulose nitrates.

pyrrhotite (pir′ə tīt), *n. Mineralogy.* a native iron sulfide having a reddish-brown to bronze color and a metallic luster, occurring in crystals and masses. It often contains nickel and is usually slightly magnetic. [from Greek *pyrrhotēs* redness, from *pyrrhos* fiery red, from *pyr* fire]

pyrrole (pir′ōl), *n. Chemistry.* a colorless liquid that smells like chloroform, obtained mostly from coal tar. It is the parent compound of chlorophyll, hemin, various proteins, and other important natural substances. *Formula:* C_4H_5N [from German *Pyrrol,* from Greek *pyrrhos* fiery red + German *-ol* -ole]

pyruvate (pī rü′vāt), *n. Chemistry.* a salt or ester of pyruvic acid: *It has been known for some time that the heart burns carbohydrates, particularly the sugar glucose and some of its breakdown products, such as lactate and pyruvate* (R. J. Bing).

pyruvic acid (pī rü′vik), *Chemistry.* a colorless acid that smells like acetic acid. It is an important intermediate product in carbohydrate and protein metabolism: *The enzymes enter into a series of 12 chemical reactions with body sugars to form pyruvic acid, a building block of fat* (Science News Letter). *Formula:* $CH_3COCOOH$ [from Greek *pyr* fire + Latin *uva* grape]

pyruvic oxidase, *Biochemistry.* any enzyme which catalyzes the oxidation of pyruvic acid.

pythagorean theorem, *Geometry.* the theorem that the square of the hypotenuse of a right triangle equals the sum of the squares of the other two sides. [named after *Pythagoras,* 582?–500? B.C., Greek philosopher and mathematician]

pyxidium (pik sid′ē əm), *n., pl.* **-ia** (-ē ə). *Botany.* a seed vessel that bursts open transversely into a top and bottom part, the top part acting as a lid. [from New Latin, from Greek *pyxidion,* diminutive of *pyxis, pyxidos* box]

pyxis (pik′sis), *n., pl.* **pyxides** (pik′sə dēz). = pyxidium.

---

cap, fāce; fäther; best, bē; tèrm; pin, five;
rock, gō, ôrder; oil, out; cup, pùt, rüle,
yü in use, *yu̇* in uric;
ng in bring; *sh* in rush; *th* in thin; ᵗʜ in then;
zh in seizure.
ə = *a* in about, *e* in taken, *i* in pencil, *o* in lemon, *u* in circus

Q

QCD, *abbrev.* quantum chromodynamics: *In the QCD theory quarks interact by exchanging a gluon* (New Scientist).

QED, *abbrev.* quantum electrodynamics.

Q scale, *Geology.* a measure of duration of vibrations in the surface of the earth: *On earth, such rubble would be a terrible conductor of tremors; on the geologist's "Q" scale ... it would rate about 10. By contrast the moon rubble scored at least 2,000* (Science News). [*Q,* abbreviation of German *Querwellen* transverse waves]

QSO (kyü′es′ō′), *n.* = quasar. [from the abbreviation of *quasi-stellar object*]

quad, *n.* a unit of energy equal to one quadrillion (10¹⁵) British thermal units, used for expressing national or global energy use and supply.

quad., *abbrev.* **1** quadrangle. **2** quadrant.

quadrangle, *n.* **1** = quadrilateral. **2** the rectangular area represented by one of the United States Geological Survey topographic and geological maps. The two common sizes are tracts about 13 miles wide by 17 miles north to south and 6 1/2 miles wide by 8 1/2 miles north to south. [from Late Latin *quadrangulum,* from Latin *quadr-* four + *angulus* angle]
—**quadrangular,** *adj.* consisting of or resembling a quadrangle; having four angles.

quadrant, *n.* **1** *Mathematics.* **a** a quarter of the circumference of a circle; arc of 90 degrees. **b** the area contained by such an arc and two radii drawn perpendicular to each other.
2 *Geometry.* one of the four parts into which a plane is divided by two straight lines crossing at right angles. The upper right-hand section is the first quadrant.
3 *Anatomy.* any one of four regions into which the abdomen is divided for diagnostic purposes.
4 *Embryology.* one of the four blastomeres in the four-cell stage of the ovum.
[from Latin *quadrantem* a fourth]
—**quadrantal** (kwod ran′tl), *adj. Mathematics.* of or having to do with a quadrant; included in the fourth part of the surface of a circle.

QUADRANTS

2nd	1st
3rd	4th

(def. 1, 2)

Quadrantid (kwod ran′tid), *n. Astronomy.* one of a shower of meteors occurring early in January: *Shower meteors come from well-defined streams, of which the Perseids, the Leonids, and the Quadrantids are among the most celebrated* (New Scientist). [from New Latin *Quadrantem,* name of a former constellation from which the showers seem to radiate]

quadrat (kwod′rət), *n. Ecology.* one of a group of small areas or plots of land, usually rectangular, laid off for close study of the relative abundance of species or of other questions involving the geographical distribution of plants and animals. [from Latin *quadratum* a square, cube; see QUADRATIC]

quadrate (kwod′rit or kwod′rāt), *n.* = quadrate bone.

quadrate bone, *Zoology.* one of the pair of bones in birds, fishes, amphibians, and reptiles that joins the lower jaw to the skull.

quadratic, *Mathematics.* —*adj.* of or involving the second power of an unknown quantity or of a variable; involving expressions of the second degree. A **quadratic equation** is one in which one or more of the terms is squared, but raised to no higher power; it can be written in the form $ax^2 + bx + c = 0$, where $a \neq 0$. A **quadratic function** is one described by the quadratic equation $f(x) = ax^2 + bx + c$; $a \neq 0$. The graph of a quadratic function is a parabola. Compare **quadric.**
—*n.* a quadratic equation.
[from Latin *quadratum* a square, cube, from *quadrus* square, related to *quattuor* four] —**quadratically,** *adv.*

quadratic form, *Mathematics.* a homogeneous polynomial in two or more variables. Compare **quantic.**

quadratics, *n.* the branch of algebra that deals with quadratic equations.

quadrature (kwod′rə chùr or kwod′rə chər), *n.* **1** *Mathematics.* the finding of a square equal in area to a given surface, especially one bounded by a curve.
2 *Astronomy.* **a** the position of a celestial body that is 90 degrees away from another: *When one of these planets attains an elongation of 90°, it is said to be at quadrature. When the planet's elongation becomes 180°, it is said to be at opposition; then the planet rises as the sun sets and vice versa* (Krogdahl, *The Astronomical Universe*). **b** either of the two points in the orbit of a celestial body halfway between the points of conjunction and opposition: *A half moon is visible at the quadratures of the moon.*

quadri-, *combining form.* four, having four; four times, as in *quadrilateral = having four sides.* [from Latin *quadri-,* related to *quattuor* four]

quadric (kwod′rik), *Mathematics.* —*adj.* of the second degree; quadratic (applied especially to functions or equations with more than two variables). A **quadric surface** is one having an equation of the second degree.
—*n.* a quantity, equation, or function of the second degree. Compare **quartic, quintic.**

quadriceps (kwod′rə seps), *n. Anatomy.* the large muscle of the front of the thigh, which extends the leg, and has four heads or origins. [from New Latin, from Latin *quadri-* four + *caput* head]

quadricipital (kwod′rə sip′ə təl), *adj. Anatomy.* having to do with the quadriceps.

quadrifoliate (kwod′rə fō′lē it), *adj. Botany.* having four leaves or leaflets; having leaves in whorls of four.

quadrilateral, *Geometry.* —*adj.* having four sides and four angles.
—*n.* **1** a plane figure having four sides and four angles. **2** a figure formed by four straight lines which, if extended, intersect at six points.
[from Latin *quadrilaterus,* from *quadri-* four + *latus* side]

quadrillion, *n., adj.* **1** (in the United States, Canada, and France) 1 followed by 15 zeros. **2** (in Great Britain and Germany) 1 followed by 24 zeros. [alteration of French *quadrillon,* from *quadri-* four + *million* million]

quadrinomial (kwod′rə nō′mē əl), *adj.* **1** *Mathematics.* consisting of four terms, such as $a^2 - ab + 4a + b^2$. **2** *Statistics.* having or involving four distinct probabilities or outcomes: *a quadrinomial distribution.* [from *quadri-* + (*bi*)*nomial*]

quadrivalence (kwod′rə vā′ləns), *n.* the quality or condition of being quadrivalent.

quadrivalent (kwod′rə vā′lənt), *adj.* **1** *Chemistry.* having a valence of four; tetravalent.
2 *Biology.* of or designating a group of four chromatids of two homologous chromosomes joined together in synapsis during the prophase of meiosis.
—*n.* **1** *Chemistry.* a quadrivalent atom or element.
2 *Biology.* a quadrivalent group of chromosomes.
—**quadrivalently,** *adv.*

quadru-, *combining form.* a variant of **quadri-,** as in *quadruped.*

quadrumane (kwod′rù mān), *n. Zoology.* a quadrumanous animal.

quadrumanous (kwod rü′mə nəs), *adj. Zoology.* **1** four-handed; using all four feet as hands, as primates other than human beings do. **2** of or belonging to a former grouping of primates that included monkeys, apes, and lemurs. [from New Latin *quadrumanus,* from Latin *quadru-* four + *manus* hand]

quadruped, *Zoology.* —*n.* an animal that has four feet: *Cattle, cats, dogs, lizards, and frogs are all quadrupeds.*
—*adj.* four-footed: *a quadruped mammal.*
[from Latin *quadrupedem,* from *quadru-* four + *pedem* foot]
—**quadrupedal** (kwo drü′pə dəl or kwod′rə ped′l), *adj.* of, having to do with, or like a quadruped; four-footed: *In comparing the hand of monkeys with the hand of man one must bear in mind an obvious fact that is all too often overlooked: monkeys are largely quadrupedal, whereas man is fully bipedal* (John Napier).

quadrupole (kwod′rə pōl′), *n. Physics.* a set or combination of two dipoles: *The current is made to flow along four rods, forming a quadrupole* (Science Journal).

qualitative analysis, *Chemistry.* a testing of a substance or mixture to find out what its chemical constituents are: *Analytical Chemistry is devoted to discovering means for separating samples of matter into their constituent parts for purposes of identification. When the purpose is to determine the exact nature, but not the amounts, of the constituents, the work is known as Qualitative Analysis* (Offner, *Fundamentals of Chemistry*).

quality, *n. Acoustics.* the character of sounds aside from pitch and volume or intensity: *We may define sound quality as the psychological effect produced by the relative intensities of the overtones present in a sound* (Shortley and Williams, *Elements of Physics*).

quanta (kwon′tə), *n.* plural of **quantum:** *Quantum theory tells us that light comes in little packages of energy called quanta* (John R. Pierce).

quantal (kwon′tl), *adj.* **1** *Physics.* of or having to do with quanta or the quantum theory: *quantal units.*
2 *Biology.* (of an effect or response) either positive or negative; all-or-none.

quantasome (kwän′tə sōm), *n. Botany.* one of the granules containing chlorophyll found inside the chloroplast of plant cells.

quantic (kwon′tik), *n. Mathematics.* any homogeneous function having two or more variables. Compare **quadratic form.**

quantitative (kwon′tə tā′tiv), *adj.* **1** *Mathematics.* concerned with quantity or quantities.
2 *Chemistry.* involving or acting on the whole quantity of a substance. —**quantitatively,** *adv.*

quantitative analysis, *Chemistry.* a testing of a substance or mixture to find out the amounts and proportions of its chemical constituents.

quantity, *n. Mathematics.* **1** something having magnitude, or size, extent, amount, etc. **2** a figure or symbol representing this. [from Latin *quantitatem,* from *quantus* how much]

quantize, *v. Physics.* **1** to apply quantum mechanics or the quantum theory to; measure (energy) in quanta: ... *rotatory energy is limited to certain discrete values, or, as the physicist says, is "quantized"* (Scientific American).
2 to restrict the magnitude of (an observable quantity) in all or some of its range to a set of distinct values, especially to multiples of a definite unit: *The electron density is, in a sense, quantized, occurring in charge-packets which we call electrons* (H. R. Paneth).
—**quantization,** *n.* the fact or process of quantizing or the state of being quantized: *The "quantization" of general relativity ... involves treating gravitational radiation not as being continuous but as consisting of tiny packets, much as light, which was once thought continuous radiation, is now known to consist of tiny light packets, or photons* (Science News Letter).

quantum (kwon′təm), *n., pl.* **-ta** (-tə). *Physics.* **1** the smallest amount of a physical quantity which can exist independently, especially a discrete quantity of electromagnetic energy. Light and heat are given off or absorbed in quanta. *One radiation quantum can modify the nucleus of a germ cell sufficiently to cause a mutation, an inheritable change of character; but only one*

---

cap, fāce, fäther; best, bē, tèrm; pin, five;
rock, gō, ôrder; oil, out; cup, pùt, rüle,
yü in use, *yù* in uric;
ng in bring; sh in rush; th in thin, ᴛʜ in then;
zh in seizure.
ə = *a* in about, *e* in taken, *i* in pencil, *o* in lemon, *u* in circus

in many million quanta will score that kind of bull's eye (Otto Frisch).

2 this amount of energy regarded as a unit. The photon is a quantum of radiant energy. The phonon is a quantum of thermal energy in the form of sound or vibration.

[from Latin, neuter of *quantus* how much]

quantum chemistry, the application of the principles of quantum mechanics to the study of chemical phenomena, especially the interaction between molecules.

quantum chromodynamics, = chromodynamics: *All this is pulled together in a theory called quantum chromodynamics (QCD) analogous to the thoroughly proved quantum electrodynamics which describes the interaction of particles through electromagnetic forces* (Technology Review).

quantum electrodynamics, *Physics.* the quantum theory as applied to electrodynamics: *Quantum electrodynamics ... is concerned with the interaction of the electron, the particle of electricity, with radiation* (Science News Letter). *The most complete theory of electromagnetic interactions is the theory of quantum electrodynamics (QED), in which not only are the electrons in atomic systems described by quantum equations, but also the behavior of the electromagnetic field under the stimulus of electron motions is quantized as well* (Science).

quantum field theory, *Physics.* any theory treating fundamental forces or fields as quantized, or transmitted by quanta. Quantum electrodynamics was the first successful quantum field theory.

quantum jump, *Physics.* an abrupt change from one energy level to another, especially such a change in the orbit of an electron with the loss or gain of a quantum of energy: *The action of the solid-state maser also depends on quantum jumps, but they are jumps of electrons within individual atoms rather than energy transitions of whole molecules* (Scientific American).

quantum-mechanical, *adj.* of or having to do with quantum mechanics: *Computed quantum-mechanics models of the electronic structure of atoms and molecules can provide a reliable and comprehensive alternative to the traditional experimental approach to chemistry* (Arnold C. Wahl). —**quantum-mechanically,** *adv.*

quantum mechanics, the branch of physics dealing with the interaction of matter and radiation, the structure of the atom and the motion of atomic particles, and with related phenomena. Quantum mechanics explains the process through which atoms emit and absorb photons, which it describes as having the characteristics of both particles and waves. Quantum mechanics developed as an expanded and systematic application of the quantum theory. *The quantum mechanics of Schrödinger and Heisenberg enables us to predict correctly the characteristic energy levels of an atom or molecule from certain perfectly general equations. When quantum mechanics is applied to a system involving macroscopic bodies, the number of characteristic energy levels is so great and the levels are so close together that the results become equivalent to those obtained by classical or Newtonian mechanics* (Shortley and Williams, *Elements of Physics*).

quantum number, *Physics.* one of a set of numbers assigned to a physical system, specifying the exact state of the system among those allowed by the quantum theory.

quantum statistics, *Physics.* the statistics of the distribution of nuclear particles, either in the form of Bose-Einstein statistics or in the form of Fermi-Dirac statistics.

quantum theory, *Physics.* the theory that radiant energy is transmitted in the form of discrete units or quanta, especially as developed by Albert Einstein from the ideas of Max Planck and extended by Niels Bohr to atomic structure. Application of the principles of quantum theory led to the development of quantum mechanics, quantum electrodynamics, quantum statistics, etc.

quaquaversal (kwä′kwə vėr′səl), *adj. Geology.* dipping away in all directions from a center, as the strata of a domed structure. [from Late Latin *quaquaversus* where toward]
—**quaquaversally,** *adv.* in all directions from a central point or area.

quark, *n. Nuclear Physics.* one of a hypothetical set of subatomic particles, each with an electric charge less than that of the electron, regarded as constituents of all hadrons. [coined by the American physicist Murray Gell-Mann from the phrase *"three quarks"* in *Finnegans Wake,* a novel by James Joyce]
ASSOCIATED TERMS: The original quark theory postulated three types or varieties (*flavors*) of quarks, now called the *down quark,* the *up quark,* and the *strange quark.* However, the discovery of new classes of subatomic particles suggested the possibility of additional flavors, which are now called *charm, bottom,* and *top.* As many as eighteen types of quark are possible according to the theory of *chromodynamics,* which holds that each quark possesses a quantum property called *color.* It is the *color force* which binds quarks together, and the carrier of this force is the *gluon.*

quart, *n.* **1** a customary unit of capacity for liquids, equal to one fourth of a gallon or 32 fluid ounces. **2** a customary unit of capacity for dry things, equal to one eighth of a peck. [from Old French *quarte,* from Latin *quarta,* feminine of *quartus* fourth]

quarter, *n.* **1** one of four equal or corresponding parts into which a thing may be, or is, divided; half of a half; one fourth.

2 *Astronomy.* one of the four periods of the moon, lasting 7 days each. SYN: phase.

3 a a point of the compass; direction. **b** one fourth of the distance between any two adjacent points of the 32 marked on a compass; 2 degrees 48 minutes 45 seconds.

quartic (kwôr′tik), *Mathematics.* —*adj.* of the fourth degree: *a quartic equation.*
—*n.* a quantity, equation, or function of the fourth degree. Compare **quadric, quintic.**

quartile (kwôr′tīl *or* kwôr′tl), *Mathematics, Statistics.*
—*n.* **1** one of the points or marks dividing a set of data into four parts, each having the same frequency. The upper quartile is the point reached or exceeded by 25 per cent of the cases plotted on the frequency scale.
2 any of the four parts thus formed. Compare **quintile.**
—*adj.* of or having to do with quartiles; being a quartile.
[from Medieval Latin *quartilis,* from Latin *quartus* fourth]

quartz, *n. Mineralogy.* a very hard mineral composed of silica and found in many different types of rocks, such as sandstone and granite. Quartzes vary according to the size and purity of their crystals. Those with submicroscopic crystals are divided into fibrous varieties, or chalcedonies, and granular varieties. Crystals of pure quartz are coarse, colorless, and transparent, and quartz in this form is commonly called *rock crystal.* Impure colored varieties of quartz include flint, agate, and amethyst. *Quartz, the commonest mineral and the one that is found most often in the form of distinct crystals, is six-sided and always comes to a point* (Frederick H. Pough). *Formula:* SiO_2 [from German *Quarz*]

quartz-diorite, *n.* = tonalite.

quartz glass, *Chemistry, Mineralogy.* a clear, vitreous solid formed from pure quartz that has been melted, characterized by an ability to withstand large and quick temperature changes, chemical inertness, high melting point, and a special transparency to infrared, visible, and ultraviolet radiations. Also called **fused quartz.**

quartzite (kwôrt′sīt), *n. Geology.* a granular rock composed essentially of quartz, and formed by the metamorphism of sandstone: *A sandstone metamorphosed becomes a quartzite, a rock of extreme hardness and resistance to erosion* (Finch and Trewartha, *Elements of Geography*).
—**quartzitic** (kwôrt sit′ik), *adj.* of the nature of or consisting of quartzite: *The majority of the rocks encountered, such as mica schist, vary from hard quartzitic and heavily jointed rocks to softer and decayed varieties* (L. H. Dickerson).

quasar (kwā′zär *or* kwā′sär), *n. Astronomy.* any of a number of celestial objects larger than stars but smaller than galaxies, that emit powerful blue light and often radio waves: *The term quasar ... was originally applied only to the starlike counterparts of certain strong radio sources whose optical spectra exhibit red shifts much larger than those of galaxies. Before long, however, a class of quasi-stellar objects was discovered with large red shifts that have little or no emission at radio wavelengths. "Quasar" is now commonly applied to starlike objects with large red shifts regardless of their radio emissivity* (Scientific American). Also called **QSO, quasi-stellar object.** [from *quas(i)-* (*stell*)*ar* (radio source)]

quasi-atom (kwā′zī at′əm), *n. Physics.* a short-lived nuclear particle resembling an atom: *Quasi-atoms are systems in which particles not normally found in atoms become bound together ... Quasi-atoms are generally unstable structures either because they are subject to matter-antimatter annihilation (positronium) or because one or more of their constituents is radioactively unstable (muonium)* (Science News).

quasifission, *n. Nuclear Physics.* a type of nuclear fission in which the target and projectile nuclei fuse briefly, exchanging large numbers of nucleons, before they separate.

quasimolecule, *n. Physics.* a structure formed by a combination of quasi-atoms: *Basically, electron ejection from an atom or a quasimolecule is due to the time-varying electric field acting on the electron as the collision partners pass each other. If the collision partners have comparable atomic numbers, it must usually be assumed that the electrons in both adjust their orbits so as to form quasimolecular states from which an electron will be removed* (W. E. Meyerhof).
—**quasimolecular,** *adj.* of or having to do with quasimolecules: *quasimolecular transitions.*

quasiparticle, *n. Physics.* a unit or quantum, as of sound, light, or heat, that has some of the properties of a particle, such as mass or momentum: *... the first quasiparticles, the phonons (sound quanta in a crystal lattice), and the concepts of two others: the exciton, a specially excited state of electrons in a crystal lattice; and the polaron, a conducting electron in an ionic lattice* (Mikhail D. Millionshchikov).

quasi-stellar, *adj. Astronomy.* of, having to do with, or being various types of quasars: *a quasi-stellar radio source, a quasi-stellar blue galaxy.*

quasi-stellar object, = quasar: *The extraordinary properties of quasi-stellar objects, or quasars, were not recognized until 1963, when Maarten Schmidt discovered the red shift of 3C 273.* (Science).

Quaternary (kwə tèr′nər ē), *Geology.* —*n.* **1** the present geological period, the later of the two periods making up the Cenozoic era, and including the Pleistocene and Recent epochs: *The Cenozoic is the last and shortest era, beginning 75 million years ago and composed of only two periods, the Tertiary and the Quaternary.* (R. Beals and H. Hoijer).
2 the deposits made in this period.
—*adj.* of or having to do with the Quaternary or its deposits: *The beginning of the Quaternary Period dates back a mere million years or so, and therefore the record is so recently inscribed, that multitudinous details can be read clearly* (Moore, *Introduction to Historical Geology*).
▶ See the note under **Tertiary.**

quaternary ammonium compound, *Chemistry.* a type of organic compound whose structure consists of a central nitrogen atom joined to four organic radicals and one acid radical.

quaternion (kwə tèr′nē ən), *n. Mathematics.* **1** the quotient of two vectors considered as depending on four geometrical elements and as expressible by an algebraic quadrinomial. Also called **hypercomplex number.**
2 quaternions, the calculus of vectors in which a quaternion is employed: *... the value of quaternions for pursuing researches in physics* (Herbert Spencer).

queen, *n. Zoology.* a fully developed female in a colony of insects, such as bees or ants, that lays eggs. There is usually only one queen in a hive of bees.

queen substance, *Biochemistry.* a pheromone secreted by queen bees and given to worker bees to inhibit the normal development of their ovaries and thereby prevent the birth of other queens: *The queen substance of the honey bee ... will inhibit the normal development*

---

cap, fāce, fäther; best, bē, tèrm; pin, fīve;
rock, gō, ôrder; oil, out; cup, pùt, rüle,
yü in use, *yu* in uric;
ng in bring; sh in rush; th in thin, ̴H in then;
zh in seizure.
ə = *a* in about, *e* in taken, *i* in pencil, *o* in
lemon, *u* in circus

of the ovaries in other insects, including ants, flies and termites, and even in prawns (New Scientist).

quicksand, *n. Geology.* a shifting mass of very deep, soft, wet sand that yields easily to pressure and will not hold up a heavy object but slowly engulfs it.

quill, *n. Zoology.* **1** Also called **quill feather.** one of the large, strong feathers of the wings or tail of a bird.
2 the hollow, horny stem of such a feather: *Each main shaft or quill sprouts forth some 600 dowls or barbs on either side to form the familiar vane of the feather* (Atlantic). See the picture at **feather.**
3 a stiff, sharp hair or spine that is like the end of a feather. A porcupine or hedgehog has quills on its back.

quincunx (kwin′kungks), *n. Botany.* an overlapping arrangement of five petals or leaves, in which two are interior, two are exterior, and one is partly interior and partly exterior. [from Latin *quincunx* five-twelfths, from *quinque* five + *uncia* a twelfth]

quinone (kwi nōn′), *n. Chemistry.* **1** a yellowish, crystalline compound with an irritating odor, obtained by the oxidation of aniline and regarded as a benzene with two hydrogen atoms replaced by two oxygen atoms. It is used in tanning and making dyes. *Quinone will, for example, oxidise another material and be itself reduced to hydroquinone* (New Scientist). *Formula:* $C_6H_4O_2$
2 any of several related compounds, some of which serve as coenzymes or vitamins.

quinque-, *combining form.* five; having five; five times, as in *quinquefoliate = having five leaves.*

quinquefoliate (kwin′kwə fō′lē it), *adj. Botany.* having five leaves or leaflets; having leaves in whorls of five.

quinquevalence (kwin′kwə vā′ləns), *n. Chemistry.* the quality or condition of being quinquevalent.

quinquevalent (kwin′kwə vā′lənt), *adj. Chemistry.* having a valence of five; pentavalent.

quintal (kwin′tl), *n.* a unit of mass in the metric system, equal to 100 kilograms. [from Medieval Latin *quintale,* from Arabic *qintār,* from Latin *centenarius,* from *centum* hundred]

quintic (kwin′tik), *Mathematics.* —*adj.* of the fifth degree: *a quintic equation.*
—*n.* a quantity, equation, or function of the fifth degree. Compare **quadric, quartic.**

quintile (kwin′tīl *or* kwin′tl), *Mathematics, Statistics.*
—*n.* **1** one of the points or marks dividing a set of data into five parts, each having the same frequency.
2 any of the five parts thus formed. Compare **quartile.**
—*adj.* of or having to do with quintiles; being a quintile. [from Latin *quintus* fifth + English (*quart*)*ile*]

quintillion, *n.* **1** (in the United States, Canada, and France) 1 followed by 18 zeros. **2** (in Great Britain and Germany) 1 followed by 30 zeros. [from Latin *quintus* + English (*m*)*illion*]

quotient (kwō′shənt), *n. Mathematics.* a number obtained by dividing one number by another. In $26 \div 2 = 13$, 13 is the quotient. [from Latin *quotiens* how many times, from *quot* how many]

R

r or **r.,** *abbrev.* or *symbol.* **1** radius. **2** rod *or* rods. **3** roentgen. **4** *Statistics.* correlation coefficient.

R or **R.,** *abbrev.* or *symbol.* **1** (hydrocarbon) radical. **2** radius. **3** Rankine. **4** rare earth. **5** ratio. **6** resistance. **7** roentgen. **8** Rydberg constant.

Ra, *symbol.* radium.

race, *n. Biology.* any interbreeding group of organisms within or below a species, such as a breed, strain, or variety: *A race, in this technical sense of the term, is a variety which is perpetuated with considerable certainty by sexual propagation* (Asa Gray).

raceme (rā sēm′ *or* rə sēm′), *n. Botany.* a simple flower cluster having its flowers on nearly equal stalks along a stem, the lower flowers blooming first. The lily of the valley, currant, and chokecherry have racemes. [from Latin *racemus* cluster (of grapes or berries)]

raceme

lily-of-the-valley

racemic (rā sē′mik *or* rə sem′ik), *adj. Chemistry.* of or designating an optically inactive mixture formed by the combination of dextrorotatory and levorotatory forms in equal molecular proportions: *L-leucine has a bitter taste, D-leucine has a sweet taste and the racemic mixture* (*both*) *is tasteless* (Scientific American). [from Latin *racemus* cluster]

racemic acid, *Chemistry.* an optically inactive form of tartaric acid, occurring in the juice of grapes along with ordinary tartaric acid. *Formula:* $C_4H_6O_6$

racemization (ras′ə mə zā′shən), *n. Chemistry.* the act or process of producing a racemic mixture.

racemize (ras′ə mīz), *v. Chemistry.* to make or become optically inactive by the mixing of dextrorotatory and levorotatory forms: *In fossil material ... the free or bound amino acids slowly racemize* (Nature).

racemose (ras′ə mōs), *adj.* **1** *Botany.* arranged in the form of a raceme; characteristic of racemes: *Indeterminate, racemose, ascending, or centripetal inflorescence is that form in which the terminal bud of the flower cluster continues to develop and increase the length of the stem indefinitely* (Youngken, *Pharmaceutical Botany*).
2 *Anatomy.* shaped like or resembling a bunch of grapes. A **racemose gland** is a compound gland formed of a system of ducts which branch into sacs and resemble clusters of grapes, as the pancreas.

rachis (rā′kis), *n., pl.* **rachises, rachides** (rak′ə dēz *or* rā′kə dēz). **1** *Botany.* **a** a stem, as those of grasses. **b** a stalk, as of a pinnately compound leaf or frond.
2 *Zoology.* the terminal part of the shaft of a feather. See the picture at **feather.**
3 *Anatomy.* the spinal column.
[from Greek *rhachis* backbone]

rad, *n.* a unit for measuring absorbed doses of ionizing radiation, equal to 100 ergs of energy per gram of absorbing material, or 0.01 joule per kilogram. The SI unit intended to replace the rad is the *gray.* See also **rem, rep.** [from *r*(*adiation*) *a*(*bsorbed*) *d*(*ose*)]

rad, *abbrev.* or *symbol.* **1** radian. **2** radical. **3** radius. **4** radix.

radar, *n. Physics.* **1** an instrument for determining the distance, direction, and speed of unseen objects by the reflection of radio waves: *Forecasters can look into a storm with radar and learn its size, shape, speed, direction of travel, and rate of development* (Robert C. Guthrie).
2 a process or system by which the reflection of radio waves is measured: *The miracle underlying all radar is that men have learned ... to measure time in such infinitesimal amounts that radio echo ranges of objects miles away can be read with accuracy* (James Phinney Baxter).
[from *ra*(*dio*) *d*(*etecting*) *a*(*nd*) *r*(*anging*)]

radar astronomy, a branch of astronomy that studies planets and other bodies in the solar system by analyzing the reflections of radar signals sent from the earth at specific targets: *Radar astronomy ... derives information that passive watching cannot* (Science News Letter).

radar telescope, a radio telescope used in radar astronomy as a radar receiver to study and measure the distance of objects in outer space.

radial, *adj.* **1** *Biology.* diverging from a common point; arranged like or in radii or rays; radiating: *a radial organ.*
2 *Anatomy.* of or near the radius bone: *the radial artery.*
3 *Geometry.* of the radius of a circle.
4 *Zoology.* of or having to do with the arms of a starfish or other echinoderm; characterized by radial symmetry: *Radial animals either drift with the water currents most of the time or live a sessile life* (Buchsbaum, *Animals Without Backbones*).

---

cap, fāce, fäther; best, bē, tėrm; pin, fīve;
rock, gō, ôrder; oil, out; cup, pùt, rüle,
yü in use, *yu* in uric;
ng in bring; *sh* in rush; *th* in thin, ᴛʜ in then;
zh in seizure.
ə = *a* in about, *e* in taken, *i* in pencil, *o* in lemon, *u* in circus

541

radial symmetry

—*n.* **1** a part that radiates or is arranged in radii. **2** *Biology.* a radial nerve, artery, or organ. —**radially,** *adv.*

radial symmetry, *Biology.* a condition in which like parts are arranged about an axis, from which they radiate like the parts of a flower, as in many echinoderms: *Both the polyps and the medusas of coelenterates exhibit radial symmetry, in which all radii are alike at any particular level* (Buchsbaum, *Animals Without Backbones*). Compare **biradial** and **bilateral symmetry.**

radial symmetry:

wild rose   starfish

radial velocity, *Astronomy.* the velocity of a star along the line of sight of an observer, determined by measuring the Doppler shift of lines in the star's spectrum, usually with a spectroscope: *By the radial velocity of a star or other object is meant the rate at which the distance between the object and the observer is changing.* (Duncan, *Astronomy*).

radian (rā′dē ən), *n. Mathematics.* a unit of angular measure equal to the angle subtended at the center of a circle by an arc equal in length to the radius of the circle. One radian is equivalent to about 57.2958 degrees. *Since the radius is contained 2π times (2π = 6.28 ...) in the circumference, there are 2π or 6.28 ... radians in one complete revolution or 360°* (Sears and Zemansky, *University Physics*). *Symbol:* rad

radiance, *n. Optics.* the radiant flux emitted by a unit area of a source into a unit solid angle.

radiant, *adj. Physics.* **1** emitting rays or waves of light, heat, etc.; radiating: *The sun is a radiant body.* **2** of or having to do with the radiation of light, heat, etc., from a common point.

—*n.* **1** *Physics.* a point or object from which light or heat radiates. **2** *Astronomy.* the point in the heavens from which the meteors in a shower seem to have come: *The radiant of a meteoric shower is the vanishing point in the perspective of the parallel trails* (Baker, *Astronomy*).

radiant energy, *Physics.* energy in the form of waves, especially electromagnetic waves. X rays, radio waves, and visible light are forms of radiant energy. *When electrons in an atom in the excited state return to lower energy levels, the energy is emitted as radiant energy of specific frequency, producing characteristic spectral lines which can be used to identify the element* (Chemistry Regents Syllabus).

radiant flux, *Physics.* the rate of emission of radiant energy, in watts or in ergs per second: *Time rate of flow of radiant energy is called radiant flux or radiant power, and when it is considered for each part of the spec-*

trum separately, it is called spectral radiant flux (Deane B. Judd).

radiant heat, *Physics.* heat transmitted by electromagnetic radiation: *The degree and characteristics of wood charring when exposed to radiant heat may be affected by the type of coating* (New Scientist).

radiate, *v.* **1** *Physics.* to emit rays or waves of (light, heat, etc.); emit (energy of any kind) in the form of rays or waves: *The sun radiates light. Stars radiate their heat away, and must burn nuclear fuel to keep going* (Nature). **2** *Biology.* (of a group of organisms) to spread out into new environments and thereby diverge or diversify: *The Australian marsupials ... radiated into a great variety of habitats* (W. C. Allee). [from Latin *radiatum* radiated, from *radius* ray]

radiation, *n.* **1** *Physics.* **a** the act or process of emitting energy in the form of rays or waves, especially electromagnetic waves: *the radiation of light, heat, sound, etc.* **b** energy radiated or transmitted in the form of rays, waves, or particles: *Radiation is wave motion, known to us as the electromagnetic waves used for radio transmission, heat, light, X rays and cosmic rays. Large numbers of extremely tiny particles in motion together act like waves* (Helen M. Davis). **c** a stream of particles or electromagnetic waves emitted by the atoms and molecules of a radioactive substance as a result of nuclear decay. Some radiations from atoms are alpha particles, beta particles, gamma rays, and neutrons. Radiation is often harmful to living tissue. *Radiobiologists have concluded that exposure to low levels of radiation accelerates normal aging* (Nathan W. Shock).

ASSOCIATED TERMS: see **conduction.**

2 *Biology.* **a** a radial arrangement or structure: *the radiation of a bird's feathers.* **b** the spread of a group of animals or plants into new environments, thereby diverging or diversifying. Compare **adaptive radiation.**

radical, *adj.* **1** *Botany.* of, having to do with, or arising from the root or base of the stem: *radical leaves.* SYN: basal.

2 *Mathematics.* having to do with or forming the root of a number or quantity: *The most important of the radical functions are those which are square roots of rational functions* (H. F. Baker).

—*n.* **1** *Chemistry.* an atom or group of atoms acting as a unit in chemical reactions: *Such radicals may be produced by the high-temperature decomposition of hydrocarbons such as methane, CH_4, which forms the methyl radical, CH_3* (M. P. Barnett).

► The word *group* is more commonly used than *radical* to refer to several connected atoms that are considered as a unit within a molecule: *Diethylbenzine contains two ethyl groups. Radical (or free radical)* often refers to a group that has one or more unpaired electrons and is not part of a larger molecule. Some writers use *radical* to refer specifically to ionic groups. **2** *Mathematics.* **a** an expression indicating the root of a quantity. EXAMPLE: $\sqrt{5}$, $\sqrt{x-3}$. **b** = radical sign. [from Late Latin *radicalis,* from Latin *radix* root]

radical axis, *Geometry.* the straight line joining the points of intersection of two circles, or between two circles that do not intersect, from which tangents at any point to both circles will be of equal length.

radical expression, *Mathematics.* an expression, especially an irrational number or quantity, involving a radical sign.

radical sign, *Mathematics.* a mathematical sign ($\sqrt{\ }$) put before a number or expression to show that a root of it is to be extracted. 16 = the square root of $16 = 4$; $\sqrt[3]{27}$ = the cube root of $27 = 3$.

radicand (rad′ə kand′), *n. Mathematics.* the quantity placed under a radical sign. [from Latin *radicandus,* gerundive of *radicare* take root, from *radix* root]

radicle (rad′ə kəl), *n.* **1** *Botany.* the part of a plant embryo that develops into the root.
2 *Anatomy.* a rootlike part, such as one of the fibrils of a nerve fiber. [from Latin *radicula,* diminutive of *radix* root]

radii (rā′dē ī), *n.* a plural of **radius.**

radio, *n. Physics.* **1** the transmission of electric impulses through space by means of electromagnetic waves. **2** = radio wave.
—*adj. Astronomy.* of or involving the emission of radio waves: *radio bursts, radio noise, radio sources.* [ultimately from Latin *radius* ray, beam]

radio-, *combining form.* **1** radio; radio waves, as in *radioacoustic, radiotelephony.*
2 radiant energy, as in *radiometry = the measurement of radiant energy.*
3 radioactive, as in *radioisotope = radioactive isotope.*
4 of the radius and, as in *radioulnar = of the radius and ulna.*
5 radial; radially, as in *radiosymmetrical = radially symmetrical.*
[ultimately from Latin *radius* ray]

radioactive, *adj. Physics.* of, having, or caused by radioactivity. Radium, uranium, and thorium are radioactive metallic elements. *Of approximately 200 radioactive nuclides found in fallout, the following are considered most likely to represent a significant hazard to living things: iodine-131, strontium-89, strontium-90, cesium-137, and carbon-14* (Renee Alexander). —**radioactively,** *adv.*

radioactive decay, *Physics.* the spontaneous disintegration of a radioactive substance, at the characteristic rate of the particular radioisotope, and accompanied by the emission of ionizing radiation in the form of particles and gamma rays: *Rutherford and Frederick Soddy (1902) explained the radioactive decay process in the case of uranium and paved the way for the identification of the elements in the radioactive series. It was recognized that radioactive decay involved the emission of charged particles—alpha and beta particles—and penetrating radiation or gamma rays* (Ralph E. Lapp).

radioactive series, *Physics.* one of several series of isotopes of certain elements representing various stages of disintegration of a radioactive substance. Three of these series occur naturally (the actinium, thorium, and uranium series) and others, such as the neptunium series, are artificially produced.

radioactivity, *n. Physics.* **1** the property exhibited by certain elements, of emitting radiation in the form of alpha particles, beta particles, or gamma rays as the result of spontaneous nuclear decay: *The intense radioactivity of radium and its compounds made it possible to investigate the nature of the radiation and the effects that accompany its emission ... for, as the investigators*

later recognized, radioactivity is a property of the atomic nucleus (Furry, *Physics*).
2 the radiation emitted by a radioactive substance: *Radioactivity is measured in curies: one curie is equal to the radioactivity from one gram of radium (37 billion atoms disintegrating per second). Thus a millionth of a curie, acting on the body over a period, is a dangerous dose* (Scientific American).

radio-astronomical, *adj.* of or having to do with radio astronomy: *radio-astronomical observations.*

radio astronomy, the branch of astronomy dealing with the detection of objects in space by means of radio waves that these objects emit. Radio astronomy enables observers to study distant galaxies and the hydrogen gas clouds of the Milky Way. *The most obvious role of radio astronomy in cosmology has been the attempt to choose between two main theories: the evolutionary theory that the universe is of limited extent and expanding, and the steady-state theory that the universe is infinite in space and time and effectively unchanging* (New Scientist).

radiobiological, *adj.* **1** of or having to do with radiobiology: *Radiobiological discussions have often taken the spontaneous mutation rate as a reference base, as an unavoidable evil which could not be averted and ought not be aggravated* (Bulletin of Atomic Scientists).
2 having to do with or resulting from the effects of radiation on living cells or organisms: *Removal of oxygen before irradiation was found to protect cells against many kinds of radiobiological damage* (Science News Letter).

radiobiology, *n.* the branch of biology dealing with the effects of radiation on living cells and organisms.

radiocarbon, *n. Physics.* a radioactive isotope of carbon, especially carbon 14: *Radiocarbon is present in the atmosphere as carbon dioxide from two sources: the action of cosmic rays on the nitrogen of the atmosphere, and the neutrons produced in the same gas by molecular explosions* (Edward S. Gilfillan).

radiocarbon dating, = carbon dating: *Radiocarbon dating is based on the fact that radioactive carbon-14 decays at a fairly regular rate. By measuring the amount of radioactive carbon-14 remaining in preserved organic material from prehistoric sites, physicists can tell within a statistically expressed margin of error when the organism died* (Frank Hole).

radiochemical, *adj.* of or having to do with radiochemistry: *One of the most important points in radiochemical work is the necessity of using an amount of radioactive element sufficient to obtain quantitative results of reliable accuracy.* (D. G. Tuck).
—*n.* a chemical to which radioactive isotopes have been added: *It is offering a full range of radiochemicals with the addition of radioisotopes to its line of laboratory reagents* (N. Y. Times).

cap, fāce, fäther; best, bē, tèrm; pin, five;
rock, gō, ôrder; oil, out; cup, pùt, rüle,
yü in use, *yu̇* in uric;
ng in bring; *sh* in rush; *th* in thin; ⊤H in then;
zh in seizure.
ə = *a* in about, *e* in taken, *i* in pencil, *o* in lemon, *u* in circus

radiochemistry, *n.* the branch of chemistry dealing with radioactive substances.

radioecology, *n.* a branch of ecology dealing with the effects of radioactivity upon living organisms, especially in a particular locality: *Radioecology concerns the interaction of radiation and radioisotopes with organisms at the population or community level of organization* (R. G. Wiegert).

radio frequency, *Physics.* a frequency at which electromagnetic waves can transmit impulses or signals, ranging from extremely low frequency (*ELF*, below 300 hertz) to extremely high frequency (*EHF*, between 30 and 300 gigahertz): *It seems that the radio frequencies emitted by atoms and molecules at low pressures are fixed, and will not change with time, so that if the frequency of a spectral line has been accurately measured, the line itself can then be made the standard of frequency* (J. Little). *Abbreviation:* R.F.

radio galaxy, *Astronomy.* a galaxy that emits radio waves: *It seemed likely ... that most radio sources in space ... were remote radio galaxies which emitted vast amounts of energy by unknown mechanisms* (London Times).

radiogenic, *adj.* formed as a product of radioactivity: *Leakage of radiogenic helium through the ocean floor leads to increases in the concentration of helium in deep ocean water* (Hans E. Seuss).

radiogeology, *n.* the branch of geology dealing with the effects of radioactivity on the earth, ocean, and the earth's atmosphere.

radioimmunoassay, *n.* *Immunology.* a method of detecting and identifying the amount or characteristics of a protein, hormone, or other substance by labeling it with a radioactive chemical and combining it with an antiserum to induce an immunological reaction. The extent of the reaction is measured by the amount of radiation emitted. *Radioimmunoassay ... is capable of detecting a 100,000th part of a millionth of a gram of steroid* (Manchester Guardian Weekly).

radioimmunological, *adj.* involving the use of radioimmunoassay: *Estimation of the concentration of peptide hormones in plasma can be made by radioimmunological techniques* (John Watt McLaren).

radioiodine, *n.* = iodine 131.

radioisotope, *n.* *Physics.* a radioactive isotope, especially one produced artificially: *Radioisotopes are chemical elements such as cobalt or strontium that have been made radioactive in atomic reactors ... The chemical structure of the elements is not altered* (Wall Street Journal).

radiolabel, *Chemistry.* —*v.* to label or tag (a substance) by adding a radioactive isotope to the substance and tracing it through one or more chemical reactions: *Variations in the composition of the antisera would make each antiserum a unique target for radiolabelling with consequent variations in results* (Nature).
—*n.* a radioactive isotope used to radiolabel a substance: *The use of a radiolabel makes the technique expensive, Purcell concedes, but the expense is justified by the increased sensitivity* (Science).

radiological, *adj.* of or having to do with radiology: *the radiological effects of exposure to X rays.*
—**radiologically,** *adv.*

radiology, *n.* **1** the branch of medicine dealing with X rays or the rays from radioactive substances, especially for diagnosis or treatment: *Imaging ... is now a major force which has profoundly influenced the practice of radiology* (D. Sutton).
2 the use of X rays to examine, photograph, or treat bones, organs, or certain manufactured products: *The selective destruction of tissues forms the basis of therapeutic radiology* (W. T. Moss).

radiolysis (rā′dē ol′ə sis), *n.* *Chemistry.* the molecular decomposition of a substance resulting from the action of radiation: *Organic fluids in the radioactive environment of a reactor are subject to radiolysis* (Bulletin of Atomic Scientists).
—**radiolytic** (rā′dē ō lit′ik), *adj.* having to do with or producing radiolysis.

radiometric, *adj.* **1** of or having to do with radiometry: *radiometric studies of infrared radiation.*
2 of or having to do with the measurement of radioactivity or ionizing radiation: *Radiometric assay of the uranium found in bones can be used in the same way as fluorine and nitrogen analyses to establish whether two bones from the same deposit are contemporary* (Frank Hole). —**radiometrically,** *adv.*

radiometry (rā′dē om′ə trē), *n.* *Physics.* the detection and measurement of radiant energy, especially that of infrared radiation.

radiomimetic (rā′dē ō mi met′ik), *adj.* *Biology.* having an effect on living tissue almost identical with that of ionizing radiation: *Most biological effects of radiation involve the formation and the integrity of the chromosomes. Chemicals that duplicate these effects are called radiomimetic* (Scientific American).

radionuclide (rā′dē ō nü′klīd), *n.* *Physics.* a nuclide that exhibits radioactivity: *Naturally occurring radionuclides may reach the body from soil and water through the food chain and thereby cause internal irradiation* (Renee Alexander).

radiopaque (rā′dē ō pāk′), *adj.* *Medicine.* not transparent to X rays or other radioactive substances; not readily penetrated by radiation. *The velocity of blood flow is easily determined by injecting a radiopaque substance into the bloodstream and measuring the rate of travel of its shadow on a film* (Scientific American).

radioprotection, *n.* *Chemistry.* protection by means of chemicals against the harmful effects of ionizing radiation.
—**radioprotective,** *adj.* providing radioprotection: *Dopamine, an intermediate hormone formed by the adrenal glands in the synthesis of norepinephrine, is a good radioprotective agent* (J. Richard Thomson).

radiosensitive, *adj.* sensitive to the action of ionizing radiation: *Another weed, Anisantha sterilis, was found remarkably radiosensitive since only 5,000 rad was lethal* (Science News Letter).
—**radiosensitivity,** *n.* the sensitivity of a substance to the action of ionizing radiation: *The increased radiosensitivity of a child's body must be considered in contrast to that of the adult* (Bulletin of Atomic Scientists).

radio spectrum, *Physics.* **1** the entire range of radio frequencies in the spectrum of electromagnetic radiation.
2 the spectrum of any source at these frequencies: *Samples of QSOs with flat radio spectra were chosen* (Nature).

radio telescope, a device used in radio astronomy for detecting and recording radio waves coming from stars and other objects in outer space. It usually consists of a radio receiver with an antenna fixed on a wide bowl-shaped reflector which collects and focuses the waves. *Radio telescopes show the structure of our galaxy much as an X-ray photograph reveals the skeleton of an animal* (J. D. Kraus). Compare **radar telescope.**

radiothorium, *n. Physics.* a radioactive isotope of thorium, having mass number 228.

radiotoxic, *adj.* of or having to do with radiotoxins: *radiotoxic effects of irradiation.*

radiotoxin, *n.* a radioactive poison: *The radiotoxins in the [irradiated] extracts penetrate the cell nucleus and combine with certain proteins such as histones ... thus affecting the mechanism controlling the synthesis of nuclei acid* (Science Journal).

radioulnar (rā′dē ō ul′nər), *adj. Anatomy.* of the radius and the ulna of the forearm: *the radioulnar joint.*

radio wave, *Physics, Astronomy.* an electromagnetic wave within the radio frequencies: *Radio waves travel at the speed of light* (A. H. Hayes). *Radio waves are a million times longer than light waves and easily penetrate clouds that are completely opaque to light* (J. D. Kraus). See the diagram at **electromagnetic spectrum.**

radium (rā′dē əm), *n. Chemistry.* a radioactive metallic element found in very small amounts in uranium ores such as uraninite. Radium is very unstable and gives off alpha particles and gamma rays as it decays in successive forms into radon, polonium, and, finally, lead. *Radium is never found free, but it is found combined in almost all rocks in extremely minute quantities* (Offner, *Fundamentals of Chemistry*). *A small block of radium is always one or two degrees higher in temperature than its environment, and this is true whether such a piece of radium is kept at room temperature, placed in boiling water, or put in a deep freeze* (I. Bernard Cohen). *Symbol:* Ra; *atomic number* 88; *mass number* (of the most stable isotope) 226; *melting point* 700°C; *boiling point* 1737°C; *oxidation state* 2. [from New Latin *radium,* from Latin *radius* ray]

radium emanation, = radon.

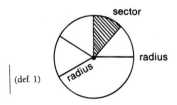

(def. 1)

sector

radius

radius

radius (rā′dē əs), *n., pl.* **radii** or **radiuses. 1** *Geometry.* **a** any line segment going straight from the center to the outside of a circle or a sphere: *Radii of the same circle or of equal circles are equal* (Herberg, *A New Geometry*). **b** the length of such a line segment: *If a circle has radius r, we shall speak of the number 2r as the diameter of the circle. Of course, the number 2r is the length of every chord through the center* (Moise and Downs, *Geometry*).

2 *Anatomy.* **a** that one of the two bones of the forearm which is on the thumb side: *The skeleton of each arm consists of the humerus or upper arm bone, the radius (on the same side as the thumb) and ulna in the lower arm ...* (Harbaugh and Goodrich, *Fundamentals of*

Biology). See the picture at **arm. b** a corresponding bone in the forelimb of other vertebrates. See the pictures at **homology, wing.**

3 *Zoology.* a line thought of as dividing an animal having radial symmetry into like parts.
[from Latin *radius* ray, spoke of a wheel]

radius vector, *pl.* **radius vectors, radii vectores** (rā′dē ī vek tō′rās). **1** *Mathematics.* a line segment, or its length, joining the fixed point and a variable point in a system of polar coordinates: *It is customary to call the distance of a point in space from the origin its radius vector* (Schaaf, *Analytic Geometry*).

2 *Astronomy.* a straight line connecting the center of the sun or other body (taken as a fixed point) with a planet, comet, or other body (taken as the variable point) orbiting around it: *The line from the center of the sun to the center of a planet (called the radius vector) sweeps out equal areas in equal periods of time* (H. E. Newell and James J. Haggerty).

radix (rā′diks), *n., pl.* **radices** (rad′ə sēz′ or rā′də sēz), **radixes. 1** *Mathematics.* a number taken as the base of a system of numbers, logarithms, or the like. The radix of the decimal system is ten.

2 *Botany.* the root of a plant, especially a plant used in the preparation of a medicine.
[from Latin *radix* root]

radon (rā′don), *n. Chemistry.* a gaseous, inert, radioactive element, formed by the radioactive decay of radium. Radon has at least 17 known isotopes. Traces of it are found in the air in various amounts. *Radon, an alpha-emitting daughter of radium, is easily measured by its radioactivity* (Scientific American). *Symbol:* Rn; *atomic number* 86; *mass number* (of the most stable isotope) 222; *half-life* 3.8 days. [from *radium*]

radula (raj′ù lə), *n., pl.* **-lae** (-lē). *Zoology.* a horny band in the mouth of a mollusk, set with tiny teeth: *[The limpet] gathers food into its mouth with a long tongue, or radula, which looks like a ribbon and is covered with rows of teeth* (William J. Clench).

rain, *n. Meteorology.* precipitation in the form of water falling in drops from the clouds. Rain is formed from moisture condensed from water vapor in the atmosphere. *The rain that is produced when clouds yield up their moisture is almost pure water; the dust particles in the air and various gases are the only contaminating substances* (M. F. Vessel).

rainbow, *n. Meteorology.* a bow or arch of seven colors seen sometimes in the sky, or in mist or spray when the sun shines on it from behind the observer. A rainbow shows all the colors of the spectrum: violet, indigo, blue, green, yellow, orange, and red. *The rainbow, one of the most beautiful of natural phenomena, is a spectrum produced by the dispersion of sunlight by spherical raindrops* (John Charles Duncan). Also called **iris.**

---

cap, fāce, fäther; best, bē, tèrm; pin, five;
rock, gō, ôrder; oil, out; cup, pùt, rüle;
yü in use, *yù* in uric;
ng in bring; *sh* in rush; *th* in thin, ᴛʜ in then;
zh in seizure.
ə = *a* in about, *e* in taken, *i* in pencil, *o* in lemon, *u* in circus

rainbow

rain forest, *Geography, Ecology.* a large, very dense forest composed chiefly of evergreen trees in a region where rain is very heavy throughout the year. Rain forests are usually in tropical areas, but sometimes in northern areas, such as southeastern Alaska. *The ... dense, steamy rain forests of Australia and New Guinea, where the sun seldom penetrates* (Scientific American). *Considering their environment, arctic plants are probably more efficient than those of the equatorial rain forests* (John C. Reed).

rain shadow, *Geography, Meterology.* a condition of relatively little rainfall, occurring on the downwind side of mountainous areas. Rain shadows are due to the warming of air as it moves downslope, which usually prevents atmospheric condensation.

ramal (rā′məl), *adj. Biology.* of, having to do with, or resembling a branch or ramus: *A ramal leaf is one which is affixed directly to a branch* (Youngken, *Pharmaceutical Botany*).

Raman effect (rä′mən), *Physics.* the scattering of incident light by the molecules of a transparent substance, in such a way that the wavelengths of the scattered light are lengthened or shortened: *In the Raman effect a quantum of light gives up some of its energy to a molecule and reappears as a scattered quantum with a lower frequency* (J. A. Giordmaine). [named after C. V. Raman, 1888–1970, Indian physicist who discovered this effect]

ramentum (rə men′təm), *n., pl.* **-ta** (-tə). *Botany.* a thin, membranous scale formed on the leaves or shoots of some ferns: *... ramentum, the chaffy scales which usually clothe the base of the petiole...* (Emerson, *Basic Botany*). [from Latin *ramentum* piece scraped off]

ramus (rā′məs), *n., pl.* **-mi** (-mī). **1** *Biology.* a branch, as of a plant, a vein, or a bone: *The spinal nerves distal to the root ganglia divide ... into anterior and posterior rami* (W. Keiller).
2 *Anatomy.* a bony process extending like a branch from a larger bone, especially the ascending part of the lower jaw that makes a joint at the temple: *The horizontal ramus—that portion of the jaw that holds the teeth—is deep and massive, as is also true in many fossil and living apes* (Elwyn L. Simons).
[from Latin *ramus* branch]

random, *adj. Statistics.* having to do with or involving the same or equal chances or probability of occurrence for each member of a group: *A random process is the sampling at random from a population of potential events* (J. B. Carroll). **—randomness,** *n.*

randomization, *n. Statistics.* the act, process, or result of randomizing: *Any test of statistical significance depends on the "randomization of the set of observations"* (Scientific American).

randomize, *v. Statistics.* to make or become random in arrangement, especially in order to control the variables; to use random sampling in (a statistical procedure, etc.): *Shuffling randomized a deck of ordered cards* (Scientific American). *If he tries to randomize in his head, unconscious biases creep in* (Martin Gardner).

—randomizer, *n.* a device that generates a random output: *Brown shows convincingly that even mechanical randomizers (e.g., chance-machines used to select lists of random digits) must inescapably display some form of bias* (Ernest Nagel).

random sample, *Statistics.* a sample so drawn from the total group that every item in the group has an equal chance of being chosen: *The simple, pure, or unrestricted random sample ... is drawn in such a manner that each item, as well as each combination of items, has an equal probability of being included in the sample* (Parl, *Basic Statistics*).
—random sampling, the process of selecting a random sample: *Choosing every fifth item, for example, is random sampling. It is representative sampling* (Emory S. Bogardus).

random variable, *Statistics.* a variable whose values are distributed according to a probability distribution: *We can think of most discrete random variables as counts (how many heads, children, spades, or accidents?) and most continuous random variables as measures (how tall, long, heavy, or intelligent?)* (F. E. Fischer). Also called **variate.**

random walk, *Statistics.* any movement or process whose individual steps are determined at random or by chance, independently of the preceding steps: *A radioactive particle, while being carried downwind at the mean windspeed, is also thrown about randomly by the turbulent forces acting on it—the movement known as a "random walk"* (New Scientist).

range, *n.* **1** *Mathematics.* **a** the set of all the values a given function may take on. **b** = domain.
2 *Statistics.* the difference between the smallest and the greatest values which a variable bears in frequency distribution: *Since the range is sensitive only to the two extreme scores in a distribution, it is not considered a very satisfactory measure of dispersion* (F. J. Crosswhite).
3 *Geology.* a row or chain of mountains: *Ordinarily the term range is applied to mountains that have a general unity of form, structure, and geologic age* (Finch and Trewartha, *Elements of Geography*).

Rankine (rang′kən), *adj.* of, based on, or according to a scale (the **Rankine scale**) in which temperatures are measured as roughly 460 degrees plus the Fahrenheit value, so that zero degrees Rankine means absolute zero (zero kelvins, or −459.67 degrees Fahrenheit). *Abbreviation:* R or Rank. [named after William J. M. Rankine, 1820–1872, Scottish scientist and engineer who devised this scale]

Raoult's law (rä ülz′), *Chemistry.* the principle that the vapor pressure of a substance in solution is proportional to the mole fraction of that substance: *Raoult's law, which is based upon experimental observations, deals especially with the effect of a solute upon the vapor pressure of a liquid* (Jones, *Inorganic Chemistry*). [named after François M. Raoult, 1830–1901, French chemist]

raphe (rā′fē), *n.* **1** *Anatomy.* a seamlike union between two parts of an organ of the body: *The scrotum is marked at its midline by a raphe.*
2 *Botany.* **a** (in certain ovules) the vascular tissue connecting the hilum with the chalaza. **b** a median line or rib on a valve of a diatom. [from Greek *rhaphē* suture, seam]

raphide (raf′īd), *n., pl.* **raphides** (raf′idz *or* raf′ə dēz). *Botany.* one of the minute needle-shaped crystals, usually composed of calcium oxalate, that occur in the cells of many plants. [from New Latin *raphides,* from Greek *rhaphides,* plural of *rhaphis* needle]

rapid eye movement, *Physiology.* the frequent and rapid, usually lateral, movement of the eyes which normally occurs during the dreaming period or state. *Abbreviation:* REM

rapids, *n. pl. Geography.* the part of a river's course where the water rushes quickly, often over rocks near the surface.

raptor (rap′tər *or* rap′tôr), *n. Zoology.* any raptorial bird; bird of prey: *Ornithologists have long suspected that the almost catastrophic decline of the peregrine falcon and other birds of prey in Britain has been matched by a corresponding decrease of avian raptors and scavengers in the rest of Europe* (New Scientist). [from Latin *raptor* robber, from *rapere* seize]

raptorial (rap tôr′ē əl), *adj. Zoology.* **1** adapted for seizing prey; having a hooked beak and sharp claws suited for seizing prey. **2** belonging to or having to do with birds of prey, such as the eagles and hawks.

rare earth, *Chemistry.* **1** an oxide of a rare-earth element: *Rare earths can be used in control rods for nuclear reactors, and also as a radiation shielding ingredient in concrete* (Science News Letter).
2 = rare-earth element: *The rare earths are ... 15 elements ... evolved from a role of interesting chemical oddities to a position of exciting importance in scientific and industrial technology* (Scientific American).

rare-earth, *adj.* of or having to do with rare earth: *The principal rare-earth mineral, monazite, is a mixture of phosphates* (Offner, *Fundamentals of Chemistry*).

rare-earth element, *Chemistry.* any of a series of metallic elements that have similar properties, ranging from lanthanum (atomic number 57) through lutetium (atomic number 71). Sometimes lanthanum is excluded from the rare-earth elements, and sometimes yttrium is included. *Rare-earth elements are not really very rare, however; the chief cost in their preparation is in separating them from one another.* (Joseph J. Becker). *Thorium occurs chiefly in association with the rare-earth elements in monazite ore* (Optima).

rarefaction (rãr′ə fak′shən), *n. Physics.* **1** a decrease in density and pressure in a medium, such as air, due to the passing of a sound wave or other compression wave. **2** the region in which this occurs. Compare **condensation.**

rasorial (rə sôr′ē əl), *adj. Zoology.* of or having to do with birds that scratch the ground for food, such as the chicken. [from New Latin *Rasores* the former order name of such birds, from Late Latin *rasor* scraper, from Latin *radere* to scrape]

rate, *n. Mathematics.* the quantity or degree of something measured in proportion or relation to something else; a quantity calculated per unit of a standard of reference: *If a jet plane covers a distance of 2000 miles in four hours, it travels at the rate of 500 miles per hour. ... To find the rate of travel, divide the distance traveled by the time required to cover the distance. If we let r represent the rate, d the distance, and t the time, we may express this rule by the formula* $r = d/t$ (Pearson and Allen, *Modern Algebra*).

ratio (rā′she ō *or* rā′shō), *n. Mathematics.* **1** the relation between two numbers or quantities expressed as a quotient; a relative magnitude. Sheep and cows in the ratio of 10 to 3 means ten sheep for every three cows, or 3 1/3 times as many sheep as cows.
2 a quotient expressing this magnitude. The ratio between two quantities is the number of times one contains the other. The ratio of 10 to 6 is 10:6, 10/6, $10 \div 6$, or $\frac{10}{6}$. The ratios of 3 to 5 and 6 to 10 are the same.
[from Latin *ratio* reckoning]

rational, *Mathematics.* —*adj.* **1** of or having to do with a rational number: *a rational fraction.*
2 involving no root that cannot be extracted.
—*n.* = rational number: *Subtraction of rationals is neither associative nor commutative* (Mathematics Regents Syllabus).

rational number, *Mathematics.* any number that can be expressed as an integer or as a ratio between two integers, excluding zero as a denominator. 2, 5, and $-1/2$ are rational numbers.

ratite (rat′īt), *Zoology.* —*adj.* **1** having a flat breastbone with no keel: *Ostriches and emus are ratite birds.* **2** of or having to do with ratite birds.
—*n.* a ratite bird.
[from New Latin *Ratitae* the order name of these birds, from Latin *ratis* raft, timber]

ravine (rə vēn′), *n. Geology, Geography.* a narrow valley with steep sides, eroded by running water. A ravine is larger than a gully and smaller than a canyon. [from French *ravine* violent rush]

ray, *n.* **1** *Physics.* a line or beam of light, electricity, or other radiant energy: *X rays, the invisible rays of the spectrum. From the wave viewpoint, a ray is an imaginary line drawn in the direction in which the wave is traveling* (Sears and Zemansky, *University Physics*).
► See the note under **beam.**
2 *Biology.* a part like a ray, such as: **a** one of the arms or branches of a starfish. **b** one of the processes which support and extend the fin of a fish. **c** the marginal portion of the flower head of certain composite plants, composed of ray flowers or petals. **d** = ray flower. **e** a branch of an umbel. **f** one of the vertical bands of tissue between the pith and the bark of a tree or other plant; medullary ray.
3 *Astronomy.* any of the bright streaks that are seen radiating from some lunar craters: *Best seen when the sun is high above them, the rays are prominent features of the full moon. The most conspicuous and longest ray system radiates from Tycho near the south pole, causing the full moon through a small telescope to appear something like an orange* (Baker, *Astronomy*).
4 *Geometry.* a set of points on a line, consisting of a given point (the *end point* of the ray) and all points on one side of it.

---

cap, fāce, fäther; best, bē, tėrm; pin, five;
rock, gō, ôrder; oil, out; cup, pùt, rüle,
y*ü* in use, y*u* in uric;
ng in bring; sh in rush; th in thin, ᴛʜ in then;
zh in seizure.
ə = *a* in about, *e* in taken, *i* in pencil, *o* in lemon, *u* in circus

547

ray flower or **ray floret,** *Botany.* one of the marginal flowers or florets of a daisy, aster, or other composite flower head, resembling a petal.

Rayleigh wave (rā′lē), *Geology.* a vertical, ripple-like seismic wave on or just below the surface of the earth: *Measurement of crustal thickness over the entire U.S. was made by noting dispersion in phase velocity of earthquake Rayleigh waves* (Science News Letter). Compare **Love wave.** [named after J. W. S. *Rayleigh,* 1842–1919, British physicist who first described the wave in 1900]

Rb, *symbol.* rubidium.

Re, *symbol.* rhenium.

react, *v. Chemistry.* **1** to undergo a reaction; produce a chemical change: *Acids react with metals to dissolve them.*
2 to cause (a substance) to form a chemical change or reaction: *One way to make formaldehyde is to react methane with hydroxyl* (Science News).

reactance (rē ak′təns), *n. Electricity.* that part, expressed in ohms, of the impedance of an alternating-current circuit which is due to inductance and capacitance, rather than resistance: *Pure reactance does not consume power* (Roy F. Allison).

reactant (rē ak′tənt), *n. Chemistry.* an element or compound that enters into a chemical reaction: *Changes in the spatial distribution of matter may also be caused by chemical reactions between the molecules of a mixture, and the rate of reaction is dependent on both the concentrations of the reactants and the number of different kinds of molecules which take part* (J. F. Pearson).

reaction, *n.* **1** *Chemistry.* a change involving the rearrangement of the atoms, molecules, etc., of one or more substances and resulting in the formation of one or more additional substances that often have different properties: *The reaction between nitrogen and hydrogen produces ammonia. It is now recognized that reactions take place in a series of steps or stages, each of which is extremely simple, involving the reaction of only very few—rarely more than two—molecules or atoms or radicals, the breaking or formation of a single bond, or the transfer of a single atom or electron from one reacting partner to another* (K. D. Wadsworth). Also called **chemical change, chemical reaction.**
2 *Physics.* an equal and opposite force which a body exerts against a force that acts upon it: *... in no case is it of importance which of the equal and opposite forces is considered the action and which the reaction* (Shortley and Williams, *Elements of Physics*).
3 *Nuclear Physics.* any process in which the nucleus of an atom undergoes a change or transformation, such as fission, fusion, or radioactive decay: *There is a difference between fusion and thermonuclear reactions. If two ions collide and fuse we call this a fusion reaction. If two ions in a uniformly hot gas collide and fuse we call this a thermonuclear reaction* (Electrical World). Also called **nuclear reaction.**
4 *Physiology.* **a** the response of a nerve, muscle, or organ to a stimulus: *The stimulus initiates, within the protoplasm, a series of changes constituting the impulse. These changes usually progress from the site of stimulation to more distant regions, where they initiate an adjustment to the stimulus. Such an adjustment is named a reaction* (Harbaugh and Goodrich, *Fundamentals of Biology*). **b** the response of the body to a test for immunization or the like: *A more general distinction depends on the extent of reaction shown by antiserum prepared against one strain of the virus, with virus preparations of a different strain* (A. W. Haslett).

reactive, *adj.* **1** *Chemistry.* readily given to or involving a reaction: *Alkynes are more reactive than alkenes.* **2** *Physiology.* readily responding to a stimulus. **3** *Electricity.* having to do with or possessing reactance.

reactivity (rē′ak tiv′ə tē), *n.* **1** *Chemistry.* the power or condition of being reactive: *The existence of such ... ions as intermediates in organic reactions is well established, and their reactivity can be readily measured by the rate at which they react with powerful acidic (negative) ions* (R. F. Homer).
2 *Nuclear Physics.* a measure of the degree to which a reactor departs from the critical condition: *If a reactor is supercritical, each generation contains more neutrons than the last by a ratio known as the reproduction factor, usually called k. The degree of criticality is usually expressed as $(k - l)/k$, which is called the reactivity* (H. M. McTaggart).

reactor, *n. Nuclear Physics.* an apparatus for the release of atomic energy by a controlled chain reaction. Reactors usually consist of a core of fissionable material, such as uranium, a moderator, such as graphite or heavy water, control rods to regulate the rate of the chain reaction, a gas or liquid coolant, and a pressure vessel that holds the core. *In a reactor that uses uranium fuel, fission occurs when the nucleus of a U-235 atom captures a neutron [and] splits into two smaller nuclei called fission fragments, releasing energy and several neutrons* (Francis T. Cole). Also called **nuclear reactor.**

reagent (rē ā′jənt), *n. Chemistry.* a substance involved in a reaction, especially a substance used to detect the presence of other substances by the chemical reactions it causes: *The addition to the blood of small amounts of reagents such as citrates, which can bind chemically and inactivate calcium ions ... prevents coagulation quite effectively* (K. S. Spiegler).

reagin (rē′ə jin), *n. Immunology.* any of a group of antibodies that react with allergens, as of hay fever or asthma: *The strange cellular disturbances that constitute hay-fever and the other allergies are known to be prompted by a reaction between macromolecules (antigens) in pollen or dust, and proteins (antibodies) which are carried in the blood serum and which are known as reagins* (New Scientist).
—**reaginic** (rē′ə jin′ik), *adj.* having to do with or being a reagin: *reaginic activity, a reaginic antibody.* —**reaginically,** *adv.*

real, *adj.* **1** *Mathematics.* either rational or irrational, not imaginary.
2 *Optics.* **a** of or having to do with an image formed by actual convergence of rays: *It is clear that whether an object is real or virtual depends merely on whether the light is diverging or converging when it enters the lens* (Hardy and Perrin, *Principles of Optics*). *A real image can be caught on a screen, a virtual image cannot* (Shortley and Williams, *Elements of Physics*). **b** having to do with or designating a focus forming such an image.

real number, *Mathematics.* any rational or irrational number: *The set of real numbers consists of all the positive numbers, all the negative numbers, and zero. ... The set of real numbers may be pictured as the set of points on a line* (Robert W. Wirtz and Hyman Gabai).

Réaumur (rā′ə myûr), *adj.* of or in accordance with the thermometric scale introduced about 1730 by Réne Antoine Ferchault de Réaumur, 1683–1757, a French physicist, in which the freezing point of water is 0 degrees and the boiling point 80 degrees: *For a while the Réaumur scale held its own against the superior thermometers of Fahrenheit and of Celsius, but slowly it lost ground and is now virtually out of use* (Isaac Asimov).

recapitulation, *n. Biology.* the supposed repetition in the development of an embryo of stages in the evolution of the species. The **recapitulation theory** holds that an embryo passes in its development from the ovum through a series of stages that resemble (and hence "recapitulate") the ancestral forms through which the species that the embryo belongs to passed in its evolutionary history. The theory is summed up in the maxim, "Ontogeny recapitulates phylogeny."

receiver, *n. Chemistry.* a vessel or container for receiving and condensing the product of distillation, receiving and containing gases, etc.

Recent, *Geology. n., adj.* = Holocene.

receptacle, *n. Botany.* the stalklike part of a flower that bears the petals, stamens, and pistils: *All these parts—calyx, corolla, stamens, and carpels—are attached to the receptacle, the somewhat specialized summit of the pedicel* (Emerson, *Basic Botany*).

receptor, *n.* **1** *Physiology.* a cell or group of cells sensitive to stimuli, such as a sense organ or the terminal portion of a sensory or afferent neuron: *The eye has two kinds of visual receptors, rods and cones* (Science News Letter). *Like the organs of sight and hearing the olfactory membrane has receptor cells which fire when stimulated and generate electrical pulses that travel to the brain* (Scientific American).
ASSOCIATED TERMS: see **neuron.**
2 *Biochemistry.* a molecular structure or site on a cell which is capable of combining with a hormone, antigen, globulin, or other chemical substances: *Receptors, or at least the notion of highly specific recognition patches on the cell surface, now pervade the thinking of many non-pharmacologists: indeed, "receptorology" (as it has inelegantly been called) is the real motive force behind the new interest in the cell surface* (New Scientist).

recessional moraine, *Geology, Geography.* a terminal moraine formed when a receding glacier halts temporarily. See the picture at **moraine.**

recessive, *Genetics.* —*adj.* of, having to do with, or designating the allele in a pair of contrasting alleles that is latent or subordinate in an organism when both alleles are present in the germ plasma. EXAMPLE: If a guinea pig inherits a gene for black fur from one parent and a gene for white fur from the other, it will have black fur, as black fur is dominant and white fur is recessive. *Recessive characters, even in the presence of dominant ones, are quite easily recognized* (G. Fulton Roberts).

—*n.* **1** a recessive allele or gene: *In a heterozygote the genes are of two kinds, dominants and recessives* (Youngken, *Pharmaceutical Botany*). ▶ See the note under **dominant.**
2 an individual possessing or transmitting a recessive trait: *Mendel discovered that in this generation the numerical proportion of dominants to recessives is approximately constant, being in fact as three to one* (William Bateson). —**recessiveness,** *n.*

recipient, *n. Medicine.* a person who receives a transfusion, graft, or transplant from another. Contrasted with **donor.**

reciprocal, *adj.* **1** *Mathematics.* **a** inversely corresponding or related; inverse: *a reciprocal spiral.* **b** based on an inverse relationship: *a reciprocal equation.*
2 *Physiology.* complementary; mutual. In the state of **reciprocal innervation,** one set of antagonistic muscles relaxes as the other contracts.
3 *Genetics.* of or designating a pair of crosses in which the male parent in one cross is of the same kind as the female parent in the other.

—*n.* **1** *Mathematics.* a number so related to another that when multiplied together they give 1: *3 is the reciprocal of 1/3, and 1/3 is the reciprocal of 3.*
2 *Physics.* a unit that is in an inverse or complementary relationship with another: *The mho is the reciprocal of the ohm.*
[from Latin *reciprocus* returning] —**reciprocally,** *adv.*

reciprocal translocation, *Genetics.* an interchange of parts between two nonhomologous chromosomes.

reclinate (rek′lə nāt), *adj. Botany.* bent or curved downward, as a leaf in a bud.

recombinant, *n.* **1** *Genetics.* an organism, cell, or other structure showing characteristics resulting from recombination.
2 = recombinant DNA: *Foreign genes are inserted into the bacterium by splicing them into a plasmid ... and reintroducing the recombinant into the bacterium* (Nature).
—*adj.* **1** *Genetics.* formed by or showing recombination: *a recombinant chromosome, recombinant inbred strains.*
2 *Molecular Biology.* having to do with or formed from recombinant DNA: *recombinant gene experimentation, recombinant genetic techniques.*

recombinant DNA, *Molecular Biology.* genetic material produced in the laboratory by breaking up and then combining DNA molecules obtained from different organisms. The recombinant DNA is introduced into a host cell where it becomes part of the cell's permanent genetic complement and is able to perform specific genetic functions. *The work is called recombinant DNA research because it involves breaking apart chains of deoxyribonucleic acid (DNA), the chemical*

cap, fāce, fäther; best, bē, tėrm; pin, five;
rock, gō, ôrder; oil, out; cup, pùt, rüle;
yü in use, *yu* in uric;
ng in bring; *sh* in rush; *th* in thin, ᴛʜ in then;
zh in seizure.
ə = *a* in about, *e* in taken, *i* in pencil, *o* in lemon, *u* in circus

that carries genetic data, and recombining them in various ways (Daniel L. Hartl).

▶ Work with recombinant DNA, often called *gene-splicing*, is generally regarded as the first major step in *genetic engineering*. Genes transplanted into bacteria by gene-splicing can cause the bacteria to produce cheaply an endless supply of scarce hormones and other body chemicals. The technique has many variations. The most widely used one consists of joining DNA molecules of any organism to bacterial molecules called *plasmids* which carry genes but reproduce independently of the chromosomes. In order to splice the donor DNA to the plasmid DNA, a *restriction enzyme* is used to cleave the molecules. When the plasmid is reintroduced into bacteria, the new genes it carries begin to function as part of the bacteria's natural genetic makeup. Besides plasmids, viruses can also be combined with the DNA of bacteria and other organisms to serve as carriers of new genes.

recombinase, *n. Biochemistry.* an enzyme that catalyzes genetic recombination.

recombination, *n.* **1** *Genetics.* the formation in offspring, either by crossing-over or by independent assortment, of new genetic combinations that are not in either parent: *Through genetic recombination, the progeny can inherit the defects of both parents or neither* (Francis H. C. Crick).

ASSOCIATED TERMS: see **evolution.**

2 *Physics.* the combining of positive and negative ions, or of holes and electrons, to form neutral atoms.

—**recombinational,** *adj. Genetics.* of or having to do with recombination: *Almost every gene in which a series of multiple alleles is known can be subdivided, through crossing over, into separate recombinational genetic units* (William K. Baker).

recombine, *v. Genetics.* to form new genetic combinations; undergo or cause recombination: *As a result of fertilization, alleles recombine. As a consequence, new allelic gene combinations are likely to be produced* (Biology Regents Syllabus).

recon (rē′kon), *n. Genetics.* the smallest molecular unit of genetic material out of which the larger units, the muton and cistron, are built. *The ultimate unit of molecular structure, the "recon," was found equal to one base pair of nucleic acid. The muton was approximately ten base pairs long* (Science News Letter). [from *rec(ombination)* + *-on*]

rect., *abbrev.* rectangle.

rectangle, *n. Geometry.* a four-sided plane figure with four right angles: *A rectangle is a kind of parallelogram. ... However, since a rectangle has four right angles, its diagonals have a property which the general parallelogram does not have. This property is commonly used as a test for rectangularity. ... If the diagonals of a parallelogram are equal, the figure is a rectangle* (Herberg, *A New Geometry*). [from Medieval Latin *rectangulus,* rectangular, ultimately from Latin *rectus* right + *angulus* angle]

rectangular (rek tang′gyə lər), *adj. Geometry.* **1** shaped like a rectangle: *a rectangular figure, a rectangular space.*

2 having one or more right angles: *a rectangular hexahedron.*

3 placed or lying at right angles: *rectangular coordinates.*

—**rectangularity,** *n.* rectangular quality or condition.

—**rectangularly,** *adv.*

rectification, *n.* the act or process of changing an alternating current into a direct current.

RECTIFICATION

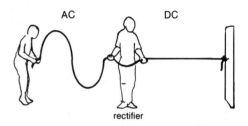

The alternating current is changed to direct current as it passes through the rectifier.

rectifier (rek′tə fī′ər), *n.* a device for changing alternating current into direct current: *The commutator of a generator acts as a rectifier, keeping a direct current in the outside line by changing the brush contacts at the same time that the current in the armature reverses* (Tracy, *Modern Physical Science*).

rectify, *v. Geometry.* to find out the length of (an arc, curve, etc.).

rectilinear (rek′tə lin′ē ər), *adj.* **1** in a straight line; moving in a straight line: *the rectilinear propagation of light.*

2 forming or lying in a straight line: *rectilinear rays of light.*

3 bounded or formed by straight lines: *A triangle is that plane rectilinear figure which has three sides* (E. V. Neale).

4 characterized by straight lines: *a rectilinear map.*

—**rectilinearly,** *adv.*

rectrix (rek′triks), *n., pl.* **rectrices** (rek trī′sēz). *Zoology.* one of the strong feathers in the tail of a bird. The rectrices serve as a rudder in flight. [from Latin *rectrix,* from *regere* to direct]

rectum (rek′təm), *n. Anatomy.* the lowest part of the large intestine, extending from the last curve of the colon to the anus. See the picture at **alimentary.** [from New Latin, from Latin *(intestinum) rectum* straight (intestine)]

rectus (rek′təs), *n., pl.* **-ti.** (-tī). *Anatomy.* any of several very straight muscles (as in the abdomen, thigh, and eye). [from New Latin *(musculus) rectus* straight (muscle)]

recumbent (ri kum′bənt), *adj. Botany.* leaning or resting on anything: *a recumbent plant, recumbent leaves.*

red alga, *Biology.* any member of a division (Rhodophyta) of characteristically red or purplish, mostly marine, algae: *Now an electron microscope study of the chloroplasts of a red alga, Laurencia spectabilis ... has located DNA here too, and has given support to a recent speculation about the origin of chloroplasts* (New Scientist).

red beds, *Geology.* a series of deep-red sedimentary strata, of a sandy, silty, or shale-like composition, mainly of the Permian or Triassic periods. Red beds often contain gypsum or salt deposits and are a conspicuous formation in the Rocky Mountains. *Triassic red beds ... are in the dry valleys of Antarctica, the red marls of Worcestershire, the hills of Alsace-Lorraine* (New Yorker).

red blood cell, *Physiology.* a cell in the blood, formed in bone marrow and containing hemoglobin, that carries oxygen from the lungs to various parts of the body: *Normal men have some 5,000,000 red blood cells [per cubic millimeter of blood]; normal women, 4,500,000. When in either sex the count drops below 4,000,000 cells, the individual is anemic* (Marguerite Clark). Also called **red cell, red corpuscle, erythrocyte.**

red cell or **red corpuscle,** = red blood cell: *Multiplication of cells and their mutant DNA occurs throughout life, and nearly 4 million new red cells pour into the bloodstream every second* (Francis H. C. Crick).

red dwarf, *Astronomy.* a star of the main-sequence group having a cooler temperature and lower luminosity than the others. Red dwarfs are generally smaller and fainter than the sun. Compare **white dwarf.**

red giant, *Astronomy.* a star of great size and brightness which has a comparatively cool surface temperature: *With the further passage of time, when a star has burned up about 10 per cent of its hydrogen, it swiftly becomes brighter and redder and moves off the main sequence on its way to becoming a red giant.* (Robert P. Kraft). Compare **white giant.**

redox (rē′doks), *Chemistry.* —*adj.* producing or containing the processes of reduction and oxidation: *A redox reaction may be considered in two parts, one representing a loss of electrons (oxidation), and the other representing a gain of electrons (reduction) Each reaction is known as a half-reaction ... Redox reactions that occur spontaneously may be employed to provide a source of electrical energy* (Chemistry Regents Syllabus).
—*n.* = oxidation-reduction: *An internal redox apparently occurs, and the electron moves from cation to anion* (Nature).

red shift, *Astronomy.* a shift of the light of stars, galaxies, etc., toward the red end or longer wavelengths of the spectrum, indicating movement outward at increasing speed, and leading to the belief that the universe is constantly expanding at an ever increasing rate of speed: *This red shift, which is a function of the velocity with which the light source appears to be receding from the observer, is most frequently interpreted as arising from the expansion of the universe following an initial "big bang" of creation; it allows measurement of extragalactic distances by means of a proportional relationship, called Hubble's law, between the recessional velocity of a light source and its distance from the Earth* (Kenneth Brecher). Compare **blue shift.**
—**red-shift,** *v.* to shift toward the red end of the spectrum: *The emission lines are red-shifted by an amount corresponding to a velocity of expansion of 700 km/s* (Robin Scagell).

red snow, *Biology.* **1** a growth of red-pigmented green algae in snow that makes the white surface seem lightly brushed with red. It is common in arctic and alpine regions. *Red snow ... has been discovered spreading over decayed leaves and mosses on the borders of small lakes* (H. Macmillan).
2 snow containing such a growth: *The phenomenon of red snow is due to the presence of some genus of algae, scientifically known as Protococcus nivalis* (E. Hawks).

red tide, *Biology.* a reddish discoloration on the surface of seawater produced at times by the sudden clustering together of billions of one-celled organisms that are toxic to fish. Red tide appears in most waters of the tropics and semitropics around Africa, Asia, South America, and California, Texas, Florida and in the Gulf of Mexico. *Similar in action to botulinus toxin and almost as deadly is the toxin secreted by the poisonous dinoflagellates. These microscopic marine organisms can kill fish by the millions when they multiply explosively in "red tides"; the concentration of their toxin in the digestive tracts of mollusks accounts for many cases of shellfish poisoning in man* (Elijah Adams).

reduce, *v.* **1** *Chemistry.* **a** to remove oxygen from (a compound): *When hydrogen is passed over hot cupric oxide, it takes oxygen from it (reduces it), so that only elementary copper remains* (Dull, *Modern Chemistry*). **b** to donate one or more electrons to (an atom or ion); change to a lower valence by the gain of electrons: *The metallic elements can be arranged in a table showing their ability to reduce ions of other metals* (Pauling, *College Chemistry*). **c** to combine with hydrogen; add hydrogen to (a compound). Contrasted with **oxidize.**
2 *Mathematics.* to simplify (an expression, a fraction, formula, etc.): *Reduce a fraction means to convert a fraction to an equal fraction with a smaller numerator and denominator. But the new fraction has the same value* (Robert L. Swain).
3 *Biology.* to undergo meiosis.

reducing agent, *Chemistry.* any substance that reduces a compound, especially by donating an electron or electrons: *An oxidizing agent is a substance that takes up electrons during an oxidation-reduction reaction. A reducing agent is the substance that furnishes the electrons in this action. It is obvious then that the substance oxidized is also the reducing agent, and the substance reduced is the oxidizing agent* (Dull, *Modern Chemistry*).

reducing sugar, *Chemistry.* a saccharide, such as glucose or fructose, that causes the reduction of copper or silver salts in alkaline solutions.

---

cap, fāce, fäther; best, bē, tèrm; pin, five;
rock, gō, ôrder; oil, out; cup, pùt, rüle,
yü in use, *yu̇* in uric;
ng in bring; *sh* in rush; *th* in thin, ᴛʜ in then;
zh in seizure.
ə = *a* in about, *e* in taken, *i* in pencil, *o* in lemon, *u* in circus

reductant

reductant (ri duk′tənt), *n.* = reducing agent: *Chlorophyll is thus a very peculiar substance: it can act both as an oxidant and as a reductant* (Scientific American).

reductase (ri duk′tās), *n. Biochemistry.* any enzyme that promotes reduction of an organic compound: *Reductase of yeast decomposes lactic acid to pyrotartaric acid and hydrogen. It also decomposes formic acid into carbon dioxide and hydrogen* (Youngken, *Pharmaceutical Botany*).

reduction, *n.* **1** *Chemistry.* **a** a reaction in which oxygen is removed from a compound: *The operation of the biosphere depends on the utilization of solar energy for the photosynthetic reduction of carbon dioxide (CO_2) from the atmosphere to form organic compounds on the one hand $(CH_2O)_n$ and molecular oxygen (O_2) on the other* (G. Evelyn Hutchinson). **b** a reaction in which an atom or ion gains one or more electrons and decreases in valence: *Oxidation may properly be defined as the process occurring when an element loses electrons or increases in valence number—becomes more positive. Reduction occurs when an element gains electrons or decreases in valence—becomes less positive or more negative* (Parks and Steinbach, *Systematic College Chemistry*). **c** a reaction in which hydrogen is combined with a compound. **2** = meiosis.

reduction division, = meiosis: *Meiosis is ... called reduction division, because the cells formed contain only half as many chromosomes as other cells* (E. V. Cowdry).

reflect, *v. Physics.* to turn back or throw back (waves or particles of light, heat, sound, etc.) from surfaces upon which they fall or strike: *In general, dark materials absorb much light, and light-colored materials reflect much of it* (Beauchamp, *Everyday Problems in Science*). Compare **refract.** [from Latin *reflectere* bend back, from *re-* back + *flectere* bend]

reflectance, *n. Optics.* the amount of light reflected by a surface in proportion to the amount of light falling on the surface: *It is evident that reflectance is related in a general way to the lightness of the color perceived* (Deane B. Judd).

reflecting telescope, a telescope in which light from the object is gathered and focused by a concave mirror and the resulting image is magnified by the eyepiece: *In the reflecting telescope, the paraboloidal mirror is mounted in a cell at the lower end of the tube (which in large reflectors is a skeleton tube composed of steel members braced for rigidity), and its focus is at the upper end where it is somewhat difficult of access. Different methods of overcoming this difficulty produce different forms of the telescope* (Duncan, *Astronomy*). Compare **refracting telescope.**

reflection, *n.* **1** *Physics.* the act of reflecting or the condition of being reflected; the turning back or throwing back of a ray of light, sound wave, a stream of electrons, etc., falling upon or striking a surface while traveling within a given medium: *A plane sound wave is reflected from a plane surface with the angle of reflection equal to the angle of incidence. It will be recognized that this is the same law governing the reflection of light waves from a plane mirror* (Sears and Zemansky, *University Physics*). Compare **refraction.** See also

angle of reflection, law of reflection. See the picture at **incidence.**

ASSOCIATED TERMS: see **diffusion.**

2 *Anatomy.* **a** the bending of a part back upon itself, especially the folding of a membrane from the wall of a cavity over an organ and back to the wall. **b** the part bent back.

—**reflective,** *adj.* that reflects; reflecting: *the reflective surface of polished metal.*

—**reflectivity,** *n.* the quality or state of being reflective.

reflector, *n. Physics.* **1** anything that reflects light rays, sound waves, etc., especially a piece of glass for reflecting light in a particular direction. Reflectors are usually concave. **2** = reflecting telescope.

reflex, *n. Physiology.* **1** an involuntary action in direct response to a stimulation of some nerve cells. Sneezing, vomiting, and shivering are reflexes. *Simple reflexes are considered to be independent of any learning process; the form of the response is always the same, that is, the same set of muscles or glands is always involved; and only one type of stimulus produces the response.* (S. A. Barnett). **2** = reflex arc.

—*adj.* **1** *Physiology.* not voluntary; coming as a direct response to a stimulation of some sensory nerve cells. Yawning is a reflex action. *Reflex behavior involves a pathway (reflex arc) over which impulses travel. In a spinal reflex there is a pathway from receptors to a sensory neuron to interneurons in the spinal cord to a motor neuron to an effector* (Biology Regents Syllabus). **2** *Mathematics.* (of an angle) more than 180 degrees and less than 360 degrees.

[from Latin *reflexum* bent back, from *re-* again + *flectere* to bend]

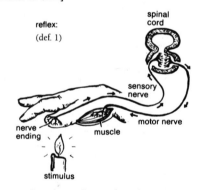

reflex:
(def. 1)

spinal cord

sensory nerve

motor nerve

nerve ending

muscle

stimulus

Heat from the candle stimulates the sensory nerves that produce the sensation of pain, which causes a motor reflex, stimulating the muscles to contract and withdraw the thumb.

reflex arc, *Physiology.* the nerve path in the body leading from stimulus to reflex action. The impulse travels inward to a nerve center and the response outward to the organ or part where the action takes place. *In the operations of a nervous system, the functional unit is a group of neurons called a reflex arc* (Shull, *Principles of Animal Biology*).

refract (ri frakt′), *v. Physics.* to bend (a ray of light, sound waves, etc.) from a straight course. Water refracts light shining through it. Compare **reflect.** [from Latin *refractum* broken back, refracted, from *re-* back + *frangere* to break]

552

refracting telescope, a telescope in which light from the object is gathered and focused by a lens (the objective) and the resulting image is magnified by the eyepiece: *In combination, two double convex lenses can form a refracting telescope* (Baker, *Astronomy*). Compare **reflecting telescope.**

refraction (ri frak′shən), *n.* **1** *Physics.* the turning or bending of a ray of light, sound waves, a stream of electrons, etc., when passing obliquely from one medium into another of different density: *Light travels at different velocities through different materials, and when passing from one medium to another, e.g., from air into water, is slowed or speeded, a process technically known as refraction.* (Science News). Compare **diffraction.**

ASSOCIATED TERMS: see **diffusion.**

2 *Optics.* measurement of the refractive power of the eye, a lens, or the like: *In optometry, the determination of the refractive errors of an eye is termed "refraction."* (Hardy and Perrin, *Principles of Optics*).

—**refractive** (ri frak′tiv), *adj.* **1** able to refract; refracting. The **refractive power** of a lens or other substance is a measure of its ability to refract light. **2** of or having to do with refraction.

refraction:
(def. 1)

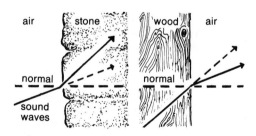

air   stone   wood   air

normal   normal

sound waves

refractive index, = index of refraction.

refractivity (rē′frak tiv′ə tē), *n. Physics.* the quality or condition of being refractive.

refractor (ri frak′tər), *n. Physics.* **1** anything that refracts light rays, sound waves, etc. **2** = refracting telescope.

refractory (ri frak′tər ē), *adj. Physiology.* **1** (of a muscle, nerve, etc.) temporarily unresponsive or not readily responsive to stimuli.

2 of or characterized by such temporary unresponsiveness to stimuli. The **refractory period** is a temporary period of reduced responsiveness immediately following a response. *In skeletal muscle, the refractory peri-*

od *is much shorter than in cardiac muscle* (L. V. Heilbrunn). —**refractorily,** *adv.* —**refractoriness,** *n.*

refrangibility (ri fran′jə bil′ə tē), *n. Physics.* **1** the property of being refrangible. **2** the amount of refraction (of light rays, etc.) that is possible.

refrangible (ri fran′jə bəl), *adj. Physics.* able to be refracted: *Rays of light are refrangible.*

reg, *n. Geography.* a desert with a stony surface and little or no sand. Compare **erg**[2]. [probably from Amharic (language of Ethiopia) *araga* rise]

regenerate, *v.* (ri jen′ə rāt) **1** *Biology.* to grow again or form (new tissue, a new organ, etc.) to replace that which has been lost or injured: *Blood vessels regenerate very quickly* (Bernard Donovan).

2 *Chemistry.* to make (a substance) reusable, especially by returning it to its original condition: *They were testing a ... system for regenerating oxygen for space crews* (New Scientist). *The cation exchanger is regenerated by passing through it a solution of an acid, such as sulphuric or hydrochloric* (K. S. Spiegler).

3 *Electronics.* to increase the amplification of (a current or signal) by transferring a portion of the power from the output circuit to the input circuit.

—*n.* (ri jen′ər it) *Biology.* a regenerated part, organ, or organism.

regeneration, *n.* **1** *Biology.* the formation of a new tissue, part, or organ to replace that which has been lost or injured; the formation of an entire organism from a part: *While asexual reproduction is not common among the echinoderms, a high power of regeneration is present and a single arm can regenerate a whole starfish* (W. Andrew).

ASSOCIATED TERMS: see **asexual.**

2 *Electronics.* the amplification of the strength of a current or signal by transferring a portion of the power from the output circuit to the input circuit.

—**regenerative** (ri jen′ə rā′tiv), *adj. Biology.* of or involving regeneration; regenerating: *Invertebrates exhibit a higher degree of regenerative ability than most vertebrates* (Biology Regents Syllabus).

region, *n.* **1** *Geography.* a part of the earth's surface having certain characteristics: *a mountainous region.*

2 *Ecology.* a division of the earth according to plant or animal life: *The purpose of biogeography is to trace out the reasons why particular species occupy the regions where they are now found* (H. R. Mill).

3 *Geophysics.* a division of the atmosphere according to height: *the E region of the ionosphere, the stratospheric region.*

4 *Oceanography.* a division of the sea according to depth: *the mid-oceanic region, the region of seamounts.*

5 *Anatomy.* a generalized division or part of the body: *the region of the heart, the gastrointestinal region.*

cap, fāce, fäther; best, bē; tèrm; pin, fīve; rock, gō, ôrder; oil, out; cup, pùt, rüle, yü in use, yu̇ in uric; ng in bring; sh in rush; th in thin; ᵺ in then; zh in seizure.

ə = a in about, e in taken, i in pencil, o in lemon, u in circus

—**regional,** *adj.* of or in a particular region: *a regional storm, the regional flora and fauna.* —**regionally,** *adv.*

regional metamorphism, *Geology.* metamorphism of very large volumes of rock: *Many geologists apply the term "regional" metamorphism to processes that have changed the rocks of great areas such as the Appalachians and northeastern Canada. Actually these processes seem to be no more than a combination of dynamic and igneous metamorphism on a large scale and hardly require a special name* (Fenton, *The Rock Book*).

regma (reg'mə), *n., pl.* **-mata** (-mə tə). *Botany.* a dry fruit of three or more carpels that separate from the axis at maturity, as in various geraniums. [from New Latin, from Greek *rhegma* a break]

regolith (reg'ə lith), *n. Geology.* the layer of soil and loose rock fragments overlying solid rock: *The regolith generally has a loose consistency; ... cemented together to form a kind of compound, secondary rock* (Paul D. Lowman, Jr.). Also called **mantle rock.** [from Greek *rhēgos* blanket + *lithos* stone]

regression, *n.* **1** *Biology.* reversion to a less developed state or form or to an average type: *In summary, the record of evolution we have shows no steady progression from simple to complex, from lower to higher except in the most general way. Regression ... and extinction are quite as much a part of the total record as is progression* (R. Beals and H. Hoijer).
2 *Genetics.* **a** the tendency of offspring to exhibit physical characteristics that are nearer those of the general population than are those exhibited by their parents; reversion to the average type: *The amount of the regression affords a useful measure of the intensity of inheritance. If the regression is slight, it means that the intensity of the inheritance is high* (J. A. Thomson). **b** the tendency of offspring to exhibit physical characteristics of an earlier type than those exhibited by their parents; reversion to an earlier type.
3 *Statistics.* the relationship between the mean value of a random variable and the corresponding values of one or more independent variables. A **regression equation** is one that gives the probable value of the dependent variable as a function of the value of the independent variable, usually by using the method of least squares. A **regression line** or **curve** is a graph of the probable value of the dependent variable plotted against the value of the independent variable.

regular, *adj.* **1** *Geometry.* having all its angles equal and all its sides equal: *a regular polygon.*
2 *Botany.* having all the same parts of a flower alike in shape and size.
3 *Crystallography.* (of a crystal) isometric.
[from Latin *regularis,* from *regula* rule]

regulator, *n.* or **regulator gene,** = regulatory gene.

regulatory gene, *Molecular Biology.* any gene that directs the synthesis of a molecule which interacts with an operator gene to control the function of an operon: *The first known class consists of structural genes, which determine the amino acid sequence and three-dimensional shape of proteins; the second is regulatory genes, which specify whether structural genes will function and therefore control the rate of enzyme synthesis* (Science News).

regurgitation (rē gėr'je tā'shən), *n.* **1** *Physiology.* **a** the return of undigested food from the stomach to the mouth. **b** the food so returned: *Since they cannot carry things well in their claws, the young are fed on a regurgitation of partly digested flesh* (Winchester, *Zoology*). **2** *Medicine.* the flow of blood back into the heart through a defective heart valve: *Heart murmurs usually arise from defective operations of the valves—either a backflow due to leaks (often called "regurgitation") or an obstruction of forward flow* (Victor A. McKusick).

reject, *v. Biology.* (of the body) to resist the introduction of (foreign tissue) by the mechanism of immunity: *He is now entering the crucial period where his body could begin to reject the implanted heart* (Observer). *A graft that will be ultimately rejected at first appears to be accepted by the host tissues* (R. M. Kirk).
—**rejection,** *n.* immunological resistance of the body to the grafting or implantation of foreign tissue: *Tissue rejection, which has led to the failure of other transplant operations, has yet to prove a problem in eardrum homografts* (Time).

rejuvenate, *v. Geology.* to restore to a condition characteristic of a younger landform or landscape, as by a drop in the sea level or an uplift of the land that causes a renewal of erosion: *Many mountain ranges show evidence of having been eroded to almost featureless plains, and then "rejuvenated" by the warping or faulting upward of the mountain stumps to form a new high mountain range. In such "second cycle" mountains the proportion of metamorphic rocks increases* (Garrels, *A Textbook of Geology*).
—**rejuvenation,** *n.* the act or process of rejuvenating a stream, region, etc.

rejuvenescence (ri jü'və nes'əns), *n. Biology.* **1** the process by which the contents of a cell break the cell wall and form a new cell with a new wall.
2 the renewal of vitality by the exchange of material between two distinct cells, as during conjugation.

relation, *n. Mathematics.* **1** a set of ordered pairs. **2** a property that holds between ordered pairs, such as is greater than, is inversely proportional to, etc.

relative frequency, *Statistics.* the ratio of the number of actual occurrences to the possible occurrences. EXAMPLE: If E occurs x times in N trials, then x/N is the relative frequency of E.

relative humidity, *Meteorology.* the ratio between the amount of water vapor present in the air and the greatest amount the air could contain at the same temperature: *If the air contains only half the amount of vapor that it can hold when saturated, the relative humidity is 50 per cent. In a good natural farming area the average relative humidity is about 65 per cent* (Walter J. Saucier).

relative maximum, *Mathematics.* a value of a variable greater than any values close to it.

relativistic, *adj. Physics.* **1** having to do with or based on the theory of relativity: *the relativistic concept of space-time, relativistic quantum mechanics.*
2 having a speed so great relative to the speed of light that the values of mass and other properties are significantly altered: *Synchrotron radiation ... is produced when beams of relativistic electrons (moving at speeds near that of light) are bent around the circular paths*

characteristic of many accelerators (William E. Spicer).

—**relativistically,** *adv.* **1** in a relativistic manner: *Most of the mass of the three quarks in a proton is relativistically converted into the tremendous energy that binds them together* (Time). **2** from a relativistic point of view.

relativity, *n. Physics.* a theory concerning the physical laws which govern matter, space, time, and motion, expressed in certain equations by Albert Einstein. According to the **special theory of relativity,** if two systems are moving uniformly in relation to each other, it is impossible to determine anything about their motion except that it is relative, for the velocity of light in a vacuum is constant, independent of either the velocity of its source or an observer, and all the laws of physics are the same in all inertial frames of reference. Thus it can be mathematically derived that mass and energy are interchangeable, as expressed in the equation $E = mc^2$, where c = the velocity of light; that a moving object appears to be shortened in the direction of the motion to an observer at rest; that a clock in motion appears to run slower than a stationary clock to an observer at rest; and that the mass of an object increases with its velocity. The **general theory of relativity** is an extension of the above theory, and deals with frames of reference that are accelerated rather than inertial, and with gravitation.

relaxation time, *Physics.* the time required, in many physical phenomena involving change or disturbance, for the elements of the process to recover equilibrium, or to effect some desired or expected result: *Damadian ... measured the "relaxation time" of these protons—that is, the time required for a proton to give up the added electro-magnetic energy* (Michael E. DeBakey). *The second is the relaxation time: the time in which the star settles down and becomes a stable member of the cluster under the influence of its gravitational interactions with nearby stars* (Scientific American).

relaxin (ri lak′sin), *n. Biochemistry.* a hormone produced by the corpus luteum during the later stages of pregnancy which facilitates birth by relaxing the pelvic ligaments.

releaser, *n. Zoology.* a stimulus that releases a specific response in another animal: *The baby waxbill's special food-begging notes are what we call 'releasers'—the signals, in other words, that make the mother waxbill feed her young* (Konrad Z. Lorenz).

releasing factor, *Physiology.* a substance produced by one gland that stimulates the release of a hormone by another gland, especially any of the hormonal substances secreted by the hypothalamus which stimulates the anterior part of the pituitary gland to release certain specific hormones: *There appears to be a chemically distinct releasing factor for each of the six anterior pituitary hormones* (New Scientist). *Abbreviation:* R.F.

reluctance, *n. Physics.* the resistance offered to the pas-

sage of magnetic lines of force, equivalent to the ratio of the magnetomotive force to the magnetic flux. [from Latin *reluctantem* struggling against]

reluctivity, *n. Physics.* the ratio of the intensity of the magnetic field to the magnetic induction of a substance; the reciprocal of permeability. [from *reluct(ance)* + *-ivity,* as in *reflectivity*]

rem, *n., pl.* **rem** or **rems.** a unit for measuring absorbed doses of radiation, equivalent to one roentgen of X rays or gamma rays: *The rem is a unit of biological damage per gram of exposed tissue; thus the implication that a chest x-ray of 25 millirems is equivalent to the (presumably whole-body) exposure received in 25 hours by a person who, standing in a radiation field, receives 1 millirem per hour, is misleading* (Science). Compare **rep.** [from *r(oentgen) e(quivalent) m(an)*]

REM, *abbrev.* rapid eye movement: *There are two distinct phases of sleep: One is accompanied by rapid eye movements and is called REM sleep. The other is non-rapid-eye-movement, or NREM, sleep. The periods of REM sleep are short early in the night and lengthen toward morning.* (V. L. Parsegian).

remainder, *n. Mathematics.* **1** the number that is left over when one integer is divided by another. The remainder plus the product of the quotient and the divisor equals the dividend.
2 the number obtained when one number is subtracted from another; difference.

remanence (rem′ə nəns), *n. Physics.* the flux density remaining in a substance after the magnetizing force has ceased. [from Latin *remanentem* staying back, remaining]

remex (rē′meks), *n., pl.* **remiges** (rem′ə jēz′). *Zoology.* any of the feathers of a bird's wing that enable it to fly; a flight feather. [from Latin *remex* oarsman, from *remus* oar]

remicle (rem′ə kəl), *n. Zoology.* a small flight feather, especially the outermost feather attached to the second phalanx of the middle finger of a bird's wing.

remigial (ri mij′ē əl), *adj.* of or having to do with a remex or remiges.

renal (rē′nl), *adj. Anatomy.* of, having to do with, or located near the kidneys: *renal arteries.* [from Latin *renalis,* from *ren* kidney]

renal corpuscle, *Anatomy.* one of the filtering structures in the cortex of the kidney, composed of a glomerulus and Bowman's capsule: *The renal corpuscles with the uriniferous tubules are the essential excretory units in the vertebrate animals generally* (Shull, *Principles of Animal Biology*). Also called **Malphigian body.**

renal gland or **renal capsule,** = adrenal gland.

---

cap, fāce, fäther; best, bē, tėrm; pin, fīve;
rock, gō, ôrder; oil, out; cup, pùt, rüle,
yü in use, *yù* in uric;
ng in bring; *sh* in rush; *th* in thin, ᴛн in then;
zh in seizure.
ə = *a* in about, *e* in taken, *i* in pencil, *o* in lemon, *u* in circus

555

reniform (ren′ə fôrm *or* rē′nə fôrm), *adj. Biology.* kidney-shaped: *a reniform leaf. The corneal substance of the eye is reniform* (T. H. Huxley). [from Latin *ren* kidney]

reniform leaf

renin (rē′nin *or* ren′in), *n. Biochemistry.* an enzyme in the kidney, capable of breaking down protein, and in- volved in the release of a peptide that affects the caliber of blood vessels: *Page made important discoveries on the workings of renin, an enzyme secreted by the kid- ney when it is starved of blood* (Time). [from German *Renin,* from Latin *renes* kidneys]

rennet (ren′it), *n. Biology.* a material obtained from the stomach mucus of calves. It contains rennin and is sometimes used as a coagulant in cheesemaking. [Mid- dle English, from *rennen* to run, Old English *rinnan*]

rennin (ren′ən), *n. Biochemistry.* an enzyme in the gas- tric juice that coagulates or curdles milk. It occurs in young infants, in calves, and also in certain other or- ganisms. Rennin is important in cheesemaking, and is now commercially produced by processes involving bacterial fermentation.[from *rennet*]

renormalization, *n. Physics.* the systematic replacement of unwanted infinities with experimentally observed values, used in quantum field theories as a method of eliminating obstacles to theoretical formulations: *This technique of hiding the mathematically undesirable in- finities is called renormalisation, and the resulting the- ory of quantum electrodynamics (QED), properly renormalised, turns out to be brilliantly successful. The infinities which arise in QED are symptomatic of nearly all quantum field theories* (New Scientist).
—**renormalize,** *v.* to apply renormalization to.

reovirus (rē′ō vī′rəs), *n. Biology.* an echovirus associated with respiratory and intestinal infections and found also in certain animal and human tumors: *The human virus is a reovirus, formerly known as the ECHO Type 10 virus, an agent of infection that causes upper respi- ratory and gastrointestinal diseases* (Scientific Ameri- can). [from *r*(espiratory) *e*(nteric) *o*(rphan) *virus*]

rep, *n., pl.* **rep** *or* **reps.** a unit of radiation used especially to measure beta rays, equal to the amount of ionizing radiation that will transfer to living tissue 93 ergs of energy per gram. Compare **rem.** [from *r*(oentgen) *e*(quivalent) *p*(hysical)]

repand (ri pand′), *adj. Botany.* having the margin slight- ly uneven or wavy: *a repand leaf.* [from Latin *repandus* bend back]

repeating decimal, *Mathematics.* a decimal in which the same figure or series of figures is repeated indefinitely. EXAMPLE: .3333+, .2323+.

repetend (rep′ə tend), *n. Mathematics.* that part of a re- peating decimal that is repeated indefinitely. The repe- tend in 6.010101 ... is 01.

replacement, *n. Geology.* the process by which a new mineral grows in the body of an old mineral by under- going simultaneous capillary solution and deposition.
replacement set, = domain.

replicase (rep′lə kās), *n. Biochemistry.* an enzyme that promotes synthesis of a complementary RNA mole- cule on an RNA template: *All that is required to pro- duce more of this RNA artificially is an enzyme, or biological catalyst, called replicase* (Robert Reinhold).

replicate, *v. Biology.* to duplicate or reproduce exactly by genetic processes: *A basic property of genetic mate- rial is its ability to replicate in a very precise way. Thus in man it is believed that at cell division the genetic material is exactly replicated in such a way that every cell of the body receives in its nucleus the total infor- mation originally present in the fertilized egg* (George W. Beadle).
—**replication,** *n.* the process by which a genetic sub- stance replicates or is replicated: *The process of mito- sis involves replication of each single-stranded chromosome during the nondividing period, resulting in double-stranded chromosomes* (Biology Regents Syllabus).
—**replicative,** *adj.* of, having to do with, or involved in replication: *DNA is the replicative molecule of the cell* (Scientific American).

replicon (rep′lə kon), *n. Molecular Biology.* a sequence of DNA nucleotide pairs that undergoes replication as a unit. The circular bacterial chromosome constitutes a single replicon, in contrast to eukaryotic DNA that may contain numerous replicons that replicate simul- taneously: *This group ... constitutes a replicon, of which there is just one in the E. coli chromosome and many on mammalian chromosomes. The bigger the chromosome the more replicons it has* (New Scientist). [from *replic*(ation) + *-on*]

representative fraction, a geographical scale expressed in the form of a fraction. EXAMPLE: The representative fraction of 1:62,500 or 1/62,500 shows that one unit of measurement on the map represents 62,500 of the same units on the earth's surface. *Abbreviation:* R.F.

repressor, *n. Molecular Biology.* a component of the operon whose function is to repress the action of the operator: *Genetic experiments some time ago showed that there are groups of genes controlled by the prod- ucts of other genes called regulators. The latter pro- duce substances called repressors, which neutralize the protein-forming activities of the genes under their con- trol* (Science News). Compare **derepressor.**

reproduce, *v. Biology.* to produce offspring; undergo re- production: *One function of all animals is to reproduce their own kind.*

reproduction, *n. Biology.* the sexual or asexual process by which living organisms produce other organisms like themselves: *The process or processes by which all living things produce others of a more or less similar nature is known as reproduction. This capacity is com- mon to microscopic single-celled organisms as well as to the complex multicellular plants and animals. The* *universality of the occurrence of this process indicates that all life comes from preexisting life* (Harbaugh and Goodrich, *Fundamentals of Biology*).

ASSOCIATED TERMS: In *asexual reproduction* new organisms develop from a cell or cells of the parent without the fusion of nuclei in processes such as *binary fission, budding,* and *sporulation. Sexual reproduction* involves the production of specialized sex cells (*gametes*) and the fusion of their nuclei (*fertilization*) to produce a fertilized egg cell (*zygote*). Sexual reproduction in flowering plants takes place through *pollination.* Reproduction in many aquatic vertebrate animals such as fish and amphibians occurs through *external fertilization;* in most terrestrial vertebrate animals, reproduction takes place through *internal fertilization.*

Most mammals are *viviparous;* most reptiles and birds are *oviparous;* some fish, amphibians, and reptiles are *ovoviviparous.* Animals in which both eggs and sperm are produced by *self-fertilization* are called *hermaphrodites.* Reproduction from an unfertilized egg is called *parthenogenesis.* Special asexual reproductive processes include *fragmentation, regeneration,* and *vegetative propagation.* See also the associated terms under **asexual** and **sexual.**

reproductive, *Biology.* —*adj.* **1** that reproduces: *sexually reproductive organisms.*
2 for or concerned with reproduction: *the human reproductive system, the reproductive tract of the female.*
—*n.* a reproductive organism, especially among the social insects: *At certain seasons winged male and female reproductives (future kings and queens) are also present* (Scientific American).

reptile, *Zoology.* —*n.* any of a class (Reptilia) of cold-blooded vertebrates that breathe by means of lungs and usually have skin covered with dry horny plates or scales. Snakes, lizards, turtles, alligators, and crocodiles are reptiles. Reptiles such as the dinosaurs were the dominant form of life during the Mesozoic.
—*adj.* of, resembling, or belonging to the reptiles.
—**reptilian,** *adj.* of or having to do with reptiles. —*n.* = reptile.

repulsion, *n. Physics.* the tendency of particles to increase their distance from one another: *Protons exert forces of repulsion on other protons, electrons exert forces of repulsion on other electrons* (Sears and Zemansky, *University of Physics*). Contrasted with **attraction.**
—**repulsive,** *adj.* of the nature of or characterized by repulsion.

RES, *abbrev.* reticuloendothelial system.

research, *n.* organized scientific investigation to solve problems, test hypotheses, or develop or invent new products: *recombinant DNA research, a marine research laboratory.*

research reactor, *Nuclear Physics.* a reactor for the study of fission and nuclear energy, or for research in radioactivity, radiology, etc.

réseau (rā zō′ *or* ri zō′), *n. pl.* -**seaux** (-zōz). *Astronomy.* a crisscross of squares photographed on the same plate as a star for the purpose of measurement. [from French]

resequent (ri sē′kwənt), *adj. Geology, Geography.* (of a stream) flowing in the same direction as a consequent stream but developing at a lower level than the original slope. Compare **insequent, obsequent, subsequent.**

reservoir, *n. Biology.* **1** a part of an animal or plant in which some fluid or secretion is collected or retained.

2 an organism that carries a disease germ or virus to which it is immune: *They hope that the laboratories will soon identify the animal carriers—the reservoirs—and the insects—vectors—that transmit the disease* (N.Y. Times).

residual, *adj. Geology.* of or designating material left in place when rock is weathered: *residual clay soil, a residual deposit. Soils formed by the weathering of rocks and the addition of humus in situ are known as residual* (White and Renner, *Human Geography*).

residual quantity, *Mathematics.* a binomial having one of its terms negative, as 2a − b.

residue, *n.* **1** *Chemistry.* a molecule incorporated in another, especially one incorporated in a protein, carbohydrate, or other organic polymer.
2 *Mathematics.* **a** the remainder left after a given number is divided into the square, cube, etc., of some integer. **b** a number congruent to a given number with respect to (modulo) a third number. A **residue class** is the class of integers congruent to each other with respect to (modulo) a given number.

resilin (rez′ə lin), *n. Biology.* an elastic substance consisting of cross-linked protein chains, found in the cuticles of many insects: *The flea's leap is powered not by muscle alone but is assisted by the elastic protein resilin. This protein, which is present in the wing-hinge ligament of dragonflies and locusts, can store and release energy more efficiently than any known rubber and can deliver power faster than most actively contracting muscles* (Miriam Rothschild). [from Latin *resilire* spring back]

resin, *n. Chemistry.* any of a class of vegetable substances, composed chiefly of esters and ethers of organic acids, that occur as a sticky yellow or brown substance exuded on the bark of various plants and trees, such as the pine and fir. Resin is transparent or translucent, does not conduct electricity, and is used in medicine, varnish, plastics, inks, and adhesives. When pine resin is heated it yields turpentine; the hard yellow substance that remains is called rosin. Copal and amber are types of resin. [from Latin *resina*]

resin duct or **resin canal,** *Botany.* a canal in the wood of trees, especially conifers, through which resin is secreted and conducted.

resinous (rez′ə nəs), *adj.* of, resembling, or containing resin: *resinous materials.*

resistance, *n. Physics.* **1** an opposing force, especially one tending to prevent motion.

2 the opposition offered by a body or substance to the passage of an electric current through it, resulting in a change of electrical energy into heat or other forms

---

cap, fāce, fäther; best, bē, tėrm; pin, fīve;
rock, gō, ôrder; oil, out; cup, pùt, rüle,
yü in use, *yù* in uric;
ng in bring; sh in rush; th in thin, ŦH in then;
zh in seizure.
ə = *a* in about, *e* in taken, *i* in pencil, *o* in lemon, *u* in circus

of energy. The resistance of a conductor in a circuit equals the potential difference across the conductor divided by the current. The SI unit of resistance is the ohm. *Resistance is the electrical counterpart of friction, and can serve the same damping function* (Roy F. Allison).

resistance:
| (def. 1)

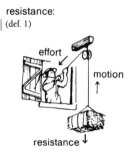

effort

motion ↑

resistance ↓

resistivity, *n. Electricity.* the characteristic resistance of a given substance; the reciprocal of conductivity: *Some materials whose resistivities lie between metals and insulators are called semiconductors* (Physics Regents Syllabus).

resistor, *n. Electricity.* a conductor used in an electronic device, such as a radio or computer, to control current or voltage in an electric circuit because of its resistance: *A resistor is simply a poor conductor of electricity* (John R. Pierce).

resolution, *n.* **1** *Physics.* the act, process, or capability of making component parts distinguishable, as of an object, closely adjoining images, or a quantity of nearly equal values: *A 1,000 kilohertz ... sonar would provide a resolution of 30 centimeters on a target 200 meters away* (Scientific American).
2 = resolving power.
3 *Chemistry.* the separation of a racemic compound or mixture into its optically active components.

resolve, *v.* **1** to break up into parts; separate into components: *white light resolves into a spectrum by a prism. Some chemical compounds can be resolved by heat.*
2 to produce separate images of; make distinguishable, by some device such as a microscope or radar, or by some scientific process: *to resolve a double star by spectroscopy.*
3 to solve or simplify by arithmetic or other mathematical operations: *to resolve an equation.*
[from Latin *resolvere,* from *re-* un- + *solvere* to loosen]

resolving power, *Optics.* the ability of an optical lens or system to produce separate images of objects very close together: *The device has a resolving power of 600,000 which means it can separate one part in 600,000* (Science News Letter). Also called **resolution.**

resonance (rez'n ǝns), *n.* **1** *Chemistry.* the oscillation of molecules between two or more structures, each possessing identical atoms, but different arrangements of electrons. It is the property of a compound having simultaneously the characteristics of two or more structural forms that differ only in the distribution of electrons. Resonance is used to describe the structure

of compounds which cannot be properly represented by a single structural formula. Resonance confers greater stability on the compound than it would possess with only one structural form. Also called **mesomerism.**
2 *Physics.* a marked increase occurring in the amplitude of oscillation of a system when the system is subjected to an oscillating force whose frequency is the same as or very close to the natural frequency of the system. It is the condition in an electric circuit adjusted to allow the greatest flow of current at a certain frequency. Resonance occurs in acoustical, atomic, electrical, magnetic, mechanical, optical, and other systems, and is applied in various types of analyses, such as nuclear magnetic resonance.
3 *Nuclear Physics.* an unstable particle of extremely short duration, or an excited state of a particle, detected as an increase, at certain energy levels, in the probability of interaction of various other particles: *The rho meson is a particle in its own right although it is also frequently referred to as a resonance* (Thomas Groves).
4 *Astronomy.* a synchronous relationship between two motions of a celestial body (as the axial and orbital motions of the moon) or between the motions of two bodies orbiting a third: *The motion of Toro [an asteroid] is controlled by a complicated resonance with the motions of the earth and Venus* (Science News).
—**resonant,** *adj.* of, involving, or exhibiting resonance.
—**resonate,** *v.* to exhibit resonance: *Both groups took particular interest in nuclear magnetic resonance (nmr) studies of their respective molecules. By this technique, the frequency with which the nucleus of a hydrogen atom resonates in an applied magnetic field can be determined, giving a measure of the local magnetic environment of that proton* (New Scientist).

resonance (def. 2)

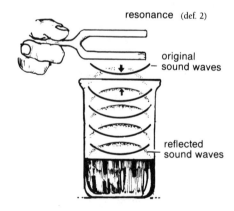

original sound waves

reflected sound waves

resorb, *v.* to absorb again; undergo resorption.
resorption, *n.* **1** *Biology.* the act or process of resorbing, especially of taking into the circulatory system food material that has been digested or broken down or of dissolving and absorbing material already in the body, such as soft tissue or bone.
2 *Geology.* the remelting and resorbing of a crystal by the molten magma in the formation of igneous rocks.
—**resorptive,** *adj.* having to do with or characterized by resorption.

respiration, *n.* **1** *Biology.* the energy-producing process by which a living organism or cell obtains oxygen from the air or water, distributes it, uses it for oxidation of food materials, and gives off carbon dioxide: *External respiration ... is the exchange of gases between the environment and the blood. It takes place in the specialized organ, such as the gills or the lungs* (Otto and Towle, *Modern Biology*). *All of the chemical and physical activities associated with anabolic processes of living organisms are made possible by the utilization of energy. In general, this energy is a by-product of a chemical change in which oxygen is consumed and carbon dioxide and water are liberated. These changes occur constantly in every living cell. The entire process is variously referred to as oxidation, physiologic burning, internal respiration, or simply respiration* (Harbaugh and Goodrich, *Fundamentals of Biology*). Compare **aerobic respiration, anaerobic respiration, cellular respiration, internal respiration.**
2 *Physiology.* the act of inhaling and exhaling; the process of breathing air through the lungs: *It is your breathing, or respiration, that gets the used air out of the air sacs in your lungs and takes in fresh air* (Beauchamp, *Everyday Problems in Science*).

respiratory, *adj.* of or having to do with respiration: *The lungs are respiratory organs.*

respiratory enzyme, *Biochemistry.* an enzyme that helps to bring about respiration in cells. Respiratory enzymes, such as oxidase, act on oxygen and the foods in the cells to produce energy. *Living cells contain substances that speed up chemical processes in the cell. Scientists call these substances respiratory enzymes* (Andrew C. Ivy).

respiratory pigment, *Biochemistry.* any of various proteins that carry oxygen in cellular respiration: *It was further discovered that ... the respiratory pigment of blood is not identical with the haemoglobin in muscle cells that makes meat red* (H. Munro Fox).

respiratory quotient, *Physiology.* the ratio of the volume of carbon dioxide released to that of oxygen consumed in a given period or interval: *The factor secreted by the pituitary gland has been found to depress the normal ratio between the oxygen inhaled in respiration and the carbon dioxide exhaled. This ratio is technically known as the Respiratory Quotient, or R.Q. This ratio is an index of the utilization by the body of its energy-giving fuels, particularly its carbohydrates* (N. Y. Times).

respiratory system, *Anatomy.* the system of organs and passages by which air is taken into the body and carbon dioxide and oxygen are exchanged. In the human body the respiratory system includes the nasal cavities, pharynx, larynx, trachea, and the lungs.

respirometric (res′pə rō met′rik), *adj.* of or having to do with respirometry: *respirometric studies of fungi.*

respirometry (res′pə rom′ə trē), *n.* *Biology.* the measurement of the degree and nature of respiration.

response, *n.* *Physiology.* a change in an organism, organ, tissue, etc., resulting from application of a stimulus; reaction to a stimulus or stimuli: *The stimulus initiates ... a series of changes constituting the impulse. These changes usually progress from the site of stimulation to more distant regions, where they initiate an adjustment to the stimulus. Such an adjustment is named a reaction or response* (Harbaugh and Goodrich, *Fundamentals of Biology*).

rest, *n.* **at rest,** *Physics.* having a net flow of zero in any given direction: *Static electricity deals with electrical charges at rest.*
► When referring to charges, the phrase *at rest* refers only to their net flow in any given direction; it does not imply that the charges themselves are not in motion.

restiform (res′tə fôrm), *adj.* *Anatomy.* cordlike; corded. The **restiform body** is one of a pair of large, cordlike bundles of nerve fibers lying one on each side of the medulla oblongata and connecting it with the cerebellum. [from New Latin *restiformis,* from Latin *restis* cord + *forma* form]

resting, *adj.* *Biology.* **1** (of spores) dormant. **2** (of cells) not dividing or preparing to divide.

restitution, *n.* *Physics.* the return of an elastic body to its original form or position when released from strain.

rest mass, *Physics.* the mass of an atom, electron, or other particle when it is regarded as being at rest.

restriction enzyme or **restriction endonuclease,** *Biochemistry.* an enzyme that separates DNA strands of an organism at a specific site which matches the DNA fragment of another organism cut by the same enzyme, enabling segments of DNA from different sources to be joined in new combinations: *A ... class of enzymes called "restriction enzymes" has made gene insertions possible* (Science News). *At least 80 enzymes, called restriction endonucleases are known to cut DNA in different specific parts of the molecule* (N. Y. Times).

resultant, *n.* **1** *Mathematics.* a single vector that is the equivalent of a set of vectors: *The resultant of two or more concurrent forces acting on a body is the single force producing the same effect. The resultant may be found by the vector addition of the individual forces* (Physics Regents Syllabus). See the picture at **vector.**
2 *Chemistry.* the substance or substances produced in a chemical reaction.

resupinate (ri sü′pə nāt), *adj.* *Botany.* bent backward; inverted; appearing as if upside down.
—resupination, *n.* a resupinate condition; inversion of parts.

retene (rē′tēn or ret′ēn), *n.* *Chemistry.* a white, crystalline hydrocarbon obtained especially from the tar of resinous woods and certain fossil resins. *Formula:* $C_{18}H_{18}$ [from Greek *rhētinē* resin]

retentivity, *n.* *Physics.* the capacity of a substance to retain induced magnetic force after the source of the magnetization has been removed.

reticle (ret′ə kəl), *n.* *Optics.* a line, or a network of fine lines, wires, or the like, placed in the focus of the objective of a telescope or other optical instrument to make

cap, fāce, fäther; best, bē, tèrm; pin, five;
rock, gō, ôrder; oil, out; cup, pùt, rüle,
yü in use, *yú* in uric;
ng in bring; *sh* in rush; *th* in thin, ₮H in then;
zh in seizure.
ə = *a* in about, *e* in taken, *i* in pencil, *o* in lemon, *u* in circus

accurate observation easier. [from Latin *reticulum* network]

reticular (ri tik′yə lər), *adj. Biology.* having the form of a net; netlike: *reticular tissue, reticular fibers.*

reticular cell, = reticulocyte.

reticular formation, *Anatomy.* a network of small nerve cells extending from the diencephalon downward through the spinal cord, with distinct formations in the medulla oblongata and mesencephalon. It exerts control over the body's motor activities. *Conditioned reflexes take place in the "wakefulness" area, a small structure deep in the brain called the reticular formation* (Science News Letter).

reticulate (ri tik′yə lit *or* ri tik′yə lāt), *adj.* **1** *Biology.* covered with a network; netlike. Reticulate leaves have the veins arranged like the threads of a net.
2 *Genetics.* (of evolution or species formation) depending on repeated crossing between several lines. **Reticulate evolution** is evolutionary change resulting from recombination between various interbreeding populations.—**reticulately,** *adv.*
—**reticulation,** *n.* a reticulated formation, arrangement, or appearance; network.

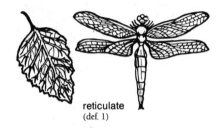

reticulate
(def. 1)

reticulocyte (ri tik′yə lə sīt), *n. Biology.* **1** a red blood cell that is not fully developed. Reticulocytes comprise from 0.1 to 1 percent of the red blood cells. **2** a cell of connective tissue, made up chiefly of a network of delicate branching fibers.

reticuloendothelial (ri tik′yə lō en′dō thē′lē əl), *adj. Anatomy.* of or having to do with the reticuloendothelial system: *a reticuloendothelial cell or tissue.*

reticuloendothelial system, *Anatomy.* a system of cells in the body, especially in the spleen, lymph nodes, bone marrow, and liver, that function in freeing the body of foreign matter and disease germs, in the formation of certain blood cells, and in the storing of fatty substances: *The reticuloendothelial system of the body ... includes cells of the spleen, lymph glands, bone marrow and liver. The cells act as scavengers to rid the body of dyestuffs and other foreign matter, such as disease germs and their poisons* (Science News Letter). *Abbreviation:* RES

reticulum (ri tik′yə ləm), *n., pl.* **-la** (-lə). **1** *Biology.* any reticulated system or structure; network.
2 *Zoology.* the second stomach of animals that chew the cud: *When first eaten the food passes into the rumen, later balls of this food, called the cud, are passed into the reticulum* (Winchester, *Zoology*). Compare **abomasum, omasum.**
[from Latin *reticulum* network, diminutive of *rete* net]

retina (ret′n ə), *n., pl.* **-inas, -inae** (-n ē′). *Anatomy.* the structure in the eyeball that is sensitive to light and receives optical images. It lines the inside back of the chamber filled with vitreous humor and contains the rods and cones near its outer surface. Its inner surface is continuous with the optic nerve. *The sensitive elements in the retina of the vertebrate eye are of two kinds: rods and cones ... In man, the central part of the retina, the fovea centralis or 'yellow spot' on which objects are normally focused, consists almost entirely of cones. These are responsible for colour-vision and for detailed resolution.* (J. L. Cloudsley-Thompson). [from Medieval Latin *retina*, probably from Latin *rete* net]

retinacular (ret′n ak′yə lər), *adj.* of or having to do with a retinaculum.

retinaculum (ret′n ak′yə ləm), *n., pl.* **-ula** (yə lə). **1** *Botany.* a viscid gland for holding the pollen masses together on the stigma of orchids and plants of the milkweed family.
2 *Zoology.* **a** a small scale or plate which in some insects checks an overprotrusion of the sting. **b** an arrangement of hooks, or hooks and bristles, which interlocks the fore and hind wings of insects when in flight.
3 *Anatomy.* a fibrous structure binding down tendons of muscles.
[from Latin *retinaculum* band, halter, from *retinere* hold back, retain]

retinal (ret′n əl), *adj. Anatomy.* of or on the retina.
—*n.* (ret′n al) = retinene.

retinene (ret′n ēn′), *n. Biochemistry.* a yellow pigment in the retina of the eye, formed when rhodopsin in the rods of the eye is decomposed by light: *In all cases the pigments which absorb the light that stimulates vision are made of vitamin A, in the form of its aldehyde, retinene, joined with specific retinal proteins called opsins.* (Scientific American). *Formula:* $C_{20}H_{28}O$ Also called **retinal.** Compare **visual yellow.**

retinite (ret′n īt), *n. Mineralogy.* any one of various fossil resins, especially one of those derived from brown coal and peat. [from Greek *rhētinē* resin]

retinol (ret′n ōl), *n.* = vitamin A.

retinotectal (ret′n ō tek′təl), *adj. Anatomy.* of or involving the network of nerve fibers connecting the retina with the dorsal part of the midbrain: *The mechanisms responsible for the formation of specific nerve connections have been analysed, most extensively in the retinotectal system of lower vertebrates: the axons of retinal ganglion cells connect selectively at local tectal sites to produce a map of the retina across the surface of the optic tectum* (Nature). [from *retina* + *tectum*]

retinula (re tin′yə lə), *n., pl.* **-lae** (-lē). *Zoology.* a cluster of pigmented cells in the compound eyes of arthropods, from which the rhabdom arises: *Each retinula [of a nocturnal moth] is composed of seven or eight photoreceptor cells* (H. R. Steeves and J. S. Vande-Berg).
—**retinular,** *adj.* of or constituting a retinula: *retinular cells.*

retractor, *n. Anatomy.* a muscle that retracts an organ, protruded part, or other process.

retral (rē'trəl), *adj. Biology.* at the back; posterior: *Beneath the retral ethmoidal spike is seen the olfactory groove* (Journal of Microscopic Sciences). —**retrally**, *adv.*

retrogradation, *n. Astronomy.* an apparent backward motion of a planet or asteroid from east to west: *What one realizes first about the planets is their appearance and disappearance in the nightly sky, their stations and retrogradations* (I. Bernard Cohen).

retrograde, *Astronomy.* —*adj.* characterized by retrogradation: *The eighth moon's path through the heavens is remarkable because, along with three other satellites of Jupiter, it moves in a clockwise direction. This motion is known as retrograde, and is a direction opposite that of most objects in the solar system* (Science News Letter).
—*v.* (of a planet or asteroid) to appear to move backward from east to west: *Once during each synodic period the planet turns and moves westward, or retrogrades, for a time before resuming the eastward motion* (Baker, Astronomy). —**retrogradely**, *adv.*

retrograde metamorphism, *Geology.* metamorphism in which relatively low-temperature minerals replace older high-temperature minerals.

retrorse (ri trôrs'), *adj. Botany, Zoology.* turned backward; turned in a direction opposite to the usual one. Contrasted with **antrorse.** [from Latin *retrorsus*, from *retro-* back + *versus* turned]
—**retrorsely**, *adv.*

retrovirus, *n. Biology.* any of a group of tumor-producing viruses containing RNA and reverse transcriptase: *The RNA tumor viruses ... are often called "retroviruses" because they reverse a step in one central dogma of biology: DNA makes RNA makes protein* (Science News) [from *re(verse)* *tr(anscriptase)* + connective *-o-* + *virus*]

retuse (ri tyüs' *or* ri tüs'), *adj. Botany.* having an obtuse or rounded apex with a shallow notch in the center: *a retuse leaf.* [from Latin *retusum* beaten back, from *re-* back + *tundere* to beat]

retuse leaf

revehent (rev'ə hənt), *adj. Anatomy.* conveying back: *revehent blood vessels.* [from Latin *revehentem* carrying back, from *re-* back + *vehere* carry]

reverse fault or **reversed fault,** *Geology.* a fault in which one side has moved above and over the other side, as in a thrust fault.

reverse shell, *Zoology.* a spiral shell, as of a gastropod, in which the whorl rises from right to left, in reverse of the usual direction.

reverse transcriptase, *Biochemistry.* an enzyme that stimulates the formation of DNA on a template of RNA in certain tumor-producing viruses. *In recent years an unanticipated type of polymerase was discovered: it uses RNA as a template and catalyzes the synthesis of complementary DNA. It is called ... reverse transcriptase, or ... RNA-dependent DNA polymer-*

revolution

ase, and is found in cells infected by certain viruses composed of RNA (Arnold W. Ravin). Also called **transcriptase.** Compare **retrovirus.**

reversible reaction, *Chemistry.* a reaction whose progress may be halted and reversed or, after change has taken place, be made to occur again in reverse, so as to leave the substances involved once again in their original state: *Thus ethyl alcohol and acetic acid react together to form ethyl acetate and water. But ethyl acetate and water also react together forming ethyl alcohol and acetic acid. In such reversible reactions, it is found that after a certain time, depending on the conditions, no further change appears to occur even though each pair of reactants is still present. In fact, a 'dynamic equilibrium' is set up, the concentrations of the components being such that both reactions proceed at the same rate* (K. D. Wadsworth).

reversing layer, *Astronomy.* the thin innermost layer of the atmosphere of the sun and other stars, where spectral lines are formed: *The surface of the sun ... is known as photosphere. It's the first of four outer layers, and directly above it is the reversing layer, surrounded by the chromosphere, which in turn is enveloped by the sun's corona* (Barnhard, Handbook of Heavens).

reversion, *n. Biology.* a return to an earlier type or average condition; the reappearance of an ancestral character or average type not exhibited by the parents or for several generations.

revertant (ri vėr'tənt), *Genetics.* —*n.* a mutant that has reverted to the earlier or normal condition, usually by a second mutation: *Quite frequently ... one finds a revertant in which the second mutation can be shown to occur in a gene different from that of the defective messenger RNA. Somehow this second mutation suppresses the effect of the original one; it enables the mutant cell to produce an active enzyme even though the gene for that enzyme is producing a defective messenger RNA* (Luigi Gorini).
—*adj.* that has reverted to the normal phenotype or earlier condition: *revertant strains, revertant cells.*

rev/min, *symbol.* revolutions per minute: *a speed of 15,500 rev/min.*

revolution, *n.* **1** *Astronomy.* **a** the motion of a heavenly body around another: *One revolution of the earth around the sun takes a year.* **b** the motion of a heavenly body as it turns on its axis. SYN: rotation. **c** the time or distance of one revolution.
2 *Geometry.* the motion of a figure about a center or axis.
3 *Physics.* the act or fact of turning around a center or axis; rotation: *The wheel of the motor turns at a rate of 1650 revolutions a minute.*

---

cap, fāce, fä̀ther; best, bē, tėrm; pin, fīve;
rock, gō, ôrder; oil, out; cup, pùt, rüle,
yü in use, *yu̇* in uric;
ng in bring; sh in rush; th in thin, ᴛʜ in then;
zh in seizure.

ə = *a* in about, *e* in taken, *i* in pencil, *o* in lemon, *u* in circus

revolve, *v.* **1** to move in a circle; move in a curve around a point: *Our Earth and Mars, Venus, and Mercury ... revolve nearest to the Sun* (David Brewster).
2 to turn around a center or axis; rotate: *The wheels of a moving car revolve.*

Reynold's number, *Physics.* a mathematical factor used to express the relation between the velocity, viscosity, density, and dimensions of a fluid in any system of flow. It is used in aerodynamics to correct the results of tests of scale-model airplanes in wind tunnels. [named after Osborne *Reynolds,* 1842–1912, Irish engineer and physicist]

R.F., *abbrev.* **1** radio frequency. **2** releasing factor. **3** representative fraction.

R factor, *Genetics.* any of a group of plasmids in various bacteria that gives them resistance to antibiotics and other drugs and that can be transmitted from one bacterium to another by conjugation: *Bacteria that have survived exposure to an antibiotic by developing resistance to it pass on this resistance in the form of genetic material—the R factor—to other species or strains* (London *Times*). [from *R(esistance) factor*]

r.h. or **RH,** *abbrev.* relative humidity: *Bacterial cells lose their viability when they are held in aerosols at various levels of humidity (RH)* (Nature).

Rh, *symbol.* **1** Rh factor: *The so-called rhesus (Rh) blood factor is another important blood antigen found in man and monkeys. About 87 per cent of the white population of the United States have this antigen, hence are called Rh-positive (designated Rh); the remaining 13 per cent are without it, that is, they are Rh-negative (indicated rh)* (Hegner and Stiles, *College Zoology*).
2 rhodium.

rhabdom (rab'dəm), *n. Zoology.* any of the rods supporting the crystalline lenses in a compound eye: *In many nocturnal insects such as Noctuid moths, glow-worm beetles and the like, there is a special kind of compound eye that forms images by superposition. Each sensory rhabdom receives light rays not only through its own facet, but through neighbouring facets also* (Science News). [from Late Greek *rhabdōma* group of rods, from Greek *rhabdos* rod]

rhabdovirus (rab'dō vī'rəs), *n. Biology.* any of a group of rod-shaped viruses containing RNA and associated with various diseases transmitted by animal or insect bites: *One of the things which is known about rabies virus is that its genetic material is ribonucleic acid (RNA). Because of this property and its morphology it has been classified with the rhabdovirus group (from the Greek rhábdos meaning a rod)* (Science Journal).

rhamnose (ram'nōs), *n. Biochemistry.* a sugar occurring in combination with a glycoside in many plants: *The possible non-nutritive sugars include sorbose (originally found in Mountain Ash berries), xylose (from oat husks), arabinose (from sugar-beet pulp), galactose (from flaxseed), rhamnose (from cherry gum) and fucose (from seaweed)* (New Scientist). *Formula:* $C_6H_{12}O_5$ [from Greek *rhamnos* the buckthorn (one of the plants in which rhamnose occurs)]

rhenium (rē'nē əm), *n. Chemistry.* a rare, heavy, silver-white, metallic element which occurs in molybdenum ore. It has chemical properties similar to those of manganese and a very high melting point that makes it valuable as an ingredient in alloys. *Symbol:* Re; *atomic number* 75; *atomic weight* 186.2; *melting point* 3180°C; *boiling point* 5627°C; *oxidation state* 4, 6, 7 [from New Latin, from Latin *Rhenus* the Rhine river]

rheo-, *combining form.* stream; flow; current, as in *rheology, rheotropism.* [from Greek *rheos* a flowing, stream, from *rhein* to flow]

rheological, *adj.* of or having to do with rheology: *First there were rheological studies on coal-oil mixtures, which led to the development of a stable slurry containing 40–60 per cent of coal; they also established the limits on the size of particle and the viscosity of the oil* (New Scientist). **—rheologically,** *adv.*

rheology (rē ol'ə jē), *n.* the science that deals with the flow and alteration of matter.

rheotactic (rē'ə tak'tik), *adj.* of or having to do with rheotaxis.

rheotaxis (rē'ə tak'sis), *n. Biology.* the movement of a cell or organism in response to the stimulus of a current of water or air.

rheotropism (rē ot'rə piz əm), *n. Biology.* the orientation in growth of plants or sessile animals in response to a current of water or air.

Rhesus factor (rē'səs), = Rh factor: *Another system of blood groups was discovered which was called the Rhesus factor* (G. Fulton Roberts).

Rh factor, *Immunology.* an antigen found on the surface of the red blood cells of most human beings. Persons who have the Rh factor are known as *Rh-positive.* Such a person's blood cells may agglutinate if they come into contact with an antibody (*anti-Rh*) produced in the body of an *Rh-negative* person who receives a transfusion of Rh-positive blood. *The Rh factor is another kind of protein found in the blood. The gene for the Rh factor is a dominant gene. About 85 percent of the people in the United States have the Rh factor in their blood* (Otto and Towle, *Modern Biology*). [from *Rh(esus)* a small monkey of India used in medical research, in the blood of which the Rh factor was first discovered]

rhin-, *combining form.* the form of *rhino-* before vowels, as in *rhinal, rhinencephalon.*

rhinal (rī'nl), *adj.* of or having to do with the nose. SYN: nasal.

rhinarium (rī när'ē əm), *n., pl.* **-naria** (-när'ē ə). *Zoology.* **1** the front part of the clypeus, as of neuropteran insects and certain beetles. **2** the area of bare skin around the nostrils of ruminants. [from New Latin, from Greek *rhinos* nose]

rhinencephalic (rī'nen sə fal'ik), *adj.* of or having to do with the rhinencephalon.

rhinencephalon (rī'nen sef'ə lon), *n., pl.* **-la** (-lə). *Anatomy.* the part of the brain most closely connected with the olfactory nerves: *All the brain damage occurred in a region of the cerebrum known as the rhinencephalon* (Science News Letter). [from New Latin, from *rhin-* nose + *encephalon*]

rhino-, *combining form.* nose; of the nose, as in *rhinovirus.* Also spelled **rhin-** before vowels. [from Greek *rhinos*]

rhinovirus (rī'nō vī'rəs), *n. Biology.* any of a group of picornaviruses associated with the common cold and other respiratory diseases.

rhizobic (rī zō′bik), or **rhizobial** (rī zō′bē əl) *adj.* of or having to do with the rhizobia.

rhizobium (rī zō′bē əm), *n., pl.* **-bia** (-bē ə). *Bacteriology.* any member of a genus (*Rhizobium*) of rod-shaped, nitrogen-fixing bacteria that live symbiotically in nodules on the roots of leguminous plants. [from New Latin, from Greek *rhiza* root + *bios* life]

rhizocarpous (rī′zə kär′pəs), *adj. Botany.* having an annual stem but perennial roots. The perennial herbs are rhizocarpous.

rhizocaul (rī′zə kôl), *n. Zoology.* the rootlike part of a polyp, used for attachment to some support. [from Greek *rhiza* root + *kaulos* stalk]

rhizogenic (rī′zə jen′ik), *adj. Botany.* producing roots: *rhizogenic tissue.*

rhizoid (rī′zoid), *Biology.* —*adj.* rootlike.
—*n.* one of the rootlike filaments by which a moss, fern gametophyte, liverwort, or fungus is attached to the substratum: *At its [a moss's] base, and sometimes elsewhere, fine, brown, hairlike threads, called rhizoids ... can be found. Under the microscope these are seen to be only one cell in thickness, and to be peculiar in that the transverse walls extend diagonally across them* (Emerson, *Basic Botany*). [from Greek *rhiza* root]

rhizomatic (rī′zə mat′ik), *adj.* = rhizomatous.

rhizomatous (rī zō′mə təs), *adj. Botany.* **1** of, having to do with, or resembling a rhizome.
2 having rhizomes: *But if more plants of the rhizomatous type are wanted, a leaf is pulled off and inserted in regular soil* (N.Y. Times).

rhizome (rī′zōm), *n. Botany.* a rootlike, creeping, underground stem of some plants, which usually sends out roots below and leafy shoots above, stores food to be used by the new plant the following year, and has nodes, buds, and small, scalelike leaves. Commonly called **rootstock.** [from Greek *rhizōma*, from *rhiza* root]

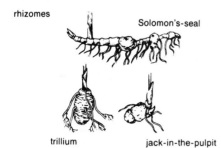

rhizomes

Solomon's-seal

trillium   jack-in-the-pulpit

rhizomic (rī zō′mik), *adj. Biology.* belonging to or consisting or rhizomes.

rhizomorph (rī′zə môrf), *n. Biology.* a rootlike filament or hypha in various fungi.

rhizomorphous (rī′zə môr′fəs), *adj. Botany.* rootlike in form.

rhizophagous (rī zof′ə gəs), *adj. Zoology.* feeding on roots.

rhizosphere (rī′zə sfir), *n. Botany, Geology.* the soil immediately surrounding the roots of a plant: *At the heart of soil-plant relationships are the events ... that*

occur on the surfaces of roots and in their immediate vicinity in the rhizosphere (Science).
—**rhizospheric** (rī′zə sfer′ik), *adj.* of or having to do with the rhizosphere: *rhizospheric compounds.*

Rh negative, *Immunology.* lacking the Rh factor: *Rh negative blood.*

rhodamine (rō′də mēn), *n. Chemistry.* **1** a red dye having brilliant fluorescent qualities, obtained by heating an amino derivative of phenol with phthalic anhydride and hydrochloric acid. *Formula:* $C_{23}H_{31}ClN_2O_3$
2 any of a group of related dyes. [from Greek *rhodon* rose + English *amine*]

rhodic (rō′dik), *adj. Chemistry.* of or containing rhodium, especially with a high valence.

rhodium (rō′dē əm), *n. Chemistry.* a silver-white metallic element found chiefly in platinum ores. It is similar to aluminum, but harder, and forms salts that give rose-colored solutions. It is resistant to acid, and is used for plating silverware, jewelry, etc. *Symbol:* Rh; *atomic number* 45; *atomic weight* 102.905; *melting point* 1966°C; *boiling point* 3725 $\pm$ 100°C; *oxidation state* 3. [from New Latin, from Greek *rhodon* rose]

rhodochrosite (rō′dō krō′sit), *n. Mineralogy.* a mineral consisting mainly of manganese carbonate and usually occurring in rose-red crystals. *Formula:* $MnCO_3$ [from Greek *rhodochrōs* rose-colored]

rhodonite (rō′də nit), *n. Mineralogy.* a rose-red mineral, consisting mainly of manganese silicate. It is sometimes used as an ornamental stone. *Formula:* $MnSiO_3$ [from Greek *rhodon* rose]

rhodoplast (rō′də plast), *n. Biology.* one of the chromatophores which bear the red coloring matter in the cells of red algae: *The concept that the photosynthetic plastids—e.g., green chloroplasts, red rhodoplasts, and other colored, light-absorbing organelles of algae and plants—derive from once-independent photosynthetic monads has even more support* (Lynn Margulis). [from Greek *rhodon* rose + *plastos* something formed]

rhodopsin (rō dop′sən), *n. Biochemistry.* a purplish-red protein pigment that is sensitive to light, found in the rods of the retina of the eye in all vertebrates. Also called **visual purple.** Compare **retinene.** [from Greek *rhodon* rose + *opsis* sight]

rhombencephalic (rom′ben sə fal′ik), *adj. Anatomy.* of or having to do with the hindbrain.

rhombencephalon (rom′ben sef′ə lon), *n.* = hindbrain.

rhombic (rom′bik), *adj.* **1** *Geometry.* **a** having the form of a rhombus. **b** having a rhombus as a base or cross section. **c** bounded by rhombuses.
2 *Chemistry.* having to do with a system of crystallization characterized by three unequal axes intersecting at right angles. SYN: orthorhombic.

rhombic

---

cap, fāce, fäther; best, bē, tèrm; pin, fīve;
rock, gō, ôrder; oil, out; cup, pùt, rüle,
yü in use, *yù* in uric;
ng in bring; *sh* in rush; *th* in thin, ᴛʜ in then;
zh in seizure.
ə = *a* in about, *e* in taken, *i* in pencil, *o* in lemon, *u* in circus

rhombohedral (rom'bə hē'drəl), *adj. Geometry.* having to do with or in the form of a rhombohedron. See the picture at **crystal.**

rhombohedron (rom'bə hē'drən), *n., pl.* **-drons, -dra** (-drə). *Geometry.* a solid bounded by six rhombic planes: *Calcite crystals belong to the hexagonal system. They cleave in three directions to form perfect rhombohedrons* (A. Pabst). See the picture at **crystal.**

rhomboid (rom'boid), *Geometry.* —*n.* a parallelogram with equal opposite sides, unequal adjacent sides, and oblique angles. —*adj.* shaped like a rhombus or rhomboid. [from Late Latin *rhomboides,* from Greek *rhombos* + *eidos* form]
—**rhomboidal** (rom boi'dl), *adj.* = rhomboid.

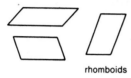

rhomboids

rhomboideus (rom boi'dē əs), *n., pl.* **-dei** (-dē ī). *Anatomy.* a pair of muscles that help to hold and manipulate the upper arms. [from New Latin, from Late Latin *rhomboides* rhomboid]

rhombus (rom'bəs), *n., pl.* **-buses, -bi** (-bī). *Geometry.* an equilateral parallelogram, usually having two obtuse angles and two acute angles. [from Latin, from Greek *rhombos*]

rhombuses

rho meson (rō), *Nuclear Physics.* a highly unstable and short-lived subatomic particle with a mass about 1400 times that of the electron, produced in high-energy collisions between particles: *There is a class of unstable particles, the neutral vector mesons, whose members resemble photons in many ways, with two important exceptions: they have mass and they exhibit the strong force. The most prominent is the rho meson* (Scientific American).

Rh positive, *Immunology.* containing the Rh factor: *Rh positive blood.*

rhumb (rum *or* rumb), *n.* **1** *Geography.* any of the 32 points of the compass. **2** = rhumb line.

rhumb line, *Geography, Astronomy.* a line on the surface of a sphere cutting all meridians at the same oblique angle: *The rhumb line or loxodrome . . . has the same azimuth throughout its length. On a Mercator chart it is represented by a straight line, but because of the convergence of the meridians it is in fact a spiral, and a craft that followed such a course indefinitely would (unless the angle with the meridian were 0 or 90°) approach the pole as a limit, making around it an infinite number of turns* (Duncan, *Astronomy*). [ultimately from Latin *rhombus*]

rhynchocoel (ring'kō sēl), *n., adj.* = nemertean.

rhyolite (rī'ə līt), *n. Geology.* a volcanic rock containing quartz and alkalil feldspar, with texture often showing the lines of flow: *The color of rhyolite ranges widely, but generally is white or light shades of yellow, brown, or red. Most rhyolites are flow banded; that is, they show streaky irregular layers as in taffy, formed by the flowing of the sticky, almost congealed magma* (Gilluly, *Principles of Geology*).
[from German *Rhyolit,* from Greek *rhyax* lava flow, torrent, from *rhein* to flow]
—**rhyolitic** (rī'ə lit'ik), *adj.* of, having to do with, or resembling rhyolite: *rhyolitic volcanism.*

rib, *n.* **1** *Anatomy.* one of the curved bones extending around the chest from the backbone to the front of the body. They are joined to the vertebrae in pairs. Human beings and most vertebrates have ribs, but the number of ribs varies considerably, with 24 ribs in humans (12 on each side of the body) and as many as 48 in some other mammals. *Certain parts of the skeleton, such as the skull and ribs, protect the delicate parts of the body from injury* (Beauchamp, *Everyday Problems in Science*).
2 *Botany.* a thick vein of a leaf.

ribcage, *n. Anatomy.* the barrel-shaped enclosure formed by the ribs of the chest.

riboflavin (rī'bō flā'vən), *n. Biochemistry.* an orange-red crystalline substance, a constituent of the vitamin B complex, that promotes growth and is present in liver, eggs, milk, spinach, etc. *Deficiency of riboflavin manifests itself by inflammatory changes of the skin and mucous membrane of the lip and the corners of the mouth* (Time). *Formula:* $C_{17}H_{20}N_4O_6$ Also called **lactoflavin** and **vitamin B₂.** [from *ribose* + Latin *flavus* yellow]

ribonuclease (rī'bō nü'klē ās), *n. Biochemistry.* an enzyme that promotes the breakdown by hydrolysis of ribonucleic acid. Also called **RNAase.**

ribonucleic acid (rī'bō nü klē'ik), *Biochemistry.* a nucleic acid important in protein synthesis and genetic transmission, found in the cytoplasm and nucleolus of all living cells and in many viruses. Ribonucleic acid consists of long, usually single-stranded chains of repeating units of ribose combined with phosphoric acid and the purine bases, adenine and guanine, and the pyrimidine bases, cytosine and uracil. It usually appears in the abbreviated form RNA.
► Ribonucleic acid (RNA), like deoxyribonucleic acid (DNA), is composed of nucleotide sequences. However, there are three major differences between the structure of DNA and RNA molecules. In RNA, the sugar part (ribose) of the molecule is substituted for deoxyribose; uracil is substituted for the thymine base in DNA; there is a single chain of nucleotides instead of the two complementary chains of the DNA molecule.

ribonucleoprotein, *n. Biochemistry.* a nucleoprotein containing a portion of ribonucleic acid: *Ribonucleoproteins . . . are known to be involved in the synthesis of proteins in living cells* (New Scientist).

ribonucleoside, *n. Biochemistry.* a nucleoside containing ribose as its sugar: *The substances are ribonucleosides, nucleic acids that occur in all living cells. These acids have been found to be helpful in checking the amount of damage that alcohol produces in a certain strain of bacteria* (Science News Letter).

ribonucleotide, *n. Biochemistry.* a nucleotide containing ribose as its sugar: *The immediate building blocks of RNA are the ribonucleotides of the four bases. A ribonucleotide is a base linked to a ribose sugar, which is linked in turn to at least one phosphate group* (Jerard Hurwitz and J. J. Furth).

ribose (rī'bōs), *n. Biochemistry.* a pentose sugar present in ribonucleic acid, riboflavin, diphosphopyridine nucleotide, and other nucleotides and nucleic acids. It is obtained in the dextrorotatory form chiefly from nucleic acids contained in plants. *Formula:* $C_5H_{10}O_5$ [alteration of *arabinose* a sugar derived from gum *arabic*]

ribose nucleic acid, = ribonucleic acid.

ribosomal (rī'bə sō'məl), *adj. Molecular Biology.* of or having to do with a ribosome or ribosomes: *ribosomal proteins; . . . the role of ribosomal nucleic acids in protein synthesis* (London *Times*).

ribosomal RNA, *Molecular Biology.* the form of RNA (ribonucleic acid) found in ribosomes: *Ribosomal RNA accounts for approximately 80 per cent of the RNA content of the bacterial cell* (Watson, *Molecular Biology of the Gene*). *It is thought that ribosomal RNA temporarily binds messenger RNA to the ribosomes* (Biology Regents Syllabus). *Abbreviation:* rRNA

ribosome (rī'bə sōm), *n. Molecular Biology.* any of the particles of ribonucleic acid and protein found in the cytoplasm of living cells where they are actively involved in the synthesis of proteins: *From the nucleus the RNA moves into the surrounding cytoplasm and makes up about half of the substance of the particles called ribosomes, which are the principal site of protein synthesis* (Scientific American). See the picture at **cell.** [from *ribo*(*nucleic acid*) + *-some*]

Richter scale (rik'tər), *Geology.* a scale for indicating the magnitude of earthquakes. [named after Charles F. *Richter,* born 1900, American seismologist who devised the scale] See below.

from this high over the northern Mississippi Valley and the Northern Plains States (N. Y. Times).

riebeckite (rē'bek īt), *n. Mineralogy.* a hydrous sodium iron silicate of the monoclinic amphibole group, found as dark blue or black prismatic crystals and often containing magnesium. [from German *Riebeckit,* from Emil *Riebeck,* 19th-century German explorer + *-it* -ite]

Riemannian (rē'män'ē ən), *adj. Mathematics.* of or having to do with the theories and equations of the German mathematician Georg Riemann (1826–1866): *the Riemannian metric, a Riemannian surface.*

Riemannian geometry, = elliptic geometry.

rift valley, *Geology, Geography.* a valley formed by the lowering of an area of land between two nearly parallel faults: *East Africa has a number of deep, long, narrow valleys called rift valleys. These rift valleys contain many long, narrow lakes* (G. H. T. Kimble). Also called **graben.** See the pictures at **divergence fault.**

rift zone, *Geology.* a large area of the earth in which a rift occurs when plates of the earth's crust move away from one another: *Recent measurements of magnetic and gravitational fields in the region also have found evidence for a very broad rift zone—an area where the plate structure may have been weakened by ancient volcanic activity. The fault zone appears to be related and parallel to the rift zone . . . and may prove to be the means by which the weakened rift zone adjusts to stress* (Science News).

right, *adj. Geometry.* having a line or axis perpendicular to another line or surface: *a right cone.*

right angle, *Mathematics.* an angle that is formed by a line perpendicular to another line; angle of 90 degrees. See the picture at **angle.**

—**right-angled,** *adj.* containing a right angle or right angles; rectangular.

Approximate Magnitude	Characteristic Effects	Number of Earthquakes per Year
2.0 to 3.4	barely felt	above 100,000
3.5 to 4.2	felt as a rumble	about 50,000
4.3 to 4.9	shakes furniture; can break dishes	about 5000
5.0 to 5.9	dislodges heavy objects; cracks walls	more than 1000
6.0 to 6.9	considerable damage to buildings	about 100
7.0 to 7.3	major damage to buildings; breaks underground pipes	15 to 18
7.4 to 7.9	great damage; destroys masonry and frame buildings	2 to 4
above 8.0	complete destruction; ground moves in waves	1 to 2

rictus (rik'təs), *n.* **1** *Zoology.* the expanse or aperture of the mouth of birds or fishes.

2 *Botany.* the orifice of a bilabiate corolla. [from Latin *rictus* a mouth opened wide, from *ringi* to gape]

ridge, *n.* **1** *Geography, Geology.* **a** a long, narrow chain of hills or mountains: *the Blue Ridge of the Appalachian Mountains.* **b** a long, narrow elevation on the ocean floor: *The oceanic rift belts form the Mid-Atlantic Ridge, the East Pacific Rise, the Carlsberg Ridge of the Indian Ocean, and the connecting ridges between* (Robert W. Decker).

2 *Meteorology.* a linear zone of high barometric pressure: *A rather weak ridge of high pressure will extend*

right ascension, *Astronomy.* the arc of the celestial equator between the vernal equinox and the great circle of the celestial sphere passing through the celestial poles and a celestial body whose position is being considered, reckoned toward the east and expressed in

cap, fāce, fäther; best, bē, tėrm; pin, fīve;
rock, gō, ôrder; oil, out; cup, půt, rüle,
yü in use, yù in uric;
ng in bring; sh in rush; th in thin, ᴛʜ in then;
zh in seizure.
ə = a in about, e in taken, i in pencil, o in lemon, u in circus

565

hours: *Unlike terrestrial longitude, which is measured both east and west from the prime meridian, right ascension is always measured toward the east* (Baker, *Astronomy*).

right atrioventricular valve, = tricuspid valve.

right-brain, *n. Anatomy.* the right hemisphere of the cerebrum. Compare **left-brain.**

right triangle, *Geometry.* a triangle with one right angle.

rill or **rille** (ril), *n.* **1** *Astronomy.* a long, narrow valley on the surface of the moon: *There are many rills, or clefts—cracks of the order of half a mile wide and of unknown depth* (Baker, *Astronomy*).
2 *Geography, Geology.* a very small stream or channel.
[from German *Rille* furrow]

ring, *n.* **1** *Chemistry.* a closed chain of atoms linked by bonds that may be represented graphically in circular form.
2 *Mathematics.* a set of elements subject to the operations of addition and multiplication, in which the set is commutative under addition and associative under multiplication, and in which the two operations are related by distributive laws.
3 = annual ring.

ring compound, *Chemistry.* a compound of several atoms united to form a ring; a cyclic or closed-chain compound: *The six carbon atoms in benzene and its derivatives are thought to be arranged in a ring or closed chain. For this reason they are sometimes called "Ring Compounds" or "Closed Chain Compounds"* (Parks and Steinbach, *Systematic College Chemistry*).

Ringer's solution or **Ringer solution,** *Biology, Medicine.* the chlorides of sodium, potassium, and calcium in the approximate proportion in which they are found in blood and tissue fluid. It is used to preserve mammalian tissue in experiments. *Ringer's solution is used in large quantities for intravenous maintenance of salt and water* (Leland C. Clark). [named after Sydney *Ringer*, 1835–1910, English physician who introduced this solution]

ring nebula, *Astronomy.* **1** a kind of planetary nebula that looks like a ring when viewed through a telescope or on a photographic plate: *Among them are a few ring nebulae, hollow spheres of gaseous material; diffuse nebulae, looking almost like fluffs of cotton set against the blackness of space; and the dark nebulae* (Bernhard, *Handbook of Heavens*).
2 Ring Nebula, the ring nebula in the constellation Lyra.

rings of Saturn
(def. 1)

crepe ring
ring B
Cassini's division
ring A

ring-porous, *adj. Botany.* having pores which differ in size and distribution, being larger and more numerous in the spring than in the summer, such as the wood of the oak, ash, and hackberry. Compare **diffuse-porous.**

rip current or **riptide,** *n. Oceanography.* a strong, narrow surface current which flows rapidly away from the shore, usually at a right angle to it, due to the return flow of water driven by the wind.

Riss (ris), *n. Geology.* the third glaciation of the Pleistocene in Europe: *It indicated four cold spells, which they promptly identified with the four glaciations that every geological and archaeological student has long been taught to recite—Günz, Mindel, Riss, and Würm* (Sunday *Times*). [named after the *Riss* river, in southwestern Germany, the Alpine site of the glaciation]

river, *n. Geography.* a large natural stream of water that flows into a lake, an ocean, or another river. The Nile is the longest river in the world (6,671 kilometers). The longest river in the United States is the Mississippi (3,779 kilometers).
ASSOCIATED TERMS: The source of a river is its *headwaters*. These flow to the *mouth* or to an *estuary*, where the river empties into a larger river, lake, or sea. The headwaters flow in *rills* and *streams*, forming branches called *tributaries*. A branch that flows away from the main stream is a *distributary*. An area drained by a river and its tributaries is a *drainage basin*. The *channel*, which is the deepest portion of a river, consists of the underwater *bed* and the *banks* on either side. The upper course of many rivers have *rapids* and *waterfalls*, whose fast currents can cut a *canyon* or a *valley*, while the lower course often forms a *flood plain* or *meander*.

RMS, *abbrev.* root-mean-square.

RNA, *n. Biochemistry.* ribonucleic acid, a genetic material found in all cells and many viruses: *RNA differs chemically from DNA in two respects: (1) the sugar part (ribose) of the backbone of the RNA molecule has an extra oxygen atom in it; and (2) the thymine base of DNA is replaced by uracil (U), which is chemically very similar* (Gordon M. Tomkins).

RNAase (är′en′ā′ās), *n.* = ribonuclease.

RNA polymerase, *Biochemistry.* an enzyme that acts upon DNA to synthesize ribonucleic acid: *Scientists have wondered how this one enzyme, RNA polymerase, could catalyze the synthesis of such differing RNA molecules* (Earl A. Evans). Compare **DNA polymerase.**

Roche limit (rōsh), *Astronomy.* a distance of 2.44 times a planet's radius, measured from the planet's center, within which a satellite or celestial body might be pulled apart by the gravitational force of the planet. Also called **Roche's limit.** [named after Edouard A. *Roche*, 1820–1883, French mathematician]

Roche lobe, *Astronomy.* a gaseous bulge formed in companion stars by each star's gravitational pull on the other: *The most widely held theory for the nova phenomenon is that the prenova is a close binary system consisting of a red dwarf, which is filling its Roche Lobe, spilling matter over towards a white dwarf companion* (Nature). [named after Edouard *Roche*]

roche moutonnée (rôsh′mü′tə nā′), *pl.* **roches moutonées** (rôsh′mü′tə nā′). *Geology.* an elongate mound of bedrock, rounded and smoothed by glacial action. Compare **moutonnée.** [from French (literally) fleecy rock]

Roche's limit, = Roche limit.

rock, *n. Geology.* **1** the mass of mineral matter of which the earth's crust is composed: *Rock is sometimes defined simply as 'an aggregate of minerals', but that again would not satisfy the geologist, for some rocks, like certain limestones, are composed almost entirely of one mineral; ... Clay and loose sand, it may be added, are to the geologist as truly rock as are the hardest* *of slates, sandstones, granites and so forth* (Jones, *Minerals in Industry*).

2 any part or layer of such matter: *Rocks comprise igneous, sedimentary, and metamorphic types. We know that igneous rocks originate both as intrusions and extrusions, that sedimentary rocks are formed by the breaking down and deposition of preexisting rock material, and that metamorphic rocks are produced mainly by heat and pressure affecting other rocks* (Moore, *Introduction to Historical Geology*).

[from Old French *roque,* from Popular Latin *rocca*] ASSOCIATED TERMS: Common igneous rocks include *basalt, gabbro, granite, obsidian,* and *pumice.* Some common sedimentary rocks are *breccia, clay, coal, flint, limestone, sandstone,* and *shale.* Common metamorphic rocks are *gneiss, marble, quartzite, schist,* and *slate.* Igneous rocks form by solidification of *magma,* and may be *extrusive* (formed from *lava*) or *intrusive.* Sedimentary rocks are formed by *deposition* of *sediments* that make up *strata.* Metamorphic rocks result from *metamorphism,* which may be *dynamic* or *contact metamorphism.* See also the associated terms under **erosion.**

rock crystal, *Mineralogy.* a colorless, transparent variety of quartz, used in optical instruments and as a semiprecious gemstone.

rod, *n.* **1** *Biology.* one of the microscopic sense organs in the retina of the eye that are sensitive to dim light: *In this dim light between daylight and darkness, either rod (night) or cone (daylight) vision may be in use. Different individuals vary in the extent to which they use either rods or cones* (Science News Letter).

2 *Nuclear Physics.* a metal bar or tube containing neutron-absorbing material such as boron or cadmium, used in a reactor to regulate the rate or type of reaction.

3 a unit of length equal to 5 1/2 yards or 16 1/2 feet (5.0292 meters). A square rod is 30 1/4 square yards or 272 1/4 square feet (25.293 meters).

rodent, *n.* *Zoology.* any of an order (Rodentia) of mammals having two continually growing incisor teeth in each jaw especially suitable for gnawing wood and similar material. Rats, mice, squirrels, porcupines, and beavers are rodents. *Rodents are omnivorous in their diet and are man's chief competitor among the mammals for food. . . . Rodents have no canine teeth, but the incisors are long and curved. This is the most easily recognized characteristic of the group* (Winchester, *Zoology*). [from Latin *rodentem* gnawing]

—**rodential** (rō den′shəl), *adj.* of or having to do with rodents.

roe, *n.* *Zoology.* **1** fish eggs, especially when contained in the ovarian membrane of the female fish.

2 the eggs or spawn of various crustaceans.

roentgen or **röntgen** (rent′gən), *Physics.* —*n.* the international unit of the intensity of X rays or gamma rays. It is equal to the quantity of radiation required to produce one electrostatic unit of electrical charge in one cubic centimeter of dry air under normal temperature and pressure. *Symbol: r.*

—*adj.* having to do with X rays or gamma rays: *The smaller part of this energy reaches the earth in the form of ultraviolet or roentgen radiation, gamma quanta, and corpuscular currents consisting of atomic nuclei and the separate elementary particles* (Bulletin of Atomic Scientists).

[named after Wilhelm K. *Röntgen,* 1845–1923, German physicist who discovered X rays in 1895]

roentgenology, *n.* the branch of radiology having to do with X rays, especially as used in medical diagnosis and treatment.

roentgen ray, = X ray.

rogue, *n.* *Biology.* an individual, usually a plant, that varies from the standard: *Strains of "rogue" tomatoes appear in certain varieties, and themselves yield more rogues in a way not typical of "Mendelian" inheritance* (Harold W. Rickett).

room temperature, an average indoor temperature of 20 to 25°C (68 to 77°F).

root, *n.* **1** *Botany.* **a** the part of a plant that grows downward usually into the ground, to hold the plant in place, absorb water and mineral foods from the soil, and often to store food material: *The chief functions of a root are absorption, storage and support. Its principal function is the absorption of nutriment and to this end it generally has . . . root hairs which largely increase the absorbing surface* (Youngken, *Pharmaceutical Botany*). **b** any underground part of a plant, especially when fleshy, as the carrot or turnip.

2 *Anatomy.* a part, as of an organ, that is embedded in tissue: *the root of a tooth, the roots of the hair.* See the picture at **tooth.**

3 *Mathematics.* **a** a quantity that produces another quantity when multiplied by itself a certain number of times: *2 is the square root of 4 and the cube root of 8* ($2 \times 2 = 4, 2 \times 2 \times 2 = 8$). **b** a quantity that satisfies an equation when substituted for an unknown quantity: *In the equation* $x^2 + 2x - 3 = 0$, *1 and* -3 *are the roots.*

root:
(def. 1)

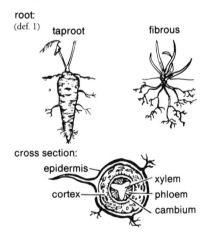

taproot · fibrous

cross section:

epidermis · xylem

cortex · phloem · cambium

root canal, *Anatomy.* a passage in the root of a tooth through which nerves and vessels reach the pulp. See the picture at **tooth.**

cap, fāce, fäther; best, bē, tèrm; pin, fīve;
rock, gō, ôrder; oil, out; cup, pùt, rüle,
yü in use, *yu̇* in uric;
ng in bring; *sh* in rush; *th* in thin, ŦH in then;
zh in seizure.
ə = *a* in about, *e* in taken, *i* in pencil, *o* in
lemon, *u* in circus

root cap, *Botany.* a mass of cells at the tip of growing roots that protects the active growing point immediately behind it: *The root cap grows from the inside and its old collapsing outer cells wear off as it is pushed through the soil. By this means the tender apical meristem immediately behind the cap is protected from damage* (Emerson, *Basic Botany*).

root collar, *Botany.* place at the base of a tree where the swelling and spreading of the roots begin.

root fungus, *Botany.* a fungus growing in symbiotic association with the roots of plants.

root hair, *Botany.* a hairlike outgrowth from the root of a plant: *Root hairs are elongated epidermal cells which increase the surface area of the root for the absorption of water and minerals. The movement of materials through the semipermeable membrane of root hairs involves both diffusion . . . and active transport* (Biology Regents Syllabus).

root-mean-square, *n.* **1** *Statistics.* the square root of the average of the squares of a set of related values.
2 *Physics.* the value of an alternating current or voltage: *It is found more convenient to describe alternating currents and voltages by their root-mean-square (abbreviated rms) values than by their maximum values. The rms value of a varying current is defined as that value of a steady current which would develop the same quantity of heat in the same time in the same resistance, and for that reason the rms value is called the effective value of the varying current* (Sears and Zemansky, *University Physics*).

root-mean-square deviation, *Mathematics, Statistics.* an average measure of the difference between two sets of numbers. Compare **standard deviation.**

rootstock, *n. Botany.* **1** = rhizome.
2 a root that serves as a stock for propagating plants.

root symbiosis, *Botany.* a symbiotic relationship between certain bacteria and fungi and the roots of the higher plants.

root tubercle, *Botany.* one of the small rootlike growths or nodules produced on the roots of leguminous plants by nitrogen-fixing bacteria.

rotate, *v. Physics.* to move around a center or axis; turn in a circle; revolve: *The Sun, like the Earth or a top when spinning, turns round, or rotates, on an axis* (J. N. Lockyer).
—*adj. Botany.* spreading out nearly flat like a wheel; wheel-shaped, as a corolla with a short tube and spreading limb.

rotation, *n. Physics.* **1** the act or process of turning around a center or axis; a turning in a circle: *Rotation is motion around an axis within the body. Thus the earth rotates daily* (Baker, *Astronomy*).
2 the time required for one such movement.
3 one such movement.
► A *rotation* refers to a turning or spinning of a body around an axis that passes through the body itself, whereas a *revolution* usually refers to motion around an external point.

rotator (rō′tā tər), *n., pl.* **rotators, rotatores** (rō′tə tôr′ēz). *Anatomy.* a muscle that turns a part of the body: *Muscles that rotate a body part are called rotators, some of which move the structure clockwise and others move it counterclockwise* (Winchester, *Zoology*).

rotatory (rō′tə tôr′ē), *adj.* **1** *Physics.* turning like a top or wheel; rotating; rotary: *rotatory motion.*
2 *Anatomy.* causing rotation: *a rotatory muscle.*

rotavirus, *n. Biology.* any of a group of circular viruses with spokelike projections that are related to the reoviruses and cause acute gastroenteritis in infants and newborn animals. [from Latin *rota* wheel + English *virus*]

rotifer (rō′tə fər), *n. Zoology.* any of a phylum (Rotifera) of microscopic water animals that have one or more rings of cilia on a disk at the head of the body, which aid in locomotion and drawing in food: *Small ponds appear choked with gelatinous algae and microscopic animals such as rotifers* (Gabriele Rabel). [from Latin *rota* wheel + *ferre* to carry]
—**rotiferal** (rō tif′ər əl), *adj.* **1** of or having to do with the rotifers. **2** having a wheel or wheellike organ.

roton (rō′ton), *n. Physics.* a quantum of energy that describes the rotating motion of liquid helium atoms, analogous to the phonon: *A roton is a kind of vortex or ring of helium atoms that can form in superfluid helium at temperatures near absolute zero* (Science News). [from Latin *rotare* rotate + English *-on*]

rotor, *n. Meteorology.* a turbulent mass of air that rotates horizontally about its axis, found especially in the area of some large mountain barriers: *Because mountain waves can be dangerous to aircraft, and because on occasion they generate intense "rotors" that can raise roofs from houses or topsoil from farms, it came to be recognized that atmospheric gravity waves were of direct significance to man* (Nature).

roughage, *n.* the coarser parts or kinds of food, such as bran and fruit skins, which stimulate the movement of food and waste products through the intestines. Also called **fiber.**

round angle, *Geometry.* a complete circle; an angle of 360 degrees.

round number, *Mathematics.* **1** a number resulting when a number is rounded off. 3874 in round numbers would be 3900 or 4000. **2** a whole number without a fraction.

royal, *adj. Chemistry.* inert; noble: *royal metals.*

royal jelly, *Zoology.* a creamy, jellylike substance, rich in vitamins and proteins, that is formed by glands in the heads of young worker bees and fed throughout the larval stage to a queen bee to give the queen a longer life and greater fertility: *Structural characteristics in a few animals are determined or modified by their food. In the honeybee, for example, any fertilized egg may develop into a queen bee; but to attain that end the larvae must be fed on "royal jelly," which is predigested pollen prepared for and given to them by the workers* (Shull, *Principles of Animal Biology*).

r.p.m. or **rpm,** revolutions per minute.

r.p.s. or **rps,** revolutions per second.

rRNA, *abbrev.* ribosomal RNA: *It was generally agreed that the genetic information in DNA is first transferred to three different types of ribonucleic acids (RNAs), designated as transfer RNA (t-RNA), messenger RNA (m-RNA), and ribosomal RNA (r-RNA)* (Robert L. Hill).

Ru, *symbol.* ruthenium.

rubber, *n.* an elastic substance obtained from the milky juice of certain tropical plants, or made synthetically by chemical processes. Rubber will not let air or water through. *Natural rubber is made up of long molecules consisting of simple hydrocarbons strung together end to end like beads in a necklace* (Edith Goldman).

rubidium (rü bid′ē əm), *n. Chemistry.* a soft, silver-white metallic element that decomposes water and ignites spontaneously when exposed to air, used in photoelectric cells: *Rubidium and cesium, like the other alkali metals, never occur free in nature . . . only minute quantities of their salts are to be found, and these are usually found associated with lepidolite, carnallite, and others of the minerals of the more plentiful elements of the family* (Jones, *Inorganic Chemistry*). *Symbol:* Rb; *atomic number* 37; *atomic weight* 85.47; *melting point* 38.89°C; *oxidation state* 1. [from New Latin, from Latin *rubidus* red, from *rubere* be red]

rubredoxin (rü′brə dok′sən), *n. Biochemistry.* a protein found in anaerobic bacteria and associated with oxidation-reduction reactions in cells. *Also included in the iron-sulfur protein class are the rubredoxins, in which single iron atoms are attached to cysteine sulfurs.* (Richard Cammack). [from Latin *ruber* red + English *redox*]

ruby, *n. Mineralogy.* a clear, hard, red precious stone. It is a variety of corundum. *Formula:* Al_2O_3 [ultimately from Latin *rubeus* red]

ruderal (rü′dər əl), *Botany.* —*adj.* growing in rubbish or waste places: *a ruderal plant, ruderal vegetation.* —*n.* a ruderal plant: *The more permanent colonisers . . . ruderals of open ground* (New Scientist). [from Latin *rudera,* plural of *rudus* broken stone]

ruff, *n. Zoology.* a collar of specially marked feathers or hairs on the neck of a bird or animal.

ruga (rü′gə), *n., pl.* **-gae** (-jē). *Biology.* a wrinkle, fold, or ridge: *the rugae of the contracted scrotum.* [from Latin]

rugose (rü′gōs *or* rü gōs′), *adj. Biology.* having rugae or wrinkles; wrinkled; ridged: *a rugose leaf.*

ruled surface, *Geometry.* a surface generated by the motion of a straight line so that the surface may be wholly covered by a group of lines. A cone or a cylinder is a ruled surface.

rule of three, a method of finding the fourth term in a mathematical proportion when three are given. The product of the means equals the product of the extremes.

rumen (rü′mən), *n., pl.* **-mina** (-mə nə). *Zoology.* **1** the first stomach of an animal that chews the cud, in which most food collects immediately after being swallowed: *Sheep, as cattle and goats do, have an "extra stomach," the rumen, where rough feed is predigested with the help of bacteria* (Science News Letter). Compare **abomasum, omasum, reticulum.** **2** the cud of such an animal. [from Latin *rumen* gullet]

ruminal (rü′mə nəl), *adj. Zoology.* of the rumen: *Portions of the ruminal contents are regurgitated, thoroughly chewed . . ., and returned to the rumen* (Science News).

ruminant (rü′mə nənt), *n. Zoology.* an animal that chews the cud. Cows, sheep, and camels are ruminants. The ruminants comprise a suborder of even-toed, hoofed, herbivorous mammals, having a

stomach with four or three separate cavities. *Ruminants—cattle, sheep, deer, goats and others—are animals that digest plant cellulose in their complex stomachs and turn it into protein* (Herbert Kondo). [from Latin *ruminantem* chewing the cud, from *rumen* gullet]

rump, *n. Zoology.* the hind part of the body of an animal, where the legs join the back.

runcinate (run′sə nit *or* run′sə nāt), *adj. Botany.* having coarse, toothlike notches or lobes pointing backward. Dandelion leaves are runcinate. [from Latin *runcina* plane (but taken as "saw")]

runner, *n. Botany.* **1** a slender stem that takes root along the ground, thus producing new plants. Strawberry plants spread by runners. **2** a plant that spreads by such stems.

runner (def. 1)

runoff, *n. Geology.* water that runs off the land in streams; that portion of precipitating water that flows overland to the sea.

runt, *n. Zoology.* **1** an animal, person, or plant that is smaller than the usual size. SYN: dwarf. **2** the smallest animal of a litter, especially a small pig.

rupestrine (rü pes′trin) *or* **rupestral** (rü pes′trəl), *Biology.* —*adj.* living or growing among rocks: *a rupestrine plant.* —*n.* a respestrine plant. [from Latin *rupes* rock]

rupicoline (rü pik′ə lin) *or* **rupicolous** (rü pik′ə ləs), *adj.* = rupestrine. [from Latin *rupes* rock + *colere* inhabit]

ruptive marking, *Zoology.* patterns of light and dark color that break up the outline of an animal to give it protective coloration. [from Latin *ruptum* broken]

Russell diagram, = Hertzsprung-Russell diagram: *a Russell diagram, plotting spectral type horizontally and luminosity vertically and representing each star by a dot* (Duncan, *Astronomy*).

rust, *n.* **1** *Chemistry.* **a** the reddish-brown or orange coating that forms on iron or steel by oxidation of the surface metal when exposed to air or moisture, made up principally of hydrated ferric oxide: *Rust is formed by the galvanic action of the iron and small particles of other metals contained in it in contact with the acidic solutions formed by the dissolution of carbon dioxide in rain water and dew* (Jones. *Inorganic Chemistry*). **b** any film or coating on any other metal

---

cap, fāce, fäther; best, bē, tėrm; pin, fīve;
rock, gō, ôrder; oil, out; cup, pút, rüle,
yü in use, *yu* in uric;
ng in bring; *sh* in rush; *th* in thin, ᵀH in then;
zh in seizure.
ə = *a* in about, *e* in taken, *i* in pencil, *o* in lemon, *u* in circus

due to oxidation or corrosion: *Aluminum has built-in protection against rust* (Newsweek).
2 *Botany.* **a** Also called **rust fungus.** any of various parasitic fungi that produce a plant disease characterized by the appearance of usually reddish spores on the leaves or stems: *If they were of no economic significance, the rusts would still hold a prominent place in botany because of their peculiarly complicated life histories, their special relationships to their hosts, and the ways in which they are thought to be related to other fungi* (Emerson, *Basic Botany*). **b** the disease caused by any of these fungi.
[Old English *rūst*]

ruthenic (rü then'ik), *adj. Chemistry.* of or having to do with ruthenium, especially with a high valence.

ruthenious (rü thē'nē əs), *adj. Chemistry.* of or having to do with ruthenium, especially with a low valence.

ruthenium (rü thē'nē əm), *n. Chemistry.* a hard, brittle, silver-white metallic element, found in platinum ores and used in alloys. Ruthenium is a rare metal similar to platinum. *Symbol:* Ru; *atomic number* 44; *atomic weight* 101.07; *melting point* 2250°C; *boiling point* 3900°C; *oxidation state* 3. [from New Latin, from Medieval Latin *Ruthenia* Russia; because it was discovered in the Ural Mountains]

Rutherford atom, *Physics.* the atom as described in the atomic theory proposed by Lord Ernest Rutherford in 1911. The mass of this atom is located in its center, or nucleus, with electrons revolving about the nucleus much like planets revolve around the sun.

rutherfordium (ruTH'ər fôr'dē əm), *n. Chemistry.* an artificially produced radioactive element having up to ten isotopes. *Atomic number* 104. Also called **kurchatovium.** [named after Lord Ernest *Rutherford,* 1871–1937, British physicist] ▶ See **element.**

rutilated (rü'tə lā'tid), *adj. Mineralogy.* containing needles of rutile: *rutilated quartz.*

rutile (rü'tēl *or* rü'til), *n. Mineralogy.* a reddish-brown or black mineral with a metallic or diamond-like luster, consisting of titanium oxide and often containing some iron. Rutile is polymorphous with anatase and brookite and is a common ore of titanium. *Formula:* TiO_2

Rydberg constant (rid'bèrg), *Physics.* an atomic constant which appears in the formulas of wave numbers in all atomic spectra. For a hypothetical atom of infinite mass, the Rydberg constant is equal to $2\pi^2 me^4/ch^3$, where m is the rest mass of the electron, e is the charge of the electron, c is the speed of light, and h is Planck's constant. *Symbol:* R. *Physicists . . . announced in November 1978 that they had used lasers and light polarizers to measure the Rydberg constant, a fundamental constant of atomic physics, to an accuracy of 3 parts per billion* (Karl G. Kessler). [named after Johannes R. *Rydberg,* 1854–1919, Swedish physicist]

S

s or **s.**, *abbrev.* or *symbol.* **1** sacral. **2** second *or* seconds. **3** specific heat. **4** strangeness. **5** stratus. **6** svedberg *or* svedbergs. **7** secondary.

S, *symbol.* **1** siemens. **2** sulfur.

sabin (sā′bin), *n. Physics.* a unit to measure the sound absorption qualities of a surface. It is equivalent to one square foot of a completely absorptive surface. [named after Wallace C. *Sabine,* 1868–1919, American physicist]

sac, *n. Biology.* a baglike part in an animal or plant, often one containing liquids. The human bladder is a sac. The honeybee carries honey in a honey sac. The octopus has an ink sac. SYN: cyst, vesicle. [from French *sac,* from Latin *saccus* sack]

saccade (sə käd′), *n. Physiology.* the small, jerking movement of the eye as it shifts from one fixed position to another: *The most common major eye movement is the saccade. Saccades usually take less than a twentieth of a second, but they happen several times each second in reading* (Scientific American). [from French *saccade* jerk, tic]
—**saccadic** (sə kä′dik), *adj.* of or characterized by a saccade or saccades: *The eyeball does not follow a continuous path but jumps discontinuously, in so-called saccadic movement* (New Scientist).

saccate (sak′āt), *adj. Botany.* **1** having the form of a sac or pouch: *a saccate corolla or calyx.* **2** having a sac or pouch. [from Latin *saccus* sack]

saccharase (sak′ə rās), *n.* = invertase. [from Latin *saccharum* sugar, from Greek *saccharon;* see SACCHARIN]

saccharate (sak′ə rāt), *n. Chemistry.* **1** a salt or ester of saccharic acid.
2 a compound of a metallic oxide with a sugar: *calcium saccharate.*

saccharic (sə kar′ik), *adj. Chemistry.* having to do with or obtained from a sugar. **Saccharic acid** is a dibasic acid, occurring in three optically different forms, produced by oxidizing glucose and various other hexose sugars. *Formula:* $C_6H_{10}O_8$

saccharide (sak′ə rīd), *n. Chemistry.* a compound consisting of one or more simple sugars; a monosaccharide or oligosaccharide.

saccharin (sak′ər ən), *n. Chemistry.* a crystalline substance obtained from coal tar, several hundred times sweeter than cane sugar, formerly used as a substitute for sugar but since 1977 regarded as potentially carcinogenic to humans. *Formula:* $C_7H_5NO_3S$ [from Latin *saccharum* sugar, from Greek *sakcharon,* ultimately from Sanskrit *śarkarā* (originally) gravel, grit]

saccharoidal (sak′ə roi′dəl), or **saccharoid** (sak′ə roid), *adj. Geology.* having a granular texture like that of loaf sugar: *saccharoidal limestone, saccharoidal crystallization.*

saccharomyces (sak′ə rō mī′sēz), *n., pl.* -ces. *Biology.* any member of a genus (*Saccharomyces*) of single-celled yeasts that lack a true mycelium and reproduce by budding: *Saccharomyces ... are strongly fermenting.* [from New Latin, from Latin *saccharum* sugar + Greek *mykēs* fungus (so called because the yeast produces alcoholic fermentation in sugary fluids)]

sacculate (sak′yə lāt) or **sacculated** (sak′yə lā′tid), *adj. Biology.* formed of little sacs; divided into saclike dilations: *In the kangaroos the whole extent of the stomach is sacculated* (Edward P. Wright).

sacculation (sak′yə lā′shən), *n. Biology.* **1** the formation of a saccule or saccules: *sacculation of the colon.* **2** a sacculate part: *a sacculation of a bronchus.*

saccule (sak′yül), *n. Anatomy.* a little sac, especially the smaller of the two membranous sacs in the labyrinth of the internal ear. [from Latin *sacculus,* diminutive of *saccus* sack]

sac fungus, = ascomycete.

sack cloud, *Meteorology.* a form of mammato-cumulus in which the pocket hanging from the cloud becomes so deep as to resemble a sack or bag.

sacral (sā′krəl), *adj. Anatomy.* of, having to do with, or in the region of the sacrum: *the sacral nerves.*

sacrococcyx (sā′krō kok′siks), *n. Anatomy.* the sacrum and coccyx regarded as one bone.

sacroiliac (sā′krō il′ē ak), *Anatomy.* —*adj.* **1** of or having to do with the sacrum and the ilium: *sacroiliac articulation.* **2** designating the joint between the sacrum and the ilium.
—*n.* the sacroiliac joint: *Strictly speaking, the sacroiliac is an area at the base of the spine* (Atlantic).

sacrosciatic (sā′krō sī at′ik), *adj. Anatomy.* of or having to do with the sacrum and the ischium: *the sacrosciatic notch.*

sacrum (sā′krəm), *n., pl.* -cra (-krə), -crums. *Anatomy.* a compound triangular bone at the lower end of the spine, made by the joining of several vertebrae, and forming the back of the pelvis: *In the sacral region the vertebrae in some animals are considerably thickened without great change in form, while in others they are much flattened and more or less fused into a platelike structure, the sacrum* (Shull, *Principles of Animal Biology*). [from Late Latin *(os) sacrum* sacred (bone) (probably so called from its being offered as a dainty in sacrifices)]

---

cap, fāce, fäther; best, bē, tèrm; pin, fīve;
rock, gō, ôrder; oil, out; cup, pùt, rüle,
yü in use, *yu* in uric;
ng in bring; *sh* in rush; *th* in thin; ᴛʜ in then;
zh in seizure.

ə = *a* in about, *e* in taken, *i* in pencil, *o* in lemon, *u* in circus

sagittal (saj′ə təl), *adj.* **1** *Anatomy.* of or having to do with a suture between the parietal bones of the skull: *... a sagittal crest running along the top of the skull* (New Scientist).
2 *Zoology.* of, having to do with, or situated in the vertical plane that divides a bilaterally symmetrical animal into halves that are approximate mirror images of each other, or any plane parallel to this plane: *the skull of the gorilla has a sagittal crest.* SYN: median.
[from Latin *sagitta* arrow]
sagittate (saj′ə tāt), *adj.* *Botany, Zoology.* shaped like an arrowhead: *Calla lilies have sagittate leaves. Some birds have sagittate markings.*

sagittate
leaf

sal, *n.* salt (used especially in pharmaceutical terms, as *sal ammoniac, sal soda, sal volatile*). [from Latin]
sal ammoniac (ə mō′nē ak), = ammonium chloride.
salicylaldehyde (sal′ə sə lal′də hīd), *n.* *Chemistry.* a fragrant, colorless to dark-red oil produced from phenol and chloroform, used in analytical chemistry. It is the aldehyde of salicylic acid. *Formula:* $C_6H_4(OH)COH$
salicylate (sə lis′ə lāt), *n.* *Chemistry.* any salt or ester of salicylic acid.
salicylic acid (sal′ə sil′ik), *Chemistry.* a white, crystalline or powdery acid used as a mild antiseptic and preservative, and in making aspirin: *To turn salicylic acid into acetylsalicylic acid (aspirin), a compound related to acetic acid is used* (Time). *Formula:* $C_6H_4(OH)CO_2H$ [from Latin *salicem* willow (so called because it was first obtained from an extract of willow bark)]
salientian (sā′lē en′shē ən), *Zoology.* —*adj.* of or belonging to an order (Salientia) of amphibians that includes the frogs and toads.
—*n.* a salientian animal.
[from Latin *salientem* leaping]
salina (sə li′nə *or* sə lē′nə), *n.* *Geology.* **1** a playa encrusted with salt.
2 a salt marsh, salt lake, or the like.
[from Spanish, from Latin *salinae* saltworks, from *sal* salt]
saline (sā′lēn′ *or* sā′līn), *adj.* **1** of salt; like salt; salty: *saline particles, a saline substance.*
2 containing common salt or any other salts: *a saline solution.*
3 *Chemistry.* **a** of or having to do with chemical salts. **b** (of medicines) consisting of or based on salts of the alkaline metals or magnesium: *a saline laxative.*
—*n.* **1** a salt spring, well, or marsh.
2 a salt of an alkali or magnesium, used as a cathartic.
3 a solution with a high concentration of salt, especially one with a concentration similar to that of the blood, used in medical treatment.
[from Latin *sal* salt]

salinity (sə lin′ə tē), *n.* **1** saline quality or condition: *The salinity of the ocean is the most significant feature in determining the structure of marine plants and animals* (Clarence J. Hylander).
2 a saline concentration: *At very low salinities the electrical resistance of the solutions becomes very high* (New Scientist).
saliva, *n.* *Physiology.* a colorless, watery fluid that the salivary glands secrete into the mouth to keep it moist, aid in chewing and swallowing of food, and start digestion: *The water, which makes up over 99 per cent of normal saliva, softens dry foods, thus facilitating swallowing* (Harbaugh and Goodrich, *Fundamentals of Biology*). [from Latin]
—**salivary** (sal′ə ver′ē), *adj.* of or producing saliva: *a salivary enzyme.*
salivary amylase, = ptyalin.
salivary gland, *Physiology.* any of various glands that empty their secretions into the mouth. The salivary glands of human beings and certain other vertebrates are digestive glands that secrete saliva containing the digestive enzyme ptyalin, salts, mucus, etc.
salivate (sal′ə vāt), *v.* *Physiology.* to secrete saliva, especially in excessive amounts: *He [Pavlov] found that if he rang a bell at the same time as he offered food to dogs, they would quite quickly associate the sound of the bell with food and upon the bell alone would cause them to salivate* (J. A. V. Butler).
—**salivation** (sal′ə vā′shən), *n.* the act or process of salivating; secretion of saliva: *It is supposed that in the simple reflex of salivation the presence of food in the mouth activates a particular clump of nerve cells* (S. A. Barnett).
salmon cloud, *Meteorology.* a band of parallel cirrostratus clouds stretching almost entirely across the sky and appearing to taper at the ends.
salmonella (sal′mə nel′ə), *n., pl.* **-nellas, -nellae** (-nel′ē), or **-nella.** *Bacteriology.* any of a genus of gram-negative, rod-shaped bacteria that cause food poisoning, typhoid, and paratyphoid fever in humans, and other infectious diseases in domestic animals. [from New Latin *Salmonella* the genus, named after Daniel E. *Salmon*, 1850–1914, American pathologist]
salpingian (sal pin′jē ən), *adj.* *Anatomy.* of or having to do with a salpinx or salpinges.
salpinx (sal′pingks), *n., pl.* **salpinges** (sal pin′jēz). *Anatomy.* **1** = Eustachian tube. **2** = Fallopian tube. [from Greek *salpinx* trumpet]
sal soda, *Chemistry.* a white, crystalline sodium carbonate containing ten water molecules, used for washing textiles, bleaching linen, etc. *Formula:* $Na_2CO_3 \cdot 10H_2O$ Also called **washing soda.** Compare **soda ash.**
salt, *n.* **1** the common name of sodium chloride (NaCl), a white substance found in the earth and in sea water, used as a seasoning, a preservative for food and hides, and in many industrial processes: *Salt is one of the common rock minerals of the earth ... Inexhaustible supplies are available for human use* (Finch and Trewartha, *Elements of Geography*). *Formula:* NaCl
2 *Chemistry.* a compound derived from an acid by replacing the hydrogen wholly or partly with a metal or an electropositive radical. A salt is formed when an acid and a base neutralize each other. Sodium bicarbonate is a salt. *Salts [are] usually defined as ionic*

compounds which in water solution yield a positive ion other than hydrogen and a negative ion other than hydroxyl. The process by which an acid and a base unite to form water and a salt is termed neutralization (Jones, *Inorganic Chemistry*). *Salts which contain a polyvalent ion combined with two different ions, are called double salts. Acid salts and basic salts are, therefore, double salts* (Offner, *Fundamentals of Chemistry*).

▶ The nomenclature of chemical salts has reference to the acids from which they are derived. For example, sulfates, nitrates, and carbonates imply salts of sulfuric, nitric, and carbonic acids. The suffix *-ate* implies a large amount of oxygen in the acids; the suffix *-ite* implies a smaller amount.

—*v. Chemistry.* **1** to treat or impregnate with a salt. **2** to add a salt to (a solution in order to precipitate a dissolved substance). **3** Usually, **salt out,** to precipitate (a dissolved substance) in this manner.

—*adj.* **1** containing salt: *salt water.* **2** flooded with or growing in salt water: *salt weeds.* **3** consisting of or containing salt water: *a salt pond.*

salt dome, *Geology.* a circular structure of sedimentary rocks resulting from the upward movement of a subterranean mass of salt; a diapiric structure: *In the northeastern area of the Caspian Sea, salt domes and subsalt formations have been discovered* (New Scientist). Also called **plug.**

salt lake, *Geology.* a lake having a high concentration of salts, found in areas of large salt deposits. Salt lakes are often saltier than oceans.

salt marsh, 1 *Geology.* a marsh periodically flooded by salt water through the actions of winds or tides. **2** *Ecology.* a marsh or swamp having a high concentration of salt, chiefly sodium chloride, found especially in tropical regions.

saltpeter, *n.* **1** naturally occurring potassium nitrate; niter. **2** sodium nitrate, especially when occurring naturally. [from Old French *salpetre,* from Medieval Latin *sal petrae* salt of rock]

saltwater, *adj.* **1** consisting of or containing salt water: *a saltwater solution.* **2** living in the sea or in water like seawater: *saltwater fish, the saltwater crocodile.*

sal volatile (vō lat′ə lē), *Chemistry.* a colorless or white crystalline salt of ammonium, used in smelling salts. *Formula:* NH$_4$Cl [from New Latin, volatile salt]

samara (sam′ər ə *or* sə mär′ə), *n. Botany.* any dry fruit that has a winglike extension and does not split open when ripe. The fruit of the maple tree is a double samara with one seed in each half. [from Latin *samara* elm seed]

samarium (sə mãr′ē əm), *n. Chemistry.* a hard, brittle, grayish-white, metallic element of the cerium group, discovered in 1879 in samarskite. Samarium is one of the rare-earth elements. It is used in control rods in nuclear reactors. *Symbol:* Sm; *atomic number* 62; *atomic weight* 150.35; *melting point* about 1072°C; *boiling point* about 1900°C; *oxidation state* 2, 3. [from *samar* (*skite*) + New Latin *-ium* suffix of chemical elements]

samarskite (sə mär′skīt), *n. Mineralogy.* a black, lustrous, orthorhombic mineral, an oxide containing niobium, uranium, cerium, samarium, and other

elements. [named after Colonel *Samarski,* a 19th-century Russian mine official]

sand, *n. Geology.* earth material consisting of grains of worn-down or disintegrated rock, mainly of silica, larger than silt but much smaller and finer than gravel.

sandbank, *n. Geography, Geology.* a ridge of sand forming a mound, shoal, or hillside.

sandbar, *n. Geography, Geology.* a ridge of sand formed in a river or along a shore by the action of tides or currents: *Many heavily-laden streams deposit sand bars in their channels thus making their courses shallower and rendering them more likely to flood their valleys* (White and Renner, *Human Geography*). Also called **sand reef.**

sand plain, *Geography, Geology.* a small sandy plain, usually a flat-topped hill originally formed as a delta by water running out of a glacier.

sand reef, = sandbar.

sandstone, *n. Geology.* a sedimentary rock formed by the consolidation of sand and held together by a natural cement such as silica or iron oxide.

sandstorm, *n. Meteorology.* a strong wind that carries along clouds of dust, especially in a desert region.

sand wave, 1 *Geology.* a large wavelike structure formed by rapidly moving currents of water. **2** *Oceanography.* a moving ridge of sand on the surface of the sea floor: *The recently-discovered phenomenon of North Sea "sand waves," moving "dunes" on the sea-floor, might present disconcerting problems, changing a level location into one with an awkward ridge in the middle of it* (New Scientist).

sanidine (san′ə dēn), *n. Mineralogy.* a mineral of the alkali feldspar family, formed at high temperature, and polymorphous with orthoclase and microcline. *Formula:* (K,Na)AlSi$_3$O$_8$ [from Greek *sanidos* plank (so called in reference to its flat crystals)]

S-A node, = sinoatrial node.

Santa Ana, *Meteorology.* a strong, hot, dry wind occurring in southern California usually in the winter. [named after the *Santa Ana* mountain range in southern California]

sap, *n. Botany.* **1** the liquid that moves through a vascular plant, carrying water and dissolved minerals upward through the plant, and water and dissolved food downward. Maple sugar is made from the sap of some maple trees. **2** = sapwood.

saphena (sə fē′nə), *n., pl.* **-nae** (-nē). *Anatomy.* either of two large superficial veins of the leg, one extending along the inner side of the leg from the foot to the groin, and the other extending along the outer and posterior side from the foot to the knee. [from Medieval Latin, from Arabic *ṣāfin*]

---

cap, fāce, fäther; best, bē, tèrm; pin, fīve;
rock, gō, ôrder; oil, out; cup, pùt, rüle,
yü in use, yù in uric;
ng in bring; sh in rush; th in thin, ŦH in then;
zh in seizure.
ə = a in about, e in taken, i in pencil, o in lemon, u in circus

—**saphenous** (sə fē′nəs), *adj.* of, designating, or having to do with a saphena or the saphenae.

sapling, *n. Botany.* a young tree, especially a young forest tree with a trunk from 1 to 4 inches in diameter.

saponification (sə pon′ə fə kā′shən), *n. Chemistry.* a reaction in which an ester is heated with an alkali, such as sodium hydroxide, to form an alcohol and a salt of an acid; alkaline hydrolysis of a fat or oil to form soap. [from New Latin *saponificare,* from Latin *saponem* soap + *facere* to make]

saponify (sə pon′ə fī), *v. Chemistry.* to subject (a fat or an oil) to saponification: *Considerable preliminary concentration can be effected by saponifying the oil, that is by turning the glycerides of fatty acids which constitute the bulk of the oil into soaps, which are sodium salts of the fatty acids* (Science News).

saponin (sap′ə nin), *n. Chemistry.* any of the glucosides obtained from the soapwort, soapbark, soapberry, and many other plants, and forming (in solution) a soapy lather when shaken. The commercial substance, a mixture of saponins, is used to produce foam in beverages, as a detergent, and in fire extinguishers. [from French *saponine,* from Latin *saponem* soap]

saponite (sap′ə nīt), *n. Mineralogy.* a clay mineral with a soapy feel that occurs in soft, amorphous masses, filling veins and cavities, as in serpentine and traprock. It is a hydrous silicate of aluminum and magnesium. [from Swedish *saponit,* from Latin *saponem* soap]

sapphire (saf′īr), *n. Mineralogy.* a clear, hard, bright-blue variety of corundum. Sapphires are valued as semi-precious stones. [from Greek *sappheiros,* from Hebrew *sappīr),* from Sanskrit *sánipriya* sapphire; (originally) dear to the planet Saturn]

sapphirine (saf′ə rin *or* saf′ī rēn), *n. Mineralogy.* a pale-blue or greenish metamorphic mineral, a silicate of aluminum, magnesium, and iron.

sapro-, *combining form.* decay; decomposition; putrefaction, as in *saprogenic, saprolite, saprophyte.* [from Greek *sapros* rotten]

saprogenic (sap′rə jen′ik), *adj. Biology.* producing or formed by decay: *saprogenic bacteria.*

saprogenous (sə proj′ə nəs), *adj.* = saprogenic.

saprolite (sap′rə līt), *n. Geology.* soft, partly decomposed rock rich in clay, and remaining in its original place.
—**saprolitic** (sap rə lit′ik), *adj.* of, having to do with, or like saprolite.

saprophagous (sə prof′ə gəs), *adj.* = saprophytic.

saprophyte (sap′rō fīt), *n. Biology.* any fungus or bacterium that lives on decaying organic matter. ▶ See the note under **heterotroph.**
—**saprophytic** (sap′rō fit′ik), *adj.* of or like a saprophyte; living on decaying organic matter: *Saprophytic fungi are the greatest destroyers of wood* (Harbaugh and Goodrich, *Fundamentals of Biology).* —**saprophytically,** *adv.*

sapwood, *n. Botany.* the soft, new, living wood between the cambium and the heartwood of most trees. See the picture at **wood.**

sarco-, *combining form.* **1** flesh; fleshy, as in *sarcocarp.* **2** muscle tissue, as in *sarcosome.* [from Greek *sarkos* flesh]

sarcocarp (sär′kō kärp), *n. Botany.* **1** the fleshy mesocarp of certain fruits, such as the peach and plum. It is the part usually eaten.
2 any fruit that is fleshy.

sarcodinian (sär′kō din′ē ən), *Biology.* —*adj.* of or belonging to a superclass (Sarcodina) of amebalike protozoans that move and take in food by means of pseudopodia. The rhizopods are sarcodinian organisms.
—*n.* a protozoan belonging to this group; rhizopod. [from Greek *sarkōdēs* fleshy, from *sarkos* flesh]

sarcolactic acid (sär′kō lak′tik), *Biochemistry.* the dextrorotatory form of lactic acid found in the blood and muscles as a result of the metabolism of glucose: *Sarcolactic acid ... has been called the "acid of fatigue" ... because a greater amount of sarcolactic acid is present in the muscles when a person is tired* (George L. Bush).

sarcolemma (sär′kō lem′ə), *n. Histology.* an elastic transparent membrane enclosing each of the fibers of muscle tissue, especially skeletal muscle tissue. [from *sarco-* muscle + Greek *lemma* husk]
—**sarcolemmal,** *adj.* of or having to do with a sarcolemma or sarcolemmas: *Each giant fiber is surrounded by a stout sarcolemmal complex with associated nuclei* (Graham Hoyle).

sarcoma (sär kō′mə), *n., pl.* **-mas, -mata** (-mə tə). *Medicine.* any of a class of cancers originating in tissue that is not epithelial, but chiefly connective tissue: *It is possible to have tumors of the mesoderm, these being usually known as sarcomata* (Science News). Compare **carcinoma, lymphoma.**
—**sarcomatous** (sar kō′mə təs), *adj.* having to do with or of the nature of a sarcoma.

sarcomere (sär′kə mir), *n. Histology.* one of the segments into which a fibril of striated muscle is divided by thin dark bands; one of the segments of a sarcostyle.

sarcoplasm (sär′kō plaz əm), *n. Histology.* the clear, hyaline substance which separates the fibrillae, or sarcostyles, in a striated muscle fiber: *Imbedded in the sarcoplasm, and forming a large part of the bulk of the cell, are numerous slender strands, the contractile myofibrils ... Collectively they give the whole cell the appearance of being marked by light and dark transverse bands ... These are the marks to which the term "striated" refers* (Shull, *Principles of Animal Biology).*
—**sarcoplasmic,** *adj.* having to do with or of the nature of sarcoplasm. The **sarcoplasmic reticulum** is a specialized form of endoplasmic reticulum found in muscle cells. *Inside muscle fibres, calcium ions are stored in the sarcoplasmic reticulum—an intracellular network of fine tubules responsible for transmitting stimuli from the cell surface to the centre of the fibre* (New Scientist).

sarcosome (sär′kō sōm), *n. Histology.* a large specialized mitochondrion found in striated muscle.

sarcostyle (sär′kō stīl), *n. Histology.* one of the delicate fibrillae which make up the fiber of a striated muscle.

sarmentose (sar men′tōs), *adj. Botany.* having or producing slender runners: *sarmentose stems, sarmentose shrubs.* [from Latin *sarmentum* twig]

saros (sār′os), *n. Astronomy.* a cycle of 18 years and 11 1/3 days, in which solar and lunar eclipses repeat themselves. If the interval contains five leap years instead of four, it is 18 years and 10 1/3 days. [from Greek]

sartorius (sär tôr′ē əs), *n., pl.* **-torii** (-tôr′ē ī). *Anatomy.* a flat, narrow muscle, the longest in the human body, running from the ilium to the top of the tibia, and crossing the thigh obliquely in front. [from New Latin, from Latin *sartor* tailor, patcher]

sastruga (sas trü′gə), *n.* = zastruga.

satellite, *n.* **1** *Astronomy.* **a** a celestial body that revolves around a planet, especially around one of the nine major planets of the solar system. The moon is a satellite of the earth. *Jupiter's eleven known satellites are sharply divided into three groups: the inner satellites and two groups of outer satellites. The five inner satellites revolve from west to east in orbits which are nearly circular and nearly in the planes of the planet's equator and orbit* (Baker, *Astronomy*). **b** any object launched into an orbit around the earth or other celestial body; an artificial satellite. Such satellites are used to send weather or other scientific information back to earth; they also aid in transmitting microwave communications, such as telephone calls and television programs, to distant points on earth. *The magnetic field of the Earth induces sufficiently large eddy currents in the metal hull of a satellite to slow down its rotation* (New Scientist).
2 *Anatomy.* Usually, **satellite cell.** any of the cells that encapsulate the bodies of nerve cells in ganglia, analogous to the Schwann cells surrounding the axons.
3 *Genetics.* a short segment of a chromosome separated from the rest by a constriction or constrictions, typically associated with the formation of a nucleolus.
4 = satellite DNA.
—**satellitic,** *adj.* of, having to do with, or of the nature of a satellite.

satellite (def. 1b)

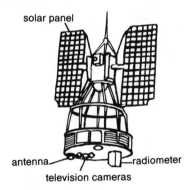

solar panel

antenna — radiometer

television cameras

satellite DNA, *Molecular Biology.* a form of DNA in animal cells which is of a different density from that of other DNA and consists of repeating sequences of nucleotide pairs: *A large fraction of the DNA in animal cells is a satellite DNA, which consists of a repeating sequence of six nucleotide pairs. In some animal species more than 10% of the total DNA is of this type* (Stephen N. Kreitzman).

saturate, *v.* **1** *Chemistry.* to cause (a substance) to unite with the greatest possible amount of another substance. A saturated solution of sugar or salt is one that cannot dissolve any more sugar or salt.

2 *Physics.* to magnetize (a substance) so that the intensity of its magnetization reaches its maximum value. [from Latin *saturatum* filled, from *satur* full]
—**saturated,** *adj.* **1** *Chemistry.* **a** that has combined with or taken up in solution the largest possible proportion of some other substance: *When a gas is confined above the surface of a liquid at a definite pressure and the system is held at a fixed temperature, the gas passes into solution until a state of equilibrium is established, and the gas molecules are leaving the solution at the same rate as an equal number of like molecules are passing into solution. Under these conditions, the liquid is said to be saturated with the gas, and the solution is called a saturated solution* (Parks and Steinbach, *Systematic College Chemistry*). **b** (of an organic compound) lacking double or triple bonds and having no free valence, as methane, ethane, propane, etc. The series of saturated hydrocarbons having the general formula $C_nH_{2n} + _2$ is called the methane or alkane series.
2 *Physics.* charged to the full extent of its capacity: *Further increase of plate potential does not increase the plate current, which is then said to become saturated* (Sears and Zemansky, *University Physics*).
3 *Optics.* (of colors) containing no white or black; of the greatest intensity.
4 *Mineralogy.* containing the greatest proportion of silica possible.
—**saturation,** *n.* **1** the act or process of saturating.
2 the fact of being saturated; saturated condition. The saturation of a color increases as the amount of white in it is decreased.

Saturn, *n. Astronomy.* the second largest planet in the solar system, and the sixth in distance from the sun. Its orbit lies between those of Jupiter and Uranus and takes 29.46 years to complete, at a mean distance from the sun of 1,500,000,000 kilometers. Saturn is encircled by a system of rings that are composed of tiny particles of ice. See the pictures at **ring, solar system.**
—**Saturnian,** *adj.* of or having to do with the planet Saturn: *Each Saturnian season is ... 7 1/2 years long, or 6300 Saturnian days* (Krogdahl, *The Astronomical Universe*).

saurian (sôr′ē ən), *Zoology.* —*n.* **1** any of the group of reptiles that includes the lizards. **2** any similar reptile, such as a crocodile or dinosaur.
—*adj.* belonging to or having to do with the saurians. [from Greek *sauros* lizard]

saussurite (sôs′yə rīt *or* sô sûr′īt), *n. Mineralogy.* a very compact mixture of albite, zoisite, and other minerals, formed by the alteration of feldspar. [from Horace B. de *Saussure*, 1740–1799, Swiss naturalist who described this mineral + *-ite*]

cap, fāce, fäther; best, bē, tėrm; pin, five;
rock, gō, ôrder; oil, out; cup, pùt, rüle;
yü in use, yú in uric;
ng in bring; sh in rush; th in thin; ᴛʜ in then;
zh in seizure.
ə = a in about, e in taken, i in pencil, o in lemon, u in circus

savanna or **savannah** (sə van′ə), *n. Geography, Ecology.*
1 a tropical or subtropical region of grassland, typically with scattered trees or shrubs: *The savanna is a mixed plant association of grass and trees, which develops as a response to marked seasons of rain and drought. But since the intensity and length of the rainy season differ strikingly from place to place, the character of the savanna formation likewise differs. It varies from fairly dense monsoon forest (with teak, deciduous mahogany or bamboo thickets) through park-like woodland, to dry thorn-forest* (White and Renner, *Human Geography*). **2** a grassy plain with few or no trees, especially in the southeastern United States. [from Spanish *sabana, zavana,* from Arawak (Indian language of Haiti)]

saxicavous (sak sik′ə vəs), *adj. Zoology.* rock-boring, as certain mollusks: *saxicavous seashells.* [from Latin *saxum* rock + *cavare* to hollow]

saxicoline (sak sik′ə lin), *adj. Zoology, Botany.* living on or among rocks. [from Latin *saxum* rock + *colere* inhabit]

Sb, *symbol.* antimony (for Latin *stibium*).

sc., *abbrev.* **1** scale. **2** science.

Sc, *symbol.* scandium.

scalar (skā′lər), *Mathematics.* —*adj.* having or involving only magnitude. Scalar numbers or quantities are used to represent length, mass, speed, etc: *Distance is a scalar quantity that represents the length of a path from one point to another* (Physics Regents Syllabus). Contrasted with **vector.**
—*n.* a scalar number or quantity: *A scalar is completely specified by the statement of a numeral and a unit—for example, a volume of ten gallons. On the other hand, a vector requires for its specification a numeral, a unit, and a direction—for example, a displacement of two feet eastward* (Shortley and Williams, *Elements of Physics*).
[from Latin *scalaris* like a ladder, from *scalae* ladder]

scalar field, *Mathematics.* a set of scalars assigned to points composing a region of space, the value of each scalar depending on the location of the point with which that scalar is associated. Compare **vector field.**

scalariform (skə lar′ə fôrm), *adj. Botany.* ladderlike: *scalariform plant cells, scalariform conjugation.*

scalar product, = dot product.

scalation, *n. Zoology.* the nature and form of the scales, as in fishes or snakes.

scale¹, *n.* **1** *Zoology.* **a** one of the thin, flat, hard plates forming the outer covering of many fishes, snakes, and lizards: *The bodies of most fish are covered by scales which overlap like shingles on a roof* (Winchester, *Zoology*). **b** a part like this in other animals, such as one of the very small plates covering the wings of moths and butterflies or one of the plates covering the tails of certain mammals.
2 *Botany.* **a** one of the parts that unite to cover a bud in winter. Scales are modified rudimentary leaves found on the leaf buds of most perennial deciduous plants. Also called **scale leaf. b** a part like those of the leaf bud, such as one of the layers of an onion bulb, or one of the scalelike leaves of a rhizome or pine cone.
3 *Chemistry.* an oxide formed on metals when heated. [from Scandinavian (Old Icelandic) *skál*]

scale², *n.* **1 a** a series of marks made along a line or curve at regular distances to use in measuring: *A thermometer has a scale.* **b** an instrument marked in this way, used for measuring.
2 a the size of a map, plan, chart, or model compared with what it represents: *Maps may have large scales or small. A ratio of 1 to 10,000,000 indicates a small scale because one unit of map distance represents 10,000,000 units of earth distance, and the map is, by comparison, extremely small. A ratio of 1 to 100,000 indicates a map of much larger scale, and a ratio of 1 to 1 would indicate a map as large as the area mapped* (Finch and Trewartha, *Elements of Geography*). **b** the equally divided line, as on a map or chart, that indicates this relationship.
3 a system of numbering: *The decimal scale counts by tens.*
—*v.* to make according to scale; alter or vary in a determined proportion: *The radio waves ... can be scaled several orders of magnitude in frequency to predict the true intensity of the optical radiation* (Scientific American).
[from Latin *scalae* ladder, steps, from *scandere* to climb]

scale leaf, *Botany.* one of the parts that unite to cover a bud in winter: *Scale leaves are reduced foliage leaves. They are found on certain rhizomes; above ground stems ... on bulbs; and forming the protective scale of scaly buds* (Youngken, *Pharmaceutical Botany*).

scalene (skā lēn′ or skā′lēn′), *adj. Geometry.* **1** (of a triangle) having three unequal sides. **2** (of a cone or other solid figure) having the axis inclined to the base. [from Late Latin *scalenus,* from Greek *skalēnos* uneven]

scalene muscle, *Anatomy.* any one of several triangular muscles that connect the upper ribs with vertebrae.

scalp, *n. Anatomy.* the skin and hair covering the skull. Beneath the scalp are layers of fat, connective tissue, and blood vessels.

scandent (skan′dənt), *adj. Botany.* climbing: *a scandent vine, a scandent stem.* [from Latin *scandentem*]

scandium (skan′dē əm), *n. Chemistry.* a gray metallic element, occurring in rare minerals that are found in Scandinavia: *Observations show that scandium is rather abundant in some of the stars* (Scientific American). *Symbol: Sc; atomic number 21; atomic weight 44.956; melting point 1539°C; boiling point 2832°C; oxidation state 3.* [from New Latin, from Latin *Scandia* Scandinavia]

scansorial (skan sôr′ē əl), *adj. Zoology.* **1** having to do with or adapted for climbing: *Woodpeckers have scansorial feet.*
2 habitually climbing: *a scansorial bird.*
[from Latin *scansorius* used for climbing, from *scandere* to climb]

scape, *n.* **1** *Botany.* a leafless flower stalk rising from the ground, as that of the dandelion.
2 *Zoology.* something like a stalk, such as the shaft of a feather.
[from Latin *scapus* stalk]

scaphoid (skaf′oid), *adj., n.* = navicular. [from Greek *skaphoidēs* boat-shaped]

scapolite (skap′ə lit), *n. Mineralogy.* any of a group of minerals of variable composition, essentially silicates of aluminum, calcium, and sodium, occurring in tetragonal crystals and also massive. [from Greek *skapos* shaft]

scapose (skă'pōs), *adj. Botany.* **1** having scapes. **2** of or like a scape.

scapula (skap'yə lə), *n., pl.* **-lae** (-lē), **-las.** *Anatomy.* **1** = shoulder blade. **2** one of the other bones of the pectoral arch of some vertebrates. [from Latin *scapulae* shoulders]
—**scapular,** *adj.* of the shoulder or shoulder blade.
—*n.* **1** *Zoology.* a bird's feather growing in the shoulder region where the wing joins the body. **2** = scapula.

scapulary (skap'yə ler'ē), *adj., n.* = scapular.

scatter, *v. Physics.* **1** to throw back or deflect (as rays of light or radioactive particles) randomly or in all directions.
2 to undergo scattering: *The majority of the energy is carried by phonons which inelastically scatter at the interface* (Nature).
—*n.* **1** *Physics.* the scattering of light or other radiation, such as radio waves.
2 *Statistics.* the degree to which repeated measurements or observations vary: *A commonly used measure of the dispersion or scatter of a number of observed values about the central values is the standard deviation* (B. Fozard).

scatter diagram, *Statistics.* a graph showing the relationship between two variables, in which one variable is marked off on the x-axis and the other on the y-axis, and points representing the values of the variables in a distribution are plotted along the lines of the graph. When the relationship is close, the points tend to form a diagonal line or a curve. *Scatter diagrams were drawn to show mean concentrations of albumin ... fibrinogen and cholesterol against age* (Nature).

scattering layer, *Biology.* a layer of plankton or other organisms in the ocean that reflects and scatters sound waves.

scavenger, *n. Zoology.* an animal that feeds on the remains of other animals which it has not killed. Compare **predator.**

scent gland, *Zoology.* a gland which secretes an odoriferous substance: *the scent glands of female moths.*

scheelite (shā'līt or shē'līt), *n. Mineralogy.* calcium tungstate, an ore of tungsten, found chiefly in brilliant crystals of various colors. *Formula:* $CaWO_4$ [from Karl W. *Scheele,* 1742–1786, Swedish chemist + *-ite*]

schiller (shil'ər), *n. Mineralogy.* a peculiar, almost metallic luster, sometimes with iridescence, occurring on certain minerals. [from German *Schiller* play of colors]

schist (schist), *n. Geology.* a medium-grained metamorphic rock usually composed essentially of mica, that splits easily into layers: *In a few places the rocks are less coarsely crystalline, and are varieties of green slate and schist* (E. F. Roots). [from Greek *schistos* cleft, from *schizein* to split]

schistocyte (skis'tə sīt), *n. Biology.* a fragmenting red blood cell.

schistocytosis (skis'tə sī tō'sis), *n. Biology.* fragmentation of a red blood cell.

schistose or **schistous** (shis'tōs), *adj. Geology.* of or resembling schist; having the structure of schist: *Schistose rock composed chiefly of muscovite, quartz,*

and biotite in varying proportions (Gilluly, *Principles of Geology*).
—**schistosity** (shis tos'ə tē), *n.* schistose structure or formation; ability to be split into layers: *a regional schistosity.*

schizo-, *combining form.* split; division; cleavage, as in *schizocarp, schizogenesis.* [from Greek *schizein* to divide, split]

schizocarp (skiz'ō kärp), *n. Botany.* any dry fruit that divides, when ripe, into two or more seed vessels that contain one seed each and do not split open, as in the carrot and celery.
—**schizocarpous** (skiz'ə kär'pəs), *adj.* of or resembling a schizocarp.

schizogamy (ski zog'ə mē), *n. Biology.* reproduction in which a sexual form is produced by fission or by budding from a sexless one, as in some worms.

schizogenesis (skiz'ə jen'ə sis), *n.* = schizogony.

schizogenous (ski zoj'ə nəs), *adj. Biology.* reproducing by schizogony.

schizogonic or **schizogonous,** *adj.* having to do with or exhibiting schizogony.

schizogony (ski zog'ə nē), *n. Biology.* reproduction by fission, as in the schizophytes: *The* [*schizont*] *when fully grown, reproduces by a process known as schizogony ... In this process the whole of the animal divides up simultaneously into a considerable number of minute individuals, all alike. Each of these develops into a sexual individual which in its turn divides up completely into sexual reproductive cells* (New Biology).

schizomycetes (skiz'ō mī'sēts), *n. pl.* = bacteria. [from New Latin *Schizomycetes* the class name, from *schizo-* + Greek *mykētos* fungi]

schizont (skiz'ont), *n. Biology.* a protozoan cell that divides asexually to form daughter cells, especially a multinucleate sporozoan cell that divides by fission into a number of new cells. Compare **gamont.** [from German *Schizont,* from Greek *schizein* to split + *ontos* being]

schizopelmous (skiz'ə pel'məs), *adj. Zoology.* (of birds) having two flexor tendons for the toes. [from *schizo-* + Greek *pelma* sole of the foot]

schizophyte (skiz'ə fīt), *n. Biology.* any of the bacteria and blue-green algae that reproduce by simple fission or by spores and that are sometimes classified as a group.

schlieren (shlir'ən), *n.pl.* **1** *Geology.* irregular, dark or light streaks or clots occurring in plutonic igneous rock because of varying proportions of the minerals present: *Many schlieren—perhaps nearly all—are stretched-out, half-assimilated inclusions of hornblende schist or similar refractory ferromagnesian-rich rocks, but some may be segregations of clots of ferro-*

---

cap, fāce, fäther; best, bē, tèrm; pin, fīve;
rock, gō, ôrder; oil, out; cup, pùt, rüle,
yü in use, *yu* in uric;
ng in bring; *sh* in rush; *th* in thin; ᴛʜ in then;
zh in seizure.
ə = *a* in about, *e* in taken, *i* in pencil, *o* in lemon, *u* in circus

school

magnesian minerals crystallized out of the granite magma itself (Gilluly, *Principles of Geology*). **2** *Physics.* **a** areas in a medium where refraction varies as a result of differences in density. **b** the shadows cast on a screen when light is refracted by these areas so that it cannot hit the screen: *Tiselius equipped his apparatus with an optical system which rendered the boundaries visible. They showed up as shadows or schlieren, from which the system is known as the "schlieren method."* (Scientific American). —*adj. Physics.* of, having to do with, or using schlieren, especially to detect irregularities in heat convection, shock wave patterns, etc.: *Focused shadowgraphs were taken through windows. They are known as schlieren photographs and show the shock waves about the model* (Science News). [from German *Schlieren*, from *Schlier* marl]

school, *Zoology.* —*n.* a large number of the same kind of fish or water animals swimming together: *A true school of fish is a group in which all of the individuals are facing in a common direction, parallel to one another and regularly spaced.* (Science News Letter). —*v.* to swim together in a school: *Fishes that schooled near the surface died and were cast ashore, and birds that ate the poisoned fishes also died* (Lorus J. and Margery Milne). [from Dutch; related to *shoal*]

schorl (shôrl), *n. Mineralogy.* tourmaline, especially iron-rich black tourmaline: *Schorl ... is opaque, has a glassy luster, no cleavage, irregular fracture and a colorless streak* (Fenton, *The Rock Book*). [from German *Schörl*] —**schorlaceous** (shôr lā′shəs), *adj.* of the nature of, resembling, or containing schorl.

schreibersite (shrī′bər sīt *or* shrī′bər zīt), *n. Mineralogy.* a phosphide of iron and nickel occurring only in meteoritic iron. *Formula:* $(Fe,Ni)_3P$ [from German *Schreibersit*, named after Karl von *Schreibers*, Austrian museum director]

Schrödinger (wave) equation (shrœ′ding ər), *Physics.* a general equation of wave mechanics describing the behavior of atomic particles passing through a field of force: *The formulation of the famous "Schrödinger equation" put quantum theory on a strict mathematical basis, and provided the foundation for its further rapid expansion* (George Gamow).

Schwann cell (shwän *or* shvän), *Biology.* any of the cells that form a myelin sheath around the nerve fibers or axons in the peripheral nervous system: *The membrane of each Schwann cell compacts on itself to form a flattened sheet that wraps around the axon repeatedly* (Scientific American). [named after Theodor *Schwann*, 1810–1882, German anatomist]

Schwarzschild radius (shvärts′shilt *or* shwôrts′shilt), *Astronomy.* the size at which the gravitational forces of a collapsing body in space become so strong that they prevent the escape of any matter or radiation: *Eventually an object whose collapse continues reaches a limiting size that depends on its mass. The size is called the Schwarzschild radius ... When the object shrinks to less than its Schwarzschild radius, it becomes a black hole* (Science News). [named after Karl *Schwartzschild*, 20th-century German astrophysicist]

sci., *abbrev.* **1** science. **2** scientific.

sciaenoid (sī ē′noid), *Zoology.* —*adj.* of or belonging to a large group of spiny-finned carnivorous fishes usually with air bladders that make a drumming sound. It includes the drumfishes and some kingfishes. —*n.* a sciaenoid fish. [from Latin *sciaena* a kind of fish]

sciatic (sī at′ik), *adj. Anatomy.* of or in the region of the hip: *the sciatic artery.* The **sciatic nerve** is a large, branching nerve which extends from the lower back down the back part of the thigh and leg. [from Medieval Latin *sciaticus*, alteration of Latin *ischiadicus*, from Greek *ischion* hip joint]

science, *n.* **1** a systematized body of knowledge based on observation and experimentation: *biological or physical sciences, pure science.* **2** a branch of such knowledge dealing with the phenomena of the universe and their laws; a physical or natural science. **3** a particular branch of knowledge or study, especially as distinguished from art: *the science of climatology.* [from Old French, from Latin *scientia* knowledge, from *scire* know] —**scientific,** *adj.* **1** based on, regulated by, or done according to the methods of science: *a scientific arrangement of fossils, scientific research.* **2** of or having to do with science; used in science: *scientific instruments.* —**scientifically,** *adv.* in a scientific manner; according to the facts and methods of science.

scientific method, an orderly method used in scientific research, generally consisting of identifying a problem, gathering all the pertinent data, formulating a hypothesis, performing experiments, interpreting the results, and drawing a conclusion.

scientific notation, a short form of mathematical notation used by scientists, in which a number is expressed as a decimal number between 1 and 10 multiplied by a power of 10. The scientific notation for 500 is 5×10^2; for 800,000,000 it is 8×10^8; for 6,570,000,000,000 it is 6.57×10^{12}. Also called **standard notation.**

scientist, *n.* person who studies and has expert knowledge of some branch of science, especially a physical or natural science.

scintillation, *n. Physics.* a flash or spark of visible or ultraviolet light produced by ions in a phosphor when it is struck by a charged particle or high-energy photon.

scintillation counter, *Physics.* a device for counting radioactive particles by detecting the number of scintillations when ionizing radiation strikes a fluorescent material: *Unlike Geiger counters, which detect radiation through ionization in a gas ... a scintillation counter records radiations by means of tiny sparklike scintillations produced when the radiation strikes a crystal* (Scientific American).

scintillometer (sin′tə lom′ə tər), *n.* = scintillation counter.

scion (sī′ən), *n. Botany.* a bud, stem, or other section of a plant which is attached to a stock in grafting: *By using root fragments, named stocks, from a Siberian crab and stem-cuttings, termed scions, from a Jonathan apple, hundreds of grafts, each essentially a potential Jonathan tree, can be obtained* (Harbaugh and Goodrich, *Fundamentals of Biology*). [from Old French *cion*]

sciophilous (sī of′ə ləs), *adj. Botany.* growing or living by preference in the shade; shade-loving. [from Greek *skia* shadow + *philos* loving]

scissile (sis′əl), *adj. Chemistry.* capable of being broken: *a scissile bond.* [from Latin *scissilis,* from *scindere* to split]

scission (sizh′ən), *n. Chemistry.* the breakage of a bond, especially in a long-chain polymer so that two smaller chains result. [from Late Latin *scissionem* a splitting, division, from Latin *scindere* to split]

scissure (sizh′ər), *n. Anatomy.* a natural opening in an organ or part. SYN: fissure. [from Latin *scissura* a split, from *scindere* to split]

scler-, *combining form.* the form of **sclero-** used before vowels, as in *scleroid, sclerite.*

sclera (sklir′ə), *n. Anatomy.* the tough, white outer membrane which covers the eyeball, except for the part covered by the cornea: *The human eye ... is surrounded by a tough, white, outer skin, called the sclera ... At the front of the eye, the sclera is replaced by tough transparent membrane, called the cornea* (Hardy and Perrin, *Principles of Optics*). Also called **sclerotic coat** and **sclerotic.** [from New Latin, from Greek *sklēros* hard]
—**scleral,** *adj.* of or having to do with the sclera: *scleral tissue.*

sclereid (sklir′ē id), *n. Botany.* a thick-walled or lignified sclerenchyma cell, usually pitted and not elongate. Also called **stone cell.**

sclerenchyma (skli reng′kə mə), *n. Botany.* nonliving plant tissue composed of thickened and hardened cells from which the protoplasm has disappeared. It is found chiefly as a strengthening and protecting tissue in the stem and in such hard parts of plants as nut shells: *Sclerenchyma or stony tissue comprises a variety of supporting elements having thickened cell walls usually composed of lignocellulose* (Youngken, *Pharmaceutical Botany*). [from Greek *sklēros* hard + *en-* in + *chyma* what is poured]
—**sclerenchymatous** (sklir′eng kim′ə təs), *adj.* of or resembling sclerenchyma: *sclerenchymatous tissue, a sclerenchymatous polyp.*

sclerite (sklir′īt), *n. Zoology.* a hard plate, spicule, or the like, of an invertebrate animal, especially one of the plates of the exoskeleton of a grasshopper or similar arthropod, composed of chitin or calcium.
—**scleritic** (skli rit′ik), *adj.* of or resembling a sclerite; hardened or chitinized.

sclero-, *combining form.* **1 a** hard, as in *sclerodermatous = having a hard body covering.* **b** hardness, as in *sclerometer = instrument for measuring the hardness of a substance.* **2** having to do with the sclera, as in *scleroiritis = inflammation of the sclera and iris of the eye.* Also spelled **scler-** before vowels. [from Greek *sklēros* hard]

sclerocauly (sklir′ə kô′lē), *n. Botany.* a condition of plant stems in which they become slender, hard, and dry. [from *sclero-* + Greek *kaulos* stem]

sclerodermatous (sklir′ə dèr′mə təs), *adj. Zoology.* having a hard body covering, as of plates or scales.

scleroid (sklir′oid), *adj. Biology.* hard; indurated.

sclerometer (sklə rom′ə tər), *n.* an instrument for measuring the hardness of a substance, especially a mineral.

sclerophyllous (sklir′ə fil′əs), *adj. Botany.* having hard, leathery leaves which resist loss of moisture, as a result of a well-developed sclerenchyma: *True mediterranean vegetation is a dry evergreen woodland type of small size, the individual species of which are sparse, scrubby, and sclerophyllous* (White and Renner, *Human Geography*).

scleroprotein, *n. Biochemistry.* any of a group of fibrous proteins, such as collagen and keratin, obtained from bones, cartilage, and other animal tissues. Also called **albuminoid.**

sclerosed (skli rōst′), *adj. Biology.* affected with sclerosis; thickened, hardened: *sclerosed tissue. At times a clot, or thrombus, develops in a sclerosed artery and completely obstructs the flow of blood* (Harry L. Jaffe).

sclerosis (sklə rō′sis), *n., pl.* **-ses** (-sēz). *Biology.* **1** a hardening of a tissue or part of the body by an increase of connective tissue or the like at the expense of more active tissue.
2 a hardening of a tissue or cell wall of a plant by thickening or the formation of wood.
[from Medieval Latin, from Greek *sklērōsis,* from *sklēros* hard]

sclerotic (sklə rot′ik), *adj.* **1** *Anatomy.* of or having to do with the sclera.
2 *Biology.* of, with, or having sclerosis; hardened; stony: *sclerotic tissue.*
—*n.* = sclera.

sclerotic coat, = sclera.

sclerotin (skler′ə tin *or* sklə rō′tən), *n. Biochemistry.* any of a group of proteins which form the outer cuticles of insects. Sclerotins become hard and dark by a natural tanning process.

sclerotium (skli rō′shē əm), *n., pl.* **-tia** (-shē ə). *Biology.* a tubelike body of compact hyphae or protoplasmic material formed in certain fungi and slime molds, capable of becoming dormant. [from New Latin, from Greek *sklēros* hard]
—**sclerotial,** *adj.* of or having to do with the sclerotium or sclerotia.

sclerotize, *v. Biology.* to harden; become sclerosed; turn to sclerotin.
—**sclerotization,** *n.* **1** the process of sclerotizing: *In millipedes the cuticle is hardened both by sclerotization and by calcification and is extremely strong, an adaptation perhaps correlated with burrowing and pushing* (Science News). **2** = sclerosis.

scobiform (skō′bə fôrm), *adj. Botany.* having the form of or resembling sawdust: *scobiform seeds.* [from Latin *scobis* sawdust + English *-form*]

cap, fāce, fäther; best, bē, tèrm; pin, fīve;
rock, gō, ôrder; oil, out; cup, pùt, rüle,
yü in use, *yu* in uric;
ng in bring; *sh* in rush; *th* in thin; ŦH in then;
zh in seizure.
ə = *a* in about, *e* in taken, *i* in pencil, *o* in
lemon, *u* in circus

scolecite (skol′ə sīt *or* skō′lə sīt), *n. Mineralogy.* a hydrated silicate of calcium and aluminum, belonging to the zeolite family. Scolecite is found in needle-shaped crystals and fibrous or radiated masses. *Formula:* $CaAl_2Si_3O_{10} \cdot 3H_2O$ [from German *Skolezit,* from Greek *skōlēx* worm + German *-it* -ite]

scolex (skō′leks), *n., pl.* **scolices** (skō′lə sēz). *Zoology.* **1** the larva of a tapeworm or similar parasitic worm. **2** the head of the adult form: *... a series of many segments, or proglottids, which come off from the neck of the scolex and trail out into the intestine, usually for several feet* (Winchester, *Zoology*). [from New Latin, from Greek *skōlēx* worm]

-scope, *combining form.* a means or instrument for viewing, observing, or examining as in *microscope, telescope, stethoscope.* [from New Latin *-scopium,* from Greek *-skopion,* from *skopein* look at, examine]

scopula (skop′yə lə), *n., pl.* **-las** or **-lae** (-lē). *Zoology.* a small brushlike pad of stiff hairs as on the tarsi of bees and spiders. [from New Latin, from Latin *scopula* broom twig]

scopulate (skop′yə lāt), *adj. Zoology.* shaped like a broom or brush; brushlike.

-scopy, *combining form.* examination; observation, as in *cranioscopy = examination of the cranium.*

scoria (skôr′ē ə *or* skō′rē ə), *n., pl.* **scoriae** (skôr′ē ē *or* skō′rē ē). *Geology.* porous, cinderlike fragments of basaltic lava, usually darker and denser than pumice: *Scoria ... has been observed to form when rising bubbles of steam are caught by the congealing of the sticky lava around them* (Gilluly, *Principles of Geology*). [from Greek *skōria,* from *skōr* excrement]
—**scoriaceous** (skôr′ē ā′shəs *or* skōr′ē ā′shəs), *adj.* having to do with or consisting of scoria: *If the bubbles are larger, the structure is scoriaceous* (Fenton, *The Rock Book*).

scorodite (skôr′ə dīt), *n. Mineralogy.* a hydrous ferric arsenate, occurring in orthorhombic crystals and in earthy form, and usually of a greenish or brown color. *Formula:* $FeAsO_4 \cdot 2H_2O$ [from German *Skorodit,* from Greek *skorodon* garlic (so called because of its smell when heated) + German *-it* -ite]

scorzalite (skôr′zə līt), *n. Mineralogy.* a phosphate of iron and aluminum, occurring in crystalline or massive form. It is a blue mineral, isomorphous with lazulite. *Formula:* $FeAl_2(PO_4)_2(OH)_2$ [from E. P. *Scorza,* born in 1899, Brazilian mineralogist + *-lite*]

scotophil (skō′tə fil) or **scotophilic** (skō′tə fil′ik), *adj. Biology.* having an affinity for or requiring darkness: *Actually the circadian rhythm was imagined to comprise two half cycles each of approximately twelve-hours duration, one of which was reckoned to be dark-requiring (scotophil), the other light-requiring (photophil)* (New Scientist and Science Journal). [from Greek *skotos* darkness + English *-phil* or *-philic* attracted to, loving]

scotopia (skō tō′pē ə), *n. Physiology.* **1** adaptation of the eyes to darkness. **2** the ability to see in darkness. [from New Latin, from Greek *skotos* darkness + *ōpos* eye]

scotopic (skō top′ik), *adj. Physiology.* of, having to do with, or designating vision in darkness or dim light, involving only the rods of the retina: *The 5,000 angstrom pigment could be correlated with scotopic (low*

intensity) or rod sensitivity of the chicken (New Scientist).

scour, *Geology.* —*v.* cut or carve out by grinding and carrying away sediment: *The stream had scoured a channel.*
—*n.* scouring action: *The movement of the tides sets up a "scour" that cuts channels where they are not wanted and fills them where man has struggled to dig them* (N. Y. Times Magazine).

scree (skrē), *n. Geology.* a steep mass of loose rocky fragments lying at the base of a cliff or on the side of a mountain. Also called **talus.** [from Scandinavian]

screw, *n. Physics.* a cylinder with an inclined plane wound around it, and fitting into or making a threaded cylindrical hole. It is a simple machine, used to raise a load over the threads by applying a small force. Screws also produce motion. A **screw propeller** is a revolving hub with rotating blades that propels by acting as a screw in a fluid.

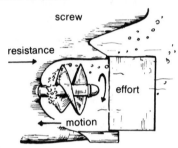

screw

resistance

effort

motion

scrotal (skrō′tl), *adj.* of or having to do with the scrotum.

scrotum (skrō′təm), *n., pl.* **-ta** (-tə), **-tums.** *Anatomy.* the pouch that contains the testicles: *The testes ... originate in the wall of the body cavity and descend through an opening in the lower abdominal wall to a sac-like scrotum* (Harbaugh and Goodrich, *Fundamentals of Biology*). [from Latin]

scrub, *n. Ecology.* land overgrown with stunted trees or shrubs, such as the Australian bush.

scutate (skyü′tāt), *adj.* **1** *Zoology.* having shieldlike plates or large scales of bone, shell, etc.
2 *Botany.* shaped like a shield: *Nasturtiums have scutate leaves.* [from Latin *scutatus* having a shield, from *scutum* shield]

scute (skyüt), *n.* = scutum: *The snake's bottom surface is covered by a series of scutes: wide scales whose free trailing edges overlap like plates of armor* (Scientific American).

scutellate (skyü′tl āt *or* skyü tel′āt), *adj. Zoology, Botany.* **1** having a scutellum or scutella.
2 formed into a scutellum.

scutellum (skyü tel′əm), *n., pl.* **-tella** (-tel′ə). *Zoology, Botany.* a small plate, scale, or other shieldlike part, such as on the feet of certain birds, the bodies of certain insects, or a cotyledon of some grasses. [from New Latin, diminutive of Latin *scutum* shield]

scutum (skyü′təm), *n., pl.* **-ta** (-tə). *Zoology.* **1** a shieldlike part of a bone, shell, etc., as on a turtle or armadillo. **2** any of the hard scales on the underside of a snake. [from Latin *scutum* shield]

scyphistoma (sī fis′tə mə), *n., pl.* **-mae** (-mē) or **-mas.** *Zoology.* a scyphozoan embryo that multiplies by budding, and gives rise to permanent colonies of scyphozoans: *The scyphistoma becomes divided into disks ... Each disk develops tentacles; and ... swims away as a minute medusa called an ephyra. The ephyra gradually develops into an adult jellyfish* (Hegner and Stiles, *College Zoology*). [from New Latin, from Latin *scyphus* cup + Greek *stoma* mouth]

scyphozoan (sī′fə zō′ən), *Zoology.* —*n.* any of a class (Scyphozoa) of marine cnidarians that include large jellyfishes having a bell-shaped, gelatinous body and long, trailing tentacles, and lacking a true polyp stage. —*adj.* of or belonging to the scyphozoans: *a scyphozoan polyp.* Compare **anthozoan, hydrozoan.**
[from Latin *scyphus* cup + Greek *zoion* animal]

scyphus (sī′fəs), *n., pl.* **-phi** (-fī). *Botany.* a cup-shaped part, such as the end of a lichen's fruit stalk or the corolla of a flower. [from Latin *scyphus* cup]

s.d. or **S.D.,** *abbrev.* standard deviation.

Se, *symbol.* selenium.

sea, *n.* **1** *Geography.* **a** the great body of salt water that covers almost three fourths of the earth's surface; the ocean. **b** any large body of salt water, smaller than an ocean, partly or wholly enclosed by land: *the North Sea.* **c** any of various relatively large landlocked bodies of fresh or salt water: *the Sea of Galilee, the Black Sea.*
2 *Oceanography.* a swell of the ocean due to wind: *a heavy sea.*
3 *Astronomy.* Often spelled **Sea.** one of the dark, flat plains on the moon once thought to be seas; mare: *The first astronaut on the moon . . . landed in the crater Alphonsus bordering on the lunar sea called Mare Nubium* (Science News Letter).
—*adj.* of, on, or from the sea; marine: *a sea animal, sea flora.*

seabed, *n.* = sea floor.

sea breeze, *Meteorology.* a breeze blowing from the sea toward the land: *Land and sea breezes are due to an unequal rate of heating and cooling of air across a shoreline* (Neuberger and Stephens, *Weather and Man*).

sea cave, *Geology.* a cave formed by the action of waves against a rocky shore.

seacoast, *n. Geography.* the land adjacent to the sea; shore or coast: *the seacoast of North America.* SYN: seashore.

sea fire, *Biology.* bioluminescence at sea, as that produced by certain luminescent marine flagellates.

sea floor, *Geology.* the floor or bottom of a sea or ocean.

sea-floor spreading, *Geology.* (in plate tectonics) the process by which the sea floor is being continuously formed and spread by upwellings from the earth's mantle along the mid-ocean ridges when crustal plates move apart, and continuously destroyed along the subduction zones, where plates push against each other and the sea floor sinks into the mantle: *In the study of the sea-floor spreading and plate tectonics, continental margins assume special significance, for it is there that plate interactions or early stages of spreading occur* (Science News).

sea gate, *Geography.* an entrance from the sea into a bay, harbor, or the like.

sea ice, *Oceanography.* frozen seawater. Sea ice has many saltwater pockets. *Both the theory and tests disprove the long-held rule that sea ice is only one-third as strong as fresh-water ice* (Science News Letter).

sea level, *Geography, Geology.* the surface of the sea, especially when halfway between mean high and low water. Mountains, plains, and ocean beds are measured as so many meters or feet above or below sea level. *Changes in mean sea level of the oceans possibly give the best indication that no drastic variations of the ice mass are taking place at present, but so many smaller factors affect sea level that it is relatively useless for measuring small changes of the Antarctic ice* (New Scientist).

seam, *n. Geology.* a layer or stratum, especially of a mineral; bed: *a seam of coal.*

seamount, *n. Geology.* a tall, cone-shaped hill or mountain arising from the sea bottom with its summit beneath the surface of the sea, and often having a flat top, as a guyot does: *Unlike the land variety of mountains—such as the Rockies or Alps which sprang largely from sediments folded by lateral pressure of landmasses, mid-ocean mountains, or seamounts, are built from basalt, an igneous rock that emerged as molten magma from within the earth, then cooled and solidified* (Igor Lobanov-Rostovsky).

seaquake, *n. Geology.* an earthquake originating at the bottom of the sea.

seashore, *n.* **1** = seacoast. **2** *Oceanography.* the area between the lines of ordinary high tide and ordinary low tide.

season, *n.* one of various regularly recurring periods or divisions of the year, especially spring, summer, autumn, or winter, characterized by specific meteorological or climatic conditions. Each season begins astronomically at an equinox or solstice, but popularly at various dates in different climates. In certain tropical regions there is only a rainy season and a dry season. —**seasonal,** *adj.* **1** having to do with the seasons: *seasonal variations in the weather.* **2** happening at regular intervals of the year: *Monsoon rains are seasonal in Asia and Africa.*

sea turn, *Meteorology.* a gale or breeze coming from the sea, generally accompanied by mists.

seawater, *n. Oceanography.* the saltwater of the sea or ocean: *Seawater can get as cold as 28 degrees without freezing because of the salt in solution* (Science News Letter).

seaweed, *n. Biology.* a plantlike organism or organisms growing in the sea, especially any of various marine algae.

sebaceous (si bā′shəs), *adj. Physiology.* having to do with or secreting a fatty or oily substance. **Sebaceous glands** are glands in the inner layer of the skin that se-

---

cap, fāce, fäther; best, bē, tèrm; pin, fīve;
rock, gō, ôrder; oil, out; cup, pút, rüle,
yü in use, *yu* in uric;
ng in bring; *sh* in rush; *th* in thin, ᴛʜ in then;
zh in seizure.
ə = *a* in about, *e* in taken, *i* in pencil, *o* in lemon, *u* in circus

crete sebum, the material that supplies oil to the skin and hair: *Sebaceous glands are distributed very unevenly . . . . For every gland on your leg, there are about 20 on your trunk and 100 on your face* (Marion B. Sulzberger). [from Latin *sebum* tallow, grease + English *-aceous*]

sebum (sē'bəm), *n. Physiology.* the secretion of the sebaceous glands, composed chiefly of fat, keratin, and cellular material: *Tiny glands at the sides of the hair follicles secrete sebum, an oil that lubricates the hair as it grows, and makes it soft and flexible* (W. V. Mayer and S. R. Friedman). [from Latin *sebum* tallow, grease]

sec or **sec.**, *abbrev.* **1** secant. **2** second *or* seconds.

secant (sē'kənt *or* sē'kant), *n.* **1** *Geometry.* a straight line that intersects a curve at two or more points.
2 *Trigonometry.* **a** the ratio of the length of the hypotenuse in a right triangle to the length of the side adjacent to an acute angle. It is the reciprocal of the cosine. **b** a straight line drawn from the center of a circle through one extremity of an arc to the tangent from the other extremity of the same arc. **c** the ratio of the length of this line to the length of the radius of the circle. *Abbreviation:* sec
—adj. *Geometry.* intersecting: *a secant plane.*
[from Latin *secantem* cutting]

geometry:

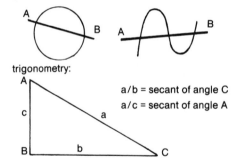

trigonometry:

a/b = secant of angle C
a/c = secant of angle A

second, *n.* **1** the SI, MKS, and CGS unit of time, a fundamental unit equal to 9,192,631,770 periods of the radiation corresponding to the transition between two levels of the ground state of the cesium-133 atom. *Symbol:* s
► The present measurement of the second, adopted by the International Committee on Weights and Measures in 1964, is accurate to 6 parts in 10^{12} or 1 second in 5000 years, and is several hundred times more accurate than the astronomical definitions formerly in use, in which the second was defined as 1/86,400 of a mean solar day (1/60 of a minute, 1/3600 of an hour) or 1/31556925.9747 of the solar year 1900 (ephemeris second).
2 a unit of angular measurement, equal to 1/60th of a minute, or 1/3600 of a degree of an angle. 12°10'30" means 12 degrees, 10 minutes, 30 seconds. *Two stars that are only a second apart require a fairly good telescope to show them separately, and yet the results of modern astronomical measurements are reliable to the hundredth of a second—the angle subtended by the di-*

ameter of a ten-cent piece at a distance of about two hundred miles (Duncan, Astronomy).
[from Old French *seconde*, from Medieval Latin *secunda* (*minuta*) second (minute), i.e., the result of the second division of the hour into sixty parts]

secondary, *adj.* **1** *Chemistry.* characterized or formed by the replacement of two atoms or groups. For instance, in secondary monohydric alcohol (CH_3CH_3CHOH), two of the hydrogen atoms of methyl alcohol (CH_3OH) have been replaced by alkyl groups ($-CH_3$). All secondary alcohols have the functional group -CHOH. In secondary sodium phosphate (Na_2HPO_4), two of the hydrogen atoms of phosphoric acid (H_3PO_4) have been replaced by sodium atoms. Compare **primary, tertiary.**
2 *Geology.* produced from another mineral by decay, alteration, or the like: *Under secondary ore deposits we group those that were developed by processes of erosion, weathering and sedimentation* (Fenton, *The Rock Book*). Compare **primary.**
—n. 1 *Astronomy.* **a** a celestial body that revolves around another, especially a satellite of a planet: *The moon is a secondary that revolves in an orbit around the earth, a primary. The earth, in turn, is a secondary that travels in an orbit around the sun* (Eric D. Carlson). **b** the dimmer star of a binary star system. Compare **primary.**
2 *Physics.* a subatomic particle produced by a cosmic ray as it enters the earth's atmosphere. Compare **primary.** See also **shower.**
3 = secondary feather.
4 = secondary color.

secondary color, *Optics.* a color produced by combining two primary colors. Also called **binary color.**

secondary consumer, *Ecology.* an animal, such as a fox or a hawk, that feeds on smaller, plant-eating animals. Compare **primary consumer.**

secondary electron, *Physics.* an electron given off by secondary emission. An anode struck by electrons flowing from a cathode will give off secondary electrons.

secondary emission, *Physics.* the freeing of electrons from the surface of a metal by bombardment with other electrons or ions: *Electrons can be knocked out of a metal or other substance by electrons as well as by positive ions. This is called secondary emission, the "secondary" electrons being the electrons knocked out, as distinguished from the primary electrons which do the knocking* (John R. Pierce).

secondary feather, *Zoology.* one of the flight feathers on the forearm of a bird's wing. Compare **primary feather.**

secondary radiation, *Physics.* radiation of cosmic rays, electrons, etc., produced as a result of the action of an earlier radiation on matter.

secondary root, *Botany.* a branch of a primary root: *This first root is called, appropriately, the primary root. Those which arise from it are sometimes known as secondary roots or simply as root branches* (Emerson, *Basic Botany*).

secondary sex character or **secondary sex** (or **sexual**) **characteristic,** *Biology.* any of the characteristics developed by males and females at puberty or at the breeding season as a result of hormonal secretions by certain cells in the reproductive glands. Secondary sex characters include differences in the size and shape of many male and female animals, the growth of breasts in females and of facial hair in human males, the growth of antlers in stages and of nuptial plumage in male birds, etc. These characteristics are called secondary because they are not directly concerned with reproduction.

secondary wave, *Geology.* an earthquake wave in which rock particles vibrate at right angles to the direction of travel. Also called **S wave.** Compare **primary wave, surface wave.**

second growth, *Ecology.* a growth or crop of vegetation replacing one previously cut or destroyed.

second quarter, *Astronomy.* **1** the period of time between the first half moon and the full moon. **2** the phase of the moon when it has revolved far enough to reveal its entire face; the full moon.

second sound, *Physics.* the rapid heat transfer of helium at very low temperatures by a motion like that of sound waves.

secreta (si krē′tə), *n. pl. Physiology.* the products of secretion; the substances secreted by a gland, follicle, etc. [from New Latin, from the neuter plural form of Latin *secretum* set apart, separated]

secretagogue (si krē′tə gôg), *n. Physiology.* a hormone or other substance that causes or stimulates secretion. [from *secreta* + Greek *agōgos* leading]

secrete, *v. Physiology.* to produce and discharge (a secretion): *to secrete digestive juices. Most people secrete the appropriate blood group substances (antigens) in bodily secretions such as saliva and tears* (Science News).

secretin (si krē′tin), *n. Biochemistry.* an intestinal polypeptide hormone that stimulates secretion of the pancreatic juice by the pancreas. Secretin is produced in the epithelial cells of the duodenum upon contact with acid.

secretion, *n. Physiology.* **1** a substance that is separated from the blood by a gland or organ, or is formed from materials supplied by the blood, in order to perform certain functions or to be excreted from the body. Secretions from various glands include saliva, mucus, tears, sweat, and bile. **Internal secretions** are the hormones formed by ductless (endocrine) glands and returned to the blood rather than discharged into a body part by a duct. **2** the act or process of producing and discharging such a substance: *the secretion of milk by the mammary gland, a decrease in the rate of secretion.*

secretory (si krē′tər ē), *adj. Physiology.* of or causing secretion; secreting: *Toward the end of pregnancy some secretory activity of the epithelial cells of the mammary glands becomes evident but secretion of liquid does not occur until parturition has been effected* (Harbaugh and Goodrich, *Fundamentals of Biology*).

secs., 1 seconds. **2** sections.

sectile (sek′tl or sek′tīl), *adj. Mineralogy.* that can be cut smoothly by a knife but cannot withstand pulverization: *If the specimen can be cut by a knife, like a piece of hard tar, and yet shatters under a sharp blow, it is sectile* (Scientific American).
—**sectility** (sek til′ə tē), *n.* the property of being sectile.

section, *n.* a thin slice of a tissue, mineral, or the like, cut off for microscopic examination: *There should be no need to cut the specimen up into thin sections in or-*

der to examine it, and this contrasts strongly with the case of the electron microscope, where sections of 1/10,000 mm. thickness are needed to avoid absorption of electrons (J. G. Thomas).

sector, *n. Geometry.* **1** the part of a circle between two radii and the included arc.
2 an instrument consisting of two rulers connected by a joint, used in measuring or drawing angles. [from Late Latin, from Latin *sector* cutter, from *secare* to cut]
—**sectorial,** *adj.* of, having to do with, or resembling a sector.

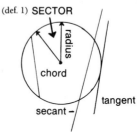

(def. 1) SECTOR

secund (sē′kund or sek′und), *adj. Botany.* arranged on one side only; unilateral: *The flowers of the lily of the valley and the false wintergreen are secund.* [from Latin *secundus* following, second (of a series)]

secundine (sek′ən dīn), *n. Botany.* the second or inner coat or integument of an ovule.

sedentary, *adj. Zoology.* **1** living in one place, as English sparrows do; not migratory or moving far: *Pigeons are sedentary birds.*
2 attached to one spot; not moving: *a sedentary mollusk.*
3 spinning a web and lying in wait: *I discovered that this was no web-spinning, sedentary spider, but a wandering hunter* (W. H. Hudson).
[from Latin *sedentarius* used to sitting, from *sedere* sit, settle]

sediment, *Geology.* —*n.* earth, stones, and other matter deposited by water, wind, or ice: *Not all sediments are of marine origin. Some appear to have been deposited in shallow coastal bays or marshes. Among these are organic deposits such as coal, iron-bearing sediments such as bog iron ore (limonite), and other forms of iron deposits. Some clearly have been deposited by streams or the wind, and still others are rocks believed to have resulted from deposits in the evaporating waters of interior basins or coastal lagoons in arid climates. Such are rock salt and gypsum* (Finch and Trewartha, *Elements of Geography*).
—*v.* to deposit as or form sediment.
[from Latin *sedimentum* matter that settles, dregs, from *sedere* settle]

cap, fāce, fäther; best, bē, tėrm; pin, five;
rock, gō, ôrder; oil, out; cup, pùt, rüle,
yü in use, yu̇ in uric;
ng in bring; sh in rush; th in thin, ŦH in then;
zh in seizure.
ə = a in about, e in taken, i in pencil, o in lemon, u in circus

sedimentological

—**sedimentary,** *adj.* formed by the deposition of sediment. Shale is a sedimentary rock. *Sedimentary rocks are derived not only from rock fragments but also from plant and animal remains* (Frederick H. Pough).
—**sedimentation,** *n.* a depositing of sediment: *The theory is that as the continents drifted apart, they deflected established oceanic currents, sending cooler water into equatorial regions where sedimentation is ordinarily heaviest* (N. Y. Times).

sedimentological, *adj.* of or having to do with sedimentology: *sedimentological studies of Tertiary strata.*

sedimentology, *n.* the branch of geology that studies the formation and structure of sediments and sedimentary rocks: *Sedimentology is the science of rocks that have accumulated from fragments of preexistent formations or from solutions through the agencies of wind, water, ice and chemistry* (New Scientist).

seed, *Botany.* —*n.* **1** the part of a plant from which a flower, vegetable, or other plant grows; a fertilized and mature ovule capable of germination into a plant similar to that from which it came: *A seed consists of a seed coat, which develops from the outer coverings of the ovule, and an embryo . . . The development of a seed into a mature plant capable of reproduction involves cell division, differentiation, and growth* (Biology Regents Syllabus). See the picture at **naked.** **2** any seedlike part of certain plants, such as the fruit of the strawberry. **3** a bulb, sprout, or any part of a plant from which a new plant will grow.
—*v.* **1** to sow (seeds): *Dandelions seed themselves.* **2** to produce seed; shed seeds: *Some plants will not seed in a cold climate.*

seed bud, *Botany.* **1** the part of a plant that develops into a seed; ovule. **2** the bud of a plant still in the seed; plumule.

seed capsule, = pericarp.

seedcase, *n. Botany.* any pod, capsule, or other dry, hollow fruit that contains seeds.

seed coat, *Botany.* the outer covering of a seed: *The seed has three important parts—a protective outer skin, or seed coat; an embryo, which will become the new plant; and a food supply, or endosperm, usually in the form of one, two, or many cotyledons, or seed leaves* (Vernon Quinn). See the pictures at **hypocotyl, monocotyledon.**

seed leaf, = cotyledon.

seedling, *n. Botany.* **1** a young plant grown from a seed. **2** a young tree less than three feet high or smaller than a sapling.

seed plant, *Botany.* any plant that bears seeds. Most seed plants have flowers and produce seeds in fruits; some, such as the pines, form seeds on cones. The spermatophytes are seed plants.

seedpod, *n. Botany.* a pod containing seeds, such as a legume.

seed vessel, = pericarp.

segment, *n.* **1** *Geometry.* **a** a part of a circle, etc., cut off by a line, especially a part bounded by an arc and its chord. **b** a part of a sphere cut off by two parallel planes. **c** = line segment.

2 *Biology.* one of the sections, parts, or divisions into which a body or organism is divided: *The body [in arthropods] is composed of a series of segments or 'somites', arranged along a longitudinal axis* (H. A. Nicholson).
—*v.* to separate or divide into segments: *The caudal muscles therefore consist of a 'dorsal' and 'ventral' muscle on each side, . . . transversely segmented by the membranous septa* (George M. Humphry).
[from Latin *segmentum,* from *secare* to cut]
—**segmental,** *adj.* **1** of, having to do with, or composed of segments: *a segmental organ, duct, etc., the segmental structure of an annelid.*
2 having the form of a segment of a circle: *a segmental arch.* —**segmentally,** *adv.*
—**segmentation,** *n.* **1** division into segments: *The leaves undergo a segmentation similar to that of the stem* (Sydney H. Vines).
2 *Biology.* the growth and division of a cell into two, four, eight cells, and so on: *Shortly after fertilization, within a time measured by minutes or hours in most animals, the fertilized egg begins to divide. This division, which is repeated in rapid succession until the egg is converted into many cells, is called cleavage or segmentation* (Shull, *Principles of Animal Biology*).

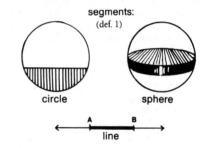

segments:
(def. 1)

circle   sphere

line

segmentation cavity, *Embryology.* the cavity of the blastula; blastocoele.

segregable, *adj. Genetics.* able to undergo segregation: *segregable characters.*

segregant, *n. Genetics.* an organism resulting from segregation: *After prolonged growth and division in culture, the scientists isolated segregants that contained the single parental number of chromosomes* (Frank B. Rothman).

segregate, *v. Genetics.* to undergo segregation: *In the progeny of hybrids characteristics derived from grandparents "segregate" or reappear* (S. M. Gruenberg).

segregation, *n. Genetics.* the separation of opposing (allelic) pairs of genes, occurring during meiosis: *One of Mendel's laws is that of segregation. It is now possible to restate this law in modern terms by saying that in the production of gametes only one gene of an allelic pair can appear in a given egg or sperm. Hence, when the parent carrying the determiners TT produces gametes, only one gene T, can be present in a given egg or sperm* (Emerson, *Basic Botany*).

seif (sif), *n. Geology.* a long, narrow dune found in sandy deserts in which the winds are variable: *Individual seifs as much as 60 miles long are known, and chains of seifs extend for more than 200 miles in western Egypt* (Gilluly, *Principles of Geology*). [from

Arabic *saif* sword (so called from the shape of the dune)]

seism-, *combining form.* the form of **seismo-** before vowels, as in *seismic.*

seismic (sīz'mik *or* sī'smik), *adj.* **1** of or having to do with earthquakes or other movements of the earth's crust: *seismic waves.*
2 caused by an earthquake or other movement of the earth's crust: *a seismic disturbance.* —**seismically,** *adv.*

seismicity (sīz mis'ə tē *or* sīs mis'ə tē), *n. Geology.* susceptibility to earthquakes; relative frequency and distribution of earthquakes in a given area: *One involves finding the number of earthquakes of this size in two places of well-measured seismicity, Southern California and New Zealand* (Bulletin of Atomic Scientists).

seismo-, *combining form.* earthquake, as in *seismology = the scientific study of earthquakes.* Also spelled **seism-** before vowels. [from Greek *seismos,* from *seiein* to shake]

seismogram, *n.* the record made by a seismograph.

seismograph, *n.* an instrument for recording the direction, intensity, and duration of earthquakes or other movements of the earth's crust.
—**seismographic,** *adj.* of or having to do with seismography or a seismograph.

seismography, *n.* **1** the art of using a seismograph.
2 the branch of seismology dealing especially with the mapping and description of earthquakes.

seismol., *abbrev.* **1** seismological. **2** seismology.

seismological, *adj.* of or having to do with seismology.
—**seismologically,** *adv.*

seismology (sīz mol'ə jē), *n.* the scientific study of earthquakes and other movements of the earth's crust: *Seismology has provided evidence that supports the plate tectonics theory* (C. E. Keen).

seismometer (sīz mom'ə tər), *n.* an instrument that detects and measures movements in the earth: *The heart of the seismometer is a pendulum, which serves as the sensing element. Its movement records the strength and character of each tremor of the earth* (Scientific American).

seismometric, *adj.* of or having to do with seismometry.

seismometry, *n.* the scientific recording and study of earthquake phenomena, especially by means of the seismometer.

selection, *n. Biology.* any natural or artificial process by which certain animals or plants survive and reproduce their kind, while others die or are prevented from breeding. Compare **natural selection, sexual selection.**

selenic (si lē'nik *or* si len'ik), *adj. Chemistry.* of or containing selenium, especially with a valence of six.

seleniferous (sel'ə nif'ər əs), adj. containing or yielding selenium, as ore.

selenious (si lē'nē əs), *adj. Chemistry.* of or containing selenium, especially with a valence of four.

selenite (sel'ə nīt), *n. Mineralogy.* a variety of gypsum, found in transparent crystals and foliated masses. [from Greek *selēnītēs* (*lithos*) (stone) of the moon, from *selēnē* moon (so called because its brightness was supposed to wax and wane with the moon)]

selenium (sə lē'nē əm), *n. Chemistry.* a nonmetallic element which exists in several forms that differ in physical and chemical properties, and is found with sulfur in various ores. Because its electrical conductivity increases with the intensity of light striking it, selenium is used in photoelectric cells. *Selenium is present in the earth's crust . . . Its presence in the soil . . . accounts for the fatal animal "alkali disease" whose cause remained a mystery for seventy-five years* (Offner, *Fundamentals of Chemistry*). *Symbol:* Se; *atomic number* 34; *atomic weight* 78.96; *melting point* 217°C; *boiling point* 684.9°C; *oxidation state* -2, $+4$, $+6$. [from New Latin, from Greek *selēnē* moon (so called because its properties follow those of *tellurium,* whose name derives from Latin *tellus* earth)]

selenocentric, *adj. Astronomy.* **1** having to do with the center of the moon.
2 with the moon as center: *The main purpose of the probe is to test a system for placing a spacecraft into a selenocentric or near-lunar orbit* (London *Times*).

selenodesy (sel'ə nod'ə sē), *n. Astronomy.* the study of the dimensions, gravity, and other physical characteristics of the moon: *A rapidly developing field appears to be lunar geodesy, a specialty the geodesists call "selenodesy"* (Science). [from Greek *selēnē* moon + English (geo)*desy*]

selenographic (si lē'nə graf'ik), *adj.* of or having to do with selenography. A **selenographic chart** is a map of the moon.

selenography (sel'ə nog'rə fē), *n. Astronomy.* **1** the science dealing with physical features of the moon: *Selenography, the name given to the branch of astronomy that is concerned with the study and mapping of the lunar features, dates back to early in the seventeenth century, when the newly invented telescope was first turned on the moon* (London *Times Literary Supplement*).
2 a map of the physical features of the moon: *In order to provide a basis for a discussion of the spots on the Moon we reproduce a selenography or drawing of the Moon* (Science Journal).
3 the geography of the moon: *Perhaps the best impression of the Moon's geography (or . . . selenography) . . . shows the general distribution of the craters and mountain ranges* (Marguerite Clarke).

selenological (si lē'nə loj'ə kəl), *adj.* of or having to do with selenology: *Tests with these [models] simulate the action of meteorites striking the Moon's surface in some ancient selenological period when it was still semi-molten* (New Scientist).

selenology (sel'ə nol'ə jē), *n.* the branch of astronomy that deals with the moon.

self, *Botany.* —*adj.* (of a flower) having the same color throughout.

---

cap, fāce, fäther; best, bē, tėrm; pin, fīve;
rock, gō, ôrder; oil, out; cup, pùt, rüle,
yü in use, *yu* in uric;
ng in bring; *sh* in rush; *th* in thin, ₮H in then;
zh in seizure.
ə = *a* in about, *e* in taken, *i* in pencil, *o* in lemon, *u* in circus

—*n.* a flower having the same color throughout.

—*v.* to fertilize (a flower) with its own pollen: *Several inbred lines from a parent rose are being produced by "selfing." Each of these lines will be further selfed* (Science News Letter).

self-adaptation, *n. Biology.* the adaptation of an organism in response to new conditions.

self-differentiation, *n. Biology.* the differentiation shown in organs, tissues, etc., that are relatively independent of neighboring structures in their development.

self-diffusion, *n. Physics.* the diffusion of atoms within a metal that is composed of identical atoms, especially at high temperature.

self-digestion, *n. Physiology.* digestion or breakdown of cells or tissue by enzymes produced by these cells. Compare **autolysis.**

self-fertility, *n. Botany.* the ability to fertilize itself.

self-fertilization, *n.* **1** *Botany.* the fertilization of the ovules of a flower by pollen from the same flower. Also called **autogamy.**
2 *Zoology.* the fertilization of two gametes from the same animal among some hermaphrodites: *Monoecious or hermaphroditic animals are provided with both male and female reproductive organs, and produce both eggs and spermatozoa; actually they are double-sexed. In some species the eggs of an individual are fertilized by spermatozoa from the same individual; this is called self-fertilization* (Hegner and Stiles, *College Zoology*). Compare **cross-fertilization.** —**self-fertile,** *adj.*

self-fertilized, *adj. Botany.* fertilized by its own pollen; autogamous: *a self-fertilized flower.*

self-fertilizing, *adj.* **1** *Botany.* fertilizing by the pollen of its own flowers alone: *Tobacco is normally a self-fertilizing species* (Science News).
2 *Biology.* breeding from closely related individuals; inbreeding: *populations of self-fertilizing organisms.*

self-luminous, *adj. Physics.* having in itself the property of giving off light, as the sun or flames: *The necessity for an outward flow of energy explains why a star must be self-luminous* (Scientific American). —**self-luminosity** *n. Bodies like radium that exhibit self-luminosity in the dark . . .* (Nature).

self-pollinate, *v. Botany.* to fertilize by transfer of pollen from the anther to the stigma of the same flower, or to the stigma of another flower or clove of the same plant. *Inbred lines [are] produced by continually self-pollinating the offspring originating from a single plant* (London *Times*). —**self-pollination,** *n.* See the picture at **pollination.**

self-replicating, *adj. Biology.* replicating oneself or itself: *Every molecular biologist believes that DNA is a self-replicating molecule* (New Scientist).

self-replication, *n. Biology.* replication of itself or by itself: *These organisms [mycoplasmas] which are even smaller than some viruses, are thought to possess only the minimum number of structures needed for self-replication and independent existence* (Scientific American).

self-sterile, *adj. Botany.* unable to fertilize itself, as certain flowers or plants: *The sweet cherry is called self-sterile, because the pollen from any one tree can-*

not pollinate the flowers on that tree or any tree of the same variety (Reid M. Brooks). —**self-sterility,** *n.*

sematic (si mat'ik), *adj. Biology.* serving as a sign or warning of danger: *sematic coloration.* Compare **aposematic.** [from Greek *sematos* sign]

semen, *n. Physiology.* the fluid produced in the male reproductive organs, normally containing the male reproductive cells (spermatozoa) that fertilize the ova. Also called **seminal fluid.** [from Latin *semen* seed]

semi-, *prefix.* **1** half, as in *semicircle = half circle.* **2** partly, as in *semiaquatic = partly aquatic.* [from Latin]

semiangle, *n. Geometry.* the half of a given or measuring angle.

semiaquatic, *adj. Botany, Zoology.* partly aquatic; growing or living close to water, and sometimes found in water.

semiarc, *n. Geometry.* half an arc.

semiarid, *adj. Ecology.* not completely arid; having sufficient rainfall to support short grasses or shrubs: *a semiarid region. Areas receiving between 10 and 20 inches of rain a year are semiarid. They are suitable for grazing and dry farming, but not for intensive agriculture except under irrigation* (Blair, *Weather Elements*). —**semiaridity,** *n.* semiarid nature or condition.

semiaxis, *n. Geometry.* half an axis, as of a hyperbola: *Superimposed on the precession of the earth's rotation axis in space are small irregularities called nutations. The principal one being an elliptical motion with semiaxes of nine and seven seconds of arc and a period of 18.6 years, which is caused by changes in the position of the moon's orbit* (Scientific American).

semicircle, *n. Geometry.* **1** either of the two identical figures formed by bisection of a circle.
2 the arc formed by half the circumference of a circle. —**semicircular,** *adj.* having the form of half a circle. —**semicircularly,** *adv.*

semicircular canal, *Anatomy.* any of three curved, tubelike canals in the inner ear that help to maintain balance. See the picture at **ear.**

semiconducting, *adj. Physics.* of, having to do with, or being a semiconductor: *a semiconducting crystal.*

semiconductor, *n. Physics.* a crystalline substance, such as germanium or silicon, that conducts electricity with an efficiency between that of metals and insulators. Semiconductors can convert alternating current into direct current and sunlight into electricity. They are the basic material of transistors and integrated circuits, which are used to build computers and other electronic devices. *Semiconductors are materials which can be tailor-made to conduct electrical current somewhat less efficiently than such a conductor as copper but much more efficiently than many other materials; it's this property that enables semiconductors to control or amplify electrical impulses* (Wall Street Journal).

semiconservative, *adj. Molecular Biology.* designating a form of replication in which the original molecular strands are conserved individually rather than together: *Watson and Crick . . . suggested that accurate duplication of DNA could occur if the chains separated and each then acted as a template on which a new complementary chain was laid down. This form of duplication was later called "semiconservative" because* it supposed that although the individual parental chains were conserved during duplication . . . their association ended as part of the act of duplication (John Cairns).

semicylindrical, *adj. Geometry.* shaped like or resembling a cylinder divided lengthwise.

semidouble, *adj. Botany.* having the outermost stamens converted into petals, while the inner ones remain perfect: *semidouble roses or poppies.*

semiellipse, *n. Geometry.* the half of an ellipse bisected by one of its axes, especially the major axis.

semiflexion, *n. Anatomy.* the posture of a limb or joint halfway between extension and complete flexion.

semifluid, *Physics.* —*adj.* partly fluid or liquid; having characteristics of both a fluid and a solid; flowing but very thick or viscous. SYN: semiliquid. —*n.* a semifluid substance.

semigroup, *n. Mathematics.* a closed set under an associative binary operation.

semiliquid, *adj., n.* = semifluid.

semilog, *adj.* = semilogarithmic.

semilogarithmic, *adj. Mathematics.* having or using a logarithmic scale or coordinates on one axis and a standard scale on the other: *a semilogarithmic graph.*

semilunar, *adj. Anatomy.* shaped like a half moon. The **semilunar cartilage** is either of the crescent-shaped cartilages of the knee joint. The **semilunar valves** are (1) a set of three crescent-shaped flaps or valves at the opening of the aorta that prevent the blood from flowing back into the ventricle. (2) a similar set of valves at the opening of the pulmonary artery. See the picture at **valve.**

semimajor axis, *Mathematics.* one half of the major axis of an ellipse or ellipsoid.

semimembranosus (sem'ē mem'brə nō'səs), *n., pl.* **-si** (-sī). *Anatomy.* one of the hamstring muscles, a long, semimembranous muscle of the back of the thigh. It flexes and rotates the leg and extends the thigh.

semimembranous (sem'ē mem'brə nəs), *adj. Anatomy.* (of a muscle) partly membranous; intersected by several broad, flat tendinous intervals.

semimetal, *n. Chemistry.* an element that is only partly metallic, as arsenic or tellurium: *An insulating crystal differs from a semiconductor only in having a large energy gap. A semiconductor with a very small energy gap (or a large amount of impurity) becomes a "semimetal"* (New Scientist).
—**semimetallic,** *adj.* partly metallic: *A compound of zinc and the semimetallic element germanium* (Harper's).

semimicro, *adj.* of, having to do with, or used in semimicrochemistry: *a semimicro centrifuge.*

semimicrochemistry, *n.* microchemistry that deals with centigram quantities.

semiminor axis, *Mathematics.* one half of the minor axis of an ellipse or ellipsoid.

seminal, *adj. Anatomy, Physiology.* of, having to do with, containing, or conveying semen. The **seminal duct** is the duct of the testis that carries semen outward, especially the part of the duct which runs from the epididymis to the ejaculatory duct. The **seminal vesicle** is either of two receptacles for holding semen, situated on each side of the base of the bladder. [from Latin *seminalis,* from *semen, seminis* seed] —**seminally,** *adv.*

seminal fluid, = semen.

seminiferous (sem'ə nif'ər əs), *adj. Anatomy, Physiology.* conveying, containing, or producing semen: *Spermatozoa are produced . . . inside the long coiled seminiferous tubules of the testis* (Richard J. Goss).

seminivorous (sem'ə niv'ər əs), *adj. Zoology.* eating or feeding on seeds: *seminivorous birds.*

semioviparous (sem'ē ō vip'ər əs), *adj. Zoology.* bearing living young so incompletely developed that they remain within the mother's pouch for a time. Kangaroos are semioviparous.

semipalmate or **semipalmated,** *adj. Zoology.* incompletely webbed; having the toes webbed only partway, as the feet of certain birds: *a semipalmate goose.*

semiparasite, *n.* = hemiparasite.

semiparasitic, *adj. Biology.* incompletely or partially parasitic: **a** usually living on a host but able to live on decaying or dead organic matter. **b** (of a plant) living on a host, but containing chlorophyll and therefore able to manufacture carbohydrates from the air, as mistletoe.

semipermeable (sem'i pėr'mē ə bəl), *adj.* partly permeable; permeable to some substances but not to others: *The plasma membrane is semipermeable.* Also called **differentially permeable.**

semiterrestrial, *adj. Biology.* partly terrestrial; not growing or living entirely on land: *Another report tended to dispel the common belief that there are no semiterrestrial shrimps and prawns* (Ronald R. Novales).

semitropical or **semitropic,** *adj.* = subtropical.
—**semitropics,** *n.pl.* = subtropics.

senary (sen'ər ē), *adj. Mathematics.* having six for the base: *the senary scale.* [from Latin *senarius* of six each, from *seni* six each]

senile, *adj. Geology.* having reached an advanced stage of erosion; made flat or level, as by the action of water or wind: *a senile valley.*

sensation, *n. Physiology.* a sensing or perceiving something by stimulation of a sense organ: *The visual sensation has three attributes: hue, saturation, and brightness. Taken together, these three are said to make up the sensation of color* (Sears and Zemansky, University Physics).

sense, *n.* **1** *Biology.* **a** any of the faculties by which stimuli from outside or inside the body are received and recognized by parts of the body. Sight, hearing, touch, taste, and smell are the five principal senses. *All our knowledge (or, at least, that part which is the concern of science) reaches us through the senses, and science, true to its first belief, attempts to construct a 'picture' of something responsible for our sensations* (R. Shil-

---

cap, fāce, fäther; best, bē, tėrm; pin, five;
rock, gō, ôrder; oil, out; cup, pùt, rüle,
yü in use, *yu* in uric;
ng in bring; *sh* in rush; *th* in thin, ᴛʜ in then;
zh in seizure.
ə = *a* in about, *e* in taken, *i* in pencil, *o* in lemon, *u* in circus

ton). **b** a receptor, or group of receptors, by which the body receives and responds to external or internal stimuli.

2 *Mathematics.* either of two opposite directions in which motion takes place: *Although, strictly speaking, emf [electromotive force] is not a vector quantity, it is useful to assign to it a direction . . . We shall arbitrarily consider the direction or sense of an emf to be from the − toward the + terminal of the seat, within the seat* (Sears and Zemansky, *University Physics*).

—*v.* **1** to perceive by the senses; experience as a sensation.

2 to detect or react to (signals, stimuli, coded data, or the like) automatically: *A photoelectric cell senses objects that a person cannot see.*

[from Latin *sensus,* from *sentire* perceive, know, feel]

sense organ, *Anatomy.* a specialized structure where sensory neurons are concentrated and which functions as a receptor. The eyes, ears, tongue, nose, and skin are sense organs.

sensible, *adj.* **1** that can be perceived by the senses: *the sensible horizon, the sensible atmosphere.*

2 capable of feeling or perceiving, as organs, tissues, or parts of the body.

3 sensitive to external influences, as a balance or thermometer.

sensillum, *n., pl.* **-la** (-lə). *Biology.* an elementary sense organ, as a single epithelial cell at the end of a sensory nerve fiber. [from New Latin, diminutive of Latin *sensus* sense]

sensitive, *adj.* **1** *Biology.* able to respond to stimulation from various external agents, such as light or gravity: *The human eye and certain photographic film are approximately 20 times more sensitive to green light than to red* (Gerald L. Pearson).

2 *Botany.* responding to external stimuli by moving, as the leaves of the sensitive plant.

3 a easily affected or influenced: *The mercury in the thermometer is sensitive to changes in temperature.* **b** easily affected by certain agents: *paper sensitive to white light.*

—**sensitivity, 1** the capacity of an organism or part to respond to stimuli. SYN: irritability.

2 the degree of this capacity: *The first step in discriminating between different wavelengths must take place in the retina, colour vision being based on differences of sensitivity between the retinal end-organs, linked to a cerebral ability to recognize which end-organs are responding and to what extent* (K. Tansely and R. A. Weale).

sensitize, *v. Immunology.* to make unusually sensitive to an antigen: *A single bee sting may sensitize a susceptible individual.*

—**sensitization,** *n.* the act, process, or result of sensitizing.

sensor, (sen′sər *or* sen′sôr), *n.* **1** a device for receiving and transmitting a physical stimulus such as heat, light, or pressure: *His final series of medical and psychological tests . . . included the attaching of medical sensors to his body for recording data on his pulse rate, respiration, and body temperature* (Chicago Sun Tribune). **2** = sense organ.

sensorimotor (sen′sər ē mō′tər), *adj. Physiology.* of or having to do with both the sensory and motor activity of the body: *The study . . . of sensorimotor performance came into prominence during the second World War as a result of the problems met in the human control of service equipment such as high-speed aircraft* (New Scientist).

sensorineural (sen′sər ē nùr′əl), *adj. Physiology.* of or having to do with the areas of sense perception controlled by sensory nerves: *Tests showed that the hearing loss is due to sensorineural impairment rather than to a middle ear lesion* (Science News Letter).

sensorium (sen sôr′ē əm), *n., pl.* **-soriums, -soria** (-sôr′ē-ə). *Physiology.* **1** a center in the brain that receives and interprets sensory stimuli: *The sensorium was after all the terminus, whereas he was immediately concerned with the pathway leading to that terminus* (London Times).

2 the whole sensory apparatus of the body.

[from Late Latin *sensorium* sense organ, from Latin *sensus* sense]

sensory, *adj. Biology.* **1** of or having to do with sensation or the senses. The eyes and ears are sensory organs. A **sensory area** or **center** is any area of the cerebral cortex associated with the receiving of sensations. **2** (of nerves, ganglia, etc.) conveying an impulse from the sense organs to the nerve center: *Thus we see that some of the nerves are sensory and pick up sensations from sense organs to carry them to the main nerve cords and brain* (Winchester, *Zoology*). Compare **motor.** See the picture at **reflex.**

sentence, *n. Mathematics.* a group of symbols that expresses a complete idea or a requirement; an equation or inequality. $4 + 2 = 6$ is a closed sentence expressing a complete idea. $x + 2 = 6$ is an open sentence expressing a requirement.

sep., *abbrev.* sepal.

sepal (sē′pəl), *n. Botany.* one of the leaflike parts which make up the calyx of a flower. The sepals are usually green and cover the unopened bud. In a carnation, the sepals make up a green cup at the base of the flower. In a tulip, the sepals are colored in the same way as the petals. [from New Latin *sepalum*]

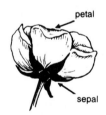

petal

sepal

sepaloid, *adj. Botany.* of or resembling a sepal.

sepalous, *adj. Botany.* having sepals.

sepsis (sep′sis), *n., pl.* **sepses** (sep′sēz). *Medicine.* **1 a** a poisoning of the system by disease-producing bacteria and their toxins absorbed into the bloodstream, as from festering wounds or putrefaction. **b** = septicemia.

2 the condition of being infected with pus-forming bacteria, such as streptococci or staphylococci.

[from Greek *sepsis* putrefaction]

ASSOCIATED TERMS: see **antisepsis.**

septal (sep′tl), *adj. Biology.* of or having to do with a septum: *The free inner margin of each septum is a thick convoluted septal filament, continued below as a threadlike acontium; both parts bear nematocysts and gland cells* (Storer, *General Zoology*).

septarian (sep tãr′ē ən), *Geology.* of, having to do with, or containing a septarium.

septarium (sep tãr′ē əm), *n., pl.* **-ia** (-ē ə). *Geology.* a roughly spherical concretion of minerals occurring in layers, as in sand or clay. A septarium is usually of calcium carbonate or carbonate of iron, having a network of cracks filled with calcite and other minerals. [from New Latin, from Latin *septum* divider, septum]

septate, *adj. Biology.* divided by a septum or septa: *All other filamentous fungi are more or less completely septate (that is, the cells are separated from each other by cross walls)* (Emerson, *Basic Botany*).

septic (sep′tik), *adj. Medicine.* of or having to do with sepsis; causing infection or putrefaction: *We did, of course, practice rigid asepsis (avoidance of infection) but septic surgery could be a nightmare* (Harper's).

septicidal (sep′tə sī′dl), *adj. Botany.* (of dehiscent seed capsules) bursting along the septa: *There are three kinds of valvular dehiscence: septicidal, loculicidal, and septifragal* (Youngken, *Pharmaceutical Botany*). [from *septum* + Latin *-cidere* to cut]

septicity, *n. Medicine.* septic character or quality: *... septicity in sewage* (New Scientist).

septifragal (sep′tə frā′gəl), *adj. Botany.* (of dehiscent seed capsules) characterized by the breaking away of the valves from the septa. Compare **septicidal, loculicidal.** [from *septum* + Latin *frag-*, stem of *frangere* to break]

septum (sep′təm), *n., pl.* **-ta** (-tə). *Biology.* a dividing wall; partition. There is a septum of bone and cartilage between the nostrils. The ventricles of the heart are separated by a septum. The inside of a green pepper is divided into chambers by septa. [from Latin *saeptum* a fence, from *saepire* hedge in]

sequence, *n. Mathematics.* a set of numbers, arranged in a specific order, that can be put into a one-to-one correspondence with the set of whole numbers in their natural order. The *n*th term of a sequence can be expressed as a function of *n*. EXAMPLE: $1/2, 1/4, 1/8, 1/16 ...$

—v. *Biochemistry.* to work out or determine the sequence of bases or other subunits in a molecule, compound, etc.: *Sophisticated new biochemical techniques for sequencing, or "reading," the genetic code stored in the DNA ... of the cell nucleus enable scientists to isolate and manipulate the genes controlling all aspects of an organism's structure and function* (Stephen M. Head).

sequential, *adj. Statistics.* of or based upon a method of sampling that is a continuous process, repeated with each increase in the number of samples: *sequential analysis. A further step in the same direction would lead to so-called sequential sampling in its ultimate form, where decisions will be made after each observation* (Parl, *Basic Statistics*). **—sequentially,** *adv.*

sequester, *v. Chemistry.* to prevent precipitation of (metallic ions) in solution by the addition of a chemical compound.

—sequestration, *n. There are substances called chelating agents which inactivate metal atoms by encircling them in large ring-shaped molecules, this phenomenon bearing the technical name of sequestration* (W. T. Read).

sequestrant, *n. Chemistry.* a compound that prevents the usual precipitation reactions of metallic ions in solution: *Many of the most common and most easily obtainable sequestrants are designed to soften water by pre-empting calcium and magnesium salts* (Newsweek).

Ser, *abbrev.* serine.

sera (sir′ə), *n.* a plural of **serum.**

serac (sā räk′), *n. Geology.* a large block or pinnaclelike mass of ice on a glacier, formed by the intersection of two or more crevasses: *Owing to their great speed, the surfaces of some of the large glaciers in Greenland and in the Himalayas are riddled with crevasses and broken up into seracs ... to a degree that makes human progress over them exceedingly slow and tedious* (M. F. Perutz). [from Swiss French *sérac*]

seral (sir′əl), *adj. Ecology.* of or having to do with a sere: *seral stages.*

sere (sir), *n. Ecology.* the complete series of ecological communities occupying a given area over hundreds or thousands of years from the initial to the final or climax stage. See also **succession.** [from *series,* perhaps influenced by Latin *serere* join, connect]

serein (sə ran′), *n. Meteorology.* a fine rain falling from an apparently clear sky, especially after sunset. The clouds may be too thin to see, or may be to the windward side. [from French, ultimately from Latin *serum* evening]

sericeous (sə rish′əs), *adj. Botany.* **1** consisting of or resembling silk. **2** covered with fine, silky hairs: *sericeous leaves.* [from Late Latin *sericeus,* from Latin *sericus* silken]

sericin (ser′ə sin), *n. Biochemistry.* a gelatinous protein that holds together the fibroin in raw silk. [from Latin *sericum* silk]

sericterium (ser′ik tir′ē əm), *n., pl.* **-teria** (-tir′ē ə). *Zoology.* a glandular apparatus in insects, especially silkworms, for the secretion of silk. [from New Latin, from Latin *sericum* silk]

series, *n., pl.* **series. 1** *Mathematics.* the sum of the terms of a sequence. EXAMPLES: $1/2 + 1/4 + 1/8 + 1/16 + ...$ is an infinite series. The series $2 + 4 + 6 + 8 + 10$ is a finite arithmetic series of the sequence 2, 4, 6, 8, 10. The series $2 + 4 + 8 + 16 + 32$ is a finite geometric series of the sequence 4, 8, 16, 32.
2 *Chemistry.* **a** a class of similar or related compounds, especially homologous organic compounds: *the methane series, the ethylene series.* **b** a sequence of elements of increasing atomic number; period or part of a period

---

cap, fāce, fäther; best, bē, tèrm; pin, fīve;
rock, gō, ôrder; oil, out; cup, pùt, rüle,
*y*ü in use, *y*ù in uric;
ng in bring; *sh* in rush; *th* in thin; ᴛʜ in then;
zh in seizure.
ə = *a* in about, *e* in taken, *i* in pencil, *o* in lemon, *u* in circus

in the periodic table: *These "Periods" are sometimes called "series" and correspond to the designations of the ... energy levels in which the first Period is the K energy level, the second Period the L energy level, the third is M, the fourth is N, and so on through the periods* (Parks and Steinbach, *Systematic College Chemistry*).

3 *Physics.* **a** a group of lines occurring in the emission spectrum of certain elements, especially hydrogen: *Under the proper conditions of excitation, atomic hydrogen may be caused to emit the sequence of lines ... called a series* (Sears and Zemansky, *University Physics*). **b** a number of circuit elements, connected to provide a single conducting path: *a group of resistors in series.*

4 *Geology.* **a** a division of rocks ranking below a system, containing the rocks formed during a geological epoch: *Two main divisions are recognized in the section located in western Texas that is taken as the standard of reference for study of American Permian formations. These divisions, which are classed as series, are separated from one another by unconformities* (Moore, *Introduction to Historical Geology*). **b** a division of soil groups: *The soil scientist ... divides each soil group into families, the families into series ... The series are divided into types ... They distinguish the soil differences of small areas, parts of farms, or even parts of fields* (Finch and Trewartha, *Elements of Geography*). [from Latin, from *serere* join]

serine (ser′ēn), *n. Biochemistry.* a colorless, crystalline compound present in many proteins or produced synthetically: *Serine is 'an amino-acid of particular significance in bacterial metabolism'* (J. A. Barnett). *Abbreviation:* Ser [from *ser(icin)* + *-ine*]

sero-, *combining form.* **1** serum, as in *seriodiagnosis = diagnosis by means of serums.*
2 serum and, as in *seromucous = containing serum and mucus.*
[from Latin *serum*]

serological or **serologic,** *adj.* of or having to do with serology: *Serological tests and examination by electron microscope have established that two viruses are distinct* (New Scientist). —**serologically,** *adv.*

serology (si rol′ə jē), *n.* **1** the scientific study of the properties and immunological reactions of serums.
2 a test performed on serum to diagnose a condition or detect some irregularity: *Food purveyors who attempt to dilute expensive meat with cheap meat ... cannot expect to get away with it if the meat is raw, because the various components are easily distinguished by serology* (New Scientist).

seromuscular, *adj. Anatomy.* having to do with both the serous and muscular coats of the intestine.

serosa (si rō′sə), *n. Biology.* **1** a serous membrane that lines the pericardial, pleural, and peritoneal cavities, enclosing their contents.
2 the embryonic envelope or chorion of reptiles and birds.
[from New Latin, from Latin *serum*]
—**serosal,** *adj.* of or having to do with a serous membrane: *a serosal lining.*

serotinous (si rot′ə nəs), *adj. Botany.* flowering or developing late in the season; late to blossom or fruit. [from Latin *serotinus,* from *sero* late]

serotonergic (sir′ə tə nėr′jik), *adj. Biochemistry.* producing or activated by serotonin, especially in transmitting nerve impulses: *... serotonergic pathways in the central nervous system* (Science). [from *seroton(in)* + *-ergic,* as in *adrenergic, cholinergic*]

serotonin (sir′ō tō′nən), *n. Biochemistry.* a compound found in various mammalian tissues, especially blood and nerve tissue. It causes blood vessels to constrict, stimulates smooth muscles, and is involved in the transmission of impulses between nerve cells: *Serotonin is formed, within some brain neurons, from tryptophan, an essential amino acid which the body cannot manufacture by itself. The brain obtains this amino acid from the bloodstream, which in turn gets it from the diet* (Richard J. Wurtman). *Formula:* $C_{10}H_{12}N_2O$ [from *sero-* + *toni(c)* + *-in*]

serotype, *Biology.* —*n.* a group of closely related organisms that have the same set of antigens: *We knew it was salmonella, but we didn't know the serotype—the species* (Berton Roueché). *The types were given numbers ... to eliminate confusion that might arise from the fact that a single serotype could produce clinically different diseases* (Leon J. LeBeau).
—*v.* to type organisms by the set of antigens they have in common; classify according to serotype: *[Of] 331 human cases of a particular Salmonella infection ... 271 have been traced by serotyping through the abattoir and back to the farm* (Manchester Guardian Weekly).

serous (sir′əs), *adj. Biology.* **1** of, having to do with, or resembling serum. A **serous fluid** is one that resembles blood serum.
2 producing or containing serum: *a serous inflammation.*

serous membrane, *Biology.* a thin membrane of connective tissue lining certain cavities of the body and moistened with a serous fluid. Compare **serosa.** ► See the note under **membrane.**

serpentine (ser′pən tēn), *n. Mineralogy.* a greenish rock-forming mineral with an oily luster, consisting chiefly of magnesium. It is sometimes spotted like a serpent's skin. Serpentine is a soft, waxy substance. A fibrous variety of serpentine (chrysotile) is the most important type of asbestos. *Formula:* $Mg_3Si_2O_5(OH)_4$

serpentinite (sėr′pən tē′nīt), *n. Geology.* a rock composed chiefly of the mineral serpentine.

serrate (ser′āt *or* ser′it), *adj. Botany, Zoology.* notched like the edge of a saw; toothed: *a serrate leaf.* [from Latin *serratus,* from *serra* a saw]

serrate leaf

serration (se rā′shən), *n. Botany, Zoology.* **1** a serrate edge or formation. **2** one of its series of notches. **3** a serrate formation.

serrula (ser'yə lə), *n., pl.* **-lae** (-lē), *Zoology.* a comblike ridge found on the appendages of arachnids, such as spiders, scorpions, and mites. [from Latin *serrula,* diminutive of *serra* a saw]

serrulate (ser'yə lit), *adj. Botany, Zoology.* very finely notched: *a serrulate leaf.*

serum (sir'əm), *n., pl.* **serums** or **sera** (sir'ə). *Biology.* **1** a clear, pale-yellow, watery part of the blood that separates from the clot when blood coagulates: *This liquid portion of the blood, known as serum, is the plasma minus the clotting proteins which were used in the formation of the clot* (Harbaugh and Goodrich, *Fundamentals of Biology*).
2 this substance, or any similar serous liquid, used to prevent or cure a disease, and usually obtained from the blood of an animal that has been made immune to the disease. Diphtheria antitoxin is a serum. *Serums containing substances that will fight the particular diseases are used for immunization among other things* (S. M. Gruenberg).
3 any watery liquid in animals. Lymph is a serum.
4 the watery substance of plants.
[from Latin *serum* whey, liquid]

serum albumin, *Biochemistry.* the albumin found in blood serum. It is the largest component of blood plasma and is a substitute for plasma. *Serum Albumin ... concentrates, as high as 25 percent, of the same viscosity as natural blood, are used regularly as plasma substitutes in cases of shock and severe burns* (Marguerite Clark).

serum globulin, *Biochemistry.* the globulin fraction found in blood serum.

sesamoid (ses'ə moid), *Anatomy.* —*adj.* of or having to do with certain small, oval, nodular bones or cartilages, as in the kneecap: *A sesamoid bone is not actually attached to a tendon as most other bones are, but rather is enclosed within the tendon itself* (Marshall R. Urist).
—*n.* a sesamoid bone or cartilage.
[from Greek *sēsamoiedēs* like sesame seed in form, from *sēsamon* sesame]

sesqui- (ses'kwi-), *combining form. Chemistry.* one and a half (usually applied to salts in which the proportions of one radical or element to another are 2:3), as in *sesquioxide.* [from Latin *sesqui-* one and a half]

sesquicarbonate, *n. Chemistry.* a salt whose composition is between a carbonate and a bicarbonate: *sodium sesquicarbonate.*

sesquioxide, *n. Chemistry.* a compound of oxygen and another element in the proportion of three atoms of oxygen to two of the other.

sesquiterpene, *n. Chemistry.* a terpene having one and a half times as many atoms in the molecule as a normal terpene. *Formula:* $C_{15}H_{24}$

sessile (ses'əl), *adj.* **1** *Botany.* attached by the base instead of by a stem: *The flower stalk of a single flower of an inflorescence is called a pedicel. When borne without such support the flower is sessile* (Youngken, *Pharmaceutical Botany*).
2 *Zoology.* sedentary; fixed to one spot. Some barnacles are sessile. *The majority of animals are free-living, able to move about independently, but some water dwellers are fixed in place, or sessile* (Storer, *General Zoology*).

[from Latin *sessilis* sitting, from *sedere* sit]
—**sessility** (se sil'ə tē), *n.* the condition of being sessile.

sessile leaves
(def. 1)

set, *n.* **1** *Mathematics.* a collection of numbers, points, or other elements which are distinguished from all other elements by specific common properties. The numbers from 0 to 10 form a set, and any number in this set is a member of the set. *Lines and planes are sets of points. In the language of set theory, combinatorial analysis is concerned with the arrangement of elements (discrete things) into sets, subject to specified conditions* (Martin Gardner).
2 *Meteorology.* the direction toward which a wind blows.
3 *Oceanography.* a series of large waves following closely a group of small waves: *Sets are the result of several factors. Chaotic wind gusts in the storm which raised the waves are one factor. The sorting process that converts a sea to swell is another* (Steve Lissau).
ASSOCIATED TERMS: Mathematical sets (def. 1) are classified according to the number of their members and the relationship between the sets. *Empty* or *null* sets have no members. *Equivalent* sets have the same number of members, and *equal* sets have identical members. *Disjoint* sets have no members in common, while *overlapping* sets have at least one member in common. A *finite* set has a limited number of members, as opposed to an *infinite* set with an unlimited number of members. A *universal* set or *universe* is the set of all members being considered at any one time. A *subset* is a set contained within another set. An *intersection* of sets includes only those members that belong to two or more sets. A *union* of sets includes all the members that belong to either or both of two sets. The *complement* of a set includes those members of a set that do not belong to a subset. A *solution set* is the set which contains all the solutions of an *open sentence.* A *replacement set* or *domain* is the set whose members are considered as possible replacements for a variable. Sets are the subject matter of *set theory.*

seta (sē'tə), *n., pl.* **-tae** (-tē). *Botany, Zoology.* a slender, stiff, bristlelike structure on an organism. Earthworms have four pairs of setae on each segment; locomotion is accomplished through the interactions of muscles and setae. [from Latin *seta, saeta* bristle]

cap, fāce, fäther; best, bē, tèrm; pin, five;
rock, gō, ôrder; oil, out; cup, pùt, rüle,
yü in use, *yù* in uric;
ng in bring; *sh* in rush; *th* in thin, ᴛʜ in then;
zh in seizure.
ə = *a* in about, *e* in taken, *i* in pencil, *o* in lemon, *u* in circus

591

setaceous (si tā'shəs), *adj. Botany, Zoology.* **1** bristlelike; bristle-shaped: *The cats ... and the seals, in which animals the long elastic setaceous whiskers are so useful as feelers* (T. Bell).
2 furnished with bristles; bristly: *a setaceous moth.*

setose (sē'tōs), *adj. Botany, Zoology.* bristly; setaceous.

set theory, the branch of mathematics that deals with sets, their properties, and their relationships: *Over the past century the study of sets, and of the relations between sets, has evolved into a theoretical structure that is one of the main branches of mathematics. Set theory is widely used in teaching to demonstrate how things can be grouped and how groups are related to one another* (Scientific American).

sex, *Biology.* —*n.* **1** either of the two divisions, female and male, into which the members of most species of animals are divided on the basis of their reproductive organs and functions: *In some animals, ... only differences in the gonads distinguish the sexes* (High School Biology). *Sex-limited traits are those due to certain genes which are capable of phenotypic expression in one sex but not in the other* (Hegner and Stiles, *College Zoology*).
2 either of a similar division in plants between organisms having female or male organs and functions.
3 the sum of the differences between female and male; maleness or femaleness: *While sex is an inherited character, it is not determined by a single gene in the same way the eye color is. Sex is determined by whole chromosomes rather than by single genes, at least in diploid organisms* (McElroy, *Biology and Man*).
4 the union of male and female organisms or gametes: *Sex is the method of recombining genes that occurs in cellular organisms with fully evolved chromosomal and mitotic equipment for transmission of the genetic material* (Simpson, *Life: An Introduction to Biology*).
—*adj.* of sex; having to do with sex: *sex distinctions, the sex organs, the sex attractant of an insect.*
—*v.* to determine the sex of: *We can sex infants before birth* (Donald Gould). *Crocodiles are very hard to sex* (New Scientist).
[from Latin *sexus*]

sexagesimal (sek'sə jes'ə məl), *Mathematics.* —*adj.* having to do with or based upon the number 60. A sexagesimal fraction is one whose denominator is 60 or a power of 60. *Each table is transcribed in the now standard adaptation of Hindu-Arabic numerals to the sexagesimal notation of the Babylonians* (Science).
—*n.* a sexagesimal fraction.
[from Latin *sexagesimus* sixtieth, from *sexaginta* sixty, from *sex* six] —**sexagesimally,** *adv.*

sex cell, *Biology.* an egg cell or sperm cell. Also called **gamete, germ cell.**

sex chromatin, = Barr body: *Female cells possess a characteristic material called sex chromatin, which can be recognized when the cells are stained and studied under the microscope* (London Times).

sex chromosome, *Biology.* either of a pair of chromosomes which in combination with each other determine sex and sex-linked characteristics: *In each human cell, there are 22 pairs of autosomes and one pair of sex chromosomes. The sex chromosomes are designated as "X" and "Y"* (Biology Regents Syllabus).

sexed, *adj. Biology.* having sex or the characteristics of sex: *a sexed animal or plant.*

sex gland, = gonad.

sex hormone, *Biochemistry.* a hormone that influences the development or stimulates the function of reproductive organs and secondary sexual characteristics: *The sex hormones are steroid chemicals. They are produced by the ovaries and the testes ... Scientists call the female sex hormones estrogens and the male hormones androgens* (Science News Letter).

sexless, *adj. Biology.* without sex; neuter: *Between the sexless, female and hermaphrodite states of these latter flowers, the finest gradations may be traced* (Darwin).

sex-limited, *adj. Genetics.* capable of phenotypic expression in one sex but not in the other: *Sex-limited characters are sometimes classed as secondary sexual characters. Examples are: beard in man; bright-colored plumage in certain male birds* (Harbaugh and Goodrich, *Fundamentals of Biology*).

sex-linkage, *n. Genetics.* the condition of being sex-linked: *Any animal or plant whose sex is determined by chromosomes, and in which, as a consequence of this chromosome relation, the male produces two kinds of spermatozoa, may be expected to show sex-linkage ... Man is one of these animals* (Shull, *Principles of Animal Biology*).

sex-linked, *adj. Genetics.* **1** located in a sex chromosome: *Traits such as color blindness are said to be sex-linked, since their genes are borne by the X chromosome that also bears the genes influencing sex* (Hegner and Stiles, *College Zoology*).
2 of, having to do with, or designating a character, such as hemophilia, that is transmitted by genes located in the sex chromosomes: *Transmission of the defect is not sexlinked but women are more likely to have acute attacks and men more likely to show skin effects* (Geoffrey Dean).

sex ratio, *Biology.* the proportion of males to females in a given population, usually stated as the number of males per 100 females. In humans, the sex ratio is 103 to 107 (that is, 103 to 107 males are born to every 100 females born).

sexual, *adj. Biology.* **1** of or having to do with sex. In sexual reproduction, animals and plants reproduce their own kind by the union of the male and female sex cells or gametes.
2 possessing sex; sexed; separated into two sexes: *Plants are naturally and primarily divided into two great divisions, called Sexual and Asexual* (John Lindley).
ASSOCIATED TERMS: Sexual reproduction in animals involves *gametogenesis* (consisting of *meiosis, oogenesis,* and *spermatogenesis*), *fertilization* (including *external* and *internal fertilization*), and *embryogenesis* (consisting of *cleavage, gastrulation, differentiation,* and *growth*). Sexual reproduction in flowering plants involves *pollination, fertilization,* and *germination.* See also the associated terms under **gametogenesis.**
—**sexuality,** *n.* the quality of being sexual; possession of the male or female reproductive characteristics.

sexual generation, *Biology.* the sexual phase in the alternation of generations; gametophyte.

sexual selection, *Biology.* natural selection perpetuating certain characteristics that attract one sex to the other, such as bright feathers in birds: *I conclude that of all* *the causes which have led to the differences in external appearance between the races of men ... sexual selection has been by far the most efficient* (Darwin, *Descent of Man*).

Seyfert galaxy (sē'fərt *or* sī'fərt), *Astronomy.* any of a group of spiral galaxies having very small, starlike centers which exhibit broad emission lines indicative of a high state of atomic excitation: *A possible link between normal galaxies and quasars may be provided by the Seyfert galaxies, which have unusually bright nuclei similar in many ways to the sharply defined quasars* (Science Journal). [named after Carl K. Seyfert, 1911–1962, American astronomer who listed and described ten of these galaxies in the 1940's]

s.g., *abbrev.* specific gravity.

shadow band, *Astronomy.* any of a series of roughly parallel broken bands, alternatively bright and dark, that move with an irregular, flickering motion over light-colored surfaces during a solar eclipse. It is seen just before and after the period of total eclipse. *The fact that shadow bands are not visible at all eclipses proves satisfactorily their atmospheric origin* (New Scientist).

shale, *n. Geology.* a fine-grained sedimentary rock, formed from hardened clay, mud, or silt in thin layers which split easily: *Shales accumulate in many different environments. As the main load brought down to the sea by great rivers is mud, it is not surprising that shale is the most abundant marine sedimentary rock ... Many shales are black, perhaps because of large amounts of carbon-rich organic matter in various stages of decomposition or due to the precipitation of black iron sulfide ... by sulfur bacteria* (Gilluly, *Principles of Geology*). [Old English *scealu* shell]

shamal (shə mäl'), *n. Meteorology.* a cold northwest wind that periodically blows across parts of central Asia and the Persian Gulf. [from Arabic *shamāl* north, north wind]

shatter cone, *Geology.* a cone-shaped rock fragment with distinctive ridges, produced by intense shock forces from the impact of a meteor: *It would be necessary to see in more detail the occurrence of tektites in relation to that of shatter cones before the two could be linked in one theory* (New Scientist).

—shatter-coned, *adj.* having or characterized by shatter cones: *shatter-coned structure.*

shear, *n.* **1** *Physics.* **a** a force causing two parts or pieces to slide on each other in opposite directions. **b** the strain or deformation resulting from this; shearing stress: *The third type of elasticity ... is one in which the shape of a body is changed without change in the volume of the body. The type of deformation involved is called a shear* (Shortley and Williams, *Elements of Physics*).

2 *Meteorology.* a change in the direction and speed of wind due to differences in temperature, altitude, etc.: *They have revealed regions of intense turbulence and strong wind shears, which are layers of winds blowing in different directions at different altitudes* (Science News Letter).

shearing plane, *Geology.* the plane along which rupture from shearing stress takes place in rocks.

shearing stress or **shear stress,** *Physics.* the stress in a body caused by shear: *With present technology shear stresses close to ultimate theoretical limits can be*

achieved for very short times in many solids (Ronald K. Linde).

shear plane, = shearing plane.

shear wave, *Physics.* a wave in an elastic medium which causes movement of the medium but no change in its volume. A secondary wave is a shear wave. *The mantle, unlike the core, was found to be solid; it could transmit shear waves—waves of torsion* (Listener).

sheath, *n.* **1** *Botany.* the part of an organ rolled around a stem or other body to form a tube: *In the grasses, the blade arises, not from a petiole, but from a sheath* (Emerson, *Basic Botany*).

2 *Zoology.* the elytron of a beetle.

sheet, *n. Geology.* a nearly horizontal layer of igneous or sedimentary rock: *Sheets, or extrusive sheets, are hardened lava flows that spread widely, most commonly from cracks or fissures. If the lava was fluid and flowed easily, it made sheets that were thin and smooth* (Fenton, *The Rock Book*).

sheet erosion, *Geology.* the washing away of soil in layers from barren, sloping land by rainfall.

sheet lightning, *Meteorology.* lightning in broad flashes. Sheet lightning is actually a reflection of lightning that occurs beyond the horizon.

shelf, *n. Geology.* **1** a ledge of land or rock, especially a submerged ledge or bedrock: *In the Gulf the shelf—the offshore sea bottom—is muddy* (Scientific American). **2** = continental shelf.

shelf fungus, *Biology.* any basidiomycete fungus, either saprophytic or parasitic, that grows in a perpendicular plane on a tree trunk or limb, where it promotes wood rot.

shelf ice, *Geology.* a ledge of ice sticking out into the sea from an ice sheet: *Some species of marine plants even manage to thrive in the lightless waters beneath the shelf ice* (Scientific American).

shelf sea, *Oceanography.* the part of the sea that covers a continental shelf.

shell, *n.* **1** *Zoology.* **a** the hard outer covering of certain animals. Oysters and other mollusks, beetles and some other insects, and turtles have shells. *The largest shell is that of the giant clam of the South Pacific Ocean* (R. Tucker Abbott). SYN: carapace. **b** the hard outside covering of an egg: *The shell is produced by lime-producing glands in the lower part of the oviduct. The egg is laid soon after the shell forms* (Otto and Towle, *Modern Biology*).

2 *Botany.* the hard outside covering of a nut, seed, or fruit.

3 *Geology.* **a** a hard, thin layer of rock. **b** the earth's crust.

4 *Physics, Chemistry.* **a** a pattern of electrons surrounding a nucleus. All of the electrons in a shell have the same principal quantum number and approximate-

cap, fāce, fäther; best, bē, tèrm; pin, fīve;
rock, gō, ôrder; oil, out; cup, pùt, rüle,
yü in use, *yủ* in uric;
ng in bring; *sh* in rush; *th* in thin, ŦH in then;
zh in seizure.
ə = *a* in about, *e* in taken, *i* in pencil, *o* in lemon, *u* in circus

ly the same energy. *The electrons group themselves in various "shells" about the nucleus. Each shell seems able to hold a certain maximum number of electrons.* (Sears and Zemansky, *University Physics*). Also called **principal energy level. b** an analogous pattern of protons and neutrons within the nucleus. See also **shell model, magic number.**

shellfish, *n., pl.* **-fishes** or (collectively) **-fish.** *Zoology.* a water animal with a shell, especially a mollusk or a crustacean that is used for food. Clams, oysters, crabs, and lobsters are shellfish. Shellfish are not true fish; they lack a backbone and look very different from regular fish.

shell model, *Nuclear Physics.* a theory of nuclear structure used to explain the relative stability of various nuclei. See also **magic number.**

shield, *n.* **1** *Geology, Geography.* a large area of exposed igneous and metamorphic rocks, usually Precambrian, in a craton: *the Canadian Shield. The broad lowland outcrop areas constitute so-called "shields," which are the nuclear parts of the continents; these are bordered by nearly flat-lying sedimentary formations.* (Moore, *Introduction to Historical Geology*).
2 *Zoology.* a protective plate covering a part on the body of an animal, such as a scute or carapace.

shield volcano, *Geology.* a low, broad, dome-shaped volcano built up of overlapping basaltic lava flows and often covering a very large area: *Mauna Loa, on the island of Hawaii, is a shield volcano.*

shift, *n. Geology.* a slight fault or dislocation in a seam or stratum.

shinbone, *n.* = tibia.

shoal, *n. Geology.* a sandbank, sand bar, or ledge of rock, coral, or the like, that makes the water shallow especially one that can be seen at low tide: *It was found that the rounded hills and promontories within and bordering the ice shelf are composed entirely of ice, much thicker than the floating shelf, that is grounded on shoals* (E. F. Roots).

shock front, 1 *Physics.* the outer part of a shock wave, at which pressure reaches the highest point: *It can be seen that at the advancing front of the wave, called the shock front, there is a very sudden drop of pressure, to that of the surrounding atmosphere. The shock front thus behaves like a moving wall of highly compressed air or water* (F. Reines).
2 *Astronomy.* a region in which a moving ionized gas medium meets a magnetic field, resulting in a shock wave: *The position of the shock front varies with the velocity of the solar wind, moving closer to the Sun during solar quiescent times and closer to the Earth during solar active periods. This in-out movement of the shock front is associated with the amounts of the terrestrial field which can advance with the rotating Earth to the sunward side* (L. O. Quam).

shock wave, *Physics.* a compression wave produced by a sudden change in pressure and particle velocity: *If the disturbance is small the result is an ordinary sound wave, but if the blow is severe, such as occurs with a massive body rushing rapidly through the air or with an explosion, the result is a definite atmospheric discontinuity called a shock wave which travels initially with a speed somewhat greater than that of sound* (O.

G. Sutton). *The passage of a shock wave through a solid can redistribute atoms, change the levels of electron energy, and alter the internal energy balance* (Ronald K. Linde).

shooting star, *Astronomy.* a common term for a meteor, especially one seen falling or darting through the sky at night: *Meteors are small, solid celestial bodies which are invisible, except those that enter the earth's atmosphere and are heated to incandescence by impact of the air molecules. Then they appear momentarily as the streaks of light across the night sky that have long been known as shooting stars* (Baker, *Astronomy*).

shore, *n. Geography.* land at the edge of a sea, lake, river, or other body of water.

shoreline, *n. Geography.* the line where land and water meet: *The shorelines of hill lands are different in many significant ways from those which characterize the meeting of plains with the sea. The steep and often serrate shores of hill lands commonly are bordered by sea bottom which slopes less gently than the continental shelf that borders plains* (Finch and Trewartha, *Elements of Geography*).

short-day plant, *Botany.* a plant that blooms only when its daily exposure to light is relatively short, as in the spring or late fall when the days are short and the nights are long. Sugar cane, wild strawberries, violets, and poinsettias are some short-day plants. *In contrast to corn, teosinte is a "short-day" plant, blooming only in seasons when the days are appreciably shorter than the nights* (Paul C. Mangelsdorf). Compare **long-day plant, day-neutral plant.**

short division, *Mathematics.* a method of dividing numbers in which each step of the division is worked out mentally. It is used to divide small numbers. Compare **long division.**

shoulder, *n. Anatomy.* **1** the part of the body to which an arm, foreleg, or wing is attached. **2** the joint by which the arm or foreleg is connected to the trunk.

shoulder blade, *Anatomy.* the flat, triangular bone of either shoulder, in the upper back. Also called **scapula.**

shower, *n.* **1** *Meteorology.* **a** a short fall of rain, characterized by a sudden start and stop: *Relatively abrupt and rapid fluctuations in intensity are characteristic of showers. Generally, individual showers are of comparatively short duration, although successive showers may extend over a period of many hours* (Neuberger and Stephens, *Weather and Man*). **b** a similar fall of snow, sleet, etc.: *a shower of hail.*
2 *Physics.* a sudden burst of subatomic particles, produced by the interaction of a cosmic ray with a nucleus high in the atmosphere. See also **primary, secondary.** [Old English *scur*]

shrub, *n. Botany.* a perennial woody plant smaller than a tree, usually with many separate stems starting from or near the ground: *Such well-known plants as roses, lilacs, blackberries, hazelnuts, sumacs, the smaller willows, and many of the foundation plantings about buildings are shrubs. This form of woody plant often makes up the chief undergrowth in forests, along stream courses, and even extends far beyond the forests into the grasslands and deserts* (Emerson, *Basic Botany*). [Old English *scrybb* brush]
ASSOCIATED TERMS: see **HERB.**

Si, *symbol.* silicon.

SI, *abbrev.* Système International d'Unités (International System of Units). See **SI unit.**

sial (sī'al), *n. Geology.* granitelike rock rich in silicon and aluminum. Sial is the chief rock underlying the land masses, as distinct from the ocean basins. *Evidence from these and other sources suggests that all the continents have foundations of relatively light rock ('sial') rich in silica and alumina which are, as it were, floating on an underlying layer of denser rock ('sima')* (New Biology). [from *si(lica)* + *al(uminum)*] —**sialic** (sī al'ik), *adj.* **1** composed largely of silicon and aluminum: *Granite is the chief sialic rock.*
2 consisting of sial: *a sialic land basin.* Compare **simatic.**

sialic acid, *Biochemistry.* any of a group of amino sugars that are components of polysaccharides and mucoproteins and are present in many bacteria and animal tissues: *The glucose is first transformed into the simple sugars from which the carbohydrate chains in glycoprotein are formed, such as galactose ... and sialic acid* (Scientific American). [from Greek *sialos* saliva + *-ic*]

sialogogic (sī al'ə goj'ik), *adj. Physiology.* stimulating or provoking an increased flow of saliva. [from Greek *sialon* saliva + *agogos* leading]

sialomucin (sī'ə lō myü'sin), *n. Biochemistry.* a mucopolysaccharide having sialic acid as its acid component.

sibling species, *Biology.* any of various species of animals that appear to be morphologically identical but do not interbreed with one another: *The genetic isolation undoubtedly is the first step in divergent evolution, and these [nearby] populations represent distinct species at a very early stage. They are referred to as sibling species* (Weier, Botany).

sickle cell, *Biology.* a red blood cell that is sickle-shaped instead of round because of an abnormality in the hemoglobin. It is characteristic of various chronic genetic diseases, especially sickle-cell anemia. *The sickle cell is a genetic mutation that occurred long ago in malarial regions of Africa* (Scientific American).

side chain, *Chemistry.* a chain of atoms attached to the principal chain in the structure of a molecule: *One approach is to "graft" synthetic side-chains on to a cellulose backbone* (New Scientist). Compare **lateral chain.**

sidereal (sī dir'ē əl), *adj. Astronomy.* **1** of or having to do with the stars or constellations: *a sidereal system.* SYN: astral, stellar.
2 measured by the apparent daily motion of the stars; based on sidereal time: *A sidereal day is shorter than a mean solar day by 3 minutes and 56 seconds* (Eric D. Carlson). *A sidereal month—27 1/3 days—is the time the moon takes to make one trip around the earth in relation to the stars* (Eugene M. Shoemaker). ▶ See the note under **day.**
[from Latin *sidereus* of the stars, astral, from *sidus* star]

sidereal time, *Astronomy.* time measured by the stars, or the hour angle of the vernal equinox. A sidereal day is one rotation of the earth in reference to any star or the vernal equinox at the meridian, and is about 4 minutes shorter than a mean solar day. A sidereal day consists of 23 hours, 56 minutes, and 4.09 seconds of mean solar time. *Although sidereal time is suited to many activities of the observatory, it is not useful for civil purposes, because our daily affairs are governed by the*

sun's position in the sky, and not by the vernal equinox. Sidereal noon, for example, comes at night during a part of the year (Baker, Astronomy).

sidereal time (def. 2)

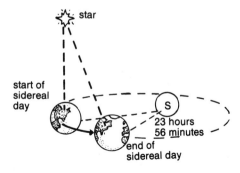

siderite (sid'ə rīt), *n.* **1** *Mineralogy.* an iron ore occurring in various forms and colors and crystalline with perfect rhombohedral cleavage. *Formula:* $FeCO_3$
2 *Astronomy.* a meteorite consisting mainly of iron and nickel. Compare **siderolite.**
[from Greek *sidēros* iron] —**sideritic** (sid'ə rit'ik), *adj.* of the nature of siderite.

siderochrome (sid'ə rə krōm'), *n. Biochemistry.* a compound that transports iron across a cell membrane into the cell, where it is metabolized. [from Greek *sidēros* iron + English *chrome*]

siderolite (sid'ər ə līt), *n. Astronomy.* a meteorite composed of a mixed mass of iron and stone.

siderophile (sid'ər ə fīl), *adj. Geology.* having an affinity for metallic iron: *siderophile elements.* Compare **atmophile.**

siemens (sē'mənz), *n.* the SI unit of electrical conductance, equal to the conductance of a body through which one ampere of current flows when the potential difference is one volt. It is the reciprocal of the ohm. *Symbol:* S [named after Ernst Werner von *Siemens*, 1816–1892, German electrical inventor]

sierozem (syer'ə zem *or* syer'ə zyôm'), *n. Geology.* a type of soil found in temperate or cool arid regions, characterized by a brownish-gray surface covering a lighter layer and supporting sparse, shrubby vegetation. It is low in humus and its lime is near or on the surface. *The gray sierozem soils of many semi-deserts in the West ... include much residual clay* (Fenton, The Rock Book). Compare **chernozem.** [from Russian *serozyom,* from *seryi* gray + *zemlya* earth, soil]

sierra (sē er'ə), *n. Geography.* a chain of hills or mountains whose peaks suggest the teeth of a saw: *The road wound up the bold sierra which separates the great*

cap, fāce, fäther; best, bē, tėrm; pin, fīve;
rock, gō, ôrder; oil, out; cup, pùt, rüle,
yü in use, *yù* in uric;
ng in bring; sh in rush; th in thin, ŦH in then;
zh in seizure.
ə = a in about, e in taken, i in pencil, o in
lemon, u in circus

sieve cell

plateaus of Mexico and Puebla (W. H. Prescott). [from Spanish *sierra* (literally) a saw]
—sierran, *adj.* of or having to do with a sierra or sierras.

sieve cell, *Botany.* an elongated cell whose thin walls have perforations which allow communication between adjacent cells of a similar nature. Sieve cells form the essential element of the phloem of gymnosperms and lower vascular plants. *Sieve cells are the only type of food-conducting cell in most lower vascular plants and gymnosperms* (Raven, *Biology of Plants*).

sieve of Eratosthenes (er'ə tos'thə nēz'), *Mathematics.* a method for finding prime numbers by writing down a series of whole numbers, beginning with 2, crossing out all the second numbers except 2, all the third except 3, and so on until all but the prime numbers remain. The method was invented by Eratosthenes, 276?–194? B.C., a Greek mathematician.

sieve plate, *Botany.* one of the thin walls, usually an end wall, of a sieve-tube member.

sieve tissue, *Botany.* tissue composed of sieve cells or sieve tubes.

sieve tube, *Botany.* a tubelike structure composed of thin, elongated cells (sieve-tube members) connected through perforations in their end walls, forming the essential element of the phloem of higher vascular plants.

sieve-tube member, *Botany.* one of the elongated, tubelike cells that make up a sieve tube. As a sieve-tube member matures, its nucleus decays, but the sieve-tube member remains functional. *Sieve-tube members are characteristically associated with specialized parenchyma cells called companion cells, which contain all of the components commonly found in living plant cells, including a nucleus* (Raven, *Biology of Plants*).

sigma (sig'mə), *n.* **1** *Nuclear Physics.* an unstable subatomic particle, a form of hyperon, having a neutral, positive, or negative charge, a mass approximately 2330 times that of the electron, and a spin of 1/2. Also called **sigma particle.** Compare **lambda, omega, xi.**
2 = sigma factor. [from Greek *sigma* the 18th letter of the Greek alphabet, equivalent to the English letter S]

sigma factor, *Biochemistry.* a protein that stimulates the synthesis of chains of ribonucleic acid: *The function of the sigma factor is to give the core enzyme its specificity to transcribing from a fixed point on a strand of DNA and producing RNA of defined length* (Science Journal).

sigma particle, = sigma (def. 1).

sigmic, *adj. Nuclear Physics.* of, having to do with, or containing a sigma or sigmas: *Kaonic and sigmic atoms tend to go together, since experiments aimed at making kaonic atoms make sigmic ones too* (Science News).

sigmoid (sig'moid) or **sigmoidal** (sig moi'dəl), *adj.* **1** resembling the Greek letter sigma; S-shaped: *a sigmoid curve, a sigmoidal fold.*
2 *Anatomy.* of or having to do with the sigmoid flexure of the colon.
[from Greek *sigmoeidēs* shaped like the letter sigma, from *sigma* + *eidos* form]

sigmoid flexure, 1 *Anatomy.* the S-shaped bend of the colon just above the rectum. **2** *Zoology.* an S-shaped curve.

sign, *n. Mathematics.* **1** any of the arithmetical and algebraic symbols used in performing operations, expressing functions, etc., such as the equal sign =, the identity sign ≡, the plus sign +, the multiplication sign ×, etc. To indicate that something is not equal to something else, the sign ≠ is used.
2 a unit of plane angle equal to 30 degrees.

signed number, *Mathematics.* a number with a plus sign or minus sign in front; positive or negative number. EXAMPLE: The equation $(+5) + (-8) = (-3)$ consists of three signed numbers. Also called **directed number.**

significant, *adj.* (of a figure or digit in a number representing a measured quantity) meaningful with respect to the accuracy of the measurement. Zeros which appear to the left of other figures are not significant; the number 0.083 has two significant figures, since it can be written 8.3×10^{-2}. Zeros which appear between other figures are always significant; the number 803 has three significant figures. Zeros which appear to the right of other figures are significant only if there is a decimal point; the number 1800 has two significant figures, while the numbers 1800. and 18.00 each have four significant figures. *In standard notation it is possible to indicate any desired number of significant figures. For example, if the figure 186,000 were known to four significant figures, it would be written 1.860 × 10^5* (Physics Regents Syllabus).

signless, *adj. Mathematics.* having no algebraic sign, or being essentially positive, like the modulus of an imaginary or a tensor.

silica (sil'ə kə), *n. Mineralogy.* a common hard, white or colorless compound, silicon dioxide, that resembles glass. Quartz, flint, opal, and sand are forms of silica. *The dioxide of silicon in its various forms is plentiful. It is estimated, in fact, that at least 12 per cent of the earth's crust is composed of this substance, which is usually referred to generically as silica* (Jones, *Inorganic Chemistry*). Formula: SiO_2 [from New Latin, from Latin *silex* flint]

silicate (sil'ə kit or sil'ə kāt), *n. Mineralogy.* a compound containing silicon with oxygen and a metal; a salt of silicic acid. Silicates constitute the greater number of the minerals that compose the crust of the earth. Mica, garnet, talc, asbestos, and feldspar are silicates. *By far the greater percentage of [minerals] fall into the group known as silicates ... The aggregates of these minerals are called rocks* (Garrels, *A Textbook of Geology*).

siliceous or **silicious** (sə lish'əs), *adj. Mineralogy.* containing or consisting of silica; resembling silica: *Paleolithic stone tools were made by chipping or flaking hard siliceous (glasslike) materials like flint, quartzite, and obsidian* (R. Beals and H. Hoijer).

silicic (sə lis'ik), *adj. Chemistry, Mineralogy.* of, obtained from, or containing silicon or silica: *The more volatile or acidic or silicic constituents of the fluid rock beneath the surface may have concentrated in the original uplifts* (Scientific American).

silicic acid, *Chemistry.* any one of various weak acids obtained from silicon. The formula of a common kind is H_4SiO_4

silicide (sil′ə sīd), *n. Chemistry.* a compound of silicon and another element or radical.

silicification (sə lis′ə fə kā′shən), *n. Geology.* conversion into or impregnation with silica: *Shells composed of calcium carbonate may be made more dense by infiltration of calcite deposited by ground water. Commonly, also, there has been replacement of the original hard parts by some other mineral in submicroscopic particles; such replacement by calcium carbonate is termed calcification; by silica, silicification; and iron pyrite, pyritization* (Moore, *Introduction to Historical Geology*).

silicify, *v. Geology.* to convert to or impregnate with silica. **Silicified wood** is wood so impregnated or replaced by silica that it has become quartz or opal.

silicle (sil′ə kəl), *n. Botany.* a short, broad silique, as in shepherd's-purse. [from Latin *silicula,* diminutive of *siliqua* seed pod]

silicon (sil′ə kən), *n.* a nonmetallic chemical element found only combined with other elements, chiefly combined with oxygen in silica and silicates. Next to oxygen, silicon is the most abundant element in nature, occurring in amorphous and crystalline forms. The crystalline form is much used in steel as a deoxidizing and hardening agent. *Small, almost microscopic chips of semiconductor material, usually the crystalline, brittle element silicon, make possible electronic functions that were impractical with the "old" technology using vacuum tubes, or even with the transistor* (Simon Ramo and Max Weiss). *Symbol:* Si; *atomic number* 14; *atomic weight* 28.086; *melting point* 1410°C; *boiling point* 2355°C; *oxidation state* 2, ± 4. [from *silica*]

silicon dioxide, = silica.

silicone (sil′ə kōn), *n. Chemistry.* any of a large group of organic compounds based on a structure in which organic groups are attached to silicon, and obtained as oils, greases, plastics, and resins. Silicones are noted for their stability and their ability to resist extremes of heat and cold, and are used for lubricants, varnishes, and insulators. *Silicones are synthetic polymers containing silicon, carbon, hydrogen, and oxygen; they may be viscous liquids or rubbery solids* (William L. Benedict).

siliconize, *v. Chemistry.* to combine, or cause to combine, with silicon.

silicular (sə lik′yə lər), *adj. Botany.* having the shape or appearance of a silicle.

silique (sə lēk′ *or* sil′ik), *n. Botany.* the characteristic podlike fruit of plants of the mustard family, a long, narrow, capsule with two valves, which splits open from the bottom upward, exposing the seeds attached to two placentae. [from Latin *siliqua* seed pod]

silk, *n. Zoology.* a fine, soft, strong protein fiber composed mainly of fibroin. It is secreted by various insects and arachnids to spin cocoons, cobwebs, etc. [Old English *sioloc*]

sill, *n. Geology.* an approximately horizontal sheet of intrusive igneous rock, found between older rock beds: *Sills are prominent features in many plateau countries. One of the best examples is the Karroo region of South Africa where ... dolerite intrusions, mostly of sill character, penetrate the strata over an area of 220,000 square miles* (Tyrrell, *Principles of Petrology*). See the picture at **volcano.** [Old English *syll*]

sillimanite (sil′ə mə nīt), *n. Mineralogy.* a silicate of aluminum occurring in orthorhombic crystals, trimorphous with andalusite and kyanite. *Formula:* Al₂SiO₅ [from Benjamin *Silliman,* 1779–1864, American chemist + *-ite*]

siloxane (sə lok′sān), *n. Chemistry.* any hydride of silicon in which silicon atoms alternate with atoms of oxygen. Siloxanes are very water-repellent and cannot be dissolved except by strong acids and alkalis. [from *sil(icon) + ox(ygen) + -ane*]

silt, *n. Geology.* **1** very fine particles of earth, sand, clay, etc., carried by moving water and deposited as sediment. Particles of silt measure from .01 to 1 mm in diameter; they are larger than clay particles and smaller than sand particles. *In the absence of the annual deposit of silt from the upper reaches of the Nile, heavy application of fertilizer would be needed on farmlands along the banks* (T. M. Schad and H. L. Edlin). **2** a deposit of such sediment occurring as a stratum in soil.
[perhaps from Scandinavian]
—**siltation,** *n.* the formation or deposition of silt.

siltstone, *n. Geology.* a fine-grained rock formed of consolidated silt: *Eventually, layers build up and a lithification, or stone-forming process, takes place. Sometimes pressure compacts, or squeezes, the water from the deposits. This locks the particles together and forms rocks called siltstone from silt, and shale from clay* (David E. Jensen).

Silurian (sə lùr′ ē ən), *or* sī lùr′ē ən), *Geology.* —*n.* **1** the third geological period of the Paleozoic era, after the Ordovician and before the Devonian, characterized by the development of early invertebrate land animals and land plants. The Silurian formerly included what is now Ordovician and what is now Silurian. **2** the rocks formed during this period.
—*adj.* of or having to do with this period or its rocks: *The total vegetation of the earth by the end of the Silurian period ... appears to have been mostly algae floating in water and growing on wet soil* (Emerson, *Basic Botany*).
[named after *Silures,* ancient people of Wales in whose region such rocks were formed]

silver, *n. Chemistry.* a shining white, metallic element occurring both natively and in combination, characterized in a pure state by its great malleability and ductility. Silver is superior to any other substance in its ability to conduct heat and electricity. *Pure silver, like pure gold, is too soft for coinage, ornaments, plate and jewellery, and the metal is usually alloyed with copper to harden it* (Jones, *Minerals in Industry*). *Symbol:* Ag; *atomic number* 47; *atomic weight* 107.868; *melting point* 961°C; *boiling point* 2212°C; *oxidation state:* 1. [Old English *siolfor*]

cap, face, father; best, bē, tėrm; pin, five;
rock, gō, ôrder; oil, out; cup, pùt, rüle,
yü in use, yu̇ in uric;
ng in bring; sh in rush; th in thin, ᴛʜ in then;
zh in seizure.
ə = a in about, e in taken, i in pencil, o in lemon, u in circus

597

silver bromide, *Chemistry.* a compound noted for its sensitivity to light, formed by the action of a bromide on an aqueous solution of silver nitrate. It is used in photography. *Formula:* AgBr

silver chloride, *Chemistry.* a compound noted for its sensitivity to light, used especially in photography for sensitizing paper. *Formula:* AgCl

silver nitrate, *Chemistry.* a colorless, crystalline, poisonous salt that becomes gray or black in the presence of light and organic matter, obtained by treating silver with nitric acid. It is used as a reagent in photography, in dyeing, to silver mirrors, as an antiseptic, etc. *Formula:* $AgNO_3$

sima (sī'mə), *n. Geology.* basaltic rock rich in iron and magnesium and low in silica. Sima is the chief constituent of the ocean floors. *While the continents are composed of a granitic crust called sial, the oceans are underlain by sima, a chemically altered surface of the earth's mantle* (C. P. Idyll). [from *si(lica)* + *ma(gnesium)*]

—**simatic** (sī mat'ik), *adj.* **1** composed largely of silicates of iron and magnesium: *Basalt is the chief simatic rock.* **2** consisting of sima: *a simatic ocean basin.* Compare **sialic.**

similar, *adj. Geometry.* having the same shape but not necessarily the same size or position; having corresponding angles equal and corresponding sides proportional: *similar triangles.* Compare **congruent.**

simoom (sə müm') or **simoon** (sə mün'), *n. Meteorology.* a hot, dry, suffocating wind carrying much sand, and occurring in the deserts of Arabia, Syria, and northern Africa. [from Arabic *simūm*]

simple closed curve, = Jordan curve.

simple equation, = linear equation.

simple fraction, *Mathematics.* a fraction in which both the numerator and the denominator are whole numbers. EXAMPLE: 1/3, 3/4, $^{219}/125$

simple fruit, *Botany.* a fruit developed from a single matured ovary, as the tomato, apple, and acorn. Simple fruits are classified as either dry or fleshy. Compare **aggregate fruit, multiple fruit.**

simple leaf, *Botany.* a leaf that is a single blade. An oak leaf is a simple leaf.

simple machine, *Physics.* any of the elementary devices or mechanical powers which multiply or change the direction of force and on which more complex machines are based. The lever, wedge, pulley, wheel and axle, inclined plane, and screw are the six simple machines.

simple sugar, *Chemistry.* any monosaccharide, such as glucose.

simplex (sim'pleks), *n., pl.* **simplexes, simplices** (sim'plə sēz), **simplicia** (sim plish'ē ə). *Geometry.* a figure or element having the minimum number of boundary points in a Euclidean space of a specified number of dimensions. The simplex of one-dimensional space is the line segment, of two-dimensional space the triangle, and of three-dimensional space the tetrahedron. [from Latin *simplex* single]

simultaneous equations, *Mathematics.* two or more equations or inequalities, with two or more unknowns, for which a set of values of the unknowns is sought that is a solution of all the equations or inequalities.

sin, *abbrev.* sine.

sinciput (sin'sə pət), *n. Astronomy.* **1** the front part of the head. **2** the upper part of the skull. Compare **centriciput, occiput.** [from Latin, from *semi* half + *caput* head]

—**sincipital** (sin sip'ə təl), *adj.* of or having to do with the sinciput.

sine (sīn), *n. Trigonometry.* the ratio of the length of the side opposite an acute angle in a right triangle to the length of the hypotenuse. The sine, secant, and tangent are the three fundamental trigonometric functions. [from Latin *sinus* bend, bosom; in Medieval Latin, translation of Arabic *jaib* sine, bosom]

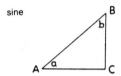

In the right triangle ABC, AB is the hypotenuse, and the sine of angle a = BC/AB, the sine of angle b = AC/AB.

sine curve, *Trigonometry.* a curve showing the relationship between the size of an angle and its sine, plotted by using the successive values as coordinates. Also called **sinusoid.**

sine wave, *Physics.* a simple wave, such as an electromagnetic or sound wave, that can be graphically represented as a sine curve. Also called **sinusoidal wave.**

single, *adj. Botany.* having only one set of petals. Most cultivated roses have double flowers with many petals; wild roses have single flowers with five petals.

single-blind, *adj.* having to do with or designating a test or experiment whose exact makeup is known to the investigators but not to the subjects: *In a single-blind study, the physician knows whether he is giving a patient drug or placebo but the patient does not know which he is receiving* (Science). Contrasted with **double-blind.**

single bond, *Chemistry.* a bond in which one electron pair is shared between two atoms; a bond involving one valence. Compare **double bond, triple bond.**

single cross or **single-cross hybrid,** *Genetics.* the hybrid produced by a single crossing of two inbred lines. Compare **double cross** or **double-cross hybrid.**

singularity, *n.* **1** *Astronomy, Physics.* a hypothetical point in space at which an object becomes compressed to infinite density and infinitesimal volume: *In practice, a singularity is surrounded by a region of space so distorted by gravity that nothing can escape; it is this region that constitutes the black hole* (New Scientist).

2 *Mathematics.* a point at which the continuity of a surface, function, etc., is broken: *The functions studied in ordinary calculus ... offer very awkward and inadequate descriptions of the more complex surfaces common among soap films. Such surfaces have what mathematicians call "singularities"—that is, edges* and vortices caused by the surface branching or intersecting with itself (Lynn A. Steen).

sinh, abbrev. hyperbolic sine.

sinistral, adj. Zoology. (of a gastropod shell) having the spire or whorl rising from right to left as viewed from the outside: In most species the shell is right-handed (dextral), being coiled clockwise as seen from the spire, but some are left-handed (sinistral) (Storer, General Zoology). [from Latin sinister left]

sinistrorse (sin'ə strôrs), adj. Botany. rising spirally from right to left: the sinistrorse stem of a vine. [from Latin sinistrorsus turned left, from sinister left + versus turned, from vertere to turn] —**sinistrorsely,** adv.

sink, v. Physics. any natural or artificial means of absorbing or removing a substance or a form of energy from a system: Gases spend varying periods there [in the troposphere] depending on the "sinks" by which each is removed from the atmosphere: incorporation into cloud droplets, reactions with other gases, loss to finely divided liquid or solid particles or the earth's surface and so on (Scientific American). Compare **heat sink.**

sinkhole, n. Geology. a funnel-shaped cavity formed in limestone regions by the removal of the rock through action of rain, ground water, or running water: As the caverns enlarge, their roofs become incapable of supporting themselves and collapse, forming sinkholes at the surface (Garrels, A Textbook of Geology).

sinoatrial (sī'nō ā'trē əl), adj. Anatomy. of or having to do with the area between the sinus venosus and the right atrium of the heart: the sinoatrial pacemaker.

sinoatrial node, Anatomy. a mass of tissue in the right atrium of the heart, near the point where the major veins enter, that originates the heartbeat: In the case of long and lingering illnesses ... the heart is the last organ to fail, and the pacemaker of the heart—the sinoatrial node—is the last to die (Paul D. White). Also called **S-A node.**

sinoauricular node (sī'nō ô rik'yə lər), = sinoatrial node.

sinter, n. Geology. a hard chemical incrustation or deposit formed on rocks by evaporation of hot or cold mineral waters. Sinter may be siliceous (consisting of silica) or calcareous (consisting of calcium carbonate). [from German Sinter dross, slag]

sinuate (sin'yù āt or sin'yù it), adj. Botany. having its margin strongly or distinctly wavy: a sinuate leaf. [from Latin sinuatum bent, from sinus a curve] —**sinuately,** adv.

sinus (sī'nəs), n. 1 Anatomy. **a** a cavity in a bone, especially one of the cavities in the bones of the skull that connect with the nasal cavity. See the picture at **uvula.** **b** a reservoir or channel containing venous blood.
2 Botany. a curve or bend, especially a curve between two projecting lobes of a leaf.
[from Latin sinus any bend or curve]

sinusoid (sī'nə soid), n. **1** = sine curve: When unrolled, each half of the paper will have a cut edge in the form of a sine curve, or sinusoid, one of the fundamental wave forms of physics (Scientific American).
2 Anatomy. one of the spaces or tubes through which blood passes in various organs such as the suprarenal gland and the liver: Blood from the digestive tract flows in the hepatic portal vein to be filtered through capillary-like sinusoids in the liver, then is collected in the hepatic veins connected to the sinus venosus (Storer, General Zoology).

sinusoidal (sī'nə soi'dəl), adj. **1** Mathematics. of or having to do with a sine curve: A sinusoidal variation can be understood in terms of a crank on a shaft which rotates at a constant rate, f turns per second ... The height, h, of the end of the crank above the level of center of the shaft varies sinusoidally with time. If we plot height or amplitude vs. time in seconds we get a sine curve or sine wave (Pierce, Electrons, Waves and Messages).
2 resembling or flowing in the wavelike course of a sine curve or curves: It is a different mechanism of swimming from that of the fish, in which a sinusoidal oscillation of the body and tail takes place in a side-to-side direction (New Scientist).
3 Anatomy. of or having to do with a sinus.
—**sinusoidally,** adv. in the manner of a sine curve: The steady state current, like the terminal voltage, is seen to vary sinusoidally with the time (Sears and Zemansky, University Physics).

sinusoidal projection, Geography. a type of map projection in which the central meridian and the equator are shown as straight lines, but the other meridians are shown as curved lines: A sinusoidal projection does not have lines of longitude of equal length. It squeezes shapes near the top and bottom, and bends them at the left and right (E. B. Espenshade, Jr.).

sinusoidal wave, = sine wave.

sinus venosus (vi nō'səs), Anatomy. **1** the chamber in the heart of lower vertebrates to which the blood is returned by the veins. **2** the cavity in the embryonic atrium of mammals in which the various venous systems are joined.

siphon (sī'fən), n. **1** Zoology. a tube-shaped organ of some animals, such as certain shellfish, for drawing in and expelling water, etc.: The squid sucks water up into its mantle cavity ... There is a little tube leading from the cavity called the siphon which can expel this water rather forcibly and propel the animal in the opposite direction using the principles of jet propulsion (Winchester, Zoology).
2 Physics. a bent tube through which a fluid can be drawn over the edge of one container into a lower container by means of air pressure. A siphon stops working when the fluid in both containers reaches the same level.
[from Greek siphōn pipe]

---

cap, fāce, fäther; best, bē, tėrm; pin, five;
rock, gō, ôrder; oil, out; cup, pùt, rüle,
yü in use, yù in uric;
ng in bring; sh in rush; th in thin, ŦH in then;
zh in seizure.
ə = a in about, e in taken, i in pencil, o in lemon, u in circus

siphonostele

—**siphonate** (sī'fə nāt), *adj. Zoology.* having a siphon or siphons.

siphon
(def. 2)

siphonostele (sī'fə nə stēl' *or* sī'fə nə stē'lē), *n. Botany.* vascular tissue in the form of a hollow tube with the pith in the center. [from Greek *siphōn* pipe + English *stele*]

siphuncle (sī'fung kəl), *n. Zoology.* **1** a small tube passing through the partitions in the shell of certain cephalopods: *The siphuncle, a coiled fleshy tube enclosed in a limy covering, extends through all the chambers, connecting them with the body of the nautilus* (William J. Clench). **2** either of two small tubular organs on the abdomen of an aphid, through which a waxy secretion is exuded. Also called **cornicle** [from Latin *siphunculus,* diminutive of *siphon* pipe, from Greek *siphōn*]
—**siphuncular** (sī fung'kyə lər), *adj.* of, having to do with, or like a siphuncle: *the siphuncular canal.*
—**siphunculate** (sī fung'kyə lāt), *adj.* having a siphuncle: *a siphunculate shell.*

sirocco (sə rok'ō), *n., pl.* **-cos.** *Meteorology.* **1** a very hot, dry, and dust-laden wind blowing from the northern coast of Africa across the Mediterranean and part of southern Europe: *Rain is followed usually by the sirocco—the hot wind from Africa—and the ground is quickly parched and cracked* (Atlantic). **2** a moist, warm, south or southeast wind in these same regions: *Sometimes the sirocco extends to the northern shore of the Mediterranean where it becomes a warm and moist wind* (Blair, *Weather Elements*).

sister chromatid, *Biology.* either of a pair of genetically identical chromatids. A prophasic or metaphasic chromosome consists of two sister chromatids.

sitosterol (sī tos'tə rōl), *n. Biochemistry.* any of several crystalline alcohols or sterols, similar to cholesterol, present in wheat, corn, bran, Calabar beans, and other plants. [from Greek *sitos* grain + English *sterol*]

sitotoxin (sī'tō tok'sin), *n. Microbiology.* a toxin or poison generated by a microorganism in vegetable food. [from Greek *sitos* grain + English *toxin*]

situs (sī'təs), *n. Anatomy.* position, situation, or location, especially the proper or original position of a part or organ: *situs inversus ... internal organs being transposed or on the wrong side.* [from Latin]

SI unit, any of the units of measurement in the International System of Units (SI), the system generally used by scientists. The seven fundamental SI units are the kilogram, second, meter, ampere, kelvin, mole, and candela. All other SI units are derived from these. For instance, the newton, the derived unit of force, is equal to a meter multiplied by a kilogram divided by a second squared ($m \cdot kg/s^2$). Multiples of all SI units are designated by a single set of combining forms; a millisecond equals one thousandth of a second, a milliampere equals one thousandth of an ampere, etc.
[*SI,* abbreviation of French *Système International (d'Unités)* International System (of Units)]
► The system of SI units is intended to replace earlier metric systems such as the MKS system (which is very similar, since it also has the meter, kilogram, second, and ampere as fundamental units) and the CGS system (which is based on the centimeter, gram, and second).

skeletal, *adj.* **1** *Anatomy.* **a** of, forming, or resembling a skeleton: *the skeletal framework. The general functions of the skeletal system are support, protection, and muscle attachment* (Winchester, *Zoology*). **b** attached to or formed by a skeleton; *The movement of your body is made possible by the skeletal muscles; that is, the muscles that are attached to the bones* (Beauchamp, *Everyday Problems in Science*). See the picture at **muscle.**
2 *Mineralogy.* of or designating a skeleton crystal.
—**skeletally,** adv.

skeletogenous (skel ə toj'ə nəs), *adj. Biology.* producing a skeleton; giving rise to a skeleton: *skeletogenous tissue.*

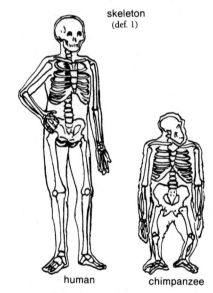

skeleton
(def. 1)

human   chimpanzee

skeleton, *n. Anatomy.* **1** the framework of bones and cartilage in vertebrates that supports the muscles and organs of the body and protects the viscera: *Vertebrate animals have bony skeletons within their bodies. These animals can grow by a continuous increase in size and not by a series of ecdyses [shedding of exoskeletons] ... There are many invertebrate animals which have no skeleton* (Mackean, *Introduction to Biology*). **2** the hard supporting or covering part of an invertebrate animal, such as the shell of a mollusk or crustacean: *In two genera, Tubipora (the organ-pipe coral) and Heliopora (the blue coral), which are widely distributed on coral reefs, a continuous calcareous skele-*

ton is developed resembling that of reef corals (L. A. Borradaile and F. A. Potts).

[from Greek *skeleton* (*sōma*) dried (body)]

ASSOCIATED TERMS: The skeleton of vertebrate animals is an *endoskeleton*, as distinguished from the chitinous *exoskeleton* of invertebrates. The human skeleton consists of the *axial skeleton*, made up of the bones of the head, neck, and trunk, and the *appendicular skeleton*, made up of the bones of the arms and legs and their supporting bones. The axial skeleton is composed of the *cranial* bones, the *cervical, thoracic,* and *lumbar vertebrae*, the *ilium, pubis,* and *ischium* of the *pelvic girdle*, the *sacrum*, and the *coccyx*. The appendicular skeleton is composed of the *clavicle, scapula*, and *sternum;* the *humerus, radius,* and *ulna* of each arm; the *carpus, metacarpus,* and *phalanges* of the wrist, hand, and fingers; the *femur, patella, tibia,* and *fibula* of the thigh and leg; and the *tarsus, metatarsus,* and *phalanges* of the ankle, foot, and toes.

skeleton crystal, *Mineralogy.* an incompletely developed crystal, having an axial framework, but with faces only partially filled in.

skew (skyü), *adj. Geometry.* not lying in the same plane in three-dimensional space: *a skew curve. Skew lines do not intersect and are not parallel.*

skin, *n. Anatomy.* the outer layer of tissue of the human or animal body, especially when soft and flexible. The skin envelops the body and consists of the dermis and epidermis. *The skin is the largest organ of the body and, next to the brain, the most complicated* (Science News Letter).

ASSOCIATED TERMS: The skin consists of the *epidermis, dermis,* and *subcutaneous* tissue. The epidermis consists of *keratinous, granular, spinous,* and *basal* cells, as well as *melanocytes.* The innermost layer of the epidermis is lined by *epithelial* cells. The dermis consists mainly of *connective tissue,* and *papillae* containing nerve endings. The subcutaneous part of the skin contains *adipose* tissue. The skin contains *sebaceous glands* and two kinds of *sweat glands,* called *apocrine* and *eccrine.* See also the associated terms under **hair.**

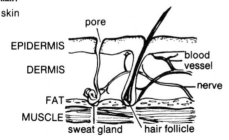

skin

skin friction, *Physics.* the friction developed between a solid and a fluid, especially the friction that occurs in the thin layer of air over the surface of a body moving at very high speeds, causing a sharp rise in temperature: *Even a well-streamlined airplane has drag caused by skin friction. Skin friction occurs in the thin layer of air next to the surface of the airplane. This portion of air is called the boundary layer. The friction results when one layer of air slides over another* (A. Wiley Sherwood).

skull, *n. Anatomy.* the bony or cartilaginous framework of the head in humans and other vertebrates, enclosing and protecting the brain: *The human skull may conveniently be divided into three principal portions: the cranium or cranial vault, the face, and the lower jaw or mandible* (R. Beals and H. Hoijer).

slate, *n. Geology.* a fine-grained, usually bluish-gray metamorphic rock that splits easily into thin, smooth layers: *A shale subjected to great pressures has its particles further flattened and arranged in more perfect*

parallelism so that the rock readily splits or cleaves. It is called slate, a rock of considerable economic value (Finch and Trewartha, *Elements of Geography*).

sleet, *n. Meteorology.* **1** clear particles of ice formed by the freezing of raindrops before they reach the ground: *Sleet is more easily removed when it has a sublayer of snow than when it is frozen directly to the walks* (Neuberger and Stephens, *Weather and Man*). **2** a thin coating of ice formed by frozen rain; glaze.

slickenside (slick′ən sīd′), *n. Geology.* a rock surface that has become polished and striated from the sliding or grinding motion of an adjacent mass of rock. Slickensides commonly occur along faults. *Many slickensides in limestone are covered with quartz, others in marble bear serpentine, while faces of dark diorite are made glossy by green epidote* (Fenton, *The Rock Book*).

slide, *n. Geology.* **1** the downward falling or sliding of a mass of earth, snow, or rock.
2 the mass of earth, snow, or rock falling or sliding down from a hill or mountainside. SYN: landslide, avalanche.

slide rule, a device consisting of a ruler with a sliding section in the center, both parts being marked with logarithmic scales, used for making rapid mathematical calculations. Slide rules have now been largely replaced by pocket digital calculators.

slip, *n. Geology.* **1** a fault in rock due to the sinking of one section.
2 a movement producing such a fault.
3 the amount of such movement, measured by the amount of displacement along the fault plane.

slip face, the steep, concave, unstable slope on the lee side of a crescent dune. See the picture at **dune.**

slope, *n.* **1** *Geology.* **a** the inclined surface of a hill, mountain, plain, etc.; any part of the earth's surface that goes up or down at an angle. **b** the angle at which such a surface deviates from the horizontal.

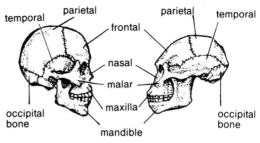

Human   Australopithecine

cap, fāce, fäther; best, bē, tėrm; pin, fīve;
rock, gō, ôrder; oil, out; cup, pút, rüle,
yü in use, *yu̇* in uric;
ng in bring; *sh* in rush; *th* in thin; ŦH in then;
zh in seizure.
ə = *a* in about, *e* in taken, *i* in pencil, *o* in
lemon, *u* in circus

601

2 *Mathematics.* the tangent of the angle formed by the intersection of a straight line with the horizontal axis of a pair of Cartesian coordinates.
[Old English *aslopen* slipped away]

slow neutron, *Nuclear Physics.* a neutron of relatively low energy, as compared with a fast neutron: *When U²³⁸ atoms capture slow neutrons, they form atoms of U²³⁹, but these atoms do not fission* (Robert L. Thornton).

slow virus, *Biology.* any of a class of viruses that are present in the body of an infected individual for a long time before they become active or infectious: *So far, scientists have convincingly linked only four chronic human diseases to slow viruses ... It would seem unlikely that only these diseases are caused by slow infections* (Richard T. Johnson). *The risks involved from the so-called "slow viruses," particularly measles viruses, ... may hide for years in human tissue and emerge later in the form of encephalitis or possibly even as a factor in multiple sclerosis* (N. Y. Times).

slug, *n.* **1** *Zoology.* any of various slow-moving, elongated mollusks that resemble snails but have no shell or only a rudimentary one. Slugs live mostly in forests, gardens, and damp places. *A study of the embryonic development of the slug reveals that a shell is formed in the embryo just as it is in the snail, but fails to continue its development to a functional size* (Winchester, *Zoology*).
2 *Physics.* a unit of mass in the foot-pound-second system, equal to about 32.17 pounds. The slug is defined as the mass which will acquire an acceleration of one foot per second per second when acted upon by a force of one pound.

slump, *Geology.* —*n.* **1** the slipping or falling of a mass of rock or unconsolidated material, especially the vertical drop of a mass of rock along a cliff face after the supporting material at the base of the cliff has been undercut by erosion. **2** the mass of fallen material.
—*v.* (of a mass of material) to slip or fall; to undergo a slump.

slush, *n.* *Meteorology.* snow or ice that is partly melted or turned into a watery mixture by rain, warm temperature, or application of a chemical substance.

Sm, *symbol.* samarium.

small calorie, *Chemistry, Physics.* = calorie: *The heat unit most commonly used in the metric system is the calorie. The small calorie is the quantity of heat which will raise the temperature of 1 gram of water from 3.5° to 4.5° ... The large calorie or "kilogram-calorie" is equal to 1000 small calories* (Parks and Steinbach, *Systematic College Chemistry*).

small circle, *Geometry, Astronomy.* a circle on the surface of a sphere whose plane does not pass through the center of the sphere.

small intestine, *Anatomy.* the narrow, winding, upper part of the intestines, between the stomach and the large intestine, where digestion is completed and nutrients are absorbed by the blood. The human small intestine is about twenty feet long and consists of the duodenum, the jejunum, and the ileum. See the picture at **alimentary.**

smaltite (smôl'tīt), *n.* *Mineralogy.* a tin-white to steel-gray mineral consisting essentially of an arsenide of cobalt, but usually also containing nickel, and occurring in crystals or in compact or granular masses. [earlier *smaltine,* from French, from *smalt* a kind of pigment]

smear, *n.* *Biology.* a thin layer of blood, tissue, etc., applied to a slide for microscopic examination or on the surface of a culture medium: *The fixed smear is covered with stain ... and allowed to dry in air. Bacteria are usually examined under the oil immersion objective, and the oil can be placed directly on top of the smear* (Burrows, *Textbook of Microbiology*).

smectic (smek'tik), *adj.* *Chemistry.* (of liquid crystals) consisting of a series of layers in which the molecules are arranged either in rows or at random: *A common example of a smectic liquid crystal is the layers forming the inner and outer surface of a soap bubble* (Science News). Compare **cholesteric** and **nematic.**
[from Latin *smecticus* cleansing, from Greek *smēktikos,* from *smēchein* to cleanse]

smectite (smek'tīt), *n.* = montmorillonite.

smegma (smeg'mə), *n.* *Physiology.* sebaceous secretion, especially that found under the prepuce: *The connection between lack of circumcision in the male and the incidence of cervical cancer in the female is believed to be effected by the smegma—a soap-like secretion in the uncircumcised male—but so far little is known about the cancer-producing properties, if any, of smegma* (Observer). [from Latin *smegma,* soap, from Greek *smēgma,* from *smēchein* to cleanse]

smithsonite (smith'sə nīt), *n.* *Mineralogy.* native carbonate of zinc: *Smithsonite is a zinc ore ... usually found with zinc deposits lying in limestone rocks* (Dana's Manual of Mineralogy). *Formula:* $ZnCO_3$ [named after James *Smithson,* 1765–1829, English chemist and mineralogist, founder of the Smithsonian Institution]

smog, *n.* *Meteorology.* **1** a combination of smoke, chemical pollutants, and fog in the air, usually present in large urban or industrial areas: *Three principal effects are caused by smog. The first is a marked reduction in visibility; the second is an effect on the upper respiratory tract—the nose and throat—and possibly on the lower respiratory tract or the deeper recesses of the lungs, and on the eyes; the third is the effect on vegetation* (Morris B. Jacobs).
2 a form of air pollution produced when sunlight causes hydrocarbons and nitrogen oxides from automobile emissions to combine in a photochemical reaction: *The reactions of hydrocarbons with nitrogen oxides in the presence of ultraviolet radiation produce the photochemical smog that appears so often over Los Angeles and other cities* (S. Fred Singer).
[blend of *smoke* and *fog*]

smoke, *n.* *Chemistry.* a dispersion of solid particles in a gas: *coal smoke, wood smoke, cigarette smoke. Strictly speaking, there is little difference between dusts and smokes, though it is customary to think of particles suspended in the air as the result of combustion or other direct chemical action as constituting a smoke* (Jones, *Inorganic Chemistry*).

smooth muscle, *Anatomy.* a type of muscle not contracted by voluntary action, with fibers in smooth layers or sheets. The muscles of the stomach, intestine, and other viscera (except the heart) are smooth muscles. Also called **involuntary muscle.** See the picture at **muscle.**

Sn, *symbol.* tin. [for Latin *stannum*].

sneeze, *n. Physiology.* a sudden, violent expulsion of air through the nose and mouth. It is a reflex caused chiefly by irritation of the nasal nerves.

Snell's law, *Optics.* the fundamental principle that the sines of the angles of incidence and refraction stand in a constant ratio to one another: *It is evident from Snell's law that the angle of refraction is always less than the angle of incidence for a ray passing from a medium of smaller into one of larger index, as from air into glass* (Sears and Zemansky, *University Physics*). [named after Willebrord van Roijen *Snell,* 1591–1626, Dutch mathematician]

snout, *n. Zoology.* **1** the projecting part of an animal's head that contains the nose, mouth, and jaws. Pigs, dogs, and crocodiles have snouts. **2** a similar projection in certain insects, such as the snout beetle and snout butterfly.

snow, *n.* **1** *Meteorology.* water vapor frozen into hexagonal crystals that fall to earth in soft, white flakes and spread often upon it as a white layer.
2 *Chemistry.* any of various substances having a snowlike appearance: *carbon-dioxide snow.*

snow blindness, *Optics.* temporary or partial blindness caused by the reflection of ultraviolet rays from snow or ice: *Snow blindness ... usually occurs at altitudes over 1,000 feet and may occur in overcast weather as well as in bright sunshine* (Science News Letter).

snowdrift, *n. Meteorology.* a mass or bank of snow piled up by the wind.

snowflake, *n. Meteorology.* a small, feathery, hexagonal crystal of snow, or any aggregation of such crystals or fragments of crystals: *Snowflakes are crystals of many beautiful, lacy patterns. The fundamental form is hexagonal, but this is subject to much intricate elaboration, apparently influenced by the temperature, and perhaps also by the rapidity of condensation. Large snowflakes are formed by the combination of many small crystals, usually at temperatures not much below freezing, never at very low temperatures* (Blair, *Weather Elements*).

soap, *n. Chemistry.* any metallic salt of an acid derived from a fat; the reaction product of an ester or fatty acid and an alkali: *If we treat stearin with an alkali, such as sodium hydroxide, they react to form soap and glycerol* (Dull, *Modern Chemistry*). Compare **saponification.**

soapstone, *n. Geology.* a soft metamorphic rock composed mostly of talc.

social, *adj.* **1** *Zoology.* living together in organized communities: *Ants, bees, and termites are social insects.*
2 *Botany.* growing in patches or clumps.
—**sociality** (sō'shē al'ə tē), *n.* the condition of being social.

society, *n.* **1** *Zoology.* an organized community of animals: *a society of wasps.*
2 *Ecology.* a group of plants, especially of a single species, considered as a unit in an ecological community.

sociobiology, *n.* the study of the biological basis of social behavior in all kinds of organisms: *The central principle of sociobiology is that living things tend to behave in a manner that maximizes their inclusive fitness; that is, they behave so as to project a maximum number of*

genes into future generations (Barash, *Sociobiology and Behavior*).

socket, *n. Anatomy.* **1** a hollow place in some part of the body in which another part moves. A person's eyes are set in sockets. The joint of the hip consists of a ball or knob fitting in a socket (the acetabulum).
2 the space in the jawbone holding the root of a tooth.

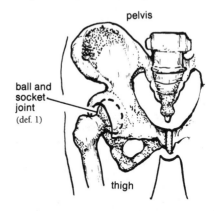

pelvis

ball and socket joint (def. 1)

thigh

soda, *n. Chemistry.* any of various substances containing sodium, such as sodium bicarbonate or baking soda, sodium carbonate or sal soda, and sodium hydroxide or caustic soda. [from Italian, from Arabic *sauda,* a plant yielding sodium carbonate]

soda ash, *Chemistry.* sodium carbonate in a powdery white form, partly purified for commercial use: *Soda ash ... takes from 30% to 40% of the total output of salt for the manufacture of glass, soap, various sodium chemicals, and in the preparation of washing-soda and so forth* (Jones, *Minerals in Industry*).

sodic (sō'dic), *adj. Chemistry.* of or containing sodium: *sodic sulfur, sodic chlorides.*

sodium, *n. Chemistry.* a soft, silver-white metallic element which occurs in nature only in compounds and reacts violently with water to form sodium hydroxide and hydrogen gas. Soda and common table salt contain sodium. Sodium is one of the alkali metals which oxidize rapidly in the presence of air and makes up about 2.8 per cent of the earth's crust. Its most familiar compound is sodium chloride, which is common table salt. *Symbol:* Na; *atomic number* 11; *atomic weight* 22.9898; *melting point* 97.8°C; *boiling point* 892°C; *oxidation state* 1. [from *soda*]

sodium benzoate, = benzoate of soda.

sodium bicarbonate, *Chemistry.* a powdery, white crystalline salt with a somewhat alkaline taste, used in cooking, medicine, manufacturing, etc. Sodium bicarbonate is a source of carbon dioxide. *Formula:*

---

cap, fāce, fäther; best, bē, tėrm; pin, fīve;
rock, gō, ôrder; oil, out; cup, pùt, rüle,
yü in use, *yu* in uric;
ng in bring; sh in rush; th in thin, ᴛʜ in then;
zh in seizure.
ə = *a* in about, *e* in taken, *i* in pencil, *o* in lemon, *u* in circus

sodium borate

$NaHCO_3$ Also called **baking soda** and **bicarbonate of soda.**

sodium borate, = borax.

sodium carbonate, *Chemistry.* a salt that occurs in a powdery white form (soda ash) and in a hydrated crystalline form (sal soda or washing soda). It is used for softening water, making soap and glass, neutralizing acids, and in medicine and photography. *Formula:* Na_2CO_3

sodium chloride, *Chemistry.* a white, crystalline substance, the chloride of sodium, that is the ordinary table salt. *Formula:* $NaCl$

sodium cyanide, *Chemistry.* a very poisonous, white crystalline salt, used in the cyanide process of extracting gold and silver from ores, in fumigating, etc. *Formula:* $NaCN$

sodium fluoride, *Chemistry.* a poisonous crystalline salt, used as an insecticide, a disinfectant, in the fluoridation of water, in treating certain forms of tooth decay, etc. *Formula:* NaF

sodium hydroxide, *Chemistry.* a white solid that is a strong, corrosive alkali. It is used in making hard soaps, rayon, and paper, in tanning, and as a bleaching agent. *Formula:* $NaOH$ Also called **caustic soda.**

sodium nitrate, *Chemistry.* a colorless, crystalline compound used to produce other nitrates, such as potassium nitrate, and in making fertilizers, explosives, etc. *Formula:* $NaNO_3$ Also called **niter.**

sodium pump, *Biochemistry.* the cellular mechanism or means by which sodium ions are moved out of cells and replaced with potassium ions: *The sodium pump ... gets hold of any sodium that gets into the cell and turns it out again* (Listener). *The "sodium pump" ... uses metabolic energy in the form of ATP to extrude sodium ions from the axon* (Scientific American).

sodium silicate, *Chemistry.* any of various colorless, white, or grayish-white crystallike substances, used in preserving eggs, in soap powders, as adhesives, etc. Also called **water glass.**

soft coal, = bituminous coal.

soft rays or **soft radiation,** *Physics.* radiation of low energy or penetrating power, as that of X rays.

softwood, *n. Botany.* **1** a coniferous tree. Pines and firs are softwoods; oaks and maples are hardwoods. *The terms hardwood and softwood, while in general use, are definately misleading, since no definite degree of hardness divides woods falling into the two groups* (White and Renner, *Human Geography*). **2** the wood of a coniferous tree: *Wood is generally called softwood if it comes from a tree which has needle-like leaves, such as pine. The broad-leaved trees, such as the oaks, provide what is commonly called hardwood* (Morris Lieff).

▶ See the note under **hardwood.**

soil, *n. Geology.* the top layer of the earth's surface, composed of rock and mineral particles mixed with animal and vegetable matter: *Soil is more common than rock at the earth's surface. Almost all outcrops of rock are less firm—more easily crumbled and broken—than is the same rock at a depth of 20 or 100 feet* (Gilluly, *Principles of Geology*). *The significance of the soil lies in its ability to furnish water and nutrients to plants* (Colby and Foster, *Economic Geography*).

ASSOCIATED TERMS: Soil contains mineral particles of *sand, silt,* and *clay,* and organic particles called *humus.* A *chernozem* is a black soil rich in humus, while a *sierozem* is a grayish soil low in *humus.* Layers of soil are called *horizons,* and most soils include three horizons: the *A horizon* or *topsoil,* the *B horizon,* and the *C horizon* or *subsoil.* Soil science, or *pedology,* uses such terms as *ped* (a natural soil aggregate), *pedon* (a small, representative volume of soil), and *polypedon* (a group of pedons that form the characteristic soil in a geographic area) to describe the structure of soils and to study *pedogenesis* (the origin and formation of soils). Pedologists of the United States classify soils according to the characteristics of a polypedon. Every polypedon falls within one of 10 orders of soils in the Comprehensive Soil Classification System (CSCS) of the Soil Survey Staff of the U.S. Soil Conservation Service. The 10 orders are:

Alfisols	Soils of humid climates.
Aridisols	Soils of dry climates.
Entisols	Soils lacking horizons.
Histosols	Soils with a large proportion of organic matter.
Inceptisols	Soils with weakly developed horizons.
Mollisols	Soils with thick, organically rich horizons.
Oxisols	Very old, highly weathered soils.
Spodosols	Soils with iron and aluminum in the B horizons.
Ultisols	Soils of warm and humid climates.
Vertisols	Soils of subtropical and arid warm climates.

soil creep, *Geology.* the slow movement or settling of surface soil down a slope: *Under these conditions rapidly eroded gullies dissect the upland minutely, but there is little softening of the contours of the features by soil creep or continuous rainwash. The result is a surface of the type known as badlands* (Finch and Trewartha, *Elements of Geography*).

soil profile, *Geology.* the succession of layers or horizons in a vertical section of soil: *There is a tendency for the crumb structure of the soil to disintegrate, and for the fine particles of the sodium-clay to run down the soil profile, sealing up all spaces and killing the roots of plants by asphyxiation* (H. G. Chippindale). Compare **solum.**

soil science, = pedology. Compare **earth science.**

sol[1] (sol *or* sōl), *n. Chemistry.* a colloidal solution: *Parts of the protoplasm may have a sol structure in which their consistency resembles that of water; then these same parts will quickly switch to a gel structure resembling strongly set gelatin* (Douglas Marsland). [short for *solution*]

sol[2] (sol), *n. Astronomy.* a Martian day, consisting of 24 hours, 37 minutes, and 22 seconds: *On sol 8—the eighth sol or day, after the first of the Viking landers had touched down on Chryse Planitia, a great basin in the northern hemisphere of Mars—the craft's sampler arm extended straight out and then dropped to the ground* (New Yorker). [probably from Latin *sol* sun, since a Martian day represents one rotation of Mars with respect to the sun]

sol., *abbrev.* **1** soluble. **2** solution.

solano (sō lä′nō), *n. Meteorology.* a dry, very warm easterly wind that blows in the southeastern coastal region of Spain in the summer. [from Spanish, from Latin *solanus* the east wind, from *sol* sun]

solar, *adj. Astronomy.* **1** of the sun: *a solar eclipse.* **2** having to do with the sun: *solar research.* **3** coming from the sun: *solar heat, solar power.*

4 measured or determined by the earth's motion in relation to the sun; based on solar time: *a solar day, the solar calendar.*

5 working by means of the sun's light or heat. A solar battery converts sunlight into electrical energy.
[from Latin *solaris*, from *sol* sun]

solar apex, *Astronomy.* the point in space, situated in the constellation Hercules, toward which the sun is moving.

solar constant, *Astronomy.* the average intensity of solar radiation, as would be measured outside the earth's atmosphere and at the earth's mean distance from the sun. It is equal to approximately 1.94 calories per square centimeter per minute. *The solar constant is the standard measure of the heat from the sun beating down on the top of the earth's atmosphere. Smithsonian scientists have measured it nearly every day for [65] years* (Science News Letter).

solar day, = mean solar day: *Night succeeds day as the earth rotates on its axis, turning first one side and then the other toward the sun. The day-night cycle represents the solar day of 24 hours* (Frank A. Brown, Jr.).
► See the note under **day.**

solar energy, *Physics.* the radiant energy transmitted by the sun; the energy of the sun's electromagnetic radiation: *The interest in solar energy is certainly understandable. The Sun keeps the Earth warm, allows vegetation to grow, and nurtures all life. Significantly it is now also recognized that the Sun's radiation can be harnessed for much—perhaps all—of man's energy needs* (Paul Rappaport).

solar flare, *Astronomy.* a sudden eruption of hydrogen gas on the surface of the sun, usually associated with sunspots, and accompanied by a burst of ultraviolet radiation, often followed by a magnetic disturbance: *Solar flares emitting blasts of gas and particles were quickly followed by more intense auroral displays and by radio blackouts* (Richard S. Lewis).

solar physics, the branch of astrophysics dealing with the physical characteristics and phenomena of the sun: *Already our concepts of the interplanetary medium, the magnetosphere of the Earth, the solar wind, basic knowledge of solar physics, and solar-atmospheric effects are evolving rapidly as space experiments are completed* (Bulletin of Atomic Scientists).

solar radiation, *Astronomy.* the electromagnetic radiation emitted by the sun: *The earth's atmosphere is relatively transparent to direct and reflected solar radiation, which is short-wave energy, only about 15 per cent being absorbed, and that chiefly by small amounts of water vapor* (Finch and Trewartha, *Elements of Geography*). *Solar radiation in the desert frequently is intense because the cloud cover is irregular, negligible, or thin* (Brooks D. Church).

solar system, *Astronomy.* the sun and all the planets, satellites, comets, asteroids, and other celestial bodies that revolve around it: *The solar system consists mainly of empty space. What material it does contain belongs almost entirely to a huge dominating incandescent globe—the sun ... and it is in fact this very dominance by the sun which gives the system its name, the solar system* (Krogdahl, *The Astronomical Universe*).

solar time, = apparent solar time: *Sidereal time is divided similarly; but the minutes and seconds of solar time are each a little longer than the corresponding units of sidereal time* (Bernhard, *Handbook of Heavens*).

solar wind, *Astronomy.* a continuous stream of ionized particles ejected by the sun, extending in all directions through interplanetary space: *The solar wind is not like a wind at earth's surface, but a vast mass of charged particles, called a plasma ... When the solar wind of charged, or ionized, particles is rushing past earth, it has a velocity of 200 miles a second* (Science News Letter). See the picture at **magnetosphere.**

solar year, the period of time required for the earth to make one complete revolution around the sun, from one vernal equinox to the next. The solar year is 365 days, 5 hours, 48 minutes, 45.51 seconds. Also called **astronomical year** and **tropical year.**

soldier, *n. Zoology.* **1** a type of worker with a large head and powerful jaws in certain ant colonies.
2 a kind of large-headed individual in a colony of termites: *The soldiers, which are also wingless and blind, are larger than the workers* (Carl D. Duncan).

solenoid (sō′lə noid), *n. Electricity.* a current-carrying coil surrounding a movable iron core, especially a spiral or cylindrical coil of wire that acts like a magnet when a current passes through it. A solenoid is used to convert electrical into mechanical energy. *A single loop or a solenoid carrying a current experiences a torque in a magnetic field* (Physics Regents Syllabus). [from French *solénoïde*, from Greek *sōlēn* channel + *eidos* form] **—solenoidal,** *adj.*

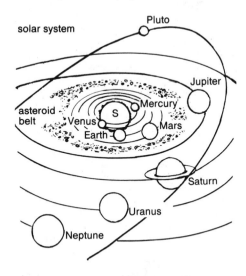

solar system

soleus (sō'lē us), *n., pl.* **solei** (sō'lē ī). *Anatomy.* a broad, flat muscle of the calf of the leg, situated immediately in front of and deeper than the gastrocnemius. [from New Latin, from Latin *solea* sandal, shoe]

solfatara (sōl'fä tä'rä), *n. Geology.* a volcanic vent or area that gives off only sulfurous gases, steam, and the like. [from Italian *Solfatara,* a volcano near Naples, from *solfo* sulfur]

solid, *adj.* **1** *Physics.* being a solid; not liquid or gaseous. Most solid substances can be melted by heat and changed to liquids; conversely, a liquid becomes solid when it freezes. *The separation of a fluid from the solid or undissolved particles which it contains ...* (J. Smith). **2** *Geometry.* having length, breadth, and thickness. A sphere is a solid figure.
—*n.* **1** *Physics.* a substance consisting of densely packed atoms. Solids are much more resistant to deformation of shape and volume than are liquids or gases. Most solids have a crystalline structure. *Solids which compose the great known part of the globe ...* (Sir Humphry Davy). *The use of the terms "solid" and "fluid" in connection with the huge pressures prevailing in the earth's interior is sometimes questioned. What a geophysicist means by the term "solid" in this context is simply that the elastic behavior of the material in question can be described by equations which match those applying to ordinary solids in normal laboratory conditions* (Scientific American). **2** *Geometry.* a body that has length, breadth, and thickness. A cube is a solid. [from Latin *solidus*]

solid angle, *Geometry.* an angle formed by the three or more planes intersecting at a common point.

solid geometry, the branch of mathematics that deals with objects having the three dimensions of length, breadth, and thickness.

solidification, *n. Physics.* the act or process of solidifying; alteration of a fluid to a solid.

solidify, *v. Physics.* to make or become solid; change from a fluid to a solid: *Extreme cold will solidify water by freezing.*

solid-state, *adj.* of or having to do with solid-state physics: *Undoubtedly, many solid-state phenomena besides superconductivity will be studied with the shortest radio waves* (Scientific American). *There are at least several solid-state electron processes beside those we have mentioned (luminescence and the flow of impurity-atom electrons)* (Harper's).

solid-state physics, the branch of physics that deals with the physical properties of solid materials, such as mechanical strength, the movement of electrons, the nature of crystals, etc. Research in solid-state physics has produced the transistor and other semiconductor devices. *From the beginning of solid-state physics, infrared, ultraviolet, and X-rays provided important methods of probing the properties in solids* (William E. Spicer).

solifluction (sol'ə fluk'shən), *n. Geology.* the downward movement of soil and rock on the face of the earth, caused by the action of the weather: *Heat can exacerbate the natural process of solifluction, the slow creep of earth down a slope* (Science News). [from Latin *solum* soil + *fluctionem* act of flowing, from *fluere* to flow]

solitary, *n.* **1** *Zoology.* living alone, rather than in colonies: *Solitary wasps are predators of a rather special sort. Only a few take prey as food for themselves; for the most part the adults of all species feed on sugars in solution, which they find in the nectar of flowers, in ripe fruit or in the honeydew secreted by aphids and other plant-sucking insects* (Scientific American). **2** *Botany.* growing separately; not forming clusters: *solitary inflorescence.*

soliton (sol'ə ton), *n. Physics.* a solitary wave of energy caused by a single disturbance: *Solitons have actually been around in solid state physics for some years, in the guise of nondispersive waves ... A soliton ripple would always remain the same height and width: it would progress over the surface maintaining its initial form* (New Scientist). [from *solit(ary)* + -*on*]

solstice (sol'stis), *n. Astronomy.* **1** either of the two times in the year when the sun is at its greatest distance from the celestial equator. In the Northern Hemisphere, June 21 or 22, the **summer solstice,** is the longest day of the year and December 21 or 22, the **winter solstice,** is the shortest. *On June 21 or 22, the sun is at its most northern point. Then it takes the northern half of the earth longer to turn through sunlight than through the night. The northern nights are short; days are long. This is the time of the solstice—summer for the north but winter for the south* (Bernhard, *Handbook of the Heavens*). **2** either of the two points reached by the sun at these times: *Four equidistant points on the ecliptic are the two equinoxes, where this circle intersects the celestial equator, and the two solstices, where it is farthest away from the equator* (Baker, *Astronomy*). [from Old French, from Latin *solstitium,* from *sol* sun + *sistere* stand still]

solstice

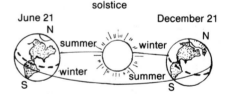

June 21
N
summer
winter
winter
summer
S

December 21
N
S

solstitial (sol stish'əl), *adj. Astronomy.* **1** of or having to do with a solstice: *The hour circles that pass through the equinoxes and solstices are known as the equinoctial and solstitial colures, respectively* (Duncan, *Astronomy*). **2** happening at or near a solstice: *solstitial rains.* **3** resembling the climate of the summer solstice.

solubility (sol'yə bil'ə tē), *n. Chemistry.* the quality that substances have of dissolving or being dissolved easily: *the solubility of sugar in water. Solubility is another property which varies with temperature, since most substances are more soluble in hot liquids than in cold. It is usually defined as the number of units of mass of the substance which will dissolve in a fixed number of units of mass of the solvent at a specified temperature* (Jones, *Inorganic Chemistry*).

soluble (sol'yə bəl), *adj. Chemistry.* that can be dissolved. Soap is soluble in water. *The product obtained by this latter method remains easily soluble in water,* whereas denatured proteins tend to become insoluble (D. L. Woodhouse and H. S. A. Sherratt). [from Latin *solubilis*, from *solvere* dissolve] —**solubleness**, *n.* —**solubly**, *adv.*

soluble RNA, = transfer RNA.

solum (sō'ləm), *n., pl.* **sola** (sō'lə), **solums**. *Geology.* the upper part of a soil profile, comprising the A and B horizons: *Thus by a complex process extending back through the ages the surface material was so changed that it became entirely different from the material beneath it. This upper changed material is the solum, or true soil. It is our most precious heritage from the past* (Colby and Foster, *Economic Geography*). [from Latin *solum* soil]

solute (sol'yüt), *n. Chemistry.* a solid, gas, or liquid that is dissolved, usually in a liquid, to form a solution: *When a solution is composed of two substances it is customary to designate as the solvent that component which is present in the larger proportion, and to call the other component the solute. This, however, is a matter of convenience, for there is no fundamental distinction between the two, and in some cases it is not clear which component is to be regarded as solvent, and which as solute* (Parks and Steinbach, *Systematic College Chemistry*). [from Latin *solutum* dissolved]

solution, *n.* **1** *Chemistry.* **a** the act or process by which a solid, liquid, or gas is uniformly mixed with another solid, liquid, or gas: *Weathering includes several processes. One of these is solution, for any kind of rock will dissolve at least to a slight degree whenever it is in contact with water* (Emerson, *Basic Botany*). **b** the homogeneous mixture formed by this process: *The components of a solution cannot be distinguished with the aid of a microscope or an ultramicroscope, and they cannot be separated by filtration or by sedimentation. The proportions of the components of a solution can be varied continuously within certain limits without producing an abrupt change in any of its properties. In this respect, a solution differs from a chemical compound whose composition cannot be varied* (Parks and Steinbach, *Systematic College Chemistry*). **c** the condition of being dissolved: *Water that has been filtered through much sand and soil may be perfectly clear and yet be carrying minerals in solution* (Beauchamp, *Everyday Problems in Science*). **2** *Mathematics.* **a** the answer to a problem. **b** a value of a variable that satisfies an equation; any number which makes an open sentence a true statement. EXAMPLE: The number 4 is a solution of $3y + 2 = 14$ because $3 \times 4 + 2 = 14$. [from Latin *solutionem* a loosing, from *solvere* loosen]

solution set, *Mathematics.* the set which contains all the solutions of an open sentence: *Depending on the problem, a solution set may be an empty set, or it may have any number of members* (C. W. Junge). *The members of a replacement set of an open sentence which make the sentence true constitute the solution set of the open sentence. If every such replacement makes the open sentence false, the solution set is empty* (Mathematics Regents Syllabus). Also called **truth set.**

solvate (sol'vāt), *Chemistry.* —*n.* a chemical substance produced by the combination of the ions or molecules of a solvent and a solute.

—*v.* to become or cause to become a solvate: *In water, the hydrogen ions combine with the atoms of the water molecules (H_2O) to form the solvated hydrogen ion (H_3O^+)* (James S. Fritz).

—**solvation** (sol vā'shən), *n.* the combination of a solute with its solvent: *The concept of the "solvation" of ions in solution, that is, their collecting a surrounding layer of bound solvent molecules, is inherently rather vague; while such a process undoubtedly occurs, different methods of measuring the number of solvated molecules give results varying from a few per ion up to values requiring more solvent than is actually present in the whole solution* (New Scientist).

solvent, *n. Chemistry.* **1** a substance, usually a liquid, that can dissolve other substances: *A great range of solvents has now been explored for selectively dissolving specific chemicals out of coal paste* (Scientific American). **2** a substance in which another is dissolved, forming a solution: *Water is the most common solvent in electrolytic solutions* (Physics Regents Syllabus). Compare **solute.**

solvolysis (sol vol'ə sis), *n. Chemistry.* a reaction in which a solvent reacts with the solute to form a new, usually intermediate, compound: *Dilute sodium hydroxide was added to the mixture. This is a common procedure in many such "solvolysis" reactions* (New Scientist).

—**solvolytic** (sol'və lit'ik), *adj.* having to do with or producing solvolysis.

soma (sō'mə), *n., pl.* **somata** (sō'mə tə). *Biology.* all the tissues of an organism except the germ cells. [from Greek *sōma* body]

somat-, *combining form.* the form of **somato-** before vowels, as in *somatic.*

somatic (sō mat'ik), *adj. Biology.* **1** of or having to do with the body: *Somatic protoplasm, or somatoplasm, is that protoplasm comprising the body tissues in general* (Harbaugh and Goodrich, *Fundamentals of Biology*). **2** having to do with or consisting of somatic cells: *somatic tissues. Many geneticists ... consider malignancies as consequences of somatic gene mutations, and believe therefore that their numbers should be strictly proportional to radiation exposure, as is the number of mutations in the reproductive cells, which cause genetic deviations* (Eugene Rabinowitch). **3** having to do with the soma, especially as contrasted with the mind or nervous system: *the somatic effects of a disease.*

—**somatically**, *adv.*

somatic cell, *Biology.* any cell of an organism except a germ cell: *The cells in a multicellular animal may be divided into (1) somatic cells or body cells (and their products), constituting the individual animal throughout its life; and (2) germ cells having to do only with*

cap, fāce, fäther; best, bē, tèrm; pin, five; rock, gō, ôrder; oil, out; cup, pùt, rüle, *yü* in use, *yu* in uric; *ng* in bring; *sh* in rush; *th* in thin; ŦH in then; *zh* in seizure. ə = *a* in about, *e* in taken, *i* in pencil, *o* in lemon, *u* in circus

607

reproduction and continuance of the species (Storer, *General Zoology*).

somatic nervous system or **somatic system**, *Physiology.* the part of the peripheral nervous system of vertebrates which controls the voluntary movements of the skeletal muscles. Compare **autonomic nervous system.**

somato-, *combining form.* body; of the body; somatic, as in *somatochrome, somatosensory.* [from Greek *soma, somatos* body]

somatochrome (sō′mə tə krōm), *n. Biology.* a nerve cell which possesses a well-marked cell body surrounding the nucleus on all sides and staining deeply in basic aniline dyes.

somatogenic, *adj. Biology.* of somatic origin; developing from the somatic cells.

somatomedin (sō′mə tə mē′din), *n. Biochemistry.* a hormone secreted by the liver which promotes the action of somatotropin. [from *somato(tropin)* + Late Latin *mediari* to intervene, mediate]

somatoplasm (sō′mə tə plaz′əm), *n. Biology.* the protoplasm of the somatic cells: *The somatoplasm dies periodically, but the germ plasm is potentially immortal* (Hegner and Stiles, *College Zoology*).
—**somatoplastic**, *adj.* of or having to do with the somatoplasm.

somatopleural (sō′mə tə plür′əl), *adj. Embryology.* of, having to do with, or forming the somatopleure.

somatopleure (sō′mə tə plür), *n. Embryology.* the outer layer of the two layers of mesoderm tissue of craniate vertebrates which forms the body wall. [from *somato-* + Greek *pleura* side]

somatosensory, *adj. Biology.* of or having to do with sensations involving the parts of the body that are not associated with the eyes, tongue, ears, and other primary sense organs: *the somatosensory cortex, the somatosensory system of the crayfish.*

somatostatin (sō′mə tə stat′in or sō′mə tə stā′tin), *n. Biochemistry.* a hormone produced by the hypothalamus that inhibits the secretion of various other hormones, especially those regulating body growth and glucagon and insulin production: *Somatostatin consists of a long peptide chain of 14 amino acids and can now be synthesized in the laboratory* (Earl A. Evans, Jr.).

somatotrophin (sō′mə tə trō′fin), *n.* = somatotropin.

somatotropic hormone, = somatotropin.

somatotropin (sō′mə tə trō′pin), *n. Biochemistry.* the pituitary hormone that regulates the growth of the body: *The growth hormone, somatotropin, which stimulates growth and the synthesis of proteins in all parts of the body, is a product of the anterior lobe* (Bradley T. Scheer). [from *somato-* + Greek *tropē* a turning]

-some, *combining form.* body, as in *chromosome = color body, monosome = single body.* [from Greek *soma*]

somite (sō′mīt), *n.* **1** *Zoology.* any of a longitudinal series of more or less similar parts or segments composing the body of certain animals, such as the earthworm. Also called **metamere.**

2 *Embryology.* a segment of the body of an embryo, such as the myotome: *There are many blocks of tissue, called somites, along the back of the embryo. These will develop into muscles* (Stephen C. Williams). [from Greek *soma* body + English *-ite*]
—**somitic** (sō mit′ik), *adj.* of, having to do with, or resembling a somite or somites: *a somitic ring or joint.*

sone (sōn), *n. Physics.* a unit of loudness. One sone is equivalent to a simple tone having a frequency of 1000 cycles per second, 40 decibels above the listener's threshold. [from Latin *sonus* sound]

sonic, *adj. Physics.* **1** of, having to do with, or using sound waves: *In the same class is the ... "sonic" drill that sets up vibrations with sound waves* (Wall Street Journal).
2 having to do with the rate at which sound travels in air (about 331 meters per second at sea level): *sonic speed.*
[from Latin *sonus* sound] —**sonically**, *adv.*

sonicate, *Physics.* —*v.* to break up or disperse (a substance) by means of high-frequency sound waves: *25 to 50 mg of DNA, were placed in a rosette flask cooled by an ice-brine bath and were sonicated* (Science).
—*n.* a particle of a substance that has been sonicated: *The sonicates were centrifuged for 10 min at 750 r.p.m.* (Nature).
—**sonication**, *n.* the act or process of sonicating: *Very small, single membraned liposomes produced by sonication of a smetic mesophase* (New Scientist).

sonic barrier, *Physics.* a sudden increase in air resistance met by an aircraft or projectile as it nears the speed of sound. Also called **sound barrier.**

sonoluminescence, *n. Physics.* luminescence produced by high-frequency sound waves. [from Latin *sonus* sound + English *luminescence*]
—**sonoluminescent**, *adj.* of or having to do with sonoluminescence: *On a sonoluminescent viewing screen, ... each part of the image will fluctuate in brightness according to its sound output* (New Scientist).

sorb, *v. Chemistry.* to absorb or adsorb: *At 78°C (the temperature of solid carbon dioxide) both oxygen and nitrogen are sorbed readily* (New Scientist).

sorbent, *n. Chemistry.* anything that absorbs or adsorbs: *A mixture of chemical compounds is applied to a stationary sorbent (blotting paper demonstrates the effect) and is then made to migrate along the sorbent* (L. J. Morris).

soredial (sə rē′dē əl), *adj. Biology.* of or resembling a soredium or soredia.

soredium (sə rē′dē əm), *n., pl.* **-dia** (-dē ə). *Biology.* a fragment of a lichen that is able to develop into a new thallus when detached from the surface of the thallus: *A great many produce reproductive bodies called soredia; these are knots of hyphae with a few algal cells entangled with them, and they are usually formed in the gonidial layer of the thallus and reach the surface through cracks in the cortex* (C. L. Duddington). [from New Latin, from Greek *sōros* a heap]

sorosis (sə rō′sis), *n., pl.* **-ses** (-sēz). *Botany.* a fleshy multiple fruit composed of the ovaries, receptacles, and associated parts of an entire cluster of flowers, as in the pineapple and mulberry. [from New Latin, from Greek *sōros* a heap]

608

sorption, *n. Chemistry.* the act or process of sorbing, or the state of being sorbed; absorption or adsorption.

—**sorptive,** *adj.* capable of sorbing; tending to sorb; characterized by sorption: *In total sorptive capacity the zeolites are quite equal to the commercial gel sorbents such as charcoal, silica gel or activated alumina* (R. M. Barrer).

sorus (sôr′əs *or* sōr′əs), *n., pl.* **sori** (sôr′ī *or* sōr′ī). *Botany.* **1** any of the dotlike clusters of sporangia on the underside of the frond of a fern: *On the under surface or on the margin of the laminae, pinnae, or pinnules of most ferns may be seen small brown patches, each of which is called a sorus* (Youngken, *Pharmaceutical Botany*).

2 a cluster of spores in various lower plants: *Those amoebae that do not become stalk cells are eventually lifted in a body to the top of the stalk, where they are transformed into capsule-shaped spores encased in a globular mass of slime. This body constitutes the sorus* (A. C. Lonert).

[from New Latin, from Greek *sōros* heap]

sound, *n.* **1** *Physiology.* the sensation produced in the organs of hearing by vibrations transmitted through air or some other medium: *A sound that is barely audible to a person with good hearing is given a value of 0 decibels (db). This is called the threshold of hearing* (Tracy, *Modern Physical Science*).

2 *Physics.* the mechanical vibrations causing this sensation. Sound travels through air in waves at a rate of about 331 meters per second at sea level under normal conditions of pressure and temperature. *All sounds are produced by a vibrating object, either solid, liquid, or gas. To create a sound a force must be applied to the solid, liquid, or gas, thus causing it to move. Thus, work is done; and since energy is the ability to do work, energy is always required to produce a sound* (Obourn, *Investigating the World of Science*). See also **sound wave.**

[Middle English *soun* (the *d* is a later addition), from Old French *son,* from Latin *sonus*]

ASSOCIATED TERMS: Sound travels in *waves* as it moves through a *medium.* Sound waves are *compression waves* produced by the *vibration* of an object. The waves are transmitted by the *condensations* and *rarefactions* generated by the vibrating object. The number of vibrations (condensations and rarefactions) produced each second is the *frequency* of the waves. As the frequency increases, the *wavelength* decreases. Frequency also determines the *pitch* of a sound. The *intensity* of a sound is determined by the *amplitude* of the vibrations that produce the sound waves. Sound waves may be *reflected, refracted, diffracted,* or reinforced by *resonance.* The intensity level of sound is measured in *decibels;* the loudness level is measured in *phons.* Sound with frequencies below the audible range is called *infrasound,* and that above the range of hearing is called *ultrasound.* The science of sound is *acoustics.*

sound barrier, = sonic barrier.

sound spectrogram, a record or diagram made by a sound spectrograph. Also called **spectrogram.**

sound spectrograph, an electronic instrument which produces a graphic representation of sound, with vertical marks representing frequency and horizontal marks representing time. Also called **spectrograph.**

sound wave, *Physics.* any of the longitudinal progressive vibrations of a material medium by which sounds are transmitted: *Sound waves in the ideal sense are mechanical disturbances propagating in a continuous medium. The basic fact of interest from a physical standpoint is that these waves carry momentum and*

energy without a net transport of mass (Henry Eyring).

sour, *adj. Chemistry.* **1** acid as a result of fermentation: *sour milk.* **2** unusually acid: *sour soil.*

south, *n. Geography.* the direction to the right as one faces the rising sun; direction just opposite north.

southeaster, *n. Meteorology.* a wind or storm from the southeast.

southern, *adj. Astronomy.* of or in the southern half of the celestial sphere: *Sirius and Canopus are southern stars.*

Southern Hemisphere, *Geography.* the half of the earth that is south of the equator. See the picture at **hemisphere.**

southern lights, = aurora australis.

south geographic pole, *Geography.* the southern end of the earth's axis. Also called **South Pole.**

southing (sou′thing *or* sou′ᴛʜing), *n. Astronomy.* declination measured southward.

south magnetic pole, *Geography.* the point on the earth's surface south of the equator toward which one end of a magnetic needle points. Its location varies but is approximately 2570 kilometers from the south geographic pole, in Wilkes Land.

south pole, 1 *Physics.* the pole of a magnet that would attract that end of a compass needle that normally points south.

2 = south magnetic pole.

3 South Pole = south geographic pole.

southwester, *n. Meteorology.* a wind or storm from the southwest.

sp., *abbrev.* **1** species. **2** specimen.

space, *n.* **1** *Physics.* the three-dimensional volume which extends in all directions and in which all matter exists: *A vacuum is empty space.*

2 *Astronomy.* **a** the region of the solar system beyond the earth's atmosphere: *translunar space, interplanetary space.* **b** the part of the universe beyond the solar system: *intergalactic space, interstellar space.*

3 *Mathematics.* **a** the set of all points. **b** a set of points that fulfills certain conditions: *topological space, Euclidean space.*

space charge, *Physics.* the electric charge contained in a given volume of space: *The escape of electrons from the surface of an isolated metallic conductor leaves the conductor positively charged. The electrons which have escaped are therefore attracted by the conductor and form a "cloud" of negative charge outside its surface. This cloud is called a space charge* (Sears and Zemansky, *University Physics*).

space lattice, *Physics.* a pattern consisting of a series of points formed by the intersections of three systems of parallel and equally spaced lines, as in the arrangement of the atoms or ions in a crystal: *The application*

cap, fāce, fäther; best, bē, tèrm; pin, fīve; rock, gō, ôrder; oil, out; cup, pùt, rüle, *yü* in use, *yu* in uric; ng in bring; sh in rush; th in thin, ᴛʜ in then; zh in seizure.

ə = *a* in about, *e* in taken, *i* in pencil, *o* in lemon, *u* in circus

of heat increases the amplitude of atomic vibrations on the space lattice of the metal, till at a certain temperature the binding forces in the crystal structure give way and the metal melts. (F. A. Fox).

spacer, *n. Molecular Biology.* a segment of DNA that separates specific genes from one another: *For the production of a small molecule of RNA that resides in ribosomes . . . there are no less than 20,000 identical genes separated from each other by an inert DNA "spacer"* (New Scientist). Compare **intervening sequence.**

space-time, *n.,* or **space-time continuum,** *Physics.* space conceived as a continuum of four dimensions (length, width, height, and time), within which physical events can be located exactly: *Ever since Einstein based relativity on a four-dimensional geometry of space-time, scientists have found the fourth and higher dimensions increasingly useful and even necessary to describe the world in which we live* (Kenneth Engel). *In his special theory of relativity, Einstein showed that time can be considered as the "fourth coordinate," supplementing the three spatial coordinates and thus forming a four-dimensional space-time continuum* (George Gamow).

spacial, *adj.* = spatial.

spadiceous (spā dish'əs), *adj. Botany.* having the form of, bearing, or resembling a spadix: *a spadiceous flower.*

spadix (spā'diks), *n., pl.* **spadixes, spadices** (spā dī'sez'). *Botany.* a spike composed of minute flowers on a fleshy stem, usually enclosed in a spathe, as in the jack-in-the-pulpit and the calla lily: *When a lily is ripe for fertilizing the spadix begins to heat up, reaching temperatures 25 degrees F. or more above the surrounding air* (Science News Letter). [from Latin, from Greek *spadix* palm branch]

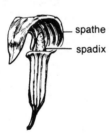

spathe — spadix

spallation (spə lā'shən), *n. Nuclear Physics.* the ejection from a nucleus of protons or neutrons, especially in the form of light nuclei, as a result of intense bombardment or excitation by foreign particles: *The little A1²⁶, seemingly identified, could have been produced by cosmic-ray spallation of argon in the atmosphere* (David W. Parkin).

spar, *Mineralogy.* any of various crystalline, shiny minerals that split easily into angular blocks, such as calcspar and fluorspar. [from Middle Low German]

spark chamber, *Nuclear Physics.* a gas-filled chamber containing charged metal plates through which subatomic particles pass, producing a visible trail of sparks by means of which the particles may be examined and identified: *In the standard spark chamber,*

bright sparks are produced by ion multiplication between the successive plates through which the original particle has passed (D. E. Yount). Compare **bubble chamber** and **cloud chamber.**

spasm (spaz'əm), *n. Physiology, Medicine.* a sudden involuntary contraction of a muscle or muscles. A *clonic spasm* is characterized by alternate contraction and relaxation of the muscles, and a *tonic spasm* by prolonged contraction without relaxation for some time. [from Greek *spasmos,* from *span* draw up, tear away] —**spasmodic** (spas mod'ik), *adj.* having to do with, resembling, or characterized by a spasm or spasms: *a spasmodic cough.* —**spasmodically,** *adv.*

spathe (spāᴛʜ), *n. Botany.* a large bract or pair of bracts, often colored, that encloses a flower cluster. The calla lily has a white spathe around a yellow flower cluster. See the picture at **spadix.** [from Greek *spathē* blade (of a sword or oar)] —**spathed,** *adj.* having or surrounded by a spathe.

spathulate (spath'yə lit), *adj.* = spatulate.

spatial, *adj. Physics.* of, having to do with, or occupying space: *New galaxies are born; others are lost to view, but their average spatial density does not change* (Science News). *'Other things' being equal, electrons tend to be in different spatial states as long as possible owing to their mutual repulsion* (H. R. Paneth). Also spelled **spacial.** —**spatiality,** *n.* spatial quality or character.

spatter cone, *Geology.* a clot or hardened mass of lava in the form of a steep-sloped cone. *Spatter cones . . . are small structures built around unimportant vents. Explosions throw lumps or sticky lava a few feet into the air; falling, they build cones or domes. Some spatter cones form on top of lava flows, where gas gathers in pockets and makes tiny explosions* (Fenton, *The Rock Book*).

spatulate (spach'ə lit), *adj.* **1** *Botany.* having a broad, rounded end and a long, narrow base: *a spatulate leaf.* **2** *Zoology.* broad and rounded (organ or part): *a spatulate joint, a spatulate process.* [from Late Latin *spatula* spoon, broad blade]

spatule (spach'ùl), *n. Zoology.* a spatulate formation or part, as the end of the tail feathers of a bird.

spawn, *n.* **1** *Zoology.* **a** the eggs or sperm deposited directly into water by aquatic animals such as fish, frogs, and shellfish. **b** the young newly hatched from such eggs. **2** *Botany.* mycelium from which mushrooms grow. —*v.* (of aquatic animals) to produce or deposit spawn; discharge eggs or sperm directly into water: *The Chinook salmon of the U.S. Northwest is born in a small stream, migrates downriver to the Pacific Ocean as a young smolt, and, after living in the sea for as long as five years, swims back unerringly to the stream of its birth to spawn* (Scientific American). [from Old French *espandre* spread out, from Latin *expandere* expand]

spec., *abbrev.* **1** specimen. **2** spectrum.

specialization, *n. Biology.* the adaptation of an organ, organism, etc., by structure or function to a particular mode of life: *Bacterial spore formation and spore germination are examples of cellular differentiation (specialization) and morphogenesis (growth and development)* (Robert G. Eagon).

specialize, *v. Biology.* to adapt by structure or function to a particular mode of life; undergo or cause to undergo specialization: *A specialized animal is one that has developed organs which fit it particularly for one specific environment . . . Hoofed animals have acquired a special kind of foot adapted to running and walking* (R. Beals and H. Hoijer). *Gall growth always originates in unspecialized tissue; the insect causes this to differentiate into specialized cells and promotes continued growth where no further growth would ordinarily occur* (William Hovanitz).

special theory of relativity. See **relativity.**

speciate (spē′shē āt), *v. Biology.* to form new species by evolutionary processes: *Hybridization prevents races speciating and provides a diverse gene pool from which a great variety of forms can ultimately be obtained* (New Scientist).
—**speciation,** *n.* the formation of new species: *Study of these chromosomal aberrations which varied from one species to the next provided great insight into incipient speciation and information as to how evolution proceeds through the ages* (Sheldon C. Reed).
ASSOCIATED TERMS: see **evolution.**

species (spē′shēz), *n., pl.* **-cies. 1** *Biology.* a group of related organisms which have certain characteristics in common and which, if they reproduce sexually, are able to interbreed freely in nature and produce fully fertile offspring. A species ranks next below a genus and may be divided into several varieties, races, or breeds. Wheat is a species of grass. *The basic group in biological classification is the species (not specie; the plural is also species) . . . Examples of common species are the house fly, yellow perch, bullfrog, and American robin. The individuals comprising a species often can be subdivided into smaller groups known as geographic races or subspecies that differ from one another by average rather than absolute characters* (Storer, *General Zoology*).
ASSOCIATED TERMS: see **kingdom.**
2 *Physics.* a particular kind of atom, nucleus, molecule, etc.: *The heavier elements all put together are about one tenth as frequent as helium; the frequency of any given species falls off rapidly with increasing atomic weight* (E. P. George).
[from Latin *species* (originally) appearance]

species-specific, *adj. Biology.* limited in reaction or effect to one species: *It has become apparent that viruses, which had been thought to be species-specific, can cross from one species to another and thus spread disease* (New Scientist).

specific, *adj. Biology.* of, having to do with, or characteristic of species: *According to the binomial system which has been universally adopted, every plant belongs to a species which is given two Latin names and these together represent its specific name. The first name is the name of the genus or the generic name; the second is the name of the species or the specific epithet . . . Thus, the Wild Cherry is named Prunus serotina, Prunus representing the name of the genus, serotina the specific epithet or kind of Prunus* (Youngken, *Pharmaceutical Botany*).

specific gravity, *Physics.* the ratio of the weight or mass of a given volume of any substance to that of an equal volume of some other substance taken as a standard (usually water at four degrees Celsius for solids and liquids, and hydrogen or air for gases). The specific

gravity of platinum is 21.5 because any volume of platinum weighs 21.5 times as much as the same volume of water at 4°C. A hydrometer is often used to find the specific gravity of a liquid. *Abbreviation:* s.g., sp. gr. Compare **density.**

specific gravity

600g iron ÷ 40g water = 15g specific gravity

specific heat, *Physics, Chemistry.* **1** the number of calories of heat required to raise the temperature of one gram of a substance one degree Celsius.
2 the ratio of the quantity of heat required to raise the temperature of a unit mass of a substance one degree Celsius to the quantity of heat required to raise the same mass of some other substance, usually water, one degree Celsius. *Abbreviation:* S, sp. ht.

specimen, *n.* **1** one of a group or class taken to show what others are like: *to collect specimens of rocks.* **2** a sample of anything selected for testing, examination, or study: *a blood specimen. There should be no need to cut the specimen up into thin sections in order to examine it, and this contrasts strongly with the case of the electron microscope, where sections of 1/10,000 mm. thickness are needed to avoid absorption of electrons* (J. G. Thomas). *Abbreviation:* sp. [from Latin, from *specere* to view]

spectra (spek′trə), *n.* a plural of **spectrum.**

spectral, *adj. Physics.* **1** of or produced by the spectrum: *spectral colors, the spectral shift of stars.* **2** carried out or performed by means of the spectrum: *spectral analysis.* —**spectrally,** *adv.*

spectral class, *Astronomy.* any of the classes into which stars are divided on the basis of their spectra. In order of decreasing temperature, the classes are as follows: O, B, A, F, G, K, M. Stars of classes O and B are bluish white; M stars are red. G stars, such as the sun, are yellow. *A spectral class is given a letter designation . . . [and] graduated subdivisions within a letter class are indicated by Arabic numeral suffixes from 0 to 9, for example, A0 or M4* (Krogdahl, *The Astronomical Universe*).

cap, fāce, fäther; best, bē, tèrm; pin, fīve;
rock, gō, ôrder; oil, out; cup, pùt, rüle,
yü in use, *yu* in uric;
ng in bring; *sh* in rush; *th* in thin, ᴛʜ in then;
zh in seizure.
ə = *a* in about, *e* in taken, *i* in pencil, *o* in lemon, *u* in circus

611

spectral line

spectral line, *Physics.* any of the dark or bright lines in a spectrum caused by the transition of atoms from one energy level to another and indicating the presence of a particular chemical element; absorption line or emission line: *Because spectral lines are very narrow, it is unlikely that two molecules will have a line at the very same wavelength. In fact, a molecule can often be identified on the basis of a single spectral line* (Patrick Thaddeus).

spectral series, *Physics.* a series of spectral lines, usually related by a mathematical formula. See also **Balmer series.**

spectrin, *n. Biochemistry.* a protein found in the membrane of red blood cells: *The two heaviest polypeptide components . . . are collectively known as spectrin. Spectrin accounts for about a third of all the protein in the red-cell membrane* (Scientific American).

spectro-, *combining form.* having to do with the spectrum or with spectrum analysis, as in *spectroscope = an instrument for spectrum analysis, spectrogram = a photograph of a spectrum.* [from *spectrum*]

spectrogram, *n. Physics.* **1** a photograph or drawing of a spectrum: *a spectrogram of the sun.*
2 = sound spectrogram: *Acoustics researchers have made spectrograms in which sound is analyzed in terms of the frequencies that are present in successive time intervals* (James A. Young, Jr.).

spectrograph, *n. Physics.* **1** a spectroscope equipped with a photographic or digital device for recording the spectrum: *The vaporized material, heated further by an electric spark, rises in a plume whose light is "read" by a spectrograph.* (National Geographic).
2 = sound spectrograph.
—**spectrographic,** *adj.* of, having to do with, or obtained by use of the spectrograph: *The corona presents a quite different picture on the spectrographic plate* (Scientific American). —**spectrographically,** *adv.*

spectrometer, *n. Physics.* a device equipped with a scale for measuring wavelengths of radiation, energies of particles, or any other quantity that covers a range of values: *Absorption of infrared light, as measured by the spectrometer, reveals the kind of . . . atoms in the atmosphere through which the light waves have passed* (Science News Letter).
—**spectrometric,** *adj.* of or having to do with a spectrometer or spectrometry: *spectrometric chemical analysis.* —**spectrometrically,** *adv.*
—**spectrometry,** *n.* the science that deals with the use of the spectrometer: *infrared spectrometry, mass spectrometry.*

spectrophotometer, *n. Physics.* an instrument used to compare the intensities of two spectra, or the intensity of a given color with that of the corresponding color in a standard spectrum: *Newly developed sensitive ultraviolet spectrophotometers have permitted direct observation of oxidation and reduction . . .* (Frederick L. Crane). Compare **microspectrophotometer.**
—**spectrophotometry,** *n.* the science that deals with the use of the spectrophotometer.

spectroscope, *n. Physics.* an instrument for producing and examining the spectrum of a ray from any source. The spectrum produced by passing a light ray through a prism or grating can be examined to determine the composition of the source of the ray. *With the possible exception of the telescope itself, no astronomical instrument has excelled the simple spectroscope in importance for our knowledge of the material universe* (Harlow Shapley).
—**spectroscopic,** *adj.* of or having to do with the spectroscope or with spectroscopy: *a spectroscopic prism. Earth-based spectroscopic studies had indicated the presence of carbon dioxide in the Venusian atmosphere* (Hugh Odishaw). —**spectroscopically,** *adv.*
—**spectroscopy,** *n.* the science dealing with the examination and analysis of spectra: *The detailed study of emission and absorption spectra is called spectroscopy. This field of study is of importance not only in giving us detailed theories of the structure of atoms and molecules based on interpretation of the observed spectra but also in providing methods of chemical analysis* (Shortley and Williams, *Elements of Physics*).

spectroscopic binary, *Astronomy.* a binary star whose two components are so close that they cannot be separated with a telescope and are known to exist only through spectrum analysis, which reveals Doppler shifts due to the individual stellar motions.

spectrum (spek′trəm), *n., pl.* **-tra** (trə), **-trums.** *Physics.*
1 the band of colors formed when a beam of white light is broken up by passing through a prism or by some other means. A rainbow has all the colors of the spectrum: red, orange, yellow, green, blue, indigo, and violet. *. . . passing sunlight through a . . . prism and obtaining a . . . rainbow-like band of colors is the sun's spectrum. The sun's spectrum actually extends beyond both the red and violet regions but these infrared and ultraviolet portions of the spectrum are not visible to the human eye* (J. Gordon Vaeth).
2 the range of wavelength formed when any other form of radiant energy is broken up: *The spectrum of electromagnetic waves . . . ranges from the extremely short gamma-rays emitted by radioactive substances to the very long waves used for radio communication* (W. C. Vaughan). Compare **band spectrum, continuous spectrum, line spectrum.**
3 = mass spectrum: *If two hydrocarbons are present together in the mass spectrometer, the spectrum that is observed is simply the sum of the spectra that each would produce if present by itself* (M. P. Barnett).
[from Latin *spectrum* appearance, from *specere* to view]

specular, *adj.* **1** of or like a mirror; reflecting; lustrous. **Specular iron** or **specular hematite** is a variety of hematite with a brilliant metallic luster.
2 of, having to do with, or resembling a speculum: *Reflection of light from a smooth surface is called regular reflection or specular reflection* (Shortley and Williams, *Elements of Physics*).

speculum (spek′yə ləm), *n.* **1** *Optics.* a mirror or reflection of polished metal. A reflecting telescope contains a speculum.
2 *Zoology.* a patch of color on the wings of many ducks and certain other birds. In ducks it is often an iridescent green or blue.
[from Latin *speculum* mirror, from *specere* to view, look]

speleology (spē′lē ol′ə jē), *n. Geology.* the scientific study and exploration of caves. [from Greek *spēlaion* cave]

612

sperm, *n. Biology.* **1** a sperm cell; spermatozoon: *The gamete from the male is commonly very small and motile, being able to swim by the whipping of one or more cilia or flagella. This motile gamete is called the sperm . . ., a term which is an abbreviation of spermatozoan* (Harbaugh and Goodrich, *Fundamentals of Biology*). *In the case of angiosperms . . . the two gametes are very different from one another. The larger one, produced by the female parent, is the egg, and the smaller comes from the male and is called the sperm* (Emerson, *Basic Botany*). See the picture at **fertilization.**
2 = semen: *Today about half of the cattle born in the world are produced from previously frozen sperm* (Science News). [from Greek *sperma* sperm, seed, from *speirein* to sow]

spermagonium (spėr'mə gō'nē əm), *n., pl.* **-nia** (-nē ə). *Biology.* one of the cup-shaped cavities or receptacles in which the spermatia of certain lichens and fungi are produced. [from Greek *sperma* sperm + a root *gen-* to bear]

spermary, *n. Biology.* the organ or gland in which spermatozoa are generated in male animals. SYN: testis.

spermatheca (spėr'mə thē'kə), *n., pl.* **-cae** (-sē). *Zoology.* a receptacle in the oviduct of many female invertebrates, for receiving and holding spermatozoa: *The female ant, along with the wasp and the bee, possesses a spermatheca, a receptacle in which the male sperm is stored and kept alive for a lifetime being withdrawn in infinitesimal quantity as needed* (Alfred L. Kroeber). [New Latin, from Greek *sperma* sperm + *thēkē* receptacle]
—**spermathecal,** *adj.* of or having to do with a spermatheca: *the spermathecal duct.*

spermatic, *adj. Biology.* **1** of or having to do with sperm or with a sperm-producing gland: *a spermatic filament, the spermatic artery.*
2 containing, conveying, or producing sperm: *the spermatic duct.*

spermatic cord, *Anatomy.* the cord by which the testicle is suspended within the scrotum, enclosing the vas deferens, the blood vessels and nerves of the testicle, and extending to the groin.

spermatic fluid, = semen.

spermatid (spėr'mə tid), *n. Biology.* a haploid cell resulting from the meiotic division of a secondary spermatocyte; an immature sperm cell: *When the secondary spermatocytes divide, each produces 2 functional cells; these are called spermatids. The spermatids change without further division into spermatozoa* (Hegner and Stiles, *College Zoology*). Compare **ootid.**

spermatium (spėr mā'shē əm), *n., pl.* **-tia** (-shē ə). *Biology.* **1** the nonmotile male gamete that fuses with the carpogonium in certain algae.
2 one of the minute cylindrical or rod-shaped bodies in certain lichens and fungi that are produced like spores in spermagonia and function as male fertilizing organs: *Reproduction is asexual by spores or sexual by spermatia, functioning like sperms, and by eggs* (Youngken, *Pharmaceutical Botany*). [from New Latin, from Greek *spermation,* diminutive of *sperma* sperm]

spermato-, *combining form.* seed; sperm, as in *spermatocyte* = *a germ cell that produces spermatozoa; spermatophyte* = *a plant that produces seeds.* [from Greek *sperma, spermatos* seed, sperm]

spermatocyte (spėr mat'ə sīt), *n. Biology.* a germ cell that gives rise to spermatozoids or to spermatozoa. A primary spermatocyte divides by meiosis to form two secondary spermatocytes which, in turn, undergo meiosis to form spermatids which are converted into spermatozoa.

spermatogenesis (spėr'mə tō jen'ə sis), *n. Biology.* the formation and development of spermatozoa: *Sperm formation is termed spermatogenesis, and the formation of the egg, oogenesis. Accordingly, in the life cycle of each of the more highly developed plants and animals there are two generations, which alternatively produce each other* (Harbaugh and Goodrich, *Fundamentals of Biology*). Also called **spermiogenesis.**
—**spermatogenetic,** *adj.* of or having to do with spermatogenesis: *the spermatogenetic process.*
—**spermatogenic,** *adj.* **1** producing spermatozoa: *the spermatogenic cords, spermatogenic cells.*
2 = spermatogenetic.

spermatogonial (spėr mat'ə gō'nē əl), *adj. Biology.* of or having to do with a spermatogonium or spermatogonia: *the spermatogonial stage.*

spermatogonium (spėr mat'ə gō'nē əm), *n., pl.* **-nia** (-nē ə). *Biology.* **1** a primitive germ cell that divides and gives rise to spermatocytes: *Peripherally located spermatogonia, or germ cells, develop into the mature spermatozoa* (Richard J. Goss). **2** = spermagonium.

spermatophore (spėr mat'ə fôr), *n. Zoology.* a capsule or case containing many spermatozoa, produced by the male of many insects, mollusks, annelids, and some vertebrate animals.

spermatophyte (spėr mat'ə fīt), *n. Botany.* any of the plants that produce seeds, including both the angiosperms and the gymnosperms: *The seed plants or spermatophytes . . . have a closer affinity to the ferns than do the ferns to the club mosses* (Fuller and Tippo, *College Botany*).
—**spermatophytic** (spėr mat'ə fit'ik), *adj.* of or having to do with a spermatophyte.

spermatozoal (spėr'mə tə zō'əl), *adj. Biology.* of, having to do with, or resembling spermatozoa or a spermatozoan: *spermatozoal filaments, spermatozoal nuclei.*

spermatozoan (sper'mə tə zō'ən), *adj.* = spermatozoal.
—*n.* = spermatozoon.

spermatozoid (spėr'mə tə zō'id), *n. Botany.* one of the tiny motile male gametes produced in an antheridium by which the female organs are fertilized. Also called **antherozoid.**

---

cap, fāce, fäther; best, bē, tėrm; pin, fīve;
rock, gō, ôrder; oil, out; cup, pùt, rüle,
yü in use, *yu̇* in uric;
ng in bring; *sh* in rush; *th* in thin, ᴛʜ in then;
zh in seizure.
ə = *a* in about, *e* in taken, *i* in pencil, *o* in lemon, *u* in circus

spermatozoon

spermatozoon (spèr′mə tə zō′ən), *n., pl.* **-zoa** (-zō′ə). *Biology.* a mature male gamete or reproductive cell, especially the motile male germ cell of an animal, usually consisting of a round or cylindrical head, a short neck or midpiece, and a long, threadlike tail: *Around the 14th day of the cycle, ovulation occurs: the follicle releases the ovum, which travels down the Fallopian tube toward the uterus. If it encounters a live spermatozoon on the way, or soon after its arrival, the ovum will be fertilized* (Time). [from *spermato-* + Greek *zoion* animal]

sperm cell, *Biology.* a male gamete or germ cell; spermatozoon. A sperm cell unites with an ovum to fertilize it.

spermidine (spèr′mə dēn), *n. Biochemistry.* a crystalline basic substance found in semen and in some animal tissues: *Substances in sperm which have been reported to kill bacteria are spermine and spermidine, both of which are nitrogenous bases* (London *Times*). *Formula:* $C_7H_{19}N_3$

spermine (spèr′mēn), *n. Biochemistry.* a crystalline basic substance found in semen, sputum, yeast, and other substances. *Formula:* $C_{10}H_{26}N_4$

spermiogenesis (spèr′mē ō jen′ə sis), *n. Biology.* **1** the development of a spermatozoon from a spermatid. **2** = spermatogenesis.

sperrylite (sper′ē līt), *n. Mineralogy.* a mineral, an arsenide of platinum, occurring in minute, silvery white, isometric crystals. *Formula:* $PtAs_2$ [from Francis L. *Sperry,* a Canadian chemist of the 1800's who discovered the mineral + *-lite*]

spessartite (spes′ər tīt), *n. Mineralogy.* a species of garnet, a manganese aluminum silicate, that is red or yellowish-red in color and contains small amounts of various other elements, such as iron and magnesium. *Formula:* $Mn_3Al_2(SiO_4)_3$ [from *Spessart,* a district in Bavaria + *-ite*]

sp. gr., *abbrev.* specific gravity.

sphalerite (sfal′ə rīt′), *n. Mineralogy.* native zinc sulfide, the main ore of zinc. Sphalerite is dimorphous with wurtzite. *Formula:* ZnS Also called **blende, zincblende.** [from Greek *sphaleros* deceptive, slippery]

sphene (sfēn), *n. Mineralogy.* a silicate of calcium and titanium commonly found in igneous rocks. *Formula:* $CaTiSiO_5$ Also called **titanite.** [from French *sphène,* from Greek *sphēn* wedge (so called because of the mineral's crystalline shape)]

sphenochasm (sfē′nə kaz′əm), *n. Geology.* a deep wedge-shaped depression of the earth's crust between two faults: *Photographs provided for geologists by the manned programs have enabled them to . . . study the theory that the Arabian Sea is a sphenochasm, formed by continental drift* (Robert C. Baumann). [from Greek *sphēn* wedge + English *chasm*]

sphenoid (sfē′noid), *adj.* **1** wedge-shaped: *a sphenoid crystal, a sphenoid part or process.*
2 *Anatomy.* of, having to do with, or designating a wedge-shaped compound bone of the base of the skull.
—n. a sphenoid bone.

sphenoidal (sfi noi′dəl), *adj.* = sphenoid.

sphere, *n.* **1** *Geometry.* a round solid figure whose surface is at all points equally distant from the center: *Roughly speaking, a circle is the boundary of a round*

region in a plane; and a sphere is the surface of a round ball in space . . . The intersection of a sphere with a plane through its center is called a great circle of the sphere (Moise and Downs, Geometry).
2 *Astronomy.* **a** = celestial sphere. **b** the particular region occupied by each of the fixed stars and planets. [from Greek *sphaira* globe, ball]
—spherical, *adj.* **1** shaped like a sphere; globular: *the spherical eyeball. Only a spherical earth could explain the shape of the moon when it is eclipsed by the intervention of the earth between it and the sun* (Will Durant).
2 formed in or on a sphere: *spherical lines or figures.*
3 of or having to do with a sphere or spheres; dealing with the properties of the sphere or spherical figures: *In spherical space the sum of the angles of every triangle is greater than 180 degrees, and the excess over 180 degrees is greater the larger the triangle is in relation to the sphere* (Scientific American). **—spherically,** *adv.*

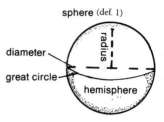

sphere (def. 1)

spherical aberration, *Optics.* aberration of rays of light resulting in a blurred or indistinct image, and arising from the spherical shape of the lens or mirror: *Spherical aberration is found in all lenses bounded by spherical surfaces. The marginal portions of the lens bring rays of light to a shorter focus than the central region* (George Wald). Compare **chromatic aberration.**

spherical angle, *Geometry.* an angle formed by two intersecting arcs of great circles of a sphere.

spherical geometry, the branch of geometry that deals with figures formed on the surface of a sphere: *Spherical geometry . . . is especially useful in navigation, because the earth is shaped approximately like a sphere* (Holmes Boynton).

spherical polygon, *Geometry.* a polygon formed on the surface of a sphere by arcs of great circles.

spherical triangle, *Geometry.* a triangle formed on the surface of a sphere by intersecting arcs of three great circles: *The sides as well as the angles of a spherical triangle are measured in degrees, and by the formulae developed in spherical trigonometry it is possible to compute any of the six parts . . . when any three are given* (Duncan, Astronomy).

spherical trigonometry, the trigonometry of spherical triangles: *Spherical trigonometry deals with triangles on the surface of a sphere* (Howard E. Eves).

spheroid (sfir′oid), *Geometry.* **—n.** a body shaped somewhat like a sphere, but not perfectly round, especially one formed by the revolution of an ellipse about one of its axes: *The earth is not a perfect sphere but an oblate spheroid, slightly flattened at the poles* (Sears and Zemansky, University Physics). **—adj.** = spheroidal.

614

spheroidal (sfi roi′dəl), *adj.* **1** shaped like a sphere: *a spheroidal galaxy.*
2 *Geometry.* having to do with a spheroid or spheroids. —**spheroidally,** *adv.*

spheroplast (sfer′ə plast), *n. Biology.* a bacterial cell that has lost most or all of its cell wall.

spherulite (sfer′ù līt), *n. Geology.* a small, spherical, concretionary mass of needlelike crystals radiating from a center, formed in certain igneous rocks: *Spherulites are the result of incipient crystallization* (Fenton, *The Rock Book*).
—**spherulitic** (sfer′ù lit′ik), *adj.* of, containing, or resembling spherulites.

sphincter (sfingk′tər), *n. Anatomy.* a ringlike muscle that surrounds an opening or passage of the body, and can contract to close it: *After the enzymes of the stomach have had time to react, the sphincter to the small intestine will open and allow the food to be squeezed into the duodenum* (Winchester, *Zoology*). [from Greek *sphinktēr*, from *sphingein* to squeeze]
—**sphincteral** (sfingk′tər əl), *adj.* of or like a sphincter.

sphingomyelin (sfing′gō mī′ə lin), *n. Biochemistry.* any of a class of phospholipids found in the brain, kidney, liver, and some other organs, consisting of choline, sphingosine, phosphoric acid, and a fatty acid. [from Greek *sphingein* to squeeze, bind + English *myelin*]

sphingosine (sfing′gə sēn), *n. Biochemistry.* a basic amino alcohol in sphingomyelin, cerebroside, and some other phosphatides. *Formula:* $C_{18}H_{37}O_2N$

sp. ht., *abbrev.* specific heat.

spicate (spī′kāt), *adj.* **1** *Botany.* having spikes; arranged in spikes: *a spicate plant or flower.*
2 *Zoology.* having the form of a spike; pointed: *a spicate appendage or process.*
[from Latin *spicatum* furnished with spikes, from *spica* spike]

spicular (spik′yə lər), *adj.* **1** having, consisting of, or covered with spicules. **2** like a spicule; slender and sharp-pointed.

spiculation (spik′yə lā′shən), *n.* formation into a spicule or spicules.

spicule (spik′yül), *n.* **1** *Anatomy.* a small, slender, sharp-pointed bone or fragment of bone.
2 *Zoology.* one of the small, spikelike, calcareous or siliceous pieces forming the skeleton of many invertebrates, especially sponges.
3 *Biology.* a small spike of flowers. SYN: spikelet.
4 *Astronomy.* one of the small, spikelike ejections of short duration from the sun's chromosphere: *Spicules are small jets of luminous material that shoot out into the atmosphere of the sun* (Scientific American).
[from Latin *spiculum* sharp point, diminutive of *spica* spike]

spiculum (spik′yə ləm), *n., pl.* **-la** (-lə). *Zoology.* **1** a sharp-pointed process or formation, as a spine of an echinoderm or a copulatory organ in a nematode. **2** = spicule. [from Latin *spiculum* sharp point; see SPICULE]

spike, *n.* **1** *Botany.* **a** an indeterminate inflorescence in which the main axis is elongated and the flowers are sessile: *A spike is like a raceme except that the flowers are sessile . . . on the main axis; that is, they have no pedicels but are attached directly to the main stalk* (Emerson, *Basic Botany*). **b** an ear of grain.

2 *Physics.* **a** a sudden, sharp uprise or peak in a motion, voltage, current, etc. **b** any tip or high point on a linear graph: *. . . brain-wave patterns characterized by six- and 14-per-second spikes in the brain-wave tracing* (Science News Letter).

spike

wheat   flower

spikelet, *n. Botany.* a small spike or flower cluster, especially a small spike in the compound inflorescence of grasses or sedges: *The spikelet is the unit of a grass inflorescence. A typical spikelet consists of a shortened zigzag axis (rachilla) bearing floral leaves* (Youngken, *Pharmaceutical Botany*).

spilite (spī′līt), *n. Geology.* a basalt altered by seawater to a mixture of low-temperature, hydrous silicates. [from French]

spilosite (spī′lə sīt), *n. Geology.* a greenish schistose rock resulting from metamorphism of slate. [from Greek *spilos* spot, speck]

spin, *n. Nuclear Physics.* **1** an internal motion of a subatomic particle analogous to rotation: *In quantum theory each particle, such as an electron, proton or neutron has a "spin," as if it were revolving about its own axis with constant speed* (R. E. Peierls).
2a the intrinsic angular momentum of a subatomic particle: *Each elementary particle of physics, such as the proton and the neutron, has an angular momentum called spin . . . The total spin, plus the . . . total linear momentum of the particles in their motion must be conserved in nuclear reactions in which particles and photons change bewilderingly into other particles and photons* (Pierce, *Electrons, Waves and Messages*). **b** a quantum number expressing spin angular momentum.
3 the total angular momentum of a nucleus (including the orbital motions of its nucleons): *Ordinary heat is motions of atoms or molecules, but when the motion has died away at 0°K, the nuclei of the atoms still have a property called "spin." Some spins have more energy than others, and the spinning nuclei can affect the spin of other nuclei near them.* (Time).

cap, fāce, fäther; best, bē, tėrm; pin, fīve;
rock, gō, ôrder; oil, out; cup, pùt, rüle,
yü in use, *yu* in uric;
ng in bring; *sh* in rush; *th* in thin, ᴛʜ in then;
zh in seizure.
ə = *a* in about, *e* in taken, *i* in pencil, *o* in lemon, *u* in circus

spinal, *adj.* **1** *Anatomy.* of, having to do with, or in the region of the spinal column or spinal cord. The **spinal marrow** is the tissue of the spinal cord.
2 *Botany, Zoology.* of, having to do with, or resembling a spine or spines.

spinal canal, the duct formed by the openings of the articulated vertebrae, containing the spinal cord.

spinal column, *Anatomy.* the flexible supporting column of bone along the middle of the back in the body of vertebrates, consisting of many vertebrae separated by intervertebral disks and held together by muscles and tendons. Commonly called **backbone, spine.**

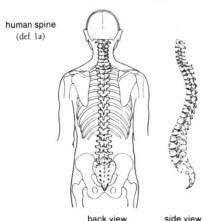

human spine
(def. 1a)

back view   side view

spinal cord, *Anatomy.* the thick, whitish cord of nerve tissue which extends from the medulla oblongata down through most of the spinal column and from which nerves to various parts of the body branch off: *The spinal cord coordinates activities between the brain and other body structures. It is a center for reflex actions* (Biology Regents Syllabus). See the pictures at **afferent, brain.**

spinal fluid, = cerebrospinal fluid.

spinal ganglion, *Anatomy.* any of the sensory ganglia situated on the dorsal root of each spinal nerve.

spinal nerve, *Anatomy.* any of the paired nerves arising from the spinal cord. There are 31 pairs of spinal nerves in the human body, 8 of which are cervical, 12 thoracic, 5 lumbar, 5 sacral, and 1 coccygeal.

spinal reflex, *Physiology.* a reflex whose pathway is through the spinal cord.

spinar (spin′är), *n. Astronomy.* a rapidly rotating galactic body: *There would be one important difference between a pulsar and a spinar. In a pulsar rotation does not contribute appreciably to the equilibrium of the object, but in the supermassive spinar centrifugal forces provide the main support against the entire object's collapsing* (Franco Pacini and Martin J. Rees). [from *spin* + *-ar,* as in *pulsar*]

spindle, *n. Biology.* the group of fibers, each composed of microtubules, along which the chromosomes are arranged during mitosis and meiosis. The spindle forms a rounded figure which tapers from the middle toward each end. See the pictures at **meiosis, mitosis.**

spindle cell, *Biology.* a spindle-shaped or fusiform cell, characteristic of some tumors: *spindle cell sarcoma.*

spine, *n.* **1** *Anatomy.* **a** a common word for the spinal column. **b** a pointed projection extending from a bone.
2 *Botany.* a stiff, sharp-pointed growth of woody tissue: *The cactus and hawthorn have spines.*
3 *Zoology.* **a** a stiff, pointed, thornlike part or projection: *A porcupine's spines are called quills.* **b** a sharp, rigid fin ray of a fish. **c** = spicule.
[from Latin *spina* (originally) thorn]

spinel (spi nel′), *n. Mineralogy.* any of a group of very hard minerals consisting chiefly of oxides of magnesium and aluminum and occurring in various colors. The magnesium may be replaced in part by ferrous iron, zinc, or manganese, and the aluminum by ferric iron or chromium. *Formula:* $MgA1_2O_4$ [ultimately, from Latin *spina* thorn]

spineless, *adj.* **1** *Zoology.* lacking a spinal column; invertebrate: *A jellyfish is a spineless animal.*
2 *Botany.* having no spines or sharp-pointed processes: *a spineless cactus.*

spin-flip, *n. Nuclear Physics.* a reversal in the direction of the spin of a particle or nucleus: *Neutral hydrogen atoms emit 21-cm radiation by a somewhat unusual "spin-flip" process in which the spins of the hydrogen nucleus and its attendant electron switch from being parallel to antiparrallel* (New Scientist).

spinneret (spin′ə ret′), *n. Zoology.* the organ by which spiders and certain insect larvae such as silkworms spin their threads for webs and cocoons. [diminutive of *spinner* one that spins]

spinnerule (spin′ər ül), *n. Zoology.* any of the tubules on the spinneret of a spider. Silk in liquid form flows through the spinnerules from silk glands in the spider's abdomen to the outside of the body. The silk then hardens into a thread.

spinode (spī′nōd), *n. Geometry.* that point on a curve where a point generating the curve has its motion precisely reversed. Also called **cusp.** [from Latin *spina* spine + English *node*]

spinor (spin′ôr *or* spin′ər), *n. Physics.* a vector having several complex numbers as components, used in quantum mechanics to describe particle spin: *Spinors representing muons, electrons and the neutrinos involved in their interaction are arranged in a certain order in a matrix* (Nature). [from *spin* + *-or,* as in *vector*]

spinous (spī′nəs), *adj.* **1** *Botany.* covered with spines; having or full of spines or thorns; thorny.
2 *Zoology.* armed with or bearing spines or sharp-pointed processes: *The porcupine has a spinous back.* **3** *Anatomy.* shaped like a spine or thorn: *the spinous process of the vertebral arch.*

spin wave, *Nuclear Physics.* a wave propagated through a crystal lattice by the deviation of magnetic moments associated with the spin angular momentum of electrons. The quantum particle of a spin wave is called a magnon.

spiracle (spir′ə kəl *or* spī′rə kəl), *n. Zoology.* a small opening for breathing, as the blowhole of a whale. Insects take in air through tiny spiracles. A shark or ray emits water through a spiracle. *The bodies of insects and related arthropods have a series of openings, or spiracles, through which air is pumped to and from an internal branched system of air ducts, or tracheae, by*

movements of the abdomen (Storer, *General Zoology*).
[from Latin *spiraculum,* from *spirare* breathe]
—**spiracular** (spī rak′yə lər), *adj.* of, having to do with, or serving as a spiracle.

spiral, *n.* **1** *Geometry.* **a** a curve on a plane which winds about a fixed point. **b** a helix or similar curve.
2 = spiral galaxy: *The Andromeda galaxy is representative only of a class of galaxies known as spirals because of their spiral or pinwheel symmetries* (Krogdahl, *The Astrononical Universe*). [from Medieval Latin *spiralis* winding, from Latin *spira* coil, from Greek *speira*]

spirals
(def. 1a)

spiral galaxy or **spiral nebula,** *Astronomy.* a galaxy or nebula appearing as one or more spiraling streams issuing from a center: *Our Milky Way system seems to be a spiral galaxy, presumably very much resembling in outline and appearance the Great Spiral in Andromeda* (Bart J. Bok). See the picture at **galaxy.**
ASSOCIATED TERM: see **galaxy.**

spiral of Archimedes, *Geometry.* a plane curve which is the locus of a point moving along a ray at a uniform rate while the ray moves at a constant angular rate. Its polar equation is $r = a\theta$, where a is a positive constant.

spiricle (spī′rə kəl), *n. Botany.* one of the delicate coiled threads on the surface of certain seeds and fruits, which uncoil when wet. [from New Latin *spiricula,* diminutive of Latin *spira* spire, coil]

spirillum (spī ril′əm), *n., pl.* **-rilla** (-ril′ə). *Bacteriology.* **1** any of a genus (*Spirillum*) of aerobic bacteria having long, rigid, spirally twisted forms and bearing a tuft of flagella. **2** any of various other spirally twisted bacteria, such as a spirochete. See the picture at **bacteria.** [from New Latin, from Latin *spira* spire, coil]
ASSOCIATED TERMS: see **coccus.**

spirochetal or **spirochaetal** (spī′rə kē′təl), *adj.* of or having to do with spirochetes.

spirochete or **spirochaete** (spī′rə kēt), *n. Bacteriology.* any of a group of motile bacteria that are slender, spiral, and able to expand and contract. One kind causes syphilis, and another causes relapsing fever. [from Greek *speira* coil + *chaitē* hair]

spirogyra (spī′rə ji′rə), *n. Biology.* any of a large genus (*Spirogyra*) of green algae that grow in scumlike masses in freshwater ponds or tanks. The cells have one or more ribbonlike chloroplasts winding spirally to the right. [from Greek *speira* coil + *gyra* circle]

spiroplasma (spī′rə plaz′mə), *n. Microbiology.* any of a genus (*Spiroplasma*) of mycoplasmas having a spiral form and lacking a cell wall, associated with various plant diseases. [from Greek *speira* coil + *plasma* something molded]

spit, *n. Geography.* **1** a narrow point of land running into the water. **2** a long, narrow reef, shoal, or sandbank extending from the shore.

splanchnic (splangk′nik), *adj. Anatomy.* of, having to do with, or in the region of the viscera: *a splanchnic nerve, the splanchnic mesoderm.* [from Greek *splanchna* viscera, from *splēn* spleen]

splanchnopleure (splangk′nə plùr), *n. Embryology.* the inner layer of the mesoderm combined with the endoderm. It develops into the connective and muscle tissue of most of the intestinal tract. [from Greek *splanchna* viscera + *pleurā* side]

spleen, *n. Anatomy.* a lymphoid organ at the left of the stomach in humans, and near the stomach or intestine in other vertebrates, that stores blood, disintegrates old red blood cells, and helps filter foreign substances from the blood. [from Greek *splēn*]

splenial (splē′nē əl), *adj. Anatomy.* of or having to do with the splenius.

splenic (splen′ik), *adj. Anatomy.* of, having to do with, or in the region of the spleen.

splenius (splē′nē əs), *n., pl.* **-nii** (-nē ī). *Anatomy.* a broad, flat muscle extending from the upper vertebrae to the neck and base of the skull. It serves to move the head and neck. [from New Latin, from Greek *splēnion* bandage, compress (used for the spleen), from *splēn* spleen]

splice, *v. Molecular Biology.* **1** to join (a gene or DNA segment of one organism) to that of another; recombine (strands of DNA molecules) from different organisms to form new genetic combinations: *The controversial research in question is a class of experiments that . . . include splicing the genes of a virus or bacterium to partially purified DNA from mammals or birds or from lower animals known to produce potent toxins or pathogens* (Science News).
2 to insert or transplant (a new or altered gene or DNA segment) into an organism such as a bacterium so as to alter its genetic makeup, especially to enable it to produce some specific chemical substance: *One valuable product has already resulted from the work: human insulin, manufactured by splicing fragments of DNA that manufacture the hormone in humans into an intestinal bacterium, causing it to start producing insulin on its own* (Newsweek).

split, *v.* **1** *Chemistry.* **a** to divide (a molecule) into two or more individual atoms or atomic groups. **b** to remove by such division.
2 *Physics.* to divide (an atom or atomic nucleus) into two portions of approximately equal mass by forcing the absorption of a neutron. Compare **fission.**

split gene, *Molecular Biology.* a segment of DNA containing one or more noncoding sectors (introns): *Molecular biologists studying other eucaryotic genes . . . showed that split genes are a common feature of eucaryotes and that introns can occur anywhere in a gene* (Daniel L. Hartl).

spodumene (spoj′ù mēn), *n. Mineralogy.* a hard, transparent or translucent silicate of aluminum and lithium, usually occurring in flat prismatic crystals. It varies in color from a light gray, yellow, or green to emerald-green and purple and is one of the main

cap, fāce, fäther; best, bē, tèrm; pin, five;
rock, gō, ôrder; oil, out; cup, pùt, rüle,
yü in use, *yù* in uric;
ng in bring; *sh* in rush; *th* in thin, ᴛʜ in then;
zh in seizure.
ə = *a* in about, *e* in taken, *i* in pencil, *o* in lemon, *u* in circus

617

sources of lithium. Some varieties are used as gems. *Formula:* LiAlSi$_2$O$_6$ [from Greek *spodoumenos* burned to ashes, from *spodos* ashes, powder (so called because it powders under the blowtorch)]

spongy cell, *Botany.* any of the cells forming the spongy layer.

spongy layer, *Botany.* a layer of loosely packed, irregularly shaped cells lying below the palisade layer in a leaf. Although most of the photosynthesis occurring in a leaf is carried out by cells of the palisade layer, those of the spongy layer are also photosynthetic. See the picture at **leaf.**

spongy parenchyma, *Botany.* the tissue formed by a layer of spongy cells: *The spongy parenchyma extends from the palisade parenchyma to the lower epidermis* (Weier, *Botany*).

spontaneous combustion, *Chemistry.* the bursting into flame of a substance as a result of heat produced by slow oxidation of the constituents of the substance itself without heat from an outside source: *Spontaneous combustion is self-initiating combustion. It occurs when a slow oxidation liberates heat which is not removed, or is removed but slowly, from the surrounding material. If the material is combustible and if it is surrounded by sufficient oxygen, combustion occurs when sufficient heat has accumulated to raise the material to its kindling temperature* (Jones, *Inorganic Chemistry*). Also called **spontaneous ignition.**

spontaneous emission, *Physics.* the release of excess energy in the form of light by an excited atom: *Light produced by the sun and by ordinary electric lights is the result of spontaneous emission caused by heat* (James P. Gordon).

spontaneous generation, *Biology.* the supposed production of living organisms from nonliving matter: *Aristotle and many of his contemporaries believed that such large animals as insects, frogs, and snakes were generated spontaneously from slime and mud. Such ideas, that life arose spontaneously, are associated with the theories of spontaneous generation or of abiogenesis* (Harbaugh and Goodrich, *Fundamentals of Biology*). Also called **abiogenesis, xenogenesis.**

spontaneous ignition, = spontaneous combustion.

sporadic E or **sporadic E layer,** *Geophysics.* an area of high ionization in the E layer of the ionosphere which interferes with the normal reflection of short-wave radio: *Sporadic E is ... sometimes only a few hundred feet thick and occurs at times now unpredictable* (Science News).

sporangial, *adj. Botany.* of or having to do with the sporangium: *sporangial cells, the sporangial layer.*

sporangiophore (spə ran′jē ə fôr), *n. Botany.* a structure or receptacle which bears sporangia: *Each sporangiophore forms a globose sporangium at its tip. As the sporangia mature, they become dark in color* (Fuller and Tippo, *College Botany*).

sporangium (spə ran′jē əm), *n., pl.* **-gia** (-jē ə). *Botany.* a receptacle or case in which asexual spores are produced; spore case. The little brown spots sometimes seen on the underside of ferns are groups of sporangia. The sporangium receives different names according to the kind of spores produced, such as *macrosporangium, microsporangium.* In mosses *sporangium*

is usually the same as *capsule. The spores of the bread mold and a good many other fungi are borne within a sporangium; other fungi ... bear their spores in chains of cells that become detached as they mature. Such spores are called conidiospores or simply conidia* (Greulach and Adams, *Plants*). [from New Latin, from Greek *spora* seed + *angeion* vessel]

spore, *Biology.* —*n.* a reproductive body, usually single-celled, which becomes free and is capable of growing into a new organism. Ferns and other flowerless plants produce spores; mold grows from spores. Some protozoans and bacteria produce spores. The spores of flowerless plants are somewhat analogous to the seeds of flowering plants. *Shortly after the mold appears, you should observe some dark objects at the ends of vertical stalks. They are spore cases; they contain asexual reproductive cells called spores* (Robert W. Menefee). *Some bacteria have the capacity to transform themselves, via an intricate sequence of organized steps, into small ovals or spheres that are highly resistant, dormant cells known as spores or endospores* (Robert G. Eagon).
—*v.* to form or produce spores.
[from New Latin, from Greek *spora* seed, spore, a sowing, from *speirein* to sow]

spore case, *Biology.* a receptacle containing spores, especially a sporangium: *As the ... body of the mold matures, many upright hyphae produce spore cases containing thousands of spores. Each spore case, or sporangium, is about the size of a pinhead* (William F. Hanna).

spore fruit, *Botany.* any part of a plant that produces spores, such as an ascocarp.

sporo-, *combining form.* spore or spores, as in *sporogenesis* = *the formation of spores.* [from Greek *spora* or *sporos* seed, spore, *speirein* to sow]

sporocarp (spôr′ə kärp), *n. Biology.* **1** a multicellular body for the formation of spores, as in red algae and ascomycetous fungi.
2 (in mosses) = sporongium.
3 a sorus in certain aquatic ferns.

sporocyst, *n. Biology.* **1** a cyst formed by sporozoans during reproduction, in which sporozoites develop.
2 an encysted sporozoan.
3 a capsule or sac containing germ cells, that develops from the embryo of trematode worms, usually within the body of a snail.
4 *Botany.* a resting cell which produces asexual spores, as in algae.
—**sporocystic,** *adj.* of or having to do with a sporocyst.

sporocyte, *n. Biology.* a cell from which a spore or spores are derived.

sporogenesis (spôr′ə jen′ə sis), *n. Biology.* **1** the formation of spores: *Meiosis occurs at various times in the life histories of primitive plants, but in the mosses, ferns, and seed plants it occurs as spores are being formed—that is, during sporogenesis* (Harbaugh and Goodrich, *Fundamentals of Biology*).
2 reproduction by means of spores.

sporogenic (spôr′ə jen′ik), *adj.* = sporogenous.

sporogenous (spə roj′ə nəs), *adj. Biology.* **1** reproducing or reproduced by means of spores: *sporogenous plants.*
2 bearing or producing spores: *There is an operculum, as usual, but the sporogenous cells form a dome-shaped mass* (Emerson, *Basic Botany*).

sporogonium (spôr′ə gō′nē əm), n., pl. -nia (-nē ə).
Botany. the sporangium of mosses and liverworts.
[from sporo- + Greek -gonos producing]

sporogony (spô roj′ə nē), n. Biology. reproduction by
means of spores, especially in certain protozoans.

sporont (spôr′ont), n. = gamont.

sporophore (spôr′ə fôr), n. Biology. 1 the branch or portion of a sporophyte which bears spores. In fungi, it is
a single hypha or branch of a hypha. The umbrella
growth, which most people call a mushroom, is really
a stalk that grows up from the mycelium. ... The umbrella is called a sporophore, which means the part
that bears the spores (William F. Hanna).
2 = sporophyte (in ferns and mosses).
[from sporo- + Greek -phoros carrying]
—sporophorous (spə rof′ə rəs), adj. 1 bearing spores.
2 of or having to do with the sporophore.

sporophyll (spôr′ə fil), n. Botany. any leaf or leaflike organ usually more or less modified, which bears spores
or sporangia: Except in a few instances the sporangia
are borne on sporophylls crowded together and forming cones or spikes (strobili) at the ends of erect
branches or as lateral branches (Youngken,
Pharmaceutical Botany). [from sporo- + Greek
phyllon leaf]

sporophyte (spôr′ə fīt), n. Botany. the individual plant
or the generation of a plant which produces spores, in
a plant which exhibits alternation of generations:
When gametes unite, a diploid zygote is formed; and
the plant developed from it is called a sporophyte because it bears spores (Miller and Leth, High School
Biology). See also alternation of generations.
—sporophytic, adj. having to do with, resembling, or
characteristic of a sporophyte.

sporopollenin (spôr′ə pol′ə nin), n. Botany. a highly resistant and durable substance that forms the outer wall
of all pollen grains except those of aquatic plants: The
sporopollenin outer wall of a pollen grain remains intact in concentrated acids and alkalis at temperatures
as high as 500 degrees Fahrenheit (Scientific American).

sporozoan (spôr′ə zō′ən), Biology. —n. any of a class
(Sporozoa) of minute parasitic protozoans which absorb food through the body wall and reproduce sexually and asexually in alternate generations. Certain
sporozoans cause diseases of man and animals, such as
malaria. Multiple fission, or sporulation, occurs in the
sporozoans ... where the nucleus divides repeatedly
and then the cytoplasm subdivides, so that a part of it
surrounds each of the many daughter nuclei (Storer,
General Zoology).
—adj. of or belonging to this class: a sporozoan parasite.
[from New Latin Sporozoa, from Greek spora spora
+ zoion animal]

sporozoite (spôr′ə zō′īt), n. Biology. any of the minute,
immature bodies, often infective, produced by the division of the spore of certain sporozoans, each of which
develops into an adult sporozoan: If a vaccinated person is bitten by an infected Anopheles mosquito and
even one sporozoite escapes the antibodies produced
by the vaccination, it can lodge in the liver and begin
the disease cycle. (Thomas H. Maugh II).

sq.

sport, Biology. —n. an animal, plant, or part of a plant
that varies suddenly or in a marked manner from the
normal type or stock. A white blackbird would be a
sport. These differences were not the slight variations
emphasized by Darwin but were wide differences,
known among plant and animal breeders of today as
sports (Harbaugh and Goodrich, Fundamentals of
Biology).
—v. 1 to vary markedly from the normal type; exhibit
spontaneous mutation: All flowers, as we know, easily
sport a little in colour (Nature).
2 to show bud variation.

sporulate (spôr′yə lāt), v. Biology. to form spores or
sporules; convert into spores: In a cell about to sporulate the nucleus is found in the centre of the cell
(Nature).
—sporulation (spôr′yə lā′shən), n. the formation of
spores or sporules: The cytosome divides, not by successive equal fissions but by many simultaneous divisions, into as many pieces as there are nuclei, thus
forming a number of small cells at the same moment.
This process is sometimes called multiple fission and
sometimes sporulation. It occurs regularly in the complicated life history of the organism of malaria (Shull,
Principles of Animal Biology).
ASSOCIATED TERMS: see asexual.

spreading factor, Biochemistry. a substance able to disperse colloids, especially the enzyme hyaluronidase.

spring, n. 1 Astronomy. the three months between the
vernal equinox and the summer solstice: The vernal
equinox [is] called "vernal" since in the northern hemisphere it marks the beginning of spring (Krogdahl,
The Astronomical Universe).
2 Geology. a small stream of water flowing naturally
from the earth: Ground water is the water that saturates the pores and cracks in soil and rock beneath the
land surface. It comes to the surface in springs, and
also swells the volume of many streams by seeping into
them from their beds and banks (Gilluly, Principles of
Geology).

spring tide, Oceanography. the exceptionally high and
low tides which come at the time of the new moon or
the full moon. Compare neap tide. See the picture at
tide.

spur, n. 1 Botany. a a slender, generally hollow, projection from a flower part, as from the calyx of colum-
bine. SYN: calcar. b a short or stunted branch or shoot
of a tree.
2 Geography. a relatively short lateral extension of a
mountain range.

sputum (spyü′təm), n., pl. -ta (-tə). Physiology. 1 saliva;
spit. 2 a mixture of saliva and mucus coughed up.
[from Latin]

sq., abbrev. square.

cap, fāce, fäther; best, bē, tèrm; pin, five;
rock, gō, ôrder; oil, out; cup, pùt, rüle,
yü in use, yu̇ in uric;
ng in bring; sh in rush; th in thin, ᴛʜ in then;
zh in seizure.
ə = a in about, e in taken, i in pencil, o in
lemon, u in circus

619

squall

squall (skwôl), *n. Meteorology.* a sudden, violent gust of wind, often with rain, snow, or sleet. [apparently related to Swedish *skval* sudden rush of water]

squall cloud, *Meteorology.* a small cloud that forms below the front edge of a thundercloud.

squall line, *Meteorology.* a line of thunderstorms preceding a cold front: *A tornado is born in the dark march across the countryside of a row of thunderstorms, called a squall line* (Science News Letter).

squama (skwā′mə), *n., pl.* **-mae** (-mē′). **1** *Zoology.* scale or scalelike part.
2 *Anatomy.* a platelike mass: *the squama of the temporal bone.*
[from Latin]
—**squamate** (skwā′māt), *adj.* having or covered with scales.
—**squamation** (skwə mā′shən), *n.* **1** the condition of being covered with scales.
2 the arrangement or pattern of the scales covering an animal.

squamosal (skwə mō′səl), *adj.* **1** *Anatomy.* of or having to do with the squama of the temporal bone. **2** = squamous.
—*n. Anatomy.* the squama of the temporal bone.

squamous (skwā′məs), *adj.* **1** *Anatomy.* **a** having to do with the squama of the temporal bone. **b** of or designating a type of epithelial tissue characterized by flat, scalelike cells: *the squamous cells of the cervix.*
2 *Zoology.* **a** consisting of scales or scalelike parts. **b** scaly; squamate.
[from Latin *squamosus,* from *squama* scale]

square, *n.* **1** *Geometry.* **a** plane figure with four equal sides and four right angles. **b** an instrument shaped like a T or an L, for drawing or testing right angles.
2 *Mathematics.* the product obtained when a number is multiplied by itself: *16 is the square of 4.*
—*adj.* **1** *Geometry.* having four equal sides and four right angles.
2 of a specified length on each side of a square: *a room ten feet square.*
3 a designating an area each of whose dimensions is that unit of length: *a square inch, a square kilometer, a square yard. A square foot measures an area one foot long and one foot wide.* SYN: squared. **b** based on such units; in square measure: *square measurement.*
—*v. Mathematics.* **a** to find or describe a square equivalent in area to: *square a circle.* **b** to multiply (a number or quantity) by itself. **c** to calculate the number of square units of measure in.
[from Old French *esquare,* ultimately from Latin *exquadrare* to make square, from *ex-* out + *quadrus* a square, from *quattuor* four]

square matrix, *Mathematics.* a matrix having an equal number of rows and columns.

square number, *Mathematics.* the product of a number multiplied by itself, as 25 (5 × 5), or 36 (6 × 6).

square root, *Mathematics.* a number that produces a given number when multiplied by itself: *The square root of 16 is 4, since 4 × 4 = 16.*

square wave, *Electronics.* a wave with a rectangular shape, which alternatively takes on two fixed values for equal periods of time.

squarrose (skwar′ōs *or* skwo rōs′), *adj.* **1** *Botany.* composed of or covered with scales, bracts, or other processes standing out at right angles or more widely, as a calyx or involucre.
2 standing out at right angles or more widely, as scales or bracts.
[from Latin *squarrosus* scurfy]

sr, *symbol.* steradian.

Sr, *symbol.* strontium.

sRNA, *abbrev.* soluble RNA.

stability, *n.* **1** *Chemistry.* the property of a compound that is not easily decomposed or easily reactive with other compounds: *The stability of water is evidence of the strength of the covalent bonds between the oxygen and hydrogen atoms* (Dull, *Modern Chemistry*).
2 *Physics.* the property of being stable.

stable, *adj.* **1** *Chemistry.* (of a compound) not easily decomposed.
2 *Physics.* **a** capable of enduring indefinitely unless disturbed by an external influence; not subject to decay: *a stable nucleus. In contrast to radioactive nuclides produced in reactors and accelerators, most naturally occurring nuclides are stable; that is, they have no tendency to decay to other nuclides* (Eicher and McAlester, *History of the Earth*). **b** tending to return to a state of equilibrium after being displaced: *a stable body, a stable system.* See the picture at **equilibrium.**

stage, *n.* **1** *Zoology.* one step or degree in a process; period of development. Frogs pass through a tadpole stage. *Insects that go through complete metamorphosis include butterflies, moths, flies, and beetles. Each goes through four stages of development: egg, larva, pupa, and adult* (Otto and Towle, *Modern Biology*).
2 *Geology.* two or more sets of related beds of stratified rocks. It is the division of rocks ranking below a series. *Seemingly abrupt changes in the nature of organic remains coincide in position with boundaries between adjoining stages* (Moore, *Introduction to Historical Geology*).

stalactites and stalagmites

stalactite (stə lak′tīt *or* stal′ək tīt), *n. Geology.* a formation of calcium carbonate, shaped like an icicle, hanging from the roof of a cave, formed by dripping water that contains calcium carbonate: *Stalactites build from the ceiling, stalagmites from the floor. In time, two such matching formations may unite into a column, but their growth is usually infinitely slow* (Jones, *Inorganic Chemistry*). [from New Latin *stalactites,* from Greek *stalaktos* dripping, from *stalassein* to trickle]

—**stalactitic** (stal′ək tit′ik), *adj.* **1** of the nature of or like a stalactite or stalactites: *the stalactitic structure of some minerals.*
2 having stalactites: *a stalactitic cave.* —**stalactitically,** *adv.*

stalagmite (stə lag′mīt *or* stal′əg mīt), *n. Geology.* a formation of calcium carbonate, shaped like a cone, built up on the floor of a cave, formed by water dripping from a stalactite: *In contrast to the thin stalactites, there are massive stalagmites, in some instances rising 40 feet from the floor of the cave* (Science News Letter). [from New Latin *stalagmites,* from Greek *stalagmos* a drop, from *stalassein* to trickle]
—**stalagmitic** (stal′əg mit′ik), *adj.* **1** of the nature of or like a stalagmite or stalagmites. **2** having stalagmites. —**stalagmitically,** *adv.*

stalk, *n.* **1** *Botany.* **a** the stem or main axis of a plant, which rises directly from the root, and which usually supports the leaves, flowers, and fruit. **b** any slender, supporting or connecting part of a plant or plantlike organism. A flower or a leaf blade may have a stalk. **2** *Zoology.* any similar part of an animal. The eyes of a crayfish are on stalks.
3 *Anatomy.* a slender part or organ resembling a stalk: *the pituitary stalk.*
[Middle English *stalke*] —**stalkless,** *adj.* —**stalklike,** *adj.*
—**stalked,** *adj.* having a stalk or stem: *a stalked barnacle or crinoid.*

stamen (stā′mən), *Botany. n., pl.* **stamens, stamina** (stam′ə nə *or* stā′mə nə). the part of a flower that contains the pollen, consisting of a slender, threadlike stem or filament which supports the anther. The stamens are surrounded by the petals. *The stamens commonly form one or more whorls between the corolla and the pistil or pistils. The number of stamens per flower is variable and may or may not bear a relation to the number of sepals or petals* (Hill, *Botany*). [from Latin *stamen* warp, thread]

STAMEN:

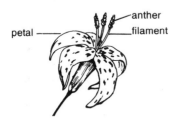

petal — anther
— filament

staminal (stam′ə nəl), *adj. Botany.* having to do with or consisting of stamens: *staminal scales, a staminal tube.*

staminate (stam′ə nit *or* stam′ə nāt), *adj. Botany.* **1** having stamens but no pistils: *If one or more pistils are present and stamens are wanting, the flower is called pistillate or female; if it possesses one or more stamens but no pistil, it is described as staminate or male* (Youngken, *Pharmaceutical Botany*).
2 having a stamen or stamens; producing stamens: *In today's cultivated corn, the staminate inflorescence or male flowering is the tassel alone and the ear carries the female flowering, the corn "silk"* (Science News Letter).

staminode (stam′ə nōde), *n.* = staminodium (def. 1): *Stamens are called staminodes if they lack anthers, or if the anthers fail to produce pollen* (Harold N. Moldenke).

staminodium (stam′ə nō′dē əm), *n., pl.* **-dia** (-dē ə). *Botany.* **1** a sterile or abortive stamen. **2** an organ resembling an abortive stamen. [from New Latin]

staminody (stam′ə nō′dē), *n. Botany.* the metamorphosis of various organs of a flower, as a sepal, petal, pistil, or bract into a stamen.

standard, *n. Botany.* **1** the large upper petal of a papilionaceous flower. Also called **vexillum.**
2 one of the three upper petallike segments of an iris: *pure white standards.*

standard deviation, *Statistics.* a measure of the dispersion or variability in a frequency distribution, equal to the square root of the arithmetic mean of the squares of the deviations from the arithmetic mean of the distribution. A distance of one standard deviation on each side of the mean of a normal curve includes 68.27 per cent of the cases in the frequency distribution. *Measures of dispersion are important for they show how much the units deviate from the mean, the mode, or the medium. There is the mean variation or deviation which shows what the average deviation of all the items is, for example, from the mean. The standard deviation is a refinement of the mean deviation; it gives greater weight to extreme deviations from the norm* (Emory S. Bogardus). Also called **root-mean-square deviation.**

standard error, *Statistics.* the standard deviation of the sample in a frequency distribution: *According to the laws of probability, given "normal distribution," 68 per cent of the cases fall within one standard error of the true value. Standard error relates to sample size* (N. Y. Times). *The standard error of the arithmetic mean is computed by dividing the standard deviation* (10.1 in the case of the control group) *by the square root of the total number of plants in the group:* $10.1 \sqrt{10} = 3.2$ (Scientific American).

standard gravity, *Physics.* an acceleration of gravity equal to about 9.80665 meters per second per second, which is approximately the value of the acceleration of gravity at the earth's surface.

standard molar enthalpy of formation, = enthalpy of formation.

standard notation, = scientific notation: *Standard notation should be used to indicate the number of significant figures and to facilitate mathematical operations with large and small numbers* (Physics Regents Syllabus).

standard temperature, = normal temperature.

cap, fāce, fäther; best, bē, tėrm; pin, fīve;
rock, gō, ôrder; oil, out; cup, pu̇t, rüle;
yü in use, yu̇ in uric;
ng in bring; sh in rush; th in thin, ŦH in then;
zh in seizure.
ə = a in about, e in taken, i in pencil, o in lemon, u in circus

standard time, *Astronomy.* the time officially adopted as the standard for a region or country, usually the nearest meridian exactly divisible by 15°. Each meridian 15° east or west of Greenwich, England, marks a time difference of one hour, making a total of 24 time zones. *Everywhere within a standard-time zone, the clocks using standard time read the same hour, minute, and second ... Between one zone and the next, there is an abrupt jump of $1^h 0^m 0^s$* (Bernhard, *Handbook of Heavens*).

standing wave, *Physics.* a wave characterized by lack of vibration at certain points (nodes) in a medium, between which areas of maximum vibration occur periodically. Standing waves are produced whenever a wave is confined within boundaries, as in a vibrating string of a musical instrument. *Whereas in a traveling wave the amplitude remains constant while the wave progresses, here the wave form remains fixed in position (longitudinally) while the amplitude fluctuates. Certain points known as the nodes remain always at rest. Midway between these points, at the loops or antinodes, the fluctuations are a maximum. The vibration as a whole is called a standing wave* (Sears and Zemansky, *University Physics*). Also called **stationary wave.** See the picture at **wave.**

stannic (stan'ik), *adj. Chemistry.* of, having to do with, or containing tin, especially with a valence of four. [from Late Latin *stannum* tin]

stannite (stan'īt), *n. Mineralogy.* a steel-gray or iron-black mineral with a metallic luster, a sulfide of tin, copper, and iron. *Formula:* Cu_2FeSnS_4 Also called **tin pyrites.**

stannous (stan'əs), *adj.* of, having to do with, or containing tin, especially with a valence of two.

stapes (stā'pēz), *n., pl.* **stapes, stapedes** (stā'pə dēz). *Anatomy.* the innermost of the three small bones of the middle ear. The other bones are the *incus* and the *malleus.* Also called **stirrup bone.** See the picture at **ear.** [from New Latin *stapes,* from Medieval Latin *stapes* stirrup]

staphylococcal (staf'ə lə kok'əl) or **staphylococcic** (staf'ə lə kok'sik), *adj.* having to do with or produced by staphylococci.

staphylococcus (staf'ə lə kok'əs), *n., pl.* **-cocci** (-kok'sī). *Bacteriology.* any of a genus (*Staphylococcus*) of gram-positive, spherical bacteria that usually bunch together, forming grapelike clusters. Several species are capable of causing serious infections in human beings and other organisms. *The penicillin-resistant staphylococci protect themselves by producing an enzyme penicillinase, which destroys penicillin* (New Scientist). [from Greek *staphylē* bunch of grapes + *kokkos* grain]

star, *n.* **1** *Astronomy.* a hot, luminous, gaseous celestial body. Stars vary in size from those slightly larger than the earth to those several million times as large as the sun. The majority of stars, including the sun, fall into the main sequence when plotted on the Hertzsprung-Russell diagram according to luminosity and spectral class.

ASSOCIATED TERMS: Stars are divided into five major types: *supergiants, giants, dwarfs, white dwarfs,* and *neutron stars* and are found in *galaxies,* of which the *Milky Way* is one. The distance between stars is measured in *light-years.* The *main-sequence stars* include most stars visible from the earth; their *luminosity* is plotted on a *Hertzsprung-Russell diagram.* A *binary system* is a pair of stars revolving about each other; an *eclipsing binary* is one in which one star periodically blocks the other's light. A *black hole* is a collapsed star. A *nova* and *supernova* are exploding stars. A *variable star* is one whose brightness varies. Other stars or starlike objects are *quasars* or *QSO's, pulsars,* and *X-ray stars.* See also the associated terms under **galaxy.**

2 *Physics.* a star-shaped pattern of lines radiating outward from the nucleus of an atom that is exploded by high-energy subatomic particles, as seen in a photograph of this effect produced in a cloud chamber. [Old English *steorra*]

starch, *n. Biochemistry.* a white, odorless, tasteless, powdery or granular substance, chemically a complex carbohydrate, found in all parts of a plant which store plant food. It is an important ingredient in the diet, reacting with certain enzymes to form glucose, maltose, and other sugars. Potatoes, wheat, rice, and corn contain much starch. [Old English *stercan* make rigid, from *stearc* stiff, strong]

star cloud, *Astronomy.* a group of stars that look like a bright, hazy area when seen without a telescope.

star cluster, *Astronomy.* a group of stars that are relatively close together, normally classified as either galactic or globular clusters.

star drift, *Astronomy.* a gradual movement of groups of stars in one direction; motion common to a group of stars.

star field, *Astronomy.* the stars in a portion of the sky seen through a telescope.

Stark effect, *Physics.* the splitting of spectral lines when an emitting atom is under the influence of a strong electric field: *If an electric field is applied to an ordinary atom, its energy levels are shifted slightly; the shift is called the Stark effect ... The nature of the shift is altered radically if the energy levels happen to be degenerate* (Scientific American). [named after Johannes Stark, 1874–1957, German physicist who discovered this effect]

starquake, *n. Astronomy.* a series of rapid changes in the shape of a star or in the distribution of its matter, detected from sudden acceleration of the star's rate of pulsation: *Starquakes may change a pulsar's pulse* (New Scientist).

stasis (stā'sis *or* stas'is), *n., pl.* **stases** (stā'sēz' *or* stas'ēz'). *Biology, Medicine.* a stoppage or stagnation of the flow of any of the fluids of the body, as of the blood in the blood vessels or of the feces in the intestines. [from Greek *stasis* a standing]

-stat, *combining form.* **1** any regulating, stabilizing, or controlling center, agent, or device, as in *rheostat* (instrument regulating the flow of electric current), *thermostat* (device for regulating temperature).
2 anything that checks or arrests, as in *bacteriostat* (agent that arrests the growth of bacteria). [from Greek *-statēs,* from *sta-,* a root of *histanai* to cause to stand]

statcoulomb (stat'kü'lōm), *n.* a CGS unit of electric charge, equal to the charge that would exert a force of one dyne on an equal charge at a distance of one centimeter in a vacuum; approximately 3.3356×10^{-10} coulomb. [from *stat(ic)* + *coulomb*]

state, *n. Physics.* **1** the condition of a substance or system with regard to composition, form, structure, or the like: *the stable state of a system, a body in a state* *of equilibrium. The general types of information arising from experiments on elastic scattering of particles are the existence of "states" ... (A state is a spatially localized system that remains compact for some time)* (Marc Ross). Compare **phase.**

2 = energy level: *In 1914 J. Frank and G. Hertz further strengthened the concept of stationary states or fixed energy levels by bombarding gas molecules with electrons* (Physics Regents Syllabus).

statement, *n. Mathematics.* an idea expressed by a closed sentence: EXAMPLE: $5 = 2 + 3$ is a statement expressing the true idea that 5 is the sum of 2 and 3.

static, *adj. Physics.* **1** having to do with bodies at rest or with forces that balance each other. Contrasted with **dynamic.**
2 acting by weight without producing motion: *static pressure.*
3 a of or having to do with stationary electrical charges that balance each other. **Static electricity** can be produced by rubbing a glass rod with a silk cloth. **b** producing such electricity: *a static discharger.*

statics, *n.* the branch of mechanics that deals with bodies at rest or forces that balance each other: *Statics and kinetics are two major branches of the science of mechanics that tells you what happens when a solid material is pushed, pulled, twisted or squashed—when forces are applied to it. Statics is at work when the particle or object doesn't move.* (Science News Letter).

stationary front, *Meteorology.* a surface between two dissimilar air masses neither of which is displacing the other and usually resulting in mild temperatures and cloudy weather: *A stationary front ... may develop from either a warm or a cold front when the fronts are retarded in their movements. In turn, one of the air masses involved may start advancing against the other, changing the stationary front into a warm or cold front, depending upon the density of the advancing air mass* (Neuberger and Stephens, *Weather and Man*).

stationary wave, = standing wave.

statistic, *n. Statistics.* **1** a measure of a characteristic of a sample: *The arithmetic mean, the standard deviation, or a measure of skewness, ... when computed from a sample, are statistics, while the same measures, when referring to the universe, are parameters* (Parl, *Basic Statistics*).
2 any value, item, etc., used in statistics: *an important statistic.*

statistical, *adj.* of or having to do with statistics; consisting of or based on statistics: *statistical information, statistical tables, statistical measures.* —**statistically,** *adv.*

statistical independence, *Statistics.* the absence of correlation between two or more ways of classifying a group.

statistical mechanics, the branch of physics in which statistical methods are used to deal with atoms, molecules, and other microscopic systems.

statistics, *n.* **1** *pl. in form and use.* numerical information collected and classified systematically: *Whenever statistics are kept, the numbers of births and deaths rise and fall in nearly parallel lines* (William R. Inge). *In working with statistics, it is often necessary to find the sum of a set of numbers before performing further calculations. A convenient shorthand using the Greek letter sigma ... has been developed to denote the sum*

of a set of numbers (Corcoran, *Algebra Two*). See **summation sign.**
2 *pl. in form, sing. in use.* the science of collecting, classifying, and interpreting such information, usually on the basis of samples and the rules of probability: *Statistics aims at supplying a sound basis for uncertain statements* (J. A. Nelder).
[from German *Statistik,* ultimately from Latin *status* state]

statoblast (stat′ə blast), *n. Zoology.* any of the horny buds developed within certain freshwater bryozoans that are set free when the parent colony dies, remain inactive throughout the winter, and develop into new individuals in the spring. [from Greek *statos* standing, static + *blastos* sprout]

statocyst (stat′ə sist), *n.* **1** *Zoology.* an organ of balance found in various crustaceans, flatworms, and other invertebrates, consisting of a sac containing particles (statoliths), as of sand or lime, suspended in fluid: *Most lugworms have so-called statocysts, consisting of a heavy particle rolling around in a sensitive vesicle, by means of which they can detect the direction of the gravitational pull* (G. P. Wells).
2 *Botany.* an organ of gravitation consisting of cells that contain starch grains and other granules (statoliths) and ectoplasm sensitive to the pressure of the grains.

statolith (stat′ə lith), *n.* **1** *Zoology.* a particle, as of sand or lime, suspended in fluid, contained in a statocyst.
2 *Botany.* a starch grain or other granule in the statocyst of a plant: *Many plants contain in the root tip large starch grains which move, under the effect of gravity, within the plant cell. It has been suggested that these starch grains, called statoliths, are the receptors of the gravitational impulse* (Science News).

staurolite (stôr′ə līt), *n. Mineralogy.* a metamorphic mineral, a silicate of aluminum and iron, yellowish-brown to dark-brown in color, found frequently twinned in the form of a cross. *Formula:* $(Fe,Mg)_2Al_8Si_4O_{23}(OH)$ [from French, from Greek *stauros* cross + French *-lite*]
—**staurolitic** (stôr′ə lit′ik), *adj.* having to do with, resembling, or characterized by the presence of staurolite.

steady state, *Physics.* the condition in which all or most changes or disturbances have been eliminated from a system: *In scientific theory, a system is in a perpetual or steady state of balance when all forces acting upon it, internal as well as external, cancel out. We see examples of the steady state in the equilibrium phase of chemical reactions, in the homeostatic mechanism of the human body, in the equilibrium state of statistical mechanics, and in the "balance of nature"* (Emilio Q. Daddario).

---

steady-state theory or **steady-state hypothesis,** *Physics, Astronomy.* the theory that the universe is in appreciably the same state as it has always been, for, although matter has been and is being lost or dispersed, other matter is continuously created in the form of hydrogen atoms to take its place: *His [Maarten Schmidt's] study suggests that a complete survey of the sky with large telescopes would reveal as many as 14 million quasars, but since they are such short-lived objects all but 35,000 or so must have burned themselves out in the time it has taken their light to reach us. Schmidt's observations provide a strong argument against the steady-state hypothesis, which holds that the universe has always looked just as it does now* (Scientific American). Contrasted with **big bang theory.**

steam, *n. Physics.* **1** water in the form of vapor or gas. Boiling water gives off steam.
2 the vapor of boiling water used, especially by confinement in special apparatus, to generate mechanical power and for heating and cooking: *The buckets change the flow of steam in such a manner as to produce a force which rotates a turbine wheel* (J. J. William Brown).

steam point, *Physics.* the temperature at which pure water boils under normal atmospheric pressure, equal to 100° Celsius. Compare **ice point.**

steapsin (stē ap′sən), *n. Biochemistry.* a digestive enzyme secreted in the pancreatic juice, which changes fats into glycerol and fatty acids. [from *stea(rin)* and *(pe)psin*]

stearic acid (stē ar′ik *or* stir′ik), *Chemistry.* a white, odorless, tasteless, saturated fatty acid, obtained chiefly from tallow and other hard fats by saponification. It exists in combination with glycerol as stearin. *Palmitic and stearic acids, containing 16 and 18 carbon atoms, respectively, are found most abundantly in the oils of plants and the fats of animals* (Harbaugh and Goodrich, *Fundamentals of Biology*). *Formula:* $CH_3(CH_2)_{16}COOH$ [from Greek *stear* fat]

stearin (stē′ər ən *or* stir′ən), *n. Chemistry.* a white, odorless, crystalline substance, an ester of stearic acid and glycerol, that is the chief constituent of many animal and vegetable fats.

steatite (stē′ə tīt), *n. Geology.* a rock composed of impure talc, with a smooth, greasy feel; soapstone. [from Latin *steatitis* soapstone, from Greek *stear, steatos* fat] —**steatitic** (stē′ə tit′ik), *adj.* of, having to do with, like, or made of steatite: *steatitic slate.*

Stefan-Boltzmann law (stef′ən bōlts′mən), *Physics.* the law stating that the total energy radiated per second by each unit area of a perfect black body is proportional to the fourth power of its absolute temperature. [named after Josef *Stefan,* 1835–1893, and Ludwig *Boltzmann,* 1844–1906, Austrian physicists]

stele (stē′lē), *n., pl.* **-lae** (-lē), **-les.** *Botany.* the central cylinder of tissue in the stems and roots of plants, consisting of the vascular or conducting tissues (xylem and phloem) together with supporting tissues: *The amount of mineral salts in the stele is maintained at a relatively low level by a rapid transport to other parts of the plant in the transpiration stream* (J. F. Sutcliffe). [from Greek *stēlē* slab, pillar]

stellar, *adj. Astronomy.* of or having to do with a star, or stars: *stellar magnitudes. The structure of the Sun's atmosphere ... is also of great interest for stellar research since the Sun is the only star whose surface we can examine in any detail* (New Scientist). SYN: astral, sidereal. [from Latin *stella* star]

stellarator (stel′ə rā′tər), *n. Physics.* a device in which highly ionized gas is confined in an endless tube by means of an externally applied magnetic field, used to produce controlled thermonuclear power: *Medium-density plasma containers ... include the stellators, originally developed at the Princeton Plasma Physics Laboratory, and the tokamaks, originally developed at the I. V. Kurchatov Institute of Atomic Energy near Moscow* (Scientific American). Compare **tokamak.** [from *stellar* (*gener*)*ator* (so called from the expectation that stellar temperatures may be generated by such a device)]

stellate (stel′āt *or* stel′it), *adj. Biology.* spreading out like the points of a star; star-shaped: *stellate ganglia. Many interneurons (that is, Golgi, basket, and stellate cells) integrate and modify the input to, and the output from, the cerebellar cortex* (Stephen T. Kitai).

stem, *n. Botany.* **1** the main part, usually above the ground, of a tree, shrub, or other plant; the firm part that supports the branches; the trunk or stalk: *As the main stem grows taller, the branches elongate ... In time the branches equal or overtake the main stem and thus produce a rounded or irregular crown* (Greulach and Adams, *Plants*).
2 the ascending axis of a plant, as distinguished from the root or descending axis: *The botanist must have some reliable means of deciding whether he is dealing with root or stem. In all but rare instances a single criterion can be used as a means of recognizing the two. It is this: The stem always has nodes and internodes wherever it grows and the root never has them under any circumstances* (Emerson, *Basic Botany*).
3 the stalk supporting a flower, a fruit, or a leaf. Specifically: **a** the peduncle of the fructification. **b** the pedicel of a flower. **c** the petiole or stalk of a leaf.
▶ Stems are either *herbaceous* (soft, slender) or *woody* (hard, thick). Herbaceous stems have only primary tissues (the epidermis, phloem, xylem, and parenchyma). Woody stems have both primary and secondary tissues (primary and secondary phloem, primary and secondary xylem, cambium, phelloderm, cork cambium, cork) which cause the formation of wood and bark.

stem cell, *Biology.* an embryonic or primitive cell that develops into specialized cells: *The marrow itself consists of fat cells, an assortment of immature blood cells, and important blood-forming cells known as stem cells. Stem cells can reproduce themselves again and again, and may develop into any type of blood cell—red cells, platelets, or white cells* (Beverly Merz).

steno-, *combining form.* narrow; small, as in *stenocephalic = narrow-headed; stenopetalous = having narrow petals.* [Greek *stenos*]

stenobathic (sten′ə bath′ik), *adj. Ecology.* living in water within a narrow range of depth: *stenobathic organisms.* [from *steno-* + Greek *bathos* depth]

624

stenohaline (sten′ō hā′lin *or* sten′ō hal′in), *adj. Ecology.* capable of living only in water whose degree of saltiness is within a narrow range: *a stenohaline animal or plant.* [from *steno-* + Greek *hals* brine]

stenomorph, *n. Ecology.* an organism that is very small due to a cramped habitat, as from crowding by other organisms.

stenopaic (sten′ə pā′ik), *adj. Optics.* having to do with or characterized by a small or narrow opening: *a stenopaic eyepiece.* [from *steno-* + Greek *opē* opening]

stenophagous (stə nof′ə gəs), *adj. Zoology.* living on a small variety of foods.

stenophyllous (sten′ə fil′əs), *adj. Botany.* having narrow leaves. [from *steno-* + Greek *phyllon* leaf]

stenotherm (sten′ə therm′), *n. Ecology.* a stenothermal organism.

stenothermal (sten′ə ther′məl), *adj. Ecology.* not able to endure a broad range of temperatures; restricted to limited temperatures: *stenothermal species.*

stenothermic (sten′ə ther′mik), *adj.* = stenothermal.

stenotopic (sten′ə top′ik), *adj. Ecology.* restricted to a single habitat: *stenotopic organisms.* [from *steno-* + Greek *topos* place]

stenotropic (sten′ə trop′ik), *n. Ecology.* having narrow limits of adaptation to varied conditions. [from *steno-* + Greek *tropē* a turning]

step fault, *Geology.* a fault in which the displacement is distributed among several parallel planes at short distances from one another instead of being confined to a single plane.

stephanite (stef′ə nīt), *n. Mineralogy.* a soft, brittle, black mineral with a metallic luster, an ore of silver. *Formula:* Ag_5SbS_4 [from German *Stephanit*]

steppe (step), *n. Geography, Ecology.* a large expanse of dry, treeless grassland: *Some writers divide arid regions into steppes, where scattered bushes and short-lived grasses furnish a scanty pasturage, and true deserts, where vegetation is sparse or absent. On this basis the deserts of North America would be mostly steppes* (Gilluly, *Principles of Geology*). [from Russian *step′*]

sterad., *abbrev.* steradian.

steradian (sti rā′dē ən), *n. Geometry.* a unit of measurement of solid angles. It is equal to the solid angle subtended at the center of a sphere by the area of the surface of the sphere equal to the square of its radius. *Symbol:* sr [from Greek *stereos* solid + English *radian*]

stereo- (ster′ē ō- *or* stir′ē ō-), *combining form.*
1 hard; solid, as in *stereotaxis, stereotropic.*
2 involving spatial arrangement, as in *stereochemistry, stereoisomerism.*
3 having or involving three dimensions; three-dimensional, as in *stereology, stereomicroscope.* [from Greek *stereos* solid]

stereochemical, *adj.* having to do with stereochemistry. —**stereochemically,** *adv.*

stereochemistry, *n.* **1** the branch of chemistry dealing with the relative position in space of atoms in relation to differences in the optical and chemical properties of the substances.

2 the spatial arrangement or position of atoms or molecules in a substance: *One essential feature of most of these chemicals is their stereochemistry—the specific shape of the molecule* (M. Frederick Hawthorne).

stereoisomer, *n. Chemistry.* one of two or more isomeric compounds that are held to differ by virtue of a difference in the spatial arrangement (not in the order of connection) of the atoms in the molecule: *Of all seed fat acids chaulmoogric is unique in having a ring-like, unsymmetrical structure. Its two forms (stereoisomers) are the same when represented in a projection formula, but different, structurally, in space* (Science News Letter).
—**stereoisomeric,** *adj.* characterized by stereoisomerism.
—**stereoisomerism,** *n.* isomerism in which the atoms are joined in the molecule in the same order but differ in their spatial arrangement.

stereology (ster′ē ol′ə jē *or* stir′ē ol′ə jē), *n.* the scientific study of the three-dimensional characteristics of objects that are normally viewed only two-dimensionally: *Up to now it has not been easy to measure brain area but the rather new science of stereology, employing statistico-geometrical methods, now makes it possible to draw conclusions concerning three-dimensionsal structures from flat images, such as cut sections* (Science Journal).
—**stereological,** *adj.* —**stereologically,** *adv.*

stereome (ster′ē ōm *or* stir′ē ōm), *n.* **1** *Botany.* the elements which give strength to a fibrovascular bundle in a plant.
2 *Zoology.* the hard tissue of the body of invertebrates. [from Greek *stereōma* solid body, from *stereos* solid]

stereometric, *adj.* having to do with or performed by stereometry. —**stereometrically,** *adv.*

stereometry (ster′ē om′ə trē *or* stir′ē om′ə trē), *n.* the measurement of solid figures; solid geometry.

stereomicroscope, *n. Optics.* a microscope with two eyepieces, used to obtain a three-dimensional image of the object viewed; a stereoscopic microscope.

stereoregular, *adj. Chemistry.* of, having to do with, or characterized by a definite and regular spatial arrangement of the atoms in the repeating units of a polymer: *stereoregular synthesis. Highly crystalline and very light fibres can now be obtained from stereoregular hydrocarbon macromolecules* (Giulio Natta).
—**stereoregularity,** *n.* the quality or state of being stereoregular.

stereoscopic, *adj. Optics.* **1** of or having to do with stereoscopy.
2 seeming to have depth as well as height and breadth; three-dimensional: *Almost every invention for improving the motion-picture art is accompanied by the claim that it produces a stereoscopic effect ... If the meaning of the term is restricted, as it should be, to the*

cap, fāce, fäther; best, bē, tėrm; pin, five;
rock, gō, ôrder; oil, out; cup, pùt, rüle,
*y*ü in use, *y*ù in uric;
ng in bring; *sh* in rush; *th* in thin, ŦH in then;
zh in seizure;
ə = *a* in about, *e* in taken, *i* in pencil, *o* in lemon, *u* in circus

reproduction of the spatial relationships by the aid of the observer's binocular sense, it is clear that such claims are absurd (Hardy and Perrin, *Principles of Optics*).

—**stereoscopy** (ster′ē os′kə pē *or* stir′ē os′kə pē), *n.* **1** the study of stereoscopic systems or effects. **2** the viewing of objects in three dimensions.

stereospecific, *adj. Chemistry.* **1** = stereoregular.

2 restricted to a particular stereoisomer; stereochemically specific: *stereospecific synthesis, a stereospecific catalyst.*

—**stereospecifically,** *adv.* in a stereospecific manner.

—**stereospecificity,** *n.* the quality or state of being stereospecific.

stereotactic, *adj.* of, having to do with, or exhibiting stereotaxis.

stereotaxis, *n. Biology.* a movement of an organism as a result of contact with a solid body. Also called **thigmotaxis.**

stereotropic, *adj. Biology.* bending or turning around under the stimulus of contact with a solid body.

—**stereotropism,** *n.* a tendency to bend or turn in response to contact with a solid body or rigid surface. Stereotropism may be positive (turning toward a solid body) or negative (turning away from a solid body).

steric (ster′ik *or* stir′ik), *adj. Chemistry.* of or having to do with the arrangement in space of the atoms in a molecule: *Steric molecular configuration affects both the intensity and character of the spectrum* (F. Degering). [from Greek *stereos* solid + English *-ic*]

—**sterically,** *adv.*

sterigma (stə rig′mə), *n., pl.* **-mata** (-mə tə), **-mas.** *Biology.* **1** a ridge extending down a stem below the point of attachment of a decurrent leaf.

2 a stalk or filament bearing a spore in a fungus.

3 a branch or outgrowth of a basidium: *At the apex of the basidia are usually four sterigmata bearing the spores.* [from Greek *stērigma* support]

—**sterigmatic** (ster′ig mat′ik), *adj.* having to do with or like a sterigma: *the apices of the sterigmatic cells.*

sterilant (ster′ə lənt), *n. Biology.* an agent that sterilizes, such as a chemical that destroys an insect's ability to reproduce: *Chemical sexual sterilants of maximum effectiveness will have to produce sexual sterility in both males and females without adverse effects on mating behavior* (Science News Letter).

sterile, *adj.* **1** *Physiology.* not able to produce offspring; incapable of reproduction: *a genetically sterile organism.* SYN: infertile.

2 *Botany.* **a** not bearing fruit or spores: *a sterile plant.* **b** producing only stamens, or producing neither stamens nor pistils: *a sterile flower.* **c** not able to germinate: *a sterile spore.*

3 *Biology.* free from living organisms, especially microorganisms: *a sterile operating area, a sterile environment.*

[from Latin *sterilis* barrenness]

—**sterility,** *n.* the condition of being sterile.

sterilize, *v. Biology.* **1** to make free from living organisms, especially microorganisms, as by heating: *to sterilize a flask by boiling.* SYN: disinfect, purify.

2 to deprive (an animal) of fertility; make incapable of producing offspring by removing the organs of reproduction or by the inhibition of their function: *to sterilize agricultural pests.*

3 to make (a plant) sterile; make unable to bear fruit or to germinate.

—**sterilization,** *n.* the act of sterilizing or the condition of being sterilized: *the sterilization by heat of organic liquids, medical instruments, etc.*

sternal, *adj. Anatomy.* of, having to do with, or in the region of the sternum: *the sternal structure of birds.*

sternite (stèr′nīt), *n. Zoology.* the ventral part of each segment of the body of an insect or other arthropod: *A remarkable feature, to which the shortness of the harvestman's body is partly due, is the forward displacement of the sternites, or ventral plates, of the abdomen* (Theodore H. Savory). [from Greek *sternon* chest]

sternoclavicular, *adj. Anatomy.* of, having to do with, or connecting the sternum and clavicle.

sternocleidomastoid (stèr′nō klī′dō mas′toid), *n. Anatomy.* either of two muscles of the neck that serve to turn and nod the head by connecting the sternum, the clavicle, and the mastoid process of the temporal bone.

—*adj.* of this muscle: *the sternocleidomastoid artery.* [from New Latin *sternum* breastbone + Greek *kleidos* clavicle + English *mastoid*]

sternocostal, *adj. Anatomy.* of, having to do with, or connecting the sternum and the ribs.

sternoscapular, *adj. Anatomy.* of or having to do with the sternum and the scapula: *a sternoscapular muscle.*

sternum (stèr′nəm), *n., pl.* **-na** (-nə), **-nums.** *Anatomy.* a long bone or series of bones, occurring in most vertebrates except snakes and fishes, extending along the middle line of the front or ventral aspect of the trunk, usually articulating with some of the ribs (in human beings, with the true ribs), and with them completing the wall of the thorax: *Birds have a keel-shaped sternum ... in order to accommodate the large amount of muscle necessary for flight* (Winchester, *Zoology*). Commonly called **breastbone.** [from Greek *sternon* chest, breastbone]

steroid, *Biochemistry.* —*n.* any of a large class of structurally related compounds containing the carbon ring of the sterols, and including the sterols, various hormones, saponins, and acids found in bile: *As many as twenty-nine different chemicals of the family known as steroids had been isolated from the adrenal cortex. Among these were six biologically active substances, including cortisone and hydrocortisone* (N. Y. Times).

—*adj.* of, resembling, or having to do with a steroid or a sterol: *The adrenal cortex secretes two types of steroid hormones, aldosterone and corticosterone.* [from *ster(ol)* + *-oid*]

steroidal, *adj. Biochemistry.* = steroid: *The effects of the steroidal hormones upon other hormone-producing systems in the body ...* (Science News Letter).

steroidogenesis (stə roi′də jen′ə sis), *n. Biochemistry.* the formation of steroids in the body: *Increased steroidogenesis by rat adrenal quarters ... normally occurs in response to adrenocorticotropic hormone* (Science).

—**steroidogenic,** *adj.* of, having to do with, or inducing steroidogenesis: *steroidogenic enzymes.*

sterol (ster′ōl or ster′ol), n. Biochemistry. any of a group of solid, chiefly unsaturated alcohols, such as ergosterol or cholesterol, present in animal and plant tissues: The sterols have sometimes been grouped with lipoids because of their solubilities, but in reality they are solid alcohols. The best known of these is cholesterol ... Gallstones are concretions of almost pure cholesterol (Harbaugh and Goodrich, Fundamentals of Biology). [shortened from (chole)sterol, (ergo)sterol)]

STH, abbrev. somatotropic hormone.

stibnite (stib′nīt), n. Mineralogy. native antimony sulfide, a lead-gray mineral occurring in orthorhombic crystals and in massive forms. It is the most important ore of antimony. Formula: Sb_2S_3 [ultimately from Latin stibium antimony]

stickslip, Geology. —n. the rapid, jerky displacement of rock along a fracture line, as distinguished from gradual, stable sliding: Theoretical and experimental work on the mechanisms of generation of earthquakes indicate that the presence of hot water along a rock fracture system tends to cause rock to move by creep and not by the process known as stickslip, which results in seismic shocks (Robert W. Rex).
—v. to undergo stickslip: The moon is aseismic because its temperature-pressure curve lies in the ... field where rock fails by stable sliding rather than by stick-slipping (Nature).

sticky ends, Molecular Biology. ends of single strands of DNA molecules that complement each other in the sequence of their nucleotides and can be linked up or reconnected in the presence of the enzyme ligase. Restriction enzymes are used to produce sticky ends. Sticky ends ... have turned out to be of utmost importance for genetic engineering research (Robert Cooke).

stigma (stig′mə), n., pl. **stigmata** (stig mä′tə or stig′mə-tə), **stigmas. 1** Botany. the part of the pistil of a flowering plant that receives the pollen: Externally, a pistil is ordinarily differentiated into three parts: (1) A terminal structure, usually rough or sticky at maturity, called the stigma; (2) a neck-like portion, the style; and (3) an enlarged spherical or cylindrical portion known as the ovary (Harbaugh and Goodrich, Fundamentals of Biology). Following pollination, the pollen grain germinates on the stigma and forms a pollen tube which extends into the ovule (Biology Regents Syllabus).
2 Zoology. a spiracle of an insect.
3 Biology. the pigmented eyespot of a protozoan or algal cell. [from Latin, from Greek stigma mark, puncture, from stizein to mark, tattoo]
—**stigmal,** adj. of or having to do with a stigma.

stigmatic (stig mat′ik), adj. **1** Botany. of a stigma: In some species, the stigma produces a sticky secretion, the stigmatic fluid. (Weier, Botany).
2 Optics. = anastigmatic (applied especially to light rays that converge to a single point).

stilb, n. Optics. a unit of brightness or luminance equal to one candela per square centimeter of a surface. [from Greek stilbein to glitter]

stilbite (stil′bīt), n. Mineralogy. a zeolite mineral, a hydrated silicate of aluminum, calcium, and sodium, usually occurring in radiating or sheaflike tufts of crystals having a pearly luster. It varies in color from white to brown or red. Formula: $NaCa_2Al_5Si_{13}O_{36} \cdot 14H_2O$ [from Greek stilbein to glitter]

stimulate, v. Physiology. to induce a stimulus in; increase temporarily the functional activity of (the body or some part of the body, especially a nerve): The muscle fibers, being excitable, may ... be directly stimulated by heat, light, chemicals, pressure, or electricity (Giese, Cell Physiology). SYN: excite, irritate.
—**stimulation,** n. the act of stimulating; the inducing of a stimulus or stimuli: It makes no difference whether the stimulation comes from a floodlight or a spark—the nerve impulse is of the same magnitude (Martin E. Spencer).

stimulated emission, Physics. the emission of radiation of a given frequency, stimulated by the presence of radiation of the same frequency. Lasers and masers operate by stimulated emission.

stimulus (stim′yə ləs), n., pl. **-li** (-lī). Physiology. something that excites the body or some part of the body (as an organ, tissue, or cell) to produce a specific response, such as transmitting an impulse along a nerve, moving a muscle, or increasing the rate of heartbeat: Some stimuli act directly upon cells or tissues and elicit a direct response (e.g., sunburn), but most animals have various kinds of specialized receptors (sense organs) to receive stimuli (Storer, General Zoology). The intensity of the stimulus will determine how many axons are made active. The more intense the stimulus, the more axons will be sending in impulses (McElroy, Biology and Man). See the picture at **reflex.** [from Latin stimulus (originally) a goad]

sting, n. **1** Zoology. the sharp-pointed part of an insect or other animal that pricks or wounds and often poisons: A wasp's sting is not left in the wound. Also called **stinger.**
2 Botany. a stiff, sharp-pointed glandular hair that secretes an irritating fluid when touched, as on the nettle.

stinger, n. Zoology. the stinging organ of an insect or other animal; sting (def. 1).

stipe (stip), n. **1** Botany. a stalk or stem, especially: **a** the stalk in flowering plants formed by the receptacle or some part of it, or by a carpel. **b** the stalk or petiole of a frond, especially of a fern or seaweed. **c** the stalk or stem in certain fungi which supports the pileus or cap: the stipe of a mushroom.
2 Zoology. a stalklike part or organ. [from Latin stipes tree trunk]

stipel (stī′pəl), n. Botany. a secondary stipule situated at the base of the leaflets of a compound leaf. [from New Latin stipella, diminutive of Latin stipula stem, stalk]

stipes (stī′pēz), n., pl. **stipites** (stip′ə tēz), **1** Zoology. a part or organ resembling a stalk, especially the second section of one of the maxillae of a crustacean or insect.
2 Botany. a stipe. [from Latin stipes tree trunk]

---

cap, fāce, fäther; best, bē, tèrm; pin, fīve;
rock, gō, ôrder; oil, out; cup, pùt, rüle,
yü in use, yu̇ in uric;
ng in bring; sh in rush; th in thin, ⊤H in then;
zh in seizure.
ə = a in about, e in taken, i in pencil, o in lemon, u in circus

stipular (stip′yə lər), *adj. Botany.* **1** of, having to do with, or resembling stipules.
2 situated on, near, or in the place of a stipule.
stipulate (stip′yə lit), *adj. Botany.* having stipules.
stipule (stip′yül), *n. Botany.* one of the pair of little leaf-like parts at the base of a leaf stem. [from Latin *stipula* stem, stalk, related to *stipes* trunk of a tree]

STIPULE

Stirling's formula, *Mathematics.* a formula for approximating the value of higher factorials, using the transcendental numbers pi and *e.* [named after James *Stirling,* 1692–1770, Scottish mathematician]
stishovite (stish′ə vīt), *n. Mineralogy.* an extremely dense form of silica produced under very high pressure (as by the impact of a meteorite): *The relatively high density of stishovite ... and its formation from pyrophillite ... suggest that it may exist in appreciable amounts in the deeper layers of the Earth's crust* (New Scientist). *Formula:* SiO_2 [from S.M. *Stishov,* a Russian mineralogist who produced it in a laboratory in 1961 + *-ite*]
stochastic (stō kas′tik), *adj. Mathematics, Statistics.* having to do with a random variable or variables; involving chance or probability: *The simplest and most celebrated example of a stochastic process is the Brownian motion of a particle* (Scientific American). [from Greek *stochastikos* proceeding by guesswork, ultimately from *stochos* aim, guess] **—stochastically,** *adv.*
stock, *n.* **1** *Botany.* **a** the main stem of a plant or tree, as distinguished from the root and branches. **b** an underground stem like a root; rhizome.
2 *Zoology.* a compound organism consisting of a colony of zooids.
3 *Geology.* a large, domelike mass of intrusive rock that has worked its way close to the surface before solidifying. Stocks are smaller than batholiths.
stoichiometric (stoi′kē ə met′rik), *adj.* having to do with stoichiometry: *the stoichiometric quantity of nitrogen needed for ammonia synthesis.* **—stoichiometrically,** *adv.*
stoichiometry (stoi′kē om′ə trē), *n. Chemistry.* **1** the quantitative relationship between the elements making up a substance, or between the substances taking part in a chemical reaction: *The problem in the formation of thin films of inorganic compounds is to assure that the stoichiometry of the film is as desired, and this, in some cases, can be controlled by the conditions of evaporation, such as residual gas pressure, heating sources and rate of evaporation* (J. Klerer).
2 the branch of chemistry that deals with such quantitative relationships: *Stoichiometry is the study of the quantitative relationships implied by chemical formulas and by chemical equations. In stoichiometry it is frequently convenient to use the mole interpretations and mole relationships in the solving of problems* (Chemistry Regents Syllabus).
[from Greek *stoicheion* element + English *-metry*]
stoke (stōk), *n.* a unit in the centimeter-gram-second system for measuring the kinematic viscosity of a fluid (the viscosity of a fluid divided by its density). [named after George *Stokes,* 1819–1903, British mathematician and physicist]
Stokes' law, *Physics.* **1** a formula that expresses the rate of settling of spherical particles in a fluid.
2 the law that the frequency of luminescence excited by radiation is usually not higher than the frequency of the exciting radiation: *Stokes' law states that fluorescence is always of a longer wavelength than that of the radiation which causes it* (George Savage). [named after George *Stokes;* see **stoke**]
stolon (stō′lon), *n.* **1** *Botany.* **a** a slender stem along or beneath the surface of the ground that takes root at the tip and grows into a new plant. A very slender, naked stolon with a bud at the end is a runner. *Bermuda grass has above-ground horizontal stems called stolons. These stems creep along the ground, and at each node, shoots and roots arise* (Weier, *Botany*). **b** a rhizome or rootstock of certain grasses, used for propagation.
2 *Zoology.* a rootlike growth in a compound organism. It is a process of the soft tissue joining a bud or zooid to the main part.
3 *Biology.* a horizontal hypha of a fungal mycelium.
[from Latin *stolonem* a shoot]

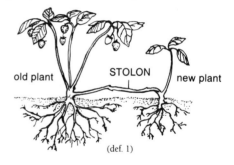

old plant   STOLON   new plant

(def. 1)

stoloniferous (stō′lə nif′ər əs), *adj.* producing or bearing stolons: *stoloniferous grasses. Their growth pattern varied from giant forms to small-leaved, creeping stoloniferous types* (New Scientist). **—stoloniferously,** *adv.*
stoma (stō′mə), *n., pl.* **-mata** (-mə tə) or **-mas. 1** *Botany.* = stomate: *Chlorophyll is not as abundant in plants with few stomata (breathing pores) as in plants with many stomata. This helped explain why photosynthesis is restricted in plants with few or defective stomata. These plants usually have pale leaves* (William P. Schenk).
2 *Zoology.* a mouthlike opening in an animal body, especially a small or simple aperture in an invertebrate. [from Greek *stoma* mouth]
stomach, *n. Anatomy.* **1** the organ of a vertebrate body that serves as a receptacle for food and in which early states of digestion occur. In human beings, it is a large, muscular, saclike part of the alimentary canal, occupying the upper part of the left side of the abdomen. The stomach's lining contains gastric glands which secrete

the enzymes and hydrochloric acid necessary for digestion. See the pictures at **alimentary, crustacean.**

2 (in invertebrates) any portion of the body capable of digesting food.

3 the part of the body containing the stomach; abdomen; belly.

[from Greek *stomachos*, from *stoma* mouth]

—**stomachic,** *adj.* of or having to do with the stomach; gastric.

stomatal (stō′mə tl), *adj. Biology.* having to do with or connected with a stomate or stomates; of the nature of a stomate: *The slitlike opening, taken with the guard cells, constitutes what is known as the stomatal apparatus* (Youngken, *Pharmaceutical Botany*). *Stomatal action constitutes an important control over water loss* (Emerson, *Basic Botany*).

stomate (stō′māt), *n. Biology.* one of the small openings in the epidermis of plants, especially of the leaves. Stomates permit the passage of water and gases into and out of the plant. Also called **stoma.** See the picture at **leaf.** [from Greek *stoma* mouth]

-stome, *combining form.* mouth; mouthlike part; opening, as in *pneumostome = an opening for the passage of air.* [from Greek *stoma*]

stomodeal or **stomodaeal** (stō′mə dē′əl), *adj. Embryology.* having to do with or having the character of a stomodeum.

stomodeum or **stomodaeum** (stō′mə dē′əm), *n., pl.* **-dea** or **-daea** (-dē ə). *Embryology.* the anterior or oral portion of the alimentary canal of the embryo: *The opening of the mouth is produced when the ectoderm at the anterior end of the embryo invaginates and pushes through the endoderm to form an opening, the stomodeum, that becomes the mouth* (Winchester, *Zoology*). [from New Latin, from Greek *stoma* mouth + *hodaios* on the way (to), from *hodos* way]

stone, *n.* **1** *Geology.* a small piece or fragment of rock.
▶ In industrial usage, *stone* refers to small or quarried masses of rock, such as limestone or granite, used to construct buildings or to form roadbeds. In geology, the word is restricted in meaning to a small fragment of rock, and is rarely used. Limestone and sandstone are not stones but sedimentary rocks.

2 *Botany.* the hard covering of a seed, especially that of a soft, pulpy fruit: *peach stones, plum stones.*

stone canal, *Zoology.* a duct in the water-vascular system of an echinoderm, usually with calcareous walls, leading from the madreporite to a vessel around the mouth.

stone cell, *Botany.* a short, hardened cell that serves to support other tissues, found especially in seeds and fruit. Stone cells are constituents of sclerenchyma. Also called **sclereid.**

storage ring, *Nuclear Physics.* a device for storing a beam of accelerated particles in a circular track and causing it to collide with an opposing beam to produce the necessary energy for the creation of new particles: *Like the circular accelerators, the storage ring uses strong magnetic fields to deflect the particles into a circular path, and also to prevent them from spreading out and being lost in the walls of the vacuum chamber in which they travel* (New Scientist).

storm, *n. Meteorology.* a strong wind with rain, snow, or hail, often accompanied by thunder and lightning, especially a wind having a velocity of 103 to 117 kilometers per hour (force 11 on the Beaufort scale).

storm center, *Meteorology.* the center of a cyclone, where there is very low air pressure and comparative calm.

storm surge, *Meteorology.* the rapid rise in water level along a coast as a tropical cyclone approaches the land.

STP, *abbrev.* standard (conditions of) temperature and pressure: *Standard temperature and pressure (STP) of a gas are defined as 0°C (273 K) and 760 mm of mercury (760 torr), or 1 atmosphere pressure* (Chemistry Regents Syllabus).

str., *abbrev.* strait.

stragulum (strag′yə ləm), *n. Zoology.* the back and folded wings of a bird taken together as a distinguishing feature. Also called **mantle, pallium.** [from Latin *stragulum* covering]

straight angle, *Geometry.* an angle of 180 degrees. A straight angle equals two right angles. See the picture at **angle.**

straight chain, *Chemistry.* an arrangement of atoms in an organic molecule having no branches or loops; an unbranched open chain: *In propane three carbon atoms are joined together to form a "straight chain"* (Parks and Steinbach, *Systematic College Chemistry*).
—**straight-chain,** *adj.* of or consisting of a straight chain: *It is used to convert straight-chain gasoline with poor antiknock properties into branched-chain, high-octane gasolines* (Scientific American).

straightedge, *n.* a strip or bar of wood, metal, or plastic having one edge accurately straight, used for drawing or testing straight lines and level surfaces.

strain¹, *Physics.* —*n.* alteration of form, shape, or volume caused by external forces: *The bending of a beam, the twisting of a rod, the compression of a liquid or gas into a smaller volume are strains* (Furry, *Physics*). *The elastic properties of materials are described in terms of two concepts known as stress and strain ... Associated with each type of stress ... is a corresponding type of strain* (Sears and Zemansky, *University Physics*). *Stress is related to the force causing deformation; strain is related to the amount of deformation* (Shortley and Williams, *Elements of Physics*).
—*v.* to cause alteration of form, shape, or volume in (a solid): *If the region in the vicinity of a stress concentration is locally strained, and the structure for some reason is exposed to a temperature high enough to allow carbon and nitrogen atoms to diffuse to the strained region, then the onset of cleavage becomes easier* (New Scientist).

[from Old French *estreindre* bind tightly, from Latin *stringere* draw tight]

cap, fāce, fäther; best, bē, tèrm; pin, five;
rock, gō, ôrder; oil, out; cup, pùt, rüle;
yü in use, yù in uric;
ng in bring; sh in rush; th in thin, ᴛʜ in then;
zh in seizure.
ə = *a* in about, *e* in taken, *i* in pencil, *o* in lemon, *u* in circus

629

strain², *n. Biology.* **1** a group of organisms that form part of a race, breed, or variety and are distinguished from related organisms by some feature: *There are thousands of wheat strains (all plants of a strain are genetically identical), and each has advantages and disadvantages in size, disease resistance, growing period and the like* (Israel Shenker).
2 an artificial variety of a domestic animal.
[Old English *strēon* a gain, begetting]

strait, *n. Geography.* a narrow channel connecting two larger bodies of water: *The Strait of Magellan ... is the only strait between the Atlantic and Pacific oceans* (S. D. Diettrich). *Abbreviation:* str.

strange, *adj. Nuclear Physics.* of, having to do with, or characterized by the property of strangeness: *Experimeters began to find evidence for particles which, although apparently produced in strong nuclear reactions seemed to decay ... consistent with a transmutation by the weak nuclear force ... Puzzled by these effects, the physicists called the new particles "strange" particles, and endowed them with a property called strangeness, the value of which depended on the number of strange quarks that a particle contained* (New Scientist).

strangeness, *n. Nuclear Physics.* a property or quantum number of certain subatomic particles that is conserved in strong interactions: *The lambda particle, the lightest one possessing the quantum property called strangeness, is only 20% heavier than a neutron or proton and differs from the neutron, for example, only in having this strangeness.* (D. Allan Bromley). *Abbreviation:* s

strata (strā′tə *or* strat′ə), *n.* plural of **stratum.**

stratification, *n. Geology.* **1** the formation of strata; deposition or occurrence in strata; bedding: *Most sedimentary rocks form distinct layers or strata. This stratification ... generally results from variations in the supply of sedimentary detritus during deposition, or from changes in the velocity of currents that are laying down the material ...* (Gilluly, *Principles of Geology*).
2 = stratum.

stratiform (strat′ə fôrm), *adj. Geology.* arranged in strata; forming a stratum or layer.

stratify, *v. Geology.* to deposit (rock) in strata; divide into strata; form strata in: *Both sandstones and shales are divided into layers or beds, and are said to be stratified* (A. H. Green). *Hardened sediments often are called stratified rock* (Finch and Trewartha, *Elements of Geography*).

stratigraphy (strə tig′rə fē), *n.* **1** the branch of geology that deals with the order and position of strata: *Stratigraphy gives more conclusive evidence. This method is used when excavation reveals several different layers of occupation. Archaeologists can usually assume that objects found in the lower layers are older than those in the upper layers* (Grahame Clark).
2 the order and position of the strata (of a country or region): *The stratigraphy of river deposits can pose obdurate problems* (G. H. Dury).

strato-, *combining form.* horizontal layer; stratum, as in *stratocumulus, stratovolcano.* [from Latin *stratus* a spreading out]

stratocumulus (strā′tō kyü′myə ləs *or* strat′ō kyü′myə-ləs), *n. Meteorology.* a cloud formation consisting of large, globular masses of dark clouds above a flat, horizontal base, usually seen in winter and occurring at heights under 2000 meters: *Low, dense clouds, such as stratocumulus and nimbostratus, reduce the natural illumination ...* (Neuberger and Stephens, *Weather and Man*). See the picture at **front.**

stratopause (strat′ə pôz′), *n. Meteorology.* an area of atmospheric demarcation between the stratosphere and the mesosphere: *The top of the stratosphere is the stratopause at a height of about 30 miles* (C. G. Knudsen and J. K. McGuire).

stratosphere (strat′ə sfir), *n. Meteorology.* the region of the atmosphere between the troposphere and the mesosphere. In the stratosphere, temperature varies little with changes in altitude, and the winds are chiefly horizontal. See the picture at **atmosphere.**
—**stratospheric**, *adj.* of or having to do with the stratosphere: *When the temperature difference across this Arctic stratospheric jet stream reached a critical value, the planetary flow pattern became unstable* (Science News Letter).

stratovolcano, *n. Geology.* a large volcano made up of alternating layers of lava and pyroclastic matter: *The history of Mount St. Helens has been one of alternating eruptions of lava and of airborne material, ranging in size from dust particles to ejected boulders or "bombs". The result: the beautifully symmetrical steep-sided cone typical of "stratovolcanoes"—other examples being Fuji in Japan and Vesuvius in Italy* (N. Y. Times).

stratum (strā′təm *or* strat′əm), *n., pl.* **-ta** (-tə), **-tums.** **1** *Geology.* a bed or formation of sedimentary rock consisting throughout of approximately the same kind of material: *These four different kinds of rock—limestone, sandstone, shale, and conglomerate—are formed from materials that have been eroded and transported by water, wind, and ice. Because they are found in layers, or strata, they are sometimes called stratified rocks* (Beauchamp, *Everyday Problems in Science*).
2 *Meteorology, Oceanography.* a region of the atmosphere or of the sea assumed as bounded by horizontal planes for purposes of calculation: *the temperature of the lower strata of the air, a stratum of constant oceanic temperature.*
3 *Biology.* a layer of tissue: *a normal stratum of epithelium.*
4 = subpopulation.
[from Latin *stratum* something spread out, from *sternere* to spread]

stratum corneum (kôr′nē əm), *Histology.* the outer layer of the skin: *There are several layers of cells in the epidermis. ... That on the outside, the stratum corneum, is horny and consists of broad, thin cells, the squamous epithelium* (Hegner and Stiles, *College Zoology*). [from New Latin]

stratus (strā′təs), *n., pl.* **-ti** (-tī). *Meteorology.* a cloud formation consisting of a horizontal layer of gray clouds that spread over a large area, occurring at heights under 2000 meters: *Cold fog is a supercooled cloud touching the ground, while cold stratus is a layer of supercooled cloud above the ground* (E. G. Droessler). See the picture at **cloud.** [from Latin *stratus* a spreading out, from *sternere* to spread]

streak, *n.* **1** *Mineralogy.* the color of the fine powder produced when a mineral is rubbed upon a hard surface. Hematite has a red streak, magnetite a black streak. *The streak is frequently used in the identification of minerals, for, although the color of a mineral may vary within wide limits, the streak is usually constant* (Dana's Manual of Mineralogy).
2 *Bacteriology.* the distribution over the surface of a medium in a line or stripe of material to be inoculated.

stream, *n.* **1** *Geography, Geology.* **a** a flow of water in a channel or bed. Small rivers and large brooks are both called streams. *Fluvial deposits are produced by sediment-laden streams. The water comes from rain or snow, and the sediment comes from the weathering of surficial materials in the stream's drainage basin* (Eicher and McAlester, *History of the Earth*). **b** a steady current of water, as in a river or in the ocean. **2 a** any current or flow: *the blood stream,* a *stream of air, gas, or electricity.* **b** a ray or beam: *a stream of light. The earth is subject to continuous bombardment by a stream of assorted atomic particles some of which reach it with very great energy. These particles, are known as the primary cosmic rays* (E. P. George).

streamer, *n. Astronomy.* **1** a ribbonlike column of light shooting across the heavens in the aurora borealis.
2 a stream of rays emitted by the sun's corona: *Whereas at maximum activity the general outline of the corona appears roughly symmetrical, though with individual streamers irregular, at times of minimum activity there are relatively short streamers above the poles and longer ones which stretch outwards more or less at right angles to the sun's axis* (A. W. Haslett).

strengite (streng'īt), *n. Mineralogy.* a purplish or deep pink mineral, a hydrated phosphate of iron, occurring mainly in botryoidal form. *Formula:* $FePO_4 \cdot 2H_2O$ [from German *Strengit*]

strepogenin (strep'ə jen'in), *n. Biochemistry.* a peptide found in insulin and other proteins that is essential to the growth of mice and certain microorganisms. Also spelled **streptogenin.**

strepto-, *combining form.* **1** twisted or linked; resembling chains, as in *streptococcus = a group of bacteria that occur in chains.*
2 streptococcus, as in *streptokinase = an enzyme derived from streptococci.*
[from Greek *streptos* curved, twisted, from *strephein* to twist]

streptobacillus (strep'tə bə sil'əs), *n., pl.* **-cilli** (-sil'ī). *Bacteriology.* a bacillus that occurs as part of a chain of bacilli. One kind of streptobacillus causes an infection transmitted by the bite of a rat.

streptococcal (strep'tə kok'əl) or **streptococcic** (strep'tə-kok'sik), *adj.* having to do with or caused by streptococci: *a streptococcal infection.*

streptococcus (strep'tə kok'əs), *n., pl.* **-cocci** (-kok'sī). *Bacteriology.* any of a genus (*Streptococcus*) of spherical, gram-positive bacteria that multiply by dividing in only one plane, usually occurring in chains or as paired cells, and causing many serious infections and diseases such as scarlet fever and erysipelas: *Examination with a microscope shows two kinds of lactic acid bacteria: chains of round (or ovoid) streptococci, called lactic streptococci, and rod-shaped lactobacilli which are also often in chains* (J. A. Barnett). [from New Latin, from Greek *streptos* curved + *kokkos* grain]

streptokinase (strep'tō kī nās'), *n. Biochemistry.* an enzyme, a component of fibrinolysin, that dissolves blood clots and waste matter associated with infections: *The most important non-human activator of profibrinolysin is streptokinase, a protein isolated from the culture broth of streptococcal bacteria* (Saturday Review).

streptolysin (strep'tə li'sin), *n. Biochemistry.* any of various hemolysins derived from streptococcal bacteria: *Streptolysin 0 ... is a cytolytic toxin of some practical importance because it is an essential reagent in the measurement of a specific antibody, antistreptolysin.* (Science).

streptomyces (strep'tə mī'sēz), *n. sing.* and *pl. Microbiology.* any of a genus (*Streptomyces*) of aerobic microorganisms found in the soil and regarded as intermediate between bacteria and fungi. Several species are sources of important antibiotics such as streptomycin. [from New Latin, from Greek *streptos* curved, twisted + *mykēs* fungus]

stress, *n. Physics.* the internal resistance to an external force, such as tension or shear, which tends to cause a change in the shape or volume of a body, expressed as the force per unit area acting on the body: *The term strain refers to the relative change in dimensions or shape of a body which is subjected to stress* (Sears and Zemansky, *University Physics*). *Stress is proportional to strain, the proportionality constant depending only on the material and not on the particular body* (Shortley and Williams, *Elements of Physics*).

stretch receptor, *Physiology.* a sense organ that is sensitive to any stretching of the tissue in which it is found. The muscle spindle is a stretch receptor. *From this work they learned that fibers called stretch receptors send out a code that is picked up and translated by ganglia, or nerve centers, and then relayed back via nerves as orders to activate muscles* (Sergei Lenormand).

stria (strī'ə), *n., pl.* **striae** (strī'ē). **1** *Geology, Mineralogy.* a slight furrow or ridge; small groove or channel, as produced on a rock by moving ice, or on the surface of a crystal or mineral by its structure: *Noteworthy features among the sedimentary rocks of the Upper Pre-Cambrian in the Great Lakes region are ... the presence, northeast of Lake Huron, of thick boulder conglomerates containing what seemed to be well-marked glacial striae and showing the characteristic heterogeneous mixture of materials seen in glacial till* (Moore, *Introduction to Historical Geology*).
2 *Biology.* a streak, stripe, or narrow band: *striae of fat lying among the muscle.*
[from Latin *stria*]

—striated (strī'ā tid), *adj.* marked with striae; striped; streaked; furrowed: *striated rock.*

---

cap, fāce, fàther; best, bē, tèrm; pin, fīve;
rock, gō, ôrder; oil, out; cup, pùt, rüle,
yü in use, yu̇ in uric;
ng in bring; sh in rush; th in thin, ŦH in then;
zh in seizure.
ə = a in about, e in taken, i in pencil, o in lemon, u in circus

striated muscle, *Anatomy.* a type of muscle with fibers of cross bands usually contracted by voluntary action, such as the muscles that move the arms, legs, and neck: *Cardiac muscle is "involuntary" striated muscle that occurs in the vertebrate heart only. Striated muscle of the skeletal type—that is, the muscles of the external muscular system—is voluntary muscle* (Hegner and Stiles, *College Zoology*).

striation (strī ā'shən), *n.* **1** striated condition or appearance: *Striation is characteristic of muscles whose contraction is rapid* (M. Foster).
2 one of a number of parallel striae; stria: *The most remarkable occurrences of glacial deposits and striations of this age are those in India. There ... the striations underlying the glacial beds show that the ice was moving from the south, from or across the present equator* (New Biology).

strict, *adj. Botany.* close or narrow and upright: *a strict stem or inflorescence.*

strigose (strī'gōs), *adj.* **1** *Botany.* covered with stiff and straight bristles or hairs: *strigose leaves.* SYN: hispid.
2 *Zoology.* marked with fine, closely set ridges, grooves, or points.
[from New Latin *strigosus,* from *striga* stiff bristle, from Latin *striga* swath, furrow]

strike, *n. Geology.* the horizontal direction of a stratum, vein, or other feature or structure of rock; direction with regard to the points of the compass. The strike of a vein is perpendicular to the direction of the dip.

strike-slip fault, *Geology.* a fault in which the displacement is largely in the direction of the strike; fault with mostly horizontal displacement.

strobila (strə bī'lə *or* strō'bə lə), *n., pl.* **-lae** (-lē). *Zoology.* **1** the body of a tapeworm, as distinct from the head, consisting of a chain of segments.
2 a stage in the development of certain jellyfish in which a series of disk-shaped bodies split off to form new individuals.
[from New Latin, ultimately from Greek *strobilos* pine cone]

strobilation (strob'ə lā'shən), *n. Zoology.* an asexual form of reproduction in which segments of the body separate to form new individuals, as in tapeworms and scyphozoans.

strobile (strob'əl *or* strō'bīl), *n.* = strobilus: *Sporangia containing spores are usually borne on terminal cones or strobiles at the tips of branches* (Greulach and Adams, *Plants*).

strobilus (strob'ə ləs), *n., pl.* **-li** (-lī). any seed-producing cone, such as a pine cone, or a compact mass of scalelike leaves that produce spores, such as the cone of the club moss: *Microspores develop on catkin-like strobili on male trees* (Weier, *Botany*). [from Greek *strobilos* pine cone, from *strobos* a whirling]

stroma (strō'mə), *n., pl.* **-mata** (-mə tə). **1** *Anatomy.* **a** the connective tissue, nerves, and vessels that form the framelike support of an organ or part: *A tumour developed in the stroma of a fibrous structure will probably be fibrous* (T. Bryant). *The ova are imbedded in a stroma of delicate and yielding cellular substance* (R. Owen). **b** the spongy, colorless framework of a red-blood corpuscle or other cell.

2 *Botany.* the dense, granular background substance of chloroplasts: *Particulates having the properties of cytoplasmic ribosomes may be seen in the stroma* (Weier, *Botany*).
[from Greek *stroma, stromatos* a spread to lie or sit on]
—**stromal,** *adj.* of or having to do with the stroma or with stromata.

stromatolite (strō mat'ə līt), *n. Geology.* a calcareous rock structure consisting of deposits of blue-green algae: *Stromatolites have been found widely distributed in both modern and fossil marine environments, chiefly in intertidal areas, and sometimes even in very salty conditions* (New Scientist). [from Greek *stromatos* a spread + English *-lite*]
—**stromatolitic** (strō mat'ə lit'ik), *adj.* of or having to do with stromatolites: *stromatolitic dolomite.*

strombuliform (strom'byə lə fôrm), *adj. Geology, Botany.* twisted or coiled into the form of a screw, helix, or spiral. [from New Latin *strombuliformis,* from Latin *strombus* spiral shell]

stromeyerite (strō'mī'ə rīt), *n. Mineralogy.* a steel-gray mineral, a sulfide of silver and copper, with a metallic luster, occurring massive and in crystals. *Formula:* $CuAgS$ [named after Friedrich *Stromeyer,* German chemist of the 1800's]

strong force, = strong interaction: *Baryons (including protons and neutrons) are the particles subject to the strong force, leptons those (including electrons) subject to the weak force* (Science News).

strong interaction, *Nuclear Physics.* an interaction between subatomic particles that is stronger than any other known force. The strong interaction causes neutrons and protons to bind in the nuclei of atoms. *Physicists know four different kinds of force fields or interactions. Two of these, the gravitational and the electromagnetic, are familiar from the macroscopic world. The other two are confined to the microscopic domain of nuclear and particle physics. These are the strong interaction (which holds atomic nuclei together) and the weak interaction (which comes into play in many forms of radioactive decay)* (Dietrick E. Thomsen).

strontianite (stron'shē ə nīt), *n. Mineralogy.* a mineral consisting of strontium carbonate, occurring in massive, fibrous forms and varying color from white to yellow and pale green. *Formula:* $SrSO_3$

strontium (stron'shē əm *or* stron'tē əm), *n. Chemistry.* a soft, silver-white metallic element which occurs only in combination with other elements, used in making alloys and in fireworks, signal flares, etc. Strontium is one of the alkaline-earth metals and resembles calcium. *Symbol:* Sr; *atomic number* 38; *atomic weight* 87.62; *melting point* 769°C; *boiling point* 1384°C; *oxidation state* 2. [from New Latin, from *Strontian,* mining locality in Argyllshire, Scotland, where the element was first found]

strontium 90, *Physics.* a radioactive isotope of strontium that occurs in the fallout from nuclear explosions. It is easily absorbed by the bones and tissues and may eventually replace calcium in the body. *Strontium 90 is of particular interest because, being chemically similar to calcium, it may be deposited in human bone. Its half-life, the time required for its radioactivity to drop to one-half of the original value, is 25 years* (Science

News Letter). *Symbol:* [90]Sr; *atomic number 38; mass number 90.* Also called **radiostrontium.**

strontium carbonate, a white, odorless powder, in its natural state the constituent of strontianite, used in sugar refining and in making salts of strontium. *Formula:* $SrCO_3$

strophiole (strof′ē ōl *or* strō′fē ōl), *n. Botany.* a cellular outgrowth near the hilum in certain seeds. [Latin *strophiolum,* diminutive of *strophium* band of cloth, from Greek *strophos* twisted band]

strophoid (strof′oid *or* strō′foid), *n. Geometry.* a curve that is the locus of intersections of two lines rotating uniformly about two fixed points in a plane. [from French *strophoïde,* from Greek *strophos* twisted band]

structural, *adj.* **1** *Biology.* of or having to do with the organic structure of an organism: *... the structural differences which separate Man from the Gorilla and the Chimpanzee* (Thomas H. Huxley). SYN: morphological.
2 *Geology.* of or having to do with the structure of rocks and the earth's crust: *structural features such as basins, crests, valleys, etc.*

structural chemistry, a branch of chemistry dealing with the structure of molecules: *Structural chemistry is concerned ... also with the relation of each individual atom in a molecule and how the atoms are bonded together. This latter concern includes the disposition of the electrons and the shape and size of the individual nuclei* (William J. Bailey).

structural formula, a chemical formula that differs from a molecular formula in that it shows how the atoms in a molecule are arranged: *[Molecular] formulas ... do not adequately show the internal arrangement of the atoms in organic molecules. Instead, the organic chemist draws a diagram ... which he calls a structural formula. He represents the atoms by their symbols, and connects them with lines to represent covalent (non-polar) bonds* (Offner, *Fundamentals of Chemistry*). EXAMPLE:

$$H-\!\!\underset{\displaystyle\overset{\displaystyle H}{|}}{\overset{\displaystyle\overset{\displaystyle H}{|}}{C}}\!\!-H \qquad or \qquad H : \overset{..}{\underset{..}{C}} : H$$

See the picture at **formula.**

structural gene, *Molecular Biology.* a sequence of nucleotides that determines the sequence of amino acids and the structure of a protein: *[The] regulatory genes specify whether structural genes will function, and therefore control the rate of enzyme synthesis* (Science News).

structural geology, a branch of geology dealing with the positions and shapes of geological features.

structure, *n.* **1** *Biology.* the arrangement of tissues, parts, or organs of a whole organism: *the structure of a flower. Variations of structure arising in the young or larvae naturally tend to affect the structure of the mature animal* (Darwin, *Origin of Species*).
2 *Geology.* **a** the character of rocks as determined by stratification and faults. **b** the features of rocks that are due to fracture or to texture.
3 *Chemistry.* the manner in which the atoms making up a particular molecule are attached to one another.

stufa (stü′fä), *n., pl.* **-fas, -fe** (-fā). *Geology.* a jet of steam issuing from a fissure in the earth in a volcanic region. [from Italian]

stupeous (stü′pē əs), *adj.* **1** *Zoology.* covered with long, loose scales, like tow: *the stupeous palpi of an insect.* **2** *Botany.* woolly.
[from Latin *stupeus* made of tow, from *stupa* course flax]

stupose (stü′pōs), *adj. Biology.* bearing tufts or mats of long hairs; composed of matted filaments like tow. [from Latin *stupa* coarse flax]

stylate (stī′lit), *adj.* **1** *Zoology.* **a** having a style. **b** styloid; styliform.
2 *Botany.* having a persistent style.

style, *n.* **1** *Botany.* the stemlike part of the pistil of a flower having the stigma at its top: *Pollen grains on the stigma ... form pollen tubes, which grow downward through the style to the ovule* (Fuller and Tippo, *College Botany*).
2 *Zoology.* a small, slender, pointed process or part: *a cartilaginous style.*
[from Old French *estile,* from Latin *stilus,* originally, pointed writing instrument]

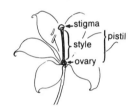

styliform, *adj.* shaped like a style; slender and pointed: *a styliform bone or appendage, a styliform projection or rock.*

stylohyoid (stī′lō hī′oid), *adj. Anatomy.* of or having to do with the styloid process of the temporal bone and the hyoid bone.

styloid (stī′loid), *adj.* **1** *Zoology.* like a style; slender and pointed: *the styloid muscles in carnivores.*
2 *Anatomy.* of or having to do with a styloid process: *the styloid nucleus.*

styloid process, *Anatomy.* **1** a sharp spine pointing down at the base of the temporal bone in man.
2 the pointed projection at the lower extremity of the ulna, on the inner and posterior side.

stylolite (stī′lə līt), *n. Geology.* a columnar structure, often with grooved sides, occurring along a thin seam in limestone or certain other rocks, usually at right angles to the plane of stratification: *The irregular marking on the limestone face is a "suture-joint" (stylolite), formed by differential solution and compaction of the*

cap, fāce, fäther; best, bē, tèrm; pin, fīve;
rock, gō, ôrder; oil, out; cup, pùt, rüle,
yü in use, *yu* in uric;
ng in bring; *sh* in rush; *th* in thin; ᴛʜ in then;
zh in seizure.
ə = *a* in about, *e* in taken, *i* in pencil, *o* in lemon, *u* in circus

633

stylomandibular

rock (Moore, *Introduction to Historical Geology*). [from Greek *stylos* pillar + English *-lite*]

stylomandibular, *adj. Anatomy.* connecting the styloid process of the temporal bone and the lower jawbone: *a stylomandibular ligament.*

stylomastoid, *adj. Anatomy.* common to the styloid and mastoid processes of the temporal bone.

stylomaxillary, *adj.* = stylomandibular.

sub-, *prefix.* **1** under; below, beneath, as in *subglacial, subcortical.*
2 resulting from further division; branch; subdivision, as in *subset, subgroup.*
3 smaller than, as in *submolecule, submicron.*
4 constituent of, as in *subnucleon, subparticle.*
5 subordinate; secondary, as in *subplate, subsere.*
[from Latin *sub* underneath, beneath]

subacetate, *n. Chemistry.* an acetate in which there is an excess of the base or metallic oxide beyond the amount that reacts with the acid to form a normal salt; basic acetate.

subaerial, *adj. Geology, Ecology.* taking place, existing, or formed in the open air or on the earth's surface: *The field evidence so far obtained is enough to show that the valley meanders were still being eroded after the last glacial maximum. In S. W. Sweden they are cut in moraine which was not exposed to subaerial denudation until some 10,000 years ago* (G. H. Dury). **—subaerially,** *adv.*

subalpine, *adj. Geography.* **1** of, having to do with, or characteristic of mountain regions next in elevation below those called alpine, usually between 1300 and 1800 meters in most parts of the North or South Temperate Zones: *a subalpine climate, a subalpine tree or plant.*
2 of or having to do with regions at the foot of the Alps.

subalternate, *adj. Biology.* alternate, but with a tendency to become opposite: *subalternate polyps, subalternate arrangement of the upper and lower teeth.*

subaqueous, *adj.* **1** *Geology.* existing, formed, or occurring under water: *subaqueous rocks.*
2 *Biology.* living or growing under water: *subaqueous plants.*

subarctic (sub ärk′tik), *adj.* of, designating, or resembling regions just south of the Arctic Circle: *a subarctic zone, a subarctic climate.*

subatom, *n. Physics.* a constituent of an atom: *Protons and electrons are subatoms.*

subatomic (sub′ə tom′ik), *adj. Physics.* **1** of or having to do with the constituents of an atom or atoms: *Like all other subatomic bits of matter, electrons have wave-like as well as particle-like properties* (R. Hofstadter). **2** of or having to do with the interior of an atom or any phenomenon occurring there: *subatomic forces, subatomic interaction of particles.*

subatomic particle, *Physics.* one of the fundamental units of which atoms and all matter are composed; elementary particle: *Protons, quarks, photons, and weakons are all subatomic particles.*

subaxillary, *adj.* **1** *Anatomy.* situated beneath the axilla: *subaxillary glands, subaxillary feathers.*
2 *Botany.* situated or placed beneath an axil.

subcellular, *adj. Biology.* smaller in size than ordinary cells: *Contemporary cell biology has come to view the cell as an organized structure containing organized subcellular structures* (Daniel Mazia). *The smaller viruses are subcellular* (Bulletin of Atomic Scientists).

subclass, *n.* **1** *Biology.* a group of organisms ranking above an order and below a class: *Linnaeus recognized groups of four different values—the class, the order, the genus* (plural, *genera*), *and the species* (plural, *species*). *To these categories have been added the phylum* (plural, *phyla*) *and subphylum* (assemblies greater *than the class*), *the subclass ... and others* (Shull, *Principles of Animal Biology*). **2** = subset.

subclavian (sub klā′vē ən), *Anatomy.* **—adj.** **1** beneath the clavicle or collarbone. The **subclavian muscle** is a small muscle extending from the first rib to the clavicle.
2 of or having to do with the subclavian artery, vein, or groove.
—n. a subclavian part, such as an artery.

subclavian artery, *Anatomy.* the large artery forming the trunk of the arterial system of the arm or forelimb: *Blood flows through the right and left subclavian arteries to the shoulders and arms* (John B. Miale).

subclavian groove, *Anatomy.* either of the two shallow depressions on the first rib for the subclavian artery and vein.

subclavian vein, *Anatomy.* the part of the main vein of the arm lying under the clavicle.

subclimax, *n. Ecology.* a stage below or preceding a climax, especially a stage in which further development has been arrested because of some factor other than climate: *Similarly in the eastern U. S. deer favor the second-growth lands where palatable young trees grow in place of the original forest. This kind of vegetation is called "subclimax." A climax* (mature) *forest generally supports few deer* (Scientific American).

subcontinent, *n. Geography.* **1** a land mass that is very large but smaller than a continent; very large island: *the subcontinent of New Guinea.*
2 a large section of a continent having a certain geographical or political independence: *the subcontinent of India.*
—subcontinental, *adj.* of, having to do with, or characteristic of a subcontinent: *Arabia, a block of subcontinental size, ... was formerly a part of Africa* (New Scientist).

subcortex, *n. Anatomy.* the white matter beneath the cortex of the brain.

subcortical, *adj.* **1** *Anatomy.* of or having to do with the subcortex: *the subcortical regions of the brain.*
2 *Zoology, Botany.* situated beneath the cortex of a sponge or the cortex of a tree.

subcritical, *adj. Nuclear Physics.* having or using less than the amount of fissionable material necessary to sustain a chain reaction: *a subcritical reactor.*

subcrust, *n. Geology.* the under portion of the crust of the earth, from about 240 to about 640 kilometers below the surface: *An alternative mechanism is offered by the occurrence of convection currents in the subcrust, below the depth of the sima* (New Biology).
—subcrustal, *adj.* happening or located below the earth's crust: *earth's subcrustal mantle.*

subcutaneous, *adj.* **1** *Histology.* under the skin: *subcutaneous tissue.*

2 *Medicine.* living under the skin: *a subcutaneous parasite.*

subdivision, *n.* *Botany.* a group of related plants ranking below a division. Also called **subphylum.**

► Current botanical usage tends to eliminate the terms *subdivision* or *subphylum,* using instead the following system of classification: *division, class, subclass, order, family, genus, species.*

subduct, *v.* *Geology.* to sink under the margin of a crystal plate; undergo or cause subduction: *The plate that rafted India, then migrated northward toward and subducted into the Tethyan trench ... evidently glided freely along parallel "megashears" on its eastern and western boundaries without interacting with the other crustal plates of the world* (R. S. Dietz and J. C. Holden). Compare **obduct.**

—subduction, *n.* the process by which one crustal plate sinks or is pushed under part of another: *In some cases an island arc forms offshore. The most prominent examples of this process are in South Asia, where the subduction of the Indian-Australian plate has generated the Indonesian archipelago, and in East Asia, where the sinking Pacific and Phillipine plates have produced the islands of Japan and the Phillipines* (K. C. Burke and J. T. Wilson).

subduplicate, *adj.* *Mathematics.* being that of the square roots of the quantities. EXAMPLE: 2 : 3 is the subduplicate ration of 4 : 9.

subdural (sub dür′əl), *adj.* *Anatomy, Medicine.* situated or existing below the dura mater: *the subdural spaces of the brain, a subdural hemorrhage.*

subdwarf, *n.,* or **subdwarf star,** *Astronomy.* any of a group of stars lying just below the main sequence in the Hertzsprung-Russell diagram, being relatively small and dim when compared to a main-sequence star of the same spectral class: *From the results gathered so far by rocket and satellite, astronomers conclude that the brightest objects in the ultraviolet sky are stellar: hot subdwarfs, subluminous in the visible but radiating copiously in the ultraviolet; the central stars of planetary nebulae; and emissionline stars ... with extended envelopes* (New Scientist). Compare **subgiant.**

suberin (sü′bər in), *n.* *Biochemistry.* a fatty substance present in the cork cells of many plants: *The walls of the cells formed outside the cork cambium become impregnated with suberin, a water-resistant substance much like cutin, and thus is formed the cork, which comprises most of the bark of older stems* (Harbaugh and Goodrich, *Fundamentals of Biology*). [from Latin *suber* cork]

suberization, *n.* *Botany.* the making of cell walls into cork by the formation of suberin.

suberize, *v.* *Botany.* to change (a cell wall) into cork tissue by the formation of suberin.

subfamily, *n.* *Biology.* a group of related plants or animals ranking above a genus and below a family. Compare **subclass** and **subdivision.**

subfossil, *Paleontology.* **—adj.** partly fossilized: *subfossil shells, subfossil remains.*

—n. a subfossil animal or plant.

subgenus, *n.* *Biology.* a group of related plants or animals ranking above a species and below a genus. See also **subclass** and **subdivision.**

subgiant, *n.,* or **subgiant star,** *Astronomy.* any of a group of stars lying just above the main sequence in the Hertzsprung-Russell diagram, being larger and brighter than a main-sequence star of the same spectral class, but fainter than a giant: *Prominent among these are the yellow and red giants like Capella and Arcturus (together with distinct groups christened subgiants and supergiants), so called because if they are cool and bright they must be very large and rarefied (in contrast to "dwarfs" on the main sequence such as the Sun)* (New Scientist). Compare **subdwarf.**

subglacial, *adj.* *Geology.* existing or formed beneath a glacier: *a subglacial stream.* **—subglacially,** *adv.*

subgroup, *n.* **1** a subordinate group; subdivision of a group, especially in botany and zoology: *Two subordinate types of this subgroup can be distinguished as far as the English elms are concerned* (R. H. Richens).

2 *Chemistry.* a division of a group in the periodic table.

3 *Mathematics.* a subset of a group which is itself a group with respect to the same operation.

subhumid, *adj.* *Climatology.* having sufficient rainfall to support the growth of tall grass or similar vegetation: *If the rainfall and the water needs of plants are about the same, the climate is subhumid* (C. Warren Thornthwaite).

subjacent (sub jā′sənt), *adj.* *Geology.* being in a lower situation, though not directly beneath; at or near the base, as of a mountain: *At the base of the Cretaceous [outcrop] is a profound unconformity, inasmuch as the subjacent rocks in most places are resistant, strongly deformed, metamorphosed Pre-Cambrian rocks or late Paleozoic granite* (Moore, *Introduction to Historical Geology*).

sublevel, *n.* *Physics, Chemistry.* = subshell.

sublimate (sub′lə māt), *v.* *Chemistry, Physics.* **—v.** to sublime (a solid substance).

—n. (sub′lə mit or sub′lə māt), a material obtained when a substance is sublimed. Bichloride of mercury is a very poisonous sublimate.

sublimation, *n.* *Chemistry, Physics.* **1** the process of subliming; direct transition from a solid to a vapor.

2 the result of subliming; a sublimate.

sublime (sə blīm′), *v.* *Chemistry, Physics.* to change directly from a solid to a vapor, without passing through a liquid phase: *When heated to moderate temperatures, iodine sublimes, forming a violet vapor that rapidly condenses to crystals on a cold surface* (Science News Letter). [from Latin *sublimis* (originally) sloping up (to the lintel), from *sub-* up to + *liminis* threshold]

sublingual, *Anatomy.* **—adj.** situated under or on the under side of the tongue: *a sublingual gland, artery, cyst, etc.*

—n. a sublingual gland, artery, or the like.

---

cap, fāce, fäther; best, bē, tèrm; pin, fīve;
rock, gō, ôrder; oil, out; cup, pùt, rüle,
yü in use, yù in uric;
ng in bring; sh in rush; th in thin, ᴛʜ in then;
zh in seizure.
ə = a in about, e in taken, i in pencil, o in lemon, u in circus

sublittoral, *adj.* **1** *Geography, Ecology.* of or near the seacoast: *a sublittoral plant.*
2 *Oceanography.* of or having to do with the area of an ocean from low tide to the edge of the continental shelf: *... the rich plant and animal life of the intertidal and sublittoral zones of the seashore* (New Biology).
—*n.* a sublittoral area or region.

sublunary or **sublunar,** *adj. Astronomy.* beneath the moon; terrestrial: *The Van Allen radiation belt is the chief feature of sublunary space* (New Scientist).

submarginal, *adj. Biology.* near the margin or edge of a body or organ: *submarginal sori, submarginal tentacles.* —**submarginally,** *adv.*

submarine, *adj.* **1** *Biology.* occurring or growing below the surface of the sea; underwater: *submarine plants.*
2 *Oceanography.* of or having to do with the region below the surface of the sea: *The continental terrace is one of the main subjects of investigation in submarine geology today* (Scientific American).

submaxilla (sub′mak sil′ə), *n., pl.* **-maxillae** (-mak sil′ē), **-maxillas.** *Anatomy.* the lower jaw or lower jawbone in man and other vertebrates.

submaxillary (sub mak′sə ler′ē), *Anatomy.* —*n.* **1** the lower jawbone. **2** a salivary gland situated beneath the lower jaw on either side.
—*adj.* of, having to do with, or situated beneath the lower jaw or lower jawbone: *a submaxillary fracture.* The **submaxillary glands** are a pair of salivary glands beneath the lower jaw.

submicroscopic, *adj. Optics.* too minute to be seen through an ordinary microscope. —**submicroscopically,** *adv.*

submucosa (sub′myü kō′sə), *n., pl.* **-sae** (-sē). *Histology.* the connective tissue lying beneath the mucous membrane and consisting of numerous minute mucous glands.
—**submucosal,** *adj.* of, having to do with, or in the region of the submucosa: *submucosal swelling.*

submultiple, *n. Mathematics.* a number or quantity that divides another without a remainder. SYN: factor.

subnuclear, *adj. Nuclear Physics.* **1** of or belonging within an atomic nucleus; being smaller than a nucleus: *subnuclear particles.*
2 of or having to do with the study of subnuclear particles: *subnuclear research, subnuclear physics.*

suborbital, *Anatomy.* —*adj.* situated below the orbit of the eye, or on the floor of the orbit, as a cartilage or nerve.
—*n.* a suborbital cartilage or nerve.

suborder, *n. Biology.* a group of related organisms ranking above a family and below an order: *Primates are usually divided into two suborders called Prosimii (including three shrews, lemurs and tarsiers), and Anthropoidea (man-apes and monkeys)* (R. Beals and H. Hoijer).

suboxide (sub ok′sīd), *n. Chemistry.* a compound of oxygen and another element or radical, containing a small proportion of oxygen.

subparticle, *n.* a constituent of a particle: *All ribosomes are constructed from two unequal subparticles* (Science Journal). *Especially puzzling has been the mass or masses of the quarks, the elementary subparti-* cles out of which the overwhelming majority of particles are supposed to be built (Science News).

subphylum, *n. Biology.* **1** a group of related animals ranking below a phylum. **2** = subdivision.

subplate, *n. Geology.* a small or secondary crustal plate: *There also is a North America, South America, Eurasia, Africa, and Antarctica plate. Additionally, there are several smaller subplates (e.g., the Nazca, Cocos, and Caribbean plates)* (Robert S. Dietz).

subpolar, *adj.* **1** *Geology.* below or adjoining the poles or polar seas of the earth in latitude: *But the great bulk of the hemisphere's air, from the subtropic to the subpolar regions, appears to circulate in the opposite direction* (Scientific American).
2 *Astronomy.* beneath a pole of the celestial sphere: *the subpolar passage of a star.*

subpopulation, *n. Statistics.* a subdivision of a population: *It is sometimes forgotten that such estimates actually represent average values in the population that has been sampled and they do not necessarily apply either to differences within various subpopulations or to differences between subpopulations* (Arthur R. Jensen). Also called **stratum.**

subpotency, *n. Biology.* a lessening in the power to transmit inherited characteristics.
—**subpotent,** *adj.* having or exhibiting subpotency.

subscapular, *Anatomy.* —*adj.* beneath, or on the anterior surface of, the scapula: *a subscapular gland or artery.*
—*n.* a subscapular muscle, artery, etc.

subsequence, *n. Mathematics.* a subordinate or secondary sequence: *a subsequence of numbers.*

subsequent (sub′sə kwənt), *adj. Geology, Geography.* **1** (of a stream) having developed its present course following the uplift of the area; being tributary to an antecedent stream. **2** cut by a subsequent stream: *a subsequent valley.* Compare **consequent, insequent, obsequent, resequent.**

subsere (sub′sir′), *n. Ecology.* a subordinate or secondary succession of plant communities: *... the species composition of early stages of the subsere* (New Scientist).

subset, *n. Mathematics.* a set, each of whose members is a member of a second set: *A subset of S is a set every element of which belongs to S. The set of numbers from 2 to 5 is a subset of the set of numbers 0 to 10. Symbol:* $\subset$ EXAMPLE: A $\subset$ B means that A is a subset of B.

subshell, *n. Physics, Chemistry.* a subdivision of an electron shell, consisting of one or more orbitals. The number of possible subshells in a shell is equal to the shell's principal quantum number. *The relation arises from the disposition of electrons around a nucleus in a series of shells and subshells, which are really a simplified physical representation of quantum-mechanical energy levels* (Scientific American). Also called **sublevel.**

subshrub, *n. Botany.* **1** a small shrub with partially woody stems. **2** = undershrub.

subsidence (səb sī′dəns), *n. Meteorology.* a gradual downward flow of air.

subsoil, *n. Geology.* the layer of earth that lies just under the surface soil: *Plants with long roots often sink their way through the topsoil in search of water and minerals in the subsoil* (Science News Letter).

subsolar, *adj. Astronomy.* directly underneath the sun; having the sun in the zenith: *The geographical point of a celestial object is the point on the earth's surface at which the object is in the zenith at a particular instant. This point is also known specifically as the subsolar point, the sublunar point, or the substellar point* (Baker, *Astronomy*).

subsonic, *adj. Physics.* **1** of or having to do with sound waves below the range of human audibility (below frequencies of about 10 hertz). Also called **infrasonic.** Compare **ultrasonic.**

2 of, having to do with, or moving at a speed less than the speed of sound (about 331 meters per second in air at sea level). —**subsonically,** *adv.*

subspace, *n. Mathematics.* a subset of a space. A subspace is itself a space.

subspecies (sub′spē′shēz), *n. Biology.* a group of related plants or animals ranking below a species. Compare **variety.**

subspecific, *adj.* of, having to do with, or like a subspecies: *a subspecific character, a subspecific rank.* —**subspecifically,** *adv.*

substance, *n. Chemistry.* any form of matter all specimens of which have identical properties and composition; any chemical element or compound. A substance is homogeneous. All samples of a substance have the same heat of vaporization, melting point, boiling point, and other properties relating to composition. ▶ See the note under **matter.**

substellar, *adj. Astronomy.* directly below any particular star; having the star in the zenith: *the substellar point of a celestial object.*

substituent, *Chemistry.* —***n.*** an atom or group of atoms taking the place of another atom or group in a compound: *The modern and more systematic method of naming alcohols is based on the number of carbon atoms in the longest continuous chain, the substituents being named and their positions numbered* (O. W. Nitz).

—***adj.*** having to do with such an atom or group of atoms.

substitution, *n. Chemistry.* the replacing of one or more elements or radicals in a compound by other elements or radicals: *The hydrogen atoms of saturated hydrocarbons can be replaced by active halogen family atoms. The general term for these reactions is halogen substitution and the products are called halogen derivatives* (Chemistry Regents Syllabus).

substrate (sub′strāt), *n.* **1** *Biochemistry.* the material that an enzyme or ferment acts upon: *The enzyme glucosidase when incubated with a substrate containing 4-methyl-unibelleferone produces the very highly fluorescent hydroxycoumarin* (F. Degering).

2 *Chemistry.* any surface on which a layer of a different material can be deposited: *Exploiting superlattices commercially will require specialised techniques for depositing very thin layers of semiconducting materials on to substrates* (The Economist).

3 *Biology.* a layer of material on which an organism can grow and multiply: *Pathogens usually lie in the soil in a dormant condition awaiting the arrival of a suitable substrate* (Ralph Baker). SYN: substratum.

substratum (sub strā′təm *or* sub strat′əm), *n., pl.* **-ta** (-tə), **-tums. 1** *Geology.* a layer of earth or rock lying just under the surface soil: *If it is true that the sedi-*

ments lie on a granitic substratum, then there was a definite time of beginning of the earth when the granites cooled and solidified and erosional processes began (Gilluly, *Principles of Geology*). Compare **subsoil.**

2 *Biology.* the substrate or medium on which an organism grows.

subtangent (sub tan′jənt), *n. Geometry.* the part of the axis of a curve cut off between the tangent and the ordinate of a given point in the curve.

subtemperate, *adj. Geography.* of, having to do with, or found in the colder regions of the Temperate Zone.

subtend (səb tend′), *v.* **1** *Geometry.* to stretch or extend under, or be opposite to (an angle, etc.): *The chord of an arc subtends the arc. A telescope normally operates by taking light from an object subtending a small angle and making it appear to subtend a large angle* (Scientific American).

2 *Botany.* to enclose in the angle between a leaf or bract and its stem.

[from Latin *subtendere,* from *sub-* under + *tendere* to stretch]

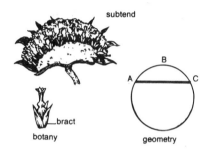

subtend

bract

botany

geometry

subtense (səb tens′), *n. Geometry.* the chord of an arc of any other subtending line.

subterrane (sub′tə rān), *n. Geology.* the bedrock under a deposit. [from Latin *subterraneus* underground]

subtilisin (sub til′ə sin), *n. Biochemistry.* an enzyme that breaks down protein, and is produced by a common species of soil bacteria, used in chemical synthesis: *Subtilisin, an enzyme which occurs in a species of bacterium, performs the task of a protease—it splits other proteins into their amino acids and has a broad preference for cleaving the substrate chain at an amino acid ester* (Science Journal). [from New Latin (*Bacillus*) *subtilis* the species of soil bacteria, from Latin *subtilis* subtle]

subtract, *v. Mathematics.* to take one number or quantity from another; find the difference between two numbers or quantities. The minus sign ($-$) means to subtract, so that $5 - 3$ means "3 subtracted from 5" or "5 minus 3": $5 - 3 = 2$. [from Latin *subtractum*

cap, fāce, fäther; best, bē, tèrm; pin, five;
rock, gō, ôrder; oil, out; cup, pùt, rüle,
yü in use, *yu* in uric;
ng in bring; sh in rush; th in thin; ᴛʜ in then;
zh in seizure.
ə = *a* in about, *e* in taken, *i* in pencil, *o* in lemon, *u* in circus

subtrahend

drawn from under, from *sub-* under + *trahere* to draw]

—subtraction, *n.* the act or process of subtracting one number or quantity from another; finding the difference between two numbers or quantities: $10 - 2 = 8$ is a simple subtraction.

—subtractive, *adj.* **1** of or having to do with subtraction. **2** tending to subtract; having power to subtract. **3** to be subtracted; having the minus sign ($-$).

subtrahend (sub'trə hend), *n. Mathematics.* a number or quantity to be subtracted from another the (*minuend*). EXAMPLE: in $10 - 2 = 8$, the subtrahend is 2. [from Latin *subtrahendus* to be subtracted, from *subtrahere* subtract]

subtriplicate, *adj. Mathematics.* being that of the cube roots of the quantities. EXAMPLE: 2 : 3 is the subtriplicate ratio of 8 : 27.

subtropical, *adj. Geography.* **1** bordering on the tropics: *a subtropical region.*
2 characteristic of subtropical regions; nearly tropical: *warm, subtropical waters.* Also called **semitropical.**

subtropics, *n.pl. Geography.* a region or regions bordering on the tropics. Also called **semitropics.**

subulate (sü'byə lit), *adj. Biology.* slender, more or less cylindrical, and tapering to a point; awl-shaped: *a subulate leaf.* [from Latin *subula* awl]

subviral, *adj. Biology.* having to do with or caused by a subvirus: *The transmissible agent of scrapie is believed by some scientists to be a subviral basic protein* (Science News).

subvirus, *n. Biology.* a viral protein or other basic constituent of a virus that has some of the properties of a virus: *For instance, attempts to remove scrapie agent from membranes should release their polysaccharide "subvirus"* (New Scientist).

subvolcanic, *adj.* = hypabyssal.

subzero, *adj.* below zero: *subzero Arctic temperatures.*

succession, *n. Ecology.* the gradual replacement of one type of community or ecosystem by another, involving a series of changes in the plant and animal life that may result in a climax: *Succession is the appearance, in a definite sequence, of one community which replaces another as environmental conditions change. Succession is as important in ecology as Mendel's laws are in genetics* (Clarence J. Hylander). *Succession may be said to begin with pioneer organisms, since these are the first plants to populate a given location ... A typical successional sequence in New York State might be: pioneer, grass, shrub, conifer, and deciduous woodland* (Biology Regents Syllabus). See also **sere.**

► Plant species (flora) dominate a community in the sense that they are the most abundant food sources. Plant succession is a major limiting factor for animal (fauna) succession. Hence communities are identified by their dominant plant species, for example Pine Barrens and Sphagnum Bog. A climax community persists until a catastrophic change, such as a forest fire or abandoned farmlands, alters or destroys it. Thereafter, succession occurs once again leading to a new climax community or to the reestablishment of the former climax community.

—successional, *adj.* of or having to do with a succession: *a successional series, a successional habitat.*

meadow

coniferous

coniferous and hardwood

hardwood

succession

succinate (suk'sə nāt), *n. Chemistry.* a salt or ester of succinic acid.

succinic acid (suk sin'ik), *Chemistry.* a colorless, odorless, crystalline dicarboxylic acid present in amber and important in the Krebs cycle. It is used as a chemical intermediate and in medicine. *Formula:* $(CH_2CO_2H)_2$ [from Latin *succinum* amber]

succinic dehydrogenase, *Biochemistry.* an enzyme that catalyzes the removal of hydrogen from succinic acid in the presence of a hydrogen acceptor.

succise (sək sīs'), *adj. Botany.* appearing as if cut or broken off at the lower end. [from Latin *succisus* cut below]

succulent, *Botany.* **—adj.** having thick or fleshy and juicy leaves or stems: *Succulent plants grow in deserts and other dry places in the world where there is little water [and] plants have large stems or leaves in which to store water* (Marcus Maxon).
—n. a succulent plant: *In desert succulents such as cacti and century plants the cells contain a large amount of colloidal material and the plants are covered with a very thick epidermis. These two factors acting together all but prevent transpiration* (Emerson, *Basic Botany*).

sucker, *n.* 1 *Botany.* **a** an adventitious shoot from the trunk or a branch of a tree or plant: *Although the individual plant bears its bunch of bananas and dies, the real stem, or trunk, is the underground rootstock; and this lives on from year to year, frequently sending up new shoots, or "suckers," which will develop into banana plants* (Colby and Foster, *Economic Geography*). **b** one of the small roots of a parasitic plant. Also called **haustorium.**

2 *Zoology.* an organ in some animals for sucking or holding fast by a sucking force.

sucrase (sü′krās), *n.* = invertase. [from French *sucre* sugar]

sucrose (sü′krōs), *n. Chemistry.* ordinary sugar obtained from sugar cane, sugar beets, etc. Sucrose is a crystalline disaccharide sugar that changes by hydrolysis to fructose and glucose. *Formula:* $C_{12}H_{22}O_{11}$ Also called **saccharose.** [from French *sucre* sugar]

sudoriferous (sü′də rif′ər əs), *adj. Physiology.* secreting or causing sweat: *The human skin differs from that of most mammals in containing sweat (sudoriferous) glands.* (Storer, *General Zoology*). [from Latin *sudor* sweat + *ferre* to bear]

suffosion (sə fyü′zhən), *n. Geology.* underground seepage of water into rock. [from Latin *suffosionem* a digging under, from *suffodere* to dig underneath]

suffrutescent (suf′rù tes′ənt), *adj. Botany.* somewhat woody or shrubby at the base.

suffrutex (suf′rù teks), *n., pl.* **suffrutices** (sə frü′tə sēz). *Botany.* 1 an undershrub, or very small shrub; a low plant with decidely woody stems, as the trailing arbutus.

2 a plant having a woody base but a herbaceous annual growth above.

[from New Latin, from Latin *sub-* under + *frutex* shoot, runner, bush]

suffruticose (sə frü′tə kōs), *adj. Botany.* having the nature of a suffrutex; small with woody stems; woody at the base but herbaceous above: *An undershrub or suffruticose stem is a stem of small size and woody only at the base, as in Bittersweet, Thyme, etc.* (Youngken, *Pharmaceutical Botany*).

sugar, *n.* 1 = sucrose. 2 *Chemistry.* any of the class of carbohydrates to which sucrose belongs, including glucose, lactose, and maltose. Sugars are soluble in water, sweet to the taste, and either directly or indirectly fermentable. According to their chemical structure, sugars are classified as monosaccharides, disaccharides, and trisaccharides. Most plants manufacture sugar.

sugar of lead, = lead acetate.

sulcate (sul′kāt), *adj. Botany, Zoology.* marked with parallel furrows or grooves, as a stem or tissue. [from Latin *sulcatum* furrowed, from *sulcus* furrow]

sulcus (sul′kəs), *n., pl.* **-ci** (-sī). *Anatomy.* 1 a groove or furrow in a body, organ, or tissue. 2 a shallow groove between two convulsions of the surface of the brain. [from Latin *sulcus* furrow, groove]

sulfatase (sul′fə tās), *n. Biochemistry.* an enzyme that catalyzes the hydrolysis of esters of sulfuric acid. One kind is commonly found in animal and plant tissues; another is present in bacteria.

sulfate, *Chemistry.* —*n.* a salt or ester of sulfuric acid. Epsom salt is a sulfate of magnesium. Gypsum is a sulfate of calcium. Ferrous sulfate is used in medicine.

—*v.* 1 to combine or treat with sulfuric acid or with a sulfate. 2 to convert into a sulfate.

sulfide, *n. Chemistry.* any compound of sulfur with another element or radical; a salt or ester of hydrogen sulfide.

sulfinyl (sul′fə nəl), *n. Chemistry.* a bivalent organic radical; -SO.

sulfite, *n. Chemistry.* a salt or ester of sulfurous acid.

sulfonate (sul′fə nāt), *Chemistry.* —*adj.* a salt or ester of a sulfonic acid.

—*v.* 1 to convert into a sulfonic acid.

2 to treat (an oil, etc.) with sulfuric acid.

—**sulfonation,** *n.* the introduction of one or more sulfonic acid radicals -SO₂OH into an organic compound.

sulfonic acid, *Chemistry.* any of a group of organic acids containing the univalent radical -SO₂OH, considered as sulfuric acid derivatives by the replacement of a hydroxyl radical (-OH). They are used in making phenols, dyes, drugs, etc.

sulfonium (sul fō′nē əm), *n. Chemistry.* a positive ion or univalent radical containing trivalent sulfur, such as H_3S^+ or $(CH_3)_3S^+$.

sulfonyl (sul′fə nəl), *n. Chemistry.* a bivalent radical, -SO₂, occurring in sulfuric acid. Also called **sulfuryl.**

sulfoxide (sulf ok′sīd), *n. Chemistry.* an organic compound containing an -SO group linked to two carbon atoms.

sulfur (sul′fər), *n. Chemistry.* a light-yellow, highly flammable, nonmetallic element that exists in several allotropic forms and burns in the air with a blue flame and a stifling odor. Sulfur is found abundantly in volcanic regions, occurring free in nature as a brittle, crystalline solid or in combination with metals and other substances, and is also a constituent of proteins. It is used in making matches and gunpowder, for vulcanizing rubber, in bleaching, and in medicine. *The biggest single use for sulfur is in making sulfuric acid to treat phosphate rock in producing phosphate fertilizers* (David M. Kiefer). *Symbol:* S; *atomic number* 16; *atomic weight* 32.064; *melting point* 112.8°C; *boiling point* 444.4°C; *oxidation state* −2, +4, +6. Also spelled **sulphur.** [from Latin *sulfur, sulphur*]

sulfurate (sul′fə rāt *or* sul′fyə rāt), *v.* = sulfurize.

—*adj.* made or consisting of sulfur.

—**sulfuration,** *n.* = sulfurization.

sulfur bacterium, *Bacteriology.* any of a group of bacteria that are able to oxidize sulfur compounds: *There are several forms of sulfur bacteria.... These are often designated as the white, red, and green sulfur bacteria* (Emerson, *Basic Botany*).

---

cap, fāce, fäther; best, bē; tėrm; pin, fīve;
rock, gō, ôrder; oil, out; cup, pùt, rüle,
*y*ü in use, *y*ù in uric;
ng in bring; sh in rush; th in thin; ŦH in then;
zh in seizure.
ə = *a* in about, *e* in taken, *i* in pencil, *o* in lemon, *u* in circus

sulfur dioxide, *Chemistry.* a heavy, colorless gas that has a sharp odor, formed in nature from volcanic activity and from the decay of organic matter. It is used to prepare sulfuric and sulfurous acids and other chemicals. *Acid rain . . . may form when sulfur dioxide in the air is converted into sulfur trioxide* (C. Frank Shaw). *Formula:* SO_2

sulfureous (sul fyùr′ē əs), *adj.* = sulfurous.

sulfuric, *adj. Chemistry.* of, having to do with, or containing sulfur, especially with a valence of six.

sulfuric acid, *Chemistry.* a heavy, colorless, oily, very corrosive acid derived from sulfur. Sulfuric acid is used in making explosives and fertilizers, in refining petroleum, and in many other industrial processes. *Formula:* H_2SO_4 Also called **oil of vitriol** or **vitriol.**

sulfurization, *n.* the act or process of sulfurizing.

sulfurize, *v. Chemistry.* to combine or impregnate with sulfur or a sulfur compound; subject to the action of sulfur.

sulfurous (sul′fər əs), *adj. Chemistry.* of, having to do with, or containing sulfur, especially with a valence of four. Also spelled **sulfureous.**

sulfurous acid, *Chemistry.* a colorless solution of sulfur dioxide in water, used as a bleach, reducing agent, etc. It is known chiefly in the form of its salts, the sulfites. *Formula:* H_2SO_3

sulfur trioxide, *Chemistry.* a chemical compound used chiefly as an intermediate in the production of sulfuric acid: *Sulfur dioxide had long been the chief suspect. Under the action of sunlight it can react with oxygen to form sulfur trioxide, which in turn can combine with water vapor to make a sulfuric acid mist* (Scientific American). *Formula:* SO_3

sulfuryl (sul′fər əl), *n.* = sulfonyl.

sulphur (sul′fər), *n.* = sulfur.

sum, *n. Mathematics.* **1** the number or quantity obtained by adding two or more numbers or quantities together; the result of addition. The sum of 2 and 3 is 5.
2 a series of two or more numbers or quantities to be added.
3 the limit of the sum of the first *n* terms of an infinite geometric progression, as *n* increases without bound. [from Latin *summa,* feminine of *summus* highest]

summation, *n. Mathematics.* the act or process of finding the sum or total. SYN: addition.

summation sign, *Mathematics, Statistics.* the Greek capital letter sigma, used to denote the sum of a series of quantities. *Symbol:* Σ

summer solstice, *Astronomy.* **1** the solstice that occurs about June 21. It is the time in the Northern Hemisphere when the noon sun appears to be farthest north. In the Southern Hemisphere this is the winter solstice.
2 the northernmost point of the ecliptic, which the sun reaches at this time. It is now in the constellation Gemini.

sun, *n. Astronomy.* **1** the brightest celestial body in the sky; a star around which the earth and other planets revolve and which supplies them with light and heat. It is a glowing ball of hot gases, chiefly hydrogen and helium. The mean distance of the sun from the earth is 150 million kilometers. Its mean diameter is 1,392,000 kilometers, its volume about 1,300,000 times that of the earth, and its mass about 333,000 times that of the earth.
2 any celestial body like the sun. Many stars are suns about which planetary systems orbit. Compare **solar, solar flare, solar radiation, solar system, solar wind.** ASSOCIATED TERMS: see **chromosphere.**

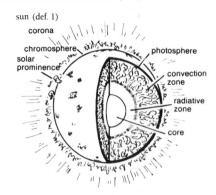

sun (def. 1)

corona
chromosphere
solar prominence
photosphere
convection zone
radiative zone
core

sunspot, *n. Astronomy.* one of the spots, darker than the rest of the photosphere, that appear periodically in certain zones of the surface of the sun, and are associated with disturbances of the earth's magnetic field: *It is everywhere recognized not only that the maximum incidence of sunspots is cyclical but that the cycle extends over a period that may vary from nine to seventeen years and that averages out to slightly more than eleven* (John Brooks). Also called **macula.** See the picture at **prominence.**

super-, *prefix.* **1** over; above; as in *superclass, superciliary.*
2 surpassing; excessive, as in *superovulation.*
3 beyond the ordinary; excessively, as in *supersaturate, superheat.*
4 very large, as in *supergiant, supercontinent.* [from Latin *super* over, above]

superactinide series (sü′pər ak′tə nid), *Chemistry.* a series of superheavy elements predicted to follow the transactinide series in the periodic table.

superciliary (sü′pər sil′ē er′ē), *adj.* **1** *Anatomy.* **a** of, having to do with, or in the region of the eyebrow. **b** designating or having to do with a prominence (**superciliary arch**) of the frontal bone over the eye.
2 *Zoology.* **a** situated over the eye: *a superciliary line or patch of color.* **b** having a marking above the eye, as various birds do.

superclass, *n. Biology.* **1** a group of related organisms ranking below a subphylum and above a class. **2** = subphylum.

supercluster, *n. Astronomy.* a group of neighboring clusters of galaxies: *Most astronomers believe that superclusters are the largest groups that can be told apart in the universe* (A. G. W. Cameron).

superconduct, *v. Physics.* to act as a superconductor; conduct electric current with no resistance at temperatures near absolute zero: *Even at lower frequencies ultrasonic waves are absorbed by the conduction electrons in various metals. That this absorption is really due to interaction with electrons is demonstrated by the fact that it disappears when the metals are cooled to the point where they become superconducting* (Klaus Dransfeld).

superconduction, *n. Physics.* **1** the conducting of electric current by means of a superconductor. **2** = superconductivity.

superconductive, *adj. Physics.* capable of or having superconductivity: *Alloys of lanthanum with yttrium and lutetium become superconductive at low temperatures* (New Scientist).

superconductivity, *n. Physics.* the ability of some metals, such as lead and tin, to conduct electric current with no resistance at temperatures near absolute zero: *Superconductivity, the strange property of metals at temperatures near absolute zero, can be eliminated by a fairly weak magnetic field* (Scientific American).

superconductor, *n. Physics.* a metal or alloy, such as lead or an alloy of niobium and tin, that can conduct electric current with no resistance at temperatures near absolute zero: *The remarkable fact about superconductors is that all the resistance vanishes at a finite temperature, different from absolute zero. Thus, in particular, although nothing happens to the impurities, they cease, for some reason, to be able to scatter the electrons* (Science).

supercontinent, *n. Geology.* any of several great landmasses, or a single great landmass, that is thought to have originally comprised the present continents and that later split up into smaller masses which drifted to form the present continents: *Dietz discounts an alternative hypothesis that all continents once were a single land mass called Pangaea. Rather he favours the notion of two supercontinents, eloquently advocated a half century ago by Wegener. Wegener's other supercontinent, called Laurasia, combined North America, Europe, and Asia* (New Scientist).

supercool, *v. Physics, Chemistry.* to cool (a liquid, especially water) below the normal freezing point without causing it to solidify or crystallize.

superfamily, *Biology.* a group of organisms ranking above a family and below an order or, according to some, below a suborder: *A common superfamily, Hominoidae, . . . sets apes and man off from monkeys* (William Howells).

superfetation (sü′pər fi tā′shən), **1** *Zoology.* the development of a second fetus before the birth of one already present in the uterus. Superfetation is a normal occurrence in some animal species.
2 *Botany.* the fertilization of the same ovule by two different kinds of pollen.
[from Latin *superfetare* to conceive during pregnancy, from *super* over, above + *fetare* to bear, from *fetus* offspring]

superfluid, *n. Physics.* a fluid, liquid helium, characterized by the complete disappearance of viscosity at temperatures near absolute zero: *Helium 4 at temperatures near absolute zero is a liquid with most remarkable properties and is the only known superfluid* (Science News Letter).
—**superfluidity,** *n.* extreme fluidity; lack of viscosity: *. . . the superfluidity of liquid helium, the frictionless flow of entire atoms, demonstrated in the liquid's ability to flow through the tiniest tubes or narrowest slits* (Scientific American).

supergalaxy, *n. Astronomy.* a very large galactic system consisting of a group of galaxies: *. . . the supergalaxy of which the Milky Way is a part* (Scientific American). Compare **local group** and **supercluster.**

supergene, *adj. Geology.* formed or originating on the earth's surface: *supergene rocks, supergene deposits.* Contrasted with **hypogene.** [from *super-* + *-gene,* as in *hypogene*]

supergiant, *n.,* or **supergiant star,** *Astronomy.* any of various extremely large and brilliant stars, ranging in luminosity from 100 to 10,000 or more times that of the sun. They are most common in the arms of spiral galaxies. *Supergiant stars are extraordinarily large and luminous giants. Examples are Rigel and Betelgeuse* (Baker, *Astronomy*).

superglacial, *adj. Geology.* situated or occurring upon a surface of ice, especially of a glacier.
—**superglacially,** *adv.*

supergranular or **supergranulated,** *adj.* of, having to do with, or forming supergranulations: *The result of the supergranular motions . . . is to spread the magnetic flux over very large areas of the solar surface, even to the poles of the sun* (Science). *Scientists care these supergranulated cells, as they are called, as the tops of convection currents bringing energy up from the center of the sun where it is generated by nuclear fusion* (William J. Cromie).

supergranulation, *n. Astronomy.* any of a large number of gaseous cells of great density and heat that extend into the chromosphere from the deeper layers of the sun: *The supergranulation "cells," unlike the small convective granulations visible on the Sun's surface, are of the order of 15000 to 30000 km across . . . with a central upflow of material which then flows outward with a velocity of about 500m/s and then down back into the body of the Sun* (New Scientist).

supergravity, *n. Physics.* any theory that unites gravity with the subatomic forces: *Since supergravity incorporates standard general relativity in a very fundamental and unique way, it was hoped that the additional constraints imposed by the added symmetry would cause the unmanageable infinities to cancel each other* (Science News).

superheat, *v. Physics.* **1** to heat (a liquid) above its normal boiling point without producing vaporization. See the picture at **geyser.**
2 to heat (steam) apart from water until it contains no suspended water droplets. The steam then resembles and will remain a dry or perfect gas at the specified pressure.

superheavy, *adj. Nuclear Physics.* **1** having a higher atomic number or greater atomic mass than those of the heaviest elements known: *Superheavy elements are generally created by accelerating heavy ions such as argon, for interactions with heavier elements such as uranium* (New Scientist).

cap, fāce, fäther; best, bē, tėrm; pin, fīve;
rock, gō, ôrder; oil, out; cup, pùt, rüle,
yü in use, *yu̇* in uric;
ng in bring; *sh* in rush; *th* in thin, ᴛʜ in then;
zh in seizure.
ə = *a* in about, *e* in taken, *i* in pencil, *o* in lemon, *u* in circus

2 of or belonging to superheavy elements: *superheavy nuclei. The beta disintegration of these 'superheavy' isotopes of uranium led naturally to the formation of einsteinium-253 and fermium-255* (Science Journal).

superhelical, *adj.* of or having to do with a superhelix: *Normal, double-helix DNA is spiral, like a loosely coiled spring. But DNA has also been isolated from cells in a superhelical form in which the helix is twisted around itself. Gellert and his co-workers found that the enzyme they discovered changed normal DNA into superhelical DNA. They named the enzyme gyrase* (Julian Davies).

superhelix, *n. Molecular Biology.* a form of DNA consisting of a double helix coiled around itself: *Recent neutron diffraction studies indicate that the DNA is wound in a superhelix around the outside of this histone core* (Nature).

superior, *adj.* **1** *Botany.* growing above some other part or organ, as: **a** the ovary when situated above or free from the (inferior) calyx. **b** the calyx when adherent to the sides of the (inferior) ovary and thus seeming to rise from its top.
2 *Astronomy.* **a** designating those planets whose orbits lie outside that of the earth (originally, according to the Ptolemaic astronomy, as having their spheres above that of the sun). **b** on the far side of the sun from the earth: *a superior conjunction of Mercury with the sun.* **c** above the horizon: *the superior passage of a star.*

superior vena cava, *Anatomy.* the large vein carrying the blood from the head, chest wall, and upper extremities to the right atrium of the heart. Compare **inferior vena cava.**

superior vocal cords, *Anatomy.* = false vocal cords.

supernatant, *n. Physics, Chemistry.* the liquid floating above or on the surface of matter deposited by centrifugation, precipitation, or other means: *Cells from the culture were found 100 % effective against the disease while the fluid or supernatant that rises on the culture offered far less protection* (Science News Letter). [from Latin *supernatare* to float on the surface]

supernova (su'pər nō'və), *n., pl.* **-vae** (-vē), **-vas.** *Astronomy.* a nova far brighter than an ordinary nova, being from 10 to 100 million times as luminous as the sun: *As an exploding star which differs from a nova both in magnitude and mechanism of the outburst, a supernova can reach the brightness of an entire galaxy. . . . These stars, which during their maximum explosive phase can emit as much energy in 1 day as the Sun does in 1,000,000 years, can be detected even in very distant galaxies* (J. Allen Hynek). *Supernovae are widely believed to result from the nearly total destruction of stars, perhaps leaving behind a neutron star or black hole as debris* (Robert P. Kirshner).

superorder, *n. Biology.* a group of organisms ranking above an order and below a class.

superorganism, *n. Biology.* a group of organisms that function as a social unit. A colony of social insects is a superorganism. *The second notion was that of the "superorganism." The collective behavior of termites in a hill . . . was an achievement only to be explained by assuming that the collection of animals, linked together, had become an intellect, a huge, crawling, thinking organism, capable of accomplishing much*

more than could be accounted for by simply summing the capacities of the individual members (Lewis Thomas).

superosculate, *v. Geometry.* to osculate at more consecutive points than usually suffice to determine the locus.
—superosculation, *n.* the act or process of superosculating.

superparasitism, *n. Biology.* the infestation of parasites by other parasites.

superphosphate, *n. Chemistry.* **1** any of various phosphates which have been treated with sulfuric acid to increase their solubility for use as fertilizers.
2 a phosphate containing an excess of phosphoric acid; acid phosphate.

superplastic, *adj. Physics.* capable of plastic deformation under very small stress at high temperature: *Superplastic materials . . . are characterized by a very small grain size . . . and an immense capacity for plastic deformation at elevated temperatures, without fracture occurring* (New Scientist).
—superplasticity, *n.* the property or quality of being superplastic.

superpose, *v. Geometry.* to place (a figure) upon another so that the two coincide.

superposed, *adj. Botany.* situated directly over some other part (used especially of a whorl of organs arranged opposite or over another instead of alternately).

superposition, *n. Geometry.* the placement of one figure upon another so that the two coincide.

supersaturate, *v. Chemistry.* to add to beyond the ordinary saturation point; saturate abnormally. A supersaturated solution is one in which more of a substance is dissolved than the solvent will hold under normal conditions.
—supersaturation, *n.* the process of supersaturating or the condition of being supersaturated: *The relative humidity of the air rose to the saturation point and even above it—a condition known as supersaturation. Some of the water molecules in the water vapor were now changed to the liquid state* (Fred R. Schlessinger).

supersonic, *adj. Physics.* **1** of or having to do with sound waves beyond the limit of human audibility (above frequencies of 20,000 hertz); ultrasonic.
2 of, having to do with, or moving at a speed greater than the speed of sound in air (about 331 meters per second at sea level in air). **—supersonically,** *adv.*

superspecies, *n. Biology.* a group of species which are related, especially on a geographical or ecological basis.

supersymmetry, *n. Physics.* a hypothetical symmetry that relates the fermions and the bosons and gravitational force to forces that operate on the subatomic level: *In supergravity, supersymmetry is extended from the global level to the local level. Remarkably, this extension leads automatically to theories that incorporate the gravitational force and suggest the possibility of unifying gravitation with the other forces* (Scientific American).

supinate (sü'pə nāt), *v. Anatomy.* to undergo or cause to undergo supination.

supination (sü'pə nā'shən), *n. Anatomy.* **1** a rotation of the hand or forelimb so that the palm faces up or forward.

2 a similar movement of the foot, hindlimb, shoulder, or the like.

3 the position which results from this rotation. [from Latin *supinatum* laid on the back, from *supinus* lying on the back, supine]

supinator (sü'pə nā'tər), *n.* a muscle, especially of the forearm, that effects or assists supination.

supp. or **suppl.,** *abbrev.* **1** supplement. **2** supplementary (angle).

supplement, *n. Geometry.* the amount needed to make an angle or arc equal 180 degrees.

supplementary angle, *Mathematics.* either of two angles which together form an angle of 180 degrees. A 60-degree angle is the supplementary angle of a 120-degree angle.

suppressor, *n. Genetics.* a mutant gene that reverses the effect of another mutant gene: *The function of a mutated gene is sometimes restored by a mutation in a different gene, called a suppressor gene, which functions by changing a part of the biochemical machinery that translates the genes* (Frank G. Rothman).

suppressor cell or **suppressor T cell,** *Immunology.* a T cell whose function is to reduce or suppress the activity of B cells so as to keep the immune system in balance: *AIDS victims develop an imbalance of those white blood cells responsible for fighting infection. Healthy persons have twice as many helper cells, which eliminate foreign organisms, as suppressor cells, which inhibit that activity; AIDS victim have more suppressor than helper cells. They are thus extremely susceptible to infection* (L. B. Berkowitz and M. J. Maslow).

supra-, *prefix.* on; above; upper, as in *supracellular, supramaxilla, suprarenal.* [from Latin *supra,* related to *super* over]

supracellular, *adj. Biology.* of or having to do with systems of biological organization above the level of the cell: *Supracellular biologists, he maintained, have made a far greater contribution to the progress of cancer therapy than molecular biologists* (Nature).

supragenic, *adj. Genetics.* of or having to do with hereditary factors above the level of the gene: *. . . supragenic functions of the chromosome, such as recombination and translocation* (H. E. Kubitschek).

supramaxilla (sü'prə mak sil'ə), *n., pl.* **-maxillae** (-maksil'ē). *Anatomy, Zoology.* the upper jaw or upper jawbone.

—**supramaxillary,** —*adj.* of or having to do with the upper jaw or upper jawbone.

—*n.* the upper jawbone.

supramolecular, *adj. Chemistry.* **1** above, or having more complexity than, a molecule.

2 made up of more than one molecule.

supraorbital, *adj. Anatomy.* situated above the orbit of the eye: *Most apes possess a marked bulge of bone, called a supraorbital ridge, which extends unbroken across the region of the skull just over the eyes. This ridge is small in the orang and very small or completely absent in man* (R. Beals and H. Hoijer).

suprarenal, *adj. Anatomy.* **1** situated above the kidney or on the kidney; adrenal.

2 of or from the adrenal glands.

—*n.* = adrenal glands.

[from *supra-* above + Latin *renes* kidneys]

supravital, *adj. Biology, Medicine.* of, having to do with, or capable of staining living cells removed from the body of an animal. Compare **intravital.**

surd (sėrd), *Mathematics.* —*n.* an irrational number, such as $\sqrt{2}$.

—*adj.* that cannot be expressed as a whole number or common factor; irrational.

[from Latin *surdus* unheard]

surface, *n. Geometry.* a portion of space extending in two dimensions; something that has length and breadth but no thickness: *a plane surface. The surface of a sphere is curved.*

surface of revolution, *Geometry.* a surface which is generated by the revolution of a curve around an axis.

surface tension, *Physics.* the tension of the surface film of a liquid that makes it contract to a minimum area. It is caused by molecular forces and measured in terms of force per unit length.

surficial (sėr fish'əl), *adj. Geology.* on or of the earth's surface: *Erosion quickly erases the surficial expression of terrestrial craters* (New Scientist).

surge, *Physics.* —*v.* to increase or oscillate suddenly or violently, as an electrical current.

—*n.* **a** a sudden or violent rush or oscillation of electrical current in a circuit. **b** a wave of pressure in a liquid system caused by a sudden stoppage of flow.

survival of the fittest, *Biology.* the survival, as a result of natural selection, of those organisms which are best adapted to their environment: *"Survival of the fittest" . . . is an expression often used to summarize the theory of natural selection. In this context it must be emphasized that "fit" refers not to health but to the production of a large number of offspring which survive to reproductive age* (Mackean, *Introduction to Biology*).

suspension, *n. Chemistry.* a mixture in which very small particles of a solid remain suspended without dissolving: *When clay is shaken up with water, in which it is insoluble, a turbid mixture is obtained, the ingredients of which slowly separate on standing. This mixture is an example of a suspension* (Offner, *Fundamentals of Chemistry*). Compare **emulsion.**

suspensor, *n. Botany.* a group of cells at the extremity of the embryo in many vascular plants, which help position the embryo in relation to its food supply: *Instead of forming a single embryo, one begins to organize from the deepest cell of each of the four columns. The cells next above elongate, forming suspensors, that is, structures that force the young, growing sporophytes deep into the tissues of the female gametophyte* (Emerson, *Basic Botany*).

suspensory ligament, *Anatomy.* a supporting ligament, especially the membrane that holds the lens of the eye in place.

---

cap, fāce, fäther; best, bē, tėrm; pin, fīve;
rock, gō, ôrder; oil, out; cup, pùt, rüle,
yü in use, *yu* in uric;
ng in bring; *sh* in rush; *th* in thin, ŦH in then;
zh in seizure.

ə = *a* in about, *e* in taken, *i* in pencil, *o* in lemon, *u* in circus

sustentacular

sustentacular (sus'ten tak'yə lər), *adj. Anatomy.* supporting: *sustentacular muscle fibers.* The **sustentacular cells** are the supporting cells of an epithelial tissue in contrast to other nearby cells with different functions. [from Latin *sustentaculum* a stay, support]

suture, *n.* **1** *Anatomy.* a line formed by a junction or closure between bones: *a cranial suture, the coronal suture.*
2 *Geology.* a principal fracture separating crustal plates of the earth: *The Vardar Zone is thought to include a line of "sutures" formed where two continental blocks converged, sweeping up an intervening ocean floor* (Walter Sullivan).

SV40, *abbrev. Biology.* Simian Virus 40 (a virus that causes cancer in monkeys, widely used in genetic and medical research): *Perhaps the most successful in vivo system for manipulating the function of mammalian genes comes from forming hybrid SV40 viruses* (Science).

svanbergite (svän'bər gīt), *n. Mineralogy.* a mineral occurring in rhombohedral crystals of a yellow, red, or brown color. It consists of sulfate and phosphate of aluminum and strontium. *Formula:* $SrAl_3(PO_4)(SO_4)(OH)_6$ [from Lars F. *Svanberg,* Swedish chemist of the 1800's + *-ite*]

svedberg (sved'bėrg *or* svä'bər ē), *n.,* or **Svedberg unit,** *Chemistry.* a measure of the rate of sedimentation of a protein or other large molecule in a centrifugal field, equal to 10^{-13} centimeter per second: *In the tadpole we find a single protein peak with a sedimentation constant of 4.3 Svedberg units* (Scientific American). *Abbreviation:* S [named after Theodor *Svedberg,* 1884–1971, Swedish chemist]

swarm, *Biology.* —*n.* a cluster of free-swimming or free-floating cells or one-celled organisms, such as zoospores, moving in company.
—*v.* to escape from the parent organism in a swarm, with characteristic movement.

swarmer, = swarm spore.

swarm spore, *Biology.* any of various motile spores produced in great abundance, especially a zoospore.

sweat, *n. Physiology.* the colorless moisture secreted by the sweat glands through the pores of the skin to help regulate the body's temperature: *Sweat is much more dilute than urine, about 99 per cent of it being water. Of its solids, sodium chloride is the most important. Urea is not very abundant; at the minimum production of sweat (600 cc. per day) only about 1.5 per cent of the total urea is lost through the skin in man* (Shull, *Principles of Animal Biology*). Also called **perspiration.** Compare **sudoriferous.**

sweat gland, *Physiology.* one of the many small coiled glands that secrete sweat. Human sweat glands are found in the deeper layer of the skin, connecting with the surface of the skin through a duct that ends in a pore. See the picture at **skin.**

swimmeret (swim'ə ret), *n. Zoology.* an abdominal limb or appendage in many crustaceans, used in respiration and for carrying the eggs in females and usually adapted for swimming. Also called **pleopod.**

sychnocarpous (sik'nə kär'pəs), *adj. Botany.* capable of bearing fruit many times without perishing, as certain trees. [from Greek *sychnos* many + *karpos* fruit]

syconium (sī kō'nē əm), *n., pl.* **-nia** (-nē ə). *Botany.* a multiple fruit developed from numerous flowers embedded in a hollow fleshy receptacle, as in the fig. [from New Latin, from Greek *sykon* fig]

syenite (sī'ə nīt), *n. Geology.* a gray, crystalline, igneous rock composed of an alkali feldspar and certain other minerals, such as hornblende: *A syenite comparatively rich in quartz is called syenite-granite; it really represents an intermediate stage between granite and normal syenite.* (Fenton, *The Rock Book*). [from Greek *Syēnítēs* (*lithos*) (stone), from *Syēnē* (now Aswan) in Egypt]
—**syenitic** (sī'ə nit'ik), *adj.* **1** of or containing syenite. **2** resembling syenite, or possessing some of its properties.

sylvanite (sil'və nīt), *n. Mineralogy.* a telluride of gold and silver, occurring in crystals or masses of gray, white, or yellow, with metallic luster. *Formula:* $(Au, Ag)Te_2$ [from (*Tran*)*sylvania,* region in Romania where it is found + *-ite*]

sylvinite (sil'və nīt), *n. Geology.* an evaporite rock consisting of a mixture of halite and sylvite, used as an ore of potassium.

sylvite (sil'vīt), *n. Mineralogy.* a salt-tasting mineral, potassium chloride, occurring in white or colorless cubes or octahedrons. It is an important source of potassium. *Formula:* KCl [from New Latin *sal digestivus sylvii* (literally) digestive salt of *Sylvius,* the Latinized name of a physician of the 1600's + *-ite*]

sym., *abbrev. Chemistry.* symmetrical.

symbiont (sim'bē ont *or* sim'bī ont), *n. Biology.* an organism living in a state of symbiosis: *While this particular symbiont dinoflagellate is not poisonous, other symbiotic zooxanthellae may be, and fish which feed on coral containing them would presumably become poisonous in turn* (Scientific American). *The aphid Myzus persicae has a requirement for only three amino acids—methionine, histidine, and isoleucine—while it is in possession of its normal intracellular symbionts, but appears to require at least ten amino acids when freed from its symbionts* (Peter W. Miles). Also called **symbiote.**

symbiosis (sim'bē ō'sis *or* sim'bī ō'sis), *n., pl.* **-ses** (-sēz'). *Biology.* any prolonged association or living together of two or more organisms of different species. [from Greek *symbiōsis,* from *symbioun* live together, from *syn-* together + *bios* life]
▶ Formerly, *symbiosis* was usually applied to associations that are beneficial to both parties—to what is known as *mutualism.* Currently *symbiosis* is applied to any special association of unlike organisms within the looser association of a community as a whole.
ASSOCIATED TERMS: Symbiosis is a general term for relationships between organisms of different species. It includes *parasitism,* in which one organism benefits at the expense of another, often injuring it; *commensalism,* in which one organism benefits while the other is unaffected; *amensalism,* in which one is inhibited while the other is unaffected; and *mutualism,* in which both organisms benefit. The term *antibiosis* refers to any kind of symbiosis which is detrimental to at least one of the organisms.

symbiote (sim'bē ōt *or* sim'bī ōt), *n.* = symbiont.

symbiotic (sim′bē ot′ik *or* sim′bī ot′ik), *adj.* of or characterized by symbiosis: *Symbiotic relationships may include nutritional, reproductive, and protective relationships* (Biology Regents Syllabus). —**symbiotically,** *adv.*

symbol, *n. Chemistry.* a letter or group of letters and numbers representing an element, atom, molecule, etc.: *A symbol not only stands for the name of the element, but it also signifies one atom of that element* (Parks and Steinbach, *Systematic College Chemistry*).

symmetric, *adj.* = symmetrical.

symmetrical (si met′rə kəl), *adj.* **1** *Geometry.* characterized by or having symmetry.
2 *Botany.* (of a flower) having the same number of parts (sepals, petals, stamens, and carpels) in each whorl. SYN: isomerous.
3 *Biology.* divisible vertically into similar halves either by one plane only (bilaterally symmetrical or zygomorphic) or by two or more planes (radially symmetrical or actinomorphic): *When the parts of a whole are balanced in respect to size, shape or position on opposite sides of a center, the object is said to be symmetrical. . . . In zoology, symmetry is the arrangement of the parts of animal bodies in relation to a central axis. Jellyfishes are an example of sea animals with radial symmetry. Some one-celled animals are asymmetrical, or not symmetrical* (Science News Letter).
4 *Chemistry.* **a** having a structure or structural formula characterized by a particular symmetry. **b** denoting a derivative of benzene in which hydrogen atoms occupying positions one, three, and five have been replaced. —**symmetrically,** *adv.*

symmetry

asymmetry

radial symmetry

bilateral symmetry

symmetry (sim′ə trē), *n.* **1** *Geometry.* a balanced arrangement; ability to be divided into two parts that are mirror images of each other. A figure has symmetry with respect to a point if that point is the midpoint of every possible line segment containing it and having endpoints on the figure. A figure has symmetry with respect to a line or plane if the line or plane bisects every possible line segment intersecting it at right angles and having endpoints on the figure. *The theorem that if, in a triangle ABC, AB=AC, then the angles at B and C are equal, is blindingly obvious (by symmetry)* (H. A. Thurston).
2 *Botany.* agreement in number of parts among the cycles of organs that compose a flower.
3 *Biology.* the arrangement of organs or other parts of an organism into similar halves around a center or axis: *Symmetry is defined as a correspondence in shape or arrangement of parts on opposite sides of a dividing line or plane, such that if the portion on one side were viewed in a mirror it would appear identical with the part on the other side* (Shull, *Principles of Animal Biology*). Compare **bilateral symmetry, radial symmetry.**
4 *Chemistry, Physics.* The arrangement of the atoms of a molecule in a definite spatial pattern or coordinate system.
5 *Physics.* a relationship between properties of subatomic particles, such as spin or electric charge, that stems from an underlying similarity.
[from Greek *symmetria,* from *syn-* together + *metron* measure]

sympathetic, *adj.* **1** *Physiology.* **a** produced in one body by transmission of vibrations of the same frequency from another body: *sympathetic vibrations.* **b** produced by responsive vibrations induced in one body by transmission of vibrations of the same frequency through the air or other medium from another: *The phenomena of resonance are examples of sympathetic sound.*
2 *Anatomy.* of or having to do with the sympathetic nervous system: *a sympathetic neuron, sympathetic ganglia.* —**sympathetically,** *adv.*

sympathetic nervous system, *Physiology.* the part of the autonomic nervous system that produces involuntary responses opposite to those produced by the parasympathetic nervous system, such as increasing the heartbeat and slowing down activity of glands and digestive and reproductive organs. It consists of two groups of ganglia connected by nerve cords, one on either side of the spinal column.

sympathomimetic (sim′pə thō′mi met′ik), *Medicine.* —*adj.* imitating or mimicking the action of the sympathic nervous system. —*n.* a sympathomimetic substance.

sympatric (sim pat′rik), *adj. Ecology.* of or having to do with sympatry; existing in the same region without interbreeding: *sympatric species.* Compare **allopatric.** —**sympatrically,** *adv.*

—**sympatry,** *n.* the existence of plant or animal species in the same area without hybridization through interbreeding: *It is known as the principle of competitive exclusion and it says, in brief, that two noninterbreeding populations that stand in precisely the same relationship to their environment cannot occupy the same territory indefinitely. They cannot, in other words, live in "sympatry" forever* (Scientific American).
[from Greek *syn-* together + *patra* fatherland]

sympetalous (sim pet′ə ləs), *adj.* = gamopetalous.

---

cap, fāce, fäther; best, bē, tėrm; pin, fīve;
rock, gō, ôrder; oil, out; cup, pùt, rüle,
yü in use, *yù* in uric;
ng in bring; *sh* in rush; *th* in thin, ṫH in then;
zh in seizure.
ə = *a* in about, *e* in taken, *i* in pencil, *o* in lemon, *u* in circus

symphysis (sim′fə sis), *n., pl.* **-ses** (-sēz′). **1** *Anatomy.* **a** the union of two bones originally separate, either by the fusion of the bony substance or by intervening cartilage, especially of two similar bones on opposite sides of the body in the median line, as that of the pubic bones or of the two halves of the lower jawbone. **b** the part or line of junction thus formed. **c** an articulation in which bones are united by cartilage without a synovial membrane. Also called **synchondrosis. d** a union or line of junction or other parts either originally or normally separate.
2 *Botany.* a fusion or coalescence of parts of a plant normally distinct.
[from Greek *symphysis* a natural growing together or articulation (of bones), from *syn-* together + *phyein* to grow]

sympodial (sim pō′dē əl), *adj. Botany.* having to do with, of the nature of, or producing a sympodium. —**sympodially,** *adv.*

sympodium (sim pō′dē əm), *n., pl.* **-dia** (-dē ə). *Botany.* an axis or stem that imitates a simple stem, but is made up of the bases of a number of axes that arise successively as branches one from another, as in the grapevine. [from New Latin, from Greek *syn-* together + *podos* foot]

syn-, *prefix.* with; together; jointly; alike, as in *synecology, syngenesis.* [from Greek *syn* together, with]

synaeresis (si ner′ə sis), *n.* = syneresis.

synantherous (si nan′thər əs), *adj. Botany.* characterized by stamens that are coalescent by means of their anthers, as a composite plant.

synanthous (si nan′thəs), *adj. Botany.* **1** characterized by the abnormal union of two or more flowers.
2 having flowers and leaves that appear at the same time.
—**synanthy** (si nan′thē), *n.* the abnormal union of two or more flowers.
[from *syn-* together + Greek *anthos* flower]

synapse (sin′aps *or* si naps′), *n.* **1** *Physiology.* the place where a nerve impulse passes from one nerve cell to another; the point at which an impulse is transmitted from the axon of one neuron to the dendrite of another: *Electron micrographs of insect muscle show that the structure of these synapses between nerve and muscle is similar to that in vertebrates: the end of the axon—the elongated portion of the nerve cell—is in close contact with the membrane of the muscle cell* (David S. Smith). See the picture at **neuron.**
2 = synapsis.
—*v.* **1** *Physiology.* to form a synapse: *An inhibitory neuron acts to inhibit all cells with which any of its terminals synapse* (Victor J. Wilson).
2 *Biology.* to unite in synapsis: *The two sets of chromosomes are so different (AA in one species and BB in another) that they do not synapse with each other during meiosis* (Greulach and Adams, *Plants*).
[from Greek *synapsis* conjuction, from *syn-* together + *haptein* fasten]

synapsis (si nap′sis), *n., pl.* **-ses** (-sēz′). **1** *Biology.* the union of homologous chromosomes during meiosis: *Zygonema is defined as the stage [of meiotic prophase] in which homologous sets of sister chromatids complete their side-to-side alignment, an association called synapsis* (Goodenough, *Genetics*). **2** = synapse.

synaptic (si nap′tik), *adj.* having to do with a synapse or a synapsis: *It has to be stressed that there is no actual physical continuity between neurons. The narrow gap separating the knobs of one neuron from the knobs or cell body of another neuron is known as the synapse. Any one neuron may have as many as a thousand or more synaptic connexions* (George M. Wyburn). —**synaptically,** *adv.*

synaptinemal complex or **synaptonemal complex** (sə nap′tə nē′məl), *Biology.* a threadlike protein structure that extends between chromosomes during the prophase of meiosis: *Electron microscopy has revealed the presence of a tripartite structure (the synaptinemal complex) associated with meiotic chromosomes. Apparently restricted to germ cells, the complex is thought by some to represent the point-to-point pairing of homologs preparing for or undergoing exchange* (Rhoda F. Grell). [from *synaptic* + Greek *nema* thread]

synaptosome (si nap′tə sōm), *n. Physiology.* a structure containing parts of nerve endings that have undergone synapse: *The most remarkable feature of synaptosomes is that the attachment of the surface membranes of the two nerve cells (the synaptic junction) remains intact even though each of these membranes has been torn from its cell during the homogenisation* (New Scientist).
—**synaptosomal,** *adj.* of or having to do with a synaptosome or synaptosomes: *synaptosomal fractions.*

synarthrodial (sin′är thrō′dē əl), *adj. Anatomy.* having to do with or of the nature of a synarthrosis. —**synarthrodially,** *adv.*

synarthrosis (sin′är thrō′sis), *n., pl.* **-ses** (-sēz′). *Anatomy.* a kind of articulation admitting of no movement, as in the sockets of the teeth. [from Greek *synarthrōsis* immovable articulation, from *syn-* together + *arthron* joint]

syncarp (sin′kärp), *n. Botany.* **1** an aggregate fruit. **2** a multiple fruit. [from *syn-* together + *-carp* fruit]
—**syncarpous** (sin kär′pəs), *adj.* **1** of or having the character of a syncarp. **2** consisting of united or coherent carpels.
—**syncarpy** (sin′kär pē), *n.* **1** the state of having united carpels. **2** the abnormal union or fusion of two or more fruits.

syncerebrum (sin ser′ə brəm), *n., pl.* **-brums, -bra** (-brə). *Zoology.* the compound brain of an insect.

synchondrosial (sing′kon drō′sē əl), *adj. Anatomy.* of or having to do with synchondrosis.

synchondrosis (sing′kon drō′sis), *n., pl.* **-ses** (-sēz′). *Anatomy.* **1** an articulation in which the bones are so fused by intervening cartilage that the joint has little or no motion. **2** = symphysis (def. 1c). [from *syn-* together + Greek *chondros* lump]

synchrocyclotron (sing′krō si′klə tron), *n. Nuclear Physics.* a cyclotron that accelerates charged particles by changing the frequency of the alternating electric field so that it synchronizes with the motion of the particles: *A ... synchrocyclotron discovered that high-energy proton beams are strongly polarized after being scattered from a hydrogen target* (Scientific American).

synchronism, *n. Physics.* the condition of being synchronous: *A way to forestall particle and applied voltage falling out of synchronism is to use as high an applied voltage as possible so that the particles have to go around as few times as possible* (E. O. Lawrence and Lloyd Smith).

synchronous, *adj. Physics.* having coincident frequency, as an alternating electric current: *. . . a synchronous compensator for an open-air hydrogen-cooling installation* (New Yorker).

synchrotron (sing'krə tron), *n. Nuclear Physics.* a particle accelerator that accelerates subatomic particles to extremely high speed by simultaneously increasing the magnetic field and changing the frequency of the electric field: *The synchrotron, requires less magnet iron than the synchrocyclotron, because it does not need a magnetic field everywhere inside it but only near the fixed orbit of the particles* (Scientific American). [from *synchro*(*nous*) + *-tron*]

synchrotron radiation, *Physics, Astronomy.* a form of electromagnetic radiation produced by the spiraling motion of high-speed electrons around a magnetic field, as in a synchrotron or in some galactic and extragalactic nebulae: *By analogy with the radio galaxies, it is tempting to suggest that this continuous spectrum is synchrotron radiation, caused by the acceleration of very high energy electrons in a magnetic field* (Science Journal).

synclinal (sin klī'nl *or* sing'klə nəl), *adj. Geology.* of, having to do with, or resembling a syncline. A **synclinal axis** is the axis toward which the slopes of a syncline converge. A **synclinal line** is the line from which a syncline slopes.

syncline (sing'klīn), *n. Geology.* a fold or folds of rock strata sloping downward from opposite directions so as to form a trough or inverted arch: *Throughout its entire length of more than 2,000 miles, the Appalachian geosyncline was greatly compressed by earth forces, which pushed previously undisturbed Paleozoic deposits northwestward against the rigid interior platform of the continent. The rocks were folded into great anticlines and synclines, and in many places they were broken by thrust faults having the upthrow side on the southeast* (Moore, *Introduction to Historical Geology*). [from *syn-* together + Greek *klinein* to lean]

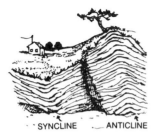

SYNCLINE   ANTICLINE

synclinorium (sing'klī nôr'ē əm), *n., pl.* **-noria** (-nôr'ē ə), **noriums.** *Geology.* a compound syncline, consisting of a series of subordinate synclines and anticlines, the whole formation having the general contour of an inverted arch. [from New Latin, from Greek *synklinein* to incline + *oros* mountain]

syncytial (sin sish'ē əl), *adj.* of or having to do with a syncytium: *syncytial fusion of cells, a syncytial structure.*

syncytium (sin sish'ē əm), *n., pl.* **-cytia** (-sish'ē ə). *Biology.* **1** a single cell containing several nuclei, formed either by fusion of a number of cells without fusion of the nuclei, or by division of the nucleus without division of the cell substance.
2 a structure composed of such cells, as one forming the outermost fetal layer of the placenta: *The body wall* [*of a rotifer*] *is a thin syncytium with nuclei, covered by a thin glassy chitin-like cuticle* (Storer, *General Zoology*).
[from New Latin, from *syn-* together + Greek *kytos* anything hollow, cell]

syndesmosis (sin'des mō'sis), *n., pl.* **-ses** (-sēz'). *Anatomy.* an articulation in which the bones are connected by ligaments, membranes, or other structures, other than those which enter into the composition of the joint. [from New Latin, from Greek *syndesmos* a fastening, binding together (as of sinews)]

syndesmotic (sin'des mot'ik), *adj. Anatomy.* bound together by a fascia, as two bones; of or having to do with syndesmosis.

synecious (si nē'shəs), *adj.* = synoecious.

synecological (sin'ek ə loj'ə kəl), *adj.* of or having to do with synecology. —**synecologically,** *adv.*

synecology (sin'ə kol'ə jē), *n.* the branch of ecology dealing with communities as distinguished from individual species: *Synecology constitutes a large part of modern ecological research, contributing to understanding of the vital interrelations of plants and animals to each other* (Clarence J. Hylander).

syneresis (si ner'ə sis), *n. Chemistry.* the loss of liquid and resulting contraction of a gel or clot. Also spelled **synaeresis.** [from Greek *synairesis* contraction, a drawing together]

synergid (si nèr'jid *or* sin'ər jid), *n. Botany.* either of the pair of small cells at the micropylar end of the embryo sac in an ovule of a flowering plant: *After fertilization, synergids and antipodal cells usually degenerate, leaving the zygote and primary endosperm nucleus within the ovule* (Weier, *Botany*). [from Greek *synergos* working together]

synergism (sin'ər jiz əm) *or* **synergy** (sin'ər jē), *n. Biology.* a combined or correlated action of a group of organs of the body, such as nerve centers or muscles, or of two or more drugs or remedies: *Synergism between auxin and gibberellin has also been demonstrated in other phases of plant development* (New Scientist). [from Greek *synergos* working together, from *syn-* together + *ergon* work]

cap, fāce, fäther; best, bē, tėrm; pin, fīve;
rock, gō, ôrder; oil, out; cup, pùt, rüle,
yü in use, yụ in uric;
ng in bring; sh in rush; th in thin, ᴛʜ in then;
zh in seizure.
ə = a in about, e in taken, i in pencil, o in
lemon, u in circus

—**synergistic,** *adj.* of or having to do with synergism: *The mixture has a synergistic effect; that is, its killing power is much greater than the sum of the power of the two poisons used separately* (N.Y. Times).

syngamic (sin gam′ik), *adj.* having to do with syngamy.

syngamous (sing′gə məs), *adj.* = syngamic.

syngamy (sing′gə mē), *n. Biology.* the union of two cells, as of gametes in fertilization. Compare **allogamy, heterogamy.**

syngeneic (sin′jə nē′ik), *adj. Immunology.* genetically identical or closely related, so as to allow tissue transplant, etc.: *Mouse recipients of transplants are called "syngeneic" when the corresponding pair of genes in each host cell are identical with those of the homozygous tumor cells* (Science News Letter). Compare **allogeneic, xenogeneic.** [from Greek *syngeneia* kinship]

syngenesious (sin′jə nē′shəs), *adj.* **1** *Botany.* (of stamens) united by the anthers so as to form a ring or tube. **2** = syngeneic.

syngenesis (sin jen′ə sis), *n. Geology.* the simultaneous formation of mineral deposits with the rock enclosing them.

—**syngenetic** (sin′jə net′ik), *adj.* of or having to do with syngenesis: *syngenetic ores.* —**syngenetically,** *adv.*

synkaryon (sin kar′ē on), *n. Biology.* a nucleus produced by the fusion of two nuclei, as in fertilization. [from *syn-* together + Greek *karyon* nut]

synkinesis (sin ki nē′sis), *n. Physiology.* a reflex or involuntary movement of muscles or limbs made simultaneously with a voluntary or deliberate movement. Swinging one's arms when walking is an example of synkinesis.

synoecious (si nē′shəs), *adj. Botany.* **1** having male and female flowers in one head, as some composite plants. **2** having male and female organs in the same receptacle, as some mosses. Also spelled **synecious.** [from *syn-* together + *-oecious,* as in *dioecious*]

synonym, *n.* **1** *Biology.* a scientific name discarded as being incorrect or out of date.
2 *Genetics.* a nucleotide triplet or codon that may be substituted for another to produce the same amino acid: *There are more than 60 translator RNA molecules, each corresponding to an RNA code word for one of the 20 amino acids. Because there are synonyms in the code, as many as six different translator RNA molecules may link to the same amino acid* (Stephen L. Wolfe).

—**synonymous,** *adj. Genetics.* consisting of or involving synonyms: *With most amino acids being designated by two or more synonyms, which typically differ only in the third position of the triplet . . . a large fraction (perhaps 70 percent) of all random nucleotide substitutions at the third position are synonymous changes and do not lead to amino acid replacements* (Scientific American).

—**synonymy,** *Biology.* a list of the several different scientific names that have been applied to a species or other group by various taxonomists.

synoptic (si nop′tik), *adj. Meteorology.* **1** of or having to do with a chart showing meteorological data from simultaneous observations at many points: *Marks were allotted on a system based on synoptic data, height of cloud-top in relation to freezing level* (A. W. Haslett).
2 of or having to do with the branch of meteorology that deals with the compilation or analysis of such data.
[from Greek *synoptikos* giving a general view, from *synopsis* a general view]

synovia (si nō′vē ə), *n. Physiology.* a viscid, clear, lubricating liquid secreted by certain membranes, such as those lining the joints. [from New Latin (coined by Paracelsus)]

—**synovial,** *adj.* consisting of, containing, or secreting synovia: *The ends of the bones are padded with cartilage at the joints and there is a lubricating liquid, the synovial fluid, that further reduces friction* (Winchester, *Zoology*).

synovium (si nō′vē əm), *n. Histology.* the membrane lining a joint of the body; a synovial membrane: *The knee joint has a lining, or synovium, which produces lubrication for the joint by secreting a fluid just slightly thicker than cooking oil. The synovium reacts defensively to injury by overproducing the fluid, which results in the well-known, if misnamed, water on the knee* (N. Y. Times Magazine). [from New Latin, from *synovia*]

synpetalous (sin pet′ə ləs), *adj.* = gamopetalous.

synsepalous (sin sep′ə ləs), *adj.* = gamosepalous.

synspermous (sin spėr′məs), *adj.* characterized by synspermy.

synspermy (sin spėr′mē), *n. Botany.* union or coalescence of two or more seeds.

syntenic (sin ten′ik), *adj. Genetics.* located on the same chromosome pair; characterized by synteny: *The demonstration that two or more genes are syntenic . . . and the assignment of genes or groups of syntenic genes to particular chromosomes can be accomplished . . . [by] following the segregation of genetic markers in informative families* (R. P. Creagan and F. H. Ruddle).

synteny (sin′tə nē), *n. Genetics.* the location of particular genes on the same chromosome pair, regardless of any hereditary linkage between them: *Genes that are on the same chromosome will therefore usually be expressed together. . . . Assaying a number of clones for various human enzymes therefore provides information on the synteny of genes* (Scientific American). [from *syn-* together + Latin *tenia* band]

syntexis (sin tek′sis), *n. Geology.* the process by which magma is formed by the melting of different types of rocks; the modification of magma composition by the assimilation of country rock. [from Greek *syntēxis* a melting together]

synthesis (sin′thə sis), *n., pl.* **-ses** (-sez′). *Chemistry.* the formation of a compound or a complex substance from its elements or simpler compounds, by one or more chemical reactions or by nuclear change. Alcohol, ammonia, and rubber can be artificially produced by synthesis. The synthesis of various new elements has been accomplished by nuclear bombardment. Protein synthesis occurs in the cells of all living organisms, but some proteins, such as insulin, have also been synthesized in the laboratory. *Generally, organic synthesis is accomplished by inducing chemical reactions to bind together basic, readily available chemicals and compounds until the desired complex molecule is created*

(Philip F. Gustafson). *Synthesis, while important in its own right as a process by which many useful materials are prepared, is often used in the laboratory in conjunction with analysis in the establishment of composition* (Jones, *Inorganic Chemistry*). [from Greek, from *syntithenai* to combine, from *syn-* together + *tithenai* to put]

—**synthesize,** *v.* to produce by chemical synthesis: *Enzymes, hormones, antibodies, genes, and viruses are all synthesized in the cells of living organisms.*

synthetase (sin'thə tās), *n.* = ligase.

synthetic, *adj. Chemistry.* of, having to do with, or involving synthesis, especially by artificial means: *synthetic resins, synthetic fibers, synthetic rubber.*

synthetic division, *Mathematics.* a short method for dividing polynomials when the divisor is a polynomial of the first degree by using only the coefficients of the polynomials.

synthetic geometry, geometry treated without algebra or coordinates, as distinguished from analytic geometry; ordinary Euclidean geometry.

syrinx (sir'ingks), *n., pl.* **syringes** (sə rin'jēz), **syrinxes.** *Zoology.* the vocal organ of birds, situated at or near the division of the trachea into the right and left bronchi: *The syrinx is partially closed with a valve, which can be tightened to produce the various notes when air is forced through it . . . but usually only the males have the beautiful songs, which are heard most during the mating season* (Winchester, *Zoology*). [from Greek *syrinx* shepherd's pipe]

system, *n.* **1** *Biology.* **a** a set of organs or parts in the body having the same or similar structure, or serving to perform some function: *the nervous system, the digestive system. In all the more complex animals the systems are everywhere made up of unlike parts, each contributing a different portion of the general process* (Shull, *Principles of Animal Biology*). **b** the organism as a whole, especially with respect to its vital processes or functions: *The living system is essentially an "open" thermodynamic system in which the cells are capable of exchanging energy with outside sources* (Atlantic). **c** each of the primary groups of tissues or parts in the higher plants: *the vascular system.*
2 *Astronomy.* **a** a group of celestial bodies forming a whole that follows certain natural laws: *the solar system.* **b** a theory of the arrangement and relationship of the celestial bodies by which their observed movements and phenomena are explained: *the Copernican system.*
3 *Geology.* a major division of rocks ranking above a series, containing the rocks formed during a geological period: *Major segments of the geologic column, which are deemed to have world-wide application, are known as systems* (Moore, *Introduction to Historical Geology*).
4 *Physics, Chemistry.* **a** a portion of matter made up of an assemblage of substances which are ordered in one or more phases and are in, or tend to approach, equilibrium. *A system is binary when it is made up of two substances, ternary if made up of three, and so on. Heterogeneous systems are made up of matter in different states of aggregation* (Parks and Steinbach, *Systematic College Chemistry*). **b** a substance, or an assemblage of substances, considered as a separate entity for the purpose of study: *Any system, such as a normal atom, containing equal numbers of protons*

and electrons, exhibits no net charge (Sears and Zemansky, *University Physics*).
5 *Mathematics.* **a** one or more sets of elements and the various operations and relations involving them: *the system of real numbers.* **b** a method of representation or notation of points, numbers, etc.: *the Cartesian coordinate system, the decimal system.*
[from Greek *systēma*, from *synistanai* bring together, from *syn-* together + *histanai* to stand, place]

systematic, *adj.* **1** of, having to do with, or according to a system: *The subject matter of all science is essentially the same: systematic observation and systematic presentation of the observations in communicable form* (F. H. George). *During the time when systematic classification was beginning to develop there was on foot a movement to give precise names to plants* (Emerson, *Basic Botany*).
2 *Statistics.* affecting all of a set of measurements; not random or chance: *a systematic error, systematic bias.*

systematics, *n. Biology.* the science of classifying groups of organisms. SYN: taxonomy.

systemic, *adj. Biology.* **1** having to do with, supplying, or affecting the organism as a whole: *systemic symptoms, a systemic vein. A systemic poison is absorbed into the plant itself, not merely sprayed on the outside* (Harper's).
2 having to do with or affecting a particular system of parts or organs of the body, especially the nervous system: *a systemic lesion.*

systemic circulation, *Physiology.* the circulation of blood throughout the body: *The customary use of the term "pulmonary circulation" for that part of the circulatory system which involves the right ventricle and the lungs, and of the term "systemic circulation" for the left ventricle and the balance of the circulatory system, seems to imply that the body has two distinct blood circuits. In actuality there is only one circuit; the systemic apparatus is one arc of it and the pulmonary apparatus is the other* (Julius H. Comroe, Jr.).

systole (sis'tə lē), *n. Physiology.* the normal, rhythmical contraction of the heart, especially that of the ventricles, when blood is pumped from the heart into the arteries. Systole alternates with diastole, the two together constituting the cardiac cycle. *After the contracting heart has forced the last drop of blood out of the ventricles, it relaxes. The contraction lasts for three-tenths of a second, the relaxation, or rest period, five-tenths of a second in a heart that beats 72 times a minute. The period of contraction is called the systole, that of relaxation the diastole* (Science News Letter). [from Greek *systolē* contraction]

—**systolic** (sis tol'ik), *adj.* having to do with, or characterized by a contraction of the heart: *The maximum or systolic pressure, the highest pressure in the artery, corresponds to the peak of the pressure-wave resulting*

---

cap, fāce, fäther; best, bē, tèrm; pin, fīve;
rock, gō, ôrder; oil, out; cup, pùt, rüle,
yü in use, yù in uric;
ng in bring; sh in rush; th in thin, ŦH in then;
zh in seizure.

ə = a in about, e in taken, i in pencil, o in
lemon, u in circus

syzygetically

from *systolic* or *ventricular contraction* (Harbaugh and Goodrich, *Fundamentals of Biology*).

syzygetically (siz′ə jet′ə klē), *adv.* with reference to a syzygy or syzygies.

syzygial (si zij′ē əl), *adj.* having to do with or of the nature of a syzygy.

syzygy (siz′ə jē), *n., pl.* **-gies.** *Astronomy.* the conjunction or opposition of two celestial bodies, or either of the points at which these occur, especially with respect to the sun and the moon. [from Latin *syzygia,* from Greek, from *syzygein* to yoke (in pairs), from *syn-* together + *zygon* yoke]

T

t., *abbrev.* **1** temperature. **2** time. **3** ton *or* tons. **4** tertiary.

T, *abbrev. or symbol.* **1** temperature. **2** tera-. **3** tesla. **4** thymine. **5** time. **6** tritium.

Ta, *symbol.* tantalum.

table, *n.* **1** an arrangement of numbers, symbols, etc. in columns and lines to indicate some relation between them: *a multiplication table, the periodic table of the elements, a table of weights and measures.* **2** *Anatomy.* either of the two large bones of the skull separated by the diploe. **3** *Geography.* = tableland.

tableland, *n. Geography.* a large, high plain; broad, elevated, nearly flat area or region. SYN: plateau, mesa, table.

tabular, *adj. Geology.* **1** tending to split into flat, thin pieces. **2** covering a large area while being relatively thin: *Small bodies with a large surface area (called tabular masses), surrounded by cool, solid rock, lose their heat more rapidly than would the same volume of material in a spherical reservoir* (Leet and Judson, *Physical Geology*).

tachylyte or **tachylite** (tak′ə līt), *n. Geology.* a black, glassy basalt of volcanic origin that is readily fusible: *Tachylites commonly occur as bombs and cinders, or scoria, thrown out by volcanoes* (Fenton, *The Rock Book*). Also called **basalt glass.** [from German *Tachylit,* from Greek *tachys* swift + *-lytos* soluble]
—**tachylytic** or **tachylitic,** *adj.* composed of or resembling tachylyte.

tachyon (tak′ē on), *n. Nuclear Physics.* any of a class of hypothetical subatomic particles that move with a speed greater than that of light: *At the velocity of light a tachyon would possess infinite energy and momentum; as the particle lost energy, it would speed up, until at zero energy its velocity would be infinite!* (New Scientist). [from Greek *tachys* swift + English *-on*]
—**tachyonic,** *adj.* having to do with or being a tachyon or tachyons: *A tachyonic electron would radiate all its energy ... in about 10^{-19} seconds* (Nature).

tachysterol (tə kis′tə rol), *n. Biochemistry.* a substance formed by irradiating ergosterol, becoming calciferol when further irradiated. *Formula:* $C_{28}H_{44}O$

taconite (tak′ə nīt), *n. Geology.* a variety of chert which contains iron, occurring especially in the Mesabi Range of northeastern Minnesota: *Taconite, one of the hardest rocks known to man, contains tiny particles of iron ore* (Wall Street Journal). [from *Taconic* mountains in western New England + *-ite*]

tactile (tak′təl), *adj. Physiology.* of or having to do with the sense of touch: *tactile stimuli.* The **tactile hairs** on animals are surrounded by nerves that respond to touch.

tactile corpuscle or **tactile bud,** *Physiology.* any of numerous minute, oval bodies which occur in sensitive parts of the skin and are associated with the sense of touch.

tadpole, *n. Zoology.* a very young frog or toad in the larval stage when it lives in water and has gills, a long tail, and no limbs. Through metamorphosis, tadpoles lose their gills and tail by absorbing them into the body and develop lungs and legs that enable them to live on land. [Middle English *taddepol,* from *tadde* toad + *pol* poll (head)]

taenia (tē′nē ə), *n., pl.* **-niae** (-nē ē), **-nias.** *Anatomy.* a ribbonlike band of tissue or muscle, such as the band of longitudinal muscles of the colon. [from Greek *tainia* band]

tag, *n. Chemistry.* a labeled substance used as a radioactive tracer: *Carbon 14 is used as a "tracer" or "tag" in research or industrial processes* (Wall Street Journal). SYN: label.

TAI, *abbrev.* International Atomic Time.

taiga (tī′gə), *n. Geography, Ecology.* **1** swampy, coniferous evergreen forest land of subarctic Siberia between the tundra and the steppes. **2** similar forest land in northeastern China and in North America. [from Russian *taĭga*] ► See the note under **biome.**

tail, *n.* **1** *Zoology.* **a** the slender extremity at the back of many vertebrates, containing the caudal vertebrae. **b** the large feathers arising from the rump of a bird: *The bird's tail ... serves as a rudder in flight, as a counterbalance in perching, and for display in courting by the males of many species. The tail may be spread or folded and elevated, depressed, or tilted to direct the course of flight* (Storer, *General Zoology*). **c** = tail fin. **2** *Astronomy.* a luminous trail of small particles from the head of a comet, extending away from the sun: *As comets near the Sun, solar heating drives off dust and gases, expanding the coma and producing the characteristic tail which may stretch for 150 million miles and be as much as ten million miles wide* (New Scientist).

tail coverts, *Zoology.* the feathers concealing the bases of a bird's tail feathers.

tail fin, *Zoology.* a fin at the posterior of the body of a fish, whale, or the like: *Their [whales'] fore-limbs have developed into fins, and a great fluke or tail fin has been evolved to propel their great mass at high speed through the ocean* (Science News Letter).

talc (talk), *n. Mineralogy.* a soft, smooth mineral, a hydrous silicate of magnesium, usually consisting of slippery, translucent sheets of white, apple-green, or gray, used in making face powder, chalk, etc.: *Talc is the softest mineral known* (Frederick H. Pough).

---

cap, fāce, fäther; best, bē, tèrm; pin, fīve;
rock, gō, ôrder; oil, out; cup, pùt, rüle,
yü in use, *yu* in uric;
ng in bring; *sh* in rush; *th* in thin, ᴛʜ in then;
zh in seizure.
ə = *a* in about, *e* in taken, *i* in pencil, *o* in
lemon, *u* in circus

tallow

Formula: $Mg_3Si_4O_{10}(OH)_2$ [from Medieval Latin *talcum,* from Arabic *talq*]

tallow, *n. Biology.* **1** the hard, white fat from around the kidneys of sheep, cows, oxen, etc.

2 the fat or adipose tissue of an animal.

3 any of various kinds of greases or greasy substances, especially those obtained from plants.

[Middle English *talowe*]

talon (tal′ən), *n. Zoology.* the claw of an animal, especially a bird of prey. [from Old French *talon* heel, ultimately from Latin *talus* ankle]

talus¹ (tā′ləs), *n., pl.* **-li** (-lī). *Anatomy.* **1** the human anklebone. Also called **astragalus. 2** the human ankle. [from Latin]

talus² (tā′ləs), *n. Geology, Geography.* a steep mass of loose rocky fragments lying at the base of a cliff or on the side of a mountain: *It is known that movement of talus, referred to as talus shift, occurs rapidly when actually in motion, and in an extremely variable manner from sector to sector and through time, as creep or as sliding, and even as overturning* (Sidney E. White). [from French]

tan or **tan.,** *abbrev.* tangent.

tangent (tan′jənt), *adj.* touching a curve or surface at one point but not intersecting. These circles are tangent: ∞
—*n.* **1** *Geometry.* **a** a tanget line, curve, or surface. **b** the part of a line tangent to a curve from the point of tangency to the horizontal axis. See the picture at **sector.**

2 *Trigonometry.* the ratio of the length of the side opposite an acute angle in a right triangle to the length of the side adjacent to the acute angle. The tangent, sine, and secant are the three fundamental trigonometric functions.

[from Latin *tangentem* touching]
—**tangential** (tan jen′shəl), *adj.* of, having to do with, or being a tangent.

tannate (tan′āt), *n. Chemistry.* a salt or ester of tannic acid.

tannic (tan′ik), *adj.* of or resembling tannin.

tannic acid, *Chemistry.* **1** the white, amorphous, strongly astringent principle derived from nutgalls. *Formula:* $C_{14}H_{10}O_9$

2 any one of various other astringent organic substances, especially the principle derived from oak bark.

tannin (tan′ən), *n. Chemistry.* any of various acid substances obtained from the bark or galls of oaks, and from certain other plants, used in tanning, dyeing, making ink, and in medicine. Tannic acid is a common tannin. [from French *tanin,* from *tanner* to tan]

tantalic (tan tal′ik), *adj. Chemistry.* of, having to do with, or derived from tantalum. **Tantalic acid** is a colorless, crystalline acid, known mainly in the form of its salts.

tantalite (tan′tə lit), *n. Mineralogy.* a heavy, black mineral with a submetallic luster, usually found with manganese or niobium. It is the chief ore of tantalum. *Formula:* $(Fe,Mn)(Ta,Nb)_2O_6$

tantalum (tan′tl əm), *n. Chemistry.* a hard, ductile, lustrous, grayish-white metallic element that occurs with niobium in tantalite and certain other minerals, and is very resistant to corrosion, used as an alloy in nuclear reactors and in surgical and dental equipment.

Symbol: Ta; *atomic number* 73; *atomic weight* 180.948; *melting point* 2996°C; *boiling point* 5425°C; *oxidation state* 5. [named after *Tantalus,* the mythological character condemned to stand in water up to his chin only to have it withdrawn from his reach whenever he tried to drink (so called because this element cannot absorb acid when immersed in it)]

tapetum (tə pē′təm), *n., pl.* **-ta** (-tə). **1** *Botany.* a cell or sheath of cells in a spore case, serving to supply nourishment to the maturing spores: *Within the innermost endothecial layer, bounding each sporangium, is the tapetum, a single-celled layer. This, near the time of dehiscence, undergoes breaking down or absorption by developing pollen or microspore cells* (Youngken, *Pharmaceutical Botany*).

2 *Anatomy.* **a** a layer of tissue in the choroid of the eye, containing reflecting crystals such as those that make cats' eyes shine at night: *Made up of tiny plates silvered with guanine crystals, the tapetum reflects incoming light back through the retina, thereby restimulating the light-sensitive rods.* (Scientific American). **b** the layer of nerve fibers forming the roof of part of the lateral ventricle of the brain.

[from New Latin, from Greek *tapēs* carpet]

taphonomy (tə fon′ə mē), *n. Paleontology.* **1** the processes and conditions to which plants and animals are subjected as they become fossilized: *The ecology of present-day organisms is of fundamental importance in paleoecological interpretations, of course, but equally valuable is ... taphonomy—the study of the death, disintegration, burials, and potential preservation of organisms as fossils* (Science).

2 the study of these processes and conditions. [from Greek *taphos* tomb + -*nomia* field of knowledge]
—**taphonomic** (taf′ə nom′ik), *adj.* of or having to do with taphonomy: *We made a taphonomic analysis to determine if they were aquatic or land animals* (Robert T. Bakker).

tardigrade (tär′də grād), *Zoology.* —*adj.* of or having to do with a class (Tardigrada) of microscopic arthropods with rudimentary circulatory and respiratory systems and four pairs of short legs, either marine or inhabiting damp places, often found as slime on ponds. —*n.* a tardigrade arthropod. [from Latin *tardigradus* slow-paced, from *tardus* slow + *gradi* to walk]

target, *n. Physics.* **1** a metallic surface, often of platinum or tungsten, opposite the cathode in an X-ray tube, upon which the cathode rays impinge and produce X rays.

2 any substance or surface subjected to bombardment by nuclear particles.

tarn (tärn), *n. Geography, Geology.* a small mountain lake or pool, especially one formed by glaciers. [from Scandinavian]

tarsal (tär′səl), *adj. Anatomy.* **1** of or having to do with the tarsus. The **tarsal bones** are the seven bones of the human tarsus. The **tarsal joint** of birds lies between the tibia and the metatarsus. *The tarsal gland, which is located on the ankle of the deer, was discovered to be of central importance in this animal's social behavior. The animals sniff each other's tarsal glands 3–4 times per hour for recognition and identity checks* (John G. Lepp).

2 of or having to do with the tarsi of the eyelids: *the tarsal ligaments, the tarsal plates.*

tar sand, *Geology.* a type of oil-yielding sandstone or sand formation containing tarry substances: *In addition to coal, oil, and natural gas, the earth's crust contains vast potential resources of oil shale and oil-bearing tar sands* (Scientific American).

tarsometatarsal (tär′sō met′ə tär′səl), *adj.* **1** *Anatomy.* having to do with the tarsus and the metatarsus. **2** *Zoology.* of or having to do with the tarsometatarsus.

tarsometatarsus (tär′sō met′ə tär′səs), *n., pl.* **-si** (-sī). *Zoology.* **1** the leg bone or shank in birds and early reptilians, consisting of united tarsal and metatarsal bones: *The tarsometatarsus ... represents the distal row of tarsal bones and the second, third, fourth, and fifth metatarsals fused together* (Hegner and Stiles, *College Zoology*). **2** the third joint of the limb of a bird.

tarsus (tär′səs), *n., pl.* **-si** (-sī). *Anatomy.* **1a** the human ankle. **b** the group of small bones composing it; collective name for the seven bones between the tibia and the metatarsus.

2 the corresponding part in most mammals, in some reptiles, and in amphibians.

3 the shank of a bird's leg; tarsometatarsus.

4 the last segment of the leg of an arthropod.

5 the thin plate of dense connective tissue that gives form to the edge of the eyelid.

[from New Latin, from Greek *tarsos*]

tartar (tär′tər), *n.* **1** *Chemistry.* **a** an acid solid, potassium bitartrate, present in grape juice and deposited as a reddish crust on the inside of wine casks. **b** this substance partly purified as cream of tartar. *Formula:* $KHC_4H_4O_6$

2 *Biology, Medicine.* a hard, yellowish substance formed by the action of saliva on food and deposited as a crust on the teeth: *Most dental plaque can easily be removed by adequate toothbrushing; dental tartar can only be removed by a dentist or dental hygienist* (Daniel A. Collins).

[from Old French *tartre,* from Medieval Latin *tartarum*]

tartaric acid, *Chemistry.* a colorless, crystalline acid found in unripe grapes but also made synthetically, used to make cream of tartar, baking powder, various esters, and in many industrial processes and products. There are three isomers of tartaric acid: one levorotatory, one dextrorotatory, and one optically inactive. Natural tartaric acid is dextrorotatory, while the synthetic form is a racemic mixture. *Formula:* $(CH(OH)CO_2H)_2$

tartrate (tär′trāt), *n. Chemistry.* a salt or ester of tartaric acid.

tassel, *Botany.* **—*n.*** **1** the cluster of long stems bearing small flowers that hangs from the top of a cornstalk: *In today's cultivated corn, the staminate inflorescence or male flowering is the tassel alone and the ear carries the female flowering, the corn "silk"* (Science News Letter).

2 a hanging catkin, blossom, flower, or bud in any tree or plant.

—*v.* to grow tassels: *Corn tassels just before the ears form.*

taste, *n. Physiology.* the sense by which the flavor of a substance is perceived by the taste buds: *The human organ of smell is inside the nose; particles of odorous substances are brought in by the air during breathing and the sense of taste is closely related to that of smell. In fact, for human beings unaided by odour, there are only four tastes, sweet, bitter, sour and salty* (N.B. Hodgson). SYN: gustation.

taste bud, *Anatomy.* one of the groups of receptor cells, chiefly in the lining of the tongue or mouth, that are organs of taste: *In the tongue, taste buds located in different regions conduct specific taste sensations* (Martin E. Spencer). *Taste buds ... are usually in or about the mouth but occur over the body in catfishes and carp and on the fins of some other fishes* (Storer, *General Zoology*).

tau (tou *or* tô), *n. Nuclear Physics.* a weakly interacting subatomic particle with a mass about three and a half thousand times that of the electron: *The tau joins the lepton group of particles. ... The electron is the best known member of the family* (New Scientist). [from *tau,* the 19th letter of the Greek alphabet and the first letter of Greek *triton* third (so called because this particle is the third charged lepton after the electron and the muon)]

taurine (tôr′ēn), *n. Chemistry.* a neutral, crystallizable substance, found in the fluids of many animals' muscles, lungs, and other organs, also resulting from the hydrolysis of taurocholic acid. *Formula:* $C_2H_7NO_3S$ [from Latin *taurus* bull (so called from its being first found in the bile of oxen)]

taurocholic acid (tô′rə kō′lik *or* tô′rə kol′ik), *Chemistry.* a crystalline acid, present as a sodium salt in human bile and the bile of oxen and most other animals, and hydrolyzing into taurine and cholic acid. *Formula:* $C_{26}H_{45}NO_7S$

tautology (tô tol′ə jē), *n. Mathematics.* a compound statement that is always true regardless of the truth value of the basic statements of which it is composed. EXAMPLE: From the statements "If x is greater than 5 then x^2 is greater than 25" and "x is greater than 5," we conclude that "x^2 is greater than 25." In a truth table this tautology is written $[(p \rightarrow q) \wedge p] \rightarrow q$. [ultimately from Greek *tautologein* repeat, from *tauto* same + *logos* saying]

tautomer (tô′tə mər), *n. Chemistry.* one of the isomeric forms exhibited in tautomerism: *Such closely similar molecular alternatives are called tautomers ... Truly tautomeric forms of a molecule interchange with each other so readily that they exist side by side in equilibrium in a sample of the substance* (New Yorker). [from Greek *tauto* same + *meros* part]

—tautomeric (tô′tə mer′ik), *adj.* of, having to do with, or exhibiting tautomerism.

tautomerism (tô tom′ə riz′əm), *n. Chemistry.* a form of isomerism in which a difference in the placement of a hydrogen atom results in two or more isomers (called tautomers) which are in equilibrium.

---

cap, fāce, fäther; best, bē, tėrm; pin, fīve;
rock, gō, ôrder; oil, out; cup, pu̇t, rüle,
yü in use, *yu̇* in uric;
ng in bring; *sh* in rush; *th* in thin, ᴛʜ in then;
zh in seizure.
ə = *a* in about, *e* in taken, *i* in pencil, *o* in lemon, *u* in circus

tautonym (tô'tə nim), *n. Biology.* a scientific name with the generic and specific names alike, now unacceptable in botany, though often found in zoology. EXAMPLE: *Cygnus cygnus,* the whooper swan. [from Greek *tautōnymos* same name]
—**tautonymic** (tô'tə nim'ik), *adj.* of or characterized by a tautonym; having the genus and species names alike.

taxis (tak'sis), *n., pl.* **taxes** (tak'sēz). **1** *Biology.* an involuntary or reflex movement in a particular direction by a free organism or a cell, such as a zoospore, in reaction to an external stimulus, such as light: *The essentially invariable type of response by which an animal orients itself toward or away from a given stimulus is termed a taxis. (The term tropism now is reserved for the turning movements of plants.) A fish that heads into a current so that the two sides of its body are stimulated equally by the passing water exhibits positive rheotaxis* (Storer, *General Zoology*). **2** = taxonomy. [from New Latin, from Greek *taxis* arrangement]

taxon (tak'son), *n., pl.* **taxa** (tak'sə). a taxonomic category, such as a family or order: *Assignment to the taxon is not on the basis of a single property but on the aggregate of properties* (Scientific American). [back formation from *taxonomy*]

taxonomic (tak'sə nom'ik), *adj.* of or having to do with taxonomy: *More impressive than disjunct genera or even larger taxonomic groups are the discontinuities in whole floras* (New Biology). —**taxonomically,** *adv.*

taxonomy (tak son'ə mē), *n.* **1** classification, especially of plants and animals on the basis of differences and similarities between them: *The primary objective of any system of taxonomy is to provide a convenient and useful filing system so that all information about any species or about any larger groups of organisms may be placed under a given category of the file. Each category is called a taxon* (Theodore L. Jahn).
2 the branch of science dealing with classification; study of the general laws and principles of classification. [from French *taxonomie,* from Greek *taxis* arrangement + *-nomia* system of laws]

Taylor's series, *Mathematics.* a power series for the expansion of a function *f*(*x*) in a neighborhood of the reference point (*a*), based on a theorem (**Taylor's theorem**) which describes approximately polynomials for the general function *f*(*x*). Compare **Maclaurin's series.** [named after Brook *Taylor,* 1685–1731, English mathematician]

Tb, *symbol.* terbium.

Tc, *symbol.* technetium.

T cell or **T-cell,** *n. Biology.* a type of lymphocyte that attacks foreign bodies, derived from the thymus gland and distinguished from a B cell by its relatively smooth surface: *Since T cells do not secrete antibodies and B cells do, it seems plausible that T cells might cause B cells to produce antibodies* (Science News). *Both the B-cells and T-cells reside primarily in the body's lymphoid tissues ... From these tissues, the cells recirculate through the body and continually monitor for the presence of potential attackers* (Time). Also called **T lymphocyte.** Compare **helper cell, suppressor cell.** [from t(*hymus*-derived) *cell*]

TDB, *abbrev.* Barycentric Dynamic Time.
TDT, *abbrev.* Terrestrial Dynamic Time.
Te, *symbol.* tellurium.

tear, *n. Physiology.* a drop of salty liquid secreted by the lachrymal glands to moisten the membrane covering the front of the eyeball and the lining of the eyelid.

tearduct, *n. Anatomy.* any of several ducts that carry tears from the lachrymal glands to the eyes or from the eyes to the nose.

technetium (tek nē'shē əm), *n. Chemistry.* a radioactive metallic element produced artificially, especially by fission of uranium or molybdenum, used to inhibit the corrosion of iron. Technetium was unknown in nature before it was produced artificially in 1937. *Technetium is an unstable element whose longest-lived known isotope has a half-life of only 216,000 years—far less than the age of the stars in which it is found* (William A. Fowler). *Symbol:* Tc; *atomic number* 43; *mass number* 99 (the most stable isotope); *melting point* ab. 2200°C; *oxidation state* 4, 6, 7. [from Greek *technētos* artificial, from *technē* art]

technology, *n.* **1** the science of the mechanical and industrial arts; applied science: *A school of technology includes courses in engineering, agriculture, electronics, and other practical fields.*
2 a particular application of technology; any method, process, or system using special tools and techniques to achieve a goal: *medical technology. To reach the ... conditions required for a net release of fusion power it is necessary to first develop many new technologies* (Scientific American).
—**technological,** *adj.* of or having to do with technology.

tectonic (tek ton'ik), *adj. Geology.* of or having to do with general changes in the structure of the earth's crust or with the forces that affect them: *Folding and faulting are major tectonic processes. Such shallow earthquakes are abundant near volcanoes ... they are now thought to originate in the same way, by movements of rocks along faults, even though the forces that bring about the faulting may be the result of a bursting magma chamber rather than tectonic forces* (Gilluly, *Principles of Geology*). [from Greek *tektōn* builder]

tectonics, *n.* **1** a branch of geology dealing with the forces affecting the structure of the earth's crust.
2 the features or characteristics of this structure.
3 = plate tectonics.

tectonism, *n. Geology.* tectonic activity, especially the action of forces that cause deformities in the earth's crust: *The Caribbean continued to close slowly. The climax of movement was during the middle Eocene when there was major tectonism on the northern and southern margins. The islands of the Greater Antilles then attained their present proportions* (Nature).

tectorial (tek tôr'ē əl), *adj. Anatomy.* covering like a roof. The **tectorial membrane** is a rooflike membrane covering the organ of Corti in the inner ear. [from Latin *tectorium* a covering]

tectum (tek'təm), *n., pl.* **-ta** (-tə). *Anatomy.* any rooflike structure of the body, especially the dorsal part of the midbrain: *Anatomical studies had shown that not all the frog's optic nerve fibers pass to the tectum* (Scientific American). [from Latin *tectum* roof]

tegmen (teg'men), *n., pl.* **-mina** (-mə nə). **1** *Biology.* a cover, covering or coating.
2 *Botany.* the thin, soft, delicate inner coat of a seed, surrounding the embryo. SYN: integument.
3 *Zoology.* a forewing of an insect when modified to serve as a covering for the hind wings.
[from Latin *tegmen*]

tegmental (teg men'təl), *adj.* having to do with the tegmentum.

tegmentum (teg men'təm), *n., pl.* **-ta** (-tə). **1** *Botany.* a the scaly coat which covers a leaf bud. b one of the scales of such a coat.
2 *Anatomy.* a covering, especially a part of the midbrain consisting of white fibers running lengthwise through a mass of gray matter.
[from New Latin, from Latin *tegmentum,* variant of *tegumentum* a covering, from *tegere* to cover]

tegument (teg'yə mənt), *n. Biology.* a natural covering; shell, capsule, or cocoon; integument.

tektite (tek'tīt), *n. Geology.* any of various rounded, glassy objects of different shapes and weights, found in various parts of the world and thought to have come from the moon or to have resulted from the impacts of large meteorites on terrestrial rocks: *Tektites differ chemically from meteorites, but some of them contain small bits of meteoritic iron* (John A. O'keefe). *Tektites are usually considered to be of acidic or intermediate material; however, they are somewhat lower in aluminum and alkali oxides and richer in iron oxide than granites, and, in optical properties, they are like the moon* (Science). [from Greek *tēktos* molten, from *tēkein* to melt]
—**tektitic** (tek tit'ik), *adj.* of or having to do with tektites: *If the glass chemist were asked to formulate a material which would survive both hypersonic entry through the atmosphere and prolonged geological weathering, he could not improve upon the tektitic formula* (New Scientist).

telemeter (tel'ə mē'tər), *n.* a device for measuring heat, radiation, pressure, speed, etc., and transmitting the information to a distant receiving station. [from *tele-* far off, distant + *-meter*]
—**telemetric,** *adj.* having to do with a telemeter or telemetry. —**telemetrically,** *adv.*
—**telemetry** (tə lem'ə trē), *n.* the use of telemeters for measuring and transmitting information.

telencephalon (tel'en sef'ə lon), *n. Anatomy.* the anterior part of the forebrain in vertebrates, comprising mainly the cerebral hemispheres. Also called **end brain.** [from *tel-* far off, tele- + *encephalon*]
—**telencephalic** (tel'en sə fal'ik), *adj.* of or having to do with the telencephalon.

teleost (tel'ē ost *or* tē'lē ost), *Zoology.* —*adj.* of or having to do with a large group of fishes with bony skeletons, including most common fishes, such as the perch or flounder, but not the sharks, rays, and lampreys: *Most teleost fishes have a swim bladder, or a bag of gas in the body cavity* (R. McNeill Alexander) —*n.* a teleost fish. [from New Latin *Teleostei* the group name, from Greek *teleios* finished + *osteon* bone]

telescope, *n.* **1** an optical instrument for making distant objects appear nearer and larger, consisting of an arrangement of lenses, and sometimes mirrors, in one or more tubes: *The most essential part of a telescope is the objective, a lens or mirror which, by refraction or* reflection, concentrates a beam of light from a luminous point such as a star into a tiny dot, an image of the star. Star images are formed in a nearly flat geometrical surface called the focal plane, and the distance of the center of the objective from this surface is its focal length* (Duncan, *Astronomy*). See also **reflecting telescope, refracting telescope.**
2 any device that is analogous to a telescope but uses forms of energy other than light. See also **radar telescope, radio telescope.**
3 *Nuclear Physics.* an arrangement of particle counters or detectors used in studies of subatomic particle reactions.
—**telescopic,** *adj.* **1** of or having to do with a telescope: *a telescopic lens.*
2 obtained or seen by means of a telescope: *a telescopic view of the moon.*
3 visible only through a telescope. —**telescopically,** *adv.*

teleseism (tel'ə sī'zəm), *n. Geology.* an earth tremor remote from a place where it is recorded or indicated by a seismograph.
—**teleseismic,** *adj.* of or having to do with a teleseism: *a network of 20 to 30 teleseismic recording stations.*

telial (tē'lē əl), *adj. Biology.* of, having to do with, or resembling a telium: *It is the telial stage that gives this fungus the name black stem rust. Under the microscope the spores appear as dark brown two-celled structures. When first formed, each of these cells has two nuclei which soon fuse, making uninucleate cells* (Emerson, *Basic Botany*).

teliospore (tē'lē ə spôr), *n. Biology.* a thick-walled spore produced by certain rust fungi generally in autumn, remaining in the tissues of the host during winter, and germinating in the spring to produce basidia. [from Greek *teleios* complete + English *spore*]
—**teliosporic,** *adj.* having to do with or characterized by teliospores.

telium (tē'lē əm), *n., pl.* **telia** (tē'lē ə). *Biology.* a sorus bearing teliospores, formed by certain rust fungi. [from New Latin, from Greek *teleios* complete]

telluric (te lùr'ik), *adj. Chemistry.* of, having to do with, or containing tellurium, especially with a valence of six.

tellurite (tel'yə rīt), *n.* a rare mineral, a native dioxide of tellurium, usually found in clusters of minute, whitish or yellowish crystals. *Formula:* TeO_2

tellurium (te lùr'ē əm), *n. Chemistry.* a silver-white nonmetallic element with some metallic properties, usually occurring in nature combined with gold, silver, or various other metals. Tellurium is poisonous, brittle unless extremely pure, and chemically similar to sulfur and selenium. *Symbol:* Te; *atomic number* 52; *atomic weight* 127.60; *melting point* ab. 450°C; *oxidation state* −2, +4, +6.

---

cap, fāce, fäther; best, bē, tèrm; pin, fīve;
rock, gō, ôrder; oil, out; cup, pùt, rüle,
yü in use, *yu* in uric;
ng in bring; *sh* in rush; *th* in thin, ɫH in then;
zh in seizure.
ə = *a* in about, *e* in taken, *i* in pencil, *o* in lemon, *u* in circus

tellurous (tel′yər əs *or* te lùr′əs), *adj. Chemistry.* of, having to do with, or containing tellurium, especially with a valence of four.

telo-, *combining form.* end; terminal; final, as in *telomere, telocentric, telophase.* [from Greek *telos* end]

telocentric (tel′ə sen′trik), *adj. Genetics.* —*adj.* having the centromere in a terminal position.
　—*n.* a telocentric chromosome. Compare **acrocentric, metacentric.**

telolecithal (tel′ə les′ə thəl), *adj. Embryology.* (of eggs) having the yolk in one end and the cytoplasm in the other end: *Eggs with little yolk are said to be alecithal or, from the uniform distribution of the yolk, homolecithal. Eggs with much yolk aggregated toward the vegetative pole are telolecithal* (Shull, *Principles of Animal Biology*). [from *telo-* + Greek *lekithos* yolk]

telomer (tel′ə mər), *n. Chemistry.* a reduced polymer formed by the reaction between a substance capable of being polymerized and an agent that arrests the growth of the chain of atoms: *Chemists found how to limit the number of atoms that link together to form a long-chain molecule, creating telomers in contrast to the long-chain polymers* (Science News Letter). [from *telo-* + (*poly*)*mer*]

telomere (tel′ə mēr), *n. Biology.* a tip or end of a chromosome. [from *telo-* + Greek *meros* part]

telophase (tel′ə fāz), *n. Biology.* the fourth and final stage of mitosis or meiosis, when the chromosomes become less visible and the nuclear membrane reappears. Telophase usually occurs in both the first and second meiotic divisions, but in certain organisms it does not occur in the first division. *Meiosis displays the same four phases characteristic of mitosis, but the prophases and metaphases of the two types of nuclear division show some definite differences. ... The anaphase and telophase are essentially the same as in mitosis. In the telophase, the nuclear material with the reduced number of chromosomes is reorganized into discrete nuclei* (Harbaugh and Goodrich, *Fundamentals of Biology*). See the picture at **mitosis.**
　—**telophasic** (tel′ə fā′zik), *adj.* of or having to do with the telophase.

telotaxis, *n. Biology.* the movement or orientation of an organism toward a particular stimulus, to the exclusion of all others. Contrasted with **tropotaxis.**

telson (tel′sən), *n. Zoology.* the rearmost segment of the abdomen in certain crustaceans and arachnids, such as the middle flipper of a lobster's tail or the sting of a scorpion. [from Greek *telson* limit]

temp., *abbrev.* temperature.

temperate, *adj.* **1** *Geography, Meteorology.* of or belonging to either of the temperate zones; neither tropical nor polar; moderate: *a temperate climate, the temperate belt. Dynamical stability is an invariable feature of the broad wind-pattern of temperate regions* (Science News). See the picture at **zone.**
2 *Microbiology.* having its genetic material fused with that of the host; not virulent: *a temperate phage.*

temperate zone or **Temperate Zone**, *Geography.* the part of the earth's surface between the tropic of Cancer and the arctic circle in the Northern Hemisphere, or the part between the tropic of Capricorn and the ant-

arctic circle in the Southern Hemisphere: *In the temperate zones the sun never appears in the zenith, nor does it become circumpolar* (Duncan, *Astronomy*).

temperature, *n.* **1** *Physics.* **a** the degree of heat or cold of any substance, measured by any of various instruments and using such scales as Celsius, Fahrenheit, Réaumur, or Kelvin. On the Celsius scale the temperature of freezing water is 0 degrees and of boiling water 100 degrees. All temperatures are based on their position above or below 273.16 kelvins, which is the triple point of water. **b** a measure of the kinetic energy of the molecules that make up a substance, usually expressed in kelvins: *The concept of temperature describes a relation between two quantities, energy and disorder. In fundamental terms what is measured by temperature is the change in the disorder of a system as the energy of the system changes. In a crystalline solid the major contribution to disorder is the vibratory motion of the atomic nuclei* (Warren G. Proctor).
2 *Physiology.* the degree of heat contained in the body of a living organism, usually measured by a thermometer: *Normal body temperature is about 98.6 degrees Fahrenheit, but a healthy child's temperature usually varies with the time of day (lower in the morning, higher toward evening) and with his activity (rising as high as 100 degrees or more after intensive exercise)* (S. M. Gruenberg).

(def. 1a)

CELSIUS		FAHRENHEIT
boiling point of water	100 —	— 212
	50 —	
		— 100
	0 —	— 32 freezing point of water
	18 —	— 0

temperature gradient, *Physics.* the rate of change in temperature: *A temperature gradient leads in a familiar way to a flow of energy from hot to cold regions by heat conduction. However, this means of energy transport is not of great importance in stars since, at such high temperatures, radiation transports much more energy than conduction* (C. M. and H. Bondi).

temperature-humidity index, *Meteorology.* a combined measurement of temperature and humidity arrived at by adding degrees of temperature to percentage of relative humidity and dividing by two. It is used as a measure of discomfort during the summer months. *Abbreviation:* THI

temperature inversion, *Meteorology.* an atmospheric condition in which a stationary layer of warm air settles over a layer of cool air near the ground, often causing a heavy concentration of pollutants in the air and mirages of light in the night sky: *The new weather station will form part of an early warning system to guard against temperature inversion, a weather phenomenon quite common in the Bantry area which causes sul-*

phur dioxide to lie "umbrella fashion" over the area (Irish Press).

template (tem′plit), *n. Molecular Biology.* a macromolecule that serves as a mold for the synthesis of another macromolecule: *DNA molecules are not the direct templates for protein synthesis. The genetic information of DNA is first transferred to RNA molecules. In turn, RNA molecules act as the primary templates that order amino acid sequences in proteins* (Watson, *Molecular Biology of the Gene*).

temporal, *adj. Anatomy.* of the temples or sides of the forehead. The **temporal bone** is a compound bone that forms part of the side and base of the skull. The **temporal lobe** is the part of each cerebral hemisphere, in front of the occipital lobe, which contains the center of hearing in the brain. See the picture at **skull.** [from Latin *temporalis,* from *tempora* the temples]

tendinous (ten′də nəs), *adj. Anatomy.* of, resembling, or consisting of a tendon or tendons: *a tendinous cord. Closely associated [with bursitis] in its symptoms is the so-called fibrositis, affecting the tendinous sheaths of the muscles* (Ralph R. Mellon).

tendon, *n. Anatomy.* a tough, strong band or cord of fibrous tissue that joins a muscle to a bone or some other part and transmits the force of the muscle to that part: *The other end of the muscle may blend with an elongated bundle of connective tissue fibers, forming a tendon, which is attached to the periosteum of another bone* (Harbaugh and Goodrich, *Fundamentals of Biology*). Compare **ligament.** [from Medieval Latin *tendonem,* from Greek *tenōn;* influenced by Latin *tendere* to stretch]

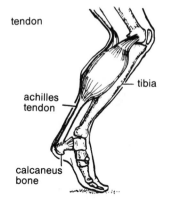

tendon

tibia

achilles
tendon

calcaneus
bone

tendon of Achilles, = Achilles' tendon.

tendril (ten′drəl), *n. Botany.* a threadlike part of a climbing plant that attaches itself to something and helps support the plant: *The tendrils of the Grape Vine are modified inflorescence branches, those of Sarsaparilla are modified stipules, and those of the Pea are modified leaflets. The ends of the tendrils of the Japan Ivy become swollen and flattened, forming adhesive disks which cling to objects with which they come into contact* (Youngken, *Pharmaceutical Botany*). [from Middle French *tendrillon*]

—**tendrilous** (ten′drə ləs), *adj.* full of or resembling a tendril or tendrils: *tendrilous appendages.*

tennantite (ten′ən tīt), *n. Mineralogy.* a gray to black mineral, a sulfide of arsenic with copper and usually iron. It is an ore of copper. *Formula:* Cu_3AsS_3 [named

after Smithson *Tennant,* 1761–1815, English chemist]

tenorite (ten′ə rīt), *n. Mineralogy.* a black oxide of copper, found in thin, iron-black scales in oxidized veins and lodes. *Formula:* CuO [from Italian]

tensile, *adj. Physics.* that can be stretched; ductile: *tensile metals.*

tensile strength, *Physics.* the maximum stress that a material can withstand before it breaks, expressed in pounds per square inch: *For iron alloys, the highest strengths are about ten times that of the pure metal, while for aluminium alloys, the corresponding figure is about eight. If one applied this lower figure to a value of 19 tons per square inch (which is perhaps reasonable for the tensile strength of pure titanium), one arrives at a figure of about 150 tons per square inch as the order of tensile strength one might hope to achieve with the strongest titanium alloys* (F. A. Fox).

tensility (ten sil′ə tē), *n.* tensile quality; ductility.

tensiometer (ten′sē om′ə tər), *n.* **1** an instrument for measuring the tautness or tension of wire, fabric, and other material.
2 an instrument for measuring the surface tension of a liquid.
—**tensiometric,** *adj.* of or having to do with a tensiometer or with tensiometry: *A tensiometric method utilizes a porous cup filled with water connected by a tube to a vacuum indicator. This approach measures the capillary potential or suction of soil water* (P. R. Nixon).
—**tensiometry,** *n.* **1** the measurement of tension or tensile strength.
2 the study of tension or tensile strength.

tension, *n. Physics.* **1** the stress caused by the action of a pulling force; the condition of being stretched between two points: *Muscles are sensitive to stretch and automatically (reflexly) adjust their activity to changes in tension* (W. F. Floyd and P. H. S. Silver).

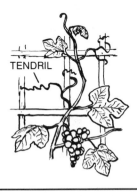

TENDRIL

---

cap, fāce, fäther; best, bē, tėrm; pin, fīve;
rock, gō, ôrder; oil, out; cup, pùt, rüle,
yü in use, yù in uric;
ng in bring; sh in rush; th in thin, ŦH in then;
zh in seizure.
ə = a in about, e in taken, i in pencil, o in lemon, u in circus

tensor

2 the force exerted by the body undergoing stress: *A common type of structure in which pushes, as well as pulls, are involved, is ... a street lamp or a sign. We wish to compute the tension in the supporting cable and the compression in the strut when the weight of the suspended body is known* (Sears and Zemansky, *University Physics*).

—**tensional**, *adj.* of or having to do with tension.

tensor (ten'sər *or* ten'sôr), *n.* **1** *Anatomy.* a muscle that stretches or tightens some part of the body.

2 *Mathematics.* a vector that can be defined only by reference to more than three components: *The gravitational equations of the general theory of relativity are written in a form of mathematics that deals with quantities called tensors* (Harper's).

tentacle, *n.* **1** *Zoology.* one of the long, slender, flexible growths, usually occurring on the head or around the mouth of an animal, used to touch, hold, or move: *the tentacles of an octopus.*

2 *Botany.* a sensitive, hairlike growth on a plant: *the tentacles of the sundew.*

[from New Latin *tentaculum*, from Latin *tentare* to try]

—**tentacled**, *adj.* having a tentacle or tentacles: *The hydra has a wormlike trunk and a tentacled head.*

—**tentacular**, *adj.* of, forming, or resembling tentacles: *tentacular papillae or outgrowths.*

tephra (tef'rə), *n.* *Geology.* fragmentary volcanic material, such as ash, dust, cinders, and volcanic bombs, given off in an eruption: *Super-hot gas-charged magma burst up from under the earth through a solid cap of island, throwing huge boulders fifty kilometers in the air and scattering hot ash, or tephra, over the whole eastern Mediterranean* (Atlantic). [from Greek *tephra* ashes]

tephrite (tef'rīt), *n.* *Geology.* any of a group of mafic volcanic rocks related to basalt, but containing one or more feldspathoids.

—**tephritic** (tef rit'ik), *adj.* of the nature of tephrite; having to do with tephrite: *a tephritic variety of rock.*

tera- (ter'ə-), *combining form.* one trillion (10^{12}) of any SI unit, as in *terahertz = one trillion hertz, terawatt = one trillion watts.* Symbol: T [from Greek *teras* monster]

terahertz, *n.* a unit of frequency equal to one trillion hertz: *Scientists at the National Bureau of Standards' Boulder, Colo, laboratories found the absolute frequency of an emission from a helium-neon laser to be 88.376245 terahertz* (Science News).

terbium (ter'bē əm), *n.* *Chemistry.* a silver-gray, rare-earth metallic element which occurs in certain minerals with yttrium and ytterbium. *Symbol:* Tb; *atomic number* 65; *atomic weight* 158.924; *melting point* ab. 1360°C; *boiling point* ab. 3041°C; *oxidation state* 3. [from New Latin, from (*Yt*)*terby*, town in Sweden]

terbium metals, *Chemistry.* a group of closely related rare-earth elements, consisting of terbium, dysprosium, europium, and gadolinium.

terbium oxide, *Chemistry.* a dark brown powder, one of the rare earths. *Formula:* Tb_2O_3

tergal (ter'gəl), *adj.* **1** *Anatomy.* of or on the back; dorsal: *the tergal wall of the thorax.*

2 *Zoology.* of or having to do with a tergum or tergite.

tergeminate (ter jem'ə nit), *adj. Botany.* (of a compound leaf) having at the base a pair of leaflets and then forking, with a pair on each branch. [from Latin *tergeminus* triple]

tergite (ter'jit), *n. Zoology.* the upper or dorsal plate of each segment or somite of an arthropod or other articulated animal: *On the opposite interior surface of the last tergite are chitinous points* (Athenaeum). [from Latin *tergum* the back]

tergum (ter'gəm), *n., pl.* **-ga** (-gə). *Zoology.* the back or dorsal portion of an arthropod or other articulated animal: *The sclerites on each somite form a dorsal tergum of four fused plates, a pleuron of three plates on each side, and a single ventral sternum* (Storer, *General Zoology*). [from Latin *tergum* the back]

term, *n. Mathematics.* **1** one of the elements composing a ratio or a fraction: *Term refers to either the numerator or the denominator of a fraction. Lower terms means a smaller numerator and denominator. In lowest terms means that the numerator and denominator of a fraction cannot be made smaller* (Robert L. Swain).

2 any part of a compound algebraic expression separated from the other parts by a plus or minus sign. In $(13ax^2 - 2bxy + y)$, $13ax^2$, $2bxy$ and *y* are the terms.

3 a point, line, or surface that limits in geometry.

terminal, *adj. Botany.* growing at the end of a stem, branch, etc., as a bud or flower. [from Latin *terminalis*, from *terminus* end] —**terminally**, *adv.*

terminal moraine, *Geology, Geography.* a moraine deposited at the end of a glacier: *A terminal moraine ... consists of a jumble of large and small particles, from great blocks the size of a house to clay resulting from the grinding of the rocks to flour by the erosive processes* (Gilluly, *Principles of Geology*). See the picture at **drumlin.**

terminal side, *Trigonometry.* the line formed by the rotation of the initial side of an angle.

terminal velocity, *Physics.* the maximum velocity of a falling body, attained when the resistance of air, water, or other surrounding fluid has become equal to the force of gravity acting upon the body: *[The] terminal velocity for a man is about 120 m.p.h. (e.g. for a parachutist delaying the opening of his chute); the corresponding velocity for a cat is probably 40 m.p.h. at the most* (New Scientist).

terminator, *n.* **1** *Astronomy.* the line separating the light and dark parts of the disk of the moon or a planet: *The spacecraft flipped over on its back and snapped the earth from a distance of 240,000 miles, giving a good picture of the terminator—the division of the sunlit and shadowed areas of our planet* (Science News Letter).

2 *Molecular Biology.* the site on a segment of DNA where the formation of messenger RNA is terminated; a sequence of nucleotides that signal the end of transcription: *Total success moved within reach only when Khorana cleared yet another formidable hurdle—the determination and synthesis of additional stretches of nucleotides, known as the promotor and terminator, that constitute the start and stop signals for the larger gene* (Charles M. Cegielski).

ternary (tèr'nər ē), *adj.* **1** *Mathematics.* **a** having three for the base: *a ternary scale.* **b** involving three variables.

2 *Chemistry, Physics.* involving or consisting of three elements or components: *a ternary alloy. Kinetic Theory considerations show that ternary collisions are at the most one thousand times less frequent than binary collisions and simultaneous collisions of four or more molecules must be relatively even more uncommon* (K. D. Wadsworth).
[from Latin *ternarius,* from *terni* three each, from *ter* thrice]

ternate (tèr'nit *or* tèr'nāt), *adj. Botany.* **1** having or consisting of three leaflets. **2** having leaves in whorls of three. [from Latin *terni* three each, from *ter* thrice]
—ternately, *adv.*

terpene (tèr'pēn'), *n. Chemistry.* **1** any of a group of isomeric hydrocarbons produced by distilling the volatile oils of plants, especially conifers: *The amounts of terpenes given off by plants are certainly sufficient to account for all the petroleum formed in previous geological periods* (Scientific American). *Formula:* $C_{10}H_{16}$
2 any of various alcohols derived from or related to terpene.
[from German *Terpentin* turpentine]

terpenoid (tèr'pə noid), *Chemistry.* **—n.** any of a group of compounds having a chemical structure similar to that of terpene.
—adj. of or having to do with a terpenoid: *Certain cucumbers produce a terpenoid substance that acts as a specific feeding attractant for cucumber beetles* (David E. Fairbrothers).

terpolymer (tèr pol'i mər), *n. Chemistry.* a compound formed by the polymerization of three different compounds, each of which usually is able to polymerize alone. [from Latin *ter* thrice + English *polymer*]

terrace, *n. Geology.* a flat, raised level of land with vertical or sloping sides, especially one of a series of such levels placed one above the other: *More common than structural terraces are valley terraces ... made up entirely of river deposits. These terraces are the remnants of old floodplains, now incised by the streams that once made them* (Gilluly, *Principles of Geology*).

terrain, *n. Geography.* any tract of land, especially considered with respect to its extent and natural features: *the hilly, rocky terrain of parts of New England.* [from French *terrain* land, ultimately from Latin *terra* earth, land]

terraqueous (tə rā'kwē əs), *adj. Biology.* living in land and water, as a plant. [from Latin *terra* land + English *aqueous*]

terrestrial, *adj.* **1** *Geology.* of the land, not water: *Islands and continents make up the terrestrial parts of the earth.*
2 *Zoology.* living on the ground, not in the air, water, or trees: *terrestrial animals. These animals are now able to crawl out on land and take their place as terrestrial vertebrates* (Winchester, *Zoology*).
3 *Botany.* living or growing on land; growing in the soil: *terrestrial plants. The terrestrial form is much less easily damaged and more suitable for study than the aquatic species* (Emerson, *Basic Botany*).

Terrestrial Dynamic Time, a time scale widely employed in astronomical tables, replacing the earlier *Ephemeris Time.* It is ahead of *International Atomic Time* (TAI) by 32.184 seconds. *Abbreviation:* TDT

terrestrial magnetism, = geomagnetism.

terrestrial planet, *Astronomy.* any of the four small planets nearest the sun and somewhat similar to each other in size: *Any theory of the origin and development of the solar system has to explain logically the extreme apparent differences between the Jovian planets (Jupiter, Saturn, Uranus and Neptune) and the terrestrial planets (Mars, earth, Venus and Mercury)* (Science News).

terrestrial space, *Astronomy.* a zone from the earth's surface to about 6400 kilometers into space, within which the earth's magnetic and electric influences are strongest.

terricolous (tə rik'ə ləs), *adj. Biology.* living on or in the ground. [from Latin *terricola* inhabitant of the earth]

terrigenous (tə rij'ə nəs), *adj. Geology.* of or having to do with shallow marine sediments deposited from neighboring land: *The original contours of submarine topography are continually being reduced and rounded by the deposition of ... Terrigenous deposits consisting of gravels, sands, silts, muds, and coral fragments eroded by wave action from the shore or washed out from the land by rivers* (White and Renner, *Human Geography*). [from Latin *terrigena* one born from the earth]

territorial, *adj. Zoology.* having to do with or exhibiting territoriality: *[Some] species become territorial only during the breeding season, when they defend a small nesting area ... Some mammals signal their territorial claims by leaving scents* (J. A. Wiens).

territoriality, *n. Zoology.* a form of behavior in which an animal establishes an area as its own and defends it from encroachments by others, usually of its own species. Territoriality is commonly found among birds, fish, lizards, and certain amphibians and mammals. *No room was left for doubting ... the territoriality of the owls* (Scientific American).

territory, *n. Zoology.* an area within definite boundaries, such as a nesting ground, in which an animal lives and from which it keeps out others of its species: *Still, the cichlids' territoriality—defense of its territory—offers an interesting analogy to the institution of private property in the higher vertebrates* (New Yorker).

tert., *abbrev.* tertiary.

tertiary, *adj. Chemistry.* characterized or formed by the replacement of three atoms or groups. For instance, in tertiary monohydric alcohol ($CH_3CH_3CH_3COH$), three of the hydrogen atoms of methyl alcohol (CH_3OH) have been replaced by alkyl groups (-CH_3). All tertiary alcohols have the functional group -COH.

---

cap, fāce, fäther; best, bē, tèrm; pin, fīve;
rock, gō, ôrder; oil, out; cup, pùt, rüle;
yü in use, *yù* in uric;
ng in bring; sh in rush; th in thin, ᴛʜ in then;
zh in seizure.
ə = *a* in about, *e* in taken, *i* in pencil, *o* in lemon, *u* in circus

Tertiary

In tertiary sodium phosphate (Na$_3$PO$_4$), three of the hydrogen atoms of phosphoric acid (H$_3$PO$_4$) have been replaced by sodium atoms. Compare **primary, secondary.**

Tertiary, *Geology.* —*n.* **1** the earlier of the two periods making up the Cenozoic era, immediately following the Mesozoic. During this time the great mountain systems, such as the Alps, Himalayas, Rockies, and Andes, appeared, and rapid development of mammals occurred. *The first period of the Cenozoic, the Tertiary, is usually divided into five subperiods or epochs: Paleocene ("ancient recent types [i.e., mammals]"), Eocene ("dawn of recent types"), Oligocene ("few recent types"), Miocene ("minority of recent types"), and Pliocene ("majority of recent types")* (R. Beals and H. Hoijer).
2 the rocks formed during this period.
—*adj.* of or having to do with this period or its rocks.
► The Cenozoic era has traditionally been divided at the Pliocene-Pleistocene boundary (about 2.5 million years ago) into the *Tertiary* and *Quaternary* periods, but many geologists now divide it instead at the Oligocene-Miocene boundary (about 26 million years ago) into the *Paleogene* (or *Eogene*) and *Neogene* periods.

tertiary consumer, *Ecology.* an animal that eats a secondary consumer in a food chain. In a food chain in which a grass-eating mouse (primary consumer) is devoured by a weasel (secondary consumer) and the weasel in turn is consumed by a hawk, the hawk is a tertiary consumer.

tertiary root, *Botany.* a branch of a secondary root.

tervalent (tėr vā′lənt), *adj.* = trivalent.

tesla (tes′lə), *n. Physics.* the SI or MKS unit of magnetic induction, equal to one weber per square meter. *Symbol:* T [named after Nikola *Tesla,* 1856–1943, Croatian-born American electrical engineer]

tesseract (tes′ə rakt), *n. Mathematics.* a four-dimensional equivalent of a cube: *A tesseract of side x has a hypervolume of x⁴. The volume of its hypersurface is 8x³* (Martin Gardner). [from Greek *tessera* four + *aktis* ray]

test[1], *Chemistry.* —*n.* **1** the examination of a substance to see what it is or what it contains: *A test for boric acid depends on the fact that boric acid colors an alcohol flame green* (Dull, *Modern Chemistry*).
2 the process or substance used in such an examination: *Cobaltous nitrate provides a simple test for the identification of aluminum.*
—*v.* to give a test (for): *to test for oxygen, to test for proteins.*
[from Old French *test* a vessel used in assaying, from Latin *testum* earthen vessel]

test[2], *n.* **1** *Zoology.* the hard covering of certain animals: *Among the sea urchins and their close relatives, the ossicles are so fused or united to each other as to form a rigid, more or less hemispherical, skeleton termed a test, covered externally by ectoderm and containing internally the viscera of the body* (Harbaugh and Goodrich, *Fundamentals of Biology*). SYN: shell.
2 = testa (def. 1). [from Latin *testa* shell]

testa (tes′tə), *n., pl.* **-tae** (-tē or -tī). **1** *Botany.* the hard outside coat of a seed: *The testa, or outer seed shell, differs greatly in form and texture. If thick and hard,* it is crustaceous; if smooth and glossy, it is polished; if roughened, it may be pitted, furrowed, hairy, reticulate, etc. (Youngken, *Pharmaceutical Botany*). SYN: integument. **2** = test[2] (def. 1). [from Latin *testa* shell]

testaceous (tes tā′shəs), *adj. Biology.* **1** of the nature or substance of a shell or shells: *a testaceous operculum.* **2** having a hard shell: *testaceous marine animals.* **3** of a dull brownish-red, brownish-yellow, or reddish-brown color: *a testaceous lark. The upper part of the body is testaceous, or potsherd colour* (R. Holme).

test cross, *Genetics.* a cross between a heterozygote and a recessive homozygote: *To determine the genotype of an individual showing the dominant phenotype, it is crossed with a homozygous-recessive individual. Recessive phenotypes among the offspring indicate a heterozygous parent. This procedure is known as a test cross* (Biology Regents Syllabus).

testes (tes′tēz), *n.* plural of **testis:** *Testes, the male sex glands, secrete testosterone.*

testicle (tes′tə kəl), *n. Anatomy.* a testis, especially the testis of a mammal, enclosed in a scrotum. [from Latin *testiculus,* diminutive of *testis* testis]
—**testicular** (tes tik′yə lər), *adj.* of or having to do with a testicle or testis: *An advance in the knowledge of testicular physiology was made when sperm agglutination in sterile men was shown to be caused by the presence of agglutinins in the seminal fluid* (Robert B. Greenblatt).

testis (tes′tis), *n., pl.* **-tes** (-tēz). *Anatomy.* **1** either of the two reproductive glands of a male vertebrate animal that secrete sperm; a mature male gonad: *The [human] testes consist of thousands of small tubules, which lead through a series of tubes to the urethra and to the outside. The tubules, called seminiferous tubules, produce the sperm* (McElroy, *Biology and Man*).
2 a similar gland in an invertebrate animal: *The [hydra's] male gonads, or testes, are formed near the oral end and release the sperm through little nipples at the tip of the organs out into the surrounding water* (Winchester, *Zoology*).
[from Latin *testis* witness (of virility)]

testosterone (tes tos′tə rōn), *n. Biochemistry.* a hormone secreted by the testes, or produced synthetically, that is responsible for the secondary sex characteristics of males. Chemically, it is a white, crystalline steroid. *Formula:* C$_{19}$H$_{28}$O$_2$ [from *testis* + *ster(ol)* + *-one*]

test tube, a thin glass tube closed at one end, used in making chemical or biological tests: *The chemist uses test tubes of various sizes to hold small amounts of liquid and solid substances* (Paul R. Frey).
—**test-tube,** *adj.* **1** of, having to do with, or contained in a test tube: *a test-tube experiment, a test-tube specimen.*
2 produced by in vitro fertilization; fertilized in laboratory apparatus: *a test-tube birth, a test-tube baby.*
3 produced by artificial insemination: *test-tube cattle.*

tetartohedral (ti tär′tō hē′drəl), *adj. Crystallography.* having one fourth of the number of faces required by the highest degree of symmetry belonging to its system: *a tetartohedral crystal.* [from Greek *tetartos* fourth + *hedra* side, base]

Tethys (teth′ēz), *n. Geology.* a large triangular sea that hypothetically separated Africa and Eurasia before the drifting of continents: *Long before the existence of the present Mediterranean, perhaps a quarter of a billion* years ago, a giant waterway, the Tethys, extended across southern Europe, Iran and southern China to the Pacific Ocean. The Mediterranean is generally considered to be a reminant of the Tethys (Science News). Compare **Pangaea.**

—**Tethyan** (teth'ē ən), adj. of or having to do with the Tethys: *The Tethyan system of ocean trenches —the trough into which were tipped the huge volumes of sediments later compressed into the Himalayas— formed part of the old sea floor lying to the north* (New Scientist).

tetra-, *combining form.* four; having four, as in *tetrachloride, tetravalent.* [from Greek, related to *tettares* four]

tetrabasic, *adj. Chemistry.* **1** (of an acid) having four hydrogen atoms that can be replaced by basic atoms or radicals.
2 having four atoms or radicals of a univalent metal.
3 containing four basic hydroxyl (-OH) radicals.

tetrachloride, *n. Chemistry.* a compound containing four atoms of chlorine combined with another element or radical.

tetracyclic, *adj.* **1** *Chemistry.* having four rings in the molecules: *tetracyclic diterpenes.*
2 *Botany.* having four whorls of floral organs.

tetrad (tet'rad), *n.* **1** *Biology.* a group of four chromatids formed in various organisms when a pair of chromosomes splits longitudinally during meiosis: *Each pair thus comes to consist of four half chromosomes, and the quadruple body formed is called a tetrad. Owing to its origin, two of the parts of each tetrad are maternal, the other two paternal* (Shull, *Principles of Animal Biology*). Also called **bivalent.**
2 *Botany.* a group of four cells, as of spores or pollen grains: *At this stage the chromosomes are beginning to split, preparatory to the next step, which is separation and migration of split chromosomes; and finally, formation of tetrad of spores, each of which has the reduced number of chromosomes* (Emerson, *Basic Botany*).
3 *Chemistry.* an element, atom, or radical with a valence of four.
[from Greek *tetrados* four, from *tetra-*]
—**tetradic** (te trad'ik), *adj.* of or having to do with a tetrad: *a tetradic cell, a tetradic element.*

tetradactyl (tet'rə dak'təl) or **tetradactylous** (tet'rə-dak'tə ləs), *adj. Zoology.* having four fingers, claws, toes, etc., on each limb: *a tetradactyl bird, tetradactylous feet.*

tetradymite (te trad'ə mīt), *n. Mineralogy.* a telluride of bismuth, found in pale, steel-gray laminae with a bright, metallic luster. *Formula:* Bi_2Te_2S [from Greek *tetradymos* fourfold]

tetradynamous (tet'rə di'nə məs), *adj. Botany.* having six stamens, four longer, arranged in opposite pairs, and two shorter, inserted lower down (characteristic of flowers of the mustard family). [from *tetra-* + Greek *dynamis* power]

tetragon (tet'rə gon), *n. Geometry.* a figure, especially a plane figure, having four angles and four sides, as a square or diamond. [from Greek *tetragōnon* quadrangle]

tetragonal (te trag'ə nəl), *adj.* **1** *Geometry.* of or having to do with a tetragon; having four angles; quadrangular: *a tetragonal prism.*
2 *Crystallography.* having three axes at right angles to each other, the two lateral axes being equal and the vertical of a different length. See the picture at **crystal.**
ASSOCIATED TERMS: see **crystal.**
—**tetragonally,** *adv.*

tetrahedral (tet'rə hē'drəl), *adj.* of or having to do with a tetrahedron; having four faces: *a tetrahedral pyramid, a tetrahedral granule.*
—**tetrahedrally,** *adv.* in a tetrahedral manner or form: *In the complete haemoglobin molecule the subunits are arranged tetrahedrally, and it is remarkable how four irregular objects can fit so closely together, projections on one being dovetailed into hollows on its neighbor, and form a nearly spherical whole* (New Scientist).

tetrahedrite (tet'rə hē'drīt), *n. Mineralogy.* native sulfide of antimony and copper, with various elements sometimes replacing one or the other of these, often occurring in tetrahedral crystals. *Formula:* $(Cu,Fe)_{12}Sb_4S_{13}$

tetrahedron (tet'rə hē'drən), *n., pl.* **-drons, -dra** (-drə). *Geometry.* a solid figure having four faces. A regular tetrahedron is a pyramid whose base and three sides are equilateral triangles. *When spheres are packed together as tightly as possible they become nests of tetrahedrons* (solid figures with four triangular sides). *Moreover, nested within every group of four tetrahedrons is an octahedron* (eight-sided figure) (Robert W. Marks). [from Greek *tetra-* + *hedra* base]

tetramer (tet'rə mər), *n. Chemistry.* a compound in which four molecules of the same substance are produced by polymerization.
—**tetrameric,** *adj.* of or having to do with a tetramer: *... having an eight-membered ring structure with bridging fluorines, based on its tetrameric nature in benzene solution* (J. S. Thayer).

tetramerous (te tram'ər əs), *adj.* **1** *Botany.* (of a flower) having four members in each whorl (generally written 4-*merous*).
2 *Zoology.* having tarsi with four joints: *tetramerous insects.*
[from Greek *tetrameres* having four parts, from *tetra-* four + *meros* part]

tetraploid (tet'rə ploid), *Genetics.* —*adj.* having four times the haploid number of chromosomes characteristic of the species: *Tetraploid snaps [snapdragons] are those that have been treated with colchicine to give them extremely large flowers, husky growth habit and good foliage* (N. Y. Times).
—*n.* a tetraploid organism or cell: *Instead of the usual two sets of heredity-bearing chromosomes, plants with four sets, or tetraploids, were produced* (Science News

cap, fāce, fäther; best, bē, tėrm; pin, five;
rock, gō, ôrder; oil, out; cup, pùt, rüle;
yü in use, *yu̇* in uric;
ng in bring; sh in rush; th in thin, ᴛH in then;
zh in seizure.
ə = *a* in about, *e* in taken, *i* in pencil, *o* in lemon, *u* in circus

Letter). *Tetraploids are the result of recent genetic research and practical flower breeding* (Gordon Morrison).
[from *tetra-* four + *-ploid*, as in *haploid, diploid*]
—**tetraploidy** (tet′rə ploi′dē), *n.* the state of having four times the haploid number of chromosomes characteristic of the species.

tetrapod (tet′rə pod), *Zoology.* —*n.* **1** = quadruped. **2** any of a superclass (Tetrapoda) of vertebrates whose members are characterized by four limbs, including amphibians, reptiles, birds, and mammals but not fishes: *Evidence accumulated during recent years has revealed close relationships between Triassic tetrapods, especially reptiles, found in Africa and South America* (Edwin H. Colbert).
—*adj.* **1** = quadruped. **2** of or designating a tetrapod or tetrapods; having four limbs: *The bone is part of a labyrinthodont amphibian of the Triassic age and is the first fossil record of tetrapod life from Antarctica* (Martin Prinz).
[from Greek *tetrapodos* four-footed]

tetrapterous (te trap′tər əs), *adj.* **1** *Zoology.* having four wings or fins: *tetrapterous flies, tetrapterous fishes.* **2** *Botany.* having four winglike appendages, as certain fruits.
[from Greek *tetrapteros* four-winged]

tetraspore, *n. Botany.* one of a group of four haploid asexual spores in red algae, resulting from the division of a mother cell: *The tetraspores give rise to the haploid plants on which antheridia and oogonia are borne* (Emerson, *Basic Botany*).

tetrastichous (te tras′tə kəs), *adj. Botany.* **1** (of flowers) arranged in a spike having four vertical rows.
2 (of a spike) having the flowers arranged in four vertical rows. [from Greek *tetrastichos* having four rows]

tetratomic (tet′trə tom′ik), *adj. Chemistry.* **1** having four atoms in each molecule: *Phosphorus does not exist as a diatomic molecule at room temperature, but exists as a tetratomic molecule, P_4* (Chemistry Regents Syllabus).
2 having four replaceable atoms or groups of atoms.

tetravalent (tet′rə vā′lənt), *adj. Chemistry.* having a valence of four; quadrivalent: *An element like carbon whose atom can hold four monovalent atoms in combination is called a tetravalent element* (Parks and Steinbach, *Systematic College Chemistry*).

tetroxide (te trok′sīd), *n. Chemistry.* an oxide containing four atoms of oxygen in each molecule with another element or radical.

TeV, *abbrev.* teraelectron volt (one trillion electron volts, used as a unit for measuring energy in particle physics).

texture, *n. Geology.* the physical structure of a rock; the spatial relationships between the mineral grains making up a rock, including their size, shape, and arrangement.

Th, *symbol.* thorium.

thalamus (thal′ə məs), *n., pl.* **-mi** (-mī). *Anatomy.* a large, oblong mass of gray matter in the posterior part of the forebrain, from which nerve fibers pass to the sensory parts of the cortex and which is connected with the optic nerve: *Strictly speaking, there are no brain centres for thinking and feeling, though there is*

an anatomically distant region, the thalamus, deep in the brain-stem which has something to do with the perception of pain and other sensations and the judgment of their quality (Science News). Compare **hypothalamus.** [from Latin *thalamus* inside room, from Greek *thalamos*]

thalassic (thə las′ik), *adj. Oceanography.* of or having to do with the smaller or inland seas, as distinct from the pelagic waters or oceans. [from Greek *thalassa* sea]

thallic (thal′ik), *adj. Chemistry.* of or containing thallium, especially with a valence of three.

thallium (thal′ē əm), *n. Chemistry.* a soft, malleable, bluish-white, metallic element that occurs in small quantities in iron and zinc ores and in various minerals. Its compounds are extremely poisonous and are used to kill insects and rodents. Thallium is also used in making glass of high refractive power. *Symbol:* T1; *atomic number* 81; *atomic weight* 204.37; *melting point* 303.5°C; *boiling point* ab. 1457°C; *oxidation state* 1,3. [from New Latin, from Greek *thallos* green shoot (so called because the element's spectrum is marked by a green band)]

thallophyte (thal′ə fit), *n. Biology.* any of a group of plantlike organisms showing no differentiation into stem, leaf, and root. The simpler unicellular forms reproduce by cell division or by asexual spores; the higher forms reproduce both asexually and sexually. Bacteria, algae, fungi, and lichens are thallophytes. See also **thallus.** [from Greek *thallos* green shoot + *phyton* plant]
—**thallophytic** (thal′ə fit′ik), *adj.* of or having to do with the thallophytes: *the thallophytic plants.*
► The thallophytes were formerly considered to comprise a major division (Thallophyta) of the plant kingdom. Most biologists no longer classify them as plants.

thallous (thal′əs), *adj. Chemistry.* of or containing thallium, especially with a valence of one.

thallus (thal′əs), *n., pl.* **thalli** (thal′ī), **thalluses.** *Biology.* a plantlike organism not divided into leaves, stem, and root; the body characteristic of thallophytes. Mushrooms, toadstools, and lichens are thalli. *A spore of the common bread mold alighting on a piece of moist bread germinates and gives rise to a branching, filamentous plant body (thallus) which produces, on special filaments, hundreds of globular sporangia* (Harbaugh and Goodrich, *Fundamentals of Biology*). [from New Latin, from Greek *thallos* green shoot]

thaw, *n. Meteorology.* a condition or period of weather above the freezing point (0 degrees Celsius) warm enough to melt ice and snow.

theca (thē′kə), *n., pl.* **-cae** (-sē′). **1** *Botany.* **a** a sac, saclike cell, or capsule in a plant. **b** a vessel containing spores in various lower plants.
2 *Anatomy, Zoology.* a case or sheath enclosing an animal, such as an insect pupa, or some part of an animal, such as a tendon.
[from New Latin, from Greek *thēkē* case, cover]
—**thecal** (thē′kəl), *adj.* having to do with or of the nature of a theca.

theorem (thē′ər əm *or* thir′əm), *n. Mathematics.* a statement or proposition that has been or is to be proved, often on the basis of certain axioms or assumptions, and that can be usually expressed by an equation or formula: *Every theorem is a statement that if a certain thing is true, then something else is also true. For ex-*

ample, *Theorem 4–9 says that if two intersecting lines form one right angle, then they form four right angles. The* if *part of a theorem is called the hypothesis; it states what is given. The* then *part is called the conclusion; it states what is to be proved on the basis of what is given* (Moise and Downs, *Geometry*). [from Greek *theōrēma,* from *theōrein* to consider, look at]

▶ See the note under **axiom.**

theoretical arithmetic, = theory of numbers.

theory, *n.* **1** an explanation or model based on observation, experimentation, and reasoning, especially one that has been tested and confirmed as a general principle helping to explain and predict natural phenomena: *the theory of evolution. Einstein's theory of relativity explains the motion of moving objects. The merit of a theory lies not necessarily in 'prediction' in the temporal sense. That would depend on chronological order which may be fortuitous. It lies in the automatic inclusion of data, which may already be known, but which have not been put explicitly into the theory in the first place. We must judge any theory by the amount of data it covers over and above those it had to be fitted to* (H. R. Paneth).

2 *Mathematics.* a set of theorems which constitute a connected systematic view of some branch of mathematics: *the theory of probabilities, group theory, set theory, the theory of numbers.*

[from Greek *theōriā* a looking at, thing looked at, from *theōrein* to consider, look at]

▶ **Theory, hypothesis** as terms in science mean a generalization reached by inference from observed particulars and proposed as an explanation of their cause, relations, or the like. **Theory** implies a larger body of tested evidence and a greater degree of probability: *The red shift in the spectra of galaxies supports the theory that the universe is continuously expanding.* **Hypothesis** designates a merely tentative explanation of the data, advanced or adopted provisionally, often as the basis of a theory or as a guide to further observation or experiment: *Animal experiments have strengthened the hypothesis that some cancers in human beings are caused by viruses.*

theory of games, = game theory.

theory of numbers, = number theory.

therapsid (thə rap′sid), *n. Zoology.* any of a group of reptiles first appearing in the Permian period, that had differentiated teeth and skulls much like the mammals: *During this period* [*the Permian*] *reptiles called therapsids developed. These reptiles resembled mammals in some ways, and scientists believe the therapsids were ancestors of the mammals* (Samuel P. Welles). [from New Latin *Therapsida* the order name, from Greek *theraps* attendant]

therm (thėrm), *n. Physics.* any of various units of heat: **a** = calorie (def. 1a). **b** = Calorie (def. 1b). **c** a unit equivalent to 1000 Calories. **d** a unit equivalent to 100,000 British thermal units. [from Greek *thermē* heat]

therm-, *combining form.* the form of **thermo-** before vowels, as in *thermal, thermion.*

thermal (thėr′məl), *adj. Physics.* of or having to do with heat; generated, measured, or operated by heat: *Thermal energy is mechanical potential and kinetic energy of random motion on a microscopic scale* (Shortley and Williams, *Elements of Physics*). The

atoms in a solid metal are always involved in thermal agitation, and ... this thermal agitation also destroys the perfect periodicity of the crystal, leading to scattering of the electrons, hence to electrical resistance (H. W. Lewis).

—*n. Meteorology.* a rising current of warm air: *Like gliding, ballooning depends for movement on luck with thermals, which are air currents rising off sun-warmed fields and hills* (New Yorker). —**thermally,** *adv.*

thermal equator, = heat equator.

thermalize, *v. Physics.* to lower the kinetic energy of (an atom or other particle) until it reaches the energy characteristic of the temperature of a particular medium: *These "hot" atoms have such kinetic energy that any molecule they may form by chemical reaction is likely to be instantly fragmented; but if they are "thermalized", that is allowed to cool by collisions with an inert gas, they will then react in a chemically intelligible manner, particularly if the reaction product can also be deexcited by rapid cooling* (D. E. H. Jones). —**thermalization,** *n.* the act or process of thermalizing: *Thermalization of the neutrons takes place and the neutron density is then measured by detectors mounted in the moderating material* (J. L. Phillips).

thermal neutron, *Physics.* a neutron moving with an energy characteristic of the temperature of the medium in which it is moving. Thermal neutrons maintain the chain reaction in a nuclear reactor. *Thermal neutrons are neutrons with kinetic energies approximating those of molecules of substances at ordinary temperatures* (Physics Regents Syllabus).

thermal pollution, 1 the discharge of artificially heated water into a natural body of water, causing a rise in natural water's temperature that is harmful to plant and animal life in it. **2** the discharge of heated gases into the surrounding air, causing a rise in air temperature that sometimes affects local weather conditions.

thermal spring, = hot spring: *As soil algae have adapted to high and low extremes in the desert, so have they developed the ability to live and carry out photosynthesis in thermal springs* (Brooks D. Church).

thermal unit, a unit adopted for measuring and comparing quantities of heat.

thermic (thėr′mik), *adj.* = thermal.

thermion (thėr′mē on′), *n. Physics.* an electrically charged particle, usually an electron, given off by a heated body. [from *therm-* + *ion*] —**thermionic** (thėr′mē on′ik), *adj.* of or having to do with thermions: *Experiment shows that the electrons emitted from hot filaments can be deflected by electric and magnetic fields. The results of experiments on these 'thermionic' electrons show that these electrons*

cap, fāce, fäther; best, bē, tėrm; pin, five;
rock, gō, ôrder; oil, out; cup, pùt, rüle;
yü in use, *yù* in uric;
ng in bring; *sh* in rush; *th* in thin, ᴛʜ in then;
zh in seizure.
ə = *a* in about, *e* in taken, *i* in pencil, *o* in lemon, *u* in circus

have a characteristic charge (Shortley and Williams, *Elements of Physics*).

thermionic current, *Physics.* **1** an electric current produced by movements of thermions. **2** a movement of thermions.

thermionic emission, *Physics.* the freeing of electrons by heat, as from a metal: *Thermionic emission was first disclosed in 1883 by Thomas A. Edison* (J. J. William Brown).

thermionics, *n.* the science of thermionic phenomena.

thermo-, *combining form.* heat; temperature, as in *thermoelectricity = electricity produced by heat.* Also spelled **therm-** before vowels. [from Greek *thermē*]

thermochemical, *adj.* of or having to do with thermochemistry.

thermochemistry, *n.* **1** the branch of chemistry dealing with the relations between chemical action and heat: *Lavoisier, in collaboration with the great physicist Pierre Simon de Laplace, made studies of the heat evolved in combustion which laid the foundation of thermochemistry* (Denis I. Duveen).
2 thermochemical properties or characteristics: *the thermochemistry of organic ions. Much of the research in mineralogy ... was concentrated on the crystal chemistry and thermochemistry of rock-forming silicates* (Robert E. Boyer).

thermocline (thèr′mō klīn), *n. Geology.* a layer within a large body of water sharply separating parts of it that differ in temperature, so that the temperature gradient through the layer is very abrupt. [from *thermo-* + Greek *klinein* to lean]

thermocouple, *n. Physics.* two dissimilar metallic conductors joined end to end, whose junction, when heated, produces a thermoelectric current in the circuit of which they form a part. Thermocouples are used as thermometers, to generate electricity, to make refrigeration devices, etc. *If two different metals are placed in contact, and the junction is warmed, an electromotive force is generated. The metals form what is called a "thermocouple," and a measurement of the electromotive force can be used to determine the temperature of the metal contact* (F. P. Bowden). Also called **thermoelectric couple.** Compare **thermojunction, thermopile.**

thermodiffusion, *n. Physics.* diffusion of heat.

thermoduric, *adj. Bacteriology.* capable of surviving the high temperature of pasteurization: *As a result of a two years' test at the National Institute for Research in Dairying, University of Reading, it was shown that the number of thermophilic organisms was greatly reduced and the thermoduric organisms were destroyed, in a typical plant used in the milk industry* (J. L. Burn). [from *thermo-* + Latin *durare* to last, endure]

thermodynamic, *adj. Physics.* of or having to do with thermodynamics; using force due to heat or to the conversion of heat into other forms of energy: *High thermodynamic efficiency is of little use for practical commercial purposes unless a reasonably cheap and readily available fuel can be burned* (J. Hodge). —**thermodynamically,** *adv.*

thermodynamics, *n.* **1** the branch of physics that deals with the relations between heat and other forms of energy, and of the conversion of one into the other. Ther-

modynamics is based on two laws, the first of which (essentially the law of conservation of energy but called the **first law of thermodynamics**) is that energy cannot be created or destroyed, only converted from one form into another, as from mechanical energy into heat, or heat into mechanical energy. The **second law of thermodynamics** is that heat will, of its own accord, flow only from a hot body to a cooler body. In addition to these, some writers refer to the following statement by Planck as the **third law of thermodynamics:** Every system has a finite, positive entropy, but at a temperature of absolute zero the entropy may become zero, as it does in perfect crystalline substances.
2 thermodynamic properties or processes: *the thermodynamics of convection and conduction.*

thermoelectric, *adj.* of or having to do with thermoelectricity: *The voltage output of a single thermoelectric junction is extremely low ...* (J. J. William Brown). —**thermoelectrically,** *adv.*

thermoelectric couple, = thermocouple

thermoelectricity, *n. Physics.* **1** electricity produced directly by heat, especially that produced by a temperature difference between two different metals used as conductors in a circuit.
2 the branch of physics that deals with such electricity.

thermoelectromotive, *adj. Physics.* of or having to do with electromotive force produced by heat.

thermoelectron, *n. Physics.* a negatively charged particle given off by a heated body.

thermogram, *n.* a measurement or record made by a thermograph: *In medicine, infrared is a supplement to X-ray diagnosis, providing doctors with a thermogram—a photograph that shows not what our eyes would see but what hundreds of thermometers would sense* (Science News Letter).

thermograph, *n.* **1** a thermometer that automatically records temperature: *Permanent continuous records of the air temperature are obtained by means of a thermograph ... [that] utilizes a flat, curved metallic tube filled with a liquid whose expansion and contraction change the curvature of the tube* (Neuberger and Stephens, *Weather and Man*).
2 the photographic or other apparatus used in thermography: *The thermograph films a fine record of the body temperature. Troubled tissues or "hot spots" show up white, while healthy, "cool" areas of the body register black* (Science News Letter).

thermographic, *adj.* of, having to do with, or obtained by a thermograph or thermography: *The thermographic method ... uses the heat of infrared rays, instead of light, for exposure* (R. F. Beckwith). —**thermographically,** *adv.*

thermography, *n.* **1** the measurement of temperature by means of a thermograph.
2 a method of recording photographically or displaying visually infrared rays emitted by the body which show differences in temperature between healthy and unhealthy tissue: *In the treatment of burns ... thermography has been hailed as a tool as essential as the scalpel in healing large-area burns* (New Scientist).

thermohaline, *adj. Oceanography.* having to do with both the temperature and salt content of the ocean: *thermohaline circulation. Because of a peculiar thermohaline situation set up by the cooling and evaporation of northward-flowing surface water, a shallow* density-driven outflow is produced that carries saline water back to the straits where it spills over the sill and into the Gulf of Aden (A. Conrad Neumann).

thermojunction, n. Physics. the point of union of the two metallic conductors of a thermocouple: The freezing and cooling operations are accomplished by tiny thermojunctions which absorb and remove heat from the cooling compartments when electric current is applied (Chester L. Dawes).

thermokinematics, n. Physics. the study of motion caused by heat.

thermolabile (thėr′mō lā′bəl), adj. Biochemistry. susceptible of destruction or loss of characteristic properties at moderately high temperatures, as certain toxins.

thermoluminescence, n. Physics. the emission of light produced by heat or exposure to high temperature, widely used to monitor the radiation dose a substance has been subjected to: A pottery sample is rapidly heated in the laboratory, the photons are released, giving off thermoluminescence or glow. The longer ago a sample was fired, the more glow is recorded, and the older the pottery is determined to be (N. Y. Times).
—**thermoluminescent,** adj. having or showing thermoluminescence: a thermoluminescent apparatus, the thermoluminescent capacity of crystals.

thermolysis (thər mol′ə sis), n. Chemistry. decomposition or dissociation by heat: The heat ... has the effect of throwing the molecule into such agitation that the mutual affinity of the atoms cannot retain them in union. This is the process of ... thermolysis (A. Daniell).
—**thermolytic** (thėr′mə lit′ik), adj. having to do with or producing thermolysis.

thermomagnetic, adj. Physics. of or having to do with the heat as a modifying effect on the magnetic properties of bodies. A **thermomagnetic effect** is any of various effects occurring when a conductor or semiconductor is placed at the same time in a temperature gradient and a magnetic field.

thermometer, n. an instrument for measuring temperature, usually by means of the expansion and contraction of mercury or alcohol in a capillary tube and bulb with a graduated scale.
—**thermometric,** adj. **1** of or having to do with a thermometer: the thermometric scale.
2 made by means of a thermometer: thermometric observations. —**thermometrically,** adv.

thermometry (thər mom′ə trē), n. the measurement of temperature: The whole science of heat is founded on thermometry and calorimetry (James C. Maxwell).

thermonuclear, adj. Physics. of or having to do with the fusion of atoms through very high temperature: a thermonuclear reaction. Thermonuclear energy differs from presently-attainable atomic fission power in that atoms are fused together as in an H-bomb instead of being split as in an atomic bomb (Wall Street Journal).

thermoperiod, n. Botany. the length of time during which a plant is exposed to a particular temperature each day: Although the thermoperiod is still under investigation, experiments indicate that for most houseplants the night temperature should be allowed to fall about eight to 14 degrees F. below the daytime temperature (Scientific American).
—**thermoperiodism,** n. the response of a plant to daily changes in temperature.

thermophile (thėr′mə fil), n. Microbiology. a bacterium, fungus, or other microorganism that is able to grow at temperatures above 40 degrees Celsius: Thermophiles—algae, bacteria or fungi which prefer a heated habitat—were predictably found in greatest numbers in the crater where there are steam vents (Science News).
—**thermophilic** (thėr′mō fil′ik), adj. requiring high temperatures for development: Some Actinomycetes are thermophilic: they are able to live and thrive at high temperatures that few other organisms could tolerate (C. L. Duddington).

thermopile, n. Physics. a device consisting of several thermocouples acting together for the production of a combined effect, as for generating currents or for ascertaining minute temperature differences: One of the most sensitive types of thermopile consists of one or more junctions of fine bismuth and tellurium wires with thin blackened gold disks to absorb the radiation (Furry, Physics).

thermoreceptor, n. Physiology. a sense organ stimulated by a change in the temperature: This appreciation of temperature conditions at the body surface, or "thermal sensitivity," is a cutaneous sense modality like touch and pain, and has its own sensory apparatus made up of sense organs or thermoreceptors, with their own communication lines or nerve relays to the brain (New Scientist).

thermoregulate, v. Biology. to regulate the temperature of the body; undergo or cause to undergo thermoregulation: While some tunas, some insects, and some reptiles thermoregulate, they do not maintain body temperature at such a specific level as mammals and birds do (E. Don Stevens).
—**thermoregulation,** n. regulation of body temperature: mammalian thermoregulation. Most thermoregulation is accomplished by selection of microhabitats with thermally distinct microclimates. ... The animals keep their temperatures in their active range by moving between light and shade, orienting their bodies parallel or perpendicular to the light rays and by seeking warmer or cooler surfaces (Science News).
—**thermoregulator,** n. an internal mechanism that regulates body temperature: Our juncos, chickadees, and kinglets appear to use different thermoregulators than the ones ... in the Sudan (New Scientist).
—**thermoregulatory,** adj. of or having to do with thermoregulation: Even in deep hibernation a mammal does not lose all control of its thermoregulatory system (Scientific American).

thermoremanence (thėr′mō rem′ə nəns), n. Physics, Geology. **1** the magnetism remaining in a rock or other substance that was once molten and then cooled to the Curie point.

---

cap, fāce, fäther; best, bē, tėrm; pin, fīve;
rock, gō, ôrder; oil, out; cup, pùt, rüle,
yü in use, yù in uric;
ng in bring; sh in rush; th in thin, ŦH in then;
zh in seizure.
ə = a in about, e in taken, i in pencil, o in lemon, u in circus

2 the process of acquiring such magnetism: *In some samples the magnetism was very stable and seemed to indicate that they had acquired their magnetism by thermoremanence—cooling through the Curie temperature—in a weak field of about 35 gammas* (New Scientist).

—**thermoremanent,** *adj.* of or having to do with thermoremanence: *Most lunar rocks have some remanent magnetization. It is widely accepted that the igneous rocks contain a component of stable thermoremanent magnetization, but the mode of magnetization cannot be defined uniquely* (Science).

thermosphere, *n. Geophysics.* the region of the atmosphere above the mesopause, in which temperature increases with height: *Auroral phenomena and magnetic activity occur in the thermosphere. Its outermost part ... is the fringe region where the earth's atmosphere merges with interplanetary gases* (Charles J. Knudsen and James K. McGuire).

thermostability, *n.* the quality of being thermostable: *the thermostability of an enzyme, cell, etc.*

thermostable, *adj. Biochemistry.* remaining stable when heated; able to undergo heating without loss of characteristic properties, as certain ferments or toxins.

thermotaxis, *n. Biology.* **1** the movement of an organism in response to changes in temperature: *Cases of directive stimulation ... have been designated Chemotaxis, Phototaxis, Thermotaxis, Galvanotaxis, and so forth* (Max Verworn).
2 the regulation of body temperature; thermoregulation.

thermotropic, *adj.* having to do with or exhibiting thermotropism: *thermotropic curvatures in plants.*

thermotropism, *n. Biology.* the tendency to bend or turn toward or away from the sun or other source of heat.

theta pinch (thā'tə *or* thē'tə), *Physics.* the rapid compression of a magnetic field surrounding highly ionized gas in order to produce a controlled fusion reaction: *Shocks are one of the practical ways to heat high-beta plasmas, and they have been used for years in many experiments. The most common method is the so-called theta pinch, in which a magnetic field surrounding a plasma is suddenly compressed* (Science News).

theta rhythm or **theta waves,** *Physiology.* a pattern of electrical oscillations in the brain in the frequency of 4 to 8 hertz: *Although they are most prominent in the hippocampus, theta waves are readily recorded from many regions of the brain, including cortex, when rodents and carnivores are alert or aroused, and during the early stages of conditioning* (Science). Compare **alpha rhythm, beta rhythm.**

THI, *abbrev.* temperature-humidity index.

thi-, *combining form.* a form of **thio-** used sometimes before vowels, as in *thiamine, thiazine.*

thiamin (thī'ə mən), *n.* = thiamine.

thiamine (thī'ə mən *or* thī'ə mēn), *n. Biochemistry.* a vitamin, a crystalline organic compound that promotes growth and aids in preventing beriberi, found in yeast, meats, whole-grain cereals, and certain vegetables, or prepared synthetically. *Formula:* $C_{12}H_{17}ClN_4OS$ Also called **vitamin B$_1$.**

thiazine (thī'ə zin), *n. Chemistry.* any of a group of compounds, each having a ring composed of four carbon atoms, one sulfur atom, and one nitrogen atom. The thiazines are the parent substances of certain dyes.

thigh, *n. Anatomy.* **1** the part of the leg between the hip and the knee. See the picture at **socket.**
2 the similar but not corresponding part of a four-legged vertebrate animal, such as the horse; upper part of the hind leg.
3 the second segment of the leg of a bird, containing the tibia and the fibia.
4 the third segment of the leg of an insect.

thighbone, *n. Anatomy.* the bone of the leg between the hip and the knee. Technical name, **femur.**

thigmotactic (thig'mə tak'tik), *adj.* = stereotactic.

thigmotaxis (thig'mə tak'sis), *n.* = stereotaxis.

thigmotropic (thig'mə trop'ik), *adj.* = stereotropic.

thigmotropism (thig mot'rə piz əm), *n.* = stereotropism.

thio- (thī'ō-), *combining form. Chemistry.* sulfur (replacing oxygen atoms in the designated oxygen compound), as in *thioarsenate, thiocyanate.* Also spelled sometimes **thi-** before vowels. [from Greek *theion* sulfur]

thioacetic acid, *Chemistry.* a liquid produced by heating glacial acetic acid with a sulfide of phosphorus, used as a reagent. *Formula:* C_2H_4OS

thioacid, *n. Chemistry.* an acid in which sulfur partly or wholly takes the place of oxygen.

thioaldehyde, *n. Chemistry.* any of a group of compounds containing the group (-CHS), and considered as aldehydes with sulfur substituted for the oxygen.

thioctic acid, (thī ok'tik), = lipoic acid.

thiol (thī'ōl *or* thī'ol), *n.* **1** = mercaptan.
2 the univalent radical -SH of the mercaptans.

thionate (thī'ə nāt), *n.* a salt or ester of a thionic acid.

thionic (thī on'ik), *adj. Chemistry.* **1** of or containing sulfur. **2** having oxygen replaced by sulfur.

thionic acid, *Chemistry.* any of a group of unstable acids represented by the formula $H_2S_nO_6$, where n equals two, four, five, and perhaps six.

thionine (thī'ə nēn), *n. Chemistry.* a dark crystalline, basic compound derived from thiazine, used as a violet stain in microscopy. *Formula:* $C_{12}H_9N_3S$

thionyl (thī'ə nəl), *n.* = sulfinyl.

thiophene (thī'ə fēn), *n. Chemistry.* a colorless liquid compound, present in coal tar, with an odor like that of benzene, and with properties similar to those of benzene. *Formula:* C_4H_4S

thiophenol (thī'ə fē'nōl), *n. Chemistry.* a colorless mobile liquid with the odor of garlic, regarded as phenol with the oxygen replaced by sulfur. *Formula:* C_6H_5SH

thiophosphate (thī'ō fos'fāt), *n. Chemistry.* a salt or ester of thiophosphoric acid.

thiophosphoric acid, *Chemistry.* an acid produced from phosphoric acid by replacing one or more oxygen atoms with an atom of sulfur.

thiosulfate, *n.* a salt or ester of thiosulfuric acid.

thiosulfuric acid, *Chemistry.* an unstable acid, considered as sulfuric acid in which one atom of oxygen is replaced by sulfur. It occurs only in solution or in the form of its salts (thiosulfates). *Formula:* $H_2S_2O_3$

third dimension, *Physics.* the dimension of depth or thickness in a figure, object, or system.

third proportional, *Mathematics.* the final term of a proportion having three variables, the second and third terms being the mean proportional. EXAMPLE: In the proportion, 2 is to 4 as 4 is to 8, the third proportional is 8.

third quarter, *Astronomy.* **1** the period between full moon and second half moon. **2** the phase of the moon represented by the second half moon, after full moon: *The moon keeps revolving, and, seven days after full moon, we again see another half circle, the third quarter* (Fred L. Whipple).

thistle tube, a funnel-shaped glass tube used by chemists, having a large bulb like the head of a thistle between the conical flaring part and the rest of the tube: *The thistle tube serves both as a funnel through which acidulated water (dilute sulfuric acid) can be added, and also as a reservoir to hold the liquid that is displaced by the gases that are collected* (Offner, *Fundamentals of Chemistry*).

thixotropic (thik′sə trop′ik), *adj.* exhibiting or characterized by thixotropy: *Certain clays suspended in solutions of alkali chloride of molar strength or greater will give gels that are thixotropic, i.e. gels that become fluid when stirred or agitated but set again when stirring is discontinued* (Science News).

thixotropy (thik sot′rə pē), *n. Chemistry.* the property of becoming fluid when stirred or agitated, as exhibited by gels. [from Greek *thixis* a touching + *-tropos* a turning]

tholeiite (thō′lē īt), *n. Geology.* a type of basalt, usually containing quartz and alkalic feldspar: *The tholeiites represent the primary magma generated in the upper mantle beneath the oceans* (Lawrence Ogden). [from *Tholey,* village in Germany + *-ite*]

—**tholeiitic** (thō′lē it′ik), *adj.* of or containing tholeiite: *In plate tectonics, magma is generated at ocean ridges and at subduction zones, the former producing alkalic and tholeiitic flood basalts and the latter basalts and calc-alkaline igneous suites* (Nature).

thoracic (thô ras′ik), *adj.* of, having to do with, or in the region of the thorax. The **thoracic cavity** contains the heart and lungs.

thoracic duct, the main trunk of the lymphatic system, that passes through the thoracic cavity in front of the spinal column and empties lymph and chyle into the blood through the left subclavian vein.

thoracolumbar (thôr′ə kō lum′bər), *adj. Anatomy.* of or having to do with the thoracic and lumbar regions, especially of the spine or the sympathetic nervous system: *The autonomic system is divided functionally into two major regions. One centers in the middle portion of the spinal cord (chest and small of back) and may be called the thoracolumbar system. The other has its center partly in the brain, partly in the lower end of the spinal cord, and is called the craniosacral system* (Shull, *Principles of Animal Biology*).

thorax (thôr′aks), *n., pl.* **thoraxes, thoraces** (thôr′ə sēz′).
1 *Anatomy.* the part of the body between the neck and the abdomen. The thorax contains the chief organs of circulation and respiration.
2 *Zoology.* the second of the three main divisions of an arthropod's body, between the head and the abdomen.
[from Latin, from Greek *thōrax* chest]

thoria (thôr′ē ə), *n. Chemistry.* a heavy white powder, an oxide of thorium, obtained from monazite. Thoria is used as a catalyst, and is also used in optical glass, crucibles, cathodes, etc. *Formula:* ThO$_2$ [from *thorium*]

thorianite (thôr′ē ə nīt), *n. Mineralogy.* a radioactive mineral consisting of an oxide of thorium, found in small brownish-black crystals having a resinous luster. Thorianite is isomorphous with uraninite. *Formula:* ThO$_2$

thorite (thôr′īt), *n.* a mineral, a silicate of thorium, occurring in crystals or in massive form. Thorite varies in color from orange-yellow to brownish-black or black. *Formula:* ThSiO$_4$ [from Swedish *thorit,* from *Thor* the Scandinavian god of thunder + *-it* -ite]

thorium (thôr′ē əm), *n. Chemistry.* a dark-gray, radioactive, metallic element present in thorite, monazite, and certain other rare minerals. When thorium is bombarded with neutrons, it changes into an isotope of uranium (U-233) which is used as a nuclear fuel. Thorium is also used to strengthen alloys. *Symbol:* Th; *atomic number* 90; *mass number* 232; *melting point* 1750°C; *boiling point* ab. 4000°C; *oxidation state* 4. [from New Latin, from Swedish *Thor* the Scandinavian god of thunder]

thorn, *n.* **1** *Botany.* a sharp point on a stem or branch of a tree or other plant. A thorn usually grows from a bud. Blackberries, roses, hawthorns, and cacti have thorns. SYN: spine, prickle. **b** a tree or other plant with thorns, especially any shrub or small tree of the hawthorns, such as the Washington thorn, a species native to the southern United States.
2 a spine or spiny process in an animal.

thoron (thôr′on), *n. Chemistry.* a radioactive, gaseous isotope of radon, formed by the disintegration of thorium.

thortveitite (thôrt vī′tīt), *n.* a very rare grayish-green mineral, a silicate of scandium and yttrium. *Formula:* (Sc,Y)$_2$Si$_2$O$_7$ [from Olaus *Thortveit,* a 20th-century Norwegian mineralogist + *-ite*]

Thr, *abbrev.* threonine.

(def. 2)

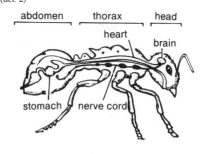

three-phase

three-phase, *adj. Physics.* **1** of or denoting any system in which three distinct phases, such as liquid, solid, or gas, exist together or in combination.
2 of or having to do with a combination of three electric circuits caused by alternating electromotive forces differing in phase by one third of a cycle (120 degrees): *a three-phase rectifier, amplifier, etc.*

threonine (thrē′ə nēn′ *or* thrē′ə nən), *n. Biochemistry.* a crystalline amino acid, a product of the hydrolysis of proteins, considered essential to human nutrition. *Formula:* $C_4H_9NO_3$ *Abbreviation:* Thr

threshold, *n.* **1** *Physiology.* the point at which a given stimulus begins to be perceptible, or the point at which two stimuli can be differentiated: *We apply a brief electric shock to the nerve from our stimulating electrodes and gradually increase its strength. When it reaches a certain strength, known as the "threshold," we suddenly see an electric wave recorded on the oscilloscope* (Scientific American).
2 *Physics.* the minimum energy required to initiate a physical process, especially a nuclear reaction.

threshold frequency, *Physics.* the minimum frequency required to eject an electron from the surface of a material: *The energy associated with the threshold frequency is called the work function of the material* (Physics Regents Syllabus).

throat, *n. Anatomy.* the part of the vertebrate body passing from the mouth through the neck to the stomach or the lungs, especially the area of the fauces and pharynx.

thrombin (throm′bən), *n. Biochemistry.* an enzyme in blood serum which reacts with fibrinogen to form fibrin, causing blood to clot: *The substance that converts fibrinogen into the insoluble fibrin is thrombin. If thrombin were normally present in the blood, clotting would result, but instead it is in the inactive form of prothrombin. Prothrombin in the presence of calcium ions is converted into thrombin* (Hegner and Stiles, College Zoology). [from Greek *thrombos* clot]

thrombocyte (throm′bə sīt), *n.* = blood platelet.

thrombokinase (throm′bō ki′nās), *n. Biochemistry.* an enzyme that promotes the conversion of prothrombin into thrombin and is active in the clotting of blood.

thromboplastic, *adj. Biochemistry.* promoting or having to do with the clotting of blood: *a thromboplastic protein.*

thromboplastin (throm′bō plas′tən), *n. Biochemistry.* **1** a protein substance found in the blood and in tissues, which promotes the conversion of prothrombin into thrombin: *Like thrombin, thromboplastin is not a normal constituent of the circulating blood, but is formed from precursor substances in what is sometimes referred to as the first stage of clotting* (Saturday Review). **2** = thrombokinase.

thrombosthenin (throm bos′thə nin), *n. Biochemistry.* a contractile protein found in human blood platelets: *Investigators ... at the National Institutes of Health in Bethesda, Md., isolated actin and myosinlike proteins from thrombosthenin—a group of contractile proteins in human blood platelets. It is probably involved in coagulation* (Earl A. Evans). [from Greek *thrombos* clot + *sthenos* strength]

thromboxane (throm bok′sān), *n. Biochemistry.* a substance produced by enzymes in the blood platelets that stimulates the aggregation of blood platelets and constricts blood vessels: *The molecular family of the prostagladins, a group of naturally occurring substances of marked physiological activity and of great therapeutic interest, was shown to include the important thromboxanes and prostacyclins* (Frederick D. Greene). [from Greek *thrombos* clot + English *ox(ygen)* + *ane*]

thrombus (throm′bəs), *n., pl.* -bi (-bī). *Physiology, Medicine.* a fibrous clot which forms in a blood vessel and obstructs the circulation. [from New Latin, from Greek *thrombos* clot]

throw, *n. Geology.* **1** a fault. **2** the extent of vertical displacement produced by a fault.

thrust, *n.* **1** *Mechanics.* the force of one thing pushing on another.
2 *Geology.* **a** a movement of one crustal unit over another. **b** = thrust fault.

thrust fault, *Geology.* a low-angle, reversed fault, produced by horizontal compression: *The rocks were folded into great anticlines and synclines, and in many places they were broken by thrust faults having the upthrow side on the southeast* (Moore, Introduction to Historical Geology).

thulium (thü′lē əm), *n. Chemistry.* a silver-white, rare-earth, metallic element of the yttrium group, found in gadolinite and various other minerals. An isotope of thulium is used as the radiating element in portable X-ray units. *Symbol:* Tm; *atomic number* 69; *atomic weight* 168.934; *melting point* 1545°C; *boiling point* 1727°C; *oxidation state* 3. [from New Latin, from Latin *Thule*, the northernmost part of the habitable world]

thumb, *n.* **1** *Anatomy.* the short, thick finger of the human hand, next to the forefinger, which can be used in opposition to the other fingers; pollex. See the picture at **muscle.**
2 *Zoology.* the corresponding digit of other animals, as in the bat. See the picture at **wing.**

thunder, *n. Meteorology.* **1** the loud noise that accompanies or follows a flash of lightning, caused by a disturbance of the air resulting from the discharge of electricity: *What about thunder? As the lightning flash cuts its narrow path through the air "the temperature rises in a few ten-millionths of a second to about 15,000 degrees Centigrade, causing the air in the channel to expand explosively and so create very powerful sound waves." Such factors as the number of separate strokes, echoes and the like produce the rumbling noises that can be heard miles away* (Scientific American). **2** = thunderbolt.

thunderbolt, *n. Meteorology.* a flash of lightning and the thunder that follows it.

thundercloud, *n. Meteorology.* a dark, electrically charged cloud that produces lightning and thunder.

thunder egg, *Geology.* a common name for a small, geodelike body formed in tuff or lava and consisting of chalcedony, opal, or agate.

thunderhead, *n. Meteorology.* one of the round, swelling masses of cumulus clouds often appearing before thunderstorms and frequently developing into thunderclouds.

thundersquall, *n. Meteorology.* a squall with lightning and thunder.

thunderstorm, *n. Meteorology.* a storm with lightning, thunder, rain, and sometimes hail: *When the air moving against a mountain slope is conditionally or convectively unstable, mountain thunderstorms may be severe and continuous and extend to great heights ...* (Blair, *Weather Elements*).

thylakoid (thī'lə koid), *n. Botany.* a saclike membrane that is the structural unit of the grana in the chloroplasts of plant cells and the site of photochemical reactions essential in photosynthesis: *The chloroplast is single-lobed and peripheral, and its lamellae exhibit a three-thylakoid arrangement.* (Leonard Muscatine). [from Greek *thylakos* pouch, sack + English *-oid*]

thymic (thī'mik), *adj. Anatomy.* of or having to do with the thymus: *Thymic nucleic acid is a source of thymine.*

thymine (thī'mēn), *n. Biochemistry.* a substance present in nucleic acid in cells. It is one of the pyrimidine bases of DNA, corresponding to uracil in RNA. *In chemical ground-plan, RNA is like DNA except in that ribose takes the place of deoxyribose and that one of the bases, thymine, is replaced by a different pyrimidine, uracil* (F. R. Jevons). *Abbreviation:* T *Formula:* $C_5H_6N_2O_2$ Compare **adenine, cytosine, guanine.** See the picture at **DNA.** [from *thymus* (so called because the substance was originally extracted from the thymus)]

thymocyte (thī'mə sīt), *n. Biology.* a lymphocyte found in the thymus gland: *All the cells originate in the bone marrow, but some of them undergo a process of maturation in the thymus, a large lymph gland at the base of the neck. Those cells are known as thymocytes, and it is the thymocytes that seem to be responsible for most of the damage done during tissue rejection* (London *Times*).

thymol (thī'mōl *or* thī'mol), *n. Chemistry.* an aromatic, white or colorless, crystalline substance obtained from thyme and other plants or made synthetically, used as an antiseptic, etc. *Formula:* $C_{10}H_{14}O$

thymosin (thī'mə sin), *n. Biochemistry.* a hormone of the thymus gland, associated with the development of T cells: *The thymus gland is a central organ in immunity. Located between the breastbone and the heart in humans and most other mammals, it triggers the development of certain cells into white blood cells that are called thymus-dependent lymphocytes, or T cells. Thymosin is a key hormone in this development* (Allan L. Goldstein).

thymus (thī'məs), *n.,* or **thymus gland,** *Anatomy.* a small glandlike organ consisting of lymphoid tissue, found in young vertebrates near the base of the neck and disappearing or becoming rudimentary in the adult. The thymus is the site of development of T cells and is important in controlling immunological reactions. *An intact thymus is required for the normal development and differentiation of immunological competence and ability to produce antibodies to a wide variety of antigens* (Hugh O. McDevitt). [from New Latin, from Greek *thymos*]

thyroactive (thī'rō ak'tiv), *adj. Biochemistry.* stimulating the activity of thyroxine and other secretions of the thyroid gland: *The iodine treatment followed large doses of thyroactive substances. The purpose of the in-vestigation was to determine whether the conditions of the ... treatment by a single dosage merely served as a vehicle for iodine or acted as a whole* (Science News Letter).

thyrocalcitonin (thī'rō kal'sə tō'nən), *n.* = calcitonin: *Experimental evidence has indicated that a hormone produced by the thyroid gland, thyrocalcitonin, plays a role in regulating calcium levels in the body, preventing the accumulation of an overdose* (Science News).

thyroglobulin, *n. Biochemistry.* a protein of the thyroid gland that contains iodine, found in the colloid substance of the gland: *All of these substances are stored in the colloid of the thyroid-follicle lumen. Their molecules are attached to the giant molecules of the protein thyroglobulin, which are so big that they cannot normally penetrate the wall of the follicle and escape into the bloodstream* (Scientific American).

thyroid, *Anatomy.* —*n.* **1** = thyroid gland: *If the thyroid is removed in a young animal, skeletal growth is arrested and sexual maturity fails* (Storer, *General Zoology*). **2** = thyroid cartilage. **3** an artery, vein, etc., near or associated with the thyroid gland.
—*adj.* of or having to do with the thyroid gland or thyroid cartilage: *Thyroid function has far-reaching effects in man's activities. The overactive gland will increase metabolism and nervousness and reduce weight.* (James H. Leathem).
[from Greek *thyreoeides* shieldlike, from *thyreos* door-shaped shield, from *thyra* door]

thyroid cartilage, *Anatomy.* the principal cartilage of the larynx, which forms the Adam's apple in human beings: *Covering the front and sides of the larynx is a large angular cartilage, the two connected sides of which suggested, to the Greek anatomists, a pair of shields fastened together at an acute angle: hence the name thyroid, or "shield-like" cartilage* (Charles K. Thomas).

thyroid gland, *Anatomy.* a large endocrine gland in the neck of vertebrates, near the larynx and upper windpipe, that secretes thyroxine and affects growth and metabolism: *Hyperthyroidism is a metabolic disorder caused by overactivity of the thyroid gland. This sits below the Adam's apple and encircles the front of the windpipe like a small horseshoe* (David Seegal).

thyroid-stimulating hormone, = thyrotropin: *The average pituitary gland of an adult, the scientists found, stores ... about half the amount excreted per day by some patients with underactive thyroid glands* (Science News Letter). *Abbreviation:* TSH

thyrotropic hormone = thyrotropin.

thyrotropin (thī'rō trō'pin), *n. Biochemistry.* a hormone produced by the pituitary gland that regulates the activity of the thyroid gland. Also called **thyrotropic hormone, thyroid-stimulating hormone.**

---

cap, fāce, fäther; best, bē, tèrm; pin, five;
rock, gō, ôrder; oil, out; cup, pùt, rüle,
*y*ü in use, *y*ù in uric;
ng in bring; *sh* in rush; *th* in thin; ᴛʜ in then;
zh in seizure.
ə = *a* in about, *e* in taken, *i* in pencil, *o* in lemon, *u* in circus

669

thyrotropin-releasing factor or **thyrotropin-releasing hormone,** *Biochemistry.* a hormone produced in the hypothalamus that causes the release of thyrotropin. *Abbreviation:* TRF or TRH

thyroxin (thī rok′sən), *n.* = thyroxine.

thyroxine (thī rok′sēn *or* thī rok′sən), *n. Biochemistry.* a hormone secreted by the thyroid gland that regulates the rate of metabolism and, in children, affects growth. It is a white, crystalline amino acid containing iodine. A synthetic form is used to treat goiter and other thyroid disorders. *Formula:* $C_{15}H_{11}I_4NO_4$

thyrse (thėrs), *n.* = thyrsus.

thyrsus (thėr′səs), *n., pl.* **-si** (-sī). *Botany.* a form of mixed inflorescence, a contracted panicle, in which the main ramification is indeterminate and the secondary or ultimate is determinate, as in the lilac and horse chestnut. [from New Latin, ultimately from Greek *thyrsos* staff, stem]

Ti, *symbol.* titanium.

tibia (tib′ē ə), *n., pl.* **tibiae** (tib′ē ē). *Anatomy.* 1 the inner and thicker of the two bones of the leg from the knee to the ankle. Commonly called **shinbone.** See the pictures at **knee, leg, tendon.**
2 a corresponding bone in amphibians, birds, reptiles, and mammals.
3 the fourth section (from the body) of the leg of an insect. [from Latin]
—**tibial** (tib′ē əl), *adj.* of or having to do with the tibia: *the tibial ligament.*

tibiofibular (tib′ē ō fib′yə lər), *adj. Anatomy.* of or having to do with both the tibia and the fibula: *the posterior tibiofibular ligament.*

tidal, *adj.* 1 *Oceanography.* of or having to do with tides. 2 *Physics, Astronomy.* of or having to do with any effect caused by differences in the strength of gravitational forces at different places: *tidal heating of a satellite's interior.*

tidal air, *Physiology.* the air that a person ordinarily inhales and exhales at each breath. Tidal air amounts to about 500 milliliters in adults.

tidal bore, *Oceanography.* a wave which moves inland or upriver as the incoming tidal current surges against the flow of river water. Tidal bores occur in long, narrow estuaries or in the rivers that flow into such estuaries.

tidal current, *Oceanography.* the movement of water toward and away from the coast as the tide rises and falls: *Tidal currents shift the sand back and forth in a more leisurely rhythm. Ocean currents do not usually brush the bottom with sufficient force to carry sand* (Scientific American).

tidal volume, = tidal air.

tidal wave, *Oceanography.* 1 a large wave or sudden increase in the level of water along a shore, caused by unusually strong winds.
2 = tsunami: *Seismic sea waves are commonly and incorrectly called "tidal waves." They are not related to tides, nor are they single huge waves. They roll across the ocean in series up to 20 minutes apart, and the first wave is seldom the largest. Crests, six or seven feet at the highest, race unobserved past ships in deep water, then pile up in the shallow water and crash against the* shore *as waves as much as 100 feet high* (Science News Letter).

3 either of two great swellings of the ocean surface (caused by the attraction of the moon and sun) that move around the globe on opposite sides and cause the tides: *True tidal waves, or tides, are not generally thought of as wave motion ... To understand the tide as a wave phenomenon we must know something of its origin. The regular surge of the waters is caused by the gravitational pull of the sun and moon on the earth. The more important effect is that of the moon* (Garrels, *A Textbook of Geology*).

tide, *n. Oceanography.* 1 the rise and fall of the surface level of the ocean, usually taking place about every twelve hours: *Tides are caused by the gravitational attraction of the sun and the moon on the waters of the earth. When all three of these bodies are in line, as they are both at new and full moon, their attraction is exerted together, and then we have what are termed "spring" tides, with a relatively large range between high and low.* (Science News Letter).
2 the inward or outward flow or current resulting from this on a coast, in a river, etc.: *Most sea-shore animals have a marked rhythm correlated with the state of the tide. At low tide they are inactive, and they are activated by the force of the waves. In the laboratory a rhythm of sensitivity to mechanical shaking persists in periwinkles ... for some days* (J. L. Cloudsley-Thompson).

NEAP TIDE:

SPRING TIDE:

Neap tide occurs during the period of weak gravitational pull of the sun and moon when they are not aligned.
Spring tide occurs during the period of strong gravitational pull of the sun and moon when they are aligned.

tidemark, *n. Oceanography.* a mark left or reached by the tide at high or low water; high-water mark or low-water mark.

tiderace, *n. Oceanography.* a strong tidal current, especially one that flows in a tideway.

tide table, *Oceanography.* a table that lists the times of the high and low tides at a place or places on each day during a particular period.

tidewater, *n. Oceanography.* 1 water in rivers, streams, etc., affected by the rise and fall of the tides. 2 low-lying land along a seacoast through which such water flows.

tideway, *n. Oceanography.* **1** a channel in which a tidal current runs. **2** a strong current running in such a channel.

tiemannite (tē′mə nīt), *n. Mineralogy.* a rare mineral, a selenide of mercury, occurring in dark-gray masses or granules with a metallic luster. *Formula:* HgSe

tiger's-eye, *n. Mineralogy.* a golden-brown mineral with a changeable luster, composed chiefly of quartz colored by iron oxide, and often used as a gemstone.

till, *n. Geology.* glacial drift or deposit of clay, sand, gravel, and boulders: *Some till is largely coarse boulders, but that from thin ice caps eroding shale and limestone may be chiefly clay and silt, with only scattered boulders. Rock fragments found in till are ... subrounded or sharply angular* (Gilluly, *Principles of Geology*). Also called **boulder clay.**

tillite, *n. Geology.* a sedimentary rock composed of compacted or cemented glacial till: *Beds of tillite—old, consolidated glacial rubble—have been studied in known glaciated regions and are unquestioned evidence of the action of deep ice cover* (Patrick M. Hurley).

timberline, *n. Geography, Ecology.* the line on mountains and in polar regions beyond which trees will not grow because of the cold: *Slashed across the tops of tall mountains of the world lies a cruel no-man's land, where howling winds make dwarfs of trees and ice beats grass and flowers to a mat. The sun punishes, ice and rocks avalanche. This is the land above "timberline"—beautiful, but harsh, dangerous, where seemingly nothing can live* (Maclean's). Also called **tree line.**

time, *n.* **1** *Physics.* **a** the indefinite and continuous duration in which events succeed one another: *the absolute, asymmetric time of Newtonian mechanics. It is generally acknowledged that the very concept of time depends upon the possibility of the repetition of events that may be considered identical* (Jacques E. Romain). Compare **arrow of time.** See also **space-time.** **b** duration measured by quanta corresponding to the highest radiation frequency or by the frequency of vibrations of an atom of cesium or the like: *quantized time, atomic time.*
2 *Astronomy.* any system of measuring or reckoning the passage of time based on the rotation of the earth on its axis and its revolution about the sun or in reference to any star: *solar time, sidereal time, Greenwich mean time.*
3 a unit of geological chronology: *Divisions of geologic time are based on the duration of time that corresponds to making of the time-rock units [stage, series, system]. The time unit corresponding to a stage is designated as age, that of a series of an epoch, and that of a system as a period. An era is a long time division comprised of two or more periods* (Moore, *Introduction to Historical Geology*).

time dilation, *Physics.* the apparent slowing down or stretching out of time for a speeding object as its velocity increases relative to another object traveling at a different velocity: *One consequence of the special theory of relativity is that rate of time flow is not the same in two coordinate systems moving relative to one another. This phenomenon, known as time dilation, accounts for the slowing down of high-velocity natural*

clocks as recorded by a stationary observer (New Scientist).

time reversal, *Physics.* the principle that if any physical process is allowed by the laws of nature, the same process with all events reversed must also be allowed. Time reversal is violated by some subatomic processes. *Time reversal is one of the basic symmetries of particle physics. If a movie could be made of a particle process, such as a photon producing a positron and an electron, and shown reversed, a physicist would see an electron and a positron coming together to form a photon, a perfectly plausible occurrence. From what he saw, he would have no way of knowing that the film had been reversed* (Science News).
—time-reverse, *v.* to reverse the order of (a sequence of events): *Because gravity is a one-way force, always attracting and never repelling, it might be supposed that the motions of bodies under the influence of gravity could not be time-reversed without violating basic laws* (Martin Gardner).

time-rock unit, *Geology.* a unit of classification of rocks based on the span of time during which they were deposited. The basic time-rock units are the *stage,* the *series,* and the *system.*

time-symmetric, *adj. Physics.* symmetric with respect to time: *Einstein's general theory of relativity is mathematically time-symmetric: The theory indicates that relativistic processes operate in two opposite directions, like a movie which can be shown running backward as well as forward. This means that while black holes implode and absorb matter and energy, there must also be stars that simultaneously explode and emit matter and energy somewhere in the universe. Color-conscious physicists have named these theoretical stars white holes* (Mort LaBrecque). Compare **arrow of time.**

time zone, a geographical region within which the same standard time is used, especially any one of 24 zones, beginning and ending at the International Date Line, into which the world is divided: *As one moves westward from one time zone to the other, one must move one's watch back an hour* (Norman D. Anderson).

tin, *n. Chemistry.* a soft, silver-white metallic element, highly malleable and having a low melting point, used in plating metals to prevent corrosion and in making alloys such as bronze and pewter: *All the world's tin is obtained from ore-bodies that were formed originally in or in the near neighborhood of rocks of the granite family, and the mineralizing agents which deposited the tin minerals came from an acid igneous magma* (Jones, *Minerals in Industry*). Symbol: Sn; atomic number 50; atomic weight 118.69; melting point 231.9°C; boiling point 2270°C; oxidation state 2,4.

---

cap, fāce, fäther; best, bē, tèrm; pin, five;
rock, gō, ôrder; oil, out; cup, pùt, rüle,
yü in use, *yù* in uric;
ng in bring; *sh* in rush; *th* in thin, ᴛʜ in then;
zh in seizure.
ə = *a* in about, *e* in taken, *i* in pencil, *o* in lemon, *u* in circus

tincal (ting′kəl), *n. Mineralogy.* a former name for crude native borax. [from Malay *tingkal*]

tincalconite (tin kal′kə nīt), *n. Mineralogy.* a colorless to white borate mineral, one of the main sources of boron. *Formula:* $Na_2B_4O_7 \cdot 5H_2O$

tissue, *n. Biology.* a mass of similar cells and their intercellular substance, working together to perform a particular function, and forming the organs and other structural parts of an organism: *muscle tissue, nerve tissue, connective tissue. Plant tissues include the epidermis, parenchyma, xylem, and phloem. Most* [*animal*] *tissues have pain receptors, with the exception of the brain, the cartilage of joints and the dense part of bone* (Martin E. Spencer). [from Old French *tissu,* originally past participle of *tistre* to weave, from Latin *texere*]

tissue culture, *Biology.* **1** the technique or process of keeping bits of animal tissue alive and growing in a sterile, nutrient medium.
2 the culture or medium in which tissue is grown: *Recently, scientists reported that cells from 12 human sarcoma-type cancers were grown in tissue cultures* (*mammalian cells cultured "in the test tube" outside of the body of the animal or human from which they were derived*) (Robert G. Eagon).

tissue respiration, = internal respiration: *The second part of respiration is called tissue respiration. In this process, the cells use the oxygen to burn the food fuels. In addition to supplying energy for tissue building, tissue respiration also supplies the heat that man and warm-blooded animals need to maintain body temperature* (Fritz Lipmann).

titanic (tī tan′ik), *adj.* of or containing titanium, especially with a valence of four: *titanic oxide, titanic iron.*

titaniferous (tī′tə nif′ər əs), *adj.* containing or yielding titanium: *Titaniferous ore is usually a black, granular mixture of ilmenite and magnetite.*

titanite (tī′tə nīt), *n.* = sphene. [from German *Titanit,* from *Titanium* titanium + *-it* -ite]

titanium (tī tā′nē əm *or* tī tā′nē əm), *n. Chemistry.* a strong, lightweight, silver-gray metallic element occurring in rutile, ilmenite, brookite, and various other minerals. It is highly resistant to corrosion and is used widely in making alloys because it unites with nearly every metal except copper and aluminum. *Symbol:* Ti; *atomic number* 22; *atomic weight* 47.90; *melting point* ab. 1660°C; *boiling point* 3287°C; *oxidation state* 2, 3, 4. [named after *Titan,* one of a family of giants in Greek mythology]

titanous (tī tan′əs *or* tī′tə nəs), *adj.* of or containing titanium, especially with a valence of three: *titanous oxide, titanous fluoride.*

titer (tī′tər), *n. Chemistry.* **1** the amount of a standard solution necessary to produce a certain result in titration.
2 the weight of a pure substance which is contained in, would react with, or would be equivalent to, a unit volume of a reagent solution, usually expressed in milligrams of solute per milliliter of solution. Also spelled **titre.**
[from French *titre* proportions in alloyed metal, quality, (originally) title, from Latin *titulus*]

titrant (tī′trənt), *n.* the substance added in titration: *With this apparatus, titrant can be added at rates of up to 10 cc a minute; thus, for the quantities involved, titrations can be completed in 20 seconds, whereas, previously they took about 2 minutes* (New Scientist).

titrate (tī′trāt), *Chemistry.* —*v.* to analyze or be analyzed by titration: *In titrating iron solutions, the ferrocyanide is not used* (G. E. Davis).
—*n.* a solution to be analyzed by titration.

titration (tī trā′shən *or* ti trā′shən), *n. Chemistry.* the process of determining the amount of some substance present in a solution by measuring the amount of a different solution of known strength that must be added to complete a chemical change: *One of the most common methods of determining the concentration of a solution of unknown value is that of titration. This method consists of measuring carefully the volume of a solution of known concentration which is required to react with an equally carefully measured volume of a solution of unknown concentration* (Jones, *Inorganic Chemistry*).

titre (tī′tər), *n.* = titer.

titrimetric (tī′trə met′rik), *adj.* of or having to do with measuring by titration: *titrimetric analysis.* —**titrimetrically,** *adv.*

titrimetry (tī trim′ə trē), *n.* measured by titration.

Tl, *symbol.* thallium.

T lymphocyte, = T cell.

Tm, *symbol.* thulium.

tn., *abbrev.* ton.

tocopherol (tō kof′ə rōl), *n. Chemistry.* any of four closely related alcohols associated with, or one of the components of, vitamin E, important as an antisterility factor in the diet, and present in wheat germ and certain other vegetable oils, milk, and lettuce and other plant leaves. The most potent tocopherol is the alpha-form, $C_{29}H_{50}O_2$. [from Greek *tokos* offspring + *pherein* to bear, produce + English *-ol*]

toe, *n. Anatomy.* one of the end parts or digits of the foot of a vertebrate.

tokamak (tō′kə mak), *n. Physics.* a device for producing controlled thermonuclear power, in which highly ionized gas is confined in an endless tube by magnetic fields generated by electric currents outside the tube and inside the gas itself: *Recently the tokamaks have produced plasmas nearer to fusion conditions than any other devices have been able to do* (Science News). Compare **stellarator.** [from Russian]

toluene (tol′yü ēn), *n. Chemistry.* a colorless, flammable, aromatic liquid hydrocarbon with a smell like that of benzene, obtained from coal tar and petroleum and used as a solvent and for making explosives, dyes, etc. *Formula:* $C_6H_5CH_3$ [from *tolu* a South American balsam]

toluic acid (tə lü′ik), *Chemistry.* a colorless carboxylic acid homologous with benzoic acid, found in four isomeric forms, derived from toluene or xylene. *Formula:* $CH_3C_6H_4CO_2H$

toluidine (tə lü′ə dēn), *n. Chemistry.* a compound analogous to aniline, found in three isomeric forms. Toluidine is a toluene derivative used especially in making dyes. *Formula:* $CH_3C_6H_4NH_2$

tombolo (tom′bə lō), *n. Geology.* a sand bar which connects an island to the mainland or to another island: *Wave transport frequently links a small offshore island to the mainland by a sand bar, known as a tombolo. The transported sand may come from the mainland or from the island, depending on the wave action and the nature of the terrain* (Scientific American). [from Italian, from Latin *tumulus* mound of earth]

tomentum (tō men′təm), *n., pl.* **-ta** (-tə). **1** *Botany.* a soft down consisting of long, soft, entangled hairs, pressed close to the surface: *leaves covered with white, silky tomentum.*
2 *Anatomy.* a downy covering, especially the flocculent inner surface of the pia mater, consisting of numerous minute vessels entering the brain and spinal cord.
[from New Latin, from Latin *tomentum* cushion stuffing]

-tomy, *combining form.* **1** surgical incision or operation, as in *tracheotomy, lobotomy.* **2** a cutting or casting off, as in *autotomy = a casting off of part of the body.* [from Greek *-tomia* a cutting]

ton, *n.* **1** a customary unit of weight equal to 2000 pounds or 907.18 kilograms (**short ton**) in the United States and Canada, 2240 pounds or 1,016.05 kilograms (**long ton**) in Great Britain.
2 = metric ton.
3 a measure of volume that varies with the thing measured; it is about equal to the space occupied by a ton's weight of the particular stuff. Thus a ton of stone is 16 cubic feet; a ton of lumber is 40 cubic feet.
4 a unit of refrigeration equal to the amount of heat needed (200 B.T.U.'s per minute) to melt one short ton of ice in 24 hours: *A one-ton conditioner can remove 288,000 B.T.U.'s of heat in 24 hours from the air it is conditioning* (Clarence L. Ringquist). *Abbreviation:* tn.
[from Old English *tunne* a measure of capacity]

tonalite (tō′nə līt), *n. Geology.* a granular, plutonic igneous rock, a variety of diorite, containing quartz, plagioclase, and biotite or hornblende. Also called **quartz-diorite.** [from *Tonale,* a pass in the Tyrol, where it is found + *-ite*]

tone, *n. Physiology.* the degree of firmness or tension normal to the organ or tissue when healthy; the normal state of tension in the body or an organ. See **tonic, tonus.**

tongue, *n.* **1** *Anatomy.* **a** the movable fleshy organ in the oral cavity of most vertebrates, having taste buds that make it the chief organ of taste, used in taking in and swallowing food, and serving in humans as an essential organ of speech: *The tongue is a complex combination of muscles which is attached in front to the chin bone and in back to the hyoid bone in the upper part of the neck* (Charles K. Thomas). See the pictures at **mouth, uvula. b** a similar organ or part in an invertebrate animal.
2 *Geography, Geology.* **a** a narrow strip of land jutting out into water. **b** a long, narrow extension of a glacier or lava flow.

tonic, *adj. Physiology, Medicine.* **1** of or having to do with the normal tone or tension of an organ or body. A **tonic reflex** is a reflex involved in maintaining the tone of an organ or body.
2 characterized by continuous contraction of the muscles: *a tonic convulsion or spasm.*

tonne (tun), *n. Especially British.* = metric ton.

tonometer (tō nom′ə tər), *n.* an instrument for measuring tension or pressure: *A tonometer is often used to detect glaucoma. The instrument measures intraocular pressure* (Richard Martin).
—**tonometric,** *adj.* of or having to do with a tonometer or tonometry.
—**tonometry,** *n.* measurement with a tonometer: *The measurement of the pressure within the eyeball, tonometry, is determined by placing the tonometer directly upon the cornea which has been made insensitive by the application of one or two drops of a local anesthetic. The procedure is routine practice in the determination of glaucoma* (Eugene M. Blake).

tonoplast (tō′nə plast), *n. Botany.* a thin membrane that surrounds the vacuole of a plant cell: *The tonoplasts separate cytoplasm and vacuole* (Weier, *Botany*). Also called **vacuolar membrane.** [from Greek *tonos* tension + English *-plast*]

tonsil, *n. Anatomy.* **1** either of the two oval masses of lymphoid tissue on the sides of the throat, just back of the oral cavity: *No one really knows the purpose of tonsils, but many medical scientists believe they aid in protecting the respiratory and digestive systems from infection* (James M. Toomey).
2 any of several similar masses at the back of the throat: *The* **pharyngeal tonsils** *are situated near the nasal passage and are better known as the adenoids. The* **lingual tonsils** *are at the back of the tongue.*
[from Latin *tonsillae* tonsils]
—**tonsillar** (ton′sə lər), *adj.* of or having to do with tonsils: *the tonsillar arches, the tonsillar plexus.*

tonus (tō′nəs), *n. Physiology.* bodily or muscular tone: *Tonus is maintained involuntarily through the contraction of only a few cells of the entire muscle. Eventually these cells fatigue, and tonus is lost. The tonus of an opposing muscle causes the muscle whose tonus is lost to be stretched. The stretching stimulates nerves ... to carry impulses which stimulate another group of cells, and tonus is restored. The nodding of and backward jerking of the head as one becomes drowsy illustrates the loss and restoration of tonus in certain neck muscles* (Harbaugh and Goodrich, *Fundamentals of Biology*). [from Latin, from Greek *tonos* tension, tone]

tooth, *n.* **1** *Anatomy.* one of the hard, bonelike parts in the mouth of vertebrates, attached in a row to each jaw, and used for biting and chewing food and as weapons of attack or defense.
2 *Zoology.* any of certain hard parts or processes in the mouth or digestive tract of invertebrates.

---

cap, fāce, fäther; best, bē, tèrm; pin, fīve;
rock, gō, ôrder; oil, out; cup, pùt, rüle,
yü in use, *yù* in uric;
ng in bring; *sh* in rush; *th* in thin, ᴛʜ in then;
zh in seizure.
ə = *a* in about, *e* in taken, *i* in pencil, *o* in lemon, *u* in circus

3 *Botany*. one of the projections at the margin of certain leaves or petals.

tooth:

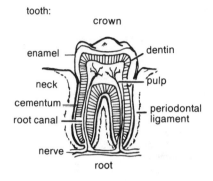

top, *Chemistry*. —*n*. the part in a distillation that volatilizes first.
—*v*. (in a distillation) to remove the part that volatilizes first.

topaz, *n. Mineralogy*. 1 a hard, transparent or translucent mineral that occurs in crystals of various forms and colors. Transparent yellow, pink, or brown varieties are used as gems. *Formula:* $Al_2SiO_4(F,OH)_2$
2 a yellow variety of sapphire or quartz, used as a gem. [from Greek *topazos*]

top cross, *Genetics*. the offspring produced when a variety is crossed with one inbred line.

topographic, *adj*. of or having to do with topography. A topographic map shows elevations as well as the positions of mountains, rivers, etc., often in color and with contour lines: *Topographic maps are used in the laboratory to observe and analyze the effects of the several geologic processes which are constantly changing the face of the earth* (Leet and Judson, *Physical Geology*).

topographical, *adj*. = topographic. —topographically, *adv*.

topography (tə pog′rə fē), *n*. 1 *Geography*. a the surface features of a place or region, including hills, valleys, streams, lakes, bridges, tunnels, roads, etc.: *The topography of the continents has been shaped in large part by river and wind erosion* (Scientific American). *Arctic topography is distinctive and varied. The absence of forest cover also contributes to unusual geomorphological processes* (John C. Reed). b a detailed description or drawing of the surface features of a place or region: *a topography of the lunar maria*. c the science of making an accurate and detailed description or drawing of places or their surface features: *Surveying is part of topography*.
2 the surface features of any object or structure: *the topography of a quartz crystal*.
3 *Anatomy*. the study of the structural features or regions of the body or its parts: *cerebral topography*. [from Greek *topos* place + English -*graphy*]

topological, *adj*. of or having to do with topology. A topological transformation in geometry is a one-to-one correspondence between the points of two figures. If one figure can be transformed into another by such a one-to-one correspondence, the two figures are considered topologically equivalent. —topologically, *adv*.

topology (tə pol′ə jē), *n*. 1 *Geometry*. the study of those properties of figures or solids that are not normally affected by changes in size or shape: *Topology makes no distinction between a sphere and a cube, because these figures can be deformed or molded into one another. Topology makes a distinction between a sphere and a torus (a doughnut-shaped figure) because these cannot be deformed into one another* (Howard W. Eves).
2 *Geography*. the topographical study of a particular locality in order to learn the relationship between its topography and its history.
3 *Anatomy*. the structure of a certain area or part of the body in relation to other areas or parts.
4 the surface structure or arrangement of parts of any object: *the topology of a plant cell*.
[from Greek *topos* place + English -*logy*]

top quark, *Nuclear Physics*. a hypothetical quark having a very large mass: *So far the sixth quark (sometimes called the top quark) has not been found despite some expensive and high-powered accelerator experiments designed to reveal it, and its absence is puzzling* (Kendrick Frazier). Also called t quark. Compare bottom quark.

topsoil, *n. Geology*. 1 the upper part of the soil; surface soil, as opposed to subsoil: *Topsoil containing humus is much desired by the farmer and the gardener, for humus soil holds more water and air than soil without it* (Science News Letter).
2 loam or other earth from or in this part of the soil, usually consisting of sand, clay, and decayed organic matter.

tor (tôr), *n. Geography*. a high, rocky, or craggy hill, knoll, etc. [Old English *torr*]

torbernite (tôr′bər nīt), *n. Mineralogy*. a radioactive mineral, a hydrated phosphate of uranium and copper, found in bright-green, tabular crystals. It is a minor ore of uranium. *Formula:* $Cu(UO_2)(PO_4)_2 \cdot 8\text{-}12H_2O$ [from *Torbern* Bergman, 1735–1784, Swedish chemist + -*ite*]

toric (tôr′ik), *adj*. of, having to do with, or shaped like a torus. A toric lens is an optical lens with a surface forming a part of a geometrical torus, used in eyeglasses because it refracts differently in different meridians. *Astigmatism ... can be corrected by a toric lens, which compensates for such variations in curvature* (Harry S. Gradle).

tornadic, *adj. Meteorology*. of, having to do with, or resembling a tornado.

tornado, *n., pl*. -does or -dos. *Meteorology*. 1 an extremely violent and destructive whirlwind extending down from a mass of dark clouds as a twisting funnel and moving over the land in a narrow path: *In an average year 145 tornadoes occur in the United States, the greatest concentration of them in the Midwest and Southeast. Although their exact cause is unknown, they are usually formed when cold, dry air, moving down from the north, overruns abnormally warm moist air moving up from the Gulf of Mexico. Under certain conditions the comparatively lighter warm air is sucked upward in great rushes. Cooling and condensation of the moist air follows, and this in turn releases latent heat and energy.* (N. Y. Times).
2 a violent, whirling squall occurring during the summer on the west coast of Africa.

[alteration of Spanish *tronada,* from *tronar* to thunder, from Latin *tonare*]

toroid (tôr′oid), *n. Geometry.* **1** a surface described by the revolution of any closed plane curve about an axis in its own plane. Compare **torus.**

2 a solid enclosed by such a surface: *The external field of a toroid is represented by the field of no poles, and is zero* (Shortley and Williams, *Elements of Physics*). [from *torus*]

—**toroidal** (tô roi′dəl), *adj.* of, having to do with, or characteristic of a toroid or torus: *The inside of a toroidal, or doughnut-shaped, vacuum chamber is flooded with electrons of modest energy* (Scientific American). —**toroidally,** *adv.*

torose (tôr′ōs), *adj.* **1** *Botany.* cylindrical, with bulges or constrictions at intervals; swelling in knobs at intervals: *a torose pericarp.*

2 *Anatomy.* protuberant; knobbed: *torose muscles.* [from Latin *torosus* bulging, from *torus* a bulge]

torpid (tôr′pid), *adj. Zoology.* characterized by torpor: *the torpid state of hibernation.*

torpor (tôr′pər), *n. Zoology.* the dormant, inactive condition of hibernating animals. [from Latin, from *torpere* be numb]

torque (tôrk), *n. Physics.* **1** a force or system of forces causing rotation or torsion: *A single loop or a solenoid carrying a current experiences a torque in a magnetic field. This torque applied to a coil provides the basis for the operation of the galvanometer and the electric motor* (Physics Regents Syllabus).

2 a measure of the effect of such a force or forces, equal to the product of the force and the perpendicular distance from the axis of rotation or torsion to the line of action of the force: *If forces are expressed in pounds and distances in feet, the unit of torque is one pound-foot. One pound-foot is the torque produced by a force of one pound at a perpendicular distance of one foot from an axis* (Sears and Zemansky, *University Physics*).
[from Latin *torquere* to twist]

torr (tôr), *n.* a unit of pressure approximately equivalent to the amount of pressure that will support a column of mercury one millimeter high; 1/760 of an atmosphere or 1333.32 pascals: *Using the best pumps and the most exquisite care in the design and operation of the systems, they could reach pressures approaching a hundred-millionth of a torr* (Scientific American). [from Evangelista *Torr(icelli*), 1608–1647, Italian mathematician and physicist]

Torricelli's law or **Torricelli's theorem** (tôr′ə sel′ēz *or* tôr′ə chel′ēz), *Physics.* a principle in hydrodynamics which states that the velocity with which a liquid flows through an opening in a container equals the velocity of a body falling freely from the surface of the liquid to the opening. [named after E. *Torricelli* (see **torr**), who developed the principle]

Torrid Zone, *Geography.* (in former climatic classifications) the very warm region between the tropic of Cancer and the tropic of Capricorn. The equator divides the Torrid Zone. Compare **Frigid Zone, Temperate Zone.** See the picture at **zone.**

torsion, *n. Physics.* the twisting or turning of a body by two equal and opposite forces; the deformation of a body by a twisting force about the body's axis of symmetry: *The ... torsion of a straight cylindrical rod or*

wire is a strain that consists of pure, though inhomogeneous, shear (Furry, et al., *Physics*). [from Late Latin *torsionem* torture, ultimately from Latin *torquere* to twist]

—**torsional,** *adj.* of, having to do with, or resulting from torsion: *When the vibrations are in a curve whose plane is at right angles to the direction of wave propagation, the vibrations are called torsional vibrations and the wave a torsional wave* (Glathart, *College Physics*).—**torsionally,** *adv.*

torso, *n. Anatomy.* the trunk of the human body; the body without head and limbs. [from Italian (originally) stalk, from Latin *thyrsus*]

torus (tôr′əs *or* tō′rəs), *n., pl.* **toruses, tori** (tôr′ī *or* tō′rī).

1 *Geometry.* **a** a doughnut-shaped surface described by the revolution of a conic section: *All the properties of knots are extrinsic properties of toruses, (or, if you prefer, one-dimensional curves that may be thought of as toruses whose meridians have shrunk to points) that are embedded in* [three-dimensional space] (Martin Gardner). **b** the solid enclosed by such a surface: *The torus is rotated around its axis which is in a horizontal plane. The material to be purified is melted in the torus so that it fills the lower part of the tube* (New Scientist). **c** = toroid.

2 *Anatomy.* a smooth, rounded swelling or protuberant part. SYN: ridge.

3 *Botany.* **a** the receptacle of a flower. **b** a thickened membrane which closes the bordered pit of a wood cell.

[from Latin *torus* a bulge]

total eclipse, *Astronomy.* an eclipse of the sun or moon in which the whole of the disk is obscured: *A total eclipse of the sun occurs when the umbra of the moon's shadow falls on the earth* (Baker, *Astronomy*). Compare **annular eclipse.**

total heat, = enthalpy.

totality, *n. Astronomy.* **1** total obscuration of the sun or moon in an eclipse; total eclipse.

2 the time or duration of this.

totipalmate (tō′tə pal′māt), *adj. Zoology.* having all four toes completely webbed, as a pelican does. [from Latin *totus* whole + English *palmate*]

—**totipalmation,** *n.* the condition of being totipalmate.

totipotence (tō tip′ə təns) or **totipotency** (tō tip′ə tən sē), *n. Biology.* the quality of being totipotent, especially the capacity of a cell to develop into a complete organ or organism: *All these facts emphasize the totipotency of the freely suspended cells derived from carrot phloem. These vegetative diploid cells behave like zygotes, and remarkable parallels have been drawn between the way free-cell clusters grow and the way in which embryos form from zygotes* (F. C. Steward).

---

cap, fāce, fäther; best, bē, tèrm; pin, five;
rock, gō, ôrder; oil, out; cup, pùt, rüle,
yü in use, yủ in uric;
ng in bring; sh in rush; th in thin; ᴛʜ in then;
zh in seizure.

ə = *a* in about, *e* in taken, *i* in pencil, *o* in lemon, *u* in circus

totipotent, *adj. Biology.* capable of developing into a complete organism, or of differentiating into any kind of cell or tissue: *Teratomas contain many different tissues, including skin, muscles, fat, glands, hair and teeth, all derived from "totipotent" cells: cells that by definition are capable of developing into all types of body tissue* (Scientific American). *It is thought by some that the interstitial cells represent a sort of embryonic tissue carried over into the adult, and that the cells are totipotent, that is, they can differentiate into any of the specialized cells of the hydra* (Hegner and Stiles, *College Zoology*). [from Latin *totus* whole + English *potent*]

tourmaline (tùr′mə lin *or* tùr′mə lēn), *n. Mineralogy.* a chemically complex mineral found in granite and pegmatite. Tourmaline is chiefly a silicate of boron and aluminum with varying amounts of other elements, especially iron, calcium, and sodium. It may be black, brown, red, pink, green, blue, or yellow, and some varieties are used as semiprecious gemstones. [from French, from Singhalese *tormalli* carnelian]

toxic, *adj.* **1** caused by a poison or toxin: *toxic anemia.* **2** producing a poison or toxin: *toxic bacteria.*

toxicity, *n.* **1** toxic or poisonous quality. **2** the amount of toxin or poison found in a substance or produced by an organism. **3** the potency of a toxic substance.

toxicology, *n.* **1** the science that deals with poisons and their effects, antidotes, detection, etc. **2** toxicological properties or characteristics: *The practical determination of the hazard index is by no means simple—the researcher must know a good deal about the toxicology and the environmental behaviour of the chemical* (New Scientist). **—toxicological,** *adj.* of or having to do with toxicology.

toxigenic, *adj.* producing toxin: *Temperatures between 44° and 50°F. are safe for short periods if the food is not grossly contaminated with toxigenic or infectious microorganisms* (Isabel Noble). **—toxigenicity,** *n.* the degree to which an organism produces toxin.

toxin (tok′sən), *n. Biochemistry.* any poison formed by a living organism as a result of its metabolism. Toxins formed by bacteria cause diseases such as diphtheria and scarlet fever. The body reacts to some toxins by producing antitoxins. *The microbes occur in milk right from the udder and in warm milk they multiply rapidly and form a toxin so resistant to heat that it cannot be neutralized by boiling* (Maclean's). Compare **endotoxin, exotoxin, phytoxtoxin, zootoxin.** [from *tox(ic)* + *-in*]

toxin-antitoxin, *n. Immunology.* a mixture of a toxin with a corresponding antitoxin, used formerly to immunize against a disease such as diphtheria.

toxoid (tok′soid), *n. Immunology.* a toxin chemically treated so that it will lose its toxic quality but retain its antigenic properties. Toxoids are used for immunization against diphtheria and tetanus.

TPN, *abbrev.* triphosphopyridine nucleotide.

t quark, = top quark: *Theory right now envisions six, and to those four it adds a t quark (prosaically called "top," but the more philosophically inclined say "truth") and a b quark ("bottom" or "beauty")* (Science News).

tr., *abbrev.* trace.

trabecula (trə bek′yə lə), *n., pl.* **-lae** (-lē *or* -lī). **1** *Anatomy, Zoology.* a small beam, bar, fibrous band, etc., in the structure of a bodily part or organ: *Medullary bone ... occurs in the form of trabeculae, or fine spicules, which grow out into the marrow cavity from the inner surface of the structural bone* (T. G. Taylor). **2** *Botany.* a projection extending across the cell cavity in the ducts of some plants, or across the cavity of the sporangium in mosses. [from Latin, diminutive of *trabs* beam] **—trabecular,** *adj.* of or having to do with a trabecula; forming or formed by trabeculae: *trabecular tissue.* **—trabeculate** (trə bek′yə lit), *adj.* having a trabecula or trabeculae: *a trabeculate organ or part.*

trace, *n.* **1** *Chemistry.* an amount of some constituent in a compound, etc., usually too small to be measured: *traces of sulfur and nitrogen oxides in pollutants. Micronutrients ... are sometimes added in traces to fertilizers* (Scientific American). **2** *Botany.* a vascular bundle leading to a leaf or bud: *Xylem of traces is continuous with xylem in leaf and stem, and phloem of the trace is continuous with phloem in leaf and stem* (Weier, *Botany*). **—adj.** *Chemistry.* consisting of a trace; too small to be measured: *a trace compound, trace gases, a chemical found in trace amounts in the body. Uranium is the most highly concentrated trace metal found in the miners' lungs* (Franklin J. Tobey, Jr.). See also **trace element.**

trace element, *Chemistry.* an element present in small amounts, especially one used by an organism and considered necessary to its proper functioning: *Tiny amounts of metals such as iron, copper, zinc, manganese and molybdenum, called trace elements, can upset the delicate balance between health and sickness* (Science News Letter). *Plants need large amounts of 10 chemical elements—... Plants also need five other chemicals, but in such tiny amounts that they are called trace elements. These are copper, boron, zinc, manganese, and cobalt* (Joseph E. Howland).

tracer, *n. Chemistry.* an element or atom, usually radioactive, that can be traced and observed as it passes through a body, plant, or other system in order to study biological processes or chemical reactions within the system: *Scientists are employing tracers, or so-called "tagged" molecules, to determine exactly how the living cell uses its food and energy to grow and to carry out all its other vital processes* (Norbert Dernbach). Also called **label.**

trachea (trā′kē ə), *n., pl.* **tracheae** (trā′kē ē′), **tracheas.** **1** *Anatomy.* the tube in air-breathing vertebrates extending from the larynx to the bronchi, by which air is carried to and from the lungs: *The trachea is kept open by rings of cartilage. The ciliated mucous membrane which lines the trachea traps microscopic particles and sweeps them toward the pharynx* (Biology Regents Syllabus). Commonly called **windpipe.** See the pictures at **bronchi, mouth.** **2** *Zoology.* one of the air-carrying tubes of the respiratory system of insects and other arthropods: *The bodies of insects and related arthropods have a series of openings, or spiracles, through which air is pumped*

to and from an internal branched system of air ducts, or tracheae, by movements of the abdomen (Storer, *General Zoology*).
3 *Botany.* **a** a tube in the xylem of some vascular plants, especially the angiosperms, formed by modified tracheids with open end walls connected end to end, permitting the passage of water and dissolved minerals. Tracheae are covered with various markings or thickenings, the spiral being the common type. Also called **vessel, duct. b** = tracheid.
[from Late Latin *trachia* windpipe, from Greek *tracheia* (*arteria*) rough (artery)]
—**tracheal** (trā′kē əl), *adj.* of, having to do with, or using the trachea: *the tracheal cartilages, tracheal respiration.*

tracheary (trā′kē er′ē), *adj. Botany.* of or having to do with tracheae or tracheids: *At the structural level, permanganate has been shown to act as an electron-dense stain for the lignin component of the cell wall of tracheary elements* (Donald E. Fosket).

tracheate (trā′kē it), *Zoology.* —*adj.* having tracheae: *tracheate arachnids.*
—*n.* a tracheate arthropod: *The insects and other tracheates* (*centipedes, millipedes, certain spiders, and so on*) *accomplish this by an amazingly efficient "tracheal system" of tubes and tubules* (Scientific American).

tracheid (trā′kē əd), *n. Botany.* an elongated cell with thick, perforated walls, that serves to carry water and dissolved minerals through a plant, and provides support. Tracheids form an essential element of the xylem of vascular plants. *Ducts and tracheids are essentially thin-walled, thus facilitating diffusion, but portions of the side walls are thickened by bands laid down in the form of spirals or rings. The pattern made by these reinforcing bands is responsible for such distinguishing terms as spiral or annular ducts or tracheids* (Harbaugh and Goodrich, *Fundamentals of Biology*).

tracheobronchial, *adj. Anatomy.* having to do with the trachea and the bronchi: *Breathing heavily polluted fog may start acute inflammation in the lining of the tracheobronchial tree and it is likely to cause exacerbations in persons whose bronchitis is chronic* (Manchester Guardian).

tracheole (trā′kē ōl), *n. Zoology.* one of the tiny branches of the trachea of an insect: *The finest branches of this ramifying system are called tracheoles; they are present most abundantly in the organs that have the highest oxygen requirement—notably the flight muscles* (David S. Smith).

tracheophyte (trā′kē ə fīt), *n. Botany.* any of a large division (Tracheophyta) of plants whose stems, leaves, and roots have well-developed vascular tissues; a vascular plant. Compare **bryophyte, pteriolophyte.** [from *trache*(a) + *-phyte*]

trachyte (trā′kīt or trak′īt), *n. Geology.* a light-colored volcanic rock consisting chiefly of alkali feldspar, and also containing augite, biotite, or hornblende. [from French, from Greek *trachys* rough]
—**trachytic** (trə kit′ik), *adj.* (of rock) having densely packed prisms of alkali feldspar lying parallel to each other.

trade wind, *Meteorology.* a wind blowing toward the equator from about 30 degrees north latitude and about 30 degrees south latitude. North of the equator,

it blows from the northeast; south of the equator, from the southeast.

tragus (trā′gəs), *n., pl.* **-gi** (-jī). *Anatomy.* the bulge that partially conceals the external opening of the ear. [from New Latin, from Greek *tragos* (originally) goat]

trajectory, *n.* **1** *Physics.* the curve described by a projectile or other body moving through space: *The trajectory is affected to a large extent by air resistance, which makes an exact analysis of the motion extremely complex* (Richards, *Modern College Physics*).
2 *Geometry.* a curve or surface that passes through a given set of points or intersects a given series of curves or surfaces at a constant angle.

tramontane (trə mon′tān or tram′ən tān), *Meteorology.*
—*adj.* (of the wind) coming across or from beyond the mountains, especially blowing over Italy from beyond the Alps.
—*n.* any cold wind from a mountain range.
[from Italian *tramontana* north wind, from Latin *transmontanus* beyond the mountains]

trans (trans or tranz), *adj. Chemistry.* of or having to do with an isomeric compound that has certain atoms on the opposite sides of a plane: *a trans configuration or structure.* Compare **cis.** [from *trans-*]

trans-, *prefix.* **1** across or beyond, as in *translunar, transfinite, transmedian.*
2 *Chemistry.* having certain atoms on opposite sides of the plane: *a trans-isomeric compound.* Contrasted with **cis-.** [from Latin *trans* across]

transactinide (trans ak′tə nīd), *adj. Chemistry.* of or belonging to the series of elements whose atomic numbers extend beyond the actinide series: *the transactinide elements 104 through 109. Element 104 is the first element in a region of the periodic table that Glenn T. Seaborg, the codiscoverer of elements 94 through 102 ... , has called the transactinide region* (Scientific American). *The transactinide series ... should extend from 104 to 112* (Science News).

transaminase, *n. Biochemistry.* an enzyme that catalyzes transamination.

transaminate, *v. Chemistry.* to cause the reversible transfer of (an amino group) from one compound to another: *The group that was selectively altered included amino-acids, for example, valine, that are not transaminated or oxidized in myocardium and those such as alanine that can be transaminated and used for energy production* (Nature).
—**transamination**, *n.* the reversible transfer of an amino group from one compound to another.

transcendental, *Mathematics.* —*adj.* not capable of being the solution of a polynomial equation with rational coefficients. π is a transcendental number. *A number is called algebraic if it is a solution of an equation such as $a + bx = 0$ ... where a, b, c, d, etc. stand for ordi-*

---

cap, fāce, fäther; best, bē, tèrm; pin, five;
rock, gō, ôrder; oil, out; cup, pùt, rüle;
yü in use, *yù* in uric;
ng in bring; *sh* in rush; *th* in thin, ᴛʜ in then;
zh in seizure.
ə = *a* in about, *e* in taken, *i* in pencil, *o* in lemon, *u* in circus

nary whole numbers ... *A number that is not algebraic is called transcendental* (Paul R. Halmos).
—*n.* a transcendental term, quantity, or number.
—**transcendentally,** *adv.*

transcortical, *adj. Anatomy.* crossing the cortex of the brain.

transcribe, *v. Molecular Biology.* to form or synthesize (a nucleic acid molecule or molecules) by transferring genetic information from a template: *to transcribe DNA into messenger RNA, to transcribe RNA from DNA.*

transcriptase (tran skrip′tās), *n.* = reverse transcriptase.

transcription, *n. Molecular Biology.* the process of forming a nucleic acid molecule by using a template of another molecule: *Gene transcription, whereby enzyme reactions mediate the synthesis of RNA molecules from DNA templates, has been investigated mostly in microbial organisms* (Nature).
—**transcriptional,** *adj.* of, having to do with, or occurring in transcription: *a transcriptional error.*
—**transcriptionally,** *adv.*

transduce, *v.* **1** *Biology.* to cause (a gene or genetic material) to be transferred from one bacterial cell to another by bacteriophages: *This analysis has been facilitated by ... the existence of transducing phages and plasmids which have the natural capability for moving chromosomal segments from one bacterium to another of the same or closely related species* (David M. Glover).
2 *Physics.* to convert (energy) from one form to another, as from heat energy to electric energy, especially for purposes of measurement: *Industrial measurement and control systems ... integrate fluctuating light, velocity, flow, and other factors which can be transduced to electric current* (Science).
[from Latin *transducere* to lead across]

transducer, *n. Physics.* any device for converting energy from one form to another, as from acoustic energy to electric or mechanical energy. Loudspeakers and microphones are examples of transducers.

transduction, *n.* **1** *Biology.* the transfer of a gene or genetic material from one bacterial cell to another by bacteriophages: *A bacterial virus can carry the bacteria's own genes from cell to cell. Like a disease carrier, it infects one bacterium with hereditary material picked up from another. This "transduction" of bacterial heredity was discovered by accident* (Scientific American). Compare **transfection, transformation.** See also **transposon.**
2 *Physics.* the conversion of energy from one form to another: *Transduction occurs when a loudspeaker changes electrical into acoustical energy.*

transect (tran′sekt), *n. Botany.* **1** a cross section of the vegetation of an area, usually that part growing along a long, narrow strip: *For the botanist a brisk 15 miles between, say, Mumbles Head to Rhossili Bay in Glamorganshire is an eventful transect, abounding in rare flora* (New Scientist).
2 a representation of such a cross section.

transfect (trans fekt′), *v. Microbiology.* to infect (a bacterial cell) with the nucleic acid of a virus so as to cause the cell to produce the virus: *Scientists employed the DNA from an animal virus, vaccinia, to transfect com-*

petent cells of B[acillus] subtilis. They obtained evidence for the production of new infective viruses within the cytoplasm of the bacterial cells (P. Sypherd). [from *trans-* + (*in*)*fect*]

transfection, *n.* the process of transfecting or the condition of being transfected; the production of viruses in bacterial cells infected with viral nucleic acid. Compare **transduction, transformation.**

transferase (trans′fǝr ās), *n. Biochemistry.* any enzyme that catalyzes the transfer of a phosphate, amine, methyl, or other group from one molecule to another.

transfer cell, *Botany.* a specialized plant cell which exchanges dissolved substances with its surroundings and transfers them across the plant membranes: *Transfer cells ... are widespread among flowering plants, conifers, ferns, mosses, and even lichens. Their distinguishing structural feature consists of irregular and extensive ingrowths of the material of their walls, so giving the cells a very high surface-to-volume ratio. They are associated with the transfer of solutes across membranes* (New Scientist).

transfer factor, *Biochemistry.* a polypeptide substance secreted by lymphocytes that is capable of transferring immunity from one cell to another.

transferrin (trans fer′in), *n. Biochemistry.* a beta globulin in blood serum that combines with iron and transports it to bone marrow and other parts of the body.

transfer RNA, *Molecular Biology.* a form of RNA (ribonucleic acid) that delivers amino acids to the ribosomes during protein synthesis: *The molecules of transfer RNA form a class of small globular polynucleotide chains ... They act as vehicles for transferring amino acids from the free state inside the cell into the assembled chain of the protein. This vital function as an intermediary between the nucleic acid language of the genetic code and the amino acid language of the working cell has made transfer RNA a major subject of research in molecular biology* (Scientific American). *Abbreviation:* tRNA Also called **soluble RNA.** Compare **messenger RNA, ribosomal RNA.**

transfinite, *Mathematics.* —*adj.* beyond or surpassing any finite number or magnitude: *a transfinite cardinal number.*
—*n.* a transfinite number.

transform, *v.* **1** *Physics.* **a** to change (one form of energy) into another. A generator transforms mechanical energy into electricity. **b** to change (an electric current) into one of higher or lower voltage, from alternating to direct current, or from direct to alternating current.
2 *Mathematics.* to change (a figure, term, etc.) to another differing in form but having the same value or quantity.
—*n. Mathematics.* an expression derived from another by changing a figure, term, etc., without changing its quantity or value.

transformation, *n.* **1** *Mathematics.* **a** the changing of the form of an expression or figure into another form. **b** the result of such an operation; transform.
2 *Biology.* a genetic change produced by the incorporation into a cell of DNA from another cell or from a virus: *In bacterial transformation a bit of DNA penetrates the boundary of a bacterial cell and becomes incorporated into the cell's genetic apparatus* (Scientific American). Compare **transduction, transfection.** See also **transposon.**

transformer, *n. Physics.* a device for changing an alternating electric current into one of higher or lower voltage by electromagnetic induction. A transformer usually consists of two coils of insulated wire wound around a hollow metal core. One of the coils (the *primary winding*) supplies the input voltage, while the other (the *secondary winding*) supplies the output voltage. *Step-up transformers* increase the voltage as necessary; *step-down transformers* decrease the voltage.

transform fault, *Geology.* one of numerous strike-slip faults between segments of mid-oceanic ridges and rises along which lateral movement of crustal plates occurs: *Global analysis has established that the big shears called transform faults are the zones along which crustal plates glide as they separate* (New Scientist and Science Journal).

transfusion, *n.* = blood transfusion.

transfusion cell, *Botany.* a thin-walled plant cell that permits the passage of water to adjacent tissues.

transgression, *n. Geology.* the spread of the sea over the land along a subsiding shoreline, producing an overlap by deposition of new strata upon old: *According to P. F. Fedorov, every transgression of the Caspian Sea which occurred during glacial advances of Pleistocene time coincides, without exception, with a regression of the Black Sea* (Science).

transient, *n. Physics.* a temporary oscillation or other disturbance occurring in a system just before it attains a steady-state condition: *For a mechanical system a transient may be discerned in the particle displacement, the particle velocity, and the pressure at any given point; the corresponding quantities in an electrical system are the charge, the current, and the voltage at a particular point* (H. J. Gray).

transistor (tran zis′tər), *n. Physics.* a small electronic device containing semiconductors such as germanium or silicon, used to amplify or control the flow of electrons in a circuit: *One of the largest applications of transistors has been as a switch in computers. The electrical characteristics of a transistor are nearly ideal for this type of application in that they closely simulate a relay in its open and closed positions* (L. B. Valdes). [from *tran*(*sfer*) + (*re*)*sistor*]

transit, *n. Astronomy.* **1** the apparent passage of a celestial body across the meridian of a place, or through the field of a telescope: *A star is at upper transit when it crosses the upper branch of the celestial meridian, and at lower transit when it crosses the lower branch* (Baker, *Astronomy*).
2 the passage of a smaller celestial body across the disk of a larger one: *Sometimes there are transits, when a satellite passes in front of the disk: note that the meaning of the word 'transit' in this connection is quite different from that of a surface-feature transit* (Listener).

transit instrument, *Astronomy.* a telescope mounted with its east-west axis fixed so that it moves in the plane of the local meridian, used to find the time of transit of a celestial body: *Observations of the times of upper transit of stars are made by means of a transit instrument ... A set of cross-hairs in its focal plane permits one to gauge the precise instant when any star is on the celestial meridian* (Krogdahl, *The Astronomical Universe*).

transition element or **transition metal,** *Chemistry.* any of a number of metallic elements with an incomplete inner electron shell. Transition elements tend to show a variety of oxidation states and to form many complex ions. Silver and gold are transition elements. *Transition elements ... owe their uniqueness to their ability to form strong complexes with ligands, or molecular groups, of the type present in the side chains of proteins. Enzymes in which transition metals are tightly incorporated are called metalloenzymes, since the metal is usually embedded deep inside the structure of the protein* (Earl Frieden).

transitive, *adj. Mathematics.* having to do with or characterized by transitivity: *We sometimes state the transitive property by saying: If two numbers are equal to the same number, they are equal* (Pearson and Allen, *Modern Algebra*).
ASSOCIATED TERMS: see **associative.**

transitivity, *n. Mathematics.* a relation such that if it holds between A and B and between B and C, it must also hold between A and C. EXAMPLE: If 12 is greater than 6 and 6 is greater than 3, then 12 is greater than 3.

translate, *v.* **1** *Physics.* to move from one point or place to another without rotation.
2 *Molecular Biology.* to cause (genetic information in messenger RNA) to direct the specific sequencing of amino acids in protein synthesis: *In this way, the transfer-RNA molecules act as "translating devices," translating a codon at one end into an amino acid at the other* (Isaac Asimov).
3 *Mathematics.* to subject (the axes of coordinates) to translation: *A common reason for translating axes is to simplify the equation of a curve* (William Karush).

translation, *n.* **1** *Physics.* motion in which there is no rotation; onward movement that is not rotary or reciprocating: *In translation, the motion of one point of a body may be taken as representing the motion of the body as a whole. The body of an automobile or trolley car moving on a level track is an example of translation* (John E. Hoyt).
2 *Molecular Biology.* the specific sequencing of amino acids in protein synthesis on the basis of information contained in messenger RNA: *Translation is the process whereby the genetic information contained in m-RNA determines the linear sequence of amino acids in the course of protein synthesis. Translation occurs on ribosomes, complex particles in the cytoplasm* (Robert G. Eagon).
3 *Mathematics.* a changing of the coordinates of a point to other coordinates which refer to new axes parallel to the old. In a **translation of axes,** the new x-axis is parallel to the old x-axis and the new y-axis is parallel to the old y-axis.

---

cap, fāce, fäther; best, bē, tėrm; pin, five;
rock, gō, ôrder; oil, out; cup, pùt, rüle,
yü in use, *yù* in uric;
ng in bring; *sh* in rush; *th* in thin, ᴛʜ in then;
zh in seizure.
ə = *a* in about, *e* in taken, *i* in pencil, *o* in lemon, *u* in circus

—**translational,** *adj.* of or having to do with translation: *The condition for translational equilibrium ensures that the body will have no tendency to change its translational speed or direction of motion* (Shortley and Williams, *Elements of Physics*).

translatory, *adj. Physics.* consisting of onward motion, as distinct from rotation: *rotatory and translatory motion.*

translocate, *v.* to cause to undergo translocation: *to translocate cobalt in a leaf, to translocate chromosomes in mice.*

translocation, *n.* **1** *Botany.* the movement or conduction of water, minerals, etc., from one part of a plant to another: *The bulk of the salts are translocated upward through the xylem, although translocation in both directions occurs through the phloem* (Greulach and Adams, *Plants*).
2 *Biology.* the transfer of part of a chromosome from one position to another in the same or in another chromosome: *If it [sterility] arises as a result of translocation of the chromosomes (in which the chromosomes are first ruptured, and then reunited in various ways so that there is a large-scale rearrangement of the genes) we call it translocation sterility* (New Scientist). ASSOCIATED TERMS: see **deletion.**

translucent, *adj. Optics.* letting light through without being transparent: *Materials like clear glass or air, which transmit light so well that a luminous object may be clearly seen through them, are called transparent; other materials such as frosted glass or paraffin, which transmit light but not clear, are called translucent* (Shortley and Williams, *Elements of Physics*). [from Latin *translucentem,* from *trans-* through + *lucere* to shine] —**translucently,** *adv.*

translunar, *adj. Astronomy.* extending beyond the moon or the moon's orbit around the earth: *translunar space.*

transmedian, *adj. Anatomy.* passing or lying across a median line, as of the body: *a transmedian muscle.*

transmit, *v. Physics.* **1** to cause (light, heat, sound, or other form of energy) to pass through a medium: *When light strikes a material, several different things may happen. It may be reflected, it may disappear (be absorbed), or it may be transmitted through the material. If it is transmitted so that we can see through the material, we say that the material is transparent; if not, we say that it is translucent* (Beauchamp, *Everyday Problems in Science*).
2 to convey (force or movement) from one part of a body or mechanism to another: *The motion is transmitted from particle to particle, to a great distance* (Thomas H. Huxley).
3 (of a medium) to allow (light, heat, etc.) to pass through: *Glass transmits light.*

transmittance, *n. Physics.* the ratio of the radiant energy transmitted by a medium to the total radiant energy incident upon the medium.

transmutation, *n. Chemistry, Physics.* a change from one atom into another atom of a different element through a gain or loss of protons by the nucleus. Transmutation may occur naturally, as by radioactive decay, or artificially, as by bombardment with neutrons. Transmutation may also refer to a change of an isotope into another isotope of the same element by a gain or loss of neutrons. *The radium atom decays first to radon, then to other unstable descendants and finally to lead. In this spontaneous transmutation from radium to lead a total of five alpha particles (plus several beta particles) is emitted* (Herman Yagoda).

transmute, *v. Chemistry, Physics.* to subject (an atom or element) to transmutation: *Thorium can be transmuted into fissionable uranium 233 which is a possible fuel for atomic power plants* (Wall Street Journal).

transonic (tran son'ik), *adj. Physics.* having to do with or moving at speeds immediately below or above the speed of sound in air (about 331 meters per second). Also spelled **transsonic.** Compare **subsonic, supersonic.**

transparency, *n. Optics.* (of a material) the ratio of the intensity of the transmitted light to the intensity of the incident light.

transparent, *adj. Optics.* transmitting light so that bodies beyond or behind can be distinctly seen: *Most window glass is transparent.* Compare **translucent.** [from Medieval Latin *transparentem* showing light through, from Latin *trans-* through + *parere* appear] —**transparently,** *adv.*

transpiration (tran'spə rā'shən), *n. Biology.* the act or process of transpiring; passage of a liquid or gas in the form of vapor through a membrane or surface, such as the passage of waste matter through the skin or moisture through the leaves of a plant: *Evaporation from the soil and transpiration of vegetation are responsible for the direct return to the atmosphere of more than half of the water which falls on the land* (Michel Batisse). Compare **evapotranspiration.**
—**transpirational,** *adj.* of or having to do with transpiration: *The mechanism by which water is transported through the xylem is best explained by the hypothesis of transpirational pull. Transpirational pull involves the transpiration of water vapor through the stomates. This exerts a pulling force on the column of water in the xylem. Because of the cohesive and adhesive properties of water, the water column does not break and is drawn up from the roots* (Biology Regents Syllabus).

transpire (tran spīr'), *v. Biology.* to pass off or send off a liquid or gas in the form of vapor through a membrane or surface, as from the living body or from the leaves or other parts of plants: *Plants which are transpiring rapidly do not necessarily absorb more mineral salts than do others in which the rate of transpiration is slow* (J. F. Sutcliffe).

transplant (trans plant'), *v.* **1** *Botany.* to remove a plant from one place and plant it in another: *Proper root pruning a few months previous to transplanting, a reduction of the amount of transpiring surface by the removal of a considerable number of leaves or leafy branches when the plant is reset, and protection of the root tips and smaller branches from drying, usually result in success in transplanting* (Emerson, *Basic Botany*).
2 *Biology, Medicine.* to transfer (an organ, skin, etc.) from one person, animal, or part of the body to another: *to transplant skin in a skin graft.*
—*n.* (trans'plant'). **1** *Biology, Medicine.* **a** the transfer of an organ, skin, etc., from one person, animal, or part of the body to another: *a heart transplant, a bone-marrow transplant. Eye banks testify to the success that has attended ... surgical transplants of eye*

corneas (N. Y. Times). **b** the part so transferred: *Transplants are often rejected by the body's immune system.*

2 *Botany.* a seedling transplanted once or several times.

3 *Microbiology.* the transfer of microorganisms from one medium to another.

—**transplantation,** *n.* **1** the act or process of transplanting or condition of being transplanted: *Failures of corneal transplantation, the scientists think might be due to the fact that the donor and recipient do not belong to the same specific blood group* (Science News Letter).

2 something that has been transplanted. SYN: graft. ASSOCIATED TERMS: see **allogeneic, allograft.**

transport, *n. Biology.* the process by which materials essential to life are absorbed and distributed throughout an organism. Compare **active transport, passive transport.**

transposable element, = transposon.

transpose, *Mathematics.* —*v.* (trans pōz′) to transfer (a term) to the other side of an algebraic equation, changing plus to minus or minus to plus.

—*n.* (trans′pōz′) a matrix formed by transposing the rows and columns of a given matrix.

transposition, *n. Mathematics.* **1** the act of transposing or condition of being transposed.

2 a permutation of a set involving only two symbols.

transposon (trans pō′zon), *n. Biology.* a segment of bacterial DNA capable of transferring its genetic properties from one bacterium to another, or from one site in a cell to another site: *Many plasmid genes which confer resistance to antimicrobial agents—including penicillin, sulfonamide, streptomycin, chloramphenicol, and tetracycline reside upon transposons* (Stanley Falkow). Also called **transposable element.** Informally called **jumping gene.** Compare **transformation, transduction.** [from *transpose* + *-on*]

transsonic (trans son′ik), *adj.* = transonic.

transuranic (trans′yù ran′ik), *n.* = transuranic element.

transuranic element or **transuranium element,** *Chemistry.* any of a group of radioactive elements whose atomic numbers are higher than that of uranium (92). The group includes neptunium, plutonium, americium, curium, berkelium, californium, einsteinium, fermium, mendelevium, nobelium, and lawrencium. *The transuranic elements, one of which is plutonium, the fuel of the atomic bomb ... are of greater atomic weight than uranium, are all unstable and decay by radioactive disintegration. They are therefore not found in any appreciable quantity in nature* (K. S. Spiegler).

transversals

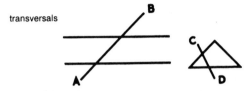

Lines AB and CD are transversals.

transversal (trans vėr′səl *or* tranz vėr′səl), *adj.* = transverse.

—*n. Geometry.* a line intersecting two or more other lines. —**transversally,** *adv.*

transverse, *Geometry.* —*adj.* of or having to do with the axis of a conic section that passes through the foci.

—*n.* a transverse axis; the longer axis of an ellipse.

transverse colon, *Anatomy.* the part of the large intestine that crosses under the liver: *The section of the large intestine which extends up toward the liver is called the ascending colon. The next part, which crosses the abdomen to the left side, is the transverse colon. The descending colon is the section that passes downward. The last part of the intestine is the rectum* (Andrew C. Ivy).

transverse process, *Anatomy.* a process projecting laterally from a vertebra: *Each vertebra is made up of a spool-like centrum surmounted by a neutral arch to house the nerve cord. Above the arch is a low neural spine, at either side is a broad transverse process, and at either end is a pair of articular processes that fit to similar processes on adjacent vertebrae* (Storer, General Zoology).

transverse vibrations, *Physics.* periodic disturbances in which the particles of the medium move at right angles to the direction of propagation.

transverse wave, *Physics.* a wave in which the individual particles of the medium move at right angles to the direction of the wave's propagation. Contrasted with **longitudinal wave.** See the picture at **wave.**

trap, *n. Geology.* a former term for basalt or other fine-grained, dark, igneous rock usually having a sheetlike or steplike structure: *Most of this igneous rock, called "trap," is in the form of sheets lying parallel to the sedimentary strata* (Moore, *Introduction to Historical Geology*). Also called **trap rock.** [earlier *trapp,* from Swedish, from *trappa* stair]

trapeziform (trə pē′zə fôrm), *adj. Botany, Zoology.* shaped like a trapezium or trapezoid.

trapezium (trə pē′zē əm), *n., pl.* **-ziums, -zia** (-zē ə). **1** *Geometry.* **a** a four-sided plane figure having no sides parallel. **b** *British.* = trapezoid (def. 1).

2 *Anatomy.* a bone of the wrist at the base of the thumb. Compare **trapezoid.**

[from Late Latin, from Greek *trapezion* (originally) little table, diminutive of *trapeza* table, from *tra-* four + *peza* foot]

(def. 1)

trapezium

trapezius (trə pē′zē əs), *n., pl.* **-zii** (-zē ī). *Anatomy.* each of a pair of large, flat, triangular muscles of the back of the neck and the upper part of the back and shoul-

---

cap, fāce, fäther; best, bē, tėrm; pin, five;
rock, gō, ôrder; oil, out; cup, pùt, rüle;
yü in use, *yù* in uric;
ng in bring; *sh* in rush; *th* in thin, ᴛʜ in then;
zh in seizure.
ə = *a* in about, *e* in taken, *i* in pencil, *o* in lemon, *u* in circus

681

ders, together forming a somewhat diamond-shaped figure. [from New Latin, from Late Latin *trapezium*]

trapezoid (trap′ə zoid), *n.* **1** *Geometry.* **a** a four-sided plane figure having two sides parallel and two sides not parallel. **b** *British.* = trapezium (def. 1).
2 *Anatomy.* the bone of the wrist articulating with the metacarpal bone that forms the base of the forefinger. Compare **trapezium.**
[from Greek *trapezoidēs* shaped like a trapezium]
—**trapezoidal,** *adj.* in the form of a trapezoid.

trapezoids

trap rock, = trap.
traveling wave, *Physics.* a wave in which the particles of the medium move with the wave and gain energy from the wave, so that they continuously overtake each other.
travertine (trav′ər tin *or* trav′ər tēn), *n. Geology.* a white or light-colored form of limestone deposited by springs, especially hot springs: *Deposits of calcium carbonate formed in a similar way around springs are known as travertine. Sometimes travertine is used as a building stone* (Jones, *Inorganic Chemistry*). [from Italian *travertino*]
tree, *n. Botany.* **1** a large perennial plant with a single woody stem (the *trunk*) and usually having branches and leaves at some distance from the ground. Most trees grow to be at least 5 to 7 meters tall and their trunks grow at least 7.5 to 8 centimeters thick. *Trees are classified as either (a) broadleaf or (b) needle leaf (conifers); (a) deciduous or (b) evergreen. Evergreens are those which retain some foliage throughout the year, while deciduous trees periodically lose their leaves and are therefore bare for a portion of the year* (Finch and Trewartha, *Elements of Geography*). **2 a** shrub or herb that resembles a tree in form or size, such as the banana and plantain.
ASSOCIATED TERMS: see **herb.**
tree line, = timberline.
trematode (trem′ə tōd), *Zoology.* —*n.* any of a class (Trematoda) of parasitic flatworms that have suckers and sometimes hooks. The flukes are trematodes.
—*adj.* of or belonging to the trematodes.
[from Greek *trēmatōdēs* with holes, from *trēmatos* hole]
tremolite (trem′ə lit), *n. Mineralogy.* a white or gray mineral, a variety of amphibole, consisting chiefly of a silicate of calcium and magnesium, occurring in fibrous masses or thin-bladed crystals. *Formula:* $Ca_2Mg_5Si_8O_{22}(OH)_2$ [from *Tremola,* valley in the Swiss Alps + *-ite*]
tremor, *n. Geology.* a small shaking or vibrating movement of the earth, often occurring before or after an earthquake.
trench, *n. Geology.* a long, narrow depression in the ocean floor: *Areas of underthrust have produced the great ocean trenches, such as the Tonga Trench in the southwestern Pacific; the Kuril, Japan, and Mariana*

Trenches in the northwestern Pacific; and the Peru-Chile Trench off the western edge of South America (Science News).
treponema (trep′ə nē′mə), *n., pl.* **-mas, -mata** (-mə tə). *Bacteriology.* any of a genus (*Treponema*) of spirochetes parasitic in humans and other warm-blooded animals, including the species that cause syphilis, relapsing fever, and yaws. [from New Latin, from Greek *trepein* to turn + *nēma* thread]
—**treponemal,** *adj.* of or having to do with treponemas: *a treponemal infection or disease.*
TRF or **TRH,** *abbrev.* thyrotropin-releasing factor; thyrotropin-releasing hormone.
tri-, *combining form.* **1** three; having three, as in *trisect, triangle.*
2 *Chemistry.* containing three atoms, radicals, or other constituents of the substance specified, as in *trioxide, trisulfate.*
[from Latin or Greek *tri-,* from *tria* three]
triacid (trī as′id), *Chemistry.* —*adj.* **1** (of a base or alcohol) having three hydroxyl (-OH) groups which may replace the hydrogen of an acid to form a salt or ester: *Bases like aluminum hydroxide and iron (III) hydroxide, which contain three hydroxyl ions in one molecule of the base, are called triacid bases* (Parks and Steinbach, *Systematic College Chemistry*).
2 having three replaceable acid atoms of hydrogen per molecule.
—*n.* an acid of which one molecule contains three hydrogen atoms which may be replaced by basic atoms or groups.
triad (trī′ad *or* trī′əd), *n. Chemistry.* an element, atom, or radical with a valence of three.
—**triadic,** *adj.* that is a triad; trivalent.
triangle, *n. Geometry.* a closed plane figure having three sides and three angles. A triangle is a polygon the sum of whose three angles is always 180°. Triangles may be classified according to their angles as acute, obtuse, or right triangles, or according to the relationships of their sides as scalene, isosceles, or equilateral.
—**triangular,** *adj.* of or having to do with a triangle, shaped like a triangle. A **triangular region** is the union of a triangle and its interior. A **triangular prism** is one whose base is a triangular region. —**triangularly,** *adv.*
triangulate, *Geometry.* —*v.* (trī ang′gyə lāt) **1** to mark out or divide into triangles: *to triangulate a rectangle.*
2 to find out by trigonometry: *to triangulate the height of a mountain. Ceilometers are the vertical beams used to triangulate cloud elevations* (Newsweek).
—*adj.* (trī ang′gyə lit) **1** composed of or marked with triangles. **2** = triangular. —**triangulately,** *adv.*
—**triangulation,** *n.* **1a** the process of triangulating; measurement done by means of trigonometry: *Maps are usually made by selecting arbitrary control points and measuring the distances and directions of other points by triangulation* (Scientific American). **b** the series or network of triangles laid out for such measurement.
2 division into triangles.
Triassic (trī as′ik), *Geology.* —*n.* **1** the earliest period of the Mesozoic era, before the Jurassic, characterized by the appearance of dinosaurs and primitive mammals, the domination of the earth by reptiles, and much volcanic activity.
2 the rocks formed during this period: *The Jurassic of the Rocky Mountain area is somewhat similar to the*

Triassic, but with more evidence of marine deposition (Garrels, *A Textbook of Geology*).

—adj. of or having to do with this period or its rocks: *Triassic rocks in North America are conveniently classed in three main parts, respectively designated as the Lower, Middle, and Upper Triassic Series. These do not correspond precisely to the threefold division of Triassic in Germany, correlation being based, instead, on the marine section of the Mediterranean area* (Moore, *Introduction to Historical Geology*). [from German *Trias*, a series of strata containing three types of deposit, from Late Latin *trias* three]

triatomic, *adj. Chemistry.* **1** containing three atoms; consisting of molecules each containing three atoms. **2** having three atoms or groups which can be replaced.

tribasic (trī bā′sik), *adj. Chemistry.* **1** (of an acid) having three hydrogen atoms which can be replaced by basic atoms or radicals. Phosphoric acid, H_3PO_4, is tribasic. **2** having three atoms or radicals of a univalent metal. **3** containing three basic hydroxyl (-OH) radicals.

tribe, *n. Biology.* a group of related organisms ranking below a subfamily or suborder and usually containing several genera: *a tribe of birds, a tribe of herbs.*

tribo- (trī′bō- or trib′ō-), *combining form.* friction; produced by friction, as in *triboelectricity.* [from Greek *tribos* a rubbing, from *tribein* to rub]

triboelectric, *adj.* of or having to do with triboelectricity: *triboelectric effects.*

triboelectricity, *n. Physics.* electricity produced by friction; static electricity: *Frictional electricity ... was supposedly known to the ancient Greeks, particularly Thales of Miletus, who observed about 600 B.C. that when amber was rubbed, it would attract small bits of matter. The term "frictional electricity" gave way to "triboelectricity," although since "tribo" means "to rub," the newer term does little to change the concept* (A. D. Moore).

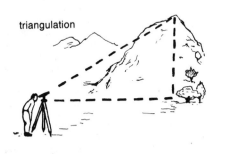

triangulation

tribological, *adj.* of or having to do with tribology. **—tribologically,** *adv.*

tribology (trī bol′ə jē), *n.* the study of friction, wear, and lubrication: *Tribology is a vital element in the science, technology and practice of engineering; it plays an important part whenever interacting surfaces meet in relative motion* (H. Peter Jost).

triboluminescence, *n. Physics.* the emission of light under friction or mechanical pressure: *Crystals of saccharin which, when freshly prepared, flash brilliantly*

on crushing, ... after a few weeks' preservation show no appreciable triboluminescence (Nature).

—triboluminescent, *adj.* having or characterized by triboluminescence.

tribophysics, *n.* the physical properties or phenomena associated with friction.

tributary (trib′yə ter′ē), *Geography.* **—n.** a stream that flows into a larger stream or body of water: *The Ohio River is a tributary of the Mississippi River.* Compare **distributary.**

—adj. flowing into a larger stream or body of water: *A river or trunk stream is formed by the confluence of tributary streams, themselves formed by smaller tributaries, and so on to the individual contributions of springs, rills, and rainwash* (Gilluly, *Principles of Geology*).

tricarboxylic acid (trī kär′bok sil′ik), *Chemistry.* an organic compound containing three carboxyl groups.

tricarboxylic acid cycle, = Krebs cycle.

tricarpellary (trī kär′pə ler ē), *adj. Botany.* having or consisting of three carpels.

triceps (trī′seps), *n. Anatomy.* the large muscle at the back of the upper arm which extends or straightens the arm. See the pictures at **arm, biceps.** [from Latin *triceps* three-headed, from *tri-* three + *caput* head]

trichite (trik′īt), *n. Geology.* a very minute, dark-colored, hairlike embryonic crystal occurring in some vitreous rocks. Trichites may be straight or curved. [from Greek *trichos* hair + English *-ite*]

trichloride (trī klôr′īd), *n. Chemistry.* a compound whose molecules contain three atoms of chlorine combined with another element or radical.

trichocyst (trik′ə sist), *n. Zoology.* one of the tiny stinging or grasping organs on the body of protozoans, consisting of a hairlike filament in a small sac: *The ectoplasm [of the paramecium] contains many spindle-shaped trichocysts, alternating between the bases of the cilia, that may be discharged as long threads to serve perhaps in attachment or defense* (Storer, *General Zoology*). [from Greek *trichos* hair + English *cyst*]

—trichocystic, *adj.* having to do with or of the nature of trichocysts: *the trichocystic formation of ciliates.*

trichogyne (trik′ə jīn), *n. Biology.* a hairlike process forming the receptive part of the female reproductive organ in certain algae and fungi. [from Greek *trichos* hair + *gynē* woman, female]

trichome (trī′kōm or trik′ōm), *n. Botany.* an outgrowth from the epidermis of plants (a general term including hairs and prickles): *During the formation of the leaf epidermis of certain plant species, there occur modifications resulting in the formation of epidermal hairs or trichomes* (Harbaugh and Goodrich, *Fundamentals*

cap, fāce, fäther; best, bē, tèrm; pin, fīve;
rock, gō, ôrder; oil, out; cup, pùt, rüle,
yü in use, yù in uric;
ng in bring; sh in rush; th in thin, ᴛʜ in then;
zh in seizure.
ə = *a* in about, *e* in taken, *i* in pencil, *o* in lemon, *u* in circus

of Biology). [from Greek *trichōma* growth of hair, from *trichos* hair]

trichotomous (trī kot′ə məs), *n. Botany.* branching into three parts; giving off shoots by threes. [from Greek *tricha* triple, triply + *-tomia* a cutting, division]

trichroic (trī krō′ik), *adj.* having or exhibiting trichroism: *trichroic crystals.*

trichroism (trī′krō iz əm), *n. Optics.* the property of some crystals of exhibiting three different colors when viewed in three different directions. [from Greek *tri-* three + *chroia* skin, color of skin]

trichromat (trī′krō mat), *n. Optics.* a person who perceives a mixture of three primary colors, such as red, green, and blue; a person with normal color vision. —**trichromatic,** *adj.* having to do with or characterized by vision requiring a mixture of three primary colors to match the color sensation experienced: *Since the normal eye has trichromatic vision it is capable of three kinds of color discrimination* (Deane B. Judd). —**trichromatism,** *n.* trichromatic vision.

triclinic (trī klin′ik), *adj. Crystallography.* having the three axes unequal and obliquely inclined. See the picture at **crystal.**

ASSOCIATED TERMS: see **crystal.**

tricrotic (trī krot′ik), *adj. Physiology, Medicine.* (of a pulse or a tracing of it) having or showing a three-fold beat. [from Greek *trikrotos* rowed with a triple stroke]

tricuspid (trī kus′pid), *Anatomy.* —*adj.* 1 having three points or cusps: *a tricuspid tooth.* Compare **tridentate.** 2 of or having to do with the tricuspid valve. —*n.* 1 a tricuspid tooth. 2 = tricuspid valve.

tricuspid valve, *Anatomy.* a valve of three segments opening from the right atrium into the right ventricle of the heart. It prevents blood from being forced back into the right atrium during contraction of the ventricles. Also called **right atrioventricular valve.** Compare **mitral valve.** See the picture at **valve.**

tricyclic, *adj. Chemistry.* containing three rings of atoms in the molecule: *Anthracene is a tricyclic organic compound.*

trident (trī′dənt), *adj.* = tridentate.

tridentate (trī den′tāt), *adj.* 1 *Anatomy.* having three teeth or toothlike points; three-pronged: *tridentate teeth.* Compare **tricuspid.** 2 *Chemistry.* of or designating a molecule three of whose atoms are joined to a metal atom or ion to form a chelate: *a tridentate ligand.*

tridymite (trid′ə mīt), *n. Mineralogy.* a crystallized form of silica found in igneous rocks, usually in twinned groups of three crystals. Tridymite is a high-temperature polymorph of quartz. *Tridymite occurs in small hexagonal tables, colorless and transparent* (James Dwight Dana). *Formula:* SiO_2 [from Greek *tridymos* threefold]

triecious (trī ē′shəs), *adj.* = trioecious. —**trieciously,** *adv.*

triene (trī′ēn), *n. Chemistry.* a compound having three double bonds.

triethyl (trī eth′əl), *adj. Chemistry.* containing three ethyl groups: *triethyl borate, triethyl citrate.*

trifluoride (trī flü′ə rīd), *n. Chemistry.* a compound containing three atoms of fluorine combined with another element or radical.

trifoliate (trī fō′lē it), *adj. Botany.* having three leaves, or three parts like leaves. Clover is trifoliate.

trifoliolate (trī fō′lē ə lāt), *adj. Botany.* consisting of three leaflets. The wood sorrel is trifoliolate.

trifurcate (trī fėr′kit *or* trī fėr′kāt), *adj. Zoology.* divided into three branches like the prongs of a fork: *a trifurcate antenna, trifurcate ribs.* [from Latin *trifurcus,* from *tri-* three + *furca* fork] —**trifurcation,** *n.* 1 the condition of being trifurcate. 2 a trifurcate formation or arrangement.

trig., *abbrev.* 1 trigonometric. 2 trigonometry.

trigamous (trig′ə məs), *adj. Botany.* having staminate, pistillate, and perfect flowers in the same head.

trigastric (trī gas′trik), *adj. Anatomy.* having three fleshy bellies, as certain muscles.

trigeminal (trī jem′ə nəl), *Anatomy.* —*adj.* of or denoting the fifth pair of cranial nerves, each of which divides into three branches having sensory and motor functions in the face: *They used the terms trigeminal neuropathy and neuritis because pain in the face is felt through the fifth, or trigeminal, nerve* (Atlanta Constitution). —*n.* a trigeminal nerve. [from Latin *trigeminus* born three, from *tri-* three + *geminus* born together]

triglyceride (trī glis′ə rīd), *n. Chemistry.* any fatty compound formed when three acid radicals replace the three hydrogen atoms of the -OH (hydroxyl) groups in glycerol: *Some researchers believe the level of other substances, the triglycerides, is more important than that of cholesterol* (Science News Letter).

trigon., *abbrev.* trigonometry.

trigonal (trī gō′nəl), *adj.* 1 = triangular. 2 *Crystallography.* having three faces or a three-fold symmetry. [from Greek *trigōnon,* neuter of *trigōnos* triangular]

trigonometric (trig′ə nə met′rik), *adj.* of, having to do with, or based on trigonometry: *Foucault demonstrated that the apparent rotation of the pendulum varies with the trigonometric sine of the latitude at which it is installed* (C. L. Stong). *The direct, or trigonometric, method of determining stellar parallaxes diminishes in accuracy as more distant stars are observed* (Baker, Astronomy). —**trigonometrically,** *adv.* by means of trigonometry: *Most stars are too far away to be surveyed trigonometrically and the distance estimates then depend on a comparison of apparent and intrinsic (absolute) magnitudes* (Bernard E. J. Pagel).

trigonometrical (trig′ə nə met′rə kəl), *adj.* = trigonometric.

trigonometric function, *Mathematics.* any of the fundamental functions of an angle or arc that may be defined as ratios of the sides of a right triangle: the sine, cosine, tangent, cotangent, secant, or cosecant. Also called **circular function.**

trigonometric series, *Mathematics.* an infinite series whose terms contain sines and cosines of angles.

trigonometry (trig′ə nom′ə trē), *n.* the branch of mathematics that deals with the relations between the sides and angles of triangles and the calculations based on these, particularly with certain functions, such as the sine, secant, and tangent: *Trigonometry in the broader sense has become a branch of the theory of functions. We should therefore expect to find that trigonometry involves considerable use of algebra. This aspect of trigonometry is sometimes spoken of as analytic trigonometry, to distinguish it from the practical trigonometry used by surveyors, engineers, navigators and astronomers* (William L. Schaaf). *Abbreviation:* trig., trigon. [ultimately, from Greek *tri-* three + *gōnia* angle + *metron* measure]

trigonous (trig′ə nəs), *adj. Biology.* having three prominent angles, as a plant stem or ovary. [from Greek *trigōnos* triangular, from *tri-* three + *gōnia* angle]

trihedral (trī hē′drəl), *adj. Geometry.* 1 having, or formed by, three planes meeting at a point: *a trihedral angle.*
2 having, or formed by, three lateral planes: *a trihedral prism.*

trihedron (trī hē′drən), *n., pl.* **-drons, -dra** (-drə). *Geometry.* a figure formed by three planes meeting at a point. [from *tri-* + Greek *hedra* seat, base]

trihydric (trī hī′drik), *adj.* = trihydroxy.

trihydroxy (trī′hī drok′sē), *adj. Chemistry.* having three hydroxyl (-OH) groups: *The most important trihydroxy alcohol is 1,2,3-propanetriol (glycerol)* (Chemistry Regents Syllabus).

triiodothyronine (trī ī′ə dō thī′rə nēn), *n. Biochemistry.* an amino acid secreted by the thyroid gland or prepared synthetically, used in treating hypothyroid conditions: *Growth hormone (GH) and the thyroid hormone triiodothyronine (T3) alter the activity of RNA polymerase enzymes in target cells* (Science News). *Formula:* $C_{15}H_{12}I_3NO_4$

trijugate (trī′jù gāt), *adj. Botany.* (of a pinnate leaf) having three pairs of leaflets. [from Latin *trijugus* threefold]

trilinear, *adj. Mathematics.* of, involving, or contained by three lines.

trilobate (trī lō′bāt), *adj. Botany.* having or divided into three lobes: *a trilobate leaf.*

trilobed, *adj.* = trilobate.

trilobite (trī′lə bīt), *n. Zoology.* any of a class (Trilobita) of small, extinct, marine arthropods of the Paleozoic era, with jointed legs and a body divided into three vertical lobes and many horizontal segments. Small fossil trilobites are widely found in various rocks. They are believed related to the crustaceans. [ultimately, from Greek *tri-* three + *lobos* lobe]

trilocular (trī lok′yə lər), *adj. Biology.* having three compartments or chambers, as the capsule of a plant, or the heart of a reptile.

trimer (trī′mər), *n. Chemistry.* 1 a molecule formed by combining three identical smaller molecules. EXAMPLE: C_6H_6 or $(C_2H_2)_3$ is a trimer formed by combining three molecules of C_2H_2
2 a compound consisting of trimers.
[from *tri-* three + *-mer,* as in *polymer*]

trimerous (trim′ər əs), *adj.* 1 *Botany.* (of a flower) having three members in each whorl (generally written 3-merous). Compare **monomerous.**
2 *Zoology.* having three segments to each tarsus, as certain insects. Compare **dimerous.**
[from *tri-* three + Greek *meros* part]

trimethyl (trī meth′əl), *adj. Chemistry.* having three methyl (-CH₃) radicals: *trimethyl aluminum.*

trimetric projection, *Geometry.* the projection of a solid using different scales at arbitrarily chosen angles for its three dimensions.

trimolecular, *adj. Chemistry.* of or made out of three molecules.

trimorph (trī′môrf), *n. Crystallography.* 1 a trimorphic substance. 2 any of its three different forms.

trimorphic, *adj.* existing in or assuming three distinct forms; exhibiting trimorphism.

trimorphism (trī môr′fiz əm), *n.* 1 *Crystallography.* the occurrence in three different forms of a crystalline substance.
2 *Zoology.* the occurrence of three forms distinct in color, size, structure, etc., in different individuals of a species.
3 *Botany.* the occurrence of three distinct forms of flowers, leaves, etc., on the same plant or in the same species.

trimorphous, *adj.* = trimorphic.

trinomial (trī nō′mē əl), *n.* 1 *Mathematics.* an expression in algebra consisting of three terms connected by plus or minus signs. $ax^2 + bx + c$ *is a trinomial.*
2 *Biology.* a scientific name of an animal or plant consisting of three terms, the first indicating the genus, the second the species, and the third the subspecies or variety. *Malus prunifolia robusta* is a trinomial.
—*adj. Mathematics.* consisting of three terms.
[from *tri-* + *-nomial,* as in *binomial*]
—**trinomialism,** *n. Biology.* the trinomial system of nomenclature; the use of trinomial names.

trinucleotide (trī nu′klē ə tīd), *n. Biochemistry.* a nucleotide consisting of a combination of three mononucleotides: *Each amino acid can be coded by a trinucleotide containing three specific bases* (Albert L. Lehninger). Compare **dinucleotide, mononucleotide.**

trioecious (trī ē′shəs), *adj. Botany.* designating a genus or other group in which staminate, pistillate, and perfect flowers occur on different plants. Also spelled **triecious.** [from *tri-* three + *-oecious,* as in *dioecious*]

triol (trī′ôl), *n. Chemistry.* a trihydroxy alcohol, such as glycerol.

triose (trī′ōs), *n. Biochemistry.* any of a class of sugars containing three atoms of carbon, and produced from glycerol by oxidation.

trioxide (trī ok′sīd), *n. Chemistry.* an oxide having three atoms of oxygen in each molecule.

tripartite (trī pär′tīt), *adj. Botany.* divided into three parts nearly to the base: *a tripartite leaf.*

tripetalous (trī pet′ə ləs), *adj. Botany.* having three petals.

triphosphate (trī fos′fāt), *n.* a substance whose molecule contains three phosphate radicals: *The system that synthesized Kornberg's DNA-type material requires the triphosphate instead of the diphosphate of the nu-*

---

cap, fāce; fäther; best, bē, tèrm; pin, fīve;
rock, gō, ôrder; oil, out; cup, pùt, rüle,
yü in use, *yu* in uric;
ng in bring; *sh* in rush; *th* in thin, ₮H in then;
zh in seizure.

ə = *a* in about, *e* in taken, *i* in pencil, *o* in lemon, *u* in circus

685

cleotides, and of course deoxyribose rather than ribose (Francis H. C. Crick).

triphosphopyridine nucleotide (trī fos'fō pir'ə dēn), *Biochemistry.* a coenzyme that serves as a hydrogen acceptor in intermediary metabolism. It is similar in function to diphosphopyridine nucleotide. *Formula:* $C_{21}H_{28}N_7O_{17}P_3$ *Abbreviation:* TPN

triphyllous (trī fil'əs), *adj. Botany.* having or consisting of three leaves. [from *tri-* + Greek *phyllon* leaf]

tripinnate (trī pin'āt), *adj. Botany.* triply pinnate (applied to a bipinnate leaf whose divisions are also pinnate). **—tripinnately,** *adv.*

triple bond, *Chemistry.* a bond in which three electron pairs are shared between two atoms, occurring in some unsaturated compounds; a bond involving three valences. Compare **single bond, double bond.**

triple point, *Physics.* the point at which the three phases (gas, liquid, and solid) of a substance exist in equilibrium: *It is very easy in principle to determine the triple point of water. Take a thermally insulated vessel containing only ice and liquid water. Then increase the volume somewhat to make space for vapor. Evaporation will take place—some slight adjustment of the quantity of ice and that of liquid water will occur in arriving at thermal equilibrium, and the system will settle down at the temperature and pressure of the triple point* (Shortley and Williams, *Elements of Physics*). See diagram at **phase.**
▶ The temperature of the triple point of water, used as the standard fixed point of thermometry, is defined as 273.16 kelvins or .01 degrees Celsius.

triplet, *n.* **1** *Physics.* **a** a group of three subatomic particles with similar characteristics but different charge states. **b** a quantum state having an internal degree of freedom with three possible values, such as the state of a system of two particles having a total spin equal to one unit.
2 *Chemistry.* a molecule having an even number of electrons, with the electrons of one pair having parallel spins.
3 *Genetics.* any combination of three bases in the genetic code: *nonsense triplets. Since three are 64 possible triplets, common amino acids are often specified by several different combinations* (Stephen M. Head). SYN: codon. Compare **trinucleotide.**

triplite (trip'līt), *n. Mineralogy.* a monoclinic mineral, a phosphate of iron and manganese containing magnesium and calcium, having a brown or blackish color. *Formula:* $(Mn,Fe,Mg)_2(PO_4)(F,OH)$ [from Greek *triploos* triple + English *-ite* (so called because of the mineral's cleavage in three directions)]

triploblastic (trip'lə blas'tik), *adj. Zoology.* having three germ layers, the ectoderm, the endoderm, and the mesoderm, as the embryos of vertebrates do. Compare **diploblastic.** [from Greek *triploos* triple + *blastos* germ, sprout]

triploid (trip'loid), *Genetics.* **—adj.** having three times the haploid number characteristic of the species: *Some females [of Drosophila flies] were triploid, that is, had three sets of chromosomes, including three X's (a total of 12 chromosomes). Triploid females are normal in appearance and behavior* (Theodosius Dobzhansky).

—n. a triploid organism.
[from New Latin *triploides,* from Greek *triploos* triple]

—triploidy (trip'loi dē), *n.* the condition of being triploid.

tripoli (trip'ə lē), *n. Geology.* any of several soft, porous, siliceous earths or rocks, often composed of the shells of diatoms, radiolarians, infusorians, etc., or of chalcedonic chert. [named after *Tripoli,* a region in northern Africa, where it was originally found]

tripolite (trip'ə līt), *n. Geology.* tripoli that is made up of the fossil remains of diatoms. SYN: diatomaceous earth, kieselguhr.

triquetrous (trī kwē'trəs *or* trī kwet'rəs), *adj. Botany, Zoology.* three-edged; having three salient angles: *triquetrous nutlets, triquetrous mandibles.* [from Latin *triquetrus* three-cornered]

triquetrum (trī kwē'trəm), *n., pl.* **-tra** (-trə). *Anatomy.* a bone of the human carpus (wrist) in the proximal row of carpal bones. [from New Latin, from Latin *triquetrum,* neuter of *triquetrous* three-cornered]

triradiate, *adj. Botany, Zoology.* having or consisting of three rays; radiating in three directions from a central point: *a triradiate spore, a triradiate bone or cartilage.* **—triradiately,** *adv.*

trisaccharide (trī sak'ə rīd), *n. Biochemistry.* any of a class of carbohydrates which on hydrolysis yield three molecules of simple sugars (monosaccharides). ASSOCIATED TERMS: see **disaccharide.**

trisect, *v. Geometry.* to divide into three parts, especially three equal parts. [from *tri-* + Latin *sectum* cut]
—trisection, *n.* division into three parts, especially the division of a line segment or an angle into three equal parts.

trisepalous (trī sep'ə ləs), *adj. Botany.* having three sepals.

triseptate (trī sep'tāt), *adj. Biology.* having three septa (partitions).

trisoctahedron (tris ok' tə hē'drən), *n., pl.* **-dra** (-drə). *Geometry.* a solid bounded by twenty-four equal faces, every three of which correspond to one face of an octahedron. [from Greek *tris-* thrice + English *octahedron*]

trisomic (trī sō'mik), *Genetics.* **—adj.** having three of a given chromosome instead of the normal set of two; being diploid except for one chromosome which is triploid; characterized by trisomy.
—n. a trisomic individual.
[from *tri-* three + Greek *soma* body]

trisomy (trī'sō mē), *n. Genetics.* the condition of having one or more chromosomes present three times in the somatic cells instead of being present only twice. Down's syndrome is usually due to trisomy of chromosome number 21.

tritium (trit'ē əm *or* trish'ē əm), *n. Chemistry.* a radioactive isotope of hydrogen, three times as heavy as ordinary hydrogen. It is used to release nuclear energy through fusion, as a bombarding particle in accelerators, and as a tracer in biomedical research. Tritium decays to form helium. *Since tritium is radioactive, with a half-life of 12.5 years, it disappears after a time from water which has been out of contact with the atmosphere. Wines, and water in wells, can be dated by their tritium content* (Scientific American). *Symbol:* T or ^3H; *atomic number* 1; *mass number* 3. Compare

686

[from New Latin, from Greek *tritos* third]

triton (trī′ton), *n. Chemistry.* the nucleus of a tritium atom, consisting of two neutrons and one proton.

trivalent (trī vā′lənt), *adj. Chemistry.* having a valence of three: *An element like aluminum or nitrogen whose atom can hold three monovalent atoms in combination is called a trivalent element* (Parks and Steinbach, *Systematic College Chemistry*). Also called **tervalent.**
—**trivalence,** *n.* the quality of being trivalent.

trivial, *adj. Biology.* of or designating a species; specific.

trivial name, 1 *Biology.* **a** the name added to a generic name to distinguish the species; specific name or epithet. **b** the common or vernacular name of an organism.
2 *Chemistry.* the common or unscientific name of a chemical compound obtained from a natural source, used before the molecular structure of the compound is known.

tRNA, *abbrev.* transfer RNA.

trochal (trō kəl), *adj. Zoology.* resembling a wheel, as the ciliated disk of a rotifer. [from Greek *trochos* wheel]

trochanter (trō kan′tər), *n.* **1** *Anatomy.* a protuberance or process on the upper part of the femur of many vertebrates.
2 *Zoology.* the second section of the leg of an insect. [from Greek *trochantēr,* related to *trechein* to run]
—**trochanteric,** *adj.* of or having to do with a trochanter: *the trochanteric bursa.*

trochlea (trok′lē ə), *n., pl.* **-leae** (-lē ē). *Anatomy.* a pulleylike structure or arrangement of parts with a smooth surface upon which another part glides, as the part of the humerus with which the ulna articulates. [from New Latin, from Latin *trochlea* pulley block, from Greek *trochilia,* diminutive of *trochos* wheel]

trochlear, *adj.* **1** *Anatomy.* **a** of, connected with, or forming a trochlea. **b** of or having to do with the trochlear muscle or trochlear nerve.
2 *Botany.* circular and narrowed in the middle, like the wheel of a pulley.

trochlear muscle, *Anatomy.* the superior oblique muscle of each eye.

trochlear nerve, *Anatomy.* either one of the fourth pair of cranial nerves that activate the trochlear muscle.

trochoid (trō′koid), *n. Geometry.* a curve traced by a point on or connected with a circle which rolls on a straight line, or on the inside or outside of another circle.
—*adj.* **1** *Anatomy.* (of a pivot joint) having one bone turning upon another with a rotary motion.
2 *Zoology.* shaped like a top, as certain shells. [from Greek *trochoeidēs* round, wheel-shaped]
—**trochoidal,** *adj.* **1** having the form or nature of a trochoid.
2 of or having to do with trochoids.
3 = trochoid. —**trochoidally,** *adv.*

trochophore (trok′ə fôr), *n. Zoology.* a free-swimming, ciliate larval form of most mollusks and of certain bryozoans, brachiopods, and marine worms. [from Greek *trochos* wheel + English *-phore*]

troilite (trō′ə līt or troi′līt), *n. Mineralogy.* a sulfide of iron found in meteorites and regarded as a variety of pyrrhotite: *Troilite is a good conductor of heat* (New Scientist). *Formula:* FeS [named after Dominico

Troili, Italian scientist who in 1766 described a meteorite containing this mineral]

Trojan or **trojan,** *n.,* or **Trojan asteroid,** *Astronomy.* any of a group of asteroids that revolve to form an equilateral triangle with Jupiter and the sun. Most of them are named after heroes of the Trojan war. *Hektor is a trojan: it lies on Jupiter's orbit at one of the four points that are an equal distance from Jupiter and the Sun, the lagrangian points* (New Scientist).

-tron, *suffix.* **1** having to do with electrons, as in *cryotron, magnetron.*
2 a device for directing the movement of subatomic particles, as in *cyclotron, synchrotron.*
[from Greek *-tron* device, instrument]

trona (trō′nə), *n. Mineralogy.* a white, gray, or yellow monoclinic mineral, a native hydrated sodium carbonate, used as a source of various sodium compounds. *Formula:* $Na_2(CO_3) \cdot Na(HCO_3) \cdot 2H_2O$ [from Swedish]

troostite (trüs′tīt), *n. Mineralogy.* a variety of willemite containing manganese, occurring in reddish hexagonal crystals. *Formula:* $(Zn,Mn)_2SiO_4$ [from Gerard Troost, 19th-century American geologist + *-ite*]

trophallaxis (trō′fə lak′sis), *n. Biology.* the reciprocal exchange of food between organisms, especially among social insects: [*In*] *some groups of termites ... an adult worker caste takes charge of nest building and trophallaxis, or mutual feeding* (New Yorker). [from New Latin, from Greek *trophē* nourishment + *allaxis* exchange]

trophic (trō′fik), *adj. Biology.* of or having to do with nutrition: *trophic diseases, the trophic capacity of an ecosystem.* [from Greek *trophē* nourishment]
—**trophically,** *adv.*

trophic level, *Ecology.* any of the stages in the flow of food from one population of organisms to another: *The sequence of trophic levels in any ecosystem forms a food chain* (Clarence J. Hylander).

trophoblast (trō′fə blast), *n. Embryology.* a layer of cells external to the embryo in many mammals, having the function of supplying it with nourishment. [from Greek *trophē* nourishment + *blastos* germ, sprout]
—**trophoblastic,** *adj.* of or having to do with the trophoblast: *There are reports of the finding of trophoblastic fragments in the general circulation in postpartum* (Science).

trophozoite (trō′fə zō′īt), *n. Biology.* a vegetative protozoan, especially a sporozoan during its growing stage: *The life cycle of Monocystis is briefly outlined as follows. ... The animals are in some unknown way transferred from one earthworm to another as spores, each containing 8 elongated bodies called sporozoites. ... Each sporozoite ... penetrates a bundle of sperm-mother cells of the earthworm, and is then termed a trophozoite* (Hegner and Stiles, *College*

---

cap, fāce, fäther; best, bē, tèrm; pin, fīve;
rock, gō, ôrder; oil, out; cup, pùt, rüle,
yü in use, *yu* in uric;
ng in bring; *sh* in rush; *th* in thin, ₮H in then;
zh in seizure.
ə = *a* in about, *e* in taken, *i* in pencil, *o* in lemon, *u* in circus

tropic

Zoology). [from Greek *trophē* nourishment + *zōion* animal]

tropic, *n.* **1** *Geography.* **a** either of two parallels of latitude on the earth's surface, one 23 degrees 27 minutes north (**tropic of Cancer**) and one 23 degrees 27 minutes south of the equator (**tropic of Capricorn**), representing the points farthest north and south at which the sun ever shines directly overhead. See the picture at **zone. b tropics** or **Tropics,** *pl.* the regions between and near these parallels of latitude. The tropics are generally the warmest regions of the earth. *The humid tropics comprise a somewhat interrupted and irregular "belt" 20 to 40° wide, around the earth and straddling the equator. This region ... lacks a winter* (Finch and Trewartha, *Elements of Geography*). **2** *Astronomy.* either of two parallels of celestial latitude, the limits reached by the sun in its apparent journey north and south.
—*adj.* (trō′pik) *Biology.* having to do with or characterized by tropism.
[from Latin *tropicus,* from Greek *tropikos* pertaining to a turn, from *tropē* a turn, a change, from *trepein* to turn]
—**tropical,** *adj.* **1** of, having to do with, or occurring in the tropics: *tropical diseases. Disturbances officially become tropical storms when the wind exceeds 39 miles an hour* (N. Y. Times). **2** native to the tropics: *Bananas are tropical fruit.* **3** of or having to do with the tropics of Cancer and Capricorn, or either one: *the tropical Pacific.*

tropical cyclone, *Meteorology.* a cyclone originating in tropical oceans, characterized by violent rainstorms and winds with velocities of up to 320 kilometers per hour. Hurricanes are tropical cyclones occurring in the West Indies. *Tropical cyclones, as their name implies, achieve their most destructive effects in low latitudes. They ... are usually born in latitudes 5–15° North and South, over the sea* (R. W. James).

tropical rain forest, *Geography, Ecology.* a rain forest in a region near the equator, characterized by year-round warmth and very heavy rainfall. Tropical rain forests are green throughout the year. The largest tropical rain forest is that of the Amazon, covering a large portion of South America.

tropical year, = solar year.

tropic of Cancer. See under **tropic.**

tropic of Capricorn. See under **tropic.**

tropism (trō′piz′əm), *n. Biology.* the turning of a plant or sessile animal, or a part of one, in a particular direction in response to some external stimulus, as that of light (*phototropism, heliotropism*), heat (*thermotropism*), gravity (*geotropism*), etc.: *In the tropisms of common plants all changes in form take place only in the younger growing parts. Old leaves and the older stems and roots remain unchanged in position* (Emerson, *Basic Botany*). [from Greek *tropē* a turning, from *trepein* to turn]
—**tropistic,** *adj.* of or having to do with a tropism: *We may use the term tropistic to describe the reactions of both fixed and free organisms to directive stimuli* (F. Keeble). —**tropistically,** *adv.*

tropo-, *combining form.* **1** a turn or turning; alternation; change, as in *tropopause, tropophyte, tropocollagen.* **2** tendency; orientation; affinity, as in *tropotaxis.* [from Greek *tropos* a turn, from *trepein* to turn]

tropocollagen (trō′pō kol′ə jən), *n. Biochemistry.* a basic protein substance from which the collagen fibers of connective tissue, bone, etc., are formed.

tropoelastin (trō′pō i las′tin), *n. Biochemistry.* a basic protein substance from which elastin is formed.

tropomyosin (trō′pə mī′ə sin *or* trop′ə mī′ə sin), *n. Biochemistry.* one of the major protein elements of muscle tissue. Tropomyosin binds to molecules of actin and helps to regulate the interaction of actin and myosin.

troponin (trō′pə nin *or* trop′ə nin), *n. Biochemistry.* a protein in muscle tissue regulated by calcium ions, important in muscle contraction. Each molecule of troponin is attached to a molecule of tropomyosin and a molecule of actin.

tropopause (trō′pə pôz′), *n. Meteorology.* the area of the atmosphere between the troposphere and the stratosphere: *The height of the tropopause varies with latitude. The height is about 10.6 miles (17 km) in equatorial regions, from which it gradually decreases toward the poles, both north and south, descending in polar regions to an elevation of only 4 or 5 miles (6–8 km), and possibly less* (Blair, *Weather Elements*).

tropophilous (trō pof′ə ləs), *adj. Botany.* adapted to a climate which is alternately dry and moist or cold and hot: *tropophilous vegetation.*

tropophyte (trop′ə fit), *n. Botany.* a plant which is adapted for growth in a climate which is alternately dry and moist or cold and hot, as the deciduous trees: *The deciduous trees and shrubs, also known as the broad-leaved plants and the summer green plants, form the principal tropophytes* (Youngken, *Pharmaceutical Botany*).
—**tropophytic** (trop′ə fit′ik), *adj.* of, having to do with, or resembling a tropophyte: *tropophytic flora.*

troposphere (trō′pə sfir), *n. Meteorology.* the lowest region of the atmosphere, below the stratosphere, within which there is a steady fall of temperature with increasing altitude. Turbulence of the air and most cloud formations occur in the troposphere. *At the top of the troposphere, called the tropopause, the temperature may be as low as −112°F (−80°C) ... The temperature drop at increasing altitudes in the troposphere plays an important part in changes in the weather* (F. Sechrist). See the picture at **atmosphere.**
—**tropospheric,** *adj.* of or having to do with the troposphere: *tropospheric turbulence.*

tropotaxis (trō′pə tak′sis), *n. Ecology.* a movement or orientation of an organism in response to two stimuli, especially lights, by means of different sense organs. Contrasted with **telotaxis.**

trough (trôf), *n.* **1** *Geology, Oceanography.* a long, narrow channel or depression between ridges on land or in the sea, or between the crests of waves: *On the ocean floor itself, there is a diversity of topographic forms ... some are depressions in the sea floor ... the elongated troughs with gentle side slopes, and the similar but steeper-sided trenches* (Gilluly, *Principles of Geology*). *The instrument is designed to count a sample of 300 waves, and to measure the height of each from trough to succeeding crest* (A. W. Haslett).

688

2 *Meteorology.* a long, narrow area of relatively low barometric pressure: *The path followed by a hurricane after it turns northward depends on the orientation and speed of movement of the trough it has selected.* (W. G. Osmun).

troy, *adj.* in, of, or by a customary system of weights (**troy weight**) used for gems and precious metals. One pound troy equals a little over four fifths of an ordinary pound. [from Middle French *Troyes,* city in France, former site of a fair at which this weight may have been used]
—*n.* a troy weight.

Trp, *abbrev.* tryptophan.

true, *adj. Biology.* conforming to or representing the type named; typical: *the true fungi. A sweet potato is not a true potato.*

true ribs, *Anatomy.* the ribs which articulate with the sternum or breatbone. In human beings the true ribs are the seven upper ribs on each side of the thorax. Compare **false ribs.**

true vocal cords, *Anatomy.* the lower pair of vocal cords, which produce the voice. Also called **inferior vocal cords.** Compare **false vocal cords.**

truncate, *adj.* **1** cut off; blunt, as if cut off: *the truncate leaf of the tulip tree.* **2** having no apex, as some spiral shells.

truncate leaf

truncated, *adj. Crystallography.* having the angles or edges of a crystal cut off or replaced by a plane face.

truncated cone or **truncated pyramid,** *Geometry.* a cone or pyramid whose apex or vertex is cut off by a plane.

truncation, *n. Crystallography.* the replacement of an angle or edge by a crystalline face.

trunk, *n.* **1** *Botany.* the main stem of a tree, as distinct from the branches and the roots.
2 *Anatomy.* a human or animal body without the head, arms, and legs. SYN: torso.
3 *Zoology.* the thorax of an insect.
4 *Biology.* the main stem of a blood vessel, nerve, or similar structure, as distinct from its branches.

truth, *n.* an informal name for the property possessed by a top quark. Compare **beauty.**

truth set, = solution set.

truth table, *Mathematics.* a table that lists all possible combinations of true and false values that can be assigned to a proposition. Truth tables are designed to show logical relationships and are used in mathematics, logic, and computer engineering. *For each symbolic circuit one can construct a "truth table," in which are listed all possible input states and the corresponding output states* (David C. Evans).

truth-value, *n. Mathematics.* the truth or falseness of a proposition or statement: *A system of logic is concerned essentially with the investigation and analysis of the truth-value of a given statement. Such a truth-value can be assigned only to a declarative sen-*

tence free of any ambiguous or vague words ... sentences such as: *Tomorrow is Tuesday. Albany is the capital of New York State ... have objective truth-values which are generally accepted.* (Mathematics Regents Syllabus).

trypanosome (tri pan′ə sōm), *n.* any of a genus (*Trypanosoma*) of parasitic flagellate protozoans inhabiting the blood of vertebrates, usually transmitted by blood-sucking insects and causing serious diseases, such as sleeping sickness. [from Greek *trypanon* a borer, anger + *sōma* body]

trypsin (trip′sən), *n. Biochemistry.* **1** an enzyme in the juice secreted by the pancreas, that aids in digestion by changing proteins to peptones by hydrolysis: *The trypsin splits proteins, proteoses, and peptones from the stomach into simpler and simpler compounds* (Shull, *Principles of Animal Biology*).
2 any enzyme having a similar function.
[from Greek *tryein* to wear down by rubbing + English (*pe*)*psin* (so called because the enzyme was originally obtained by rubbing the pancreas with glycerin)]

trypsinogen (trip sin′ə jən), *n. Biochemistry.* an inactive form of the pancreatic enzyme trypsin, converted into trypsin by enterokinase in the small intestine.

tryptamine (trip′tə mēn), *n. Biochemistry.* a crystalline substance closely related to serotonin, formed in the tissues from tryptophan. It is an intermediate in various metabolic processes in animals and plants. *Formula:* $C_{10}H_{12}N_2$

Truth Table

of the conditional p —→ q, representing the proposition, "If a person lives in Chicago, then he lives in Illinois." EXAMPLE:

p	q	p —→ q
John lives in Chicago	John lives in Illinois	True
John lives in Chicago	John doesn't live in Illinois	False
John doesn't live in Chicago	John lives in Illinois	True
John doesn't live in Chicago	John doesn't live in Illinois	True

See another example under **conjunction.**

tryptic, *adj.* having to do with, produced by, or of the nature of trypsin: *It seems probable ... that proteolytic digestion in plants is always tryptic* (Athanaeum).

cap, fāce, fäther; best, bē, tèrm; pin, fīve;
rock, gō, ôrder; oil, out; cup, pùt, rüle;
yü in use, *yü* in uric;
ng in bring; sh in rush; th in thin, ᴛн in then;
zh in seizure.
ə = *a* in about, *e* in taken, *i* in pencil, *o* in lemon, *u* in circus

tryptophan (trip′tə fan), *n. Biochemistry.* a colorless, solid, essential amino acid formed from proteins by the digestive action of proteolytic enzymes: *Researchers have been experimenting with tryptophan, the chemical from which the body makes serotonin. Only a small dose of tryptophan—which is found in many foods, notably milk—seems to ease the insomniac to sleep* (Time). *Formula:* $C_{11}H_{12}N_2O_2$ *Abbreviation:* Trp [from *tryptic* + Greek *phainein* to appear]

TSH, *abbrev.* thyroid-stimulating hormone.

tsunami (tsü nä′mē), *n. Oceanography.* a large, destructive ocean wave caused by an underwater earthquake or volcanic eruption: *The final remarkable feature of tsunamis is that the height of the waves in the deep ocean is only about three to five metres. However, the energy locked up in these small metre-high waves ... becomes all too apparent as the water shoals and land approaches ... a five-metre-high water crest travelling at 600 km/h in deep water becomes a devastating 30-metre-high wave travelling at 50 km/h as it approaches the shore* (New Scientist). Commonly called **tidal wave.** [from Japanese]
—**tsunamic,** *adj.* of or having to do with a tsunami: *tsunamic waves.*

tubal, *adj. Anatomy.* of, having to do with, or involving a tube, such as the Fallopian tube or the Eustachian tube. **Tubal ligation** is a form of female sterilization in which the Fallopian tubes (oviducts) are cut and the severed ends are tied. **Tubal tonsils** are small aggregations of lymphatic tissue at the entrance of the Eustachian tube.

tube, *n.* **1** *Botany.* the lower united portion of a gamopetalous corolla or a gamosepalous calyx.
2 *Anatomy.* any of various tubular or cylindrical organs or parts: *the Eustachian tube, the bronchial tubes.*
3 = tube of force.

tube foot, *Zoology.* one of the many tubular organs of locomotion of a starfish, sea urchin, or other echinoderm.

tuber (tü′bər *or* tyü′bər), *n.* **1** *Botany.* a solid, thickened portion or outgrowth of an underground stem, of a more or less rounded form, bearing modified axillary buds from which new plants may arise. A potato is a tuber.
2 *Zoology.* a rounded swelling or projecting part in an animal body.
[from Latin *tuber* lump]

tubercle (tü′bər kəl *or* tyü′bər kəl), *n.* **1** *Botany.* a small, rounded knob or projection on a plant, such as a nodule on the root of a legume.
2 *Zoology.* a small, rounded projection or protuberance, as on a bone or on the surface of the body in various animals.
3 *Anatomy.* a knob near the head of a rib at the point of articulation with the transverse process of a vertebra.
4 *Medicine.* a small swelling on the surface of the body or in a part or organ, especially a mass of lymphocytic and other cells characteristic of tuberculosis.
[from Latin *tuberculum,* diminutive of *tuber* lump]

tuberculate *or* **tuberculated,** *adj.* having or affected with tubercles; characterized by tubercles: *tuberculate grain, tuberculate skin.*

tuberosity, *n. Botany, Zoology.* **1** the quality or condition of being tuberous.
2 *Anatomy.* a large, irregular protuberance of a bone, especially for the attachment of a tendon or ligament.

tuberous, *adj. Botany, Zoology.* **1** bearing tubers: *a tuberous plant, a carp with a tuberous body.* **2** of the nature of tubers or abnormal swellings: *vascular tuberous growth.* **3** covered with rounded knobs or swellings: *tuberous flesh.*

tuberous root, *Botany.* a true root (usually one of a cluster) thickened so as to resemble a tuber, but bearing no buds, as in the lesser celandine and the dahlia.
—**tuberous-rooted,** *adj.* having a tuberous root or roots: *tuberous-rooted irisis and anemones.*

tubicolous (tu bik′ə ləs), *adj. Zoology.* inhabiting a tube, as a mollusk with a tubular shell, an annelid with a tubular case, or a spider which spins a tubular web: *Unlike the majority of tubicolous marine worms, Chaetopterus lacks any kind of protrusible tentacular apparatus for obtaining food and oxygen from the surrounding waters. Instead, it possesses three highly modified segments in the middle of its body which act to pump a current of nutrient-laden water through the tube* (S. C. Brown). [from Latin *tubus* tube + *colere* inhabit]

tubular, *adj. Botany, Zoology.* **1** shaped like a tube: round, hollow, and open at one or both ends: *When the secreting cells thus indented form a channel of nearly uniform diameter, the gland is said to be tubular* (Shull, *Principles of Animal Biology*). **2** that is a tube; consisting of a tube: *a tubular corolla or calyx.*

tubule, *n. Anatomy.* a small tube, especially a minute tubular structure in an animal body: *the uriniferous tubules of the kidney.*

tubulin (tü′byə lin), *n. Biochemistry.* a globular protein that is the basic structural unit of microtubules: *Reports of ... research on microtubules have been concerned with the role of calcium in regulating the assembly of microtubules* (Richard Weisenberg).

tufa (tü′fə *or* tyü′fə), *n. Geology.* **1** any of various porous rocks formed of powdery matter consolidated and often stratified, especially a form of limestone deposited by springs, lakes, etc.: *Tufa, or calcareous tufa, as it is sometimes called to distinguish it from volcanic tuff, also forms in springs and lime-saturated lakes, although to some degree its deposition seems to be fostered by the work of lime-secreting algae* (Birkeland and Larson, *Putnam's Geology*). **2** = tuff. [from Italian *tufo,* from Latin *tofus*]

tufaceous (tü fā′shəs *or* tyü fā′shəs), *adj.* of, having to do with, or resembling tufa: *tufaceous limestone, tufaceous accumulations around the craters of volcanoes.*

tuff, *n. Geology.* a rock produced by the consolidation of volcanic ash and other volcanic fragments: *Where grains of clay, sand or gravel were plentifully mingled with ash, tuff grades into shale, sandstone, or conglomerate. It is these bedded tuffs that contain fossils, often both abundant and good* (Fenton, *The Rock Book*). See the picture at **volcano.** [from Middle French *tuf,* from Italian *tufo* tufa]

tundra (tun′drə), *n. Geography, Ecology.* **1** a vast, level, treeless plain in the arctic regions. The ground beneath the surface of the tundra is frozen even in summer. Much of Siberia and northern North America is tundra.
2 any of various tracts and areas similar to this, as certain plateaus in the Andes and elsewhere. [from Russian]
▶ See the note under **biome.**

tungsten (tung′stən), *n. Chemistry.* a heavy, hard, steel-gray metallic element found only in certain rare minerals. Its melting point is higher than that of any other metal. Tungsten is added to steel to make it harder, stronger, and more elastic. It is also used in making electric light-bulb filaments, surgical instruments, solar energy devices, etc. *Symbol:* W; *atomic number* 74; *atomic weight* 183.85; *melting point* 3410°C; *boiling point* 5660°C; *oxidation state* 6. Also called **wolfram.** [from Swedish, from *tung* heavy + *sten* stove]

tungstic (tung′stik), *adj. Chemistry.* of or containing tungsten, especially with a valence of six.

tungstite (tung′stīt), *n. Mineralogy.* a rare, yellow or yellowish-green mineral, tungsten trioxide, usually occurring in powdery form. *Formula:* $WO_3 \cdot H_2O$

tunic, *n.* **1** *Biology.* a natural covering of a plant, animal, part of an animal, etc.: *the tunic of a seed, the tunics of the onion. A simple ascidian is cylindrical or globase, attached by a base or stalk. It is covered with a tough elastic layer, the ... tunic, of cellulose-like material* (which is rare in animals) (Storer, *General Zoology*). **2** = tunica. [from Latin *tunica*]

tunica (tü′nə kə), *n., pl.* **-cae** (-kē, -kī, or -sē). *Anatomy.* a covering or enveloping membrane or fold of tissue; a sheath that envelops or lines an organ or other part of the body. Also called **tunic.** [from Latin *tunica* tunic]

tunicate (tü′nə kit *or* tü′nə kāt), *adj.* **1** *Botany.* made up of concentric layers. An onion is a tunicate bulb.
2 *Zoology.* **a** having a tunic or outer covering: *tunicate sea animals.* **b** belonging to the tunicates: *a tunicate mollusk.*
—*n. Zoology.* any of a subphylum (Tunicata *or* Urochordata) of small sea chordates, including the ascidians and allied forms, characterized by a saclike body enclosed in a tough, leathery membrane, with a single or double opening through which the water enters and leaves the pharynx: *The particular group of protochordates ... are called tunicates because their bodies are encased in a little tunic or jacket looking like a small bag. Sea-squirts, which send forth a couple of jets of water if touched, are tunicates* (Science News Letter).

turbellarian (tėr′bə lãr′ē ən), *Zoology.* —*n.* any of a class (Turbellaria) of flatworms, including the planarians, which inhabit fresh or salt water or damp earth. They have external cilia which produce minute whirls in the water.
—*adj.* of or belonging to the turbellarians.
[from Latin *turbellae* bustle, stir, plural of *turba* crowd, disturbance]

turbidimetric, *adj.* of or having to do with turbidimetry.
—**turbidimetrically,** *adv.*

turbidimetry (tėr′bə dim′ə trē), *n. Chemistry.* the process of measuring or determining the turbidity of liquids.

turbidite (tėr′bə dīt), *n. Geology.* **1** sediment deposited by turbidity currents.
2 a rock formed from this sediment: *Many of the essential characteristics of normal turbidites reappear in these carbonate turbidites, such as graded bedding and alternation with fine-grained normal sediments* (A. Brouwer).

turbidity (tėr bid′ə tē), *n. Chemistry.* cloudiness or opacity of a normally clear liquid due to a suspension of solid particles or colloidal droplets.

turbidity current, *Geology.* an underwater stream of silt, mud, or the like, usually along the bottom of a slower-moving body of water: *A turbidity current is a type of density current which owes its relatively great density, as compared to the surrounding water, to solids in suspension. These currents are generated in the sea by muddy river water or by slumps, landslips, or mud flows on any submarine slope* (Robert S. Dietz).

turbinate (tėr′bə nit *or* tėr′bə nāt), —*adj. Biology.* **1** shaped like a spinning top or inverted cone: *a turbinate fruit, a turbinate shell.*
2 *Anatomy.* of, having to do with, or denoting certain scroll-like, spongy bones of the nasal passages in higher vertebrates: *The surface of the turbinate bones in front of the organ is supplied with sensory fibres from a different nerve; and these, too, respond to many of the substances which can be smelt* (A. W. Haslett).
—*n.* **1** a turbinate shell. **2** a turbinate bone.

turbopause, *n. Geophysics.* an area in which atmospheric turbulence ceases, especially such an area at the base of the thermosphere: *At the so-called turbopause this mixing [of gases] effectively ceases and the helium is released. Thus, if the turbopause falls in altitude, more helium rises to the upper atmosphere* (New Scientist).

turbulence, *n.* **1** *Meteorology.* an eddying motion of the atmosphere, interrupting the flow of wind: *Other evidence of upper air turbulence can be found in the behavior of noctilucent clouds, meteor trails, and sounding rockets. Noctilucent clouds are often wavy in appearance and continually change their shape in a manner that would be logically caused by vertical air movement ... All this leads to the inevitable conclusion that the upper air, primarily the stratosphere, is a region of turbulence and strong vertical as well as strong horizontal winds* (J. Gordon Vaeth).
2 *Physics.* the state of any fluid in which turbulent flow is taking place.

turbulent flow, *Physics.* flow in a fluid characterized by constant changes in direction and velocity at any particular point: *Air can flow in the boundary layer in two ways. The air particles can move (1) in orderly paths ... or (2) in irregular paths. The parallel flow is called*

cap, fāce, fäther; best, bē, tėrm; pin, fīve;
rock, gō, ôrder; oil, out; cup, pùt, rüle,
yü in use, *yù* in uric;
ng in bring; *sh* in rush; *th* in thin, ᴛʜ in then;
zh in seizure.
ə = *a* in about, *e* in taken, *i* in pencil, *o* in lemon, *u* in circus

691

tusk

laminar flow. The irregular flow is called turbulent flow (A. Wiley Sherwood). See the picture at **fluid flow.**

tusk, n. Zoology. a very long, pointed, projecting tooth. Elephants, walruses, and wild boars have two tusks; narwhals usually have only one tusk. *Hogs have canine teeth (eyeteeth) that often develop into sharp tusks, particularly in adult males. These tusks serve as tools for digging and as weapons for fighting* (Tony J. Cunha).

twin, n. **1** Biology. one of two offspring born at the same time to the same mother: *There are two types of twins: 'identical' (monozygotic, uniovular or one-egg twins) and 'fraternal' (dizygotic, diovular or two-egg twins). Twins of the first type are the result of the splitting of a single fertilized ovum and are thus genetically identical. Those of the second type develop from separate ova, and are no more similar genetically than are full sibs (that is, than pairs of brothers and/or sisters)* (Science News).
2 Crystallography. a composite crystal consisting of two crystals, usually equal and similar, united in reversed positions with respect to each other: *One part of the twin may appear to have been derived from the other by a revolution about some crystal direction common to both* (Dana's Manual of Mineralogy).
—**v.** Crystallography. to unite so as to form a twin: *two crystals twinned round an axis, the twinning of feldspars.*

tylosis (tī lō′sis), n., pl. **-ses** (-sēz′). Botany. a growth from a cell wall into the cavity of woody tissue: *In some woods the parenchyma cells, before they die, push bladderlike outgrowths called tyloses into the vessels, blocking them and making water conduction impossible* (New Scientist). [from Greek *tylōsis* formation of a callus, from *tylē* callus]

tympana (tim′pə nə), n. a plural of **tympanum:** *The grasshopper possesses sensory organs which include eyes, tympana, and antennae* (Biology Regents Syllabus).

tympanic, adj. Anatomy. of or having to do with the eardrum or the middle ear: *the tympanic muscle, nerve, etc.* The **tympanic cavity** is the large, irregularly shaped cavity of the middle ear.

tympanic bone, Anatomy. **1** a bone in mammals supporting the eardrum and enclosing the passage of the external ear.
2 a similar bone in lower vertebrates, such as a bone supporting the lower jaw in fishes.

tympanic membrane, Anatomy. the membrane which separates the external ear from the middle ear; the eardrum: *The tympanic membrane picks up vibrations so that the frog may have a sense of hearing* (Winchester, Zoology). *The tympanic membrane of crickets is mechanically tuned to a rather narrow spectrum of frequencies between four and six kilohertz* (John G. Lepp). Also called **tympanum.**

tympanum (tim′pə nəm), n., pl. **-na** (-nə) or **-nums. 1** = tympanic membrane. **2** = middle ear. [from Medieval Latin, from Latin *tympanum* drum, from Greek *tympanon*]

Tyndall beam, Optics, Chemistry. the visible path of a light beam that enters a colloid and is scattered by colloidal particles in the Tyndall effect. [named after John Tyndall, 1820–1893, English physicist]

Tyndall effect, Optics, Chemistry. the scattering of light in different colors by the particles of a colloid: *The bluish appearance of a light beam passing through something like a soap solution is called the "Tyndall effect"* (George Gamow). See also **ultramicroscope.**

type, n. Biology. **1** a general plan or structure characterizing a group of animals, plants, or other organisms: *Such types or common plans as those of the Arthropoda, the Annelida, the Mollusca ...* (Thomas H. Huxley).
2 a type species, type genus, or other taxon for which a broader taxon is named.
3 = type specimen: *It is common practice now for the author of a species to designate one particular specimen as the type of the species, and to indicate the museum or other collection in which it is placed* (Shull, Principles of Animal Biology).

type genus, Biology. the genus from which the name of the family or subfamily is taken; theoretically, the genus most perfectly exhibiting the family characteristics.

type species, Biology. the species from which the name of the genus is taken; theoretically, the species most perfectly exhibiting the generic characteristics.

type specimen, Biology. an individual or specimen from which the description of the species or subspecies has been prepared and upon which the specific name has been based.

typhlosole (tif′lə sōl), n. Zoology. a ridge or fold extending along the inner wall of the intestine and partly dividing the intestinal cavity in various animals, such as lampreys and certain ascidians, mollusks, and worms: *In order to increase the area where absorption of food can take place the intestine folds downward from the dorsal surface to form a projection into its interior, which is called a typhlosole. This makes the intestine much more efficient for its length than it would be otherwise* (Winchester, Zoology). [from Greek *typhlos* blind + *sōlēn* channel]

typhoon (tī fün′), n. Meteorology. a violent tropical cyclone occurring in the western Pacific or the Indian Ocean, chiefly during the period from July to October: *The western Pacific typhoons cut a storm path 50 to 100 miles (80 to 160 kilometers) wide. They travel slowly, but the violent gusty winds within the circle of the typhoon cause great destruction* (James E. Miller). [from Chinese (Cantonese) *tai fung* big wind]
ASSOCIATED TERMS: see **cyclone.**

typical, adj. Biology. that is the type of the species, family, or other group: *the typical genus.*

Tyr, abbrev. tyrosine.

tyramine (tī′rə mēn), n. Biochemistry. a colorless, crystalline amine produced by bacterial action or by the decarboxylation of tyrosine, found in citrus fruits, ripe cheese, putrefied animal tissue, etc. It has been used as an agent that imitates certain effects of the sympathetic nervous system. *Formula:* $HOC_6H_4CH_2NH_2$ [from *tyr(osine)* + *amine*]

tyrosinase (tī′rō sə nās), n. Biochemistry. an enzyme present in vegetable and animal tissues. It converts tyrosine into melanin and similar pigments by oxidation.

tyrosine (tī′rə sēn), *n. Biochemistry.* a white, crystalline amino acid produced by the hydrolysis of a number of proteins, such as casein. It is a precursor of adrenalin, thyroxine, and melanin. *The synthesis of melanin from ... tyrosine has been known for a long time and some insect melanins can be synthesized with better yield in laboratory experiments by incubating small pieces of the integument with compounds such as di-hydroxy-phenyl-alanine, commonly abbreviated to 'Dopa'* (B. Nickerson). *Formula:* $C_9H_{11}NO_3$ *Abbreviation:* Tyr [from Greek *tyros* cheese]

cap, fāce, fäther; best, bē, tèrm; pin, fīve; rock, gō, ôrder; oil, out; cup, pùt, rüle, *yü* in use, *yu̇* in uric; *ng* in bring; *sh* in rush; *th* in thin, ᴛʜ in then; *zh* in seizure.
ə = *a* in about, *e* in taken, *i* in pencil, *o* in lemon, *u* in circus

U

U, *symbol* or *abbrev.* **1** uranium. **2** universe. **3** uracil.

U-235, = uranium 235.

U-238, = uranium 238.

ubiquinone (yü′bə kwi nōn′ *or* yü′bə kwin′ōn), *n. Biochemistry.* any of a group of quinones that assist oxidation and reduction within cells, especially by serving as electron or hydrogen carriers between the flavoproteins and the cytochromes in the mitochondria of various cells: *Ubiquinone is an integral unit of the electron-transport system, and is synthesised by a long, complex series of reactions* (New Scientist). Also called **coenzyme Q.** [blend of Latin *ubique* everywhere and English *quinone*]

uintaite or **uintahite** (yü in′tə īt), *n. Mineralogy.* a variety of asphalt having a black color with a brown streak and a brilliant luster. [from the *Uinta* Mountains, Utah, where deposits of the mineral are found + *-ite*]

ulexite (yü′lek sīt), *n. Mineralogy.* a mineral, a hydrous sodium and calcium borate, occurring in loose, rounded masses of white, needle-shaped crystals that transmit light from one end to the other without being transparent. *Formula:* $NaCaB_5O_9 \cdot 8H_2O$ [from G. L. *Ulex,* 19th-century German chemist + *-ite*]

ulna (ul′nə), *n., pl.* **-nae** (-nē), **-nas.** *Anatomy.* **1** the thinner, longer bone of the forearm, on the side opposite the thumb: *The skeleton of each arm consists of the humerus or upper arm bone, the radius (on the same side as the thumb) and ulna in the lower arm, eight carpal or wrist bones, and the bones of the hand* (Harbaugh and Goodrich, *Fundamentals of Biology*). See the picture at **arm.**
2 a corresponding bone in the forelimb of a vertebrate animal: *The forearm [in bats] consists of a rudimentary ulna and a long, curved radius* (E. P. Wright). See the picture at **wing.**
[from New Latin, from Latin *ulna* elbow]
—**ulnar,** *adj.* **1** of or having to do with the ulna; towards the side on which the ulna is situated: *the ulnar border of the hand, the ulnar portion of the forearm.*
2 in or supplying the part of the forearm near the ulna: *the ulnar artery, nerve, or vein.*

ulnocarpal, *adj. Anatomy.* of or having to do with the ulna and the wrist (carpus).

ulnoradial, *adj. Anatomy.* of or having to do with the ulna and the thicker and shorter bone of the forearm (radius).

ultimate analysis, *Chemistry.* a form of analysis in which the percentage of each element in a compound or mixture is determined. Compare **proximate analysis.**

ultimate strength, *Physics.* the inherent resistance in a piece of material equal but opposed to the ultimate stress.

ultimate stress, *Physics.* the stress necessary to break or crush a piece of material.

ultimobranchial body or **ultimobranchial gland,** *Anatomy.* a small structure near the thyroid gland of most vertebrates, derived from the last of several embryonic pouches of the gills or pharynx and the source of the cells in the thyroid gland that produce the hormone calcitonin. [from Latin *ultimus* last + English *branchial*]

ultra-, *prefix.* **1** beyond; on the other side of, as in *ultraviolet = beyond the violet.*
2 beyond the limits of; transcending, as in *ultrasonic.*
3 extremely, as in *ultrabasic.*
[from Latin, from *ultra* beyond]

ultrabasic, *adj. Geology.* extremely rich in base-forming elements and poor in silica: *Underlying the crust, geophysical studies have shown, the Earth's mantle is made of different, denser material—probably an ultrabasic rock like peridotite, strongly deficient in silica* (Peter Stubbs). Also called **ultramafic.**

ultracentrifuge (ul′trə sen′trə fyüj), *n. Chemistry.* a centrifuge that can spin at very high speed (up to 80,000 revolutions per minute) and attain centrifugal fields up to 500,000 times the force of gravity, used for rapidly breaking up macromolecules, determining particle sizes, analyzing the contents of body fluids, etc.
—*v.* to subject to the action of an ultracentrifuge: *The microsomes can be isolated by ultracentrifuging a cell brei (or mush), and when in a concentrated form appear as an amber-coloured jelly* (New Biology).
—**ultracentrifugal,** *adj.* of or by means of an ultracentrifuge. —**ultracentrifugally,** *adv.*
—**ultracentrifugation,** *n.* the act or process of subjecting to the action of an ultracentrifuge: *These particles, just barely visible under ordinary microscopes, can be separated from the rest of the cell by ultracentrifugation* (Scientific American).

ultradian (ul trā′dē ən), *adj. Biology.* of or having to do with biological rhythms or cycles that recur more than once per day: *Ultradian rhythms were first suspected more than 20 years ago when researchers discovered that rapid eye movement or REM sleep occurs in cycles of 90 to 100 minutes* (Science News). Compare **circadian, infradian.** [from *ultra-* beyond + Latin *dies* day + English *-an*]

ultramafic, *adj.* = ultrabasic.

ultramicro, *adj.* = ultramicroscopic.

ultramicroscope, *n. Optics.* an instrument for making visible particles too small to be seen in an ordinary microscope. It consists of a compound microscope with several lenses and a powerful arc lamp whose light is brought to an intense focus. To study colloidal particles, the light is scattered according to the Tyndall effect, illuminating the particles so that they appear as bright points of light against a dark background. *Objects as small as 6/1,000,000 of a millimeter can be seen with an ultramicroscope* (Joseph Valasek).

ultramicroscopic, *adj.* **1** too small to be seen with an ordinary microscope: *Viruses are ultramicroscopic organisms that cause a variety of diseases ranging from wilt in tomato and tobacco plants to influenza, yellow fever and poliomyelitis* (N. Y. Times).
2 having to do with an ultramicroscope: *ultramicroscopic examination of bacterial viruses.*
—**ultramicroscopically,** *adv.*

ultrasonic, *adj. Physics.* of or having to do with sound waves or vibrations beyond the range of human hearing or above frequencies of 20,000 hertz. Ultrasonic waves are short and easily reflected, forming echoes, and at certain frequencies can produce high energy. *In ultrasonic techniques (echo sounding or sonar) echoes are reflected back to the receiving transducer and plotted on a cathode-ray tube in a fashion which geometrically represents the location of their source within the body* (I. Donald). Compare **subsonic.** See also **ultrasound.**

ultrasonics, *n.* **1** the science that deals with ultrasonic waves, especially their application in science and industry.
2 ultrasonic waves; ultrasound: *In cleaning delicate and intricate items such as watch mechanisms, small gears, surgical instruments or parts with blind holes, ultrasonics can be used to remove dirt, swarf, grease or grinding paste quickly and efficiently* (New Scientist).

ultrasound, *n. Physics.* sound waves or vibrations above a frequency of 20,000 hertz; ultrasonic waves: *The most spectacular use of ultrasound in medicine was in visualization of internal tissues, such as the eye or the brain, by a technique reminiscent of sonar (the underwater equivalent of radar): pulses of sound energy are sent into the body and their reflections are mapped on a picture tube* (Charles Süsskind).

ultrastructural, *adj.* of or having to do with ultrastructure: *The cells, even in the heart-shaped embryo of cotton and Capsella, are virtually indistinguishable in terms of ultrastructural characteristics* (William A. Jensen).

ultrastructure, *n. Biology.* a structure with extremely fine details, invisible to an ordinary microscope; ultramicroscopic structure: *Particular attention is being paid to the ultrastructure of neurosecretory systems in crustaceans and in vertebrates, using the electron microscope* (New Scientist).

ultraviolet, *Physics.* —*adj.* of or having to do with the invisible part of the spectrum whose rays have wavelengths shorter than those of the violet end of the visible spectrum and longer than those of the X rays. The ultraviolet spectrum is usually considered to extend from about 50 to 380 nanometers. *ultraviolet light. Ultraviolet rays shorter than 320 nm are effective in killing bacteria and viruses . . . Direct exposure of the skin to ultraviolet rays from the sun or from other sources produces vitamin D in the body* (James A. R. Samson). See the picture at **electromagnetic spectrum.**
—*n.* ultraviolet radiation; the part of the spectrum comprising ultraviolet radiation: *Camera film is very sensitive to ultraviolet* (Tracy, *Modern Physical Science*). Abbreviation: **UV**

ultravirus, *n. Biology.* a virus that can go through the finest filters; an extremely small virus.

umbel (um′bəl), *n. Botany.* a flower cluster in which stalks nearly equal in length spring from a common center and form a level or slightly curved surface, as in parsley. A simple umbel has only one set of rays, as in the ginseng; in a compound umbel the pedicels or rays each bear an umbel, as in the carrot and dill. [from Latin *umbella* parasol, diminutive of *umbra* shade]
—**umbellate** (um′bə lit *or* um′bə lāt), *adj.* **1** of or having umbels; forming an umbel or umbels: *an umbellate flower.*
2 having flowers in umbels: *an umbellate plant.*

umbel

umbelliferous (um′bə lif′ər əs), *adj. Botany.* **1** bearing an umbel or umbels: *The parsley and carrot are umbelliferous.* **2** belonging to the parsley family. [from New Latin *Umbelliferae* the parsley family, from Latin *umbella* parasol + *ferre* to bear]

umbellule (um′bəl yül), *n. Botany.* a small or partial umbel; an umbel formed at the end of one of the primary pedicels or rays of a compound umbel.

umbilical (um bil′ə kəl), *adj. Anatomy.* of, having to do with, or situated near the umbilical cord: *the umbilical region, the umbilical vein.*

umbilical cord, *Anatomy.* a cordlike structure that connects the navel of an embryo or fetus with the placenta of the mother. It carries nourishment to the fetus and carries away waste. *The mammalian embryo is attached to the placenta by the umbilical cord, an organ of rope-like appearance which forms as the embryo develops. The umbilical cord contains large blood vessels which carry the embryo's blood to and from the placenta* (Harbaugh and Goodrich, *Fundamentals of Biology*).

umbilicus (um bil′ə kəs *or* um′bə li′kəs), *n., pl.* **-ci** (-kī *or* -sī). **1** *Anatomy.* the round, usually depressed and puckered scar in the middle of the surface of the abdomen, where the umbilical cord was attached before birth; the navel.
2 *Botany, Zoology.* a small depression or hollow suggestive of a navel, such as the hilum of a seed. [from Latin *umbilicus* navel]

umbo (um′bō), *n., pl.* **umbones** (um bō′nēz). **1** *Zoology.* the most protuberant portion of a bivalve mollusk shell, located near the hinge: *Thus, the age of a clam can be estimated. Near the hinge and slightly anterior is a little hump in the shell called the umbo. The rings*

cap, fāce, fäther; best, bē, tėrm; pin, fīve;
rock, gō, ôrder; oil, out; cup, pùt, rüle;
yü in use, *yu̇* in uric;
ng in bring; *sh* in rush; *th* in thin, ᴛʜ in then;
zh in seizure.
ə = *a* in about, *e* in taken, *i* in pencil, *o* in lemon, *u* in circus

radiate out from this region, so this is the oldest portion of the shell (Winchester, *Zoology*).

2 *Anatomy.* a rounded projection of the surface of the tympanic membrane.

[from Latin *umbo* boss of a shield, knob, related to *umbilicus* navel]

umbra (um′brə), *n., pl.* **-brae** (-brē), **-bras.** *Astronomy.*
1 the completely dark portion of the shadow cast by the earth, moon, etc., during an eclipse: *This umbra is the full shadow and is quite dark. The unaided eye from the earth sees the umbra with sharp clear edges; but with low-power glasses the shadow appears somewhat diffuse, and high magnification prevents us entirely from defining the exact edge of the dark shadow* (Bernhard, *Handbook of Heavens*). See the picture at **eclipse.**

2 the dark inner or central part of a sunspot: *A sunspot consists of two distinct parts: the umbra, the more or less central dark part, and the penumbra, the lighter border which is three fourths as bright as the photosphere* (Baker, *Astronomy*).

[from Latin *umbra* shade, shadow]

—**umbral,** *adj.* of or having to do with an umbra: *the umbral shadow of an eclipse.*

unary (yü′nə rē), *adj. Mathematics.* consisting of, using, or applied to only one object or numeral: *Eight cells on the paper tape are marked 1111 + 111, signifying the addition of 4 and 3 in the "unary" system in which an integer n is symbolized by n 1's* (Martin Gardner).

[from Latin *unus* one + English *-ary;* patterned on *binary*]

uncertainty principle, *Physics.* the principle in quantum mechanics that certain coordinates of a single physical object, such as the position and velocity of an electron, can never be accurately determined simultaneously: *Among the many important experimental phenomena illustrating the uncertainty principle is the Compton effect. Here, a photon is scattered by an electron; the momentum of the photon is rendered uncertain as a result of its scattering by the electron and the electron is moved from its original position by the impact received from the photon* (A. L. Loeb). Also called **Heisenberg uncertainty principle, indeterminacy principle.**

unciform (un′sə fôrm), *Anatomy.* —*adj.* denoting or having to do with the hamate bone, its hooklike process, or any similar hooklike process.

—*n.* the hamate bone.

[from New Latin *unciformis,* from Latin *uncus* hook + *forma* form]

unciform process, *Anatomy.* **1** the process projecting from the palmar surface of the hamate bone. **2** a hook-shaped process of the ethmoid bone.

uncinate (un′sə nāt), *adj. Anatomy, Zoology.* bent at the end like a hook; hooked; unciform: *the uncinate process of the pancreas.*

unconformity, *n. Geology.* a surface representing a break in the continuity of strata, with the rock immediately above that surface being younger than the rock beneath: *Unconformities are classified into three types—disconformities, angular unconformities, and nonconformities—with each type having important implications for the geologic history of the area in* which it occurs (Plummer and McGeary, *Physical Geology*).

uncus (ung′kəs), *n., pl.* **unci** (ung′kī *or* un′sī). *Anatomy, Zoology.* a hooklike part or process: *the unci of a locust.* [from Latin *uncus* hook]

underbrush, *n. Botany.* bushes, shrubs, and small trees growing under large trees in woods and forests.

undergrowth, *n.* **1** = underbrush. **2** *Zoology.* the shorter, finer coat of hair underlying the outer coat of any of various animals: *the undergrowth of sheep and goats.*

undershrub, *n. Botany.* a plant having a shrubby base: *Such low-growing shrubby plants as heather . . . are termed undershrubs* (J. E. Willis). Also called **subshrub.**

undertow, *n. Oceanography.* any strong current below the surface of the water, moving in a direction different from that of the surface current: *The water thrown forward by breaking waves rushes upon the shore only to lose its velocity and run back, under the pull of gravity, beneath other oncoming waves. The returning water is called the undertow, and it has sufficient force to be an important factor in erosion* (Finch and Trewartha, *Elements of Geography*).

undifferentiated, *adj. Biology.* not differentiated; not modified and specialized in the course of development: *Generally, invertebrate animals possess more undifferentiated cells than do vertebrate animals. As a result, invertebrates exhibit a higher degree of regenerative ability than most vertebrates* (Biology Regents Syllabus).

undulation, *n. Physics.* a wavelike motion in air or in another medium, as in the propagation of sound or light; a vibration or wave.

undulatory theory, = wave theory.

unguiculate (ung gwik′yə lit), *adj.* **1** *Zoology.* **a** having nails or claws, as distinguished from hoofed animals and cetaceans: *an unguiculate mammal, unguiculate quadrupeds.* **b** ending in, or taking the form of, a nail or claw: *unguiculate limbs or digits, unguiculate mandibles.*

2 *Botany.* having a clawlike base: *unguiculate petals.*

—*n. Zoology.* an unguiculate animal, especially a quadruped.

[from New Latin *unguiculatus,* from Latin *unguiculus,* diminutive of *unguis* nail, claw, hoof]

unguis (ung′gwis), *n., pl.* **-gues** (-gwēz). **1** *Zoology.* a nail, claw, or hoof, especially on a quadruped. Also called **ungula.**

2 *Botany.* the narrow, clawlike base of certain petals, by which they are attached to the receptacle.

[from Latin *unguis* nail, claw, hoof]

696

ungula (ung′gyə lə), *n., pl.* **-lae** (-lē). **1** = unguis (def. 1).
2 *Geometry.* a cylinder, cone, or other solid figure, the
top part of which has been cut off by a plane oblique
to the base. [from Latin *ungula* hoof, from *unguis* nail,
claw]

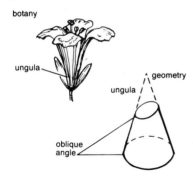

botany

ungula

geometry

ungula /

oblique
angle

—**ungular,** *adj.* of or having to do with an ungula.
ungulate (ung′gyə lit *or* ung′gyə lāt), *Zoology.* —*adj.* **1**
having hoofs: *The competition . . . must be between giraffe and giraffe, and not with the other ungulate animals* (Darwin, *Origin of Species*).
2 having the form of a hoof; hoof-shaped: *the ungulate
feet of a gazelle.*
—*n.* any of a group of mammals that have hoofs, including the ruminants, horses, rhinoceroses, elephants, and pigs; an ungulate mammal: *Bison, cattle,
sheep, horses, and other ungulates, both wild and domestic, gather into herds and flocks* (Simpson, *Life:
An Introduction to Biology*). [from Latin *ungulatus,*
from *ungula* hoof]
unguligrade (un′gyə lə grād), *Zoology.* —*adj.* walking
on hoofs, with the rest of the foot off the ground: *The
primitive type of foot posture is called plantigrade.
The entire palm or sole rests on the ground . . . This
type gave way to the digitigrade posture, the animals
walking upon their digits . . . The third type of foot
posture is the unguligrade . . . These ungulates walk
on modified nails or hoofs, such as those of the horse*
(Hegner and Stiles, *College Zoology*).
—*n.* an unguligrade animal.
[from New Latin *unguligrada,* from Latin *ungula* hoof
+ *gradus* step, grade, from *gradi* to step]
uni- (yü′nə- *or* yü′nē-), *prefix.* one; a single; having, or
made of, only one, as in *unicellular = having one cell.*
[from Latin, from *unus* one]
uniarticular or **uniarticulate,** *adj. Anatomy, Zoology.*
having a single joint or articulation: *uniarticular tarsi,
a uniarticulate snake.*
uniaxial, *adj.* **1** *Optics.* having a single axis along which
light can propagate without double refraction: *There
is one direction in a quartz crystal . . . in which both
beams travel with the same velocity. This is known as
the optic axis and, since quartz contains but one such
axis, it is known as a uniaxial crystal* (Hardy and Perrin, *Principles of Optics*).
2 *Botany.* having only one axis, as when the primary
stem of a plant does not branch and terminates in a
flower.
unicellular, *adj. Biology.* having one cell only; single-celled: *Aging and senescence cannot be observed
in unicellular organisms, since they divide continuous-*

*ly, and it is not possible to distinguish the daughter
cells from their parents* (Nathan W. Shock).
—**unicellularity,** *n.* the condition of being single-celled.
unicostate, *adj. Botany.* having one large vein, the midrib, running down the center: *a unicostate leaf.*
unicuspid, *Anatomy.* —*adj.* having only one cusp, as an
incisor or canine tooth. —*n.* a unicuspid tooth.
unidirectional, *adj. Physics.* **1** flowing, moving, or radiating in only one direction: *unidirectional pulses, unidirectional radio waves.*
2 of, involving, or responding in only one direction:
*Unidirectional microphones are sensitive to sounds
from one direction* (Samuel Seely).
unifactorial, *adj. Genetics.* involving, dependent on, or
controlled by one gene only. Contrasted with
multifactorial.
unified field theory, *Physics.* any theory seeking to account for several fundamental forces by demonstrating
mathematical or physical connections between them:
*In 1949, Einstein, hoping that a unified field theory
would remove the need for a statistical interpretation
of quantum physics and make matter a property of the
field, published "A generalized theory of gravitation,"
a theory of such mathematical complexity that no
practicable method of experimentally testing its conclusions has yet been found* (J. G. May). Compare
field theory. See also **Weinberg-Salam theory.**
unifilar (yü′nə fi′lər), *adj.* having, suspended by, or using a single thread or fiber: *a unifilar magnetometer.*
[from *uni-* + Latin *filum* thread]
uniflorous, *adj. Botany.* having or bearing only one flower.
unifoliate, *adj. Botany.* **1** having only one leaf: *a unifoliate stem.* **2** = unifoliolate.
unifoliolate, *adj. Botany.* **1** (of a leaf) compound in
structure but having only one leaflet, as in the orange.
2 (of a plant) having such leaves.
uniformitarianism, *n. Geology.* the school of thought,
associated with the 19th-century geologist Charles Lyell, which assumes that geological phenomena are the
product of natural forces operating over long periods
of time with considerable, though not necessarily uniform, uniformity: *Lyell was, if anything, too extreme in his
uniformitarianism and was unready even to admit the
possibility of mild and limited catastrophes* (Isaac Asimov). Contrasted with **catastrophism.**
unijugate, *adj. Botany.* (of a pinnacle leaf) having only
one pair of leaflets.
unilateral, *adj.* **1** *Botany.* having all the parts arranged
on one side of an axis; turned to one side: *unilateral
flowers, as those of the forget-me-not.*

cap, fāce, fäther; best, bē, tėrm; pin, fīve;
rock, gō, ôrder; oil, out; cup, pùt, rüle,
yü in use, yu̇ in uric;
ng in bring; sh in rush; th in thin, ᴛʜ in then;
zh in seizure.
ə = a in about, e in taken, i in pencil, o in
lemon, u in circus

2 *Anatomy, Medicine.* occurring on or affecting only one side of the body or one side of an organ or part: *a unilateral section of the medulla, unilateral sciatica.* **3** *Geometry.* having only one side; one-sided: *a unilateral surface.* —**unilaterally,** *adv.*

unilobed or **unilobar,** *adj. Anatomy.* having or consisting of only one lobe: *a unilobed lung, a unilobar liver.*

unilocular, *adj. Botany, Zoology.* having or consisting of only one cavity or cell: *One form of sporangium originates as a single, enlarged, rounded cell at the end of a short branch (unilocular sporangium)* (Emerson, *Basic Botany*). *The Paper Nautilus, the Cowries, . . . etc. are unilocular shells* (W. Wood).

unimolecular, *adj. Chemistry.* **1** involving only one type of molecule: *a unimolecular reaction.*
2 having the thickness of a single molecule; monomolecular: *Available as an item of mass production . . . was a filtering medium with a uniform cell structure that approaches unimolecular dimensions which may be likened to a submicroscopic honeycomb section* (Leon J. Le Beau).

uninucleate, *adj. Biology.* having a single nucleus: *a uninucleate cell. In the excysted amoeba, each nucleus divides and as it does so the body too divides, separating off the daughter nucleus, so that eight uninucleate amoebae develop from the product of a single cyst* (R. S. J. Hawes).

union, *n. Mathematics.* a set including all the members which belong to either or both of two sets without repeating any members. EXAMPLE: If set A = { 1, 2, 3, 4 } and set B = { 4, 5, 6 }, then the union of the two sets is { 1, 2, 3, 4, 5, 6 }. The union of two sets A and B is denoted by A ∪ B, the symbol for union being ∪.

uniovular, *adj. Biology.* = monozygotic.

uniparous (yü nip′ər əs), *adj.* **1** *Zoology.* bearing or producing one egg or one offspring at a birth: *Elephants are uniparous.* **2** *Botany.* developing a single axis at each branching: *a uniparous cyme.* [from New Latin *uniparus,* from Latin *unus* one + *parere* to bear]

unipolar, *adj.* **1** *Physics.* **a** produced by or proceeding from one electric or magnetic pole: *unipolar conductivity.* **b** having or operating by means of one electric or magnetic pole: *In unipolar devices conduction is due to a single type of carrier: either holes or electrons, but not both* (New Scientist).
2 *Anatomy.* having or confined to only one fibrous process: *the unipolar nerve cells of the spine and cranium.*
—**unipolarity,** *n.* unipolar quality or condition.

uniseptate, *adj. Biology.* having only one septum or partition: *uniseptate spores.*

unisexual, *adj.* **1** *Biology.* having the essential reproduction organs of one sex only; not hermaphroditic: *Marine snails are generally unisexual, or of one sex only* (W. J. Clench). **2** = diclinous (def. 2).

unit, *n.* **1** *Mathematics.* a single magnitude or number considered as the base of all numbers, especially the smallest whole number; one; 1: *to count in units, tens, and hundreds.*
2 a quantity taken as a standard of measurement, especially: **a** a standard physical quantity in any system of measurement: *a unit of mass, an angstrom unit, the electromagnetic unit of charge. An inch, a degree, and*

a radian are standard units. **b** the quantity of a vaccine, serum, drug, or other agent necessary to produce a specific effect: *international units of vitamins, a unit of blood in transfusions.*
—**adj. 1** of, having to do with, or equivalent to a unit: *a unit measure, the unit rate of velocity, a circuit of unit resistance.* **2** consisting of, containing, or forming a unit or units: *a unit dose of radiation.*

unitarity, *n. Nuclear Physics.* the principle that if a particle can decay by several modes, the sum of the fractions decaying by each mode should add up to one: *Among all the processes particle physicists have to deal with, the decay of K into two muons may seem a small thing to cause consternation, but its nonappearance indicated a possible violation of a fundamental law, the principle of unitarity* (Science News).

unit cell, *Crystallography.* the basic building block of a crystal, consisting of the smallest group of molecules, atoms, or ions repeated over and over to make up the crystal's geometric pattern: *In a crystal all the atoms are distributed in an orderly arrangement throughout the space occupied by the crystal. This is called a lattice . . . These patterns are repeated throughout the crystal, and each is called a unit cell* (W. D. Corner).

unit circle, *Mathematics.* a circle whose radius is one unit of distance; a circle of radius 1.

unit magnetic pole, *Physics.* a magnetic pole which repels a like pole at a unit distance of one centimeter with a force of one dyne.

unity, *n. Mathematics.* **1** a quantity or magnitude regarded as equivalent to 1 in calculation, measurement, or comparison: *As the sum of the coefficients is unity, only two need be known and then the third can be found by simple arithmetic. This fact makes possible the use of a two-dimensional diagram on which only two of the coefficients are plotted* (C. L. Boltz). **2** = identity element.

univalence (yü′nə vā′ləns), *n. Chemistry.* a univalent quality or condition.

univalent (yü′nə vā′lənt), *adj.* **1** *Chemistry.* having a valence of one; monovalent.
2 *Biology.* single (applied to a chromosome which lacks, or does not unite with, its homologous chromosome during synapsis).
—**n.** *Biology.* a univalent chromosome: *There was also an increase with age in the number of chromosomes that had separated from their partner. Such separated chromosomes are called univalents, and they can arise through the failure of chiasmata association* (R. G. Edwards and Ruth E. Fowler).

univalve, *Zoology.* —**n. 1** any mollusk having a shell consisting of one piece: *These shells demonstrate the two main classes of mollusks. About three of every four shells are univalves, with a single spiral shell. The snails and their relatives build univalve shells. The bivalve mollusk has two pieces, joined by a hinge, and is typified by clams, mussels, and oysters* (Science News Letter).
2 the shell of such a mollusk.
—**adj. 1** having a shell consisting of one piece. **2** composed of a single piece: *a univalve shell.*

universal, *adj. Mathematics.* having to do with a universe of objects or numbers; comprising all the elements under consideration: *a universal set.*

Universal Coordinated Time, = Coordinated Universal Time.

universal donor, *Immunology.* a blood donor who has type O blood, which can be given fairly safely in transfusions to a person of any blood type: *In emergency group O blood, having no antigens, may be given to anybody irrespective of group (hence the term 'universal donor' applied to this group)* (G. Fulton Roberts).

universal gas constant, = gas constant.

universal gravitation, *Physics.* gravitation conceived as a property of all matter in the universe. *The gravitational force or gravitational pull exerted by the earth on objects near its surface . . . is merely an example of a property possessed by all material bodies, called universal gravitation* (Shortley and Williams, *Elements of Physics*).

Universal Time, a generic designation for several international time standards, the two most widely used being Coordinated Universal Time and UT1.

universe, *n.* **1** *Astronomy.* the totality of matter and energy in space and time, including all galaxies and other celestial bodies. Cosmology is the study of the universe: *The astrophysicists tell us that there is practically no lithium, beryllium, or boron in the universe. This might be due to the fact that these light elements are consumed very rapidly in the nuclear chemistry occurring in stellar interiors* (E. P. George).
2 *Mathematics.* the set of all objects being considered at one time. The universe might be the set of all natural numbers, the numbers from 0 through 10, all animals, or the animals on a farm. *Abbreviation:* U
3 *Statistics.* the total number of items from which a sample is selected: *The process of drawing general conclusions about a universe on the basis of the characteristics of a sample from the universe is known as sampling, or statistical inference* (Parl, *Basic Statistics*). SYN: population.

unknown, *n.,* or **unknown quantity,** *Mathematics.* a quantity whose value is to be found, usually represented by a letter, such as x, y, or z. In the equation $3x - 1 = 2x + 1$ the unknown is x. *We are usually interested in finding values of the unknown or unknowns which will make the two sides of an equation equal* (Paul R. Rider).

unsaturated (un sach′ə rā′tid), *adj.* **1** *Chemistry.* **a** (of a solvent or a solution) able to absorb or dissolve an additional quantity of a substance: *A solution which has not yet reached such a state of equilibrium and in which dissolution of the solute may still take place is described as unsaturated* (Jones, *Inorganic Chemistry*). **b** (of an organic compound) having a double or triple bond so that additional atoms or groups may be taken on without the liberation of other atoms, radicals, or compounds: *Propylene is unsaturated; it has unused chemical bonds that can bind the incoming tritium atom to make a short-lived intermediate, a propyl radical (H_2C-CHT-CH_3)* (Richard Wolfgang). *The unsaturated fats, including so-called polyunsaturates, generally are found in vegetable oils and usually are liquid at room temperature* (Wall Street Journal).
2 *Mineralogy.* incapable of developing in the presence of excess silica; not containing all of the silica that could be present: *The feldspathoids are unsaturated minerals. A nepheline syenite is an unsaturated rock.*

unsaturated radical, *Chemistry.* an organic radical having a double or triple bond which joins two atoms of carbon.

upheaval, *n.* *Geology.* the act of raising, or fact of being raised, above the original level, especially by volcanic action: *Very extensive regions . . . have been undergoing slow and gradual upheaval* (Charles Lyell). Compare **upthrow.**

uplift, *Geology.* —*n.* an elevation or rise of a part of the earth's surface: *Uplift . . . turns sea bottoms into land. Large parts of our continent consist of old sea bottoms which were raised until they formed land. This movement upward of the earth is called diastrophism* (Carroll L. Fenton).
—*v.* to cause (a part of the earth's surface) to rise above the surrounding area: *A few islands in the main ocean basins have been uplifted at irregular rates* (William M. Merrill).

upper, *adj.* *Geology.* **1** Usually, **Upper.** being or relating to a later division of a period, system, or the like: *Upper Cambrian.*
2 lying nearer the surface and formed later than others of its group, type, or class: *an upper stratum.*

upper air or **upper atmosphere,** *Meteorology.* the region of the atmosphere above the troposphere: *Determination of the composition, characteristics, and processes of the upper atmosphere . . . is accomplished by . . . sampling of atmospheric conditions and constituents by such research vehicles as rockets and balloons [and] . . . application of physical laws to the data obtained, and deduction of the conditions, constituents, and processes which produce these phenomena* (J. Gordon Vaeth).

upper bound, *Mathematics.* a number greater than or equal to every number in a given set of real numbers. Contrasted with **lower bound.** Compare **least upper bound.**

Upper Carboniferous, *Geology.* **1** the name outside of North America for the Pennsylvanian period: *Generally, the distinctness in character of the Lower and Upper Carboniferous is as strongly marked in Europe and Asia, for example, as that separating Mississippian from Pennsylvanian* (Moore, *Introduction to Historical Geology*).
2 the rocks formed during this period.

up quark, *Nuclear Physics.* a type of quark having a charge of $+ 2/3$ and a spin of $+ 1/2$: *It was possible to distinguish between two types of quarks on the basis of their different masses and electric charges, and these were called "up quarks" and "down quarks." The two kinds of quarks were sufficient to construct neutrons and protons* (Sheldon L. Glashow). Also called **u quark.**

---

cap, fāce, fäther; best, bē, tèrm; pin, five;
rock, gō, ôrder; oil, out; cup, pùt, rüle,
yü in use, *yu* in uric;
ng in bring; sh in rush; th in thin, ᴛн in then;
zh in seizure.
ə = *a* in about, *e* in taken, *i* in pencil, *o* in lemon, *u* in circus

upsilon (yüp′sə lon), *n.*, or **upsilon particle,** *Nuclear Physics.* any of a group of extremely heavy, short-lived, neutral particles, with a mass about ten times as large as that of the proton. Upsilons consist of bottom quarks paired with their own antiquarks. *Two groups . . . have been studying the upsilon particle, a b quark combined with a b antiquark. They confirmed the existence of two excited upsilon particles whose quarks and antiquarks move more rapidly than the quarks and antiquarks of normal upsilons. Physicists analyze the energy differences among the various versions of the upsilon to measure the force that binds the quark to its antiquark* (Robert H. March). [from *upsilon,* name of the 20th letter of the Greek alphabet (υ)]

upslope fog, *Meteorology.* a fog formed by the cooling of stable moist air as it moves from lower to higher elevations: *When air moves upslope against a mountain side, or even up a gradually sloping plain, the adiabatic cooling due to ascent may result in saturation and the development of an upslope fog* (Blair, *Weather Elements*).

uptake, *n. Physiology.* absorption; ingestion: *Patient research work has been carried out . . . on all aspects of the subject—the effect of diet, of mineral uptake during the period in which the teeth are being formed* (New Scientist). *Iodine is used by the thyroid gland, and the amount and rate of uptake of iodine varies with the functional state of the thyroid* (Shields Warren).

upthrow, *n. Geology.* an upward dislocation of a mass of rock, generally caused by faulting. Compare **upheaval.**

upthrust, *Geology.* —*n.* an upward movement of part of the earth's crust: *The most significant discovery was that the Transantarctic Mountain chain, of which the Queen Maud Range is a part, was caused not by a single upthrust but at least partly by the sinking of adjacent land* (Laurence M. Gould). See also **uplift.**
—*v.* to push up, as in an upthrust: *The region as a whole consists of a geological anticline, . . . upthrust by an outer ripple of those same Miocene foldings that formed the Alps and the Himalayas* (New Scientist).

upwarp, *Geology.* —*v.* to fold or bend in an anticlinal or upward manner: *In the latter part of the [Tertiary] period there was a change. The whole region was moved vertically upward several hundred feet, and parts were upwarped more than 2,000 feet. This caused rejuvenation of the streams and relatively rapid erosion of weak rock belts, but hard rocks were worn down little* (Moore, *Introduction to Historical Geology*).
—*n.* an upwarped condition or area.

upwelling, *n. Oceanography.* a process, common along continental coastlines, in which nutrient-laden waters from the ocean depths rise to the surface: *Upwelling typically occurs during local summer, when prevailing winds force warm surface waters away from the coast. Cold, deep water then rises or "upwells" to replace the wind-displaced surface water, bringing with it rich supplies of nutrients and causing an increase in the production of living organisms* (Myrl C. Hendershott).

u quark, = up quark: *A proton is made up of two u quarks and one d quark. Its electrical charge is the sum of the quarks' charges,* +1. *The neutron's combination is udd, so it contains equal amounts of positive and negative charges* (Robert H. March).

uracil (yùr′ə səl), *n. Biochemistry.* a pyrimidine base present in ribonucleic acid, corresponding to thymine in deoxyribonucleic acid: *RNA is a giant molecule built up of nucleotides containing ribose sugar and any one of four bases: adenine, uracil, guanine, and cytosine* (Scientific American). *Formula:* $C_4H_4N_2O_2$ Abbreviation: U [from *ur(ea)* + *ac(etic)* + *-il,* variant of *-yl*]

uralite (yùr′ə līt), *n. Mineralogy.* a green secondary amphibole formed by alteration of pyroxene. [from German *Uralit,* from the *Ural* Mountains + *-it* -ite]
—**uralitic** (yùr′ə lit′ik), *adj.* of, having to do with, containing, or consisting of uralite: *uralitic porphyry.*

Uranian, *adj. Astronomy.* of or having to do with the planet Uranus: *The Uranian rings are far too faint and too close to the planet to be detected by direct observation* (Joseph Ashbrook).

uranic, *adj. Chemistry.* of or containing uranium, especially with a valence of six.

uraniferous, *adj. Geology.* containing or yielding uranium: *uraniferous shale.*

uraninite (yù ran′ə nīt), *n. Mineralogy.* a mineral consisting largely of an oxide of uranium, found usually in blackish, pitchlike masses. It is a source of radium, uranium, and actinium. *Highest amounts of uranium in minerals are the uraninites, with about 85%* (Science News Letter). *Formula:* UO_2 Also called **pitchblende.**

uranium, *n. Chemistry.* a very heavy, silvery-white, highly radioactive metallic element occurring in uraninite and certain other minerals, and containing three natural isotopes of mass numbers 234, 235, and 238. Uranium is the main source of nuclear energy. It combines readily with most elements to form compounds that are extremely poisonous. *Symbol:* U; *atomic number* 92; *mass number* 238 (the most abundant and stable isotope); *melting point* 1132°C; *boiling point* 3818°C; *oxidation state* 3, 4, 5, 6. See also **uranium 235, uranium 238.** [from New Latin, named after *Uranus,* the planet]

uranium 235, *Physics.* a radioactive isotope of uranium that makes up about 0.7 per cent of naturally occurring uranium. It is fissionable by slow neutrons and capable of sustaining a rapid chain reaction. *Symbol:* ^{235}U; *atomic number* 92; *mass number* 235; *half-life* 7.13×10^8 years. Also called **U-235.**

uranium 238, *Physics.* a radioactive isotope of uranium that makes up about 99 per cent of all naturally occurring uranium. Though it is not fissionable, it can capture neutrons to form ^{239}P, a fissionable isotope of plutonium which is used as a source of nuclear energy. *Symbol:* ^{238}U; *atomic number* 92; *mass number* 238; *half-life* 4.51×10^9 years. Also called **U-238.**

uranous, *adj. Chemistry.* of or containing uranium, especially with a valence of four.

Uranus (yù rā′nəs *or* yùr′ə nəs), *n. Astronomy.* one of the largest planets in the solar system and the seventh in distance from the sun. The diameter of Uranus is 50,800 kilometers, more than four times that of the earth. It has fifteen moons or satellites and a system of thin rings around it. *The mere existence of radio emissions virtually confirmed that Uranus has a magnetic field. But Voyager's magnetometer soon added the unexpected fact that the axis of the field was tilted some 55° away from the planet's own rotation axis, compared with less than 20° for the fields of other worlds known to have them* (Science News). See the picture at **solar system.** [from *Uranus* the god of the sky or heaven in Greek mythology, from Greek *Ouranos* (literally) the heavens, sky]

uranyl (yur′ə nil), *n. Chemistry.* a bivalent radical, $UO_2{}^{2+}$, existing in many compounds of uranium and forming salts with acids.

—**uranylic,** *adj.* of or containing uranyl.

urate (yur′āt), *n. Chemistry.* a salt of uric acid: *A bladder is present in lizards, but the excretory wastes are semisolid as in birds and most reptiles, being passed from the cloaca as whitish material (urates) with the feces* (Storer, *General Zoology*).

urea (yu rē′ə *or* yur′ē ə), *n. Biochemistry.* a crystalline, nitrogenous solid present in solution in the urine of mammals and in other body fluids as a product of protein metabolism. It is also produced synthetically for use in fertilizers, in medicine, etc. *The use of urea as a dietary supplement for cattle to provide nitrogen for protein manufacture . . . promises to allow the conversion of such nitrogen-poor carbohydrates as wood-pulp and wastepaper to valuable protein* (Science News). *Formula:* $CO(NH_2)_2$ [from New Latin, from French *urée,* from *urine* urine]

urease (yur′ē ās), *n. Biochemistry.* an enzyme present in various bacteria, fungi, beans, etc., which promotes the decomposition of urea into ammonium carbonate.

uredinial (yur′ə din′ē əl), *adj. Biology.* of or having to do with the uredinium: *the uredinial stage of rust fungi.*

uredinium (yur′ə din′ē əm), *n., pl.* **-ia** (-ē ə). *Biology.* a pustule bearing uredospores, formed by certain rust fungi. [from Latin *uredinem* itch, blight, from *urere* to burn]

urediospore (yu rē′dē ə spôr), *n.* = uredospore.

uredospore (yu rē′də spôr), *n. Biology.* a one-celled, orange or brownish spore produced by certain rust fungi. Uredospores appear in the summer after the aeciospores and before the teliospores, and reproduce and extend the fungus rapidly.

ureotelic (yur′ē ō tel′ik), *adj. Biology.* having urea as the main constituent of nitrogenous waste. Mammals are ureotelic. Compare **uricotelic.** [from *urea* + Greek *telos* end]

ureter (yu rē′tər *or* yur′ə tər), *n. Anatomy.* a duct that carries urine from a kidney to the bladder or the cloaca: *The ureters empty into the base of the urinary bladder and the bladder empties the urine to the outside through a tube not found in the frog at all, the urethra* (Winchester, *Zoology*). [from Greek *oureter,* ultimately from *ouron* urine]

urethra (yu rē′thrə), *n., pl.* **-thrae** (-thrē), **-thras.** *Anatomy.* a duct in most mammals through which urine is discharged from the bladder, and, in males, through which semen is also discharged: *The two vasa deferentia enter the base of the urethra, which is a common urinogenital canal through the male copula-*

tory organ, or penis (Storer, *General Zoology*). [from Greek *ourethra,* ultimately from *ouron* urine]

—**urethral** (yu rē′thrəl), *adj.* of or having to do with the urethra: *the urethral canal, prostatic and other urethral glands.*

uric, *adj. Physiology.* of, having to do with, or found in urine: *uric calculi.*

uric acid, *Biochemistry.* a white, crystalline acid, only slightly soluble in water, formed as a waste product of the metabolism of purines and found in the urine of reptiles, birds, humans, etc.: *Chemically uric acid is a member of a very common class of substances —namely, the purines. Among the purines are adenine and guanine, which are building stones of the nucleic acids (DNA and RNA) and other essential constituents of all living cells* (Scientific American). *Formula:* $C_5H_4N_4O_3$

uricase (yur′ə kās), *n. Biochemistry.* a liver enzyme in lower mammals that breaks down uric acid into allantoin before excretion: *The enzyme chemical is called uricase. Lack of it prevents the human body from breaking down uric acid crystals that form in the joints of certain susceptible persons, making them subject to attacks of gout* (Science News Letter).

uricotelic (yur′ə kō tel′ik), *adj. Biology.* having uric acid as the main constituent of nitrogenous waste. Birds and snakes are uricotelic. Compare **ureotelic.** [from *uric* (acid) + Greek *telos* end]

uridine (yur′ə dēn), *n. Biochemistry.* a white powder, a nucleoside of uracil, present in ribonucleic acid: *Uridine nucleotide is used by the cell to bind and carry sugar fragments to the points where they are needed for growth processes* (New Scientist). *Formula:* $C_9H_{12}N_2O_6$

urinary, *adj. Anatomy.* **1** of, having to do with, or resembling urine: *a urinary secretion.*
2 of, occurring in, or affecting the organs that secrete and discharge urine: *urinary organs, the urinary duct.*

urine, *n. Physiology.* a waste product of the body of vertebrates that is excreted by the kidneys. Human urine is a fluid that passes through the ureters into the bladder, and is then discharged from the body through the urethra; it is normally amber in color and slightly acid, and has a specific gravity of about 1.02. It is made up of water, urea, creatinine, uric acid, and various inorganic salts including sodium, potassium, ammonia, calcium, and magnesium. The urine of birds and reptiles is excreted in the form of a semisolid white paste. *Urine consists of about 96 per cent water, 2 per cent urea, 0.5 per cent uric acid, and 1.5 per cent inorganic salts* (Shull, *Principles of Animal Biology*). [from Latin *urina*]

uriniferous (yur′ə nif′ər əs), *adj. Anatomy.* conveying urine: *The parenchyma of the kidney consists of closely packed uriniferous tubules, between which are*

---

cap, fāce, fäther; best, bē, tėrm; pin, five;
rock, gō, ôrder; oil, out; cup, put, rüle,
yü in use, *yu* in uric;
ng in bring; sh in rush; th in thin, ᴛʜ in then;
zh in seizure.

ə = *a* in about, *e* in taken, *i* in pencil, *o* in lemon, *u* in circus

blood vessels and a scanty amount of interstitial tissue (Copenhaver, *Bailey's Textbook of Histology*).

urinogenital, *adj.* = urogenital: *In vertebrate animals the reproductive and excretory systems are intimately connected and together they comprise the urinogenital system* (Shull, *Principles of Animal Biology*).

urinous, *adj. Physiology.* of, having to do with, or contained in urine: *urinous salts.*

urn, *n. Botany.* the theca in which the spores of a moss are produced.

uro-¹, *combining form.* urine; urinary, as in *urogenous* = *secreting urine.* [from Greek *ouron* urine]

uro-², *combining form.* tail; posterior part, as in *urochord, uropod.* [from Greek *oura* tail]

urocanic acid (yùr'ə kan'ik), *Biochemistry.* an acid that is a product of the metabolism of histidine, found in the urine of dogs and also in the human epidermis, where it is thought to absorb and dissipate harmful ultraviolet rays. *Formula:* $C_6H_6N_2O_2$ [from *uro-¹* + Latin *canis* dog]

urochord (yùr'ə kôrd), *n. Zoology.* the notochord of an ascidian larva, usually limited to the caudal region: *Among the sea squirts, the urochord persists throughout life* (A. Wilson). [from *uro-²* + *chord*]
—**urochordal,** *adj.* of or having to do with the urochord: *urochordal larvae.*

urochrome, *n. Biochemistry.* the yellow pigment which colors urine.

urodele (yùr'ə dēl), *Zoology.* —*adj.* of or belonging to an order (formerly Urodela, now Caudata) of amphibians which retain the tail throughout life, including the salamanders and newts: *a urodele amphibian.*
—*n.* a uredele amphibian: *No urodele . . . has more than four digits in the manus* (Thomas H. Huxley). [from Greek *oura* tail + *dēlos* visible]

urogenital, *adj. Anatomy.* having to do with the urinary and genital organs: *The anus of man expels the feces only, since there is no cloaca in the adult human being. However, there is a cloaca in the human embryo, but during embryonic development tissues separate the urogenital ducts from the termination of the intestine so they have separate openings before a child is born* (Winchester, *Zoology*).

urokinase (yùr'ō ki'nās), *n. Biochemistry.* a protein enzyme that dissolves blood clots: *Urokinase [is] found as a trace in human urine* (N. Y. Times).

urolith (yùr'ə lith), *n. Physiology, Medicine.* a calculus formed in any part of the urinary passages.

urology (yù rol'ə jē), *n.* the branch of medicine dealing with the urogenital tract in the male or the urinary tract in the female and their diseases.

uronic acid (yù ron'ik), *Chemistry.* any of a group of compounds found in urine and in mucopolysaccharides, formed by oxidation of the primary alcohol groups of sugars. Galacturonic acid and glucuronic acid are common uronic acids. [from Greek *ouron* urine + English *-ic*]

uropod (yùr'ə pod), *n. Zoology.* an abdominal appendage of an arthropod, especially one of the last pair of paddlelike appendages of a crustacean: *The tail appendages, or uropods, are held together side by side, and their tips are placed in contact with water which*

then flows up the capillary channel between them (New Biology). See the picture at **crustacean.** [from *uro-²* + Greek *podos* foot]

uropygial (yùr'ə pij'ē əl), *adj. Zoology.* of or having to do with the uropygium. The **uropygial gland** is a large gland opening on the backs of many birds at the base of the tail, with an oily secretion used in preening the feathers.

uropygium (yùr'ə pij'ē əm), *n. Zoology.* the rump of a bird, which bears the tail feathers. [from Greek *ouropygion,* from *oura* tail + *pygē* rump]

urostyle, *n. Zoology.* the posterior unsegmented portion of the vertebral column in certain fishes and amphibians: *The vertebral column or backbone [of the frog] consists of 9 vertebrae and a bladelike posterior extension, the urostyle* (Hegner and Stiles, *College Zoology*). [from *uro-²* + Greek *stylos* pillar]

Ursa Major (ėr'sə), *Astronomy.* the most prominent northern constellation, shaped somewhat like a bear with an enormous tail, and including the seven stars of the Big Dipper, two of which (the Pointers) point toward the North Star (Polaris).

Ursa Minor, *Astronomy.* the northern constellation, shaped somewhat like a bear, that includes the seven stars of the Little Dipper, with the North Star (Polaris) at the end of its handle.

urushiol (ü rü'shē ol), *n. Biochemistry.* a poisonous, pale, oily liquid derived from catechol. It is the active irritant principle in poison ivy, and is used in making lacquer and for tests as an allergen. [from Japanese *urushi* lacquer + English *-ol*]

UT1 (yü'tē'wun'), *n.* an international time standard based on the Earth's rotation, equivalent to the Greenwich Time that formerly served as the primary standard of time. Though primary time standards are now based on atomic clocks, the national time services of many nations broadcast corrections that enable one to determine UT1 to within 0.1 second. Because it is based on the Earth's rotation, UT1 is useful for celestial navigation. [*from U(niversal) T(ime) 1*]

UTC, *abbrev.* Coordinated Universal Time.

uterine (yü'tər ən *or* yü'tə rīn'), *adj.* of, having to do with, or in the region of the uterus: *the uterine canal, uterine contractions at birth.* The **uterine appendages** are the ovaries and oviducts. The **uterine cycle** is a series of changes occurring in the uterus about every 28 days, resulting in ovulation and menstruation.

uterus (yü'tər əs), *n., pl.* **uteri** (yü'tə rī'), **uteruses.** *Anatomy.* **1** the organ of the body in most female mammals that holds and nourishes the young till birth: *Changes taking place in the uterus are coordinated with changes taking place in the ovaries. This must be so for pregnancy to proceed . . . Ovulation occurs when the endometrium is just about midway in its development. Since it takes a few days for the fertilized egg to reach the uterus, the egg enters the uterus when the endometrium is just approaching its peak of development* (McElroy, *Biology and Man*). See the picture at **embryo. 2** a corresponding part in other animals: *the uterus of fishes, leeches, and worms.* [from Latin womb, belly]

utricle (yü'trə kəl), *n.* **1** *Botany, Zoology.* a small sac or baglike body in an animal or plant, such as a thin seed vessel resembling a bladder: *If the pericarp is papery and surrounds the seed in a bladdery manner, as in the* *goosefoot, the achene is termed a utricle* (Richard S. Cowan).

2 *Anatomy.* the larger of the two membranous sacs in the labyrinth of the inner ear, the saccule being the smaller: *There are two other organs in the inner ear which are believed to register gravity. These are the tiny cavities known as the "utricle" and "saccule" which contain microscopic solid particles called "otoliths" or, literally, "ear-stones". The otoliths may be regarded as . . . detecting the direction of gravity and also measuring its magnitude* (Arthur C. Clarke). Also called **utriculus.**
[from Latin *utriculus,* diminutive of *uter, utris,* skin bag]

—**utricular** (yü trik′yə lər), *adj.* **1** of, having to do with, or resembling a utricle: *utricular vessels or bodies.*
2 having or composed of a utricle or utricles: *cellular, vesicular, and utricular tissue.*

utriculus (yü trik′yə ləs), *n., pl.* **-li** (-lī). = utricle (def. 2): *The macula is the sensor organ of the utriculus—the part of the ear that responds to gravity* (Science News).

UV, *abbrev.* ultraviolet.

uvarovite (ü vä′rə fit), *n. Mineralogy.* an emerald-green variety of garnet containing chromium. *Formula:* $Ca_3Cr_2Si_3O_{12}$ [from Count S. S. *Uvarov,* 1786–1855, president of St. Petersburg Academy + *-ite*]

uvea (yü′vē ə), *n.* **1** *Anatomy.* the posterior, colored surface of the iris of the eye. **2** the middle, vascular coat of the eye, composed of the iris, choroid membrane, and the ciliary body. Also called **uveal tract.** [from Medieval Latin, from Latin *uva* grape]
—**uveal** (yü′vē əl), *adj.* of or having to do with the uvea. The **uveal tract** consists of the iris, the ciliary body, and the choroid.

uvula (yü′vyə lə), *n., pl.* **-las, -lae** (-lē′). *Anatomy.* the small piece of flesh hanging down from the soft palate in the back of the mouth: *The soft palate can be clearly seen without a laryngoscope as well as that curious cone-shaped appendage known as the uvula which dangles from the velum* (Simeon Potter). [from Late Latin, diminutive of Latin *uva* grape]
—**uvular,** *adj.* of or having to do with the uvula. The **uvular glands** are small follicles of the mucous membrane covering the uvula.

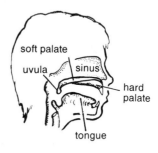

cap, fāce, fäther; best, bē, tèrm; pin, five;
rock, gō, ôrder; oil, out; cup, pùt, rüle,
yü in use, *yù* in uric;
ng in bring; *sh* in rush; *th* in thin, ᴛʜ in then;
zh in seizure.
ə = *a* in about, *e* in taken, *i* in pencil, *o* in lemon, *u* in circus

V

v. or **v,** *abbrev.* or *symbol.* **1** vector. **2** velocity. **3** volt. **4** voltage. **5** volume.

V, *abbrev.* or *symbol.* **1** vanadium. **2** volt. **3** voltage.

vac., *abbrev.* vaccum.

vaccine (vak sēn′), *Immunology.* —*n.* a mildly toxic preparation of weakened or killed bacteria or viruses of a particular disease, administered to an individual to induce immunity against that disease. Vaccines work by causing the body to develop antibodies against the disease-causing microorganisms. *Most vaccines are injected into the body, but Sabin polio vaccine is taken by mouth. A single dose of some vaccines provides lifelong protection against infection. Other vaccines require several doses to produce immunity and then must be reinforced at regular intervals by booster doses* (Alan R. Hinman). *Isolation and purification of a virus is necessary before a vaccine can be made* (Boyce Rensberger).
—*adj.* of or having to do with a vaccine: *vaccine therapy, attenuated vaccine virus.*
[from Latin *vaccinus* of or from cows, from *vacca* cow (so called because the original preparation was a suspension of cowpox virus used to produce immunity against smallpox)]
—**vaccination,** *n.* the inoculation or administration of a vaccine, usually by injection. *Active immunization* is another term for vaccination.

vacuolar (vak′yu̇ ə lər), *adj. Biology.* of, having to do with, or of the nature of a vacuole or vacuoles: *the vacuolar wall. The vacuolar limits are defined by a one-layered vacuolar membrane* (Greulach and Adams, *Plants*).

vacuolar membrane, = tonoplast.

vacuolate (vak′yu̇ ə lāt) or **vacuolated** (vak′yu̇ ə lā′tid), *adj. Biology.* (of a cell or cytoplasm) containing one or more vacuoles: *The egg in cotton is a highly vacuolate cell, and the cytoplasm is spread thinly between the wall and the large vacuole* (William A. Jensen).

vacuolation (vak′yu̇ ə lā′shən), *n. Biology.* the formation of vacuoles; change to a vacuolar condition: *The subsequent stages of development, cell elongation and vacuolation, do not strictly speaking belong to germination proper. It may be assumed that they are no different from the similar processes occurring in any growing tissue of the adult plant* (A. M. Mayer).

vacuole (vak′yu̇ ōl), *n. Biology.* a cavity in the cytoplasm of a cell, bound by a single membrane and containing water and waste products of cell metabolism. Vacuoles are typically found in plant cells. *In plant cells there are often found chloroplasts and other pigment-containing plastids, as well as large vacuoles which contain a solution less viscous but usually more acid than the rest of the cytoplasm* (D. L. Woodhouse and H. S. A. Sherratt). *In the protozoa, temporary vacuoles are common. They usually either enclose bodies of food in process of digestion, in which case*

they are called food vacuoles, or disappear at intervals by ejecting their liquid contents through the surface layer of protoplasm into the surrounding medium. The latter kind is called a pulsating or contractile vacuole (Shull, *Principles of Animal Biology*). See the pictures at **cell, centrosome.** [from French, from Latin *vacuus* empty]
► See the note under **organelle.**

vacuum, *n., pl.* **vacuums, vacua** (vak′yu̇ ə). *Physics.* **1** a space completely devoid of matter: *In empty space where there is no material (a vacuum), there can be no molecules* (Beauchamp, *Everyday Problems in Science*). *It was natural to conclude that radiant energy would travel fastest in a vacuum where the density is zero.* (Barnard, *Science: A Key to the Future*).
2 an enclosed space from which almost all the air or other gas has been removed, especially to permit experimentation without atmospheric distortion: *The rapid and reliable production of a really good vacuum ... despite tiny leaks that cannot be altogether eliminated, depend usually on [a device] ... producing a vacuum from which all but perhaps one hundred-thousandth of the air is removed* (Furry, *Physics*).
[from Latin, neuter of *vacuus* empty]

vacuum tube, an electron tube from which almost all the air has been pumped out, and in which the movement of electrons is controlled by electric and magnetic fields. Cathode-ray tubes, which include television picture tubes and other video display tubes, are the most widely employed vacuum tubes.

vadose (vā′dōs), *adj. Geology.* having to do with or occurring in the unsaturated area between the earth's surface and the water table: *the vadose zone. The descending vadose water dissolves the soluble minerals and carries them in solution down into the zone of saturation* (White and Renner, *Human Geography*). [from Latin *vadosus* shallow, from *vadum* ford]

vagal (vā′gəl), *adj. Anatomy.* of or having to do with the vagus nerve: *In 20 more or less sedentary men a 6- to 12-week period of vigorous physical retraining restored the vagal tone toward normal* (Science News Letter).

vagile (vaj′əl or vaj′īl), *adj. Zoology.* characterized by vagility; able to move around: *vagile organisms.* [from Latin *vagus* wandering + English *-ile,* as in *sessile*]

vagility (və jil′ə tē), *n. Zoology.* the condition of being able or free to move around; mobility: *The ability of insects to cross oceans depends on their habits and vagility in immature and adult stages* (Science).

vagina (və jī′nə), *n., pl.* **-nas, -nae** (-nē). *Anatomy.* **1** the membranous passage in female mammals that leads from the uterus to the vulva or external genital organs: *In a woman the single uterus opens to the vagina, which leads to the outside, while in the frog the two uteri open into the cloaca* (Winchester, *Zoology*).

2 a similar passage in certain animals other than mammals: *In the intestinal roundworm, the two uteri are joined in a single short vagina.*

[from Latin *vagina* sheath]

—**vaginal** (vaj'ə nəl), *adj.* of or having to do with the vagina: *the vaginal canal, vaginal arteries.*

vagus (vā'gəs), *n., pl.* **vagi** (vā'jī). = vagus nerve. [from New Latin, from Latin *vagus* wandering]

vagus nerve, *Anatomy.* either of the tenth pair of cranial nerves, extending from the brain to the heart, lungs, stomach, and other organs: *Production of gastric juices for digesting food results from stimulation of the vagus nerves* (Science News Letter).

Val, *abbrev.* valine.

valence (vā'ləns), *n.* **1** *Chemistry.* **a** the combining capacity of an atom or radical, determined by the number of electrons that an atom will lose, add, or share when it reacts with other atoms. Elements whose atoms lose electrons, such as hydrogen and the metals, have a positive valence. Elements whose atoms gain electrons, such as oxygen and other nonmetals, have a negative valence. Oxygen has a negative valence of 2; hydrogen has a positive valence of 1; one atom of oxygen combines with two of hydrogen to form a molecule of water. Compare **oxidation state. b** a unit of valence: *Oxygen has two valences.*

2 *Biology, Immunology.* the ability of certain substances, such as serums or antibodies, to interact or to produce a certain effect, especially the capacity of an antibody molecule to combine with antigen molecules: *The valence number of IgG antibody molecules is 2; the valence number of IgM antibodies is 10.*

[from Latin *valentia* strength, from *valere* be strong]

valence electron, *Chemistry.* an electron in the outer shell of an atom. In a chemical change, the atom gains, loses, or shares such an electron in combining with another atom or atoms to form a molecule. *Since they [the electrons in the outer shell] determine the combining capacity of the atom, they are sometimes spoken of as the valence electrons of the atom, and the shell to which they belong is called the valence shell* (Briscoe, *College Chemistry*).

-valent, *combining form.* having valence, as in *trivalent.* [from Latin *valentem* being strong]

valine (val'ēn), *n.* *Biochemistry.* an amino acid constituent of protein, essential to growth. It can be isolated by hydrolysis of fish proteins for use as a nutrient, in medication, and in biochemical research. *Peptide-bond formation proceeds much like a zipper, starting with the amino acid valine at one end of the chain and closing bond after bond until the protein molecule is finished* (Scientific American). *Formula:* $C_5H_{11}NO_2$ *Abbreviation:* Val [from *val(eric acid)*, $C_5H_{10}O_2$, present in the herb valerian + *-ine*]

vallate (val'āt), *adj. Anatomy.* surrounded by a ridge or elevation; having a surrounding ridge or elevation. The **vallate papillae** are large, flat papillae in front of the V-shaped groove at the back of the tongue. [from Latin *vallatum* surrounded with a rampart, from *vallum* rampart]

vallecula (və lek'yə lə), *n., pl.* **-lae** (-lē). *Anatomy.* a shallow groove or depression, such as the depression between the hemispheres of the cerebellum or the depression between the back of the tongue and the epi-

valve

glottis. [from Late Latin *vallicula* depression, diminutive of Latin *vallis* valley, furrow]

valley, *n. Geography, Geology.* **1** a broad area of low-lying land situated between hills or mountains and usually having a river or stream flowing along its bottom: *If conditions permit, the many agents working toward valley-widening ultimately produce a broad valley with a wide, level floor* (Leet and Judson, *Physical Geology*). **2** a wide region drained by a river and its tributaries: *the Mississippi valley.* [from Old French *valée,* from *val* vale, from Latin *vallis*]

valley flat, *Geology.* a low, level deposit of sediment in the channel of a stream: *Gradually, by undermining and caving the valley wall on one side and depositing on the other, the stream produces the broad valley flat of old age* (Finch and Trewartha, *Elements of Geography*).

valley glacier, *Geology.* a glacier that occupies a mountain valley, as in the Alps and the mountains of the western United States: *A valley glacier ... is far wider and thicker than the corresponding stream of meltwater, so that valley glaciers extending below the snowline obstruct the drainage of lateral tributary streams* (G. H. Dury). Also called **alpine glacier.**

valley train, *Geology, Geography.* the alluvium deposited by glacial outwash, partially filling the valley below the retreating front of the glacier.

valvate (val'vāt), *adj. Botany.* **1** meeting without overlapping, as the parts of certain buds do: *valvate sepals or petals.*

2 united by the margins only and opening as if by valves, as the capsules of regularly dehiscent fruits and certain anthers do.

3 composed of or characterized by such an arrangement of parts: *a valvate structure, valvate estivation or vernation.*

[from Latin *valvatus* having folding doors, from *valvae,* plural of *valva* one of a pair of folding doors]

valve, *n.* **1** *Anatomy.* **a** a part of the body that controls the flow of a fluid, especially by preventing it from flowing back to its source. The valves of the heart are membranes that control the flow of blood into and out of the heart. *At the pyloric orifice the circular layer forms a sphincter valve—the pyloric valve—which regulates the exit of food (chyme) from the stomach into the small intestine* (Edwards, *Concise Anatomy*). **b** a similar part or structure serving to close a passage for other reasons: *The ears are short, and have each a very small inner valve* (William Bingley).

2 *Zoology.* one of the two or more parts of hinged shells like those of oysters and clams, or the whole shell when it is in one piece, as in snails.

3 *Botany.* **a** one of the sections formed when a seed vessel bursts open. **b** a section that opens like a lid when an anther opens. **c** either of the halves of the shell of

cap, fāce, fäther; best, bē, tėrm; pin, five;
rock, gō, ôrder; oil, out; cup, pùt, rüle;
*y*ü in use, *y*ù in uric;
ng in bring; sh in rush; th in thin, ŦH in then;
zh in seizure.
ə = *a* in about, *e* in taken, *i* in pencil, *o* in lemon, *u* in circus

705

valvular

a diatom: *The glassy wall does not make a complete, impervious covering over the cell but is really in two parts, known as valves, which fit together like a box and its lid* (Emerson, *Basic Botany*).
[from Latin *valva* one of a pair of folding doors]

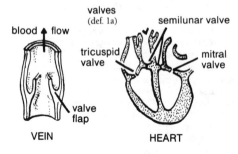

valves
(def. 1a)

blood ↑ flow

semilunar valve

tricuspid valve

mitral valve

valve flap

VEIN

HEART

valvular (val'vyə lər), *adj. Anatomy.* **1** of, having to do with, or affecting a valve or valves, especially the valves of the heart: *the valvular apparatus between the atria and ventricles.*
2 having the form or function of a valve: *a valvular capsule, a valvular calyx.*

vanadate (van'ə dāt), *n. Chemistry.* a salt or ester of vanadic acid.

vanadic (və nā'dik), *adj. Chemistry.* of or containing vanadium, especially with a valence of three or of five.

vanadic acid, *Chemistry.* any of a group of acids known only in the form of their salts (vanadates), especially vanadium pentoxide, V_2O_5.

vanadinite (və nad'ə nīt), *n. Mineralogy.* a mineral consisting of a vanadate and chloride of lead, occurring in brilliant red or yellow crystals: *Vanadium is found in complex minerals the most important of which are vanadinite ... and carnotite* (Offner, *Fundamentals of Chemistry*). Formula: $Pb_5(VO_4)_3Cl$

vanadium (və nā'dē əm), *n. Chemistry.* a very hard, silver-white, malleable metallic element occurring in certain iron, lead, and uranium ores, such as vanadinite, and used especially in making various strong alloys of steel: *Vanadium in combination with other elements is widely distributed in igneous and sedimentary rocks, but only in small quantities, although it may, as the result of the decomposition of the minerals containing it, become more concentrated in iron-rich soils and bauxite clays* (Jones, *Minerals in Industry*). *Symbol:* V; *atomic number* 23; *atomic weight* 50.942; *melting point* ab. 1900°C; *boiling point* ab. 3000°C; *oxidation state* 2, 3, 4, 5 [from New Latin, from *Vanadis,* Scandinavian goddess of love (so called because it was discovered in Sweden)]

vanadous (və nā'dəs), *adj. Chemistry.* of or containing vanadium, especially with a valence of two or of three.

Van Allen belt or **Van Allen radiation belt,** *Geophysics.* either of two broad bands of intense radiation surrounding the earth, consisting of charged particles from the solar wind that have been captured by the earth's magnetic field: *Inside the current sheet is the inner magnetosphere, which includes the region of the Van Allen belts, where the trapped electrons and protons gyrate within the field* (Von Braun and Ordway,

New Worlds). See the picture at **magnetosphere.** [named after James A. *Van Allen,* 20th-century American physicist who discovered the belts in 1958]

Van de Graaff generator (van'də graf'), *Physics.* a generator that produces electric potentials of very high voltages, used to accelerate charged particles to energies of about 10 million electron volts: *The Van de Graaff generator builds up a large electric charge ... inside the top sphere. The charge on the sphere is continuously built up to very high voltages. The Van de Graaff generators are used for experimental work in nuclear physics* (Obourn, *Investigating the World of Science*). Also called **electrostatic generator.** [named after Robert *Van de Graaff,* 1901–1967, American physicist]

van der Waals forces or **van der Waals force** (van'dər wôlz'), *Chemistry.* the relatively weak forces or force of attraction existing between nearby atoms or molecules that are not bound to each other: *Even in the absence of dipole attraction and hydrogen bonding, as in nonpolar molecules, weak attractive forces exist between molecules. These forces are called van der Waals forces. Van der Waals forces make it possible for species of small nonpolar molecules (such as hydrogen, helium, oxygen, etc.) to exist in the liquid and solid phases under conditions of low temperature and high pressure* (Regents Chemistry Syllabus). [named after Johannes *van der Waals,* 1837–1923, Dutch physicist]

vane, *n.* the flat, soft part of a feather on the shaft, attached to the central quill by the rachis. See the picture at **feather.**

vanish, *v. Mathematics.* to take on the value of zero: *The axis must vanish, before P and P¹ would reach 180°* (H. J. Brooke).

vapor, *n. Physics.* a gas formed from a substance that is usually a liquid or a solid; the gaseous form of a liquid or solid. Water vapor is water in a gaseous state that is always present in the air, forming clouds, rain, and snow when condensed.
▶ A vapor resumes its liquid or solid state under high pressures at ordinary temperatures, while a gas compressed at ordinary temperatures remains in the gaseous state.
—vaporization, *n.* the process of changing, or of being changed, into vapor: *Evaporation and boiling of liquids are forms of vaporization* (Louis Marick). Compare **sublimation.**
—vaporize, *v.* to change into vapor; evaporate: *Magnetic fields are used to contain the plasma since all materials vaporize long before the required temperature of some 100 million degrees is reached* (Science News Letter). *Many molecular species are stable only at temperatures below the point where they would liquefy or vaporize* (Alan Holden).

vapor pressure or **vapor tension,** *Physics.* the pressure exerted by a vapor in an enclosed space when the vapor is in equilibrium with its liquid or solid at any specified temperature: *The vapor pressure in the container of sea water is lower than in the flask of pure water, because its water molecules, being bound to salt ions, do not evaporate as easily* (Scientific American).

var., *abbrev.* **1** variable. **2** variation. **3** variety.

variable, *n.* **1** *Mathematics.* **a** a quantity that can assume any of the values in a given set of values. *Example:* In the equation $x + y = 7$, x and y are variables; the equation is true for any values of *x* and *y* whose sum is seven. **b** a symbol representing such a quantity.

2 *Statistics.* anything that varies in magnitude; any quality or characteristic that may assume a large number of values in different individual cases. Compare **dependent variable, independent variable, random variable.**

3 = variable star.

—*adj. Biology.* deviating, as from the normal or recognized type: *a variable species, a variable character or form.* SYN: aberrant. —**variableness,** *n.* —**variably,** *adv.*

variable star, *Astronomy.* a star that varies periodically in brightness or magnitude: *Variable stars fall chiefly into three classes with respect to the cause of their variability: (1) Pulsating stars, comprising Cepheid, long-period, and irregular variables; (2) novae, supernovae, and also nova-like stars; (3) eclipsing stars which vary in light owing to periodic eclipses of mutually revolving stars* (Baker, *Astronomy*).

variance (vãr′ē əns), *n.* **1** *Statistics.* the square of the standard deviation: *The variance of the distribution ... is the average value of the square of the deviations of x from its central value or mean* (J. A. Nelder).

2 *Chemistry.* the number of conditions, such as temperature or pressure, which must be fixed in order that the state of the system may be defined; degree of freedom of a system.

variate (vãr′ē āt), *n.* = random variable.

variation, *n.* **1** *Biology.* **a** a deviation of an animal or plant from the normal or recognized type: *Variation ... is quite as universal a fact in life as is heredity, and both are significant in understanding man. Variations must never be confused with modifications. The latter are changes which take place in the organism as a result of temporary or special environmental conditions and which are not inherited* (Ralph Beals and Harry Hoijer). **b** an animal or plant showing such deviation or divergence: *If a new variation or mutation were to appear in nature, and if it were adapted for survival, genetics could account for the transmission of the new characters; but how could these new characters have appeared in the first place if hereditary is governed by genes?* (I. Bernard Cohen). **c** a difference existing between the individuals of a species: *Along with the tendency for offspring to resemble their parents there is thus a parallel tendency of offspring to differ from their parents. Such differences, known as variations, appear repeatedly throughout the plant and animal kingdom* (Harbaugh and Goodrich, *Fundamentals of Biology*).

ASSOCIATED TERMS: see **evolution.**

2 *Astronomy.* the deviation of a celestial body, such as the moon, from its average orbit or motion.

3 *Mathematics.* one of the different ways in which the members of any group or set may be combined.

4 *Geography.* the angular difference between geographic north and magnetic north from any point: *The angle between the horizontal component and the true north-south direction is called the variation or declination* (Sears and Zemansky, *University Physics*).

varied, *adj. Zoology.* having different colors (used especially in the names of animals): *the varied bunting.*

variegated, *adj. Botany, Zoology.* varied in appearance; marked with patches or spots of different colors; many-colored: *Pansies are usually variegated.*

variegation, *n. Botany, Zoology.* the condition or quality of being variegated; varied coloring: *leaf and flower variegation.*

varietal (və rī′ə təl), *adj. Biology.* of, having to do with, or constituting a variety: *In order to avoid introducing any dangerous malady from the East, e.g. the Bunchy Top virus disease, varietal collections, carefully selected, are first of all sent to Kew and maintained there under observation for a suitable period of time in a specially prepared greenhouse* (C. W. Wardlaw).

variety, *n. Biology.* **1** a division of a species: *Names of varieties are applied by giving the name of the species followed by the prefix "var." before the varietal epithet, as Chenopodium ambrosioides var. anthelminticum* (Youngken, *Pharmaceutical Botany*). SYN: subspecies.

2 a plant or animal differing from those of the species to which it belongs in some minor but permanent or transmissible particular: *a variety of corn, varieties of birds.*

varistor (və ris′tər), *n. Electronics.* a semiconducting resistor whose resistance varies with the applied voltage: *A ... varistor protects a device by short circuiting surges. In operation the varistor is usually connected in parallel with the component or circuit it is protecting. When a high voltage transient comes along the varistor stops being a good insulator and becomes a conductor, thus clamping the line voltage to a safe level and blocking off the transient* (New Scientist). [from *var(iable res)istor*]

varix (vãr′iks), *n., pl.* **varices** (vãr′ə sēz′). *Zoology.* a longitudinal elevation or swelling on the surface of a gastropod shell. [from Latin *varix* dilated vein]

varve (värv), *n. Geology.* a layer or series of layers of sediment deposited in a lake or other body of still water within one year. Varves are used to determine the age of sediments in dating geological phenomena. *Radiocarbon dating of tree rings and clay varves on lake bottoms will permit scientists to determine solar cycles even before 220 B.C.* (Science News Letter). [from Swedish *varv* layer]

—**varved,** *adj.* deposited in varves: *Varved (banded) clay consists of a succession of double laminae, interpreted as resulting from one year's sedimentation for each band* (G. H. Dury).

vary, *v.* **1** *Mathematics.* to undergo or be subject to a change in value according to some law: *to vary inversely as the cube of y.*

2 *Biology.* to exhibit or be subject to variation, as by natural or artificial selection.

vas (vas), *n., pl.* **vasa** (vā′sə *or* vā′zə). **1** *Anatomy.* a duct or vessel conveying blood, lymph, or other fluid through the body (usually found in Latin phrases, such as *vas breve* a short gastric artery, plural *vasa brevia*): *The blood flows down into the medulla in the long*

---

cap, fāce, fäther; best, bē, tèrm; pin, fīve;
rock, gō, ôrder; oil, out; cup, pùt, rüle;
yü in use, *yu̇* in uric;
ng in bring; *sh* in rush; *th* in thin, ᵀH in then;
zh in seizure.
ə = *a* in about, *e* in taken, *i* in pencil, *o* in lemon, *u* in circus

707

vasal

straight vessels (the vasa recta) for variable distances and then breaks up into capillaries (H. E. de Wardener). See also **vas deferens, vas efferens.**

2 Botany. a tube or conduit in a plant (usually found in Latin plural phrases, such as vasa propria sieve tubes).

[from Latin vas vessel]

vasal (vā′səl or vā′zəl), adj. Anatomy. having to do with or connected with a vas or vasa: a vasal structure, a vasal fluid.

vascular (vas′kyə lər), adj. Biology. having to do with, made of, or provided with vessels that carry a body fluid, such as the blood or lymph in an animal or the sap in a plant. The **vascular system** of an animal or plant consists of the vessels and organs that convey fluids through the body of the organism: The third of the tissue systems is the vascular, a complex system, consisting chiefly of phloem and xylem and used for the conduction of water, mineral salts, and foods as well as for strengthening and support (Hill, Botany). [from Latin vasculum, diminutive of vas vessel]

vascular bundle, Botany. a strand of vascular tissue consisting mostly of xylem and phloem. It is the structural unit of the stele in vascular plants. In the leaf, a vascular bundle is made up of three parts: an outer sheathing layer, usually one cell thick, the bundle sheath, and two masses of conductive tissue, the xylem and phloem (Emerson, Basic Botany).

vascular cambium, Botany. cambium that gives rise to secondary phloem and secondary xylem: A continuous layer of vascular cambium is formed from precambrial cells that have not lost their meristematic character (Weier, Botany).

vascular cylinder, = stele.

vascularity, n. Biology. the quality or condition of being provided with vessels to circulate body fluid; vascular form or condition: The quantity of blood a leech is capable of drawing varies ... according to the vascularity of the part (Robert T. Hulme).

vascularize, v. Biology. to make or become vascular; to form vessels, especially blood vessels: [The] mucous membrane of the stomach becomes highly vascularized (J. H. Bennet).

—**vascularization,** n. **1** Biology. the process of vascularizing; formation of vessels, especially blood vessels: Estrogen initiates vascularization of the uterus lining (Biology Regents Syllabus).

2 Medicine. an abnormal or pathological formation of blood vessels: Corneal vascularization is a problem in many corneal diseases and may also occur after a corneal transplant (R. I. Pritikin and M. L. Duchon).

vascular plant, Biology. a plant in which the structure is made up in part of vascular tissue. Vascular plants comprise the spermatophytes (seed-producing plants) and pteridophytes (ferns and fernlike plants). Also called **tracheophyte.**

vascular ray, Botany. a sheet or band of conducting vascular tissue which extends radially from the xylem through the cambium into the phloem: Vascular Rays ... are formed as a result of the activity of the cambium during secondary growth of the plant axis (Youngken, Pharmaceutical Botany).

vascular tissue, Botany. the tissue in a vascular plant, consisting essentially of phloem and xylem, which carries the sap throughout the plant: Simple multicellular plants, bryophytes, lack vascular tissue. ... Higher plants, tracheophytes, possess vascular tissue for intercellular transport (Biology Regents Syllabus).

vasculature (vas′kyə lə chúr), n. Anatomy. the system or arrangement of blood vessels in the body or any part of the body: The migration of the germ cells takes place at first through the vasculature (Science Journal). One concept of the pathogenesis of cholera holds that the major area of action of the toxin is to the vasculature supplying the mucosa of the small bowel (H. T. Norris).

vas deferens (def′ə rənz), pl. **vasa deferentia** (def′ə-ren′shē ə). Anatomy. the excretory duct that conveys sperm from the testis to the urethra: Sperm [in planarian worms] are produced in the numerous testes and make their way through the slender vasa efferentia (sing. vas efferens) into one of the two vasa deferentia (sing. vas deferens) and thence down into the seminal vesicles which act as a storage chamber for the sperm until the time comes for their discharge (Winchester, Zoology). Also called **ductus deferens.** [from New Latin, deferent vessel]

vas efferens (ef′ə rənz), pl. **vasa efferentia** (ef′ə ren′shē-ə). Anatomy. any of a number of small ducts conveying sperm from the testis to the epididymis. [from New Latin, efferent vessel]

vaso- (vā′sō- or vā′zō-), combining form. blood vessel, as in vasoactive, vasomotor. [from Latin vas vessel]

vasoactive, adj. Physiology. acting on the blood vessels, especially by constricting or dilating them: A vasoactive peptide, perhaps responsible for the local swellings, has been isolated from plasma from patients during attacks (Chester A. Alper).

vasoconstriction, n. Physiology. constriction of the blood vessels, especially by the action of a nerve or a drug: Noise causes vasoconstriction of the cochlear vessels in the ear, accompanied by a significant reduction in the flow and number of red blood cells reaching the organ of Corti, the site of hearing (Samuel Rosen).

—**vasoconstrictive,** adj. serving to constrict blood vessels: a vasoconstrictive nerve, a vasoconstrictive drug.

—**vasoconstrictor,** n. something that constricts blood vessels, such as a nerve or a drug: Serotonin is ... a vasoconstrictor isolated first from beef serum in 1948 (Frank P. Mathews).

vasodilation or **vasodilatation,** n. Physiology. dilation of the blood vessels, especially by the action of a nerve or a drug.

—**vasodilative,** adj. serving to dilate blood vessels: a vasodilative nerve, a vasdilative drug.

—**vasodilator,** n. something that dilates blood vessels, such as a nerve or a drug: Histamine belongs to the group of substances known chemically as amines and physiologically as vasodilators (New Biology).

vasomotor, adj. Physiology. of or having to do with the nerves and nerve centers that regulate the contraction and expansion of blood vessels: The vasomotor system ... has three functions: (1) it regulates the distribution of the blood ... ; (2) it regulates blood pressure; and (3) it regulates body temperature in a warm-blooded animal through controlling heat loss at the surface of the body (Harbaugh and Goodrich, *Fundamentals of Biology*).

vasopressin, *n. Biochemistry.* a pituitary hormone that contracts small blood vessels, raises blood pressure, and reduces the excretion of urine by the kidneys: *Vasopressin is the antidiuretic hormone of mammals which, by promoting the reabsorption of water from the kidney tubules, concentrates the urine and decreases the volume formed* (Bradley T. Scheer).

vasopressor, *Physiology.* —*adj.* of, having to do with, or causing constriction of blood vessels: *The life-saving vasopressor drugs, used to combact shock after a heart attack, conserve the blood supply to heart and brain at the expense of other less sensitive organs in the body* (Science News Letter).
—*n.* a substance that constricts blood vessels; vasoconstrictor.

vaterite (vat′ə rīt), *n. Mineralogy.* one of the three crystalline forms of calcium carbonate, the others being calcite and aragonite. *Formula:* $CaCO_3$ [from H. Vater, 20th-century German mineralogist + *-ite*]

vault, *n. Anatomy.* an arched structure, especially that of the skull: *This vault is formed by the nasal bones and the nasal processes of the maxillary bones* (R. Knox). ... *the upper and middle portions of the cranial vault* (H. Miller).

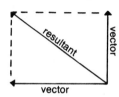

(def. 1a)

The vector lines represent the magnitude and direction of force, the resultant line represents the result of the force exerted by the vectors at the point where they meet. Compare the picture at **composition of forces.**

vector (vek′tər), *n.* **1** *Mathematics.* **a** a quantity having direction as well as magnitude, usually represented by a directed line segment whose direction indicates the direction of the vector and whose length indicates the magnitude of the vector: *By a vector we mean an ordered pair of real numbers. Thus a vector is simply a point of a coordinate plane. One way to draw a picture of a vector (x,y) is to draw the axes and then plot the point P = (x,y). Another way to draw a picture of a vector is to draw the directed segment from the origin O to the point P = (x,y).* (Moise and Downs, *Geometry*). Contrasted with **scalar. b** = radius vector. **2** *Biology.* **a** an organism, such as a mosquito or tick, that transmits disease-causing microorganisms: *The more frequently and the more exclusively a species of mosquito feeds on rabbits, the more efficient will it be as a vector* (F. Fenner and M. F. Day). *Aedes aegypti, a vector for yellow fever too, and Culex tarsalis both transmit the virus that causes the equine encephalitis* (Science News Letter). **b** any agent that acts as a carrier or transmitter: *Molecular cloning calls for the insertion of a fragment of foreign DNA into a vector (in this case a plasmid, a small circle of bacterial DNA) that serves to introduce the fragment into bacteria, where it replicates* (Scientific American).
—*adj. Mathematics.* of or involving a vector; having

both direction and magnitude: *a vector equation. Mathematically, velocity is a vector quantity, because it has both speed and direction* (Robert F. Paton). [from Latin *vector* carrier, from *vehere* to carry]

vector field, *Mathematics.* a set of vectors composing a region of space, the value of each vector depending upon the location of the point from which that vector radiates. Compare **scalar field.**

vectorial (vek tôr′ē əl), *adj.* of or having to do with a vector or vectors: *Another new instrument being installed at Palomar is a vectorial recorder which photographs a pattern of the earth's surface motion in two dimensions* (Science News Letter).

vector meson, *Nuclear Physics.* any of a class of composite subatomic particles having one unit of spin, including the omega, phi, and rho mesons: *Vector mesons are particles with the same spin as a photon, but they are hadrons* (Dietrick E. Thomsen). *The name "vector meson" is derived from the intrinsic angular momentum, or spin, of these particles, which is unity and which transforms from one spatial coordinate system to another as a vector quantity* (F. V. Murphy and D. E. Yount).

vector product, = cross product.

vector space, *Mathematics.* a system consisting of a set of generalized vectors and a field of scalars, having the same rules for vector addition and scalar multiplication as those for physical vectors and scalars.

vegetable, *adj. Botany.* **1** of plants; having to do with plants: *vegetable substances, vegetable life. Lignite is formed from vegetable matter in a manner similar yet midway between that forming peat and coal* (Science News Letter).
2 obtained from the fruit or seeds of plants: *vegetable fats, vegetable wax. Vegetable oils include olive oil, corn oil, and linseed oil.*
[from Medieval Latin *vegetare* to grow, flourish, ultimately from Latin *vegetus*, lively, vigorous]
► In botany, the word *vegetable* is acceptable only as a descriptive adjective, not as a noun. In popular usage, such parts of plants as heads of lettuce, spinach, potatoes, corn, tomatoes, and string beans are called vegetables. In botany, lettuce and spinach are *leaves,* potatoes are *tubers,* and corn, tomatoes, and string beans are *fruits* (ripened ovaries). Note that the popular usage of *fruit* also differs from the botanical usage.

vegetable kingdom, = plant kingdom.

vegetal, *adj. Biology.* of or having to do with the vegetal pole: *The two halves ("animal" and "vegetal") do not contain a complete sample of the field of differentiation; consequently, neither can develop into a complete embryo* (Simpson, *Life: An Introduction to Biology*).

---

cap, fāce, fäther; best, bē, tėrm; pin, fīve;
rock, gō, ôrder; oil, out; cup, pùt, rüle;
yü in use, *yù* in uric;
ng in bring; *sh* in rush; *th* in thin, ŦH in then;
zh in seizure.
ə = *a* in about, *e* in taken, *i* in pencil, *o* in lemon, *u* in circus

709

vegetal pole, *Biology.* the part of an egg's surface located opposite to the animal pole and usually containing the principal mass of yolk: *In another type of egg, the telolecithal egg, the yolk is abundant as compared to the amount of cytoplasm, and is located at the vegetal pole ... The very small amount of cytoplasm is located, cap-like, at the opposite or animal pole* (Harbaugh and Goodrich, *Fundamentals of Biology*). Also called **vegetative pole.**

vegetation, *n. Botany.* plant life, especially the total mass of plants growing in a particular area: *A desert is an area which produces insufficient vegetation to support habitation under natural conditions: it is not merely a region short of water* (R. N. Elston).
—**vegetational,** *adj.* of or having to do with vegetation: *vegetational areas, vegetational types.*

vegetative, *adj.* **1** *Botany.* **a** concerned with growth and development rather than reproduction: *The ordinary vegetative organs such as roots, stems, or leaves of many species readily form adventitious roots and shoots* (Emerson, *Basic Botany*). **b** causing or promoting growth in plants; inducing vegetable growth: *vegetative mold, a vegetative matrix.*
2 *Physiology.* of or having to do with the involuntary functions controlled by the autonomic nervous system: *the vegetative processes of the body, such as growth and repair.*

vegetative pole, = vegetal pole.

vegetative propagation, *Botany.* a form of asexual reproduction in multicellular plants, in which new plants develop from the roots, stems, or leaves of the parent plant.
ASSOCIATED TERMS: see **asexual.**

vegetative reproduction, *Botany.* = vegetative propagation: *Vegetative reproduction, or reproduction without fertilization and seed formation, takes place in some plants* (Harold N. Moldenke).

vein, *n.* **1** *Anatomy.* any of the membranous, branching tubes forming part of a system of vessels that carry the blood to the heart from all parts of the body: *Blood flows out from the heart to the tissues through an ever-branching system of vessels. Arteries divide into arterioles ... and the latter give rise to the capillaries. At the other end, capillaries join to form venules, venules join to form veins, and eventually, the veins unite to form the great veins that return blood to the heart* (Harper's).
2 *Botany.* one of the strands or bundles of vascular tissue forming the principal framework of a leaf: *The vein is commonly surrounded by a mass of parenchyma, sometimes with small amounts of collenchyma or fibrous tissue above and below* (Greulach and Adams, *Plants*). SYN: nervure. See the picture at **leaf.**
3 *Zoology.* one of the ribs that strengthen the wing of an insect.
4 *Geology.* **a** a continuous crack or fissure in rock filled with one or more minerals, especially a metallic ore, different from the containing rock: *The so-called "true fissure veins" are veins that differ sharply in mineral content and structure from their wall rocks and generally break away from them cleanly* (Gilluly, *Principles of Geology*). **b** a deposit, as of ore or coal, having a more or less regular development in length,

width, and depth: *a vein of copper.* SYN: lode. **c** a small natural channel within the earth or ice through which water trickles or flows.
—**veined,** *adj.* having or showing veins: *a veined leaf.*
—**veining,** *n.* the formation or arrangement of veins or veinlike markings on or in something: *the veining of a leaf.* SYN: venation.

veinlet, *n.* **1** *Anatomy.* = venule. **2** *Botany.* a branch or subdivision of a vein: *The veins divide and subdivide, making a network, and are so distributed that no part of the mesophyll is far removed from one or more veinlets* (Emerson, *Basic Botany*).

veiny, *adj. Botany.* having many veins: *veiny leaves.*

velamen (və lā′mən), *n., pl.* **-lamina** (-lam′ə nə). **1** *Botany.* the thick, outer, spongy tissue covering the aerial roots of epiphytic orchids.
2 *Anatomy.* a membranous covering or partition. SYN: velum.
[from Latin *velamen* covering, from *velare* to cover, from *velum* veil]

velar (vē′lər), *adj. Anatomy.* of or having to do with a velum, especially the soft palate: *Velar closure ... is the closure of the oral passage by the tongue against the lower surface of the velum* (Henry A. Gleason, Jr.).

veld or **veldt** (velt *or* felt), *n. Geography, Ecology.* the open, grass-covered plains of southern Africa, often with bushes but with very few trees: *In the case of the high veldt ... , the whole is an elevated plateau steppe. In many districts this is a thick carpet of fine, soft, red grass; in others it is typical bush veldt* (White and Renner, *Human Geography*). [from Afrikaans *veld,* from Dutch *veld* field]
ASSOCIATED TERMS: see **grassland.**

veliger (vē′lə jər), *n. Zoology.* the larva of a mollusk when it has a ciliate swimming membrane or velum: *The development of the eggs of most mollusks includes a trochophore stage ... which becomes a veliger larva, so called because of the presence of a band of cilia, the velum, in front of the mouth* (Hegner and Stiles, *College Zoology*). [from Latin *velum* veil + *-ger* bearing, from *gerere* to bear]

velocity, *n. Physics.* the rate at which a body moves in a given direction. Velocity is a vector quantity whose magnitude is expressed in units of distance and time, such as meters per second: *When physicists speak of the velocity of a body, they refer not only to its speed at a point in its path but to its direction of motion as well* (Pierce, *Electrons, Waves, and Messages*).
▶ *Speed* is not the same as *velocity:* speed is the rate of motion of a body irrespective of direction, whereas velocity is the rate of motion in a particular direction.

velum (vē′ləm), *n., pl.* **-la** (-lə). **1** *Anatomy.* **a** a membrane or membranous partition resembling a veil, especially the soft palate: *Above and behind the velum is the nasal pharynx opening into the cavity of the nose which acts as a resonance chamber* (Simeon Potter). **b** any one of several membranes connected to or in the brain: *The two medullary vela are inclined obliquely towards each other* (G. V. Ellis).
2 *Zoology.* a ciliated membrane which covers the veliger of a mollusk, serving as an organ of swimming or locomotion: *The free-swimming young teredo hangs in the water from a velum, an extraordinary mobile structure that looks like an animated umbrella and*

functions as an organ of locomotion (Scientific American).

3 *Biology.* a membranous structure or covering in certain fungi.
[from Latin *velum* covering, veil]

vena cava (vē′nə kā′və), *pl.* **venae cavae** (vē′nē kā′vē). *Anatomy.* either of two large veins that empty blood from the upper and lower halves of the body into the right atrium of the heart. Compare **inferior vena cava** and **superior vena cava.** [from Latin *vena cava* hollow vein]

venation (ve nā′shən), *n. Biology.* **1** the arrangement of veins in the blade of a leaf, in an insect's wing, etc. SYN: nervation. **2** these veins collectively. [from Latin *vena* vein]

venin (ven′in), *n. Biochemistry.* any of a group of poisonous substances present in the venom of snakes, toads, and scorpions.

Venn diagram (ven), *Mathematics.* a diagram using circles and rectangles to represent various types of sets and to show the relationship between them. In a Venn diagram, separate sets may be represented by two or more separate circles, and overlapping sets by two or more overlapping circles. [named after John *Venn,* 1834–1923, English logician]

venom, *n. Zoology, Physiology.* a poisonous substance secreted by special glands of some snakes, spiders, scorpions, lizards, and similar animals, who inject it into their prey or enemy by biting or stinging: *Venom is a complex of organic materials having various physiological effects, and each kind of venom has distinctive characteristics and different toxicity* (Storer, *General Zoology*). Compare **toxin.**
—**venomless**, *adj.* not producing venom: *Boa constrictors are venomless snakes.*
—**venomous**, *adj.* producing venom: *Rattlesnakes are venomous.*

venous (vē′nəs), *adj.* **1** *Physiology.* of, having to do with, or contained in the veins: *venous circulation. Venous blood is dark-red after giving up oxygen and becoming charged with carbon dioxide.*
2 *Botany, Zoology.* having numerous veins: *a venous leaf, the venous wings of insects.*
[from Latin *venosus,* from *vena* vein]

vent, *n.* **1** *Zoology.* the excretory opening at the end of the digestive tract, especially in birds, reptiles, amphibians, and fishes.
2 *Geology.* the opening of a volcano in the earth's crust. Compare **fumarole.**
[from Old French *vent* wind and *évent* vent, blowhole, both ultimately from Latin *ventus* wind]

venter (ven′tər), *n.* **1** *Anatomy.* **a** the abdomen. **b** the cavity of the abdomen. **c** the hollowed surface of a bone. **d** the prominent central portion of a muscle.
2 *Zoology.* the part of lower forms of animal life corresponding in function or position to the abdomen of mammals.
3 *Botany.* the thickened basal portion of an archegonium: *Each archegonium contains an egg imbedded in its thick, fleshy base, called the venter* (Emerson, *Basic Botany*).
[from Latin *venter* belly]

ventral, *adj.* **1** of, having to do with, or situated on or near the surface or part opposite the back: *The lower or ventral surface [of a frog] comprises the broad throat beneath the head; the chest, thorax, or pectoral region adjacent to the forelegs; the belly, or abdomen, behind; and the pelvic region between the hind legs* (Storer, *General Zoology*).
ASSOCIATED TERMS: see **dorsal.**
2 *Botany.* of or belonging to the anterior or lower surface, as of a carpel: *a ventral style. Above the egg, in the upper part of the venter, is a smaller, cone-shaped cell, the ventral canal cell* (Emerson, *Basic Botany*).
[from Late Latin *ventralis,* from Latin *venter* belly]
—**ventrally,** *adv.* in a ventral position or direction: *The anterior abdominal vein ... runs ventrally and forward* (H. N. Martin and W. A. Moale).

ventral fin, = pelvic fin.

ventricle (ven′trə kəl), *n. Anatomy.* **1** either of the two lower chambers of the heart that receive blood from the atria and force it into the arteries: *The muscular wall of the left ventricle is about a half-inch in thickness in normal people, and that of the right ventricle is about two-thirds as thick. The left ventricle has to pump blood into strong elastic arteries against a pressure of about 110–145 mm. of mercury, while the right ventricle pumps blood into the lungs against a pressure of only about 25–30 mm.* (H. M. Marvin). See the picture at **heart.**
2 any of a series of connecting cavities in the brain, normally numbering four in the adult human being, formed by enlargement of the neural canal.
3 any hollow organ or cavity of the body, such as the space between the true and false vocal cords.
[from Latin *ventriculus,* diminutive of *venter* belly]

ventricose (ven′trə kōs), *adj. Biology.* swollen or protuberant, especially on one side: *a ventricose calyx or corolla, the ventricose shell of a mollusk. The [insect's] gullet ... is ventricose when it dilates into a large bag or crop* (E. Newman).

ventricular (ven trik′yə lər), *adj. Anatomy.* of, having to do with, or resembling a ventricle: *In the beating cycle the ventricular septum muscle tends to contract first; thus it provides a stiff member against which the surrounding ventricular muscles contract to squeeze the blood out of the ventricles* (Scientific American).

ventriculus (ven trik′yə ləs), *n., pl.* **-li** (-lī). *Zoology.* **1a** the stomach or digestive cavity of certain insects, fish, and reptiles. **b** the gizzard in birds.
2 the body cavity of a sponge.
[from Latin, diminutive of *venter* belly]

ventrolateral, *adj. Anatomy.* of, having to do with, or affecting both ventral and lateral parts: *the ventrolateral area of the thalamus. By a different type of specialization for locomotion, members of another branch from the ancestral stock developed ventrolateral lobelike outgrowths of the body segments, and thus became walking animals* (Science News Letter).

---

cap, fāce, fäther; best, bē, tèrm; pin, fīve;
rock, gō, ôrder; oil, out; cup, pùt, rüle;
yü in use, *yù* in uric;
ng in bring; *sh* in rush; *th* in thin; ᴛʜ in then;
zh in seizure.
ə = *a* in about, *e* in taken, *i* in pencil, *o* in lemon, *u* in circus

ventromedial, *adj. Anatomy.* of, having to do with, or affecting both ventral and medial parts: *Ground squirrels with lesions in the ventromedial nucleus of the hypothalamus overeat and become obese, just as do rats, cats, and humans with similar damage* (Evelyn Satinoff).

venule (ven'yül), *n. Anatomy.* a small vein, especially one that begins at the capillaries and connects them with the larger veins: *Capillaries join to form venules, venules join to form veins, and eventually, the veins unite to form the great veins that return blood to the heart* (Harper's). [from Latin *venula,* diminutive of *vena* vein]

Venus, *n. Astronomy.* the sixth largest planet in the solar system and the second in distance from the sun. Venus is the planet that comes nearest to the earth, and the brightest planet as seen from the earth. It appears in the evening sky when moving toward the earth, and in the morning sky when moving away from the earth. *Venus proved to be a relatively dry planet, with an atmosphere almost wholly composed of carbon dioxide, and, as such, is hostile to life as we know it on earth* (John B. Adams). See the picture at **solar system.** [named after the Roman goddess of love and beauty]
—**Venusian,** *adj.* of or having to do with the planet Venus: *The Venusian atmosphere is about one hundred times as massive as that of the earth* (Von Braun and Ordway, *New Worlds*).

vermiculite (vər mik'yə lit), *n. Mineralogy.* a clay mineral that resembles mica and consists of a hydrous silicate of aluminum, iron, and magnesium occurring in small foliated scales. When expanded by heat, vermiculite is used for insulation, soundproofing, etc. *Vermiculite ... has the property of expanding by heat-treatment to a permanent volume some 15 times its volume as mined. It is this expanded mineral that is used in the nailable concrete* (Science News Letter). [from Latin *vermiculus,* diminutive of *vermis* worm (so called from the appearance of the mineral when heated)]

vermiform (vèr'mə fôrm), *adj. Zoology, Anatomy.* shaped like a worm; long, thin, and more or less cylindrical: *vermiform larvae, vermiform fishes such as the eel and lamprey.* [from New Latin *vermiformis,* from Latin *vermis* worm + *forma* form]

vermiform appendix or **vermiform process,** *Anatomy.* a slender tube, closed at one end, growing out of the large intestine in the lower right-hand part of the abdomen: *The vermiform process or appendix ... is extremely variable [in length], being anywhere from 1 to 10 inches, with an average of about 4 inches. It is commonly the site of inflammation (appendicitis), necessitating its removal (appendectomy). It is functionless and represents the vestigial remains of a more extended cecum* (Edwards, *Concise Anatomy*). Commonly called **appendix.**

vernacular, *Biology.* —*n.* the common name of a plant or animal, not its scientific New Latin name, such as *black-eyed Susan* for *Rudbeckia serotina.*
—*adj.* of or designating the common name given to a plant or animal. The vernacular name of plants of the genus *Equisetum* is horsetail.
—**vernacularly,** *adv.* in or according to the vernacular.

vernal equinox, *Astronomy.* the equinox that occurs in the Northern Hemisphere about March 21: *The point and time (March 21) of the sun's crossing the celestial equator northward are both called the vernal equinox—called "equinox" (from the Latin meaning equal night) because at such times day and night are of equal length* (Krogdahl, *The Astronomical Universe*). Compare **autumnal equinox.** [from Latin *vernalis* of spring, from *ver* spring]

vernalize, *v. Botany.* to cause (a plant) to bloom and bear fruit early by subjecting the seed or bulb to a very low temperature.
—**vernalization,** *n.* the act or process of vernalizing: *Vernalization consists of the transformation of winter cereals into spring varieties by chilling and soaking the seeds* (Laurence H. Snyder).

vernation, *n. Botany.* the arrangement or formation of the leaves of plants or fronds of ferns in the bud, with reference to their folding, coiling, and other such characteristics. Compare **foliation.** [from Latin *vernare* to bloom, as in spring, from *vernus* of spring, from *ver* spring]

vernier or **Vernier** (vèr'nē ər), *n.* a small movable scale for measuring a fractional part of one of the divisions of the fixed scale to which it is attached, used, for example, in astronomical and surveying instruments: *Experiments were conducted in a greenhouse during the summer in reservoirs ... containing 20 cm of pot soil and filled up with tap water. The circumference of the floats and the length of the petioles were measured with a sliding gauge with Vernier scale (accuracy 0.1 mm)* (Nature).
—*adj.* furnished with a vernier: *a vernier caliper or compass.*
[named after Pierre *Vernier,* 1580–1637, French mathematician]

vernix (vèr'niks), *n.,* or **vernix caseosa** (kā'sē ō'sə), *Embryology.* a waxy white substance covering the skin of a fetus to prevent its softening by the amniotic fluid. [from Medieval Latin *vernix* varnish + Latin *caseus* cheese]

verruca (ve rü'kə), *n., pl.* **-cae** (-kē *or* -sē). *Biology.* a wartlike growth or prominence on the body of a plant or animal. [from Latin *verruca* wart]
—**verrucose** (ver'ù kōs), *adj.* covered with small wartlike swellings or protuberances.

versatile, *adj.* **1** *Zoology.* **a** turning forward or backward: *the versatile toe of an owl.* **b** moving freely up and down, and laterally: *versatile antennae.*
2 *Botany.* attached at or near the middle so as to swing or turn freely: *a versatile anther.*
[from Latin *versatilis* turning, from *versare,* frequentative of *vertere* to turn]

vertebra (ver'tə brə), *n., pl.* **-brae** (-brā *or* -brē), **-bras.** *Anatomy.* **1** one of the cartilaginous or bony members of the spinal column in vertebrates: *Each piece of the backbone is called a vertebra ... Each vertebra has several projections to which muscles are attached* (Beauchamp, *Everyday Problems in Science*).
2 the vertebrae, *pl.* the spinal column; backbone: *the vertebrae of a whale.*
[from Latin *vertebra* (originally) joint, turning place, from *vertere* to turn]

—**vertebral,** *adj.* **1** of, having to do with, or being a vertebra or the vertebrae: *the vertebral arteries.*

2 composed of vertebrae; spinal: *the vertebral chain of bones.*

3 having a spinal column; vertebrate: *vertebral animals, vertebral life. Cuvier ... was the first to divide animals into vertebral and invertebral* (J. Scott). —**vertebrally,** *adv.*

vertebral column, = spinal column: *In the higher chordates the notochord is present in the embryo only, since it is replaced by the bony vertebral column in the adult form* (Winchester, *Zoology*).

vertebrate (ver′tə brit *or* ver′tə brāt), *Zoology.* —*n.* an animal that has a backbone; any of the large subphylum (Vertebrata) of chordates having a segmented spinal column and a brain case or cranium, including fishes, amphibians, reptiles, birds, and mammals: *All vertebrates have a vertebral column, or backbone, as their structural axis. This is a flexible, usually bony support that develops around the notochord, supplanting it entirely in most species* (Curtis, *Biology*).

▶ According to the commonest classification, there are seven living classes of vertebrates, including the *Agnatha* (lampreys and hagfishes), the *Chondrichthyes* (cartilagenous fishes, such as the sharks and rays), the *Osteichthyes* (sturgeon, trout, salmon, and other bony fishes), the *Amphibia* (salamander, frog, etc.), the *Reptilia* (turtle, snake, lizard, etc.), *Aves* (birds), and *Mammalia* (duckbill, kangaroo, opossum, rat, dog, horse, human being, etc.).

—*adj.* **1** having a spinal column or backbone: *vertebrate animals.*

2 of or having to do with vertebrates: *the vertebrate skeleton. The earliest known vertebrate remains are those of Fishes* (Herbert Spencer).

vertex (ver′teks), *n., pl.* **vertices** (ver′tə sēz′), **vertexes.**
1 *Geometry.* **a** the point opposite to and farthest from the base of a triangle, pyramid, or other figure having a base. A triangle consists of three vertices and three sides joining pairs of the vertices. **b** the point where the two sides of an angle meet. **c** any point of intersection of the sides of a polygon or the edges of a polyhedron: *A polyhedral angle of a polyhedron is one made up of the edges meeting at a vertex. A diagonal is a line segment joining two vertices not in the same face* (William Karush). **d** the point in a curve or surface at which the axis meets it: *The point V, where the axis intersects the surface, is called the vertex* (Sears and Zemansky, *University Physics*).

2 *Astronomy.* = zenith.

3 *Optics.* the point, at the center of a lens, where the axis intersects the surface.
[from Latin *vertex* (originally) whirl, whirling, from *vertere* to turn]

vertical, *adj.* **1** *Geometry.* (of a straight line or plane surface) extending at right angles to the plane of the horizon; perpendicular; upright: *The x-axis and all lines parallel to it are called horizontal; the y-axis and all lines parallel to it are called vertical* (Moise and Downs, *Geometry*).

2 *Astronomy.* of, situated at, or passing through the vertex or zenith; directly overhead: *At each equinox the sun appears vertical over the equator* (Archibald Geikie). See also **vertical circle.**

3 *Botany.* lying in the direction of the stem or axis; lengthwise: *vertical leaves.*
—*n.* **1** a vertical line, plane, position, direction, or part.
2 = vertical angle. —**vertically,** *adv.*

vertical angle, *Geometry.* either of two angles formed by two intersecting lines and lying on opposite sides of the point of intersection: *The angles formed by the blades of [open] shears are called vertical angles ... Vertical angles are not adjacent, but are rather opposite angles at intersecting lines* (Herberg, *A New Geometry for Secondary Schools*).

vertical circle, *Astronomy.* any great circle of the celestial sphere perpendicular to the plane of the horizon, passing through the zenith and nadir. Also called **azimuth circle.** See the picture at **azimuth.**

verticality, *n.* the quality or condition of being vertical; vertical position; perpendicularity.

verticil (ver′tə sil), *n. Botany, Zoology.* a whorl or circle of leaves, hairs, or other parts, growing around a stem or central point. [from Latin *verticillus* whorl, diminutive of *vertex* vertex]
—**verticillate** (ver tis′ə lāt *or* ver′tə sil′āt), *adj.* forming verticils or whorls: *verticillate leaves or flowers.*

vesical (ves′ə kəl), *adj. Anatomy.* or or having to do with the bladder; formed in the urinary bladder. [from Latin *vesica* bladder]

vesicle (ves′ə kəl), *n.* **1** *Biology.* a small bladder, cavity, sac, or cyst, especially one filled with fluid: *The vesicles on the axon side of the synapse which apparently contain the transmitter substance are so tiny that, in accordance with the Heisenberg uncertainty principle, there is a relatively large uncertainty as to their location over a period as brief as a millisecond* (John C. Eccles).

2 *Geology.* a small spherical or oval cavity in a volcanic igneous rock, produced by the expansion of bubbles of gas or vapor during solidification of the rock.
[from Latin *vesicula,* diminutive of *vesica* bladder, blister]

vesicular (və sik′yə lər), *adj. Biology.* **1** of, having to do with, or resembling a vesicle or vesicles: *vesicular cells.*

2 characterized by the presence of vesicles: *vesicular lungs. Even some small vesicular rocks were completely coated by the bright condensate, suggesting at least some direct condensation of water vapor on these rocks* (Science). —**vesicularly,** *adv.*

vespertine (ves′pər tin), *adj.* **1** *Zoology.* flying or appearing in the early evening. Bats and owls are vespertine animals.

2 *Botany.* opening in the evening, as some flowers do.

3 *Astronomy.* descending toward the horizon in the evening: *a vespertine planet or star.*
[from Latin *vesper* evening]

---

cap, fāce, fäther; best, bē, tėrm; pin, five;
rock, gō, ôrder; oil, out; cup, pùt, rüle,
yü in use, *yu* in uric;
ng in bring; *sh* in rush; *th* in thin, ŦH in then;
zh in seizure.
ə = *a* in about, *e* in taken, *i* in pencil, *o* in lemon, *u* in circus

vessel, *n.* **1** *Biology.* a tube carrying blood or other fluid. Veins and arteries are blood vessels. Compare **vascular. 2** = trachea (def. 3).

vestibular, *adj.* of or having to do with a vestibule: *The frogs, whose inner ear is similar to man's, were launched into space ... with microelectrodes surgically implanted in the vestibular (inner ear) nerves leading from the sensor cells in the frogs' otoliths* (Science News).

vestibule, *n.* *Anatomy.* a cavity that leads to another (usually larger or more important) cavity. The vestibule of the ear is the central cavity of the inner ear. The vestibule of the oral cavity is the space anterior to the teeth and gums. See the picture at **ear.**

vestige, *n.* *Biology.* a part, organ, etc., that is no longer fully developed or useful but performed a definite function in an earlier stage of the existence of the same organism or in ancestral organisms: *Rudimentary organs, ... as ... the vestige of an ear in earless breeds* (Darwin, *Origin of Species*). *The aquatic reptile ... which retains only the vestiges or rudiments of eyes* (Charles Lyell). [from French, from Latin *vestigium* footprint]
—**vestigial** (ve stij'ē əl), *adj.* (of a part, organ, etc.) no longer fully developed or useful. Vestigial organs or structures in the human body include the vermiform appendix, the coccyx, the wisdom teeth, hair on the body, nipples on males, and muscles to move the ears.
—**vestigially,** *adv.*

vesuvianite (və sü'vē ə nīt), *n.* *Mineralogy.* a metamorphic mineral, a silicate of aluminum, calcium, iron, and magnesium, sometimes with other elements, occurring in massive or, more frequently, in square crystals of various colors. *Formula:* $Ca_{10}Mg_2Al_4(SiO_4)_5(SiO_7)_2(OH)_4$ Also called **idocrase.** [from *Vesuvius,* the volcano near Naples, Italy + *-ite*]

vexillary (vek'sə ler'ē), *adj.* of or having to do with a vexillum. **Vexillary estivation** is a mode of estivation in which the exterior petal, as in the case of the vexillum, is largest and encloses and folds over the other petals.

vexillum (vek sil'əm), *n., pl.* **vexilla** (vek sil'ə).
1 *Botany.* the large upper external petal of a papilionaceous flower. **2** *Zoology.* the web or vane of a feather. [from Latin *vexillum* flag or banner]

viability, *n.* *Biology.* the condition of being viable; the ability to live, grow, or develop, as in certain climates or environments: *The viability of the seed, that is to say, its ability to germinate ... can nowadays be determined by staining the seed with tetrazolium chloride, a dye which gives a characteristic red colour with live seeds but not with dead ones. It is a useful method for sampling seeds of doubtful viability* (A. M. Mayer).

viable, *adj.* *Biology.* capable of living, growing, or developing, as a spore, seed, egg, embryo, etc.: *Some kinds of vegetable and flower seed are viable (capable of germination) for many years* (N. Y. Times). *They point out that the grafts do not become viable after implantation, but instead they serve as a framework to hold the bud which is under pressure* (John C. Vanatta). [from French, from *vie* life, from Latin *vita*]
—**viably,** *adv.*

vibraculum (vī brak'yə ləm), *n., pl.* **-la** (-lə). *Zoology.* one of the long, slender, whiplike, movable appendages or organs of certain bryozoans. [from New Latin, from Latin *vibrare* to shake, oscillate]

vibrate, *v.* *Physics.* **1** to swing to and fro or otherwise move in an alternating or reciprocating motion (especially in reference to an elastic body or medium whose equilibrium has been disturbed): *At low temperatures the atoms [of a solid] vibrate a little about their mean position in the lattice, but as the temperature is raised the average energy of the atoms increases and on the whole they vibrate more violently, until at the melting point the lattice structure breaks up and the material becomes a liquid with comparative freedom of atomic movement* (R. S. Barnes).
2 to measure by moving to and fro: *A pendulum vibrates seconds.*
[from Latin *vibrare* to shake] ► See note at **vibration.**

vibratile (vī'brə təl *or* vī'brə tīl), *adj.* **1** having to do with or characterized by vibration: *the vibratile action of spermatozoa.*
2 having a vibratory motion: *vibratile cilia.*

vibration, *n.* *Physics.* **1** an alternating or reciprocating motion, especially that produced in an elastic body or medium by the disturbance of equilibrium.
2 a single movement of this kind: *The wavelength is the distance travelled by the wave front in the period of one vibration. Since the frequency is the number of vibrations per second, frequency multiplied by wavelength gives the distance the wave front travels in one second. This is, of course, the velocity of propagation* (W. C. Vaughan).
► When the reciprocating movement is comparatively slow, such as that of a pendulum, the term *oscillation* is commonly used, while the term *vibration* is generally confined to a motion with rapid reciprocations or revolutions.
—**vibrational,** *adj.* of, having to do with, or of the nature of vibration: *The lower the musical tone, the smaller is the number of vibrational cycles a second* (Simeon Potter). *As the vibrational energy of a molecule increases, the atoms vibrate more and more about their normal positions* (K. D. Wadsworth).
—**vibrationally,** *adv.* in a manner characterized by or resulting in vibration: *Many chemical reactions occur with changes of bond lengths, and such reactions will almost surely leave the products in vibrationally excited states* (George C. Pimentel).

vibratory, *adj.* *Physics.* of the nature of, characterized by, or consisting of vibration: *Vibratory motion results when a body hanging from a spring is pulled downward and then released; other common examples of this type of motion are the vibrations of strings and air columns of musical instruments* (Shortley and Williams, *Elements of Physics*). *Singing and speaking ... produce different vibratory patterns in the vocal cords* (Science News Letter).

vibrio (vib'rē ō), *n., pl.* **-rios.** any of a genus (*Vibrio*) of short, curved, motile bacteria, often shaped like a comma, spiral, or S, such as the species that causes Asiatic cholera. Also called **comma bacillus.** See the picture at **bacteria.** [from Latin *vibrare* to vibrate]
—**vibrionic,** *adj.* having to do with or caused by vibrios.

vibrissa (vī bris'ə), *n., pl.* **-brissae** (-bris'ē).

1 *Anatomy.* one of the short hairs growing in a nostril.
2 *Zoology.* **a** one of the long, bristlelike, tactile organs growing upon the upper lip and elsewhere on the head of most mammals; a whisker, as of a mouse. **b** one of the special set of long, slender, bristlelike feathers that grow in a series along each side of the rictus (gape of the mouth) of many birds, such as flycatchers and goatsuckers. The vibrissae entangle the legs and wings of insects, and thus diminish their struggling when caught. [from Late Latin *vibrissae* hairs in the nostrils, related to Latin *vibrare* to vibrate]

vicariance (vī kãr′ē əns), *n. Biology.* the geographical separation of similar species of plants and animals by a barrier such as a mountain range or an ocean resulting from massive displacements of the earth's crust: *The debate here is whether a certain pattern of species distribution was caused by dispersal or vicariance* (Niles Eldridge). [from Latin *vicarius* (originally) substituted, from *vicis* change, alteration]

vicarious, *adj. Physiology.* occurring in some part or organ other than the normal one.

vicenary (vis′ə ner′ē), *adj. Mathematics.* having 20 for the base: *a vicenary scale.* [from Latin *viceni* twenty each, related to *viginti* twenty]

vicinal plane (vis′ə nəl), *Mineralogy.* a subordinate plane in a crystal, whose position varies little from that of the fundamental plane which it replaces. [from Latin *vicinalis* neighboring, from *vicinus* neighbor]

villi (vil′ī), *n., pl.* of **villus.** **1** *Anatomy.* **a** tiny, hairlike parts growing out of the mucous membrane of the small intestine, that increase the surface area to facilitate the absorption of substances: *Fatty acids and glycerol are absorbed through the villi into the lacteals and are transported in the lymph. Monosaccharides and amino acids are absorbed through the villi and enter the capillaries to be transported to the liver* (Biology Regents Syllabus). **b** fingerlike projections of the chorion that contribute to the formation of the placenta of mammals.
2 *Botany.* the long, straight, soft hairs that cover the fruit, flowers, and other parts of certain plants. [from Latin, plural of *villus* tuft of hair]

villiform (vil′ə fôrm), *adj. Zoology.* having the form of villi; numerous, slender, and closely set, as the teeth of certain fishes.

villous (vil′əs), *adj.* **1** having villi; covered with villi: *the villous internal surface of the small intestine, villous plants.*
2 of the nature of or resembling villi: *villous processes, villous formations.* —**villously,** *adv.*

villus (vil′əs), *n.* singular of villi: *The nutrient passes through the epithelial cell into the core of the villus and enters either the blood or the lymph circulation without ever having crossed the muscle layer* (Darrell R. Van Campen).

vimen (vī′mən), *n., pl.* **vimina** (vim′ə nə). *Botany.* a long, flexible shoot of a plant. SYN: twig. [from Latin *vimen* twig]
—**vimineous** (vi min′ē əs), *adj. Botany.* producing vimina: *a vimineous plant or tree.*

vinculum (ving′kyə ləm), *n., pl.* **-la** (-lə). *Mathematics.* a line drawn over several terms to show that they are to be considered together. EXAMPLE: $\overline{4 + 5} \times 6$ means 9×6. [from Latin *vinculum* bond, from *vincire* to bind]

violet rays, *Physics.* the shortest rays of the spectrum that can be seen, having wavelengths of about 380 nanometers.

viral, *adj.* of, having to do with, or caused by a virus: *a viral infection. When bacteria that harbor the viral chromosome in a latent state are exposed to certain chemicals or to physical agents such as ultraviolet light or X rays ... the virus breaks off from the chromosome and begins to replicate* (Gordon M. Tomkins).

virga (vėr′gə), *n. Meteorology.* streamers of rain or snow falling from a cloud, but dissipated before they reach the ground. [from Latin *virga* twig, streak (in the sky)]

virgate (vėr′git *or* vėr′gāt), *adj. Botany.* producing a large number of small twigs: *virgate branches, a virgate stem.* [from Latin *virga* twig]

virgin, *n. Zoology.* **1** an animal that has never mated. **2** a female insect, such as a bee or wasp, that produces fertile eggs by parthenogenesis.
—*adj.* **1** *Zoology.* **a** being a virgin: *a virgin female.* **b** occurring by parthenogenesis: *virgin birth.*
2 *Botany.* in a natural state; never cleared or cut by humans: *a virgin forest.*

virion (vī′rē on *or* vir′ē on), *n. Biology.* a mature virus particle, consisting of RNA or DNA enclosed in a protein shell and constituting the infectious form of the virus: *Somewhat earlier other workers had fractionated virions—the actual virus particles as distinct from the forms assumed by the virus inside cells—and had found RNA polymerases, enzymes that catalyze the synthesis of RNA from its building blocks* (Howard M. Temin). [from *virus* + *-on*]

virogene (vī′rō jēn′), *n. Biology.* a virus-producing gene; a gene capable of specifying the synthesis of a virus: *Tricks for getting the virogenes to express themselves as infectious viruses in cultured cells were discovered some time ago. But the viruses produced in this way are generally not very good at reproducing themselves in the animal* (New Scientist).

viroid (vī′roid), *n. Biology.* any of various infectious particles, smaller than known viruses, that consist entirely of single-stranded RNA and cause disease in plants: *Viroids have been identified by T. O. Diener as the smallest known agents of infectious disease. The molecular weight of viroids is estimated to be as little as $7.5–8.5 \times 10^4$ daltons, in marked contrast to the conventional plant virus genomes, which have molecular weights of approximately 2×10^6 daltons* (R. K. Horst).

virtual, *adj.* **1** *Optics.* **a** of or having to do with an image formed when the rays from each point of the object diverge as if from a point beyond the reflecting or refracting surface: *It is clear that whether an object is real or virtual depends merely on whether the light is diverging or converging when it enters the lens* (Hardy and Perrin, *Principles of Optics*). *A virtual image is*

cap, fāce, fäther; best, bē, tėrm; pin, fīve;
rock, gō, ôrder; oil, out; cup, pùt, rüle,
*y*ü in use, *y*ù in uric;
ng in bring; sh in rush; th in thin, ᴛʜ in then;
zh in seizure.
ə = a in about, e in taken, i in pencil, o in lemon, u in circus

715

subjective in that it appears to form where an image could not possibly exist because rays do not actually intersect at the image point (Physics Regents Syllabus). **b** having to do with or designating a focus forming such an image.

2 *Nuclear Physics.* being too transient to be detected or observed directly: *A virtual photon can have very different properties from a real one. A real photon is one that is flying free and can be detected, in a light beam or an X-ray beam, for instance. A virtual photon is one that is emitted and absorbed so quickly that its existence cannot be detected* (Dietrick E. Thomsen).

virulent (vir′yə lənt *or* vir′ə lənt), *adj. Biology.* able to cause disease by breaking down the protective mechanisms of the host: *Both rust and smut resistance in host plants frequently comes to be of little avail when new, more virulent strains of the parasite arise* (Emerson, *Basic Botany*). [from Latin *virulentus,* from *virus* poison] —**virulently,** *adv.*
—**virulence,** *n.* the degree to which an infectious microorganism is able to break down the protective mechanisms of its host.

virus, *n. Biology.* **1** any of a large group of disease-producing agents that are smaller than bacteria, are composed of a core of RNA or DNA and an outer coat of protein, and are dependent upon living cells for their reproduction and growth. Viruses are shaped like rods or spheres and range in size from about 10 to 250 nanometers. They are able to infect almost all types of organisms, including bacteria, and cause such diseases as rabies, measles, polio, influenza, chicken pox, and the common cold. *When we consider the reproduction of viruses we are in quite a dilemma. We know that viruses can form crystals and that objects that form crystals (such as table salt, ordinary sugar, dry ice, asbestos and diamond, to mention a few) are not considered to be examples of life. Yet we also know that in certain situations (usually when they are inside living cells) viruses behave very much like living things; certainly they reproduce like living things* (Robert W. Menefee). *The likelihood that viruses are fundamental causative agents in many mammalian cancers has implications for diagnosis, for immunological treatment, and for cancer prevention* (Michael J. Brennan). Compare **echovirus, oncornavirus, papovavirus, slow virus.** See also **bacteriophage.**

2 any disease caused by a virus, such as tobacco mosaic in plants, distemper in dogs, foot-and-mouth disease in cattle, and hepatitis and yellow fever in human beings: *The place of grippe can never be filled by ... the new viruses, which last a mere thirty-six hours and are common in every sense* (Harper's). [from Latin *virus* poison]

viscera (vis′ə rə), *n., pl.* of **viscus.** *Anatomy.* the soft internal organs of the body, especially of the abdominal cavity, including the heart, stomach, liver, intestines, kidneys, etc.: *Most of the abdominal and pelvic viscera ... are covered with a serous membrane called the peritoneum* (Edwards, *Concise Anatomy*). [from Latin]
—**visceral,** *adj.* of, having to do with, or in the region of the viscera: *The second type of muscle in the frog's body is the smooth or visceral muscle. ... The name visceral is given because it is found primarily in the visceral organs* (Winchester, *Zoology*). —**viscerally,** *adv.*

viscid (vis′id), *adj. Botany.* covered with a sticky secretion, as leaves. [from Late Latin *viscidus,* from Latin *viscum* birdlime]

viscoelastic (vis′kō i las′tik), *adj. Physics.* having the properties of viscosity and elasticity: *Silicone putty is ... viscoelastic. A ball of it will bounce, but when left on a table for a few hours the same ball will flow under the force of gravity into a pancake* (Scientific American).
—**viscoelasticity,** *n.* the quality or condition of being viscoelastic: *All substances show a combination of elastic and fluid behavior that is termed viscoelasticity* (Arthur V. Tobolsky).

viscosity (vis kos′ə tē), *n. Physics.* a property of fluids that causes them to resist flowing as a result of internal friction from the fluid's molecules moving against each other. All fluids have some degree of viscosity. *Helium II must be extraordinary fluid; in the terms of physics, it must have an extremely low viscosity, meaning an extremely small internal frictional resistance to flow. The viscosity of a liquid is usually measured by letting it flow through a narrow capillary tube* (Eugene M. Lifshitz). [from Late Latin *viscosus* viscous]

viscous (vis′kəs), *adj. Physics.* having or marked by viscosity: *A liquid which resists flowing, or resists the action of any other deforming force upon it, is said to be viscous* (Jones, *Inorganic Chemistry*). [from Late Latin *viscosus,* from *viscum* birdlime]

viscus (vis′kəs), *n.* singular of **viscera.** *Anatomy.* any visceral organ, especially one within the abdominal cavity: *Imperfection of any viscus, as lungs, heart or liver ...* (Herbert Spencer).

visible light, *Physics.* light consisting of electromagnetic waves that can be seen, as contrasted with ultraviolet and infrared waves that are invisible. Visible light ranges in wave-length from about 380 to about 710 nanometers. *Visible light occupies less than one octave of the spectrum of electromagnetic waves* (W. C. Vaughan). See the picture at **electromagnetic spectrum.**

visible spectrum, *Physics.* the part of the spectrum that can be seen, appearing as a band of colors merging through continuous hues into each other from red to violet: *... hence the only interstellar absorption lines of appreciable strength in the visible spectrum are produced by the relatively scarce elements, sodium and calcium* (Lyman Spitzer, Jr.).

visual binary or **visual double,** *Astronomy.* a binary or double star that can be seen as two stars with a telescope and sometimes with the unaided eye: *Most double stars appear as one to the unaided eye. If they can be seen as two stars, or if a telescope reveals them as two stars, they are called visual doubles or visual binaries* (Charles A. Federer, Jr.).

visual field, = field of vision.

visual purple, *Biochemistry.* a photosensitive, purplish-red protein present in the rods of the retina, that, in the presence of light, is bleached to form visual yellow: *If it takes a long time for the visual purple to form, the individual cannot see well in dim light. This condition is known as night blindness* (Matthew Luckiesh). Also called **rhodopsin.**

visual yellow, *Biochemistry.* a substance formed in the rods of the retina from rhodopsin after exposure to light and converted to a yellow pigment before it is broken down into retinene and vitamin A.

vital capacity, *Physiology.* the amount of air that the lungs can hold when breathing in as deeply as possible and then exhaling: *A change in the vital capacity may be significant. Thus, the vital capacity of a patient suffering from active lung disease falls as the disease progresses* (Winton and Bayliss, *Human Physiology*).

vitalism, *n. Biology.* the former theory that the behavior of a living organism is, at least in part, due to a vital principle or force that has none of the characteristics of matter or energy as defined by physics and chemistry: *Du Bois-Reymond [1818–1896] disproved vitalism where the electrical portions of the body were involved. Vitalism maintained itself chiefly in connection with the chemical aspects of the body and was not laid to rest until Buchner's work a century later* (Isaac Asimov).
—**vitalistic,** *adj.* of or having to do with vitalism.

vitamin, *n. Biochemistry.* any of a group of organic substances necessary for the normal growth and nourishment of the body, found in small amounts in various plant and animal foods such as milk, butter, raw fruits and vegetables, and also prepared medicinally in the form of tablets, injections, etc. Lack of essential vitamins causes such diseases as rickets and scurvy, as well as general poor health. Several vitamins, such as vitamin D and vitamin K, are produced by the body itself, but not always in sufficient quantities to meet the body's needs. [from Latin *vita* life + English *amin(e)* (so called because it was originally thought to be an amine derivative)]

vitamin A, *Biochemistry.* a fat-soluble vitamin found in milk, butter, cod-liver oil, egg yolk, liver, green and yellow vegetables, etc., that increases the resistance of the body to infection and prevents night blindness. It exists in two known forms, A_1 (*Formula:* $C_{20}H_{30}O$) and A_2 (*Formula:* $C_{20}H_{28}O$).

vitamin B_1, = thiamine.

vitamin B_2, = riboflavin.

vitamin B_6, = pyridoxine.

vitamin B_{12}, *Biochemistry.* a vitamin containing cobalt, found especially in liver, milk, and eggs and active against pernicious anemia. *Formula:* $C_{63}H_{90}N_{14}O_{14}PCo$ Also called **extrinsic factor.**

vitamin B_c, = folic acid.

vitamin B complex, *Biochemistry.* a group of water-soluble vitamins including thiamine (vitamin B_1), riboflavin (vitamin B_2), nicotinic acid, pyridoxine (vitamin B_6), pantothenic acid, inositol, para-aminobenzoic acid (vitamin B_x), biotin (vitamin H), choline, and folic acid (vitamin B_c), which are found in high concentration in yeast and liver.

vitamin B_t, *Biochemistry.* a vitamin found chiefly in meat, liver, and milk, essential to the growth of the meal worm and certain insects but not to higher animals or humans. Formula: $C_7H_{15}NO_3$ Also called **carnitine.**

vitamin B_x, = para-aminobenzoic acid.

vitamin C, *Biochemistry.* a water-soluble vitamin found in citrus fruits, tomatoes, leafy green vegetables, etc., and also made synthetically, used especially in the prevention and cure of scurvy. *Formula:* $C_6H_8O_6$ Also called **ascorbic acid, cevitamic acid.**

vitamin D, *Biochemistry.* a fat-soluble vitamin found in cod-liver oil, milk, egg yolk, etc., and produced by irradiating ergosterol and other sterols. Vitamin D prevents rickets and is necessary for the growth and health of bones and teeth. It exists in several related forms, including D_2 (calciferol), D_3, and D_4.

vitamin D_2, = calciferol.

vitamin D_3, *Biochemistry.* the natural form of vitamin D, found in fish-liver oils, irradiated milk, and all irradiated animal foodstuffs. *Formula:* $C_{27}H_{44}O$

vitamin D_4, *Biochemistry.* a vitamin produced by irradiating a form of ergosterol. *Formula:* $C_{28}H_{46}O$

vitamin E, *Biochemistry.* a fat-soluble vitamin found in lettuce and other plant leaves, wheat germ oil, and milk, that is necessary for some reproductive processes and aids in preventing abortions. Lack of vitamin E is associated with sterility. *Formula:* $C_{29}H_{50}O_2$ See also **tocopherol.**

vitamin G, = riboflavin.

vitamin H, = biotin.

vitamin K, *Biochemistry.* a fat-soluble vitamin found in green leafy vegetables, alfalfa, fish meal, egg yolk, tomatoes, etc., that promotes clotting of the blood and prevents hemorrhaging. It exists in several related forms, the best known ones being vitamin K_1 and K_2.

vitamin K_1, *Biochemistry.* a vitamin present in green plants, used in the formation of prothrombin. *Formula:* $C_{31}H_{46}O_2$

vitamin K_2, *Biochemistry.* a vitamin found in putrified fish meal and in microorganisms. *Formula:* $C_{41}H_{56}O_2$

vitamin P, *Biochemistry.* a water-soluble crystalline substance found in citrus fruits, paprika, etc., that promotes capillary resistance to hemorrhaging.

vitamin P complex, = bioflavonoid.

vitamin PP, = nicotinic acid.

vitellarium (vit′ə lãr′ē əm), *n., pl.* **-iums, -ia** (-ē ə). *Zoology.* the gland of the ovary which secretes the vitellus of the egg in certain invertebrates, especially worms. [from New Latin, from Latin *vitellus* egg]

vitellin (vi tel′ən *or* vī tel′ən), *n. Biology.* a protein contained in the yolk of eggs. [from Latin *vitellus* egg]

vitelline (vi tel′ən *or* vī tel′ən), *adj. Biology.* of or having to do with the vitellus or egg yolk. The **vitelline membrane** is the transparent membrane enclosing an egg yolk.

vitellogenesis, *n. Biology.* the formation of vitellus or yolk.

vitellus (vi tel′əs *or* vī tel′əs), *n. Biology.* the yolk of an egg: *The vitellus is limited by a cell membrane and surrounded by a moderately thick, transparent coat called the zona pellucida* (Science Journal). [from Latin]

---

cap, fāce, fäther; best, bē, tèrm; pin, fīve;
rock, gō, ôrder; oil, out; cup, pùt, rüle,
yü in use, yu̇ in uric;
ng in bring; sh in rush; th in thin, ₮H in then;
zh in seizure.
ə = a in about, e in taken, i in pencil, o in lemon, u in circus

vitrain (vit'rān), *n. Geology.* a lithotype of coal having a vitreous luster. Compare **clarain, durain, fusain.** [from Latin *vitrum* glass + English *-ain,* as in *fusain*]

vitreous humor (vit'rē əs), *Physiology.* the transparent, jellylike substance that fills the eyeball behind the lens: *The eye ... is filled with liquid, the humors of the eye. That between the translucent cornea and the lens is the aqueous humor; the semisolid material between lens and retina is the vitreous humor* (H. V. Neal and H. W. Rand). [from Latin *vitreus* of glass, from *vitrum* glass]

vitrify (vit'rə fī), *v. Chemistry, Geology.* to change into a glass or glassy substance, especially by fusion due to heat: *... lumps of vitrified matter which were tentatively identified ... as volcanic scoriae* (Emily Vermule). [from Latin *vitrum* glass]
—**vitrification,** *n.* the process of vitrifying.

vitriol (vit'rē əl), *n. Chemistry.* 1 sulfuric acid.
2 any of certain sulfates of metals, such as **blue vitriol,** a sulfate of copper, **green vitriol,** a sulfate of iron, or **white vitriol,** a sulfate of zinc. Vitriols are characterized by a glassy appearance. [from Medieval Latin *vitriolum,* from *vitrum* glass]

vitta (vit'ə), *n., pl.* **vittae** (vit'ē). 1 *Botany.* one of a number of elongated, club-shaped canals or tubes for oil, occurring in the fruit of most plants of the parsley family.
2 *Zoology, Botany.* a band or stripe of color, as on the bill of a bird or the leaf of a plant. [from Latin *vitta* fillet, chaplet]
—**vittate** (vit'āt), *adj.* marked or striped with vittae: *vittate mite.*

vivianite (viv'ē ə nīt), *n. Mineralogy.* a phosphate of iron, usually occurring in crystals of blue and green color. *Formula:* $Fe_3(PO_4)_2 \cdot 8H_2O$ [named after J. G. Vivian, 20th-century mineralogist]

viviparous (vī vip'ər əs), *adj.* 1 *Biology.* bringing forth live young, rather than eggs. Most mammals and some other animals are viviparous. *This is true viviparous reproduction, with the young being born alive, in contrast to the oviparous reproduction found in most other animals, where eggs are laid* (Winchester, *Zoology*).
2 *Botany.* reproducing from seeds or bulbs that germinate while still attached to the parent plant. [from Latin *viviparus,* from *vivus* alive + *parere* give birth to, bear]
—**viviparity** (viv'ə par'ə tē), *n.* the condition of being viviparous.

VLDL, *abbrev. Biochemistry.* very-low density lipoprotein (a lipoprotein containing a very large proportion of lipids to protein and carrying most cholesterol from the liver to the tissues). Compare **HDL, LDL.**

vocal cords or **vocal folds,** *Anatomy.* two pairs of folds of mucous membranes in the throat, projecting into the cavity of the larynx. The lower pair (inferior or true vocal cords) can be pulled tight and the passage of breath between them then causes them to vibrate, which produces the sound of the voice. The upper pair (superior or false vocal cords) do not directly aid in producing voice.

vol., 1 volcano. **2** volume.

volar (vō'lər), *adj. Anatomy.* of or having to do with the palm of the hand or the sole of the foot. Compare **palmar.** [from Latin *vola* hollow of the palm or sole]

volcanic, *Geology.* —*adj.* **1** of or caused by a volcano; having to do with volcanoes: *a volcanic eruption, volcanic activity.*
2 discharged from or ejected by a volcano or volcanoes; consisting of materials produced by igneous action: *volcanic cinders, volcanic dust.*
3 characterized by the presence of volcanoes: *volcanic regions.*
—*n.* a volcanic rock: *lunar volcanics. The larger scale features of lavas ... are aids not only in distinguishing volcanics from sediments but in giving clues to the conditions under which the volcanics were formed* (Garrels, A Textbook of Geology). —**volcanically,** *adv.*

volcanic ash, *Geology.* finely pulverized pyroclastic material ejected from a volcano in eruption: *Volcanic ash is not really ash, since it has not been burned. It consists of particles roughly the size of peas or shot, which do look like coarse ashes. They are solid or porous fragments of obsidian ... When partly decomposed, volcanic ash makes very rich soil* (Fenton, *The Rock Book*).

volcanic bomb, *Geology.* a piece of lava, often large and hollow, ejected by a volcano in eruption: *Volcanic bombs ... are fragments that range from an inch to several feet in diameter. They include both blocks of solidified lava that were broken and tossed out and lumps of molten material that became round, pear-shaped, or irregularly massive during their ascent, hardening as they did so* (Fenton, *The Rock Book*).

volcanic cone, *Geology.* a hill around the rim of a volcano, consisting chiefly of matter ejected during eruptions: *Nearly all volcanic cones are characterized by a relatively small funnel-shaped depression—called a crater—marking the top of the conduit through which the eruptive products were channeled* (Birkeland and Larson, *Putnam's Geology*).

volcanic glass, *Geology.* a natural glass produced by the very rapid cooling of lava: *Extrusive rocks ordinarily are cooled quickly, giving little time for the collection of molecules of similar sorts into crystals of appreciable size. Such are, therefore, finely crystalline, or they may have no crystals at all and are then classed as volcanic glass* (Finch and Trewartha, *Elements of Geography*). Compare **obsidian.**

volcaniclastic, *Geology.* —*adj.* consisting of volcanic fragments or sediments: *volcaniclastic rocks.*
—*n.* a volcaniclastic rock: *The major detrital sources were nearby granitic batholiths and andesitic lava flows, all of late Mesozoic age, rhyolitic volcaniclastics of Cenozoic age, and uplifted sedimentary rocks of all ages* (Science).

volcanic rock, *Geology.* a crystalline or glassy igneous rock formed by volcanic action near the earth's surface: *The lavas and solid fragments erupted from volcanoes are called volcanic rocks. They are composed in large part of microscopic mineral crystals and glass* (Gilluly, *Principles of Geology*).

volcanism, *n. Geology.* the phenomena connected with volcanoes and volcanic activity: *Some of the chains of crater pits must have their origin in a type of vol-*

canism, *perhaps in lava flows of matter liquefied by the impact* (Atlantic). Also spelled **vulcanism.**

volcano, *n., pl.* **-noes** or **-nos.** *Geology.* **1** an opening in the earth's crust through which steam, ashes, and lava are expelled in periods of activity: *There is abundant evidence that local heating and melting of the earth's solid crust and upper mantle do take place, for these processes are responsible for present-day volcanoes—all of which release large quantities of volatile elements at the earth's surface* (Eicher and McAlester, *History of the Earth*). **2** a cone-shaped hill or mountain around this opening, built up of the material thus expelled. Compare **shield volcano, composite volcano.** [from Italian, from Latin *Vulcanus* Vulcan, the Roman god of fire]

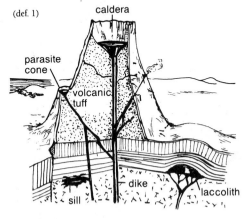

(def. 1)
caldera
parasite cone
volcanic tuff
dike
sill
laccolith

volcanogenic, *adj. Geology.* originating in or produced by volcanoes: *In the Indian Ocean gaps in the sedimentary record have been encountered in a broad spectrum of terrigenous, pelagic, biogenic and volcanogenic sediments encompassing late Mesozoic and Cainozoic time* (Nature).

volcanological, *adj.* of or having to do with volcanology: *a volcanological laboratory or observatory.*

volcanology, *n.* the scientific study of volcanoes and volcanic phenomena.

volt, *n.* the SI or MKS unit of potential difference or electromotive force, equal to the difference in potential needed to cause a current of one ampere to flow through a resistance of one ohm. *Symbol:* V [named after Alessandro *Volta,* 1745–1827, Italian physicist]

voltage, *n. Electricity.* potential difference or electromotive force expressed in volts. A current of high voltage is used in transmitting electric power over long distances. *Abbreviation: v.*

voltaic, *adj. Electricity.* of, having to do with, or producing an electric current by chemical action. A **voltaic cell** is an electric cell consisting of two electrodes, each of a different metal, connected by a wire and immersed in an electrolyte; galvanic cell. An electric current is produced in the wire by a chemical reaction between the electrolyte and one of the electrodes. SYN: galvanic.

volume, *n.* **1** *Mathematics.* the amount of space occupied by a body as measured in three dimensions, expressed in cubic units; cubic capacity: *The volume of an open-necked balloon is proportional to the cube of its diameter ... It is therefore of advantage to make the*

balloons as large as possible, *for then the density of the balloons is least, and the attainable height greatest* (A. J. Herz and R. M. Tennent). **2** *Physics.* the amount of loudness of a sound or the magnitude of the current transmitting it. *Abbreviation:* v., vol.

volumetric (vol′yə met′rik), *adj.* of or having to do with measurement by volume. Volumetric analysis is quantitative chemical analysis of a substance by titration with a standard solution. **—volumetrically,** *adv.*

voluntary muscle, = striated muscle.

volute (və lüt′), *n. Zoology.* **1** a turn or whorl of a spiral shell. **2** any of a group of gastropods that have a spiral shell. [from French, from Latin *voluta,* feminine past participle of *volvere* to roll] **—voluted,** *adj.* having a coil, whorl, or volute: *a voluted shell.*

volutin (vol′yə tin), *n. Microbiology.* a granular substance that stains with basic dyes and is sometimes found in certain bacteria and yeasts. [from New Latin *volutans* the species name of the bacteria in which this substance was first found]

volva (vol′və), *n. Biology.* the membranous covering that completely encloses many fungi, especially mushrooms, in the early stage of growth. [from Latin, variant of *vulva* womb]

volvent, *n. Zoology.* one of a series of pear-shaped cells on the tentacles of a hydra. The volvents release short, thick threads to capture small swimming animals by coiling about them. [Latin *volventem* rolling (so called because of the cell's coiled structure)]

vomer (vō′mər), *n. Anatomy.* a small, thin bone of the skull forming a large part of the nasal septum and having the shape of a plowshare. The vomer divides the nostrils in most vertebrates. [from Latin *vomer* plowshare] **—vomerine** (vō′mər in), *adj.* of or having to do with the vomer: *the vomerine region of the skull.*

vortex (vôr′teks), *n., pl.* **vortexes, vortices** (vôr′tə sēz′). *Physics.* a spiral motion of a fluid within a limited area, especially a whirling mass of water, air, etc., that sucks everything near it toward its center: *Sometimes, for reasons still unknown, a great mass of air rises, instead of a small cell. The wind rushing in to fill the gap is deflected by the earth's spin, producing a vortex. All elements of the storm form in a spiral configuration: the clouds, the cascading rains and the winds* (N. Y. Times). SYN: whirlpool. [from Latin, variant of *vertex.* See VERTEX.]

vortical (vôr′tə kəl), *adj. Physics.* of, having to do with, or moving in a vortex: *vortical currents.*

vorticity (vôr tis′ə tē), *n. Physics.* **1** the condition of a fluid with respect to its vortical motion: *The effect of compressibility is to make the wave spread out in all directions and to attenuate, until there is a disappear-*

---

cap, fāce, fäther; best, bē, tèrm; pin, fīve;
rock, gō, ôrder; oil, out; cup, pùt, rüle,
yü in use, *yu̇* in uric;
ng in bring; sh in rush; th in thin, ᴛʜ in then;
zh in seizure.
ə = *a* in about, *e* in taken, *i* in pencil, *o* in lemon, *u* in circus

ance of all features except a vortex (*spiralling*) motion along the lines of magnetic force: this "vorticity", in fact, travels linearly and without attenuation, as does the whole wave in the ideal case (New Scientist).

2 a measure of the rate of vortical motion or rotation in a fluid: *The vorticity is equal to twice the angular velocity around an axis through the point* (Scientific American).

vulcanism, *n.* = volcanism.

vulva (vul′və), *n., pl.* **-vae** (-vē), **-vas.** *Anatomy.* the external genital organs of the female. [from Latin *vulva* womb]
—**vulval,** *adj.* of or having to do with the vulva.

vulviform, *adj.* **1** *Zoology.* shaped like a vulva. **2** *Botany.* shaped like a cleft with projecting edges.

W

w or w., *abbrev.* 1 weight. 2 work.

W, *symbol* or *abbrev.* 1 watt *or* watts. 2 tungsten (German, *Wolfram*). 3 W particle.

wacke (wak′ə), *n. Geology.* a rock similar to sandstone, consisting of unsorted angular fragments of rocks and minerals. [from German *Wacke* pebbles and gravel in riverbeds]

wad, *n. Mineralogy.* an impure, earthy ore of manganese oxide minerals. It is generally soft and of a dark-brown or black color. [from a local name in Great Britain for black lead or plumbago]

wadi (wä′dē), *n. Geology.* a usually dry valley or ravine in Arabia, northern Africa, etc., through which a stream flows during the rainy season: *The wadis, entirely without water during most of the year, may become torrents of muddy water filled with much debris after one of these flooding rains* (Finch and Trewartha, *Elements of Geography*). [from Arabic *wādī*]

waist, *n.* 1 *Anatomy.* the part of the human body between the ribs and the hips.
2 *Zoology.* the slender part of the abdomen of various insects, such as wasps, ants, and some flies.

wall, *n. Biology.* the side part of any hollow thing: *the wall of the stomach, the wall of a cell, the abdominal wall.*

waning moon, *Astronomy.* the moon between the full moon and new moon, when its light appears to be diminishing because a smaller area is visible. Compare **waxing moon.**

warm-blooded, *adj. Biology.* having blood whose temperature stays about the same regardless of the surroundings. Birds and mammals are warm-blooded; reptiles, amphibians, and fishes are cold-blooded. *A layer of fat beneath the skin, plus a covering of hair, fur, or feathers, helps keep a warm-blooded animal warm* (James E. Heath). *Warm-blooded animals must eat more than cold-blooded ones to fuel their high rate of metabolism* (Vincent J. Maglio). Also called **endothermic, homoiothermic.**

warning coloration, *Zoology.* natural coloring or marking of an animal that serves to warn or alarm enemies: *Warning coloration is ascribed to some butterflies and other insects, considered to be distasteful to their enemies—they "advertise" their unpalatability. Bees and wasps with stout stings are often marked conspicuously with black and yellow* (Storer, *General Zoology*).

waste, *n.* 1 *Biology.* **a** unusable or excess material left over from the breakdown of tissue: *The liver shares in the excretion of urea, since it helps convert protein wastes into urea. When proteins are broken down, ammonium salts are among the products. These salts are converted into urea partly in the liver, but the actual excretion is elsewhere* (Shull, *Principles of Animal Biology*). **b** such material discharged from the body; excrement.

2 *Geography.* material derived by mechanical and chemical erosion from the land, carried by streams to the sea.

water, *n.* 1 *Chemistry.* a colorless, tasteless, odorless liquid, a compound of hydrogen and oxygen, freezing at 0 degrees Celsius (32° Fahrenheit) and boiling at 100 degrees Celsius (212° Fahrenheit). Water is essential to most plant and animal life and in the form of oceans it covers about 70 per cent of the earth's surface. *Water has the rare property of being denser as a liquid than as a solid, and it is probably the only substance that attains its greatest density at a few degrees above the freezing point (four degrees centigrade)* (Scientific American). *Formula:* H_2O
2 *Physiology.* any of various watery substances or secretions occurring in or discharged from the body, such as sweat, saliva, urine, serum, etc.: *Water loss occurs from the skin, the respiratory tract, the faeces, and the kidney. On occasions it may also be lost in expectoration, tears, haemorrhage, etc. Normally the water loss is balanced by the intake and that manufactured from foods* (McDowall, *Physiology and Biochemistry*).

water cycle, 1 *Geology.* the cycle by which water evaporates from oceans, lakes, and other bodies of water, forms clouds, and is returned to those bodies of water in the form of rain and snow, the runoff from rain and snow, or ground water: *The oceans do not dry up because when the rain falls, it either drops back into the ocean or sinks into the ground, feeds the streams, and returns to the oceans. We call this process the water cycle* (Gerhard Neumann). Also called **hydrologic cycle.**
2 *Biology.* the cycle by which living organisms absorb or consume water and return it to the atmosphere by transpiration, respiration, and other metabolic processes: *The use of water in photosynthesis and its production in respiration is also part of the water cycle* (Greulach and Adams, *Plants*).

waterfall, *n. Geology, Geography.* any sudden descent of a stream or river. A waterfall usually occurs from a great height and is perpendicular or nearly perpendicular.
► If the volume of the water is small, the waterfall is often called a *cascade;* if it is large, it is called a *cataract.* A waterfall is often called *falls,* as in Niagara Falls, Victoria Falls.

---

cap, fāce, fäther; best, bē, tèrm; pin, five;
rock, gō, ôrder; oil, out; cup, pùt, rüle,
yü in use, *yù* in uric;
ng in bring; sh in rush; th in thin, ᴛH in then;
zh in seizure.
ə = a in about, e in taken, i in pencil, o in lemon, u in circus

water hole, *Astronomy.* a part of the radio region of the electromagnetic spectrum that is comparatively free of noise: *This comparatively quiet region of the electromagnetic spectrum, now generally accepted as the prime spectral band to be searched for interstellar signals, has ... been dubbed the "water hole," since it is bounded at the low-frequency end by the hydrogen line at 1,420 megahertz and at the high-frequency end by the hydroxyl lines between 1,612 and 1,720 megahertz* (Scientific American).

water level, *Geography, Geology.* **1** the surface level of a stream, lake, or other body of water.
2 the plane below which the ground is saturated with water; water table.

water of crystallization or water of hydration, *Chemistry.* water that is combined with certain crystalline substances to form a hydrate. When the water is removed by heating, it leaves an anhydrous salt. *A crystallized substance that contains water of crystallization is a hydrate. Each hydrate holds the definite amount of water that is necessary for formation of its crystal* (Tracy, *Modern Physical Science*).

watershed, *n. Geography, Geology.* **1** the ridge between the regions drained by two different river systems. On one side of a watershed, rivers and streams flow in one direction; on the other side, they flow in the opposite direction.
2 the area drained by one river system: *The Tennessee River drains a watershed of 40,000 square miles* (Science News Letter).

water-soluble, *adj. Chemistry.* that will dissolve in water: *water-soluble vitamins. Their work dealt with water-soluble fertilizer salts, and they found that such salts are held by the soil and are not easily leached out by rainwater* (K. S. Spiegler).

waterspout, *n. Meteorology.* a rapidly spinning column of air and water, produced by the action of a whirling cloud mass over an ocean or lake; a tornado over water, occurring usually in tropical regions: *Whirling winds and sea spray form the funnel clouds of waterspouts, the seafaring relatives of tornadoes. A ... study of the life cycles of waterspouts may help scientists understand how tornadoes, the more violent storms, develop* (Richard H. Chesher).

water table, *Geology.* the level below which the ground is saturated with water: *After a prolonged rain the ground may be completely soaked; the water sinks into the ground, and if a dry spell follows it may leave the upper ground dry. But somewhere below the surface the tiny spaces between the grains of sand or other materials are filled with water. The top of this saturated part of the ground is the water table* (Colby and Foster, *Economic Geography*). Also called **water level.**

water vapor, *Physics.* water in a gaseous state, especially when fairly diffused as it is in the air, and below the boiling point, as distinguished from steam: *The water vapor in the atmosphere is primarily supplied by the oceans* (K. K. Turekian). *Water vapor absorbs solar and terrestrial radiation, and provides much of the energy for the development of storms* (C. G. Knudsen and J. K. McGuire).

water-vascular, *adj. Zoology.* of or having to do with the circulation of water in the vessels of certain animals, especially the echinoderms. The **water-vascular system** is the system of water-filled canals connecting the tube feet of echinoderms: *The water-vascular system ... is a division of the coelom, peculiar to echinoderms* (Hegner and Stiles, *College Zoology*).

Watson-Crick, *adj. Molecular Biology.* of, having to do with, or derived from the double-helical model (the **Watson-Crick model**) of the molecular structure of DNA devised by the American biologist James D. Watson (born 1928) and the British biophysicist Francis H. C. Crick (born 1916): *Chromosomes are composed of tremendously large DNA molecules that, when stretched out into individual Watson-Crick duplexes, are thousands of times longer than their containing cells* (H. E. Kubitschek). Compare **double helix.**

watt, *n.* the SI or MKS unit of power, equal to one joule per second. In the case of electrical power, the number of watts is equal to the current in amperes multiplied by the electrical potential in volts. *The watt and the* $ft \cdot lbf/sec$ *are both inconveniently small units for many practical power measurements, and hence it has been found desirable to define larger units. In the MKS system two multiples of the watt are commonly used: the kilowatt (1 kw = 1000 watts) and the megawatt (1 megawatt = 1000 kw = 1,000,000 watts* (Shortley and Williams, *Elements of Physics*). Symbol: W [named after James *Watt,* 1736–1819, Scottish engineer]

wattage, *n.* power expressed in watts, especially kilowatts.

watt-hour, *n.* a measure of electrical energy or work, equal to the work done by one watt acting for one hour; 3600 joules.

wattle, *n. Zoology.* **1** the bright-red flesh hanging down from the throat on the males of chickens, turkeys, or other domestic fowls and certain other birds.
2 a fleshy appendage below the throat of certain reptiles, such as the iguana.
3 the barbel of a fish.

wave, *n.* **1** *Oceanography.* a moving ridge or swell of water caused by the wind, tides, and currents: *The greater part of marine erosion is accomplished by waves ... In small waves the motion is confined to surface waters, but in great ones there is sufficient agitation to cause some churning of the bottom at considerable depths* (Finch and Trewartha, *Elements of Geography*).
2 *Physics.* any disturbance traveling through a medium by which energy is transferred from one particle of the medium to another without causing any permanent displacement of the medium itself. Seismic waves travel through the earth. Sound waves can travel through the air. Light and heat waves can travel through a vacuum. Waves are usually measured by their length, amplitude, velocity, and frequency. In a longitudinal wave, such as a sound wave, the motion of the particles is parallel to the direction in which the wave travels; in a transverse wave, such as a light wave, the motion is perpendicular to the direction in which the wave travels. Compare **vibration.** See also **compression wave, standing wave, traveling wave, wave motion.**

wave equation, *Physics.* **1** a partial differential equation that is used to describe wave motion. **2** = Schrödinger wave equation: *In formulating his wave equation Schrödinger cast in rigorous mathematical form the earlier body of quantum-mechanical knowledge* (Arnold C. Wahl).

wave form, or **waveform,** *n. Physics.* the form assumed by a wave, especially the form obtained by plotting a characteristic of the wave against time: *Each individual mode can be described mathematically as a damped sinusoid . . . That is, they are waveforms that gradually die away as time passes* (Science News).

wave front, *Physics.* the continuous line or surface including all the points in space reached by a wave or vibration at the same instant as it travels through a medium: *When waves spread out from a small source in a uniform medium, the wave fronts are spheres concentric with the source* (Sears and Zemansky, *University Physics*).

wave function, *Physics.* a mathematical function in quantum mechanics describing the propagation of the wave associated with any particle or group of particles: *An important quantity in wave mechanics is the wave function* Ψ *. . . The quantity* Ψ^2 *at any particular point is a measure of the probability that the particle will be near that point* (Halliday and Resnick, *Physics*).

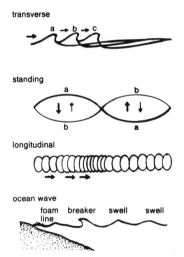

transverse

a → b → c

standing

a   b

b   a

longitudinal

ocean wave

foam breaker swell swell
line

wavelength, *n. Physics.* the distance between one peak or crest of a wave of light, heat, or other electromagnetic energy, and the next corresponding peak or crest; the distance between successive particles that are in the same phase at the same time, measured in the direction in which the wave is traveling. The wavelengths of radio waves are measured in meters; the wavelengths of X rays are measured in nanometers. *By choosing different wavelengths of light, scientists can observe the distribution of various elements, such as hydrogen and calcium, on the surface of the sun* (Robert H. Baker). *Laser light . . . obtains its special properties because it consists of only one wavelength, and its electromagnetic waves are all in step with each other, reinforcing each other* (Harry Schwartz). *Abbreviation:* w.l. *Symbol:* λ

wavellite (wā′və līt), *n. Mineralogy.* hydrous phosphate of aluminum, found in globular aggregates with a radiating structure. *Formula:* $Al_3(OH)_3(PO_4)_2 \cdot 5H_2O$ [from William *Wavell,* died 1829, English physician who discovered it + -ite]

wave-mechanical, *adj.* of or having to do with wave mechanics: *Each atom consists of a comparatively small central core, or nucleus, surrounded by a number of electrons. In the old atomic theory these were considered to move in orbits round the nucleus. In the newer wave-mechanical picture of the atom the orbits are replaced by a probability distribution of electrons. One particular electron cannot be definitely located at a particular point at a given time; all that can be said is that there is a certain probability of its being there* (W. D. Corner).

wave mechanics, *Physics.* a theory that ascribes characteristics of waves to subatomic particles and attempts to interpret physical phenomena on this basis: *Wave mechanics gives a far better account of atomic structure than the planetary atom conception, but it is highly mathematical and gives us little to visualize* (Glathart, *Foley's College Physics*). *While explaining all the atomic phenomena for which Bohr's theory already worked, wave mechanics also explained those phenomena for which Bohr's theory failed (such as the intensities of spectral lines, etc.)* (Ganow, *Thirty Years that Shook Physics*).

wave motion, *Physics.* the process by which waves propagate through a medium, characterized by a forward undulating or vibrational motion. Sound waves and electromagnetic waves have different kinds of wave motion.

wave number, *Physics.* the number of waves per unit distance in a series of waves of a given wavelength; the reciprocal of the wavelength of a wave: *Over this limited region it was found that the attenuation of starlight on a logarithmic scale of intensity varied linearly with wave number (reciprocal wavelength), and the same relationship appeared to hold over a fairly wide area of the sky* (Science Journal).

wave theory, *Physics.* the theory that light is propagated in undulatory movement or waves: *The wave theory of light supposes that the energy which causes it radiates in waves from a source such as a lamp or the sun in something like the way that ripples spread over the surface of a pond when a stone is dropped into the water* (Baker, *Astronomy*). Also called **undulatory theory.**

wave train, *Physics.* a group of waves sent out at successive intervals along the same path from a vibrating body: *If a wave train, such as that of light, strikes the edge of an object, the direction of the wave is changed. This is known as diffraction* (Robert F. Paton).

cap, fāce, fäther; best, bē, tèrm; pin, five;
rock, gō, ôrder; oil, out; cup, pùt, rüle;
yü in use, yủ in uric;
ng in bring; sh in rush; th in thin, ŦH in then;
zh in seizure.
ə = a in about, e in taken, i in pencil, o in
lemon, u in circus

wavy, *adj. Botany.* wavelike; undulating: *a leaf with a wavy margin.*

wax, *n.* **1** *Biochemistry.* any of various organic compounds or mixtures similar to fats and oils but containing no glycerides, secreted by certain plants and animals. Waxes include beeswax, lanolin, and carnauba wax. *Lipids are responsible for the familiar water-repellent character of the surfaces of plants, animals, and insects. In chemical composition the surface lipids differ from the internal lipids. Collectively they are called waxes because of their peculiar physical properties, although in strict chemical terms, wax refers to esters of long-chain alcohols with long-chain acids* (Science).
2 any similar substance found in various minerals, such as ozocerite and paraffin.

waxing moon, *Astronomy.* the moon between the new moon and full moon, when its light appears to be increasing because a larger area is visible. Compare **waning moon.**

Wb, *symbol.* weber.

weak force, = weak interaction: *The W and Z particles ... are the fundamental particles that transmit the weak force* (J. W. Rohlf).

weak interaction, *Nuclear Physics.* an interaction between subatomic particles that is closely related to electromagnetism and that causes radioactive decay and other subatomic reactions. Though fundamentally equal in strength to electromagnetism, it appears weaker because it only acts between particles that are very close to each other. Its carriers or quanta are the W and Z particles. *Weak interactions do not manifest themselves in everyday life, and the only trace of their existence is the beta decay of radioactive nuclei* (Marek Demianski). See also **Weinberg-Salam.**

weakon (wē′kon), *n. Physics.* a particle that is a carrier of the weak interaction; W particle or Z particle. Also called **intermediate boson** or **intermediate vector boson.** [from *weak* (*interaction*) + *-on*]

weapon, *n. Biology.* any organ of an animal used for fighting or for protection, such as claws, horns, teeth, and stings: *... typical modern monkeys, with the canine teeth enlarged into weapons* (A. S. Woodward).

weather, *n. Meteorology.* **1** the condition of the atmosphere at a particular time or place with respect to temperature, moisture, violence or gentleness of winds, clearness or cloudiness, etc.: *Arctic weather apparently follows the same pattern as air masses elsewhere on the earth and is susceptible, therefore, to the same forecasting techniques* (John C. Reed).
2 windy, rainy, or stormy weather: *damage done by the weather.*

weathering, *n. Geology.* the chemical and physical processes acting to break down materials at the earth's surface.

weather map, *Meteorology.* a map or chart showing conditions of temperature, barometric pressure, precipitation, direction and velocity of winds, etc., over a wide area for a given time or period: *A weather map is a map of a large area showing the weather conditions existing at a given time. The distribution of pressure and temperature over the area is shown by isobars and isotherms. On modern maps fronts and air masses are lo-*cated and named . . . *The barometric depression, or cyclone, is a prominent feature of weather maps, outside the tropics* (Blair, *Weather Elements*).

web, *n. Zoology.* **1** the fabric of delicate silken threads spun by a spider. Webs are used to ensnare prey and and as nests or shelters.
2 a membrane or skin joining the toes of swimming birds and certain other water animals.
3 a series of barbs on each side of the shaft of a bird's feather. SYN: vexillum.

weber (web′ər *or* vā′bər), *n.* the SI or MKS unit of magnetic flux, equal to one volt multiplied by one second or to 10^8 maxwells. *Symbol:* Wb [named after Wilhelm E. *Weber*, 1804–1891, German physicist]

webfoot, *n. Zoology.* **1** a foot in which the toes are joined by a web.
2 a bird or animal having webbed feet.
—web-footed, *adj.* having the toes joined by a web: *a web-footed alligator.*

wedge, *n.* **1** *Physics.* a device that tapers to a sharp edge or to a point, used to split wood and other materials or to adjust the positions of heavy objects. It is a simple machine. *The wedge may be looked upon as two inclined planes back to back . . . Illustrations of the wedge are the axe, the hatchet, a knife, etc.* (Hoyt, *Concise Physics*).
2 *Meteorology.* a long narrow area of high pressure between two cyclonic systems. SYN: ridge.

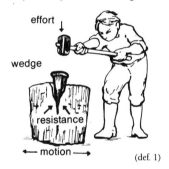

(def. 1)

weed, *n. Botany.* a useless or troublesome plant, especially one that is damaging to crops or poisonous to animals: *A well-established stand of grass exerts a powerful controlling influence that almost entirely excludes annual weeds, but when it is weakened or destroyed, the controls it exerts are relaxed and the weeds develop* (Emerson, *Basic Botany*).

weight, *n.* **1** *Physics.* the force with which a body is attracted to the earth or some other field of gravitation. The weight of a body is the product of the mass of the body and the acceleration of gravity, and is expressed in such units as the newton and dyne. *The force with which an object is pulled vertically downward toward the earth (the force of gravity) is called the weight of that object. Weight is an ever-present force on all objects near the earth. The vertical is defined as the direction of the force of gravity; the horizontal is the plane normal to the vertical* (Shortley and Williams, *Elements of Physics*). ▶ See the note under **mass.**

2 a a system of standard units used for expressing weight, such as avoirdupois weight or troy weight. **b** a unit of such a system: *Pounds and ounces are customary weights.*

3 *Statistics.* **a** a factor assigned to a number in a computation, as in determining an average, to make the number's effect on the computation reflect its importance: *The weights used in calculating the weighted arithmetic mean need not always be frequencies. In many cases, the weights are assigned according to the relative importance of the group they represent* (Parl, *Basic Statistics*). **b** the frequency of an item in a statistical compilation. *Abbreviation:* w., wt.

—*v. Statistics.* to give a weight to: *a weighted average.*

weight density, *Physics.* the weight of a substance per unit volume.

weightless, *adj. Physics.* not experiencing the effects of gravity; being in a state of free fall: *Astronauts are weightless in space, and so they must anchor themselves with straps and other devices to keep from floating about* (William J. Cromie).

—**weightlessness,** *n.* a condition in which the effects of gravity are not felt: *The astronauts [are] given calcium to make up for the deficiency of it in their bones caused by weightlessness* (D. E. Fink and G. C. Wilson).

▶ The term *weightlessness* is technically misleading since an individual orbiting beyond the earth's atmosphere is still attracted by the earth's gravitational field. His condition is due to the fact that the gravitational pull on him is neutralized by centrifugal force. See **free fall.**

Weinberg-Salam (wīn′bėrg sä′läm), *adj. Physics.* of or having to do with a unified field theory which treats the weak interaction and electromagnetism as parts or phases of the same phenomenon: *The Weinberg-Salam theory predicts that there should be a weak interaction between the electrons in an atom and the particles in the nucleus of the atom and that weak interaction should not conserve parity* (Nature). *The Weinberg-Salam model includes a fairly arbitrary "recipe" of quarks and leptons . . . The recipe of the model—four quarks (one of them the predicted charmed quark) and four leptons (electron, muon, and two neutrinos) became "folklore"* (New Scientist). [named after Steven *Weinberg,* born 1933, American physicist, and Abdus *Salam,* born 1926, Pakistani physicist, who independently developed the theory]

wernerite (wėr′nə rīt), *n.* = scapolite. [from Abraham G. *Werner,* 1749?–1817, German mineralogist + -*ite*]

westerlies, *Meteorology.* the prevailing westerly winds found in certain latitudes: *Between the westerlies and the trades are the horse latitudes* (Scientific American).

Western Hemisphere, *Geography.* the half of the world that includes North and South America.

wet cell, an electric cell having a liquid electrolyte. Compare **dry cell.**

wet way, *Chemistry.* the method of analysis in which the reactions are produced mostly in solutions and by the use of liquid reagents.

whalebone, *n.* = baleen.

Wharton's jelly, *Anatomy.* mucoid connective tissue which constitutes most of the bulk of the umbilical cord: *It has been reported that the female may be im-* *munized against fetal tissue through injection of a tissue (Wharton's jelly) from the umbilical cord of fetuses* (Elmer B. Harvey). [named after Thomas *Wharton,* 1614–1673, English anatomist]

wheel and axle, *Physics.* an axle on which a wheel is fastened. As a simple machine one of its uses is to lift weights by winding a rope or chain onto the axle as the wheel is turned. *Most of the mechanisms of watches, clocks, . . . and automobile transmissions consists of wheels and axles. Every mechanical toy that must be "wound up" to make it go has one, usually more than one, wheel and axle* (Glathart, *College Physics*).

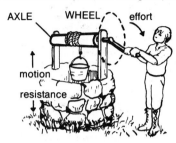

whirlpool, *n. Oceanography.* a current of water whirling round and round rapidly and violently; eddy or vortex of water: *Where the tide is thrown from side to side against sunken rocks, or where two opposing currents meet . . ., the water forms whirlpools* (Archibald Geikie).

whirlwind, *n. Meteorology.* a current of air whirling violently round and round; a whirling windstorm; vortex of air: *Whirlwinds are small, shallow whirls of upflowing and inflowing air, beginning at the ground level* (Blair, *Weather Elements*).

white blood cell, *Biology.* any colorless blood cell with a nucleus: *Several types of white blood cells exist. Phagocytic white blood cells engulf and destroy bacteria at the site of injection by the process of phagocytosis . . . Lymphocytes are another type of white blood cell that is associated with the immune response. These white blood cells produce specific antibodies which act against foreign molecules known as antigens* (Biology Regents Syllabus). Also called **white cell, white corpuscle, leucocyte.** Compare **red blood cell.**

white cell or **white corpuscle,** = white blood cell.

white dwarf, *Astronomy.* a white star of low luminosity, small size, and very great density: *White dwarfs are the normal end-product of low-mass stellar evolution. Most live quiet lives in this state. A few can, with "help," go to a supernova* (Hyron Spinrad). *A moderate-sized star that has collapsed into a dense, or "white dwarf," state sucks material from a companion star until . . . Its internal structure can then no longer resist*

---

cap, fāce, fäther; best, bē, tèrm; pin, fīve;
rock, gō, ôrder; oil, out; cup, pùt, rüle,
yü in use, *yu̇* in uric;
ng in bring; *sh* in rush; *th* in thin, ᴛʜ in then;
zh in seizure.
ə = *a* in about, *e* in taken, *i* in pencil, *o* in lemon, *u* in circus

the compression of its own mass and it collapses, initiating explosives reactions in its carbon-oxygen core (Walter Sullivan). Compare **red dwarf.**

white frost, = hoarfrost.

white giant, *Astronomy.* a star of the main-sequence group having a higher temperature and luminosity than the others: *The most conspicuous region is the "main sequence", a fairly narrow band ranging from hot and luminous stars (white giants) to cool and faint stars (red dwarfs). ... The majority of the stars of our galaxy are white dwarfs and main sequence stars less luminous than the sun. The most luminous stars form a small fraction of a galaxy, but are of course visible over larger distances* (Annual Review of Nuclear Science). Compare **red giant.**

white hole, *Astronomy, Physics.* a hypothetical hole in outer space from which energy and stars and other celestial matter emerge or explode: *Narlikar and others thought it likely that quasars could be white holes, colossally powerful explosions in space, in which matter and radiation would suddenly appear and be spewed out, apparently out of nothing, since white holes, if they existed, were the precise opposite of black holes ... white holes could not only solve the mystery of the quasars; they could also explain the enormously powerful X-rays coming from the centres of ordinary galaxies, and provide a source for the most powerful of cosmic rays* (John Newell). Compare **black hole.**

white iron pyrites, = marcasite.

white light, *Physics.* **1** the light which comes directly from the sun, and which has not been decomposed, as by refraction: *The sun as a whole . . . is composed of all elements, each of which contributes one or more colors to the visible spectrum. White light is a mixture of all these colors* (Scientific American). See the picture at **dispersion.**
2 any light producing the same color or color sensation as direct sunlight.

white matter, *Anatomy.* whitish nerve tissue, especially in the brain and spinal cord, that consists chiefly of nerve fibers with myelin sheaths: *The outer or white matter of the spinal cord consists of bundles of medullated fibers connecting between various parts of the brain and the nuclei of spinal nerves and adjustor neurons* (Storer, *General Zoology*). Compare **gray matter.**

white noise, *Physics.* the sound heard when the entire range of audible frequencies is produced at once, as in the operation of a jet engine.

whiteout, *n. Meteorology.* a condition in arctic and antarctic regions in which the sky, the horizon, and the ground become a solid mass of dazzling reflected light, obliterating all shadows and distinction: *A whiteout can be more devastating than a blizzard, for the snow on the ground merges with a solid white overcast of clouds, with no visible point of junction* (John Brooks).

whole blood, *Immunology.* natural blood with none of the essential components, such as plasma and platelets, removed.

whole gale, *Meteorology.* a wind with a velocity of 89 to 102 kilometers per hour on the Beaufort scale.

whole number, *Mathematics.* **1** a positive integer or zero. The set of whole numbers is usually $\{ 0, 1, 2, 3, \ldots \}$, 0 sometimes being excluded.
2 a number denoting one or more whole things or units.

whorl (hwėrl *or* hwôrl), *n.* **1** *Botany.* a circle of leaves or flowers round a single node or point on the stem of a plant: *The flower consists of four whorls, or sets, of parts. These whorls are (1) the calyx, (2) the corolla, (3) the stamens, and (4) the pistils* (Harold N. Moldenke). *The developing leaves at that stage of the plant's growth are tightly rolled in a whorl* (Stanley D. Beck). SYN: verticil.
2 *Zoology.* one of the turns of a spiral shell.
3 *Anatomy.* one of the spiral curves in the cochlea of the ear.
[Middle English *whorle,* apparently variant of *whirl*]
—**whorled,** *adj.* **1** having a whorl or whorls: *a whorled plant, a whorled shell.*
2 arranged in a whorl: *whorled petals.*

wild, *adj.* **1** *Zoology.* living or growing in a natural habitat; not tamed: *The great majority of animals are wild. Individual wild animals can usually be tamed, but they easily become wild again* (Lorus and Margery Milne). SYN: undomesticated.
2 *Botany.* produced or yielded naturally, without the aid of man; uncultivated: *wild honey, wild cherries. Wild forms of apples, strawberries, blackberries, wheat, and many other cultivated crops are known, but they are far inferior to the cultivated fruits and crops that have been developed from them* (Emerson, *Basic Botany*).

wild type, *Genetics.* a normal strain or ordinary type of an organism, as distinguished from a mutant strain or type.
—**wild-type,** *adj.* of or belonging to a wild type: *a wild-type gene, a wild-type virus. In its simplest form, the theoretical model of a closed population consists essentially of a pool of genes, the genes carried by the individuals of which the population is composed. Most of them are "wild-type" alleles, that is to say those alleles which have become characteristic of the population through the action of natural selection in the past* (T. C. Carter).

willemite (wil'ə mīt), *n. Mineralogy.* a mineral, a silicate of zinc, found in masses or hexagonal prisms of various colors from light greenish-yellow to red and black, all of which exhibit bright yellow fluorescence under ultraviolet light. It is a minor zinc ore. *Formula:* Zn_2SiO_4 [from German *Willemit,* from *Willem I,* 1772–1843, king of the Netherlands + *-it* -ite]

wind, *n. Meteorology.* **1** air in motion, varying in force from a slight breeze to a strong gale: *The two most important climatic functions of wind are (a) the maintenance of a heat balance between the higher and lower latitudes . . . and (b) the transportation of water vapor from the oceans to the lands* (Finch and Trewartha, *Elements of Geography*).
2 a strong wind; gale.

wind chill, *or* **windchill,** *n. Meteorology.* the combined cooling effect on the human body of air temperature and wind speed. If the air temperature is 10°F and the wind speed is 10 mph, the estimated wind chill is −9°F. *Wind chill, which is actually another name for the dry convective cooling power of the atmosphere,* is a term descriptive of the cooling effect of air movement and low temperature (New Scientist).

wind erosion, *Geology.* the removal of topsoil by dust storms.

windpipe, *n.* = trachea (def. 1). See the picture at **bronchi.**

wind-pollinated, *adj. Botany.* fertilized by pollen carried by the wind; anemophilous: *Many grasses and many trees are wind-pollinated . . . The pollen grains of some wind-pollinated plants average around 0.025 mm. in diameter, and are sometimes, as in pines, provided with wings, which greatly facilitate their transport in air currents* (Hill, *Botany*).

wind pollination, *Botany.* fertilization with pollen carried by the wind: *Wind pollination is the common type in plants with inconspicuous flowers, as in grasses, poplars, walnuts, alders, birches, oaks, ragweeds, and sage* (Weier, *Botany*).

wind rose, *Meteorology.* a diagram indicating the relative frequency, force, and other factors of the winds from various directions at some given place.

wind scale, *Meteorology.* a system of numbers or symbols used to record the speed of the wind.

windstorm, *n. Meteorology.* a storm with much wind but little or no precipitation.

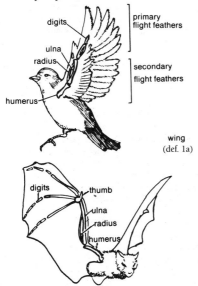

wing
(def. 1a)

wing, *n.* **1** *Zoology.* **a** one of the movable parts of a bird, insect, or bat used in flying, or a corresponding part in a bird or insect that does not fly. Birds have one pair of wings; insects usually have two pairs. **b** any similar structure, such as the folds of skin of a flying squirrel or one of the enlarged fins of a flying fish.
2 *Botany.* **a** either of the two side petals of a pealike flower. Also called **ala. b** a leafy or membranous expansion or thin extension, as of a samara: *The wings of the samaras of such trees as maple, ash, and elm are broad extensions of the ovary wall which greatly enlarge the surface on which the wind can act without materially increasing the weight of the fruit* (Emerson, *Basic Botany*). See the picture at **naked.**
3 *Anatomy.* the ala of the nose.

---

wingspread, *n. Zoology.* the distance between the tips of the wings of a bird, bat, insect or other such animal, when they are spread.

winter solstice, *Astronomy.* **1** the solstice that in the Northern Hemisphere occurs about December 21 or 22, when the noon sun appears to be farthest south. **2** (for the Northern Hemisphere) the point on the ecliptic farthest south of the celestial equator, which the sun reaches at this time. *The summer solstice is the most northern point of the ecliptic, the sun's position on June 22; the winter solstice is the most southern point, the sun's position on December 22. Owing to the plan of leap years, these dates vary slightly* (Baker, *Astronomy*).

Wisconsin, *adj. Geology.* of or having to do with the most recent period of glaciation in North America, beginning about 115,000 years ago and lasting about 95,000 years. Compare **Illinoian, Kansan, Nebraskan.**

wisdom tooth, *Anatomy.* the third molar tooth in humans. It is the back tooth on either side of each jaw, usually appearing between the ages of 17 and 25.

witherite (wiᴛʜ′ə rīt), *n. Mineralogy.* native barium carbonate, a rare white, gray, or yellowish mineral. *Formula:* $BaCO_3$ [from William *Withering,* 1741–1799, English physician who first described and analyzed the mineral]

w.l., *abbrev.* wavelength.

Wolffian body (wúl′fē ən), = mesonephros. [named after Kaspar Friedrich *Wolff,* 1733–1794, German anatomist]

Wolffian duct, = mesonephric duct: *The male hormone testosterone acts on the reproductive system, causing the Wolffian (male) duct to survive and the Müllerian (female) duct to retrogress* (Barbara Ford).

Wolfram (wúl′frəm), *n.* **1** = tungsten. **2** = wolframite. [from German *Wolfram*]

wolframite (wúl′frə mīt), *n. Mineralogy.* a mineral containing tungsten, iron, and manganese, occurring in crystals or in massive form. Wolframite is an important ore of tungsten. *Formula:* $(Fe,Mn)WO_4$

wollastonite (wúl′ə stə nīt), *n. Mineralogy.* a native silicate of calcium, occurring as crystals or in massive form. It is a common product of the metamorphism of limestone by intrusive igneous bodies. *Formula:* $CaSiO_3$ [from William H. *Wollaston,* 1766–1828, British physicist and chemist + *-ite*]

wood, *n. Botany.* the hard, fibrous substance that makes up the trunk and branches beneath the bark of trees and shrubs: *Wood is composed of cellulose fibers and a substance called lignin. The cellulose is the main body and lignin is the bond that holds the cellulose together. Without lignin, wood is a loose bundle of fibers. Without cellulose, it is a porous sponge of lignin* (Science News Letter). *From a practical . . . standpoint*

---

cap, fāce, fàther; best, bē, tèrm; pin, five;
rock, gō, ôrder; oil, out; cup, pùt, rüle,
yü in use, *yu* in uric;
ng in bring; *sh* in rush; *th* in thin, ᴛʜ in then;
zh in seizure.
ə = *a* in about, *e* in taken, *i* in pencil, *o* in
lemon, *u* in circus

. . . wood is all that portion of woody exogenous plant axis inside of the cambium line. It is the principal strengthening and water-conducting tissue of stems and roots (Youngken, *Pharmaceutical Botany*).

▶ See the note under **hardwood.**

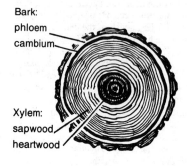

Bark:
phloem
cambium

Xylem:
sapwood
heartwood

wood ray, = xylem ray.

wool, *n.* **1** *Zoology.* **a** the fine, soft undercoat of sheep and various other mammals such as the goat, alpaca, angora, camel, and vicuña. Wool consists of nearly cylindrical fibers of keratin covered by minute overlapping scales. **b** the furry hair of some insect larvae, such as the caterpillar.
2 *Botany.* a downy substance found on certain plants. Also called **tomentum.**

work, *n.* *Physics.* the transfer of energy occurring when a force is applied to a body, causing the body to move in the direction of the force producing it. It is equal to the product of the force and the distance through which the force moves and is commonly expressed in joules, ergs, or foot-pounds. *In order for physical work to be done, it is necessary for a force to act on a body and for the body to experience a displacement that has a component parallel to the direction in which the force is acting* (Shortley and Williams, *Elements of Physics*). *In mechanics, work is defined as force exerted in direction of motion times distance of motion* (Pierce, *Electrons, Waves, and Messages*). *Abbreviation:* w or w.

work function, *Physics.* the energy required to release an electron as it passes through the surface of a metal: *The minimum energy required is called the work function, and it is designated by the Greek letter phi. This energy can be supplied by heat . . . light . . . or bombarding particles, usually other electrons or positive ions* (W. P. Dyke).

world ocean, *Geography.* the continuous body of water that covers most of the earth's surface, made up of all the oceans—the Pacific, the Atlantic, the Indian, and the Arctic, and all the seas. The continents lie like islands in the world ocean. *The actual ocean basins are more than filled by water, and consequently the water of the world ocean overlaps some 10 million square miles of the lower edges of the continents* (White and Renner, *Human Geography*).

worm, *n.* *Zoology.* any of numerous small, slender invertebrates, usually soft-bodied and lacking legs, including annelids, nematodes, platyhelminths, and nemerteans.

wormhole, *n.* *Astronomy, Physics.* a hypothetical passageway in space connecting a black hole and a white hole: *Particularly appealing . . . is the concept of "wormholes," which tunnel through the contorted space-time geometry of black holes into other universes—or emerge into our own universe at some other time and place. If a star went through such a wormhole it might, according to one hypothesis, burst forth far away, in space-time dimensions, radiating intense energy* (Walter Sullivan).

W particle, *Nuclear Physics.* a subatomic particle with either a positive or negative charge, held to be a carrier, or quantum unit, of the weak interaction. It has a mass about 86 times that of a proton. Compare **Z particle, weakon.**

wrist, *n.* *Anatomy.* **1** the joint that connects the hand with the arm.
2 a corresponding joint or part of the forelimb of an animal.
3 the bones of this part; carpus.

wt., *abbrev.* weight.

Würm (vûrm *or* wûrm), *n.* *Geology.* the fourth and most recent glaciation of the Pleistocene in Europe: *The last major glaciation, called the Wisconsin in North America and the Würm . . . in Europe, was marked by the growth of several large ice sheets in the Northern Hemisphere and of smaller glaciers in alpine and polar areas* (G. H. Denton and S. C. Porter). Compare **Günz, Mindel, Riss.** [named after *Würm,* a lake in southern Germany, a locality of the glaciation]

wurtzite (wèrt′sīt), *n.* *Mineralogy.* a native sulfide of zinc, a crystalline dimorph of sphalerite. *Formula:* (Zn,Fe)S [from Charles *Würtz,* 1817–1884, French chemist + *-ite*]

X

x, *symbol.* **1** an unknown quantity (especially in algebraic equations, along with *y* and *z*).
2 abscissa.

xanth-, *combining form.* the form of **xantho-** before vowels, as in *xanthic, xanthin.*

xanthate (zan′thāt), *n. Chemistry.* a salt of xanthic acid, usually potassium or sodium, used especially as a flotation agent to keep water from wetting sulfide ores.

xanthein (zan′thē in), *n. Chemistry.* the yellow water-soluble coloring matter of flowers.

xanthene (zan′thēn), *n. Chemistry.* a crystalline compound that is the basis of a group of mainly yellow dyes. It is formed by the reduction of xanthone. *Formula:* $C_{13}H_{10}O$

xanthic (zan′thik), *adj.* **1** *Botany.* yellow (applied especially to a series of colors in flowers passing from yellow through orange to red).
2 *Biochemistry.* of or having to do with xanthin or xanthine.

xanthic acid, *Chemistry.* any of a group of unstable acids having the general formula $ROCS_2H$ (in which R denotes a hydrocarbon radical), especially a colorless, oily liquid, $C_3H_6OS_2$, with a strong odor.

xanthin (zan′thin), *n. Biochemistry.* **1** the yellow non-water-soluble coloring matter of flowers.
2 = xanthine.

xanthine (zan′thēn′ *or* zan′thən), *n. Biochemistry.* a crystalline, nitrogenous substance, a purine, present in the urine, blood, liver, and muscle tissue, and also in various plants. *Formula:* $C_5H_4N_4O_2$

xanthine oxidase, *Biochemistry.* an enzyme in animal tissues that catalyzes the oxidation of xanthine and other purines: *Xanthine oxidase ... contains eight iron atoms, two molybdenum atoms and two molecules incorporating riboflavin (one of the B vitamins) in a giant molecule more than 25 times the size of cytochrome c* (Earl Frieden).

xantho-, *combining form.* yellow, as in *xanthophyll* = a yellow pigment. Also spelled **xanth-** before vowels. [from Greek *xanthos* yellow]

xanthone (zan′thōn), *n. Chemistry.* a crystalline compound that is the basis of a group of naturally occurring yellow dyes. *Formula:* $C_{13}H_8O$

xanthophore (zan′thə fôr), *n. Biology.* a chromatophore containing a yellow pigment: *Unlike melanophores, the cells that carry the yellow pigment (xanthophores) do not increase or decrease in number in response to outside stimulation* (Scientific American).

xanthophyll (zan′thō fil), *n. Biochemistry.* **1** a yellow pigment related to carotene, present in plant cells and thought to be a product of the decomposition of chlorophyll. *Formula:* $C_{40}H_{56}O_2$
2 any of various related yellow pigments.
[from Greek *xanthos* yellow + *phyllon* leaf]
—**xanthophyllic,** *adj.* having to do with or containing xanthophyll.

xanthoproteic acid (zan′thō prō tē′ik), = xanthoprotein: *A very characteristic property of nitric acid is that of staining wool, silk, and the skin a bright-yellow color. This is caused by the formation of xanthoproteic acids and is used as a test for proteins* (Parks and Steinbach, *Systematic College Chemistry*).

xanthoprotein (zan′thō prō′tēn), *n. Biochemistry.* a yellowish substance formed by the action of concentrated nitric acid on protein.

xanthopterin (zan thop′tər in), *n. Biochemistry.* a yellow pigment found in the wings of butterflies. *Formula:* $C_6H_5N_5O_2$ [from *xantho-* + Greek *pteron* wing]

x-axis, *n. Mathematics.* the horizontal axis in a system of rectangular coordinates, as on a chart or graph. Compare **y-axis.**

X-chromosome, *n.,* or **X chromosome,** *Genetics.* one of the two chromosomes that determine sex in many animals; a sex chromosome. A fertilized egg cell containing two X chromosomes, one from each parent, develops into a female. *Every man receives an X-chromosome from his mother and does not transmit it to his sons. Every woman receives an X-chromosome from each of her parents* (Curt Stern). Compare **Y-chromosome.**

Xe, *symbol.* xenon.

xenia (zē′nē ə), *n. Botany, Genetics.* the genetic effects produced by pollen on the embryo and endosperm or on the maternal tissues of the fruit. [from Greek *xenia* being a guest, from *xenos* guest]

xenic, *adj. Chemistry.* having to do with or derived from a compound of xenon and another element or radical: *xenic acid.*

xeno- (zen′ə- *or* zē′nə-), *combining form.* foreign; strange, as in *xenolith, xenograft.* [from Greek *xenos* guest]

xenobiotic, *Biology, Chemistry.* —*n.* a drug or other foreign substance capable of harming or affecting a living organism.
—*adj.* foreign to a living organism: *a xenobiotic chemical.*

xenogamous (zə nog′ə məs), *adj. Botany.* of or produced by cross-fertilization.
—**xenogamy,** *n.* cross-fertilization between the flowers of different plants.

xenogeneic (zen′ō jə nē′ik), *adj. Immunology.* (of transplanted tissue) deriving from an organism of a different species: *Although the efficiency of entry of*

---

cap, fāce, fäther; best, bē, tèrm; pin, fīve;
rock, gō, ôrder; oil, out; cup, pùt, rüle,
yü in use, *yu̇* in uric;
ng in bring; sh in rush; th in thin, ᴛʜ in then;
zh in seizure.
ə = *a* in about, *e* in taken, *i* in pencil, *o* in
lemon, *u* in circus

xenogeneic immunoblasts into the lamina propria of the small gut was somewhat less than that of syngeneic cells, the phenomenon appears to be qualitatively similar (Nature).
ASSOCIATED TERMS: see **allogeneic.**

xenogenesis, *n. Biology.* **1** = alternation of generations. **2** the supposed production of offspring wholly and permanently unlike the parent.
3 = spontaneous generation.
—**xenogenetic,** *adj.* of the nature of or having to do with xenogenesis.
—**xenogenic,** *adj.* = xenogenetic.

xenograft, *n.* = heterograft: *If the donor and recipient are non-identical members of the same species, the graft is called an allograft or homograft; if they belong to different species, the graft is called a xenograft or heterograft* (New Scientist).

xenolith (zen′l ith *or* zēn′l ith), *n. Geology.* a fragment of older rock embedded in an igneous mass. [from Greek *xenos* strange + *lithos* rock]
—**xenolithic,** *adj.* of or having to do with a xenolith: *With the exception of xenolithic (literally "foreign rock") fragments of mantle that are occasionally brought to the surface by kimberlite pipes and some basalt formations, direct sampling of the upper mantle is impossible* (Scientific American).

xenon (zē′non *or* zen′on), *n. Chemistry.* a heavy, colorless, odorless, gaseous element, present in very small quantities in the air and used in bubble chambers and in filling flashbulbs, electron tubes, etc. Xenon is obtained from liquid air. It is largely inert, but it does combine with fluorine. *Xenon is only one tenth as abundant as krypton, making up roughly 1 part in 20,000,000 parts of air. It is more than four times as dense as air (130:29), liquefies at −108.1°C, freezes at 112.0°C, and has its critical temperature at 16.6°C* (Offner, *Fundamentals of Chemistry*). *Symbol:* Xe; *atomic number* 54; *atomic weight* 131.30; *oxidation state* 2, 4, 6, 8. [from Greek *xenon,* neuter of *xenos* strange]

xenotime (zen′ə tīm), *n. Mineralogy.* a yellowish-brown, natural phosphate of yttrium, which resembles zircon in form but is not as hard. Xenotime is used as a source of yttrium. *Formula:* YPO_4 [from *xeno-* + Greek *timē* honor]

xer-, *combining form.* the form of **xero-** before vowels, as in *xerarch, xeric.*

xeric (zir′ik), *adj. Botany.* **1** lacking moisture: *There are certain varieties of wheat that are very xeric when compared with others* (Emerson, *Basic Botany*). Compare **hydric, mesic. 2** = xerophytic.

xero- (zir′ə-), *combining form.* dry, as in *xerophilous, xerothermic.* Also spelled **xer-** before vowels. [from Greek *xeros* dry]

xeromorph, *n. Botany.* a plant adapted to saltwater marshes or highly alkaline soils: *The term 'xerophyte' is now limited to those plants able to endure conditions of drought, while salt-marsh plants are known as 'xeromorphs'* (R. N. Elston). [from *xero-* + Greek *morphē* form]

xerophilous (zi rof′ə ləs), *adj. Biology.* adapted to or thriving in a dry climate: *The cactus is a natural xerophilous plant.* [from *xero-* + Greek *philos* loving]
—**xerophily,** *n.* the condition or character of being xerophilous.

xerophyte (zir′ə fit), *n. Botany.* a plant that loses very little water and can grow in deserts or very dry ground. Cactuses, sagebrush, century plants, etc., are xerophytes. [from *xero-* + *-phyte*]
ASSOCIATED TERMS: see **hydrophyte.**
—**xerophytic** (zir′ə fit′ik), *adj. Botany.* having the character of a xerophyte; xerophilous: *Although ... xerophytic (literally "dry plant") organisms may be able to conserve internal supplies of water for a long time, however, they still need some occasional dew or rain* (G. Evelyn Hutchinson).
—**xerophytically,** *adv.* in the manner of a xerophyte.
—**xerophytism** (zir′ə fi tiz′əm), *n.* the quality or condition of being xerophytic: *They studied the physiological conditions governing xerophytism in plants growing in dry or salty habitats in temperate regions, such as coastal marshes and sand-dunes* (R. N. Elston).

xerothermic, *adj. Biology.* **1** being both dry and hot: *a xerothermic climate or period.*
2 adapted to a climate or environment that is both dry and hot: *xerothermic organisms.*

xi (zī *or* ksē), *n.* Also called **xi particle** *or* **xi hyperon.** *Nuclear Physics.* a hyperon, either neutral or negative, present in cosmic rays and having a mass approximately 2580 times the mass of an electron: *Decay products of the xi are a lambda-zero and a neutral pion which decays instantaneously into two gamma rays* (New Scientist). Compare **lambda, sigma, omega.** [from the name of the 14th letter of the Greek alphabet (Ξ)]

xiphisternal cartilage (zif′ə ster′nəl), = xiphisternum.

xiphisternum (zif′ə stėr′nəm), *n., pl.* **-na** (-nə). *Anatomy.* the posterior or lower part of the sternum of mammals (in man usually called *xiphoid cartilage*). [New Latin, from Greek *xiphos* sword + New Latin *sternum* sternum]

xiphoid (zif′oid), *adj. Anatomy.* **1** shaped like or resembling a sword: *a xiphoid bone.* SYN: ensiform.
2 of or having to do with the xiphisternum or the xiphoid cartilage.

xiphoid cartilage *or* **xiphoid process,** *Anatomy.* the cartilaginous lower end of the sternum in humans.

xiphophyllous (zif′ə fil′əs), *adj. Botany.* having sword-shaped leaves. [from Greek *xiphos* sword + *phyllon* leaf]

X-radiation, *n. Physics.* radiation made up of X rays: *X-radiation is emitted from the anode surface as a consequence of its bombardment by the electron stream* (Sears and Zemansky, *University Physics*).

X ray, *Physics.* **1** an electromagnetic ray having an extremely short wavelength that ranges from about 0.1 to 10 nanometers, formed by the bombardment of a metal target with high-speed electrons in a vacuum tube or by the transfer of electrons to the inner shells of heavy atoms. X rays penetrate solids such as body tissue, affect photographic films, plates, and fluorescent screens, ionize gases, and produce secondary radiations. X rays are used in medicine to locate breaks in bones, foreign objects lodged in the body, etc., and to diagnose and treat diseases. In nature, X rays are generated by the high temperature inside the sun and other stars. *When a beam of X rays is directed through a crystal, the rays are reflected from the layers of atoms, and these rays form a pattern that indicates the spacing and arrangement of the atoms* (Scientific American). See the picture at **electromagnetic spectrum.**

2 a photograph made by means of X rays. Also called **roentgen ray.** See also **roentgen.**

[half-translation of obsolete German *X-Strahlen* X rays, from *X,* in sense of "unknown" + *Strahl* ray, beam]

▶ X ray is usually written with a capital *X.* It is not hyphenated as a noun, but it is as an adjective or as a verb: *an X-ray examination, to X-ray the chest.*

—X-ray, *v.* to examine, photograph, or treat with X rays.

—X-ray, *adj.* **1** of or having to do with X rays: *An X-ray tube is a vacuum tube for generating X rays.* **2** done by means of X rays: *an X-ray photograph, an X-ray examination of the teeth.*

X-ray astronomy, a branch of astronomy dealing with the nature and sources of X rays in space: *X-ray astronomy ... can be carried out only from rockets and satellites sent above the atmosphere, since the atmosphere screens out the radiation* (N. Y. Times).

X-ray crystallography, the study of the arrangement of atoms, ions, or molecules in crystals and chemical substances by X rays: *X-ray crystallography has determined the structures of many compounds ... and some pictures show the position of every atom to within a hundredth of an angstrom.* (New Scientist).

X-ray diffraction, *Physics.* the scattering of X rays on contact with matter, with changes in radiation intensity as a result of differences in atomic structure within the matter. It is an important method of studying atomic and molecular structure and is used in X-ray crystallography. *... the techniques of X-ray diffraction, which have contributed so much to the understanding of the inner structure of metals and alloys* (F. A. Fox).

X-ray pulsar, *Astronomy.* a pulsar that is the source of powerful X-ray emissions: *One of the most remarkable developments of X-ray astronomy was the ... discovery of an X-ray pulsar in the Crab Nebula* (Richard B. Hoover).

X-ray source, = X-ray star: *Most of the known discrete X-ray sources lie close to the Milky Way and therefore very probably belong to the local galaxy* (Science Journal).

X-ray spectrometer, a spectrometer using X rays by which the chemical constituents of a substance are separated into their characteristic spectral lines for identification and determination of their concentration: *An X-ray spectrometer ... automatically (under computer control) performs the analysis and provides a printout of the analytical results* (Max Tochner).

—X-ray spectrometry, the use of an X-ray spectrometer; chemical analysis by means of an X-ray spectrometer.

X-ray spectroscopy, = X-ray spectrometry.

X-ray star, *Astronomy.* a celestial object, especially a star, that radiates a relatively large amount of energy as X rays: *The X-ray star is approximately 1,000 light-years away and ... radiates about 500 times more energy as X rays than as visible light* (E. L. Schücking). *New discoveries would shed light on the relationship between pulsars and other kinds of X-ray stars* (New Scientist).

X-ray telescope, a telescope designed to detect X rays, used in X-ray astronomy: *Because X rays cannot be focused by ordinary lenses or reflected by conventional telescopes, X-ray telescopes have an unusual shape. They are built in the form of a slightly tapered cylinder, because the manner in which X rays are reflected is not the same as the reflection of light rays. To reflect X rays efficiently, the diameter of the telescope must be large and its reflecting section must be long* (Herbert Friedman).

xyl-, *combining form.* the form of xylo- before vowels, as in *xylan, xylene.*

xylan (zī′lan), *n. Biochemistry.* a yellow, gelatinous compound, a pentosan, found in woody tissue. It yields xylose when hydrolyzed. *One disadvantage of hardwoods over soft is that hardwood contains a higher percentage of xylan, a gummy substance that must be removed* (Science News Letter).

xylem (zī′lem), *n. Botany.* the tissue in a vascular plant, consisting essentially of woody fibers, tracheids, parenchymatous cells, and (in angiosperms) tracheae, through which water and dissolved minerals pass upward from the roots. The xylem also provides support for the plant. *Xylem contrasts markedly with all other tissues in that more than half of its divisions are longitudinal* (Frank Cusick). *Xylem elements differ from other plant cells by their thick inner walls of lignin and cellulose which give wood its characteristic toughness* (Science News). Compare **phloem.** See the pictures at **root, wood.** [from German *Xylem,* from Greek *xylon* wood]

xylem ray, *Botany.* a ray or plate of parenchyma cells extending from the pith to the cambium in a woody stem. Also called **wood ray.** Compare **phloem ray.**

xylene (zī′lēn′), *n. Chemistry.* any of three isomeric, colorless, liquid hydrocarbons present in coal and wood tar, naphtha, etc. Commercial xylene is a mixture of all three, and is used in making dyes, as a raw material for polyester fibers, etc. *Formula:* $C_6H_4(CH_3)_2$ Also called **xylol.**

xylitol (zī′lə tôl), *n. Chemistry.* a crystalline alcohol derived from xylose, used as a sugar substitute. *Formula:* $C_5H_{12}O_5$

xylo- (zī′lō), *combining form.* wood; woody, as in *xylocarp, xylophage.* Also spelled **xyl-** before vowels. [from Greek *xylon* wood]

xylocarp, *n. Botany.* a hard and woody fruit.

—xylocarpous, *adj.* having fruit which becomes hard and woody.

---

cap, fāce, fäther; best, bē, tèrm; pin, five;
rock, gō, ôrder; oil, out; cup, pùt, rüle,
yü in use, *yù* in uric;
ng in bring; *sh* in rush; *th* in thin, ᴛH in then;
zh in seizure.
ə = *a* in about, *e* in taken, *i* in pencil, *o* in lemon, *u* in circus

xylogen (zī′lə jən), *n. Botany.* **1** wood or xylem in a formative state. **2** = lignin.

xyloid (zī′loid), *adj. Botany.* **1** of or having to do with wood. **2** like wood; ligneous.

xylol (zī′lōl), *n.* = xylene.

xylology (zī lol′ə jē), *n.* the study of the structure of wood.

xylophage (zī′lə fāj), *n. Zoology.* a xylophagous organism; an insect, mollusk, or crustacean that eats or destroys wood.

xylophagous (zī lof′ə gəs), *adj. Zoology.* **1** feeding on wood, as some insect larvae. **2** boring into or destroying wood, as some mollusks and crustaceans. [from New Latin *xylophagus,* from Greek *xylon* wood + *phagein* to eat]

xylose (zī′lōs), *n. Biochemistry.* a crystalline, pentose sugar present in woody plants: *While some simple sugars may contain as few as three carbon atoms, those occurring most commonly contain five or six ... Xylose, an example of such a sugar, is ... a constituent of such complex carbohydrates as pentosans, gums, and hemicelluloses* (Harbaugh and Goodrich, *Fundamentals of Biology*). *Formula:* $C_5H_{10}O_5$

xylyl (zī′ləl), *n. Chemistry.* a univalent radical, part of xylene. *Formula:* C_8H_9-

xylylene (zī′lə lēn), *n. Chemistry.* a bivalent radical, part of xylene. *Formula:* -C_8H_8-

Y

y, *symbol.* **1** an unknown quantity (especially in algebraic equations, along with *x* and *z*). **2** ordinate.

y., *abbrev.* yard or yards.

Y, *symbol.* yttrium.

yard, *n.* a customary measure of length, equal to 36 inches (3 feet) or 0.9144 meter. *Abbreviation:* yd., y. [Old English *gerd* rod]

y-axis, *n. Mathematics.* the vertical axis in a system of rectangular coordinates, as on a chart or graph. Compare x-axis.

Yb, *symbol.* ytterbium.

Y-chromosome, *n.,* or Y chromosome, *Genetics.* one of the two chromosomes that determine sex in many animals; a sex chromosome. A fertilized egg cell containing a Y chromosome develops into a male. *The Y chromosome does not carry genes for ordinary sex-linked characteristics, but in the case of a few characteristics ... genes are located on the Y chromosome* (Colin, *Elements of Genetics*). *The Y-chromosome is transmitted from father to son only in the male line* (Theodosius Dobzhansky). Compare X chromosome.

yd., *pl.* yd. or yds. *abbrev.* yard.

year, *n. Astronomy.* **1** the period of the earth's revolution around the sun: *The solar or astronomical year is 365 days, 5 hours, 48 minutes, 45.51 seconds.*
2 the time in which any planet completes its revolution around the sun: *Among those 26 [asteroids] there are only three whose ... mean distance from the sun—are less than the earth's ... As a result, it takes less time to circle the sun than does any other known asteroid ... only 0.759 earth-years long—about nine months* (Science News).
3 the time it takes for the sun to make an apparent journey from a given star back to it again: *The sidereal year is 20 minutes, 23 seconds longer than the solar year.*

yeast, *n. Botany.* any of a genus (*Saccharomyces*) of minute, one-celled, ascomycetous fungi which produce alcoholic fermentation in saccharine fluids. The commercial yeast used in the production of alcoholic beverages consists of masses of yeast cells and spores. *Some yeasts divide by fission, as do bacteria. In the majority of yeasts, however, new cells grow out of the mother cell, much as a small bubble would form ... The small "bubbles" formed from the mother yeast cell are called buds* (Weier, *Botany*).

yellow enzyme, = flavoprotein.

yellow-green algae, *Botany.* a division (Chrysophyta) of algae having a yellow-green to green pigment masking the chlorophyll.

-yl, *combining form. Chemistry.* a radical composed of two or more elements (with one usually designated by the base word) acting like a simple element and forming the foundation of a series of compounds, as in alkyl, acetyl, carbonyl. [from French *-yle,* from Greek *hylē* substance, matter, wood]

yolk, *n. Biology.* **1** the yellow internal part of an egg of a bird or reptile, surrounded by the albumen or white, and serving as nourishment for the young before it is hatched.
2 the corresponding part in any animal ovum or egg cell, which serves for the nutrition of the embryo, together with the protoplasmic substances from which the embryo is developed.
[Old English *geolaca,* from *geolu* yellow]

yolk sac, *Embryology.* a membranous sac filled with yolk, attached to and providing food for the embryo. In cephalopods and lower vertebrates, it is the only source of food for the embryo, for these animals do not develop a placenta. *The developing embryo at first is only a small ball of cells called a blastula. Later the ball of cells develops a cavity within it near one side. Two more cavities then form within the thick inner cell mass. One is the yolk sac, the other is the amniotic sac* (Stephen C. Williams).

yolk stalk, *Embryology.* a narrow, ductlike part that unites the yolk sac to the middle of the embryo's digestive tract.

ytterbia (i tèr′bē ə), *n. Chemistry.* a heavy, white powder, an oxide of ytterbium, which forms colorless salts. *Formula:* Yb_2O_3

ytterbic (i tèr′bik), *adj. Chemistry.* of or containing ytterbium, especially with a valence of three.

ytterbium (i tèr′bē əm), *n. Chemistry.* a rare-earth metallic element whose compounds resemble those of yttrium. It occurs with yttrium in gadolinite and various other minerals, and is used in making special alloys. *Ytterbium ... was discovered by Marignac in 1878. In 1907, G. Urbain of France found that what Marignac had called ytterbium was actually composed of two elements, which he named lutetium and neoytterbium; the latter was eventually listed as ytterbium* (Scientific American). *Symbol:* Yb; *atomic number* 70; *atomic weight* 173.04; *melting point* 824°C; *boiling point* 1466°C; *oxidation state* 2,3. [from New Latin, from *Ytterby,* town in Sweden, where it was first discovered]

ytterbium metals, = yttrium metals.

yttria (it′rē ə), *n. Chemistry.* a heavy, white powder, an oxide of yttrium, obtained from gadolinite and other rare minerals. *Formula:* Y_2O_3

yttrium (it′rē əm), *n. Chemistry.* a dark-gray metallic element resembling and associated with the rare-earth elements, occurring in gadolinite, xenotime, and vari-

cap, fāce, fàther; best, bē, tèrm; pin, five;
rock, gō, ôrder; oil, out; cup, pùt, rüle,
yü in use, *yù* in uric;
ng in bring; *sh* in rush; *th* in thin; ᵀH in then;
zh in seizure.
ə = *a* in about, *e* in taken, *i* in pencil, *o* in
lemon, *u* in circus

ous minerals. It is used in making iron alloys and to remove impurities from metals. Yttrium compounds are used to make incandescent gas mantles. *Symbol:* Y; *atomic number* 39; *atomic weight* 88.905; *melting point* 1523°C; *boiling point* 3337°C; *oxidation state* 3. [from New Latin, ultimately from *Ytterby.* See YTTERBIUM.]

yttrium metals, *Chemistry.* a group of metals that include yttrium and the rare-earth metals dysprosium, erbium, holmium, lutetium, thulium, and ytterbium. Also called **ytterbium metals.**

Z

z, *symbol.* an unknown quantity (especially in algebraic equations, along with *x* and *y*).

Z, *symbol* or *abbrev.* **1** atomic number. **2** zenith. **3** zenith distance.

zaratite (zär′ə tīt), *n. Mineralogy.* a hydrous carbonate of nickel, of a green color, found as an incrustation and in stalactites. *Formula:* $Ni_3(CO_3) (OH_4) \cdot 4H_2O$ [from Spanish *zaratita*]

zastruga (zäs trü′gə), *n., pl.* **-gi** (-gē). *Geology.* one of a series of wavelike ridges formed in snow by the action of the wind, and running in the direction of the wind. Also spelled **sastruga.** [from Russian *zastruga* groove]

z-axis, *n. Mathematics.* the third axis in a three-dimensional system of rectangular coordinates, the other two axes being the x-axis and y-axis.

Z disk or **Z disc,** *Biochemistry.* a thin, dark disk of fibrous protein that passes through striated muscle fiber and marks the boundaries of contiguous contractile units: *Skeletal muscles contract by way of thick protein rods sliding between thin rods. The muscle is given strength, and its speed of contraction increased, by intermittent cross-walls known as Z discs ... But because of those Z-discs, contraction is limited; the contracted length of a skeletal muscle is only 40 per cent of the extended length* (New Scientist). Also called **intermediate disk.** Compare **Z line.** [half-translation of German *Z-Scheibe,* abbreviation of *Zwischenscheibe* intermediate disk]

zeatin (zē′ə tin), *n. Biochemistry.* a cytokinin originally isolated from young maize kernels. [from New Latin *Zea* the maize plant genus + (*kine*)*tin*]

Zeeman effect (zā′män), *Physics.* the separation of lines of the spectrum that occurs in light emanating from a source in a magnetic field: *The method involves the Zeeman effect, which occurs when a single emission line or absorption line of the electromagnetic spectrum is produced in the presence of a magnetic field. The Zeeman effect will usually split the line into three components: one unshifted in frequency, one shifted slightly higher in frequency and one shifted slightly lower* (G. L. Berge and G. A. Seielstad). [named after Pieter Zeeman, 1865–1943, Dutch physicist]

zein (zē′in), *n. Biochemistry.* a protein contained in corn, used in plastics, coatings, and adhesives: *By-products of wet milling include zein, a protein used as a source of many plastics* (G. F. Sprague). [from New Latin *Zea* the maize genus, from Latin *zea* spelt, from Greek *zeia*]

zeitgeber or **Zeitgeber** (tsīt′gā′bər or zīt′gā′bər), *n. Biology.* any time indicator, such as light, dark, or temperature, that influences the workings of the biological clock: *For many animals the photic zeitgebers are believed to involve the twilight periods. Many animals customarily begin or cease activity during twilights, and simulated twilights influence the activity of numerous captive mammals* (Science). [from German *Zeitgeber* (literally) time giver]

zenith (zē′nith), *n. Astronomy.* **1** the point in the heavens directly overhead; a point where a vertical line intersects the celestial sphere. It is opposite to nadir. For an observer at the North Pole, the North Star would be about at the zenith. *Every different point on the earth has a different zenith. The point directly opposite the zenith on the celestial sphere, i.e., the imaginary point directly under the observer's feet is called the nadir. The imaginary circle midway between zenith and nadir, dividing the sky into two hemispheres, is called the astronomical horizon* (Krogdahl, *The Astronomical Universe*). See the picture at **nadir.**
2 the point of highest altitude of a celestial body, relative to a particular observer or place. *Abbreviation:* Z See the picture at **azimuth.**
[from Old French or Medieval Latin *cenith,* from Arabic *samt(ar-rās)* the way (over the head)]
—**zenithal** (zē′nə thəl), *adj.* of, having to do with, or occurring at the zenith: *a zenithal projection of the lands around the Poles.*

zenith distance, *Astronomy.* the angular distance of a celestial body from the zenith of a particular observer or place; the complement of a body's altitude, measured in degrees along the vertical circle passing through the zenith and the body: *At the North Pole, the zenith distance of the North Star is about 0 degrees. Abbreviation:* Z Also called **co-altitude.**

zeolite (zē′ə līt), *n. Mineralogy.* any of a large group of minerals consisting of hydrated silicates of aluminum with alkali metals, commonly found in the cavities of igneous rocks and in various altered volcanogenic sediments. Zeolites have their atoms arranged in an open crystal framework that can hold other atoms or molecules much as a sponge holds water. [from Swedish *zeolit,* from Greek *zein* to boil + Swedish *-lit* -lite (so called because it boils or swells under the blowpipe)]
—**zeolitic** (zē′ə lit′ik), *adj.* having to do with or consisting of zeolite: *zeolitic aluminosilicates.*

zero, *n., pl.* **zeros** or **zeroes.** *Mathematics.* **1** the symbol or digit 0, denoting the complete abscence of quantity or magnitude. In arithmetic, zero is the identity element of addition, since 0 added to any number gives the same number as the sum: $2 + 0 = 0 + 2 = 2$, etc. Zero is also the cardinal number of the empty set.

---

cap, fāce, fäther; best, bē, tèrm; pin, five;
rock, gō, ôrder; oil, out; cup, pùt, rüle,
yü in use, *yû* in uric;
ng in bring; *sh* in rush; *th* in thin, ⱦн in then;
zh in seizure.
ə = *a* in about, *e* in taken, *i* in pencil, *o* in lemon, *u* in circus

A number multiplied by zero gives zero. Zero is often symbolized $\emptyset$ to avoid confusion with the letter O.
2 a the starting point or neutral position on a scale; a point from which the graduation of a scale begins. **b** the temperature that corresponds to zero on the scale of a thermometer.
—*adj. Meteorology.* **1** denoting a ceiling not more than 16 meters high.
2 denoting visibility of not more than 55 meters in a horizontal direction.
[from Italian, from Arabic *sifr* empty]

zero g or **zero gravity,** *Physics.* a condition in which gravitational force cannot be felt or observed, characteristic of a free-falling body: *The zero gravity attained in an orbiting satellite also can be achieved experimentally for a few seconds in an airplane flying along a path such that the plane's speed and direction are temporarily what they would be in a natural orbit at that altitude* (Science News Letter). Compare **weightlessness.**

zero magnitude, *Astronomy.* a measure of brillance of certain stars, being 2 1/2 times as bright as first magnitude.

zero-point energy, *Physics.* the kinetic energy remaining in a substance at the temperature of absolute zero: *Quantum theory ... implies that every physical system must retain a finite "zero-point energy," which means that however low the temperature, the system will continue to fluctuate about an equilibrium position* (R. Furth).

zero-sum, *adj. Mathematics.* having to do with or characterized by a situation in game theory in which the total winnings of one side equal the total losses of the other side: *Most game theory work has been on what are called two-person zero-sum games. This means that the conflict is between two players ... and whatever one player wins the other loses* (Martin Gardner). See also **maximin, minimax.**

zeroth (zir′ōth), *adj. Mathematics.* of or at zero; being zero: $x^0 = x$ to the zeroth power.

zero-zero, *adj. Meteorology.* denoting conditions of severely limited visibility in both the horizontal and vertical directions: *zero-zero weather.*

zinc (zingk), *n. Chemistry.* a shiny, bluish-white metallic element that is hard, brittle, and little affected by air and moisture at ordinary temperatures, but at high temperatures burns in air with a bright, blue-green flame. Zinc occurs in nature combined with other elements in sphalerite and other minerals. It is widely used in industry as a coating for iron, in alloys such as brass, as a roofing material, in electric batteries, in paint, etc. *Symbol:* Zn; *atomic number* 30; *atomic weight* 65.37; *melting point* 419.58°C; *boiling point* 907°C; *oxidation state* 2. [from German *Zink*]

zincate (zing′kāt), *n. Chemistry.* a salt of zinc hydroxide, such as $Zn(OH)_2$, when it acts as a weak acid.

zincblende, = sphalerite.

zinc chloride, *Chemistry.* a water-soluble crystal or crystalline powder, used in galvanizing, electroplating, and as a wood preservative, disinfectant, etc.: *Anhydrous zinc chloride, $ZnCl_2$, may be made by the direct union of chlorine with the metal. The anhydrous salt is used as a caustic in surgery and as a dehydrating agent in organic reactions ... A solution of zinc chloride is used to dissolve the metal oxides on the surface of metals before soldering* (Parks and Steinbach, *Systematic College Chemistry*). *Formula:* $ZnCl_2$

zincic (zing′kik), *adj. Chemistry.* having to do with, consisting of, or resembling zinc.

zinciferous (zing kif′ər əs), *adj. Geology.* containing or yielding zinc.

zincite (zing′kīt), *n. Mineralogy.* native zinc oxide, of a deep-red or orange-yellow color. *Formula:* ZnO

zinc sulfide, *Chemistry.* a yellowish or white powder occurring naturally as sphalerite, used as a pigment and as a phosphor in television screens and on watch faces. *Formula:* ZnS

zinkenite (zing′kə nīt), *n. Mineralogy.* a steel-gray sulfide of antimony and lead. *Formula:* $Pb_6Sb_{14}S_{27}$ [from German *Zinkenit*, from J. K. L. *Zinken,* 19th-century German mineralogist + *-it* -ite]

zircon (zèr′kon), *n. Mineralogy.* a native silicate of zirconium that occurs in tetragonal crystals, variously colored. Transparent zircon is used as a gem. Zircon is composed chiefly of silicon, oxygen, and zirconium, but contains small amounts of other elements, such as hafnium, iron, and the rare earths. *Formula:* $ZrSiO_4$ [ultimately probably from Arabic *zargūn,* from Persian *zargūn* (literally) golden, from *zar* gold]

zirconia (zèr kō′nē ə), *n. Chemistry.* a dioxide of zirconium, usually obtained as a white, amorphous powder, used in making incandescent gas mantles and refractory utensils: *Technology continued to work on producing a better imitation diamond. Eventually a material known as "zirconia" (cubic zirconium oxide) was developed. This material has the hardness of YAG [Yttrium Aluminum Garnet] plus the dispersion of diamond. It is by far the best diamond imitation ever produced* (Joel E. Arem). *Formula:* ZrO_2 [from New Latin, from *zircon*]

zirconic (zèr kon′ik), *adj. Chemistry.* of, having to do with, or containing zirconia or zirconium.

zirconium (zèr kō′nē əm), *n. Chemistry.* a grayish-white metallic element commonly found as a black powder or as a grayish crystalline substance. Zirconium has both acidic and basic properties and is found in many minerals, such as zircon and baddeleyite. It is used in alloys for wires, filaments, etc., in making steel, and in nuclear reactors. *Zirconium has high resistance to corrosion and is used in atomic reactors to protect uranium fuel from water carrying heat out of the reactor core. It is considered ideal for this purpose because it absorbs few of the neutrons necessary to carry on a chain reaction* (Wall Street Journal). *Symbol:* Zr; *atomic number* 40; *atomic weight* 91.22; *melting point* ab. 1852°C; *boiling point* 3578°C; *oxidation state* 4. [from New Latin, from *zircon*]

zirconium oxide, = zirconia.

zirconyl (zèr′kə nəl), *n. Chemistry.* a bivalent radical, ZrO-.

Z line, *Biochemistry.* a Z disk as it appears longitudinally: *One difference between muscle cells and nonmuscle cells is that in muscle fibers the actin filaments are anchored to flat protein structures called Z lines, which are emplaced between every two contractile units. The actin filaments extend perpendicularly from both surfaces of the Z line and run in opposite directions on opposite sides. In nonmuscle cells, however, the actin* filaments are anchored directly to the cell membrane (Elias Lasarides and Jean Paul Rivel).

Zn, *symbol.* zinc.

zo-, *combining form.* the form of **zoo-** before vowels, as in *zoic, zooid.*

zoarium (zō ār′ē əm), *n., pl.* **-aria** (-ār′ē ə). *Zoology.* the colony or aggregate of individuals of a compound animal. [from New Latin, from *zoon* animal, from Greek *zōion*]

zodiac (zō′dē ak), *n. Astronomy.* an imaginary belt of the heavens, extending about 8 degrees on both sides of the apparent yearly path of the sun and including the apparent paths of the major planets and the moon. The zodiac is divided into 12 equal parts, called *signs,* named after 12 constellations. *The zodiac is the concept of dividing the path of sun, moon, and planets around the heavens into twelve equal parts, each named after a constellation of fixed stars ... The series runs: ram, bull, twins, crab, lion, virgin, scales, scorpion, archer, goat, water-carrier, fishes* (Alfred L. Kroeber). [from Greek *zōidiakos* (*kyklos*) (circle) of the figures, in reference to the figures of the ram, archer, scales, etc., representing the signs of the zodiac, from *zōidion* sculptured figure, sign of the zodiac, diminutive of *zōion* figure, animal]

zodiacal light (zō dī′ə kəl), *Astronomy.* a cone-shaped glow of faint light in the sky, seen near the ecliptic at certain seasons of the year, either in the west after sunset or in the east before sunrise. Zodiacal light is caused by sunlight reflecting on meteoric dust particles concentrated in the plane of the ecliptic. *The zodiacal light is so named because it is seen against the zodiacal constellations which lie along the ecliptic* (Oliver J. Lee). Compare **Gegenschein.**

zoea (zō ē′ə), *n., pl.* **zoeae** (zō ē′ē) or **zoeas.** *Zoology.* a larval stage of development in crustaceans, especially in decapods such as crabs, characterized by spines on the carapace and rudimentary thoracic and abdominal limbs: *In the first stage, the young crab, called a zoea, is about 1/25 of an inch in length* (Science News Letter). [from New Latin, from Greek *zōē* life]
—**zoeal,** *adj.* of or having to do with a zoea or zoeae.

zoic (zō′ik), *adj.* **1** *Zoology.* of or having to do with living animals; characterized by animal life. **2** *Geology.* containing animal fossils.

zoisite (zoi′sīt), *n. Mineralogy.* an orthorhombic mineral of the epidote group, a silicate of aluminum and calcium, sometimes containing iron instead of aluminum. *Formula:* $Ca_2Al_3Si_3O_{12}(OH)$

zonal, *adj.* **1** of, having to do with, or divided into zones; characterized by or arranged in zones: *Because they are characteristic of large regions or zones, which are more or less similar to the great climatic and vegetational regions, these are called the zonal soils* (Finch and Trewartha, *Elements of Geography*).
2 *Meteorology.* latitudinal; easterly or westerly: *December through March ... the easterly circulation in the stratosphere follows a pattern known as zonal flow—air at 80,000 feet moves around the globe at a constant latitude, veering neither to the north or south* (Science News).
3 *Botany.* marked with circular bands of color: *The leaves of certain varieties of geranium are zonal.*

—**zonality,** *n.* zonal character or arrangement: *Characterization of the zonality of soil is of particular value in landscape descriptions (soil maps and data) which have specific practical value in helping to guide land-use decisions for the future* (Gerald W. Olson). —**zonally,** *adv.*

zona pellucida (zō′nə pe lü′sə də), *Embryology.* a thick, tough, transparent membrane surrounding the yolk of a developed mammalian ovum: *At last, one—and only one—sperm that penetrates the tough zona pellucida, the glass-clear membranous shell of the ovum, joins its pronucleus with that of the egg. At that moment conception takes place and, scientists generally agree, a new life begins* (James C. G. Conniff). [from New Latin (literally) pellucid zone]

zonate or **zonated,** *adj.* marked with or divided into zones: *The coral snake is zonate, with alternating rings of color.*

zonation, *n. Biology.* **1** distribution of plants or animals in zones or regions of definite character: *Light penetration limits the distribution of plants, resulting in a zonation of the seaweeds: the green algae live in the uppermost, well-lighted zones; the brown algae in the intermediate zone; and the red algae at greatest depths* (Clarence J. Hylander).
2 formation of zones or concentric layers, as in the growing plant cell.
3 any arrangement in zones: *an indistinct zonation of minerals in rocks.*

zonda (zon′də), *n. Meteorology.* a wind of the foehn type in the Argentine pampas. [from American Spanish]

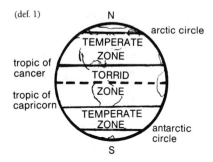

(def. 1)

zone, *n.* **1** *Geography.* any of the five great divisions of the earth's surface, bounded by imaginary lines going around the earth parallel to the equator. The zones are distinguished by differences of climate (the *torrid zone,* the two *temperate zones,* and the two *frigid zones*). *The division of the earth into the five classic zones bounded by the Tropics of Capricorn and Cancer and by the polar circles is of ancient origin and is*

cap, fāce, fäther; best, bē, tėrm; pin, five;
rock, gō, ôrder; oil, out; cup, pùt, rüle,
yü in use, *yu* in uric;
ng in bring; *sh* in rush; *th* in thin, *ᴛʜ* in then;
zh in seizure.
ə = *a* in about, *e* in taken, *i* in pencil, *o* in lemon, *u* in circus

purely on a solar climate basis. These are zones of possible sunshine rather than of actual climate ... The latitudinal zones, notwithstanding their old names of torrid, temperate, and frigid, merely mark differences in the elevation of the sun (Blair, Weather Elements).
2 Biology. a region or area characterized by certain forms of animal or plant life which are in turn determined by certain environmental conditions.
3 Ecology. an area characterized by more or less uniform vegetation. SYN: belt.
4 Anatomy. a growth or structure surrounding some part in the form of a ring or cylinder: a ciliary zone.
5 Geology. = horizon (defs. 2a, b).
6 Mathematics. a part of the surface of a sphere contained between two parallel planes.
[from Greek zōnē (originally) girdle]

zonule (zō′nyül), n. Anatomy. a small zone or band, as of a ligament.

zoo- (zō′ə-), combining form. animal or animals, as in zoology = the science of animals. Also spelled **zo-** before vowels. [from Greek zōion animal, related to zōē life]

zoogenic, adj. produced from animals; of animal origin: Limestones formed from shells are zoogenic.

zoogeographic or **zoogeographical,** adj. of or having to do with zoogeography: As far as land mammals are concerned the world's major zoogeographic provinces are at present four in number: the Holarctic-Indian, which consists of North America and Eurasia and also northern Africa; the Neotropical, ... the Ethiopian, ... and the Australian (Björn Kurtén). —**zoogeographically,** adv.

zoogeography, n. Zoology. the study of the geographical distribution of animals, especially the causes and effects of such distribution and of the relationships between certain areas and the groups of animals inhabiting them: Somewhat related to ecology is the geographic distribution of animals, or zoogeography. Ecology relates partly to local distribution of organisms, as determined by environmental conditions. Zoogeography also involves these questions of local distribution, since no species can live where the conditions are not suitable, and wrong conditions constitute barriers to distribution (Shull, Principles of Animal Biology).

zoogloea (zō′ə glē′ə), n. Bacteriology. a jellylike cluster of bacteria swollen by the absorption of fluids from the medium in which they are grown: The bacteria gradually swell up into zoogloea masses, until finally their bodies break down into soluble nitrogenous substances which are partly absorbed and assimilated and partly stored as reserve nitrogenous food by the green leguminous plant (Youngken, Pharmaceutical Botany). [from New Latin, from Greek zōos alive + gloios gelatinous substance]
—**zoogloeal,** adj. of or having to do with a zoogloea or zoogloeas: the zoogloeal form of bacteria.

zooid (zō′oid), n. Zoology. each of the distinct individuals which make up a colonial or compound animal organism; a polyp, hydroid, or the like.

zool., abbrev. **1** zoological. **2** zoology.

zoologic, adj. = zoological.

zoological, adj. **1** of animals and animal life: a zoological region, a zoological exhibition.
2 having to do with zoology: zoological works, a zoological experiment.
—**zoologically,** adv. in relation to zoology: Borneo and New Guinea ... are zoologically wide as the poles asunder (A. R. Wallace).

zoology, n. **1** the branch of biology that deals with animals and animal life; study of the structure, physiology, development, classification, etc., of animals: Botany and zoology, to state it in a highly oversimplified manner, arose from the 16th century on as applied sciences, attached to medicine. Botany started as a broadened study of medicinal herbs and early botanical gardens. ... Zoology arose in connection with human anatomy and physiology (Ernst Mayr).
2 the animals living in a particular area or period: the zoology of Alaska, the zoology of the Pleistocene. SYN: fauna.
3 zoological facts or characteristics concerning a particular animal or group of animals: the zoology of vertebrates.

zoonosis (zō on′ə sis or zō′ə nō′sis), n., pl. **-ses** (-sēz′). Biology. any disease of animals that can be transmitted to humans, such as parrot fever (psittacosis): Dog saliva may contain viral zoonoses that include rabies and other infections (Science News Letter). One of the main functions of the veterinarian is to prevent and control zoonoses (D. W. Bruner).

zoophagous (zō of′ə gəs), adj. Biology. feeding on animals: The countless host of animals that inhabit the depths of the ocean ... are necessarily zoophagous (Nature). SYN: carnivorous.

zoophile (zō′ə fil), n. Botany. a zoophilous plant or its seed.

zoophilic (zō′ə fil′ik), adj. = zoophilous.

zoophilous (zō of′ə ləs), adj. **1** Botany. (of plants) adapted for being pollinated by animals.
2 Zoology. (of insects) showing a preference for feeding on animals other than humans.

zoophyte (zō′ə fit), n. Zoology. any of various invertebrate animals resembling plants, such as sea anemones, corals, sponges, etc.: Sponges are just barely animals. In fact, they are such a borderline case that until the nineteenth century they were called zoophytes, the animal-plants. Sponges are among the most primitive forms of multicellular animal life (Shirley A. Pomponi). [from Greek zōiophyton, from zōion animal + phyton plant]
—**zoophytic** (zō′ō fit′ik), adj. of, having to do with, or resembling zoophytes: zoophytic structures.

zooplankter (zō′ə plangk′tər), n. Biology. any of the animals or animal-like organisms in a zooplankton: The zoologist studies the nutritional requirements and efficiencies of food conversion of marine animals (zooplankters) feeding on marine phytoplankters or on other small animals (John D. H. Strickland).

zooplankton (zō′ə plangk′tən), n. Biology. the part of the plankton of any body of water which consists of animals: To the zooplankton belong the protozoa, the sea anemones, the corals and the incredibly shaped jellyfishes (Scientific American). Compare **phytoplankton.**
—**zooplanktonic** (zō′ə plangk′ton′ik), adj. of or having to do with zooplankton: zooplanktonic crops.

zoosporangium (zō′ə spə ran′jē əm), *n., pl.* **-gia** (-jē ə). *Botany.* a receptacle or sporangium in which zoospores are produced: *The zoospores are set free through an opening formed at the apex of the zoosporangium* (Thomas H. Huxley).
—**zoosporangial,** *adj.* of or having to do with a zoosporangium or zoosporangia.

zoospore, *n.* **1** *Botany.* an asexual spore that can move about by means of cilia or flagella, produced by some algae and fungi: *Spores produced by water-inhabiting organisms usually develop one or more hair-like cilia which enable them to swim. Such spores are called zoospores* (Harbaugh and Goodrich, *Fundamentals of Biology*).
2 *Zoology.* a minute, freely moving, flagellate or ameboid organism released by the sporocyst of various protozoans.
[from Greek *zōos* alive + English *spore*]
—**zoosporic,** *adj.* of, having to do with, or resembling a zoospore or zoospores: *zoosporic organisms, zoosporic reproduction.*

zoosterol (zō os′tə rōl), *n. Biochemistry.* any sterol originating in animals, as distinguished from a phytosterol.

zootoxin, *n. Biology.* a toxin or poison of animal origin. Zootoxins include the venoms of spiders, snakes, and scorpions. Compare **phytotoxin.**

Zorn's lemma (zôrnz), *Mathematics.* the principle that if a set is partially ordered and each completely ordered subset has an upper bound, then the set has at least one element greater than any other element in the set. [named after Max A. *Zorn*, born 1906, German mathematician]

Z particle, *Nuclear Physics.* a subatomic particle with a neutral electric charge and a mass about 99 times that of the proton, held to be a carrier, or quantum unit, of the weak interaction: *The theory predicting that the weak force is carried by a Z particle of zero electric charge, or a W particle of either positive or negative charge, depending on the reaction, was so firmly established that the actual discovery was more a cause for congratulation than for surprise. All three predicted particles have now been produced by the giant colliding-beam accelerator at CERN, the European Laboratory for Particle Physics near Geneva* (Walter Sullivan).

Zr, *symbol.* zirconium.

Zugunruhe (tsuk′ûn rü′ə), *n. Biology.* the migratory drive in animals, especially birds: *Birds, like insects, emphasize locomotory as opposed to vegetative functions during long-distance flight; the well-known Zugunruhe or migratory restlessness is a case in point* (Science). [from German, from *Zug* travel + *Unruhe* unrest]

Zwicky galaxy (zwik′ē), *Astronomy.* any of a class of galaxies that have relatively small masses and have most of their luminosity concentrated in a small area. [named after Fritz *Zwicky*, 1898–1974, Swiss-born American astronomer, who catalogued this class of galaxies]

zwitterion (zwit′ər ī′ən *or* swit′ər ī′ən), *n. Chemistry.* a molecule which has both a positive and a negative charge, on opposite sides, as in neutral amino acids and certain protein molecules; a dipolar molecule.

[from German *Zwitterion,* from *Zwitter* hybrid + *Ion* ion]
—**zwitterionic,** *adj.* of or having to do with a zwitterion or zwitterions: *a zwitterionic form or structure.*

zyg-, *combining form.* the form of **zygo-** before vowels, as in *zygapophysis.*

zygapophyseal or **zygapophysial** (zī′gə pə fiz′ē əl), *adj. Anatomy.* of or having to do with a zygapophysis; articular, as a vertebral process.

zygapophysis (zī′gə pof′ə sis), *n., pl.* **-ses** (-sēz′), *Anatomy.* one of the articular processes on the neural arch of a vertebra which interlock each vertebra with the one above and below. Each vertebra normally has four, two anterior and two posterior. [New Latin, from *zyg-* yoke + *apophysis*]

zygo- (zī′gə-), *combining form.* **1** yoke; yoked or paired, as in *zygodactyl = having the toes in pairs.*
2 union; joining, as in *zygospore.* Also spelled **zyg-** before vowels. [from Greek *zygon* yoke]

zygodactyl (zī′gə dak′təl), *Zoology.* —*adj.* having the toes arranged in pairs, with two before and two behind, as the feet of a climbing bird.
—*n.* a zygodactyl bird, such as a parrot.
[from *zygo-* + Greek *daktylos* finger, toe]

zygodactylous (zī′gə dak′tə ləs), *adj.* = zygodactyl.

zygoma (zī gō′mə), *n., pl.* **-mata** (-mə tə) or **-mas.** *Anatomy.* **1** the bony arch below the socket of the eye in vertebrates, formed by the zygomatic bone (cheekbone) and the zygomatic process of the temporal bone. Also called **zygomatic arch.**
2 = zygomatic process.
3 = zygomatic bone.
[from New Latin, from Greek *zygōma* bolt, bar, from *zygon* yoke]
—**zygomatic** (zī′gə mat′ik), *adj.* of, having to do with, or situated in the region of the zygoma: *a zygomatic muscle.*

zygomatic arch, = zygoma.

zygomatic bone, *Anatomy.* the three-sided bone forming the lower boundary of the socket of the eye. Also called **jugal bone, zygoma,** and, commonly, **cheekbone.**

zygomatic process, *Anatomy.* a process of the temporal bone which articulates with the zygomatic bone to form the zygoma.

zygomorphic, *adj. Botany.* (of a flower) symmetrical about a single plane; divisible vertically into similar halves in only one direction, as the sweet pea. Compare **actinomorphic.**
—**zygomorphism** or **zygomorphy,** *n.* the condition of being zygomorphic.

zygonema (zī′gə nē′mə), *n.* = zygotene.

zygophyte, *n. Botany.* a plant which reproduces by the fusion of two similar gametes (zygospores).

---

cap, fāce, fäther; best, bē, tėrm; pin, fīve;
rock, gō, ôrder; oil, out; cup, pùt, rüle,
yü in use, *yù* in uric;
ng in bring; *sh* in rush; *th* in thin, ᴛʜ in then;
zh in seizure.
ə = *a* in about, *e* in taken, *i* in pencil, *o* in lemon, *u* in circus

zygosity (zī gos′ə tē), *n. Genetics.* the condition of a zygote, especially with respect to homozygosity and heterozygosity: *The zygosity of twins can now be determined with better than 90 percent accuracy by means of a brief questionnaire* (Arthur R. Jensen).

zygospore, *n. Biology.* a spore formed by the union of two similar gametes, as in various algae and fungi.
—**zygosporic,** *adj.* of or having to do with a zygospore or zygospores.

zygote (zī′gōt), *n. Biology.* **1** the cell formed by the union of two germ cells or gametes. A fertilized egg is a zygote. *In a multicellular organism a zygote is merely a beginning. From this beginning a new individual develops by repeated mitotic divisions. But these divisions may not occur immediately. In some species a zygote produces a thick covering that is resistant to heat and drying; in this form it may remain dormant for months or even years* (Miller and Leth, *High School Biology*).
2 the individual which develops from this cell: *The zygote goes through a series of mitotic divisions. ... During the early stages of its development, the zygote's energy comes from the nutrients stored in the large ovum* (Otto and Towle, *Modern Biology*). Compare **embryo, fetus.**
[from Greek *zygōtos* yoked, from *zygoun* to join together, from *zygon* yoke]

zygotene (zī′gə tēn), *n. Biology.* the stage of the prophase of meiosis in which homologous chromosomes become paired: *The prophase of meiosis I is highly specialized and occupies a comparatively long period of time. For these reasons it is subdivided into five stages: leptotene, zygotene, pachytene, diplotene and diakinesis* (R. G. Edwards). [from *zygo-* + Greek *tainia* band]

zygotic (zī got′ik), *adj. Biology.* of, having to do with, or resembling a zygote: *zygotic division, a zygotic form.*

zym-, *combining form.* the form of **zymo-** before vowels, as in *zymase.*

zymase (zī′mās), *n. Biochemistry.* an enzyme complex in yeast which, in the absence of oxygen, changes sugar into alcohol and carbon dioxide or, into lactic acid. In the presence of oxygen zymase changes sugar into carbon dioxide and water: *In making wine, use is made of an enzyme, zymase, produced by the microorganisms called yeasts. Zymase acts as a catalyst in decomposing sugar glucose to ethyl alcohol and carbon dioxide* (Anthony Standen).

zyme (zīm), *n. Biochemistry.* a substance causing fermentation; ferment. [from Greek *zymē* leaven]

zymo-, *combining form.* **1** fermentation; ferment, as in *zymology, zymolysis.*
2 enzyme, as in *zymogen, zymogram.*
3 yeast, as in *zymosterol.* Also spelled **zym-** before vowels.
[from Greek *zymē* leaven]

zymogen (zī′mə jən), *n. Biochemistry.* the inactive precursor of an enzyme; a proenzyme: *Many enzymes are formed in the cells where they are produced as inactive zymogens; only in the presence of another substance, an activator, does the enzyme become functional* (Storer, *General Zoology*). [from *zymo-* + *-gen*]

zymogen granule, *Biochemistry.* a proenzyme in a gland cell, especially a cell of the pancreas or the stomach.

zymogenic, *adj. Biochemistry.* **1** of or having to do with a zymogen or zymogens.
2 causing fermentation: *Yeasts are zymogenic organisms.*

zymogram (zī′mə gram), *n. Biochemistry.* a diagram or other representation of the different molecular forms of an enzyme, obtained by electrophoresis: *Most varieties have consistent patterns of isozymes (variants of the esterase enzymes) ... The isozyme distribution can be summarized in a zymogram, which provides the basis for identifying the variety of the seedling. But the identification is not a "finger-printing"—some varieties cannot be distinguished from each other* (New Scientist).

zymology (zī mol′ə jē), *n.* the study of fermentation and of ferments and their action.

zymolysis (zī mol′ə sis), *n.* = fermentation.

zymophore (zī′mə fôr), *n. Biochemistry.* the active part of an enzyme. Compare **active site.**

zymosan (zī′mə san), *n. Biochemistry.* an insoluble residue from the cell walls of yeast, used experimentally in immunological and cancer research. [from Greek *zymōsis* fermentation]

zymosis (zī mō′sis), *n.* **1** *Biochemistry.* fermentation. **2** *Biology.* the development of an infectious disease, formerly thought to be analogous to fermentation. [from Greek *zymōsis,* from *zymoun* to leaven, ferment, from *zymē* leaven]

zymosterol (zī mos′tə rōl), *n. Biochemistry.* a sterol obtained from yeast.

zymotic (zī mot′ik), *adj. Biochemistry.* of, having to do with, or causing zymosis or fermentation: *the zymotic process.*

zymurgy (zī′mer jē), *n.* the branch of chemistry dealing with the process of fermentation. [from *zym-* + Greek *-ourgos* making, from *ergon* work]

PERIODIC
TABLE

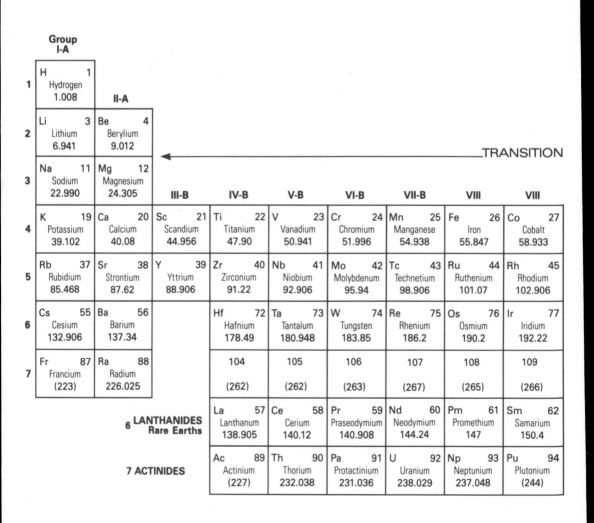

**Group
I-A**

TRANSITION

	I-A	II-A	III-B	IV-B	V-B	VI-B	VII-B	VIII	VIII
1	H 1 Hydrogen 1.008								
2	Li 3 Lithium 6.941	Be 4 Berylium 9.012							
3	Na 11 Sodium 22.990	Mg 12 Magnesium 24.305							
4	K 19 Potassium 39.102	Ca 20 Calcium 40.08	Sc 21 Scandium 44.956	Ti 22 Titanium 47.90	V 23 Vanadium 50.941	Cr 24 Chromium 51.996	Mn 25 Manganese 54.938	Fe 26 Iron 55.847	Co 27 Cobalt 58.933
5	Rb 37 Rubidium 85.468	Sr 38 Strontium 87.62	Y 39 Yttrium 88.906	Zr 40 Zirconium 91.22	Nb 41 Niobium 92.906	Mo 42 Molybdenum 95.94	Tc 43 Technetium 98.906	Ru 44 Ruthenium 101.07	Rh 45 Rhodium 102.906
6	Cs 55 Cesium 132.906	Ba 56 Barium 137.34		Hf 72 Hafnium 178.49	Ta 73 Tantalum 180.948	W 74 Tungsten 183.85	Re 75 Rhenium 186.2	Os 76 Osmium 190.2	Ir 77 Iridium 192.22
7	Fr 87 Francium (223)	Ra 88 Radium 226.025		104 (262)	105 (262)	106 (263)	107 (267)	108 (265)	109 (266)

**6 LANTHANIDES
Rare Earths**

La 57 Lanthanum 138.905	Ce 58 Cerium 140.12	Pr 59 Praseodymium 140.908	Nd 60 Neodymium 144.24	Pm 61 Promethium 147	Sm 62 Samarium 150.4

7 ACTINIDES

Ac 89 Actinium (227)	Th 90 Thorium 232.038	Pa 91 Protactinium 231.036	U 92 Uranium 238.029	Np 93 Neptunium 237.048	Pu 94 Plutonium (244)